Springer Japan KK

T. Yamashita   K. Tanabe (Eds.)

Advances in Superconductivity XII

Proceedings of the 12th International Symposium on
Superconductivity (ISS '99), October 17–19, 1999, Morioka

With 1,199 Figures

 Springer

TSUTOMU YAMASHITA, Professor
New Industry Creation Hatchery Center
Research Institute of Electrical Communication
Tohoku University
2-1-1 Katahira, Aoba-ku, Sendai 980-8577, Japan

KEI-ICHI TANABE
Director, Division 6
Superconductivity Research Laboratory
ISTEC
1-10-13 Shinonome, Koto-ku, Tokyo 135-0062, Japan

ISBN 978-4-431-70270-2   ISBN 978-4-431-66877-0 (eBook)
DOI 10.1007/978-4-431-66877-0

Printed on acid-free paper

Typesetting: Camera-ready by the authors

SPIN: 10743731

Foreword

The International Symposium on Superconductivity, which has been held annually since 1988, is a forum for presenting the most up-to-date information about a broad range of research and development in superconductivity, from fundamental aspects to applications.

More than 10 years have passed since the discovery of high-temperature oxide superconductors and various developments of their applications began. It can be said that the prospects for application of oxide superconductors have recently opened up. Great progress has been made toward practical use, for example, of the flywheel, which uses bulk materials; the high-performance cryo-cooled magnet made of bismuth wire; SQUIDs; and microwave devices. These were the results of persistent efforts to develop materials from the viewpoint of materials science and engineering. Nevertheless, the development of second-generation wires and integrated digital devices seems urgent. I believe these developments would provide solutions to future energy and environmental problems.

Also important is progress in comprehensive understanding of high-temperature superconductivity. At the symposium, special attention was focused on the two-dimensional character, or the c-axis transport, resulting from the unique structural and electronic properties of cuprates. There was also much discussion on the newest results in the field of vortex physics and device physics, and on some fundamental issues in bulk materials, wires, and thin films. These papers in the proceedings will be instructive for many researchers and for students who are to enter this field.

I sincerely hope that this proceedings will not be simply a record of the symposium, but will play an important role in inspiring research and development from fundamental physics to a wide variety of applications for the next decade.

SHOJI TANAKA
Vice President, ISTEC
Chairperson, Steering Committee

Preface

This volume contains the proceedings of the 12th International Symposium on Superconductivity (ISS '99), which was held at the Hotel Metropolitan Morioka in Morioka, Japan, October 17–19, 1999, under the cosponsorship of the International Superconductivity Technology Center (ISTEC), the Government of Iwate Prefecture, Morioka City, and the Industrial Vitalization Center for Tohoku. The ISS has been held annually since 1988 to discuss the latest topics in superconducting science and technology. It aims to promote worldwide cooperation and information exchange among the scientists, engineers, and business administrators involved in this field.

ISS '99 was attended by 596 participants from 19 countries. The total number of presentations was 366, including 1 special plenary lecture by S. Tanaka (ISTEC-SRL); 3 plenary lectures; 45 invited talks; 88 ordinary oral presentations; and 229 posters. These contributions were arranged into the following sections: (a) Plenary Lectures, (b) Physics and Chemistry, (c) Bulks, (d) Wires and Tapes, (e) System Applications, (f) Films and Junctions, and (g) Electronic Devices. The programs were summarized with a closing address given by D. Gubser. The majority of the papers presented at ISS '99 were included in the proceedings after being reviewed by selected specialists who are engaged in superconductivity research.

We are indebted to many people for their valuable contributions to all stages of the symposium. In particular, we wish to express our sincere gratitude to the International Advisory Committee for its assistance in choosing speakers and also to the Program Committee, which selected the papers and arranged the program. Once again, the publication of these proceedings was greatly facilitated by the diligence of the authors in preparing their manuscripts and by the unstinting support of the referees.

Tsutomu Yamashita
Kei-ichi Tanabe

Organization of ISS'99

Sponsored by:
International Superconductivity Technology Center

Cosponsored by:
Iwate Prefecture
Morioka City
Industrial Vitalization Center for Tohoku

Supported by:
Ministry of International Trade and Industry, Agency of Industrial Science and
    Technology
Science and Technology Agency (Prime Minister's Office)
Ministry of Education, Science, Sports and Culture
Ministry of Transport
Ministry of Posts and Telecommunications
New Energy and Industrial Technology Development Organization (NEDO)
Tohoku Bureau of International Trade & Industry

Approved by:
Cryogenic Association of Japan
Japan Chemical Industry Association
Japan Fine Ceramics Association
Japan Electronic Industry Development Association
Japan Mining Industry Association
Research and Development Association for Future Electron Devices
The Federation of Electric Power Companies
The Institute of Electronics, Information and Communication Engineers (IEICE)
The Iron and Steel Institute of Japan (ISUJ)
The Japan Iron and Steel Federation (JISF)
The Japan Electrical Manufacturers' Association
The Japanese Electric Wire & Cable Makers' Association
The Institute of Electrical Engineers of Japan
The Japan Society of Applied Physics
The Japan Institute of Metals
The Physical Society of Japan
The Chemical Society of Japan
The Ceramic Society of Japan
The Magnetics Society of Japan

Organizing Committee

Chairperson:

S. Nasu	ISTEC

Members:

T. Satou	Ministry of International Trade and Industry
H. Masuda	Iwate Prefectural Government
H. Kuwashima	Morioka City
T. Yashima	Industrial Vitalization Center for Tohoku
K. Inaba	Tohoku Bureau of International Trade and Industry
K. Kajimura	Electrotechnical Laboratory
N. Okayama	Sumitomo Electric Industries, Ltd.
E. Shoyama	Hitachi, Ltd.
S. Tanaka	ISTEC-SRL
S. Nakajima	Emeritus Professor of the University of Tokyo
M. Tachiki	National Research Institute for Metals

Steering Committee

Chairperson:

S. Tanaka	ISTEC-SRL

Members:

I. Watanabe	Iwate Prefectural Government
N. Ohta	Morioka City
T. Abe	Tohoku Bureau of International Trade and Industry
M. Murayama	Industrial Vitalization Center for Tohoku
M. Suwa	Electrotechnical Laboratory
T. Yamashita	Tohoku University
K. Noto	Iwate University
E. Kikuchi	Tohoku Electric Power Co., Inc.

Program Committee

Co-chairpersons:

T. Yamashita	Tohoku University
D . Gubser	Naval Research Laboratory

Members:

H. Akoh	Electrotechnical Laboratory
H. Ikuta	Nagoya University
T. Ishida	Osaka Prefecture University
Y. Ishimaru	Fujitsu Laboratories, Ltd.
Y. Endoh	Tohoku University
H. Ohsaki	The University of Tokyo
A. Oota	Toyohashi University of Technology
M. Kato	Tohoku University
K. Kadowaki	University of Tsukuba
Y. Kitaoka	Osaka University
H. Kumakura	National Research Institute for Metals
K. Sawada	Railway Technical Research Institute, JR
J. Shimoyama	The University of Tokyo
S. Tajima	ISTEC-SRL
Y. Tanaka	ISTEC
K. Tanabe	ISTEC-SRL
S. Tahara	NEC Corporation
I. Tsukada	CRIEPI
M. Naito	NTT
M. Nagano	Tohoku Electric Power Co. , Inc.
K. Noto	Iwate University
K. Hayashi	Sumitomo Electric Industries, Ltd.
A. Fujimaki	Nagoya University
A. Maeda	The University of Tokyo
Y. Matsuda	The University of Tokyo
K. Matsumoto	ISTEC-SRL
M. Murakami	ISTEC-SRL
H. Yamauchi	Tokyo Institute of Technology
H. Wada	National Research Institute for Metals

General Affairs Committee

Chairperson:

K. Noto	Iwate University

Members:

J. Kuroda	Iwate Prefectural Government
H. Hareyama	Morioka City
J. Yanai	Industrial Vitalization Center for Tohoku
M. Nagano	Tohoku Electric Power Co., Inc.

Contents

* Invited papers

1 Plenary Lectures

2 Physics and Chemistry

2.1 Mini-Symposium:
What Is the Two-Dimensionality in High-Temperature Superconductors?

2.2 Solid State Chemistry

2.3 Normal and Superconducting State Properties

2.4 Vortex Physics

XVIII

3 Bulk Materials

3.1 RE-Ba-Cu-O

3.2 Characterization

3.3 Bi-Sr-Ca-Cu-O

4 Wires and Tapes

4.1 YBCO Type Superconductors

XXIV

4.3 Applications

5 System Applications

5.1 Magnetic Levitation, Bearings, and Actuators

XXVIII

6 Films and Junctions

6.1 Film Preparation and Characterization

6.2 Film Growth and Mechanisms

6.3 Junction Fabrication and Characterization

7 Electronic Devices

7.1 SQUID Applications

7.2 Microwave Devices

7.3 Digital Devices

7.4 Novel Devices

1 Plenary Lectures

The Second Industrial Revolution and Superconductivity Technology

Shoji Tanaka
Superconductivity Research Laboratory / ISTEC
1-10-13 Shinonome, Koto-ku, Tokyo 135-0062, Japan

Abstract: Progress in information technology has been moving with increasing swiftness since 1995 and society has been changing rapidly through the global information network. This seems to indicate that we are now in the process of a second industrial revolution. In the near future, nearly a billion PCs will be distributed in homes and offices around the world and information will move at high speed of more than 10 megabits per second. Furthermore large numbers of equipment, routers, telecom switches, servers will be connected to the network system for information processing. They will require an enormous amount of electricity, which is an environmental problem that cannot be ignored for the future of the world. New superconductivity technology will be able to provide excellent solutions to these problems in the near future, by serving new electronic devices which will be operated on very small amounts of electricity, at very high speeds, with new electricity storage systems and new electricity transmission systems etc. In this paper the outlook of future superconductivity technology will be presented, from the standpoint of the new industrial revolution.

Keywords: superconductivity, electronic devices, SFQ devices, network

INTRODUCTION

Recently, the development of the information technology has been very quick and its effects have covered the entire world. The structures of many kinds of industry; business, production and so on, are changing rapidly and the life styles of people are also drastically changing. Thus I believe that we are now in the process of a second industrial revolution and the core of this revolution is the Internet.

The Internet is a world wide network system, consisting of numerous numbers of information equipment. Now, more than hundred millions of personal computers, millions of telecom switches, routers and servers are connected to the Internet through optical fiber network systems and wireless communication systems.

Furthermore, the Internet of the next generation is going to be developed, where picture communication will play dominant role in addition to the communication by vocal and literature, and it will happen before the year 2010. Such new systems will all require information equipment (1) of very high performance, 100 times to 1000 times faster, (2) low power consumption of not exceeding current systems, (3) downsizing as small as possible. An expected scheme of such future system is shown in Fig. 1. It will be difficult to construct such new systems by using conventional technology, and therefore it is expected that superconducting electronics will play an important role in very near future.

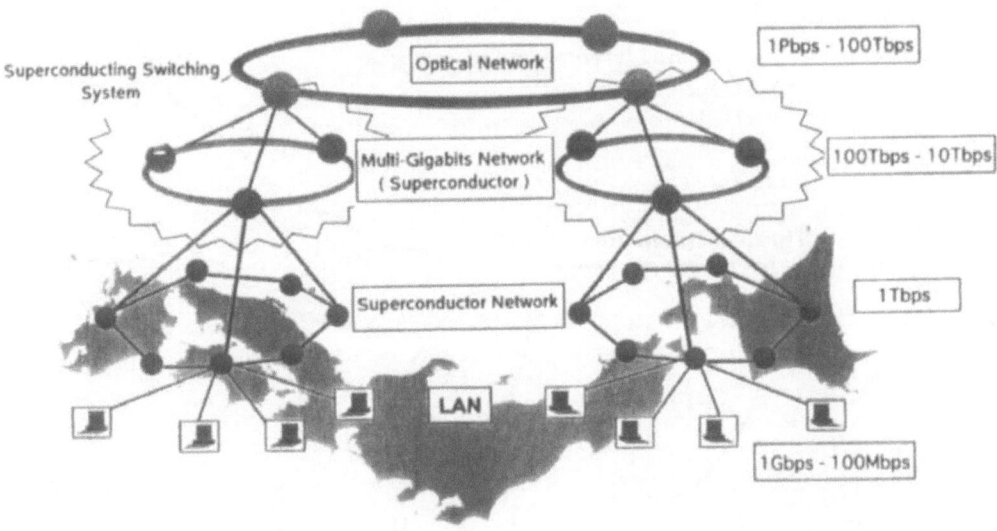

Fig. 1. Hybrid network system.

[1] SINGLE FLUX QUANTUM DEVICE AND ITS CIRCUITS

The principle of the SFQ is already established as is shown in Fig. 2.   It is just the same as SQUID device, in which the magnetic field is quantized as magnetic quantum fluxes.   When the number of flux is 0 or 1, it corresponds to an information signal of 0 or 1.   The operation characteristics of this device are shown in Fig. 3.   Here it is shown that the operation speed reaches 1ps and the consuming power is as small as 1nW.   Then it is possible in principle, to construct circuits, which are almost 100 times faster and 1/100 times smaller in power consumption than with conventional CMOS circuits, respectively.

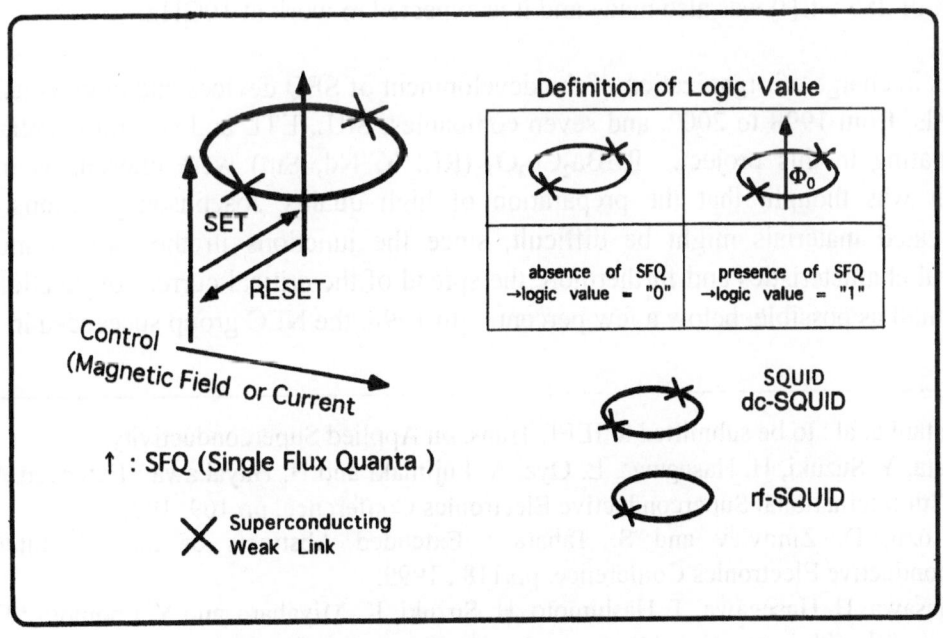

Fig. 2. Controlling SFQ with use of SQUID.

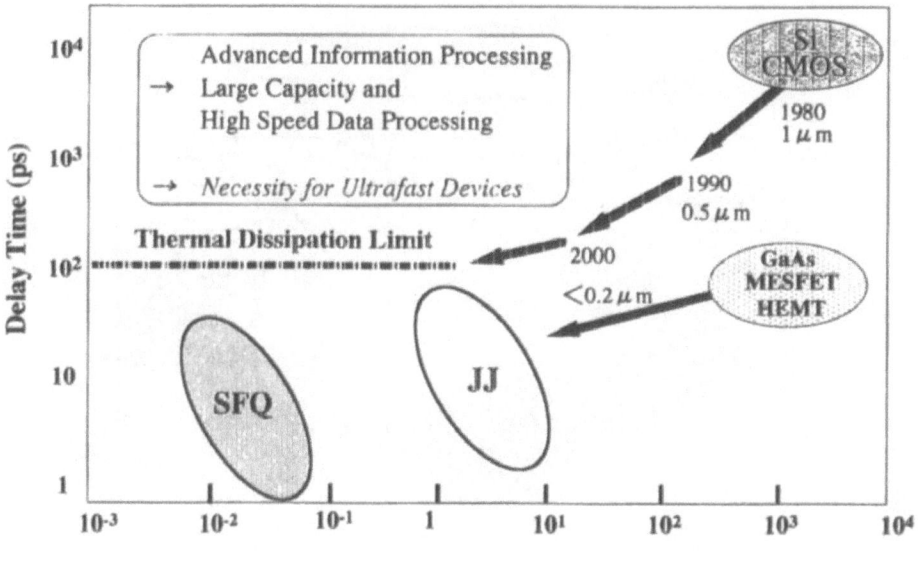

Power Consumption (μ W)

Fig. 3. Overview of electronic devices.

However, the history of the SFQ circuits is so short, that we first have to confirm what kinds of circuits are possible to make and also how high a performance is obtained. Then we started two national projects.

1) The Science and Technology Agency is in charge of the organization of the basic research on the characteristics of SFQ circuits and also to create new types of circuits, from 1997 to 2001, and three companies, SRL, ETL and several universities are participating in this project. The fabrication of LTS SFQ circuits is by the Fundamental Research Laboratory of NEC. High speed ring oscillator[1], shift resister[2] and arbiter circuits[3] have already been made and have provoken to work at high speed performances of around 50GHz. High speed DRAM of 256 bits[4] was also made and it is expected to work at 10GHz.

2) MITI is in charge of organization of the development of SFQ devices and circuits using HTS materials, from 1998 to 2002, and seven companies, SRL, ETL and several universities are participating in this project. $REBa_2Cu_3O_7$ (RE: Y, Nd, Sm) were chosen, as materials. First, it was thought that the preparation of high quality Josephson junctions in such complicated materials might be difficult, since the junctions in the circuits must have beautiful characteristics and furthermore the spread of the critical current of junctions might be as small as possible, below a few percent. In 1998, the NEC group succeeded in

[1] Y. Tarutani et al : to be submitted to IEEE Trans. on Applied Superconductivity.
[2] F. Furuta, Y. Suzuki, H. Hasegawa, E. Oya, A. Fujimaki and H. Hayakawa : Extended Abstracts of the 7th International Superconductive Electronics Conference, pp.109, 1999.
[3] S. Yorozu, D. Zinoviev and S. Tahara : Extended Abstracts of the 7th International Superconductive Electronics Conference, pp.118, 1999.
[4] S. Nagasawa, H. Hasegawa, T. Hashimoto, H. Suzuki, K. Miyahara and Y. Enomoto : Extended Abstract of the 7th International Superconductive Electronics Conference, pp.111, 1999.

making high quality junctions of YBCO[5]. The method employed by them is shown in Fig.4. In order to make ramp edge structure, the YBCO film (1) made by laser ablation method on the substrate is bombarded by argon ions and after heat treatment in oxygen atmosphere YBCO film (2) is made by the same laser ablation method. The very thin barrier is formed between both films and the junction shows beautiful characteristics as is shown in Fig.5. The structure of the barrier was investigated by TEM method in SRL and it was found that the barrier has the thickness of about 20A and structure was clearly different from YBCO as is shown in Fig.6[6]. Furthermore, the spread 1 σ of the critical current in 100 junctions made on one chip is 8%.

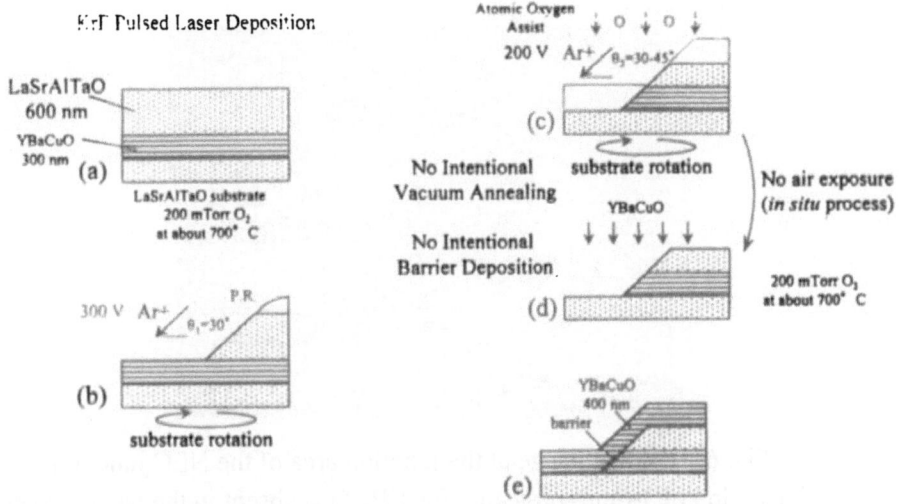

Fig. 4. Fabrication process of ramp edge junction.

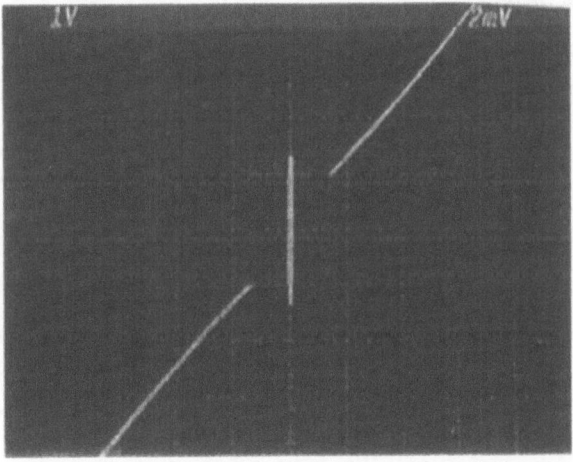

Fig. 5. Typical I-V characteristics at 4.2K for the edge junctions.
(The vertical scale is 1.0mA/div., and the horizontal scale is 2.0mV/div.)

[5]   T. Satoh, M. Hidaka and S. Tahara : IEEE Trans. on Applied Superconductivity 9 (2), pp.3141, 1999.
[6]   J.G. Wen, N. Koshizuka, S. Tanaka, T. Satoh, M. Hidaka and S. Tahara : Applied Physics Letters, 75 (16), pp.2470, 1999.

The Hitachi group[7], the Toshiba group[8] and SRL immediately tried to make junctions by using their own equipments and all of them obtained almost similar results. The relations between the critical current I_c and I_cR_n product obtained in the three groups are summarized in Fig.7 and this indicates that the junctions obtained by the three groups have the same characteristics and that the fabrication method employed is suitable in making large scale integration of SFQ devices. This year the NEC group tried to integrate 1000 junctions and the obtained spread of the critical current was less than 10%[9]. These results mentioned above give us hope in obtaining in a few years 10000 junctions integration having spread of the critical current of 5%.

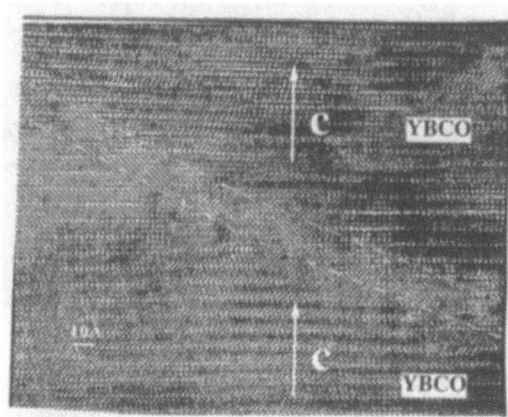

Fig. 6. HRTEM image at the junction area of the NEC junction.
The period of 3-times provskite for YBCO is absent in the barrier layer.
The 2nm barrier layer is identified as a cubic phase (a~4.1A).

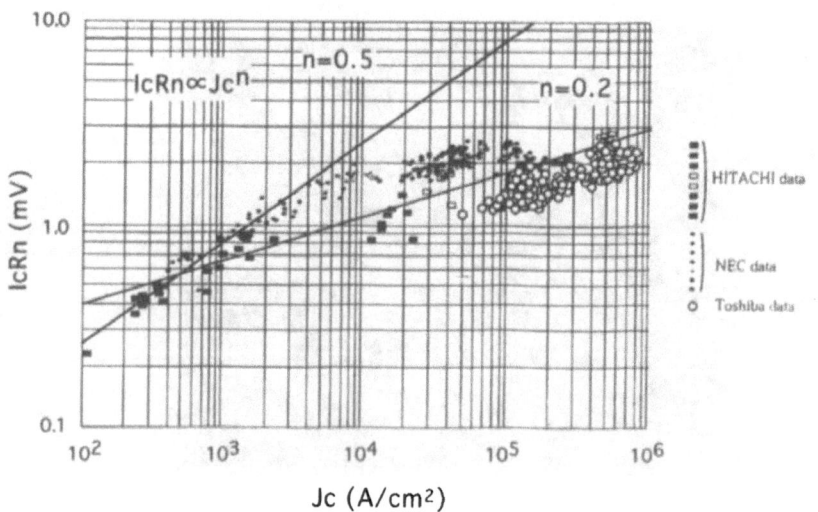

Fig. 7. Relation between I_c and I_cR_n for surface modified junctions.

[7] Y. Soutome, T. Fukazawa, A. Tsukamoto and Y. Tarutani : submitted to Applied Physics Letters.
[8] J. Yoshida, S. Inoue, H. Sugiyama and T. Nagano: to be published in Physica C.
[9] T. Satoh, J.G. Wen, M. Hidaka, S. Tahara, N. Koshizuka and S.Tanaka : to be published in Superconductor Science and Technology, 1999.

In order to reach real SFQ circuits, we have to solve other difficulties. And one of them is the fabrication of high quality ground plane. The ideal case of the structure of the SFQ circuits is shown in Fig. 8. In this figure, the ground plane is single crystal, in which the superconductivity characteristics shows the highest quality. Then the density of pinning centers for magnetic flux must be very small compared with superconducting films made by other methods. Furthermore, it is known that the surface of the single crystal can be made very flat, the roughness of which is about 10A. However, as the growth of single crystal of large diameter is difficult, we, in SRL, make the ground planes by using thick films made by the Liquid Phase Epitaxy method. The films are grown on the substrates of MgO single crystals and their characteristics are almost the same as that of YBCO single crystals. We succeeded in polishing the surface of these films and their roughness is about 10A. We are going to grow some insulating films on the ground plane and further superconducting films, in which many Josephson junctions will be integrated as is shown in Fig. 8.

Basic Technologies

(1) Single Crystal Growth

(2) Homoepitaxial Growth of High Quality Thin Film

(3) Multilayer Structure

(4) Junction Fabrication

(5) Passivation

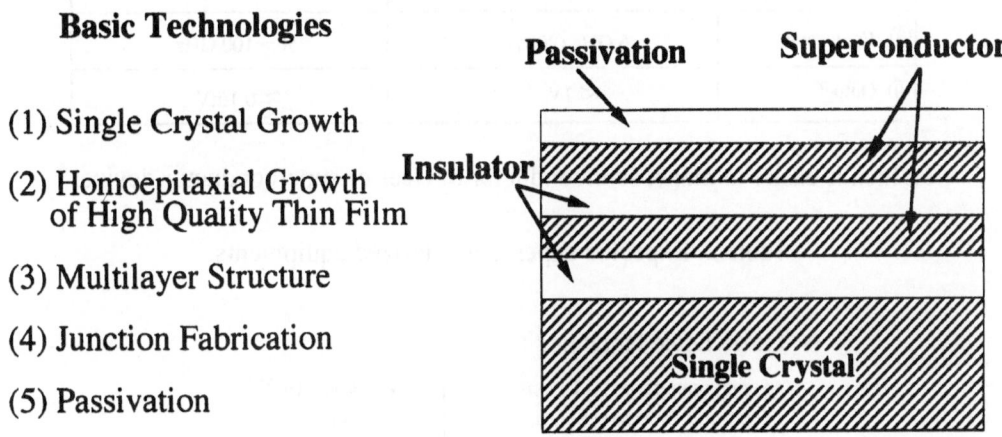

Fig. 8. Image of future SFQ circuit.

[2] FABRICATION PRECESSES OF SFQ CIRCUITS

When we discuss the possibility of the SFQ circuits of large scale integration, it becomes very important to investigate the future fabrication processes. As the fabrication processes of very large scale integrated circuits in the semiconductor industry is very well developed, it is the easiest way to compare the fabrication processes in superconductors and semiconductors, as is shown in Table 1. In this table, the biggest difference in both cases is the line width of the circuits. In the well developed semiconductor chips, the line width approaches to 0.1 μ m. On the other hand, in the case of LTS SFQ circuits, it was already proved that circuits of 1 μ m line width and of containing more than 10,000 junctions still show the high frequency characteristics of higher than 50GHz, and it is expected that such high frequency characteristics will expand to circuits of more than 100,000 junctions. Therefore, in the fabrication of the SFQ circuits, we can use the rather old fashioned lithography equipment in the semiconductor industry. And other processes will be also rather easier than semiconductor fabrications. The biggest inferior in the SFQ circuits fabrication is the design technology of the very primitive stage. We have no suitable CAD system for large scale integration of SFQ devices yet, and so it is necessary to develop CAD system as quick as possible.

8

Table 1. Comparison of fabrication process in semiconductor and superconductor.

	Semiconductor	Superconductor
① Line Width	$< 0.1 \mu m$	$0.8 \mu m$
② Lithography	after 2005	completed
③ Structure	Three Dimensional	Two Dimensional
④ Number of Masks	> 20	≈ 10
⑤ Wiring	Multi-layered ; high resistance heating	Multi-layered ; no resistance no heating
⑥ Wafer	8 inchs ; surface roughness several nm's	3 inchs ; surface roughness several nm's
⑦ Design	Efficient Tool ; Margin. large	Small Tool ; Margin. small
⑧ Frequency	5 GHz (2005)	50~100 GHz
⑨ Output	$\approx 1V$	$\approx 0.1mV$

Which is easier to make, semiconductor devices or superconductor devices?

Table 2. Expected superconducterized equipments.

● Large Scale Machines :

· Supercomputer > 10 T Flops
· Electronic Switching Board
· Superserver

⌐ Low Tc Chips (4.2K)
  or
⌐ High Tc Chips (40-50K)

● Medium Scale Machines

· Supercomputer < 10 T Flops
· ....
· ....

⌐ High Tc Chips (40-50K)

Refrigeration Efficiency is 20 times lager at 50K than that at 4.2K.

[3] THE FUTURE SFQ CIRCUITS AND THE INTERNET OF THE NEXT GENERATION

The possibility of large scale SFQ circuits is now going to be opened and very high performance information equipment will come to the market before the year 2010, if suitable funds and suitable number of engineers are provided for development. The expected superconducterized equipmemnts are shown in Table 2. They must have very high speed performances and must have very small power consumptions and also be very small in size. For instance, the Super Router will have a very high speed of 100Tbps, and the Super Server of 1Tflops will consume about 15kW of electricity and their size will be in one rack. This information equipment will be necessary in constructing the very high performance Internet systems of the next generation and such Internet systems will accelerate the industrial revolution of the world more intensively.

Key Insights from Structural Studies of High-Temperature Superconductors: Is There a Path to Higher T_c?

James D. Jorgensen

Materials Science Division and Science and Technology Center for Superconductivity, Argonne National Laboratory, Argonne, IL 60439 USA

Abstract: Structural studies have allowed the development of a model for the "ideal" high-temperature superconductor. For a given compound, the maximum T_c is traditionally achieved by using a chemical variable to adjust the carrier concentration to the optimum value. When comparing different compounds at their optimum doping, the highest T_c is observed for compounds with flat CuO_2 planes. T_c can also be enhanced if the charge reservoir region, or blocking layer, is metallic. In general, these three criteria cannot simultaneously be met by adjusting a single chemical/structural variable. Additionally, recent work on $HgBa_2CuO_{4+x}$ and 123 compounds as a function of doping suggest that electronically-driven structural distortions may hinder attempts to produce higher T_c's by chemical substitutions. The ideal high-T_c compound has not yet been discovered. But, in spite of these challenges, the search should continue.

Keywords: Crystal structure, Maximum T_c, Doping, Electronic Structure, Electronically-driven structural instability

INTRODUCTION

Although over fifty distinct copper-oxide superconductors have been discovered since 1986 [1], the superconducting transition temperature, T_c, (at ambient pressure) has not been raised above 135 K, which was achieved in the three-layer $HgBa_2Ca_2Cu_3O_{8+x}$ compound in 1993.[2] Nevertheless, the wide variety of structural features manifest by these compounds has led to a consensus concerning what chemical/structural features give rise to the highest T_c. In this paper, I review these conclusions and discuss the challenge of finding new compounds with higher T_c's.

OPTIMUM DOPING

The first critical insight from structural studies was that chemical modification of the charge reservoir layer (also called the blocking layer) could be used to create carriers in the metallic CuO_2 conduction layer which is responsible for the superconductivity. Perhaps the most convincing demonstration of this principle was that the bond valence sum for the Cu atoms in the CuO_2 planes in $YBa_2Cu_3O_{6+x}$ scaled with the oxygen content in the CuO_x chains in the charge reservoir layer.[3] This bond valence sum is an estimate of the charge at the plane Cu site calculated from the lengths of the Cu-O bonds around this site. This concept of creating carriers in the CuO_2 planes through chemical modification of the charge reservoir layer came to be known as the charge transfer model.

Consistent with this hypothesis, it was found that a variety of defects could be used in the charge reservoir layer to control the concentration of charge carriers. For example, in the insulating La_2CuO_4 compound, superconductivity can be created by substitution of a 2+ cation (e.g., Ba, Sr, or Ca) on the La^{3+} site or by insertion of interstitial oxygen defects in the La_2O_2 charge reservoir layer.[4] However, the relationship between the charge transfer and the defect chemistry can be rather complex, such as the case where defect ordering in the chain region of $YBa_2Cu_3O_{6+x}$, at constant oxygen content, dramatically affects the charge transfer and the T_c.[5]

DEFECTS IN THE CuO_2 LAYERS

Whereas defects in the charge reservoir layer may be required to achieve the desired carrier concentration, defects in or near the CuO_2 layers are clearly detrimental to superconductivity. This is especially true for the substitution of metal atoms on the plane Cu site.[6] Small concentrations of such defects destroy superconductivity. $(La,Sr,Ca)_3Cu_2O_{6+x}$ is a particularly interesting system for the study of defects near the CuO_2 layers. This is the two-layer compound in the series beginning with La_2CuO_4. When first discovered, $La_2SrCu_2O_6$ was found to be metallic, but, mysteriously, not superconducting. It was subsequently shown that substitution of a small Ca cation at the metal site between the two CuO_2 planes (analogous to the Y site in $YBa_2Cu_3O_7$) produced superconductivity by reducing the dimensions of the structure in this region and eliminating the formation of an interstitial oxygen defect within the double CuO_2 layer.[7] The formation of defects in or near the CuO_2 layers may explain why some metallic layered copper-oxide compounds do not exhibit superconductivity.

OPTIMUM STRUCTURE FOR HIGH T_c

The observation of a wide range of maximum T_c's in the various copper-oxide superconductors, after carrier concentration is optimized and detrimental defects are eliminated in or near the CuO_2 layers, argues that there is an optimum structure for achieving the highest T_c. Conclusions about the features of this optimum structure come from comparing different compounds. The highest T_c's are observed for compounds with flat CuO_2 planes. The $HgBa_2Ca_{n-1}Cu_nO_{2n+2+x}$ compounds satisfy this criteria. The n=1, 2, and 3 members of this series exhibit the highest T_c's reported for any one-, two-, or three-layer compounds (Table 1). The important feature of the structures of these three compounds is that the strong bonding of the apical oxygen atom to Hg results in a weak, and unusually long, bond of this oxygen atom to the plane Cu atom. This long apical Cu-O bond reduces the repulsion between the apical oxygen atom and the oxygen atoms in the CuO_2 plane, allowing flat CuO_2 planes.

Table 1. Buckling angles of the CuO_2 planes (Cu-O-Cu) and copper-oxygen apical bond lengths (Cu-O) for $HgBa_2Ca_{n-1}Cu_nO_{2n+2+x}$ compounds. Numbers in parentheses are standard deviations of the last significant digit. (from Refs. 8, 9, 10)

No. of layers, n	T_c (K)	Cu-O-Cu (°)	Cu-O (Å)
1	95	180	2.780(1)
2	126	179.4(2)	2.775(3)
3	135	178.4(4)	2.741(6)

The importance of the buckling angle of the CuO_2 planes can be seen by comparing the T_c's and buckling angles for several compounds, as shown in Fig. 1.[11] Only compounds with two CuO_2 layers are shown in this figure because the most accurate structural data are available for these compounds. The same behavior is observed for three-layer compounds, but with fewer data. (For one-layer compounds, the occurrence of different structural distortions, such as the coordinated tilting of CuO_6 octahedra in La_2CuO_4, make such a comparison difficult.) For the two-layer compounds with insulating charge reservoir layers ($La_2CaCu_2O_6$, Tl-1212, Bi-2212, Tl-2212, and Hg-1212), the correspondence between T_c and CuO_2 plane buckling is remarkable. Flat planes lead to higher T_c's. A few degrees of buckling lowers T_c substantially. The $HgBa_2CaCu_2O_{6+x}$ compound has a buckling angle of 179.4°, implying that a small increase in T_c (perhaps 10 K) would be achieved if the plane could be made perfectly flat.

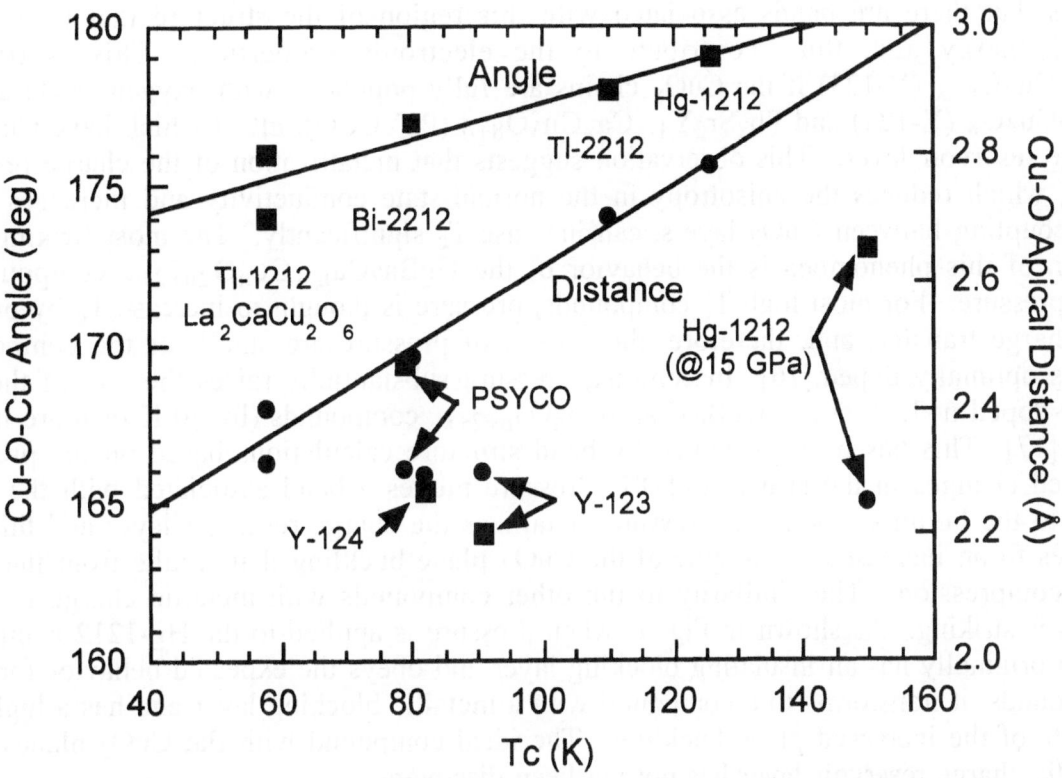

Fig. 1. Relationship between T_c, the Cu-O-Cu bucking angle of the CuO_2 planes (square symbols), and the Cu-O apical bond distance (round symbols) for compounds with two CuO_2 layers.

A number of studies where the buckling angle has been varied while holding the doping level constant have confirmed this relationship between T_c and buckling angle. The simplest and most direct confirmation comes from the studies of $La_{2-x}Sr_xCuO_4$ vs. pressure by Yamada and Ido.[12] Pressure reduces the buckling angle of the CuO_2 planes in the orthorhombic structure until a continuous transformation to a tetragonal structure with flat planes occurs. T_c increases linearly with pressure until the planes become flat and then remains constant with pressure. Experiments by Dabrowski et al., where the buckling angle is varied through chemical substitution in $La_{2-x}M_xCuO_4$ (M=Nd,Ca,Sr) lead to the same conclusion.[13] The same behavior is observed in other high-T_c compounds, such as the 123 materials, and there are several electronic structure calculations that provide an explanation in terms of how buckling affects the electronic density of states at the Fermi energy. This topic is reviewed in Ref. 14.

A recent report of the effect of structural disorder on T_c may also be explained in terms of the effects of disorder on local buckling angle. Attfield et al. [15] showed that T_c decreases systematically with increasing variance of the size of the charge reservoir cation in $Ln_{1.85}M_{0.15}CuO_4$, where the combination of Ln (La or Nd) and M (Sr, Ba, or Ca) is chosen to conserve the doping level while introducing structural disorder because of the different sizes of the cations. They offered no fundamental explanation for why the resulting lattice strains lower T_c. I suggest that the suppression of T_c is caused by increases in the buckling angle (which was not measured) in the local structure resulting from the strain.

Fig. 1 also illustrates the behavior of several two-layer compounds that violate the expected relationship between T_c and buckling angle. These are all compounds where the charge reservoir layer is metallic. In this context, the concept of a metallic charge reservoir layer means that there are bands associated with this region of the structure that lie near the Fermi energy and, thus, contribute to the electronic properties. This is true for $YBa_2Cu_3O_{6+x}$ (Y-123) if the CuO_x chains are fully populated with oxygen ($x \approx 1$) and for $YBa_2Cu_4O_8$ (Y-124) and $Pb_2Sr_2Y_{1-x}Ca_xCu_3O_{8+\delta}$ (PSYCCO), all of which have Cu in the charge reservoir layer. This observation suggests that metallization of the charge reservoir layer, which reduces the anisotropy in the normal-state conductivity and increases the c-axis coupling between CuO_2 layers, can increase T_c significantly. The most striking illustration of this phenomena is the behavior of the $HgBa_2Ca_{n-1}Cu_nO_{2n+2+x}$ compounds at high pressure. For most high-T_c compounds, pressure is thought to increase T_c by promoting charge transfer; and, therefore, the effects of pressure are small for the composition that is optimally doped.[16] In contrast, pressure substantially raises the T_c's of the optimally-doped n=1, 2, and 3 $HgBa_2Ca_{n-1}Cu_nO_{2n+2+x}$ compounds (by 30 K or more in each case).[17] This has been explained by band structure calculations based on the pressure-induced changes in the structures.[18] Pressure moves a band associated with the HgO_x layer to the Fermi energy; i.e., pressure metallizes the charge reservoir layer and this contributes to an increased T_c in spite of the CuO_2 plane buckling that results from the structural compression. The similarity to the other compounds with metallic charge reservoir layers is striking. As shown in Fig. 1, when pressure is applied to the Hg-1212 compound, which originally has an insulating blocking layer and obeys the expected behavior for those compounds, it transforms to a compound with a metallic blocking layer and has a higher T_c in spite of the increased plane buckling. The ideal compound with flat CuO_2 planes and a metallic charge reservoir layer has not yet been discovered.

ELECTRONIC BARRIER TO ACHIEVING HIGHER T_c

These observations for the $HgBa_2Ca_{n-1}Cu_nO_{2n+2+x}$ compounds provide critical insight into the challenge that must be met to increase the (ambient pressure) T_c of layered copper-oxide compounds beyond the present record of 135 K. One must achieve a structure with flat CuO_2 planes; the chemistry of the charge reservoir layer must be adjusted to achieve optimal doping of the CuO_2 planes; and the charge reservoir layer must also be metallic. In terms of the electronic structure, I assume that the latter two criteria mean that bands associated with the CuO_2 planes and the charge reservoir layer must simultaneously be at the Fermi energy. These criteria cannot, a priori, simultaneously be achieved by adjusting a single chemical variable. Hence, it is clear why both chemical doping and pressure must be used to achieve the highest T_c in the $HgBa_2Ca_{n-1}Cu_nO_{2n+2+x}$ compounds -- two variables are needed to adjust two bands to lie at the Fermi energy.

Recent work on the $HgBa_2CuO_{4+x}$ compound suggests that electronically-driven structural distortions may increase the difficulty of making the ideal high-T_c material. Fig. 2 shows the T_c and unit cell volume of $HgBa_2CuO_{4+x}$ as a function of the internal structural parameter $[z(Ba)-z(O2)]$.[19] This structural parameter, which generally decreases with increasing oxygen content, is a measure of the charge transfer as oxygen is added to the compound. At the maximum T_c, as the material passes from the under-doped to the overdoped state, there is a region where T_c remains constant and the unit cell volume increases while oxygen is added. This anomalous increase in the unit cell volume changes the structure so that the band associated with the Hg-containing charge reservoir layer does not lie at the Fermi energy; i.e., the structure distorts, by increasing its cell volume, to avoid placing this band at the Fermi energy. The application of pressure, which reduces the cell volume, can be viewed as removing this electronically-driven structural distortion and moving the charge-reservoir band to the Fermi energy.

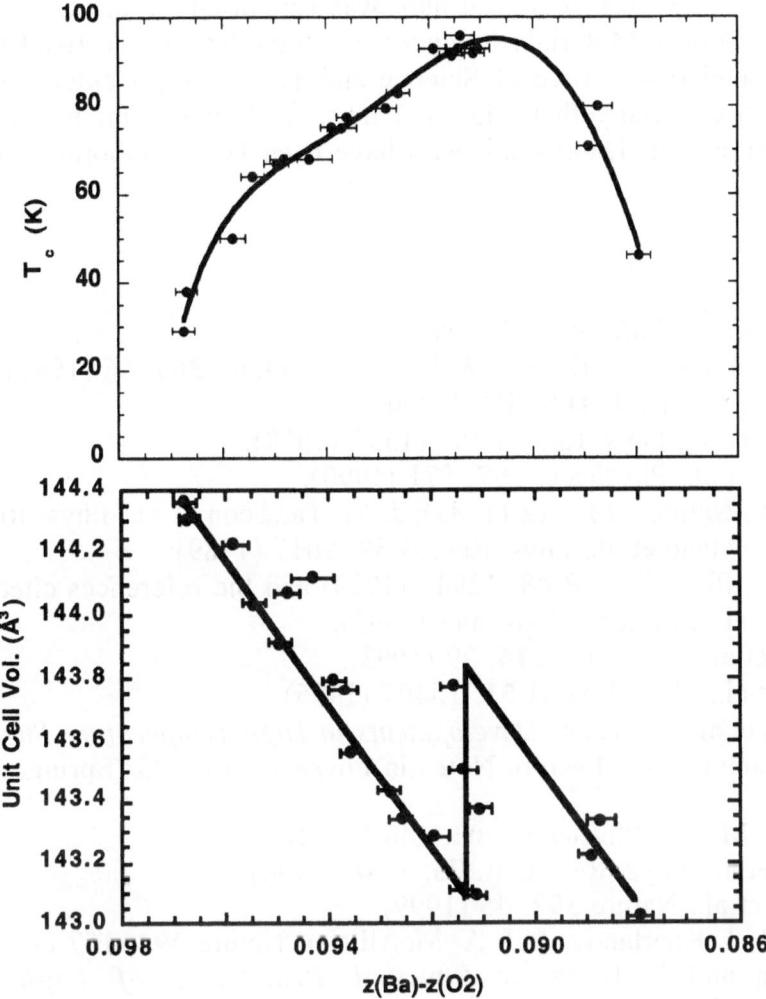

Fig. 2. T_c (top frame) and unit cell volume (bottom frame) of $HgBa_2CuO_{4+x}$ vs. the structural parameter $[z(Ba)-z(O2)]$, which is a measure of the charge transfer (see text).

Such behavior (structural distortion that lowers the density of states at the Fermi energy) is common in conventional superconductors, but has seldom been explicitly studied in the high-T_c materials. However, this behavior is not unique to the $HgBa_2CuO_{4+x}$ compound. A recent study of a chemically-substituted 123 compound in which both the under-doped and over-doped states can be accessed by changing the oxygen content shows a similar electronically-driven structural distortion at the maximum T_c.[14] In the 123 compound, the structural distortion manifests itself as an increased buckling of the CuO_2 planes, which we have already argued will lower T_c.

The challenge of achieving a higher T_c in the layered copper-oxide compounds is, thus, not an easy task. The desired criteria are well understood, but more than one chemical and/or structural variable is needed to simultaneously achieve the ideal structural and electronic properties. Recent work suggests that, as this ideal compound is approached, electronically-driven structural distortions can suppress T_c by distorting the structure to avoid the ideal structural and/or electronic properties. In spite of these challenges, the T_c of over 160 K achieved in the $HgBa_2Ca_2Cu_3O_{8+x}$ compound at high pressure (in a compound with buckled planes) argues that further increases in T_c at ambient pressure should be possible.

Acknowledgments: The work reviewed here was supported by the U. S. Department of Energy, Energy Research - Materials Sciences, contract No. W-31-109-ENG-38 and the National Science Foundation, Office of Science and Technology Centers, grant No. DMR 91-20000. I wish to especially thank D. G. Hinks, O. Chmaissem, P. G. Radaelli, J. L. Wagner, H . Shaked, and B. Dabrowski, who have been key collaborators in much of the work.

REFERENCES

1. R. J. Cava, Physica C **282-287**, 27 (1997).
2. A. Schilling, M. Cantoni, J. D. Guo, & H. R. Ott, Nature **363**, 56 (1993).
3. R. J. Cava et al., Physica C **165**, 419 (1990).
4. J. D. Jorgensen et al., Phys. Rev. B **38**, 11337 (1988).
5. J. D. Jorgensen et al., Physica C **167**, 571 (1990).
6. Y. Maeno et al., Nature **328**, 512 (1987); J. M. Tarascon et al., Phys. Rev. B **37**, 7458 (1988); R. S. Howland et al., Phys. Rev. B **39**, 9017 (1989).
7. H. Shaked et al., Phys. Rev. B **48**, 12941 (1993) and the references cited therein.
8. J. L. Wagner et al., Physica C **210**, 447 (1993).
9. P. G. Radaelli et al., Physica C **216**, 29 (1993).
10. J. L. Wagner et al., Phys. Rev. B **51**, 15407 (1995).
11. J. D. Jorgensen et al., in *Recent Developments in High Temperature Superconductivity*, edited by J. Klamut et al., Lecture Notes in Physics, Vol. 475 (Springer-Verlag, 1996) p. 1.
12. Y. Yamada and M. Ido, Physica C **203**, 240 (1992).
13. B. Dabrowski et al., Phys. Rev. Lett. **76**, 1348 (1996).
14. O. Chmaissem et al., Nature **397**, 45 (1999).
15. J. P. Attfield, A. L Kharlanov, & J. A. McAllister, Nature **394**, 157 (1998).
16. J. S. Schilling and S. Klotz, in *Physical Properties of High Temperature Superconductors, Vol. III*, edited by D. M. Ginsberg (World Scientific Publ., Singapore, 1992) p. 59.
17. L. Gao et al., Phys. Rev. B **50**, 4260 (1994).
18. D. L. Novikov, O. N. Myrasov, & A. J. Freeman, Physica C **222**, 38 (1994); D. J. Singh & W. E. Pickett, Physica C **233**, 237 (1994).
19. J. D. Jorgensen et al., in *High Temperature Superconductors and Novel Materials*, edited by G. Van Tendeloo et al. (NATO Book Series, Kluwer Academic Publ., B. V., 1999) p. 109.

FLUX PINNING AND CRITICAL CURRENTS IN LTS AND HTS CONDUCTORS

David Dew-Hughes

Oxford University, Department of Engineering Science, Parks Road, Oxford OX1 3PJ, UK

Abstract: What determines J_c in superconducting wires and tapes? The relationship between microstructure, flux pinning and J_c for low temperature superconductors is believed to be well understood. Simple flux-pinning theory leads to the observed scaling law for total pinning force $\propto$ b(1 - b) in NbTi conductors. In bronze-process Nb_3Sn conductors the microstructure contains paths which can accommodate flux shear and a theory of shear by flux lattice dislocations gives good agreement with experiment. The major differences between low temperature superconductors and high temperature superconductors are that the latter have much smaller coherence lengths and are highly anisotropic. The consequence of short coherence lengths are materials in which grain boundaries and other defects act as weak links. The anisotropy requires that high temperature superconductors must be highly textured in order to carry significant transport currents. The processing necessary to produce textured conductor, thermo-mechanical in the case of powder-in-tube conductor, or deposition in the case of coated conductor, results in an expensive product.

Keywords: Superconductivity, conductors, flux pinning, critical currents, low temperature superconductors, NbTi, Nb_3Sn, high temperature superconductors, BiSCCO, YBCO.

LOW TEMPERATURE CONDUCTORS.

A frequent complaint heard in the superconductivity community is "when are we going to have some real commercial applications for high temperature superconductors". It is worth bearing in mind that it took 50 years after the original discovery of superconductivity before the first commercial applications were realised. This was because it was not until about 1960 that alloys and compounds, based upon the element Niobium, capable of carrying high supercurrent densities, were developed. The pioneers in this work were the groups at Westinghouse and Bell Labs [1]. In the niobium-based ductile BCC alloys, it was found that cold deformation significantly enhanced the critical current density, J_c. It was thought that dislocations might somehow be connected with the elements of the "Mendelssohn Sponge [2]", and in 1962 I was asked by A H Cottrell to look at this problem. My student, A V Narlikar, began a transmission electron microscopical study of a series of niobium alloys. It was at the Colgate conference on the Physics of Superconductivity [3], held the following year, that Abrikosov's ideas on the flux line lattice [4] became accepted by the superconductivity community, and the connections between flux pinning and critical currents were established.

Meanwhile we had chosen to carry out experiments on a Nb25at%Ta alloy, as, at 4.2K this material had an upper critical field well within the 4 Tesla capability of our newly delivered NbZr magnet constructed by the fledgling Oxford Instrument Company. J_c was increased by cold deformation, as expected, but annealing at about 1000°C caused an even greater increase in J_c. The effect was further enhanced by more cold deformation after annealing. This result was unexpected; annealing should have reduced the dislocation density, with a concomitant reduction in flux pinning and J_c. Transmission electron microscopy showed that the deformed structure was heavily dislocated, but that the dislocations were fairly uniformly distributed throughout the material. Annealing at 1000°C lead to a redistribution of dislocations into cell-walls; further deformation sharpening the cell walls and enhancing the difference in dislocation density between walls and interiors of the cells [5]. It was clear that pinning was due to an interaction between flux lines and tangles of dislocations or cell walls, and not individual dislocations [6]. This lead to the idea of ΔK pinning [7,8], the theory for which was developed by Hampshire and Taylor [9].

Some 90% of all superconductor sold commercially is in the form of multifilamentary NbTi. Rods of the alloy are inserted in a copper matrix, and drawn down, often with repeated bundling, drawing and annealing schedules, to produce a multifilamentary composite wire. The superconducting filaments have a heavily deformed microstructure, with grains, sub-grains and non-superconducting αTi particles elongated in the direction of drawing. Pinning occurs at the sub-grain boundaries and αTi interfaces, and is a mixture of normal particle and ΔK pinning, with a pinning function in which the critical Lorentz force $J_c \wedge B$ is proportional to $b(1 - b)$, b being the reduced induction B/B_{c2} [10]. The microstructure is elongated in the direction of the current flow, and the Lorentz force acting on the flux vortices is such as to drive them across the sub-grain and normal particle boundaries. The critical current is associated with the unpinning of flux vortices from these boundaries. Theory and experiment are well-matched [11]. The above expression seems to hold whenever the critical current is determined by flux pinning with a density of pins less than the density of flux lines. The b term arises because, as the density of flux lines increases, so increases the total length of line pinned. The (1 - b) term represents the decrease in superconducting order parameter with increasing induction.

The other commercial conductor is based on the intermetallic A15 type compound Nb_3Sn. Multifilamentary conductor is fabricated by some variant of the bronze process. In this process rods of niobium are inserted in a copper/tin bronze ingot as matrix, and drawn, again with rebundling, to form a composite of niobium filaments in the bronze matrix. Reaction between the tin content of the bronze and the niobium at an elevated temperature converts the latter into Nb_3Sn filaments. This procedure is necessary, as the intermetallic compound is brittle and non-deformable. The critical Lorentz force in these materials is found to obey a scaling law similar to that postulated by Kramer [12], namely $b^{1/2}(1 - b^2)$. The critical current density increases as the grain size decreases, as would be expected if the pinning occurred at the grain boundaries, and as it does in NbTi. The $(1 - b^2)$ term may be taken as indicative of some flux shearing process, as the C_{66} modulus of the flux line lattice varies as $(1 - b^2)$ at high values of b.

It is not immediately obvious as to why these two types of material should behave in such different fashion. As can be seen in Table 1, their superconducting parameters and scale of microstructure are not vastly different. However, when one comes to examine the microstructure of Nb_3Sn, it is found to be very different from that of NbTi. This is not at all unexpected, due to the very different ways in which both microstructures are generated. That of bronze-processed Nb_3Sn consists of columnar grains whose axes are normal to the axes of the filaments [13]. The Lorentz force will act parallel to some of these boundaries,

Table 1: Comparison between NbTi and Nb_3Sn.

Property	NbTi	Nb_3Sn
T_c	~9K	~18K
H_{c2}	12-14T	~25T
ξ_0	5nm	3-4nm
G-L kappa	~35	~20
grain size	20-25nm	20-25nm
scaling law	$b(1-b)$	$\sim b^{1/2}(1-b)^2$

driving the flux lines along them rather than across them. A path is thus provided down which flux can shear, and the author has put forward a mechanism of flux lattice dislocation assisted shear [11]. Values of critical Lorentz force predicted on this model are close both to the Kramer law and to observation; in addition the model predicts an inverse dependence of J_c on grain size, as experimentally observed, but not predicted on the Kramer theory.

The relation between production processes and microstructure, and the relation between microstructure, flux pinning, and critical current density, are now well established and understood for the commercial low temperature conductors based on NbTi and Nb_3Sn. There are two further points to be made. One is to point out an interesting consequence of the two scaling laws. The slope of the function $b(1 - b)$ approaches 1 as $b \rightarrow 1$, whereas the slope of the $b^{1/2}(1 - b^2)$ function approaches 0 as $b \rightarrow 1$. This means that NbTi conductors still have appreciable values of critical current at fields close to H_{c2}, but that for bronze-processed A15 conductors the critical current density becomes vanishingly small close to H_{c2}. If the microstructure of these latter materials could be manipulated into something similar to that of the ductile alloys, high critical currents close to the upper critical field of the material would be possible. Magnets generating close to 25 Tesla could be manufactured from Nb_3Sn without having to contemplate inserts of more exotic materials.

The second point to be made is in respect of bronze-processed V_3Ga. The critical Lorentz force in this material does not appear to obey any simple scaling law. However V_3Ga is very strongly paramagnetically limited, i.e. the actual, experimental, upper critical field is much reduced from that deduced theoretically from its basic superconducting parameters. Strong paramagnetism in the normal state reduces the free energy of the normal state with respect to the superconducting state in high magnetic fields, and superconductivity ceases prematurely [14]. If the critical Lorentz force is plotted versus a reduced induction derived in respect of the predicted non-limited B_{c2}, the low induction part of the curve, before the onset of the paramagnetic limitation, can be fitted to the usual A15 $b^{1/2}(1 - b^2)$ scaling law [15]. This suggests that the paramagnetic limitation comes in suddenly, and the flux line lattice has no prior warning of this effect. This point will be seen to be of relevance to the later discussion of high temperature superconductors.

High Temperature Conductors.

The immediate expectation from the discovery of the high temperature, mixed copper oxide superconductors, was the "magnet builders' dream". It was thought that the high temperature aspect of these materials could be exploited at 77K to build electromagnets that would compete with permanent magnets, offering inductions in excess of 2 Tesla. At low temperatures, the high critical fields would allow of competition with low temperature superconductors and the 21 Tesla maximum field available from existing A15 conductor would be exceeded. These goals have proven to be difficult of realisation. The production of long lengths of flexible conductor, able to carry high current densities in high magnetic fields, has turned out to be a very serious challenge. The behaviour of HTS materials is very different to that of LTS materials, especially in respect of critical currents in magnetic fields.

Figure 1 shows schematically a J_c versus B curve typical of most high temperature superconductors. The curve consists of three regions: an initially region in which the critical current decreases rapidly as soon as the field is turned on; a region, which can be linear, falling slowly with increasing field, and a third region in which the critical current falls to zero. The middle region may appear to be perfectly horizontal, indicating no dependence of critical current on applied field. It may also extend to very high fields, especially in Bi-2212 at temperatures below 20K. An extreme example is a sample of spray-pyrolised Tl-1223 in which the critical current density at 4.2K is constant with field

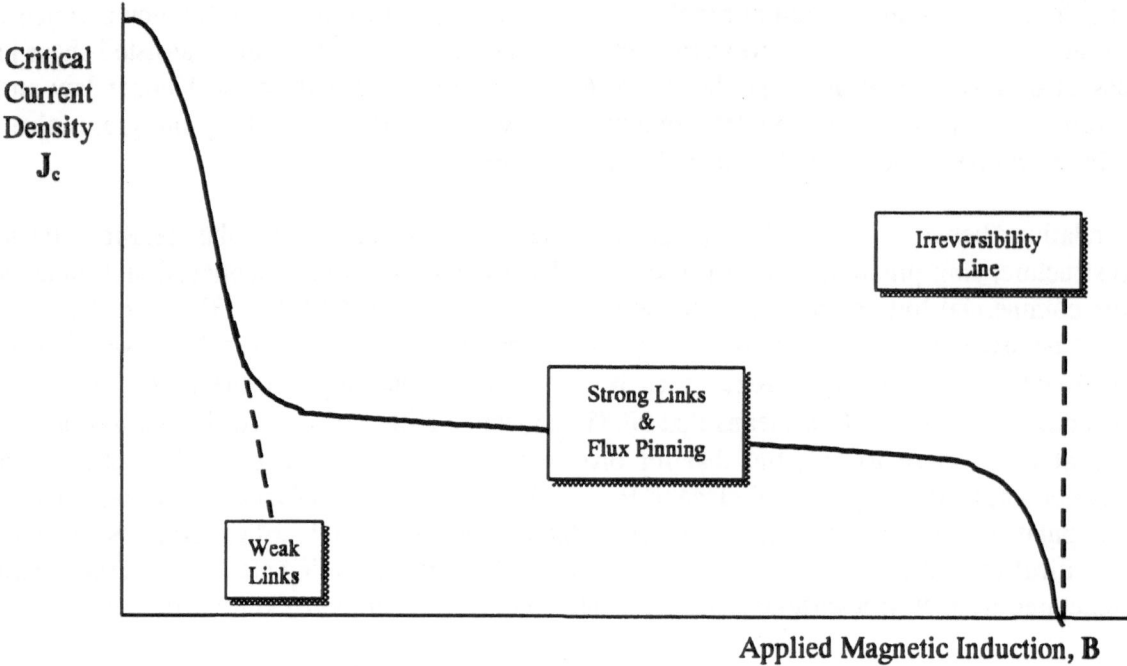

Figure 1: Jc versus B, schematic, for typical HTS conductor.

up to inductions of 40T [16]. As the temperature is increased, all regions of the curve move to lower values of field and critical current density. In particular the cut-off field decreases, and the (negative) slope of the middle region increases.

The significant fundamental differences between LTS and HTS are that the latter are anisotropic and have rather small coherence lengths. Structurally the mixed oxide superconductors are tetragonal, or near tetragonal, with a,b lattice parameters lying in the range 0.375-0.395 nm, and c-axis parameter 3-12 times greater. This structural anisotropy leads to anisotropy in the physical properties of the compounds. In single crystals, the critical current density in the ab plane is many times greater than that in the c direction, normal to the ab plane. The superconducting coherence length, ξ, is small in these compounds; that in the c-direction is just a few tenths of a nanometre in length, of similar magnitude to the region of crystallographic disturbance in the boundary between two grains. The consequence of this small range of coherence is that grain boundaries in high temperature superconductors act as weak links, i.e. the superconducting wave functions in adjacent grains are only weakly coupled to one another. The overall critical transport current density in a superconductor is determined by whichever is the lesser of the *intragrain* or the *intergrain* current densities. The *intragrain* current density is controlled by flux pinning, the *intergrain* current density is a measure of the ability of current to flow from one grain to an adjacent grain. This latter depends upon the strength of the superconducting link across the boundary, and in the case of anisotropic superconductors, upon the relative orientation between the two grains [17]. The initial rapid drop in J_c with field is due to many weak links between grains being progressively switched off as the field is increased [18].

The current that is left is now being carried by the few strong links that exist between the grains, and the number of these is relatively insensitive to magnetic field. The strength of supercurrent depends upon the proportion of grain boundaries that are strong links, and improvements in conductor performance require that the number of intergrain strong links be increased at the expense of the weak links. Many models have been proposed to account for the manner in which current is transferred from grain to grain in anisotropic mixed oxide superconductors [19]. It is not possible in the time at my disposal to discuss the relative merits of the various models, but the conclusions from these models, confirmed by experience, is that the proportion of strong links between grains, and hence the intergrain current, is maximised by grain alignment. The material is textured so that the c-axis of the grains is close to being normal to the direction of current flow, and that the ab planes of the grains are in near parallelism to one another. In effect, the conductor is as close to being a single crystal as possible.

Once a degree of texture has been established, the current density is further determined by flux pinning. A fully texture material will carry no appreciable current density if pinning is weak. Conversely, a material with strong pinning will also have a low critical current density if there is no texture. In anisotropic materials, the pinning of flux is also anisotropic [20]. The pinning strength is a function of the direction of an external magnetic field relative to the ab planes of the superconductor. The critical current density is much higher with the field parallel to the ab planes, than when it is perpendicular to them. The HTS compounds consist of groups of one, two or three copper oxide layers, which are responsible for the superconductivity, separated by layers of other oxides that are essentially insulating. With the field, and hence the flux lines, lying parallel to the ab planes, the flux lines will tend to place themselves in the insulating layers. The pinning mechanism, known as intrinsic pinning, is similar to that by normal particles in LTS. The density of pins is, however, much greater than the density of flux lines. This explains the relative insensitivity of the current density to external magnetic field in the middle region of the J_c versus B curve. When the applied field is normal to the ab planes, the intrinsic pinning no longer acts to hinder flux line motion; critical current densities are much lower than when the field is parallel to the planes. The situation is made worse by the fact that flux lines normal to the ab planes tend to split into "pancakes" [21]. This tendency is greater the greater the ratio of non-superconducting oxide layer thickness to superconducting oxide layer thickness, and hence the degree of anisotropy in the material. The anisotropy can be reduced, and flux pinning can be enhanced, by chemical substitution that distorts the crystal structure, by the addition of non-superconducting phases, and by irradiation. The latter is discussed by Weinstock later in this conference. A detailed investigation of the effect of other phases upon the phase evolution, microstructure, and superconducting properties of Bi-2212 tape has been reported by Huang and Dew-Hughes [22].

As the applied field continues to increase, a value is reached at which the critical current falls to zero. This is the irreversibility field, above which it becomes impossible to pin flux. Irreversibility in magnetisation experiments also disappears. The magnitude of the irreversibility field decreases as the anisotropy and tendency to form pancake vortices increases. There is controversy as to the origin of the irreversibility field. Arguments persist as to whether it is caused by flux lattice melting, or by thermally activated depinning. What is interesting is that the critical Lorentz force in HTS materials in many cases follows scaling laws similar to those found for LTS. The one difference is that the reduced induction used in the scaling laws is that relative to the irreversibility field rather than the upper critical field. There are many examples of this in the literature. The point can be illustrated by reference to a recent example of single crystal Hg-1223 [23]. The significance of this becomes obvious when it is compared with the behaviour of the paramagnetically limited V_3Ga mentioned above. Scaling with the irreversibility field in HTS indicates that this field is an intrinsic property of the flux line lattice. Its advent is anticipated; it is not something that is suddenly imposed on the material by an external agency, as is the paramagnetic limit in V_3Ga.

COMMERCIAL HTS CONDUCTOR

The necessity for texture, maintained over long lengths in the case of conductor, has to date restricted the commercial development of HTS conductor to three materials and three processes.

The greater the anisotropy of the HTS phase, the more readily will it form as a textured product. It is ironic that the intrinsic material parameter that aids texture formation also acts to reduce pinning. The two requirements for high current density conductor, texture and flux pinning, are to some extent in competition.

Bi-2212 is coated onto a silver or silver alloy tape substrate. Melt-processing and annealing result in a textured product [24]. T_c of this material is only a few degrees above 77K, and at this temperature the irreversibility field is very small. This material is mainly proposed for high field, greater than 20T, applications at liquid helium or liquid hydrogen temperatures.

Bi-2223 conductor is fabricated by the OPIT (oxide powder in tube) process. The HTS powder is packed into silver, or silver alloy, tubes and drawn down into fine wire. It is interesting to note that the first use of powder in tube for superconductor was a mixture of Nb and Sn powders in a Nb tube to produce the first Nb_3Sn conductor [24]. A silver tube was first used in the production of conductor from the $PbMoS_6$ Chevrel phase superconductor [25]. An early attempt at HTS conductor based on YBCO also followed the powder in silver tube approach [26]. Multifilamentary conductor is fabricated by rebundling and redrawing. The wire is finally flattened to a tape form and then subjected to a partial melting and annealing process which results in the development of a textured product. This material has an acceptable critical current density at 77K in low magnetic fields.

YBCO conductor is prepared by deposition of a thin film of YBCO on a textured substrate. In the IBAD (ion-beam assisted deposition) process a metallic substrate is coated with a biaxially textured zirconia film as the template for the epitaxial YBCO film. In the alternative, RABiTS™ (roll-assisted biaxially textured substrate) process, the metallic substrate itself is textured by a schedule of controlled rolling and annealing. There are variations on all of the above processes, many of which will be described in subsequent presentations in this conference.

Of the above processes, only one, OPIT Bi-2223, is in commercial production. It does, however, result in an expensive product. The most useful measure of the cost of an electrical conductor is the price per kiloamp meter, i.e. the cost of a meter length of sufficient cross-section of conductor to enable it to carry one kiloamp. Two years ago the cost of Bi-2223 conductor was about $US 1500/kAm, according to quotations from one US and three European manufacturers. Of this amount, roughly 30% is raw material costs, split equally between the superconducting powder and the silver or silver alloy tube, 30% is the cost of the mechanical working, wire drawing and rolling, and 40% is the cost, presumably largely the energy costs, of the prolonged annealing treatments [28]. The price is dominated by the costs of the extensive and complicated thermomechanical processing that must be carried out in order to create the textured microstructure necessary for maximum current density. The price should be compared with that of copper conductor at about $US 16 per kAm, and that for multifilamentary NbTi conductor which varies from less than $US 1 per kAm for MRI magnet conductor produced in large quantities, to about $US 5/kAm for the more specialised conductor for small 10 Tesla laboratory magnets. It is clearly economically advantageous to use low temperature superconductor in place of copper in many applications. This lesser cost results from the much higher current densities achievable with low temperature superconductors. For HTS materials to compete, the cost of conductor per kAm must undergo a considerable reduction.

An examination of the various factors involved in the conductor cost, as quoted above, suggests that the only way a significant improvement can be achieved is by an order of magnitude or more increase in the critical current density. In the two years since the above figures were collected the cost has fallen. Malozemoff, in his talk later in this conference, will quote a price of $300/kAm for a large quantity of Bi-2223 tape. This lowering of cost has been brought about by developments in processing leading to higher current densities.

CONCLUSIONS

The critical current density in both low temperature and high temperature superconductors is controlled by their microstructure. Flux pinning in the ductile alloys based on Niobium occurs at dislocation tangles, sub-grain boundaries and interfaces with non-superconducting second phases (αTi). Flux shear along columnar grain boundaries seems to be the controlling mechanism in the bronze-route A15 materials. There is scope here for improving the performance at high field by the manipulation of microstructure. The incorporation of a fine dispersion of an inert second phase, could it be achieved, would both refine the grain structure, prevent the columnar grain formation, and provide normal particle pinning. How this is to be achieved is not at all clear. In the high temperature superconductors microstructural control must provide both a high degree of texture and flux pinning. Progress here is needed to bring costs down to level at which HTS can compete with LTS in electric power applications.

1.   Hulm, J.K., Kunzler, J.E. and Matthias, B. T. (1981) *Physics Today* **34**, 34.
2.   Mendelssohn, K. and Moore, J. R. (1935) *Nature* **135**, 826.
3.   Proceedings published (1964) in *Rev. Mod. Phys.* **36**.
4.   Abrikosov, A. A. (19557) *Soviet Phys. JETP*, **5**, 1174.
5.   Narlikar, A. V. and Dew-Hughes, D. (1966) *J. Materials Science* **1**, 317.
6.   Narlikar, A. V. and Dew-Hughes, D (1964) *Physica Stat. Sol.* **6**, 383.
7.   Dew-Hughes, D (1966) *Mat. Sci Engg.* **1**, 2.
8.   Dew-Hughes, D. and Witcomb, M. J. (1972) *Phil. Mag.* **26**, 73.
9.   Hampshire, R. G. and Taylor, M. T. (1972) *J. Phys. F.* **2**, 89.
10. Dew-Hughes, D. (1974) *Phil. Mag.* **30**, 293; *IEEE Trans Magnetics* **MAG-23**, 1172.
11. Dew-Hughes, D. (1987) *Phil. Mag.* **B55**, 459.
12. Kramer, E. J. (1973) *J. Appl. Phys.* **44**, 1360.
13. Pande, C. S. (1979) in Luhman and Dew-Hughes, (eds.), *Metallurgy of Superconducting Materials*, Academic Press, New York, p 171.
14. Clogston, A. M. (1962) *Phys Rev. Lett.* **9**, 266; Chandrasekhar, B. S. (1962) *Appl. Phys. Lett.* **1**, 7.
15. Dew-Hughes, D. (1978) *J. Appl. Phys.* **49**, 327.
16. Ryan, D. T. *et al.* (1996) *IEEE Trans. Magnetics* **32**, 2803.
17. Dimos, D., Chaudhari, P,. Mannhart, J. and LeGoues, F.K. (1988) *Phys. Rev Lett.* **61**, 219.
18. Peterson, R. L. and Ekin, J. W. (1990) *Phys. Rev.* **B42**, 8014.
19. Bulaevskii, L.N., Clem, J.R., Glazman, L.I. and Malozemoff, A.P. (1992) *Phy.s Rev.* **B45**, 2545; Hensel, B., Grivel, J.C., Jeremie, J., Perin, A., Pollini, A. and Flukiger, R. (1993) *Physica* **C205**, 329.
20. Kes, P. H., Aarts, J., Vinokur, V. and van der Beek, C.J. (1990) *Phys. Rev. Lett.* **64**, 1063.
21. Clem, J.R. (1991), *Phys. Rev.* **B43**, 7837.
22. Huang, S-l and Dew-Hughes, D (1999) *Physica* **C319**, 104.
23. Karpinski, J. *et al.* (1999) *Superconductor Sci. Tech.* **12**, R153.
24. Burgoyne, J. W. *et al.* (1996) in U. Balachandran, P.J. McGinn and J.S. Abell (eds.), High Temperature superconductors: Synthesis, Processing and Large Scale Applications, The Minerals, Metals and Materials Society, Warrendale, Pa., p 167.
25. Kunzler, J. E. (1961) *Rev. Mod. Phys.* **33**, 501.
26. Luhman, T. S. and Dew-Hughes, D. (1978) *J. Appl. Phys.* **49**, 936.
27. Cowey, L., Jones, H. and Dew-Hughes, D. (1988) *Cryogenics* **28**, 181.
28. Krauth, H., (1997) contribution to discussion at SCENET Workshop on Materials for Power Applications, Kassel, Germany.

2 Physics and Chemistry

c-axis Superfluid Response and Quasiparticle Conductivity in $Bi_2Sr_2CaCu_2O_{8+\delta}$ and $Bi_2Sr_2CuO_{6+\delta}$

[1]M.B. Gaifullin, [1]Yuji Matsuda, [2]N. Chikumoto, [3]J. Shimoyama,
[3]K. Kishio, and [4]R. Yoshizaki

[1]Institute for Solid State Physics, University of Tokyo, Roppongi 7-22-1,Minato-ku, Tokyo 106, Japan and CREST, Japan Science and Technology Corporation.
[2]Superconductivity Research Laboratory, ISTEC,Shibaura 1-16-25, Minato-ku, Tokyo 105, Japan
[3]Department of Superconductivity, University of Tokyo, Bunkyo-ku, Tokyo 113, Japan
[4]Institute of Materials Science, University of Tsukuba, Tsukuba, Ibaraki 305-8573, Japan

Abstract: Josephson plasma resonance for underdoped $Bi_2Sr_2CaCu_2O_{8+\delta}$ and $Bi_2Sr_2CuO_{6+\delta}$ have been measured by sweeping the microwave frequency continuously. The resonance enables us to determine the superfluid density and quasiparticle conductivity in the c-axis accurately. We show that the superfluid response and the low energy excitations out of the condensate in the c-axis of these materials are very different from those in the ordinary Josephson multilayer tunnel junctions.

Keywords: Josephson Plasma, Superfluid, Quasiparticle, Penetration Depth

INTRODUCTION

The interlayer electron transport is one of the most important subject for understanding the mechanism of high-T_c superconductors (HTSC). In the superconducting state, the electrodymanics within the ab-planes have been studied extensively [1]. On the other hand, the c-axis transport of Cooper pairs and quasiparticles is not well understood due to the high anisotropy of HTSC. Josephson plasma resonance (JPR) is a novel tool to obtain the information on both the superfluid and the low-energy excitations out of the condensate [2, 3, 4]. In this paper, by performing the JPR measurements, we present new data about the superfluid density and quasiparticle conductivity for underdoped $Bi_2Sr_2CaCu_2O_{8+\delta}$ and $Bi_2Sr_2CuO_{6+\delta}$ and discuss the difference between HTSC and ordinary Josephson multilayer tunnel junctions.

EXPERIMENT

The microwave frequency was swept continuously from 20 GHz to 150 GHz using backward-wave oscillators. We used a bolometric technique to detect a very small microwave absorption by the sample. To keep the microwave power constant, we employed a leveling loop technique. Underdoped $Bi_2Sr_2CaCu_2O_{8+\delta}$ and $Bi_2Sr_2CuO_{6+\delta}$ single crystals were grown by the traveling floating zone method.

JOSEPHSON PLASMA RESONANCE

The plasma frequency in the c-axis $\omega_{pl}(= c/\sqrt{\varepsilon_0}\lambda_c$, λ_c is the c-axis penetration depth and ε_0 is the dielectric constant) provides a very direct measurement of the c-axis superfluid density n_c via $\omega_{pl}^2 = 4\pi n_c e^2/\varepsilon_0 m^*$, where m^* is the effective mass of the electron. In extremely anisotropic $Bi_2Sr_2CaCu_2O_{8+\delta}$ and $Bi_2Sr_2CuO_{6+\delta}$ ω_{pl} fall into the microwave window. The microwave absorption P_{abs} is determined by the imaginary part of the dielectric function $\varepsilon_c(\omega)$; $P_{abs} \propto \mathrm{Im}1/\varepsilon_c(\omega)$ [4]. When $\hbar\omega_{pl} \ll \Delta$ (Δ is the superconducting energy gap), $\varepsilon_c(\omega)$ can be expressed as

$$\varepsilon_c(\omega) = \varepsilon_0\{1 - \frac{\omega_{pl}^2}{\omega^2} - \frac{\omega_{qp}^2}{\omega(\omega + i/\tau)}\}, \tag{1}$$

where ω_{qp} and τ are the plasma frequency and the scattering time of the quasiparticles, respectively. When $\omega\tau \ll 1$, P_{abs} can be written as

$$P_{abs}(\omega, T) \propto \frac{4\pi\sigma_{qp}^c(T)/\varepsilon_0}{\{1 - \omega_{pl}^2(T)/\omega^2\}^2 + \{4\pi\sigma_{qp}^c(T)/\varepsilon_0\omega\}^2}, \tag{2}$$

where $\sigma_{qp}^c = \varepsilon_0\omega_{pl}^2 e^2\tau/4\pi$ is the quasiparticle conductivity. The resonance occurs at $\omega = \omega_{pl}$ and the line width is proportional to σ_{qp}^c. Thus the resonance enables us to determine the superfluid density and quasiparticle conductivity.

SUPERFLUID RESPONSE

Figure 1 depicts the T-dependence of ω_{pl} for underdoped $Bi_2Sr_2CaCu_2O_{8+\delta}$ (T_c=82.5 K,

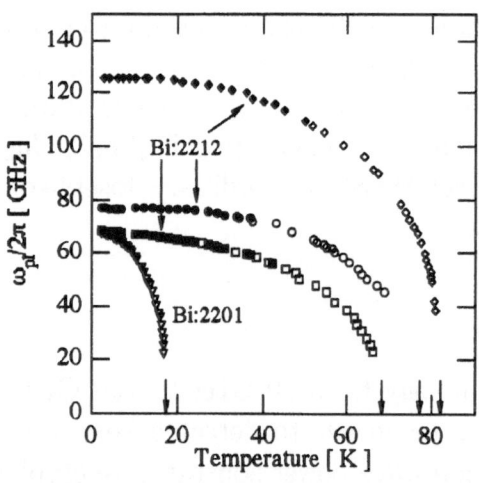

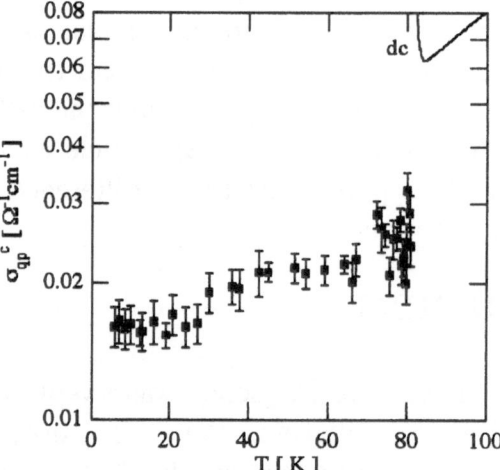

Figure 1: T-dependence of Josephson plasma frequency of underdoped $Bi_2Sr_2CaCu_2O_{8+\delta}$ and underdoped $Bi_2Sr_2CuO_{6+\delta}$. The solid (open) symbols represent ω_{pl} determined by sweeping frequency (temperature).

Figure 2: Quasiparticle conductivity for slightly underdoped $Bi_2Sr_2CaCu_2O_{8+\delta}$ (T_c=82.5 K).

77.2 K, and 68.0 K) and underdoped Bi$_2$Sr$_2$CuO$_{6+\delta}$ (T_c=16.5 K). The detailed analysis for the T-dependence of ω_{pl} is given in Ref.[5]. When going from slightly to strongly underdoped Bi$_2$Sr$_2$CaCu$_2$O$_{8+\delta}$, $\omega_{pl}/2\pi$ at T=0 falls from 125 GHz to 68 GHz. The c-axis critical current density j_c is also obtained through the relation

$$\omega_{pl}^2 = \frac{8\pi^2 cdj_c}{\varepsilon_0 \Phi_0},\tag{3}$$

where d is the interlayer distance (d=1.2 nm) and Φ_0 is the flux quantum. The above ω_{pl} corresponds to j_c from 900 A/cm^2 to 270 A/cm^2, using ε_0=6. According to the simplest tunneling model which assumes Fermi liquid and fully incoherent tunneling (parallel momentum of Cooper pairs not conserved), $j_c(0)$ is given by $j_c(0) = \pi\Delta(0)/2ed\rho_c$, where ρ_c is the normal state tunneling resistivity. If we apply the above expression to Bi$_2$Sr$_2$CaCu$_2$O$_{8+\delta}$ with T_c=82.5 K ($j_c(0) = 900$ A/cm^2) using $\Delta(0)$=25 meV from STM measurement, ρ_c is estimated to be 370 Ωcm. This value is approximately 25 times larger than ρ_c just above T_c (16Ω cm). This fact strongly indicates that the transport mechanism through the Josephson junction in these materials is quite different from those in ordinary junctions.

QUASIPARTICLE CONDUCTIVITY

Figure 2 shows σ_{qp}^c obtained from the line width for underdoped Bi$_2$Sr$_2$CaCu$_2$O$_{8+\delta}$. The T-dependence is very different from that reported in Ref.[6]. Below T_c, σ_{qp}^c falls to low values, then decreases gradually with T. At low temperatures, σ_{qp}^c remains finite. This is because in d-wave superconductors the impurity scattering gives rise to a finite quasiparticle density of states at the Fermi level. The monotonic decrease of σ_{qp}^c below T_c is in contrast to the quasiparticle conductivity in the ab-plane which shows a broad peak below T_c due to the suppression of the quasiparticle scattering. This suggests that the quasiparticle transport in the c-axis is not influenced by that in the ab-plane.

References

[†] Corresponding author, e-mail:ym@issp.u-tokyo.ac.jp

[1] D.A. Bonn*et al.* (1996) Czech. J. Phys. **46**, 3195.

[2] Y. Matsuda *et al.*, (1995) Phys. Rev. Lett., **75**, 4512 , Y. Matsuda *et al.* (1997) *ibid*, **78**, 1972, M.B. Gaifullin *et al.* (1998) *ibid.* **81**, 3551.

[3] L.N. Bulaevskii *et al.* (1995) Phys. Rev. Lett. **74**, 801.

[4] Y. Ohhashi and S. Takada (1999) Phys. Rev. B **59**, 4404, S.E. Shafranjuk and M. Tachiki (1998) Europhys. Lett. **44**, 348, S.N. Artemenko *et al.* Phys. Rev. B in press. T. Koyama, J. Phys. Soc. Jpn in press.

[5] M.B. Gaifullin *et al.*,(1999) Phys. Rev. Lett. **83**, 3928.

[6] H. Kitano *et al.* (1998) Phys. Rev. B **57**, 10946.

Study of Highly Anisotropic Conductivity and Penetration Depth of $Bi_2Sr_2CaCu_2O_y$ by Using a Cavity Perturbation Technique

H. Kitano[1], K. Kinoshita[1], Y. Tsuchiya[1], K. Iwaya[1], R. Abiru[1], and A. Maeda[1,2]

[1]Department of Basic Science, University of Tokyo, 3-8-1, Komaba, Meguro-ku, Tokyo, 153-8902, Japan
[2]CREST, Japan Science and Technology Corporation (JST), 4-1-8, Honcho, Kawaguchi, 332-0012, Japan

Abstract: The highly anisotropic properties of the superconducting $Bi_2Sr_2CaCu_2O_y$ (BSCCO) were studied by a cavity perturbation technique with the microwave magnetic and electric fields, which were parallel or perpendicular to the CuO_2 planes. Through careful measurements and analysis, we succeeded in obtaining the quantitatively reliable results of microwave conductivity σ_1 and penetration depth λ both in the ab-plane and along the c-axis, except the detailed behavior of σ_1^c in the superconducting state. Temperature evolution of $\lambda^2(0)/\lambda^2(T)$ was found to be quite anisotropic, suggesting that the quasiparticle conduction remained to be anisotropic even in the superconducting state.

Keywords: Microwave conductivity, Penetration depth, Anisotropy, $Bi_2Sr_2CaCu_2O_y$

INTRODUCTION

The highly anisotropic properties of the high-T_c cuprates have been discussed for many years, in terms of the mechanism of superconductivity. In particular, the c-axis motions of quasiparticles (QPs) and Cooper pairs in the Meissner state is one of basic problems which are still far from a complete understanding. In order to study them by using a microwave (MW) cavity perturbation technique, very careful measurements and analysis are required, since the in-plane highly conductive nature may give serious effects to the c-axis property [1, 2]. Here, we present a study of the highly anisotropic nature of the superconducting $Bi_2Sr_2CaCu_2O_y$ (BSCCO). We succeeded in obtaining the quantitatively reliable results of the microwave conductivity σ_1 and penetration depth λ both in the ab-plane and along the c-axis, except the detailed behavior of σ_1^c in the superconducting state.

EXPERIMENTAL

The high-quality single crystals of BSCCO were grown by the floating zone method. T_c was determined as 87K by magnetization measurements. The MW properties were investigated by using a cavity perturbation technique with a cylindrical normal Cu cavity resonator operated at 50 GHz in the TE_{011} mode. As for the MW properties in the

ab-plane, we measured the surface impedance $Z_s(= R_s + iX_s)$ in the configuration of $H_\omega \parallel c$. On the other hand, the c-axis properties were studied by measuring the complex dielectric constant $\epsilon(= \epsilon_1^c + i\epsilon_2^c)$ in the configuration of $E_\omega \parallel c$ [3, 4].

RESULTS AND DISCUSSION

Figure 1(a) shows the temperature dependence of the inverse square of the normalized penetration depth $\lambda^2(0)/\lambda^2(T)$ in the ab-plane and along the c-axis. We confirmed that our results of $\lambda^2(0)/\lambda^2(T)$ along each direction quantitatively agreed with reported results by other measurements [4]. Our results strongly suggest that the temperature evolution of $\lambda^2(0)/\lambda^2(T)$, which is proportional to the superfluid density, is quite anisotropic below T_c. In a simple d-wave BCS model with the coherent QP motion in all directions, the temperature dependnce of $\lambda^2(0)/\lambda^2(T)$ in the ab-plane and along the c-axis must be the same with each other [5, 6]. Three possibilities are considered as explanations for the anisotropic evolution. First possibility is the coherent motion with the c-axis transfer t_c which strongly depends on $k_{\parallel}$ [5]. Second one is the incoherent motion assisted by impurities [6]. Third one is the pair tunneling of Cooper pairs[7]. In the first model, it was expected that $\Delta\lambda_c \sim T^5$ at low temperatures, while it was expected that $\Delta\lambda_c \sim T^2$ in other two models. Our results indicated that $\Delta\lambda_c \sim T^2$ at low temperatures, showing a sharp contrast to previous results on the same material [8]. We also performed the measurement in the same configuration of $H_\omega \perp c$ as Jacobs et $al.$ and Shibauchi et $al.$ [8]. Surprisingly, the frequency shift $\Delta f(= f_s - f_0)$ was found to be negative both in the normal and superconducting states. This strongly suggests that the sample is not in the skin depth regime, which was assumed in previous reports [8]. Thus, we can conclude that the change of $\Delta\lambda_c$ can hardly be detected in the configuration of $H_\omega \perp c$.

Figure 1(b) shows the temperature dependence of σ_1 in the ab-plane and along the c-axis. The whole temperature dependence of σ_1^{ab} below T_c agreed with the ordinary behavior of high-quality crystals of BSCCO [9]. On the other hand, a non-metallc ($d\sigma_1^c/dT > 0$) behavior, which was expected from dc resistivity in the normal state, was clearly observed in σ_1^c above T_c. In addition, the obtained values of $\sigma_1^c(T_c)$ ($\sim 0.06 \ \Omega^{-1}cm^{-1}$) and $\lambda_c(0)$ ($\sim 170 \ \mu m$) was found to satisfy roughly the modified Basov correlation [10]. In the superconducting state, σ_1^c showed a small drop just below T_c. We found that the magnitude of its drop strongly correlated to the slope of $\lambda_c^2(0)/\lambda_c^2(T)$ just below T_c [4]. Thus, a small drop in σ_1^c just below T_c seemed to suggest the same rapid drop in the density of QPs moving along the c-axis, which was proportional to $1 - \lambda_c^2(0)/\lambda_c^2(T)$. We also observed an anomalous increase in σ_1^c below ~ 80 K, as shown in Fig. 1(b). Evidently, this increase in σ_1^c with decreasing temperatures cannot be explained by the coherent motion of QPs along the c-axis, because such coherent motion was never expected from the anisotropic evolution of $\lambda^2(0)/\lambda^2(T)$, as shown in Fig. 1(a). If this increase is intrinsic to σ_1^c of BSCCO, it could be explained by only a special mechanism such as the interaction between QPs and Cooper pairs. The $\cos\phi$ term of Josephson current might be a possible candidate [11]. However, at present, we cannot also exclude the possibility that this behavior was extrinsic. We should note that other experiments by using the Josephson plasma resonance reported a different behavior [12]. More careful measure-

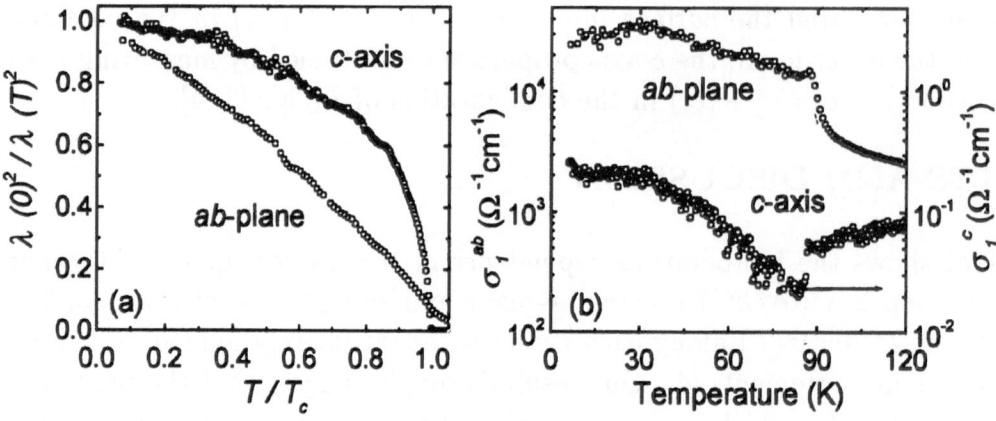

Figure 1: (a)The temperature dependence of $\lambda^2(0)/\lambda^2(T)$ in the ab-plane and along the c-axis. (b)The temperature dependence of the microwave conductivity σ_1.

ment is required to obtain the detailed behavior of σ_1^c in the superconducting state.

CONCLUSION

We succeeded in obtaining the results of σ_1 and λ of BSCCO both in the ab-plane and along the c-axis, except the detailed behavior of σ_1^c in the superconducting state. The temperature evolution of $\lambda^2(0)/\lambda^2(T)$ was found to be quite anisotropic below T_c. The change of $\Delta\lambda_c$ was $\sim T^2$ at low temperatures, which could be explained by impurity assited incoherent hoppoing [6] or Cooper pair tunneling model [7]. To discuss more detailed properties of QPs moving along the c-axis, more careful measurement on σ_1^c is required.

1. H. Kitano, T. Hanaguri, and A. Maeda, Phys. Rev. B **57**, 10946 (1998).
2. A. Hosseini et al., Phys. Rev. Lett. **81**, 1298 (1998).
3. A. Maeda, T. Hanaguri, and H. Kitano, in Advances in Superconductivity XI, edited by N. Koshizuka and S. Tajima (Springer-Verlag, Tokyo, 1999), pp.193-198.
4. H. Kitano et al., J. Low Temp. Phys. in press.
5. T. Xiang and J. M. Wheatley, Phys. Rev. Lett. **77**, 4632 (1996).
6. R. J. Radtke, V. N Kostur, and K. Levin, Phys. Rev. B **53**, R522 (1996).
7. T. Xiang and J. M. Wheatley, Phys. Rev. Lett. **76**, 134 (1996).
8. T. Jacobs et al., Phys. Rev. Lett. **75**, 4516 (1995); T. Shibauchi et al., Physica C **264**, 227 (1996).
9. S-. F. Lee et al., Phys. Rev. Lett. **77**, 735 (1996); J. Corson et al., Physica B in press.
10. S. Chakravarty et al., Phys. Rev. Lett. **82**, 2366 (1999).
11. R. E. Harris, Phys. Rev. B **10**, 84 (1974); ibid. **11**, 3329 (1975).
12. M. B. Gaifullin et al., Phys. Rev. Lett. to be published.

Interlayer Tunneling Spectroscopy for $Bi_2Sr_2CaCu_2O_{8+\delta}$ Using Intrinsic Junctions

Minoru Suzuki[1], Takao Watanabe[2], and Azusa Matsuda[2]

[1]Department of Electronic Science and Engineering, Kyoto University, Kyoto 606-8501, Japan
[2]NTT Basic Research Laboratories, 3-1 Morinosato, Wakamiya, Atsugi, Kanagawa 243-0198, Japan

Abstract: By means of the interlayer tunneling spectroscopy, which employs very thin mesas comprised of approximately 10 series-connected intrinsic tunnel Josephson junctions of $Bi_2Sr_2CaCu_2O_{8+\delta}$, we have obtained the following important findings. The normal tunneling resistance exhibits linear temperature dependence above T_c, and shows a sharp drop below T_c, implying that the major in-plane scattering of quasiparticles is due to electron-electron interaction. The superconducting gap is clearly discriminated from the pseudogap, which is at variance with a previously accepted picture for the relationship between the superconducting gap and the pseudogap.

Keywords: interlayer tunneling spectroscopy, normal tunneling resistance, superconducting gap, pseudogap

INTRODUCTION

Tunneling spectroscopy is one of the most effective probes into the superconducting properties and the electronic states of superconductors. This method usually employs tunnel junctions, of which the superconductor-insulator-superconductor (SIS) junctions are preferable to the superconductor-insulator-normal metal (SIN) junctions from the energy-resolution point of view. However, the SIS tunnel junctions of high-T_c superconductors (HTS) have been proved to be very difficult to fabricate even based on the present state of the art technology. Therefore, many of the tunneling spectroscopy measurements have employed scanning tunneling microscopes (STM) or point contact junctions, all of which are based on SIN junctions.

Although artificial HTS tunnel junctions are still unrealized despite the enormous effort made thus far, the SIS structure itself can be found in a layered crystal structure of HTS. In particular, the $Bi_2Sr_2CaCu_2O_{8+\delta}$ system represents a typical crystal structure of this type. Kleiner el al. [1, 2] found that a piece of $Bi_2Sr_2CaCu_2O_{8+\delta}$ crystal functions just as a stack of SIS tunnel Josephson junctions. These Josephson junctions, naturally built in a crystal structure of layered superconductors, are called intrinsic Josephson junctions (IJJs). The interlayer tunneling spectroscopy (ITS) makes use of a stack of these IJJs [3]. Since the interface of the IJJ is ideally sharp and flat on an atomic scale, the data obtained by ITS technique are most likely to provide essentially physical properties of the material employed with least influence from the interface irregularity or gradation.

There are, however, merits and demerits in ITS. The major merit comes from the use of SIS junctions, from which high energy resolution and clear spectra are basically attained. Another merit is that the c-axis resistivity of the very portion probed by ITS can also be measured almost concomitantly. This is very important in the case of materials whose surfaces are quite subject to change in composition, such as oxygen content in the case of HTS in particular. To be strict, there is no assurance in scanning tunneling spectroscopy (STS) or angle-resolved photoemission spectroscopy (ARPES) experiments that the surface composition is the same

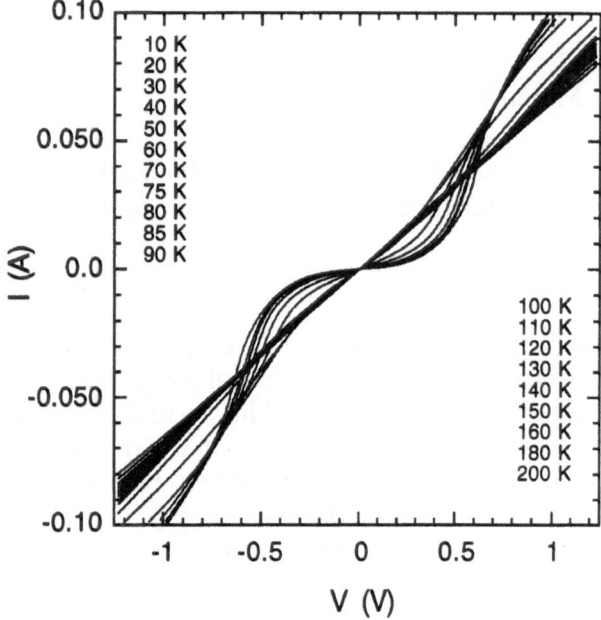

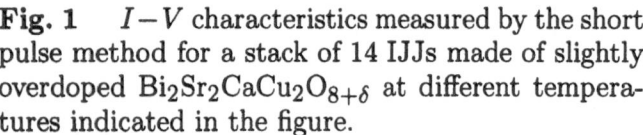

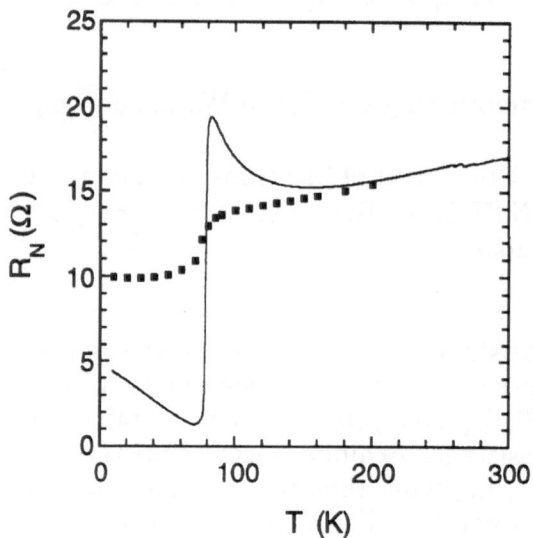

Fig. 1 $I-V$ characteristics measured by the short pulse method for a stack of 14 IJJs made of slightly overdoped $Bi_2Sr_2CaCu_2O_{8+\delta}$ at different temperatures indicated in the figure.

Fig. 2 T-dependence of the stack resistance R_c (solid line) and the normal state tunneling resistance R_N (filled squares) for the same sample in Fig. 1. T_c is 77.2 K. The dc current for the measurement of R_c is 100 μA. The semiconductive behavior of R_c below T_c is due to the voltage state of one junction whose critical current is small and suppressed by the digital voltmeter.

as what is expected. They can be different for a number of reasons. Namely, these techniques probe only crystal surfaces, which are prone to change, instead of bulk of material. In the case of ITS, however, the concomitant measurement of the the c-axis resistivity ρ_c and its temperature dependence reinforces the composition of the portion to be probed. Thus ITS is solider than other surface spectroscopy techniques with regard to the doping dependence.

The major demerits of ITS are twofold. Since a stack of ITSs contains a number of junctions connected in series, current injection causes significant self-heating. Because the temperature rise due to the self-heating often causes quantitative ambiguity in the physical properties and their interpretation, it should be reduced to a very small and almost negligible proportion. The second demerit comes from the necessity to apply a high voltage on a specimen, which frequently causes a puncture of specimens. We adopted the following two measures to eliminate these demerits.

First, we have fabricated very small and thin mesas on top of the $Bi_2Sr_2CaCu_2O_{8+\delta}$ crystal surface [4]. This sort of miniaturization was significantly effective in reducing the self-heating effect [5]. However, this measure alone is not sufficient to eliminate the heating effect completely. Therefore, we further adopted short pulse method in the measurement. The combination of these two measures led to measurements of tunneling characteristics with an influence of the heating effect less than 3 % on the voltage scale in the case of 20 μm square specimens [6]. With these measures, the ITS technique has been applied to the $Bi_2Sr_2CaCu_2O_{8+\delta}$ system to provide an important result concerning the superconducting gap and the pseudogap in this system.

SAMPLE PREPARATION AND MEASUREMENTS

Specimens used for ITS consist of a thin mesa with a lateral dimension of $10\mu m$ and a thickness of approximately 15 nm fabricated on the surface of a $Bi_2Sr_2CaCu_2O_{8+\delta}$ single crystal, which was grown by the traveling-solvent floating-zone method [7]. In fabrication, the mesa thickness was controlled by the Ar ion-milling time. Before the fine-patterning process, the Au/Ag electrode was evaporated on the cleaved surface of a $Bi_2Sr_2CaCu_2O_{8+\delta}$ crystal and annealed at 430 to 450 °C in oxygen atmosphere or in vacuum depending on the aimed carrier doping level for the specimen. The sides of the mesa were filled with an SiO film, which also forms an insulating layer at the same time for the upper electrode wiring. A schematic cross section of a specimen is described in previous publications [4, 6].

The short pulse method in the present ITS measurements is basically the same as the dc four-probe method except for the following pulse-related portion [6]. Current pulses with a width of 1 μs were supplied from an arbitrary waveform generator. The voltage responses from the specimen were detected with a four-channel digital oscilloscope. The voltage output was obtained by averaging 50 voltage pulse responses at a time of 500 ns from the triggered point. Even in this situation of the short pulse method, the measurement is still not completely free from the self-heating. When the current height exceeds 40 mA for a specimen whose junction lateral size is 20 μm, the influence of the heating becomes discernible. The voltage decay due to the heating effect becomes largest when the current height is approximately 80 mA, where the influence on the voltage is about 3%. Thus the influence of the self-heating effect is approximately 3% or less on the voltage scale in the 20 μm junction case. In the present case, the junction lateral size is 10 μm and the influence of the heating effect is much less.

NORMAL TUNNELING RESISTANCE

Figure 1 shows a set of current-voltage $(I - V)$ characteristics at various temperatures T from 10 to 200 K for a 13-IJJ 20 μm square mesa made of a slightly overdoped Bi_2Sr_2Ca-$Cu_2O_{8+\delta}$ crystal observed. It is clearly seen that each $I - V$ curve evolves a linear portion above the superconducting gap voltage V_g, i.e., the normal state tunneling resistance part. In usual tunnel junctions made of conventional superconductors, the normal state tunneling resistance R_N represents a tunneling probability of the junction barrier and has little relation to the physical properties of the superconductor. In the present case, however, R_N or the stack resistance R_c represents the resistivity in the c-axis, i.e., ρ_c. Therefore, it is possible to measure ρ_c simultaneously by this ITS method as a function of bias voltage. In this sense, it is particularly important to observe that R_N exhibits a significant T-dependence. Since this behavior is not basically seen in conventional superconductors, it is regarded as peculiar to high-T_c superconductors.

Figure 2 shows the T-dependence of R_N (filled squares) determined from the $I - V$ curves in Fig. 1. There are two significant features in this T-dependent behavior. The first feature is that R_N exhibits linear T-dependence above T_c and coincides with the linear part of R_c (solid line). When the scattering time associated with the c-axis tunneling is much smaller than that within the CuO_2 planes, the tunneling is mostly incoherent, while it is coherent when the tunneling interval is greater than the scattering time within the CuO_2 planes. In the latter case, the T-dependence of in-plane transport scattering is reflected on the T-dependence of ρ_c and R_N[8]. Therefore, the linear T-dependence observed in R_N implies that the transport along the c-axis is partly conveyed via the coherent tunneling. This T-linear dependence of ρ_c is observed only when the density of states (DOS) at the Fermi level is almost T-independent. When the $Bi_2Sr_2CaCu_2O_{8+\delta}$ system is under doped, the strongly T-dependent change in DOS, i. e., the evolution of a pseudogap, masks the T-dependence of the scattering time. This leads to a semiconductive T-dependence for R_c below a temperature at which the pseudogap starts

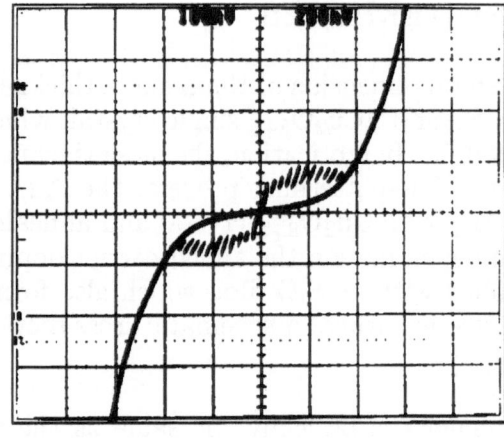

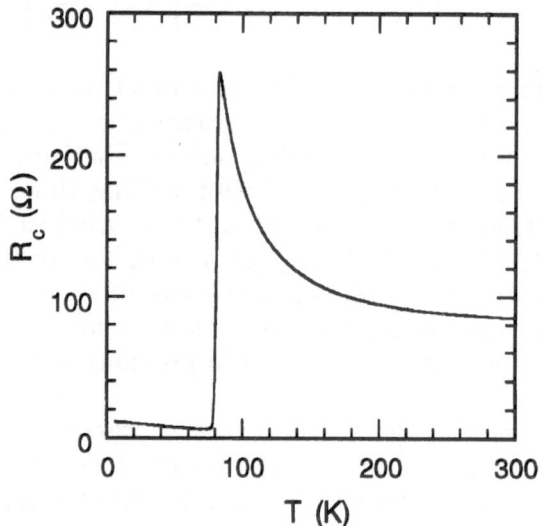

Fig. 3 An oscilloscope image of current-voltage characteristics for a stack of 12 intrinsic junctions of slightly under doped $Bi_2Sr_2CaCu_2O_{8+\delta}$ at 7.1 K. Y-axis: 1mA/div. X-axis: 200 mV/div.

Fig. 4 Temperature dependence of a stack resistance R_c for the same sample in Fig. 3. The stack contains 12 intrinsic junctions. The significantly semiconductive R_c behavior reinforces its under doped composition.

to evolve. The semiconductive T-dependence is due to the nonlinear $I - V$ curve associated with the evolving pseudogap. Since R_N is measured at a bias voltage higher than both the pseudogap edge and the superconducting gap edge, it gives a resistance free from the DOS effect and reflects the in-plane transport. The difference of R_c and R_N in Fig. 2 reflects this situation.

Another feature observed in Fig. 2 is the sharp drop in R_N at T_c. This sudden decrease in R_N implies that the number of scattering bosons decreases below T_c. Since this decrease is dominantly correlated with the decrease in the electron density due to the condensation of the superconducting electron pairs, the scattering bosons are most likely of electronic origin. This means that the major scattering in the CuO_2 planes is caused by the electron-electron interaction. More importantly, this consequence gives rise to a further implication that the occurrence of the high-T_c superconductivity is caused by a non-phononic origin.

PSEUDOGAP OF UNDERDOPED SAMPLES

Figure 3 shows an oscilloscope image of $I - V$ characteristics for a 12-junction 10 μm square IJJ stack of slightly under doped $Bi_2Sr_2CaCu_2O_{8+\delta}$. In this image, 12 resistive branches are clearly seen, indicating that the stack is composed of 12 IJJs. Therefore, the stack thickness is exactly determined from this observation. Figure 4 shows the T-dependence of the stack resistance R_c, which is proportional to ρ_c in the usual transport experiments. The semiconductive T-dependence reinforces that the specimen is in the under doped region. The ρ_c curve also gives a value of T_c=79 K for this specimen as determined resistively. This value is consistent with the fact that the specimen is slightly under doped.

Figure 5 shows the T-dependence of the $I - V$ characteristics at 10 K observed for the same specimen in Fig. 4. At a first glance, the $I - V$ curve appears to have neither gap structure nor the normal state tunneling resistance region. However, a structure related to the superconducting gap shows up in $dI/dV - V$ curves. The normal tunneling part in missing in the

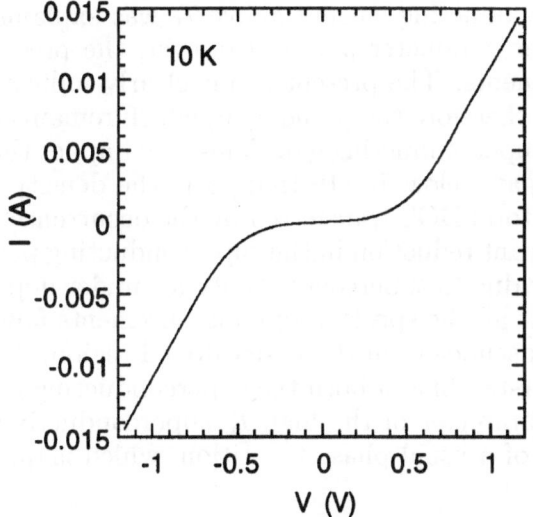

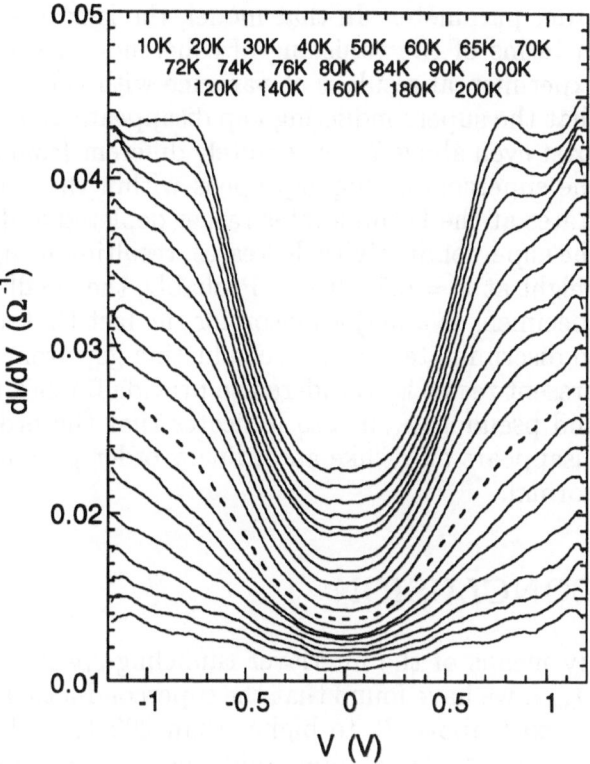

Fig. 5 $I-V$ characteristics for a stack of 12 intrinsic Josephson junctions of slightly under doped $Bi_2Sr_2CaCu_2O_{8+\delta}$ measured at 10 K by the short pulse method.

Fig. 6 A set of dI/dV curves for a stack of 12 intrinsic Josephson junctions of slightly underdoped $Bi_2Sr_2CaCu_2O_{8+\delta}$ at various temperatures indicated inside. Each curve is vertically shifted for an appropriate value. The dashed curve was observed at 80 K, very close to T_c.

$I-V$ curve in Fig. 5. This is due to the existence of the pseudogap, which is expected to lie in much higher voltage range than in Fig. 5. The observation of the normal tunneling part and the determination of R_N need the application of much higher voltage, which is difficult at present due to the resultant puncture destruction of specimens.

Figure 6 shows a set of $dI/dV - V$ curves obtained numerically from the $I - V$ data, a part of which is shown in Fig. 5. In Fig. 6, the dashed line indicates the curve observed at 80 K, which is very close to T_c. The small peak at 0.75 V decreases in its height when T increases, changes to a cusp, and then disappears at T_c. Furthermore, the peak position seems to shift to lower energies as T increases, although the data is not sufficient to conclude definitely. These behaviors strongly imply that the small peak at 0.8 to 0.9 V is due to the superconducting gap. In that case, the peak position gives a superconducting energy gap of $2\Delta = 72$ meV for a single junction. This value is larger than that of slightly overdoped $Bi_2Sr_2CaCu_2O_{8+\delta}$, as reflected by the under doped level of this specimen. It is also seen that there is a significant gap structure in the $I - V$ curves even above T_c. This gap structure is due to the pseudogap whose peak is expected to lie outside the voltage range in this experiment. (We limited the maximum applied voltage to 1.2 V in this case in order not to destroy samples by a puncture, which frequently took place.) This structure was observed in a number of spectroscopy techniques, including STS [9, 10] and ARPES [11]. In those experimental results, it is argued that the superconducting gap continues to exist even above T_c as a pseudogap. In its interpretation, the superconducting gap and the pseudogap are of the same origin and represent the same order parameter. In that model, the lack of superconducting current above T_c was explained in terms of the vanishing of coherency in the order parameter phase. However, the present experimental result is at variance with these arguments. The present result clearly indicates that the superconducting gap disappears at T_c and therefore the pseudogap, which remains to exist even above T_c, is definitely different from the superconducting gap. This also implies that the superconducting gap opens within the pseudogap below T_c. In that case, the density of states at the Fermi level is rather depleted and the total DOS transferred by the occurrence of the superconductivity decreases, resulting in significant reduction in the superconducting peak height at $V = 0.8 - 0.9$ V. Probably, the small peak due to superconductivity for under doped specimens is a major reason for the fact that almost all the spectroscopy measurements failed to discriminate the superconducting gap from the pseudogap in the under doped region. The present result is considered to provide further understanding of both the superconducting gap and pseudogap. it also indicates that the order parameter of the high-T_c superconductivity disappears at T_c like an ordinary order parameter of a usual phase transition, which is quite normal.

CONCLUSION

By means of the interlayer tunneling spectroscopy for a slightly under doped $Bi_2Sr_2CaCu_2O_{8+\delta}$, we have found that the superconducting gap disappears at T_c but the pseudogap remains to exist above T_c to higher than 200 K. It is found that the superconducting gap and the pseudogap are definitely different. It is also found that the normal state tunneling resistance exhibits a significant temperature dependence. The characteristic linear T-dependence of R_N above T_c implies that the coherent tunneling takes place along the c-axis. The sudden decrease in R_N at T_c implies that the major scattering in the CuO_2 planes is due to electron-electron interaction.

ACKNOWLEDGMENTS

The authors would like to express their gratitude to Profs. M Tachiki, Y. Matsuda and Yu. I. Latyshev for illuminating discussions.

1. R. Kleiner, F. Steinmeyer, G. Kunkel, and P. Müller, Phys. Rev. Lett. **68**, 2394 (1992).
2. R. Kleiner and P. Müller, Phys. Rev. B **49**, 1327 (1994)
3. M. Suzuki, T. Watanabe, and A. Matsuda, Phys. Rev. Lett. **82**, 5361 (1999).
4. K. Tanabe, S. Karimoto, Y. Hidaka, and M. Suzuki, Phys. Rev. B **53**, 9348 (1996).
5. Yu. I. Latyshev, T. Yamashita, L. N. Bulaevskii, M. J. Graf, A. V. Balatsky, and M. P. Maley, Phys. Rev. Lett. **82**, 5345 (1999).
6. M. Suzuki, T. Watanabe, and A. Matsuda, IEEE Trans. Appl. Supercond. **9**, 4507 (1999).
7. T. Watanabe, T. Fujii, and A. Matsuda, Phys. Rev. Lett. **79**, 2113 (1997).
8. N. Kumar and A. M. Jayannavar, Phys. Rev. B **45**, 5001 (1992).
9. Ch. Renner, B. Revaz, J. -Y. Genoud, K. Kadowaki, and Ø. Fischer, Phys. Rev. Lett. **80**, 149 (1998).
10. A. Matsuda, S. Sugita, and T. Watanabe, Phys. Rev. B **60**, 1377 (1999).
11. H. Ding, A. F. Bellman, J. C. Campuzano, M. Randeria, M. R. Norman, T. Yokoya, T. Takahashi, H. Katayama-Yoshida, T. Mochiku, K. Kadowaki, G. Jennings, and G. P. Brivio, Phys. Rev. Lett. **76**, 1533 (1996).

Charge Transport along the c-Axis in High-T_c Cuprates[*]

Yoichi Ando

Central Research Institute of Electric Power Industry, Komae, Tokyo 201-8511, Japan

Abstract: Using 61-T pulsed magnetic fields, the normal-state ρ_{ab} and ρ_c are measured in Bi-2201 system down to 0.66 K, and the coexistence of the "metallic" ρ_{ab} and the "semiconducting" ρ_c, usually called the charge confinement behavior, was confirmed to extend far below T_c. Recent measurement of the c-axis magnetoresistance under 16 T dc magnetic field in heavily underdoped Y-123 crystals revealed that the peculiar c-axis charge transport, and thus the charge confinement, is fundamentally related to the antiferromagnetic spin fluctuations.

Keywords: anisotropy, resistivity, transport properties, normal state, magnetoresistance

INTRODUCTION

One of the most unusual normal-state properties of the high-T_c cuprates is the temperature dependence of the anisotropic resistivity: in samples near optimum doping, the in-plane resistivity ρ_{ab} decreases linearly with decreasing T over a wide temperature range, while the out-of-plane resistivity ρ_c increases rapidly at low temperatures. To elucidate whether this contrasting behavior extends far below T_c and thus is truly a "ground state" property of the normal state in the absence of superconductivity, we measured the normal-state ρ_{ab} and ρ_c in La-doped $Bi_2Sr_2CuO_y$ (Bi-2201) crystals by suppressing superconductivity with 61-T pulsed magnetic fields [1]. In sufficiently clean samples, we confirmed that the metallic ρ_{ab} coexists with a "semiconducting" ρ_c down to the lowest experimental temperature, 0.66 K, giving evidence of the non-Fermi-liquid nature of the cuprates. This result also evidences the existence of a charge confinement mechanism that persists to low temperatures. To further investigate the charge confinement, we have recently measured the c-axis magnetoresistance in heavily underdoped $YBa_2Cu_3O_{6+x}$ (Y-123) crystals that show Néel order. It was found that the peculiar c-axis transport is strongly related to the antiferromagnetism, because the c-axis magnetoresistance shows a step-like increase across the Néel temperature [2]. Such results strongly suggest that the charge transport along the c-axis, which takes place as a hopping process across the charge-confining CuO_2 planes, is fundamentally related to the antiferromagnetic spin fluctuations.

CHARGE CONFINEMENT IN THE ZERO TEMPERATURE LIMIT

The contrasting behavior of ρ_{ab} and ρ_c has been the subject of intense study, both theoretically and experimentally [3], and is often counted as an evidence for the non-Fermi-liquid nature of the cuprates [4]. However, there have been discussions that the contrasting behavior might be a finite temperature effect in a Fermi liquid, where, for example, c-axis transport is due to phonon-assisted hopping [5]. Clearly, such question must be settled by measurements of the normal-state resistivity at low temperatures.

The most straightforward way to measure the normal-state resistivity below T_c is to suppress superconductivity with an intense magnetic field. We suppressed superconductivity in La-doped

Bi-2201 single crystals using a 61 T pulsed magnetic field and measured both ρ_{ab} and ρ_c in the normal-state down to 0.66 K. The Bi-2201 system we used is a particularly suitable compound for studying the contrasting behavior: T_c is relatively low among the high-T_c cuprates; ρ_c shows a strong divergence [6]; and the linear-T behavior in ρ_{ab} is quite robust (extends up to 700 K [6]). The $Bi_2Sr_{2-x}La_xCuO_y$ single crystals we report here are slightly overdoped, with nominal $x=0.05$ and the midpoint T_c of about 13 K. The crystals were grown by the flux method and the cleanest ones show the in-plane resistivity ρ_{ab} of about 130 $\mu\Omega$cm at 100 K. To measure the anisotropic resistivity, we employ a six-terminal method: two current contacts are located on the top ab face of the crystal with a pair of voltage contacts placed in between; an additional pair of voltage contacts is placed on the bottom face directly beneath the top-face voltage pair. Analyzing the top- and bottom-face voltages using the linear anisotropic resistivity model of Busch et $al.$ [7] gives ρ_{ab} and ρ_c.

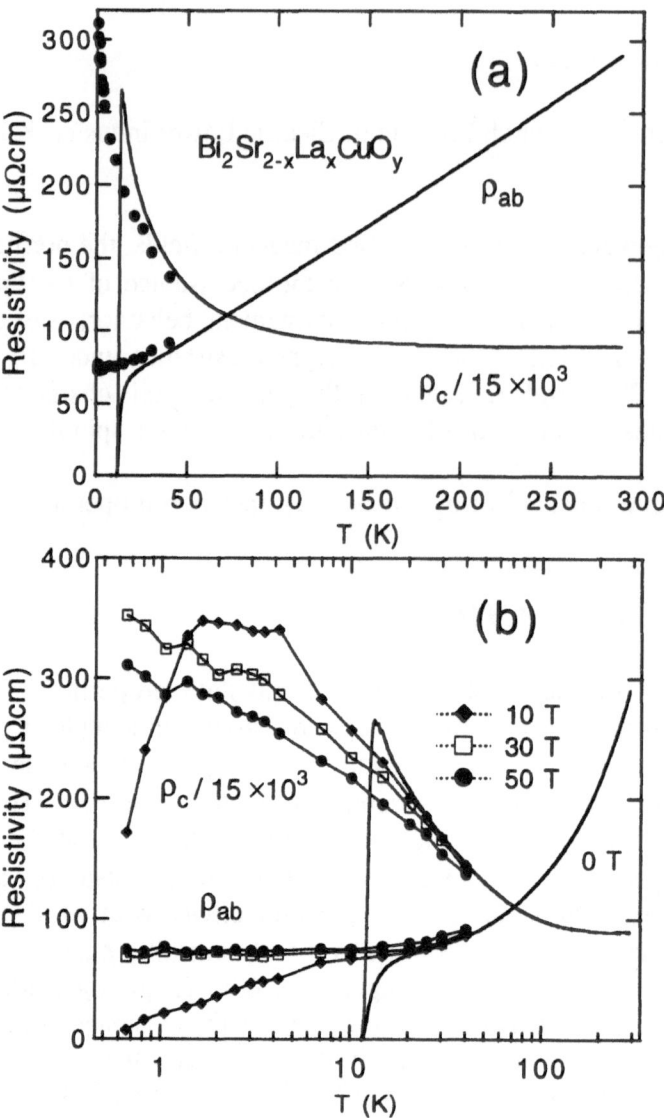

Fig. 1. (a) T dependence of ρ_{ab} and ρ_c of one of the cleanest Bi-2201 samples in 0 and 60 T. (b) log$T$ plot of ρ_{ab} and ρ_c for fixed values of magnetic field, emphasizing the metallic ρ_{ab} and diverging ρ_c in the zero temperature limit. ρ_c data are divided by 15×10^3.

The result of the high magnetic field measurement of one of the cleanest samples is shown in Fig. 1(a). In this sample, ρ_{ab} stays metallic down to the lowest experimental temperature, 0.66 K, and ρ_c continues to diverge. Therefore, the contrasting behavior of ρ_{ab} and ρ_c persists down to $T/T_c = 0.05$. This strongly suggests that the metallic in-plane conduction and "semiconducting" out-of-plane conduction can indeed coexist in the zero-temperature limit when the in-plane disorder is sufficiently small [1]. Note that ρ_{ab} in this Bi-2201 sample becomes as small as 74 $\mu\Omega$cm, which corresponds to $k_Fl \approx 42$ in the 2D model ($k_Fl =hc_0/\rho_{ab}e^2$, where $c_0 = 12$ Å is the interlayer distance). The data suggest that ρ_{ab} is saturating to a residual resistivity, although we cannot exclude the possibility that ρ_{ab} will cross over to an insulating behavior below our experimental temperature range. Ordinarily, however, such a large value of k_Fl would assure metallic behavior to extremely low temperatures. Figure 1(b) shows a log$T$ plot of ρ_{ab} and ρ_c for various fixed magnetic fields in order to emphasize the low-temperature behavior. The large negative MR in ρ_c is evident and ρ_c diverges roughly logarithmically at low temperatures, comparable to the behavior reported in $La_{2-x}Sr_xCuO_4$ [8]. On the other hand, ρ_{ab} below T_c shows almost no temperature dependence and little magnetoresistance in this clean sample. This suggests that the c-axis transport is uncorrelated with the in-plane transport. This behavior of Bi-2201 contrasts strongly with that of $La_{2-x}Sr_xCuO_4$, where ρ_{ab} becomes insulating whenever ρ_c is diverging at low temperatures [8,9].

Let us discuss the implication of the above result. A clear indication is that the c-axis transport is incoherent even at very low temperatures. Within the framework of the Fermi-liquid theory, several models have been proposed [3] to explain the incoherence, including renormalization of the interlayer hopping rate [10], dynamical dephasing [11], and interlayer scattering [5,12]. Accounting for a "semiconducting" ρ_c in these Fermi-liquid models in addition to the incoherence is more difficult and two possibilities have been discussed: phonon-assisted hopping [5,10] and temperature-dependent suppression of the density of states at the Fermi energy, $N(0)$ [13]. Rojo and Levin have found that phonon-assisted hopping can give a "semiconducting" ρ_c over a broad temperature range, as long as the phonon energies are sufficiently large in helping electrons to hop across the planes. In this mechanism, however, ρ_c must eventually become proportional to ρ_{ab} as the characteristic phonon energy becomes small at low temperatures, roughly $T \leq 20$ K (an energy scale which is approximately 1/40 of the highest energy c-axis phonons [5]). In our data, ρ_c continues to diverge at our lowest experimental temperature, 0.66 K, which clearly speaks against the phonon-assisted hopping model. The data pose an additional difficulty for Fermi-liquid models: the fact that ρ_{ab} shows almost no temperature dependence and little magnetoresistance at low temperatures, while ρ_c continues to show strong temperature and magnetic-field dependences. In the interlayer scattering models [5,12], the temperature dependence of ρ_c is determined by the scattering time from the interplane (off-diagonal) disorder, τ_1, and also by the scattering time from the in-plane (diagonal) impurities, τ_2. Since the same scattering times enter into the expression of ρ_{ab} in these interlayer scattering models, it is difficult to account for the temperature-independent ρ_{ab}. Alternatively, Zha, Cooper and Pines have proposed [13] that the low-temperature upturn in ρ_c is due to a reduction in $N(0)$. Although there is no microscopic calculation of ρ_{ab} in this model, one would expect ρ_{ab} to be temperature dependent when $N(0)$ changes with temperature.

On the other hand, in non-Fermi-liquid theories of the high-T_c cuprates, the incoherence of the c-axis conduction results from the in-plane quasiparticle confinement [4,14]. Both the resonating valence bond theory [15,16] and the Luttinger liquid theory [17] give metallic ρ_{ab} accompanied by "semiconducting" ρ_c in the zero-temperature limit. To summarize, the observation of coexistence of metallic ρ_{ab} and "semiconducting" ρ_c down to the lowest experimental temperature, 0.66 K, points toward a non-Fermi-liquid ground state in the high-T_c cuprates.

c-AXIS CHARGE TRANSPORT AND THE SPIN FLUCTUATIONS

A fundamental role of the magnetic interactions in the c-axis transport was recently found in a study of heavily underdoped $RBa_2Cu_3O_{6+x}$ (R=Tm, Lu) in the vicinity of the Néel temperature T_N [18]; it was found that the c-axis transport occurs through two conduction channels, one of which is essentially blocked by the AF ordering. As a result of such blocking, a steep increase in both ρ_c and the anisotropy ratio ρ_c/ρ_{ab} was observed upon cooling below T_N. In a study of the ab-plane and c-axis magnetoresistance (MR) of a series of heavily underdoped $YBa_2Cu_3O_{6+x}$ (Y-123) crystals, we found that the c-axis MR undergoes a drastic change in the vicinity of the Néel temperature [2]. Also, quite unexpectedly, no feature associated with the AF ordering was found in the transverse ab-plane MR [$I \parallel a(b)$; $H \parallel c$]. These new data indicate that the development of the AF correlations and the formation of the long-range Néel order have a profound influence only on the charge transport across the CuO_2 planes, leaving the in-plane transport unchanged. Moreover, the data show that the longitudinal c-axis MR (H // c) is apparently governed by the AF fluctuations even in the temperature range *above* T_N, indicating that the spin fluctuations are playing a major role in the c-axis transport regardless of the presence of the Néel order.

The MR measurements are performed either by sweeping temperature (controlled by a Cernox resistance sensor) under constant magnetic fields up to 16 T, or by sweeping the field at a fixed temperature stabilized by a capacitance sensor to an accuracy of about 1 mK. The latter method allows measurements of $\Delta\rho/\rho$ as small as 10^{-5} at 10 T. In Fig. 2 we present a set of ρ_c curves obtained for the same Y-123 single crystal at slightly different oxygen contents in the AF region. The rise in ρ_c induced by the AF transition becomes more and more evident as T_N is lowered. Note that both the zero-field and the 16 T data are shown in the main panel of Fig. 2, where the difference between the two becomes noticeable below T_N. The inset of Fig. 2 demonstrates an unusual behavior of the c-axis MR, in which a step-like increase in $\Delta\rho_c/\rho_c$ is observed upon cooling through T_N. Except for a small difference in the MR step width, which is obviously related to the width of the AF transition, this striking feature is very

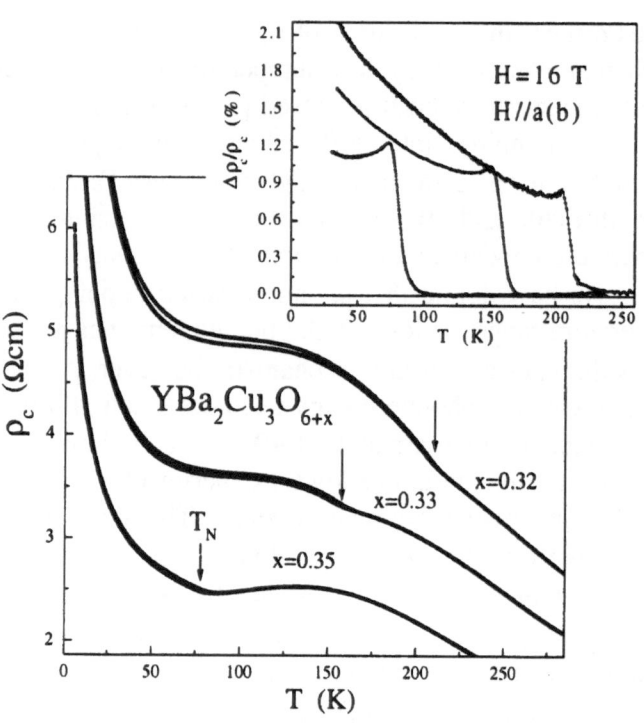

Fig. 2. ρ_c (T) of an Y-123 single crystal at three different oxygen contents. Curves for H=0 and 16 T [H // $a(b)$] are shown. The kink on the resistivity curves marks the AF transition. Inset: T dependences of the c-axis transverse MR [H // $a(b)$] at 16 T .

reproducible within a set of Y-123 crystals. One would expect that the Néel transition in Y-123, like other phase transitions associated with the magnetic subsystem, is considerably affected by the application of magnetic fields. If a strong magnetic field suppresses AF order and lowers T_N, the c-axis MR should become negative, because ρ_c is enhanced below T_N; however, we have found that the c-axis MR is *positive*, which is opposite to such a naive expectation. Also surprisingly, when we look at the in-plane transport, we do not find any anomaly that can be associated with the Néel transition, and the ab-plane MR, $\Delta\rho_{ab}/\rho_{ab}$, is always smooth in the vicinity of T_N.

One would naturally ask whether the out-of-plane transport is sensitive exclusively to the long-range order arising below T_N; if the short-range AF correlations above T_N also contribute to ρ_c and its MR, we may expect that the AF fluctuations play an essential role not only in the AF compositions but also in the superconducting compositions. To clarify this point, we obtained more precise MR data with the field-sweeping measurements to investigate the behavior above T_N, where the MR becomes very small. The T dependences of $\gamma_\perp H^2$ and $\gamma_{//} H^2$ components of $\Delta\rho_c/\rho_c$ (for $H \perp c$ and H // c, respectively) presented in Fig. 3 depict the qualitative difference in the MR behavior for the two directions of the magnetic field. For H // ab [Fig. 3(a)], the c-axis MR changes at T_N in a step-like manner by up to two orders of magnitude. The step separates regions below and above T_N with relatively weak dependence of the MR on temperature. This behavior implies that the sensitivity to the magnetic field appears abruptly with the onset of the long-range AF order. On the other hand, for $H /\!/ c$ [Fig. 3(b)] we observe a MR peak at T_N, which is accompanied by a tail spreading to far above T_N. The MR as a function of temperature has no discontinuity at T_N and one can infer from Fig. 3(b) that the c-axis MR grows as T^{-k} with decreasing temperature until the Néel transition interrupts this tendency. However, the right-hand side of the MR peak for $H /\!/ c$ (the T^{-k} behavior) apparently shifts with T_N when the x is changed, which indicates its relation to the AF ordering. Therefore, we can conclude that a mechanism associated obviously with the AF fluctuations dominates the c-axis MR in a wide temperature range above T_N as well. This observation clearly demonstrates that the short-range AF correlations play an essential role in the out-of-plane transport.

The contrasting behavior of the ab- and c-axis MR in the heavily underdoped Y-123 indicates that changes that occur in the spin subsystem at T_N are influential only on the electron transport

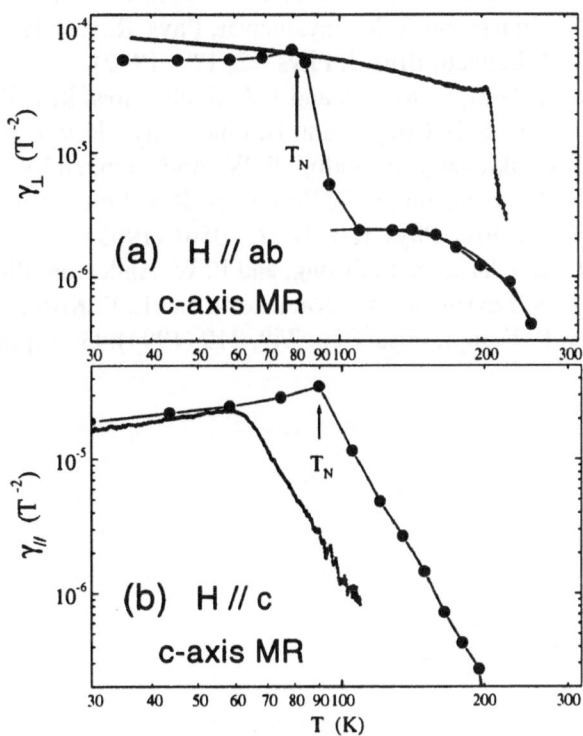

Fig. 3. T dependences of the $\gamma_\perp H^2$ term in the c-axis MR with (a) $H /\!/ ab$ and (b) $H /\!/ c$. A Lu-123 crystal was used for the field sweep measurements (solid circles). The MR data for Y-123 under sweeping temperature are shown for comparison (dots).

across the CuO_2 planes and apparently not on the in-plane one. It is known that the heavily underdoped Y-123 above T_N possesses well-developed dynamic AF correlations in the CuO_2 planes [19] and the Néel temperature actually corresponds to the establishment of the AF order along the c-axis. It is possible that the freezing of the spin degrees of freedom below T_N causes an increase in ρ_c, if the spin fluctuations assist the electron hopping between the CuO_2 planes. Since an increase in ρ_c also takes place when the magnetic field is applied, one can infer that the field suppression of the spin fluctuations is likely to be the main source of the positive c-axis MR in our heavily underdoped Y-123. Also, the dramatic changes in the out-of-plane transport associated with the evolution of the magnetic state might suggest that it is the spin subsystem that is responsible for the charge confinement within the CuO_2 planes.

*Work done in collaboration with A. N. Lavrov, Kouji Segawa, J. Takeya, G. S. Boebinger, A. Passner, N. L. Wang, C. Geibel, and F. Steglich.
1. Y. Ando et al., Phys. Rev. Lett. 77, 2065 (1996); ibid. 79, 2595(E) (1997).
2. A. N. Lavrov, Y. Ando, K. Segawa, and J. Takeya, Phys. Rev. Lett. 83, 1419 (1999).
3. For a review, see S. L. Cooper and K. E. Gray, in Physical Properties of High Temperature Superconductors IV, ed. by D. M. Ginsberg (World Scientific, Singapore, 1994).
4. P. W. Anderson, Science 256, 1526 (1992).
5. A. G. Rojo and K. Levin, Phys. Rev. B 48, 16861 (1993).
6. S. Martin et al., Phys. Rev. B 41, 846 (1990).
7. R. Busch et al., Phys. Rev. Lett. 69, 522 (1992).

8.   Y. Ando *et al.*, Phys. Rev. Lett. **75**, 4662 (1995).

9.   G. S. Boebinger *et al.*, Phys. Rev. Lett. **77**, 5417 (1996).

10. N. Kumar and A. M. Jayannavar, Phys. Rev. B **45**, 5001 (1992).

11. A. J. Leggett, Braz. J. Phys. **22**, 129 (1992).

12. M. J. Graf, D. Rainer, and J. A. Sauls, Phys. Rev. B **47**, 12089 (1993).

13. Y. Zha, S. L. Cooper, and D. Pines, Phys. Rev. B **53**, 8353 (1996).

14. S. Chakravarty, A. Sudbø, P. W. Anderson, and S. Strong, Science **261**, 337 (1993).

15. P. W. Anderson and Z. Zou, Phys. Rev. Lett. **60**, 132 (1988).

16. N. Nagaosa, Phys. Rev. B **52**, 10561 (1995).

17. D. G. Clarke, S. P. Strong, and P. W. Anderson, Phys. Rev. Lett. **74**, 4499 (1995).

18. A. N. Lavrov, M. Yu. Kameneva, and L. P. Kozeeva, Phys. Rev. Lett. **81**, 5636 (1998).

19. A. P. Kampf, Phys. Rep. **249**, 219 (1994); G. Aeppli *et al.*, Science **278**, 1432 (1997).

C-Axis Infrared Spectra of Bilayer High-T_c Cuprates: The Effect of Intrabilayer Josephson Coupling

V. Zelezny[1*], S. Tajima[1], T. Motohashi[2], J.Shimoyama[2], K. Kishio[2] and D. van der Marel[3]

[1]Superconductivity Research Laboratory/ISTEC, Tokyo 135-0082, Japan
[2]Dept. of Applied Chem., The University of Tokyo, Tokyo 113, Japan
[3]Lab. of Sol. St. Phys., University of Groningen, Tne Netherland

Abstract: The c-axis optical spectra of single crystals of $YBa_2(Cu_{1-x}R_x)_3O_{6.6}$, R=Ni and Zn, and $Bi_2Sr_2CaCu_2O_{8+\delta}$ have been measured at temperatures between 7 and 300 K. The complex conductivity shows several interesting features known from pure YBCO – an anomalous interaction of optical phonons with the electronic background and appearance of a bump at 420 cm^{-1}. It was found that these features are present in the conductivity spectra of all bilayer high-T_c cuprates. The observed bump can be interpreted as an intrabilayer plasmon predicted by the multilayer model, which strongly interacts with optical phonons. The concentration dependence of the optical conductivity shows that the primary parameter, which determines the intensity of the mentioned effect, is the critical temperature (T_c) or superfluid density.

Keywords: c-axis infrared spectra, Josephson plasmon, temperature and substituent dependence

INTRODUCTION

A lot of experimental and theoretical work has been performed, but we are still very far from understanding the unusual behavior of the c-axis infrared properties in high-T_c superconducting cuprates. The most striking and difficult understandable property of the cuprates is their high anisotropy. The anisotropy is often so high that the coherence length along c-axis $\xi_c < d_c$ the lattice constant. In this case it is much better to consider high-T_c superconductor as a stacked array of two-dimensional superconducting layers than an anisotropic medium. When the adjacent layers are coupled together by Josephson tunneling we obtain the model which was suggested by Lawrence and Doniach[1]. There are several important consequences of this model. This approach strongly modifies the charge dynamics by Josephson coupling. Many of these materials exhibit dc and ac intrinsic Josephson effects. The c-axis polarized infrared reflectance shows an abrupt rise at low frequencies below T_c, which is called the Josephson plasmon.

Recently van der Marel and Tsvetkov [2] generalized this model for multilayer materials and showed that more than one plasma frequency is possible. When the CuO_2 layers are separated along the c-axis by two different types of blocking layers, as in the T^* 214 compounds, two plasma resonances should exist[3]. This effect has been demonstrated for a powder sample of T^* phase $SmLa_{1-x}Sr_xCuO_{4-\delta}$. According to this model, in multilayer materials with a crystal form, such as YBCO and $Bi_2Sr_2CaCu_2O_z$,, the coupling between intracellular layers leads to the appearance of an additional excitation, which corresponds to antiphase tunneling between the layers. In this case, the dielectric function besides two zeros shows a pole. The additional pole and zero characterize the transverse and longitudinal frequency of the excitation, which is manifested in the reflectivity spectrum as an additional *reststrahlen* band. This electronic excitation always strongly interacts with the CuO_2 plane bond-bending oxygen vibration at 320 cm^{-1}.

In this paper we report the c-axis infrared reflectance of underdoped Bi2212 and Y123 substituted by Zn and Ni. The Bi2212 samples show in agreement with other bilayer materials a bump (400 – 550 cm^{-1})

below T_c. The appearance of the bump for Zn and Ni doped sample correlates with T_c. The experimental data are well reproduced by the van der Marel and Tsvetkov model.

RESULTS AND DISCUSSION

The infrared reflectance along the c-axis has been measured in a broad spectral ($30 - 4000$ cm^{-1}) and temperature region ($7 - 300$ K). The measured samples were YBCO single crystals and Bi2212 mosaics. The optical conductivity and dielectric function were evaluated using Kramers-Kronig analysis.

It is well known that the anomalous behavior of the c-axis reflectivity and also conductivity (bump, c-axis resonance) has been observed in most underdoped cuprates (Y123, Y124, PSYCO) but it has not been found in single layer materials (LSCO, $Tl_2Ba_2CuO_6$). This has been recently confirmed by our observation of this effect on two Bi2212 mosaics with T_c 85 and 60 K. There have been no reports on the temperature dependence of the Bi2212 reflectance because of difficulty in preparing a large enough crystal for infrared measurements. Bi2212 is much more anisotropic than YBCO and its large carrier effective mass results in very low conductivity along c-axis. The absence of free carriers substantially modifies the spectrum in Fig.1, which is much more semiconductive than YBCO. The dominant contribution to the spectrum is from the infrared active phonons. The resonance at 450 cm^{-1} is much less pronounced and strongly broadened. In this case, instead of a real bump appearance, it is possible only to observe an increase in intensity in the valley between phonon peaks. The Josephson plasma frequency in Bi2212 is very low ≈ 5 cm^{-1} [4] for the same reasons as the c-axis conductivity depression and we cannot observe it in the far-infrared region. We cannot therefore decide if both anomalies appear here at the same temperature as was observed also for YBCO.

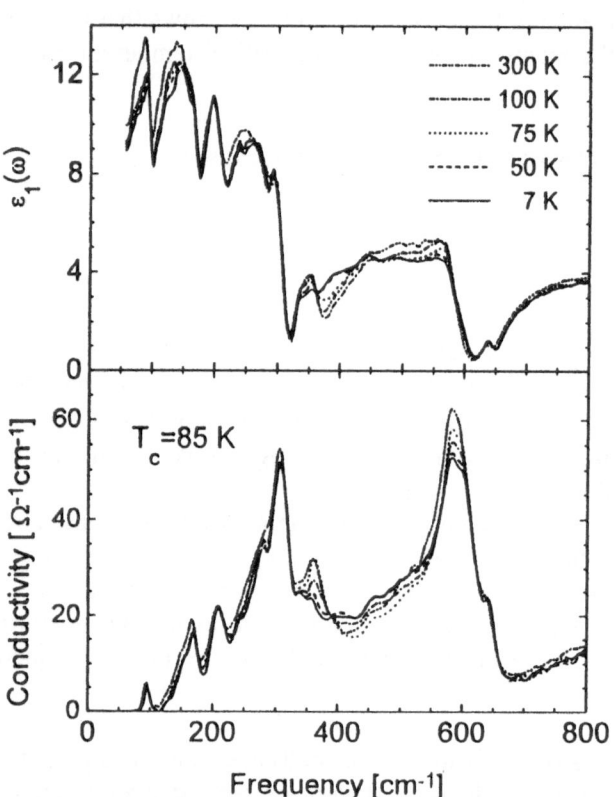

Fig.1. Temperature dependence of the c-axis conductivity and dielectric function for the underdoped Bi2212.

The influence of the Zn doping on the reflectivity spectrum is well established. The spectrum of pure YBCO below T_c in Fig.2 shows a sharp low-frequency plasma edge and pronounced bump around $400 - 500$ cm^{-1} and strong anomaly of the 320 cm^{-1} phonon. These features are strongly suppressed in the spectra of Zn-doped samples. It was originally thought that the Zn substitution also reduces the antiferromagnetic fluctuations at $Q=(\pi,\pi)$ which are responsible for deepening the pseudogap[5]. The correlation between this phenomenon and the appearance of a bump at 450 cm^{-1} allows us to draw a conclusion about its origin and spin-phonon coupling anomaly at 320 cm^{-1}. Some recent experiments show that the problem of the pseudogap is more complex. A study of the $\sigma_c(\omega)$ spectrum in Y123 and La214 shows that Zn substitution has no influence on the pseudogap [6]. Other experiments (μSR., T_1^{-1} and thermoelectric coefficient) indicate that the superconducting pairs are localized around each Zn atom. This dramatically reduces the condensate density and T_c and seriously changes the charge dynamics of the samples with concentration x=0.006 and 0.012. This provides a basis for the conclusion that both the low-frequency plasma edge and the bump are correlated in the same way to T_c but not to the pseudogap temperature T^*.

The reflectivity of Ni doped YBCO at 10 K is illustrated also in Fig.2. The low temperature spectra of the Ni-doped sample are more smeared than those of pure YBCO due to additional scattering by Ni atoms. On cooling the sample, both the low-frequency plasma edge and the bump at 450 cm^{-1} can be easily observed below T_c, which for this sample is 54 K. This also demonstrates that the mere presence of substituents in sample, which do not affect T_c, also do not suppress both spectral features.

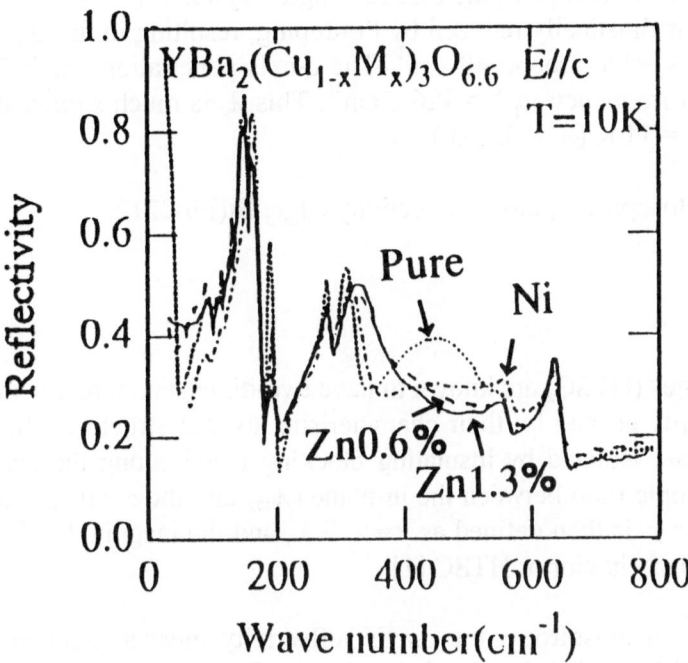

Fig.2. The c-axis reflectivity of the substituted and underdoped YBCO at 10 K.

When we compare the c-axis spectra of underdoped pure and substituted YBCO ($YBa_2(Cu_{1-x}R_x)_3O_{6.6}$; R=Zn, x=0.0 ($T_c$=59K), 0.006 ($T_c$=39K), 0.013 ($T_c$=18K) and Ni, x=0.01 ($T_c$=54K)) and Bi2212 we can make several interesting conclusions. The bump at 450 cm^{-1} has been observed in all bilayer cuprates, whose c-axis infrared spectra have been studied. There are no report of its appearance in single layer cuprates. It turns out to be an intrinsic property of the double pyramid structure in these materials. It is always coupled to the anomalous behavior of the oxygen bending phonon at 320 cm^{-1}. The Josephson plasma edge and the bump at 450 cm^{-1} are suppressed by the substitution for Cu. The experiments show that they are correlated to T_c but practically insensitive to T^*. On basis of these considerations we conclude that the bump at 450 cm^{-1} has properties which were predicted by van der Marel and Tsvetkov [2] for transverse (antiphase) optical Josephson plasma. The model of the stacked layers is a good approximation for understanding the c-axis charge response.

Acknowledgments. This work was supported by the New Energy and Industrial Technology Development Organization (NEDO) as Collaborative Research and Development of Fundamental Technologies for Superconductivity Applications.

* On leave from Institute of Physics, ASCR, Prague, Czech Republic
1.   W.E. Lawrence, S. Doniach, Proc. 12th Int. Conf.Low Temp.Phys., Kyoto (1970).
2.   D. van der Marel and A. Tsvetkov, Czech.Journ. of Phys. **46**, 3165 (1996).
3.   H. Shibata and T. Yamada, Phys.Rev.Lett. **81**, 3519 (1998).
4.   O.K.C. Tsui et al., Phys.Rev.Lett. **73**, 724 (1994).
5.   R. Hauff et al., Phys.Rev.Lett. **77**, 4620 (1996).
6.   K. Mizuhashi et al., Phys.Rev.B **52**, R3884 (1995).

Josephson Plasma Reflectivity Edge in Pb-doped Bi2212 Single Crystals: Direct Evidence for Reduced Anisotropy of Bi2212 by Pb-doping

Teruki Motohashi,[1] Jun-ichi Shimoyama,[1] Kohji Kishio,[1] Kenji M. Kojima,[1] Shin-ichi Uchida,[1] and Setsuko Tajima[2]

[1]Dept. of Superconductivity, Univ. of Tokyo, 7-3-1 Hongo, Bunkyo-ku, Tokyo 113-8656, Japan
[2]SRL-ISTEC, 1-10-13 Shinonome, Koto-ku, Tokyo 135-0062, Japan

Abstract: We have successfully observed, for the first time, the Josephson plasma reflectivity edge of Bi-based high-T_c cuprate using carrier overdoped Bi(Pb)2212 single crystal (T_c = 65 K). An extremely large anisotropy of Bi2212 was drastically reduced by Pb-doping, resulting in the appearance of the Josephson plasma reflectivity edge around 40 cm^{-1}. The c-axis penetration depth λ_c = 12.6 μm was estimated from the plasma frequency $\omega_{ps}{}^c$ = 126.2 cm^{-1}. This λ_c is much shorter than that of carrier overdoped Bi2212 with T_c = 71 K (λ_c = 35 μm).

Keywords: electromagnetic anisotropy, Josephson plasma reflectivity edge, Bi(Pb)2212

INTRODUCTION

High temperature superconducting cuprates (HTSC) are known to have two-dimensional nature in its superconducting properties. This is mainly caused by their characteristic layered structures. In the crystals, superconducting CuO_2 planes are isolated by insulating blocking layers along the crystal c-axis, giving an extremely large anisotropic ratio between the in-plane (λ_{ab}) and the c-axis penetration depth (λ_c). The anisotropy parameter γ is then defined as $\gamma \equiv \lambda_c/\lambda_{ab}$, and the magnitude of γ is essential in understanding the flux pinning behavior of HTSC [1].

Among various methods for evaluation of anisotropy, the c-axis reflectivity measurement in the superconducting state is the most powerful one. It is known that a sharp reflectivity edge associated with the Josephson plasma appears in the far-infrared region in several HTSC compounds [2,3]. The edge frequency $\omega_p{}^c$ and the Josephson plasma frequency $\omega_{ps}{}^c$ [$\equiv (\varepsilon_\infty)^{1/2}\omega_p{}^c$], where ε_∞ is the dielectric constant, are connected with the c-axis penetration depth $\lambda_c \equiv c/\omega_p{}^c = c/(\varepsilon_\infty)^{1/2}\omega_p{}^c$, and give direct information on the anisotropy in the superconducting state.

Recently, we have found that the Pb doping in Bi2212 systematically and largely decreases the resistivity anisotropy ρ_c/ρ_{ab} in the normal state [4]. In a carrier overdoped $Bi_{1.6}Pb_{0.6}Sr_{1.8}CaCu_2O_y$ single crystal with T_c = 65 K, $\rho_c/\rho_{ab} \sim 1200$ was obtained at 100 K; this value is comparable to that of less anisotropic HTSC compounds (LSCO and YBCO) in their carrier underdoped state. On the other hand, anisotropy of Bi(Pb)2212 in the superconducting state had been still unclear. In the present study, c-axis reflectivity measurements were performed at various temperatures in order to evaluate the anisotropy of Bi(Pb)2212 in the superconducting state.

EXPERIMENTAL

Single crystals with a nominal composition of $Bi_{1.6}Pb_{0.6}Sr_{1.8}CaCu_2O_y$ were grown by the FZ method. Crystals were post-annealed in oxygen atmosphere to produce carrier-overdoped samples with T_c = 65 K. Polarized reflectivity measurements were carried out using a fast-scan Fourier transform spectrometer (Bruker IFS113v) in the far- and mid-infrared range (30 ~ 4000 cm^{-1}) at temperatures between 8 and 295 K. Spectra were taken by detecting the reflection from a mosaic sample which consists of ten pieces of crystals piled up with the aid of an epoxy resin and polished along a face parallel to the c-axis. The contribution of the epoxy resin in the original reflectivity spectra was subtracted to obtain the intrinsic spectra of Bi(Pb)2212 single crystal. Details of the experimental procedures are described elsewhere [5].

RESULTS AND DISCUSSION

Figure 1 shows the c-axis reflectivity spectra of the Bi(Pb)2212 single crystal at various temperatures. At 295 K, the spectrum shows a semiconducting feature with relatively low reflectivity down to 30 cm^{-1}, indicating a lack of any electronic contribution along the c-axis. No remarkable change is observed in the spectrum down to 75 K. Below T_c (= 65 K), however, a sharp reflectivity edge appears around 40 cm^{-1}, making a dip structure at higher energy side of the edge. From the similarity of the behaviors reported in the c-axis spectra of LSCO [2] and YBCO [3], the observed reflectivity edge here is associated with the Josephson plasma. For comparison, we also measured the c-axis reflectivity spectra (inset of Fig. 1) of a carrier overdoped Pb-free Bi2212 single crystal (T_c = 79 K) annealed under the same conditions. As clearly seen in the inset, no appreciable spectral change is observed even in the superconducting state down to 40 cm^{-1}: this is quite contrastive to the case of Bi(Pb)2212.

The reflectivity spectra of Bi(Pb)2212 were transformed into the complex dielectric function [$\varepsilon(\omega) = \varepsilon_1(\omega) + i\varepsilon_2(\omega)$] through Kramers-Kronig analysis. In the normal state, $\varepsilon_1(\omega)$ was positive and almost constant (~ 10) in the lowest energy region, whereas at 8 K it rapidly decreased and then changed its sign showing a characteristic ω-dependence in the superconducting state: $\varepsilon_1^s(\omega) \propto \omega^{-2}$. The Josephson plasma frequency ω_{ps}^c = 126.2 cm^{-1} was obtained from a slope of the $\varepsilon_1^s(\omega)$ vs. ω^{-2} plot, and λ_c was then calculated to be 12.6 μm.

This is the first observation of the Josephson plasma edge in Bi-based HTSC in the far-infrared reflectivity measurements. We emphasize that the present result is the most convincing evidence for the remarkably reduced anisotropy of Bi(Pb)2212 in the superconducting state. Up to date, there has been no report for observations of the Josephson plasma reflectivity edge in Bi-based HTSC due to its extremely large electromagnetic anisotropy, which pushes the Josephson plasma edge out of the experimental window in the infrared reflectivity measurements (~ 30 cm^{-1}) [6].

It is known that the magnitude of λ_c intrinsically correlates with the c-axis conductivity (σ_c) in the normal state. Basov et al. [7] have proposed a universal line in the σ_c vs. λ_c plot where λ_c^2 is inversely proportional to σ_c just above T_c for a variety of HTSC compounds. In Fig. 2, we plot σ_c and λ_c of Bi(Pb)2212 together with data for other HTSC compounds. The present result lies just on the universal line. By using the relationship between σ_c and λ_c, we can estimate λ_c-value of the Pb-free Bi2212 (T_c = 79 K) in the present study. Since σ_c was obtained to be ~ 0.33 Ω^{-1} cm^{-1} at 100 K by the transport measurement [4] and corresponding λ_c is approximately 60 μm, the edge frequency is expected to be ~ 15 cm^{-1}, which is out of the experimental window in our spectrometer.

Figure 2 also demonstrates a strong dependence of λ_c on carrier doping level. The magnitude of λ_c systematically decreases with increasing carrier concentration in each compound. Since it is known that Pb doping in Bi2212 gives the increase of the carrier concentration in the CuO$_2$ planes, quite short λ_c in Bi(Pb)2212 may be partly caused by the overdoping state of the crystal. However, the overdoped Pb-free Bi2212 with nearly the same T_c (= 71 K) in Ref. 9 has much longer λ_c (= 35 μm) compared to our Bi(Pb)2212, λ_c = 12.6 μm. It seems to be difficult to explain the difference of λ_c only taking the carrier doping level for each crystal into consideration. We thus speculate that Pb substitution would intrinsically reduce the anisotropy of Bi2212, in addition to the carrier doping effect.

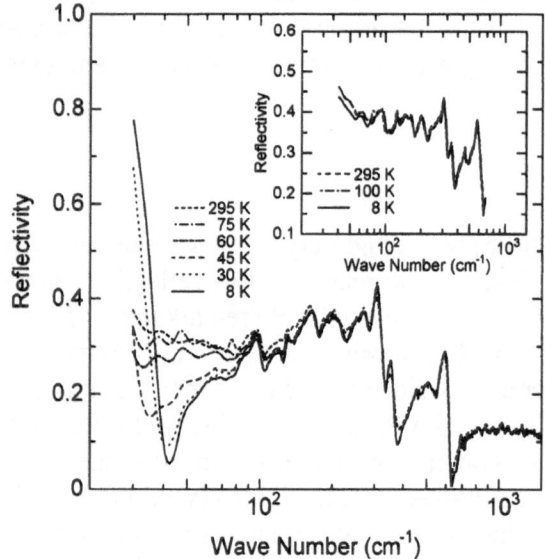

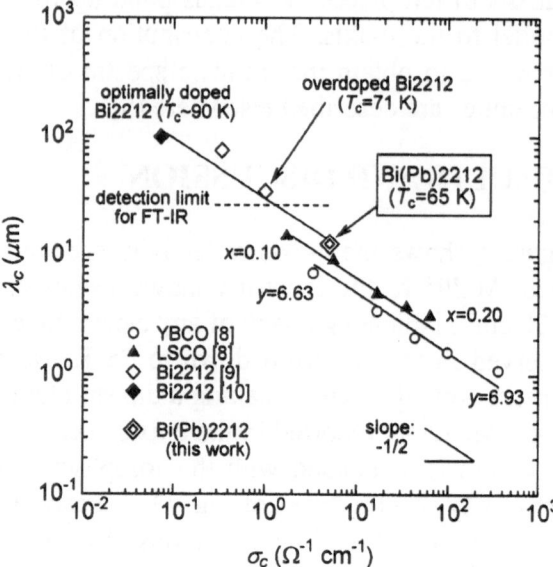

Fig. 1. c-axis reflectivity spectra of $Bi_{1.6}Pb_{0.6}Sr_{1.8}CaCu_2O_y$ single crystal at various temperatures. Inset represents the c-axis spectra of overdoped Bi2212 single crystal. For the spectra of Bi2212, the contribution of epoxy resin in the spectra has not been subtracted.

Fig. 2. Correlation between σ_c and λ_c of Bi(Pb)2212 together with the data for YBCO [8], LSCO [8], and Bi2212 [9,10]. Solid lines are eyeguides with a slope of -1/2. A broken line represents the detection limit for the Josephson plasma edge in the FT-IR reflectivity measurements.

CONCLUSION

In order to evaluate the electromagnetic anisotropy of Bi(Pb)2212 in the superconducting state, c-axis reflectivity measurements were carried out on the overdoped Bi(Pb)2212 single crystal (T_c = 65 K) at various temperatures. Below T_c, a sharp reflectivity edge was observed around 40 cm^{-1}, corresponding to the appearance of the Josephson plasma in the superconducting state. This is the first observation of the Josephson plasma reflectivity edge in Bi2212, which is the most anisotropic compound in many high-T_c cuprates. The c-axis penetration depth λ_c = 12.6 μm was estimated from the plasma frequency $\omega_{ps}{}^c$ = 126.2 cm^{-1}. This λ_c value is shorter than that of the overdoped Pb-free Bi2212 with T_c = 71 K (λ_c = 35 μm). These results indicate that the anisotropy of Bi2212 is drastically reduced by Pb-doping both in superconducting and normal states.

1. K. Kishio, *Coherence in High Temperature Superconductors* (World Scientific, Singapore, 1996) pp. 212-225.
2. K. Tamasaku, Y. Nakayama, and S. Uchida, Phys. Rev. Lett. **69**, 1455 (1992).
3. C.C. Homes, T. Timusk, R. Liang, D.A. Bonn, and W.N. Hardy, Phys. Rev. Lett. **71**, 1645 (1993).
4. T. Motohashi, Y. Nakayama, T. Fujita, K. Kitazawa, J. Shimoyama, and K. Kishio, Phys. Rev. B **59**, 14080 (1999).
5. T. Motohashi, K.M. Kojima, J. Shimoyama, S. Tajima, K. Kitazawa, S. Uchida, and K. Kishio, submitted to Phys. Rev. Lett.
6. S. Tajima, G.D. Gu, S. Miyamoto, A. Odagawa, and N. Koshizuka, Phys. Rev. B **48**, 16164 (1993).
7. D.N. Basov, T. Timusk, B. Dabrowski, and J.D. Jorgensen, Phys. Rev. B **50**, 3511 (1994).
8. S. Uchida and K. Tamasaku, Physica C **293**, 1 (1997).
9. H. Shibata and A. Matsuda, Phys. Rev. B **59**, R11672 (1999).
10. J.R. Cooper, L. Forró, and B. Keszei, Nature **343**, 444 (1990).

Interlayer Tunneling in Bi-2212: Coherency and Charging Effects

Yu. I. Latyshev[1,2], V. N. Pavlenko[1,2], S.-J. Kim[2], and T. Yamashita[2]

[1] Institute of Radio-Engineering and Electronics Russian Acad. of Sciences, 11 Mokhovaya str., 103907 Moscow, Russia
[2] Research Institute of Electrical Communication, Tohoku University, 2-1-1, Katahira, Aoba-ku, Sendai 980-8577, Japan

Abstract: The interlayer tunneling has been studied on high quality Bi-2212 stacks of micron to the submicron lateral size. We found that low temperature and low voltage tunneling *I-V* characteristics can be self-consistently described by Fermi-liquid model for a *d*-wave superconductor with a significant contribution from coherent interlayer tunneling. The gap and pseudogap interplay with variation of temperature and magnetic field has been extracted from the *I-V* characteristics. We consider also the role of charging effects for submicron stacks and the role of the trapped flux for stacks of annular geometry.

Keywords: Bi-2212, stacked junctions, interlayer tunneling coherency, gap symmetry, pseudogap, charging effects, annular stacks.

INTRODUCTION

Low temperature quasiparticle transport properties can provide important information about low temperature normal state and superconducting order parameter symmetry in layered high-T_c cuprates. Naturally, those studies are difficult because they require very high magnetic field to suppress strong shunting by superconducting currents. One of the possibilities proposed in [1] is to study interlayer tunneling *I-V* characteristics of small enough stacked junction. On that way one needs to avoid masking effects on the *I-V* characteristics due to Joule selfheating or quasiparticle injection. It may be achieved with pulse technique [2] or using small, submicron lateral size stacks [3]. We review here our recent measurements of interlayer tunneling characteristics in small Bi-2212 stacks with emphasis on significant contribution of coherent processes to the interlayer tunneling and on d-wave symmetry of the order parameter.

EXPERIMENTAL

As a base material for junction fabrication we used Bi-2212 single crystal whiskers grown by impurity free method [4]. Recently they have been characterized [4] as a very perfect crystalline object. For the fabrication of stacked junction we used the conventional FIB machine of Seiko Instruments Corp., SMI 9800 (SP) with Ga$^+$-ion beam. The junctions have been fabricated by double-sided processing of whisker with FIB. The details of fabrication steps described in [5]. For fabrication of annular type junctions a small hole at the center of the stack has been etched. Four Ag-contact pads have been evaporated and annealed before FIB processing to avoid Ga-ions diffusion into the junction body. Parameters of the stacks are listed in the Table 1.

Low temperature and voltage *I-V* characteristics. Figures 1 (b) and 1 (c) show the *I-V* charactrristics of samples #2 and #4. The fully superconducting overlap geometry of the stack (Fig. 1a) was used to suppress the effects of quasiparticle injection on the tunneling characteristics, usually occuring in junctions of the mesa type with a normal metal top electrode [1,6]. We also substantially reduced the effects of selfheating in our submicron mesa junctions. Self-heating manifests itself in the form of an S-shaped *I-V* curve near the gap voltage V_g [3]. The measured

Table 1. Parameters of Bi-2212 stacked junctions.

No	S (μm^2)	N	V_g (V)	I_c (μA)	Notes
1	0.4	70	2.4	0.1	
2	2.0	65	1.3	12	
3	1.5	50	1.1	6	
4	0.6	35	1.7	0.25	
5	0.3	50	2.2	0.07	
6	36x0.5	50	3.0	14	Array 6x6 junctions
7	400	40	0.8	1000-2000	Data taken from Ref. [1]
8	0.4	70	2.4	~0.05	
H-1	2.0	35	1.4	0.6	Contains a hole D=0.2μm

temperature dependence of the c-axis resistivity of the stack was typical for slightly overdoped Bi-2212 crystals with $T_c \approx 77$ K [7]. The critical current I_c was determined from the *I-V* characteristics as the current of switching from the superconducting to the resistive state, averaged over the stack. The variation of the critical current along the stack is not large (usually within 15%), indicating a good uniformity of our structures. The *c*-axis critical current density, J_c, for the junctions with in-plane area $S > 2$ μm^2 was typically 600 A/cm^2 at T = 4.2 K [3]. The superconducting gap (pseudogap) voltage of the stack, V_g, was determined from the *I-V* characteristics as the voltage at the maximum of the *dI/dV*. The gap of the intrinsic junction, $2\Delta_0 \approx eV_g / N$, where N is the number of elementary layers in the stack, reaches value as high as 50 meV (see Ref. [3]). The multibranched structure, which is clearly seen in Fig. 1(b) corresonds to subsequent transition of the intrinsic junctions into the resistive state for increasing voltage [8]. At voltages $V > V_g$ all junctions are resistive. In downsweep of voltage, starting from $V > V_g$ the *I-V* curve is observed in the all junction resistive state. Here, only quasiparticles contribute to the c-axis transport. Corresponding quasiparticle conductivity, σ_q, thus can be defined directly from that part of the *I-V* curve. The Ohmic resistance, R_n, at $V > V_g$ is well defined (Fig. 1c). This resistance is nearly temperature independent (Fig. 2b) and corresponds to the conductivity $\sigma_n(V > V_g) \approx 80$ (kΩ cm)$^{-1}$ for energies above superconducting gap and pseudogap.

We found out that interlayer tunneling *I-V* characteristics at low temperatures essentially differ from those of conventional Josephson junctions between s-wave superconductors. We specify [9]: (1) strong disagreement with Ambegaokar-Baratoff (A-B) relation, $J_c^{AB}(0) = \pi \sigma_n \Delta_0 / 2 e s$, with s the spacing between intrinsic superconducting layers (15.6 Å); (2) quadratic dependences of quasiparticle conductivity $\sigma_q(V, T)$ on V and T (Fig. 2): $\sigma_q(V, 0) = \sigma_q(0, 0) (1 + \alpha V^2)$, $\sigma_q(0, T) = \sigma_q(0, 0) (1 + \beta T^2)$ with $\alpha = 0.014 \pm 0.003$ (meV)$^{-2}$, $\beta = (8 \pm 2) \times 10^{-4}$ K^{-2}, (3) nonzero and universal value of $\sigma_q(0, 0) \approx 2$ (kΩ cm)$^{-1}$.

We found empirically the modified relation of the A-B type,

$$J_c(0) \approx \pi \sigma_q(0, 0) \Delta_0 / e s \tag{1}$$

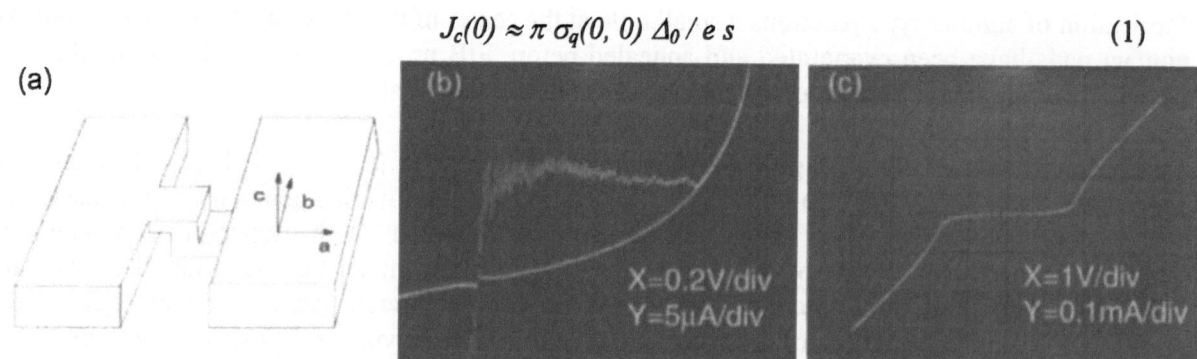

Fig. 1. Schematic view of the junction (a); *I-V* characteristics of the Bi-2212 stacks in (b) enlarged scale for sample #2 and (c) extended scale for sample #4. *T*=4.2 K.

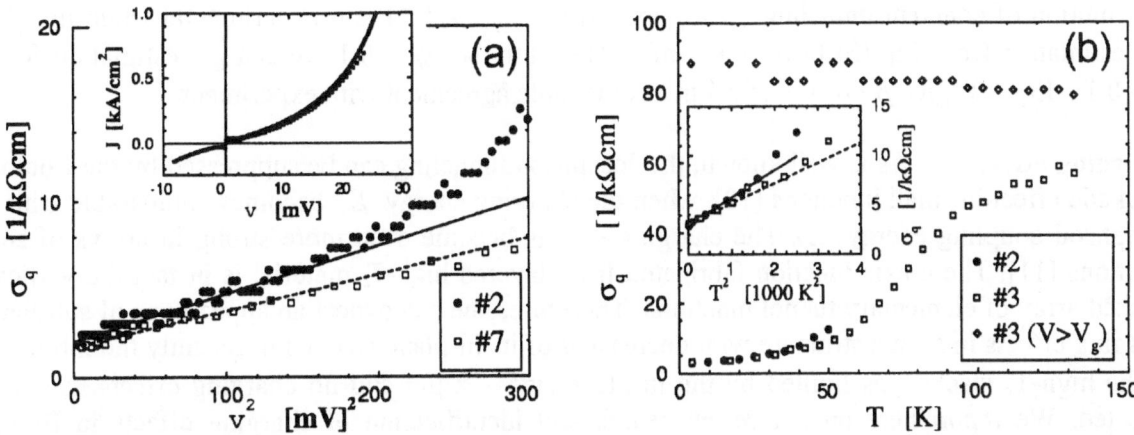

Fig. 2. Quasiparticle dynamic conductivity (a) vs $v^2 = V^2/N^2$ at $T=4.2$ K and (b) vs T for $v > V_g/N$ and $v \to 0$. Inset in (a): corresponding J-v curves; (b): σ_q vs T^2 at $v \to 0$. Lines are fits for $v < 10$ mV and $T^2 < 1000$ K^2, correspondingly.

and scaling relation between α and β, $\beta/\alpha = \text{const} \approx 15 \pm 4$. It was shown that all these features can be described self-consistently by Fermi-liquid model for quasiparticles in clean d-wave superconductor with resonant scattering [9]. Impurity scattering leads to the formation of gapless state near the node directions φ_g at angles $\varphi_g \pm \varphi_0/2$, with $\varphi_0 \sim \gamma/\Delta_0$ where γ is the impurity bandwidth of quasiparticles. That results in a nonzero density of states at zero energy, $N(0)\,\varphi_0$, where $N(0)$ is the 2D density of states per spin at the Fermi level, and leads to a universal quasiparticle interlayer conductivity $\sigma_q(0, 0)$. It was shown also [9] that the values of $\sigma_q(0, 0)$, $J_c(0)$ and coefficients α and β are strongly dependent on the coherency of interlayer tunneling. If to denote the weight for in-plane momentum conserving (coherent) tunneling as a and for incoherent as $(1-a)$, the ratio $J_c(0)/\sigma_q(0,0)$ is expressed as follows [9]

$$J_c(0)/\sigma_q(0,0) = (\pi\Delta_0/e\,s)\,[a + (1-a)\,\Delta_0/\varepsilon_F]\,/\,[a + (1-a)\,\gamma/\varepsilon_F] \qquad (2)$$

with ε_F the Fermi energy. Correspondingly for coefficients α and β it was obtained [9]

$$\alpha \approx 1/24\,\gamma^2\,[1 + (1/a - 1)\,\gamma/\varepsilon_F] \qquad \beta \approx \pi^2/(18\,\gamma^2)\,[1 + (1/a) - 1)\,\gamma/\varepsilon_F] \qquad (3)$$

One can see that Eq (2) turns to the experimentally found relation (1) only for significant

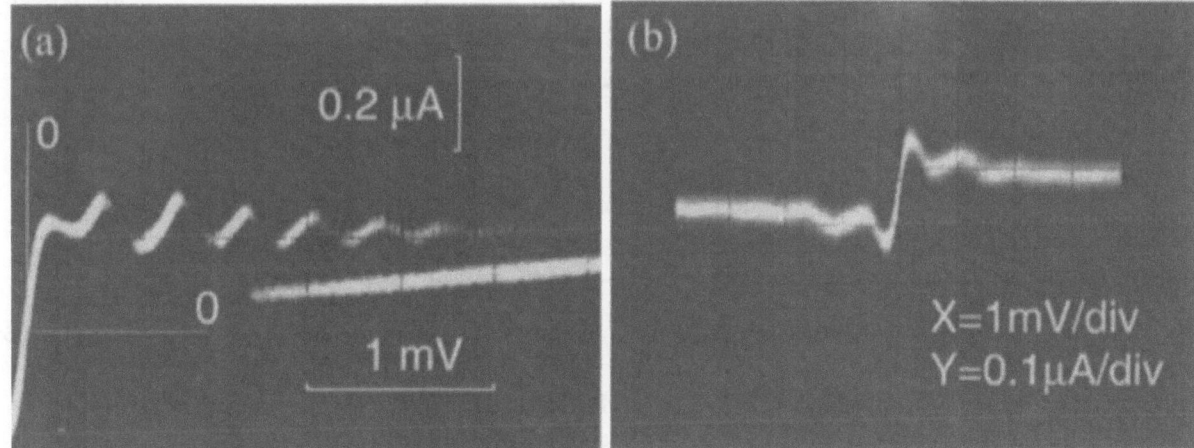

Fig. 3. Extended scale I-V characteristics of submicron Bi-2212 stacks #4 (a), #5 (b). The period of structure ΔV corresponds to the charging energy for single Cooper pair transfer through the stack.

contribution of coherent tunneling, $a \gg \max \{\Delta_0 / \varepsilon_F, \gamma / \varepsilon_F\}$. From experimental value for β we can estimate γ from Eq. (3) to be $\gamma \approx 3$ mV. Then for $\Delta_0 / \varepsilon_F \approx 0.1$ we can get estimation for a, $a \gg 0.1$. Eqs. (3) give $\beta / \alpha = 4 \pi^2 / 3$ in a reasonable agreement with experiment.

Charging effects. As is well known, the Josephson tunneling can be suppressed by the Coulomb blockade effect in small junctions [10], when the charging energy E_c becomes comparable with the Josephson coupling energy E_J. The charging effects become even more strong in arrays of small junctions [11]. The c-axis junction fabricated from layered high-T_c material is in fact the vertically stacked array of elementary tunnel junctions. Therefore, one can expect an appearance of substantial charging effects in such a structure with decreasing of its in-plane size. Until recently the fabrication of the high-T_c stacks was limited by the in-plane size ~ 2 μm and no charging effects have been detected. We report here on our recent search and identification of charging effects in Bi-2212 submicron stacks.

For the submicron junctions we clearly observed well defined tunneling charcteristics with superconducting gap (40-50 meV per elementary junction) (Fig. 1c). We found also a number of new features [3]. The I-V characteristics have no hysteresis and multibranched structure, the critical current has a finite slope increasing with a temperature, the periodic structure of current peaks develops on the I-V curves at low temperatures (Fig.3), $J_c(0)$ is reduced. Formally, all these details resemble the features typical for the single Cooper-pair tunneling in small tunnel junctions [12].

However, the period of observed Coulomb staircase ΔV was much higher than the charge energy for a single pair transfer through a single elementary junction E_{c0}. To explain that, we have analyzed the possibility of existence of charge soliton in our stacks, previously introduced for planar arrays of tunnel junctions [13]. The energy of charge soliton E_s is proportional to the number of junctions located within soliton length L_s. The latter, can be defined as [14]: $L_s = 2(C_0/C_g)^{1/2}$ with C_0 the capacitance of single junction and C_g the stray capacitance. Our estimations showed that in Bi-2212 stacks the charge soliton length is surprisingly high, ~ 400 elementary junctions, because of the very small value of stray capacitance. It implies that for a stack containing ~50 junctions all of them will be included in the single Cooper pair soliton length. Therefore, the whole stack will respond to the single pair transfer as a single unit. That can explain the big charge energy and a reason of disappearance of the multibranched structure in our submicron stacks. The value of E_s estimated as $E_s = N E_{c0}$ quite well corresponds to ΔV [3].

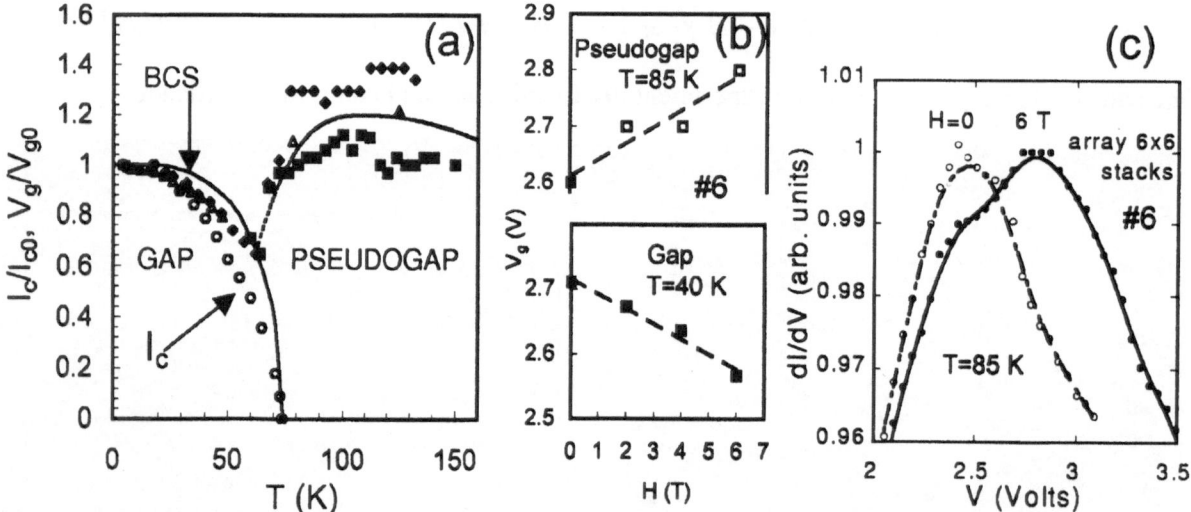

Fig. 4. Gap and pseudogap dependences on T (a) and magnetic field $H//c$ (b). Denotes of the panel (a) correspond to the following samples: ♦ - #3, Δ - #8, ■ - #1. (c) shows shift of the dI/dV spectrum of sample #6 under magnetic field in the pseudogap region. Line for pseudogap temperature dependence is guide to eye.

With an increase of the temperature above 4.2 K the Coulomb staircase structure of current peaks gradually washes out and disappears above 12 K (for the stack #5) when the charge soliton energy $E_s = \Delta V$ becomes less than kT. The behaviour of the sloped critical current under conditions $E_J << kT$, $E_J << E_c$, which are roughly valid for our case, has been analysed in [15]. It was shown that in the case, when the resistance of the environment Z_1 is less than $R_Q = h/(4e^2)$,, the junction will have classical phase diffusion behaviour and the zero bias resistance, R_0, should be proportional to $(kT)^2$: $R_0 = 2Z_1 (kT / E_J)^2$. That is in a qualitative agreement with our experiment.

Gap and pseudogap spectroscopy. The interlayer tunneling I-V characteristics in small stacked junctions provide an important possibility of gap and pseudogap spectroscopy in layered high-T_c cuprates. In comparison with other widely used methods for studies gap and pseudogap like ARPES or STM (see as a review Ref.[16]) which are in fact surface methods, the interlayer tunneling spectroscopy gets an information from the body of the single crystal. Our submicron Bi-2212 stacks show high quality tunneling characteristics. Both the gap and the pseudogap are clearly seen in the I-V characteristics (see e. g. Fig. 1c). Fig. 4a shows temperature dependence of gap and pseudogap. Starting from low temperatures a gap goes down more rapidly than usual BCS dependence (solid line). We defined T_c at a point (77 K) where $I_c(T)$ turns to zero. At $T > 65$ K the gap evolves into the pseudogap that was observed up to 160 K. The maximum pseudogap value $2\Delta_p$ at ~ 110 K is about 20% higher than superconducting gap $2\Delta_0$. We have never observed simultaneously a gap and a pseudogap at the interval 65 K $< T <$ 77 K. The most interesting observed feature is that the pseudogap abruptly goes down approaching T_c from high temperatures. It may be an indication of the anti-coexistence of the gap and the pseudogap. That type of temperature behaviour has been reproduced recently on break Bi-2212 junctions [17], but is in conflict with STM measurements on Bi-2212 [18], where neither gap nor pseudogap variation with temperature has been reported. To clarify the point we undertaken studies of the gap and pseudogap dependences on magnetic field H // c up to 6T (see Fig. 4b). We found that the gap is suppressed by magnetic field. To the contrast the pseudogap enhances by H (Fig. 4b,c). That observation excludes pseudogap origin due to superconducting fluctuations [19] or due to existence of preformed Cooper-pairs [20] above T_c and points out to the different origin or ordering for the gap and for the pseudogap formation.

Annular type stacks. Small annular junctions were the subjects of particular interest last decade **Fig. 5.** I_c vs $H_{//}$ dependences for the annular type junction. (a) without trapped flux; (b) with the flux

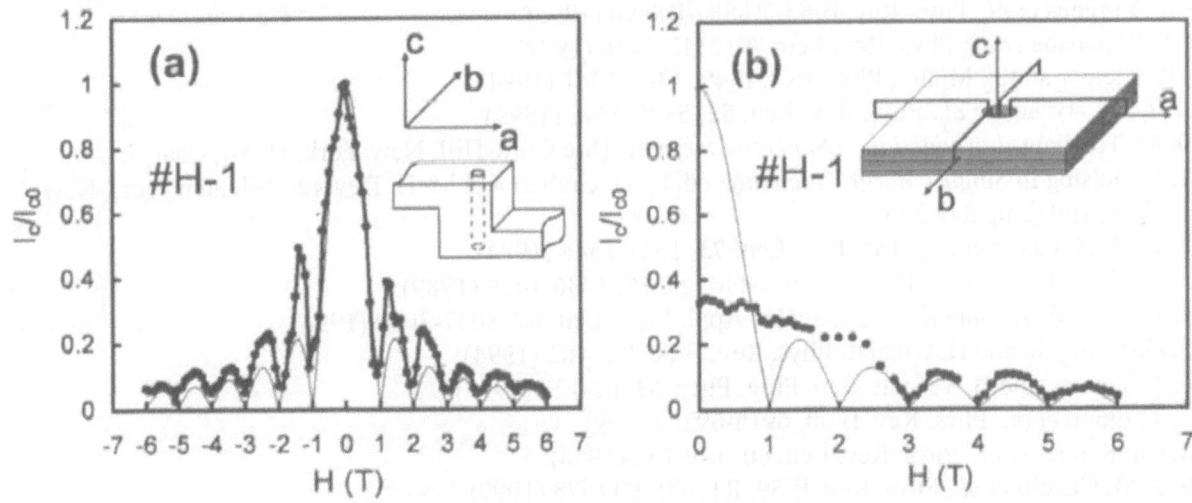

trapped in the hole at $H_\perp = 1$T. Insets in (a): junction geometry; (b) schematically shown magnetic field lines of a trapped flux. Thin solid line corresponds to Fraunhofer-type dependence.

because of possibility of flux trapping (see, e.g. [21]). Related magnetic field can contain radial component affecting Josephson critical current. Here we report on our recent studies of intrinsic dc Josephson effect in small annular type Bi-2212 mesas and its sensitivity to the trapped flux. The stacks were of squared geometry with lateral size $L = 1$-1.5 μm and contained the small hole at the center (with diameter 0.2 μm) threaded the whole thickness of the stack (insert to Fig.5a). We found that perpendicular field H > 0.1 T induces long living state with flux trapped in a hole. That state is characterized by suppressed value of the critical current due to an appearance of internal radial field parallel to the layers (see insert to Fig. 5 b). Fig.5 shows dependence of the critical current on parallel magnetic field for the annular type stack for both cases of zero (a) and of nonzero (b) trapped flux. For the case of zero trapped flux $I_c(H_{//})$ well follows expected Fraunhofer-type dependence $I_c(H_{//}) \propto |sin\, x / x|$ with $x = (\pi\, L\, s\, H / \Phi_0)$, where Φ_0 is flux quantum [22]. The period of oscillations ΔH exactly corresponds to the flux quantum per elementary junction for the measured stack with $L = 1.4$ μm. The flux trapping leads to suppression of critical current and to suppression of the first oscillations of $I_c(H_{//})$ (Fig.5b), the bigger trapped flux the more oscillations are disturbed.

CONCLUSIONS

We studied interlayer tunneling on small stacked junctions of different size and geometry. The results obtained point out to the d-wave symmetry of order parameter, to the significant contribution from coherent interlayer tunneling, to the "non-superconducting" origin of the pseudogap. We uncovered also the influence of charging effects and flux trapping on interlayer tunneling.

Acknowledgments.   The work has been done in collaboration with Los Alamos National Lab. We are thankful to L. N. Bulaevskii, M. J. Graf, A. V. Balatsky, M. Maley and N. Morozov for cooperation. We thank A. M. Nikitina for providing us with Bi-2212 single crystal whiskers. This work was supported by CREST, the Japan Science and Technology Corporation, and the Russian State Program on HTS under grant No. 99016.

1. K. Tanabe *et al.*, Phys. Rev. B **53**, 9348-9352 (1996).
2. M. Suzuki *et al.*, Phys. Rev. Lett. **82**, 5361-5364 (1999).
3. Yu .I. Latyshev *et al.*, JETP Letters **69**, 84-90 (1999), *ibid.* E. **69**, 640-643 (1999).
4. Yu .I. Latyshev *et al.*, Physica **C 216**, 471-477 (1993).
5. Yu. I. Latyshev, S.-J. Kim, T. Yamashita, IEEE Trans. on Appl. Supercond., **9**, 4312-4315 (1999).
6. A. Yurgens *et al.*, Phys. Rev. B **53**, R8887-R8890 (1996).
7. T. Watanabe *et al.*, Phys. Rev. Lett. **79**, 2113-2116 (1997).
8. R. Kleiner and P. Müller, Phys. Rev. B **49**, 1327-1341 (1994).
9. Yu .I. Latyshev *et al.*, Phys. Rev. Lett. **82**, 5345-5348 (1999).
10. M. Tinkham, *Introduction to Superconductivity*, (Mc Graw-Hill, New York, 1996), Chap.7.
11. P. Delsing in *Single Charge Tunneling* ed. by H. Grabert and M. H. Devoret, (Plenum Press, New York, 1992), pp.249-274.
12. D. B. Haviland *et al.*, Phys. Rev. Lett. **73**, 1541-1544 (1994).
13. K. K. Likharev *et al.*, IEEE Trans. on Magn. **25**, 1436-1439 (1989).
14. K. K. Likharev and K. A. Matsuoka, Appl. Phys. Lett. **67**, 3037-3039 (1994).
15. G.-L. Ingold and H. Grabert, Phys. Rev. B **50** 395-402 (1994).
16. T. Timusk and B.W. Statt, Rep. Prog. Phys. **62**, 61-122 (1999).
17. T. Ekino *et al.*, Phys. Rev. B **60**, 6916-6922 (1999).
18. Ch. Renner *et al.*, Phys. Rev. Lett. **80**, 149-152 (1998).
19. A.M. Cucolo *et al.*, Phys. Rev. B **59**, R11675-R11678 (1999).
20. Q. Chen *et al.*, Phys. Rev. Lett. **81**, 4708-4711 (1998).
21. C. Nappi, R.Cristiano and M.P. Lisitskii, Phys. Rev. B **58**, 11685-11691 (1998).
22. L. N. Bulaevskii, J. R. Clemm, L. I. Glazman, Phys. Rev. B **46**, 350-355 (1992).

Anomalous charge transport of the parent antiferromagnet $Bi_2Sr_2RCu_2O_8$

I. Terasaki, T. Takemura, T. Takayanagi and T. Kitajima

Department of Applied Physics, Waseda University, Tokyo 169-8555, JAPAN

Abstract: The resistivity and the thermopower for $Bi_2Sr_2Ca_{1-x}R_xCu_2O_8$ single crystals are measured and analyzed with a special interest in the parent antiferromagnet insulators. Above room temperature, the parent insulators show an electric conduction confined in the CuO_2 plane and a decreasing thermopower with temperature, which are very similar to those for high-T_c cuprates. These data strongly suggest that essential anomalies of high-T_c cuprates inheres in the parent insulators.

Keywords: parent antiferromagnet insulator, confinement, pseudo-gap

INTRODUCTION

The anisotropic charge transport of high-T_c cuprates (HTSC) has been one of the biggest anomalies that cannot be explained by conventional solid-state theories. In this context the parent antiferromagnetic (AF) insulators are to be carefully investigated as a limit of underdoping. Since they exhibit a characteristic two-dimensional (2D) AF fluctuation above the Néel temperature T_N, their transport properties are expected to show a qualitative change below and above T_N. There appears increasing evidence of close resemblance between the AF order and the d-wave superconductivity (SC) [1, 2], which again implies the anomalous properties of the parent insulators.

In spite of the above prospects, the transport properties of the parent insulators have attracted less interest. We have been studying the charge transport of the parent insulator $Bi_2Sr_2RCu_2O_8$ (R=Y and rare-earth) [3] and here we report on the recent results of the resistivity and the thermopower from 4 to 500 K.

EXPERIMENTAL

Single crystals of $Bi_2Sr_2Ca_{1-x}R_xCu_2O_8$ were grown by a flux technique. The growth conditions and the characterization of the samples were described elsewhere [3]. The resistivity was measured from 4.2 to 500 K through a four-probe method. The thermopower was measured using a steady-state technique from 4.2 to 450 K with a temperature gradient of 0.5-1.0 K. A thermopower of voltage leads was carefully subtracted.

RESULTS AND DISCUSSION

Figure 1(a) shows the in-plane resistivity ρ_{ab} and the out-of-plane resistivity ρ_c of $Bi_2Sr_2ErCu_2O_8$ single crystals. Although ρ_{ab} and ρ_c are insulating, their temperature dependence is significantly different, which is clearly seen in the resistivity ratio ρ_c/ρ_{ab}. As shown in Fig. 1(b), ρ_c/ρ_{ab} for $Bi_2Sr_2Ca_{1-x}Er_xCu_2O_8$ systematically evolves with x [3]. We should emphasize that ρ_c/ρ_{ab} smoothly changes with x above room temperature. If one looked at ρ_c/ρ_{ab} only above room temperature, one could not distinguish the parent insulators (x=0.5 and 1) from the superconductors (x=0 and 0.1). Thus we may say that the holes are confined in the parent insulator as well as in HTSC. Furthermore, ρ_c/ρ_{ab} for x=1.0 and 0.5 takes a broad maximum at a certain temperature T_{max}, as indicated by arrows in Fig. 1(a). By comparing the μSR data for $Bi_2Sr_2Ca_{1-x}Y_xCu_2O_8$ [4], we find that T_{max} is very close to T_N. These data suggest that the 2D AF fluctuation above

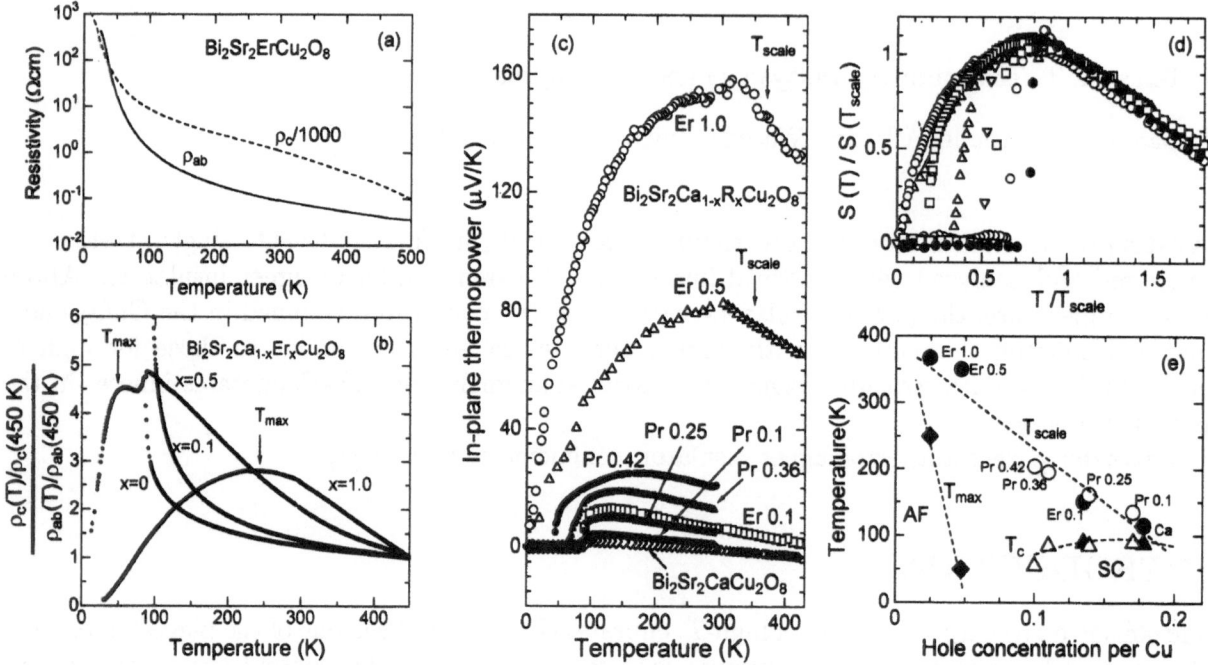

Fig. 1 (a) The in-plane resistivity (ρ_{ab}) and the out-of-plane resistivity (ρ_c) of $Bi_2Sr_2ErCu_2O_8$ single crystals, (b) ρ_c/ρ_{ab} for $Bi_2Sr_2Ca_{1-x}Er_xCu_2O_8$ normalized at 450 K, (c) The in-plane thermopower of $Bi_2Sr_2Ca_{1-x}R_xCu_2O_8$, (d) The same data in Fig. 1(c) normalized at T_{scale}, and (e) T_{max}, T_{scale} and T_c plotted as a function of the hole concentration per Cu estimated from the room-temperature thermopower.

T_N causes the confinement in the CuO_2 plane, and that the 3D AF order below T_N releases the holes from the confinement. In other words, the confinement is effective only in a spin liquid. Very recently, Hanke et al. [5] have pointed out that the AF order in the parent insulator can be regarded as a Bose-Einstein condensation (BEC) of $S = 1$ magnons in the resonating-valence-bond background. In this context, the charge transport for the parent insulator may probe some anomalies associated with BEC of magnons.

The in-plane thermopower S of the parent insulators is also anomalous. In Fig. 1(c), the thermopowers for $Bi_2Sr_2Ca_{1-x}Er_xCu_2O_8$ and $Bi_2Sr_2Ca_{1-x}Pr_xCu_2O_8$ are plotted as a function of temperature. S at room temperature monotonically increases with x, which is consistent with the literature [6]. This indicates that doping levels are controlled by the substitution of Er or Pr for Ca. A significant anomaly is that S for the parent insulators is roughly expressed by $A - BT$ above a temperature T_{scale}, where A and B are positive constants. Note that S for HTSC also obeys the relation of $A - BT$, while the diffusive part of the thermopower of conventional conductors will be independent of temperature at high temperatures for $E_F \ll k_BT$. Thus B and T_{scale} are another hallmark that discriminates the parent insulator and HTSC from the other solids. In addition, B and T_{scale} is larger for larger x, implying a certain relationship between them. To see the relationship more clearly we plot $S(T)/S(T_{scale})$ as a function of T/T_{scale} in Fig. 1(d). Most unexpectedly, all the curves of $S(T)/S(T_{scale})$ fall on a single curve, which clearly indicates a good scaling relation of the thermopower.

Let us compare T_{scale} with other experiments. In Fig. 1(e), T_c, T_{max} and T_{scale} are plotted as a function of hole concentration per Cu (p) that is estimated through an empirical relation to the room-temperature thermopower [6]. T_{scale} linearly decreases with p, which reminds us of the pseudo-gap temperature T^*. In fact, we have found T_{scale} for our superconducting samples to be close to T^* observed in the STM/STS or photoemission experiments [7]. Although the physical meaning of T^* is still controversial, many physical parameters follow a similar scaling relation. In particular, the Hall coefficient [8] and the susceptibility [9] can be scaled in terms of T/T^*, which strongly suggests that the carrier density or the density of states is a function of T/T^*. Accordingly the thermopower may also be expressed as a function of T/T^* in the sense that it gives a measure of the carrier density or the density of states.

The scaling behavior is consistent with a recent photoemission study on a parent insulator $Ca_2CuO_2Cl_2$ [1, 2], where a d-wave-like charge gap is opened in the k space. This means that the holes are doped initially near the node $(\pi/2, \pi/2)$ at low temperatures, and the doped region would be somewhat spread around the node at high temperature, because the electron correlation weakened by thermal excitation decreases or smears the charge gap. This situation indeed resembles what is observed in the underdoped HTSC. At $T=0$, only the node has the zero-energy excitation, and above T^* the Fermi surface grows from $(\pi/2, \pi/2)$ to $(\pi/2, 0)$ and $(0, \pi/2)$.

As mentioned above, we can capture a gross feature of the phase diagram of $Bi_2Sr_2Ca_{1-x}R_xCu_2O_8$ by measuring ρ_{ab}, ρ_c and S, which give T_c, T_{max} ($\sim T_N$) and T_{scale} ($\sim T^*$) as a function of p. Hence we would like to emphasize that the resistivity and the thermopower are a simple and powerful tool to explore the phase diagram. We further claim that the high-temperature transport is essentially the same between the parent insulator and HTSC. Above all, T_{scale} is indicative of the existence of the pseudo-gap in the parent insulator, which should be further examined by different probes.

SUMMARY

We prepared a set of single crystals of $Bi_2Sr_2Ca_{1-x}R_xCu_2O_8$, and measured the resistivity and the thermopower. We have found that the resistivities of the parent insulators exhibit a confinement behavior above T_{max} (near the Néel temperature) and the thermopower shows a good scaling behavior for a certain temperature T_{scale} (near the pseudo-gap temperature). These results strongly suggest a close similarity between the antiferromagnetic insulators and the high-temperature superconductors.

Acknowledgments. This work was partially supported by KAWASAKI STEEL 21st Century Foundation. The authors would like to thank T. Itoh, T. Kawata, K. Takahata, Y. Iguchi and T. Sugaya for collaboration.

1. R. B. Laughlin, Phys. Rev. Lett. **79**, 1726-1729 (1997).

2. F. Ronning, C. Kim, D. L. Feng, D. S. Marshall, A. G, Loeser, L. L. Miller, J. N. Eckstein, I. Bozobic and Z. X. Shen, Science **282** 2067 (1998).

3. T. Kitajima, T. Takayanagi, T. Takemura and I. Terasaki, J. Phys. Condens. Matter **11**, 3169-3174 (1999).

4. N. Nishida, S. Okuma, H. Miyatake, T. Tamegai, Y. Iye, R. Yoshizaki, K. Nishihara, K. Nagamine, R. Kadono and J. H. Brewer, Physica C**168**, 23-28 (1990).

5. W. Hanke, M. G. Zacher, E. Arrigoni and S. C. Zhang, cond-mat/9908175.

6. S. D. Obertelli, J. R. Cooper and J. L. Tallon, Phys. Rev. B**46**, 14928-14931 (1992).

7. T. Takemura, T. Sugaya and I. Terasaki, in the JPS fall meeting (1999).

8. H. Y. Hwang, B. Batlogg, H. Takagi, H. L. Kao, R. J. Cava, J. J. Krajewski and W. F. Peck, Jr, Phys. Rev. Leet **72** 2636-2639 (1994).

9. T. Nakano, M. Oda, C. Manabe, N. Momono, Y. Miura and M. Ido, Phys. Rev B**49** 16000-16008 (1994).

Theory for c-Axis Transport in HTSC

Kosaku Yamada and Youichi Yanase

Department of Physics, Graduate School of Science, Kyoto University, Kyoto, 606-8502, Japan

Abstract: We show that the anomalous transport phenomena in high-T_c cuprates can be understood on the basis of the Fermi liquid theory by taking into account strong anti-ferromagnetic spin fluctuations. To understand the transport phenomena in high-T_c cuprates the following two points are important. First, there exist two regions, called hot spots and cold spots on the quasi-two dimensional Fermi surface. The incoherent carriers around hot spots give little contributions to the conductivity; the conduction of the current owes to the carrier around cold spots. Secondly, the c-axis transfer matrix element between layers is small for cold spots and large for hot spots, according to band calculations. By the combined effect of the above two points, the c-axis conductivity becomes incoherent even when the inplane conductivity is coherent owing to the cold spots. The anisotropic behavior on the Fermi surface also appears in the pseudogap phenomena, which are inherent in the quasi-two dimensional system. We show that superconducting fluctuations above T_c give rise to the pseudogap.

Keywords: Resistivity, Paseudogap, Cuprate, Superconducting Fluctuation

INTRODUCTION

Since the discovery of high temperature superconductors (HTSC), nature of the normal state above T_c has been explained on the basis of various standpoints. In this paper we show that the transport phenomena in the normal state of cuprates can be naturally explained by taking anti-ferromagnetic spin fluctuations into account on the basis of the Fermi liquid theory [1].

In the latter part of this paper the pseudogap phenomena observed in the underdoped systems are explained by the selfenergy correction due to the superconducting fluctuation in the quasi-two dimensional strong coupling superconductors [2].

DAMPING RATE OF QUASIPARTICLES

We calculate the electron selfenergy due to the interaction with spin fluctuations, which are written by the dynamical susceptibility,

$$\chi(q,\omega) = \chi_Q/(1 + \xi^2(q - Q)^2 - i\omega/\omega_s). \tag{1}$$

Here, ξ and ω_s are the anti-ferromagnetic correlation length in the lattice space unit and the characteristic frequency of the anti-ferromagnetic spin fluctuations, respectively. The static susceptibility χ_Q at $Q = (\pi, \pi)$ is written as $\chi_Q = \alpha\xi^2$, α being the scale factor of the order of 10 (eV^{-1}). Usually, ξ is large, $4 \sim 6$ and ω_s is small, around 10 meV. These features strongly enhance the anti-ferromagnetic fluctuation.

We consider the selfenergy correction due to the interaction with the magnetic fluctuation,

$$\Gamma(p, \varepsilon_n; p', \varepsilon'_n : q, \omega_m) = \Gamma(q, \omega_m)$$
$$= g^2 \chi(q, \omega_m) = g^2 \chi_Q / [1 + \xi^2 (q - Q)^2 + |\omega_m|/\omega_s]. \tag{2}$$

The energy spectrum ε_k is given by the tight-binding model as

$$\varepsilon_k = -2t(\cos k_x + \cos k_y) + 4t' \cos k_x \cos k_y - \mu, \tag{3}$$

where t and t' are the nearest and next nearest neighbor transfer integrals, respectively. We set $t=0.5$eV and $t' = 0.45t$. The Fermi surface is shown in Fig. 1.

In this paper we calculate the selfenergy part by using a one-loop approximation and obtain the following result for the imaginary part of it,

$$\text{Im}\Sigma^R(k, \varepsilon) = -\frac{g^2}{2} \chi_Q \omega_s \int_{\text{FS}} \frac{dq}{(2\pi)^2 |v_{k-q}|} \left\{ \log \left[1 + \frac{\varepsilon^2}{\omega_q^2} \right] + \frac{4T}{\omega_q} \tan^{-1} \left[\frac{\pi^2 T}{4\omega_q} \right] \right\}, \tag{4}$$

where $\omega_q = \omega_s(1 + \xi^2(q - Q)^2)$ and the momentum integration should be carried out on the Fermi surface $\varepsilon_{k-q} = 0$. Here it is noted that Eq.(4) reduces to $\text{Im}\Sigma^R(k, \varepsilon) \propto \varepsilon^2 + (\pi T)^2$ in the low temperature region, $\varepsilon, T \ll \omega_s$, and $\text{Im}\Sigma^R(k, \varepsilon) \propto \log[\varepsilon/\omega_q] + 2T/\omega_q$ in the high temperature region.

The Eq.(4) possesses the momentum dependence which specifies the important regions called 'hot spot' and 'cold spot' on the Fermi surface.

As shown in Fig. 2, the hot spot is located in the vicinity of anti-ferromagnetic Brillouin zone, while the cold spot is located far from that zone.

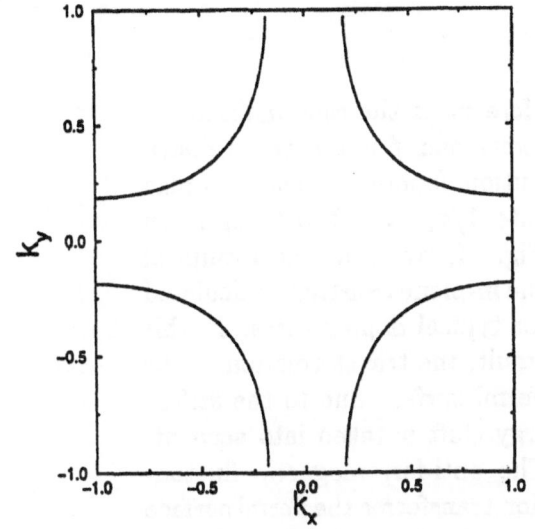

Fig. 1 The non-interacting Fermi surface of high-T_c cuprates.

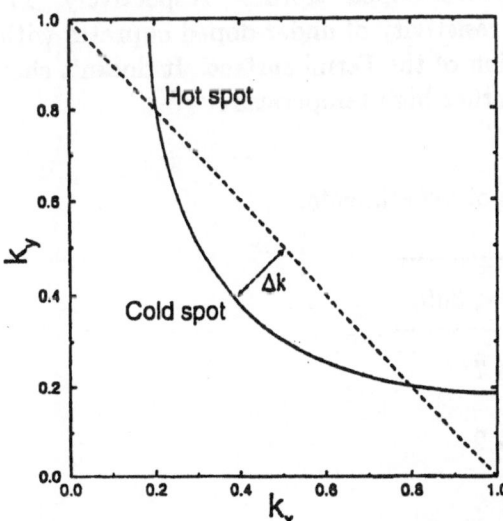

Fig. 2. The 'hot spot' and 'cold spot' of the Fermi surface.

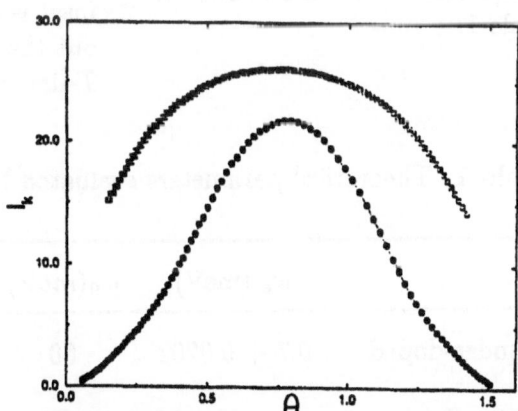

Fig. 3 The momentum dependence of the mean free path. The marks of ●, and □ correspond to under-doped and over-doped cuprates, respectively. Here, $\theta = \text{Arctan}(k_y/k_x)$. The under-doped case shows the strong anisotropy.

The momentum k around the hot spot on the Fermi surface satisfies $\varepsilon_k = \varepsilon_{k-Q} = 0$, and quasi-particles at the hot spots can exchange the spin fluctuations with momentum $Q = (\pi, \pi)$, while quasi-particles at the cold spots can't do it. Thus, the quasi-particles at hot spots possess a large damping rate, while those at cold spots possess a small damping rate. We show the mean free path as a function of $\theta = \tan^{-1}(k_y/k_x)$ in Fig. 3. We can see that in the underdoped case the mean free path shows the strong anisotropy.

TRANSPORT COEFFICIENT

By using the Kubo formura, we can derive the expression for the conductivity [3],

$$\sigma_{\mu\nu} = e^2 \int \frac{dk}{(2\pi)^2} \left\{ -\tau_k \left(\frac{df}{d\varepsilon} \right)_{\varepsilon = \varepsilon_k} \right\} v_\mu J_\nu, \tag{5}$$

$$J_\nu = v_\nu + \int \frac{dk'}{(2\pi)^2} \frac{a^2 \Gamma_{22}(k - k', 0) \tau_k'}{4i} v_\nu'. \tag{6}$$

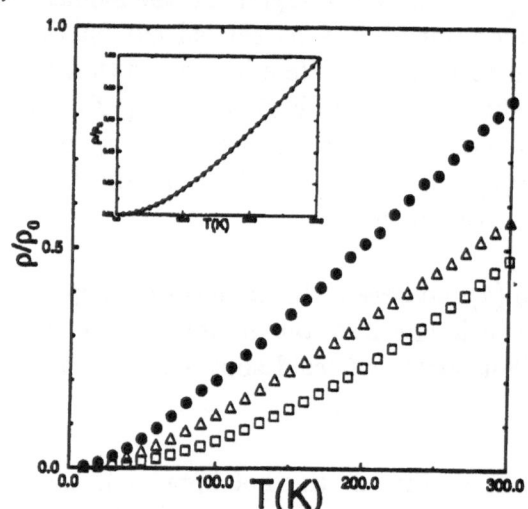

Here v_μ is the renormalized velocity and f is the Fermi distribution function. The damping rate $1/\tau_k = -\mathrm{Im}\Sigma(k, \varepsilon_k)$. In Fig. 4, we show the results of the in-plane resistivity calculated for typical doping cases. In this result, the transformation of the Fermi surface due to the selfenergy shift is taken into account. The anti-ferromagnetic fluctuation transforms the Fermi surface towards the anti-ferromagnetic Brillouin zone. Our theoretical parameters estimated from experimental results are shown in Table 1.

Fig. 4. The temperature dependence of the in-plane resistivity. The marks of ●, △ and □ correspond to under-doped, optimally-doped and over-doped cuprates, respectively. The inset is the in-plane resistivity of under-doped cuprates without the transformation of the Fermi surface. It doesn't show T-linear law up to rather high temperature.

Table 1. Theoretical parameters evaluated by the results of experiments.

	ω_s (meV)	ω_0 (meV)	$\xi(100 < T < 300)$
under-doped	$0.7 + 0.020T$	60	$3.0 \sim 4.7$
optimally-doped	$3.4 + 0.048T$	40	$1.5 \sim 2.2$
over-doped	30.0	10.8	0.6

Here, we add a comment on the Hall coefficient. The H-linear term of $\sigma_{\mu\nu}$ is given by

$$\sigma_{\mu\nu} = 2\frac{e^3}{c}H \int \frac{d\mathbf{k}}{(2\pi)^2} \left[J_\mu \frac{\partial J_\nu}{\partial k_\nu} - J_\nu \frac{\partial J_\mu}{\partial k_\nu} \right]$$
$$\times \left\{ \tau_k{}^2 \left(-\frac{\partial f}{\partial \varepsilon} \right)_{\varepsilon = \varepsilon_k} \right\} v_\mu, \tag{7}$$

where J_μ is given by Eq.(6) with Γ_{22} being the vertex correction [4].

To explain the temperature dependence of the Hall coefficient, the vertex correction given by Eq.(6), which is not derived by the Boltzmann equation, is essential, when the anti-ferromagnetic fluctuation is strong. This work has been done by Kanki, Kontani and Ueda [5].

RESISTIVITY ALONG c-AXIS

According to the band calculation [6], the transfer matrix element along c-axis possesses the following momentum dependence,

$$t_\perp \propto \left(\frac{\cos k_x - \cos k_y}{2} \right)^2. \tag{8}$$

This expression means that the interplane transfer between cold spots near $k_x = k_y$ vanishes, while that between hot spots is finite. From this fact the ratio ρ_c/ρ_{ab} shown in Fig. 5 is obtained. The mean free path along c-axis is shown in Fig. 6 in the unit of interplane distance c. With decreasing doping, the quasi-particles near the hot spots become incoherent owing to large damping rate and the coherent motion along c-axis is destroyed. In this case, although the quasi-particles near the cold spots are coherent, the transfer matrix between cold spots on the neighboring layers is absent. These facts explain the coherent in-plane transport and incoherent out-of-plane transport.

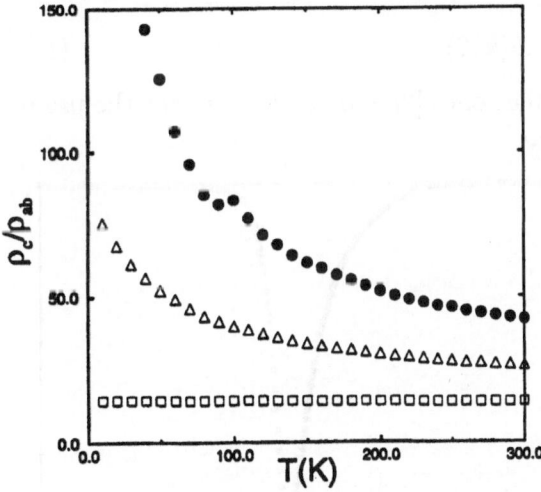

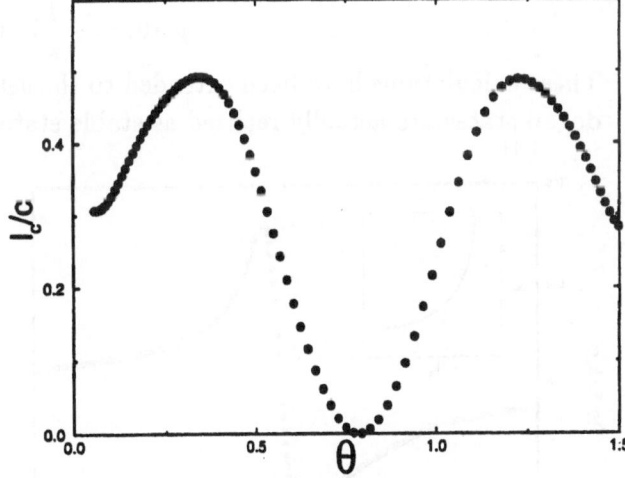

Fig. 5 The anisotropy of the resistivity ρ_c/ρ_{ab}. ρ_c and ρ_{ab} are the c-axis resistivity and the in-plane resistivity, respectively. The marks are the same as Fig. 4. The leap around 100 K in the under-doped case is caused by the numerical error.

Fig. 6. The mean free path along the c-axis for the under-doped cuprates. Here, $\theta = \mathrm{Arctan}(k_y/k_x)$, and c is a lattice space along the c-axis.

PSEUDOGAP IN UNDERDOPED SYSTEM

The pseudogap is observed by the nuclear magnetic resonance (NMR), tunneling experiments and angle resolved photo-emission spectroscopy (ARPES). The pseudogap arises from superconducting fluctuations in quasi-two-dimensional systems. By using the one-loop approximation for the superconducting fluctuation, we can obtain the spectrum possessing pseudogap near the Fermi surface. The t-matrix near and above T_c is given by

$$t^{-1}(q,\omega) = t_0 + bq^2 - \omega(a_1 + ia_2). \tag{9}$$

Here, t_0 tends to 0 as $T \to T_c$ and b is given by the square of coherence length, which becomes short in case of strong coupling superconductors. The electron selfenergy due to the interaction with superconducting fluctuations is given by

$$\Sigma(k,\varepsilon_n) = -T \sum_{q,\omega_n} G(k-q,\varepsilon_n-\omega_n)t(q,\omega_n). \tag{10}$$

For the strong attractive interaction, the selfenergy part behaves oppositely to the ordinary Fermi liquid; the imaginary part of the selfenergy possesses a peak at the Fermi energy and the slope of the real part of selfenergy becomes positive in contrast to negative slope in the Fermi liquid. This effect arises from the resonant scattering of quasi-particles corresponding to the pair formation and reduce the spectral density at the Fermi surface. In addition to the large damping effect, the real part reduces the quasi-particle density near the Fermi surface, and the coefficient γ of the T-linear specific heat becomes small as observed in the pseudogap region. We show the selfenergy part in Fig. 7 and the spectrum in Fig. 8, respectively.

From these results, we can obtain the T-linear specific coefficient in the pseudogap region,

$$\gamma = \frac{2\pi^2 k_B 2}{3} \sum_k \rho_k(0) \left[1 - \frac{\partial \Sigma_k(\omega)}{\partial \omega}\Big|_{\omega=0} \right]. \tag{11}$$

$$\rho_k(0) = -\frac{1}{\pi} \mathrm{Im} G_k(0) = A(k,0). \tag{12}$$

These calculations have been extended to the self-consistent ones [2] and confirmed that the pseudogap states are actually realized as stable states.

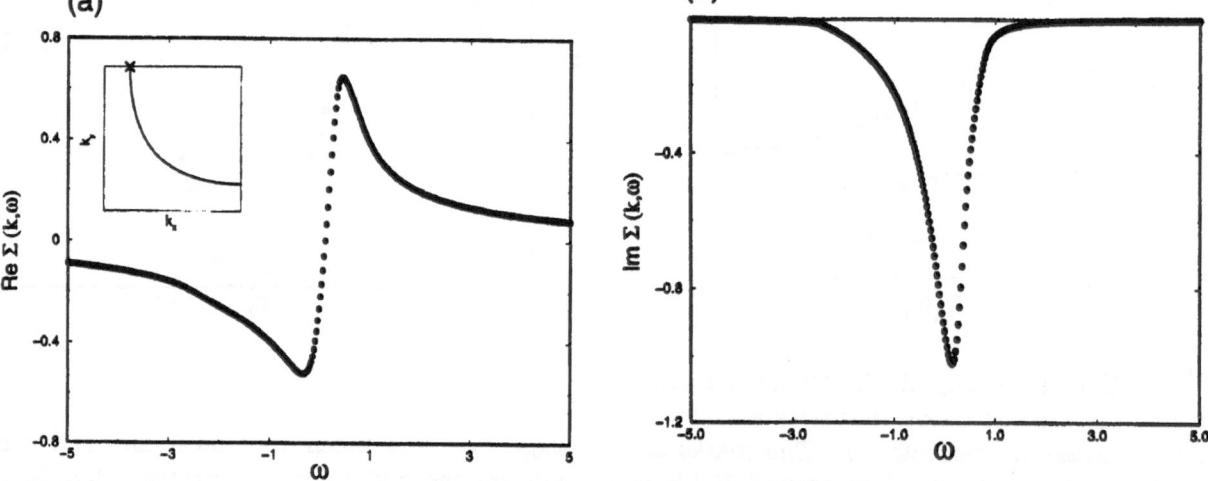

Fig. 7. The single particle self-energy on the Fermi surface near $(0,\pi)$ obtained by the lowest order calculation. (a) The real part. (b) The imaginary part. Here, $k = (0.589, \pi)$, attractive interaction $g_s = -1.0$, and $T = 0.21$. The k-point is shown in the inset. Here, $T_c = T_{\mathrm{MF}} = 0.185$.

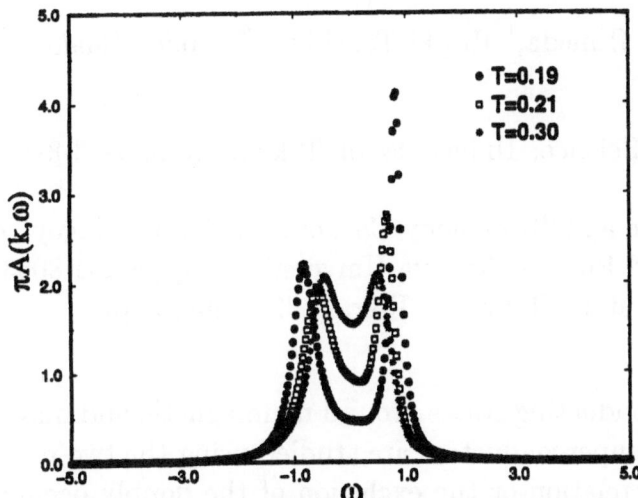

Fig. 8. The spectral weight for various temperatures $T = 0.19$ (circles), $T = 0.21$ (squares), $T = 0.30$ (stars). The other parameters are the same as those in Fig. 7.

TRANSPORT IN PSEUDOGAP REGION

According to the experimental results, with opening pseudogap the in-plane resistivity decreases, while the c-axis resistivity increases. These results can be explained as follows [1, 2]. With developing pseudogap near the hot spots, anti-ferromagnetic fluctuations are suppressed. Owing to this fact, the in-plane resistivity decreases rapidly with decreasing temperature. On the other hand, the c-axis conductivity which arises from the transfer between hot spots on the neighboring layers decreases, since the spectral weight near the hot spots decreases owing to the formation of pseudogap. The Hall coefficient decreases with decreasing temperature owing to the suppression of antiferromagnetic fluctuations.

CONCLUSION

On the basis of the Fermi liquid theory, we have explained the transport phenomena in the caprate superconductor taking into account the anitiferromagnetic fluctuation. Moreover, taking into account the superconducting fluctuation we have explained also the pseudogap phenomena, where the instability of the Fermi liquid state begins.

1. Y. Yanase and K. Yamada, J. Phys. Soc. Jpn. **68**, 548-560 (1999).
2. Y. Yanase and K. Yamada, J. Phys. Soc. Jpn. **68**, 2999-3015 (1999).
3. K. Yamada, K. Yosida and K. Hanzawa, Prog. Theor. Phys. Sup. No.108, 141-171 (1992).
4. H. Kohno and K. Yamada, Prog. Theor. Phys. **80**, 623-643 (1988).
5. H. Kontani, K. Kanki and K. Ueda, Phys. Rev. B59, 14723-14739 (1999). K. Kanki and H. Kontani, J. Phys. Soc. Jpn. **68**, 1614-1624 (1999).
6. O.K. Anderson, A.I. Liechtensteln, O. Jepsen and F. Paulsen, J. Chem. Solids, **55**, 1573-1591 (1995).

Impurities and Vortex Cores in the t-J Model

Masao Ogata,[1] Akihiro Himeda,[1] Hiroki Tsuchiura,[2] Yukio Tanaka,[2,3] and Satoshi Kashiwaya[2,4]

[1]Department of Basic Science, University of Tokyo, Komaba 3-8-1, Meguro-ku, Tokyo 153-8902, Japan
[2]CREST, Japan Science and Technology Corporation (JST), Nagoya 464-8063, Japan
[3]Department of Applied Physics, Nagoya University, Nagoya 464-8063, Japan
[4]Electrotechnical Laboratory, Tsukuba, Ibaraki 305-8568, Japan

Abstract: The superconducting states around nonmagnetic and magnetic impurities and vortex cores in high-T_c superconductors are studied using the two-dimensional t-J model. The effect of strong correlation or the exclusion of the doubly occupancies is taken into account via Gutzwiller approximation. The spatial dependence of the order parameters and the local density of states are obtained from the numerical diagonalization of the Bogoliubov-de Gennes equation derived in the Gutzwiller approximation. Around a non-magnetic impurity, zero-energy states are found which can be compared with the recent STS experiments. It is shown that the degree of localization of the zero-energy state depends on the doping. On the other hand, around a magnetic impurity, we find a splitting of the zero-energy peak in the local density of states. Around the vortex core, it is found that the antiferromagnetic correlation develops inside the core when the system is close to the boundary to the antiferromagnetic region.

Keywords: nonmagnetic impurity, magnetic impurity, vortex core, t-J model, zero-energy state, scanning tunneling spectroscopy

INTRODUCTION

The bulk properties of the in-plain superconducting state in high-T_c materials are consistent with the BCS superconductivity with the pair potential of the two-dimensional $d_{x^2-y^2}$-wave symmetry. However in several cases of inhomogenity, such as interfaces and in the vicinity of impurities and vortices, novel interference effects are expected due to the sign change of the pair potential [1,2]. Basically this sign change leads to a formation of a zero-energy state. Recently scanning tunneling spectroscopy (STS) has developed intensively so that the direct observation of local density of states becomes possible with atomical spatial resolution. If the zero-energy state is formed, it shows up as a zero-energy peak in the STS experiments which have actually succeeded near the vortex core [3] and surfaces [4]. Furthermore very recently the zero-energy peak is clearly observed in the vicinity of nonmagnetic impurity in BSCCO [5]. However some experiments have shown that there are unexpected behaviors [3,6]. These anomalies will be due to the strong two-dimensionality and/or strong correlation characteristic to the high-T_c superconductors. Thus we study the d-wave superconducting states using the two-dimensional t-J model in which we can elucidate the effect of strong correlation as well as the doping dependences [7-10]. Since the main interaction in the t-J model is the superexchange interaction, there appear interesting phenomena relating to the magnetism in the superconducting state.

One of the problems is the electronic states around nonmagnetic and magnetic impurities, and the other is the antiferromagnetic vortex core.

In this paper we first study the interference effect in the nonmagnetic impurity problem. It has been shown that the lowest eigenvalue approaches zero as the impurity scattering approaches the unitary limit [11,12]. We show that this zero-energy state actually gives the zero-energy peak, which can be observed in STS around impurities [10]. Here we also study the doping dependence of the zero-energy state to show that in the underdoped region this state is localized around the impurity, while it has a slow decay ($\sim 1/r$) along the nodes of the gap in the overdoped region. Next we study the *magnetic* impurity case and find a splitting of the zero-energy peak in the local density of states. This can be directly checked with STS experiments.

The microscopic structure of vortex cores in the high-T_c superconductors turns out to be a very interesting subject because recent STS observations of vortex cores in YBCO[3] and BSCCO[6] show unexpected behaviors. There are several discrepancies between the previous theories and experiments. Theoretically a zero-energy peak in the STS near the vortex core is expected [13-15], but experimentally a splitting of the zero-energy peak is observed for YBCO and the density of states inside the core is strongly suppressed in BSCCO. These clearly contradict the previous theories for d-wave superconductivity. We showed that it is possible to resolve the discrepancy for YBCO if we use the t-J model and consider the induced s-wave component [8]. In this paper we discuss another possibility of the antiferromagnetic vortex core which can be an explanation for the discrepancy for the STS experiments for BSCCO.

t-J MODEL AND GUTZWILLER APPROXIMATION

The Hamiltonian of the t-J model is written as

$$\mathcal{H} = -t \sum_{\langle i,j \rangle \sigma} (c_{i\sigma}^{\dagger} c_{j\sigma} + \text{h.c.}) - t' \sum_{\langle i,j \rangle' \sigma} (c_{i\sigma}^{\dagger} c_{j\sigma} + \text{h.c.}) + J \sum_{\langle i,j \rangle} \boldsymbol{S}_i \cdot \boldsymbol{S}_j - \mu \sum_{i,\sigma} c_{i\sigma}^{\dagger} c_{i\sigma}, \quad (1)$$

where $\langle i,j \rangle$ ($\langle i,j \rangle'$) means the summation over (next) nearest-neighbor pairs. In order to study the electronic states in inhomogeneous systems, we assume site-dependent order parameters, $\Delta_{ij} = \langle c_{i\uparrow} c_{j\downarrow} \rangle$. The variational state is given by $P_G|\text{BCS}(\Delta_{ij})\rangle$ with $P_G = \prod_i (1 - \hat{n}_{i\uparrow} \hat{n}_{i\downarrow})$ being a Gutzwiller projection operator. Δ_{ij} are determined so as to minimize the variational energy $E_{\text{var}} = \langle \text{BCS}(\Delta_{ij})|P_G \mathcal{H}_{tJ} P_G|\text{BCS}(\Delta_{ij})\rangle$. It is usually difficult to estimate the variational energy, E_{var}, due to the Gutzwiller projection. Here we use a Gutzwiller approximation [16,7], in which the constraint is taken into account as a statistical average. This leads to

$$E_{\text{var}} = \langle \text{BCS}(\Delta_{ij})|\mathcal{H}_{\text{eff}}|\text{BCS}(\Delta_{ij})\rangle,$$

where the parameters t and J are replaced with t_{ij}^{eff} and J_{ij}^{eff} in the effective Hamiltonian $\mathcal{H}_{\text{eff}}$ [16-18]. Minimizing E_{var}, we obtain a Bogoliubov-de Gennes equation together with a self-consistent equation for Δ_{ij}. We solve numerically the Bogoliubov-de Gennes equation and carry out an iteration until the self-consistent equation for Δ_{ij} is satisfied. Thus we obtain the fully quantum results. The effect of strong correlation is taken into account by Gutzwiller approximation. The local density of states are also obtained for various J/t and the doping rate δ.

NONMAGNETIC IMPURITY CASE

For studying the effect of nonmagnetic impurity, we add an impurity potential

$$\sum_{i\sigma} V_i^{\text{imp}} n_{i\sigma}, \qquad (2)$$

in addition to the t-J model. The impurity potentials V_i^{imp} are nonzero only on the impurity sites and we assume $V_i^{\text{imp}} = 1000t$ to consider the unitary scattering limit which is relevant to the Zn impurity. We regard the $N_L \times N_L$ square lattice as a unit cell of which the impurity is located at the center. We find that it is necessary to take the size of the unit cell large enough, otherwise the numerical calculation gives a fictitious peaks as in ref. [19].

Figure 1(a) shows the obtained order parameters for $\delta = 0.15$, which are decomposed into extended s-wave and d-wave components. From Fig. 1(a), one can see that the d-wave order-parameter is suppressed within an about 5 lattice constants around the impurity site. Simultaneously we find an s-wave component induced around the impurity but its amplitude is less than 1% relative to the bulk d-wave component.

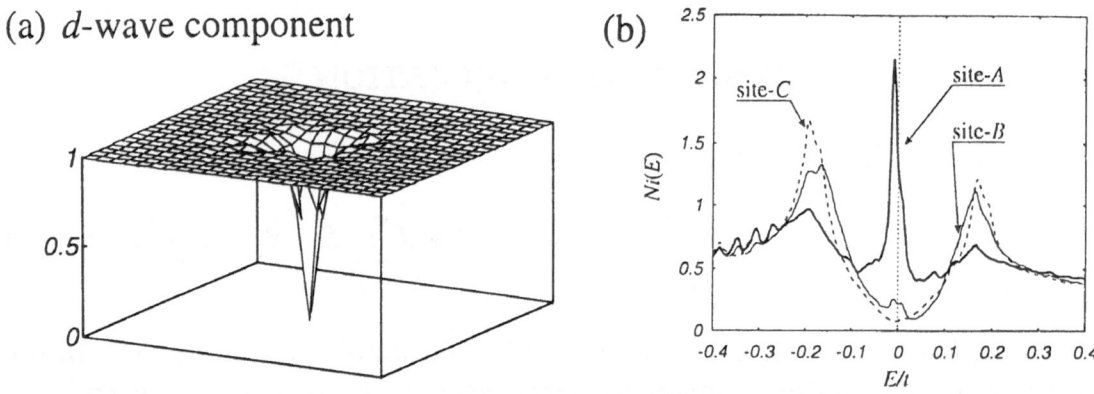

Fig. 1: (a) Spatial dependence of the d-wave component of the pair potential normalized by the bulk value and (b) the local density of states at three sites. See the text for the positions of the sites, A, B and C. The doping rate is $\delta = 0.15$ and $t'/t = -0.4$, $J/t = 0.2$. t' is chosen to reproduce the Fermi surface observed experimentally. The size of the unit cell is 27×27 sites.

In the same approximation we calculate the local density of states around the impurity:

$$N_i(E) = \frac{1}{N_c} \sum_{k\alpha} \{|u_i^\alpha(\boldsymbol{k})|^2 \delta(E^\alpha(\boldsymbol{k}) - E) + |v_i^\alpha(\boldsymbol{k})|^2 \delta(E^\alpha(\boldsymbol{k}) + E)\}, \qquad (3)$$

with i representing a site and N_c being the number of the magnetic unit cells. This quantity is directly related to the tunnel conductance of the STS experiment. Figure 1(b) shows the LDOS obtained on various sites, A, B and C. On the sites located on the edge of the unit cell (site-C), we actually find the typical DOS for d-wave superconductivity. On the nearest-neighbor site of the impurity (site-A), the LDOS detects the zero-energy states as the zero-energy peak. This is due to the fact that, in the unitary scattering limit, there always exist some impurity-scattering processes in which the quasiparticles feel the sign-change of the pair potential. Different from the case with vortex or surface, the splitting of the zero-energy peak does not appear since the induced s-wave component is real and rather small. It is interesting that the local DOS at site B, which is next to the site A, does not have a zero-energy peak. This means that the zero-energy state has a strong spatial variation, which is a Friedel oscillation in the vicinity of half-filling. It is under way to compare our results with the recent STS experiments near the Zn impurities observed in BSCCO [5].

The spatial dependence of the zero-energy state shows that it is approximately localized around the impurity and there are no $1/r$-tails along the diagonal direction. This behavior is consistent with the results in refs. [20] and [21]. An interesting question is how the zero-energy state behaves in the overdoped region. To examine the doping dependence of the zero-energy state, attention should be given to the size of the unit cell since the coherence length becomes longer with increasing δ. Taking account of this point, we study the doping dependence of the zero-energy state. We find that in the overdoped region the zero-energy state has a slow decay ($\sim 1/r$) along the nodes of the gap, which is consistent with the results in refs. [11] and [22]. This indicates that there is an interesting doping dependence.

MAGNETIC IMPURITY CASE

In the same approximation we study the case of magnetic impurity. In order to take account the effect of the magnetism of the impurity, we assume an effective magnetic field on the nearest-neighbor sites of the impurity by regarding the fixed impurity moment.

$$H_{\text{eff}} = J'\langle S_i^z \rangle. \tag{4}$$

The result is shown in Fig. 2. We find that the zero-energy peak splits into two peaks, which is to be checked in STS experiments. We confirm that the splitting is due to the effective magnetic field because, its amplitude is roughly proportional to the amplitude of J'.

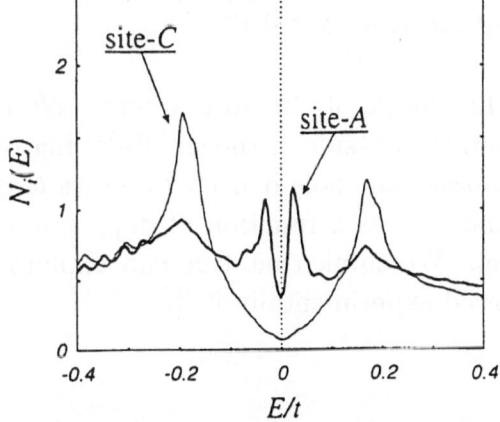

Fig. 2: The local density of states at the nearest-neighbor site A of the *magnetic* impurity, compared with the bulk density of states for $\delta = 0.15$ and $t'/t = -0.4$, $J/t = 0.2$.

VORTEX CORE

The uniform magnetic field is introduced in terms of Peierls phase of the hopping term as $t_{ij} = t_{eff} \exp(i \frac{e}{\hbar c} \int_i^j \boldsymbol{A} \cdot \boldsymbol{dr})$. We assume a square vortex lattice and each $N \times N$ site has one vortex; typically we use $N = 18$. We find that that an extended s-wave component is induced around the core with a four-fold symmetry. This is characteristic in the low-doping regime [8]. As a result of this induced s-wave component, the zero-energy peak in the local density of states splits. We expect that this results may explain the recent experiments of scanning tunneling spectroscopy in $YBa_2Cu_3O_{7-\delta}$.

Another interesting possibility for the vortex states in the t-J model is an antiferromagnetic vortex core which was expected in the SO(5)-symmetric theory [23]. Since the d-wave superconducting order parameter is suppressed inside the core, the antiferromagnetic local moment may show up in the core. We studied this possibility using the modified Gutzwiller approximation where the antiferromagnetic local moment is taken into account [17,18]. For example, in Fig. 3 we show the local density of states near the vortex core when some antiferromagnetic moment is induced in the case of the doping rate $\delta = 0.10$. This indicates that the zero-energy states are strongly suppressed due to the presence of the local antiferromagnetic excitation gap. We speculate that the BSCCO has a larger value of J/t than YBCO, and thus BSCCO is close to the boundary to the antiferromagnetically ordered state in the phase diagram and the induced antiferromagnetic short-range correlation is larger. This can be the origin of the difference between the BSCCO and YBCO observed experimentally.

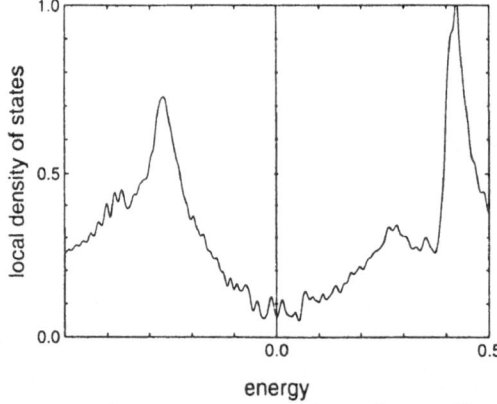

Fig. 3: Local density of states near the vortex core when the antiferromagnetic moment is induced inside the core found in the t-J model for $\delta = 0.10$.

Finally we briefly discuss the charge of the vortex core. We find that the hole density inside the core is *smaller* than the outside in the small-doping region. On the other hand, when the doping rate is increased, the hole density becomes *larger* than the outside. This change of the charging of the core as a function of doping is unusual and characteristic to the underdope t-J model. We think that this can explain the Hall anomaly in the superconducting state observed experimentally [24].

Summary

We have shown that the zero-energy state gives zero-energy peak in the LDOS around a nonmagnetic impurity in the d-wave superconducting state of the underdoped t-J model.

The induced *s*-wave component is real but small and thus there is no splitting of the zero-energy peak. Furthermore we found that the degree of localization of the zero-energy state depends on the doping rate; in the underdoped region it is approximately localized (radius $5a$) around the impurity, while in the overdoped region it has a slow $1/r$ decay. As for the magnetic impurity, we predict the splitting of the zero-energy paek which is ready to be checked experimentally. We also study the microscopic structure of vortex cores and electronic states around the core in the two-dimensional t-J model. Especially in the low-doping region, we find the possible antiferromagnetic vortex core. In this case the local density of states inside the core is strongly suppressed. These results especially in the underdoped region may explain the anomalies in the STS experiments. Thirdly, the sign of charge of the vortex core changes as a function of doping. This can be related to the Hall anomaly.

Acknowledgements. The author wishes to thank H. Fukuyama, T. M. Rice, F. C. Zhang, Ø. Fischer, Ch. Renner, Y. Matsuda, K. Machida, M. Sigrist, and M. Ichioka for their useful discussions.

1. C. R. Hu, Phys. Rev. Lett. **72**, 1526 (1994).
2. Y. Tanaka and S. Kashiwaya, Phys. Rev. Lett. **74**, 3451 (1995); Phys. Rev. B **53**, 9371 (1996); Phys. Rev. B **53**, 11957 (1996).
3. I. Maggio-Aprile, Ch. Renner, A. Erb, E. Walker and O. Fischer, Phys. Rev. Lett. **75**, 2200 (1995).
4. L. Alff, H. Takashima, S. Kashiwaya, N. Terada, H. Ihara, Y. Tanaka, M. Koyanagi and K. Kajimura, Phys. Rev. B **55**, 14757 (1997).
5. S. H. Pan, E. W. Hudson, K. M. Lang, H. Eisaki, S. Uchida, and J. C. Davis, cond-mat/9909365.
6. Ch. Renner, B. Revaz, K. Kadowaki, I. Maggio-Aprile, and Ø. Fischer, Phys. Rev. Lett. **80**, 3606 (1998).
7. H. Yokoyama and M. Ogata, J. Phys. Soc. Jpn, **65**, 3615 (1996).
8. A. Himeda, M. Ogata, Y. Tanaka and S. Kashiwaya, J. Phys. Soc. Jpn. **66**, 3367 (1997).
9. Y. Tanuma, Y. Tanaka, M. Ogata and S. Kashiwaya, J. Phys. Soc. Jpn. **67**, 1118 (1998).
10. H. Tsuchiura, Y. Tanaka, M. Ogata and S. Kashiwaya, J. Phys. Soc. Jpn. **68**, 2510 (1999).
11. Y. Onishi, Y. Ohashi, Y. Shingaki and K. Miyake, J. Phys. Soc. Jpn. **65** 675 (1996).
12. A. V. Balatsky, M. I. Salkola and A. Rosengren, Phys. Rev. B **56** 15547 (1995).
13. Y. Wang and A. H. MacDonald, Phys. Rev. B **52**, R3876 (1995).
14. N. Schopohl and K. Maki, Phys. Rev. B **52**, 490 (1995).
15. M. Ichioka, N. Hayashi, N. Enomoto, and K. Machida, Phys. Rev. B **53**, 15316 (1996).
16. F. C Zhang, C. Gros, T. M. Rice, and H. Shiba, Supercond. Sci. Technol. **1**, 36 (1988).
17. A. Himeda and M. Ogata, to appear in Phys. Rev. B.
18. A detailed estimation of t_{eff} and J_{eff} will be published elsewhere.
19. H. Tanaka, K. Kuboki and M. Sigrist, Int. J. Mod. Phys. B **12** 2447 (1998).
20. P. A. Lee, Phys. Rev. Lett. **71** 1887 (1993).
21. M. Franz, C. Kallin and A. J. Berlinsky, Phys. Rev. B **54** 6897 (1996).
22. A. V. Balatsky and M. I. Salkola, Phys. Rev. Lett. **76** 2386 (1996).
23. D. P. Arovas, A. J. Berlinsky, C. Kallin and S. C. Zhang, Phys. Rev. Lett. **79**, 2871 (1997).
24. T. Nagaoka, Y. Matsuda, H. Obara, A. Sawa, T. Terashima, I. Chong, M. Takano and M. Suzuki, Phys. Rev. Lett. **80**, 3594 (1998).

Inplane and Interplane Charge Dynamics in Layered Manganite Crystals

T. Kimura [1], T. Okuda [1], and Y. Tokura [1,2]

[1] Joint Research Center for Atom Technology (JRCAT), Tsukuba 305-0046, Japan
[2] Department of Applied Physics, University of Tokyo, Tokyo 113-0033, Japan

Abstract: We have investigated the inplane and interplane charge dynamics in bilayered manganite crystals, $La_{2-2x}Sr_{1+2x}Mn_2O_7$ ($0.3 \leq x \leq 0.5$), which show the highly anisotropic charge-transport and magnetic properties, and the colossal magnetoresistance. In the layered magnetic system, the interplane as well as inplane charge dynamics is expected to critically depend on the interlayer magnetic coupling within and between the ferromagnetic-metallic (FM) MnO_2 bilayers. The interlayer magnetic coupling shows the strong doping-level dependence, and is related to not only the magneto-transport properties but also the magneto-structural ones. We have also studied the low-temperature transport properties. The weak localization effect as evidenced by $T^{1/2}$-dependence of the inplane conductivity was observed for the ferromagnetic bilayered manganite, which indicates the layered magnetic system is a disordered three-dimensional metal in the ground state.

Keyword: Layered manganite, Anisotropy, Magnetoresistance

INTRODUCTION

Although recent observations of the large negative magnetoresistance (MR) effect have shed renewed light on the study of perovskite manganites, a great deal of underlying physics attracted the broad interest in the study of these systems. Recently extensive studies have been performed in the so-called Ruddlesden-Popper (RP) structure series for manganese oxides which are characterized by the expression $(R, A)_{n+1}Mn_nO_{3n+1}$ (R and A being trivalent rare earth or divalent alkaline earth ions, respectively). The basic structure in this homologous series appears to be based on alternate stacking of rocksalt-type block layers $(R, A)_2O_2$ and n MnO_2-sheets along the c-axis. The $n=2$ member of the RP series for manganites, $La_{2-2x}Sr_{1+2x}Mn_2O_7$ show a wide variety of field-induced phenomena as well as $n = \infty$ pseudo-cubic compounds, including a colossal magnetoresistance (CMR) effect [1] related to paramagnetic insulator−to−ferromagnetic metallic transition. The charge-ordering transition has also been observed in the bilayered manganite with the carrier concentration $x=0.5(=1/2)$ [2]. Another remarkable feature for bilayered manganite is the interplane tunneling magnetoresistance (TMR) effect which provides a novel approach to the large MR attainable at low magnetic fields [3]. Several kinds of the ferromagnetic tunneling junctions with use of perovskite manganites have so far been investigated, such as trilayer junctions [4], granular polycrystalline samples [5], and thin-films on bicrystal substrates [6]. The bilayered compound, however, is composed of ferromagnetic-metallic (FM) MnO_2 bilayers with intervening insulating (I) $(La,Sr)_2O_2$ blocks. In other word, the bilayered manganite intrinsically contains the infinite arrays of FM/I/FM tunneling junctions in its crystal structure. Here, we present the systematic study on the magneto-transport, magnetic, and magneto-elastic properties of $La_{2-2x}Sr_{1+2x}Mn_2O_7$ ($0.3 \leq x \leq 0.5$) crystals.

EXPERIMENTAL

A series of $La_{2-2x}Sr_{1+2x}Mn_2O_7$ ($0.3 \leq x \leq 0.5$) crystals was grown by the floating zone method. The grown crystals were characterized by the powder and single-crystal x-ray diffraction (XRD) measurements to ensure single phase of the bilayer structure. The grown crystals were oriented using x-ray back-reflection Laue technique, and cut out into rectangular slab specimens. Resistivity measurements were made by the conventional four-probe technique with current parallel (ρ_{ab}) and perpendicular (ρ_c) to MnO_2 bilayers. ^3He-^4He dilution refrigerator was used for the measurement in a low temperature region down to 30 mK. Measurements of specific heat were done by using the relaxation method. The striction was measured using a uniaxial strain gauge which was attach to the widest face of a specimen.

CRYSTAL AND MAGNETIC STRUCTURE

We display in Fig. 1 the lattice parameters (a), the schematic phase diagram (b), and lattice distortion and Mn-O bond lengths (c) of x=0.3, 0.4, and 0.5 crystals at room temperature [7]. The lattice parameter of the $a(b)$-axis slightly increases with increase of x, whereas that of the c-axis decreases more rapidly. Variation of the lattice parameters is closely connected with that of the Mn-O bond lengths. The Mn-O(1) bond length along the $a(b)$-axis inconsiderably elongates with increasing x, while the Mn-O(2) bond length along the c-axis steeply shrinks. Since the decrease of x corresponds to the e_g-electron doping, the systematic change of the Mn-O bond lengths reflects the orbital-state occupancy of the e_g electrons. A remarkable expansion of the out-of-plane Mn-O(1) bond length with lower hole concentration x (or higher electron concentration) implies that the doped electrons prefer to occupy the $3z^2 - r^2$ orbital-state. Thick arrows on the right hand of the respective schematic crystal structures in Fig. 1(c) illustrate a variation of the low-temperature spin structure within a bilayer unit as a function of the doping level, which has been revealed by recent neutron diffraction studies [8, 9, 10, 11, 12]. The magnetic moments in the ground state always couple ferromagnetically within the constituent *single* MnO_2 layer. However, the coupling between the respective MnO_2 layers within a bilayer unit depends on the doping level. In a low-doped region around $x \approx 0.3$, the magnetic moments of respective *single* MnO_2 layers couple ferromagnetically within a bilayer, and align along the c-axis. As x is increased, the magnetic moments direct along the ab-plane and show canted antiferromagnetism beyond $x \approx 0.4$. With further increasing x toward x=0.5, the A-type AF order at approximately 210 K. The magnetic structure of the antiferromagnetic (AF) phase can be viewed as the layered antiferromagnetism, in which the magnetic moments lie in the ab-plane, couple ferromagnetically within the constituent single MnO_2 layer but show AF order between the respective MnO_2 layers within a bilayer unit. Such a evolution of the magnetic structure which strongly depends on the doping level may have a close connection to the change in orbital state.

In the light of the charge/orbital-ordering at the doping level with the commensurate value of x=0.5(=1/2), the bilayered manganite shows a complicated temperature dependence of charge/orbital ordered state [13]. We display in Fig. 2 the temperature profiles of (a) the resistivity, (b) the intensity of superlattice reflections, $(\frac{7}{4}, \frac{1}{4}, 0)$ and $(\frac{7}{4}, -\frac{1}{4}, 0)$ normalized by that of (2, 0, 0) reflection which was obtained by single crystal XRD measurements. The XRD measurements confirmed the successive reentrant structural changes: Weak superlattice reflections arising from the orbital-ordering show up at $(h\pm\frac{1}{4}, k\pm\frac{1}{4}, 0)$ below $\sim$210 K, as shown in Fig. 2(b). With further decreasing temperature below $\sim$100 K, the superlattice spots disappear again in our experimental resolution. Figure 2(a) shows a steep rise of ρ_{ab} toward lower temperatures around 210$\sim$220 K where the superlattice reflections appear [Fig. 2(b)]. With further decreasing temperature, both ρ_{ab} and ρ_c exhibit a broad peak centered at $\sim$170 K, and decrease toward lower temperature in accordance with the suppression of the intensity of the superlattice reflections. Another noteworthy feature in Fig. 2(a) is presence of a remarkable hysteresis between the cooling and warming runs in the temperature region of $\sim$70 K$\leq T \leq \sim$210 K. Below $\sim$70 K where the superlattice reflections is not observed, the thermal hysteresis also disappears, and the both ρ_{ab} and ρ_c increase with decreasing temperature although the temperature dependence is much weaker than the Arrhenius-type thermal activation. The hysteretic behavior and the peak structure centered at $\sim$170 K in the resistivity appears to be with the charge/orbital-ordering whose pattern is illustrated in the inset of Fig. 2. The reentrant disappearance of the superlattice reflections may be attributed to the switching of the orbital-ordered state from the staggered $d_{3x^2-r^2}/d_{3y^2-r^2}$ to the uniform $d_{x^2-y^2}$. The charge/orbital ordered state in $LaSr_2Mn_2O_7$ appears to be altered to the layered antiferromagnetic (AF) state with the uniform ordering of $d_{x^2-y^2}$ orbitals by lowering temperature.

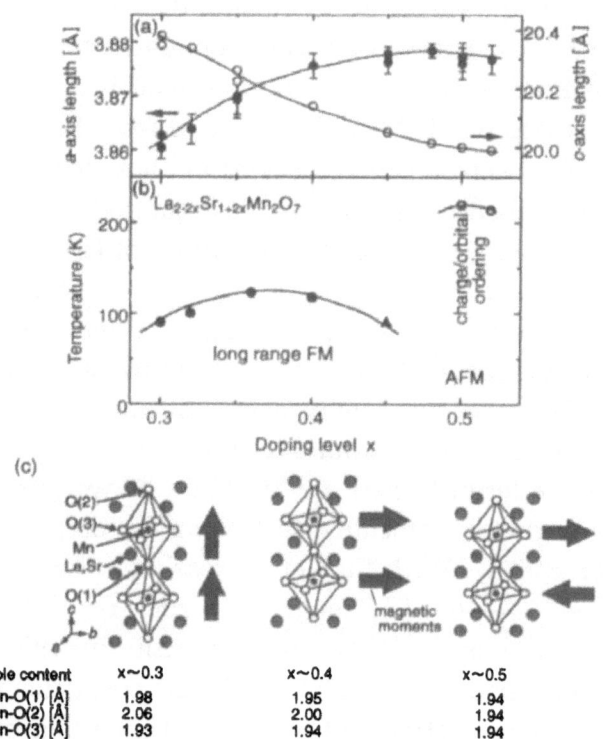

Fig. 1: The schematic phase diagram (b), doping-level dependence of the lattice parameters (a) and lattice distortion (c) at room temerature of $La_{2-2x}Sr_{1+2x}Mn_2O_7$. Thick arrows on the right hand of respective crstal structures indicate the spin structures within a bilayer unit at low temperatures.

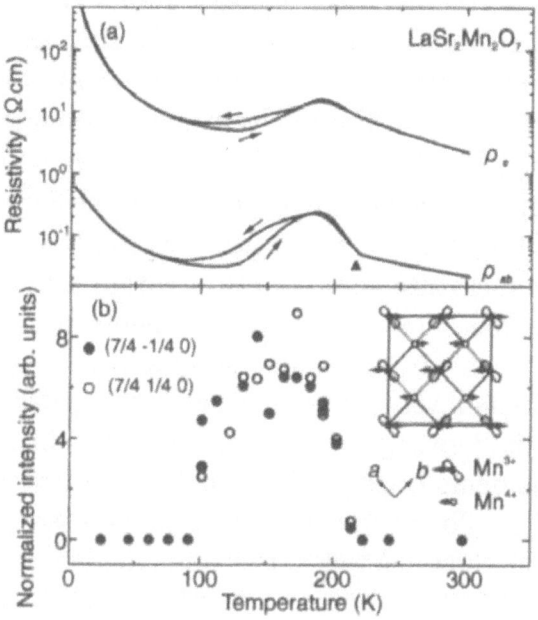

Fig. 2: Temperature dependence of (a) inplane (ρ_{ab}) and interplane (ρ_c) resistivity, (b) integrated intensity of ($\frac{7}{4}, -\frac{1}{4}, 0$) and ($\frac{7}{4}, \frac{1}{4}, 0$) superlattice reflection [normalized by fundamental (2,0,0) reflection] characteristics of the orbital-ordered state, measured with an x-ray diffraction imaging plate for a $LaSr_2Mn_2O_7$ crystal. Inset: Schematic view of possible charge and orbital ordering in $LaSr_2Mn_2O_7$.

MAGNETOTRANSPORT AND MAGNETOSTRUCTURAL PROPERTIES

We show the strong correlation between magnetotransport and magnetoelastic properties in the bilayered manganite. Figure 3 displays temperature profiles of the resistivity and the striction along the ab-plane ($\Delta L_{ab}/L_{ab}$) and the c-axis ($\Delta L_c/L_c$) in several magnetic fields with $H\|ab$ for the crystals with $x=0.3$ and 0.4 [7]. Both the crystals show gigantic magnetostriction which is closely correlated with the CMR effect, as clearly seen in Fig. 3. A steep drop in resistivity (or the CMR effect) is observed at T_c for $x=0.3$ and $x=0.4$ crystals. For both the crystals, the specific heat measurement in a high H shows a finite electronic specific-heat constant γ ($\sim 3\pm1$ mJ/K^2) at the FM phase [Fig. 4(a)], which is suggestive of a finite density of state at the Fermi level [14]. Furthermore, as shown in Fig. 4(b) the $T^{1/2}$ dependence of the inplane conductivity and the finite value of the inplane zero conductivity indicates that in a strict sense the ferromagnetic bilayered manganite is a disordered 3D metal in the ground state, accompanying a weak localization effect [15].

For the $x=0.3$ crystal which shows a positive change in spontaneous lattice striction along the ab-plane toward the spin-ordered state, $\Delta L_{ab}/L_{ab}$ considerably increases by applying H above T_c [Fig. 3(a)], whereas $\Delta L_c/L_c$ decreases [Fig. 3(c)]. On the other hand, in the $x=0.4$ crystal with negative spontaneous lattice striction along the ab-plane, $\Delta L_{ab}/L_{ab}$ decreases, yet $\Delta L_c/L_c$ increases by applying H [Figs. 3(b) and 3(d)]. In both the $x=0.3$ and $x=0.4$ crystals, the magnetostriction as well as the MR is enhanced around T_c, and becomes small in going away from T_c. However, the sign and magnitude of magnetostriction strongly depend on the doping level.

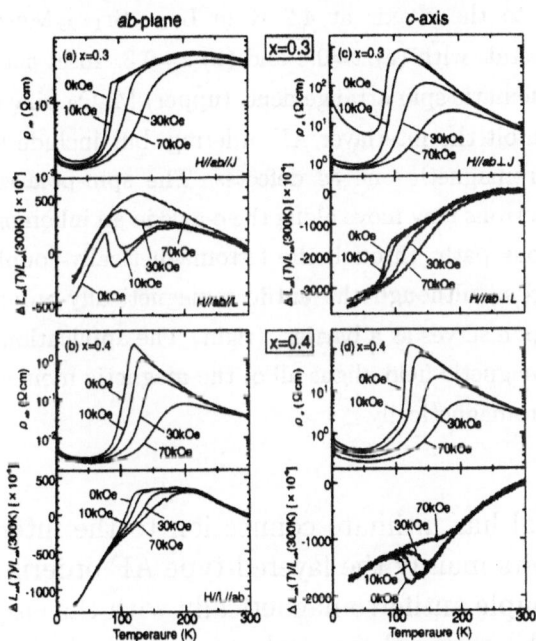

Fig. 3: Temperature dependence of inplane resistivity (ρ_{ab}) and striction [$\Delta L_{ab}(T)/L_{ab}(300$ K)] and interplane resistivity (ρ_c) and striction [$\Delta L_c(T)/L_c(300$ K)] in several magnetic fields for La$_{2-2x}$Sr$_{1+2x}$Mn$_2$O$_7$ ($x=0.3$ and 0.4) crystals.

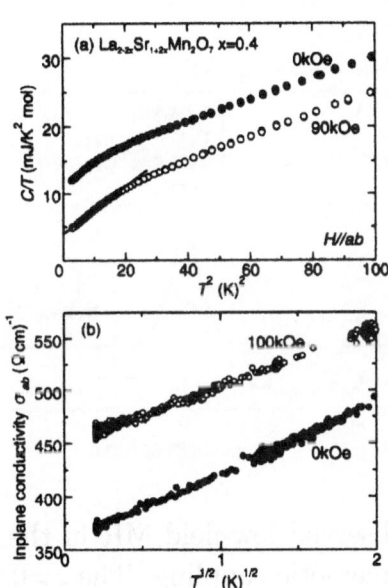

Fig. 4: (a) Temperature dependence of specific heat C as shown in C/T vs T^2 plot in 0 and 90 kOe. The exptrapolated (by a solid line) C/T value at 0 K in 90 kOe expresses the γ value. (b) Inplane conductivity σ_{ab} for an $x=0.4$ crystal in the low-temperature region down to 30 mK, plotted as a function of $T^{1/2}$.

To understand the origin of the field-induced striction in the bilayered manganite, we should take account of the orbital degrees of freedom of the e_g-like conduction electrons of Mn^{3+}. Since the orbital character is strongly affected by a lattice form, the observed structural distortion by changing temperature, magnetic field and doping level should reflect sensitively a change in the orbital state.

INTERPLANE TUNNELING MAGNETORESISTANCE

Recent progress of the giant magnetoresistance study in magnetic multilayers has shed light on the study of the TMR effect in the FM/I/FM junctions. A recent finding of the interplane TMR effect in a layered-AF $La_{2-2x}Sr_{1+2x}Mn_2O_7$ (x=0.3) crystal provides another novel approach to the large MR attainable at low magnetic fields [3]. The isothermal MR curve of an x=0.3 crystal (b) at 4.2 K is shown in Fig. 5 for both ρ_{ab} and ρ_c together with those of an x=0.4 crystal (a), for comparison. In the x=0.4 crystal with the ferromagnetic coupling between the bilayers, the MR ratio is very small in both ρ_{ab} and ρ_c. In the x=0.3 crystal, by contrast, the ρ_c drastically decreases in a low-field region in accordance with the magnetization process, and becomes constant when the magnetization is saturated at above an external field of $H_{sat} \approx 5$ kOe. Although a similar field dependence is also observed in the inplane MR, the interplane MR $[\rho_c(0) - \rho_c(H_{sat})]/\rho_c(H_{sat})$ as large as several hundreds % is much greater than the inplane MR.

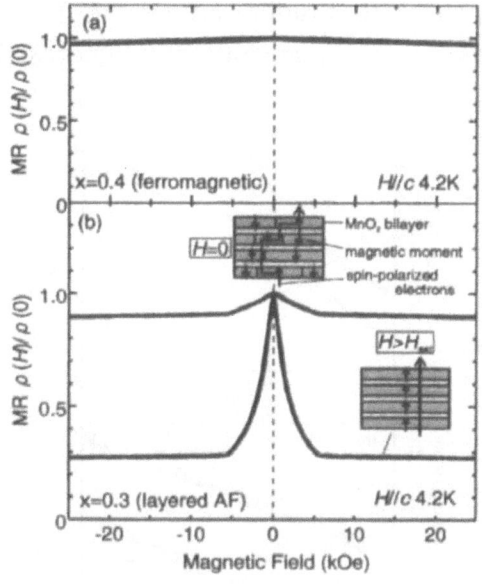

Fig. 5: Normalized inplane (ρ_{ab}) and interplane (ρ_c) resistivity as a function of a magnetic field parallel to the c axis at 4.2 K in $La_{2-x}Sr_{1+2x}Mn_2O_7$ crystals with (a) x=0.4 and (b) x=0.3. Inset shows schematic spin arrangement; (upper) MnO_2 bilayers exhibit the interlayer AF ordering, but include the ferromagnetic one as defects. The spin-polarized electrons may move along the c-axis in an inhomogeneous path, through the ferromagnetically-coupled region, although the antiferromagnetically-ordered region serves as a barrier. (right) The application of a magnetic field aligns all of the magnetic moments ferromagnetically.

The observed low-field MR in the x=0.3 crystal has intimate connection to the inter-bilayer magnetic coupling. The x=0.3 crystal shows mainly the layered-type AF ordering in which the ferromagnetically ordered bilayers couple antiferromagnetically with the easy axis along the c-axis. The c-axis transport of the spin-polarized electron is blocked at the insulating block due to the AF-type coupling between the adjacent MnO_2 bilayers, although the inter-bilayer coupling mode contains a high density of domain wall, i.e. the ferromagnetic interlayer coupling mode. Such a ferromagnetically interlayer coupled part may serve as a leaky current path along the c-axis (illustrated in the inset of Fig. 5). By applying a magnetic field, however, a sort of metamagnetic transition from the layered AF to the ferromagnetic state takes place at a low field due to an extremely weak AF inter-bilayer coupling. The magnetization process removes such AF-coupled carrier-blocking boundaries and allows the inter-bilayer tunneling of spin-polarized electrons, as illustrated in the right inset of Fig. 5. Such a field-switching of the magnetic structure in coincidence with the tunneling-type MR has been confirmed by a neutron diffraction measurements [10]. The bilayered manganite can be viewed as an infinite array of FM/I/FM junctions and/or an infinite stack of spin valves.

SUMMARY

We have performed systematic measurements of the inplane and interplane magneto-transport and magnetostructural properties for $La_{2-2x}Sr_{1+2x}Mn_2O_7$ ($0.3 \leq x \leq 0.5$) crystals which are viewed as composed of ferromagnetic-metallic MnO_2 bilayers with intervening non-magnetic insulating $(La,Sr)_2O_2$ blocks. Most of characteristic features observed in the pseudo-cubic perovskite manganite are reproduced in the bilayered manganite, such as the colossal magnetoresistance and the charge-ordering by the control of the doping level. However, novel phenomena such as the interlayer tunneling magnetoresistance have been found in the bilayered manganite, which is attributed to the variation of the intra-bilayer and inter-bilayer magnetic coupling.

Acknowledgments. We would like to thank K. Yamamoto, T. Ishikawa, Y. Tomioka, R. Kumai, and Y. Okimoto for helpful discussions. This work, supported in part by NEDO, was performed in JRCAT under the joint research agreement between NAIR and ATP.

[1] Y. Moritomo, A. Asamitsu, H. Kuwahara, and Y. Tokura, Nature **380**, 141 (1996).

[2] J. Q. Li, Y. Matsui, T. Kimura, and Y. Tokura, Phys. Rev. B **57**, R3205 (1998).

[3] T. Kimura, Y. Tomioka, H. Kuwahara, A. Asamitsu, M. Tamura, and Y. Tokura, Science **274**, 1698 (1996).

[4] J. Z. Sun, W. J. Gallagher, P. R. Duncombe, L. Krusin-Elbaum, R. A. Altman, A. Gupta, Yu Lu, G. Q. Gong, and Gang Xiao, Appl. Phys. Lett. **69**, 3266 (1996).

[5] H. Y. Hwang, S-W. Cheong, N. P. Ong, and B. Batlogg, Phys. Rev. Lett. **77**, 2041 (1996).

[6] N. D. Mathur, G. Burnell, S. P. Isaac, T. J. Jackson, B.-S. Teo, O. J. L. MacManus-Driscoll, L. F. Cohen, J. E. Evetts, and M. G. Blamire, Nature **387**, 266 (1997).

[7] T. Kimura, Y. Tomioka, A. Asamitsu, and Y. Tokura, Phys. Rev. Lett. **81**, 5920 (1998).

[8] P. D. Battle, D. E. Cox, M. A. Green, J. E. Millburn, L. E. Spring, P. G. Radaelli, M. J. Rosseinsky, J. F. Vente, Chem. Mater. **9**, 1042 (1997).

[9] D. N. Argyriou, J. F. Mitchell, C. D. Potter, S. D. Bader, R. Kleb, and J. D. Jorgensen, Phys. Rev. B **55**, R11965 (1997).

[10] T. G. Perring, G. Aeppli, T. Kimura, Y. Tokura, and M. A. Adams, Phys. Rev. B **58**, R14693 (1998).

[11] K. Hirota, Y. Moritomo, H. Fujioka, M. Kubota, H. Yoshizawa, and Y. Endoh, J. Phys. Soc. Jpn. **67**, 3380 (1998).

[12] M. Kubota, H. Fujioka, K. Ohoyama, K. Hirota, Y. Moritomo, H. Yoshizawa, and Y. Endoh, J. Phys. Chem. Solids (in press).

[13] T. Kimura, R. Kumai, Y. Tokura, J. Q. Li, Y. Matsui, Phys. Rev. B **58**, 11081 (1998).

[14] T. Okuda, T. Kimura, and Y. Tokura, Phys. Rev. B **60**, 3370 (1999).

[15] P. A. Lee and V. Ramakrishnan, Rev. Mod. Phys. **57**, 287 (1985).

Magnetic Correlations in Lightly-doped La$_{2-x}$Sr$_x$CuO$_4$

Masaaki Matsuda

The Institute of Physical and Chemical Research (RIKEN), Wako, Saitama 351-0198, Japan

Abstract: A review is given of the neutron scattering studies in the lightly-doped La$_{2-x}$Sr$_x$CuO$_4$, which exhibits insulating spin-glass behavior. Most remarkable feature is that the static spin corrrelations are incommensurate at low temperature across the entire insulating spin-glass region. The incommensurate positions imply a one-dimensional spin modulation which is rotated by 45° from that in the superconducting phase.

Keywords: La$_{2-x}$Sr$_x$CuO$_4$, spin-glass, stripe ordering, neutron scattering

INTRODUCTION

Extensive studies on the high-T_c superconducting copper oxides have revealed an intimate connection between magnetism and superconductivity [1]. The phase diagram of La$_{2-x}$Sr$_x$CuO$_4$ shows that the magnetic state exhibits a drastic change with Sr doping. The parent material La$_2$CuO$_4$ shows three-dimensional (3D) long-range antiferromagnetic (AF) ordering below ~325 K [2]. When Sr is doped in the material, the 3D AF ordering quickly disappears. However, the low temperature magnetic phase is replaced by a spin-glass phase in which elastic magnetic Bragg rods, originating from two-dimensional AF correlations, develop gradually [3,4].

In the superconducting samples, a remarkable feature is that magnetic correlations become incommensurate (IC) [5-7]. Detailed studies on hole concentration dependence of the magnetic excitations were performed by Yamada *et al.* [8] They found that the incommensurability (δ) is almost linear with hole concentration (x) below x~0.12 as shown in Fig. 1. Recently, static magnetic ordering is observed in superconducting La$_{1.88}$Sr$_{0.12}$CuO$_4$ [9,10]. The elastic magnetic peaks are observed at IC positions where magnetic excitation peaks are found. This is probably related with the stripe ordering of spin density wave and charge (hole) density wave found in La$_{2-y-x}$Nd$_y$Sr$_x$CuO$_4$, in which the stripe runs along the a_{tetra} or b_{tetra} axis (collinear stripe) [11].

INCOMMENSURATE SPIN CORRELATIONS IN THE CuO$_2$ PLANE

Very recently, Wakimoto *et al.* reported detailed hole concentration dependence of the spin correlations in the spin-glass phase ($0.03 \leq x \leq 0.06$) [12]. Most remarkable finding is new type of IC spin correlations in La$_{1.95}$Sr$_{0.05}$CuO$_4$. This was first interpreted as the rotation of 45° from the parallel satellites in the superconducting phase. Then the careful examination of the intensity profiles revealed [13] that there are only 2 satellite peaks along b_{ortho} while in superconducting compounds the IC peaks are located parallel to both the a_{tetra} and b_{tetra} axes. The magnetic peaks are observed at the IC positions $(1,\pm\varepsilon,0)$ and $(0,1\pm\varepsilon,0)$ with ε~0.064. The magnetic correlations strongly suggest the diagonal stripe, in which the stripe runs along the b_{ortho} axis. These results suggest that static magnetic correlations change from diagonal to collinear stripe at x=0.055±0.005 where the insulator-to-metal transition occurs.

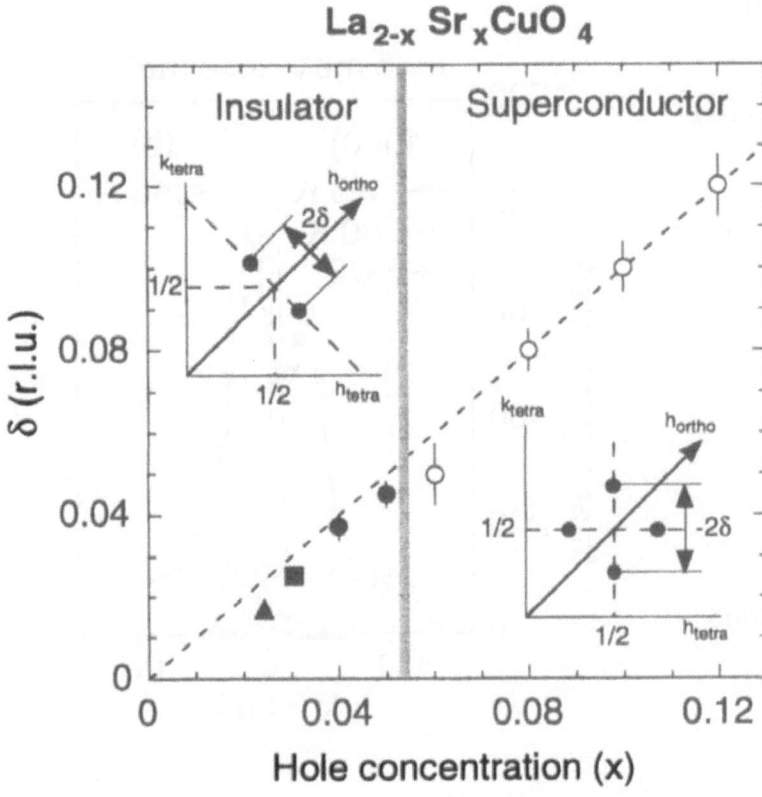

Fig. 1 Hole concentration (x) dependence of the splitting of the IC peaks δ in tetragonal reciprocal lattice units [15]. Open circles indicate the data for the inelastic IC peaks reported by Yamada et al. [8]. Filled circles and square are the data for the elastic IC peaks reported by Wakimoto et al. [13,19]. The filled triangle is obtained by Matsuda et al. [15]. The broken line corresponds to $\delta = x$. The insets show the configuration of the IC peaks in the insulating phase (diagonal stripe) and the superconducting phase (collinear stripe).

The next step is to clarify whether the IC magnetic correlations persist throughout the spin-glass phase down to the critical concentration of $x=0.02$ for 3D Néel ordering. Matsuda et al. first tried to check this with $La_{1.98}Sr_{0.02}CuO_4$ sample [14], which was grown by flux method in a platinum crucible. Due to a small magnetic impurity phase with a long-range AF ordering, it was difficult to perform measurements in the ($HK0$) scattering plane, which is necessary to determine whether the magnetic correlations are IC or not. Matsuda et al. have then performed neutron scattering experiments in $La_{1.976}Sr_{0.024}CuO_4$ [15], which was grown by travelling solvent floating zone technique. Below ~40 K elastic magnetic peaks develop and at low temperatures the peaks are clearly resolved at the IC positions $(1,\pm\varepsilon,0)$ and $(0,1\pm\varepsilon,0)$ with ε~0.0232. This corresponds to the same diagonal one-dimensional spin modulation observed in $La_{1.95}Sr_{0.05}CuO_4$ which has ε~0.064 [13]. The open and filled circles in Fig. 2(a) correspond to the IC magnetic peaks from the two domains in the ($HK0$) zone. Figures 2(b)-(d) show transverse and longitudinal elastic scans around $(1,0,0)$ and $(0,1,0)$. Two peaks are observed in the transverse scan A while one intense peak together with a weak shoulder on the low-h side is observed in the longitudinal scans B and C. The instrumental resolution at $(1,0,0)$ can be estimated from higher-order reflections, which in turn are measured by removing the Be filters. As illustrated in Figs. 2(b) and 2(c), the magnetic peaks are much broader than the resolution along both h and k.

The solid lines in Figs. 2(b)-(d) are the results of fits to a convolution of the resolution function with 3D squared Lorentzians. The two intense peaks in Fig. 2(b) originate primarily from the magnetic signals at $(1,1\pm\varepsilon,0)$ in domain A while the weak shoulder in Fig. 2(c) originates from magnetic signals at $(0,1-\varepsilon,\pm1)$ in domain B. The relatively intense peaks observed at $(0,1\pm\varepsilon,0)$ occur because of the short correlation length along the c axis, which in turn makes the $(0,1\pm\varepsilon,L)$ with L odd magnetic peaks broad along the c axis. The instrumental resolution function is also elongated along

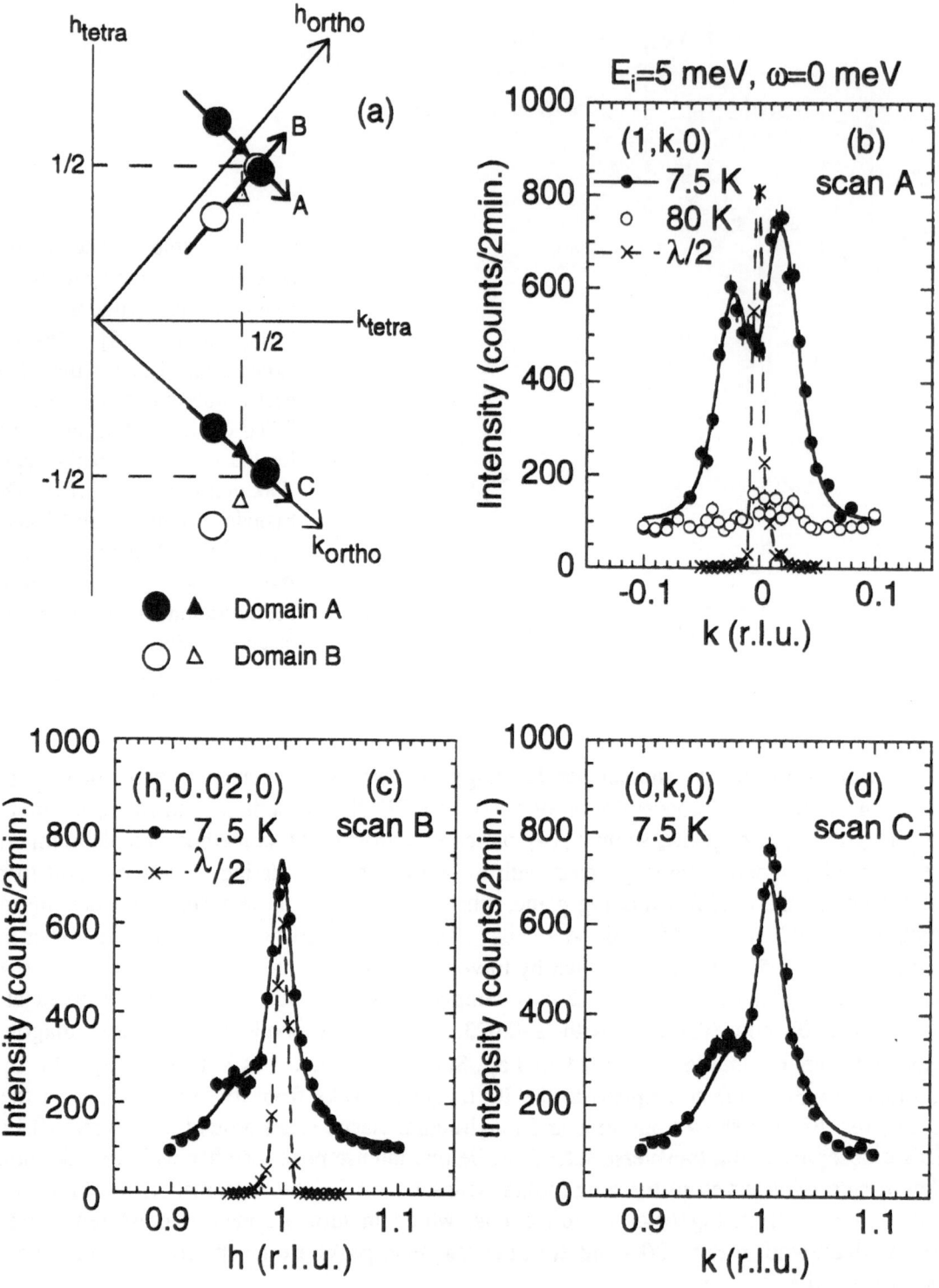

Fig. 2 (a) Diagram of the reciprocal lattice in the (HK0) scattering zone [15]. Filled and open symbols are for domains A and B, respectively. The triangles and circles correspond to nuclear and magnetic Bragg peaks, respectively. The thick arrows show scan trajectories. Transverse (b) and longitudinal elastic scans (c) and (d) around (1,0,0) and (0,1,0) at 7.5 K (filled circles) and 80 K (open circles) in $La_{1.976}Sr_{0.024}CuO_4$ [15]. The crosses represent the higher-order Bragg peaks observed at (1,0,0) by removing the Be filters. The broken lines are guides to the eyes. The peak width represents the instrumental resolution. The solid lines are the results of fits to a convolution of the resolution function with 3D squared Lorentzians with ξ_a=94.9 Å, ξ_b=39.9 Å, ξ_c=3.15 Å, and ε=0.0232 r.l.u.

the c axis so that the magnetic signals are effectively integrated. The observed data are fitted with ξ'_a=94.9±4.0 Å, ξ'_b=49.9±1.3 Å, ξ'_c=3.15±0.08 Å, and ε=0.0232±0.0004 r.l.u., where ξ'_a, ξ'_b, and ξ'_c represent the inverse elastic peak widths in Q along the a, b, and c axes, respectively. The calculation reproduces the observed profiles quite well.

The incommensurability ε corresponds to the inverse modulation period of the spin density wave. Here, ε is defined in orthorhombic notation so that $\varepsilon=\sqrt{2}\times\delta$ where δ is defined in tetragonal units. As shown in Fig. 1, δ follows reasonably well the linear relation $\delta\approx x$ over the range $0.03\leq x\leq 0.12$ which spans the insulator-superconductor transition. In a charge stripe model this corresponds to a constant charge per unit length in both the diagonal and collinear stripe phases, or equivalently, 0.7 and 0.5 holes per Cu respectively because of the $\sqrt{2}$ difference in Cu spacings in the diagonal and collinear geometries. Our value for x=0.024 falls slightly below the $\delta\approx x$ line and indeed it corresponds to ~1 hole/Cu as in $La_{2-x}Sr_xNiO_4$ where there is ~1 hole/Ni [16]. We note that Machida and Ichioka predict 1 hole/Cu throughout the diagonal stripe phase [17].

There are at least two possible origins of the finite correlation lengths for the static order in $La_{1.976}Sr_{0.024}CuO_4$. The first is that the lengths simply measure the spin decoherence distance of the AF spin clusters. The second is that the disorder originates primarily from a random distribution of stripe spacings and orientations as discussed by Tranquada $et\ al.$ [18] Further experiments and theoretical calculations will be required to choose between these possibilities.

SPIN STRUCTURE

The magnetic structure in the spin-glass phase is now considered. Matsuda $et\ al.$ performed a detailed study on the magnetic structure in the spin-glass $La_{1.98}Sr_{0.02}CuO_4$ [14]. Figures 3(a) and 3(b) show the L dependence of the magnetic elastic peaks in $La_{1.98}Sr_{0.02}CuO_4$ measured at $(1,0,L)$ and $(0,1,L)$ at 1.6 K, respectively. The background estimated from the high temperature data (60 K) was subtracted so that the remaining signal is purely magnetic. These scans probe the between-plane magnetic correlations. Broad peaks are observed at $(1,0,even)$ where magnetic Bragg peaks exist in La_2CuO_4. There are some characteristic features. Firstly the peaks are broad, indicating that the spin correlations along the c axis are short-ranged. Secondly, the magnetic intensity at $(1,0,even)$ initially increases with increasing L, in contrast to the behavior found for the magnetic Bragg intensities in pure La_2CuO_4. Lastly, the magnetic intensities at $(1,0,L)$ are much larger than those at $(0,1,L)$. From these results, one can deduce that spin clusters are formed in $La_{1.98}Sr_{0.02}CuO_4$. However, the spin clusters have a different geometrical structure from that in pure La_2CuO_4.

The simplest model to explain the increase of the intensity with increasing L along both $(1,0,L)$ and $(0,1,L)$ is that the cluster AF spin is randomly directed within the ab plane. In this case, the intensity would vary with L like

$$1/2[1+\sin^2\theta(L)]\,f(Q)^2 \qquad (1)$$

where $\theta(L)$ is the angle that the Q-vector of the $(1,0,L)$ or $(0,1,L)$ reflection makes with the ab plane and $f(Q)$ is the magnetic form factor, which is approximately constant for the range of L's considered here. The solid lines in Figs. 3(a) and 3(b) are the calculated profiles using as the intrinsic line shape 3D squared Lorentzians convoluted with the instrumental resolution function. The parameters used are ξ_a=160 Å, ξ_b=25 Å, and ξ_c=4.7 Å. We should note that a result equivalent to Eq. (1) is obtained by fixing the spin direction along $(H,H,0)$ or by assuming equal admixtures of 3D correlated phases where the spin vector s is along or perpendicular to the AF propagation vector τ/a_{ortho}. In the Néel state of pure La_2CuO_4 the spin is along b_{ortho} while just above $T_N=325$ K, that is, at 328 K when the correlation length is about 800 Å [20] the spin is randomly oriented in the ab plane. Because of the latter result it seems physically plausible that in the frozen spin clusters below 40 K in $La_{1.98}Sr_{0.02}CuO_4$ the spin direction would also be random. This is consistent with the fact that the net Ising anisotropy favoring the b_{ortho} axis from the Dzyaloshinsky-Moriya interaction is only about 0.1 K in energy. We should note that in all cases we assume that the propagation vector of the AF order is along a_{ortho} in order to account for the pronounced peaks at (1,0,L) for L even alone.

With increasing hole concentration, the correlation length perpendicular to the CuO_2 plane becomes rapidly reduced. In $La_{1.95}Sr_{0.05}CuO_4$, the result of L-scans shows that the spin correlations perpendicular to the CuO_2 plane are almost negligible [13].

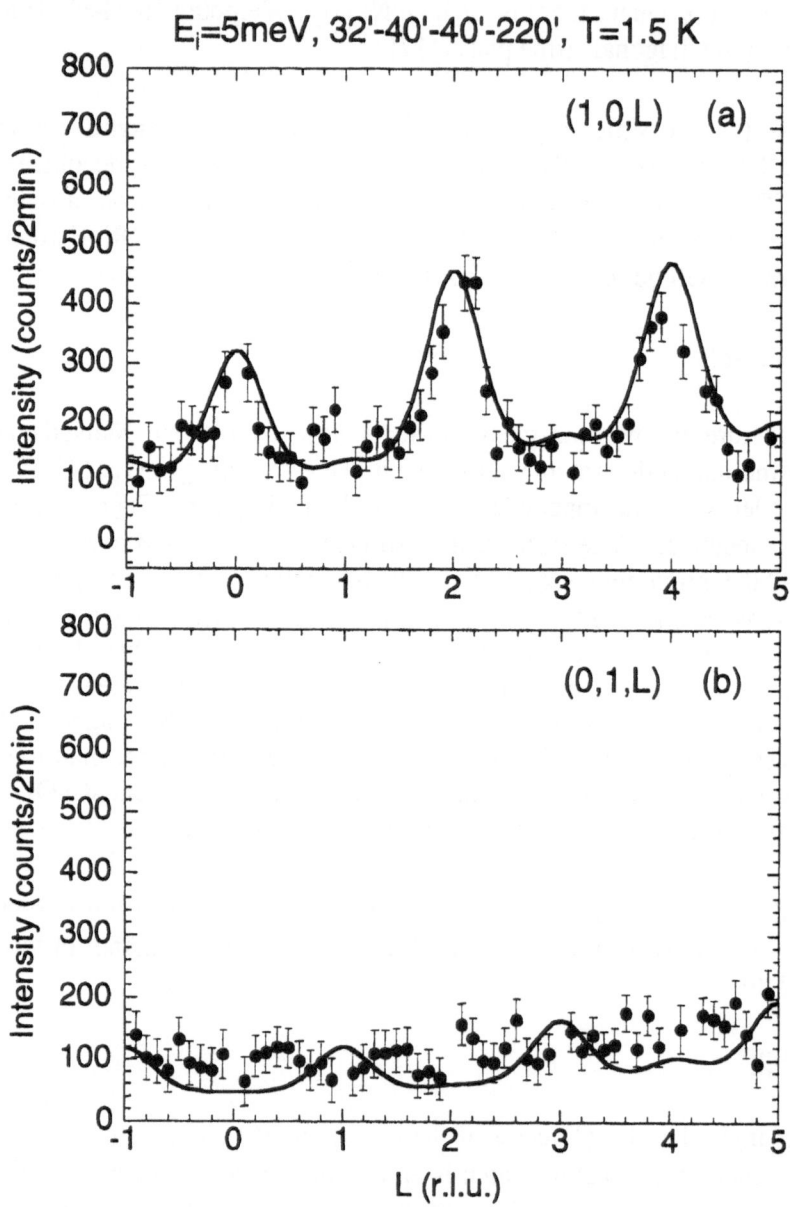

Fig. 3   Elastic scans along (1,0,L) (a) and along (0,1,L) (b) at 1.6 K in $La_{1.98}Sr_{0.02}CuO_4$ [14]. The background intensities measured at 60 K are subtracted. The solid lines shows the results of calculations with 3D squared Lorentzians with $\xi_a=160$ Å, $\xi_b=25$ Å, and $\xi_c=4.7$ Å.

Acnowledgments. This work has been performed in collaboration with R. J. Birgeneau, P. Böni, Y. Endoh, M. Fujita, M. Greven, M. A. Kastner, S.-H. Lee, Y. S. Lee, G. Shirane, S. Wakimoto, and K. Yamada. This study was supported in part by the U.S.-Japan Cooperative Program on Neutron Scattering operated by the United States Department of Energy and the Japanese Ministry of Education, Science, Sports and Culture and by a Grant-in-Aid for Scientific Research from the Japanese Ministry of Education, Science, Sports and Culture.

1. M. A. Kastner, R. J. Birgeneau, G. Shirane, and Y. Endoh, Rev. Mod. Phys. **70**, 897 (1998).
2. D. Vaknin, S. K. Shinha, D. E. Moncton, D. C. Johnston, J. Newsam, C. R. Safinya, and H. King, Phys. Rev. Lett. **58**, 2802 (1987).
3. B. J. Sternlieb, G. M. Luke, Y. J. Uemura, T. M. Riseman, J. H. Brewer, P. M. Gehring, K. Yamada, Y. Hidaka, T. Murakami, T. R. Thurston, and R. J. Birgeneau, Phys. Rev. B **41**, 8866 (1990).
4. B. Keimer, N. Belk, R. J. Birgeneau, A. Cassanho, C. Y. Chen, M. Greven, M. A. Kastner, A. Aharony, Y. Endoh, R. W. Erwin, and G. Shirane, Phys. Rev. B **46**, 14034 (1992).
5. H. Yoshizawa, S. Mitsuda, H. Kitazawa, and K. Katsumata, J. Phys. Soc. Jpn. **57**, 3686 (1988).
6. R. J. Birgeneau, Y. Endoh, Y. Hidaka, K. Kakurai, M. A. Kastner, T. Murakami, G. Shirane, T. R. Thurston, and K. Yamada, Phys. Rev. B **39**, 2868 (1989).
7. S.-W. Cheong, G. Aeppli, T. E. Mason, H. A. Mook, S. M. Hayden, P. C. Canfield, Z. Fisk, K. N. Klausen, and, J. L. Martinez, Phys. Rev. Lett. **67**, 1791 (1991).
8. K. Yamada, C. H. Lee, K. Kurahashi, J. Wada, S. Wakimoto, S. Ueki, H. Kimura, Y. Endoh, S. Hosoya, G. Shirane, R. J. Birgeneau, M. Greven, M. A. Kastner, and Y. J. Kim, Phys. Rev. B **57**, 6165 (1998).
9. T. Suzuki, T. Goto, K. Chiba, T. Shinoda, T. Fukase, H. Kimura, K. Yamada, M. Ohashi, and Y. Yamaguchi, Phys. Rev. B **57**, 3229 (1998).
10. H. Kimura, K. Hirota, H. Matsushita, K. Yamada, Y. Endoh, S.-H. Lee, C. F. Majkrzak, R. Erwin, G. Shirane, M. Greven, Y. S. Lee, M. A. Kastner, and R. J. Birgeneau, Phys. Rev. B **59**, 6517 (1999).
11. J. M. Tranquada, J. D. Axe, N. Ichikawa, Y. Nakamura, S. Uchida, and B. Nachumi, Phys. Rev. B **54**, 7489 (1996).
12. S. Wakimoto, R. J. Birgeneau, Y. Endoh, P. M. Gehring, K. Hirota, M. A. Kastner, S. H. Lee, Y. S. Lee, G. Shirane, S. Ueki, and K. Yamada, Phys. Rev. B **60**, R769 (1999).
13. S. Wakimoto, R. J. Birgeneau, M. A. Kastner, Y. S. Lee, R. Erwin, P. M. Gehring, S. H. Lee, M. Fujita, K. Yamada, Y. Endoh, K. Hirota, and G. Shirane, cond-mat/9908115, Phys. Rev. B (submitted).
14. M. Matsuda, R. J. Birgeneau, P. Böni, Y. Endoh, M. Greven, M. A. Kastner, S.-H. Lee, Y. S. Lee, G. Shirane, S. Wakimoto, K. Yamada, cond-mat/9907435, Phys. Rev. B (submitted).
15. M. Matsuda, R. J. Birgeneau, Y. Endoh, M. Fujita, M. A. Kastner, G. Shirane, S. Wakimoto, and K. Yamada, Nature (submitted).
16. J. M. Tranquada, D. J. Buttrey, and V. Sachan, Phys. Rev. B **54**, 12318-12323 (1996).
17. K. Machida and M. Ichioka, cond-mat/9812398.
18. J. M. Tranquada, N. Ichikawa, and S. Uchida, Phys. Rev. B **59**, 14712 (1999).
19. S. Wakimoto and S.-H. Lee, (private communication).
20. R. J. Birgeneau, M. Greven, M. A. Kastner, Y. S. Lee, B. O. Wells, Y. Endoh, K. Yamada, and G. Shirane, Phys. Rev. B **59**, 13788 (1999).

Growth of Thin-film $RuSr_2GdCu_2O_8$ by Pulsed-laser Deposition and Ex-situ Annealing

James E. McCrone[1], Gary Gibson[2], Jeffrey L. Tallon[3], John R. Cooper[1] and Zoe Barber[2]

[1] IRC in Superconductivity, University of Cambridge, Madingley Road, Cambridge, UK
[2] Dept. of Materials Science and Metallurgy, University of Cambridge, Cambridge, UK
[3] New Zealand Institute for Industrial Research Ltd., P.O. Box 31310, Lower Hutt, N.Z.

Abstract: We have prepared epitaxial films of the rutheno-cuprate $RuSr_2GdCu_2O_8$ (Ru-1212). Our films, though not phase-pure as yet, display ferromagnetic ordering and signs of the coexisting superconductivity that has drawn so much attention to the bulk material, where the two types of order (which are usually incompatible) uniformly coexist on a microscopic scale. In the absence of single crystals, study of polycrystalline films will help elucidate the electronic structure of Ru-1212, its anisotropy and the interplay between ferromagnetism and superconductivity.

Keywords: Superconductivity, Thin films

INTRODUCTION

The successful synthesis[1] and the subsequent discovery of microscopically coexisting superconductivity and electronic ferromagnetism[2, 3], have made $RuSr_2GdCu_2O_8$ (Ru-1212) a topical material. Structurally it is a triple perovskite, with one RuO_2 and two CuO_2 layers per unit cell. There is evidence suggesting that the RuO_2 layer is an itinerant ferromagnet, as it is in $SrRuO_3$. In support of this, neutron diffraction studies have so far shown no magnetic lattice[4], and band-structure calculations suggest a conducting RuO_2 layer[5]. In contrast, transport properties such as the thermo-electric power and Hall effect may be understood purely in terms of conducting, under-doped CuO_2 layers[6], favouring a local-moment picture. Possible scenarios for coexistence have been explored by Pickett *et al*[7], who postulate the existence of an unusual Fulde-Ferrell-Larkin-Ovchinnikov (FFLO) superconducting phase or possibly a "π phase". In the former case the magnetic field is accommodated by spatial modulation of the superconducting order parameter, and in the latter the parameter alternates in phase between CuO_2 layers, producing nodes in the ferromagnetic layers. Experimental work to test these theories has so far been limited to sintered material, where experiments see an average of anisotropic crystal properties. In this article we describe the production of epitaxial thin films of Ru-1212, which will allow studies of the true in-plane properties to be carried out.

FILM GROWTH

Films were grown by pulsed laser deposition (PLD), using a KrF excimer laser ($\lambda = 248$ nm) with a pulse energy density at the target of approximately $1.4 Jcm^{-2}$ and a frequency of 10Hz. The target was a sintered pellet of phase-pure $RuSr_2GdCu_2O_8$ (Ru-1212) of diameter 25mm, whose synthesis is described elsewhere[1, 2]. Strontium titanate ($SrTiO_3$) substrates were mounted on a heated stainless-steel block in the conventional on-axis geometry. The deposition chamber was evacuated to at least 1×10^{-5}mbar before each run, after which oxygen was bled into the system to maintain a constant pressure during growth.

A typical X-ray spectrum for an as-deposited film shows a set of (00*l*) reflections corresponding to a *c*-axis lattice parameter of 5.94Å, with a number of other impurity peaks. We experimented with a range of deposition conditions, varying substrate temperature between 750 and 850°C, the target to substrate distance from 28 to 60mm, and the oxygen pressure from 0.1 to 0.3mbar, but no major change was observed in the X-ray spectra of the films. As yet, the major as-deposited phase is unidentified.

Our lack of success with *in-situ* growth led us to experiment with various *ex-situ* post-annealing strategies, initially at 1060°C, the temperature at which long-term annealing has been shown to improve the crystal quality of sintered Ru-1212[3]. The films were placed in an alumina crucible and annealed in a clean quartz furnace tube in flowing high-purity oxygen for periods of up to five hours. The crucible was pushed into the hot furnace and removed quickly after the anneal using a quartz rod, then cooled rapidly on a copper block.

Production of the desired 1212 phase was achieved by annealing at temperatures of between 1000 and 1100°C, though the purity of the resultant Ru-1212 phase varied with annealing temperature and time. Figure 1 shows the X-ray spectrum for film Z379, which was deposited at 750°C in

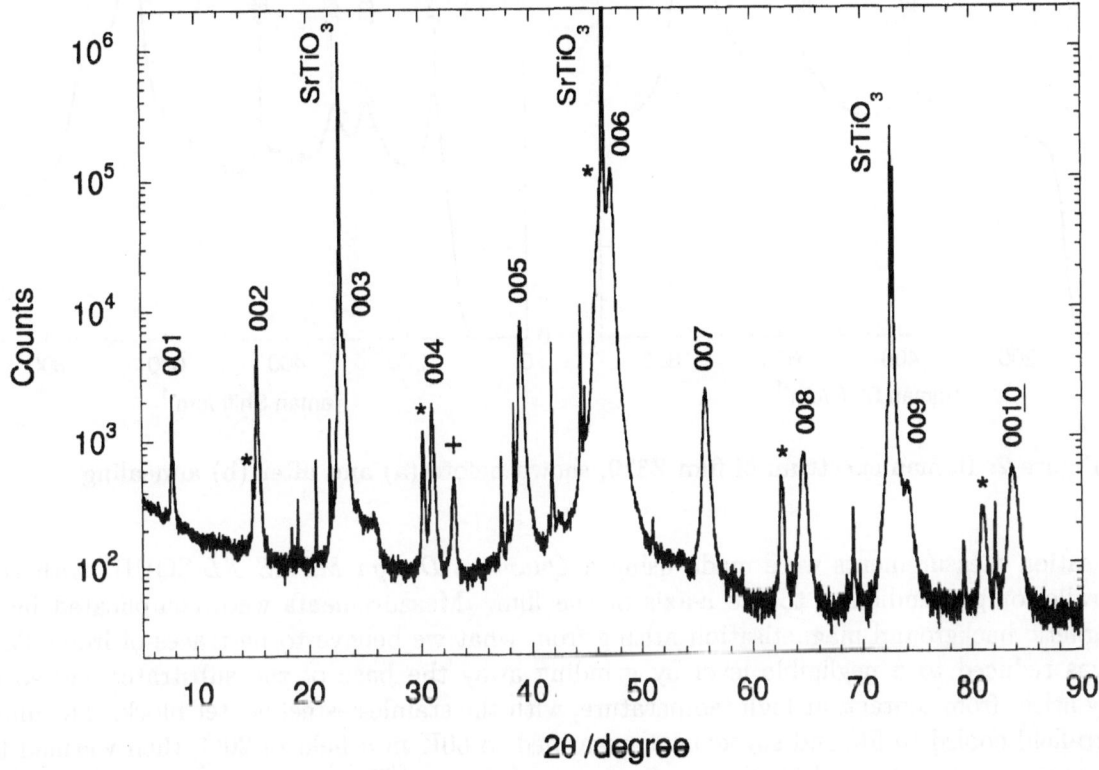

Figure 1: X-ray spectrum for film Z379, after annealing at 1050°Cfor one hour in flowing O₂. The data were taken using Cu-kα radiation.

a 0.22mbar O₂ atmosphere and subsequently annealed for one hour at 1050°C. The film now shows strong (00*l*) reflections, with a *c*-axis lattice parameter of 11.54(8)Å (after correction for height errors from the raw spectrum[8]). This value agrees well with the 11.56Å reported from neutron diffraction studies on sintered Ru-1212[4]. We estimate crudely from the relative areas of neighbouring peaks (e.g. the (004) and nearby peak marked with an asterisk) that the total impurity content is ~35% in this film: the significant impurity peaks in the spectrum come from unconverted as-deposited material, marked with an asterisk, and one other phase, marked with a cross. All the measurements reported here refer to this film, but we have achieved almost complete elimination of the as-deposited phase in other films by annealing for the same time at the higher temperature of 1080°C. This increased the amount of the new impurity phase (marked with a cross in Fig. 1), but the overall phase purity is higher. We believe that this new impurity is the result of non-stoichiometry in the as-deposited material, which may be eliminated by adjusting the deposition conditions.

CHARACTERISATION

We performed Raman spectroscopy on the films as a further check of their structure. The spectrum of the as-grown film, shown in Fig. 2a, is very different from that of the annealed film (Fig. 2b), illustrating the fact that a complete change of structure has taken place. For the annealed film the shifts of all the Raman peaks agree extremely well with those which Pringle et al [9] attributed to the apical, CuO_2 and RuO_2 oxygen vibrations in sintered Ru-1212.

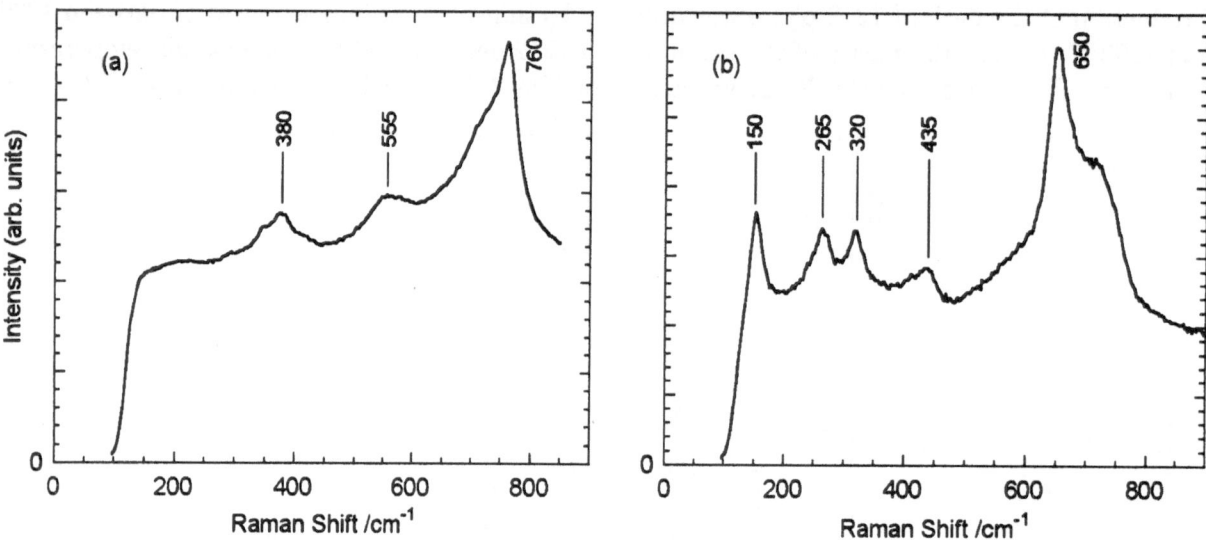

Figure 2: Raman spectrum of film Z379, shown before (a) and after (b) annealing.

Magnetisation measurements were made using a *Quantum Design MPMS XL* SQUID with the field parallel or perpendicular to the c-axis of the film. Measurements were complicated by a ferromagnetic background magnetisation arising from what we believe to be traces of iron. This signal was reduced to a negligible level by grinding away the base of the substrate, and so it probably arises from contact, at high temperature, with the stainless-steel heater block. The films were zero-field cooled to 5K and subsequently warmed to 60K in a field of 20G, then warmed to room temperature in a field of 10kG in order to probe T_c and T_{Curie}, respectively. The loss of ferromagnetic order can be seen clearly in Fig. 3a by the sharp decrease in magnetic moment as the sample is warmed through T_{Curie}, which is ~150K for this film. This value is obtained by fitting the magnetisation to Eqn. 1 for T>160K.

$$M = C_1 + \frac{C_2}{T} + \frac{C_3}{T - T_{Curie}}$$

(1)

C_1 is the diamagnetic substrate signal (~constant for this range of temperature), C_2 the paramagnetic Gd signal and C_3 is that due to the Ru ferromagnetism. Deriving T_{Curie} in the same way for a sintered sample (Fig. 3b) gives a value of 141K.

Magnetic observation of the superconducting transition at T_c is hampered severely by the background diamagnetism of the substrate and the remanent field of the Ru moments. Insulating

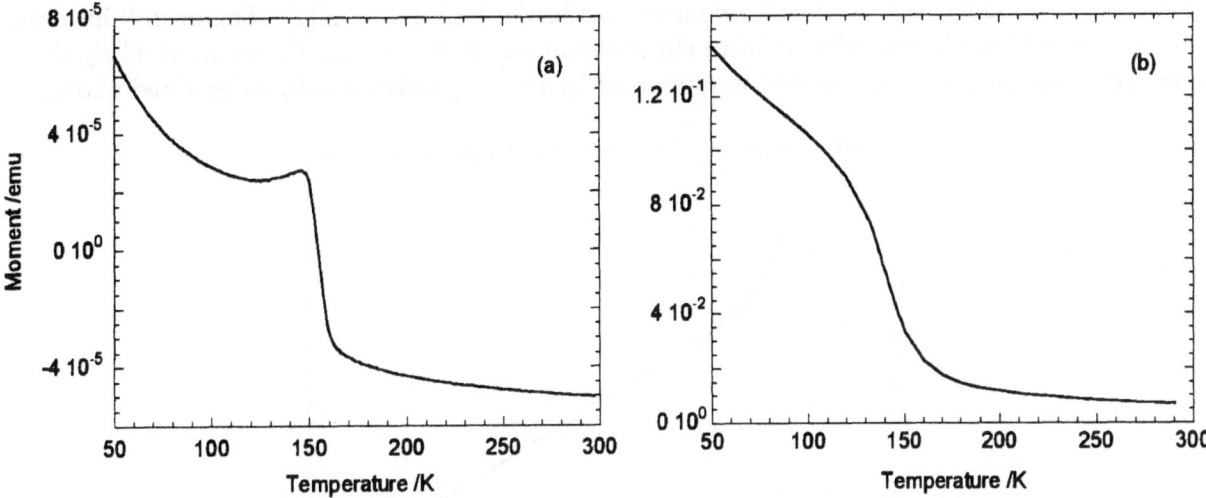

Figure 3: (a) Magnetisation of film Z379 as a function of temperature and (b) data taken in the same way for a sintered sample of mass 0.0143g.

boundaries between grains lead to a very small diamagnetic shielding fraction, an effect also observed in sintered material, where the shielding fraction just below the thermodynamic transition (46K)[10] is of order 1% and does not approach 100% until the grain boundaries become superconducting at lower temperatures. Figure 4 shows our best evidence for a superconducting transition with H∥c in an applied field of 20G. The onset of superconductivity is at ∼36K, compared with the 46K seen in the bulk material; the rapidly increasing moment at low temperatures is due to the paramagnetic Gd atoms. We have observed similar behaviour to that shown in Fig. 4 in another piece of the same film, but find a lack of reproducibility after field and temperature cycling which is being investigated.

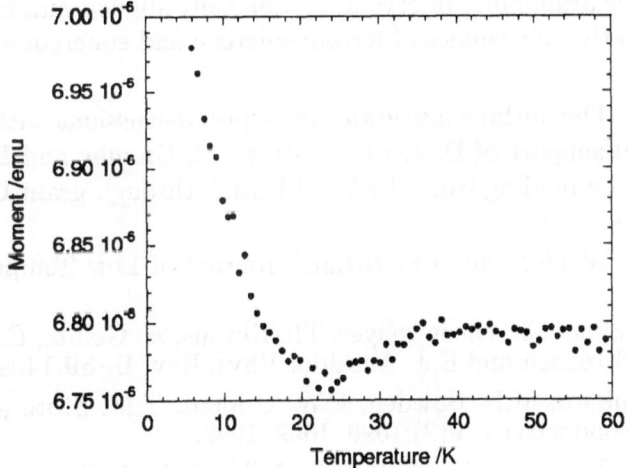

Figure 4: Magnetisation data for film Z379 taken with H∥c. The film was zero-field cooled to 5K and then measured while warming in a field of 20G.

Four-point AC resistivity measurements were made down to 5K using a continuous-flow cryostat. To date, all our films have been macroscopically non-metallic, but they do display a broad feature in the resistance at T_{Curie} similar to that of the sintered material. Figure 5 shows resistance data plotted on a semi-logarithmic scale. We ascribe the increased resistance near T_{Curie} to spin-dependent scattering by the Ru moments, as in the bulk material[6]. The overall insulating behaviour and the absence of a resistive superconducting transition at T_c are most likely due to poor grain boundaries. This picture is supported by the magnetisation data described above.

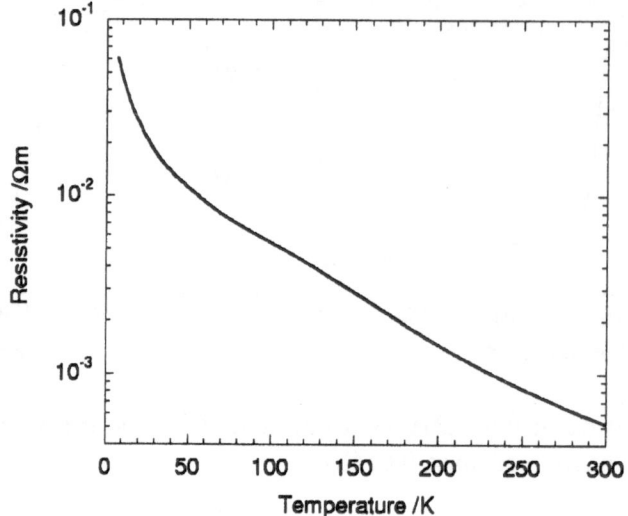

Figure 5: Resistivity of film Z379 as a function of temperature. The film thickness is estimated to be 400nm by comparison with known YBCO growth rates.

SUMMARY

We have succeeded in producing epitaxially aligned films of Ru-1212 with impurity inclusions of ~10-35% by PLD growth and *ex-situ* annealing. Work is continuing to improve film quality. The existing films show a T_{Curie} of 150K and a T_c of 36K, in broad agreement with previous results for sintered samples. The availability of crystallographically aligned material should allow some of the questions relating to the coexistence of ferromagnetism and superconductivity to be answered.

Acknowledgements. The authors acknowledge helpful discussions with E. Tarte and F. Kahlmann, and the technical support of D. Astill, S. Brown, I. Ganney and E. Robinson. This work was partially supported by funding from the UK EPSRC, through grant GR/M205941.

1. L. Bauernfeind, W. Widder and H.F. Braun. Journal of Low Temperature Physics, **105**(5-6):1605–1610, 1996.

2. C. Bernhard, J.L. Tallon, Ch. Niedermayer, Th. Blasius, A. Golnik, E. Brucher, R.K. Kremer, D.R. Noakes, C.E. Stronach and E.J. Ansaldo. Phys. Rev. B, **59**:14099, 1999.

3. J.L. Tallon, C. Bernhard, M.E. Bowden, P.W. Gilberd, T.M. Stoto and D.J. Pringle. IEEE Trans. Appl. Superconductivity, **9**(2):1696–1699, 1999.

4. O. Chmaissem, J.D. Jorgensen, H. Shaked, P. Dollar and J.L. Tallon. Sub. to Phys. Rev. B.

5. R. Weht, A.B. Shick and W.E. Pickett. High Temperature Supercon.: Proc. HTS '99, 1999.

6. J.E. McCrone, J.R. Cooper and J.L. Tallon. J. Low Temp. Phys.: Proc. of MOS '99, in press.

7. W.E. Pickett, R. Weht and A.B. Shick. Phys. Rev. Lett., in press.

8. B.D. Cullity. *Elements of X-ray Diffraction.* (Addison-Wesley, 1978).

9. D.J. Pringle, J.L. Tallon, B.G. Walker and H.J. Trodahl. Phys Rev B, **59**(18):R11679–11682, 1999.

10. J.L. Tallon, C. Bernhard and J.W. Loram. J. Low Temp. Phys.: Proc. of MOS '99, in press.

The Metal-Insulator Transition and High-Temperature Superconductivity

Peter P. Edwards

School of Chemistry, The University of Birmingham, Edgbaston, Birmingham B15 2TT, England

Abstract: A discussion is given of the metal-insulator transition in high-T_c layered cuprates for temperatures above T_c. We attempt to place this composition-induced metal-insulator transition in relation to that known for systems such as doped semiconductors, expanded fluid metals, compressed fluid hydrogen, etc. When viewed from this global perspective, the location of the metal-insulator transition in layered cuprates appears consistent with that found in these other systems, when issues relating to carrier identify and concentration, etc. are taken into consideration. However, several features of the metal-insulator transition in cuprates are *qualitatively* different from that in the range of systems reviewed here ; in particular, it is suggested that the quasi two-dimensional nature of the layered cuprates is particularly relevant to their remarkable electronic behaviour. Thus, the insulator-to-metal transition, the progenitor of high-temperature superconductivity in cuprates, is two-dimensional rather than three-dimensional, in nature. The pronounced localization of carriers, large lattice polarizability, and extant disorder in this two-dimensional system all appear to be key factors governing the evolution of the metallic state - and presumably also the high-T_c superconducting state - in the layered cuprates.

Keywords: The Metal-Insulator Transition / High-Temperature Superconductivity.

INTRODUCTION

The issue of the metal-insulator transition is now recognised as a key element in any discussion of high-temperature superconductivity in layered cuprates [1-3]. In materials such as $La_{2-x}Sr_xCuO_4$, one starts with the antiferromagnetic insulator La_2CuO_4 and introduces p-type carriers via the controlled chemical substitution of La by Sr. The phenomenon of high-temperature superconductivity then evolves directly from the introduction of excess holes into this antiferromagnetic insulating state. The generic phase diagram of the prototypical $La_{2-x}Sr_xCuO_4$ is now well established [1]. Perhaps less well-known is a most remarkable, complementary example of p-type doping in a series of metal-superconductor-(antiferromagnetic) insulator transitions [4,5] within the so-called septenary system $(Ca_{1-x}Y_x)Sr_2(Tl_{1-y}Pb_y)Cu_2O_7$. Taking as an example a particular branch of the compositional phase space, *viz* $(Ca_{1-x}Y_x)(Tl_{0.5}Pb_{0.5})Sr_2Cu_2O_7$, a predominantly single-phase material can be obtained across the entire homogeneity range, spanning the 78K superconductor $(Ca)Sr_2(Tl_{0.5}Pb_{0.5})Cu_2O_7$, up to the highest T_c (108K for $x = 0.2$) and finally through to the antiferromagnetic insulator at Cu^{2+} (Figure 1). A central issue to now emerge thus centres on the relationship between high-temperature superconductivity and the composition-induced insulator-to-metal transition in the layered cuprates.

It is our belief that a deeper insight into the phenomenon of high-temperature suprconductivity may ultimately be gained by viewing the layered 2D-cuprates within the broader perspective of other electronic materials and systems undergoing compositionally-induced metal-insulator transitions.

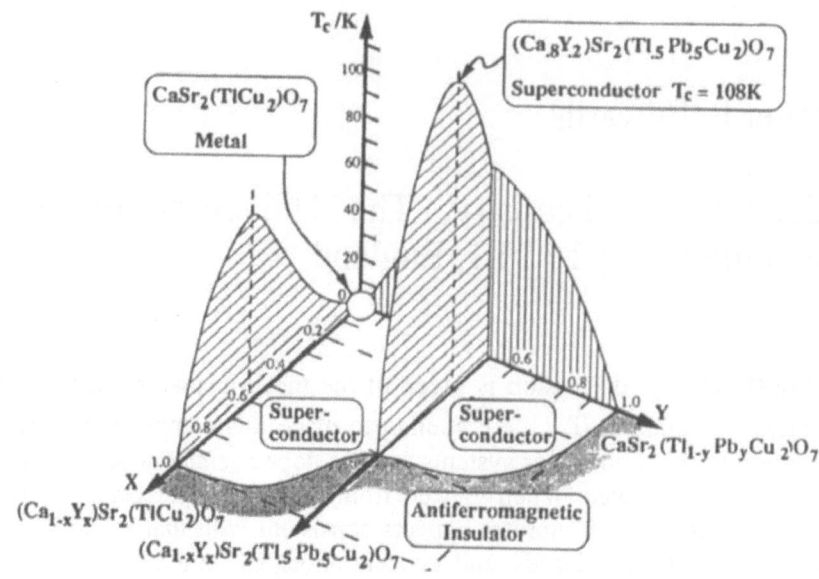

Figure 1

Compositionally-induced
superconductor-insulator
phase transition in
$(Ca_{1-x}Y_x)Sr_2(Tl_{1-y}Pb_y)Cu_2O_7$.
Original experimental data
from refs. 4,5 and redrawn
by W. Y. Liang.

THE METAL-INSULATOR TRANSITION

Background

The electronic phase transition signifying the transformation of metal-to-insulator, and vice-versa, has proved to be surprisingly recalcitrant to a complete theoretical analysis [6-8]. The current intense effort in trying to understand the 'anomalous' metallic state (for $T > T_c$) of layered cuprates will now have to take cognisance [1] of " ... *the time-honoured (which means very hard) problem of the metal-insulator transition*".

Mott first posed the question as to whether insulating materials could, under appropriate conditions, become metals [9]. His conclusion - that under certain circumstances they would - was perhaps not surprising; what was remarkable was Mott's proposal that at the transition from insulator to metal, *all* the available valence electrons would become free (or itinerant) at once - not just a few of them. Mott presented arguments [9] to show that the insulator-metal transition must be discontinuous at $T = 0K$. A sketch of the proposed Mott transition from metal-to-insulator at $T = 0K$ is given in Figure 2. Here, as the average distance, d, between one-electron centres decreases, a first-order electronic phase transition from insulator-to-metal occurs. Mott predicted that the metal-insulator transition would occur at some critical average separation, d_c, given by [9,10]

$$d_c \sim 4a_H{}^* ,$$

where $a_H{}^*$ is the effective Bohr-orbit radius of the localized, one-electron centre at low carrier concentration, deep into the insulating regime. Clearly, d_c can be related to the critical density (n_c) of electronic centres giving the venerable Mott criterion,

$$n_c{}^{1/3}a_H{}^* \sim 0.25 \qquad\qquad .... \quad (1)$$

This criterion is known to be a highly effective indicator of the location of the metal-insulator transition [6-8].

Figure 2

The Mott metal-insulator transition at T = 0K. The parameters d_c (equivalently n_c) and σ_{min} are key indicators of the metal-insulator transition. This cartoon shows a first-order metal-insulator transition at a critical value of d_c (n_c). Modified from Edwards et al [7].

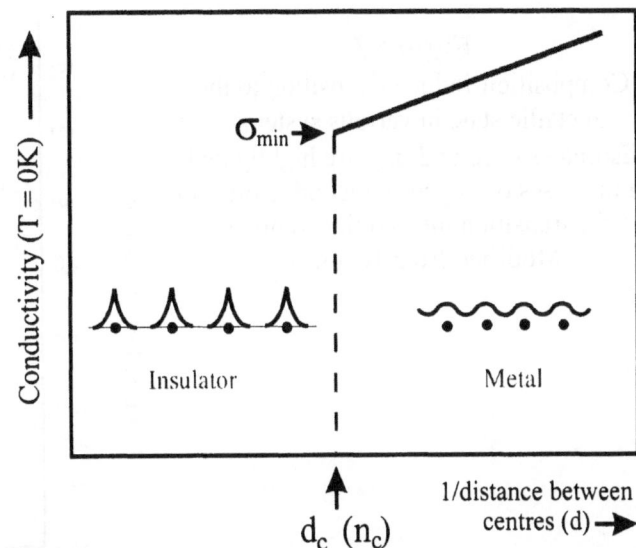

According to Mott [11], the d.c. electrical conductivity would jump discontinuously from a *minimum metallic conductivity*, σ_{min}, to zero at d_c (equivalently n_c). The concept was based on the Ioffe-Regel criterion for implicit metallic behaviour [12]; this reasons that materials only become metallic - and remain metallic - when the mean-free-path of the valence (conduction) electron becomes comparable to, or exceeds, the mean distance between the centres supplying the electrons. Mott coupled this notion with Anderson's prediction that for a certain value of the disorder, electronic states become localized [13], to give an estimate for σ_{min} [11,14], *viz*

$$\sigma_{min} = C(e^2/\hbar)d_c^{-1} = C(e^2/\hbar)n_c^{1/3} \qquad \text{........} \quad (2)$$

where C is a constant, typically between 0.05 and 0.025.

The introduction of excess holes into the layered cuprates leads to a gradual increase in the room temperature conductivity [15] and a presumed composition-induced insulator-to-metal transition at these 'high' temperatures (*c.f.* T = 0K!; Figure 2). It is therefore instructive to place such generic behaviour for the layered cuprates in the context of other materials traversing a compositionally-induced metal-insulator transition [6-8]. In Figure 3 we show the measured d.c. electrical conductivity as a function of the effective number density of carriers for various fluid and solid state materials [16,17]; these encompass fluid hydrogen at very high elemental (carrier) density, expanded fluid metals (e.g. Hg), through to the prototypical (low-carrier-density) doped semiconductors Si : P and Ge : Sb. We also include two representative oxide systems, Na_xWO_3 (WO_3:Na) and $LaNi_{1-x}M_xO_3$; note that the sodium tungsten bronzes are highly conducting in their metallic state. Obviously there are underlying, differentiating features characteristic of each system which need to be addressed within any detailed examination. Nevertheless, in terms of such global considerations, the layered cuprate materials are indeed representative of other relatively high (carrier) density (*ca.* 10^{21} - 10^{22} cm^{-3}) systems traversing the metal-insulator transition. What sets the cuprates apart from the other materials, of course, is the presence of high-temperature superconductivity in the vicinity of the metal-insulator transition. Clearly, to advance any meaningful discussion, one must therefore attempt to locate that metal-insulator transition in cuprate materials for which the T → 0K behaviour of the electrical conductivity (Figure 2) is obviously not accessible because of the occurrence of high-temperature superconductivity.

High-T_c Cuprates: Doped 2D Charge-Transfer Insulators

Jarrell and co-workers [18] have outlined a phenomenological model for the metal-insulator transition (T > T_c) in $La_{2-x}Sr_xCuO_4$ in which electronic transfer is taken to be predominantly intralayer. At low

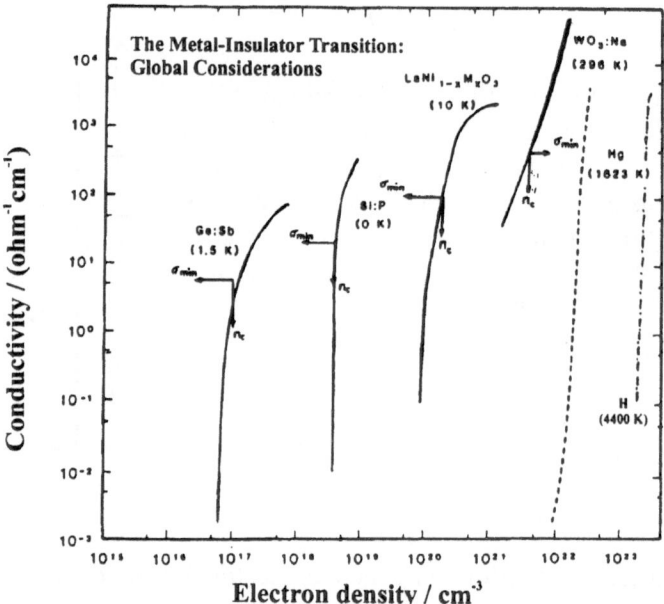

Figure 3

Composition-induced transition to the metallic state in various systems.
Estimates of n_c and σ_{min} are highlighted in the cases of doped semiconductors and the transition metal oxide systems.
Modified from 16 and 17.

doping levels, excess holes in the CuO_2 layers are coulombically bound to their accompanying dopant ion (Sr^{2+}) in a localized 2D hydrogenic orbital, having a_H^* of between 4-8Å, which typically encompasses 1-2 unit cells. Many authors have attributed the temperature dependence of the electrical resistivity in the insulating regime of the layered cuprates as variable-range-hopping conduction between spatially-located states [19,20]. The likely origin of this localization will be the random electronic potential experienced by excess holes in the CuO_2 plane because of their coupling to the statistically distributed dopant ions. Such disorder gives rise to a mobility gap and a concomitant mobility edge (E_c) in the electronic density-of-states.

Thomas [21] has argued that in the layered cuprates as the density of holes is continually increased, a transition from insulator-to-metal occurs at a critical composition prescribed by the Mott criterion (Eq.1). It has previously been found from other systems that the Mott criterion appears to be a realistic and accurate indicator for an Anderson or a Mott-Hubbard transition [6]. Estimates for a_H^* and n_c for the layered cuprates are included in Figure 4, which is a representation of the several classes of systems [7,8] undergoing compositionally-induced metal-insulator transitions (Figure 3); these again range from doped semiconductors through the high-T_c cuprates and highly conducting sodium tungsten bronzes, through finally to expanded fluid metals and compressed fluid hydrogen.

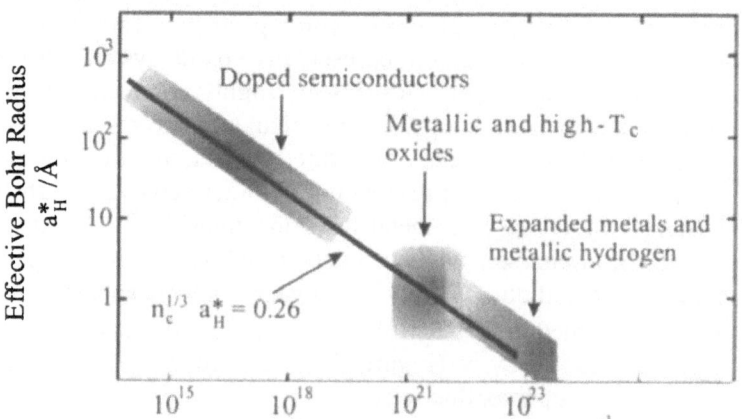

Figure 4

Correlation of a_H^* with n_c for a range of systems. The shaded regions indicate the approximate range of parameters for each system; the solid line is the Mott critrion.

The only rigorous criterion [6-9] for differentiating between a metal and an insulator is the value of the electrical conductivity at T = 0K; there, metals have a finite conductivity (with the exception of superconductors, which show infinite conductivity) whilst insulators have zero electrical conductivity (Figure 2). Clearly, in the case of a high-T_c material, any attempts at extrapolating toward T = 0K are ruled out by the appearance of superconductivity at high temperatures. However, in the vast majority of systems reviewed here, n_c for the metal-insulator transition at *finite* temperatures is estimated by two, 'course-grained' approaches. The first assumes that metallic behaviour is recognised from the temperature coefficient of electrical resistivity ; with metallic behaviour as $d\rho / dT > 0$, transforming to insulating behaviour, $d\rho / dT < 0$ at some critical carrier concentration. The second approach relies on the Mott-Ioffe-Regel (MIR) criterion (Eq.2) for an estimate of the value of the d.c. electrical conductivity at the boundary between metal and insulator (Figure 2).

Although the metal-insulator transition at $T = 0K$ may not be discontinuous, it is widely recognised that the MIR criterion gives a realistic value of σ_{min} for a wide range of systems undergoing the metal-insulator transition [22]. Furthermore, the value of σ_{min} is known to scale with the critical carrier concentration, via Eq.2 [23]. In Figure 5 we show the variation of σ_{min} with n_c for our range of materials, including the layered cuprates. The region corresponding to the high-T_c cuprates exhibiting metal-insulator transitions $(T > T_c)$ is clearly in line with the behaviour of other electronic materials *at* the very threshold of the metal-insulator transition [22,23]. These correlations serve to emphasise a key aspect of the layered high-T_c cuprates; namely, that these materials undergo a continuous, composition-induced transformation from the insulating to the superconducting regimes, without any classical *"metallic"* (i.e. highly conducting) ground state in between [1,24,25].

By and large, therefore, the gross features and characteristics of the metal-insulator transition for $T > T_c$ in layered cuprates can be understood within the framework of models used for what one might term 'conventional' systems, but with several important differences. Amongst these are :- (i) The undoped material is invariably an antiferromagnetic Mott insulator, highlighting the key role of electron-electron interactions. (ii) Excess holes, when bound to their accompanying dopant ions, are to a good approximation confined to a single CuO_2 layer [17,26]. The subsequent growth of the metallic phase with hole density is therefore almost purely two-dimensional. Importantly, this may imply that there is no true (discontinuous) insulator-to-metal transition at $T = 0K$ in these systems, but rather a crossover from strong-to-weak electron localization [26]. A potent feature of the layered cuprates may simply be that the low-dimensionality very effectively inhibits, or suppresses any potential first-order electronic phase transition to a *"genuinely metallic"*, highly conducting state. Recall, that extremely high conductivities are indeed possible in the 3D perovskite $NaWO_3$ and ReO_3 (Figure 3). The presence of a mobility gap - reflecting the undoubted presence and importance of Anderson-localization effects - will also enhance the spatial localization of electronic states in the 2D CuO_2 layers. (iii) The large value of the static dielectric constant, as compared to that at optical frequencies, also sets the layered cuprates apart from their conventional doped semiconductor counterparts. We note that for systems with such large differences in dielectric constants, one anticipates the formation of polaronic states in the insulating regime [26-29].

It appears that such a related confluence of circumstances in the layered cuprates may also enhance the possibility of bipolaron formation [27,29]. Notably, a 2D crystal (and electronic) structure for the cuprates, coupled with (iii), will lead to strong attractive forces between the two (hole) polaronic carriers in the CuO_2 planes, producing a bound, bipolaronic state. Several authors have advanced the viewpoint that these bipolaronic states, being doubly-charged bosons, act as (real space) precursors for the high-temperature superconducting state. Similarly, the belief is that in the insulating (hopping) regime, the participating carriers are already bipolarons (bosons) [27-29].

Within semiconducting terminology the layered 2D cuprate systems are heavily compensated, since the number of charge carriers is precisely half the number of hole states. Mott has pointed out that charged bosons in such a disordered system will behave in certain respects as fermions. Here mutual (charge) repulsion, instead of the Pauli principle, will limit the occupation number of charged bipolaronic particles in each localized state [30,31].

Figure 5

Correlation of σ_{min} with n_c for a range of experimental systems. The shaded regions indicate the approximate range of parameters for each system. The dotted lines represent values of C = 0.05 and 0.025 in Equation 2, expressing the magnitude of σ_{min} for various systems having different values of n_c.

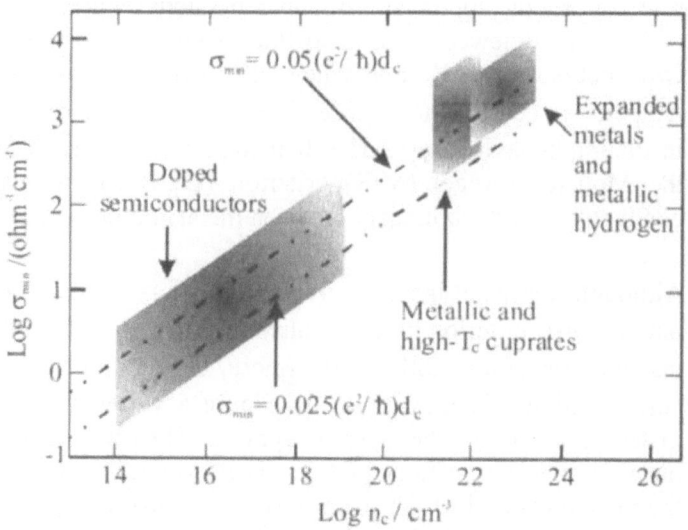

We expect, therefore, the system to show a mobility edge, and a metal-insulator transition (not necessarily discontinuous) would occur when the Fermi energy drops below the mobility edge and at a value of n_c given by the Mott criterion (Eq. 1). The Mott criterion appears to be valid for an Anderson or Mott-Hubbard transition, and the constant is found empirically to have the value ~0.26 for a wide variety of systems [6,7,22]. Recent experiments [32] on $La_{2-x}Sr_xCuO_4$ in a pulsed 61T magnetic field allow one to probe the metal-insulator transition in the absence of superconductivity. Importantly, under these conditions, the location of the metal-insulator transition is placed at an Sr composition corresponding to the optimally-doped T_c and, incidentally, at a composition close to the Mott criterion (Figure 4).

The intriguing possibility arises, therefore, that the transition to the metallic state at T = 0K, also recognised at T > 0K by two key indicators (Figures 4, 5), may also be coincident with the Bose transition (condensation) to the superconducting state [3]. The Mott criterion, highly successful for describing the metal-insulator transition in a wide range of doped semiconductors and related fermion systems (expanded metals, *etc*) may now also be appropriate for bosons. A particularly fruitful area of future research activity should therefore centre on the nature of the metal-insulator transition, but applied to the case when the current carriers are charged *bosons*, rather than *fermions*.

CONCLUDING REMARKS

In this brief review we have discussed the composition-induced metal-insulator transition in the layered, high-T_c cuprates in relation to that known in other systems; notably, doped semiconductors, expanded fluid metals and compressed fluid hydrogen. At any temperature above absolute zero, there is no clear dividing line between metals and insulators. Furthermore, the very occurrence of superconductivity at high temperatures further exaggerates the difficulties of locating the metal-insulator transition in the cuprates. Here we have attempted to apply two, coarse-grained prescriptions for identifying the metal-insulator transition, based on the Mott criterion and the Mott-Ioffe-Regel concept of a minimum metallic conductivity. These, and other, indicators highlight a key feature; namely, that the high-T_c electronic state of the layered cuprates prevails in the immediate vicinity of the metal-insulator transition. Just how closely linked are the two composition-induced electronic phase transitions now remains to be fully investigated. However, it remains abundantly clear that looking for superconductivity - *and potential high-temperature superconductivity* - in materials close to a metal-insulator transition will remain a highly-profitable activity for future investigations.

1. Y. Iye in *Metal-Insulator Transitions Revisited,* edited by P.P. Edwards and C.N.R. Rao (Taylor and Francis, London, 1995), pp. 211-230.
2. T. Ito, H. Takagi, S. Ishibashi, T. Ido and S. Uchida, Nature **350,** 596-598 (1991)
3. P.P. Edwards, N.F. Mott and A.S. Alexandrov, J. of Supercond. **11,** 151-154 (1998).
4. For a review, see R.S. Liu and P.P. Edwards, in *Materials Science Forum,* edited by J.J. Pouch, S.A. Alterovitz, R.R. Romanofsky and A.F. Hepp, (Trans. Tech. Publications, Switzerland, 1993), pp. 435-464.
5. R.S. Liu, P.P. Edwards, Y.T. Huang, S.F. Wu and P.T. Wu, J. Solid State Chem. **86,** 334-339.
6. N.F. Mott, *Metal-Insulator Transitions,* (Taylor and Francis, London, 1990).
7. P.P. Edwards, R.L. Johnston, F. Hensel, C.N.R. Rao and D.P. Tunstall, Solid State Physics **52,** 229-338 (1999).
8. P.P. Edwards, R.L. Johnston, C.N.R. Rao and D.P. Tunstall, Philos. Trans. of the Roy. Soc., **A356,** 1-278 (1998).
9. N.F. Mott, Philos. Mag. **6,** 287-309 (1961).
10. N.F. Mott, Can. J. Phys. **34,** 1356-1367 (1956).
11. N.F. Mott, Philos. Mag. **26,** 1015-1026 (1972).
12. A.I. Ioffe and A.R. Regel, Prog. Semicond. **4,** 239-291 (1960).
13. P.W. Anderson, Rev. Mod. Phys. **50,** 191-201 (1978).
14. N.F. Mott and E.A. Davis, *Electronic Processes in Non-Crystalline Solids,* (Clarendon Press, Oxford, 1979).
15. B. Batlogg, H.Y. Hwang, H. Takagi, R.J. Cava, H.L. Kao and J. Kwo, Physica C **235-240,** 130-133 (1994).
16. P.P. Edwards in *Crystal Engineering,* edited by K.R. Seddon and M. Zaworotko (Kluwer Academic, The Netherlands, 1999) pp. 409-431.
17. P. Ganguly, N.Y. Vasanthacharya, C.N.R. Rao and P.P. Edwards, J. Sol. State Chem. **54,** 400-406 (1984).
18. M. Jarrell, D.L. Cox, C. Jayaprakash and H.R. Krishnamurthy, Phys. Rev. B. **40,** 8899-8907 (1989).
19. P. Mandal, A. Poddar, B. Ghosh and P. Choudhury, Phys. Rev. B. **43,** 13102-13111 (1991).
20. See, for example, C. Quitmann, D. Andrich, C. Jarchow, M. Fleuster, B. Beschoten, G. Guntherodt, V.V. Moshchalkov, G. Mante and R. Manzke, Phys. Rev. B. **46,** 11813-11825 (1992).
21. G.A., Thomas in *High Temperature Superconductivity,* Proc. 39th Scottish Universities Summer School in Physics, St. Andrews, June 1991, edited by D.P. Tunstall and W. Barford (SUSSP, Edinburgh, 1991), pp.169-206.
22. P.P. Edwards, T.V. Ramakrishnan and C.N.R. Rao, J. Phys. Chem. **99,** 5228-5239 (1995).
23. H. Fritzsche in *The Metal Non-Metal Transition in Disordered Systems,* Proc. 19th Scottish Universities Summer School in Physics, St. Andrews, August, 1978, edited by L.R. Friedman and D.P. Tunstall (SUSSP, Edinburgh, 1978). pp.193-238.
24. Y. Iye in *Physical Properties of High Temperature Superconductors III,* ed. D.M. Ginsbery (World Scientific, Singapore, 1992), pp.1-76.
25. C.N.R. Rao, J. Chem. Soc. Chem. Commun. **19,** 2217-2221 (1996).
26. C.Y. Chen, R.J. Birgeneau, M.A. Kastner, N.W. Preyer and T. Thio, Phys. Rev. **B43,** 392-401.
27. D. Emin and M.S. Hillery, Phys. Rev. **B39,** 6575-6593 (1989).
28. G. Verbist, F.M. Peters and J.T. Devreese, Phys. Rev. **B43,** 2712-2720 (1991).
29. A.S. Alexandrov and N.F. Mott, *High Temperature Superconductors and other Superfluids* (Taylor and Francis, London, 1994).
30. N.F. Mott, Contemp. Phys. **31,** 373-385 (1990).
31. N.F. Mott in *Polarons and Bipolarons in High-T_c Superconductors and Related Materials,* edited by E.K.H. Salje, A.S. Alexandrov and W.Y. Liang (Cambridge University Press, 1995) pp.1-10.
32. G.S. Boebinger, Y. Ando, A. Passner, T. Kimura, M. Okuya, J. Shimoyama, K. Kishio, K. Tamasaku, N. Ichikawa and S. Uchida, Phys. Rev. Letts. **77,** 5417-5420 (1996).

CONTROL OF Ni-SUBSTITUTION SITE IN YBa$_2$(Cu,Ni)$_3$O$_{7-\delta}$

S. ADACHI, Y. ITOH , T. MACHI, E. KANDYEL, S. TAJIMA, and N. KOSHIZUKA

SRL-ISTEC, 1-10-13 Shinonome, Koto-ku, Tokyo 135-0062, Japan

ABSTRACT

We have developed a new preparation method to make Ni-substituted YBa$_2$Cu$_3$O$_{7-\delta}$ samples with different distributions of Ni at Cu(1)("*chain*") and Cu(2)("*plane*") sites. We found that the high-temperature annealing procedure can change the substitution site of Ni. Ceramic samples of YBa$_2$Cu$_{3-x}$Ni$_x$O$_{7-\delta}$ (x = 0.00, 0.02, 0.05 and 0.10) prepared in 1-atm oxygen atmosphere were annealed at 800 - 1000 °C in various atmospheres with different oxygen partial pressures, and subsequently oxygenated below 550 °C. The lower the oxygen partial pressure was, the lower the T_c became. For the samples with the same x, no appreciable difference in lattice parameters and oxygen contents was observed. This result indicates that the distribution of Ni atoms between the chain and plane sites could be altered by the annealing treatment.

KEYWORDS: Y-123, Ni, substitution, annealing effects

INTRODUCTION

Study of transition metal substitution effects for high-T_c cuprate superconductors is one way to understand mechanism of the superconductivity [1,2]. Especially, the Ni and Zn substitution effects are very interesting because of the contrast in magnetism of the substituents [3-5]. The previous reports on the YBa$_2$Cu$_3$O$_{7-\delta}$ superconductor inform that Zn suppresses T_c most significantly among the examined transition-metals and the effect of T_c suppression by Ni-substitution is weaker. YBa$_2$Cu$_3$O$_{7-\delta}$ has two Cu sites, which are the Cu(1) ("*chain*") and Cu(2) ("*planes*") sites. It is thought that the 2-dimensional CuO$_2$ plane containing the Cu(2) site is responsible for high-T_c superconductivity. It is possible that the observed weak T_c-suppression effect by Ni-substitution results from preferential substitution of Ni into the Cu(1) site. In actual, several researchers served experimental results suggesting Ni-substitution on the Cu(1) site [6,7]. To study transition metal substitution effects for high-T_c superconductors using YBa$_2$Cu$_3$O$_{7-\delta}$, information about the location of the Ni substituents is needed. For an ideal way to study the impurity effects, it is required to produce samples in which Ni is preferentially substituted for Cu(2) site in the plane.

Coordination number of oxygen for Cu(1) is changeable in a range of 2~4. Jorgensen *et al.* [8] investigated *in situ* neutron powder diffraction at elevated temperatures under various oxygen partial pressures, $p(O_2)$. Their results indicate that the occupancy of the oxygen sites next to Cu(1) at a fixed temperature in the range of 400 - 900 °C can be controlled by changing $p(O_2)$. These results suggest us a way to control the chemical potential at Cu(1) site by tuning the condition of heat-treatments. Assuming that the site selection of substituted Ni is linked with the chemical potential at Cu(1), we may be able to control the ratio of Ni occupancies at the two sites by changing the heat-treatment condition. The chemical potential of Ni with 4-fold coordination is lower than that for 2-fold coordination [9,10]. Therefore, Ni atoms prefer the Cu(2) site to the Cu(1) in large δ composition, which is achieved under low-$p(O_2)$ atmospheres at high temperatures.

In this work, we try to make YBa$_2$(Cu,Ni)$_3$O$_{7-\delta}$ samples through different heat-treatments. The obtained samples have the same cation composition, oxygen content, and unit cell dimension, but appreciably different T_c values. This provides a new synthetic method for YBa$_2$(Cu,Ni)$_3$O$_{7-\delta}$ in which Ni atoms preferentially occupy the Cu(2) site.

EXPERIMENTAL

Samples were prepared by a solid-state reaction method. Y$_2$O$_3$, BaCO$_3$, CuO and NiO powders were

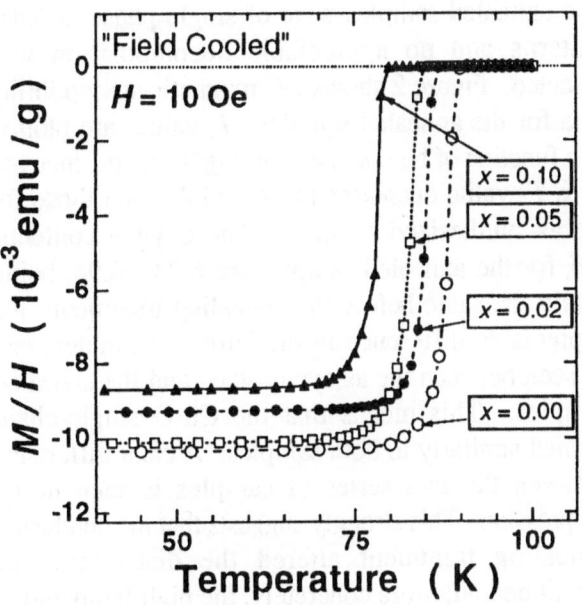

Fig.1. Temperature dependence of DC magnetic susceptibility for as-synthesized samples.

weighed out at the nominal composition of $YBa_2Cu_{3-x}Ni_xO_{7-\delta}$ ($x = 0.00, 0.02, 0.05$ and 0.10) and thoroughly mixed. The powder mixture was calcined at 860 - 920 °C for 3 days in air with intermediate grindings. The calcined powders were pelletized and sintered at 970 °C for 2 days in flowing oxygen gas and then slowly cooled at a rate of 100 °C/h. The obtained "as-synthesized" samples were annealed at 800 °C for 3 days in flowing nitrogen gas, and then quenched, The annealed samples were finally oxygenated by heat-treatment at 550 °C for 18 days in flowing oxygen gas, and then slowly cooled at a rate of 50 °C/h. Different annealing conditions were also attempted.

Structural properties were examined by powder X-ray diffraction (XRD) using $CuK\alpha$ radiation. DC magnetic susceptibility was measured by cooling samples in an external field of 10 or 20 Oe using a superconducting quantum interference device (SQUID) magnetometer. The oxygen contents were determined by a conventional iodometric titration method, assuming valences for Y, Ba, Ni [11] and O are +3, +2, +2 and -2, respectively.

RESULTS AND DISCUSSION

The as-synthesized samples were single-phase in XRD. Figure 1 shows the temperature dependences of DC magnetic susceptibility. T_c values are lowered with increasing x. These values are in agreement with the results in the previous reports [2,4]. The oxygen contents, $7-\delta$, for the as-synthesized samples are 6.92 - 6.95, indicating that almost fully oxygenation was attained for the samples. The obtained results of lattice parameters, T_c and oxygen content are summarized in Table I. It is clearly indicated that Ni substitution brings about changes in the lattice parameters and T_c.

Table I   Summary of lattice parameters, T_c and oxygen content for $YBa_2Cu_{3-x}Ni_xO_{7-\delta}$.

Sample processing condition	x	Lattice parameters (Å)			T_c (K)	Oxygen content $7-\delta$
		a	b	c		
(as-synthesized)	0.00	3.819(1)	3.888(1)	11.693(3)	90.5	6.93
	0.02	3.819(1)	3.887(1)	11.690(3)	87.5	6.95
	0.05	3.822(1)	3.885(1)	11.683(4)	85	6.94
	0.10	3.821(1)	3.885(3)	11.680(4)	80	6.92
800°C, N₂	0.00	3.820(1)	3.888(1)	11.692(3)	91	6.93
	0.02	3.819(1)	3.887(1)	11.689(2)	86	6.94
	0.05	3.821(1)	3.886(1)	11.685(4)	80	6.93
	0.10	3.821(2)	3.884(2)	11.678(6)	70	6.93
(14 days)	0.10	3.822(1)	3.884(1)	11.678(5)	70	6.95
900°C, air	0.10	3.821(1)	3.885(1)	11.679(3)	75	6.93
1000°C, $p(O_2)$ = 400 atm	0.10	3.823(2)	3.884(2)	11.679(7)	81	6.92

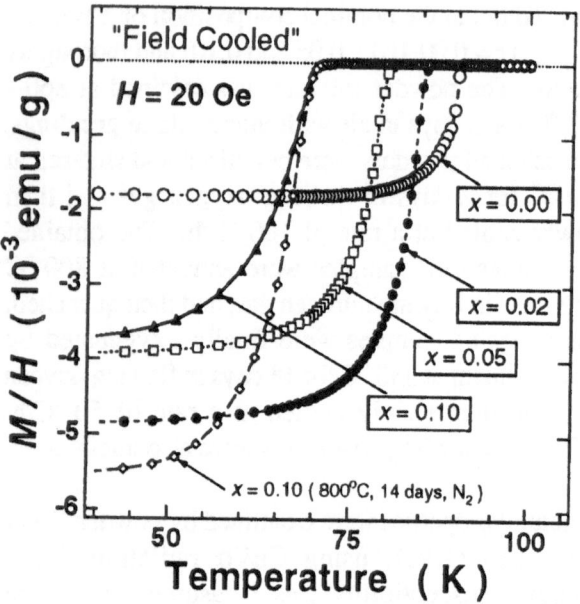

Fig.2. Temperature dependence of DC magnetic susceptibility for annealed samples.

The annealed samples were of single-phase in XRD patterns and no appreciable decomposition was detected. Figure 2 shows DC magnetic susceptibility data for the annealed samples. T_c values are plotted as a function of Ni content, x, in Fig.3. As the increase of x, T_c value decreases more rapidly than those for the as-synthesized samples. The oxygen contents, $7-\delta$, for the annealed samples are 6.93 - 6.94, being nearly the same before the annealing treatment. No appreciable difference in the lattice parameters can be seen between the as-synthesized and the annealed samples. This means that the Cu-O single-chain formed similarly in both samples. A clear difference between the two series of samples is seen in T_c-suppression. This strongly suggests that the conducted annealing treatment altered the distribution of substituted Ni; more concretely, the high-temperature annealing in the reducing atmosphere moved Ni atoms from the Cu(1) site to the Cu(2) one.

To check whether the 3-days annealing is enough to attain an equilibrium condition or not, a longer annealing duration of 14 days was examined for the sample with $x = 0.10$. No appreciable difference in lattice parameters, T_c and oxygen content was observed between the 3-days and 14-days-annealed samples (Table I and Fig. 2). The annealing treatments in different $p(O_2)$'s, which were in air and in a mixture of 80% Ar - 20% O_2 gases at 2000 atm (i.e. $p(O_2)$ = 400 atm), were performed for the sample with $x = 0.10$. The observed T_c's decrease with decreasing $p(O_2)$ during high-temperature annealing, as plotted in Fig. 3. This indicates that we can control the chemical potential of the Cu(1) site and the preference of Ni-substitution site by changing $p(O_2)$ during high-temperature annealing.

ACKNOWLEDGMENT

This work was supported by the New Energy and Industrial Technology Development Organization (NEDO) as Collaborative Research and Development of Fundamental Technologies for Superconductivity Applications.

REFERENCES

1. Xiao G et al. (1987) Phys. Rev. B 35: 8782
2. Markert JT et al. (1989) MRS Bull. 14: 37
3. Liang R et al. (1990) Physica C 170: 307
4. Ishida K et al. (1993) J. Phys. Soc. Jpn. 62: 2803
5. Mendels P et al. (1999) Europhys. Lett. 46: 678
6. Howland RS et al. (1989) Phys. Rev. B 39: 9017
7. Bridges F et al. (1990) Phys. Rev. B 42: 2137
8. Jorgensen JD et al. (1987) Phys. Rev. B 36: 3608
9. Shannon RD (1976) Acta Crystallogr. A 32: 751
10. Navrotsky A, Kleppa OJ (1967) J. Inorg. Nucl. Chem. 29: 2701
11. Yoshimuma K et al. (1997) Advances in Superconductivity IX (Springer Verlag, Tokyo) p 377

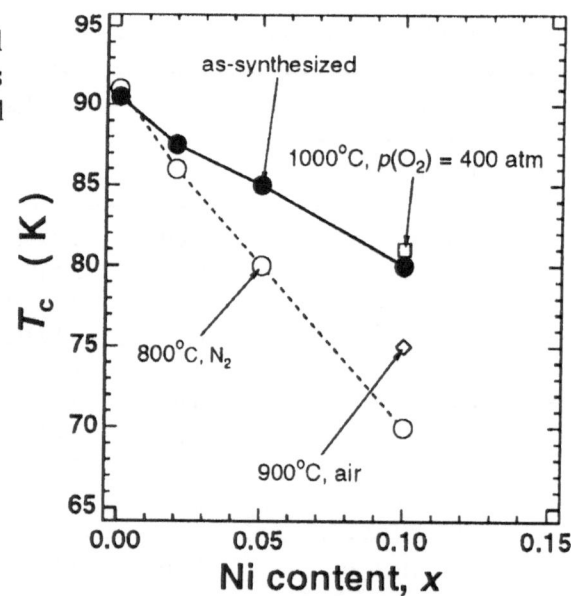

Fig.3. T_c value as a function of Ni content, x.

Observation of Mixed-Valence State in the BaSm($Cu_{1-x}Fe_x$)$_2$O$_{5+\delta}$ Double-Perovskite Phase

Johan Lindén[1*], Jin Nakamura[1], Pavel Karen[2], Arne Kjekshus[2], Maarit Karppinen[1†], and Hisao Yamauchi[1]

[1]Materials and Structures Laboratory, Tokyo Institute of Technology, Yokohama 226-8503, Japan
[2]Department of Chemistry, University of Oslo, N-0315, Norway

Abstract: A set of Fe-based, double-perovskite solid solutions BaSm($Cu_{1-x}Fe_x$)$_2$O$_{5+\delta}$ ($x = 1.0$, 0.95, 0.90 and 0.50) was synthesized. The samples were characterized by ^{57}Fe Mössbauer spectroscopy at 77 and 300 K. All samples were magnetically ordered at 300 K. In reduced samples, a single component, formally denoted as $Fe^{2.5+}$, dominated the spectra. Upon oxidizing or doping the samples with Cu the intensity of this component decreases dramatically. The overall amounts of di-, tri-, and mixed-valent Fe corresponded well to the values of x and δ. Below the Verwey-type transition temperature ($T_V \approx 200$ K) the mixed-valence state separates into pure high-spin Fe^{3+} and Fe^{2+}, as verified by Mössbauer spectroscopy. Resistivity measurements showed that the samples are semiconducting at 300 K. Upon decreasing temperature, the resistivity increases exponentially, but at T_V there is a change in activation energy.

Keywords: Mössbauer spectroscopy, Double perovskite, Mixed valence

INTRODUCTION

The antiferromagnetic double-perovskite cuprate BaYCuFeO$_{5+\delta}$, first synthesized by Er-Rakho *et al.* [1], has so far occupied a less prominent position in material science than, *e.g.*, the Ba$_2$YCu$_3$O$_{7-x}$ triple-perovskite, due to the fact that the former lacks the superconducting properties. In the double-perovskite structure, the Fe (Cu) atoms reside inside square-pyramidal coordination polyhedra whose basal planes face each other, not unlike in the BaYCu$_3$O$_{7-x}$ phase. The Cu-free version of the double-perovskite was first synthesized by Karen and Woodward with Sm and Nd at the rare-earth site [2]. Excess oxygen atoms may enter the site between the planes [2], thus increasing locally the coordination number of Fe from 5 to 6 [2,3].

Recently a Verwey-type mixed-valence state has been observed in the BaSmFe$_2$O$_{5+\delta}$ double-perovskite phase [4]. In samples with reduced oxygen content ($\delta \approx 0$), the Fe atoms, which occupy a single lattice site, have an overall valence in between +3 and +2. Their ^{57}Fe Mössbauer spectra recorded at room temperature are nevertheless dominated by a single antiferromagnetic component, which is assigned to the valence state $Fe^{2.5+}$ [4]. Upon cooling of the samples below the Verwey-type transition temperature $T_V \approx 200$ K, a charge separation occurs, and the expected amounts of divalent and trivalent iron are found in the Mössbauer spectra [4]. For BaYCuFeO$_{5+\delta}$, Caignaert *et al.* [5] observed a change in the magnetic structure at $T \approx 200$ K which is well below the Néel temperature of 450 K. Whether this rearrangement of spins is related to the Verwey-type transition in BaSmFe$_2$O$_{5+\delta}$ is yet unconfirmed.

EXPERIMENTAL

The preparation of the Cu-free samples $BaSmFe_2O_{5+\delta}$ is reported elsewhere [4]. Cu-doped samples were synthesized using high-purity $BaCO_3$, Sm_2O_3, CuO and Fe_2O_3 as starting materials. After thorough mixing, the powders were reacted at 985 °C for 15 h in flowing Ar. Then the samples were placed in vacuum capsules containing a Fe/FeO getter and sintered at 985 °C for 40 h. Nominal compositions corresponding to $x = 1, 0.95, 0.9$ were prepared in this manner. Additionally a sample having $x = 0.5$ was synthesized by using an air atmosphere and a temperature of 1000 °C (24 h). Reduction of the samples was achieved by annealing in a thermobalance under a 5 % H_2/Ar atmosphere at 600 °C. The phase composition of the samples was determined by powder x-ray diffraction (XRD) and structural details of the present phases were investigated by transmission-electron microscopy. ^{57}Fe Mössbauer spectra were recorded at 77 and 300 K in a transmission geometry. Details on the analysis of the spectra are given in Ref. [4]. Energy-dispersive x-ray analysis (EDX) was applied for checking the concentration of elements in the unit cell. For resistivity measurements the as-synthesized $BaSm(Cu_{0.05}Fe_{0.95})_2O_{5+\delta}$ powder was pelletized and sintered at 985 °C in the vacuum capsule (final δ value was approx. 0.2). The resistivity curve was measured down to 77 K using a simple two-point DC set up. Thin Ag electrodes were attached using Ag paste. Samples were also characterized by magnetic susceptibility measurements.

RESULTS AND DISCUSSION

All samples were found to crystallize in the double-perovskite structure. The phase purity was more than 90 % according to XRD. According to EDX, the expected amounts of copper were present in the samples. A change in the magnetic susceptibility of the most Fe-rich samples was observed in the temperature range of 180 – 210 K, indicating the occurrence of the Verwey-type transition. A corresponding change occurred in the resistivity of the $BaSm(Cu_{0.05}Fe_{0.95})_2O_{5.2}$ sample, manifested on a semi-logarithmic scale with a change in slope of the otherwise nearly linear dependences. The resistivity undergoes a profound change between 300 and 160 K, from some 30 Ωm to $3 \cdot 10^5$ Ωm. Due to these high values of resistance, the contact contributions could be neglected in the 2-point method used.

Fig. 1 shows the Mössbauer spectra for the $BaSm(Cu_{0.10}Fe_{0.90})_2O_{5.1}$ sample recorded at 77 and 300 K. The component assigned to $Fe^{2.5+}$ is highlighted. It has an isomer shift of 0.55 mm/s, $i.e.$, it lies between that of high-spin Fe^{2+} and high-spin Fe^{3+} [6]. Bond-valence calculations also give a formal valence of +2.5 [2]. The internal field is around 30 T. Below T_V this component is split into pure Fe^{2+} and Fe^{3+} as shown by the highlighted components of the 77 K spectrum. The internal field of the Fe^{2+} component has an extremely low value of 8 T, which causes the resonance lines to be concentrated to the center of the Mössbauer spectra, but the isomer shift of 1.27 mm/s is typical for high-spin Fe^{2+}. The high-spin Fe^{3+} component has an internal field of 53 T and an isomer shift of 0.39 mm/s, thus confirming the assignment. For a detailed discussion on the component assignment see Ref. 4.

When the oxygen content of the samples increases, the intensity of the $Fe^{2.5+}$ component in the 300 K data and the intensity of the Fe^{2+} component in the 77 K data decreases rapidly [4]. The same is true for the 77 K data of the Cu-doped samples when the Cu content $(1-x)$ is increased. However, the 300 K spectrum of $BaSm(Cu_{0.50}Fe_{0.50})_2O_{5.0}$ was dominated by a component resembling the one assigned to $Fe^{2.5+}$, among other by its internal field being approximately 30 T. Its isomer shift of 0.22 mm/s, however, unambiguosly proves it to be a Fe^{3+} state. The open question is whether the low value of the internal field value indicates a certain inclination towards formation of the mixed-valence between Fe^{3+} and Cu^{2+}. The Néel temperature of the similar $BaY(Cu_{0.50}Fe_{0.50})_2O_{5.0}$ phase is as high as 446 K [7] and therefore one would expect a higher internal field value at 300 K. Future endeavors will include attempts to synthesize the $BaSm(Cu_{1-x}Fe_x)_2O_{5+\delta}$ solid solution with $0.5 < x < 0.9$ in order to check in detail the influence of Cu doping on the $Fe^{2.5+}$ state.

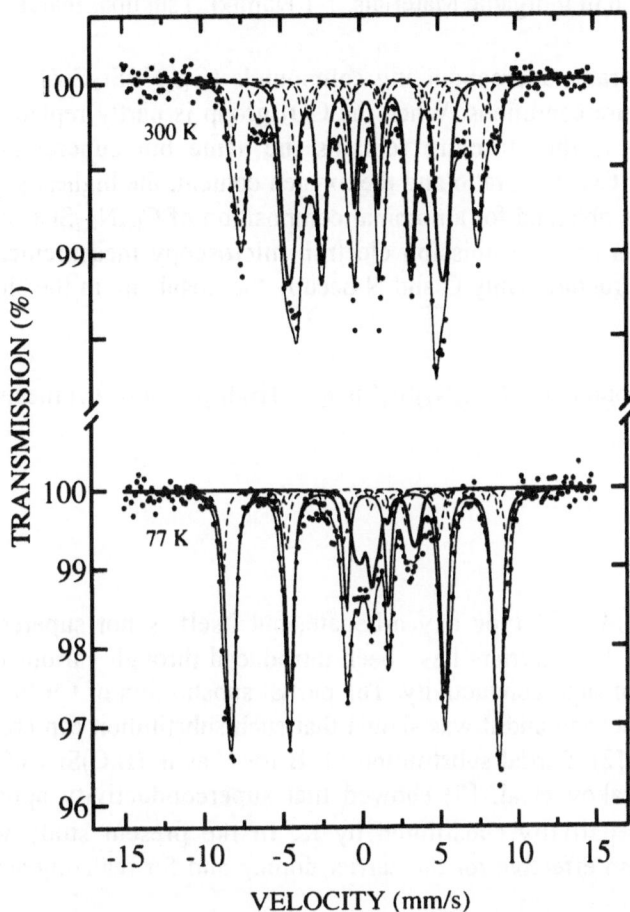

Fig.1. Mössbauer spectra from the $BaSm(Cu_{0.10}Fe_{0.90})_2O_{5.12}$ measured at the indicated temperatures.

Acknowledgments. The kind support from the Japan Society for the Promotion of Science is acknowledged (J. L.). The authors are grateful to Dr. M. Lippmaa for performing the resistivity measurements.

* On leave from Physics Department, Åbo Akademi, FI-20500 Turku, Finland.

† Permanent address: Lab. of Inorg. Anal. Chem. Helsinki Univ. Technol., FI-02150 Espoo, Finland.

1. L. Er-Rakho, C. Michel, Ph. Lacorre, and B. Raveau, J. Solid State Chem. **73**, 531 (1988).
2. P. Karen and P.M. Woodward, J. Mater. Chem. **9**, 789 (1999).
3. M. Nagase, J. Lindén, J. Miettinen, M, Karppinen, and H. Yamauchi, Phys. Rev. B **58**, 3371 (1998).
4. J. Lindén, P. Karen, A. Kjekshus, T. Pietari, and M. Karppinen, Phys. Rev. B, Dec. 1999
5. V. Caignaert, I. Mirebeau, F. Bourée, N. Nguyen, A. Ducouret, J.-M. Greneche, and B. Raveau, J.Solid State Chem. **114**, 24 (1995).
6. N.N. Greenwood and T.C. Gibb, "Mössbauer Spectroscopy", p. 91, Chapman and Hall, London, 1971.
7. C. Meyer, F. Hartmann-Boutron, Y. Gros, and P. Strobel, Solid State Commun. **76**, 163 (1990).

New oxycarbonitrate superconductors $(C_{1-x}N_x)Sr_2CuO_{5+y}$ with a 1201-type structure

Nikolai Zhigadlo, Yoshihiro Anan, Toru Asaka, Yoshio Matsui, and Eiji Takayama-Muromachi

National Institute for Research in Inorganic Materials, 1-1 Namiki, Tsukuba, Ibaraki, 305-0044, Japan

Abstract: New oxycarbonitrate superconducting compounds $(C_{1-x}N_x)Sr_2CuO_{5+y}$ were obtained under high pressure/high temperature conditions. When the CO_3 group is partly replaced by NO_3 group as in the $(CO_3)_{1-x}(NO_3)_xSr_2CuO_{2+y}$, the structure remains the same but superconductivity appears. By controlling both the nominal CO_3/NO_3 ratio and the oxygen content, the highest superconducting volume fraction with T_c of 33 K was obtained for a nominal composition of $C_{0.8}N_{0.2}Sr_2CuO_{5.3}$. X-ray diffraction, electron-probe microanalysis and transmission electron microscopy measurements indicated that this phase has an M-1201 type structure. Only C and N occupy the metal site in the charge reservoir without partial replacement of Cu.

Keywords: High-T_c superconductor, $(C_{1-x}N_x)Sr_2CuO_{5+y}$, High-pressure synthesis, HRTEM, Magnetic susceptibility

INTRODUCTION

The CSr_2CuO_5 is a typical M-1201 type oxycarbonate, but itself is not superconducting because no carriers are present [1]. The hole carriers have been introduced through various chemical substitutions resulting in the appearance of superconductivity. The partial substitution of Cu for C and Ba for Sr as in $(Cu,C)(Ba,Sr)_2CuO_5$ was first tried and it was shown that such substitution can create enough number of holes for superconductivity [2]. Partial substitution of B for C as in $(B,C)Sr_2CuO_5$ is also effective for the hole doping [3,4]. Kazakov et al. [5] showed that superconductivity appears in $CSr_{2-x}K_xCuO_5$ $(0.25 \leq x \leq 0.7)$ where Sr is partially substituted by K. In the present study we found that partial substitution of N for C is also effective for the carrier doping and for realizing superconductivity in the M-1201 type structure.

EXPERIMENTAL

The powders of CuO (99.9 %), $Sr(NO_3)_2$ (99.9 %), Sr_2CuO_3, SrO_2, $SrCO_3$ (99.9 %), were used to obtain nominal mixtures for high-pressure synthesis. Sr_2CuO_3, was prepared in advance through solid state reaction of CuO and $SrCO_3$ (99.9 %) at 980° C for 5 days with several intermediate grindings. SrO_2 was prepared through a solution route [6]. The initial reagents were thoroughly grounded and sealed in gold capsules for high-pressure treatments. Ag_2O was put into the mixture as an oxidizing agent. Samples were allowed to react at 5.5 GPa and at 1270° C for 1 h using a belt-type high-pressure apparatus, then quenched to room temperature. Prepared specimens were examined by X-ray powder diffraction analysis with CuK_α radiation. The lattice parameters were determined by the least squares method. Electron diffraction (ED) patterns and lattice images were taken using a Hitachi H-1500 high-resolution transmission electron microscope (HRTEM) operated at 800 kV. The electron probe micro analysis (EPMA) was carried out using a JEOL JXA-8600MX analyzer. The magnetic susceptibilities were measured for pulverized samples under zero-field-cooling (ZFC) and field-cooling (FC) conditions in an applied field of 20 Oe by a DC SQUID magnetometer (Quantum Design, MPMS).

RESULTS AND DISCUSSION

Starting compositions, cell parameters of the M-1201 phases and superconducting properties for some of the samples prepared are listed in Table 1. The superconducting transition temperature, T_c in the table is the onset value determined from the magnetic susceptibility data collected by the zero-field-cooling method. The magnetic susceptibility at 5 K by the zero-field-cooling method is given as a rough measure of the superconducting volume fraction. As seen in the table, the superconducting volume fraction depends strongly on the nominal CO_3/NO_3 ratio and the nominal oxygen content. The highest superconducting volume fraction which was large enough to assume bulk nature of superconductivity was obtained for the nominal composition, $C_{0.8}N_{0.2}Sr_2CuO_{5.3}$, with $T_c = 33$ K (see below).

Table 1. Chemical compositions of starting mixtures, cell parameters and superconducting properties of the high-pressure samples.

No.	Starting composition	a (Å)	c (Å)	V (Å^3)	T_c (K)	χ (10^{-3} emu/g)[a]
1	$CSr_2CuO_{5.35}$	3.8980(3)	7.4871(7)	113.76(2)	-	-
2	$C_{0.9}N_{0.1}Sr_2CuO_{5.35}$	3.8990(3)	7.4792(6)	113.70(2)	13	-0.59
3	$C_{0.8}N_{0.2}Sr_2CuO_{5.3}$	3.8974(2)	7.4724(5)	113.50(1)	33	-5.59
4	$C_{0.8}N_{0.2}Sr_2CuO_{5.35}$	3.8983(4)	7.4716(8)	113.54(2)	31	-5.15
5	$C_{0.8}N_{0.2}Sr_2CuO_{5.45}$	3.8984(4)	7.4692(8)	113.52(2)	31	-5.05
6	$C_{0.7}N_{0.3}Sr_2CuO_{5.35}$	3.8984(2)	7.4676(5)	113.49(1)	35	-2.57
7	$C_{0.7}N_{0.3}Sr_2CuO_{5.45}$	3.8976(3)	7.4719(6)	113.51(2)	31	-3.16
8	$C_{0.7}N_{0.3}Sr_2CuO_{5.55}$	3.8978(2)	7.4728(6)	113.53(1)	31	-3.02

[a]Magnetic susceptibility at 5 K.

Figure 1 shows the powder X-ray diffraction patterns of high-pressure products with nominal compositions $(C_{1-x}N_x)Sr_2CuO_{5.35}$, x=0.0, 0.1, 0.2, 0.3 (sample No. 1, 2, 4, 6, respectively).

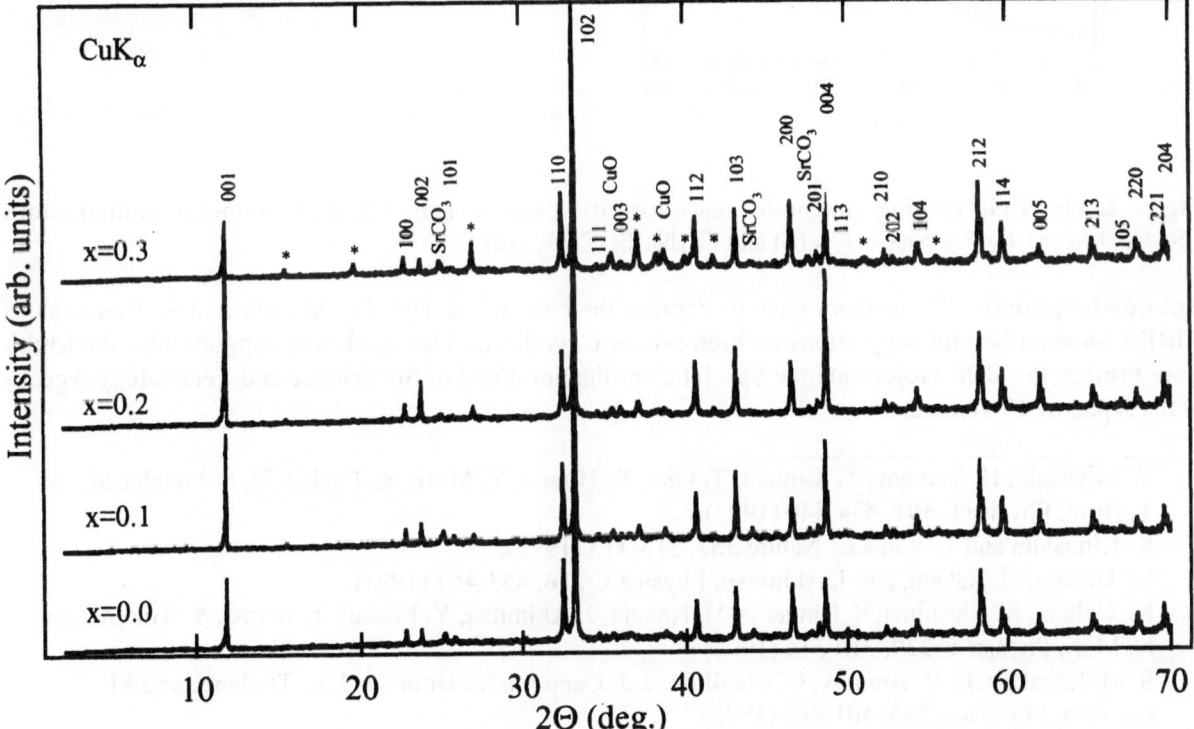

Fig. 1. The powder X-ray diffraction patterns of samples with nominal compositions $(C_{1-x}N_x)Sr_2CuO_{5.35}$, x=0.0, 0.1, 0.2, 0.3. Peaks marked "*" belong to unidentified impurity phase(s).

Most of the X-ray peaks of each sample can be indexed assuming a tetragonal cell. Some peaks can be assigned to unreacted CuO and $SrCO_3$. Although presence of metal Ag was confirmed by EPMA, X-ray peaks corresponding to it are not seen in Fig. 1 probably because of its amorphous-like structure. Peaks denoted by "*" are due to a small amount of unidentified impurity phase(s). The tetragonal phase in each sample was identified by the ED and HRTEM measurements to have an M-1201 type layered structure. To know the composition of the present M-1201 phase, we performed EPMA measurements for the sample with the $(C_{0.8}N_{0.2})Sr_2CuO_{5.35}$ nominal composition. Although, the N and C contents could not be determined by EPMA, we obtained the Sr and Cu contents to be 0.50±0.01. This ratio of 0.50(1), i.e. 1/2 is expected from the composition of $(C,N)Sr_2CuO_y$ and it was concluded that the M-site is occupied only by C and N, without partial substitution of Cu. The field-cooling magnetic susceptibilities data are given in Fig. 2 for the samples with nominal compositions of $CSr_2CuO_{5.35}$ (a), $C_{0.9}N_{0.1}Sr_2CuO_{5.35}$ (b) and $C_{0.8}N_{0.2}Sr_2CuO_{5.3}$ (c). Superconducting transition was not found in the nitrogen-free $CSr_2CuO_{5.35}$ sample. On other hand, superconductivity appeared in the two nitrogen-containing samples. In particular, the $C_{0.8}N_{0.2}Sr_2CuO_{5.3}$ sample showed large enough diamagnetism below 33 K to assume bulk superconductivity. These results indicate that the substitution of N for C creates enough carriers for superconductivity.

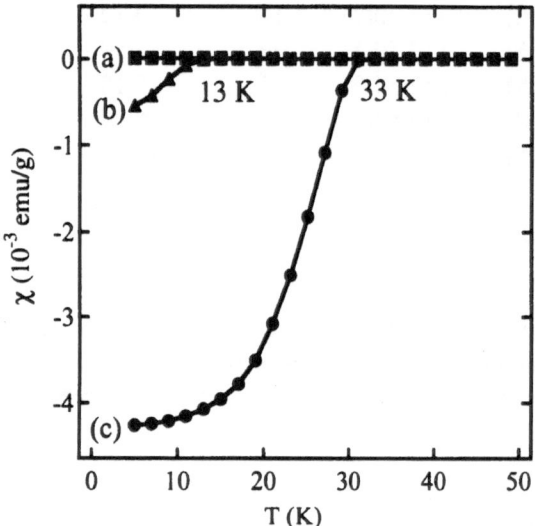

We can assume the pentavalent state for N because of the present high oxygen-pressure synthesis condition. Electron doping rather than hole doping is expected from the simple substitution of NO_3^{-1} for CO_3^{-2}, but the electron doping is highly unlikely because of the octahedral coordination of the Cu atom. It seems that excess oxygen atoms are incorporated into the (C,N) plane, simultaneously with the N substitution. In other words, the CO_3^{-2} group seems to be substituted by NO_4^{-3} or NO_6^{-7} group, resulting in the creation of the hole carriers.

Fig. 2. The field-cooling magnetic susceptibility data of samples with nominal compositions $CSr_2CuO_{5.35}$ (a), $C_{0.9}N_{0.1}Sr_2CuO_{5.35}$ (b) and $C_{0.8}N_{0.2}Sr_2CuO_{5.3}$ (c).

Acknowledgments. The authors wish to express their thanks to Drs. M. Akaishi and S. Yamaoka of NIRIM for their helpful suggestions on high-pressure synthesis. This work was supported by the Multi-Core Project, the COE Project and the Special Coordination Fund of the Science and Technology Agency of the Japanese Government.

1. Y. Miyazaki, H. Yamane, T. Kajitani, T. Oku, K. Hiraga, Y. Morii, K. Fuchizaki, S. Funahashi, and T. Hirai, Physica C **191**, 434-440 (1992).
2. K. Kinoshita and T. Yamada, Nature **357**, 313-315 (1992).
3. M. Uehara, H. Nakata, and J. Akimitsu, Physica C **216**, 453-457 (1993).
4. M. Uehara, M. Uoshima, S. Ishiyama, H. Nakata, J. Akimitsu, Y. Matsui, T. Arima, Y. Tokura, and N. Mori, Physica C **229**, 310-314 (1994).
5. S. M. Kazakov, E. V. Antipov, C. Chaillout, J. J. Capponi, M. Brunner, J. L. Tholence and M. Marezio, Physica C **253**, 401-406 (1995).
6. M. Isobe, T. Kawashima, K. Kosuda, Y. Matsui, and E. Takayama-Muromachi, Physica C **234**, 120-126 (1994).

High pressure synthesis of new superconductors MoNiP and MoRuP with T_c of above 15 K

Ichimin Shirotani[1], Mitsuru Takaya[1], Isamu Kaneko[1], Chihiro Sekine[1] and Takehiko Yagi[2]

[1]Faculty of Engineering, Muroran Institute of Technology, Mizumoto, Muroran-shi 050-8585, Japan
[2]The Institute for Solid State Physics, University of Tokyo, Roppongi, Tokyo 106-8666, Japan

Abstract: Ternary molybdenum phosphides MoNiP with a Fe_2P-type structure and MoRuP with a Co_2P-type structure have been prepared at around 1600 °C and 4 GPa. The resistivity of both compounds decreases with decreasing temperature and sharply drops at around 15.5 K. These compounds are new superconductors with the highest T_c in the metal phosphides.

Keywords: Superconductivity, High pressure synthesis, New superconductor

INTRODUCTION

Ternary metal compounds ZrRuX(X = P and Si) crystallize in two modifications, a Fe_2P-type hexagonal structure(h-ZrRuX)[1,2] and a Co_2P-type orthorhombic structure(o-ZrRuX)[3,4]. Each layer in the Fe_2P-type hexagonal structure is occupied by either Zr and X atoms or Ru and X atoms. The two-dimensional triangular clusters of Ru_3 are formed, and linked with each other through the Ru-X bonds in the basal plane. In contrast, the orthorhombic structure has layers which are filled with Zr, Ru and X atoms and these layers are all equivalent. The orthorhombic phase of ZrRuP transforms to the hexagonal one at around 3.5 GPa and 1100 °C[5]. Superconducting transition temperatures(T_c's) of the several ternary metal compounds with the Fe_2P-type structure are considerably high, above 10 K[1,4,6,]. On the other hand, the T_c's of the superconductors with the Co_2P-type structure are between 2 and 5 K[2,4,7]. The T_c's of the compounds with the Fe_2P-type structure are much higher than those of the compounds with the Co_2P-type one. However, we have found that the T_c of ZrRhSi with the Co_2P-type structure is above 10 K[8].

Various binary molybdenum phosphides have been prepared at high temperatures[9,10]. Several metal-rich molybdenum phosphides show the superconductivity at low temperatures ; Mo_3P with the T_c of 7K has the highest T_c in the binary molybdenum phosphides[10]. Ternary molybdenum phosphide MoNiP has the Fe_2P-type structure[11]. Thus, the electrical behavior of the compound is very interest. Recently, we have prepared ternary molybdenum phosphides MoNiP and MoRuP with layer structures at high temperatures and high pressures. These compounds show superconducting transition at around 15.5 K. In this report the properties of the new superconductors prepared at high pressure are discussed.

EXPERIMENTAL

Using a wedge-type cubic-anvil high pressure apparatus, ternary molybdenum compounds were pre-

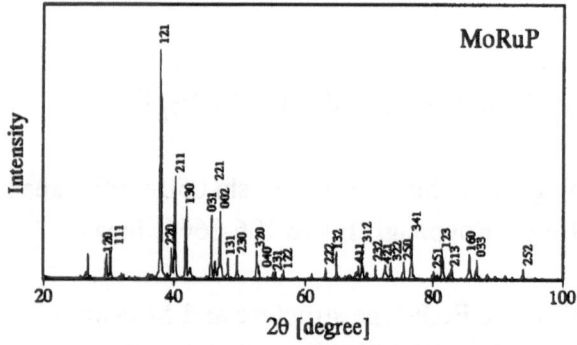
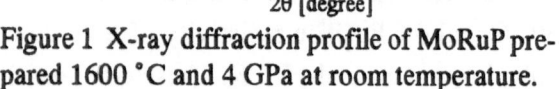

Figure 1 X-ray diffraction profile of MoRuP prepared 1600 °C and 4 GPa at room temperature.

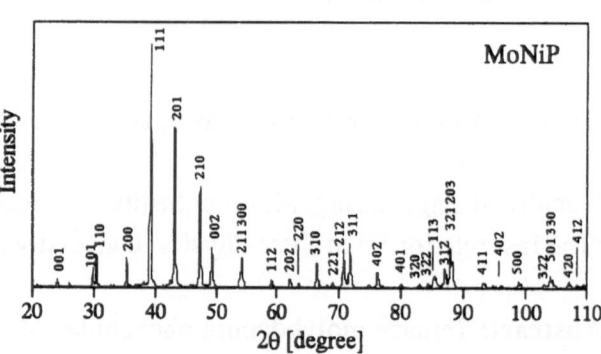

Figure 2 X-ray diffraction profile of MoNiP prepared 1600 °C and 4 GPa at room temperature.

pared at high temperatures and high pressures[4]. The sample container made of pyrophyllite is formed into a cube of 21 mm on an edge. A sample assembly for the preparation of these compounds is similar to that used for the synthesis of black phosphorous[12]. The starting materials are put into a crucible made of BN. The crucible with a graphite heater is inserted into the pyrophyllite cube. The ternary molybdenum phosphides were prepared by reaction of stoichiometric amounts of each metal and P powders at around 4 GPa. The reaction temperatures were between 1500 and 1700 °C. The samples were characterized by powder x-ray diffraction using CuK_α radiation and silicon as a standard. Figure 1 and 2 show the x-ray diffraction profiles of MoRuP and MoNiP prepared at around 1600 °C and 4 GPa. The diffraction lines in MoRuP are assigned by the index of the Co_2P (TiNiSi)-type structure. On the other hand, MoNiP has the Fe_2P-type hexagonal structure. The lattice constants obtained from these data are given in table 1.

RESULTS AND DISCUSSION

Figure 3 shows electrical resistivity - temperature curve for MoNiP prepared at high pressure. The resistivity decreases with decreasing temperature and sharply drops at around 15.5 K. The magnetic element Ni in MoNiP is contained. However, the compound has the considerably high T_c. Figure 4 shows electrical resistivity-temperature curve for MoRuP prepared at high pressure. The sharp decrease of the resistivity is observed at around 15.5 K. The dc magnetic susceptibility of both com-

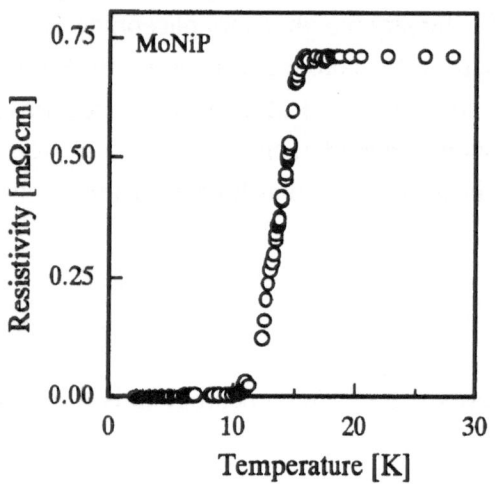

Figure 3 Resistivity of MoNiP at low temperatures.

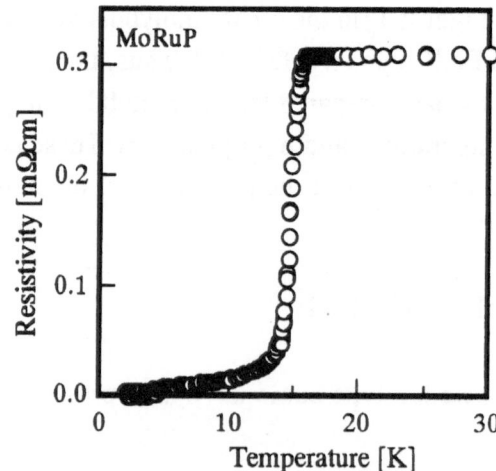

Figure 4 Resistivity of MoRuP at low temperatures.

Table 1 T_c's and crystal data of MoNiP, MoRuP and ternary zirconium compounds

	h-ZrRuP	h-ZrRuSi	MoNiP	o-ZrRuP	ZrRhSi	MoRuP
Structure	Hexagonal, Fe_2P-type			Orthorhombic, Co_2P-type		
a(Å)	6.459	6.684	5.858	6.417	6.549	6.035
b(Å)	-	-	-	7.323	7.459	6.944
c(Å)	3.778	3.672	3.668	3.862	3.918	3.857
c/a	0.577	0.550	0.626	0.602	0.598	0.639
V/fomula unit (Å^3)	45.50	47.67	36.35	45.37	47.85	40.40
T_c(K)	13	12.2	15.5	4	10.5	15.5

pounds rapidly decreases above 15 K. These are due to the superconducting transition. MoNiP with the Fe_2P-type structure and MoRuP with the Co_2P-type structure are new superconductors with the T_c of above 15 K. The T_c of both compounds is highest in the metal phosphides. Binary molybdenum phosphides Mo_3P, Mo_8P_5 and Mo_4P_3 have been prepared at high temperatures and high pressures. The superconductivity is found at around 7 K for Mo_3P, 5.8 K for Mo_8P_5 and 3 K for Mo_4P_3[10]. Mo_3P has the highest T_c in the binary metal phosphides. The phosphide superconductors contained molybdenum atoms are very interest.

The T_c's and crystal data of the superconductors with Fe_2P and Co_2P-type structures are summarized in table 1. The cell volumes / formula unit of MoNiP and MoRuP are significantly smaller than those of the zirconium compounds. The c/a of both molybdenum compounds is considerably larger than that of the zirconium compounds. MoNiP and MoRuP have stronger two-dimensional character compared with the zirconium compounds. These may be closely related to the superconducting properties of ternary molybdenum phosphides.

1. H. Barz, H.C. Ku, G.P. Meisner, Z. Fisk and B.T. Mattias, Proc. Natl. Acad. Sci., U.S.A., **77**, 3132-3134(1980).

2. V. Johnson and W. Jeitschko, J. Solid State Chem., **4**, 123-130(1972).

3. R. Muller, R.N. Shelton, J.W. Richardson and R.A. Jacobson, J. Less Common Met., **92**, 177-183(1983).

4. I. Shirotani, K. Tachi, K. Takeda, S. Todo, T. Yagi and K. Kanoda, Phys. Rev., **B52**, 6197-6200(1995).

5. I. Shirotani, K. Tachi, N. Ichihashi, T. Adachi, T. Kikegawa and O. Shimomura, Phys. Lett., **A205**, 77-80(1995).

6. I. Shirotani, K. Tachi, Y. Konno, S. Todo and T. Yagi, Phylosophical Magazine, **B79**, 767-776(1999).

7. W. Xian-Zhong, B. Chevalier, J. Etourneau and P. Hagenmuller, Mat. Res. Bull., **20**, 517(1985).

8. I. Shirotani, Y. Konno, Y. Okada, C. Sekine, S. Todo and T. Yagi, Solid State Commun., **108**, 967-970(1998).

9. S. Rundqvist and T. Lundström, Acta chem. Scand., **17**, 37-46(1963).

10. I. Shirotani, I. Kaneko, M. Takaya, C. Sekine and T. Yagi, Physica B, in press.

11. P. R. Guerin and M. Sergent, Acta Cryst., **B33**, 2820-2823(1977).

12. I. Shirotani, Mol. Cryst. Liq. Cryst., **86**, 1943-1952(1982).

Epitaxial growth of single-crystalline thin film of two-legged spin ladder compound $Ca_{14}Cu_{24}O_{41}$

Yutaka Furubayashi[1*], Takahito Terashima[1], Mikio Takano[1], Ken-ichiro Yagi[2], Shigetada Shima[2], and Hikaru Terauchi[2]

[1]Institute for Chemical Research, Kyoto University, Uji, Kyoto-fu 611-0011, Japan
[2]Advanced Research Center of Science, Kwansei-Gakuin University, Sanda, Hyogo 669-1337, Japan

*email: furubaya@scl.kyoto-u.ac.jp

Abstract: We report the growth of single-crystalline thin film of Sr-free $Ca_{14}Cu_{24}O_{41}$. RHEED and XRD measurement revealed that the structure of the film is identical to that of bulk $Sr_{14-x}Ca_xCu_{24}O_{41}$ and Ca-content x is 14. Electric behavior of ρ_c is T-linear down to 20K and semiconductive at lower temperature, while ρ_a is semiconductive at whole temperature range. This behavior is similar to that of ρ_{ab} and ρ_c for underdoped high-T_c superconductors. But no sign of superconductivity emerged.

Key words: spin ladder, $(Sr,Ca)_{14}Cu_{24}O_{41}$ compounds, thin film growth, superconductivity

1 Introduction

Spin ladder systems of $S=1/2$ antiferromagnet have attracted much attention. According to theoretical studies [1,2], even-legged spin ladders have energy gap in the spin excitation and the possibility of d-wave superconductivity with a high temperature scale of $\Delta/2$ (Δ is spin gap). Especially the superconductivity was observed through a transport study under high pressure >3GPa for $(Sr,Ca)_{14}Cu_{24}O_{41}$[3,4].

$Sr_{14-x}Ca_xCu_{24}O_{41}$ contains a Cu_2O_3 two-legged ladder sheet and a CuO_2 1D-chain sheet alternately stacked along b-axis. The c-axis is parallel to ladders and chains and the a-axis is perpendicular to them (shown in fig 1.). It is known that the density of holes in ladder sheet increases as Ca-content increases [5] and for $x > 10$ the resistivity along c-axis becomes metallic in wide range of temperature.

The Sr-free $Ca_{14}Cu_{24}O_{41}$, is desired for the extended study of spin-charge coupled dynamics in spin ladder systems as well as the elucidation of the possibility of superconductivity under ambient pressure, because of the highest carrier density. For bulky single crystals the Ca-content x is limited to 12 because of requirement of high oxygen pressure. One promising way to grow metastable compounds is the epitaxial. In this paper we report the epitaxial growth of Sr-free $Ca_{14}Cu_{24}O_{41}$ thin film and the results of electric resistivity and anisotropy.

2 Experimental

$Ca_{14}Cu_{24}O_{41}$ thin film was grown by the pulsed laser deposition . The substrate is the single crystal $Sr_{14}Cu_{24}O_{41}$ grown by the floating zone method. The target is the mixture of $CaCO_3$ and CuO at a ratio of 14:26 sintered at 780 °C in oxygen atmosphere. The film was grown with substrate temperature of 600 °C, an oxygen pressure of 1.6×10^2 Pa, and deposition rate of 0.05 nm/s. For oxydation, the

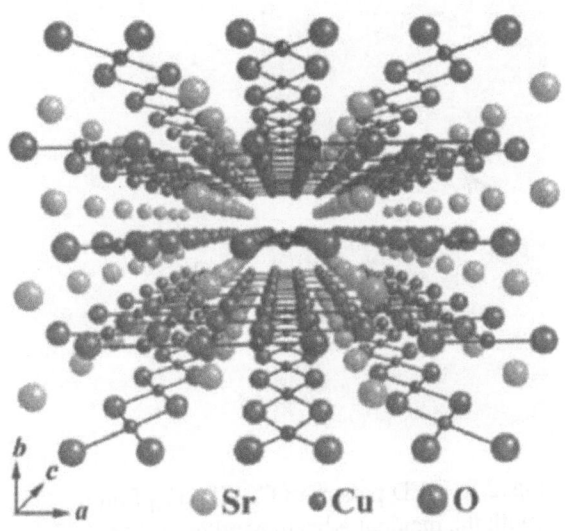

Fig. 1 Perspective view of the crystal structure of $Sr_{14}Cu_{24}O_{41}$ showing the geometrical arrangement of the Cu_2O_3 ladders and the CuO_2 chains.

grown film is treated with partially ozonized oxygen gas in cooling.

The identification of the structure is done by the reflection high energy electron diffraction (RHEED) and the X-ray diffraction (XRD). Electric resistivity under ambient pressure is measured by conventional 4-probe method.

3 Results and Discussion

Figure 2 is the RHEED patterns of a film with the thickness of 20 nm. The upper pattern for the incident e-beam parallel to a-axis consists of two sets of streaks. The streaks indexed as $(0\ 7n)$ are by the c-axis of ladders and the streaks indexed as $(0\ 10n)$ by the c-axis of chains, respectively. The lower pattern for the incident e-beam parallel to c-axis consists of only a set of streaks common to ladders and chains. Therefore the structure of this film is identical to that of 14-24 phase.

In fig. 3 the lattice spacing a, b, and c and unit cell volume V are plotted as a function of Ca content x [7,8] . Because of the in-plane lattice mismatch, the in-plane lattice a and c are expanded and the out-of-plane b shrunk. But the volume can be fitted with Vegard's Law. Therefore, we can conclude that the Ca content x in this film is 14.

Figure 4-(a) shows electric resistivity for the $Ca_{14}Cu_{24}O_{41}$ film with the current parallel to c-axis (ρ_c) and parallel to a-axis (ρ_a) as a function of temperature (film thickness is 100 nm) . The behavior of ρ_c is T-linear down to 20 K and semiconductive at lower temperature. On the other hand, ρ_a is semiconductive at all range of temperature. This behavior of ρ_c and ρ_a is similar to that of ρ_{ab} and ρ_c for the underdoped high-T_c superconductors. Anisotropy ratio, ρ_a/ρ_c, for the $Ca_{14}Cu_{24}O_{41}$ film is shown in fig. 4-(b). The value of ρ_a/ρ_c is 200 in wide range of temperature and increases monotonously as temperature decreases.

As compared to bulky $(Sr,Ca)_{14}Cu_{24}O_{41}$, we can mention as follows: (1) the range of metallic resistivity is wider and (2) the in-plane anisotropy ρ_a/ρ_c is larger than for the bulky $(Sr,Ca)_{14}Cu_{24}O_{41}$ [4]. Therefore, $Ca_{14}Cu_{24}O_{41}$ in this work contains the most heavily hole-doped two-legged ladders preserving quasi-1D nature.

Acknowledgements

This work has been partly supported by CREST (Core Research for Evolution Science and Technology) of the Japan Science and Technology Corporation (JST).

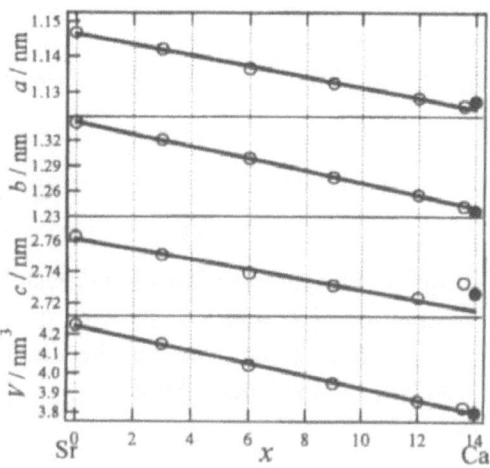

Fig. 3 The lattice spacing a, b, c and unit cell volume V are plotted as a function of Ca content x.

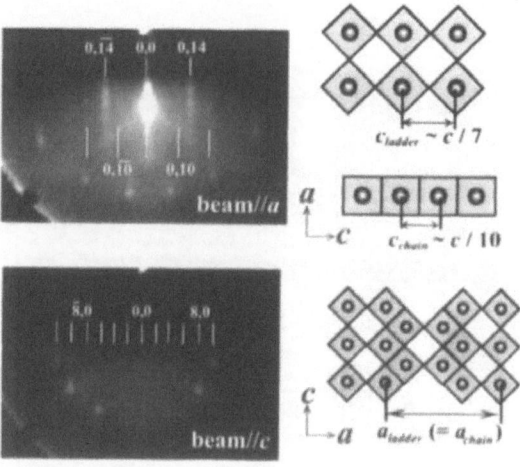

Fig. 2 RHEED pattern of $Ca_{14}Cu_{24}O_{41}$ film with the incident e-beam parallel to a-axis (upper) and parallel to c-axis (downer).

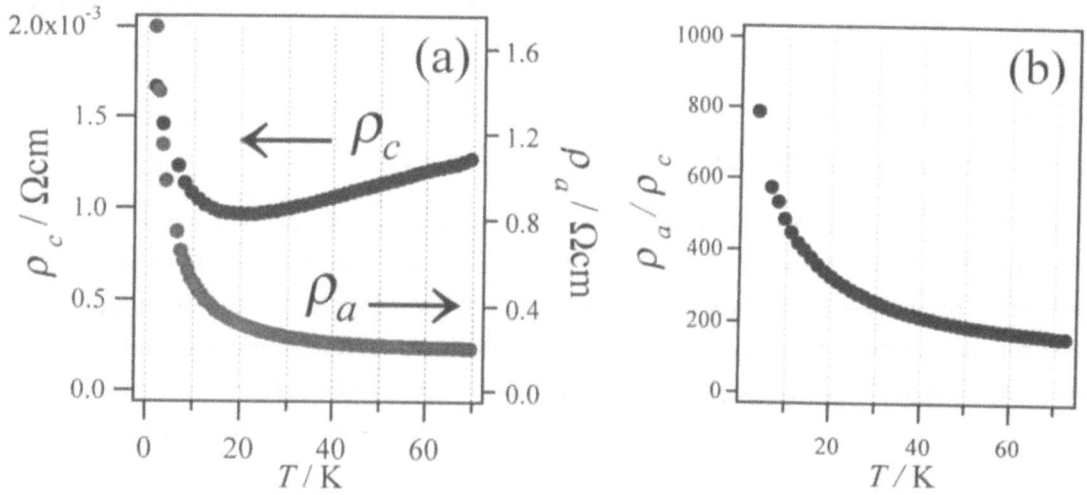

Fig 4 (a) Resistivity for $Ca_{14}Cu_{24}O_{41}$ film with the current paralell to a-axis (ρ_a) and c-axis (ρ_c) as a function of temperature. (b)Anisotropy ratio ρ_a / ρ_c as a function of temperature.

References

[1] E. Dagotto et al., Phys. Rev. B **45**, 5744 (1992)

[2] T. M. Rice et al., Europhys. Lett. **23**, 445 (1993)

[3] M. Uehara et al., J. Phys. Soc. Jpn. **65**, 2764 (1996)

[4] T. Nagata et al., Physica C **282-287**, 153 (1997)

[5] T. Osafune et al., Phys. Rev. Lett. **78**, 1980 (1997)

[6] X. Li et al., Physica C **229**, 251 (1994)

[7] M. Kato et al., Physica C **258**, 284 (1996)

[8] M. Isobe et al., Phys. Rev. B **57**, 613 (1998)

Growth and Characterization of Single Crystals of Tl-based Cuprate Superconductors

Masashi Hasegawa[1], Humihiko Takei[2], Kouich Izawa[1], and Yuji Matsuda[1]

[1]Institute for Solid State Physics, University of Tokyo, Roppongi, Minato-ku, Tokyo 106-8666 Japan
[2]Graduate School of Science, Osaka University, Machikaneyama, Toyonaka-shi, Osaka 560-0043 Japan

Abstract : A new closed-atmosphere method using a gold crucible has been developed to grow single crystals of the Tl-based cuprate superconductors by a self-flux growth method. This new method can provide us mm-sized single crystals of Tl-2201. The value in the resistivity ratio $R(300K)/R(T_c)$ (RRR) of one of the as-grown overdoped Tl-2201 crystals is 18.3. It should be noted that the transition temperature width of one of the annealed Tl-2201 crystals ($T_c = 20$ K) is about 1 K. The irreversibility line and effects of the Pb-ion irradiation on the transmittivity critical field (resistive critical field) of the overdoped Tl-2201 crystal are clarified.

Keywords : Tl-based cuprate superconductors, high quality single crystal, irreversibility line, transmittivity critical field, Tl-2201

INTRODUCTION

The Tl-based cuprate superconductors, especially $Tl_2Ba_2CuO_x$ (Tl-2201) are important to clarify intrinsic physical properties of the cuprates high temperature superconductors. However, systematic studies using their high quality single crystals are limited, compared to the other cuprate superconductors, such as the Bi-based and La-based cuprate superconductors. Hasegawa et al. had reported crystal growth of the Tl-based cuprate superconductors with the CuO_2 double-layered stackings, and also clarified local structure changes and physical properties of the grown crystals [1-4]. However, there are still issues to be resolved.

In this article, growth technique by a newly developed closed-atmosphere method using a gold crucible and characterization results will be described. In addition, the irreversibility line and effects of the Pb-ion irradiation on the transmittivity superconducting transition, which is essentially same as resistive transition of overdoped Tl-2201 crystals will be also clarified.

Experimental

Nominal compositions of starting mixture is Tl_2O_3(3N) : BaO_2(2N) : CuO(4N) = 1 : 2.5 : 2. The mixture was heated at 600 °C for 10 min in an oxygen atmosphere and cooled rapidly to room temperature. Then, the mixture was ground and put into a gold crucible which was about 15 mm in diameter, 20 mm in height and 0.4 mm in thickness. The gold crucible was, then, put into an alumina crucible and also covered tightly by alumina cement to be held mechanically. After soaking at 920 °C for 10 hours, it was cooled down to 880 °C at a rate of 0.7 °C/h, then rapidly cooled to the room temperature in the furnace. After the growth procedure, as-grown crystals were annealed on a gold foil in a flow of dried Ar gas mixed with oxygen gas (Ar : O_2 = 9 : 1) at an appropriate temperature to control T_c. The details of the crystal growth was described elsewhere [5].

108

Grown single crystals were observed using optical and scanning electron microscopes. The single crystals were examined by an electron probe microscope analyzer (EPMA) to investigate the composition. Temperature dependence of the electric resistivity parallel to the c-plane were measured by the four probes method. Temperature dependence of the magnetization was measured using a SQUID magnetometer. The broadband fundamental transmittivity was measured by the local Hall probe magnetometer. Columnar defects in the crystal were introduced by the Pb-ion irradiation along the c-axis with the matching field of 1 T.

Results and Discussion

The typical size of the removed crystals clean of flux was ~1 mm x ~1 mm x ~70 μm. They showed metallic luster and often showed curved fine steps on the surface (basal-plane). These steps indicate a lateral growth of the crystals. Such steps suggest that the crystals were grown under the condition of solute enrichment. The EPMA analysis indicated that the compositions of the crystals were determined as $Tl_{1.9}Ba_2Cu_{1.0}O_x$ where the Ba content was fixed as 2 and homogeneous. The oxygen content was not determined.

All of the overdoped Tl-2201 crystals ($T_c \sim 20$ K) showed almost T^2-temperature dependence. It should be noted that the c-plane resistivity ratio R(300K)/R(T_c) (RRR) of one of the overdoped as-grown crystals ($T_c = 20$ K) is 18.3. Figure 1 shows the temperature dependence of magnetization and irreversibility behavior of one of the Tl-2201 crystal. This crystal was annealed at 673 K for 75 min. It should be noted that the transition temperature width is about 1 K (10%-90%). These results indicate extremely high quality of the crystals.

Figure 2 shows the irreversibility line determined by irreversibility behavior in the temperature dependence of magnetization. The temperature dependence of the irreversible field is well described by a power law, $(1-T/T_c)^{1.5}$ below 1 T and $(1-T/T_c)^{5.3}$ above 1 T. The power value of 5.3 above 1 T is larger than that of Y-123 crystal (~1.5 above 1 T) [6] and smaller than that of Bi-2212 crystal (~17 above 1 T) [7]. This means that the overdoped Tl-2201 crystal is more anisotropic than that of the Y-123 crystal and less anisotropic than that of the Bi-2212 crystal. This results is reasonable on the basis of their CuO_2 layer stackings in the crystal structure.

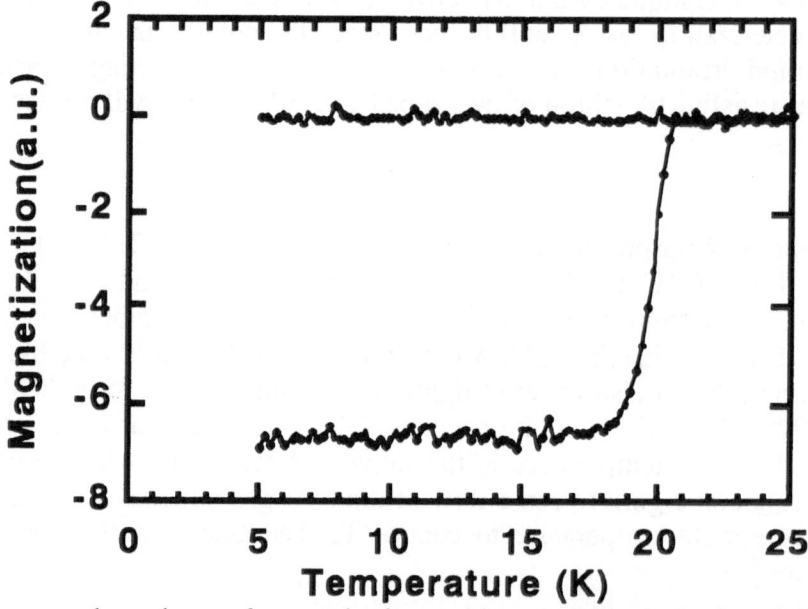

Figure 1 Temperature dependence of magnetization and irreversibility behavior of the Tl-2201 crystal. The applied magnetic field is 1 mT and parallel to the c-axis.

Effects of the Pb-ion irradiation on the transmittivity superconducting transition of another crystal in magnetic field, which is essentially same as resistive superconducting transition [8] are shown in Figure 3. It is found that the transmittivity critical field H* is increased by the irradiation. Our recent studies reported that the resistive superconducting transition at H* of the overdoped Tl-2201 was attributable to the melting of the vortex lattice [9,10]. Accordingly, it should be noted that the columnar defects induced by the irradiation are useful to improve H*, *ie.*, prevent the melting of the vortex lattice of the cuprate superconductors. The irreversibility field and resistive critical field may be concerned with each other. Thus, the temperature dependence of H* of the crystal whose irreversibility line is shown in Figure 2 will be clarified in the near future to compare with each other.

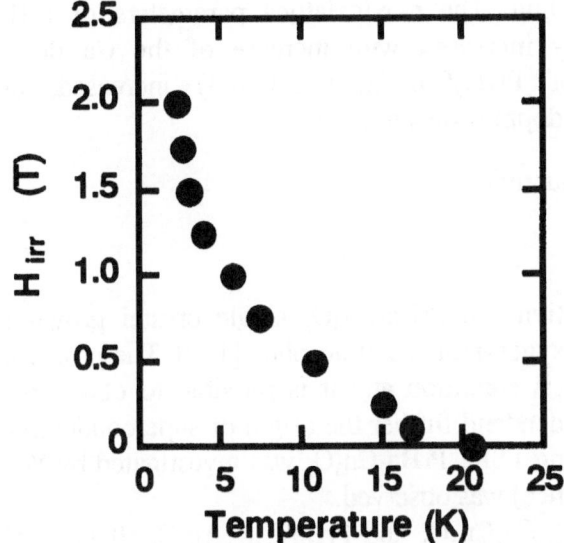

Figure 2 Irreversibility field H_{irr} line.

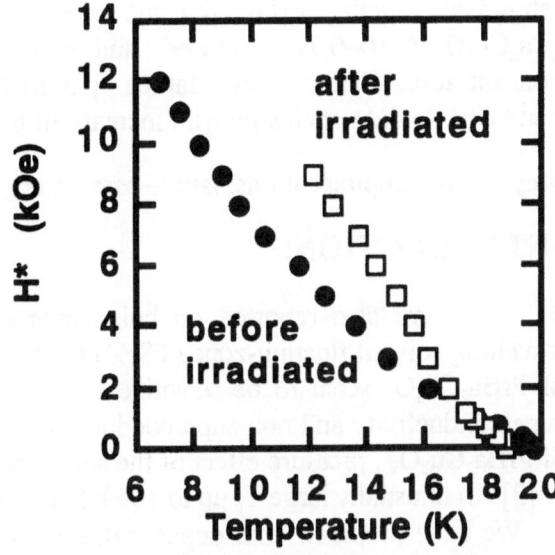

Figure 3 Irradiation effects on transmittivity critical field H*.

Acknowledgments

The authors are grateful to Mr. A. Shibata and Dr. N. Chikumoto for their help on the magnetization measurements. They also wish to thank Dr. C. J. van der Beek and Dr. M. Konczykowski for their help on the Pb-ion irradiation and transmittivity measurements. This work was supported by a Grant-in-Aid for Scientific Research from the Ministry of Education, Science and Culture, and partially supported by CREST of Japan Science and Technology Corporation.

References

1. M. Hasegawa, Y. Matsushita, Y. Iye and H. Takei, Physica C **231**, 161-166(1994).
2. M. Hasegawa, Y. Matsushita and H. Takei, *Advances in Superconductivity VII*, Eds. K.Yamafuji and T.Morishita, Springer-Vlg., 723-728(1995).
3. Y. Matsushita, M. Hasegawa, F. Sakai and H. Takei, Jpn. J. Appl. Phys. **34**, L1263-L1266(1995).
4. M. Hasegawa, Y. Matsushita and H. Takei, Physica C **267**, 31-44(1996).
5. M. Hasegawa, H. Takei, K. Izawa and Y. Matsuda J. Low Temp. Physics, accepted.
6. N. Kobayashi, K. Hirano, T. Nishizaki, H. Iwasaki, T. Sasaki, S. Awaji, K. Watanabe, H. Asaoka and H. Takei, Physica C **251**, 255-262(1995).
7. Y. Xu and M. Suenaga, Phys. Rev. B **43**, 5516-5525(1991)
8. J.Gilchrist and M. Konczykowski, Physica C **212**, 43-60(1993).
9. K. Izawa, H. Takahashi, Y. Matsuda, M. Hasegawa, N. Chikumoto, C. J. van der Beek and M. Konczykowski, J. Low Temp. Physics, accepted.
10. K. Izawa, A. Shibata, H. Takahashi, Y. Matsuda, M. Hasegawa, N. Chikumoto, C. J. van der Beek and M. konczykowski, Physica B, accepted.

Structural Properties of $PrBa_{2-x}Ca_xCu_3O_y$ and $PrBa_2Cu_{3-x}Ag_xO_y$: Correlation with Electronic and Magnetic Properties

Z. Zou[1], J. Ye[2], T. Minawa[4], H. Kawanaka[3] Y. Nishihara[3,4]

[1]National Institute of Materials and Chemical Research , 1-1 Higashi, Ibaraki 305, Japan
[2]National Research Institute for Metals, Tsukuba, Ibaraki 305, Japan
[3]Electrotechnical Laboratory,1-1-4 Umezono,Tsukuba Ibaraki 305, Japan
[4]Faculty of Science, Ibaraki University, Bunkyo 2-2-1, Mito 310, Japan

Abstract: We examined the effects of Ca doping $PrBa_{2-x}Ca_xCu_3O_y$ (x=0~0.7) and Ag doping $PrBa_2Cu_{3-x}Ag_xO_y$(x=0~0.3) on the structural, electrical and magnetic properties. Rietveld refinement of powder x-ray diffraction data indicated that both doped systems have a tendency to become a tetragonal structure even after the oxygen annealing. The c-axis lattice parameters of $PrBa_{2-x}Ca_xCu_3O_y$ (x=0~0.7) decreased, and conductivity increased with increase of the Ca doping concentration. The c-axis lattice parameter of $PrBa_2Cu_{3-x}Ag_xO_y$(x=0~0.3) increased, and conductivity did not change with increase of the Ag doping concentration.

Keywords: doping cations, lattice parameters, conductivity

INTRODUCTION:

We have also reported on bulk superconductivity in $PrBa_2Cu_3O_x$ single crystal grown by travelling solvent floating-zone (TSFZ) method in oxygen-reduced atmosphere[1, 2]. The properties of $PrBa_2Cu_3O_x$ seem to be sensitive to the synthesis condition and it is possible to obtain both superconductivity and non-superconductivity. To understand further the origin of superconductivity of $PrBa_2Cu_3O_x$, pressure effect of the superconducting TSFZ $PrBa_2Cu_3O_x$ was investigated by Ye *et al.*[3] An unusually large T_c up to 105 K(zero-resistance) was observed.

We have prepared both doped systems of $PrBa_{2-x}Ca_xCu_3O_y$ and $PrBa_2Cu_{3-x}Ag_xO_y$. It has been reported that Ca ion can occupy the R site in $RBa_{2-x}Ca_xCu_3O_y$ (R=Y, La) systems and the displaced R move to the Ba site.[4,5] The Ag ion suggests to occupy the Cu site and lead to change of c-axis parameter. The effects of Ca and Ag substitution on the structural, electrical and magnetic properties are investigated. These results are discussed in conjunction with the roles of hybridization and atomic disorder in the anomalous physical properties of $PrBa_2Cu_3O_x$.

EXPERIMENTAL

The polycrystalline samples of both $PrBa_{2-x}Ca_xCu_3O_y$(x=0~1.0) and $PrBa_2Cu_{3-x}Ag_xO_y$(x=0~0.3) were prepared by solid state reaction method using high purity(99.99), $BaCO_3$, $CaCO_3$, Ag_2O, and CuO. All as-grown samples were annealed in flowing pure oxygen gas.[6] The chemical composition of samples was examined by scanning electron microscope-X-ray energy dispersion spectrum (SEM-EDS). The average ratio of Pr:Ba(Ca):Cu(Ag) was 1.00:1.93:2.93. From these experimental results we confirmed that these samples have the Y123 composition. Powder x-ray diffraction data were collected at 295 K on a Rigaku RINT-2000 diffractometer using $CuK\alpha$ radiation (λ=1.54178Å) with condition of 40kV and 38mA. The structure refinement was carried out using the Rieveled refinement method. Electrical resistivity was measured by the standard four probe method. Magnetic susceptibility measurements were performed using a SQUID magnetometer.

RESULTS AND DISCUSSION

The powder x-ray diffraction data of annealed $PrBa_{2-x}Ca_xCu_3O_y$ (x=0~1.0) were collected with a step scan procedure in the range of $2\theta = 5$ to $100°$. The step interval was $0.024°$ and scan speed, $1°/min$. The sample with x = 0.7 showed weak reflections due to the unknown impurity phase, which indicates that it is beyond the limit of the Ca doping range. Full-profile structure refinements of the collected powder x-ray diffraction data for $PrBa_{2-x}Ca_xCu_3O_y$ (x=0~0.7) were performed using Rietveld program REITAN.[7] The variations with x of lattice parameters and unit cell volume show in Fig.1. It is obvious that a-axis lattice parameters increase with increase of the Ca concentration from x=0.0 to x=0.1. However, b, c-axis lattice parameters decrease rapidly with increase of the Ca concentration. The decrease of b-axis with increasing x indicates loss of b-site oxygen which is in agreement with substitution of the Ca for Y in Y123 system.[8]

Figure 2 shows the temperature dependence of magnetic susceptibility of $PrBa_{2-x}Ca_xCu_3O_y$. A clear magnetic ordering temperature(T_N) were observed about 16 K at x=0.0 and 0.1 samples. However, no magnetic ordering temperature was observed in $PrBa_{2-x}Ca_xCu_3O_y$ at x=0.5 and 0.7 samples. This result and the variation of effective magnetic moment[7] in the system supported strongly the fact that a part of Ca ions can occupy Pr site and the displaced Pr move to the Ba site in $PrBa_{2-x}Ca_xCu_3O_y$. Similar features were observed in $(Y_{1-x}Pr_x)Ba_2Cu_3O_y$ and $(Pr_{1-x}Ca_x)Ba_2Cu_3O_y$ systems, where the reduction of T_N caused by effect for Pr by the non-magnetic ion Ca.[11]

It is noted that the resistivity decreases when Ca doped into $PrBa_{2-x}Ca_xCu_3O_y$, for x<0.5, and increase with further x increasing. The similar features was also reported in ref.[10]. The decrease in electrical resistivity by Ca doping into $PrBa_{2-x}Ca_xCu_3O_y$ might be explained by assuming that the substitution of Pr by Ca will weaken the magnitude of the hybridization between Pr 4f electrons and O2p and then resumes conductivity.[5] However, the resistivity at x=0.7 shows a increases ($\rho_{x=0.7}>\rho_{x=0.5}$). The result that resistivity increases with decreasing Pr showed an opposite behavior with $(Y_{1-x}Pr_x)Ba_2Cu_3O_y$, where the resistivity increases with increasing Pr.[11] On the other hand, Pr in Ba sites could reduce the formal oxidation state on the Cu(1)-O(4) linkage and increase the resistivity[6].

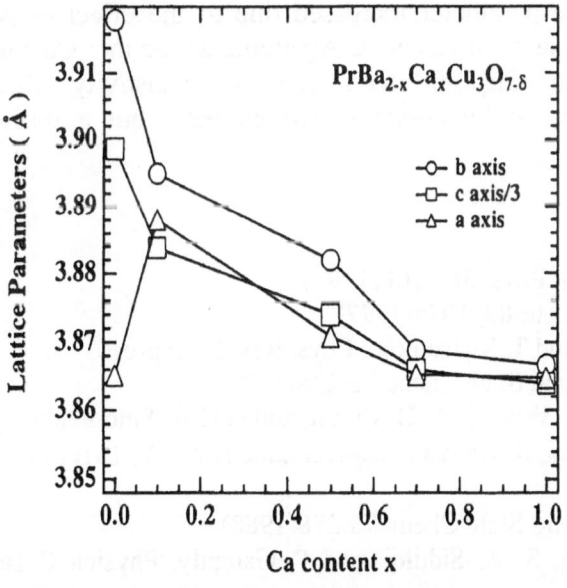
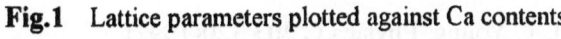

Fig.1 Lattice parameters plotted against Ca contents

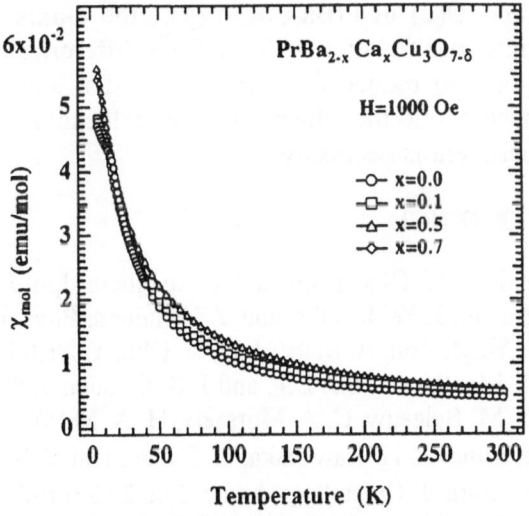

Fig.2 Temperature dependence of magnetic susceptibility of the $PrBa_{2-x}Ca_xCu_3O_y$ (x=0~0.7) at an applied magnetic field of 1000 Oe

Figure 3 shows the variations with x of lattice parameters of annealed $PrBa_2Cu_{3-x}Ag_xO_y$ (x=0~0.3) samples. A fierce changes occur as Ag content from x=0.0 to x=0.1. However, a very small changes occur as Ag content from x=0.1 to x=0.3. It is obvious that a, and c-axes lattice parameter increases rapidly with increase of the Ag content in the region of x<0.1. However, b-axis lattice parameter decreases rapidly in the region of x<0.1. The variations of a, and b-axis show a similar changes to that of Ca doped $PrBa_{2-x}Ca_xCu_3O_y(x=0\sim0.7)$. The substitution of Ag for Cu has a same tendency to become a tetragonal structure, but c-axis lattice parameter increases with x .

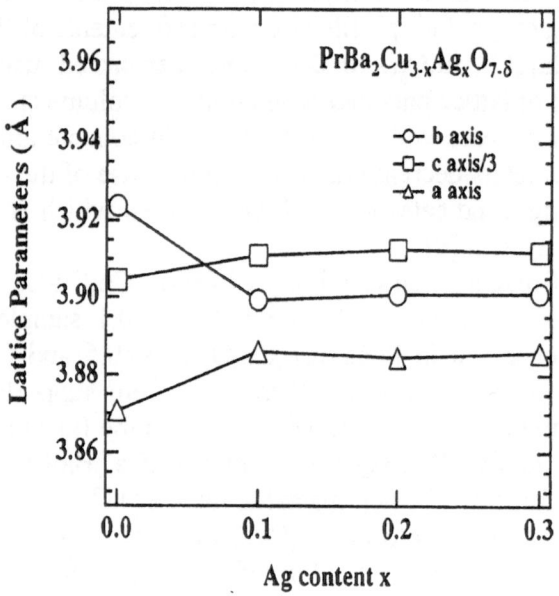

Fig.4 Lattice parameters plotted against Ag contents

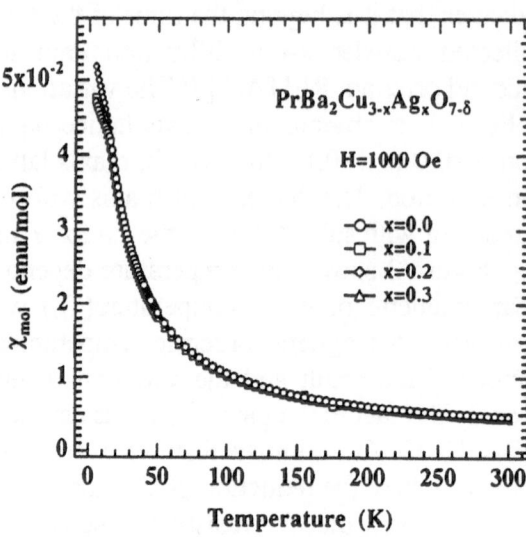

Fig.3 Temperature dependence of magnetic susceptibility of the $PrBa_2Cu_{3-x}Ag_xO_y(x=0\sim0.3)$ at an applied magnetic field of 1000 Oe

The temperature dependence of magnetic susceptibility of $PrBa_2Cu_{3-x}Ag_xO_y$ is shown in Fig. 4. The magnetic ordering temperature did not change in $PrBa_2Cu_{3-x}Ag_xO_y$ with increasing Ag. The magnetic ordering temperature in $PrBa_2Cu_3O_7$ was found to be strongly related to the Cu-O chain, but the magnetic interaction mechanism of $PrBa_2Cu_3O_7$ is not yet clear.[12] It is noted that the resistivity did not change with x. Although we can not confirm expansion of atoms distance between Pr and O(2) in $PrBa_2Cu_{3-x}Ag_xO_y$, the c-axis lattice parameter increased due to the effect of Ag doping into Cu sites. Judging from difference of Ag and Cu valences, Ag atoms doped into Cu sites might also reduce the formal oxidation state on the Cu(1)-O and increase the resistivity, which canceled out the effect of hole delocalization caused by c-axis lattice increase, but a further experiment is necessary.

REFERENCE

1. Z. Zou, K. Oka, T. Ito, and Y. Nishihara, Jpn. J. Appl. Phys. **36**, L18 (1997)
2. Z. Zou, J. Ye, K. Oka, and Y. Nishihara, Phys. Rev. Lett. **80**, 1074(1997)
3. J. Ye, Z. Zou, A. Matsushita, K. Oka, Y. Nishihara, and T. Matsumoto, Phys. Rev. B, in press
4. A. Manthiram, S.J. Lee, and J. B. Goodenough, J. Solid State Chem. 73 278(1988)
5. D. M. Beleeuw, C. A. Mutsaers, H. A. M. Vanhal, H. Verweij, A. H. Carim, and H.C.A. Smoorenburg,
6. T. Minawa, H. Kawanaka, H. Bando, and Y. Nishihara, *Advances in superconductivity X*, 131(1998)
7. F. Izumi, J. Crystallogr Assoc. Jpn. **27** 23(1985)
8. A. Manthiram, S.J. Lee, and J. B. Goodenough, J. Solid State Chem. 73 278(1988)
9. C. Infante, M. K. Elmously, R. Dayal, M. Husain, S. A. Siddiqi, and P. Ganguly, Physica C **167** 640(1990)
10. H. D. Yang, M. W. Lin, C. H. Luo, H. L. Tsay, and T. F. Young, Physica C **203** 320(1992)
11. H. D. Yang, M. W. Lin, C. H. Luo, H. L. Tsay, and T. F. Young, Physica C **203** 320(1992)
12. D. M. Beleeuw, C. A. Mutsaers, H. A. M. Vanhal, H. Verweij, A. H. Carim, and H.C.A. Smoorenburg, Physica C **156** 126(1988)

Substitution Effects of Ba by Sr and La in Physical and Structural Properties of $Ba_2Cu_3O_4Cl_2$ Compounds

Jinhua Ye[1], Zhigang Zou[2], and Akiyuki Matsushita[1]

[1]National Research Institute for Metals, 1-2-1 Sengen, Tsukuba, Ibaraki 305, Japan
[2]National Institute of Materials and Chemical Research, 1-1 Higashi, Tsukuba, Ibaraki 305, Japan

Abstract: Doping effects of Ba by Sr and La on electrical, magnetic and structural properties of $Ba_2Cu_3O_4Cl_2$ compounds were investigated. Ceramic samples of $Ba_{2-x-y}Sr_xLa_yCu_3O_4Cl_2$ were synthesized by the solid-state reaction and characterized by powder X-ray diffraction and Rietveld structure refinement. No apparent structural variation has been recognized with the increase of doping amount, except a smooth change in lattice parameters. Resistivity and magnetic susceptibility measurements indicated that $Ba_{2-x}Sr_xCu_3O_4Cl_2$ exhibits antiferromagnetic and semiconducting behavior. On the other hand, with the increase of trivalent La content, the compound tends to behave metallic. Moreover, a transition to diamagnetic state at about 20K was observed.

Keywords: Substitution effect, Electrical property, Magnetic property

INTRODUCTION

$Ba_2Cu_3O_4Cl_2$ is an insulator with a layered structure composed of Cu_3O_4 planes and blocking layers. The compound has received much attention, because its similar structure to those of the high-T_c superconductors, and its complicated magnetic properties. The symmetry of the crystal structure is tetragonal (space group: $I4/mmm$) at room temperature [1]. For Cu atoms in the Cu_3O_4 plane, there are two sites, i.e., CuI and CuII sites. It is known that the CuI spin is antiferromagnetically ordered and CuII spin is paramagnetically ordered at room temperature [2-4]. The competition between the CuI and CuII spins is considered to result in the multi-steps magnetic transitions of the compound.

In order to investigate the effects of carrier density and atomic size on physical properties of the compound, we prepared a series of samples with Ba being substituted by La^{3+} and Sr^{2+}. Here we report the electrical, magnetic and structural properties of $Ba_{2-x-y}Sr_xLa_y Cu_3O_4Cl_2$ (x=0~1, y=0~0.4) compounds.

EXPERIMENTAL

Poly-crystalline samples of $Ba_{2-x-y}Sr_xLa_y Cu_3O_4Cl_2$ were prepared by solid state reaction method using high purity (99.99%) $BaCO_3$, $BaCl_2$, $SrCO_3$, La_2O_3, and CuO under ambient pressure. The starting materials were mixed and pressed into pellets, and sintered inside an alumina crucible at 900 °C for 50 hours using an electric furnace in air. All the samples were annealed at 800 °C for 24 hours in a flowing of pure oxygen gas. Powder X-ray diffraction data were collected at room temperature on a Rigaku RINT-2000 diffractometer using CuKα radiation (=1.54178Å). The structure refinement was carried out using the Rietveld refinement method. Magnetic susceptibility measurements were performed in the temperature range 5-300 K using a Quantum Design Superconducting Quantum Interference device (SQUID) magnetometer. Temperature dependence of electrical resistivity of the compounds was also measured with the conventional four-probe dc method.

RESULTS AND DISCUSSION

1. Ba substitution by Sr^{2+}

Fig.1 shows the X-ray diffraction patterns of the Sr^{2+} substituted $Ba_{2-x}Sr_xCu_3O_4Cl_2$ (x=0, 0.5, 1.0) compounds. We can see that the Bragg peak distributions of the doped compounds remain unchanged, while peak positions shift to the higher 2θ angles with the increase of the Sr contents. Reitveld structural refinement of the diffraction patterns revealed a liner decrease of the lattice parameters from a= 5.5141(2)Å and c=13.824(5) Å in x=0 compound to a=5.4968(6) Å and c=13.230(8)Å in x=1.0 compound. The remarkable contraction of the a- and c- lattice parameters are understandable from the difference in the atomic radius of the Ba and Sr ions (Ba^{2+}: 1.44 Å, Sr^{2+}: 1.21 Å).

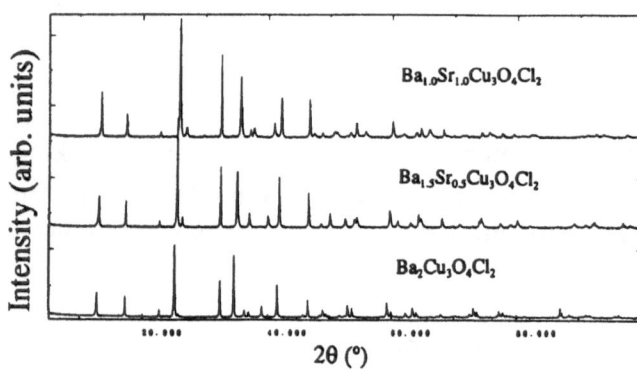

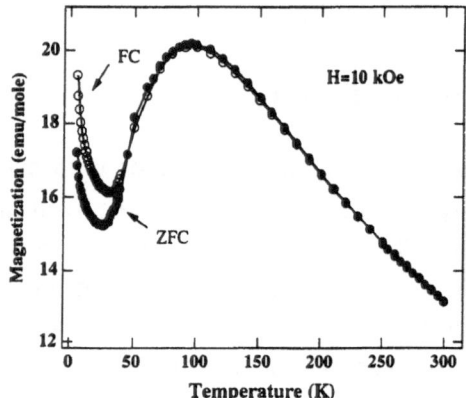

Fig. 1 X-ray diffraction patterns of the Sr substituted $Ba_{2-x}Sr_xCu_3O_4Cl_2$ (x=0, 0.5, 1.0) compound

Fig. 2 Temperature dependent magnetization M(T) curves of $BaSrCu_3O_4Cl_2$

Fig. 2 shows the temperature dependence of the susceptibility of $BaSrCu_3O_4Cl_2$. The magnetic field applied was 10 KOe. A broad peak around 100 K and a minimum at 32 K were observed. If compared Fig. 2 to those reported by Ruck et al, [3] for $Ba_2Cu_3O_4Cl_2$, it is not difficult to find out that these two compounds, Sr-doped or not, have a very similar temperature dependence of the susceptibility. Only the T_N's differ slightly. The complicated magnetic properties of the Sr substituted $Ba_{2-x}Sr_xCu_3O_4Cl_2$ compound are considered caused by the similar origin as that reported in $Ba_2Cu_3O_4Cl_2$ compound. Below T ~330 K, CuI and CuII spins order antiferromagnetically and paramagnetically, respectively. At ~30 K the CuII spins turn to order antiferromagnetically too, corresponding to a minimum in the M(T) curve. The competition of the antiferromagnetically ordered CuI spins and paramagnetically ordered CuII spins seems to result in the complicated magnetic properties of the compound. Our results showed that although Sr^{2+} doping into $Ba_2Cu_3O_4Cl_2$ compound have led to a significant contraction in the lattice parameters of the compound, the magnetic properties of the compound has not been affected severely by the Sr^{2+} doping.

2. Ba substitution by La^{3+}

Fig. 3 shows the X-ray diffraction patterns of the La substituted $Ba_{2-y}La_yCu_3O_4Cl_2$ (y=0, 0.2, 0.4) compounds. In contrary to Sr substituted compounds, no meaningful change in lattice parameters were observed in La-doped compounds.

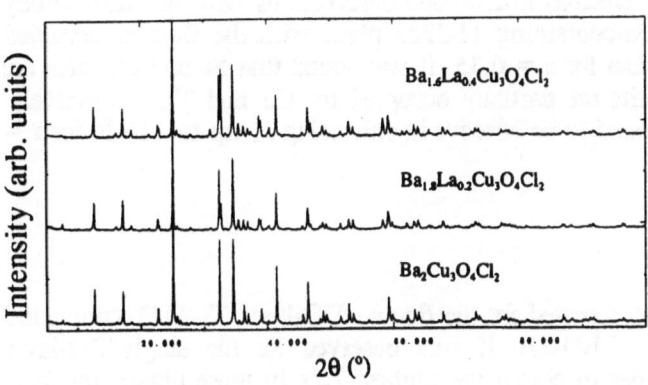

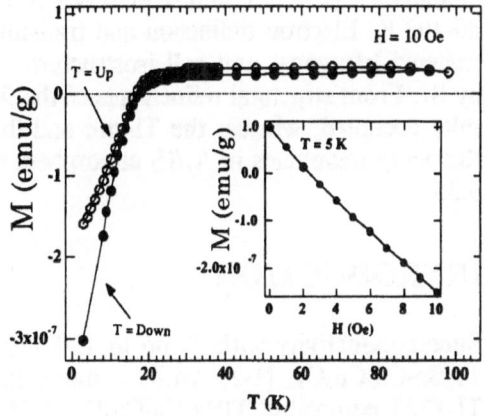

Fig. 3 X-ray diffraction patterns of the La substituted $Ba_{2-y}La_yCu_3O_4Cl_2$ (y=0, 0.2, 0.4) compounds.

Fig. 4 Temperature dependent magnetization M(T) curves of $Ba_{1.8}La_{0.2}Cu_3O_4Cl_2$

Temperature dependent magnetization M(T) curves of $Ba_{1.8}La_{0.2}Cu_3O_4Cl_2$ under a magnetic field of 10 Oe is shown in Fig. 4. The inset shows field dependence of the magnetization at 5 K. A transition into diamagnetic state at ~20 K is observed. The Meissner volume fraction is about 15%. From X-ray diffraction pattern shown in Fig. 3, a small amount of impurity phase, whose peak intensity increase with the increase of La content, can be recognized and determined to be non-supuerconducting $La_4BaCu_5O_{11}$ compound [5]. Measurements of electrical resistivity of the $Ba_{1.8}La_{0.2}Cu_3O_4Cl_2$ compound revealed a metallic behavior of the compound. However, no obvious anomaly could be observed in the temperature range where diamagnetic transition was observed from SQUID measurements. The results suggest that the observed diamagnetic transition at ~20 K, as well as metallic behavior of the compound is most likely caused by the increase in carrier density along with the La^{3+} doping. However, more systematic study is needed to make clear the origin of the diamagnetic transition of the La doped compound and its relation to occurrence of superconductivity.

1. Von. R. Kipka and HK. Muller-Buschbaum, Z. Anorg. Allg. Chem. **419**, 58-62 (1976).
2. Y. Yamada, N. Suzuki, and J. Akimitsu, Physica B **213&214**, 191-193 (1995).
3. K. Ruck, D. Eckert, G. Krabbes, M. Wolf, and K. –H. Muller, J. Solid State Chem. **141**, 378-384 (1998).
4. T. Ito, H. Yamaguchi, and K. Oka, Phys. Rev. B **55**, R684-687 (1997).
5. F. Herman, R. V. Kasowski, and W. Y. Hsu, Phys. Rev. B **37**, 2309-2312 (1988).

A New Family of 2223-Type Superconductors, $(Tl_{1-x}Hg_x)_2Sr_2Ca_2Cu_3O_y$ Synthesized Under High Pressure

E. Kandyel, X.-J. Wu, S. Adachi and S. Tajima

SRL/ISTEC, 10-13 Shinonome, 1-Chome, Koto-ku, Tokyo 135-0062 Japan

ABSTRACT: We have synthesized new Ba-free Tl-based superconductor $(Tl_{1-x}Hg_x)_2Sr_2Ca_2Cu_3O_y$ using high-pressure high-temperature technique. These compounds are chemically and structurally stabilized by partial substitution of Hg for Tl. While Hg-free compound is multi-phased, Hg-doped compounds ($x = 0.35$-0.65) were nearly single phased and found to be superconducting with T_c's of 80-103 K. Electron diffraction and transmission electron microscope observations have revealed a body centered tetragonal unit cell isostructural with Ba-containing Tl-2223 phase with the Ba-sites occupied by Sr. From structural refinements of the XRD data for $x = 0.35$, it was found that Sr and Cu sites are fully occupied, whereas the Tl-site and the Ca-site are partially occupied by Cu and Tl, respectively. Reducing treatments in Ar/H_2 atmosphere were found to be effective in enhancing T_c up to 116 K for $x = 0.35$.

INTRODUCTION

Superconductivity with T_c up to 120-130 K was reported for the double-TlO-layer Tl-2223 compound $Tl_2Ba_2Ca_2Cu_3O_{10}$ [1-2], while a rather low T_c of 100-110 K was observed for the single-TlO-layer Tl-1223 compound $TlBa_2Ca_2Cu_3O_9$ [3,4]. In order to obtain the highest T_c's in these phases the hole density needs to be optimized, which can be achieved through cationic substitution and/or through control of the oxygen nonstoichiometry. In the Tl-1223 system, T_c can be enhanced up to 124 K when Ba was totally replaced by Sr and Tl was partially substituted by Pb in the composition $(Tl_{0.5}Pb_{0.5})Sr_2Ca_2Cu_3O_9$ [5]. Thus, if a Ba-free Tl-2223, $(Tl,M)_2Sr_2Ca_2Cu_3O_y$ could be synthesized, their superconducting properties are of particular interest. Recently, we have synthesized $(Tl_{1-x}Hg_x)_2Sr_2Ca_2Cu_3O_y$, ($x = 0.35$-$0.65$) using a high-pressure technique [6]. Incorporation of Hg appears to stabilize thallium strontium cuprates with double-TlO-layer in a manner similar to the role that Pb plays in the Bi/Sr system. The solubility limit of Hg in the compound was found to be around $x = 0.5$. In this paper, we report on the synthesis, structural and superconducting properties of thallium strontium cuprates, $(Tl_{1-x}Hg_x)_2Sr_2Ca_2Cu_3O_y$, with 2223-type structure.

EXPERIMENTAL

The samples were prepared using a high-pressure technique. Tl_2O_3, HgO, $SrCuO_2$, Ca_2CuO_3 and CuO were used as starting materials. These powders were mixed to a nominal composition $(Tl_{1-x}Hg_x)_2Sr_2Ca_2Cu_3O_y$ ($0 \leq x \leq 1$). Furthermore, the oxygen content, y, was varied, keeping the cation composition constant and changing the ratios between $SrCuO_2$, SrO_2 and CuO. High-pressure synthesis was performed using a cubic-anvil-type apparatus. Details of the experimental setup have already been described elsewhere [6]. First, pressure was gradually increased to 3 GPa over a time span of 1 h, and then heat treatment was performed at 1000°C for 1 h. Next, the sample was quenched by reducing the electrical power to zero, and finally the pressure was released in 1 h. The pulverised samples were then annealed at 300-350°C in reducing atmosphere (Ar/H_2). The crystal structure of the samples were examined by a powder X-ray diffraction (XRD) using Cu K_α radiation. Step-scanned data were collected at 0.02° intervals for 8s over a 2θ range from 3-120°. The intensity data were refined by using the RIETAN-94 program. Electron diffraction (ED) and transmission electron microscopy (TEM) were performed using a JEOL-4000EX instrument operating at 400 kV. The magnetic properties were investigated in the "field-cooling" mode using a superconducting quantum interference device (SQUID) magnetometer in an external field of 10 Oe.

116

RESULTS AND DISCUSSION

The compounds starting from $(Tl_{1-x}Hg_x)_2Sr_2Ca_2Cu_3O_y$ $(x = 0.35-0.50)$ gave a much better phase quality. It is especially emphasized that the desired 2223 phase did not form at all without Hg. There is no doubt that the Hg substitution was effective in stabilizing Ba-free Tl-2223 structure. The effect of oxidizing atmosphere on the synthesis of the Ba-free Tl-2223 phase was investigated using sample with $x = 0.35$. The nominal oxygen content, y, was varied from 9.65 to 10.25. The main products and T_c are summarized in Table 1. A nearly single phase (Tl,Hg)-2223 can be synthesized for a very narrow range of y around 9.65. When the nominal oxygen content became larger than 9.65, (Tl,Hg)-2212 started to appear and its amount increased with a further increase in y-value. Moreover, the samples with $y > 10.0$ are compraised of (Tl,Hg)-2212 as dominant phase with a trace of (Tl,Hg)-2223. The oxygen atmosphere dependence of the phase formation of the Ba-free $(Tl_{0.65}Hg_{0.35})_2Sr_2Ca_2Cu_3O_y$ is different from that previously observed for Ba-containing $(Tl_{0.3}Hg_{0.7})_2Ba_2Ca_2Cu_3O_y$, which was synthesized using a high-pressure technique. The latter compound could be obtained for a wider range of the oxygen content, y, from 9.3 to 10.5 [7].

Table 1. Effects of oxidizing atmosphere on the phase and T_c (onset) of $(Tl_{0.65}Hg_{0.35})_2Sr_2Ca_2Cu_3O_y$ samples.

y	main phase(s)	minor phases	T_c (K)
9.65	2223	CaSrCuO, CuO	103
9.75	2223	2212, CaSrCuO, CuO	100
9.85	2223, 2212	CaSrCuO, CuO	98
10.00	2212, 2223	CaSrCuO, CuO	94
10.25	2212	2223, CaSrCuO, CuO	90

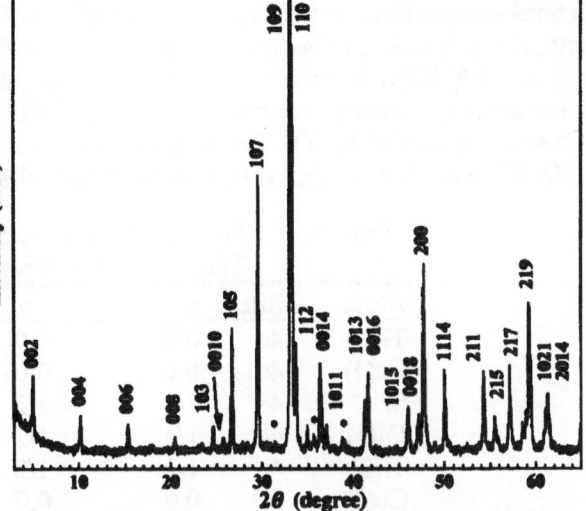

Fig. 1 XRD pattern for nominal composition $(Tl_{0.65}Hg_{0.35})_2Sr_2Ca_2Cu_3O_{9.65}$.

Figure 1 shows the XRD pattern for the sample with nominal composition $(Tl_{0.65}Hg_{0.35})_2Sr_2Ca_2Cu_3O_{9.65}$. Almost all the diffraction peaks can be indexed with a tetragonal unit lattice $a = 3.8130(5)$ and $c = 34.7230(4)$ Å. All the reflections for this phase satisfy a diffraction condition of $h+k+l=2n$, i.e., being consistent with a body-centered lattice. A few peaks from impurities ((Ca,Sr)CuO$_2$ and CuO) are labeled by asterisks. Since the number of peaks from the impurity phases was not large, and the intensities of these peaks were weak, this sample could be considered as a nearly single phase. ED and HRTEM measurements were taken along the [100] direction for $(Tl_{0.65}Hg_{0.35})_2Sr_2Ca_2Cu_3O_{9.65}$. Symmetry and lattice parameters deduced from the ED patterns are in agreement with those determined by the XRD data. The atomic image clearly shows the existence of double (Tl,Hg)-O sheets. In the HRTEM images of the present $(Tl,Hg)_2Sr_2Ca_2Cu_3O_y$, no superstructure can be visible.

Figure 2 shows the temperature dependence of DC magnetic susceptibility for an as-synthesized sample with nominal composition $(Tl_{0.65}Hg_{0.35})_2Sr_2Ca_2Cu_3O_{9.65}$. As-prepared sample shows a large Meissner volume, indicating bulk nature of superconductivity with T_c of 103 K which is 20 K lower than that reported for Ba-containing $(Hg,Tl)_2Ba_2Ca_2Cu_3O_y$ [8]. It has been shown that the critical temperature of the latter compound can be improved dramatically by annealing in a reducing atmosphere. So, small amounts of the sample were post annealed in reducing conditions Ar/H$_2$ (3% H$_2$) at 300-350°C for different periods. The T_c increases monotonically with increasing the strength of reducing condition, i.e. longer annealing time, reaching a critical temperature of 116 K after 40 min. The increase in T_c is also accompanied by a small weight loss and an increase in the room temperature c-axis length, with both phenomena resulting from oxygen evolution. The main peaks, in the XRD patterns of the annealed samples, remain those from the 2223 structure and no additional peaks from other superconductors, such as (Tl,Hg)-12$(n-1)n$ are observed, reinforcing no substantial decomposition of the main phase.

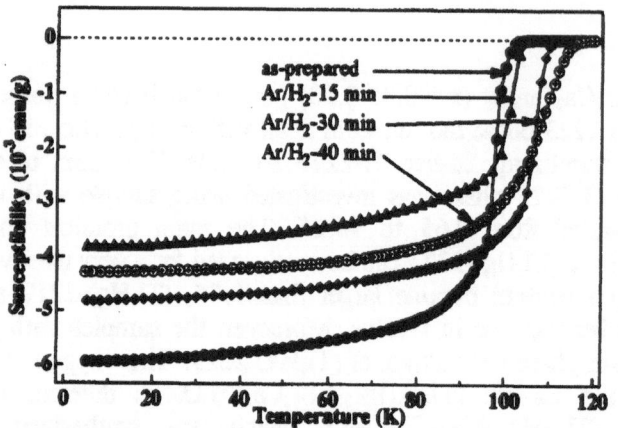

Fig. 2. Temperature dependence of FC magnetic susceptibility for as-prepared and annealed samples with nominal composition $(Tl_{0.65}Hg_{0.35})_2Sr_2Ca_2Cu_3O_{9.65}$.

Using the X-ray data of as-prepared sample with nominal composition $(Tl_{0.65}Hg_{0.35})_2Sr_2Ca_2Cu_3O_{9.65}$, we refined the structure of (Tl,Hg)-2223 phase. Table 2 lists the final refined structural parameters. The refined composition was $(Tl_{0.891}Cu_{0.109})_2Sr_2(Ca_{0.925}Tl_{0.075})_2Cu_3O$, with satisfactory $B = 0.88(7)$ $Å^2$ for Tl site, $B = 0.3(2)$ $Å^2$ for Ca-site, $R_P = 3.40\%$, $R_{WP} = 4.53\%$ and $R_1 = 3.30\%$. The refined lattice parameters are $a = 3.8130(5)$ Å and $c = 34.7230(4)$ Å. From the structural refinements, it was found that Sr and Cu sites are fully occupied whereas about 11% of the Tl-site is occupied by Cu and more than 7% of the Ca-site is occupied by Tl. The oxygen in the TlO layer (O_4) is shifted from its ideal position 4e site $(1/2,1/2,z)$ to 16m site (x,x,z) to achieve a more desirable bond length.

Table 2.   Refined structural parameters for $(Tl_{0.65}Hg_{0.35})_2Sr_2Ca_2Cu_3O_{9.65}$.
The structure is tetragonal (space group $I4/mmm$).

Atom	Site	x	y	z	B ($Å^2$)	g
Tl(1)	4e	0.0	0.0	0.2193(1)	0.88(7)	0.89(1)
Cu(1)	4e	0.0	0.0	0.2193	0.88	0.11
Sr	4e	1/2	1/2	0.1452(1)	0.6(1)	1.0
Ca(1)	4e	1/2	1/2	0.0477(8)	0.3(2)	0.925(6)
Tl(2)	4e	1/2	1/2	0.0477	0.3	0.075
Cu(2)	2a	0.0	0.0	0.0	0.4(2)	1.0
Cu(3)	4e	0.0	0.0	0.0937(2)	0.2(2)	1.0
O(1)	4e	1/2	0.0	0.0	1.0	1.0
O(2)	8g	0.0	1/2	0.0900(5)	1.0	1.0
O(3)	4e	0.0	0.0	0.1589(8)	1.0	1.0
O(4)	4e	0.578(5)	0.578	0.2191(8)	1.0	0.25

Summary

$(Tl_{1-x}Hg_x)_2Sr_2Ca_2Cu_3O$, with 2223-type have been stabilized by Hg under high pressure. As-prepared samples were highly over doped exhibiting T_c of 80-103 K and a large enhancement in T_c up to 116-118 K is observed after annealing in Ar/H_2 atmosphere.

Acknowledgments
This work was supported by the New Energy and Industrial Technology Development Organization (NEDO) and the Special Coordination Fund of the Science and Technology Agency of Japan

1. Z.Z. Sheng, A.M. Hermann, Nature 332 (1988) 138.
2. H. Yamauchi, T. Kaneko, Phase Transitions 41 (1993) 21.
3. S.S.P. Parkin, V. Lee, A. Nazzal, R. Savoy, R. Beyers, S. La Placa, Phys. Rev. Lett. 61 (1988) 750.
4. R. Sugise, M. Hirabayashi, N. Terada, T. Shimomura, H. Ihara, Jpn. J. Appl. Phys. 27 (1988) 1709.
5. R.S Liu, S.H. Hu, D.A. Jefferson, P.P. Edwards, Physica C 198 (1992) 318.
6. E. Kandyel, X.-J. Wu, S. Adachi, S. Tajima, Physica C (1999) in press.
7. T. Tatsuki et al., Adv. in Supercond. IX, Proc. of the 9th Inter. Symp. on Supercond., Springer-Verlag, Tokyo, (1995) 399.
8. T. Tatsuki, A. Tokiwa-Yamamoto, T. Tamura, S. Adachi, K. Tanabe, Physica C 278 (1997) 160.

New Series of Hg-Based, $(Hg,Cr)Sr_2Ca_{n-1}Cu_nO_y$ $(n = 2-5)$ Superconductors Synthesized Under High Pressure

E. Kandyel, X.-J. Wu, S. Adachi and S. Tajima

SRL/ISTEC, 10-13 Shinonome, 1-Chome, Koto-ku, Tokyo 135-0062 Japan

ABSTRACT: We succeeded in preparing several members of a new family of $(Hg,Cr)Sr_2Ca_{n-1}Cu_nO_y$ superconductors under high-pressure at 6 GPa and 1000°C. The phase formation was confirmed by X-ray and electron diffraction analyses as well as high-resolution electron microscopy. These compounds are tetragonal and analogs to $HgBa_2Ca_{n-1}Cu_nO_y$ layerd cuprates with Ba completely substituted by Sr. The first important feature deals with the synthesis of 1212 phase, $(Hg,Cr)Sr_2CaCu_2O_y$, without additional elements like Y with T_c of 80 K. The second important point deals with the successful synthesis of higher members (n = 3-5) of this family for the first time. The highest T_c of 98 K was observed for (Hg,Cr)-1223 phase.

INTRODUCTION

Following the discovery of superconductivity at 90 K in $HgBa_2CuO_{4+\delta}$ [1], two series of Ba-free Hg-based cuprates, $(Hg,M)Sr_2(Ca,R)_{n-1}Cu_nO_y$ (n = 1,2) have been synthesized in an evacuated quartz tube by introducing both high-valence cations (M) into the Hg-site and Y or rare-earth element (R) into the Ca-site. Among all these Ba-free Hg-based cuprates, Cr-doped compounds show a peculiar behavior, such as high-T_c of 60 K in (Hg,Cr)-1201 phase [2,3] and no superconductivity in the 1212 one [4]. The nonsuperconductivity of $(Hg,Cr)Sr_2(Ca,Y)Cu_2O_y$ is attributed to the lack of hole in the conducting CuO_2 planes due to the presence of Y^{3+} at the Ca-site. If a (Hg,Cr)-1212 compound without Y is synthesized there will be more hope of its being superconducting. However, it is very difficult to isolate Y-free HgSr-1212 at ambient pressure. High-pressure is sometimes quite effective in extending a solubility limit of elemental replacement in cuprate superconductors. This is related to the fact that most cuprate superconductors have perovskite-related dense structures which are stable at high-pressure. In the present study, we carried out phase search experiments under high-pressure of 6 GPa for the Hg-Cr-Sr-Ca-Cu-O system. We were able to stabilize the first five members of $(Hg,Cr)Sr_2Ca_{n-1}Cu_nO_y$ series of superconductors (n = 1-5). Their structural and superconducting properties are reported.

EXPERIMENTAL

HgO, Cr_2O_3, $SrCuO_2$, Ca_2CuO_3, SrO_2, CuO and Cu were used to obtain mixtures with nominal compositions of $(Hg_{0.7}Cr_{0.3})Sr_2Ca_{n-1}Cu_nO_y$ (n = 1-6). $SrCuO_2$ and Ca_2CuO_3 were prepared in advance from CuO and $SrCO_3$ or $CaCO_3$ by the conventional solid state reaction at 950°C for 48 h in flowing oxygen gas with intermediate grinding. The oxygen content was varied, keeping the cation composition constant and changing the ratios between $SrCuO_2$, SrO_2, CuO and Cu. The powder mixtures was pressed into a pellet and charged into a gold capsule. The sample was heat treated at 1000°C for 1 h under a pressure of 6 GPa using a cubic-anvil-type apparatus. Crystal structure was analyzed by means of X-ray powder diffraction (XRD) using Cu Kα radiation, electron diffraction (ED) and transmission electron microscopy (TEM). The compositional analyses of crystal grains were carried out using an energy-dispersive X-ray analyzer attached to a TEM (TEM-EDX). The DC susceptibility was measured using SQUID magnetometer in an external magnetic field of 10 Oe.

RESULTS AND DISCUSSION

We tested nominal compositions $(Hg_{0.7}Cr_{0.3})Sr_2Ca_{n-1}Cu_nO_y$ with various n-values (n = 1-6) and oxygen contents, y. Table I summarizes starting compositions and superconducting properties of the high-pressure products. The superconducting transition temperature, T_c in the table is the onset value determined from the magnetic susceptibility data collected by the field-cooling method.

Table 1: Chemical compositions of starting mixture, observed superconducting phases and superconducting properties of the high-pressure products.

Sample No.	Starting Compositions	Observed Phases	T_c (K)
1	$(Hg_{0.7}Cr_{0.3})Sr_2CuO_{4.8}$	1201	58
2	$(Hg_{0.7}Cr_{0.3})Sr_2CuO_{4.6}$	1201	60
3	$(Hg_{0.7}Cr_{0.3})Sr_2CaCu_2O_{7.0}$	1201	60
4	$(Hg_{0.7}Cr_{0.3})Sr_2CaCu_2O_{6.6}$	1201,1212	84
5	$(Hg_{0.7}Cr_{0.3})Sr_2CaCu_2O_{6.2}$	1212	80
6	$(Hg_{0.7}Cr_{0.3})Sr_2CaCu_2O_{6.0}$	1212,1223	95
7	$(Hg_{0.7}Cr_{0.3})Sr_2Ca_2Cu_3O_{8.30}$	1212,1223	93
8	$(Hg_{0.7}Cr_{0.3})Sr_2Ca_2Cu_3O_{8.15}$	1223	98
9	$(Hg_{0.7}Cr_{0.3})Sr_2Ca_2Cu_3O_{8.00}$	1223,1234	90
10	$(Hg_{0.7}Cr_{0.3})Sr_2Ca_3Cu_4O_{10.30}$	1223,1234	64
11	$(Hg_{0.7}Cr_{0.3})Sr_2Ca_3Cu_4O_{10.15}$	1234	64
12	$(Hg_{0.7}Cr_{0.3})Sr_2Ca_3Cu_4O_{10.00}$	1234,1245	62
13	$(Hg_{0.7}Cr_{0.3})Sr_2Ca_4Cu_5O_{12.30}$	1234,1245	60
14	$(Hg_{0.7}Cr_{0.3})Sr_2Ca_4Cu_5O_{12.15}$	1245	45
15	$(Hg_{0.7}Cr_{0.3})Sr_2Ca_4Cu_5O_{12.00}$	1245	45

Figure 1 shows the XRD patterns for samples # 2, 5, 8, 11 and 14. From the similarity of these patterns to the previous data of M-12$(n-1)n$ [5] it was considered that every sample contains a similar M-12$(n-1)n$-type of compound with a tetragonal structure. The lattice parameters shown in Fig. 1 indicate that these samples contain n = 1, 2, 3, 4, 5 members of the (Hg,Cr)-12$(n-1)n$ series, respectively. The c-dimension increases linearly with increasing n with an increment of 3.2 Å. This increment corresponds to thickness of a composite Ca-CuO$_2$ plane and is consistent with the values obtained for other series of superconductors synthesized under high-pressure. The a-dimensions also increases monotonically with n. The a-dimensions of the present Y-free (Hg,Cr)-1212 compound (3.8452 Å) is shorter than that reported for the nonsuperconducting (Hg,Cr)Sr$_2$(Ca,Y)Cu$_2$O$_y$ (3.8526 Å) [4], suggesting an increase in the hole concentration in the conducting CuO$_2$ planes. We carried out similar identifications for other samples. The detected members of the homologous series are listed in Table 1 with simplified notation, 12$(n-1)n$.

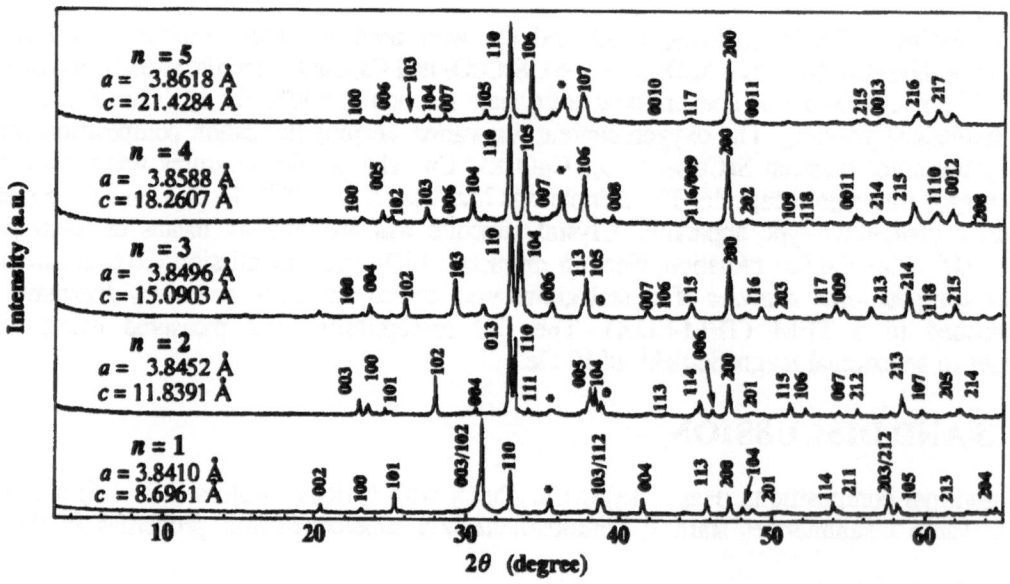

Fig. 1. XRD patterns of the members with n = 1 to 5 of the $(Hg_{0.7}Cr_{0.3})Sr_2Ca_{n-1}Cu_nO_y$ series.

ED patterns and corresponding high-resolution TEM images for a crystal grains of samples # 2, 5, 8, 11 and 14 showed that the observed phases had a primitive tetragonal unit cell with lattice parameters of $a = 3.84$ Å and $c = 8.6+3.2(n-1)$ Å with $n = 1$-5. Almost perfect stacking of layer is observed for phases with $n = 1$-5 whereas samples with $n > 5$, as nominal composition, show an intergrowth of phases with different n-values. TEM-EDX compositional analysis for sample # 8 gave the composition $Hg_{0.62}Cr_{0.38}Sr_{2.11}Ca_{1.89}Cu_{2.93}O_y$. The Cu/(Sr+Ca) and Sr/Ca ratios were very close to 3/4 and 1.0, respectively. The charge reservoir layer contains only Hg and Cr, the chromium content was, however, slightly larger than the nominal one. All these results support with the 1223 structure of this sample.

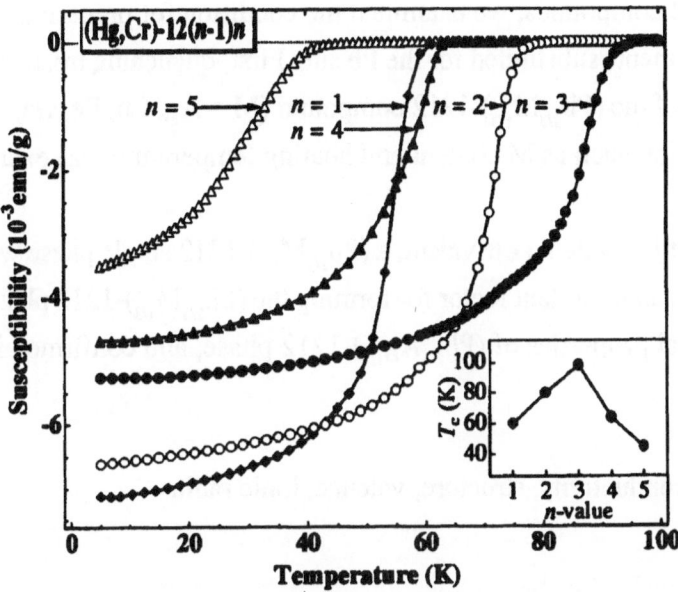

Fig. 2. FC magnetic susceptibility data for samples with # 2, 5, 8, 11 and 14.

Fig. 2 shows typical field-cooling DC susceptibility data for the samples # 2, 5, 8, 11 and 14. Every sample show large enough diamagnetic susceptibility at 5 K indicating bulk superconductivity. Only one transition is observed for each sample indicating the presence of only one superconducting phase in full agreement with XRD data (Fig. 1). T_c of (Hg,Cr)-1201 ($n = 1$) is 60 K, being consistent with results of Shimoyama et al. [2] and Chmaissem et al. [3]. Superconductivity is also observed in Y-free (Hg,Cr)-1212 ($n = 2$) phase at 80 K. The onset T_c of this series increases from 60 K for $n = 1$ to a maximum value of 98 K for $n = 3$ and decreases thereafter down to 64 and 45 K for $n = 4$ and 5, respectively. Samples with more or less oxygen than the optimized ones, as in the samples # 5, 8, 11, or 14, show the formation of lower or higher members of this series besides the main phase and exhibit two transitions in susceptibility data.

Summary

A new series of high-T_c superconductors $(Hg_{0.7}Cr_{0.3})Sr_2Ca_{n-1}Cu_nO_y$ ($n = 1$-5) has been synthesized under high-pressure at 6 GPa and 1000°C. The compounds are tetragonal with lattice parameters $a = 3.84$ Å and $c = 8.6+3.2(n-1)$ Å with $n = 1$-5. The onset T_c increases from 60 K for $n = 1$ to a maximum value of 98 K for $n = 3$ and decreases thereafter down to 64 and 45 K for $n = 4$ and 5, respectively.

Acknowledgments
This work was supported by the New Energy and Industrial Technology Development Organization (NEDO) and the Special Coordination Fund of the Science and Technology Agency of Japan

[1] S.N. Putilin, E.V. Antipov, O. Chmaissem and M. Marezio, Nature **362**, 226 (1993).

[2] J. Shimoyama, S. Hahakura, K. Kitazawa, K. Yamafuji and K. Kishio, Physica C **224**, 1 (1994).

[3] O. Chmaissem, T.Z. Deng and Z.Z. Sheng, Physica C **242**, 17 (1995).

[4] O. Chmaissem and Z.Z. Sheng, Physica C **242**, 23 (1995).

[5] S.M. Loureiro, Y. Matsui and E. Takayama-Muromachi, Physica C **302**, 244 (1998).

Structure and Physical Properties of $(Pb_{2/3}M_{1/3})$-1212 Phase (II)

Y. Ichimaru, K.Satoh, S. Kambe, K. Yamaguchi, O. Ishii

Graduate School of Science and Engineering, Yamagata Univ, 4-3-16, Jonan, Yonezawa, 992-8510, Japan

Abstract: It is well-known that $(Pb_{1-x}M_x)Sr_2(Y_{1-y}Ca_y)Cu_2O_z$ (M = Bi, Cu, V, Cd, Mg, Sr and Ca) compounds show superconductivity whose Tc ranges from 40K to 108K[1].

For exploring novel Pb-contained 1212 compounds, we examined the condition for preparing the $(Pb_{2/3}M_{1/3})$-1212 phase and the kind of elements substituted for the Pb site. First, quenching treatment was essential for forming a single phase of the $(Pb_{2/3}M_{1/3})$-1212 compound (M = Ag, Cu, Fe, Ga, In, and Ni). Particularly, preparation condition such as M content and heating temperature was examined in detail.

It was revealed that when the M element is divalent or trivalent, a $(Pb_{2/3}M_{1/3})$-1212 single phase was formed, indicating that a charge balance is an important factor for forming the $(Pb_{2/3}M_{1/3})$-1212 phase.

Especially, we investigated the physical properties of $(Pb_{2/3}Ag_{1/3})$-1212 phase, and confirmed its non-superconductivity.

Keywords: $(Pb_{2/3}M_{1/3})Sr_2YCu_2O_z$, superconductivity, structure, valence, ionic radius

1. Introduction

Pb-contained oxide superconductors are classified into Pb-3212 phase[2] discovered in 1988 and 1212 phase[3] done in 1989. While the block layer of Pb-3212 phase is composed of PbO-Cu-PbO layer , that of Pb-1212 phase is composed of Pb(M)O layer, respectively.

It is well-known that $(Pb_{1-x}M_x)Sr_2(Y_{1-y}Ca_y)Cu_2O_z$ (M = Bi, Cu, V, Cd, Mg, Sr and Ca) compounds show superconductivity whose Tc ranges from 40K to 108K.

In this experiment, substitution of Ag, Cu, In, V, Co, Ni, Fe, Ga, Al, Cr, Pb, Mn, Sn, Ti, Ge, Ze, Ce, Nb, Ta and W for Pb was examined for clarifying the conditions of the ionic radius and the valence of the substituted ion M to form a single phase of the $(Pb_{2/3}M_{1/3})Sr_2YCu_2O_z$.

2. Experiment

Ceramic samples were prepared by a solid-state reaction. PbO, $SrCO_3$, Y_2O_3, CuO and MO_x (oxide M) powders were mixed in a nominal composition of $(Pb_{2/3}M_{1/3})Sr_2YCu_2O_z$[4]. The powder mixture was calcined at 800 ˚C for 10h in air, powdered and pressed. They were sintered at 1000 ˚C for 10h, and cooled by quenching (10 minutes) or slow cooling (4 hours). They were powdered again before measuring powder X-ray diffraction (XRD). Impurity phases were identified by the XRD experiment. We calculated the r, the ratio of a maximum peak height of the impurity vs. that of the 1212-phase. We call the $0 \leq r < 5\%$, $5 \leq r < 10\%$ and $r \geq 10\%$ samples single phase, impurity-contained phase and multi phase, respectively.

3. Results and Discussion

3.1 $(Pb_{2/3}M_{1/3})Sr_2YCu_2O_z$

In Fig.2, XRD patterns of the $(Pb_{2/3}M_{1/3})Sr_2YCu_2O_z$ phase with M = Ag, Cu, Fe, Ga, In, Ni and Pb are shown. The peaks are homologous with those of the $PbSr_2YCu_2O_z$, revealing that a structure of the $(Pb_{2/3}M_{1/3})Sr_2YCu_2O_z$ is similar to the $PbSr_2YCu_2O_z$.

In contrast, M = Ti, Co and Zr substitution made impurity-contained phase.

It was also revealed that M = V, Cr, Mn, Sn, Ge, Ce, Nb, Ta, W and Al substitution was not favorite for Pb substitution in this system. In Fig.3, a phase diagram of the $(Pb_{2/3}M_{1/3})Sr_2YCu_2O_z$ is shown. It was found that divalent, trivalent and tetravalent valences are necessary for forming the (Pb,M)-1212 single phase. In Fig.4, a relation between the valence of M, Pb and the number of oxygen, z. It is found that the valence of Pb is stable between Pb^{3+} and Pb^{4+}.

3.2 $(Pb_{2/3}Ag_{1/3})Sr_2YCu_2O_z$

Among the $(Pb_{2/3}M_{1/3})$-1212 single phase, the M = Ag phase had a smallest resistance as shown in Fig.5 (a). So we also measured the magnetic susceptibility at low temperature. In Fig.5 (b), magnetic susceptibility curve of M = Ag sample is shown.

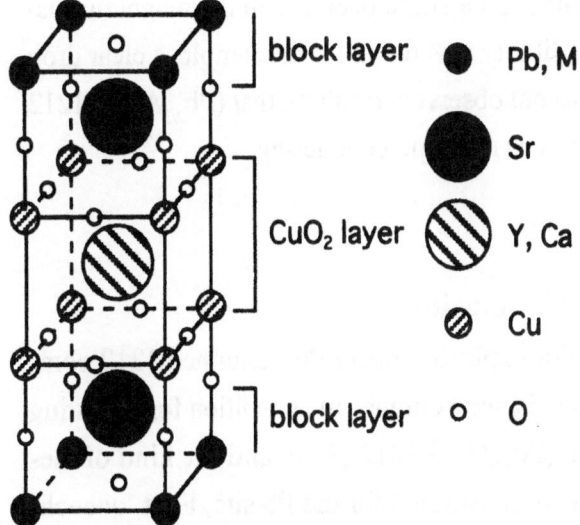

Fig.1 Schematic representation of the crystal structure of (Pb,M)1212-phase.

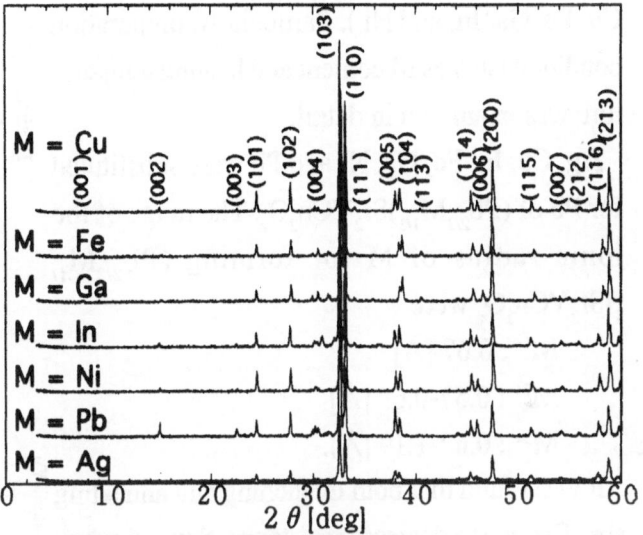

Fig.2 Powder XRD patterns of single phase $(Pb_{2/3}M_{1/3})Sr_2YCu_2O_z$ samples.

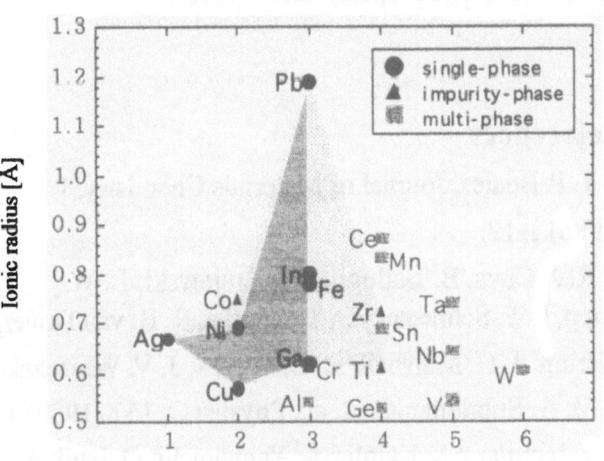

Fig.3 Phase diagram of $(Pb_{2/3}M_{1/3})Sr_2YCu_2O_z$.

Although a slight decrease in the magnetic susceptibility was found for the sample, a clear drop was not observed, resulting that $(Pb_{2/3}Ag_{1/3})$-1212 phase is not superconducting.

4. Conclusion

For exploring novel Pb-contained 1212 compounds, we examined the condition for preparing the $(Pb_{2/3}M_{1/3})$-1212 phase and the kind of elements substituted for the Pb site. First, quenching treatment was essential for forming a single phase of the $(Pb_{2/3}M_{1/3})$-1212 compound (M = Ag, Cu, Fe, Ga, In, and Ni). Particularly, preparation condition such as M content and heating temperature was examined in detail.

Ag, Cu, In, Fe, Ga, Ni and Pb were substituted for Pb of $(Pb_{2/3}M_{1/3})Sr_2YCu_2O_z$. The range of the ionic radius of M for forming $(Pb_{2/3}M_{1/3})Sr_2YCu_2O_z$ were

M^{1+} : 0.67 [Å]

M^{2+} : 0.57-0.69 [Å]

M^{3+} : 0.62-1.19 [Å].

It was found that both quenching and annealing are effective to decrease resistance. Superconductivity did not occur for these samples though they showed metallic property below 70K.

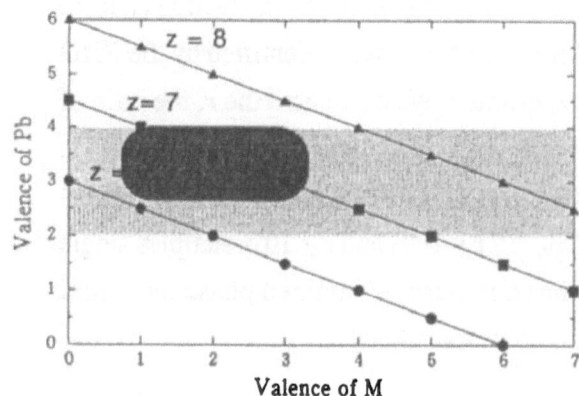

Fig.4 Relation between the valence of M, Pb and the number of oxygen, z.

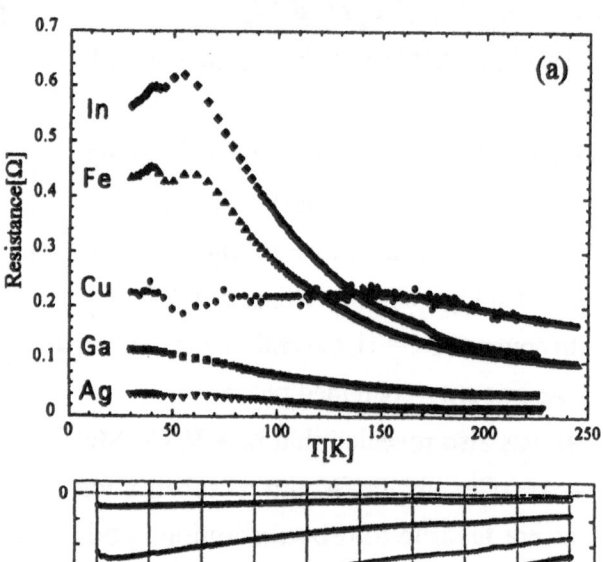

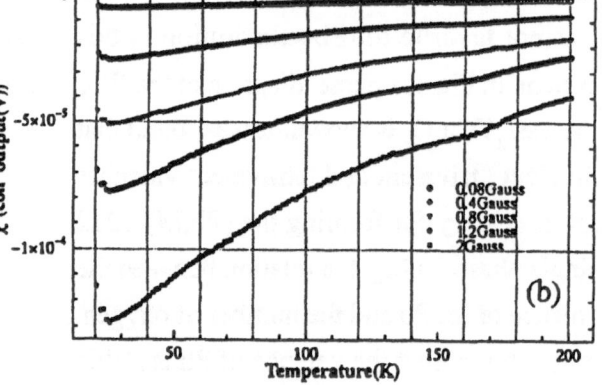

Fig.5 Dependence of (a) resistance and (b)magnetic susceptibility on temperature (M = Ag)

References

1) T. P. Beales, Journal of Materials Chemistry, 1 (1998) 1-12.

2) R. J. Cava, B. Batlogg, J. J. Krajewski, L. W. Rupp, L. F. Schneemeyer, T. Siegrist, R. B. van Dover, P, Marsh, W. F. Peck, Jr., P. K. Gallagher, S. H. Glarum, J. H. Marshall, R. C. Farrow, J. V. Waszczak, R. Hull and P. Trevor: Nature 336 (1988) 211.

3) M. A. Subramanian et. al., Physica C, 157 (1989) 124.J. Y. Lee et. al., J. Mater. Res., 4 (1989) 763.

4) Y. Ichimaru, S. Kambe, K. Yamaguchi, O. Ishii, Advance in Superconductivity XI (Springer-Verlag Tokyo, 1999) 419-422.

Diffusion of Water Vapor in YBa$_2$Cu$_3$O$_x$

SHOICHI EDO[1] and TOSHIHIKO TAKAMA[2]

[1]Department of Mechanical System Technology, Hokkaido Polytechnic College, Zenibako 3-190, Otaru 047-0292, Japan.
[2]Department of Applied Physics, Faculty of Engineering, Hokkaido University, Kita-ku, Sapporo 060-8628, Japan.

Abstract: Diffusion of water vapor in YBa$_2$Cu$_3$O$_x$ was studied by isothermal thermogravimetry. The weight gain during isothermal heating was measured under the vapor pressure of 2.3 kPa. Thermal activation energy for the diffusion was determined to be 0.51 eV. Diffusion coefficients obtained by assuming that the samples consist of sphere powders were 0.2×10^{-13}, 0.7×10^{-13} and 1.1×10^{-13} cm^{2}s^{-1} at temperatures of 131, 151 and 173 °C, respectively. The values are consistent with those reported for oxygen diffusion near the temperatures.

Keywords: superconductor, YBa$_2$Cu$_3$O$_x$, water vapor, diffusion, activation energy

INTRODUCTION

The superconductor YBa$_2$Cu$_3$O$_x$ ($6 < x < 7$) absorbs water vapor at temperatures up to 200 °C without chemical decomposition [1,2]. During the absorption, the compound transforms to normal conductor with the structure having a doubled unit cell along the c-axis [2-4]. The lattice spacing along the c-axis increases by about 16 % in comparison with that in the original structure [4]. We have observed that the weight of the compound increases in proportion to the square root of isothermal heating time by thermogravimetry (TG) [1,2]. It is thought that the gain is originated from diffusion of OH$^-$ ions into the oxygen vacant sites in the Cu-O basal plane.

The diffusion of water vapor has been studied for the compound with x of 6.78 by Ikuma et al. [5]. They obtained the diffusion coefficients of OH$^-$ ions of 10^{-19} to 10^{-18} cm^{2}s^{-1} at temperatures from 100 to 150 °C. The measurements were performed under the vapor pressure of 270 Pa. They excluded the data at the vapor pressure of 2.7 kPa, because they considered that the weight change takes place along with chemical decomposition. In the present paper, the diffusion accompanying the structural change is studied for the compound with x of 6.4.

EXPERIMENTAL

Polycrystalline sample of YBa$_2$Cu$_3$O$_{6.95}$ was prepared by conventional solid-state reaction at 940 °C for 24 h in air and by cooling to 350 °C at a rate of 1 K/min. It was annealed at 725 °C for 6 h in air and quenched into liquid N$_2$ to obtain oxygen-deficient sample with x of 6.4. The weight gain due to water vapor absorption was measured by TG system (Rigaku TAS100) accurate to 10 μg. The powders of 200 mg in a platinum container were heated to three temperatures of 131, 151 and 173 °C at a rate of 10 K/min in Ar gas flow. Then they were annealed in a flow of a mixture of Ar gas and water vapor at the given temperatures. The partial pressure of vapor was kept at 2.3 kPa which is the saturation pressure at room temperature.

RESULTS AND DISCUSSION

Figure 1 shows the weight gains ΔW at temperatures of 131, 151 and 173 °C as a function of the square root of isothermal heating time t. The gain at given time increases with an increase in temperature. It is seen that the gain at 173 °C increases linearly up to about 4×10^4 second, although the others slightly deviate from straight line. If the gain is generated by diffusion and the process obeys Fick's second low, the gain ΔW at time of t is written as

$$\Delta W_t = A\sqrt{Dt} ,\tag{1}$$

where A is a constant. The diffusion coefficient D is given by

$$D = D_0 e^{-\frac{E}{kT}},\tag{2}$$

where E is the thermal activation energy and D_0 is the frequency factor. Eqs. (1) and (2) lead to

$$\ln t = \ln \frac{\Delta W_t^2}{A^2 D_0} + \frac{E}{kT}.\tag{3}$$

Figure 2 shows plots of ln t against $1/T$, where t is time at which the weight gain reaches 0.5, 1.0, or 1.5 mg. It is seen that ln t is proportional to $1/T$. The values of E obtained from the slope were 0.47, 0.53 and 0.53 eV at the weight gains of 0.5, 1.0 and 1.5 mg, respectively. The mean value of 0.51 eV is slightly small in comparison with 0.6 eV obtained at the vapor pressure of 270 Pa [5].

Following the treatment by Ikuma *et al.* [5], we assumed that OH$^-$ ions diffuse into the oxygen vacant sites in the Cu-O basal plane and that the particles of the powder sample are spherical in shape. According to them, the diffusion coefficient is given by

$$\frac{\Delta W_t}{\Delta W_\infty} = 6\sqrt{\frac{Dt}{\pi r^2}} ,\tag{4}$$

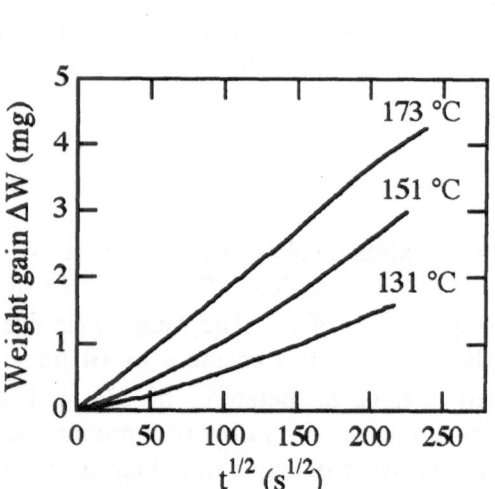

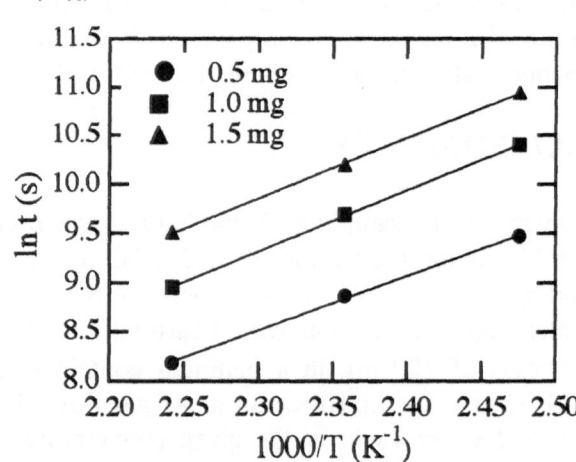

Fig. 1   Isothermal weight gain, ΔW, measured for powder samples of YBa$_2$Cu$_3$O$_{6.4}$ (200 mg) at temperatures of 131, 151, and 173 °C as a function of the square root of time (sec).

Fig. 2   Ln t as a function of $1/T$, where t denotes time (sec) at which the weight gain ΔW reaches 1.5, 1.0, and 1.5 mg at temperatures of 131, 151, and 173 °C.

where ΔW_∞ is the weight gain at infinite time at which all of the vacant sites are occupied by OH⁻ ions and r means the radius of the particle [5,6]. Scanning electron microscope gave the mean radius of 5 μm for the present powders. From Eqs. (1) and (4), one obtains the relation,

$$A = \frac{6\Delta W_\infty}{\sqrt{\pi r^2}}. \tag{5}$$

At infinite time, the compound should be expressed by $YBa_2Cu_3O_x(OH)_{8-x}$. For the composition, ΔW_∞ is 8.2 mg for the sample of 200 mg with x of 6.4. The coefficients $AD^{1/2}$ in Eq. (1) were calculated from the plot shown in Fig. 1 by the least squares method. By using these values and the constant A in Eq. (5), the diffusion coefficients D were determined to be 0.2×10^{-13}, 0.7×10^{-13} and 1.1×10^{-13} cm²s⁻¹ at temperatures of 131, 151 and 173 °C, respectively. They are 4~5 orders in magnitude larger than that observed at the vapor pressure of 270 Pa [5]. The difference in D is thought to result from difference in crystal structure. In our sample, the lattice spacing along the c-axis largely increases with the structural change as mentioned in the introduction. The increase should accelerate the diffusion. The mean frequency factor D_0 was 6.5×10^{-8} cm²s⁻¹.

Figure 3 is a comparison of temperature dependence of the diffusion coefficient for water vapor and oxygen [7]. It is noted that the present results lie on the extrapolated line from the oxygen data to lower temperatures. This indicates that the oxygen-deficient compound absorbs water vapor at almost the same rate as that for oxygen near the temperatures.

In summary, diffusion of water vapor into oxygen-reduced $YBa_2Cu_3O_{6.4}$ has been studied by means of TG at temperatures of 131, 151 and 173 °C. Thermal activation energy E was 0.51 eV when we assume that OH⁻ ions diffuse into all of the oxygen vacant sites in the Cu-O basal planes. Diffusion coefficient varied from 0.2×10^{-13} to 1.1×10^{-13} cm²s⁻¹ between 131 and 173 °C.

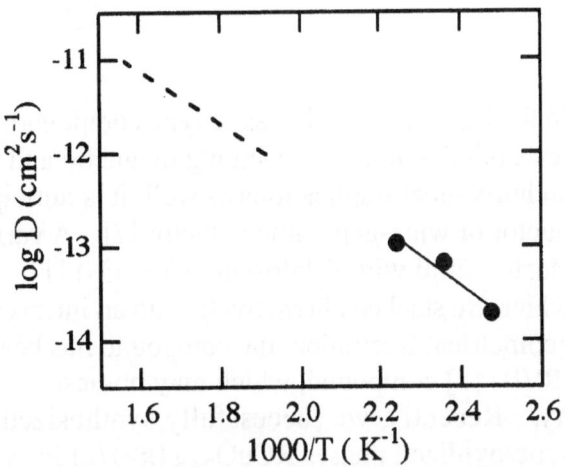

Fig. 3 Comparison of temperature dependence of diffusion coefficient D between the present result (closed circles) and that for oxygen diffusion (broken line) [7].

Acknowledgment

The authors wish to acknowledge Dr. K. Tsuchiya of Toyohashi University of Technology for his suggestions of the experiment.

1. S. Edo and T. Takama, *Advances in Superconductivity VIII*, edited by H. Hayakawa and Y. Enomoto (Springer-Verlag Tokyo, 1996), 401.
2. S. Edo and T. Takama, Jpn. J. Appl. Phys. **37**, 3956 (1998).
3. T. Takama and S. Edo, Physica C **235-240**, 401 (1994).
4. S. Edo and T. Takama, *Advances in Superconductivity XI*, edited by N. Koshizuka and S. Tajima (Springer-Verlag Tokyo, 1999), 443.
5. Y. Ikuma, M. Yoshimura, and S. Kabe, J. Mater. Res. **5**, 17 (1990).
6. J. Crank, *The Mathematics of Diffusion*, 2nd ed. (Clarendon Press, Oxford, 1975).
7. S. R. Rothman, J. L. Routbort, and J. E. Baker, Phys. Rev. B **40**, 8852 (1989).

Synthesis of $ACuO_2$ (A=Al, Ga, Fe and Rh) delafossites containing two-dimensional triangular lattices

K. Isawa and M. Nagano

R & D Center, Tohoku Electric Power Co., Inc., Sendai 981-0952, JAPAN

Abstract: The structural and physical properties of layered cuprates of $ACuO_2$ (A=Al, Ga, Fe and Rh) delafossites have been investigated.   The oxygen content in each $ACuO_2$ sample was found to be nearly stoichiometric.   Since all the $ACuO_2$ compounds decomposed easily by thermal oxidation, the delafossite-derived superoxide $RCuO_{2+\delta}$ ($\delta \geq 0.5$), as previously reported for R= Y, La, Pr, Nd, Sm and Eu, hardly seemed to be obtained for all the elements presently studied. On the basis of the Arrhenius plots of the ρ-vs-T curves, the electrical conduction in $ACuO_2$ was likely due to thermally activated carriers.   The magnetic susceptibilities for all the samples except for $FeCuO_2$ were nonmagnetic.   On the other hand, $FeCuO_2$ exhibited antiferromagnetic transitions at 16K and 11K, as previously reported.

Keywords: delafossite, oxygen stoichiometry, electrical resistivity, antiferromagnet

In the last decade, delafossite-type compounds have attracted attention in the field of fundamental research due to their intriguing magnetic and transport properties [1-6].   From the viewpoint of technological implications as well, it is anticipated to utilize the compounds as transparent conductor or wide-gap semiconductor [7]. A variety of $ACuO_2$ ternary compounds are known to be isostructural with delafossite ($CuFeO_2$) [1].   The cations A^{3+} and Cu^+ form triangular lattices which are stacked alternatively with an intervening oxygen layer along the c-axis.   Because of the geometrical frustration, the compound has been presumed to be a model resonating valence bond (RVB) [8] compound, which may possess a novel quantum characteristic, e. g., superconductivity.   Recently, we successfully synthesized [2] the rare earth delafossites $RCuO_2$ and their superoxidized phases $RCuO_{2+\delta}$ ($\delta > 1/2$) for various rare earth elements, following the work by Cava et al. [4].   Furthermore, we observed [2,3] peculiar behavior in the magnetic susceptibilities for $(Y,Ca)CuO_{2+\delta}$ ($\delta \sim 1/2$) and unusual enhancement in thermoelectric power at low temperature regime for almost all $RCuO_{2+\delta}$ samples.   In this work, we studied the phase stability and the oxygen stoichiometry of $ACuO_2$ (A=Al, Ga, Fe and Rh) delafossites to clarify whether new superoxide, i. e. $ACuO_{2+\delta}$, could be obtained or not.   The electrical and magnetic properties of the $ACuO_2$ samples were also examined.

Polycrystalline samples of $ACuO_2$ delafossites were prepared by a solid state reaction technique as described in detail in Ref.2.   The phase stability of the samples was studied by thermogravimetry (TG) up to 1000°C in N_2/O_2 gas mixtures with various O_2 concentration. The phase content and the lattice parameters were determined by powder x-ray diffraction (XRD) analysis using $CuK\alpha$ radiation.   The electrical resistivity of the samples was measured by a conventional dc four-probe method.   The dc magnetic susceptibility was measured using a SQUID magnetometer.

The optimized sintering conditions are represented in Table 1.   It was revealed for the synthesized $ACuO_2$ samples that nearly all the diffraction peaks in their powder XRD patterns were assigned to the rhombohedral unit cell, although very small peaks from unreacted starting materials were still detected. Note that $FeCuO_2$ was obtained quite easily by firing in a vacuum furnace with a molybdenum heater, in comparison with a complex sintering procedure proposed by Dumere et al. [5]. Since the excess oxygen (δ) in the as-sintered samples is determined by TG analysis at δ=0.04-0.02±0.03 (see Table 1), the oxygen content of each as-sintered sample is nearly stoichiometric. Figure 1 shows the lattice parameters of the as-sintered $ACuO_2$ samples: a and c, with respect to the ionic radius r_A. In the figure, a and c for $RCuO_2$ (R: rare earth elements) are also

Table 1   Starting materials, sintering conditions and analytical oxygen content for $ACuO_2$.

	starting materials	sintering conditions	atmosphere	oxygen content
$AlCuO_2$	Al_2O_3, Cu_2O	1100°C, 48hr (x 2)	in flowing N_2	2.04±0.03
$GaCuO_2$	Ga_2O_3, Cu_2O	1100°C, 48hr	3×10^{-6} Torr	2.02±0.03
$FeCuO_2$	Fe_2O_3, Cu_2O	1000°C, 12hr	4×10^{-6} Torr	2.02±0.03
$RhCuO_2$	Rh_2O_3, CuO	900°C, 96hr	air	2.02±0.03

replotted [2]. The value of a decreases pseudolinearly as a function of r_A descending from La^{3+} (r_A=1.032Å) to Al^{3+} (r_A=0.535Å)[9].   On the other hand, the c value is almost independent of r_A. The typical TG curve, which was obtained for a $FeCuO_2$ sample under various atmosphere are shown in Fig. 2. Upon heating at 1°C/min under the pure O_2 atmosphere, the sample weight reaches a plateau above 690°C. The powder XRD pattern of the final product indicates that the sample decomposed into Fe_2CuO_4 and CuO. As clearly seen in Fig.2, the decomposition is likely inevitable by thermal oxidation because the sample tends to decompose even at a very low oxygen concentration such as 500ppm O_2 in N_2. The decomposition was commonly observed for all the other samples. Therefore, the following possibilities are deduced: (i) the $ACuO_2$ (A=Al, Ga, Fe and Rh) phase does not prefer to load or intercalate oxygen atoms anymore, or (ii) superoxide $ACuO_{2+\delta}$ is metastable phase, so that it is hard to isolate the phase by thermal oxidation before the decomposition. Figure 3 shows the resistivities (ln ρ) as a function of reciprocal temperature (T^{-1}) for the $ACuO_2$ samples. The obtained ln ρ -vs-T^{-1} curves are pseudo-linear, and the slope of the curves, i. e., $d(\ln \rho)/d(T^{-1})$, is nearly independent of A. According to the Arrhenius plots shown in Fig. 3, the electrical conduction in $ACuO_2$ is likely caused by thermally activated carriers. The values of activation energies (E_a) was estimated as ~0.24 eV for all the samples. These values of E_a were in good agreement with that of P-doped $AlCuO_2$ thin film (E_a ~0.2 eV [7]). Figure 4 shows the temperature dependence of the dc magnetic susceptibility (χ) at H=1T for the $ACuO_2$ samples.

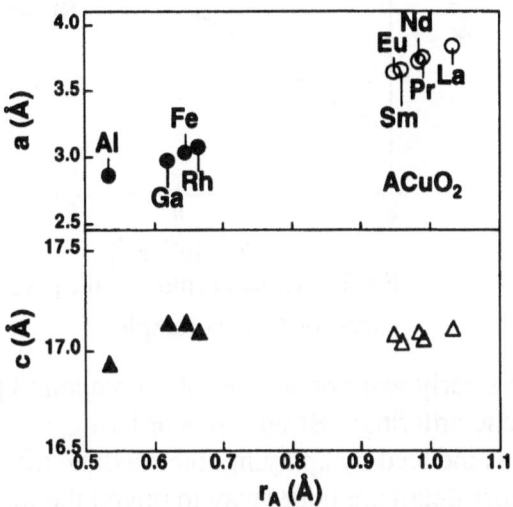

Fig.1   Lattice constants a and c as a function of ionic radius r_A.

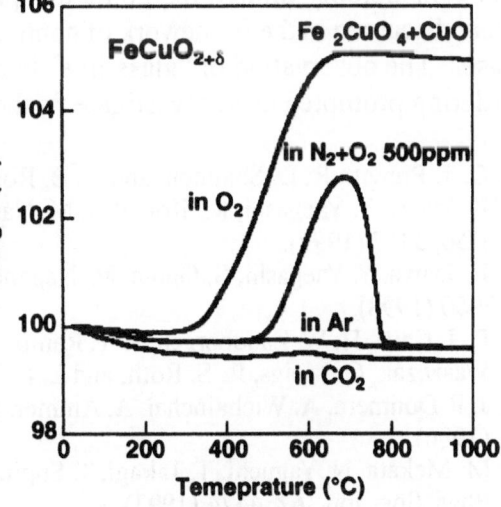

Fig.2   Change in weight of powered $FeCuO_2$ sample in a various atmosphere.

Except for FeCuO$_2$, χ of each as-sintered sample is nearly independent of temperature above 30K, and below 30K it slightly increases with decreasing temperature. The former behavior is consistent with that expected for a nonmagnetic system (3d^{10} for Cu$^+$ ions), whereas the latter is supposed to be attributed to small magnetic impurities which are not detected by XRD analysis and /or lattice imperfections. On the other hand, the temperature dependence of χ in FeCuO$_2$ exhibited basically Curie-Weiss type behavior above ~20K, and then χ decreased rapidly on cooling down to 2K. Two sharp magnetic transitions were detected at around 16K and 11K.

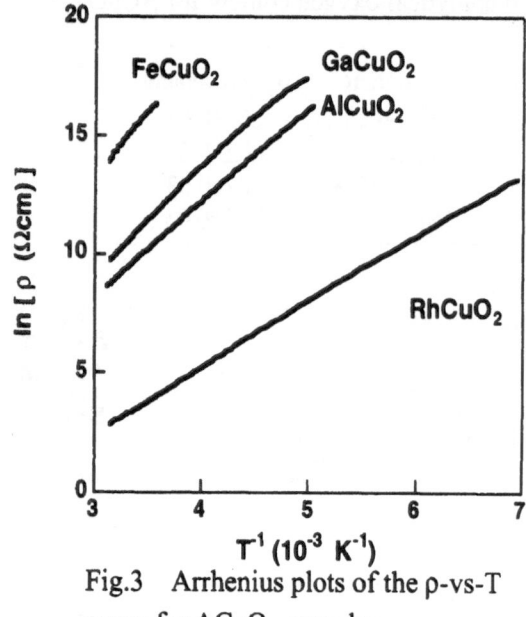

Fig.3 Arrhenius plots of the ρ-vs-T curve for ACuO$_2$ samples.

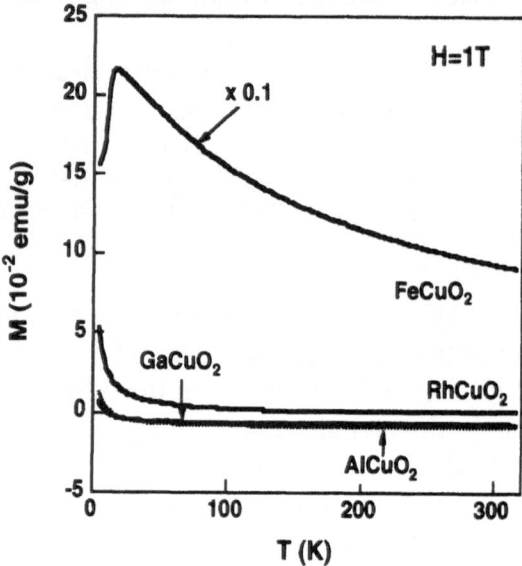

Fig.4 χ-vs-T curves for ACuO$_2$ samples.

In a early work on the FeCuO$_2$ compound [6], the transitions were identified as an antiferromagnetic ordering. Besides this ordering, it was recently found for FeCuO$_2$ that a "glass-like" state was induced by applying intermediate magnetic field (H<1T) below ~100K. Measurements in more detail are under way to unveil the magnetism.

In conclusion, nearly stoichiometric ACuO$_2$ compounds were successfully synthesized for A=Al, Ga, Fe and Rh. All the ACuO$_2$ samples appeared to decompose by thermal oxidation even under the considerably diluted oxygen atmosphere. The physical properties of ACuO$_2$ compounds were discussed in the framework of nonmagnetic semiconductor with the exception of the FeCuO$_2$ case. The observation of "glass-like" behavior for FeCuO$_2$ in addition to the antiferromagnetic ordering prompted us to investigate further into the magnetic property of the compound.

1. C. T. Prewitt, R. D. Shannon, and D. B. Rogers, Inorg. Chem. **10**, 719 (1971).
2. K. Isawa, Y. Yaegashi, M. Komatsu, M. Nagano, S. Sudo, M. Karppinen, and H. Yamauchi, Phys. Rev. **B56**, 3457 (1997).
3. K. Isawa, Y. Yaegashi, S. Ogota, M. Nagano, S. Sudo, K. Yamada, and H. Yamauchi, Phys. Rev. **B57**, 7950 (1998).
4. R. J. Cava, H. W. Zandbergen, A. P. Ramirez, H. Takagi, C. T. Chen, J.J. Krajewski, W.F. Peck, Jr., J. V. Waszczak, G. Meigs, R. S. Roth, and L. F. Schneemeyer, J. Solid State Chem. **104**, 437 (1993).
5. J. P. Doumere, A. Wichainchai, A. Ammer, M. Pouchard, and P. Hagenmuller, Mater. Res. Bull. **21**, 745 (1986).
6. M. Mekata, N. Yaguchi, T. Takagi, T. Sugino, S. Mitsuda, H. Yoshizawa, N. Hosoito, and T. Shinjo, J. Phys. Soc. Jpn. **62**, 4474 (1993).
7. H. Kawazoe, M. Tasukawa, H. Hyodo, M. Kurita, H. Yanagi, and H. Hosono, Nature **389**, 939 (1997).
8. P. W. Anderson, Mater. Res. Bull. **8**, 153 (1973).
9. R. D. Shannon, Acta Crystallogr. **A. 32**, 751 (1976) .

Preparation of $(As_{0.33}Cu_{0.67})Sr_2YCu_2O_{6.88}$ and its physical properties

Eiji Sato[1] , Shiro Kambe[1], Takahiro Akao[1,2], Tetsu Ohsuna[3], Kenji Hiraga[3] and Osamu Ishii[1]

[1]*Graduate School of Science and Engineering, Yamagata University, 4-3-16, Jonan, Yonezawa, Yamagata 992-8510, Japan*

[2]*Materials and Structures Laboratory, Tokyo Institute of Technology, 4259, Nagatsuta, Midori-ku, Yokohama 226-0027, Japan*

[3]*Institute for Materials Research, Tohoku University, 2-1-1 Katahira, Sendai 980-0812, Japan*

Abstracts : $(As_{0.33}Cu_{0.67})Sr_2YCu_2O_y$ was newly prepared. The $(As_{0.33}Cu_{0.67})Sr_2YCu_2O_{6.88}$ has a tetragonal unit cell with a = 3.829(6), c = 11.47(1)Å. Annealing the sample in an oxygen atmosphere led to an increase in the Cu valence as large as 2.38. However, resistivity measurement showed that $(As_{0.33}Cu_{0.67})Sr_2YCu_2O_{6.88}$ was not superconducting but semiconducting. Nonsuperconductivity of $(As_{0.33}Cu_{0.67})Sr_2YCu_2O_{6.88}$ was attributed to the fact that the large number of O ions was located in O(3) site, which resulted in localization of holes. In this case, about 70% of oxygen ions in the $(As, Cu)O_{0.88}$ layer is located at O(3) site, leading to localization of holes. If the number of oxygen ions in the O(4) site is increased, superconductivity can be expected.

Keyword: $(As_{0.33}Cu_{0.67})Sr_2YCu_2O_y$, Crystal Structure, Physical Properties

INTRODUCTION

Layered copper oxides, $(M_xCu_{1-x})Sr_2YCu_2O_y$ are interesting compounds because of its variety of M ions and appearance of superconductivity. The value of x varies from 0.15 to 1.0, superconductivity is obtained at low levels of substitution of M = Ti, V, Fe, Co, Ga, Ge, Mo, W and Re. It is also found that when x is around 0.33, the largest number of single phase (M,Cu)-1212 are obtained [1]. So, we examined the requirements for preparing a single phase of $(M_{0.33}Cu_{0.67})Sr_2YCu_2O_y$ (M = Li, Al, Ti, V, Cr, Fe, Co, Ni, Zn, Ga, Ge, As, Se, Zr, Nb, Mo, Ru, Pd, In, Sn, Ta, W, Re, Os and Pt). Among these compounds, we prepared a novel layered cuprate, $(As_{0.33}Cu_{0.67})Sr_2YCu_2O_y$. We also report its structure and physical properties.

EXPERIMENTAL

The samples of (As, Cu)-1212 phase with nominal composition of $(As_{0.33}Cu_{0.67})Sr_2YCu_2O_y$ were prepared by a conventional solid state reaction [2,3]. Powders were mixed in an agate mortar, calcinated at 800°C for 10h in air. The calcinated powder were mixed, pressed and sintered at 850-1000°C for 10h in air. After sintering, the sample was quenched to room temperature on a metal plate. The crystal structure was determined by X-ray diffraction method. In order to dope enough holes to the (As, Cu)-1212 phase, all the samples were annealed at 500°C in O_2 atmosphere.

Oxygen content and the average Cu valence were determined by the coulometry method. In order to measure element composition, EPMA (electron probe microanalysis) measurements were carried out using an EDAX analytical system.

Its structural parameters were determined by using Rietveld analysis (RIETAN) [4]. By high-resolution transmission electron microscopy (HRTEM), electron diffraction patterns and high-resolution electron microscopic (HREM) atomic images were obtained.

Resistivity-temperature curves (5K≤T≤300K) were measured by the conventional four-probe method. DC susceptibility was measured on cooling for a sintered samples in an applied field of 0.08 - 2.0 Gauss.

RESULTS AND DISCUSSION

$(As_{0.33}Cu_{0.67})Sr_2YCu_2O_y$ {(As, Cu)-1212} sintered at various temperature and quenched. From XRD pattern, impurity phases were drastically decreased by sintering at 1000°C. In order to confirm the tolerance of As ratio, the x of $(As_xCu_{1-x})Sr_2YCu_2O_y$ was successively changed from 0 to 1. It turned out that the (As, Cu)-1212 single phase can be successfully prepared only within a very narrow region around x=0.33

From the coulometry experiment, the oxygen content, y of the (As, Cu)-1212 was found to be 6.88±0.02, which corresponded to 2.16+ of the Cu valence. In order to increase the hole concentration and to investigate the temperature dependence of the resistivity, the sample was annealed in an O_2 atmosphere at 500°C 20 h. After annealing, the oxygen content was increased from 6.88 to 7.17; the Cu valence was increased from 2.16 to 2.38.

The average composition determined by EPMA was As : Cu = 0.32 : 2.68, which well agreed with the nominal one. Rietveld analysis was carried out for the $(As_{0.33}Cu_{0.67})Sr_2YCu_2O_{6.88}$. Table 1 shows crystallographic data for this sample. The final reliability factors were R_{wp}=7.75% , R_e=5.28 % indicating that the refinements were successful. The crystallographic data obtained by Rietveld analysis clearly displays the atomic arrangement and shows that the lattice parameters are a=3.829(6)Å and c=11.47(1)Å. The simulated structure image agrees very well with the observed TEM image.

Table 1.
Crystallographic data for $(As_{0.33}Cu_{0.67})Sr_2YCu_2O_{6.88}$

Atom	Site	x	y	z	Occupation	B(Å)
As	1a	0	0	0	0.3284	0.5
Cu(1)	1a	0	0	0	0.6682	0.5
Sr	2h	0.5	0.5	0.1979	1.0	0.5
Y	1d	0.5	0.5	0.5	1.0	0.2
Cu(2)	2g	0	0	0.3528	1.0	0.5
O(1)	4i	0	0.5	0.3700	0.9132	0.5
O(2)	2g	0	0	0.1502	0.9945	1.5
O(3)	1c	0.5	0.5	0	0.7037	1.5
O(4)	2f	0	0.5	0	0.3568	1.0

Lattice constants: a=3.82967 Å and c=11.47110 Å : space group, tetragonal P4/mmm(Vol. A, No. 123); R_{wp}=7.75%; R_p=5.36%; Rr=16.58%; R_e=5.28%; S=1.4689; d=0.1974; R_i=2.75%; R_f=4.41%.

Temperature dependence of resistivity for (As, Cu)-1212 is shown in Fig. 1. The increase in oxygen content, y from 6.88 to 7.17 leads to a decrease in resistivity from $5.0 \times 10^4 \, \Omega \, m$ to $1.5 \times 10^{-2} \, \Omega \, m$ at 100K. However, the sample, whose Cu valence was determined to be 2.38, was not superconducting but semiconducting. In Fig. 2, temperature dependence of magnetic susceptibility is shown, but superconductivity was not observed for the (As, Cu)-1212. However, transitions of susceptibility were observed at around 50 K, 100 K and 160 K, where anomalies resistivity were also observed, although cause of these magnetic transitions is not clear. Nonsuperconductivity of the (As, Cu)-1212 is probably caused by existence of two oxygen sites, O(3) and O(4), in the (As, Cu)O$_{0.88}$ layer. We previously reported that the oxygen in the O(3) site drastically localizes holes, resulting that amount of itinerant holes in the CuO$_2$ layer is decreased [5]. In this material, about 70% of oxygen ions in the (As, Cu)O$_{0.88}$ layer are

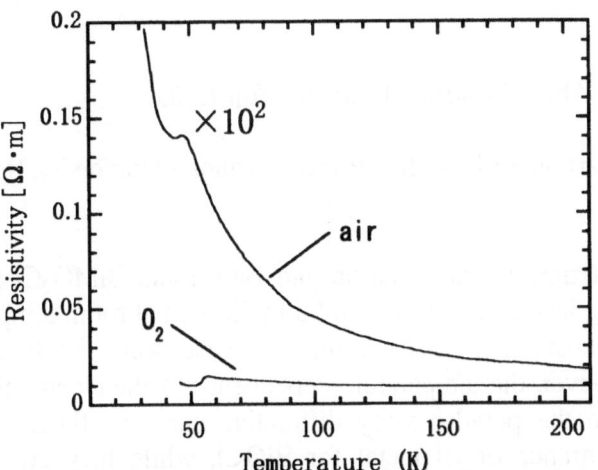

Fig. 1. Temperature dependence of resistivity for (As, Cu)-1212.

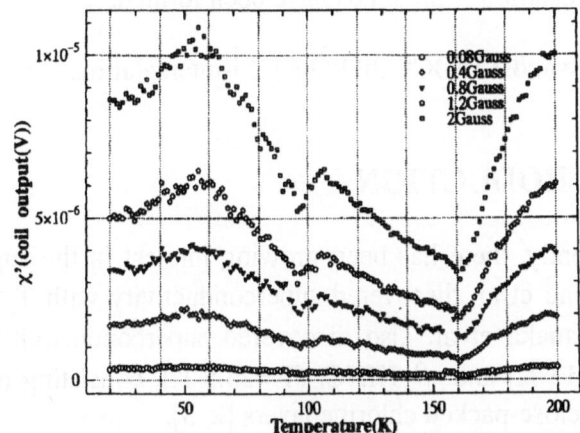

Fig. 2. Temperature dependence of susceptibility for (As, Cu)-1212.

located at O(3) site, leading to localization of the hole. In order to induce superconductivity, the increase of the number of oxygen ions in the O(4) site can be required.

CONCLUSION

A single phase of the (As, Cu)-1212 was successfully prepared for x = 0.33 in (As$_x$Cu$_{1-x}$)Sr$_2$YCu$_2$O$_y$. From the HRTEM and the Rietveld analysis, it was found that the (As$_x$Cu$_{1-x}$)Sr$_2$YCu$_2$O$_y$ has a tetragonal structure (space group P4/mmm ; a = 3.829(6)Å, c = 11.47(1)Å. Resistivity measurements showed that (As$_{0.33}$Cu$_{0.67}$)Sr$_2$YCu$_2$O$_{6.88}$ was not superconducting but semiconducting.

Reference

[1] T. Den and T. Kobayashi, *Physica C*, **196** (1992) 141-152.

[2] S. Kambe, T. Akao, I. Shime, S. Ohshima, K. Okuyama, *Adv. Supercond.* **6** (1994) 339.

[3] S. Kambe, T. Akao, I. Shime, S. Ohshima, K. Okuyama, *Mater, Sci. Eng. B* **22** (1995) 57.

[4] F. Izumi, H. Asano, H. Murata, N.Watanabe, *J.Appl. Crystallogr.* **20** (1987) 411.

[5] S.Kambe, E. Sato, T. Akao, S. Ohshima, K, Okuyama and R. Sekine, *Phys. Rev. B* **60** (1999) 687.

Li-Intercalation into the Bi-Based Oxychlorides with the Layered Structures

Yuji Abe, Masatsune Kato and Yoji Koike

Department of Applied Physics, Tohoku University, Sendai 980-8579, Japan

Abstract: Li-intercalation into BiOCl and $BiMO_2Cl$ (M = Pb, Ba, Sr and Ca) has been carried out using hexane solution of n-buthyllithium at room temperature. In the case of BiOCl and $BiPbO_2Cl$, the host samples have turned black immediately and the electrical resistivity has decreased. However, the temperature dependence of the electrical resistivity has still remained semiconductive. From the powder x-ray diffraction analysis, these changes have been guessed to be due to the appearance of Bi metal for BiOCl, while they are due to the formation of a new intercalation compound Li_xBiPbO_2Cl for $BiPbO_2Cl$. In the case of $BiMO_2Cl$ (M = Ba, Sr and Ca), on the other hand, the host samples have not changed in color and remained insulating, suggesting that no intercalation compounds have been formed.

Keywords: BiOCl, $BiPbO_2Cl$, Li-intercalation, Electrical resistivity, Powder x-ray diffraction

INTRODUCTION

Recently, there has been growing interest in the superconductivity of Li-intercalation compounds. Takano et al. discovered superconductivity with $T_c \sim$ 2 K in the Li-intercalated $KCa_2Nb_3O_{10}$ [1]. Yamanaka et al. also discovered superconductivity of Li-intercalated layered nitrides β-ZrNCl (T_c=13 K) and β-HfNCl (T_c=25.5 K), consisting of Zr(Hf)-N double layers sandwiched between two close-packed chlorine layers [2, 3].

Figure 1(a) shows the crystal structure of the Bi-based oxychloride BiOCl, which is similar to that of β-Zr(Hf)NCl. It is composed of the double chlorides layer $[Cl_2]$ and the two-dimensional $[Bi_2O_2]$ layer, instead of the $[Zr(Hf)_2N_2]$ layer in β-Zr(Hf)NCl. In addition, it is well known that $(Ba, K)BiO_3$ with the three-dimensional Bi-O network exhibits superconductivity with $T_c \sim$ 30K [4]. Therefore, the Li-intercalated BiOCl is expected as a candidate for a new superconductor with the two-dimensional Bi-O network. In this paper, we report a trial of Li-intercalation into BiOCl. Moreover, we also report a trial of Li-intercalation into $BiMO_2Cl$ (M = Pb, Ba, Sr, Ca), composed of the $[Bi_2O_2]$ layer and the single chloride layer [Cl], as shown in Fig. 1(b) [5].

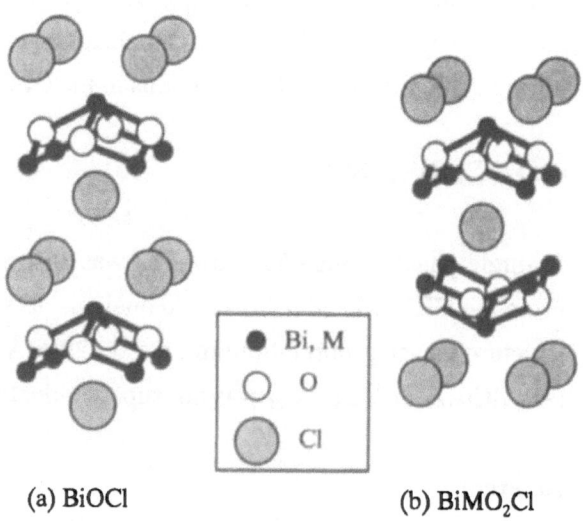

(a) BiOCl   (b) $BiMO_2Cl$

Fig. 1. Crystal structures of (a) BiOCl and (b) $BiMO_2Cl$ (M = Pb, Ba, Sr, Ca).

EXPERIMENTAL

Commercial powder of BiOCl was used as host samples of BiOCl. It was pressed into a pellet with the dimensions of 10 mm in diameter and 1 mm in thickness and sintered in air at 700 ℃ for 24 h. Host samples of $BiMO_2Cl$ (M = Pb, Ba, Sr, Ca) were prepared by the conventional solid-state reaction method, using powders of BiOCl, PbO, BaO_2, SrO and CaO. The powders weighted stoichiometrically were mixed, pelletized and heated in air at 670 ℃ for 24 h. All products were characterized to be of the single-phase of $BiMO_2Cl$ by the powder x-ray diffraction using CuKα radiation. Li-intercalation was carried out by soaking the pelletized host samples into 20 ml of hexane solution of 1.6M n-buthyllithium. The electrical resistivity was measured keeping the samples in the solution. Leads wires of copper had been attached to the host samples with silver paste before the Li-intercalation was carried out.

RESULTS AND DISCUSSION

In the case of BiOCl, the host sample, which had been white in color and insulating, turned black 2h soaking after in the solution. As shown in Fig. 2, the electrical resistivity at room temperature decreased with increasing Li-intercalation time, and was 10^3 Ω cm after 50 h. In the powder x-ray diffraction pattern of the sample which was soaked for 10 h, however, peaks due to Bi metal slightly appeared, and the intensity of the peaks due to Bi metal increased with the increase of the Li-intercalation time, as shown in Fig. 3. Thus, both of the change in color and the decrease in the electrical resistivity at room temperature are guessed to be due to the appearance of Bi metal, which was extracted through the reduction of BiOCl during the Li-intercalation process. However, there remains a possibility that a new intercalation compound Li_xBiOCl was formed within 8h soaking in the beginnings of the Li-intercalation process, because the host sample turned black and the electrical resistivity decreased in spite of no observation of peaks due to Bi metal in the x-ray diffraction pattern.

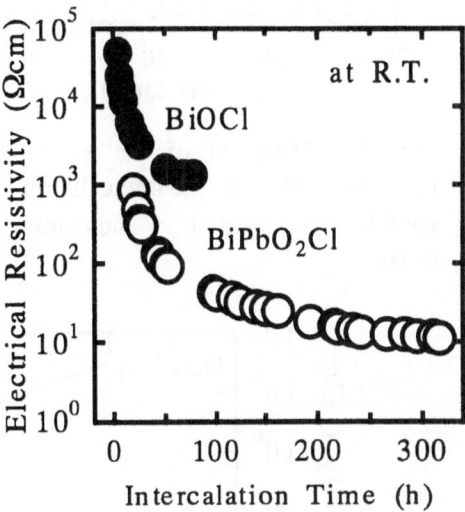

Fig. 2. Dependence of the electrical resistivity at room temperature on the Li-intercalation time for BiOCl and $BiPbO_2Cl$.

In the case of $BiPbO_2Cl$, the host sample also turned black and the electrical resistivity at room temperature decreased with increasing Li-intercalation time. It was 10 Ω cm after 300 h, for M = Pb, as shown in Fig. 2. Unlike BiOCl, from the powder x-ray diffraction analysis, it was found that Bi metal did not appear within 50 h soaking and that the structure of $BiPbO_2Cl$ was almost retained during the Li-intercalation process even after 300 h, as shown in Fig. 4. These results suggest that a new intercalation compound Li_xBiPbO_2Cl has been formed. The temperature dependence of the electrical resistivity in the Li-intercalated $BiPbO_2Cl$ has still remained semiconductive, as shown in Fig. 5. However, the constancy of the lattice parameters and the semiconductive behavior through the Li-intercalation suggest that the intercalated Li-content was small. The Li-content is not estimated and the site occupied by intercalated Li ions is not identified either. Values of the electrical resistivity at room temperature for the Li-intercalated $BiPbO_2Cl$ shown in Fig. 5 are higher than those in Fig. 2. This may be due to the fact that the Li-concentration in the hexane solution of n-buthyllithium decreases by the precipitation of LiOH during the Li-intercalation process. For

BiMO$_2$Cl (M = Ba, Sr, Ca), the host samples did not change in color and remained insulating, suggesting that Li-intercalation compounds were not formed.

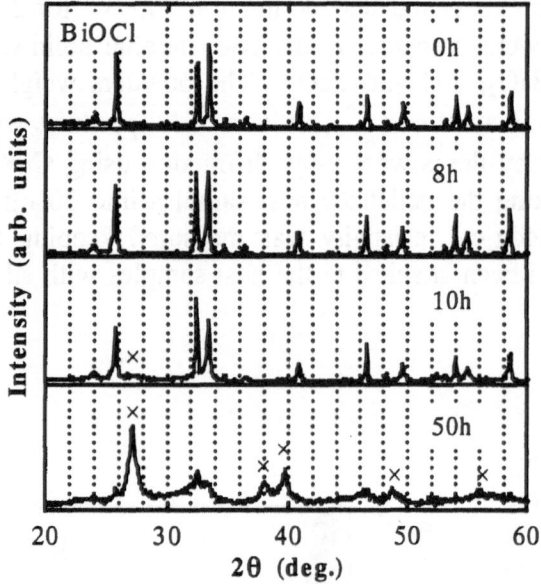

Fig. 3. Variation of the x-ray diffraction patterns with the Li-intercalation time for BiOCl. The symbol × indicates a peak due to Bi.

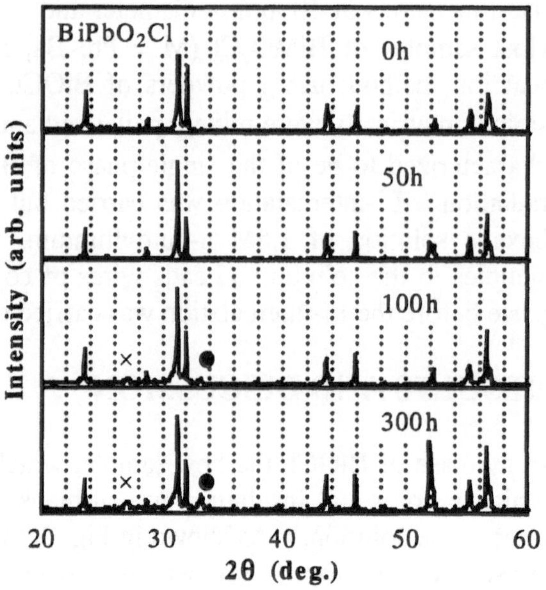

Fig. 4. Variation of the x-ray diffraction patterns with the Li-intercalation time for BiPbO$_2$Cl. Symbols × and ● indicate peaks due to Bi and Pb$_7$Bi$_3$, respectively.

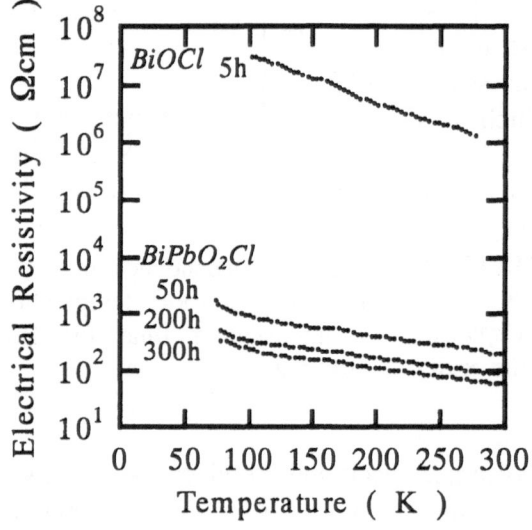

Fig. 5. Temperature dependence of the electrical resistivity for the Li-intercalated BiOCl and BiPbO$_2$Cl.

Acknowledgment This work was supported by a Grant-in-Aid for Scientific Research given by the Ministry of Education, Science and Culture, Japan and also supported by CREST of Japan Science and Technology Corporation.

1. Y. Takano, H. Taketomi, H. Tsurumi, T. Yamadaya and N. Mori, Physica B 237-238, 68 (1997).
2. S. Yamanaka, H. Kawaji, K. Hotehama, M. Ohashi, Adv. Matter. 9, 172 (1996).
3. S. Yamanaka, K. Hotehama and H. Kawaji, Nature 392, 580 (1998).
4. R. J. Cava, B. Batlogg, J. J. Krajewski, R. Farrow, L. W. Rupp, Jr., A. E. White, K. Short, W. F. Peck and T. Kometani, Nature 332, 814 (1988).
5. J. Ketterer and V. Kramer, Mat. Res. Bull. 20, 1031 (1985).

Carrier Doping into $Ba_2Cu_3O_4Cl_2$ with the Cu_3O_4 Plane through the Li-substitution for Cu

Takaaki Tanaami, Masatsune Kato and Yoji Koike

Department of Applied Physics, Tohoku University, Sendai 980-8579, Japan

Abstract: A trial of the carrier doping into $Ba_2Cu_3O_4Cl_2$ with the Cu_3O_4 plane, which is composed of $Cu(1)O_2$ and extra $Cu(2)$ sublattices, has been carried out through the partial substitution. We have succeeded in obtaining single-phase samples of $Ba_2Cu_{3-x}Li_xO_4Cl_2$ ($x \leqq 1.0$). With increasing x, the c-axis is almost constant and the a-axis tends to decrease. All samples exhibit semiconductive behavior. The electrical resistivity decreases with increasing x up to 0.5 and tends to increase slightly for $x > 0.5$ due to disorder in the Cu_3O_4 plane through the substitution of Li^+ for Cu^{2+}. Both of the antiferromagnetic transition temperatures with respect to $Cu(1)$ and $Cu(2)$ decrease with increasing x. However, the antiferromagnetic ordering in the $Cu(1)O_2$ sublattice still remains even for $x = 0.5$. Considering that the formal valence of Cu is +2.2 for $x = 0.5$, holes doped through the Li-substitution may be localized in the $Cu(2)$ site.

Keywords: Cu_3O_4 plane, $Ba_2Cu_{3-x}Li_xO_4Cl_2$, carrier doping, electrical resistivity, magnetic susceptibility, substitution effect

INTRODUCTION

The common structural feature of all high-T_c superconductors found so far is to have the two-dimensional CuO_2 plane. Recently, superconductivity was discovered under high pressures in the so-called spin-ladder compound $Sr_{0.4}Ca_{13.6}Cu_{24}O_{41}$ with the two-dimensional Cu_2O_3 plane [1]. This seems to imply that any cuprates with the two-dimensional Cu-O plane exhibit superconductivity. Therefore, we have taken notice of the cuprate $A_2Cu_3O_4X_2$ (A = Sr, Ba; X = Cl, Br) [2-4] with the two-dimensional Cu_3O_4 plane. Fig. 1(a) shows the schematic representation of the Cu_3O_4 plane. The Cu_3O_4 plane can be regarded as a $Cu(1)O_2$ sublattice with additional $Cu(2)$ ions at the center of every second plaquette in the $Cu(1)O_2$ sublattice. Fig. 1(b) shows the crystal structure of $A_2Cu_3O_4X_2$. This compound can be regarded as an alternate stack of the Cu_3O_4 plane and the flourite-type A_2X_2 layer along the c-axis, which is similar to structures of high-T_c superconductors. $Cu(1)$ ions are coordinated with two X^- ions at the apical sites, while $Cu(2)$ ions have no apical anion.

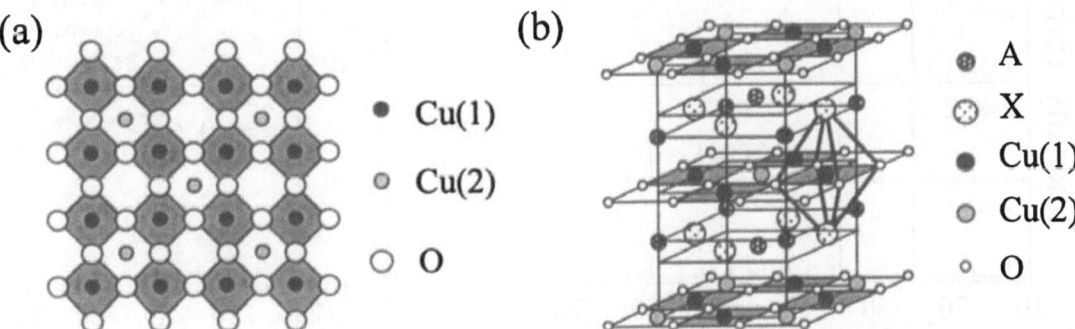

Fig. 1. (a) Schematic representation of the Cu_3O_4 plane. (b) Crystal structure of $A_2Cu_3O_4X_2$ (A = Sr, Ba; X = Cl, Br).

The mother compound $A_2Cu_3O_4X_2$ is an insulator. Cu(1) and Cu(2) spins show three-dimensional antiferromagnetic orders below T_N (1) and T_N (2), respectively. Moreover, a broad peak due to the two-dimensional antiferomagnetic order in the Cu(2) sublattice is observed around T_{2D} (2) in the temperature dependence of the magnetic susceptibility [5]. With decreasing ionic-radii of A and X, the electrical resistivity decreases and values of T_N (1) and T_{2D} (2) increase [6]. This behavior is explained as being due to the increase of the overlap between the Cu 3d orbit and the O 2p, X p ones. With decreasing ionic-radii of A and X, however, the $A_2Cu_3O_4X_2$ (2342) phase with the Cu_3O_4 plane turns to be a mixture of CuO and the $A_2CuO_2X_2$ (2122) phase with the CuO_2 plane [7]. In this report, we try carrier doping into $Ba_2Cu_3O_4Cl_2$ through the partial substitution of Li^+ for Cu^{2+} and investigate the substitution effects on the electrical and magnetic properties.

EXPERIMENTAL

A trial of the carrier doping into $Ba_2Cu_3O_4Cl_2$ was carried out through the partial substitution of Li^+ for Cu^{2+}. Polycrystalline samples were prepared by the solid-state reaction method. Powders of $BaCO_3$, CuO, $BaCl_2$ and Li_2CO_3 were used as raw materials. The phase of the products was identified by powder x-ray diffraction using CuK_α radiation. Electrical resistivity measurements were carried out by the standard DC four-point probe method. The magnetic susceptibility was measured using a SQUID magnetometer in a magnetic field of 1T.

RESULTS AND DISCUSSION

Fig. 2 shows powder x-ray diffraction patterns of $Ba_2Cu_{3-x}Li_xO_4Cl_2$ for various values of x. It is found that single-phase samples are obtained for $0 \leqq x \leqq 1.0$. Fig. 3 shows the temperature dependence of the electrical resistivity ρ of $Ba_2Cu_{3-x}Li_xO_4Cl_2$ for various values of x. Semiconductive behavior is observed for all samples. Fig. 4 shows the temperature dependence of the magnetic susceptibility χ of $Ba_2Cu_{3-x}Li_xO_4Cl_2$ for various values of x. With increasing x, both of the increase just below T_N (1) and the broad peak around T_{2D} (2) are suppressed.

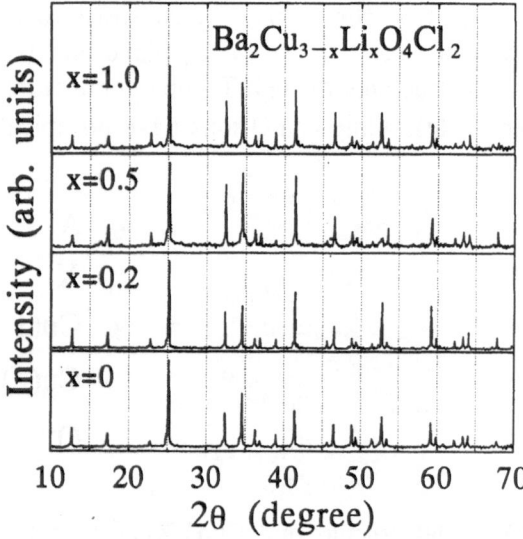

Fig. 2. Variation of the powder x-ray diffraction patterns of $Ba_2Cu_{3-x}Li_xO_4Cl_2$ for various values of x.

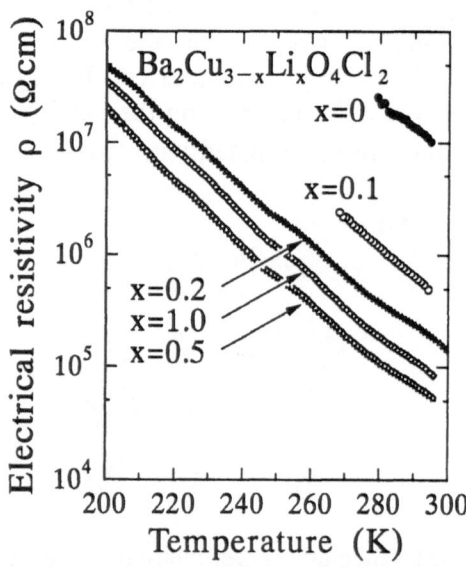

Fig. 3. Temperature dependence of the electrical resistivity ρ of $Ba_2Cu_{3-x}Li_xO_4Cl_2$ for various values of x.

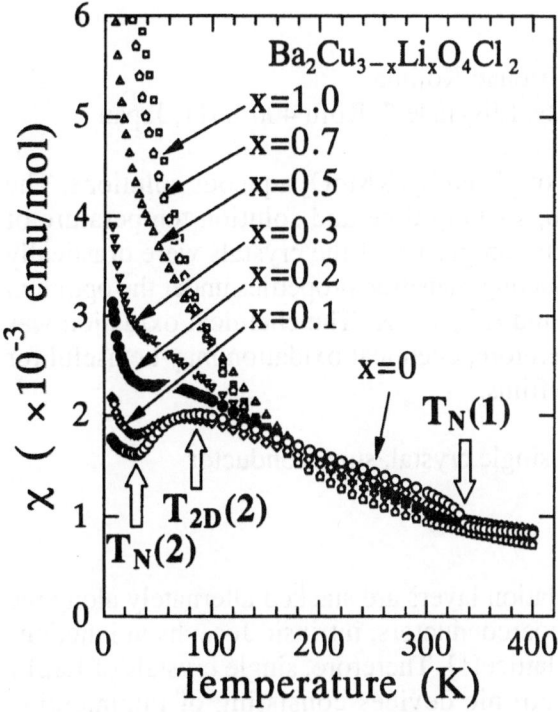

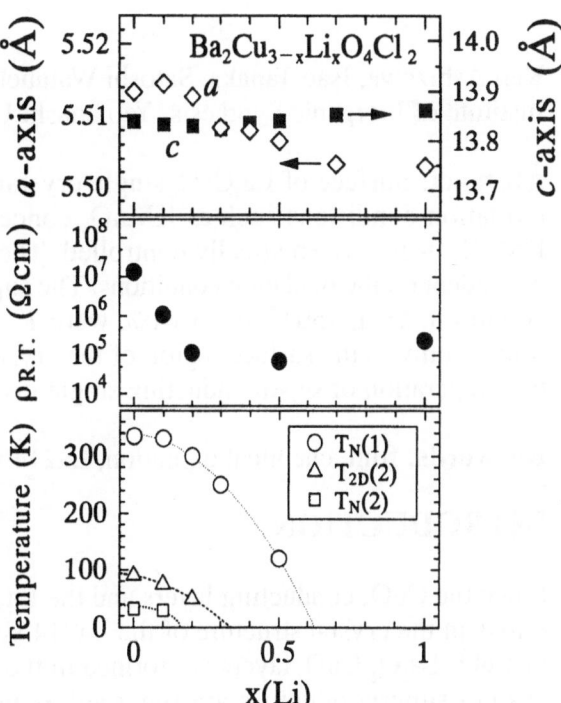

Fig. 4. Temperature dependence of the magnetic susceptibility χ of $Ba_2Cu_{3-x}Li_xO_4Cl_2$ for various values of x.

Fig. 5. Dependences on x of the lattice parameters, electrical resistivity at room temperature $\rho_{R.T.}$, $T_N(1)$, $T_N(2)$ and $T_{2D}(2)$ for $Ba_2Cu_{3-x}Li_xO_4Cl_2$.

Dependences on x of the lattice parameters, electrical resistivity at room temperature $\rho_{R.T.}$, $T_N(1)$, $T_N(2)$ and $T_{2D}(2)$ for $Ba_2Cu_{3-x}Li_xO_4Cl_2$ are summerized in Fig. 5. With increasing x, the c-axis is almost constant and the a-axis tends to decrease. The decrease in the a-axis suggests that holes are doped into the Cu_3O_4 plane through the partial substitution of Li^+ for Cu^{2+}. With increasing x up to 0.5, $\rho_{R.T.}$ decreases down to $\sim 10^4$ Ωcm. For x >0.5, $\rho_{R.T.}$ tends to increase slightly due to disorder in the Cu_3O_4 plane through the Li^+ substitution for Cu^{2+}. Moreover, it is found that values of $T_N(1)$, $T_N(2)$ and $T_{2D}(2)$ decrease with increasing x. However, the antiferromagnetic ordering in the $Cu(1)O_2$ sublattice still remains even for x = 0.5. Considering that the formal valence of Cu is +2.2 for x = 0.5, holes doped through the Li-substitution may be localized in the Cu(2) site. We have not succeeded in any other trials of substitution.

Acknowledgments: This work was supported by CREST of Japan Science and Technology Corporation, and also by Iketani Science and Technology Foundation.

1. M. Uehara, T. Nagata, J. Akimitsu, H. Takahashi, N. Mori and K. Kinoshita, J. Phys. Soc. Jpn. **65**, 2764 (1996).
2. B. Grande and Hk. Müller-Buschbaum, Z. Naturforsch. B **31**, 405 (1975).
3. R. Kipka and Hk. Müller-Buschbaum, Z. Anorg. Allg. Chem. **419**, 588 (1976).
4. W. J. Zhu, F. Wu, Y. Z. Huang, C. Dong, H. Chen and Z. X. Zhao, Mater. Res. Bull. **29**, 219 (1994).
5. T. Ito, H. Yamaguchi and K. Oka, Phys. Rev. B **55**, R684 (1997).
6. M. Kato, T. Tanaami and Y. Koike, Proceedings of LT22, Physica B, in press.
7. T. Tanaami, M. Kato and Y. Koike, Proceedings of MOS99, J. Low Temp. Phys., in press.

CHEMICAL OXIDATION ON THE SURFACE OF La$_2$CuO$_4$ SINGLE CRYSTALS

Ken Ashizawa, Isao Tanaka, Satoshi Watauchi and Hironao Kojima
Institute of Inorganic Synthesis, Yamanashi University. Miyamae 7, Kofu 400-8511, Japan

Abstract: Surface of La$_2$CuO$_4$ single crystals were oxidized by KMnO$_4$ aqueous solutions. The oxidation conditions such as KMnO$_4$ concentration, soaking time and solution temperature of KMnO$_4$, were systematically controlled. The magnetic properties of the crystals were drastically dependent on the oxidation conditions. The superconducting transition properties under the optimum condition, 48 h, 150℃ and 5 wt%, were T_c = 28 K and ΔT_c = 2 K. The chemical oxidation was limited only to the surface region of the crystals. Therefore, chemical oxidation may be useful for the preparation of superconducting single-crystalline films.

Keywords: film, chemical oxidation, La214 system, single crystal, superconductor

INTRODUCTION

Since the CuO$_2$ conducting layers and the La$_2$O$_2$ insulation layers are stacked alternately along the c-axis in the crystal structure of the La214 system superconductors, intrinsic Josephson junctions of CuO$_2$-La$_2$O$_2$-CuO$_2$ layers are formed in the crystal lattice[1]. Therefore, single crystals of La214 system superconductors are regarded as microelectronic devices consisting of innumerable Josephson junctions. The Josephson plasma is generated by the coupling between the Josephson current flowing along the c-axis and the electromagnetic field. The intrinsic Josephson junctions of HTSC are superior to SIS Josephson junctions. Single-crystalline films and layers of HTSC with the thickness larger than several micrometers are needed to be applied for superconducting devices with intrinsic Josephson junctions in the HTSC single crystals.

Although the stoichiometric La$_2$CuO$_4$ is an antiferromagnetic insulator[2], the oxygen-excess La$_2$CuO$_{4+\delta}$ becomes a superconductor. Chemical oxidation[3,4], annealing in high-pressure oxygen gas and electrochemical oxidation were well known as oxygen insertion methods. In this study, the superconducting single-crystalline layers of oxygen-excess La$_2$CuO$_{4+\delta}$ were formed by direct chemical oxidation into the surface region of the non-superconducting La$_2$CuO$_4$ single crystals using aqueous solutions of KMnO$_4$. The single crystal films were obtained easily by this process, which has no problems encountered in general thin films preparation such as lattice consistence and thermoequibrium conditions.

EXPERIMENTAL

The La$_2$CuO$_4$ single crystals to be used for oxidation experiments were prepared as follows; The La$_2$CuO$_4$ polycrystalline feeds and solvents were prepared in stoichiometric amount with high purity starting materials (La$_2$O$_3$ and CuO). The feeds and solvent were calcined at 850℃ for 24 h in oxygen. The heated powder was formed into a cylindrical shape of 6 mm in diameter and about 60 mm in length, sintered at 1250℃ for the feed rods and at 950℃ for the solvents for 12 h in oxygen. The La$_2$CuO$_4$ single crystals were grown by the TSFZ method using an infrared heating furnace at the growth rate of 1.0 mm/h under oxygen pressure of 0.2 MPa. The single crystals of 5 mm in diameter and about 50 mm in length were obtained. The grown crystals were sliced into sections of 1mm thickness perpendicular to the a-axis. The crystals were annealed at 900℃ for 40h in nitrogen to remove excess oxygen which makes the as-grown crystals superconducting. The tips of the single crystals were oxidized into aqueous solutions of 5-15 wt % KMnO$_4$ at 50-150℃ for 48-96h, using the special airtight instrument with Teflon crucible (FLON INDUSTRY, model F-1029-04S) to suppress water evapolation.

The surface conditions and erosion of the single crystals was observed using a polarizing microscope. The treated crystals were characterized by DC magnetization measurement using a SQUID magnetometer (QUANTUM DESIGN, model MPMS-5S). The zero-field cooled and field-cooled magnetization were measured in temperature range between 5 and 40 K under an external magnetic field of 1 Oe parallel to the a-axis of the tips.

RESULTS AND DISCUSSION

The high quality single crystals were grown in 5 mm diameter and in 50 mm length by TSFZ method (Fig. 1). Figure 2 shows the results of magnetization measurements in the La_2CuO_4 single crystals. Very weak Shielding signal was observed in the as-grown crystals. By annealing in nitrogen, shielding signal disappeared. This result indicated that the single crystals become nonsuperconductors by N_2-annealing. The Shielding signal of chemical oxidized crystals was about 40 times larger than the as-grown crystals.

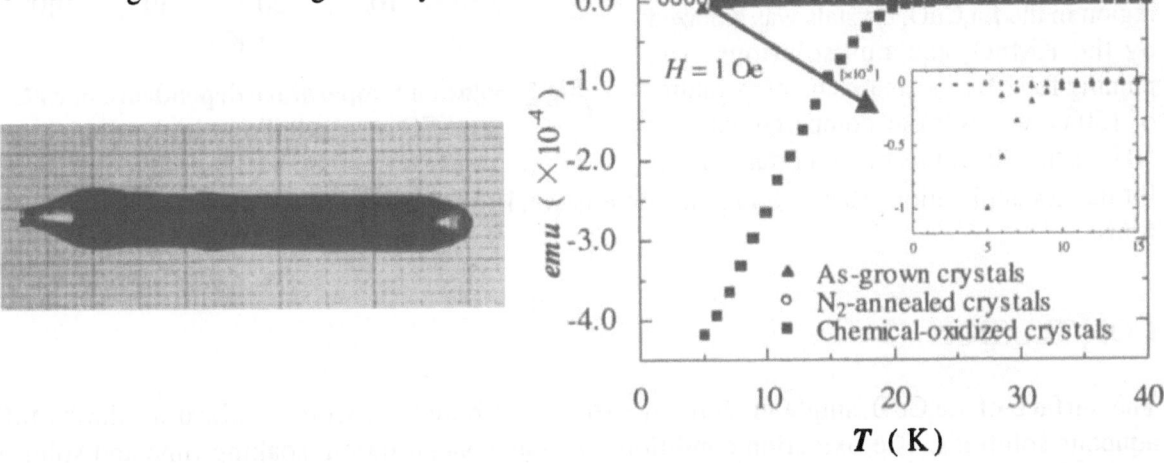

Fig.1 single crystal of La_2CuO_4.

Fig.2 The comparison with magnetization between as-grown, N_2-annealed and chemical-oxidized (5 wt%, 48h, 50℃) crystals.

Figure 3 shows temperature dependence of the zero-field cooled (ZFC) and field cooled (FC) susceptibilities along the a-axis for the La_2CuO_4 single crystals treated in 5 wt% $KMnO_4$ aqueous solutions at 50℃ for varied soaking time. The superconducting transition properties of the crystals becomes superior with increasing the soaking time. The superconducting transition properties were almost constant for the soaking time more than 96h. As shown in figure 4, the superconducting

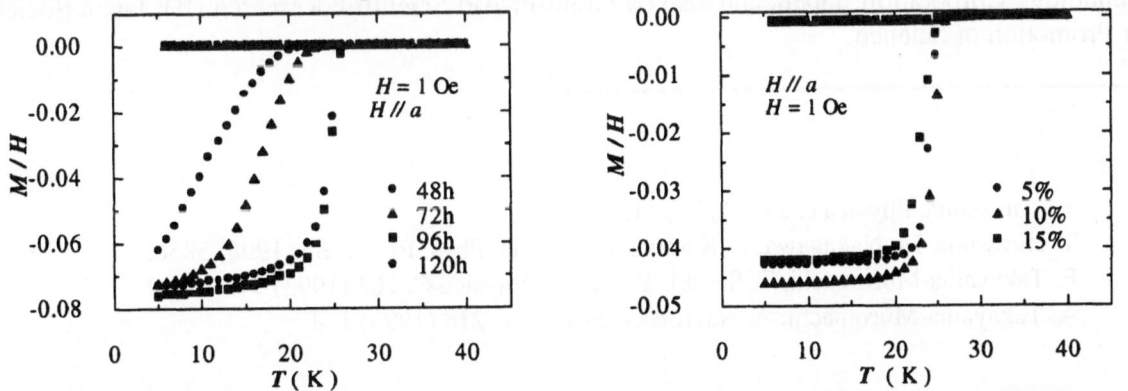

Fig.3 Soaking-time dependence of ZFC and FC susceptibilities in the single crystals treated in 5 wt% $KMnO_4$ at 50℃.

Fig.4 Dependence on $KMnO_4$ concentration in the solutions of ZFC and FC susceptibilities in the single crystals treated at 100℃ for 96h.

properties of the crystals were almost independent on the KMnO$_4$ concentration in the aqueous solutions. However, the superconducting transition temperature increases remarkably about 5 K with increasing the treated temperature, as shown in figure 5.

$T_{c\,onset}$ and ΔT_c were determined to be about 28 K and 2 K, respectively under the optimum condition. The volume of the superconducting phase was estimated to be about 7% from the Shielding fractions. This result suggests that the only surface region in the La$_2$CuO$_4$ crystals was oxidized by the KMnO$_4$ aqueous solutions. The pinning force for the treatment temperature of 150℃ was weak as compared with for 50℃ and 100℃. It is the important factor for devices applications that the pinning force is weak, because the magnetic flux move into the films.

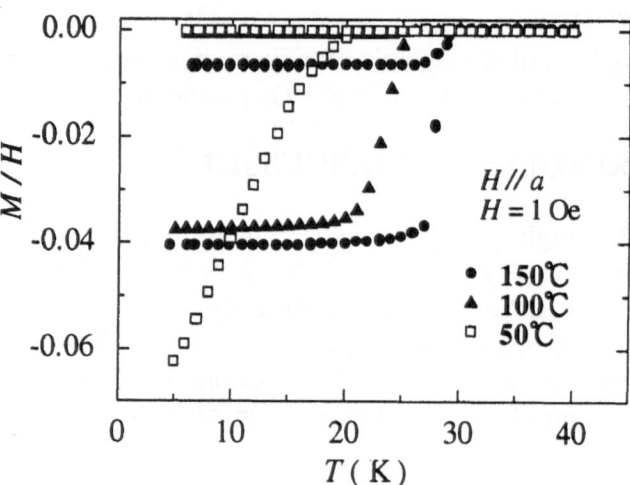

Fig.5 Solution temperature dependence of ZFC and FC susceptibilities in the single crystals treated 5wt% for 48 h.

CONCLUSION

The surface of La$_2$CuO$_4$ single crystals by grown TSFZ method were oxidized in the KMnO$_4$ aqueous solutions. The oxidation conditions such as concentration, soaking time and solution temperature of KMnO$_4$ were controlled systematically. The magnetic properties of crystals drastically depended on the oxidation conditions. The superconducting transition temperature under the optimum condition, 48 h, 150℃ and 5 wt%, is about 28 K and ΔT_c 2 K. The volume of the superconducting phase was estimated to be about 7% from the Shielding fractions. This result sugests that chemical oxidation was limited only to the surface region of the crystals, and that chemical oxidation may be useful for the preparation of superconducting films.

Acknowledgments. This work was supported partly by CREST Project, Japan Science and Technology Corporation, Japan, and also by Grant-in-Aid Scientific Research (B), Japan Society for Promotion of Science.

1.   T. Yamashita, Physica C **293** (1997) 31.
2.   T. Hirayama, M. Nakagawa, A. Sumiyama, Y. Oda, Phys. Rev. B **58** (1998) 5856.
3.   E. Takayama-Muromachi, T. Sasaki, Y. Matsui, Physica C. **207** (1993) 97.
4.   A. Takayama-Muromachi, A. Navrotsky, Physica C **218** (1993) 164.

Study of $Bi_2Sr_2CaCu_2O_y$ Single Crystal Surfaces by X-ray Photoelectron Spectroscopy

Yasuhiro Yamauchi, Satoru Kishida and Heizo Tokutaka

Department of Electrical and Electronic Engineering, Tottori University, Koyama, Tottori 680-8552, Japan

Abstract: We heated $Bi_2Sr_2CaCu_2O_y$(BSCCO) single crystals at various temperatures in air, Ar gas and O_2 gas, and investigated their surface states by X-ray photoelectron spectroscopy(XPS). From the results, we found that the Cu valence and the contents of Sr occupying Sr and Ca sites in BSCCO single crystals were affected by the atmosphere of heat-treatments.

Key words: BSCCO, single crystal, XPS, clean surface, heat-treatments

INTRODUCTION

In order to fabricate the devices using $Bi_2Sr_2CaCu_2O_y$(BSCCO) superconductors and to carry out their surface analysis exactly, it is necessary to obtain their clean surface. The heat-treatment of oxides in air is known to be one of surface cleaning methods. However, it is not clarified what occurs on the surface BSCCO single crystals when they were heat-treated in air. Since X-ray photoelectron spectroscopy(XPS) is available for understanding the chemical bond natures of constituent elements on the surfaces of BSCCO single crystals, we used XPS in order to analyze them.

In this study, we heated the BSCCO single crystals in air, Ar gas and O_2 gas, and investigated chemical bond natures of the elements Cu and Sr on the surfaces by XPS.

EXPERIMENTAL

BSCCO single crystals were prepared by a self flux method[1]. After cleaving the samples in air using an adhesive tape, they were heated at the temperatures from 300°C to 800°C for 1h in air, Ar gas and O_2 gas. Then, the samples were quickly transferred into an XPS chamber. The XPS measurements were carried out with Shimadzu ESCA750 spectrometer in a base pressure less than 3×10^{-5} Pa using X-ray source of MgK_α. The XPS peak positions of all constituent elements were corrected by the C-1s XPS peak position at about 285 eV.

RESULTS AND DISCUSSION

We measured the XPS spectra of Cu-2p core levels from the surfaces of the BSCCO single crystals heated at various temperatures in various atmospheres. The Cu-2p 1/2 main and satellite peaks were observed at about 954 eV and 962 eV, respectively. It was reported that the satellite peak was always observed when the valence of Cu was 2+, and that Cu^{2+} was dominant on the clean surface of BSCCO superconductors[2]. Therefore, we expect that the degree of a clean surface is able to be estimated from the XPS intensity ratios of the satellite to main peaks.

Figure 1 shows the relative XPS intensity ratios of Cu-2p 1/2 satellite to main peaks as a function of heat-treatment temperature. The surfaces of BSCCO single crystals were heated in various atmospheres. In Fig.1, the intensity ratios increased up to the temperature of 400℃ in the atmospheres of air and O_2 gas, and then were saturated at the temperature of more than 400℃. This indicated that Cu^{2+} was dominant, and that the clean surfaces were obtained by heating the BSCCO single crystals at the temperatures of more than 400℃ in air and O_2 gas. In Ar gas, the intensity ratios decreased with increasing the temperatures of heat-treatments. This indicated that the valence of Cu on the BSCCO single crystal surfaces was reduced from 2+ to 1+ by the heat-treatment in Ar gas, namely, in the atmosphere of little oxygen. It was reported that the carbon and the impurity oxygen on the surface of BSCCO single crystals were removed and Cu^{2+} was dominant by heat-treating at the temperature of more than 400℃ in air and O_2 gas[3]. Our results were consistent with the report[3]. Therefore, the surface of BSCCO single crystals is thought to be clean. In addition, Cu valence was affected by the atmosphere of heat-treatments.

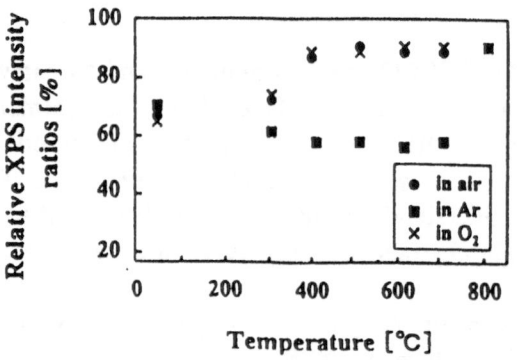

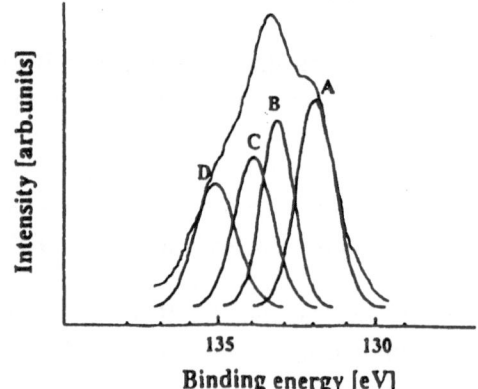

Fig. 1. The relative XPS intensity ratios of Cu-2p 1/2 satellite to main peaks from the surfaces of BSCCO single crystals heated in various atmospheres as a function of heat-treatment temperature.

Fig. 2. The typical Sr-3d XPS spectrum separated into four peaks.

Figure 2 shows the typical Sr-3d XPS spectrum which was separated into four peaks by using the parameters of the reference[4]. In general, Sr and Ca are known to be easily substituted each other in BSCCO superconductors. In Fig.2, the Sr-3d XPS spectrum consisted of four peaks which are due to Sr peak at Sr site (A and C), and at Ca site (B and D). Although not represented in the figure, we found that Ca-2p XPS spectrum also consisted of four peaks as well as that of Sr. By separating the Ca-2p and Sr-3d XPS spectra, we can estimate the degree of substitution between Sr and Ca in the BSCCO single crystals.

Figure 3 shows the Sr-3d 5/2 relative XPS intensities from the surfaces of BSCCO single crystals heated at various temperatures in (a)air and (b)Ar gas. The Sr intensity in the XPS spectra was normalized by a maximum value. Hereafter, the intensities of A,B,C and D peaks are called a relative intensity. In Fig.3(a), the contents of Sr occupying Sr and Ca sites were independent of temperatures of heat-treatments in air and O_2 gas. Although not represented in the figure, the dependence of annealing temperature in O_2 gas on Sr contents was approximately equal to that in air. From the results, we suggest that the Sr contents were not affected by the temperatures of heat-treatments in air and O_2 gas. In Fig.3(b), the Sr intensities at Sr and Ca sites increased and decreased, respectively with the temperature of heat-treatments. In Figs.1 and 3(b), the heat-treatment at the temperature of 400℃ in Ar gas changed the valence of Cu and the Sr contents. In addition, the

145

behavior of Fig.1 at the temperatures of more than 400°C was approximately equal to that of Fig.3(b). Therefore, the valence of Cu which changes by the atmosphere of heat-treatments is thought to affect the contents of Sr occupying Sr and Ca sites.

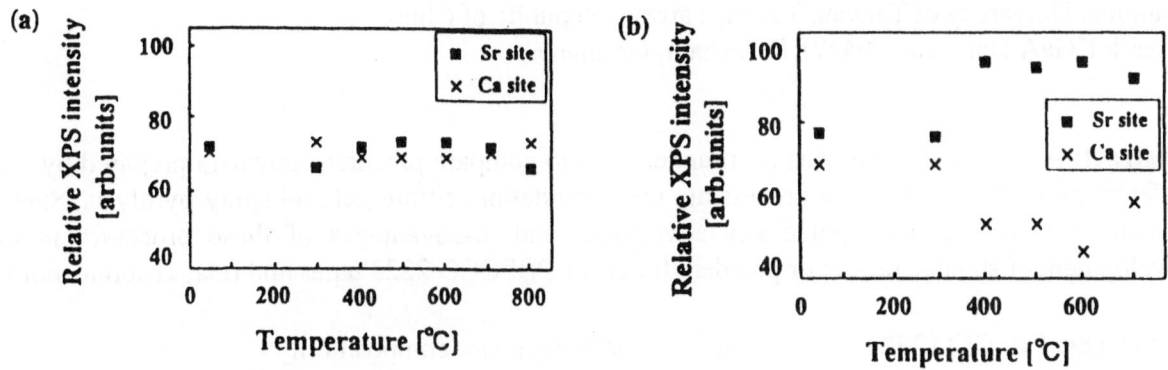

Fig. 3. The Sr-3d 5/2 relative XPS intensities from the surfaces of BSCCO single crystals heated at various temperatures in (a)air and (b)Ar gas.

CONCLUSIONS

We investigated the surface compositions and the chemical bond natures of the BSCCO single crystals heated at various temperatures in the atmospheres of air, Ar gas and O_2 gas. From the results, we found that the contents of Sr occupying Sr and Ca sites were affected by the atmosphere of heat-treatments, and were related with the valence of Cu.

REFERENCES

1. Kishida S, Yumoto T, Nakanishi S, Tokutaka H, Fujimura K (1995) J Cryst Growth 153: 17
2. Kishida S, Tokutaka H, Toda F, Fujimoto H, Futo W, Nishimori K, Ishihara N (1990) Jpn J Appl Phys 28: L438
3. Yamauchi Y, Kishida S, Tokutaka H (1998) J. Surf. Anal 5: 312
4. Kohiki S, Wada T, Kawashima S (1988) Phys Rev Lett 38: 7051

A study of the reactivity of PBSCCO 2223 precursor powders prepared by solid-state processing, co-precipitation, citrate gel and spray pyrolysis

Ya Wei Hsueh[1], Ru Shi Liu[1*], Lee Woodall[2] and Michael Gerards[2]

[1]National University of Taiwan, Taipei, Taiwan, Republic of China
[2]Merck KGaA Darmstadt, 64271 Darmstadt, Germany

Abstract: A systematic study has been undertaken to compare precursor powders prepared by four different processes - solid state processing, co-precipitation, citrate gel and spray pyrolysis. Such a systematic study has highlighted key advantages and disadvantages of these processes to the development of suitable precursor powders for OPIT PBSCCO 2223 tapes and related applications.

Keywords: PBSCCO 2223, spray pyrolysis, reactivity, phase compositiom

INTRODUCTION

Development of PBSCCO 2223 conductors using the OPIT (Oxide Powder in Tube) process have resulted in J_c (77 K, 0 T) of short Bi-2223/Ag tapes approaching 7×10^4 A/cm^2 [1]. However, such encouraging values are only attained with very painstaking control of all processing parameters, relating both to the powder precursors and to the tape manufacturing process. Control of the powder composition and morphology, as well as the mixture of crystalline phases present, will determine the choice of processing parameters to be used in tape manufacture.

EXPERIMENTAL

Powders synthesised from the solution routes (co-precipitated, citrate gel and spray pyrolysis) were prepared from a mixed acidic solution of $Bi(NO_3)_3 \cdot 5H_2O$, $Sr(NO_3)_2$, $Pb(NO_3)_2$, $Ca(NO_3)_2 \cdot 4H_2O$ and $Cu(NO_3)_2 \cdot 3H_2O$ (purity 99%), corresponding to the nominal composition $Bi_{1.7}Pb_{0.4}Sr_{1.8}Ca_{2.2}Cu_{3.2}O_x$.

Blue, co-precipitated powder was prepared by titrating a mixed-metal nitrate solution into a triethylamine/oxalic acid mix (ratio 1.8~2.2); precipitation complete at a pH of 4.2~4.3. The precipitant was subsequently filtered and then dehydrated at 80°C for 24 hours. The powder was then heated at 200°C for 2 hours. Citrate gel powder was prepared by first adding one gram equivalent of citric acid (for each gram equivalent of metals) to a mixed nitrate solution with subsequent addition of ethylenediamine, dropwise into the solution with constant stirring, until the pH reached 6. The solution was gelled at 100°C and then heated to 200°C for 2 hours, during which swelling and then combustion occurred. Spray pyrolysis powder was prepared by *Merck KGaA* in which the nitrate solution was sprayed as a fine, sub-micron mist into a heated reaction chamber. Solid State powder was prepared from an intimate, stoichiometeric mix of oxide and carbonates, PbO, Bi_2O_3, $SrCO_3$, $CaCO_3$ and CuO.

All of the powders where calcined together in air, first at 730°C for 24 hours, then at 800°C for 48 hours (with intermediate grinding after 24 hour) and finally sintered at 842°C for 96 hours (with intermediate grinding every 24 hours).

X-ray powder diffraction measurements were carried out with a SCINTAG (X1) diffractometer (Cu $K\alpha$ radiation, $\lambda=1.5406$ Å) and with a Siemens D5000 Diffractometer. The mass content of the different phases and lattice constants of the 2212-phase (space group Cccm) were also determined by the Rietveld method (program FullProf [2]).

RESULTS

$Bi_{1.7}Pb_{0.4}Sr_{1.8}Ca_{2.2}Cu_{3.2}O_x$ Precursor Phase development. The X–ray diffraction traces for samples calcined up to 800°C are shown in figure 1 with the phase percentages for a number of samples, determined by Rietveld analysis shown in figure 2.

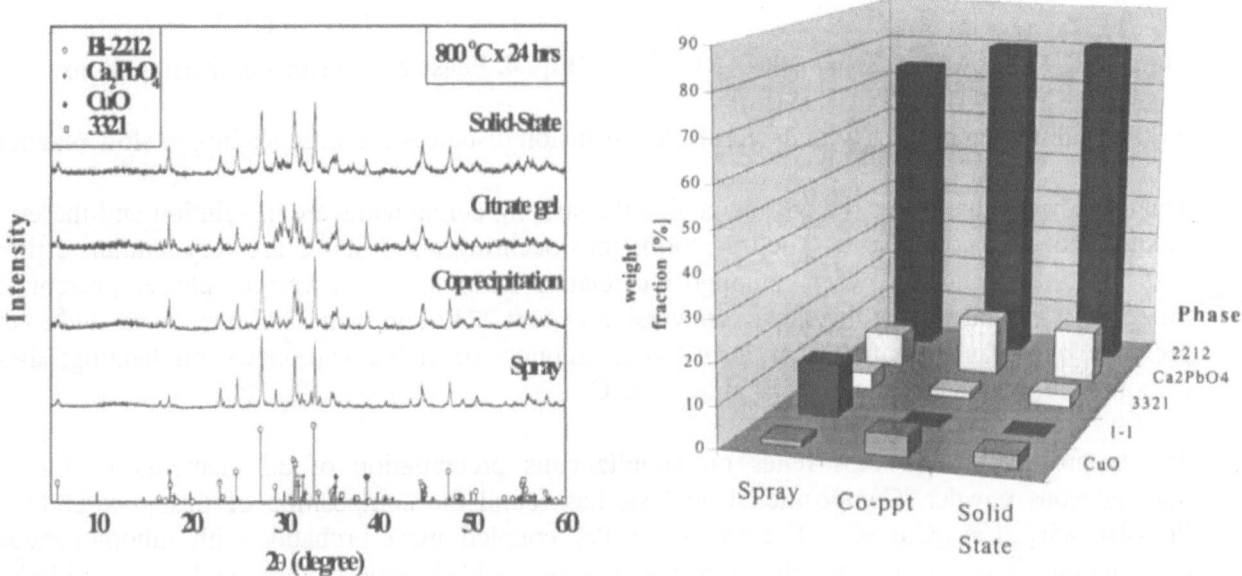

Fig. 1. XRD of powders calcined at 800°C   **Fig. 2.** Phase % for various powders

There are clear differences between the powders prepared using the different processes, co-precipitated powder having the most CuO and Ca_2PbO_4, spray pyrolysis the most $(Ca_{1-x}Sr_x)_yCuO_2$ phases (x ~0.5, y~0.85) [1-1 phase]. Citrate gel powder is thought to contain some B-Sr-Ca-oxide phases as minor components.

The development of the Bi-2223 phase was followed by sintering the samples at 842°C, as can be seen in figure 3 and figure 4. Reaction to the 2223 phase was followed, for simplicity, by using the ratio of the. XRD peak areas, Bi-2223 (115) / [Bi-2223 (115) + Bi2212 (113)]. Bi-2223 phase development clearly occurs far faster for the spray pyrolysis powder than for all the others, forming at least twice as much 2223 in the first 24 hours than co-precipitated powder.

DISCUSSION AND CONCLUSIONS

The route to forming the 2223 phase is highly dependent upon the method of preparing the precursor. Although the solid state route offers the simplest route to 2223, the reaction times required are very long in sintered samples, over 100 hours. The primary particle size of the raw materials is relatively large, the homogeneity of the powder critically dependent upon the

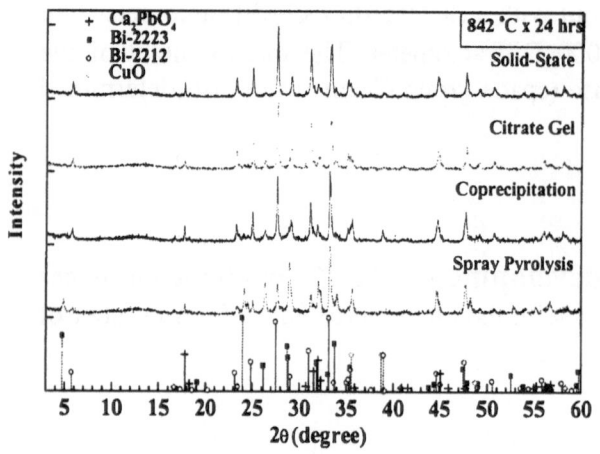

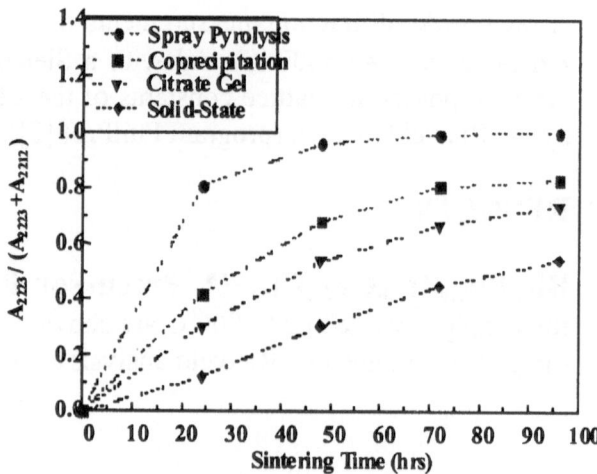

Fig. 3. XRD of powders calcined at 842°C

Fig. 4. Phase conversion comparing all powders

mechanical mixing process with the result that diffusion distances are large leading to slow reaction.

The citrate gel route offers the advantage that the starting components are in solution and therefore mixing is on the atomic scale. The transformations occuring up to 800°C are very similar to those seen in co-precipitated powder, although the relative amounts of the various phases present are different. It is thought that the slower conversion to first 2212 and then 2223 may be related to the presence of organic moitites that form small amounts of stable carbonates on heating; these carbonates only slowly decomposing above 800°C.

The co-precipitation process relies on simultaneous precipitation of all elements to form a homogeneous powder. Compositional analysis has found the composition of this powder to be $Pb_{0.42}Bi_{1.79}Sr_{1.65}Ca_{1.88}Cu_{3.57}O_x$. The excess of Pb, coupled most probably with inhomogeneous precipitation, is responsible for the large amounts of Ca_2PbO_4 seen in the powders at and below 800°C and may also be responsible for the relatively fast conversion of 2212 to 2223.

The spray pyrolysis process produces powder that transforms to the 2223 phase far faster than the others, reacting almost twice as fast as co-precipitated powder in the first 24 hours. This is due to the highly homogeneous, fine grain (primary particle) nature of this powder. The precursor formed at 800°C, containing 1-1 and smaller amounts of Ca_2PbO_4 may be more suitable for faster 2223 formation because of small amounts of Pb within the 2212 grains.

In conclusion, powders prepared by the spray pyrolysis process are more reactive than those produced by the other routes. Assuming that the faster transformation can be controlled in OPIT tape manufacture, significant manufacturing cost savings could be achieved using powders made by this process.

Acknowledgements. The Authors are grateful to Prof. W.W Schmahl and S. Räth (Ruhr-Universität, Bochum, Germany) for the Rietveld assessment of these powders.

---

* Author to whom correspondance should be addressed

[1]   Rupich M W *1998 International Workshop on Superconductivity*, Okinawa, Japan

[2]   FullProf Version 3.5b (1998), J. Rodriguez-Carvajal Laboratoire Leon Brillouin (CEA-CNRS)

ESR Study of the Quasi One-Dimensional Cuprate $(Ca_{1-x}Y_x)_{0.82}CuO_2$

Y. Miyazaki,[*] N. C. Hyatt, P. A. Anderson, and P. P. Edwards

School of Chemistry, University of Birmingham, Edgbaston, Birmingham B15 2TT, United Kingdom

Abstract: An electron spin resonance (ESR) study has been performed on polycrystalline samples of the quasi one-dimensional cuprate $(Ca_{1-x}Y_x)_{0.82}CuO_2$ with $0 \leq x \leq 0.435$, in which the formal Cu valence varies from +2.36 to +2.00. The samples with $0 \leq x \leq 0.20$ exhibit broad peaks at around 30 K in their DC magnetic susceptibility, while the samples with $0.20 < x \leq 0.435$ show an antiferromagnetic transition below 29 K. We have found that the magnetic behavior of the compounds with $0 \leq x \leq 0.20$ is accurately described by a contribution from alternating Heisenberg chains (ACH), and a Curie-Weiss (CW) component. After subtracting the CW contribution, the compounds show typical one-dimensional behavior with spin symmetric exchange interactions. Most of the doped hole carriers appear to be localized and dilute the long-range spin correlations along the chain.

Keywords: ESR, Cuprates, Low-dimensional magnetism

INTRODUCTION

Compounds of composition $(Ca_{1-x}Y_x)_{0.82}CuO_2$ comprise solely edge-shared one-dimensional (1D) CuO_2 chains and accommodate a wide range of formal Cu valence (Cu^{n+} = 2.36 - 2.00) dependent on the yttrium concentration x ($0 \leq x \leq 0.435$). Recent studies [1,2] have shown that the compounds undergo a distinctive evolution of magnetic behavior, from a 1D alternating Heisenberg state, to a 3D antiferromagnetically ordered state, upon increasing x (equivalent to decreasing Cu^{n+}). Since the structural differences are very small throughout the doping range, the hole concentration evidently plays a key role in determining the electronic structure of the compound system. However, no detailed magnetic study of the solid solution has been reported. We have investigated the magnetic properties of samples over the whole range of x. Here we report the results of ESR and DC susceptibility measurements performed through the whole doping-region, from the 1D to the 3D ordered regimes.

EXPERIMENTAL

Samples were prepared using standard ceramic methods from high-purity $CaCO_3$, Y_2O_3 and CuO. Details of the preparation procedure have been described in a previous report [2]. The DC magnetic susceptibility of the powdered samples was measured using a Cryogenics S100 SQUID magnetometer under a magnetic field of 1000 Oe. ESR measurements were performed on the powdered samples using a Bruker ESP 300 spectrometer operating at a frequency of 9 GHz and temperatures from 296 to 4 K.

RESULTS AND DISCUSSION

At 296 K, we observed broad isotropic ESR spectra, with a typical g-value of g_{iso} = 2.06(2) and a peak-to-peak line width, ΔH_{pp}, of 870 ± 40 Oe for all the samples. Both the $g_{iso}(T)$ and $\Delta H_{pp}(T)$ values were almost temperature independent from 296 K to 100 K. Below 50 K, all the samples showed a negative shift in g_{iso} values depending on their composition. Their behavior can be well described in terms of spin symmetric exchange interactions. A detailed study of the $g_{iso}(T)$ and $\Delta H_{pp}(T)$ behavior will be presented in a separate paper. In Figs. 1 (a) and (b), we show the integrated ESR intensity, I_{ESR}, against temperature, together with the DC susceptibility, $\chi_{DC}(T)$ for the samples with $x = 0$ ($Cu^{+2.36}$) and $x = 0.20$ ($Cu^{+2.20}$). Following the analysis widely adopted for low-dimensional cuprates [3], we assume that $\chi_{DC}(T)$ consists of three contributions:

$$\chi_{DC}(T) = \chi_0 + \chi_{CW}(T) + \chi_{ACH}(T), \quad (1)$$

where χ_0 is a temperature independent term, and $\chi_{CW}(T)$ is the Curie-Weiss (CW) contribution and $\chi_{ACH}(T)$ is the component derived from alternating Heisenberg chains (ACH) [4]. The expression of the $\chi_{CW}(T)$ and $\chi_{ACH}(T)$ terms is presented in ref [5]. As is clearly seen, the $I_{ESR}(T)$ value is not proportional to the DC susceptibility, $\chi_{DC}(T)$ for both samples. However, after subtracting the $\chi_{CW}(T)$ (dot and dashed line) and χ_0 contributions, the remainder $\chi_{ACH}(T)$ curves (dotted line) are satisfactorily superimposed on the I_{ESR} values (normalized at 296 K), except at very low temperatures. We have extended this analysis to the other samples lying in the 1D region and confirmed that all the samples show a similar relationship between the $\chi_{ACH}(T)$ and $I_{ESR}(T)$ curves.

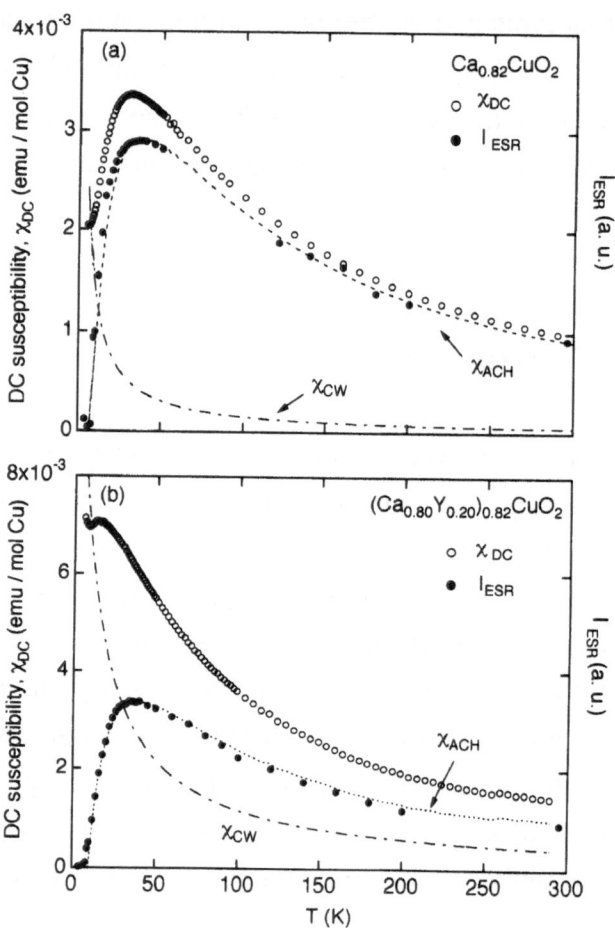

Fig. 1 Temperature dependence of the ESR integrated intensity, I_{ESR}, DC magnetic susceptibility, χ_{DC}, and the fitted curves for $(Ca_{1-x}Y_x)_{0.82}CuO_2$ with (a) $x = 0$ and (b) $x = 0.20$.

Next, we determined the spin numbers contributing to the $\chi_{CW}(T)$ and $\chi_{ACH}(T)$ terms, N_{CW} and N_{ACH}, per formula unit. These spin numbers have been obtained from the fitted parameters on each $\chi_{DC}(T)$ curve [5]. In Fig. 2, we show the N_{CW} and N_{ACH} values against Cu^{n+}. Over the 1D region, the sum of the N_{ACH} and N_{CW} reasonably corresponds to the total spin numbers. For example, the sample with $x = 0.20$ ($Cu^{n+} = 2.20$) has 0.12(1) mol/f.u. of CW spins and 0.65(1) mol/f.u. of ACH spins. The obtained total spin number, $N_{ACH} + N_{CW} = 0.77(2)$ mol/f.u., is very close to N_{total} of 0.80. Since all the samples are electrical insulators, the doped hole carriers would form localized $S = 0$ spin singlets and appear to dilute the spin interactions along the chain. Most of the CW contribution to the $\chi_{DC}(T)$ is thus considered to derive from spin fragments, presumably located next to these singlets, which are not detected in our ESR experiment. In order to rationalize the spin dynamics of this system, a further detailed study using high-quality single crystals is currently underway.

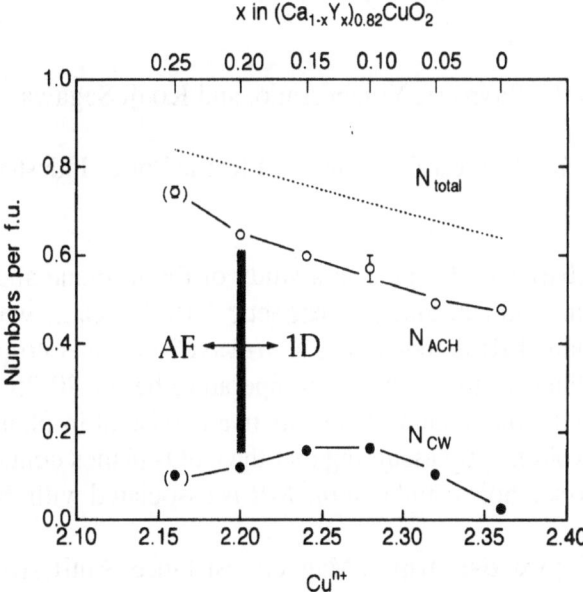

Fig. 2 The derived spin numbers per formula unit (f.u.) of the Curie-Weiss (CW) and alternating Heisenberg chain (ACH) contribution, N_{CW} and N_{ACH}, plotted against the nominal Cu valence.

Acknowledgment

The authors thank T. Green for his technical support and assistance during the ESR measurements. This work was supported, in part, by the JSPS (Japan Society for the Promotion of Science) Post Doctoral Fellowships for Research Abroad to Y. M.

---

* Present address: Department of Applied Physics, Graduate School of Engineering, Tohoku University, Aoba 08, Sendai 980-8579, Japan

1. A. Hayashi *et al.*, Phys. Rev. B **58**, 2678 (1998).
2. Y. Miyazaki *et al.*, Chem. Eur. J. **5**, 2265 (1999).
3. M. Kato et al., Physica C **258**, 284 (1996).
4. J. W. Hall *et al.*, Inorg. Chem. **20**, 1033 (1981).
5. Y. Miyazaki *et al.*, submitted to Phys. Rev. Lett.

Manifestations of the Charged Stripes in the Magnetoresistance of Heavily Underdoped YBa$_2$Cu$_3$O$_{6+x}$

A. N. Lavrov*, Yoichi Ando, and Kouji Segawa

Central Research Institute of Electric Power Industry, 2-11-1 Iwado-kita, Komae, Tokyo 201-8511, Japan

Abstract: We present a study of the in-plane and out-of-plane magnetoresistance (MR) in heavily-underdoped, antiferromagnetic YBa$_2$Cu$_3$O$_{6+x}$, which reveals a variety of striking features. The in-plane MR demonstrates a "d-wave"-like anisotropy upon rotating the magnetic field H within the ab plane. With decreasing temperature below 20-25 K the system acquires memory: exposing a crystal to the magnetic field results in a persistent in-plane resistivity anisotropy. The overall features can be explained by assuming that the CuO$_2$ planes contain a developed array of stripes accommodating the doped holes, and that the MR is associated with the field-induced topological ordering of the stripes.

Keywords: Stripes, Magnetoresistance, Antiferromagnetic, YBaCuO

INTRODUCTION

In high-T_c cuprates the conducting state appears as a result of hole or electron doping of the parent antiferromagnetic (AF) insulator. The tendency of doped holes to segregate may give rise to an intriguing microscopic state with carriers gathered within an array of quasi-1D "stripes" separating AF domains [1-3]. An ordered striped structure has been observed [4] in La$_{1.6-x}$Nd$_{0.4}$Sr$_x$CuO$_4$ and in La$_2$NiO$_{4.125}$, while most superconducting cuprates demonstrate incommensurate magnetic fluctuations [5] which can be considered as dynamical stripe correlations [1]. Dynamical or static stripes might be responsible for the peculiar normal state of cuprates as well as for the occurrence of superconductivity [1], but still very little is known about the electron dynamics in the stripes.

In this paper we report an extraodinary behavior of the magnetoresistance (MR) in antiferromagnetic YBa$_2$Cu$_3$O$_{6+x}$, which provides evidences that conducting stripes actually exist in CuO$_2$ planes and have a considerable impact on the electron transport.

EXPERIMENTAL METHODS

The high-quality YBa$_2$Cu$_3$O$_{6+x}$ single crystals were grown by the flux method in Y$_2$O$_3$ crucibles, and a high-temperature annealing was used to reduce their oxygen content. The MR was measured by sweeping the magnetic field at fixed temperatures stabilized by a capacitance sensor with an accuracy of ~1 mK. The angular dependence of the MR was determined by rotating the sample within a 100° range under constant magnetic fields up to 16 T.

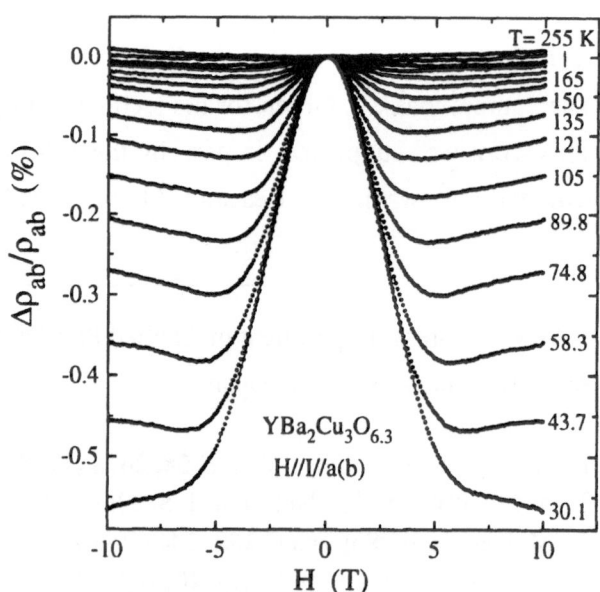

Fig. 1. Longitudinal in-plane MR of YBa$_2$Cu$_3$O$_{6.3}$. The data are averaged over several field sweeps.

RESULTS AND DISCUSSION

The $YBa_2Cu_3O_{6+x}$ crystals even being located deep in the AF range of the phase diagram ($x \sim$ 0.3) are far from conventional insulators: the in-plane resistivity ρ_{ab} remains "metallic" at high T and at low T it grows slower than expected for the hopping electron transport [6,7]. These AF crystals demonstrate an unusual behavior of the in-plane MR, $\Delta\rho_{ab}/\rho_{ab}$, when the magnetic field H is applied along the CuO_2 planes, Fig.1. At weak fields, the longitudinal in-plane MR $[H//I//ab]$ is negative and follows roughly a T-independent ζH^2 curve, but abruptly saturates above some threshold field. The threshold field and the saturated MR value gradually increase with decreasing temperature. The MR anomaly becomes noticeable near the Néel temperature $T_N \approx 230$ K, but evolves rather smoothly through T_N, which indicates that the long-range AF order itself is not responsible for its origin.

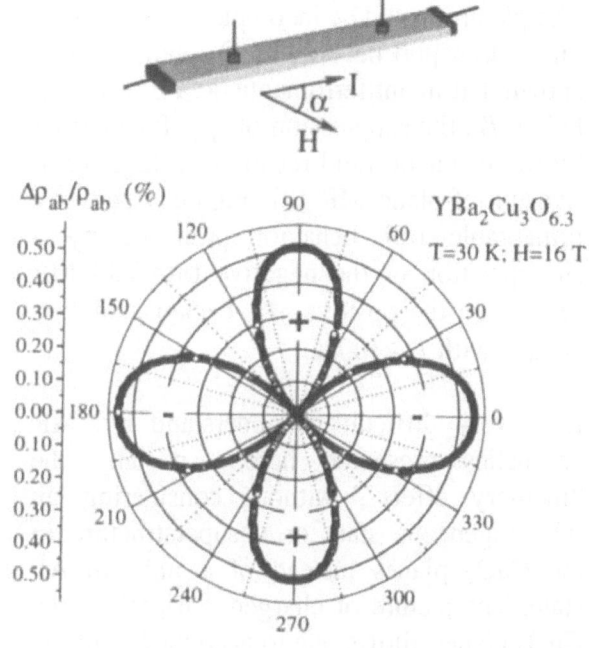

Fig. 2. Angular dependence of the MR (H//ab; H=16 T); the sign of MR is indicated.

When the magnetic field is turned in the plane to become perpendicular to the current $[H//ab; H \perp I]$, the low-field MR term just switches its sign, retaining its magnitude and the threshold-field value [7]. MR measurements performed upon rotating H within the ab plane reveal a striking anisotropy with a "d-wave"-like symmetry, i.e. $\Delta\rho_{ab}/\rho_{ab}$ changes from negative at $\alpha=0$ to positive at $\alpha=90°$, being zero at about 45°, Fig.2. It is worth noting that the low-field MR feature is not observed at all when the magnetic field is applied along the c-axis.

The most intriguing peculiarity of the low-field MR appears at temperatures below ~25 K, where the H-dependence of ρ_{ab} becomes irreversible. Figure 3 shows the low-field MR term measured for $H \perp I$ (for clarity, the background MR, γH^2, determined at high fields is subtracted: $\Delta\rho_{ab}/\rho_{ab}=(\Delta\rho_{ab}/\rho_{ab})^*+\gamma H^2$). Initially the irreversibility appears as a small hysteresis on the MR curve, but upon cooling to 10 K it becomes much more pronounced (the MR peaks are shifted from $H=0$ and strongly suppressed). We note that the first field sweep which starts at $\Delta\rho_{ab}/\rho_{ab}=0$ differs significantly from the subsequent ones. The salient point here is that the resistivity does not return to its initial value after removing the magnetic field; hence, the system acquires a memory. The application of the magnetic field at low T introduces a persistent resistivity anisotropy to the CuO_2 planes.

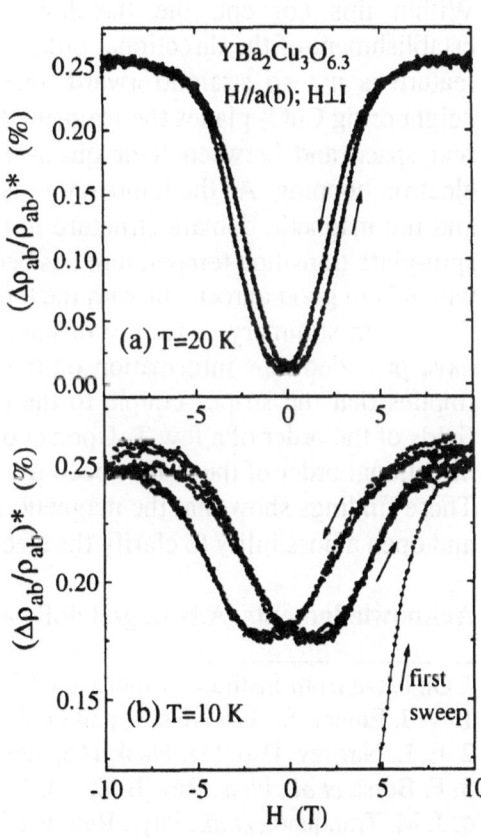

Fig. 3. The low-field MR component. Each curve contains data of 4 field sweeps performed at a rate of 1 T/min.

The picture would be incomplete without data on the transport between CuO_2 planes. It was shown that in antiferromagnetic $YBa_2Cu_3O_{6+x}$ below T_N, the suppression of spin fluctuations by the magnetic field results in a large positive out-of-plane MR [8]. Figure 4 shows a remarkable MR behavior produced by a superposition of the negative low-field MR feature on the positive γH^2 background in a sample with $T_N \geq 300$ K.

It is very difficult to understand the MR anomalies presented here, especially the "memory effect", without considering an inhomogeneous state or a superstructure in the CuO_2 planes instead of a uniform AF state. The picture of charged "stripes" in the CuO_2 planes allows one to account for all the

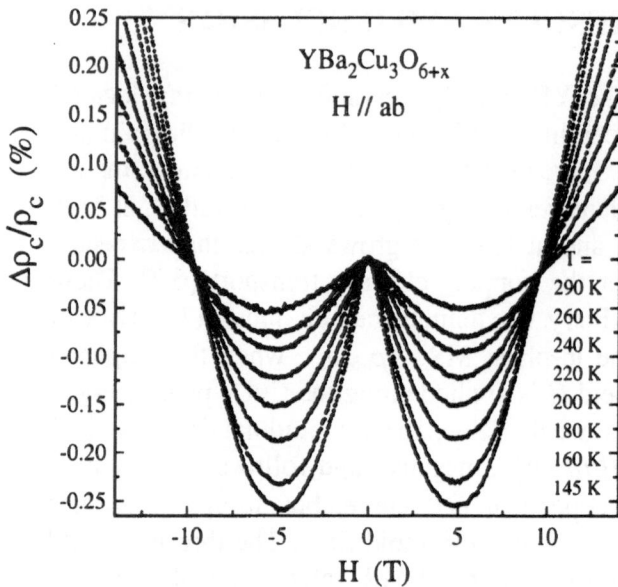

Fig. 4. Transverse out-of-plane MR of $YBa_2Cu_3O_{6+x}$.

observed MR peculiarities, by assuming that the magnetic field gives rise to a directional ordering of the stripes [7]. Actually, the aligning of stripes with confined carriers moving along would change the current paths and introduce the in-plane anisotropy. The rotation of stripes by the magnetic field gives an excellent explanation for the in-plane MR with the "d-wave"-shaped angular dependence. Within this concept, the threshold field of several Tesla is presumably coming from the establishment of the directional order of the stripes. Though an explanation of the out-of-plane MR feature is not so straightforward, one can imagine that by adjusting the direction of stripes in neighboring CuO_2 planes the magnetic field increases the overlapping both between the stripes in the real space and between their quasi-1D carriers in the k-space, enhancing the probability of the electron hopping. As the temperature is lowered, it is expected that the stripe dynamics slows down and the magnetic domain structure in the CuO_2 planes is frozen, forming a cluster spin glass. The spin-glass transition temperature has been reported to be about 20-25 K for the AF compositions [9], which is in good agreement with the temperature where the hysteretic MR behavior is found.

In summary, a variety of unusual MR features found in heavily underdoped $YBa_2Cu_3O_{6+x}$ have provided new information on the conducting "stripes" in the CuO_2 planes. The MR behavior implies that the stripes couple to the external magnetic field and undergo topological ordering at fields of the order of a few T. Upon cooling below ~20 K the dynamics of stripes gets slower and the directional order of the stripes becomes persistent, giving rise to a "memory effect" in the resistivity. These findings show that the magnetic field can be used as a tool to manipulate the striped structure and open a possibility to clarify the electron dynamics within the stripes.

Acknowledgments. A.N.L. gratefully acknowledges the support from JISTEC.

* On leave from Institute of Inorganic Chemistry, Lavrentyeva-3, 630090 Novosibirsk, Russia.

1. V. J. Emery, S. A. Kivelson, and O. Zachar, Phys. Rev. B **56**, 6120 (1997).
2. E. L. Nagaev, Usp. Fiz. Nauk **165**, 529 (1995).
3. F. Borsa *et al.*, Phys. Rev. B **52**, 7334 (1995).
4. J. M. Tranquada *et al.*, Phys. Rev. B **54**, 7489 (1996). *ibid.* **52**, 3581 (1995).
5. K. Yamada *et al.*, Phys. Rev. B **57**, 6165 (1998).
6. A. N. Lavrov, M. Yu. Kameneva, and L. P. Kozeeva, Phys. Rev. Lett. **81**, 5636 (1998).
7. Y. Ando, A. N. Lavrov, and K. Segawa, Phys. Rev. Lett. **83**, 2813 (1999).
8. A. N. Lavrov, Y. Ando, K. Segawa and J. Takeya, Phys. Rev. Lett. **83**, 1419 (1999).
9. Ch. Niedermayer *et al.*, Phys. Rev. Lett. **80**, 3843 (1998).

Evidence for One-Dimensional Charge Transport in the Stripe Ordered Phase of La$_{2-x-y}$Nd$_y$Sr$_x$CuO$_4$

Takuya Noda, Kaya Kobayashi, Hiroshi Eisaki and Shin-ichi Uchida

Department of Superconductivity, University of Tokyo, Hongo 7-3-1, Bunkyo-ku, Tokyo 113-8656, Japan

Abstract: Doping dependence of the Hall coefficient (R_H) is presented for La$_{1.4-x}$Nd$_{0.6}$Sr$_x$CuO$_4$ in the spin-charge stripe ordered phase. For $x \leq 1/8$, a remarkable decrease in the magnitude of R_H at low temperatures provides evidence for one-dimensional charge transport. For $x > 1/8$, the R_H remains relatively large in the stripe ordered phase. The results indicate a crossover from one- to two-dimensional charge transport taking place at $x = 1/8$. Zn-substitution for Cu recovers the large values of R_H, indicating that disorder caused by Zn-substitution in the stripe ordered phase is destructive of one-dimensional charge transport.

Keywords: Stripe, Hall effect, One-dimensionality, Crossover at 1/8, Zn impurity

INTRODUCTION

Stripe, the one-dimensional (1D) spin and charge density modulations in the two-dimensional (2D) CuO$_2$ planes, was proposed by Tranquada *et al.* to account for the anomalous behavior in La$_{2-x-y}$Nd$_y$Sr$_x$CuO$_4$ (LNSCO) [1], in which slight changes in the crystal structure leads to suppression of the superconductivity. There are stripe fluctuations in La$_{2-x}$Sr$_x$CuO$_4$ (LSCO) over a wide x range [2] and, even the static stripes have been observed in LSCO with $x < 1/8$ [3]. Furthermore, a possibility is suggested that Zn might stabilize the stripe ordered phase. It is reported that Zn enhances the suppression of superconductivity for carrier concentration ~ 1/8 in many cuprates [4].

Because of the lack of direct evidence for 1D, full consensus on the existence of spin-charge stripes as well as Zn-stabilized stripe order has not been established. The difficulty lies in that the charge stripes in the adjacent CuO$_2$ plane are directed by 90° as suggested from the neutron and x-ray scattering measurements [1] [5] and what we can see is the average of the two orientations.

The Hall effect measurement has the ability to detect the 1D charge dynamics. Suppose a bundle of independent 1D strings, then we should not expect any Hall effect, as the carriers cannot hop to the adjacent strings. Therefore, we might be able to obtain the direct evidence for or against 1D charge dynamics from the behavior of R_H, which is connected to the off-diagonal conductivity σ_{xy}. To investigate diagonal conductivity (σ_{xx}) and σ_{xy}, the in-plane resistivity and the Hall coefficient measurements were performed for La$_{1.4-x}$Nd$_{0.6}$Sr$_x$CuO$_4$ with $x = 0.10$, 0.12, 0.13 and 0.15 and for La$_{1.48}$Nd$_{0.4}$Sr$_{0.12}$Cu$_{1-z}$Zn$_z$O$_4$ with $z = 0$ and 0.04.

RESULTS AND DISCUSSIONS

R_H shows dramatic a change of doping and temperature (T)-dependences below the structural phase transition temperature, (T_0) (low temperature orthorhombic (LTO) - low temperature tetragonal (LTT)) (Fig.1), although in-plane resistivity does not show any remarkable change. For $x \leq 1/8$, R_H shows a rapid decrease, after showing a small discontinuous drop at T_0, approaching zero at low temperatures. This decrease has nothing to do with superconductivity, since it takes place well above resistive T_c, and the values are independent of magnetic fields. For $x=0.13$, with only 1% increase of doping, R_H does not show such a remarkable change below T_0. The decrease of R_H below T_0 is gradual and R_H preserves finite values even at the lowest temperature. For $x=0.15$, R_H continues to increase across T_0, ending up at the fairly large value, comparable with the value at 200K.

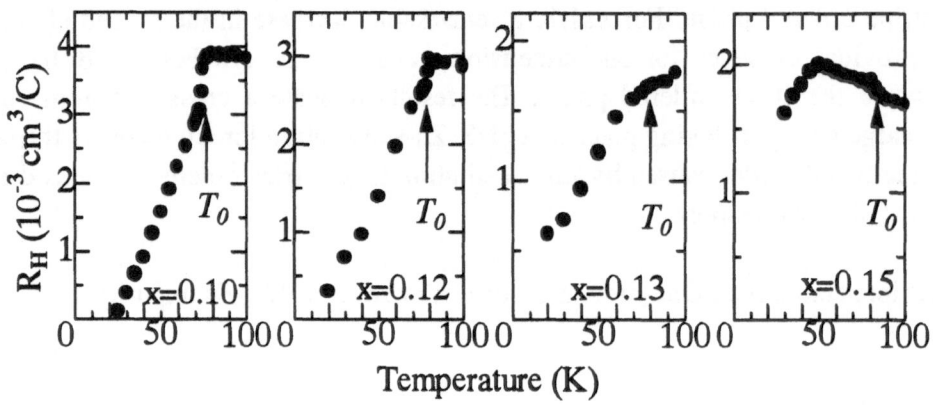

Fig. 1. Temperature dependence of R_H for LNSCO with y=0.6, x=0.10, 0.12, 0.13 and 0.15 (from right to left) measured at a magnetic field of 5T parallel to the c-axis with the current along the CuO_2 plane. Arrows indicate T_0.

The T-dependence of σ_{xx}, σ_{xy} for x=0.12, y=0.6 and σ_{xy} for x=0.12, y=0.4 are shown in Fig.2, which illustrates that σ_{xy} decreases approximately linearly in temperature below T_0, approaching to zero as T→0. The rapid decrease of R_H (and hence σ_{xy}) is robust against the out-of-plane disorder and thus an intrinsic property in the stripe ordered phase for $x \leq 1/8$. In the same figure, we overlay the x-ray intensity of charge order peaks (I_c) for x=0.12, y=0.40 [5]. $\sigma_{xy}(T)$ follows the T-dependence of I_c, which demonstrates that the charge order causes the suppression of the transverse motion of the carriers.

Apparently the radical reduction of R_H or σ_{xy} seen for $x \leq 1/8$ is associated with the suppression of the cyclotron motion caused by confinement of carriers within 1D charge stripes. By contrast, σ_{xy} and R_H for x=0.13 and 0.15 has finite values even at the lowest temperatures well below T_0. This indicates that the charge transport for $x > 1/8$ sustains 2D nature. In this sense, the observed change when we go through 1/8 is a 1D-2D dimensional crossover.

To the contrary, the reduction of R_H does not take place for the Zn-substituted compound (Fig.3), indicating that Zn destroys the confinement of carriers within static 1D charge stripes and recovers 2D nature. One may suppose that the recovery of R_H is because of complete disappearance of stripe order. This is unlikely because the resistivity of Zn-substituted LNSCO shows a small jump at a temperature close to T_0 of LNSCO, evidencing a robustness of the LTT phase and hence of the stripe order. From neutron scattering measurements on $La_{1.88}Sr_{0.12}Cu_{1-z}Zn_zO_4$, the magnetic correlation length and the onset-temperature of the static spin modulation for z=0.03 is lower than those for z=0, indicating that Zn-substitution degrades the coherence of stripe order [3]. This result is consistent with our results of R_H. In addition it is theoretically supported that holes tend to be accumulated close to the Zn sites so as to minimize the number of broken antiferromagnetic bonds. In such a case, each stripes could no longer regular, because the randomness of Zn-substitution is destructive of long range stripe order, even if Zn-substitution serve to stabilize the short range order. Then there might arise extra stripe path perpendicular to direction of the charge stripe.

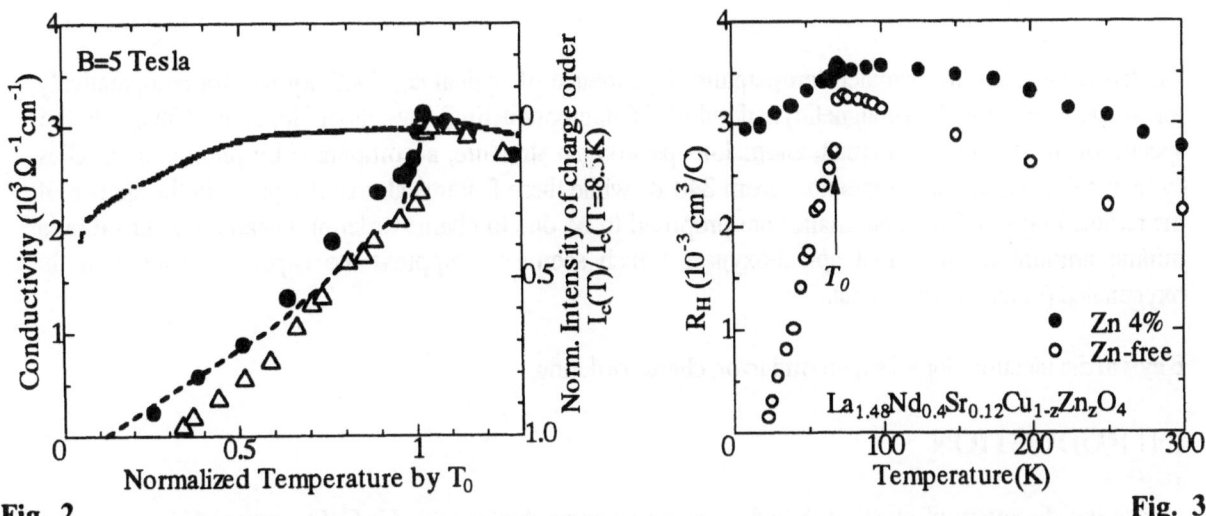

Fig. 2   **Fig. 3**

Fig. 2. Temperature dependence of σ_{xy} (closed circle), σ_{xx} (dots) for x=0.12, y=0.6 and σ_{xy} (open triangle) for x=0.12, y=0.4. In this plot σ_{xy} for y=0.6 and 0.4 is multiplied by 200 and 120, respectively. The intensity of the charge satellite peaks (I_c) for x=0.12, y=0.4 (dashed line) is superposed on the σ_{xy} data. Note that I_c is plotted with vertical axis upside down and temperature is normalized by T_0 (T_0=78K for x=0.12, y=0.6 and T_0=68K for x=0.12, y=0.4).

Fig. 3. Temperature dependence of R_H for $La_{1.48}Nd_{0.4}Sr_{0.12}Cu_{1-y}Zn_yO_4$ with z=0 and 0.04. Arrows indicate T_0.

CONCLUSION

We have investigated the doping and temperature dependences of the Hall coefficient in the stripe ordered phase of Zn-free and Zn-substituted LNSCO. We have demonstrated that the suppression of σ_{xy} for Zn-free LNSCO is correlated with the development of charge order for $x \leq 1/8$. This gives evidence for one-dimensional charge transport and thus for the presence of the charge stripes. We have observed the recovery of R_H for Zn-substituted LNSCO, indicating that the in-plane disorder in the stripe ordered phase is destructive of one-dimensional charge transport.

1. J. M. Tranquada *et al.*, *Nature* **375**, 561 (1995); J. M. Tranquada *et al.*, Phys. Rev. **B54**, 7489 (1996); J. M. Tranquada *et al.*, Phys. Rev. Lett. **78**, 338 (1997); J. M. Tranquada, N. Ichikawa and S. Uchida, to be published.

2. S. -W. Cheong *et al.*, Phys. Rev. Lett. **67**, 1791 (1991); T. E. Mason *et al.*, Phys. Rev. Lett. **71**, 919 (1993); K. Yamada *et al.*, Phys. Rev. **B57**, 6165 (1998).

3. T. Suzuki *et al.*, Phys. Rev. **B57**, 3229 (1998); H. Kimura *et al.*, Phys. Rev. **B59**, 6517 (1999).

4. M. Akoshima *et al.*, Phys. Rev. **B57**, 7491 (1998).

5. M. v. Zimmermann *et al.*, Europhys. Lett. **41**, 629 (1998); T. Niemöller *et al.*, cond-mat/9904383.

Optical Spectra in $Nd_{1.85}Ce_{0.15}CuO_{4+y}$ crystal: Implication of charge ordering

Y. Onose, Y. Taguchi, T. Ishikawa*, S. Shinomori, K. Ishizaka, and Y. Tokura

Department of Applied Physics, University of Tokyo, Tokyo 113-8656, Japan

Abstract: We have investigated temperature dependence of optical conductivity spectra comparatively for oxygenated (antiferromagnetic) and reduced (superconducting) crystals of $Nd_{1.85}Ce_{0.15}CuO_{4+y}$. In the spectra of the oxygenated crystal, anomalous pseudogap structure, accompanied by phonon anomalies, evolves with decreasing temperature from 340 K, while these features almost disappear in the spectra of the reduced crystal. These anomalies are proposed to be due to charge ordering instability induced by a minute amount of interstitial apical-oxygen, which seems to suppress the superconductivity in the oxygenated (or as-grown) crystal.

Keywords: electron-doped superconductor, charge ordering

INTRODUCTION

Since the discovery of electron-doped cuparate superconductor, $Nd_{2-x}Ce_xCuO_{4+y}$ system[1], an effect of oxygen-reducing procedure has long been a mystery. Superconductivity appears only for the appropriately reduced crystal with $0.14 < x < 0.17$, but never shows up in as-grown or oxygenated crystals[1]. In the as-grown or oxygenated crystal with $x=0.15$, antiferromagnetic order emerges below T_N=120-160 K [2-4]. It has been revealed by a recent neutron diffraction experiment[5] that a minute amount ($\Delta y \sim 0.02$) of interstitial apical oxygens, which should not be present in the ideal T'-structure, are removed by the reducing procedure. Therefore, apical oxygens as "impurities" seem to suppress the superconductivity and to induce the antiferromagnetism instead. To study the role of the apical oxygen, we have investigated temperature variation of optical spectra comparatively for both oxygenated and reduced single crystals of $Nd_{1.85}Ce_{0.15}CuO_{4+y}$. Only for the oxygenated crystal, we have found anomalous pseudogap formation accompanied by lowering of local lattice symmetry, which is similar to the anomalies observed in typical charge ordering systems.[6-8]

EXPERIMENTAL

$Nd_{1.85}Ce_{0.15}CuO_{4+y}$ single crystals studied in this work were grown by the traveling solvent floating-zone method in O_2 atmosphere of 4atm. To obtain superconducting samples, some pieces of crystal were annealed in flowing Ar gas for 100 hours at 1000 °C and subsequently in O_2 for 50 hours at 500 °C. Reflectivity spectra were measured on the polished *ab* face of the nonsuperconducting and the superconducting crystals with typical size of $6 \times 4 \times 2$ mm^3. In order to remove possible stress at the polished surface, we annealed the as-grown crystal in O_2 for 100 hours at 1000 °C and the reduced crystal at 500 °C in Ar for 50 hours after the polishing. In measurements of reflectivity spectra, temperature dependence of the reflectivity spectra was measured for 0.01-3 eV over the range of 10-390 K. The room temperature data for above 3-36 eV were used to perform the Kramers-Kronig analysis and deduce optical-conductivity spectra at respective temperatures. For the analysis, we assumed constant reflectivity or Hagen-Rubens relation below 0.01 eV and ω^{-4} extrapolation above 36 eV.

RESULTS AND DISCUSSION

We plot in the Fig. 1 the temperature dependence of resistivity for both the oxygenated and the reduced crystals. The resistivity for the oxygenated crystal decreases with lowering temperature at around 300K but shows an gradual upturn at around 200 K, while the resistivity for the reduced crystal monotonically decreases from 300 K to the superconducting transition temperature (T_c=25 K). It is worth noting that the resistivity for the oxygenated crystal is fairly low ($\rho \sim$2 m Ω cm) even at the lowest temperature (4.2 K), which suggests that there are still itinerant charge carriers in the antiferromagnetic phase.

We show reflectivity spectra of the oxygenated and the reduced crystals at 10 K and 290K in Fig. 2. In the spectrum of the oxygenated crystal, a broad peak is observed at around 2 eV, which is ascribed to a subtle trace of charge-transfer (CT) gap excitation from O $2p$-like states to Cu 3d-like states. Below 1 eV, a high-reflectance band arises, although the structures of the optical phonon are clearly discernible below 0.07 eV, implying poor dielectric screening. In the spectrum at 10K, a hollow structure at around 0.3 eV and also a broad peak structure at around 0.05 eV are present, which are responsible for a pseudogap-like structure (0.2-0.3 eV) and activated phonon structure (around 0.042 eV), respectively, in the optical conductivity spectrum (see Figs. 3). As for the reduced crystal, the remnant of CT excitations become less clear, and the high-reflectance metallic band below 1 eV is increased. Most importantly, both the hollow structure at around 0.3 eV and the broad peak structure at around 0.05 eV no longer exist at 10 K as well as at 290 K. The difference of reflectivity spectra at 290 K for the oxygenated and reduced crystals is ascribed partly to the increase of doping level induced by reducing procedure, as previously illustrated by the optical study at room temperature on reduced and unreduced Pr$_{2-x}$Ce$_x$CuO$_4$ crystal[9]. However, the fact that the hollow structure at around 0.3 eV and the broad peak structure at around 0.05 eV appear only in the spectrum of the oxygenated crystal at 10 K cannot be explained by the difference of doping level alone. Therefore, we should consider that a minute amount of interstitial apical oxygen drastically modifies electronic and local lattice structure in addition to changing of doping level.

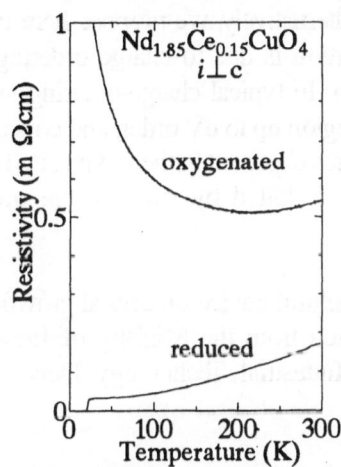

Fig. 1. The temperature dependence of in-plane resistivity for an oxygenated and a reduced crystal of Nd$_{1.85}$Ce$_{0.15}$CuO$_{4+y}$.

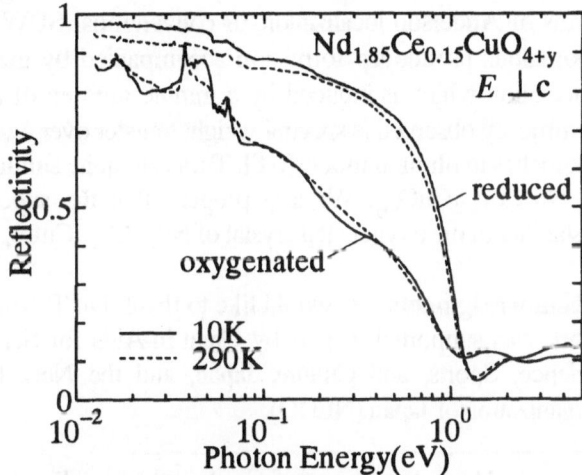

Fig. 2. Reflectivity spectra for an oxygenated and a reduced crystal at 10K and 290K.

Temperature dependence of the optical conductivity below 0.8 eV for the oxygenated crystal is shown in Fig. 3(b). Spectral weight between 0.1 eV and 0.3 eV is gradually transferred across the isosbetic (equal-absorption) point at 0.3 eV to a higher energy region ($\geqq$ 0.3 eV) with decreasing temperature below 340 K, making a gap-like structure. The spectral shape shows no more temperature-variation above 340 K. Therefore, it is likely that the pseudogap formation begins at around 340 K and gradually evolves below this temperature, although the gap does not open completely even at the lowest temperature.

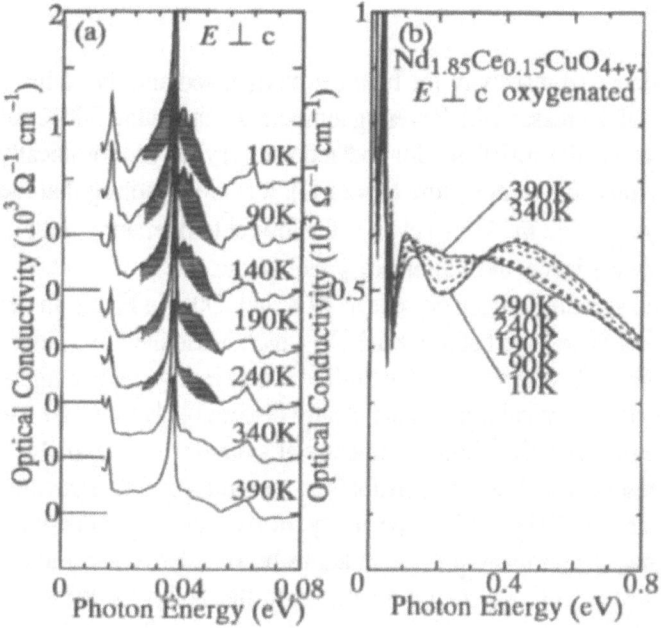

Fig. 3. Temperature dependence of optical conductivity spectra for an oxygenated crystal of $Nd_{1.85}Ce_{0.15}CuO_{4+y}$ (a) in a low energy region below 0.8 eV, showing the phonon anomaly, and (b) in a mid-infrared region, showing the pseudogap feature.

We plot in Fig. 3(a) the temperature variation of the optical conductivity spectra below 0.08 eV for the oxygenated crystal. In the in-plane polarization spectra of the ideal T'-structure, four phonon modes would be expected from the factor group analysis. Four phonon modes are actually observed, around 0.016, 0.037, 0.042, and 0.063 eV in the conductivity spectra above 340K. However, the broad band centered around 0.042 eV (hatched structure in Fig. 3(a)) grows in intensity with decreasing temperature from 340 K in this region, where some sharp modes are buried in, or interfere with, this broad band. Such an emergence of additional phonon modes implies a lowering of local lattice symmetry. These activated phonon modes are nearly absent above 340K and remarkably grow in intensity with decreasing temperature from 340K. The temperature dependence is quite parallel with the evolution of pseudogap.

The temperature variation of the spectra over an energy scale as large as 1 eV cannot be accounted for in terms of Anderson localization, or conventional SDW transition. Alternatively, we propose here that the anomalous pseudogap formation accompanied by the lattice distortion is due to charge ordering or its fluctuation which is induced by a minute amount of apical oxygen. In typical charge-ordering systems, commonly observed is spectral weight transfer over a wide energy region up to eV order, and concomitant anomalies in phonon modes[6-8]. These are quite similar to the features observed in the oxygenated crystal of $Nd_{1.85}Ce_{0.15}CuO_{4+y}$. We also propose that the superconductivity is killed by such charge ordering instability in the oxygenated crystal of $Nd_{1.85}Ce_{0.15}CuO_{4+y}$.

Acknowledgments. We would like to thank Dr. T. Kimura for his helpful advice on crystal growth. This work was supported in part by Grant-in-Aids for Scientific Research from the Ministry of Education, Science, Sports, and Culture, Japan, and the New Energy and Industrial Technology Development Organization of Japan (NEDO).

*present address: Department of Condensed Matter Physics, Tokyo Institute of Technology, Tokyo 152-8551, Japan

1. Y. Tokura, H. Takagi, and S. Uchida, Nature (London) **337**, 345 (1989).
2. G. M. Luke *et. al.*, Nature(London) **338**, 49 (1989).
3. S. Kambe, H. Yasuoka, H. Takagi, S. Uchida, and Y. Tokura, J. Phys. Soc. Jpn. **60**, 400 (1991).
4. M. Matsuda *et. al.*, Phys. Rev. B **45**, 12548 (1992).
5. A. J. Schultz, J. D. Jorgensen, J. L. Peng, and R. L. Greene, Phys. Rev. B **53**, 5157 (1996).
6. T. Katsufuji, T. Tanabe, T. Ishikawa, Y. Fukuda, T. Arima, and Y. Tokura, Phy Rev B **54**, R14230 (1996).
7. T. Ishikawa, S. K. Park, T. Katsufuji, T. Arima, and Y. Tokura, Phys. Rev. B **58**, R13326 (1998).
8. Y. Okimoto, Y. Tomioka, Y. Onose, Y. Otsuka, and Y. Tokura, Phys. Rev. B **57**, R9377 (1998).
9. T. Arima, Y. Tokura, and S. Uchida, Phys. Rev. B **48**, 6597 (1993).

Incommensurate Spin Fluctuations in YBCO*
- Charge and Spin Stripe Scenario -

Yasuo Endoh [1], Masatoshi Arai [2], Setsuko Tajima [3]

[1]Institute for Materials Research, Tohoku University, Katahira, Aoba-ku, Sendai 980-8577, Japan and CREST
[2]Institute of Materials Structure Science, KEK, Oho, Tsukuba, Ibaraki, 305-0011, Japan
[3]Superconducting Research Laboratory, ISTEC, Shinonome, Koto-ku, Tokyo 135-0062, Japan

Abstract : A brief summary of the recent inelastic neutron scattering studies on spin dynamics of the superconducting YBCO is given. Spin fluctuations in CuO_2 plane was experimentally shown to be of the incommensurate (IC) spin density wave state as those in LSCO. This new result suggests that the IC spin fluctuations is generic in the superconducting Cu oxides. The inherent relation to the high temperature superconductivity and the IC spin state is discussed on the STRIPE scenario.

Keywords : YBCO, incommensurate spin fluctuations, stripe, neutron scattering

INTRODUCTION

During the past decade extensive neutron scattering studies have been performed for the $YBa_2Cu_3O_{7-y}$ (YBCO) system of the best known high temperature superconductors. Among numerous experimental results from this YBCO system, essentially two important facts are revealed. Namely a prominent peak at 41 meV develops in the superconducting state for an optimally-doped crystal (y=0) at the (π,π) reciprocal position of 2 dimensional (2D) CuO_2 plane [1,2]. It was defined as the resonant peak. Then for an under-doped crystal with Tc=62 K shows incommensurate (IC) magnetic peaks at $(\pi, \pi\pm\delta)$ and $(\pi\pm\delta,\pi)$ [3],which is the same IC spin density wave (ISDW) structure in the $La_{2-x}Sr_xCuO_4$ (LSCO) superconductors [4]. This ISDW state has been interpreted by the 1D spin stripe model existed in the 2D CuO_2 lattice in the LSCO system [6]. Since the incommensurability, δ or the inverse spacing of the spin stripe is proportional to both the doping concentration of holes, n_h, and Tc in LSCO [7], the ISDW suggesting the stripe realization must be robust to the high temperature superconductivity (HTSC) at least in Cu oxide materials [8].

In this brief report, we describe the recent inelastic neutron scattering studies of spin dynamics in YBCO [8] and the results will be discussed in comparison with those of LSCO.

INELASTIC NEUTRON MAGNETIC SCATTERING FROM YBCO

Inelastic neutron scattering experiments were carried out on the chopper spectrometer, MARI installed at ISIS in the Rutherford Appleton Laboratory in UK. High quality single crystals were synthesized by the SRL-CR method at the Superconductivity Research Laboratory of ISTEC. The excellent crystal is of all single phase of YBCO. Then, we tried to control the oxygen concentration, y by annealing at high temperatures for an order of a month or more. The total volume of the sample for a series of the present experiments was about 30 g consisting of several pieces of the single crystalline slabs which have the mosaicness of 0.5 - 1.0 deg. The samples were characterized by the bulk measurement to determine Tc, which is 67 K($\pm$ 2.2 K) and 92 K($\pm$ 1 K), respectively for under- and optimally-doped crystals. Note that Tc was determined as the midpoint and ΔTc in the parenthesis is the temperature width of the transition.

The results for the under-doped sample were reported in the recent publication [9]. The results for the optimally-doped sample is reported here, though the data analysis is not completed yet at this stage. At 10 K we could identify a sharp resonance peak at 41 meV at (π,π) position, just below the strong scattering of phonon branch at around 43 meV. Since the resonant peak evolves below Tc, the peak is more clearly visible by subtracting 100 K data from the 10 K data as shown in Fig.1 of the scattering intensity contour map. The intensities in the lower energies below the resonant energy are negative, which essentially accords with the first report by Rossat-Mignod [1]. However the new result is shown that a double peak feature in q at lower energies clearly gives rise to the ISDW state as shown in Fig.2. The incommensurability, δ is larger than that of under-doped (Tc=67 K) crystal. The resonant peak at 100 K above Tc almost diminishes but it remains as weak IC peak structure as well.

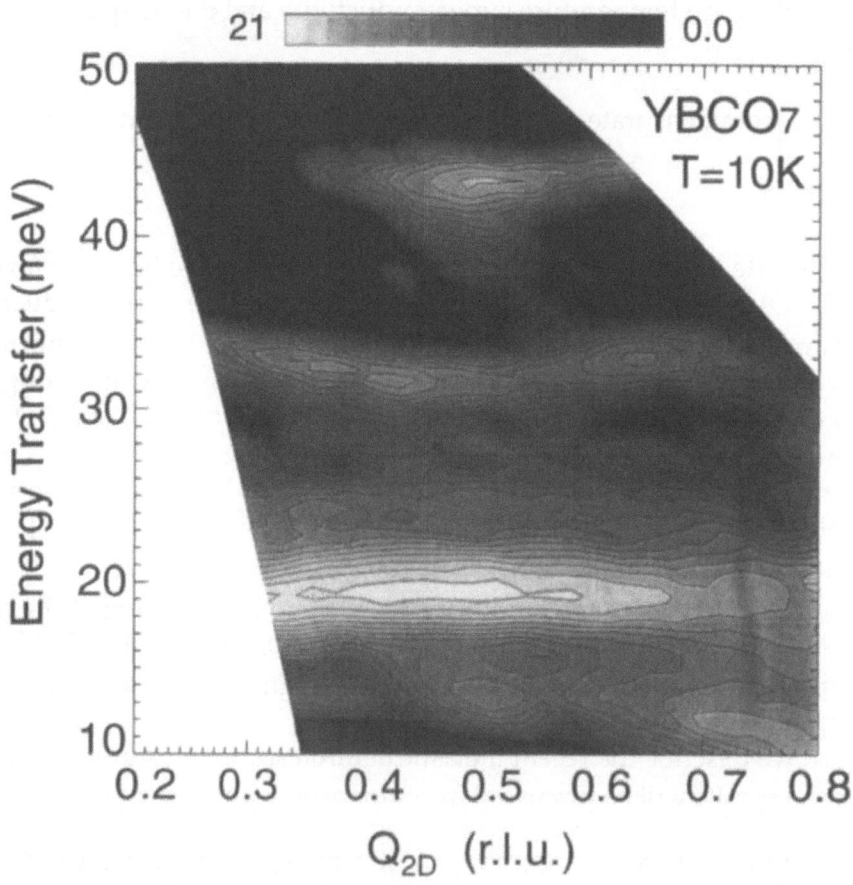

Fig.1   Scattering contour map spanned in (q_{2D},ω) space at 10 K. The data were taken by those of scan at 10 K subtracted from the 100 K scan. Note that scattering intensities around 28 meV are negative.

SPIN STRIPES

The results are discussed on the basis of the stripe scenario [6] providing a big issue for the elucidation of HTSC mechanism. The doping dependence of δ, or the spacing of the spin stripes strongly suggests that doped charges are confined in the 1D charge stripes forming a tweed pattern transformed from the IC structure. If the same anti phase magnetic structure across the charge stripe is formed, the spacing of the charge stripe should be a half of the spin stripes, or 2δ which corresponds to the half filling of charge ($\delta=x$). Spin excitations were found to be a quite generic feature in both LSCO and YBCO. Namely the IC structure maintains in high energies without a significant change of δ as well as q width. Thus we extend to apply an important conclusion from the spin dynamics in LSCO [7] that a linear relation between δ and Tc holds even in YBCO. We argue, then that the spacing of the spin stripes in the optimally doped YBCO is shorter than that of the optimally doped LSCO. This important result suggests that the number of the confined charges in the stripes is larger in YBCO, which may give rise to the higher Tc. It also gives a new phenomenological scale factor of 3/2 by direct comparison of the linear relationship of δ vs Tc in two systems.

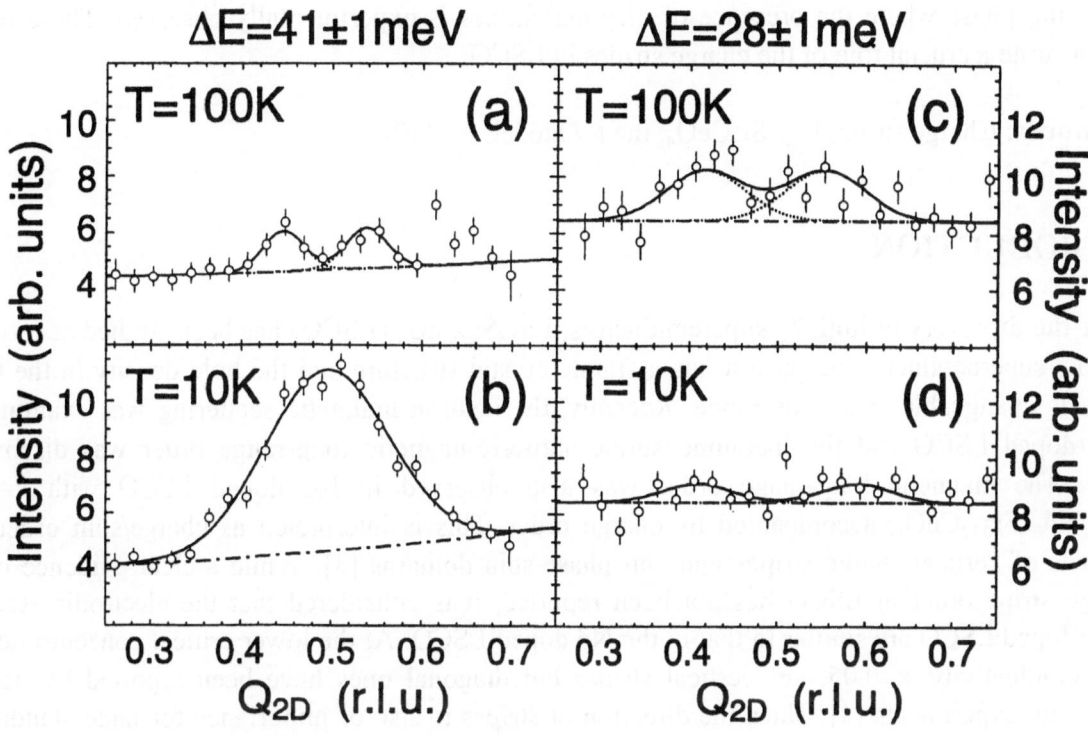

Fig.2 Typical q scan data at the fixed transferred energy of 28, 41 meV at 10 and 100 K.

Acknowledgement : The authors thank A.Garrett, T.Nishijima, K.Tokumoto, Y.Shiohara, C.Frost and S.Bennington for their great efforts in our collaboration. They also acknowledge H.Fukuyama, S.Maekawa, N.Nagaosa and T.Tohyama for stimulated discussions.

----

*This work has been supported by the CREST project sponsored by JST, and the Grant in Aid of the Scientific Research Project sponsored by Monbusho.

1. J.Rossat-Mignod et al., Physica C 185 - 189, 86 (1991)
2. H.F.Fong et al., Phys. Rev.Lett. 75, 316 (1995)
3. H.Mook et al., Nature. 395, 580 (1998)
4. S.Hayden et al., Phys. Rev.Lett. 66, 821 (1991)
5. V.J.Emery, S.Kivelson and H.Q.Lin, Phys. Rev.Lett. 64, 475 (1990)
6. K.Yamada et al., Phys.Rev.B 57, 6165 (1998)
7. S.Wakimoto et al, Phys.Rev.B to appear
8. M.Arai et al., Phys. Rev.Lett. 82, 821 (1999)

Charge Stripes and Electronic States in Underdoped La$_{2-x}$Sr$_x$CuO$_4$

Yasumasa Shibata, Takami Tohyama, Susumu Nagai,* and Sadamichi Maekawa

Institute for Materials Research, Tohoku University, Sendai 980-8577, Japan

Abstract: We investigate the electronic states of underdoped La$_{2-x}$Sr$_x$CuO$_4$ (LSCO) by the *t-t'-t''-J* model containing charge stripes. The numerically exact diagonalization calculation is employed on small clusters. The model with vertical stripes consistently explains the physical properties observed in the angle-resolved photoemission and the optical conductivity experiments near $x=0.12$, while the diagonal-stripe model does not explain them. Rather the latter seems to be applicable to the insulating phase where the presence of diagonal stripes is experimentally observed. These results demonstrate a crucial role of the charge stripes in LSCO.

Keywords: Charge stripe, La$_{2-x}$Sr$_x$CuO$_4$, the *t-J* model, ARPES

INTRODUCTION

Since the discovery of high T_C superconductors, La$_{2-x}$Sr$_x$CuO$_4$ (LSCO) has been studied as a typical cuprate superconductor, because it has a simple crystal structure and the hole density in the CuO$_2$ plane is changeable in a wide range. Recently, the neutron magnetic scattering was examined in underdoped LSCO and the incommensurate antiferromagnetic long-range order was discovered [1,2]. The magnetic long-range order was also observed in Nd doped LSCO with $x=0.12$, La$_{1.48}$Nd$_{0.4}$Sr$_{0.12}$CuO$_4$, accompanied by charge order. This is interpreted as charge/spin order that consists of vertical charge stripes and anti-phase spin domains [3]. While a clear evidence of the charge stripe order in LSCO has not been reported, it is considered that the electronic states in underdoped LSCO are similar to that in the Nd doped LSCO. At the lower critical concentration for superconductivity, x=0.05, not vertical stripes but diagonal ones have been reported by neutron scattering experiments [4]. Thus, the direction of stripes is also of importance for understanding of the electronic states of LSCO. Recent angle-resolved photoemission spectroscopy (ARPES) experiment on underdoped LSCO has also shown anomalous doping dependence [5]: The spectrum near $(\pi/2,\pi/2)$ along the $(0,0)$-(π,π) direction shows a broad peak in insulating and overdoped samples, while the spectrum in underdoped samples near $x=0.12$ has no peak structure near the Fermi level. This is in contrast with underdoped Bi$_2$Sr$_2$CaCu$_2$O$_{8+\delta}$ (Bi2212) where a sharp peak appears near $(\pi/2,\pi/2)$ [6].

In this paper, the electronic states of LSCO are examined in terms of the effect of stripes on various excitation spectra. By using a microscopic model with realistic parameters for LSCO, we find that vertical stripes consistently explain characteristic features in the ARPES spectrum as well as the optical conductivity of underdoped LSCO [7]. In contrast, the model with diagonal stripes do not explain the underdoped ARPES spectrum, but is qualitatively consistent with the spectrum in insulating phase where the diagonal stripes are experimentally observed.

VERTICAL STRIPES

As a microscopic model describing the CuO_2 plane, we employ the t-t'-t''-J model, where t' and t'' are second and third nearest-neighbor hoppings, respectively. In LSCO, we estimated the ratio t'/t and t''/t to be -0.12 and 0.08, respectively, and $J/t=0.4$ [7]. To obtain the excitation spectra, the exact diagonalization calculation is performed on small clusters. The tendency toward the stripe instability is modeled by introducing a configuration-dependent stripe potential V_s into the clusters. The magnitude of V_s is assumed to depend on the number of holes n_h in each column along the y-direction of the lattice. $V_s(n_h)$ in each column is assumed that $V_s(0)= V_s(1)=0$, $V_s(2)=-2V$, and $V_s(3)=-3V$, being $V>0$.

Figures 1 shows the single-particle spectral function $A(\mathbf{k},\omega)$ on a 18-site cluster with 2 holes. When there is no stripe potential [Fig. 1(a)], quasi-particle (QP) peaks with large weight are seen below and above the Fermi level at $(\pi/3,\pi/3)$ and $(2\pi/3,2\pi/3)$, respectively. For $V/t=1$ [Fig. 1(b)], however, there is no distinct QP peak and the spectra become more incoherent. This is consistent with the ARPES results for $x=0.12$ [5]. In contrast to the spectra along $(0,0)$ to (π,π), the $(2\pi/3,0)$ spectrum remains sharp with the

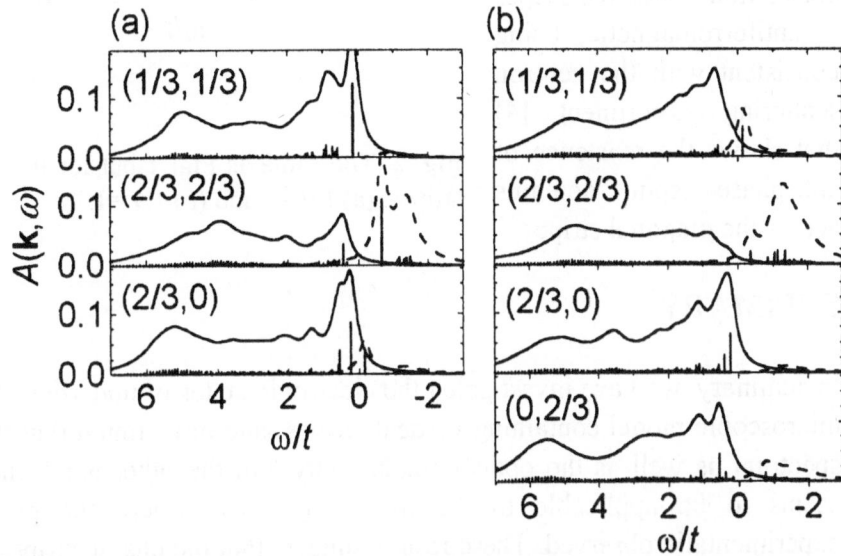

Fig. 1. Single–particle spectral function $A(\mathbf{k},\omega)$ for the 18-site t-t'-t''-J model without (a) and with (b) vertical stripes ($V/t=1$). The solid and dashed curves represent the electron-removal and addition spectra, respectively. The momentum is measured in units of π.

lowest electron-removal energy. This feature is explained as a precursor of the localization of carriers along the direction perpendicular to the stripes [7,8]. Correspondingly, the Drude weight along this direction is strongly suppressed. However, the optical conductivity in the midinfrared region is enhanced [7]. This is consistent with the fact that the midinfrared absorption at around ω~0.27 eV in LSCO is enhanced as compared with those of $YBa_2Cu_3O_{6.6}$ and Bi2212 [9]. Along the stripes, large Drude weight indicating one-dimensional metallic behavior is obtained [7]. We note that the vertical stripe model also explains anti-phase spin domains observed by the neutron scattering experiments.

DIAGONAL STRIPES

In order to see whether the broad spectral feature along the $(0,0)$-(π,π) direction is intrinsic to the vertical stripes, we examine the effect of diagonal stripes as shown in Fig. 2. Here, a 4x4 cluster and the t-J model with $J/t=0.4$ are used to preserve the ground state with total momentum of $(0,0)$. A diagonal stripe potential is introduced, similar to Fig. 1 but with diagonal direction of hole pairs. Even if the stripe potential is turned on, the $(\pi/2,\pi/2)$ spectrum remains sharp. This is because the potential does not change the nature of diagonal hole-pair configuration dominated in the ground state. Therefore, the diagonal charge stripes do not explain the broadness along the $(0,0)$-(π,π) direction. Rather, they may be associated with enhanced spectral weight along the direction at the critical concentration of $x=0.05$ [5]. We note that the spin correlation across the stripes is antiferromagnetic, being consistent with the neutron scattering experiment [4] that shows the existence of anti-phase spin domains across the diagonal stripes.

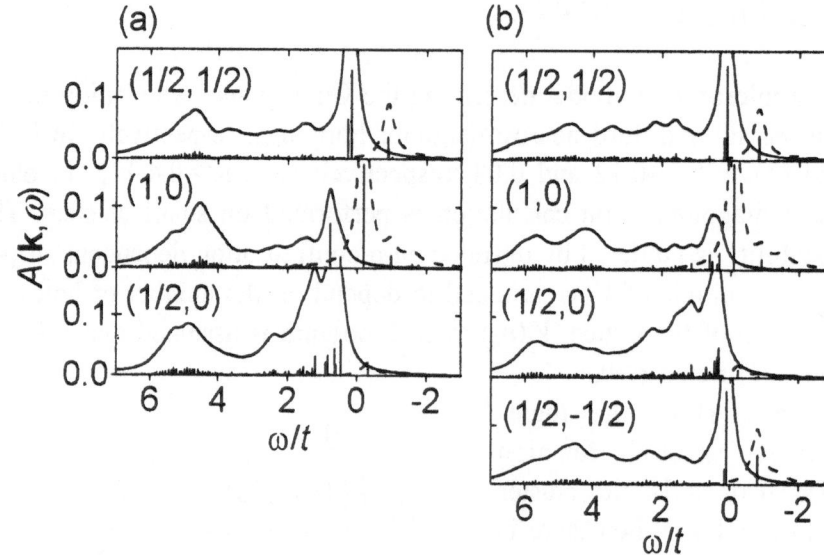

Fig. 2. The same as Fig. 1 but for the 16-site t-J model with diagonal stripes. (a) $V/t =0$ and (b) $V/t =1$.

SUMMARY

In summary, we have investigated the electronic states of underdoped LSCO. We have introduced a microscopic model containing vertical stripes, and have found that the model explains the ARPES spectrum as well as the optical conductivity. On the other hand, the model with diagonal stripes seems to be applicable to the insulating phase where the presence of diagonal stripes is experimentally observed. These results suggest that the charge stripes are an essential ingredient for the explanation of the physical properties of LSCO, although the d-wave superconductivity is suppressed by the vertical stripes [7].

Acknowledgements. This work was supported by CREST, NEDO, and Priority-Areas Grants from the Ministry of Education, Science, Sports and Culture of Japan. The parts of the numerical calculation were performed in the Supercomputer Center in ISSP, University of Tokyo, and the supercomputing facilities in IMR, Tohoku University.

* Present address: NTT DATA corp., 2-11-17 Tsukiji, Chuo-ku, Tokyo 104-0045, Japan
1. T. Suzuki *et al.*, Phys. Rev. B **57**, R3229 (1998).
2. H. Kimura *et al.*, Phys. Rev. B **59**, 6517 (1999).
3. J. M. Tranquada *et al.*, Nature **375**, 561 (1995).
4. S. Wakimoto et al., Phys. Rev. B **60**, R769 (1999).
5. A. Ino *et al.*, J. Phys. Soc. Jpn. **68**, 1496 (1999); cond-mat/9902048
6. C. Kim *et al.*, Phys. Rev. Lett. **80**, 4245 (1998).
7. T. Tohyama *et al.*, Phys. Rev. Lett. **82**, 4910 (1999).
8. M. I. Salkola *et al.*, Phys. Rev. Lett. **77**, 155 (1996).
9. S. Tajima *et al.*, Europhys. Lett. **47**, 715 (1999).

SO(5) Multicritical Phenomena of Superconductivity and Antiferromagnetism in Organic Superconductors

Shuichi Murakami* and Naoto Nagaosa

Department of Applied Physics, University of Tokyo, Bunkyo-ku, Tokyo 113-8656, Japan

Abstract: We study theoretically the multicritical phenomena of the superconductivity (SC) and antiferromagnetism (AF) in κ-BEDT salts. This system is in the strong coupling regime both for the SC and the AF, and the critical fluctuation is observed within $T - T_c \leq 10K$. The phase diagram and the NMR relaxation rate $1/T_1$ is analysed in terms of the renormalization group method, and the following results are obtained; (i) the bicritical phenomenon observed experimentally indicates the rotational symmetry, i.e., SO(5) symmetry, within the 5-dimensional order paramter space of the SC and the AF. (ii) the critical exponent x for the divergence of $1/T_1$ is well explained by $x = \nu(z - 1 - \eta)$ with the dynamical exponent $z = 3/2$ for the AF region while $z = \phi/\nu \sim 1.84$ at the bicritical point. These results strongly suggest that the origin of the SC is common with the AF and its symmetry is d-wave.

Keywords: Organic conductor, antiferromagnetism, bicritical point, SO(5)

INTRODUCTION

In κ-(BEDT-TTF)$_2$X (abbreviated as κ-(ET)$_2$X), recent detailed experiments revealed the interesting phase diagram including the bicritical phenomenon between the antiferromagnetism (AF) and the singlet superconductivity (SC) in the plane of temperature and a parameter p controlling the ratio of the Coulomb repulsion U and the bandwidth W [1,2]. Above a characteristic temperature $T^*(U/W)$, the NMR relaxation rate T_1^{-1} increases as the temperature is lowered in a manner independent of p. Below $T^*(U/W)$, on the other hand, T_1^{-1} diverges towards the AF transition temperature, while it shows a spin-gap behavior in the SC side.

MODEL AND ITS CRITICAL PROPERTIES

Let us consider a generic Ginzburg-Landau model for these compounds;

$$H = \int d^d r \left[\frac{1}{2} r_{\parallel} |\vec{\sigma}|^2 + \frac{1}{2} |\vec{\nabla}\vec{\sigma}|^2 + \frac{1}{2} r_{\perp} |\vec{s}|^2 + \frac{1}{2} |\vec{\nabla}\vec{s}|^2 + u|\vec{\sigma}|^4 + 2w|\vec{\sigma}|^2|\vec{s}|^2 + v|\vec{s}|^4 \right], \quad (1)$$

where $\vec{\sigma}$ and $\vec{s}$ are the order parameters of the SC and the AF, respectively. The RG recursion relations for u, v, w up to order ϵ are written as [3]

$$\frac{du}{dl} = \epsilon u - \frac{10u^2 + 3w^2}{2\pi^2}, \quad \frac{dv}{dl} = \epsilon v - \frac{11v^2 + 2w^2}{2\pi^2}, \quad \frac{dw}{dl} = \epsilon w - \frac{4u + 5v + 4w}{2\pi^2} w. \quad (2)$$

The only stable fixed point is the biconical one with $(u_B^*, v_B^*, w_B^*) = 2\pi^2 \epsilon(0.0905, 0.0847, 0.0536)$. However, not all the points in this space will flow to this fixed point through the RG, as shown in Fig.2. If $uv - w^2 > 0$, the RG flow converges to the biconical fixed point (u_B^*, v_B^*, w_B^*) and the phase diagram shows a tetracritical behavior (Fig.1(a)). The critical exponents are $\nu_B = 0.5 + 0.132\epsilon$ and and $\alpha_B = -0.0278\epsilon$. The four second-order phase boundaries are written as

$|g| \sim |t|^{\phi_B}$ with the crossover exponent $\phi_B = 1 + 0.135\epsilon$, where $t \sim (2r_\parallel + 3r_\perp)/5$ and $g \sim (r_\parallel - r_\perp)$ are scaling fields corresponding to the temperature and the anisotropy between the AF and SC phases, respectively. If $uv - w^2 < 0$, on the other hand, the RG flow runs into an unstable region, which indicates a first-order transition between the ordered and the disordered phases near the triple point, as shown in Fig.1(c). Two branches of the first-order transition lines terminate at tricritical points.

Only in the case $uv = w^2$ a bicritical behavior (Fig.1(b)) is predicted by the RG analysis. This bicritical behavior is governed by the Heisenberg fixed point $u_H^* = v_H^* = w_H^* = 2\pi^2\epsilon/13$, which is stable only on the surface $uv = w^2$, and the corresponding exponents are $\nu_H = 0.5 + 0.135\epsilon$, $\phi_H = 1 + 0.192\epsilon$, $\alpha_H = -0.0385\epsilon$. The experimentally observed bicritical behavior strongly suggests $uv = w^2$, which corresponds to the rotational symmetry in the 5-dimensional order parameter space of $(\vec{\sigma}, \vec{s})$. It is interesting to note that only if $uv = w^2$ is the model (1) smoothly related to the SO(5) NLσ model in [4].

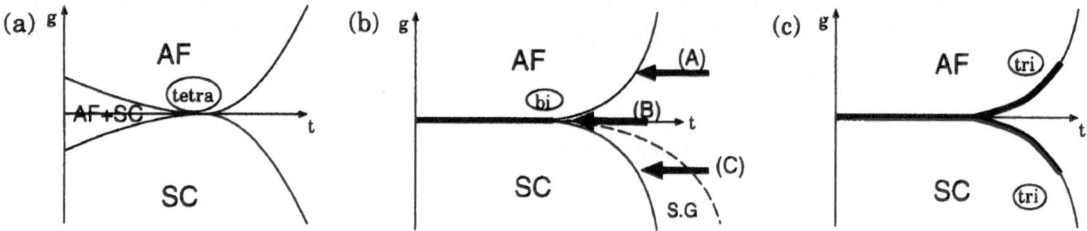

Fig. 1. Schematic phase diagrams of the model (1). The thick lines represent first-order lines, and the thin ones are the second-order lines. (a) $uv > w^2$:tetracritical, (b) $uv = w^2$:bicritical, (c) $uv < w^2$:tricritical.

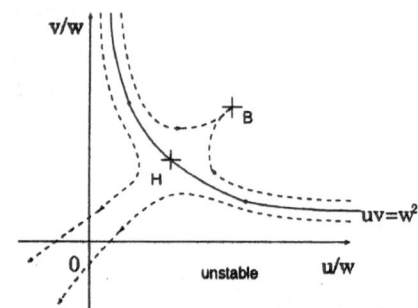

Fig. 2. RG flow of the model (1). The points "B" and "H" are the biconical and the Heisenberg fixed points, respectively.

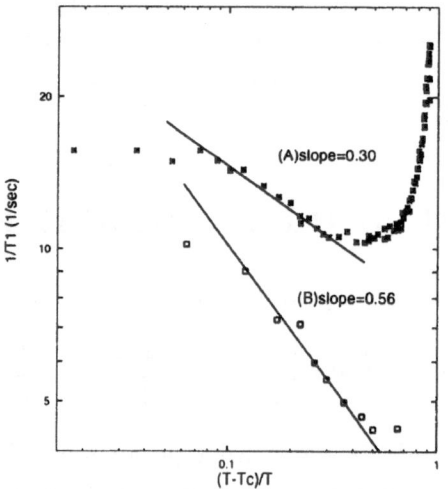

Fig. 3. Log-log plot of T_1^{-1} v.s. $\frac{T - T_{c,AF}}{T}$ for (A) κ-(ET)$_2$Cu[N(CN)$_2$]Cl (open squares), and (B) deuterated κ-(ET)$_2$Cu[N(CN)$_2$]Br (solid squares). (Data from [2].)

DYNAMICS

As for the dynamic critical phenomena. the NMR linewidth is proportional to $(T - T_{c,AF})^{-x}$ on the AF side of the normal phase with $x = \nu(z - 1 - \eta)$, where z is a dynamic critical exponent. When the system is in the vicinity of $T_{c,AF}$, but not so near the bicritical point, z is equal to

$d/2 = 3/2$. Thus, the exponent is $x_{AF} = 0.315$ up to $O(\epsilon^3)$. On the other hand, in the bicritical region, we get $z = \phi_H/\nu_H \sim 1.84$, resulting in $x_{BP} = 0.589$, up to $O(\epsilon^2)$. In approaching the SC phase it does not diverge but shows a spin-gap behavior below a characteristic temperature $t^*_{S.G.} \sim |g|^{1/\phi_H}$ because of the singlet formation. In the whole critical region of the normal phase, T_1^{-1} has a scaling form $T_1^{-1} = t^{-x_{BP}} f(g/t^{\phi_H})$. In Fig.3 is shown the log-log plot of T_1^{-1} v.s. $\frac{T-T_{c,AF}}{T}$ for (A) κ-(ET)$_2$Cu[N(CN)$_2$]Cl, and (B) deuterated κ-(ET)$_2$Cu[N(CN)$_2$]Br. The former is located in the AF region a bit away from the bicritical point while the latter nearly at the bicritical point as shown in Fig.1(b). The critical exponent x in the region $1K \leq T - T_{c,AF} \leq 10K$ gives 0.30 ± 0.04 for (A) and 0.56 ± 0.04 for (B). These values are in reasonably good agreement with the above theoretical values, which supports that the critical region is rather wide ($\sim 10K$) and $\xi_{AF} \sim 1$. As for experimental estimations of the SC coherence length ξ_{SC}, there is still some controversy. However, the rotational symmetry between the AF and SC suggests that ξ_{SC} is also short (~ 1). It is also inferred that the observed spin gap behavior is due to the large fluctuation of the SC order parameter in the SC compound κ-(ET)$_2$Cu(NCS)$_2$ ((C) in Fig.1(b)).

The rotational symmetry between AF and SC order parameters suggests that the mechanism of the AF and SC is common and the underlying microscopic quantum model has the enhanced dymanical symmetry, i.e., SO(5) [4]. Therefore it is likely that the symmetry is d-wave. We believe that the (a) half-filling, (b) intermediate Coulomb interaction, and (c) nearly 2-dimensional Fermi surface in κ-(ET)$_2$X makes the SO(5) symmetric model a promising candidate. When these conditions are violated, e.g., by doping carriers and/or applying external magnetic field, the SO(5) symmetry would be broken and the bicritical phenomenon will turn into tetra- or tricritical behavior. The nearly one-dimensional compound (TMTSF)$_2$X would not have the SO(5) symmetry, and considering the fact that the coexistence of the AF and the SC has never been observed, the normal-AF and normal-SC transition is also expected to be first order (Fig.1(c)).

SUMMARY

We have studied theoretically the critical phenomena of the AF and the SC in κ-(ET)$_2$X from the viewpoint that this system is in the strong coupling regime and nontrivial critical phenomena are observed. The fluctuation of both the AF and the SC order parameters are treated in terms of the RG method, employing $\epsilon = 4 - d$ expansion. Classification of the scaling trajectories leads to three types of multicritical phenomena (Fig.1), i.e., (a) tetracritical, (b) bicritical, and (c) tricritical phenomena. The bicritical phenomenon (b) is unstable towards tetra- and tricritical ones, and is realized only when the rotational symmetry is there within the 5-dimensional order paramter space of SC and AF. Dynamic critical phenomena are also studied especially for the NMR relaxation rate $1/T_1$, and its critical exponent x is in reasonable agreement with the ϵ-expansion result both for the AF region and at the bicritical point.

Acknowledgments: The authors acknowledge K. Kanoda, M. Kardar, and K. Miyagawa for fruitful discussions. This work is supported by Grant-in-Aid for COE Research No. 08CE2003 from the Ministry of Education, Science, Culture and Sports of Japan.

---

* e-mail: murakami@appi.t.u-tokyo.ac.jp.
1. K. Kanoda, Hyperfine Interactions **104** 235 (1997).
2. A. Kawamoto et al., Phys. Rev. **B52** 15522 (1995); unpublished.
3. J. M. Kosterlitz, D. R. Nelson, and M. E. Fisher, Phys. Rev. **B13** 412 (1974).
4. S.-C. Zhang, Science **275** 1089 (1997).
5. S. Murakami and N. Nagaosa, cond-mat/9910001.

Evolution of Spectral Function from Insulator to Superconductor: What do We Learn from Angle-Resolved Photoemission?

Takami Tohyama and Sadamichi Maekawa

Institute for Materials Research, Tohoku University, Sendai 980-8577, Japan

Abstract:

The angle-resolved photoemission spectroscopy provides unique information for the dramatic change of high-T_c cuprates from insulator to superconductor upon carrier-doping. To extract the underlying physics, we investigate the single-particle spectral function of the extended t-J model with long-range hopping terms. In undoped system, a broad spectrum at $k=(\pi,0)$ is found to be due to a novel spin liquid state around a photo-doped hole, accompanied by the separation of the spin and charge degree of freedom. With doping, the $(\pi,0)$ spectrum is systematically evolved from broad line shape in underdoped system to sharp one in overdoped system. We argue that the broadness in the underdoped system is an indication of the spin liquid state.

Keywords: ARPES, the t-J model, spin liquid, spectral function

INTRODUCTION

The most dramatic features of high-T_c superconducting cuprates are that undoped insulating antiferromagnets change into superconductors upon doping of carriers and that superconductivity exists in a certain range of the carrier density. The angle-resolved photoemission spectroscopy (ARPES) is one of the primary sources of such changes of the electronic structure due to doping. Systematic changes of the ARPES spectra with doping from the undoped insulators to underdoped and overdoped $Bi_2Sr_2CaCu_2O_{8+\delta}$ superconductors have been reported [1-3], focusing mainly on the dispersion and spectrum at around momentum $k=(\pi,0)$ or $(0,\pi)$ where the superconducting gap with $d_{x^2-y^2}$ symmetry shows maximum. An interesting observation is that a broad spectrum at $(\pi,0)$ in undoped samples [4] is continuously evolved into a sharp peak in overdoped samples: The spectrum in underdoped samples remains broad, and then it becomes sharper and sharper with further doping. In addition, maximum positions of peaks along $(\pi/2,\pi/2)$ to $(\pi,0)$ show similar dispersion between undoped and underdoped samples, which follows a $d_{x^2-y^2}$ gap function [5]. These results indicate a close relation between antiferromagnetism and superconductivity.

To elucidate the problem and extract the underlying physics, we perform theoretical study of the single-particle spectral function in two-dimensional cuprates by using numerically exact diagonalization method on small clusters of the extended t-J model with long-range hopping terms [3,6]. In undoped system, a novel spin liquid state is realized around a photo-doped hole with momentum $k=(\pi,0)$. The $(\pi,0)$ spectrum thus becomes very broad accompanied by the separation of the spin and charge degree of freedom [6]. This naturally explains a d-wave-like dispersion observed in the undoped insulator [5]. With hole doping, the $(\pi,0)$ spectrum shifts to the Fermi level keeping its broadness [3], indicating the spin liquid state. Since a high-energy pseudogap with order of J is determined by the position of the broad spectrum, the spin liquid picture is the most probable explanation of the origin of the pseudogap.

SINGLE HOLE IN ANTIFERROMAGNETIC INSULATOR

The first ARPES experiment on undoped insulators was performed by Wells *et al.* [4] for $Sr_2CuO_2Cl_2$. From the comparison between the ARPES data and the *t-J* results, it is found that the *t-J* model can quantitatively explain the observed dispersion from $(0,0)$ to (π,π) with a width of $2.2J$. However, the model does not explain the dispersion with a width of about $2J$ along the $(0,\pi)$-$(\pi,0)$ direction: It predicts a nearly flat dispersion along this direction. This discrepancy has led to intense theoretical studies, from which it was found that the discrepancy may be resolved by introducing second and third neighbor hopping matrix elements, t' and t'' [3]. Not only the dispersion but also the ARPES line shape contains important information. The quasiparticle (QP) peak at $(\pi/2,\pi/2)$ is sharp, while at the $(\pi,0)$ point the peak is strongly suppressed. Such a suppression is also reproduced very well by the *t-t'-t''-J* model but not by the *t-J* model [3].

Ronning *et al.* pointed out that the dispersion follows the form of a d_{x2-y2} gap [5]. Recent high resolution data, however, show that, while the dispersion near $(\pi,0)$ follows the d-wave function, the deviation from the function is remarkable near $(\pi/2,\pi/2)$ [7]. We found that the deviation is consistent with a dispersion obtained from the *t-t'-t''-J* model [6]. Although the model describes the observed dispersion very well, a frequently arising question is about the applicability of t' and t'' terms on the real system: The experimental fact that the dispersion in the $(0,\pi)$-$(\pi,0)$ direction has the almost same one as in the $(0,0)$-(π,π) direction is hard to be ascribed to the t' and t'' terms, because the $(0,0)$-(π,π) dispersion is regulated by J [8,9]. This question is resolved by examining the dependence of the QP energy difference $E(\pi,0)-E(\pi/2,\pi/2)$ on J [6]. The difference in the *t'-t'-t''-J* model is found to have the same J dependence as $E(0,0)-E(\pi/2,\pi/2)$, with the magnitude of about $2J$. Therefore, even in the *t-t'-t''-J* model, the width of the dispersion is governed by the spin degree of freedom, as is the case of the *d*-wave RVB picture [8,10].

The above facts indicate that the excited state with momentum $(\pi,0)$ has peculiar features in contrast to other momentum states. In fact, from examinations of the spin correlation around a doped hole of the *t-t'-t''-J* model, it was concluded that a novel spin liquid state is realized around the hole with momentum $(\pi,0)$ in contrast to a Néel-like state at around $(\pi/2,\pi/2)$ [6,11]. In addition, the dynamical properties of spin and charge degrees of freedom also provide us useful information on the QP state with $(\pi,0)$ [6]. We found that, while in the *t-J* model both spin and charge components are carried by the QP state, only spin component is involved in the *t-t'-t''-J* model. This is because of the separation of the spin and charge degrees of freedom, consistent with the spin liquid picture mentioned above. Finally, we note that this spin liquid state should be distinguished from the *d*-wave RVB state [8,10]: The former is seen in the one-hole state with momentum $(\pi,0)$ and the excitation energy of $\sim 2J$, while the RVB theory predicts a spin liquid state which is independent of the momentum of a doped hole.

DOPING DEPENDENCE OF ARPES LINE SHAPE

In this section, we discuss the systematic evolution of the line shape of the ARPES spectra upon carrier-doping in the normal state above T_c. As is the case of undoped materials, the peak at the $(0,0)$-(π,π) Fermi crossing in the underdoped $Bi_2Sr_2CaCu_2O_{8+\delta}$ is sharp, while the peak at $(\pi,0)$ is broad and suppressed. In the overdoped sample, the peak at $(\pi,0)$ moves closer to the Fermi energy and becomes sharp with increased intensity while the line shape on the $(0,0)$-(π,π) cut remains more or less the same as in the underdoped sample [3]. The shift of the broad QP peak at $(\pi,0)$ toward the Fermi energy is also summarized in [5]. The QP position, whose energy is about 100 to 200 meV, is sometimes called high-energy pseudogap. A similar doping dependence was observed in single plane $Bi_2Sr_{2-x}La_xCuO_{6+\delta}$ [12].

The t-t'-t''-J model reproduces the experimental evolution of the ARPES line shape on hole-doping [3]. The peak at around $(\pi/2,\pi/2)$ is sharp, which is independent of doping, while the spectrum at $(\pi,0)$ is broad in the underdoped case and becomes sharp in the overdoped one. The peak position of the $(\pi,0)$ spectra shifts to near the Fermi level with doping. The great breadth of the QP peak at $(\pi,0)$ in the underdoped case is caused by the reduction of the QP weight, which may come from two alternative sources. (i) The same mechanism of the coupling between charge motion and spin background discussed above for the undoped case could be still effective in the lightly doped material: A spin liquid state around a hole with momentum $(\pi,0)$ induced by t' and t'' survives even in the underdoped region. This may be a probable explanation of the origin of the high-energy pseudogap [5]. (ii) Alternatively, there could be larger phase space for decay of QPs because of strong coupling of the photo-hole to collective magnetic excitations near $q=(\pi,\pi)$ [13]. The Fermi surface topology is changed by the inclusion of t' and t'' in a way that enhances this coupling. Both of these mechanisms are likely to have less importance for the over-doped case.

SUMMARY

The fact that the line shape of the $(\pi,0)$ spectrum is continuously evolved from insulating to underdoped samples demonstrates the continuity of the electronic states from the antiferromagnetic insulators to the superconductors. We found that the spin liquid state around a photo-doped hole with momentum $(\pi,0)$ is the origin of the broad line shape in the insulator. The spin liquid state is supposed to remain even in the underdoped system, playing a crucial role in the formation of the pseudogap.

Acknowledgements. We would like to thank Y. Shibata, C. Kim, Z.-X. Shen, and N. Nagaosa for collaboration and discussion on this subject. This work was supported by CREST, NEDO, and Priority-Areas Grants from the Ministry of Education, Science, Sports and Culture of Japan. The numerical calculations were performed in the Supercomputer Center in ISSP, University of Tokyo, the supercomputing facilities in IMR, Tohoku University.

1. D. M. King *et al.*, J. Phys. Chem. Solids 56, 1865 (1995).
2. S. LaRosa *et al.*, Phys. Rev. B **56**, R525 (1997).
3. C. Kim *et al.*, Phys. Rev. Lett. **80**, 4245 (1998).
4. B. O. Wells *et al.*, Phys. Rev. Lett. **74**, 964 (1995).
5. F. Ronning *et al.*, Science **282**, 2067 (1998).
6. T. Tohyama, Y. Shibata, S. Maekawa, Z.-X. Shen, and N. Nagaosa, cond-mat/9904231.
7. C. Kim, private communication.
8. R. B. Laughlin, Phys. Rev. Lett. **79**, 1726 (1997).
9. W. Hanke, M. G. Zacher, E. Arrigoni, and S. C. Zhang, cond-mat/9908175.
10. X.-G. Wen and P. A. Lee, Phys. Rev. Lett.**79**, 1726 (1996)
11. G. Martins, R. Eder, and E. Dagotto, Phys. Rev. B **60**, R3716 (1999).
12. J. M. Harris *et al.*, Phys. Rev. Lett.**79**, 143 (1997).
13. Z.-X. Shen and J. R. Schrieffer, Phys. Rev. Lett.**78**, 1771 (1997).

Angle-resolved photoemission study of $PrBa_2Cu_3O_7$ and $PrBa_2Cu_4O_8$

T. Mizokawa[1], C. Kim[2], Z-X. Shen[2], A. Ino[3], T. Yoshida[3], A. Fujimori[1,3], M. Goto[4],
H. Eisaki[4], S. Uchida[4], M. Tagami[5], K. Yoshida[5], A. I. Rykov[5], Y. Siohara[5], K. Tomimoto[5],
S. Tajima[5], Yuh Yamada[6], S. Horii[4], N. Yamada[7], Yasuji Yamada[5], I. Hirabayashi[5]

[1]Department of Complexity Science and Engineering, University of Tokyo, Tokyo 113-0033, Japan
[2]Department of Applied Physics and Stanford Synchrotron Radiation Laboratory, Stanford University, Stanford, CA94305 , U.S.A.
[3]Department of Physics, University of Tokyo, Tokyo 113-0033, Japan
[4]Department of Advanced Materials Science, University of Tokyo, Tokyo 113-0033, Japan
[5]Superconductivity Research Laboratory, International Superconductivity Technology Center, Tokyo 135-0062, Japan
[6]Faculty of Science and Engineering, Shimane University, Matsue 690-0823, Japan
[7]Department of Applied Physics and Chemistry, University of Electro-Communications, Chofu, Tokyo 182-8585, Japan

Abstract: We have performed an angle-resolved photoemission study on $PrBa_2Cu_3O_7$ (Pr123) and $PrBa_2Cu_4O_8$ (Pr124) which have hole-doped Cu-O chains. Dispersive features with one-dimensional (1D) character are observed in Pr123 and Pr124 and are attributed to signals from the hole-doped Cu-O chains which are approximately 1/4-filled. While, in Pr123, the 1D feature loses its spectral weight near the Fermi level (E_F), it crosses E_F in Pr124. These results suggest that, while the hole-doped Cu-O single chains in Pr123 have instability to charge ordering, the hole-doped Cu-O double chains in Pr124 can get rid of it.

Keywords: ARPES, Cu-O chain, spin-charge separation, charge ordering

INTRODUCTION

In $PrBa_2Cu_3O_7$ (Pr123) and $PrBa_2Cu_4O_8$ (Pr124), the CuO_2 planes do not have enough carriers to cause superconductivity [1]. On the other hand, the Cu-O single chains in Pr123 and the Cu-O double chains in Pr124 are heavily hole-doped and show semiconducting and metallic behaviors, respectively [1,2]. Therefore, Pr123 and Pr124 give us a unique opportunity to study the electronic structures of the hole-doped Cu-O chains. In this work, we study the hole-doped Cu-O chains in Pr123 and Pr124 using angle-resolved photoemission spectroscopy (ARPES).

EXPERIMENTAL

Single crystals of Pr123 were grown in a MgO crucible by a pulling technique and were detwinned by annealing at 500°C in oxygen atmosphere under uniaxial stress. Naturally-untwinned single crystals of Pr124 were grown by a flux method under oxygen pressure of 11 atm [3]. The ARPES measurements were performed at the beamlines 5-3 and 5-4 of Stanford Synchrotron Radiation Laboratory (SSRL). The chamber pressure during the measurements was less than 5×10^{-11} Torr. The

173

samples were cooled to 10 K and cleaved *in situ*. The cleaved surfaces were the *ab*-plane, where the *b*-axis is in the Cu-O chain direction. The cleanliness of the surfaces was checked by the absence of a hump at ~ 9.5 eV. The position of the Fermi level (E_F) was calibrated with gold spectra.

RESULTS AND DISCUSSION

The ARPES spectra of Pr123 and Pr124 along the Cu-O chain direction are shown in Fig. 1(a). Here, k_a and k_b are the momemta perpendicular to the chain and along the chain in units of $1/a$ ($a = 3.87$ Å for Pr123, 3.88 Å for Pr124) and $1/b$ ($b = 3.93$ Å for Pr123, 3.90 Å for Pr124), respectively. In Pr123, the dispersive feature takes a band maximum at $k_b/\pi \sim 0.25$ and rapidly loses its weight for $k_b/\pi > 0.25$ without reaching E_F. On the other hand, in Pr124, the dispersive feature reaches E_F at $k_b/\pi \sim 0.25$. Although only the ARPES data for $k_a/\pi \sim 1$ are shown in Fig. 1, the dispersions hardly depend on k_a and have good one-dimensionality. These facts indicate that the Cu-O chains in Pr123 and Pr124 are approximately 1/4-filled, namely, 50% hole-doped. For Pr123, the suppression of the intensity at E_F would be a manifestation of the charge ordering or charge density wave in the 1/4-filled Cu-O single chain [4]. Actually, charge instability in the Cu-O chain has been observed by NMR and NQR measurements of Pr123 [5]. On the other hand, in the Cu-O double chain, charge ordering would be frustrated because of the interaction between the two chains within the double chain. Consequently, Pr124 can get rid of the instability to charge ordering and can have finite spectral weight at E_F.

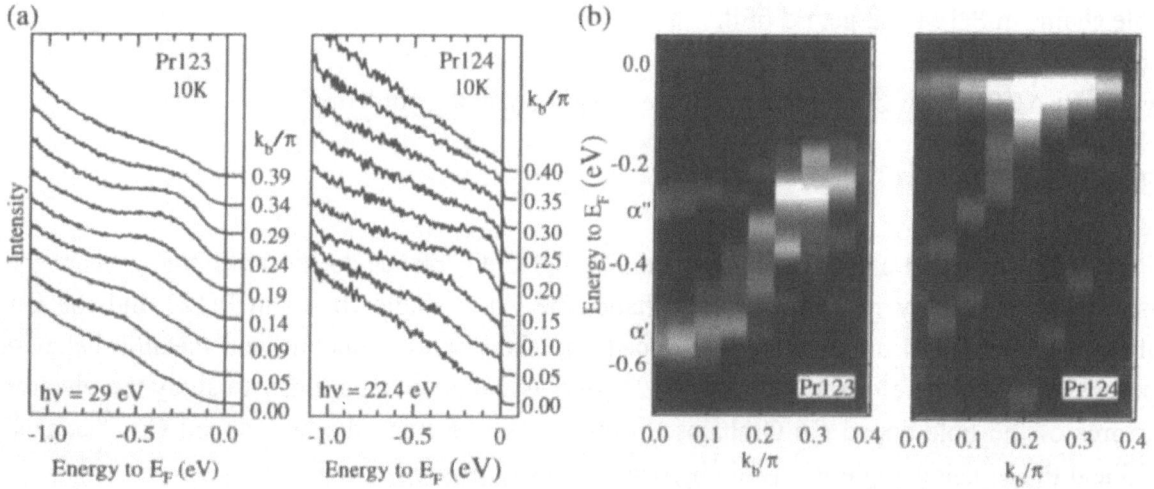

FIG. 1. (a) ARPES spectra taken along the Cu-O chain in Pr123 and Pr124. (b) Second derivatives of the ARPES spectra along the chain direction. k_b is the momemtum along the chain. The energy resolutions were approximately 40 meV for Pr123 and 20 meV for Pr124.

In order to show the dispersions clearly, the second derivatives of the ARPES spectra are displayed in Fig. 1(b). In Pr123, two dispersive features labled as α' and α'' are visible as two bright belts which can be attributed to holon and spinon dispersions and are manifestations of the spin-charge separation in one-dimensional (1D) systems as observed in $SrCuO_2$ [6]. The holon and spinon dispersions have the width of ~ 0.5 and ~ 0.1 eV, respectively, which approximately agree with t and J in the t-J model for the cuprates [7]. On the other hand, Pr124 has no separate spinon and holon features expected for a Tomonaga-Luttinger (TL) liquid [8]. This result indicates that the interaction between the double chains is relevant and that the Cu-O chains in Pr124 might be regarded as a two-dimensional (2D) system. Actually, the overall band dispersion of Pr124 shows some deviations from the ideal 1D behavior realized in Pr123. This is consistent with the fact that Pr124 has large Hall coefficient at low temperature [2]. Since the distance between the neighboring chains in Pr124 is as long as that in Pr123, the interchain hopping term in Pr124 should be as small as that in Pr123. Probably, when each chain is metallic, the interchain coupling becomes relevant and, consequently, the chains form a 2D system with strong anisotropy. Since the interchain hopping term is expected to be small compared to t or J, the present data raise a question whether such a weak interchain coupling is enough to wipe out holon and spinon features expected in ARPES spectra of Cu-O chains. It should also be noted that the ARPES spectra of Pr124 are different from those expected for a conventional Fermi liquid. The dispersive feature is very broad and cannot be interpreted as a quasi-particle peak of a Fermi liquid. In addition, the ARPES spectra for $k_b/\pi < 0.25$ have substantial spectral weight at E_F although the dispersive feature is located well below E_F. This behavior is similar to that observed in the hole-doped CuO_2 plane of the high-T_c cuprates [9].

CONCLUSION

We have studied the hole-doped Cu-O chains in Pr123 and Pr124 using ARPES. While, in Pr123, the 1D features from the Cu-O chain lose their weight near E_F, the dispersive feature crosses E_F in Pr124. These facts indicate that the charge ordering occurs in the Cu-O single chain and is suppressed in the Cu-O double chain. The lack of spinon and holon features in Pr124 indicates that the interchain coupling is relevant in Pr124 compared to Pr123. Neither the TL-liquid picture nor the Fermi-liquid picture explain the ARPES spectra of Pr124. It would be interesting to study further the difference between the metallic Cu-O chains in Pr124 and the CuO_2 plane in the high-T_c cuprates.

Acknowledgements. The authors would like to thank I. Terasaki, K. Penc, T. Thoyama, and S. Maekawa for valuable comments and the staff of SSRL for technical support. Crystal gorwth and characterization of Pr123 were supported by NEDO. SSRL is operated by the U. S. DOE Office of Basic Energy Scinece and Division of Chemical Science.

1. J. L. Peng *et al.*, Phys. Rev. B 40, 4517 (1989); K. Takenaka *et al.*, Phys. Rev. B 46, 5833 (1992);   Y. Yamada *et al.*, Physica C 231, 131 (1994).
2. I. Terasaki *et al.*, Phys. Rev. B 54, 11993 (1996).
3. S. Horii *et al.*, submitted to Phys. Rev. B.
4. P. A. Lee, T. M. Rice, and P. W. Anderson, Phys. Rev. Lett. 31, 462 (1973).
5. B. Grévin *et al.*, Phys. Rev. Lett. 80, 2045 (1998).
6. C. Kim *et al.*, Phys. Rev. Lett. 74, 964 (1996); Phys. Rev. B 56, 15589 (1997).
7. S. Maekawa, T. Thoyama, and S. Yunoki, Physica C 263, 61 (1996).
8. J. Solyom, Adv. Phys. 28, 201 (1979); V. Meden and K. Schonhammer, Phys. Rev. B 46, 15753 (1992); J. Voit, Phys. Rev. B 47, 6740 (1993); K. Penc *et al.*, Phys. Rev. Lett. 77, 1390 (1996).
9. D. Dessau *et al.*, Phys. Rev. Lett. 71, 2781 (1993); Z-X. Shen and D. Dessau, Phys. Rep. 253, 1 (1995); A. Ino *et al.*, J. Phys. Soc. Jpn. 68, 1496 (1999).

Elastic magnetic signals in the superconducting $La_{2-x}Sr_xCuO_4$

Hiroyuki Kimura[1]*, Hiroki Matsushita[1], Kazuma Hirota[1], Yasuo Endoh[1], Masaki Fujita[2], Kazuyoshi Yamada[2], Gen Shirane[3], Seung-Hun Lee[4], Young Sang Lee[5], Marc A. Kastner[5], and Robert J. Birgeneau[5]

[1]Department of Physics, Tohoku University, Sendai 980-8578
[2]Institute for Chemical Research, Kyoto University, Uji 610-0011
[3]Department of Physics, Brookhaven National Laboratory, Upton, NY 11973-5000
[4]National Institute of Standards and Technology, Gathersburg, MD 20899
[5]Department of Physics, Massachusetts Institute of Technology, Cambridge, MA 02139

Abstract: Incommensurate elastic magnetic signals in the superconducting $La_{2-x}Sr_xCuO_4$ have been extensively studied on neutron scattering. Kimura *et al.* clarified that the elastic signal for $x = 0.12$ starts developing below 31 K where the superconductivity appears. The peak width are very sharp and reaches almost resolution limit, indicating a long-range static magnetic correlation. In the present work, we report a systematic study of the incommensurate elastic peaks for a wide hole-doping range ($0.08 \leq x \leq 0.15$). The elastic peaks were observed for the $0.10 \leq x \leq 0.135$ samples. It was also found that the static magnetic correlation is most developed at $x = 0.12$. Our results is qualitatively consistent with μSR and NMR measurements.

Keywords: $La_{2-x}Sr_xCuO_4$, Incommensurate elastic peak, Neutron scattering

In La-based high-T_c cuprates, it is well established that the low-energy spin excitation at incommensurate wave vectors, $(\frac{1}{2} \pm \varepsilon \, \frac{1}{2} \, 0)$ and $(\frac{1}{2} \, \frac{1}{2} \pm \varepsilon \, 0)$ plays an important role for superconductivity. Yamada *et al.* have carried out systematic neutron scattering study for $La_{2-x}Sr_xCuO_4$ (LSCO) and clarified that there exists a linear relationship between the incommensurability ε and T_c in a wide hole-doping range ($0.06 \leq x \leq \frac{1}{8}$) [1]. Furthermore, in $YBa_2Cu_3O_{6+x}$ (YBCO), incommensurate spin fluctuation was also observed [2] and the value of ε is consistent with the "T_c vs. ε curve" established in LSCO. The results strongly suggests that the incommensurate spin fluctuations have some universal features among high-T_c cuprates.

It has been believed for a long time that in the superconducting state, antiferromagnetic correlations is purely dynamical, namely the spin fluctuation is dominant in the CuO_2 plane. However, recently Suzuki *et al.* found the elastic magnetic signals, implying the static antiferromagnetic correlations, in the superconducting $La_{1.88}Sr_{0.12}CuO_4$ [3]. These signals are incommensurate with the lattice and the ε is almost consistent with the hole concentration, which is quite similar to that of the inelastic signals due to spin fluctuations observed at the same hole concentration [1]. Furthermore, Kimura *et al.* made comprehensive studies of LSCO for $x = 0.10$, 0.12, 0.15 and established that the elastic peaks were observed at both $x = 0.10$ and 0.12 in the superconducting state, while for $x = 0.15$ no signals were observed [4]. Since the static antiferromagnetic correlation is mostly enhanced at $x = 0.12$, there have been controversial discussions on the origin of the static correlation with respect to the suppression of high-T_c superconductivity around $x = \frac{1}{8}$ ($\frac{1}{8}$-problem). On the other hands, as shown in Figure 1 (a), the appearing temperature of the elastic peak for $x = 0.12$ is identical within the errors to the onset of T_c (= 31.5 K). One can easily associate from this result the cooperative interplay between the static correlation and the superconductivity. Actually, in excessively oxygenated La_2CuO_{4+y}, the static antiferromagnetic correlation and the superconductivity emerge simultaneously [5].

In order to elucidate whether the static correlation and the superconductivity are competitive or cooperative, we have carried out systematic investigations of incommensurate elastic peaks for a wide hole-doping range in LSCO. In this report, we present the experimental results for $x = 0.08$, 0.10, 0.12, 0.13, 0.135, 0.15, showing that the elastic magnetic peaks exist in $0.10 \leq x \leq 0.135$.

Single crystals of LSCO were grown by TSFZ method [6]. Sample characterizations for all the sample were shown in Ref. [1,4,7], where it is shown that all the crystals are homogeneous for Sr doping and their concentrations are well controlled. Neutron scattering measurements were performed by using triple-axis-spectrometer HER at a cold neutron guide in JRR-3M (JAERI) and SPINS at NIST. The incident neutron beam was fixed at 5 meV. In the present experimental condition, the energy resolutions for both the spectrometer were estimated to be 0.2 meV with FWHM.

Elastic magnetic peaks were observed for $x = 0.10, 0.12, 0.13, 0.135$ while for $x = 0.08$ and 0.15 any elastic signals were below the limit of detectability. Figures 1 (a), (b) show the temperature dependences of the elastic peaks for $x = 0.12$ [4], 0.13 [7], respectively. Each peak spectrum at lowest temperature is shown in the inset of each figure. It is clearly seen that the appearing temperature of the elastic peak for $x = 0.13$ ($\equiv T_{0.13}^*$) is lower than that for $x = 0.12$ ($T_{0.12}^*$). More importantly, $T_{0.13}^*$ ($\sim 20K$) is inconsistent with T_c. The similar behavior was already observed

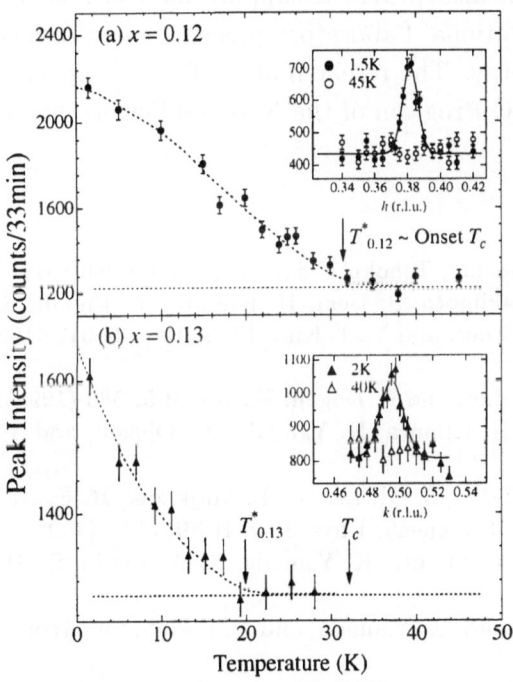

Figure 1: Temperature dependences of the elastic peaks for (a); $x = 0.12$ and (b); 0.13. Each inset shows the peak spectrum at the lowest temperature.

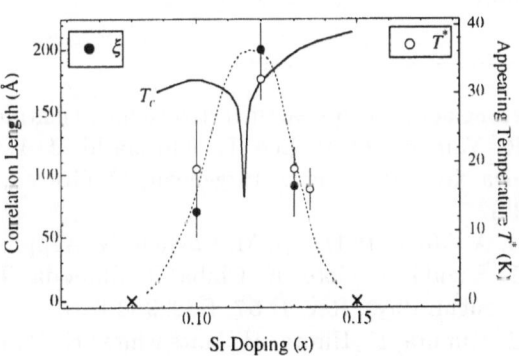

Figure 2: In-plane correlation length (filled circle, left axis) and appearing temperature of the elastic peaks (open circle, right axis) as a function of hole doping x.

for the $x = 0.10$ sample [4], implying that the elastic peaks are indeed enhanced at the vicinity of $x = \frac{1}{8}$. Static antiferromagnetic correlation length in the CuO_2 plane was estimated by the analyses of the intrinsic peak width taking into account of instrumental resolutions. The results show that the in-plane correlation is mostly extended at $x = 0.12$, of which length ξ was estimated to be at least 200 Å. On the other hand, ξ's for $x = 0.10$ and 0.13 were also estimated to be about 90 Å, which are shorter than that for $x = 0.12$. As for $x = 0.12$, much shorter out-of-plane correlation (~ 15 Å) comparing with the in-plane one was also observed, suggesting the 2D-like antiferromagnetic correlation.

Figure 2 shows the summarized data of our present work; the doping dependence of the appearing temperature T^* (right vertical axis) and the in-plane correlation length ξ (left vertical axis). It is now clear that the static antiferromagnetic correlation is mostly stabilized at $x \sim \frac{1}{8}$, where the superconductivity is a little suppressed. Our present results indicate that the static correlation competes with the superconductivity, which is qualitatively consistent with μSR and NMR study in LSCO [8,9]. In addition to LSCO system, recently, evolution of quite low frequency component of spin fluctuations are observed by μSR studies in Zn-doped YBCO and Bi-2212 system at $\frac{1}{8}$ hole-doping [10]. These results imply that the evolution of static or quasi static antiferromagnetic correlation around $x \sim \frac{1}{8}$ degrades their superconductivity, which might be a common feature in the high-T_c superconductors. However, it should be noted that in La_2CuO_{4+y}, which has very high $T_c \sim T^*$ value, the interplay between the static correlation and the superconductivity is not yet clear. More systematic studies are required in excess-oxygen system.

Present work was supported by CREST sponsored by the Japan Science & Technology Cooperation. The US-Japan cooperative research program also provided support for the neutron scattering experiment at NIST. Work at Brookhaven National Laboratory was carried out under Division of Material Science, U. S. Department of Energy. The research at MIT was supported by the National Science Foundation and by the MRSEC Program of the National Science Foundation.

---

*Present address: Research Institute for Scientific Measurements, Tohoku University, Sendai 980-8577

[1] K. Yamada, C. H. Lee, K. Kurahashi, J. Wada, S. Wakimoto, S. Ueki, H. Kimura, Y. Endoh, S. Hosoya, G. Shirane, R. J. Birgeneau, M. Greven, M. A. Kastner, and Y. J. Kim, Phys. Rev. B **57**, 6165 (1998).

[2] H. A. Mook, P. Dai, S. M. Hayden, G. Aeppli, T. G. Perring, and F. Dogan, Nature **395**, 580 (1998).

[3] T. Suzuki, T. Goto, K. Chiba, T. Shinoda, T. Fukase, H. Kimura, K. Yamada, M. Ohashi, and Y. Yamaguchi, Phys. Rev. B **57**, R3229 (1998).

[4] H. Kimura, K. Hirota, H. Matsushita, K. Yamada, Y. Endoh, S. -H. Lee, C. F. Majkrzak, R. Erwin, G. Shirane, M. Greven, Y. S. Lee, M. A. Kastner, and R. J. Birgeneau, Phys. Rev. B **59**, 6517 (1999).

[5] Y. S. Lee, R. J. Birgeneau, M. A. Kastner, Y. Endoh, S. Wakimoto, K. Yamada, R. W. Erwin, S. -H. Lee, and G. Shirane, Phys. Rev. B in press.

[6] C. H. Lee N. Kaneko, S. Hosoya, K. Kurahashi, S. Wakimoto, K. Yamada, and Y. Endoh, Supercond. Sci. Technol. **11** (1998) 891.

[7] H. Matsushita, H. Kimura, M. Fujita, K. Yamada, K. Hirota, and Y. Endoh, J. Phys. Chem. Solids. **60** (1999) 1071.

[8] K. Kumagai, K. Kawano, I. Watanabe, K. Nishiyama, and K. Nagamine, J. Supercond. **7** (1994) 63.

[9] T. Goto, S. Kazama, K. Miyagawa, and T. Fukase, J. Phys. Soc. Jpn. **63** (1994) 3494.

[10] I. Watanabe, M. Akojima, M. Aoyama, Y. Koike, S. Ohira, W. Higemoto, and K. Nagamine, September Meetings of Phys. Soc. Jpn. (1999).

Charge Dynamics of $A_{14}Cu_{24}O_{41}$ Ladder Compounds in Two Ends: Sr14 and Ca14

Kenji M. Kojima,[1*] Makoto Someya,[1] Naoki Motoyama,[1] Hiroshi Eisaki,[1] Shin-ichi Uchida,[1] Yutaka Furubayashi,[2] Takahito Terashima,[2] and Mikio Takano[2]

[1]Department of Superconductivity, University of Tokyo, Hongo, Bunkyo-ku, Tokyo 113-8656, Japan
[2]Institute for Chemical Research, Kyoto University, Uji, Kyoto-fu 611-0011, Japan

Abstract: We report optical reflectivity measurement of the spin ladder compound $A_{14}Cu_{24}O_{41}$. Reflectivity of the purely A=Sr system (Sr14) exhibited a 'metallic' behavior in the E//c (//leg) polarization at room temperature, while below ~200K, there was a shift of the spectral weight from infrared to visible region. The increase of the chain Cu^{3+} signal (ω~3eV) suggests that some of the holes on the ladder move to the chain sites at low temperatures. We present a preliminary result of optical reflectivity of A=Ca compound (Ca14) as well, which is available only in a form of thin film evaporated onto Sr14 substrate.

Keywords: spin ladder, $A_{14}Cu_{24}O_{41}$ compounds, optical reflectivity, charge-ordering

INTRODUCTION

Since discovery of the high Tc cuprates, there has been vital effort to understand the anomalously high Tc's. Spin ladder system, which has localized S=1/2 spins coupled by antiferromagnetic interactions on the ladder structure, has drawn much attention, because of its similarities to the high Tc cuprates; existence of a spin gap and d-wave superconductivity upon hole doping [1,2]. As a realization of the 2-leg spin ladder, $A_{14}Cu_{24}O_{41}$ compounds has been actively investigated, since single crystal is available and holes are naturally doped with the site A occupied by di-valent ions, such as Sr^{2+} and Ca^{2+}. It is noteworthy that superconductivity was actually found with Tc=12K in highly hole-doped $Sr_{0.4}Ca_{13.6}Cu_{24}O_{41.84}$ under an applied hydrostatic pressure of 3GPa [3].

One of the complications that this series of compounds may have is the CuO_2 chain layer stacking alternately with the Cu_2O_3 ladder layer. Doped holes are distributed to the ladders *and* to the chains, sometimes making interpretation of experimental data difficult. Optical reflectivity measurement is one of the rare experimental techniques available to distinguish holes on the chain and the ladder sites [4]. At room temperature, the number of doped holes on the ladder increases as the A=Sr^{2+} site is substituted by Ca^{2+} ions in $(Sr_{14-x}Ca_x)Cu_{24}O_{41}$ [4]. At the x=8 and 11 composition where the ladders have relatively large number of holes, charge ordering (CDW) of paired holes have been observed [4]. In the purely A=Sr compound, magnetic and charge anomalies at T~200K have been detected by various probes [5-7]. In order to identify whether the anomalies correspond to a charge order, or some other effects, we have measured optical reflectivity.

Only recently, the other end compound A=Ca became available in a form of thin film on Sr14 substrate [8]. DC resistivity exhibited a metallic behavior down to ~20K; comparing to Ca12, the metallic behavior extends to a lower temperature in Ca14. Although evaluation of optical constants in thin film is not straight-forward [9], we present a preliminary measurement of Ca14.

RESULTS AND DISCUSSION

We have measured optical reflectivity using a Fourier transform spectrometer (Bruker IFS113v; 4meV<ω<0.5 eV) and a grating monochrometer (Jasco CT-25C; 0.5eV<ω<5eV). In Fig.1a, we show the optical reflectivity of Sr14 compound. At room temperature, the reflectivity exhibits a Drude-like behavior in the infrared regime with a plasma edge at ω_p~0.6eV. However, the temperature dependence is different from that of a regular metal; below T~200 K, the plasma edge diminishes so that the reflectivity behaves as that of an insulator. Instead, in the visible region is seen an increase of the peak intensity at energy ω~2eV and 3eV. From the measurements of high Tc cuprates [10], CuO_2 chain compound (Li_2CuO_2) and $(Sr_{14-x}Ca_x)Cu_{24}O_{41}$ [4], the 2eV peak has been identified as the charge-transfer (CT) gap and the 3eV peak as an excitation of Cu^{3+} on the CuO_2 chains. The transfer of the spectral weight is evident from the optical conductivity (Fig.1b), obtained by the standard Kramers-Kronig analysis. The spectral weight of far- to mid-infrared decreases with lowering temperature, being transferred to the excitation centered at 2 and 3eV. Below 200K, only the spectral weight in the mid-infrared region is remaining with a broad peak at ω~0.4 eV. This feature is common to lightly doped $La_{2-x}Sr_xCuO_4$ and the peak is known as the 'mid-infrared' absorption [10]. We have integrated the optical conductivity from 4meV to the iso-absorption point (1.2eV), and extracted the number of carriers, electrons (holes) involved in the low-energy excitations on the ladder (Fig.1c). The number of carriers on the ladder shows a rapid decrease and then saturates at T~200K.

The decrease of the effective carrier number on the ladder may allow two interpretations; (1) the holes remain on the ladder, but they are localized by forming a charge order, and (2) the holes move to the chain at low temperatures, and the number of holes on the ladder decreases. So far, interpretation (1) has been favored to explain the anomalies at T~200K [6,7]. However, in optical measurements of Sr14 compound, there has been no direct observation of charge order, such as a CDW gap or a collective excitation which are present at higher Ca content [4]. The increase of the

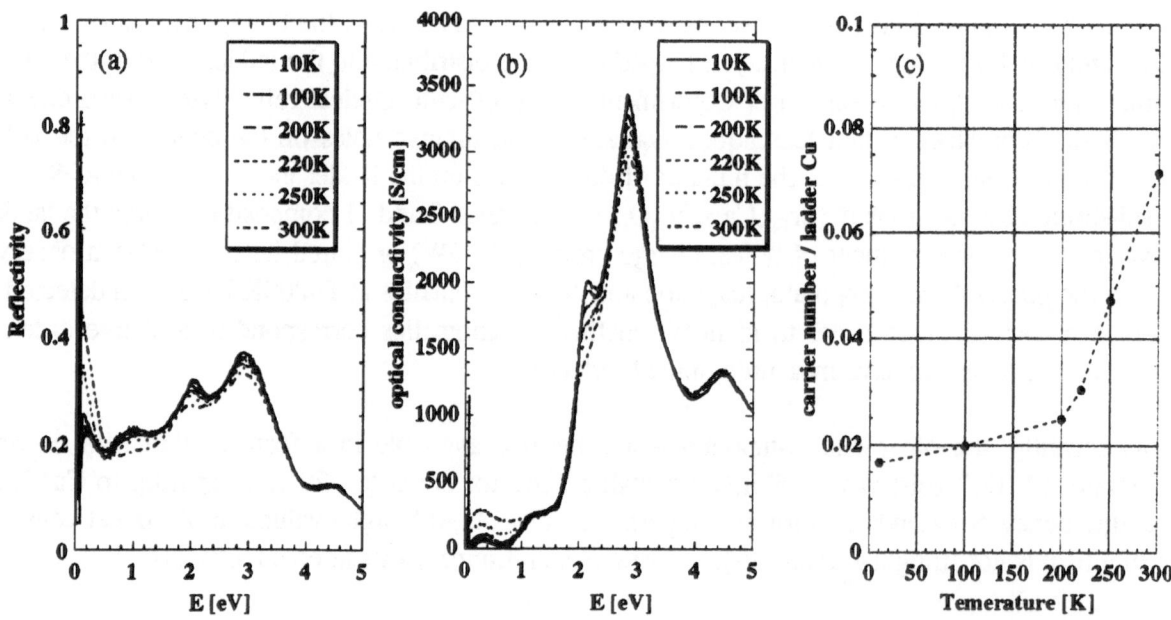

Fig.1 Optical reflectivity (a), conductivity (b) and number of carriers on the ladder (c) of Sr14 compound.

Cu^{3+} signal of the chain ($\omega \sim 3eV$ peak) at low temperatures suggests that the situation (2) is not a negligible factor to explain the anomalies.

The other end compound Ca14 has the most metallic DC conductivity in the $A_{14}Cu_{24}O_{41}$ family [8]. In Fig.2, we show the optical reflection of Ca14 thin film evaporated onto the Sr14 substrate with a thickness of nominally 100nm. Since the skin depth of infrared light ($\sim 1.8 \mu m$ at $\omega = 1eV$) is longer than the film thickness, the substrate is visible through the Ca14 thin film. However, the enhanced reflection of the sample at 10K is due to the high conductivity of Ca14, since the substrate is insulating, and exhibits relatively low reflectance. If the film thickness is reliable, the conductivity of Ca14 is in the same order with Ca11, since the calculated reflection from Ca11 thin layer exhibits similar reflectance.

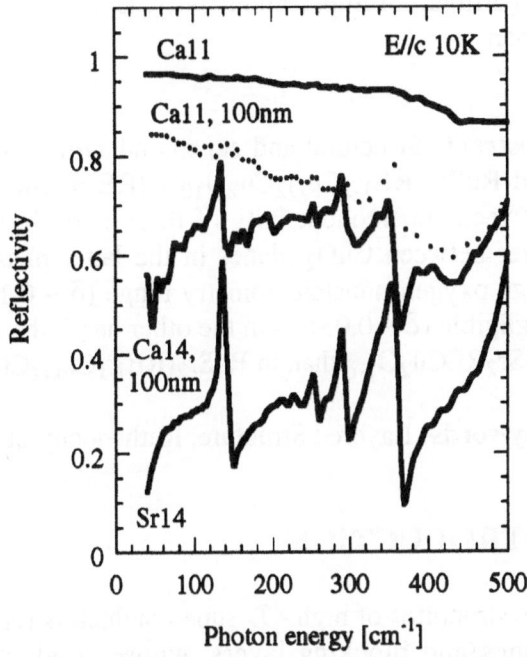

Fig.2 optical reflectivity of bulk (Ca11 and Sr14) and thin film on Sr14 samples (Ca14). Ca11 thin film data is a simulation.

CONCLUSIONS

We have measured optical reflection of Sr14 compound, where various anomalies have been reported at T~200 K. The number of carriers on the ladder was found to decrease and saturate at this characteristic temperature. The spectral weight was found to shift to the CT excitation on the ladder *and* the Cu^{3+} signal on the CuO_2 chain. The latter suggests that the doped hole on the ladder moves to the chain, requesting us to re-consider the simple charge density wave picture favored to interpret the T~200 K anomalies.

Acknowledgement: The research has been partly supported by COE grant from Monbusho.

---

* e-mail: kenji@lyra.t.u-tokyo.ac.jp

1. T. M. Rice S. Gopalan, and M. Sigrist, Europhys. Lett. **23**, 445 (1993); S. Gopalan, T. M. Rice and M. Sigrist, Phys. Rev. B**49**, 8901 (1994).
2. M. Sigrist, T. M. Rice and F. C. Zhang, *et al.*, Phys. Rev. B**49**, 12058 (1994); M. Troyer, H. Tsunetsugu and T.M. Rice, Phys. Rev. B**53**, 251 (1996).
3. M. Uehara *et al.*, J. Phys. Soc. J pn. **65**, 2764 (1996).
4. T. Osafune *et al*, Phys. Rev. Lett. **78**, 1980 (1997); Phys. Rev. Lett. **82**, 1313 (1999).
5. N. Motoyama *et al.*, Phys. Rev. B**55**, R3386 (1997).
6. T. Imai *et al.*, Phys. Rev. Lett. **81**, 220 (1998); M. Takigawa *et al*, Phys. Rev. B**57**, 1124 (1998).
7. H. Kitano *et al.*, private communication (1999).
8. Y. Furubayashi *et al.*, Phys. Rev. B**60**, R3720 (1999).
9. K.E. Kornelsen *et al.*, Phys. Rev. B**44**, 11882 (1991).
10. Y. Tokura *et al.*, Phys. Rev. B**41**, 11657 (1990); S. Uchida *et al.*, Phys. Rev. B**43**, 7942 (1992).

Structural and Superconducting Properties of Ruthenocuprates

Kenji D. Otzschi, Toru Hinouchi, Tomonori Mizukami, Jun-ichi Shimoyama, and Kohji Kishio

Department of Superconductivity, University of Tokyo, 7-3-1 Hongo, Bunkyo-ku, Tokyo 113-8656, Japan

Abstract: Structural and superconducting properties of layered ruthenocuprates, $RuSr_2RECu_2O_{8-\delta}$ and $RuSr_2(RE_{1-x}Ce_x)_2Cu_2O_{10-\delta}$ (RE = Sm, Eu, Gd), have been investigated. The differences between superconductivity of these two phases are due to their structural differences, *i.e.* fluorite layer between CuO_2 planes in the latter phase. Because of it, $RuSr_2(RE_{1-x}Ce_x)_2Cu_2O_{10-\delta}$ has a large oxygen-nonstoichiometry range ($\delta \sim 0.15$), while the oxygen defects in $RuSr_2RECu_2O_{8-\delta}$ are negligible ($\delta \sim 0.03$). On the other hand, the RE/Sr substitution is considered to be more feasible in $RuSr_2RECu_2O_{8-\delta}$ than in $RuSr_2(RE_{1-x}Ce_x)_2Cu_2O_{10-\delta}$.

Keywords: Layered Structure, Ruthenocuprate, Fluorite, Perovskite Blocking Layer

INTRODUCTION

The structures of high - T_c superconductors (HTSC) are considered to be alternate stackings of CuO_2 planes and blocking layers, whose conductivity is known to affect on their electromagnetic anisotropy, defined as $\gamma^2 = \rho_c/\rho_{ab}$, and the flux pinning strength. Developments of low-anisotropic HTSC are crucial for their practical applications, because of their high J_c in a high magnetic field. Among them, $YBa_2Cu_3O_{7-\delta}$ is known to have the lowest anisotropy ($\gamma^2 \approx 50$). $RuSr_2RECu_2O_{8-\delta}$ (Ru1212(RE), RE = Sm, Eu, Gd) has a Y123-related structure [1], where Cu-O chains are replaced by highly conductive $SrRuO_3$ perovskite-block ; The reported resistivity of $SrRuO_3$ single crystal is 2.8×10^{-4} Ωcm (at 300K) [2], making this phase a useful candidate for high-J_c HTSC. Another structurally related $RuSr_2(RE_{1-x}Ce_x)_2Cu_2O_{10-\delta}$ (Ru1222(RE), RE = Sm, Eu, Gd) phase is also found to have this blocking layer, where the $(RE,Ce)_2O_2$-fluorite layer substitutes for the RE layer. Moreover, both Ru1212 (RE = Gd) [3] and Ru1222 (RE = Eu, Gd) [4] phases have been recently reported to exhibit coexistence of superconductivity in the CuO_2 planes ($T_{c, onset} = 45K \sim 40K$) and ferromagnetism in RuO_2 layers ($T_{Curie} = 135K$ and 80 K, respectively). In this paper, we describe structural and superconducting properties of these two ruthenocuprates, reporting differences of their oxygen nonstoichiometry and cations' substitution.

EXPERIMENTAL

Polycrystalline samples with nominal compositions, $RuSr_2RECu_2O_8$ and $RuSr_2(RE_{0.7}Ce_{0.3})_2Cu_2O_{10}$, were prepared by solid state reaction from high purity powders of RuO_2, $SrCO_3$, RE_2O_3 (RE = Sm, Eu, Gd, Dy), CeO_2, and CuO. The optimized calcination and sintering conditions [5] were applied for each phase, and some samples were post-annealed under various atmosphere as described later. Phase compositions were analyzed by powder X-ray diffraction collected by a Mac Science MXP18 assembly with a step of 0.02° (Cu-K_α). Oxygen contents were measured by hydrogen-reduction method using a Mac Science TG-DTA 2000. The resistivity measurements were carried out by standard four-probe method in the temperature range from 4.2 to 280K. Oxygen nonstoichiometry was measured using a electro-microbalance with a sensitivity of $\sim 50\mu g$, which corresponds to $\Delta\delta \sim 0.001$ when a 3g of the sample is used.

RESULTS AND DISCUSSION

SrRuO$_3$ and Sr(Ru$_{0.5}$RE$_{0.5}$)O$_3$ are known to easily form as a second phase impurity during the synthesis of Ru1212 and Ru1222 phases, respectively [1, 6], and bother further measurements. So we have optimized the synthesis conditions [5], and found that the low temperature calcination at 800°C in air drastically improves sample purity. Samples were then sintered at 1000 ~ 1020°C (for Ru1212) or 1070°C (for Ru1222) several times with intermediate grindings.

Figure 1(a) shows temperature dependence of resistivity for the Ru1212(Gd) and Ru1212(Eu) phases. The as-prepared Ru1212(Gd) sample exhibited a metallic temperature-dependence down to about 100K and showed superconducting transition ($T_{c, onset}$ = 43K, $T_{c, zero}$ = 17K) after a slight upturn. Annealing under high-O$_2$ pressure (350atm) did not make much difference in its resistivity, but only resulted in a slight increase of $T_{c, onset}$ to 45K and steepen the slope of its normal state resistivity against temperature, which is due to improvements of grain connectivity. Meanwhile, a long-time sintering at 1020°C (6 ~ 8 days) was necessary to obtain superconducting Ru1212(Eu) sample. Post-annealing under flowing O$_2$ did not affect its T_c, but only changed its normal state resistivity. It is noticed that the superconducting transition temperature of Ru1212(Eu) ($T_{c, onset}$ = 25K and $T_{c, zero}$ = 7K) is much lower than that of Ru1212(Gd).

Figure 1(b) and 1(c) respectively show temperature dependence of resistivity for the Ru1222(Gd) and the Ru1222(Eu) phases. The as-prepared sample of Ru1222(Gd) (sintered in air at 1070°C, then quenched) did not show any resistive transitions down to the lowest temperature, only exhibiting semiconductive behavior. By annealing in air at lower temperatures, e.g. 500°C for 1 day, the Ru1222(Gd) phase became superconducting with $T_{c, onset}$ = 35K and $T_{c, zero}$ = 16K. It should be noticed that further annealing under flowing O$_2$ resulted in a large increase of $T_{c, onset}$ to 43K and $T_{c, zero}$ to 27K, while its normal state resistivity became metallic. As clearly seen in the figure, Ru1222(Eu) showed the similar annealing-condition dependence to Ru1222(Gd), i.e. semiconducting to metallic/superconducting. And it has the max

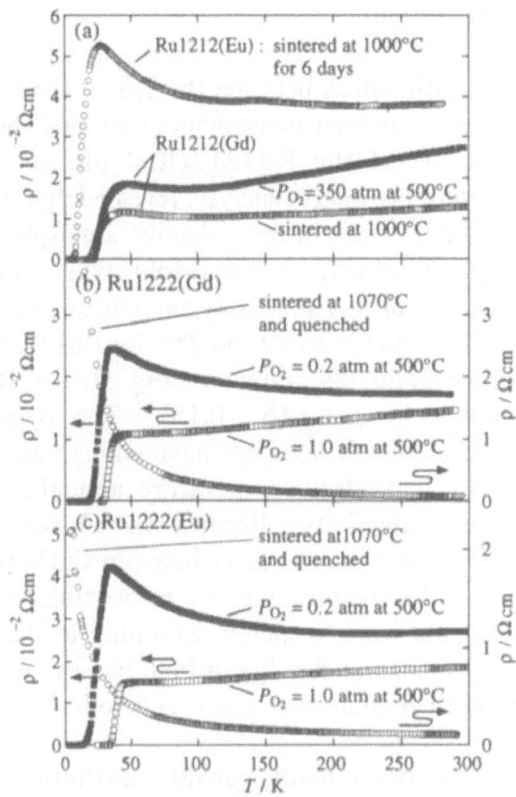

Fig.1. Temperature dependence of resistivity of (a) RuSr$_2$RECu$_2$O$_{8-\delta}$ (RE=Gd, Eu), (b) RuSr$_2$Gd$_{1.4}$Ce$_{0.6}$Cu$_2$O$_{10-\delta}$, and (c) RuSr$_2$Eu$_{1.4}$Ce$_{0.6}$Cu$_2$O$_{10-\delta}$.

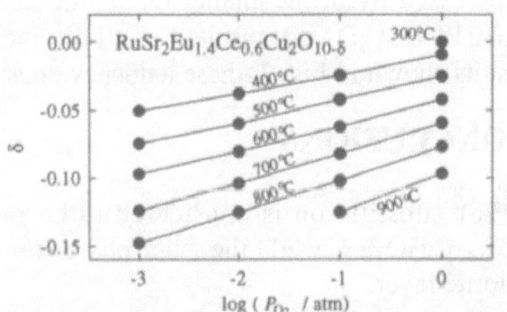

Fig.2. Oxygen nonstoichiometry of Ru1222(Eu).

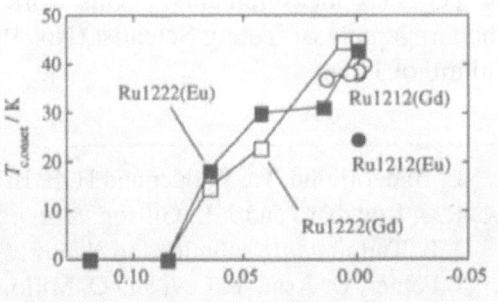

Fig.3. Oxgen-content dependence of T_c's of Ru1212 and Ru1222 phases.

transition temperature of $T_{c, onset} = 40K$ and $T_{c, zero} = 28K$, which is almost the same as that of Ru1222(Gd).

The differences between the T_c's of the Ru1212(RE) and the Ru1222(RE) phases can be explained by their oxygen nonstoichiometry and the cations' (RE/Sr) substitution. First of all, the oxygen content of the Ru1212(RE) phases was measured to be $y = 8.00\pm0.01$, and detailed thermogravimetric analysis revealed that the oxygen deficiency, δ, in the Ru1212(RE) phases is less than 0.03 even in the reductive atmosphere ($P_{O_2} = 10^{-3}$ atm at 880°C). Thus, $SrRuO_3$ blocking-layer is robust, i.e. the 6-fold oxygen coordination around Ru^{5+} ion is considered to be rigid. Because of this small oxygen nonstoichiometry, the T_c's of the Ru1212(RE) phases ($T_{c, onset} = 43 \sim$ 46K for Ru1212(Gd) and 25K for Ru1212(Eu)) are not sensitive to the annealing conditions. On the other hand, as shown in Fig. 2, the Ru1222(RE) phase shows a relatively large oxygen-nonstoichiometry ($\delta \sim 0.15$). The as-prepared Ru1222(RE) contains a large amount of oxygen defects, and it does not have enough amount of carrier to show superconductivity (Fig. 1(b) and 1(c)). By low-temperature annealing, the amount of doped holes increases to evoke its superconductivity. Because $SrRuO_3$ blocking-layer is considered to accommodate very few oxygen defects, the fluorite layer between CuO_2 planes, which differentiates the Ru1222 structure from the Ru1212 structure, must be responsible for this nonstoichiometry. Samples with various oxygen contents were prepared according to the T-P-y phase-diagram (Fig. 2), and the oxygen-content dependence of the T_c's of both phases are summarized in Fig. 3. The T_c increases with increasing oxygen content in all cases, and both phases are considered to be still in the underdoped regime.

On the other hand, partial substitution of RE for Sr is possibly responsible for lower T_c of Ru1212(Eu) than that of Ru1212(Gd). Eu^{3+}, which is larger than Gd^{3+} and has closer ionic radius to Sr^{2+}, tends to substitute for Sr^{2+}-site to decrease the amount of carrier holes, just like lager lanthanoide ions replace Ba^{2+} in the RE123 systems. The driven out Sr^{2+} can easily form $SrRuO_3$ impurity in the reaction, which hampers superconductivity of Ru1212(Eu). This substitution can be suppressed by using smaller rare earth cations, and actually partial substitution of Dy for Gd, i.e. Ru1212($Gd_{0.8}Dy_{0.2}$), resulted in a little increase of $T_{c, onset}$ to 49K. Meanwhile, according to the results shown in Fig. 3, these tendency does not seem remarkable in the Ru1222(RE) systems.

CONCLUSIONS

RE/Sr substitution is considered to be more feasible in $RuSr_2RECu_2O_{8-\delta}$ than in $RuSr_2(RE_{1-x}Ce_x)_2Cu_2O_{10-\delta}$, while the latter phase shows much larger oxygen nonstoichiometry because of the fluorite layer.

Acknowledgments. Authors are grateful to Prof. Y. Ueda and Dr. M. Isobe (ISSP, Univ. of Tokyo) for TG-DTA measurements. This work was supported by CREST, JST, and a Grant-in-Aid for Encouragement of Young Scientist (No. 10750603) given by the Ministry of Education, Science and Culture of Japan.

1. L. Bauernfeind, W. Widder and H. F. Braun, *Physica* **C254** 151 - 158 (1995).
2. R. J. Bouchard and J. L. Gillson, *Mater. Res. Bull.* **7**, 873 (1972).
3. J. L. Tallon *et al.*, submitted to *Nature*; C. Bernhard *et al.*, submitted to *Phys. Rev. B*.
4. I. Felner, U. Asaf, Y. Levi and O. Millo, *Phys. Rev. B* **55**, R3374 - R3377 (1997).
5. K. Otzschi, T. Mizukami, T. Hinouchi, J. Shimoyama and K. Kishio, to be published in *J. Low Temp. Phys.* as proceedings of PC-MOS99 (Stockholm, 1999).
6. T. Kaibin, Q. Yitai, Z. Yadun, Y. Li, C. Zuyao and Z. Yuheng, *Physica* **C259** 168 - 172 (1996).

Antiferromagnetic Correlations and the Pseudogap in HTS Cuprates

Jeffery L. Tallon

Industrial Research Ltd., P.O. Box 31310, Lower Hutt, New Zealand. E-mail: J.Tallon@irl.cri.nz

Abstract: Evidence is presented from $1/T_1T$ NMR and ARPES data for the sudden disappearance of 2D antiferromagnetic (AF) correlations in the lightly overdoped region ($p \approx 0.19$) at the T=0 metal-insulator transition where the pseudogap energy falls to zero. AF fluctuations thus appear to be intimately associated with the pseudogap and serve primarily to weaken superconductivity, strongly reducing the condensation energy and superfluid density.

Keywords: antiferromagnetism, pseudogap, superconductivity, doping

INTRODUCTION

The high-T_c superconducting (HTS) cuprates exhibit a generic phase behaviour as a function of hole concentration, p, ranging from an antiferromagnetic (AF) insulator at zero doping to a metallic Fermi liquid at high doping with the appearance of superconductivity (SC) between. In spite of the disappearance of the 3D Néel state prior to the onset of SC 2D AF correlations are observed to persist well out into SC compositions [1]. In addition, underdoped cuprates exhibit normal state (NS) correlations above T_c which result in a depletion of the density of states (DOS) referred to as the *pseudogap* [2]. While many models have been proposed for understanding the pseudogap a widely favoured scenario is that it arises from incoherent pairing fluctuations above T_c [3], an outlook not shared by the present author. In view of the lack of consensus as to the nature of the pseudogap and the origin of the pairing interaction a number of key questions may be asked: (i) is the progression from a strongly AF-correlated underdoped phase to a Fermi liquid smooth or discontinuous? (ii) is the disappearance of the pseudogap gradual (as might occur in a phase fluctuation model) or sudden (as for a competing correlation)? (iii) what is the relation, if any, between the pseudogap and AF correlations? The lack of such basic comprehension of the experimental situation goes a long way to understanding why so many mutually exclusive theoretical models of cuprate superconductivity are still extant [4]. We address these questions here.

AF CORRELATIONS AND THE PSEUDOGAP

It is widely believed that $1/^{63}T_1T$ provides a clear measure of AF correlations, where $1/^{63}T_1$ is the copper spin-lattice relaxation rate given by

$$1/T_1T \sim \sum_{\mathbf{q}} |A_{\mathbf{q}}|^2 \ \chi''(\mathbf{q},\omega_o) / \omega_o. \tag{1}$$

$A_{\mathbf{q}}$ are the hyperfine coupling form factors, ω_o is the NMR frequency and $\chi''(\mathbf{q},\omega_o)$ is the imaginary part of the dynamic spin susceptibility. Typical data for underdoped cuprates show that $1/^{63}T_1T$ has a Curie-like $1/T$ dependence at high T, generally associated with AF correlations, but then falls at lower T due to the opening of the pseudogap [5]. The maximum occurs at T* which is widely viewed as the temperature at which the spin gap opens. Millis, Monien and Pines [6] introduced a phenomeno-logical expression for $\chi(\mathbf{q},\omega_o)$ which is enhanced at the AF wave vector $\mathbf{q} = \mathbf{Q}_{AF} \equiv (\pi,\pi)$. Inserting this in eqn. (1) and assuming that the AF correlation length satisfies $\xi^2 \gg 1$ one finds $1/^{63}T_1T \approx a_1\chi_s \, \tau_{SF}$ where χ_s is the static spin susceptibility and τ_{SF} is the AF spin fluctuation lifetime. The Curie-like T-dependence of $1/^{63}T_1T$ at high T thus suggests that $\tau_{SF} \sim 1/T$ and thus $1/^{63}T_1 \approx a_2\chi_s$. These two relations actually are found to be well satisfied in the case of Y-124 [7]. The same approximations yield $1/^{17}T_1T \approx a_3\chi_s$. Finally, the Knight shift, $K_s \approx a_3 \chi_s + \sigma$ where σ is the chemical shift. Thus

$$1/^{63}T_1 \sim 1/^{17}T_1T \sim (K_s - \sigma) \sim \chi_s. \tag{2}$$

Again, these relationships are well satisfied for Y-124 [7]. The characteristic T-dependence of $1/^{63}T_1T$ can thus be seen to derive from the $1/T$ dependence of τ_{SF} and the T-dependence of χ_s which, like S^{el}/T (with S^{el} = electronic entropy), is progressively and smoothly depressed with decreasing T due to the pseudogap. T* thus loses its meaning as a well-defined point at which a spin gap opens.

Experimentally, $1/^{63}T_1T$ maintains its high-T Curie-like T-dependence across the entire overdoped region in La-214 [5] thus suggesting that AF correlations also persist across the overdoped region. However, we argue that this inference is not justified. In La-214 (and less markedly in other cuprates) χ_s itself is found to develop an increasing Curie-like dependence in the overdoped region [8] possibly due to the proximity of the van Hove singularity. By reference to equ. (2) this could account for the persisting $1/T$ dependence of $1/^{63}T_1T$, i.e this dependence in the overdoped region could derive from χ_s rather than τ_{SF}. A more robust measure of AF correlations is the ratio $^{17}T_1/^{63}T_1$ in which the effect of χ_s is divided out. From the above considerations $^{17}T_1/^{63}T_1 \sim \tau_{SF}$ but more generally,

$$^{17}T_1 / ^{63}T_1 \sim \langle 1+f_q^2 \rangle \approx 1 + a_3 \xi^2 = 1 + C_{AF} T^{-1} \tag{3}$$

where the average is over q. Here f_q is the ratio of the enhanced AF susceptibility to the bare FL susceptibility and C_{AF} is a measure of the AF correlations. The ratio $^{17}T_1/^{63}T_1$ has been determined [9] by Takigawa et al. for Y-123 at two doping levels and by Tomeno et al. for Y-124 and the T-dependence of eqn. (3) is found. We have fitted the data to obtain the p-dependent parameter $C_{AF}(p)$ and this is plotted in Fig. 1 (diamonds). C_{AF} falls sharply towards zero at the critical doping point of p=0.19 just where the pseudogap energy, E_g, determined from NMR and heat capacity falls to zero, as shown by the inset. Values of p are determined from δ values in $YBa_2Cu_3O_{7-\delta}$ or from the roughly parabolic variation of T_c with p which may be approximated by $T_c = T_{c,max} \times [1 - 82.6(p-0.16)^2]$ [10].

One major effect of AF correlations is to heavily reduce quasiparticle (QP) lifetimes near the zone boundary at $k=(\pi,0)$ due to scattering from spin fluctuations. This may be seen in the suppression of the NS QP peaks in ARPES spectra at the FS crossing near $(\pi,0)$ but not at the FS crossing on the zone diagonal near $(0,0)$ [11]. If one focuses on the spectra near $(\pi,0)$ at about 100K, i.e. above T_c, then underdoped samples show the suppression of the QP peak as well as the pushing back of the leading edge due to the NS pseudogap. In contrast, overdoped samples exhibit a closure of the pseudogap (the mid-point of the leading edge coincides with the Fermi energy) and the recovery of the QP peak. We have examined the ARPES spectra at 100K of 11 Bi-2212 samples with different doping states from the Stanford and Chicago groups and summarise the data in Fig. 2. This, again, shows the pseudogap energy falling to zero [12] near p=0.19 and the abrupt recovery of the QP peak at the same point. The T_c values are plotted as open squares for all spectra with suppressed QP peaks and as solid squares where the QP is fully recovered. There is a sudden recovery at p=0.19 as indicated by the spectra shown in the figure either side of this point [11,12]. Also shown in Fig. 1 are the values of T_{min} where the ab-plane resistivity of La-214 (open circles) and Y-123 (filled circles) cross over from a metallic to semiconducting T-dependence. The point where $T_{min} \rightarrow 0$ is the metal/insulator transition and it clearly coincides with the disappearance of both the pseudogap and AF correlations.

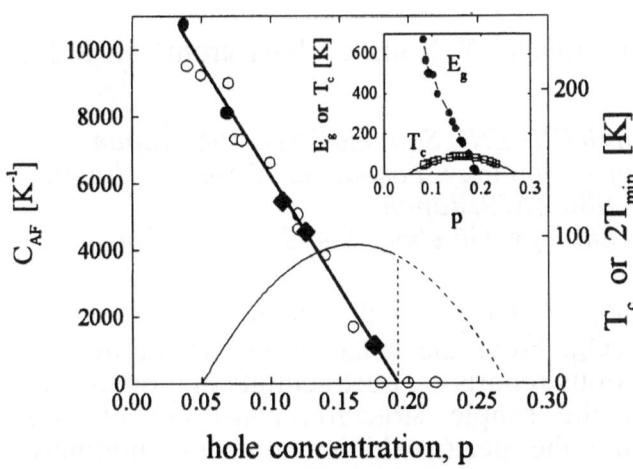
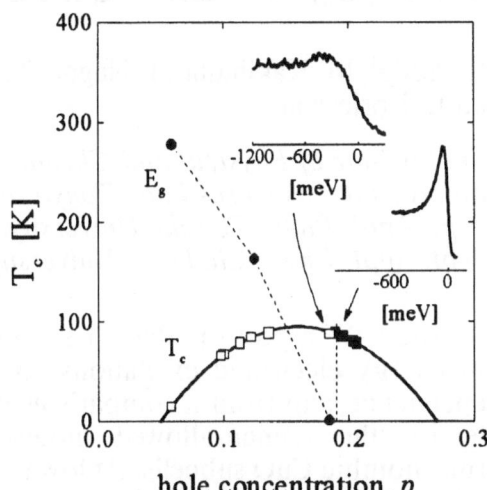

Fig. 1. The p-dependence of the AF parameter C_{AF} for Y-123 and Y-124 (♦) and crossover temperature T_{min} for La-214 (O) and Y-123 (●). Inset: the p-dependence of T_c and the pseudogap energy, E_g, for $Y_{0.8}Ca_{0.2}$-123.

Fig. 2. The p-dependence of E_g and T_c for for Bi-2212 from ARPES. Open squares: no quasi-particle (QP) peak. Filled squares: full QP peak as shown in the two insets.

The above results reveal the *sudden* disappearance of AF correlations at the critical doping state p=0.19, just where the pseudogap disappears and at the location of the metal-insulator transition at T=0. At this point the T=0 condensation energy, superfluid density and critical currents all pass through a sharp maximum and this has been interpreted within a quantum critical point scenario [13].The sudden loss of AF correlation is further borne out by inelastic neutron scattering which, for fully oxygenated Y-123, shows only a weak enhancement in susceptibility at $q=Q_{AF}$ that barely rises above background [1]. Moreover, Rübhausen et al [14] have observed a sudden loss of the 2000 cm⁻¹ two-magnon Raman scattering peak in the lightly overdoped region (at p≈0.20). This peak, observed in underdoped and optimally-doped samples, is attributed to a photon-induced two-magnon excitation in an AF background. Its demise indicates the destruction of the AF background. These results indicate that AF correlations are intimately associated with the pseudogap which however is more than just spin correlations. It is known from heat capacity that both spin and charge degrees of freedom freeze out equally with the establishment of the pseudogap state [2]. Spin- and charge-ordered stripes are then a possible scenario. Antiferromagnetism would appear only to weaken and suppress SC with progressive underdoping and the question has to asked whether spin fluctuations can be responsible for pairing if they are substantially suppressed over much of the overdoped side.

1. P. Bourges, in *Gap Symmetry and Fluctuations in High Temperature Superconductors* ed. by J. Bok, G. Deutscher, D. Pavuna and S.A. Wolf (Plenum, 1998).
2. J.W. Loram et al., Physica C **235-240**, 134 (1994).
3. V.J. Emery and S.A. Kivelsen, Nature **374**, 434 (1995).
4. T. Timusk and B. Statt, Rep. Prog. Phys., **62**, 61 (1999).
5. C. Berthier, M.H. Julien, M. Horvatic and Y. Berthier, J. Phys. I France **6**, 2205 (1996).
6. A.J. Millis, H. Monien and D. Pines, Phys. Rev. B **42**, 167 (1990).
7. G.V.M. Williams, J.L. Tallon and J.W. Loram, Phys. Rev. B **58**, 15053 (1998).
8. J.W. Loram et al., *10th Anniversary HTS Workshop* (World Scientific, Singapore, 1996), p. 341.
9. M. Takigawa et al., Phys. Rev. B **43**, 247 (1991); I. Tomeno et al., Phys. Rev. B **49**, 15327 (1994).
10. J.L. Tallon et al., Phys. Rev. B **51**, 12911 (1995).
11. C. Kim et al., Phys. Rev. Lett., **80**, 4245 (1998).
12. H. Ding et al., Nature **382**, 51 (1996); M.R Norman et al., Phys. Rev. Lett., **79**, 3506 (1997).
13. J.L. Tallon et al., phys. stat. sol. **215**, 531 (1999).
14. M. Rübhausen et al., Phys. Rev. Lett., **82**, 5349 (1999).

CHARGE AND SPIN DYNAMICS IN SPIN-LADDER $Sr_{14}Cu_{24}O_{41}$ INVESTIGATED BY RAMAN SCATTERING

M. Osada[1], M. Kakihana[2], I. Nagai[2], T. Noji[3], T. Adachi[3], Y. Koike[3], J. Bäckström[4], M. Käll[4], and L. Börjesson[4]

[1]*The Institute of Physical and Chemical Research (RIKEN), Saitama 351-0198, Japan*
[2]*Materials & Structures Lab., Tokyo Institute of Technology, Yokohama 226-8503, Japan*
[3]*Dept. Appl. Phys., Tohoku University, Sendai 980-8579, Japan*
[4]*Dept. Appl. Phys., Chalmers University of Technology, Göteborg S-412 96, Sweden*

Abstract: We report on Raman scattering study of changes in the phonon spectrum and low-energy electronic excitations for $Sr_{14}Cu_{24}O_{41}$. From the polarization and Ca-doping dependence, and from a comparison with previous reports on isostructural compounds, we identify the Raman-allowed modes within the simple structure, consisting of two orthorhombic CuO subcells. At low temperatures, the spectra exhibit a number of normally forbidden phonons as well as additional low-energy excitations. For $T < \sim 200$ K, we observe a large loss of low-energy spectral weight that linked to changes in the linewidth of the 246-cm^{-1} phonon. These effects, analogous to the pseudogap effect in the high-T_c cuprates, can be correlated with recently reported resistivity experiments that show the pair formations in the ladders for $T < \sim 200$ K.

Keywords: $Sr_{14-x}Ca_xCu_{24}O_{41}$, Spin-ladder system, Raman scattering, Pseudogap

INTRODUCTION

Recently, spin-ladder $Sr_{14}Cu_{24}O_{41}$ material has attracted much interest owing to the discovery of superconductivity in the Ca-doped materials under high-pressure [1]. Apart from the practical importance as a non-CuO_2-plane superconductivity, the fundamental understanding of the charge and spin dynamics in this system becomes important testing ground for high-T_c superconductivity (HTS), and the so-called pseudogap phenomenon is an interesting topic [2].

Raman scattering has yielded important information about the physical properties of strongly correlated systems. In HTS cases for example, Raman scattering can act as a tool to probe the low-lying elementary excitations associated with the CuO_2 planes, which are essential to investigate the HTS mechanism [3]. Although a similar study is expected for spin-ladder $Sr_{14}Cu_{24}O_{41}$ material, the previous Raman studies have been performed only for polycrystalline samples [4] and there is no systematic study on temperature dependence of low-energy excitation spectra. Towards a better understanding of the charge and spin dynamics in this system, we investigate Raman scattering study of changes in the phonon spectrum and low energy electronic excitations for $Sr_{14-x}Ca_xCu_{24}O_{41}$.

EXPERIMENTAL

Single crystals of $Sr_{14-x}Ca_xCu_{24}O_{41}$ with $x = 0$ and 7 were grown by the NaCl flux method [5]. Raman measurements were performed in a backward micro-configuration, using the 514.5 nm line from an Ar$^+$ laser (~ 1 mW). The scattered light was dispersed by a subtractive triple spectrometer (Atago Bussan R64000-COE) and collected with a liquid-nitrogen-cooled CCD detector. For low-temperature measurements ($T = 10 - 400$ K), the crystals were mounted in a continuous helium flow optical cryostat with a temperature stability better than ± 1 K, which was confirmed by recording anti-Stokes spectra.

RESULTS AND DISCUSSION

Fig. 1 shows polarized Raman spectra of $Sr_{14}Cu_{24}O_{41}$ single crystal at room temperature. Strong peaks show up in the zz and xx polarizations while the phonon features are considerably weaker in the yy polarization. The Ca-doped sample shares with the same polarization behavior, although phonon peaks become broad even at low temperatures and fine structures are lost. The main reason of less-structural spectra in the Ca-doped sample is randomness of crystal structures.

In the zz and xx polarizations, four prominent peaks at 246, 301, 545, and 580 cm^{-1} can be observed. These four A_g modes stem from the atoms of Sr, ladder-Cu, ladder-O, and chain-O which are Raman-active within the ideal structure, consisting of $[CuO_2]_\infty$ and $[Sr_2Cu_2O_3]_\infty$ orthorhombic (Fmmm) subcells [4]. From simple mass considerations, we assign two low-frequency modes at 246 and 301 cm^{-1} to Sr

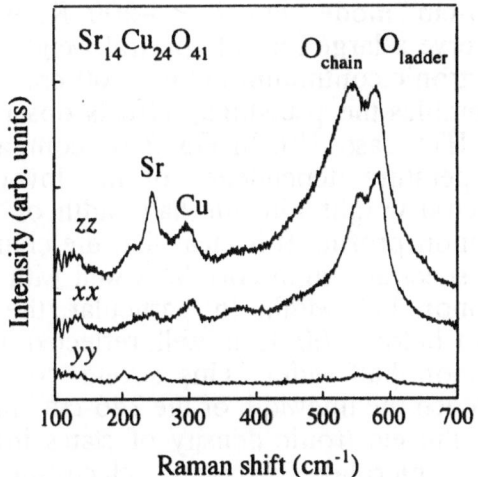

Fig. 1. Polarized Raman spectra of $Sr_{14}Cu_{24}O_{41}$ single crystal under three different geometries, together with our proposed assignment.

and ladder-Cu modes, respectively. Further support of this assignment comes from the Ca-doping dependence. In the Ca-doped sample, the (Sr/Ca) mode is expected at a higher frequency due to the decreasing average mass and ionic radius of the (Sr/Ca) ions, which is in accordance with the observed hardening of the 246-cm^{-1} mode. The identification of the two high-frequency oxygen modes is troublesome because their frequencies are very close. From a comparison with $Ca_{1-x}Sr_xCuO_2$ containing similar CuO chains [6], however, we assign the lower frequency mode to the chain-O mode and the higher one to the ladder-O mode.

We now turn to the temperature dependence in $Sr_{14}Cu_{24}O_{41}$. Fig. 2 shows the temperature dependence of zz-polarized spectra. At low temperatures, the spectra exhibit a number of normally forbidden phonons as well as additional low-energy excitations. These features up to 3000 cm^{-1}, which are much pronounced in the zz polarization (along the chains), do not have a Raman-allowed one-phonon origin and will be discussed elsewhere.

When altering the temperature, we find some interesting changes in the low-frequency modes. The strongest effect concerns the Sr mode at ~ 246 cm^{-1}, which shows a complex behavior (Fig. 3). It is clearly seen that the linewidth of the Sr phonon decreases rapidly below 200 K but again increase at further low temperatures. We note that the sharpening occurs at about the same temperature as recent resistivity experiments report anomalies that can be attributed to the hole pairing in the ladders for $T < \sim 200$ K. [7].

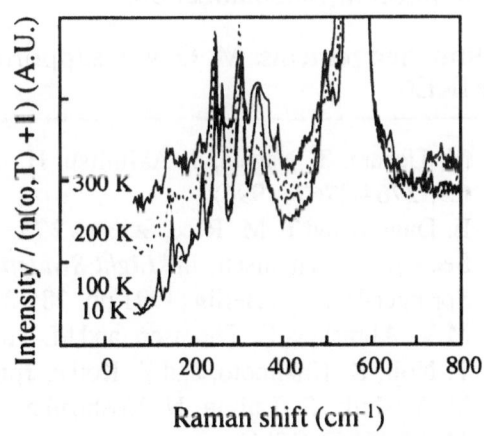

Fig. 2. Temperature dependent Raman spectra of $Sr_{14}Cu_{24}O_{41}$ for zz polarization.

Obviously, the electronic continuum is connected with the anomalous changes of the 246-cm^{-1} mode. For $T < \sim 200$ K, we also observe a large loss of spectral weight in the electronic continuum below ~ 700 cm^{-1}, which resembles the psuedogap effects observed in the HTS cases [8]. In Fig. 3 we compare the temperature dependence of the low-energy spectral weight with the half-width of the Sr phonon profile. It is clear that the change in the spectral weight correlates well with the Sr phonon half-width. In particular, the rapid drop below 200 K is well reflected by the phonon half-width. This strong correlation between the linewidth of the 246-cm^{-1} phonon and the electronic density of states indicates that interplay between electronic and vibrational degrees of freedom is affected at ~ 200 K.

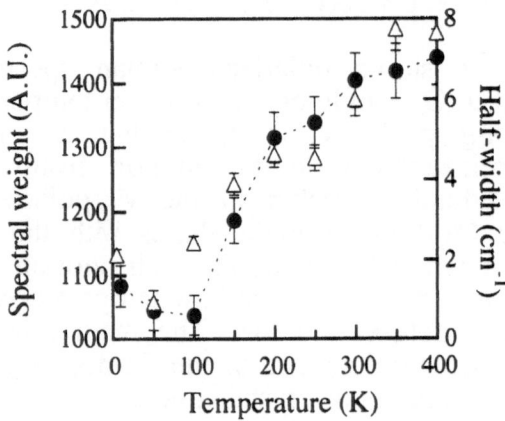

Fig. 3. Spectral weight below 700 cm^{-1} (closed circles) and the 246-cm^{-1} phonon half-width (open triangles) versus temperature.

The similarity between $Sr_{14}Cu_{24}O_{41}$ and HTS points to that the suppression of the low-energy spectral weight is indicative of some kind of pair formation. One common element connecting $Sr_{14}Cu_{24}O_{41}$ and HTS is the presence of antiferromagnetic correlations. Hence, if part of the pairing interaction is related to these correlations, it is not surprising that one can observe similar effects in both systems. It is, however, not clear how the anomaly of the Sr mode should be interpreted within the above framework. A related hypothesis is that the phonon anomaly is partly due to an interaction with magnon excitations, which have been discussed in Y124 case [9]. In this context, we note that the mode at 265 cm^{-1} appears at ~ 200 K and that its frequency approximately fits to the value of the spin-gap energy $\Delta_{SP} = \sim 260$ cm^{-1} found in neutron scattering [10]. This suggests that the spin-gap energy is of the right order of magnitude to allow for a coupling to the phonon system. It is then possible that the low-frequency tail of spin excitation couples to the 246-cm^{-1} phonon and gives rise to the phonon renomalization.

Acknowledgements. M. O. was supported by Special Researcher's Basic Science Program at RIKEN.

1. M. Uehara, T. Nagata, J. Akimitsu, H. Takahashi, N. Mori, and K. Kinoshita, J. Phys. Soc. Jpn. **65**, 2764-2767 (1996).

2. E. Dagoto and T. M. Rice, Science **271**, 618-623 (1996).

3. See *e.g.*, C. Thomsen, in *"Light Scattering in Solids VI"* edited by M. Cardona and G. Güntherodt, Springer-Verlage, Berlin (1991 pp. 285-359.

4. M.V. Abrashev, C. Thomsen, and M. Surtchev, Physica C **280**, 297-303 (1997).

5. T. Noji, K. Kakimoto, and Y. Koike, Jpn. J. Appl. Phys. **37**, 100-101 (1998).

6. M. Yoshida, S. Tajima, N. Koshizuka, S. Tanaka, S. Uchida, and S. Ishibashi, Phys. Rev. B **44**, 11997-12002 (1991).

7. T. Adachi, K. Shiota, M. Kato, T. Noji, Y. Koike, Solid State Commun. **105**, 639-642 (1998) 639.

8. R. Nemetschek, M. Opel, C. Hoffmann, P. F. Müller, R. Hackl, H. Berger, L. Forró, A. Erb, and E. Walker, Phys. Rev. Lett. **78**, 4837-4840 (1997).

9. M. Käll, A. P. Litvinchuk, L. Börjesson, P. Berastegui, and L.-G. Johansson, Phys. Rev. B **53**, 3566-3572 (1996).

10. R. S. Eccleston, M. Uehara, J. Akimitsu, H. Eisaki, N. Motoyama, and S. Uchida, Phys. Rev. Lett. **81**, 1702-1705 (1998).

Single Crystal Growth and Characterization of the 4-leg Spin-Ladder Compound $La_2Cu_2O_5$

Chinnathambi Sekar, Takao Watanabe, and Azusa Matsuda

NTT Basic Research Laboratories, 3-1 Wakamiya, Morinosato, Atsugi-shi, Kanagawa 243-0198, Japan.

Abstract: Crystallization of $La_2Cu_2O_5$ in the La_2O_3-CuO system has been investigated in air, flowing-oxygen, and under high-pressure oxygen $[P(O_2)=3atm]$ atmospheres. Large $La_2Cu_2O_5$ single crystals $(0.2x7.0x0.05\ mm^3)$ have been grown using modified slow cooling (MSC) method under flowing-oxygen atmosphere. Electrical resistivity parallel to the ladder direction (//b) in as-grown and HIP-treated $La_2Cu_2O_5$ crystals was measured using the standard four-probe method.

Keywords: Spin-ladder, $La_2Cu_2O_5$, Single crystal, Flux method, Hot Isostatic Pressing (HIP)

INTRODUCTION

The recent proposal [1] of possible occurrence of superconductivity in even-leg spin-ladder systems and the subsequent discovery [2] of superconductivity under pressure in the two-leg spin ladder compound $Sr_{14-x}Ca_xCu_{24}O_{41}$ have sparked renewed interest in the study of ladders. $La_2Cu_2O_5$ and $La_8Cu_7O_{19}$, n=2 and 3 members of the new homologous series of lanthanum cuprates $La_{4+4n}Cu_{8+2n}O_{14+8n}$, have 4-leg and 5-leg ladder structures respectively in the lateral direction [3]. Large single crystals are indispensable for precise measurement of their physical properties using sophisticated characterization methods.

A close observation of the phase diagram of the La_2O_3-CuO system reveals that the crystallization of La-225 involves a narrow liquidus line in terms of both temperature and composition [4]. Hence, growth of large La-225 single crystals by standard slow cooling (SSC) method was thought to be difficult [3,4]. Recently, we found that the liquidus line widens in O_2 atmosphere compared to air [5]. By utilizing this, we grew good-quality La-225 single crystals by SSC. The La-225 crystal sizes have been increased using a modified slow cooling (MSC) in which a flux-poor starting composition is used to attain enough supersaturation and a rapid cooling from high temperature to the appropriate crystallization temperature is adopted to avoid the formation of high-temperature-phase La_2CuO_4. In the present work, we report the growth of La-225 crystals in air, flowing O_2, and under high-pressure oxygen atmospheres and the results are compared.

EXPERIMENT

The starting materials for crystal growth were prepared by mixing La-225 powder (sintered under flowing O_2 atmosphere at 1015°C), and CuO-flux in different molar ratios [referred to in terms of La_2O_3-CuO(molar ratio) hereafter]. Crystal growth was carried out using platinum crucibles. Both SSC and MSC were used. In flowing O_2 atmosphere, the starting material (1:5.25) was heated to 1115°C, soaked at this temperature for 10 hr, and then slow-cooled at the rate of 1.7°C/h to 1020°C (SSC). For the starting composition in the range of 1:4.75-1:3.75, the charge was heated to 1115°C (or more) and soaked at this temperature for 5 hr. Later the melt was fast-cooled to Ts (1065°C) in 10 min. and then slow-cooled from Ts to 1020°C at the rate of 1°C/h (MSC). At the end of the growth runs, the charge was furnace-cooled to temperature at a faster rate. In air, single crystals were grown from different starting compositions. For the excess-CuO range (1:9-1:7), the charge was heated to 1100°C and soaked for 10 hr. Slow cooling was carried out from 1100°C-1010°C at the rate of 1-2°C/h (SSC). For poor-CuO range (1:5.25-1:3.75), the charge was heated to about 150°C (or more) higher than the crystallization temperature (Ts) and soaked at this temperature for 3-5 hr. Later the melt was fast-cooled to Ts (1050°C) in 10 min. and then slow-cooled from Ts to 1010°C at the rate of 0.7-1°C/h (MSC). Finally, the charge was furnace-cooled to room temperature at a faster rate. Crystals were harvested by mechanically breaking the charge. Attempts were also made to grow La-225 crystals from poor-CuO starting composition under high pressure oxygen atmosphere [P(O$_2$)=3atm] by SSC. X-ray diffraction (XRD) studies were carried out on as-grown single crystals and the length of the c-axis was calculated from (002n) peaks.

RESULTS AND DISCUSSION

Crystal growth in flowing O$_2$ atmosphere: A systematic investigation was carried out on the crystallization of La-225 in the La$_2$O$_3$-CuO system for different starting compositions. For a given starting composition, soaking temperature dependence was also studied.

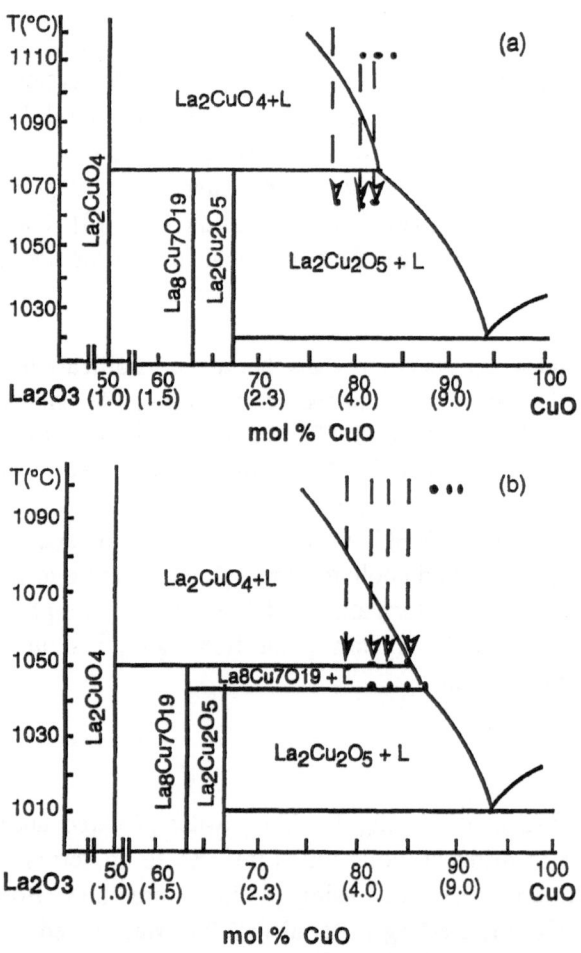

Fig. 1 Schematic representation of the quasi-binary phase diagram of the La$_2$O$_3$-CuO system in a) flowing oxygen and b) air. The numbers in paranthesis show starting composition in molar ratio (CuO/La$_2$O$_3$)

The results indicate that the liquidus line of crystallization widens in O$_2$ atmosphere compared to air. The results are presented in Fig. 1. For the starting composition of 1:5.25, crystals grew like druses and hence are free from growth defects. However, the length of the crystals couldn't be increased beyond 2-3 mm. In the poor-CuO region, large La-225 (0.2x7.0x0.05 mm^3) crystals were grown by MSC. A dendritic growth pattern was observed on the surface of the as-grown crystals. For the still-poor-CuO region (1:3.75), large square platelets made of a number of tiny La-225 crystals were grown. It should be mentioned that it was not possible to grow crystals from the excess-CuO range (< 5.5 mol % CuO) due to extensive creeping. *Crystal growth in air*: Unlike in flowing O$_2$ atmosphere, it became possible in air to grow pure La-225 crystals in the excess-CuO range by SSC method. Long, thin, needle-like crystals grew inside the cavities for the starting composition in the range of 1:9 - 1:7. For the poor- CuO range, mixed crystals containing both La-225 and La-8719 phases were obtained (1:5 - 1:4.5). By choosing the growth conditions carefully, we have succeeded in growing pure, large La-8719 single crystals for the first time by MSC in air.

Crystal growth under high oxygen pressure: Pure La-225 single crystals were grown under high-pressure oxygen [P(O$_2$) = 3 atm] with the starting composition of 1:3.75. Further experiments are needed to find the optimum starting composition, pressure, and growth temperature.

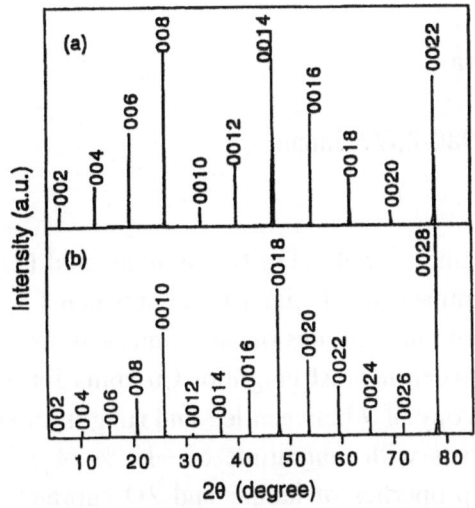

Fig. 2 XRD pattern of pure a) La$_2$Cu$_2$O$_5$ and
b) La$_8$Cu$_7$O$_{19}$ single crystals

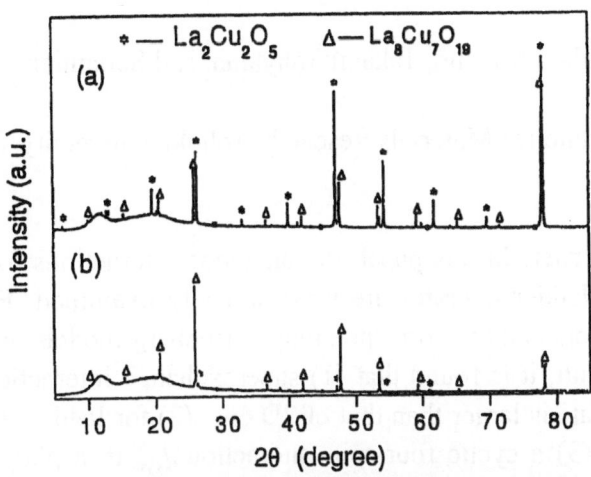

Fig. 3. XRD pattern of mixed crystals grown in air
from the starting composition of a) 1:4.5 and
b) 1:5.0 [La$_2$O$_3$:CuO (molar ratio)]

Characterization: Figs. 2a and 2b show the XRD pattern of pure La-225 and La-8719 single crystals, whose "c" lattice parameters were estimated as 27.978 and 34.684Å respectively. Fig. 3 shows the XRD pattern of mixed crystals grown in air. Temperature-dependent electrical resistivity measurements were carried out parallel to the ladder axis (//b) by the standard four-probe method. The result indicates that the as-grown pure La-225 crystals are insulating. The crystals were then hole-doped by means of high-oxygen- pressure annealing in a HIP furnace for 210 h at 600°C under 400 atm. The estimated room-temperature resistivities of the as-grown and HIP-treated La-225 crystals are 1.7x10^3 Ω cm and 384 m Ωcm.

CONCLUSIONS

Large La-225 single crystals with a maximum size of 0.2x7.0x0.05 mm^3 have been grown in flowing O$_2$ atmosphere by a modified slow cooling method. High-pressure-oxygen crystal growth with poor-CuO starting composition also leads to the growth of pure La-225 single crystals. Contrary to this, mixed crystals of La-225 and La-8719 and also pure La-8719 crystals grew in air for the corresponding starting composition. Temperature-dependent electrical resistivity measurements parallel to the ladder axis (//b) showed that the as-grown La-225 crystals are insulating. The resistivity decreases significantly upon hole doping by means of high-oxygen-pressure annealing in a HIP furnace, though no transition from the semiconducting behavior was observed.

Acknowledgment: The authors wish to thank Dr. Hiroyuki Shibata for extending the use of the HIP facility.

1. E. Dagoto and T. M. Rice, Science 271 (1996) 618.
2. M. Uehara, T. Nagata, J. Akimitsu, H. Takahashi, N. Mori and K. Kinoshita, J. Phys. Soc. Japan 65 (1996) 2764.
3. R. J. Cava, T. Siegrist, B. Hessen, J. J. Krajewski, W. F. Peck, Jr., B. Batlogg, H. Takagi, J.V. Waszczak, L.F. Schneemeyer and H. W. Zandbergen, J. Solid. State Chem. 94 (1991) 170; Physica C 177 (1991) 115.
4. A. N. Malyuk, A. A. Zhokhov, I. I. Zver'kova, A. M. Ionov, G. A. Emel'chenko and V. Sh. Shekhtman, Superconductivity 5(12) (1992) 2202.
5. C. Sekar, T. Watanabe and A. Matsuda (submitted to J. Crystal Growth).

Magnetic interaction in 1D, 2D and ladder cuprates

Yoshiaki Mizuno, Takami Tohyama, and Sadamichi Maekawa

Institute for Materials Research, Tohoku University, Sendai 980-8577, Japan

Abstract: In this paper, the magnetic interactions for one-dimensional (1D), two-dimensional (2D) and ladder cuprates are systematically examined. By a comparison of eigenstates between Cu-O clusters and the corresponding Heisenberg models, we evaluate magnitudes of these interactions. As a result, it is found that (1) superexchange interaction J between nearest neighbor Cu spins for 1D cuprate is larger than that of 2D one, (2) for ladder an anisotropy of J between leg and rung is small, and (3) a cyclic four-spin interaction J_{cyc} in a plaquette exists with magnitude of ~10 % of J for ladder and 2D cuprates. The effect of J_{cyc} on the magnetic properties of ladder and 2D cuprates is briefly discussed.

Keyword: superexchange interaction, four-spin interaction, Cu-O cluster, exact diagonalization

INTRODUCTION

A variety of insulating cuprates, including the parent compounds of high-T_c superconductors, afford us an opportunity to study magnetic properties of low-dimensional systems. Recent experiments for insulating cuprates have shown that the superexchange interaction between nearest neighbor (NN) Cu spins J remarkably depends on Cu-O network structure [1], i.e., (i) J in 1D cuprates is larger than that in 2D ones, and (ii) J along leg of ladder cuprates is larger than that along rung. Moreover, very recently, it has been reported that additional interactions such as a four-spin (4S) interaction are important for explaining the spin excitation spectra for ladder cuprates [2]. These experimental results indicate the necessity to examine a relation between magnetic interaction and Cu-O network structure, and to establish proper magnetic descriptions for the cuprates. In this paper we evaluate systematically magnetic interactions for 1D, 2D and ladder cuprates from theoretical viewpoint.

HOW TO CALCULATE MAGNETIC INTERACTIONS

A general model to describe the electronic states of cuprates is a d-p model, where Cu$3d_{x2-y2}$ and O$2p_\sigma$ orbitals are taken into account. The Hamiltonian contains hopping interactions between the $3d$ and $2p$ orbitals (T_{pd}) and between the $2p$ orbitals (T_{pp}), an energy-level separation between the d and p orbitals (Δ), on-site Coulomb interactions at Cu site (U_d) and O one (U_p), and direct exchange interaction between Cu$3d$ and O$2p$ orbitals (K_{pd}). These physical parameters are determined by the procedure as a previous study [1,3]. Considering these interactions, we adopt small Cu-O clusters shown in Fig. 1(a-1, 2 and 3), which are simulating 1D, 2D and ladder cuprates, respectively.

T_{pd} and T_{pp} are obtained by considering the bond length dependence and local environment via the Madelung potential around Cu and O ions [1]. The latter is essential for the difference of J among 1D, 2D and ladder cuprates. Δ is determined from the difference in the Madelung potential between Cu and O sites, and the dielectric constant [3]. We take below La_2CuO_4, $SrCu_2O_3$, and Sr_2CuO_3 as

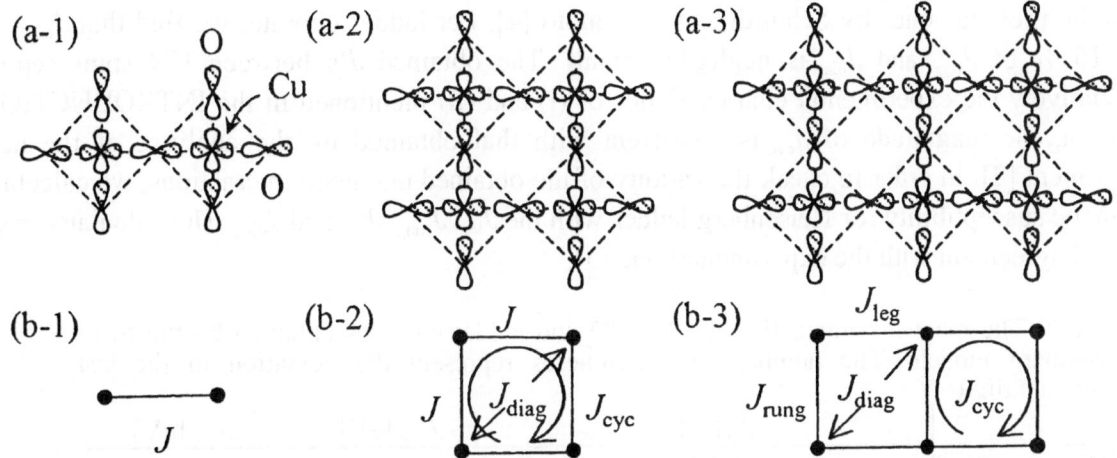

Fig. 1. (a) The Cu-O clusters for 1D, 2D and ladder cuprates and (b) the corresponding Heisenberg-type models.

typical insulators of 1D, 2D and ladder cuprates, respectively. The obtained parameters are listed in Table 1. Since U_d and U_p are independent of material, they are set to be 8.5 eV and 4.1 eV, respectively, and K_{pd} is taken to be 0.05 eV [3]

Table 1. The parameters for the typical cuprates used in the clusters.

	T_{pd} [eV]	T_{pp} [eV]	Δ [eV]
1D cuprate	1.16 [a], 1.09 [b]	0.49	3.1 [a], 2.5 [b]
2D cuprate	1.24	0.38	3.3
Ladder cuprate	1.08 [c], 1.15 [d], 1.29 [e]	0.42, 0.53 [f]	2.6 [c], 2.6 [d]

a: Parallel to chain, b: Perpendicular to chain
c: Leg direction, d: Rung direction, e: Interladder direction

We evaluate the magnetic interactions by associating the eigenstates of the Cu-O clusters with those of the corresponding Heisenberg-type models, which are shown in Fig. 1(b-1, 2 and 3). Not only the superexchange interaction

J between NN spins, but also a diagonal interaction J_{diag} and 4S interaction J_{cyc} are considered for 2D and ladder cuprates. The Heisenberg-type Hamiltonian for 2D cuprate is, for example, written by

$$H = J \sum_{<i,j>} \mathbf{S}_i \cdot \mathbf{S}_j + J_{diag} \sum_{<<i,j>>} \mathbf{S}_i \cdot \mathbf{S}_j + J_{cyc} \sum_{plaquette} \left(P_{ijkl} + P_{ijkl}^{-1} \right), \tag{1}$$

where $\mathbf{S}_j$ is a spin operator at i site, and J_{cyc} is defined as a coefficient of 4S cyclic permutation operators P_{ijkl} and P_{ijkl}^{-1}, which can be rewritten by using two- and four-spin interactions as

$$P_{ijkl} + P_{ijkl}^{-1} = 4\left[(\mathbf{S}_i \cdot \mathbf{S}_j)(\mathbf{S}_k \cdot \mathbf{S}_l) + (\mathbf{S}_i \cdot \mathbf{S}_l)(\mathbf{S}_j \cdot \mathbf{S}_k) - (\mathbf{S}_i \cdot \mathbf{S}_k)(\mathbf{S}_j \cdot \mathbf{S}_l) \right] \tag{2}$$

$$+ (\mathbf{S}_i \cdot \mathbf{S}_j) + (\mathbf{S}_j \cdot \mathbf{S}_k) + (\mathbf{S}_k \cdot \mathbf{S}_l) + (\mathbf{S}_l \cdot \mathbf{S}_i) + (\mathbf{S}_i \cdot \mathbf{S}_k) + (\mathbf{S}_j \cdot \mathbf{S}_l) + 1/4,$$

where $ijkl$ denotes a set of 4 spins in a plaquette. The J, J_{diag}, and J_{cyc} are determined by associating the lowest several eigenvalues of the Cu-O clusters with those of the Heisenberg-type models. The eigenvalues are calculated by using the numerically diagonalization method.

RESULTS AND DISCUSSION

The results are summarized in Table 2. For 2D cuprate, J is ~0.15 eV, similar to the previous study [1], where Cu_2O_7 cluster is used. J_{cyc} is ~ 7 % of the J, and J_{diag} is zero. These results are consistent with the previous ones by Schmidt and Kuramoto [4]. For ladder cuprate, we find that $J_{leg} > J_{rung}$, $J_{cyc} \sim 10$ % of J_{leg}, and J_{diag} is negligibly small. The obtained J's between NN spins reproduce qualitatively the experimental characteristics of (i) and (ii) mentioned in the INTRODUCTION. In addition, the magnitude of J_{cyc} is consistent with that obtained by the analyses of the neutron experiment [2]. In order to check the validity of the obtained magnetic interactions, we calculate the magnetic susceptibility for Heisenberg ladder with the J_{leg}, J_{rung}, J_{cyc} and J_{diag}. The calculated result is in good agreement with the experimental one [5].

Table 2. The magnetic interaction for 1D, 2D and ladder cuprates obtained by the fit to Heisenberg models. The numbers in parentheses represent the deviation in the last significant digit.

	J [eV]	J_{diag} [eV]	J_{cyc} [eV]
1D cuprate	0.17	–	–
2D cuprate	0.146 (1)	0.000 (0)	0.011 (1)
Ladder cuprate	J_{leg}: 0.195 (5), J_{rung}: 0.15 (2)	0.003 (2)	0.018 (2)

Let us consider the effect of J_{cyc} on the magnetic properties in 2D and ladder cuprates. The J_{cyc} reduces the superexchange interaction between NN spins effectively because it works as magnetic frustration to antiferromagnetic arrangement in a quadratic plaquette. For 2D cuprate, the frustration works all J's equivalently, while for ladder, due to the structural anisotropy, J_{rung} is influenced twice more strongly than J_{leg}. This leads to an effective anisotropy between leg and rung. In the previous analysis of the magnetic susceptibility by the simple Heisenberg ladder with only NN interactions, it has been pointed out that $J_{leg}^{eff} \sim 2 J_{rung}^{eff}$ [6]. Therefore, it is considered that this anisotropy results from the renormalization of J_{leg} and J_{rung} by J_{cyc}.

SUMMARY

We have examined the magnetic interactions in various cuprates systematically and obtained the values of them, consistent with the experimental results. It has been found that there is four-spin interaction with considerable magnitude. Further study on the effect of J_{cyc} on the several physical quantities is necessary for a quantitative investigation between theory and experiment.

Acknowledgement: This work was supported by a Grant-in-Aid for Scientific Research on Priority Areas from the Ministry of Education, Science, Sports and Culture of Japan, CREST and NEDO. The parts of the numerical calculation were performed in the Supercomputer Center in ISSP, University of Tokyo, and the supercomputing facilities in IMR, Tohoku University.

1. Y. Mizuno, T. Tohyama, and S. Maekawa, Phys. Rev. B **58**, R14713-R14717 (1999) and references therein.
2. M. Matsuda, K. Katsumata, R. S. Eccleston, S. Brehmer and H. –J. Mikeska, unpublished.
3. Y. Mizuno, T. Tohyama, S. Maekawa, T. Osafune, N. Motoyama, H. Eisaki, and S. Uchida, Phys. Rev. B **57**, 5326-5335 (1998).
4. Y. Mizuno, T. Tohyama, and S. Maekawa, cond-mat/9906444, and to be published in J. Low Temp. Phys.
5. H. J. Schmidt and Y. Kuramoto, Physica C **167**, 263-266 (1990).
6. J. C. Johnston, Phys. Rev. B **54**, 13009-13016 (1996).

Resonant Inelastic X-Ray Scattering in Insulating Cuprates: A Numerical Study of the Hubbard Model

Kenji Tsutsui, Hiroshi Kondo, Takami Tohyama, and Sadamichi Maekawa

Institute for Materials Research, Tohoku University, Sendai 980-8577, Japan

Abstract: The momentum dependence of the RIXS spectrum in insulating cuprates is examined theoretically. Cu K-edge RIXS spectrum is calculated by using the exact diagonalization technique on small clusters in the Hubbard model with Cu 1s-core band. We find that the RIXS spectrum related to charge-transfer excitations has the characteristic momentum-transfer dependence. The dependence is explained by the particle-hole excitations from occupied lower Hubbard band (LHB) to unoccupied upper Hubbard band (UHB). The results suggest that the RIXS is very useful for extracting the feature of the unoccupied UHB, and provides an opportunity for the understanding of the different behavior of hole- and electron-doped superconductors.

Keywords: RIXS, insulating cuprates, Hubbard Model, SDW mean-field approximation

Resonant inelastic x-ray scattering (RIXS) has recently received much attention as a powerful technique to investigate elementary excitations in the strongly correlated electron systems. In particular, it has been demonstrated that, by using the high-resolution experiments, the momentum-dependent measurement of the charge transfer gap in insulating cuprates is possible [1,2]. From the theoretical side, the present authors have predicted the dependence based on the two-dimensional Hubbard model with the second and third neighbor hoppings [3]. The dependence has been explained by the particle-hole excitations from occupied lower Hubbard band (LHB) to unoccupied upper Hubbard band (UHB) by numerically calculating the spectral function of the two-body Green's function [3]. In this paper, we discuss the excitations by using the antiferromagnetic (AF) spin-density-wave (SDW) approximation and suggest that the interference effect arising from the AF long-range order plays a crucial role in the excitations.

A minimal model that can describe the LHB and UHB is the Hubbard model as used in the analysis of O 1s x-ray absorption spectrum [4]. Hereafter, we use a term 3d-electron system to stand for LHB and UHB. The Hubbard Hamiltonian with second and third neighbor hoppings is written as

$$H_{3d} = -t \sum_{\langle i,j \rangle_{1st},\sigma} d^{+}_{i,\sigma} d_{j,\sigma} - t' \sum_{\langle i,j \rangle_{2nd},\sigma} d^{+}_{i,\sigma} d_{j,\sigma} - t'' \sum_{\langle i,j \rangle_{3rd},\sigma} d^{+}_{i,\sigma} d_{j,\sigma} + \text{H.c.} + U \sum_{i} n^{d}_{i,\uparrow} n^{d}_{i,\downarrow} , \quad (1)$$

where $d_{i,\sigma}$ is the annihilation operator of 3d electron with spin σ at site i, $n^{d}_{i,\sigma} = d^{+}_{i,\sigma} d_{i,\sigma}$, the summations $\langle i,j \rangle_{1st}$, $\langle i,j \rangle_{2nd}$, and $\langle i,j \rangle_{3rd}$ run over first, second, and third nearest-neighbor pairs, respectively, and the rest of the notation is standard. The on-site Coulomb energy U corresponds to the charge transfer energy of cuprates [3].

In the intermediate states of Cu K-edge RIXS process, 3d electrons interact with a 1s-core hole created by the dipole transition of an 1s electron to 4p orbital due to an absorption of an incident photon with energy ω_i and momentum $\mathbf{K}_i$. This interaction is written as

$$H_{1s-3d} = -V_c \sum_{i,\sigma,\sigma'} n^{d}_{i,\sigma} n^{s}_{i,\sigma'} , \quad (2)$$

where $n^s_{i,\sigma}$ is the number operator of $1s$-core hole with spin σ at site i, and V_c is taken to be positive. This interaction, V_c, causes excitations of the $3d$ electrons across the gap. The photo-excited $4p$ electron is assumed to go into the bottom of the $4p$ band with momentum $\mathbf{k}_0$ and not to interact with either the $3d$ electrons or the $1s$-core hole due to delocalized nature of the $4p$ orbital. In the final state, the $4p$ electron goes back to the $1s$ orbital emitting a photon with energy ω_f and momentum $\mathbf{K}_f$. The RIXS spectrum is then given by

$$I(\Delta\mathbf{K},\Delta\omega)=\sum_{\alpha}\left|\langle\alpha|\sum_{\sigma}s_{\mathbf{k}_0-\mathbf{K}_f,\sigma}p_{\mathbf{k}_0,\sigma}\frac{1}{H+\varepsilon_{1s-4p}-E_0-\omega_i-i\Gamma}p^+_{\mathbf{k}_0,\sigma}s^+_{\mathbf{k}_0-\mathbf{K}_i,\sigma}|0\rangle\right|^2\delta(\Delta\omega-E_\alpha+E_0),(3)$$

where $H = H_{3d} + H_{1s\text{-}3d}$, $\Delta\mathbf{K} = \mathbf{K}_i - \mathbf{K}_f$, $\Delta\omega = \omega_i - \omega_f$, $s_{\mathbf{k},\sigma}$ ($p_{\mathbf{k},\sigma}$) is the annihilation operator of the $1s$-core hole ($4p$ electron) with momentum $\mathbf{k}$ and spin σ, $|0\rangle$ is the ground state of the half-filled system with energy E_0, $|\alpha\rangle$ is the final state of the RIXS process with energy E_α, Γ is the inverse of the relaxation time in the intermediate state, and ε_{1s-4p} is the energy difference between the $1s$ level and the bottom of the $4p_z$ band. The values of the parameters are set to be $U/t = 10$, $V_c/t = 15$, and $\Gamma = 1$ [3]. Figure 1 shows the RIXS spectrum in Eq. (3) calculated on 4×4-site cluster by using the modified version of the conjugate gradient method as well as the Lanczös technique [3]. The spectrum strongly depend on the momentum showing a feature that the weight shifts to higher energy region with increasing $|\Delta\mathbf{K}|$. In addition, the threshold of the spectrum at $\Delta\mathbf{K} = (\pi/2,0)$ is lower in

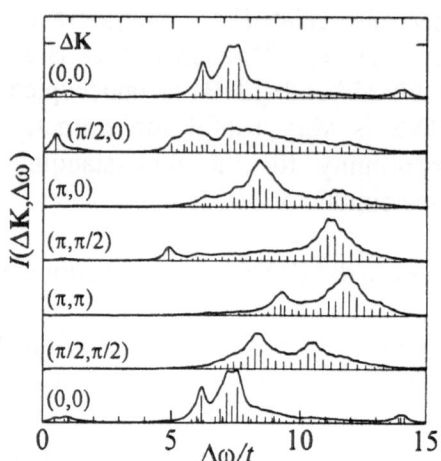

Fig. 1. Momentun dependence of the resonant inelastic x-ray scattering in the Hubbard model [3].

energy than that at $(0,0)$. At $\Delta\mathbf{K} = (\pi/2,\pi/2)$, however, the spectrum appears above the threshold at $(0,0)$, resulting in an anisotropic momentum dependence between the spectra along $(0,0)$ to $(\pi/2,0)$ and along $(0,0)$ to $(\pi/2,\pi/2)$. Note that it is insufficient to explain the dependence by using the convoluted spectrum of the single-particle excitation spectra [3].

As discussed in Ref. 3, the dependence has been explained by the spectral function of the two-body Green's function $B(\mathbf{k}, \Delta\mathbf{K}; \Delta\omega)$ defined by

$$B(\mathbf{k},\Delta\mathbf{K};\Delta\omega)=\sum_{\alpha'}\left|\langle\alpha'|\sum_{\sigma}d^+_{\mathbf{k}+\Delta\mathbf{K},\sigma}d_{\mathbf{k},\sigma}|0\rangle\right|^2\delta(\Delta\omega-E_\alpha+E_0),\qquad(4)$$

where the states $|\alpha'\rangle$ have the same point-group symmetry as that of the final states of the RIXS process. In particular, we have found that the intensity of $B(\mathbf{k}, \Delta\mathbf{K}; \Delta\omega)$ with $[\mathbf{k}, \Delta\mathbf{K}] = [(\pi/2,-\pi/2), (\pi/2,\pi/2)]$ is much small compared with that with $[\mathbf{k}, \Delta\mathbf{K}] = [(\pi/2,0), (\pi/2,0)]$, indicating that the excitation from the occupied state at $(\pi/2,-\pi/2)$ to the unoccupied state at $(\pi,0)$ is almost forbidden [3]. One of the possible factors of this fact is the AF environment around the 'quasi-particles'. We show in the following that the matrix element of the excitation from the occupied LHB state at $(\pi/2,-\pi/2)$ to the unoccupied UHB state at $(\pi,0)$ is rigorously zero in the simple AF-SDW mean-field approximation.

The AF-SDW mean-field Hamiltonian is given by

$$H_{\text{MF}} = \sum_{\mathbf{k}\in M.Z.}\left(E^-_{\mathbf{k}}a^+_{\mathbf{k},\sigma}a_{\mathbf{k},\sigma} + E^+_{\mathbf{k}}b^+_{\mathbf{k},\sigma}b_{\mathbf{k},\sigma}\right),\qquad(5)$$

where the summation of $\mathbf{k}$ runs in the half of the original Brillouin zone enclosed by the magnetic zone (M.Z.) boundary $|k_x \pm k_y| = \pi$, $E^{\pm}_{\mathbf{k}} = -4t' \cos k_x \cos k_y - 2t'' (\cos 2k_x + \cos 2k_y) \pm \sqrt{4t^2(\cos k_x + \cos k_y)^2 + U^2 m^2}$, m being the staggered magnetization which is determined self-consistently, and the operators $a_{\mathbf{k},\sigma}$ and $b_{\mathbf{k},\sigma}$ are the annihilation operators of the LHB and HUB, respectively. They are given by

$$\begin{pmatrix} d_{\mathbf{k},\sigma} \\ d_{\mathbf{k}+Q,\sigma} \end{pmatrix} = \begin{pmatrix} \cos\frac{\theta_{\mathbf{k},\sigma}}{2} & \sin\frac{\theta_{\mathbf{k},\sigma}}{2} \\ -\sin\frac{\theta_{\mathbf{k},\sigma}}{2} & \cos\frac{\theta_{\mathbf{k},\sigma}}{2} \end{pmatrix} \begin{pmatrix} a_{\mathbf{k},\sigma} \\ b_{\mathbf{k},\sigma} \end{pmatrix}, \tag{6}$$

where $\mathbf{Q} = (\pi,\pi)$, $\sin\theta_{\mathbf{k},\sigma} = \dfrac{Um\sigma}{\sqrt{4t^2(\cos k_x + \cos k_y)^2 + U^2 m^2}}$, and $\cos\theta_{\mathbf{k},\sigma} = \dfrac{-4t(\cos k_x + \cos k_y)}{\sqrt{4t^2(\cos k_x + \cos k_y)^2 + U^2 m^2}}$. At half filling, the ground state $|0\rangle_{\text{SDW}}$ is fully occupied state in the LHB ($a^{+}_{\mathbf{k},\sigma}$).

We consider the spectral function $B(\mathbf{k},\Delta\mathbf{K};\Delta\omega)$ of Eq. (4) with $[\mathbf{k}, \Delta\mathbf{K}] = [(\pi/2,-\pi/2), (\pi/2,\pi/2)]$. Because the point-group symmetry of the ground state is A_1, the symmetry of $|\alpha'\rangle$ with total momentum $(\pi/2,\pi/2)$ in Eq. (4) has the even for the reflection with respect to the line of $k_x = k_y$. Instead of the restriction in $|\alpha'\rangle$ in Eq. (4), we use the irreducible tensor operator for $d^{+}_{\mathbf{k}+\Delta\mathbf{K},\sigma} d_{\mathbf{k},\sigma}$. This operator is $d^{+}_{\mathbf{k}+\Delta\mathbf{K},\sigma} d_{\mathbf{k},\sigma} + d^{+}_{\mathbf{k}'+\Delta\mathbf{K},\sigma} d_{\mathbf{k},\sigma}$ with $[\mathbf{k}, \mathbf{k}', \Delta\mathbf{K}] = [(\pi/2,-\pi/2), (-\pi/2,-\pi/2), (\pi/2,\pi/2)]$. Thus $B(\mathbf{k},\Delta\mathbf{K},\Delta\omega)$ with $[\mathbf{k}, \Delta\mathbf{K}] = [(\pi/2,-\pi/2), (\pi/2,\pi/2)]$ becomes $1/4 \times \Sigma_\alpha |\langle\alpha|(d^{+}_{\mathbf{k}+\Delta\mathbf{K},\sigma} d_{\mathbf{k},\sigma} + d^{+}_{\mathbf{k}'+\Delta\mathbf{K},\sigma} d_{\mathbf{k}',\sigma})|0\rangle|^2 \delta(\Delta\omega - E_\alpha - E_0)$ where $\mathbf{k}' = (-\pi/2,\pi/2)$ and $|\alpha\rangle$ is not restricted for the symmetry. Using the transformation of Eq. (6), one obtains that

$$\left(d^{+}_{\mathbf{k}+\Delta\mathbf{K},\sigma} d_{\mathbf{k},\sigma} + d^{+}_{\mathbf{k}'+\Delta\mathbf{K},\sigma} d_{\mathbf{k}',\sigma}\right)|0\rangle_{\text{SDW}} = \sin^2\frac{\theta_{\mathbf{k},\sigma} - \theta_{\mathbf{k}+\Delta\mathbf{K},\sigma}}{2} b^{+}_{\mathbf{k}+\Delta\mathbf{K},\sigma} a_{\mathbf{k},\sigma}|0\rangle_{\text{SDW}}, \tag{8}$$

and the coefficient $\sin\frac{\theta_{\mathbf{k},\sigma} - \theta_{\mathbf{k}+\Delta\mathbf{K},\sigma}}{2}$ is equal to zero when $\mathbf{k} = (-\pi/2,\pi/2)$ and $\mathbf{k} + \Delta\mathbf{K} = (\pi,0)$. Therefore, the excitation from the occupied LHB state at $(\pi/2,-\pi/2)$ [and $(-\pi/2, \pi/2)$] to the unoccupied UHB state at $(\pi,0)$ [and $(0,\pi)$] is forbidden in the AF-SDW approximation. This result is consistent with the small intensity of $B(\mathbf{k},\Delta\mathbf{K};\Delta\omega)$ with $[\mathbf{k}, \Delta\mathbf{K}] = [(\pi/2,-\pi/2), (\pi/2,\pi/2)]$ in the cluster calculations.

In summary, we have examined the momentum dependence of the RIXS using the numerically exact diagonalization technique on small clusters. Characteristic momentum dependence has been found in the spectrum. The dependence has been explained by the particle-hole excitations from LHB to UHB. We have also discussed the excitations by using the AF-SDW approximation and suggested that the interference effect arising from the AF long-range order plays a crucial role in the excitations.

Acknowledgments. This work was supported by Priority-Areas Grants from the Ministry of Education, Science, Sports and Culture of Japan, CREST, and NEDO. Computations were carried out in ISSP, University of Tokyo; IMR, Tohoku University; and Tohoku University.

1. P. Abbamonte, C. A. Burns, E. D. Isaacs, P. M. Platzman, L. L. Miller, S. W. Cheong, and M. V. Klein, Phys. Rev. Lett. **83**, 860-864 (1999).
2. Z. Hasan, E. D. Isaacs, and Z.-X. Shen, et al., unpublished.
3. K. Tsutsui, T. Tohyama, and S. Maekawa, cond-mat/9905372; to be published in Phys. Rev. Lett.
4. C. T. Chen, F. Sette, Y. Ma, M. S. Hybertsen, E. B. Stechel, W. M. C. Foulkes, M. Schluter, S-W. Cheong, A. S. Cooper, L. W. Rupp, Jr., B. Batlogg, Y. L. Soo, Z. H. Ming, A. Krol, and Y. H. Kao, Phys. Rev. Lett. **66**, 104-108 (1991).

Measurement of Local Oxygen Concentration in $YBa_2Cu_3O_y$ by Convergent-Beam Electron Diffraction

Zentaro Akase[1], Yuji Tanaka[1*], Yoshitsugu Tomokiyo[1] and Masashi Watanabe[2]

1) Department of Materials Science and Engineering, and 2) Research Laboratory for High Voltage Electron Microscopy, Kyushu University, Fukuoka 812-8581, Japan

Abstract: Convergent-beam electron diffraction (CBED) patterns were observed from small areas in nm-scale using a transmission electron microscope JEM-2010FEF. In order to obtain information on local oxygen concentration y, measured intensities of CBED patterns were analyzed with the dynamical theory of electron diffraction. Intensities of $00l$ systematic reflections change with the irradiation time. The observed change in the $00l$ intensities can be explained in terms of oxygen deficiency. Similarly local change in y is detected by careful observations of CBED patterns. The present approach may be applicable to the measurement of local change in ionicity or carrier density.

Keywords: Oxygen deficiency, Convergent-beam electron diffraction, Structure factor, Energy filtering

INTRODUCTION

Structures of YBa2Cu3Oy (Y-123) depend on oxygen concentration y. An orthorhombic superconducting phase transforms to a tetragonal, non-superconducting phase when y is less than 6.5 [1]. A value of y easily changes with temperature and oxygen partial pressures. In an oxygen-deficient phase, ordering of some of the oxygen sites occurs. Many stacking faults are usually observed in sintered or melt-processed bulk Y-123 material [2]. The stacking fault consists of an extra CuO layer, which indicates local deviation from Y-123 stoichiometry [3]. A local change in oxygen concentration gives rise to a local change in structures and may influence superconducting properties such as T_c, J_c, H_{irr} and the peak effect. One of the methods to measure a local oxygen concentration is the X-ray energy dispersive spectrometry in transmission electron microscopy. Unfortunately, it is not very easy to measure the concentration of light elements in compounds composed of heavy elements. Therefore, in this paper we will extract information of structure factors from CBED patterns and demonstrate the presence of local change in oxygen concentration in Y-123.

We used a new transmission electron microscope JEM-2010FEF equipped with a field-emission gun and an energy filter of omega-type (Fig.1) [4]. Large angle CBED patterns of $00l$ systematic reflections were recorded by a slow-scan CCD camera or imaging plates. Electron energy-loss spectra were also obtained to determine a thickness at the same areas where CBED patterns were taken. The sample made by the quench and melt-growth (QMG) method was used (T_c = 92K. J_c > 10^4 A/cm^2 at 1T, 77K).

RESULTS AND DISCUSSION

According to our calculations of electron-diffraction intensities based on the many-beam dynamical, it is realized that intensities of $00l$ systematic reflections strongly depend on the oxygen concentration in Y-123. In the present experiment, we have used diffraction higher than 0 0 12 since the intensities are not sensitive to a change in specimen thickness at $l \geq 12$.

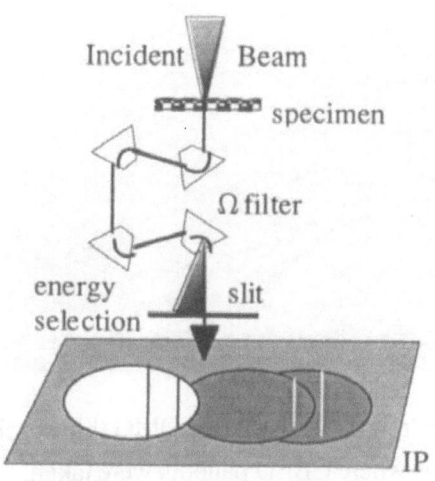

Fig.1. Schematic drawing of energy-filtering convergent-beam electron diffraction.

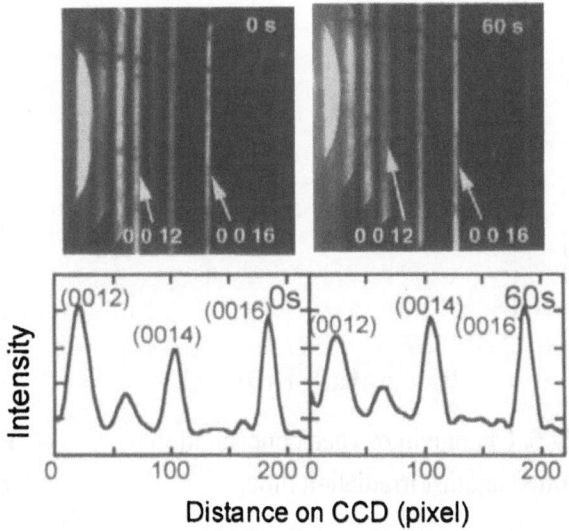

Fig.2. Observed large angle CBED patterns and intensity profiles read out from imaging plates.

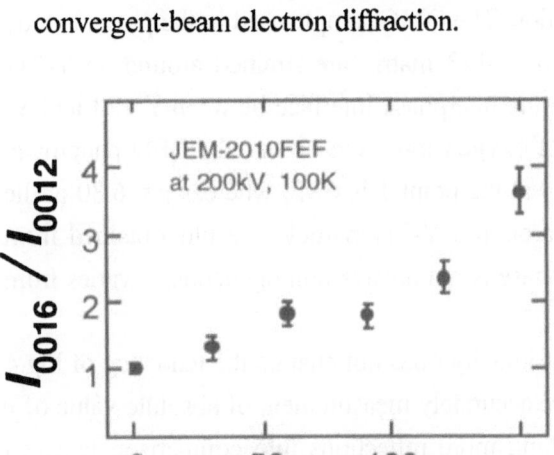

Fig.3. Measured intensity ratio I_{0016}/I_{0012} as a function of irradiation time.

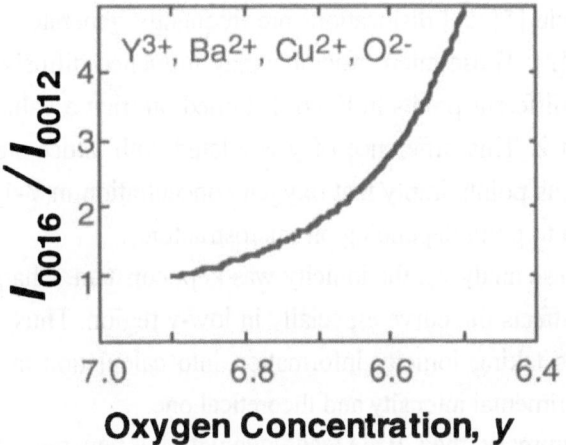

Fig.4. Calculated intensity ratio I_{0016}/I_{0012} as a function of oxygen concentration.

(1) In order to confirm detection sensitivity for the local change in oxygen concentration, an in-situ experiment was performed at first. Figure 2 shows large angle CBED patterns recorded during illumination of 2 nm focused probe at a fixed position in a Y-123 thin specimen. Intensities of 0 0 12 and 0 0 16 reflections change with an increase of irradiation time as shown in Fig.2 and Fig.3. The change in diffraction intensities is due to the oxygen deficiency caused by local beam-heating at a high degree of vacuum because electron optical conditions are kept constant during the observation. Figure 4 shows the ratio of I_{0016} to I_{0012} calculated with the Bloch-wave method assuming Y^{3+}, Ba^{2+}, Cu^{2+} and O^{2-} as a function of y. The value of y is six plus the occupancy of oxygen at CuO chain in an unit cell. Figure 4 transforms the intensity ratio in Fig.3 into the oxygen concentration y as shown in Fig.5. In Fig.4, the intensity ratio is not very sensitive to y in high-y region. So, in Fig.5, the error bar of y in high-y region is longer than one in low-y region.

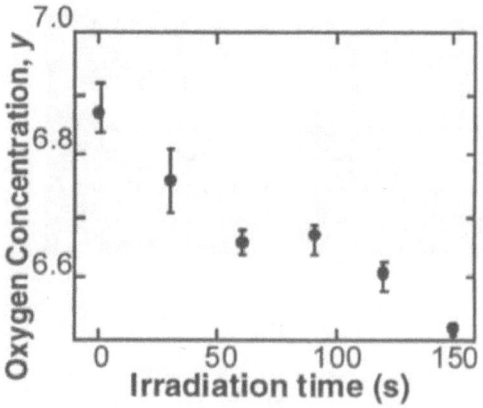

Fig.5. Change in oxygen concentration y plotted against irradiation time.

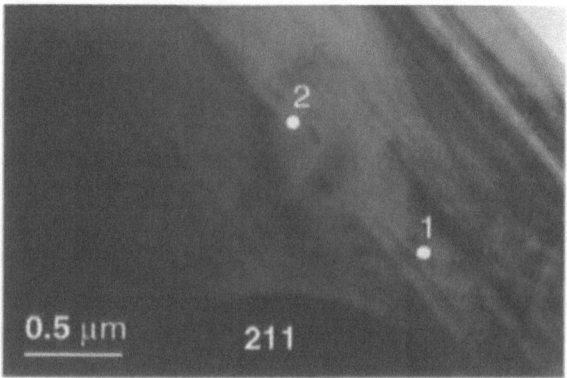

Fig. 6. Electron micrograph of a QMG thin specimen and positions where CBED patterns were taken. Measured values of y, 1: $y = 6.92$, 2: $y = 6.80$

(2) Next, we measured a local change in oxygen concentration. The QMG sample used in the present study contains particles of Y_2BaCuO_5 phase (Y-211). Lattices of Y-123 matrix are strained around an Y-211 particle [5] and dislocations are frequently generated from the interphase interface between Y-123 and Y-211 [2]. These microstructures may influence diffusivity of oxygen ions. We observed CBED patterns at two different points in Fig.6. It turned out that a value of y at the point 1 is 6.92, whereas $y = 6.80$ at the point 2. This difference of y is related with lattice strain around a Y-211 particle. Results obtained from various points imply that oxygen concentration in Y-123 phase is not always homogeneous; it varies from point to point depending on microstructures.

In these analyses, the ionicity was kept constant. Charge of ionicity does not change the tendency of Fig.4 but affects the curve especially in low-y region. Thus, more accurately measurement of absolute value of y needs taking ionicity information into calculation and taking more reflections into comparison between experimental intensity and theoretical one.

In summary, the CBED technique with the new microscope is the very powerful approach to detect a local change in oxygen concentration in High T_c superconductors. This method is applicable for a measurement of local change in ionicity of constituent elements or carrier density in high T_c superconductors.

Acknowledgments. The work was partly supported by the Proposal-Based New Industry Creative Technology R & D Promotion Program.

*Present address: Nissho Co., R & D Laboratory, 700 Nojisho, Kusatu, 525-0055, Japan

1. R.J.Cava, A.W.Hewat, E.A.Hewat, B.Batlogg, M.Marezio, K.M.Rabe, J.J.Krajewski, W.F.Tom Peck Jr, and L.W.Rupp Jr, Physica C 165, 419-433 (1990).

2. Y.Tomokiyo, S.Murakami and Y.Suyama, in Advances in Superconductivity IX, edited by S.Nakajima and M.Murakami (Springer-Verlag, Tokyo, 1997), pp.547-550.

3. Y.Tomokiyo, E.Tanaka, X.Zheng, H.Kuriyaki and K.Hirakawa, Physica C, 219, 288-296 (1994).

4. Y.Tomokiyo, S.Matsumura and T.Manabe, J. Microscopy, 194, 210-218 (1998).

5.Y.Tomokiyo, Y.Omori, E.Tanaka and Y.Suyama, in Advamces in superconductivity VI, edited by T.Fujita and Y.Shiohara (Springer-Verlag, Tokyo, 1994), pp.447-450.

Anisotropic dielectric constant of the parent antiferromagnet $Bi_2Sr_2MCu_2O_8$ (M=Dy, Y and Er) single crystals

T. Takayanagi, T. Kitajima, T. Takemura and I. Terasaki

Department of Applied Physics, Waseda University, Tokyo 169-8555, JAPAN

Abstract: The anisotropic dielectric constants of the parent antiferromagnet $Bi_2Sr_2MCu_2O_8$ (M=Dy, Y and Er) single crystals were measured from 80 to 300 K. The in-plane dielectric constant is found to be very huge (10^4-10^5). This suggests a remnant of the Fremi surface of the parent antiferromagnet. The out-of-plane dielectric constant is 50-200, which is three orders of magnitude smaller than the in-plane one. A significant anomaly is that a similar out-of-plane dielectric constant is observed in superconducting samples.

Keywords: parent antiferromagnet, dielectric constant, insulator-metal transition

INTRODUCTION

In a low hole density, the CuO_2 plane shows high resistivity and antiferromagnetic (AF) order at low temperature, which is called a parent AF insulator. With doping, an insulator-metal transition (IMT) arises, and the system changes from the AF insulator to a superconductor. For IMT, the dielectric constant ε is of great importance in the sense that it provides a measure of localization length in the insulator. Chen *et al.* [1] have first pointed out the importance of ε in the studies of high-T_c cuprates (HTSC). However, they studied ε only for $La_2CuO_{4+\delta}$, which has various structural phase transitions that might affect ε seriously. Another problem is that they studied ε only near 4.2 K, although the resistivity anisotropy was strongly dependent on temperature. Thus it should be further examined to study ε for other HTSC over a wider temperature range.

We have been studying the charge transport of the parent insulator $Bi_2Sr_2MCu_2O_8$ (M=Y and rare-earth) [2]. In this proceedings we report on measurements and analyses of the anisotropic dielectric constants from 80 to 300 K.

EXPERIMENTAL

Single crystals of $Bi_2Sr_2MCu_2O_8$ (M=Dy,Y and Er) were grown by a self-flux method. The growth conditions and the sample characterization were described in Ref [2]. The resistivity was measured using a four-probe technique, and a ring configuration was used for the out-of-plane direction. The dielectric constants were measured with a two-probe technique using a lock-in amplifier (Stanford Research SR630 and SR844). A typical contact resistance was 50-100 Ω for the in-plane direction, and 1-10 Ω for the out-of-plane direction. Thus the measurement along the in-plane direction is less accurate near room temperature, where the contact resistance becomes comparable with the sample resistance. Detailed information on the measurements will be written elsewhere.

All the samples of $Bi_2Sr_2MCu_2O_8$ were insulating, and the doping levels of the as-grown crystals were slightly different for different M. We do not yet understand the M dependence, but the melting points and/or the liquidus lines may depend on M to give a slight variation in composition. Thus crystals with different M's act as a set of parent insulators with slightly different doping levels. We estimated the hole concentration per Cu (p) by measuring the room-temperature thermopower [3]. With good reproducibility, M=Dy was nearly undoped (p=0-0.02), and M=Er and Y were slightly doped (p=0.02-0.04). The doping levels (and the measurement results) were nearly the

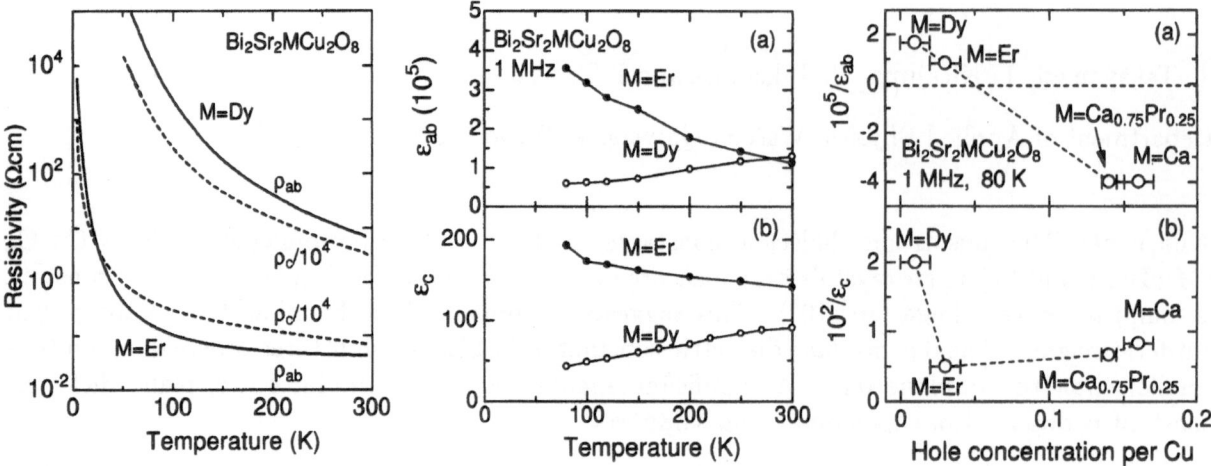

Fig. 1   The in-plane resistivity (ρ_{ab}) and the out-of-plane resistivity (ρ_c) of $Bi_2Sr_2MCu_2O_8$ single crystals. (left panel)

Fig. 2   The dielectric constant of $Bi_2Sr_2MCu_2O_8$ single crystals. (middle panel) (a) The in-plane direction (ε_{ab}) and (b) the out-of-plane direction (ε_c).

Fig. 3   The reciprocal of the dielectric constant $1/\varepsilon$ plotted as a function of hole concentration per Cu. (right panel). (a) The in-plane direction ($10^5/\varepsilon_{ab}$) and (b) the out-of-plane direction ($10^2/\varepsilon_c$). For the superconducting samples (M=Ca and $Ca_{0.75}Pr_{0.25}$), see text.

same between M=Er and Y, we discuss the data for M= Dy and Er below.

RESULTS AND DISCUSSION

Figure 1 shows the in-plane resistivity (ρ_{ab}) and the out-of-plane resistivity (ρ_c) for M=Er and Dy. Reflecting the different doping levels, both ρ_{ab} and ρ_c are larger for M=Dy than for M=Er. The temperature dependence is also different between M=Dy and Er. In particular, ρ_{ab} for M=Er is nearly independent of temperature at 300 K, which indicates that the in-plane conduction is nearly metallic. It should be noted here that ρ_c/ρ_{ab} is strongly dependent on temperature and the doping levels, which suggests the confinement behavior in the AF insulator [2].

Figure 2 shows the in-plane dielectric constant (ε_{ab}) and the out-of-plane dielectric constant (ε_c) for M=Er and Dy at 1 MHz. Both ε_{ab} and ε_c are larger for M=Er than M=Dy, which indicates that the sample for M=Er is closer to IMT boundary. It should be emphasized that ε_{ab} is as huge as 10^4-10^5. We think that the huge ε_{ab} comes from an electronic origin, because (1) ε_{ab} is very sensitive to the doping levels and (2) the dielectric loss Im $\varepsilon_{ab} \propto 1/\rho_{ab}$ is large compared with conventional ferroelectric materials. The charge order or the variable range hopping may be an origin of the huge ε_{ab}. Thus we may say that the huge ε_{ab} is a remnant of the the Fermi surface calculated by band theories.

An important feature is that ε_c remains positive and finite in the superconducting samples. Kitano *et al.* [4] found that ε_c of $Bi_2Sr_2CaCu_2O_8$ near T_c was 40-50 at 10 GHz, whereas Terasaki and Tajima [5] measured that it was 120 at 100 MHz. These values are of the same order of ε_c for the parent insulators, and we may say that the out-of-plane conductance of HTSC is a "remnant" of the parent insulator. Another feature is that the temperature dependence of ε_c is different between M=Er and Dy. Recently we have found that ε_c for all the samples, *including superconducting ones*, can be understood with the Debye description of dielectric relaxation [6], which has been used for the analyses of the dielectric response of the charge density wave [7].

The reciprocal of the dielectric constant ($1/\varepsilon_{ab}$ and $1/\varepsilon_c$) at 80 K is plotted as a function of hole concentration per Cu in Fig. 3. For comparison, the data for the superconducting samples are also plotted. ε_c is employed from Refs. [5, 6], and ε_{ab} is estimated from the Drude model as $\varepsilon_{ab}(\omega \to 0) = -(\omega_p/\gamma)^2$, where ω_p and γ are the plasma frequency and the damping factor respectively. By putting $\hbar\omega_p$=1.1 eV and $\hbar\gamma$=k_BT, we get $\varepsilon_{ab} = -2.5 \times 10^4$, which is in the same order of ε_{ab} for M=Dy and Er. As shown in Fig. 3(a), $1/\varepsilon_{ab}$ crosses zero near p=0.05, and goes negative in the metallic side. This is exactly what we see IMT in doped Si. On the other hand, although $1/\varepsilon_c$ becomes smaller for M=Er than for M=Dy, $1/\varepsilon_c$ for the superconducting samples is positive, and stay at the same order. Thus ε_c is unlikely to diverge at IMT, as Chen et $al.$ previously found that ε_c for La$_2$CuO$_{4+\delta}$ does not diverge at IMT [1]. We should note that the gross feature of Fig. 3 is not largely dependent on frequency and temparature, although the data for 1MHz at 80 K was rather arbitrarily selected. More detailed analysis is in progress.

Chen et $al.$ pointed out two possibilities for the non-divergent ε_c. One is that IMT in HTSC occurs only along the in-plane direction, and the other is that the heavy effective mass along the out-of-plane direction makes the effective Bohr radius of a hole shorter than the c-axis length. Our data favors the former scenario. According to the latter scenario, $\varepsilon_{ab}/\varepsilon_c$ would be equal to the effective mass ratio, which disagrees with our observation that ε_c for Bi$_2$Sr$_2$$MCu_2O_8$ is larger than ε_c for La$_2$CuO$_{4+\delta}$. Thus the non-divergent ε_c does not solely comes from the anisotropic effective mass, but from the anomalous conduction mechanism such as "confinement".

SUMMARY

In summary, we prepared single crystals of Bi$_2$Sr$_2$$MCu_2O_8$ (M=Y and rare-earth) and measured the anisotropic dielectric constants ε_{ab} and ε_c from 80 to 300 K. The present study has revealed that ε_{ab} (10^4-10^5) is about three orders of magnitude larger than ε_c (10^2). The huge ε_{ab} is a remnant of the Fermi surface, where the dc conductivity is suppressed by the strong correlation or localization. We have found that ε_c remains near 10^2 across the insulator-metal transition, which means that the transition occurs only along the in-plane direction. This can be a piece of evidence of the confinement behavior of the parent insulators.

Acknowledgments. This work was partially supported by The Kawakami Memorial Foundation, and by Waseda University Grant for Special Research Projects (99A-556). The authors would like to thank S. Tajima for the rf-conductivity measurements in Superconductivity Research Laboratory, International Superconductivity Technology Center. They also appreciate T. Itoh, T. Kawata, K. Takahata Y. Iguchi and T. Sugaya for collaboration.

1. C.Y. Chen, N. W. Preyer, P. J. Picone, M. A. Kastner, H. P. Jenssen, D. R. Gabbe, A. Cassanho, and R. J. Birgeneau, Phys. Rev. Lett. **63**, 2307-2310 (1989).

2. T. Kitajima, T. Takayanagi, T. Takemura and I. Terasaki, J. Phys. Condens. Matter **11**, 3169-3174 (1999).

3. S. D. Obertelli, J. R. Cooper, and J. L. Tallon, Phys. Rev. **B 46**, 14928-14931 (1992).

4. H. Kitano, T. Hanaguri and A. Maeda, Phys. Rev. **B 57**, 10946-10950 (1998).

5. I. Terasaki and S. Tajima, Meeting Abstracts of the Physical Society of Japan **52**, Issue 1, p. 651 (1997) (in Japanese).

6. T. Kitajima, T. Takayanagi and I. Terasaki, Meeting Abstracts of the Physical Society of Japan **54**, Issue 2, p. 584 (1999) (in Japanese).

7. R. J. Cava, R. M. Fleming, P. Littlewood, E. A. Rietman, L. F. Schneemeyer and R. G. Dunn, Phys. Rev. B**30**, 3228-3239 (1984).

Features of Crystal Structures in the metallic phase of $Sr_{1-x}La_xTiO_3$

M. Mogi, Y. Inoue, M. Arao*, and Y. Koyama*

NISSAN ARC, LTD., Yokosuka, Kanagawa 237-0061, Japan
*Dept. of Materials Science and Engineering, Waseda Univ., Shinjuku-ku, Tokyo 169-8555, Japan

The structural properties of metallic $Sr_{1-x}La_xTiO_3$ with $0.2 \leq x \leq 0.95$ have been investigated in the help of Rietveld refinement using X-ray powder diffraction. A structural phase diagram was confirmed to have a $R\bar{3}c$ phase in $0.2 \leq x < 0.5$ and a $Pbnm$ phase in $0.5 \leq x \leq 0.95$. Particularly, the Ti-O-Ti bond angles decrease monotonously with increasing x in the $Pbnm$ phase. This suggests that a change in the band width plays a certain role in the metal-insulator transition, which takes place around $x=0.95$.

Keywords:$Sr_{1-x}La_xTiO_3$, X-ray powder diffraction, Ti-O-Ti bond angle, metal-insulator transition

INTRODUCTION

In the $Sr_{1-x}La_xTiO_3$ system, $LaTiO_3$ (with $x=1.0$) is one of the Mott-type insulators, and a metal-insulator(M-I) transition takes place around $x=0.95$ [1,2]. That is, $Sr_{1-x}La_xTiO_3$ exhibits a metallic behavior in the composition range of $0<x\leq0.95$. A systematic change of physical properties in the metallic phase suggests that the M-I transition can be correlated with 3d band filling. Because substitution of La for Sr decreases an effective ionic radius, however, some structural change can be expected and plays a certain role in the M-I transition. We have investigated structural features by transmission electron microscopy and found that the crystal structure changes $Pm3m$ to $Pbnm$ via $R\bar{3}c$ with increasing La content [3,4]. Note that these changes are closely related to a rotation of TiO_6 octahedra, which results in a change of band width via the Ti-O-Ti bond angle. These structural changes should influence the electronic structure of $Sr_{1-x}La_xTiO_3$. In this work, therefore, X-ray powder diffraction profiles of metallic $Sr_{1-x}La_xTiO_3$ with $0.2\leq x\leq0.95$ were collected at room temperature to determine Ti-O-Ti bond angles at each La content. The bond angles were calculated from atomic coordinates of Ti and O ions, which were obtained by Rietveld refinements.

EXPERIMENTAL PROCEDURE

$Sr_{1-x}La_xTiO_3$ samples with $0.2\leq x\leq0.95$ were prepared by an arc-melting technique in an $20\%H_2$ /Ar atmosphere. X-ray powder diffraction profiles were collected on a diffractometer operated at 40kV and 300mA, using monochromatic Cu $K\alpha$ radiation. We measured profiles in a range of $5\leq2\theta\leq120°$. To obtain more than 10,000 counts for an intensity of the highest peak, the counting time per step was set to be 1s. The range of 2θ and step width were 5-120° and 0.02° , respectively.

The Rietveld refinement was made for each powder diffraction profile in $20\leq2\theta\leq120°$ by the Rietveld-analysis program RIETAN 94[5,6]. The refinement started with the $R\bar{3}c$ space group for $0.2\leq x<0.5$, and the *Pbnm* space group for $0.5 \leqq x \leqq 0.95$. The detailed procedure of the refinement is as follows. Lattice parameters were first refined while other profile-function parameters were fixed. After refining atomic coordinates of ions, we get preferred-orientation and isotropic thermal parameters. Finally all the parameters including profile-function parameters were optimized simultaneously.

RESULT AND DISCUSSION

Figure 1 shows a measured X-ray powder diffraction profiles of $Sr_{0.8}La_{0.2}TiO_3$ at room temperature, together with a calculated diffraction profile. In the inset of Fig. 1, the details of diffraction peaks at $2\theta=38°$ are also shown. In addition to diffraction peaks due to the *Pm3m* structure of cubic symmetry, there are small peaks in the profile, such as shown in the inset. The peak should reflect the deviation from the cubic symmetry. Based on our previous electron diffraction studies, we indexed the peaks in terms of the $R\bar{3}c$ structure, and found that the small peak at 38° can be indexed as the 113 peak. From similar analysis of other samples, it was confirmed that samples with $0.2 \leqq x < 0.5$ have the $R\bar{3}c$ structure, which is characterized by the $R_{25}{}^x + R_{25}{}^y + R_{25}{}^z$ rotational

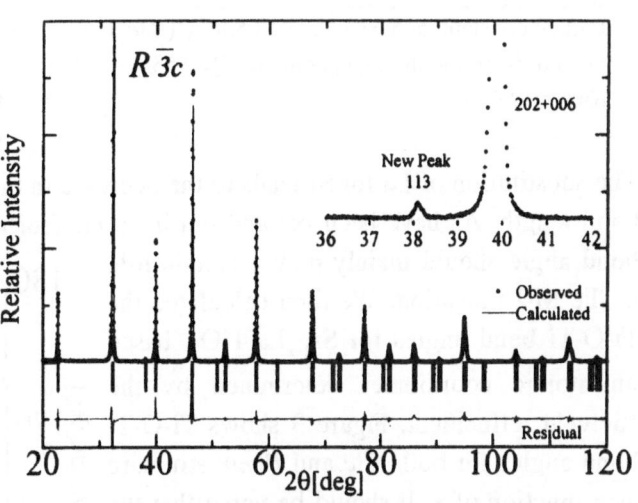

Fig.1 X-ray powder diffraction profile of $Sr_{1-x}La_xTiO_3$ with $x=0.2$ at room temperature.

displacement. Note that in the R_{25} displacement the rotation directions of the two neighboring TiO_6 octahedra along the rotation axis are opposite to each other and the superscript denotes the rotation axis. The obtained structure parameters of $Sr_{0.8}La_{0.2}TiO_3$ were shown in Table 1.

$Sr_{1-x}La_xTiO_3$ with $0.5 \leqq x \leqq 0.95$ was, on the other hand, confirmed to have the *Pbnm* structure. For example, an X-ray powder diffraction profile of the $x=0.95$ sample at room temperature is shown in Fig. 2. A detailed profile in $30 \leqq 2\theta \leqq 55°$ is shown in the inset. Note that the M-I transition occurs around $x=0.95$. Many peaks with weak intensities are observed in the profile, as indicated by open circles in the inset, in addition to diffraction peaks due to cubic perovskite structure. A Rietveld-refinement analysis of this profile indicated that the peaks are entirely consistent with these due to the *Pbnm* structure. The structure parameters

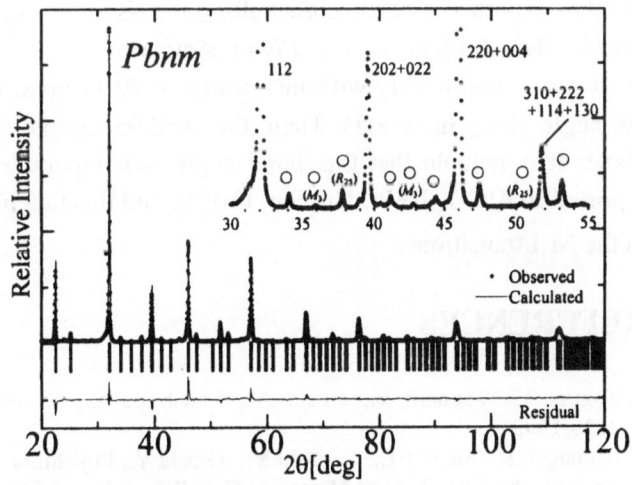

Fig.2 X-ray powder diffraction profile of $Sr_{1-x}La_xTiO_3$ with $x=0.95$ at room temperature.

of $Sr_{0.95}La_{0.05}TiO_3$ are listed in Table 2. It should be noted that the $R_{25}{}^x + R_{25}{}^y + R_{25}$ displacement is involved in the *Pbnm* structure and that two neighboring octahedra along the rotation axis are rotated in the same directions in the M_3 displacement. That is, the peaks at $2\theta \fallingdotseq 38°$ and 50° in the inset are produced by the R_{25} displacement, while the peaks at $2\theta \doteqdot 36°$ and $43°$ comes from the M_3 displacement. Eventually, these experimental data confirmed our previous result obtained by electron diffraction[4].

Table 1. Structure parameters of La$_{0.2}$Sr$_{0.8}$TiO$_3$.

Atom	Wyckoff symbol	Atomic coordinates		
		x	y	z
La/Sr	6a	0	0	0.25
Ti	6b	0	0	0
O	18e	0.479	0	0.25

Lattice constant a=5.536 Å, c=13.538 Å ($R\bar{3}c$)
Overall isotropic thermal parameter Q=0.55
Rwp =12.44%

Table 2. Structure parameters of La$_{0.95}$Sr$_{0.05}$TiO$_3$.

Atom	Wyckoff symbol	Atomic coordinates		
		x	y	z
La/Sr	4c	0.284	0.250	0.993
Ti	4a	0.5	0	0
O	4c	0.482	0.250	0.073
O	8d	0.277	0.023	0.713

Lattice constant a=5.585 Å, b=5.576 Å ,c=7.887 Å
Overall isotropic thermal parameter Q=0.40($Pbnm$)
Rwp =13.77%

The substitution of La for Sr leads to the decrease in the Ti-O-Ti bond angle and the increase in the Ti-O bond length. As have been pointed out in a number of previous studies in perovskite oxides[7, 8], the bond angle should mainly play a certain role in the M-I transition. We then calculated the Ti-O-Ti bond angles for Sr$_{1-x}$La$_x$TiO$_3$, based on atomic coordinates determined by the Rietveld refinement. Figure 3 shows Ti-O-Ti bond angles for both $R\bar{3}c$ and $Pbnm$ structure as a function of x. It should be noted that the $R\bar{3}c$ structure is characterized by only one bond angle, while there are two types of angles in the $Pbnm$ structure. One is along the c axis and the other are in the ab-plane. The bond angle in the $R\bar{3}c$ structure decrease from $173.4°$ to $169.3°$ with increasing x. Similarly, both the bond angles along c axis and in the ab-plane in the $Pbnm$ structure

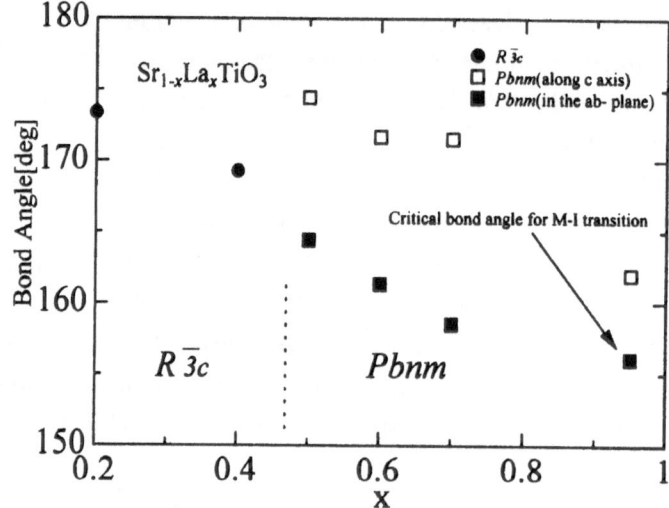

Fig.3 Ti-O-Ti bond angles as a function of the La content.

decrease monotonously with increasing x. At each x, the angles in the ab-plane is always smaller than the angle along the c axis. Then, the x=0.95 sample has the smallest bond angle of $156.1°$ in the ab-plane. It is notable that this bond angle is comparable with the critical bond angle for M-I transition reported in RNiO$_3$(R=Pr, Nd, Sm, Eu)[8], and this fact proved that the structural factor plays a certain role in the M-I transition.

REFERENCES

1. Tokura Y., Taguchi Y., Okada Y., Fujishima Y., Arima T., Kumagai K., and Iye Y. *Phys. Rev. Lett.* **70**, 2126(1993).
2. Kumagai K., Suzuki T., Taguchi Y., Okada Y., Fujishima Y., and Tokura Y. *Phys. Rev.* **48**, 7636(1993).
3. Arao M., Koyama Y., and Munakata F. in Proceedings of *the 9th International Symposium on superconductivity(ISS '96)*, edited by S. Nakajima and M. Murakami(Springer-Verlag, Tokyo, 1997), p.289.
4. Arao M., Munakata F. ,and Koyama Y. *Physica C* **282**, 1111(1997).
5. Izumi F., "*The Rietveld Method*", edited by R. A. Young, Oxford University Press, Oxford, Char. 13(1993).
6. Kim Y. I., and Izumi F. *J. Ceram. Soc. Jpn.* **102**, 401(1994).
7. for example, Crandles D. A., Timusk T., Garrett J. D., and Greedan J. E., Physica C **201**, 407(1992)., and Okimoto Y. , Katsufuji T., Okada Y., Arima T., and Tokura Y., *Phys. Rev.* B**51**, 9581(1995).
8. Torrance J. B., Lacorre P., and Nazzal A. I. *Phys. Rev.* **45**, 8209(1992).

Features of Microstructure in $Ca_2MnO_{4-\delta}$

Yasuhide Inoue[1], Yoichi Horibe[2], and Yasumasa Koyama[2]

[1]NISSAN ARC, LTD., 1 Natsushima-cho, Yokosuka, Kanagawa 237-0061, Japan
[2]Dept. of Materials Science and Engineering, Waseda Univ., 3-4-1 Ookubo, Shinjuku-ku, Tokyo 169-8555, Japan

Features of crystal structures and a related microstructure in $Ca_2MnO_{4-\delta}$ ceramic sample have been investigated by transmission electron microscopy. In electron diffraction patterns at room temperature, there exist $h/2\ k/2\ l$-type and $h/2\ k/2\ l/2$-type superlattice reflection spots, in addition to fundamental spots due to the tetragonal K_2NiF_4-type structure. That is, there coexist the Bbcm and I41/acd structures in our samples. In dark field images taken by the superlattice spots, further there are three types of regions arranged along the c axis. A careful analysis indicated that one of the regions has the I41/acd structure and the others are two variants of the Bbcm one. The microstructure was therefore understood to be a layered structure, which is composed of the regions with the I41/acd and Bbcm structures.

Keywords: $Ca_2MnO_{4-\delta}$ oxide , transmission electron microscopy, microstructure

INTRODUCTION

It is known that the two-dimensional CuO_2 plane involved in crystal structures of cuprates is one of the most important factors in understanding of superconductivity. Based on a t-J-V model, it was suggested that electronic phase separation occur in the CuO_2 plane. The phase separation is thought to be directly related to a low-temperature structural phase transition from a low-temperature orthorhombic (LTO)phase to a low-temperature tetragonal (LTT) one found in $La_{2-x}Ba_xCuO_4$[1-2]. Elucidation of the low-temperature structural transition is therefore needed for understanding of physical properties in cuprates.

$Ca_{2-x}Sm_xMnO_{4-\delta}$ also exhibits the same LTO-to-LTT phase transition around x～0.3. According to the previous works, there are I41/acd and Bbcm phases in 0<x<0.25, in addition to the LTO and LTT phases[3-5]. Both structures are produced by an introduction of the rotational displacement of the oxygen octahedra about the [001] axis into the normal K_2NiF_4-type structure. Note that the <110> and <100> rotational displacements are, respectively, involved in the LTO and LTT structures. A striking feature of the [001] rotation is that the rotational displacement can not propagate along the c axis. That is, the structural transition characterized by the [001] rotation has a strong two-dimensional nature. As a result of the two-demensional nature, structural disorder along the c axis is easily induced. The coexistence of the I41/acd and Bbcm structures must have some relation with such structural disorder. As the first step of elucidation of the LTO-to-LTT structural transition, therefore we have examined features of the crystal structures and a related microstructure in $Ca_2MnO_{4-\delta}$ by transmission electron microscopy.

EXPERIMENTAL PROCEDURE

Ca2MnO4-δ ceramic samples were prepared by a conventional solid state reaction. Starting powders of CaCO3 and Mn2O3 were mixed mechanically and pressed into pellets. The pellets were sintered at 1523 K for 72 h in an air. Characteristic features of the microstructure were examined, using H-8100A and H-9000UHR transmission electron microscopes. Specimens for electron microscopy observation were prepared by crushing the ceramic samples.

RESULT AND DISCUSSION

Figure 1 shows four electron diffraction patterns of Ca2MnO4-δ at room temperature. Electron incidences of Figs. 1(a), 1(b), 1(c), and 1(d) are parallel to the [001], [100], [111], and [130] directions, respectively. Diffraction spots are indexed in terms of the tetragonal K2NiF4-type structure with no atomic displacement. In these diffraction patterns, two types of superlattice reflection spots were observed in addition to fundamental spots due to the tetragonal K2NiF4-type structure. One is $h/2\ k/2\ l$-type superlattice reflection spots, as is indicated by arrows A, B, and C. The other is $h/2\ k/2\ l/2$-type superlattice reflection spot, as is indicated by an arrow D in Fig. 1(d). As is seen in Fig. 1(d), an intensity of the $h/2\ k/2\ l/2$-type spot is stronger than that of the $h/2\ k/2\ l$-type one. From a careful examination of the electron

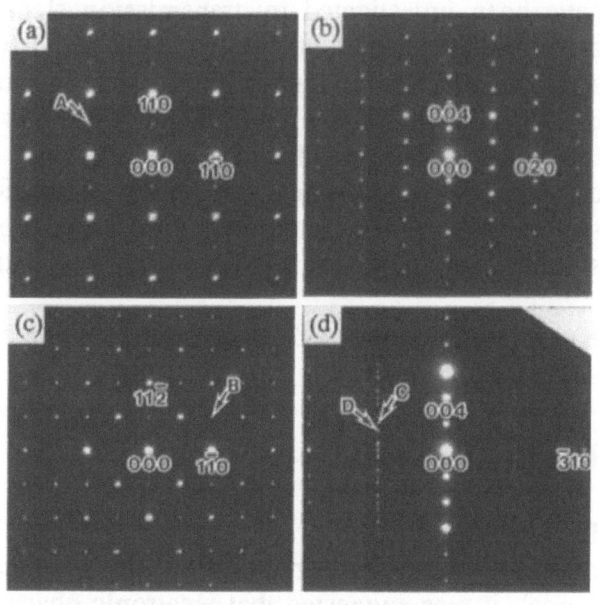

Fig. 1.Electron diffraction patterns of Ca2MnO4-δ. Electron incidences are parallel to the (a) [100], (b) [001], (c) [111], and (d) [130] directions, respectively.

diffraction patterns, superlattice spots along the [110] and [1$\bar{1}$0] directions through an origin are due to double diffraction. That is, the $h/2\ k/2\ l$-type spots have extinction rule of $h=2n$ and $k=2n$ and $h+k=4n$, $l=2n+1$ and $h+k=4n+2$, $l=2n$ and $|h|=|k|$, on the other hand, $h/2\ k/2\ l/2$-type spots have that of $h=2n$ and $k=2n$ and $l=2n$ and $|h|=|k|$ (n; integer).

In order to understand the details of a microstructure, we took dark field images of these superlattice reflection spots. Figure 2 shows a dark field image taken by the 3/2 1/2 2 and the 3/2 $\bar{1}$/2 3/2 superlattice reflection spots indicated by the arrows C and D in Fig. 1(d), respectively. An electron incidence is parallel to the [130] direction. In the dark field image, there are three types of regions indicated by arrows A, B, and C, respectively, which are randomly distributed along the [001] direction. The regions A, B, and C are observed as dark-line contrast, faint bright-band contrast, and bright-line contrast, respectively. When the specimen was slightly tilted about the [001] direction, the contrasts of the regions A and C are reversed each other, but the contrast is not changed in the region B. The most important feature is that the regions A and C give rise to the $h/2\ k/2\ l$-type superlattice spots in diffraction patterns, while the $h/2\ k/2\ l/2$-type spots come from the region B. This implies that the crystal structure of the regions A and C is different from that of the region B. The dark field image indicates that three types of regions are arranged along the c axis. According to the previous works, two types of the crystal structures with space groups of I4₁/acd and Bbcm have been reported at room temperature. Calculated extinction rules of these structures clearly showed that the I4₁/acd and Bbcm structures give rise to the $h/2$ $k/2$ l-type and $h/2$ $k/2$ $l/2$-type superlattice reflection spots, respectively. Based on this result, it is understood that the region B has the I4₁/acd structure and the regions A and C are variants of the Bbcm structure. The latter is consistent with the contrast reverse of the regions A and C, which was mentioned before.

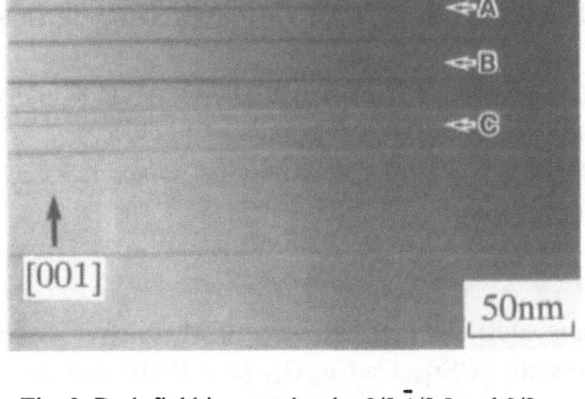

Fig. 2. Dark field image taken by 3/2 $\bar{1}$/2 2 and 3/2 $\bar{1}$/2 3/2 superlattice spots shown in Fig. 1(d).

Let us briefly explain the existence of two variants of the Bbcm structure. Figure 3 shows rotational displacements of the MnO₆ octahedra in the Bbcm structure. When two octahedra in two neighboring MnO₂ layers are specified by A₁ and A₂, as is indicated in Fig. 3(a), there are two types of rotational displacements of the octahedra. One is that the octahedra A₁ and A₂ are rotated in the same direction, Fig. 3(a). In the other, the rotation directions of these octahedra (B₁ and B₂) are opposite to each other, Fig. 3(b). That is, a difference between the two rotational displacements results in two variants of the Bbcm structure.

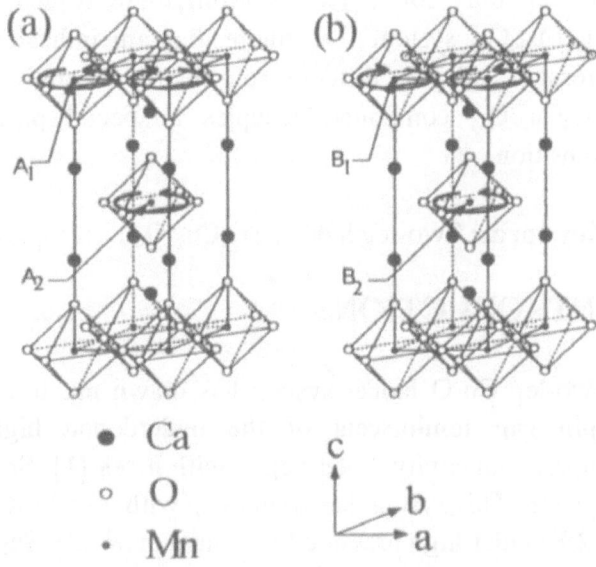

Fig. 3. The rotational displacement of the oxygen octahedra in the Bbcm structure. Two types of the rotational displacements shown in (a) and (b).

CONCLUSION

The present experiment data showed that the present $Ca_2MnO_{4-\delta}$ ceramic samples has the composite microstructure, which is mixture of the I4₁/acd region and two variants of the Bbcm structure. The I4₁/acd and Bbcm structures are characterized by the rotation of the MnO₆ octahedron about the [001] direction. It is therefore understood that the rotational displacement can not easily propagate along the rotational axis.

REFERENCES

1. J. D. Axe, A. H. Moudden, D. Hohlwein, D. E. Cox, K. M. Mohanty, A. R. Moodenbaugh, and Y. Xu, Phys. Rev. Lett. 62, 2751 (1989).
2. R. M. Fleming, B. Batlagg, R. J. Cava, and E. A. Rietman, Phys. Rev. B 35, 7191 (1987).
3. C. Chaumont, A. Daoudi, G. Le Flem, and P. Hagenmuller, J. Solid State Chem., 14, 335 (1975).
4. M. E. Leonowicz, R. Poeppelmeier, and M. J. Longo, J. Solid State Chem., 59, 71 (1985).
5. J. Takahashi, T. Kikuchi, H. Sato, and N. Kamegashira, J. Alloys and Comps., 192, 99 (1993).

Pressure Induced Phase Transition on $Sr_{14}Cu_{24}O_{41}$ with Doped Two-Leg Cu-O Ladders

Naoki Motoyama[1], Hiroshi Eisaki[1], Shin-ichi Uchida[1], Nao Takeshita[2], and Nobuo Mori[2]

[1]Department of Superconductivity, The University of Tokyo, Yayoi 2-11-16, Bunkyo-ku, Tokyo 113-8656, Japan
[2]Institute for Solid State Physics, University of Tokyo, Roppongi 7-22-1, Minato-ku, Tokyo,108-8666, Japan

Abstract: We measured electrical resistivity under hydrostatic pressure ($P \leqq 8.5$ GPa) for single crystals of $Sr_{14-x}Ca_xCu_{24}O_{41}$ ($x = 0$-10) and $Sr_{13}Y_1Cu_{24}O_{41}$. The resistivity for $x = 0$, which shows activation-type insulating behavior at ambient pressure, becomes metallic above 8 GPa. At 5 GPa, the temperature dependence of this resistivity shows a feature suggestive of a CDW transition. This feature is not observed for both Ca- and Y-substituted compound. Superconductivity under pressure is seen only for $x \geqq 10$. From these results, we have determined an x-P phase diagram of $Sr_{14}Cu_{24}O_{41}$ system. The phase diagram indicates that Ca-substitution (x), regarded as "chemical pressure", plays a different role than hydrostatic pressure (P). This diagram also indicates that the $Sr_{14}Cu_{24}O_{41}$ compound occupies a special position, such as CDW and an insulator-to-metal transition.

Keywords: Two-leg ladder, $Sr_{14}Cu_{24}O_{41}$, Transport properties, M-I transition, Hydrostatic pressure

INTRODUCTION

Two-leg Cu-O ladder system has drawn much attention, since theoretical works predicted a finite spin gap reminiscent of the underdoped high-T_c cuprates, and suggested a possibility of superconductivity when doped with holes [1]. $Sr_{14}Cu_{24}O_{41}$ system is one of the realizations of such system. This system has a spin gap with ~ 500 K [2], and a superconductivity for $x = 13.6$ with $T_c = 12$ K under high pressure by Uehara *et al.* [3]. Superconductivity is also observed for $x = 11.5$ single crystal at 3.5 GPa with $T_c = 9$ K [4].

Considering that the superconductivity so far occurs only in heavily Ca-substituted compound under high pressure, it is indispensable to make the x-P phase diagram and to understand the respective role of Ca-substitution and hydrostatic pressure. In this study, we measured electrical resistivity under hydrostatic pressure ($P \leqq 8.5$ GPa) for the single crystals of $Sr_{14-x}Ca_xCu_{24}O_{41}$ ($x = 0$-10) and $Sr_{13}Y_1Cu_{24}O_{41}$. To generate hydrostatic pressure, a cubic-anvil-type apparatus was used.

RESULTS AND DISCUSSIONS

In Fig. 1, we show the temperature dependence of the electrical resistivity for $Sr_{14}Cu_{24}O_{41}$ under high pressure along the (a) c-axis (along the ladders) (ρ_c) and (b) a-axis (across the ladders) (ρ_a). What we notice first is that an insulator-to-metal transition occurs at pressure of ~ 6.5 GPa. At 5 GPa, the resistivity shows a steep increase at $T \sim 80$ K, suggestive of a CDW transition. This charge-

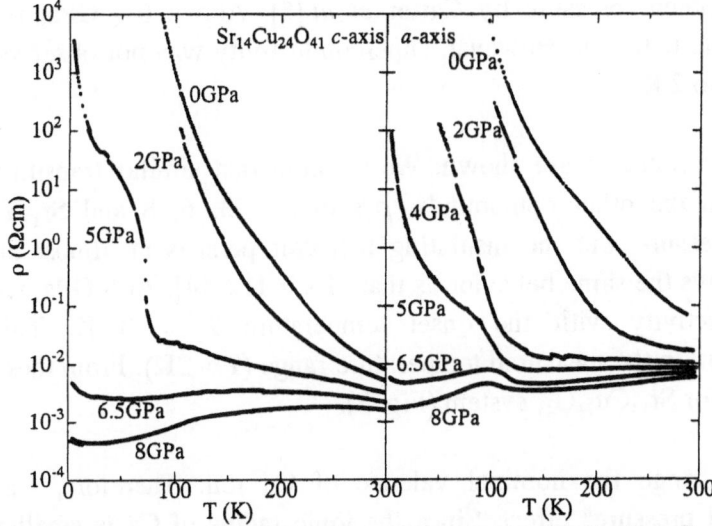

Fig. 1. The pressure dependence of ρ_c and ρ_a for $Sr_{14}Cu_{24}O_{41}$.

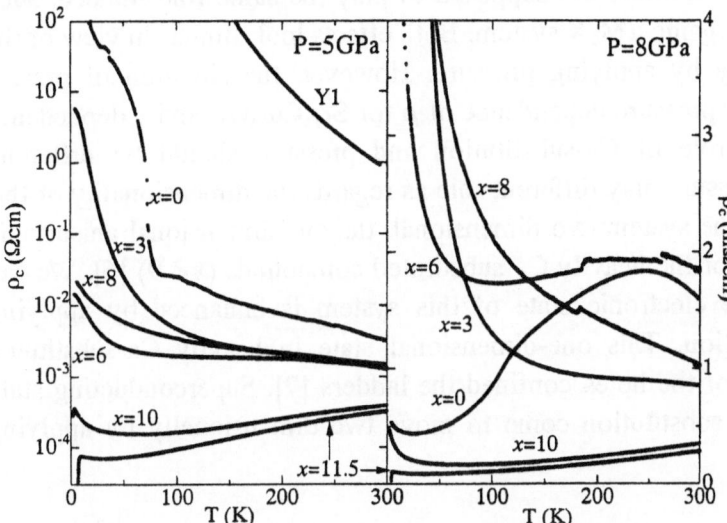

Fig. 2. The x dependence of ρ_c at 5 GPa and 8 GPa.

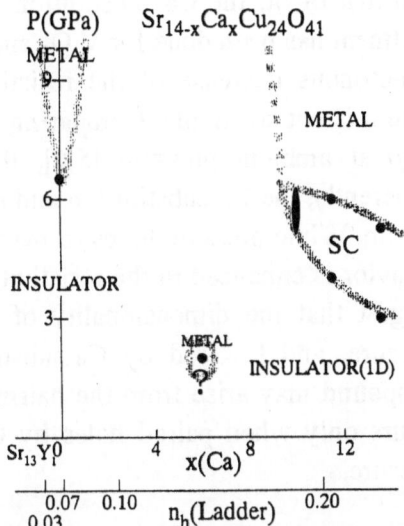

Fig. 3. The phase diagram of $Sr_{14}Cu_{24}O_{41}$ system expected by our results. n_h(Ladder) indicates the hole density of the ladder estimated by the optical measurements [8].

ordered state has been suggested at ambient pressure by Carter *et al.*[5]. Above 6.5 GPa, the resistivity, both ρ_c and ρ_a, shows metallic behavior. However, superconductivity was not observed on this compound even at 8.5 GPa down to 2 K.

In Fig. 2, ρ_c at 5 GPa and 8 GPa for various x's are shown. We mention that similar transition observed for $Sr_{14}Cu_{24}O_{41}$ is not seen in the other compounds. ρ_c's of x = 3, 6, 8 and $Sr_{13}Y_1$ compounds decrease with increasing pressure, but the insulating behavior persists at almost all pressure range. For x = 10, the result shows the same behavior as that of x = 11.5 [4]. At 5 GPa, we observed the sign of the superconductivity with the onset temperature T_c = 3 K. This superconductivity persists up to 6 GPa within the measured temperature range ($T > 2K$). From these results, we obtain the x-P phase diagram for $Sr_{14}Cu_{24}O_{41}$ system (Fig. 3).

The Ca-substitution for Sr does not change the nominal valence of Cu ion. Therefore, Ca-substitution can be considered "chemical pressure" effect. Since the ionic radius of Ca is smaller than that of Sr, the Ca-substitution and pressure are supposed to play the same role. Indeed, such treatment has been done for V_2O_3 and organic TM_2X system. Both effects look similar in view of the monotonous decrease of the resistivity by applying pressure. However, the situation of present system is not so simple. Comparing the pressure dependence of ρ for $Sr_{14}Cu_{24}O_{41}$ and x dependence of ρ at ambient pressure [5,6], the role of Ca-substitution and pressure should be different. Apparently, the Ca-substitution and pressure play different role as regards the dimensionality of this system. While pressure tends to make the system two-dimensional, the one-dimensional anisotropic behavior is enhanced in the resistivity for the heavily Ca-substituted compounds ($x \geqq 9$) [6]. We can suggest that the dimensionality of the electronic state of this system is enhanced by applying pressure, and lowered by Ca-substitution. This one-dimensional state in heavily Ca-substituted compound may arise from the pairing of the holes confined the ladders [7]. Superconducting state occurs only when paired holes by Ca-substitution come to move two-dimensionally by applying pressure.

For lower hole density, the pair formation becomes difficult due possibly to stronger repulsive interaction between holes, and the holes are strongly localized on the ladders, making the insulating phase robust against pressure. $Sr_{14}Cu_{24}O_{41}$ is unique in showing an insulator-to-metal transition. Considering that the metallic phase is difficult to be realized for other compounds, randomness should play a role in the localization of single holes in the Ca-substituted compounds.

1. T. M. Rice, S. Gopalan, and M. Sigrist, Europhys. Lett. **23**, 445-448 (1993), E. Dagotto, J. Riera, and D. Scalapino, Phys. Rev. **B45**, 5744-5747 (1992).
2. S. Tsuji *et al.*, J. Phys. Soc. Jpn., **65**, 3474-3477 (1996), R. S. Eccleston *et al.*, Phys. Rev. Lett. **81**, 1702-1705(1998).
3. M. Uehara *et al.*, J. Phys. Soc. Jpn., **65**, 2764-2767 (1996).
4. T. Nagata *et al.*, Phys. Rev. Lett. **81**, 1090-1093(1998).
5. S. A. Carter *et al.*, Phys. Rev. Lett. **77**, 1378-1381(1996).
6. N. Motoyama *et al.*, Phys. Rev. **B55**, 3386-3389(1997).
7. T. Osafune *et al.*, Phys. Rev. Lett. **82**, 1313-1316(1999).
8. T. Osafune *et al.*, Phys. Rev. Lett. **78**, 1980-1983(1997).

Impurity Effect in High T_c Superconductors

Masahiro Goto and Shin-ichi Uchida

Department of Superconductivity, University of Tokyo, Hongo 7-3-1, Bunkyo-ku,
Tokyo 113-8656, Japan

Abstract: In-plane resistivity and magnetic susceptibility measurements have been carried out on the single crystals of $YBa_2Cu_3O_{7-y}$ and $La_{2-x}Sr_xCuO_4$ with various kinds of impurities. In the underdoped region, the divalent impurity effect is characterized by an extremely large residual resistivity due to scattering in the unitality limit and an induced magnetic moment as large as a free spin, whatever kind of impurity is substituted. On the other hand, the residual resistivity from the monovalent impurity Li, and the trivalent impurity Al, is smaller and the induced magnetic moment is larger than that from the divalent impurity. These results indicate that a monovalent/trivalent impurity in the high-T_c material introduces an additional hole/electron which is localized on the impurity and thus the impurity behaves like a magnetic impurity.

Keywords: impurity effect, residual resistivity, induced magnetic moment

INTRODUCTION

Non-magnetic impurities, especially Zn, in copper oxide superconductors have a dramatic influence on the critical temperature T_c, the in-plane resistivity ρ_{ab}, and the magnetic susceptibility χ.[1] In the underdoped region, the residual resistivity, ρ_o is anomalously large. The equation of the ρ_o in the two dimensional system is given by

$$\rho_o = \frac{4\hbar}{e^2} \frac{n_{imp}}{n} \cdot \sin^2 \theta_o$$

To explain a large ρ_o, according to eq. (1), we have to assume the phase shift θ_o equals $\pi/2$, indicating that the scattering due to the Zn ion is in the unitality limit, and the carrier concentration n equals x, the doped hole concentration. χ is also influenced largely and the induced magnetic moment is as large as that of a S=1/2 free spin. These two quantities decrease in magnitude, as the carrier concentration increases. In this study, we extend the variety of impurities, Mg, Li, and Al. Mg is ionized to 2+, the same valence as Zn. Al is ionized to 3+ and Li is ionized to 1+. These impurities have no 3d electrons and are non-magnetic. We focus on the correlation between the valence of the impurity and the transport properties and magnetic properties.

RESULTS AND DISCUSSIONS

In the underdoped region, the large ρ_o is observed in the Mg substituted system. This large ρ_o is comparable to that in the Zn substituted system. It is concluded that Mg is also a potential scatterer in the unitality limit. In the system Al and Li doped, the temperature dependence of resistivity is the same as the Zn substituted system, the parallel shifts as the impurity concentration increases, indicating the carrier concentration in CuO_2 plane does not change. However, a difference from the Zn substituted system is a smaller residual resistivity. The results suggest that the ρ_o is determined not by the presence or otherwise of the 3d core but by the valence of impurities.

In the divalent impurity substituted systems, the magnetic moment is induced as large as that of S=1/2 free spins. Compared to that, in the Al and Li substituted system, larger magnetic moment is induced. It is considered that this is due to the excess holes/electrons which are localized near the impurities and contribute to the magnetic moment. In this regard the monovalent and trivalent impurity behave as a magnetic impurity. From this consideration, it is consistent that the residual resistivity is smaller as in the Ni substituted system.

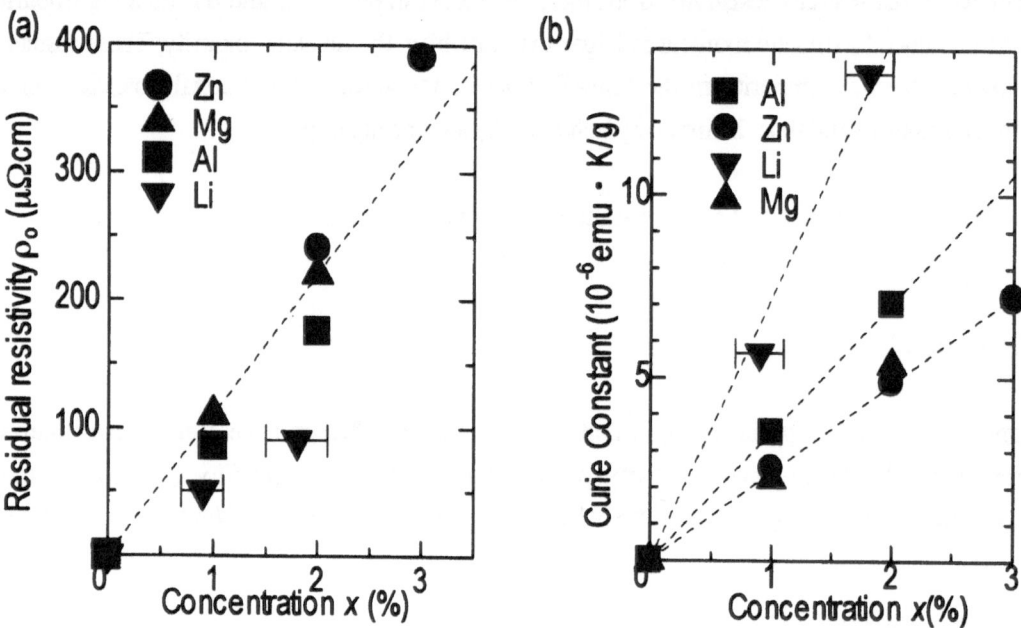

Fig.1 (a) Impurity concentration dependence of the ρ_o in the various impurity substituted system. The dashed line corresponds to in the unitality limit and n=x. The residual resistivity in the monovalent/trivalent impurity substituted system is significantly smaller than that in the divalent impurity substituted system. (b) Impurity concentration dependence of Curie constant. The Curie constant in the monovalent/trivalent impurity substituted system is larger than that in the divalent impurity substituted system.

In Fig.2, we show normalized critical temperature T_c/T_{c0} plotted as a function of the in-plane sheet residual resistance per CuO_2 plane. Fig.2 displays a universal T_c depression in the underdoped regime. The T_c suppression is controlled by the magnitude of two dimensional residual resistivity not by the species of impurities. The residual resistivity shows the magnitude of the scattering by the impurity, therefore in the same way in the superconducting state, the impurities affect charge pairs in CuO_2 plane. This result is consistent with the muon spin rotation experiment, the n_s/m^* decreases as the impurity increases.[2]

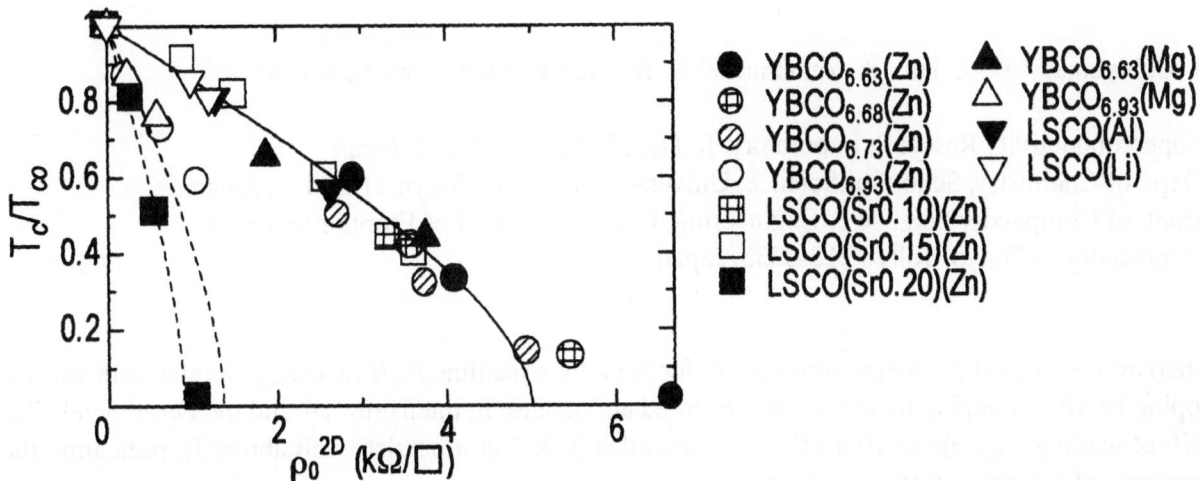

Fig.2 Normalized critical temperature T_c/T_∞ plotted as a function of the in-plane 2D residual resistance (per CuO_2 plane). The Solid curve is a line onto which all the date in the underdoped regime merge. Theoretical estimates by Radtke *et al* [3]. for a d-wave superconductor with nonmagnetic impurities are in the region between the dashed curves.

Conclusion

We have investigated the charge transport and the magnetic properties of underdoped high-T_c compound with various impurities. The monovalent/trivalent impurity is an weaker scatterer but induces larger magnetic moment than the divalent impurities. This can be considered that the monovalent and trivalent impurity behave as a magnetic impurity.

[1] Y. Fukuzumi *et. al*, Phys. Rev. Lett **76**, 684 (1996)

[2] B. Nachumi *et. al*, Phys. Rev. Lett. **77**,5421(1996)

[3] R. J. Radtke *et. al*, Phys. Rev. B **48**, 653 (1993)

Photoemission Study of Electronic State near the Fermi Level in $HgBa_2CuO_{4+\delta}$

H. Uchiyama,[1,2] W.-Z. Hu,[1] A. Yamamoto,[1] S. Tajima,[1] K. Saiki,[3] and A. Koma[2]

[1]Superconductivity Research Laboratory, ISTEC, Tokyo 135-0062, Japan
[2]Dept. of Chemistry, School of Science, University of Tokyo, Tokyo 113-0033, Japan
[3]Dept. of Complexity Science & Engineering, Graduate School of Frontier Sciences,
  University of Tokyo, Tokyo 113-0033, Japan

Abstract: We report photoemission spectra for the polycrystalline $HgBa_2CuO_{4+\delta}$ samples with various doping levels. As doping increases, we observed an increase in intensities around the Fermi level. The shift of leading edge about 10 meV was observed at 20K, but it persists well above T_c, indicating the existence of a normal state pseudogap

Keywords: $HgBa_2CuO_{4+\delta}$, photoemission spectroscopy, superconducting gap

INTRODUCTION

From the early stage of high-T_c study, the photoemission spectroscopy of $Bi_2Sr_2CaCu_2O_8$ (Bi2212) has provided a lot of information about its electronic state such as topology of the Fermi surface, magnitude and symmetry of the superconducting gap as well as the pseudogap. However, it is not clear whether all these electronic features are common in all high-T_c cuprates or not. In fact, the photoemission spectra of $La_{2-x}Sr_xCuO_4$ (La214) show some differences from the results for Bi2212. In order to know the origin of these differences, it is of great interest to study the Hg1201 system, because it contains a single Cu-O octahedron in a formula cell as is similar to La214 but shows a very high critical temperature (T_c=98 K). In this paper, we report the angle-integrated photoemission study on five samples of Hg1201 with various oxygen contents (doping levels). We investigated the intensities around E_F and the shift of the leading edge, which gives a value of the superconducting and pseudo- gaps.

EXPERIMENTS

Polycrystalline samples of Hg1201 were prepared by the solid state reaction method [1]. A doping level was controlled by temperature and atmosphere of the post annealing. In the X-ray diffraction, there was no sign of impurities, indicating that these crystals are high quality. T_c was determined by magnetization measurements. Photoemission studies were carried out with a helium discharge lamp (He I ($h\nu$=21.2 eV)). The total resolution is about 30 meV at 20 K. Backpressure was less than 1×10^{-10} Torr. All samples were filed *in situ* every 40 minutes during the measurements to avoid the surface contamination. Binding energy of spectra was calibrated using the leading edge of gold. The references were electrically contacted to the samples. The spectra were corrected for the He I* satellite subtraction.

RESULTS AND DISCUSSIONS

Figure 1 shows the photoemission spectra near E_F for various doping levels at 20 K ($< T_c$). The spectra are normalized to the intensity at 0.5 eV, above which the spectrum is less sensitive to doping. Increase

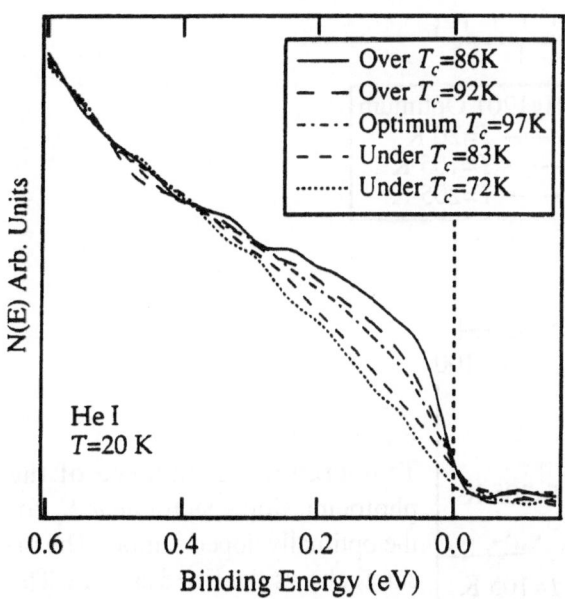

Fig.1
Photoemission spectra for the different carrier doped samples of Hg1201 at 20 K with He I. The spectra are normalized to the intensity at 0.5 eV.

of the intensity around E_F with doping is a similar behavior to the former data of the other high-T_c materials (Bi2212 [2] and La214 [3]). While the spectral intensity is strongly suppressed in the underdoped samples, it grows up with hole doping, becoming close to the Fermi-Dirac distribution. The spectral shape does not change in the higher binding energy region above 150 meV with temperature. This spectral change with doping may be linked with the development of Fermi surface, which has been detected in the angle-resolved photoemission spectroscopy [4].

Figure 2 shows temperature dependence of the spectrum near E_F for the optimally doped Hg1201 ($T_c \sim 97$ K). The spectra are normalized at 150 meV. By the spectral fitting to the Fermi-Dirac distribution function convoluted with the gaussian for arbitrary FWHM (not instrumental resolution), the shift of the leading edge midpoint in the spectrum at 20 K was estimated to be about 10 meV. Moreover, the shift does not appreciably change with temperature. As shown in the inset a), the fitted lines are crossing at one point below E_F, which is different from the behavior of gold shown in the inset b). If we evaluate the shift of the leading edge from this crossing point, the shift is about 8 meV. The spectra of the overdoped samples show the similar shifts above and below T_c, while the shift cannot be observed in the underdoped samples because of the strong suppression of the density of state near E_F.

In Fig.2, there is no peak feature below T_c, which is different from the data of Bi2212 [5]. We cannot rule out the possibility that the peak feature is diminished by the imperfect surface of our samples, because there is a clear peak feature near the gap energy in the tunnel data [6]. On the other hand, it is also possible that, since there is no peak feature in the photoemission spectra of La214 [3,7], absence or weakness of the coherence peak might be a common property in the monolayer compounds with single CuO-octahedron in a formula unit.

The estimated leading edge shift must be an average gap value in the case of a d-wave gap. The ratio of the leading edge shift (8~10 meV) to the gap from the tunneling data (33 meV) is about the same as that of La214 (the shift from the photoemission data [3,7] (~3 meV) to the gap from the tunneling data (9 meV)). Because the shifts in Hg1201 are insensitive to temperature, the gap is supposed to open even at 205 K. This suggests that the superconducting and pseudo- gaps smoothly connect with each other and the pseudogap persists above the spin gap temperature ($T^* = 140$ K) determined from NMR results [8].

Finally we point out two important features in the spectra of Hg1201. One is that with the increase of

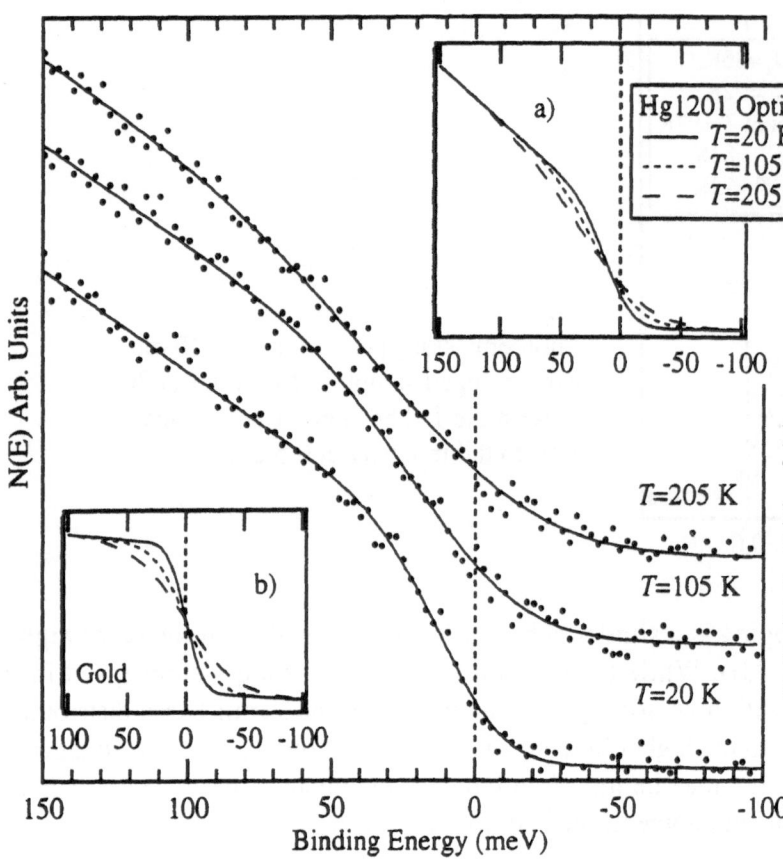

Fig. 2
Temperature dependence of the photoemission spectra near E_F for the optimally doped sample. The inset a) shows the fitted spectra. The inset b) shows the gold spectra. All spectra are normalized to the intensity at 150 meV.

carrier doping, the spectral intensities increase below ~500 meV. This may be a common feature in the high-T_c materials. The other is that the gap feature of ~10 meV remaining above T_c is similar to that in the spectra of La214 except for the gap value, whereas this feature is far from that of Bi2212. The similarity between the spectra of Hg1201 and La214 may be caused by the structural similarity with a CuO_2 layer in a formula unit, resulting in similar Fermi surfaces. The difference of the leading edge shift between them is consistent with the difference in T_c.

SUMMARY

We measured the angle-integrated photoemission spectra of Hg1201 with various doping levels. The intensities in the vicinity of E_F increase with the increase of carrier doping. The shift of the leading edge is estimated to be about 10 meV, being almost independent of temperature (20-205 K) .

Acknowledgments. This work was supported by the New Energy and Industrial Technology Development Organization (NEDO) as Collaborative Research and Development of Fundamental Technologies for Superconductivity Applications.

1. A. Yamamoto *et al.*, J. Mater. Res. **14** 644 (1999).

2. M.A. van Veenendaal *et al.*, Phys. Rev. B **47** 446 (1993).

3. A. Ino *et al.*, Phys. Rev. Lett. **81** 2124 (1998).

4. A. Ino *et al.*, J. Phys. Soc. Jpn. **68** 1496 (1999); F. Ronning *et al.*, Science **282** 2067 (1998).

5. J.-M. Imer *et al.*, Phys. Rev. Lett. **62** 336 (1989).

6. H. Murakami *et al.*, J. Phys. Soc. Jpn. **63** 2653 (1994); J.Y.T. Wei *et al.*, Phys. Rev. B **57** 3650 (1998).

7. T. Sato *et al.*, Phys. Rev. Lett. **83** 2254 (1999).

8. Y. Itoh *et al.*, J. Phys. Soc. Jpn. **67** 312 (1998).

Zn/Ni-Substitution Effects on out-of-plane Resistivity and Coherence Length in $YBa_2Cu_3O_{7-\delta}$

T. Masui, K. Tomimoto and S. Tajima

Superconductivity Research Laboratory, ISTEC, 1-10-13 Shinonome, Tokyo, 135-0062, Japan

Abstract: We measured out-of-plane resistivity (ρ_c) for pure and Zn/Ni substituted $YBa_2Cu_3O_{7-\delta}$ single crystals in the underdoped regime where the incoherent carrier transport is observed. In the heavily underdoped regime, Zn/Ni substitution to the Cu sites suppresses the increase of ρ_c at low temperatures, resulting in the reduction of anisotropy in the normal state. For highly doped crystals, we also report Zn substitution effect on the anisotropy of coherence length as a function of oxygen content.

Keywords: out-of-plane resistivity, impurity, underdoped region, anisotropy, coherence length

INTRODUCTION

In the high-Tc cuprates, the effect of Zn substitution on electronic properties has been intensively studied, but many ploblems remain open questions. In $YBa_2Cu_3O_{7-\delta}$, Zn/Ni substitution for Cu sites reduces anisotropy of coherence length (ξ) in the superconducting state, which cannot be explained by assuming anisotropic effective mass [1]. In the normal state incoherent charge transport has been established, at least in the underdoped region. It is expected that the out-of-plane coherence length is strongly affected by the normal state resistivity via a modification of Josephson coupling. In this work, we measured out-of-plane resistivity (ρ_c) and in-plane and out-of-plane coherence length (ξ_{ab}, ξ_c) of $YBa_2(Cu_{1-x}M_x)_3O_{7-\delta}$ (M=Zn, Ni) single crystals in the underdoped region and discuss the effect of Zn/Ni substitution on the electronic anisotropy.

EXPERIMENTAL

$YBa_2Cu_3O_{7-\delta}$ single crystals were grown by a crystal pulling method. The amount of impurity (Zn,Ni) were determined by ICP analysis. Crystals were annealed in O_2-flow condition, and were quenched rapidly to liquid nitrogen. The oxygen content was estimated from annealing temperature. Resistivity measurements were carried out by four-terminal method. ξ_{ab} and ξ_c were determined by magnetization measurements [1,2]. The magnetization was measured by a commercial SQUID magnetometer. Magnetic field up to 7 Tesla was applied parallel and perpendicular to the CuO_2 layers.

RESULTS

Figure 1 shows the temperature dependence of normalized resistivity along the c-axis for several underdoped crystals of $YBa_2Cu_3O_{6.5}$. All the samples were annealed in the same furnace under the same condition. The temperature dependence of dc magnetization for each samples show single superconducting transition temperature. The absolute values of ρ_c at room temperature are not much different between the pure and the impurity substituted crystals. At low temperatures,

221

ρ_c shows a semiconductor-like increase, and drops at T_c. The over-all $\rho_c(T)$ are similar to the reported one with oxygen content $7-\delta \sim 6.63$ [3]. It should be noted that the value of ρ_c at just above T_c seems to be affected by impurity substitution. In the study, we have measured ρ_c of twenty crystals in the heavily underdoped regime. For the pure YBCO crystals, we found the difference in $\rho_c(T)$ between the samples is small, if they are annealed in the same condition. By contrast, for Zn/Ni substituted crystals, the reproducibility of $\rho_c(T)$ shown in Fig. 1 is not enough for quantitative analysis. Nevertheless, we state that any steeper increase of ρ_c was not found for the Zn/Ni substituted crystals than that for the pure YBCO crystal. That is, the ρ_c for Zn/Ni substituted samples are less enhanced at low temperatures. For the higher oxygen contents, our measurements do not yet give a clear conclusion.

Figures 2 show ξ_{ab} and ξ_c for the pure and Zn substituted crystals, with oxygen content near the optimal. The ξ_{ab} is not sensitive to oxygen content for both cases, and has almost constant values of $\sim$14 Å. On the other hand, ξ_c shows a distinct contrast between the pure and Zn substituted crystals. For the pure YBCO crystals, ξ_c decreases as the oxygen content is reduced, while ξ_c seems to increase for the Zn substituted ones.

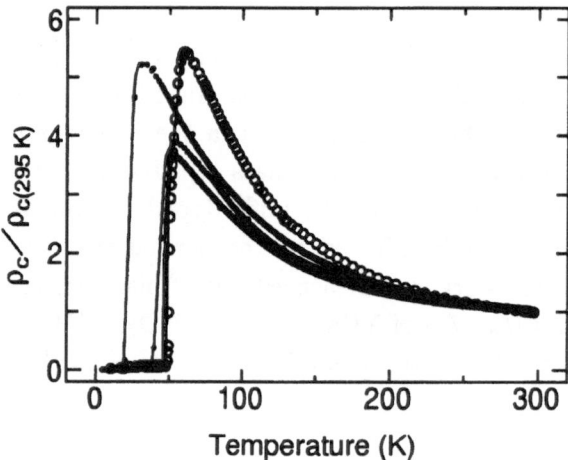

Fig. 1. Temperature dependence of ρ_c for pure and Zn/Ni substituted $YBa_2(Cu_{1-x}M_x)_3O_{7-\delta}$ crystals, with oxygen content $7-\delta \sim 6.5$. From top, pure, Zn 1.6 %, Zn 0.6 %, and Ni 1 % substituted ones. The room temperature resistivities for all the samples are within the range of 60 ± 10 mΩ. Lines are guides for the eyes.

DISCUSSION

Although we cannot completely exclude the possibility that unexpected conduction path may exist in crystals, incoherent transport of YBCO along c-axis seems to be suppressed by replacing Cu sites with Zn/Ni ions. This corresponds to the effect of substitution on the c-axis coherence length in superconducting state.

In high T_c cuprates, the superconducting state is realized on CuO_2 plane. In underdoped region, the bulk properties of superconduting state can be explained as stacks of Josephson coupling layers. Although all cuprate superconductors possess very similar structure, the electronic properties in the c-direction are quite different among the materials. This means a strong effect of the property of the blocking layers and the distance between CuO_2 layers on the c-direction properties and the

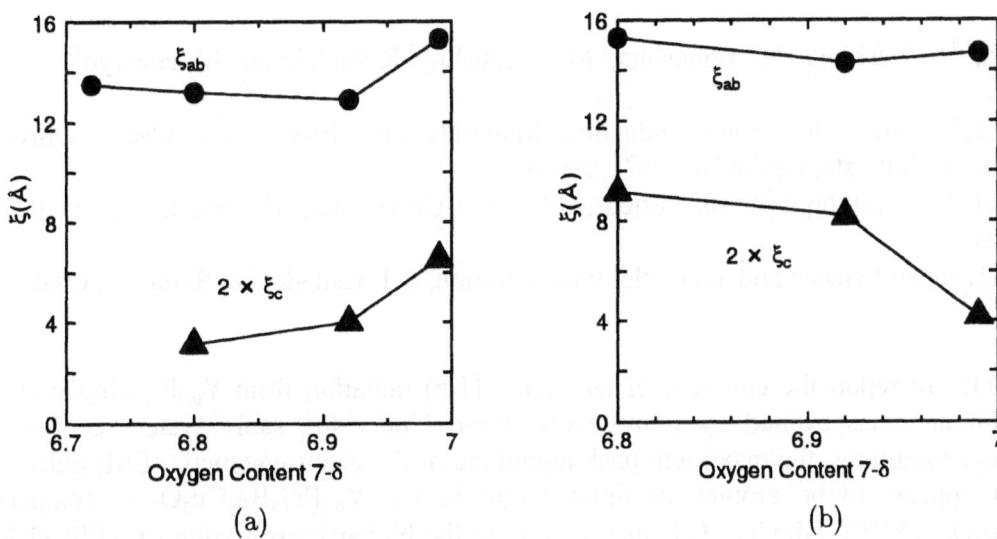

Fig. 2. The in-plane and out-of-plane coherence lengths near the optimum for (a)pure and (b)Zn 0.6 % substituted YBCO crystals.

anisotropy ratio. In the present case, impurity changes neither the property of blocking layer nor the interplane distance. We should consider an alternative factor to decrease the anisotropy.

In the normal state for the optimally doped $YBa_2Cu_3O_{7-\delta}$ where the coherent transport is dominant in all the crystal directions, it has been revealed that impurity ions act as scattering center in the unitary limit for current not only in the ab-plane but also in the c-direction [4]. The present results imply that in underdoped region impurity ions do not only act as simple scattering centers but reduce incoherency of the out-of-plane transport. In other words, impurity ions weaken the carrier confinement within the CuO_2 plane. The difference in the change of ξ_c found in Fig. 2(a) and Fig. 2(b) also supports the view, reflecting some contribution of the incoherent transport along the c-axis to supercurrent even near the optimally doped region.

SUMMARY

We measured out-of-plane resistivity for underdoped pure and Zn/Ni substituted $YBa_2Cu_3O_{7-\delta}$ crystals. The results reveal that Zn/Ni sustitution suppresses incoherent transport along c-axis. Zn substitution also affects on ξ_c when crystals are underdoped. The present results imply that Zn/Ni substitution weakens the confinement of carriers within CuO_2 plane.

Acknowledgements. This work is supported by New Energy and Industrial Technology Development Organization (NEDO) as Collaborative Reseach and Development of Fundamental Technologies for Superconductivity Applications.

1. K. Tomimoto *et al*, Phys. Rev. B **60**, 114-117 (1999).

2. N. R. Werthamer *et al*, Phys. Rev. **147**, 295 (1966).

3. K. Mizuhashi *et al*, Phys. Rev. B **52**, R3884-R3887 (1995).

4. K. Semba *et al*, Phys. Rev. B **49**, 10043-10046 (1994).

Terahertz radiation properties from YBCO and YPBCO thin films

H. Wald,[1,2] S. Nashima,[1] M. Yamashita,[1] M. Tonouchi,[1,3] P. Seidel,[2] and M. Hangyo[1]

[1] Research Center for Superconducting Materials and Electronics, Osaka University, 2-1 Yamadaoka, Suita-shi, Osaka 565-0871, Japan
[2] Institut für Festkörperphysik, Friedrich-Schiller-Universität, Helmholtzweg 5, 07743, Jena, Germany
[3] CREST, Japan Science and Technology Corporation, 2-1 Yamadaoka, Suita-shi, Osaka 565-0871, Japan

Abstract: We report the emission of terahertz (THz) radiation from $Y_{0.7}Pr_{0.3}Ba_2Cu_3O_7$ (YPBCO) thin film antennas excited by femtosecond laser. Under the same system configuration and excitation conditions, the maximum peak amplitude of the electromagnetic (EM) pulse in the time domain appears to be around 10 times larger in the $Y_{0.7}Pr_{0.3}Ba_2Cu_3O_7$ as compared to the $YBa_2Cu_3O_{7-\delta}$ (YBCO) device. This may be due to the higher transmission of YPBCO in the THz frequency range. We employ time domain spectroscopy (TDS) to determine the electromagnetic field transmission of the film/substrate interface of films made of the above mentioned materials in the frequency range from 0.2 to 2 THz. We compare the difference in the transmission measured by TDS with the difference in the signal efficiency of the emitted electromagnetic subpicosecond pulse.

Keywords: terahertz radiation, femtosecond laser, time domain spectroscopy, $Y_{0.7}Pr_{0.3}Ba_2Cu_3O_7$

INTRODUCTION

Y-Pr Substitution changes the critical temperature of the superconducting alloy $Y_xPr_{1-x}Ba_2Cu_3O_7$ that was found to be dependent on the Pr content [1]. Previous studies on TDS transmission measurements of $Y_xPr_{1-x}Ba_2Cu_3O_7$ [2] reported an increasing transmission in the frequency region up to 1.5 THz with an increasing Pr content. This suggests that the emission ratio of the electromagnetic pulse amplitude generated by the transient supercurrent change due to laser irradiation at the same excitation conditions such as laser power, pump power, bias current, and device specifications of Pr doped YBCO is higher than that of pure YBCO. In this report, we observe the THz emission from YPBCO and YBCO, measure the complex transmission of the film by TDS, and calculate the transmission change of only the output interface between film and substrate. We also compare the bias current dependence of the peak intensity for both materials.

EXPERIMENTAL

The antenna devices are c-axis-oriented YPBCO and YBCO film bridges with a film thickness between 80 ~ 160 nm and were deposited on 0.5-mm thick MgO substrates either by pulsed laser deposition or electron-beam vaporization. The typical properties of the YPBCO film are $j_c=5 \cdot 10^6 A/cm^2$ (at T=16K) and the $T_c=(51\pm2)$ K while that of the YBCO film are $j_c=8 \cdot 10^6 A/cm^2$ (at T=16K) and the $T_c=(84\pm2)$ K. The pattern of the dipole antenna (20 μm wide and 20 μm long) was made by standard photolithography. The laser spot size is around 25 μm in diameter. The femtosecond beam is produced by a mode locked Ti:sapphire laser and a low-temperature grown GaAs antenna was used to detect the signal. The windows of the optical cryostat cut the transmitting frequencies at a value above 2 THz. In the TDS measurements, we used an InAs(111) wafer to produce the THz beam. The YBCO and YPBCO antennas and the InAs wafer were all subjected to the femtosecond infrared laser beam (700 nm) to generate ultrashort electromagnetic pulses.

RESULTS AND DISCUSSION

Figure 1 shows the terahertz waveforms of YPBCO while figure 2 shows the bias current dependence of the peak amplitude of a YPBCO and a YBCO dipole antennas with the same configuration (film thickness, dipole pattern, etc.) and excited under the same conditions (laser probe power, detector power, etc.). The shape and the peak intensity of the EM pulse depend on the excitation conditions, the system alignment and the device properties. The signal amplitude increases with increasing bias current (fig.2) without significant distortion in the waveform (fig.1). This suggests no heating effect on the device even for higher currents of up to 150 mA at 5 mW excitation laser power. The FWHM in all bias currents is around 0.7 ps. No voltage drop due to flux flow could be measured. At a given temperature (T=17K) and laser power (P=5mW), the peak

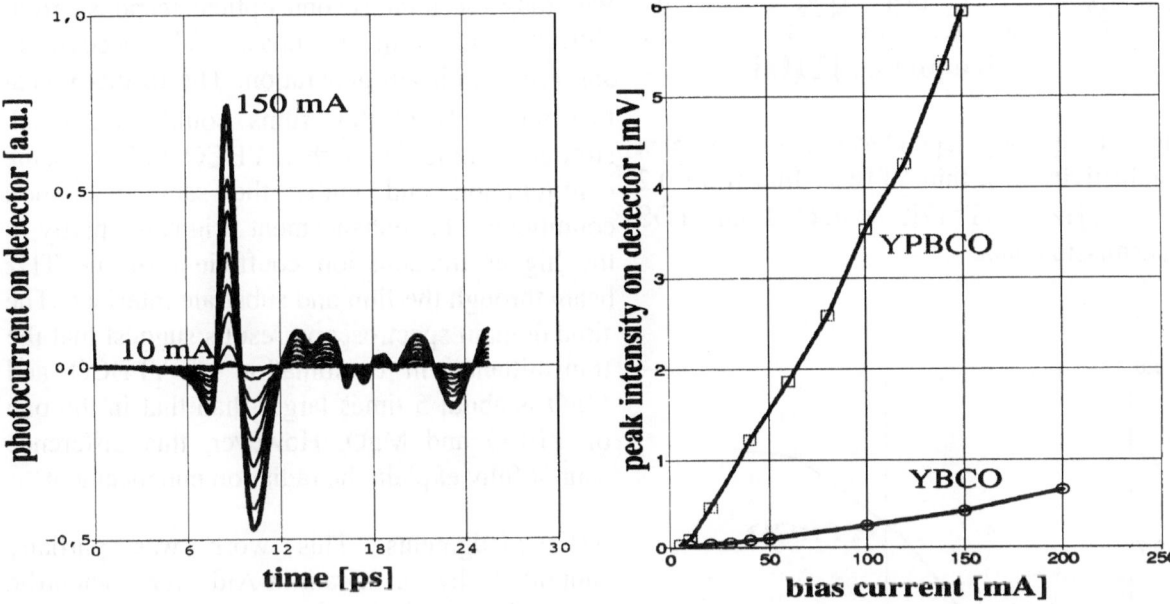

Fig. 1 Terahertz radiation waveforms emitted from YPBCO thin film for various bias currents at a probe power of 5mW.

Fig.2: Bias current dependence of THz radiation from YBCO and YPBCO, excited with a laser power of 5 mW

intensity for YPBCO depends nearly linear on the bias current. The same dependence was earlier reported for YBCO [3]. The maximum peak amplitude of the signal emitted from YPBCO seems to be about 10 times larger as that of pure YBCO which indicates that YPBCO has a higher THz radiation efficiency than the YBCO. This might be partly attributed to a higher transmission against the generate pulse at the film/substrate interface resulting from the lower conductivity of YPBCO.
To investigate the influence of the transmission, we apply TDS-THz spectroscopy to thin films of both materials. Our results were found to be in good agreement with previously reported literature [2]. From the measured transmission coefficients on an air/film/substrate interface, we calculate the complex conductivity (fig.3) and the complex refraction index of the superconducting film. Furthermore we determine the refraction index of the MgO substrate in the whole usable frequency range. From this data we extract the transmission coefficient values of the interface between the

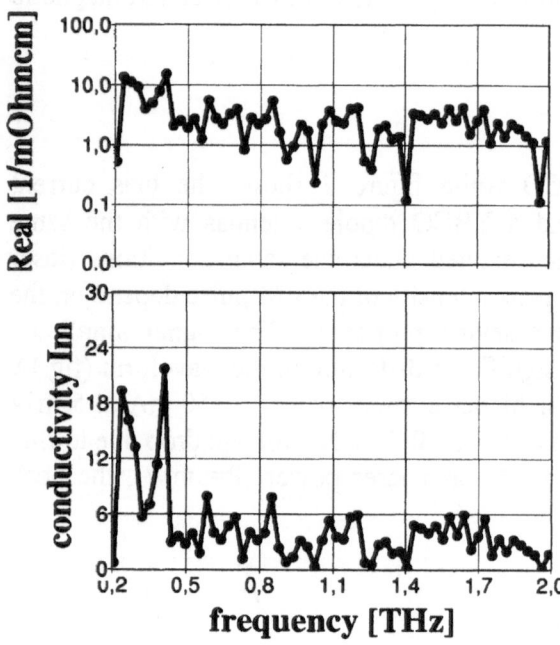

Fig.3: Frequency dependence of the complex conductivity of a thin YPBCO film from 0.2 to 2 THz at T=17K extract from TDS transmission data.

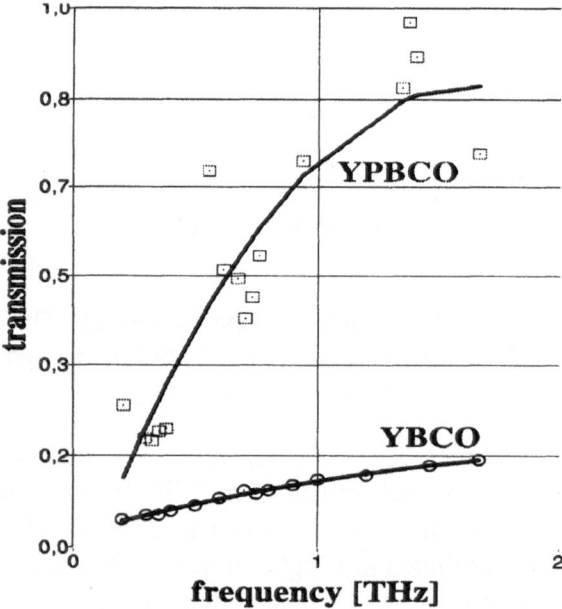

Fig.4: Comparison of the transmission on the film / substrate interface calculated from TDS transmission data, at T=17K.

superconducting film and the substrate (fig.4). The results suggest that the THz beam transmission of the YPBCO/MgO interface is around 5 times larger in the region between 0.2 and 2 THz compared to the YBCO/MgO interface. A possible reason for this is the lower number of carriers or lower carrier mobility in YPBCO. However, the difference in the maximum peak amplitude of the emitted THz signal between the YPBCO and the YBCO is at least 10 times (fig.2). The difference in energy band [4] and carrier dynamics [5] of YPBCO might be responsible for such enhancement.

SUMMARY

We study the femtosecond optical response from YPBCO thin film antennas, and successfully observe THz beam generation. The results reveal that the YPBCO thin films could emit much stronger THz radiation than YBCO with the same configuration and under the same excitation conditions. This enhancement originates partly in the higher transmission coefficient of the THz beam through the film and substrate interface. The time domain spectroscopy results suggest that the transmittance in combination of YPBCO and MgO is about 5 times larger than that in the one of YBCO and MgO. However, this difference cannot fully explain the radiation enhancement.

Acknowledgments. This work was partially supported by a Grant-in-Aid for Scientific Research on Priority Area (A), No. 10142101, from the Ministry of Education, Science, Sports, and Culture, Japan and the Deutsche Akademische Austauschdienst (DAAD).

REFERENCES
1. Radousky H.B., J.Mater.Res. 7(1992) 1918.
2. Buhlmeier R., Brorson S.D., Trofimo I.E., White J.O., Habermeier H.-U., Kuhl J., Phys. Rev. B, Vol.50, 13 (1994).
3. Tonouchi M, et al., Jpn. J. Appl. Phys., 35, 2624 (1996).
4. Neumeier J.J., Maple M.B., Torikachvili, Physica C 156 (1988) 574.
5. Nashima S, Tonouchi M, Hangyo M, Barholz K-U, and Seidel P, Advances in Superconductivity XI, eds. N. Kashizuka and Tajima S, Springer-Verlag Tokyo, 1999, pp.93-96.

TERAHERTZ RADIATION FROM YBCO THIN FILMS

EXCITED WITH 1.55 μm FEMTOSECOND LASER PULSES

Takashi Kondo,[1]* Hitoshi Saijyo,[1] Masayoshi Tonouchi,[1,2] and Masanori Hangyo[1]

[1]Research Center for Superconducting Materials and Electronics, Osaka University, Yamadaoka 2-1, Suita, Osaka 565-0871, Japan
[2]CREST, Japan Science and Technology Corporation, Yamadaoka 2-1, Suita, Osaka 565-0871, Japan

Abstract: Previously, we reported terahertz (THz) radiation effects from YBCO thin films using a femtosecond laser with a wavelength of 800 nm. In this work, a 1.55 μm femtosecond laser was used to excite YBCO thin films and THz radiation was successfully observed.

Keywords: THz radiation, 1.55 μm femtosecond laser, YBCO antenna, supercurrent modulation, interband transition, intraband transition

INTRODUCTION

Previously, we reported strong THz radiation effects from high-Tc superconductors (HTS's) excited by a femtosecond laser with a wavelength of 800 nm [1]. This effect has stimulated interest in the field of microwave photonics. From the view point of the application, it is of technological importance to realize THz beam generation using lasers with a wavelength of 1.55 μm. We explained the mechanism of THz radiation from HTS's by ultrafast supercurrent modulation due to the optical excitation. If the hypothesis that the supercurrent can be modulated by the optical pairbreaking is correct, the femtosecond laser with a wavelength much longer than 800 nm could be used for THz beams excitation. Thus HTS's could be a potential material for THz radiation using the 1.55 μm femtosecond laser. In the present work, we demonstrate THz radiation from YBCO thin films excited with the 1.55 μm femtosecond laser.

EXPERIMENTAL PROCEDURE

Figure 1 shows a schematic illustration of the prepared samples. 100-nm-thick YBCO films grown on an MgO substrate were patterned into a bow-tie antenna structure. The typical critical current of the device was ~ 350 mA at 10 K and Tc was ~ 83 K. An MgO hemispherical lens with a diameter of 3 mm was attached to the backside of the MgO substrate in order to enhance the collection efficiency of THz radiation [2]. Figure 2 shows a schematic diagram of a THz beam generation and detection system. An optical parametric oscillator (OPO) employs a lithium triborate (LBO) nonlinear optical crystal to generate infrared frequencies pumped by a mode-locked Ti:sapphire laser with a center wavelength of 810 nm. The OPO's output signal operating with a pulse width of 120 fs in full width at half maximum (FWHM), the center wavelength of 1.55 μm and a repetition rate of 82 MHz was used for the excitation. The excitation laser was focused on the center of the YBCO antenna by a lens. THz beam emitted from the opposite side through the MgO hemispherical lens was focused onto a detector. A bow-tie shaped photoconductive antenna made of Au/Ge/Ni alloy on a low-temperature-grown GaAs (LT-GaAs antenna) was used as a detector. The OPO's residual output signal (810 nm) was used to trigger the LT-GaAs antenna. The excitation laser was chopped mechanically at 2 kHz, and the current induced by the electric field of THz radiation was amplified by a current amplifier and lock-in detected. A THz waveform in the time domain was obtained by changing the delay time between the excitation and the trigger pulses.

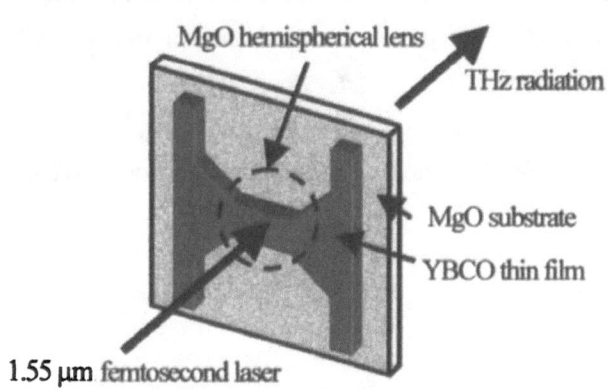

Fig. 1. A structure of a bow-tie shaped YBCO antenna.

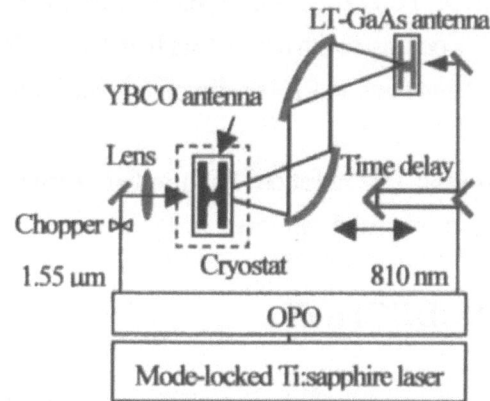

Fig. 2. A schematic diagram of a THz beam generation and detection system.

RESULTS AND DISCUSSION

Figure 3 shows a waveform of an electromagnetic pulse radiated from the current biased YBCO antenna excited by the 1.55 μm femtosecond laser. The FWHM of the first positive pulse is less than 1 ps. The small pulse 8 ps after the main pulse is due to the multiple reflection in the MgO substrate. The inset of Fig. 3 shows the Fourier component, which extends from 0.1 to ~ 0.8 THz. Since we used bow-tie shaped antennas for the emitter and the detector, THz radiation components with low frequencies are enhanced [3]. The intrinsic bandwidth was similar to the radiation excited by the 800 nm laser. Figure 4 shows waveforms of THz radiation from the vortex trapped YBCO antenna. They were measured

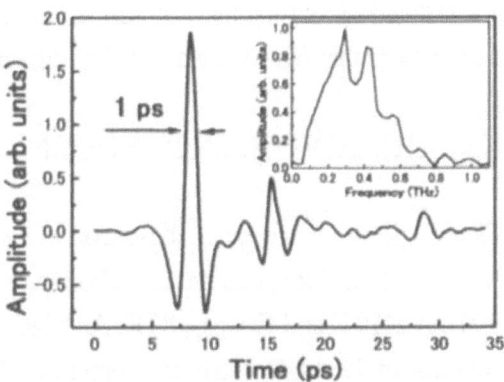

Fig. 3. A THz waveform and the Fourier component (inset). The 200 mA biased YBCO antenna was excited with power of 30 mW at 11 K.

after applying about 500 gauss magnetic field **B** for a few seconds. The sign of the waveform was reversed when the applied magnetic field was reversed. These results suggest that the ultrafast optical modulation of the supercurrent is the origin of the THz radiation.

The photon energies of the 800 nm laser and the 1.55 μm laser are about 1.55 eV and 0.8 eV, respectively. Since a charge transfer gap (CT-gap) energy of YBCO is larger than 0.8 eV [4], the excitation with the 1.55 μm laser induces the intraband transition of the carriers, whereas that with the 800 nm laser brings about the interband transition from O (2p) orbit to upper Cu (3d) one. The fact that the THz radiation is similarly generated regardless of the wavelength in the present work suggests that the femtosecond laser can modulate

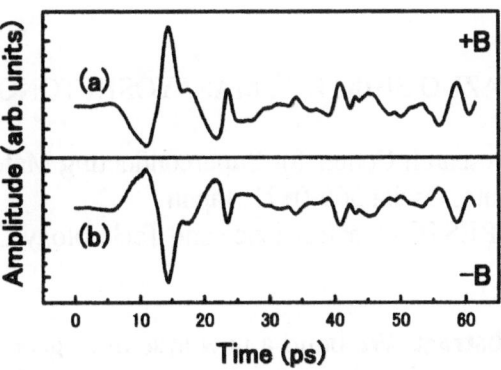

Fig. 4. Waveforms of THz radiation from a vortex trapped YBCO antenna: (a) measured after applying a positive magnetic field and (b) measured after applying a negative one. The YBCO antenna was excited with power of 130 mW at 14 K.

the supercurrent even in the case of the intraband excitation. The further comparison of THz radiation power and waveforms is now under way.

SUMMARY

We examined the 1.55 μm femtosecond laser excitation of the YBCO thin film antennas and successfully observed THz radiation from the current-biased antennas and the vortex trapped ones. The radiation properties were similar to that excited with the 800 nm femtosecond laser. The results suggest that the intraband carrier excitation by the femtosecond laser also induce the ultrafast supercurrent modulation.

Acknowledgments. We would like to thank Dr. M. Morikawa of Osaka University for technical assistance. This work was partially supported by a Grant-in-Aid for Scientific Research from the Ministry of Education, Science, Sports and Culture, and the public participation program for the promotion of creative info-communications technology R&D of Telecommunications Advanced Organization of Japan (TAO).

---

* E-mail: kondo@rcsuper.osaka-u.ac.jp

1. M. Hangyo, S. Tomozawa, Y. Murakami, M. Tonouchi, M. Tani, Z. Wang, K. Sakai, and S. Nakashima, Appl. Phys. Lett. **69**, 2122-2124 (1996).

2. M. Tonouchi, M. Tani, Z. Wang, K. Sakai, M. Hangyo, N. Wada, and Y. Murakami, IEEE Trans. Appl. Supercond. **7**, 2913-2916 (1997).

3. M. V. Exter, Ch. Fattinger, and D. Grischkowsky, Appl. Phys. Lett. **55**, 337-339 (1989).

4. S. L. Cooper, D. Reznik, A. Kotz, M. A. Karlow, R. Liu, M. V. Klein, W. C. Lee, J. Giapintzakis, D. M. Ginsberg, B. W. Veal, and A. P. Paulikas, Phys. Rev. **B**, 8233-8248 (1993).

Optical Magnetic Flux Generation by Selected Femtosecond Laser Pulses

KAZUO SHIKITA[1*], MASAYOSHI TONOUCHI[1, 2], AND MASANORI HANGYO[1]

[1]Research Center for Superconducting Materials and Electronics, Osaka University, 2-1 Yamadaoka Suita, Osaka 565-0871, Japan
[2]PRESTO, Japan Science and Technology Corporation, Osaka 565-0871, Japan

Abstract: We build a new system to generate magnetic flux quanta in a superconductive thin film loop under the control of femtosecond laser pulse numbers, and examine the pulse-number dependence of the magnetic flux generation. The pulses are controlled by an Acousto-Optic Modulator (AOM) driven by an external pulse generator. The results indicate that the magnetic flux can be generated in the superconductive loop by a single shot fs laser pulse operation.

Keywords: magnetic flux generation, femtosecond laser, pulse-select, YBCO thin film loop

INTRODUCTION

Previously, we reported that magnetic flux in superconductive loops can be controlled by a femtosecond (fs) laser [1-5]. The fs laser generally produces a pulse train at a frequency of about several tens of MHz. From the viewpoint of application, it is of technological importance to investigate the optical magnetic flux generation under the control of the pulse numbers. In the present work, we build a new system to control the fs laser pulse number while generating the magnetic flux quanta in the superconductive YBCO thin film loop, and examine the pulse-number dependence of the magnetic flux generation.

EXPERIMENTAL

Figure 1 shows the schematic structure of the measured device near the center. The hole is fabricated at the center of 100 μm-width YBCO strip-line and has a diameter of about 50 μm. The experimental procedures are the same as previously reported [2] except for the optical pulse selection. We used a pulse selector (Spectra-Physics Model 3980) to pick out a single pulse in an 82 MHz pulse train. The efficient, Bragg-angle Acousto-Optic Moderator (AOM) in a pulse selector, is synchronized to the 82 MHz pulse train and selects pulses from the train. The acoustic wave in AOM changes the direction of propagation of the selected pulse by diffracting the beam about 3°. The unselected pulses go straight and are captured by a beam block. Thus we can select pulses. However, the repetition rate of the selected pulses is limited to the range of 400 Hz ~ 4 MHz.

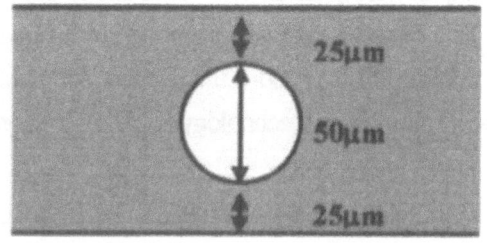

Fig. 1. The schematic of the center of the experimental device. The device is made of YBCO thin film on MgO substrate.

We then employ an external pulse generator, which controls the frequency of acoustic waves and make it possible to select optional pulses. Figure 2 shows the schematic of the magnetic flux generation system with the pulse selector and the external pulse generator. The energy of one pulse is 0.6 nJ, the diameter of laser spot is 25 μm (FWHM) at focal point and the pulse width is 80 fs (FWHM). The magnetic flux is generated under the external magnetic filed of 0.9 mT. The other details of the experimental system for generating magnetic flux in superconductive loop has been reported in our previous work [2].

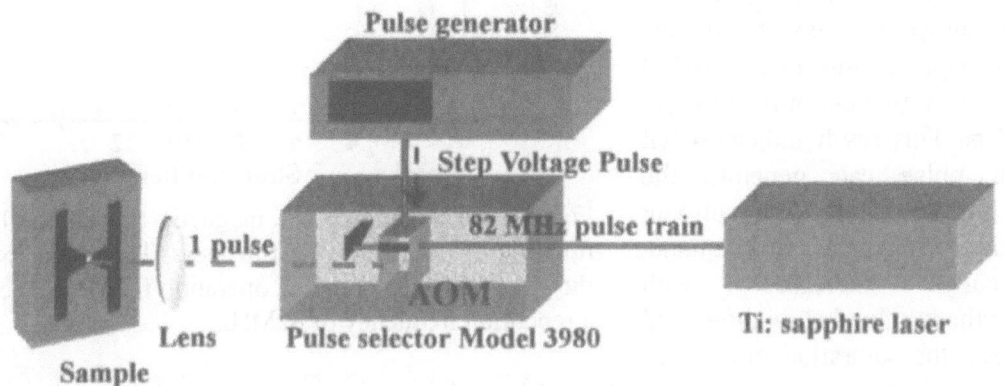

Fig. 2. The schematic of magnetic flux generation using one fs pulse.

RESULTS AND DISCUSSION

Figure 3 shows an example of the supercurrent distribution image after magnetic flux generation. We irradiated the five shots of the selected pulse on the edge of the bridge under the external magnetic field. After the removal of the external field, the x component of the supercurrent distribution was visualized by terahertz radiation mapping [2,6]. The clear persisting supercurrent circulating around the hole is seen in Fig.3, and this distribution is similar to the one generated by the unselected pulses (82MHz).

Figure 4 shows the supercurrent profile along X = 150 μm in Fig. 3. One can quantitatively estimate the supercurrent density distribution using the calibration coefficient from the THz radiation

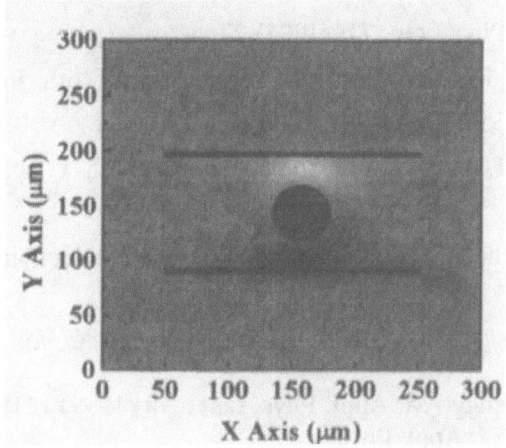

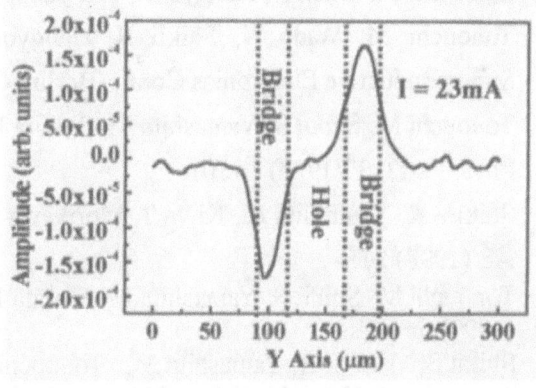

Fig. 3. The current distribution image after magnetic flux generation by 5 selected pulses.

Fig. 4. The current distribution of X = 150 μm. The loop current is about 23 mA.

amplitude to the current density [7]. The distribution in Fig.4 indicates that a persistent current of about 23 mA flows along the loop. This corresponds to the magnetic flux quanta of about 700 ϕ_0 in the hole on the assumption of a loop inductance of about 70 pH, where ϕ_0 is a single flux quantum.

Figure 5 shows the amount of the generated magnetic flux quanta (ϕ/ϕ_0) in the loop as a function of the number of the shots. This result indicates that one shot pulse can generate the magnetic flux in the loop. The amount of generated magnetic flux quanta monotonically increases with increasing the number below three, and then we see the saturation effect. We

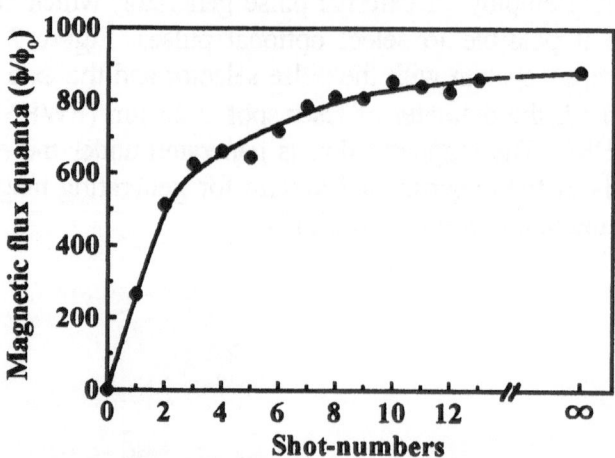

Fig. 5. The amount of the magnetic flux in the loop as a function of the number of the shots. The value "∞" means the results of the continuos operation for a few second with a repetition frequency of 82MHz.

obtained similar results from the current-biased loop. Thus we can expect that the optical magnetic flux generation effects can be utilized for the femtosecond optical signal detectors.

CONCLUSION

We built a new system to control fs pulse-number with a pulse selector and examined the magnetic flux generation using the selected pulses. The supercurrent distribution images observed by the THz radiation mapping indicated that the magnetic flux can be generated in the superconductive loop without phase transition by even a single shot. It was also found that the amount of magnetic flux increases with increasing shot-number below three shots, and then saturates at a finite value.

ACKNOWLEDGMENTS. This work was partially supported by a Grant-in-Aid for Scientific Research on Priority Area (A), No. 10142101, from the Ministry of Education, Science, Sports, and Culture, Japan.

*E-mail: shikita@rcsuper.osaka-u.ac.jp

1.   Tonouchi M, Wada N, Hangyo M, and Sakai K,   Appl Phys Lett., 71, (1997) 2364
2.   Tonouchi M, Wada N, Shikii S, Hangyo M, Tani M, and Sakai K, Ext. Abs. of 6th Int. Superconductive Electronics Conf., (Berlin, Germany, 1997), pp.248.
3.   Tonouchi M, Shikii S, Yamashita M, Shikita K, Kondo T, Morikawa O, and Hangyo M:, Jpn. J. Appl. Phys. Part2, 37(1998) L1301.
4.   Shikita K, Yamashita M, Kiwa T, Morikawa O, Tonouchi M, Hangyo M, Adv. in Superconductivity XI, (1998) 215
5.   Tonouchi M, Shikii S, Yamashita M, Shikita K, and Hangyo M, IEEE trans. Appl. Supercond. Vol. 9 (1999), in press.
6.   Shikii S,   Kondo T, Yamashita M,   Tonouchi M, and Hangyo M, Appl. Phys. Lett. , 74 (1999) 1317.
7.   Tonouchi M, Yamashita M, and Hangyo M, submitted to J. Appl. Phys.

Dependence of Magnetic Flux Generation in YBCO Thin Film Loop on Femtosecond Laser Excitation Position

KAZUO SHIKITA[1*], MASAYOSHI TONOUCHI[1,2], AND MASANORI HANGYO[1]

[1]Research Center for Superconducting Materials and Electronics, Osaka University, 2-1 Yamadaoka Suita, Osaka 565-0871, Japan
[2]PRESTO, Japan Science and Technology Corporation, Osaka 565-0871, Japan

Abstract: We have reported that magnetic flux quanta are controllable in a YBCO thin film loop with femtosecond (fs) laser pulses. However, the generation mechanism is still an open question. In the present work, after characterizing the fs laser beam profile, we study how the optical magnetic flux generation depends on laser excitation position. The supercurrent distribution images observed by terahertz radiation mapping method reveal that the magnetic flux generation in the thin film loop cannot be explained by the simple vortex penetration through the optically excited area.

Keywords: magnetic flux generation, superconductive loop, fs laser pulse

INTRODUCTION

Recently we discovered that femtosecond optical pulses can control magnetic flux quanta in a high-Tc superconductive loop [1-5]. Although the magnetic flux generation is speculatively explained by partial supercurrent modulation, the physics underlying the phenomena is unclear. In this work, we control the position of a femtosecond laser beam to generate the magnetic flux, and examine how the magnetic flux generation depends on the excitation position.

EXPERIMENTAL

Experimental procedures are the same as reported so far [2]. Prior to the magnetic flux generation experiments, we analyzed a fs laser beam profile. Figure 1 shows the observed profile of the beam at the focal point, which indicates that a FWHM is about 25 μm. The magnetic flux generation is carried out with a bias current of 100 mA and a laser power of 30 mW at 16 K. The generated magnetic flux are visualized by terahertz radiation imaging of the supercurrent distribution in the loop.

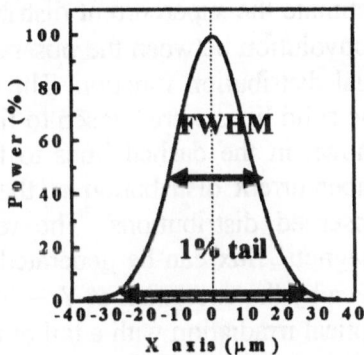

Fig. 1 The observed fs laser beam profile at the focal point.

RESULTS AND DISCUSSION

Figure 2 shows the supercurrent distribution images near the center of the bridge after magnetic flux generation. The black and white circles, respectively, represent the hole in the loop and the position where the fs laser is focused during magnetic flux generation. The surrounding bright and dark areas represent the positive and negative x-components of the supercurrent distribution, respectively. These images indicate that the persistent supercurrent flows after the flux-generation-process was done at the position 25-μm-away from the bridge edge (Y = 75 μm) or at the edge (Y = 100 μm)

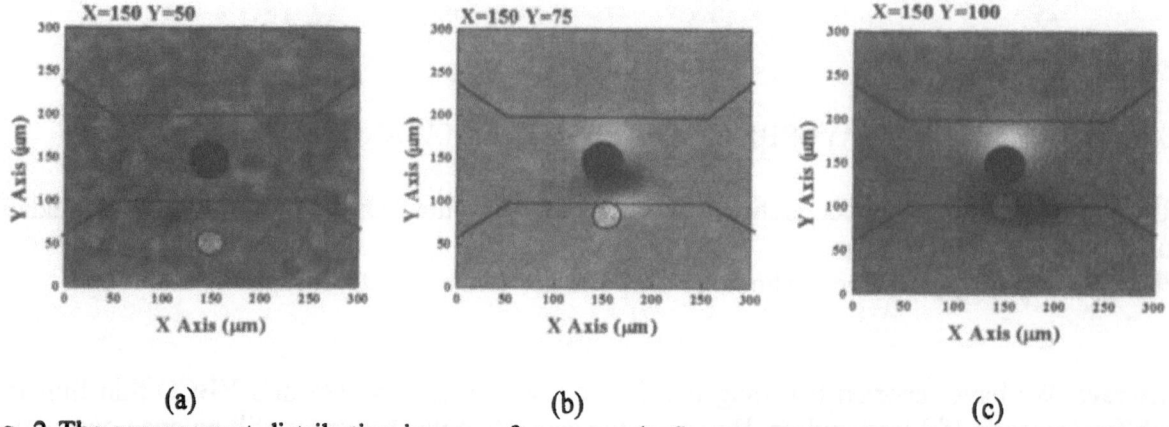

(a)   (b)   (c)

Fig. 2 The supercurrent distribution images after magnetic flux generation. The lines show the edge of superconductive bridge. The black circle at the center of the bridge represents the hole. The bright and white parts represent the positive or negative supercurrent distribution, respectively. The laser excitation positions (represented by the white circle) are at (a) X = 150 μm, Y = 50 μm, (b) X = 150 μm, Y = 75 μm and (c) X = 150 μm, Y = 100 μm.

whereas no flux was observed after the process at the position 50-μm-away from the bridge edge (Y = 50 μm). Optical magnetic flux generation are confirmed in Fig.2 (b) and (c). However, both have different supercurrent distribution. At the position Y= 100 μm, positive current flows in Fig.2 (b); negative one flows in Fig.2 (c).

Figure 3 shows the supercurrent distributions along X = 150 μm after optical excitation at various positions. The distributions given by the closed symbols represent the experimental data, which originated in the convolution between the laser beam profile and the real distribution. We qualitatively estimate the supercurrent distribution by calculating the convolution between the observed beam profile and the trial distribution function. The trial functions given by the solid lines were chosen to fit the convolution curves shown in the dashed lines to the data. The spike-like supercurrent distribution at the edges can explain the observed distributions. The result indicates that the magnetic flux can be generated in the hole of the loop even by the excitation at Y = 75 μm, which allows the optical irradiation with a tail of the laser beam. It is also suggested that the direction of the supercurrent flow at Y = 100 μm after the magnetic flux is generated by the laser irradiation at Y = 75 μm is different from other cases. If the magnetic flux penetrated from outside, the polarity of the magnetic flux would not be affected by changing the excited positions. Therefore, the simple vortex penetration at the optically excited area cannot explain the present phenomena.

We calculated the magnetic flux distribution based on a Biot Savart' law. Magnetic field **H** is given by

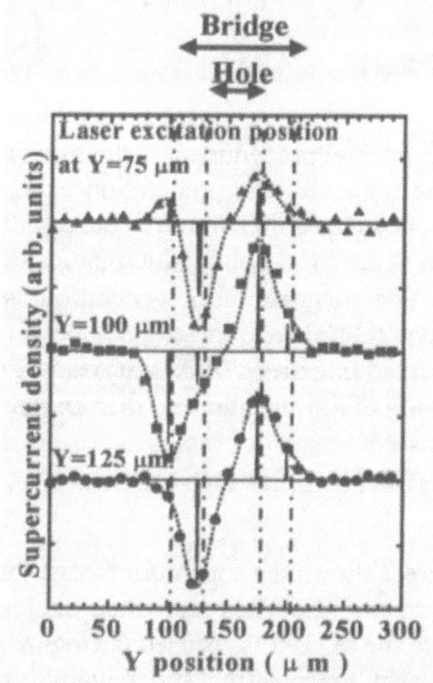

Fig. 3 The supercurrent distributions along X = 150 μm for the optical excitation at various positions.

$$H = \frac{1}{4\pi} \int_L \frac{\mathbf{I} \times \mathbf{i_r}}{r^2} dl$$

,

where the vector **I** represents supercurrent along loop L, r is the distance from a point in line L to at a point of observation, and **i$_r$** is the unit vector of the component of the direction from Q to P.

Figure 4 shows the magnetic flux distributions at X = 150 μm. Positive magnetic flux exists in the hole. Although finite magnetic flux is estimated to exist inside the films, this is simply due to estimation error. In the real system, the supercurrent distributes wider than that of the trial function to prevent the magnetic flux penetration into the films.

CONCLUSION

We studied the dependence of the optical magnetic flux generation on the laser irradiating position. The magnetic flux can be generated even by focusing the laser beam at the position 25 μm away from edge of the bridge. This indicates that the generation mechanism cannot be simply explained by the flux penetration from the outside of the superconductive loop to the inside. We also estimated the supercurrent and magnetic flux distribution, and we found that the optical magnetic generation depends on the fs laser focusing position.

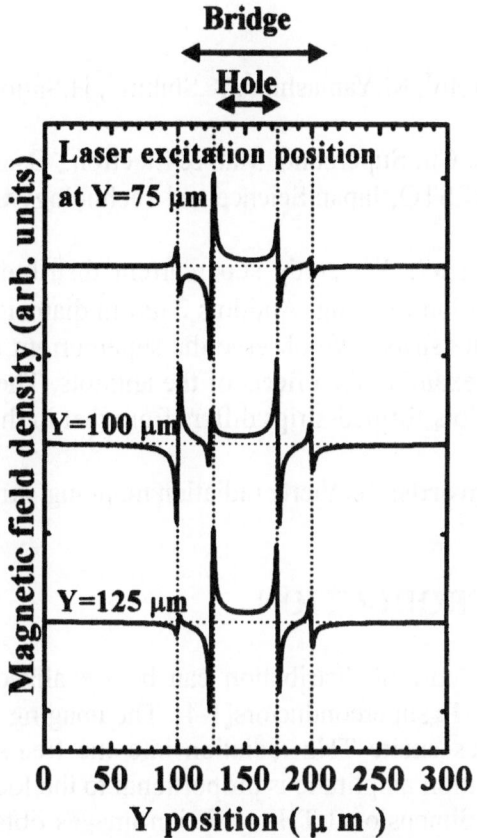

Fig. 4 The estimated magnetic flux distributions corresponding to the supercurrent distributions in Fig. 3.

ACKNOWLEDGMENTS. This work was partially supported by a Grant-in-Aid for Scientific Research on Priority Area (A), No. 10142101, from the Ministry of Education, Science, Sports, and Culture, Japan.

*E-mail: shikita@rcsuper.osaka-u.ac.jp

1. Tonouchi M, Wada N, Hangyo M, and Sakai K,   Appl Phys Lett., 71, (1997) 2364

2. Tonouchi M, Wada N, Shikii S, Hangyo M, Tani M, and Sakai K, Ext. Abs. of 6th Int. Superconductive Electronics Conf., (Berlin, Germany, 1997), pp.248.

3. Tonouchi M, Shikii S, Yamashita M, Shikita K, Kondo T, Morikawa O, and Hangyo M: Jpn. J. Appl. Phys. Part2, 37(1998) L1301.

4. Shikita K, Yamashita M, Kiwa T, Morikawa O, Tonouchi M, Hangyo M, Adv. in Superconductivity XI, (1998)215

5. Tonouchi M, Shikii S, Yamashita M, Shikita K, and Hangyo M, IEEE trans. Appl. Supercond. Vol. 9 (1999), in press.

Observation of Supercurrent Distribution
In Antidot-Formed $YBa_2Cu_3O_{7-\delta}$ Thin Film Strips

A.Moto[*], M.Yamashita[*], K.Shikita[*], H.Saijo[*], M.Tonouchi[*,**], and M.Hangyo[*]

[*]Res. Ctr. Supercond. Mat. & Electron., Osaka University, Osaka 565-0871, Japan
[**]PRESTO, Japan Science and Technology Corporation, Osaka 565-0871, Japan

Abstract: We study supercurrent distributions in YBCO thin film strips by means of terahertz radiation imaging. Antidots 2 μm in diameter are fabricated by Ar ion milling at an interval of 4 μm on the strips. We observe the supercurrent distributions in the strips with and without the antidots, and examine the effects of the antidots. The results indicate that the supercurrent distribution in the antidots-formed-strips differs from that in the strips without the antidots.

Keywords: Terahertz radiation mapping, Antidots, YBCO thin film, Supercurrent distribution

INTRODUCTION

Supercurrent distribution can be visualized by observing terahertz (THz) radiation emitted from high-Tc superconductors[1-4]. The imaging system utilizes the principle that the femtosecond optical pulses excite THz radiation into the free-space by optical supercurrent modulation, and that the radiation amplitude is proportional to the local supercurrent density at the optically excited area. The two-dimensional THz radiation images obtained by scanning the laser beam on the sample can be quantitatively transferred into the supercurrent density distribution without any significant deformation of the trapped-vortex-distribution[5]. In this work, we apply this technique to study the bias current dependence of supercurrent distributions in YBCO thin film strips with and without the antidots.

EXPERIMENTAL PROCEDURE

We have observed supercurrent distribution using Terahertz radiation imaging system [1]. YBCO thin films were patterned into 100-nm thick by 120-μm wide strips. On a part of the strip, the antidots with a diameter of 2 μm are fabricated at an interval of 4 μm by Ar ion milling, as illustrated in Figure 1. Laser power of 5mW for the THz radiation imaging is used, so that no significant distortion is induced in the vortex distribution trapped in the strips due to laser exposure.

Assuming the equations bellow, we quantitatively estimate the supercurrent density distribution in the strips from the THz radiation amplitude.

$$I_x = K \times d \times z \times \sum R, \quad (1)$$

$$J_x = K \times R, \quad (2)$$

where I_x is the bias current, K is a constant to convert the radiation amplitude to the current density, d is film thickness, z is the resolution

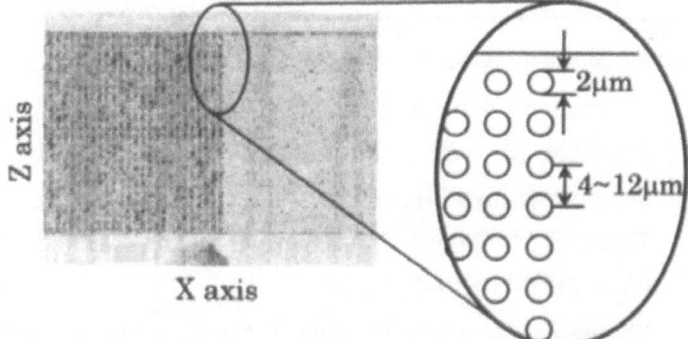

Fig. 1. Photograph and schematic illustration of the prepared sample.

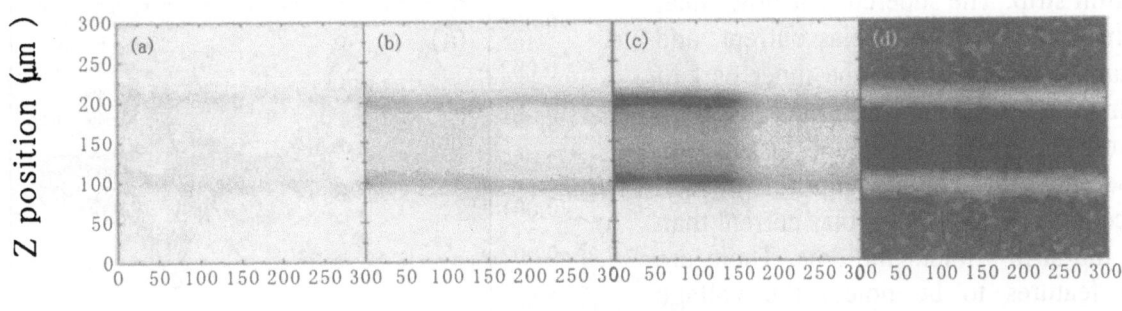

Fig. 2. Images of supercurrent distributions at T = 17K with a bias current of (a) 50 mA, (b) 250 mA, and (c) 460 mA, and (d) after removal of the bias current. There are antidots in the left-hand-side of the strip.

along Z axis, R is the maximum radiation amplitude, J_x is the supercurrent density. We also calculated the magnetic flux distribution along Y axis using Biot-Savart's law from the supercurrent density distribution along the Z axis. The y component of the magnetic flux density $B_y(z)$ is expressed by the following formula,

$$B_y(z) = \frac{\mu_0}{2\pi} \int_{-\infty}^{\infty} \frac{J_x(z')d}{z - z'} dz' + B_{EX},$$

(3)

where B_y is the magnetic flux density along Y axis, μ_0 is the magnetic permeability at vacuum, B_{EX} is the external magnetic field along Z axis[5].

RESULTS AND DISCUSSION

Figure 2 shows the images of THz radiation amplitude distribution observed by the THz mapping at various bias currents I_B. The voltage is observed at I_B larger than 400 mA, and reaches to be 1.5 mV at I_B of 460 mA. Since the antidot-formed area should have a smaller critical current Ic than that in the strip without antidots, the critical current density Jc is roughly estimated to be 7MA/cm² at 17K. The images indicate that even with a bias current much smaller than Ic, the distribution in the strip with the antidots differs from that in the strip without the dots as shown in Fig.2 (a). On the other hand, we see no significant difference in the supercurrent distributions in the vortex trapped state as given in Fig.2 (d). We also observed radiation anomaly in the flux-flow state; the radiation intensity suddenly increased at I_B above Ic with increasing bias current. We will discuss this phenomenon separately.

Figure 3 shows the supercurrent density distributions along the Z axis at various bias current. The distributions are qualitatively explained by the Bean's critical state model for

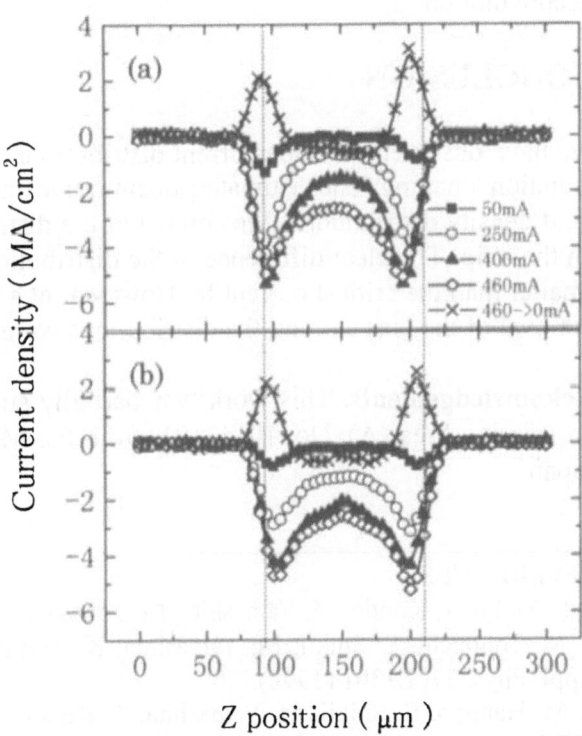

Fig.3. Bias current dependence of supercurrent desnity distributions in a strip (a) without the antidots and (b) with the antidots.

a thin film strip. The supercurrent flows near the strip edges at a small bias current, and then start to penetrate into the inner part of the strip with increasing bias current. However, the broadening of the distribution into the inner part in the anti-dots-formed strip occurs at a much lower bias current than that in the strip without the dots. There are several features to be noted; the voltage appears in the antidot- formed strip before the supercurrent distribution becomes uniform; the distribution in the trapped state is not affected by the existence of the antidots; and so on.

Figure 4 shows the calculated magnetic field density distributions corresponding to Fig.3. This indicates that the magnetic field is shielded near the center of the strip with a bias current lager than Ic. Contrary to our expectation, the flux penetration is observed at the edges of strips when bias current is small. At present, we have no clear explanation. Since the observed distribution is explained by the convolution of the laser beam profile and real supercurrent distribution, the further discussion requires the analysis based on the deconvolution.

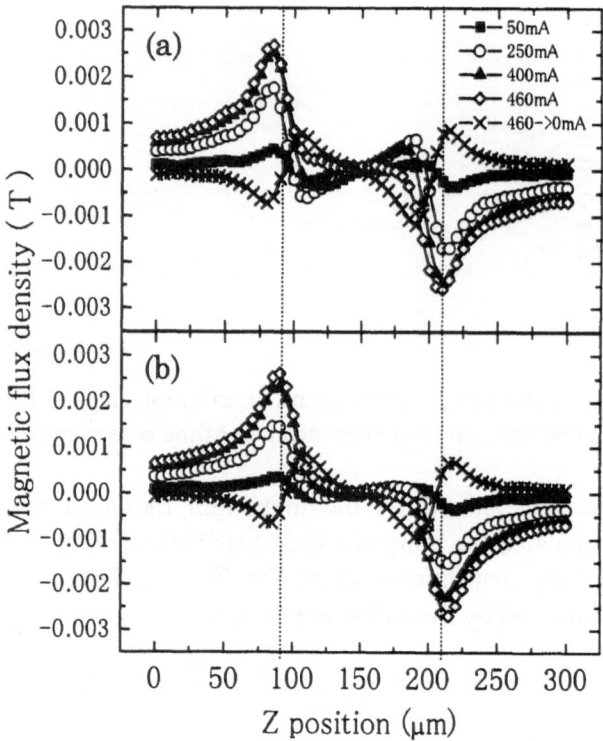

Fig. 4. Bias current dependence of magnetic flux desnity distributions in a strip (a) without the antidots and (b) with the antidots.

CONCLUSION

We have observed the supercurrent distributions in the YBCO thin film strips by means of terahertz radiation imaging, and estimated quantitative the supercurrent density distributions and magnetic field density distributions. We compared the distributions in the strips with and without the antidots on the strip. The clear difference in the distributions were observed when the bias current was much smaller than the critical current Ic. However, at a bias current above Ic and a remanent state after the removal of the bias current, the distributions were observed similar to the each other.

Acknowledgements. This work was partially supported by a Grant-in-Aid for Scientific Research on Priority Area (A), No. 10142101, from the Ministry of Education, Science, Sports, and Culture, Japan.

REFERENCES

1. S. Shikii, T. Kondo, M. Yamashita, M. Tonouchi, and M. Hangyo, Appl. Phys. Lett. 74, 1317 (1999).
2. M. Tonouchi, S. Shikii, M. Yamashita, K. Shikita, T. Kondo, O. Morikawa, and M. Hangyo, Jpn. J. Appl. Phys. 37, L1301 (1998).
3. M. Hangyo, S. Shikii, M. Yamashita, T. Kondo, M. Tonouchi, M.Tani, and K. Sakai, to be published IEEE Trans. Appl. Supercond. (1999)
4. O. Morikawa, H. Saijo, M. Yamashita, M. Tonouchi, and M. Hangyo, to be published in A.P.L.
5. M. Tonouchi, M. Yamashita, and M. Hangyo, J.A.P. (During a contribution)
6. E. Zeldov, J. R. Clem, M. McElfreshand. Darwin, Phys. Rev. B 49, 9802 (1994).

Sensitive Detection of Liquid Nitrogen Temperature Fluctuation Induced by Bubbling Using HTS Microwave Absorption

Ken-ichi Itoh, Akinori Hashizume, Masaki Tada, Jiro Yamada, V.V. Srinivasu, Masanori Ashida, Tatsuo Itoh, and Tamio Endo*

Faculty of Engineering, Mie University, Kamihama, Tsu, Mie 514-8507, Japan.

Abstract: Large fluctuating signals of microwave absorption in high temperature super-conductors (HTS) induced by liquid nitrogen bubbling, were observed for the first time. This fluctuation reflects temperature fluctuations of the liquid nitrogen. A complicated process of the liquid nitrogen from small bubbling up to evaporation was elucidated by means of the microwave absorption. This is a new bolometric function of HTS.

Keywords: Microwave absorption in HTS, Large fluctuating signal, Liquid nitrogen bubbling, Temperature fluctuation

INTRODUCTION

We have frequently observed characteristic large fluctuations of microwave absorption signals in high temperature superconductors (HTS) [1]. To elucidate an origin of this curious behavior, we conducted experiments of the microwave absorption by sweeping magnetic field and measurement time with monitoring temperatures of coolant and HTS samples. It was clarified that the origin of this signal fluctuations was liquid nitrogen bubbling. We report in this paper that the complicated temperature changes of liquid nitrogen caused by the bubbling can be monitored with fidelity by recording the microwave absorption signal of HTS.

EXPERIMENTAL

A bulk HTS sample of Bi2212 was used in this experiment. The sample was zero-field cooled to the liquid nitrogen temperature and it was put in a TE_{102} cavity as shown in Fig.1. The field modulated microwave absorption was measured on this sample at 9.3 GHz [2-4]. We employed two dc-field (H_a) applications on the sample. One is to sweep H_a upward and downward. The other is to fix H_a at 50 G. The signal was recorded with time in the latter case. The modulation field (H_m) was superimposed upon H_a in the same direction at 100 kHz. In the case of field sweep experiment, the temperature of the sample was measured before the sweep using a Cu-Constantan thermocouple. In the case of fixed H_a, the Cu-Constantan thermocouple was

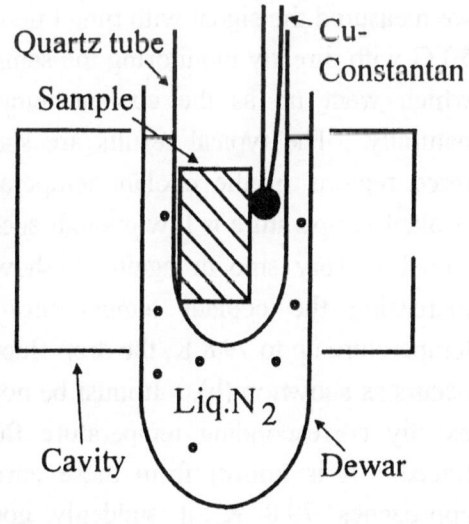

Fig.1. Experimental apparatus.

attached to the sample then the precise sample temperature (T) could be directly monitored during the absorption measurement. The coolant temperature naturally increased with time as atmospheric oxygen gradually dissolved into the liquid nitrogen.

RESULTS AND DISCUSSION

The field modulated microwave absorption signal S is shown in Fig.2 for the field sweep up and down at various sample temperatures. The signal is usually very smooth as shown in (a) at sufficiently low temperatures such as 77.3 K. However, when the temperature is increased to around 79 K, the signal shows a very curious behavior, i.e., repeated sudden drop and following large fluctuation as shown in (b). We call it a "drop-fluctuation" henceforth. Further increase in the temperature to around 80.5 K results in a disappearance of the drop-fluctuation as shown in (c). Thus it turned out that the characteristic fluctuation is observed only in a small interval of the temperature around 79 K. We examined the temperature dependence of S and elucidated that S decreases drastically with increasing T around 79 K. Therefore it can be expected that the large fluctuation of S reflects changes in the sample temperature caused by some field-dependent super-conducting properties such as flux jump.

In other to confirm the changes in sample temperature, we measured the signal with time t under the fixed H_a of 50 G with directly monitoring the sample temperature T which went up as the coolant temperature went up naturally. The typical results are shown in Fig.3 for three regions of the coolant temperature. When the coolant temperature is low enough such as T<78 K, the signal is very smooth again as shown in (a). With increasing the coolant temperature and the sample temperature up to 79.8 K, the drop-fluctuation frequently occurs as shown in (b). It must be noted in (b) that the exactly corresponding temperature fluctuation can be traced. It is known from these curves that when T approaches 79.8 K, it suddenly goes down and S suddenly goes up correspondingly. After showing the minimum of T, it turns up gradually and the corresponding S also turns down gradually to the original magnitude. Further increase in the coolant temperature

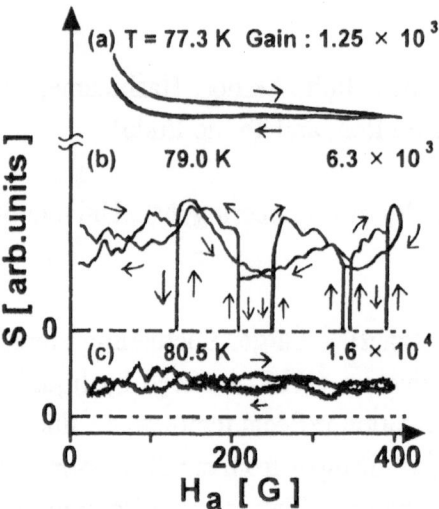

Fig.2. S vs H_a for various sample temperatures.

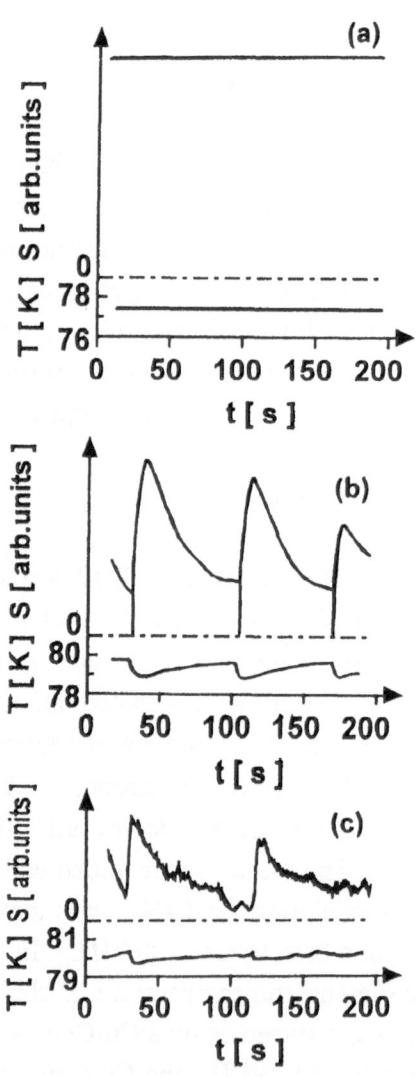

Fig.3. S and T vs time t for various coolant temperatures.

241

and the sample temperature up to 80.4 K results in the middle-size fluctuations of S without the drop signal and in the small oscillating signals. The temperature drops corresponding to the "nondrop-fluctuations" are not as large as those shown in (b). The temperature changes corresponding to the small signal oscillations are very small. Using these data, we plotted the signal magnitudes as a function of the corresponding sample temperature. We obtained an universal linear relation between them. Thus it was clarified that the fluctuating signal exactly reflects the sample temperature fluctuation.

In order to clarify an origin of the sample temperature fluctuation, we measured just the sample temperature with time without applying the microwave power under zero field. We obtained the same sample temperature fluctuations as shown in Fig.3 (b). This result surely indicates that the origin for the signal fluctuations is not, say, the flux jump but the liquid nitrogen bubblings. The signal drop is then caused by unlocking of the cavity resonance induced by vibrating motion of the sample due to the bubbling.

When the bubbling occurs, it takes out the latent heat from ambient liquid nitrogen, then the sample temperature is reduced. The drop of this sample temperature is larger for the more intensive bubbling. After each bubbling, the heat is introduced to the liquid nitrogen through the sample tube from ambient atmosphere. Therefore the sample temperature is raised again to the original level. If the bubbling is not intensive, the unlocking of the resonance is not induced. Then the signal drop is not caused as shown in Fig.3 (c). The small signal oscillation indicates the small continuous bubbling. Thus we could elucidate the detailed bubbling sequence by utilizing the microwave absorption in HTS for the first time.

CONCLUSION

We showed a new bolometric function based on the new mechanism using the microwave absorption in HTS material.

Acknowledgements. This work was partially supported by Grant-in-Aid for Scientific Research from Ministry of Education.

---

*Corresponding author, E-mail: endo@cm.elec.mie-u.ac.jp

1. T.Endo, S. Yamada, M. Horie, N. Hirate, KT. Itoh, and Y. Tsutsumi: *Advances in Super-conductivity X*, eds. K. Osamura and I. Hirabayashi (Springer, Tokyo, 1998) pp.179-182.
2. T. Endo, Solid State Physics, **33-1**, 35-45 (1998), (in Japanese).
3. T. Endo and H. Yan, *Studies of High Temperature Superconductors*, vol. **14**, edited by Anant Narlikar (Nova Science Publishers, New York, 1995), pp.65-106.
4. T. Endo, H. Yan, S. Nagase, and H. Shibata, J. Supercond. **8**, 259-269 (1995).

FREQUENCY DEPENDENCE OF AC MAGNETIZATION IN RE-Ba-Cu-O SUPERCONDUCTORS NEAR THE VORTEX-GLASS TRANSITION

TOMOKAZU FUKUZAKI[1] , NORIKO CHIKUMOTO[2] , KAZUO INOUE[2] ,
HIROYASU OGIWARA[1] , MASATO MURAKAMI[2]

[1]Shonan Institute of Technology, 1-1-25 Nishikaigan, Tujido, Fujisawa, Kanagawa 251-8511, Japan
[2]Superconductivity Research Laboratory, ISTEC, 1-16-25 Shibaura, Minato-ku, Tokyo 105-0023, Japan

ABSTRACT

The measurement of AC susceptibility was performed for RE-Ba-Cu-O superconductors to study the vortex glass liquid transition. Critical scaling relation was derived from the frequency dependence of susceptibility (χ). The vortex glass-liquid transition temperature T_g and the critical exponent (ν and z) was determined from two different models : the vortex glass-liquid transition model and the depinning model. We performed scaling analysis using these two models. By comparing the critical exponent obtained from these models, we found that the phase transition line exists nearby the irreversibility line. The analysis based on the vortex glass-liquid transition model suggest that the fluxoid behave as of two-dimensional system.

KEYWORDS: AC magnetization, vortex glass-liquid transition, (Nd,Eu,Gd)-Ba-Cu-O

INTRODUCTION

It is known that there is the second order vortex-glass/liquid (VG/VL) transition in the magnetic phase diagram of high temperature superconductors. It is also discussed that this phase transition may be the origin of the irreversibility line. However, Matsushita et. al. [1] suggested that the temperature dependence of the irreversibility line can be explained by flux creep and flux flow model (depinning model) without taking the phase transition into account. If a second order transition exists, it is expected that critical scaling relation by critical exponents(ν and z) which can be derived from analysis of AC susceptibility ($\chi = \chi '+i \chi ''$) is applicable. Therefore, we have measured the frequency dependence of AC magnetization by an AC susceptometer and tried to apply scaling with critical exponent. The critical exponent can be calculated by two kinds of methods, the vortex glass-liquid transition model and the depinning model. In the depinning model, the temperature of $J_c=0$ is determined to be the transition temperature T_g. On the other hand, in the VG/VL transition model, T_g is determined from frequency independent phase which is given by |($\pi - \theta$)|$_{T=Tg} = \pi$ -arctan($\chi ''/ \chi '$)[2].

EXPERIMENTAL

$Nd_{0.33}Eu_{0.33}Gd_{0.33}Ba_2Cu_3O_7$ (NEG123) sample prepared by the OCMG process in 0.1% O_2 in Ar was used in the present study. NEG123 is known to exhibit high J_c values[3]. We used the sample which contains the 20 mol % of Gd211 second phase and the dimension of sample is 0.99 x 0.91 x 0.71 mm^3. The sample exhibit critical temperature T_c of 92K. AC Magnetization measurement were performed with an AC magnetometer (Lake Shore 7000 Series). We measured AC susceptibility ($\chi '$,

χ") near the peak of χ" for the DC field range from 2.5T to 7T and the AC frequency range between 100Hz and 2kHz. All of the measurement was performed with the magnetic field applied parallel to the c axis of the sample.

RESULTS AND DISCUSSION

Figure 1 shows the plots of the phase (π - θ) versus temperature for B=4T. At the transition temperature T_g, the phase (π - θ) take a certain value which is independent of frequency. The phase angle at T_g were 175.78° . It is known that the phase angle at T_g is given by:

$$|(\pi - \theta)|_{T=Tg} = \{\pi - (\pi/2)[1/(2z)]\}$$

where z is a dynamic critical exponent. Using this equation, z = 10.66 is obtained.
As shown in Fig. 2, the critical regime are scaled into the universal functions $|\chi_{sc\pm}(f_{sc})|$ for B=4T. The insets shows frequency dependence of susceptibility amplitude $|\chi|$. where, χ_{sc} and f_{sc} were defined by the following equation based on the VG/VL transition model:

$$\chi_{sc\pm}(f_{sc}) \equiv \chi |1-(T/T_g)|^{-\nu/2} \quad , \quad f_{sc} \equiv f|1-(T/T_g)|^{-\nu z}$$

where χ_{sc+} and χ_{sc-} are universal scaling function at $T>T_g$ and $T<T_g$, respectively, and ν is a static critical exponent. We used the values T_g=81.83, ν=1 in the present analysis.

Figure 3 shows the temperature dependence of critical current density J_c, which was determined from the DC magnetization measurement performed with a SQUID magnetometer. In the depinning model, temperature dependence of J_c [4] is given by :

$$J_c \propto (T_g-T)^{2\nu}$$

According to this equation, the temperature of J_c=0 determins T_g. From the data shown in Fig.3, we obtained T_g=82.5K at B=4T. As it can be seen from Fig.3, $J_c^{(1/2\nu)}$ versus T became linear in the vicinity of zero J_c. From this plot ν=1.3 was obtained. The scaling with this static critical exponent is shown in Figure 4.

Critical exponent which was obtained from the analysis based on the VG/VL transition model was ν=1 and z=10.66. On the other hand, for the depinning model, ν=1.3 and z=5 was obtained. The static critical exponent were similar in both models, while the dynamic critical exponent were largely different. In the calculation of the critical exponent using the depinning model, it was supposed that the irreversibility line is the phase transition line. However, if the phase transition line exist under the irreversibility line, the smaller value of the critical exponent is expected. So, the observed difference may be explained in terms of the over estimate of T_g value in the depinning model.
As for the value of z , it is known to be strongly influenced by the flux pinning strength ; sample with a wider distribution of flux pinning strength has a smaller z value [5]. According to the microstructural study [3], the individual pinning centers of NEG123 superconductors are distributed randomly. The value of critical exponent obtained in the present study using the VG/VL transition model is similar to that calculate by the Monte-Carlo simulation [6]. Particularly, the value of z is close to that obtained for the two-dimensional (2D) case. However, in the present system, it is unlikely that the 2D phase transition will occur because of small anisotropy. On the other hand, the

value obtained the depinning model is close to the 3D case, if we assume the irreversibility line exhibit the phase transition.

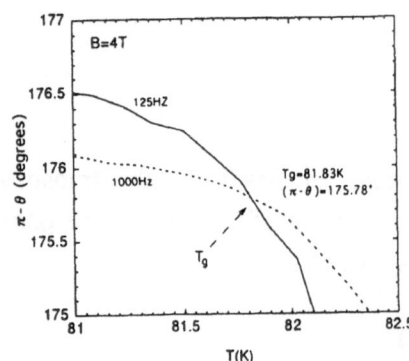

Fig.1 Plot of the phase ($\pi - \theta$) of AC
susceptibility versus temperature.
at 4T

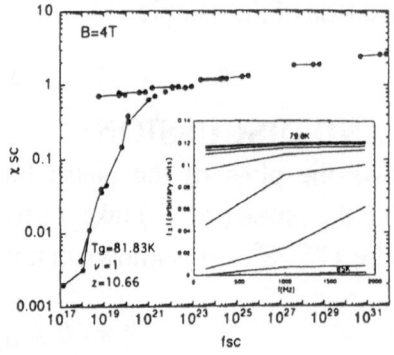

Fig.2 Critical regime which are scaled into the
universal functions $|\chi_{sc\pm}(f_{sc})|$ for B=4T.
Inset: Frequency dependence of susceptibility

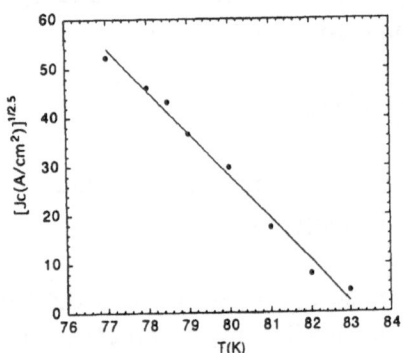

Fig.3 Plot of $J_c^{(1/2\nu)}$ versus T at 4T.

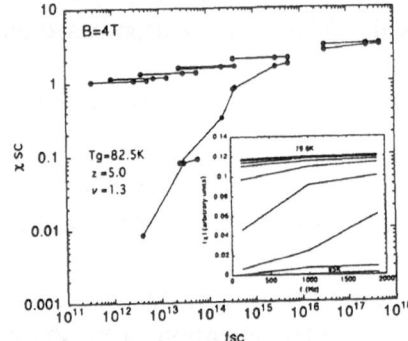

Fig. 4   Scaling of frequency- susceptibility
amplitude curves by critical exponent
which is given from   temperature
dependence of J_c at 4T

ACKNOWLEDGEMENT
This work was supported by New Energy and Industrial Technology Development Organization(NEDO) as Collaborative Research and Development of Fundamental Technologies for Superconductivity Applications under the New Sunshine Program administered by the Agency of Industrial Science and Technology (AIST) of the Ministry of International Trade and Industry (MITI) of Japan.

REFERENCES
[1] T. Matsushita, N. Ihara and T. Tohdoh : preprint of CEC /ICMC 1995, Columbus.
[2] D. S. Reed, N. –C. Teh, W. Jiang, U. Kriplani, D. A. Beam, and F. Holtzberg : Phys. Rev. B 49, 4384(1994)
[3] For example, M. Muralidhar et. al. : Physica C 313, 232(1999)
[4] T. Matsushita, M. Kiuchi, K. Noguchi : Physica C 290, 38(1997)
[5] T. Matsushita, T. Tohdo, N. Ihara : Physca C 259, 321(1996)
[6] K. Yamafuji, T.Kiss : Physica C 290, 9(1997)

Magnetic properties of polycrystalline KClO$_3$-doped Y$_{(1-0.2x)}$Ba$_{(2-0.2x)}$K$_x$Cu$_3$O$_y$(x = 0 ÷ 0.40) superconductors

Michael R. Koblischka, Anjela Veneva, and Masato Murakami

Superconductivity Research Laboratory, Internarional Superconductivity Technology Center
16-25 Shibaura 1-chome, Minato-ku, Tokyo 105-0023, Japan

Abstract: The effect of KClO$_3$-addition on the magnetic properties of polycrystalline high-temperature superconductors with nominal composition Y$_{(1-0.2x)}$Ba$_{(2-0.2x)}$K$_x$Cu$_3$O$_y$ ($x = 0 ÷ 0.40$) was investigated. DC susceptibility as function of temperature was measured in field-cooled cooling and field-cooled warming modes in magnetic fields in the range 10 mT $\leq \mu_0 H_a \leq$ 7 T on KClO$_3$-doped and pure YBCO samples. From the Meissner curves measured at a field of 10 mT, we deduce that the samples are single-phase. However, the low-temperature behaviour of the samples with KClO$_3$ additions at large applied fields is different from pure YBCO.

Key words: YBCO HTSC, KClO$_3$-addition, magnetic properties, flux pinning

INTRODUCTION

Investigating the influence of the different additions on the magnetic and electrical properties and microstructure of the high-T_c superconductors (HTSC) is of interest, as much for the understanding of the intergranular phenomena, as for solving technological problems. A special interest is associated with doping of YBa$_2$Cu$_3$O$_{7-\delta}$ (YBCO) superconducting materials by alkali-chlorine-containing additions, which improve the superconducting properties [1–4]. A T_c of 96 K has been observed for Y$_1$Ba$_{2-x}$K$_x$Cu$_3$O$_y$ samples (x = 0.2) with KCl [4]. The aim of this work is to investigate the effect of KClO$_3$-addition on the magnetic properties of polycrystalline YBCO HTSC. Here, we present field-cooled cooling (FCC) and -warming (FCW) measurements of KClO$_3$-doped and pure YBCO HTSC in magnetic fields up to 7 T.

EXPERIMENTAL

Polycrystalline KClO$_3$-doped YBCO high-temperature superconductors (HTSC) with nominal composition Y$_{(1-0.2x)}$Ba$_{(2-0.2x)}$K$_x$Cu$_3$O$_y$ ($x = 0 ÷ 0.40$) were synthesized by a solid state reaction in air [3]. Powders of Y$_2$O$_3$, BaCO$_3$, CuO, all with a purity 99.99 % and KClO$_3$ (more than 99.9 %), were mixed, pressed into pellets and heated at 880 °C for 24 h. The reacted pellets were reground, pressed again and sintered at 915 °C for 20 h. They were annealed at 600 °C for 14 h, before being cooled to room temperature. A detailed study on the thermal behavior of the samples was performed by differential thermal analysis (DTA) on the starting mixtures. The phase content, microstructure and morphology of the crystalline grains were characterized using X-ray powder diffraction (XRD) and scanning electron microscopy (SEM) coupled with an electron probe microanalyzer (EPMA).

Temperature scans in both FCC and FCW modes were carried out in various fields between 10 mT and 7 T. The measurements were performed in a commercial SQUID magnetometer (Quantum Design model XL), enabling to measure in a *continuous* temperature sweep mode with a controlled temperature sweep rate dT/dt = 35 mK/min; the datapoints are recorded in steps of 50 mK in

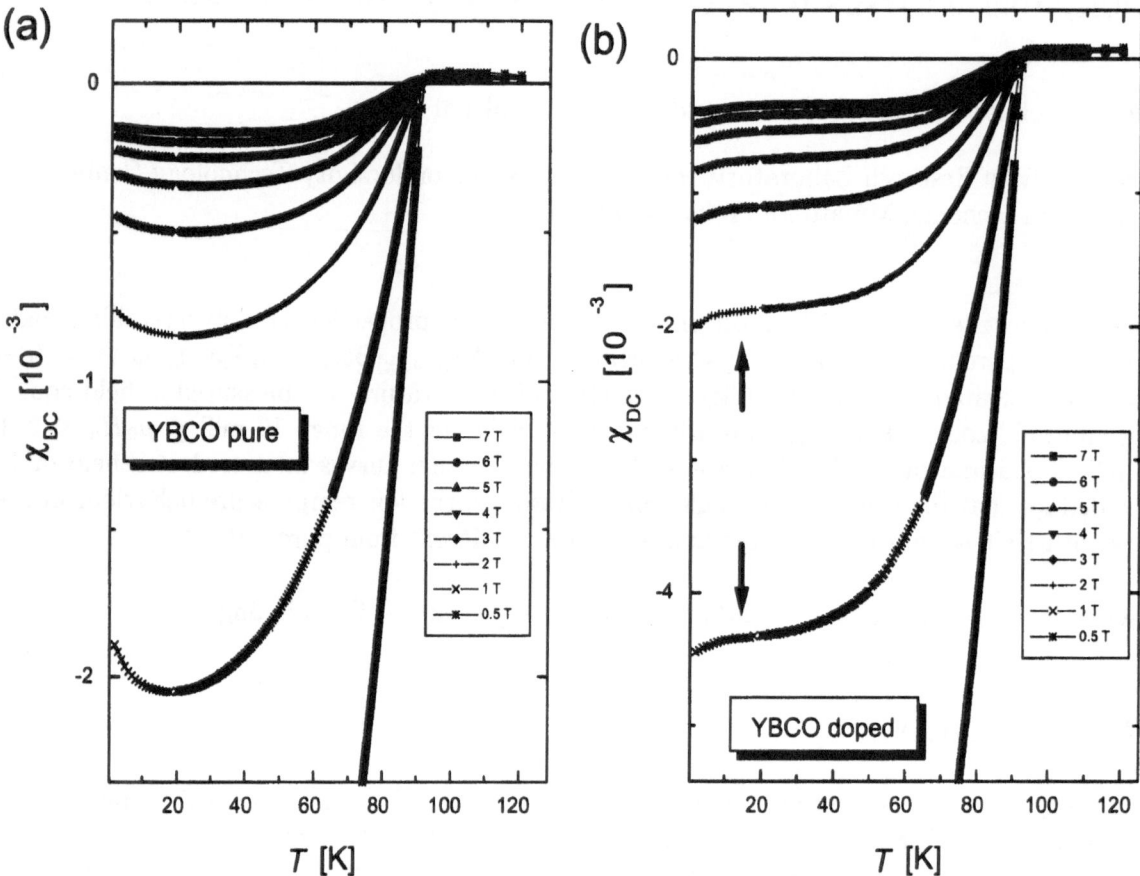

Fig. 1. (a): Temperature scans (FCC) of the DC susceptibility χ_{DC} of a pure YBCO sample in applied fields between 0.1 T and 7 T; the FCW data are omitted for clarity. Note that all curves are monotonuous in contrast to $NdBa_2Cu_3O_{7-\delta}$ crystals [5] in all applied fields. Further, at low temperatures there is an upturn of χ_{DC} due to the paramagnetic contribution of Cu. (b): The same experiments on the $KClO_3$-doped sample with $x = 0.30$. Again, all curves are monotonous, but at low T, their behaviour is different from the pure sample as marked by arrows.

the transition region. All $\chi_{DC}(T)$ curves are measured between 1.7 K and 120 K. Note that the temperature sweep is *not* halted for taking a datapoint as in a conventional SQUID magnetometer. No averaging of the signal is performed, and the scan length is 1 cm. This procedure ensures a large number of datapoints even in a sharp superconducting transition. More details of the measurement procedure are given by Koblischka et al. [5].

RESULTS AND DISCUSSION

After the second heat treatment the XRD data showed that in all cases the $KClO_3$-doped samples are pure 123 phase materials with only an indication of traces of CuO for the sample with $x = 0.40$. We would like to emphasize that the X-ray diffraction patterns of the doped samples do not contain any such extra lines within the detection limit of the X-ray method (normally the detection limit of XRD for impurities is estimated around 4-5 %, so it is not possible to detect minor impurities). And, no chlorine or other impurity phases related with chlorine or potassium are found in the $KClO_3$-doped samples. We have carried out detailed microstructural studies and SEM-EPMA revealed traces of potassium and chlorine in the grain boundaries of the samples with addition [3]. The values of the lattice parameters of $KClO_3$-doped samples are comparable to the data published for the desired YBCO orthorhombic structure [6]. The influence of $KClO_3$ on the superconducting properties depends on the concentration of the addition in the initial batch. AC susceptibility data of these samples are given by Veneva et al. [3]. The grain size of the pure YBCO sample is ≈ 28 μm; that of the $KClO_3$-doped samples ≈ 40 μm.

In Fig. 1 (a), χ_{DC} of the pure YBCO sample is plotted for various applied fields; in (b) the same experiment is shown for the $KClO_3$-doped sample with x = 0.30. All χ_{DC} curves are monotonous, and no secondary step in the transitions can be observed in fields above 4 T as in some YBCO single crystals and all $NdBa_2Cu_3O_{7-\delta}$ samples investigated in a similar way [5]. This demonstrates that the polycrystalline YBCO samples with their relatively small grains do not contain oxygen vacancy clusters providing a spatial distribution of T_c, in contrast to bulk samples. At low temperatures ($T < 15$ K), all FCC curves of the pure sample exhibit a clear upturn of the χ_{DC} curves, which is due to the paramagnetic moment of Cu^{2+}. These are located in the Cu-O-chains [7], and their number depends on the oxygenation state. In stark contrast to this behaviour, the doped sample in (b) exhibits a *downturn* of the FCC/FCW curves towards more negative (= diamagnetic) values, which develops with increasing applied field. This may point to a change in the copper valence, or, more likely, to the fact that the grain boundary regions become more strongly superconducting. In case that the coupling strength of the grains increases, shielding currents can flow on a larger length scale [8], which results in an increased diamagnetic signal. This latter explanation is in accordance with previous results obtained from AC susceptibility studies [3], where it was concluded that the $KClO_3$ addition leads to a modification of the coupling strength between the grains and to an alteration of the pinning effectiveness for the intergranular flux lines. It was shown in Ref. [3] that the grain connectivity is indeed improved by the $KClO_3$-addition, if small amounts of $KClO_3$-additions ($x = 0.20 \div 0.30$) are employed. This stronger superconducting signal may then completely mask the paramagnetic contribution of the copper.

CONCLUSIONS

We have presented FCC/FCW measurements on pure YBCO and $KClO_3$-doped samples in fields between 10 mT and 7 T. All χ_{DC} curves are found to be monotonous, but the doped samples exhibit a clear downturn to more diamagnetic values at low temperatures. This indicates that the increasing $KClO_3$-addition leads to a modification of the coupling strength of the grains.

Acknowlegdments. This work is partially supported by New Energy and Industrial Technology Development Organization (NEDO). AV and MRK gratefully acknowledge support from the Japanese Science and Technology Agency (STA).

1. B. Okai, Jpn. J. Appl. Phys. **29**, L2193-L2195 (1990).
2. A. Veneva, I. Iordanov, L. Toshev, A. Stoyanova and D. Gogova, Physica C **308**, 175-184 (1998).
3. A. Veneva, M. R. Koblischka, N. Sakai and M. Murakami, presented at the MOS'99 conference, 28.7.-2.8.99, Stockholm, Sweden, to be published in J. Low Temp. Phys.
4. K. H. Yoon and S. S. Chang, J. Appl. Phys. **67**, 2516-2519 (1990).
5. M. R. Koblischka, M. Muralidhar, T. Higuchi, K. Waki, N. Chikumoto and M. Murakami, Supercond. Sci. Technol. **12**, 288-292 (1999).
6. JCPDS powder diffraction, Alphabetical Indexes (1997) file no. 38-1433; 39-486; 39-1434
7. T. R. McGuire, T. R. Dinger, P. J. P. Freitas, W. J. Gallagher, T. S. Plaskett, R. L. Sandstrom and T. M. Shaw, Phys. Rev. B **36**, 4032-4035 (1987).
8. M. R. Koblischka, Th. Schuster and H. Kronmüller, Physica C **219**, 205-212 (1994).

Magnetic properties of superconducting and non-superconducting (Nd,Eu,Gd)-123

Michael R. Koblischka, Miryala Muralidhar, and Masato Murakami

Superconductivity Research Laboratory, Internarional Superconductivity Technology Center
16-25 Shibaura 1-chome, Minato-ku, Tokyo 105-0023, Japan

Abstract: The temperature dependence of the DC susceptibility $\chi(T)$ is measured between 1.7 K and 350 K on superconducting and non-superconducting bulk melt-processed $(Nd_{0.33}Eu_{0.33}Gd_{0.33})Ba_2Cu_3O_y$ by means of SQUID magnetometry. A strong superconducting contribution is found to coexist with a large paramagnetic moment provided by the Nd and Gd ions. The paramagnetic contributions measured on both types of samples follow a Curie-Weiss law, however, an antiferromagnetic ordering is not observed down to 1.7 K.

Key words: Light rare earth-123 superconductors, magnetic properties, flux pinning

INTRODUCTION

The quest to increase the current densities j_c of bulk, melt-processed superconductors led to the preparation of ternary compounds comprising three light rare earth elements Nd, Eu, and Gd on the rare earth site of the 123 structure. The samples with the nominal composition $(Nd_{0.33}Eu_{0.33}Gd_{0.33})Ba_2Cu_3O_y$ ("NEG") yield superconductors with a significantly larger j_c and irreversibility fields, H_{irr} [1, 2]. This strong superconducting contribution (exhibiting a record high current density for bulk samples at 77 K, 3T: 68000 A/cm^2), however, coexists with large paramagnetic moments provided by the Nd and Gd ions, as shown earlier in field-cooling (FCC) or -warming (FCW) measurements [3].

In the present paper, we focus on the magnetic properties of normalconducting as well as superconducting NEG, in order to understand the characteristic behaviour of the FCC/FCW curves.

EXPERIMENTAL

Bulk NEG samples were prepared using the oxygen-controlled melt growth (OCMG) process in 0.1% partial pressure of oxygen. In this process, the substitution of Nd, Eu or Gd on the Ba site is nearly suppressed; the remaining substitution provides additional flux pinning [2]. Details of the heat treatment schedules are described in detail elsewhere [1]; the normal-conducting samples did not receive the final oxygen treatment.

Temperature scans in both field-cooling (FCC) and field-warming (FCW) modes were carried out in various fields between 10 mT and 7 T. The measurements were performed in a commercial SQUID magnetometer (Quantum Design model XL), enabling to measure in a *continuous* temperature sweep mode with a controlled temperature sweep rate $dT/dt = 35$ mK/min; the datapoints are recorded in steps of 50 mK. Note that the temperature sweep is *not* halted for taking a datapoint as in a conventional SQUID magnetometer. No averaging of the signal is performed, and the scan length is only minimal. This procedure ensures a large number of datapoints even in a sharp superconducting transition. All $M(T)$ curves of the superconducting samples are measured between 1.7 K and 120 K; the $M(T)$ curves of the normal-conducting samples are recorded between 1.7 K and 350 K; the sweep rate for this experiment is 1 K/min and data are collected every 100 mK. More details of the measurement procedure are given by Koblischka et al. [3].

RESULTS AND DISCUSSION

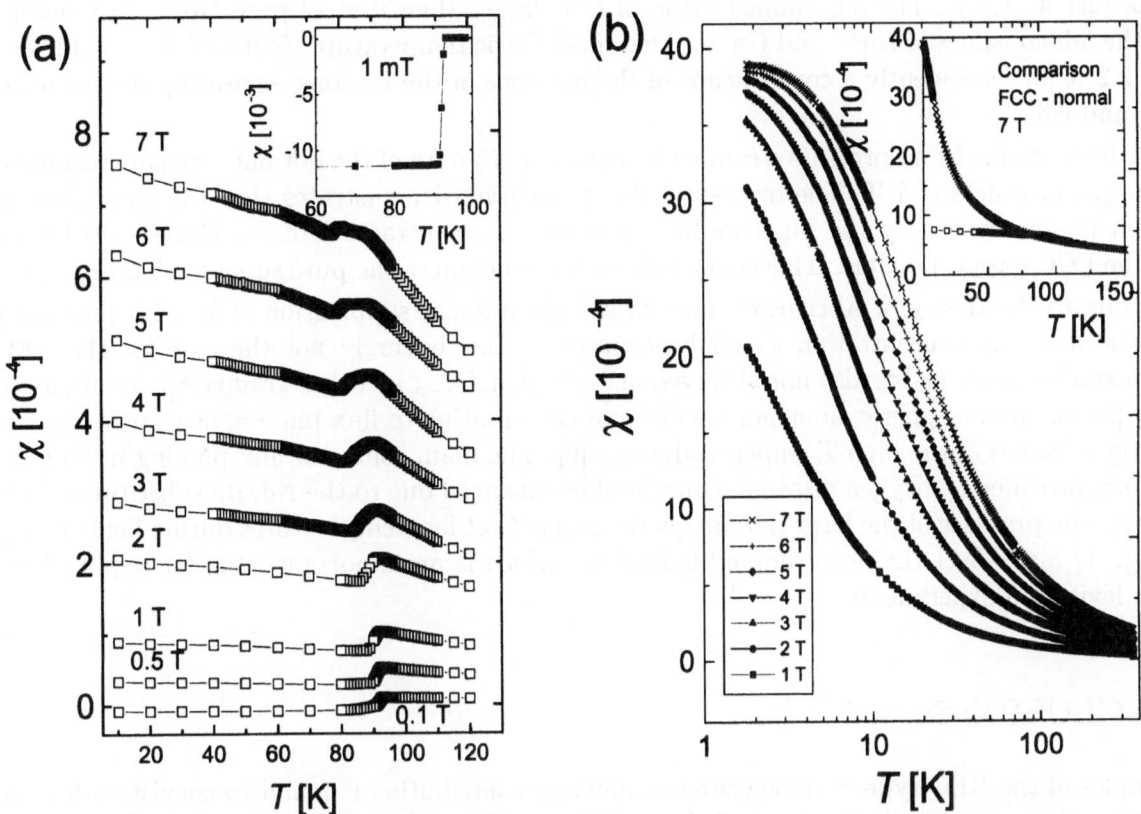

Fig. 1. (a): Temperature scans (FCC) of the magnetic moment $M(T)$ of the NEG sample in applied fields between 0.1 T and 7 T; the FCW data are omitted for clarity. Note the large paramagnetic moment caused by the presence of Nd and Gd. The inset shows a measurement in an applied field of 1 mT after zero-field cooling ("standard measurement"). (b): Temperature dependence of normal-conducting NEG (y = 6.1) between 1.7 and 350 K. The inset shows a comparison of the FCC data of superconducting NEG with the $\chi(T)$ of the normal-conducting sample at $\mu_0 H_a = 7$ T.

In Fig. 1 (left side), $\chi(T)$ is plotted for various applied fields measured during FCC; the FCW data are omitted for clarity. The inset presents the superconducting transition in an applied field of 1 mT after zero-field cooling ("standard measurement"). This shows the sample to be single-phase, with $T_{c,\text{onset}} = 93.1$ K and $\delta T_c \approx 1.5$ K. The $\chi(T)$ scans measured during FCC show that the magnetic moments of Nd (3.54 μ_B) and Gd (7.94 μ_B) cause a considerable increase of χ on decreasing T. Therefore, the magnetic moment is *positive* even in the superconducting state. In fields between 1 mT and 2 T, the superconducting transitions are quite sharp; T_c is found to reduce slightly as a function of field to 91.5 K at 5 T. Note the kink in the transitions in fields above 2 T, which is attributed to a finely dispersed rare earth-rich phase with weaker superconducting properties providing additional flux pinning of the δT_c-type. This increased δT_c-pinning is essential for the improved performance of the NEG samples at high fields, i.e. the so-called fishtail effect [4]. Note the behaviour of $\chi(T)$ above T_c, which is following a Curie-Weiss law,

$$\chi(T) = C/(T - \Theta_c) \tag{1}$$

In contrast to earlier work on $GdBa_2Cu_3O_{7-\delta}$ (GdBCO) polycrystalline samples [5, 6], the superconducting contribution in the NEG samples cannot be suppressed within our field range. To illustrate the low temperature behaviour, $\chi(T)$ was measured on a normal-conducting sample from the same batch (i.e. without oxygen annealing, y = 6.1), as shown in fig. 1 (right side). At 1 T, we obtain from fits to eq. (1) $\Theta_c = (-1.83 \pm 0.01)$ K and $C = (0.947 \pm 0.002)$ K. However, in our data no antiferromagnetic ordering is observed down to 1.7 K; the curve at 7 T clearly approaches a peak just at 1.7 K. The determined value of C is larger than that of pure Gd^{3+}, but smaller than the added values of Nd^{3+} and Gd^{3+}. This small Curie temperature (GdBCO: $\Theta_c \approx 3.3$ K and $T_N = 2.2$ K [5]) is evidently a consequence of the presence of the Eu ions separating the moments of Gd and Nd.

The inset to this figure presents a comparison of the $\chi(T)$ data of the normal- and superconducting sample at a field of 7 T. The overlap of this two curves demonstrates that the paramagnetic moment is not affected by the superconducting state, so a subtraction of the Curie contribution from the FCC data is possible. This enables the reconstruction of the pure superconducting signal. According to the theory of Abrikosov and Gor'kov [7], a linear suppression of T_c with increasing magnetic impurity concentration would be expected. This is surely not the case for the NEG system studied here, and is also not observed in $NdBa_2Cu_3O_{7-\delta}$ by other groups [8]. As discussed earlier [2], the paramagnetic moments are also not contributing to flux pinning as a conseqence of the large κ values of the high-T_c superconductors [9]. The main source of flux pinning in NEG is, therefore, provided by oxygen disorder, considerably enhanced due to the Nd/Ba substitution [10]. However, the presence of the large paramagnetic moments at low temperatures during field-cooling (see fig. 1) may affect the flux trapping behaviour, which is an important issue for applications, e.g. in levitation experiments.

CONCLUSIONS

In samples of the NEG type, a strong superconducting contribution is found to coexist with large paramagnetic moments provided by the Nd and Gd ions. No antiferromagnetic ordering is observed in normal-conducting samples down to 1.7 K, due to an increased spacing of the Nd and Gd ions by the presence of Eu.

Acknowlegdments. This work was partially supported by NEDO. MRK and MMD acknowledge support from STA and NEDO, respectively.

1. M. Muralidhar, M. R. Koblischka, T. Saitoh and M. Murakami, Supercond. Sci. Technol. **11**, 1349-1358 (1998).
2. M. R. Koblischka, M. Muralidhar and M. Murakami, Appl. Phys. Lett. **73**, 2351-2353 (1998).
3. M. R. Koblischka, M. Muralidhar, T. Higuchi, K. Waki, N. Chikumoto and M. Murakami, Supercond. Sci. Technol. **12**, 288-292 (1999).
4. M. R. Koblischka, M. Muralidhar and M. Murakami, presented at the 4th ICAM-IUMRS conference, June 14 – 17, 1999, Beijing, P.R.China, to be published in Physica C.
5. G. Hilscher, T. Holubar, G. Schaudy, J. Dumschat, M. Strecker, G. Wortmann, X. Z. Wang, B. Hellebrand and D. Bäuerle, Physica C **224**, 330-344 (1994).
6. H. Theuss and H. Kronmüller, Physica C **242**, 155-163 (1995).
7. A. A. Abrikosov and L. P. Gor'kov, Sov. Phys. JETP **12**, 1243-1252 (1961).
8. W. H. Tang and J. Gao, IEEE Trans. Appl. Supercond. **9**, 2113-2116 (1999).
9. R. P. Hübener, *Magnetic Flux Structures in Superconductors*, (Springer, Berlin, 1979).
10. M. R. Koblischka, A. J. J. van Dalen, T. Higuchi, S. I. Yoo and M. Murakami, Phys. Rev. B **58**, 2683-2867 (1998).

PREPARATION OF (Nd,Eu,Gd)-Ba-Cu-O SINGLE CRYSTALS

Muralidhar Miryala, Michael R Koblischka and Masato Murakami

Superconductivity Research Laboratory, ISTEC-SRL, Division 3, 1-16-25 Shibaura, Minato-ku, Tokyo 105, Japan

Abstract: Single crystals of (Nd,Eu,Gd)-Ba-Cu-O type were grown in Pt crucible by flux method. High purity sintered $(Nd_{0.33}Eu_{0.28}Gd_{0.38})Ba_2Cu_3O_y$ "NEG-123" powders and CuO flux were added in 1:1 ratio at room temperature and the growth was performed in Ar-0.1% O_2 atmosphere. Several crystals of approximately rectangular shape were found in the loose material; the size of 800 x 400 μm in the a-b plane and around 20 μm in the c-axis direction was successfully obtained using a crucible 50 mm in diameter. These crystals show growth steps on the surface and no twin boundaries are observed in polarized light in the as-grown samples. Chemical analysis on the crystals revealed that the average matrix atomic ratio is Nd:Eu:Gd $\approx$ 2:1:1. Measurements of DC susceptibility demonstrate that these samples exhibit the onset of superconductivity at 92 K.

Key words: NEG-123 single crystals, flux method, microstructure, critical temperature.

INTRODUCTION

The preparation, microstructure and characterization of melt-processed (Nd, Eu, Gd)-Ba-Cu-O have attracted a great deal of attention in the last two years [1-3]. The combination of different RE elements in the RE site with an appropriate mixing ratio and synthesis conditions allows one to develop samples with a high performance flux pinning capability as compared to Nd-123 or Y-123 [2]. These mixed materials show superior critical current density's in high magnetic fields at 77K and microstructural results reveal that sub-micron Gd-211 secondary phase particles are dispersed in the crystal matrix [3]. To further understand more details about the flux pinning mechanism and crystal properties, the growth of single crystal of this composition is important. In this paper, we attempt for the first time to synthesize NEG-123 type single crystals using a flux method under a controlled oxygen atmosphere.

EXPERIMENTAL

Single crystals of the NEG-123 type were grown by a flux method from sintered NEG-123 powders and CuO flux with 1:1 ratio. The high purity Nd_2O_3, Eu_2O_3, Gd_2O_3, $BaCO_3$ and CuO commercial powders were weighed to have a nominal composition of $(Nd_{0.33}Eu_{0.28}Gd_{0.38})Ba_2Cu_3O_y$. The precursor powders were ground thoroughly and calcinated at 880 °C for 24 h with intermediate grinding, and pressed into pellets, which were further sintered at 1020 °C for 48h. Sintered NEG-123 and CuO powders were mixed in a Pt crucible and the crystal growth was performed in the Ar-0.1% oxygen atmosphere. After a complete melt was formed, the furnace was slowly cooled. After the crystal growth we could not detect green phase particles. The crystals grown at top layer were CuO crystals. In the cavities we found many small crystals that were removed mechanically with care. The selected crystals were further annealed in a tube furnace under flowing oxygen gas of 400 ml/min in different steps. The oxidation conditions are summarized in Table 1. Optical microscopic observations were carried out for grown crystals, and the sizes of the crystals were measured to evaluate the morphology of crystals. The chemical composition of the as-grown crystals were determined by an TEM by EDX analyzer. Magnetic measurements of single crystals were performed using a Quantum Design MPMS-7 SQUID magnetometer equipped with a 7 T superconducting magnet for $H_a \parallel c$.

Table 1. Oxidation conditions and SQUID results on NEG-123 type single crystals.

Sl. No.	Sample	Oxygenation details	$T_{c, onset}$	ΔT_c
1	Step 1	240 h, 600 °C – 300 °C	92 K	32 K
			81 K	22 K
2	Step 2	225 h, 400 °C – 300 °C	88 K	18 K
3	Step 3	820 h, 600 °C – 300 °C	92 K	13 K

RESULTS AND DISCUSSION

Figure 1 shows the typical crystals obtained from the flux growth method and mostly were rectangular in shape. Crystals show growth steps on the surface plus remnants of molten material. Otherwise the surfaces are clean. As-grown samples had no twins when observed under polarized light. The thickness of the crystals was approximately 20 μm like YBCO crystals prepared for the first time in a similar way.

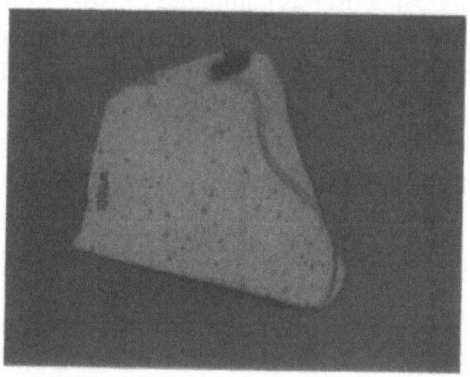 

Fig. 1 (Nd, Eu, Gd)-123 single crystals obtained from the flux method.

Compositional analyses with TEM-EDX showed that the single crystals are NEG-123 type with the average composition of Nd:Eu:Gd being close to 2:1:1.

Figure 2 shows the temperature dependence of the magnetization of a NEG-123 single crystals after oxygen annealing with different steps with the field 1 mT applied parallel to the c-axis. Onset T_c was defined by the temperature at which the Meissner signal begins to appear. It is interesting to note that the sample with the first step oxygenation show a two step transition at around 92 K and 81 K respectively. Only single step was observed after second step oxygenation. The results of

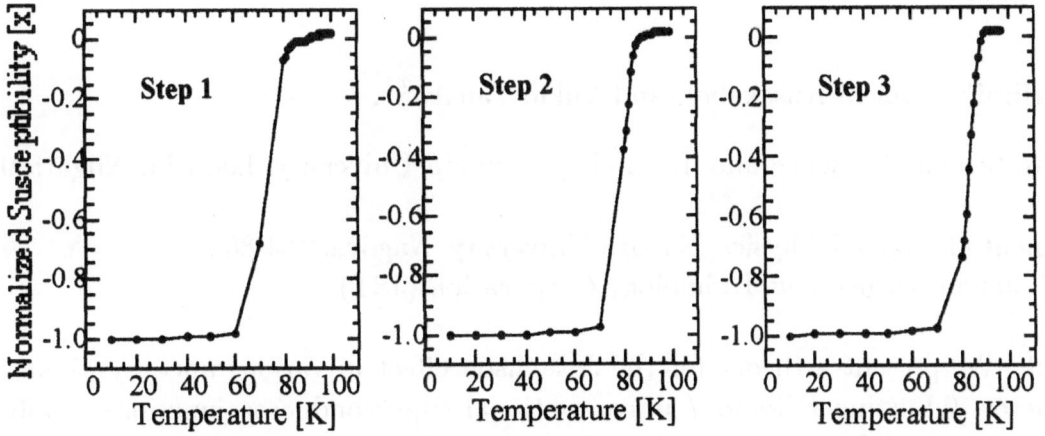

Figure 2. Temperature dependence of magnetization of O₂-annealed NEG-123 type single crystals (H ∥ c axis and applied magnetic field is 1 mT).

T_c (onset) and ΔT_c values are summarized in Table 1. It is evident that the superconducting transition widths are decreased progressively with increasing the oxygenation process. For full oxygen annealing of NEG123 more time is required like the case of Nd-123 [4]. Further experiments are under progress for the optimization of oxygen annealing conditions.

SUMMARY

Single crystals of NEG-123 type were successfully grown by the flux method under 0.1% oxygen partial pressure atmosphere. TEM by EDX analysis showed that as grown single crystals are NEG123 and the chemical ratio of the Nd:Eu:Gd matrix is 2:1:1. DC magnetization measurements $T_{c, onset}$ around 92 K.

Acknowledgements

This work was supported by New Energy and Industrial Technology Development Organization (NEDO) as a part of its Research and Development of Fundamental Technologies for Superconductor Applications Project under the New Sunshine Program administrated by the Agency of Industrial Science and Technologies M.I.T.I of Japan.

References

1. M. Muralidhar, M. R. Koblischka, T. Saitoh, and M. Murakami, Supercond. Sci. Technol. **11**, pp. 1349-1358 (1998).
2. M. R. Koblischka, M. Muralidhar and M. Murakami, Appl. Phys. Lett. **73**, pp. 2351-2353 (1998).
3. M. Muralidhar, M. R. Koblischka, and M. Murakami, Physica C **313**, pp. 232-240 (1999).
4. Th. Wolf, A-C. Bornarel, H. Kupfer, R. Meier-Hirmer and B. Obst, Phs. Rev. B, 56, pp. 6308-6319 (1997).

Josephson Effect in Luttinger Liquid between Anisotropic Superconductors

Takashi Hirai[1,2], Koichi Kusakabe[1], and Yukio Tanaka[2,3],

[1]Graduate School of Science and Technology, Niigata University, Ikarashi, Niigata 950-2181
[2]Department of Applied Physics, Nagoya University, Nagoya, 464-8063
[3]CREST, Japan Science and Technology Corporation (JST)

Abstract: We present a theory for the Josephson effect in various unconventional superconductor / Luttinger liquid / unconventional superconductor junctions, applying the bosonization method by Haldane and Maslov. Depending on the parity of superconductors and appearance of the zero-energy state, the temperature dependence of the maximum Josephson current (J_c) shows different properties. The interaction effect on the Josephson current appears in the renormarized velocity in our formulation, but essensial temperature dependence are the same as that in non-interacting cases.

Keywords: Josephson effect, Luttinger liquid, triplet superconductor, zero-energy states

INTRODUCTION

Josephson effects in superconductor / Luttinger liquid / superconductor ($s/LL/s$) junction are studied theoretically [1-3] pursuing Josephson effects through low-dimensional correlated electon systems. Since a decisive factor of Josephson current is macroscopic phase of the superconductor, the Josephson current is very sensitive to the symmetry of the superconductor. An important concept is formation of the zero-energy states (ZES) formed near surfaces of the superconductor[4-8]. ZES appears when sign of the pair potential changes on the Fermi surface. If we consider unconventional superconductor/ normal metal /unconventional superconductor ($us/n/us$) junction, ZES exists irrespective of symmetry of the pair potential (d-wave or p-wave). This could be a disadvantage, when one wants to determine the parity of the unconvensional superconductor using Josephson junctions [9]. In case of junctions with one-dimentional systems, however, the effect of ZES appears only when triplet superconductor (ts) is used. This is due to geometry of the junction, where quasi-particles are injected parpendicular to the interface. For the singlet superconductor (ss), ZES does not affect the Josephson current [10]. In the one-dimensional electron gas (1DEG), the interaction effect should be important, so we have to consider $us/LL/us$ junctions. We can actually expect several interesting characteristics in $us/LL/us$ junctions by changing parity of the unconvensional superconductor.

FORMULA and RESULTS

We consider $us/LL/us$ junctions with perfectly flat interfaces in the clean limit. The interface is perpendicular to the x-axis and is located at $x = 0$ and $x = d$, where d is the length of the LL region. We model the insulator located between the superconductors and the 1DEG by a delta functions, namely $H\delta(x)$ and $H\delta(x - d)$, where H denotes the strength of the barrier. We assume that the superconductors are two-dimensional. The Fermi wave number k_F and the effective mass m are assumed to be equal in the left- and right superconductors. In the LL, the magnitude of the Fermi wave number and the effective mass are also chosen as k_F and m, respectively.

In 1DEG, there exist right- and left-going electron- and hole-quasiparticles. We define fermion field operators using a solution of the Bogoliubov-de Gennes equation (BdGE) for an $s/1DEG/s$ junction. Geometry of the junction is shown in Fig. 1. At low temperatures, only low-lying excitations with energy $|E_n| \ll \Delta$ will contribute to the Josephson effect, where Δ is the superconducting gap. So, the fermion fields should be expanded by these low-lying excitations. Due to the Andreev- and normal reflections at interfaces, two boundary conditions are enough to describe boundary condition for every fields: $\Psi_{+,s}(x + L) = e^{i\pi\theta(\varphi)}\Psi_{+,s}(x)$ where s takes the values +1(-1) for the $\uparrow(\downarrow)$ spin projection and $L = 2d$, respectively. The twist phase, $\theta(\varphi)$ is defined using a phase factor $\gamma(\varphi)$ as $\theta = \frac{1}{\pi}\tan^{-1}(Im\gamma/Re\gamma)$. Here $\varphi \equiv \varphi_L - \varphi_R$ is difference in the macroscopic phase of two superconductors. Omitting weak energy dependence for modes with $E_n \simeq 0$, γ is determined for several junctions as follows.

1. $s(d)$-wave/1D electron gas/$s(d)$-wave (singlet):

$$\gamma = \frac{-\{\sigma_N^2\cos\varphi + 2(1-\sigma_N)(1/\delta+\delta)\} + \sqrt{\{\sigma_N^2\cos\varphi + 2(1-\sigma_N)(1/\delta+\delta)\}^2 - (\sigma_N-2)^4}}{(\sigma_N-2)^2}$$

2. p-wave/1D electron gas/p-wave (triplet): $\gamma = -e^{-i\varphi}$

3. $s(d)$-wave (singlet)/1D electron gas/p-wave (triplet): $\gamma = \frac{-i\sigma_N\sin\varphi + \sqrt{(\sigma_N-2)^2 - \sigma_N^2\sin^2\varphi}}{\sigma_N-2}$

where $\sigma_N = \frac{Z^2}{4+Z^2}$, $Z = \frac{2mH}{\hbar^2 k_F}$, $\delta = \frac{2+iZ}{2-iZ}$, respectively. Now, an important observation is that H does not appear in the formula for the p-wave, so that the Josephson effect is independent of the amount of H for the p-wave.

We can derive bosonized representation of the Luttinger liquid Hamiltonian:

$$H = H_{non-ZM} + H_{ZM}(\varphi)$$
$$H_{ZM}(\varphi) = \frac{1}{L}(\pi v_F + \frac{g_{2\perp}+g_{2\parallel}+g_{4\perp}+g_{4\parallel}}{4})(2N - \theta(\varphi))^2 \tag{1}$$

where $g_{i\parallel(\perp)}, i = 2, 4$ are coupling constants of forward scatterings between parallel (perpendicular) spins, and N is the winding number. Details of the derivation are discussed elsewhere [11]. The Josephson current is obtained as $J = -\frac{2ek_BT}{\hbar}\frac{\partial}{\partial\varphi}\log Z(\varphi)$.
The interaction effect on the Josephson current appears in the renormarized velocity in our formulation. Note that φ-dependence exists only in the zero-mode part. Since the current has to be zero when $\varphi = 0$, the winding number should satisfy $2N = 2n + 1$. Thus, we can easily obtain the current-phase relation and the maximum Josephson current for $us/LL/us$ junctions.

In the following, we will consider three cases; Temperature dependence of the maximum Josephson current J_c of (1) singlet superconductor / 1DEG / singlet superconductor

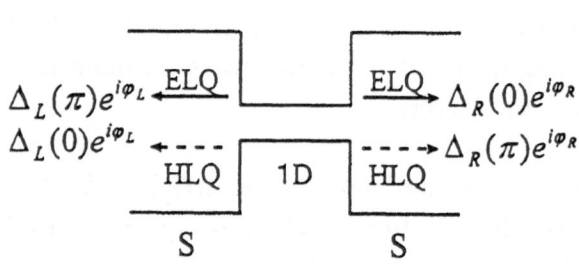

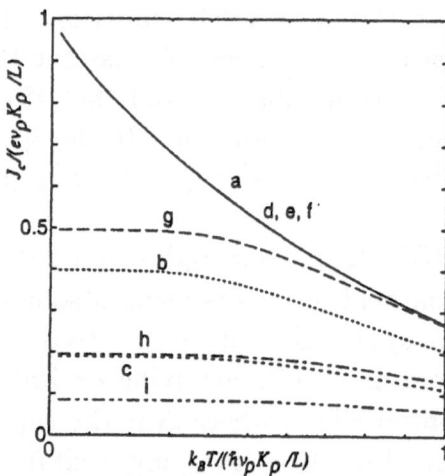

Figure 1: A schematic illustration of the $s/1DEG/s$ junction. The effective pair potentials for the reflected ELQ, the reflected HLQ, the transmitted ELQ and the transmitted HLQ are $\Delta_L(\pi)exp(i\varphi_L)$, $\Delta_L(0)exp(i\varphi_L)$, $\Delta_R(0)exp(i\varphi_R)$, $\Delta_R(\pi)exp(i\varphi_R)$, respectively.

Figure 2: Maximus Josephson current J_c plotted in $ss/o/ss$ for a.$Z = 0$, b.$Z = 0.5$, c.$Z = 1.0$, in $ts/o/ts$ for d.$Z = 0$, e.$Z = 0.5$, f.$Z = 1.0$, and in $ss/o/ts$ for g.$Z = 0$, h.$Z = 0.5$, i.$Z = 1.0$.

$(ss/o/ss)$ junction, (2) triplet superconductor / 1DEG / triplet superconductor $(ts/o/ts)$ junction and (3) singlet superconductor / 1DEG / triplet superconductor $(ss/o/ts)$ junction. Temperature dependence of the maximum Josephson current J_c for three cases is plotted in Fig. 2. For $ss/o/ss$ junction, magnitude of J_c is reduced with increasing Z. On the other hand, for $ts/o/ts$ case, J_c never changes. This result indicates the formation of ZES. For $ss/o/ts$ case, J_c takes small value compared with $ss/o/ss$ and $ts/o/ts$ cases in $Z = 0$ at low temperature, and is reduced with increasing Z.

The reason why renormalized factor $(1/k_F d)^{K_\rho^{-1}-1}$ obtained Fazio [1] does not appear in our formulation [10] is that we assume that energy dependence of γ is negligible. If we could introduce this effect, renormalization on the barrier potential may appear. However, essential feature of J_c, i.e. dependence on the parity of the pair potential is the same as the results obtained in previous paper considering 1DEG being non-interacting system [10].

Acknowledgements: This work has been partially supported by the Core Research for Evolutional Science and Technology (CREST) of the Japan Science and Technology Corporation (JST).

1. R. Fazio, *et al.* Phys. Rev. Lett. **74** , 1843 (1995).
2. D. L. Maslov, *et al.* Phys. Rev. B. **53**, 1548 (1996).
3. Y. Takane, J. Phys. Soc. Jpn. **66**, 537 (1997).
4. C. R. Hu, Phys. Rev. Lett. **72**, 1526 (1994).
5. L. J. Buchholtz and G. Zwicknagl, Phys. Rev. B. **23**, 5788 (1986).
6. J. Hara and K. Nagai, Prog. Theor. Phys. **74**, 1237 (1986)
7. Y. Tanaka and S. Kashiwaya, Phys. Rev. Lett. **74**, 3451 (1995).
8. Yu. S. Barash, *et al.* Phys. Rev. B. **55**, 15282 (1997).
9. L. Alff, *et al.* Phys. Rev. B. **55**, R14757 (1997).
10. Y. Tanaka, *et al.* Phys. Rev. B. **60**, 6308 (1999).
11. T. Hirai, *et al.* preprint.

LOWER CRITICAL FIELDS AND HOLE DENSITIES IN THE CuO$_2$ PLANE OF OXIDE SUPERCONDUCTORS

Masami Mashino[1], Shinjiro Tochihara[1], Hiroshi Yasuoka[1], Hiromasa Mazaki[1], Minoru Osada[2], and Masato Kakihana[3]

[1]Dept.Math.&Phys., National Defense Academy, 1-10-20 Hashirimizu,
Yokosuka-shi, Kanagawa 239-8686
[2]RIKEN, 2-1 Hirosawa, Wako-shi, Saitama 351-0198
[3]Materials & Structures Lab., Tokyo Inst. of Tech., 4259 Nagatsuta-cho, Midori-ku, Yokohama-shi, Kanagawa 226-8503

Abstract: The lower critical field H_{c1} is one of the important quantities for understandings of the fundamental behavior of type-II superconductors. We evaluated H_{c1} for several oxide superconductors with various hole densities within the framework of the modified Kim-Anderson critical-state model. The correlation between H_{c1} and hole density seems to be consistent with the results of muon spin relaxation (μSR), i.e., H_{c1} varies in proportion to the hole density in the CuO$_2$ plane.

Key Words: lower critical field, modified Kim-Anderson critical-state model, hole density

INTRODUCTION

In cuprate based superconductors, substitution of elements with different valence electrons brings about a change in the hole density of the CuO$_2$ plane, and this results in variation of the critical temperature T_c [1, 2]. The change in the hole density is expected to cause any effect on the magnetic properties of the samples. To interpret the magnetic behaviors of type-II superconductors, the lower critical field H_{c1} is an important quantity and has been studied by employing a variety of methods. However, the results of previous works have been still conflicting. This inconsistency is possibility caused by other effects (demagnetization, surface barriers, etc.) which would be involved in the evaluation of H_{c1}.

Recently, we have fully derived the magnetization equations of type-II superconductors within the framework of the modified Kim-Anderson (KA) critical-state model [3], where H_{c1} and surface barriers ΔH are explicitly taken into account. From the calculated initial magnetization curves and full-hysteresis loops, we have found that the magnetization curve is merely expanded upward and downward with ΔH, while H_{c1} introduces a step-like feature into the hysteresis loop. Besides, it has been revealed that this step width precisely corresponds to $2H_{c1}$ without the effect of surface barriers.

In order to reveal the correlation between the magnetic response and the carrier concentration, we prepared YBa$_2$Cu$_{3-x}$Co$_x$O$_{7-d}$ ($x = 0.00, 0.10, 0.20$) and Bi$_2$Sr$_{2-y}$La$_y$CuO$_{8+d}$ ($y = 0.20, 0.40$) single-crystal superconductors and measured the magnetization curves using a SQUID magnetometer. Analyses based on this theoretical prediction have revealed H_{c1} is closely correlated to the carrier concentration.

THEORETICAL

In the modified KA model, the critical-current density J_c and the effective applied field H_{eff} are assumed as [3, 4]

$$J_c = \frac{k}{B_0 + |B_i|},$$ (1)

$$H_{eff} = H - \frac{H}{|H|}H_{c1} - \frac{dH/dt}{|dH/dt|}\Delta H.$$ (2)

In Eq.(1), k and B_0 are material parameters, and B_i is the local magnetic-flux density inside the specimen. We consider an infinitely long cylinder with radius a and the applied field along the cylinder axis. A

257

sample is located in an external field $H = H_{dc} + H_{ac}\cos(\omega t)$, where H_{dc} is a dc magnetic field and H_{ac} is an ac field amplitude. To complete hysteresis loops for any applied field H, it is necessary to consider 113 stages of H, and all the magnetization equations have been derived [3].

In Fig.1, we show typical initial and hysteresis $M(H)$ curves calculated using the appropriate magnetization equations for $H_{dc} = 0$ and $H_{ac} = 4H_p$, where M and H are normalized to the full-penetration field H_p. The solid loop is for $\Delta H = 0$, $H_{c1} = 0.15H_p$ and the dashed loop is for $\Delta H = 0$, $H_{c1} = 0$. As shown in the figure, we find the solid line traced at the center of the saturated magnetization curves for the magnetization process shifts from the zero-magnetization line. This shift is caused by a nonzero value of H_{c1}, and this deviation has been revealed to correspond to H_{c1}. Besides, the model calculation has shown that the loop is only expanded upward and downward when ΔH has a nonzero value (not shown in the figure). This means that the effect of ΔH and H_{c1} appears independently on the feature of $M(H)$ curves. In this method, we can determine H_{c1} without any effect of surface barriers.

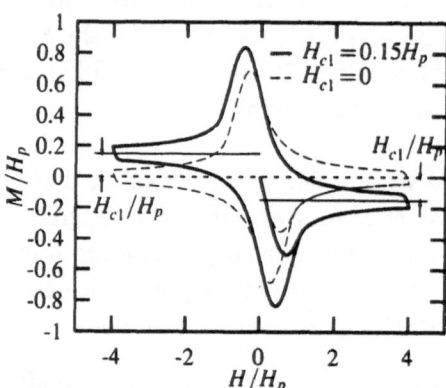

Fig. 1. Theoretical $M(H)$ curves, scaled by H_p, for $H_{dc} = 0$ and $H_{ac} = 4H_p$. Solid loop is for $\Delta H = 0$, $H_{c1} = 0.15H_p$ and dashed loop is $\Delta H = 0$, $H_{c1} = 0$.

EXPERIMENTAL

Five samples (A-E) were prepared. The composition and T_c(onset) of each sample are listed in Table 1. These samples were grown by the self-flux method. The axes of these crystals were determined by the X-ray diffraction measurement. A high crystallinity was confirmed by Raman scattering and electron probe microanalysis, and no impurity phase was detected [1, 2].

Table 1. Composition and T_c(onset) of the present samples.

Sample	Composition	doping state	T_c(onset)(K)
A	$YBa_2Cu_3O_{7-d}$		89.0
B	$YBa_2Cu_{2.90}Co_{0.10}O_{7-d}$	underdoped	71.8
C	$YBa_2Cu_{2.80}Co_{0.20}O_{7-d}$	underdoped	62.0
D	$Bi_2Sr_{1.8}La_{0.20}CuO_{6+d}$	overdoped	10.9
E	$Bi_2Sr_{1.6}La_{0.40}CuO_{6+d}$	underdoped	9.74

For each sample, the doping state in the CuO_2 plane was carefully confirmed from the Raman spectroscopy [1, 2] as listed in Table 1.

The magnetization curves at various temperatures below T_c were measured using a SQUID magnetometer, where the applied field was parallel to the c-axis of the specimen. In Fig.2, we show the typical initial and hysteresis curves of sample D at 5.0K, where the downward shift of the loop appears. As discussed before, the observed downward shift H'_{c1} from the base line $\mu_0 M = 0$, certainly corresponds to the lower critical field, where $\mu_0 = 4\pi \times 10^{-7}$ H/m. Since the samples are not infinite cylinder treated in the calculation, observed H'_{c1} should be corrected for the effect of demagnetization, i.e., H'_{c1} should be corrected using the demagnetization coefficient N which was estimated from the sample geometry. Using $H_{c1} = H'_{c1}/(1 - N)$, we can obtain the intrinsic lower critical field H_{c1}.

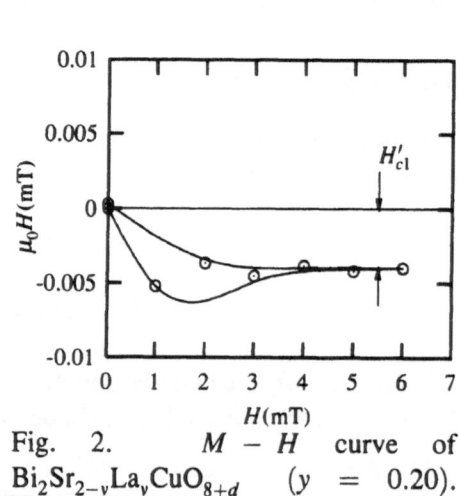

Fig. 2. $M - H$ curve of $Bi_2Sr_{2-y}La_yCuO_{8+d}$ ($y = 0.20$). $T = 5.0$K.

RESULTS AND DISCUSSION

Figure 3 shows the temperature dependence of the lower critical fields of samples A-C together with our previous data [5]. By a least-squares fitting of $\mu_0 H_{c1}$, we obtained $\mu_0 H_{c1}(0)$ for each sample. In Fig.4, we show $\mu_0 H_{c1}(0)$ as a function of the hole density in the CuO_2 plane for samples A-C. $H_{c1}(0)$ appeared to be proportional to p, where the hole density p of each sample was estimated from the bond valence sum [1, 2]. The results show a similar proportional relation to our previous data [5]. For samples D and E, we cannot refer to the proportional relation between p and $H_{c1}(0)$, because we have only two data points, but the results show an enhancement with increasing of the hole density.

In the Ginzburg-Landau (GL) theory, H_{c1} is related to the London penetration depth λ_L as $H_{c1} \propto \lambda_L^{-2}$, while λ_L is given as

$$\lambda_L^2 = \frac{m^*}{n_s e^2}, \qquad (3)$$

where m^* and n_s are respectively the effective mass and the number density of the superconducting pair.

From the GL theory treated under the assumption that the condensed pairs are in the superfluid state with uniform velocity, we can derive the relation $n_s = n$ at $T = 0$ K, where n is the number density of the electrons (or holes). From these relations, we obtain

$$H_{c1}(0) \propto n_s. \qquad (4)$$

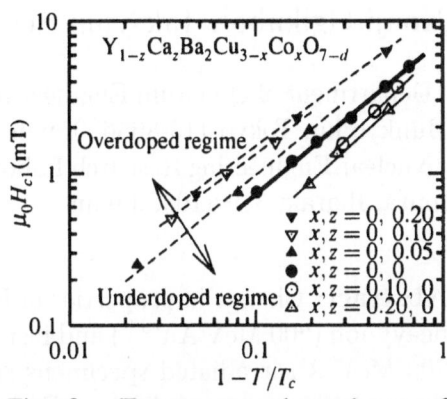

Fig. 3. Temperature dependence of lower critical fields of samples A-C. Upper three break lines are previous data from Ref. 5.

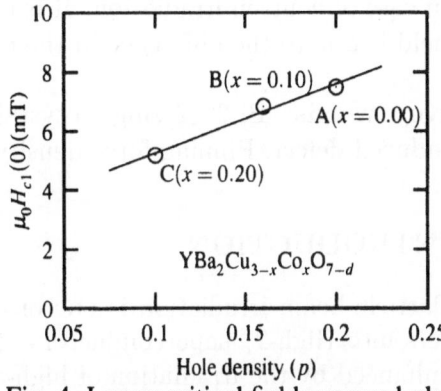

Fig. 4. Lower critical fields versus hole concentrations in the CuO_2 plane of samples A-C.

Comparing our results to those from μSR measurements [6], the relation of n_s and λ in the underdoped region has the same tendency. However, μSR experiments on $Tl_2Ba_2CuO_{6+d}$ systems which are strongly overdoped by changing the value of d show that $\lambda^{-2}(0) \propto n_s/m^*$ decreases to zero as T_c suppresses in overdoped samples.

Measurements of the magnetization curves for the cuprate based single crystals have revealed that H_{c1} is closely correlated to the carrier concentration. $H_{c1}(0)$ has been found to be in proportion to n_s for underdoped and overdoped samples. Measurements of H_{c1} for strongly overdoped samples are desired.

[1] P. Berastegui et al., J. Solid State Chem. **127**, 56 (1996).

[2] M.Kakihana et al., Phys.Rev.B **47**, 5359 (1993).

[3] S.Tochihara, H.Yasuoka, and H.Mazaki, Physica C **295**, 101 (1998).

[4] I. W. Dunn and P. H. Brit, J.Appl.Phys. **1**, 1469 (1968).

[5] S. Tochihara, et al., in *Advances in Superconductivity XI*, edited by N. Koshizuka ans S. Tajima (Springer-Verlag, Tokyo, 1999),pp. 267-270.

[6] Y.J.Uemura et al., Nature **364**, 605 (1993).

Change in Pinning Properties of $Bi_2Sr_2CaCu_2O_{8+x}$ Single Crystals due to Particle-Beam Irradiation followed by Thermal Annealing

Kouji Ogikubo[1], Takayuki Terai[1], Kenji Yamaguchi[2], and Michio Yamawaki[1]

[1]Department of Quantum Engineering and Systems Science, The University of Tokyo, 7-3-1 Hongo, Bunkyo-ku, Tokyo 113-8656, Japan
[2]Nuclear Engineering Research Laboratory, The University of Tokyo, 2-22 Shirakatashirane, Tokai-mura, Ibaraki 319-1106, Japan

Abstract: Change in properties of $Bi_2Sr_2CaCu_2O_{8+x}$(Bi-2212) single crystals due to high-energy heavy-ion (200 MeV Au^{13+}) and fast neutron irradiation followed by thermal annealing is reported. 200 MeV Au irradiated specimens show higher F_p values than neutron irradiated specimens. In the case of neutron irradiation, F_p decreased by the thermal annealing treatment at 673 K. On the other hand, F_p did not change even by thermal annealing up to 12 h in the case of 200 MeV Au irradiation. The relationships between pinning force density $F_p/F_{p,\max}$ and the reduced magnetic field $b = B/B_{irr}$ were compared among heavy-ion and neutron irradiated and thermally annealed specimens. F_p took the maximum value at $b \approx 0.3$ in case of heavy-ion irradiation and at $b \approx 0.1$ in case of neutron irradiation. It is considered that such a different dependence of F_p on magnetic field is due to the difference in the defect type produced by each irradiation.

Key words: Bi-2212 single crystal, Particle-beam irradiation, Thermal annealing, Radiation-induced defect, Pinning force density

INTRODUCTION

Particle-beam irradiation is one of the most effective methods to introduce strong pinning centers into High-T_c superconductors. It has been reported that the critical current density (J_c) is enhanced by the irradiation of high-energy particles such as electrons, neutrons and ions[1-4]. In particular, high-energy heavy-ion irradiation and fast neutron irradiation are hopeful methods. High-energy heavy-ions are expected to produce columnar defects near the target surface, which gives a very large pinning force for the vortices parallel to the defects. Fast neutron irradiation, on the other hand, can introduce defect clusters in all the area of the specimen isotropically. The size and the number of radiation-induced defects as pinning centers can be changed by thermal annealing treatment after irradiation. This treatment is quite effective for the improvement of the superconducting properties in some cases[4], because superconducting properties are strongly affected by the defect structure in materials. In this study, we irradiated Bi-2212 single crystal specimens with high-energy heavy-ions (200 MeV Au^{13+}) or fast neutrons (> 1 MeV) and annealed them at high temperature to understand the effects of particle-beam irradiation and thermal annealing on the pinning properties.

EXPERIMENTAL

Bi-2212 single crystal specimens used were prepared by the floating-zone method. Their size was 1 mm × 1 mm × 30-50 μm(c-axis). All of the specimens were annealed at 1073 K for 72 h in air and quenched to room temperature in advance. In high-energy heavy-ion irradiation, the specimens were irradiated with 200 MeV Au^{13+} ions parallel to the c-axis of the specimen using a tandem accelerator at JAERI (Japan Atomic Energy Research Institute). The fluence was 5×10^{10} cm^{-2}, at which the enhancement of J_c due to irradiation is maximized[3]. In fast neutron irradiation, the specimens sealed in a quartz capsule of 10^{-4} Pa were irradiated with fast neutrons (> 1 MeV) using JMTR reactor at JAERI. The fluence was 6×10^{17} cm^{-2}, at which the enhancement of J_c is maximized[4]. The irradiation temperature was room temperature (< 310 K) for all case of irradiation. After irradiation, 200 MeV Au^{13+} irradiated specimens and neutron irradiated specimens were annealed in air at 673 K for 1 h to 12 h and 1 h to 24 h, respectively. Magnetization hysteresis curves of the specimens after irradiation and annealing were measured with a vibrating sample magnetometer (VSM) at 20 K and 40 K as a function of applied magnetic field (B) parallel to the c-axis. The macroscopic pinning force density (F_p) was obtained from $J_c \times B$, where J_c is the critical current density in a-b plane of the specimen calculated from the magnetization curves using the modified Bean's model[5]. In the case of heavy-ion irradiation, the incident ion did not penetrate through the specimen. Therefore, in order to evaluate the effect of columnar defects on the property improvement, the value of F_p is normalized with the penetration depth of the incident ion calculated by TRIM code. The irreversibility field (B_{irr}) was obtained from the magnetization curves by determining the magnetic field at which the hysteresis width was lower than the experimental error.

RESULTS AND DISCUSSION

The macroscopic pinning force density F_p obtained from the magnetization measurement at 40 K is shown in Fig.1. 200 MeV Au irradiated specimens show higher F_p values than neutron irradiated specimens. The annealing time dependence of F_p at 40 K, 0.03 T is shown in Fig.2, where neutron irradiation, F_p decreased by the thermal annealing treatment at 673 K. This means that the defect clusters in several nm size were reduced in size and the pinning force was decreased by thermal annealing. On the other hand, F_p did not change even by thermal annealing up to 12 h in the case of 200 MeV Au irradiation. This result indicates that the size of columnar defects produced by 200 MeV Au irradiation did not change by thermal annealing at 673 K.

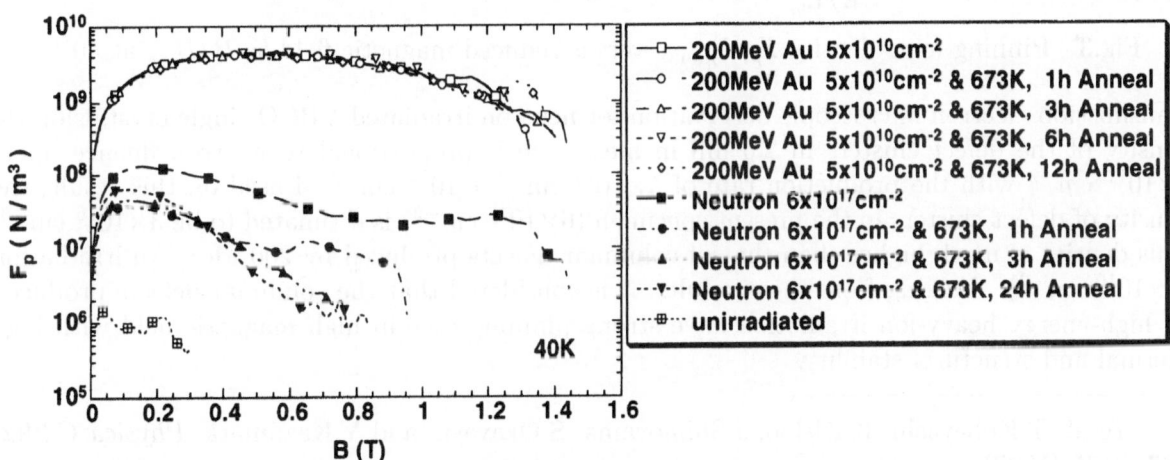

Fig.1 Macroscopic pinning force density F_p versus magnetic field at 40 K

The relationships between the pinning force density ($F_p/F_{p,\max}$) and the reduced magnetic field ($b = B/B_{irr}$) for 200 MeV Au and neutron irradiated specimens are shown in Fig.3. For 200 MeV Au irradiated specimens and annealed specimens, F_p took the maximum value at $b \approx 0.3$. For fast neutron irradiated specimens and annealed specimens, on the other hand, F_p took the maximum value at $b \approx 0.1$. This means that the columnar defects produced by 200 MeV Au irradiation can trap vortices up to higher magnetic field than the defect clusters produced by neutron irradiation. This different dependence of F_p on magnetic field is considered to be due to the difference in the pinning properties between each columnar defect and each defect cluster. According to the

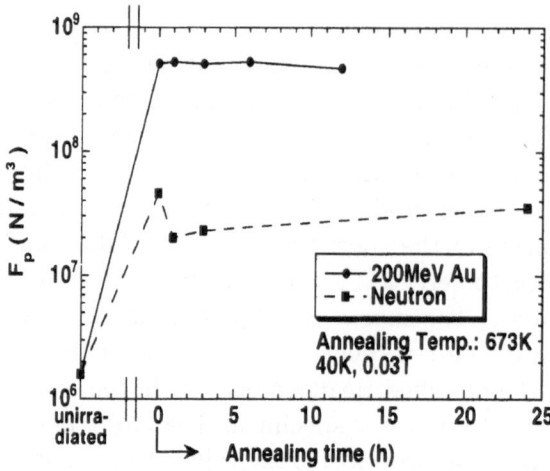

Fig.2　Annealing time dependence of F_p at 40 K, 0.03 T

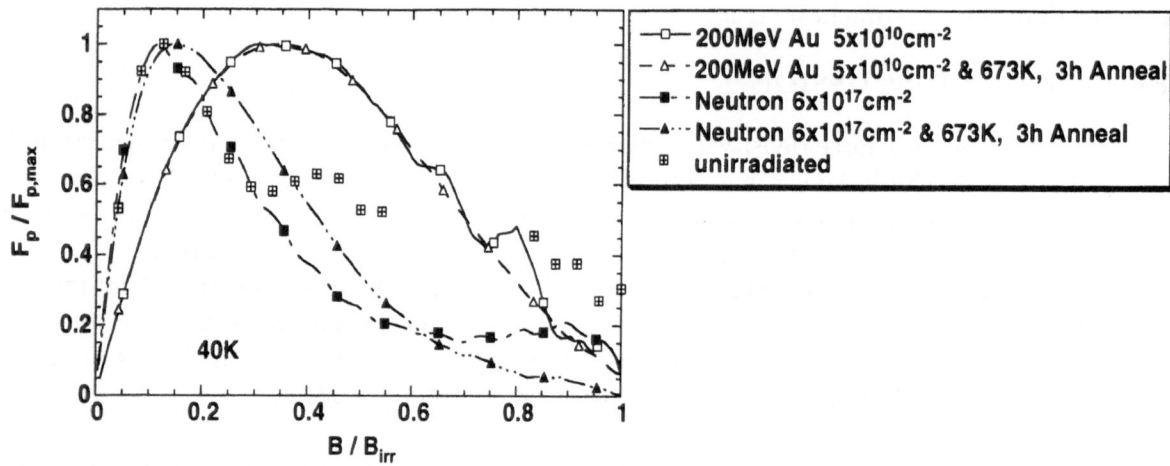

Fig.3　Pinning force density $F_p/F_{p,\max}$ versus reduced magnetic field $b=B/B_{irr}$ at 40 K

transmission electron microscopic observation for neutron irradiated YBCO single crystals[6], the density of the defect clusters in 2.5 nm in mean size is proportional to neutron fluence up to 8×10^{17} cm^{-2} with the production rate of 5×10^{15} cm^{-3} / 10^{17} cm^{-2}. Based on this result, the density of defect clusters in the present specimen (6×10^{17} cm^{-2}) is estimated to be 3×10^{16} cm^{-3}. This density is much higher than that of columnar defects produced by 200 MeV Au irradiation (5×10^{10} cm^{-2}). Judging from these results, it is considered that the columnar defects produced by high-energy heavy-ion irradiation have strong pinning force in high magnetic field with high thermal and structural stability.

1. T.Terai, T.Kobayashi, K.Kishio, J.Shimoyama, S.Okayasu, and Y.Kazumata, *Physica C* **282-287**, 2135 (1997).

2. T.Terai, T.Kobayashi, Y.Ito, K.Kishio, and J.Shimoyama, *Physica C* **282-287**, 2285 (1997).

3. K.Ogikubo, T.Kobayashi, T.Terai, S.Tanaka, K.Kishio, and J.Shimoyama, in *Advances in Superconductivity X*, edited by K.Osamura and I.Hirabayashi (Springer-Verlag, Tokyo,1998), p.481.

4. K.Ogikubo, T.Kobayashi, T.Terai, S.Tanaka, and K.Kishio, in *Advances in Superconductivity XI*, edited by N.Koshizuka and S.Tajima (Springer-Verlag, Tokyo, 1999), p.541.

5. E.M.Gyorgy, R.B.van Dover, K.A.Jackson, L.F.Schneemeyer, and J.V.Waszczak, *Appl. Phys. Lett.* **55**, 283 (1989).

6. M.C.Frischherz, M.A.Kirk, J.Farmer, L.R.Greenwood, and H.W.Weber, *Physica C* **232**, 309 (1994).

Analysis of Thermal Stability in a Small HTS Tape Coil by Using Finite Element Method

Kentaro Ohya,[1] Kensuke Ogata,[1] Takanobu Kiss,[1] Teruo Matsushita,[1,2] and Masakatu Takeo[1]

[1]Graduate School of ISEE, Kyushu University, Fukuoka, 812-8581, Japan
[2]Faculty of CSSE, Kyushu Institute of Technology, 820-8502, Japan

Abstract: Because of the large heat capacity, the influence of local disturbances on the stability is quite small in high Tc superconductors (HTSC). In addition, smooth voltage-current characteristics make resistive state in HTSC very stable. However, a design of HTSC coils using the critical current density conventionally defined by usual criterions is difficult, because these characteristics change nonlinearly with temperature and/or magnetic field. Therefore, it is necessary to estimate the thermal stability of HTSC coil based on the transport property and the quench property. In this study, we analyze the thermal stability in a Bi-2223 superconducting solenoid coil by using the finite element method with taking account of the nonlinear resistance.

Keywords: HTSC coil, Thermal stability, Quench propagation velocity

INTRODUCTION

Bi-based high T_c superconductor is one of the most promising materials because of the steady progress in a fabrication technology for long size tapes and the possibility of high magnetic field magnets operating at relatively high temperatures. However, it is difficult to distinguish clear boundary between superconducting and resistive states in the HTSC coil, so the critical current density in HTSC determined by the conventional electric filed criterion, 1μ V/cm, or resistivity criterion, 10Ω m, is no longer a critical material parameter. For example, the resistive state is very stable even though the applied current is considerably larger than the critical current because of smooth voltage-current characteristics. In this paper, we measured the heat propagation in a small HTSC coil, and analyzed the thermal stability in the coil by using a finite element method.

EXPERIMENT

The test coil was prepared using a Bi-2212 superconducting tape. The tape of cross section 3.4 × 0.24 mm² was produced by a powder-in-tube method and 66 superconducting filaments were imbedded in the silver matrix. The Ag : Bi ratio was 71 : 29. The tape as wound in a single layer on a coil former of diameter 55 mm and height 51 mm. The coil former was made of copper and covered by a thin insulating Teflon layer. A sketch of the coil is shown in Fig.1.The coil was instrumented by potential taps, resistive thermometers to measure the tape temperature and heaters to initiate the quench, as shown in Fig. 1. Two kinds of experiments were performed under cooling by flowing helium gas. One was the measurements of temperature and voltage at each part of the coil under a constant transport current only, and the other was that of temperature when the heater power was also input.

ANALYSIS

The temperature variation is calculated using the three-dimensional heat balance equation:

$$C\frac{\partial T}{\partial t}=\nabla\cdot[k\nabla T]+\dot{Q}+\dot{Q}_J-G(T-T_0),$$

(1)

where C is the heat capacity per unit volume, k is the thermal conductivity, Q is the heating power density by the heater, Q_J is the Joule heating power density, G is the thermal conductance per unit volume to the surroundings, T_0 is the bias temperature and t is time. We assumed that only the silver matrix contribute to the calculation thermal conduction in the tape. In the above, Q_J is given by

$$\dot{Q}_J = E \times J$$

(2)

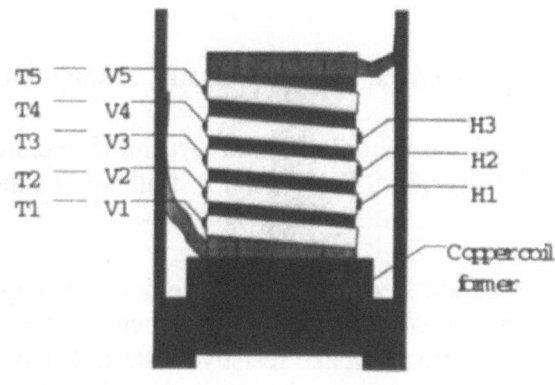

Fig. 1. Winding of measured coil and arrangements of potential tape, thermometers and heater.

using the electric field, E. Because of widely distributed flux pinning strength, the transport $E(J)$ characteristics are strongly nonlinear and can be described in the analytical form as:

$$E(J)=\rho_{FF}\int_0^J\left(\frac{J'-J_{cm}(B,T)}{J_0(B,T)}\right)^m dJ',$$

(3)

where ρ_{FF} is a uniform flux flow resistively, J_{cm} is the minimum value of J_c, J_0 is the width of J_c distribution, and m is a numerical parameter which determines the shape of J_c distribution. These parameters for the E-J curve were determined by the DC measurement in the low electric field region.

G can not be determined only by materials but is strongly influenced by the arrangement of the coil. So, it is necessary to experimentally estimate the G-value. For this purpose, we simply assume that the temperature is uniform over the thickness of the tape and the small temperature gradient along the length is neglected. As a result, (1) is reduced to

$$C\frac{\partial T}{\partial t}=\dot{Q}_J-G(T-T_0)$$

(4)

in the absence of the heating power. The Q_J-T relationship of the coil was estimated by (2). Using the relationship and (4), the time evolution of temperature was estimated under an assumption of the G-value. The G-value determined so that the estimated temperature trace agrees with the experimental one. In the case of bias temperature 40 K, it was 4042 W/Km^3 [1].

RESULT AND DISCUSSION

Fig. 2(a) shows observed variations of temperature at T2, T3, T5 at various values of transport current at $T_0 = 40$ K in the absence of the heater input. The corresponding results of numerical analysis are shown in Fig. 2(b). It is seen these results agreed very well, suggesting that (1) describes correctly the thermal behavior of the present coil. Then, this equation was used for the analysis of the thermal quench characteristics with the transport current and the heating power. Fig.3 shows observed temperature variations at each point at $T_0 = 40$ K under a transport current 80 A and a heating power 1.4 W at T$_3$. The temperature at the heated section immediately rose to 64 K after starting the heating, but then went down to 58 K because of good heat conduction to the copper coil former. The temperatures at other sections were much lower than that at the heated section and slowly went up together. This behavior is different from that of LTSC coil in which the temperature of only the heated point rises catastrophically. The critical heating power to reach 77 K

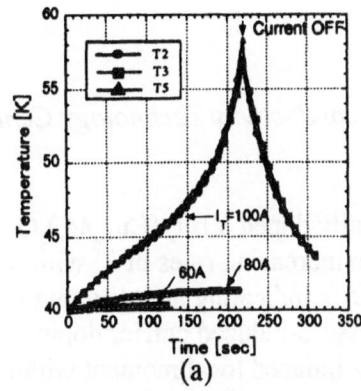
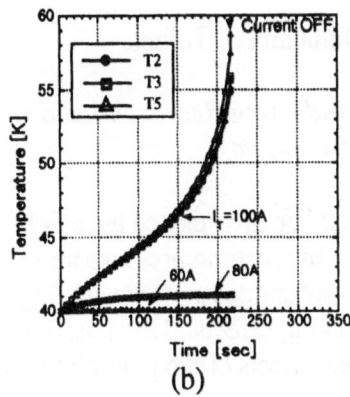

Fig. 2. (a) Measured and (b) calculated temperature traces in the absence of heater input at $T_0 = 40$ K. The temperature at each part is approximately the same.

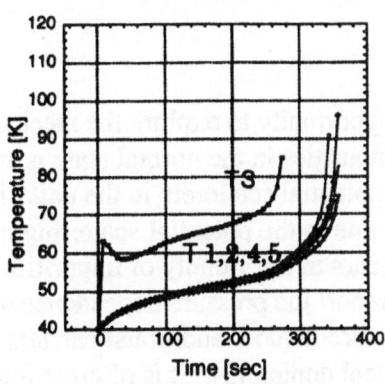
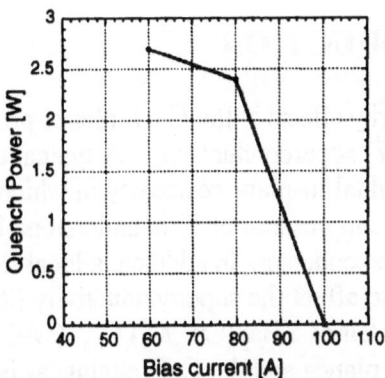

Fig. 3. Calculated temperature traces with heating at T_3, at $T_0 = 40$ K.

Fig. 4. Stability margin of bias temperature, $T_0 = 40$ K

at the heated section was determined. The obtained result is shown in Fig. 4. This critical heating power to the quench gives a measure for the stability margin. The quench propagation velocity to reach 77 K was estimated as a function of the heater power. It was of the order of 0.05-0.3 mm/sec and was much smaller than that of LTSC. This warns strongly that HTSC coils might be locally damaged at a heated region when a large disturbance takes place. Hence, the estimation of the exact stability margin is very useful in designing HTSC coils.

CONCLUSIONS

We analyzed the dynamics of quench in HTSC coil using the finite element method with taking account of the non linear resistive characteristics. The result clarified that the quench propagation velocity of HTSC is much smaller than that of LTSC, and warns that HTSC coils might be locally damaged due to a large disturbance.

ACKNOWLEDGMENTS

This study was partly funded by NEDO, Japan, under the proposal-Based New Industry Creative Technology R&D Promotion Program.

REFERENCES

[1] T. Kiss, M. Inoue, K. Hasegawa, K. Ogata, V.S. Vysotsky, Yu. Ilyin, M. Takeo, H. Okamoto and F. Irie, *IEEE Trans. Supercond.*, vol. **9**, pp. 1073-1076, 1999.

Suppression of Pair Breaking under High Pressure in YBa$_2$Cu$_3$O$_{7-\delta}$

K. Yoshida and S. Tajima

Superconducting Research Laboratory, International Superconductivity Technology Center, Tokyo 135-0062, Japan

Abstract: We investigated the pair breaking effects in the optimally doped YBa$_2$(Cu$_{1-x}$M$_x$)$_3$O$_{7-\delta}$ (M=Zn, Ni) under hydrostatic pressure up to 2 GPa. We found that the increasing rates of T_c with pressure for impurity-substituted systems are larger than that for YBa$_2$Cu$_3$O$_{7-\delta}$, indicating an intriguing reduction of pair-breaking effects. One possible origin is based on the pressure-induced carrier doping which may change the effects of the potential scattering and/or an impurity-induced local moment within the CuO$_2$ planes.

Key words: pressure effect, pair-breaking effect, YBa$_2$(Cu$_{1-x}$M$_x$)$_3$O$_{7-\delta}$ (M=Zn, Ni)

INTRODUCTION

The impurity effects in the CuO$_2$ planes provide a challenging opportunity to explore the mechanism of the high-T_c superconductivity. A typical change induced by impurities in the normal state is seen in a large residual in-plane resistivity in which impurities work as potential scatterers in the unitarity limit [1]. The suppression of T_c is understood by the pair breaking due to the potential scattering in the d-wave superconductor. In addition, a local magnetic moment appears in the vicinity of impurities, which should also affect the superconductivity [2]. In this work, we report the pressure dependence of T_c for single crystals of YBa$_2$(Cu$_{1-x}$M$_x$)$_3$O$_{7-\delta}$ (M=Zn, Ni). Applied pressure induces additional carriers in both the CuO$_2$ planes and the CuO chains, as is similar to the chemical doping [3]. It is of great interest to investigate the effects of the modification of the electronic state by pressure on the pair breaking.

EXPERIMENTAL

We grew high-quality large single crystals of YBa$_2$Cu$_3$O$_{7-\delta}$ (Y123) with substitution of 0.35% Zn or 1.0% Ni for Cu using a crystal pulling technique. The composition of the crystals were analyzed by the inductively coupled plasma (ICP) analysis. Crystals were annealed in a flowing O$_2$ atmosphere for two weeks at 480 °C to obtain the optimally-doped samples. T_c determined by the resistivity drop below the detection limit is 92.9 K (pure-Y123), 84.0 K (0.35% Zn-substituted Y123, Y123:Zn) and 87.1 K (1.0% Ni-substituted Y123, Y123:Ni) at ambient pressure. The out-of-plane resistivity ρ_c was measured using a four probe method under hydrostatic pressure P up to 2 GPa. Pressure was generated by using a piston-cylinder device with a pressure transmitting medium of an equal mixture of Fluorinert FC70 and FC77. Resistivity measurements at each pressure were carried out during heating run, after the applied pressure was set at room temperature. The constant applied pressure was actively maintained by controlling the load during the temperature excursion.

RESULTS and DISCUSSIONS

Figure 1 shows the temperature dependences of ρ_c below 300 K at several pressures for Y123 and Y123:Zn. $\rho_c(T)$ decreases with P in both cases. The decreasing rate for Y123:Zn is somewhat smaller than that for Y123, as shown in the insets of Fig.1. As was indicated in our previous study [3], one characteristic pressure effect is an enhancement of the carrier doping level which is estimated as ~+0.01 hole/GPa. The semiconductor-like upturn of ρ_c at low temperatures is a sensitive probe for the doping

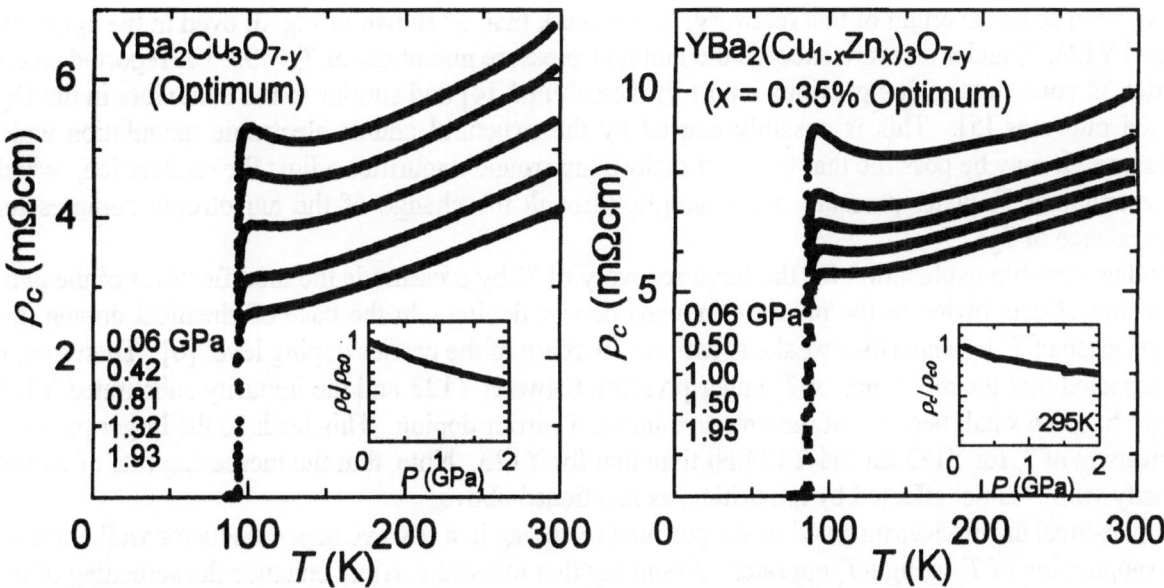

Fig.1 The T dependence of ρ_c at several pressures for pure-Y123 and 0.35% Zn-substituted Y123. The inset shows the pressure dependence of ρ_c at 295 K.

level. The reduction of the upturn with P in Y123:Zn is qualitatively the same with the behavior in Y123, as shown in Fig. 1. This suggests that the increasing rate of the carrier doping level with P is independent of the impurity substitution, which is consistent with the fact that impurities do not appreciably change the carrier density. It should be noted here that the above pressure effects are qualitatively shared with Y123:Ni.

Figure 2 shows the superconducting transition in ρ_c under high pressure for Y123:Zn. T_c is plotted in Fig. 3, together with the data of Y123 and Y123:Ni. T_c increases with P in the three materials. The rates for the impurity-substituted systems, $dT_c/dP = +1.5$ K/GPa for Y123:Zn and $+1.1$ K/GPa for Y123:Ni, are larger than $+0.8$ K/GPa for pure Y123, indicating an intriguing recovery of T_c by pressure.

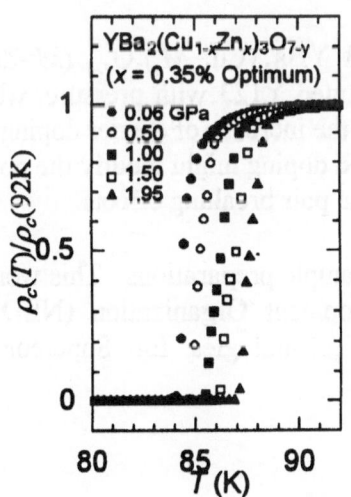

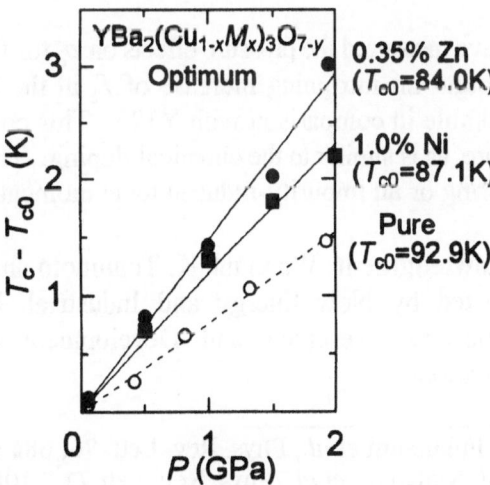

Fig.2 The temperature dependence of ρ_c for 0.35% Zn-substituted Y123 below 92 K at several pressures.

Fig.3 The variation of T_c with P for the optimally doped YBa$_2$(Cu$_{1-x}$M$_x$)$_3$O$_{7-\delta}$ (M=Zn, Ni).

Now, we discuss the origin of this recovery. We remark that, as shown in Fig. 3, even in the optimally doped Y123, T_c under pressure exceeds the ambient-pressure maximum of $T_c \sim 93.5$ K reported so far, which is consistent with a previous report by Tozer *et al.* [4] and similar to the behaviors in the Hg-based cuprates [5]. This is possibly caused by the structural and/or electronic modulation under pressure. It may be possible that the local distortions around impurities adjust the modulation, which may enhance T_c under pressure, for example, through the change of the anisotropic compression dependence of T_c.

Another possible explanation for the large recovery of T_c by pressure is the modifications of the pair-breaking effects owing to the pressure-induced carrier doping. In the case of chemical doping, the suppression of T_c by impurities weakens with the increase of the carrier doping level [6]. Therefore, it is expected that the difference of T_c under pressure between Y123 and the impurity substituted Y123 might become small because of the pressure-induced carrier doping. This leads to the larger pressure derivatives of T_c for Y123:Zn and Y123:Ni than that for Y123. Note that the increasing rate of carrier density with P is not affected by impurities, as mentioned above.

The pair-breaking mechanism based on the potential scattering in a d-wave superconductor well explains the suppression of T_c in high-T_c cuprates. Assuming that induced carriers enhance the screening of the local distorted potential around the impurities, it possibly causes the recovery of superconductivity. However, in the high-T_c cuprates, it is apparently difficult to screen because of the reduced density of states at the Fermi energy due to the opening of the psudogap. It should be noted here that, even in this case, the screening can occur due to quasiparticles in the node lines [7]. In another way, the carrier doping promotes the three dimensionality of the electronic state. The change of the anisotropy modifies the criterion of the unitarity limit, which affects the pair breaking. In order to clarify this potential scattering mechanism, it is important to measure the in-plane resistivity under high pressure.

The decrease of a local moment with the increase of carriers observed in the magnetic measurements for the chemical doping [2] is also a possible candidate for the suppression of the pair breaking. A model of the appearance of the local moment originates from a destruction of spinon singlet by impurities. The screening of the localized spin by the induced carriers possibly reduces the local moment. In addition, the weakening of the antiferromagnetic correlation must contribute to the reduction of the local moment.

CONCLUSION

We have measured the pressure effects on ρ_c for the optimally doped $YBa_2(Cu_{1-x}M_x)_3O_{7-\delta}$ (M=Zn, Ni). We found an intriguing increase of T_c in the Zn- and Ni-substituted Y123 with pressure, which is remarkable in comparison with Y123. This could originate from the increase of carrier doping under pressure, as is similar to the chemical doping. This pressure-induced doping might modify the potential scattering or an impurity-induced local moment, contributing to the pair breaking effects.

Acknowledgment. We thank K. Tomimoto and A.I. Rykov for sample preparations. This work was supported by New Energy and Industrial Technology Development Organization (NEDO) as Collaborative Research and Development of Fundamental Technologies for Superconductor Applications.

1. Y. Fukuzumi *et al.*, Phys. Rev. Lett. **76**, 684 (1996).
2. A.V. Mahajan. *et al.*, Phys. Rev. Lett **72**, 3100 (1994).
3. K. Yoshida *et al.*, in *Advances in Superconductivity XI*, edited by N. Koshizuka and S. Tajima (Springer-Verlag, Tokyo, 1999) p57.; K. Yoshida *et al.*, Phys. Rev. B, in press.
4. S.W. Tozer *et al.*, Phys. Rev. B **47**, 8089 (1993).
5. L. Gao *et al.*, Phys. Rev. B **50**, 4260 (1994).
6. H. Alloul *et al.*, Phys Rev. Lett. **67**, 3140 (1991).
7. G. Khaliullin *et al.*,, Phys. Rev. B **56**, 11882 (1997).

Resonant Raman Scattering in Twin-Free YBa$_2$(Cu$_{1-x}$Zn$_x$)$_3$O$_{7-\delta}$ Single Crystals

Mikhail Limonov,[1]* Setsuko Tajima,[1] and Akio Yamanaka[2]

[1] Superconductivity Research Laboratory, ISTEC, 10-13, Shinonome 1-Chome, Koto-ku, Tokyo 135-0062, Japan
[2] Chitose Institute of Science and Technology, Chitose, Hokkaido 066-8655, Japan

Abstract: The Resonant Raman scattering spectra of twin-free YBa$_2$(Cu$_{1-x}$Zn$_x$)$_3$O$_{7-\delta}$ single crystals with optimal oxygen content have been investigated in a temperature range from 300 to 10K in the xx- and yy-polarizations. To extract the Zn-doping dependence of the gap, we analyzed the temperature and spectral changes in the electronic response function $\rho(\omega)$. The fitting results suggest a broadening and low-frequency shift of the 2Δ peak in the Zn-doped crystals. This picture for the electronic response explains the phonon renormalization effects in the Raman spectra below T$_c$.

Keywords: Resonant Raman scattering, Superconducting gap, Phonons, Electrons

Raman scattering has been widely used as a tool to investigate electronic band structures of different materials. In this paper, precise temperature and polarization dependencies of Raman scattering spectra (RSS) have been investigated for optimally doped YBa$_2$(Cu$_{1-x}$Zn$_x$)$_3$O$_{7-\delta}$ single crystals. These crystals were grown by a top-seeded pulling technique and oxygenated under uniaxial pressure in order to obtain the orthorhombic twin-free samples [1]. The pure crystals have optimal T$_c$ = 93K. The Zn-doped sample (x = 0.4%) with optimal oxygen content has T$_c$ = 85K. RSS were studied using a T64000 Jobin-Ivon triple spectrometer with a liquid-nitrogen cooled CCD detector. The typical spectral resolution was 3 cm^{-1}. The RSS were obtained in the xx- and yy-polarizations using several lines from an Ar$^+$-Kr$^+$ laser ranging from 1.9 eV to 2.5 eV.

Figure 1 presents yy-polarized RSS of the YBa$_2$(Cu$_{1-x}$Zn$_x$)$_3$O$_{7-\delta}$ crystals at T = 10K. The A$_g$ Raman modes of the YBa$_2$Cu$_3$O$_7$ lattice are dominating in the RSS at λ_{ext} = 501 nm, while at λ_{ext} = 568 nm the most intense are the "forbidden" modes, which appear in the Raman spectra of YBa$_2$Cu$_3$O$_{7-x}$ crystals owing to the oxygen defects in the CuO chains [2]. The resonance profiles of several modes at T=10K are shown in Fig.2.

At T < T$_c$, remarkable superconductivity induced changes take place in the RSS. To extract the phononic and electronic parameters correctly, we have performed a treatment of the RSS using formula on a basis of the Green's function approach [3,4]:

$$I(\omega) = AB(\omega)\,\mathrm{g}(\omega)\left\{1 + \frac{\mathrm{g}(\omega)}{1+\mathrm{g}(\omega)}\left(\frac{\left[S\frac{1+\mathrm{g}(\omega)}{\mathrm{g}(\omega)}+(\omega-\Omega)\right]^2}{\Gamma_0^2\left(1+\mathrm{g}(\omega)\right)^2+(\omega-\Omega)^2}-1\right)\right\} \tag{1}$$

where $B(\omega)$ is the Bose factor, $\Gamma = \Gamma_0 + V^2\rho(\omega) = \Gamma_0(1+g(\omega))$ and $\Omega = \Omega_0 + V^2R(\omega)$ are renormalized linewidth and frequency, V is the electron-phonon coupling constant, $\rho(\omega)$ and $R(\omega)$ are the imaginary and real parts of the electronic response $\chi(\omega) = -R(\omega) + i\rho(\omega)$. In our fitting we have used the following formula for the $\rho(\omega)$ - function:

$$\rho(\omega) = C_0 + (C_1\,\omega + C_2\,\omega^2)/[\,\Gamma_e^{\,2} + (\omega - D)^2]. \tag{2}$$

To extract the Zn-doping dependence of the superconducting gap, we analyzed the temperature dependence of the electronic response function $\rho(\omega)$ in terms of the parameters D and Γ_e. We found a pronounced difference in these parameters for the pure and Zn-doped crystals as shown in Fig.3. This difference suggests a broadening and low-frequency shift of the 2Δ peak in the Zn-doped crystal in comparison with the pure crystal. This allows us to explain phonon renormalization effects in RSS of the $YBa_2(Cu_{1-x}Zn_x)_3O_{7-\delta}$ crystals below T_c.

Among the all phonon lines, the superconductivity induced effect on the 340 cm^{-1} line is most pronounced. The temperature dependence of the renormalized linewidth Γ is plotted in Fig.4. Below T_c the 340 cm^{-1} line exhibits broadening. For the pure sample, the linewidth Γ exhibits a bump with a maximum at T ~ 75K. For the Zn-doped sample, the linewidth Γ strongly increases just below T_c and at lower T it becomes less sensitive to temperature. The difference in Γ at T=10K can be explained by difference in $\rho(\omega)$ in pure and Zn-doped compounds. Namely, in the case of the Zn-doped crystal the maximum of the $\rho(\omega)$ function is closer to the energy of the 340 cm^{-1} line (Fig.3). This leads to a larger broadening $\Gamma = \Gamma_0 + V^2\rho(\omega)$ at low temperatures in comparison with the case of the pure crystal.

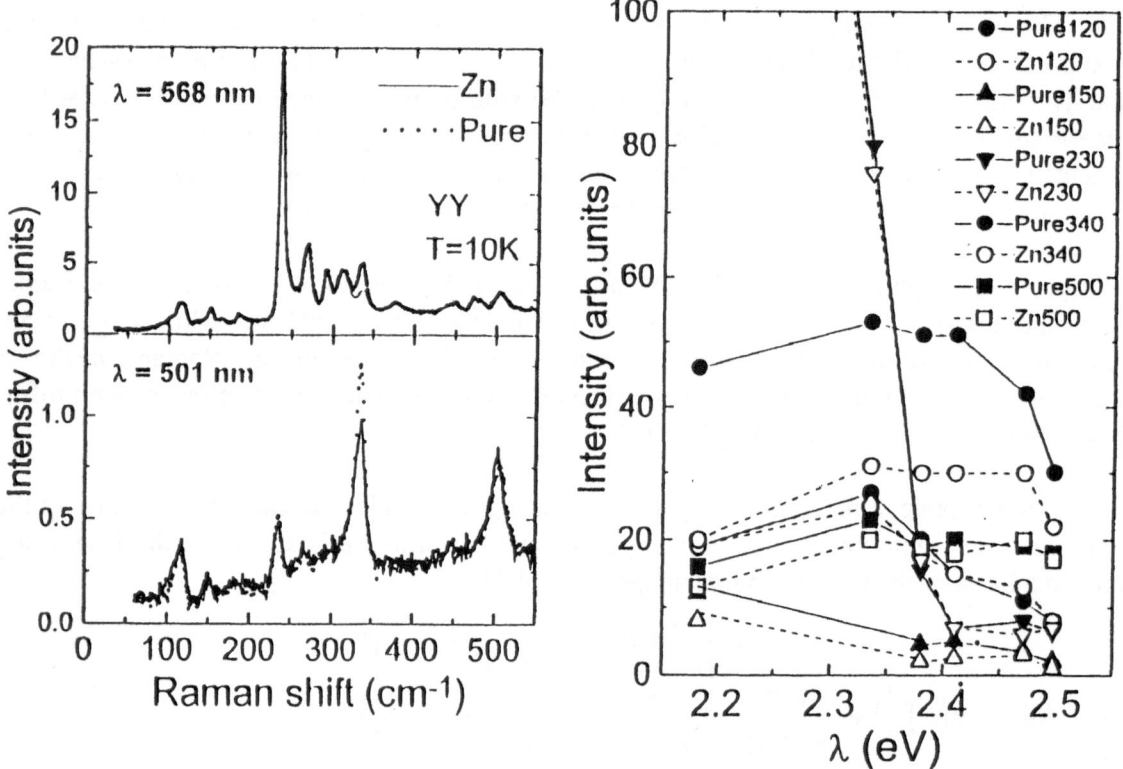

Fig.1 The RSS of the $YBa_2(Cu_{1-x}Zn_x)_3O_{7-\delta}$ crystals in yy-polarization at 10K. λ_{ext} = 568 and 501 nm.

Fig.2 The resonance profiles of the selected lines in the RSS of the pure and Zn-doped $YBa_2(Cu_{1-x}Zn_x)_3O_{7-\delta}$ crystals at T=10K.

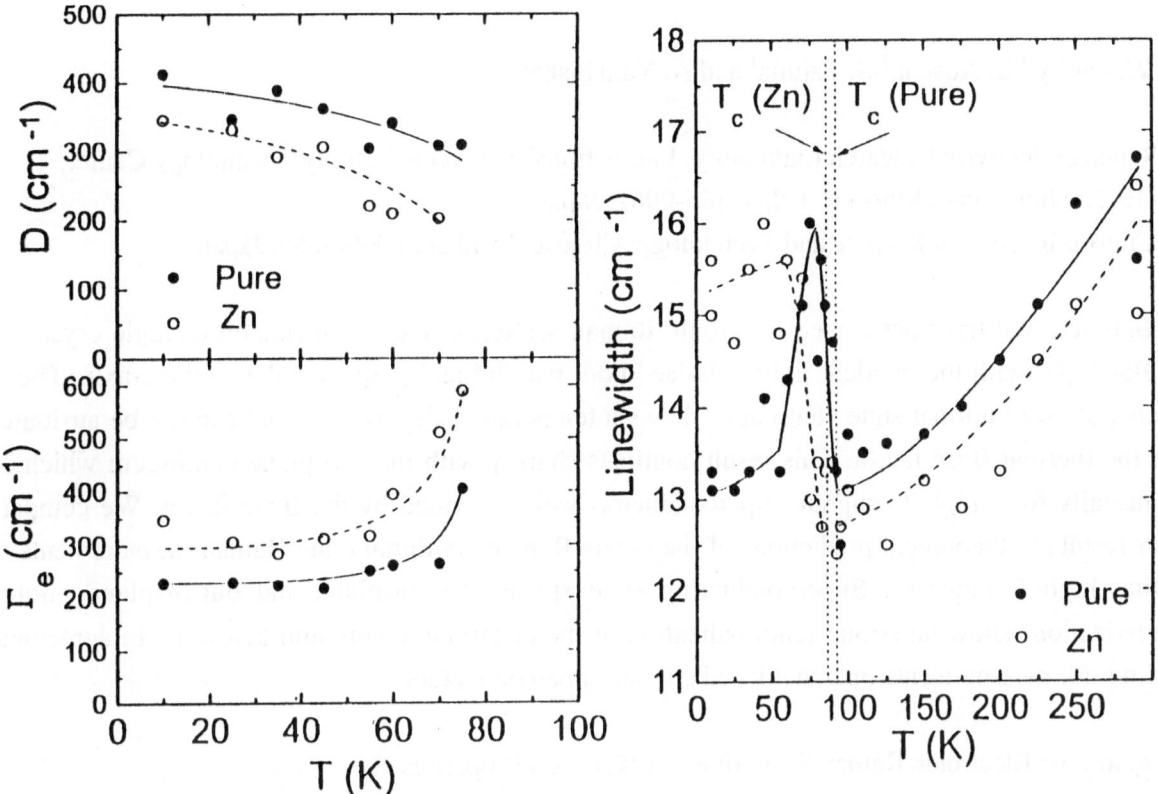

Fig.3 Temperature dependences of D and Γ_e fitting parameters of the $\rho(\omega)$ function.

Fig.4 Temperature dependences of the renormalized linewidths $2\Gamma = 2(\Gamma_0 + V^2\rho)$ for the 340 cm^{-1} line in the RSS of the pure and Zn-doped crystals.

Another interesting Zn-effect on the superconductivity induced features is a remarkable decrease in the intensity of the 340 cm^{-1} line (see Fig.1). Figure 2 shows that this effect is observed at all excitation laser lines. Our analysis of the Raman intensity based on formula (1) shows that this difference originates from the change in linewidth $\Gamma = \Gamma_0(1+g(\omega))$ of this line by Zn-doping.

Acknowledgments. The authors are thankful to A.I.Rykov for preparation of the samples.
This work is supported by New Energy and Industrial Technology Development Organization (NEDO) as Collaborative Research and Development of Fundamental Technologies for Superconductivity Applications.

---

[*] Permanent address: A.F.Ioffe Physical-Technical Institute, 194021 St.Petersburg, Russia.

1. A.I.Rykov, W.J.Jang, H.Unoki, S.Tajima, in *Advances in Superconductivity VIII*, Edited by H. Hayakawa and Y. Enomoto, (Springer-Verlag, Tokyo, 1996) pp.341-344.

2. A.G.Panfilov, M.F.Limonov, A.I.Rykov, S.Tajima, and A.Yamanaka, Phys. Rev. B **57**, R5634-5637 (1998).

3. X.K.Chen, E.Altendorf, J.C.Irwin, R.Liang, and W.N.Hardy, Phys. Rev. B **48**, 10530-10540 (1993).

4. A.G.Panfilov, S.Tajima, and A.Yamanaka, in *Advances in Superconductivity XI*, Edited by N.Koshizuka and S.Tajima, (Springer-Verlag, Tokyo, 1999) pp.77-80.

Electronic Raman Scattering in YBa$_2$Cu$_4$O$_8$

J.W. Quilty,[1] S. Adachi,[1] S. Tajima[1] and A. Yamanaka[2]

[1] Superconductivity Research Laboratory, International Superconductivity Technology Center,
1-10-13 Shinonome, Koto-ku, Tokyo 135-0062, Japan
[2] Chitose Institute of Science and Technology, Chitose, Hokkaido 066-8655, Japan

Abstract: We have performed electronic Raman scattering measurements on a single crystal of YBa$_2$Cu$_4$O$_8$ with the incident light polarised both parallel and perpendicular to the c-axis. The c-axis polarized normal-state continuum shows a temperature dependence that can not be attributed to the thermal Bose factor. This result contrasts sharply with the a-b plane continuum which is, unusually for a high-T_c cuprate superconductor, well described by the Bose factor. We compare our results to theoretical predictions of the c-axis Raman continuum and Raman measurements of other high-T_c cuprates. Superconducting state spectra for in-plane and out-of-plane photon polarizations show no strong renormalization of the electronic continuum below T_c, in agreement with prior measurements on underdoped cuprate superconductors.

Keywords: Electronic Raman Scattering, Y124, c-axis Properties

INTRODUCTION

The a-b plane electronic Raman continuum of YBa$_2$Cu$_4$O$_8$ (Y124) is unusual because it exhibits a strong temperature dependence in the normal state that is well described by the thermal Bose factor while most other high-T_c cuprate superconductors exhibit a temperature independent normal state continuum [1]. Recent years have seen an increasing interest in the use of Raman scattering to probe the electronic states of high-T_c superconductors and the peculiar c-axis properties seen via resistivity, optical and IR conductivity, and penetration depth measurements. Theoretical calculations of the normal-state c-axis Raman response in YBCO superconductors, based on the plane-chain coupling model, have recently been performed [2] which predict the appearance of features within the electronic Raman continuum directly related to the pseudogap. The c-axis Raman spectrum of Bi$_2$Sr$_2$CaCu$_2$O$_{8+\delta}$ (Bi2212) has also recently been measured [3] and a peak observed in the A_{1g} symmetry electronic continuum below T_c identified with the superconducting gap. In the context of these recent results, we were motivated to perform c-axis polarization-resolved Raman measurements on Y124 single crystals to determine whether the effects of the pseudogap and superconducting gap can be observed in the electronic continuum. The results of c-axis continuum measurements are also of interest because of the unusual in-plane Raman continuum exhibited by Y124, which includes a Bose-like temperature dependence in the normal-state and only a very weak superconductivity-induced renormalization below T_c [1].

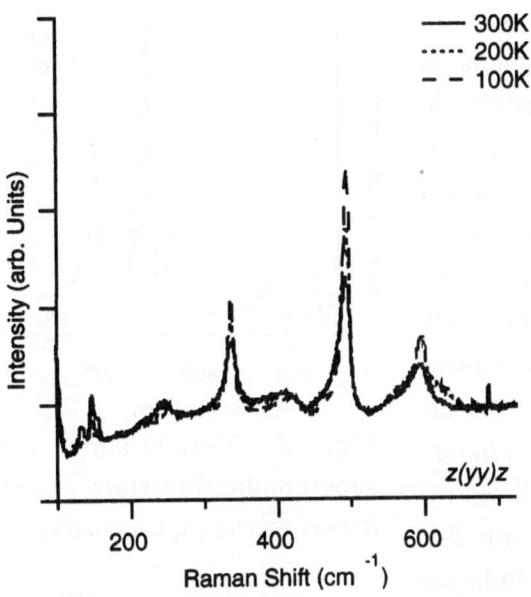

Fig. 1. Normal-state *a-b* plane Raman spectra (A_g symmetry).

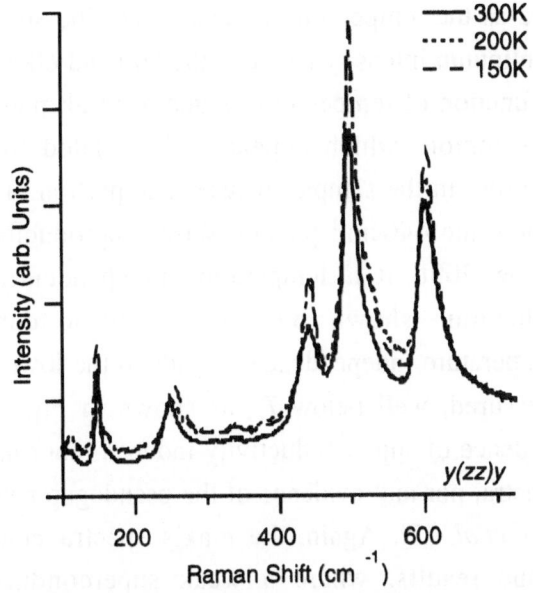

Fig. 2. Normal-state *c*-axis Raman spectra (A_g symmetry).

EXPERIMENTAL

Single crystals of $YBa_2Cu_4O_8$ (Y124) were grown in a Y_2O_3 crucible using the self-flux method with high-pressure Ar-O_2 gas mixture described elsewhere [4]. All spectra presented here were taken from a crystal of dimensions 0.4×0.3×0.1 mm^3, with a T_c (onset) of 80K determined via magnetic susceptibility. The crystal was mounted on the cold finger of a high-vacuum closed-cycle cryostat and Raman spectra measured with the 514.5nm line of an argon-krypton-ion laser, at an incident laser power of 5mW. The laser was focussed to a spot of approximate area 0.05mm^2 yielding an incident power density of around 10W/cm^2. Measurements of the Stokes:Anti-Stokes phonon intensity ratio indicated no significant sample heating and we estimate local heating of less than 10K for our cryostat configuration. Scattered light was collected in the pseudo-backscattering geometry, with a Jobin-Yvon T64000 triple monochromator and a charge coupled device used to measure the Raman spectrum.

RESULTS

Figure 1 shows in-plane normal state Raman spectra for temperatures of 300, 200, and 100K and Fig. 2 shows *c*-axis polarized spectra at 300, 200, and 150K. The thermal Bose contribution has been removed from the raw data by division, and the spectra have been normalized to unity at high frequency. The spectra in Fig. 1 show the previously observed temperature independence of the *a-b* plane electronic continuum [1] but Fig. 2 reveals a temperature dependent *c*-axis continuum in the normal state. The behaviour seen in Fig. 2 can be related to the *c*-axis continuum calculations of Wu *et al.* [2] where the continuum intensity in the range of wavenumbers shown here is seen to rise as the temperature is decreased. The strong variation in continuum intensity between the 500 and 600cm⁻¹ phonons as a function of temperature is due to weak phonon features in this region, which appear to be related to impurities or disorder in the sample surface. The portion of the continuum below the 440cm⁻¹ phonon shows reproducible behaviour – below 200K it is temperature independent, similar to the behaviour shown by the *a-b* plane continuum. This temperature independence extends to the lowest temperatures measured, well below T_c, as shown in Fig. 3. We find no evidence of superconductivity-induced renormalization in our spectra, nor any evidence of the pseudogap peak predicted by Wu *et al.* [2]. Again, the *c*-axis spectra contrast with *a-b* plane results, where a weak superconductivity-induced

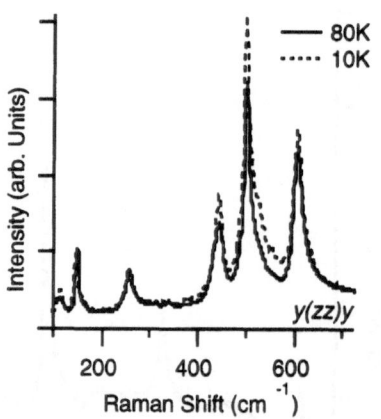

Fig. 3. Normal-state and superconducting-state *c*-axis Raman spectra (A_g symmetry).

renormalization is observed [1]. The absence of a superconductivity induced renormalization in our *c*-axis measurements may well be due to the weakness of such a renormalization compared to that seen in the *a-b* plane, as recently shown in *c*-axis Raman measurements on Bi2212 [3].

CONCLUSIONS

Our results show a temperature-dependent normal state continuum in the *c*-axis Raman spectra of a Y124 single crystal, which becomes temperature-independent below around 200K. We find no firm evidence of a temperature dependent redistribution of the *c*-axis electronic Raman continuum that could be attributed to the opening of a pseudogap or superconducting gap in our crystal. Contrasting behaviour between in-plane and out-of-plane electronic continua in the high-T_c cuprates is not surprising – they potentially arise from different physical processes [2] – but the temperature-dependent to temperature-independent crossover in *c*-axis continuum behaviour is not well explained by existing calculations of the *c*-axis electronic Raman response.

Acknowledgments. Thanks are due to Dr. H. Mori for assistance with the SQUID magnetometer. This work is supported by the New Energy and Industrial Technology Development Organization (NEDO) as collaborative research and development of fundamental technologies for superconductivity applications.

[1] S. Donovan, J. Kircher, J. Karpinski, E. Kaldis, and M. Cardona, J. Supercond. **8**, 417-420 (1995)

[2] W.C. Wu and J.P. Carbotte, Phys. Rev. B **56**, 6327-6334 (1997)

[3] H.L. Liu, G. Blumberg, M.V. Klein, P. Guptasarma, and D.G. Hinks, Phys. Rev. Lett. **82**, 3524-3527 (1999)

[4] S. Adachi, K. Nakanishi, K. Tanabe, K. Nozawa, H. Takagi, W.-Z. Hu, and M. Izumi, Physica C **301**, 123-128 (1998)

The Strong Electron Correlation Effect in Quasi-two dimensional organic superconductors

I. Kanazawa

Department of Physics, Tokyo Gakugei University,Koganei-shi, Tokyo 184-8501,Japan

Abstract:We have discussed the relationship between the superconducting critical temperature T_c and the effective mass of holes in the quasi-two dimensional organic superconductors by means of GIC model and the Eliashberg equation, and present one of possible models, which might explain this relationship.

Keywords: Organic superconductors, Electron correlation, Quasi-two dimension

INTRODUCTION

Organic system, the BEDT-TTF($\kappa - ET_2X$ superconducting salts), have layered structures with quasi-two dimensional electronic bands like cuprates. Systematic NMR, electrodynamic, thermodynamic and transport measurements under pressure have provided an opportunity to obtain unified understanding of the mechanisms of the M-I transition in $\kappa - (ET)_2X$ family with a parameter of X and superconductivity enhanced near the M-I transition [1]. Especially the presence of antiferromagnetic phases around the superconducting phase in the BEDT-TTF salts strongly suggests the importance of the electron correlation effect. The investigation for the quasi-two dimensional organic superconductors is very important in relation to the high-T_c cuprates, which can be approximatery a quasi-two dimensional superconductor. Caulfield et.al. [2] have performed high pressure magnetotransport measurements on the organic superconductor $\kappa - (BEDT - TTF)_2Cu(NCS)_2$ and found out that the enhancement of the effective mass of the carrier and the superconducting behavior are directly connected. In this study, we will discuss the relationship between the effective mass and the critical temperature T_c by the gauge-invariant carrieron model(GIC) [3-7], which is based on the gauge-invariant effective Lagrangian density in the quasi-two dimensional strongly-correlated electron system, and the Eliashberg equation, and will propose one of possible models, which might explain this relationship.

A MODEL SYSTEM

In the quasi-two dimensional organic salts, the carrier in the α-orbit in the metallic phase near the antiferromagnetic insulator phase is regarded as a kind of quasi-particle "carrieron", which is composed of the hole and the cloud of the massive gauge fields B_μ^1, B_μ^2, and the massless gauge field B_μ^3 around the hole [3-7]. In other words, we can think that a "carrieron" is a complex particle composed of the hedgehog-like (monopole-like) soliton and the hole trapped into the mobile soliton. Taking into account that the symmetry in the undoped (2+1) dimensional quantum antiferromagnet is invariant under local SU(2) [8], it is assumed that the perturbing gauge fields B_μ introduced by the hole has a local SU(2) symmetry. Then it is suspected that SU(2) gauge fields B_μ^a are spontaneously broken through the Anderson-Higgs mechanism in a way similar to the antiferromagnetic symmetry breaking around the hole. We set the symmetry breaking $\langle 0|\phi_a|0\rangle = (0,0,\mu)$ of the Bose field ϕ_a in the Lagrangian density as follows.

$$
\begin{aligned}
L \;=\; & \frac{1}{2}(\partial_i N_c^j - g_1\epsilon_{abc}\epsilon_{ijk}B_i^b N_a^k)^2 \\
& + \psi^\dagger(i\partial_0 - g_2 T_a B_0^a)\psi \\
& - \frac{1}{2m}\psi^\dagger(i\nabla - g_2 T_a B_{\mu\neq 0}^a)\psi \\
& - \frac{1}{4}(\partial_\nu B_\mu^a - \partial_\mu B_\nu^a + g_3\epsilon_{abc}B_\mu^b B_\nu^c)^2 \\
& + \frac{1}{2}(\partial_\mu\phi_a - g_4\epsilon_{abc}B_\mu^b\phi_c)^2 \\
& - \lambda^2(\phi_a\phi_a - \mu^2)^2
\end{aligned}
\tag{1}
$$

After the symmetry breaking, we can obtain the effective Lagrangian density.

$$
\begin{aligned}
\mathcal{L}_{eff} \;=\; & \frac{1}{2}(\partial_i N_c^j - g_1\epsilon_{abc}\epsilon_{jik}B_i^b N_a^k)^2 \\
& + \psi^\dagger(i\partial_0 - g_2 T_a B_0^a)\psi \\
& - \frac{1}{2m}\psi^\dagger(i\nabla - g_2 T_a B_{(\mu\neq 0)}^a)^2\psi \\
& - \frac{1}{4}(\partial_\nu B_\mu^a - \partial_\mu B_\nu^a + g_3\epsilon_{abc}B_\mu^b B_\nu^c)^2 \\
& + \frac{1}{2}(\partial_\mu\phi_a - g_4\epsilon_{abc}B_\mu^b\phi_c)^2 \\
& + \frac{1}{2}m_1^2[(B_\mu^1)^2 + (B_\mu^2)^2] + m_1[B_\mu^1\partial_\mu\phi_2 - B_\mu^2\partial_\mu\phi_1] \\
& + g_4 m_1\{\phi_3[(B_\mu^1)^2 + (B_\mu^2)^2] - B_\mu^3[\phi_1 B_\mu^1 + \phi_2 B_\mu^2]\} \\
& - \frac{m_2^2}{2}(\phi_3)^2 - \frac{m_2^2 g_4}{2m_1}\phi_3(\phi_a)^2 - \frac{m_2^2 g_4^2}{8m_1^2}(\phi_a\phi_a)^2
\end{aligned}
\tag{2}
$$

Where N_c^j is the spin parameter, ψ is Fermi field of the hole, $m_1 = \mu g_4$, $m_2 = 2\sqrt{2}\lambda\mu$. The effective Lagrangian describes two massive vector fields B_μ^1 and B_μ^2, and one massless U(1) gauge field B_μ^3. Because these masses are formed through the Higgs mechanism by introducing the hole, the field B_μ^1 and B_μ^2 exist around the hole within the length of $\sim 1/m_1 \equiv R_c$. From the first term in eq.(2), the spin N_c^j is much distorted from the antiferromagnet state within the length of $\sim R_c$ around the hole. Furthermore, the spin order parameter will be distorted in the long-range by the massless U(1) gauge field B_μ^3 [6]. When N(i),N(j), and N(k) are spins on triangle sites i,j,and k within $\pi R_c^2(\tilde{i}\,)$ around the hole at the site $\tilde{i}$, the spin liquid parameter $q_{\tilde{i}}$ is introduced as follows, $q_{\tilde{i}} \equiv \displaystyle\sum_{(ijk)\in\pi R_c^2(\tilde{i})} N(i)\cdot(N(j)\times N(k))$. (ijk) are local triplet sites of spins. Because the hole trapped into the hedgehog-like mobile soliton is thought as that the instanton-like fluctuation [9] is stabilized by the hole [10], we can assume that $|q_{\tilde{i}}|$ is approximately proportional to the topological number of the instanton. The quantized gauge fields B_μ^a are expressed in $B_\mu^a = (2\pi)^{-3/2}\sum_a \int [a_i^a(p)e_\mu^i(p)exp(ipr) + a_i^{a\dagger}(p)e_\mu^i(p)exp(-ipr)]d^3p/\sqrt{2\omega_p^a}$. Where $\omega_p^a = \sqrt{p^2 + m_1^2}$, (a =1,2) and $\omega_p^a = \sqrt{p^2}$,(a = 3), $a_i^{a\dagger}(p)$ and $a_i^a(p)$ are the creation and annihilation operators of the gauge particle B_μ^a with momentum p, respectively, $e_\mu^i(p)$ are the polarization vectors. The effective interaction H_{int} between two holes can be introduced approximately through the summation of B_μ operators as follows, $H_{int} \sim g_2^2 \sum_p \sum_{k,k'} \sum_a \frac{\omega_p^a}{(\epsilon_k - \epsilon_{k-p}) - \omega_p^{a2}} c_{k'+p}^\dagger c_{k'} c_{k-p}^\dagger c_k$,

where $c_k^\dagger$ and c_k are the creation and annihilation operators of the hole with momentum k, respectively. In the case of $|\epsilon_{k\pm p} - \epsilon_k| < \omega_p^a$,the effective interaction between two holes becomes the attractive one. Because of $k + k' \sim 0$, ω_p^a can approximate to $\sim \omega_{2k_F} = \sqrt{(2k_F)^2 + m_1^2}$. By using the Eliashberg equation, the gap function is introduced approximately as follows,

$$\begin{aligned}
\Delta(\omega(k,k')) = & \frac{\lambda}{1+\lambda}\sqrt{(2k_F)^2 + m_1^2}\int_0^\infty dz Re \frac{\Delta(z)}{\sqrt{z^2 - \Delta^2(z)}} \cdot \\
& \frac{1}{2}\{[tanh \quad z/2 + f(z) + n(\omega_{2k_F})] \cdot \\
& \left(\frac{1}{\omega + z + \omega_{2k_F} - i\delta} - \frac{1}{\omega - z - \omega_{2k_F} + i\delta}\right) \cdot \\
& -[f(z) + n(\omega_{2k_F})] \cdot \\
& \left(\frac{1}{\omega - z + \omega_{2k_F} - i\delta} - \frac{1}{\omega + z - \omega_{2k_F} + i\delta}\right)\}. \qquad (3)
\end{aligned}$$

Where λ is $\sim \frac{\partial \sum}{\partial \omega}|_{\omega=0}$, and $\sum$ is the self energy of the hole. $f(z) = (exp(\beta z) + 1)^{-1}$ and $n(\omega_{2k_F}) = (exp(\beta\omega_{2k_F}) - 1)^{-1}$. The effective mass of the hole is thought of as composing of the effective mass introduced in perturbation approximation and the effective mass m_1 of the gauge field B_μ around the hole. From eq.(3), we can get the relationship that the critical temperature T_c becomes higher as the effective mass m_1 increases. This is consistent with the experimental result [2].

CONCLUSION

We have discussed the quasi-two dimensional organic superconductors by GIC model, which is based on the gauge-invariant effective Lagrangian density in the strongly-correlated electron system, and by using the Eliashberg equation. We have proposed one of possible models, which might explain the relationship between the critical temperature T_c and the effective mass of the holes.

1. K.Kanoda,K.Miyagawa,A.Kawamoto, and Y.Nakazawa, Phys.Rev.B54,76(1996).
2. J.Caulfield,W.Lubczynski,F.L.Pratt,J.Singleton,D.Y.K.Ko,W.Hayers,M.Kurmoo, and P.Day,J.Phys.C6,2911(1994).
3. I.Kanazawa, in The Physics and Chemistry of Oxide Superconductors, (Springer-Verlag,1992)p481.
4. I.Kanazawa, Physica C185-189,1703(1991).
5. I.Kanazawa, Synth.Met.71,1641(1995).
6. I.Kanazawa, Superlattice and Microst.21,279(1997).
7. I.Kanazawa, Physica B, in press.
8. I.Affleck,Z.Zou,T.Hsu, and P.W.Anderson, Phys.Rev.B38,754(1988).
9. F.D.M.Haldane, Phys.Rev.Lett.,61,1029(1988).
10. P.Wiegmann, Prog.Theor.Phys.Suppl.107,243(1992).

TUNNELING STUDIES OF SUPERCONDUCTING MECHANISM IN BKBO

A. Suzuki, M. Suzuki, Y. Aso, X. G. Zheng, S. Tanaka

Faculty of Science and Engineering, Saga University, Saga, 840-8502, Japan

Abstract : The tunneling measurement was performed for the planar MIS junction fabricated on $Ba_{0.6}K_{0.4}BiO_3$ single crystal at various temperature. Two clear peaks due to superconducting state were observed in tunnel conductance characteristic $G(V)$ at 4.2 K. From $G(V)$ characteristic, order-parameter was determined as $\Delta_0 = 4.6$ meV, this resulted in the coupling strength factor $2\Delta_0 / k_BT_c = 3.7$. The excitation with an energy of 0.4 eV, which was predicted from the analysis of the infrared reflectivity, was not observed in the $G(V)$ characteristic.

Key words : BKBO, tunneling spectroscopy, excitation

INTRODUCTION

The crystal structure of $Ba_{1-x}K_xBiO_3$ (BKBO) is a perovskite structure and does not contain Cu. This indicates that the superconducting mechanism in BKBO is different from the one in high T_c cuprate with an electron-spin interaction via a spin of Cu. In spite of relative low carrier concentration, the superconducting transition temperature T_c of BKBO is near 30 K for $x = 0.4$ which is the upper limit value of BCS theory. The high T_c of BKBO has been believed to be explained by the strong coupling BCS theory due to the breathing mode. However, the analysis of the far-infrared data based on simple BCS model requires $\lambda \approx 0.7$ for $T < T_c$ and $\lambda \leq 0.2$ for $T > T_c$, where λ is a measure of coupling strength ($\lambda = 2 \int_0^\infty \alpha^2 F(\omega)/\omega d\omega$). To remove the discrepancy in λ, Kaufmann et al. [1] proposed to add an high energy excitation mode with an energy of 0.4 eV to the electron-phonon coupling. The introduction of high energy excitation mode enables BKBO to have $T_c = 30$ K in a weak coupling model. Kaufmann et al. [1] asserts that the excitation can arise from a charge fluctuation through a structural fluctuation, because the phase of $T_c = 30$ K is just near the metal-insulator transition and is unstable. In present experiment, we tried to find the excitation by the tunneling spectroscopy.

EXPERIMENTAL RESULT AND DISCUSSION

The single crystal of BKBO was grown by the electro-chemical method. The tunnel junction was fabricated on the cleaved surface of BKBO by successively depositing SiO as a tunnel barrier and Ag as a top electrode. The tunnel conductance $G(V)$ characteristic was measured by the standard modulation method.

As seen in Fig. 1, The $G(V)$ curve has a clear gap centered at zero-bias which is responsible to the superconducting state, and has several structures in the bias region from 20 mV (and -20 mV) to 100 mV (and -100 mV) which arise from phonons. A clear phonon structure does not usually appear in $G(V)$, because of degradation in the surface of the single crystal. The positive-bias corresponds to a positive polarity of BKBO relative to a metal electrode. Therefore, the $G(V)$ characteristic suggests that the electronic density of states is high in the conduction band in comparison with the valence band. From the analysis of *G(V)* using the Dynes' equation in BCS theory, we get $\Delta_0 = 4.6$ meV as an order parameter. Using $T_c = 29$ K determined by the temperature dependence of the resistivity, the coupling strength factor results in $2\Delta_0/k_B T_c = 3.7$. This means that the electron-phonon coupling is between a weak coupling and a strong coupling, and is consistent with the result of analysis of the imaginary part of the optical conductivity [2].

Next, the excitation mode with an energy of 0.4 eV is discussed. There are two processes through which the excitation appears in the *G(V)* ; 1) the one through the electronic density of states modulated by the electron-phonon interaction under superconducting state, and 2) another through the indirect tunneling with a conservation of energy and momentum

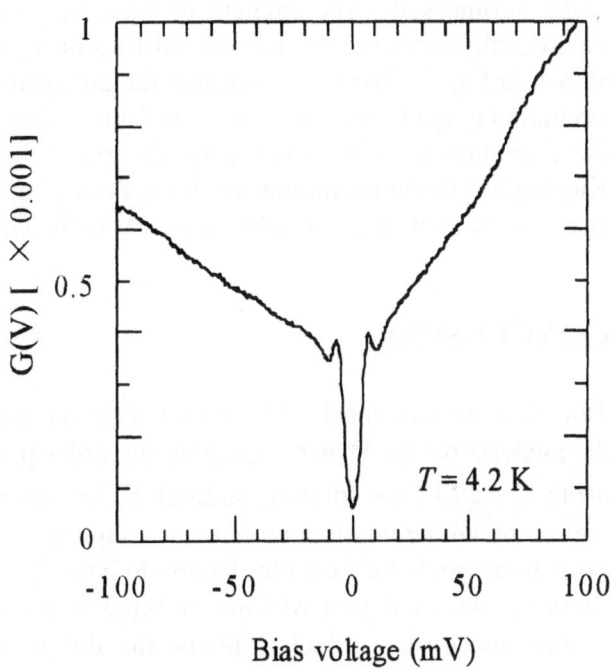

Fig. 1.

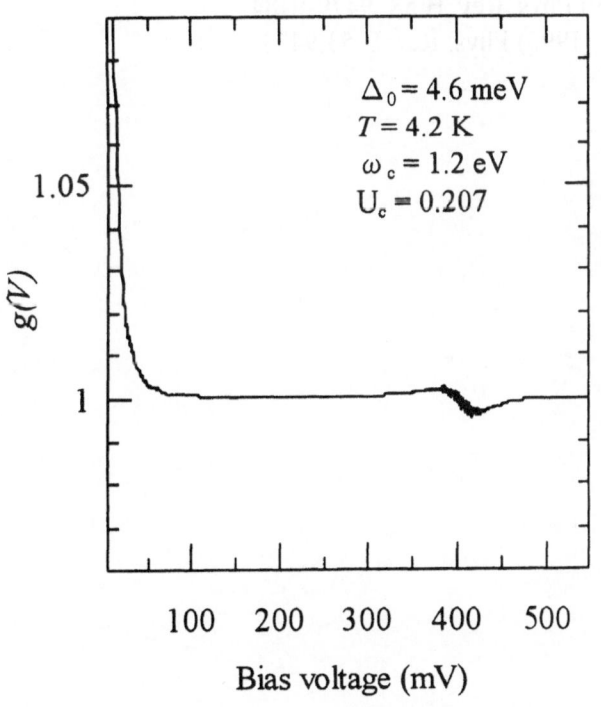

Fig. 2. The strength of excitation structure based on Migdal-Eliashberg equation

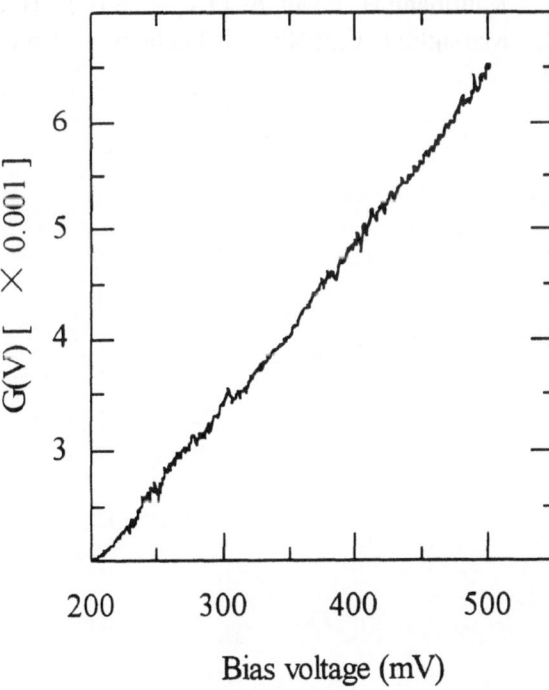

Fig. 3.

under normal state. The strength of excitation structure in the $G(V)$ by the former process can be theoretically estimated by the Migdal-Eliashberg equation. The theoretical structure of excitation is shown in Fig. 2. Taking account into the enlarged scale in $g(V)$, the estimation demonstrates that the excitation in the former process is too faint to detect by tunneling spectroscopy. But we expected that the excitation is easily detected by the tunneling spectroscopy, because in the model proposed by Kaufmann [1] the excitation has large density of states at 0.4 eV. However, as seen in Fig. 3 any structure can not be observed near 0.4 eV in the $G(V)$ characteristic.

CONCLUSION

Tunneling measurement was performed for the planar tunnel junction fabricated on BKBO. From the analysis by the Dynes' equation, the order-parameter was estimated as $\Delta_0 = 4.6$ meV. Further using $T_c = 29$ K, the coupling strength factor was determined as $2\Delta_0/k_B T_c = 3.7$, whose value is in a regime of moderate electron-phonon coupling. Although as pointed out by Kaufmann et al. the excitation with 0.4 eV enables BKBO to have $T_c = 30$ K in a regime of weak coupling and moderate coupling, the excitation was not detected in the $G(V)$ characteristic. As Kaufmann [1] suggested, another mechanism which explains the difference in the values of λ above and below T_c is ascribed to the strong momentum dependence of electron-phonon interaction. This is achieved by carrying out a vertex correction.

1. Kaufmann H. J, Dolgov O. V, Salje E. K. H, (1999) Phys. Rev. B 58: 9479-9484
2. Marsiglio F, Carbotte J. P, Puchkov A, Timusk T, (1996) Phys. Rev. B 53:9433

PSEUDO-GAP STUDIES IN $Bi_2Sr_2Ca_{1-x}Na_xCu_2O_8$ BY TUNNELING SPECTROSCOPY

M. Suzuki, A. Suzuki, Y. Aso, X. G. Zheng, S. Tanaka ,M. Kuno

Department of Physics, Faculty of Science and Engineering, Saga University, Saga, 840-8502, Japan

Abstract: The tunneling conductance $G(V)$ at various temperatures was measured for the planar junction fabricated on $Bi_2Sr_2Ca_{1-x}Na_xCu_2O_8$ (BSCCO). The optimal doped and the over-doped BSCCO were prepared by doping oxygen and / or by substituting Na for Ca. The pseudo-gap was observed in BSCCO ranging from the optimal doped region to the over doped one for both doping methods. The gap structure due to superconducting state was clear at 4.2 K, weakened with increasing temperature, disappear near T_c, and was followed by pseudo-gap above T_c. The pseudo-gap was observable up to 180 K whose thermal energy is near the spin gap energy.
Key words: NIS planar junction, tunneling spectroscopy, pseudo-gap

INTRODUCTION

The superconducting mechanism has been investigated longer than ten years.   In this period, many experiments of NMR, the inelastic neutron scattering, the penetration depth and the angle resolved photoelectron spectroscopy [1] have clarified that high T_c cuprate is d wave-superconductor.   Next step is in making clear the origin of pairing force. The high T_c cuprate shows the anomalous properties at a temperature higher than T_c that an electrical resistivity increases linearly with temperature (T), Hall coefficient dramatically increases with decreasing T and $1/TT_1$ obeys to the Curie-Weiss law [1].   These anomalous properties are thought to be brought by a spin fluctuation. After a spin gap was found to open in high T_c cuprate by NMR measurement at a temperature (T^*) remarkably higher than T_c, a similar gap has been detected in the angle resolved photoelectron spectroscopy, the temperature dependence of resistivity [1] and the tunneling conductance, and is referred to as a pseudo-gap. The pseudo-gap seems to intimately relate to a pairing force.   In present experiment, to get the information of pairing force the hole doping dependence of the pseudo-gap was investigated.

EXPERIMENTAL RESULT AND DISCUSSION

The single crystal of BSCCO was grown by the flux method. Holes were introduced by doping oxygen and / or by substituting Na for Ca. From the x-ray diffraction pattern, the crystal structure of BSCCO was found to not change with doping oxygen and / or substituting Ca by Na except the decrease of lattice constant c. The hole concentration increased with doping and substituting. The tunneling spectrum was measured for the planar junctions fabricated on optimal doped single crystal and over doped one.

The $G(V)$ characteristics are respectively shown in Figs. 1 and 2 for the optimal oxygen doped

BSCCO (P=0.172, p=9.8x10^{20} cm^{-3}) and the over oxygen doped one (P=0.186, p=8.3x10^{21} cm^{-3}), where P is a deviation value from +2 in a copper valence charge and p is a hole concentration. Two clear peaks due to superconducting state are observed in both $G(V)$ characteristics at 4.2 K. The

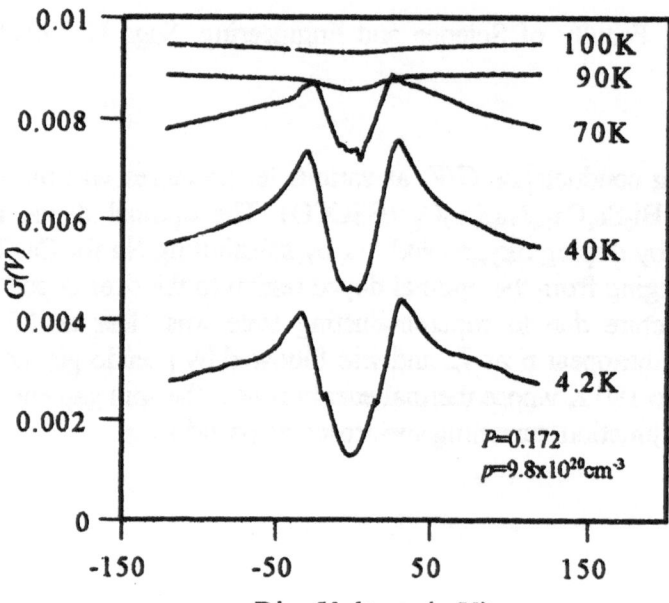

Fig.1 $G(V)$characteristics in optimal doped BSCCO

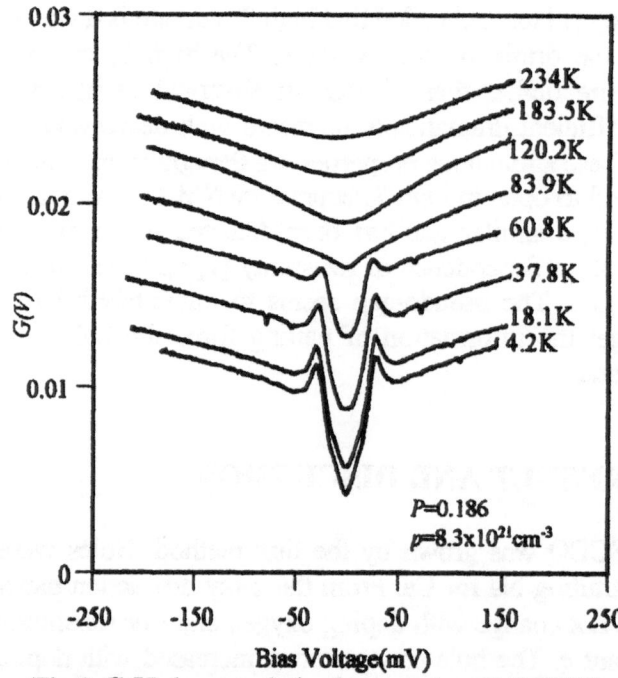

Fig.2 $G(V)$characteristics in over doped BCSSO

pseudo-gap is observed up to 100 K for optimal doped BSCCO and up to 183 K for over doped one. The background conductance in superconducting state is Λ type for optimal doped BSCCO but V type for over-doped one. This seems to reflect the special electronic state in optimal doped one and to relate to experimental fact that the conductivity is maximum at optimal doped one [2].

Next, the $G(V)$ characteristics for substitution of Na for Ca are discussed. As seen in Fig. 3, two peaks due to superconducting state are observed in the $G(V)$ characteristics at 4.2 K for the 15 % Na doped BSCCO ($x=0.15$, $p=2.9 \times 10^{21}$ cm^{-3}) and disappear near 49 K. However, the pseudo-gap is observed up to 186 K.

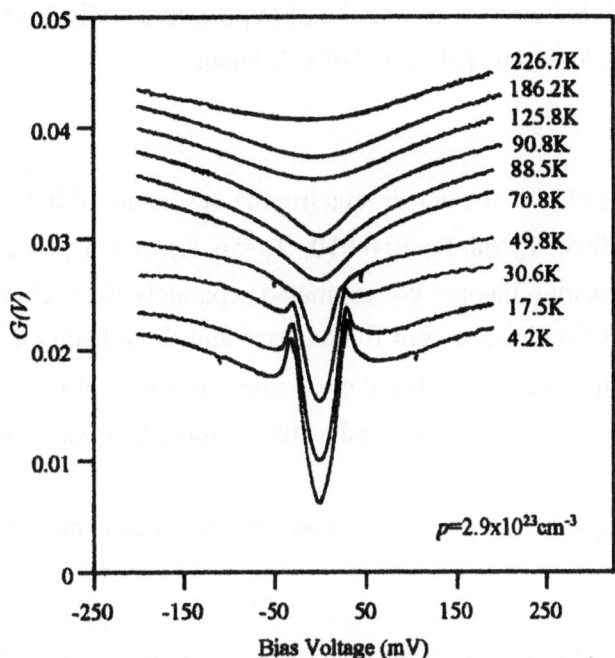

Fig.3 $G(V)$ charactristics in 15% Na doped BSCCO

In both hole doping methods, the pseudo-gap is found to appear for over doped BSCCO as well as optimal doped one and to continuously evolve to the superconducting gap with decreasing T.

CONCLUSION

Holes were introduced to BSCCO single crystal by doping oxygen and / or substituting Ca for Na. From the Hall effect measurement, the hole concentration was confirmed to increase with doping and / or substituting. In both methods of introducing hole, the pseudo-gap was observed in $G(V)$ characteristic for not only the optimal doped BSCCO but also the over doped one , and was found to continuously connect to the superconducting gap with decreasing temperature. The temperature at which the pseudo-gap appears is near the temperature at which the spin-gap opens.

1. Sato M, (1997) BUTSURI 52: 174-176; Fukuyama H, (1997) BUTSURI 52: 176-180; Yasuoka H, (1997) BUTSURI 52: 197-200: Endo Y, (1997) BUTSURI 52: 201-297
2. Idemoto Y,Toda T,Fueki K, (1995) PhysicaC 249:123-132;The experimental result by present authors also supports that the conductivity in BSCCO is maximum at the optimal doped one.

Zn-DOPING EFFECT ON THE PSEUDO SPIN-GAP OF $YBa_2(Cu_{1-x}Zn_x)_4O_8$ VIA Cu NQR

Y. Itoh, T. Machi and N. Koshizuka

Superconductivity Research Laboratory, International Superconductivity Technology Center,
10-13 Shinonome 1-chome, Koto-ku, Tokyo 135-0062, Japan

Abstract: We report zero-field ^{63}Cu nuclear quadrupole resonance (NQR) measurements for $YBa_2(Cu_{1-x}Zn_x)_4O_8$ (x=0.005, T_c=68 K; x=0.010, T_c=56 K) and a possible application of impurity-induced NQR relaxation theory. We estimated separately the ^{63}Cu NQR relaxation rate $(1/T_1)_{HOST}$ due to the host Cu electron spin fluctuations and the relaxation rate $1/\tau_1$ due to the Zn-induced spin fluctuations. We found that the Zn impurities do not destroy the pseudo spin-gap in the low-lying magnetic excitations around the antiferromagnetic wave vector.

Keywords: NQR, Zn doping effect, high-T_c superconductors, pseudo spin-gap

Effect of nonmagnetic impurity Zn substitution for Cu in high-T_c superconductors has attracted much interests [1]. The inelastic neutron scattering [2], the ^{89}Y NMR spectra [3], and the Cu NMR linewidth measurements [4] reveal that Zn impurities induce the low frequency staggered spin fluctuations (~10 meV) for $YBa_2Cu_3O_{7-\delta}$. But, the controversy surrounds the experiments on the Zn effect for the pseudo spin-gap. The previous Cu NQR relaxation study reports that the pseudo spin-gap is removed by Zn [5], but the other studies report the robust gap feature [2-4]. The detailed analysis of the Cu nuclear spin-lattice relaxation curves can provide us with microscopic spatial information on low-frequency electron spin dynamics with antiferromagnetic correlation [6].

We present a possible application of the impurity-induced Cu nuclear spin-lattice relaxation theory [7] for the Zn-substituted $YBa_2(Cu_{1-x}Zn_x)_4O_8$ (x=0.005, T_c=68 K; x=0.010, T_c=56 K) to estimate the planar $^{63}Cu(2)$ nuclear spin-lattice relaxation rate $(1/T_1)_{HOST}$ due to the host Cu electron spin fluctuations and the $^{63}Cu(2)$ relaxation rate $1/\tau_1$ due to the Zn-induced spin fluctuations separately.

We have carried out the planar Cu(2) NQR measurements with the coherent type pulsed spectrometer for the powdered samples [8]. We fit the following function to the experimental recovery curves of the $^{63}Cu(2)$ NQR spin-echo amplitude M(t), measured by an inversion

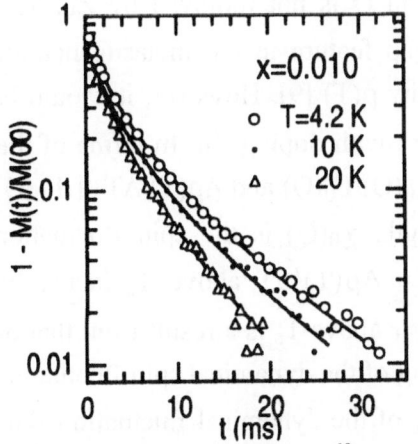

Fig. 1. T dependence of the ^{63}Cu(2) nuclear spin-echo recovery curve for the Zn x=0.010.

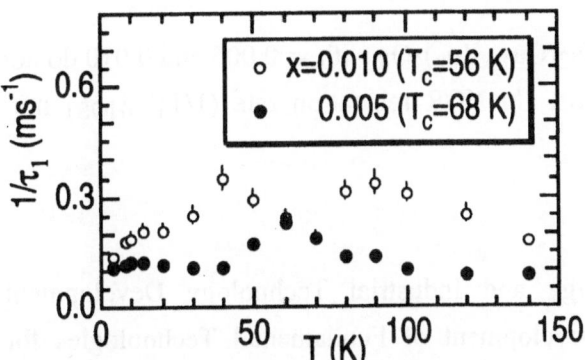

Fig. 2. The ^{63}Cu(2) NQR relaxation rate $1/\tau_1$ due to the Zn-induced electron spin fluctuations.

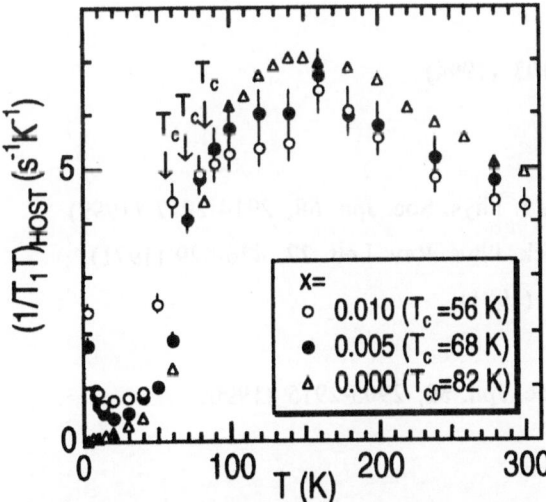

Fig. 3. The host ^{63}Cu(2) NQR relaxation rate $1/T_1T$ due to the host electron spin fluctuations. The pseudo spin-gap temperature T_s is nearly independent of the Zn content.

recovery spin-echo technique as a function of t (the time interval between the inversion and the accompanying pulses),

$$1-M(t)/M(\infty)=E_0\exp[-3t/(T_1)_{HOST}-\sqrt{3t}/\tau_1], \quad (1)$$

where E_0 is a fraction of initial inversion. Here, we assume a model in which the enhanced staggered spin fluctuations are induced around Zn [4]. The locally enhanced staggered spin fluctuations can yield the distance(r)-dependent $T_1(r)$ at each Cu site around Zn and then the term of $-\sqrt{3t}/\tau_1$ [6]. Equation (1) derived from $T_1(r)\sim r^{2D}$ in D dimensions is the same form as the stretched exponential curve due to the magnetic impurity Ni [6].

Figure 1 shows the T dependence of the ^{63}Cu NQR recovery curves $[1-M(t)/M(\infty)]$ for x=0.010. The recovery curves are nonexponential functions, described by the solid curves of eq. (1) with the fitting parameters E_0, $(T_1)_{HOST}$ and τ_1.

Figure 2 shows the estimated $1/\tau_1$ as a function of T for x=0.005 and 0.010. The T dependences of $1/\tau_1$ are similar to those for the Ni doping [6]. It is noticeable that the low temperature $1/\tau_1$ for x=0.005 below T_c levels off, which suggests the local spin susceptibility around Zn of $\chi_{local}\sim1/T$.

Figure 3 shows the T dependence of $(1/T_1T)_{HOST}$ for x=0.005 and 0.010. The pseudo spin-gap temperature $T_s(\sim160$ K), at which $1/T_1T$ takes a maximum, is not changed by the Zn doping. $(1/T_1T)_{HOST}$ is a wave-vector averaged low frequency dynamical spin susceptibility $\chi''(q, \omega\sim0)$ enhanced around q=Q=$[\pi, \pi]$. Thus, the pseudo spin-gap in the low frequency spin excitations around Q is not removed by Zn. This result is consistent with the Zn effect on the pseudo spin-gap feature in the inelastic neutron scattering [2], the ^{89}Y NMR [3, 4], and the electrical resistivity $\rho(T)$ [9]. However, it should be noted that the magnitude of $(1/T_1T)_{HOST}$ above T_{c0} decreases with doping Zn. In terms of the antiferromagnetic spin fluctuation theory, we have $1/T_1T \sim \xi^2 \chi_0(Q)/\Gamma_0(Q)$ and $\Delta\rho(T)/\Delta T \sim 1/\Gamma_0(Q)$ above T_s, where ξ is the antiferromagnetic correlation length, $\chi_0(Q)$ is the spin fluctuation amplitude and $\Gamma_0(Q)$ is the fluctuation energy [10]. Since $\Delta\rho(T)/\Delta T$ above T_s is nearly independent of the Zn content [9], the decrease of $(1/T_1T)_{HOST}$ above T_s is a result from that of the amplitude $\chi_0(Q)$. Thus, the Zn does not induce the softening of the dynamical spin fluctuation peaked around $\Gamma_0(Q)$, i. e. the Zn does not act as the pinning of the dynamical fluctuation. The rapid increase of $(1/T_1T)_{HOST}$ far below T_c as well as $1/\tau_1T$ is also noticeable. The Zn makes a drastic change in the electron states around Zn [1, 3] but also a long-range effect over the host both above and below T_c.

In conclusion, we found that Zn impurities in $YBa_2(Cu_{1-x}Zn_x)_4O_8$ with x=0.005 and 0.010 do not remove the pseudo spin-gap behavior in the host Cu NQR relaxation rate $(1/T_1T)_{HOST}$ nor decrease the pseudo spin-gap temperature T_s.

Acknowledgments

This work was supported by the New Energy and Industrial Technology Development Organization as Collaborative Research and Development of Fundamental Technologies for Superconductivity Applications.

1. H. Alloul et al., cond-mat / 9905424.

2. Y. Sidis et al., Phys. Rev. B53, 6811-6818 (1996).

3. A. V. Mahajan et al., Phys. Rev. Lett. 72, 3100-3103 (1994).

4. M.-H. Julien et al., preprint (1999).

5. G.-q. Zheng et al., Physica C263, 367-370 (1996).

6. Y. Itoh, T. Machi, N. Watanabe and N. Koshizuka, J. Phys. Soc. Jpn. 68, 2914-2917 (1999).

7. M. R. McHenry, B. G. Silbernagel and J. H. Wernick, Phys. Rev. Lett. 27, 426-429 (1971).

8. T. Miyatake et al., Phys. Rev. B 44, 10139-10145 (1991).

9. K. Mizuhashi et al., B52, R3884-3887 (1995).

10. T. Moriya, Y. Takahashi and K. Ueda, J. Phys. Soc. Jpn. 59, 2905-2915 (1990).

Observation of Vortices by Lorentz Microscopy

Ken Harada[+*], Hiroto Kasai[+], Osamu Kamimura[+], Tsuyoshi Matsuda[+], Takaho Yoshida[+*] Akira Tonomura[+*], Jun-ichi Shimoyama[#*], Kohji Kishio[#*] and Koichi Kitazawa[#*]

[+]Advanced Research Laboratory, Hitachi Ltd., Hatoyama, Saitama 350-0395,
[#]Department of Chemistry, University of Tokyo, Bunkyou -ku, Tokyo 113-8656,
[*]CREST, Japan Science and Technology Corporation (JST), Kawaguchi, Saitama 332-0012

Abstract: Dynamic behaviors of vortices (magnetic-flux-lines) in the high-temperature superconductor $Bi_2Sr_2CaCu_2O_{8+\delta}$ were observed by Lorentz microscopy with a 300-kV field-emission electron microscope. When a drive force was exerted on vortices by changing the magnetic field H, some vortices began to move with forming a plastic flow. We found various regimes of vortex motion: at low H, the vortex motion which was slow migration below a specimen temperature T of about 25 K, gradually changed to hopping above 25 K. At higher H and T the vortex-vortex interaction became appreciable and changed the forms of flow.

Keywords: Vortex dynamics, Vortex lattice, Lorentz microscopy, Electron diffraction, Field emission electron microscope

INTRODUCTION

Dissipation-free current in a superconductor can be achieved only when the vortices (magnetic-flux-lines) are pinned down against the current-induced force. Much effort has been expended on attaining high critical current density J_c without enough knowledge about microscopic mechanism of vortex pinning and its depinning. Although simulations and macroscopic measurements tried to predict vortex depinning phenomena, the behavior of vortices, especially in high-temperature superconductors, is complicated and not yet fully understood [1]. Here we report the real-time observation of vortex motion in thin films of a high-temperature $Bi_2Sr_2CaCu_2O_{8+\delta}$ (BSCCO(2212)) by Lorentz microscopy [2,3] with a 300-kV field-emission electron microscope. We found various dynamic regimes of vortex motion [4], some of which have been predicted by simulations and macroscopic measurements, *i.e.* filamentary flows [5], lattice flow [6].

EXPERIMENTAL

A single crystalline of $Bi_2Sr_2CaCu_2O_{8+\delta}$ (BSCCO(2212)) was grown by the floating-zone technique and annealed to be oxygen over-doped whose critical temperature T_c was 85 K [7]. The BSCCO(2212) thin films of thickness about 200 nm with a 100×100-μm^2 uniform area were prepared by cleaving and selected for the observations [3].

The experimental arrangement is shown in Fig. 1. A prepared specimen was set on a low-temperature stage tilted 45 degrees to the incident electron beam and the external magnetic field which was applied parallel to the electron beam for the purpose not to deflect the beam. This

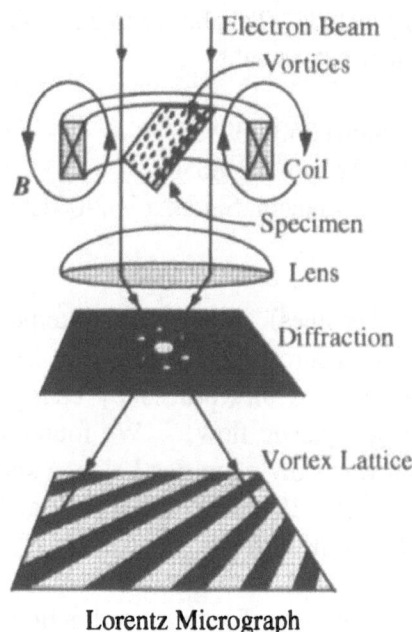

Lorentz Micrograph

Fig. 1. Experimental setup.

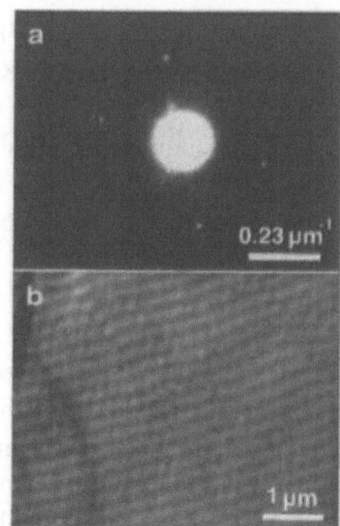

Fig. 2. (a) Electron diffraction from a Nb vortex lattice under 500 G at 7 K. Two diffraction spots were missed by weak scattering. (b) Nb vortex lattice observed by Fresnel (out-of-focus) method in Lorentz microscopy.

experimental setup enabled us to observe the vortex motion during the magnetic field change and also to observe the vortex lattice within higher magnetic field up to 1,400 G and its electron diffraction. These details of this setup and of another kind of Lorentz microscopy, *i.e.* Foucault method, are described elsewhere [8,9]. Figure 2 shows, as an example, a Nb vortex lattice and its electron diffraction.

The vortex motion was investigated through a TV system [10,11] within a range of magnetic field H, $0 \leq H \leq 45$ G, and at a temperature T, $15 \leq T \leq 50$ K. Within the ranges individual vortices could be observed and their manner of movement changed in ways depending on H and T.

RESULTS

We found various regimes of vortex motion. They are summarized on H-T diagram in Fig. 3 and are described below in detail with figures. When vortices were sparse ($H < 1$ G) and there was little interaction between them, their dynamic behavior below $T = 25$ K (region I) was very different from that above $T = 25$ K (region II). In the region I ($H < 1$ G, $T < 25$ K) in Fig. 3, all the vortices migrated at about 1.8 μm/s while maintaining the positions relative to one another. Their dynamics at 20 K are reproduced in Fig. 4 from video frames [13].

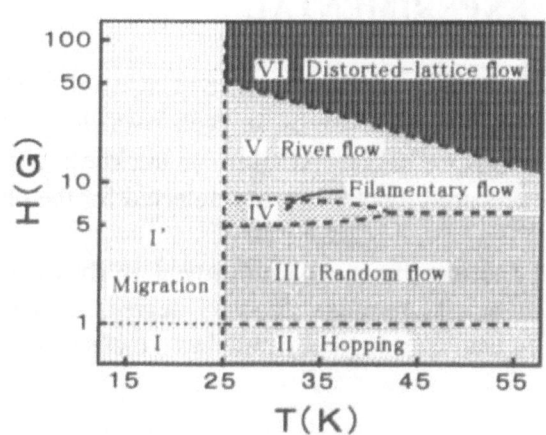

Fig. 3. *H-T* diagram of vortex motion.

In the region II ($H < 1$ G, $T > 25$ K) new pinning centers appeared and prevented the vortices from migrating. The vortices were trapped at the centers and suddenly left by hopping whose distances were greater than 50 μm. The vacant centers were soon occupied by newly arrived vortices, see, for example vortex 'A' hopped away then its vacant site was occupied by vortex 'C' in Fig. 5. Because of the high hopping speed, the vortex image seemed to blink [12].

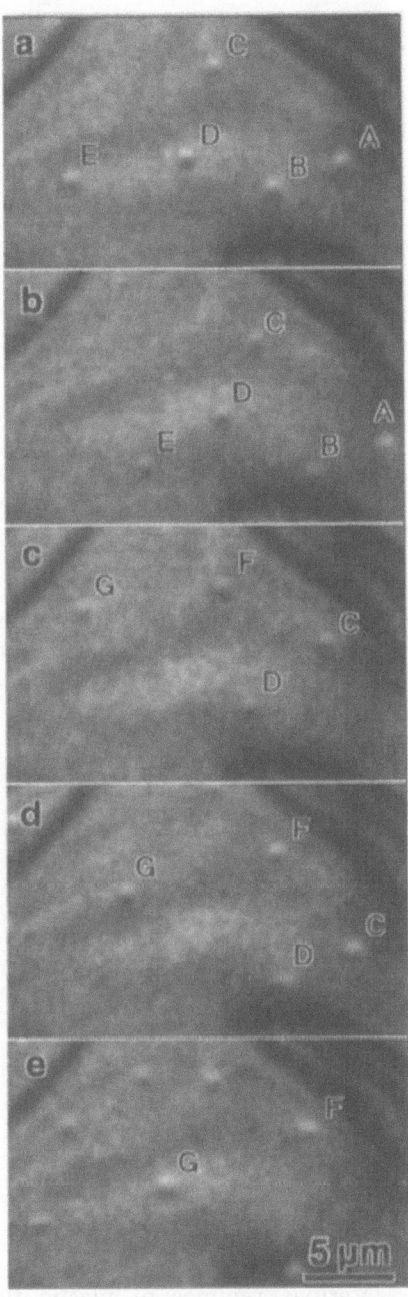

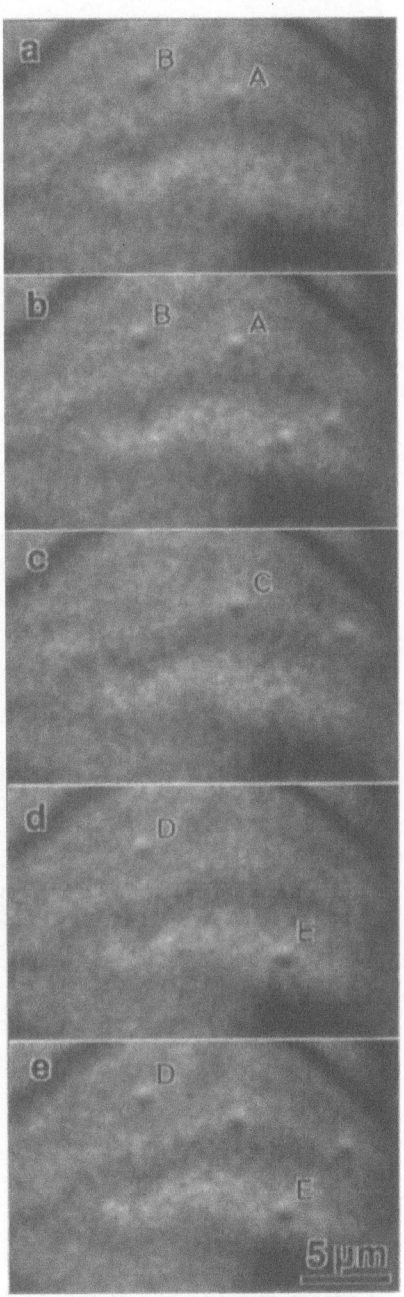

Fig. 4. Lorentz micrographs from video frames showing the migration at $H < 1$G and 20 K: (a) t=0 s, (b) t=1.4 s, (c) t=3.0 s, (d) t=4.4 s, (e) t=6.0 s. Each vortex is identified by a capital letter.

Fig. 5. Lorentz micrographs at 30 K showing the sudden hopping over more than 50 μm: (a) t=0 s, (b) t=0.6 s, (c) t=2.4 s, (d) t=3.6 s, (e) t=5.0 s. Pinning centers must be located at the position of vortex 'A', vortex 'B' and vortex 'E'.

When vortices became denser in region I' ($H > 1$ G, $T < 25$ K), the vortex-vortex interaction became stronger: the vortex motion was not essentially different from the migration occurred in region I except that moving vortices were slow down to about 0.7 μm/s and formed a more regular lattice (Fig. 6).

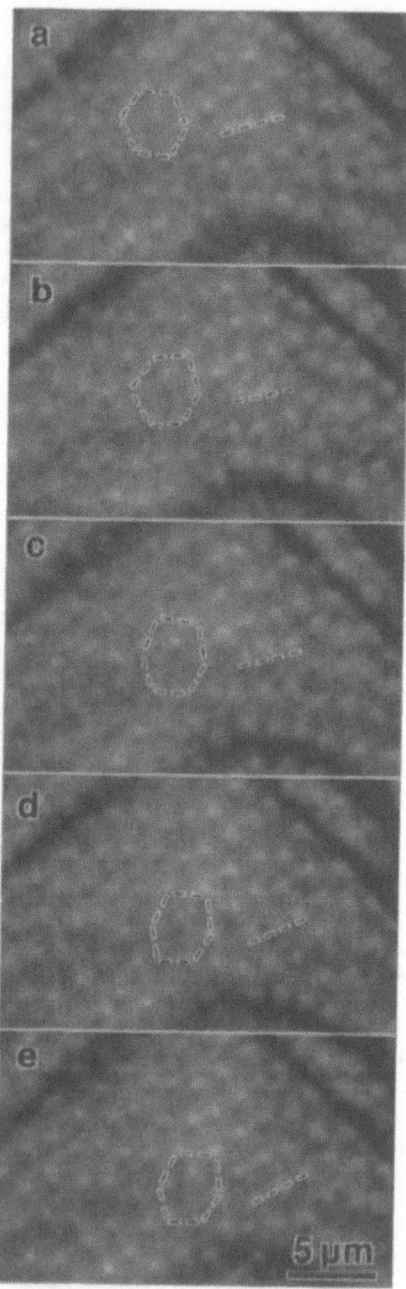

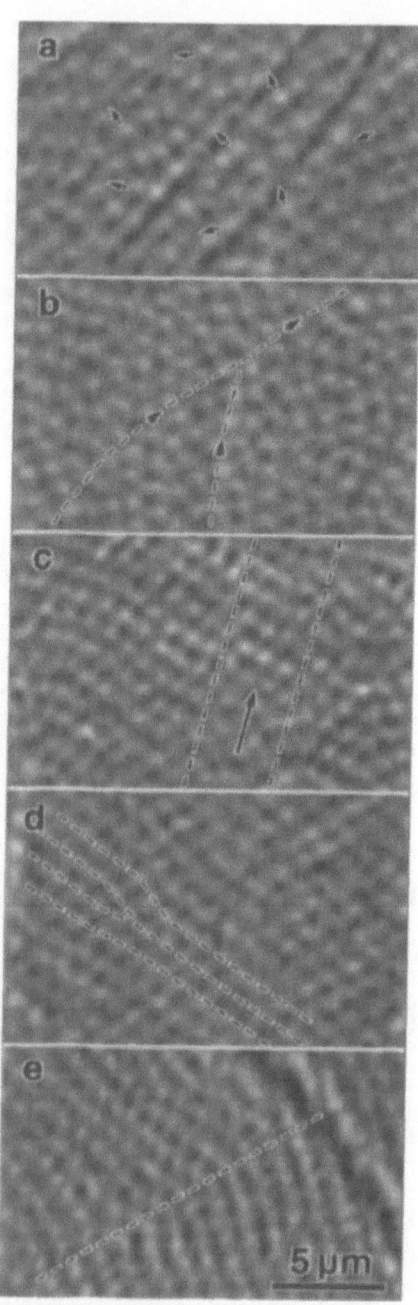

Fig. 6. Lorentz micrographs at 5 G and 20 K: (a) t=0s, (b) t=1.4 s, (c) t=3.0 s, (d) t=4.4 s, (e) t=6.0 s.   Vortex-vortex interaction makes them a regular lattice structure.

Fig. 7. Various regimes of plastic flow of vortices: (a) random flow (4 G, 30 K), (b) filamentary flow (6 G, 30 K), (c) river flow (13 G, 30 K), (d) phase slip (20 G, 50 K), (e) recrystallization (20 G, 50 K).

We think that this T-dependence of vortex motion reflects two kinds of pinning centers: at $T < 25$ K vortices move smoothly due to the existence of extremely small and many pinning centers, and at $T > 25$ K vortices hop among some effective pinning centers. High-resolution electron microscopy was used to identify what the effective pinning centers were, but very few defects were found at locations where vortices were often trapped.

In the region III ($1 < H < 5$ G, $T > 25$ K), the hopping distance was reduced to the intervortex distance because of the interruption due to surrounding vortices. Vortices sometimes pushed other vortices to move and sometimes squeezed themselves between pinned vortices. As a whole, many vortices here and there in field of view moved randomly (Fig. 7(a)).

In the region IV ($5 < H < 7$ G, $25 \leq T \leq 40$ K), hopping vortices knocked other vortices out in succession forming a filamentary flow [5]. Several vortices in a filament hopped from one site to the adjacent site simultaneously (Fig. 7(b)). Sometimes such a synchronous flow was frozen for a while and reappeared again, not only along the same path but also along different paths. There were cases where two filaments joined or intersected.

In the region V ($7 < H < 45$ G, $T > 25$ K), the vortex-vortex interaction became stronger, and vortices at rest formed domains of lattices approximately 10 μm in size. When vortices began to move, however, they did not maintain the lattice form: some of them flowed in rivers (Fig. 7(c)). The widths and locations of the rivers were not fixed but intermittently changed.

In the region VI ($H > 15$ G, $T > 25$ K), vortices tended to form a single and rigid lattice extending over the entire field of view. When a drive force was exerted on the lattice by changing the magnetic field, however, the lattice seemed to divide into domains due to the spatial difference in pinning. The lattice was deformed, changing its form in various ways with time: For example, two domains were suddenly displaced by a short distance with respect to each other (phase slips) along the flow direction. Edge dislocations were produced between them at one time and disappeared at another (Fig. 7(d)). Another way the lattice form changed is that a lattice orientation in one domain suddenly changed against other domains, and the whole vortex configuration was recrystallized as shown in Fig. 7(e). But after a few seconds the other domains also moved, returning to the original perfect crystal form. Such process was repeated and the vortex lattice lines flowed undulating with time.

CONCLUSION

The microscopic motion of vortices in high-temperature superconductors was observed systematically and then several dynamical regimes on H and T were found including new behavior of vortices on temperature.

Acknowledgments. We express sincere thanks to Profs. F. Nori of Michigan University, A. Maeda of Tokyo University, G. W. Crabtree of Argonne National Laboratory, M. Tachiki of National Research Institute of Metals, A. R. Bishop and N. G. Jensen of Los Alamos National Laboratory, E. Bodenschatz of Cornell Univ., Dr. M. Yamasaki of Hitachi Systems Development Laboratory and Drs. Y. Tarutani, N. Osakabe, T. Onogi and R. Sugano of Hitachi Advanced Research Laboratory for their helpful discussions on the present experiments. Thanks are also

292

due to Messrs S. Saitou, S. Matsunami and N. Moriya of Hitachi Advanced Research Laboratory for their technical assistance.

[1]   G. W. Crabtree and D. R. Nelson : *Physics Today* **50** (1997) 38-45.
[2]   K. Harada, T. Matsuda, J. E. Bonevich, M. Igarashi, S. Kondo, G. Pozzi, U. Kawabe and A. Tonomura : *Nature* **360** (1992) 51-53.
[3]   K. Harada, T. Matsuda, H. Kasai, J. E. Bonevich, T. Yoshida, U. Kawabe and A. Tonomura : *Phys. Rev. Lett.* **71** (1993) 3371-3374..
[4]   A. Tonomura, H. Kasai, O. Kamimura, T. Matsuda, K. Harada, J. Shimoyama, K. Kishio and K. Kitazawa : *Nature* **397** (1999) 308-309.
[5]   N. Grønbech-Jensen, A. R. Bishop and D. Dominguez : *Phys. Rev. Lett.* **76** (1996) 2985-2988.
[6]   A. E. Koshelev and V. M. Vinokur : *Phys. Rev. Lett.* **73** (1994) 3580-3583.
[7]   Y. Kotaka, T. Kimura, H. Ikuta, J. Shimoyama, K. Kitazawa, K. Yamafuji, K. Kishio and D. Pooke : *Physica C* **235-240** (1994) 1529-1530.
[8]   T. Yoshida, J. Endo, H. Kasai, K. Harada, N. Osakabe, A. Tonomura and G. Pozzi : *J. Appl. Phys.* **85** (1999) 1228-1230.
[9]   T. Yoshida, J. Endo, K. Harada, H. Kasai, T. Matsuda, O. Kamimura, A. Tonomura, M. Beleggia, R. Patti and G. Pozzi : *J. Appl. Phys.* **85** (1999) 4096-4103.
[10] T. Matsuda, K. Harada, H. Kasai, O. Kamimura and A. Tonomura : *Science* **271** (1996) 1393-1395.
[11] K. Harada, O. Kamimura, H. Kasai, T. Matsuda, A. Tonomura and V. V. Moshchalkov : *Science* **274** (1996) 1167-1170.
[12] Some of the video scenes are available on the web site (http://www.nature.com/cgi-bin/SupplData.cgi/author/Tonomura,+A.) and other informations are also at (http://www.hatoyama.hitachi.co.jp/).

Anomalous Microwave Response and Upper Critical Field in Overdoped $Tl_2Ba_2CuO_{6+\delta}$

K. Izawa[1,2], A. Shibata[1] H. Takahashi[1], Y. Matsuda[1,2], M. Hasegawa[1], N. Chikumoto[3], C.J. van der Beek[4], and M. Konczykowski[4]

[1]Institute for Solid State Physics, University of Tokyo, Minato-ku, Tokyo 106-8666 Japan
[2]CREST, Japan Science and Technology Corporation.
[3]Superconductivity Research Laboratory, ISTEC, Minato-ku, Tokyo 105-0023 Japan
[4]Laboratoire des Solides Irradies, Ecole Polytechnique, Palaiseau, 91128, France

Abstract The transmittivity T_H and microwave surface impedance $Z_s = R_s + iX_s$ have been measured in order to clarify the unusual dc resistive transition of overdoped $Tl_2Ba_2CuO_{6+\delta}$ in magnetic field. The transition field H^* is enhanced by the columnar defect, indicating that H^* is not the mean-field upper critical field but the depinning point of the vortices. R_s shows unusual field dependence which is quite different from that expected from the result of the dc resistivity.

Keywords: Overdoped $Tl_2Ba_2CuO_{6+\delta}$, Transmittivity, Surface impedance, Upper critical field

INTRODUCTION

In overdoped $Tl_2Ba_2CuO_{6+\delta}$, the dc resistivity suddenly jumps at a certain magnetic field H^* [1]. Although this behavior seems to be the mean-field upper critical field H_{c2}, temperature dependence of H^* shows upward curvature with decreasing temperature T. Many theoretical efforts [2, 3, 4, 5, 6] have been made to explain this unusual transition in terms of H_{c2}. On the other hand, a pairing onset temperature estimated by specific heat [7] and electronic Raman scattering [8] has been reported to be much higher than H_{c2}. These results suggest that the superconducting order parameter exists even above H^*. Furthermore, it is theoretically pointed out that the resistive transition is caused by the depinning of the vortices [9]. To uncover the origin of this anomalous resistive transition, we have measured the transmittivity T_H and microwave surface impedance $Z_s = R_s + iX_s$ up to 7 T.

EXPERIMENT

The high-quality single crystals were grown by using a gold crucible in closed atmosphere. Details of the sample preparation were described elsewhere[10]. The broadband fundamental and the third harmonic transmittivity measurements were performed by using the local Hall probe magnetometer at 7.753 Hz before and after Pb-ion irradiation. The columnar defects (CDs) were introduced by the irradiation in the crystal along the c

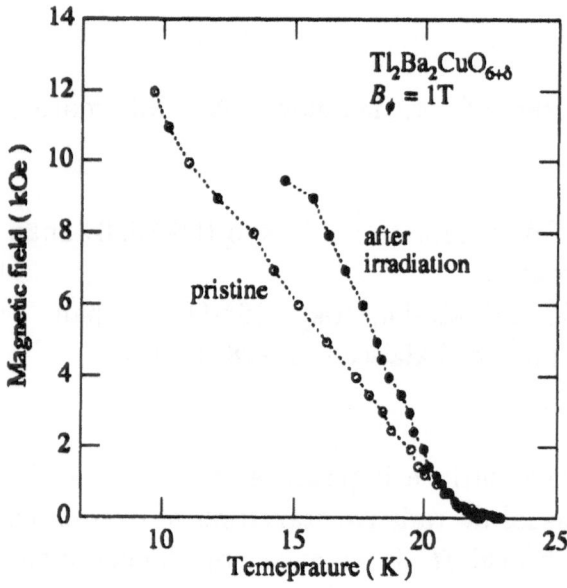

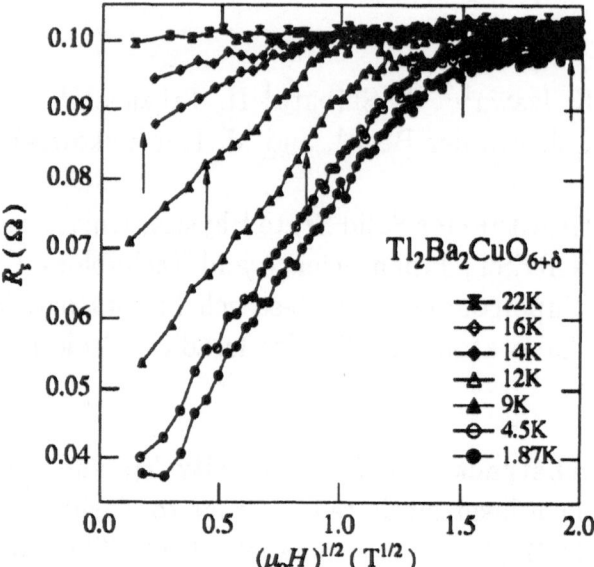

Figure 1: Effect of the Pb-ion irradiation on the transition point T^*. T^* is determined by the onset of $|T_{H3}|$. The open and solid circle indicate T^* before and after irradiation, respectively.

Figure 2: The surface resistance at the selected temperature plotted as a function of $\sqrt{H}$. The arrows indicate the dc resistive-transition points at 14, 12, 9, 4.5 and 1.87 K.

axis with the matching field $B_\phi = 1$ T. The surface impedance was measured at 28.5 GHz by using a oxygen-free Cu cavity resonator operated in the TE_{011} mode. The crystal was located at the center of the cavity at which the microwave magnetic field and dc magnetic field up to 7 T were applied along to the c axis.

RESULTS AND DISCUSSIONS

With decreasing T, the in-phase component of the fundamental transmittivity T'_H, which is related to the shielding current in the crystal, rapidly drops at T^*, at which dc resistivity also shows the sharp transition. At the same time, the third harmonics $|T_{H3}|$ steeply increases from zero to finite value at T^*, followed by a peak. The non-zero $|T_{H3}|$ below T^* implies that the presents of a nonlinear $I(V)$ response due to the flux pinning. As H is increased, these transition behavior shifts to lower T without broadening. With decreasing T, the transition point determined by the onset of $|T_{H3}|$ is found to decrease with upward curvature as shown in Fig. 1. The upward curvature of the transition point is consistent with that of dc resistivity while this transition point is slightly higher than that of the dc resistivity. Introduction of the CD enhances H^* in magnetic field although it slightly reduces T_c in zero field. It is noteworthy that the effect of the CD on the sharpness of the transition is too small. Since the CD acts as a strong pinning center, this result leads to the conclusion that H^* is not H_{c2} but the depinning point of vortices.

Figure 2 shows H dependence of R_s at the selected T. The arrows indicate the dc resistive-transition points determined by the field at which dc resistivity shows 50 % of the normal value. With increasing H, R_s increases as $\sqrt{H}$, indicating that the flux flow resistance $\rho_v \propto R_s^2$ is proportional to H. The increase of R_s is terminated at a certain field H_k and then R_s exhibits a field independent value above H_k. In general, according to the Bardeen-Stephen formula $\rho_v = (H/H_{c2})\rho_n$, ρ_v increases linearly up to the normal value ρ_n at H_{c2}. Therefore, H_k is associated with H_{c2}. However, above 5.5 K, R_s increases without anomaly at H^* up to H_k; the result of surface resistance suggests that the system is in the mixed state even above H^* while the dc resistivity exhibits the system is in the normal state. This result is consistent with the conclusion that $H^*(< H_{c2})$ is the depinning point of the vortices. On the other hand, below 5.5 K, R_s shows a field independent value even below H^* where the dc resistivity shows zero, indicating that H_k is not H_{c2}. Furthermore, H_k increases with a downward curvature with decreasing T. This T dependence of H_k is quite different from that of H^*. These unusual transport properties may be reflected in anomalous microwave response of the quasiparticles. An understanding of this unusual behavior requires further experiments.

CONCLUSION

We have measured the transmittivity and microwave surface impedance of overdoped $Tl_2Ba_2CuO_{6+\delta}$ in magnetic field up to 7 T. By introducing columnar defects, we revealed that the dc resistive transition is caused by the depinning of the vortices. The surface resistance increases as $\sqrt{H}$ and shows a kink at H_k, followed by a field independent value. The kink field H_k is not related to the dc resistive transition field H^*.

References

[1] A.P. Mackenzie, et al., Phys. Rev. Lett. **71**, 1238 (1993).

[2] A.S. Alexandrov, Phys. Rev. **B48**, 10571 (1993).

[3] A.J. Schofield, Phys. Rev. B **51**, 11733 (1995).

[4] Yu.N. Ovchinnikov and V.Z. Kresin, Phys. Rev. B **52**, 3075 (1995).

[5] T.Koyama and M. Tachiki, Phys. Rev. B **53**, 2662 (1995).

[6] G. Kontliar and C.M. Varma, Phys. Rev. Lett. **77**, 2296 (1996).

[7] A. Carrington, et al., Phys. Rev. B **54**, R3788 (1996).

[8] G. Blumberg, et al., Phys. Rev. Lett. **78**, 2461 (1997).

[9] V.B. Geshkenbein, et al., Phys. Rev. Lett. **80**, 5778 (1999).

[10] M. Hasegawa, et al., Proc. of MOS 99.

Vortex Dynamics below the Kosterlitz-Thouless Transition Detected by Voltage Noise

Satoshi Okuma, Nobuhito Kokubo, and Mikio Kamada

Research Center for Very Low Temperature System, Tokyo Institute of Technology, 2-12-1, Ohokayama, Meguro-ku, Tokyo 152-8551, Japan

Abstract: We have made simultaneous current-voltage $(I - V)$ and voltage noise S_V measurements of thin (6-nm thick) amorphous Mo_xSi_{1-x} films to study vortex dynamics below the Kosterlitz-Thouless transition temperature in the presence of the current. In the current region where the $I - V$ characteristics show strong nonlinearity, the large $1/f$-type noise is observed. At high frequency f the spectrum $S_V(f)/V$ ($S_V(f)$ divided by V) for different I seems to collapse onto a single straight line. Below the characteristic frequency f_C, $S_V(f)/V$ deviates downward from the straight line. The characteristic time $\tau_C = 1/f_C$ is much larger than the travelling time of each free vortex across the sample, implying the presence of the long-time correlation of dissociated vortices.

Keywords: Kosterlitz-Thouless Transition, Noise, Vortex dynamics, Amorphous films

INTRODUCTION

It is well established that the superconducting transition in two dimensions (2D) is described by the Kosterlitz-Thouless vortex-antivortex unbinding theory [1]. Although numerous studies have been performed on the KT transition [1-6], very little is known about the dynamics of dissociated vortices [6]. In particular, vortex dynamics well below the KT transition temperature T_{KT} has not been studied yet. In this work we have made current-voltage $(I - V)$ and current-induced voltage noise S_V measurements for the thin amorphous Mo_xSi_{1-x} film far below T_{KT}, where the dissociation of the bound vortex pairs occurs in the presence of a large current. We have observed the large $1/f$ like noise, which cannot be explained by a simple shot-noise (SN) model. The possible vortex dynamics will be discussed. Preliminary data and more detailed discussion concerning the present work are reported elsewhere [7].

EXPERIMENTAL

The amorphous Mo_xSi_{1-x} film [4,7-10], 6 nm in thickness, 5.0 mm in length, and 0.55 mm in width w, was prepared in UHV by coevaporation of pure Si and Mo onto a rotating glass substrate held at room temperature. The mean-field superconducting transition temperature T_{C0} and critical field $\sim H_{C2}$ are 3.69K and 7.5T, respectively. To cancel the ambient magnetic field, a small perpendicular field ($\sim 10^{-5}$ T) was applied using a superconducting magnet. The voltage-noise spectrum $S_V(f)$ and $I - V$ characteristics were measured by the standard four-terminal method [7-10]. To reduce the flicker noise at low frequency, the film was directly immersed in superfluid of liquid ^{4}He.

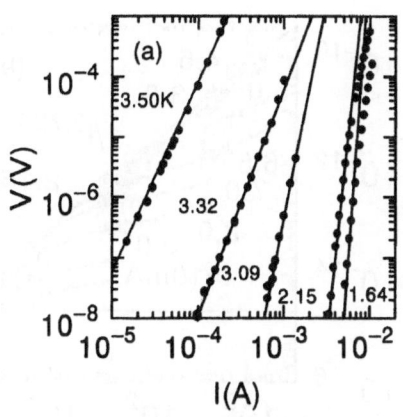 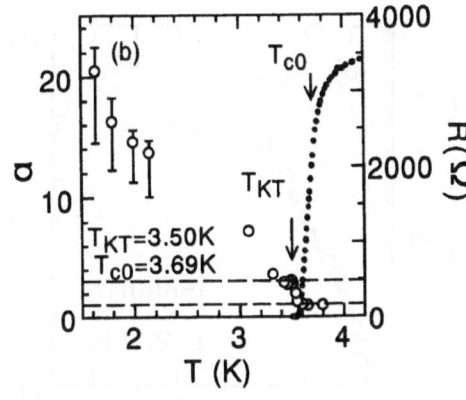

Fig.1 (a) $I - V$ characteristics of the 6-nm thick Mo_xSi_{1-x} film for selected temperatures. (b) The resistive transition and the exponent $\alpha(T)$ obtained from the power-law relation, $V \propto I^{\alpha(T)}$, in the small current limit.

RESULTS AND DISCUSSION

Shown in Fig.1(a) are the $I - V$ characteristics at different temperatures. Around the zero-resistance temperature ($\sim$3.5 K) they follow a power-law relation expressed as $V \propto I^{\alpha(T)}$. Upon cooling, $\alpha(T)$ shows a step-like increase from 1 to $\sim$3 at around 3.5K and increases monotonically, such as shown in Fig.1(b). Here, T_{KT} is defined as 3.50K where $\alpha = 3$. At low temperatures ($T \leq 2.15K$) well below T_{KT} the $I - V$ characteristics show negative curvature on a log-log scale, which is not explained with the simple KT theory. In this current-voltage range we have observed substantial broad-band noise $S_V(f)$. In Fig.2(a), S_V at fixed f as well as $I - V$ characteristics at 2.15 K is shown. S_V appears near the onset of V and rises with increasing I (or V). With further increasing I, S_V shows a broad peak. As will be shown below, this behavior is related to the change in the spectral form as a function of I. Figure 2(b) displays the noise spectra $S_V(f)/V$ ($S_V(f)$ divided by V) at 2.15 K for different I. Interestingly, these spectra at high f seem to collapse onto a single $1/f^\beta$ ($\beta \approx 0.75$) line shown with a dashed line. Physical origin responsible for this collapse is not specified at present. At low f, $S_V(f)/V$ deviates downward from the straight line. A characteristic frequency f_C at which $S_V(f)/V$ shows a deviation is strongly current dependent, which is approximately expressed as $f_C \propto I^\beta$ with $\beta \sim 20$.

These spectra are markedly different from the spectrum predicted by the SN model for free vortices. In Fig.2(b) we illustrate the spectral form expected from the SN model. Here, we assume that the vortex mean free path l is half of the sample width w. We note the following points: (i) The transit time τ_l for a vortex traveling ($l = w/2$) along the sample width is estimated to be less than $\sim \mu$s in the current region studied. However, the observed characteristic time $\tau_C = 1/f_C$ is even longer than $\sim \mu$s. (ii) The SN model predicts $1/\tau_l \propto I$, while f_C exhibits the much stronger current dependence as mentioned above. (iii) The flux-bundle sizes $n(\sim 10 - 10^5)$ estimated from the magnitud of $S_V(f)/V$ at low f are much larger than unity. All of these results imply that the observed voltage noise cannot be explained by the simple SN model. It is also important to point out that one can see a trend for $S_V(f)/V$ to approach the SN spectrum with increasing I, although we are currently not able to observe the complete SN spectrum at high I due to some experimental reasons.

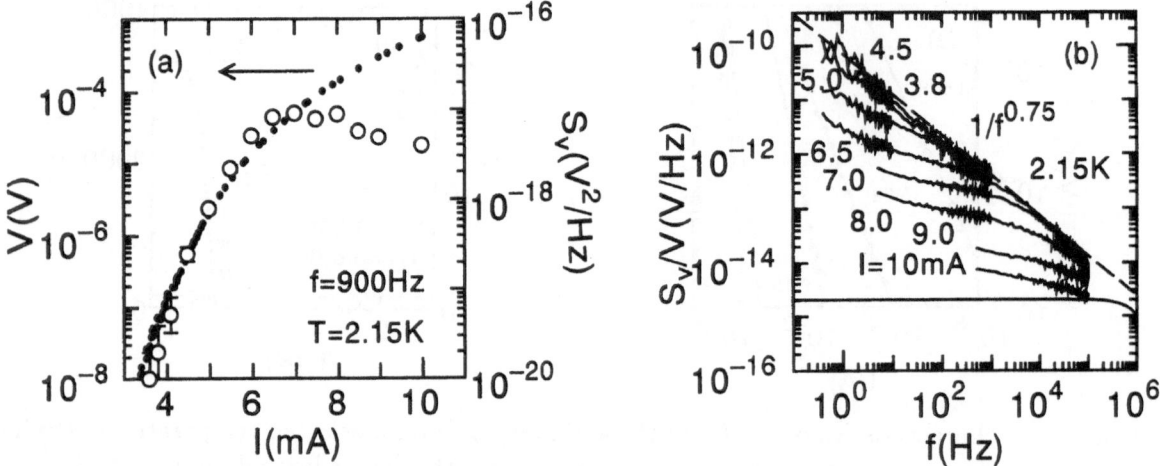

Fig.2 (a) $I - V$ characteristics and the current dependence of S_V at $f=$ 900 Hz at $T=$ 2.15 K. (b) The noise spectra $S_V(f)/V$ ($S_V(f)$ divided by V) at 2.15 K for different I. The dashed line corresponds to the $1/f^\beta$ ($\beta \approx 0.75$) line and the full line represents the spectrum predicted by the shot-noise model.

The observed $1/f$-type $S_V(f)$ (or longer τ_C than τ_l) indicates the presence of long time correlation in the motion of dissociated vortices. Here, we propose a possible model to explain the present results qualitatively. A basic idea is that motion of a small number of free vortices can trigger the successive dissociation process of bound pairs: A single vortex dissociated and driven by the current has a chance to cross between a vortex and an antivortex of a bound pair, leading to the dissociation of the bound pair into free vortices. This dissociation process can persist until the vortex annihilates with another antivortex or arrives at the side of the film. The dissociated free vortices can also dissociate other bound pairs. Thus the number of bound pairs decreases (the number of free vortices increases) rapidly and finally the successive dissociation process ceases. This successive dissociation process generates an elementary voltage pulse with a sharp peak and long time width $\tau \sim \tau_C$ ($\gg \tau_l$), giving rise to the observed voltage noise $S_V(f)$. More detailed discussions will be given elsewhere [7]. To prove this model convincingly, further measurements of $S_V(f)$ involving the field dependence as well as temperature dependence are in progress.

Acknowledgments. The authors would like to thank A. Maeda, T. Onogi, and R. Sugano for useful discussions and M. Muto for technical assistance.

1. J. M. Kosterlitz and D. J. Thouless, J. Phys. **C6**, 1181 (1973); V. L. Brezinskii, Sov. Phys. JETP **34**, 610 (1972).

2. B. I. Halperin and D. R. Nelson, J. Low Temp. Phys. **36**, 599 (1979).

3. A. M. Kadin, K. Epstein, and A. M. Goldman, Phys. Rev. **B27**, 6691 (1983).

4. S. Okuma, T. Terashima, and N. Kokubo, Solid State Commun. **106**, 529 (1998); S. Okuma, Mater. Sci. Eng. **B25**, 187 (1994).

5. R. Sugano, T. Onogi, and Y. Murayama, Phys. Rev. **B45**, 10789 (1992).

6. T. J. Shaw et al., Phys. Rev. Lett. **76**, 2551 (1996).

7. N. Kokubo, M. Kamada, and S. Okuma, preprint; N. Kokubo et al., in Proceedings of 22nd. Int. Conf. on Low Temperature Physics (Helsinki, 1999), to be published.

8. S. Okuma, T. Terashima, and N. Kokubo, Phys. Rev. **B58**, 2816 (1998).

9. N. Kokubo, T. Terashima, and S. Okuma, J. Phys. Soc. Jpn., **67**, 725 (1998).

10. S. Okuma and N. Kokubo, Phys. Rev. **B**, to be published.

The Conduction Noise Observed in $Bi_2Sr_2CaCu_2O_y$ Single Crystal

Yoshihiko Togawa*, Ryuichi Abiru*, Katsuya Iwaya*, Haruhisa Kitano* and Atsutaka Maeda*†

*Department of Basic Science, The University of Tokyo,
3-8-1, Komaba, Meguro-ku, Tokyo, 153-8902, Japan
†CREST, Japan Science and Technology Corporation (JST),
4-1-8, Honcho, Kawaguchi, 332-0012, Japan

Abstract: The vortex dynamics was investigated by the measurement of the conduction noise in $Bi_2Sr_2CaCu_2O_y$. The narrow-band noise (NBN) was observed in the vortex-solid phase. The shift of the NBN frequency was found to scaled well with the so-called washboard frequency, which is a characterisitic frequency of the velocity modulation of the coherently moving periodic structure under random pinning. Our finding shows that the coherency of the moving vortex lattice is well developed around the first-order vortex lattice phase transition (FOT).

Keywords: Vortex dynamics, Conduction noise, Washboard modulation

INTRODUCTION

The dynamics of vortices has attracted much attention. Under a driven-current, where Lorentz force is exerted on vortices, new dynamic vortex states could appear. One of the most fascinating features is a reordering of moving vortices from plastic flow to coherent flow [1-3]. The uniform flow of a periodic structure in randomly distributed pinning centers shows the periodic velocity modulation with the so-called washboard frequency, f_w, which is given by $f_w = \langle v \rangle / a$, where $\langle v \rangle$ and a is the averaged velocity and the spacing of the periodic lattice, respectively. No direct observation, however, has been reported in the vortex lattice (VL) of any superconductors, except for two interference experiments in Al thin-film [4] and $YBa_2Cu_3O_y$ [5]. In the interference experiments, a large ac driving current was applied to enhance a coherent motion of the VL and the nature of the moving VL is largely changed. Therefore, the direct observation of the washboard modulation without the external ac driving force is needed to investigate the dynamic vortex states.

EXPERIMENTAL

Single crystals of $Bi_2Sr_2CaCu_2O_y$ were grown by the floating-zone method and annealed to an optimally doped state. The critical temperature T_c, defined as the zero resistivity temperature, was 92.2 K. The dimensions of the sample was $1.5 \times 0.5 \times 0.015$

mm^3. The conduction noise spectra were taken around the FOT in a swept magnetic field under various currents. Magnetic field was applied perpendicular to the CuO_2-plane, while the current was within the plane. The fluctuating voltage between the potential electrodes was magnified 100 times using an SR554 preamplifier and fed into an HP-35670A FFT analyzer, which can analyze the noise spectrum up to 1600 Hz by a resolution of 1600 points. Each spectrum data was typically averaged 100 times. The background noise level was as low as 10^{-18} V^2/Hz.

RESULTS AND DISCUSSION

The anomaly of the magnetization, which is associated with the FOT, was found at 70 Oe at 80 K in this sample. The characteristic noise features were observed in the vortex solid phase below the FOT. Figure 1 shows the conduction noise spectra for specific fields under a current of 133 A/cm^2. In lower-field region, which are not shown in Fig. 1, a broadband noise (BBN) was observed as well as that of the local density fluctuation [6]. The behavior of the BBN observed in the conduction noise is quite similar to that observed in the local density noise. Therefore we can safely ascribe the origin of the BBN to a plastic-flow like motion of the VL, as was already discussed [6,7]. Another feature is that a narrow-band noise (NBN) with a peak at a special frequency appeared in each conduction noise specrtum as shown in Fig. 1. The NBN frequency, f_{NBN}, defined at the peak position, shifted to higher frequencies with incresing field. The noise spectrum became featureless again with further increasing field.

Figure 2 shows the magnetic-field dependence of f_{NBN}. The observed NBN suggests the periodic modulation of the translational velocity of the VL. The most possible candidate as its origin is the washboard modulation. We also plot the expected f_w, estimated from the dc resistivity measurement in Fig. 2 as follows. $f_w = \langle v \rangle / a = (\sqrt{3}/2)^{\frac{1}{2}} \rho j / (B\Phi_0)^{\frac{1}{2}}$, where ρ, j, B, Φ_0 are the resistivity, current density, magnetic field and flux quantum, respectively. Since the NBN was observed in the field region where the resistivity was below our measurement senstivity, the direct comparison between f_{NBN} and f_w was impossible except the data at 70 Oe at 66.7 A/cm^2. However, the data points of f_{NBN} connect with that of f_w quite smoothly. This is strong evidence that the NBN observed in the conduction noise resulted from the velocity modulation of the VL with the washboard frequency. On the other hand, the inverse of the transit time, τ_{tr}^{-1}, which scaled the NBN obseved in the local-density fluctuation of the VL well [6], is approximately three orders of the magnitude smaller than f_w.

Our observation of the BBN before the resistivity onset and that of the washboard noise in a slightly higher-field region are quite consistent with the picture obtained in a numerical simulation [3], where the pinned VL changes into the coherently moving VL with washboard velocity modulation, via the plastically deformed VL with increasing driving force. The detailed features, however, are inconsistent with the theoretical predictions. The data in Fig. 2 show the width of the washboard noise became broader and the height of the peak intensity decreased with increasing magnetic field. These

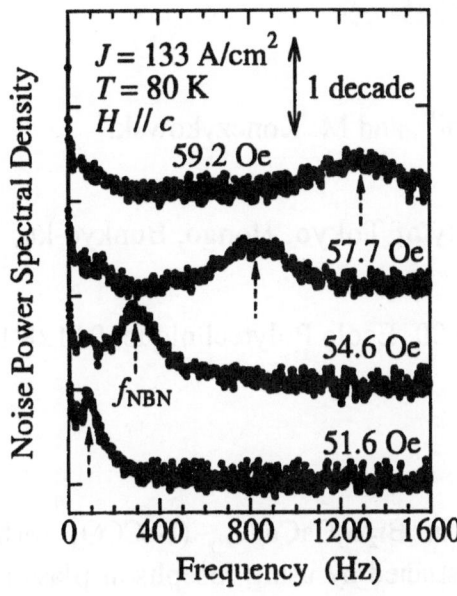
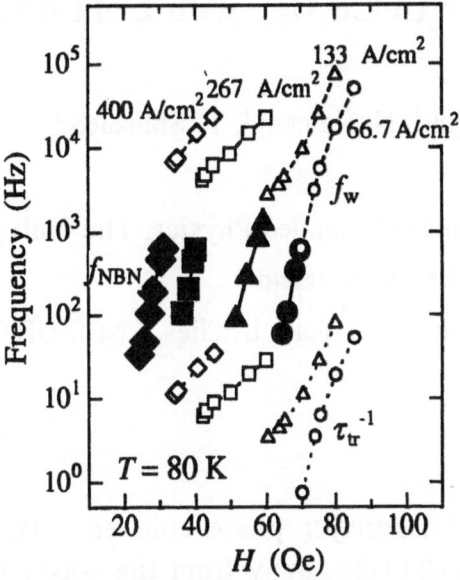

Fig. 1 Conduction noise specrta of $Bi_2Sr_2CaCu_2O_y$ single crystal

Fig. 2 Frequency dependence of f_{NBN}, f_w and τ_{tr}^{-1}. See texts for details.

results suggest that the coherence of the VL represented by the washboard noise deteriorates at a higher driving force. This tendency is on the contrary of the numerical simulation and theories, which show a very coherent nature even in the high diriving force limit. The disagreement should be a problem to be solved in future.

CONCLUSION

We have observed the direct washboard modulation of the VL in superconductors for the first time. Our experimental findings tell that the coherency of the moving vortex lattice is well developed around the vortex lattice phase transition, via plastic flow and will provide much information on new physical aspects of dynamics of vortices.

Acknowledgements: We would like to thank for T. Tsuboi, T. Hanaguri and H. Matsukawa for fruitful discussions. This work was, in part, supported by Grant-in-Aid for Scientific Research on Priority Area "Vortex Electronics". One of the authors (Y.T.) is supported by the JSPS Research Fellowships for Young Scientists.

1. Koshelev, A. E. & Vinokur, V. M. *Phys. Rev. Lett.* **73**, 3580(1994).
2. Doussal, P. Le. & Giamarchi, T. *Phys. Rev. B.* **57**, 11356(1998).
3. Olson, C. J., Reichhardt, C. & Nori, F. *Phys. Rev. Lett.* **81**, 3757(1998).
4. Fiory, A. T. *Phys. Rev. Lett.* **27**, 501(1971).
5. Harris, J. M. *et al. Phys. Rev. Lett.* **74**, 3684(1995).
6. Tsuboi, T., Hanaguri, T & Maeda, A. *Phys. Rev. Lett.* **80**, 4550(1998).
7. Maeda, A. *et al. J. Low Temp. Phys.* in press.

Angular Dependence of Josephson Plasma Resonance in $Bi_2Sr_2CaCu_2O_{8+y}$ with Columnar Defects

T. Tamegai[1], N. Kameda[1], T. Shibauchi[1*], S. Ooi[1+], and M. Konczykowski[2]

[1]Department of Applied Physics, The University of Tokyo, Hongo, Bunkyo-ku,
Tokyo 113-8656, Japan

[2]Laboratoire des Solides Irradies, CNRS URA-1380, Ecole Polytechnique, 91128 Palaiseau,
France

Abstract: Interlayer phase coherence (IPC) in $Bi_2Sr_2CaCu_2O_{8+y}$ (BSCCO) with columnar defects (CDs) tilted away from the c-axis was studied by using Josephson plasma resonance (JPR). IPC in the vortex liquid state shows nonmonotonic field dependence, which we call "recoupling" of decoupled vortex liquid. When CDs are introduced along the c-axis, recoupling occurs at a characteristic field of $H^* = (1/5 \sim 1/3)B_\Phi$, where B_Φ is the matching field. When CDs are introduced at an angle, θ_{CD}, the density of CDs in each CuO_2 layer decreases to $B_\Phi^{eff} = B_\Phi \cos \theta_{CD}$, while the pinning potential is enhanced due to the increase in the size of CDs. At larger θ_{CD}, recoupling occurs at $H^* = (1/5 \sim 1/3)B_\Phi^{eff}$. This means that the recoupling is controlled only by the density of CDs in each CuO_2 layer. Angular dependence of JPR field, $H_p(\theta_H)$, in BSCCO with $\theta_{CD} = 0°$ is smooth near $\theta_H = 0°$. As θ_{CD} increases, $H_p(\theta_H)$ starts to show a pronounced peak near $\theta_H = \theta_{CD}$. These features can be qualitatively understood by rescaling the angles in anisotropic superconductors when they are converted into isotropized system.

KEYWORDS: Josephson plasma resonance, $Bi_2Sr_2CaCu_2O_{8+y}$, columnar defect, angular scaling

INTRODUCTION

Vortex phase diagram in high temperature superconductors has been extensively studied using various techniques. Discovery of vortex lattice melting transition using local Hall probe magnetometry has renewed interest on this issue [1]. Josephson plasma resonance (JPR), which probes the interlayer phase coherence, has been shown to be a powerful tool to study the vortex states in high temperature superconductors [2,3]. Recently, decoupling nature of the melting transition has been demonstrated by using JPR [4]. Vortex phases are affected by the introduction of pinning centers. When we introduce correlated disorder such as columnar defects (CDs) into superconductors, a well-defined "Bose glass" phase is expected [5]. When the number of vortices is equal to the number of CDs at the matching field B_Φ, "Mott insulator" phase is also predicted. Low temperature relaxation measurements in $YBa_2Cu_3O_7$ shows an anomaly in the relaxation rate near B_Φ [6]. However, whether it actually exists or not is still unsettled both theoretically and experimentally.

302

Recent JPR measurements in BSCCO with CDs have shown that the vortex liquid phase is divided into two phases, one with almost decoupled and another with coupled vortex liquid [7,8]. This is also confirmed by c-axis transport measurements [9]. Monte Carlo (MC) simulations by Sugano et al., based on Lawrence-Doniach model has shown the presence of field induced transition at $B_\Phi/3$ accompanied by the enhancement of c-axis coherence, consistent with experimental observations [10]. In addition to this, they find another anomaly at B_Φ, which they ascribe to the remanent of the Mott insulator phase. On the other hand, Wengel and Tauber have concluded from their MC simulation that the Mott insulator phase survives only in a situation where vortex-vortex interaction is negligible, namely $\lambda/d \ll 1$ (d; spacing between CDs) [11]. The effect of CDs on the vortex phases can be controlled not only by changing the density of CDs but also by changing the pinning potential of them. This can be done by tilting the CDs with respect to the c-axis [12,13].

Here we report on the angular dependence of JPR in BSCCO with tilted CDs, and address the issue of the presence of Mott insulator phase. We also give a qualitative explanation to the angular dependence of JPR in these systems by rescaling the angles.

EXPERIMENTAL

BSCCO single crystals used in the present study were grown by the floating zone method [14]. Optimally-doped crystals with T_c of about 90 K were irradiated by 6 GeV Pb ion to introduce CDs at GANIL (Caen, France). The dose-equivalent matching field B_Φ, was fixed to 20 kG. The angles between the CDs and the c-axis (θ_{CD}) were chosen to be 0, 20, 45, 56, 70, 80°. The thickness of the sample was chosen so that the Pb ions can pass through the sample even at $\theta_{CD} = 80°$. JPR was measured using cavity perturbation method at 56 GHz. The sample was placed in a cylindrical copper cavity operated at TE011 mode at the position of maximum microwave electric field. For angular dependent studies, we used a superconducting magnet system with two orthogonal coils (horizontal: 50 kOe, vertical: 30 kOe), which enabled us to apply magnetic field in arbitrary directions without moving the microwave cavity. The irreversible magnetic properties were characterized using a commercial SQUID magnetometer (Quantum Design; MPMS-XL5).

RESULTS AND DISCUSSION

Figure 1(a) shows temperature dependence of JPR field H_p for BSCCO with CDs introduced at various angles (θ_{CD}) from the c-axis when the field is applied parallel to the c-axis. Reflecting the fact that the density of CDs is higher in samples with smaller θ_{CD}, H_p at lower temperature in samples with smaller θ_{CD} is higher than that with higher θ_{CD}. H_p shows a sharp enhancement at θ_{CD} dependent field at high temperatures. In addition to this, weak but temperature independent resonances are observed in all samples as shown in Fig. 1(b) [15]. When the field is normalized by the effective matching field of $B_\Phi \cos\theta_{CD}$, the jump of the phase coherence occurs between almost constant fraction (1/5 - 1/3) of $B_\Phi\cos\theta_{CD}$. This suggests that the recoupling of vortex liquid in BSCCO occurs in a broad range of tilting angle of CDs, and the recoupling field is determined by the number of defects in each CuO_2 layer. When θ_{CD} is larger than 70°, temperature dependence of $H_p/B_\Phi\cos\theta_{CD}$ becomes steeper

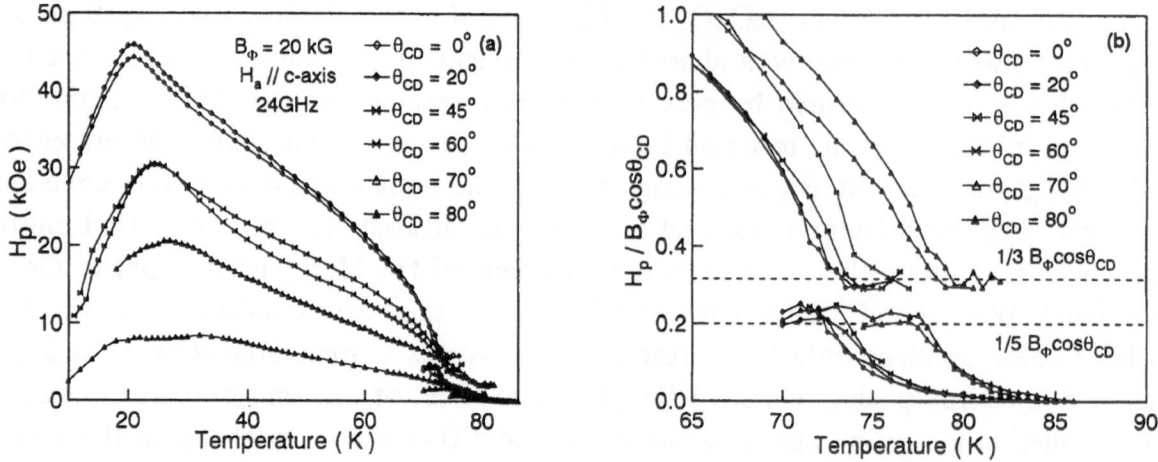

Fig. 1 (a) $H_p(T)$ measured at 56 GHz in BSCCO with θ_{CD} = 0, 20, 45, 56, 70, and 80 ° and with B_Φ = 20kG when H // c. (b) H_p normalized by $B_\Phi \cos\theta_{CD}$ at high temperatures.

especially at high temperatures. This is consistent with the increase in the pinning potential, which is confirmed by reversible magnetization and critical current measurements in BSCCO with CDs at large angles from the c-axis [12,13]. Increase in the pinning potential due to the increase in the size of CDs by a factor of $1/\cos\theta_{CD}$ explains this trend.

Figure 2 shows angular dependence of H_p in the decoupled vortex liquid region in BSCCO with θ_{CD} = 56 °. Although angular dependence of H_p is almost symmetric with respect to θ_H = 0 °, a slight enhancement of H_p is discernible when $\theta_{CD} = \theta_H$. As we have already reported [15], the enhancement of the phase coherence occurs not as a discontinuous jump, but as a continuous crossover from decoupled to recoupled liquid. Even at a field a little bit lower than $1/5 B_\Phi \cos\theta$, there is a region where local matching field is slightly higher than the average and shows recoupling. The asymmetry of angular dependence becomes stronger as we approach $1/5 B_\Phi \cos\theta_{CD}$.

Fig. 3(a) shows temperature dependence of JPR in BSCCO with θ_{CD} = 80 ° when the field is aligned almost parallel to CDs. At the highest temperature of 80 K, there is only one JPR resonance. As temperature is lowered, a broad dissipation peak start to emerge near 7 kOe. In a temperature range between 74 and 72 K, triple resonance peaks are clearly observed.

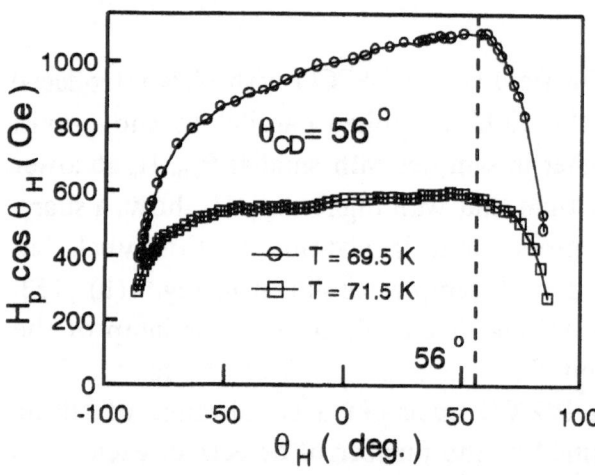

Fig. 2 (a) Angular dependence of H_p in BSCCO with θ_{CD} = 56 ° at T = 69.5 and 71.5 K in the decoupled vortex liquid regime. Slightly asymmetric $H_p(\theta_H)$ with respect to θ_H = 0 indicates the presence of the effect of columnar defects even in this temperature range.

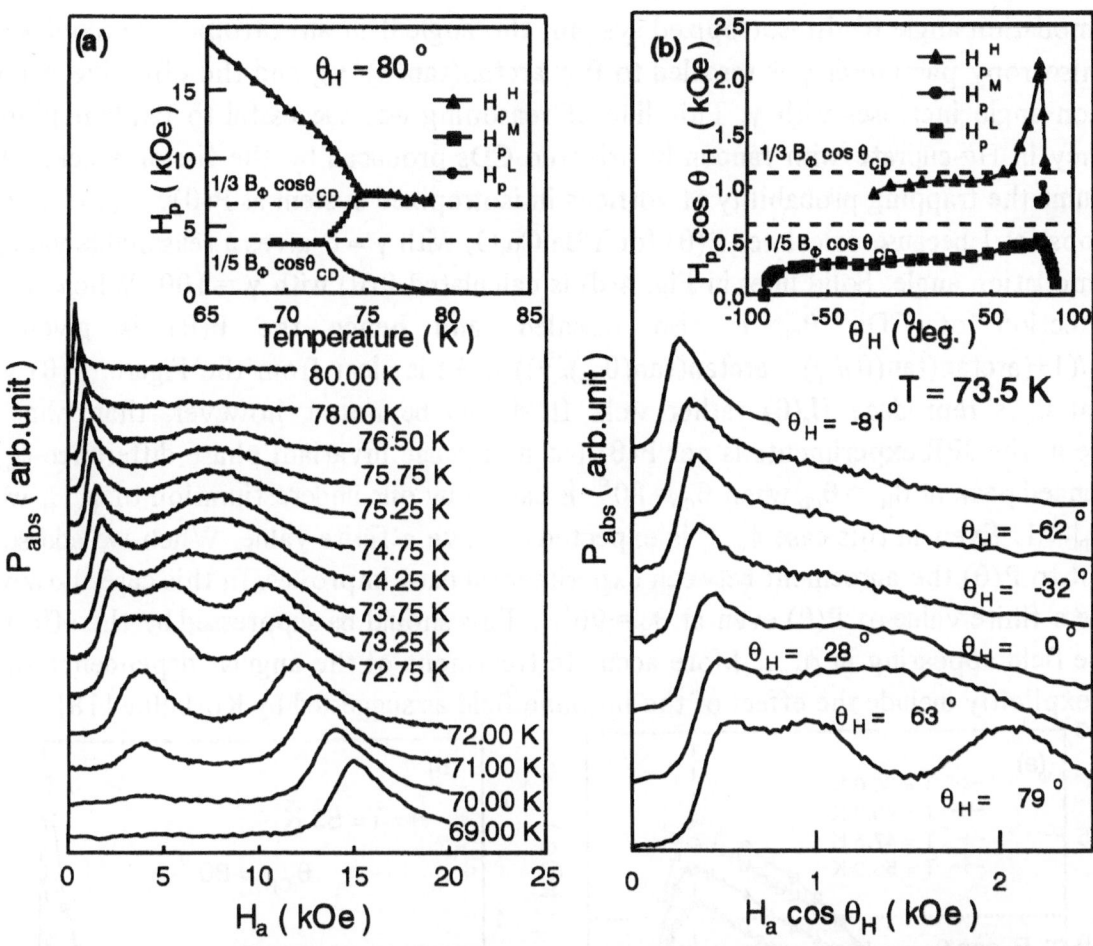

Fig. 3 Microwave dissipation as functions of magnetic field at 56 GHz in BSCCO with CDs. (a) Temperature dependence; inset shows H_p as a function of temperature. (b) Angular dependence; inset shows $H_p(\theta_H)$ at T = 73.5 K.

At lower temperatures, the lower resonance peak is suppressed and the higher resonance peak survives down to the lowest temperatures. The intermediate resonance peak connects two temperature independent lines as shown in the inset of Fig. 2(a). Microwave dissipation as functions of field at various field angles (θ_H) from the c-axis at T = 73.5 K are shown in Fig. 3(b). Triple resonance peaks are observed in a narrow angle region near H // CDs. Below $\theta_H = 73°$, only two resonance peaks are observed. Below about $\theta_H = -30°$, only one resonance is resolved as shown in Fig. 3(c).

Angular dependences of $H_p\cos\theta_H$ at low temperatures in BSCCO with $\theta_{CD} = 56$ and $80°$ are shown in Fig. 4(a) and (b), respectively. $H_p\cos\theta_H$ takes a maximum at $\theta_H = \theta_{CD}$. It should be noted that the angular dependence does not show any anomaly at $H = B_\Phi\cos\theta_{CD}$ (dotted lines). This clearly indicates that Mott insulator phase is not realized in the explored temperature range. Fig. 4(c) summarizes angular dependence of $H_p\cos\theta_H$ in BSCCO with $\theta_{CD} = 0, 20,$ and $45°$. In all samples, $H_p\cos\theta_H$ takes its maximum when $\theta_H = \theta_{CD}$. However, the peak in $H_p\cos\theta_H$ becomes broader when θ_{CD} is lower. Overall features of angular dependence can be understood by considering the angular scaling of physical quantities in anisotropic superconductors predicted by Blatter et al. [16]. If we introduce CDs in an isotropic superconductor, vortices are expected to be trapped in the CDs within a characteristic accommodation angle θ^{*}. In isotropized system, the angle θ in anisotropic superconductors with anisotropy parameter γ is rescaled to $\tilde{\theta} = \arctan(\tan(\theta)/\gamma)$, and the effective accommodation angle increases with γ. This line of reasoning was successful to explain pinning efficiency in Hg-cuprate with randomly oriented CDs produced by the fission tracks [17]. We assume the trapping probability of vortices in isotropized system as $P(\theta)= 1/(1+(\theta/a)^{2})$. We choose $a=1$ because calculated $P(\theta)$ for $YBa_2Cu_3O_7$ with $\gamma = 6$ gives a reasonable value of accommodation angle. Solid lines in Fig. 4(d) is calculated $P(\theta)$ with $\gamma = 100$. When we tilt the direction of CDs, θ_{CD} is also rescaled, and hence the $P(\theta)$ is given by $P(\theta)=1/(1+(\arctan(\tan(\theta)/\gamma) - \arctan(\tan(\theta_{CD})/\gamma))^{2})$. As is clear from the Figure, $P(\theta)$ with different θ_{CD}'s reproduce $H_p(\theta)$ rather well. It should be noted, however, that what we measure in the JPR experiments is not $P(\theta)$ but it is gauge invariant phase difference $\phi_{n,n+1}$. Pronounced peak at $\theta_H = \theta_{CD}$ when $\theta_{CD}= 80°$ is caused by our underestimation of $\phi_{n,n+1}$ when $|\theta_H - \theta_{CD}|\gg 1$. Even in this case $\phi_{n,n+1}$ is expected to have a finite value. When we add some constant to $P(\theta)$ the agreement between experimental data improves. In this case, however, we have a finite value of $P(\theta)$ even at $\theta_H = 90°$. This should be suppressed by the effect of in-plane field appearing in $\phi_{n,n+1}$. More accurate treatment of the angular dependence of H_p should explicitly include the effect of the in-plane field as suggested by Koshelev [18].

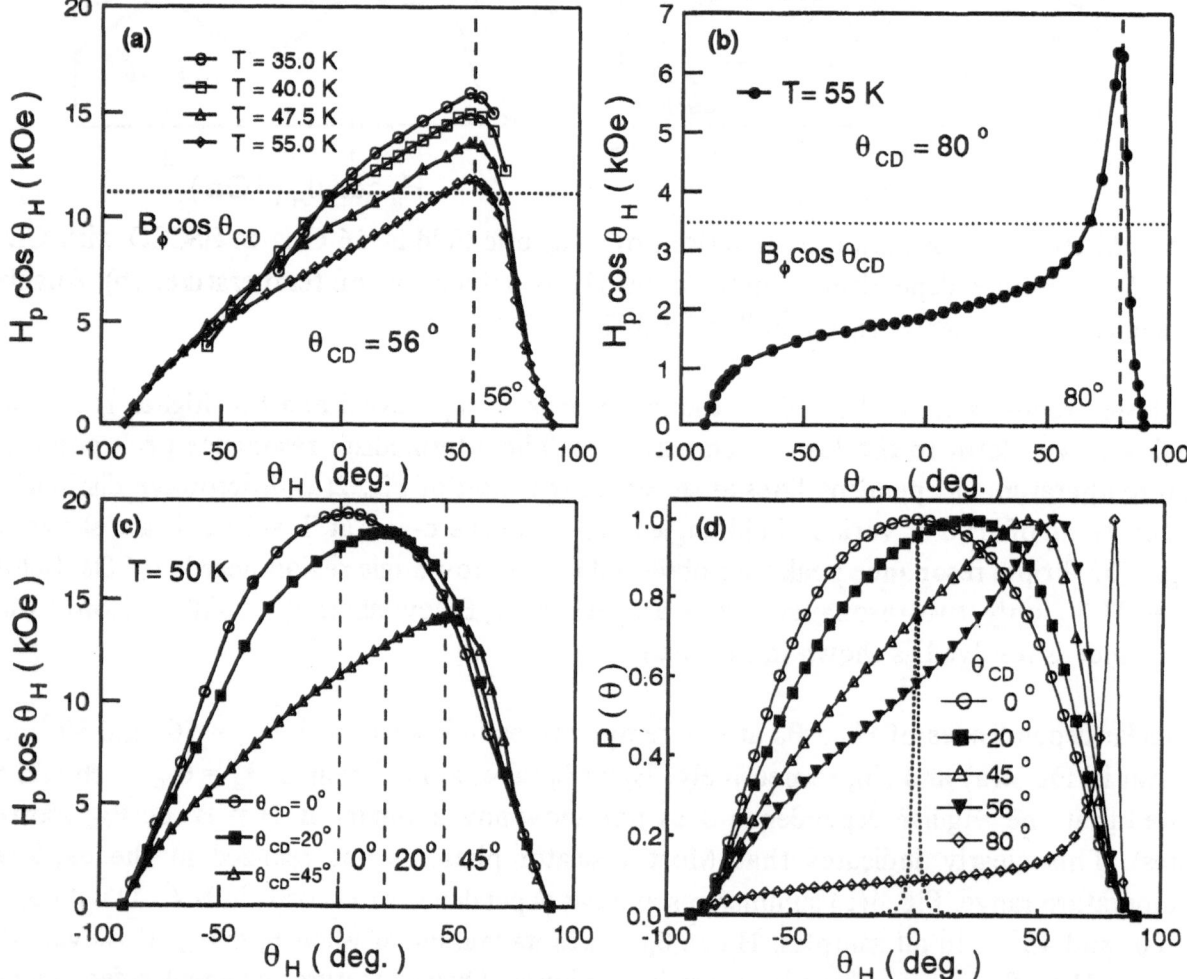

Fig. 4 Angular dependence of H_p in BSCCO with (a) $\theta_{CD} = 56°$, (b) $\theta_{CD} = 80°$, and (c) $\theta_{CD} = 0, 20,$ and $45°$. Dotted lines in (a) and (b) shows the effective matching field $B_\phi\cos\theta_{CD}$. (d) Angular dependence of trapping probability of vortices $P(\theta)$. Dotted line is for a sample with $\gamma =1$, and solid lines are for samples with $\gamma =100$ and with indicated θ_{CD}'s.

CONCLUSION

Angular dependence of Josephson plasma resonance in $Bi_2Sr_2CaCu_2O_{8+y}$ with columnar defects tilted away from the c-axis was studied. Recoupling of the decoupled vortex liquid in this system is found to occur at a characteristic field of $H^* = (1/5 \sim 1/3)B_\phi \cos\theta_{CD}$. This means that the recoupling field is controlled only by the density of CDs in each CuO_2 layer. Angular dependence of JPR field H_p in $Bi_2Sr_2CaCu_2O_{8+y}$ with $\theta_{CD} = 0°$ is smooth near $\theta_H = 0°$, while it shows a pronounced peak near $\theta_H = \theta_{CD}$ when θ_{CD} is large. These features of angular dependence of H_p is qualitatively understood by rescaling the angles in anisotropic superconductors when they are converted into isotropized system.

Acknowledgements. This work is partly supported by CREST and Grant-in-Aid for Scientific Research from Ministry of Education, Science, Sports, and Culture, Japan.

*Present address: IBM T. J. Watson Research Center Box 218, Yorktown Heights, NY 10598, USA

+Present address: National Research Institute for Metals, Sengen 1-2-1, Tsukuba 305-0047, Japan

1, E. Zeldov, D. Majer, M. Konczykowski, V. B. Geshkenbein, V. M. Vinokur, and H. Shtrikman, Nature **375**, 373-376 (1995).

2. O. K. C. Tsui, N. P. Ong, Y. Matsuda, Y. F. Yan, and J. B. Peterson JB, Phys. Rev. Lett. **73**, 724-727 (1994).

3. Y. Matsuda, M. B. Gaifullin, K. Kumagai, T. Mochiku, K. Kadowaki, and K. Hirata, Phys. Rev. Lett. **75**, 4512-4515 (1995).

4. T. Shibauchi, T. Nakano, M. Sato, T. Kisu, N. Kameda, N. Okuda, S. Ooi, and T. Tamegai, Phys. Rev. Lett. **83** (1999) 1010-1013 (1999).

5. D. R. Nelson, and V. M. Vinokur, Phys. Rev. Lett. **68**, 2398-2401 (1992).

6. K. M. Beauchamp, T. F. Rosenbaum, U. Welp, G. W. Crabtree, and V. M. Vinokur, Phys. Rev. Lett. **75**, 3942-3945 (1995).

7. M. Sato, T. Shibauchi, S. Ooi, T. Tamegai and M. Konczykowski, Phys. Rev. Lett. **79**, 3759-3762 (1997).

8. M. Kosugi, Y. Matsuda, M. B. Gaifullin, L. N. Bulaevskii, N. Chikumoto, M. Konczykowski, J. Shimoyama, K. Kishio, K. Hirata, and K. Kumagai, Phys. Rev. Lett. **79**, 3763-3766 (1997).

9. N. Morozov, L. N. Bulaevskii, M. P. Maley, and J. Sarrao, Phys. Rev. **B57**, R8146-8149 (1998).

10. R. Sugano, T. Onogi, K. Hirata, and M. Tachiki, Phys. Rev. Lett. **80**, 2925-2928 (1998).

11. C. Wengel, and U. C. Tauber, Phys. Rev. **B58**: 6565-6579 (1998).

12. R. J. Drost, C. J. van der Beek, H. W. Zandbergen, M. Konczykowski, A. A. Menovsky, and P. H. Kes, Phys. Rev. **B59**: 13612-13615 (1998).

13. S. Hebert, V. Hardy, G. Villard, M. Hervieu, Ch. Simon, and J. Provost, Physica **C299**, 259-266 (1998).

14. S. Ooi, T. Shibauchi, and T. Tamegai, Physica **C302**, 339-345 (1998).

15. T. Tamegai, M. Sato, T. Shibauchi, S. Ooi, and M. Konczykowski, Advances in Superconductivity X, 473-476 (1998).

16. G. Blatter, V. Geshkenbein, and A. I. Larkin, Phys. Rev. Lett. **68**, 875-878 (1992).

17. L. Krusin-Elbaum, G. Blatter, J. R. Thompson, D. K. Petrov, R. Wheeler, J. Ullmann, and C. W. Chu, Phys. Rev. Lett. **81**, 3948-3951 (1998).

18. A. E. Koshelev, L. N. Bulaevskii, and M. P. Maley, Phys. Rev. Lett. **81**, 902-905 (1998).

Depinning and Phase Transformation in the Vortex Solid Phase in $Bi_2Sr_2CaCu_2O_{8+\delta}$

R. Sugano[1], T. Onogi[1], K. Hirata[2], and M. Tachiki[2]

[1]Advanced Research Laboratory, Hitachi, Ltd., Hatoyama, Saitama 350-0395, Japan
[2]National Research Institute for Metals, 1-2-1 Sengen, Tsukuba, Ibaraki 305-0047, Japan

Abstract: We investigated the vortex phase diagram of a model $Bi_2Sr_2CaCu_2O_{8+\delta}$ via a Monte Carlo simulation based on the Lawrence-Doniach model. We find that a strong point-like disorder newly induces a depinning transition as pre-melting, from a strongly-pinned vortex glass (or low-field Bragg glass) to a weakly-pinned soft glass well below the melting line when the temperature is increased. This phase boundary makes an almost vertical line near 30 K. Subsequently, at temperatures above 30 K, the intermediate glass melts with interlayer decoupling of the vortex lines, implying a loss of interlayer long-range coherence. These two phase boundaries are compared to recent experimental data.

Key words: Computer simulation, $Bi_2Sr_2CaCu_2O_{8+\delta}$, vortex glass, depinning

INTRODUCTION AND MODEL

The inhomogeneity inherent in a single crystal (including oxygen defects) makes a vortex phase diagram much more complicated in comparison to an ideal disorder-free system. For example, recent experimental study of the vortex penetration for $Bi_2Sr_2CaCu_2O_{8+\delta}$ (BSCCO) crystals through a surface barrier reported that a phase-separating line exists well below the melting line[1], suggesting a depinning by the point disorder associated with the oxygen defects. In this paper we numerically explore the effects of strong point-like disorders on the vortex-matter phase diagram of BSCCO over a wide range of magnetic fields (B) and temperatures (T).

Based on the Lawrence-Doniach model, we performed a finite-temperature Monte Carlo simulation of N_v Josephson-coupled vortex lines penetrating through N_z layers in the presence of randomly distributed point-pins, by using a method similar to that in Ref [2]. With pancake-vortex coordinates labeled by $\{r_i(z)\}$ of the i-th vortex line $(i = 1, 2, \cdots, N_v)$, the in-plane intervortex interaction is given by a modified Bessel function $\epsilon_0 d K_0(|r_{ij}|/\lambda_{ab})$ with magnetic penetration depth $\lambda_{ab}(T) = \lambda_{ab}(0)\sqrt{1 - (T/T_c)^4}$. We set $\lambda_{ab}(0) = 2000$ Å, coherence length $\xi_{ab}(0)=10$ Å, layer thickness $d=10$ Å, and effective mass anisotropy $\gamma = \sqrt{M_c/M_{ab}} = 100$. We also set $N_v = 16$ and $N_z = 40$, where periodic-unit cells with image potentials are used in each layer. A point pin is modeled [3] by a cylindrical potential well with radius $c_0 = \xi_{ab}(0)$ and depth $U_0(T) = (\epsilon_0 d/2)\ln[1 + (c_0/\sqrt{2}\xi_{ab}(T))^2]$. We used four random pin configurations for the sample average, with a pin density of $7 \times 10^{11}/cm^2$.

RESULTS AND DISCUSSION

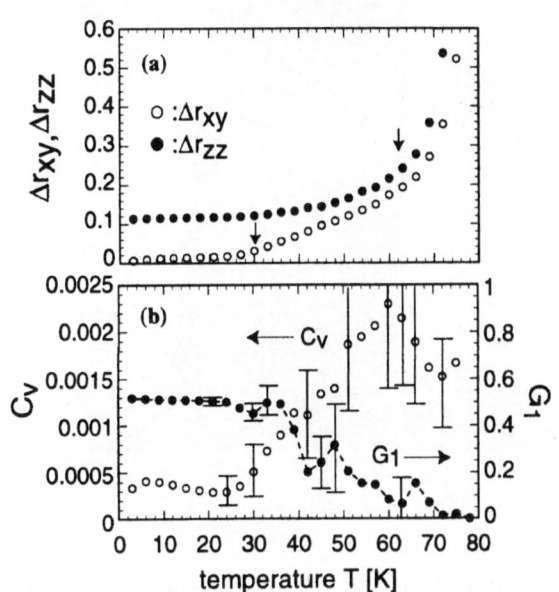

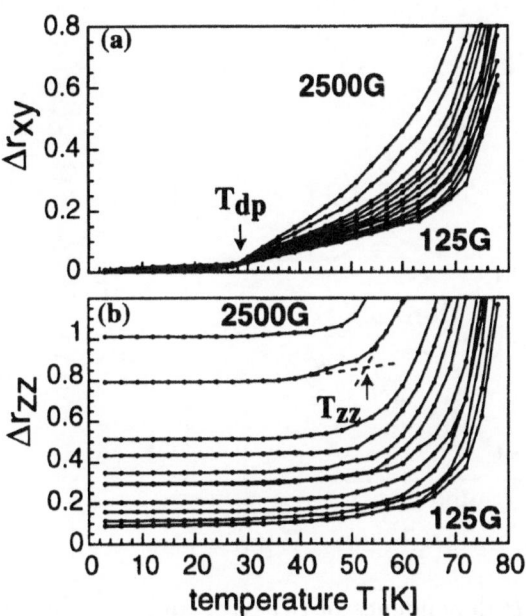

Fig. 1. T-dependence of (a) Δr_{xy}, Δr_{zz}, (b) G_1, and C_v, at $B = 200$ G.

Fig. 2. T-dependence of (a) Δr_{xy}, and (b) Δr_{zz}, for $B = 125, 200, 250, 320, 400, 640, 800, 1000, 1600, 2500$ G.

Figure 1(a) shows the typical T-dependent behavior of the in-plane vortex fluctuation, $\Delta r_{xy} = \langle |r_{i,z} - r_{i,z}^{\text{ave}}|^2 \rangle_{i,z}^{1/2} / a_0$ ($a_0 = \sqrt{\Phi_0/B}$), and the out-of-plane fluctuation, $\Delta r_{zz} = \langle |r_{i,z} - \langle r_{i,z}^{\text{ave}} \rangle_z|^2 \rangle_{i,z}^{1/2} / a_0$ at 200 G. We can see that Δr_{xy} bends twice at around 30 and 65 K. Below 30 K, where $\Delta r_{xy} \simeq 0$, almost all vortices are trapped into defects and frozen. Above 30 K, Δr_{xy} increases linearly up to $\simeq 0.25$ where $T \simeq 65$ K. In this intermediate regime, as shown in Fig. 1(b), the first Bragg peak intensity G_1 becomes unstable in contrast to its constant behavior at lower temperatures. This suggests that vortices begin to wander by depinning from the defects above 30 K. trapped into point pins and are frozen below 30 K. The Δr_{zz} does not show any distinct change around 30 K. Sharp change in the slope of Δr_{zz} occurs only at around 65 K, where we can see a prominent peak in the T-dependence of the specific heat, C_v. Above 65 K, both Δr_{xy} and Δr_{zz} proliferate and G_1 almost vanishes. This suggests that the vortices form quasi-triangular lattice (Bragg glass) below 65 K, and melt into vortex liquid (or pancake gas).

Figure 2 shows the T-dependent profiles of Δr_{xy} and Δr_{zz} for the various field B. Δr_{xy} sharply bend and cross each other at a depinning temperature $T_{dp} \simeq 30$ K for the whole field range. The T_{dp} seems insensitive to the magnetic field. Around this T_{dp}, Δr_{xy} takes a much smaller value (below 0.03) compared to the standard Lindemann melting criteria of 0.2-0.3. This suggests the occurrence of a depinning transition before melting. Well above T_{dp}, Δr_{zz} shows an abrupt change in slope at $T_{zz}(B)$, then proliferates rapidly above T_{zz}, implying the loss of the c-axis coherence. At T_{zz}, Δr_{xy} satisfies the Lindemann melting criteria, suggesting the vortices melting into the decoupled vortex liquid.

While Bragg glass appears below T_{zz} in the low-field regime, G_1 almost vanishes in the higher field. Figure 3(a) shows G_1 as a function of B below 36 K. G_1 decreases and vanishes at $B_{G1}(T)$. The B-dependence of Δr_{zz} also shows an abrupt change in slope near B_{G1},

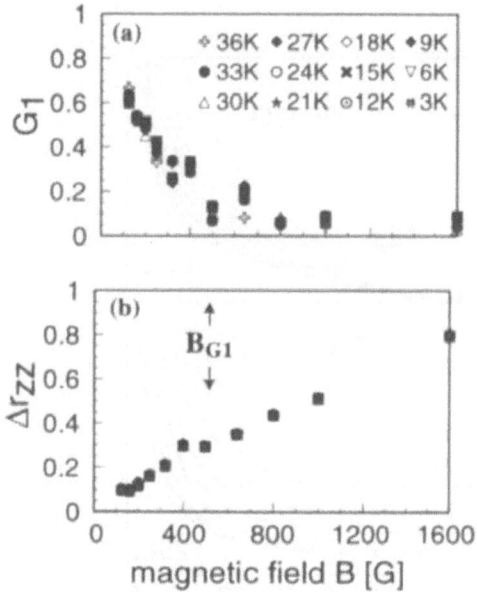

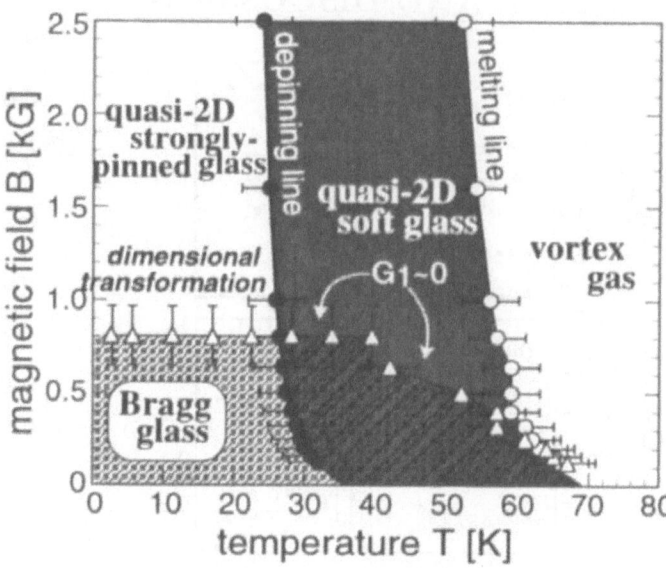

Fig. 3. B-dependence of (a) Δr_{zz} and (b) G_1 below $T = 36$ K.

Fig. 4. Calculated vortex phase diagram. T_{dp}: solid circles, T_{zz}: open circles, $B_{G1}(T)$: open triangles.

suggesting a reduction of c-axis coherence. B_{G_1} seems independent of T, therefore the $B_{G1}(T)$ line parallel to the c-axis may correspond to the 2D-3D crossover field observed in the BSCCO single crystals[4]. Above B_{G1}, a quasi-2D strongly-pinned glass phase may appear at low temperatures, distinguished from the low-field Bragg glass. And this strongly-pinned glass is expected to be transformed into weakly-pinned (soft) glass at T_{dp}, with increasing T.

Figure 4 shows the global B-T phase diagram obtained from our calculations. An almost vertical depinning line T_{dp} divides the vortex solid phase into two phases: quasi-2D strongly-pinned glass and quasi-2D soft-glass in the higher field. At lower fields, an increase in T, Bragg glass depins at T_{dp} and melts into the liquid phase at T_{zz} via the soft-glass phase, in the intermediate temperature regime between T_{dp} and T_{zz}. The simulated phase diagram also sheds light on another novel phase boundary called "T_x-line", which was experimentally pointed by Fuchs, et al. [1] and is still under debate[5]. We suggest that the T_x-line corresponds to melting from the quasi-2D soft glass to pancake vortex gas.

Acknowledgments. This research was supported by Joint Research Promotion System on Computational Science and Technology (Science and Technology Agency, Japan).

REFERENCES

1. D. T. Fuchs, *et al.*, Phys. Rev. Lett. **80**, 4971-4974 (1998).
2. R. Sugano, *et al.*, Phys. Rev. Lett. **80**, 2925-2928 (1998).
3. D. R. Nelson and V. M. Vinokur, Phys. Rev. B **48**, 13060-13097 (1993).
4. T. Tamegai, *et al.*, Physica C **213**, 33-42 (1993).
5. T. Shibauchi, *et al.*, Phys. Rev. Lett. **83**, 1010-1013 (1998).

Magnetization Behavior of BSCCO with Various Configuration of Columnar Defects

Noriko Chikumoto[1], Marcin Konczykowski[2], Yuji Matsuda[3], Jun-ichi Shimoyama[4], Kohji Kishio[4], and Masato Murakami[1]

[1]Superconductivity Research Lab., ISTEC, Shibaura, Minato-ku, Tokyo 105-0023, Japan
[2]Laboratoire des Solides Irradiés, Ecole Polytechnique, 91128 Palaiseau, France
[3]Inst. for Solid State Physics, The Univ. of Tokyo, Roppongi, Minato-ku, Tokyo 106-0032, Japan
[4]Dept. of Superconductivity. The Univ. of Tokyo, Hongo, Bunkyo-ku, Tokyo 113-0033, Japan

Abstract : $Bi_2Sr_2CaCu_2O_{8+\delta}$ single crystals were irradiated with 5.8 GeV Pb-ions to introduce columnar defects (CD). The orientation of the defect structure was altered by changing the direction of incident beam. The peak effect was observed in the sample with CD // c-axis // B_a, and it smeared gradually as CD was tilted away from the field direction (c-axis). The irreversibility field B_{irr} was scaled with the effective defect density, except for the region where the "re-coupling" transition occurs.

Keywords : BSCCO, Columnar Defects, Bose-glass, Peak Effect

INTRODUCTION

Recently it has been revealed both experimentally [1-3] and numerically [4] that the strong pinning force and the linear geometry of the columnar defects (CD) causes an increase of the c-axis correlation of vortices in the liquid phase, which results in the "re-coupling" of pancake vortices at an intermediate field. It was also reported that, in the low-temperature Bose-glass phase, the sample exhibits non-monotonic field dependence of J_c ('peak effect') which is accompanied by the re-entrant behavior of the irreversibility line [5]. Furthermore, recent computer simulation study showed that the peak effect occurs simultaneously with a rapid change in the interlayer-coupling of vortices [6]. Since all of these events occur at the same specific field, $B \approx 0.3$ T for $B_\Phi = 1$ T sample, where B_Φ is the matching field, the underlying mechanism might be the same. For the moment, we consider these anomalies are caused by the reduction of vortex mobility due to the increased inter-vortex repulsive force, which results in the increase of interlayer coupling. In order to acquire a further insight into the importance of the c-axis correlation of vortices, we investigated the effect of the defect configuration on the peak effect and the irreversibility line (B_{irr}, T_{irr}).

EXPERIMENTAL

$Bi_2Sr_2CaCu_2O_{8+\delta}$ (BSCCO) single crystalline samples were grown with a floating zone method, the details of which were described elsewhere [7]. The samples were cut from a large single crystal with $T_c = 85.7$ K (a slightly under-doped state). All samples were

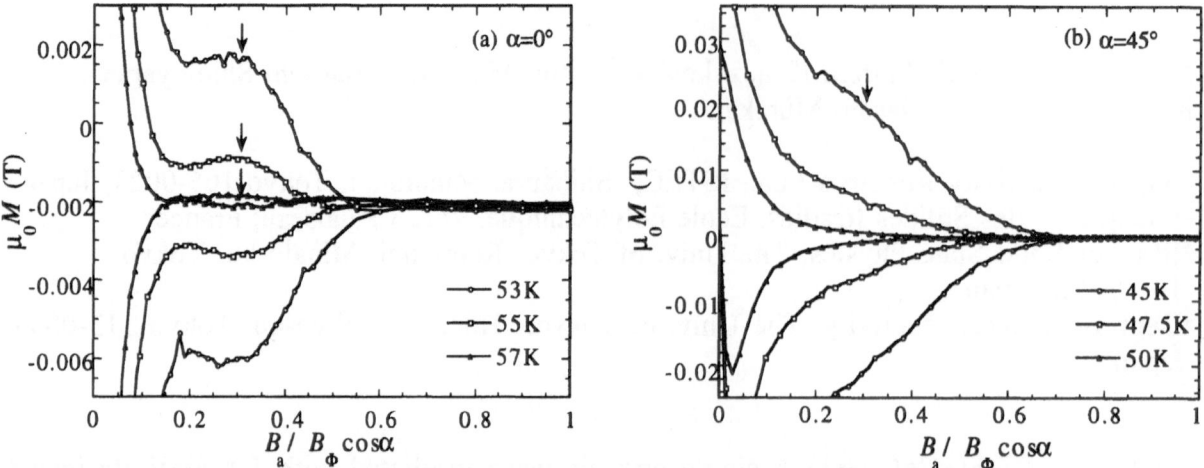

Fig.1 Part of magnetic hysteresis loops for BSCCO crystals irradiated at angles (a) $\alpha = 0°$ and (b) $45°$ with respect to c-axis ($B_\Phi = 1$ T). The magnetic field was applied parallel to c-axis.

irradiated with 5.8 GeV Pb-ions to the same total ion fluence (measured perpendicular to the ion beam) of 5×10^{10} cm^{-2} ($B_\Phi = 1$ T) at Grand Accélérateur National d'Ions Lourds (GANIL) (Caen, France), and the orientation of the ion beam with respect to the c-axis was changed by tilt angle α. It means that the density of CD in the ab-plane $B_\Phi{}^{ab}$ decreases with α as $B_\Phi{}^{ab} = B_\Phi \cos \alpha$, but the larger size of the track projections on the ab-plane for larger α. Magnetization measurements were performed using a SQUID Magnetometer (Quantum Design), with magnetic fields applied parallel to the c-axis.

RESULTS AND DISCUSSION

First we discuss the effect of the CD orientation on the peak effect. Figure 1(a) shows a typical peak effect observed in the hysteresis loop of the sample with CD // c-axis. The peak appears in a relatively high temperature range close to the irreversibility line (IL) and at a temperature independent peak field, $B_{pk} \sim 0.3$ T. Such a clear peak effect is only observed in the sample with CD // c-axis and is smeared when the CD is tilted away from c-axis, as shown in Fig.2 (b) for $\alpha = 45°$. However, if we look carefully, a slight change can be observed in the slope dM/dB at around $B_a/B_\Phi{}^{ab} = 0.3$ (shown by arrow). This result indicates that the inter-vortex force starts to affect the vortex behavior at almost the same filling factor, $B_a/B_\Phi{}^{ab} \sim 1/3$, however for the appearance of the peak effect, the pancake vortices should be lined up toward the c-direction (or the field direction). In other words, an increase of J_c was achieved by cooperative work of inter- and intra-plane forces.

Next we show the effect of CD orientation on IL. In Fig.2 (a), dependence of IL on CD orientation is summarized. IL shifts toward low field with increasing α. However, as shown in Fig. 2 (b), all curves are roughly scaled by $B_\Phi{}^{ab}$, suggesting that the shift of IL with α is mainly due to the decrease of $B_\Phi{}^{ab}$. The effect of CD orientation on IL of 're-coupling' region is shown in the inset of Fig.2 (b). We can see that CD orientation strongly affects the re-entrant behavior of IL and as a result larger T_{irr} for smaller α. This result may reflect the reduction of the peak effect by tilting CD. We also notice that IL of $\alpha = 70°$ sample deviates somehow from the master curve. In order to discuss the effect of CD with large tilting angle, we compare IL of the samples with different α ($\alpha = 0°$ and $70°$) but with

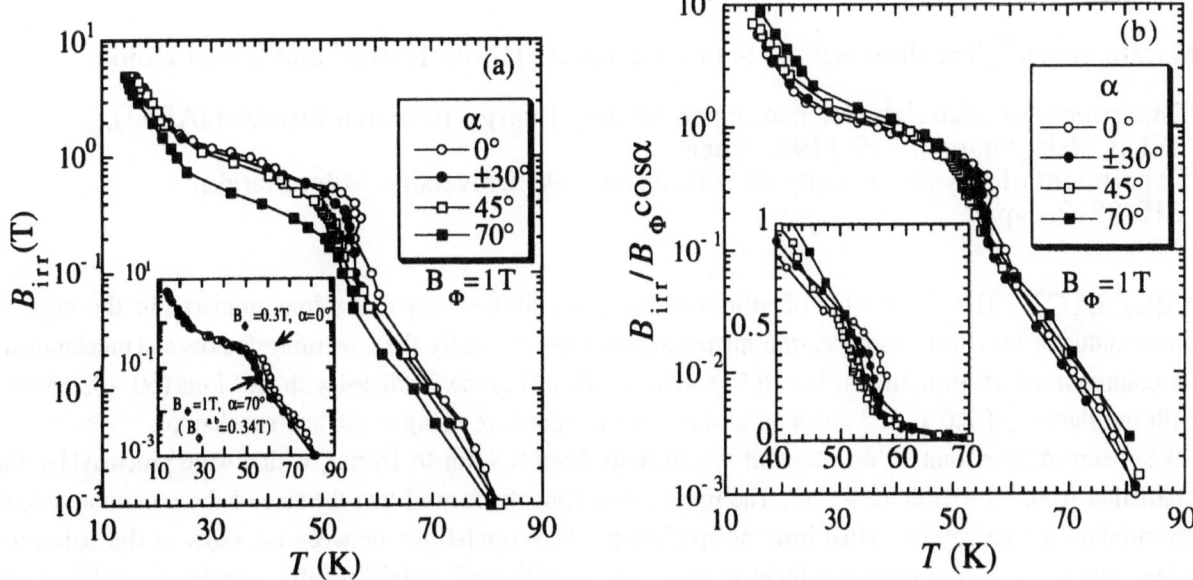

Fig. 2. (a) Temperature dependence of B_{irr} for heavy-ion irradiated BSCCO crystals with the same $B_\Phi =$ 1 T but with different irradiation angles, $\alpha = 0°, \pm30°, 45°$ and $70°$ with respect to c-axis. Inset compares the IL for the samples having similar effective matching field, $B_\Phi^{ab} \sim$ 0.3 T. (b) Plots of IL normalized by B_Φ^{ab}

similar $B_\Phi^{ab} \sim$ 0.3 T in the inset of Fig. 2 (a). At high fields, $B > B_\Phi^{ab}$, IL perfectly coincides, indicating that IL in this region only depends on B_Φ^{ab}. On the other hand, IL at low field was largely shifted toward low temperature, which also suggests that the inter-plane coherence of pancake vortex is important in the low field regime.

Acknowledgments. This work was supported by the New Energy and Industrial Technology Development Organization (NEDO) as Collaborative Research and Development of Fundamental Technologies for Superconductivity Applications.

1. M. Kosugi, Y. Matsuda, M.B. Gaifullin, L.N. Bulaevskii, N. Chikumoto, M. Konczykowski, J. Shimoyama, K. Kishio, K. Hirata, and K. Kumagai, Phys. Rev. Lett. **79**, 3763 (1997) ; Phys. Rev. **B 59**, 8970 (1999).
2. M. Sato, T. Shibauchi, S. Ooi, T. Tamegai, and M. Konczykowski, Phys. Rev. Lett. **79**, 3759 (1997).
3. N. Morozov, M.P. Maley, L.N. Bulaevskii and J. Sarrao, Phys. Rev. **B 57**, R8146 (1998).
4. R. Sugano, T. Onogi, K. Hirata and M. Tachiki, Phys. Rev. Lett. **80**, 2925 (1998)
5. N. Chikumoto, M. Kosugi, Y. Matsuda, M. Konczykowski and K. Kishio, Phys. Rev. **B 57**, 14507 (1998).
6. R. Sugano, T. Onogi, K. Hirata and M. Tachiki, Phys. Rev. **B 60**, 9734 (1999).
7. N. Motohira, K. Kuwahara, T. Hasegawa, K. Kishio and K. Kitazawa, J. Ceram. Soc. Jpn **97**, 994 (1989).

Defect Structure of High-T$_c$ Superconductor by High-Energy Heavy Ion Irradiation

Masato Sasase[1*], Takahiro Satou[2], Satoru Okayasu[1], Hiroki Kurata[1] and Kiichi Hojou[1]

[1]Department of Materials Science, Japan Atomic Energy Research Institute(JAERI), Tokai, Naka, Ibaraki, 319-1195, Japan
[2]Department of Physics, Faculty of Science, Ibaraki University, Mito, Ibaraki, 310-8512, Japan

ABSTRACT : The observation of columnar defects for high-energy heavy ions produced in the high-T$_c$ superconductor was made with electron microscope in order to clarify the governing factors and mechanisms for columnar defects formations. The $Bi_2Sr_2CaCu_2O_8$ (Bi-2212) was irradiated with Au^+ ions (60 ~ 300 MeV) with the fluence of 2.0×10^{10} ions/cm^2 at the room temperature using a Tandem accelerator.

We observed the columnar defects with the diameter from 8.4 nm to 16 nm, which were produced by the irradiation from 60 to 300 MeV Au^+. The present experiments showed that the size of the columnar defects depended on the energy deposited into the specimens. The correlations between the sizes of the columnar defects and the deposited energy of incident ions were investigated, and the results were interpreted in terms of a time dependent line source model of thermal spike. It was found that a part of the deposited energy contributes to the columnar defects formation.

KEYWORDS : high-energy heavy ion, columnar defects, thermal spike

INTRODUCTION

The enhancement of the critical current density (J_c) of the high-T$_c$ superconductor(HTCS's) is very important in practical application. However, the J_c limitation arises from a high rate of thermal creep associated with a lack of strong pinning centers. On the other hand, when new pinning centers such as twin boundaries, stacking faults and precipitates etc. are introduced to the material in an attempt to improve performance, the J_c in an applied field is great enhanced.

Heavy ions with an energy range of several hundreds MeV create amorphous regions, which are called columnar defects, along their linear tracks in the crystal of HTCS's. It is well known that the such defects enhance the J_c, since these defects act as strong pinning centers of flux lines[1, 2]. Previous results show that the diameter of the columnar defects should be comparable to the coherence length[3] in order to obtain the optimum results for J_c. Therefore, it is of primary importance to predict the conditions of ion irradiation which is suitable for the creation of columnar defects with optimum size.

We examined the correlations between the sizes of the columnar defect produced with Au^+ ions irradiation and the deposited energy in the Bi-2212. The experimental results were interpreted in terms of a time dependence line source (TDLS) model [4] of thermal spike.

EXPERIMENTAL

Specimens used in this study were c-axis oriented $Bi_2Sr_2CaCu_2O_8$ single crystals(T_c=90 K). The specimens were irradiated at room temperature with 60~300 MeV Au^+ ions with a fluence of 2.0×10^{10} ions/cm^2 using Tandem accelerator at JAERI. The charge number was from 8 to 24. The range and deposited energy (dE/dx)$_e$ of these ions in the Bi-2212 was calculated by using TRIM codes[5]. Irradiation defects were observed by using Transmission Electron Microscopy (TEM : JEOL JEM-2000F type).

RESULTS AND DISCUSSIONS

In the present experiments, the diameters of the columnar defects have been observed as a function of the energies deposited into the specimens. Fig.1(a) and (b) show the TEM images of columnar defects for 60 MeV and 300 MeV Au$^+$ ions irradiation, respectively. These lattice images were observed along the direction perpendicular to the c-plane. The diameter of the columnar defects were 8.4 and 16.0 nm, respectively. At these experimental conditions, the deposited energy $(dE/dx)_e$ were calculated[5] to 11.6 and 29.0 keV/nm, respectively. These value for each ion energy were listed in Table 1.

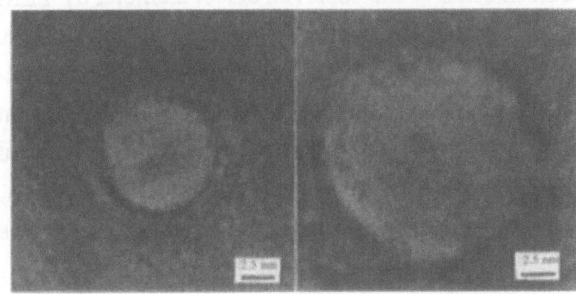

Fig. 1 The TEM images of columnar defects produced at Au$^+$ ion energy of (a) 60 MeV and (b) 300 MeV in HTCS's.

Table 1 The relationship between irradiated ion energy and diameter of columnar defect. Calculated deposited energy is also listed.

Ion energy (MeV)	diameter of defect (nm)	deposited energy $(dE/dx)_e$ (keV/nm)
60	8.4	11.6
120	9.0	19.0
180	10.5	23.8
240	12.5	26.8
300	16.0	29.0

The simplest model ever used is the thermal spike model in order to explain the columnar defects formation under the various conditions. In this model, it is assumed that instantaneous transfer of the whole energy of excited electrons to the lattice, and liberation of heat source obeying the macroscopic equation of lattice thermal conduction[6]. However, this model cannot be applicable for all materials, since it neglects the relaxation time for electron-phonon interactions. In the case of semiconductor, the cylindrical region along the ion track will be heated far above the melting point during several 10^{-10} sec [4]. The produced defect along the track will be two or three order larger than the observed defect in the present experiments.

Based on these consideration, Izui at al. proposed a modified model of columnar defect formation in semiconductor and ionic crystal instead of the thermal spike model. This model is experimentally proved in semiconductor[4, 6]. It can be considered that semiconductor and the HTCS's have the same property to analyze the defect formation process, since the carrier concentration in these materials are low enough. The TDLS model is feasible to apply in the HTCS's.

In this model, it is assumed that the material melts as a cylindrical shape during the time of 5 x 10^{-12} sec, then the amorphous phase is formed if the temperature exceeds the melting point T_m. The basic idea of this model which is proposed by Izui is as follows ; secondary electrons are produced with energies ranging from a few eV to 1 keV in narrow cylindrical region of about 1 nm in diameter along the trajectory of an ion. Among these excited electrons, electrons with relatively high energy quickly run away from this region leaving a row of positively charged atoms. On the other hand, electrons with low energy are bound in this narrow region by Coulomb attraction of these positive ions. These electrons with low energy transfer their energies to the lattice by electron-phonon interaction. Taking into account of the relaxation time of electron-phonon interaction, it is assumed that the thermal energy is released to lattice time dependent at the rate Q(t), then the temperature T(r, t) with normal distance r from the line source at the time t is given

$$Q(t) = -n\frac{d\varepsilon}{dt} = nA\left(\frac{A}{2}t + \varepsilon_o^{-\frac{1}{2}}\right)^{-3}$$

$$T(r,t) = \frac{1}{4\pi K}\int_o^t \frac{Q(t')}{t - t'}\exp\left\{-\frac{r^2}{4D(t - t')}\right\}dt'$$

$$w = \int Q(t)dt = n\varepsilon_o = \left(\frac{dE}{dx}\right)_e$$

where K and D are the thermal conductivity and the thermal diffusion coefficient, respectively, and w and $(dE/dt)_e$ are same value for the energy loss rate of electron due to electron-phonon collisions[4], ε_o is the initial energy which an electron receives from incident ion and n is the number of electrons contributing to the thermal spike per unit length along this line source.

This model is applied to the present experimental conditions. The calculated results for 60 MeV and 300 MeV Au$^+$ ion irradiation are shown in Fig. 2 and 3, respectively. In these figures, the temperature variation of the produced columnar defect are shown as a function of time t. The deposited energy, w , is taken as a parameter. This value w is varied 2.9 to 11.6 keV/nm, and from 7.3 to 29 keV/nm for 60 MeV and 300 MeV, respectively. The distance from the line source, r , is taken from a half of the observed defect diameter by TEM images. The dotted line in the figure at the 1000 °C shows the melting point (T_m) of the present sample. According to the pico-second laser pulse irradiation experiments[7], the melting or consequent recrystall-ization takes place just above the T_m within the several 10^{-12} sec. It is assumed that melting of the present sample take place when the sample is heated above 1000 °C for 5 x 10^{-12} sec. In the basis of the calculated results for 60 MeV and 300 MeV, the required deposited energies , w, are 3.9 keV/nm and 9.7 keV/nm, respectively. These values are apparently lower than the deposited energy in Table 1. This difference shows that only a part of the deposited energy effectively contributes to the columnar defects formation.

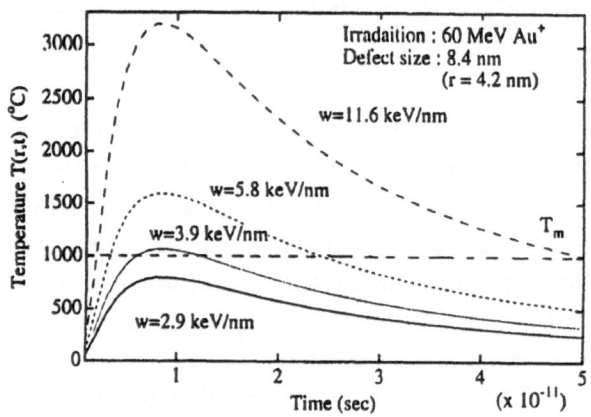

Fig. 2 Temperature variation at r=4.2 nm for each w as a function of t.

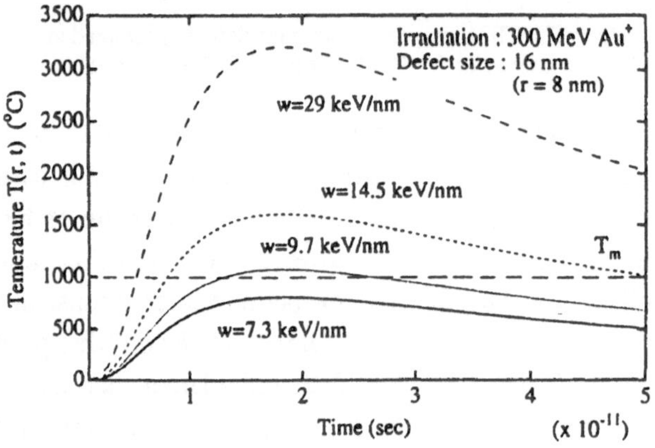

Fig.3 Temperature variation at r=8 nm for each w as a function of t .

CONCLUSION

The correlations between the sizes of the columnar defect produced with Au$^+$ ions irradiation and the deposited energy have been investigated in the Bi-2212. The present results shows that around a third of the deposited energy is consumed for defect formation even though there are some uncertainty in the calculation.

REFERENCE

[1] J. R. Thompson, Y. R. Sun, H. R. Kerchner, D. K. Christen, B. C. Sales, B. C. Chakoumakos, A. D. Marwick, L. Civale and J. O. Thomson, Appl. Phys. Lett., 60(1992)2306.

[2] W. Jiang, N.-C. Yeh, S. Reed, U. Kriplani, D. A. Bean, M. Konczykowski, T. A. Tombrello and F. Holtzberg, Phys. Rev. Lett. 72(1994)550.

[3] L. Civale, A. D. Marwick, T. K. Worthigton, M. A. Kirk, J. R. Thompson, L. Krusin-Elbaum, Y. Sun, J. R. Clem and F. Holtzderg, Phys. Rev. Lett. 67(1991) 648.

[4] K. Izui, J. Phys. Soc. Jpn. 20(1965)915.

[5] J.F. Ziegler, "Handbook of Stopping Cross Section for Energetic Ions in All Elements" (Pergamon, New York, 1980).

[6] S. Furuno, H. Otsu, K. Hojou and K. Izui, Nucl. Instrm. and Meth. B107(1996)223.

[7] G. A. Rozgonyi, H. Baumgart, F. Phillipp, R. Vebbing and H. Oppolzer, "Laser and Electron-Beam Interactions with Solids" (North-Holland, New York, 1982)177.

NONLOCAL ELECTRODYNAMICS IN $Bi_2Sr_2CaCu_2O_{8+\delta}$ SINGLE CRYSTAL IN A CORBINO DISK GEOMETRY

Yuri Eltsev, Koichi Nakao, Susumu Shibata and Naoki Koshizuka

Superconductivity Research Laboratory, ISTEC, 10-13 Shinonome 1-chome Koto-ku, Tokyo, 135, Japan

Abstract: Electrical transport measurements in $Bi_2Sr_2CaCu_2O_{8+\delta}$ single crystal have been made using a Corbino disk geometry with current applied between center of disk and its perimeter and voltage response measured across two pairs of equally spaced potential contacts located along sample radius. In the normal state radial potential distribution is well discribed by a local conductivity model. Below T_c at $0.02T \leq B \leq 0.2T$ contrary to $1/r$ current density distribution voltage response measured near the sample perimeter exceeds that taken closer to the center. This nonlocal behavior was observed above vortex melting transition indicating strong transverse vortex-vortex correlation in the vortex liquid state.

Keywords: $Bi_2Sr_2CaCu_2O_{8+\delta}$, superconducting disk, vortex lattice melting, transverse vortex correlation

Among the many aspects of vortex dynamics in high-T_c superconductors (HTSC) one of the current questions is whether local or non-local electrodynamics is required to describe transport properties of the vortex state. In a local conductor like a normal metal electrical field at any point of sample is determined by a local value of a (non-uniform) transport current while in a type-II superconductor correlated motion of vortices due to finite vortex line tension or transverse vortex-vortex interaction may result in a non-local voltage response induced distantly from the position where current acts on the flux line. Possible non-local effects due to the *longitudinal* correlation of vortices have been extensively studied in $Bi_2Sr_2CaCu_2O_{8+\delta}$ (Bi-2212) and $YBa_2Cu_3O_{7-\delta}$ (Y-123) single crystals using transport measurements in a flux transformer (FT) contact configuration with magnetic field applied perpendicular to CuO_2 layers [1]. However, different groups reported controversial results with arguments advanced both in favor and against applicability of the local picture to explain FT experiments. On the other hand substantial non-local contribution to the transport properties due to the *transverse* vortex interaction has been recently found in Y-123 in a Corbino disk geometry [2] and in FT contact configuration with B//*ab*-planes [3].

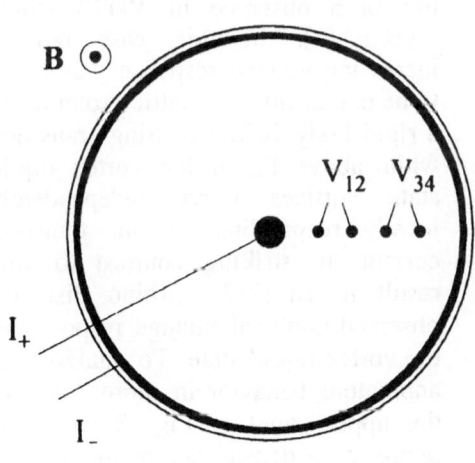

Fig. 1. Corbino disk contact configuration.

We report on the observation of nonlocal voltage response in Bi-2212 single crystals in the Corbino disk geometry (see Fig. 1). In this contact configuration current entering the sample at the center of disk and leaving it at the perimeter generates current density decreasing as $1/r$ with radius. In the mixed state the Lorentz force on the vortices//*c*-axis induced by this current pushes them along closed circular orbits and vortex velocity distribution may be resolved by measuring the radial potential distribution. We show that in the normal state in agreement with a local conductivity model the voltage profile follows diminishing as $1/r$ current density while below T_c contrary to the driving force distribution vortices near the sample perimeter move faster than those close to the center indicating strong deviation from the local picture. This nonlocal behavior was found above flux line melting transition suggesting high transverse vortex-vortex correlation in the vortex liquid state.

$Bi_2Sr_2CaCu_2O_{8+\delta}$ single crystal was grown by floating zone technique [4] and annealed at $400^{\circ}C$ in air for

20h. Thin platelet crystals of about 1.5 x 1.5 x 0.015mm^3 were carefully cut into a disk shape. Electrical contacts in the Corbino disk geometry as shown in Fig. 1 were made by silver paint and attached by 10µm gold wires. After annealing at 400°C in air for 2 h contact resistances were below 1 Ω. The voltage was measured with subnanovolt resolution using lock-in-amplifier and low noise transformer at 16 Hz. Two voltage responses as indicated in Fig. 1 were measured simultaneously during one temperature sweep using scanner. In total five samples in the Corbino disk contact configuration have been studied.

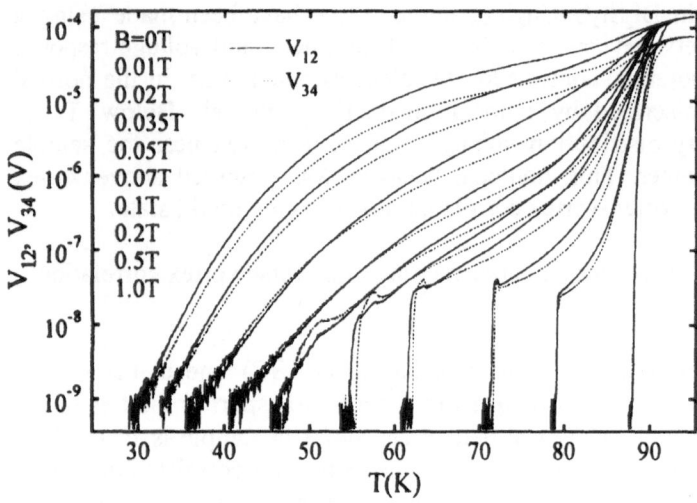

Fig. 2. Tempereature dependence of V_{12} and V_{34} with current I=2 mA at a few different fields.

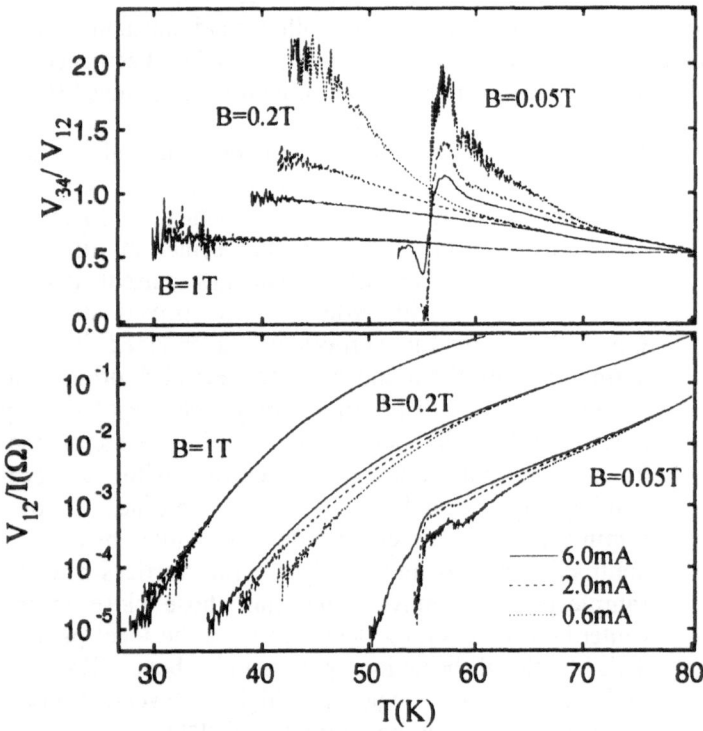

Fig. 3. Temperature dependence of the voltage ratio V_{34}/V_{12} (upper panel) and voltage response V_{12} (lower panel) at B=0.05T, 0.2T, 1T and various currents.

In Fig. 2 we show temperature dependence of voltage response in the inner part of disk (V_{12}) and outer one (V_{34}) taken at a few magnetic fields as indicated in the figure. Similar to results obtained with usual linear contact arrangement in a strip sample [5] at B≤0.05T just above onset of dissipation we observed sharp voltage step associated with flux line lattice melting transition while at higher fields with decreasing temperature both voltages continiously diminish to zero resembling glass-like transition to the vortex solid state. Above T_c and in the upper part of superconducting transition $V_{12}>V_{34}$ reflecting decreasing with radius current density while with the further decrease of temperature in the field range 0.02T≤B≤0.2T opposite to 1/r current density distribution we found $V_{34}>V_{12}$ indicating strongly nonlocal behavior.

Recently nonlocal voltage response in the Corbino disk contact configuration has been observed in Y-123 single crystal [2]. In this case radially increasing voltage response was found to be due to flux line lattice rotation as a rigid body below melting transition while above T_m in the vortex liquid state vortices move independently locally responding to the transport current. In striking contrast to this result in Bi-2212 Corbino disk we observed nonlocal voltage response in the vortex liquid state. To analyse this anomalous behavior in more detail in the upper panel in Fig. 3 we show temperature dependence of the voltage ratio V_{34}/V_{12} which reflects transverse vortex-vortex velocity correlation. At B≤0.2T approaching melting transition from above voltage ratio V_{34}/V_{12} monotonously increases suggesting high transverse vortex interaction in the vortex liquid state. Transverse vortex correlation strongly depends on the value of applied transport current decreasing with the increase of I. At higher fields B>0.2T V_{34}/V_{12} does not depend on temperature and value of transport current through the sample indicating absense of correlation in the vortex motion. As one can see from the lower panel in Fig. 3 at fields B≤0.2T where we observed nonlocal behavior the voltage response displays strongly non-linear dependence on value of current. Furthermore region of magnetic phase diagram where we observed non-Ohmic voltage response exactly coincides with the B-T range of nonlocal behavior. This result suggests the same origin for both non-linearity and nonlocality.

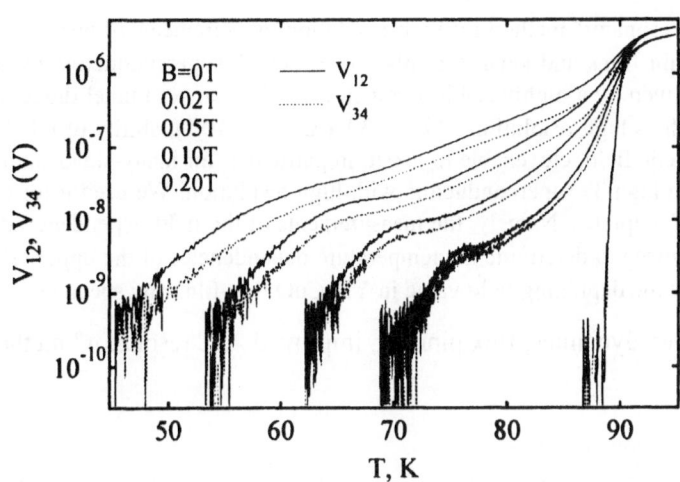

Fig. 4. Temperature dependence of V_{12} and V_{34} with I=2 mA for sample with broadened melting transition.

Decribed nonlocal behavior is extremely sensitive to the crystal quality. Strongly nonlocal voltage response with $V_{34}>V_{12}$ was observed in "clean" samples displaying sharp melting voltage step while in crystals with broadened melting transition the radial potential distribution is found to be in qualitative agreement with local conductivity model. This result is illustrated in Fig. 4 where we show temperature dependence of two voltage responses V_{12} and V_{34} for the crystal of lower quality.

In summary, we have studied transport properties of Bi-2212 single crystal in the Corbino disk configuration. Strongly nonlocal voltage response due to transverse vortex-vortex interaction was found in magnetic field range 0.02T≤B≤0.2T in the vortex liquid state. The region of magnetic phase diagram where we observed this nonlocal behavior exactly coinsides with B-T range of non-Ohmic voltage response. Nonlocal behavior has been observed in clean samples clearly exhibiting resistive melting transition while in samples with higher level of defects where we did not find sharp melting step the nonlocal voltage response is also absent.

Acknowledgments. This work is supported by New Energy and Industrial Technology Development Organization (NEDO).

1. R. Busch, G. Ries, H. Werthner, G. Kreiselmeyer, and G. Saemann-Ischenko, Phys. Rev. Lett. **69**, 522 (1992); H. Safar, E. Rodriguez, F. de la Cruz, P. L. Gammel, L. F. Schneemeyer, and D. J. Bishop, Phys. Rev. B **46**, 14238 (1992); H. Safar, P. L. Gammel, D. A. Huse, S. N. Majumdar, L. F. Schneemeyer, D. J. Bishop, D. Lopez, G. Nieva, and F. de la Cruz, Phys. Rev. Lett. **72**, 1272 (1994); Yu. Eltsev and Ö. Rapp, Phys. Rev. B **51**, 9419 (1995); Yu. Eltsev and Ö. Rapp, Phys. Rev. Lett. **75**, 2446 (1995); D. Lopez, E. F. Righi, G. Nieva, and F. de la Cruz, Phys. Rev. Lett. **76**, 4034 (1996); R. A. Doyle, W. S. Seow, Y. Yan, A. M. Campbell, T. Mochiku, K. Kadowaki, and G. Wirth, Phys. Rev. Lett **77**, 1155 (1996); C. D. Keener, M. L. Trawick, S. M. Ammirata, S. E. Hebboul, and J. C. Garland, Phys. Rev. B **55**, R708 (1997); K. Nakao, Yu. Eltsev, J. G. Wen, S. Shibata, and N. Koshizuka, Physica **C** (*in press*).
2. D. Lòpez, W. K. Kwok, H. Safar, R. J. Olsson, A. M. Petrean, L. Paulius, and G. W. Crabtree, Phys. Rev. Lett. **82**, 1277 (1999).
3. Yu. Eltsev and Ö. Rapp, Phys. Rev. **B** (*in press*).
4. G. D. Gu, T. Egi, N. Koshizuka, P. A. Miles, G. J. Russel, and S. J. Kennedy, Physica C **263**, 180 (1996).
5. S. F. W. R. Rycroft, R. A. Doyle, D. T. Fuchs, E. Zeldov, R. J. Drost, P. H. Kes, T. Tamegai, S. Ooi, and D. T. Foord, Phys. Rev. B **60**, R757 (1999).

The Possibility of Determination of the Material Parameters in HTS by Means of the Improved "LC Resonator" Method

Samvel Gevorgyan[1-2], Takanobu Kiss[1], Hovsep Shirinyan[2], Hiroshi Katsube[1], Tomokazu Ohyama[1], Teruo Matsushita[1], Masakatsu Takeo[1], and Kazuo Funaki[1]

[1] Graduate School of ISEE, Kyushu University, Fukuoka, 812-8581, Japan
[2] Institute for Physical Research of NAS, Ashtarak-2, 378410, Armenia

Abstract: We have essentially improved the "LC resonator" method for high-resolution measurements of penetration depth λ of radio frequency magnetic field into thin films and small-size plate-like high-T_c superconductors by replacing the solenoid testing coil by the flat one driven by a highly stable-frequency and low-power tunnel diode oscillator. The advantages of the new technique such as high resolution $-\Delta\lambda \sim$ 1-3Å ($\Delta\lambda/\lambda \sim 2 \cdot 10^{-6}$), ability of reliable operation in wide ranges of temperature and magnetic field, etc. enable to use it, in particular, for many-sided studies of the peculiarities of vortex dynamics in thin-film high-T_c superconductors with high resolution. We used it for determination of some material parameters in high-T_c cuprates. Namely, the measurements of the field-dependent radio frequency ($\sim$23MHz) penetration depth $\lambda(H,T)$ allowed to determine the temperature dependencies of the upper critical field, $H_{C2}(T)$, the Labusch parameter, $\alpha(T)$, and the depinning field value in YBaCuO thin-film ring specimen.

Key words: material parameters in HTS, vortex dynamics, flux pinning, improved "LC resonator" method

INTRODUCTION

For practical applications of thin HTS it is important to determine the material parameters in films such as the critical fields, the pinning force constant (the Labusch parameter), the depinning field values, etc. over wide ranges of temperature and magnetic field. In this connection the penetration depth, λ, is the convenient measurand which characterizes the superconductive state from many sides. It is sensitive to changes induced by the external fields. At magnetic field-dependent studies of λ the experiments are performed by use of the resonant methods, because the changes in λ are small and hence, high resolution is required [1]. The "LC resonator" method [2], used for tests at a MHz range, has advantages among the methods for λ measurement. Due to the low frequency it enables to avoid the quasiparticle excitation by the testing field. But, it is good only for bulk samples. Applied for thin flat HTS specimens, it gives the resolution at most $\Delta\lambda \sim$ 20Å, because of very low filling factor of solenoid coils arising from the small sample volume. An improved "LC resonator" technique [3], developed by us, has $\Delta\lambda \sim$ 1Å resolution, attainable only with the SQUID's, enough to solve this task. Besides, it can operate at high magnetic fields. Its advantages become evident at tests close to T_c. These enables to study the peculiarities of the vortex motion in thin flat HTS at the beginning of the formation of superconducting state. We used it for definition of quantities of the mixed state, which are important as material parameters, and determine flux pinning properties in HTS.

EXPERIMENTAL METHOD

We essentially improved the traditional "LC resonator" method for high-resolution detection of changes of λ at the transition of thin flat HTS materials by replacing the solenoid coil by a flat one driven by a stable frequency tunnel diode (TD) oscillator [3-4]. In new technique the measuring effect is determined by a distortion of the coil's field configuration near its face by a flat sample, due to the external influences, resulting in the final shift of the oscillator's frequency. In contrast to the traditional method, in improved one the testing field is densely distributed near coil's face. Besides, owing to flat geometry (providing higher filling factor) the thin HTS sample at its S/N transition distorts it much strongly due to the penetration of the flux of testing field, generated by a coil, into the specimen. The changes, $\delta\lambda_S$, of the shielding length λ_S of radio frequency (rf) magnetic field in the sample are detected as the resonant frequency shifts, δF, of the oscillator by $\delta\lambda_S = -G \cdot \delta F$, where G is a geometric factor. It depends strongly on the density of testing field's distribution near coil and the shape of the sample. It depends exponentially also on the position of the sample from the coil [3]. We tested various designs for flat coil to maximize the density of the field configuration near it. We optimized also the distance between the sample and the coil. In these senses the circular coils with dense winding provide better G-factors for the flat objects, pressed to theirs face, than the square ones [5]. Stability of TD oscillators with circular coils is better also [4]. All these enabled to reach the resolution $\Delta\lambda$~1-3Å ($\Delta\lambda/\lambda$~2·10^{-6}) for the first tested devices in λ measurements. Note that in general the measured magnetic shielding length, λ_S, we deal at the experiments, is not the same as the London penetration depth, λ_L, in Meissner state or the pinning penetration depth, λ_{pin}, in a mixed state. But, the effective shielding length λ_S can be related directly to the λ_L or λ_{pin} and the flux flow resistivity [6]. However, how these lengths are related to each other in fact, can be finally determined by a direct calibration of the testing device by use of the specimens with the known physical quantities of λ_L and λ_{pin}.

RESULTS AND DISCUSSION

Below we present the results of study of the magnetic transition curves of 3mm in diameter and 200μm in line width thin-film YBCO ring specimen at different temperatures. It was patterned by the typical chemical etching method from the c-axis oriented and 0.2μm thick film [3]. Actually, we measured the shielding length, λ_S, at 23MHz (or its changes) and the obtained data enabled to determine the temperature dependencies of the upper critical field $H_{C2}(T)$ and the pinning force constant $\alpha(T)$. We could estimate also the temperature dependence of the depinning field, H_{dep}, for our ring over the range 70K to T_c. Typical curves for $\delta\lambda_{pin}$ (or it's the same for δF) vs. $H(\|c)$ at different temperatures are shown in Fig.1. At small fields (at the beginning of the mixed state) the measured data obey the square root dependence of $\delta\lambda_{pin}$ vs. field $\{[\delta\lambda_{pin}(H,T)]^2 \cong [\Phi_0/\mu_0\alpha(T)]\cdot B(H)\}$, similar to that found in [7]. These data allowed to get the pinning force constant $\alpha(T)$. The temperature dependence of α is shown in Fig.2. As is seen the $\alpha(T)$ data are in a good agreement with the $[1-(T/T_c^*)^2]^2$ temperature dependence close to T_c with T_c^*=86.8K. It is also seen that the pinning forces become insignificant for tested specimen slightly below the transition temperature determined by the linear extrapolation to T_c of the corresponding $H_{C2}(T)$ dependence shown in Fig.3. The curves in Fig.3 are plotted by use of the data presented in Fig.1 by the methods described below. At higher fields the $\delta\lambda_{pin}(H)$ dependence changes from the square root to the linear, then quadratic and then the curves become much sharper compared with quadratic low, close to the end of the S/N transition. Finally, the oscillator's frequency (or the $\delta\lambda_{pin}(H)$) is saturated at the same constant value if the magnetic field becomes high enough. Besides, no magnetic hysteresis was observed at the further decreasing of the field value at any tested temperature. That is why, one can determine the value of the H_{C2}, by the field value at the saturation point. Fig.3 shows the measured $H_{C2}(T)$ curve for tested sample. As was mentioned, it allows to determine the T_c=87.4K and the value of $dH_{C2}/dT(T_c)$=−0.69±0.02 T/K by the linear extrapolation to T_c of the $H_{C2}(T)$ curve. The $H_{C2}(T)$ data obey the $H_{C2}(0)\cdot[1-(T/T_0)^2]^\beta$ dependence with β=1.22±0.03.

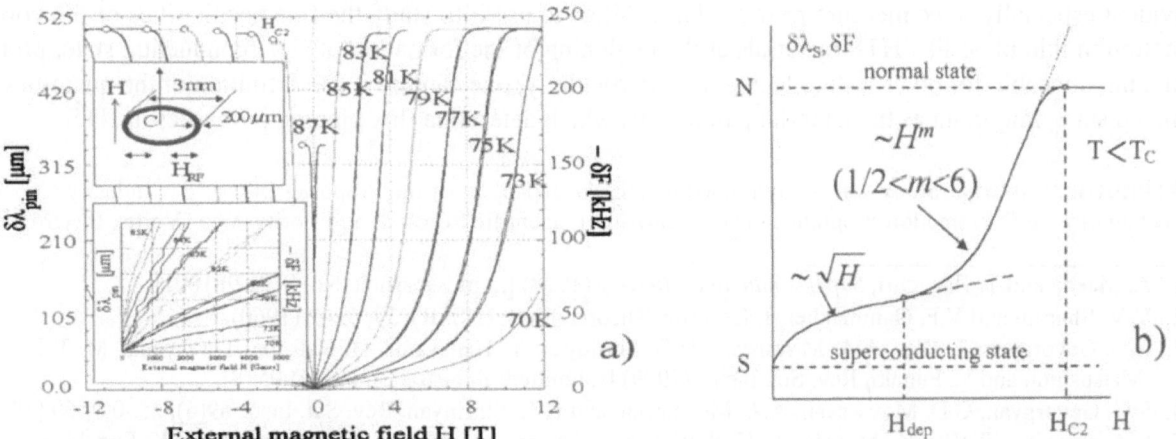

Fig.1. Magnetic transition curves at different temperatures: **a)** Experiment: top inset – field configuration; bottom inset – behavior of the curves for small magnetic field values. **b)** Schematic diagram.

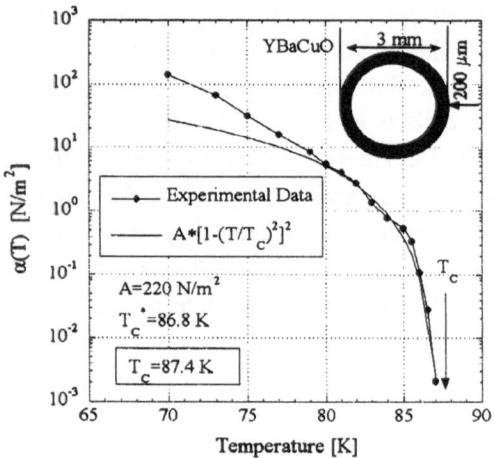

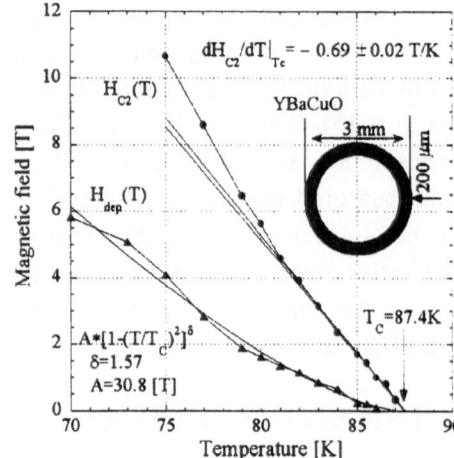

Fig.2. Temperature dependence of the pinning force constant $\alpha(T)$ (Labusch parameter) for the tested sample

Fig.3. Temperature dependencies of the $H_{C2}(T)$ and the $H_{dep}(T)$, for the tested YBaCuO thin-film ring sample.

The achieved large signal-to-noise ratio ($\sim 10^5$) for the tested small sample shows that one can use such technique for study of the fine peculiarities of the vortex dynamics in HTS thin films with high resolution. In particular, the registered square root behavior of the transition curves with subsequent deviation from it at H_{dep} enables to study the vortex depinning in the mixed state. In fact, the onset point of the deviation from the square root dependence might be used as a signal that the depinning process starts. That is, H_{dep} gives the depinning field. Fig.3 shows the determined temperature dependence of $H_{dep}(T)$ for the tested ring sample. The $H_{dep}(T)$ data obey the $H_{dep}(0)\cdot[1-(T/T_c)^2]^\delta$ temperature dependence with $\delta=1.57$ and $H_{dep}(0)=30.8T$. Additionally, the sharp shape of the curves $\delta\lambda_{pin}(H)$ in Fig.1 at the end of the transition permits also, in principle, to reveal and may permit also to separate the possible mechanisms responsible for the final destruction of the superconductivity in HTS materials at a fields close to H_{C2}. However, this item needs further additional analysis of the experimental data (probably, obtained on various composition HTS materials) and therefore, we will discuss it in a separate paper.

In conclusion, we improved the "LC resonator" method for high-resolution measurements of the penetration depth, $\lambda(H,T)$, of rf magnetic field into thin flat HTS materials by replacing the solenoid testing coil by the flat one driven by a tunnel diode oscillator. The new technique has an excellent resolution ($\Delta\lambda \sim 1$-3Å; $\Delta\lambda/\lambda \sim 2\cdot10^{-6}$), attainable only with SQUID's. It can operate even in high magnetic fields (it was tested up to 12T) with almost the same high resolution, which is superior to SQUID's. The new technique can measure the linear changes of λ in dynamic range of about 1mm with almost the same high resolution, which is also superior to SQUID's and other techniques. Its better resolution compared to others become evident especially if we measure near T_c. These allow to use it to study the fine peculiarities of the vortex motion in thin plate-like HTS materials at the beginning of the formation of superconducting state, probably, unnoticeable for other testing techniques. Particularly, one can use it for definition of the quantities of mixed state, important as the material parameters, which determine flux pinning properties in HTS.

Acknowledgments. Study was partly funded by NEDO, Japan, under the Proposal–Based New Industry Creative Technology R&D Promotion Program and Grant-in-Aid of Scientific Research on Priority Area "Vortex Electronics".

1. A. Maeda and T. Hanaguri, *Superconductivity Review* (1998) [in press: article No. SCR-981001].

2. Y.V. Sharvin and V.F. Gantmakher, J. Experim. Theoret. Phys. (USSR), **39**, 1242 (1960).

3. S.G. Gevorgyan, T. Kiss, A.A. Movsisyan, H.G. Shirinyan, Y. Hanayama, H. Katsube, T. Ohyama, M. Takeo, T. Matsushita, and K. Funaki, Rev. Sci. Instr., (1999) [submitted: Registration No. A99518].

4. S.G. Gevorgyan, G.D. Movsesyan, A.A. Movsisyan, and H.G. Shirinyan, Rev. Sci. Instr. **69**(6), 2550-2560 (1998)

5. S. Gevorgyan, T. Kiss, A. Movsisyan, H. Shirinyan, T. Ohyama, M. Takeo, T. Matsushita, and K. Funaki, Proc. of the 8-th Workshop on Low Temperature Detectors (LTD-8, 1999) [to be published in NIM-A (2000)].

6. M.W. Coffey and J.R. Clem, Phys. Rev. B, **45**(17), 9872-9881 (1992.)

7. D-H Wu and S. Sridher, Phys. Rev. Lett. **65**(16), 2074-2077 (1990).

Vortex Lattice Melting in Underdoped $YBa_2Cu_4O_8$

Takekazu Ishida, Kentaro Kitamura, Naoki Kagawa, and Kiichi Okuda
Department of Physics and Electronics, Osaka Prefecture University, Sakai, Osaka 599-8531, Japan

Seiji Adachi and Setsuko Tajima
SRL-ISTEC, 10-13 Shinonome 1-chome, Koto-ku, Tokyo 135-0062, Japan

Abstract: Vortex lattice melting of $YBa_2Cu_4O_8$ (Y124) crystals has been investigated by means of the bulk measurements. The vortex lattice softening as a precursor to the first-order melting transition can clearly be seen in χ' as well as in χ'' when the field is applied in parallel to the c axis. The melting field of Y124, approximated as $H_m = 217(1 - T/T_c)^{1.48}$ (kG), is appreciably lower than that of Y123. This suggests a substantial reduction of dimensionality in Y124.

Keywords: $YBa_2Cu_4O_8$, melting transition, ac susceptibility, dc magnetization

INTRODUCTION

Vortex lattice melting has exclusively been studied using the $Bi_2Sr_2Ca_1Cu_2O_{8+\delta}$ (Bi2212) crystal [1] and the $YBa_2Cu_3O_{7-\delta}$ (Y123) crystal [2, 3]. The former is a two-dimensional system while the latter is an anisotropic three-dimensional system. It is useful for understanding this phenomena if another cuprate of intermediate dimensionality is available for studies in this field.

The Y124 crystal has several features. (1) The oxygen stoichiometry is exactly eight. (2) The hole concentration is classified to the uderdoped regime in an as-grown state. (3) The crystal has no twin planes. (4) The CuO layer of a double-chain structure is highly metallic. (5) The mass anisotropy might be in between Y123 and Bi2212 [4]. (6) The occurrence of the first-order melting transition ($H \parallel c$) in Y124 was suggested by a resistivity drop as a function of temperature [5], but the drop was not convincingly sharp as was often observed for clean Y123 crystals.

In the present paper we describe the first *bulk* observation of the vortex lattice melting in Y124.

EXPERIMENTAL

The Y124 single crystals were grown by a self-flux method under a high-pressure gas mixture of 80% Ar−20% O_2 [6]. The starting mixture in a Y_2O_3 crucible was heated at 1140°C for 1h under 2000 atm and cooled down to 1040°C in 2 days. Two tiny crystals (denoted as YD#1 and YD#2) were used for the ac and dc magnetic measurements.

The ac magnetization $M_{ac} = (\chi'_1 - i\chi''_1)H_{ac}V$ (V; sample volume) and the dc magnetization M_{dc} were measured using a SQUID (Superconducting Quantum Interference

Device) magnetometer (Quantum Design MPMS-XL). Both dc and ac fields were applied in parallel to the c axis.

We monitored the SQUID output during a constant-rate temperature sweep by fixing a sample at the center of the SQUID coils to see a subtle change in M_{dc} expected for the first-order transition.

RESULTS AND DISCUSSION

We reported the sequential vortex states upon lattice melting ($H||c$) using an untwinned Y123 crystal [2, 3]. The melting phase transition was evident in magnetization jump as well as in the softening phenomena in χ' and χ''. However, the ac susceptibility is extremely sensitive compared to the dc magnetization, and hence is useful to see the melting phenomena using a tiny sample. As shown in Fig. 1, a similar profile of χ' and χ'' to Y123 was observed using YD#1 in the $H \parallel c$ configuration. The melting temperature T_m of YD#1 was determined by means of

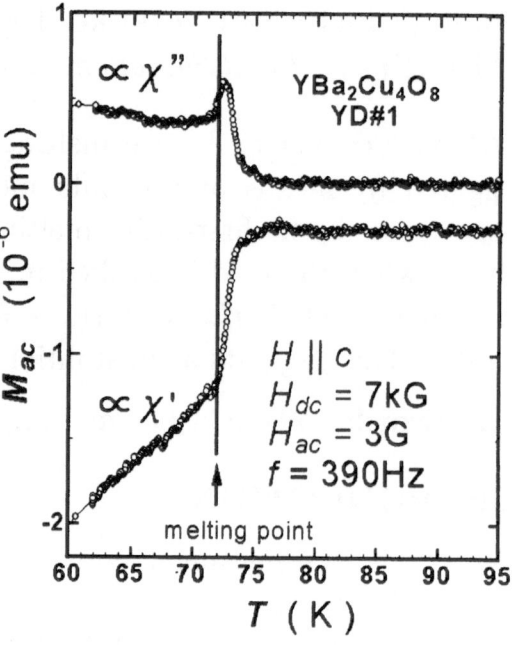

Figure 1: The ac susceptibility as a function of decreasing T in $H||c$.

the χ' shoulder (see a vertical line in Fig. 1). An enhancement in χ'' at temperatures below T_m can be regarded as a lattice softening.

In Fig. 2, we show the melting field of YD#1 determined by the χ' shoulder as a function of temperature. The data fitting to the $H = H_0(1-T/T_c)^n$ form yields $H = 217(1-T/T_c)^{1.48}$. This line is appreciably lower than $H = 331(1-T/T_c)^{1.5}$ (see dashed line) reported by Hussey *et al.* [5]. We observed a softening behavior in YD#2 as well, and T_c and the melting field of YD#2 are almost the same to those of YD#1. Presumably, the superconducting properties are not so altered from sample to sample due to a well-established stoichiometry of oxygen. In Fig. 2, we compare the Y124 lines with the melting line $H = 1270(1 - T/T_c)^{1.5}$ of Y123 [3].

We also carried out the dc magnetization measurements of YD#1 as a function of decreasing

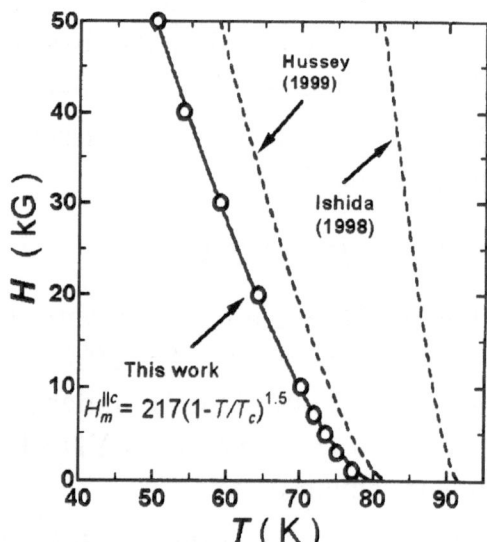

Figure 2: The melting lines of Y124 and Y123 in $H \parallel c$.

325

temperature. The external field was applied at 92 K and subsequently the temperature was decreased at the constant rate of 0.2 K/min. The sample position was always at the center of three SQUID coils. This method has advantages, i.e., no influence by field inhomogeneity, quick data acquisition, and high resolution in M_{dc}. However, this method does not give the absolute magnetization and is influenced from the signal drift, the external noise, and the flux jump in a sample.

As shown in Fig. 3, the relative dc moment ΔM_{dc} shows a discontinuous change at 65.4 K under 20 kG. This temperature is in accordance with that determined by the ac susceptibility (see Fig. 2). We checked the reproducibility of the anomaly temperature in ΔM_{dc} of Fig. 3. The magnetization jump was in the order of 10^{-6} emu. The evaluation of the magnetization jump per unit volume $4\pi\Delta m$ and the entropy change ΔS per vortex segment requires the exact determination of sample volume.

In conclusion, we confirmed the first-order melting transition of Y124 in $H \parallel c$ in terms of bulk quantities. The melting line is fairly lower than that of Y123.

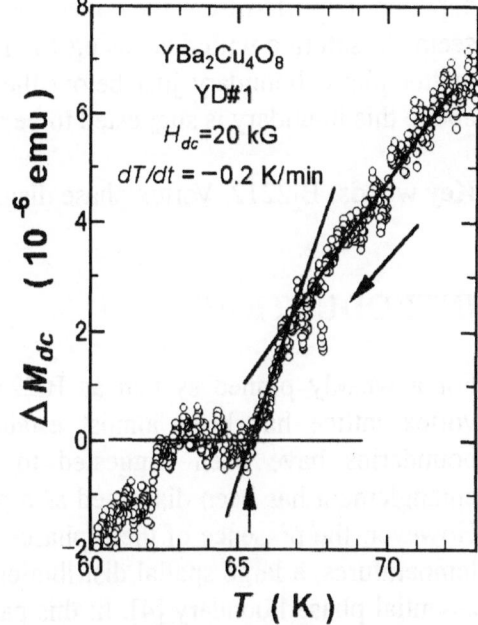

Figure 3: The dc magnetization (relative) as a function of T.

Acknowledgments: This work is partially supported by a Grant-in-Aid for Scientific Research on Priority Area "Vortex Electronics" and by the Iwatani Naoji Foundation. The work at ISTEC-SRL is partially supported by New Energy and Industrial Technology Development Organization (NEDO).

References

[1] E. Zeldov, D. Majer, M. Konczykowski, V.B. Geshkenbein, V.M. Vinokur, and H. Shtrikman, Nature **375**, 373−376 (1995).

[2] T. Ishida, K. Okuda, and H. Asaoka, Phys. Rev. B**56**, 5128−5131 (1997).

[3] T. Ishida, K. Okuda, A.I. Rykov, S. Tajima, and I. Terasaki, Phys. Rev. B**58**, 5222−5225 (1998).

[4] D. Zech, C. Rossel, L. Lesne, H. Keller, S.L. Lee, and J. Karpinski, Phys. Rev. B**54**, 12535−12542 (1996).

[5] N.E. Hussey, H. Takagi, N. Takeshita, N. More, Y. Iye, S. Adachi, and K. Tanabe, Phys. Rev. B**59**, R11668−11671 (1999).

[6] S. Adachi, K. Nakanishi, K. Tanabe, K. Nozawa, H. Takagi, W.-Z. Hu, and M. Izumi, Physica C **301**, 123−128 (1998).

Magnetic Relaxation and Vortex-Lattice Phase Boundary of Bi-2212

Y. Yamaguchi, G. Rajaram*, N. Shirakawa , A. Mumtaz**, H.Obara, T. Nakagawa and H. Bando

Electrotechnical Laboratory, Tsukuba, Ibaraki 305-0045, Japan

Abstract: Magnetic relaxation of Bi2212 has been measured by use of a micro Hall sensor array. The activation energy (U) of vortex below the second peak increases with decreasing M with an exponent expected from the collective creep theory. On the contrary the U above the second peak seems to saturate with decreasing M. These observations conclude a presence of a definite vortex-matter phase boundary just below the second-peak field. The vortex phase at high field region above this boundary is suggested to be a 'viscous vortex liquid'.

Key words: Bi2212, Vortex phase diagram, Magnetic relaxation, Activation energy, Second peak

INTRODUCTION

For a weakly pinned system as Bi2212 ($Bi_2Sr_2CaCu_2O_8$), first-order phase transition (FOT) of vortex lattice has been almost established at high temperature side [1], and various phase boundaries have been suggested to connect with the FOT boundary [2]. Recently, vortex entanglement has been discussed as a primarily origin of the second peak at low temperatures [3]. However, the presence of these phases and the continuity among them are less understood. At low temperatures, a large spatial distribution of magnetization and a large magnetic relaxation blur out essential phase boundary [4]. In this paper we report magnetic relaxation at low temperatures of a nearly optimally doped Bi2212 crystal using a micro Hall sensor array.

EXPERIMENT AND RESULTS

The sample was an as-grown $Bi_{2.1}Sr_{1.9}CaCu_2O_{8+\delta}$ crystal that was used in a previous study [5]. The doping level was nearly optimum but slightly shifted to the over-doped side and Tc was 90.5 K. A Hall sensor array was attached to the sample with the dimensions of $0.9 \times 0.8 \times 0.03$ mm^3, as shown in the inset to Fig. 1. The array was fabricated from a GaAs/AlGaAs multi-layer film consisting of seven sensors (six sensors to measure B and one for H) with each active area of 15×15 μm^2. External field H, produced by a superconducting solenoid, was applied parallel to the crystal c axis (perpendicular to the widest sample plane). Magnetic induction B_i (i=1-6 is the sensor number) at the sample surface was measured using four two-channel DC voltmeters simultaneously.

Magnetization curves ($4\pi M_i$-vs-H curves, where $4\pi M_i = B_i - H$) were measured with a ramp-and-pause type of change of H. The data shown in Fig. 1 were measured within two minutes of the pausing time. The values, measured simultaneously with different sensors, give a spatial profile of the magnetization over the sample [6]. The profile at the region of reversible magnetization (high temperature or high field) was a dome shaped profile due to surface barrier, while the profile near the second peak was almost V shape in accordance with the Bean's model.

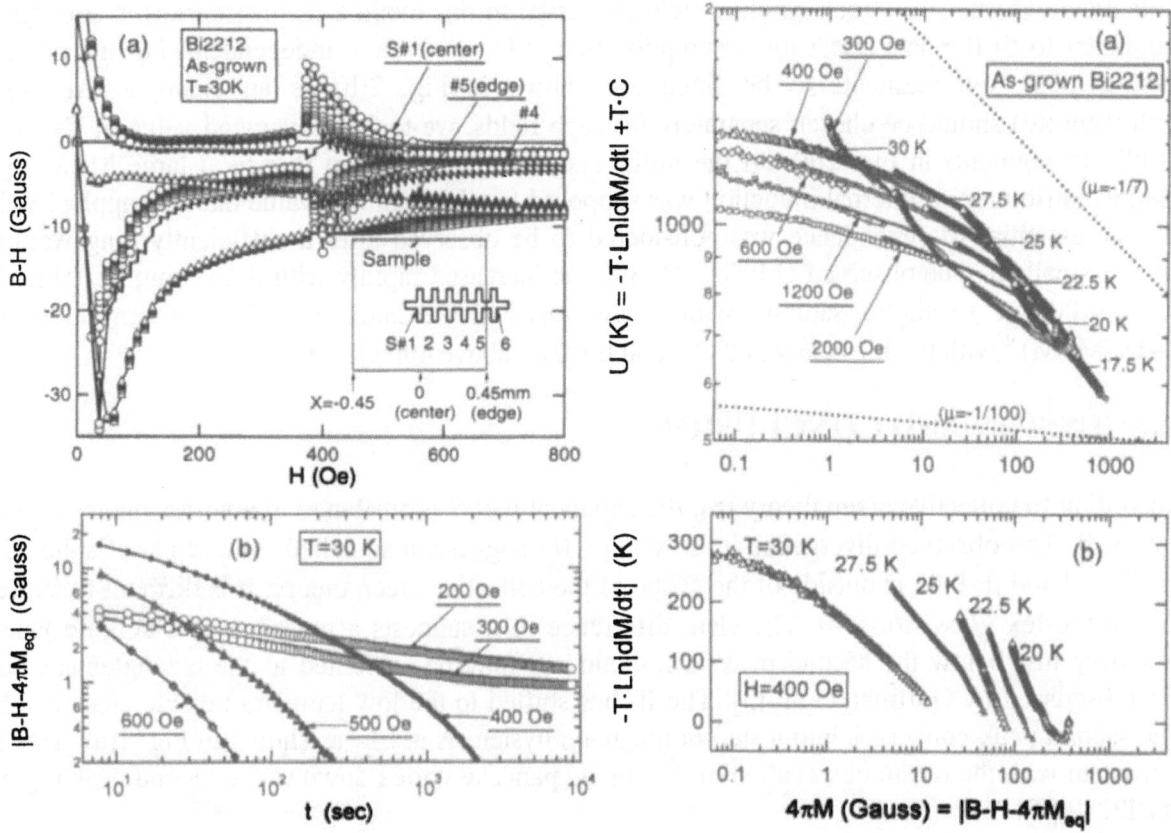

Fig. 1. (a) M-vs-H curves (where $4\pi M = B-H$) at 30 K. The field (H) was swept with pausing a few minutes at a fixed field, where several data were taken. The inset shows the position of the Hall sensor on the sample. (b) M-vs-t curves at the sample center (S#1).

Fig. 2. (a) Activation energy (U) evaluated by Maley's method from relaxation data, partly shown in Fig. 1(b). The constant C (=35 for all the data sets) is an adjustable parameter for the raw data as shown in (b).

Detailed relaxation curves (M-vs-t curves) were measured at several selected fields (Fig. 1(b)) and various temperatures. We analyzed the magnetization at the sample center (S#1) only, as the effect of the dome shape was assumed to be smallest. The equilibrium magnetization (M_{eq}) was determined as a middle value of the ascending and the descending branch of the magnetization curves. The magnetization M above the second peak was observed to relax into the M_{eq} after a long time. On the contrary, the M below the second peak approached to a different value from M_{eq}. The irreversibility line (IL) seemed to shift towards the low field and low temperature side without limit.

To see quantitative behavior of the relaxation, activation energy U(B,T,j) for various fields was calculated from Maley's method [7]. The relation between U and M is given by a rate equation of thermally activated motion of flux as,

$$d\,|M|\,/\,d\,t = (B\omega r/\pi d)\,\exp(-U/k_B T),$$

i.e. $$U/k_B = -T\,\ln|dM/dt| + T\,\ln(B\omega r/\pi d),$$

where ω is the attempt frequency, r the hop distance, d (=1 mm) the sample dimension and k_B the

Boltzmann constant. We took external field (H) as B. In this method, $C=\ln(B\omega r/\pi d)$ is adjustable parameter to fit the data under the assumption that U is temperature independent. For the present measurement, the segments to be fitted are shown in Fig. 2(b) as an example. Although $C=\ln(B\omega r/\pi d)$ should be chosen separately for each fields, we took an averaged value of C=35 to fit all the segments in Fig. 2(a). At the initial stage of the relaxation (hence at large M) a large spatial distribution of internal induction was supposed to take a smaller value than the applied field H. The essential M-dependence was considered to be observed after a sufficiently long waiting time at small M. The observed U below H=300 Oe increased rapidly with decreasing M, while U above 400 Oe seems to saturate with decreasing M. We had an analytical form close to $U=U_0(M_0/M)^\mu$ with $\mu \sim 1/7$ below 300 Oe, and $\mu <0.03$ above 400 Oe.

DISCUSSION AND CONCLUSION

According to collective creep theory [8], the exponent $\mu=1/7$ is attributed to a vortex line system at low field. The observed divergence in U with $\mu\sim 1/7$ suggests a so-called 'vortex glass' solid. On the other hand $\mu<0.03$ is outside of the scope of the collective creep theory. It is difficult to expect such a 'vortex glass' for $\mu\sim0$. The clear difference in μ suggests a presence of a definite phase boundary just below the second peak. The boundary may be attributed to the entanglement line (EL) discussed by Goffman et al. [3]. The IL has shifted to the low temperature side close to EL. The second peak converges into a step in magnetic hysteresis at EL, as shown in Fig. 1(a). This is consistent with the recent observation of decoupled pancake vortex down to the second peak region in Bi2212 [9].

In conclusion the second peak is a result of a sharp change in magnetic relaxation at entanglement line EL, and the second peak is expected to become a step in magnetic hysteresis at EL after a sufficiently long waiting time. The low temperature glass region above EL is concluded to be a 'viscous vortex liquid', which becomes gradually into the pancake gas with increasing temperature. The EL is probably a vortex phase transition in a weak pinning system like Bi2212.

Acknowledgements. One of the author (G.R.) acknowledges a support by NEDO for his stay at ETL as a fellow in a collaborated work with ISTEC. Another author (A.M.) thanks JST for his stay at ETL as an STA fellow.

* Present address. University of Hyderabad, Hyderabad 500 046, India

** Present address. Quaid-i-Azam University , Isramabad , Pakistan

1. E. Zeldov, D. Majer, M. Konczykovski, V. B. Geshkenbein, V. M. Vinokur and H. Shtrikman, Nature (London) **375** (1995) 373.

2. D.T. Fuchs, E.Zeldov, T.Tamegai, S.Ooi, M.Rappaport, H.Shtrikman, Phys Rev Letts. **80**(1998)4971.

3. M.F. Goffman, J.A. Herbsommer, F. de la Cruz, T.W. Li and P.H. Kes, Phys. Rev. B **57** (1998) 3663.

4. S. Anders, R. Parthasarathy, H.M. Jaeger, P. Guptasarma, D.G. Hinks and R. van Veen, Phys. Rev. B **58** (1998) 6639.

5. A. Mumtaz, Y. Yamaguchi, K. Oka, G. Rajaram, Physica C **302** (1998)331.

6. Y. Yamaguchi, G.Rajaram, N.Shirakawa, H.Obara, T.Nakagawa, A.Mumtaz and H.Bando, Advances in Supercond. XI (Proc. ISS'98), eds. N.Koshizuka and S.Tajima (Springer-Verlag, 1999) p. 223.

7. M.P. Maley, J. O. Willis, H. Lessure and M. E. McHenry, Phys. Rev. B **42** (1990) 2639.

8. M.V. Feigel'man, V.B. Geshkenbein, A.I. Larkin and V.M. Vinokur, Phys. Rev. Letts. **63**(1989) 2303.

9. T. Shibauchi, T. Nakao, M. Sato, T. Kisu, N. Kameda, N. Okuda, S. Ooi and T. Tamegai, Phys. Rev. Letts. **83** (1999) 1010.

EXPERIMENTAL EVIDENCE FOR PINNING-INDUCED TRANSITIONS IN NdBa$_2$Cu$_3$O$_{7-y}$ CRYSTALS

A.K. PRADHAN, S. SHIBATA, K. NAKAO AND N. KOSHIZUKA

Superconductivity Research Laboratory, ISTEC, 10-13, Shinonome, 1-chome, Koto-ku, Tokyo 135-0062, Japan

Abstract: We report on magnetotransport measurements of the effects of systematic introduction of point defects by a novel substitution technique on the clean and twinned single crystals of NdBa$_2$Cu$_3$O$_{7-y}$. The vortex lattice melts for a minimum field of 100 G for H //ab. Intrinsic pinning is found to dominate the low temperature phase below the melting transition only. A crossover from the first-order melting transition to the vortex-glass transition is detected for the first time with a precise and controlled introduction of point disorders by substitution for H//c.

KEY WORDS: melting transition, magnetoresistance, substitution, glass transition

INTRODUCTION

The observation of first-order vortex lattice melting transition is now established in clean and untwinned YBa$_2$Cu$_3$O$_x$ (YBCO) [1-3], Bi$_2$Sr$_2$CaCu$_2$O$_{8+y}$ (BSCCO) and more recently in twinned NdBa$_2$Cu$_3$O$_{7-y}$ (NBCO) [4] single crystals. However, a Bose-Glass (BG) phase is predicted very recently in twinned YBCO crystal [5]. There are enough controversy regarding the actual nature of the transition in the presence of twin planes, isotropic point defects and the presence of both of them. Point-like disorders induced by either electron [6] or proton irradiation suppress the first-order melting transition in YBCO crystals. The electron irradiated sample shows a polymer-like glass transition and on the contrary a Vortex-Glass (VG) transition is predicted in proton irradiated sample. Apart from irradiation, oxygen related defects suppress the melting transition. Another intriguing feature is the effect of intrinsic pinning on the melting transition when the magnetic field is aligned parallel to the ab plane. The influence of intrinsic pinning on the melting transition could not be probed by the transport measurement directly.

We report the evidence for a crossover from a first-order vortex lattice melting to a vortex-glass transition in twinned NBCO crystal with increasing defect density due to Nd-Ba substitution in a controlled way. We have also shown that intrinsic pinning does not influence the melting transition, however plays a role only below it. The results suggest that there is a threshold of pinning disorder which must be exceeded in order to observe the VG transition.

EXPERIMENTALS

The crystals were grown by self-flux method under the precise control of varying reduced O$_2$ partial pressure in Ar-atmosphere (PO$_2$) [7]. Crystals were annealed at 325 °C for 300h in flowing O$_2$. All crystals show twin planes with average twin spacing of ~1-2 μm under polarizing microscope. We have shown earlier [4] that the defect density in NBCO crystals increases when PO$_2$ increases during growth. High sensitivity resistance measurements were carried out by a standard four probe ac technique and *I-V* characteristics were measured using dc method as described earlier [4].

RESULTS AND DISCUSSION

Fig.1 shows the temperature dependence of resistance for H//ab plane at various field values for PO$_2$=0.05%. The high quality of the crystal is indicated by its sharp superconducting transition, T_c=95.3K, and width ΔT_c [(10-90)%]≈250 mK. The resistivity displays a *"kink"* at the melting transition where it drops sharply to zero. The inset of Fig.1 shows *I-V* characteristics for *H* //c for

PO_2=0.03%. The influence of intrinsic pinning on the melting transition for H //ab plane is insignificant.

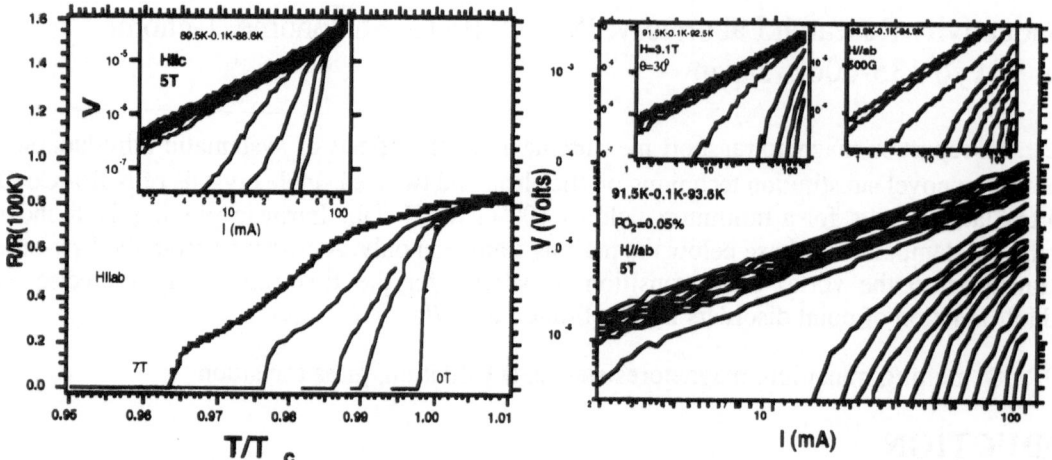

FIG.1 Normalized resistance of NBCO crystal with PO_2=0.05% at H =0, 1, 2, 5, and 7T for H //ab. Inset shows I-V curves at H =5T for H //c.

FIG.2 Isothermal I-V curves of PO_2=0.05% at H =5T for H //ab. Inset (left) shows the isothermal I-V curves at H =500G for H//ab. Inset (right) shows the isothermal I-V curves at H =3.1T for θ =30° away from H //ab.

In Fig.2, we show I-V characteristics for H =5T and 500G (right inset) for H //ab in a temperature interval of 0.1K. The I-V curves exhibit Ohmic behavior above T_m, and a non-Ohmic behavior below. Similar behavior was also observed down to 100G. However, the down turn behavior in I-V curves below T_m (showing a rounded feature) down to the lowest temperature is characteristically different from when the magnetic field is oriented either 30° away from H // ab as shown in the inset (left) of Fig.2 or for H// c (inset of Fig.1). We attribute this behavior of I-V curves below T_m for H// ab plane to the intrinsic pinning [8] that becomes active only below T_m as temperature is lowered. Our recent magnetization measurements on NBCO crystal with PO_2=0.03% for H//ab plane show the periodic oscillations in M-H curves above 70K only. This corroborates our results from the transport measurement that the intrinsic pinning is effective only at low temperature, presumably below 70K and does not exhibit any significant influence on the melting line.

Fig.3 shows the temperature dependence of the resistivity in crystal with PO_2=0.07% for H// c. Here, the kink in resistivity has completely been disappeared, indicating that the first-order melting transition has been suppressed and rather a continuous transition has taken place. Similar behavior was also observed for H // ab. However, it becomes important to study the nature of transition with increasing pinning in this system.

In the presence of weak random point defects a vortex-glass phase was proposed and seen in thin films and single crystals of YBCO. In the VG prediction, the resistance just above the transition should follow the dynamic scaling [9] $R \sim (T-T_g)^s$ where s is the field independent constant and T_g is the second-order glass transition temperature. The critical exponents, v and z are related to s through s = $v(z-d+2)$ and becomes $v(z-1)$ for dimension d=3. In the inset (left) of Fig.3, we show $R(T)$ for H// c for PO_2=0.07%, where the solid lines are fits using the above equation. T_g was determined from the intercept of the linear fit of $(\partial lnR/dT)^{-1}$ Vs. T giving a slope $1/v(z-1)$. The scaling was in excellent agreement with the experimental results. We obtained a field independent value of s =6.2±0.4 for H// c and s =6.0±0.4 for H// ab in the field range of 1 to 7T and shown in the inset of Fig.3 (right). These values are also in excellent agreement with the predicted values of the VG model.

In order to test the validity of the VG transition in NBCO crystal with PO_2=0.1%, we show I-V curves for H =1T in Fig. 4. The I-V curves show the change of curvature at the vortex-glass transition line T_g. The I-V curves were scaled using the scaling for vortex-glass model and shown in the inset of Fig.4. The I-V curves scaled very well with glass exponent v =1.6±0.3 havig T_g=91.3K. However, similar scaling was not possible for 0.07% becuse of linear behavior of I-V curves at all temperatures.

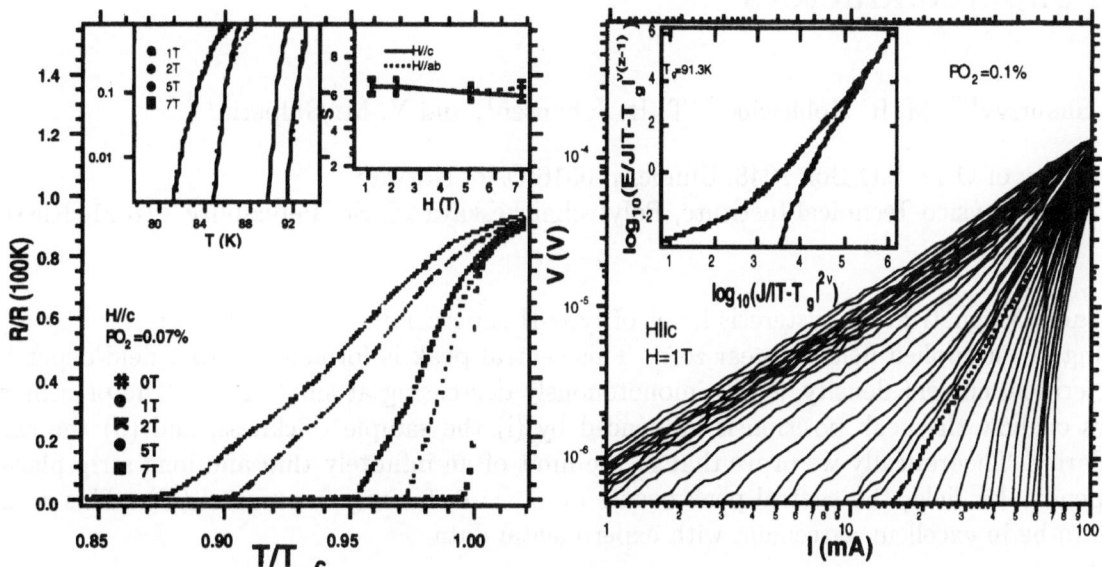

FIG.3 Normalized resistance of NBCO crystal with PO_2=0.07% for H //c at several field values. Inset (left) shows vortex-glass fits of the resistance data for H //c. Inset shows the scaling parameter s as a function of H //c and //ab.

FIG.4 Isothermal I-V curves of PO_2=0.1% at H =1T for H //c. Inset shows the scaling of I-V curves at H =1T using vortex-glass.

We note that when PO_2 is increased to 0.1%, the I-V curves show an excellent scaling for VG transition [10] confirming that a threshold pinning is necessary to observe this transition. It may be noted that the irreversibility line shifts progressively upward suggesting the enhancement in pinning with increasing PO_2. Our results do not support the evidence for either a Bose-glass [5] or a polymer-glass like transition [6] as observed in YBCO crystal. However, the downward shift of the melting line may be a consequence of increased anisotropy due to increasing disorder in the crystal [11].

CONCLUSION

In summary, we present the first evidence for a systematic and distinct crossover from the first-order vortex lattice melting transition to a continuous vortex-glass transition in $NdBa_2Cu_3O_{7-y}$ crystal with increasing point disorder by substitution. The first-order melting transition is suppressed systematically with increasing defect density and for a threshold substitution of Nd-Ba ions, a vortex-glass transition appears. The first-order vortex lattice melting is noticed even at a field of 100G for H //ab plane and is not affected by the intrinsic pinning although it becomes prominent below T_m.

This work is supported by New Energy and Industrial Technology Development Organization.

REFERENCES

1. H. Safar et al. Phys Rev Lett. 69(1992) 824; W.K. Kwok et al.Phys Rev Lett 69(1992) 3370.
2. D.E. Farrell et al. Phys Rev Lett 67(1991) 1165.
3. U. Welp et al. Phys Rev Lett. 76 (1996) 48091.
4. A.K. Pradhan et al, Europhys. Lett. 46 (1999) 787; Phys. Rev.B 59(1999) 11563.
5. S.A. Grigera et al. Phys. Rev. Lett. 81(1998) 2348.
6. J.A. Fendric et al. Phys. Rev. Lett. 74(1995) 1210.
7. S. Shibata et al Jap. J. Appl. Phys. Lett. 38(1999)L1171.
8. L. Balents L and D. R. Nelson, Phys. Rev. Lett. 73(1994) 2618.
9. M.P.A. Fisher, Phys. Rev. Lett. 62(1989) 1415.
10. A.K. Pradhan et al. (submitted).
11. A.E. Koshelev and V.M. Vinokur, Phys. Rev. B57 (1998) 8026.

Granularity and the central peak in magnetization loops of thin superconductors

D. V. Shantsev[1,2], M. R. Koblischka[1], T. H. Johansen[1], and Y. M. Galperin[1,2]

[1] University of Oslo, P.O.Box 1048, Blindern, 0316 Oslo, Norway
[2] A.F. Ioffe Physico-Technical Institute, Polytechnicheskaya 26, St. Petersburg 194021, Russia

Abstract: Magnetization hysteresis loops of type-II superconductors are characterized by a peak occuring at an applied field B_p near zero. This central peak is formed due to a field-dependence of the critical current density, $J_c(B)$, monotonously decreasing at small fields. The present work focuses on how the peak position is influenced by (i) the sample thickness, and (ii) the sample granularity. Theoretically we prove that in the limit of an infinitely thin and long strip placed in a perpendicular field, the central peak occurs at $B_p = 0$, for any function $J_c(B)$. This result is shown to be in excellent agreement with experimental data.

Key words: magnetization loops, central peak, perpendicular geometry, granularity

INTRODUCTION

The investigation of magnetic hysteresis loops (MHLs) is a widely used tool to characterize superconducting samples, and in particular, to estimate the critical current density and its dependence on magnetic field. An ever-present feature of the MHLs is a peak in the magnetization located at an applied field B_p near zero. This so-called central peak is formed due to a field dependence of the critical current density, $J_c(B)$, monotonously decreasing at small fields. When the sample is a long cylindrical body placed in a parallel applied field, the peak position can be calculated analytically within the critical-state model for several $J_c(B)$ dependences [1]. One always finds on the descending field branch that $B_p < 0$, and similarly $B_p > 0$ on the ascending branch of large-field loops. In simple terms the shift is a consequence of the local flux density, B, lagging behind the applied field. However, in some experiments one finds B_p very close to zero or even shifted to the positive side so that the peak occurs before the remanent state is reached. This latter case is, e.g., commonly observed in mono- and multifilamentary Bi-2223 tapes [2,3]. Two factors are known to cause a shift of B_p toward positive values: decreasing sample thickness [5] and increasing its granularity [3,4]. However, quantitative predictions for the peak position with account of these effects are lacking and the present work aims to fill this gap.

RESULTS AND DISCUSSION

Consider a long thin uniform superconducting strip with edges located at $x = \pm w$, the y-axis pointing along the strip, and the z-axis normal to the strip plane. The magnetic field, B_a, is applied along the z-axis, so screening currents flow in the y-direction. Assume that the strip is in a fully penetrated state, i.e., the sheet current is everywhere equal to the critical one, $J(x) = \text{sign}(x) J_c[B(x)]$. This state can be reached after applying a very large field, and then reducing it to some much smaller value. The field distribution, $B_z(x)$ then satisfies the integral equation,

$$B_z(x) - B_a = -\frac{\mu_0}{\pi} \int_0^w \frac{J_c[B_z(u)]}{x^2 - u^2} \, u \, du, \qquad (1)$$

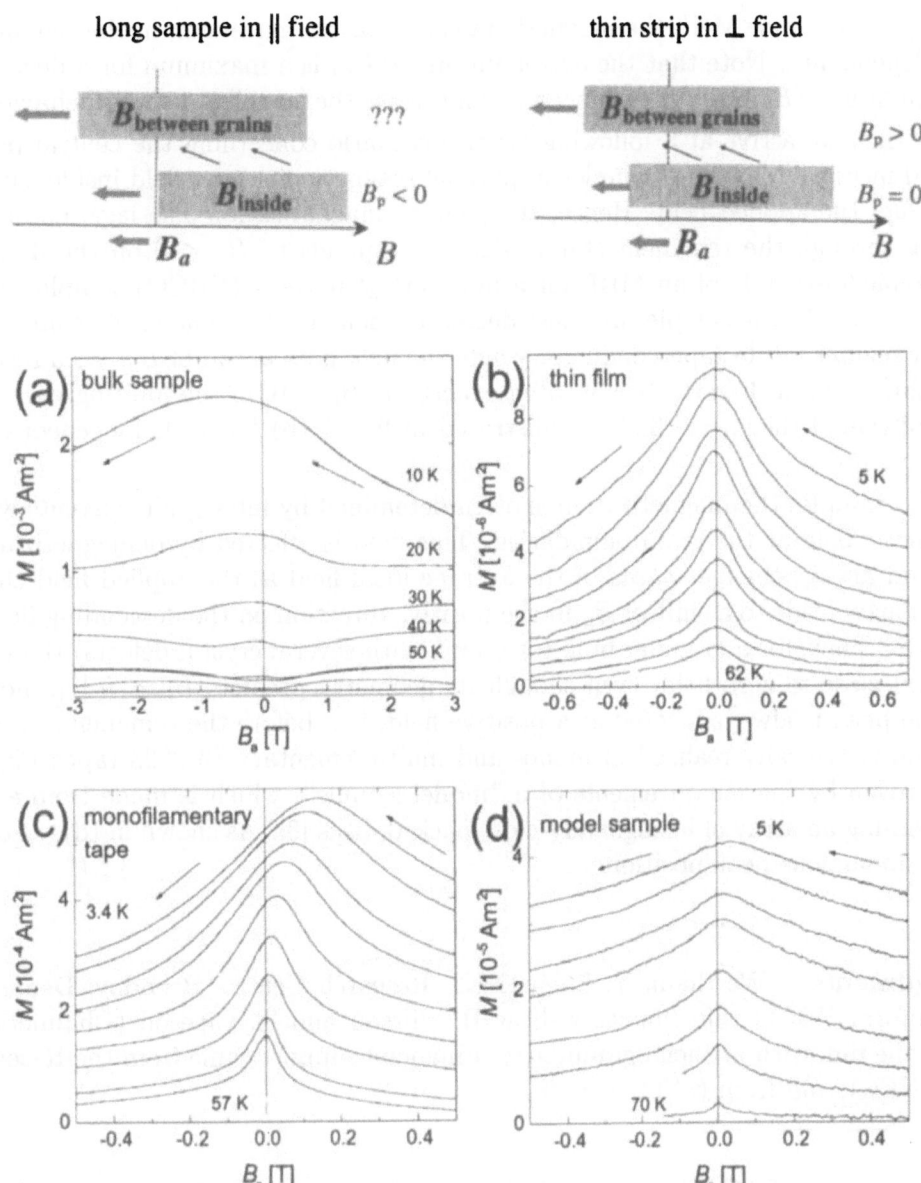

Fig. 1. Upper: A drawing illustrating the main conclusion of the paper. Lower: The descending branches of the MHLs (arrows indicate the direction of the field sweep): (a) bulk sample (thickness 300 μm), temperatures are between 10 and 40 K. (b) a very homogeneous YBCO thin film. The peak is located exactly at zero field; (c) monofilamentary Bi-2223 tape with a very pronounced *anomalous* peak position indicated by the dashed line; (d) model sample, the peak is found here at positive fields as well.

which follows immediately from the Biot-Savart law. This equation does not have an analytical solution for $B_z(x)$ in case of general $J_c(B)$. However, in the remanent state, $B_a = 0$, its solution always has an interesting symmetry [6]

$$B_z(x) = -B_z(\sqrt{w^2 - x^2}) .\qquad(2)$$

Moreover, a similar symmetry relation holds for the function $\partial B_z(x)/\partial B_a$. This symmetry leads to an important result for the magnetic moment of the strip [6]:

$$dM/dB_a = 0 \quad \text{at} \quad B_a = 0 .\qquad(3)$$

Consequently, on a major MHL the magnetic moment has an extremum in the remanent state for *any* $J_c(B)$-dependence. Note that the extremum in $M(B_a)$ is a maximum for a decreasing $J_c(B)$, and a minimum if $J_c(B)$ has a pronounced second peak, the so-called fishtail behavior.

Based on this, we arrive at a following general scenario concerning the central peak position, as illustrated in Fig. 1. For long samples in parallel magnetic field the field inside superconductor everywhere *lags* the applied field. Hence, the peak in magnetization also lags, i.e., it is observed *after* passing through the remanent state. This corresponds to $B_p < 0$ on the descending field branch. A typical example of an MHL for a bulk $YBa_2Cu_3O_{7-\delta}$ (YBCO) sample with fishtail is shown in Fig. 1(a). As the sample thickness decreases, demagnetization effects come into play and there appear regions inside superconductor where the field goes ahead of B_a. As a result, the peak position is shifted towards zero. It is located exactly at $B_p = 0$, in the limiting case of a uniform strip of infinitesimal thickness. This is illustrated in fig. 1 (b) for an homogeneous YBCO thin film.

In granular samples the magnetization is often detemined by inter-grain currents which depend on the magnetic field at the grain boundaries. This field is affected by demagnetization effect of the grains and always changes *ahead* of the average local field as the applied field changes. As a result, granularity leads to a shift of B_p in the positive direction on the descending field branch for all thicknesses. Thus, for a granular bulk (or a bulk with several crystal defects) the peak position may be found close to zero field, even though its geometry predicts $B_p < 0$ [7]. For a granular thin strip the peak is always located at a positive field, i.e., before the remanent state is reached. This situation is typically realized in mono- and multifilamentary Bi-2223 tapes Fig. 1(c). This concept is proven by the measurements of a "model sample", which is made from a YBCO thin film by patterning an array of hexagonally-close packed disks [8]. As shown in (d), also this sample exhibits the anomalous peak position.

Acknowlegdments. We thank Y. Shen (NKT Research Centre, Brøndby, Denmark) for the excellent uniform YBCO thin film as well as B. Nilsson and T. Claeson (Chalmers, Gøteborg, Sweden) for the film with artificial granularity. Financial support came from the Research Council of Norway (NFR), and from NATO via NFR.

1. D.-X. Chen and R. B. Goldfarb, J. Appl. Phys. **66**, 2489 (1989); T. H. Johansen and H. Bratsberg, J. Appl. Phys. **77**, 3945 (1995).

2. M. E. McHenry, M. P. Maley and J. O. Willis, Phys. Rev. B **40**, 2666 (1989); M. R. Koblischka, L. Pŭst, A. Galkin and P. Nalevka, Appl. Phys. Lett. **70**, 514-517 (1997); M. R. Koblischka et al, J. Appl. Phys. **83**, 6798 (1998).

3. K.-H. Müller, C. Andrikidis and Y. C. Guo, Phys. Rev. B **55**, 630 (1997).

4. J. E. Evetts and B. A. Glowacki, Cryogenics **28**, 641 (1988).

5. M. Däumling, and W. Goldacker, Z. Phys. B **102**, 331 (1997).

6. D. V. Shantsev, M. R. Koblischka, Y. M. Galperin, T. H. Johansen, L. Pŭst and M. Jirsa, Phys. Rev. Lett. **82**, 2947-2950 (1999).

7. M. R. Koblischka, L. Pŭst, M. Jirsa and T. H. Johansen, Physica C **320**, 101-114 (1999).

8. M. R. Koblischka, L. Pŭst, A. Galkin, P. Nalevka, M. Jirsa, T. H. Johansen, H. Bratsberg, B. Nilsson and T. Claeson, Phys. Rev. B **59**, 12114-12120 (1999); M. R. Koblischka, L. Pŭst, P. Nalevka, M. Jirsa, T. H. Johansen, H. Bratsberg, B. Nilsson and T. Claeson, Advances in Superconductivity XI, Springer, Tokyo, pp. 693-696 (1999).

The Dimensional Crossover Triggered by Vortex-Lattice Dislocations

Ken SUGAWARA

Ibaraki Polytechnic College, 864-4 Suifu-cho, Mito, Ibaraki 310-0005, Japan

Abstract: Collective pinning of a vortex lattice with screw dislocations is investigated. The energy of the screw dislocations is calculated taking account of dispersion of the tilt modulus of vortex lattices, and this calculation is applied to the collective-pinning model proposed by Mullock and Evetts. Comparison of the revised Mullock-Evetts model with the dimensional crossover observed in the amorphous superconductor Nb_3Ge is much more successful than that of the conventional collective-pinning models.

Keywords: Vortex pinning, Vortex lattice, Nb compounds

INTRODUCTION

The collective-pinning theory proposed by Larkin and Ovchinnikov [1] (LO) represents summation of weak elementary pinning forces f_p. The LO theory for two-dimensional (2-D) pinning (*i.e.*, pinning of straight vortex lines) is verified by many experiments on superconducting films [2]. However, in the case of three-dimensional (3-D) pinning (*i.e.*, pinning of curved vortex lines), comparison between the theory and experiments on superconducting bulks is unsuccessful [3, 4]. A possible explanation is that the discrepancies in the 3-D case are due to omitting effects of screw dislocations of vortex lattices (VLs) [3-6]. Although the plastic approach in collective-pinning theory was proposed by Mullock and Evetts [7] (ME), there still remains the discrepancies. In the present paper, a VL-dislocation model slightly different from the ME model is proposed, taking account of dispersion of the tilt modulus c_{44} of a VL.

THEORY

According to the LO theory [1], the volume pinning-force density for the number density n of the pinning centers is given by $F_p = (n\langle f_p^2 \rangle / V_c)^{1/2}$, where V_c is the correlation volume. Following the ME method [7], let us suppose that a VL domain with V_c is separated by the parallel slip planes spaced D_c apart, where each slip plane is enclosed by two adjacent screw-dislocation lines with the spacing L_c. When the VL has no edge dislocations, the correlation length R_c along the screw-dislocation lines is determined by elastic shear deformation. Then V_c is given by $L_c D_c R_c$. Since the VL-domain displacement arising from pinning approximately equals the half length of the VL constant a_0, the volume energy density of the pinned VL can be expressed as

$$U = (1/2)c_{66}(a_0/2R_c)^2 + (E_{SD}/D_cL_c) - (a_0/2)(n\langle f_p^2 \rangle / R_cD_cL_c)^{1/2}, \qquad (1)$$

where c_{66} is the shear modulus of the VL, and E_{SD} is the line energy density of a VL screw dislocation. Minimizing U gives the correlation volume for equilibrium states. The line energy density E_{SD} depends on c_{44}. Note that, when $D_c \sim a_0 \ll (c_{66}/c_{44})^{1/2}L_c \sim R_c$, dispersion of c_{44} [8] considerably reduces E_{SD} for the large Ginzburg-Landau (GL) parameter κ. Then we can put approximately [9]

$$E_{SD} = (a_0^2/4\pi)[c_{66} \cdot c_{44}(D_c)]^{1/2} \cdot 2.12(2D_c/\sqrt{3}a_0)^{1/2} \qquad (2)$$

with

$$c_{44}(D_c) \simeq (B/\kappa\xi)^2(1 - B/\mu_0H_{c2})\{[(\pi/D_c)^2 + (1 - B/\mu_0H_{c2})/\kappa^2\xi^2]^{-1} + \xi^2\mu_0H_{c2}/2B\}/\mu_0 \qquad (3)$$

335

for a superconductor with the upper critical field H_{c2} and the coherence length ξ under the magnetic-flux density B.

Substituting Eq. (2) into Eq. (1) and minimizing U give

$$D_c = \sqrt{3}a_0/2, \quad R_c = 4c_{66}E_{\mathrm{SD}}/n\langle f_p^2\rangle, \quad \text{and} \quad L_c = 128c_{66}E_{\mathrm{SD}}^3/\sqrt{3}a_0^3(n\langle f_p^2\rangle)^2. \tag{4}$$

Thus F_p for the 3-D case is

$$F_p(\text{3-D}) = (n\langle f_p^2\rangle/D_c R_c L_c)^{1/2} = a_0(n\langle f_p^2\rangle)^2/16c_{66}E_{\mathrm{SD}}^2. \tag{5}$$

For the amorphous limit of a VL, R_c in Eq. (1) is replaced by a_0, and minimizing U gives

$$F_p(\text{AL}) = (n\langle f_p^2\rangle/D_c\, a_0\, L_c)^{1/2} = n\langle f_p^2\rangle/4E_{\mathrm{SD}}. \tag{6}$$

For a film superconductor with the thickness d, if pinning is 3-D, the volume energy density for the 3-D case is lower than that for the 2-D case, $i.e.$, $U(\text{3-D}) < U(\text{2-D})$, and $L_c < d$. On the other hand, if $U(\text{3-D}) > U(\text{2-D})$, pinning is 2-D even though $L_c < d$. In this case, the term $E_{\mathrm{SD}}/D_c L_c$ is excluded from Eq. (1) and $R_c D_c L_c$ is replaced by $R_c^2 d$. Then minimizing U gives [2]

$$F_p(\text{2-D}) = (n\langle f_p^2\rangle/R_c^2 d)^{1/2} = 2n\langle f_p^2\rangle/a_0 c_{66}d. \tag{7}$$

If the minimum value of Eq. (1) in which L_c is replaced by d is smaller than the value of $U(\text{2-D})$, even though L_c given by Eq. (4) is larger than d, the condition $L_c = d$ represents the most stable state, namely, the *intermediate state* between 2-D pinning and 3-D pinning [9]. Then we can obtain

$$F_p(\text{IM}) = (n\langle f_p^2\rangle/D_c R_c d)^{1/2} = [4(n\langle f_p^2\rangle)^2/3c_{66}a_0^3 d^2]^{1/3}. \tag{8}$$

The inequality $U(\text{IM}) < U(\text{2-D})$, where $U(\text{IM})$ is the volume energy density for the intermediate state, is transformed into $d > L_{\mathrm{IM}}$ with

$$L_{\mathrm{IM}} = (2/9)\{[8c_{66}^{1/3}E_{\mathrm{SD}}/\sqrt{3}a_0(n\langle f_p^2\rangle)^{2/3}] + [2(n\langle f_p^2\rangle)^{1/3}/c_{66}^{2/3}]\}^3. \tag{9}$$

COMPARISON WITH EXPERIMENTS

In $\mathrm{Nb_3Ge}$ films, the crossover from 2-D pinning to 3-D pinning was confirmed by measuring dependence of F_p on d [5], and the discontinuous abrupt jump of F_p was observed.

Figure 1(a) shows L_c/d, L_{IM}/d, and R_c/a_0 versus $b \equiv B/\mu_0 H_{c2}$ obtained from Eqs. (4) and (9), where parameters are fixed at $\mu_0 H_{c2} = 3.5\,\text{Tesla}$, $\kappa = 70$, $d = 10\,\mu\text{m}$, and $n\langle f_p^2\rangle/b(1-b)^2 = 1.0 \times 10^{-5}\,\text{N}^2/\text{m}^3$; c_{66} is given by $\mu_0 H_{c2}^2 b(1-b)^2(1 - 0.58b + 0.29b^2)/8\kappa^2$ [8]. At $b = 0.85$ and $b = 0.89$, there occur the crossover from 2-D pinning to the intermediate state and the crossover from the intermediate state to 3-D pinning, respectively. At $b = 0.97$, the VL reaches the amorphous limit. When $L_{\mathrm{IM}} = d$ ($i.e.$, $b = b_{\mathrm{on}}$), the replacement of the correlation volume $R_c^2 d \to R_c D_c d$ leads to discontinuous increase in F_p because $D_c \sim a_0 \ll R_c$. As a result, we can obtain F_p versus b as shown in Fig. 1(b). The F_p-versus-b curve is similar to that observed in $\mathrm{Nb_3Ge}$ [5, 6]. On the other hand, the conventional collective-pinning models lead to no dimensional crossover for the same values of the parameters. However, the value of $n\langle f_p^2\rangle/b(1-b)^2 = 1.0 \times 10^{-5}\,\text{N}^2/\text{m}^3$ is at least ten times larger than the reasonable values of $n\langle f_p^2\rangle/b(1-b)^2$ reported in Refs. 2 and 4. Possibly the present estimation of E_{SD} is too rough to obtain satisfactory agreement.

At b_{on}, using Eqs. (7) and (8), we can obtain dependence of the ratio $F_p(\text{IM})/F_p(\text{2-D})$ on the temperature T as follows:

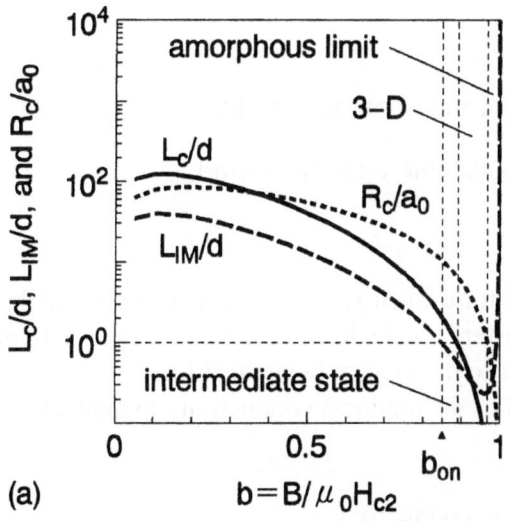 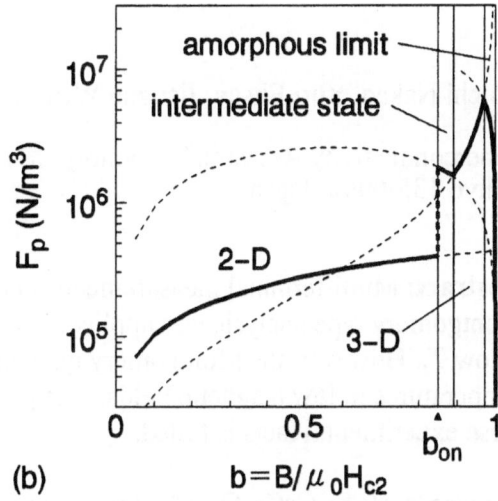

Fig. 1. The dimensional crossover triggered by the screw dislocations of a VL. (a) The thick solid line, the thick dashed line, and the thick doted line represent L_c/d, L_{IM}/d, and R_c/a_0 versus b, respectively. (b) F_p versus b (the thick solid line). The thin dashed lines represent Eqs. (7), (8), (5), and (6).

$$F_p(\text{IM})/F_p(\text{2-D}) \propto [(1 - T/T_c)b_{on}(1 - b_{on})^2]^{1/3}, \tag{10}$$

where T_c is the critical temperature; the relations of $c_{66} \propto H_{c2}^2 b(1 - b)^2$, $n\langle f_p^2\rangle \propto H_{c2}^3 b(1 - b)^2$, and $H_{c2} \propto 1 - T/T_c$ are used. Note that the above ratio is independent of estimation of E_{SD}. As shown in Fig. 2, the results for Nb$_3$Ge agree well with expectation of Eq (10). This supports the explanation that the discontinuous abrupt jump of F_p in Nb$_3$Ge is due to screw-dislocation lines penetrating into the VL.

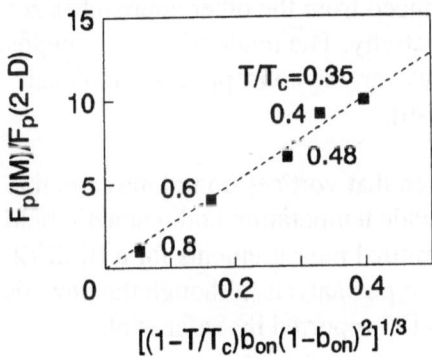

Fig. 2. $F_p(\text{IM})/F_p(\text{2-D})$ at b_{on} for several temperatures in Nb$_3$Ge. The experimental plots are extracted from the data of Wördenweber and Kes (*i.e.*, Fig. 1(b) in Ref. 6). The dashed line is a fit of Eq. (10). For $T/T_a = 0.8$, since the crossover was not observed, b_{on} is adopted as b maximizing F_p, and then we can put $F_p(\text{IM})/F_p(\text{2-D}) = 1$.

1. A. I. Larkin and Yu. N. Ovchinnikov, J. Low Temp. Phys. **34**, 409-428 (1979).
2. P. H. Kes and C. C. Tsuei, Phys. Rev. B **28**, 5126-5139 (1983); R. Wördenweber, A. Pruymboom, and P. H. Kes, J. Low Temp. Phys. **70**, 253-277 (1988).
3. E. H. Brandt and U. Essmann, Phys. Status Solidi B **144**, 13-38 (1987).
4. P. H. Kes and R. Wördenweber, J. Low Temp. Phys. **67**, 1-15 (1987).
5. R. Wördenweber and P. H. Kes, Phys. Rev. B **34**, 494-497 (1986).
6. R. Wördenweber and P. H. Kes, Cryogenics **29**, 321-327 (1989).
7. S. J. Mullock and J. E. Evetts, J. Appl. Phys. **57**, 2588-2592 (1985).
8. E. H. Brandt, J. Low Temp. Phys. **26**, 709-733, 735-753 (1977).
9. K. Sugawara, J. Low Temp. Phys. **117**, 127-147 (1999).

Montgomery Type Analysis for the Anisotropic Resistivity of $Bi_2Sr_2CaCu_2O_x$ Below and Above T_c

Koichi Nakao, Yuri Eltsev, Jianguo Wen, Susumu Shibata, and Naoki Koshizuka

Superconductivity Research Laboratory, ISTEC, 10-3 Shinonome 1-chome, Koto-ku, Tokyo 135-0062, Japan

Abstract: Multi-terminal measurements were performed for a $Bi_2Sr_2CaCu_2O$ single crystal, and the Montgomery type analysis was applied. It gave the consistent results in the normal state, but did not below T_c. However, the Montgomery type analysis seems to be applicable again at lower temperatures in low magnetic fields. One possible model of the nonlocal conductivity to explain these experimental facts is tested.

Keywords: $Bi_2Sr_2CaCu_2O_x$, Montgomery method, nonlocal conductivity

INTRODUCTION

An established technique to analyze a conductivity in a highly anisotropic material, such as $Bi_2Sr_2CaCu_2O$ (Bi-2212), is the Montgomery[1] technique or its modified versions[2]. In the Montgomery type analysis, the potential distributions are calculated numerically and compared with experimental results. Then the resistivity components are inversely solved. These procedure should be valid provided that the electric conductivity is linear and local. The breakdown of the Montgomery type analysis means that at least one of these conditions is not fulfilled.

Montgomery type analysis has been applied to high T_c superconductors. In the case of $YBa_2Cu_3O_x$ (Y-123), Safar et al[3] performed a multi-terminal measurements and deduced apparent anisotropy ρ_c/ρ_{ab} from two different current distributions and found that this ratio deduced from one configuration diverges as temperature decreases while that deduced from the other approaches zero. They attributed this breakdown to the nonlocality of the conductivity. The nonlocality was supposed to be due to the fact that a flux line moves as a line object. However, the concept of the nonlocal conductivity has not been accepted by all the researchers yet[4-6].

In the case of Bi-2212, in contrast to Y-123, it has been expected that vortices do not move as line objects but behave as uncorrelated pancake vortices[2,7] in a wide temperature and magnetic field range. Recently, however, Keener et al.[8] performed multi-terminal measurements for a Bi-2212 single crystal and observed the breakdown of the Montgomery type analysis, although the deviation of the apparent anisotropy was not so large as in the case of Y-123 reported by Safar et al.

In the present work, we systematically tested the applicability of the Montgomery type analysis to a Bi-2212 single crystal below and above T_c, by performing full numerical calculations using the formalism proposed by Levin[9].

EXPERIMENTAL

The sample was grown by a floating zone technique, and cleaved into thin platelets. The size of the sample was 3.6 x 0.85 x 0.007mm^3 and the contact configuration is illustrated in Fig.1. The superconducting transition temperature was 87K and there was no trace of Bi-2223 phase.

Measurements were performed using a low frequency (83Hz) ac technique, and the measuring current was 0.25mA. Magnetic fields were generated by a superconducting solenoid and applied along the c-axis. We measured resistances in 8 different contact configurations defined in Fig.1. Every set of two resistances can deduce the resistivity components ρ_{ab} and ρ_c. We used 20 combinations (R_i, R_j)'s among 28 possible combinations excluding 8 combinations which are very sensitive to experimental errors.

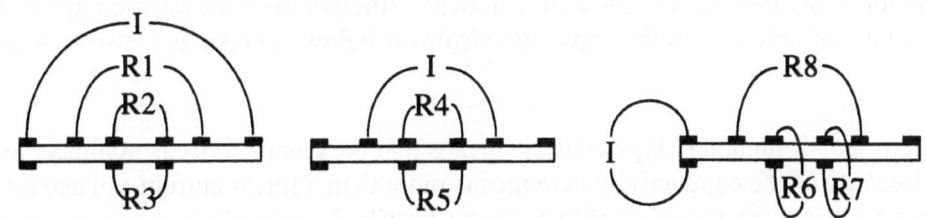

Fig.1. Contact configuration and definition of 8 resistances.

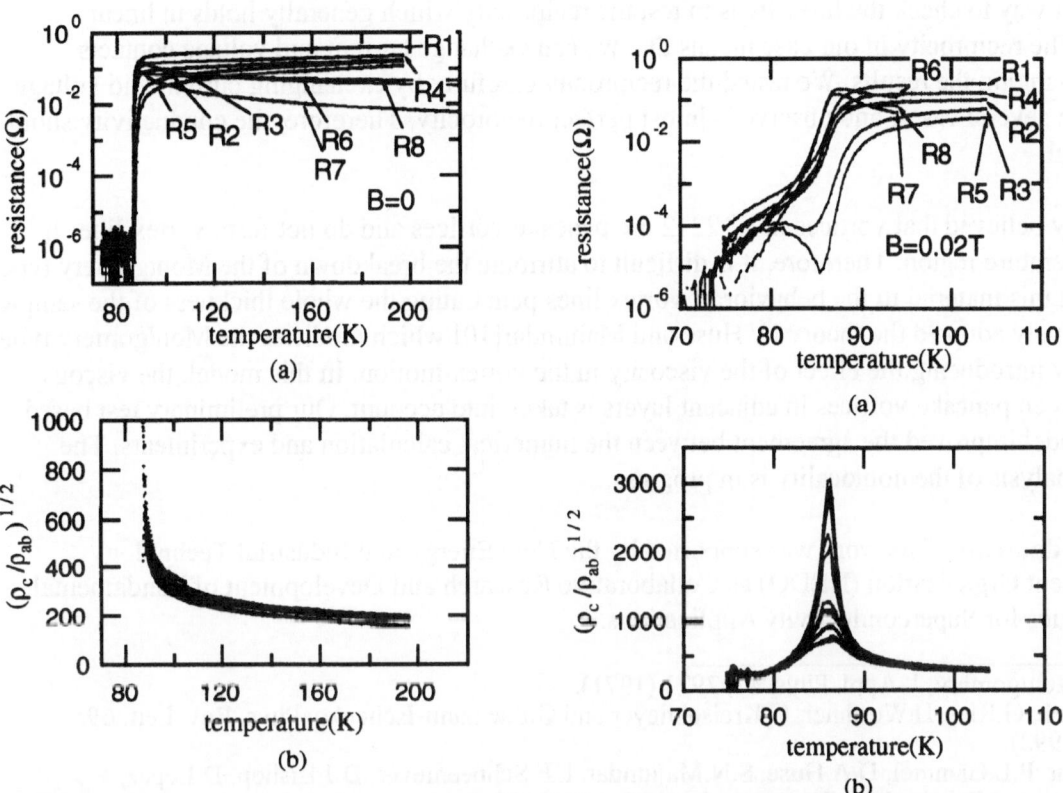

Fig.2. (a) Temperature dependence of eight resistances, and (b) apparent anisotropy based on the Montgomery type analysis in zero magnetic field.

Fig.3. (a)Temperature dependence of eight resistances, and (b) apparent anisotropy based on the Montgomery type analysis in a magnetic field of 0.02T.

DISCUSSION

The coincidence of the apparent anisotropy deduced from 20 different combinations above T_C is fairly good as shown in Fig.2(b). Although there is a discrepancy of about 20%, it is not unreasonable if the neglect of the contact width and the finite accuracy in the estimation of the sample thickness are taken into account. All the curves show a steep increase as temperature approaches T_C from above. The situation drastically changes when a small magnetic field is applied.

The apparent anisotropy shows a peak, and at the low temperature side of the peak, it is even smaller than that above T_C as shown in Fig.3(b). In the peak region, the discrepancy between values deduced from different contact configurations is large and cannot be explained by possible small errors in the estimation of the sample size or contact positions. This qualitatively agrees with the result of Keener et al.[8], and means that the Montgomery type analysis cannot be applied in this region. An interesting feature in Fig.3 is that at the lowest temperatures (75K~80K), the coincidence is recovered and it seems that the Montgomery type analysis is applicable again. In larger magnetic fields, the behavior is qualitatively the same (not shown), although the peak position moves to lower temperatures and the coincidence of the apparent anisotropy below the peak is not so clear as that in Fig.3.

The breakdown of the Montgomery type analysis means that the conductivity is nonlinear and/or nonlocal. The linearity of the conductivity here means more than a linear current-voltage relation. The linearity means that a superposition of two current distributions results in a superposition of potential distributions each component of which corresponds to each current distribution. A convenient way to check the linearity is to test the reciprocity which generally holds in linear systems. The reciprocity in our case means that we can exchange current and voltage contacts without changing the results. We tested the reciprocity carefully by exchanging current and voltage contacts in several cases, and observed almost perfect reciprocity. Therefore, the conductivity should be nonlocal.

It is widely believed that vortices in Bi-2212 are pancake vortices and do not form vortex lines in high temperature region. Therefore, it is difficult to attribute the breakdown of the Montgomery type analysis in this material to the behavior of vortex lines penetrating the whole thickness of the sample. We tentatively adopted the theory by Huse and Majumdar[10] which modifies the Montgomery type analysis by introducing the effect of the viscosity in the vortex motion. In this model, the viscous force between pancake vortices in adjacent layers is taken into account. Our preliminary test based on this model improved the agreement between the numerical calculation and experiments. The detailed analysis of the nonlocality is in progress.

Acknowledgments. This work was supported by the New Energy and Industrial Technology Development Organization (NEDO) as Collaborative Research and Development of Fundamental Technologies for Superconductivity Applications.

1. H.C.Montgomery, J. Appl. Phys. **42**, 2971 (1971).
2. R.Busch, G.Ries, H.Werthner, G.Kreiselmeyer and G.Saemann-Ischenko, Phys. Rev. Lett. **69**, 522 (1992).
3. H.Safar. P.L.Gammel, D.A.Huse, S.N.Majumdar, L.F.Schneemeyer, D.J.Bishop, D.López, G.Nieva and F.de la Cruz, Phys. Rev. Lett. **72**, 1272 (1994).
4. E.H.Brandt, Rep.Prog.Phys. **58**, 1465 (1995).
5. Yu.Eltsev and Ö.Rapp, Phys. Rev. **B51**, 9419 (1995).
6. Yu.Eltsev and Ö.Rapp, Phys. Rev. Lett. **75**, 2446 (1995).
7. H.Safar, E.Rodriguez, F.de la Cruz P.L.Gamel, L.F.Schneemeyer and D.J.Bishop, Phys. Rev. **B46**, 14238 (1992).
8. C.D.Keener, M.L.Trawick, S.M.Ammirata, S.E.Hebboul and J.C.Garland, Phys. Rev. **B55**, R708 (1997).
9. G.A.Levin, J. Appl. Phys. **81**, 714 (1997).
10. D.A.Huse and S.N.Majumdar, Phys. Rev. Lett **71**, 2473 (1993)

Temperature and Field Dependence of Magnetization Measured on a Grain Aligned Hg-1223 Samples

Yasukuni Matsumoto*, Yuusuke Nagafuji*, Chizu Taka,** and Akihiko Nishida**

*Department of Electrical Engineering Fukuoka University, 8-19-1 Nanakuma, Fukuoka 814-0180
 Japan
**Department of Applied Physics Fukuoka University, 8-19-1 Nanakuma, Fukuoka 814-0180 Japan

Abstract: Temperature and field dependence of magnetization have been measured on grain aligned Hg-1223 samples when a magnetic field is parallel or perpendicular to c-axis, and also on non grain aligned Hg-1223 samples. The critical current density J_c estimated from the hysteresis width ΔM of the magnetization loop does not show a scaling law. The J_c, and hence the ΔM, decays exponentially with increasing temperature for both magnetic field directions of $B//c$-axis and $B//ab$-plane for grain aligned samples and also non grain aligned one.

Keywords: Hg-1223, Magnetization, Field and temperature dependence of J_c, Flux pinning

INTRODUCTION

It is well known that reduced pinning force F_p/F_{pmax} measured at each temperature is scaled into a universal curve in low T_c superconducting materials, where F_{pmax} means the maximum pinning force. It turns that the critical current density J_c's are also scaled into a universal curve represented as

$$J_c = A\{H_{c2}(0)\}^m (1-t^2)^m b^\gamma (1-b)^\delta , \qquad (1)$$

where $t=T/T_c$ and $b=B/B_{c2}$[1]. Equation (1) means that a log-log plot of J_c versus $1-t^2$ shows a straight line. In case of high T_c superconductors, there are many reports that the scaling law is valid only in higher temperature region, and the data in low temperature region remarkably deviate from the universal curve described by eq.(1). This characteristic was explained as (i) effect of flux creep or (ii) existence of more than two pinning mechanisms. Recently, Hirano et al. showed that such characteristic is also observed in Bi-2212 single crystals and comes from the effect of flux creep because of weak pinning in their specimen [2]. But, the flux creep may not be responsible for such characteristic because the deviation from eq.(1) is widely observed in many HTSC's having different pinning force. The critical current density J_c is occasionally determined by the magnetization measurement. Burlachkov et al. showed a result that ΔM is proportional to $\exp(-2T/T_o)$ when a 2D pancake vortex surmounts the surface barrier (Bean-Livingston barrier)[3]. Their result shows that the J_c does not behave as eq.(1) because ΔM is proportional to J_c in usual case. We are interested in the temperature and field dependence of J_c, and report here the experimental results obtained from the magnetization measurements on powdered Hg-1223 (Re-doped) samples.

EXPERIMENTAL

Re-doped Hg-1223 samples were prepared by a sintering method. Mercury oxide powder was weighed to a non stoichiometric ratio of $Hg:Ba:Ca:Cu_{0.9}Re_{0.1} = 1.7:2:2:3$ because mercury will vaporize in the evacuated quartz tube. The X-ray diffraction pattern of the samples shows that no trace of $CaHgO_2$ is observed but there are very weak peaks from Ba-Cu-O impurity phase(s).

A half quantity of pulverized sample with mean diameter of 10μ m was mixed with epoxy resin and solidified under a magnetic field of 10T. (grain aligned sample). The remaining powder was solidified under zero magnetic field. (non aligned sample). Grain alignment was confirmed by the X-ray profile. The magnetization M of these samples was measured by using VSM at temperatures of 6.7 - 110K and under magnetic field of 0 - 2T. The T_c of both samples is 128K.

RESULTS AND DISCUSSION

Hysteresis curves for three cases of $B//c$-axis and B in ab-plane of grain aligned sample and non aligned sample are resemble to each other except the magnitude. In comparing the magnitude of magnetization at the same field, it is largest for the case of $B//c$-axis and is smallest for the case of B in ab-plane, while the value of non grain aligned sample lies between them. Anisotropy factor defined by $M(B//c)/M(B$ in ab-plane$)$ is $3\sim4$ which is slightly smaller than the reported values.

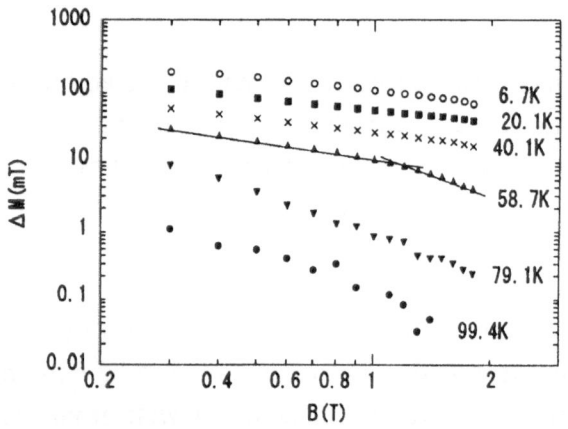

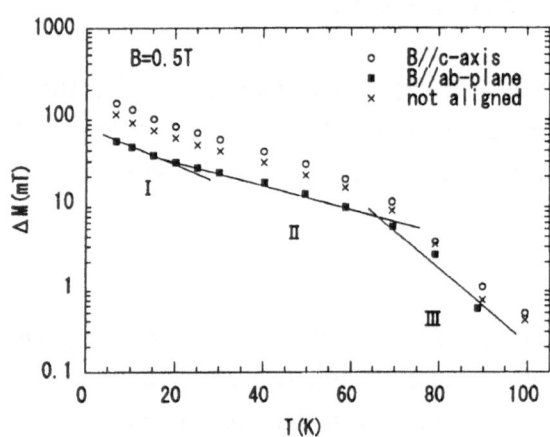

Fig1. Field dependence of ΔM for $B//c$-axis.

Fig.2 Temperature dependence of ΔM for three cases

Magnetic field dependence of the hysteresis width ΔM in the case of $B//c$-axis is shown in Fig.1. From this figure, it is obvious that ΔM is represented as $\Delta M \propto B^{-\mu}$ and μ is approximately single value at low temperatures. But, at higher temperatures, μ consists of two values. For example, we estimate $\mu_1 = 0.81$ in low field region and $\mu_2 = 2.0$ in high field region at $T=58.7K$. Field dependence of ΔM for B in ab-plane and for non aligned sample are essentially the same as Fig.1. Temperature dependence of ΔM for each case at $B=0.5T$ is shown in Fig.2. It is obvious that their features are qualitatively the same to each other. The ΔM can be represented as $\Delta M \propto \exp(-T/T_0)$. This temperature dependence is essentially the same as a result by Burlachkov et al.[3]. But, it is not possible that all experimental points are fitted on a straight line with a certain T_0 value. Experimental

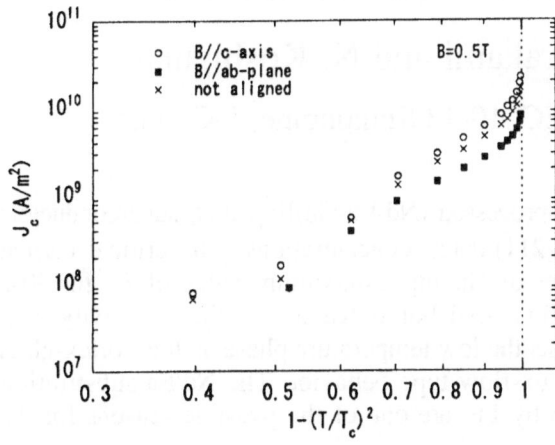

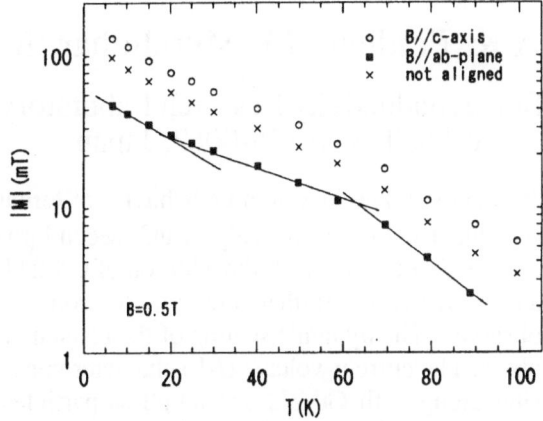

Fig.3 Temperature dependence of Jc's .   Fig.4. Temperature dependence of magnetization

points are divided into three regions of I, II and III indicated in the figure. In case of $B//c$-axis, we obtain the T_0 values of T_0=19.5K for region I, T_0=28.4K for region II and T_0=9.7K for region III, respectively. According to Burlachkov *et al.*, much contribution of the surface barrier to the ΔM may be dominant in higher temperature region near irreversibility temperature T_{irr}. For this reason, the region III shown in Fig.3 may reflect the surface barrier effect. However, another exponential decay named region I or II does not attribute to the surface effect, because these regions are far from the T_{irr}. Recently, Krelaus *et al.* showed that the temperature dependence of bulk and surface hysteresis are essentially the same to each other[4]. According to them, the temperature dependence of bulk hysteresis is also divided into three parts, which is very resemble to our case shown in Fig.2. Figure 3 shows the J_c versus $1-t^2$ plots at B=0.5T for three cases. The J_c's are determined by applying the Bean model to measured ΔM values. (We used 10μm as the mean diameter of the particle size.) The J_c sharply drops in low temperature region and this characteristic is common with three cases. In higher temperature region of 20~70K, J_c's lie on a straight line with a slope of m~5.5, but J_c's at higher temperatures than 70K deviate from this line downward. Such behavior is observed at another field. It is concluded from this figure that we can not represent the $\log J_c$ versus $\log(1-t^2)$ plots by one straight line. This feature comes from the temperature dependence of the ΔM.

Temperature dependence of absolute value of the magnetization at B=0.5T for three cases are shown in Fig.4. These were measured in the first increment of field. It is clear that the magnetization itself has essentially the same characteristic as the hysteresis width ΔM.

As a conclusion, the key point for the problem is why the magnetization has such characteristic. The magnetization M of a superconductor in mixed state is determined by the magnetic induction B inside the material, which is strongly dependent on the flux pining. Temperature dependence of the flux pinning force in high T_c superconductors is not settled yet.

1. A.M.Campbell and J.E.Evetts, Adv. Phys. **21** 372 (1972).

2. T.Hirano, T.Matsushita, Y.Nakayama, J.Shimoyama and K.Kishio, in *Advances in Supercon.-ductivity XI* 489-492 (1999).

3. L.Burlachkov, V.B.Geshkenbein, A.E.Koshelev, A.I.Larkin and V.M.Vinokur Phys. Rev. B**50** 16770-16773 (1994).

4. J.Krelaus, M.Reder, J.Hoffmann and H.C.Freyhardt, Physica C**314** 81-92(1999).

Flux pinning in melt-processed ternary (Nd-Eu-Gd) Ba$_2$Cu$_3$O$_y$ superconductors with second-phase addition

A.K. Pradhan, M. Muralidhar, M. Murakami and N. Koshizuka

Superconductivity Research Laboratory, ISTEC, 10-13 Shinonome, 1-Chome, Koto-ku, Tokyo 135-0062, Japan

Abstract: The flux pinning behavior of ternary melt-processed (Nd-Eu-Gd)Ba$_2$Cu$_3$O$_y$ superconductors is studied with varying Gd$_2$BaCuO$_5$ second-phase (Gd-211) defect concentrations. The critical current density J_c increases with the addition of Gd-211 particles displaying a maximum value of J_c for 30% and decreases on further addition. A pronounced field-induced bump feature in the resistivity was observed. The dynamic scaling of the resistance suggests the low temperature phase as the vortex-glass phase. The current-voltage (*I-V*) characteristics show flux-flow type behavior. The Nd/Ba substitution sites along with Gd-211 second-phase particles refined by Pt are one of the possible reasons for the vortex entanglement in the liquid phase.

KEY WORDS: flux pinning, magnetoresistance, second-phase, glass transition

INTRODUCTION

The critical current density in high-T_c superconductors is still below the required value for high current applications. Tremendous efforts have been put forward to enhance the critical current density, J_c in layered high-T_c superconductors for their applications at 77K. The recent development [1,2] of NdBa$_2$Cu$_3$O$_y$ (NdBCO) materials under controlled atmosphere and oxygen-controlled melt-growth (OCMG) processed ternary (LRE)BaCuO (where LRE=light rare-earth elements, i.e., Nd, Sm, Eu, Gd) superconductors [3-6] has opened up a new insight for power applications because of their high J_c (≥75,000 A/cm^2) at 77K and T_c ~94 K. A solid solution between the LRE atoms and Ba provides additional flux pinning due to the composition fluctuation in the sample leading to a spatial variations of T_c. Further flux pinning can be provided by Gd$_2$BaCuO$_5$ (Gd-211) particles which can be added to the starting powders as in YBCO. This increased pinning can be explained due to δT_c or $\Delta \kappa$- pinning.

The high flux pinning and their actual behavior in magnetic field are not known yet. The study of the dynamic response of the flux vortices is necessary to understand the actual pinning mechanism. In this paper, we report the resistive transitions in magnetic fields, the current-voltage (*I-V*) characteristics and magnetization J_c for several NEG samples with varying degree of Gd-211 as second-phase addition.

EXPERIMENTALS

Bulk samples of (Nd$_{0.33}$Eu$_{0.33}$Gd$_{0.33}$)Ba$_2$Cu$_3$O$_y$ + Gd-211 were prepared by the OCMG process using 0.1% O$_2$ partial pressure in Ar atmosphere. A series of samples with addition of 0 to 40 mol% Gd-211 second-phase particles (Gd$_2$BaCuO$_5$) was prepared. In order to refine the resulting Gd-211 particle size, 0.5 mol% Pt was added. The details of the preparation and heat treatment schedules are described elsewhere [4]. Standard ac technique (with I =1mA, f =17 Hz) was used to measure the resistivity for H //c using a low-noise ratio-transformer. Isothermal *I-V* characteristics were measured using dc method. The magnetization measurements were performed in a commercial SQUID magnetometer.

RESULTS AND DISCUSSION

Fig.1 shows the magnetic field dependence of J_c at 77 K for 0 to 40% Gd-211 second-phase addition in NEG. J_c (0) and J_c ^{peak} increases remarkably with increasing Gd-211% addition, reaches maximum for 30%, and then decreases for further increase of Gd-211 addition such as at 40%. Similar trend was observed for J_c (*H*) at the peak field in the intermediate field range. H_{irr} increases as Gd-211 addition is increased from 0 to 20%. However, H_{irr} is reduced for 30% Gd-211 addition.

Figs.2 (a) to (e) show resistive transition for NEG samples containing Gd-211 second phase from 0 to 40%. NEG-Gd0% shows a foot feature in the zero-field transition and a larger transition width compared to NEG-Gd10 and 20 and 30% samples which show a much sharper transition. However, NEG-Gd40% shows a larger transition width with a rounded foot feature. The most distinguishing

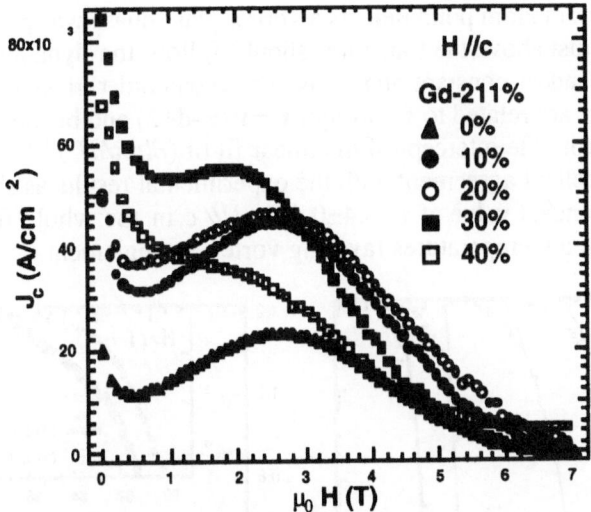

FIG.1 Field dependence of the critical current density, J_c at 77K for NEG samples containing 0 to 40 mol% second-phase and 0.05% Pt.

difference among these samples is that NEG-Gd20, 30 and 40% display a huge and broad bump feature in resistance for higher magnetic fields. NEG-Gd20% sample exhibits the most distinguishing bump feature at about 70% of the resistive transition. Although a very small bump feature is observed for NEG-Gd0 and 10% in high fields, it is not observed for $H \leq$ 2T. In order to test the influence of sample thickness on the bump feature, we show the resistive transition of NEG-Gd20% thinned down to almost less than 50% of the initial thickness of the sample in the inset of Fig.2 (c) for H =7T. We observed the similar feature at about 40% of the resistive transition in thinner sample that rules out the effect of inhomogeneity in current distribution in the sample to cause the bump feature. The distinguished feature of all these curves is that the bump feature is a strong function of field. The down

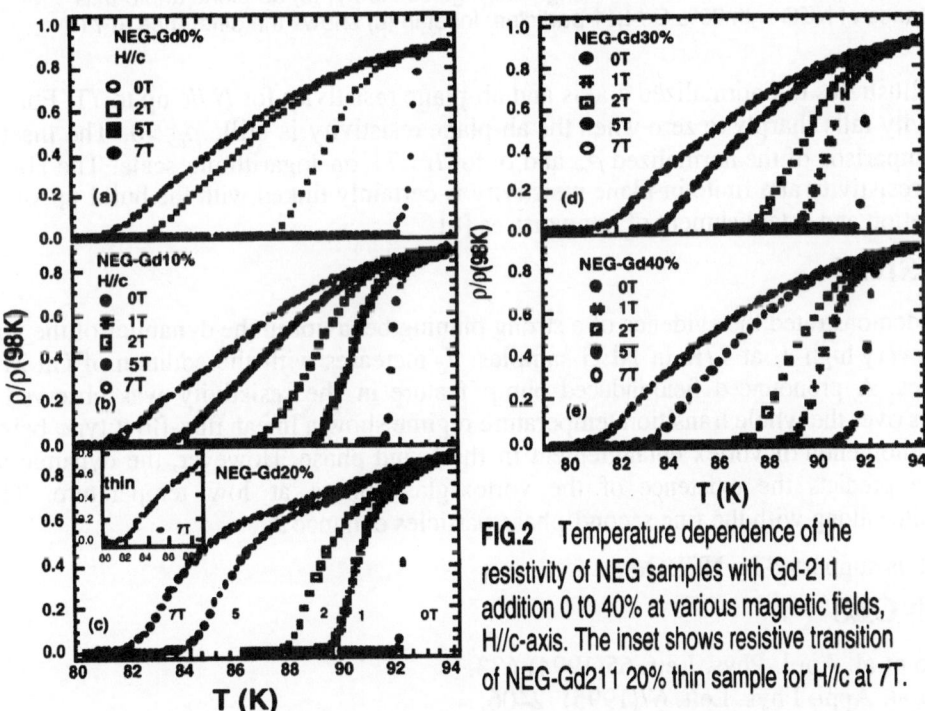

FIG.2 Temperature dependence of the resistivity of NEG samples with Gd-211 addition 0 t0 40% at various magnetic fields, H//c-axis. The inset shows resistive transition of NEG-Gd211 20% thin sample for H//c at 7T.

turn in resistance below T_b is due to strong pinning. The large reduction in normal state resistivity in NEG-Gd10 to 40% compared to NEG-Gd0% is due to Pt addition that creates percolating paths because of the proximity effect. Furthermore, Pt refines 211 particles to submicron size that contribute to the bump feature. Similar bump feature was also reported very recently in NEG samples with the addition of NEG -211 particles and attributed to the vortex entanglement in the liquid phase [6].

In the presence of weak random point defects a vortex-glass phase was proposed. In the VG prediction, the resistance just above the transition should follow the dynamic scaling [7] $R \sim (T-T_g)^s$ where s is the field independent constant and T_g is the second-order glass transition temperature. The critical exponents, ν and z are related to s through $s = \nu(z-d+2)$ and becomes $\nu(z-1)$ for dimension d=3. T_g was determined from the intercept of the linear fit of $(\partial lnR/dT)^{-1}$ Vs. T giving a slope $1/\nu(z-1)$. The scaling was in excellent agreement with the experimental results as shown in Fig.3 (a) and (b). We obtained a field independent value of $s = 5.4 \pm 0.4$ for $H// c$ in the whole field range. The IV curves show linear behavior [8] at all temperatures favoring vortex entanglement.

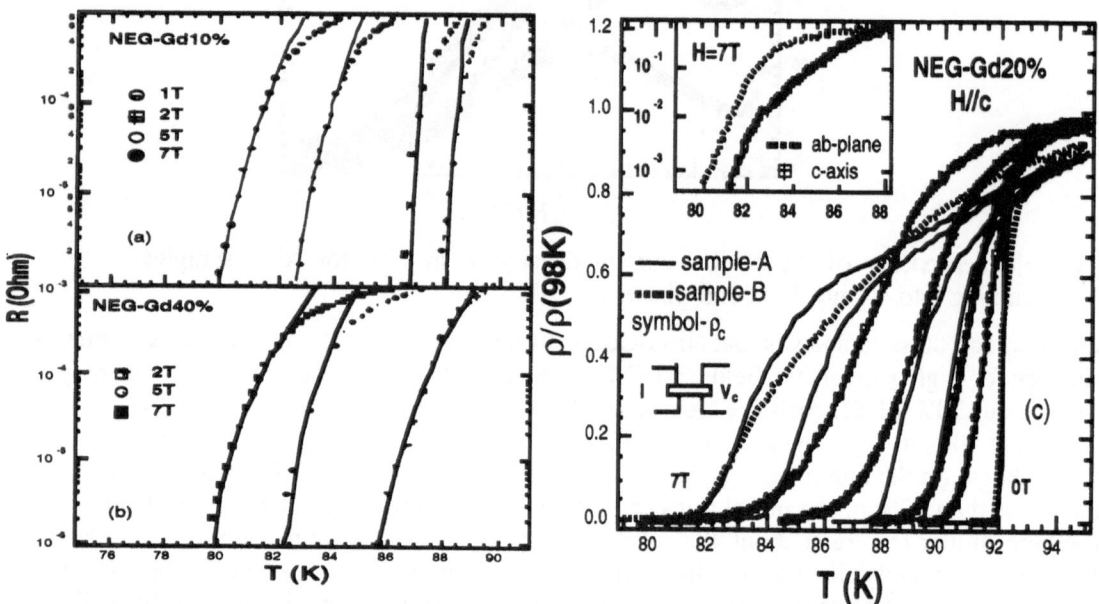

FIG.3 Scaling of resistance curves using vortex-glass theory, (c) ab-plane and c-axis resistance of NEG with 20% Gd-211 particles. Inset in (c) shows the difference at 7T

Fig.3(c) illustrates the normalized c-axis and ab-plane resistivity for $H //c$ up to 7T. For $H > 0$, the c-axis resistivity falls sharply to zero when the ab-plane resistivity is still, $\rho_{ab} > 0$. The inset of Fig.3 shows the comparison of the normalized ρ_{ab} and ρ_c for $H = 7T$ on logarithmic scale. The abrupt fall of out-of-plane resistivity at a finite in-plane resistivity is certainly linked with the build up of the c-axis vortex correlation and establishment of line vortices [8].

CONCLUSION

We have demonstrated the evidence of a strong pinning behavior in the dynamics of the vortex state that renders a very high J_c at 77K in NEG samples. J_c increases with the addition of Gd-211 second-phase particles. A pronounced field-induced bump feature in the resistivity was observed. The IV characteristics over the whole transition temperature regime show a linear flux-flow type behavior that favors the phenomenon of vortex entanglement in the liquid phase. However, the dynamic scaling of the resistance predicts the existence of the vortex-glass phase at low temperature. The Nd/Ba substitution sites along with the fine second-phase particles enhance J_c .

This work is supported by NEDO.

REFERENCES

1. S. I. Yoo et al. Appl. Phys. Lett. 65(1994) 633.
2. T. Egi et al. Appl. Phys. Lett. 67(1995) 2406.
3. M. Murakami et al. Supercond. Sci. Technol.9 (1996) 1015.
4. M. Muralidhar et al. Supercond. Sci. Technol.10 (1997) 663.
5. M. R. Koblischka et al. Appl. Phys. Lett.73(1998) 2351.
6. A. K. Pradhan et al. Appl. Phys. Lett. 75 (1999) 253.
7. M. P. A. Fisher, Phys. Rev. Lett. 62(1989) 1415.
8. A. K. Pradhan et al. J. Appl. Phys. 86(1999) 5705.

Orientation Dependence of Magnetization Near *ab* Plane in Micron-thick YBCO Films

Amit Rastogi, Hirofumi Yamasaki and Akihito Sawa

Electrotechnical Laboratory, 1-1-4 Umezono, Tsukuba, Ibaraki 305-8568, Japan

Abstract: We have studied the orientation dependence of magnetization for the applied magnetic field oriented close to the *ab* plane in YBCO films of micron thickness. Using a SQUID magnetometer, the longitudinal (parallel to the applied field direction) and the transverse (orthogonal to the applied field) components of the magnetization were measured. We have observed that most of the magnetization at any given orientation comes from that along the *c*-axis, due to the screening currents in the *ab* plane. However, magnetization along the *ab* plane cannot be neglected when the field is oriented very close to the *ab* plane.

Keywords: YBCO films, *ab*-plane magnetization, orientation dependence.

INTRODUCTION

The geometric effects dominate the magnetic behavior of superconducting samples such as thin films. The extent of these effects depends upon the shape anisotropy, *i.e.*, the ratio of their width to their thickness. Therefore, in case of the *c*-axis oriented films, we expect that magnetization would be dominated by its component along the *c*-axis (M_c) due to the screening currents in the *ab* plane [1]. Consequently, in inclined fields, the magnetization would be scaled by $\sin\theta$, θ being the angle between the *ab* plane and the applied field. There have been reports of such observations in YBCO thin films [2,3]. We have shown earlier [4] that in micron-thick films of YBCO, there is a significant magnetization along the *ab* plane (M_{ab}). Due to the thickness of the films, it is possible that the vortices can be localized along the *ab* plane. The critical current density (J_c) values obtained from magnetization curves along the *ab* plane were comparable with the values obtained by transport measurements [4].

This report investigates the orientation dependence of magnetization for the applied field oriented close to the *ab* plane in micron-thick films of YBCO. We investigate the validity of the equation

$$M_{Long} = M_{ab}\cos\theta + M_c\sin\theta, \tag{1}$$

where M_{Long} is the longitudinal magnetization parallel to the applied field. We have studied two samples with their J_c values about an order of magnitude apart.

EXPERIMENTAL

Micron-thick films of YBCO were deposited on SrTiO$_3$ (STO) and Y-ZrO$_2$ (YSZ) single crystal substrates by pulsed laser ablation method. The *c*-axis oriented films were subjected to dc magnetization studies by a SQUID magnetometer (Quantum Design). A horizontal sample rotator with an angular resolution of 0.1° was used to study the orientation dependence of magnetization around the *ab* plane. Both the longitudinal (along the applied field) and the transverse (orthogonal to the applied field) components of the magnetization were recorded while the dc field was swept in ±100mT and ±1T ranges at a constant temperature.

RESULTS and DISCUSSION

We report here our observations from two samples, A (deposited on YSZ, film thickness (t) = 2μm and size 3.5×3.5 mm^2) and B (deposited on STO, t = 1μm and size 2×5 mm^2). First, we analyzed the quality of these two samples by estimating their J_c. J_c values can be calculated from the magnetic hysteresis curves by using the Bean's model. $J_c = 2\Delta M/t$ ($H \parallel ab$ plane) and $J_c = 2\Delta M/[w(1-w/3l)]$ ($H \parallel c$-axis). Where, w is the width, l is the length of the film and ΔM is the difference between the magnetization for increasing and decreasing fields. At 60K, in sample A, the J_c values were 5×10^9A/m^2 ($H \parallel ab$ plane) and 0.6×10^9A/m^2 ($H \parallel c$-axis). In sample B the corresponding values were 4.3×10^{10}A/m^2 and 3.2×10^{10}A/m^2, respectively. Sample A had lower J_c and its value along the ab plane was eight times larger than that along the c-axis. The J_c of sample B on the other hand was well within the expected range.

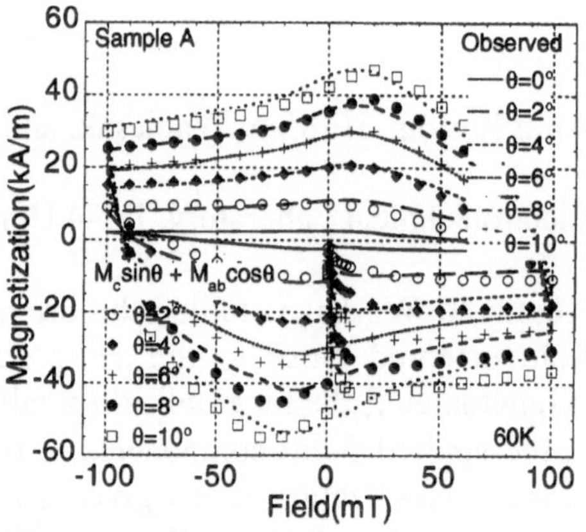

Fig. 1 shows the (longitudinal) magnetization (M) *vs.* applied field (H) curves (hitherto referred as M-H curves) from sample A at various orientations of the applied field at 60K and scan range of $\pm H_o = \pm100$mT. The solid lines

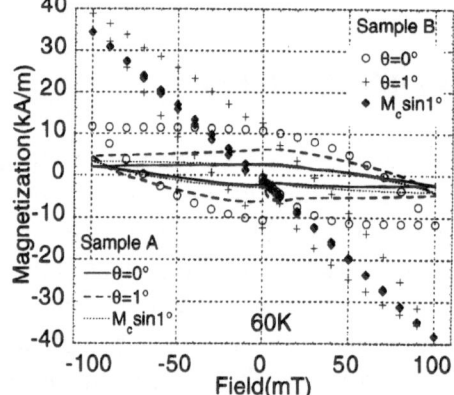

Fig. 1. M-H curves from Sample A at 60K for a field scan of $\pm H_0 = \pm100$mT and $0 \leq \theta \leq 10°$.

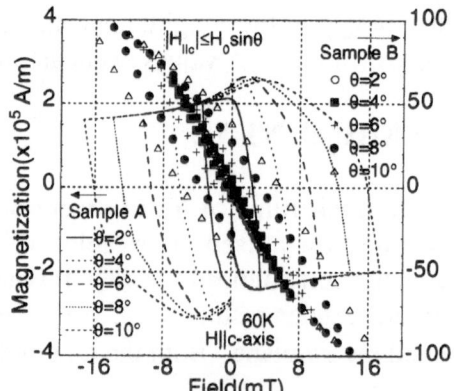

Fig. 2. M-H curves from samples A and B at 60K for $\theta = 0°$ and $1°$.

represent the M-H curves at various θ values while the markers represent the data calculated in accordance with Eq.(1). We obtained M_{ab} and M_c from the measurements in $H \parallel ab$ plane and in $H \parallel c$-axis, respectively. For the M-H curves along the c-axis, the field scans correspond to $\pm H\sin\theta$ so as to compare them with the M-H curves at various orientations, since we expect that at various θ values, the field component along the c-axis is $H\sin\theta$. These curves (along the c-axis) are also scaled by $\sin\theta$ to retrieve M_{Long} [Eq.(1)]. We observe that the two curves agree quite well. However, the

interesting point here is that M_{ab} cannot be neglected, especially for smaller θ. Fig. 2 shows the M-H curves from sample A and B at $\theta = 0°$ and $1°$. For comparison, $M_c\sin\theta$ is also shown. In both samples, there is significant contribution from M_{ab}.

Fig. 3 compares the M-H curves along the c-axis from samples A and B. The field scans correspond to $\pm H\sin\theta$. In sample A, the flux penetrates at lower fields, the magnetization is hysteretic even at low field scans of ±4mT and the absolute value of magnetization is also low. In sample B, the magnetization is almost reversible and non-hysteretic in the same scan range for $\theta\leq4°$. It also has a larger M. Therefore, we expect sample B to

Fig. 3. M-H curves ($H \parallel c$-axis) from samples A and B at 60K. The field scans correspond to $\pm H\sin\theta$.

have better quality and hence a higher J_c.

Fig. 4 shows the M-H curves from sample B at various orientations of the applied field at 60K and scan range of ±100mT. The solid lines represent the M-H curves at various θ values while the markers represent the $M_c \sin\theta$ data calculated from the M-H curves along the c-axis. As in Fig. 1, here also, for the M-H curves along the c-axis the field scans correspond to ±$H\sin\theta$. The M-H curves obtained at various θ values agree quite well with $M_c \sin\theta$. In Fig. 5, we compare the longitudinal and transverse magnetization at $\theta = 2°$, $5°$, $10°$ and also the c-axis magnetization (field scans correspond to ±$H\sin\theta$ with H_0 = 1T). M_{Trans} is expressed as $M_{Trans} = M_{ab}\sin\theta - M_c\cos\theta$,

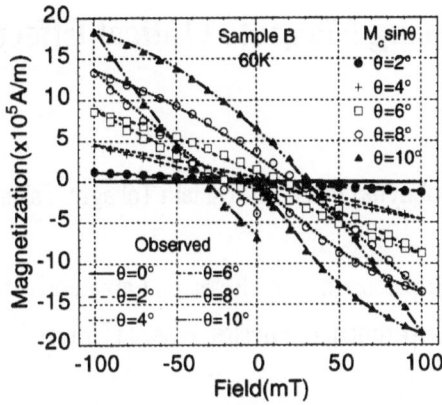

Fig 4. M-H curves from Sample B at 60K for a field scan of ±H_0 = ±100mT and $0 \le \theta \le 10°$.

and for small θ, $M_{Trans} \approx M_c$ (because $M_c \gg M_{ab}$). Also, $M_{Long} = M_{ab}\cos\theta + M_c\sin\theta \approx M_c\sin\theta$ (except for $\theta \approx 0°$). Therefore $M_{Long} = M_{Trans}\sin\theta$. In Fig. 5, the initial slopes of all the three curves are similar but after the full flux penetration, there are deviations. While the transverse and longitudinal magnetizations agree well, the c-axis magnetization is slightly larger than M_{Trans}, or $M_{Long}/\sin\theta$. This deviation can be explained if we consider the flux penetration along the ab plane. We have already observed a significantly hysteretic magnetization for the field oriented along the ab plane. This would reduce the J_c along the ab plane [5]. Therefore for the inclined fields, the dominant c-axis magnetic moment (coming from the screening currents in the ab plane) would also be reduced.

In another sample, we have observed a complete deviation from the expected behavior i.e. $M_{Long} < M_c\sin\theta$. Details of the study would be published elsewhere.

In conclusion, we have analyzed the orientation dependence of magnetization in micron-thick films of YBCO. In the two samples (their J_cs being an order of magnitude apart), the magnetization at any given θ follows Eq.(1) for the field scan of ±100mT. In both samples, the in-plane magnetization (M_{ab}) has a significant value when the applied field is oriented very close to the ab plane. In sample B with higher J_c, at higher field scans and various θ values, M_{Long} and $M_c\sin\theta$ begin to disagree after the full flux penetration has been achieved. This would be due to the flux penetration along the ab plane which results in a reduction of the ab plane J_c and hence a smaller than expected magnetization.

Acknowledgments

The authors are grateful to Y. Mawatari for useful discussions. AR would like to thank New Energy Development Organization, Japan for a fellowship based on a cooperative research agreement between Electrotechnical Laboratory, Japan and International Superconductivity Technology Center, Japan.

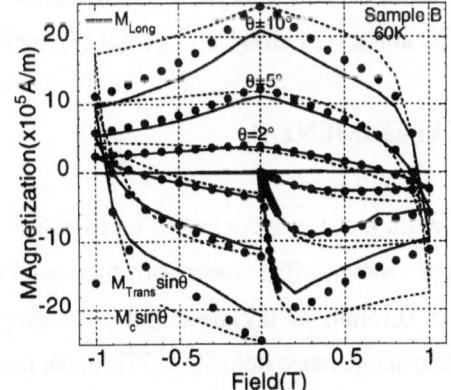

Fig 5. M-H curves from Sample B at 60K for a field scan of ±H_0 = ±1T and $0 \le \theta \le 10°$. Note that M_{Long} and $M_{Trans}\sin\theta$ agree well while $M_c\sin\theta$ is slightly larger.

1. E.H. Brandt, Rep. Prog. Phys. **58**, 1465 (1995).
2. H. Teshima *et al.*, Appl. Phys. Lett. **58**, 2833 (1991).
3. B.P. Thrane *et al.*, Phys. Rev. B **54**, 15518 (1996).
4. A. Rastogi, H. Yamasaki and A. Sawa, IEEE Trans Appl. Supercond. **9**, 1986 (1999).
5. M.V. Indenbom, *et al.,* Physica C**226**, 325 (1994).

Observations of Flux Penetration in High-Temperature Superconductors Using Magneto-Optical Effect of EuO Layer

Nobuyuki Iwata[A,B], Toshiaki Takagi[A], Takato Machi[A], Tadataka Morishita[A], and Kay Kohn[B]

[A]Superconductivity Research Laboratory, ISTEC, 1-10-13, Shinonome, Koto-ku, Tokyo 135-0062, Japan
[B]Department of Physics, Waseda University, Shinjuku-ku, Tokyo 169-8555, Japan

Abstract: Ferromagnetic oxide EuO is a promising candidate for magneto-optical layer to investigate the flux penetration in high-temperature superconductors. EuO has the Curie temperature of 69K. Au film of $1\mu m$ was deposited on EuO film as a reflection and protection layer. The magnetic moments of Eu^{2+} were ordered in plane. We have succeeded for the first time in observing the magneto-optical images of distribution of magnetic flux in $NdBa_2Cu_3O_{7-\delta}$ single crystal using EuO film, at 5K up to 5.5T and at 60K up to 2.2T.

Keywords: EuO, magneto-optical, Faraday rotation, $NdBa_2Cu_3O_{7-\delta}$ single crystal, magnetic flux

INTRODUCTION

For applications of high temperature superconductor, it is important to get the information of the flux distribution and local critical current densities (J_C). To obtain those information, Lorentz microscopy, Bitter technique, scanning tunneling microscopy, and magneto-optical (MO) technique have been investigated. Among these techniques, the MO technique has advantages of real time and macroscopic resolution. For MO investigation EuSe is used as magneto-optic layer (MOL) at high magnetic fields at lower temperatures than 40K[1,2]. EuSe is impossible to observe MO images at higher temperatures, where so-called peak effect in J_C appears. EuO is a promising candidate as MOL to cover a wide temperature and magnetic field range. X. J. Yu et al. have not succeeded to observe MO image using EuO film due to poor quality of EuO films[3]. In this paper we compare MO images of $NdBa_2Cu_3O_{7-\delta}$ (Nd123) single crystal using a single phase EuO film and using a EuSe film.

EXPERIMENTAL

The Eu metal was evaporated on MgO(001) substrate in a partial oxygen pressure of 4.5×10^{-8} Torr in an ultra-high vacuum MBE system, of which basal pressure was 2.0×10^{-9} Torr. MgO substrate was polished both sides and 0.15mm in thickness. Substrate temperature was 300℃. Magnetization curves were measured in the temperature range from 5K to 77K under magnetic fields up to 7T with a SQUID magnetometer.

The magnetic flux distribution of Nd123 single crystal was visualized by detecting the Faraday rotation through 2500Å EuO layer which was located immediately above the Nd123 sample exposed by a xenon lamp. Band pass filter of 550nm and 602nm was used for images detected by EuO and EuSe, respectively. The magnetic field was applied up to 7T perpendicular to the film plane, i.e. along the c-axis of the Nd123.

RESULTS AND DISCUSSION

EuO epitaxially grew on a MgO(001) substrate with an orientation of EuO[100]//MgO[100] and EuO(001)//MgO(001). The full width at half maximum (FWHM) of the rocking curve for EuO(002) was 0.41 degrees. Figure 1 shows magnetization curves of a 2300Å EuO film at 5K, 40K, 60K and 77K with a magnetic field applied parallel (filled symbols) and normal (open symbols) to the film plane. The magnetic moment per Eu atom was 6.88 ± 0.34 μ_B. The ferromagnetic Curie temperature was determined to be 69K from the temperature dependence of remanent magnetization. That the values of the open symbols is larger than those of the filled symbols at high magnetic fields is due to a difficulty in the correction of diamagnetism of the sample holder. The magnetic moment of EuO film had in-plane anisotropy. Below 40K the magnetization normal to the film was saturated around 3.5T. Above 60K the magnetization remained unsaturated even at 7T, which was available to observe MO images at such high temperatures and magnetic fields.

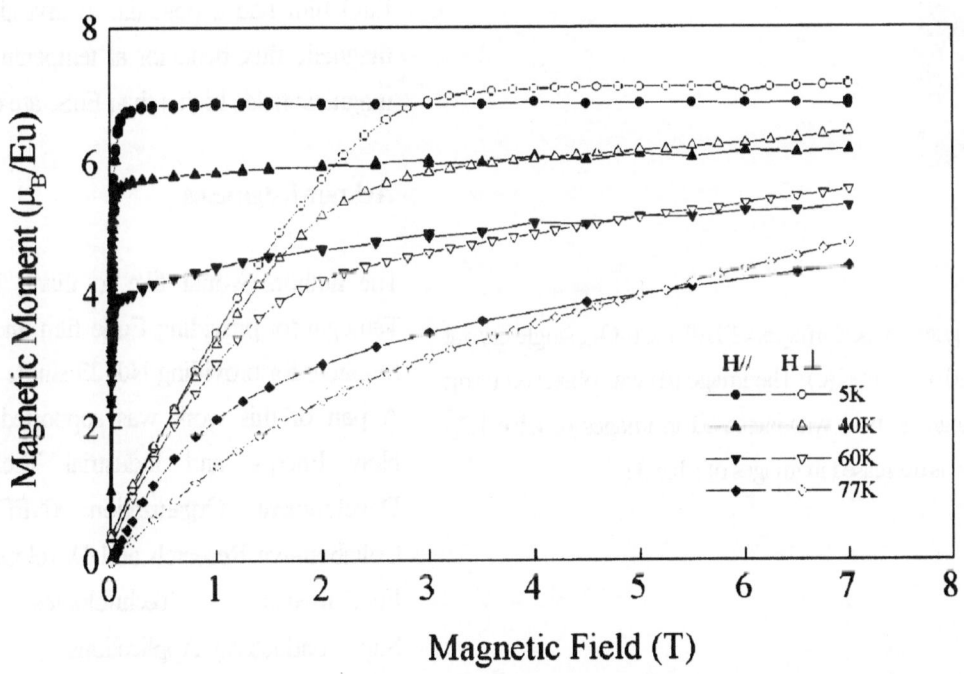

Figure 1 : Magnetization curves of 2300Å EuO film at 5K, 40K, 60K and 77K with magnetic field applied parallel (filled symbols) and normal (open symbols) to the film plane.

at 5K
a-1)2.5T   a-2)5.5T   a-3)1.0T

at 40K
b-1)1.0T   b-2)3.0T   b-3)0.5T

at 60K
c-1)1.0T   c-2)2.2T   c-3)0.5T

at 40K
d-1)1.0T

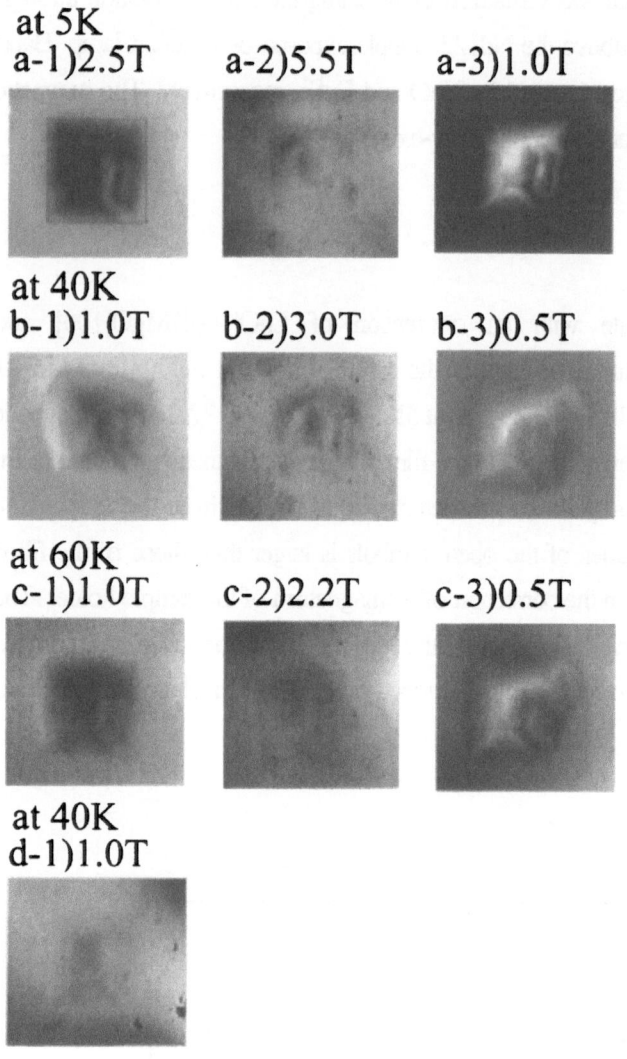

Figure 2 : Magneto-optical images of NdBa$_2$Cu$_3$O$_{7-\delta}$ single crystal at 5K(a), 40K(b) and 60K(c). The image (d) was observed using EuSe. The magnetic field was increased in images of a,b,c-1,2,) up to 7T and was decreased in images of a,b,c-3).

The MO images were observed at 5K 40K and 60K. The sample shape (0.83mm × 0.87mm) is indicated by a dashed line in fig.a-1). The Nd123 sample excludes the magnetic flux even at high magnetic fields as shown in figs. a-1), b-1), and c-1). At 40K images with EuO and with EuSe are compared, see figs.b-1) and d-1). At 60K the image was no longer detected by EuSe. As magnetic field is decreased, the magnetic flux is trapped near the center of the sample, as shown in figs. a-3), b-3) and c-3).

CONCLUSION

We have succeeded in observing MO images of Nd123 single crystal at 5K up to 5.5T and at 60K up to 2.2T using a single phase EuO film. It was demonstrated that the EuO film had a potential to investigate the magnetic flux behavior at temperatures and magnetic fields higher than EuSe are used.

Acknowledgments

The authors would like to thank Prof. T. Tamegai for providing EuSe film and Dr. Y. Shiohara for providing Nd123 single crystals. A part of this work was supported by the New Energy and Industrial Technology Development Organization (NEDO) as Collaborative Research and Development of Fundamental Technologies for Superconductivity Applications.

[1]M. R. Koblischka and R. J. Wijngaarden : Supercond. Sci. Technol. **8** (1995) 199.

[2]M. Zamboni, S. I. Yoo, T. Higuchi, K. Waki, S. Koishikawa, and M. Murakami : Physica C **281** (1997) 218.

[3]X. J. Yu, E. Batalla, and L. S. Wright : J. Mater. Res. **8** (1993) 462.

Influence of DC Field Sweep Rate on Non-Resonant Microwave Absorption in a YBa$_2$Cu$_3$O$_x$ Superconductor

Jiro Yamada, V.V. Srinivasu, Masaki Tada, Ken-ichi Itoh, Akinori Hashizume, Ikutaro Kometani, Khairil Anwar, and Tamio Endo*

Faculty of Engineering, Mie University, Kamihama, Tsu, Mie 514-8507, Japan

Abstract: Non-resonant microwave absorption intensity and reverse-sweep hysteresis were measured as a function of dc field sweep rate at 77 K in a YBa$_2$Cu$_3$O$_x$ bulk sample. Intensity of the non-resonant microwave absorption increases with increasing the sweep rate and saturates at higher sweep rates. The sweep rate influence is stronger at lower fields than at higher fields. We interpret the sweep rate dependence of the non-resonant microwave absorption in terms of much different flux profiles for slower (equilibrium) and faster (transient) sweep rates. Sweep rate dependence of the hysteresis qualitatively follows the dynamical model calculation of Portis, which takes into account the Anderson type of time decay of currents at sample surface.

Keywords: Microwave absorption, YBa$_2$Cu$_3$O$_x$ superconductor, Vortex dynamics

INTRODUCTION

Microwave absorption in High-T$_c$ Superconductors (HTSC) is very sensitive to applied magnetic fields [1-6]. Non-resonant microwave absorption (NMA) technique [1] is a convenient and very sensitive way to measure applied field dependence of the microwave absorption. In this technique, a weak ac field is superimposed on the sweeping dc field and the microwave absorption is recorded as a derivative signal vs dc field using a phase sensitive detection. The microwave absorption now depends on an equilibrium or a transient flux dynamics depending upon the sweep rate and flux relaxation times involved. NMA thus allows to study the flux dynamics and relaxation effects in HTSC. For example, Erhart et al. [7] have taken advantage of NMA technique to study the physics involved in various dynamic models such as minimal glassy model and Anderson model, by measuring "reverse-sweep hysteresis" in NMA as a function of the sweep rate of applied field. In this paper, we revisit the sweep rate dependence of NMA in a YBa$_2$Cu$_3$O$_x$ (YBCO) bulk sample. We show that NMA signal intensity strongly depends on the sweep rate, as flux profiles are much different for slower and faster dc field sweeps. We also show that the reverse-sweep hysteresis (difference between forward and reverse sweep NMA signal intensities measured at the same field) qualitatively fits to the Anderson type of relaxation model extended to NMA by Portis [8].

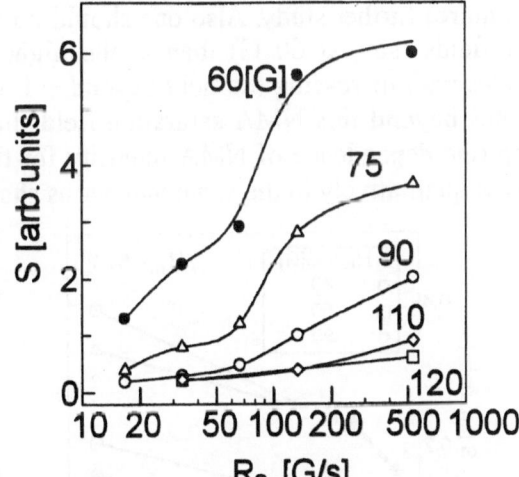

Fig. 1. NMA signal intensity (S) vs sweep rate (R$_s$), measured at different applied fields shown for a YBCO bulk sample. Solid lines are guide to eye.

EXPERIMENTAL

NMA measurements on a YBCO bulk sample at 77 K were carried out using the conventional Varian EPR Spectrometer. The microwave frequency and power were fixed at 9.4 GHz and 0.1 mW respectively. Modulation field amplitude and frequency were fixed at 5 G and 100 kHz respectively. The sample was zero-field cooled to 77 K. The sweep rate (R_s) of the dc field H_a was varied between 0.2 to 533 G/s. The sample was kept in maximal H_{rf} position in the cavity. A detailed description of the experimental set up is given elsewhere [3-6].

RESULTS AND DISCUSSION

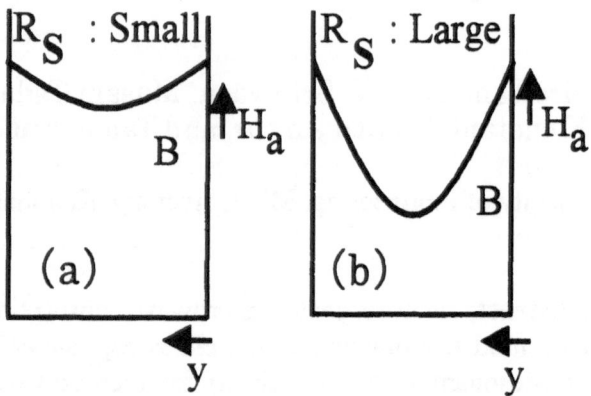

Fig.2. Flux profiles for (a) slow (equilibrium) and (b) rapid (transient) sweep rates of the applied field H_a. One can note that transient flux density gradients are steeper than equilibrium ones.

(a) *NMA intensity*: Figure 1 shows NMA signal intensity (S) vs sweep rate (R_s) for the YBCO sample, measured at 77 K in a range of R_s from 16 G/s to 533 G/s. A general increase in NMA signal intensity (which is more rapid at lower fields than at higher fields) as increasing the sweep rate from 16 to 150 G/s should be noted. However beyond this, further increase in sweep rate results in a kind of saturation in NMA intensity at the lower fields. Also there is a step like behaviour around 75 G/s. The general increase in NMA signal intensity as the sweep rate is increased can be understood in the following way. The state of affairs for the transient and equilibrium flux profiles are depicted in Fig.2. The flux profile at the higher sweep rate is much steeper than at the slower one. The steeper flux profile means higher current density at the surface and higher magnetic pressure, which leads to higher intensity of NMA [4]. The step like behaviour could be due to a transition from inter-grain to intra-grain response, which is not clear at present and requires further study. Also one should note that the influence of R_s on S is much higher at the lower fields (say, at 60 G) than at the higher fields (say, at 120 G). Incidentally, NMA itself saturates with increasing the field beyond ~ 150 G and simultaneously all the sweep rate influence vanishes beyond this NMA saturation field. In a lower range of sweep rate (0.2 to 1.7 G/s), the sweep rate dependence of NMA intensity for the forward field sweep and the reverse field sweep behaves qualitatively in the same manner as shown in Figs. 3 (a) and (b).

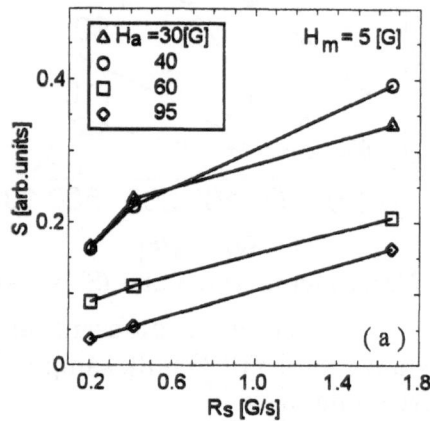
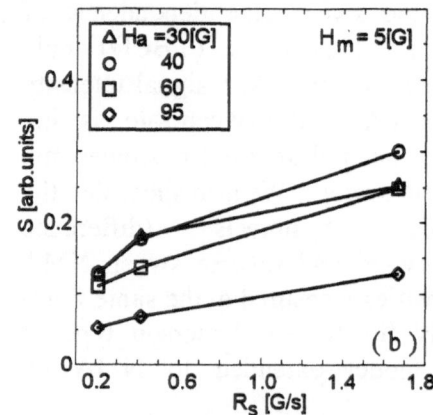

Fig. 3. NMA signal intensity (S) vs sweep rate (R_s) at a very low R_s range for the YBCO sample at 77 K. (a) Forward sweep and (b) reverse sweep. Solid lines are guide to eye.

355

(b) *Reverse-sweep hysteresis:* Erhart et al. [7] measured the reverse-sweep hysteresis in the non-resonant microwave absorption in YBCO granular samples. Both Erhart et al. [7] and Portis [8] noted the importance of plotting the reverse-sweep hysteresis vs the sweep rate, as such a plot gives important physics of the flux relaxation in these samples. Erhart et al.[7] were able to qualitatively fit their data to the minimal glassy model. Portis[8] suggests to use a more physical dynamic model based on the Anderson type of decay or relaxation. In Fig.4 we plot inverse of reverse-sweep hysteresis $(1/\Delta)$ vs inverse of sweep rate $(1/R_s)$. One can observe an initial linear increase and then a reduction of slope with increasing $1/R_s$. This behaviour qualitatively matches well with the prediction of the

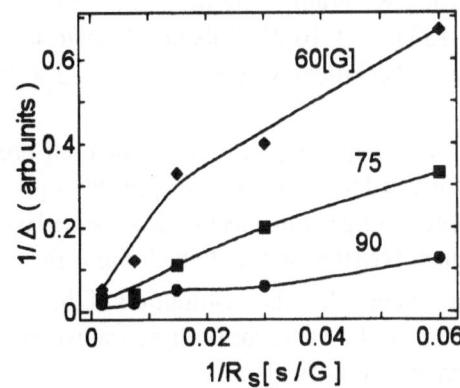

Fig.4. Inverse of reverse-sweep hysteresis $(1/\Delta)$ in NMA vs $1/R_s$ at 77 K. Solid lines are guide to eye.

dynamical model discussed by Portis[8] which relates the reverse-sweep hysteresis and the sweep rate directly for NMA measurements. This model is based on the assumption that flux penetration is associated with the time decay of currents at the surface of sample and that the time decay itself is of the Anderson type. We believe that the dynamical model suggested by Portis carries more physical sense than the minimal glassy model because the dynamical model essentially depends on the suf_ace flux density gradients or currents, upon which the microwave absorption strongly depends.

CONCLUSIONS

We measured the sweep rate dependence of non-resonant microwave absorption on YBCO bulk sample at 77 K. NMA signal intensity depends strongly on the sweep rate, and it increases with increasing sweep rate, saturating at higher sweep rates. The influence of sweep rate is more prominent at lower fields than at higher fields. The reverse-sweep hysteresis in NMA qualitatively follows the dynamical model given by Portis[8], with the Anderson type relaxation.

Acknowledgements. Grant-in-Aid for Scientific Research from the Japanese Ministry of Education, Science and Culture, and JSPS fellowship for one of the authors (VVS) are acknowledged.

*Corresponding author. E-mail: endo@cm.elec.mie-u.ac.jp

1. S.V. Bhat, P. Ganguly, T.V. Ramakrishnan, and C.N.R. Rao, J. Phys. C **20**, L559 (1997).
2. C.S. Krafft and C.F. Beckner, J. Appl. Phys. **69**, 4907 (1991).
3. T. Endo, H. Yan, S. Nagase, and H. Shibata, J. Supercond. **8**, 259 (1995).
4. T. Endo and H. Yan, *Studies of High Temperature Superconductors*, vol. **14**, edited by Anant Narlikar (Nova Science Publishers, New York, 1995), pp.65-106.
5. T. Endo and T. Wada, Jpn. J. Appl. Phys. **31**, 3303 (1992).
6. T. Endo and H. Yan, Jpn. J. Appl. Phys. **33**, 103 (1994).
7. P. Erhart, B. Senning, S. Mini, L. Fransioli, F. Waldner, J.E. Drumheller, A.M. Portis, E. Kaldis, and S. Rusiecki, Physica C **185-189**, 2233 (1991).
8. A.M. Portis, *Lecture notes in Physics, vol.48, Electrodynamics of High-Temperature Superconductors*, (World Scientific, Singapore, 1992), pp.199-204.

Ac Susceptibility in Melt-Processed Nd-Ba-Cu-O Superconductor

K. Inoue[1], K. Waki[1,2] and M. Murakami[1]

[1]SRL-ISTEC, 1-16-25 Shibaura, Minato-ku, Tokyo 105-0023, Japan
[2]Railway Technical Research Institute, 2-8-38 Hikari-cho, Kokubunji 185-8540, Japan

Abstract: Magnetic properties of the melt-processed Nd-Ba-Cu-O superconductor have been investigated with the third harmonic ac susceptibility (χ_3). The critical current density (J_c) evaluated from χ_3 decreased with decreasing the frequency accompanied by the shift of the secondary peak towards the lower field. The E-J characteristics are obtained from the frequency dependence of χ_3 and the magnetic relaxation measurement. On the assumption of the power law $E \sim J^n$, we found that the field which gives the peak n value was the same for both measurements. This suggests that the field-induced pinning is the most effective near this field.

Keywords: third harmonic ac susceptibility, secondary peak effect, field-induced pinning, n value

INTRODUCTION

LRE-Ba-Cu-O (LRE: light rare-earth elements of Nd, Sm, Eu, Gd) superconductors, which were melt processed in a reduced oxygen atmosphere, exhibit high J_c accompanied by the secondary peak effect [1]. LRE-Ba-Cu-O has a $LRE_{1+x}Ba_{2-x}Cu_3O_y$ type solid solution. When melt processed in a reduced oxygen atmosphere, LRE-rich $LRE_{1+x}Ba_{2-x}Cu_3O_y$ clusters 10-50 nm in diameter are dispersed in the $LREBa_2Cu_3O_y$ matrix [2]. These clusters with depressed T_c are weak-superconducting and are driven normal with increasing magnetic fields and thus can act as field-induced pinning centers, which is the source of the secondary peak effect.

Magnetic properties of Nd-Ba-Cu-O have extensively been studied by the dc magnetization, magnetic relaxation [3, 4], and ac susceptibility [5]. Ac susceptibility measurements allow us to study magnetic properties of superconductors in a high electric field region. According to the Bean model [6], the imaginary part of the third harmonic ac susceptibility (χ_3'') is inversely proportional to J_c. Waki *et al.* [5] found that an anomalous decrease in the χ_3'' as a function of the bias dc field is associated with the secondary peak effect in Nd-Ba-Cu-O samples.

In the present paper, we report the frequency dependence of χ_3 and its relation to J_c. E-J characteristics deduced from the χ_3 will also be compared with the results of magnetic relaxation measurements.

EXPERIMENTAL

Nd-Ba-Cu-O samples with the nominal composition of Nd: Ba: Cu = 1.8: 2.4: 3.4 were grown by the oxygen-controlled-melt-growth (OCMG) method. The melt-growth process was performed in a 0.1%O_2 - Ar atmosphere. The grown sample was cut into small specimens with dimensions of 0.37 × 0.41 × 2.37 mm such that the longest axis is parallel to the c axis, and then the samples were subjected to oxygen annealing at 300 ℃ for 260 h. Ac susceptibility measurements were performed with a Lakeshore 7229 ac susceptometer and dc magnetization measurements were carried out with Quantum Design MPMS-7 SQUID magnetometer. For both measurements, the fields were applied parallel to the c axis.

RESULTS AND DISCUSSION

Figure 1 shows the temperature dependence of the fundamental ac susceptibility (χ_1) with an ac field amplitude of 0.14 mT at a frequency of 125 Hz. The sample exhibits the critical temperature (T_c) of 94.2 K with a fairly sharp transition. The real part of the χ_1 is almost -1, showing that the demagnetization effect is negligible.

Figure 2 shows the dc bias field dependence of the χ_3 with an ac field amplitude of 2.83 mT at various frequencies (56 - 1000 Hz) under 83 K. A decrease in the χ_3'' near 1.5 T is ascribed to the secondary peak effect [5]. When the Bean critical state model [6] is valid, for an infinite slab of thickness $2d$, χ_3'' is expressed by

$$\chi_3'' = 2h_{ac}/15\pi J_c d, \tag{1}$$

under the condition that the ac field does not reach the center, where h_{ac} is the ac field amplitude. We applied this model to the present χ_3'' data and estimated J_c values. The results are shown in Fig. 3. Thus obtained J_c values decrease as the frequency decreases accompanied by the shift of the secondary peak towards the lower field. It is also notable that both the J_c values and the peak fields determined from χ_3'' are higher than those obtained from magnetization hysteresis loops. These results suggest that the J_c values are strongly affected by flux creep such that J_c values are lower as the time scale of the experiments are increased. The shift of the secondary peak field towards lower fields with increasing the time scale of the measurements reflects the competition between the field-induced pinning and the flux creep.

Next, we analyze the electric field (E) during ac measurements. On the basis of the Bean critical state model, E is expressed by

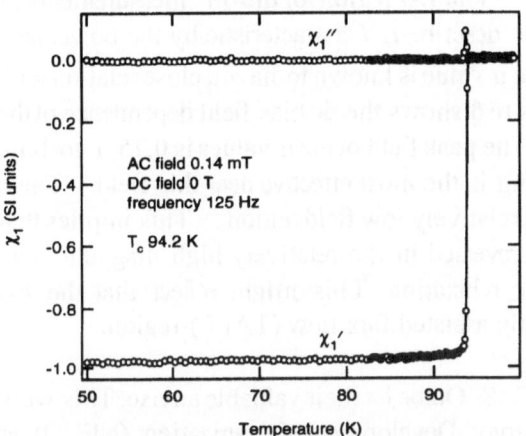

Fig. 1. Temperature dependence of the fundamental ac susceptibility with an ac field amplitude of 0.14 mT at a frequency of 125 Hz.

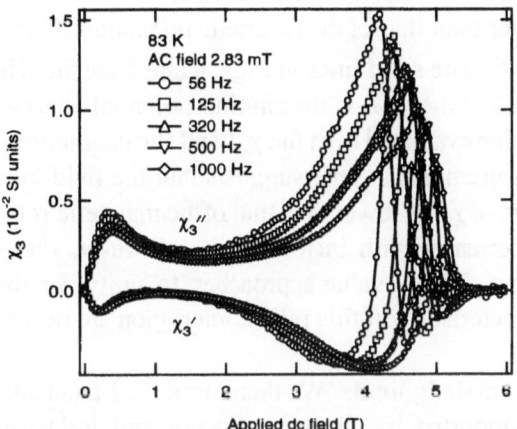

Fig. 2. Dc bias field dependence of χ_3'' with an ac field amplitude of 2.83 mT at various frequencies (56 - 1000 Hz) at 83 K.

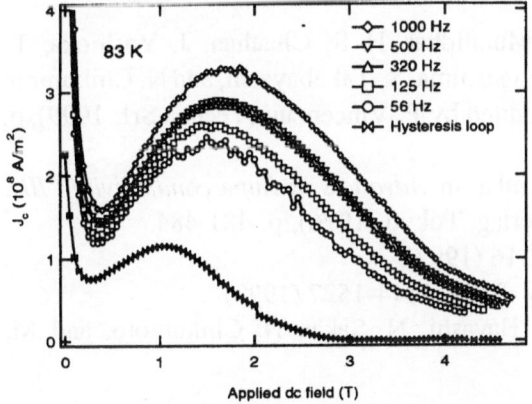

Fig. 3. Field dependence of J_c estimated from the χ_3'' and magnetization hysteresis loops.

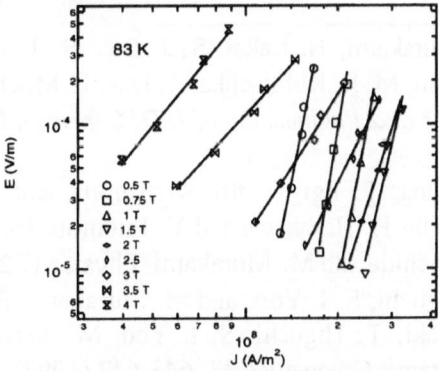

Fig. 4. E-J characteristics obtained from the χ_3'' at different dc magnetic fields. The solid lines are the fitted lines assuming the power law: $E \sim J^n$.

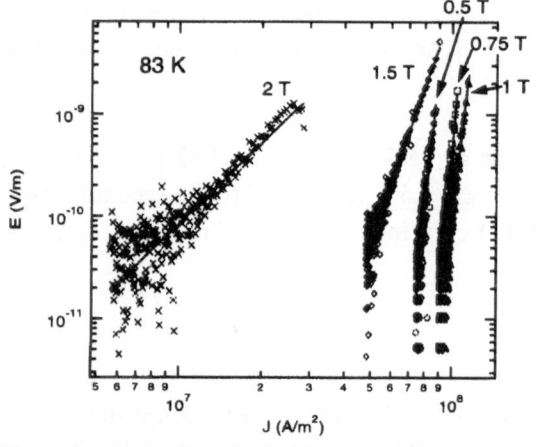

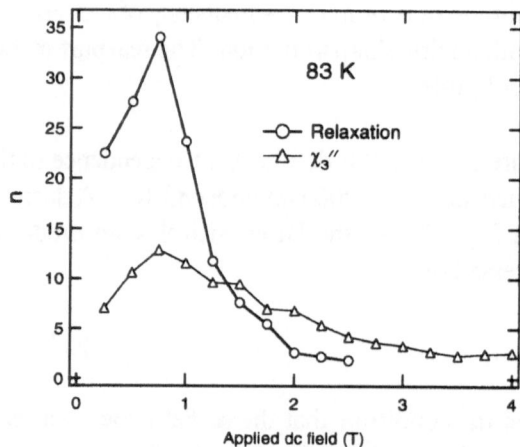

Fig. 5. *E-J* characteristics obtained from the magnetic relaxation at different dc magnetic fields. The solid lines are the fitted lines assuming the power law: $E \sim J^n$.

Fig. 6. Applied dc bias field dependence of the n value evaluated from the χ_3'' and the magnetic relaxation.

$$E = 2\pi f h_{ac} \lambda, \tag{2}$$

where $\lambda = h_{ac}/J_c$ is the ac field penetration depth, f is the frequency. Figure 4 shows the *E-J* characteristics obtained from χ_3'' at different dc magnetic fields. For comparison, we also deduced *E-J* properties from relaxation measurements, which are presented in Figure 5. The *E-J* region of the χ_3'' measurements is higher than that of the magnetic relaxation. It is common to describe *E-J* characteristic by the power law: $E \sim J^n$. The solid lines in Figs. 4 and 5 are fitted lines. The n value is known to have a close relation with $U_c/k_B T$, where U_c is the pinning potential energy [7]. Figure 6 shows the dc bias field dependence of the n value evaluated from the χ_3'' and the magnetic relaxation. The peak field of the n values is 0.75 T for both measurements. This suggests that the field-induced pinning is the most effective near this field. The n value of χ_3'' is lower than that of the magnetic relaxation in a relatively low field region. This implies that U_c decreases with increasing *J*. However, the n value is reversed in the relatively high magnetic field region. The n value approaches to unity for the magnetic relaxation. This might reflect that the *E-J* characteristics on this relaxation region are near the thermally assisted flux flow (TAFF) region.

Acknowledgments. We thank Prof. T. Matsushita and Dr. E. S. Otabe for their valuable advise. This work is supported by the New Energy and Industrial Technology Development Organization (NEDO) as Collaborative Research and Development of Fundamental Technologies for Superconductivity Applications under the New Sunshine Program administered by the Agency of Industrial Science and Technology (AIST) of the Ministry of International Trade and Industry (MITI) of Japan.

1. M. Murakami, N. Sakai, S. J. Seo, S. I. Yoo, M. Muralidhar, H. S. Chauhan, J. Yoshioka, T. Higuchi, M. R. Koblischka, A. Das, T. Mochida, K. Nagashima, S. Takebayashi, and N. Chikumoto, *Science and Engineering of HTC Superconductivity*, edited by P. Vincenzini (Techna Srl, 1999), p. 39-50.

2. Wu Ting, T. Egi, R. Itti, K. Kuroda, and N. Koshizuka, in *Advances in Superconductivity VIII*, edited by H. Hayakawa and Y. Enomoto (Springer-Verlag, Tokyo, 1996), p. 481-484.

3. T. Mochida and M. Murakami, Physica C **290**, 311-316 (1997).

4. T. Higuchi, S. I. Yoo, and M. Murakami, Phys. Rev. B **59**, 1514-1527 (1999).

5. K. Waki, T. Higuchi, S. I. Yoo, M. Watahiki, N. Hayashi, N. Sakai, N. Chikumoto, and M. Murakami, Cryogenics **37**, 643-647 (1997).

6. C. P. Bean, Rev. Mod. Phys. **36**, 31-39 (1964).

7. E. Zeldov, N. M. Amer, G. Koren, A. Gupta, M. W. McElfresh, and R. J. Gambino, Appl. Phys. Lett. **70**, 680-682 (680).

Pinning force diagram of the ternary superconductor (Nd,Eu,Gd)-123 with secondary phase additions

Michael R. Koblischka, Miryala Muralidhar, and Masato Murakami

Superconductivity Research Laboratory, Internarional Superconductivity Technology Center
16-25 Shibaura 1-chome, Minato-ku, Tokyo 105-0023, Japan

Abstract: $(Nd_{0.33}Eu_{0.33}Gd_{0.33})Ba_2Cu_3O_y$ ("NEG") samples are prepared containing extremely fine insulating 211 particles, which are uniformly dispersed in the superconducting matrix. As a function of the concentration of the 211 particles, the shape of the $j_c(H_a)$ curves at $T = 77$ K can be changed from a very pronounced fishtail behaviour to a plateau-like behaviour. Based on these observations and on the scaling of the volume pinning forces, we construct a pinning force diagram for the NEG system.

Key words: Light rare earth-123 superconductors, magnetic properties, flux pinning

INTRODUCTION

Flux pinning is one of the crucial problems in the development of technical high-T_c superconductors, especially because of the high operation temperature (77 K) aimed for in applications. The development of the light rare earth (LRE) superconductors provided samples with an increased critical current density, j_c, as compared to YBCO. Characteristic for these samples is the strongly developed secondary peak or fishtail effect, thus yielding a large j_c at fields of about 2.5 T. Recently, we have prepared samples of the type $(Nd_{0.33}Eu_{0.33}Gd_{0.33})Ba_2Cu_3O_y$ ("NEG"), which leads to an even further increase of j_c. Furthermore, we could successfully embed 211 particles of submicron size into the superconducting matrix [1]. The presence of these 211 particles influences the $j_c(H_a)$ behaviour drastically; but the position of the secondary peak remains unchanged. The scaling of the volume pinning forces, F_p, of the LRE superconductors leads to peak positions $h_0 > 0.4$ [2]; in the case of pure NEG even $h_0 \approx 0.5$ [3]. The high peak position is maintained until the shape of the $j_c(B)$ curves changes; then peak is shifted towards lower values indicating pinning at normal conducting pinning sites. This suggests that the peak effect is an unique property of the superconducting matrix (i.e. oxygen vacancy clusters), whereas the 211 particles provide effective pinning in the entire temperature range acting quasi as a "background" pinning mechanism for the peak effect. Based on these observations, we construct a pinning force diagram for the NEG system.

EXPERIMENTAL

Bulk samples of NEG were prepared using the OCMG process in 0.1% partial pressure of oxygen. A series of samples with additions of 10 to 50 mol% NEG-211 (that is, a mixture of Nd-422, Eu-211, and Gd-211 in the same ratio as the NEG matrix) is prepared. In order to refine the resulting 211 particle size, 0.5 mol% Pt is added to the samples. Details of the heat treatment schedules are described in detail elsewhere [1]. Magnetization loops (MHLs) are measured using a SQUID magnetometer (Quantum Design models MPMS 7 and XL) with a maximum field of $\pm$ 7 T; $H_a \parallel$ c axis. To minimize field inhomogeneities, the scan length is set to 1 cm.

RESULTS AND DISCUSSION

Figure 1 shows the field dependence of the critical current densities, $j_c(H_a)$, of the NEG samples. As measure for the 211 concentration, we use the initially added amount of NEG-211; this is justified by an analysis of polarization images. At an addition of 50 mol% NEG-211, j_c at zero field (self-field) increases even further, reaching 100000 A/cm^2. However, the peak effect is destroyed, and therefore, j_c in larger fields is severly reduced.

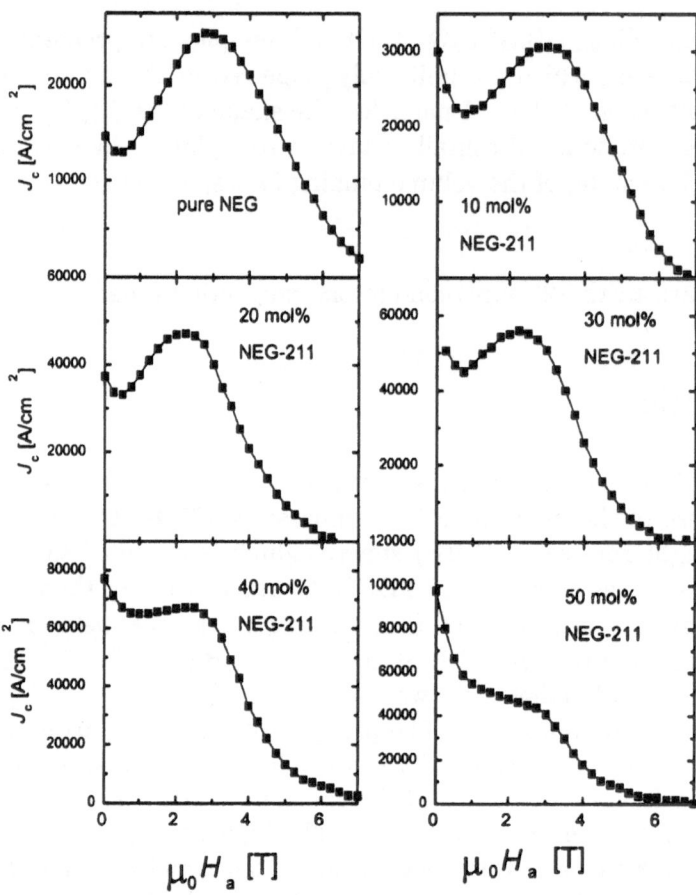

Fig. 1. (a): $j_c(H_a)$ of all NEG samples with different initial amounts of NEG-211; at $T = 77$ K and $H_a \parallel c$. Note the change of j_c, but also the considerable change of shape of the $j_c(H_a)$ curves.

Let us now discuss the basic underlying pinning mechanisms. Griessen et al. [4] presented strong evidence for a dominating δl-pinning in various YBCO thin films. In contrast to this, several authors found evidence for the δT_c-pinning in Pr-doped Y-123 and (K,Ba)BiO$_3$ single crystals [5, 6]. In general, the peak effect in $j_c(H_a)$ is due to oxygen vacancy clusters in conjunction with metal impurities as demonstrated by Erb et al. [7] using ultrapure YBCO single crystals. However, such oxygen vacancy clusters are, strictly spoken, nothing else than δT_c-pinning sites providing locally a reduction of T_c [8]. The presence of the LRE/Ba solid solution in the LRE superconductors leads to an increase of the disorder in the oxygen sublattice, and hence to an increase of the δT_c-pinning. This is indicated in the $j_c(H_a)$ curves by larger values of H_{peak}, and in the pinning force scaling by the increased h_0 [3]. Note that the contribution of the δT_c-pinning is only weak as compared to the pinning provided by the insulating inclusions. Therefore, in thin films with their much higher j_c, the pinning is only provided by the δl-pinning type as found by Griessen et al. [4]. Further, the δT_c-pinning is only effective at elevated temperatures [8]. The

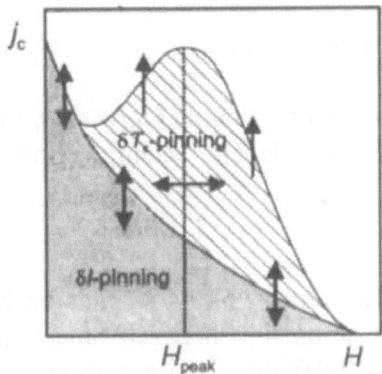

Fig. 2. Pinning force diagram, deduced from the NEG data with various NEG-211 additions. The borderline between the two pinning mechanisms can be influenced by, e.g., neutron irradiation or addition of 211 particles. H_{peak} can be influenced by e.g changing the matrix composition. The peak height can be influenced by oxygenation, or as in the case of NEG, by increased disorder within the matrix.

importance of the δT_c-pinning mechanism in *bulk* superconductors is further illustrated by the possibility to compose the $j_c(H_a)$ curves of a sample with secondary peak effect from two different contributions, which was demonstrated by Jirsa et al. [9]. One contribution is responsible for the central peak [$j_c(0\text{T})$], and is quickly decaying with increasing field. The other contribution is negligible at low fields, but increasing with increasing field and responsible for the formation of the secondary peak. Therefore, we may state that the secondary peak effect (and hence, the δT_c-pinning) is a property of the superconducting matrix. The 211 particles do not affect the fishtail shape, as long as their concentration is below a certain limit. The formation of an homogeneous NEG matrix is essential for the high j_c values exhibited by the ternary compounds at elevated temperatures, i.e. 77 K. Note the considerable change of shape of the field dependence of the critical current densities with increasing concentration of NEG-211 particles. This demonstrates the effectivity of the submicron-sized pinning sites achieved here. These particles provide a very effective pinning, forming quasi the "background" for the peak effect. In low fields, pinning is only due to these insulating 211 particles. This leads to the pinning diagram presented in Fig. 2.

In conclusion, we may state that the peak effect is due to δT_c-pinning, provided by oxygen vacancy clusters and, more pronounced, by the influence of the LRE/Ba solid solution.

Acknowledgements. This work was partially supported by NEDO. MRK and MMD are supported by STA and NEDO, respectively.

1. M. Muralidhar, M. R. Koblischka, T. Saitoh and M. Murakami, Supercond. Sci. Technol. **11**, 1349-1358 (1998).

2. M. R. Koblischka, A. J. J. van Dalen, T. Higuchi, S. I. Yoo and M. Murakami, Phys. Rev. B **58**, 2683-2867 (1998).

3. M. R. Koblischka, M. Muralidhar and M. Murakami, Appl. Phys. Lett. **73**, 2351-2353 (1998).

4. R. Griessen, H. H. Wen, A. J. J. van Dalen, B. Dam, J. Rector, H. G. Schnack, S. Libbrecht, E. Osquiguil and Y. Bruynseraede, Phys. Rev. Lett. **72**, 1910-1913 (1994).

5. H. H. Wen, Z. X. Zhao, Y. G. Xiao, B. Yin and J. W. Li, Physica C **251**, 371-378 (1995).

6. W. Harneit, T. Klein, L. Baril and C. Escribe-Filippini, Europhys. Lett. **36**, 141-145 (1996).

7. A. Erb, J.-Y. Genoud, F. Marti, M. Däumling, E. Walker and R. Flükiger, J. Low Temp. Phys. **105**, 1023-1027 (1996).

8. G. Blatter, M. V. Feigel'man, V. B. Geshkenbein, A. I. Larkin and V. M. Vinokur, Rev. Mod. Phys. **66**, 1125-1425 (1994).

9. M. Jirsa, L. Pŭst, D. Dlouhý and M. R. Koblischka, Phys. Rev. B **55**, 3276-3284 (1997).

Vortex Pinning on Low Energy Light Ions Irradiation

T.Satou[1]*, M.Sasase[2], S.Okayasu[2] and K.Hojou[2]

[1]Department of Physics, Faculty of Science, Ibaraki University, Mito 310-8512, Japan
[2]Department of Materials Science, Japan Atomic Energy Research Institute, Tokai-mura,Ibaraki,319-1195, Japan

Abstract: Irradiations of 45keV He$^+$ and 22.5keV H$^+$ ions on high-T$_c$ superconductor thin films were accomplished to investigate relationships between superconducting properties and irradiation defects. Low fluence region below 2×10^{-4}dpa, lattice parameter of c-axis and T$_c$ did not change, indicating that the irradiation does not affect the crystal structure and T$_c$. On the other hand, increase of J$_c$ was observed below 2×10^{-4}dpa on He$^+$ irradiation. This can be considered that point defect acts as effective pinning centers. The reason of increase of J$_c$ is considered that the density of irradiation defect is approximately equal to the density of vortex below 2×10^{-4}dpa.

Keywords: EuBa$_2$Cu$_3$O$_{7-x}$ thin film, vortex pinning, point defect, dpa, ion irradiation, He$^+$, H$^+$,

INTRODUCTION

It is well known that pinning properties on high-Tc superconductors are strongly affected with irradiation defects, depending their sizes, distributions, densities and their shapes [1]. Generally, it is known that columnar defects formed by high energy heavy ion irradiation act as effective pinning centers. On the other hand, pinning properties of the point defects has been investigated, but it has not clearly been understood yet [2]. In this paper, the relationships between point defects formed by the irradiation of low energy light ions and superconducting properties have been investigated.

EXPERIMENTAL

In the present study, we used c-axis oriented EuBa$_2$Cu$_3$O$_{7-x}$ (EBCO) thin films with the thickness of about 250nm on the MgO (100) substrate deposited by the reactive sputtering method. Before irradiation, lattice parameter of c-axis (C) was 1.173nm, superconducting transition temperature (T$_c$) was 87K and critical current density (J$_c$) was 2.04×10^6 (A/cm^2) at 5K under 0.1T. The sample size was 2.0mm x 2.0mm. To clarify the effect of ion species by irradiation, two different ions, 45keV He$^+$ and 22.5keV H$^+$, were used for the irradiation with the fluence of 5×10^{11} to 1×10^{16} ions/cm^2 at room temperature. For the H$^+$ irradiation, 45keV H$_2^+$ ions were used. It was assumed that the most of individual H$^+$ ions have a half of the energy of H$_2^+$: 45keV H$_2^+ \rightarrow$ 22.5keV H$^+$. Accelerating voltage was decided to make ion pass through superconductor. The irradiation was performed along the c-axis direction. X-ray diffraction (XRD) patterns were measured before and after the irradiations. The peak positions of (003), (005), (006) and (007) were used to calculate the lattice parameter of c-axis by least square fitting method. Before and after irradiation, magnetizations of specimen were measured with a commercial SQUID magnetometer (MPMS, QUANTUM DESIGN). Magnetic field was applied parallel to the c-axis. J$_c$ was calculated from the hysterisis loop at 5K and 40K using modified Bean's model [3]. T$_c$ was determined from low field magnetization curve.

The displacement per atom (dpa) is the amount of damage and is corresponding to the fluence linearly, then it is convenient for the comparison of different ion irradiations. Dpa was used instead of fluence. Distribution of the both ions inside the samples and ratio of dpa were estimated from TRIM code calculation [4], as shown in Fig.1 and Fig.2. It is seen from Fig.1 that 59.2% of He$^+$ and 93.1% of H$^+$ ions were implanted into EBCO species with the range of 250nm. Peak dpa which is the maximum of dpa was used.

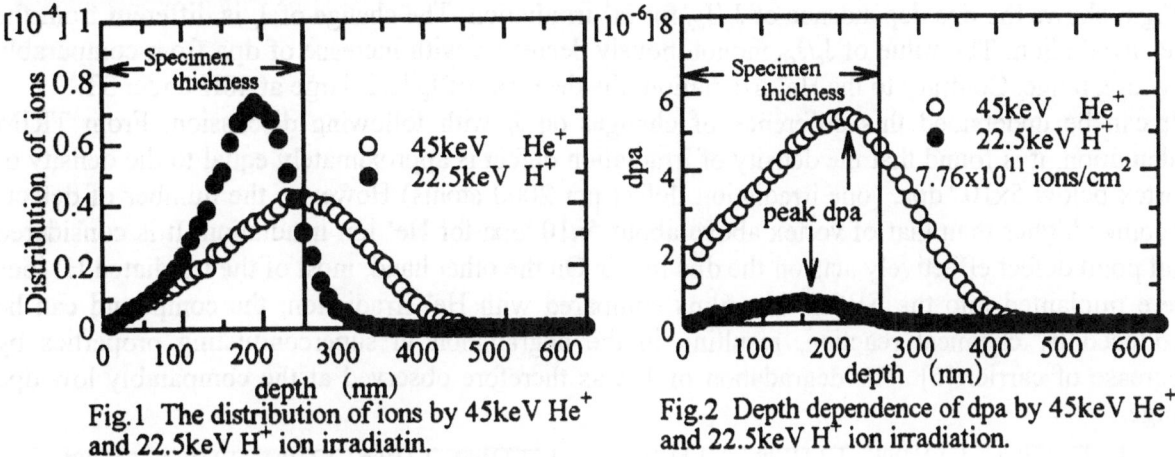

Fig.1 The distribution of ions by 45keV He$^+$ and 22.5keV H$^+$ ion irradiatin.

Fig.2 Depth dependence of dpa by 45keV He$^+$ and 22.5keV H$^+$ ion irradiation.

RUSULTS AND DISCUSSION

The change of crystal structure by ion irradiation was investigated by XRD. Peak positions shifted to low angle with increase of dpa, full width at half maximum was increased and peak intensity decreased for both He$^+$ and H$^+$ irradiation. In Fig.3, the value of C/C_0, the increase ratio of lattice parameter of c-axis, is plotted as a function of peak dpa, where C_0 is lattice parameter of c-axis before irradiation. The value of C/C_0 is almost steady below about 2×10^{-4}dpa, but increased above about 2×10^{-4}dpa. It is found that the change of lattice parameter of c-axis depends on dpa only regardless of ion species.

In Fig.4, the change of T_c/T_{c0} is plotted as a function of peak dpa, where T_{c0} is the superconducting transition temperature before irradiation. T_c/T_{c0} is almost steady at low dpa range below about 2×10^{-4}dpa, then turns to decrease above about 2×10^{-4}dpa. Finally, superconductivity has been destroyed. The change of T_c is associated to the change of lattice parameter of c-axis. This indicates that the irradiation did not affect the crystal structure and T_c at low fluence region below 2×10^{-4}dpa.

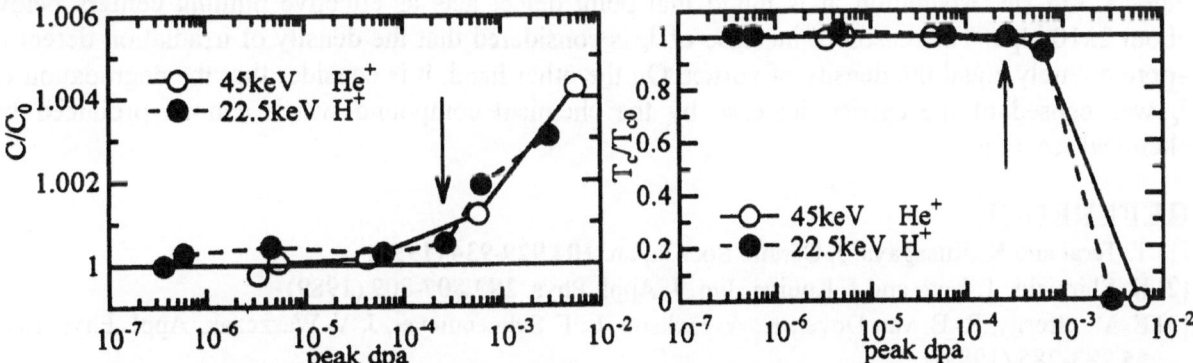

Fig.3 Dpa dependence of the change of C/C_0 by ion irradiation.

Fig.4 Peak dpa dependence of T_c/T_{c0} by ion irradiation.

The change of J_c on each ion irradiation was investigated by the SQUID magnetometer. Fig.5

shows the dpa dependence of J_c/J_{c0} for He$^+$ ion irradiation, where J_{c0} is the critical current density before irradiation. It is found that J_c/J_{c0} shows a plateau structure below 5×10^{-4}dpa, but turn to decrease above about 5×10^{-4}dpa, which is corresponding to the change of c-axis lattice parameter and T_c. It was observed two times of increase of J_c at 40K under 3T. It is consider that point defect acts as effective pinning centers at relatively high temperature and high magnetic field range for He$^+$ ion irradiation [5].

Fig.6 shows the dpa dependence of J_c/J_{c0} for H$^+$ irradiation. The change of J_c is different from the He$^+$ irradiation. The value of J_c/J_{c0} monotonously decreases with increase of dpa from comparably low dpa range. Contrary to the He$^+$ irradiation, the decrease of J_c/J_{c0} is large at 40K under 3T.

It can be understood this difference of changes on J_c with following discussion. From TRIM calculation, it is found that the density of irradiation defect is approximately equal to the density of vortex below 5×10^{-4} dpa. (one irradiation defect per 2000 atoms) However, the number of defects becomes higher than that of vortex above about 5×10^{-4}dpa for He$^+$ ion irradiation. It is considered that point defect effectively acts on the dpa range. On the other hand, most of the irradiated H$^+$ ions were implanted into the EBCO thin films compared with He$^+$ irradiation, the compound can be produced by chemical reaction, resulting in the degradation in superconducting properties by decrease of carrier [6]. The degradation of J_c was therefore observed at the comparably low dpa range.

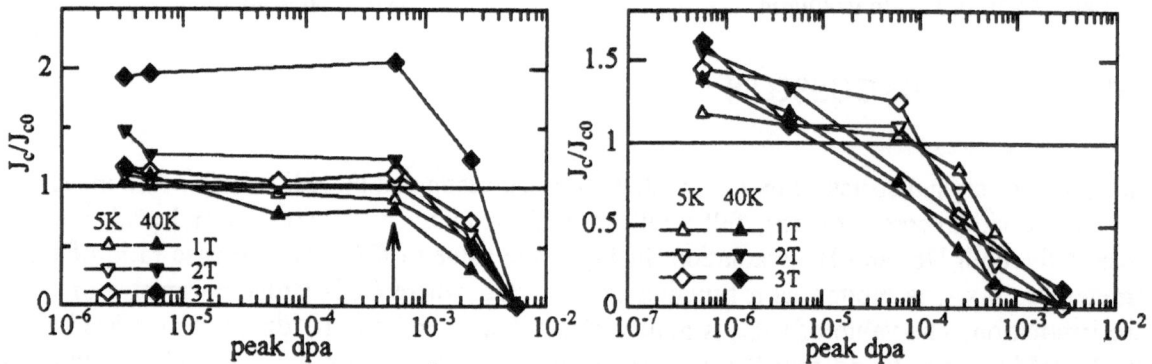

Fig5　Peak dpa dependence of J_c/J_{c0} by He$^+$ irradiation.　　Fig.6　Peak dpa dependence of J_c/J_{c0} by H$^+$ irradiation.

CONCLUSION

Irradiations of 45keV He$^+$ and 22.5keV H$^+$ ions on superconducting thin film EuBa$_2$Cu$_3$O$_{7-x}$ were accomplished to investigate the relationships between superconducting properties and irradiation defects. On He$^+$ irradiation, it is found that point defect acts as effective pinning centers below about 2×10^{-4}dpa. The reason of increase of J_c is considered that the density of irradiation defect is approximately equal the density of vortex. On the other hand, it is consider that the degradation of J_c was caused of the carrier decrease by the chemical compound which can be produced by chemical reaction.

REFERENCE

[1] T. Terai and K. Kusagaya, J. Ceram. Soc. Japan. **104** 929-934 (1996)

[2] K. Shiraishi, T. Kato and J. Kuniya, Jpn. J. Appl. Phys. **28** L807-809 (1989)

[3] E. M. Grorgy, R. B. van Dover, K. A. Jackson, L. F. Schneemeyer, J. V. Waszczak, Appl. Phys. Lett. **55** 283-285 (1989).

[4] J. P. Biersack, L.G. Haggmark, Nucl. Instr. and Meth. **174** 257 (1980)

[5] S. Okayasu and Y. Kazumata, in Advances in Superconductivity IX edited by S. Nakajima and M. Murakami (Springer-Verlag, Tokyo, 1997), pp.507-510

[6] K. Shiraishi, T. Kato, J. Kuniya, T. Kamo and S. Matuda, Jpn. J. Appl. Phys. **27** L564-566 (1989)

Dynamical Vortex Behaviors in Optimally Doped $Bi_2Sr_2CaCu_2O_{8+x}$ Crystal and ESR Anomalies

AKIHIKO NISHIDA,[1] CHIHIRO TAKA,[1] TAKASHI YASUDA,[2] AND KAZUMI HORAI[1]

[1]Department of Applied Physics, Faculty of Science,
Fukuoka University, Fukuoka 814-0180, Japan
[2]Department of Computer Science and Electronics,
Kyushu Institute of Technology, 680-4 Kawazu, Iizuka 820-8502, Japan

Abstract: Fluctuating properties of flux lines in the mixed state of optimally doped $Bi_2Sr_2CaCu_2O_{8+x}$ are studied with anomalous ESR results of free radical spins coated on the surface of the crystal. Measurements of ESR have been performed with the static magnetic field parallel to the c-axis under the field-cooling condition. The linewidth does not increase until the crystal is cooled down 10 K lower than $T_c = 90$ K, indicating that the local field variation due to flux lines is motionally averaged out by the thermal fluctuation in the vortex liquid phase. On the other hand, ESR intensity indicates clear reduction at T_c due to viscous flux motion with field modulation for ESR observations. Thermal fluctuation and forced vibration of vortices at respective characteristic frequencies are discussed to be responsible for ESR anomalies.

Key words: local field distribution, vortex liquid, motional narrowing, dynamics

INTRODUCTION

Much interest has been focussed on the vortex phase diagram in high-T_c superconductors with melting and glass transitions [1-3]. Although many of the vortex matters have been investigated with a kind of static method, transition of vortex states may in principle depend on the characteristic time of measurements or events. Fendrich *et al.* [4] studied dynamical vortex phases with simultaneous magnetization, resistivity and *I-V* measurements under the driving Lorentz force, which might reflect relatively slow dynamics. Higher-frequency dynamics are possibly investigated by the resonance method, such as μSR [5], NMR [6], ESR [7] and JPR [8].

In an effort for higher-frequency approach to the exotic vortex behaviors, we are applying spin-probe ESR method where external spins adhered on the surface of the superconductor monitor local field variations in a microscopic scale. In our previous experiment of ESR in as-grown $Bi_2Sr_2CaCu_2O_{8+x}$ single crystal [7], we observed rather small linewidth broadening just below T_c, corresponding to fairly "soft" vortex nature in this system. However, the as-grown crystal is in the overdoped state without annealing and may still include some pinning centers. In addition, different vortex matters have been pointed out depending on the degree of carrier doping [2]. Therefore, it is very interesting to study effects of different doping and annealing on the vortex states. In this work, we report on the spin-probe ESR measurements of the optimally doped $Bi_2Sr_2CaCu_2O_{8+x}$ crystal and examine characteristic vortex behaviors in different doping.

EXPERIMENTAL

The single crystalline $Bi_2Sr_2CaCu_2O_{8+x}$ was prepared by the KCl flux method as in the previous report [7]. The prepared as-grown crystal was annealed for 15 hours at 400°C under oxygen partial pressure of 0.4 Torr in order to obtain optimum doping with $T_c = 90.0$ K. The surface of the crystal was coated by a free radical DPPH (2,2-diphenyl-1-picrylhydrazyl) as the spin-probe with thickness of about 250Å. DPPH ESR was observed at 9.0 GHz of microwave by the 100 kHz field modulation with the static magnetic field parallel to the c-axis of the crystal. The static field was swept at most 40 G around the resonance field of $ca.$ 3219 G, under which the sample was field-cooled, and ESR signals were obtained by warming the sample step by step. ESR signals were taken with two directions of the field sweep (*i.e.* increasing and decreasing the static field) at respective temperatures.

RESULTS AND DISCUSSION

The irreversibility field H_{irr} and the 2nd peak field H_p were estimated with the vibrating sample magnetometer with the static field parallel to the c-axis, the results being shown in Fig. 1. The area above H_{irr} corresponds to vortex liquid or fluctuating solid phase. The region below H_{irr} and above H_p is considered to be the vortex glass phase, while the region below H_p and below H_{irr} is considered to be the Bragg glass or ordered solid phase. Since ESR is observed around 3.2 kG of the static field (as indicated by the double-sided arrow), we expect to observe local field variations due to fluctuating and steady vortices above and below $T_{irr} = 34.2$ K (solid arrow). It is noted that the area between T_{irr} and T_c is wider for the optimally doped crystal than the as-grown case, providing extended range of "soft" and fluctuating vortices.

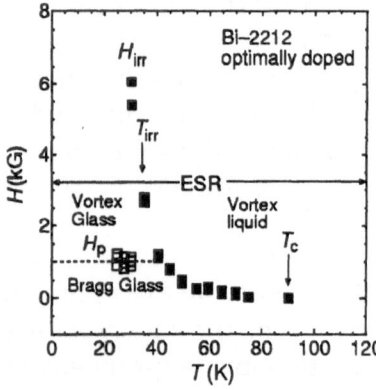
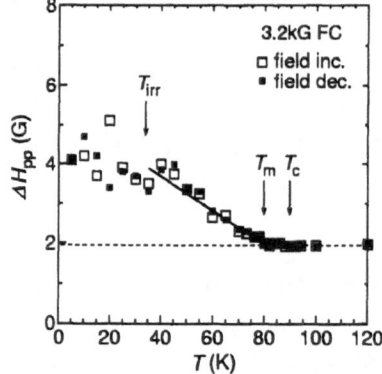
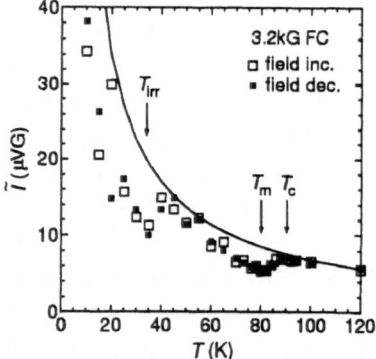

Figure 1: Irreversibility H_{irr} and second peak H_p fields as a function of temperature.

Figure 2: Temperature dependence of the peak-to-peak ESR linewidth ΔH_{pp}.

Figure 3: Temperature dependence of the integral intensity $\tilde{I}$.

Figure 2 represents temperature dependence of the peak-to-peak ESR linewidth, ΔH_{pp}, for the optimally doped crystal. In contrast to the as-grown sample [7], the linewidth for the optimally doped crystal does not increase just below T_c of 90 K, but starts to increase at 80 K which is 10 K lower than T_c. Absence of the broadening above 80 K is considered to be due to motional narrowing effect of thermally melted vortices. Thus, we denote 80 K as the melting temperature T_m, above which vortices are in the liquid phase. However, this

T_{m} should be considered as dynamical one and may be different from the static estimation such as resistivity anomaly. This is because our spin-probe captures an instantaneous field distribution with the time window of the Larmor precession (about 10^{-10} s). Similar situation is also pointed out as the "snapshot" effect in the JPR experiment [8].

We next find that the coefficient of the linear broadening (solid line in Fig. 2) for the optimum crystal is smaller than that for the as-grown case, resulting in evaluation of the zero-temperature penetration depth $\lambda(0) = 0.85$ μm in contrast to 0.63 μm for the as-grown case [7]. The value of $\lambda(0)$ which is not only larger than the literature (static) value but also larger than the as-grown crystal suggests considerable narrowing mechanism remains even in the *solid* state below T_{m} for the optimally doped crystal. Thus, we name it as the "soft" solid. We further note that the hysteresis between field-up (open squares) and -down (solid squares) sweeps appears not just at T_{irr} but appears at somewhat lower temperatures with smaller degree, suggesting steady or "ordered" solid rather than the frozen glass. These results suggest that the optimally doped crystal more closely represents intrinsic vortex matters with weaker flux pinning than the as-grown sample.

Finally, the temperature dependence of the integral intensity $\tilde{I}$, as shown in Fig. 3, indicates deviation from the Curie-law (solid line) at T_{c} for the optimum crystal, again in contrast to the as-grown case. This decrease in the intensity is owing to viscous flux motion (forced vibration) at 100 kHz modulation and energy dissipation in the melted vortex phase. Remaining deviation even below T_{m} infers that vortex "solid" is abnormal, allowing flux motion survives in the state. Recovery of the intensity around 60 K may be explained by impedance mismatch between ESR detection system and flux system whose viscosity is varying with temperature. Below the irreversibility temperature T_{irr} the intensity shows significant drop again, which might be interpreted as energy dissipation by pinned vortices yet moving within the spatial size of the pinning center.

In summary, various ESR anomalies were observed and revealed exotic flux behaviors, especially fluctuating and viscous nature of vortices. When observed at high frequency, the vortex system in the optimally doped crystal was considered to be in the liquid state at $T_{\mathrm{m}} < T < T_{\mathrm{c}}$, in the anomalous "soft" solid at $T_{\mathrm{irr}} < T < T_{\mathrm{m}}$, and in the "ordered" solid at $T < T_{\mathrm{irr}}$. Presence of higher-field (abnormal) tails and lower-field (rather normal) tails in the ESR profiles also support this conjecture of "soft" and "ordered" solid state, respectively. Fluctuations at different characteristic times (10^{-10} s and 10^{-5} s) were thought to be important for the motional narrowing effect and the dissipation phenomena.

1. E. Zeldov, D. Majer, M. Konczykowski, V. B. Geshkenbein and V. M. Vinokur: Nature **375** (1995) 373.
2. T. Nishizaki, T. Naito, N. Kobayashi: Phys. Rev. B **58** (1998) 11169.
3. D. T. Fuchs, E. Zeldov, T. Tamegai, S. Ooi, M. Rappaport, H. Shtrikman: Phys. Rev. Lett. **80** (1998) 4971.
4. J. A. Fendrich, U. Welp, W. K. Kwok, A. E. Koshelev, G. W. Crabtree, and B. W. Veal: Phys. Rev. Lett. **77** (1996) 2073.
5. C. M. Aegerter, S. H. Lloyd, C. Ager, S. L. Lee, S. Romer, H. Keller, and E. M. Forgan: J. Phys. Condens. Matter **10** (1998) 7445.
6. Y. Maniwa, T. Mituhashi, K. Mizoguchi and K. Kume: Physica C **175** (1991) 401.
7. A. Nishida, C. Taka, T. Yasuda, and K. Horai: Advances in Superconductivity **XI** (1999) 477.
8. Y. Matsuda, M. B. Gaifullin, K. Kumagai, M. Kosugi, K. Hirata: Phys. Rev. Lett. **78** (1997) 1972.

A PINNING MECHANISM OF YBaCuO THIN FILM DISPERSED WITH DIFFERENTLY-ORIENTED CRYSTAL GRAINS

Y.Sasaki, , K.Michishita, Y. Higashida Y.Kubo, K.Nakamura*, and K.Takeda**

Japan Fine Ceramics Center, 2-4-1 Mutsuno, Atsuta-ku, Nagoya 456-8587 ,Japan
*Department of Systems Management and Engineering, Nagoya Institute of Technology, Gokiso, Showa-ku, Nagoya 466-0061, Japan
**Super-GM, Umeda UN Bldg., 5-14-10 Nishitenma, Kita-ku, Osaka 530-0047, Japan

Abstract:
On the surface of YBaCuO thin film grown on LaAlO$_3$ substrate using laser deposition method with Jc value of about 1×10^6 A/cm^2, had many crystal grains of about 1μm$\times 0.1 \mu$m. These crystal grains had the same crystal structure as the matrix, but had different crystal axis. The magnetization at 77 K had a small peak when the magnetic field were applied at an angle of 30° from c-axis of the matrix, which miay be explained by the flux pinning by the boundaries between the crystal grains and the matrix..

Keywords:
YBaCuO thin film, Bitter decoration technique, Flux pinning, High resolution transmission electron microscopy

INTRODUCTION

It is well known that, in high Tc oxide superconductor, the single crystal thin film or poly-crystal thin film has very high critical current density (Jc). However, it is not clear at present why such high Jc value can be attained in the case of the thin film and what kinds of defects are responsible for the pinning.   According to the Bitter decoration experiments, the magnetic flux distribution is disordered possibly due to the strong pinning centers. The YBaCuO thin film grown on LaAlO$_3$ (001) substrate, is characterized by including many crystal grains dispersed in the matrix [1]. It is pointed out that the boundaries between these crystal grains and the matrix may act as strong pinning centers.

To clarify the effect of these crystal defects upon the pinning, the micro-structural analysis was performed by high resolution transmission electron microscopy (HRTEM). Further, the magnetization was measured as a function of magnetic field direction. A part of the pinning was interpreted by the pinning of fluxoids by the grain boundaries between the dispersed crystal grains and the matrix.

EXPERIMENTAL RESULTS

The YBaCuO thin film was manufactured by Toshiba Co. Ltd. on LaAlO$_3$ substrate using laser deposition method, which had Jc value of about 1×10^6 A/cm^2. As shown in Fig.1, the film included many crystal grains of about 1μm$\times 0. 1 \mu$m.

The electron beam diffraction showed that the small crystal grains have also the same crystal structure as the matrix. Fig.2 shows the HRTEM image of a boundary between the small grain and the matrix. It was constituted by the combination of two interfaces: The one where the (001) plane of the small grain is connected directly to the (100) or (010) plane of the matrix. The other is a twin boundary of each (103) plane. Further, it was also confirmed that the (001) plane of matrix YBaCuO crystal is connected directly to the (001) plane of LaAlO$_3$ substrate without any amorphous layer.

Fig.1 SEM photograph of YBaCuO thin film. A lot of rectangular grains are observed.

Fig.2 High resolution TEM image of the boundary
between the small grain and the matrix. It is composed of two patrs: 90° interface where (001) plane of the grain connect to (100) plane or (010) plane of the matrix, and that connected each other by (103) twin boundary .

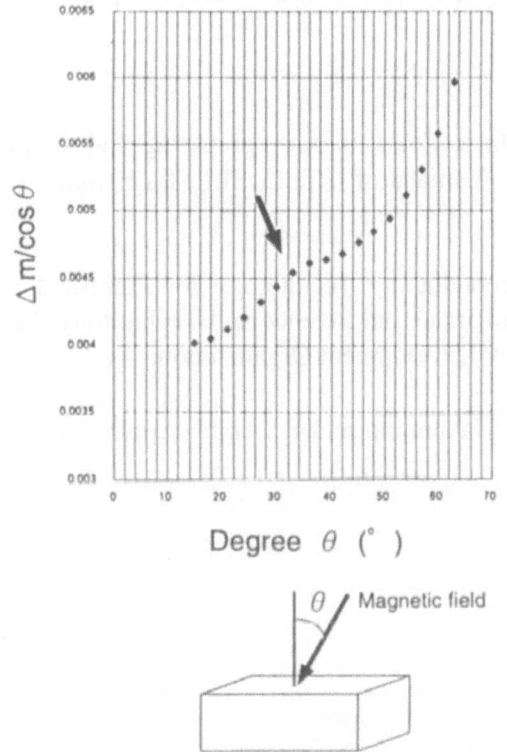

Fig.3 Magnetization hysteresis versus the azimuth angle of applied magnetic field direction

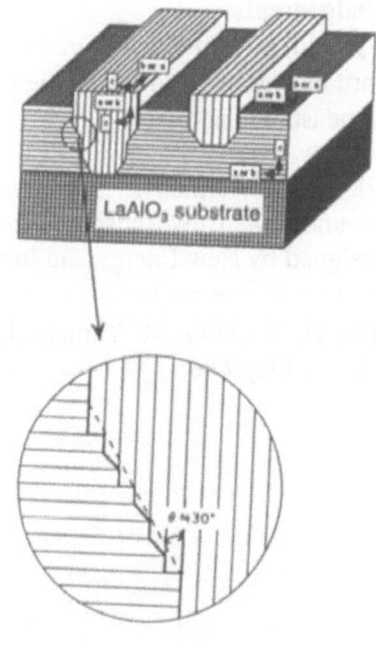

Fig.4 Schematic illustration of the 30° interface between the small grain and the matrix.

Let θ be the angle between the applied magnetic field and the normal direction of the film surface. Fig.3 is θ-dependence on the magnetization of the film at 0 Oe, and 77 K. The small peak in the magnetization was observed at $\theta = 30°$.

DISCUSSIONS

From HRTEM analysis, it was clarified that these crystal grains have the same composition and crystal structure as the matrix, but their crystal axis is simply different from that of the matrix. Fig.4 is a schematic illustration of structure of the boundary proposed. The c axes are oriented parallel to the film surface, or ab plane. Then the interface between the crystal grains and the matrix may perpendicular to the surface, i.e. $\theta = 0°$. The other interface make an angle of about 30°, i.e. $\theta = 30°$ from normal direction of the film surface..

The high Tc superconductor has so strong anisotropy that the shielding current flows most easily when magnetic field is applied along c-axis, in other words, perpendicularly to the film surface. So that, the magnetization hysteresis varies as $\cos\theta$. Then the value of magnetization in Fig.3 is divided by $\cos\theta$. The curve increased monotonically as θ increased, which is due to the intrinsic pinning of high Tc superconductor with high anisotropy along c axis. It had a small peak at around 30°, which may be related to the flux pinning.

CONCLUSION

The peak in the magnetization measurement at $\theta = 30°$ may result from the flux pinning by small crystal grains whose axis has different crystal direction of the matrix. Further, the flux pinning may be occurred at the boundaries where $\theta = 30°$.

Acknowledgements:
Authors would like to thank Dr. H. Yoshino of Toshiba Co. Ltd. for preparing the samples. Authors are also grateful to Dr. I. Hirabayashi and Dr. Y. Yamada of ISTEC, SRL, for helping us by magnetic measurements.

This work was performed as a part of R&D on Superconducting Technology for Electric Power Apparatus under the New Sunshine Program of Agency of Industrial Science and Technology, MITI, being consigned by New Energy and Industrial Technology Development Organization.

1. H. Fuke, H. Yoshino, M. Yamazaki, T. D. Thanh, S. Nakamura, and K. Ando, Appl. Phys. Lett. 60(21), (1992)2686.

High Frequency Surface Impedance Measurement in the mixed state of $Bi_2Sr_2CaCu_2O_y$

Yoshishige Tsuchiya[1], Katsuya Iwaya[1], Tetsuo Hanaguri[2], Haruhisa Kitano[1], Atsutaka Maeda[1,3], Jun-ichi Takeya[4], Kei Nakamura[4,5], and Yoichi Ando[4]

[1]Department of Basic Science, The University of Tokyo, 3-8-1, Komaba, Meguro-ku, Tokyo 153-8902, Japan

[2]Department of Advanced Materials Science, The University of Tokyo, 7-3-1, Hongo, Bunkyo-ku, Tokyo 113-8656, Japan

[3]CREST, Japan Science and Technology Corporation, 4-1-8, Honcho, Kawaguchi 332-0012, Japan

[4]Central Research Institute of Electric Power Industry, 2-11-1, Iwato-kita, Komae, Tokyo 201-8511, Japan

[5]Department of Energy Science, Tokyo Institute of Technology, Nagatsuta, Yokohama 226-8502, Japan

Abstract: Magnetic-field (H) dependence of high-frequency surface impedance, Z_s, was measured in exactly the same sample of $Bi_2Sr_2CaCu_2O_y$, where the anomalous "plateau behavior" was observed in the magnetic-field dependence of thermal conductivity, $\kappa(H)$. No drastic change was found in $Z_s(H)$ at the magnetic-field where the plateau started to show up in κ. This result suggests that a phase transition of the condensate is unlikely to be an origin of the plateau.

Keywords: quasiparticle excitation, high-frequency response, $Bi_2Sr_2CaCu_2O_y$

INTRODUCTION

The problem of quasiparticle excitation in the superconducting state of high-T_c cuprates has attracted much attention because of their expected unusual properties. Recently, a sharp kink and subsequent plateau structure in the magnetic field dependence of thermal conductivity, κ, were reported by Krishana *et al.* [1] in $Bi_2Sr_2CaCu_2O_y$ (BSCCO). Based on the data, they suggested the occurrence of a new phase transition in the condensate from the $d_{x^2-y^2}$ state to the fully-gapped state [1]. Many efforts have been made in order to clarify this remarkable feature both theoretically and experimentally. In particular, a recent study by Ando *et al.* suggested that the observed plateau feature was quite sensitive to the impurity scattering rate of samples [2]. However, to obtain more information which is related to this phenomenon, it would be necessary to compare the $\kappa(H)$ result with other measurements. High-frequency electromagnetic response also provides useful information on the quasiparticle excitation. If such a sudden change takes place in the electronic state, a similar anomaly will be expected in the high-frequency surface impedance. In this paper, we report the magnetic-field dependence of the surface impedance $Z_s = R_s + iX_s$ in exactly the same samples of BSCCO whose thermal conductivity were measured prior to the Z_s measurement, and discussed the data in a comparative manner.

EXPERIMENTAL

$Bi_2Sr_2CaCu_2O_y$ single crystals were grown by the floating zone method and were carefully annealed and quenched to obtain uniform oxygen content in the sample. The T_c's determined by the dc susceptibility were 91 K for two crystals used in this study. Details for the thermal conductivity measurement were described elsewhere [2]. Surface impedance was measured by the cavity perturbation technique with a cylindrical Cu cavity operated at 96.0 GHz in the TE_{013} mode. The surface resistance R_s and the surface reactance X_s were obtained from the changes in the quality factor of the cavity and resonance frequency, respectively. The typical dimension of sample was $0.15 \times 0.15 \times 0.02$ mm^2. In all measurements, to investigate the in-plane responses, rf magnetic-field and applied dc magnetic-field directions are parallel to the c-axis of the samples.

RESULTS AND DISCUSSION

Figure 1 shows the magnetic-field dependence of Z_s at two different temperatures in sample A, together with the magnetic-field dependence of thermal conductivity $\kappa(H)$. As shown in the Fig.1, the plateau structure was observed in $\kappa(H)$ profile. Continuous lines and open circles correspond to field-swept (FS) data after zero-field cooling and field-cooled (FC) data, respectively. Except for the low-field region below 0.5T, both data coincided with each other, which suggested that an effect of the field inhomogeneity due to the pinning on the response at 96.0 GHz could be neglected in the high-field region. Therefore, we discuss the $Z_s(H)$ profile based on the FC data. In general, surface impedance Z_s of the type-II superconductor is related to the complex effective penetration depth $\tilde{\lambda}$ as $Z_s = i\mu_0\omega\tilde{\lambda}$, where μ_0 is the vacuum permeability and ω is the angular frequency. The $\tilde{\lambda}$ was described by several possible vortex dynamics parameter, such as a pinning strength, viscous force, creep factor, quasi-particle density, and penetration depth, etc [3]. Since the vortex motion was closely related to the electronic structure of vortex core [4], we expect an anomaly in the microwave response if a drastic change takes place in the electronic structure.

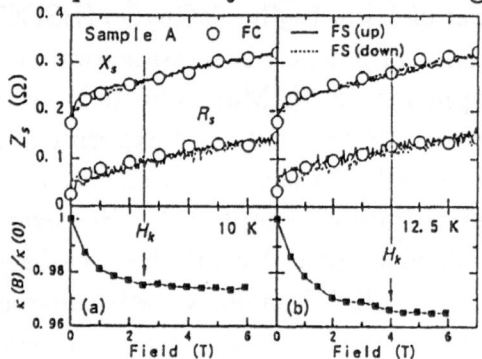

Fig. 1 $Z_s(H)$ of the sample A, together with the observed plateau structure in $\kappa(H)$.

At both temperatures, however, both R_s and X_s increase monotonically with increasing field, and at H_k's, defined as the field above which the plateau started in $\kappa(H)$ at each temperature, no distinct anomaly was observed in the $Z_s(H)$ profile.

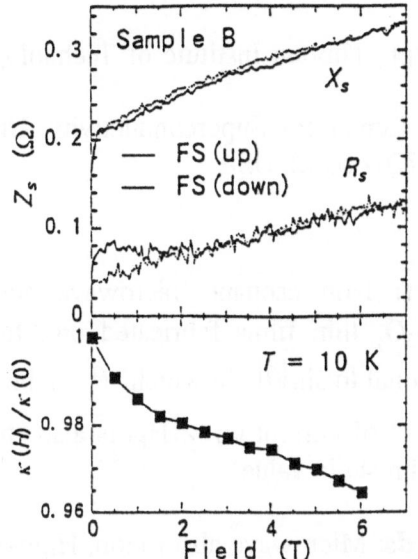

In Fig. 2, we showed the $Z_s(H)$ in sample B, together with the $\kappa(H)$ data which did not show the "plateau behavior". We could not find any qualitative difference of the $Z_s(H)$ between sample A and sample B, where the plateau was not observed in $\kappa(H)$. These results of $Z_s(H)$ implies that any kinds of phase transition are unlikely to be an origin of the plateau.

In BSCCO, we reported that the effect of additional increase of London penetration depth with increasing field played an important role around at the vortex melting transition [5]. In the high-field and the low temperature region, however, these effects would be rather small and the vortex viscosity η can be roughly estimated from the Z_s data .

Fig. 2 $Z_s(H)$ of the sample B with no plateau structure in $\kappa(H)$ profile.

The estimated value $\sim 5 \times 10^{-8}$ Ns/m^2 indicates that the core structure of vortices in BSCCO is not largely different from the conventional "normal" core. However, for more reliable numbers, measurements at several different frequencies are needed.

CONCLUSION

Magnetic-field dependence of high-frequency surface impedance was measured in exactly the same sample of BSCCO, where the anomalous "plateau behavior" was observed in the magnetic-field dependence of thermal conductivity. No drastic change was found in $Z_s(H)$ at the magnetic-field where the plateau started to show up in $\kappa(H)$. This result suggests that a phase transition of the condensate is unlikely as an origin of the plateau.

Acknowledgements: One of the authors (Y.T.) is supported by the JSPS Research Fellowships for Young Scientists.

1. K. Krishana, N. P. Ong, Q. Li, G. D. Gu, and N. Koshizuka, Science **277**, 83 (1997).
2. Yoichi. Ando, J. Takeya, K. Nakamura, Yasushi Abe, and A. Kapitulnik, *preprint.*
3. M. W. Coffey, and J. R. Clem, Phys. Rev. Lett. **67**, 386 (1991).
4. M. Golosovsky, M. Tsindlekht, and D. Davidov, Supercond. Sci. Technol. **9**, 1 (1996).
5. T. Hanaguri, *et al.*, Phys. Rev. Lett. **82**, 1273 (1999).

Anomalous Behaviors of Non-Resonant Microwave Absorptions of High-T_C Superconductors

Kazushi Sugawara[1], Nobuhito Arai[2†], Susumu Ichimura[1], Hideki Naoi[1], and Haruo Hirose[1]

[1] EE Dept., Nippon Institute of Technology, Gakuendai 4-1, Miyashiro-machi, Saitama 345-8501, Japan.
[2] Texas Center for Superconductivity, University of Houston, Houston Science Center, Houston, Texas 77204-5932, USA.

Abstract: Non-resonant microwave absorption (NRMA) measurements have been done on $YBa_2Cu_3O_y$ thin films fabricated on MgO (100) substrates. The linewidth, ΔH_{PP}, is inversely proportional to $\sin|\theta|$, in which θ is the angle between an applied magnetic field and film surface. The effect of current on ΔH_{PP} is also studied. ΔH_{PP} sharply increases at currents greater than a certain threshold value.

Keywords: Microwave absorption, High-T_C superconductor, Effect of magnetic field direction, DC current

INTRODUCTION

It is well known that superconductors absorb microwave at low magnetic fields, and this absorption is called the non-resonant microwave absorption (NRMA). The NRMA is considered to originate from vortices. However, its detailed physical origin is not well understood presently. In this study, various phenomenological aspects of NRMA have been studied on thin film samples of $YBa_2Cu_3O_y$ fabricated on MgO (100) substrates. A particular attention has been paid to the effects of direction of applied magnetic field and dc current flowing in the sample.

EXPERIMENTAL

Two thin film samples of YBCO (about 800 Å thick) were fabricated on MgO (100) substrates by sputtering. All the films are c-axis oriented. A conventional ESR spectrometer (JEOL TES-TE200T_C) with a cylindrical TE$_{011}$ mode cavity operating at about 9 GHz was employed. A magnetic field was modulated at 100kHz. The modulation amplitude is 1 Gauss. The samples are about 3mm×3mm in size. The sample was inserted into a quartz tube with ID ≃ 4mm. The NRMA

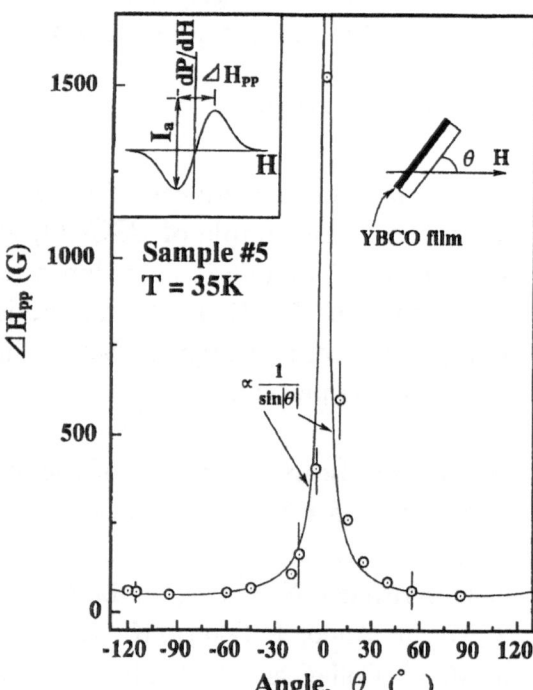

Fig. 1. ΔH_{PP} as a function of magnetic field direction, θ.

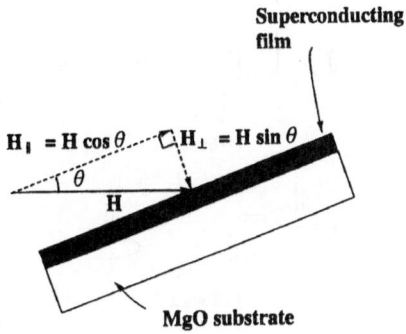

Fig. 2. A magnetic field, H, applied to a thin film sample.

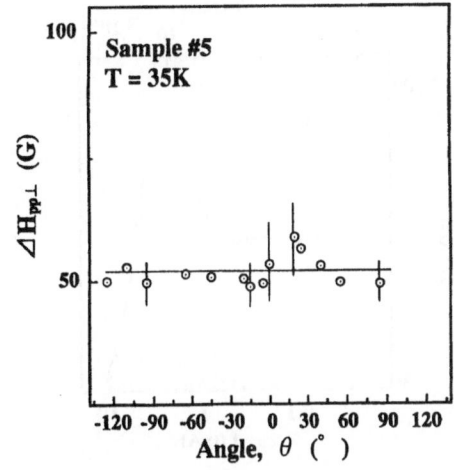

Fig. 3. $(\Delta H_{PP})_\perp$ vs. θ. $(\Delta H_{PP})_\perp$ are reproduced from ΔH_{PP} given in Fig. 1.

measurements have been done below T_C down to about 4K.

RESULTS AND DISCUSSION

The first derivative of power absorbed, dP(H) / dH, of the NRMA is qualitatively illustrated in the inset in Fig. 1, where the peak-to-peak linewidth, ΔH_{PP}, and "amplitude intensity", I_a, are defined. The NRMA signals of both samples disappeared at around 80K, which is considered to be T_C. ΔH_{PP} is related to an average vortex diameter, d, by $d^2 = 8 \phi_0/(\pi \Delta H_{PP})$[1]. Here, ϕ_0 is a quantum fluxon. The open circles in Fig. 1 represent ΔH_{PP} against the direction of magnetic field, θ, at 35K. ΔH_{PP} is nearly proportional to $1/\sin|\theta|$ (see solid curves in Fig. 1). Quite similar result has been confirmed for the NRMA of thin film of Bi-Sr-Ca-Cu-O system by Sugawara et al.[2]. These behaviors can be interpreted as follows. The applied static magnetic field can be divided into the normal ($H_\perp$) and parallel ($H_\parallel$) components to the superconducting film as is shown in Fig. 2.

In the NRMA measurements, dP(H) / dH is recorded against H, not against $H_\perp$. By reproducing $dP(H_\perp)/dH_\perp$ spectra from raw data of dP(H) / dH, the linewidth, $(\Delta H_{PP})_\perp$, measured in the $H_\perp$ axis can be obtained. Other simple way to evaluate the $(\Delta H_{PP})_\perp$ is to use the relation, $(\Delta H_{PP})_\perp = \Delta H_{PP} \times \sin\theta$. Based on this equation, ΔH_{PP} given in Fig. 1 are converted into $(\Delta H_{PP})_\perp$ and plotted against θ in Fig. 3. As is seen in Fig. 3, $(\Delta H_{PP})_\perp$ is nearly independent of θ within experimental error. Quite similar results have been confirmed for the data of aforementioned sample of Bi-Sr-Ca-Cu-O system[3]. These results indicate that the normal component of static magnetic field plays an essential role for the generation of NRMA, and presumably for the generation of vortices. However, it is noted that I_a is proportional to $\sin|\theta|$. This is reasonable since total fluxons passing through a superconducting film may be proportional to $\sin|\theta|$.

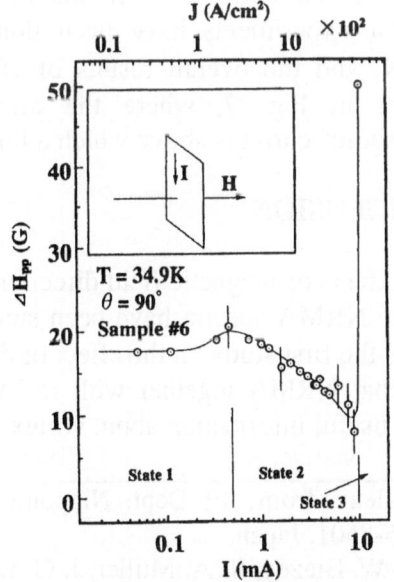

Fig. 4. ΔH_{PP} against current (I) and density of current (J).

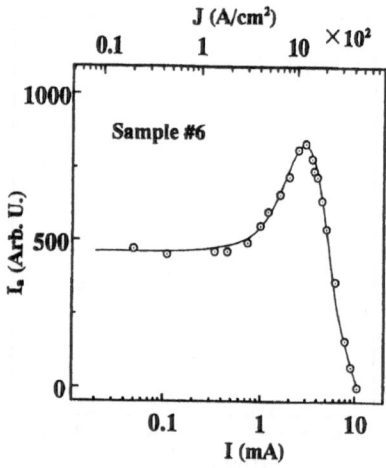

Fig. 5. Amplitude intensity, I_a, against current (I) and density of current (J).

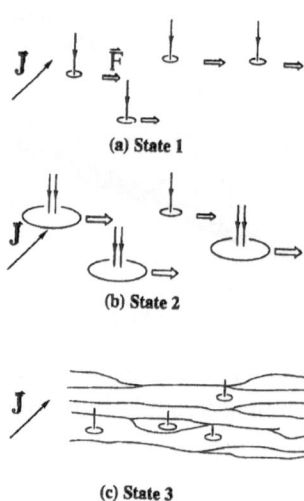

Fig. 6. Qualitative illustrations of vortex states.

Effect of flowing current. We have studied the NRMA of the YBCO film, in which a dc current, I, is flowing. The current was applied downward, perpendicularly to the static magnetic field, as is shown in the inset in Fig. 4. Figure 4 represents the current dependence of ΔH_{PP}. The upper axis in Fig. 4 represents the density of current, J. As is clearly seen in Fig. 4, ΔH_{PP} gradually decreases at currents between about 0.5mA and 10mA, and sharply increases at currents greater than about 10mA. The current dependence of the amplitude intensity, I_a, is plotted in Fig. 5. From the current dependence of ΔH_{PP}, we speculate a possible three different vortex states, States 1~3, as shown in Fig. 4. In the State 3, vortex flow may occur. These three states may reflect such states as qualitatively illustrated in Fig. 6. The States 1 and 2 are similar, but slight difference in their average vortex sizes. The average vortex size in the State 2 is slightly larger. Similar experiments have been done at 4.5K, 70.7K and 72.1K, and the overall results of ΔH_{PP} vs. I relations are shown in Fig. 7, where the arrows represent possible "threshold" currents above which a liquid state may occur.

CONCLUSION

The effects of magnetic field direction and flowing dc current on the NRMA spectra have been studied. To our knowledge, this is the first study on the effect of dc current on NRMA. We feel that NRMA together with resistivity measurements can yield useful information about vortex states and dynamics.

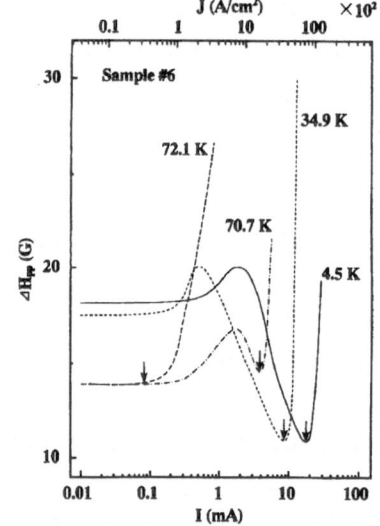

Fig. 7. ΔH_{PP} as a function of current (I) and density of current (J).

† On leave from   EE Dept., Nippon Institute of Technology, Gakuendai 4-1, Miyashiro-machi, Saitama 345-8501, Japan.

1. K. W. Blazey, K. A. Müller, J. G. Bednorz, W. Berlinger, G. Amoretti, E. Buluggiu, A. Vera and F. C. Matacotta, Phys. Rev. B36, 7241-7243 (1987)

2. K. Sugawara, T. Sugimoto, D. J. Baar, Y. Shiohara and S. Tanaka, Mod. Phys. Letters B5, 1981-1987 (1991)

3. K. Sugawara, S. Ichimura, H. Naoi, H. Hirose and N. Arai, In preparation.

Magnetic Properties of Bi-2223 Tapes Irradiated by Xe Ion

Hiroshi Ikeda[1], Naoshi Kuroda[2], Yoshiaki Tanaka[3], Tadashi Kambara[4], Kozo Yoshikawa[5] and Ryozo Yoshizaki[1]

[1]Institute of Materials Science and Cryogenics Center, University of Tsukuba, Ibaraki 305-8577, Japan
[2]Japan Atomic Energy Research Institute, Tokai-mura, Ibaraki 319-1195, Japan
[3]National Research Institute for Metals, Ibaraki 305-0047, Japan
[4]Institute of Physical and Chemical Research (RIKEN), Wako-shi, Saitama, 351-0198, Japan
[5]Takasago Research Center, Mitsubishi Heavy Industries, Ltd., Hyogo 676-0008, Japan

Abstract: We have investigated the magnetic properties of Ag-Cu alloy sheathed Bi-2223 tapes where Hf impurities are included in the pristine alloy. We introduced columnar defects in the tapes parallel to the aligned c-axis (normal to the tape surface) by Xe-ion irradiation. We found that the irreversibility line in the heavy-ion irradiated tapes showed the shift toward higher fields. Moreover, the irreversibility curve of the irradiated sample is located in the higher temperature side of the irradiated Ag-sheathed tape (without Hf impurities in the tape) even in the higher magnetic field than the matching field. In addition, the improved irreversibility line is observed in the configuration of the field parallel to the tape (i.e., perpendicular to the columnar defects).

Key words: Heavy ion irradiation, Columnar defects, Irreversibility line, Pinning Force

INTRODUCTION

Superconducting wires and tapes for practical use should have sufficient thermal and mechanical properties in addition to high critical current density (Jc). The heavy-ion irradiation has attracted much attention, since it produces columnar defects, which yield a quite large pinning energy of high-temperature superconductors for vortices parallel to the defects [1-3]. Since the irradiation can introduce the defects in a controlled manner, it can be used as a useful tool to probe the response of flux line system to known defect structure. It is not trivial question, however, whether columnar defects act cooperatively with point defects to pin vortices or not. We have investigated the effect of columnar defects on the irreversibility line for the point pin enriched sample of Bi-2223 tapes.

EXPERIMENTAL

The Ag-Cu alloy sheathed tapes were prepared by the powder-in-tube method. Ag-Cu alloy sheaths in this study were filled with the Cu-poor Bi-2223 powder, since copper atoms were expected to diffuse into inner oxide core from the outer Ag-Cu alloy sheath. We have introduced Hf elements in the Ag-Cu alloy sheathed Bi-2223 tapes, which result in the improvement of the transport Jc. The details of the preparation method, and the fundamental properties of the samples were reported elsewhere [4]. Bi-2223 tapes with typically $3.0 \times 3.0 \times 0.1$ mm^3 were irradiated with 3.5GeV ^{163}Xe^{31+} ion at the RIKEN ring cyclotron facility to introduce columnar defects along perpendicular to the tape surface direction. The total pin density was estimated to be $7.2 \times 10^{10} \text{cm}^{-2}$, which corresponded to a dose-equivalent matching field of $B\phi = 1.4 \pm 0.2$ T. This type of irradiation has produced continuous amorphous tracks with the diameter of ≈ 6 nm throughout the thickness of the Bi-2223 sample. We estimated sample qualities using X-ray diffraction measurement and high-resolution transmission electron microscope. We confirmed that the Hf atoms were substituted for $0.5 \sim 1\%$ of Sr by high-resolution analyzed electron microscopy (HRAEM) [5]. The magnetic properties were measured by using a superconducting quantum interference device (SQUID) magnetometer. The magnetic field was applied perpendicular and parallel to the wide surface of the tape.

RESULTS AND DISCUSSION

Figure 1 (a) and (b) show magnetic field versus the irreversibility temperature (Tirr) for the Ag and Ag-Cu alloy sheathed Bi-2223 tapes with columnar defects by solid symbols and for the unirradiated tapes by open symbols. The magnetic field applied (a) perpendicular (H//c) and (b) parallel (H//ab) to the tape surface direction, respectively.

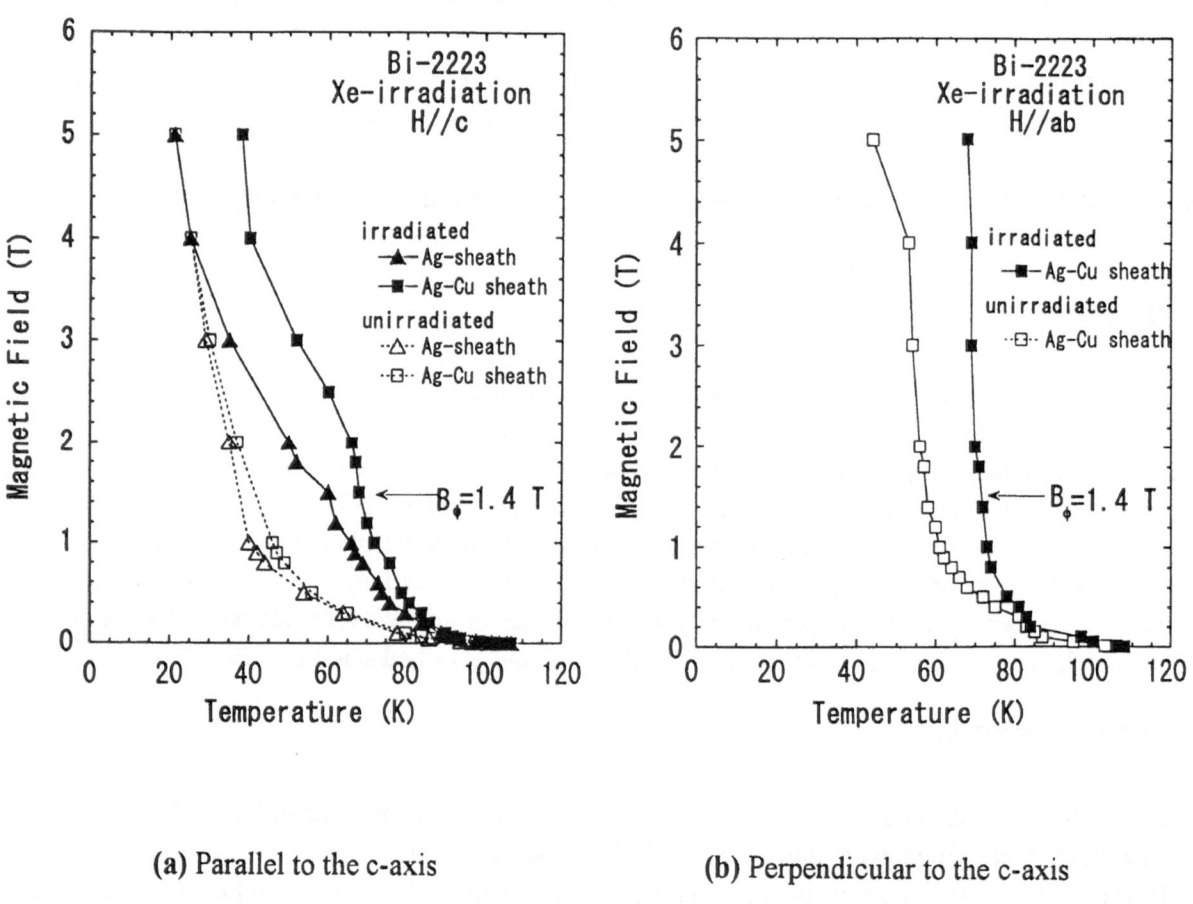

(a) Parallel to the c-axis   (b) Perpendicular to the c-axis

Fig. 1. The irreversibility temperature are plotted in H-T plane for the Ag and Ag-Cu alloy sheathed Bi-2223 tapes with columnar defects by solid symbols and for the unirradiated ones by open symbols, respectively.

In Fig.1(a) we can see that the irreversibility line curve for the irradiated Ag sheathed tapes is located in the higher temperature side of the unirradiated ones below 4 T and merges to the one for the unirradiated samples at the field of about triplicate of Bφ. We found that the irreversibility line curve for the Ag-Cu alloy sheathed tapes (with Hf impurities) was drastically shifted to the higher temperatures in the measured field up to 5 T by the Xe-ion irradiation. This behavior suggests that pinning force density (Fp) for the irradiated Ag-Cu alloy sheathed tapes is larger than that of the Ag sheathed ones. In fact, according to the results of the pinning force density measurement, the field dependence of Fp of the Bi-2223 tape samples was improved by the Hf substitution for Sr due to the collective pinning [6]. In the case of (b) the perpendicular to the c-axis, we also found that the irreversibility line curve for the Ag-Cu alloy sheathed tapes (with Hf impurities) was drastically shifted to the higher temperature side compared with parallel to the c-axis. It comes from the strong anisotropy in the flux pinning. Therefore, it is expected that the columnar defects will act as effective pinning centers in cooperation with the point defects induced by Hf doping.

CONCLUSION

We observed the effect of the columnar defects in the Bi-2223 tapes. We found the irreversibility curve of the irradiated Ag-Cu alloy sample (with Hf impurities) is located in the higher temperature side of the irradiated Ag sheathed tape even in the higher magnetic field than the triplicate of the matching field Bφ. These behaviors suggest that the columnar defects become more effective pinning centers in cooperation with the point defects induced by Hf doping in the Ag-Cu alloy sheathed tape samples.

ACKNOWLEDGMENTS

The authors wish to thank A.Iwase and N.Ishikawa at Japan Atomic Energy Research Institute for valuable discussions and preparing the columnar defect samples.

REFERENCES

1. N. Chikumoto , M. Konczykowski, M. Kosugi, Y. Matsuda, J. Shimoyama and K. Kishio   (1997) *Advances in Superconductivity X*, Osamura and Hirabayashi (Eds) 95-98
2. K. Hirata, T. Mochiku, S. Miyamoto and N. Nishida (1997) *Advances in Superconductivity X*, Osamura and Hirabayashi (Eds) 553-556
3. K. Kadowaki and K. Kimura (1997) *Advances in Superconductivity X*, Osamura and Hirabayashi (Eds) 107-110
4. M. Ishizuka, Y. Tanaka and H. Maeda (1995) Physica C **252**: 339-347.
5. H. Ikeda, Y. Tanaka, R. Yoshizaki, M. Ishizuka, K. Yoshikawa and H. Maeda (**1998**) *Advances in SuperconductivityX,* Osamura and Hirabayashi (Eds) 787-790.
6. H. Ikeda, Y. Tanaka, R. Yoshizaki, M. Ishizuka, K. Yoshikawa and H. Maeda (**1997**) *Advances in Superconductivity* **IX,** Nakajima and Murakami (Eds) 847-850.

MICROWAVE FLUX FLOW RESISTIVITY OF CLEAN LIMIT SUPERCONDUCTOR YNi$_2$B$_2$C

A.Shibata, K.Izawa, Y.Matsuda, H.Takeya[A], A.Kamimura[A], K.Hirata[A]

Institute for Solid State physics, University of Tokyo, Minato-ku, Tokyo 106-8666, Japan.
[A]National Research Institute for Metals, Tsukuba, Ibaraki 305, Japan

Abstract: We measured the microwave flux flow resistivity ρ_f of the clean limit s-wave superconductor YNi$_2$B$_2$C. We show that ρ_f increases linearly with H. This result shows that the number of the quasiparticle inside the vortex core increases in proportion to H and provides a strong evidence that the size of the vortex core does not change with H. This is inconsistent with the recent μSR and STM results which claim that the core shrinks with H. Moreover, core shrinkage with decreasing T in constant H (Kramers-Pesch effect) was not observed down to 1.5K ($\sim 0.1T_c$). The present results imply that the YNi$_2$B$_2$C is a s-wave superconductor with the anisotropic energy gap.

Keywords: YNi$_2$B$_2$C , vortex core, quasiparticle, flux flow resistivity

INTRODUCTION

Recently, interest in the vortex state of the clean limit superconductors (SCs) was renewed by the discovery of high-T_c superconductors. For example, the possibility that the size of the vortex core shrinks with H was pointed out by μSR and STM measurements for clean limit SC 2H-NbSe$_2$ [1,2]. Moreover, it has also been reported that the electronic specific heat C_e, which is proportional to the quasiparticle (QP) density of states (DOS), shows an unusual H-dependence in the clean limit SCs NbSe$_2$, YNi$_2$B$_2$C, CeRu$_2$ [3,4,5]. In these materials, C_e shows an nonlinear H-dependence, especially in YNi$_2$B$_2$C C_e increases as $C_e \propto H^{1/2}$. This behavior was discussed in terms of the shrinkage of the vortex core [4,6]. On the other hand, de Haas van Alphen effect has been observed in the mixed state of these materials, indicating the existence of the delocalized QP outside the vortex core [7]. One of the direct quantities to clarify the shrinkage of the vortex core with H is the flux flow resistivity ρ_f, because the energy dissipation due to the flux motion is mainly caused by the QP trapped inside the vortex core. Another interesting subject in the clean limit s-wave SCs is the core shrinkage with decreasing T at very low temperatures in a constant H. This effect was predicted by Kramers and Pesch for three decades ago [8], but there is no direct experimental evidence of such an effect. In this paper, to study the shrinkage of the vortex core, we performed ρ_f measurements in the microwave regime on the clean s-wave SC YNi$_2$B$_2$C.

EXPERIMENT

We have made microwave measurements (28.5 GHz) on YNi$_2$B$_2$C single crystal (T_C=14.3 K) grown by the traveling solvent floating zone method. The microwave measurements were made in a circular cylindrical Cu cavity resonator supporting TE$_{011}$ mode with Q-values of $\sim$ 25,000. The cavity perturbation technique was used to obtain the surface impedance ($Z_S=R_S+iX_S$ where R_S is the surface resistance and X_S is the surface reactance.).

According to Coffey and Clem, Z_S is given by

$$Z_S = i\mu_0\omega\lambda\sqrt{\frac{1-(i/2)\delta_v^2/\lambda^2}{1+2i\,\lambda^2/\delta_{nf}^2}} \quad , \tag{1}$$

when the microwave frequency is much higher than the pinning frequency [9]. Here λ is the penetration depth of the superfluid, δ_{nf} is the skin depth of the normal fluid and δ_v is flux flow skin depth which is given by $\delta_v^2 = 2\rho_f/\mu_0\omega$. In low field when λ is larger than δ_v, R_S can be written as $\sim \rho_f/\lambda$, while in high field when δ_v exceeds λ, R_S can be written as $\sim \rho_f/\delta_v$. Thus ρ_f is obtained from Eq.1. Usually, the flux flow resistivity is given by the Bardeen-Stephen formula, $\rho_f = (H/H_{C2})\rho_n$ (ρ_n is the resistivity in the normal state), if the DOS of the QP inside each vortex core does not depend on H. On the other hand, if the shrinkage of the core occurs, an nonlinear H-dependent flux flow resistivity, $\rho_f = (H/H_{C2})^{1/2}\rho_n$, is expected to be observed because the DOS of the QP inside the core increases as $H^{1/2}$. Thus the measurements of R_S give a crucial test for the shrinkage of the vortex core.

RESULTS AND DISCUSSIONS

Figure 1 shows R_S as a function of $H^{1/2}$. R_S increases with H and becomes constant above H_{C2}. At $H > 0.05$ T, R_S increase as $H^{1/2}$, indicating $\rho_f \propto H$. Figure 2 shows the fit of the data to Eq.1. The fit to the data is very good with only one fitting parameter λ. This H-dependence of R_S provides a strong evidence that the DOS of the QP inside the core does not change with H and rules out the possibility of the shrinkage of the vortex core.

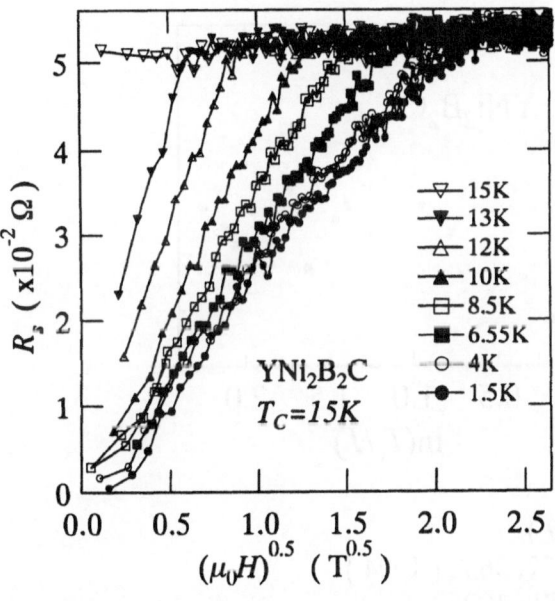
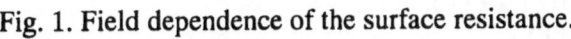
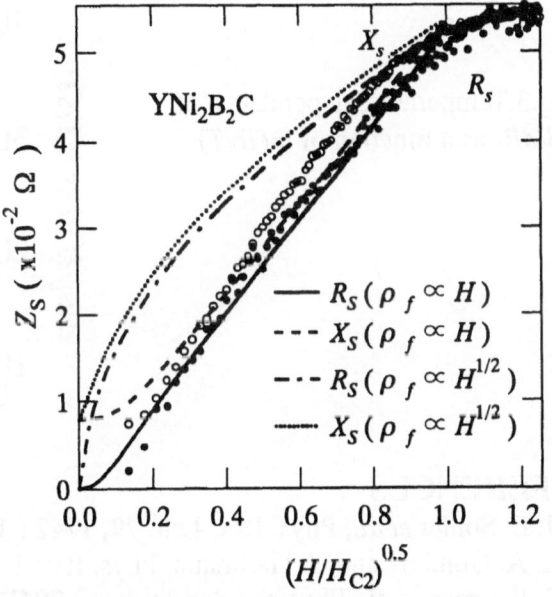

Fig. 1. Field dependence of the surface resistance.

Fig. 2. H-dependence of R_S and X_S at 1.5K. The open (solid) circles show R_S (X_S). The lines represent the results of the fitting assuming different H-dependence of ρ_f. In the fitting we used λ=350Å.

The present results contradict the results of μ SR. They also indicate that the deviation from the H-linear dependence of C_e cannot be attributed to the DOS enhancement due to the shrinkage of the vortex core. We point out here that the nonlinear H-dependence of C_e originates from the anisotropic s-wave symmetry of the Cooper pair in YNi$_2$B$_2$C. In anisotropic s-wave SCs, QP extend outside the core and overlap the QP of the neighboring vortices. This effect gives rise to $H^{1/2}$ dependent C_e, similar to the d-wave vortex state. The anisotropic s-wave state is consistent with the impurity effect of C_e observed in Ref[4], the dHvA effect which can be observed at $H \ll H_{C2}$ [7], and square lattice of the vortices [10].

We finally discuss the shrinkage of the vortex core with decreasing T in a constant H. This shrinkage is caused by the rapid change of the QP population inside the core with T. This effect gives rise to the logarithmic correction to R_S as

$$\frac{dB}{dR_S} = 0.60\lambda\mu_0 H_{C2}\frac{1}{\rho_n}ln\left(\frac{\Delta}{\kappa_B T}\right), \qquad (2)$$

Figure 3 depicts dB/dR_S as a function of $ln(T/T_c)$. dB/dR_S is obtained by the fitting of R_S by Eq.1. dB/dR_S is almost T-independent at low temperatures down to 1.5K ($\sim 0.1T_C$), indicating no logarithmic dependence. At the present stage, it is unknown why KP effect is not observed. One of the reasons may be the anisotrpic energy gap of YNi$_2$B$_2$C which smears out the discrete energy levels of QP inside the core. To clarify this point, the measurement of R_S at lower temperature is necessary.

Fig.3. Temperature dependence of dB/dR_S as a function of $ln(Tc/T)$

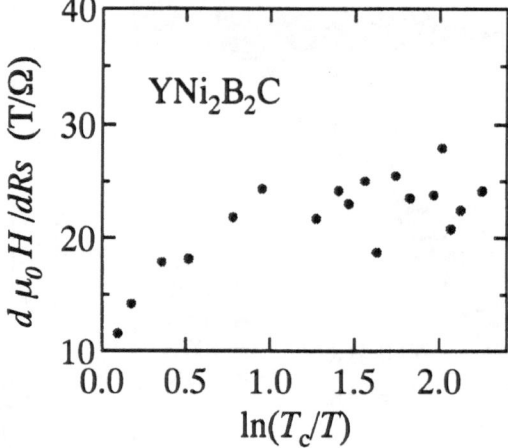

REFERENCES
[1] J. E. Sonier *et al.*, Phys. Rev. Lett. **79**, 1742 (1997).
[2] A. A. Golubov and U. Hartmann, Phys. Rev. Lett. **72**, 3602 (1994).
[3] D. Sanchez *et al.*, Physica (Amsterdam) **204B**, 167 (1995).
[4] M. Nohara *et al.*, J. Phys. Soc. Jpn. **68**, 1078 (1999).
[5] M. Hedo *et al.*, J. Phys. Soc. Jps. **67**, 272 (1998).
[6] J. E. Sonier *et al.*, Phys. Rev. Lett. **82**, 4914 (1999).
[7] T. Terashima *et al.*, Phys. Rev. B **56**, 5120 (1997).
[8] S. G. Doettinger *et al.*, Phys. Rev. B **55**, 6044 (1997).
[9] M. W. Coffey and J. R. Clem, Phys. Rev. Lett. **67**, 386 (1991).
[10] M. Yethiraj *et al.*, Phys. Rev. Lett. **78**, 4849 (1997)

MAGNETIC FIELD DEPENDENCE OF Jc IN NdBa$_2$Cu$_3$O$_{7-\delta}$ SINGLE CRYSTALS GROWN UNDER VARIOUS PARTIAL OXYGEN PRESSURES

Susumu SHIBATA,   A.K.PRADHAN,   Takato MACHI   and   Naoki KOSHIZUKA

Superconductivity Research Laboratory, International Superconductivity Technology Center,
1-10-13 Shinonome, Koto-ku, Tokyo, 135-0062, Japan

Abstract: NdBa$_2$Cu$_3$O$_{7-\delta}$ (Nd123) single crystals were grown by the horizontal-Bridgman like method under various partial O$_2$ pressures (0.03~ 3%) in Ar atmosphere. This work revealed that the peak effect disappears around 0.03% ~0.05% partial O$_2$ pressures. Furthermore, it turned out that both the peak field and irreversibility field increase with decreasing PO$_2$ from 3% to 0.1% and then decrease below 0.1%. These results are explained by the decrease of the amount of Nd/Ba substitution which is consistent with the fact that the orthorombicity increases with decreasing PO$_2$.

Key words: NdBa$_2$Cu$_3$O$_{7-\delta}$, single crystal, peak effect, substitution of Nd ions

INTRODUCTION

Among high Tc superconductors, NdBa$_2$Cu$_3$O$_{7-\delta}$ (Nd123) system is promising for practical applications because of its large peak effect in Jc observed at relatively high temperatures and high fields[1]. However, there remains various problems such as the flux pinning mechanism and identification of pinning centers. Since Nd ions are easy to substitute for Ba site partially in the Nd123 matrix, it is considered that this substitution is a possible pinning center for the peak effect in Jc[2]. In order to make clear the pinning center, it seems to be helpful to obtain the Jc-H curves for the samples with different Nd/Ba substitution grown under different conditions. In this paper, we report the substitution effect on the peak effect of the single crystals grown under various O$_2$ partial pressures.

EXPERIMENTS

The single crystals of Nd123 were prepared by the horizontal-Bridgman like method in YSZ boats(100mm length, 30mm width and 14mm height) under the following partial O$_2$ pressures of 0.03, 0.04, 0.05, 0.06, 0.07, 0.1, 0.13, 0.3, 0.65, 1 and 3% in Ar atmosphere. The starting composition was NdO$_{1.5}$:BaCO$_3$: CuO= 1:19.8:19.2[3]. The details of crystal growth is described elsewhere[4]. The as-grown single crystals were annealed at 325℃ for 14 days in flowing O$_2$ atmosphere to fully oxidize the crystals.

X-ray diffraction patterns of the grown single crystals showed that all samples are of single phase. Polarization light micrographs of these crystals showed that they have twin boundary planes present in

383

both the [110] and [1̄10] directions. The measurements of the superconducting transition temperature (T_c) and the magnetic hysteresis M-H loops of these crystals were carried out using a SQUID(QUANTUM DESIGN, MPMS) magnetometer. The orthorhombic lattice parameters of the a and b axes of these crystals were determined using the powder X-ray diffraction(XRD) method.

RESULTS AND DISCUSSION

Fig.1 shows the temperature dependencies of magnetization in the field of 50 Oe applied parallel to the c-axis for the various single crystals grown under various O_2 partial pressures. We find that the transition width $\triangle$Tc increases with raising O_2 partial pressure. As shown in Fig.2 all Tc (onset) values are higher than 94K. It turns out that the Tc increases with lowering O_2 partial pressure. The slight change in Tc may be due to the variation of the microstructures or the carrier concentration of the Nd123 matrix among these crystals.

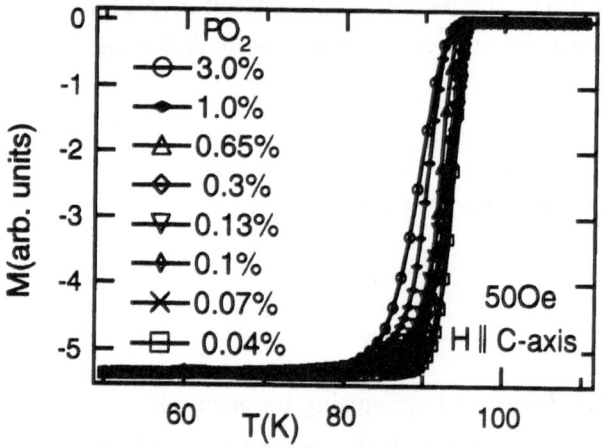

Fig.1 Temperature dependence of magnetization for Nd123 single crystals grown under different PO_2

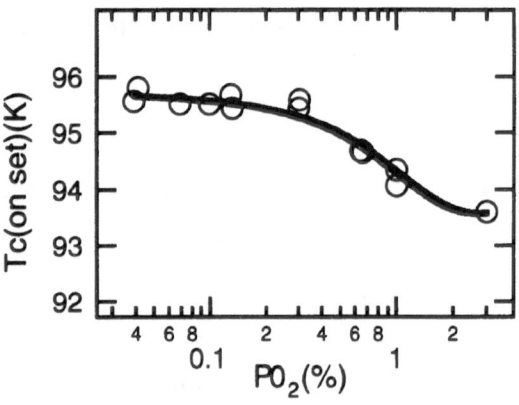

Fig.2 Tc(on set) of Nd123 single crystals vs PO_2 in Ar atmosphere

Recently, we have found that the peak effect is observed in the M-H loops for the crystals grown under the O_2 pressures down to 0.06%, but it disappears for those of 0.03, 0.04 and 0.05% at 77K in the applied magnetic fields parallel to the c-axis of the crystals[4]. Fig.3 shows the relationship between Hirr(irreversibility field), Hp(peak field) and PO_2 (%) respectively at 77K. Both Hp and Hirr increase with decreasing PO_2 from 3% to ~0.1% and decrease below ~0.1%. These behaviors below 0.1% may be ascribed to the change of the microstructures probably related to the Nd/Ba substitution[4]. It is known that with decreasing PO_2, the Nd/Ba substitution over the sample is suppressed and the localized substituted regions appear in the stoichiometric 123 matrix. A STM work by Ting et al. [5] revealed that there are compositionaly modulated circular regions with the size of about 20nm in the 123 matrix, and the Nd/Ba substitution is responsible for the finely distributed regions. These regions may have lower Tc compared with that of the matrix due to the lack of carrier concentration or due to the strain field caused by the substitution.

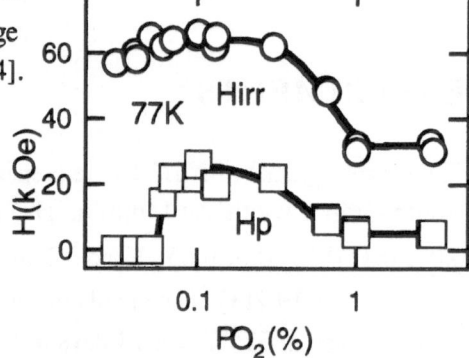

Fig.3 Relationship between Hirr, Hp and PO_2(%)

The increase of Hp is explained by the increase of Tc and Hc_2 of the pinning centers with decreasing substitution. It is clear that the disappearance of peak effect below 0.05% is caused by decreasing the

385

amount of substitution into a critical concentration required for emergence of peak effect. In fact, the first order melting transition for these vortex lattice was observed for the crystals having no peak effect, indicating that this system is significantly clean and less disordered[6]. The increase of Hirr is ascribed to the enhancement of the pinning force due to suppression of the substitution.

Fig.4 shows the relationship between the orthorombicity {(b-a)/(a+b)} x100 of Nd123 single crystals and PO$_2$. The fact that the orthorhombicity{(b-a)/(a+b)}x100 decreases with increasing PO$_2$ is consistent with the expected variation of Nd/Ba substitution. This also leads to the speculation that in the case of low PO$_2$, the stoichiometry of Nd123 in the matrix is maintained, but in relatively high PO$_2$ cases above 0.1% it slightly deviates from its stoichiometry. It is reported that the amount of oxygen vacancies in Nd123 single crystals increases with decreasing Nd/Ba substitution in fully oxygenated state[7]. In our experiment the peak effect disappeared with decreasing the substitution, where the oxygen vacancies are to increase. It is clear that our result is inconsistent with the oxygen vacancy model for the peak effect well known in Y123 crystals.

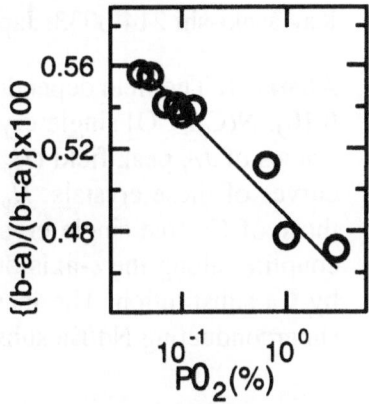

Fig.4 Relationship between the orthorombicity{ (b-a)/(a+b)}x100 of Nd123 single crystals and PO$_2$.

CONCLUSION

We studied the Jc-H characteristics of Nd123 single crystals grown under different oxygen partial pressures(0.3~3%) in Ar atmosphere. We obtained the following results. (1)Tc increases up to 95.5K with decreasing O$_2$ partial pressure. (2)The peak effect is observed at 77K in the Jc-H curves for the crystals grown under the O$_2$ partial pressure above 0.06%, but it disappeared around 0.05~0.03%. (3) Hp and Hirr increased with decreasing PO$_2$ from 3% to 0.1%. (4) The orthorombicity {(b-a)/(a+b)} x100 of Nd123 single crystals increases with decreasing PO$_2$.

These results support that the Nd/Ba substitution is responsible for the peak effect in Nd123 single crystals, although the oxygen disordering associated with the substitution is not ruled out for the origin of the peak effect.

ACKNOWLEDGMENT

This work was supported by the New Energy and Industrial Technology Development Organization (NEDO) as Collaborative Research and Development of Fundamental Technologies for Superconductivity Applications.

REFERENCES
[1] M. Murakami, N Sakai, T.Higuchi and S.I.Yoo, Sci.Technol. 9(1996)1015.
[2] S.I.Yoo and R.W.Mccallum, Physica C 210 (1993) 147.
[3] S.Shibata, H Unoki, K Kuroda and N Koshizuka, J. Materials Science Letters 16(1997)1295
[4] S.Shibata, A.K Pradhan and N. Koshizuka, J. J.Appl. Phys. 38(1999)L1169
[5]W.Ting, T.Egi, K.Kuroda, N Koshizuka and S.Tanaka, J. J.Appl. Phys. 35(1996)4034
[6] A.K.Pradhan, S.Shibata, K.Nakao and N.Koshizuka, Phys. Rev. B59(1999)11563
[7] E.Goodlin, M.Limonov, A.Panfilov, N.Khasanova, A.Oka, S.Tajima and Y.Shiohara, Physica C 300 (1998)250.

IMPROVEMENT OF IRREVERSIBILITY FIELD BY Ca SUBSTITUTION IN Nd123 SINGLE CRYSTALS

*Kazuhisa Itoi[A,B], Takato Machi[A], Kiyoshi Kuroda[A], Naoki Koshizuka[A] and Shigehiko Arai[B]

[A]Superconductivity Research Laboratory, ISTEC, 1-10-13 Shinonome, Koto-ku, Tokyo 135-0062, Japan
[B]School of Science and Technology, Meiji University, 1-1-1 Higashimita, Tama-ku, Kawasaki-shi 214-0033, Japan

Abstract: The field dependence of J_C is studied for $Nd_{1-x}Ca_xBa_2Cu_3O_{7-\delta}$ [(x =0, 0.01, 0.02, 0.05, 0.10): N(C)BCO] single crystals grown by the traveling solvent floating zone (TSFZ) method. Values of J_C, peak field (H_{peak}) and irreversibility field (H_{irr}) are evaluated from magnetization curves of these crystals. H_{peak} and H_{irr} values as a function of T / T_C increase compared with those of Ca free single crystals. The enhancement of H_{irr} is ascribed to the increase of vortex coupling along the c-axis due to decrease of anisotropy with increase of carrier concentration by Ca substitution. The increase of H_{peak} is explained by the increase of T_C and H_{C2} of weak superconducting Nd/Ba substituted regions with Ca doping.

Keywords: Nd123, irreversibility field, peak field

INTRODUCTION

(RE)123 (RE = Nd, Sm...) are promising superconducting materials because of their high T_C ~96K and large critical current densities at high fields showing "peak effect" or "fishtail effect". It is a crucial issue for developing practical materials to increase J_C, peak field H_{peak} and irreversible field H_{irr} in (RE)123 superconductors [1]. Chikumoto *et al.* [2] reported that H_{peak} and H_{irr} increased by using high-pressure oxygenation in Nd123 and Sm123 which have RE/Ba substituted regions. If the improvement of H_{irr} and H_{peak} is caused by increasing carrier concentration, we will be able to increase these fields by Ca^{2+} doping. In this study, we have tried to improve the field dependence of J_C by controlling carrier concentration with Ca substitution in Nd123 single crystals.

EXPERIMENTAL

We have grown N(C)BCO single crystals (x=0, 0.01, 0.02, 0.05, 0.10) by the TSFZ method. The details of the crystal growth are described elsewhere [3]. Then we annealed the single crystals at 320 °C for 2 weeks in a tube furnace under 1 l/min of flowing oxygen. We measured magnetic moment of oxidized single crystals by using a SQUID magnetometer (Quantum Design) and a VSM (Oxford Instruments). We obtained T_C from the diamagnetic curves with an applied field of 50Oe. J_C, H_{peak} and H_{irr} values were evaluated from the magnetization hysteresis curves measured by VSM. The magnetic field was applied parallel to the c–axis up to 14T at a sweep rate of 0.6 T/min.

---

* E-mail : itoi@istec.or.jp

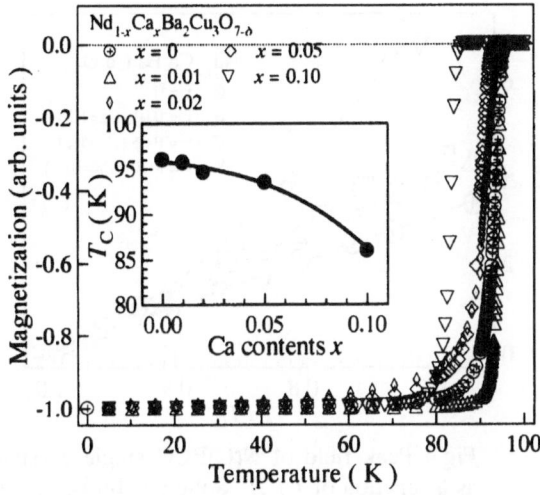

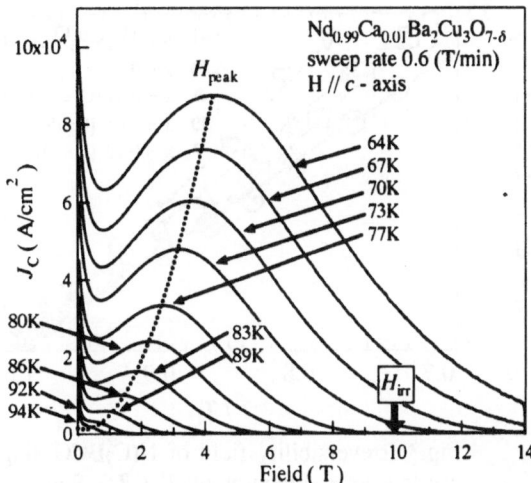

Fig.1 Temperature dependence of magnetization at $H(50Oe)$ pararell to the c-axis of the N(C)BCO single crystals. The inset is the Ca content dependence of T_C in N(C)BCO.

Fig.2 Field dependence of J_C at several temperatures (from 64K to 94K) in Ca-1%Nd123 single crystal. H_{irr} denotes the value at 77K. A dotted line is a guide for the eye.

RESULTS & DISCUSSION

Figure 1 shows the temperature dependence of magnetization for N(C)BCO single crystals measured by SQUID magnetometer. We find that T_C decreases with increasing Ca content. Lattice constants, a, b and c measured by XRD decreased, whereas the orthorhombicity defined as $(a-b)/(a+b)$ increased with increasing Ca content. It is known that the c-axis in Nd123 system increases and the orthorhombicity decreases with decreasing oxygen content [4]. Therefore, the decrease of T_C with increasing Ca content means that N(C)BCO single crystals shifted from a nearby optimum doped state to an overdoped state.

Figure 2 shows the field dependence of J_C for a Ca-1%Nd123 single crystal measured by VSM at several temperatures. It turns out that both H_{irr} and H_{peak} shift toward higher field with decreasing temperature.

Figure 3 shows Ca content dependence of H_{irr} as a function of T/T_C. The value of H_{irr} in Ca-1%Nd123 reaches about 10T at $T/T_C = 0.8$ (77K), which is about 2T higher than that of H_{irr} in Ca free Nd123 at the same temperature. H_{irr} as a function of T/T_C shifts to higher field side with increasing Ca content except for $x = 0.05$. The anomaly in Ca-5%Nd123 may be due to poor cryatallinity as shown in the broadening of ΔT_C. We can explain the improvement of H_{irr} property by the increase of vortex coupling along the c–axis due to decrease of anisotropy with increase of carrier concentration by Ca substitution.

Figure 4 shows Ca content dependence of H_{peak} as a function of T/T_C. The H_{peak} value also shift toward higher field side with increasing Ca content. Before we mention the reason for this shift of H_{peak}, we consider possible origins on the peak effect in Nd123. At present there are arguments both for the pinning centers and the mechanism of peak effect in (RE)123 systems [1,5-7].

As for the former, there are two main contributions : (1) Clusters of oxygen vacancies [5] and (2) RE/Ba substituted sites [1]. The first model, in fact, may hold for the peak effect in $YBa_2Cu_3O_{7-\delta}$ with a small amount of oxygen vacancies. However, in Nd123 system a clear disappearance of peak effect is observed with reducing the amount of Nd/Ba substitution in spite of the same amount of oxygen vacancies [6]. Thus it is unlikely that the peak effect is due to oxygen vacancies as far as Nd123 is concerned.

Concerning the peak effect in (RE)123 systems, the following two mechanisms are thought

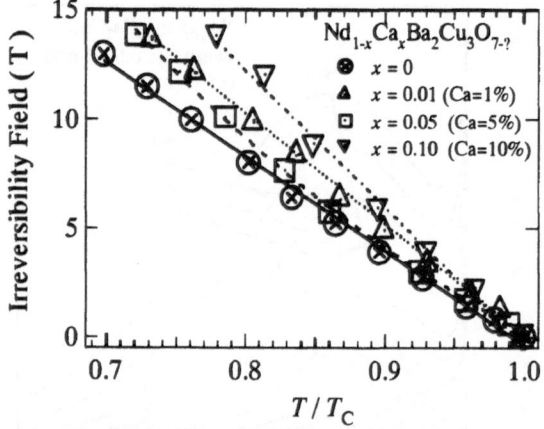
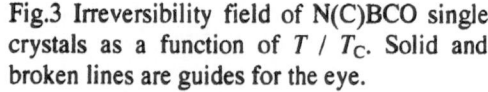

Fig.3 Irreversibility field of N(C)BCO single crystals as a function of T / T_C. Solid and broken lines are guides for the eye.

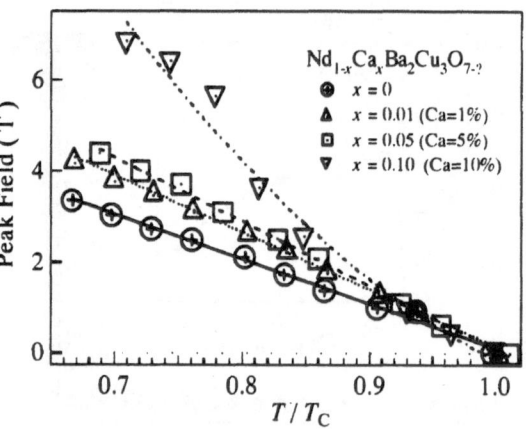

Fig.4 Peak field of N(C)BCO single crystals as a function of T / T_C. Solid and broken lines are guides for the eye.

to be dominant : (1) ΔT_C pinning: In the presence of lower T_C and H_{C2} regions in the higher T_C matrix, the lower T_C regions play a role of pinning, and the pinning force becomes maximum when the pinning centers change to a normal state with increasing field. (2) Synchronization effect [7]: The vortex lines rearrange along with pinning centers associated with the field induced disordering of vortex lattices.

According to the synchronization model, H_{peak} should decrease with lowering temperature because the field at which disordering of vortex lattice occurs decreases [7] in a relatively high temperature region. While, Fig.2 shows the opposite behavior for decreasing temperature. As a result, it is difficult to explain the enhancement of H_{peak} based on this model. The peak effect behavior in Fig.2 is rather consistent with the ΔT_C pinning model in Nd123 system. The T_C and H_{C2} of Nd/Ba substituted region will increase with Ca doping, since the carrier concentration on the substituted one also increases with the doping. Then, a larger field is necessary to make the weak superconducting Nd/Ba substituted region into a normal state. It may result in the enhancement of H_{peak} with increasing Ca content.

CONCLUSION

We succeeded in the growth of Ca doped Nd123 single crystals by TSFZ method. We found that H_{peak} and H_{irr} as a function of T / T_C increased compared with those of Ca free NdBa$_2$Cu$_3$O$_{7-\delta}$. It seems that the improvement of H_{irr} value is caused by the increase of vortex coupling along the c-axis with increase of carrier concentration by Ca substitution. The increase of H_{peak} may result from the increase of T_C and H_{C2} of Nd/Ba substituted regions.

Acknowledgment: This work is supported by the New Energy and Industrial Technology Development Organization (NEDO) as Collaborative Research and Development of Fundamental Technologies for Superconductivity Applications.

1. M. Murakami *et al.*: Supercond. Sci. Thechnol. 9 (1996) 1015
2. N. Chikumoto *et al.*: Advances in Superconductivity, vol.9 no.1 (1997) 531
3. K. Kuroda *et al.*: J. Cryst. Growth vol.173 no.1/2 (1997) 73.
4. H. Shaked *et al.*: Phys. Rev. B 41 (1990) 4173.
5. Th. Wolf *et al.*: Phys. Rev. B 56 (1997) 6305
6. S. Shibata *et al.*: to be published in JJAP.
7. A. A. Zhukov *et al.*: Phys. Rev. B 51 (1995) 12704.

ANISOTROPY OF *E-J* CHARACTERISTICS IN YBCO SUPERCONDUCTORS

Masayoshi Inoue[1], Masaru Kiuchi[1], Takanobu Kiss[1], Masakatsu Takeo[1], Teruo Matsushita[1], Satoshi Awaji[2], and Kazuo Watanabe[2]

[1]Graduate School of Information Science and Electrical Engineering, Kyushu University, Fukuoka 812-8581, Japan
[2]Institute for Material, Tohoku University, Sendai, 980-8577, Japan

Abstract: To investigate the anisotropic characteristics in YBCO superconductors, we measured the electric field (E) vs. current density (J) characteristics at various conditions such as temperature, T, magnetic field, B, and field angle, θ. We extracted the parameters related to the distribution of the critical current density, J_c and estimated the anisotropic properties with respect to the scaling characteristics of the pinning force densities and the transition fields.

Key words: Anisotropic properties, J_c distribution, Transport characteristics, High magnetic field, YBCO thin films,

INTRODUCTION

For the anisotropy of the crystal structure, the superconducting properties in high T_c superconductors (HTS) such as *E-J* characteristic, critical current density, J_c, and transition field, B_{GL}, depend on sensitively the direction of applied magnetic field. Moreover, since HTS have the extraordinary high irreversibility field at low temperature, high magnetic field is required to estimate the properties. We have already shown that a method to formulate the nonlinear *E-J* characteristics based on a study on stochastic characteristics of flux pinning and its scaling in a disordered pin medium. In this study, we estimated the anisotropy properties of YBCO superconducting thin films including the high magnetic field region. The magnetic field was applied at various directions from *ab*-plane to *c*-axis. We extracted the parameters on the statistic distribution of J_c. Then the scalings and other properties at various conditions were compared.

EXPERIMENTS

C-axis oriented YBCO thin films were deposited on $SrTiO_3(100)$ single crystal substrate by the excimer laser ablation method with a typical thickness of 200nm. The films were patterned into 100μm wide and 1mm long bridge shape by the photolithography and chemical etching process. Transport properties of the bridge were measured by the four probe method. We used not only the commercial superconducting magnet but also the hybrid magnet at Tohoku University to apply the high magnetic field up to 27T. The sample was rotated to change the direction of the applied magnetic field angle. The angle, θ, was defined by the angle from *ab*-plane of the thin films.

THEORY

We have already proposed a method to formulate the nonlinear $E(J)$ characteristics in (B, T) plane based on a study on stochastic characteristics of flux pinning and its scaling in a disordered pin medium as follows [1,2].

$$E(J) = \rho_{FF} \int_0^J \left[(J - J_{cm})/J_0 \right]^m dJ \qquad (1)$$

where, ρ_{FF} is the flux flow resistivity. On the other hand, J_{cm}, J_0 and m are related to the distribution function of J_c, i.e., J_{cm} is the minimum value of J_c, J_0 is the variance of the J_c distribution and the value of m is numerical parameter characterizing the shape of J_c distribution. The typical value of J_c denoted by J_k can be approximately estimated by $J_k=J_{cm}+J_0$. Transition field, B_{GL} is the magnetic field where $J_{cm}=0$. The parameter m is determined from the measured straight line ($J_{cm}=0$) in the Log$E$-Log$J$ characteristics. For each Log$E$-Log$J$ curves, the parameters, J_{cm} and J_0, are determined as fitting parameters. Then we can collect the parameters, J_{cm} and J_0, in a (B, T) plane from the measurements at various B, T and θ conditions.

RESULTS AND DISCUSSION

We extracted the parameters of J_c distribution such as m, J_{cm} and J_0 from the E-J characteristics by using the statistic model. The parameter m was constant to be 5.4. Using the parameters J_{cm} and J_k, the scaling characteristics of the pinning force densities $F_{pm}(\equiv J_{cm}B)$ and $F_{pk}(\equiv J_kB)$ were obtained. The result of F_{pm} was shown Fig. 1. It has been confirmed that the similar scaling also holds for F_{pk} at various conditions.

Fig. 2 indicates the temperature and magnetic field angle dependencies of B_{GL}. The field angle dependencies of the reduced fields $B_{GL}(\theta)/B_{GL}(0°)$ and $B_k(\theta)/B_k(0°)$ were shown in Fig. 3 for the

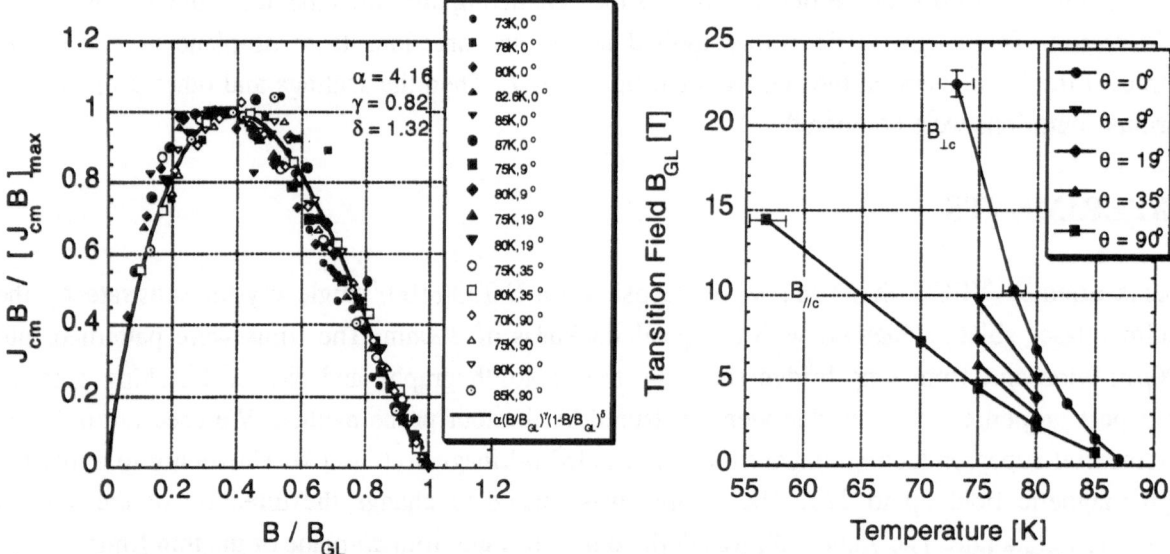

Fig. 1 Scaling characteristics of the minimum pinning force density with respect to the temperatures and magnetic field angels.

Fig. 2 Temperature and magnetic field angle dependencies of the B_{GL}.

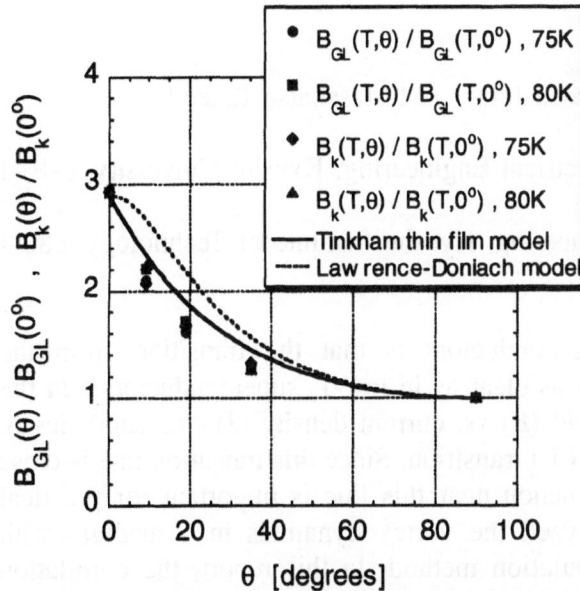

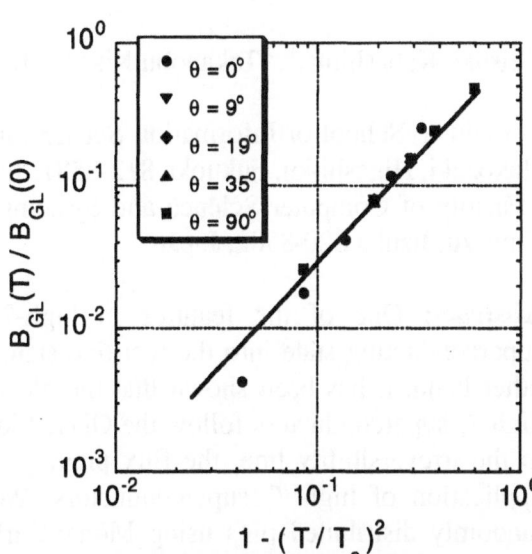

Fig. 3 Angular dependencies of the reduced fields $B_{GL}(\theta)/B_{GL}(0°)$ and $B_k(\theta)/B_k(0°)$. The solid line indicates Tinkham thin film model [3] and the broken line indicates Lawrence-Doniach model [4]

Fig. 4 Temperature dependencies of the $B_{GL}(T)$ divided by each $B_{GL}(0K)$. The $B_{GL}(T)/B_{GL}(0K)$ properties are scaled with respect to the temperature in the form of $[1-(T/T_c)^2]^{1.5}$.

temperatures of 75K and 80K. Note that the angular dependencies are the similar forme for these two quantities at different temperatures. We compared the experimental results with theoretical curves by Tinkham thin film model [3] and Lawrence-Doniach model [4]. Our results agreed rather well with that of Tinkham (2-D) model shown by the solid line, while it deviated from the Lawrence-Doniach (3-D anisotropic) model shown by the broken line. This result suggests that the coupling between CuO_2 planes in the present YBCO thin film is weak in these temperature region. The anisotropic ration, $\gamma \equiv \xi_{ab}/\xi_c$ where ξ_{ab} and ξ_c are relevant coherent lengths, is determined as 2.9 as an adjusting parameter.

Temperature dependencies of $B_{GL}(T)$ divided by each $B_{GL}(0K)$ were shown in Fig. 4. As the angler dependence for the $B_{GL}(0K)$, we used the same dependence shown in Fig. 3. It has been shown that the $B_{GL}(T)/B_{GL}(0K)$ properties are scaled with respect to the temperature in the form of $[1-(T/T_c)^2]^{1.5}$.

ACKNOWLEDGMENTS

This study was partly funded by NEDO, Japan, under the Proposal-Based New Industry Creative Technology R&D Promotion Program and Grant-in-Aid of Scientific Research on Priority Area "Vortex Electronics"

REFERENCES

1. T. Kiss, T. Nakamura, K. Hasegawa, M . Inoue, H . Okamoto, K . Funaki, M . Takeo, K. Yamafuji, F. Irie *Proceeding of ICEC17*, eds D. Dew-Huges et al. (IOP, London, 1998), pp 427-430
2. K. Yamafuji, T. Kiss,   Physica C **290**, pp 9-22 (1997)
3. M. Tinkham,   Phys. Lett. **9** 217 (1964)
4. Lawrence W E, Doniach S, Proc. 12 th Int. Conf. On Low Temperature Physics, ed Kanda E (Tokyo: Keigaku) pp 361 (1971)

Flux pinning characteristics for G-L transition with Monte-Carlo Simulation

Keisuke Kabashima,[1] Takanobu Kiss,[1] Teruo Matsushita,[1,2] and Masakatsu Takeo [1]

[1] Graduate School of Information Science and Electrical Engineering, Kyushu University, 6-10-1 Hakozaki, Higashi-ku, Fukuoka 812-8581, Japan
[2] Faculty of Computer Science and Systems Engineering, Kyushu Institute of Technology, 680-4 Kawazu, Iizuka 820-8502, Japan

Abstract: One of the features in high-T_c superconductors is that the transition from the superconducting state into the resistive state is not as clear as in low-T_c superconductors. On the other hand, it has been shown that the electric field (E) vs. current density (J) characteristics in high-T_c superconductors follow the Glass-Liquid (G-L) transition. Since this transition line is close to the irreversibility line, the flux pinning phenomenon near this line is important for practical application of high T_c superconductors. We analyzed the vortex dynamics in a medium with randomly distributed pins using Monte-Carlo simulation method. In this report, the correlation between the flux lattice and distributed pins is discussed in the vicinity of the transition line.

Keywords: Pinning, Glass-liquid transition, Monte-Carlo simulation, Cross-correlation function

INTRODUCTION

The pinning characteristics in High-T_c superconductors (HTS) are strongly influenced by various aspects which originate from their anisotropic stratified structures. These are a relatively large fluctuation, the anisotropic superconducting properties and the small condensation energy density. As a result, the pinning characteristics are significantly deteriorated by the thermal activation of flux lines at high temperatures. In addition, the pinning property is strongly affected by the superconducting property, which is speculated to fluctuate locally due to an inhomogeneous oxygen concentration. Hence, the local pinning strength is considered to distribute widely, and this seems to result in the fact that the resistive transition does not appear clearly as in Low-T_c superconductors (LTS). On the other hand, it is known that HTS obeys the so-called Glass-Liquid (G-L) transition, which is the resistive transition in the limit of zero transport current. So, the G-L transition line gives the applicable upper limit of the superconductor, and it is important to investigate the behavior of pinned flux lines in the vicinity of the G-L transition line.

Recently, Matsushita et al. theoretically investigated the thermal depinning in the vicinity of the G-L transition line using a kind of mean field theory, and showed that the thermal depinning transition itself is a phase transition of the second order. It was also clarified that the variation in the degree of disorder of the flux line lattice, S, with temperature is discontinuous at the transition point, T_g [1].
On the other hand, we have developed the Monte-Carlo simulation and investigated the behavior of pinned flux lines in detail [2]. In this paper, we examined the theoretical prediction of Matsushita et al. by calculating the correlation between pin and flux line lattice (FLL) in random pin medium using this simulation method.

MONTE-CARLO SIMULATION

The movement of flux lines in a random pin medium can described with a simple elastic model in the simulation. A two-dimensional network was composed in the condition that the interaction between a flux line and a pin was of threshold logic and that the elastic interaction was considered only between the nearest flux lines. For simplicity, default positions i.e., the positions of flux lines in the pin-free limit, are assumed to form a perfect square lattice. The displacement of the flux line from its equilibrium position is proportional to the pinning force which it suffers. This flux line network is relatively moved by a slight distance to the fixed pin medium. As a result, the distribution of the displacement from an equilibrium position can be computed, and its time-development can be calculated. The details of this simulation are given in Ref. [2].

RESULTS AND DISCUSSION

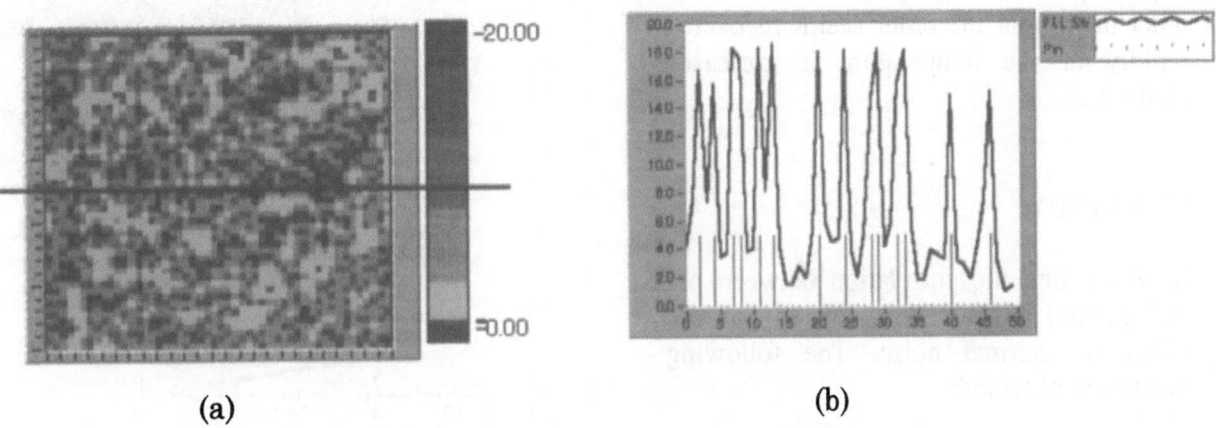

(a)   (b)

Fig. 1. (a) Spatial distribution of time-average-displacement in the FLL, and (b) Cross-sectional profile of the distortion along the line in (a).

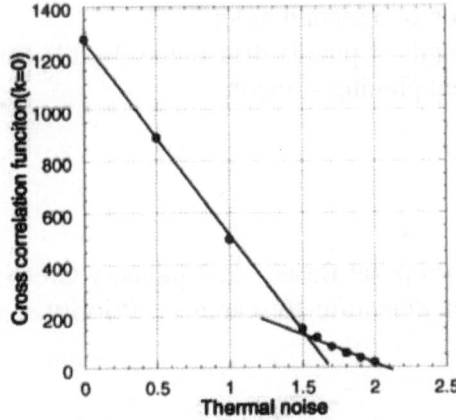

Fig. 2. Cross-correlation function for $k=0$ vs. thermal noise. This relationship corresponds to that between the degree of disorder FLL and temperature.

The results of simulation are shown in Fig. 1, where Fig. 1(a) is the spatial distribution of the time-average-displacement in FLL, and Fig. 1(b) shows the cross-sectional profile of the distribution along the normal direction to the driving force. These show that the pinned FLL is warped.

One-dimensional cross-correlation function between pin and FLL is defined by

$$R(k) = \int_{-\infty}^{\infty} P(x)F(x+k)dx ,$$

where $P(x)$ is the pinning strength at position x and $F(x)$ is a magnitude of FLL distortion. Since $F(x)$ is proportional to $P(x)$ as mentioned above, $P(0)$ is proportional to $\langle F^2(x) \rangle$ with $\langle \rangle$ denoting the statistical average. Hence, $P(0)$ gives the degree of disorder of FLL defined in [1]. The variation in

$P(0)$ with the strength of thermal noise is shown in Fig.2. This figure corresponds to the relationship between the degree of disorder and temperature, since the thermal noise is proportional to temperature. This indicates that the variation of the degree of disorder with temperature changes discontinuously at the transition point and agrees qualitatively with the theoretical prediction in [1] that the transition is of the second order.

Fig.3 shows cross-correlation function vs. temperature characteristics for $k=0$ when the inhomogeneous distribution of flux pinning strength is introduced. It is seen that $R(0)$ reduces significantly as the inhomogeneous distribution is introduced. At the same time, the break of the cross-correlation function vs. thermal noise curve becomes more gentle. At the same time, the break of the cross-correlation function vs. thermal noise curve becomes more gentle. However, it is essentially the same as result in Fig.2 without the distribution. In other words, the correlation between pin and FLL is strong in the glass state below T_g, resulting in the large degree of disorder in FLL. On the other hand, that the order seems recovered rapidly at the temperature is increased above T_g.

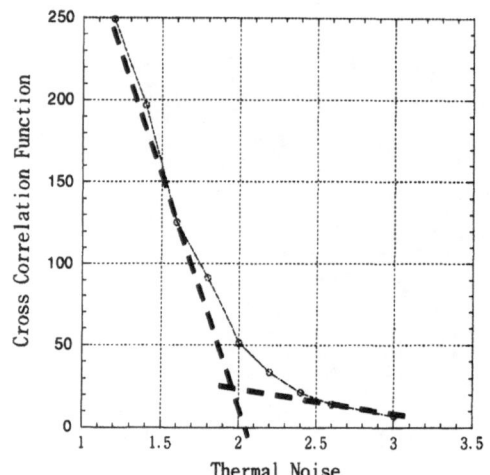

Fig. 3. Cross-correlation function for $k=0$ vs. thermal noise when the inhomogeneous distribution of flux pinning strength is introduced.

SUMMARY

In this paper, the correlation between pin and pinned FLL is investigated for the effect of thermal noise. The following results are obtained:

(1) The variation in correlation between pin and FLL with temperature changes discontinuously at a certain magnitude of thermal noise. This breaking temperature is considered as the transition temperature.
(2) It is shown that the thermal depinning is a phase transition of a second order.
(3) This behavior is essentially unchanged even if the strength of pins is distributed. In this case, the transition becomes more gradual due to the distributed pinning strength.

Acknowledgements

This study was partly funded by NEDO, Japan, under the Proposal-Based New Industry Creative Technology R&D Promotion Program and Grant-in-Aid of Scientific Research on Priority Area "Vortex Electronics".

REFERENCES

1. T. Matsushita and T. Kiss, Physica C **315**, 12 (1999).
2. K. Yamafuji and T. Kiss, Physica C **290**, 9 (1997).

Error Analysis in the Study of Transport Characteristics in HTS Using a Statistic Model

Shouichi Nishimura[1], Takanobu Kiss[1], Masakatsu Takeo[1], and Teruo Matsushita[1,2]

[1]Dept. Electrical and Electronic Systems Eng., Kyushu University, Fukuoka 812-8581, Japan
[2]Faculty of Computer Science and Systems Engineering, Kyushu Institute of Technology, 680-4, Kawazu, Iizuka 820-8502, Japan

Abstract: Based on a statistic model describing the critical current distribution, we proposed a method to describe the current-transport properties in HTS. The fitting error between the analytical and measured current-voltage characteristics was analyzed. It has been shown that the analytical form can be fitted with the measurement in wide conditions of magnetic field and temperature. These results indicate that the statistic model is suitable to analyze the transport properties in HTS. Using this method, we can estimate the homogeniety of the material as well as the pinning strength.
Key words: Current-transport properties, Critical current, Pinning, High T_c superconductor

INTRODUCTION

The rounded current-transport characteristics in HTS influence strongly the relevant properties such as critical current density, J_c, thermal stability and ac loss. Based on a statistic analysis of flux pinning, we proposed a method to characterize the transport properties in HTS [1,2]. The homogeniety of J_c and its thermodynamic properties can be estimated by using the statistic model. In this paper we carried out error analysis in the regression using the model. The fitting error between the model and the measurement will be discussed for the electric field (E) vs. current density (J) curves in various conditions of temperature and magnetic field.

E-J CURVES BASED ON THE STATISTIC MODEL

We have shown that the current-transport properties in a mixed state of HTS are goverened by the percolation process among unpinned clusters [1]. As the transport current is increased, the unpinned clusters are connected to each other. Then, the flux flow voltage will be induced when the unpinned cluster percolates. In an idealistic uniform medium where the pinning strength is constant, all pins will break at the same time at the percolation limit. In other words, the depinning probability function will be a step function. On the other hand, in a random pin medium where pinning sites and their strength distribute randomly, the pins will break with a certain distribution. The percolation limit corresponds to the minimum value of the critical current density, J_{cm}. It was shown that the depinning probability is increased in proportion to the power of ($J-J_{cm}$) in the vicinity of the percolation threshold [1]. If we assume that the elementary current-voltage characteristics can be defined with a certain J_c and a constant flux flow resistivity, i.e., that the dynamic resistance is constant when $J>J_c$, while the resistance is zero when $J \leq J_c$, the current-voltage characteristics in the random medium can be obtained by integrating the distribution function of J_c with respect to the transport current density. That is,

$$E(J) = \rho_{FF} \int_0^J [(J - J_{cm})/J_0]^m \, dJ , \tag{1}$$

where J_0 is the scale parameter indicating the half-width of the distribution, m is the numerical parameter which determines the shape of distribution and ρ_{FF} is the resistivity for uniform flux flow. The typical value of J_c in the distribution denoted by J_k is given by $J_{cm}+J_0$. Extracting the parameters J_{cm}, J_0 and m from the measured E-J curves, we can estimate the homogeniety of the material and the pinning strength [2,3].

EXTRACTION OF THE PARAMETERS AND FITTING ERROR ANALYSIS

Typical examples of E-J curves in a YBCO thin film bridge, 200nm thick, 100μm wide and 1mm long, are shown in Fig. 1. The magnetic field was applied perpendicular to the film surface. The dots indicate measurements while the solid lines are eq. (1). Note that the analytical form can be fitted quite well with the measurements even in wide range of electric field and current density, i.e., more than three decades of electric field and two decades of current density. Figure 2 indicates the averaged fitting error as a function of m, where the value of ρ_{FF} was assumed to be constant as 10μΩcm since the influence of ρ_{ff} is small in case of large m.

The parameters, J_{cm} and J_0, were determined by the nonlinear regression for each m-value based to minimize the residual error of electric field in logarithmic scale. The optimal value of m determined by the minimum error was around 5.4 at each temperature of the measurement in the range of 70 to 85K. J_{cm} and J_k extracted for the constant value of $m=5.4$ were shown in Fig. 3 (a) and 3(b), respectively. The magnetic field- and temperature-dependencies of J_{cm} and J_k are well described by the scaling of corresponding pinning force densities, i.e., the minimum strength $F_{pm} \equiv J_{cm}B$ and the typical strength $F_{pk} \equiv J_k B$. The solid lines in Fig. 3(a) and (b) are

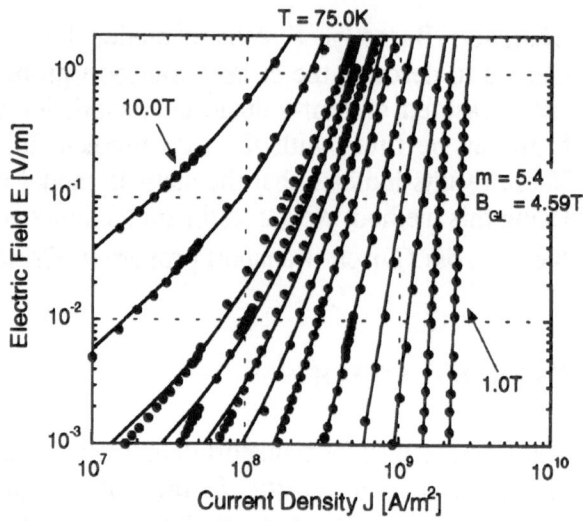

Fig. 1 Magnetic field dependent E-J curves in a c-axis oriented YBCO film measured at constant temperature of 75K. The dots are measured results while the solid lines are the analytical expression (1). The power index m was constant, 5.4. The value of ρ_{FF} was assumed to be constant as 10μΩcm. The other two parameters, J_{cm} and J_0, were determined by the nonlinear regression for each E-J curve. The thermodynamic properties of these two parameters can be attributed to the scaling of the minimum and typical values of volume pinning force densities as shown in Figs. 3(a) and 3(b).

$$F_{pm} \equiv J_{cm}B = AB_{GL}{}^{\varsigma}\left(B/B_{GL}\right)^{\gamma}\left(1-B/B_{GL}\right)^{\delta} \quad (2.a)$$

$$F_{pk} \equiv J_k B = AB_k{}^{\varsigma}\left(B/B_k\right)^{\gamma}\left(1-B/B_k\right)^{\delta} \quad (2.b)$$

with numerical parameters; (a) $A=6.53 \times 10^8$, $\varsigma = 1.7$, $\gamma = 0.82$, $\delta = 1.32$ and (b) $A=7.89 \times 10^8$, $\varsigma = 1.7$, $\gamma = 1.09$, $\delta = 1.88$. The critical fields, B_{GL} and B_k, were defined as the magnetic fields where J_{cm} and J_k were zero, respectively. Temperature dependencies of these critical fields were determined as $B_{GL}(T)=30[1-(T/T_c)^2]^{1.5}$ and $B_k(T)=75[1-(T/T_c)^2]^{1.5}$.

The mean square error (MSE) of electric field value in logarithmic scale was plotted in (B, T)-plane as shown in Fig. 4. It can be seen that the

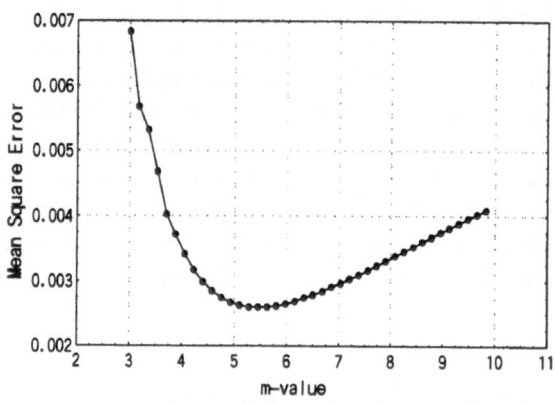

Fig. 2 Mean square error of electric field in log-scale as a function of m. The value was averaged for magnetic fields at a constant temperature of 75K.

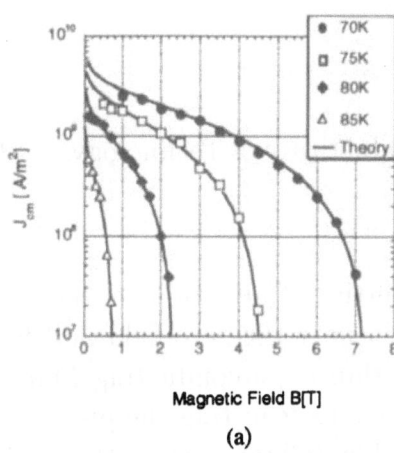

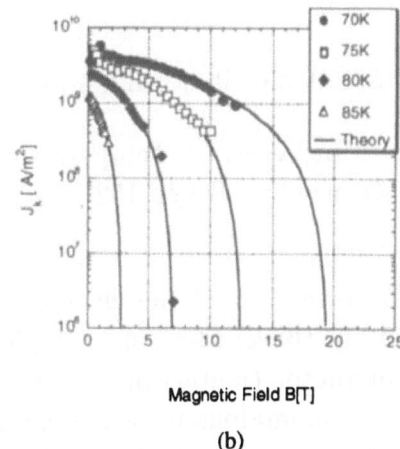

Magnetic Field B[T]

(a)

Magnetic Field B[T]

(b)

Fig. 3 Magnetic field- and temperature-dependencies of extracted J_c-related parameters (a) J_{cm}, minimum value, and (b) J_k, typical value. These can well be described by the scaling of the corresponding pinning force densities (solid lines) shown by eqs. (2.a) and (2.b), respectively.

target function can be fitted well with the measured curves in a wide magnetic field and temperature region. In most region the value of MSE was less than 4×10^{-3}.

CONCLUSION

The fitting error between the analytical form based on the statistic model and the measured $E\text{-}J$ curves was analyzed in a wide magnetic field and temperature region. The analytical form could be fitted with the measurement quite well even in wide ranges of electric field and current density. The value of MSE of electric field in logarithmic scale was less than 4×10^{-3} in the most measured region. The extracted J_c related parameters exhibit a clear scaling which is attributed to those of the pinning force densities. It was also shown that the statistic characteristic of the pin distribution was almost the same in the measured region between 70 to 85K. These results indicate that the present statistic model is reliable for the analysis of transport $E\text{-}J$ characteristics in HTS.

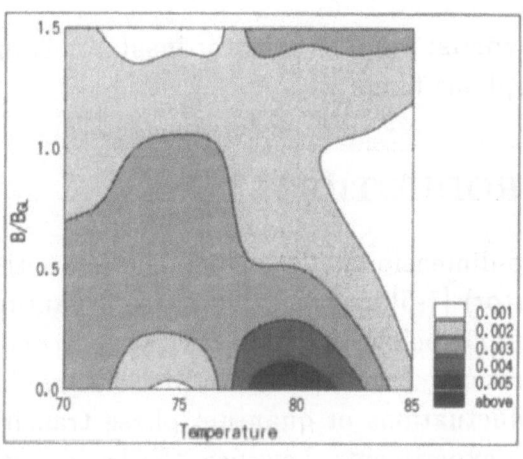

Fig. 4 Mean square error plotted in (B, T)-plane in the regression for $m=5.4$.

Acknowledgments.

This study was partly funded by NEDO, Japan, under the ProposalBased New Industry Creative Technology R&D Promotion Program and Grant-in-Aid of Scientific Research on Priority Area "Vortex Electronics".

REFERENCES

1. K. Yamafuji and T. Kiss, Physica C **290**, 9-22 (1997)
2. T. Kiss, T. Nakamura, K. Hasegawa, M. Inoue, H. Okamoto, K. Funaki, M. Takeo, K. Yamafuji and F. Irie *Proceedng of ICEC17*, eds D. Dew-Huges et al. (IOP, London, 1998), pp. 427-430
3. T. Kiss, T. Matsushita and F. Irie, Superconductor Science and Technology: in press

Magnetoresistance and Vortex States near the Superconductor-Insulator Transition in a-$\text{Mo}_x\text{Si}_{1-x}$ Films

Satoshi Okuma, Satoshi Shinozaki, and Makiko Morita

Research Center for Very Low Temperature System, Tokyo Institute of Technology, 2-12-1, Ohokayama, Meguro-ku, Tokyo 152-8551, Japan

Abstract: We have studied the electrical transport properties of ultrathin (4-nm thick) and thick (300-nm thick) films of amorphous (a-)$\text{Mo}_x\text{Si}_{1-x}$ near the zero-field and field-driven superconductor-(metal)-insulator transitions. For thin superconducting films we have observed an anomalous negative magnetoresistance (MR) suggesting the presence of the localized Cooper pairs at $T \to 0$ in perpendicular fields B higher than the critical field B_C, while for thick films the negative MR is not visible. These results indicate that the two dimensionality (2D) plays an important role in the appearance of the negative MR (or localized Copper pairs). We have also found from preliminary measurements that the negative MR is no longer visible as the field $B_\parallel$ is applied parallel to the film surface. This result may suggest the presence of mobile vortices as well as localized Cooper pairs in $B > B_C$, consistent with the view of the 2D quantum phase transition.

Keywords: Superconductor-insulator transition, Two dimensions, Bose-glass insulator, Amorphous films

INTRODUCTION

In two-dimensional (2D) superconductors the existence of the novel insulating phase (Bose insulator) [1-5] and metallic phase (quantum vortex liquid) [6-8] has been suggested just above and below the field-driven superconductor-insulator transition (SIT), respectively. Theoretically, these phases or phase transitions have been attributed to the 2D large quantum fluctuations or quantum phase transitions which occur at zero temperature (T=0) [9]. In experiments, however, the basic picture of the 2D SIT at T=0 has not been fully established. Part of reason is that the systematic studies at sufficiently low temperatures, using ideal 2D superconductors, have not been performed except for quite limited systems.

In this paper we have made systematic transport measurements near the zero-field and field-driven SIT's in a series of 4-nm thick films of amorphous (a-)$\text{Mo}_x\text{Si}_{1-x}$ at temperatures down to $\sim$0.04 K [3]. In particular, we focus on the magnetoresistance (MR) at low temperatures for superconducting films and discuss the possible vortex state in 2D just above the field-driven SIT. The thick (300 nm) film has been also measured to study the effects of dimensionality on the MR and the vortex state.

EXPERIMENTAL

The a-$\text{Mo}_x\text{Si}_{1-x}$ films used in this study were prepared by coevaporation of pure Mo and Si in UHV. To obtain a series of samples with continuously changing x at a fixed thickness, a gradient deposition technique was used [3,10]. Even for the ultrathin films with thickness of 4 nm, they were confirmed to be very uniform and viewed as nearly an ideal 2D system [3]. The resistance was measured by standard four-terminal dc and ac locking methods. The field was applied perpendicular (B) or, in some cases, parallel $(B_{\parallel})$ to the film surface. In parallel fields, the transport current was also parallel to the field direction.

RESULTS AND DISCUSSION

Figures 1(a) and 1(b) show the temperature dependence of the resistance $R(T)$ of thin(4 nm) and thick (300 nm) films, respectively. For the thin film an Arrhenius type of $R(T)$ is observed at finite B lower than the critical field B_C for the superconductor-insulator transition [11]. The activation energy $U(B)$ obtained from the slope of the straight line follows the functional form, $U(B) \propto \ln(B_0/B)$, predicted by the 2D dislocation model [12]. In contrast, for the thick film $R(T)$ clearly deviates from the activated behavior. When we attempt to extract $U(B)$ using the low-$R(T)$ data in a limited range of T, the above relation apparently breaks down. We find that $R(T)$ well obeys the power-law functional form $R(T) \sim (T - T_g)^{\nu(z-1)}$ with exponents of reasonable magnitude predicted by the 3D vortex-glass theory. These results indicate that the vortex-solid states for these films are vortex glass. We note here that $T_g = 0$ for 2D (thin films).

Shown in Figs.2(a) and 2(b) is the MR in $B > B_C$ for thin films with ρ_n=3.1 and 7.2 $\mu\Omega$ and the thick film with ρ_n=7.8 $\mu\Omega$, respectively. Here, ρ_n is the normal-state resistivity at 10 K. At $T \geq 0.3$ K the MR is always monotonic and positive for these films. For thin films, however, an anomalous peak and a subsequent decrease of MR is commonly observed below 0.1 K irrespective of disorder ρ_n. This result, together with the finding that the MR is always positive for thin insulating films [3], suggests the presence of the localized Cooper pairs even on the insulating side of the field-driven SIT. Since the negative MR is only visible at very low temperature, its origin is likely to be attributed to quantum fluctuations.

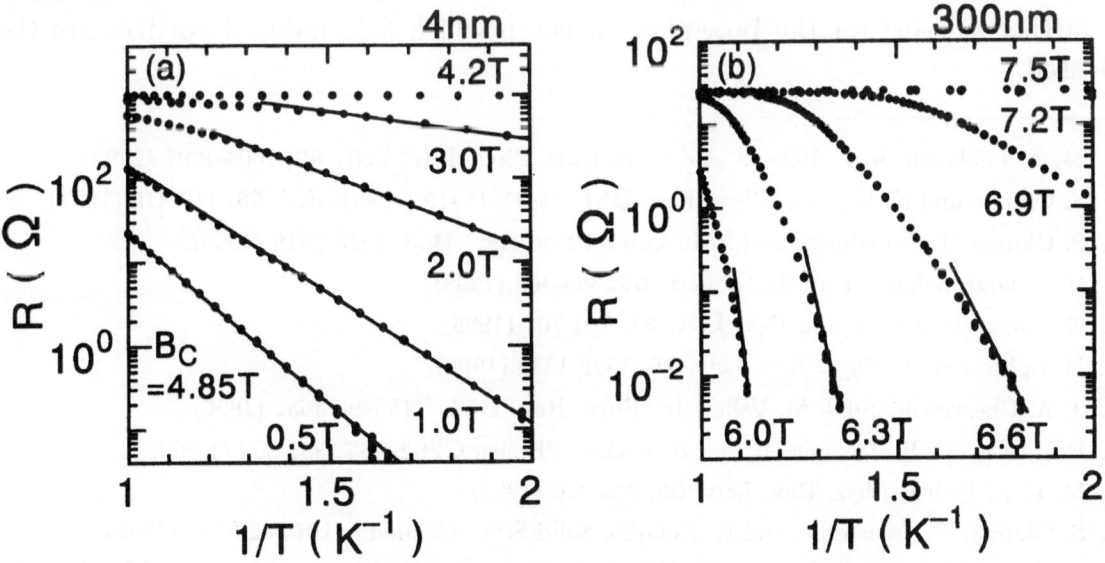

Fig.1 Arrhenius plots of the resistance $R(T)$ for the (a) thin (4-nm thick) and (b) thick (300-nm thick) films in various perpendicular fields.

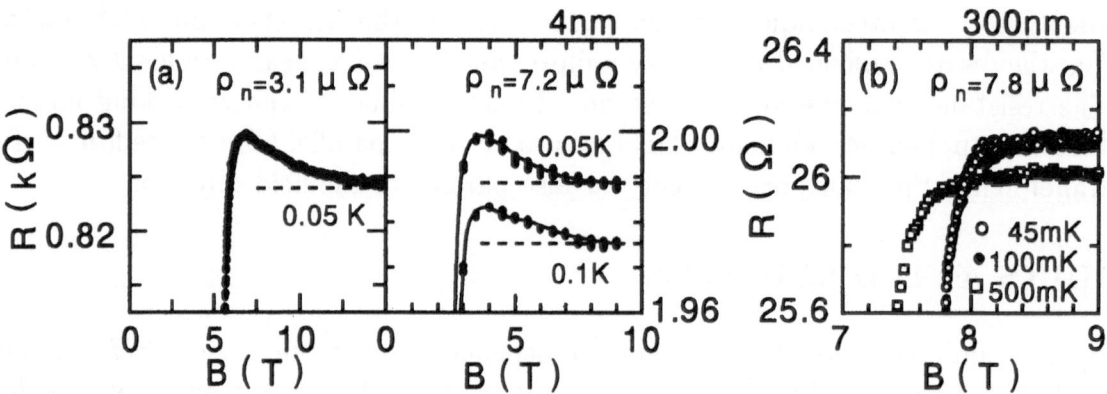

Fig.2 The MR at low temperatures for (a) the thin films with ρ_n=3.1 and 7.2 $\mu\Omega$ and (b) the thick film with ρ_n=7.8 $\mu\Omega$.

For the thick film the MR is always positive and the anomalous negative MR is not observed down to the lowest temperature ($\sim$0.04 K) measured. It is noted that the degree of disorder ρ_n=7.8 $\mu\Omega$ in this thick film is close to that ρ_n=7.2 $\mu\Omega$ in the thin film. These results indicate that the two-dimensionality plays an important role in the appearance of the negative MR (or localized Cooper pairs) at $B > B_C$.

All of these results are consistent with the dirty-boson model of Fisher, in which the quantum phase transition from the 2D vortex-glass to Bose-glass phase takes place with increasing B [9]. Within this model the field region where the negative MR is observed corresponds to the Bose-glass insulator. To prove this prediction more convincingly, detection of the quantum-mechanically-melted vortices in the Bose-glass phase is important. Thus we are conducting the MR measurements in parallel fields, in which the contribution from the motion of the field-induced vortices is negligible. Preliminary results show that the negative MR observed for the thin film in perpendicular fields is no longer visible when the field is applied parallel to the film surface. If this result is verified commonly for other thin superconducting films with different disorder, we may conclude that mobile vortices are present in the insulating region where the negative MR is observed. This will be a further support for the Bose-glass phase in which field-induced vortices are (Bose) condensed.

1. M. A. Paalanen, A. F. Hebard, and R. R. Ruel, Phys. Rev. Lett. **69**, 1604-1607 (1992).

2. S. Okuma and N. Kokubo, Phys. Rev. **B51**, 15415-15419 (1995); *ibid.* **56**, 410-415 (1997)

3. S. Okuma, T. Terashima, and N. Kokubo, Phys. Rev. **B58**, 2816-2819 (1998).

4. V. F. Gantmakher *et al.*, JETP Lett. **68**, 363-369 (1998).

5. N. Marković *et al.*, Phys. Rev. Lett. **81**, 701-704 (1998).

6. D. Ephron *et al.*, Phys. Rev. Lett. **76**, 1529-1532 (1996).

7. J. A. Chervenak and J. M. Valles, Jr., Phys. Rev. **B54**, R15649-15652 (1996).

8. P. H. Kes, M. H. Theunissen, and B. Becker, Physica **C282-287**, 331-334 (1997).

9. M. P. A. Fisher, Phys. Rev. Lett. **65**, 923-926 (1990).

10. S. Okuma, T. Terashima, and N. Kokubo, Solid State Commun. **106**, 529-533 (1998)

11. S. Okuma, S. Shinozaki, and N. Kokubo, in *Advances in Superconductivity XI*, edited by N. Koshizuka and S. Tajima (Springer-Verlag, Tokyo, 1999), pp.185-188.

12. M. V. Feigel'man, V. B. Geshkenbein, and A. I. Larkin, Physica **C167**, 177-187 (1990).

In-plane Field Contribution for Josephson Plasma Mode in Under-doped $Bi_2Sr_2CaCu_2O_{8+\delta}$

Itsuhiro Kakeya[1,2], Tomoyuki Wada[1], Ryo Nakamura[1], and Kazuo Kadowaki[1,2]

[1]Institute of Materials Science, University of Tsukuba, Tsukuba, Ibaraki, 305-8573, Japan
[2]CREST, Japan Science and Technology Cooperation (JST)

Abstract: Josephson plasma resonance (JPR) experiments under magnetic fields parallel to CuO_2 layers have been performed in $Bi_2Sr_2CaCu_2O_{8+\delta}$ single crystals with doping range being from optimal to under-doped side. The results are found to be quite unique and cannot be accounted for the conventional notion of JPR for $H \mathbin{/\!/} c$, the systematic measurements as functions of the doping, applied microwave frequency, and magnetic field suggest that the new JPR mode can be excited with strong coupling to the Josephson vortex lattice.

Keywords: Josephson plasma resonance, parallel magnetic field, Josephson vortex

INTRODUCTION

Josephson plasma resonance (JPR) has been studied for several years as a sensitive probe for vortex states[1], and has been revealed new phenomena concerning interlayer coupling of pancake vortices[2]. Up to now, the features of JPR in vortex liquid state in perpendicular field ($H \mathbin{/\!/} c$) are almost comprehended. However, JPR in parallel field ($H \mathbin{/\!/} ab$) is quite mysterious. In this orientation, Josephson vortices parallel to the ab plane are considered to be created inside the insulating block layers, and their oscillation mode can take part in the JPR resonance through strong coupling between them because the mass seems to be much smaller than the one of pancake vortices[3]. Thus intriguing plasma modes coupled with the vortex oscillations are expected to be observed. Although there are a few experimental reports[4] in the vicinity of $H \mathbin{/\!/} ab$ and the results are interpreted based on theoretical suggestions[5], neither consistency nor adequacy has been obtained.

In this report, we present experimental results of JPR in perpendicular field at which in-plane field contribution is dominant as functions of temperature, microwave frequency, and doping in under-doped region. Behaviors of the observed resonance are quite unique and cannot be explained by the conventional theories of JPR for $H \mathbin{/\!/} c$. This strongly suggests that new theory involving Josephson vortices explicitly is required for JPR in $H \mathbin{/\!/} ab$.

EXPERIMENTAL DETAILS

We have used three under-doped $Bi_2Sr_2CaCu_2O_{8+\delta}$ single crystals U1, U2, and U3, which were grown by the modified Traveling Solvent Floating Zone (TSFZ) method and annealed in a vacuum in order to change the doping level. An optimally-doped crystal OP was also used for comparison. Their T_c's obtained by magnetization measurements with a SQUID magnetometer are 70, 72, 78, and 92 K for U1, U2, U3, and OP, respectively. JPR measurements were performed at four microwave frequencies of 20.0, 25.5, 34.3, and 44.2 GHz by a reflective type of the bridge balance circuit with TE_{102} mode rectangular cavity resonators. Samples were placed inside the cavity so that the microwave electric field was parallel to the c axis, in which configuration only the longitudinal Josephson plasma mode is excited[6]. The resonance curves were obtained either by sweeping magnetic field at a fixed temperature (FS) or by sweeping temperature at a fixed external field (TS). The external field was applied by a split-pair of superconducting magnet and the direction was varied with rotating the microwave setup including the cavity with respect to the magnet by a high-precision goniometer. The direction of the ab plane was determined by looking at the symmetry of angular dependence of the JPR lines.

RESULTS AND DISCUSSION

In sample U1, two kinds of resonance lines are observed at 25.5 GHz by TS experiment as shown in Fig. 1(a). One is observed in a low field region with sharp and symmetric line-shape. This resonance shifts to lower temperatures with wider line-width as the external field increases. The resonance temperature, however, tends to increase above 1000 Oe, and the resonance finally disappears at 1500 Oe. The other one begins to appear at 800 Oe with a broad line from lower temperature. This line also shows non-monotonic field dependence: as the field increases, resonance temperature once increases then decreases above 1300 Oe. We named these lines as "higher temperature branch (HTB)" and "lower temperature branch (LTB)" for the former and the latter, respectively. Figure 1(b) presents the resonance curves obtained by TS experiment at 44.2 GHz. Only one resonance line is observed. This resonance is identified to be the same branch as HTB obtained at 25.5 GHz. We plotted all the resonance peaks, HTB and LTB, in Fig. 1(c). Both of branches shift to lower temperature as the microwave frequency increases and the re-entrant behaviors for HTB and LTB are observed at all frequencies. Similar results were obtained in sample U2 (T_c = 73 K).

It is obvious from Fig1(c) that the gap region between HTB and LTB observed at 20.0 GHz is smaller than one at 25.5 GHz. This means that the splitting is more enhanced at higher frequencies, so that LTB is pushed out from the experimental temperature range as observed at 44.2 GHz. On the other hand,

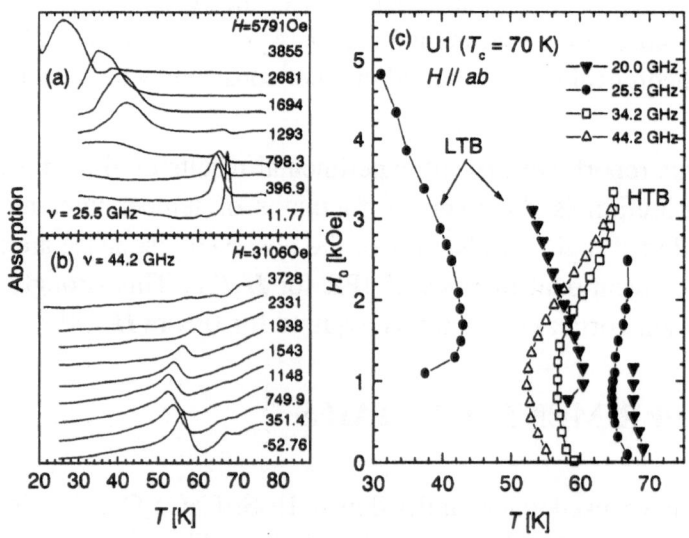

Fig. 1 Resonance curves at 25.5 (a) and 44.2 GHz (b). Plots of resonance peak for all measured frequencies are shown in (c)

LTB comes closer to HTB at lower frequency and they are expected to merge to a single resonance with and monotonous temperature dependence at frequency much lower than 20 GHz. In optimally-doped crystals, such monotonous temperature dependence in $H \parallel ab$ is actually observed even at 34 GHz[7]. These are interpreted that the feature of the resonance mode drastically depends on a relation between the incidental microwave frequency ω_0 and sample anisotropy from which zero-field plasma frequency ω_p can be extracted. Therefore, we can interpret that two split resonance branches with the re-entrant behavior are observed for larger ω_0/ω_p, whereas the monotonous temperature dependence is observed for smaller ω_0/ω_p. In fact, split resonance modes are observed at 44.2 GHz even in the optimally doped crystal as well as one at 20.0 GHz in sample U1, as shown in Fig. 2. This result strongly supports the above interpretation.

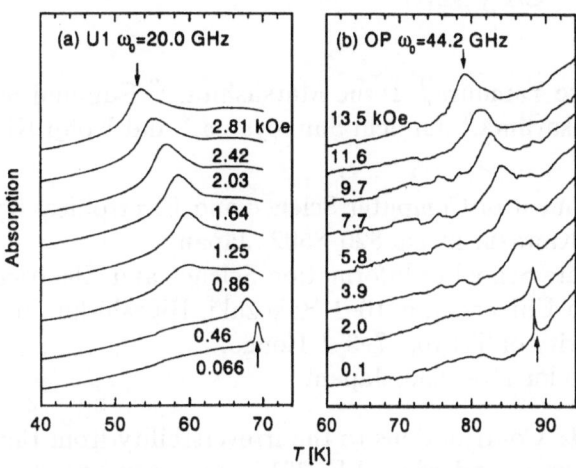

Fig. 2 Resonance curves at 20 GHz in under-doped crystal U1 (a) and at 44 GHz in optimally-doped crystal OP (b). LTB and HTB resonances, which are pointed by down and up arrows respectively, show the similar behavior qualitatively in both panels.

SUMMARY

We have observed two split Josephson plasma mode in magnetic field orientation being parallel to the ab plane. These branches shift to higher temperature and become closer at lower frequencies. This frequency dependence also depends on anisotropy of the crystal, which determines inherent plasma frequency. From these results, the ratio of incident microwave frequency and the plasma frequency is an important parameter for JPR in parallel in-plane field.

1. L. N. Bulaevskii, M. P. Maley, and M. Tachiki Phys. Rev. Lett. **74**, 801-804; Y. Matsuda, *et al.*, Phys. Rev. Lett. **75**, 4512-4515 (1995); A. E. Koshelev Phys. Rev. Lett. **77**, 3901-3904 (1996).

2. T. Shibauchi, *et al.*, Phys. Rev. Lett. **83**, 1010-1013 (1999); I. Kakeya, R. Nakamura, and K. Kadowaki, to be published in Physica B

3. L. N. Bulaevskii, M. Ledvij, and V. G. Kogan, Phys. Phys. Rev. B **46**, 366-380; J. R. Clem and M. W. Coffey, Phys. Rev. B **42**, 6209-6216.

4. O. K. C. Tsui, *et al.*, Phys. Rev. B, 2948-2951 (1997); Y. Matsuda, *et al.*, Phys. Rev. B **55**, 8685-8688 (1997).

5. L. N. Bulaevskii, *et al.*, Phys Rev B **55**, 8482-8489 (1997).

6. I. Kakeya *et al.*, Phys Rev B **57**, 3110-3115 (1998).

7. I. Kakeya, R. Nakamura, and K. Kadowaki, *Advances in Superconductivity XI*, 609-612 (1999).

Distribution of Shielding Currents in Underdoped Bi-2212 Single Crystal

Syunsuke Yamaura,[1] Teruo Matsushita,[1,2] Edmund Soji Otabe,[1] Yuri Nakayama,[3] Jun-ichi Shimoyama,[3] and Kohji Kishio[3]

[1]Department of Computer Science and Electronics, Kyushu Institute of Technology, 680–4 Kawazu, Iizuka 820–8502, Japan
[2]Graduate School of Information Science and Electrical Engineering, Kyushu University, 6–10–1 Hakozaki, Higashi-ku, Fukuoka 812–8581, Japan
[3]University of Tokyo, 7–3–1 Hongo, Bunkyo-ku 113–8656, Japan

Abstract: Contributions to the irreversibility from the bulk and surface pinning are investigated in detail for an underdoped Bi-2212 single crystal using the Campbell method. The measured flux profile showed a linear flux distribution suggesting a uniform bulk shielding current without surface pinning, as the Bean model predicts, in the field-temperature region far below the irreversibility line. Getting close to the irreversibility line, the AC penetration depth became longer and might exceed the specimen size, because the critical current density approached zero. Hence, the obtained profile was different from exact one. However, the theoretical analysis using the theoretical model of Campbell shows that there is no contribution from the surface irreversibility in this specimen.

Keywords: Underdoped Bi-2212, Campbell method, Surface pinning

INTRODUCTION

Recently two of the authors [1] found that a first-order transition line existed below the irreversibility line for underdoped Bi-2212 single crystals. This seemed to be consistent with the theoretical prediction [2]. On the other hand, the contribution from the surface irreversibility has been discussed in detail [3]. If the irreversibility line is attributed to the surface pinning in underdoped Bi-2212, the phase diagram theoretically predicted in [2] must be corrected for the bulk flux line system. Thus, the distribution of the shielding current is investigated using the Campbell method [4] for such a Bi-2212 specimen to clarify if the irreversibility is governed by bulk pinning or surface pinning. Since the observed flux profile is significantly influenced by the reversible motion of flux lines in the vicinity of the irreversibility line, the theoretical analysis was done using a model of Campbell [5] in which such a motion is taken into account.

EXPERIMENTS

A specimen was cut from an underdoped Bi-2212 single crystal prepared by a floating zone method and its dimension was 4.83 mm long, 1.65 mm wide and 0.062 mm thick. The c-axis was directed normal to the flat wide surface and the critical temperature was 85.0 K. The magnetic flux distribution was measured using the Campbell method. A DC magnetic field was applied along the c-axis and a small AC magnetic field was applied parallel to the length of the specimen. The frequency of the AC field was 35.0 Hz. Before starting the measurement, the AC field of a sufficient amplitude was applied to the specimen. This made the current to shield only the AC field, resulting in a complete penetration of the DC field. Then, the AC field amplitude was gradually reduced to zero to eliminate the shielding current. After that the penetration of AC field from the flat surface was measured.

The Campbell method gives a serious overestimation of the critical current density, J_c, in the case where J_c is so small that the AC penetration depth, λ_0', becomes comparable to or larger than the specimen size [6]. Hence, the DC magnetization measurement was also performed using a SQUID magnetometer. The critical current density was estimated from the magnetization hysteresis and the irreversibility field was determined by the field at which J_c is reduced to 1.0×10^6 A/m^2.

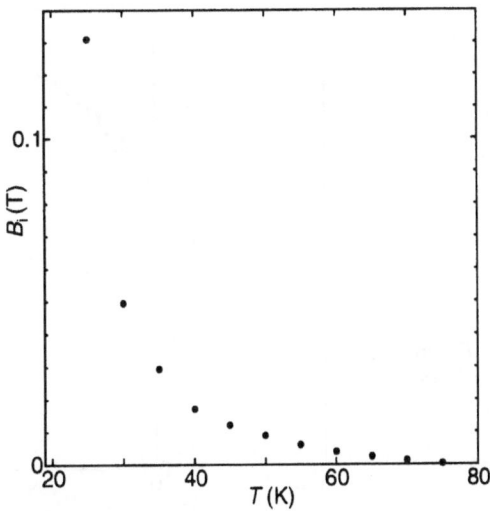

Fig. 1. Irreversibility line of underdoped Bi-2212 single crystal.

RESULTS AND DISCUSSION

Figure 1 shows the irreversibility field of the specimen. It increases sharply with decreasing temperature below 30 K. The flux profile at 25 K and 20 mT is shown in Fig. 2. The penetration depth,λ', starts at a finite value (λ'_0), increases linearly, and is saturated at the center (about 31 μm) as increasing AC field amplitude. The DC field, 20 mT, is lower enough than the irreversibility field, 131 mT, at this temperature. This agrees with the prediction of the irreversible Bean model and shows that there is no contribution from the surface. That is, if there is a finite contribution from the surface, the extrapolation of the linear part of the flux profile to zero penetration depth gives a positive AC field amplitude [7]. This gives a half of the surface irreversibility in a full critical state.

The flux profile at 30 K and 20 mT is shown in Fig. 3. This condition is fairly close to the irreversibility field of 49.5 mT at this temperature. This result seems to show that the surface irreversibility does not exist in this condition as well. However, the observed value of λ'_0 is of the same order of magnitude with a half of the specimen thickness, $t/2$. Since λ'_0 is approximately given by

$$\lambda'_0 = \left(\frac{a_f B}{2\pi \mu_0 J_c} \right)^{1/2} \tag{1}$$

with a_f denoting a flux line spacing, such a large λ'_0 is ascribed to a very low J_c. Since the practical penetration depth is limited by $t/2$, if λ'_0 estimated from Eq. (1) using the observed J_c is smaller than $t/2$, the observed flux profile is approximately correct. In this case, λ'_0 estimated from Eq. (1) is 9.6 μm and the above condition is fulfilled. Strictly speaking, however, the observed flux profile does not give a valid clarification that the surface irreversibility does not exist in the specimen, since the observed flux profile is influenced by the reversible flux motion.

The flux profile can be simulated using the Campbell model [5] even under a remarkable reversible flux motion. The details of the analysis are given in [6]. The solid line in Fig. 2 is a theoretical result of the analysis at $T = 25$ K and $B = 20$ mT with $J_c = 4.6 \times 10^7$ A/m^2 observed using a SQUID. In the above the surface irreversibility is assumed not to exist. The agreement between the theory and the experiment is satisfactory. The solid line in Fig. 3 shows the theoretical result at the 30 K and 20 mT with an assumption of $J_c = 1.6 \times 10^7$ A/m^2, where the J_c value estimated from a SQUID magnetometer is 9.3×10^6 A/m^2. The reason for the difference in J_c-values is not clear. However, the theoretical result explains the observed result well. Hence, it can be speculated that the surface irreversibility does not exist even in this case. Since the flux pinning is of a nonlocal

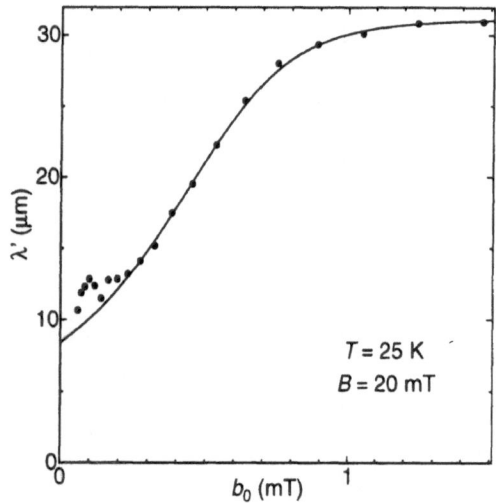

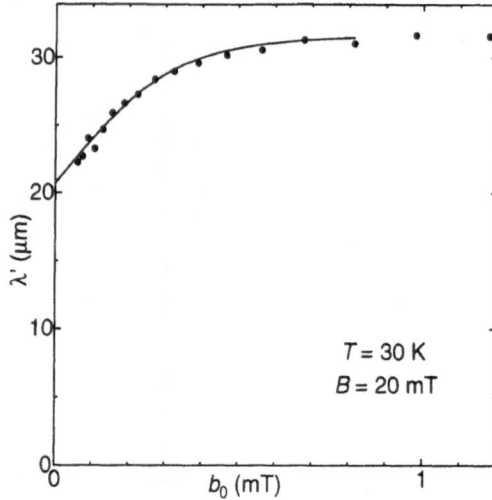

Fig. 2. Flux profile at 25 K and 20 mT. The solid line is a theoretical result of the Campbell model.

Fig. 3. Flux profile at 30 K and 20 mT. The solid line is a theoretical result of the Campbell model.

property so that all the quantities are averaged within the scale of correlation length, λ_0'. Hence, if the surface pinning is not observed in the temperature-field region where λ_0' is much shorter than $t/2$, it proves that the surface pinning does not exist in the vicinity of irreversibility line where λ_0' is very long.

The present results show that the irreversibility comes from the bulk pinning in the underdoped Bi-2212 single crystal. This suggests that the irreversibility line observed by Kishio *et al.* [1] for a similar specimen also originates from the bulk pinning, indicating that the prediction in [2] is valid.

CONCLUSION

Distribution of the shielding current in an underdoped Bi-2212 single crystal was measured using Campbell method and the flux profile was analysed using the Campbell model. The following results are obtained:

1. Any sign of strong surface pinning was not obtained in wide ranges of temperature and magnetic field. Hence, all the irreversibility properties including the irreversibility line originate from the bulk pinning.

2. This result suggests the possibility that the first order transition line appears below the vortex glass-liquid transition line for extremely two-dimensional superconductors.

REFERENCES

1. K. Kishio, J. Shimoyama, S. Watauchi and H. Ikuta, *Proc. of 8th Int. Workshop on Critical Currents in Superconductors* (World Scientific, Singapore), pp. 35–38 1998.
2. T. Matsushita and T. Kiss, Physica C **315**, 12–22, 1999.
3. See for example, D. T. Fuchs, E. Zeldov, T. Tamegai, S. Ooi, M. Rappaport and H. Shtrikman, Phys. Rev. Lett., **80**, 4971–4974, 1998, and references therein.
4. A. M. Campbell, J. Phys. C, **2**, 1492–1501, 1969.
5. A. M. Campbell, J. Phys. C, **4**, 3186–3198, 1971.
6. N. Ohtani, E. S. Otabe, T. Matsushita and B. Ni, Jpn. J. Appl. Phys., **31**, L169–171, 1992.
7. T. Matsushita, T. Honda, Y. Hasegawa and Y. Monju, J. Appl. Phys., **54**, 6526–6532, 1983.

Current-Voltage Curves of a Bi-2212/Ag Multilayered Tape Prepared by PAIR Process

Akihito Yamasaki,[1] Yuri Ueno,[1] Masaru Kiuchi,[2] Teruo Matsushita,[1,2]
Hanping Miao,[3] Hiroaki Kumakura,[3] and Hiroshi Kitaguchi[3]

[1]Department of Computer Science and Electronics, Kyushu Institute of Technology,
  680–4 Kawazu, Iizuka 820–8502, Japan
[2]Graduate School of Information Science and Electrical Engineering,
  Kyushu University, 6–10–1 Hakozaki, Higashi-ku, Fukuoka 812–8581, Japan
[3]National Research Institute for Metals , 1–2–1 Sengen,
  Tsukuba 305–0047, Japan

Abstract: The current-voltage curves of a Bi-2212 tape prepared by PAIR process was measured in a magnetic field parallel to the c-axis and their scaling was examined. It was found that the two critical indices and the transition temperatures for the scaling were approximately the same as those in a previously measured conventional tape specimen, although the observed critical current density was very much improved. The observed results are compared with the theoretical analysis using the flux-flow model.

Keywords: PAIR process, distribution of flux pinning strength

INTRODUCTION

Bi-based superconductors have a higher growth rate along the a-b plane than in the direction of the c-axis during a reaction and are suitable materials for long superconducting tapes. It is also known that superconducting grains near an interface with silver are almost perfectly aligned and the critical current density near the interface is larger than that in the superconducting region far from the interface. This indicates that the flux pinning strength is widely distributed in the tape. In addition, the current-voltage characteristic, one of the most important properties for application of superconductors, is significantly affected by this inhomogeneity. That is, the n-value, which is defined by expressing the relation between the electric field and the current density as $E \propto J^n$, is well known to depend on the distribution of the flux pinning strength.

It was recently reported that the critical current density in Bi-2212/Ag multilayer tape was improved almost two times higher at low temperatures by adopting the Pre-Annealing and Intermediate Rolling(PAIR) process [1]. It was considered that the grain alignment in the whole region of the tape was improved by PAIR process. This suggests that the distribution width of flux pinning strength becomes narrow in PAIR processed tapes. In this article, the current-voltage curves were measured for a multilayed Bi-2212 tape prepared by PAIR process and their scaling was examined.

EXPERIMENTS

The specimen used in this study was a silver-sheathed Bi-2212 superconducting tape prepared by the PAIR process and contained two superconducting layers. The length, width and thickness of the superconducting layer were about 50 mm, 3.5 mm and 10 μm, respectively. The c-axis was approximately directed normal to the flat surface of the specimen. The critical temperature, T_c,

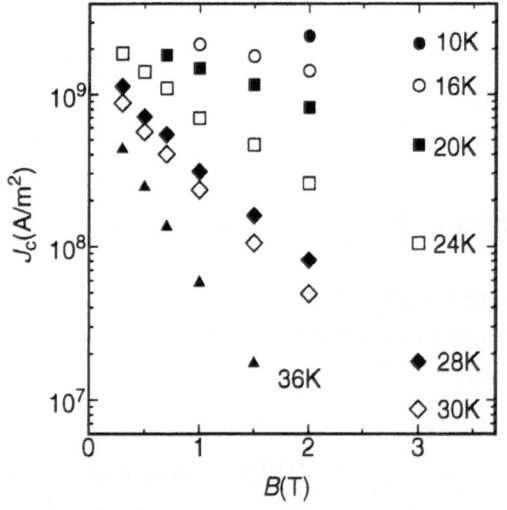

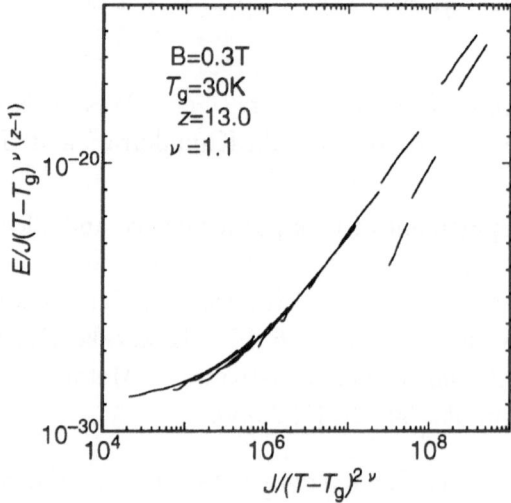

Fig. 1. Magnetic field dependence of critical current density in the temperature range 10 - 36K.

Fig. 2. Scaled current-voltage curves at $B = 0.3$ T.

measured by SQUID was 79.0 K. The current-voltage curves were measured by the four probe method under the magnetic field parallel to the c-axis. A pulsed transport current with a period of 1 s was applied to the specimen to reduce the Joule heat at current leads. The voltage was measured across the voltage terminals separated by 1.0 cm. The critical current density was determined by the electric field criterion of $E = 1.0 \times 10^{-4}$ V/m. In silver-sheathed tapes, the current-voltage curves only of the superconducting Bi-2212 layer cannot be estimated because of the current sharing to the silver sheath. Here, we approximated the silver-sheath tape by a parallel circuit of the silver-sheath and the superconducting layers. After the measurement of the characteristics of the silver-sheath tape, the superconducting layer was broken and the resistivity of the silver-sheath was measured, and then, the current-voltage characteristic of the superconducting layer was analytically estimated.

RESULTS AND DISCUSSION

Figure 1 shows the observed critical current densities of the PAIR processed tape. It amounts to 1.49×10^9 A/m² at 20K and at 1T. This value is more than ten times larger than that in a Bi-2212 tape prepared by a conventional process [2]. The scaling of the measured current-voltage curves is examined in the same manner as in the vortex glass-liquid transition model, and the dynamic and static critical indices, z and ν, are analytically estimated. Figure 2 shows scaled current-voltage curves at $B = 0.3$ T. The two critical indices are shown in Figs. 3 and 4. It is seen that z increases with increasing magnetic field, while ν decreases. These values are almost the same as those obtained for the specimen prepared by the conventional method [2]. In addition, the G-L transition line is also approximately the same between them. This indicates that the distribution width of flux pinning strength is not appreciably improved in the PAIR processed tape, although the local critical current density itself is increased.

These observed current-voltage characteristics are compared with the theoretical analysis using the flux-flow model in which the distribution of flux pinning strength was taken into account [3]. It is found that the agreement is good between the theoretical and experimental results at low

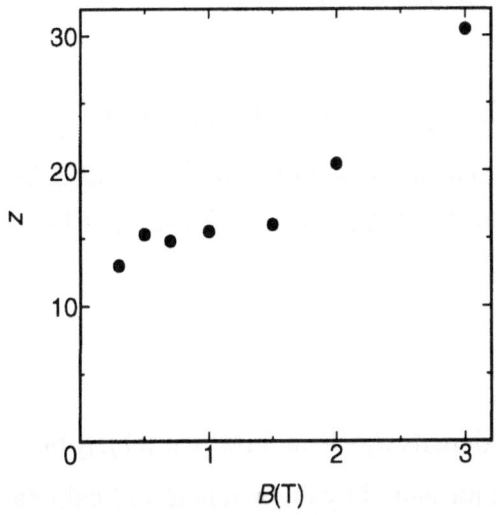

Fig. 3. Dynamic critical index of PAIR processed tape.

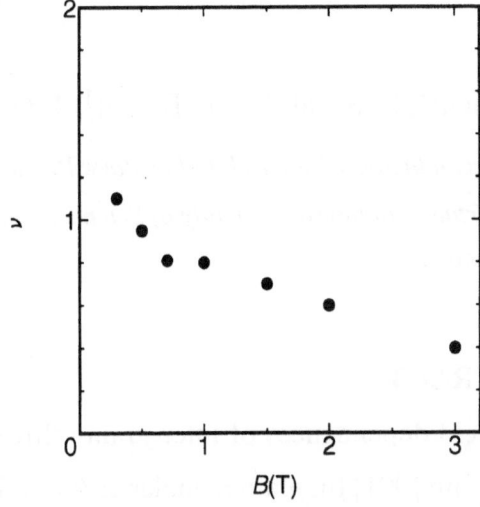

Fig. 4. Static critical index of PAIR processed tape.

temperatures. However, the current-voltage characteristics at high temperatures are significantly different between the theoretical and experimental results. That is, the observed effective flux flow resistivity was about three orders of magnitude lower than the known value. This suggests that the current sharing occurred more significantly at high temperatures than assumed in the parallel circuit model. This might be the reason for the observed relatively low z-value in spite of high critical current density in the PAIR processed tape specimen.

CONCLUSION

The current-voltage curves of a Bi-2212 tape prepared by PAIR process was measured in a magnetic field parallel to the c-axis and their scaling was examined. The observed current-voltage characteristics were compared with the theoretical analysis using the flux-flow model in which the distribution of flux pinning strength was taken into account. The following results are obtained:

1. The critical current density of a Bi-2212 tape prepared by a PAIR process is more than ten times larger than that in a Bi-2212 tape prepared by a conventional process.

2. However, the two critical indices and the G-L transition temperatures in the present PAIR processed specimen are approximately the same as those in the conventional tape specimen. This suggests that the wide distribution of the local critical current density was not improved even by the PAIR process.

3. The observed effective flow resistance was much smaller than the known value at high temperatures, suggesting that the current sharing might occur significantly at these temperatures.

REFERENCES

1. H. Miao, H. Kitaguchi, K. Togano and T. Hasegawa Physica C **301**, 116–122 (1998).
2. K. Noguchi, M. Kuchi, M. Tagomori, T. Matsushita and T. Hasegawa, *in Adv Supercond IX* (Springer-Verlag, Tokyo, 1996), pp. 625–628.
3. T. Matsushita, T. Todoh and N. Ihara, Physica C **259**, 321–325 (1996).

MAGNETO-OPTICAL OBSERVATION OF FLUX DENSITY DISTRIBUTION IN YBa$_2$Cu$_3$O$_{7-\delta}$ BICRYSTAL FILMS

T. Machi[1], S. Sasaki[1,2], T. Takagi[1], J. G. Wen[1], K. Kajita[2] and N. Koshizuka[1]

[1]*Superconductivity Research Laboratory, ISTEC, 1-10-13, Shinonome, Koto-ku, Tokyo, 135-0062 Japan*
[2]*Department of physics, Faculty of Science, Toho University, 2-2-1, Miyama, Funabashi-shi, Chiba, 274-8510 Japan*

ABSTRACT

The field dependences of inter-grain critical current density J_c^{GB} in YBa$_2$Cu$_3$O$_{7-\delta}$ bicrystal films with [001] tilt grain boundaries (GBs) have been measured by the magneto-optical imaging technique. The results on the intra-grain critical current density J_c^{IG} were consistent with those obtained by the transport measurements. It was found that a bicrystal film with 4° tilt angle has a peak around 500 Oe in J_c^{GB}/J_c^{IG}-H curve, while the higher tilt angle films such as 24° have no peak in the curves. It seems possible to explain the peak in J_c^{GB}/J_c^{IG}-H curve by the matching field effect caused by the vortex pinning on the dislocation core array along the GB observed by TEM.

KEYWORDS : YBa$_2$Cu$_3$O$_{7-\delta}$, bicrystal, film, magneto-optical imaging, J_c

INTRODUCTION

It is a crucial issue to understand the transport mechanism of grain boundaries(GBs) in high-T_c superconductors when we consider their applications to wires and tapes. It is necessary to study the tilt angle dependence of the inter-grain J_c (J_c^{GB}) in the films with well characterized GBs. Dimos *et al.*[1] firstly reported the misorientation angle dependence of J_c^{GB} in bicrystal thin films of YBa$_2$Cu$_3$O$_{7-\delta}$ (hereafter Y123) with (001) tilt geometry prepared by pulsed-laser deposition (PLD) technique. The magneto-optical imaging on this kind of samples[2] supported their transport results on the misorientation angle dependence[1]. However, TEM observation of such samples revealed that the grain boundaries have meandering microfacet structures[3]. Therefore, it has been desired to obtain bicrystal films with no meandering GBs. Recently, Takagi *et al.*[4] succeeded in growing bicrystal films of Y123 with various (001) tilt angles by liquid phase epitaxy (LPE) method. Wen *et al.*[5] showed by TEM observation that these films have a large single facet structure. In present paper, we report the results of the flux penetration behaviors and the values of the intra-grain J_c^{IG} and inter-grain J_c^{GB} for these "ideal" bicrystals obtained by using the magneto-optical imaging technique.

EXPERIMENTAL

Y123 bicrystal films (tilt angle = 4°, 12°and 24° ; thickness =2.5μm, 1.9μm and 2.5μm, respectively) were prepared by LPE method on MgO bicrystal substrates. The details of film preparation are described elsewhere[4]. An iron garnet film with in-plane magnetization (the saturation field = 800 Oe) was used as a magneto-optical image detector. All measurements were carried out under zero field cooled condition at 7K. The brightness of magneto-optical image was converted into flux density by using the flux density dependence of brightness at the outside position sufficiently far from the sample.

RESULTS AND DISCUSSION

Figure 1 shows a typical change of magneto-optical images for 4°-bicrystal film with increasing external field. In order to calculate the intra-grain J_c^{IG}, because the thickness was not sufficiently larger than the penetration depth, we used Zeldov's formula[6] instead of the simple Bean model. Figure 2 shows a magneto-optical image for the film at 400Oe at 7K and the line profiles of flux distribution at various fields from the one side to the other. In order to evaluate the maximum J_c^{IG} value of this material, we chose a plot line X-Y that is the sharpest line profile in this area. The obtained J_c^{IG} value was about 1.37×10^7 A/cm^2, which is about two times larger than that obtained by the transport measurement[4].

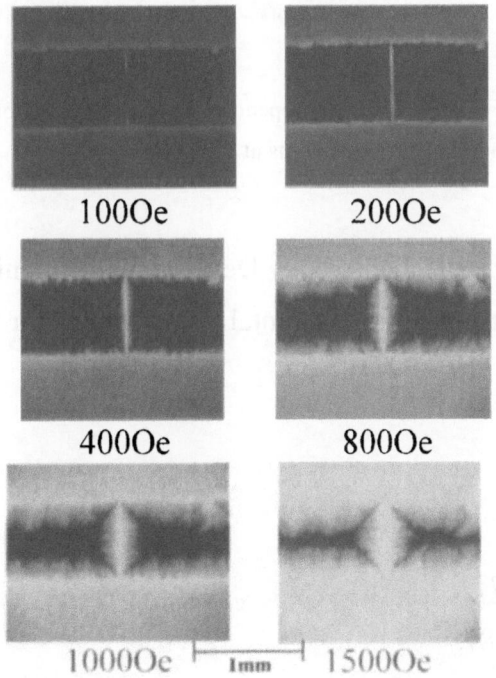

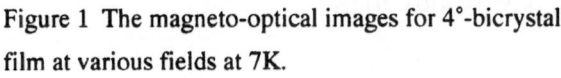

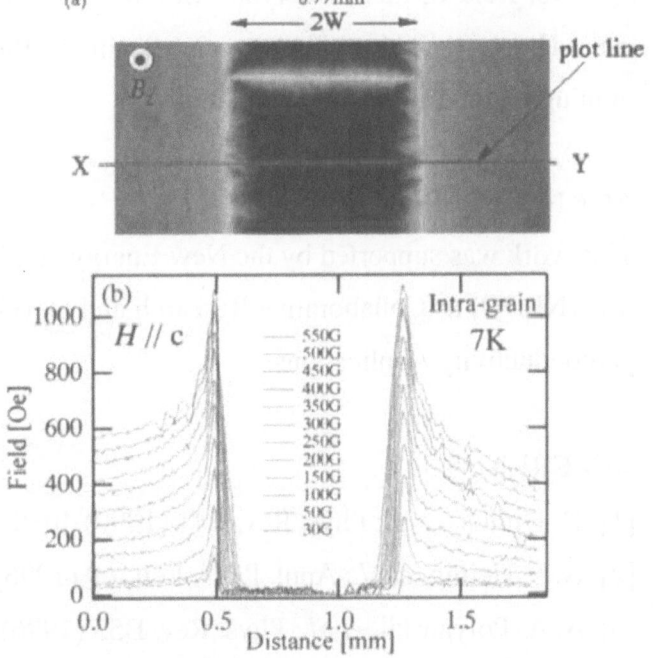

Figure 1 The magneto-optical images for 4°-bicrystal film at various fields at 7K.

Figure 2 (a) The magneto-optical image for 4°-bicrystal film at 400Oe at 7K (b) The line profiles along X-Y at various fields.

The inter-grain J_c^{GB} was obtained from the relation $J_c^{GB} = -\cos(2\alpha) J_c^{IG}$ by using α defined in Fig.3 as done by Polyanskii et al.[3]. Figure 4 shows the field dependences of J_c^{GB} / J_c^{IG} for 4°- and 24°-bicrystal films. J_c^{GB} / J_c^{IG} of 4°-bicrystal film was higher than that of 24°-bicrystal film. J_c^{GB} / J_c^{IG} of 24°-bicrystal film decreased rapidly with increasing the external field rather than that of 4°-bicrystal film. Since J_c^{GB} of 4°-bicrystal film at 1000Oe was about one order smaller than that of J_c^{IG} from the transport measurement[4], it seems that the value of J_c^{GB} / J_c^{IG} obtained from the magneto-optical imaging shows a good agreement with it. It is noted that J_c^{GB} / J_c^{IG} of 4°-bicrystal film has a peak around 500 Oe. TEM image of this film revealed that the distance between dislocation cores along GB was 5.6nm[5]. If we assume the trianglar arrangement and Gaussian distribution of flux lines around GB, we can estimate the matching field of several hundreds Oe from this distance. Therefore, the peak in J_c^{GB} / J_c^{IG} vs. field of the 4°-bicrystal film seems to be caused by the matching effect associeted with the flux pinning on the dislocation core array.

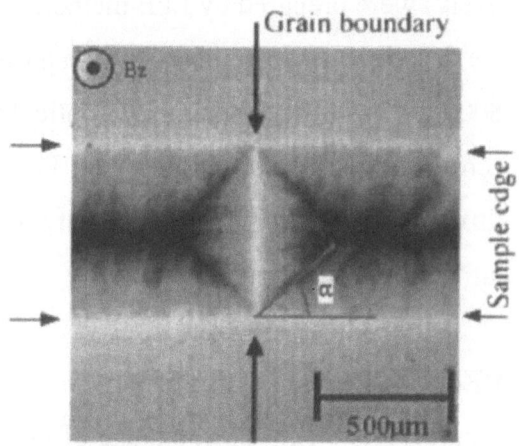

Figure 3 The definition of angle α. J_c^{GB} / J_c^{IG} value is calculated from this angle.

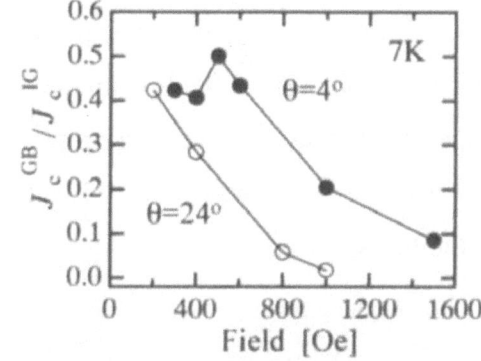

Figure 4 The field dependences of J_c^{GB} / J_c^{IG} for 4°- and 24°-bicrystal films at 7K.

ACKNOWLEDGMENT

This work was supported by the New Energy and Industrial Technology Development Organization (NEDO) as Collaborative Research and Development of Fundamental Technologies for Superconductivity Applications.

REFERENCES

[1] D. Dimos, et al.; Phys. Rev. B41 (1990) 4038.

[2] N. F. Heinig, et al.; Appl. Phys. Lett. 69 (1996) 577.

[3] A. A. Polyanskii, et al.; Phys. Rev. B53 (1996) 8687.

[4] T. Takagi, et al.; IEEE Trans. on Appl. Supercond. 9 (1999) 2328.

[5] J. G. Wen, et al.; IEEE Trans. on Appl. Supercond. 9 (1999) 2046.

[6] E. Zeldov, et al.; Phys. Rev. B49 (1994) 9802.

Anomalous Magnetization Behavior in The Vortex Liquid State in $Bi_2Sr_2CaCu_2O_{8+\delta}$

Kazuhiro Kimura[1], Ryo Koshida[1], Satoru Okayasu[3], Masao Sataka[3], Yukio Kazumata[4], Wai K. Kwok[5], George W. Crabtree[5], and Kazuo Kadowaki[1,2]

[1]Institute of Materials Science, University of Tsukuba,1-1-1 Tennodai, Tsukuba, Ibaraki 305-8573, Japan
[2]CREST, Japan Science and Technology Corporation (JST)
[3]Japan Atomic Energy Research Institute, Tokai, Naka, Ibaraki 319-1195, Japan
[4]Nihon Advanced Technology Co., Tokai, Naka, Ibaraki 319-1106, Japan
[5]Materials Science Division, Argonne National Laboratory, Argonne, Illinois 60435, USA

Abstract: Weak irreversible behavior was found in magnetization curves in the vortex liquid state above the irreversibility line, H_{irr}, but below a characteristic field, H_F, in as-grown $Bi_2Sr_2CaCu_2O_{8+\delta}$ single crystals and in columnar defected ones by using a SQUID magnetometer. H_F was also determined by temperature sweep measurements, and is strongly influenced by disorders in the samples. With increasing the number of columnar defects H_F systematically shifts higher temperature and field region. These results strongly suggest existence of a disorder-induced new phase between H_{irr} and H_F.

Keywords: the vortex liquid state, weak pinning, disorder, the second-order transition

INTRODUCTION

Vortex matter in high-T_c superconductors exhibits a variety of different types of phases due to strong thermal fluctuations, highly anisotropic layered structures, and various kinds and strength of pinning effects. In the case of pure limit it is expected that the vortex liquid region appears in a wide temperature and magnetic field region. The mean-field upper critical field, H_{c2}, does not show a sharp character of the second-order transition as found in conventional type-II superconductors and becomes a cross-over phenomenon from normal state to the fluctuating superconducting vortex liquid state. The true superconducting state can be realized only below the first-order vortex-line-lattice melting transition (VLLMT). In the case of actual single crystals, although they are clean and high quality ones grown by floating zone technique, the first-order transition persists down to the critical point, T_{cr}, below T_c and changes the character to the second-order transition. It was found that this critical point moves very sensitively to higher temperature by introducing higher disorder levels in the sample[1]. We have studied systematically this point by introducing columnar defects with $B_\Phi = 0.005, 0.01, 0.02, 0.05, 0.1, 0.2, 0.5, 1$ T. In samples with the columnar doses above $B_\Phi = 0.01$ T, the first-order VLLMT can no longer be observable as a sharp transition ($T_{cr} \rightarrow T_c$) and changes the character to the second-order transition (vortex glass transition). Moreover, it was found that the peak effect usually observed at low temperatures below the critical point also changes the character from sharp jump to broad peak effect.

In this paper we will present a systematic study of the effect of columnar defects in single crystals by measuring magnetization curves. The results show that the weak pinning or very small number of columnar defects induces a new vortex phase in the so-called vortex liquid state.

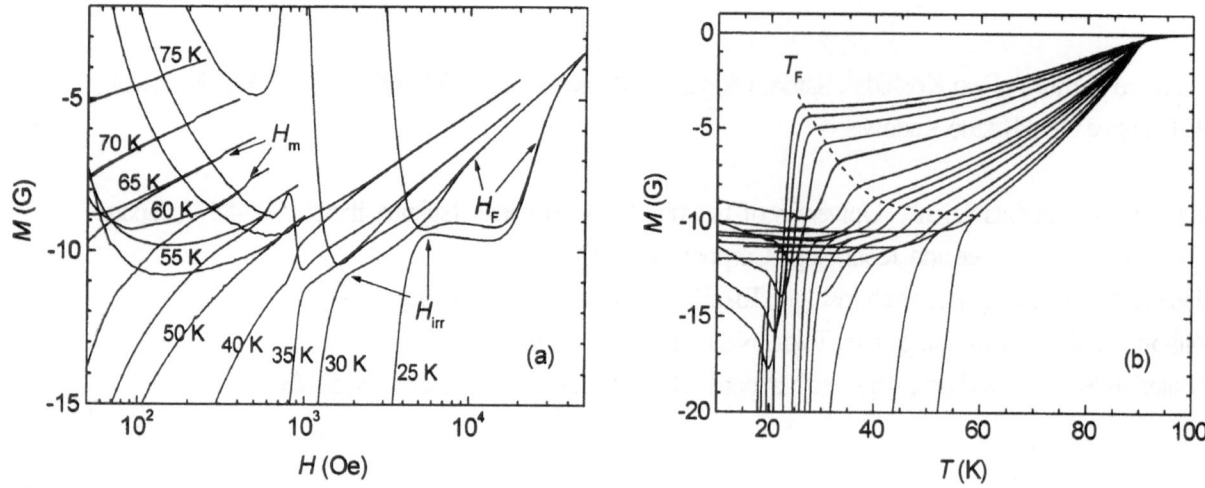

Fig. 1 (a) *M-H* curves of an as-grown $Bi_2Sr_2CaCu_2O_{8+\delta}$. (b) *M-T* curves of an as-grown $Bi_2Sr_2CaCu_2O_{8+\delta}$ (H = 0.03, 0.04, 0.07, 0.1, 0.2, 0.3, 0.5, 1, 2, 3, 4, and 5 T).

EXPERIMENTS

Magnetization measurements have been performed in $Bi_2Sr_2CaCu_2O_{8+\delta}$ single crystals with and without columnar defects by using a SQUID magnetometer (Quantum Design). RSO (Reciprocal Sample Option) method was used with parameters of the frequency 4 Hz and the amplitude 1 cm. $Bi_2Sr_2CaCu_2O_{8+\delta}$ single crystals were grown by the traveling solvent floating zone technique. Columnar defects were introduced into the crystals from the same batch by the heavy-ion irradiation at Argonne National Laboratory (B_Φ = 0.005, 0.01, 0.02, 0.05, 0.1 T) and at Japan Atomic Energy Research Institute (B_Φ = 0.2, 0.5, 1 T).

RESULTS AND DISCUSSION

Fig.1 (a) shows a set of magnetization hysteresis (*M-H*) curves of an as-grown $Bi_2Sr_2CaCu_2O_{8+\delta}$ single crystal. In the higher temperature region above 45 K magnetization jump associated with the first-order VLLMT (represented by H_m) can be observed. In the lower temperature region below 40 K ($\sim$ the critical point, T_{cr}) the magnetization jump is no longer observed, but instead, a weak irreversible behavior below H_F can clearly be seen in the vortex liquid state above the irreversibility line (the second-order vortex glass transition line), H_{irr}, where main hysteresis almost closes (see Fig. 1 (a)). Such a behavior was also observed in all as-grown crystals we measured. The magnetization relaxation was examined between H_{irr} and H_F, and only less than 0.3 % relaxation was found in 10 hours. The field-cool magnetization is identical with the zero-field one. These results indicate that there exists some kind of weak pinning mechanism between H_{irr} and H_F definitely different from the one below H_{irr}. The existence of this new vortex state between H_{irr} and H_F can be associated with the weak pinning residing in the single crystals.

H_F as well as H_{irr} can be determined more clearly by temperature-sweep measurement. A series of *M-T* curves is shown in Fig.1 (b), where magnetization steps are clearly observed in the reversible region in *M-T* curves at T_F (H), which corresponds quite well to H_F (T) determined from *M-H* curves. The height of the step grows larger with applied magnetic field increased. Enhancement of the step height with larger amplitude was not observed in our measurements in contrast to the results in Ref. 2.

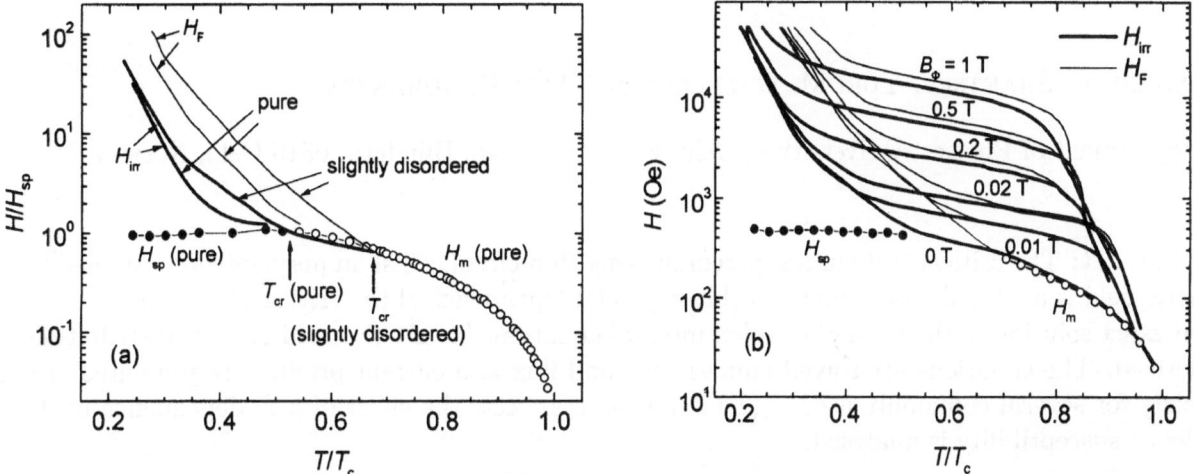

Fig. 2 (a) *H-T* phase diagram of two as-grown. T_{cr} represents the critical point. (b) *H-T* phase diagram of with columnar defects ($B_\Phi = 0, 0.01, 0.02, 0.2, 0.5$, and 1 T).

In order to study the nature of the region in the vortex liquid state between H_{irr} and H_F we have checked several as-grown samples with different disorder level and different sizes, and columnar defected ones with $B_\Phi = 0.005 - 1$ T. The weak pinning region between H_{irr} and H_F persists even in the sample with $B_\Phi = 1$ T, and the essential features of both *M-H* and *M-T* curves above H_{irr} are qualitatively the same in all samples. H_F in two as-grown crystals, pure one and slightly disordered one, are plotted together with H_{irr}, H_m and the second peak field, H_{sp}, in *H-T* phase diagram in Fig.2 (a). In the slightly disordered sample the magnetization hysteresis area is larger than that in the pure sample, and magnetization jump denoting the first-order transition is no longer detected below 60 K. It is noted that H_F strongly influenced by disorders and always lies below the critical point temperature in as-grown samples we have measured. It appears that H_m splits into H_{irr} and H_F at the critical point, T_{cr}. The weak pinning phenomena between H_{irr} and H_F may be a result of the second-order transition. However, it is not necessary for H_F to emerge at the critical point, especially in inhomogeneous samples. It is also noted that H_F as well as H_{irr} does not depend on the sample size, which supports the interpretation that these phenomena are intrinsic in the vortex system.

Fig. 2 (b) illustrates the *H-T* phase diagram of samples with columnar defects of $B_\Phi = 0, 0.01, 0.02, 0.2, 0.5$, and 1 T. H_F as well as H_{irr} shifts systematically to the higher temperature and the higher field region with increasing the number of columnar defects. This indicates that the weak pinning region between H_{irr} and H_F becomes more stable with disorders increased, and H_F may merge into the first-order transition line in the ideal crystal in the pure limit. It is intriguing to correlate these phenomena with the recent theoretical predictions, where the vortex-loop unbinding[3,4] or defects[5] cause phase transitions.

1. K. Kimura, S. Okayasu, M. Sataka, Y. Kazumata, and K. Kadowaki, preprint
2. T. Shibauchi *et al.*, Phys. Rev. Lett. **83**, 1010 (1999)
3. Z. Tesanovic, Phys. Rev. **B 59** 6449 (1999)
4. A. K. Nguyen and A. Sudbo, EuroPhys. Lett. **46**, 780 (1999)
5. M. J. W. Dodgson, V. B. Geshkenbein, and G. Blatter, cond-mat/9902244

Magnetic behavior of thin superconducting disks with a field-dependent critical current

Daniel V. Shantsev, Yuri M. Galperin, and Tom H. Johansen

Department of Physics, University of Oslo, P. O. Box 1048, Blindern, 0316 Oslo, Norway

Abstract: The critical state in a superconducting thin circular disk in perpendicular applied magnetic field is analysed. Assuming an arbitrary field dependence of the critical sheet current, $J_c(B)$, an exact solution in the form of coupled integral equations for the flux and current distributions is derived. The equations are solved numerically, and flux and current profiles are presented graphically for several commonly used $J_c(B)$ dependences. Also the effect of a B-dependence in J_c on the ac susceptibility is analysed.

Key words: thin disk, critical state model, critical current, flux distribution

INTRODUCTION

The critical state model (CSM) is widely accepted to be a powerful tool in the analysis of magnetic properties of type-II superconductors. During the last years much attention has been paid to the CSM analysis of thin samples in perpendicular magnetic fields (perpendicular geometry). Explicit analytical expressions for flux and current distributions have been obtained for a long thin strip [1, 2] and a thin circular disk [3 - 6] assuming a constant critical current (the Bean model). Theoretical results with a B-dependent j_c have so far been obtained only for long samples in a parallel field (parallel geometry) [7, 8] and for a long thin strip [9].

In this paper, we derive an exact solution for the case of a thin circular disk characterized by a *general* $j_c(B)$. We find a set of integral equations, which is solved numerically to obtain the field and current density distributions in various magnetized states as well as the ac susceptibility. Results are presented for several $j_c(B)$ models.

BASIC EQUATIONS

Consider a thin superconducting disk of radius R and thickness d, where $d \ll R$. We assume either that $d \geq \lambda$, where λ is the London penetration depth, or, if $d < \lambda$, that $\lambda^2/d \ll R$. We put the origin of the reference frame at the disk center and direct the z-axis perpendicularly to the disk plane. The external magnetic field $\mathbf{B}_a$ is applied along the z-axis, the z-component of the field in the plane $z = 0$ being denoted as B. The current flows in the azimuthal direction, with a sheet current denoted as $J(r) = \int_{-d/2}^{d/2} j(r, z) \, dz$, where j is the current density.

Consider a zero-field-cooled disk placed in the external field B_a. The disk then consists of an inner flux-free region, $r \leq a$, and of an outer region, $a < r \leq R$, penetrated by magnetic flux. To obtain expressions for the current and flux distribution we follow a procedure originally suggested in [3] and then generalized in [9] for the case of a B-dependent J_c in a thin strip. This yields a set of three coupled integral equations [10]

$$J(r) = \begin{cases} -\dfrac{2r}{\pi} \displaystyle\int_a^R dr' \sqrt{\dfrac{a^2 - r^2}{r'^2 - a^2}} \dfrac{J_c[B(r')]}{r'^2 - r^2}, & r < a \\ -J_c[B(r)], & a < r < R \end{cases} \tag{1}$$

$$B(r) = B_a + \frac{\mu_0}{2\pi} \int_0^R F(r,r')J(r')dr' , \qquad B_a = \frac{\mu_0}{2} \int_a^R \frac{J_c[B(r')]}{\sqrt{r'^2 - a^2}} \, dr'. \qquad (2)$$

Here $F(r,r') = K(k)/(r+r') - E(k)/(r-r')$, where $k(r,r') = 2\sqrt{rr'}/(r+r')$, while K and E are complete elliptic integrals. In the case of B-independent J_c, these equations are reduced to the explicit Bean-model results derived in [4] and [5]. For a given $J_c(B)$-dependence, the magnetic behavior is found by solving Eqs. (1), and (2) numerically as described in [10].

In remagnetized states, i.e., decreasing B_a from a maximum value B_{am}, the flux distribution can be found as $B(r) = B_m(r) + \tilde{B}(r)$. Here $B_m(r)$ is the flux profile at B_{am}, while $\tilde{B}(r)$ satisfies equations similar to the above, but with a different critical current $\tilde{J}_c(r) = J_c[B_m(r) + \tilde{B}(r)] + J_c[B_m(r)]$.

RESULTS

In the calculations we used the following $J_c(B)$ dependences, $J_c = J_{c0}/(1 + |B|/B_0)$ (Kim model), and $J_c = J_{c0} \exp(-|B|/B_0)$ (exponential model). Shown in Fig. 1 are the flux and current distributions for increasing field for the exponential model with different J_{c0} and B_0. These parameters are chosen so that the flux front position $a = 0.2R$, and the applied field B_a, are the same for all the profiles. Hence, one can follow the variations in the profile shape as the B-dependence of J_c changes. Contrary to the Bean model, we see that with $J_c(B)$ one has (i) a nonuniform $J(r)$ at $a < r < R$, with a minimum at the disk edge where $|B|$ has a peak; (ii) a cusped peak in $J(r)$ at $r = a$, since the magnetic field vanishes at this point with infinite derivative; (iii) a steeper $B(r)$ near the flux front, and reduced peaks at the edges.

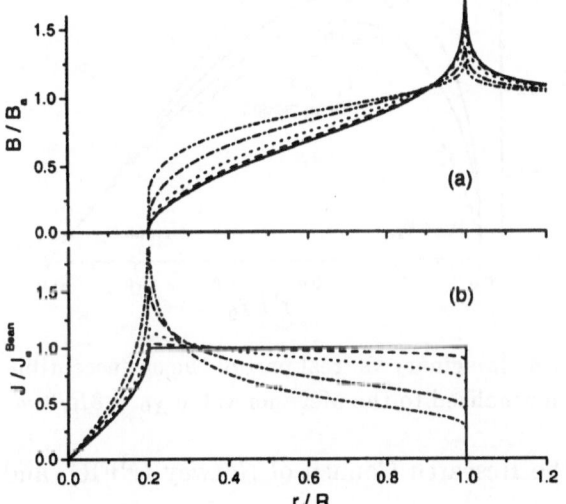

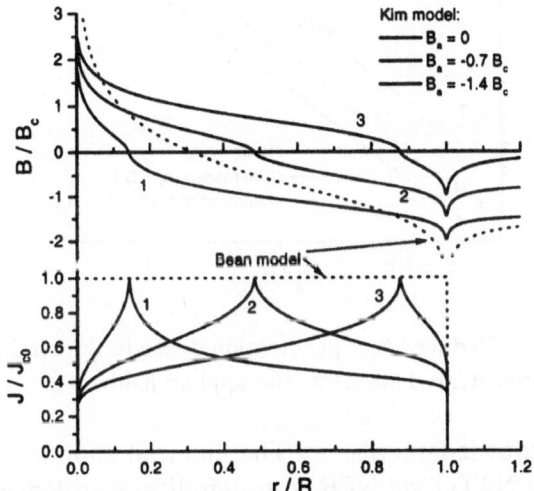

Fig. 1. Flux density (a) and current (b) profiles for flux penetration into a virgin state for the exponential model (solid lines represent the Bean model)

Fig. 2. Flux and current density profiles during field descent after first increasing B_a to a very large value; $B_0 = B_c$, where $B_c \equiv \mu_0 J_{c0}/2$.

Figure 2 shows profiles for fully-penetrated decreasing-field state for the Kim model. We see that near the points where $B = 0$ there is an abrupt increase in the slope of $B(r)$ and a corresponding peak in $J(r)$. In contrast, the Bean model has a uniform $J(r)$, while the shape of $B(r)$ is constant, although the profile is shifted according to the variation in applied field.

To analyze quantitatively the role of a B-dependent J_c let us consider the position of the flux front, a, during increasing B_a. For a long cylinder, in the limit of small parallel applied fields one has that [8], $a/R \approx 1 - B_a/B_c + (\mu_0 R j_c'(0)/2)(B_a/B_c)^2$ where here $B_c \equiv \mu_0 j_{c0}R$. We see that the B-dependence of j_c enters the expansion only in the second-order term, and hence, can be ignored at small B_a. In contrast, for a thin disk, the local field near the edge is very large even for small B_a. Thus, an expansion of $J_c(B)$ in powers of B is not everywhere convergent, and we have therefore performed numerical calculations of $a(B_a)$. From Fig. 3 one can see that for small B_a all the models yield a parabolic $a(B_a)$ dependence. A more detailed analysis shows that the overall behavior of the penetration depth can be fitted well by the Bean model result [5], $a/R = 1/\cosh(B_a/B_c)$ with an effective value of B_c. We find that the effective value is well described by $B_c^{\text{eff}} = B_c(1 - \alpha B_c/B_0)$ for $B_0/B_c > 1$. Here $\alpha = 0.42$ for the exponential model, and $\alpha = 0.36$ for the Kim model. A similar analysis of the long thin strip case shows that the same relation is valid only with different values $\alpha = 0.60$ and 0.51 for the exponential and the Kim model, respectively.

From the flux and current distributions one can calculate the magnetization hysteresis loops and, from that also the ac susceptibility $\chi = \chi' + i\chi''$. We analyse the behavior of χ' and χ'' as functions of the ac field amplitude, B_{am}. Shown in Fig. 4 are parametric plots of χ'' versus $-\chi'$ for $J_c(B) = J_{c0}\exp{-|B|/B_0}$ with different B_0. A typical feature of these plots is that χ'', which characterises the ac loss, has a maximum. The dashed curve, $B_0 \to \infty$, reproduces the Bean model result obtained in [5] with maximal value $\chi''_{\max} = 0.24$. We find that a B-dependence of J_c leads to (i) a shift of the peak position to a lower $|\chi'|$, and (ii) to an increase in the peak value to 0.32 for $B_0 = B_c$. Furthermore, a dramatic change of the large B_{am} behavior is observed due to the B-dependence of J_c; χ'' falls to zero more abruptly than in the Bean model. Interestingly, at low B_{am} the behavior of both χ' and χ'' is well described by the Bean-model expressions [5] with the same effective field, B_c^{eff}, defined above. Therefore, the slope of the $\chi''(\chi')$ curve as $\chi' \to -1$ remains the same for *any* $J_c(B)$ (the straight line on Fig. 4). This result allows a universal, $J_c(B)$-independent test for the critical-state model.

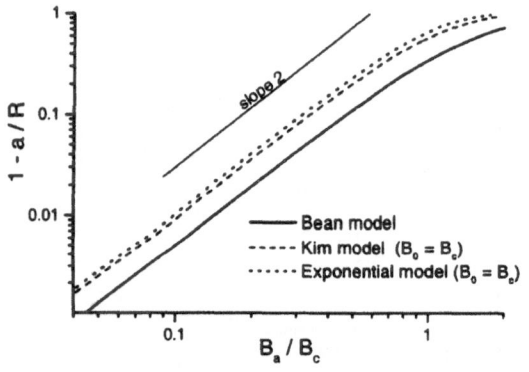

Fig. 3. Reduced flux penetration depth in the virgin flux-penetrated state vs. the applied field.

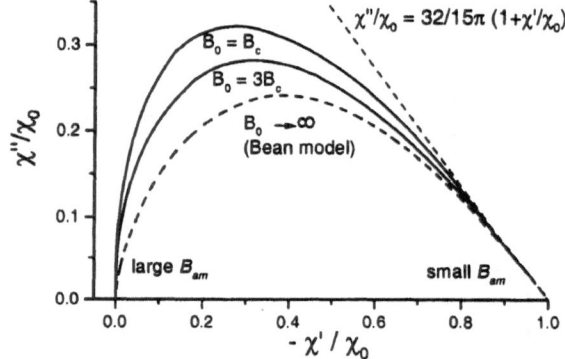

Fig. 4. Imaginary vs. real part of the ac susceptibility normalized to the Meissner value $\chi_0 = 8R/3\pi d$.

Acknowledgments. The financial support from the Research Council of Norway (NFR), and from NATO via NFR is gratefully acknowledged.

1. E. H. Brandt, and M. Indenbom, Phys. Rev. B **48**, 12893 (1993).
2. E. Zeldov, J. R. Clem, M. McElfresh, and M. Darwin, Phys. Rev. B **49**, 9802 (1994).
3. P. N. Mikheenko and Yu. E. Kuzovlev, Physica C **204**, 229 (1993).
4. J. Zhu, J. Mester, J. Lockhart, and J. Turneaure, Physica C **212**, 216 (1993).
5. J. R. Clem and A. Sanchez, Phys. Rev. B **50**, 9355 (1994).
6. Some final expressions obtained in [3] are not correct, see discussion in [4].
7. J. R. Clem, J. Appl. Phys. **50**, 3518 (1979).
8. T. H. Johansen and H. Bratsberg, J. Appl. Phys. **77**, 3945 (1995).
9. J. McDonald and J. R. Clem, Phys. Rev. B **53**, 8643 (1996).
10. D. V. Shantsev, Y. M. Galperin, T. H. Johansen, Phys. Rev. B **60**(17), in press (1999).

Pinning Potential and AC Susceptibility in Superconducting Pb-Bi Alloys

Akitoshi Matsuda

Department of Electrical Engineering, Kyushu Sangyo University,
2-3-1 Matsukadai, Higashi-ku, Fukuoka 813-8503, Japan

Abstract: In oxide superconductors, the critical current are largely influenced by thermally activated flux creep at high temperature owing to their weak pinning strength. But, it is difficult to control the pinning strength in high-temperature oxide superconductors. The pinning strengths of the superconducting Pb-Bi alloys can be controlled easily by varying the Bi content. In order to investigate the effect of reversible motion of fluxoids, the AC susceptibility was measured for Pb-Bi alloys. The apparent pinning potentials and AC susceptibilities are systematically investigated for Pb-Bi alloys with different pinning strengths.

Keywords: critical current, flux creep, AC susceptibility, Pb-Bi alloys, pinning potential

INTRODUCTION

The flux pinning characteristic and critical current are investigated for superconducting Pb-Bi alloys with different pinning strength[1]. The measurement of fundamental AC susceptibility has been used for characterization of superconductors, such as the critical temperature, irreversibility line and the critical current. Therefore, it is important at first to clarify the relation between the pinning strength and the effect of reversible motion of fluxoids. In this paper, we clarify the dependence of reversible motion of fluxoids on the flux pinning strength by measuring fundamental AC susceptibility in Pb-Bi superconducting.

EXPERIMENTS

The superconducting Pb-Bi alloy specimens were prepared as follows: Lead and bismuth with the purity of 99.99% were mixed in a desired composition and melt at 500°C for 5 hours in a vacuum of 10^{-4} Torr. In general, Pb-Bi alloys consist of the superconducting epsilon phase and the normal bismuth-rich phase. We fabricated some specimens with atomic ratios of bismuth, 20at% ~ 46at%. Magnetic moments measurement were carried out using a SQUID magnetometer. The critical current density, J_c, were evaluated from the magnetization data[2]. The temperature dependence of the AC susceptibility was measured under various AC magnetic field and pinning strength. The DC and superposed AC magnetic field were applied along the long axis of the specimen.

RESULTS AND DISCUSSION

The field dependence of the pinning strength, F_p, computed from the $F_p = J_c \times B$ is shown in Fig. 1. The B_e was applied as $0 \Rightarrow 0.4$(T) after zero-field cooling from room temperature. The values of F_p for Pb-24at%Bi specimen at $T=5.0$K are about $5 \sim 10$ times smaller than those of Pb-46at%Bi specimen. This reveals that Pb-24at%Bi specimen has the weakest pinning strength and Pb-46at%Bi specimen has the strongest pinning strength in these specimens. At all temperatures, the magnetization decayed almost logarithmically with time. Therefore, we could obtain a very reliable normalized relaxation rate from the linear reduction of M with $\ln t$ [3]. The time dependent $M(t)$ is given

$$M(t) = M_0 \left(1 - \frac{k_B T}{U_0^*}\right) \ln \frac{t}{t_0}, \tag{1}$$

where M and M_0 are magnetization at time t and t_0, U_0^* is magnitude of the apparent pinning potential and k_B is Boltzmann's constant. M_0 is the initial magnetization at the beginning of measurement. The apparent pinning potential, U_0^*, obtained from Eq. (1) is expressed as

$$U_0^* = \frac{k_B T}{R_c/M_0} \ln 10. \tag{2}$$

where R_c is the creep rate.

Figure 2 shows the temperature dependence of the apparent pinning potential, U_0^*, for specimens with different pinning strength. It is found that there are the peak of U_0^* for Pb-37at%Bi specimen at near $T=5.0$K and Pb-46at%Bi specimen at near $T=5.5$K. On the other hand, there are not the peak of U_0^* for Pb-24at%Bi, Pb-28at%Bi specimens.

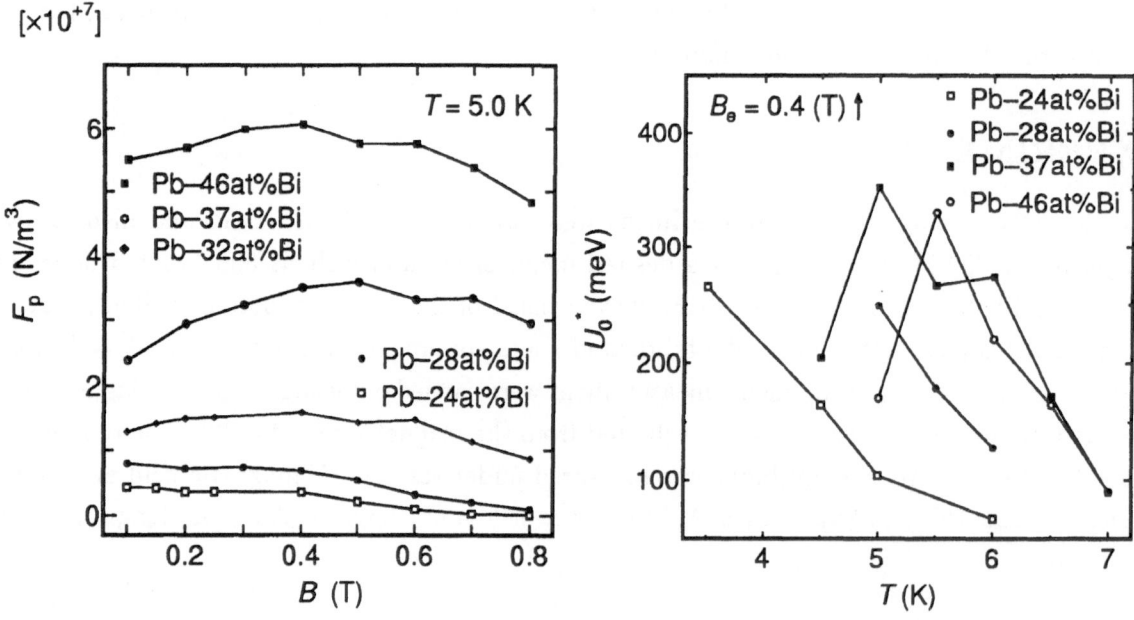

Fig. 1. Temperature dependence of pinninng strength at $T=5.0$K.

Fig. 2. Temperature dependence of U_0^* obtained from magnetization relaxation.

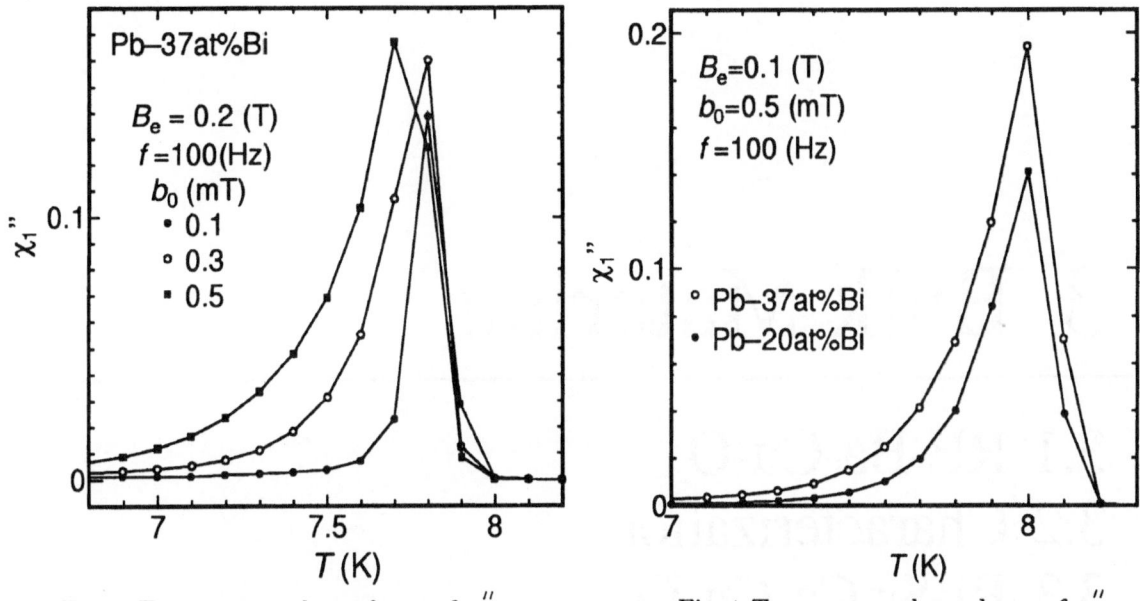

Fig. 3. Temperature dependence of χ_1''.   Fig. 4. Temperature dependence of χ_1''.

Figure 3 shows the results of temperature dependence of the fundamental AC susceptibility, χ_1'', for Pb-37at%Bi specimen in various AC magnetic fields. It is found that the peak value of χ_1'' decreases according to decreasing the AC magnetic field. Figure 4 shows the results of temperature dependence of χ_1'' for Pb-37at%Bi, Pb-20at%Bi specimens. It is found that the peak value of χ_1'' decreases according to weakening the pinning strength. The peak values of χ_1'' decrease with decreasing AC magnetic field and/or weakening pinning strength, suggesting that the reversible motion of fluxoid becomes significant by the thermally activated flux creep at high temperatures due to their weak flux pinning interactions.

CONCLUSIONS

In this article, the apparent pinning potential and fundamental AC susceptibility of superconducting Pb-Bi alloys were measured and the following results are obtained. (1)We can easily control the pinning strength of Pb-Bi specimens by varying the Bi content. (2)The values of U_0^* have the peak in strong pinning strength, but there is no peak in weak pinning strength specimens. (3)The peak value of χ_1'' decreases according to decreasing AC magnetic field and weakening pinning strength. The reversible fluxoid motion becomes prominent for weaker pinning strength and/or at higher temperature.

Acknowledgements. This work was financially supported by the Sasakawa Scientific Research Grant from The Japan Science Society.

1. T. Matsushita, A. Matsuda, and K. Yanagi (1993) Physica C **213**, 477-482.

2. A. Matsuda, T. Akune, and N. Sakamoto (1998) Advances in Superconductivity **X**, 505-508.

3. A. Matsuda, T. Akune, and N. Sakamoto (1999) IEEE Trans. Applied Superconductivity Vol.9, No.2, 2199-2202.

3 Bulk Materials

SIMULATION OF CRYSTAL GROWTH IN (RE)Ba$_2$Cu$_3$O$_x$

Georg J. Schmitz

ACCESS e.V., Intzestr 5, D-52072 Aachen, Germany

Abstract: Recent developments of the phase-field concept and its applications in modeling microstructures evolving during solidification of multicomponent and multiphase alloys are reviewed. Special emphasis is given to the qualitative description of phenomena occurring during the peritectic solidification of (RE)Ba$_2$Cu$_3$O$_x$ superconductors. Future directions of the method like e.g. coupling to thermodynamic databases or coupling between macroscopic process simulation and simulation of microstructure evolution are highlighted.

Keywords: phase-field modeling, multiphase systems , peritectic solidification, high-temperature superconductors, growth mechanisms

INTRODUCTION

The phase field approach of modeling phenomena occurring during solidification has gained more and more importance in the recent years. Being based on fundamental aspects of phase-transitions in general, research in the "phase-field" has essentially been performed by mathematicians and theoretical physicists. Recent developments - e.g. the extension of the phase field concept to the description of multiphase equilibria and multicomponent alloys - make this method a powerful tool for materials scientists trying to optimize the quality of their products by optimizing the microstructure of the materials.

THE PHASE-FIELD METHOD

The phase-field theory , which is based on Time-Dependant Ginzburg Landau theories adresses first order phase-transitions such as solidification/crystallization. An order parameter, which in solidification problems often is identified with the fraction ϕ of the solid phase, varies between 0 and 1. The equations of motion are derived by functional differentiation of a free energy formulation, which generally consists of three contributions :

- a gradient energy being related to gradients of the order parameter
- a potential energy
- a driving force term describing the deviation from equilibrium being a function of the difference ΔG of the Gibbs energies of the bulk phases.

The stationary solution $(d\phi/dt = 0)$ of the simplest phase-field equation

$$\tau\, d\phi/dt = \varepsilon^2 \nabla^2 \phi + W'(\phi) + m(\Delta G) \qquad (1)$$

in thermodynamic equilibrium $(m(\Delta G)= 0)$ yields a hyperbolic tangent profile for the order parameter in case of $W(\phi)$ being a double well potential . This solution is centered at position **x**$=0$ in space. In case of a non-vanishing driving force $(m(\Delta G)\neq 0)$ this profile moves in space with a velocity proportional to the degree of deviation from equilibrium preserving the profile shape. In solidification problems this motion is associated with a release of the latent heat of fusion appearing as a source term $(\sim Ld\phi/dt)$ in the heat conduction equation:

$$dT/dt = \lambda \nabla^2 T + Ld\phi/dt \qquad (2)$$

The non-linear coupling between the phase-field equation (1) and the heat-diffusion equation (2) via the contributions $Ld\phi/dt$ and $m(\Delta G)$ is the basis for the evolution of complex structures like dendrites when numerically iterating these equations in time [1-9].

During solidification of alloys, however, in general several phases are coexisting like e.g. the $(RE)_2BaCuO_5$ ("211"), $(RE)Ba_2Cu_3O_x$ ("123") and liquid phases during solidification high T_c superconductors. To describe equilibria of multiple phases, like e.g. a liquid phase and two solid phases in peritectic solidification, an additional order parameter is required to distinguish the two solid phases. A possible approach is the definition of an order parameter ϕ_i for each of the phases i , which may be interpreted as the fraction of phase i within a certain control volume [10,11]. This interpretation naturally leads to the constraint:

$$\Sigma\phi_i = 1 \qquad (3)$$

The equations of evolution for the different order parameters ϕ_i are obtained by functional minimization of a free energy functional essentially consisting of a summation of energy-contributions of the dual phase-boundaries liquid/solid1, liquid/solid2 and solid1/solid2 . This kind of multi-phase-field functional allows the individual treatment of the different interfaces with their specific properties like e.g. surface energy σ_{ij} or latent heat L_{ij} . It could be shown - both analytically (for the sharp interface limit) and numerically - that the respective equations of evolution lead to equilibria obeying classical laws at triple junctions like e.g. Young's law describing the angles between different phases as a funtion of the different interfacial energies [12].

Simulations have meanwhile been performed on binary alloys Fe-C [13,14], and eutectics [15] in both 2D and 3D using linearized binary phase diagrams. Especially the simulations of eutectic growth recovered well known analytical relations for the spacing between the two solid phases [16]. A review of the development of the phase-field concept for binary alloys can be found in [17].

SIMULATIONS OF GROWTH PHENOMENA IN (RE)BA$_2$CU$_3$O$_x$

Besides macroscopic process modelling of Y-Ba-Cu-O crystallization without [18,19] or with fluid flow [20] , Monte-Carlo-Simulations of diffusion [21] and different cellular automata type simulations of microstructure evolution [22,23,24] especially phase-field simulations have been applied to describe the diffusion controlled peritectic growth in the Y-Ba-Cu-O system, where a pseudo-binary phase diagram is used to describe the peritectic reaction

$$(RE)_2BaCuO_5 \text{ ("211")} + \text{Liquid} \rightarrow (RE)Ba_2Cu_3O_x \text{ ("123")}$$

The kinetics of this reaction to a large extent is controlled by the diffusion of Yttrium in the liquid .First simulations accordingly adressed the diffusive interaction of a peritectic 123 growth front with a properitectic 211 particle and recovered experimental observations of partially/totally dissolving 211 particles and their overgrowth by the 123 growth front [25].

A more detailed analysis of these simulations revealed a dissolution/reprecipitation process [26] as at least partially contributing to the pushing of 211 particles leading to a macroscopically inhomogeneous distribution of these particles and to the formation of x-like patterns. The formation of these patterns is moreover enhanced by numerically observed yttrium enrichment at the corners of a facetted, anisotropically growing 123-crystal [27].

Anisotropic ripening of the 211 particles has been observed experimentally ranging from spherical 211 particles to needle-like shapes depending on amount of dopants in the melt.This behaviour could qualitatively be reproduced numerically by varying the interfacial tension between 211 and liquid phases and/or by changing the rate of Yttrium diffusion [28].

All simulations cited above are based on pseudobinary sections of the real, quaternary phase diagram and estimates for thermophysical data like e.g. interfacial energies. Although they are only valid within these constraints, they yielded most useful information and ideas for further development of real process technology. Especially the variation of initial conditions in the numerical simulations has indicated a potential benfit for processing if the initial experimental conditions e.g.consist of a regular geometrical arrangements of precursors [29,30]

Especially a grain selection mechanism in polycrystallineYBaCuO growth being observed in two dimensional simulations has now been verified experimentally paving the way towards biaxially textured thick films on metallic substrates [31]. Future simulations in the Y-Ba-Cu-O system will support both the further development of existing processes and creation of new processes like e.g. infiltration of fabrics [32]

CONCLUSIONS AND FUTURE PERSPECTIVES

The phase-field concept and its numerical realization seem to become a powerful tool for the prediction of microstructure evolution even in complex thermodynamic systems like high temperature superconductors. Open questions relate especially to issues of computational efficiency in terms of realistic 3D simulations in direct coupling with thermodynamic/thermophysical databases. Finally, the simulated microstructure has to be related to the macroscopic process parameters leading to its formation. First attempts to reach this objective are made by coupling macroscopic and microscopic simulations for laser-surface remelting of aluminum coatings [33]. The variation of initial and boundary conditions in qualitative simulations already at present provides useful hints how to control microstructure formation and accordingly the properties of the resulting material.

Acknowledgments: This paper in large parts represents a summary of activities and projects in ACCESS' microsimulation group, which are supported by British Steel, Hoogovens, Bayer as well as by BMBF and DFG. The experimental work on high temperature superconductors being compared to the numerical results is funded by the German Federal Ministry of Higher Education ,Research and Technology (BMBF) under grant 13 N 7490.

1.) G.Fix, Phase field models for free boundary problems, in "Free boundary Problems" Vol. II, Ed. A.Fasano, M.Primicerio (Piman, Boston 1983)
2.) J.B.Collins, H.Levine, Diffusive interface model of diffusion-limited crystal growth, Phys. Rev. B, Vol. 31 No. 9 (1985) 6119-6122
3.) G. Caginalp, P. C. Fife, Phys. Rev. B 33 11 (1986)7792
4.) A.A.Wheeler, W.J.Boettinger, G.B.McFadden, Phase-field model for isothermal phase transitions in binary alloys, Phys. Rev. A, Vol. 45 No. 10 (1992) 7424-7439
5.) R. Kobayashi, Modeling and numerical simulations of dendritic crystal growth, Physica D 63 (1993) 410-423
6.) S.-L. Wang, R.F.Sekerka, A.A.Wheeler, B.T.Murray, S.R.Coriell, R.J.Braun,
7.) G.B.McFadden, Thermodynamically-consistent phase-field models for solidification, Physica D 69 (1993) 189-200
8.) T.Ihle, H.Müller-Krumbhaar, Fractal and compact growth morphologies in phase transitions with diffusion transport, Phys. Rev. E, vol. 49 No. 4 (1994) 2972-2991
9.) A.Karma, W. J.Rappel, Numerical Simulation of Three-Dimensional Dendritic Growth, Phys. Rev. Lett. Vol. 77 No. 19 (1996) 4050-4053
10.) I. Steinbach, F. Pezzolla, B. Nestler, M. Seeßelberg, R. Prieler, G.J. Schmitz, J.L.L.Rezende: A phase field concept for multiphase systems. Physica D 94(1996), pp 135-147.
11.) J. Tiaden, B. Nestler, H.J. Diepers, I. Steinbach: The Multiphase-Field Model with an Integrated Concept for Modeling Solute Diffusion. Physica D (1998)115, pp 73-86.
12.) H.Garcke, B.Nestler, B.Stoth "A multiphase concept: numerical simulation of moving phase boundaries and multiple junctions" , SIAM J. on Applied Mathematics (1999) in press
13.) J. Tiaden U. Grafe: A Phase-Field Model for Diffusion and Curvature Controlled Phase Transformations in Steels. subm. to: PTM International Conference on Solid-Solid Phase Transformations '99, Kyoto, May 1999.
14.) J. Tiaden : "Phase-Field simulations of the peritectic solidification of Fe-C" J. Crystal Growth 198/199 (1999) 1275-1280

15.) M. Seeßelberg, J. Tiaden, G.J. Schmitz, I. Steinbach: Peritectic and Eutectic solidification: Simulations of the microstructure using the multi-phase-field method. Proc. of the 4th Int. Conf. on Solidification Processing, Sheffield, 7.-10. July 1997. Ed. by J. Beeck, H. Jones, pp 440-443.

16.) M. Seeßelberg, J. Tiaden: Simulations of Binary Eutectic Microstructures Using the Multi-Phase-Field Method. Proc. 8th Conf. on Modeling of Casting, Welding and Advanced Solidification San Diego, June 1998, pp 557-564.

17.) I. Steinbach, G.J. Schmitz: Direct numerical simulation of solidification structure using the phase field method. Proc. 8th Conf. on Modeling of Casting, Welding and Advanced Solidification San Diego, June 1998, pp 521-532.

18.) M. Seeßelberg, G.J. Schmitz, B. Nestler, I. Steinbach: Macroscopic and microscopic modeling of the growth of YBaCuO bulk material. IEEE Trans. on Appl. Supercond. 7(1997)2, pp 1739-1742.

19.) M. Seeßelberg, G.J. Schmitz, T. Wilke, M. Ullrich: Simulation of Thermal Fields for the Optimization of the Melt-Texturing Process of HTSC Bulk Materials. EUCAS '97, Inst. of Phys. Conf. Ser., No 158, pp 841-844.

20.) Y. Namikawa, M. Egami, Y. Shiohara , J. Mater Res. 9(1996) 28

21.) A Pekalski, M Ausloos;Monte-Carlo Simulation of Oxygen Diffusion in Planar Model of 123 - YBCO - Low-Temperature Regime and Effect of Trapping Barrier; Physica C 226 1-2 (1994)188

22.) N.Vandewalle, R.Cloots, M.Ausloos J. Mater. Res 10(1995)268

23.) E.A.Goodilin, N.N.Oleynikov, A.N.Branov, Y.D.Tretyakov , Inorganic Materials 29(1993) 1285

24.) Ch. Wolters, J. Laakmann, S. Rex, G.J. Schmitz: Numerical simulation of the influence of Y2BaCuO5 particles on the growth morphology of peritectically solidifying YBa2Cu3O7-x (Proc. EUCAS '93), Göttingen,ed. H.C. Freyhardt, Oberursel: DGM 1993, p 353.

25.) G.J. Schmitz, B. Nestler, H.J. Diepers, F. Pezzolla, R. Prieler, M. Seeßelberg, I. Steinbach: Numerical Simulation of YBaCuO-Growth Phenomena using the Phase Field Method. Applied Superconductivity, EUCAS 1995, Institute of Physics Conference SeriesNo 148 (1995), pp 167-170.

26.) G.J. Schmitz, M. Seeßelberg, B. Nestler, R. Terborg: Modeling of RE-Ba-Cu-O growth using transparent organic analogues and numerical simulations. Physica C 282-287 (1997), pp 519-520.

27.) G.J. Schmitz, B. Nestler, M. Seeßelberg: YBCO Melt-Processing Development by Numerical Simulation. J. Low Temp. Phys. 105(1996), p 1451.

28.) G.J. Schmitz, B. Nestler: Simulation of phase transitions in multiphase systems: peritectic solidification of (RE)Ba2Cu3O7-x superconductors. Mat. Sci. and Eng. B53(1998), pp 23-27.

29.) G.J. Schmitz, O. Kugeler: Isothermal production of uniaxially textured YBCO superconductors using constitutional gradients. Physica C 275(1997), pp 205-210.

30.) G.J. Schmitz, A. Tigges, J.C. Schmidt: Texturing of (RE)Ba2Cu3O7-x thick films by geometrical arrangements of reactive precursors. Supercond. Sci. Technol. 11(1998) pp 950-953.

31.) M. Tarka, E.S. Reddy, J. Noudem, G.J.Schmitz , to be published

32.) E. Sudhakar Reddy, M. Tarka, J. Noudem, G.J.Schmitz "YBa2Cu3Ox monodomain fabrics prepared by an infiltration process", subm. Proceedings PASREG 99 , Morioka Japan, October 99

33.) G. Laschet, H.-J. Diepers, I. Steinbach: Micro-Macrosimulation of laser remelting of an aluminum coating on steel. Proc. of ECLAT '98. European Conference on Laser Treatment of Materials, Hannover 22.-23.09.1998, pp 265-270

Effect of fabrication process on Jc-H property for untwinned orthorhombic Nd123 single crystal

Akihiro Oka, Satoshi Koyama, Teruo Izumi and Yuh Shiohara

Superconductivity Research Laboratory Division 4, 1-10-13 Shinonome, Koto-ku, Tokyo 135-0062, Japan

Abstract: The critical current density(Jc)-H properties for untwinned orthorhombic Nd123 single crystals were studied. The untwinned crystals were fabricated by means of two different heat treatment processes. Firstly, the as-grown crystals were preliminarily twinned by annealing in a pure oxygen gas flow, subsequently detwinned with uni-axial stress (Detwinned sample). Secondly, the as-grown crystals were directly annealed with uni-axial stress in a pure oxygen gas flow (Twin-free sample). The Jc-H property for the twin-free crystals showed a larger peak effect than that for the detwinned crystals. The results of the Jc-H properties and EXAFS measurements suggested that the Nd/Ba ratio in the twin-free crystals was fluctuated with a shorter wavelength than that in the detwinned crystals.

Key words: Nd123, untwinned single crystal, critical current density, Nd/Ba ratio fluctuation, EXAFS

INTRODUCTION

Nd123 system has been known as a high-quality superconducting material, which has a high Tc value (96K) and anomalous peak effects in the Jc-H properties. In order to develop bulk applications for the Nd123 system, the larger Jc values in high magnetic fields are expected. It has been reported that nonsuperconducting Nd-rich phases in the Nd123 system, which are attributed to the occurrence of Nd/Ba ratio fluctuation with a wavelength of a few tens of nanometer during annealing, act as pinning centers for the peak effects [1]. Here, the Nd/Ba configurations in the Nd123 system should be affected by the oxygen configurations in Cu-O basal planes. Therefore, the Nd/Ba ratio fluctuation in the Nd123 system should be also affected by annealing with uni-axial stress such as untwinning processes, because the uni-axial stress leads to the alignment of the oxygen configurations in Cu-O basal planes [2]. In the present study, the Jc-H properties for untwinned orthorhombic Nd123 single crystals after annealing with uni-axial stress were studied. The Jc-H properties for the untwinned crystals were compared each other, which were fabricated by means of two different heat treatment processes. Furthermore, the Nd/Ba ratio fluctuation in these crystals was evaluated by using EXAFS measurements.

EXPERIMENTAL

The Nd123 single crystals were grown by the top-seeded solution-growth (TSSG) method [3]. The composition of the grown crystals was analyzed by inductively coupled plasma atomic emission spectrometry (ICP-AES) and confirmed to be Nd : Ba : Cu = 1.07 : 1.95 : 3.00. The Nd123 samples were cut from the grown crystals into rectangular plates with dimensions of $2.0 \times 2.0 \times 0.5mm^3$, whose edges of 2.0mm correspond to <110> direction of the crystal lattice. The untwinned orthorhombic samples were fabricated by means of two different heat treatment processes. Firstly, the as-grown samples were preliminarily twinned by annealing at 500°C for 200h in a pure oxygen gas flow, and then detwinned at 500°C for 100h, subsequently at 300°C for 200h with a uni-axial load with 3.92N in <100> direction from the corner of the samples (Detwinned sample). Secondly, the as-grown samples were directly annealed with a uni-axial load with 3.92N from 700°C to 300°C for 200h in a pure oxygen gas flow (Twin-free sample). These untwinning heat treatment processes are described in detail elsewhere [4]. Furthermore, two different twinned samples were also fabricated by the same heat treatment conditions as the detwinned or the twin-free samples except that the twinned samples were annealed without uni-axial stress (Twinned sample 1 and Twinned sample 2, respectively). These samples were confirmed to be untwinned or twinned by using polarized optical microscopy. The magnetization hysteresis (M-H) loops for the detwinned, the twin-free, and the twinned samples were measured at 77K with c//H on the zero field cooling (ZFC) condition by using a superconducting quantum interference de-

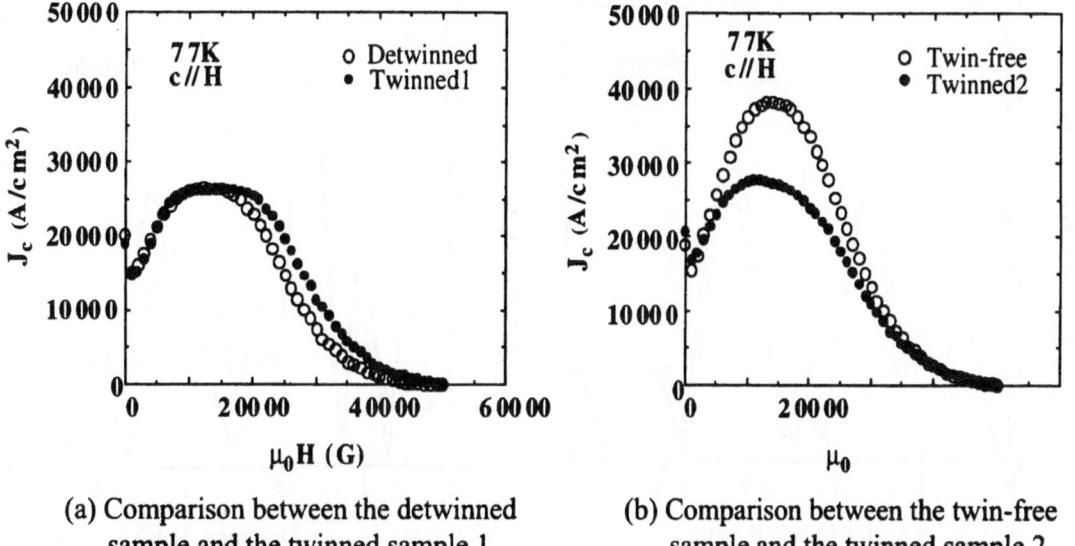

(a) Comparison between the detwinned
sample and the twinned sample 1

(b) Comparison between the twin-free
sample and the twinned sample 2

Fig. 1. Comparison of Jc-H properties at 77K with c//H among the detwinned, and the twin-free, and the twinned Nd123 samples after annealing.

vice (SQUID) magnetometer, and the Jc values for the samples were estimated from the M-H loops applying the extended Bean critical-state model [5]. Furthermore, the Nd/Ba ratio fluctuation in the detwinned, the twin-free, and the twinned samples were evaluated by using EXAFS measurements. The unpolarized Cu K-edge EXAFS spectra (Rigaku R-EXAFS-3000V) for the crushed samples were measured at room temperature in a fluorescence mode using a tungsten X-ray tube with a Si monochromater. The peaks at around 1.6Å and 2.5Å~4.5Å corresponding to the oxygen configurations in Cu-O basal planes and the Nd/Ba configurations, respectively, in the magnitude of Fourier transform of the EXAFS oscillations, were compared among the samples [6].

RESULTS AND DISCUSSION

Figure 1 shows the comparison of the Jc-H properties at 77K among the detwinned, the twin-free, and the twinned Nd123 samples after annealing. As shown in the figure, the Jc-H property for the twin-free sample showed a much larger peak effect than that for the twinned sample 2. On the other hand, the Jc-H property for the detwinned sample showed almost the same peak effect as that for the twinned sample 1. The Jc values for the peak effects correspond to the number of the Nd-rich regions in the samples, assuming that the elemental pinning force of any Nd-rich region is equal each other. Therefore, it can be suggested from these results that the wavelength of the Nd/Ba ratio fluctuation in the twin-free samples is shorter than that in the twinned samples. Furthermore, it can be suggested that the wavelength of the Nd/Ba ratio fluctuation in the detwinned samples is almost the same as that in the twinned samples. Figure 2 shows the comparison of the magnitude of Fourier transform of the EXAFS oscillations among the detwinned, the twin-free, and the twinned Nd123 samples after annealing. In the magnitude of Fourier transform for the twin-free sample, the peak at around 3.0 Å was split into two peaks, at the same time that the peak intensity at around 1.6 Å was much larger than that for the twinned sample 2. On the other hand, in the magnitude of Fourier transform for the detwinned sample, the peak at around 3.0 Å was not split into two peaks and almost the same as that for the twinned sample 1, at the same time that the peak intensity at around 1.6Å was also almost the same as that for the twinned sample 1. Here, the magnitude of Fourier transform for the twin-free sample showed the similar tendency to that for twin-free Y123 single crystals, in which the fluctuation of cation such as Y/Ba atoms does not exist [7]. Therefore, it is considered that these results show the shorter wavelength of the Nd/Ba ratio fluctuation in the twin-free sample corresponding to the prevention of the progress of this fluctuation, to add to the alignment of the oxygen configurations in Cu-O basal planes. On the other hand, it is considered that those show the similarity of the Nd/Ba ratio fluctuation and the oxygen configurations in Cu-O basal planes between the detwinned and the twinned samples. With regard to the shorter wavelength of the Nd/Ba ratio fluctuation in the twin-free samples, it is noticeable that the

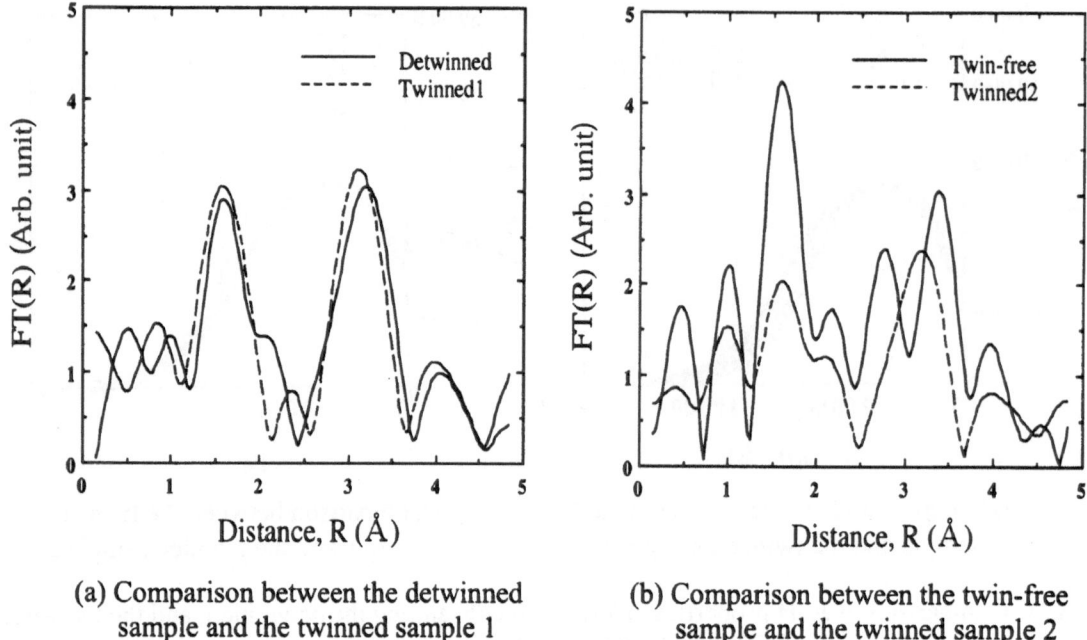

(a) Comparison between the detwinned
sample and the twinned sample 1

(b) Comparison between the twin-free
sample and the twinned sample 2

Fig. 2. Comparison of the magnitude of Fourier transform of the EXAFS oscillations among
the detwinned, the twin-free, and the twinned Nd123 samples after annealing.

distribution of oxygen atoms in Cu-O basal planes during annealing is restricted to a perpendicular
direction to the uni-axial stress. When the Nd/Ba ratio fluctuation proceeds in the samples, it is necessary for the oxygen atoms in the Cu-O basal planes to distribute along not only a parallel direction but
also a perpendicular direction to the uni-axial stress, because of the compensation for an excess of
electric charge attributed to the substitution of Nd^{3+} for Ba^{2+}. Therefore, the restriction of the oxygen
configurations in Cu-O basal planes due to the uni-axial stress should remarkably prevent the progress
of the Nd/Ba ratio fluctuation in the twin-free samples. On the other hand, it is considered that, once the
Nd/Ba ratio fluctuation in the detwinned samples occurs during annealing preliminarily without uni-
axial stress, it is quite difficult to redistribute this fluctuation on the following detwinning condition,
because of the high activation energy for the redistribution of the Nd/Ba ratio fluctuation in the samples.

CONCLUSION

The Jc-H properties for untwinned orthorhombic Nd123 single crystals were studied, which were fabricated by two different heat treatment processes such as the detwinning and the twin-free processes.
The Jc-H property for the twin-free crystals showed a larger peak effect than that for the detwinned
crystals. Furthermore, the magnitude of Fourier transform of the EXAFS oscillations was compared
among these crystals. These results suggested that the Nd/Ba ratio in the twin-free crystals was fluctuated with a shorter wavelength than that in the detwinned crystals.

Acknowledgments. This work was supported by the New Energy and Industrial Technology Development Organization (NEDO) as Collaborative Research and Development of Fundamental Technologies for Superconductivity Applications.

1. T. Hirayama and Y. Ikuhara, J. Mater. Res. **12** (1997) 293.
2. H. Schmid, E. Burkhardt, B. N. Sun and J.-P. Rivera, Physica C **157** (1989) 555.
3. M. Nakamura, H. Kutami and Y. Shiohara, Physica C **260** (1996) 297.
4. A. Oka, S. Koyama, T. Izumi and Y. Shiohara, Physica C **314** (1999) 269.
5. E. M. Gyorgy, R. B. van Dover, K. A. Jackson, L. F. Schneemeyer and J. V. Waszczak, Appl. Phys. Lett.
   55 (1989) 283.
6. J. B. Boyce, F. Bridges, T. Claeson, R. S. Howland and T. H. Geballe, Phys. Rev. B **36** (1987) 5251.
7. A. Oka, S. Koyama, T. Izumi and Y. Shiohara, Physica C **319** (1999) 249.

Synthesis of c-axis Oriented RE-Ba-Cu-O (RE=Sm and Nd) Superconductors and Performance of Superconducting Permanent Magnets Activated by Pulsed Fields

Uichiro Mizutani, Hiroshi Ikuta,* Tetsuhisa Hosokawa, Hiromasa Ishihara, Kouichi Tazoe, Tetsuo Oka,# Yoshitaka Itoh,# Yousuke Yanagi,# Masaaki Yoshikawa#

Department of Crystalline Materials Science, Nagoya University, Furo-cho, Chikusa-ku, Nagoya 464-8603, Japan
* Center for Integrated Research in Science and Engineering, Nagoya University, Furo-cho, Chikusa-ku, Nagoya 464-8603 Japan
# IMRA MATERIAL R&D Co., Ltd, Hachiken-cho, Kariya, Aichi 448-0021, Japan

Abstract: The c-axis oriented single-domain Ag-bearing Sm123 and Nd123 superconductors with 30-36 mm in diameter were synthesized by using the Nd123 (001) single crystal as a seed under Ar-gas flowing atmosphere. The maximum trapped fields are 9 Tesla at 25 K for the Sm123 and 6.3 Tesla at 50 K for the Nd123 in the static field cooling magnetization. The motion of the flux line in the Sm123 driven by the pulsed field was studied in comparison with that in the Y123. The Nd-Fe-B plate ($50 \times 50 \times 5 mm^3$) can be satisfactorily magnetized by scanning it over the Sm123 superconducting permanent magnet. We also report on the construction of a pulse field-driven face-to-face type magnetic field generator capable of producing 1.8 Tesla in an open space with the gap of 12 mm.

Keywords: superconducting permanent magnet, Sm123, Nd123, pulse field magnetization

INTRODUCTION

The trapped field of the $REBa_2Cu_3O_y$ (RE123) bulk superconductors has been enormously increased in the last few years by choosing either light rare earth element RE=Sm[1, 2], Nd [3, 4] and Gd [5] in place of Y or by irradiation with thermal neutrons [6]. A high-quality RE123 (RE=Nd, Sm and Gd) superconductor can be synthesized in a reduced oxygen gas atmosphere to suppress the Ba/RE substitution in the 123 matrix. The c-axis oriented single-domain Sm123 and Nd123 superconductors have been grown to 30-36 mm in diameter by reinforcing the mechanical strength with the addition of Ag_2O powders. It is pointed out that further increase in trapped field is limited not by the pinning force but by the mechanical strength of the sample.

We focus on three different subjects in this work. First, the trapped field ability of the c-axis oriented single-domain Ag-bearing Sm123 and Nd123 superconductors is discussed. Second, we discuss why the Sm123 can be magnetized by the practically important pulsed field magnetization (PFM) method [7, 8] to less extent than the Y123 and explain the combined use of the IMRA technique with a yoke to be very effective to overcome this difficulty. As the final subject, we show the successful magnetization of a large Nd-Fe-B plate by using the Sm123 superconducting permanent magnet and also the construction of a face-to-face magnetic field generator.

EXPERIMENTAL

The Sm123 samples containing Ag_2O up to 20 wt.% and Sm_2BaCuO_5 (Sm211) particles with the molar ratio of Sm123 to Sm211 equal to 3:1 or 3:2 were prepared under the Ar-gas flow atmosphere [1,2]. The growth of 211 particles was suppressed by adding 0.5 wt.%Pt, while the Nd123 (001) crystal was used as a seed to synthesize the c-axis oriented single domain Sm123 superconductor. The final product after post-annealing in oxygen-gas flow was 30-36 mm in diameter and 15 mm in thickness. The Nd123 superconductor containing 15 and 20 wt.%Ag$_2$O with Nd123:Nd422:BaCO$_3$=5:1:x (x=0 and 0.58) was likewise synthesized [3,4]. The Sm123 and Nd123 superconductors thus synthesized were magnetized by the static field produced by a 10 T superconducting solenoid magnet in the temperature range 25-80 K. The Sm123 samples were also magnetized by a pulsed field. A pulsed current with a rise time of about 1 ms was fed to the solenoid coil wound around the sample. The resulting magnetic field at the center of the coil is calculated from the current and coil constant. Its maximum field is called the applied field. For comparison, the Ag-free Y123 containing Y123:Y211=5:2 with 36 mm in diameter was also synthesized.

The temperature dependence of the axial component of the trapped flux density B$_z$ was measured by using the transverse-type Hall sensor mounted on the top surface of the sample (F W Bell, model BHT 921) or by scanning the Hall sensor 1.2 mm above the sample surface (F W Bell BHA 921). The measurement was made within 5-20 min after reducing the external field to zero. The dynamical motion of flux in the superconductor was studied by detecting the signal from a series of pick-up coils concentrically located on the surface of the cylindrical sample with 36 mm in diameter. The diameters of 7 coils "E2", "E1", "5" to "1" were 50, 35, 30, 24, 18, 12 and 6 mm so that coils "E1" and "E2" were located at the edge and outside of the sample, respectively. The pick-up coils were sandwiched by two similar samples with the gap of about 2 mm to minimize the straying field. The induced voltage was sampled with a rate of 1 μs.

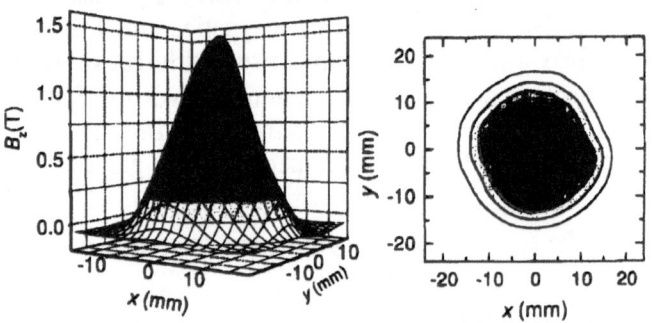

Figure 1. Trapped flux distribution at 77 K for the Sm123 sample containing 10 wt.%Ag2O with 36 mm in diameter. The axial component Bz was measured by mapping the Hall sensor at 1.2 mm above the surface of the sample after magnetizing at 4 T in the field cooling mode [2].

RESULTS AND DISCUSSION

Static field magnetization of Sm123 and Nd123 superconductors

The Sm123 and Nd123 samples were magnetized by using static magnetic field in the field cooling (FC) mode. Figure 1 shows the distribution of the axial component of the trapped flux density B$_z$ at 77 K for the Sm123 sample with the diameter of 36 mm. A single-cone distribution profile is obtained, indicating the absence of serious weak-links. Its value at 1.2 mm above the center of the sample surface reaches 1.7 Tesla (Note that 2.1 Tesla at 0.55 mm above the surface was recorded by the Hall sensor directly mounted onto the sample). Its temperature dependence is plotted in Fig.2 for Sm123 samples containing different amounts of Ag$_2$O. The value of B$_z$ increases rapidly with decreasing temperature for all samples. The maximum trapped field reaches 9.0 Tesla at 25 K for the Sm123 sample with

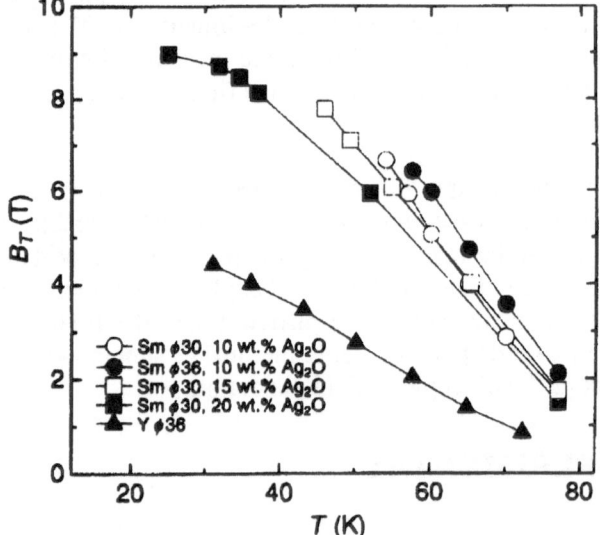

Figure 2. Temperature dependence of the trapped flux density measured at 0.55 mm above the surface of the sample by the Hall sensor 20 min after switching off the external field. The data for the Y123 are shown for comparison [2].

20 wt.% Ag_2O.

The superconducting transition temperature T_c of the Nd123 could increase to 94 K when the $BaCO_3$ content x is chosen to be 0.58. The trapped field for the Nd123 with 15 wt.%Ag_2O turned out to be 1.04 Tesla at 77 K and increased to 6.3 Tesla at 50 K.

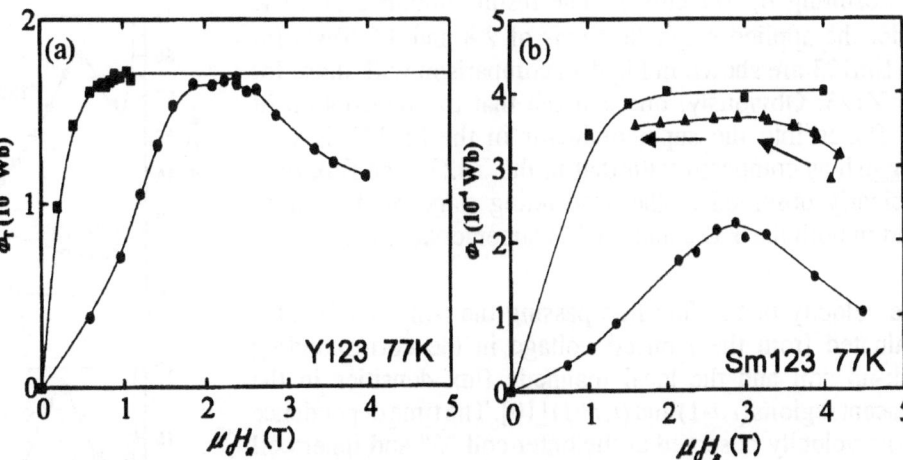

Fig.3 Applied field dependence of total magnetic fluxes trapped in (a) Y123 and (b) Sm123 samples with 36 mm in diameter. (■): static FC, (●): single pulsed field, (▲): the IMRA method coupled with soft-steel yoke.

Pulse field magnetization of the Sm123 superconductors

Figure 3 shows the applied field dependence of the total magnetic flux trapped in both Sm123 and Y123 superconductors at 77 K due to a single-pulsed field magnetization. As is clearly seen, the PFM technique can trap as many fluxes as the static FC mode for the Y123 but only 50 % of the fluxes captured by the static field for the Sm123. This poses a serious difficulty upon the practical application of the Sm123 superconductor as a superconducting permanent magnet in spite of its trapping ability superior to the Y123. As shown in Fig.3, however, the use of the soft-steel as a yoke coupled with the adoption of the "IMRA" technique could raise the trapping fluxes up to 90 % of the static FC magnetization. Here IMRA stands for the abbreviation of "Iteratively Magnetizing pulsed-field operation with Reducing Amplitudes" [9] and the method consists of applying a pulse field large enough to magnetize the central portion of the sample in the first operation and subsequent iterative PFM operations with reducing amplitudes [9]. To shed more light on the difference in the trapping ability between the Sm123 and Y123 and to elucidate the mechanism involved in the IMRA method coupled with a yoke, we have studied the motion of fluxes penetrated into the Sm123 and Y123 during the PFM by analyzing the signals measured by a series of pick-up coils.

The magnetic flux density $B_{ij}(t)$ averaged in between coils i and j at a time t is given by the difference in the total fluxes encircled by these two coils, each of which is calculated by integrating the induced voltage with respect to time up to t [10]. The spatial distribution of the magnetic flux density is then easily obtained from

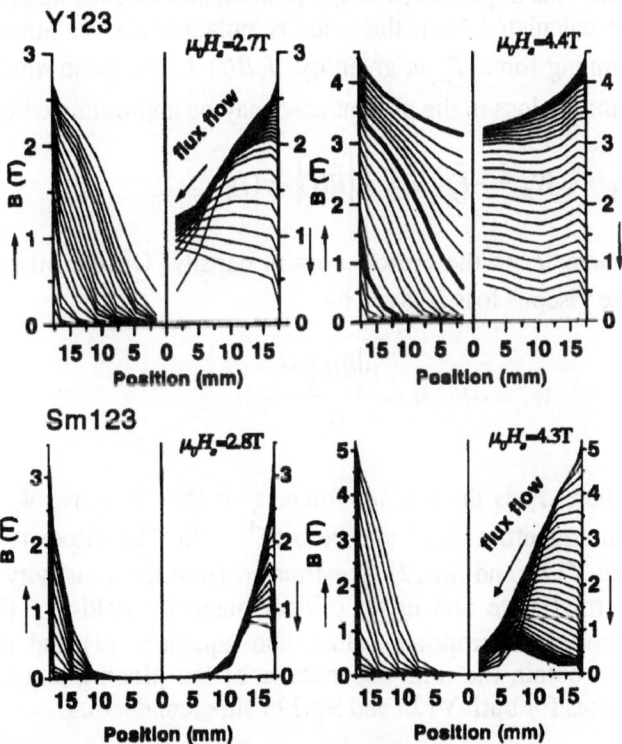

Fig.4. Flux density distribution during the PFM process in Y123 and Sm123 samples at 77 K. The center of the sample is taken as origin. The data on the left and right hand sides in each panel represent the ascending and descending stages of a single-pulse magnetization process, respectively.

the resulting $B_{ij}(t)$-t curves. The results obtained at 77 K under the applied magnetic fields of 2.8 and 4.3 Tesla for the Sm123 are shown in Fig.4 in comparison with those for the Y123. Obviously, one can see that the penetration of the fluxes into the superconductor in the Sm123 is very sluggish as compared with that in the Y123. The flux flow is clearly observed in the descending stage of the pulsed field in both the Y123 and Sm123 superconductors.

The velocity of the flux line passing the coil i can be also evaluated from the induced voltage in the corresponding pick-up coil and the local magnetic flux densities in the adjacent regions $(i, i$-1$)$ and $(i, i$+1$)$ [10]. The time dependence of the velocity observed at the outer coil "5" and inner coil "2" is plotted in Fig.5 for both the Sm123 and the Y123. The maximum velocity observed at the outer coil reaches 30 m/s at about 0.25 ms in the Y123, whereas it is only 8 m/s at 0.7 ms in the Sm123. On the other hand, the maximum velocity observed at the inner coil still maintains 18 m/s at about 0.3 ms in the Y123, whereas it is blurred and less than 7 m/s in the Sm123. The results indicate that the motion of the flux line in the Sm123 superconductor is much slower than that in the Y123.

The time dependence of the pinning and viscous losses can be calculated from the velocity obtained above. Since the pinning force $\vec{F}_p$ is given by $J_c\vec{B}(t)$ in the Bean model, the pinning loss in the present case may be approximated by

$$W_p = -\vec{F}_p \cdot \vec{v} = J_c\left|\vec{B}(t)\right| \cdot \left|\vec{v}(t)\right|, \tag{1}$$

where J_c is the critical current density. On the other hand, the viscous loss is given by

$$W_v = -\vec{F}_v \cdot \vec{v} = \frac{\eta\left|\vec{B}(t)\right|v^2}{\phi_o}, \tag{2}$$

where $\vec{F}_v$ is the viscous force, η is the viscosity, ϕ_o is the flux quantum equal to 2.067×10^{-15} Wb. The viscosity η for the Y123 and Sm123 is estimated from the resistivity in the normal state and upper critical magnetic field H_{c2} [7]. By inserting appropriate values into equations (1) and (2), we can obtain the time dependence of the pinning and viscous losses for both Y123 and Sm123 superconductors.

As shown in Fig.6, the pinning loss in the Y123 at the coil "3" reaches its maximum value of about 7×10^8 W/m^3 at 0.5 ms whereas the viscous loss 60×10^8 W/m^3 upon application of the external pulsed field of 4.3 Tesla at 77 K. In contrast, the pinning and viscous losses in the Sm123 superconductor

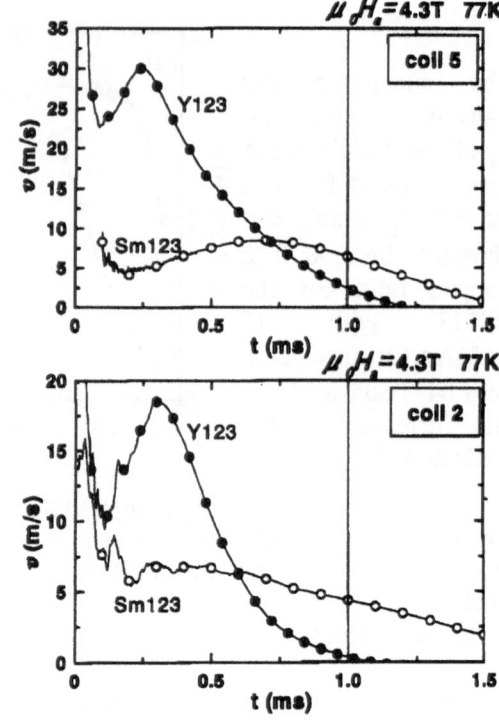

Fig.5 Time dependence of the flux velocity observed at coils "5" and "2" in Sm123 and Y123 samples at 77 K. The vertical line indicates the peak of the applied pulsed field.

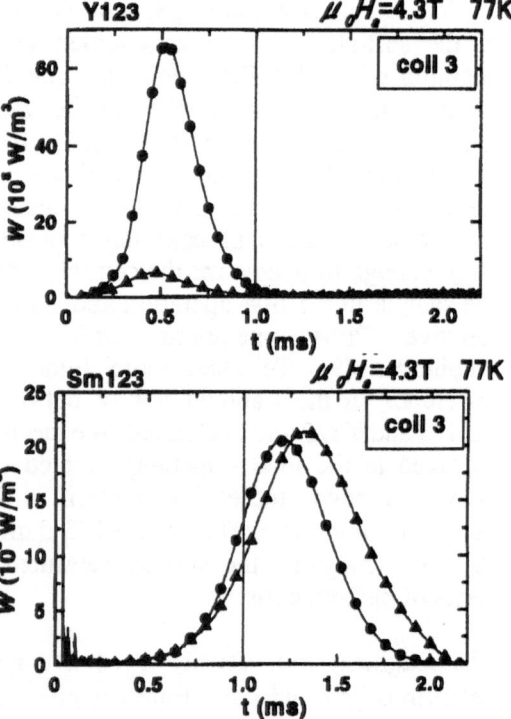

Fig.6 (▲) Pinning and (●) viscous losses in Y123 and Sm123 samples at the coil "3" at 77 K, when the applied field is 4.3 Tesla. The vertical line indicates the peak of the applied field.

are almost the same and take their maximum value of about $20x10^8$ W/m^3 at about 1.3 ms. The fact that the pinning force in the Sm123 is much stronger than that in the Y123 is well reflected in these data. Moreover, we see that the flux penetration occurs immediately after the operation of the pulsed field and the losses become maximum at about 0.5 ms in the Y123 but that the penetration is substantially retarded and the maximum losses occur at about 1.3 ms in the Sm123, at which the external pulsed field has already passed its peak. This explains the reason for the poor performance of the flux trapping ability due to the single pulse magnetization for the Sm123 samples.

We consider it to be quite effective to insert the superconductor into a pair of soft-steel yokes so that the superconductor is exposed to the external field longer than the duration of a pulsed field with 1 ms. The yoke is instantaneously magnetized by the pulse field and remains fully magnetized until the pulse field completely fades away. The magnetic field tends to be expelled from the superconductor but the long-surviving magnetic field of about 2 Tesla generated by the yoke will facilitate the penetration of fluxes into it. In addition, we believe it to be also effective to suppress the flux flow, which occurs at the descending stage of the pulsed field as shown in Fig.4.

In the case of the IMRA method, the first optimum applied field is strong enough to trap the magnetic fluxes at the center of the sample. Because of the heat generation, a large amount of fluxes escape from the periphery of the sample by passing through the region in between the mutually perpendicular facet lines, where the pinning force is believed to be relatively weak. The escape of the flux lines along the preferential paths is pronounced at low temperatures but the escape path becomes isotropic at 77 K [11]. The amplitudes of the successive pulses are progressively lowered so that the heat generation is suppressed and the escape of the fluxes is reduced. In this way, the magnetic fluxes down to the peripheral region can be gradually accumulated and the trapped field distribution approaches to a conical profile obtained by the static field as that shown in Fig.1.

3.3 performance as superconducting permanent magnet

First, we demonstrate that a rectangular Nd-Fe-B permanent magnet with the size 50x50x5 mm^3 can be satisfactorily magnetized by scanning its surface over the Sm123 superconducting permanent magnet, which is magnetized beforehand by a static field. The Sm123 superconductor with 36 mm in diameter is cooled by the Gifford-McMahon (GM) refrigerator down to 35 K in the static field of 7.2 Tesla. As shown in Fig.7, the trapped field of the Sm123 superconductor thus magnetized produces the z-component magnetic field exceeding 3 Tesla immediately above the surface of the head of the vacuum vessel. The demagnetized Nd-Fe-B plate was slowly scanned horizontally at the distance 2.5 mm above the surface along its long axis with 10 mm in pitch in every alternative direction. The resulting magnetic field trapped by the Nd-Fe-B permanent magnet is measured by scanning the Hall sensor 1.2 mm above its surface. The magnetic field distribution along the direction perpendicular to the magnetizing direction is shown in Fig.8 as a function

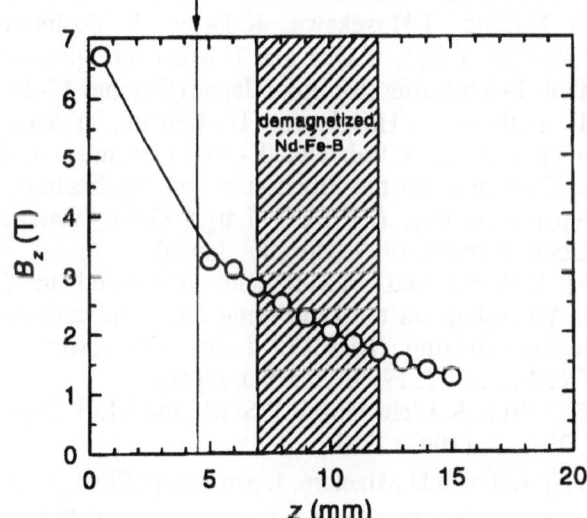

Fig.7 The z-dependence of Bz of the Sm123 superconducting magnet magnetized by the static field of 7.2 T at 35 K. The shaded area shows the region where the Nd-Fe-B plate passes. The arrow indicates the surface of the vacuum chamber.

of distance y, the origin of which is taken at the center of the plate. The data obtained by exposing the plate to the static field are included for comparison. It is clear that the Nd-Fe-B permanent magnet is magnetized by the superconducting magnet to the same extent as that obtained by exposing it to a uniform static field. It is also important to note that the distribution is fairly smooth without any signature due to the 10 mm-pitch scanning of the Nd-Fe-B plate along the direction perpendicular to the y axis.

We show in Fig.9 the face-to-face type superconducting permanent magnet system, which is driven by the pulsed field magnetization for a pair of the Sm123 superconductors mounted on the respective cold stages of the GM-refrigerators. The gap between the face-to-face two vacuum chambers can be varied. Even when the open gap is zero, the two Sm123 superconductors are 6.4 mm apart because of a finite gap between its surface and outer surface of the vacuum chamber. A pair of the Sm123 superconductors are magnetized by applying the IMRA method with the maximum applied field of 6.2 Tesla at 32 K in the presence of the soft-steel yoke. The magnetic field produced at the center of the gap is 1.8 Tesla when the gap available as an open space is chosen to be 12 mm. This field is not strong enough to magnetize the Nd-Fe-B plate at present but further increase in magnetic field beyond 3 Tesla is in progress.

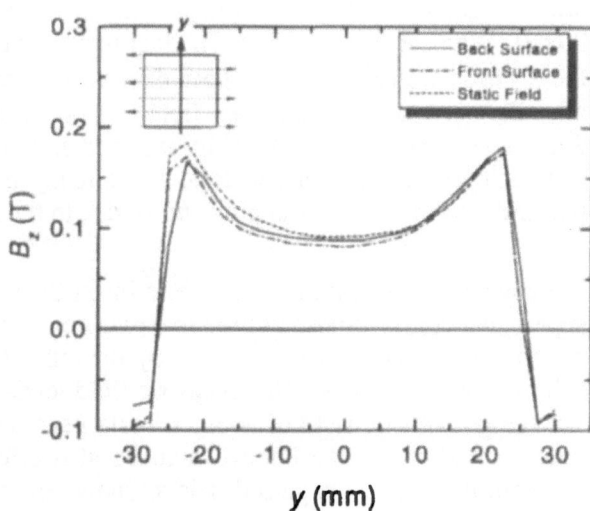

Fig.8 Trapped field distribution of the Nd-Fe-B plate magnetized by the Sm123 superconducting maget. Inset shows the direction of 10 mm-pitch scanning of the Nd-Fe-B plate.

[1] H.Ikuta, A.Mase, Y.Yanagi, M.Yoshikawa, Y.Itoh, T.Oka and U.Mizutani, Supercond.Sci.Technol. 11 (1998) 1345

[2] U.Mizutani, A.Mase, H.Ikuta, Y.Yanagi, M.Yoshikawa, Y.Itoh and T.Oka, Mat.Sci.Eng. (1999)

[3] H.Ikuta, T.Hosokawa, K.Tazoe, M.Yoshikawa, Y.Yanagi, Y.Itoh, T.Oka and U.Mizutani, presented at this ISS meeting (Morioka, Japan) October 17-21.

[4] H.IKuta, T.Hosokawa, U.Mizutani, Y.Yanagi, M.Yoshikawa, Y.Itoh and T.Oka, Presented at 2nd Int.Workshop on the Processing and Applications of Superconducting (RE)BCO Large Grain Materials, (October 19-22, 1999, Morioka, Japan)

[5] S.Nariki and M.Murakami, Presented at 2nd Int.Workshop on the Processing and Applications of Superconducting (RE)BCO Large Grain Materials, (October 19-22, 1999, Morioka, Japan)

[6] Y.Ren, R.Weinstein, R.P.Sawh and J.Liu, Physica C 282-287 (1997) 2301

Fig.9 Face-to-face pulsed field-driven magnetic field generator. The magnetic field of 1.8 Tesla is available at present, when an open space is 12 mm wide.

[7] Y.Itoh and U.Mizutani, Japan.J.Appl.Phys. 35 (1996) 2114

[8] Y.Itoh, Y.Yanagi and U.Mizutani, J.Appl.Phys. 82 (1997) 5600

[9] U.Mizutani, T.Oka, Y.Itoh, Y.Yanagi, M.Yoshikawa and H.Ikuta, Applied Superconductivity, 6 (1998) 235

[10] A.Terasaki, Y.Yanagi, Y.Itoh, M.Yoshikawa, T.Oka, H.Ikuta and U.Mizutani, Proc. of the 10th Int.Symp. on Superconductivity (ISS'97), October 27-30, 1997, Gifu, pp.945-948

[11] Y.Yanagi, Y.Itoh, M.Yoshikawa, T.Oka, T.Hosokawa, H.Ishihara, H.Ikuta and U.Mizutani, presented at this ISS meeting (Morioka, Japan) October 17-21.

Large YBCO - Monoliths with Peak Effect and High Trapped Fields

Gernot Krabbes[1], Günter Fuchs[1], Peter Schätzle[1], Stefan Gruss[1], Jai W. Park[2], Ferdinand Hardinghaus[2]

[1]IFW - Institute of Solid State and Materials Research Dresden, D-01171 Dresden
[2] Solvay Barium Strontium GmbH, D-30002 Hannover, Germany

Abstract: The present paper reports on the trapped fields of 9.0 and 11.4 T (on top of a single YBCO cylinder) which have been trapped at 42.5 and 17 K, respectively. The composite material consists of a YBCO-matrix with Ag inclusions and was reinforced by a bandage from steel. A modified melt crystallisation process was applied to grow the superconducting bulk. On the other hand, the trapped fields at 77 K benefit from the newly proposed chemical doping by Zn on Cu plane sites which results in a well pronounced peak effect between 1 and 3 T. Thus, more than 1.1 T has efficiently been trapped in a cylindrical sample of only 25 mm in diameter.

Keywords: $YBa_2Cu_3O_7$ bulk, substitution of Cu, peak effect, trapped field

INTRODUCTION

Trapped fields in bulk high T_c superconductors are limited by the critical current density j_c and the appearing weak links, both restricting the supercurrent in the critical state. The most striking factor to achieve very high fields, however, has turned out to be the limited strength to resist the strong Lorentz force appearing during magnetising the samples. The highest fields so far have been trapped in U-doped YBCO samples irradiated by thermal neutrons [1]. Damages induced by fission products improved pinning and, eventually, mechanical strength. In previous papers, admixing Ag has been proposed to improve the mechanical strength of the YBCO material [2, 3]. The investigation of parts of the phase diagram, which are relevant to the melt crystallisation process have recently been reported in ref [4], from which it becomes clear that Ag does not significantly occupy lattice sites in the Y-123 phase, and preparation has to be directed to a composite material with homogeneously distributed Ag inclusions.

On the other hand, in the region of applied fields between 0.5 and 5 T, weak links and critical current densities are the limiting factors [5]. Therefore, alternative RE-123 phases are of increasing interest due to the appearing peak in the j_c vs. H representation. The appearance of the peak effect [6, 7] as known so far is attributed to mechanisms which lower T_c or B_{irr}, respectively as it becomes obvious comparing the solid solutions in $RE_{1+y}Ba_{2-y}Cu_3O_7$ (RE = Nd, Sm ....) with the fixed composition of $YBa_2Cu_3O_7$.

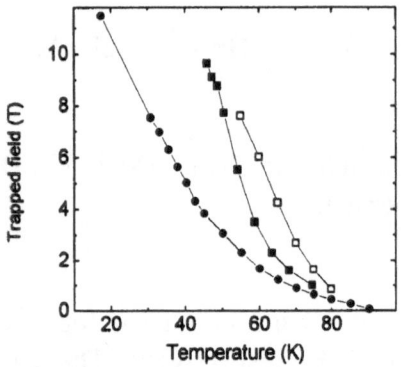

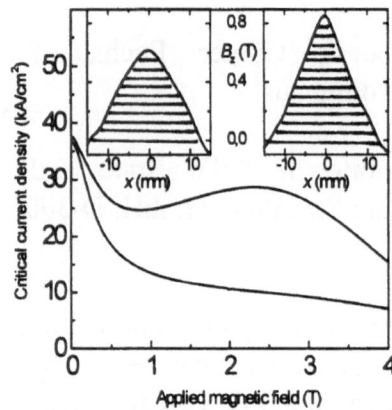

Fig. 1: Trapped field vs. temperature in one single disk of YBCO/Ag (●).and in Ag-free mini-magnets ■ without and □ with peak effect)

Fig. 2: Sections across Bean's cones and dependence of j_c on the applied magnetic field at 77K; left figure and lower line: pure, right and upper line: Zn doped

VERY HIGH TRAPPED FIELDS IN YBCO/Ag COMPOSITES

Cylindrical melt-textured YBCO bulk samples of 25 mm diameter were prepared following the modified melt crystallisation process (MCP, [8]) from a precursor which contains Ag powder admixed and using seeds from Sm-123. The dependence of the maximum remanent field on temperature was measured in the cryostat of a superconducting 18 T magnet. The disks were field cooled from 100 K to the measuring temperature. Then, the magnetic field was slowly reduced to zero at a rate of 0.1 T/min in order to avoid flux jumps due to thermal instabilities.

Very high trapped fields up to 11.4 T have been achieved by the addition of Ag and applying a bandage from stainless steel in order to strengthen the samples against the magnetic tensile stress [9]. Investigating the microstructure reveals that small Ag precipitates have been formed during the solidification which hinder the propagation of cracks, thus increasing the mechanical strength. Fig. 1 represents the trapped field on top of the enforced single cylinder versus temperature compared with two mini-magnets each consisting of a pair of silver - free YBCO cylinders. The latter reveal higher j_c, but they break under the influence of the Lorentz force already near 9.6 and 8 T, respectively.

PEAK EFFECT BY ZN DOPING

The precursor powder was an intimate mixture of $YBa_2Cu_3O_{7-\delta}$, 0.2 to 0.4 mole Y_2O_3 per 1 mole of $YBa_2Cu_3O_{7-\delta}$, 0.5 wt. % Pt and the Zn dopant, the amount of which is not larger than 1% in the experiments. The MCP was applied to cylinders of 25 and 35 mm diameter, respectively followed by oxidising in oxygen flow. Neither inhomogeneity due to enrichment of the dopant nor secondary phases nor alteration of the twin structure have been detected..

The dependence of j_c on applied field is shown in Fig. 2 for two YBCO samples. A pronounced peak effect is observed near 3 T for the Zn doped material which was prepared under equivalent annealing conditions and at the same final-oxidizing temperature (less than 400 °C)

as the sample without Zn. The advantage of the peak effect for achieving high trapped fields for temperatures above 40 K can be concluded from Fig. 1 and from the field of 1.12 T, which was trapped in a 25 mm cylinder of the new material compared with the value for 750 mT for the conventional composition. Fig. 2 also illustrates the different field profiles of both samples which are closely related to the corresponding j_c vs. B relationship.Field induced pinning is expected to have the same reason as the depression of T_c, namely the arising magnetic moment in the proximity of the nonmagnetic Zn impurity, which cannot form the singlet state as usual in the CuO_2 plane.

CONCLUSIONS

Very high fields can be trapped in YBCO/Ag composite material which is reinforced by a steel bandage. Ag-inclusions dissipate mechanical energy thus preventing crack propagation. In the present state, j_c is lower than in Ag-free materials. Therefore, cooling to 17 K is necessary to trap 11.4 T which is the highest field so far has been achieved, whereas 9.0 T at 47 K can be achieved with the modified YBCO revealing the peak effect. No irradiation treatment was applied.

Zn for Cu substitution on Cu chain sites has been developed to be a proper method to generate a peak effect at 77 K. This chemical route results in YBCO material of high efficiency for trapped field and levitation applications.

1. R. Weinstein, J. Liu, Y. Ren, R.P. Sawh, D. Parks, C. Foster , V. Obot, Proc. 10th Anniversary HTS Workshop on Physics, Materials and Applications, Houston, March, 1996
2. S. Kohayashi, S. Yoshizawa, H. Miyariri, H. Nakane, S. Nagaya, Materials and Engeneering **B 53** (1998) 70
3. P. Schätzle, G. Krabbes, S. Gruß, G. Fuchs, IEEE Transactions Appl. Supercond. **9** (1999) 2022
4. U. Wiesner, G. Krabbes, M. Ueltzen, C. Magerkurth, J. Plewa, H. Altenburg, Physica C **294** (1998) 17
5. G. Fuchs, C. Wenger, A. Gladun, S. Gruss, P. Schaetzle, G. Krabbes, J. Fink, K. H. Müller, L. Schultz, Applied Superconductivity 1999,. Inst. of Phys. Conf. Series, to be publ. 1999
6. S. Iwata, S. Nagaya, H. Ikuta, U. Mizutani, Advances in Superconductivity XI, Proceed.. ISS '98 Intern. Symp. Superconductivity, Springer publ. Tokyo 1999, p. 657
7. A. Erb, E. Walker, J. Y. Genoud, R. Flükiger, Physica C **282-287** (1997) 89
8. G. Krabbes, P. Schätzle, W. Bieger, G. Fuchs, Applied Supercond. **6** (1998) 61
9. S. Gruss, G. Fuchs, P. Schaetzle, G. Krabbes, J. Fink, K.-H. Müller, L. Schultz, Applied Superconductivity 1999,. Inst. of Phys. Conf. Series, to be publ. 1999
10. G. Krabbes, G. Fuchs, P. Schätzle, S. Gruss, J. W. Park, F. Hardinghaus, S. L. Drechsler, R. Hayn, Applied Superconductivity 1999,. Inst. of Phys. Conf. Series, to be publ. 1999

The authors are indebted to Mrs G. Stöver, B. Preuß, B. Thaut for careful technical assistance. Support by BMBF, DFG and Fonds der Chemischen Industrie is gratefully acknowledged.

MULTI-SEEDING OF YBCO SUPERCONDUCTORS

CHAN-JOONG KIM,[1] YOUNG A. JEE,[1] HO-JIN KIM,[2] JIN-HO JOO,[2] SANG-CHUL HAN,[3] HAN-YOUNG-HEE HAN,[3] TAE-HYUN SUNG,[3] SANG-JUN KIM[3] AND GYE-WON HONG[1]

[1]Superconductivity Research Project, Korea Atomic Energy Research Institute, P. O. Box 105, Yusong, Taejon 305-600, Korea
[2]Schools of Metallurgy and Materials Science Engineering, Sungkyunkwan University, Suwon, Kyounggi-do 440-756, Korea
[3]Power System Laboratory, Korea Electric Power Research Institute, 103-16, Munjidong, Yusong, Taejon 305-380, Korea

ABSTRACT: Due to the low yttrium solubility in Ba-Cu-O liquid, the growth rate of a $YBa_2Cu_3O_7$ (Y123) grain in the liquid is low. It takes a few hundred hours to fabricate a large single grain YBCO superconductor. A multi-seeding technique may shorten the processing time needed for the fabrication of large area textured YBCO superconductors. In this study, we tried a top-seeded melt growth process combined with multi-seeding. The processing variables, the growth mode of Y123 grains at the top surfaces, and the levitation forces and trapped magnetic fields of the multi-seeded sample are reported.

INTRODUCTION

Among the melt processes developed so far, the top-seeded melt growth (TSMG) process [1] is known to be an effective technique for the fabrication of large single grain YBCO superconductors. In this technique, a $NdBa_2Cu_3O_7$ or $SmBa_2Cu_3O_7$ seed is placed on the top of a YBCO compact made from a mixture of $YBa_2Cu_3O_7$ (Y123) and Y_2BaCuO_5 precursor powders, and then melt texture growth heat treatment follows. During slow cooling through a peritectic temperature (T_p) after appropriate holding above T_p a Y123 grain grows from the seed. Due to the low growth rate of the Y123 grain, however, it takes several hundred hours to grow a large size Y123 grain sample. In case of single seeding, total processing time is determined by the time required for the growth front of a YBCO grain nucleated at the central seed to reaches to the edges of the YBCO compact. If a large number (n) of seeds are placed at the top surface, the processing time can ideally be reduced to 1/n of that of the single seeding, since n numbers of Y123 grains grow at the top surface simultaneously. In this study, we tried multi-seeding for the fabrication of large area textured YBCO superconductors. We report the growth mode of Y123 grains at the top surfaces and the relevant magnetic properties.

EXPERIMENTALS

YBCO superconductors were prepared by the TSMG process combined with multi-seeding. The seeds used in this study were single grain Sm1.8 slabs with a composition of $Sm_{1.8}Ba_{2.4}Cu_{3.4}O_x$, prepared by the conventional melt process. $Y_{1.8}Ba_{2.4}Cu_{3.4}O_x$ (Y1.8) powder was pressed using a 3 cm rectangular mold into pellets. The surfaces of Y1.8 compacts were coated with Yb_2O_3 powder to prevent the undesirable Y123 nucleation at the sample surface [2]. Different numbers of seeds (one to six) were placed at the top surfaces of the Y1.8 compacts. The seeded Y1.8 compacts were melt-processed following the heating cycles of Fig. 1.

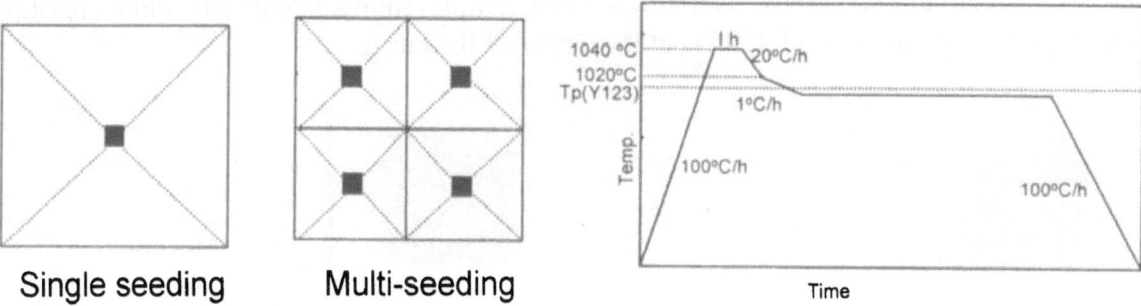

Fig. 1 Concept of multi-seeding and the heating cycles of the TSMG process.

The prepared Y1.8 samples were annealed at 500 °C for 50 h in flowing oxygen for oxygen diffusion. Magnetic levitation forces at 77 K were estimated from force-distance hysteresis curves between samples and a Nd-B-Fe permanent magnet with a surface field of 5000 Gauss and a diameter of 20 mm. Samples were cooled with and without the magnetic field of the permanent magnet. The surface magnetic flux density of the samples was measured by a Hall probe (Lakeshore model HGCT 3020).

RESULTS AND DISCUSSION

Figure 2 is photos of the top surfaces of TSMG-processed Y1.8 samples combined with multi-seeding. It can be seen that Y123 grains corresponding to the numbers of the seeds grew at the top surfaces with a rectangular growth mode.

Fig. 2. Photos of top surfaces of TSMG-processed YBCO samples with multi-seeding.

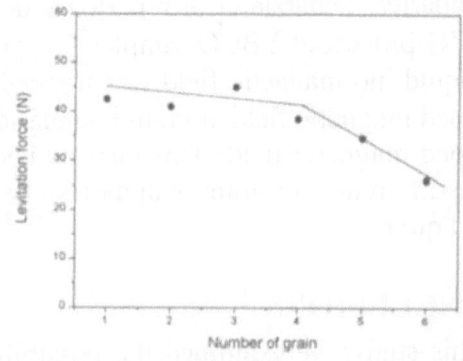

Fig. 3. Levitation force versus the number of Y123 grains of TSMG-processed YBCO samples.

Figure 3 shows the maximum levitation force versus the number of Y123 grains, estimated from force-distance curves at 77 K. The maximum levitation force of the sample prepared with single seed is 43 N. The levitation forces of the multiple seeded samples maintain similar value up to triple seeded case.   But it decreased significantly when the number of seeds exceeds 4, showing a maximum value of 25 N, which is about 60 % of that of the single seeded sample. The decrease of the levitation force is attributed to the presence of the grain boundaries at the top surface.

Figure 4 show the trapped magnetic field profiles of TSMG-processed YBCO samples with one, two, three and six seeds. The single seed sample shows symmetric field distribution showing a maximum value of 3500 G at the center of the curve.

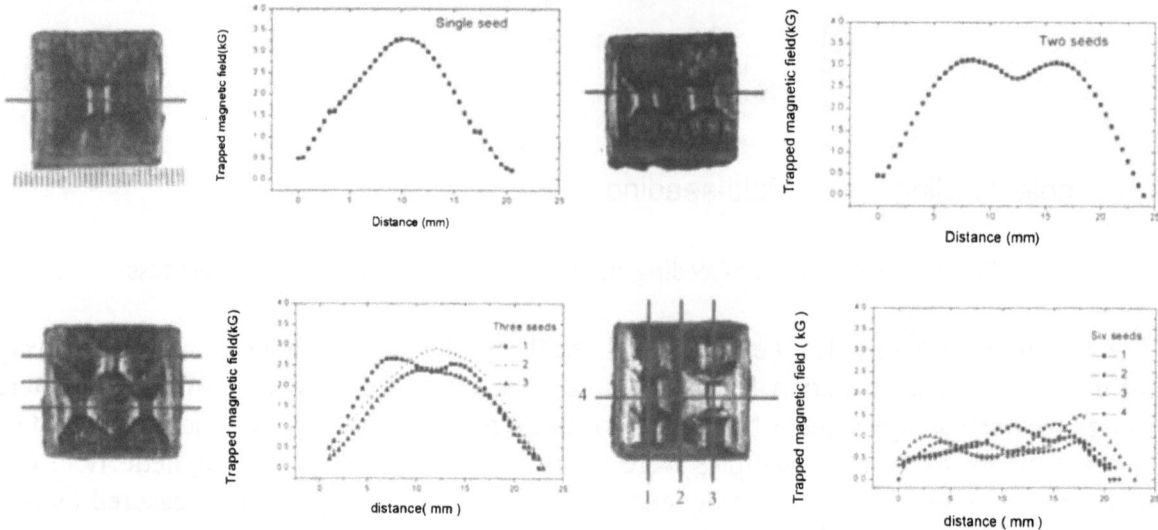

Fig. 4. Trapped magnetic fields of TSMG-processed YBCO samples with multi-seeding.

The double-seed sample shows two peak points and one deflection between the two peaks. The locations of the two peaks and the one deflection correspond to the centers of the two grains grown at the two seeds and the grain boundary, respectively. The trapped magnetic fields of the peak and the deflection are 3200 G and 2900 G, respectively. The sample prepared using six seeds show several defection points. The trapped magnetic fields are relatively low compared to those of the samples with a fewer numbers of grains, probably due to the presence of grain boundaries. Schatzle et al reported much liquid was included at the grain boundaries of the TSMG-processed YBCO samples prepared with multi-seeding [3]. As a result of the presence of liquid, no magnetic field was trapped at the grain boundary. In contrast to their result, the trapped magnetic field at grain boundaries of this study was more than 50 % of the maximum-trapped magnetic field. This may be because the fabrication conditions of the two processes were different. The grain boundaries of this study may be more strongly linked or may contain less liquid.

CONCLUSIONS

In this study, we examined the possibility of multi-seeding for the fabrication of large area textured YBCO superconductors. The levitation force and trapped magnetic field of the samples decreased with an increasing number of seeds, but the degree of the degradation was not much higher. Improvement of the grain boundary microstructure is necessary to minimize the degradation of the magnetic properties.

REFERENCES

1. M. Morita, S. Takebayashi, M. Tanaka, K. Kimura, K. Miyamoto, K. Sawano, Adv. Supercond. **III** 733 (1991).
2. C-J. Kim, Y. A. Jee, S-C. Kwon, T-H. Sung and G-W. Hong, Physica C **315** 263 (1999).
3. P. Schatzle, G. Krabbes G. Stover, G. Fuchs and D. Schlafer, Supercond. Sci. Technol. **12** 69 (1999).

EXPLORING THE TERNARY SUPERCONDUCTORS OF THE TYPE (Nd,Eu,Gd)-Ba-CuO

Muralidhar Miryala, Michael R Koblischka and Masato Murakami

Superconductivity Research Laboratory, ISTEC-SRL, Division 3, 1-16-25, Shibaura, Minato-ku, Tokyo, 105, Japan

Abstract: It has already been demonstrated that the $(Nd_{0.33}Eu_{0.33}Gd_{0.33})Ba_2Cu_3O_y$ (NEG) bulk superconductors have a superior performance as compared to YBCO or NdBCO, at all fields and temperatures. Initially we started from an even ratio of the light rare earth elements (LRE) in the matrix and in the 211 particles. After acquiring promising results with these samples, we now explore the effect of variation in LRE mixing ratio. We have prepared $(Nd_{0.33}Eu_{(0.66-x)}Gd_x)Ba_2Cu_3O_y$ bulk samples grown by the OCMG-process, in which x ranged from 0 to 0.33. The heat treatment profiles for melt processing were determined according to the peritectic decomposition temperatures obtained from the thermal analysis measurements. The sample with x = 0.25 had a large critical current density (J_c) of 90 000 A/cm^2 at 77 K and 1.7 T.

Key words: melt processing, NEG-123, LRE mixing, high critical current density, flux pinning

INTRODUCTION

Recently, the development of new $(Nd_{0.33}Eu_{0.33}Gd_{0.33})Ba_2Cu_3O_y$ bulk superconductor materials called NEG-type have demonstrated that the microstructure, as well as flux pinning properties of these new materials are strikingly different from those of the YBCO or NdBCO [1-3]. Critical current densities (J_c's) and irreversibility fields have improved significantly at liquid nitrogen temperature[3]. Our recent experimental results also indicated that the optimum configuration of the superconducting matrix in the NEG system is not necessarily Nd:Eu:Gd=1:1:1, and the composition of the matrix can be influenced by using different LRE mixtures within the matrix or by adding different concentrations of 211 particles in the matrix [4-5]. Therefore it is important to further optimise the matrix composition to improve the critical current density and strength the flux pinning. For this study, we chose a fixed amount of Nd, and changed the Eu and Gd concentrations, while keeping the oxygenation procedure constant.

EXPERIMENTAL

High-purity commercial powders (5N) of Nd_2O_3, Eu_2O_3, Gd_2O_3, $BaCO_3$ and CuO were weighed to have a nominal composition of $(Nd_{0.33}Eu_{(0.66-x)}Gd_x)Ba_2Cu_3O_y$ in which x ranged from 0 to 0.33. The starting powders were ground thoroughly and calcinated at 880 ^{o}C for 24 h with intermediate grinding, and pressed into pellets. The sintering was carried out at 925 ^{o}C for 15 h. This entire process was repeated three times under low oxygen partial pressure. Finally, the precursor powders were pressed into pellets with a diameter of 20 mm and a thickness of 15 mm, which were subjected to cold isostatic press (CIP) with a pressure of 2000 Kg/cm^2. A MgO (100) seed was placed on the top and centre of the pellet, which was subsequently OCMG-processed in 0.1% partial pressure of O_2 with a gas flow rate of about 300 ml/min. The details of the heat treatment schedule and oxygen annealing process can be found elsewhere [1]. Magnetization hysteresis loops (MHLs) were measured mainly at 77 K in applied fields up to 7 T using a commercial SQUID magnetometer (Quantum Design, model MPMS7). The scan length was 1 cm to minimize field inhomogeneties. The external magnetic field was always applied parallel to the c-axis of the samples.

RESULTS AND DISCUSSION

Figure 1 presents the current density characteristics, J_c of the NEG-123 samples measured at 77 K for fields applied parallel to the c-axis. In all cases, we observe a strongly developed secondary

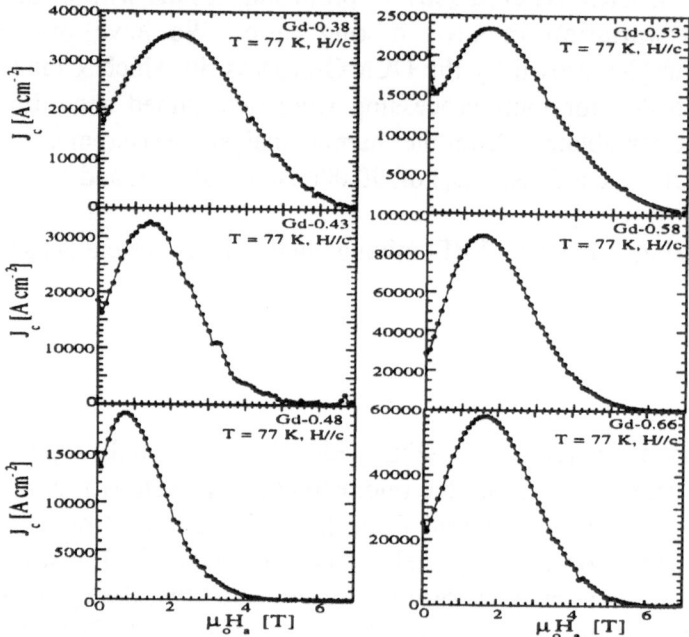

Fig. 1. J_c-B properties (77K, B $\parallel$ c-axis) for (Nd,Eu,Gd)-Ba-Cu-O superconductors with varying Gd content x, ranging between 0.38 and 0.66. Note the high critical current density around 90 000 A cm^{-2} at the peak field 1.7 T. All samples are prepared by Ar-0.1% partial pressure of O_2.

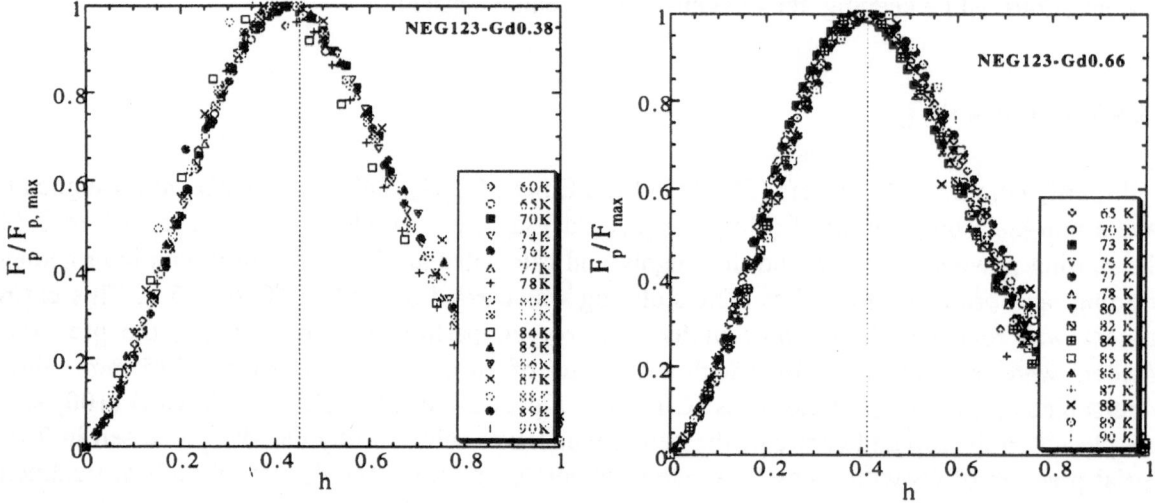

Fig. 2. Fp/Fp,max scaling of NEG-123 (Gd=0.38 and Gd=0.66) samples at temperature between 65K and 90K. The scaling is found to be excellent.

peak effect. It is evident that the Gd content has a large influence on the J_c-B properties. The peak position is shifted from 2.3T to 1.7T as the concentration of Gd varied from x=0.38 to x=0.66. The irreversibility fields also decreased as the Gd content increased. We observed a very high irreversibility field around the Nd:Eu:Gd ratio close to 1:1:1. However the high peak J_c of 90 000 A cm^{-2} at 1.7 T (77K, H//c,) in the sample with x = 0.58. These results clearly showed that NEG-123 systems offer one more degree of freedom. The physical properties can be modified to suit the end use applications, by changing the matrix ratio.

To further clarify this mechanism we performed a scaling analysis of normalized volume pinning forces, F_p, versus a reduced field h = H_a/H_{irr} for two sample: x=0.38; and x=0.66. The results are presented in fig. 2. The peaks are at h_0 = 0.45 for x=0.38 and 0.41 for x=0.66. It is evident that in the Nd:Eu:Gd ratio the Gd content influence the flux pinning strength. In our privies work, in the case of Nd:Eu:Gd, 1:1:1, we found the peak position at h_0 = 0.50. Our present results show that matrix mixing ratio close to 1:1:1 and also a matrix with higher Gd content will be promising materials for high magnetic field and as well as low field applications.

SUMMARY

Melt-textured NEG-123 samples with different concentration of Gd were prepared by the OCMG-process. Critical current density measurement analysis indicated that the NEG-123 materials can be used for various applications at 77K, since their magnetic properties can be modified by adjusting the mixing ratio of three rear earth elements. A large critical current density (J_c) of 90 000 A/cm^2 was achieved at 77K and 1.7 T for H || c axis. The volume pinning forces versus a reduced field analysis showed that the peak position is shifted from h_o=0.50 (x=0.33) to 0.41 (0.66) which indicated that the matrix mixing ratio is more important for strengthening δT_c-type pinning .

Acknowledgements

This work was supported by New Energy and Industrial Technology Development Organization (NEDO) as a part of its Research and Development of Fundamental Technologies for Superconductor Applications Project under the New Sunshine Program administrated by the Agency of Industrial Science and Technologies M.I.T.I of Japan.

References

1. M. Muralidhar, M. R. Koblischka, T. Saitoh, and M. Murakami, Supercond. Sci. Technol. **11**, pp. 1349-1358 (1998).
2. M. Muralidhar, M. R. Koblischka, and M. Murakami, Physica C **313**, pp. 232-240 (1999).
3. M. R. Koblischka, M. Muralidhar and M. Murakami, Appl. Phys. Lett. **73**, pp. 2351-2353 (1998).
4. M. Muralidhar, and M. Murakami, Physica C **309**, pp. 43-48 (1998).
5. M. Muralidhar, M. R. Koblischka, and M. Murakami, Supercond. Sci. Technol. **12**, pp.555-562 (1999).

Fe-Doped YBCO Single Crystals Grown By Crystal Pulling

Xin Yao, Akihiro Oka, Teruo Izumi and Yuh Shiohara

Superconductivity Research Laboratory, ISTEC, 1-10-13 Shinonome Koto-ku,   Tokyo 135, Japan

Abstract: Fe-doped $YBa_2Cu_3O_{7-\delta}$ (Fe-YBCO) single crystals were grown by a crystal pulling technique. Cu sites substituted by Fe were realized up to 5.83% due to varying Fe content in the liquid in the range of 0-3.73 mol%.   The increasing substitution of Fe at Cu sites resulted in decreasing superconducting transition temperature (T_c), which was strongly dependent on the annealing temperature.   A crystal growth model of Fe-YBCO solid   solution   was also discussed.

Keywords: superconductors, single crystal, substitution

INTRODUCTION

Cationic substitutions in the Y-Ba-Cu-O system have become more and more attractive.   This is because substitution studies can provide some insight into the mechanism of superconductivity and also spread the area of practical applications[1].   We earlier reported growth of Zn-doped YBCO single crystals, as a result of the cooperative work on the structural, chemical and physical properties measurements[2,3].   Recently, Fe doped YBCO materials were found to be stable with respect to diffusion of Fe, i.e. there is negligible Fe diffusion into its neighboring layers[4].   This means that the Fe doped YBCO crystal is a promising material for insulator or low T_c substrates for electronic devices application.   The aim of this work is to optimize a growth process to grow Fe-YBCO single crystals with a maximum substitution content.   Our interests will focus on the crystal growth   mechanism .

EXPERIMENTAL

The high quality $Ba_3Cu_5O_z$ powders were used as a raw material for the solvent. The Y_2O_3 crucibles were used to prevent contamination. Y_2BaCuO_5 (Y211) powders were placed at the bottom of the crucible and then the Ba-Cu-O composite was put on it. $YBa_2Cu_3O_z$ thin films were deposited on single crystals MgO by rf thermal plasma evaporation and were then used as seed crystals.   After melting of the whole powders, Fe_2O_3 pellets were charged into the melt.   Then the liquid was held at a high temperature of approximately 1030°C for 24 h to gain complete melting and a uniform distribution of Fe ions in the melt.   Crystal and flux compositions were analyzed by an ICP technique.   Crystals were cut into approximately 1.5x1.5x1.5 mm^3 and then oxygenated at different temperatures, 500°C for 3 d and 420°C for 5 d.   Magnetic measurements of single crystals were carried out using a SQUID.

RESULTS AND DISCUSSI ON

A   series of Fe doped Y123 single crystals were grown at the temperatures in the range of 1000-978°C. Substituted Fe contents varied from 0 to 2.76 mol% in crystals and accordingly x   varied in the range of 0-0.0583.   Fe content x appears to have an upper limit of 0.0583 in   Fe-YBCO solid solutions by the crystal pulling method, which is much lower than those prepared by the sintered method (x= 0.1~0.14, Ref[1]), but is very close to that achieved in the case of the flux method (x= 0.065, Ref[6]). Figure 1 illustrates the Fe substitution concentration in the crystal as a function of the Fe concentration in the

446

liquid. The Fe effective distribution coefficient of k_{eff}, which is the solid composition divided by the bulk liquid composition, were obtained to be 0.35 ~0.75. There is a higher value of k_{eff} when Cu sites were substituted more in YBCO crystals as shown in Fig. 1.

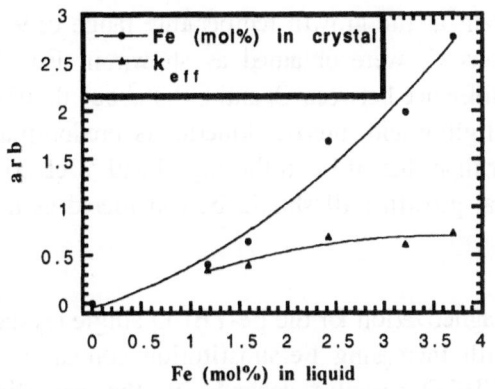

Fig. 1. The relationship of the Fe composition between liquid/crystal and k_{eff}

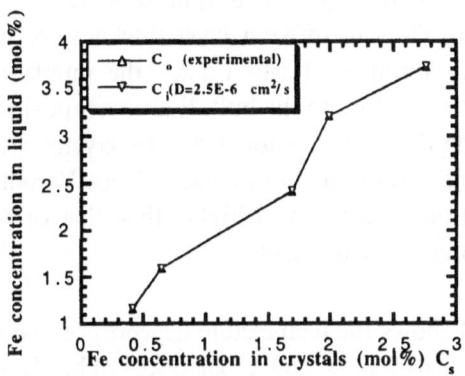

Fig. 2. The relationship between the crystal compositions and the liquid compositions.

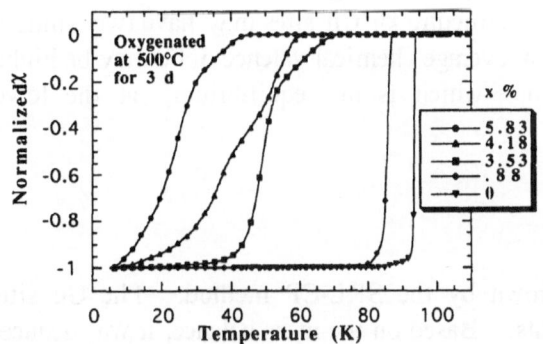

Fig. 3. The temperature dependence of normalized magnetization of Fe-YBCO samples.

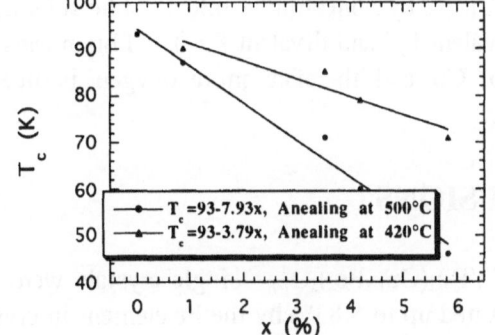

Fig. 4. The T_{co} as a function of the Fe content x in samples oxygenated at different temperatures.

In crystal growth, a thin solute diffusion boundary layer exists in front of the growing crystal interface. Three typical growth models could be considered, including: (1) local equilibrium with liquid diffusion, (2) interface kinetics with liquid diffusion and (3) interface kinetics. In the model one, a local equilibrium state is assumed at the interface, i.e. $C_i = C_L(T_s)$, whereas in the model three it is assumed that the interface liquid composition is equal to the bulk liquid composition, i.e. $C_i = C_o$. However, in most cases, crystal growth follows model two. In that case, the total driving force for crystal growth can be divided into two parts, diffusion factor $C_i - C_o$ and interface kinetics $C_L(T_s) - C_i$. Here C_o, C_i and $C_L(T_s)$ are bulk liquid composition, interface liquid composition and equilibrium liquid composition, respectively. It should be pointed out that all the compositions mentioned above are relevant to the Fe element only since the interest of this work concentrates on the correlation between Fe compositions (liquid and solid) and Fe-doping behaviors in the crystal growth. The interface liquid composition C_i can be calculated from the mass balance equation

$$R_c(C_i - C_s^*) = D_L \frac{\partial C}{\partial x} \approx D_L \frac{C_i - C_o}{\delta_c} \qquad (1)$$

$$\delta_c = 1.6 D_L^{1/3} v^{1/6} \omega^{-1/2} \qquad (2)$$

where R_c is crystal growth rate (cm/s) in the growth direction, D_L is the Fe diffusivity in the liquid (cm^2/s). δ_c is a diffusion layer thickness (cm) and can be estimated by the Cochran's analysis as shown in Eq.2[7]. Some appropriate physical properties ($D_L=2.5 \times 10^{-6}$ cm^2/s, $v = 0.14$ cm^2/s) and experimental data ($\omega = 120$ rpm=4 π rad/s) were used for calculation. It was assumed that the diffusivity D_L and diffusion layer thickness δ_c remain constant at the growth temperature range of 978-1000°C. From Eq.1 and Eq.2, the interface compositions C_i were obtained as shown in Fig. 2, which is very close to the bulk liquid composition C_o (the difference between C_i and C_o is about 0.26%), indicating that the diffusion factor for crystal growth is negligible and interface kinetics is predominant from the Fe solute point of view. Since Y content is lower than that of Fe in the liquid and Y effective distribution ratio is much higher than that of Fe, the crystal growth still should be considered as a Y diffusion controlled model.

Figure 3 shows the temperature dependence of normalized magnetization for the Fe-YBCO single crystals after oxygenation at 500°C for 3 d. The T_c decreases with increasing Fe substitution content x in crystals. It was found that the magnetization of Fe-YBCO samples depends on the annealing temperature. Figure 4 shows the correlation among the onset T_{co}, the Fe content x, and annealing temperature. It can be concluded that the lower annealing temperature of 420°C results in a higher T_{co}. The tendency was more significant in higher x samples. The Fe substitution at Cu sites causes the reduction of the onset T_{co} about 8 K/[Fe/(Fe+Cu)]% by annealing at 500°C while about 4 K/[Fe/(Fe+Cu)]% by annealing at 420°C. The substituted structure for Cu sites may have two kinds of Fe ions (trivalent Fe^{3+} and divalent Fe^{2+}). This means that average chemical valence of Fe may be higher than that of Cu and therefore more oxygen is needed, which is in equilibrium at the lower temperature.

CONCLUSIONS

A series of $YBa_2(Cu_{1-x}Fe_x)_3O_{7-\delta}$ single crystals were grown by the SRL-CP method. The Cu sites were substituted up to 5.83% by the Fe element in crystals. Based on the mass balance, it was deduced that the Fe interface composition C_i is almost the same as the Fe bulk liquid composition C_o, suggesting that interface kinetics is rate limiting and the diffusion factor is insignificant for crystal growth when considering the Fe solute only. The substitution of Fe at Cu sites results in the decrease of superconductivity of materials. The magnetization of samples strongly depends on the annealing temperature. The reduction of T_c was about 4 and 8 K/[Fe/(Fe+Cu)]% by annealing at 420 °C and 500 °C, respectively.

Acknowledgment. This work is/was supported by the New Energy and Industrial Technology Development Organization (NEDO) as Collaborative Research and Development of Fundamental Technologies for Superconductivity Applications.

1. J. T. Markert, B. D. Dunlap, and M. B. Maple, MRS Bulletin **14**, 37 (1989).
2. X. Yao, K. Ohtsu, S. Tajima, H. Zama, F. Wang, and Y. Shiohara, J. Mater. Res. **11**, 1120 (1996).
3. R. Hauff, S. Tajima, W. -J. Jang, and A. Rykov, Phys. Rev. Lett. **77**, 4620 (1996).
4. T. Usagawa, J. wen, S. Koyama, T. Utagawa, and K. Tanabe, Jpn. J. Appl. Phys. **38**, 436 (1999).
5. Y. Yamada and Y. Shiohara, Physica C **217**, 182 (1993).
6. Y. Eltse and O. Rapp, Phys. Rev. B **51**, 9419 (1995).
7. J. A. Burton, R. C. Prim, and W. P. Slichter, J. of Chem. Phys. **21**, 1987 (1953).

JOINING OF MELT-PROCESSED RE-Ba-Cu-O SUPERCONDUCTOR

Junya Maeda, Susumu Seiki, Makoto Kambara*, Yuichi Nakamura, Teruo Izumi and Yuh Shiohara

Superconductivity Research Laboratory, ISTEC, 1-10-13 Shinonome, Koto-ku, Tokyo 135-0062, Japan

Abstract: The melt-processed Y-Ba-Cu-O superconducting bulks were jointed using a Yb-Ba-Cu-O superconducting solder having a lower peritectic temperature than that of the YBCO matrix. Generally, the Yb-Ba-Cu-O superconducting solders were sandwiched between Y-Ba-Cu-O superconducting bulks. The problems of this sandwich soldering were the segregation of RE211 particles and voids formed at the final solidified regions that might interrupt the current flow through the jointed region. In this paper, we discussed the solidification behavior in this sandwich-soldering region for understanding the formation of these non-superconducting phases at the final solidification regions.

Keywords: Soldering, Joining, Yb-Ba-Cu-O, Solidification behavior, Particle pushing

INTRODUCTION

In the superconducting current leads, to obtain long length superconducting rods are effective to reduce heat leaks. Several process to joint melt-processed superconducting bulks have been investigated [1]-[3]. K. Salama et al. reported that the YBCO bulks could be jointed by the hot-press process [1]. They confirmed superconducting transitions at the jointed region by the transport Tc measurement.
On the other hand, K. Kimura et al. proposed the soldering method [2]. They used the Yb-Ba-Cu-O superconducting powder as the solder, and sandwiched it between the YBCO superconducting bulks. The peritectic temperature of the Yb-Ba-Cu-O system measured in our another study is about 1238K, that is lower than that of the Y-Ba-Cu-O system (1278K). They succeeded in melting the Yb-Ba-Cu-O solder only without decomposition of the Y-Ba-Cu-O matrix. In this paper, we also tried to apply this sandwich type joining to joint Y-Ba-Cu-O specimens, and discuss the solidification behavior at the soldered regions.

EXPERIMENTS

Single domain Y-Ba-Cu-O superconducting rods prepared for the jointed matrices were fabricated by unidirectional-solidification described in detail in our previous paper [4]. The Y-Ba-Cu-O superconducting rods were cleaved to obtain the specimens.
The Yb-Ba-Cu-O solder was prepared by the following procedure; raw materials of Yb_2O_3, $BaCO_3$, CuO were mixed thoroughly with the prescribed composition of $Yb_{1.2}Ba_{2.1}Cu_{3.1}O_x$. This mixed powder was calcined at 1173K for 12hours and grinned. This process was repeated four times.
Yb-Ba-Cu-O compound powder mixed with toluene was sandwiched between the cleavage planes of a couple of Y-Ba-Cu-O specimens. They were set into a temperature controlled muffle furnace in which the specimen temperatures were measured by the R-type thermocouple right below the specimens. Temperatures were heated up to 1263K, in order to melt the Yb-Ba-Cu-O solder without decomposing the Y-Ba-Cu-O matrix since the peritectic temperature of the Y-Ba-Cu-O material is about 1278K, while that of the Yb-Ba-Cu-O is about 1238K, as mentioned before. Subsequently, the specimens were cooled to 1183K with the cooling rates of 1.0K/h. The microstructures at the jointed region were observed by polarized optical microscopy, and scanning electron microscopy (SEM) equipped with the electro micro analyzer (EPMA).

---

*Present affiliation: IRC in supercond., Cavendish Lab., Univ. of Cambridge, UK

RESULTS

Figure 1 shows the back-scattering electron compositional images (BEI-COMPO) of the transversal sections of the jointed specimen. In these figures, thick dark and light dark region indicates the Y-Ba-Cu-O and the solidified Yb123 respectively, and white particles in the soldered region are Yb211 particles.

At the center of the soldered region, Yb123 are connected partially, but large flat bubbles and Yb211-dense-layer are also segregated at that region. These Yb211 dense layer and the large voids might interrupt the current path in the soldered region.

DISCUSSION

Firstly, It is discussed about the macro-segregation of the Yb211 particles. A. Endo et al. have been explained the macro-segregation of the high-temperature stable phase particles with the U-C-J pushing-trapping model [5][6]. RE211 particles are pushed by the RE123 growing interface in which the interfacial energy balance described in the following that is a main driving force for pushing.

$$\Delta\sigma_0 = \sigma_{123-211} - \sigma_{L-211} - \sigma_{123-L} \qquad (1)$$

where $\sigma_{123-211}$ is the interfacial energy between the RE123 matrix and a RE211 particle and the subscript of L is BaO-CuO melt. According to paper by Endo et al, the critical size (r*) of a RE211 particle which could be trapped by the growing RE123 matrix, is able to be determined roughly by the undercooling (ΔT), the interfacial energy balance ($\Delta\sigma_0$) and the viscosity of the melt (η).

$$r^* \propto \Delta\sigma_0 / \eta\Delta T \qquad (2)$$

During the soldering, the temperature of the Yb-Ba-Cu-O solder was changed from 1263K to 1183K at the cooling rate of 1.0K/h, and the undercooling (ΔT) might be increased. Therefore, the size of the Yb211 particles, which were pushed by the RE123 growing interface and accumulated at the final solidified region were larger than those of the initial solidification region. At the initial region near the Y-Ba-Cu-O matrix, Yb211 particles were pushed easily, because of the low undercooling, i.e. a low growth rate of the RE123, and it induces the macro-segregation in the soldered region.

Secondly, the voids formation in the final solidified region was discussed. Oxygen gas was generated during the initial heating by the decomposition of the sintered Yb-Ba-Cu-O powder with the following reaction:

$$Yb_1Ba_2Cu_3O_x \rightarrow Yb_2Ba_1Cu_1O_5 + Ba_3Cu_5O_y \, (melt) + O_2 \, (gas) \qquad (3)$$

Not all oxygen gas generated by the decomposition could be reabsorbed into the Yb123 matrix during solidification. Excess oxygen gas coalesced and formed bubbles in the $Ba_3Cu_5O_x$ melt. These gas bubbles were also pushed by the growing RE123 interfaces similar to that of RE211 particles, and finally segregated at the final solidification region.

Summary

The Y-Ba-Cu-O specimens were soldered with the sandwiched Yb-Ba-Cu-O powder. After the soldering, Yb211 particles and gas bubbles were segregated in to the final solidified region. These phenomena were explained by the U-C-J pushing-trapping model. It is necessary for the further R&D to consider about this macro-segregation in the final solidification region in order to keep current path through the soldered region.

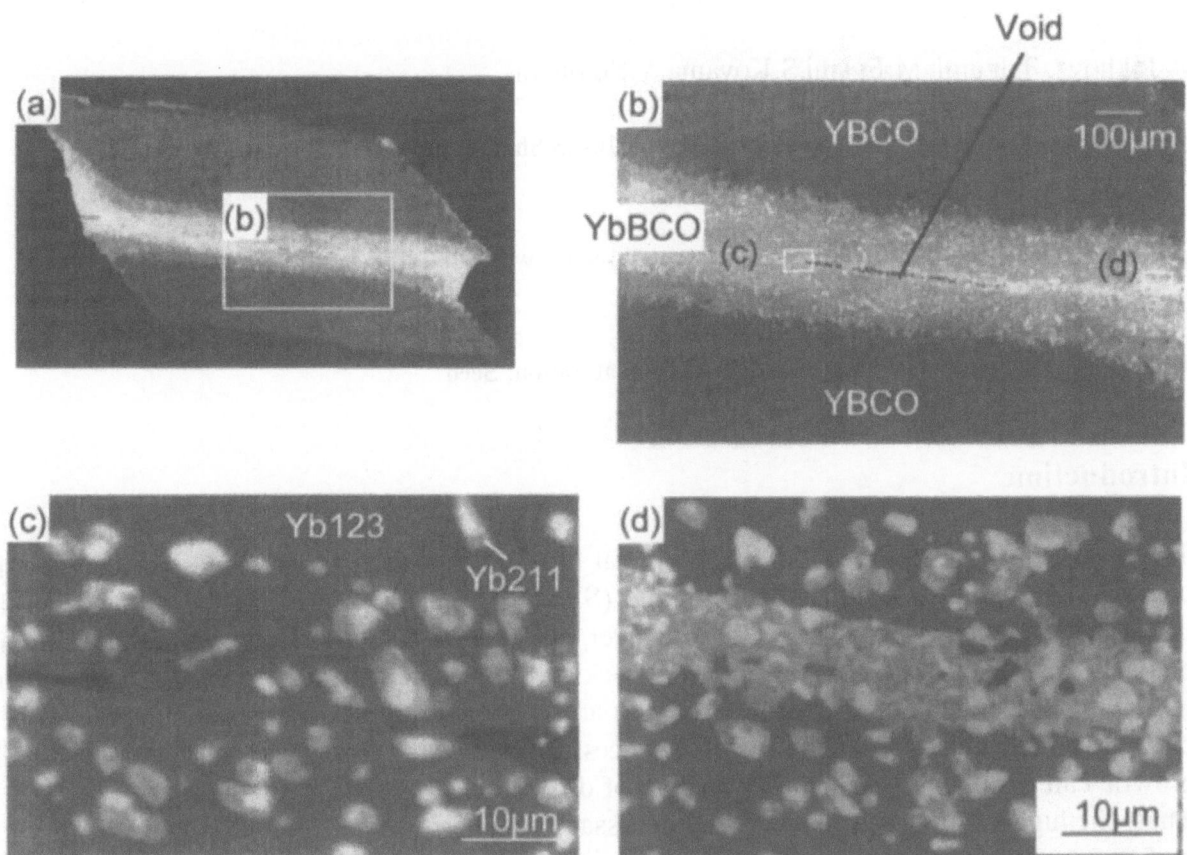

Figure 1   COMPO images showing the transversal sections of the soldered YBCO specimen. It can be found that the Yb211 particles and the voids are segregated around the final solidified regions.

Acknowledgement: This study was supported by the Proposal-Based New Industry Creative Type Technology R&D Promotion Program from the New Energy and Industrial Technology Development Organization (NEDO) Japan.

Reference
[1] K. Salama and Selvamanickam V., Appl. Phys. Lett., 60(1992),pp898
[2] K. Kimura, K. Miyamoto, M. Hashimoto, Advances in Superconductivity VII, Springer-Verlag(Tokyo), (1995), pp681
[3] N. Sakai, D. N. Matthews, R. Hedderich, H. Takaichi and M. Murakami, Advances in Superconductivity VI, Springer-Verlag(Tokyo), (1996) ,pp803
[4] J.Maeda Y.Shiohara, J. Mater. Res, 14(1999), pp2739
[5] A. Endo, Y. Watanabe, K. Miyake, T. Umeda, K. Murata, Y. Shiohara, J. Jpn. Inst. Met. 61(1997),pp963
[6] D. R. Uhlmann, B. Chalmers and K. A. Jackson, J. Appl.Phys., 35(1964),pp2986

Using a small size and high quality monocrystal as a seed material for SRL -CP growth process.

A.Jokhov*, T.Izumi, M.Egami,S.Koyama,Y.Shiohara.

Superconductivity Research Laboratory,ISTEC,1-10-13 Shinonome 1-chome,Koto-ko,Tokyo,Japan

Abstract: The high quality Y123 monocrystal seed was used in crystal growth by SRL-CP method.

Keywords: Single crystal growth of Y123; Supersaturation; Seed

Introduction

The large and high quality Y123 single crystal have been desired as substrates for the electric devices. The Solute Rich Crystal Pulling (SRL-CP) method [1] has been successfully developed to get the large crystals. However, the above crystal involves several subgrain boundaries which can be observe on X-ray rocking curves and topograph pictures. The defects inherit the imperfection of seeding material and lead to multidomain structure of obtained crystal. Unfortunately, necking effect that successfully are using in Czochralski method for melt growth can't still apply to SRL-CP, because of difficulties the crystal growth process.
Thus, the high quality seed is one of the necessary conditions to get the bulk monocrystal with perfect crystallinity. The purpose of the work is to get the good connection between the seed and crystal holder for following crystal growth process.

Experimental

Y123 single crystal seed were prepared by cutting the seed from the best possible crystalline part of separately grown Y123 single crystal to exclude subgrain boundaries. The rocking curve of the seed is shown in Fig.1. The full width at half maximum (FWHM) is about 0.2°.

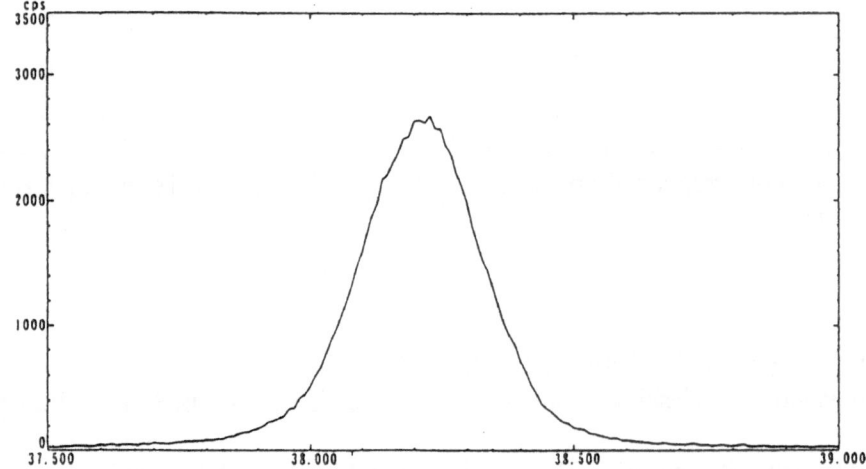

Fig.1. X-ray rocking curve of (005) diffraction of seed monocrystal.

Then the Melt Texture Growth (MTG) process [2] was applied to get the good connection between the seed crystal and crystal holder. The heat pattern of the process is shown in Fig.2.

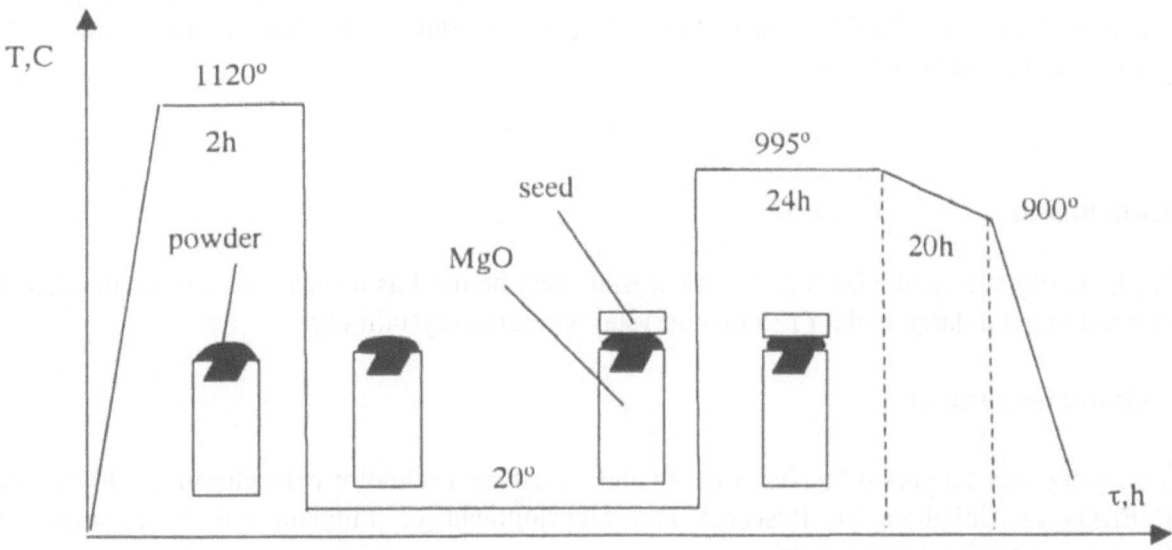

Fig.2 The heat pattern for the connecting process of the seed and MgO holder.

Small part of powder with $Y_{1.6}Ba_{2.3}Cu_{3.3}O_x$ composition was placed on the top of MgO crystal holder and the construction was heated up to 1120°C and kept at the temperature for 2 hours. Then the sample was quenched to room temperature. After that the previously prepared seed are placed on the top of freeze melt and the sample introduced into the furnace at 995°C. During 20 hours the well known MTG process take place to get the good connection between the seed and MgO bar. Slow cooling the furnace down to 900°C allows to escape the low melted composition in the sample. The section of final junction is shown in Fig.3.

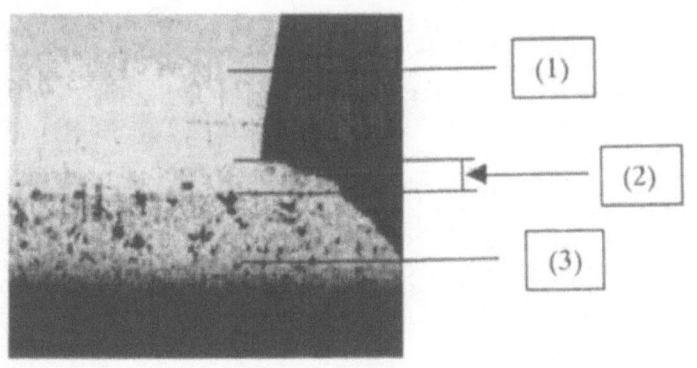

Fig.3 The section of the junction between the seed crystal and MgO bar.
(1) - Y123 seed
(2) – monodomain layer of "123" phase
(3) - ceramics material of "123" and "211" phases

Results and discussion

There is a thin layer ~ 0.1mm between the seed crystal (1) and ceramics part of the sample (3). This layer (2) of "123 "composition is monodomain and contains "211 "particles that was shown by EPMA analyses. Not any considerable amount of low temperature melting mixture of $BaCuO_2$ and CuO phases was observed in ceramics part of the sample. Thus the peritectic reaction "211" +L ="123" has finished completely during the holding and slow cooling periods of the MTG process.

Conclusion

High quality and small size Y123 monocrystal may be used as a seed material for the SRL-CP method to get a large scale Y123 monocrystal with high crystallinity.

Acknowledgement

This work was supported by New Energy and Industrial Technology development Organization (NEDO) as Collaborative Research and Development of Fundamental Technologies for Superconductivity Applications.

References

1. X.Yao and Y.Shiohara, "Large ReBCO single crystals: growth processses and superconducting properties", Supercod. Sci. Technol.10 (1997) 249-258.
2. M.Cima, M.Flemings et.al., "Semisolid solidification of high temperature superconducting oxides", J.Appl.Phys.72 (1), 1 July 1992.

Crystal Growth of CaLaBaCu3Ox Tetragonal Superconductors

Ryusuke Kita, and Tomoki Suzuki

Graduate School of Science and Engineering, Shizuoka University,

Johoku 3-5-1, Hamamatsu, Shizuoka 432-8651, Japan

Abstract: The crystal growth of the tetragonal $CaBaLaCu_3O_x$ from the molten solution prepared by the mixture of $CaBaLaCu_3O_x$ and $BaCuO_2$-CuO flux was investigated. The tetragonal films with the lattice constants of $a=3.88$ Å and $c=11.64$ Å were grown on $NdGaO_3$ (110) (NGO) by liquid phase epitaxy. The change from c-axis orientation to a-axis orientation was observed with increasing the time of keeping the temperature of the solution constant. Single crystals were also grown by the self-flux technique.

Keywords: crystal growth, tetragonal superconductor, liquid phase epitaxy, single crystal

INTRODUCTION

$YBa_2Cu_3O_{7-\delta}$ (YBCO) has been extensively studied for its application to electronics since it has high Tc of 90K above the liquid nitrogen boiling point and relatively high Jc. There are serious problems, however, such as surface roughening due to the twin defects caused by the phase transition from the orthorhombic to the tetragonal structure and the high water reactivity. It is expected to be possible to prepare the high-quality crystals without twin defects since tetragonal superconductors such as $(Y,Ca)(Ba,La)_2Cu_3O_x$ (YCBLCO) [1] or $CaBaLaCu_3O_x$ (CBLCO) [2] with Tc of 80K have no phase transition from the orthorhombic to the tetragonal structure. Furthermore it is reported for that the tetragonal superconductors exhibit high water stability and thermal stability probably due to the disorder of oxygen in the Cu-O chain layers [3]. Synthesis of single crystals or single-crystalline films of this system are very important for understanding the physical properties. There are few studies, however, for the crystal growth of the CBLCO system. In this paper we have investigated the crystal growth of tetragonal CBLCO from the melt by liquid phase epitaxy (LPE) and the self-flux technique.

EXPERIMENTAL

The present experimental setup consisted of a chamber furnace, a Y_2O_3 crucible to hold the melt and a driving device which lifts and lowers the substrates. We use $CaBaLaCu_3Ox$ as a solute and $BaCuO_2$-CuO as a flux. The starting material of two compounds were $CaCO_3$(99.99%), $BaCO_3$(99.95%), La_2O_3(99.99%) and CuO(99.9%). The nominal composition of CLBCO was Ca:Ba:La:Cu=1:1:1:3 and that of the $BaCuO_2$-CuO flux was Ba:Cu=3.0:7.0. CLBCO powder was prepared by solid state reaction of appropriate quantities of starting materials and reacted in air at temperatures between 900°C and 970°C for 48 h. The formation of CLBCO phase with a tetragonal structure for the powder was examined by x-ray diffraction (XRD) with Cu Kα radiation. Two compounds were mixed and put into a yttria crucible and heated up to 1035°C to make a solution. The growth condition was as follows: The solution was kept at 1035°C for 2h and cooled to 1026°C in 1 h and further cooled to 1016°C at a rate of 5°C/h. After the final cooling the temperature of the melt was kept constant. NdGaO₃ (110) (NGO) single crystals were used for substrates. The substrates were vertically dipped into the molten solution for 10min during the cooling or with keeping the temperature of the melt constant after the cooling. The growth temperatures were measured using Pt-Rh thermocouples near the surface of the melt. The orientation of the grown films was examined by XRD.

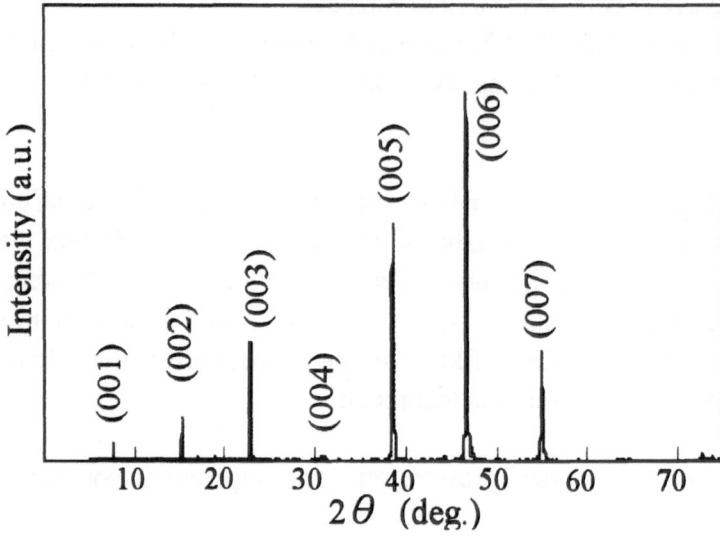

Fig. 1. XRD pattern of the c-axis oriented film on the NdGO₃(110) substrate grown at 1021°C.

RESULTS AND DISCUSSION

Fig. 1 shows the XRD pattern of the film grown on NGO at 1021°C during the second cooling. The film grown at 1021°C showed c-axis orientation with its c-axis perpendicular to the substrate. The lattice constant calculated from the XRD pattern was c=11.68Å. Fig. 2 shows the XRD patterns of the films grown from the solution kept at 1016°C for (a) 23h (film A) and (b) 47h (film B). The film A showed the mixture of a-axis and c-axis orientation. This result also shows the structure of the film is tetragonal with the lattice constants of a=3.87Å and c=11.64Å. The film B showed a-axis orientation with the lattice constant of a=3.88Å. These lattice constants correspond to those reported for the bulk CBLCO compounds [2]. These results of the XRD measurements also represent that the crystalline orientations of the films changed from c-axis to a-axis depending on the time of keeping the temperature constant. The tendency is different form that reported for the YBCO LPE films,

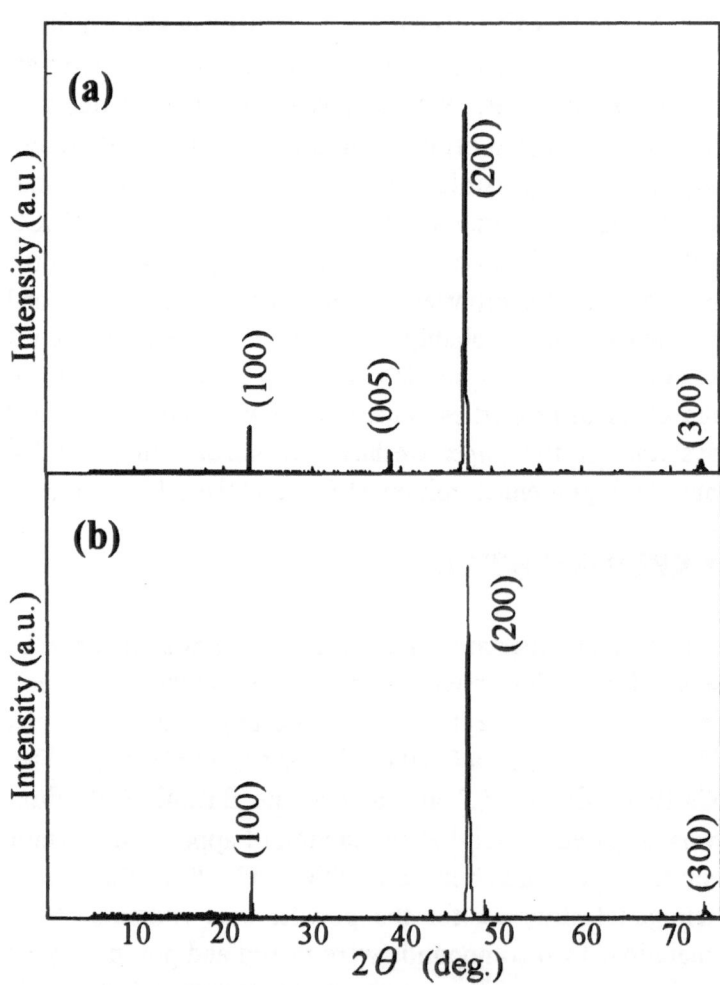

Fig. 2. XRD patterns of the films on the NdGO₃(110) substrate grown from the solution kept at 1016°C for (a) 23h and (b) 47h.

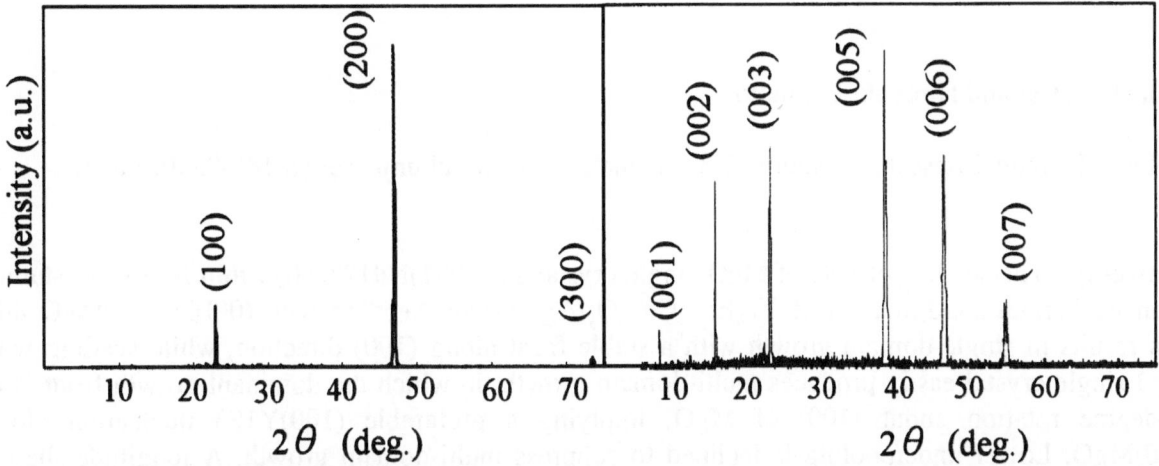

Fig. 3. XRD patterns for the two sides of the single crystal grown from the melt.

in which, the growth temperature of the a-axis oriented films is higher than that of the c-axis oriented films [4]. We also investigated the growth of CBLCO single crystals from the melt with the same composition used for the LPE growth. The growth conditions are as follows; The molten solution was kept at 1035°C for 2h and cooled to 1020°C in 1 h. The solution was further cooled to 1010°C at a rate of 0.2°C/h and finally cooled to 500°C in 50 h. The single crystal with a size of 3x3x1 mm^3 was obtained. The XRD patterns for the two sides of the single crystal were shown in Fig. 3. The lattice constants calculated from the XRD patterns were a=3.86Å and c=11.79Å.

CONCLUSIONS

We have investigated the crystal growth of the tetragonal CLBCO superconductors from the melt prepared by the mixture of CLBCO and $BaCuO_2$-CuO flux. The films with a tetragonal structure were grown on NGO substrates by LPE. The films showed the change from c-axis orientation to a-axis orientation. The change of the orientation was dependent on the keeping time of the molten solution. The lattice constants for the film were a=3.88Å and c=11.64Å. Single crystals with the lattice parameters of a= 3.86Å and c= 11.79Å were also grown by the self-flux technique.

Acknowledgement

The authors would like to acknowledge to Dr. Morishita and Dr. Yamada for helpful discussions.

References

1. A. Manthiram and J.B. Goodenough, Physica C **159**, 760-768(1989).
2. D. Goldschmidt, A. Knizhnik, Y. Direktovich, G.M. Reisner and Y. Eckstein, Phys. Rev. B **49**, 15928-15935 (1994).
3. J.P. Zhou, R-K. Lo, J.T. McDevitt, J. Talvacchio, M.G. Forrst, B.D. Hunt, Q.X. Jia and D.Reagor, J. Mater. Res. **12**, 2958-2975 (1997).
4. T. Kitamura, M. Yoshida, Y. Yamada, Y. Shiobara, I. Hirabayashi and S. Tanaka, Appl. Phys. Lett. **66**, 1421-1423 (1995).

Influence of seed crystallography on melt-textured domain of YBaCuO/Ag system

Chuanbing Cai and Hiroyuki Fujimoto

Railway Technical Research Institute, 2-8-38 Hikari-cho, Kokubunji, Tokyo 185-8540, Japan

Abstract: The seeding effects of MgO single crystal and (001)Nd123/MgO thin film were studies in an isothermal solidification of $Y_{1.8}Ba_{2.4}Cu_{3.4}O_y$/Ag system. Seeding with (001)Nd123/MgO thin film results in single-domain growth with a stable front along (100) direction, while seeding with MgO single crystal easily produces multi-domain growth, in which the dominant growth front is in 45-degree rotation about (100) of MgO, implying a preferable (100)Y123 nucleation along (110)MgO. Lower undercooling is inclined to suppress multi-domain growth. A longitude-shaped Nd123/MgO thin film is suggested as a promising seed to achieve a large-sized sample.

Keywords: Nd123/MgO thin film, seed, single domain, nucleation

INTRODUCTION

Seeding technique is widely used to obtain a large-sized single-domain $REBa_2Cu_3O_y$(RE123) bulk superconductor. A successful seed should not only ensure a single nucleation but also control the orientation of grown domain. Melt-grown or flux-grown bulk Nd(Sm)123, even Y123 itself are commonly believed as good seeds for Y123 crystal growth[1-2]. However, from practical viewpoint, there are some inconveniences for them, e.g. orientation confirmation, contamination deletion and size limitation etc., which usually result in a poor reproducibility. In addition, a cubic-shaped bulk seed easily leads to 90° tilt boundaries, implying a serious influence of geometric morphology of seed on the textured domain[2]. In present work, we compare the different seeding functions between MgO single crystal and Nd123/MgO thin film so as to understand more about the effect of seed crystallography on the textured domain. To our knowledge, few people use thin film as a seed in melt growth, although a well-quality thin film shows salient advantages, such as less impurity contamination, free high angle grain boundaries and convenient cutting in an alternative size.

EXPERIMENTAL

The ϕ 20mm pellet of $Y_{1.8}Ba_{2.4}Cu_{3.4}O_y$ sintered powder with 10 wt% Ag_2O and 0.5 wt% Pt was kept in an isothermal alumina box and confirmed by two thermocouples close to sample. The (001) Nd123 /MgO thin film or MgO single crystal was placed on the top of pellet before thermal cycle i.e. cold seeding method was employed. The samples were raised to 1040℃ for 1h and quickly decreased to 1000℃ for next 1h, and then rapidly cooled down to various holding temperature(T_h), corresponding $\Delta T = T_{m-p} - T_h$(T_{m-p} of 970℃ is the monotecto-peritectic temperature for YBCO/Ag system)[3]. After a given holding time(t_h, 30h to 60h), the samples were taken out. The grown domain was examined by polarized optical microscopy, x-ray diffraction and trapped field map.

RESULTS AND DISCUSSION

Similar to using bulk Nd123 crystal, seeding with (001) Nd123 film results in single-domain growth with a stable front parallel to (100)Nd123/MgO thin film (Fig.1(up)). XRD pattern shows only (001) peak presence, indicating that the grown domain are oriented with the c-axis normal to sample's surface(Fig.1(down)). In contrast, seeding with MgO single crystal easily produces multiple domains originated from seed, although spontaneous nucleation doesn't appear(Fig.2(A,B)). The dominant growing front is in 45-degree rotation about (100) of MgO. It implies that (100) Y123 crystal preferentially nucleates along (110)MgO. In addition, the multi-domain growth is dependent on the undercooling. Lower undercooling seems to be effective to suppress multi-domain growth. When $\Delta T=5^{\circ}C$, only a dominant domain appears(Fig.2(C)).

Fig.3 shows an image of Y123 crystallization begins from seed site. The critical spherical nucleus with a radius $r^*(\propto 1/\Delta T)$ transforms into a cap of the size[4]. It is assumed at equilibrium on the interface of seeding crystal

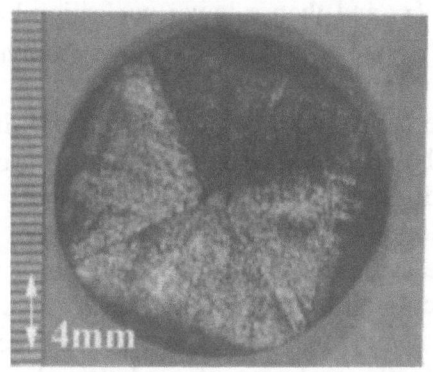

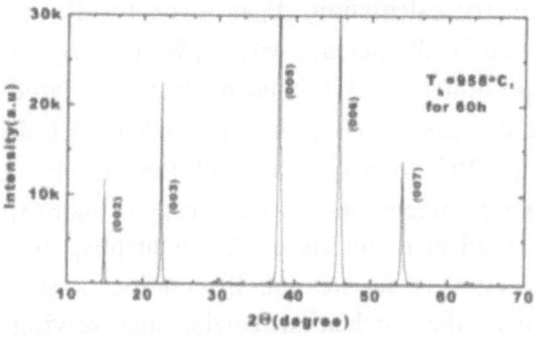

Fig.1. A typical Nd123/MgO seeded sample. (up)Optical image, (down) XRD pattern

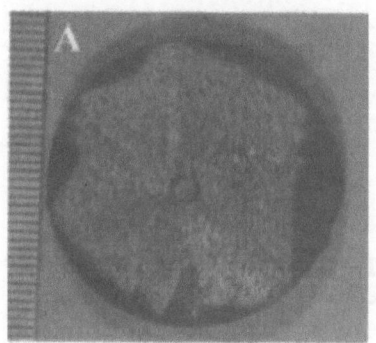
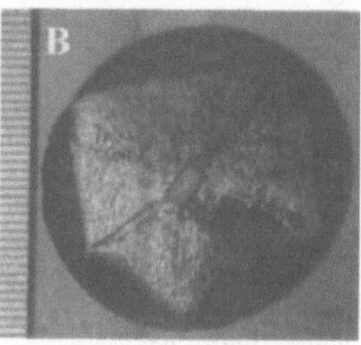
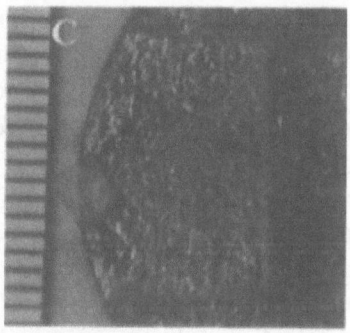

(A) $T_h=956^{\circ}C$, 30h   (B) $T_h=962^{\circ}C$, 30h   (C) $T_h=965^{\circ}C$, 30h(seed at edge)

Fig.2. MgO seeded samples, melt-processed at 1040°C, but at various undercooling
All scale resolutions are 0.5mm.

and melt(liquid and 211 particles) when $\sigma_{SM}-\sigma_{CS}=\sigma_{MC}\cos\theta$, where σ_{SM}, σ_{CS}, σ_{MC} are the surface energies of seed-melt, crstal-seed and melt-crystal interface, respectively. Correspondingly, the formation free-energy ΔG_c^* is less by a factor $f(\theta)$ than that required to form a spontaneous nucleus, i.e. $\Delta G_c^*=\Delta G^* f(\theta)$, where, θ is the seed-crystal contact angle and $f(\theta)$ decreases with θ becoming small. A small θ is deduced from large σ_{SM}, small σ_{CS} or σ_{MC}. Different from σ_{SM} and σ_{MC}, σ_{CS} is controllable by choosing seed material, it should decrease with decreasing lattice mismatch between seed and crystal. Nd123 is close to Y123 lattice parameter, so Y123 is easy to grow epitaxially on Nd123 thin film. However, for MgO seed, there exists more complex lattice matching. Two types of basic ways are: (a) (100)Y123 rotated in 45 degree to (100)MgO with 2.4% lattice misfit; (b) (100)Y123 parallel to (100)MgO with 9.1% lattice misfit[5]. Obviously, the former exhibits a good matching, which should lead to a smaller contact angle θ, and then a lower nucleation barrier. So, 45-degree rotation matching occurs preferably, which is the reason why 45-degree growth front is dominant in those MgO seeded samples.

On the other hand, compared with the first nucleus size, the seed size is large enough to provide more sites for other nucleation along the melt-seed interface(Fig.3). So, the nucleus-seed contacting size(hereafter termed as effective seeding size, d*) is a critical parameter for initial nucleation. To ensure single nucleation, an ideal case should be that d* is close to the real size of seed. Based on a simple geometry calculation, d* is given by d*=2r*sin θ. Obviously, d* increase with r*, while r* is in reverse proportional to ΔT, thus d* becomes large when ΔT decreases. As an example, when ΔT decrease from 12℃ to 5℃, d* will increase by 2.4. A larger d* means first nucleus occupies more space at the seed-melt interface. Consequently, the other nucleation is suppressed. In addition, a small ΔT implies the nucleation barrier and driving force decreasing and then nucleation frequency lowering. All of these are possible reasons why a lower undercooling is inclined to single-domain growth. However, for a single lattice matching case, even if a larg-sized seed provides more initial nucleation sites, it does not mean resulting in the multiple domains since all of nuclei are possible to joint each other to form an integrate grow front. We tentatively apply a longitude-shaped Nd123 thin film as seed to fabricate a sample. Different from normal square shape, the grown domain appears in a rectangle shape, seeming to extend from seed outline. The trapped field contour at 77K after 4,000G magnetization shows a single dome with no hollows(Fig.4), implying an uniform single domain with few high angle grain boundaries is achieved. Hereby, it is worthy to indicate that a longitude-shaped Nd123 thin film is a promising seed to achieve those large-sized samples which obtained only by multi-seeded technique at present.

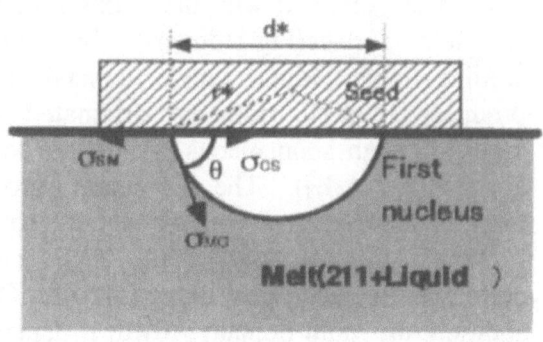

Fig.3. Illustrative equilibrium of seeding nucleation and effective seeding size

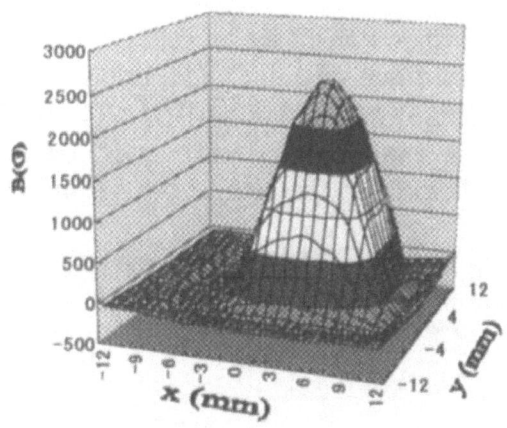

Fig.4. Trapped field contour for a longitude-shaped Nd123/MgO seeded samples.

In summary, seeding with (001) Nd123/MgO film results in single-domain growth in melt-processed Y123, while seeding with MgO single crystal easily produces multi-domain growth, in which preferable growth along (110)MgO. Lower ΔT is inclined to suppress multi-domain growth.

Acknowledgments. Authors are grateful to Dr. Y. Nakamara, Ms. K. Tachibana, Mr. K. Waki and Mr. H.Kamijo for their helps and discussion in this work.

1. M.Morita, S.Takebayashi, M.Tanaka and K.Kimura, Adv. Supercond.,3(1991)733.
2. X.F.Zhang, Supercond.Sci.Technol. 11(1998)1391
3. Y.Nakamura, K.Tachibana, H.Fujimoto, Physica C 306(1998)259
4. M.C.Flemings, Solidification Processing, McGraw-HILL, Inc.(1974)P290
5. D.M.Hwang, T.S.Ravi, et al., Appl. Phys. Lett.57(1990)1690

Melt growth of Ag doped Gd-Ba-Cu-O bulk superconductors by using Nd-Ba-Cu-O seed crystal

Takashi Saitoh[1], Koichi Kamada[1], Kazumasa Iida[2], Naomichi Sakai[3] and Masato Murakami[3]

[1]Iwate Industrial Research Institute, 3-35-2, Iiokashinden, Morioka-shi, Iwate, 020-0852 Japan
[2]SRL-ISTEC, 3-35-2, Iiokashinden, Morioka-shi, Iwate, 020-0852 Japan
[3]SRL-ISTEC, 16-25, Shibaura 1-Chome, Minato-Ku, Tokyo, 105-0023 Japan

Abstract: We have studied the melt growth of Ag doped Gd-Ba-Cu-O (Gd123 + 40mol% Gd211 + xwt% Ag, x=0-10) bulks by using a cold-seeding method with Nd-Ba-Cu-O seed crystal. Growth temperatures of the Ag doped bulks decreased with increasing Ag contents. EPMA and magnetic measurements showed that Gd element diffused into the seed crystal and formed (Nd, Gd)123 composite at the interfacial region. The formation of the composite phase raised the growth temperature. For cold-seeding, therefore it is necessary to increase the growth temperature for the fabrication of large grain samples.

Keywords: Ag doping, Gd123, melt growth, Nd123 seed crystal

INTRODUCTION

Oxygen-controlled melt-growth (OCMG) -processed Gd-Ba-Cu-O {$GdBa_2Cu_3O_y$ (Gd123)+Gd_2BaCuO_5 (Gd211)} bulk superconductor exhibits large critical current density (J_c) [1, 2], and are expected as materials for strong trapped field magnets. Such a bulk is usually fabricated by top-seeded melt-growth (TSMG) technique. There are two kinds of seeding techniques for TSMG: cold-seeding; and hot-seeding method. The cold-seeding method has an advantage that it can be performed at room temperature. For practical applications, it is necessary to increase mechanical properties of the bulks. It is reported that addition of Ag is effective in increasing the mechanical properties of the bulks [3, 4]. In this paper, we studied the melt-growth of the Gd-Ba-Cu-O bulks with various Ag contents by using cold-seeding method with Nd-Ba-Cu-O seed crystal.

EXPERIMENTAL PROCEDURE

Commercial powders of Gd123, Gd211 and Ag_2O were mixed with an automatic agate mortar for 2 hrs to have nominal compositions of $Gd_{1.8}Ba_{2.4}Cu_{3.4}O_y$ + x wt% Ag (x=0, 1, 2, 5, 10). Decomposition temperatures (T_p) of Ag doped samples were determined by differential thermal analysis (DTA). The well-mixed powders were uniaxially pressed into ϕ 20mm x 15mm pellets and then subjected to cold isostatic press (CIP) under 2ton/cm². Cold-seeding method was employed for the fabrication of single domain samples, in which a Nd123 seed crystal was placed on the center top of the pellet at room temperature. The pellets were then melt-processed in 1%O_2-Ar atmosphere. Heat schedule was determined based on the T_p as shown in Fig.1. After the melt-process, the samples were cut into two pieces from the seed crystal. Microstructures of the samples were observed by scanning electron microscope (SEM) and composi-

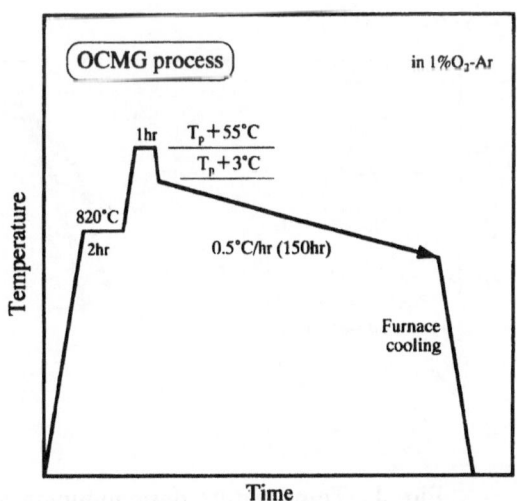

Fig. 1. Heat schedule for the OCMG-process.

tional analyses were carried out with an electron probe micro analyzer (EPMA). The melt grown samples were cut into rectangular shape with dimensions of 1.5mm x 1.5mm x 0.5mm and annealed in pure O_2, and then T_c was measured with a vibrating sample magnetometer (VSM).

RESULTS AND DISCUSSION

Fig. 2 shows the T_p as a function of Ag content. In the range from 0 to 5wt% Ag doping, the T_p is linearly decreased with increasing Ag content. The T_p of the sample with 10wt% Ag is about 40°C lower than that of the sample without Ag doping. Top view of the melt-processed samples (a) without Ag and (b) with 10wt% Ag are shown in Fig. 3. Both samples could grow from the seed crystal, even though they are not single domain. The number of the domains of the sample with 10wt% Ag is much larger than that of the sample without Ag. Fig. 4 shows temperature dependencies of normalized DC susceptibilities for the melt-processed samples with various Ag contents. Onset T_c's of the samples were 94K-95K for all the samples, and the transition was broadened with increasing Ag content. Here, the transition width of the 10wt% Ag doped sample was about 2°C larger than that of the sample without Ag doping. EPMA compositional maps at the interface between the 10wt% Ag doped sample

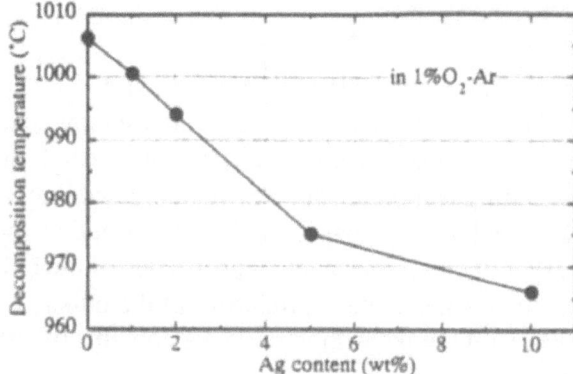

Fig. 2. Peritectic decomposition temperature of Gd-Ba-Cu-O as a function of Ag contents.

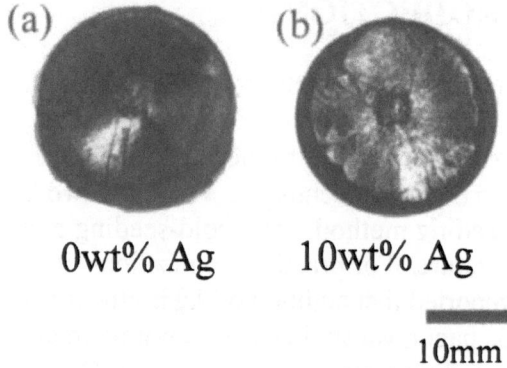

0wt% Ag   10wt% Ag

10mm

Fig. 3. Top views of the OCMG-processed Gd-Ba-Cu-O samples with and without Ag doping. (a) without Ag, (b) with 10wt%Ag.

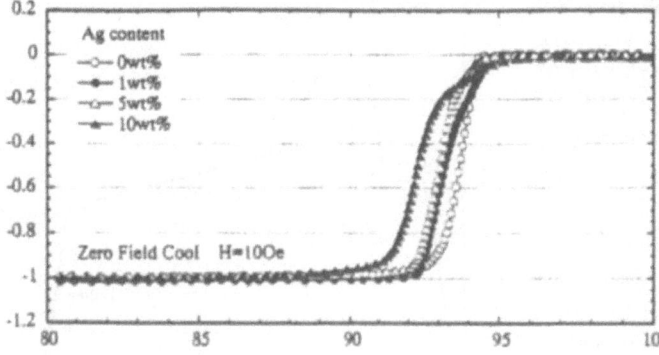

Fig. 4. Temperature dependencies of the normalized DC susceptibilities for the OCMG-processed Gd-Ba-Cu-O superconductors with various Ag contents. The samples were cut from the interfacial region between bulk and seed crystal.

and Nd-Ba-Cu-O seed crystal were shown in Fig. 5. It is observed that Nd and Gd mutually diffused into the matrix and then formed (Nd, Gd)123 composite phase at the interface. Both of the deterioration in T_c and the multi-domain behavior are probably due to the formation of the interfacial composite phase. In cold-seeding method, it takes a long time to completely decompose RE123 phase, so that interfacial reaction occurred. It is well known that the peritectic temperatures of the RE123 increase with increasing the average ionic radius of the RE ion. Thus, the decomposition temperature at the interfacial region is higher than the expected tem-

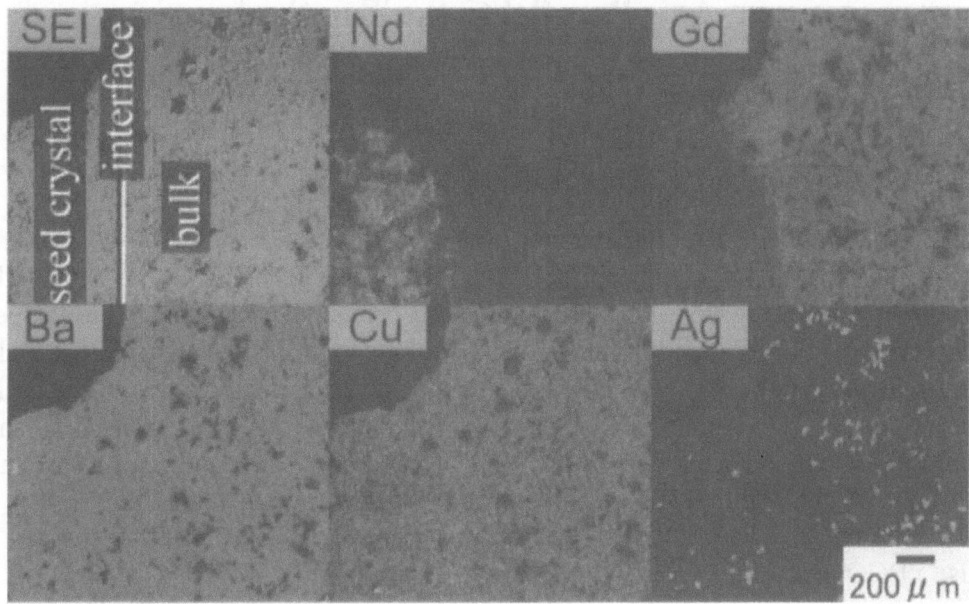

Fig. 5. EPMA map analysis of Gd-Ba-Cu-O with 10wt%Ag.

perature. In the case of the sample with 10wt% Ag, the selected growth temperature was much lower than the actual growth temperature. Therefore, the top view showed multi-domain feature. Such a problem can be solved by employing higher growth temperatures by considering the interfacial reaction.

CONCLUSIONS

We have studied the melt growth of Ag doped Gd-Ba-Cu-O (Gd123+40mol% Gd211+x wt% Ag, x=0-10) by using the cold-seeding method with Nd-Ba-Cu-O seed crystal. All the samples could grow from the seed crystal, even though they were not single domain. It is found that growth temperature of the bulk decreases with increasing Ag contents, and the EPMA observations showed that Nd and Gd mutually diffused into the matrix and then formed composite phase at the interface, which made the decomposition temperature at the interfacial region much higher than the expected temperature. Therefore, it is necessary to consider the interfacial reaction for the fabrication of large grain samples.

ACKNOWLEDGEMENTS

This works was supported by new Energy and Industrial Technology Development Organization (NEDO) as Collaborative Research and Development of Fundamental technology for Superconductivity Applications under the New Sunshine Program administered by the Agency of Industrial Science and Technology (AIST) of the Ministry of International Trade and Industry (MITI) of Japan.

REFERENCES

[1] Yoo S. I. , Sakai N, Takaichi H and Murakami M Appl Phys Lett 65 (1994) 633-635
[2] Murakami M, Sakai N, Higuchi T and Yoo S. I. Supercond Sci Technol. 9 (1996) 1015-1032
[3] Y. Nakamura, K. Tachibana, S.I. Yoo and H. Fujimoto : Advances in Superconductivity X, (1997) 649.
[4] S. Haseyama, S. Kobayashi, M. Satoh, H. Miyairi, H. Nakane and S. Nagaya : Advances in Superconductivity X , (1997) 653.

Observation and Analysis of the Growth Traces of Single-Domain YBCO Bulk Superconductors With c-Axis Parallel to the Top Surface

W.M.Yang, L.Zhou, Feng Yong, P.X.Zhang, J.R.Wang,
C.P.Zhang, Z.M.Yu, X.D.Tang, X.Z.Wu

Northwest Institute for Nonferrous Metal Research, Xi'an, 710016

Abstract: The morphology of the across traces on the top surface of single-domain YBCO bulk samples with c-axis parallel to the top surface have been clearly revealed. In the sample, the c-axis is parallel to the top surface, which indicates that YBCO crystal starts to grow from the center to the edge in all the four regions divided by across traces, but the growth style are different in the adjacent regions, the YBCO crystal growth is along a(or b) axis in one region and along c-axis in the adjacent regions, so the traces are forming a 78°angle when the four regions meet each other, and the clear traces are formed by the different growth rates and styles along c-axis and a (or b) axis and also the slightly difference in height of YBCO crystal between adjacent regions.

Keywords: YBCO bulk, single-domain, melt growth, morphology

INTRODUCTION

Many works have been done on the investigation of the microstructure of YBCO Superconductors. It is found that the microcracks lines parallel to ab-plane are low angle grain boundaries[1], or defects formed during the melt growth process and resulted in the formation of gaps filled with rejected liquid phases between the platelets[2]. The microcracks are parallel to the ab-plane commonly described by the anisotropy expansion of the YBCO crystal during the oxygen annealing process. Now large single domain YBCO bulk superconductors can be fabricated [3-6], and usually there are two types of the across traces on the top surface of single domain YBCO bulks, but up to now, the origin of the cross traces are not fully understood[7]. In this paper, the original growth microstructure of the cross traces on the top surface of single domain YBCO bulk superconductors are well revealed, and a simple growth model has been provided to interpret the growth mechanism of single-domain YBCO bulks with c-axis parallel to its top surface.

EXPERIMENTAL

The fabrication process of single-domain YBCO bulks was described elsewhere[8,9], in brief, the 123, Y_2O_3 and CeO_2 powders were mixed with proper ration and ball-milled in air and pressed into pellets(35mm in diameter). The pellets seeded with Nd-123 single crystal were heated up to1060℃ and keep some time for homogeneous melt of YBCO material. While the samples were slowly cooled from the temperature slightly above the peritectic temperature of YBCO, the melt starts to epitaxial grow from the Nd-123 seed, and finally leading to the growth of large single-domain YBCO samples. In this experiment, the Nd-123 seed was placed on the top of the sample with c-axis parallel to the top surface. There are two cross traces on the top surface of the sample, which forms a 78°angle, as shown in Fig.1. The microstructure of the top surface and cross-section has been carefully investigated by optical observation and SEM(JMS 5800,Japan).

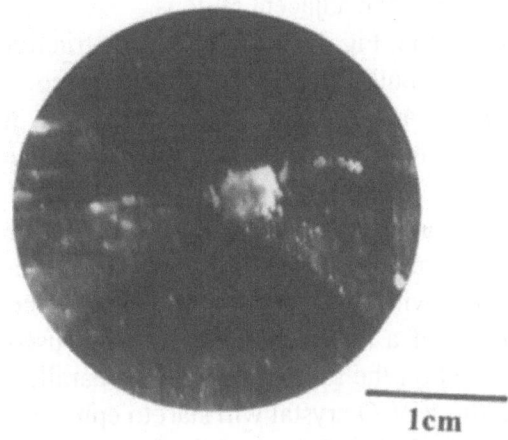

Fig. 1 The typical macrostructure of single-domain YBCO bulk superconductors with c//top surface

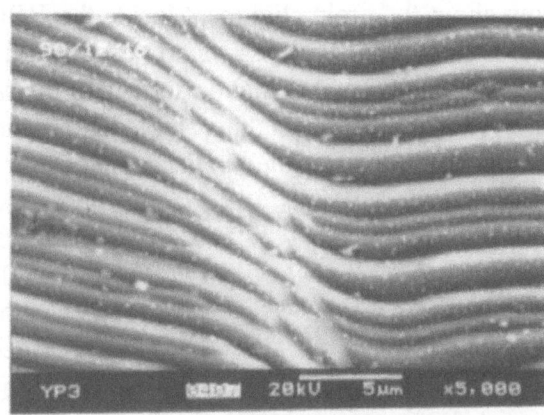

Fig.2 the growth traces of single-domain YBCO bulk on the top surface

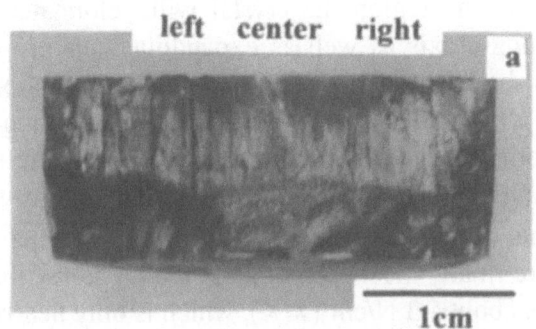

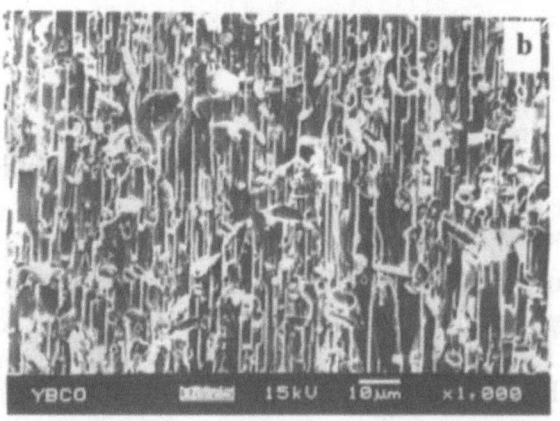

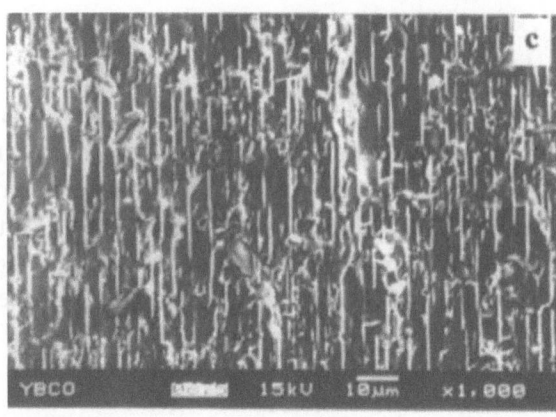

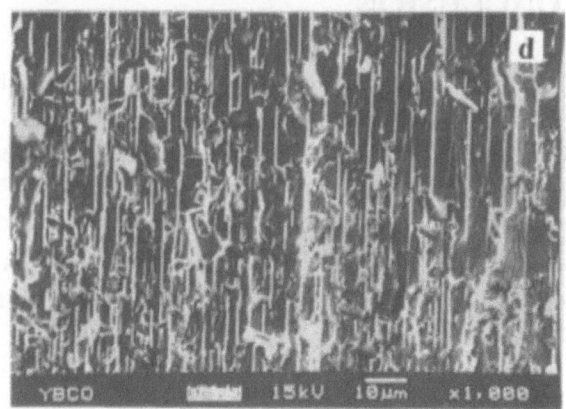

Fig.3 the macrostructure and microstructure of the cross section of YBCO samples. a). The macrostructure of the cross section. The microstructure of the across section on the left side(b), center(c), and right side(d).

RESULTS AND DISCUSSION

Fig.2 shows the growth morphology of the across traces on the top surface of single-domain YBCO bulk with c-axis parallel to the top surface. As we can see from this figure, there are many long, smooth and parallel YBCO growth stripes formed during the melt growth process, which indicates that the ab-plane is parallel to the axis of the cylinder sample in some extent. The narrow and smooth stripes are formed by the continues growth of YBCO crystal along a (or b) axis. The stairs are formed by the stair growth of YBCO along c-axis. According to the results, the across traces are of a 78° angle formed by the different growth rates and growth styles along c-axis and a (or b) axis and also the slightly difference in height of YBCO crystal between adjacent regions.

How about the grain-alignment inside the YBCO bulk sample, Fig.3 shows the macrostructure and microstructure of the cross section of single-domain YBCO bulk samples. As shown in Fig.3a, the ab-planes are normal to the top surface of the sample. These can be seen more clear in Fig.3b, 3c and 3d, which indicates that the ab-planes are normal to the top surface of the YBCO sample and homogeneously distributed in the sample.

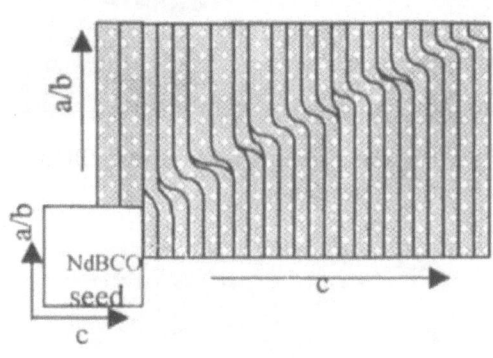

Fig.4 the formation process of the cross traces of samples with c-axis parallel to the top surfaces

According to the results and analysis above, we can deduce the formation process of the cross traces of YBCO samples with c-axis parallel the top surfaces. As shown in Fig.4, if a NdBCO single-crystal is used as a seed and placed on the sample with c-axis parallel to its top surface, the YBCO crystal will start to epitaxial grow from the sides of NdBCO seed while the temperature of YBCO bulk slightly cooled below the peritectic temperature. The YBCO crystals will grow in all the four regions but with different growth styles from the top view, in one region the crystal will elongate the growth along c-axis as well as expanding their growth along a (or b) axis on both sides; in the adjacent regions, while the YBCO is also expanding the growth along c-axis on both sides the YBCO crystals will elongate the steady growth along a (or b) axis during the melt growth

process. The different growth styles of YBCO crystals in the four regions result in the two across traces with a 78 degree angle while the four regions are finally meet each other at the diagonal lines, as shown in Fig.1. The levitation force of the sample is about 5.3 N/cm^2(77K), which is only nearly a half of that of the sample with c-axis normal to the top surface.

CONCLUSION

It is found that there are many long and smooth YBCO platelets, which are parallel stacked one another in the sample, and the sample was divided into four regions by the two traces, which means that YBCO crystal starts to grow from the center to the edge in all the four regions, but the growth style are different in the adjacent regions, the YBCO crystal growth is along a(or b) axis in one region and along c-axis in the adjacent regions, so the traces are forming a 78°angle when the four regions meet each other, and the cross traces are formed by the different growth rates and styles along c-axis and a (or b) axis, and also the slightly difference in height of YBCO crystal between adjacent regions.

REFERENCE
1.   P.McGinn et al, Appl. Phys. Lett.57(1990)1455
2.   T.IZumi and Y.Shiohara, J. Mater. Res.7(1992)16
3.   M. Morita, K. Nogashima, et al, Material science and engineering B53 (1998)159-163.
4.   W.Gawalek, et al, Appl. Supercond 2 (7-8) (1995) 465.
5.   P.Gautier-picard, et al, Materials science and engineering B 53(1998)66.
6.   W.M.Yang, L.Zhou, Y Feng, P.X.Zhang et al, Physica C 305(1998)269-274
7.   P.Diko, V.R.Tode, D.J.Miller, K.C.Goretta, Physica C 278(1997)192
8.   W.M. Yang, Lian Zhou, ,Y. Feng, P.X. Zhang et al, Physica C307(1998)217-276
9.   W.M. Yang, Lian Zhou, Y. Feng, P.X. Zhang et al, Physcia C319(1999)164-168

Observation And Analysis of the Growth Trace of Single-Domain YBCO Bulk Superconductors With c-Axis Normal to the Top Surface

W.M.Yang, L.Zhou, Y.Feng , P.X.Zhang, J.R.Wang,
C.P.Zhang,   Z.M.Yu,   X.D.Tang

Northwest Institute for Nonferrous Metal Research, Xi'an, 710016 China

Abstract: The morphology of the across traces on the top surface of single-domain YBCO bulk sample with c-axis normal to the top surface has been clearly revealed. It is found that there are many fish-scale like YBCO layers well connected each other on the top surface of the samples. In this sample the c-axis is normal to the top surface, which indicates that YBCO crystal starts to grow from the center to the edge along a(or b)axis in all the four regions divided by across traces, and so the traces form a 90°angle when the four regions meet each other, and the clear traces are formed just because of the slightly difference in height of YBCO crystal between adjacent regions.

INTRODUCTION

Melt growth process is one of the most promising method for fabricating large bulk YBCO superconductors with high levitation force. The levitation forces are closely related with grain-alignment and microstructure of the YBCO samples[1,2] and the magnetic field distribution[3]. Many works have been done on the investigation of the microstructure of YBCO Superconductors. It is found that the microcracks are parallel to the ab-plane commonly described by the anisotropy expansion of the YBCO crystal during the oxygen annealing process. Now large single domain YBCO bulk superconductors can be fabricated [4-7], and usually there are two types of the across traces on the top surface of single domain YBCO bulk Superconductors, but up to now, the origin of the cross traces are not fully understood[8]. In this paper, the original growth microstructure of the cross trace on the top surface of single domain YBCO bulk superconductors is well revealed, and a simple growth model has been provided to interpret the growth machinism of single-domain YBCO bulks with c-axis normal to its top surface.

EXPERIMENTAL

The fabrication process of single-domain YBCO bulks was described elsewhere[1,2], in brief, the 123, Y_2O_3 and CeO_2 powders were mixed with proper ration and ball-milled in air and pressed in to pellets(ϕ35mm). The pellets seeded with Nd-123 single crystal were heated up to1060℃ and keep some time for homogeneous melt of YBCO material. The samples were slowly cooled from the temperature slightly above the peritectic temperature of YBCO, the melt starts to epitaxial growth from the Nd-123 seed, and finally leading to the large single-domain YBCO samples. The grain-orientations of single-domain YBCO bulk samples are closely related with the grain-orientation of the Nd-123 seeds. In this experiment, the Nd-123 seed was placed on the top of the sample with c-axis normal to the top surface. There are two clearly cross traces on the top surface of the sample, which forms a 90°angle, as shown in Fig.1. The microstructure of the top surface and cross-section has been carefully investigated by optical microscopy and SEM(JMS 5800,Japan).

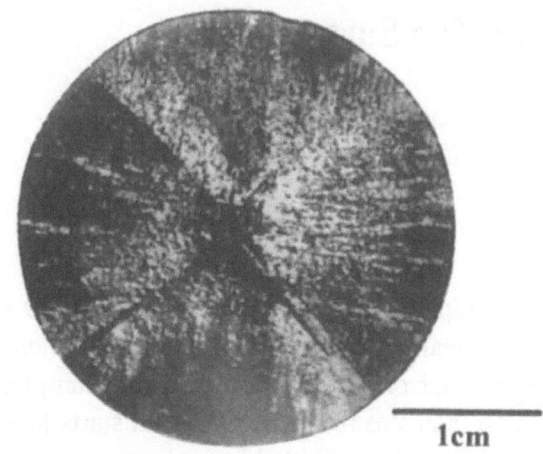

Fig. 1 The typical macrostructure of single-domain
YBCO bulk superconductors

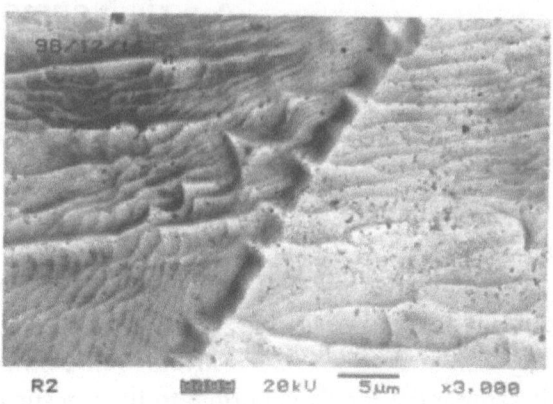

Fig.2 the growth traces of single-domain YBCO
bulk on the top surface

left   center   right

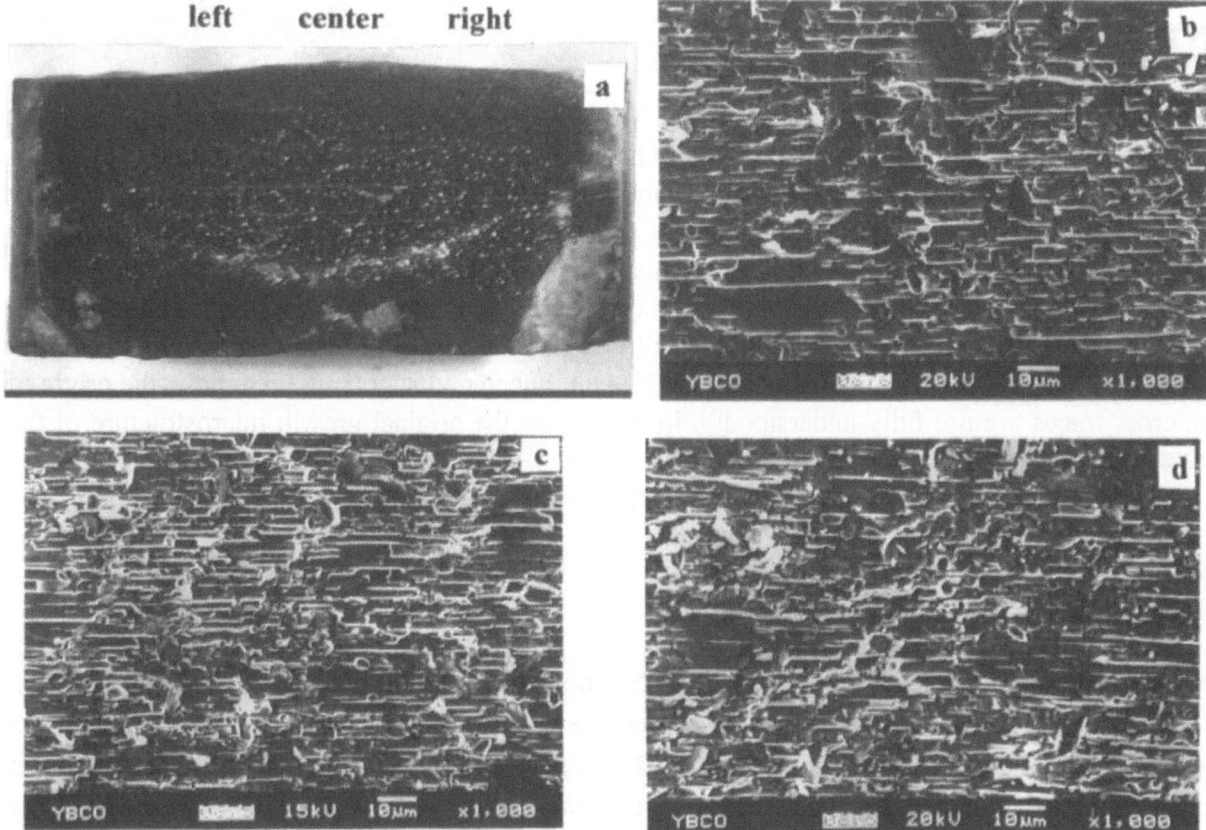

Fig.3 The   macrostructure and microstructure of the cross section of single-domian YBCO samples. a). The macrostructure of the
cross section, the microstructure of the across section on the left side(b), center(c), and right side(d).

RESULTS AND DISCUSSION

Fig.2 shows the growth morphology of the across traces on the top surface of single-domain YBCO
bulk with c-axis normal to the top surface. As we can see from this figure, There are many fish-scale
like YBCO layers stacked up each other on the top surface of the sample. The across trace is clear
shown in this figure, there are many curvature steps formed during the melt growth process. The
morphology is much different from that of single domain YBCO samples with c-axis parallel to the
top surface(submitted to ISS'99). According to the results, the across traces are formed just because
of the slightly difference in height of YBCO crystals in each adjacent region on both sides of the
trace.

How about the grain-alignment inside the YBCO bulk sample, Fig.3 shows the cross section macrostructure and microstructure of the single-domain YBCO bulk samples. As shown in Fig.3a, the ab-planes are parallel to the top surface of the sample. These can be seen more clear in Fig.3b, 3c and 3d, which means that the ab-planes are parallel to the top surface of the YBCO sample, and homogeneously distributed in the sample. All the results are in consistence with the results of XRD and Φ-scan(not shown here).

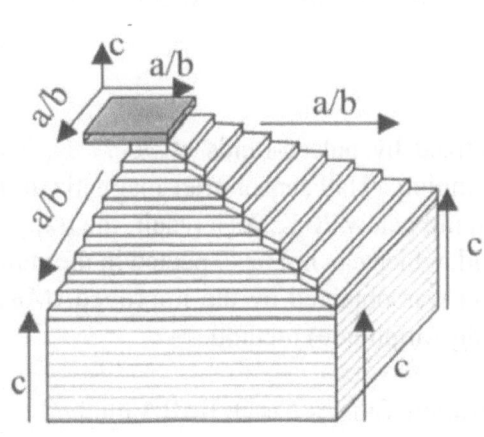

Fig.4 the formation process of the cross traces of samples with c-axis normal the top surfaces

According to the results and analysis above, we can deduce the formation process of the cross traces of samples with c-axis normal the top surfaces. As shown in Fig.4, if a NdBCO single-crystal is used as a seed and placed on the top of the sample with ab-plane parallel to its top surface, the YBCO crystal will start to epitaxial grow from the sides of NdBCO seed while the temperature of YBCO bulk slightly cooled below the peritectic temperature. The growing YBCO crystals will also expand their growth on both sides while the YBCO crystals elongate their steady growth during the melt growth process. The YBCO crystals in the four regions are all growing along a-axis or b-axis, finally meet each other at the diagonal line and form the across traces as shown in Fig.1. The morphology of these traces are different from the regions and formed just because of the slightly difference height of the adjacent regions, as indicated in Fig.4. The levitation force of the sample is up to 12 N/cm^2(77K).

CONCLUSION

It is found that for a given sample with c-axis normal to the top surface, there are many fish-scale like YBCO layers well connected each other on the top surface, which indicates that YBCO crystal starts to grow from the center to the edge along a(or b)axis in all the four regions divided by across traces, and finally the traces are forming a 90°angle when the four regions meet each other. The clear traces are formed just because of the slightly difference in height of YBCO crystal between adjacent regions.

Reference

1. Wan-min Yang, Lian Zhou, ,Yong Feng, Pingxiang Zhang, et al, Physica C307(1998)217-276
2. Wan-min Yang, Lian Zhou, Yong Feng, Pingxiang Zhang et al, Physcia C319(1999)164-168
3. Wan-min Yang, et al, The effect of magnetic field distribution on the levitation force of single - domain YBCO bulk superconductors. CEC/ICMC'99, 1999/07/12-16, In Canada
4. Mitsuru Morita et al, Material science and engineering B53 (1998)159-163.
5. W.Gawalek et al, Appl. Supercond 2 (7-8) (1995) 465.
6. P.Gautier-picard, et al, Materials science and engineering B 53(1998)66.
7. W.M.Yang, L.Zhou, Y. Feng, P.X.Zhang et al, Physica C 305(1998)269-274
8. P.Diko et al, Physica C 278(1997)192

TRAPPED FIELD DISTRIBUTION ON Sm-Ba-Cu-O BULK SUPERCONDUCTOR BY PULSED-FIELD MAGNETIZATION

Yousuke Yanagi[1], Yoshitaka Itoh[1], Masaaki Yoshikawa[1], Tetsuo Oka[1], Tetsuhisa Hosokawa[2], Hiromasa Ishihara[2], Hiroshi Ikuta[3] and Uichiro Mizutani[2]

[1]IMRA MATERIAL R&D Co., Ltd., 5-50 Hachiken-cho, Kariya, Aichi, 448-0021 Japan
[2]Department of Crystalline Materials Science, Nagoya University, Nagoya, 464-8603 Japan
[3]Center for Integrated Research in Science and Engineering, Nagoya University, Nagoya, 464-8603 Japan

ABSTRACT: A melt-processed Sm-Ba-Cu-O was magnetized by pulsed-fields at 30-77 K. The trapped field distribution on the sample was measured by scanning a Hall sensor after magnetization. We found that flux lines penetrated into the sample by flux jump below 70 K. As a result, the trapped field of the sample reached more than 70% of the applied field, which can not be expected in the static zero field cooling. The maximum trapped field of 3.8 Tesla was obtained by the improved IMRA (Iteratively Magnetizing pulsed-field operation with Reducing Amplitude) method.

Keywords: Trapped field distribution, Pulsed-field magnetization, Sm-Ba-Cu-O, IMRA method

INTRODUCTION

Recently, we have fabricated the Sm-Ba-Cu-O bulk superconductor, which can trap a strong magnetic field of 9 Tesla at 25 K [1]. We construct a magnetic field generator using the Sm-Ba-Cu-O bulk magnets cooled by refrigerators [2]. In order to use bulk superconductors in this kind of applications, we started research in a pulsed field magnetization (PFM) technique suitable for them. Then, we found that heat generation inherently occurred in the superconductors by flux motion during a PFM, caused decrease in the trapped field especially at low temperature. To overcome this, we have proposed the IMRA (Iteratively Magnetizing pulsed-field operation with Reducing Amplitude) method [3,4]. In the present work, we report the trapped field distribution on a Sm-Ba-Cu-O sample by the PFM especially at low temperature below 70 K and the optimal PFM procedure using the IMRA method.

EXPERIMENTAL

The sample was Sm-Ba-Cu-O ($SmBa_2Cu_3O_x$: Sm_2BaCuO_5=3:1 in molar ratio, with 0.5 wt.%Pt and 10 wt.%Ag_2O addition) with the size of 36 mm in diameter and 16 mm in thickness. Figure 1 shows a photograph of the sample. We confirmed that this sample had no serious weak links by measuring the trapped field distribution after field cooling (FC) in the liquid nitrogen. The sample was cooled to a temperature of 30-77 K by a refrigerator, and was magnetized by feeding the pulsed-current with the rise time of 13 ms to the magnetizing coil set outside a vacuum chamber. We define the applied field $\mu_0 H_a$ as the product of the maximum current flown in the coil and the coil constant. The trapped field distribution after each magnetization was measured by scanning a Hall sensor (F. W. Bell BHA921) 0.5 mm above the sample surface. The maximum-trapped field $B_{T\,max}$ was measured by touching the sensor to the sample surface at the position of the maximum value in the trapped field distribution measurement.

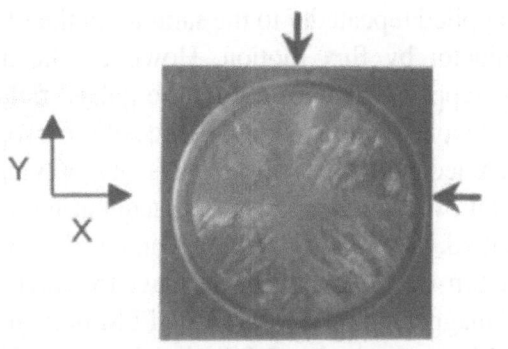

Fig. 1 A photograph of the polished surface of the sample. The sample is reinforced by a stainless steel ring. Short arrows show facet lines.

RESULTS AND DISCUSSION

Figure 2 shows the applied field $\mu_0 H_a$ dependence of $B_{T\ max}$ and trapped field distribution after applying pulsed-fields with various intensities to the sample with no trapped field (Single Pulsed-Field Magnetization: S-PFM). When $\mu_0 H_a$ is smaller than 3.5 Tesla (Fig. 2(a)), only a small field is trapped in the peripheral area of the sample. However, once $\mu_0 H_a$ exceeds 3.5 Tesla (Figs. 2(b) and 2(c)), the magnetic flux unexpectedly penetrates to the central area of the sample. As a result, the field more than 2 Tesla is trapped on the areas approximately 45-degree to the facet lines. We think this is due to flux jump caused by the extremely large gradient of the magnetic flux density in the sample. The $B_{T\ max}$ take a maximum value of 3.0 Tesla at $\mu_0 H_a = 4.2$ Tesla (Fig. 2(d)). We define this value of $\mu_0 H_a$ as the optimum applied field. In this case, the ratio of $B_{T\ max}$ to $\mu_0 H_a$ exceeds 70%, which is much larger than that by the static zero field cooling (ZFC) magnetization. This shows that the PFM technique does not require extremely large applied field. When $\mu_0 H_a$ exceeds 5 Tesla (Figs. 2(e) and 2(f)), $B_{T\ max}$ decreases with increasing $\mu_0 H_a$. In this range of $\mu_0 H_a$, magnetic field is trapped on the facet lines where no field is trapped at lower applied fields. This unique change in trapped field distribution was noticeable at low temperatures especially below 70K. We believe that the large gradient of the flux density and the difference in the local J_c in the sample caused the selective penetration by flux jump and the resulting heat generation reduces the trapped field in the 45-degree areas at high applied fields.

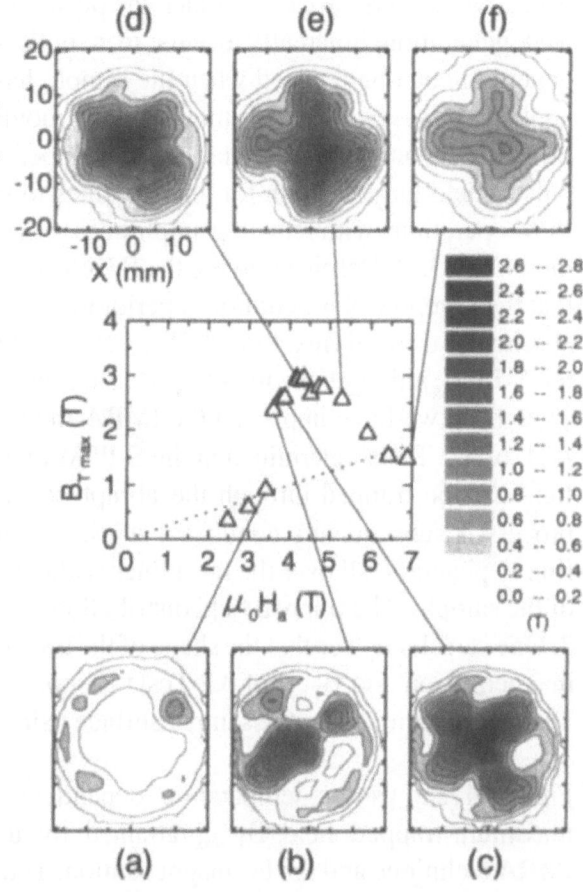

Fig. 2 Applied field dependence of the trapped field distribution by the S-PFM at 35 K. The X and Y axes in the distribution map coincide with the facet lines of the sample shown in Fig. 1. Dashed line denotes the $B_{T\ max}$ - $\mu_0 H_a$ line expected in static ZFC.

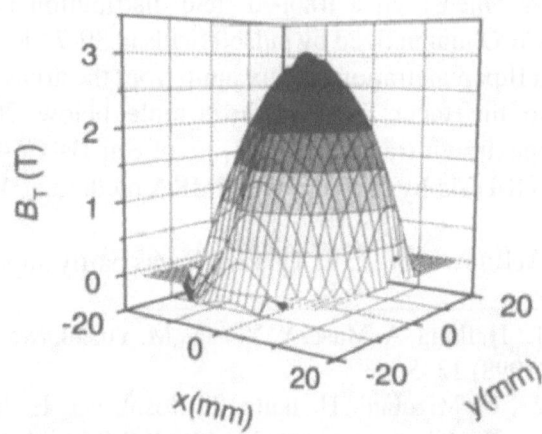

Fig. 3 The trapped field distribution 0.5mm above the sample by the improved IMRA method at 35K. $B_{T\ max}$ of 3.2 Tesla was obtained on the sample surface.

Next, we discuss the case in which the pulsed-fields are applied repeatedly to the sample. In the PFM, heat generation inherently occurs in a bulk superconductor by flux motion. However, the heat generation can be reduced when the sample has already trapped flux lines before the pulsed field is applied. This is because the amount of flux moving in the sample reduces in the presence of the trapped flux. Based on this idea, we have developed the IMRA technique, which consists of applying a pulsed-field large enough for the flux lines to penetrate all over the sample and subsequent iterative PFM operations with reducing amplitudes. Using this method, the total trapped flux can be effectively increased [3,4]. Before applying the IMRA method to the Sm-Ba-Cu-O sample, we have to remember that the abrupt flux penetration superior to the static ZFC magnetization occurs in the PFM only in the absence of trapped flux. In fact, $B_{T\,max}$ at 35 K by the IMRA was limited to 2.6 Tesla whereas 3.2T by the S-PFM, although the total trapped flux by the IMRA was 1.9 times larger than that by the S-PFM. Therefore, we have improved the IMRA method by combining with the S-PFM. First, the optimum $\mu_0 H_a$ of 4.2 Tesla determined in the S-PFM experiment is applied to the sample, where the maximum field can be trapped through the abrupt penetration by flux jump. Then, the conventional IMRA process is subsequently conducted starting with the larger $\mu_0 H_a$ of 6.2 Tesla, which is necessary for flux to penetrate all over the facet lines without reducing the intensity of the trapped field at the center of the sample. The trapped field distribution obtained by the improved IMRA method is shown in Fig. 3. One can clearly see that the shape of the trapped field distribution is similar to that attained by the FC magnetization and $B_{T\,max}$ of 3.2 Tesla is superior to that by S-PFM. At 30 K, the trapped field of 3.8 Tesla was achieved on the sample surface using this procedure.

Figure 4 shows the temperature dependence of the maximum-trapped field $B_{T\,max}$ attained by using the IMRA technique and the FC magnetization. The data for the Y-Ba-Cu-O sample is also plotted for comparison. It should be pointed out that the trapped field of 3.8 Tesla in Sm-Ba-Cu-O obtained by the improved IMRA method is comparable to that in Y-Ba-Cu-O by the FC magnetization.

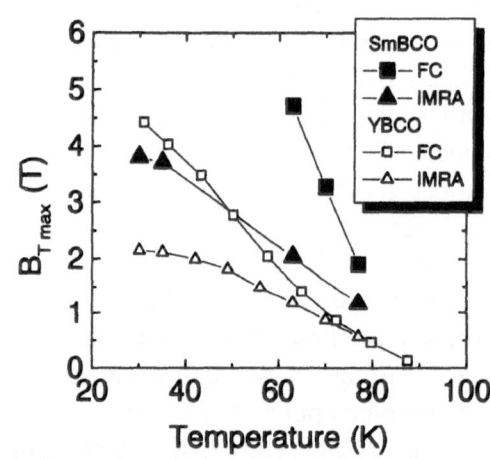

Fig. 4 Temperature dependence of $B_{T\,max}$ for the Sm-Ba-Cu-O and Y-Ba-Cu-O samples 36 mm in diameter by the FC and the IMRA method.

CONCLUSION

We measured a trapped field distribution on Sm-Ba-Cu-O magnetized by pulsed fields at 30-77 K. We found a flux penetration by flux jump from the areas 45-degree to the facet lines of the sample below 70 K. The maximum trapped field $B_{T\,max}$ of Sm-Ba-Cu-O reached 3.8 Tesla by the improved IMRA method at 30K.

Acknowledgment. This work was partly supported by Grant-in-Aid from MITI.

1. H. Ikuta, A. Mase, Y. Yanagi, M. Yoshikawa, Y. Itoh, T. Oka and U. Mizutani: Supercond. Sci. Tech. **11** (1998) 1345
2. U. Mizutani, H. Ikuta, T. Hosokawa, H. Ishihara, K. Tazoe, T. Oka, Y. Itoh, Y. Yanagi and M. Yoshikawa: presented at 12th ISS (Morioka, Japan) October 17-19, 1999
3. U. Mizutani, T. Oka, Y. Itoh, Y. Yanagi, M. Yoshikawa and H. Ikuta: Applied Superconductivity, **6** (1998) 235
4. Y. Yanagi, Y. Itoh, M. Yoshikawa, T. Oka, A. Terasaki, H. Ikuta and U. Mizutani: *Advances in Superconductivity X* (Springer-Verlag, Tokyo, 1998) p. 941

Electric-Field Criterion for the Magnetic Measurement of Critical Current Densities of Oxide Superconductors

Hirofumi Yamasaki and Yasunori Mawatari

Electrotechnical Laboratory, 1-1-4 Umezono, Tsukuba, Ibaraki 305-8568, Japan

Abstract: Due to the broad current-voltage characteristics of high-T_c oxide superconductors, a criterion is necessary to determine the critical current density J_c. In transport measurements electric-field criteria such as 1 µV/cm are often used to determine J_c. However, in the case of magnetic J_c measurements with the Bean's model, J_c criterion has been usually disregarded. We describe methods to estimate the induced electric field during magnetization measurements, and propose that we should accompany the electric-field criterion with the magnetic J_c data.

Key words: Magnetic critical current density, electric-field criterion, magnetization

INTRODUCTION

High-T_c oxide superconductors show broad current-voltage (E-J) characteristics, and therefore a criterion is necessary to determine the critical current density J_c. In transport measurements electric-field criteria such as 1 µV/cm are often used to determine J_c, and the criterion is always written with the measured J_c data. However, in the case of magnetic J_c measurements J_c criterion has been usually disregarded. In this study, we propose that we should accompany the magnetic J_c data with a criterion by which the J_c is determined. We describe the method to extract the E-J characteristics (EJC) by measuring the magnetic-field-sweep rate dependence of magnetizations, from which the electric-field criterion for J_c is easily obtained. This method is suitable for a vibrating-sample magnetometer (VSM) with which magnetization can be measured in realtime during a field scan. In the case of a SQUID magnetometer this method cannot be applied, and we propose another method in which the induced electric field can be roughly estimated by short-time flux-creep measurements.

METHODS FOR ESTIMATING THE INDUCED ELECTRIC FIELDS

Vibrating Sample Magnetometer. Let us consider a superconducting strip of length l, width w and thickness d ($l >> w > d$) that is in an applied magnetic field H_a perpendicular to the strip plane (parallel to the z-axis). When H_a is swept with a constant sweep rate $\beta = \mu_0 dH_a/dt$ and the strip is in the full-penetration condition, we can assume a steady state in which β is much larger than the time derivative of the magnetic flux density due to the shielding current. In this case, $|\beta| >> |\partial B_z/\partial t - \beta| \sim (1 - D)\mu_0|\partial M/\partial t|$, where B_z is the z component of the magnetic flux density and D is the demagnetization factor [1, 2]. The electric field E induced in the superconductor in the steady state has a simple distribution and is proportional to β. The magnetization M of the sample is nearly proportional to the shielding current density J inside the sample. Therefore, the β dependence of M reflects the E-J characteristics of the superconducting specimen. Figure 1 shows the effect of the magnetic-field-sweep rate β on magnetization M at 77.3 K in a melt-textured YBa$_2$Cu$_3$O$_{7-\delta}$ (YBCO) strip (0.5 × 0.8 × 7.8 mm), measured with a vibrating sample magnetometer. $|M|$ is larger with larger sweep rate β, because larger β leads to larger E, which in turn leads to larger J (M). The influence of β is significant at 5 T due to broad EJC in such a high field (inset).

Mawatari et al. calculated E and J at the edge of the strip [1, 2]:

$$E = -\beta w/2, \qquad (1) \qquad J = (2 + \alpha) \cdot 2M/w, \quad \alpha = \frac{\partial(\ln |M|)}{\partial(\ln |\beta|)} = \frac{\beta}{M} \frac{\partial M}{\partial \beta} . \qquad (2)$$

We can extract EJC with Eqs. (1) and (2). In this method E and J distributions in the specimen are taken into account. If $\alpha \to 0$ in Eq. (2), average current density $\langle J \rangle = 4M/w$, resulting in the well-known Bean model equation, $J_c = 4M/w$ or $2\Delta M/w$, where ΔM is the hysteresis width. The parameter α in Eq. (2) reflects the inhomogeneous J distribution in the sample. In the case of a power-law EJC with a power index n ($E \propto J^n$), we get $\alpha = 1/n$. This result demonstrates that the deviation from the Bean model calculation increases with small n-values that indicate broad EJC.

For the YBCO strip the β dependence of hysteresis widths showed clear power-law behavior at 77.3 K (not shown here). The power index directly gives the parameter α in Eq. (2), and we then calculated E and J at the edge of the strip using Eqs. (1) and (2), replacing $2M$ with ΔM. Figure 2 shows the E-J curves at various fields at 77.3 K. In this electric-field window $E = 10^{-10}$-10^{-5} V/m power-law behavior $E \propto J^n$ was observed for every field ($\mu_0 H_a$ = 0.2-5.0 T). In lower magnetic fields ($\mu_0 H_a \leq 1$ T) the power index n was large (≥ 25), which indicates that the E-J curves are steep. In this case, the parameter $\alpha = 1/n$ is small compared with 2 in Eq. (2), and the correction for the J distribution is quite small ($< 2\%$). In high magnetic fields, however, n is not as large and the correction from the Bean model equation becomes substantial, e.g., ~10% at 5 T (Fig. 2).

Now that we have obtained EJC, it is easy to show an electric-field criterion for the magnetic J_c. For example, J_c is 3.2×10^8 A/m^2 at 77.3 K and 0.2 T and 1.4×10^7 A/m^2 at 5.0 T with a criterion of $E = 1$ μV/m (Fig. 2). If we take a lower E criterion such as $E = 1$ nV/m, J_c becomes 2.6×10^8 A/m^2 at 0.2 T and 3.8×10^6 A/m^2 at 5.0 T. Substantially different J_c values are obtained with different criteria at 5.0 T, due to the broad EJC.

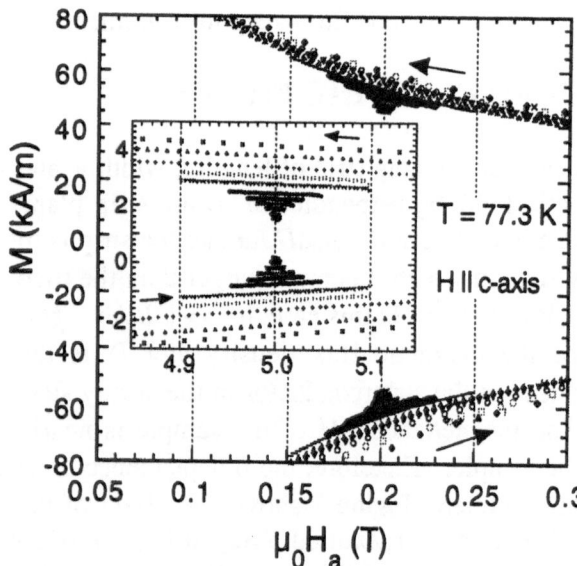

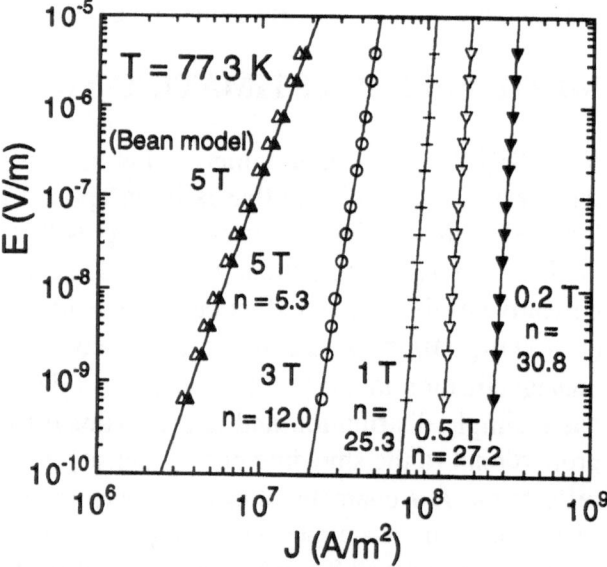

Fig. 1. Magnetic-field-sweep rate β dependence of M, measured in a YBCO strip at 77.3 K around $\mu_0 H_a$ = 0.3 T and 5.0 T (inset). The values of β were 0.0001, 0.0003, 0.0006, 0.0012, 0.003, 0.006, 0.012, 0.03, 0.06, 0.12, 0.3 and 0.6 T/min.

Fig. 2. The E-J characteristics of a YBCO strip in applied magnetic fields of $\mu_0 H_a$ = 0.2-5.0 T at T = 77.3 K. The correction from the simple Bean's model equation is substantial at 5 T.

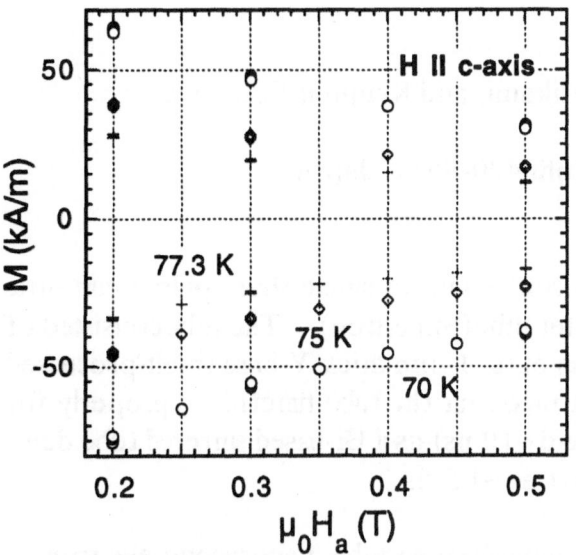

Fig. 3. Magnetic hysteresis curves measured with a SQUID magnetometer. The data at $\mu_0 H_a$ = 0.2, 0.3 and 0.5 T were measured three times with an interval of 30 seconds.

Table. 1. Induced electric fields calculated in the short-time flux creep measurement.

$T = 70$ K			
	$|M|$ (A/m)	δM (A/m)	E (nV/m)
0.2 T (up)	75403		
after 1 min	74193	1210	10.6
after 2 min	73472	721	6.3
0.2 T (down)	64043		
after 1 min	63139	904	8.0
after 2 min	62529	610	5.4
$T = 77.3$ K			
	$|M|$ (A/m)	δM (A/m)	E (nV/m)
0.5 T (up)	17305		
after 1 min	16970	335	2.9
after 2 min	16760	210	1.8
0.5 T (down)	12640		
after 1 min	12325	315	2.8
after 2 min	12119	206	1.8

SQUID Magnetometer. In the case of a SQUID magnetometer the magnetization data cannot be obtained in realtime, and the above method cannot be applied. However, it is possible to roughly estimate the induced electric-field by short-time flux-creep measurements, using Faraday's law,

$$E = -(\mu_0 w/2)(dM/dt). \tag{3}$$

Figure 3 shows the M-H curves measured in a YBCO strip (w = 0.84 mm) with a SQUID magnetometer. After the magnetic field reached the set value ($t = 0$), the first measurement was performed from $t \approx 10$ sec to $t \approx 35$ sec. After that the second measurement was done from $t \approx 66$ sec to $t \approx 91$ sec, and the third measurement was done from $t \approx 122$ sec to $t \approx 147$ sec. We regard the second measurement as the data "after 1 minute", and the third one as that "after 2 minutes". The induced electric fields calculated with Eq. (3) are shown in Table. 1. The magnetic J_c is usually calculated with the Bean's model equation $J_c = 2\Delta M/w$. It is seen from Table 1 that the J_c at 70 K and 0.2 T is 3.3×10^8 A/m^2 with a criterion of $E \sim 9$ nV/m and that J_c at 77.3 K and 0.5 T is 7.1×10^7 A/m^2 with a criterion of $E \sim 3$ nV/m. Comparing this result with Fig. 2, we found that a lower E criterion is generally applied for the measurements with a SQUID magnetometer than with a VSM.

Equations for Other Geometries. A similar method can be used for a superconducting disk (radius: R) in perpendicular magnetic fields [1, 2]. Equations corresponding to Eqs. (1)–(3) are $E = -\beta R/2$, $J = (3 + \alpha)\cdot M/R$, and $E = -(\mu_0 R/2)(dM/dt)$ with the same definition of α. For other geometries such as thin square or rectangular samples, simple analytical equations like Eqs. (1)–(3) can be used if the Bean's critical state model is assumed [3]. Moreover, numerical calculations with power-law EJC ($n = 19$) reveal that there is not much deviation from these simple calculations [3].

1. Y. Mawatari, A. Sawa, H. Obara, M. Umeda, and H. Yamasaki, Appl. Phys. Lett. **70**, 2300–2302 (1997).

2. Y. Mawatari, A. Sawa, H. Obara, M. Umeda, and H. Yamasaki, Cryogenic Eng. (in Japanese) **32**, 485–490 (1997).

3. E. H. Brandt, Phys. Rev. B **52**, 15442–15457 (1995).

A Double-Layered Supertron Consisting of Bi-Based Sintered and Y-Based Melt-Processed Pipes

Hidenori Matsuzawa, Yoshinori Watanabe, Koji Mikami, and Kenjirou Fukasawa

Faculty of Engineering, Yamanashi University, Kofu 400-8511, Japan

Abstract: To focus and guide electron beams of continuous to single short-pulsed currents, we examined a double-layered high-T_c superconductor tube (Supertron). The tube consisted of an inner 1.5-mm thick Bi-based sintered pipe and an outer 1-mm thick Y-based melt-processed pipe. Operation characteristics of the tube confirmed that the tube functioned properly for single pulsed electron beams (~340 keV, ~1 kA, and ~10 ns) as a Bi-based sintered tube does. At 80 K, the tube generated electron pulses as short as ~1.5 ns.

Keywords: Lens for electron beam, Supertron, Double-layered tube, nanosecond electron current

INTRODUCTION

Since the first demonstration [1] of a novel lens (Supertron) for charged particle beams, many experimental results [2] have been accumulated and some papers [3-5] have treated the lens theoretically. The principle of Supertrons is as follows: When an electron beam enters a superconducting tube (Supertron), the self-magnetic field of the beam is confined to the bore of the tube by the Meissner effect. The field thus confined is enhanced, and hence narrows the beam itself. The experiments revealed that Bi-based sintered tubes focus single short-pulsed electron beams well, whereas Y-based melt-processed tubes function worst. We explained these functioning of the tubes by a ferrite-core model, because there exist many resemblances between ferrite cores and Supertrons. The model led us to propose a double-layered high-T_c superconductor tube [6] for focusing and guiding all kinds of electron beams ranging from continuous to single short-pulsed currents. The inner Bi-based sintered pipe will focus single short-pulsed electron beams, because the pipe material behaves like ferrite cores at high frequencies. The outer Y-based melt-processed pipe will focus continuous electron currents, because the pipe material is composed of large grains (crystals) and has much fewer grain boundaries than the Bi-based sintered material and hence is suitable for confining the self-magnetic fields of continuous electron beams.

EXPERIMENTAL SETUP

Figure 1 shows the experimental setup used. Superconductor tubes were glued onto the inner surface of the copper heat sink with electrically conducting paste. The heat sink was cooled with a helium-gas refrigerator down to ~60 K. Primary intense electron currents (~340 keV, ~1 kA, ~10 ns) were field emitted from the cathode by applying high-voltage pulses to the diode consisting of the cathode and anode (heat sink). Neon gas was introduced into the diode space at pressures of the order of ~0.1 Torr (~13.3 Pa) to neutralize space charges of the primary electron currents. The Faraday cup detected the intense electron currents focused with the lens. The net electron currents flowing through the bore of the lens consisted of primary intense electron currents and secondary low-energy plasma electrons. The Rogowski coil received the magnetic fields that arrived at the coil after being induced by the net electron currents and then penetrating through the lens walls. Signals from both the Faraday cup and Rogowski coil were simultaneously observed on a high-speed oscilloscope (Tektronix SCD 1000, 1 GHz).

EXPERIMENTAL RESULTS

We examined the following four tubes: (1) a copper tube (heat sink itself) [6], (2) a Bi-based sintered tube [2], (3) a Y-based melt-processed tube [6], and (4) a double-layered tube. This paper mainly describes the double-layered tube.

Figure 2 shows the waveforms of the focused electron currents (first peaks) and magnetic fields (second peaks). The second-peak waveforms were delayed 26 ns from the first peaks. The results in Fig. 2 are summarized in Fig. 3, where exist notable features as follows: First, the focs

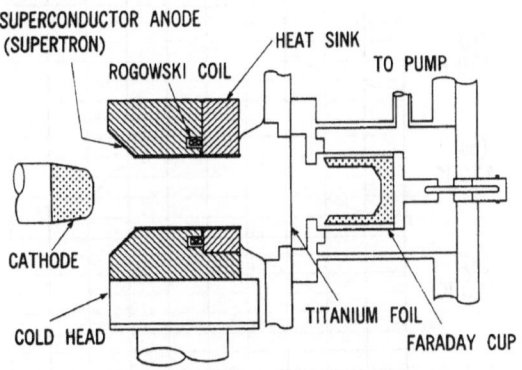

Fig. 1. Experimental setup.

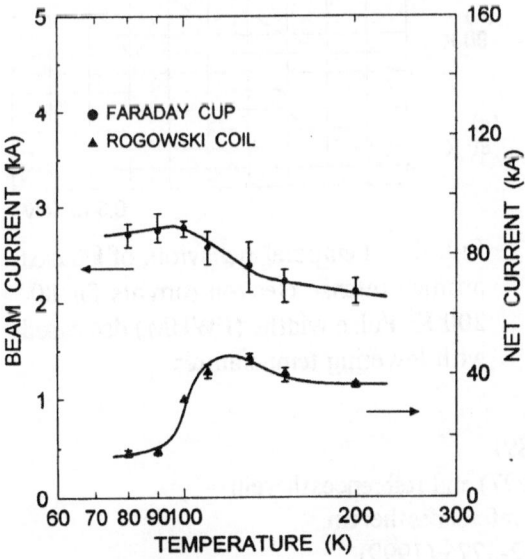

Fig. 3. Temperature dependences of focused primary electron currents detected with Faraday cup and magnetic fields received with Rogowski coil.

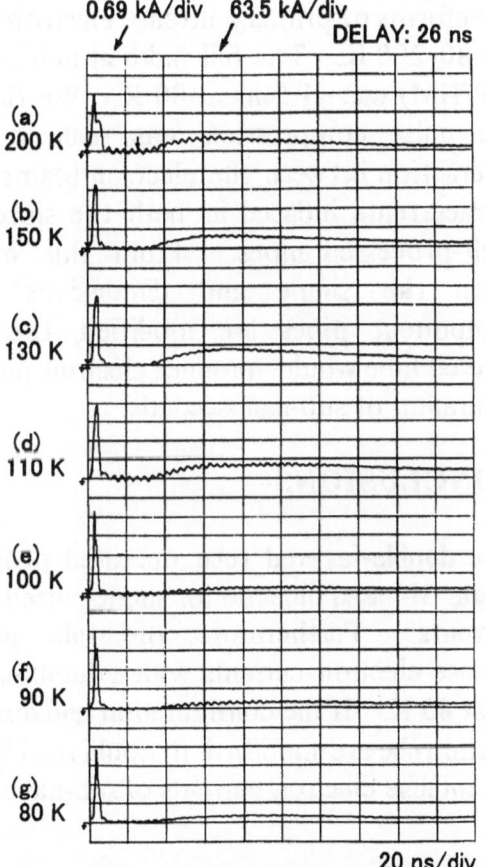

Fig. 2. Temporal behaviors of primary intense electron currents (first peaks) detected with Faraday cup and magnetic fields (second peaks) observed with Rogowski coil.

focused electron beam currents (left-hand-side ordinate) shows the temperature-dependent behaviors similar to those for single Bi-based sintered tubes. Namely, the electron currents increased with decreasing temperatures down to the critical temperature T_c of ~90 K. Below the T_c, the electron currents were almost constant, because the beam diameters were thoroughly narrowed to be detected with the Faraday cup. In another experiments, the beam diameters for the double-layered tube were certainly narrowed with decreasing temperatures as was the case of single Bi-based sintered tubes. Second, the Rogowski coil-detected magnetic fields (right-hand-side ordinate) below ~90 K were much lower than the values for another single tubes. The facts mean that the double-layered tube shielded and confined the self-magnetic fields of net electron currents quite well. Third, the detected magnetic fields gradually turned from ascending to descending at 90-110 K. The rounded turning is ascribed to the two different values of T_c's of the component-pipe materials. The experimental results thus far mentioned support that the double-layered tube functions properly for single pulsed electron beams.

The primary electron currents were compressed in duration by replacing Bi-based sintered tubes with the double-layered tube. Figure 4 shows the waveforms of primary intense electron currents for 80-200 K. The full width at half maximum (FWHM) was ~1.5 ns at 80 K. We think that the pulse compression was realized by the interaction between the electron beams and the intracurrents induced in both the sintered and melt-processed pipes. From this modeling, when the shape and dimensions of the component pipes are modified, the double-layered tube would produce electron pulses with a duration of sub-nanoseconds.

CONCLUSIONS

The double-layered tube operated properly as single Bi-based tubes do for single pulsed electron currents. Furthermore, the tube generated intense electron currents with a duration of ~1.5 ns at 80 K. If the configuration and dimensions of the tube are modified, it would readily generate intense electron currents of sub-nanoseconds.

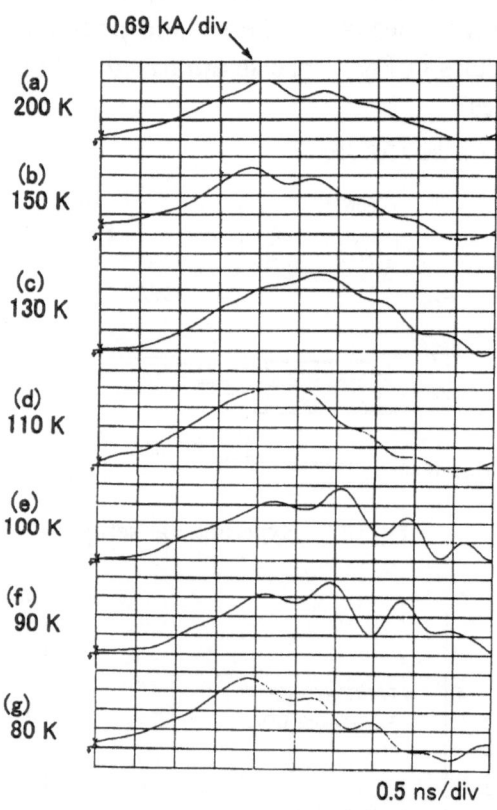

Fig. 4. Temporal behaviors of focused primary intense electron currents for 80-200 K. Pulse widths (FWHM) decreased with lowering temperatures.

1. H. Matsuzawa et al., J. Appl. Phys. **65**, 2596-2603 (1989).
2. H. Matsuzawa et al., Jpn. J. Appl. Phys. **36**, 98-104 (1997) and references therein.
3. P. Roth, Jpn. J. Appl. Phys. **36**, 4537-4538 (1997) and references therein.
4. K. Sakai and Y. Nakamura, Jpn. J. Appl. Phys. **38**, 3772-3775 (1999).
5. C. Buzea, M. Agop, and N. Rezlescu, Jpn. J. Appl. Phys. **38** (1999) 5863.
6. H. Matsuzawa et al., in *Advances in Superconductivity XI*, edited by N. Koshizuka and S. Tajima (Springer, Tokyo, 1999), pp. 1309-1312.

Fishtail Effect And Flux Pinning Characteristic In PMP $Y_{0.6}Ho_{0.4}Ba_2Cu_3O_y$

Y. Feng,[1] A.K. Pradhan, [1] J.G. Wen ,[1] N. Koshizuka,[1] and L. Zhou[2]

[1] Superconductivity Research Laboratory, ISTEC, 1-10-13 Shinonome, Koto-ku, Tokyo 135-0062, Japan
[2] Northwest Institute for Nonferrous Metal Research, P.O.Box 51, Xi'an Shaanxi, 710016, P.R.China

Abstract: The magnetization curves of Ho-doped YBCO samples , which were prepared by a powder melting process technique, were measured using a SQUID magnetometer in a wide range of temperatures . It is found that the fishtail feature is observed below 77K for H parallel to c in $Y_{0.6}Ho_{0.4}Ba_2Cu_3O_y$, while the peak effect is absent in pure YBCO.. It is concluded that the Ho doping is responsible for the fishtail feature rather than the oxygen defect pinning or matching effect.

Keywords: YBCO, Fishtail effect, Flux pinning, Element-doping

INTRODUCTION

For the applications of bulk high-Tc superconductors (HTSC) in the magnetic field, the large critical current densities are required. A peak effect or fishtail was usually observed in the Jc (H) curves of YBCO. In order to further improve Jc, the flux pinning mechanism should be clearly understood. Therefore, studies on the origin of the fishtail are of great importance both for fundamental research and practical applications of HTSC. The fishtail phenomenon has been also found in many other HTSC such as Nd-Ba-Cu-O, La-Sr-Cu-O, Bi-Sr-Ca-Cu-O and Tl-Ba-Ca-Cu-O. Several pinning mechanism have been proposed although the origin of the peak effect is not completely understood. Daeumling et al attributed the fishtail to the flux pinning induced by the oxygen deficient regions[1]. Other explanations include the vortex melting, crossover from elastic to plastic response and dimensional crossover from three dimensional (3D) to 2D of the flux line lattice. In addition, some other results clearly indicate that the peak effect is not the intrinsic property of YBCO sample and depends on the quality of the sample. Recently. Genoud et al reported that the peak effect is absent in extremely clean YBCO crystals. The fishtail should result from the defects or weak superconducting regions in the sample [2].

In this work, we present a study of the peak-effect in $Y_{0.6}Ho_{0.4}Ba_2Cu_3O_y$ sample prepared by the powder melting process method. The fishtail effect is found in this sample below 77K, whereas no peak is observed in pure YBCO. We propose that the peak effect in our material may be linked with the contribution from the paramagnetic moment of Ho or the stress field induced by the Ho doping.

EXPERIMENTAL

The samples with nominal compositions of $YBa_2Cu_3O_y$ (YBCO) and $Y_{0.6}Ho_{0.4}Ba_2Cu_3O_y$ were prepared by a powder melting process (PMP) method The pre-sintered bars were put into a tube furnace with a temperature gradient of 50-100°C/cm for the melting process in air. The highest temperature in the melting zone was below 1030°C. The moving speed of the sample was around 2mm/h. Then, these specimens were re-annealed at 900°C for 30h in order to investigate the fishtail effect. These samples were annealed at 400-550°C in a flowing oxygen for a long time in order to ensure the complete oxygenation.   The dimensions are $2.2 \times 0.92 \times 0.35 mm^3$ and $2.3 \times 0.8 \times 0.36 mm^3$ respectively for the $Y_{0.6}Ho_{0.4}Ba_2Cu_3O_y$ and YBCO samples.

The critical temperatures and magnetic hysteresis loops of the samples were measured using a SQUID magnetometer (Quantum Design MPMS) with the applied field parallel to the c-axis. The result shows that the Tc of YBCO and $Y_{0.6}Ho_{0.4}Ba_2Cu_3O_y$ is   90.8 and 91K receptively   with a sharp transition width less than 1K.

RESULTS AND DISCUSSION

The critical current density was calculated from the magnetization loops using Bean's critical state model. Figure 1 gives the field dependence of Jc at different temperatures   for the $Y_{0.6}Ho_{0.4}Ba_2Cu_3O_y$ sample in magnetic field parallel to the c-axis. It can be clearly observed that the peak effect is well developed at temperatures below 60K, while no fishtail phenomenon is found above 77K. In addition, the peak shape both in width and height changes as temperature varies. The peak becomes broader when temperature increases, which is different from other reports [3]. At temperatures above 40K, Jc decreases slowly with increasing field after Jc reaches its maximum value , which means that the pinning centers responsible for this fishtail remains effective over a field range near the peak field region. Below 40K, Jc drops quickly in the field region above the peak field (Hp) as the field increases, suggesting that the pinning centers controlling the peak rapidly become less effective above the peak field. As observed in other YBCO and TlBaCuO samples, the peak field is strongly   temperature dependent in our sample. Above 40K, Hp drops almost linearly as temperature grows, which indicates that the flux pinning in this temperature range is attributed to the defects in the sample classified as a source of correlated disorderIn addition, it can be seen from Fig.1 that the fishtail minimum (H_{min}) shifts to the higher field as temperature decreases. Also, H_{min} changes significantly with temperature around 40K. Although no scaling law for the fishtail minimum has been established it is thought that this minimum is due to a superposition of two pinning contributions with different temperature dependence.

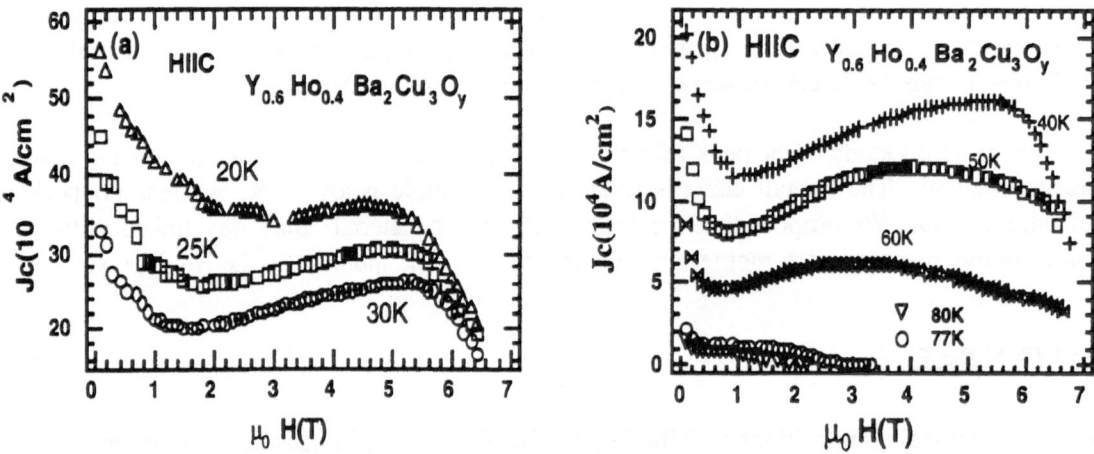

Fig. 1. Magnetic field dependence of Jc for various temperatures in the Ho-doped sample.

Although several models have been developed to explain the origin of the fishtail, it is not fully understood yet. Here, the peak effect is found below 60K, while no fishtail is observed above 77K in the Ho-added sample. It is interesting to note that no peak effect is observed even when temperature drops to 40K in the pure YBCO sample prepared under the same conditions as the Ho-doped sample (shown in Fig.2). It is believed that the oxygen content does not change in the Ho-added specimen since Ho has the same valence state as Y. Therefore,   the peak effect in our sample is   not due to the oxygen defects. Some results also indicate that the fishtail effect in YBCO can not be simply attributed to the oxygen defect pinning. We think that the peak in our sample may be related to the contribution from the paramagnetic moment of Ho or the additional stress field induced by the local substitution of Ho for Y. A theoretical calculation demonstrates that the stress field resulting from the Ho doping for Y in YBCO can introduce a new strong pinning[4]. In a recent paper, it has been found that strain field clusters created by the Sr doping are effective pinning centers in the Sr-added YBCO crystal; causing the peak effect in Jc(H) [5]. Furthermore, the change in the intrinsic parameters and the structure defect density introduced by the partial substitutions of Zn for Cu are considered to result in the strong peak effect in Zn-doped sample [6]. In order to find the possibility of the paramagnetic contribution from Ho, we carried out the field-cooling (FCC) magnetization in the temperature region from 140K to 20K for various fields. Figure 3 gives the temperature dependence of the magnetization of the Ho-added sample at different fields. These results show that the magnetic moment is always positive in both normal and superconducting state. Also, the magnetization increases on decreasing temperature and increasing field. The behavior of magnetization above Tc can be well described by the Curie-Weiss law. This result obviously indicates the contribution of the paramagnetic moment of Ho.

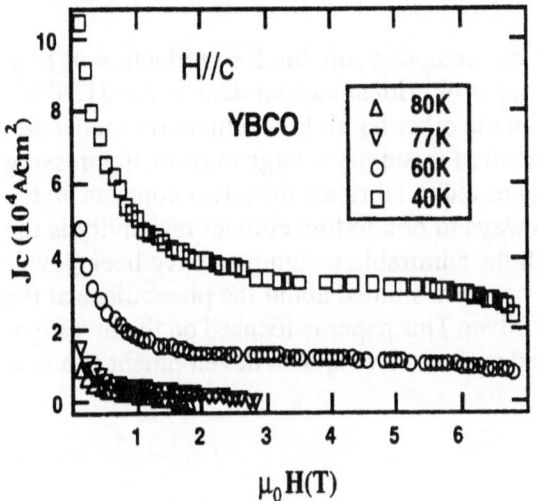

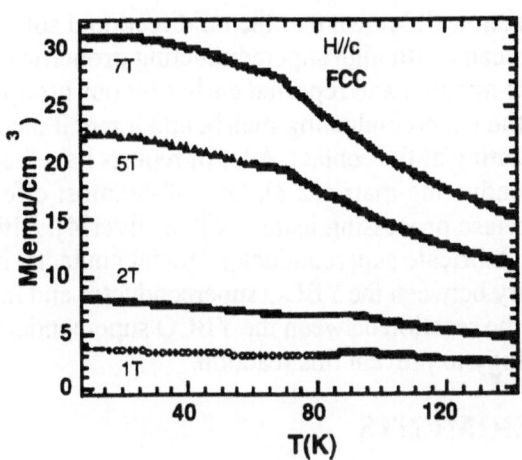

Fig.2. Field dependence of Jc in YBCO, showing no peak until 40K.

Fig.3. Temperature dependence of magnetization of Ho-doped sample.

CONCLUSION

The fishtail effect is found in a powder melting processed $Y_{0.6}Ho_{0.4}Ba_2Cu_3O_y$ bulk sample below 60K, while the peak is absent in this sample for T≥77K and in pure YBCO for T≥40K. The peak field increases linearly with decreasing temperature above 40K. The fishtail effect in our sample may be linked with the paramagnetic moment contribution of Ho or the stress field created by the Ho doping.

Acknowledgments: This work is supported by the New Energy and Industrial Technology Development Organization (NEDO) as Collaborative Research and Development of Fundamental Technologies for Superconductivity Applications.

1. M.Daemuling, J.M.Senutjens, and D.C.Larbalestier, Nature **346**, 332 (1990).
2. J.Y.Genoud, B.Revaz, A Erb, A.Mirmelsein, and A.Junod, Czech J.Phys. **47**, 1047 (1997).
3. J. L.Vargas and D.C.Larbalestier, Appl. Phys. Lett. **60,** 1741 (1992).
4. Y. Li, N.Chen, and Z.X.Zhao, Physica C **224**, 391 (1994).
5. K.Saito, H.U.Nissen, C.Beeli, T.Wolf, W.Schauer, and H.Kupfer, Phys. Rev. B **58**, 6645 (1998).
6. M.Hussain, S.Kuroda, and K.Takita, Physica C **297**, 176 (1998).

Fabrication process of low resistivity connection between YBCO superconductor and silver

Susumu Seiki, Junya Maeda, Yuichi Nakamura, Teruo Izumi and Yuh Shiohara

Superconductivity Research Laboratory, ISTEC, Koto-ku, Tokyo 135-0062, Japan

Abstract: YBCO superconducting rods prepared by the unidirectional solidification method were dipped into the molten silver in order to fabricate a metal contact on the surface of the rods. In the case of dipping into the pure molten silver, a copper poor reacted layer was formed at the interface between $Y_1Ba_2Cu_3O_x$(Y123) and silver, although the temperature of the molten silver was controlled to be lower than the decomposition temperature of the Y123 phase. This reaction could be suppressed by means of addition of Y123 powder into the molten silver. This method successfully realized the contact with the low resistivity of $4.6 \times 10^{-12} \Omega m^2$.

Keywords: YBCO, silver, connection, resistivity, dipping

INTRODUCTION

It has been well known that the unidirectional solidification method is suitable for production of long current leads with high superconducting properties. The superior critical current density Jc=71,700A/cm^2 was attained and reported earlier by our group[1]. On the other hand, low contact resistance between the superconducting matrix and a metal is also required to supply a large current suppressing joule heating at the contact. A lot of reports have been suggested to fabricate the silver contacts on the superconducting matrix[2,3]. One of the most effective ways to obtain low contact resistivity is the liquid phase processing using molten silver[4]. Although the admirable techniques have been developed to fabricate superconductor / metal contacts, it has not been studied about the phenomena at the boundary between the YBCO superconductor and melted silver. This paper is focused on the investigation of the reaction between the YBCO superconductor and molten silver and on development of a new technology to prevent this reaction.

EXPERIMENTS

The $Y_{1.8}Ba_{2.4}Cu_{3.4}O_x$(YBCO) superconducting leads(1.6mm$\phi \times$100mml) were prepared by the unidirectional solidification method with pulling rates of 1.0mm/h and temperature gradients at the growth interface of G=25°C/mm. The unidirectionally solidified YBCO superconducting leads were dipped into four different kinds of melts that were listed in Table. 1. Each of the alloyed metal and oxide was set into an MgO crucible and was melted in an electric furnace. The bottom edges of the YBCO leads were dipped into the melt at 953°C, which is lower than the peritectic temperature of the YBCO composite (1005°C), and kept in this melt for 1h. The leads were pulled up to take out immediately and cooled down to the room temperature. Then the leads were annealed in oxygen gas flow at 500°C for 400h. Additionally, the liquid compositions were evaluated by ICP-AES measurement for the samples which were obtained by dipping Al_2O_3 rods into the liquid. An optical microscope and a scanning electron microscope (SEM) equipped with a wavelength dispersive spectrometer (EPMA-WDS) were used for observation of the structure and characterizations. The contact resistivities of these connections at 77K were measured by the conventional method[5].

RESULTS AND DISCUSSION

The side view and the longitudinal cross section of the specimen dipped into A-type melt is shown in figures 1(a) and (b), respectively. In these figures, it can be seen that liquid silver adhered on the superconducting lead, but a reacted layer formed between the YBCO matrix and the adhered silver. Thickness of the reacted layer was about 50μm and the composition was analyzed as Y:Ba:Cu = 1.0:2.0:1.6. Figure 2 shows the longitudinal section of the specimen dipped into the B-type melt. YBCO superconductor was decomposed to $Y_2Cu_2O_x$(202), CuO(001), and Ba-Cu-O melt, and the 123 phase can not be found after dipping completely. Figures 3 and 4 show the longitudinal cross section of the specimen dipped into C- and D-type melts. In these figures, neither reaction at the boundary between the superconducting matrix and the molten silver nor decomposition of 123 phase can be seen. Table 2 shows the results of Cu concentration in each melt. The clear difference in the concentration was observed.The dissolution of <123> into silver melt can be described by,

$$\langle Y123 \rangle \longrightarrow [Y] + [Ba] + [Cu] \tag{1}$$

where <123> is the solid Y123 phase and [Y] , [Ba] and [Cu] are the dilute solutions of Y,Ba and Cu elements in the silver melt, respectively. Then the free energy change due to the reaction(ΔG_D) can be written when the activity of <Y123> is assumed to be unity,

$$\Delta G_D = \Delta G_D^0 + RT \ln\left(a_{[Y]} \cdot a_{[Ba]} \cdot a_{[Cu]}\right) \tag{2}$$

where ΔG_D^0 is the standard free energy change and $a_{[i]}$ is the activity of the dilute solution of i element. The equilibrium state, which can be obtained by the condition of $\Delta G_D = 0$, gives the equilibrium product of $a_{[Cu]}^{*D} \cdot a_{[Cu]}^{*D} \cdot a_{[Cu]}^{*D}$. In the case of D-type melt, product of activities are considered to be naturally achieved to the equilibrium state, and the dissolution was prevented. In the case of the A-type melt, the liquid can be recognized to be an undersaturated state, which means the negative value of ΔG_D. Consequently, the reaction of Eq.(1) proceeded in the type-A. Then the reacted layer at boundary is considered to be formed due to a kinetic effect. The no reaction in the C-type melt in spite of the difference in the Cu concentration in liquid suggests the contribution of the strong interaction of each element. Further quantitative analyses of each concentration are required to clarify the phenomena.
On the other hand, the reaction of the B-type melt can be described by,

$$\langle Y123 \rangle + [Cu] \longrightarrow \langle 202 \rangle + \langle 001 \rangle + (BCO) \tag{3}$$

where <202> and <001> are the solid phases of $Y_2Cu_2O_x$ and CuO, respectively and (BCO) is the Ba-Cu-O liquid. Then the free energy change(ΔG_B) can be given by the assumption of $a_{<123>}=a_{<202>}=a_{<001>}=1$,

$$\Delta G_B = \Delta G_B^0 + RT \ln\left(a_{(BCO)} / a_{[Cu]}\right) \tag{4}$$

where ΔG_B^0 is the standard free energy change in this reaction and $a_{(BCO)}$ is the activity of the Ba-Cu-O liquid. The equilibrium activity of [Cu] for this reaction, $a_{[Cu]}^{*D}$ can be obtained by the condition $\Delta G_B = 0$. The actual activity can be recognized to be higher than $a_{[Cu]}^{*D}$, which means the negative value of ΔG_B. Consequently the reaction in Eq.(3) was thought to be proceeded in the type B. The case of the B-type melt is considered to correspond to the above phenomenon. All relationship between the Cu-potential in the liquid and reaction was schematically shown in Figure 5.

Table 1. Dipping condition.

Composition	Temp.	Time
A Pure Ag	953°C	1h
B Ag:CuO=95:5(wt%)	953°C	1h
C Ag:Ba3Cu5Ox=95:5(wt%)	953°C	1h
D Ag:Y123=95:5(wt%)	953°C	1h

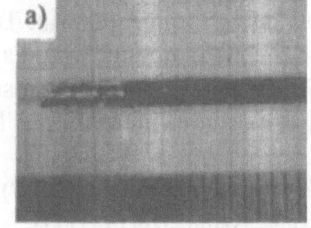 

Fig. 1 a) Side view of YBCO superconducting lead dipped into Atype melt at 953°C for 1h
b) Longitudinal cross section of this specimen

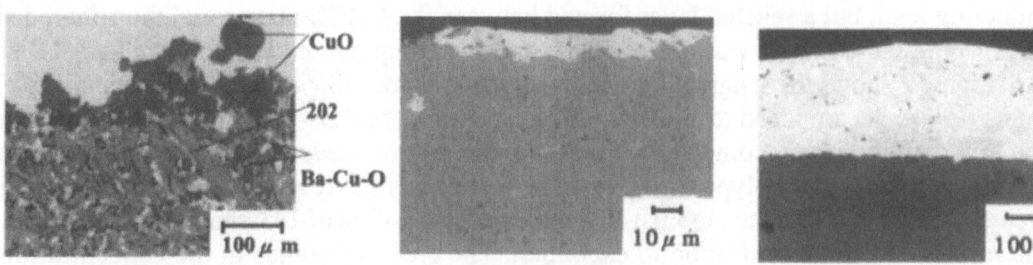

Fig. 2 Longitudinal cross section of YBCO superconductinglead dipped into Btype melt at 953℃ for 1h

Fig. 3 Longitudinal cross section of YBCO superconductinglead dipped into Ctype melt at 953℃ for 1h

Fig. 4 Longitudinal cross section of YBCO superconductinglead dipped into Dtype melt at 953℃ for 1h

Table 2 Cu concentration in each melt

Melt type	[Cu] (wt%)
B	0.83 [6]
C	0.42
D	0.18

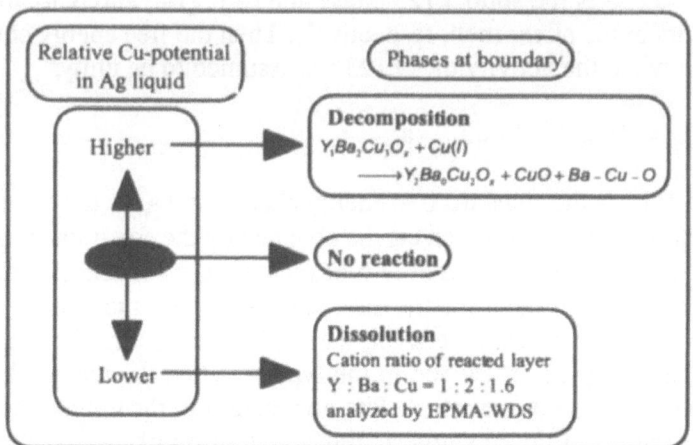

Fig. 5 The relationship of relative Cu-potential in Ag liquid

At the last, the contact resisitivity of the connection using the A-type melt was $3.6\times10^{-8}\Omega m^2$. On the other hand, the low contact resistivity of $4.6\times10^{-12}\Omega m^2$ was successfully obtained using the D-type melt.

CONCLUSION

A reacted layer was formed at the interface between YBCO superconducting matrix and molten silver in the case of the dipping using pure silver liquid. It is necessary to optimize copper concentration in the liquid silver for suppressing these reaction. We can successfully prevent these reaction with using the $Y_1Ba_2Cu_3O_x$ powder added silver melt. The contact with the low resistivity of $4.6\times10^{-12}\Omega m^2$ was successfully obtained by this method.

ACKNOWLEDGMENT

This work was supported by the New Energy and Industrial Technology Development Organization (NEDO) as Collaborative Research and Development of Fundamental Technologies for Superconductivity Applications under the New Sunshine Program administered by the Agency of Industrial Science and Technology (AIST) of the Ministry of International Trade and Industry (MITI) of Japan.

[1]   Y. Imagawa, K. Kakimoto, Y. Shiohara: Physica C **280**(1997)245
[2]   J. Mass, V. A. Gasparov, D. Pavuna: Nature **328**(1987)603
[3]   J. Joo, J. G. Kim, W. Nah: Supercond. Sci. Technol. **11**(1998)645
[4]   N. Mori, K. Kawazoe, A. Toji, K. Ogi: J. Jpn. Weld. Soc. **14**(1996)162
[5]   S. Jin et al., Appl. Phys. Lett. **54** [25] 2605-07(1989)
[6]   H. Nishiura et al., J. Am. Ceram. Soc, **81** [8] 2181-87(1998)

Microstructure and Superconducting Properties of Bulk Superconductors Prepared from Sm123-RE211 (RE = Y, Gd and Sm) and Sm123-Nd422 Precursors

Shinya Nariki, Seok-Jong Seo, Naomichi Sakai and Masato Murakami

Superconductivity Research Laboratory, ISTEC, 1-16-25 Shibaura, Minato-ku, Tokyo 105-0023, Japan

Abstract: The samples consisting of $SmBa_2Cu_3O_y$ (Sm123) precursors with RE_2BaCuO_5 (RE211, RE=Y, Gd and Sm) or $Nd_4Ba_2Cu_2O_{10}$ (Nd422) were melt-textured in a reduced oxygen atmosphere. We have studied the relationship between the size of RE211 particles and the critical current density (J_c). When Y211 and Gd211 were added to a starting material, the size of RE211 phase particles in the bulk sample was remarkably reduced. As a result, J_c values of these samples were larger than those of Sm123/Sm211 sample. The large c-axis oriented bulk sample 32mm in diameter was successfully fabricated from Sm123-Gd211precursor with 20wt% Ag_2O addition. The trapped magnetic field of this bulk reached 1.0T at 77K.

Keywords: bulk, melt-processing, Sm123, (Sm, Gd)-Ba-Cu-O, trapped magnetic field

INTRODUCTION

LRE-Ba-Cu-O (LRE = Nd, Sm, Eu and Gd) superconductors fabricated in a reduced oxygen atmosphere exhibit higher critical temperatures (T_c) than Y-Ba-Cu-O superconductor, and the magnetic field dependence of critical current density (J_c) exhibits secondary peak effect [1]. As a result, J_c values of these superconductors were higher than those of Y-Ba-Cu-O superconductor in high fields. Recently, Ikuta and coworkers [2, 3] reported that the large c-axis oriented Ag-doped Sm-Ba-Cu-O superconductor was successfully fabricated by the melt-process in a reduced oxygen atmosphere. The trapped field of this sample exceeds that of Y-Ba-Cu-O sample with a similar size.

In a melt-processed Sm-Ba-Cu-O superconductor, the size of Sm_2BaCuO_5 (Sm211) particles, which provides flux pinning, are reduced by the addition of Pt. However, the effect of Pt addition on the size reduction of Sm211 was not remarkable, compared to its effect on the Y211 [4]. If the size of Sm211 particles can be reduced, the J_c of Sm-Ba-Cu-O will be further enhanced. In this work, $SmBa_2Cu_3O_y$ (Sm123) precursors compounded with RE211 (RE = Y, Gd and Sm) or $Nd_4Ba_2Cu_2O_{10}$ (Nd422) were melt-textured in a reduced oxygen atmosphere. Relationship between the size of RE211 particles and the J_c value of bulk sample has been investigated.

EXPERIMENTAL

The melt textured samples were synthesized from the precursors with the molar ratio of Sm123 : RE211 (RE=Y, Gd and Sm) = 10 : 4 and Sm123 : Nd422 = 10 : 2. 0.5wt% Pt was added in the mixed powders of Sm123 and RE211 (or Nd422). The mixtures were first uni-axially pressed into pellets 20 mm in diameter and 15mm in thickness, and subjected to cold-isostatic pressing (CIP) under a pressure of 200MPa. An MgO (100) seed was placed on the center of the pellet, which was then OCMG (oxygen-controlled-melt-growth) processed in flowing mixture gas of $1\%O_2$ and 99%Ar. The heat treatment profiles were scheduled as follows. The peritectic decomposition temperatures (T_m) of precursor powders were determined with differential thermal analysis (DTA) measurements. For example, T_m was 1001℃ for Sm123-Y211 mixed powder. Pellets of precursors were heated to 1080℃ and held for 1 h, cooled to (T_m + 10)℃ in 30 min, and then cooled at a rate of 0.25℃/h to (T_m − 30)℃

and finally cooled to room temperature. The samples could be grown into a single domain. All the OCMG-processed samples were annealed in flowing oxygen at 300°C for 100 h.

The phase identification was performed by X-ray powder diffraction (XRD) analysis. Microstructure of the samples was observed with a scanning electron microscope (SEM). Energy dispersive X-ray (EDX) analysis with SEM was also performed to analyze the chemical composition of the matrix and the second phase. T_c values were measured with a Quantum Design SQUID magnetometer in an applied magnetic field of 1 mT. Magnetization loops were measured at 77K up to 7 T for fields applied parallel to the c-axis for J_c measurements.

RESULTS AND DISCUSSION

X-ray diffraction analyses show that the samples prepared from Sm123-RE211 (RE = Y, Gd and Sm) precursors consist of 123 and 211, while the sample prepared from Sm123-Nd422 precursor is composed of 123 phase and 422 phase. Figure 1 shows scanning electron micrographs for the microstructure of the samples. In Sm123/Sm211 sample, the size of Sm211 particles was about 2 to 3 μ m as shown in Fig.1 (a). As shown in Figs.1 (b) and (c), the size of RE211 particles in the Sm123 compounded with Gd211 and Y211 was reduced to about 1 μ m. For the Sm123 sample with Nd422 addition, the size of 422 particles was larger than those of 211 particles in the other samples as shown in Fig.1 (d). EDX analyses showed that the 123-matrix phase and 211 (or 422) second phase contain both Sm and RE ions, showing the presence of the following solid solutions: $(Sm, RE)Ba_2Cu_3O_y$ and $(Sm, RE)_2BaCuO_5$ or $(Sm, Nd)_4Ba_2Cu_2O_{10}$.

Figure 2 shows temperature dependence of dc-susceptibility in zero-field-cooled (ZFC) mode in the presence of 1 mT. The onset T_c of Sm123/Y211 decreased with the substitution of Y^{3+} for Sm^{3+} in 123 phase. The J_c was estimated by using an extended Bean model [5]. Figure 3 shows the J_c -B curves for the present samples. The J_c values of Sm123/Y211 and Sm123/Gd211 samples were higher than those of Sm123/Sm211. Fine 211 inclusions leads to a large J_c in these samples. On the other hand, J_c of Sm123/Nd422 sample was lower than those of the other samples. It is clear that the J_c values are greatly affected by the size of 211 or 422 inclusions.

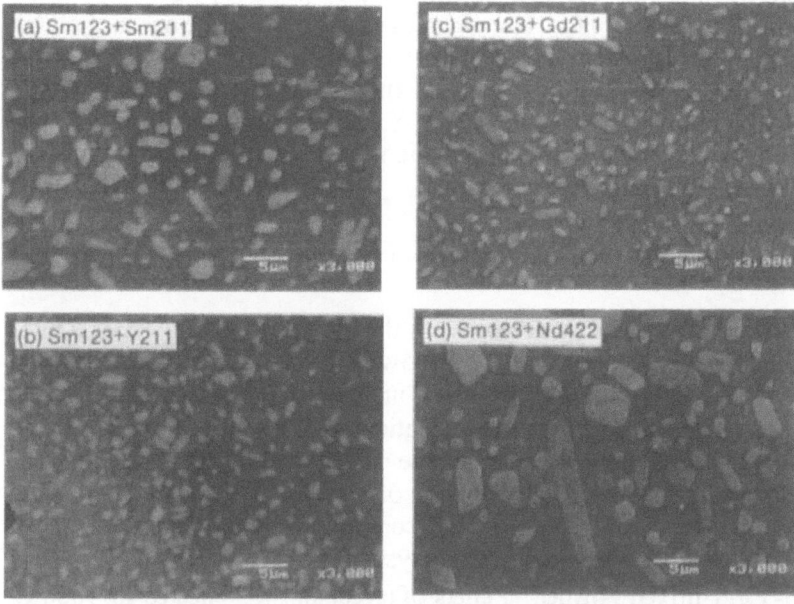

Fig.1. SEM photographs of the bulk samples prepared from the precursors of Sm123-RE211 [RE : (a) Sm, (b) Y and (c) Gd] and (d) Sm123-Nd422.

Large c-axis oriented bulk sample was also fabricated from the precursor with the composition of Sm123 : Gd211 = 10 : 3. 0.5wt% Pt and 20 wt% Ag_2O were added to the starting materials. The addition of Ag suppresses the formation of macro-cracks in large bulk samples [2]. The precursor pellet 40mm in diameter and 20mm in thickness was melt-textured in a mixture gas of 0.1%O_2 and 99.9%Ar. To obtain a large single-domain bulk, the hot seeding method [6] was employed and Nd123 (001) was used as a seed crystal. We succeeded in fabricating a single domain bulk with a diameter of 32 mm. After oxygen annealing

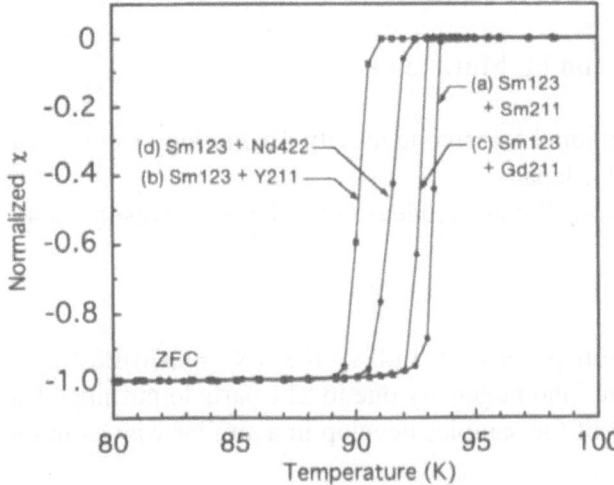

Fig.2. Temperature dependence of dc susceptibility.

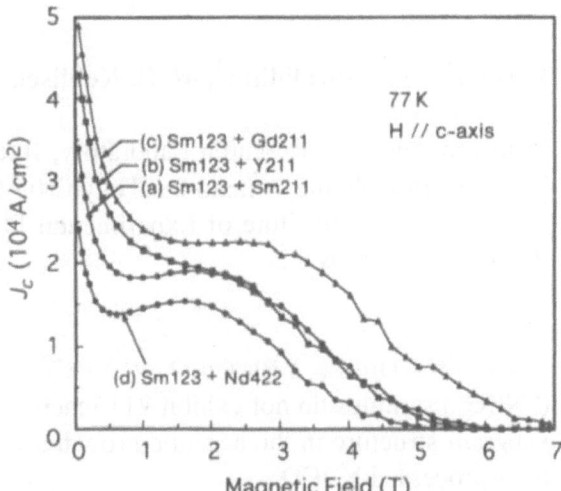

Fig.3. Field dependence of critical current density (J_c).

at 300℃, the trapped field measurements were performed. The sample was field-cooled to liquid nitrogen temperature in 5T. After the external field was removed, the trapped magnetic field was measured using a Hall sensor, which was scanned at 0.5 mm above the sample surface. As shown in Fig.4, the maximum trapped magnetic field reached 1.0T. Sm-Ba-Cu-O bulk sample with a diameter of 32 mm was also melt-textured using the precursor of Sm123 : Sm211 = 10 : 3 with 20 wt% of Ag$_2$O. The trapped magnetic field of this sample was about 0.8 T at 77K. The (Sm, Gd)-Ba-Cu-O bulk has a high trapped magnetic field compared to Sm-Ba-Cu-O bulk with 20wt% of Ag$_2$O in a similar size, because of J_c enhancement through the size reduction of 211 phase.

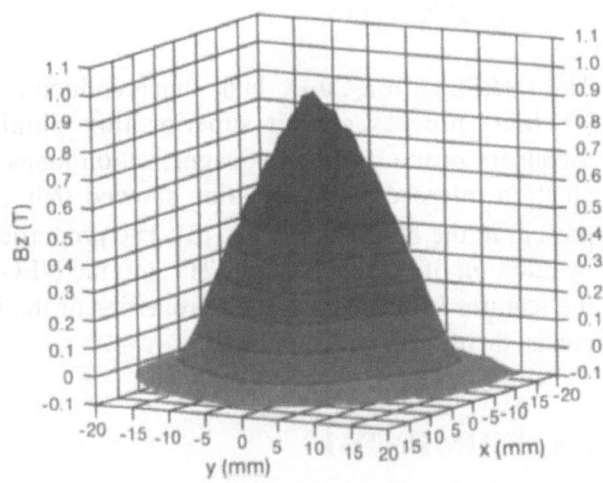

Fig.4. Trapped field distribution (77K) of the single domain Ag-added (Sm, Gd)-Ba-Cu-O sample 32mm in diameter. The maximum value of the trapped field is about 1.0T.

Acknowledgment. This work was supported by the New Energy and Industrial Technology Development Organization (NEDO) as Collaborative Research and Development of Fundamental Technologies for Superconductivity Applications.

1. M. Murakami, N. Sakai, T. Higuchi and S. I. Yoo, Supercond. Sci. Technol. **9**, 1015-1032 (1996).
2. H. Ikuta, A. Mase, Y. Yanagi, M. Yoshikawa, Y. Itoh, T. Oka and U. Mizutani, Appl. Supercond. **6**, 109-117 (1998).
3. H. Ikuta, S. Ikeda, A. Mase, M. Yoshikawa, Y. Yanagi, Y. Itoh, T. Oka and U. Mizutani, Supercond. Sci. Technol. **11**, 1345-1347 (1998).
4. C. J. Kim and G. W. Hong, Supercond. Sci. Technol. **12**, R27-R41 (1999).
5. E. M. Gyorgy, R. B. van Dover, K. A. Jackson, L. F. Schneemeyer and J. V. Waszcazk, Appl. Phys. Lett. **55**, 283-285 (1989).
6. Y. A. Jee, G. W. Hong, T. H. Sung and C. J. Kim, Physica C **314**, 211-218 (1999).

Polarized light microscopy characterization of (Nd,Eu,Gd)-123 superconductors prepared by OCMG process

P. Diko[1,2], M. Muralidhar[1], M. R. Koblischka[1] and M. Murakami[1]

[1]Superconductivity Research Laboratory, International Superconductivity Technology Center, 16-25 Shibaura 1-chome, Minato-ku, Tokyo 105-0023, Japan
[2] On leave from Institute of Experimental Physics, Slovak Academy of Sciences, Watsonova 47, 04353 Kosice, Slovakia

Abstract: Unlike $YBa_2Cu_3O_7$ ("YBCO") melt-processed bulks, the $(Nd,Eu,Gd)Ba_2Cu_3O_7$ ("NEG") samples do not exhibit 211 macroscopic inhomogeneity due to 211 particle pushing. The subgrain structure in the a- and c-growth sectors of the samples develop in a similar way as in the melt-processed YBCO.

Key words: (Nd,Eu,Gd)-123, OCMG, microstructure, bulk superconductors.

INTRODUCTION

The $(Nd,Eu,Gd)Ba_2Cu_3O_y$ bulk superconductors prepared by the oxygen controlled melt growth (OCMG) process exhibit superior flux pinning properties accompanied by a pronounced secondary peak effect in the magnetization loops as compared to YBCO or NdBCO [1]. Transition electron microscopy observation showed that Gd rich 211 particles smaller than 0.1 μm were present in the NEG-123 matrix [2]. The properties of the NEG system were then improved further by addition of 30 mol% of Gd-211 into the NEG matrix [3]. The detailed microstructural study of the features influencing overall properties of the bulk as are: 211 homogeneity, subgrain structure etc. is, therefore, fully approved.

EXPERIMENTAL

Bulk samples of NEG (Nd:Eu:Gd = 1:1:1) were prepared using the OCMG process in 0.1% partial pressure of oxygen. Samples with addition of 40 mol % NEG-211 (that is, a mixture of Nd-422, Eu-211, and Gd-211 in the same ratio as the NEG matrix) and 30 mol% of Gd-211 were prepared. In order to refine the resulting 211 particle size, 0.5 mol% Pt was added to the samples. The crystallization process was initialized by an MgO (100) seed which was placed at the center on the top of the basal pellet surface. The microstructural features of these samples were studied by optical microscopy after polishing or etching in a solution of 2 wt.% Br in ethanol. Observations in normal and polarized light were performed. The quantitative microstructural data were obtained using the image processing system Image-Pro Plus.

RESULTS AND DISCUSSION

MgO seeding lead to single grain samples (Fig. 1) with the c-axis parallel to the top surface of the sample. The central angle of the c-growth sectors (c-GS) was smaller then 90°, which implies that the growth rate was higher for c-growth than for a-growth. Quantitative measurements of 211 particle macrohomogeneity over the entire sample was performed. First, a difference in 211 particle content in the c-and a-GS-s was checked for. The growth sector boundary (GSB) region was visualized by polarized light (a- and c- subgrains can be seen when the crystal is observed in extinction position [4] (Fig. 2a). In no extinction crystal positions the 211 particles are visualized (Fig. 2b). No difference was detected in the 211 content between the a- and c-GS-s in either of the samples. Therefore, the exact position of GSB could not be defined by a sharp change in 211 content as in the case of YBCO-Y123 melt-grown bulks [5].The GSB lies in the subgrain free region, somewhere between a- and c-GS-s. The observed 211 microinhomogeneity (211 free regions) was caused by the large particle size of starting YBCO powder [6].

The 211 content (characterized by 211 volume percentage (V_{211})) and 211 size (average diameter d_{211}) are plotted in Fig. 3 as a function of the distance from the seed. Both are nearly constant for the NEG-NEG-211 sample. This result can only be explained by the nonexistence of 211 pushing by the growing crystal. In the light of this finding, the presence of 211 particles smaller then 0.1 µm in the NEG-123 matrix can be understood [1,2]. This behavior is unique - unobserved in YBCO bulks so far. This implies that there are no limitations to growing large diameter tiles due to the growth inhibition by pushed 211 particles of submicron size in this system.

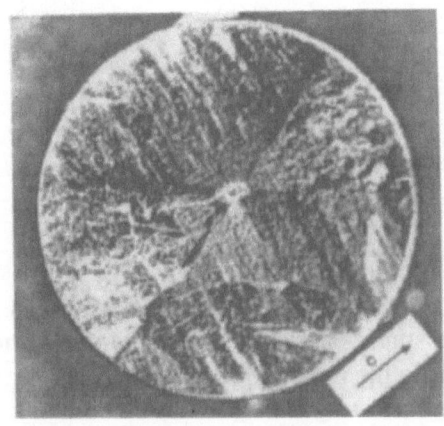

Fig. 1. Photograph of the MgO seeded NEG+NEG211 samples; diameter = 18 mm

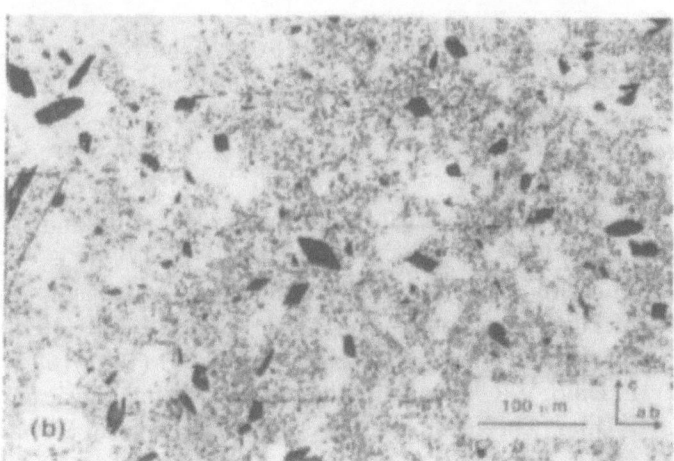

Fig. 2. a- and c-growth sectors (GS) marked by a and c-subgrains (a). Equal 211 concentration in a- and c-GS (b).

In the NEG-Gd211 sample the V_{211} and d_{211} are not constant. V_{211} increases from 12 vol. % at the seed to the 27 vol. % at the sample rim (Fig. 3). This behavior could be associated with 211 pushing. The increase of d_{211} from 0.65 μm at the seed to 0.96 μm at the rim opposes pushing phenomenon, which would imply a decrease in 211 particle size with the distance from the seed in the case of slow cooling growth [5]. The macroscopic inhomogeneity in the NEG-Gd211 sample is apparently caused by another effect, as we observed melt leaking from this sample. As the melt contains mainly Ba and Cu, the sample became rich in Gd-rich 211. The melt loss can be attributed to the composition change of the melt due to reaction of Gd-211 particles with the melt. Gd from the 211 goes to the melt and Nd and Eu from the melt to the 211 particles. Apparently, this process changes the properties of the melt (GdBaCuO has the lowest peritectic temperature of the light rare earth ions in the system [7]) and can cause melt leakang.

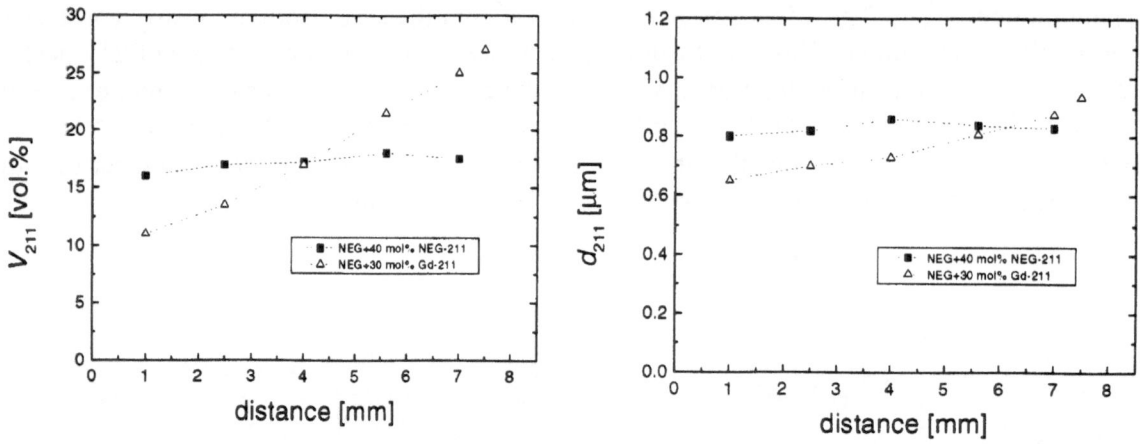

Fig.3. 211 content (V_{211}) and size (d_{211}) as function of distance from the seed.

CONCLUSIONS

Unlike YBCO bulks, the NEG samples do not exhibit 211 macroscopic inhomogeneity due to 211 particle pushing by growth front. The subgrain structure in the a- and c-growth sectors of the samples and the a-b microcracking develops in a similar way as in the melt-processed YBCO.

Acknowledgements

This work was supported by JISTEC, by the New Energy and Industrial Technology Development Organization (NEDO) and by Slovak Academy of Sciences

References

1. M. Muralidhar, M. R. Koblischka, T. Saitoh and M. Murakami, Supercond. Sci. Technol. **11** (1998) 1349.
2. M. Muralidhar, M. R. Koblischka and M. Murakami, Physica C **313** (1999) 232.
3. M. Muralidhar and M. Murakami, Physica C **309**, (1998) 43.
4. P.Diko, N.Pellerin, P.Odier, Physica C **247** (1995)169.
5. P. Diko, V.R. Todt, D.J. Miller, K.C. Goretta, Physica C **278** (1997) 192.
6. P.Diko, Ch.Wende, D.Litzkendorf, Th.Klupsch, W.Gawalek, Supercond. Sci. Technol. **11** (1998) 68.
7. M.Muralidhar, H.S.Chauhan, T.Saitoh, K.Kamada, K.Segawa and M.Murakami, Supercond. Sci. Technol. **10** (1997) 663.

FLUX PINNING IN MELT-PROCESSED $(Nd_{0.25}Sm_{0.25}Eu_{0.25}Gd_{0.25})$ $Ba_2Cu_3O_y$ SUPERCONDUCTORS

Muralidhar Miryala, Michael R Koblischka, Kazumasa Iida and Masato Murakami

Superconducting Research Laboratory, ISTEC, Division 3, 1-16-25, Shibaura, Minato-ku, Tokyo 105, Japan.

Abstract: We report on the microstructure and $J_c(B)$ properties of oxygen-controlled-melt-growth (OCMG) processed $(Nd_{0.25}Sm_{0.25}Eu_{0.25}Gd_{0.25})Ba_2Cu_3O_y$ "NSEG-123" superconductors, which contain about 0 to 40 mol% of $(Nd, Sm, Eu, Gd)_2BaCuO_5$ "NSEG-211" secondary phase and 0.5 mol% platinum. Microstructural observations clarified that the Pt addition was effective in reducing the size of NSEG-211 secondary phase like the case of $(Nd,Eu,Gd)_2BaCuO_5$ "NEG-211" in the $(Nd_{0.33}Eu_{0.33}Gd_{0.33})Ba_2Cu_3O_y$ "NEG-123" system. A large J_c of 43000 A/cm^2 was achieved at 77 K and 2.5 T. However, these values are lower than those of the NEG-123 system.

Key words: melt processing, NSEG-123, microstructure, Pt addition, critical current density.

INTRODUCTION

In order to improve the critical current density of bulk high T_c superconductors, it is common to embed small normal conducting particles into the 123 matrix. This process depends on several parameters such as the initial secondary phase, particle size, Pt and CeO_2 content. Our recent results show that J_c values of NEG-123 composite could significantly be enhanced at 77 K mainly due to the fact that the size of RE_2BaCuO_5 (RE-211; RE: rare earth elements) could be reduced [1,2]. Transmission electron microscopic observations with compositional analyses confirmed that small RE-211 particles with an average diameter smaller than 0.1 µm mainly comprise Gd-211[3]. As a result, a critical current density of 70000 A cm^{-2} at 77K and 2.5 T applied field was achieved with at 40 mol% 211 addition. These achievements motivated us to study how the secondary phase particles act if four rare earth elements are compounded at the RE site. In the present investigation, we selected the (Nd,Sm,Eu,Gd)-Ba-Cu-O system in which the ratio of four RE elements is even and studied the effect of NSEG-211 addition on flux pinning enhancement in the NSEG-123 system.

EXPERIMENTAL

High purity Nd_2O_3, Sm_2O_3, Eu_2O_3, Gd_2O_3, $BaCO_3$ and CuO commercial powders were weighed to have a nominal composition of $(Nd_{0.25}Sm_{0.25}Eu_{0.25}Gd_{0.25})Ba_2Cu_3O_y$. The precursor powders were ground thoroughly and calcined at 880 °C for 24 h with intermediate grinding, and pressed into pellets, which were further sintered at 1020 °C for 48h. Commercial high purity Nd-422, Sm-211, Eu-211 and Gd-211 powders with diameter less than 3 µm were employed to make the NSEG-211 second phase as small as possible. Finally, the NSEG-211 powders were added to NSEG-123 with the volume fractions of 10%, 20%, 30% and 40 mol%. 0.5 mol% Pt was also added to all the samples, since Pt addition was effective in reducing the size of NEG-211 [1]. The mixed powders were first pressed into pellets 20mm in diameter and 15mm in thickness, which were then subjected to CIP (cold isostatic press) under a pressure of 2000 kg/cm^2. For melt growth a MgO (100) seed was placed at the center top of the pellets which were then OCMG-processed in 0.1% partial pressure of O_2 with a gas flow rate of about 300 ml/min. The details of the heat treatment schedules can be found elsewhere [1].For oxygen annealing, samples with dimensions (a×d×c) ≈ 1.5×1.5×0.5 mm^3 were cut from the as-grown crystals and annealed in flowing O_2 gas in the trmperature range 300 to 600 °C with the following heat treatment schedule. The samples were heated to 600 °C in 2 h, held for 2 h and cooled to 500 °C in 12 h, then to 400 °C in 24 h and finally to 300 °C in 50 h and held there for 150 h followed by furnace cooling. Microstructural features of the samples were observed with an optical microscope and scanning electron microscope (SEM). The oxygenated samples were used for the T_c and J_c measurements using a Quantum Design MPMS-7 SQUID magnetometer with an applied magnetic field of 1 mT. To minimize field inhomogeneties, the scan length is restricted to 1 cm. The magnetic J_c was estimated according to the extended Bean's model by using the relation, $J_c = (20\Delta M)/[a(1-a/3b)]$, where ΔM is the hysteresis between increasing and decreasing field processes in emu/cm^3, a and b (a<b) are cross sectional dimensions (in cm) of the sample perpendicular to the applied magnetic field.

RESULTS AND DISCUSSION

Figures 1 (a) and (b) show SEM micrographs of NSEG123 sample with 20 and 40 mol% of NSEG-211 second phase. It is clear that NSEG-211 particles are finely dispersed in the NSEG123 matrix like the case of NEG123/NEG211 composites [1], showing that an addition of NSEG-211 is also effective in achieving fine dispersion of the second phases.

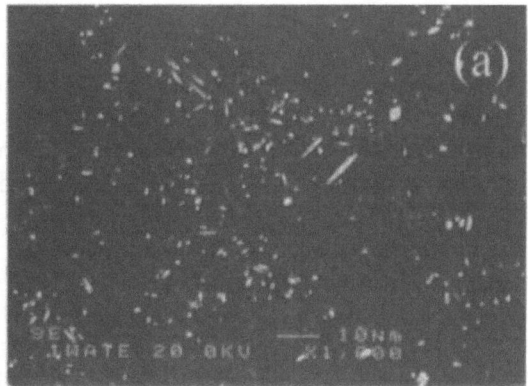 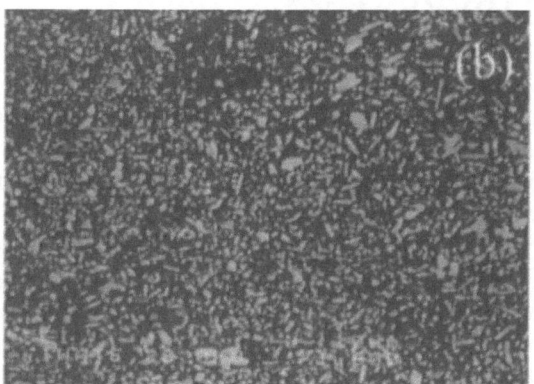

Fig. 1. SEM images for NSEG123 sample with (a) 20 and (b) 40 mol% of NSEG-211 and 0.5 mol% of Pt.

Figure 2 presents the current density characteristics of (Nd,Sm,Eu,Gd)-Ba-Cu-O samples with different amounts of NSEG-211 measured at 77 K for fields applied parallel to the c-axis. All the samples exhibit the secondary peak effect similar to that of NEG-123. The maximum J_c values for samples pure, 10, 20, 30 and 40 mol% are 39 000, 38 000, 41 000, 32 000 and 43 000 A cm^{-2} at the respective peak field of 1.6, 3, 2.6, 2.6 and 2.5 T (H ∥ c, at 77K). Due to the dispersion of fine secondary phase particles, the zero field J_c values are systematically improved with addition of NSEG-211. However, the peak values are low as compared to the NEG-123 system. These results show that the combination of rare earth elements and mixing ratio of the matrix is more important to improve the flux pinning and enhance the critical current density at 77 K. To further clarify this mechanism, we performed a scaling analysis of normalized volume pinning forces, F_p, versus a reduced field $h = H_a/H_{irr}$ for the pure NSEG-123 and the results are presented in fig. 3. The peak lies at $h_0 = 0.41$, which is similar to Nd-123 [4]. Our present results show that the mixing ratio is more important to improve the flux pinning strength at higher magnetic fields.

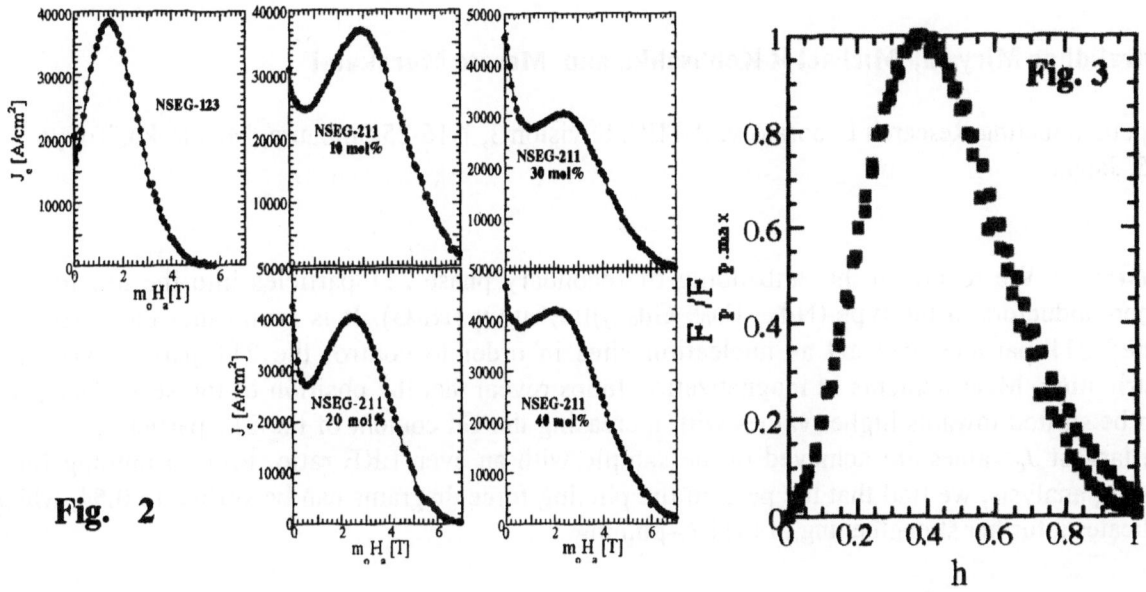

Fig. 2

Fig. 3

Fig. 2. J_c-B properties (77K, B $\parallel$ c-axis) for (Nd,Sm,Eu,Gd)-Ba-Cu-O superconductors with different volume fractions of the NSEG-211 phase. All samples are prepared by Ar-0.1% partial pressure of O_2.

Fig. 3. $F_p/F_{p,max}$ scaling of pure NSEG-123 sample at temperature between 65K and 90K. The scaling is found to be excellent.

SUMMARY

Melt-textured NSEG-123 samples with different concentration of NSEG-211 second phase were prepared by the OCMG-process. Microstructural analysis indicates that very fine 211 particles are dispersed in the 123 matrix like the case of NEG-123. The sample with 40mol% NSEG-211, a large critical current density (J_c) of 46 000 and 43 000 A cm^{-2} were achieved at 77K 0 T and 2.5 T, respectively for H $\parallel$ c axis. The volume pinning forces versus a reduced field analysis shows that the peak position is h_o=0.41, which indicates that the mixing ratio is more important for strengthening δT_c-type pinning.

Acknowledgements

This work was supported by New Energy and Industrial Technology Development Organization (NEDO) as a part of its Research and Development of Fundamental Technologies for Superconductor Applications Project under the New Sunshine Program administrated by the Agency of Industrial Science and Technologies M.I.T.I of Japan.

References

1. M. Muralidhar, M. R. Koblischka, T. Saitoh, and M. Murakami, Supercond. Sci. Technol. **11**, pp. 1349-1358 (1998).
2. M. Muralidhar, M. R. Koblischka, and M. Murakami, Physica C **313**, pp. 232-240 (1999).
3. M. Muralidhar, and M. Murakami, Physica C **309**, pp. 43-48 (1998).
4. M. R. Koblischka, A J J van Dalen, T. Higuchi, S. I. Yoo and M. Murakami, Phys. Rev. B. **56**, pp. 2863-2868 (1998).

EMBEDDING OF SECONDARY PHASE 211 PARTICLES IN MELT-TEXTURED TERNARY (Nd,Eu,Gd)-Ba-Cu-O

Muralidhar Miryala, Michael R Koblischka and Masato Murakami

Superconducting Research Laboratory, ISTEC, Division 3, 1-16-25 Shibaura, Minato-ku, Tokyo 105, Japan

Abstract: We report on the embedding of secondary phase 211 particles into the ternary 123 superconductors of the type $(Nd_{0.33}Eu_{0.33}Gd_{0.33})Ba_2Cu_3O_y$ (NEG). It is found that the externally added 211 particles can act as nucleation sites in order to control the 211 particle size and distribution. Measurements of magnetization loops reveal that the position of the secondary peak can be shifted towards higher values with increasing the Gd content of the 211 particles, whereas the largest J_c values are achieved in the sample with an even LRE ratio. From a pinning force scaling analysis, we find that the peak in the pinning force diagrams can be shifted to 0.54, which indicates a further strengthening of the δT_c-pinning.

Key words: $(Nd,Eu,Gd)Ba_2Cu_3O_y$, melt processing, critical current density, flux pinning.

INTRODUCTION

The ternary superconductors of the type $(Nd_{0.33}Eu_{0.33}Gd_{0.33})Ba_2Cu_3O_y$ (NEG) offer enhanced possibilities to engineer the flux pinning sites as compared to 123 systems with yttrium or one or two light rare earth (LRE) elements [1-5]. Our recent results reveal that the optimum configuration of the superconducting matrix in the NEG-123 system is not necessarily 1:1:1, and the composition of the matrix can be influenced by using different LRE mixtures within the matrix or within the externally added 211 particles [6,7]. Therefore, it is essential to further investigate the NEG system by varying the Gd content either in the initial superconducting matrix and/or in the 211 particles. In the present paper, we restrict ourselves to the variation in the composition of the added 211 particles, while keeping the initial matrix and the oxygenation procedure constant.

EXPERIMENTAL

High-purity commercial powders (5N) of Nd_2O_3, Eu_2O_3, Gd_2O_3, $BaCO_3$ and CuO were weighed to have a nominal composition of $(Nd_{0.33}Eu_{0.33}Gd_{0.33})Ba_2Cu_3O_y$. The nominal mixture ratio of Nd:Eu:Gd is fixed at 1:1:1. In a first step, precursor powders were prepared. The starting powders were ground thoroughly and calcinated at 880 °C for 24 h with intermediate grinding, and pressed into pellets. The sintering was carried out at 925 °C for 15 h. This entire process was repeated three times under low oxygen partial pressure. NEG-123 bulk samples with a volume fraction of 10 mol% of NEG-211 secondary phase were prepared using sintered NEG-123 and commercial Nd-422, Eu-211 and Gd-211 powders. High purity NEG-211 powders with diameter less than 1 μm were employed to make the initial NEG-211 particles as small as possible. In order to clarify the effect of NEG-211 mixing ratio on the microstructure and J_c (B) properties of the NEG-123 system, we varied the cation ratio of the NEG-211. The samples are referred to as samples A (0.40:0.40:0.20), B (0.33:0.33:0.33), C (0.30:0.30:0.40), D (0.20:0.20:0.60) and E (0.10:0.10:0.80). Finally, the precursor powders were pressed into pellets with a diameter of 20 mm and a thickness of 15 mm, which were subjected to cold isostatic press (CIP) with a pressure of 2000 kg/cm^2. A MgO (100) seed was placed on the centre top of the pellet, which was subsequently OCMG-processed in 0.1% partial pressure of O$_2$ with a gas flow rate of about 300 ml/min. The details of the heat treatment schedule and oxygen annealing process can be found elsewhere [1]. Magnetization hysteresis loops (MHLs) were measured mainly at 77 K in applied fields up to 7 T using a commercial SQUID magnetometer (Quantum Design, model MPMS7). To minimize field inhomogeneties, the scan length of 1 cm was used. The external magnetic field was always applied parallel to the c-axis of the samples.

RESULTS AND DISCUSSION

The J_c data obtained from the MHLs are presented in figs. 1 (a) and (b). From these graphs, we clearly see that the peak position, B$_{peak}$, can be shifted towards higher fields by increasing the Gd-content in the initial 211 particles. Sample D also exhibits the highest irreversibility field, H_{irr}. The overall highest J_c values are, however, obtained in the sample with an even mixing ratio. A further increase of the Gd content reduces B$_{peak}$ again to comparable values with the 0.33:0.33:0.33 samples, but the overall J_c is further reduced. The J_c values for the samples A, B, C, D and E are 48000, 68000, 58000, 40000 and 36000 A/cm^2 at the peak of 2.6, 2.6, 2.7, 3.5, and 2.8T (H $\parallel$ c, at 77K), respectively.

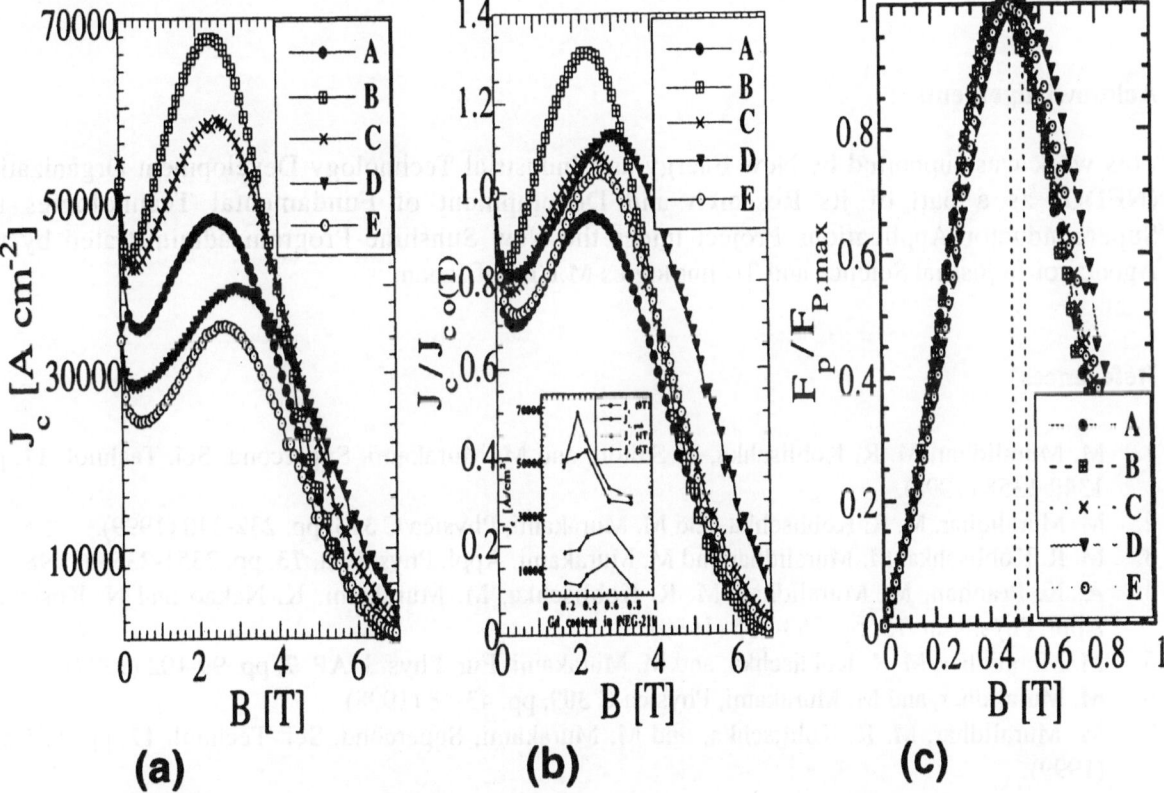

Fig. 1. Field dependence of the critical current density (T = 77 K, H$_a$ parallel to the c-axis) for samples A – E. (a) J_c data of samples A – E. (b) Normalized plot of J_c. For normalization, we used the current density at zero field, J_c (0T). The inset shows the current density in self-field, J_c (0T), the current density at the peak position, $J_{c, peak}$ and two current values at 4 T and 5 T as a function of the Gd concentration in NEG-211. (c) Scaling of the normalized pinning forces, $F_p/F_{p,max}$, vs the reduced field, $h = H_a / H_{irr}$ of the samples A – E at 77 K. The peak can be shifted up to 0.54 (sample D).

It is interesting to note that the NEG-211 secondary phase with even ratio sample exhibit a record high J_c value of 66000 A/cm^2 at 77K and 3 T (H ∥ c). The current density in self-field [$J_c(0T)$], the current density at the peak position, $J_{c,peak}$ and two selected current densities at 4 T and 5T are presented in the inset to Fig. 1 (b) as a function of the Gd concentration. We selected current densities at 4T and 5T, in order to show the improvement of the J_c in high magnetic fields. From these results, it is important to learn more about the microscopic pinning. Therefore, we performed a scaling analysis of the normalized volume pinning forces, F_p, versus a reduced field $h = H_a / H_{irr}$. H_{irr} is determined from the MHLs using a criterion of 100 A/cm^2. Since H_{irr} depends on the sweep rate of the external magnetic field H_a, it is important to compare only data with the same electric field across the sample (i.e. identical sweep rate of H_a). We use the 77 K data of all our samples comparison, as presented in Fig. 1 (c). As expected form the observed increased peak position in the magnetization curves, the data of sample D exhibit an even higher peak position, now located at about $h_0 = 0.54$.

CONCLUSIONS

The mixing ratio has proved extremely effective in increasing the pinning site density by externally fine NEG-211 particles. High critical current density values for sample B were 69000 and 66000 A/cm^2 at the peak field and 3T respectively. The volume pinning forces versus a reduced field analysis revealed that the peak position can be shifted to around $h_o \approx 0.54$, which suggest that the δT_c-type pinning was further enhanced.

Acknowledgements

This work was supported by New Energy and Industrial Technology Development Organization (NEDO) as a part of its Research and Development of Fundamental Technologies for Superconductor Applications Project under the New Sunshine Program administrated by the Agency of Industrial Science and Technologies M.I.T.I of Japan.

References

1. M. Muralidhar, M. R. Koblischka, T. Saitoh, and M. Murakami, Supercond. Sci. Technol. **11**, pp. 1349-1358 (1998).
2. M. Muralidhar, M. R. Koblischka, and M. Murakami, Physica C **313**, pp. 232-240 (1999).
3. M. R. Koblischka, M. Muralidhar and M. Murakami, Appl. Phys. Lett. **73**, pp. 2351-2353 (1998).
4. A. K. Pradhan, M. Muralidhar, M. R. Koblischka, M. Murakami, K. Nakao and N. Kosizuka, Appl. Phys. Lett. **75**, pp. 253-255 (1999).
5. M. Muralidhar, M. R. Koblischka, and M. Murakami, Eur. Phys. J. AP **7**, pp. 99-102 (1999).
6. M. Muralidhar, and M. Murakami, Physica C **309**, pp. 43-48 (1998).
7. M. Muralidhar, M. R. Koblischka, and M. Murakami, Supercond. Sci. Technol. **12**, pp. 555-562 (1999).

TEM OBSERVATION OF INTERFACES BETWEEN $(Nd,Eu,Gd)_2BaCuO_5$ SECONDARY PHASE AND THE $(Nd,Eu,Gd)Ba_2Cu_3O_y$ MATRIX IN OCMG-PROCESSED (Nd,Eu,Gd)-Ba-Cu-O

M. Muralidhar, M. R. Koblischka, A. Das, N. Sakai, and M. Murakami

Superconducting Research Laboratory, ISTEC, Division 3, 1-16-25, Shibaura, Minato-ku, Tokyo, 105, Japan

Abstract: Microstructural observations of oxygen-controlled-melt-growth (OCMG) processed $(Nd,Eu,Gd)Ba_2Cu_3O_y$ "NEG-123" /$(Nd,Eu,Gd)_2BaCuO_5$ "NEG-211" samples have been conducted using a transmission electron microscope (TEM). High-resolution TEM (HRTEM) images demonstrate that the microstructural defects that might act as pinning centres were not observed around the NEG-211 secondary phase particles. Furthermore, the 211/123 interface was found to be very clean and sharp.

Key words: $(Nd,Eu,Gd)Ba_2Cu_3O_y$, Eu_2BaCuO_5, Gd_2BaCuO_5, HRTEM, Flux pinning.

INTRODUCTION

Oxygen controlled melt-growth (OCMG) process has received a great deal of attention in the field of bulk superconductors. The interest arises from the fact that OCMG processed materials have a good flux pinning property at higher magnetic fields as compared to $YBa_2Cu_3O_y$ "Y-123" [1-3]. This behaviour is also enhanced in newly developed ternary superconductors of the type $(Nd_{0.33}Eu_{0.33}Gd_{0.33})Ba_2Cu_3O_y$ (NEG-123) and the critical current density (J_c) values are recorded in these materials are suitable for bulk applications at 77 K [1]. However, further enhancement is important for both better performance and a safety margin. A careful observation of the interfaces between the 211 secondary phase and 123 matrix is of prime importance for understanding both the flux pinning and the chemical process involved in the crystallization to improve the critical current densities at higher magnetic fields. In this paper, we investigated the 123/211 interfaces using a high-resolution transmission electron microscopy (HRTEM) and the boundary around the 211 secondary phase was observed. We also estimated chemical composition of the different shapes of 211 secondary phase particles and found that extremely fine 211 secondary phase particles contain both Eu and Gd on the rare earth site.

EXPERIMENTAL

The preparation of $(Nd_{0.33}Eu_{0.33}Gd_{0.33})Ba_2Cu_3O_y$ ("NEG") is described in [1]. Commercial Nd-211, Eu-211, and Gd-211 powders with diameter less than 1 μm were employed to make the initial NEG-211 particles as small as possible. NEG bulk samples with a volume fraction of 10 mol% of NEG-211 particles in 1:1:1 ratio were added to the starting composition. 0.5 mol% Pt and 1 mol% CeO_2 were also added to the sample, since Pt and CeO_2 additions were effective in reducing the size of LRE 211 [4], where LRE stands for the light rare earth elements such as Nd, Sm, Eu and Gd. The details of the melt growth and oxygen annealing process are described elsewhere [1]. Microstructural features and chemical composition of the samples were analyzed with a JEOL JXA-8900M WD/ED compound microanalyzer. The fine structure was studied with a H9000UHR transmission electron microscope operating at 400 kV. The crystal orientation perpendicular to the sample surface and all the TEM observations in this paper were taken with the incident electron beam along the [001] direction.

RESULTS AND DISCUSSION

TEM images of the interface between the NEG-211 secondary phase and the NEG-123 matrix are shown in Fig. 1 (a). There is no disordered phase localized in the boundary region. Figure 1 (b) is a that HRTEM image of mid part in Fig. 1 (a), which shows 123/211 interface (indicated by an arrow) is clean. These observations demonstrate that the crystal structure of the 123 matrix is not distorted, even in the vicinity of the interface. This crystal lattice fringes for the 123 and 211 secondary phase can also be seen in fig. 1 (b).

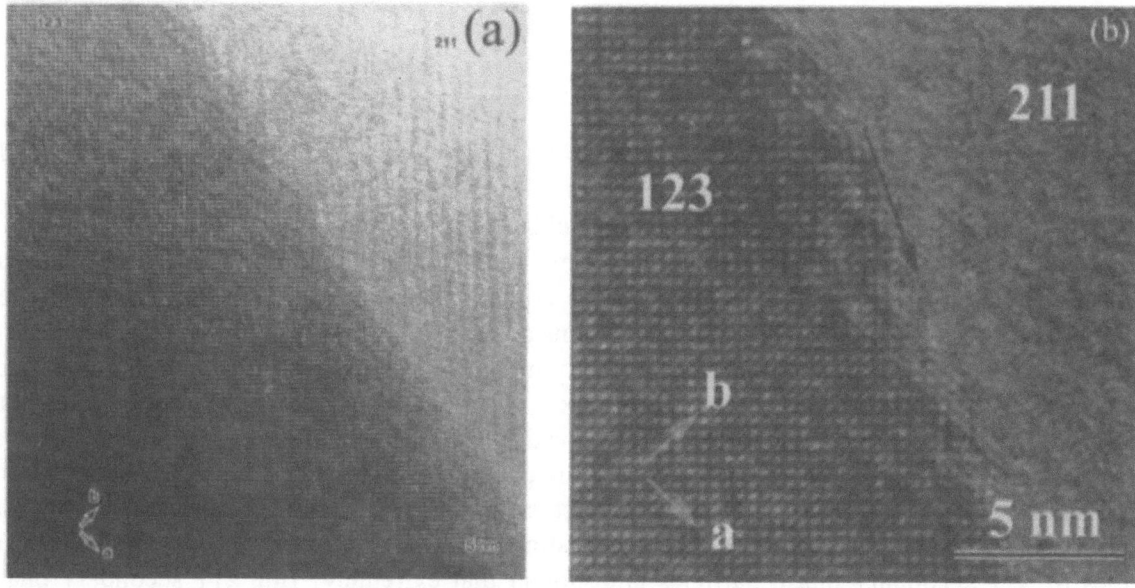

Fig. 1. HRTEM image of the interface between 211 secondary phase and 123 matrix. (b) is a magnified HRTEM image of the mid part of Fig. 1 (a). The interface is marked by a arrow and the lattice fringes of the 123 and the 211 inclusion can clearly be seen.

Figure 2 (a and b) shows a another TEM image of the 211 secondary phase in the NEG-123 matrix. It should be noted that fine secondary phase particles of the 211 phase with diameters of less than 0.1 μm are present in the microstructure and may contribute to the flux pinning as discussed in ref. [3].

TEM observations presented in this work provide direct evidence that the crystal structure of the NEG-123 matrix is not distorted even in the vicinity of the interface and no dislocation could be observed around the interface. These results support the idea that very small 211 secondary phase particles provide effective flux pinning in the (Nd,Eu,Gd)-Ba-Cu-O superconductors. Another idea is that as the interface between 211/123 is very sharp, the order parameter can vary in the distance of a coherence length at such an interface, leading to a large flux-pinning force. Murakami et al [5] have already reported that the sharpness of the 211/123 interface is critical for flux pinning enhancement in the case of MPMG processed Y-123.

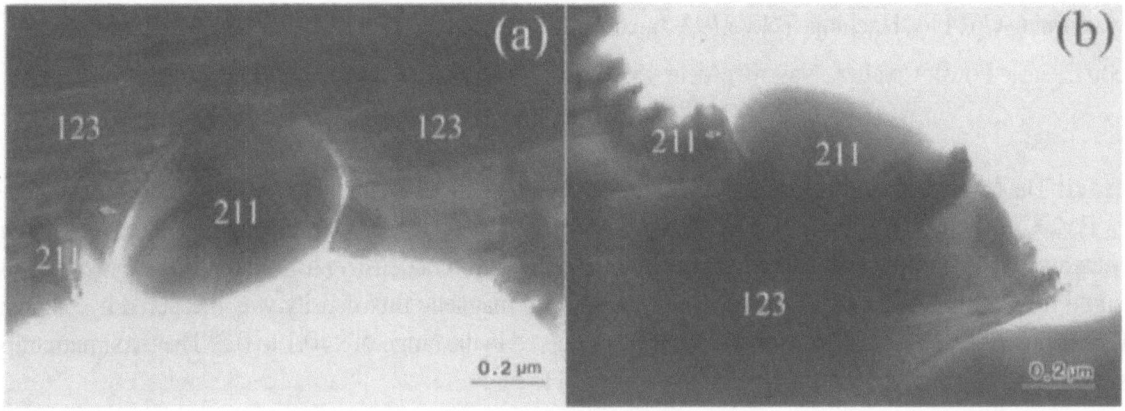

Fig. 2. Transmission electron micrograph for melt-processed $(Nd,Eu,Gd)Ba_2Cu_3O_y$ with 10 mol% NEG-211, 0.5 mol% Pt and 1 mol% of CeO_2. Very fine LRE-211 inclusions are dispersed in the 123 matrix.

CONCLUSION

We have used high-resolution transmission electron microscopy (HRTEM) to investigate 211/123 interface and found that the crystal of the 123 matrix is not distorted even at the interface. We also observed the sharpness of the 211/123 interface. TEM-EDX analysis revealed that the sub-micron sized secondary phase particles consist mainly of Gd and Eu in the rare earth site.

Acknowledgements

This work was supported by New Energy and Industrial Technology Development Organization (NEDO) as a part of its Research and Development of Fundamental Technologies for Superconductor Applications Project under the New Sunshine Program administrated by the Agency of Industrial Science and Technologies M.I.T.I of Japan.

References

1. M. Muralidhar, M. R. Koblischka, T. Saitoh, and M. Murakami, Supercond. Sci. Technol. **11**, pp. 1349-1358 (1998).
2. M. R. Koblischka, M. Muralidhar and M. Murakami, Appl. Phys. Lett. **73**, pp. 2351-2353 (1998).
3. M. Muralidhar, M. R. Koblischka, and M. Murakami, Physica C **313**, pp. 232-240 (1999).
4. C.J. Kim, K.B. Kim, I.H. Kuk and G.W. Hong, J. Mater. Res., **12**, pp. 38- (1997).
5. K. Yamaguchi, M. Murakami, H. Fujimoto, S. Gotoh, N. Koshizuka and S. Tanaka, Jap. Jurnal. Of Appl. Phys., **29**, pp. L1428-L1431 (1990).

MAGNETIC PROPERTIES OF Ag-DOPED SmBaCuO-SYSTEM SUPERCONDUCTORS

S. Kohayashi[1], S. Haseyama[1], and S. Nagaya[2]

[1]Dowa Mining Co., Ltd., Hachioji, Tokyo 192, Japan
[2]Chubu Electric Power Co., Inc., Nagoya, Aichi 459, Japan

Abstract: The large single domain of Ag-doped SmBaCuO-system superconductors with several compositions of $Sm_{1+2x}Ba_{2+x}Cu_{3+x}O_y$ (x=0, 0.1, 0.2, 0.3 and 0.4) were prepared by a melt process. The sample size was 45mm in diameter and 20mm in thickness. The dependence of the critical current density (Jc) for applied magnetic field was measured by a SQUID. The axial component of the trapped magnetic flux density was measured by scanning a Hall element sensor. The remarkable peak effect was observed in the range of x=0.1 to 0.2. The maximum trapped magnetic flux density was highest at x=0.2, then it was 1.3T.

Keywords: REBaCuO superconductor, magnetic property, trapped field, melt process

INTRODUCTION

The critical current density (Jc) of NdBaCuO-system and SmBaCuO-system superconductor prepared by a melt process is higher than that of YBaCuO-system superconductor in the high magnetic field region due to the peak effect [1]. We reported that the maximum trapped magnetic flux density of the SmBaCuO-system superconductor which has higher Jc in the high magnetic field region, was higher than that of YBaCuO-system superconductor[2]. On the other hand, it was reported the peak effect of the NdBaCuO-system superconductor depends on the composition ratio of $Nd_4Ba_2Cu_2O_{10}$ phase to $NdBa_2Cu_3O_y$ phase [3]. Therefore, it is expected that the Jc and the maximum trapped magnetic flux density of SmBaCuO-system superconductor also depends on the composition. In this paper, the large single domain of Ag-doped SmBaCuO-system superconductors with several compositions of $Sm_{1+2x}Ba_{2+x}Cu_{3+x}O_y$ (x=0, 0.1, 0.2, 0.3 and 0.4) were prepared. And the optimum composition of the SmBaCuO-system superconductor for the trapped magnetic flux density was studied.

EXPERIMENTAL

Sm_2O_3, $BaCO_3$, CuO, Pt and Ag powders were mixed with appropriate ratio as follows: The nominal compositions were Sm:Ba:Cu=1+2x:2+x:3+x (x=0, 0.1, 0.2, 0.3 and 0.4), and Pt of 0.5wt% and Ag of 10wt% were added in that. The mixed powders were calcined at $920°$C for 24 hours. Then, the calcined powders were pulverized and pressed into pellets. The pellets were partially melted at $1100°$C for 0.5h. After that, the pellets were cooled at $10°$C/min till the temperature of the top decreased to $1020°$C. Then, the temperature gradient of $5°$C/cm was applied in the furnace where the upper part of the pellets were in lower temperature. When the temperature of the top decreased to those temperature, the ab-plane of seed crystals were placed in contact to the top surface of the pellets to grow along the c-axis of the $REBa_2Cu_3O_{7-\delta}$ crystal, respectively. The seed crystals were cut out from Ag-non-doped $Sm_{1.8}Ba_{2.4}Cu_{3.4}O_y$ bulks prepared by the same melt process. They were cooled at $1°C/h$ from 1020 to $870°C$. After that, they were cooled to room temperature at $40°C/h$. The all specimens except x=0 were oriented in the direction of the c-axis, and they were composed of single domain. The size of the prepared specimens were 45mm in diameter and 20mm in thickness. On the other hand, the specimen of x=0 contained a number of second phases such as $BaCuO_2$ and CuO. The crystal orientation of x=0 did not reflect the orientation of the seed crystal and it was composed of several crystals.

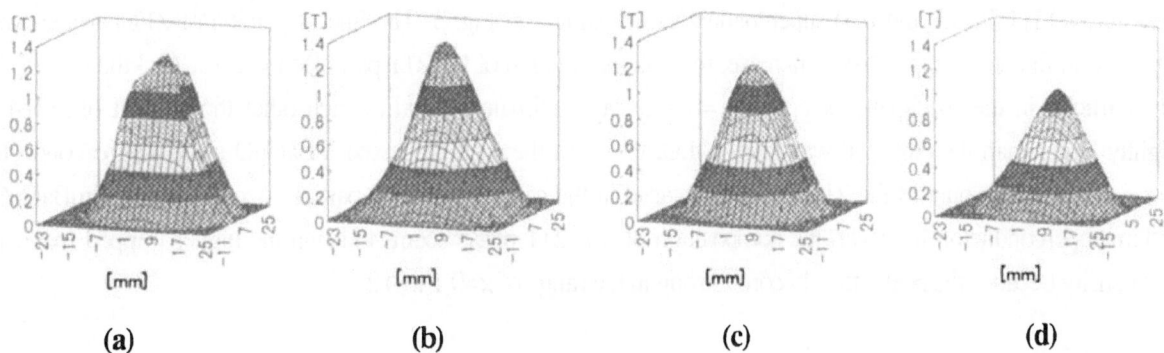

(a)   (b)   (c)   (d)

Fig. 1. Contour maps of the trapped magnetic flux density of the Ag-doped $Sm_{1+2x}Ba_{2+x}Cu_{3+x}O_y$ -system (x=(a)0.1, (b)0.2, (c)0.3 and (d)0.4) superconductors field-cooled at 77K under 2.1T.

The distribution of trapped magnetic flux density was measured as follows: The prepared specimen was field-cooled to 77K by immersing it in liquid nitrogen under 2.1T. The axial component of the trapped magnetic flux density in the specimen was measured by scanning a Hall element sensor (F. W. Bell, BHT-921). It moved step by step in an area of $50×50mm^2$ at a height of 1mm above the top surface of the specimen with a pitch of 2mm. A magnetization-hysteresis loop of the prepared specimen was measured by a SQUID. The critical current density was estimated by using the Bean model. The sample size to measure the magnetization was about $2.5×2.5×2mm^3$, where the c-axis and applied magnetic field was parallel to the 2mm edge. The size and the distribution of $Sm_2BaCuO_5(211)$ phase was observed by polarized optical micrographs.

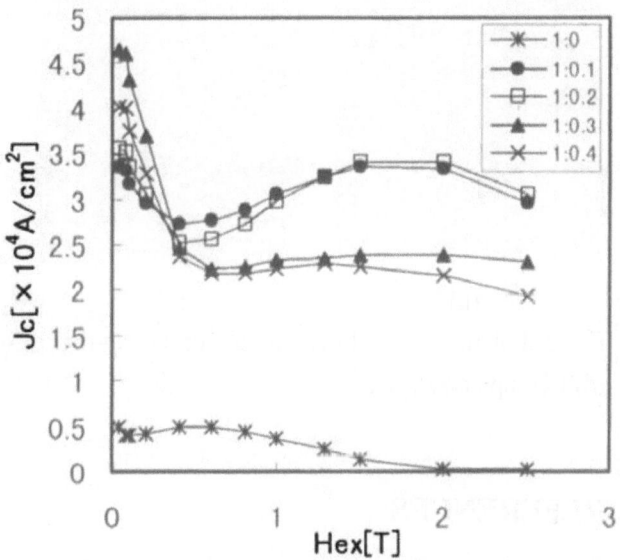

Fig. 2. Dependence of the critical current density of Ag-doped $Sm_{1+2x}Ba_{2+x}Cu_{3+x}O_y$ -system (x=0.1, 0.2, 0.3 and 0.4) superconductor for applied magnetic field Hex.

RESULTS AND DISCUSSION

The contour maps of the trapped magnetic flux density of Ag-doped $Sm_{1+2x}Ba_{2+x}Cu_{3+x}O_y$ -system (x=0.1, 0.2, 0.3 and 0.4) superconductors magnetized at 2.1T are shown in Fig. 1. The distribution of the trapped magnetic flux density formed a single dome shape for each of specimens. The maximum trapped magnetic flux density was highest at x=0.2, then it was 1.3T. The maximum trapped magnetic flux density of the specimens of x=0.1, 0.3 and 0.4 were 1.27T, 1.14T and 0.96T, respectively. On the other hand, the maximum trapped magnetic flux density of x=0 was extremely low which was 0.02T due to a number of second phases crystallized out of unreacted liquid phase. The specimens of x=0.1 and 0.2 exhibit remarkable peak effect compared with that of x=0.3 and 0.4, and the critical current density (Jc) of x=0.1 and 0.2 were higher than that of x=0.3 and 0.4 in the range of 0.5 to 2.5T. Although the origin of these peak effect is not still cleared, the relative value of the maximum trapped magnetic flux density between the each specimens approximately correspond to that of the maximum Jc in the high magnetic field above 0.5T. The polarized optical micrographs of the cross-section of the Ag-doped $Sm_{1+2x}Ba_{2+x}Cu_{3+x}O_y$-system (x=0.1, 0.2, 0.3 and 0.4) superconductors are shown in Fig. 3. The fine dispersion of 211 particles was observed in the photographs. As x increase, the volume fraction of the 211 particles increase. And the size of the 211 particles in the specimens except for x=0.4 are approximately equal to each other though that of x=0.4 is slightly larger than the others. It was known that, the Jc of the melt-processed YBaCuO-system superconductor increased as the amount of the 211particle increased in the case of the same particle size [4]. In the SmBaCuO-system superconductor, however, the composition of few 211 phase seems to bring the higher trapped magnetic flux density because the peak effect become strong in the range of x=0.1 to 0.2.

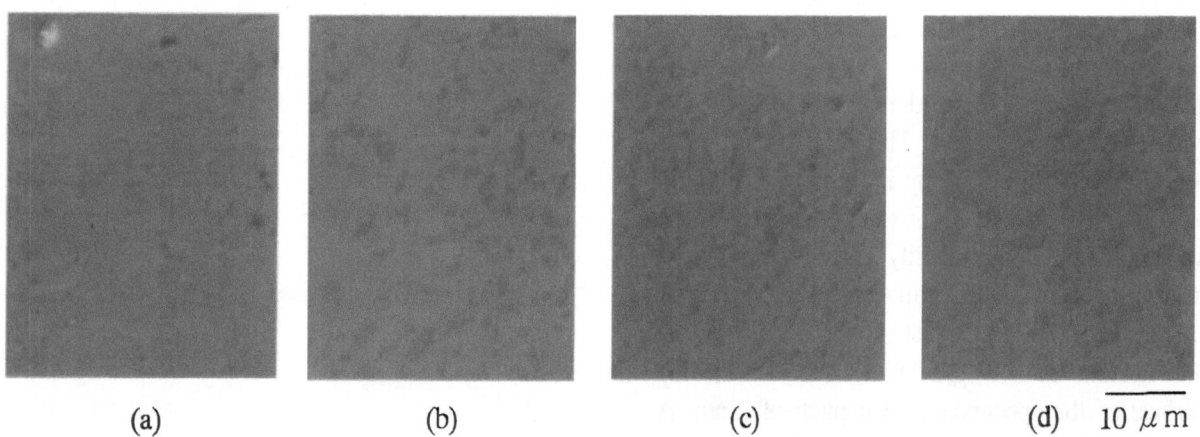

(a)   (b)   (c)   (d)   10 μ m

Fig. 3. Polarized optical micrographs of the Ag-doped $Sm_{1+2x}Ba_{2+x}Cu_{3+x}O_y$ -system (x=(a)0.1, (b)0.2, (c)0.3 and (d)0.4) superconductors

REFERENCES

1. M. Murakami, S. Yoo, T. Higuchi, N. Sakai, J. Weltz, N. Koshizuka and S. Tanaka, Jap. J. Appl. Phys. **33**, pp. L715-717 (1994).
2. S. Kohayashi, H. Miyairi, S. Yoshizawa, S. Haseyama, S. Nagaya, M. Satoh and H. Nakane, in Advances in Superconductivity 10, edited by N. Koshizuka and S. Tajima (Springer-Verlag, Tokyo, 1997), pp. 693-696.
3. S. Ikeda, T. Oka, Y. Yamada, M. Yoshikawa, Y. Yanagi, Y. Itoh and U. Mizutani, Jap. J. Appl. Phys. **36**, pp. L345-348 (1997).
4. M. Murakami, H. Fujimoto, S. Gotoh, K. Yamaguchi, N. Koshizuka and S. Tanaka, Physica C 321 (1991) 185.

The trapping characteristics of YBCO bulks using pulse magnetization

H. Ichikawa [1], A. Ninomiya [1], T. Ishigohka [1], H. Kamijo [2], H. Fujimoto [2]

[1]Seikei University, 3-3-1 Kichijoji –Kita, Musashino-shi, Tokyo 180-8633, JAPAN
[2]Railway Technical Research Institute, 2-8-38 Hikari-cho, Kokubunji-shi, Tokyo 185, JAPAN

Abstract: We have been studying magnetization characteristics of plural HTS bulks arranged in array. Previously, we presented experimental results using field-cool method in copper coils [1],a superconducting coil [2]. This time, we have carried out a pulse magnetization experiment using a capacitor bank and a pulse coil wound by copper wire. Plural HTS bulks are arranged in array in the magnetization coil. The trapped magnetic flux densities of the bulks are measured 2-dimensionally using Hall probe.

Keywords: HTS bulk, pulse-magnetization, trapped flux

INTRODUCTION

Recently, the application of bulk high Tc superconductor (HTS) as a very strong permanent magnet is gathering many interests. However, a steady-state magnetization of such a HTS bulk requires very high field usually generated by superconducting magnet. And, the use of superconducting magnet is not so easy. So, on a practical point of view, pulse magnetization using normal conducting copper coil would be preferable. Meanwhile, at present, it is very difficult to make a big YBCO superconducting bulk for various large-scale applications. We consider that a number of arranged bulks in array would behave as a big single bulk. So, the authors carried out experiments in order to study fundamentals of the pulse-magnetization characteristics of plural number of HTS bulks arranged in array.

503

PULSE MAGNETIZATION TEST

A YBCO bulk used for the experiment has a diameter of 20 mm and a thickness of 10 mm. 9 Bulks were arranged in an array of 3×3.

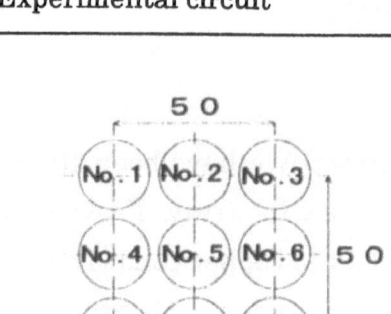

Fig. 1 Experimental circuit

A pulse magnetization coil was wound by normal conducting copper wire. For the peak currents of 262 A and 472 A, it generates magnetic fields of 0.5 T and 0.9 T at the center of the coil, respectively.

Figure 1 shows the experimental circuit. In order to obtain only the first positive half-wave, a diode was inserted in the circuit. The peak current appears at 74.5 ms and the pulse widths of the first half-wave is 194.8 ms for the capacitance of the capacitor bank C=220 mF (long pulse). Whereas, these figures become 26.8 ms and 57.8 ms for C=22 mF (short pulse), respectively.

Fig. 2 Arrangement of bulks

The magnetization process is as follows: 1) arrayed bulks are inserted into the pulse coil, 2) after they were cooled to 77 K, a pulse magnetic field is applied by magnetizing coil.

Bulks are arranged in parallel with a distance of 25 mm between the centers. An experiment was carried out also for single bulk arrangement other than the arrayed arrangements of 3×3. The positions allotted to the individual bulks are fixed throughout the all experiments, and they are numbered from 1 to 9. For the single bulk arrangement, the bulk at No. 5 position in 3×3 arrangements is used as shown in Figure 2. In the experiment, the position of No. 5 bulk is set at the center of the coil. At the position 2 mm above the surfaces of bulks, a Hall probe sensor is located. It can scan the magnetic field in axial (vertical) direction on a 2-dimensional plane. The magnetic field was measured by 2 mm pitch in case of 9-bulk arrangements, and it was measured by 0.5 mm pitch in case of single-bulk arrangement.

EXPERIMENTAL RESULTS

Figure 3 and 4 show the trapped magnetic flux distribution for the single and the 3×3 arrangements for B=0.5 T, long pulse (C=220 mF) magnetization, respectively. And, Figure 5 shows the variation of magnetic flux density to the distance from the center

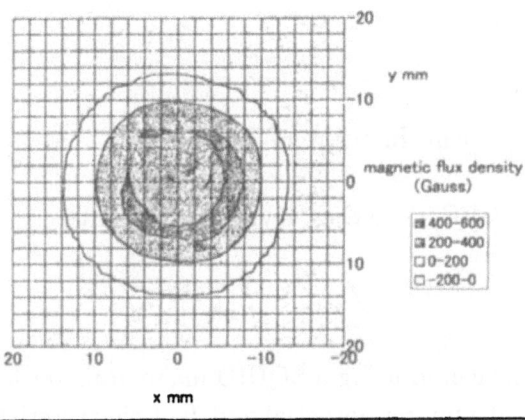

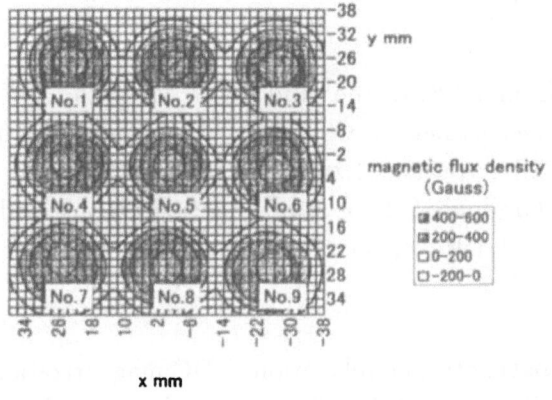

Fig. 3 Magnetic distribution on No. 5 bulk

Fig. 4 Magnetic distribution for 3×3 arrangements

of bulk in radial direction for various arrangements and locations, that is, the single and the 3×3 arrangements (No. 1, 2, 4, 5 bulks). From Figures 3, 4, and 5, it can be recognized that the flux distribution has a co-axial symmetry, and the magnetic field at the center of the bulk is lower than the outer region. The magnetic field for the 3×3 arrangements is higher than the

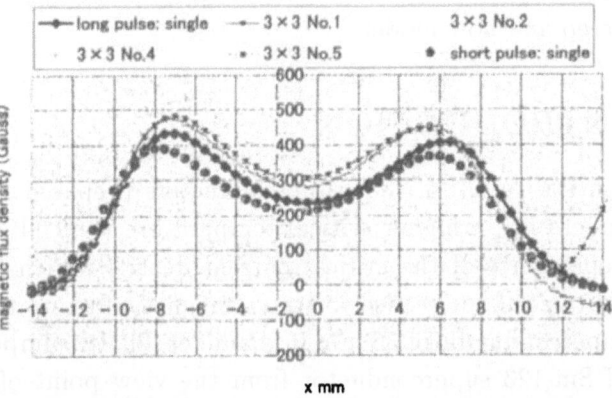

Fig. 5 Shape of trapped magnetic flux density

single arrangement. And, from Figure 5, we can understand that the longer pulse width (slower wave) brings the larger magnetized field.

As for the shapes of curves in Figure 5, it can be estimated that when the magnetic flux is going to penetrate the bulk, it begins at the outside and its motion is disturbed by the electromagnetic viscosity. This is probably the reason for the fall of the magnetic field at the center of the bulk.

1. H.Kamijo, et al. : Flux-Trapping Tests of Oxide Superconducting Bulks Arranged in Rows and Columns (1): 57th Meeting on Cryogenics and Superconductivity, C1-11, p.48, 1997.11

2. H.Ichikawa, et al. : Flux-Trapping Tests of Oxide Superconducting Bulks Arranged in Rows and Columns (2): 58th Meeting on Cryogenics and Superconductivity, F2-15, p.186, 1998.5

Irreversible Characteristics of Melt Processed Sm-123 Superconductor

Edmund S. Otabe[1] and Teruo Matsushita[1,2]

[1]Department of Computer Science and Electronics, Kyushu Institute of Technology, 680–4 Kawazu, Iizuka 820–8502, Japan

[2]Graduate School of Information Science and Electrical Engineering, Kyushu University, 6–10–1 Hakozaki, Higashi-ku, Fukuoka 812–8581, Japan

Abstract: The relaxation of DC magnetization was measured using a SQUID magnetometer for a melt processed Sm-123 superconductor and its apparent pinning potential and E-J characteristics were estimated. Both results were compared with the theoretical calculation based on the flux creep and flow model in the low field region below the peak effect.

Keywords: Sm-123, Irreversible properties, Apparent pinning potential, E-J characteristics, Flux creep and flow model

INTRODUCTION

Sm-Ba-Cu-O(Sm-123) superconductors prepared by the melt process in a reduced oxygen atmosphere has a higher critical temperature ($\sim$ 94 K) than Y-Ba-Cu-O and a better magnetic field dependence of the critical current density which frequently shows the peak effect [1]. Recently Ikuta *et al.* succeeded to trap a magnetic flux up to 9.0 T at 25 K in a Sm-123 melt single domain superconductor of 36 mm in diameter [2]. It is important to understand the irreversible properties of Sm-123 superconductor from the view point of application. The irreversible properties such as the irreversibility field and the apparent pinning potential in Bi-2212 single crystal were well explained by the flux creep and flow model [3, 4]. In this study, we focus on the apparent pinning potential and E-J characteristics of Sm-123. These characteristics were estimated from the measurement by a SQUID and the results were compared with the theoretically calculated results using the flux creep and flow model.

EXPERIMENTAL

Specimen was a Sm-123 superconductor prepared by the melt process in a controlled reduced oxygen atmosphere. The nominal composition of 123- and 211-phases of the specimen was 3:1 and Ag of 20 wt% was included. The specimen was cut from a large grain and its size was 1.09 mm × 2.01 mm × 2.56 mm. The c-axis was along the long length of the specimen. The critical temperature determined by the DC susceptibility in field cooled process was 94.0 K. All measurements were done using a SQUID magnetometer. The magnetic field was applied parallel to the c-axis.

RESULTS AND DISCUSSION

Figure 1 shows the magnetic field dependence of the critical current density, J_c, estimated from the measured hysteresis of the magnetization at various temperatures. The critical current density shows the peak effect in a wide rage of temperature. The relaxation of normalized magnetization under various magnetic fields at 77.3 K is shown in Fig. 2. The initial value of the magnetization, M_0, was estimated by extrapolating the result of the first three points to $t = 1$ s. The relaxation rate was determined from the variation in the range of 300 to 700 s. The apparent pinning potential, U_0^*, obtained from the relaxation rate is shown in Fig. 3. The peak field of U_0^* is slightly lower than that of J_c. E-J characteristics can be calculated from the relaxation of magnetization and the result at 77.3 K is shown in Fig. 4.

The apparent pinning potential and the E-J characteristic are theoretically explained by the flux creep and flow model [3, 4]. In this model, the pinning potential, U_0, can be estimated using the virtual critical current density, J_{c0}, in the flux creep free case. Here the parameters of J_{c0} which represent its magnitude and temperature and field dependencies are determined from the observed J_c value at low temperatures and magnetic fields where the flux creep is not significant. In this sense, the theoretical calculation is expected to be applicable only to the magnetic field region below peak field of J_c. However, we dare to calculated U_0^* and E-J characteristics in the entire region of the measurement. The theoretically calculated results of U_0^* and E-J characteristic using these pinning parameters are shown by thick solid lines in Fig. 3 and 4, respectively.

U_0^* starts to increases around the magnetic field (~ 0.5 T) where J_c starts to deviate from a common field dependence like $B^{\gamma-1}$, and U_0^* and J_c take a maximum and minimum approximately the same field. Hence, it is considered that the present theory is applicable to the field region below 0.5 T at 77.3 K. In this field region the observed U_0^* is less field-dependent and slightly lower than the prediction. A good agreement is obtained between the experimental and theoretical results of E-J characteristics at this field range. There results are fairly consistent with the assumption of parameter determined at low fields.

It is expected that the peak effect caused by substituted regions with lower T_c with the mechanism of repulsive kinetic energy interaction under the proximity effect. It is also expected that the usual attractive condensation energy interaction still survives even at high fields, whereas the substituted

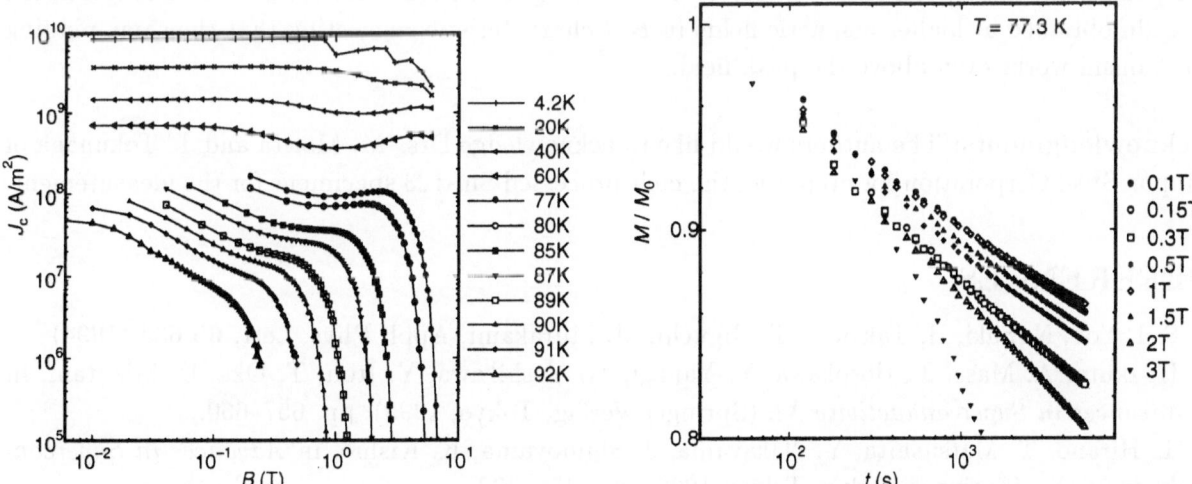

Fig. 1. Magnetic field dependence of critical current density at various temperatures.

Fig. 2. Relaxation of normalized magnetization at various magnetic fields at 77.3 K.

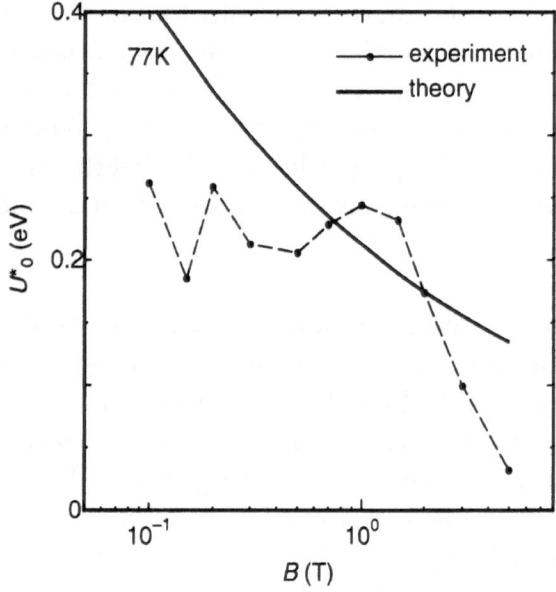

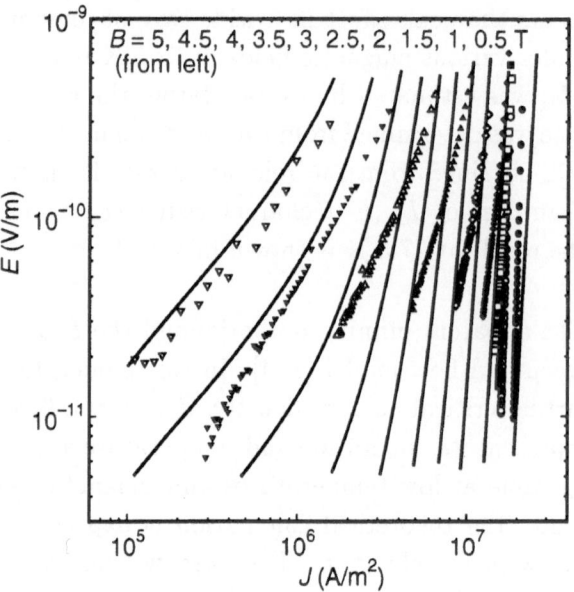

Fig. 3. Magnetic field dependence of apparent pinning potential at 77.3 K.

Fig. 4. Electric field vs current density at 77.3 K at various magnetic fields.

regions becomes inactive due to diminution of the probity effect [5]. In Fig. 4, it is found that the agreement is fairly good at higher magnetic fields again in E-J characteristic. This result seems to support the above hypothesis.

SUMMARY

In this study, the apparent pinning potential and E-J characteristics of a bulk Sm-123 superconductor were estimated and the results were compared with the theoretical results based on the flux creep and flow model. Since the parameters were determined in the low magnetic field region below the peak field, the agreement was better at lower magnetic fields. However, a better agreement is again obtained at higher magnetic fields in E-J characteristic suggesting that the same pinning mechanism works even above the peak field.

Acknowledgments: The authors would like to acknowledge Drs. M. Morita and T. Tokunaga of Nippon Steel Corporation for preparing the melt processed Sm-123 specimens for the measurement.

REFERENCES

1. S. I. Yoo, N. Saki, H. Takaichi, T. Higuchi, M. Murakami, Appl. Phys. Lett. **65** 633 (1994)
2. H. Ikuta, A. Mase, T. Hosokawa, Y. Yanagi, M. Yoshikawa, Y. Itoh, T. Oka, U. Mizutani, in *Advances in Superconductivity XI*, (Springer-Verlag, Tokyo, 1999), pp. 657–660.
3. T. Hirano, T. Matsushita, Y. Nakayama, J. Shimoyama, K. Kishio, in *Advances in Superconductivity XI*, (Springer-Verlag, Tokyo, 1999), pp. 489–492
4. T. Matsushita, T. Tohdoh N. Ihara, Physica C **259**, pp. 321–325, (1996).
5. T. Matsushita and E. S. Otabe, to be publish in Physica C (Proceedings of US-Japan Workshop, Yamanashi, Oct. 13–15, 1999)

A METHOD FOR SIMULTANEOUSLY EVALUATING THE ELECTROMAGNETIC PROPERTIES AND MEAN FREE PATH OF SUPERCONDUCTOR

Akira Taguchi[1*], Nakane Hiroshi[1], Yamazaki Sadao[1], Syuetu Haseyama[2], Soji Otabe[3]
[1]Kogakuin University,1-24-2 Nisisinjyuku, sinjyku-ku,Tokyo 163-8677 Japan
[2]Dowa Mining Co, Ltd., Hachioji, Tokyo 192,Japan.
[3]Kyushu Institute of Technology 680-4, Kawazu, Iizuka 820-8502 Japan

Abstract: The real part ΔR and the imaginary part ΔX of the difference in the impedance of a solenoid coil with and without a YBCO sample were measured by the SRPM method at 77.3K in the frequency range of 100Hz to 1MHz. The frequency dependence of ΔR and ΔX was analyzed theoretically and experimentally. From the frequency dependence of ΔR, the region of the normal and anomalous skin effect was discriminated. In the normal skin effect region, resistivity and magnetic penetration depth were obtained, and in the anoumalous skin effect region, the mean free path of electron was evaluated.

Keywords: SRPM method, YBCO superconductors, resistivity, penetration depth, mean free path

INTRODUCTION

The SRPM method is the acronym of "Simultaneous Resistivity and magnetic Penetration depth Measurement" With the SRPM method, the difference in the impedance ΔZ of a solenoid coil with a sample and another one without a sample are vectrially measured as shown in Fig.1. ΔZ is separated into the real part ΔR and the imaginary part ΔX. The resistivity ρ and magnetic penetration depth λ are simultaneously evaluated by finding out the agreement between the measured and the calculated value. The measurement of coil impedance at 77.3K was curried in the cryostat. The very low resistivity can be measured by the SRPM method, because this method is used to nullify the resistance of the lead-lines between the coil and the impedance measuring equipment (LCR meter) and the sample properties is evaluated without electrically touching the sample. When analyzing the frequency dependence of ΔR for superconductor, there is a possibility that can be distinguished whether the sample is within normal skin effect (N.S.E) or anoumalous skin effect (A.S.E). A method for obtaining ρ and λ in the region of N.S.E and ℓ (mean free path) in the region A.S.E is proposed in this paper.

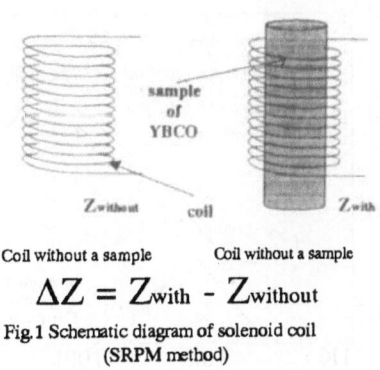

$$\Delta Z = Z_{with} - Z_{without}$$

Fig.1 Schematic diagram of solenoid coil
(SRPM method)

Table.1 Frequency dependence of ΔZ

sample	$\Delta R \propto$	$\Delta X \propto$
N.S.E.	f^2 or $f^{3/2}$	f
A.S.E.	$f^{2/3}$ or $f^{5/3}$	f

EXPERIMENT AND DISCUSSION

Considering the measurement the frequency range of 100Hz to 1MHz, it is difficult to measure ΔR and ΔX by only one coil, because of the decrease of the measuring sensitivity of coil. At the low frequency, a coil of 900 turns and radius of 6.1mm was used and at the high frequency, a coil of 25 turns and radius of 4.18mm was used. It was assumed from Table 1 that ΔX is proportional to frequency (f). The measuring results of ΔR are shown by • sign in Fig.2 and 3. In Fig 2, though the sensitivity of the measuring instrument was inadequate below 1kHz, it is assumed that the frequency dependence of ΔR below 2kHz is proportional to f^2, At above 2kHz, $\Delta R \propto f^{5/3}$. In Fig 3, $\Delta R \propto f^{5/3}$ below 50Hz, and $\Delta R \propto f^{2/3}$ above 50kHz. As a result, this method is a powerful tool for distinguishing whether the conduction mechanism lies in the region of N.S.E or A.S.E. It is found out that the border frequency between N.S.E and A.S.E is about 2kHz. In the region of N.S.E, ρ and λ can be evaluated using the measured and calculated values of ΔR and ΔX [1]. Then, the mean free path ℓ can be evaluated from the frequency dependence of ΔR and ΔX above 2kHz, by using ρ and λ. The values of ρ, λ and ℓ are shown in Table 2. In this table, the value of λ measured by the AC inductive method is shown. The values of λ evaluated by the two methods had a close correlation. The relation between the measured values of ΔR and the calculated values of ΔR using the values of ρ, λ and ℓ in the Table 2. This correlation is well. In Fig.5, when λ changes from 20 to 40 μm, the broken line(λ=20μm), solid line(λ=30μm) and the dot-dot-dash-line(λ=40μm) shows the calculated values of ΔR using the values of r and l in Table.2. At frequency above 50kHz, the measured values (• sign) are within 20~30μm. At low frequency below 50kHz, the measured values are within 30~40μm. the values of λ=40μm by this method and the one of λ=30.8μm by AC inductive method are within the region of λ=20~40μm.

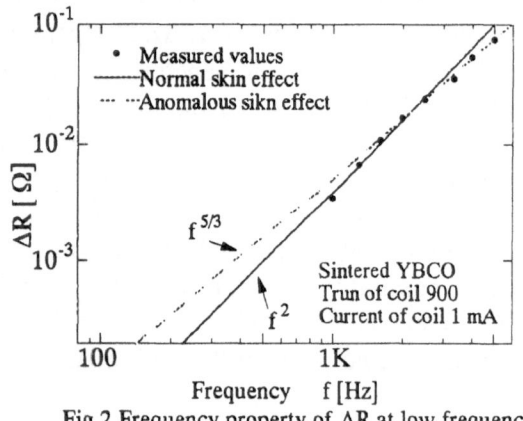

Fig.2 Frequency property of ΔR at low frequency

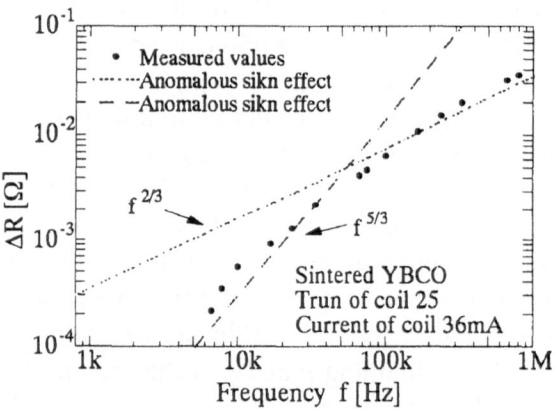

Fig.3 Frequency property of ΔR at high frequency

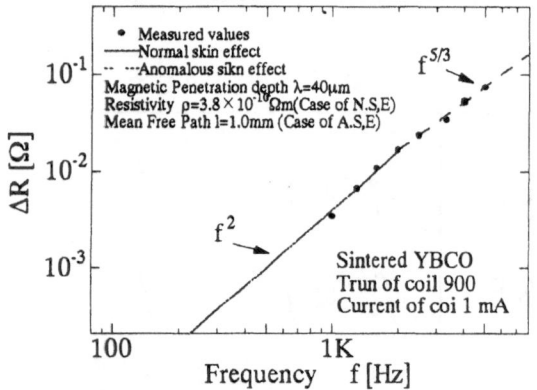

Fig.4 Frequency property of ΔR at low frequency

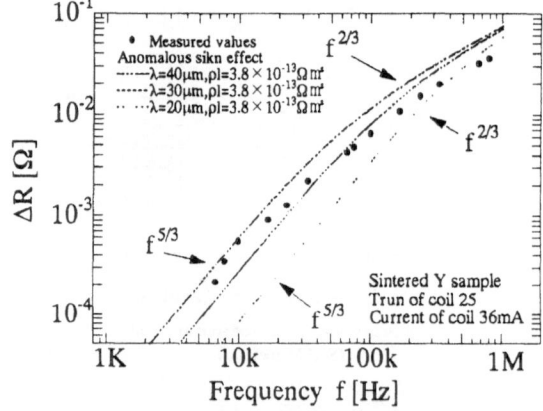

Fig.5 Frequency property of ΔR at high frequency

Table.2 Result measuring

sample of information	SRPM method		AC inductive method
	N.S.E.	**A.S.E.**	
frequency	~2kHz	2kHz~	35Hz
ρ [Ωm]	3.8×10^{-10}	3.8×10^{-10}	
λ [μm]	30	30	40
ℓ [mm]		1.0	

CONCLUSION

By using YBCO sintered rod shaped sample, a method for simultaneously evaluating the resistivity ρ, magnetic penetration depth λ and mean free path of electron ℓ was studied. The difference in the impedance of coil(real part ΔR and imaginary part ΔX) was measured at 77.3K between 100Hz and 1MHz. The frequency dependence of ΔR and ΔX was analyzed theoretically and experimentally. It is obtained from the experiments that the frequency dependence of ΔR is proportional to f^2 (normal skin effect), $f^{5/3}$ (anoumalous skin effect) and $f^{2/3}$ (anoumalous skin effect) and as the frequency (f) is increased. The values of ρ and λ can be evaluated in the region of normal skin effect. Then, ℓ can be evaluated in the region of anoumalous skin effect using the values of ρ and λ. The deviation between the values of λ by this method and by the conventional method (AC inductive method) is very small.

REFERENCE

[1] H.Nakane, T.Watanabe, M.Kobayashi and T.Hashimoto, "Frequency Dependence of Resistivity of High-Purity Copper at Low Temperature,"JJAP vol. 32(1993)pp. 3199-3203 Part 1, 1993

[2] H.Nakane, "Calculation of the Difference in Impedance for a Solenoid Coil with and without a Sample Conductor," IEEE Trans. Instrum. Meas., vol. 40, pp.544-548, 1991

[3] J.M.Ziman, Principles of the theory of solids, the Syndics of the Cambridge University Press, 1972H,Miyairi ea al."Analyses of the Electromagnetic properties of Oxide superconductors by the SRPM Method," to be published in IEEE Trans.Applied Superconductivity,June 1999

J_c-B Properties of (Nd-Y)123 System Melt Textured in Air

H.S. Chauhan[1] and M. Murakami[1,2]

[1]SRL-ISTEC, Morioka Lab., 3-35-2 Iioka-Shinden, Morioka, Iwate 020-0852, Japan
[2]SRL-ISTEC, Tamachi Lab., 1-16-25 Shibaura, Minato-ku, Tokyo 105-0023, Japan

Abstract: The substitution of Nd ion at Ba sites in the Nd123 system depends on many factors such as oxygen partial pressure, growth temperature and melt composition as evident from their J_c-B results. The survey of literature reveals that optimization of composition of precursor may be achieved mainly by changing Ba/Cu ratio, using Nd rich precursors and adding Y123. In this work we report the growth of (Nd-Y)123 precursor added with 20% (Nd-Y)211 and 0.5wt% Pt in air using a slow cooling method. The peak effect was clearly observed in the melt textured (Nd-Y)123 samples in the present study.

Keywords: (Nd-Y)123, Peak effect, Critical current density (J_c), Field-induced pinning centers.

INTRODUCTION

The substitution of Nd ion at Ba sites in the Nd123 system depends on many factors such as oxygen partial pressure[1,2], growth temperature[2,3] and melt composition[4-13] as evident from the corresponding J_c-B results. The survey of literature reveals that optimization of composition of the precursor may be achieved mainly by three different ways:- i) by changing Ba/Cu ratio[4], ii) by using Nd rich precursors[5] and iii) by adding Y123[6-13]. In all these three ways one can find suitable conditions for melt texturing in air which are preferable for better control of experiment and equipment. Matthew *et al* first reported high J_c in (Nd-Y)123 systems processed in air [6]. It was 8.77 x10^4 A/cm^2 at 77K, H=0 for a composition of $(Nd_{0.75}-Y_{0.25})123$. Cochrane *et al* used $(Nd_{0.5}-Y_{0.5})123$ precursor composition in melt texturing in air and found formation of (Nd-Y)123 and (Nd-Y)211 phases but same time ruled out any Nd/Ba substitution[7]. Varanasi *et al* reported significant improvement in magnetic properties by using precursor of Y123 mixed with Nd_2O_3 in melt texturing and also found (Nd-Y)422 inclusions[8]. Schatzle *et al* investigated the influence of addition of Nd_2O_3, Nd422, Nd_2BaO_4 and Y_2O_3 in (Nd0.5Y0.5)123 composite powder[9]. In the case of Nd_2O_3 addition broad transition with low T_c of 87K was found, whereas Nd422 or Nd_2BaO_4 additions resulted in higher T_c of 92K. The highest J_c of 4.4x10^4 A/cm^2 at 77K and 0T was recorded in case of Nd_2BaO_4 and Pt additions. In their J_c-B curve no peak effect was reported, however, there was one kink at about 1T. Recently, Mahmoud and Russell [10,11] had made detailed investigations on melt texturing in air using different precursors such as pure (Nd-Y)123 and with additions of (Nd-Y)211, Pt and CeO_2. They had not observed any peak effect and ruled out Nd/Ba substitution. Yao *et al* [12,13] have reported single crystal growth of (Nd-Y)123 systems and also found peak effect. In present investigation of melt texturing in air, we prepared Y123 and 20 mol% Y211 and Nd123 and 10 mol% Nd422 (both Ba rich) added with 0.5 wt% Pt and found evident peak effect.

EXPERIMENTAL

$Nd_{0.9}Ba_{2.1}Cu_3O_y$ +10% $Nd_{3.6}Ba_{2.4}Cu_{1.8}O_z$ and $YBa_2Cu_3O_y$ +20% Y_2BaCuO_5 were mixed in 1:1 ratio and ground after addition of 0.5wt% Pt. The well mixed powders were used for making the pellets. Once pellet was sintered at 920°C for 20 h. A small MgO(100) seed crystal was placed at the center top of the pellet before keeping in the furnace at room temperature. Then it was subjected to a heat-treatment in air environment without any flow of gas. The furnace temperature was raised to 1180°C in 4h and held for 4h. Then temperature was decreased to 1060°C in 30 minutes and held for 10 minutes. Then it was cooled at a rate of 1°/h up to 980°C and finally cooled to room temperature in 10h. The small specimens were selected from the processed pellet and oxygen annealed with the following heat pattern. The furnace temperature was raised to 600°C in 3h and held for 2h. It was lowered to 500°C in 12h, then to 400°C in 25h and to 300°C in 50h. At 300°C temperature was held for 145 h and then furnace cooled. The oxygenated samples were used for T_c and J_c measurements. The dc magnetization measurements were made with a Quantum Design MPMS SQUID magnetometer to determine the T_c and J_c. In both cases the direction of magnetic field was parallel to c-axis of the sample. The magnetic J_c was estimated according to the extended Bean's model [14].

RESULTS and DISCUSSION

Figure 1 shows the temperature dependence of normalized dc susceptibility measured for the oxygenated samples(S1-S4), which were randomly selected from the processed pellet. The results shown here are measured in zero-field cooled mode and in the presence of 10 G. The values for T_c onset, T_{cm} (mid-point) and superconducting transition (ΔT) are displayed in Table 1. The T_c onset is chosen at the point where normalized susceptibility changed to negative values, T_{cm} is taken as temperature at half of normalized dc susceptibility and ΔT is temperature difference between 10% and 90% of normalized dc susceptibility values. T_c onset varies between 91.6 K to 93.0 K. We can see that T_{cm} varies between 89.64 K to 93.02 K, whereas ΔT lies in the range of 5.23 K to 0.85 K. Figure 2 depicts the results of the field dependence of J_c of the oxygenated samples measured at 77K with magnetic field applied parallel to the c-axis of the samples. The zero-field J_c [$J_c(0)$] and peak J_c [$J_c(pk)$] and peak positions [$H(pk)$] obtained for four different samples are listed in table 1.

One can see that peak positions seem to match with increasing T_{cm} of S1 to S4. We earlier reported that the superconducting and J_c-B properties exhibit the growth temperature dependence in the OCMG processed Nd123 system [3], in that the higher growth temperature lead to the higher T_c and the lower overall J_c with higher irreversibility (Hirr) and $H(pk)$, whereas the lower growth temperature resulted in lower T_c values and lower Hirr, but overall J_c was enhanced. The J_c-B properties in the Nd-Ba-Cu-O system had been explained by field induced pinning centers[1]. In this case, the regions of low T_c phase with Nd-Ba-Cu-O solid solutions embedded in a nearly stoichiometric Nd123 matrix possessing higher

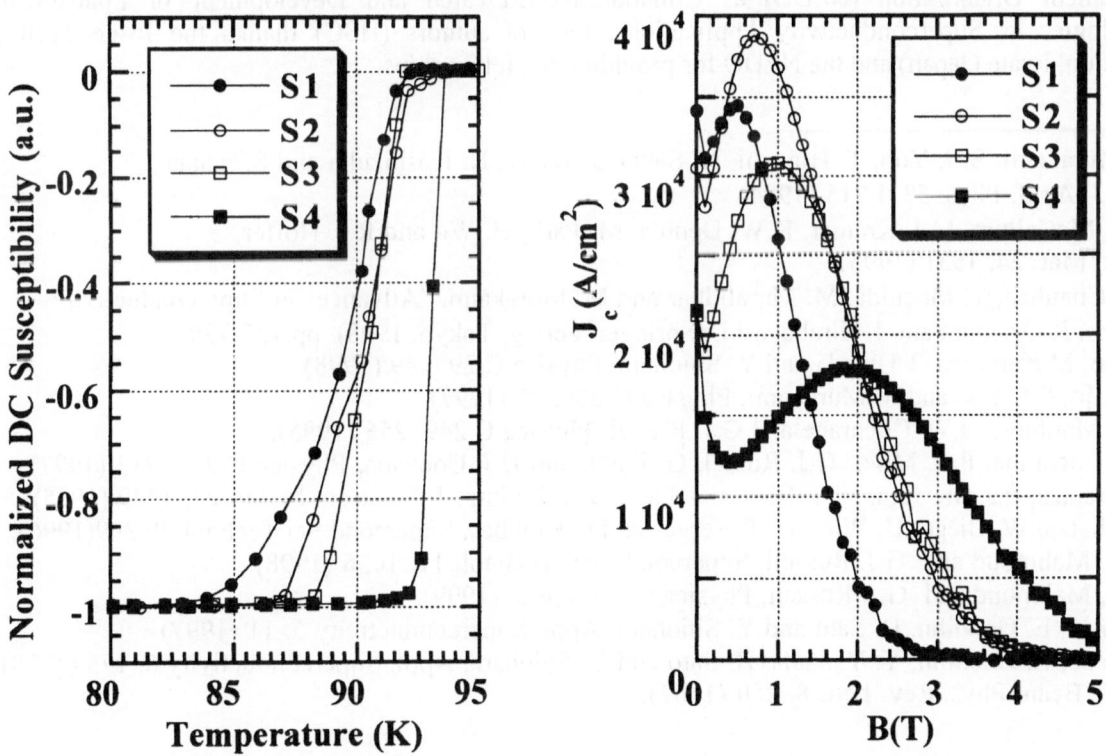

Fig. 1. (Left) The temperature dependence of dc susceptibility of oxygenated (Nd-Y)123 samples.
Fig. 2. (Right) J_c-B properties at 77K and H // c-axis of oxygenated (Nd-Y)123 samples.

Table 1 Superconducting and J_c-B properties of annealed melt textured (Nd-Y)123 samples.

Sample	T_c(onset) (K)	T_{cm} (K)	ΔT (K)	J_c(0) A/cm^2	J_c(pk) A/cm^2	H (pk) (T)
S1	92.2	89.64	5.23	34,180	34,487	0.5
S2	93.5	90.38	4.12	30,519	38,832	0.7
S3	91.6	90.56	2.66	21,258	30,824	1.0
S4	93.0	93.02	0.85	21,496	17,772	2.0

T_c, were considered to act as pinning centers when these are driven to normal state by the applied field. We earlier suggested that the chemical composition, the volume fraction and size of low T_c phase clusters may control the J_c-B properties[3]. Using this analogy we may say that the peak position is related to low T_c phase clusters which provide the maximum pinning due to proper cluster size and large volume fraction and by changing them to normal state above a certain magnetic field. Further, in similar analogy we can explain the different peak positions in different samples caused by variation of Nd/Ba substitution as a result of growth at different temperatures which is the case for a slow cooling growth experiment.

CONCLUSIONS

Our present investigations of melt textured (Nd-Y)123 indicate that the addition of Y123 can effectively be used in air processing of the Nd123 system to achieve better superconducting and J_c-B properties. It is also clear that the peak effect is observed in the present study suggesting that the low T_c clusters are formed as a result of Nd/Ba substitution and acting as field induced pinning centres. These are preliminary results and thus there is a possibility of better results by optimising the precursor composition, process and annealing conditions.

Acknowledgments. This work was supported by the New Energy and Industrial Technology Development Organization (NEDO) as Collaborative Research and Development of Fundamental Technologies for Superconductivity Applications. One of authors (HSC) thanks the Iwate Technos Foundation, Iwate (Japan) and the NEDO for providing the fellowships.

---

1. M. Murakami, S. I. Yoo, T. Higuchi, N. Sakai, J. Weltz, N. Koshizuka and S.Tanaka, Jpn. J. Appl. Phys. 33, L715 (1994).
2. R.W. McCallum, M.J. Kramer, K.W. Dennis, M. Park, H. Wu and R.J. Hoffer, Elect. Mat. 24, 1931 (1995).
3. H.S. Chauhan, T. Mochida, M. Muralidhar and M. Murakami, "Advances in Superconductivity-X", eds. K.Osamura and I. Hirabayashi, (Springer-Verlag, Tokyo, 1998) pp.725-729.
4. X.Yao, M.Kambara, T.Umeda and Y. Shiohara, Physica C 296, 69 (1998).
5. H. Kojo, S.I. Yoo and M.Murakami, Physica C 289, 85 (1997).
6. D.N. Matthews, J.W. Cochrane and G.J. Russel, Physica C 249, 255 (1995).
7. J.W. Cochrane, P.A. Miles, G.J. Russel, G. Foran and D.J. Cookson, Physica C 277, 213 (1997).
8. C.Varanasi, P.J. McGinn, H.A. Blackstead and D.B. Pulling, J. Electron. Mater. 24, 1949 (1995).
9. P. Schatzle,W. Bieger,U. Wiesner, P.Verges and G.Krabbes, Supercond.Sci.Technol. 9, 869(1996).
10. A.S. Mahmoud and G.J. Russell, Supercond. Sci. Technol. 11, 1036 (1998).
11. A.S. Mahmoud and G.J. Russell, Physica C - in press (1999).
12. X. Yao, E. Goodilin, H. Sato and Y. Shiohara, Appl. Superconductivity 5, 11 (1997).
13. X. Yao, E. Goodilin, Y. Yamada, H. Sato and Y. Shiohara, Appl. Superconductivity 6, 175 (1998).
14. C. P. Bean, Phys. Rev. Lett. 8, 250 (1962).

J_C-B PROPERTIES OF $NdBa_2Cu_3O_y$ SYSTEM PROCESSED BY A NOVEL MELT GROWTH TECHNIQUE

H.S. Chauhan[1] and M. Murakami[1,2]

[1]SRL-ISTEC, Morioka Lab., 3-35-2 Iioka-Shinden, Morioka, Iwate 020-0852, Japan
[2]SRL-ISTEC, Tamachi Lab., 1-16-25 Shibaura, Minato-ku, Tokyo 105-0023, Japan

Abstract: Based on the growth temperature dependence of J_c-B properties we earlier reported a technique in which a sample was grown alternatively at higher and lower temperatures. In this work we investigate the growth of $NdBa_2Cu_3O_y$+10 mol% $Nd_{3.6}Ba_{2.4}Cu_{1.8}O_z$ system in 0.1%O_2-Ar atmosphere, in which temperature is slowly decreased and increased between two temperatures alternatively. The highest peak J_c value obtained was 4.75×10^4 A/cm^2 at 77K and 2.61T. At 77K in all the samples reported here J_c of 5,000 A/cm^2 or more was recorded at 6T. It is evident that J_c values are improved in the higher field region by this novel growth process.

Keywords: $Nd_{3.6}Ba_{2.4}Cu_{1.8}O_z$, OCMG, Critical current density (J_c), Field-induced pinning centers.

INTRODUCTION

In RE123 systems RE/Ba substitution was a major reason for the poor superconducting and J_c-B properties and consequently there were limited activities on these materials. The advent of oxygen-controlled-melt-growth(OCMG) made this substitution controllable and resulting in better superconducting and J_c-B properties[1-3]. This has caused considerable interests in these materials considering their candidature for applications. At the same time there were attempts to grow Nd123 system in air and then anneal them at high temperature in Ar environment for getting better properties[4,5]. Further, Kojo et al [6] reported improved superconducting and J_c-B properties in air processed samples by using Ba rich (Nd123 and Nd422) precursors. We earlier reported growth temperature dependence of J_c-B properties [7] and based on this finding we developed two temperature step growth method which offered possibility of tailoring of J_c-B properties[8, 9]. In present study we use Nd123 and $Nd_{3.6}Ba_{2.4}Cu_{1.8}O_z$ precursors for the melt growth. In the present case an improvised heat pattern has been used in which during growth temperature decreases and increases slowly between two set values and this cycle is repeated many times. As sample is growing alternatively at various temperatures within a given range and thus manipulating the pinning centers which lead to improved propertics.

EXPERIMENTAL

$NdBa_2Cu_3O_y$ and 10 mol% of $Nd_{3.6}Ba_{2.4}Cu_{1.8}O_z$ powders were mixed well and used for making the pellets. A small MgO (100) seed crystal was kept at the center top of the pellet before starting the experiment. Then it was subjected to a heat-treatment in Ar-0.1%O_2 environment as described here. The furnace temperature was raised to 1070°C in 4h and held for 30 minutes. Then temperature was decreased to 1016°C in 30 minutes held for 5 hours. After that temperature was lowered to 982°C in 30 minutes. The growth was achieved by decreasing/ increasing temperature between 978°C to 982°C in 20 minutes. This cycle was repeated many times to achieve substantial growth. Finally, sample was cooled to room temperature in 10h. The small size of specimens were selected from the processed pellets and oxygen annealed with the following heat pattern. The furnace temperature was raised to 600°C in 3h and held for 2h. It was lowered to 500°C in 12h, then to 400°C in 25h and to 300°C in 50h. At 300°C temperature was held for 145h and then samples were furnace cooled. These samples were used for T_c and J_c measurements made on dc SQUID magnetometer. In all the cases the direction of magnetic field was applied parallel to the c-axis of the sample. The hysteresis measured at 77K was used in J_c calculations according to the extended Bean's model [10] same as reported earlier [9].

RESULTS AND DISCUSSION

Figure 1 shows the temperature dependence of normalized dc susceptibility measured for the oxygenated samples(A-D), which were randomly selected from the processed pellet. The results shown here are measured in zero-field cooled mode and in the presence of 10 G. The values for T_c onset, T_{cm} (mid-point) and superconducting transition (ΔT) are displayed in Table 1. The T_c onset is chosen at the point where normalized susceptibility changed to negative values, T_{cm} is taken as temperature half of normalized dc susceptibility and ΔT is temperature difference between 10% and 90% of normalized dc susceptibility values. T_c onset value of 94.0 K is same for three samples and only one sample shows value of 94.5 K. We can see that T_{cm} varies between narrow range of 93.20 to 93.70 K, whereas ΔT lies in the range of 0.75 to 1.39 K. Figure 2 displays the results of the field dependence of J_c of the oxygenated samples (A-D) measured at 77K with magnetic field kept parallel to the c-axis of the samples. The zero-field J_c [$J_c(0)$] and peak J_c [$J_c(pk)$,], $J_c(6T)$ and peak positions [H(pk)] obtained for four different samples are listed in table 1.

Let us try to see if there exists any correlation between T_c [fig. 1] and J_c [fig. 2] results of our present investigations. We earlier reported that the superconducting and J_c-B properties exhibit growth temperature dependence in the OCMG processed Nd123 systems [7]. And it was found that higher growth temperature lead to higher T_c and lower overall J_c with higher irreversibility (Hirr) and H(pk), whereas the lower growth temperature resulted in lower T_c values and so lower Hirr, but overall J_c was enhanced. But in present case we find that even T_{cm} values are slightly changing but Hpk is almost constant. Contrary to expectation based on our previous results [7] we find that increasing T_{cm} values are matching well with decreasing $J_c(pk)$ values in different samples. But as J_c-B properties are not intrinsic properties and so such a difference can take place which is caused by different growth processes in two cases. This is also supported by our previous works [8,9] which proved that tailoring of J_c-B properties may be achieved by manipulating the growth conditions. It is clearly seen that J_c-B properties are improved in higher field region which may be understood in terms of so called low T_c clusters formed in the present process. The J_c-B properties in Nd-Ba-Cu-O systems had been explained by field induced pinning centers[1]. In this case, the regions of low T_c phase Nd-Ba-Cu-O solid solution embedded in a nearly stoichiometric Nd123 possessing higher T_c, were considered to act as pinning centers when these are driven to normal state by the applied field when it is large enough to do so. We earlier suggested that the chemical composition, the volume fraction and size of low T_c phase clusters may control the J_c-B properties[7] based on our OCMG investigations on Nd123 +20 mol% Nd422 system using isothermal growth technique.

Using this analogy we may say that the peak position is caused by low T_c phase clusters which provides maximum pinning centers due to proper cluster size and large volume fraction and by changing them to normal state above a certain magnetic field. Similarly other points on J_c-B curve may be considered the function of chemical composition of cluster, and of also their size and volume fraction. Our present results show that even T_c (onset) are almost same J_c-B properties may vary. This indicates that low T_c phase clusters are playing important role in determining J_c-B properties. We can not see their signature in superconducting transition measurement due to very low field used in such a case. So better ways to probe them thoroughly for finding the relation between processing parameters and low T_c phase clusters (chemical compositions, volume fraction, size and distribution) are of utmost importance. One important point is that measurements with different techniques should be performed on the same set of samples.

Table 1:- Superconducting and J_c -B properties of annealed Nd-Ba-Cu-O samples (A-D) fabricated by a improvised growth process.

Sample	T_c(onset) (K)	T_{cm} (K)	ΔT (K)	$J_c(0)$ A/cm^2	$J_c(pk)$ A/cm^2	H (pk) (T)	$J_c(6T)$ A/cm^2
A	94.0	93.20	0.83	40,870	47,562	2.61	4986
B	94.0	93.27	0.75	41,559	42,752	2.61	5328
C	94.5	93.67	1.34	35,777	33,781	2.61	6126
D	94.0	93.70	1.39	37,103	33,183	2.71	5537

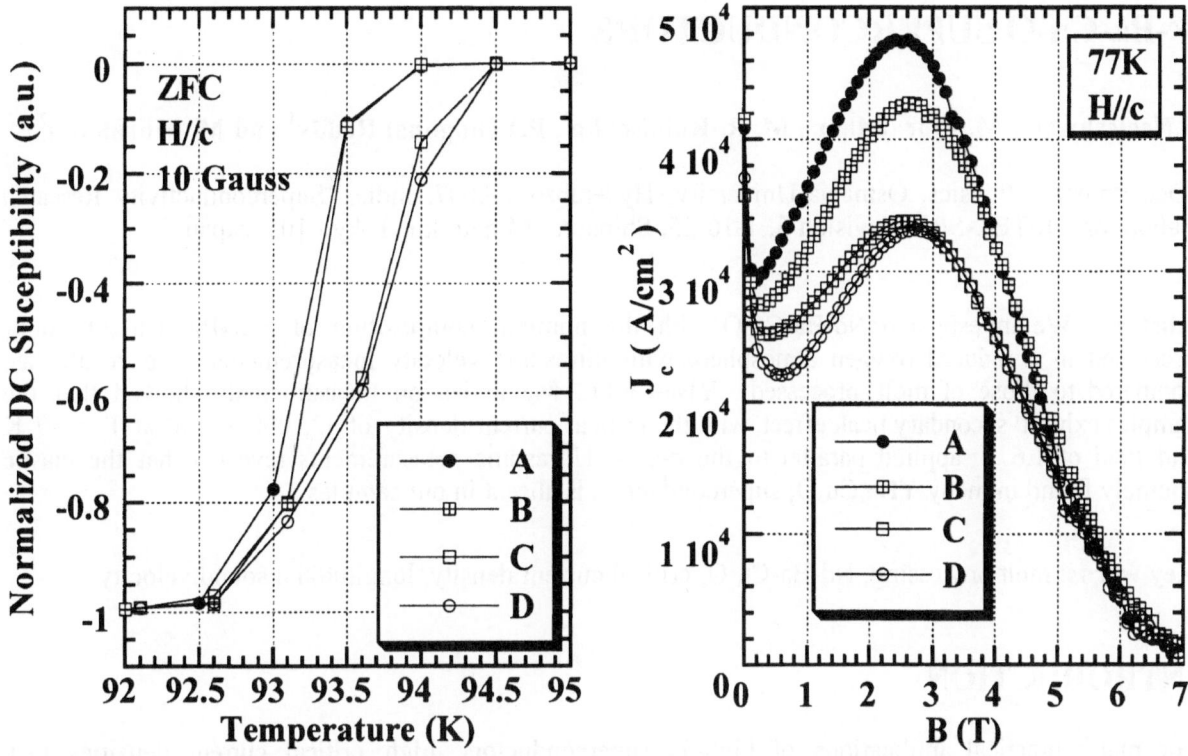

Fig. 1. (Left) The temperature dependence of dc susceptibility of oxygenated samples.
Fig. 2. (Right) J_c-B properties of oxygenated samples at 77K and H//c-axis.

CONCLUSIONS

In the present study the growth of the Nd-Ba-Cu-O system was achieved using $NdBa_2Cu_3O_y$+10 mol% $Nd_{3.6}Ba_{2.4}Cu_{1.8}O_z$ precursor by a novel OCMG process, in which temperature is decreased and increased between two temperatures alternatively. It is evident that the improvised process has enhanced J_c values in the higher field region. Further, this method may be used in large domain growth.

Acknowledgments. This work was supported by the New Energy and Industrial Technology Development Organization (NEDO) as Collaborative Research and Development of Fundamental Technologies for Superconductivity Applications. One of authors (HSC) thanks to the Iwate Technos Foundation, Iwate (Japan) and the NEDO for providing the fellowships.

----

1. M. Murakami, S. I. Yoo, T. Higuchi, N. Sakai, J. Weltz, N. Koshizuka and S.Tanaka, Jpn. J. Appl. Phys. 33, L715 (1994).

2. S. I. Yoo, M. Murakami, N. Sakai, T. Higuchi and S. Tanaka, Jpn. J. Appl. Phys. 33, 1000(1994).

3. S. I. Yoo, N. Sakai, H. Takaichi, T. Higuchi and M. Murakami, Appl. Phys. Lett. 65, 633 (1994).

4. X. Obradors, R.Yu, B. Martinez, F. Sandiumenge, N. Vilata, V. Gomis and S. Pinol, Procd. of Intl. wks. on Critical Currents in Superconductors for Practical Applications, Xian, 1997.

5. A.M. Hu, S. Jia, H. Chen and Z.X. Zhao, Physica C 272, 297 (1996).

6. H. Kojo, S.I. Yoo and M.Murakami, Physica C 289, 85 (1997).

7. H.S. Chauhan, T. Mochida, M. Muralidhar, and M. Murakami, "Advances in Superconductivity-X", eds. K.Osamura and I. Hirabayashi, (Springer-Verlag, Tokyo, 1998) pp. 725.

8. H. S. Chauhan, M. Murakami, Applied Superconductivity 6, (1998) 169.

9. H.S. Chauhan and M. Murakami,"Advances in Superconductivity-XI", Eds. N. Koshizuka and S. Tajima (Springer-Verlag, Tokyo, 1999) pp. 537.

10. C. P. Bean, Phys. Rev. Lett. 8, 250 (1962).

A COMPARISON OF LOW TEMPERATURE ELASTIC BEHAVIOUR BETWEEN MLET-PROCESSED Nd-Ba-Cu-O AND Y-Ba-Cu-O SUPERCONDUCTORS

S.Neeleshwar[1], M. Muralidhar[2], M. R. Koblischka[2], P.Venugopal Reddy[1] and M. Murakami[2]

[1]Department of Physics, Osmania University, Hyderabad-500007, India. [2]Superconductivity Research Laboratory, ISTEC-SRL, Division 3, 1-16-25, Shibaura, Minato-ku, Tokyo 105, Japan.

Abstract: We investigated Nd-Ba-Cu-O with the nominal composition of $Nd_{1.8}Ba_{2.4}Cu_{3.4}O_y$ melt-processed in a reduced oxygen atmosphere with ultrasonic velocity measurements. The results are compared to those of melt processed $YBa_2Cu_3O_y$. Magnetization measurements showed that the samples exhibit secondary peak effect, with the critical current density of 32, 000 A/cm^2 at T = 77 K and field of 1.6 T applied parallel to the c-axis. Ultrasonic measurements revealed that the elastic anomaly found in many $YBa_2Cu_3O_y$ superconductors is absent in our samples.

Key words: melt processing, Nd-Ba-Cu-O, critical current density, longitudinal sound velocity.

INTRODUCTION

For many practical applications of high-T_c superconductors, high critical current densities (J_c), especially in the presence of large magnetic fields, are required. In order to achieve high J_c values, the grain alignment as well as the introduction of effective pinning centers are essential. Furthermore, since the grain boundaries are known to act as weak links in high-T_c superconductors, the unidirectional solidification process is adopted wherein a temperature gradient is maintained. In fact, a few $YBa_2Cu_3O_y$ materials or "Y-123" were already prepared by unidirectional solidification process, which are hereafter designated as melt powder melt growth (MPMG) materials[1]. Later, as an offshoot to further development of the MPMG technique, almost perfect grain alignment was achieved in $NdBa_2Cu_3O_7$ and $SmBa_2Cu_3O_y$ materials prepared by top-seeding method using the oxygen controlled melt growth (OCMG) process [2]. As a matter of fact, a number of OCMG processed materials with the nominal composition, $(LRE)Ba_2Cu_3O_y$, where LRE stands for the light rare earth elements such as Nd, Sm, Eu and Gd, are found to have a pronounced secondary peak effect in the magnetization loops, which results in a large critical current density in a high field even at high temperatures around 77 K. Therefore, these new materials are technically very important and interesting mainly due to the fact that physical characters are admitted to be close to those of a single crystal and worth studying various aspects of OCMG crystals. Recent development of the OCMG process enabled us to grow high T_c and large J_c Nd-Ba-Cu-O bulk superconductors. Thus we compared elastic properties of Y-Ba-Cu-O materials.

EXPERIMENTAL

High purity commercial powders of Nd_2O_3, $BaCO_3$, and CuO (5N) were weighed to have a nominal composition of $Nd_{1.8}Ba_{2.4}Cu_{3.4}O_y$. The starting materials were mixed and ground for 4 h using a ball mill. The mixed powder was pressed into pellets of 6 cm in diameter by applying a load of 2 ton for 10 min and then calcinated for 24 h at 880 °C. This process was repeated three times. The calcinated powders were pressed into pellets followed by sintering at 1030 °C for 48 h in air. Finally, the sintered pellets after crushing into powders, were again pressed into pellets of 20 mm in diameter with the help of a cold isostatic press by applying a pressure of 2000 kg/cm^2. A MgO (100) seed was placed on the center top of the pellet, which was then OCMG processed in reduced oxygen atmosphere. The details of the heat treatment schedules can be found elsewhere [3]. For oxygen annealing, samples with dimensions (a×d×c) ≈ 1.5×1.5×0.5 mm^3 were cut from the as-grown crystals and annealed in flowing O$_2$ gas in the trmperature range 300 to 600 °C with the following heat treatment schedule. The samples were heated to 600 °C in 2 h, held for 2 h and cooled to 500 °C in 12 h, then to 400 °C in 24 h and finally to 300 °C in 50 h and held there for 150 h followed by furnace cooling. The oxygenated samples were used for the T_c and J_c measurements using a Quantum Design MPMS-7 SQUID magnetometer with an applied magnetic field of 1 mT. To minimize field inhomogeneties, the scan length is restricted to 1 cm. The magnetic J_c was estimated according to the extended Bean's model by using the relation, $J_c = (20\Delta M)/[a(1-a/3b)]$, where ΔM is the hysteresis between increasing and decreasing field processes in emu/cm^3, a and b (a<b) are cross sectional dimensions (in cm) of the sample perpendicular to the applied magnetic field. The sound velocity measurements were carried out by means of the ultrasonic pulse transmission technique in the temperature range from 80 to 300 K, using a sample with dimensions of 4 mm in thickness and 10 mm in diameter. X- and Y- cut PZT transducers with a fundamental frequency of 1 MHz were used for transmitting and receiving the longitudinal (V$_l$) and shear (V$_s$) waves. The transit time of 0.01 μs using a 100 MHz digital storage oscilloscope (Tektronix Model 2221).

RESULTS AND DISCUSSION

The field dependence of the critical current density is shown in Fig. 1. In the present investigation the magnetic field is applied parallel to c-axis. It is evident from the figure that an anomalous peak effect is seen at about 1.6 T and the irreversibility field, H$_{irr}$, exceeds 5.5 T at 77 K, which is commonly observed among the melt processed LRE-Ba-Cu-O materials prepared in reduced oxygen atmosphere, and ascribed to field induced pinning by LRE-rich clusters dispersed in the matrix. The peak J_c value of the sample reached 32,000 A/cm^2 at 1.6 T (77 K , H ∥ c).

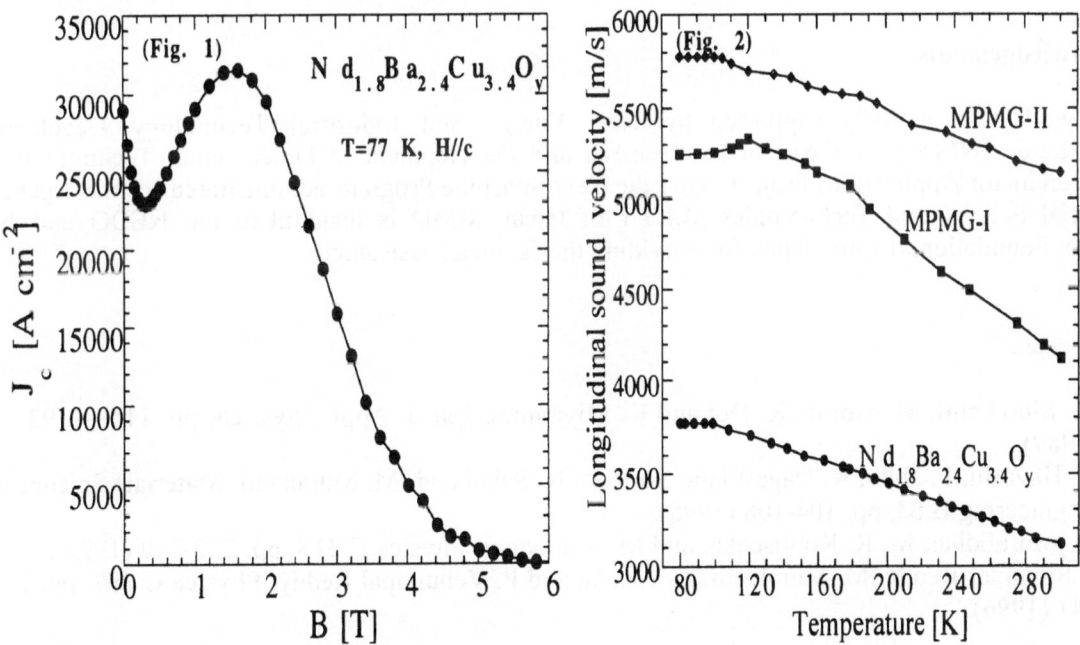

Fig. 1 Field dependence of the critical current density (T=77K, H$_a$ parallel to the c-axis) for the OCMG-processed Nd$_{1.8}$Ba$_{2.4}$Cu$_{3.4}$O$_y$.

Fig. 2 Temperature dependence of longitudinal sound velocity for the MPMG-processed Y$_1$Ba$_2$Cu$_3$O$_y$ materials (MPMG-I and MPMG-II) and OCMG-processed Nd$_{1.8}$Ba$_{2.4}$Cu$_{3.4}$O$_y$.

520

The variation of longitudinal sound velocity of $Nd_{1.8}Ba_{2.4}Cu_{3.4}O_y$ material as a function of temperature is shown in Fig. 2. It can be seen from the figure that the longitudinal sound velocity increases continuously with decreasing temperature down to about 100 K and then shows a non linear change at 95K, being the vicinity of the superconducting transition temperature (T_c). Such behaviour has a close agreement with the previous reports [4], in which the non linear changes were observed in the temperature region from 100 to 250 K. In order to compare the elastic behaviour of YBCO-123 material, we performed the longitudinal sound velocity of MPMG I and II, and the results are presented in the Fig. 2. One can see the hump (MPMG-I material) at 120 K which indicates the presence of an elastic anomaly. However, another Y-123 material, prepared by an improved version of the MPMG technique (MPMG-II material) has no such anomaly. Similarly, in the case of $Nd_{1.8}Ba_{2.4}Cu_{3.4}O_y$ sample (OCMG material), except in the vicinity of T_c, no anomaly has been observed.

It is interesting that the MPMG-II material does not show any elastic anomaly above T_c but the MPMG-I material does. Coming to the material of the present investigation ($Nd_{1.8}Ba_{2.4}Cu_{3.4}O_y$), taking the above observations into account and since the elastic anomalies are totally absent in measured temperature region, one may conclude that the material might be having better microstructure than both the MPMG materials, because under only this assumption, the absence of an elastic anomaly can be explained. Microstructure and grain size are the main controlling factors as far as the mechanical properties of ceramic superconductors are concerned.

SUMMARY

The ultrasonic sound velocity measurements show that in the superconducting phase the ultrasonic velocity is found to remain almost constant. On the basis of temperature variation of the longitudinal sound velocity of Nd-123 it was concluded that the OCMG material of the present investigation will be defect-free compared to YBCO from the viewpoints of mechanical properties.

Acknowledgements

This work was partially supported by New Energy and Industrial Technology Development Organization (NEDO) as a part of its Research and Development of Fundamental Technologies for Superconductor Applications Project under the New Sunshine Program administrated by the Agency of Industrial Science and Technologies M.I.T.I of Japan. MMD is thankful to the NEDO and Iwate Techno. Foundation, Iwate, Japan for providing the financial assistance.

References

1. M. Murakami, M. Morita, K. Doi and K. Miyamoto, Jpn. J. Appl. Phys. **28**, pp. 1189-1193 (1989).
2. N. Hayashi, P. Diko, K. Nagashima, S.I.Yoo, N. Sakai and M. Murakami, Materials Science and Engineering B **53**, pp. 104-108 (1998).
3. M. Muralidhar, M. R. Koblischka, and M. Murakami, Physica C **313**, pp. 232-240 (1999).
4. R.Ravinder Reddy, M. Murakami, S. Tanaka and P. Venugopal Reddy, Physica C **257**, pp. 137-141 (1996).

An Overview of U/n Processing

Roy Weinstein

Physics Dept. and Texas Center for Superconductivity, University of Houston, Houston, TX 77204-5506 USA

Abstract: Adding uranium to HTS powders, prior to texturing, and irradiating with thermal neutrons, n, after texturing, produces very effective pinning centers for a variety of HTS systems. Record values of J_c result. Fission of ^{235}U provides two high energy, high Z ions. These leave tracks of aligned quasi-columnar defects of length δ, which act as pinning centers. Recent experiments show $\delta \sim 2.7 \mu m$. In Y123 compounds of U, formed during texturing, are small, ~300nm. In order for U/n pinning to be most effective, the deposits of U should have a spacing, $S \leq 2\delta \sim 5.4 \mu m$. Also, for processing large objects, the penetration depth of neutrons should be large. Values of penetration depth for selected HTS are presented. Results of the U/n process in bulk Y123 and Ag-BiSCCO tape are presented. Nd123 and other HTS systems are discussed.

Keywords: Pinning Centers, High J_c, High H_{irr}, Isotropy, Fission Damage

INTRODUCTION

Critical current density, J_c, can be increased by orders of magnitude, by introducing pinning centers of various geometries. Such pinning centers can be introduced by several means: e.g., chemical [1], hot press [2], nanorods [3], irradiation [4], etc. Irradiation pinning centers are the most effective due to control over density, homogeneity and geometry of the defects.

The U/n process is a way to produce HTS with leading characteristics (J_c, H_{irr}, anisotropy), in large bulk objects, at low cost with low radioactivity [5]. In this process, an intimate mixture of ^{238}U and ^{235}U of controlled ratio is added to conventional HTS precursor powders, prior to texturing. After texturing the HTS is irradiated with isotropic thermal neutrons, n, to fission some of the ^{235}U. Each fission results in 2 heavy high Z fragments, f_1 and f_2, and a small number of neutrons (average ~2.5). The two fragments have nuclear charge of $Z_1+Z_2=92$, and are usually of unequal mass and Z (e.g., $^{130}Sn + ^{103}Mo$, or ^{149}Xe and ^{94}Sr). The energy release, 160-180MeV, depends on the fission products. Each of the energetic fragments has a range of 6-10μm, depending upon its energy. The number of fissions is proportional to F_n x $M(^{235}U)$, where F_n is the fluence of n [6]. The ionization loss per unit length of the fission fragments, dE/dx, is below that of the ions used to create amorphous columnar defects [7]. In fact it is below 2000eV/Å. See Fig. 1. Therefore, one might expect that fission fragments do not produce columnar defects. However, dE/dx is a statistical average [8], about which fluctuations occur of approximately ±75%. Because of these large fluctuations, fission fragments do create defects which are columns, broken columns, and string-of-beads. The magnitude of the fluctuations can be used to estimate the length, δ, of aligned ionization damage caused by fission fragments. In Fig. 1, this is estimated to be $2 < \delta < 4 \mu m$ [5], about 3 times shorter than the track length of the fragment. Studies of the U/n process have been done in Y123 [4,5], BiSCCO 2223 [9,10], Nd123, Sm123 and Gd123 [11]. The study of U/n-Y123 is farthest along, and will be presented first to clarify the issues.

Y123

Chemistry. Most work on U/n-Y123 has been done with HTS powders including Pt and

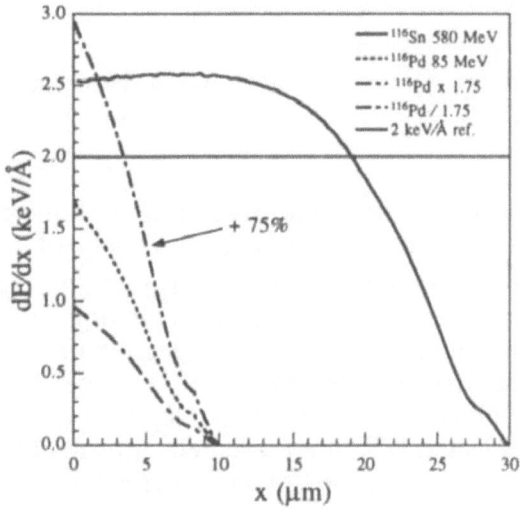

Fig. 1. Theoretical dE/dx along path of symmetric fission fragments, 85 MeV Pd. Envelopes of ±75% Landau fluctuations are shown, as is dE/dx for 580 MeV Sn.

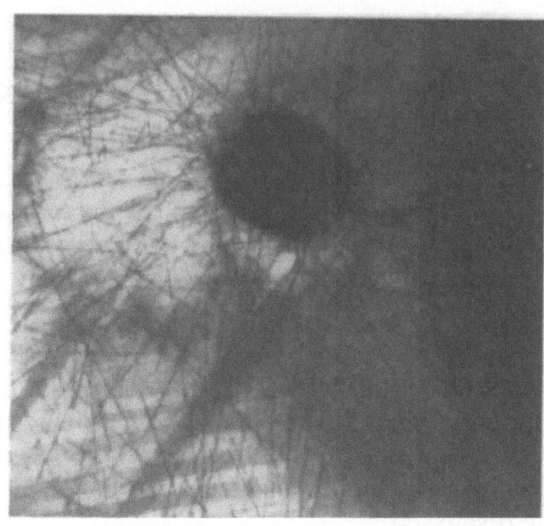

Fig. 2. Fission fragment tracks in U/n-Y13. Large deposits are Y-5, ~300nm diam., from which fragments radiate. Short tracks originate outside field of view.

excess Y211, with UO_4 added. During texturing, small (300nm), stable deposits of the U compound $(U_{0.6}Pt_{0.4})YBa_2O_6$ (called Y-5) are formed [12]. The number of these deposits is proportional to the mass of U, M_U. Thus the spacing of these deposits varies as $M_U^{-1/3}$. Spacing of $2<S<6\mu m$ has been explored. The Y-5 deposits act as chemical pinning centers, prior to irradiation. These "U-Chem" centers increase J_c by about 10% per 0.1%U(wt) [12] up to about M_U~1%(wt). Work aimed to extend this U-chemical pinning continues at Houston.

Irradiation Effects. TEM studies of U/n-Y123 clearly show the fission track damage including columns, broken columns, and string-of-beads. See Fig. 2. J_c increases markedly. Typical values of J_c achieved with 0.3%U, 250ppm ^{235}U and $8 \times 10^{16} n/cm^2$ are [5, 13, 14]: $J_c(77K,0.25T)=290kA/cm^2$; $J_c(50K,0.5T)=1.0MegA/cm^2$; $J_c(30K,14T =400kA/cm^2$. The U-Chem [12] and U/n [5, 13, 14] gains in J_c are multiplicative, and range from [14]

$$14 < \frac{J_c(B_A,T) \text{ with irradiated U}}{J_c(B_A,T) \text{ with no U}} < 40$$

Other characteristics of Y123 are also changed. Anisotropy, $A=J_c(B_A\|ab)/J_c B_A\|c)$, is reduced from ~3 to ~2. For M_U=0.15%U(wt), 250ppm ^{235}U, fluence F_n ~4 x $10^{16}n/cm^2$, $\Delta T_c \approx$ -0.7K, and creep rate is increased by 1/5. The maximum trappable field, $B_{t,max}$, for such a grain, 2cm. diam x 0.8cm, is increased by a factor of R_M=4.5 by irradiation and an added 15% by U-Chem pinning, for a total of R_M~5.2. Such a grain traps 2.2 Tesla at 77K, and 4.2T at 65K. At 6 months, the residual radioactivity is <200nc/g (7400Bq/g).

Maximum Useful Spacing of Deposits. The U added to the Y123 powders has been varied from $0.0375<M_U<1.0$%(wt). This varies the number of Y-5 deposits by ~27, and therefore varies the interdeposit spacing, S, by ~3. At the lower values of M_U the values of R_M begin to decline, indicating $S>2\delta$. (See Fig. 3.) From these experiments we find S~5.4μm (±50%) and estimate the value of δ~2.7μm. This is in good agreement with the earlier theoretical estimates [5] of $2\mu m < \delta_{theory} < 4\mu m$. We conclude that to fully exploit the U/n process, deposits of the U compound should be separated by

$$S < 5.4 \mu m$$

Penetration of Thermal Neutrons. Some n are absorbed by each isotope of each element present in the HTS. If n_i = # atoms/unit vol. of the ith isotope, and σ_i = its cross-section for n absorption, then the fraction of n remaining at depth z is [6]

$$f = e^{-\sum_i N_i \sigma_i z}$$

where the sum is over all isotopes. 1/e of the neutrons (38%) remain at $z=d_e$, where

$$d_e = 1 / \sum_i N_i \sigma_i .$$

Values of d_e for selected HTS are given in Table 1. It is desirable that neutrons increase J_c approximately uniformly in the HTS volume. Experiment shows that R_M, and hence the increase in J_c, changes $\sim \pm 5\%$ in the region $3 < F_n < 18 \times 10^{16} n/cm^2$. Thus J_c will vary little (~5%) due to fluence if the fluence at the HTS surface, compared to that at maximum depth, varies by a factor of less than 6. Objects of size $2\text{-}4 d_e$ will be very uniform, and we expect nearly uniform J_c for grains of U/n-Y123 of d>9cm.

Table 1. Absorption distance d_e at which incident n are diminished by factor $1/e \approx 0.368$.

HTS	d_e
Y123	4.44 cm
Sm123	3.67×10^{-2} cm
Nd123	2.16 cm
Gd123	4.06×10^{-3} cm
BiSCCO 2223	4.6 cm
Tl BaCCO 2223	4.55 cm

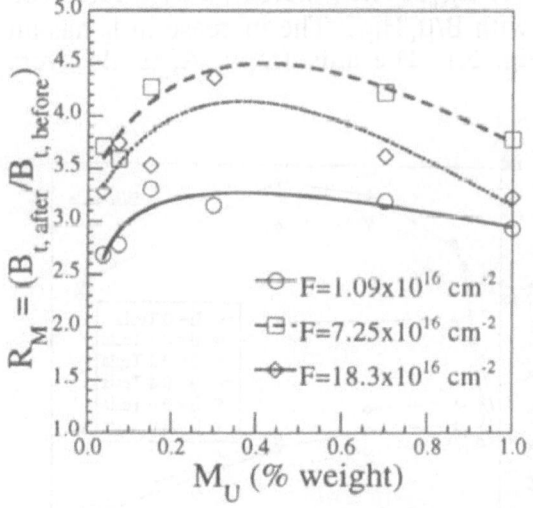

Fig. 3. Enhancement ratio of trapped magnetic field, R_M, vs. mass of U added, M_U. Note fall-off of R_M for $M_U \sim 0.0375\%$U(wt).

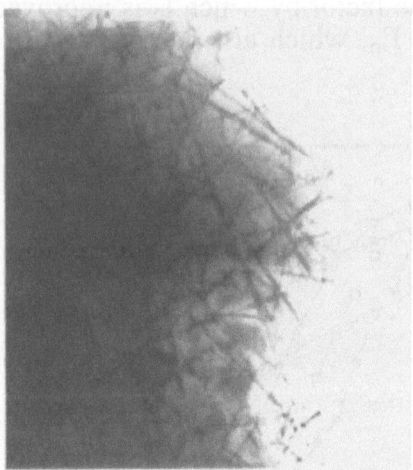

Fig. 4. Fission fragment damage in U/n-Ag-BiSCCO tape. Deposits, if any, are too small to observe.

BiSCCO 2223

Work on BiSCCO has been done by collaboration of the groups of S.X. Dou (University of

Wollongong, Australia), H.W. Weber (Atominstitut, Vienna) and R. Weinstein (Houston). Table 1 indicates that large volumes of BiSCCO can be processed uniformly by the U/n method.

Chemistry. The first experiment on Ag-BiSCCO [9] tape was done with the best mass of U as determined from the Y123 studies [5], 0.3%U, including 250ppm ^{235}U. The tape was produced in Wollongong by PIT methods. After texturing, SEM and TEM studies in Houston showed no visible U deposits. Hence, the deposits were smaller than 2nm (or were incorporated into the BiSCCO matrix). After irradiation at the Texas A&M TRIGA reactor, the expected fission fragment damage was seen (see Fig. 4). The very small deposits of U-compound mean that $S \ll 5.4\mu m$, and there is no longer need to add ^{238}U to get more deposits. On the basis of this first experiment [9] a second test was performed [10]. In the first test, 0.3% M_U with 250ppm ^{235}U was used. In the second test we used 1500ppm of ^{235}U, thus reducing the amount of U by 2, while increasing the amount of ^{235}U by 6.

Radioactivity. The radioactivity is dominated by the (n, γ) interaction with silver. The activity of the Ag is proportional to $M_{Ag} \times F_n$. The number of fissions in BiSCCO is proportional to $F_n M(^{235}U)$. The number of fissions in exp. #2 could be set equal to that of exp. #1 by using F_n(exp. #2)=1/6 x F_n(exp. #1). The advantage of the added ^{235}U is approximately a 6-fold reduction in the activity of Ag. The resulting activity in test #2 was ~3μc/cm at $F_n=1 \times 10^{16} n/cm^2$. A series of tests is now in progress at Wollongong to determine the maximum mass % of U for doping BiSCCO. It appears that $M(^{235}U)$ can be increased to ~2%(wt), and the resulting Ag radioactivity can be reduced to ~200nc/cm (7,400Bq/cm).

Effects of U/n process on BiSCCO. We report results of the second BiSCCO experiment, using 1500ppm ^{235}U(wt). Transport critical current of the Ag-BiSCCO tape was determined using a criterion of 1μV/cm. For the applied field, B_A, parallel to the c axis ($B_A \| c$) the irreversibility field was increased by a factor of ~1.9. As a result of this, I_c increased by a factor of 76 at $B_A \| c$ ~0.8 Tesla. (See Fig. 5.) H_{irr} for $B_A \| ab$ was increased by a factor of ~2.4. The factor by which I_c is improved increases with $B/\mu_o H_{irr}$. The increase in I_c has an optimum F_n, which also increases with B. (See Fig. 6.) The anisotropy, A, is also very

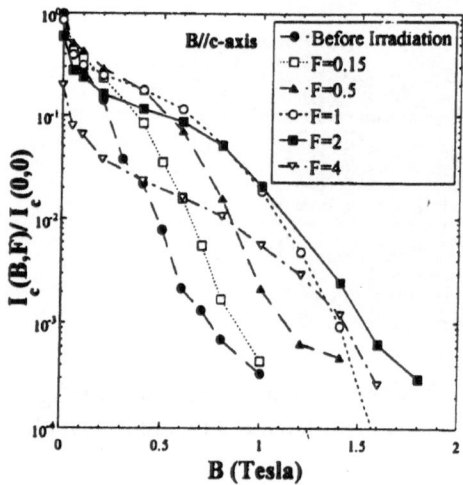

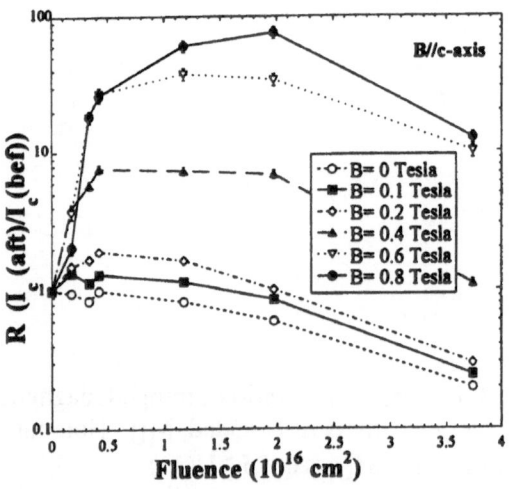

Fig. 5. Critical current (normalized) vs. $B_A \| c$, with fluence, F, as parameter. Note I_c is increased by factor of 76 at $B_A \| c$ ~0.8T, $F_n=2$.

Fig. 6. Ratio of I_c after irradiation to I_c before. Note there is an optimum fluence which increases with B.

significantly improved by the U/n process. At B_A=0.4T, A is reduced from 17 to 2. At B=0.8T (near the pre-irradiation irreversibility point), A is reduced from 300 to 3. The decrease of T_c for F=1 x 10^{16}n/cm^2 is ~3K.

Another well known process for treating Ag-BiSCCO tape is the use of high energy protons (e.g., 800MeV) to fission the Bi atoms [15], which we denote the "p$^+$/f" process. We have compared the U/n and p$^+$/f processes and conclude that the increase in I_c in U/n is a factor of 2.5 higher than p$^+$/f. The cost for p$^+$/f is a factor of 70 higher. The radioactivity for p$^+$/f is a factor of 14 higher at t=0. The irreversibility field is marginally higher and the anisotropy marginally lower for U/n than for p$^+$/f.

Sm123 AND Nd123

Work on Sm123 and Nd123 has been done by a collaboration of the groups of M. Murakami (ISTEC), G. Krabbes (IFW Dresden), H.W. Weber (Atominstitut, Vienna), and R. Weinstein (Houston).

In Sm123 the compounds of U form small deposits of Sm-5 (i.e., Y-5 with Y→Sm), suitable to the U/n method. However, the neutron penetration distance is only d_e=3.7 x 10^{-2}cm in Sm123. Thus, while U/n-Sm123 is suitable for thick film work, it is not suitable for trapped field magnets. The dominant characteristic in TEM studies of U/n-Sm123 is a very *low* number of fission tracks. The thermal neutrons are absorbed on the outer layers of the grain.

In Nd123, d_e~2.16cm. Thus the U/n process should work well on objects ~5cm in size. However, UO$_4$ mixed directly into Nd123 powders forms ~5μm deposits of Ba$_3$UO$_7$ during texturing. The spacing between these deposits greatly exceed 5.4μm, and leave large volumes of HTS without fission tracks. Also, most of the fission fragment damage occurs within the deposits, and therefore does not benefit the HTS.

To improve this situation the compound Nd-5 was fabricated in Houston, and used at Dresden in place of UO$_4$ with a modified temperature profile for texturing. The resulting size of the Nd-5 deposits is ~1.5μm. Trapped field increases of ~4, and J_c~90kA/cm^2, were achieved in a prior generation of U/n-Nd123 by an ISTEC-Houston group. We expect greater increases in trapped field when the newer grains with 1.5μm deposits are irradiated.

SUMMARY

The average rate of energy loss by fission fragments is too low to create columnar defects. However, fluctuations result in columns, broken columns and string-of-bead defects, which make very successful pinning centers. The length of these aligned defects is δ~2.7μm (±50%).

Compounds of U form during texturing. The spacing of these deposits, S, should be S<2δ ~5.4μm to avoid having large volumes of HTS without U/n pinning centers. Also, fission fragments formed in U deposits ≥3μm will create most of their useful damage within the U compound instead of within the HTS.

The spatial homogeneity of J_c in any given HTS depends upon the average penetration distance of thermal neutrons. Good homogeneity results in HTS grains a few times larger than the values of d_e, listed in Table 1.

U/n-Y123 trapped field magnets meet the criteria on S and d_e. The deposits of U compounds ~300nm in size, are themselves useful chemical pinning centers prior to irradiation. After irradiation, record values of J_c for textured materials are achieved. Anisotropy is reduced from

~3 to ~2. Trapped field, e.g., in a grain 2cm diam. x 0.8cm, exceeds 2.2 Tesla at 77K, and 4.2 Tesla at 65K.

In U/n-Ag-BiSCCO tape, H_{irr} is approximately doubled for both $B_A \| c$ and $B_A \| ab$. J_c improvement increases with $B/\mu_o H_{irr}$. e.g., for $B_A \| c$ ~0.8 Tesla, J_c increases by a factor of 76. Anisotropy is reduced dramatically. Compared to the method of high energy proton fission of Bi, our analysis indicates the U/n process provides J_c ~2.5 higher, H_{irr} and anisotropy marginally better, and cost orders of magnitude lower.

Acknowledgments. We are grateful for the free exchange of information from our collaborators at Atominstitut, Vienna; ISTEC, Japan; IFW, Dresden; and Wollongong University, NSW Australia. We especially acknowledge an informative, long visit by Mr. Tadashi Mochida of ISTEC. We thank the staff of the Texas A&M Reactor for irradiations, and for their help. We thank Dr. Joseph Kulik of the TCSUH TEM Lab for his patience and skill. The work at Houston was supported by the State of Texas via the ATP program and TCSUH, and by the Army Research Office and the Welch Foundation.

1. M.Murakami, invited paper, in *Proceedings of TCSUH International Workshop on HTS Materials, Bulk Processing, and Applications*, ed. C.W.Chu *et al.*, World Scientific, p.491, 1992.
2. D.F.Lee, A Satpathy, V.Selvamanickam, and K.Salama, *Supercond. Sci. Technol.*, **8**, 423 (1995).
3. P.Yang and C.M.Lieber, *J. Mater. Res.* **12**, 2981 (1997).
4. R.Weinstein, J.Liu, Y.Ren, I.G.Chen, V.Obot, R.P.Sawh, C. Foster and A.Crapo, invited paper, in *Proc. International Workshop on Superconductivity*, Kyoto, Japan, 1994.
5. R.Weinstein, R.P.Sawh, Y.Ren and D.Parks, *Mater. Sci. Engineer.* **B53**, 38 (1998).
6. R.Weinstein, *Interaction of Radiation with Matter*, McGraw-Hill, 1964.
7. L.Civale, A.D. Marwick, T.K. Worthington, M.A. Kirk, J.R.Thompson, L.Krusin-Elbaum, Y.Sun, J.R.Clam, and F.Holtzberg, *Phys. Rev. Lett.*, **67**, 648, (1991).
8. O.Blunck and S.Leisegang, Z. Phys. **128**, 500 (1950).
9. G.W.Schulz, C.Klein, H.W.Weber, S.Moss, R.Zeng, S.X.Dou, R.Sawh, Y. Ren, and R.Weinstein, *Appl. Phys. Lett.*, **73**, 3935 (1998).
10. R. Weinstein, 9th International Workshop on Critical Currents (IWCC-99), Madison, Wisconsin, July 7-10, 1999.
11. Chemical studies are in progress by collaboration of the research groups of M.Murakami, ISTEC, Tokyo, G.Krabbes, IFW, Dresden, and R.Weinstein, Houston.
12. R.-P. Sawh, Y.Ren, R.Weinstein, W.Hennig, and T.Nemoto, *Physica C*, **305**, 159 (1998).
13. R.Weinstein, R.Sawh, Y.Ren, M.Eisterer, and H.W.Weber, *Supercond. Sci. Technol.*, **11**, 959 (1998).
14. R.Weinstein, R.P.Sawh, Y.Ren, M.Eisterer, and H.W.Weber, in *Proc. 8th US-Japan Workshop on High Tc Superconductivity*, Tallahassee FL, ed. J.Schwartz, p.74, (1997).
15. L.Krusin-Elbaum, J.R.Thompson, R.Wheeler, A.D.Marwick, C.Li, S.Patel, D.Shaw, L.Lisowski, and J.Ullmann, *Appl. Phys. Lett.*, **64**, 3331 (1994).

BULK Bi2212 TEXTURING BY SOLIDIFICATION IN A HIGH MAGNETIC FIELD AND HOT FORGING

Robert Tournier,[1] Sybille Pavard,[1] Daniel Bourgault,[1] and Catherine Villard[2]

[1] Laboratoire de Cristallographie / Consortium de Recherche pour l'Emergence de Technologies Avancées (CRETA) - Centre National de la Recherche Scientifique – BP 166 – 38042 Grenoble cedex 9- France
[2] Centre de Recherches sur les Très Basses Températures / CRETA - Centre National de la Recherche Scientifique – BP 166 – 38042 Grenoble cedex 9- France

Abstract: Bi2212 bulk materials containing MgO are textured by solidification in a magnetic field. The maximum temperature dwell effect at $T = T_m$ on the Bi2212 processing is studied by measuring its magnetic susceptibility $\chi(T)$. T_m is varied as 877°C + ΔT, 877°C corresponding to the melting and solidification onsets. For $\Delta T \geq 15$°C, the Bi2212 phase is fully melted. For $\Delta T < 15$°C, the melting is time dependent in relation with the oxygen loss. The best orientation and critical current density Jc = 160 kA/cm² at 4K are obtained for $\Delta T = 15$°C. Hot forging slightly increases the Jc of Bi2212 processed in a magnetic field.

Keywords: Magnetic susceptibility, magnetic melt processing, Bi2212, bulk materials

INTRODUCTION

Magnetic melt processing (MMP) is known to induce a c-axis orientation when applied to bulk Bi-cuprates [1,2]. H. Maeda has orientated Bi(Pb)2212 in tapes and bulk materials processed in a high magnetic field [3]. In this paper, we study the effect on texturing of the maximum dwell temperature reached during MMP of bulk Bi2212 materials containing MgO. We also present results on applying hot forging after MMP on this material.

MAGNETIC MELT PROCESSING

Experimental. 10 wt % of MgO is added to *Hœchst* Bi2212 powder. This mixture is pressed into 20-mm diameter pellet of 10 g. The pellet placed in a zirconia crucible is processed in a tubular furnace inserted in a superconducting magnet [2]. The temperature can reach 1100°C and the maximum vertical magnetic field is 8 Tesla. The sample is placed in the maximum of magnetic force $(H_a.dH_a/dz)$: it is submitted to a 5.7 Tesla magnetic field and the gradient is 43 T/m (dH_a/dz). The use of an electronic balance allows the in-situ measurement of the magnetic susceptibility χ deduced from the magnetic force exerted on the sample: $m.\chi.H_a.dH_a/dz$ (m is the sample mass).

Ten pellets are heated up to a maximal temperature (T_m) during a two-hour dwell. The samples are then slowly cooled (15°C/h) down to 680°C where the cooling rate is increased.

The magnetic susceptibility measurement as a probe. The evolution of the magnetic susceptibility with the temperature of a Bi2212 pellet during a thermal cycle up to 960°C is plotted in fig. 1. As the liquid phase is more paramagnetic than the solid one, the onset temperature of melting can be determined to be 877°C. The T_m temperatures are chosen higher than this temperature, where a very small fraction of metastable Bi2212 crystallites are supposed to coexist with a liquid phase. For the 10 pellets, T_m is determined from its difference ΔT with the onset of melting (see fig. 1). ΔT is changed from 6 to 24 °C.

Microstructure observation. The microstructure orientation is observed by SEM in secondary electron mode on faces cleaved parallel and perpendicular to the direction of the processing magnetic field H_a. The sample processed at the lowest temperature ($\Delta T = 6°C$) is the only one having no preferred orientation. For the others, the platelets are preferably aligned perpendicular to the magnetic field direction H_a. The best texture reported in fig. 2 is obtained for $\Delta T = 15°C$.

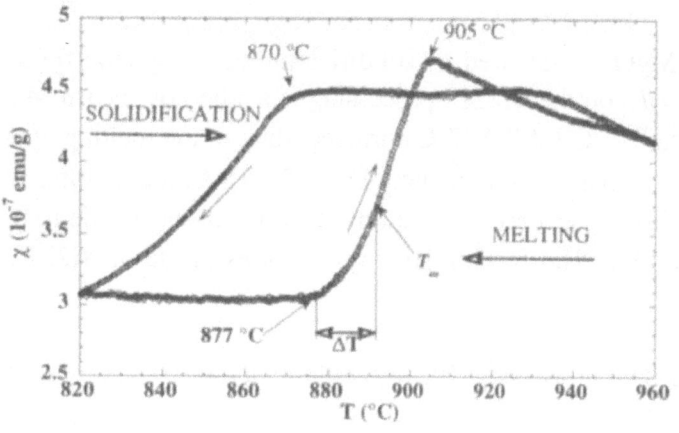

Fig. 1: Magnetic susceptibility evolution of a Bi2212 pellet during a thermal cycle up to 960°C. The melting starts at 877°C.

Fig. 2: Micrograph of a cleaved surface for the sample processed with $\Delta T=15°C$. ($T_m=892°C$)

Superconducting properties evolutions with T_m. The critical current densities J_c are determined at 77 K from DC transport measurements on 6-mm² section bars in self-field, using a criterion of 1 μV/mm. These Jc are not optimized by post-annealing in a reduced atmosphere. Magnetization loops up to 2 Tesla are performed at 4 K on 8-mm³ cubes. These measurements are used to determine an anisotropy factor A defined by $A = \Delta M_{//}/\Delta M_{\perp}$ where $\Delta M_{//}$ and $\Delta M_{\perp}$ are the maximum magnetization values obtained for an exciting field respectively applied parallel and perpendicular to H_a. Jc and A are both maximum for $\Delta T = 15°C$ (fig. 3). The magnetic melt processing performed at 15°C above the onset of melting leads to the best orientation (magnetic anisotropy A and SEM observations) and to the higher critical current density. This higher Jc is also linked to the lowest secondary phases proportion detected by SEM using EDS [5].

The in-situ susceptibility magnetic measurement as a tool. The evolution of χ versus T is shown in Fig. 4 for different temperature dwells at $T = T_m$. The χ behavior is different for $\Delta T < 15°C$ as compared to $\Delta T > 15°C$. Weighting during processing without applying magnetic field shows that the melting is accompanied by an oxygen loss which is time dependent. It explains why for $\Delta T < 15°C$, the melted part increases at a constant temperature T_m during the 2-hour dwell. Nevertheless, this time is not sufficient to completely melt the material even for $\Delta T = 14°C$ and to attain the oxygen equilibrium. When the sample is cooled, χ decreases because the material still contains a small Bi2212 solid part. The Bi2212 phase is fully melted at $T = 892°C$ during the processing time. At this temperature, the oxygen loss first produces an apparent increase of χ followed by a decrease because the oxygen loss tends to reduce the copper valence from 2+ to 1+. The susceptibility needs a minimum of 2 hours to attain the equilibrium. The equilibrium susceptibility is the same for all T = T_m as high as 900°C (ΔT = 23°C). It is important to note that at the equilibrium, χ is temperature independent down to the solidification onset at about the same temperature as the melting onset. It clearly indicates that at the equilibrium in oxygen content the melting and the solidification occur at the same temperature. The optimized overheating rate ΔT = 15°C of the melt to obtain the best orientation using the MMP process has the same value as for YBCO [6]. In addition, the sample for ΔT = 15°C has the highest susceptibility in the solid state due to good orientation along the c-axis.

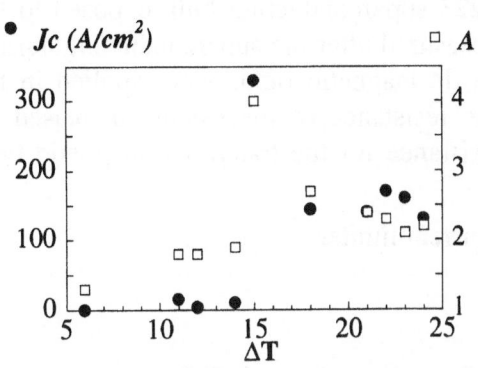

Fig. 3: Transport Jc measured at 77K in self-field for samples processed at different temperatures T_m above (ΔT) the onset of melting.

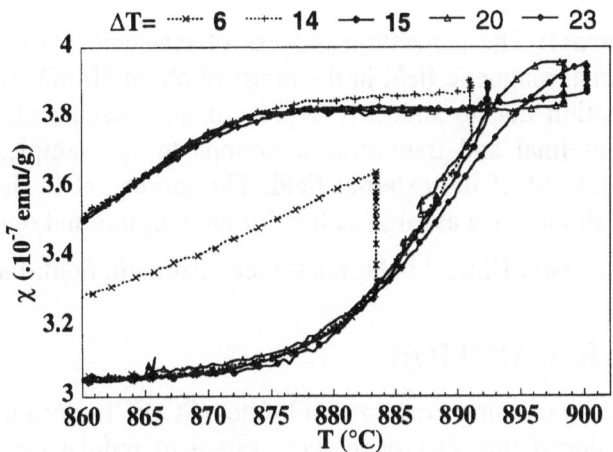

Fig. 4: χ evolution with temperature for samples processed at different temperatures (different ΔT).

Jc_m AT 4K AND COMBINATION WITH HOT FORGING

Hot forging (HF) applied after MMP slightly increases the orientation degree and the superconducting properties of Bi2212/MgO [4]. Jc_m are deduced from magnetization loops measured at 4K using the Bean model. With no thermal treatment in a reduced atmosphere applied after processing, Jc_m is 161 kA/cm² for a MMP sample and 165 kA/cm² for a sample HF after MMP.

CONCLUSION

Solidification of Bi2212/MgO in a high magnetic field leads to a textured material along the c-axis. In this process, the Bi2212 phase is fully melted at temperatures ΔT = 15°C above the melting onset temperature. Before cooling, large part of the melting window broadening is due to the slow oxygen exchange with the air. No post-annealing in a reduced atmosphere is used to increase Jc. Nevertheless, Jc higher than 160 kA/cm² are obtained at 4K by magnetic measurement. This Jc can be slightly enhanced by a subsequent hot forging.

1. J.G. Noudem, J. Beille, D. Bourgault, D. Chateigner and R. Tournier, Physica C **264**, 325 (1996)
2. S. Pavard, C. Villard, D. Bourgault and R. Tournier, Supercond. Sci. Technol. **11**, 1359 (1998)
3. H. Maeda, W.P. Chen, K. Kakimoto, P.X. Zhang, K. Watanabe, M. Motokawa, H. Kitaguchi, H. Kumakura and K. Itoh , *Advances in Superconductivity XI,* Proc. of ISS'98, p. 823 (1998)
4. S. Pavard, D. Bourgault, C. Villard and R. Tournier, Physica C **316**, 198 (1999)
5. S. Pavard, D. Bourgault, C. Villard and R. Tournier To be published
6. M.R. Lees, D. Bourgault, P. de Rango, P. Lejay, A. Sulpice and R. Tournier, Phylosophical Mag. B **65**, 1395 (1992)

GENERATING CHARACTERISTICS OF RESISTANCE IN Bi2223 SUPERCONDUCTING BULK EXPOSED TO EXTERNAL MAGNETIC FIELD

K. Kato[1], H. Shimizu[1], Y. Yokomizu[1], T. Matsumura[1] and N. Murayama[2]

[1]Nagoya University, Furo-cho, Chikusa-ku, Nagoya 464-8603, JAPAN
[2]National Industrial Research Institute of Nagoya, Hirate-cho, Kita-ku, Nagoya 462-0057, JAPAN

Abstract: The generating process of resistance of a Bi2223 superconducting bulk exposed to the external magnetic field in the range of about 80 mT was measured after the superconducting/normal transition due to suddenly supply of ac overcurrent. The dc magnetic fields were applied in the longitudinal and transverse directions to the sample. The resistance of the sample increased by application of the external field. The increment in the resistance for the transverse magnetic field was about twice as large as that for the longitudinal one.

Keywords: Bi2223 bulk, resistance, magnetic field, fault current limiter

INTRODUCTION

Superconducting fault current limiter (SC-FCL) with high T_c superconductor (HTS) is expected to be introduced into electric power system to reduce the short-circuit current [1]-[3]. It is one of the problems that the resistance of HTS is small for the SC-FCL application. In order to increase the resistance of HTS, it may be effective to apply the magnetic field to the HTS element [2][3]. In this paper, the critical current and resistance of a Bi2223 bulk exposed to magnetic field is experimentally investigated. Supposing the application to the SC-FCL, we supplied ac overcurrent of 60 Hz to the sample conductor and measured the transient aspects of normal resistance generating after superconducting/normal (S/N) transition. In the experiments, dc external magnetic fields were externally applied in the longitudinal and transverse direction to the transport current in the sample.

EXPERIMENTAL SET-UP AND PROCEDURE

The Bi2223 bulk sintered at 860℃ for 20 hours under 300 kg/cm^2 pressure was adopted as a sample conductor [4]. The sample has a highly grain-aligned microstructure as shown in Fig.1. The length and cross-sectional area of the sample are 105 mm and 2.25×1.35 mm^2, respectively. The critical current without magnetic field I_{C0} was measured to be 120 A (defined by the criterion of 0.1 mV/cm).

The dc external fields were applied to the sample in the longitudinal and transverse direction to the sample current. Fig.2 (a) and (b) illustrate the experimental set-ups for applying the longitudinal and transverse magnetic field, respectively. In the case of the longitudinal magnetic field, the sample was arranged along the center axis of the magnet coil. To apply the transverse magnetic field, we inserted the sample into the air gap of iron core with a magnetizing winding. We supplied three kinds of ac (60Hz) overcurrent having the peak value of 157, 172 and 189A which are 131, 143 and 158 % of I_{C0} to the sample. The transient

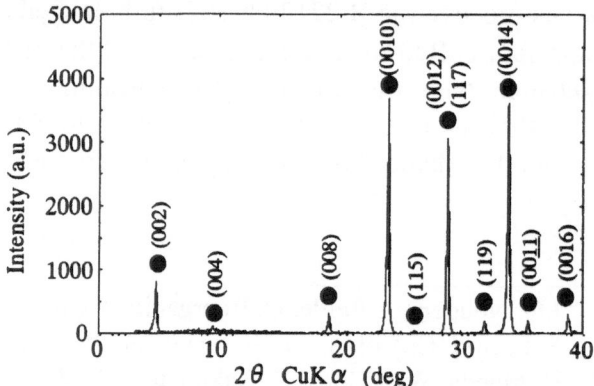

Fig. 1. XRD pattern for the sinter forged sample. X-ray was irradiated to the surface of the sample, which was perpendicular to the pressing direction.

behaviors of voltage across the sample were measured when the overcurrent was suddenly supplied for only a half cycle to the sample exposed to the magnetic flux. The flux density in the longitudinal direction is adjusted to be 0, 41.6 and 85.5 mT. In case of the transverse direction, the flux density of 0, 40.6 and 79.2 mT is applied.

VOLTAGE-CURRENT CHARACTERISTICS AND CRITICAL CURRENT

Fig.3 shows the voltage-current characteristics derived under the conditions of the longitudinal magnetic flux density B_L=85.5mT and transverse one B_T=79.2mT as well as no magnetic field. The voltage-current characteristics are hardly influenced by the peak value of overcurrent. As shown in this figure, the application of the magnetic field degrades the critical current which is the instantaneous value when the voltage begins to appear across the sample. However, the magnetic field dose not change the shape of the voltage-current characteristic curve after S/N transition. The rising rate dv/di is 0.017 Ω/m at the current of more than 170 A.

Fig.4 shows dependence of critical current on flux density. It is found that the critical current is independent of the peak value of the overcurrent. In the case of longitudinal magnetic field, the critical current is almost linearly degraded at a rate of 0.55 A/mT. On the other hand, the critical current is degraded more strongly by the application of the transverse magnetic field than the longitudinal one. For example, the magnetic field of 80 mT reduce the critical current to 76 A in the longitudinal case and 51 A in the transverse one. These values correspond to 64 % and 43 % of I_{C0}.

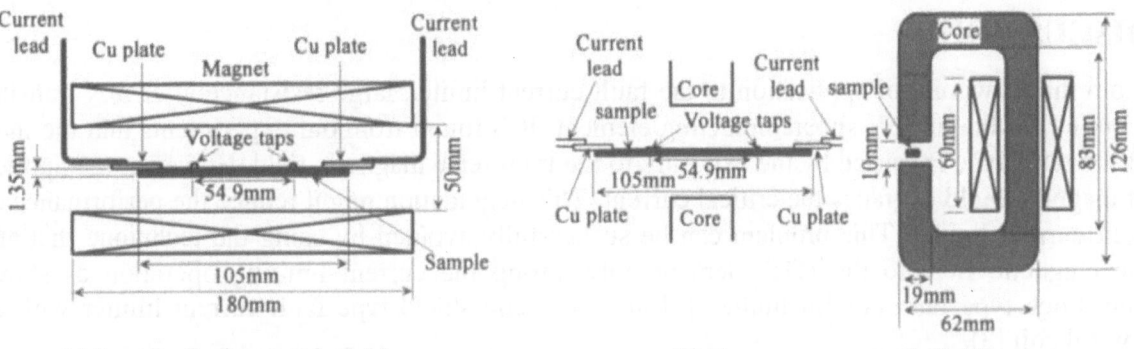

(a) Longitudinal magnetic field (b) Transverse magnetic field

Fig. 2. Experimental set-up.

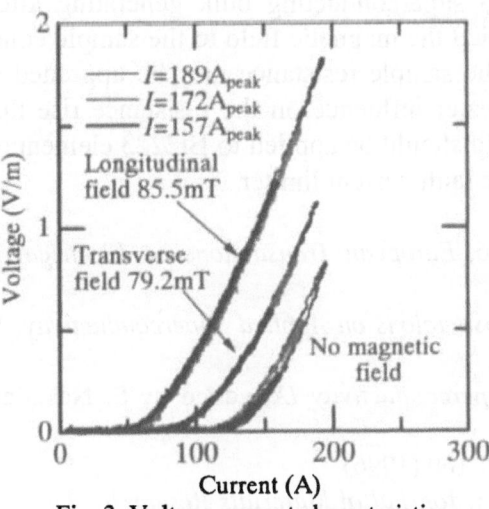

Fig. 3. Voltage-current characteristics.

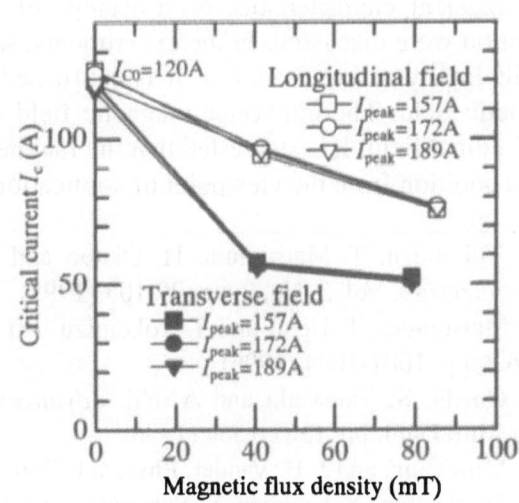

Fig. 4. Dependence of critical current on magnetic flux density.

High T_c superconductor has magnetic anisotropy and the magnetic field along c-axis of HTS is more influential to the superconductivity of Bi2223 than the magnetic field having the other direction [5]. Hence, the critical current decreases to lower level in transverse magnetic field than longitudinal one.

RESISTANCE AFTER S/N TRANSITION

Fig. 5 indicates the sample resistance per unit length as a function of magnetic flux density. The resistance was obtained by dividing the voltage with the current. The resistance increases not only with the current but also with the magnetic field. The resistance for the transverse field is larger than

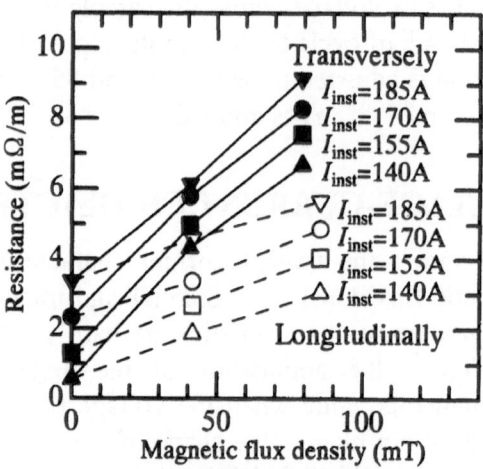

Fig. 5. Resistance as a function of magnetic flux density.
(I_{inst} : Instantaneous value of sample current)

that for longitudinal one. For example, the increments of resistance caused by the application of the magnetic field of 80 mT are measured to be 2.5mΩ/m and 6.0 mΩ/m for the longitudinal and transverse magnetic field, respectively and these values are independent of the magnitude of current. By comparing Fig. 5 with Fig. 4, it could be concluded that the resistance rise due to the magnetic field is brought about in compensation for the degradation of the critical current.

DISCUSSION

From the viewpoint of application to the fault current limiter, large resistance after S/N transition is required for the high T_c superconducting element. It is found from our experiments that the increase of the resistance is caused by the exposure to the transverse magnetic field. However, the application of magnetic field degrades the critical current. This degradation might reduce the performance of the fault current limiter. This problem can be successfully avoided by using the technique that applies the magnetic field to the HTS element only during the current limiting operation as shown in Flux-Lock type fault current limiter [2] and magnetic shield type fault current limiter with active control coil [3].

CONCLUSION

The transient characteristics of resistance of Bi2223 superconducting bulk generating after S/N transition were discussed. In the experiments, we applied the magnetic field to the sample conductor longitudinally and transversely. It is confirmed that the sample resistance may be upgraded by the magnetic field. The transverse magnetic field has greater influence on the resistance rise than the longitudinal field. It is suggested that the magnetic field should be applied to Bi2223 element only at fault condition from the viewpoint of application to the fault current limiter.

1. Y. Yokomizu, T. Matsumura, H. Okubo and Y. Kito, *European Transactions on Electrical Power Engineering*, Vol. 5, No. 2, pp. 99-105 (1995)
2. T. Matsumura, T. Uchii and Y. Yokomizu, *IEEE Transactoions on Applied Superconductivity*, Vol. 7, No.2, pp. 1001-1004 (1996)
3. T. Onishi, S. Yamazaki and A. Nii, *Advances in Superconductivity IX*, edited by S. Nakajima and M. Murakami, pp. 1361-1364 (1996)
4. N. Murayama and J. B. Vander, Physica C **256**, pp. 156-160 (1996)
5. N. Murayama, Y. Kodama. S. Sakaguchi and F. Wakai, *Journal of Materials Research*, Vol. 6, No. 7, pp. 1425-1432 (1991)

Effect of Oxygen Partial Pressure on the Microstructure and Formation of Bi-2223 Phase in the Partial-melting and Sintering Process

Xiaoye Lu[1], Akihiko Nagata[1], Naotoshi Sato[1], Kazuhisa Sugawara[1],
Shin-ichi Kamada[1] and Shuji Hanada[2]

[1]Faculty of Engineering and Resource Science, Akita University, Akita 010-8502 Japan
[2]Institute for Materials Research, Tohoku University, Sendai 980-8577 Japan

Abstract: The effect of oxygen partial pressure on the microstructure and formation of the (Bi,Pb)-2223 phase in the partial-melting and sintering process was studied. Samples with two nominal compositions $Bi_{1.84}Pb_{0.34}Sr_{1.91}Ca_{2.03}Cu_{3.06}O_y$ and $Bi_{1.8}Pb_{0.4}Sr_{1.9}Ca_{2.1}Cu_{3.5}O_y$ were prepared by following sintering conditions after partially melted at 875 ℃ for 1 h in air: (i) 760 ℃ for 120 h in N_2, (ii) 825 ℃ for 120 h in 8 % O_2+N_2, (iii) 840 ℃ for 120 h in air and (iv) 845 ℃ for 120 h in O_2. In the samples sintered in pure N_2 and in pure O_2, the Bi-2212 phase as a major phase exists without the Bi-2223 phase. The Bi-2223 phase appears in the samples sintered in air and in 8 % O_2+N_2, and the highest fraction of the Bi-2223 phase appears in the sample with the composition of $Bi_{1.8}Pb_{0.4}Sr_{1.9}Ca_{2.1}Cu_{3.5}O_y$ sintered in 8 % O_2+N_2.

Keywords: Bi-2223, Microstructure, Oxygen, Partial melting, Crystal growth

INTRODUCTION

Although several procedures have been previously proposed in order to increase the proportion of Bi-2223 phase in bismuth system[1-4], and especially it has been found that the addition of lead oxide (PbO) is successful in producing a high proportion of Bi-2223 phase[4], the synthesis of single-phase Bi-2223 superconductor is still difficult due to a poor understanding of its complex phase diagram, as well as melting and solidification behavior. Furthermore, the synthesis of Bi-2223 phase also critically depends on the atmosphere of heat-treatment[5-8]. A reduced oxygen partial pressure was found to effectively enhance the efficiency of synthesizing Bi-2223 phase[5,6]. However, these data were obtained in the usual solid state sintering method. The effect of the atmosphere of heat-treatment on the microstructure and formation of the Bi-2223 phase in sintering process after being partially melted has not yet been reported. In this study, we report the effect of oxygen partial pressure on the microstructure and formation of the (Bi,Pb)-2223 phase during sintering after partial melting.

EXPERIMENTAL

The precursor powders with two nominal compositions $Bi_{1.84}Pb_{0.34}Sr_{1.91}Ca_{2.03}Cu_{3.06}O_y$ and $Bi_{1.8}Pb_{0.4}Sr_{1.9}Ca_{2.1}Cu_{3.5}O_y$ were prepared. After being calcined at 800 ℃ for 12 hours and pulverized, the precursor powders were cold-pressed into bulk samples with a size of $2mm \times 3mm \times 15mm$. These bulk samples were heat-treated by following sintering conditions after partially melted at 875 ℃ for 1 h in air: (i) 760 ℃ for 120 h in N_2, (ii) 825 ℃ for 120 h in 8 % O_2+N_2, (iii) 840 ℃ for 120 h in air and (iv) 845 ℃ for 120 h in O_2. The microstructure and formation of the Bi-2223 phase were studied by DTA, XRD, and AC susceptibility measurements.

RESULTS AND DISCUSSION

DTA measurements were performed on the $Bi_{1.8}Pb_{0.4}Sr_{1.9}Ca_{2.1}Cu_{3.5}O_y$ precursor powders in various oxygen atmospheres between room temperature and 1000 ℃. In all these heating curves between 780 and 1000 ℃ as shown in Fig. 1, three endothermic peaks are resolved. According to the obervation of high-temperature optical microscope and analyses of XRD and EDS, these three peaks correspond to the formation of the liquid phase (first peak), the melting of (Sr,Ca) CuO_2 (second peak) and the melting of $(Sr,Ca)_2CuO_3$ (third peak). Figure 2 shows the dependence of the temperature (T_m) of the liquid phase formation on the oxygen partial pressure. It is seen that T_m decreases with decreasing oxygen partial pressure, being 865 ℃ in O_2, 860 ℃ in air, 845 ℃ in 8 % O_2+N_2, and 780 ℃ in N_2, respectively.

Figures 3 and 4 show the AC susceptibilty versus temperature plots of $Bi_{1.84}Pb_{0.34}Sr_{1.91}Ca_{2.03}Cu_{3.06}O_y$ and $Bi_{1.8}Pb_{0.4}Sr_{1.9}Ca_{2.1}Cu_{3.5}O_y$ bulk samples sintered at T_m-20 ℃ for 120 h in various oxygen atmospheres after partil melting at 875 ℃ for 1 h in air, respectively. It can be seen that the susceptibility changes at about 80 K due to the Bi-2212 phase in the samples sintered in pure N_2 and in pure O_2 occur almost in one step. On the other hand, for the samples sintered in air and in 8 % O_2+N_2, the changes in susceptibility occur in two steps, i.e. the changes at 108 K due to the Bi-2223 phase and the another changes at about 80 K due to the 2212 phase, are observed.

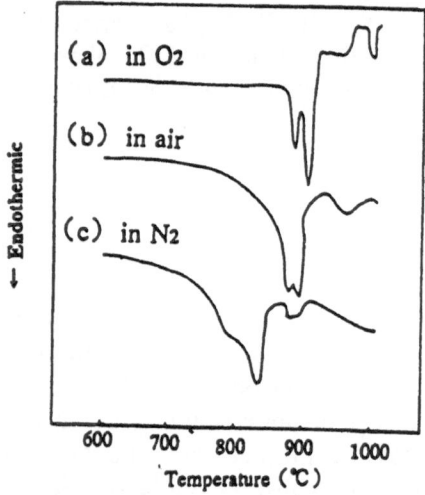

Fig. 1. DTA measurements performed on $Bi_{1.8}$
$Pb_{0.4}Sr_{1.9}Ca_{2.1}Cu_{3.5}O_y$ calcined powders
(a) in O_2, (b) in air, (c) in N_2

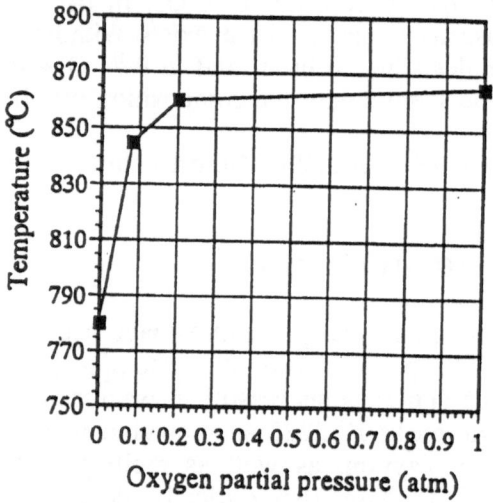

Fig. 2. The dependence of the temperature
of the liquid phase formation on the
oxygen partial pressure

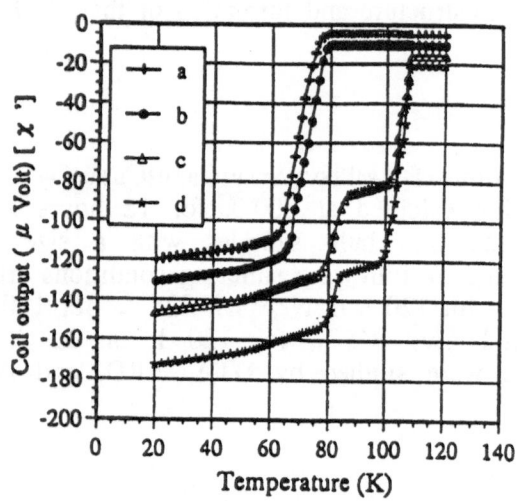

Fig.3.AC susceptibility versus temperature plots of
$Bi_{1.84}Pb_{0.34}Sr_{1.91}Ca_{2.03}Cu_{3.06}O_y$ bulk samples
sintered for 120 h in following conditions:
(a)760 ℃ in N_2, (b)825 ℃ in O_2, (c)
840 ℃ in air, (d)845 ℃ in 8%O_2+N_2

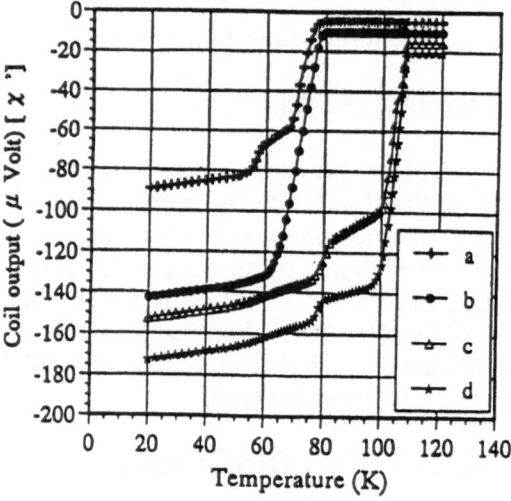

Fig.4. AC susceptibility versus temperature plots
of $Bi_{1.8}Pb_{0.4}Sr_{1.9}Ca_{2.1}Cu_{3.5}O_y$ bulk samples
sintered for 120 h in following conditions:
(a)760 ℃ in N_2, (b)825 ℃ in O_2, (c)
840 ℃ in air, (d)845 ℃ in 8%O_2+N_2

Figures 5 and 6 show the XRD patterns for $Bi_{1.84}Pb_{0.34}Sr_{1.91}Ca_{2.03}Cu_{3.06}O_y$ and $Bi_{1.8}Pb_{0.4}Sr_{1.9}Ca_{2.1}Cu_{3.5}O_y$ bulk samples sintered at T_m-20 ℃ for 120 h in various oxygen atmospheres after partil melting at 875 ℃ for 1 h in air, respectively. In the samples sintered in pure N_2 and in pure O_2, the Bi-2212 phase as a major phase exists without the Bi-2223 phase. The Bi-2223 phase appears in the samples sintered in air and in 8 % O_2+N_2, and the highest fraction of the Bi-2223 phase appears in the sample with the composition of $Bi_{1.8}Pb_{0.4}Sr_{1.9}Ca_{2.1}Cu_{3.5}O_y$ sintered in 8 % O_2+N_2.

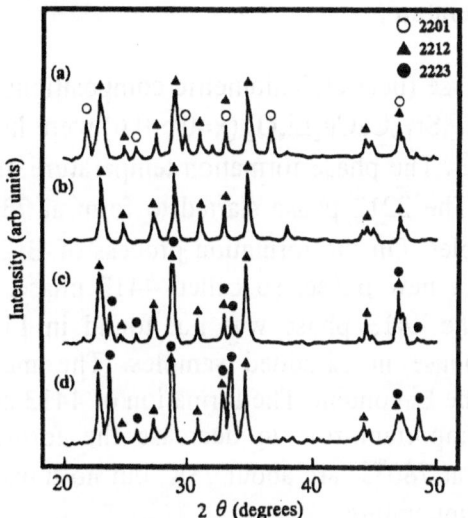

Fig. 5. XRD patterns for $Bi_{1.84}Pb_{0.34}Sr_{1.91}Ca_{2.03}Cu_{3.06}O_y$ bulk samples sintered for 120 h in following conditions:
(a)760 ℃ in N_2, (b)825 ℃ in O_2, (c) 840 ℃ in air, (d)845 ℃ in 8%O_2+N_2

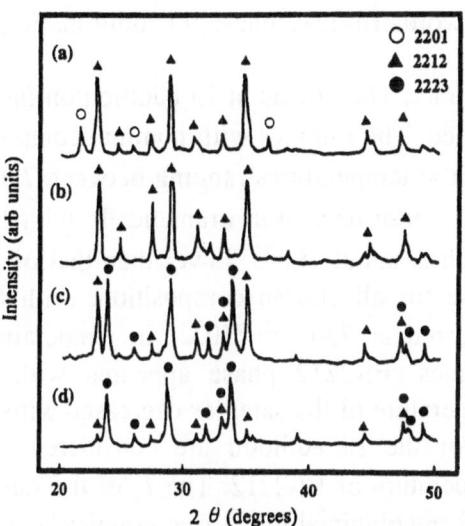

Fig. 6. XRD patterns for $Bi_{1.8}Pb_{0.4}Sr_{1.9}Ca_{2.1}Cu_{3.5}O_y$ bulk samples sintered for 120 h in following conditions:
(a)760 ℃ in N_2, (b)825 ℃ in O_2, (c) 840 ℃ in air, (d)845 ℃ in 8%O_2+N_2

CONCLUSIONS

The effect of oxygen partial pressure on the microstructure and formation of the (Bi,Pb)-2223 phase in the partial-melting and sintering process was studied. Samples with two nominal compositions $Bi_{1.84}Pb_{0.34}Sr_{1.91}Ca_{2.03}Cu_{3.06}O_y$ and $Bi_{1.8}Pb_{0.4}Sr_{1.9}Ca_{2.1}Cu_{3.5}O_y$ were prepared by following sintering conditions after partially melted at 875 ℃ for 1 h in air: (ⅰ) 760 ℃ for 120 h in N_2, (ⅱ) 825 ℃ for 120 h in 8 % O_2+N_2, (ⅲ) 840 ℃ for 120 h in air and (ⅳ) 845 ℃ for 120 h in O_2. In the samples sintered in pure N_2 and in pure O_2, the Bi-2212 phase as a major phase exists without the Bi-2223 phase. The Bi-2223 phase appears in the samples sintered in air and in 8 % O_2+N_2, and the highest fraction of the Bi-2223 phase appears in the sample with the composition of $Bi_{1.8}Pb_{0.4}Sr_{1.9}Ca_{2.1}Cu_{3.5}O_y$ sintered in 8 % O_2+N_2.

[1] D. Pandey, A. K. Singh, P. K. Srivastava, A. P. Singh, S. S. R. Inbanathan and G. Singh, Physica C **241**, 279 (1995).
[2] J. C. Toledano, D. Morin, J. Schneck, H. Faqir, O. Monnereau, G. Vacquier, P. Strobel and V. Barnole,(1995) Physica C **253**, 53 (1995).
[3] P. V. P. S. S. Sastry, J. V. Yakhmi and R. M. Iyer, Physica C **161**, 656 (1989).
[4] M. Takano, J. Takada, K. Oda, H. Kitaguchi, Y. Miura, Y. Ikeda, Y. Tomii and H. Mazaki, Japan.J.Appl.Phys. **27**, L1041 (1988).
[5] U. Endo, S. Koyama and T. Kawai, Jpn. J. Appl. Phys. **27**, L1476 (1988).
[6] K. Aoto, H. Hattori, T. Hatano, K. Nakamura and K. Ogawa, Jpn. J.Appl.Phys. **28**, L2196 (1989).
[7] Y. L. Chen and R. Stevens, J. Am. Ceram. Soc. **75**, 1160 (1992).
[8] W. Zhu and P. S. Nicholson, J. Appl. Phys. **73**, 8423 (1993).

LOW TEMPERATURE SYNTHESIS OF Bi-2212 PHASE BY Li ADDITION

I. Chong, H. Kitaguchi, and K. Togano
National Research Institute for Metals, 1-2-1, Sengen, Tsukuba, Ibaraki 305-0047, Japan

Keywords:Bi-2212 phase, Li addition, Formation temperature,

Abstract: The effects of Li addition on the Bi-2212 Phase (near stoichiometric compositions) are studied. The samples with nominal compositions of $Bi_{2.1}Sr_{1.9}CaCu_2Li_xO_z$ (x=0~0.6) were heated in air at temperatures ranging between 720℃ to 780℃. The phase formation temperature of Bi-2212 phase decreased dramatically with Li addition. The 2212 phase started to form at 730℃, which is about 100℃ lower than that of Li free samples. On the formation process of Bi-2212 phase for all chosen compositions with Li addition, a new phase, so called 4413 phase, was appeared at 730℃~750℃ in temperature range. The 4413 phase was not found in Li-free samples. Bi-2212 phase appeared with the 4413 phase in Li-added samples. The melting temperature of the samples decreased with increasing the Li content. The formation of 4413 phase and/or the Li addition are considered to play an important role to decrease the formation temperature of Bi-2212. The T_c of the samples heated at 780℃ are about 75K, but no transition has been obtained for Li-free samples heated at same temperature.

INTRODUCTION

It has been revealed that the addition or substitution of some elements, Pb, Ba, Li to Bi-Sr-Ca-Cu-O system enhances Bi-2212 phase formation and critical temperature (T_c) becomes higher than non added or non substituted phase [1-3]. Horiuchi et al. reported Li substitution to Cu site ($Bi_{2.2}Sr_{1.8}CaCu_{2.15-x}Li_xO_z$, x=0-0.7). They reported that the Li addition or substitution promotes the formation of Bi2212 phase and T_c becomes higher than that of non substituted phase and the substituted Li occupied Cu site from result of X-ray diffraction analysis [4]. They pointed out that the added Li has driven Cu out of Bi-2212 phase. Fujiwara et. al reported the Li addition to Bi-2212 ($Bi_2Sr_{1.5}Ca_{1.5}Cu_2Li_xO_y$ x=0~0.6). And they found out a new oxide phase containing Li (4413 phase) at 710℃~730℃ temperature range and the lattice parameter c increases with Li addition to Bi-2212 phase[5]. They pointed out that the new phase plays a role in promotion of formation of Bi-2212 phase. Because they studied the effect of Li addition to only one composition ($Bi_2Sr_{1.5}Ca_{1.5}Cu_2Li_xO_y$, x=0~0.6) it is not clear the relationship with the the 4413 phase and Li addition to the other composition. The 4413 phase was not found in Li substituted Bi-2212 phase ($Bi_{2.2}Sr_{1.8}CaCu_{2.15-x}Li_xO_z$, x=0-0.7) by Horiuch et al. [3]. Present study aims to clarify the relationship with the 4413 phase and Li addition for some Bi-2212 compositions, $Bi_{2.1}Sr_{1.9}CaCu_2Li_xO_y$ (x=0, 0.2, 0.4, 0.6). We report the effects of Li addition on Bi-2212 phase formation and the 4413 phase.

EXPERIMENTAL

Starting compositions were $Bi_{2.1}Sr_{1.9}CaCu_2Li_xO_z$ (x=0, 0.2, 0.4, 0.6). Starting materials Bi_2O_3, $SrCO_3$, $CaCO_3$, Li_2CO_3 and CuO were weighed at an appropriate ratio, mixed in an agate mortar and pressed into pellets, and heated in air at 700-780℃ for 12 hours. All the products were examined by powder X-ray diffraction (XRD) using Cu-Kα radiation and scanning electron microscopy (SEM). DTA were performed to determine melting temperature. AC magnetic susceptibility was measured by using an AC induction magnetometer for T_c determination.

RESULTS AND DISCUSSION

Fig. 1 shows the XRD patterns of $Bi_{2.1}Sr_{1.9}CaCu_2Li_xO_y$ (x=0~0.6) compositions samples heated at 740℃ for 12 hours. The Li-free sample was composed of Bi-2201and CuO. Bi-2212 phase was not observed. Li-added samples were composed of Bi-2201, Bi-2212, Bi-4413 (Bi-2212+Bi-2201) and CuO. The intensity of 4413 phase peak became strong with increasing Li content and the intensity of 2201 phase peaks decreased with increasing Li content. The peaks of 4413 phase were observed in the samples heated at 730℃ with Bi-2212 phase also, but the peaks were weaker than those of samples heated at 740℃ and the 4413 peaks were not found below 730℃.

The XRD patterns of $Bi_{2.1}Sr_{1.9}CaCu_2Li_xO_y$ (x=0~0.6) samples heated at 750℃ for 12 hours are shown in Fig.2. The intensity of 4413 peaks became weaker than those of samples heated 740℃. In the sample with x=0.6 the 4413 phase is almost disappeared and the intensity of Bi-2212, 2201 phase peaks became strong. The 4413 phase seems to have a relationship with decreasing phase formation temperature.

The XRD patterns of $Bi_{2.1}Sr_{1.9}CaCu_2Li_xO_y$ (x=0~0.6) compositions samples heated at 780℃ for 12 hours are shown in Fig. 3. Bi-2212 phase in Li-free sample started to form at this temperature. In Li added compositions, amount of the Bi-2201 phase increased again at this temperature though the peaks of Bi-2212 phases became sharper with increasing crystallinity, and the 4413 phase was not observed.

The DTA curves of powders with norminal compositions $Bi_{2.1}Sr_{1.9}CaCu_2Li_xO_y$ (x=0~0.6)

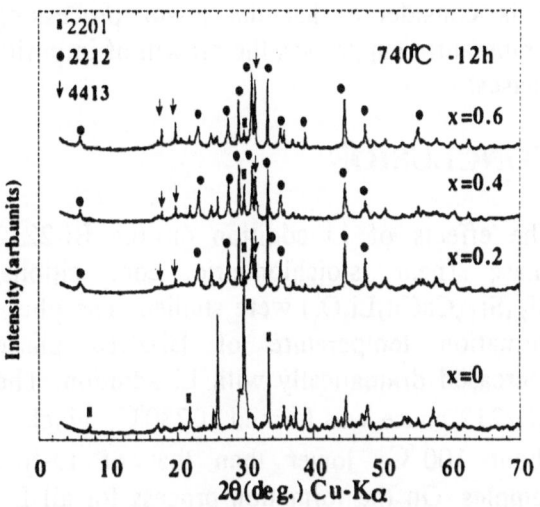

Fig.1. XRD patterns of the samples with $Bi_{2.2}Sr_{1.8}CaCu_2Li_xO_y$ compositions heated at 740 ℃

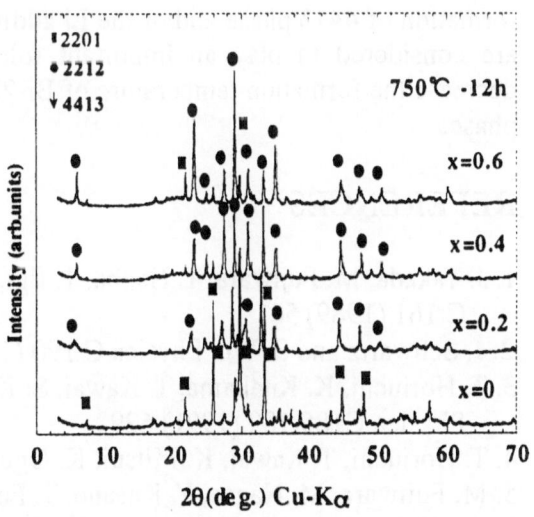

Fig.2. XRD patterns of the samples with $Bi_{2.2}Sr_{1.8}CaCu_2Li_xO_y$ compositions heated at 750 ℃

calcined at 780℃ are shown in Fig. 4. Li-added sample started to melt at 750℃ while Li-free sample started to melt above 850℃. The melting temperature decreased with increasing Li content.

The T_c of all Li-added samples were about 75K. The values of T_c are not so high because the heating temperature is so low but it is so noteworthy that the superconducting phase can be prepared at relative low temperature (750℃). It suggests the possibility to decrease processing temperature of Bi-2212 conductors. It is considered that the lower processing temperature suppresses the growth of impurity phases.

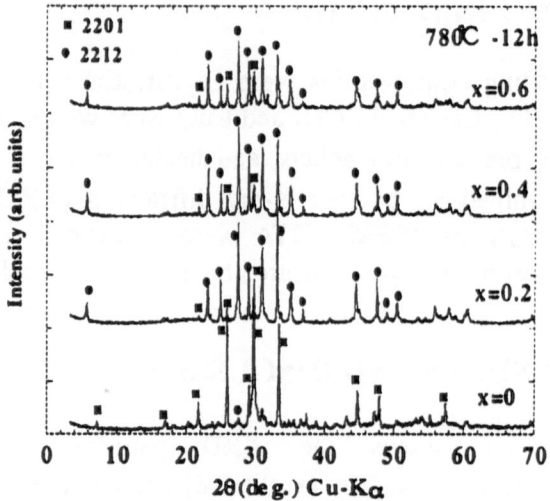

Fig.3. XRD patterns of the samples with $Bi_{2.2}Sr_{1.8}CaCu_2Li_xO_y$ compositions heated at 780℃.

CONCLUSION

The effects of Li addition on the Bi-2212 phase (near stoichiometric compositions, $Bi_{2.1}Sr_{1.9}CaCu_2Li_xO_y$) were studied. The phase formation temperature of Bi-2212 phase decreased dramatically with Li addition. The Bi-2212 phase was formed at 730℃, which is about 100 ℃ lower than that of Li-free samples. On the formation process for all Li-added compositions, an intermediate phase, so called 4413 phase, was appeared with Bi2212 phase at 730℃～750℃ temperature range, but the 4413 phase was not found in Li-free samples. The melting point of the samples decreased according to the Li content. The formation of 4413 phase and/or the Li addition are considered to play an important role to decrease the formation temperature of Bi-2212 phase.

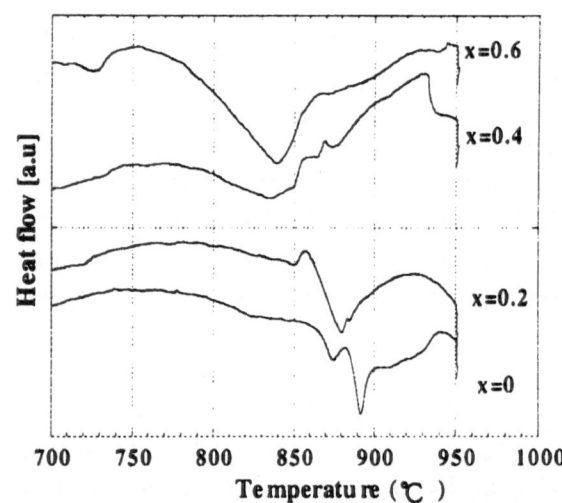

Fig.4. DTA measurements of powders calcined at 780℃ with $Bi_{2.1}Sr_{1.9}CaCu_2Li_xO$ compositions.

REFERENCES

1. J. Takada, M. Fujiwara, T. Nanba, T. Egi, Y. Ikeda, Z Hiroi, M. Takano, and H. Mazaki, Physica C 161 (1989) 561
2. J. Schwartz and S. Wu, Physica C 190 (1991) 169
3. T. Horiuchi, K. Kitahama, T. Kawai, S. Kawai, S. Hontsu, K. Ogura, I. Shigaki, and Y. Kawate, Physica C185-189 (1991) 629
4. T. Horiuchi, T. Kawai, K. Mitsui, K. Ogura, and S. Kawai, Physica C 168 (1990) 309-314
5. M. Fujiwara, M. Nagae, Y. Kusano, T. Fujii, and J. Takada, Physica C (1997) 317

THE EVALUATION OF OPTIMAL SINTERING TIME BY THE SRPM METHOD FOR OBTAINING LARGE Jc OF Bi-2223 BULK

Toru Kodaira[1]*, Hiroshi Nakane[1], Sadao Yamazaki[1],
SyujiYoshizawa[2], Isamu Tezuka[2], Syuetu Haseyama[3]

[1]Department of Electrical Engineering, Kogakuin University, Shinjuku, Tokyo 163-8677, Japan
[2]Advanced Materials Research and Development Center, Department of Chem., Meisei University, Hino, Tokyo 191-8506, Japan
[3]DOWA Mining Co. Ltd., Hachioji, Tokyo 192, Japan

Abstract: To obtain the large value of Jc, it is important to find the optimal sintering time of Bi-2223 bulk. Then, to analyze the electromagnetic properties of Bi-2223 sintered bulk, pellet samples were made by varying the sintering time and the number of Cold Isostatic Press (CIP). The properties of their samples were mainly measured by the SRPM method which could evaluate the values related to the surface impedance of samples within the frequency band of kHz from MHz. In this paper, the temperature dependence of the real part of the difference in the coil impedance was especially focused. The best conditions of the sintering time and number of CIP were investigated by SRPM method, and the results were reported.

Keywords: SRPM method, Bifilar-coil, Bi-2223, sintering time, Jc

INTRODUCTION

The SRPM method is the abbreviation of the $\underline{S}$imultaneous $\underline{R}$esistivity (ρ) and magnetic $\underline{P}$enetration depth (λ) $\underline{M}$easuring method. In the SRPM method for the pellet sample, the double coil (bifiler winding coil), which are wound at the same position, is placed on one side of the sample plate. The difference in the synthetic impedance (ΔZ) of double coils is measured when the

Table 1. Pellet Samples

Item	Sintering Time [hours]	CIP	Jc [A/cm²]
Sample 1	10	0	34
Sample 2	20	0	112
Sample 3	40	0	195
Sample 4	80	0	289
Sample 5	150	0	323
Sample 6	50	0	321
Sample 7	100	1	1,897
Sample 8	150	2	2,782

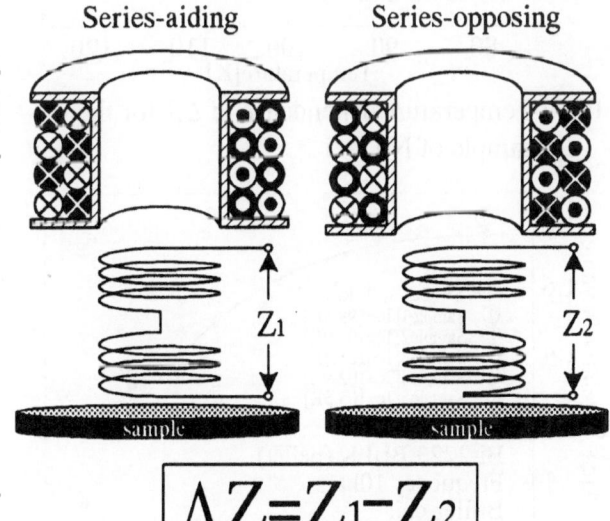

$$\Delta Z = Z_1 - Z_2$$

Fig.1 Schematic diagram of Bifiler-coil (SRPM method.)

current of the double coils flows in the same and opposite directions. The difference (ΔZ) is separated into the real part ΔR and the imaginary part ΔX. Bi-based sintered pellet samples were made under various conditions. The relations among the sintering time, the iteration of Cold Isostatic Press (CIP), the value of Jc at 77.3 K and the temperature dependence of ΔR were analyzed.

EXPERIMENT

The eight Bi-2223 sintered pellets were prepared for these measurements. The size of the pellet samples were 13 mm in diameter and 1mm in thickness. The samples were sintered at 850 ℃ in air. The sintering time and number of CIP are shown in Table 1. The sample of No.1 ~ 5 and No.6 ~ 8 were made in different lots. The ΔZ in Fig.1 for each samples was measured at the frequency of 10 kHz and in the temperature range of 77.3 ~ 120 K with the LCR meter. To measure the value of Jc, after the measurement of the SRPM method, the samples were cut out and shaped into square rods of 0.5 mm $\times$ 0.5 mm. The value of Jc at 77.3 K were estimated from the measured current gradually increasing the supplied current and 0.5 mV was detected between the two voltage terminals of 8 mm in spacing.

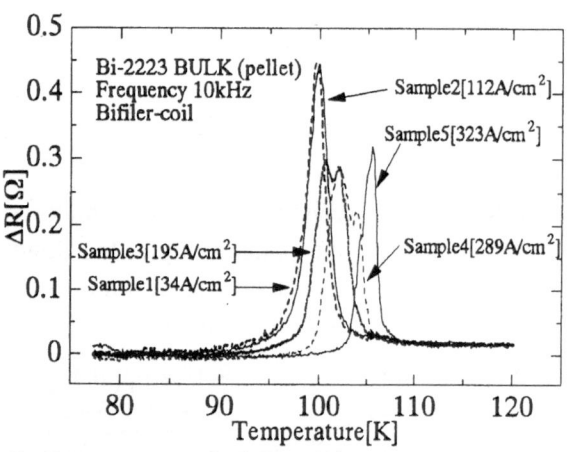

Fig.2 Temperature dependence of ΔR for the sample of No.1~5

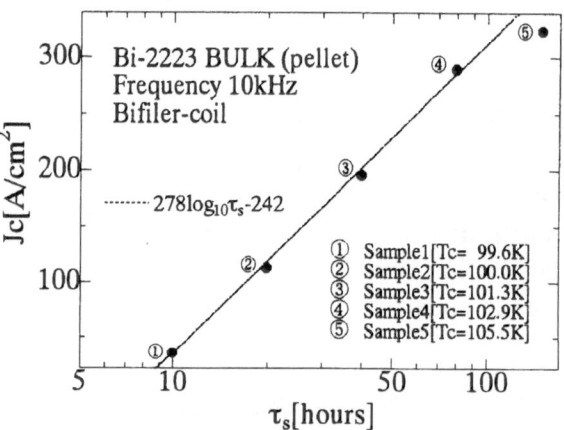

Fig.3 Jc - τs for the sample of No.1~5

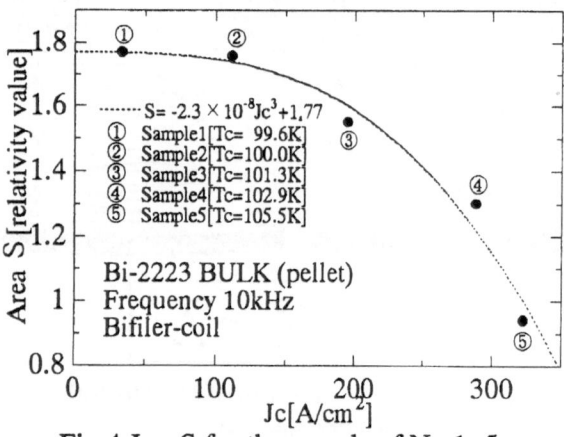

Fig.4 Jc - S for the sample of No.1~5

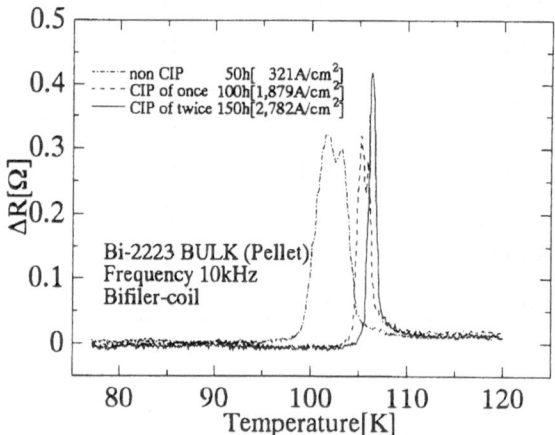

Fig.5 Temperature dependence of ΔR for the sample of No.6~8

RESULTS AND DISCUSION

1. The temperature dependence of ΔR for the samples with various sintering time.

For investigating relations among the sintering time, Jc and temperature dependence of ΔR, the temperature dependence of ΔZ was measured for the samples of No.1 ~ 5. The measuring result of ΔR is shown in Fig.2. In this figure, the temperature having the deviation of ΔR from $\Delta R=0$ at low temperature side shows Tc (offset) and the same temperature at high temperature side is Tc (onset). The presence of 2223 phase and grain junction of 2223 phase were observed in these properties. As the reason, a little quantity of 2212phase was observed by x-ray diffraction but not by the SRPM method is perhaps that the quantity was too little,

As shown in Fig.3, when the sintering time for the pellet samples was under 80 hours, the relation between the sintering time τs and Jc is obtained as follows ;

$$Jc = 278\log_{10}\tau s - 242$$

When τs is over 80 hours, Jc is rapidly saturated.

2. The relation between Jc and area of ΔR.

Fig.3 shows the relationship between Jc and S which is obtained by integrating ΔR on the temperature axis. The value of Jc increases as S becomes smaller. The value of S is proportional to the loss due to jule heating and magnetic hysteresis loss. The relation between Jc and S is given as follows ;

$$S = -2.3 \times 10^{-8} \times Jc^{3} + 1.77$$

It means that the sharpness of the curve has strong dependence on Jc.

3. The influence of the CIP on Jc.

To analyze the effect of the iteration of CIP on Jc by using the samples of No.6 ~ 8, the temperature dependence of ΔZ was measured. The result is shown in Fig.5. By iterating the CIP, the value of Jc is larger. This result says that the number of the iteration of CIP becomes the very important factor to create the superconductor with Bi-2223 phase.

CONCLUSION

Bi-based pellet samples were made by varying the sintering time and the number of iteration of CIP. The value of Jc and the temperature dependence of ΔR (surface impedance) by the SRPM method were measured and the relations between the measured values were analyzed. As for the Bi-2223 phase sample without CIP, obviously Jc is proportional to the logarithm of the length of the sintering time until it reaches the saturation point of 80 hours. For the sample without CIP, the integration of ΔR on temperature axis becomes smaller as the increase of Jc. It was confirmed that the SRPM method is an effective means to detect Bi-2212, 2223 phases and grain junction of phase which were not possible by observing the X-ray diffraction pattern.

We intend to further analyze the relations between the length of sintering time and the number of CIP using the SRPM method to obtain the best conditions to make samples with a lager Jc.

REFERENCES
[1] Nakane H, et al., European Conference on Applied Superconductivity (EUCUS'99) p76 4-7
[2] Nakane H, et al., Advances in Superconductivity X(Proceeding of ISS'97), (1997) 677-680
[3] Haseyama S, et al., The Japan Institute of Metals, Vol.61, No.9, (1997) 900-905

Growth of $Bi_2Sr_2Ca_{n-1}Cu_nO_y$ Superconducting Whiskers and Their Characterization

N.Sato[*], S.Kishida[*], T.Hirao[*], S.-J.Kim[**], T.Yamashita[**], S.F.W.R.Rycroft[***] and W.Y.Liang[***]

[*]Tottori University, 4-101 Koyama-Minami, Tottori 680, Japan
[**]Tohoku University Research Institute of Electrical Communication, Tohoku University, 2-1-1 Katahira, Aoba-ku, Sendai 980-8577, Japan
[***]IRC in Superconductivity, University of Cambridge, Madingley Road, Cambridge, CB3 OHE, UK

Abstract: We prepared $Bi_2Sr_2Ca_{n-1}Cu_nO_y$ (Bi-based) whiskers and investigated their characteristics. Bi-based whiskers with a length of 8mm were grown at the optimal growth temperature, 880°C for 240h. While the resistance-temperature characteristics showed that whiskers had mostly 2223 phase, X-ray diffraction patterns and susceptibility-temperature characteristics indicated that Bi-based whiskers were mostly Bi-2212 phase.

Keywords: Bi-based whisker, glassy plate, growth rate, XRD, R-T, DC SQUID

INTRODUCTION

In order to fabricate $Bi_2Sr_2Ca_{n-1}Cu_nO_y$ (Bi-based) high-Tc superconducting devices, it is necessary to prepare high quality superconductors. Bi-based single crystals were known to be grown by a self-flux method [1], a vertical bridgeman method [2], a KCl flux method [3] and TCFZ method [4]. However, it is still difficult to obtain high quality single crystal. As in general whiskers are expected to have a good crystallinity, we tried to make Bi-based whiskers with high quality. In this paper, we report the preparation technique of the Bi-based whiskers from quenched glassy plates and their characteristics.

EXPERIMENTAL

Bi_2O_3, $SrCO_3$, $CaCO_3$, CuO and Al_2O_3 powders were used as raw materials. The starting material was prepared by mixing powders with the composition of Bi:Sr:Ca:Cu:Al =1:1:1:2:0.75. Then, it was transferred into an alumina crucible and then, heated at 1100°C for 0.5h in air. Finally, it was poured onto a polished iron plate and pressed quickly with the other polished iron plate in order to prepare a glassy plate with high internal stress. The glassy plate was heated at the temperature of 880°C in flowing oxygen of 10-80ml/min for 48-240h. Then, the glassy plate where Bi-based whiskers were grown, was cooled to room temperature in the furnace.

Superconducting property was investigated by resistance-temperature (R-T) and susceptibility-temperature measurements. Phases of whiskers were determined mainly by X-ray diffractometer with Cu-Kα radiation. Scanning electron microscopy (SEM) was also carried out in order to observe the surface morphology.

RESULTS AND DISCUSSION

Figure 1 shows optical photograph of a glassy plate on which Bi-based whiskers are grown. The plate was heated at the temperature of 880°C for 240h in flowing oxygen of 80ml/min. The glassy plate was bent, since it was partially melted during the heating. In the figure, most of whiskers grew vertically on the both surfaces of the glassy plate.

Figure 2 shows SEM image of whiskers grown at the same growth condition. Shape of the whiskers were thin and plate-like, The size of whiskers changed from whisker to whisker. This is caused by the existence of inhomogeneous places on the glassy plate. This factor may affect the supplement of materials necessary for the growth of whisker.

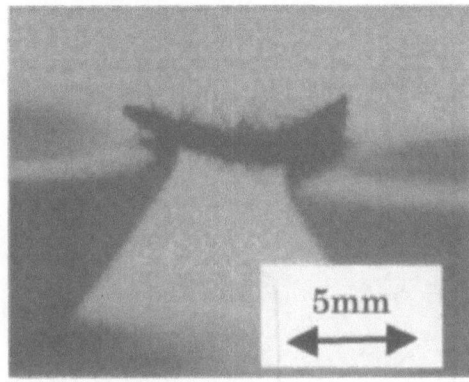

Fig.1 Optical photograph of a glassy plate after heating, which is supported by an alumina boat.

Fig.2 SEM image of Bi-based whiskers.

Figure 3 shows maximum growth rate of Bi-based whiskers as a function of growth temperature. We defined maximum growth rate as growth rate for a longest grown whisker. As shown in the figure, whiskers grew in the temperature range from 830 to 880°C and did not grow at 885°C, due to the melting of the glassy plates. The maximum growth rate was obtained at 880°C.

Figure 4 shows X-ray diffraction pattern of a Bi-based whisker. XRD peaks were well assigned to those of c-axis oriented 2212 superconductor, whereas XRD peaks due to (2223) phase were not observed. Therefore, this result indicated that the whisker was c-axis oriented and (2212) single phase. The rocking carve of (008) was also measured, and the FWHM value of the (008) peaks was about 0.08°. Since this value was limited to the resolution of our XRD instruments, it would be smaller.

Figure 5 shows R-T characteristics of a Bi-based whisker. As shown in the figure, transitions corresponding both to (2223) and (2212) phases were observed. The transition due to (2223) phase was sharp, and the transition temperature width was about 3K. The transition due to a (2212) phase was not clear. The ratio of (2212) to (2223) phases was about 0.01, when estimated from the transition height. The susceptibility-temperature characteristics measured by using a DC SQUID, which is more effective to estimate the superconductive volume, however, indicated that the (2212) phase was dominant in the whisker. This means that (2223) phase, small in amount, mainly exists in the current pass.

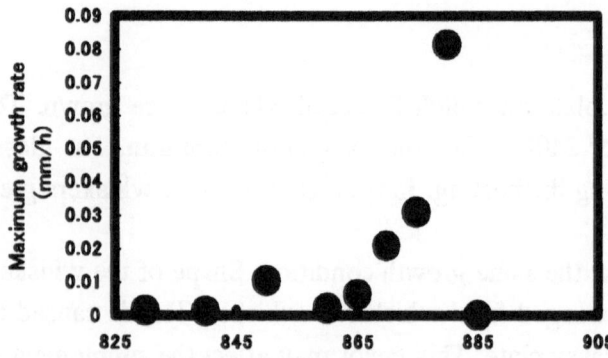

Fig.3 Maximum growth rate of the Bi-based whisker as a function of growth temperature.

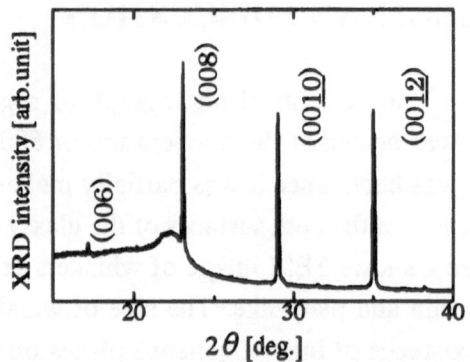

Fig.4 X-ray diffraction pattern of a Bi-based Whisker.

CONCLUSIONS

We prepared Bi-based whiskers from quenched grassy plates, and investigated their characteristics. Bi-based whisker with the length of about 8mm was grown at the temperature of 880 ℃ , which gave the maximum growth rate. From the results of XRD, R-T and susceptibility-temperature characteristics, we found that the whiskers were mostly 2212 phase with a small amount of 2223 phase.

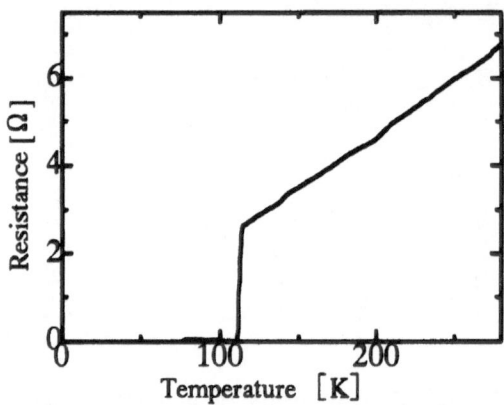

Fig.5 R-T characteristics of a Bi-based whisker.

Acknowledgment: This research is supported partly by the Electric Technology Research Foundation of Chugoku.

1. S. Kishida, S. Nakanishi, H.Tokutaka, K.Fujimura, Phys. Stat. Sol. A 151 (1995) 17.

2. S. Kishida, E. Hosokawa, J. Crystal Growth 192 (1998) 136.

3. A. Katsui, Jpn. J. Appl. Phys. 27 (1988) L844.

4. S.Takekawa, H. Nozaki, A. Umezono, K. Kosuda, M. Kobayashi, J. Crystal Growth 92 (1988) 687.

Growth of $Bi_2Sr_2CaCu_2O_y$ single crystals by the vertical Bridgman method and its modified methods

M.Nakamura[1], S.Kishida[1], H.Tanaka[1], and W.Y.Liang[2]

[1] Department of Electrical and Electronic Engineering, Tottori University, 4-101 Koyama-Minami, Tottori 680-8552, Japan

[2] IRC in Superconductivity, University of Cambridge, Madingley Road, Cambridge, CB3 OHE, UK

Abstract: We investigated annealing characteristics of as-grown $Bi_2Sr_2CaCu_2O_y$ (Bi-2212) single crystals, made by the vertical Bridgman (VB) method. We found that those as-grown crystals high quality since superconducting transition was very sharp after annealing at around 500℃. We also examined a modified VB method with rotating mechanism, and obtained plate-like single crystals of Bi-2212 single-phase and 9×3 mm^2 in size.

Keywords: Bi-2212 single crystal, vertical Bridgman method, anneal

INTRODUCTION

In order to understand physical properties and crystal structure exactly, it is necessary to obtain large size and high quality $Bi_2Sr_2CaCu_2O_y$ (Bi-2212) single crystals. It is difficult to prepare them, because the substitution of elements and the mixture of superconducting phase occur in $Bi_2Sr_2Ca_nCu_nO_y$ superconductors. Since a vertical Bridgman (VB) method is able to control the production of seed crystals, the large size Bi-2212 single crystals can be easily obtained by it [1-4]. However, the optimum growth condition for obtaining a high quality Bi-2212 single crystal is not clarified. In this study, we prepared Bi-2212 single crystals by the VB method with/without rotating a crucible during crystal growth.

EXPERIMENTAL

The starting material with the composition of Bi:Sr:Ca:Cu=2:2:1:2 was prepared by mixing and grinding powders Bi_2O_3(99.0%), $SrCO_3$(95.0%), $CaCO_3$(99.0%) and CuO(95.0%). It was pressed and calcined at 740℃ for 15 hours in order to suppress spouting out from an alumina crucible. Then, it was again ground and transferred into an alumina crucible designed by us. Figure 1 shows a schematic illustration of the equipment of the VB method. Crystal growth was carried out by pulling down the crucible at a speed of 1 mm/h in air, after it was completely melted at 985℃ in an electric furnace. The crucible could be rotated during crystal growth at 100 to 200 rpm as a modification. When the temperature of the top of the alumina crucible reached to the temperature of about 830℃, pulling the crucible down was stopped and then, it was cooled in the furnace. The single crystals formed were mechanically picked up from the broken crucible. In order to investigate the crystal structure and superconductivity, we observed the X-ray diffraction (XRD) patterns, the resistance-temperature (R-T) and AC susceptibility (ACS) characteristics.

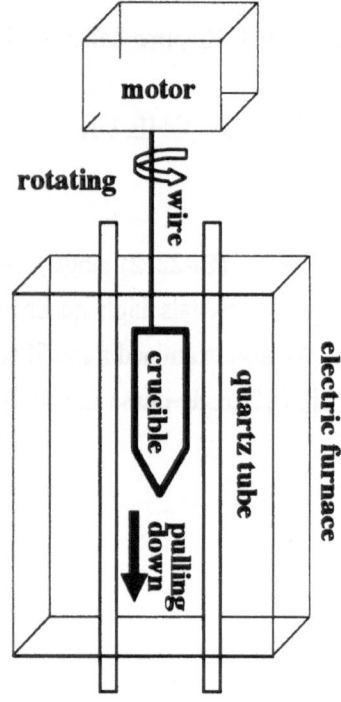

Fig.1 Crystal growth by a vertical Bridgman method.

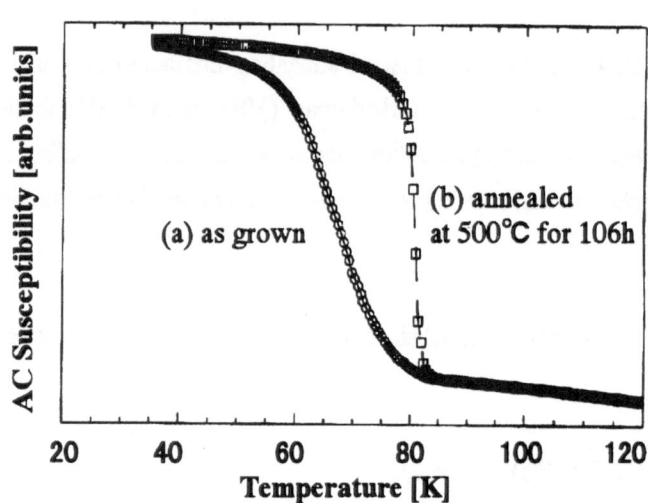

Fig.2 ACS characteristics of (a) an as grown and (b) an annealed Bi-2212 single crystals.

RESULTS AND DISCUSSION

Figure 2 shows ACS characteristics of (a) an as-grown Bi-2212 singe crystal and (b) the single crystal annealed at 500℃ in flowing oxygen gas. The plate-like Bi-2212 single crystals with a size of $25 \times 5mm^2$ were grown by the VB method without rotating the crucible [5]. However, as shown in Fig.2 (a), the $T_{C,ON}$, defined as the temperature at the 10% ACS transition, was about 80K. The transition temperature width ΔT, defined as the difference in temperature between 10% and 90% ACS transitions, was about 20K. The as-grown single crystal was then annealed in flowing oxygen gas at 500℃ for 106h in order to improve the superconductivity of the as-grown single crystal. As shown Fig.2 (b), the ΔT of the annealed single crystal decreased from 20K to 3K. This indicated that the superconductivity of the as-grown single crystals was improved by annealing at the temperature of 500℃. This may be caused by enriched and homogeneously distributed oxygens in the single crystal which is introduced during annealing in flowing oxygen gas. Therefore, the single crystal grown by the VB method is thought to be a high quality.

Figure 3 shows the ΔT of Bi-2212 single crystals as a function of annealing temperature. The ΔT of as-grown single crystal, corresponding to that at 0℃, was about 20K. With increasing annealing temperature, the ΔT decreased. The minimum ΔT was about 3K and obtained at 500℃. This indicates that high quality single crystal is able to be obtain by annealing as-grown single crystal. The annealing will produce the single crystals with homogenous distribution of oxygen in them. Therefore, we believe that as-grown single crystals have good crystallinity. The ΔT increase at the temperatures of more than 600℃ may be due to the compositional deviation from the ideal one of the Bi-2212 phase, which is caused by high temperatures of annealing.

Figure 4 shows the XRD patterns of Bi-2212 single crystals grown by the VB method (a) without and (b) with rotating mechanism. The single crystals grown by the modified (rotating crucible) VB in Fig.4(b) were plate-like and their typical size was about $9 \times 3 mm^2$. The single crystal showed a Bi-2212 single-phase as well as that grown by the VB method in Fig.4(a). From the results, we found that the Bi-2212 single crystals were also grown by using this modified VB method.

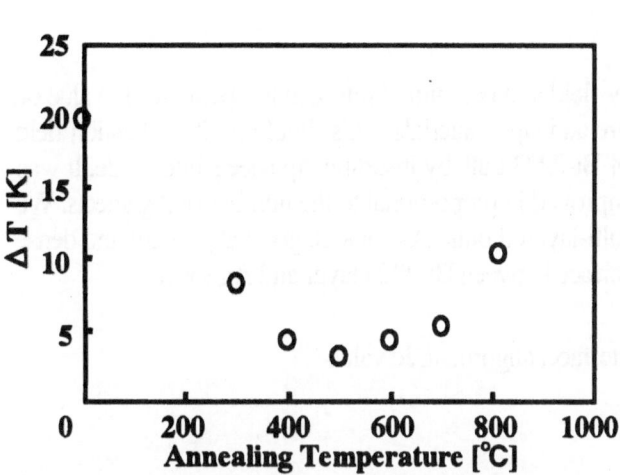

Fig.3 The transition temperature width ΔT of Bi-2212 single crystals as a function of annealing temperature.

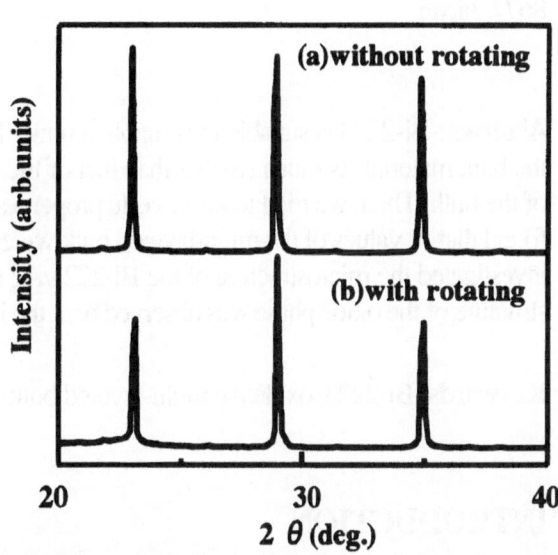

Fig.4 X-ray diffraction patterns of Bi-2212 single crystals grown by the VB method (a) without and (b) with rotating mechanism.

CONCLUSIONS

We investigated annealing characteristics of as-grown Bi-2212 single crystals, and single crystals are high quality since the ΔT is improved by annealing at 500℃. We also tried the modified VB method, and obtained plate-like single crystals of Bi-2212 single-phase and about 9×3 mm^2 in size.

Acknowledgment. This research is supported partly by the Electric Technology Research Foundation of Chugoku.

1. A.Ono, S.Sueno, F.P.Okamura, Jpn.J.Appl.Phys.27 (1988)L786.
2. A.Tanaka, S.Kishida, A.Shibasaki, H.Tokutaka, K.Fujimura, Jpn.J.Appl.Phys.36 (1997)L761.
3. A.Tanaka, S.Kishida, A.Shibasaki, H.Tokutaka, K.Fujimura, J.Crystal Growth 182 (1997)60.
4. S.Kishida, E.Hosokawa, J.Crystal Growth 192 (1998)136.
5. S.Kishida, M.Nakamura, and W.Y.Liang, (to be submitted into J.Crystal Growth)

Microstructure of Bi-2223 Phase in oxide/Ag Multi-Layered Bulk

Yoshimitsu Hishinuma,[1] Shoji Nemoto,[1] Tomokazu Shimada,[1] Isamu Tezuka,[1] Shuji Yoshizawa,[1] Shuetsu Haseyama,[2] Sadao Yamazaki[3] and Hiroshi Nakane[3]

[1]Advanced Materials Research and Development Center, Meisei University, Hodokubo 2-1-1, Hino, Tokyo 191-8506, Japan
[2]Superconductivity Development Center, Dowa Mining Co., Ltd., Tobuki-cho 277-1, Hachioji, Tokyo 192-0001, Japan
[3]Department of Electrical Engineering, Kogakuin University, Nishisinjyuku 1-24-2, Sinjuku-ku, Tokyo, 163-8677, Japan

Abstract: Bi-2223 is suitable to be applied under low field and near liquid nitrogen temperature. Jc value of the bulk materials is much smaller than that of the wire and tape materials. This should limit application field of the bulk. Then, we tried to improve Jc properties of Bi-2223 bulk by inserting Ag sheets into oxide. It was found that Jc values of the multi-layered bulk were improved in proportional to the number of Ag sheets. We investigated the microstructure of the Bi-2223/Ag multi-layered bulk. As a result, good alignment and dense structure of the oxide phase was observed near the interface between Bi-2223 layer and Ag sheet.

Keywords: Bi-2223, oxide/Ag multi-layered bulk, interface, alignment, Jc value

INTRODUCTION

Since the discovery of high-temperature superconductors, various preparation processes have been tried to improve the critical current density (Jc). Drastic improvement of Jc was attained especially for melt-processed $YBa_2Cu_3O_y$ and $Bi_2Sr_2CaCu_2O_x$ (Bi-2212). In the case of $(Bi, Pb)_2Sr_2Ca_2Cu_3O_x$ (Bi-2223) system, the other processes, adding Ag to a calcined powder and increasing the number of an intermediate pressing, were applied.[1-4] But, Jc of Bi-2223 bulks is not necessarily high enough for the applications. In the case of Bi-2223 superconducting wire and tape made by powder-in-tube (PIT) method, the Jc is one order magnitude higher than that of the bulk.[5] Since the superconducting core in PIT wire and tape was similar to the sintered bulk, it is expected that the Jc of the sintered bulk could be increased by improving the grain growth rate and the crystal orientation by introducing Ag into the superconductor. In this paper, Bi-2223/Ag multi-layered sintered bulk was prepared by inserting Ag sheets into superconductor, and the morphology of superconductor next to Ag was studied by an x-ray diffraction (XRD) and a scanning electron microscopy (SEM) method. The effect of the morphology state of superconductor on the Jc of the bulk was investigated.

EXPERIMENTS

In this study, calcined Bi-2223 powders were prepared by a coprecipitation method. The nominal composition of cations was Bi:Pb:Sr:Ca:Cu=1.85:0.35:1.90:2.05:3.05. The calcined compounds were well ground, and constant volume was picked up. The multi-layered bulks were prepared by stacking the calcined powder and Ag sheet of $50\mu m$ in thickness layer by layer. They are pressed with 1 ton/cm^2 for 1 min in metal dies of 13mm in diameter. Fig.1 illustrates the structure of multi-layered bulk. The samples were sintered by the standard profile, 840°C for 10~50 hours, respectively. In order to study on the effect of inserting Ag sheet, the microstructure of the interface of oxide/Ag was observed using scanning electron microscope (SEM) after removing Ag sheet by the Hg-amalgam method. X-ray diffraction analysis was performed to observe not only the phases

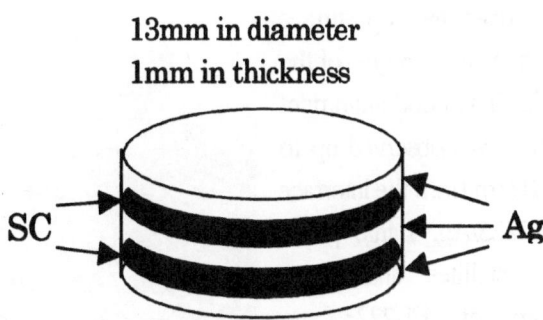

Fig. 1. Schematic drawing of the multi-layered bulk.

formed by the heat-treatment, but also c-axis alignment of plate-like grains estimated by Full Width at Half Maximum (FWHM) of (0010) peak of Bi-2223 phase after removing Ag sheet by the Hg-amalgam method. In evaluation of superconducting properties, we measured critical current density (Jc). Jc was obtained from transport critical current (Ic) and by cross-section area of the oxide layer. Transport Ic was measured by a four-probe method at 77 K in self-fields and Ic criterion was 1μV/cm.

RESULTS AND DISSCUTION

Crystalline characteristic of superconductor in the vicinity of interface between superconductor and Ag sheet was studied by XRD method. Sintering time dependence on the peak intensity and FWHM of (0010) peak at the interface of Bi-2223/Ag is shown in Fig.2. In this study, sintering temperature was 840°C, and sintering time was limited to 50 hours. This figure indicates that (0010) peak intensity of Bi-2223 phase was increased with extending the sintering time, while FWHM of (0010) peak of Bi-2223 was decreased with extending the sintering time. Furthermore, we studied minutely on microstructure near the interface of Bi-2223/Ag. Fig.3 shows relationship between the intensity and FWHM of (0010) peak of Bi-2223 phase and distance from the interface of Bi-2223/Ag. The stronger diffraction intensity and narrower rocking curve of (0010) peak were observed on the top surface of the superconductor next to Ag sheet. The intensity of (0010) peak was decreased

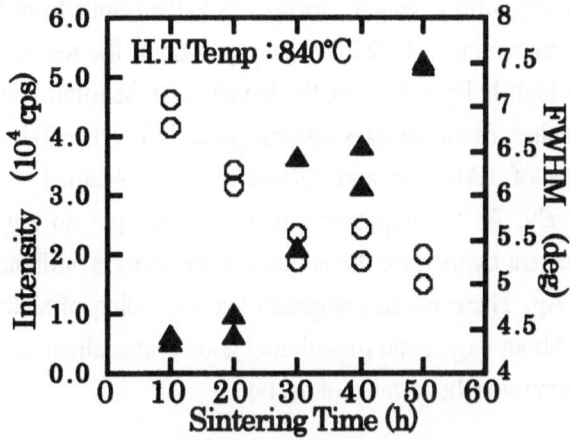

Fig. 2. Sintering time dependence on the peak intensity and FWHM value of the interface Bi-2223/Ag.
▲:Intensity, ○:FWHM

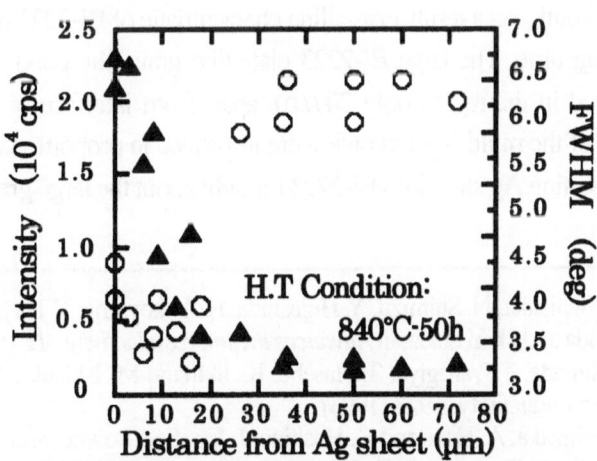

Fig. 3. Change of the intensity and FWHM of Bi-2223 (0010) vs. distance from the interface of Bi-2223/Ag.
▲:Intensity, ○:FWHM

steeply from the top surface forward the inner region of the Bi-2223. The good alignment of grains was observed up to only 20μm from the interface of Bi-2223/Ag. Large plate-like crystallites and dense structure of Bi-2223 are observed in the top surface of the superconductor next to Ag sheet in Fig. 4(a). Fig. 4(b) shows small crystallites and porous structure at the middle region of superconductor several hundred μ m apart

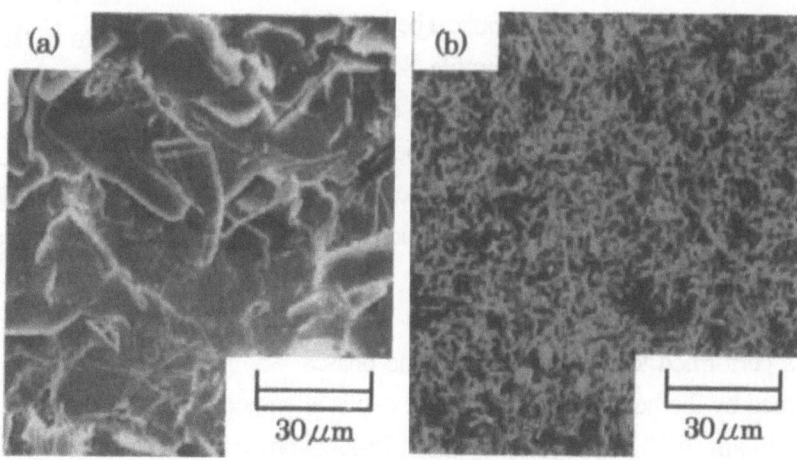

Fig. 4. SEM photographs of (a) the top surface of the superconductor next to Ag sheet and (b) the surface of middle region of superconductor several hundred μm apart from the interface of Bi-2223/Ag.

from the top surface, which corresponds to the results from the XRD. Therefore, by inserting Ag into oxide, the microstructure of Bi-2223 was improved in the region up to 20μm from the interface of Bi-2223/Ag forming high Jc layer. Further, the Jc values of the multi-layered bulk were measured at 77K and self-field.[6] The Jc values of the samples by sintered at 840°C for 50 hours increased with the number of Ag sheet. The Jc values of 180A/cm^2 and 320A/cm^2 were obtained when two and four Ag sheets were laminated, respectively. As the improvement of Jc value per an Ag sheet is constant, it is mentioned that this Jc improvement by increase the number of Ag sheet is attributed to increase of the number of the high Jc layer next to Ag. These results suggested that extending of sintering time and inserting Ag sheet into Bi-2223 brought about large grain growth and good c-axis alignment of Bi-2223 phase at the interface Bi-2223/Ag, which improved the Jc value of the bulk.

CONCLUSION

We investigated the microstructure of Bi-2223 phase and Jc properties of the prepared oxide/Ag multi-layered bulk. As a result, crystalline characteristic of Bi-2223 near the Ag sheet was improved with extending sintering time. The large Bi-2223 plate-like grain, the good c-axis alignment and the dense structure were observed in the region only 20μm apart from interface of Bi-2223 oxide/Ag. Further, we found that Jc values of the multi-layered bulk were improved in proportional to the number of Ag sheets. This is by reason that inserting Ag sheet into Bi-2223 brought about the large grain growth and the good c-axis alignment.

1. K. Michishita, N. Shimizu, Y. Higashida, H. Yokoyama, Y. Hayami, T. Tsunooka, E. Inukai, Y. Kubo, A. Saji, N. Kuroda and H. Yoshida, in *Advances in Superconductivity II* (Springer-verlag, Tokyo, 1992), pp.273-276
2. Y. Yamada, T. Yanagiya, T. Hasebe, K. Jikihara, M. Ishizuka, S. Yasuhara and M. Ishihara, IEEE trans. Appl. Superconductivity. 3 923 (1993)
3. T. Nishizaka, A. Hane and K. Hoshino, J. Jpn. Soc. Powder and powder metallurgy., **41**, 1505 (1994)
4. X.K. Fu, V. Rouessac, Y.C. Guo, P.N. Mikheenko, H.K. Liu and S.X. Dou, Physica C **320**, pp.183-188 (1999)
5. M. Satou, Y. Yamada, S. Murase, T. Kitamura and Y. Kamisada, Appl. Phys. Lett., **10**, 640 (1994)
6. S. Yoshizawa, Y. Hishinuma, S. Nemoto, I. Tezuka, S. Haseyama, S. Yamazaki and H. Nakane, Appl. Superconductivity 1999, to be accepted.

Introducing A Melt Processing into Fabrication of Bi(Pb)2223 Bulks

W. P. Chen,[1] H. Maeda,[2] K. Watanabe,[1] and M. Motokawa[1]

[1]Institute for Materials Research, Tohoku University, Sendai 980-8577, Japan
[2]Kitami Institute of Technology, Kitami 090-8507, Japan

Abstract: Melt processing is important for texture development in high-T_c superconductors. For Bi(Pb)2223, however, a melt processing usually has a negative influence on its formation. In this paper a special melt processing is studied. Using Bi(Pb)2201 and $CaCuO_2$ (mixed a ratio of 1:2) as the precursor compounds, firstly Bi(Pb)2212 is formed by a melt processing, and then a long time sintering is carried out. It is found that most Bi(Pb)2212 can react with surplus $CaCuO_2$ and the resultant Bi(Pb)2223 is of high content and high density. An obvious correlation is also observed between the grain orientation of Bi(Pb)2212 and Bi(Pb)2223, which suggests that texture development in Bi(Pb)2223 bulks can be obtained by orienting the grains of Bi(Pb)2212 during the melt processing by floating zone method or magnetic melt processing.

Keywords: Bi(Pb)2223, melt processing, bulk

INTRODUCTION

In recent years many researches have been focused on inducing texture in high-T_c superconductors and several techniques have been developed. Among these techniques, the melt-textured growth (MTG) for YBCO bulks [1], the floating-zone method for Bi2212 bulks, and the magnetic melt processing (MMP) for both YBCO and Bi2212 bulks [2] are very successful and frequently applied. These three techniques are all based on a melt processing, with the presence of a temperature gradient for floating zone method and a magnetic field for MMP. It is well-known that a slow solidification from a melt corresponding to the Bi(Pb)2223 stoichiometry leads to a mixture of very poor reactivity, from which the reformation of the Bi(Pb)2223 phase is incomplete. So up to now these techniques based on a melt processing cannot be satisfactorily applied to Bi(Pb)2223 bulks.

For Bi-based superconductors different precursor compositions give rise to different reaction sequences [3]. In our previous work it was found that the density of Bi(Pb)2212 bulks can be considerably improved when Bi(Pb)2201 and $CaCuO_2$ are used as the precursors for the partial melting-solidification process [2]. Recently we found that Bi(Pb)2212 thus formed also has a relatively high reactivity. With a higher content of $CaCuO_2$, Bi(Pb)2212 can react with surplus $CaCuO_2$ and a high content of Bi(Pb)2223 can be formed after a long time sintering. As a melt processing is included in the fabrication of Bi(Pb)2223 bulks, this result can be used for these melt processing-based techniques to introduce texture in Bi(Pb)2223 bulks.

EXPERIMENTAL

$Bi_{1.65}Pb_{0.35}Sr_{1.9}Cu_1O_x$ and $CaCuO_2$ were synthesized separately by solid state reactions and then were mixed at a ratio of 1:2. Pellets of 10 mm in diameter and 5 mm in thickness were pressed from the mixture and heat-treated in flowing 8 % O_2 + Ar gas with a temperature profile shown in Fig. 1. For comparison some pellets were also treated with only a part of the temperature profile. All the samples were analyzed by X-ray diffraction and scanning electron microscope.

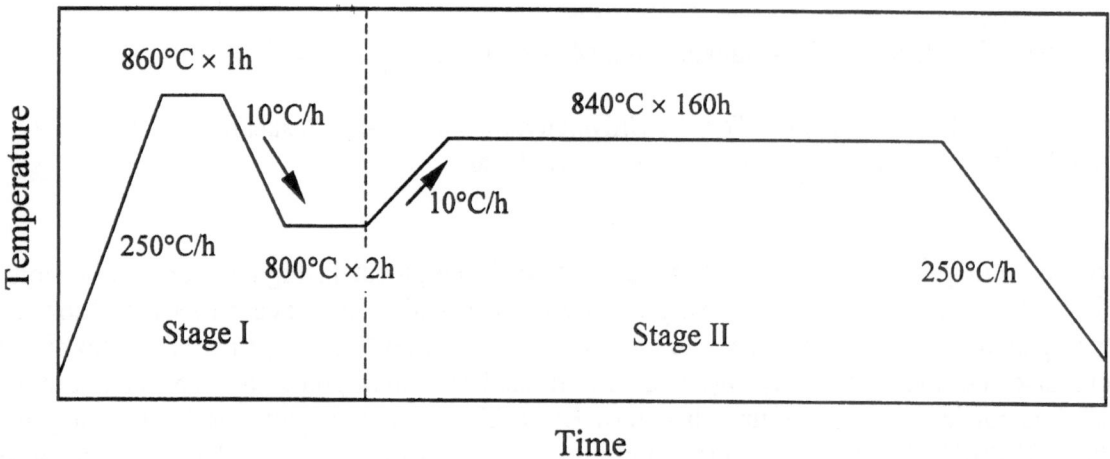

Fig. 1. Temperature profile for the heat-treatment of Bi(Pb)2223 bulks with a melt processing.

RESULTS AND DISCUSSION

The heat-treatment is composed of two stages. In the first stage a partial melting-solidification occurs. Fig. 2a shows the X-ray diffraction pattern obtained from the top surface of a sample taken out after the first stage heat-treatment. These lines are from Bi(Pb)2212 and so Bi(Pb)2212 is the main phase in the sample at this stage. It is very clear that the (00l) lines are greatly enhanced, which indicates that many grains are oriented parallel to the sample surface. It is well-known that some grain orientation proceeds from the surface inward in Bi2212/Ag tapes by partial melting-solidification [4]. This orientation in surface is the sign of the partial melting-solidification process.

The second stage is a long time sintering. Bi(Pb)2212 formed in the first stage reacts with surplus $CaCuO_2$ and Bi(Pb)2223 is formed. Fig. 2b shows the X-ray diffraction pattern recorded from the surface of a sample after the whole heat-treatment. Most lines are from Bi(Pb)2223. Only a few weak lines are from Bi(Pb)2212. This indicates that most Bi(Pb)2212 has transformed into Bi(Pb)2223 after the long time sintering and a high content of Bi(Pb)2223 has been obtained. As a partial melting-solidification process has occurred in the first stage heat-treatment, this result shows a melting process can be introduced into the fabrication of Bi(Pb)2223 bulks when Bi(Pb)2201 and $CaCuO_2$ are used as the precursor compounds. During the partial melting Bi(Pb)2201 is melted while $CaCuO_2$ is in a solid state. $CaCuO_2$ has a relatively high content in the mixture and it can act as a skeleton holding the melt from spreading. In this way a homogenous composition can be retained after the partial melting and a high content of Bi(Pb)2223 can be formed afterward. (00l) lines of Bi(Pb)2223 are greatly enhanced so the Bi(Pb)2223 grains are highly oriented in the surface. Obviously this texture has resulted from the texture of Bi(Pb)2212. It suggests that Bi(Pb)2223 grain alignment can be obtained by orienting Bi(Pb)2212 grains, which can be realized by floating-zone method or magnetic melting processing.

Some pellets were heat-treated only with the second stage long time sintering and a representative X-ray diffraction pattern is shown in Fig. 2c. No texture can be detected as that in Fig. 2b and the amount of remaining Bi(Pb)2212 is obviously higher. This indicates that a melting-solidification process is helpful for the formation of Bi(Pb)2223 when Bi(Pb)2201 and $CaCuO_2$ are used as the precursor compounds. There is a 30 % increase in diameter for the samples processed without the partial melting-solidification; while that for the samples with the partial melting-solidification is only 10 % and the density is much higher. So the microstructure can also be greatly improved by the partial melting-solidification.

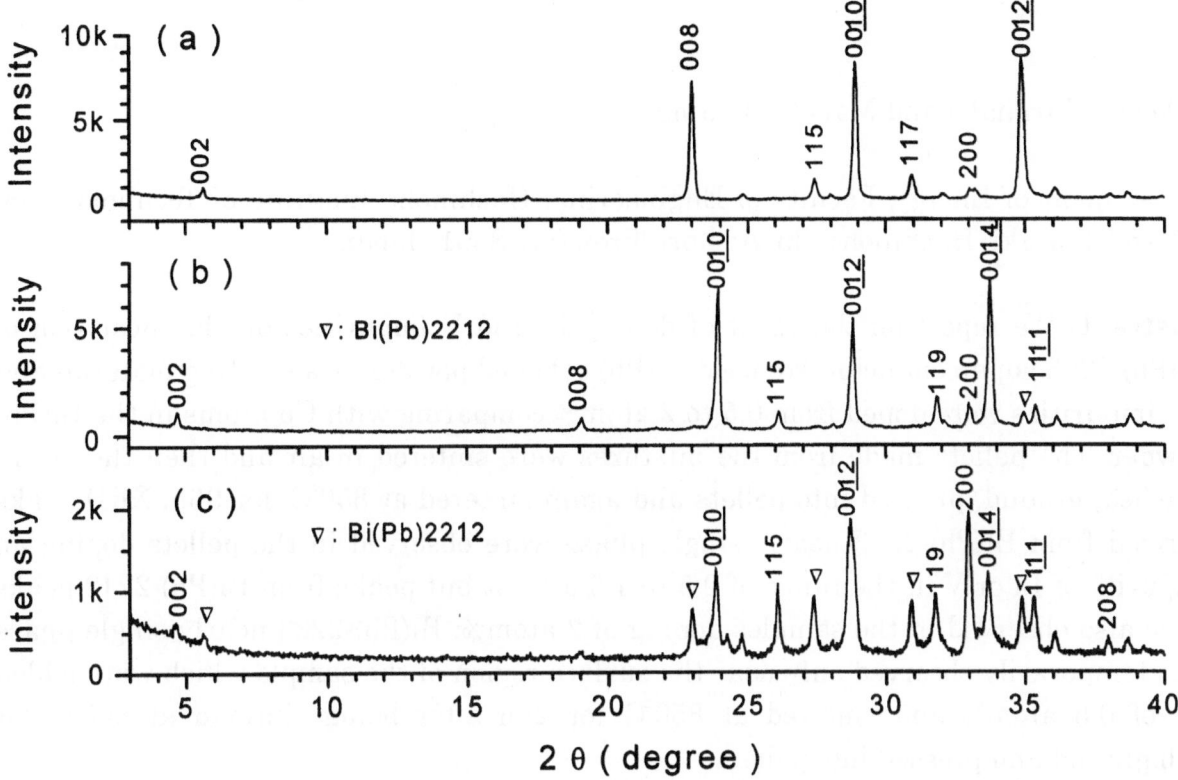

Fig. 2. X-ray diffraction pattern recorded from surface of (a) a sample with only the first stage heat-treatment; (b) a sample with the whole heat-treatment; (c) a sample with only the second stage heat-treatment

SUMMARY

With Bi(Pb)2201 and $CaCuO_2$ used as the precursor compounds, a melt processing has been successfully introduced into the fabrication of Bi(Pb)2223 bulks. Bi(Pb)2212 is firstly formed by the melt processing and then it reacts with surplus $CaCuO_2$ during a long time sintering. Comparison experiment shows that the melt processing can promote the formation of Bi(Pb)2223 phase and increase its density. As an obvious correlation is observed between the grain orientation of Bi(Pb)2212 and Bi(Pb)2223, it is proposed that texture development in Bi(Pb)2223 bulks can be obtained by orienting the grains of Bi(Pb)2212 during a melt processing.

1. G. Desgardin, I. Monot, and B. Raveau, Supercond. Sci. Technol. **12**, R115-R133 (1999).
2. W. P. Chen, H. Maeda, K. Kakimoto, P. X. Zhang, K. Watanabe, M. Motokawa, H. Kumakura, and K. Itoh, J. Cryst. Growth **204**, 69-77 (1999).
3. W. Wong-Ng, L. P. Cook, and W. Greenwood, J. Mater. Res. **14**, 1695-1706 (1999).
4. J. Kase, N. Irisawa, T. Morimoto, K. Tagano, H. Kumakura, D. R. Dietderich, and H. Maeda, Appl. Phys. Lett. **56**, 970-972 (1990).

The Effect of Metal Impurity Additions of Lithium and Vanadium on the Formation of Bi(Pb)-2223 Superconductor

Takeshi Muranaka and Makoto Hiyama

Department of Energy, Faculty of Engineering, Hachinohe Institute of Technology,88-1,Myo,Ohbiraki, Hachinohe-shi, Aomori, Pref.,031-8501, Japan

Abstract: We report on the effect of doping Li and V impurities on the formation of Bi(Pb)-2223 superconductor. We used Bi(Pb) calcined powder as a starting material and the impurities were doped from 0.5 to 2 atom% comparing with Cu atoms in the Bi(Pb) powder. The pellets made from the mixtures were sintered in air and then they were crushed, ground, pressed into pellets and again sintered at 850℃ for 96h. XRD peaks derived from Bi(Pb)-2223 nearly single phase were observed in the pellets doping an impurity of Li or V in the range of 0.5 to 1.5 atom% but peaks from Bi(Pb)-2212 peaks were also observed in the samples doping of 2 atom%. Bi(Pb)-2223 nearly single phase was temporarily observed only near the surface region of the sample which were added V of 0.5 atom% and sintered at 850℃ for 24h after being sintered at 850℃ for 48h,ground and pressed into pellets.

 Keywords: Bi(Pb)-2223 superconductor, Metal impurity addition, Lithium ,Vanadium

INTRODUCTION

Many investigators have attempted to dope impurity metals such as 3d transition metals [1-4] and a light element of Li [5] in the form of oxides to prepare high Tc 2223 phase of Bi(Pb) bulk superconductor. Among these elements, lithium and vanadium seem to have a superior character to enhance the formation of Bi(Pb)-2223 phase. The aim of this work is to study the effect of metal impurity additions of the two elements on the formation of Bi(Pb)-2223 superconductor by means of X-ray powder diffraction.

EXPERIMENTAL

To simplify the procedure and to reduce the error in the sample treatment, we chose Bi(Pb) calcined powder as a starting material which has a nominal composition of $Bi_{1.85}$ $Pb_{0.35}$ $Sr_{1.90}$ $Ca_{2.05}$ $Cu_{3.05}$. Lithium carbonate or vanadium oxide powder was added to the calcined power with the cation ratios of Cu: impurity metal atom = 1: X (X = 0, 0.005, 0.01, 0.015, 0.02). The powder mixture was pressed into pellets of 15mm diameter, about 1mm thick and the pellets was sintered at 850℃ for 48h in a programmable muffle furnace. After being cooled to room temperature, the pellets were crushed and ground in a mortar , pressed into pellets of the same diameter and again sintered at

850℃ for 24-120h. X-ray diffraction patterns of prepared samples were recorded by a Rigaku RADⅢ-C system with a graphite monochrometer in the range of $2\theta = 3\text{-}65°$.

RESULTS AND DISCUSSIONS

Fig.1 and Fig.2 show the variation of XRD patterns obtained from the pellets for the same sintering condition of 850℃ for 96h and for various Li and V contents respectively. Although XRD patterns with no impurity additions in Fig.1 and Fig.2 show some low Tc peaks, the patterns from pellets with Li or V addition of 0.5,1.0 and 1.5 atom% indicate almost high Tc peaks. The patterns with Li or V addition of 2.0 atom% again represent some low Tc peaks. The variation tendency of XRD patterns depending on the impurity contents resembles in both cases of lithium and vanadium additions.

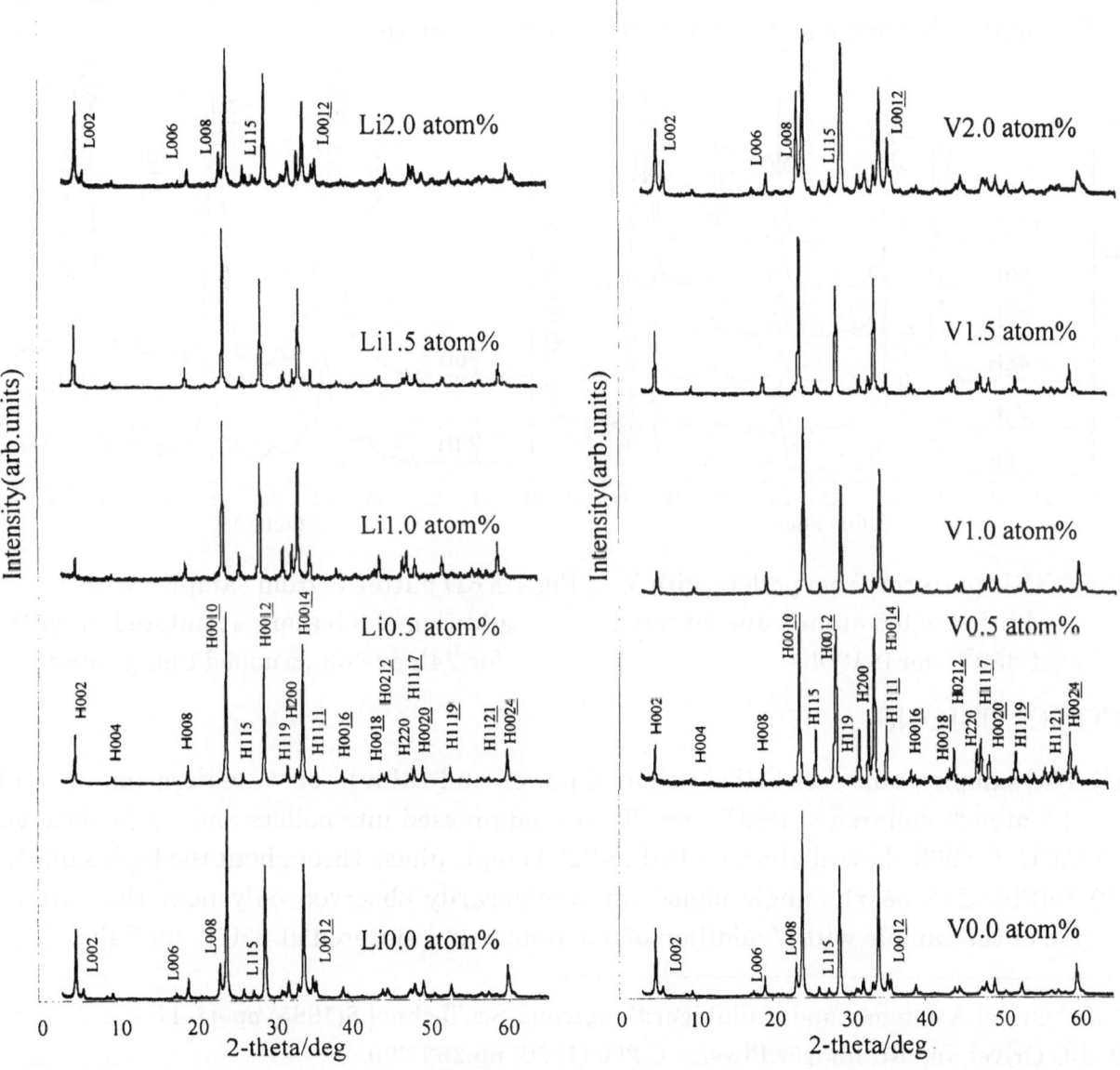

Fig.1 XRD patterns from pellets added Li impurity and sintered at 850℃ for 96h. Li contents are indicated.

Fig.2 XRD patterns from pellets added V impurity and sintered at 850℃ for 96h. V contents are indicated.

Fig.3 represents the variation of XRD patterns obtained from pellets which were added V of 0.5 atom% and sintered at 850℃ for various sintering time of 0-120h. The peak intensities derived from (008) in Bi(Pb)-2212 low Tc phase decreased at the sintering time of 24h, increased at the sintering time of 48h and 72h and again decreased in the sintering time more than 96h. It is noticeable that Bi(Pb)-2223 high Tc phase was once enriched by the shorter sintering time of 24h. Although this effect will relate to the impurity addition of vanadium, we have no idea at present to explain the variation of XRD patterns. Fig.4 shows XRD patterns of ground samples with V addition of 0.5 atom%, sintered at 850℃ for 24h and 96h. Peaks from Bi(Pb) low Tc phase were clear in the sample sintered for 24h but these peaks almost disappeared in the sample sintered for 96h.We deduced from these observations in Fig.3 and Fig.4 that Bi(Pb) 2223 single phase were formed only near the surface region of the sample under the sintering conditions of 850℃, 24h and then the single phase was formed gradually throughout the bulk sample after being sintered at 850℃ for 96h or more.

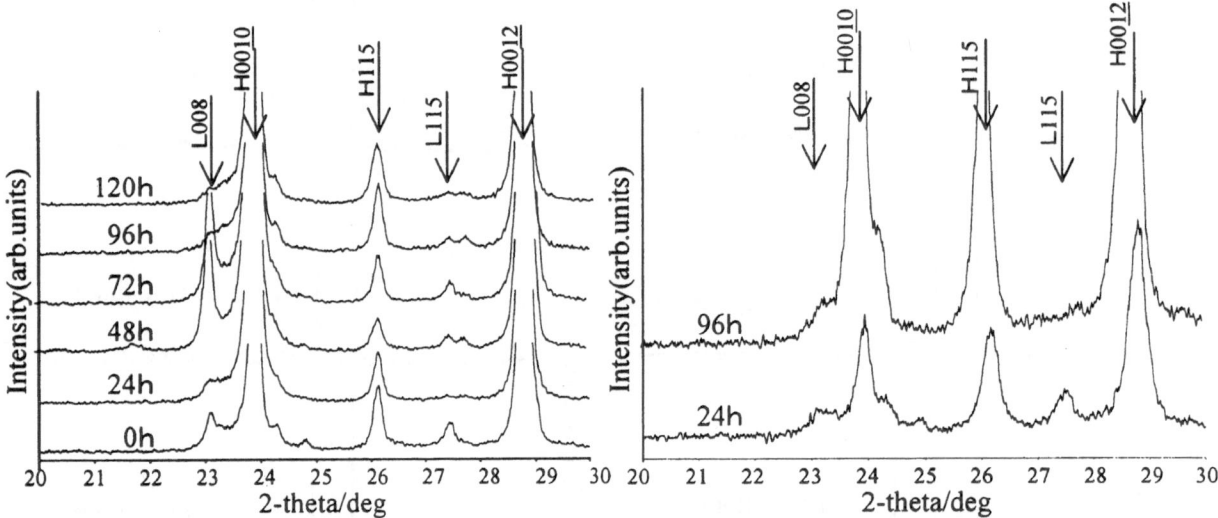

Fig.3 XRD patterns from pellets with V addition of 0.5 atom% and sintered at 850℃ for 0-120h.

Fig.4 XRD patterns from samples with V addition of 0.5 atom% ,sintered at 850℃ for 24h or 96h ,crushed and ground.

CONCLUSIONS

(1) The samples mixed of Bi(Pb) calcined power and lithium or vanadium impurity of 0.5-1.5 atom%,sintered at 850℃ for 48h,ground,pressed into pellets and again sintered at 850℃ for 96h showed almost a Bi(Pb)-2223 single phase throughout the bulk sample.
(2) Bi(Pb)-2223 nearly single phase was temporarily observed only near the surface region of the sample with V addition of 0.5 atom% and sintered at 850℃ for 24h.

1.J-C Grivel,A.Jeremie and R.Flukiger:Supercond.Sci.Technol.8(1995) pp.41-47.
2. J-C Grivel and R.Flukiger:Physica C **256** (1996) pp.283-290.
3.K.Watanabe and M.Kojima: Supercond.Sci.Technol.11(1998) pp.392-398.
4.K.kakimoto et al.:The Abstracts of the 58th Meeting of Applied Physics No.1,(1997)p.198.
5.S.Kanbe et al.: The Abstracts of the 58th Meeting of Applied Physics No.1,(1997)p197.

Performance of 1KA Class HTS Current Leads Made of Bi2223 Rods

T.C Wang[1]*, P.X.Zhang[1], Y.R Cai[1], B.Q Fu[1], J.R Wang[1], Y.Feng[1], L Zhou[1],
L.H Song[2], H.M Wen[2] and L.Z Ling[2]

[1]Northwest Institute for Nonferrous Metal Research, P.O Box51, Xi'an, Shanxi, P.R China
[2]Institute of Electronics Chinese Academy of Sciences, Beijing, P.R China

Abstract: The process and performance of High Temperature Superconducting Current Leads whose current up to 1000A were introduced. The HTS leads were made of Bi-2223 rods by a multiple process of cold-isostatic press (MCIP) and heat treatments. The transport currents were measured by a standard four-probe DC method. The results showed the current over 1000A at 77K and 0T, and 500A at 77K and 0.1T.

Keywords: current lead, Bi-2223 rod, magnetic field

INTRODUCTION

The study of High Tc Superconducting material has great advanced so far. Specifically, the HTS materials have good application characteristics for LTS technique, such as the lower heat conductivity for current leads. More and more attention have been paid to the Bi-2223 superconducting materials to be used for current leads for LTS magnets due to it's high Tc (Tc=108K). It is reported that big progress have been made for the high Tc superconducting current leads recently. Variety of standard products of HTS current leads with different shape were available, such as YBCO current leads by MTG process, Bi-2212 current leads by melt cast process [1-3], Bi-2223 current leads by cold isostatic press, Ag-sheath Bi system tape current leads by overlapping and combination [4-5], and so on. In this paper, Bi-2223 current leads had been fabricated by a multiple cold-isostatic press (MCIP) and heat treatment process. The critical currents were measured mainly at 77K and in different field up to 0.3T. The microstructure of the samples were characterized by the X-ray diffraction and SEM.

EXPERIMENT

The BSCCO precursors with a metal ratio near 2:2:2:3 were synthesized by using Bi_2O_3, PbO, $SrCO_3$, CaO and CuO powders as starting materials by the nitrate Co-deposition method. The powders were decomposed for three times at 450℃× 30h,700℃× 20h and 800℃× 10h respectively. Then, the precursor powders were obtained by mixing thoroughly, in which Bi-2212 phase was the major phase. The powders were put into rubber mode, sealed tightly and formed into rod samples by the cold isostatic press method at a pressure of 200MPa. The formed materials were sintered at 830~850℃ for 60~100h in atmosphere. Then, the samples were put into rubber mode and pressed again with the same pressure in order to obtain more tight samples. The samples were heat treated with the same temperature and time as the first treatment. The Bi2223 rod and copper wires were jointed with Sn-Pb alloy and soft braided copper wire (show in Fig.1). The Bi-2223 phase superconductors could be manufactured in a large variety of shape and dimensions by the MCIP method. Presently, they were available for the samples with dimension of Φ10× L100mm, Φ15 ×L200mm, Φ33×L150mm, even for the tubes up to Φ37mm in outside diameter and 200mm in length. The wall thickness of tubes could be chosen freely within a range of about 4 to 8mm.

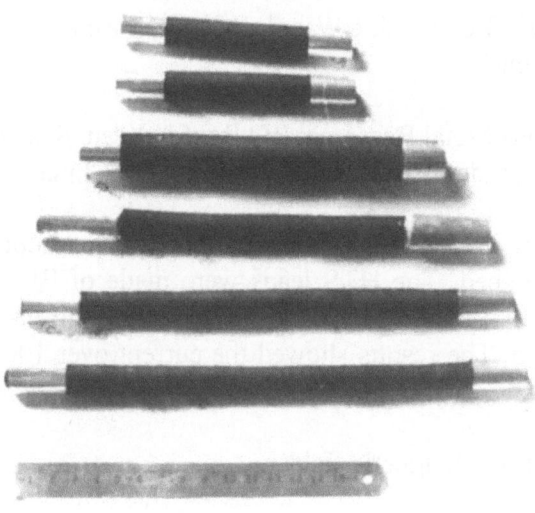

Fig.1. Bi2223 rod current leads

RESULTS AND DISCUSSION

Sample's Ic was measured by standard four-probe method with a Criterion of 1 μ V/cm. The measured samples were 17mm in diameter and 260mm in length. Fig.2. showed the results of Ic for the rods at 77K in self-field. Samples' Ic was also measured in different magnetic field. A copper coil of 50mm in diameter and 250mm in height was used as background field. The samples were fixed in the center of magnetic field. The current direction was parallel to magnetic field by using a criterion of 1 μ V/cm for Ic. Fig.3. showed the results of Ic measurement for the rods at 77K in different magnetic field.

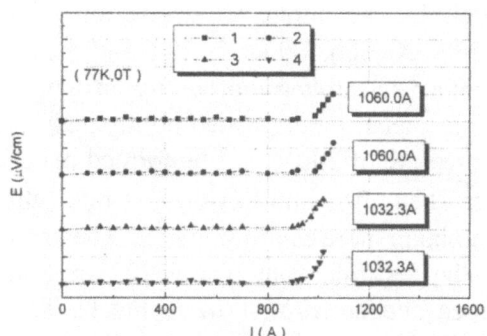
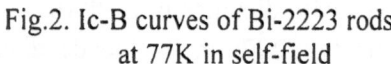

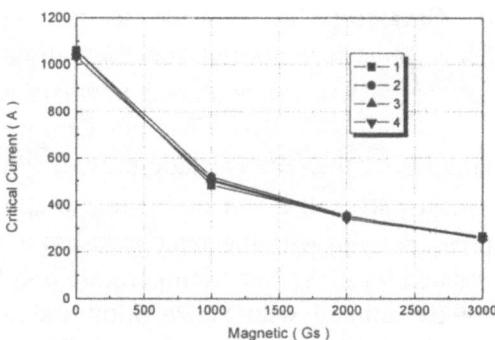

Fig.2. Ic-B curves of Bi-2223 rods at 77K in self-field

Fig.3. Ic-B curves of Bi-2223 rods at 77K in different magnetic field

In addition, the contact resistance between the Bi-2223 rod and the Ag alloy foil was measured by voltage-ampere method. It was very low to be $10^{-7}\,\Omega$ at 77K when the current of current lead was 200A.

The measurement results showed that the Ic exceeded 1000A at 77K in self-field for four rods. It is found from Fig.3 that Ic decreased similarly for all the samples. When the magnetic field changed from 0 to 0.1T, Ic of all current leads dropped to about 500A. The microstructure of samples were studied by SEM. The photograph in Fig.4 showed is the typical microstructure for the cross section. of sample. The grains seemed to be regular plate or layer and with good orientation. It was proved that orientation could be obtained in MCIP sample.

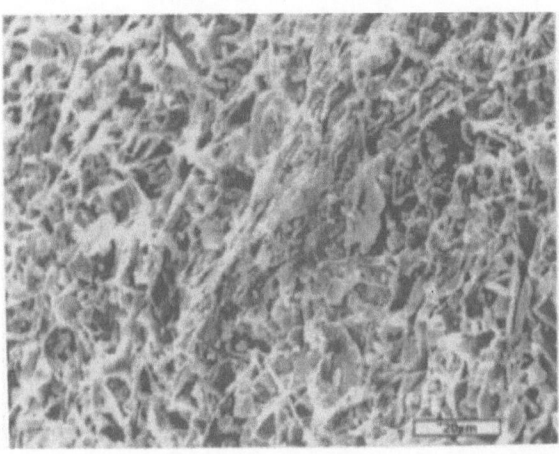

Fig.4. Microstructure of a MCIP rod in the cross section

CONCLUSION

Bulk Bi2223 material with good orientation could be obtained through MCIP method in a wide variety of shapes and dimensions. Their Ic could be up to 1000A at 77K in self-field. They had stable superconducting properties even in high magnetic field. A special fabrication of joint leads to low contact resistance between superconductors and copper wire. The design of soft braided copper wire can avoid the fracture of ceramic sample. It is believed that the improvement of Ic for current leads is because of the improvement of weak link, density of structure, good contact and low resistance between superconductor and copper wires.

1. D.R.Watson, M.Chen, D.M.Glowacka, N.Admopovos, B.Soylu, B.A.Gowackiand J.E Evance, IEEE Trons Appl. Supercond., Vol.5, No.2,P801-804(1995)
2. Joachim Back, Steffen Elschner, Peter F.Herrmann, IEEE Trans Appl. Supercond., Vol..5, No.2, P1409-1412(1995)
3. J.Bhakta, I.R Harris and J.S Abell., IEEE Trans Appl. Supercond., Vol.5, No.2, P1482-1485(1995)
4. T.Honjo, S.miyakt and T.Haseqawa., IEEE Trans Appl. Supercond., Vol.5, No.2, P1486-1489(1995)
5. T.Ando, T.Isono, H.Tsvji, T.Kato, T.Hikata and K.Sato., IEEE Trans Appl. Supercond., Vol.5, No.2, P817-820(1995)

DEVELOPMENT OF BIAXIALLY TEXTURED YBa$_2$Cu$_3$O$_7$ COATED CONDUCTORS IN THE U.S.

Robert A. Hawsey, Donald M. Kroeger, and David K. Christen

Oak Ridge National Laboratory, P. O. Box 2008, Oak Ridge, Tennessee 37831-6195, USA

Abstract: Two new processes have been under development since 1991 that promise a new, cost-effective way to manufacture flexible, high current density wires made from YBa$_2$Cu$_3$O$_7$ (YBCO). The key is to prepare a textured substrate, or "template," on which the YBCO may be deposited as a biaxially aligned thick film. Ion beam assisted deposition (IBAD) of yttria stabilized zirconia or magnesium oxide on alloy tapes enables a final superconducting layer with grain-to-grain, in-plane alignment to within 3-5 degrees. Similar results are achieved on rolling-assisted, biaxially textured substrates (RABiTS) using a variety of oxide layers on textured nickel tapes. The performance of research lengths of prototype wires in strong magnetic fields at 65 K already exceeds that of NbTi and Nb$_3$Sn in liquid helium. A scalable, *ex-situ* process for the YBCO coating has been demonstrated on both types of substrates. Consistent values of critical current density (J$_c$) greater than 1×10^6 A/cm^2 are now obtained on RABiTS, and J$_c$'s in excess of 2×10^6 A/cm^2 have been obtained on both substrates. A nonmagnetic variation of RABiTS (Ni-13% Cr) has also been shown to yield J$_c$ greater than 1.5×10^6 A/cm^2. Six private companies in the U.S. are scaling up YBCO coated conductors for power and physics applications.

Keywords: High temperature superconductors, YBCO, coated conductors, ion beam assisted deposition, RABiTS, biaxial texture, electron beam evaporation, sol gel, pulsed laser deposition

INTRODUCTION

The U.S. Department of Energy (DOE) leads the U.S. national effort to develop high-temperature superconductor (HTS) wires and to demonstrate prototype electric power applications using the best wires available today. With a year 2000 budget of nearly $32 million, the DOE is sponsoring perhaps the largest single national effort toward research and development of HTS wires.

Due in part to this government-national laboratory-university-industry partnership, long lengths of "first generation" BSCCO-2223 and -2212 HTS tapes are now commercially available. Since 1991 the performance of BSCCO-2223 tapes has increased by an order of magnitude while the price of the tapes has decreased dramatically. According to Riley [1], in 1999 kilometer lengths of BSCCO-2223 tapes were available from at least one U.S. company at a price of approximately $300/kA-m. This price represents a substantial reduction from that quoted just one year ago of $1,000/kA-m [2]. Yet it may still be said that the BSCCO-2223 tapes are in the "precommercial" stage, based upon production quantities alone. In 1998, the total annual U.S. production capability for BSCCO-2223 tapes was approximately 350 kilometers. For comparison, a number of companies worldwide each can produce nearly 20,000 km per year of NbTi for magnetic resonance imaging and research applications.

The use of BSCCO tapes at liquid nitrogen temperatures is presently limited by thermally activated flux motion to those applications involving low magnetic fields. For this reason, most of the field tests of transformers, motors, generators, and current limiters have been conducted at or below 30 K.

Previously, the utility of the YBCO compound was constrained to current leads and other "short" applications of HTS due to the lack of technology to produce biaxially textured films on practical metal-based substrates. This situation changed with the development of ion beam assisted deposition (IBAD) substrates pioneered first in Japan [3] for HTS tapes and subsequently refined significantly in 1995 by the Los Alamos group [4]. Also, in 1996 the Oak Ridge National Laboratory (ORNL) presented the first results using a template called rolling-assisted, biaxially textured substrates (RABiTS) [5,6].

The RABiTS template is formed from high-purity nickel and certain alloys using thermomechanical processing and recrystallization to form highly cube-textured tape. Appropriate buffer layers and the superconductor coating are then deposited epitaxially on the RABiTS tape. The high critical current densities obtained using this substrate may be attributed to the strong biaxial texture leading to a preponderance of low-angle grain boundaries in the structure, starting with the metal tape and ending with the superconducting film [7].

The IBAD technique starts with a polycrystalline nickel-based superalloy tape and generates a highly in-plane oriented template through deposition of yttria stabilized zirconia (YSZ) or MgO in the presence of a well-collimated "assisting" ion beam directed at an appropriate angle to the substrate. After epitaxial deposition of a thin cap layer (often CeO_2 in the case of YSZ), the template may then be used for superconductor deposition by a variety of techniques, yielding high J_c's.

HIGH CRITICAL CURRENTS

High J_c's have been produced on both IBAD and RABiTS substrates using a number of techniques for superconductor deposition. The research workhorse, pulsed laser deposition (PLD), frequently yields J_c's in excess of 2×10^6 A/cm^2 on either substrate at 77 K and in self-field, with champion numbers over 3×10^6 A/cm^2. Los Alamos can now routinely obtain critical currents of 200 A for 1-cm-wide short (<5-cm) tapes. Another method for YBCO growth has recently attracted attention. The ex-situ, or "BaF$_2$" process involves the room-temperature deposition of yttrium, barium fluoride, and copper onto RABiTS or IBAD substrates, followed by a batch post-anneal at elevated temperature in the presence of water vapor and oxygen to form the superconducting phase. Using this process, high-J_c films were grown for the first time on metal substrates by the Oak Ridge group [7,8] in 1998. A summary of the typical J_c versus temperature obtained using RABiTS/YBCO(PLD) is shown in Fig. 1 [9, 10]. Recent results for the field dependance of J_c in RABiTS/YBCO (ex-situ) are shown in Fig. 2 [8].

Using the ex-situ approach, typical results for 0.3-μm YBCO on RABiTS are $1.0-1.7 \times 10^6$ A/cm^2 (77 K, H=0) with the best value now 2.3×10^6 A/cm^2 [8]. This process works equally well for the IBAD/YSZ substrates produced by Los Alamos. For a 15-cm section of tape processed in 1-cm pieces, J_c exceeded 1×10^6 A/cm^2 over the entire section and J_c exceeded 2.5×10^6 A/cm^2 over the central 10-cm section with the best texture [8]. It is generally observed that J_c's in the best films on both RABiTS and IBAD substrates are substantially higher than would be expected from their texture as derived from x-ray diffraction (XRD). An explanation for this is found from grain orientation maps obtained from electron backscatter diffraction patterns which indicate that local or grain-to-grain misorientations are smaller than global variations determined from XRD [7,8].

The overall, or engineering current density, calculated using the critical current divided by the entire cross sectional area of the coated conductor, is also of interest for applications. The highest numbers to date have been achieved by the Los Alamos group on a 1-cm x 5-cm IBAD/YSZ substrate that was just 40-μm thick. According to Foltyn [1], a 1.5-μm YBCO film grown on this template conducted 176 A (75 K) and had an engineering current density of nearly 42,000 A/cm^2. The highest engineering current density on RABiTS using the *ex-situ* process was reported by Feenstra [1] as 28,000 A/cm^2.

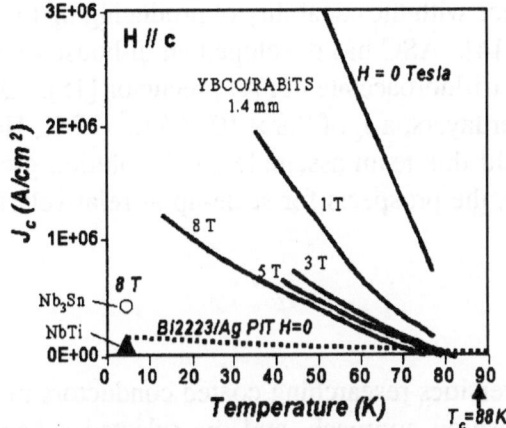

Fig. 1. Temperature dependance of the J_c of YBCO (PLD) on RABiTS as a function of applied magnetic field.

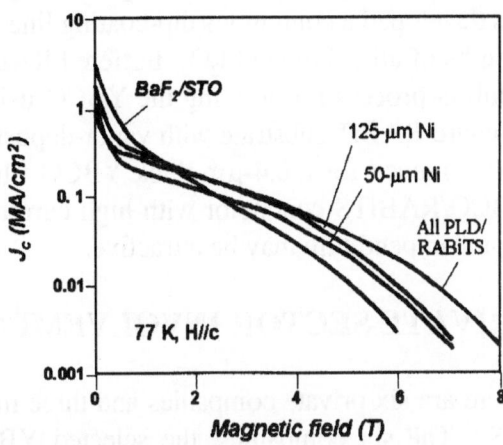

Fig. 2. Magnetic field dependance of the J_c of RABiTS/YBCO (*ex-situ*) compared with STO/YBCO (*ex-situ*) and RABiTS/YBCO (PLD).

LENGTHS

The U.S. leader in demonstration of 1-m lengths of YBCO coated conductor is Los Alamos. A number of tapes made with IBAD/YSZ templates and YBCO (PLD) that are 1-m long have exceeded 100 A critical current over the entire length. For example, one tape reported recently by Foltyn [1] had a critical current of 122 A over a 97-cm measurement length, with a critical current of over 140 A along a 78-cm section. The most uniform tape produced by Los Alamos this year had cm-to-cm variations along the length of ±20%.

Oak Ridge has focused on developing the continuous processing parameters necessary for the *ex-situ* YBCO coated conductor approach. The work is at an early stage, with end-to-end $J_c > 5 \times 10^5$ A/cm^2 (77 K, H=0) reported by List [1] for 5-cm sections of tape where the buffer layers and YBCO were deposited with moving tape.

PROCESS DEVELOPMENT

A number of critical issues that were identified at a 1998 workshop on coated conductor development are being addressed by the national laboratories [11]. Progress towards a completely nonmagnetic (77 K) stronger version of the RABiT substrate as reported by Goyal [1], for example, has enabled short samples of single-orientation, cube-textured Ni-Cr(13%) alloy with a J_c of 1.4×10^6 A/cm^2

(77 K, H=0). In addition, a YBCO/RABiTS was recently demonstrated with totally conductive buffer layers (SrRuO$_3$ by PLD and LaNiO$_3$ by sputtering) with $J_c = 0.6 \times 10^6$ A/cm^2 (77 K, H=0) and $J_c = 1.4 \times 10^6$ A/cm^2 (64 K, H=0) [1]. Such a structure could be important for electrical stability of magnets and other devices during transients. In an effort to evaluate an alternative YBCO deposition approach, Balachandran [1] reported that the team of Argonne National Laboratory, Los Alamos, and Intermagnetics General (IGC) has demonstrated a J_c in excess of 2×10^6 A/cm^2 (I_c>80 A) on a short sample of YBCO/IBAD tape using metal-organic chemical vapor deposition to grow the YBCO.

One of the interesting opportunities provided by the RABiTS approach is to apply all the layers on the metal tape by nonvacuum methods. Recently, the American Superconductor (ASC)/ORNL team announced progress towards this potentially low-cost way to make a coated conductor [12]. ORNL has developed a continuous dip-coating line and furnace with the capability of producing up to 3-m lengths of all-solution Gd$_2$O$_3$- buffered RABiTS [13,14]. ASC has developed an in-house *ex-situ* solution process for growing the YBCO using metal trifluoroacetate (TFA) precursor [15]. On a standard RABiT substrate with vapor-deposited buffer layers, a J_c of 0.8×10^6 A/cm^2 (77 K, H=0) was achieved for a 0.4-µm-thick YBCO film. Should this team assemble an all-solution grown YBCO/RABiTS conductor with high current density, the prospects for scale-up at relatively low capital expenditure may be attractive.

PRIVATE SECTOR INVOLVEMENT

There are six private companies and three major universities researching coated conductors in the U.S. The six companies, the selected YBCO deposition approach, and the selected substrate technology are shown in Table 1.

Table 1. U.S. companies developing YBCO coated conductors.

Company	HTS Deposition Approach	Substrate
3M	E-beam co-evaporation (*in-situ* and *ex-situ*)	IBAD, RABiTS
Oxford Superconducting Technology	PLD (research stage)	RABiTS
MicroCoating Technologies	CCVD (open atmosphere CVD), also TFA	RABiTS
EURUS Technologies	TFA others	RABiTS
American Superconductor	TFA	RABiTS
Intermagnetics General	PLD, MOCVD	IBAD

The work of the ASC and IGC teams was discussed earlier. 3M Company is leveraging a number of resources, including internal funding, the U.S. DOE's Superconductivity Program, and the Defense Advanced Research Project Agency to perform research that will lead to development of a continuous process for making lengths of coated conductor. They have been working with Los Alamos and Oak Ridge National Laboratories, as well as Southwire Company and Stanford University since 1997. 3M has succeeded in producing 100-m lengths of biaxially textured nickel tape compatible with buffer layer technology, including their own high-rate reactive sputter deposition for the YSZ and CeO$_2$

layers. $J_c > 10^6$ A/cm^2 has been obtained on this textured nickel. The 3M electron beam evaporation system has a 30-kW differentially pumped gun and a tape transport system capable of handling 4-cm-wide by 100-m long tape. According to O'Neill [1], short sample results obtained by 3M and ORNL using 3M nickel and CeO$_2$ vary in J_c from 0.2 to 0.5 × 10^6 A/cm^2 (77 K, H=0). Stanford University is conducting research for this team towards high-rate *in-situ* YBCO deposition.

Oxford Superconducting Technology (OST) has also succeeded in duplicating the highly textured nickel tape developed by Oak Ridge. Using Oxford Instruments buffer layers deposited by PLD, Brookhaven National Laboratory (BNL) deposited 2-µm-thick YBCO using the *ex-situ* approach with a resultant $J_c > 0.8 × 10^6$ A/cm^2 (77 K, H=0) and 0.17 × 10^6 A/cm^2 in a 1 Tesla background field [16]. OST has recently partnered with MicroCoating Technologies to scale up this technology and to research using MCT's open-atmosphere combustion chemical vapor deposition (CCVD) process for the oxide layer deposition. EURUS has also duplicated the biaxially textured nickel and has provided samples to a number of organizations worldwide. In one demonstration announced recently, the Air Force Research Laboratory used PLD to deposit buffer layers and YBCO on EURUS' nickel and measured a self-field J_c of 1 × 10^6 A/cm^2 [17].

FUTURE TRENDS

There is intense research under way in the U.S. to complete development of continuous processes for fabrication of coated conductors. In fact, an estimated $16 million will be spent in the U.S. in 2000 on this research and development. Within the year 2000, it is likely that the first 5-10 meter lengths of YBCO conductor will be produced by industry, with scale up to 100-m lengths starting in 2001. As shown in Fig. 3, by the end of 2002, 10-100 meter lengths will begin to become available for construction of simple cables, magnets, and the like.

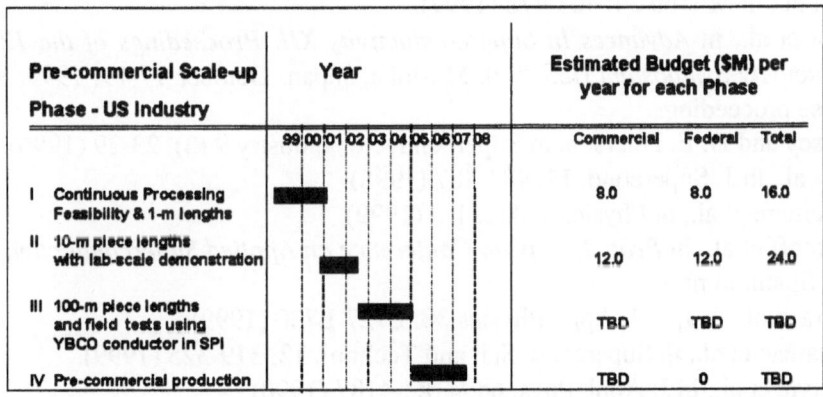

Fig. 3. Projected time line for coated conductor scale-up in the U.S.

CONCLUSION

Moving tapes of YBCO(PLD)/IBAD-YSZ on nickel alloys consistently yield critical currents of the order 100 A/cm width, and short sample results on IBAD-MgO templates are very promising. High J_c's (> 2 × 10^6 A/cm^2) have been achieved on IBAD and RABiTS templates using the *ex-situ*, or "barium fluoride" approach. One company (ASC) has demonstrated nearly 1 × 10^6 A/cm^2 (77 K, H=0) using a nonvacuum YBCO growth process on RABiTS. A total of five processes for deposition of buffer layers and YBCO on nickel tapes are being investigated by six U.S.-based companies. Given the investment by the government and private sectors, there is a high likelihood that 10-100 meter lengths of YBCO conductor will be available for applications by the end of 2001.

Acknowledgements. The authors are grateful for the assistance of the following researchers in the preparation of this manuscript: Oak Ridge National Laboratory* - Ron Feenstra, Amit Goyal, Dominic Lee, Fred List, Parans Paranthaman, and David Norton; Los Alamos National Laboratory - Paul Arendt, Steve Foltyn, and Terry Holesinger; Argonne National Laboratory - Balu Balachandran; University of Wisconsin - Susan Babcock and David Larbalestier; Stanford University - Bob Hammond and Luke S. J. Peng; 3M Company - Dave O'Neill; MicroCoating Technologies - Shara Shoup; Oxford Superconducting Technology - Ken Marken; EURUS Technologies - Mike Tomsic; American Superconductor - Marty Rupich; and Intermagnetics General - Pradeep Haldar. Research sponsored by the U.S. DOE Office of Energy Efficiency and Renewable Energy, Office of Power Technologies.

*Oak Ridge National Laboratory is managed by Lockheed Martin Energy Research for the U.S. DOE under contract No. DE-AC05-96OR22464.

1. Proceedings of the *Annual U.S. DOE Peer Review Meeting for the Superconductivity Program*, Washington, D.C., USA, July 26-27, 1999, in press.
2. R. Hammond, in *Advances in Superconductivity XI, Proceedings of the 11th International Symposium on Superconductivity (ISS'98)*, Fukuoka, Japan, November 16-19, 1998, Eds N. Koshizuka, S. Tajima, Springer-Verlag Tokyo 1999, pp. 43-48.
3. V. Iijima et al., in Appl. Phys. Lett. **60**:769-771 (1992).
4. X. D. Wu et al., in Appl. Phys. Lett. **67**:2397- 2399 (1995).
5. D. P. Norton et al., in Science **274**: 755-757(1996).
6. A. Goyal et al., in Appl. Phys. Lett **69**: 1795-1797 (1996).
7. A. Goyal et al., in Micron **30**: 463-478 (1999).
8. R. Feenstra et al., in *Advances In Superconductivity XII, Proceedings of the 12th International Superconductivity Symposium (ISS '99)*, Morioka, Japan, October 17-19, 1999, Springer-Verlag Tokyo, these proceedings.
9. R. A. Hawsey and D. E. Peterson, in Superconductor Industry **9** (3), 23-29 (1996).
10. A. Goyal et al., in J. Supercond. **11**: 481-487 (1998).
11. D. K. Finnemore et al., in Physica C **320**: 1-8 (1999).
12. A. Malozemoff et al., in *Proc. European Conference on Applied Superconductivity (EUCAS '99)*, Barcelona, Spain, in press.
13. I. Matsubara et al., in Jpn. J. Appl. Physics **38**: L727-L730 (1999).
14. M. Paranthaman et al., in Supercond. Sci. and Technol. **12**: 319-325 (1999).
15. P. C. McIntyre et al., in J. Appl. Phys. **68**: 4183-4187 (1990).
16. V. F. Solovyov et al., extended abstract, Intl. Workshop on Critical Currents, July 7-10, 1999, Univ. of Wisconsin, Madison, WI USA.
17. T. L. Peterson, private communication.

DEVELOPMENT OF COATED CONDUCTORS IN JAPAN

Yuh Shiohara and Natsuro Hobara
Superconductivity Research Lab., ISTEC, 1-10-13 Shiononome, Koto-ku, Tokyo, 135-0062 Japan

Abstract: The recent achievements of critical current densities in excess of 1MA/cm^2 at 77K in YBCO films deposited over appropriate bi-axially textured buffer layers/metal substrates have stimulated interest in the potential future applications of coated conductors at liquid nitrogen temperature and high magnetic fields, which are called as the next/second generation tape conductors. Several different processes for obtaining the bi-axially textured buffer layers/metal substrates as well as thicker film deposition processes and non-vacuum processes are reviewed. Under the Ministry of International Trade and Industry (MITI) of Japan administration through the New Energy and industrial Technology Development Organization (NEDO) as collaborative research and development of fundamental technologies for superconductivity applications, Superconductivity Research Lab. was given a responsibility of carrying out research and development of the coated conductors starting from the fisical year of 1999. This paper reviews the research goals of the project including the recent development of the related research by individual research group.

INTRODUCTION

In the last ten years, a major effort has been devoted to the Oxide-Powder-In-Tube (OPIT) process to produce long length high temperature superconducting wires and tapes. This OPIT process, however, can be applied only to BSCCO superconductors. Taking advantage of YBCO, which has its irreversibility at high temperatures, applications at the liquid nitrogen temperatures in preparing superconductive power transmission cables, coils for magnets or current limiter, etc. are highly expected, although its critical superconductive transition temperature is lower than those of BSCCO conductors. The OPIT process when applied to YBCO does not provide expected properties. The YBCO superconductive material has a characteritic critical current density of the order of 1MA/cm^2 when deposited on single crystal substrates. The critical current density of the film on a poly-crystalline substrate is drastically reduced by the presence of high angle grain boundaries[1] and is termed the weak link problem. Accordingly, for the tape applications it is strongly desired to grow the films with highly in-plane aligned crystalline structures, i.e. alignment along not only c-axis but a-axis. The intergranular critical current densities of the films so accomplished approach the film intragranular current densities.

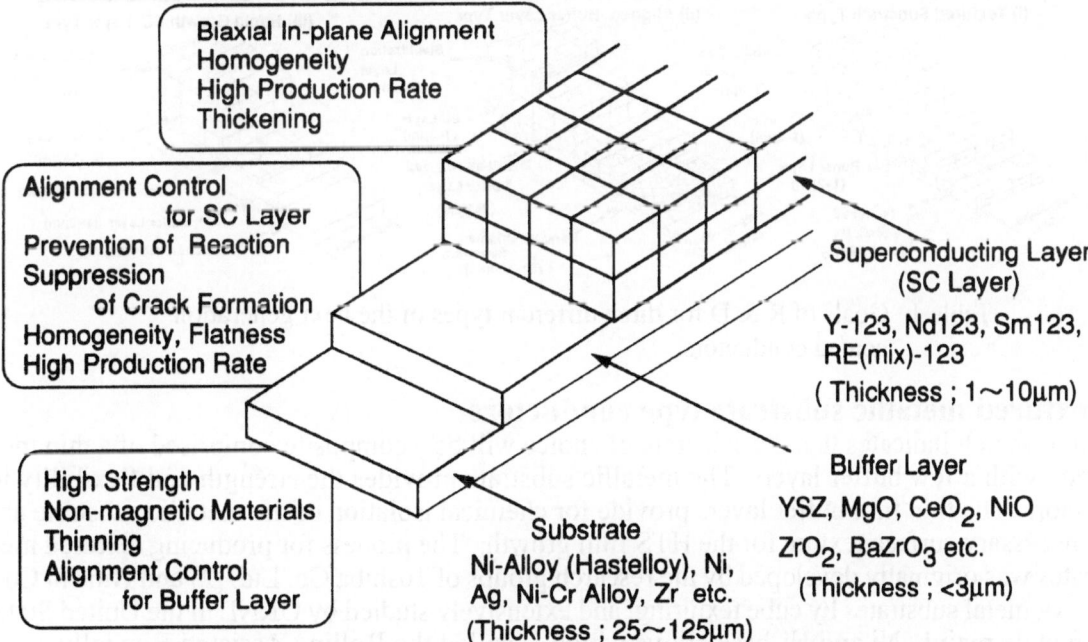

Figure 1 Materials processibng issue for the next generation coated conductors, including basic structure, key factors and candidate materials for each layer.

The YBCO coated conductors are basically constructed by the layers of buffer layers and superconducting layers on a metal substrate as shown in Fig.1. Each layer including a metallic substrate has a specific role for the long length conductors, also as described in the figure. The candidate materials and the thickness of each layer are also presented in the figure. It is noted that a thin passivation layer for stabilization, insulation and encapsulation is excluded.

CURRENT STATUS AND GOALS

The national project in Japan for development of the next generation superconductive coated conductors is carried out by developing the processes to produce three different types of conductors, as listed in Table 1, including the list of candidate materials, key processes and the goals after four years. Conceptually, YBCO coated conductors consist of epitaxially grown films on an appropriate substrate with or without a few buffer layers. Therefore, the substrate or the buffer layer must have a suitably textured surface structure to provide the required in-plane alignment growth of the HTS crystalline films, resulting in avoidance of the weak linkage due to existence of only low angle grain boundaries. The current status, objectives and future goals of the respective conductors are explained briefly in the followings.

		Substrate	Buffer Layer	SC Layer	Key Process	Target
Textured Substrate Type	Non-reactive / High Strength	Ni, Ag, Clad Materials, Ni-Cr, Ni-V, Ni-base Alloy etc.	None, NiO, ZrO$_2$, BaZrO$_2$, MgO, YSZ, Y$_2$O$_3$, CeO$_2$ etc.	Y123, Sm123, Nd123 etc.	Rolling/Annealing, Surface Polishing, Surface-oxidation Epitaxy, Buffer Layer, SC layer Evaluation(Jc etc.)	Length; 10~100m, Sub.Thickness ≦100μm, Jc≧10^5~10^{6}A/cm^2 (77K)
Aligned Buffer Layer Type	ISD / IBAD	Poly Crystalline Hastelloy, Ni-base Alloy etc.	YSZ, MgO, CeO$_2$ etc.	Y123, Sm123, Nd123 etc.	Substarate Polishing, IBAD process, ISD process, SC layer Evaluation(Jc etc.)	Length; 100~1000m, Sub.Thickness≦100μm, Jc≧10^4~10^{5}A/cm^2(77K), Production Rate; >1m/h
Rapid Growth SC Layer Type	MOD / LPE	Ni, Ag, Crad Materials, Ni-base Alloy etc.	None, MgO, YSZ, NiO, BaZrO$_3$, CeO$_2$ etc	Y123, Sm123, Nd123 etc.	Sub. Polishing, Buffer Layer, Homogeneous Seed Film, MOD process, LPE process, Evaluation(Jc etc.)	Length; 1~10m, Sub.Thickness≦100μm, SC Thickness≧5μm, Jc≧10^5~10^{6}A/cm^2(77K), Production Rate; >1m/h

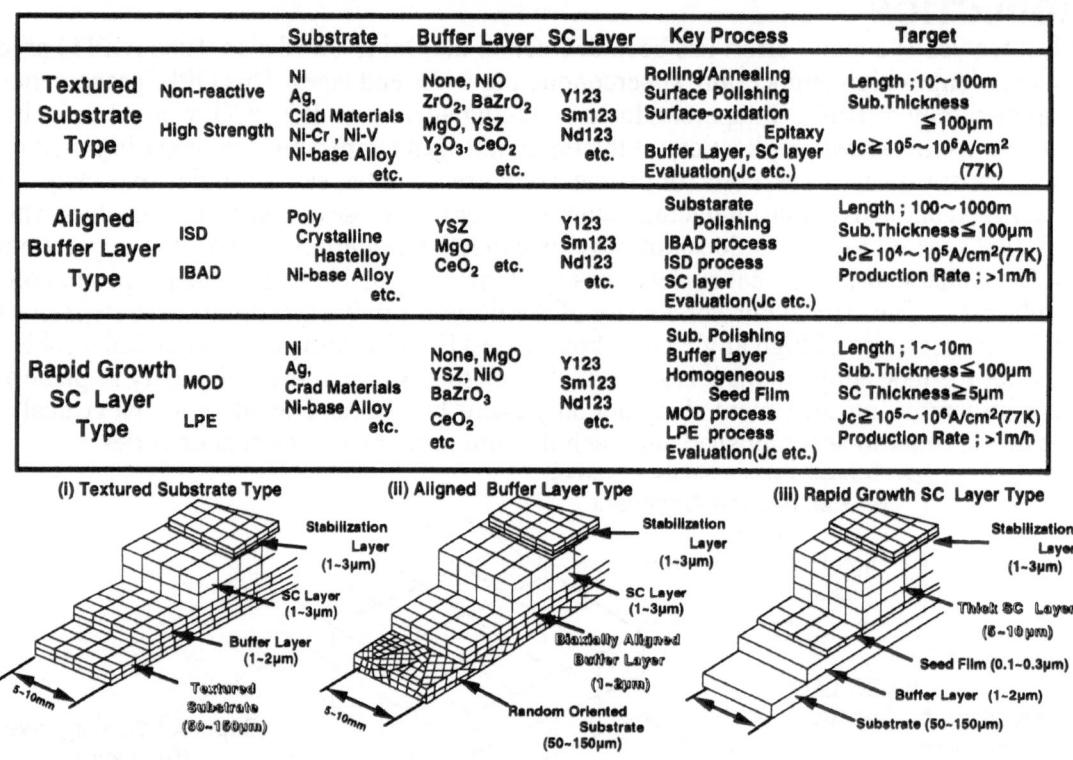

Table 1   Goals of R & D for three different types of the next generation
coated conductors

(1) Textured metallic substrate type conductors

Current research indicates that the substrate of choice will be a composite comprised of a thin metallic substrate with a few buffer layers. The metallic substrate provides the strength and flexibility to the conductors. The thin film buffer layers provide for chemical isolation of the metallic substrate as well as the necessary surface texture for the HTS film growth. The process for producing textured metallic substrates was originally developed by the research groups of Toshiba Co. Ltd. [2] and Hitachi Co. Ltd. [3] for Ag metal substrates by cube texturing, and extensively studied by ORNL in the United States for other metals mainly Ni and Ni-base alloys which is called the Rolling Assisted Biaxially Textured Substrate (RABiTS) process.[4]

In this project, two different textured metallic substrates will be investigated. The first one utilizing the cube textured Ni tape with the NiO buffer layer by means of Surface Oxidation Epitaxy (SOE)[5] together with an MgO buffer layer and depositing YBCO HTS films by Pulsed Laser Deposition(PLD) will be developed by the research group of the Furukawa Electric Co. Ltd. extensively collaborated with SRL-ISTEC. Application of the SOE buffer layer to the BaF$_2$ process and development of high deposition rate processing of the MgO buffer layer using E-beam deposition are also planned. Currently, the cube texturing process for non-magnetic metallic alloys has been investigated. Recent results of the X-ray pole figure analyses of the Ni textured substrate and the NiO buffer layer indicate a clear four fold symmetry, and the Jc-B characteristics for the YBCO coated conductors deposited over the buffer layers achieve the Jc values of $3\times10^5 A/cm^2$ at 77K and 0T and $1\times10^4 A/cm^2$ at 77K and 4T, as shown in Figure 2.

The second one utilizing the cube textured Ag tape with or without a buffer layer will be investigated by Toshiba Co. Ltd. In the case of the textured Ag tape, negligible effect of chemical reaction between Ag and HTS films is a strong advantge. However, the cost and the strength might be a diadvantage. Up to now,1m long YBCO/textured Ag tapes have been produced, in which Jc values are in the order of $10^5 A/cm^2$. It should be noted that in this process no thin buffer layer deposition process was applied which should be cost-effective. Currently, development of the process to strengthen the silver tape is under investigation by means of solution hardening such as Ag-Cu, Ag-Ni alloying and Ag cladding over high strength Ni core tapes.

Goals of this textured metallic substrate type conductors within the next four years are; 10-100 m long, thickness of the substrate ≤ 100 μm, tape production rate $\geq 10m/h$, and Jc $\geq 10^5$-10^6 A/cm^2.

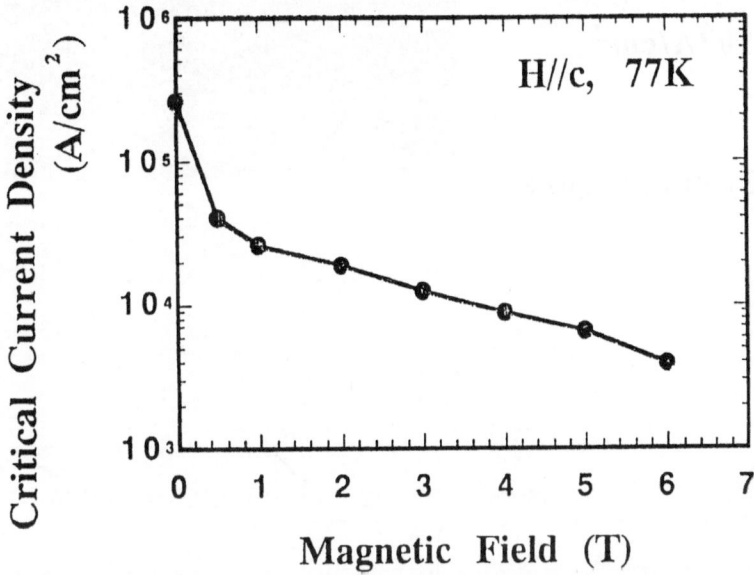

Figure 2 Jc-B characteristics for the YBCO coated conductors deposited over the buffer layers of MgO and NiO(SOE) on a textured Ni substrate.

(2) Aligned buffer layer type conductors

The process of this aligned buffer layer type conductors differs from the previous textured metallic substrate type conductors in which no texture is forced on the metallic tapes but rather the first buffer layer deposited on the metal tape is forced to have a preferred in-plane alignment independent of the underlying metal tapes, which may broaden the choice of the metallic substrate materials. Two different processes have been applied.

The first one is the Inclined Substrate Deposition (ISD) process [6] for the YSZ or CeO2 thin buffer layer in which the buffer layer was deposited by PLD onto an inclined metallic tape of Hastelloy, nickel-base alloy. Hastelloy is the choice for the metallic tape substrate which has a good thermal expansion match with the YSZ buffer layer as well as the YBCO films. In-plane alignment of the crystalline buffer layer was attained due to the shadowing effect for the crystal growth in the vapor deposition. It should be noted that this process does not need any other assisted force such as ion beam and is capable of achieving high deposition rate of about 0.7μm/min. This ISD process including high rate deposition of YBCO films was developed at first in the world by the Sumitomo Electric (SEI) and the Tokyo Electric Power Company (TEPCO). The Ic values of 37A at 77K, OT for a1m long tape and 62.4A for a 1cm long tape have been achieved. Most recently, a 3m long tape with the Jc value higher than $10^4 A/cm^2$ has been sucessfully produced as shown in Figure 3. The immediate objective of this ISD process is application of a high power industrial scale Eximer laser to the long length continuous production of YBCO conductors.

The second is the Ion Beam Assisted Deposition (IBAD) process [7] for the YSZ thin buffer layer, in which the impinging ions from the ion beam selectively cause re-emission of growth-unit particles from the growing surface remaining a certain crystal alignment due to formation of somewhat micro-channelling. This is equivalent to leaving particles which have a particular expected alignment. This IBAD process for depositing the in-plane aligned YSZ buffer layer was developed at first in the world by the research group of the Fujikura Co. Ltd. Application of the IBAD YSZ buffer layer to the YBCO coated conductors has been investigated in the Super-GM project [8] in Japan and has been followed and promoted by LANL [9] in U.S. Up to now, 2.9m long tape with homogeneous and well aligned YSZ buffer layer of 1.1mm in thickness has been produced. Most recently, 2.1m long YBCO tape on the buffer layer has been successfully produced, in which the Jc value of $2.39 \times 10^5 A/cm^2$ and the Ic value of 25.3A at 77K ,OT have been achieved as shown in Figure 4. The IBAD process is an extremely slow process and by its general nature is carried out in combination with a vapor phase deposition process such as the pulsed laser deposition (PLD) process. Accordingly, the immediate objective for this research and development is increasing the area of deposition in order to achieve a higher gross production rate to make it possible to produce a100m or longer tape.

3m Long YBCO Wire

$Jc > 10^4 A/cm^2$

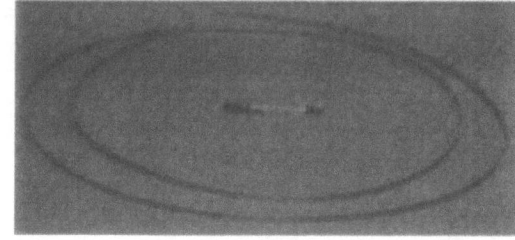

YBCO(103) Pole Figure

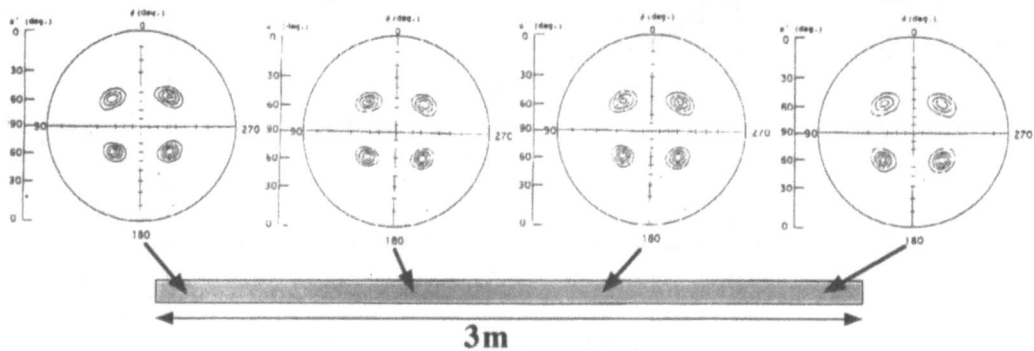

Figure 3 Photograph and results of x-ray pole figure analysis of a 3m long
YBCO tape produced by means of the ISD process

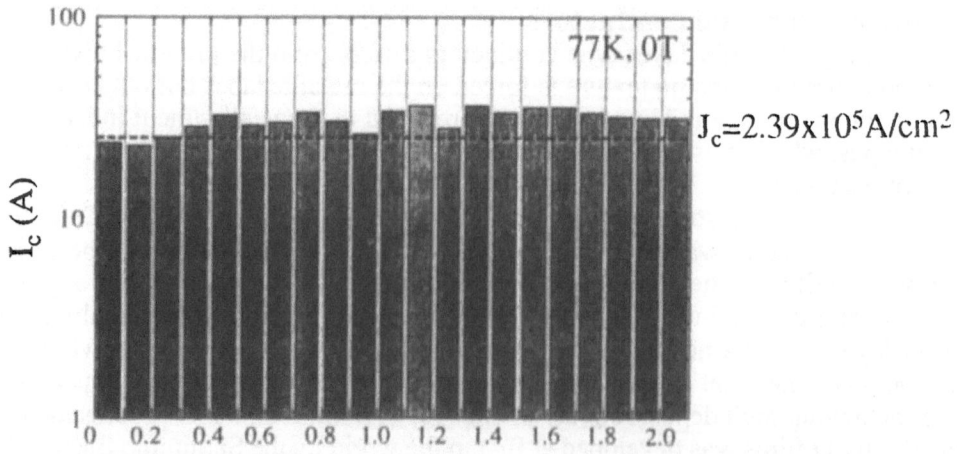

Figure 4 Longitudinal Ic distribution and Jc values of a 2m long YBCO tape
produced by means of the IBAD process. For comparison, Ic = 36.7 A
and Jc = 3.40x105A/cm2 for a 10 cm short sample.

As the above-mentioned processes for this alingned buffer layer type conductors have been well investigated compared with other processes, the two research groups, namely the Sumitomo Electric and the Fujikura, will develop the process focusing on production of long length YBCO coated conductors and investigate the further problems and key technologies for realization of long length coated conductors. Characterization of the long YBCO tapes produced by the groups will be performed by the TEPCO and the Chubu Electric Power Company, considering the future power applications. Although the cost of the coated conductors so produced may not be acceptable, it is important at least to demonstrate the capability of the YBCO coated conductors for electric power applications such as power transmission cables and for coils for magnets or current limitters. Goals of this aligned buffer layer type conductors within the next four years are; 100-1000 m long, thickness of the substrate ≤ 100 μm, tape production rate ≥ 1m/h, and Jc $\geq 10^4$-10^5 A/cm^2.

(3) Rapidly grown superconducting layer type conductors

The engineering critical current density (overall Jc) is the critical current ratioed to the superconductor cross section area divided by the total conductor cross sectional area. It is easy to see that the overall Jc is only a fraction of the superconducting material critical current density of Jc when the cross sectional overhead due to the metallic substrate, a number of buffer layers and a stabilization layer is taken into account. It is therefore required to maximize both the Jc values of the HTS films and the superconducting cross sectional area relative to the total conductor cross sectional area. Two methods of maximizing the superconducting material cross sectional area are obvious,that are increasing the superconducting film thickness and coating both sides of the metallic substrate. The former method is not straightforward either as there is evidence that the film critical current density maximizes for thicknes of up to a few microns at least by the vapor deposition processes. The important element for assessing a candidate manufacturing process is cost. Processes requiring high vacuums will be more costly than an atmospheric pressure processes, and processes having many sequential operations correspondingly have increased costs. Accordingly, development of the innovative processes is strongly expected, which includes the process to produce thick films without degradation of Jc values and non-vacuum process. Two research groups will conduct to develop this type of conductors. The first one is application of the Metal Organics Decomposition (MOD) process to the production of YBCO coated conductors, which is a kind of non-vacuum process. The Showa Electric Wire & Cable Co. Ltd. will investigate the YBCO MOD process on the textured Ag substrates using respective cationic octylate acids, considering the future application to the high strength textured metallic substrates as mentioned before. They successfully produced a 15m long YBCO coated conductors using polycrystalline randomly oriented Ag tape as a metallic substrates, in which the Jc value is the order of 10^4A/cm^2 at 77K, 0T. It is therefore expected to increase the Jc values using the textured Ag substrates. Further invesigations will be resquired, since the Jc values strongly depends on the heat treatment conditions, as shown in Figure 5. Further applications of other metal organics including trifluoroacetates [10] and naphthenic acids to the production of coated conductors and the buffer layers will be investigated with collaboration of SRL-ISTEC. Recently, the SRL-ISTEC group produced YBCO thin films on CeO$_2$ buffered MgO single crystalline substrates by the MOD process using naphthenic acids and achieved the Jc value of 1.2×10^5A/cm^2 at 77K, 0T. They also repeatedly produced YBCO thin films on LaAlO$_3$ single crystalline substrate by the MOD process using trifluoroacetates and attained the high Jc value of 3×10^6A/cm^2 at 77K, 0T.

The second is application of the Liquid Phase Epitaxy (LPE) process [11] to production of thick YBCO and NdBCO or SmBCO coated conductors, which is a non-vacuum process and is capable of achieving relatively high crystal growth rates of several μm/min, which is evident from that this process is generally used for production of bulk single crystals. SRL-ISTEC will conduct this research using two different approaches, one is the low temperature LPE process [12] by means of addition of Ag and BaF$_2$ to the BaO-CuO solutions in order to apply to the textured Ag tape substrates, the other is the high temperature LPE process [13] utilizing appropriate buffer layers such as MgO on either textured Ni substrates or Hastelloy tapes with an in-plane aligned buffer layer and an MgO saturated YBCO seed layer. Recently, YBCO LPE thick films are grown on the thin YBCO seed films over the textured Ag tape. Although the YBCO seed films deposited by MOCVD shows eight fold symmetry in crystal alignment, the LPE grown YBCO thick films show a strong tendency of the four fold symmetry, which is a new finding of the merit of the LPE process. Results are asown in Figure 6. In the case of the high temperature LPE process, the construction of the LPE coated conductors was defined as shown in

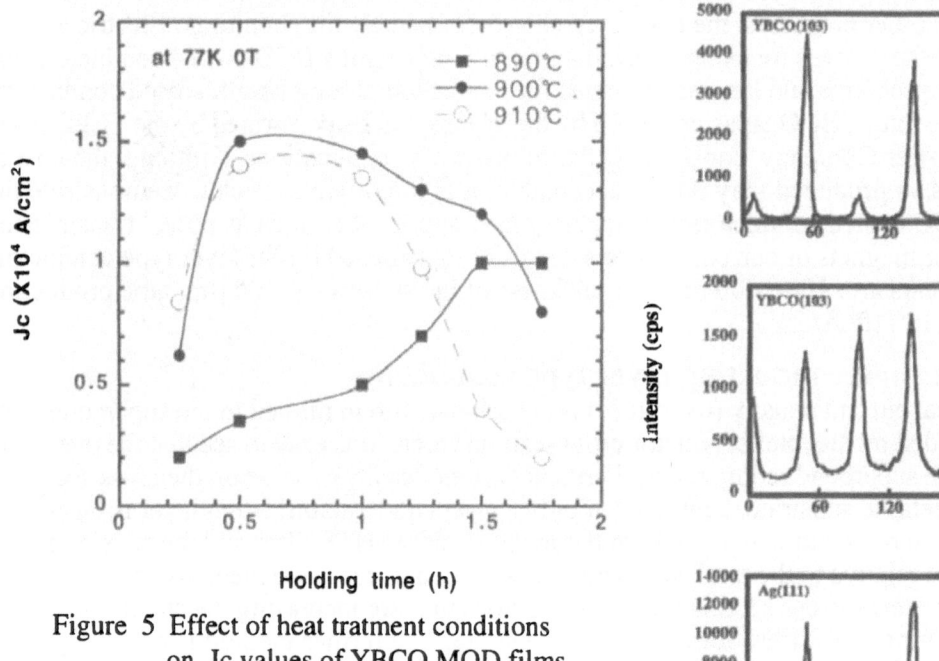

Figure 5 Effect of heat tratment conditions
on Jc values of YBCO MOD films
coated on a polycrystalline Ag tape.

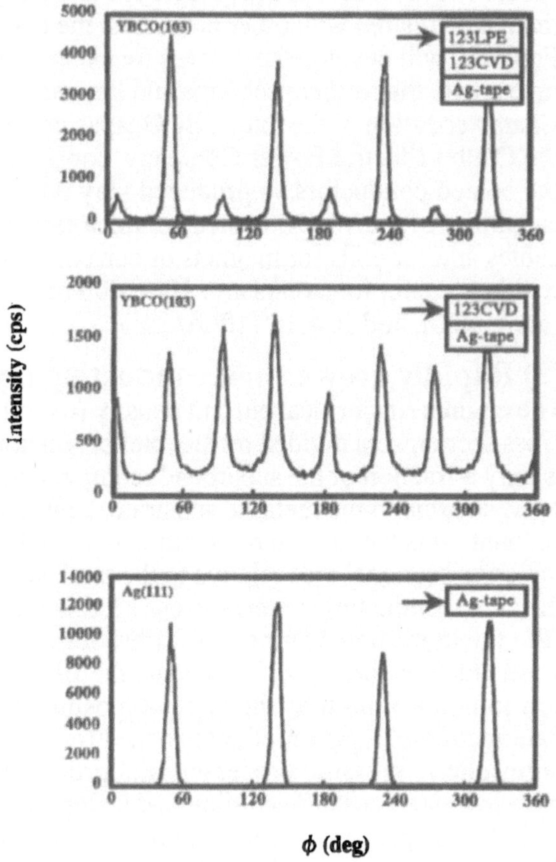

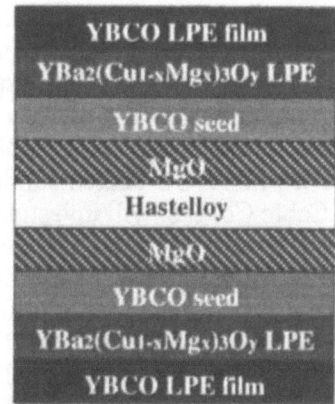

Figure 7 Construction of the YBCO coated conductors
for the high temperature LPE films.

Figure 6 X-ray ϕ-scans of the YBCO low
temperature LPE film over YBCO
seed film (MOCVD) on a textured
Ag tape

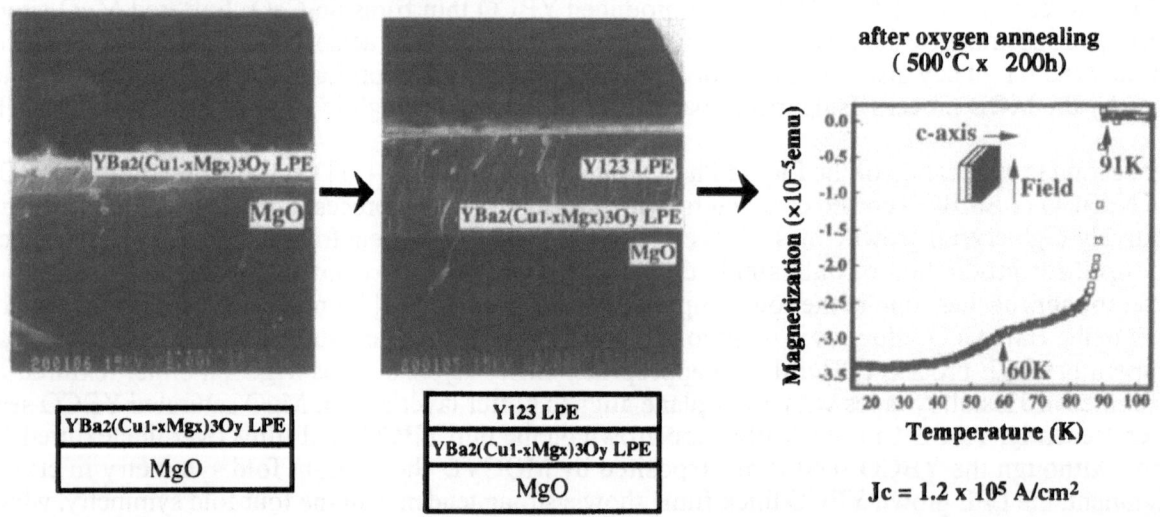

Figure 8 Cross sectional view and superconductive charcteristics of the LPE films on sinle
crystalline MgO substrates with an MgO saturated YBCO layer as a buffer layer.

Figure 7. YBCO thick films are grown on the layers of MgO saturated YBCO seed layer and an MgO buffer layer on a Hastelloy substrate. Effect of the MgO saturated YBCO seed layer on the property of the thick YBCO films was investigated, and the result indicated that the seed layer did not degrade the superconducting properties of the thick YBCO layers, as shown in Figure 8. Development of the process to achieve in-plane alignment of the MgO buffer layer by means of the ISD process using E-beam deposition is now under investigation.

Goals of this rapidly grown superconducting layer type conductors within the next four years are; 1-10 m long, thickness of the substrate ≤ 100 μm, thickness of the superconducting layer ≥ 5μm, tape production rate ≥ 1m/h, and Jc $\geq 10^5$-10^6 A/cm^2.

SUMMARY

The national project in Japan for development of the next generation coated conductors is reviewed. Realization of 100m long YBCO tapes, development of high strength textured metallic substrates, and development of innovative processes to obtain thick superconducting films without degradation of Jc properties and to produce coated conductors in the non-vacuum atmosphere are the main objectives for the next four years. After the years, combinations of the individual successful processes as well as materials including substrates, buffer layers and superconducting layers will offer the feasible coated conductors for future applications.

ACKNOWLEDGMENT

The author would like to thank to all research scientists for providing valuable research results to this paper. This paper contains the work supported by the New Energy and Industrial Technology Development ment Organization (NEDO) as Collaborative Research and Development of Fundamental Technologies for Superconductivity Applictions under the New Sunshine Program administrated by the Agency of Industrial Science and Technology (AIST) of the Ministry of International Trade and Industry (MITI) of Japan.

[1] D. Dimos et al.: Phys. Rev. Lett.(1988) 61, 219
[2] M. Yamazaki et al.: Advances in Superconductivity X, (1997), Springer-Verlag, 619
[3] N. I. Sugiyama et al.: Advances in Superconductivity IX, (1996), Springer-Verlag, 927
[4] A. Goyal et al.: Appl. Phys. Lett. (1996) 69, 1795
[5] K.Matsumoto et al.: Advances in Superconductivity XI, (1998), Springer-Verlag, 773
[6] K. Hasegawa et al.: Proc. of 16 ICEC/ICMC (1997) Elsevier Science, 1413
[7] Y. Iijima et al.: Appl. Phys. Letters, 60, (6) (1992) , 2885
[8] Y. Iijima et al.: Advances in Superconductivity X, (1997), Springer-Verlag, 599
[9] P.N. Arendt et al.: Appl. Superconductivity (1998) 4, 429
[10] P.C. McIntyre et al.: J.Appl. Phys., 71, (4) (1992) , 1868
[11] Y. Yamada et al.: Physica C 217 (1993), 182
[12] Y. Yamada et al.: J.Crystal Growth 167 (1996), 566
[13] N. Hobara et al.: in this Proceedings

Continuous Fabrication of Long and High J_C YBCO Tapes on YSZ Textured Buffer Layers by IBAD Method

Mariko Kimura[1], Yasuhiro Iijima[1], Takashi Saitoh[1] and Kaoru Takeda[2]

[1]Fujikura Ltd., 1-5-1, Kiba, Koto-ku, Tokyo 135, Japan
[2]Super-GM, 5-14-10, Nishitenma, Osaka 530, Japan

Abstract: Long and high Jc YBCO thin film tapes were continuously fabricated on flexible Ni-based alloy (width: 10mm, thickness: 0.1mm). Biaxially aligned YBCO films were deposited by pulsed laser deposition (PLD) on textured YSZ buffer layers formed by using ion-beam-assisted deposition (IBAD). Consumption of a target was controlled as a result of having examined a scanning method of the target. The Jc value of $2.4 \times 10^5 A/cm^2$, and the Ic value of 25.3A were obtained at 77K, 0T over a whole length of 2.1m with 1.06μm YBCO films on 0.7μm YSZ buffer layer.

Keywords: YBCO films, YSZ buffer layer, in-plane alignment, Pulsed Laser Deposition (PLD), Ion-Beam-Assisted Deposition (IBAD)

INTRODUCTION

High Jc values and long length are needed for applications of high-Tc superconductors to electric power technology, such as wires, magnets, and current leads. YBCO retains high Jc in high magnetic fields because of its stronger pinning property. Recently, YBCO tapes are considered to be the second generation of HTS wire, and several groups have succeeded in fabrication of YBCO tapes on metallic substrates of approximately 1m length [1]. IBAD is one of the methods to be the most promising in practical use in a future. By using this method, severe characteristics of intergranular weak links of YBCO films can be eliminated and the Jc value of $1.1 \times 10^6 A/cm^2$ (77K, 0T) was obtained in a short sample. But several factors limited the length of YBCO tapes during long time PLD process, as stability of temperature control, consumption of YBCO target, etc. We have examined a scanning method of a target. As a result, a decrease amount in thickness of the long time deposited YBCO film was reduced. The YBCO tape of 2.1m was fabricated by using this improved target scanning method.

EXPERIMENTAL

Roll-milled tapes of Hastelloy C276 (width: 10mm, thickness: 0.1mm) were used as substrates. YSZ buffer layers were continuously deposited by IBAD. Uniformly textured YSZ buffer layers were demonstrated in the length of 3m (FWHM<20 degree) [2]. For the purpose of preventing the formation of BZO at the interface and lattice mismatch between YBCO and YSZ, 0.1μm thick Y_2O_3 is deposited on YSZ using PLD. YBCO layers were continuously deposited by PLD. Ag films were deposited for the stabilizer and the protector with 10μm thickness by RF magnetron sputtering. Post-annealing in pure oxygen atmosphere was carried. The superconducting properties were measured by the DC four-probe method using a 1μV/cm criterion.

RESULTS AND DISCUSSION

Examination of scanning method of a target

As a method to scan a target, rotation scanning has been used generally. In this method, only a circumference part of a target was consumed so that an incident angle of the laser beam on a target surface was fixed in a one direction, and there were problems that plume would become small. We tried the method to combine rotation scanning and linear scanning. This operation method can control the consumption of the surface so that it can scan the entire surface of a target and let an incident angle of beam on a target surface turn over. Fig.1 shows the cross sectional SEM image of a target. The flatness nature of the surface morphology improved remarkably.

Fig.2 shows decrease ratioes of the thickness of YBCO deposited layers. A decrease ratio of the film thickness after the progress was controlled in about 10% for 9 hours.

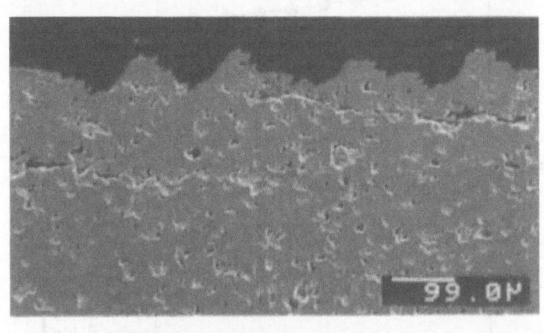

(a) Before improvement

(b) After improvement

Fig.1 Cross sectional SEM image of a target

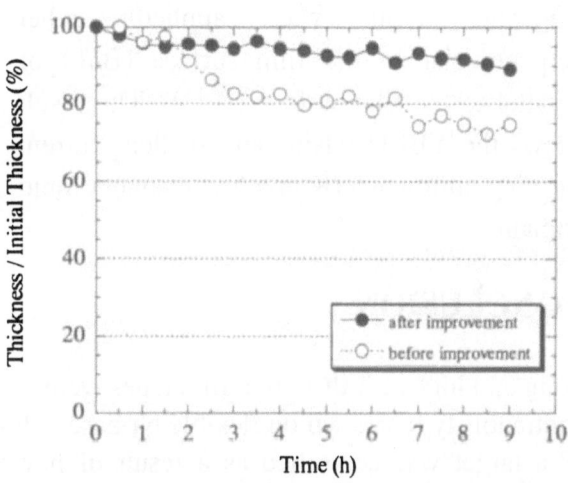

Fig.2 Decrease ratio of the thickness of YBCO layer.

Fabrication of long YBCO tapes

The YBCO tape of 2.1m length YBCO tape was fabricated by using this improved target scanning method. Fig.3 shows the longitudinal Ic distribution of the 2.1m length tape. The Jc value of $2.4 \times 10^5 A/cm^2$, and the Ic value of 25.3A were obtained at 77K, 0T over a whole length of 2.1m with 1.06μm YBCO films on 0.7μm YSZ buffer layers. The maximum value

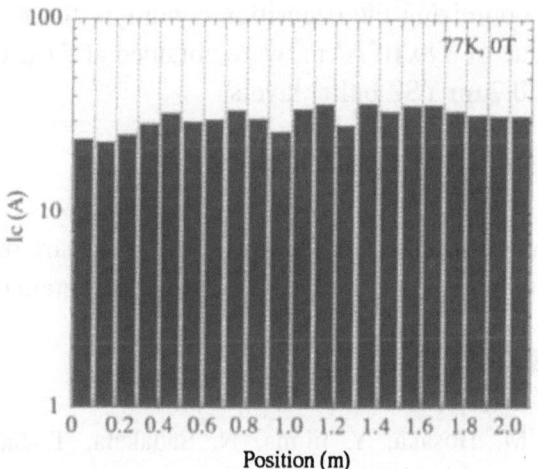

Fig.3 Longitudinal Ic distribution of 2.1m length tape.

of Jc evaluated by a voltage terminal interval of 10cm was $3.4 \times 10^5 A/cm^2$ (Ic=36.7A). We planned to increase the Ic value by optimizing the deposition condition. Fig.4 shows the longitudinal Ic distribution of 1.1m length high Ic tape. The Ic value of 44.9A, and the Jc value of $3.8 \times 10^5 A/cm^2$ were obtained at 77K, 0T over a length of 0.9m with 1.10 μm YBCO films on 0.7μm YSZ buffer layers. The FWHM of the (103) pole figure for the YBCO film of this tape was 9.9°.

Jc-B properties of a low magnetic field domain

Figure 5 shows Jc-B properties of a low magnetic field domain. The Jc value of the sample at 77K, 0T was $5.0 \times 10^5 A/cm^2$. The magnetic field was applied either perpendicular to the film surface (B//c) or parallel to the film surface (B⊥c). The result shows the YBCO retains an excellent current carrying ability at 77K in a low magnetic field domain.

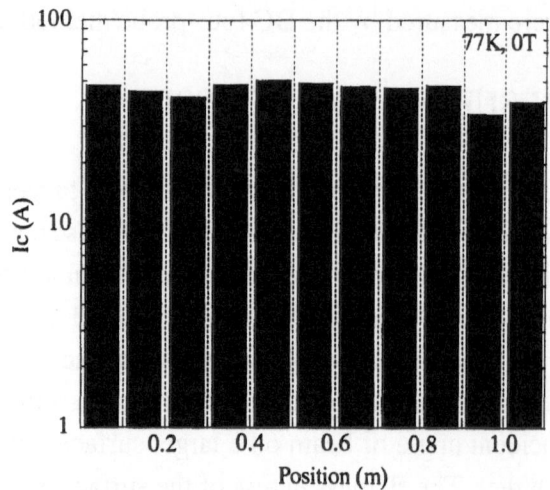

Fig.4 Longitudinal Ic distribution of 2.1m length tape

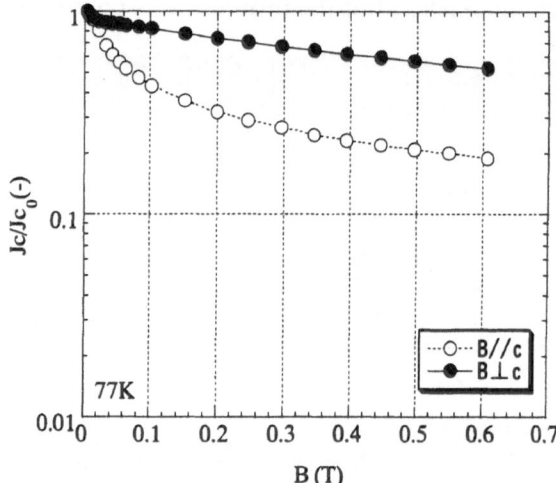

Fig.5 Jc-B properties of a low magnetic field domain.

CONCLUSION

Long and high Jc YBCO thin film tapes were continuously fabricated on flexible Ni-based alloy (width: 10mm, thickness: 0.1mm). Consumption of a target was controlled as a result of having examined a scanning method of the target. A decrease ratio of the YBCO deposited film thickness after improvement was controlled in about 10% for 9 hours. The Jc value of $2.4 \times 10^5 A/cm^2$, and the Ic value of 25.3A were obtained at 77K, 0T over a whole length of 2.1m with 1.06μm YBCO films on 0.7μm YSZ buffer layers. As a result of optimizing the deposition condition, the Ic value increased. The Ic value of 44.9A, and the Jc value of $3.8 \times 10^5 A/cm^2$ were obtained at 77K, 0T over a length of 0.9m with 1.10 μm YBCO films on 0.7μm YSZ buffer layers.

ACKNOWLEDGEMENT

This work has been carried out as a part of R&D on superconducting technology for power apparatuses under the New Sunshine Program of AIST, MITI, being consigned by NEDO.

REFERANCES

1. M. Hosaka, Y. Iijima, N. Sadakata, T. Saitoh, O. Kohno, J. Yoshitomi (1997) Advances in Superconductivity IX: 749
2. Y. Iijima, M. Kimura, T. Saitoh, K. Takeda (1999) submitted to Phisica C

High Critical Current Density YBa$_2$Cu$_3$O$_{7-\delta}$ Tapes Prepared By Surface-Oxidation Epitaxy Method

*Kaname Matsumoto, SeokBeom Kim, and Izumi Hirabayashi, [1]Tomonori Watanabe, [1]Naoki Uno, and [1]Masaru Ikeda

Superconductivity Research Laboratory, ISTEC, 2-4-1, Mutsuno, Atsuta-ku, Nagoya 456-8587, Japan
[1]*The Furukawa Electric Co., Ltd.*, 500, Kiyotaki, Nikko 321-1493, Japan

Abstract: YBa$_2$Cu$_3$O$_{7-\delta}$ (YBCO) films with high critical current density (J_c) were successfully fabricated on nickel tapes buffered with epitaxial NiO prepared by the surface-oxidation epitaxy (SOE). To enhance the superconducting properties of the YBCO films on the SOE-grown NiO, depositions of thin oxide cap layers such as YSZ, CeO$_2$, and MgO on NiO were investigated. These oxide cap layers were epitaxially grown on NiO. A substantially improved data of the critical temperature T_c = 88K and J_c = 3x10^5 A/cm^2 (77K, 0T) and 1x10^4 A/cm^2 (77K, $H//c$, 4T) were obtained for YBCO film on NiO/Ni tape, by using a MgO cap layer with a thickness of 50 nm.

Keywords: YBCO, surface-oxidation, NiO, MgO, PLD, coated conductor

INTRODUCTUION

New production methods to realize highly textured YBa$_2$Cu$_3$O$_{7-\delta}$ (YBCO) films on metallic substrates are necessary, in order to obtain YBCO coated conductors with high critical current density (J_c). The challenge seems to produce single crystalline tape of 1km length, and further technological innovation is required. We have proposed the surface-oxidation epitaxy method (SOE) to produce long YBCO coated conductors [1, 2]. The critical temperature (T_c) of 86-87K and J_c = 4-6x10^4 A/cm^2 (77K, 0T) for the YBCO films on the SOE-grown NiO were reported so far [2]. In this paper, depositions of thin oxide cap layers such as YSZ, CeO$_2$ and MgO on NiO were investigated, to enhance the superconducting properties of the YBCO films on NiO. High J_c of 3x10^5 A/cm^2 (77K, 0T) and excellent magnetic field behavior for YBCO film on the SOE-grown NiO was obtained by using a thin MgO cap layer.

EXPERIMENTAL

The biaxially oriented NiO layer was prepared by the oxidation of {100}<001> textured nickel tape at 1000-1200 °C for 3 hours in flowing Ar gas in an electric furnace. Each of the thin YSZ, CeO$_2$ and MgO cap layers, 50–200 nm thick, was deposited on NiO/Ni tape by pulsed laser deposition (PLD) using a KrF excimer laser. YBCO films were also deposited on MgO/NiO/Ni tapes by PLD. Crystalline phases and in-plane orientation of NiO, oxide cap layers, and YBCO films were studied by X-ray diffraction and by X-ray pole figure measurement. The microstructure and surface morphology were investigated by the transmission electron microscopy (TEM), the scanning electron microscopy (SEM), and the atomic force

Table 1. Texture and surface quality of oxide cap layers on NiO/Ni substrates by PLD.

	YSZ	CeO_2	MgO
Texture of cap layer	Biaxial	Biaxial	Biaxial
Surface morphology	Rough	Rough	Smooth
Roughness (Ra)	10-30 nm	10-20 nm	3-10 nm
Thickness of cap layer	200 nm	200 nm	50-100 nm
Substrate temperature	700°C	500°C	500°C
Oxygen pressure	100 mTorr	30 mTorr	1 mTorr

microscopy (AFM). J_c of the best YBCO films on MgO/NiO/Ni tapes was measured in a magnetic field parallel to the c-axis of YBCO films ($H//c$) at 77K by a dc four-probe method.

RESULTS AND DISCUSSION

The SOE-grown NiO has a flat crystal face, (100) preferred orientation, and high in-plane ordering. The full width half maximum (FWHM) of X-ray ϕ scan of (103)YBCO peak of the film deposited on the NiO layer was a degree of 10-12. However, J_c was lower than the values reported by other methods. This originates from an existence of grooves with depth of 30–150 nm on the grain boundaries of NiO. Such grain boundaries produce the superconducting weak coupling, and it becomes a cause of the J_c lowering[2].

YSZ, CeO_2 and MgO cap layers deposited on NiO/Ni tapes, to improve the surface quality of NiO, showed the biaxial orientation in any combination of YSZ/NiO/Ni, CeO_2/NiO/Ni, and MgO/NiO/Ni. The texture and surface quality of specimens were summarized in Table 1. All surfaces of the cap layers showed brightness. However, according to the observation of SEM and AFM, the roughness was large for the YSZ and CeO_2 surfaces, while the MgO surface was smooth in the present experiment. The smoothness of the surface also changes by the fabrication process of cap layers and afterwards heat treatment, etc [3, 4].

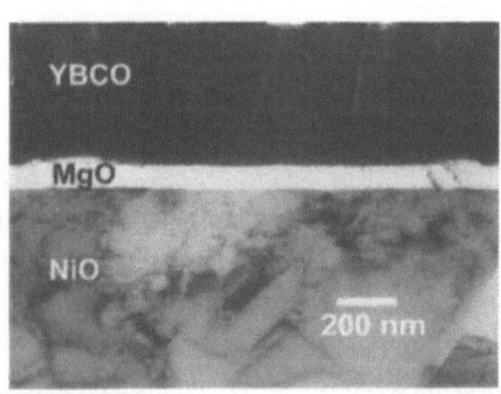

Fig.1. TEM photograph of cross section of YBCO/MgO/NiO/Ni tape.

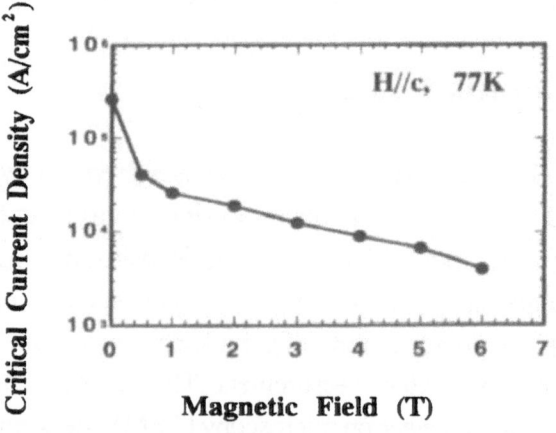

Fig.2. J_c as a function of magnetic field ($H//c$) of YBCO/MgO/NiO/Ni tape.

The YBCO film was deposited on MgO/NiO/Ni tape. By the pole figure measurement, it was proven that YBCO film epitaxially grows on MgO cap layer in cube on cube. FWHM of ϕ scan of YBCO(103) was 11 degrees and FWHM of the rocking curve for (005)YBCO was 2 degrees. A TEM cross section of the YBCO/MgO/NiO/Ni is shown in Fig.1. Epitaxial growth of MgO on NiO/Ni was also confirmed by the electron diffraction. MgO has NaCl structure and the lattice constant is 4.21Å. This value is very approximate for lattice constant 4.17Å of NiO with the same crystal structure, so that MgO grows on NiO in cube on cube and provides the template of epitaxial growth of YBCO.

The zero resistance T_c of YBCO/MgO/NiO/Ni tape reached 88K, with a transition width (10-90%) of 1.5 K, and R(300K)/R(100K)=3 has been achieved. The zero-field transport J_c of the 0.4 μm thick YBCO film was 3×10^5 A/cm^2 at 77K. Several films in the same configuration have comparable J_c values. The magnetic field dependence of J_c of the film for $H//c$ at 77 K is shown in Fig. 2. The behavior was consistent with those of high-J_c YBCO films deposited on single crystalline substrates, indicating that the current path is strongly linked. The results suggest that a thin MgO layer capped the grain boundaries of NiO, so that the formation of high-angle grain boundaries and/or the nickel contamination in the YBCO film in the vicinity of the NiO grain boundaries were suppressed effectively.

SUMMARY

A substantially improved superconducting properties in YBCO films on the SOE-grown NiO/Ni tape was obtained by using MgO cap layer. MgO prevented formation of a weak link and/or nickel contamination in YBCO film in the vicinity of NiO grain boundaries. Zero-field J_c of 3×10^5A/cm^2 and improved magnetic field behavior at 77 K was observed for YBCO film on MgO/NiO/Ni tape.

Acknowledgment. This work was supported by the New Energy and Industrial Technology Development Organization (NEDO) as Collaborative Research and Development of Fundamental Technologies for Superconductivity Applications under the New Sunshine Program Administered by the Agency of Industrial Science and Technology (AIST) of the Ministry of International Trade and Industry (MITI) of Japan.

* Present address: The Furukawa Electric Co.,Ltd.

1. K. Matsumoto, Y. Niiori, I. Hirabayashi, N. Koshizuka, T. Watanabe, Y. Tanaka, and M. Ikeda, Advances in Superconductivity X, ed. K. Osamura and I. Hirabayashi (Springer-Verlag, Tokyo, 1998), pp. 611.
2. K. Matsumoto, S. B. Kim, J. G Wen, I. Hirabayashi, T. Watanabe, N. Uno, and M. Ikeda, IEEE Trans. Appl. Supercond., **9**, 1539 (1999).
3. C. Y. Yang, S. E. Babcock, A. Goyal, M. Paranthaman, F. A. List, D. P. Norton, D. M. Kroeger, and A. Ichinose, Physica C **307**, 87 (1998).
4. M. Becht and T. Mosrishita, J. Alloys Compounds, **251**, 310 (1997).

Thin Film YBCO Tape with Over 3 meter Length Fabricated by Inclined Substrate Deposition

K. Muranaka[†], K. Fujino[†], S. Hahakura[†], K. Ohmatsu[†], H. Takei[†], Y. Sato[††], S. Honjo[††], and Y. Takahashi[††]

† Electric Power System Technology Research Laboratories, Sumitomo Electric Industries Ltd., Osaka 554-0024, Japan

†† Power Engineering R&D Center, Tokyo Electric Power Company, Yokohama 230-8510, Japan

Abstract: We have employed the inclined substrate deposition (ISD) method in order to get a biaxially textured YBCO film [1,2]. For verification of high producibility of ISD method, we have been trying to fabricate YBCO tapes with several meter lengths using the Reel-to-Reel continuous deposition system. The X-ray pole figure measurement revealed that ISD buffer layers and YBCO layers on them were in-plane aligned over 3 meter long. The YBCO tape had the high Jc value of more than 10^4 A/cm^2 at 77.3K over 3 meter length.

Keywords: YBCO thin film, 3 meter long, pulsed laser deposition, ISD, in-plane alignment

INTRODUCTION

High-Tc superconducting (HTSC) thin film tapes, which consist of metal substrates, buffer layers, and YBCO films have been developed for power applications. In view of scaling up the HTSC thin film tapes, substrate must be elongated. Single crystals were not employ as tape substrates for industrial limitation in regards of length and flexibility. Among polycrystalline materials, a Ni-based alloy was selected as a substrate because of flexible, unoxidizable and non-magnetic properties. Buffer layer was needed to prevent the interaction between substrates and HTSC films. Both the YBCO layers and the buffer layers are formed by pulsed laser deposition (PLD). For high current density of YBCO, in-plane aligned YSZ buffer layers were fabricated by ISD method. YBCO film formation was carried out by a normal PLD. YBCO films on ISD buffer layers were also biaxially textured in a substrate plane. The best Jc value of 4.3×10^5 A/cm^2 was achieved for a short sample [2], and the 1 meter long YBCO film tape with Jc of 1.5×10^5 A/cm^2 was successfully fabricated by ISD [3]. In this paper we report on the continuous deposition of over 3 meter long YBCO thin film with Reel-to-Reel substrate transfer system.

EXPERIMENTAL

The technique of continuous deposition on moving substrate is necessary for production of long length tapes. Figure 1 shows the Reel-to-Reel substrate transfer system, which we designed and constructed for continuous deposition of long length substrate a few years ago. A pair of reels transports a substrate tape through the deposition chamber. Over 3 meter length HTSC thin film tape consisted of YBCO film on flexible Ni-based alloy substrate with YSZ buffer layer. The substrate which size was 10mm wide and 0.1 to 0.2 mm thick was continuously sliding on a heater stage during deposition. Both YSZ and YBCO layers were deposited by PLD with KrF (λ=248nm) excimer laser. YSZ layers were deposited twice at the production speed of 0.7m/h with ISD method. YBCO layers were deposited on YSZ layers at the production speed of 0.6m/h. The thickness of YBCO was from 0.7 to 1.0 μm. The tape surface was protected with sputtered Ag layer. The orientation of films was measured by X-ray pole figure measurement. Jc was measured by conventional four-probe method in liquid nitrogen.

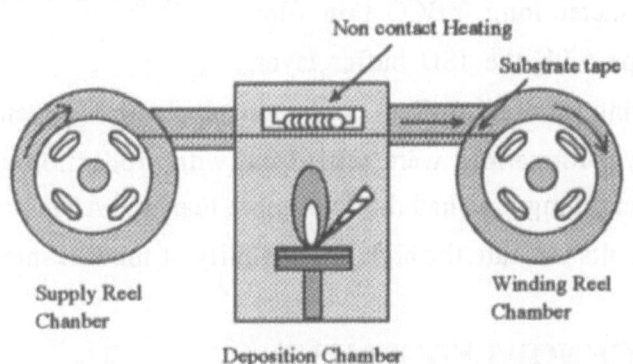

Figure 1. Schematic diagram of the Reel-to-Reel substrate transfer system

RESULTS AND DISCUSSION

Figure 2 shows the X-ray pole figure of YBCO layer observed in 1 meter apart. We can see that the YBCO films had the <001> axis normal to the substrate and the highly in-plane alignment over the whole length. Phi scans show the full width at half maximum in-plane orientation distributed between 20.5° and 23° in the whole length of 3 meter. 3 meter long YBCO tape was divided into 3 parts, and measured Jc of them in every 10 cm voltage tap spacing. Figure 3 shows the distribution of Jc of

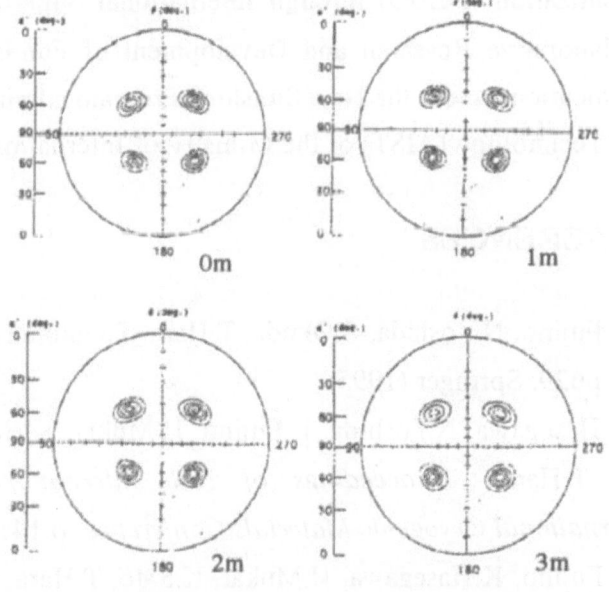

Figure 2. X-ray (103) pole figure of the YBCO observed in 1 meter apart

3 meter long tape. We obtained the maximum Jc of 1.8×10^4 A/cm^2 at 10 cm spacing. The value of Jc was more than 10^4 A/cm^2 in the whole area of 3 meter.

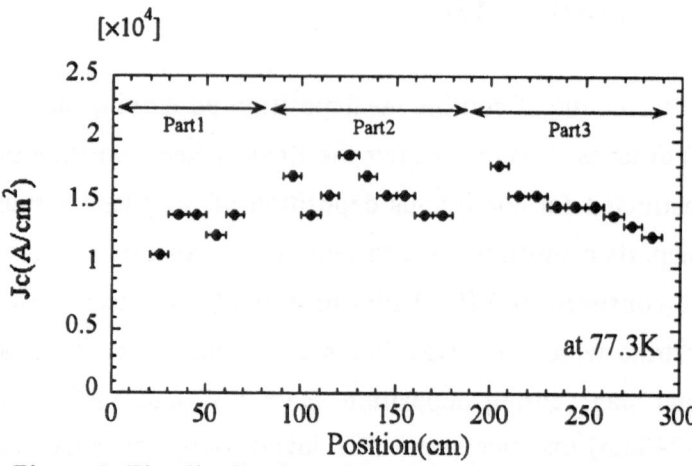

Figure 3. The distribution of Jc of 3 meter long tape

CONCLUSION

We have successfully fabricated 3 meter long YBCO thin film tape with the ISD buffer layer using the Reel-to-Reel continuous deposition system. Then the ISD buffer layer and the YBCO layer formations were carried out with production speeds of 0.7m/h and 0.6m/h respectively. 3 meter long tape had the Jc of more than 10^4 A/cm^2 over the whole length. We suppose that these results indicate the high producibility of the ISD method.

ACKNOWLEDGEMENT

This work was supported by the New Energy and Industrial Technology Development Organization (NEDO) through International Superconductivity Technology Center (ISTEC) as Collaborative Research and Development of Fundamental Technologies of Superconductivity Applications under the New Sunshine Program administered by the Agency of Industrial Science and Technology (AIST) of the Ministry of International Trade and Industry (MITI) of Japan.

REFERENCES

1 K.Fujino, N.Yoshida, S.Okuda, T.Hara, T.Ohluma and H.Ishii : *Advances in Superconductivity VII*, p629. Springer (1995)

2 K.Hasegawa, N.Yoshida, K.Fujino. H.Mukai, K.Hayashi, K.Sato, T.Ohkuma, S.Honjo, H.Ishii and T.Hara : *Proceedings of 16th International Cryogenic Engineering Conference / International Cryogenic Materials Conference*, p.1413. Elsevier Science (1997)

3 K.Fujino, K.Hasegawa, H.Mukai, K.Sato, T.Hara, T.Ohkuma, H.Ishii and S.Honjo : *Advances in Superconductivity VIII*, p675. Springer (1996)

New Metallic Substrate for YBCO Superconducting Tape

H. Yoshino, M. Yamazaki, T.D. Thanh, Y. Kudo and H. Kubota

Advanced Materials and Devices Laboratory, Corporate R&D Center, Toshiba Corp.,
1 Komukai Toshiba-cho, Saiwai-ku, Kawasaki 210-8582, Japan

Abstract: New Ag alloys (Ag-Cu alloy, Ag-Cu/Ni clad alloys) for YBCO superconducting tape were investigated with respect to the mechanical tensile strength, crystal orientation and superconducting property. These alloys were obtained by cold roll in the size of 10 mm wide and 0.1 mm thick. The crystal orientation of these as-rolled alloys showed Ag(110) parallel to the tape surface and changed to (210) after annealing below $700°C \times 30min$. The proof stress $\sigma_{0.2}$ for these clad alloys was more than 80 MPa, which is about 4 times larger than that of pure Ag. Pole figure of YBCO film deposited by PLD at $700°C$ showed in-plane alignment with a FWHM of $22°$. The Jc value of 1.4×10^5 A/cm^2 was obtained at 77 K.

Keywords: YBCO superconductor, Ag clad alloy, Mechanical strength, In-plane alignment

INTRODUCTION

The direct deposition of YBCO superconducting film on Ag tape without any buffer layer is one of the most promising methods for preparing a low-cost superconducting tape because of the simple manufacturing process. It has been clarified that when the suitable crystal plane aligned, for example Ag(110), the YBCO film deposited on it showed good in-plane alignment [1], and a high Jc value of 1.2×10^5 A/cm^2 at 77K was reported for the (110) textured Ag tape [2]. On the other hand, pure Ag tape becomes very soft after deposition of YBCO film by annealing. Therefore, it is necessary to improve the mechanical strength of Ag tape to facilitate practical handling.

In this work, some Ag alloys (Ag-Cu alloy, Ag-Cu/Ni clad alloys) were investigated for the purpose of increasing mechanical strength. Crystal orientation of these Ag alloys and YBCO films deposited on them were also studied.

EXPERIMENTAL

Ni, Ni-Cu alloy and Ni-Ag alloy were used as core metals and Ag-Cu alloy, containing a small amount of Cu, was used as a clad metal. Cu addition is effective to suppress diffusion of Cu from YBCO film. These clad alloys were prepared by cold roll without intermediate heat treatment. The reduction ratio was more than 98 %, and finally obtained tape size was 0.1mm thick and 10 mm or 20 mm wide. Crystal orientation of these alloys was identified using X-ray diffraction pattern and pole figure measurement. The mechanical tensile strength was measured by the conventional tension test method at room temperature after annealing at $700°C$ for 30 min. in N$_2$ atmosphere. The deposition of YBCO film on these Ag clad tapes has been done using a pulsed KrF excimer laser deposition method (PLD). The deposition temperature was about $700°C$, and film thickness of YBCO was controlled to be about 500 nm. After deposition of YBCO film, the sample was cooled slowly to room temperature as oxygen gas was induced into the vacuum chamber. Crystal orientation of the YBCO films was studied by X-ray diffraction pattern and pole figure. Jc value was measured by the four-probe method with dc current at 77K. The distance of the electrode was

10 mm and the criterion was 1 μ V/cm.

RESULTS AND DISCUSSION

The change of crystal orientation by heat treatment for Ag-Cu tape is shown in Fig.1. Only a strong (220) diffraction peak was observed for as-rolled tape and (420) diffraction peak increased with heat treatment. The increase of this (420) diffraction peak by heat treatment reappeared well. In the case of pure Ag tape, many peaks of (111), (200), (220) and (311) were observed and the intensity of peaks was changed randomly. Pole figures of Ag after annealing at 700℃ are shown in Fig. 2. It can be seen that Ag(420) peak exists in the center and two Ag(200) peaks are fairly sharp. This means that Ag-Cu tape is textured Ag(210) and aligned in-plane.

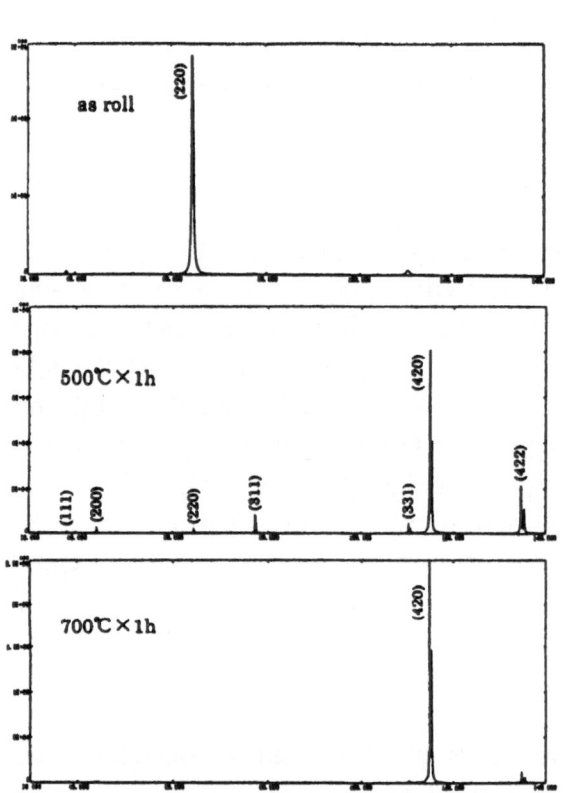

Fig. 1   Change of crystal orientation of Ag-Cu tape by heat treatment

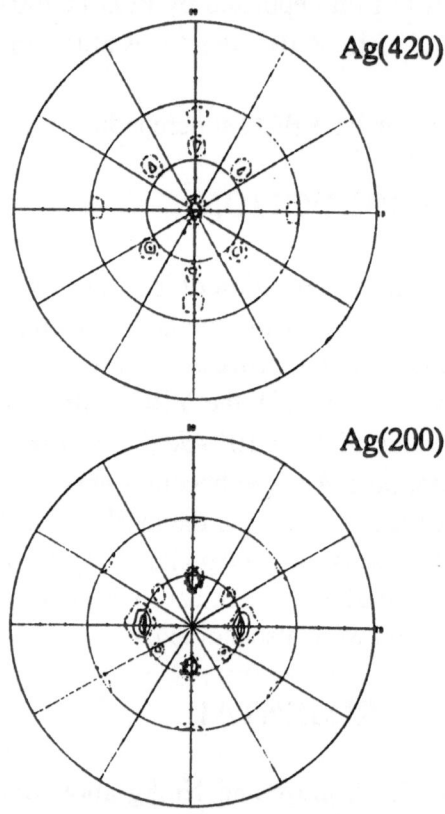

Fig. 2   Pole figures of Ag-Cu tape after annealing at 700℃ × 1h

Fig. 3 shows β scan profile of YBCO film after deposition on Ag-Cu tape. YBCO(103) diffraction pattern shows 4 times symmetry and the full-width-at-half-maximum (FWHM) is about 34 °. The FWHM is not particularly sharp but fairly good Jc of 2.0×10^5 A/cm^2 (Ic=18 A) was obtained at 77K. When in-plane alignment is improved, higher Jc will be obtained.

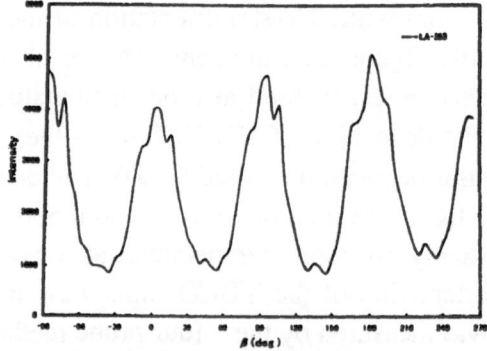

Fig. 3 β scan profile of YBCO deposited on Ag-Cu tape

Stress-strain curves were measured for the clad tapes and the proof stress $\sigma_{0.2}$ was determined at 0.2% strain. Proof stress are summarized in Fig. 4 and data on Ag and Ag-Cu tape are compared. In this figure, it is clarified that the values of $\sigma_{0.2}$ for clad alloys are about four times higher than that of Ag. In the case of the clad alloy using Ni-Cu core alloy, partial cleavage of the clad metal was observed after heat treatment in the air. It is considered that Cu in the Ni-Cu core alloy oxidized at the interface of clad by heat treatment in the air, and metallic binding energy became weak.

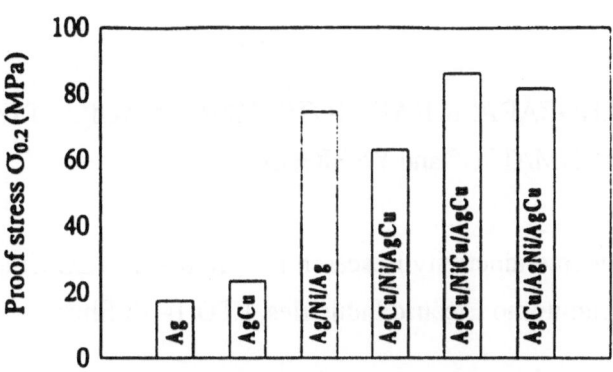

Fig. 4   Proof stress $\sigma_{0.2}$ for various Ag clad alloys

Fig. 5 shows pole figure and β scan profile of YBCO deposited on Ag-Cu/Ag-Ni clad alloy. Fairly good in-plane alignment was observed and the value of FWHM was about 22°. The Jc value of this sample was 1.4×10^4 A/cm^2 (Ic=12.2A) at 77 K.

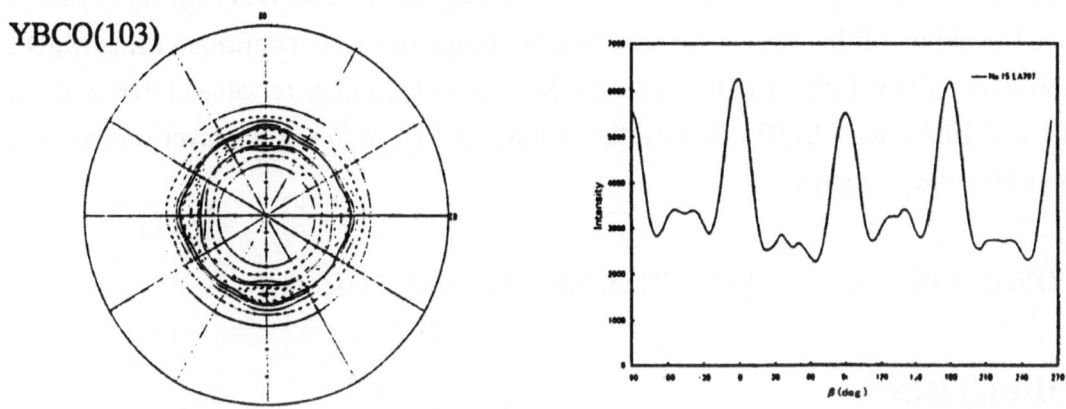

Fig. 5   Pole figure and β scan profile of YBCO deposited on Ag-Cu/Ag-Ni clad tape

CONCLUSION

New metallic substrate for YBCO superconducting tape was prepared by the clad method with Ag-Cu alloy and Ni alloys. Tensile strength was increased by the clad method and high proof stress $\sigma_{0.2}$ was obtained. These clad tapes have textured structure of Ag(210) and it is clarified that YBCO film deposited on Ag(210) has in-plane alignment. High Jc values of 2.0×10^5 A/cm^2 and 1.4×10^5 A/cm^2 were obtained for Ag-Cu tape and Ag-Cu/Ag-Ni clad alloy, respectively.

Acknowledgment. This work is supported by the New Energy and Industrial Technology Development Organization (NEDO) through ISTEC as Collaborative Research and Development of Fundamental Technologies for Superconductivity Applications.

1. J.D. Budai, R.T. Young, and B.S. Chao   Appl. Phys. Lett. 62 (1993) 1836.
2. M. Yamazaki, T.D. Thanh, Y. Kudo, H. Kubota, H. Yoshino, and K. Inoue, in Advances in Superconductivity X, edited by K. Osamura and I. Hirabayashi (Springer-Verlag, Tokyo 1998), pp.619-622.

DEVELOPMENT OF Y-SYSTEM COATED CONDUCTOR ON METAL SUBSTRATE BY LPE METHOD

N.HOBARA, K.KAKIMOTO, Y.NAKAMURA, T.IZUMI, T.YUASA, Y.TAKAHASHI, K.FUJINO*, K.OHMATSU* and Y.SHIOHARA

Superconductivity Research Laboratory, ISTEC,Koto-ku,Tokyo,135-0062,JAPAN
*Sumitomo Electric Industries, LTD.1-1-3,Shimaya,Konohana-ku,Osaka,554,JAPAN

ABSTRACT

The fabrication process of Y-system superconducting films on Hastelloy substrates by the Liquid Phase Epitaxy (LPE) method was developed. The reaction between Hastelloy and the oxide melt was prevented by an MgO buffer layer for MgO saturated oxide melt. Consequently, the suitable construction for tape processing by LPE was found as $YBa_2Cu_3O_y$<LPE>/$YBa_2(Cu,Mg)_3Oy$<LPE> /MgO/ Hastelloy. The ability of this system superconducting properties were confirmed using MgO single crystal substrates. Y123<LPE>/Y123(Mg)<LPE>/MgO revealed a high Tc value of 91K and a high Jc value of $1.2 \times 10^5 A/ cm^2$(77K,0T). Finally, the continuous $YBa_2 (Cu,Mg)_3O_y$ layer was succeeded in growing on Hastelloy substrates.

KEY WORDS: LPE, coated conductor, Y123, MgO, Hastelloy, buffer layer

INTRODUCTION

In application of the long length RE-system coated conductor, it is important to obtain three -dimensional crystal orientation alignment, high critical current, homogeneity and a high production rate. The LPE process is advantageous for fabricating thick-films at a high growth rate with high superconducting properties in comparison with the commonly used vapor processes[1]. However, In order to grow superconducting films on metal substrates by LPE, there are several problems to be solved such as the prevention of a reaction between metal and liquid oxide, realizing in-plane alignment and steady state growth for long time, etc. In this study, we have investigated at first to design the construction which is suitable for LPE tape processing.

EXPERIMENTAL

1.Prevention of reaction

The process to prevent the reaction was investigated by the following procedure using an MgO buffer layer, which is recognized to be stable for Ba-Cu-O melt. The Hastelloy substrates covered with MgO films were prepared. The substrates were dipped into the two different melts at 1000℃. One has the composition of Y:Ba:Cu=0.6:37.3:62.1 and the other was basically the same composition but saturated with MgO at the growth temperature.

2.LPE growth on MgO single crystal substrate

In order to clarify whether 123 crystal will grow from MgO saturated melt, and to evaluate its superconducting properties, MgO single crystal substrates with Y123 seed films were dipped into the MgO saturated solution. Furthermore second dipping was performed into the $Ba_3Cu_5O_x$ solution without MgO on it. The grown samples were annealed at 500℃ for 200 hours in a flowing oxygen atmosphere.

3.LPE growth on Hastelloy

To apply the MgO saturated process for metal substrate, the substrate of Y123seed/MgO/Hastelloy was dipped into the MgO saturated solution.

The sample characterizations were carried out by optical microscopy, SEM, EPMA and SQUID.

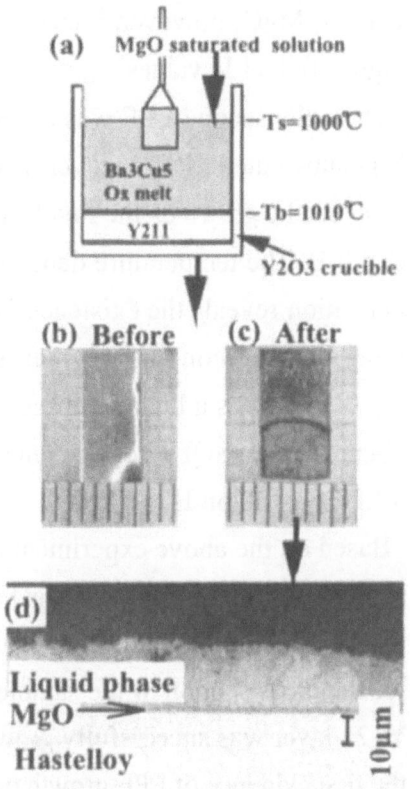

Fig.1. Experimental procedure for reaction using MgO saturated melt (a) and the typical results (b-d). (b)appearance before dipping, (c) that after dipping and (d) the cross section of the sample after dipping

RESULTS AND DISCUSSION

1.Prevention of reaction

Although an MgO buffer layer was dissolved into the melt completely in the case of the dipping into the non saturated melt, dissolution was scarcely observed in the case of the dipping into the melt saturated with MgO. Fig.1 shows appearance of the samples before and after dipping, and the cross section of the MgO/Hastelloy substrate after vertical dipping observed by optical microscopy. The reaction of the MgO buffer layer with the BaO-CuO melt was prevented using the MgO saturated melt. This result suggests that the MgO saturated melt has to be used for the LPE 123 growth.

2.LPE growth on MgO single crystal substrate

Figure.2(a) shows the cross section of the film grown from MgO saturated melt on the MgO single crystal substrate. The film revealed a Tc value of 40K after annealing in oxygen. The above results indicate that the Y123-type structure

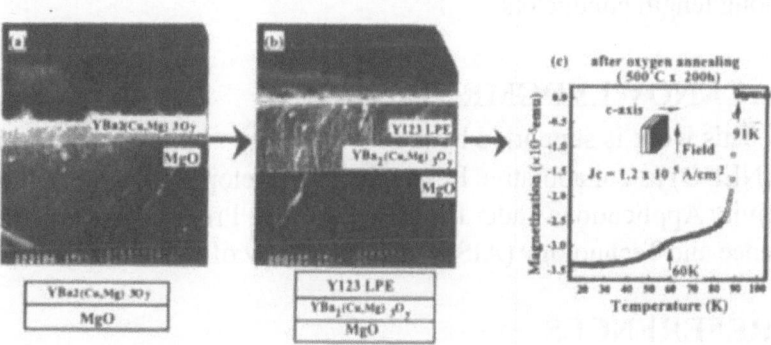

Fig. 2. SEM photographs and superconducting property of Y123/YBa$_2$(Cu, Mg)$_3$O$_y$/MgO LPE film

was formed even from the MgO saturated liquid, and the additional LPE growth form the melt without MgO, however, is necessary to obtain higher Tc and Jc values.

Then, the second LPE was carried out on the Mg- substituted LPE123 layer. The second LPE layer can be grown on the first layer as shown in Fig.2(b). The temperature dependence of magnetization reveals the existence of two difficult kinds of superconducting phases as shown in

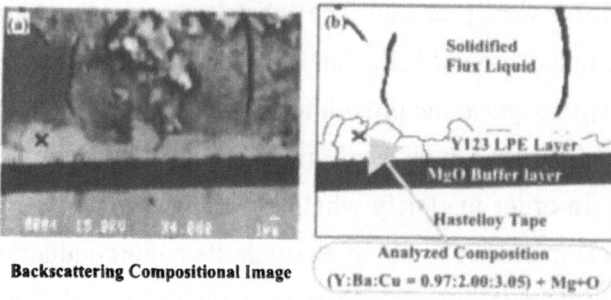

Backscattering Compositional Image

Fig.3. Backscattering compositional image (a) and its schematic image (b) of the sample grown from MgO saturated melt on Hastelloy sabstrate using an MgO buffer layer

Fig.2(c). One is a high Tc phase of 91K and the other is a low Tc phase of 60K. Furthermore, the Jc obtained from SQUID measurement was $1.2 \times 10^5 A/cm^2 (77K,0T)$.

3.LPE growth on Hastelloy

Based on the above experimental results of growth of the 123 phase on an MgO single crystal substrate, LPE growth from the MgO saturated melt on Hastelloy substrate using an MgO buffer layer was tried by the dipping method. Figure.3 shows the results of EPMA analysis for the cross section of the sample. According to the result, the continuous Y123 layer was successfully grown on the MgO buffer layer. This may be the first evidence of LPE growth on Hastelloy substrates for Y123 superconducting materials.

Through the above preliminary examinations, the suitable construction of coated conductor using Hastelloy tape could be defined for the LPE process. The construction is schematically shown in Fig.4.

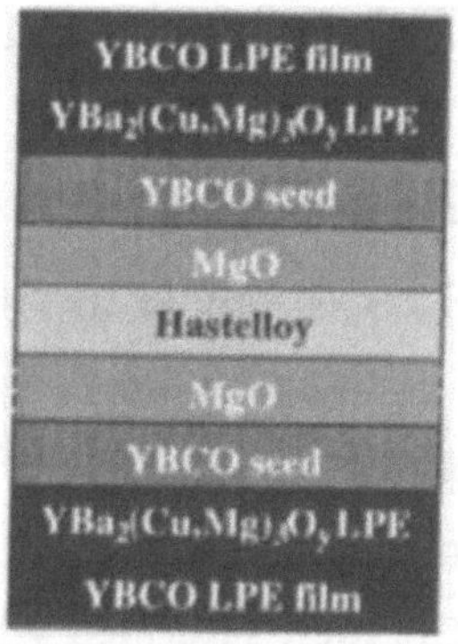

Fig.4. Schematic image of defined construction which is suitable for coated conductors using Hastelloy tape by LPE process

CONCLUSION

The high temperature LPE process was applied to fabricate Y123 conductors on Hastelloy substrates. It was found that the LPE processing is promising using two buffer layers of MgO and $YBa_2(Cu,Mg)_3Oy$. Further investigations are required, especially on optimizing the dimension of each layer as well as the fabrication of long length conductors.

ACKNOWLEDGMENT

This work is supported by the New Energy and Industrial Technology Development Organization (NEDO) as collaborative Research and Development of Fundamental Technologies for Superconductivity Applications under the New Sunshine Program administrated by the Agency of Industrial Science and Technology (AIST) of the Ministry of International Trade and Industry (MITI) of Japan.

REFERENCES

[1] Xin Yao et.al.95K NdBCO Single Crystal Grown in Air by Controlling Liquid Composition Jpn.Appl.Phys.Vol.36(1999)

Effects of the Final Heat-Treatment Conditions on Microstructures of YbBa$_2$Cu$_3$O$_{7-\delta}$ Superconducting Final Films Deposited on LaAlO$_3$(001) Substrates by the Dipping−Pyrolysis Process

Junko Shibata,[1*] Katsuya Yamagiwa,[2] Izumi Hirabayashi,[2] Tsukasa Hirayama,[1] and Yuichi Ikuhara[3]

[1]Japan Fine Ceramics Center, 2-4-1 Mutsuno, Atsuta-ku, Nagoya, 456-8587 Japan
[2]Superconductivity Research Laboratory, ISTEC, 2-4-1 Mutsuno, Atsuta-ku, Nagoya, 456-8587 Japan
[3]Department of Materials Science, The University of Tokyo, 2-11-16 Yayoi, Bunkyo-ku, Tokyo, 113-8656 Japan

Abstract: We investigated effects of final heat-treatment conditions on microstructures of YbBa$_2$Cu$_3$O$_{7-\delta}$ films formed on LaAlO$_3$(001) substrates by the dipping-pyrolysis process. First, we prepared amorphous precursor films by heating the spin-coated substrates at 425℃ in air. Subsequently, YbBa$_2$Cu$_3$O$_{7-\delta}$ films were prepared by heat-treating these precursor films at 750℃ in an Ar gas flow under various conditions of heating rate and holding time. Microstructures of the final films were studied by transmission electron microscopy. In conclusion, it was found that rapid heating rate at the final heat-treatment is necessary for the epitaxial growth of the superconducting films, and that long holding time is effective for the grain growth of the crystals.

Keywords: dipping-pyrolysis process, YbBa$_2$Cu$_3$O$_{7-\delta}$ film, transmission electron microscopy, final heat-treatment

INTRODUCTION

The dipping pyrolysis process is a promising method for producing superconducting films with high critical temperature (T$_C$). It was reported that YBa$_2$Cu$_3$O$_{7-\delta}$(Y123) films with zero resistance at 92K and with zero-field critical current densities (J$_C$) higher than 5×10^6A/cm^2 at 77K were prepared on LaAlO$_3$(LAO) substrates by using this method[1]. In this method, metal organic compounds on the substrates decompose and are crystallized to form the superconducting films during initial heat-treatment and final heat-treatment. In these heat-treatments, it is important to control the nucleation and growth of crystals for obtaining good superconducting properties. We reported effects of the initial heat-treatment conditions on microstructures of the films in the previous papers[2],[3]. In the present study, we have investigated the relationship between the final heat-treatment conditions and the microstructures of the films by transmission electron microscopy(TEM).

EXPERIMENTAL

YbBa$_2$Cu$_3$O$_{7-\delta}$(Yb123) precursor films were prepared by spin-coating LAO(001) substrates with a solution including the naphtenate salts of yttrium, barium, and copper(1:2:3 metal ratio) dissolved in toluene, and then by heating the substrates at 425℃ for 30min in air. The final films were prepared by heat-treating these precursor films in an Ar gas flow with the oxygen partial pressure p(O$_2$) of 10^{-4} atm at 750℃. More detail information about preparation of the films was described in the former paper[4]. Table 1 shows five different conditions of the final heat-treatment at 750℃ for preparing Film A,B,C,D, and E. The cooling rate from 750℃ to room temperature was 3℃ /min when Film A,C,D and E were prepared. Film B was prepared by heating the precursor film at

0.5℃/min to 750℃, and by turning the heater off to let the temperature of the film dropping to 30℃. Crystal Phases and microstructures of these final films were characterized by X-ray diffraction(XRD) using Cu-Kα radiation and by the TEM observation. In this work we used transmission electron microscope, JEM-2010, operating at 200kV with the point resolution of 0.194nm.

Table.1   Final heat-treatment conditions for preparing Film A,B,C,D, and E.

	Heating rate [℃/min]	Annealing temperature [℃]	Holding time at Annealing temperature [hours]	Cooling rate [℃/min]
Film A	20	750	10	3
Film B	0.5	750	0	(heater off)
Film C	0.5	750	10	3
Film D	3	750	2	3
Film E	3	750	10	3

RESULTS AND DISCUSSIONS

In XRD patterns of Yb123 final films, LAO(00l) and Yb123(00l) peaks were very strong. This implied that the films were c-axis oriented Yb123 films. However, results of the TEM observation did not completely agree with those of XRD.

Figure 1 shows a cross-sectional transmission electron micrograph of Film A observed from the direction of LAO[010]. An electron diffraction pattern at the top right in this micrograph was obtained from the selected area including both the film and the substrate. The film is grown into a c-axis oriented Yb123 film with a thickness of 100nm.

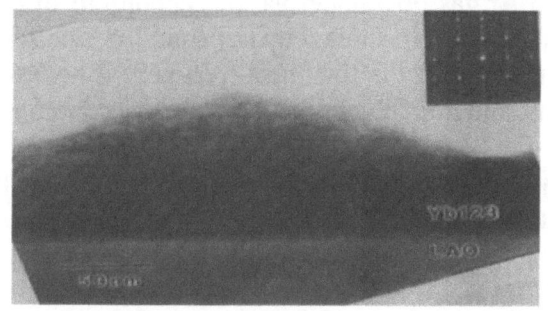

Fig.1   Cross-sectional transmission electron micrograph and an electron diffraction pattern of Film A observed from the direction of LAO[010]. A c-axis oriented Yb123 film is grown with a thickness about 100nm.

Figure 2(a) and 2(b) shows the cross-sectional micrographs of Film B. This film is polycrystalline. Nonsuperconducting phases such as $BaCuO_2$ and $Yb_2Cu_2O_5$ were found by electron diffraction patterns and energy dispersive X-ray spectroscopy (EDS) analysis.

Figure 3 shows the cross-sectional micrograph and the electron diffraction pattern of Film C. An amorphous film is seen in the vicinity of the interface between the film and the substrate. The c-axis of Yb123 crystal on the amorphous film is slightly tilted from the direction perpendicular to the surface of the LAO substrate.

Figure 4 shows the cross-sectional electron micrograph of Film D. As shown in this figure, Film D is a polycrystalline film of Yb123 and other crystals such as Yb_2O_3. In this film, a c-axis oriented Yb123 crystal in the shape of an island is seen on the surface of the substrate.

Figure 5 shows the cross-sectional electron micrograph of Film E. As shown in this figure, all the surface of the LAO substrate is covered with the c-axis oriented Yb123 film. In one half of Film E, the c-axis oriented Yb123 film was grown over 100nm in thickness. However, in the other half, a-axis oriented Yb123 crystals were also observed.

From results of the TEM observation, we suggest that when the precursor film is rapidly heated to the final annealing temperature, the heteroepitaxial growth is predominant and the random nucleation is restrained. On the other hand, slow heating facilitates the random nucleation. And at higher temperatures, the grain growth occurs. In the case that many nonsuperconducting crystals are generated, the epitaxial growth is suppressed even if the film is hold for enough time at the annealing temperature. In conclusion, high heating rate at the final heat-treatment is necessary for

591

the epitaxial growth of the superconducting films, and long holding time is effective for the grain growth of the crystals even if a small amount of nonsuperconducting crystals are formed.

(a)

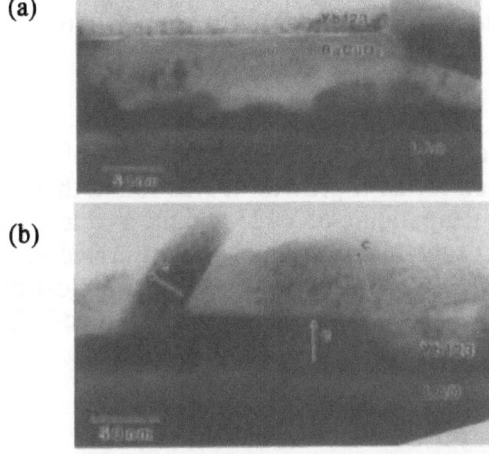

(b)

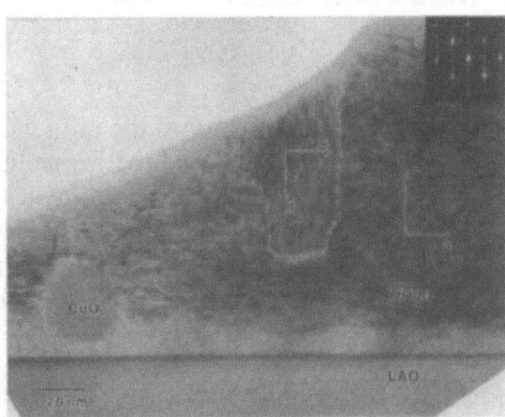

Fig.2   cross-sectional transmission electron micrographs of Film B. The film is polycrystalline: (a)BaCuO$_2$ and Yb$_2$Cu$_2$O$_5$ nonsuperconducting phases are also seen in this film; (b)Yb123 crystals are randomly oriented.

Fig.3   Cross-sectional electron micrograph of Film C. An amorphous layer is seen in the vicinity of the interface between the film and the substrate. An a-axis oriented Yb123 crystal is also visible.

Fig.4   Cross-sectional electron micrograph of Film D. It is noted that the film is polycrystalline, containing Yb123 and other crystals such as Yb$_2$O$_3$.

Fig.5   Cross-sectional electron micrograph of Film E. In this film all the surface of the LAO substrate is covered with the c-axis oriented Yb123 film.

CONCLUSIONS

By transmission electron microscopy, we observed cross sections of Yb123 films formed on LAO(001) substrates by the dipping-pyrolysis process. These films were prepared by the final heat-treatment of precursor films at various heating rate and holding time. As a result of that observation, it was found that rapid heating at the final heat-treatment is necessary for the epitaxial growth of the superconducting films. Furthermore, long holding time was found to be effective for the grain growth of the crystals.

Acknowledgments: This work was supported by the New Energy and Industrial Technology Development   Organization(NEDO).

1. P. C. McIntyre, M. J. Cima, J. A. Smith, Jr., R. B. Hallock, M. P. Siegal, and J. M. Phillips: J. Appl. Phys. **71**, 1868-1877(1992).
2. J. Shibata, K. Yamagiwa, I. Hirabayashi and T. Hirayama: Jpn. J. Appl. Phys. **37**, L1141-1143(1998).
3. J. Shibata, K. Yamagiwa, I. Hirabayashi, X. L. Ma, J. Yuan, T. Hirayama and Y. Ikuhara: Jpn. J. Appl. Phys. **38**, 5050-5053(1999).
4. K. Yamagiwa, I. Hirabayashi: Physica C **304,** 12-20(1998).

PROCESS AND CHARACTERISTICS OF $YBa_2Cu_3O_y$ / $YBa_2(Cu_{1-x}Mg_x)_3O_y$ FILMS ON MgO SUBSTRATES BY LPE

Kazuomi Kakimoto, Natsuro Hobara, Christian Krauns, Yuichi Nakamura, Teruo Izumi and Yuh Shiohara

Superconductivity Research Laboratory, ISTEC, 1-10-13 Shinonome, Koto-ku, Tokyo 135-0062, Japan

Abstract: We have investigated the influence of MgO in liquid for the LPE growth process of Y123 film. From the measurement of the Mg-solubility of solution for LPE growth, it was found that the existence of MgO in liquid scarcely influences the peritectic temperature of Y123 and the Y-solubility. The Mg-substituted Y123 LPE layer was grown on an MgO substrate from the MgO-saturated solution. Additionally, the Y123 LPE layer was grown from the solution without MgO under almost the same temperature condition as that for the Mg-substituted Y123 growth on the MgO substituted Y123 Layer. The Y123 LPE layer on Mg-substituted Y123 LPE layer revealed a Tc value of 90 K.

Keywords: YBCO, LPE film, MgO, saturated solution

INTRODUCTION

We have studied the suitability of the LPE method[1] for preparation of superconductive tapes on metallic substrates. The Y-Ba-Cu-O solution for LPE growth is very reactive with almost all metals except with silver. Therefore, it is regard to find out an effective buffer layer to prevent this reaction. Recently[2], the effective combinations for the coated tape by LPE were discovered. When the hastelloy is considered as a metallic substrate, the MgO buffer layer for the MgO-saturated solution is effective for preventing dissolution of hastelloy. However, the MgO-substituted Y123 film, which can not be avoided to grow from an MgO-saturated solution, shows lower Tc than that of Y123 without this substitution. Therefore, it is necessary to grow a Y123 LPE layer without substitution on the Mg-substituted Y123 LPE layer. However, detailed investigation of the influence of MgO for the growth has not been reported. In this paper, we have investigated the influence of MgO for the growth and its characterization of Y123/Mg-substituted Y123/MgO LPE film.

EXPERIMENTAL PROCEDURE

To investigate the influence of dissolved Mg on the Y-solubility and the preritectic temperature of Y123, liquid sampling experiments were carried out using MgO and Y_2O_3 crucibles. The $Ba_3Cu_5O_y$ melt on Y211 layer was prepared for this experiment. The sampling was carried out in the temperature range from 900 °C to 1300 °C. Then, to investigate the dependance of dissolved Mg on Y-solubility, the flux was sampled from the $Ba_3Cu_5O_x$ melt in the MgO crucible. These samples were dissolved into 1N HNO_3 solution. The concentrations of Mg, Ba, Cu and Y were analyzed by ICP. The Mg-substituted Y123 LPE film was grown from the MgO-saturated solution, which consisted of $Ba_3Cu_5O_x$ melt, Y211 powder and MgO powder in the Y_2O_3 crucible, on the MgO (100) substrate. The temperatures of the surface and the bottom of the liquid were 1000 °C and 1010 °C respectively. Futhermore, the Y123 LPE film was grown from the solution without MgO addition under the same crystal growth condition. The sample was annealed at 500 °C in 100 %O_2 for 200 hours. The Tc values were determined by SQUID.

RESULTS AND DISCUSSION

Figure 1 shows the Y-solubility as a function of temperature for $Ba_3Cu_5O_x$ melt in Y_2O_3 and MgO crucibles. From this result, no influence of dissolved Mg on either the Y-solubility or the peritectic

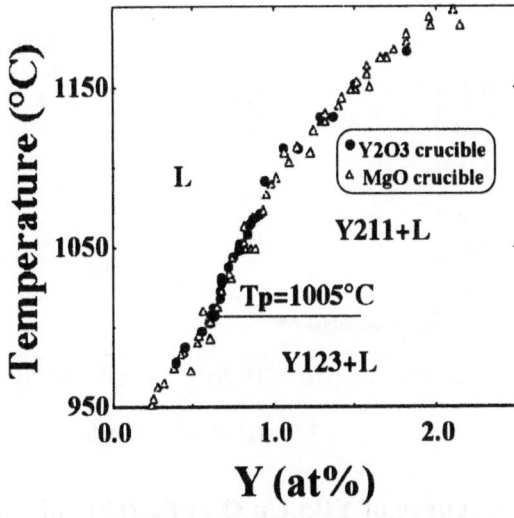

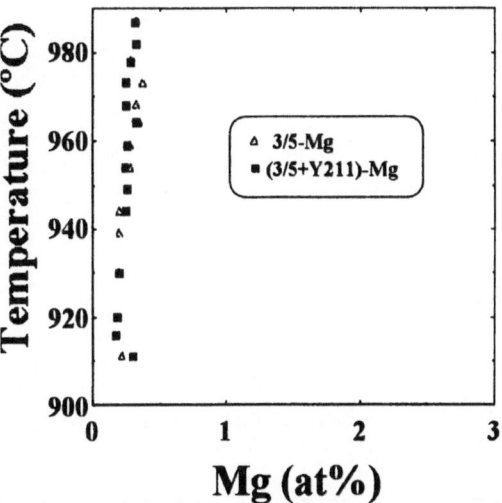

Fig.1 Y-solubility lines for $Ba_3Cu_5O_x$ melt in Y_2O_3 and MgO crucibles.

Fig.2 Mg-solubility of $Ba_3Cu_5O_x$ melt and of ($Ba_3Cu_5O_x$ melt +Y211).

temperature of Y123 could be detected at least within the accuracy of our measurements. Fig.2 shows the Mg-solubility of the $Ba_3Cu_5O_x$ melt with or without Y211. This figure shows that the influence of Y additions on the Mg solubility is negligible within the accuracy of our measurements. We should discuss these results from a thermodynamic point of view. If it is assumed that the activity of pure MgO solid is equal to unity, the free energy change by the dissolution into the Ba-Cu-O melt of the pure MgO solid, ΔG_d, is expressed by eq.1,

$$\Delta G_d = \Delta G_m + RT \ln \frac{a_{[MgO]}}{a_{\langle MgO \rangle}} = \Delta G_m + RT \ln \left(\gamma^0_{Mg} \cdot \gamma_{Mg} \cdot x_{Mg} \right) \tag{1}$$

where ΔG_m, $a_{[MgO]}$, $a_{\langle MgO \rangle}$, γ^0_{Mg}, γ_{Mg} and x_{Mg} are the free energy change by the melting of the pure MgO solid, the activity of MgO as a dilute solution for BaCuO solvent, the activity of a pure MgO solid, the exchange coefficient from the Raoult's reference to the Henry's reference, the activity coefficient of Mg-solute and the Mg-molar fraction respectively. Furthermore, γ_{Mg} is expressed by eq.2,

$$\ln \gamma_{Mg} = \varepsilon^{Mg}_{Mg} \cdot x_{Mg} + \varepsilon^{Ba}_{Mg} \cdot x_{Ba} + \varepsilon^{Cu}_{Mg} \cdot x_{Cu} \tag{2}$$

where ε^{Mg}_{Mg}, ε^{Ba}_{Mg} and ε^{Cu}_{Mg} are Mg-interaction parameters for respective elements, and x_{Mg}, x_{Ba} and x_{Cu} are the molar fraction of the elements respectively. If it can be assumed to be no substitution of Y in MgO, the free energy change by the dissolution into the Ba-Cu-O melt with existence of $YO_{1.5}$ of the pure MgO solid, $\Delta G_d'$, is expressed by eq.3,

$$\Delta G_d' = \Delta G_m + RT \ln \left(\gamma^0_{Mg} \cdot \gamma'_{Mg} \cdot x'_{Mg} \right) \tag{3}$$

where f_{Mg}' and x_{Mg}' are the activity coefficient and Mg-concentration in the case of the Y-Ba-Cu-O melt. Then, f_{Mg}' is given by eq.4,

$$\ln \gamma'_{Mg} = \varepsilon^{Mg}_{Mg} \cdot x'_{Mg} + \varepsilon^{Y}_{Mg} \cdot x'_Y + \varepsilon^{Ba}_{Mg} \cdot x'_{Ba} + \varepsilon^{Cu}_{Mg} \cdot x'_{Cu} \tag{4}$$

where e_{Mg}^Y is Y-interaction parameter on Mg ,and x_Y' x_{Ba}' and x_{Cu}' are the molar fraction of the respective elements in the case of the Ba-Cu-O melt with the $YO_{0.5}$ solution. The results of fig.1 reveal the existence of x_Y in the case of MgO crucible. Then, if the MgO mol fractions of x_{Mg} and x_{Mg}' are almost the same under the saturated condition, which means $\Delta G_d = \Delta G_d' = 0$, e_{Mg}^Y should be zero. This means that e_Y^{Mg} should also be zero. Therefore, it is considered that the Y-concentration is independent of the Mg-concentration as shown in fig.1. From these results, it is suggested that Mg-substituted Y123 film can grow under growth conditions similar to that for Y123 film. Fig.3 shows a Tc curve of a LPE film

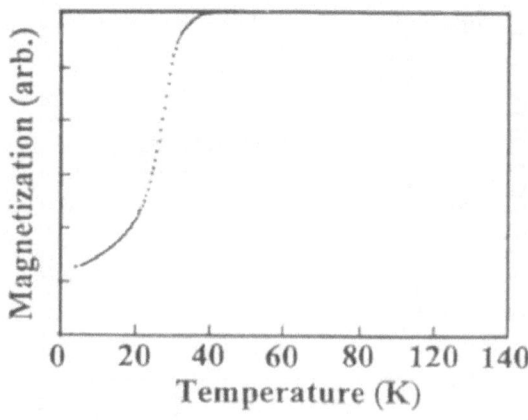
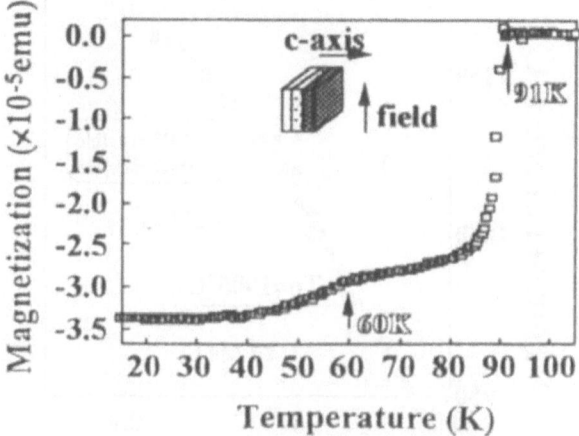

Fig.3 Tc curve of YBa$_2$(Cu$_{1-x}$Mg$_x$)$_3$O$_y$ on MgO substrate by LPE method.　　**Fig.4 Tc curve of YBa$_2$Cu$_3$O$_y$/YBa$_2$(Cu$_{3-x}$Mg$_x$)O$_y$ on MgO substrate by LPE method.**

grown from Mg-saturated solution. It shows low Tc (~40K) due to the Mg-substitution. Therefore, it is thought that an additional Y123 LPE layer on the Mg-substituted Y123 LPE layer is necessary to obtain better characteristics. Fig.4 shows a Tc curve of an Y123 LPE layer grown on Mg-substituted Y123 LPE layer at the same growth conditions as the Mg-substituted LPE layer. From this result, it was found that the Y123 LPE layer grew on the Mg-substituted Y123 LPE layer without MgO under almost the same temperature condition as that for the Mg-substituted Y123 growth and reveals high Tc (~91K). From the above mentioned results, it is shown that the MgO is a very suitable material as a buffer layer for the LPE process to prepare the coated conductor using hastelloy tape as a substrate.

CONCLUSION

Within the error of our measurements, neither the Y-solubility nor the Mg-solubility of the Ba$_3$Cu$_5$O$_x$ melt was influenced by the respective addition of Mg or Y. Accordingly, the peritectic temperature of Y123 was not affected by the presence of Mg in the melt and therefore the Y123 films with or without Mg-substitution could be obtained by the LPE method under the same conditions regardless of MgO addition. Although the Mg-substituted Y123 LPE films show low Tc (~40 K) due to the Mg-substitution, the additional LPE Y123 film without Mg substitution grown on the Mg substituted Y123 film exhibits a Tc value of 91K. Therefore, MgO is a suitable buffer layer for fabrication of coated conductors by the LPE method.

Acknowledgement: This work was supported by the New Energy and Industrial Technology Development Organization (NEDO) as Collaborative Research and Development of Fundamental Technologies for Superconductivity Applications.

1.K.Kakimoto, Y.Ishida, T.Izumi and Y.Shiohara : Proceedings of the 11th International Symposium on Superconductivity (ISS'98). 749 (1998).
2.K.Kakimoto, N.Hobara, Y.Nakamura, T.Izumi, K.Fujino, K.Ohomatu, and Y.Shiohara : Extended Abstracts (The 60th Autumn Meeting, 1999); The Japan Society of Applied Physics. 170 (1999).

FLUORITE AND RELATED TYPE OXIDE BUFFER LAYERS FOR Y-123 COATED CONDUCTORS STUDIED BY IBAD METHOD

Yasuhiro Iijima, Mariko Kimura, and Takashi Saitoh

Materials Technology Lab., Fujikura Ltd., 1-5-1, Kiba, Koto-ku, Tokyo 135-8512, JAPAN

Abstract: Biaxially aligned growth were studied for fluorite type (YSZ, CeO_2) and rare-earth C type (Y_2O_3) oxide films on polycrystalline Ni-based alloy substrates by IBAD method. Cube-textured (all axes aligned with a <100>axis substrate normal) YSZ, CeO_2 films were obtained by low energy (<300eV) ion bombardment at low temperatures (<300°C). Besides, cube textured Y_2O_3 films were obtained in far narrow conditions with a quite low energy (150eV)-ion bombardment at the temperature of 300°C.

Keywords: Ion-beam-assisted deposition (IBAD), Y-123 coated conductors, fluorite structure

INTRODUCTION

Biaxially aligned Yttria Stabilized Zirconia (YSZ) films formed by ion-beam-assisted deposition (IBAD) are reliable template layers for Y-123 coated conductors [1]. However, intercalation of thin Y_2O_3 layers between YSZ and Y-123 film is effective to compensate lattice mismatch and prevent slight interdiffusion [2]. It is worthy to form more adequate buffer materials directly on alloy tapes by IBAD.

This paper describes crystalline alignment properties for YSZ, CeO_2 (fluorite structure), and Y_2O_3 (rare-earth-C structure). Their lattice constants are shown in Table 1. Fig. 1 shows a schematic view of fluorite structure. Rare-earth C structure corresponds to one whose 1/4 of oxygen ions deleted from fluorite structure. Several fluorite type oxides were reported to be formed by IBAD, but rare-earth C ones have not be ever-reported[3-4].

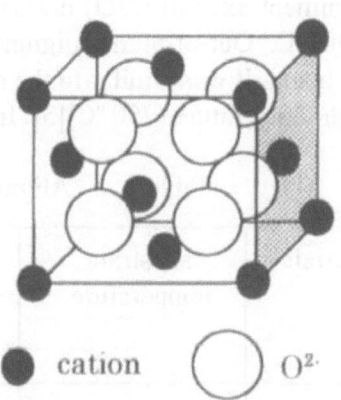

cation O^{2-}

Fig. 1. Fluorite structure

Table 1. Buffer materials for Y-123 coated conductors.

materials	YSZ	CeO_2	Y_2O_3	YBCO
structure	fluorite	fluorite	rare-earth C	deformed perovskite
lattice constant (A)	5.14	5.41	10.6	3.81
cation distance (A)	3.63	3.83	3.75	3.81
lattice mismatch (%)	4.7	0.5	1.5	-

EXPERIMENTAL

Films were formed on mirror-like polished polycrystalline Ni-based alloy substrates by dual ion beam sputtering method. Substrate temperatures were ranged from -150 to 500 °C, controlled by a thermocouple on a dummy plate next to samples. Assisting ion beam was Ar^+ or Kr^+ beam with the energy below 300 eV. The beam incident angle was set to 55 degrees from substrate normal. Oxygen gas was introduced to chamber with partial pressure of 1.0×10^{-4} Torr. Growth structures were characterized by X-ray diffraction (XRD). Thicknesses of films were 0.4-1.0 μm.

RESULTS & DISCUSSION

Biaxial alignment of off normal IBAD process is characterized by two crystalline axes simultaneously fixed during growth: an axis aligned normal to substrate, and another axis aligned to the direction of incident ions. The crystalline alignment properties for YSZ, CeO_2, and Y_2O_3 thin films are summarized in Table 2. Bold letters indicated "cube-texture", a <100> axis aligned normal, that must be held by template layers for Y-123 conductors. Cube textured YSZ, and CeO_2 films were obtained at substrate temperatures below 300°C above room temperature. Optimized temperature was near 100 °C. Higher substrate temperature was required to form cube textured Y_2O_3 films, optimized at 300°C. At higher temperature than 300°C, out-of-plane alignment axis changed to <111> for YSZ, and to random for CeO_2, and Y_2O_3, respectively.

The alignment axes of CeO_2 dramatically changed by changing the assisting ions to Kr^+ from Ar^+ below 300°C. Out-of-plane alignment axis changed to <111>, and a <100> axis aligned to the incident beam. It was similar to the results on CeO_2 by Zhu et. al. obtained with Ar^+ bombardment at very high temperature (750 °C)[3]. In all other cases for table 2, a <111>axis aligned to incident ions.

Table 2.   Alignment axes for buffer materials by IBAD.

materials	substrate temperature	Ar^+ assisting		Kr^+ assisting	
		normal axis	ion axis	normal axis	ion axis
YSZ	-150 °C	amorphous		amorphous	
	< 300 °C	**<100>**	**<111>**	**<100>**	**<111>**
	300 - 500°C	<111>	<111>	<111>	<111>
CeO_2	< 300 °C	**<100>**	**<111>**	<111>	<100>
	300 - 500°C	random	<111>	random	<111>
Y_2O_3	90-100 °C	amorphous		amorphous	
	300 °C	**<100>**	**<111>**	**<100>**	**<111>**
	500 °C	random	<111>	random	<111>

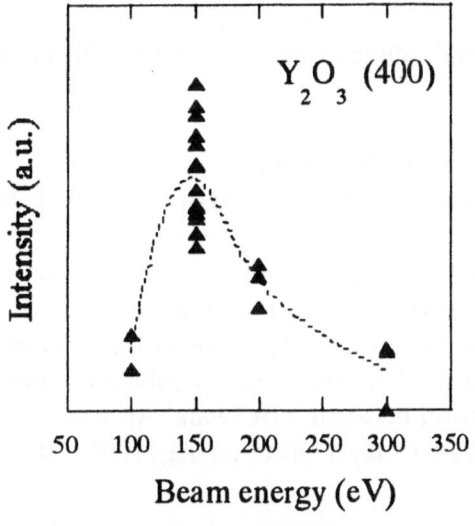
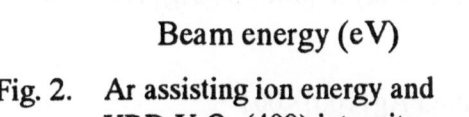

Fig. 2.   Ar assisting ion energy and XRD Y_2O_3 (400) intensity.

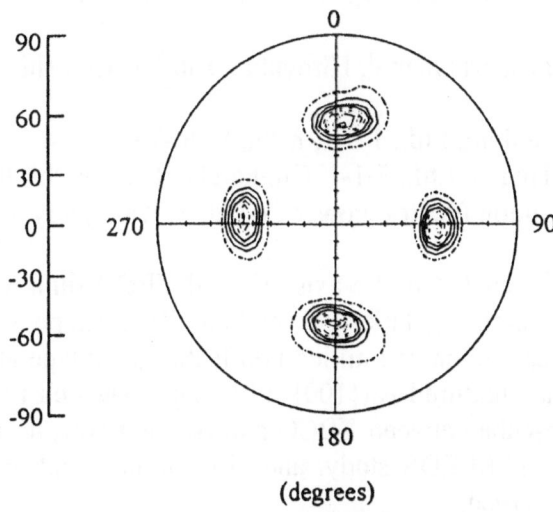

(degrees)

Fig. 3.   X-ray (222) pole figure for a Y_2O_3 film.

Y_2O_3 requires very narrow conditions of assisting ion beam energy for enough crystallization to have aligned texture.   Fig. 2 shows the optimized ion energy to be 150eV, estimated from XRD Y_2O_3 (400) intensity. In this condition cube texture was obtained as shown in Fig. 3. The azimuthal FWHM for (222) poles was 26 degrees.   It is a clear contrast to rather wide ion energy up to 300 eV allowed for crystallization of (100) aligned YSZ, and CeO_2.   It suggests that Y_2O_3 is so sensitive to Ar^+ radiation damage. Table 3 shows the density of lattice energy for buffer materials calculated by using Born-Habar cycle. Y_2O_3 has apparently lower bonding energy than YSZ, or CeO_2, that comes from lower valence number of cations and less oxygen ions of rare-earth C type structure.

Table 3.   Lattice energy for buffer materials.

materials		YSZ	CeO_2	Y_2O_3
lattice energy	(kJ/mol)	10935*	10355	12861
lattice energy density	(eV/nm³)	3344*	2715	1794

(*) for ZrO_2

CONCLUSION

Cube-textured YSZ, CeO_2, and Y_2O_3 thin films were deposited on polycrystalline Ni-based alloy substrates by IBAD at low temperatures. The alignment axes of CeO_2 films dramatically changed between Ar^+ and Kr^+ assisting ions. The optimized temperature for Y_2O_3 was 300 °C, higher than YSZ or CeO_2. Quite low energy (150eV) ion bombardment was required for crystallization of Y_2O_3 films. The calculated lattice energy suggests the possibility that Y_2O_3 is so sensitive to ion radiation damage because of insufficient bonding energy.

1. Y. Iijima, M. Hosaka, N. Tanabe, N. Sadakata, T. Saitoh, O. Kohno, and K. Takeda: J. Mater. Res. 13, 3106 (1998)
2. M. Hosaka, Y. Iijima, N. Sadakata, T. Saitoh, O. Kohno, and K. Takeda: *Advances in Superconductivity vol. 9*, ed. S. Nakajima and M. Murakami, p.749 , Springer, Tokyo, 1997
3. S. Zhu, D. H. Lowndes, J. D. Budai, and D. P. Norton: Appl. Phys. Lett. 65, 2012 (1994)
4. V. Betz, B. Holzapfel, D. Raouser, and L. Schultz: Appl. Phys. Lett. 71, 2952 (1997)

Superconducting Property of $Y_1Ba_2Cu_3O_x$ Films Formed on Silver Substrates by Continuous Chemical Vapor Deposition

Kazunori Onabe[1], Hiroyuki Akata[2], Kazutoshi Higashiyama[2], Shigeo Nagaya[3] and Takashi Saitoh[1]

1Fujikura Ltd., 1-5-1, Kiba, Koto-ku, Tokyo 135-8512, Japan
2 Hitachi Ltd., 7-1-1, Omika-cho, Hitachi-shi, Ibaraki-ken 319-1292, Japan
3Chubu Electric Power Co., 20-1, Kita-Sekiyama, Ohdaka-cho, Midori-ku, Nagoya 459-8522, Japan

Abstract: Good c-axis oriented YBCO films were formed directly on both non-textured and cube-textured Ag tapes. 11m long YBCO tape with Jc ($1\mu V/cm$) of $3.0\times10^4 A/cm^2$ at 77K, 0T was prepared on roll milled non-textured Ag tape at a forming rate of 3.0m/h. YBCO films formed on cube-textured Ag{100}<001> tapes showed two types of the location, cube-on-cube and diagonal-on-cube between YBCO and Ag, however, no inter-diffusion between YBCO and Ag was analyzed by TEM-EDS study, and also the epitaxial growth of YBCO layer grew on Ag{100}<001> was observed.

Key words: CVD, rapid formation, long YBCO tape, textured Ag{100}<001>

INTRODUCTION

Silver is one of the functional materials which can perform as the electrical stabilizer and on which YBCO films can be formed directly, so that the superconducting tape with the simplest structure is realizable by using Ag tape as the substrate. It has been reported in-plane aligned YBCO films were formed on Ag(100), (110) and (111) single crystalline substrates[1]. And cube-textured Ag{100}<001> tapes were investigated for long superconducting wires[2]. In this work, YBCO films were formed directly on non-textured and cube-textured Ag tapes by a continuous chemical vapor deposition (CVD), and the crystallinity, surface morphology, interfacial microstructure and superconducting property of YBCO/Ag tapes were studied.

EXPERIMENTAL

YBCO films were formed on non-textured and {100}<001> textured Ag tapes by a continuous CVD technique. Hot-wall type CVD system was used in this study[3]. Oxygen partial pressure and a deposition temperature were optimized at 1.3-1.4Torr and 800°C, respectively. The moving rate of the substrate was 1-3m/h. Ag layer for the stabilizer with 3-6μm in thickness was formed on YBCO layer by a rf-magnetron sputtering method. Superconducting properties were measured by a dc four-probe method, and Ic was determined at a criterion of 0.1 and 1μV/cm at 77K, 0T.

RESULTS & DISCUSSION

Figure 1 and figure 2 showed the typical X-ray pattern and the surface morphology of YBCO films on non-textured Ag substrates. X-ray pole figure measurement of YBCO(103) peak of these tapes showed no in-plane alignment, however, good c-axis orientation and smooth surface of YBCO layers were obtained by optimizing CVD conditions. The best superconducting property was given in the limited deposition condition, so an oxygen partial pressure and a deposition temperature had to be controlled accurately for long time deposition to prepare the long superconducting tape.
11m long YBCO superconducting tape was prepared continuously by using roll milled non-textured Ag tape with $8mm^W \times 0.2mm^T \times 11m^L$. The forming rate of YBCO tape was 3.0m/h, and the thickness of YBCO layer was 0.35μm. Jc distribution of this tape was shown in figure 3. Jc of

$3.0 \times 10^4 A/cm^2$ ($1\mu V/cm$) and $1.6 \times 10^4 A/cm^2$ ($0.1\mu V/cm$) at 77K, 0T was maintained through the total length. The n value was estimated at 3.7 from I-V curve.

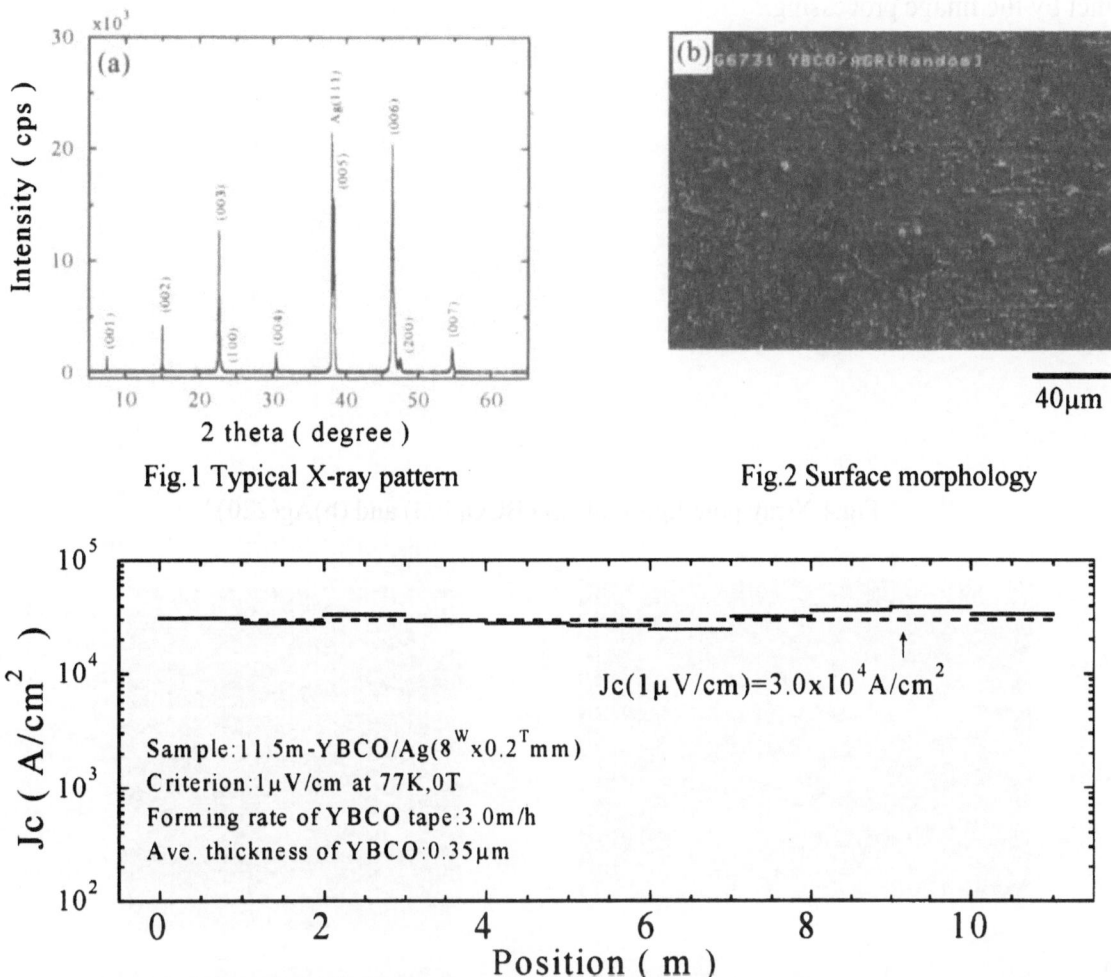

Fig.1 Typical X-ray pattern

Fig.2 Surface morphology

Fig.3 Jc distribution of 11m long YBCO superconducting tape on non-textured Ag

Cube-textured Ag{100}<001> were also employed to improve the in-plane alignment in YBCO films. The size of CUTE-Ag tape was $3^W \times 0.05^T \times 30^L$mm. YBCO film was formed continuously on it at the same deposition condition of non-textured Ag tape. Figure 4 showed the results of X-ray pole figure measurement of (a)YBCO(103) and (b)Ag(220) peaks. Two types of the location, cube-on-cube and diagonal-on-cube between YBCO and Ag, were observed, so it was necessary to texture the YBCO grains to obtain high Jc. Jc of these tapes were $1\sim2\times10^4 A/cm^2$ typically. The surface morphology and the cross-sectional TEM image were shown in figure 5(a) and (b). No inter-diffusion between YBCO and Ag was analyzed by TEM-EDS study. And also the high resolutive TEM image at the interface between YBCO and Ag{100}<001> was shown in figure 6. The epitaxial growth of YBCO layer with the miss-fit transition was made distinct by the image processing. The miss match factor of the lattice constant between YBCO and Ag was estimated at 5~6%, and it was equal to the result of the image processing, approximately.

CONCLUSIONS

Ag tape was investigated as one of the functional substrate that was realizable the superconducting tape with the simplest structure. 11m long YBCO superconducting tape with Jc ($1\mu V/cm$) of

$3.0 \times 10^4 \text{A/cm}^2$ at 77K, 0T was prepared at a forming rate of 3.0m/h on non-textured Ag tape. YBCO films on cube-textured Ag{100}<001> showed two types of the location, cube-on-cube and diagonal-on-cube between YBCO and Ag, however, no inter-diffusion between YBCO and Ag was analyzed by TEM-EDS study, and the epitaxial growth of YBCO layer on Ag{100}<001> was made distinct by the image processing.

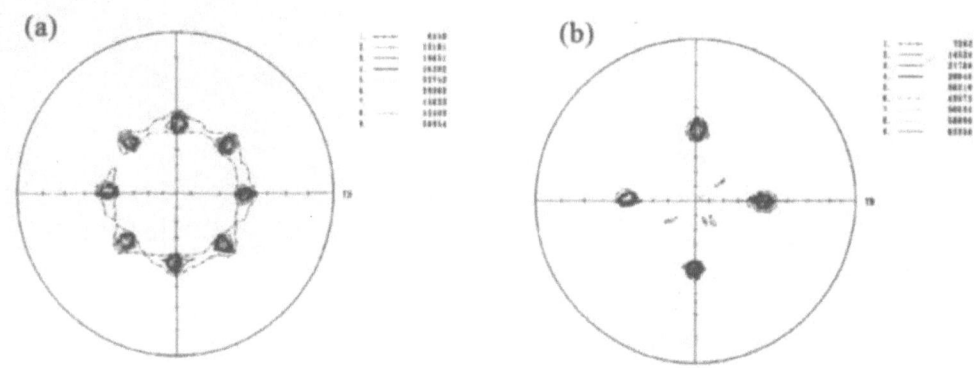

Fig.4 X-ray pole figure of (a)YBCO(103) and (b)Ag(220)

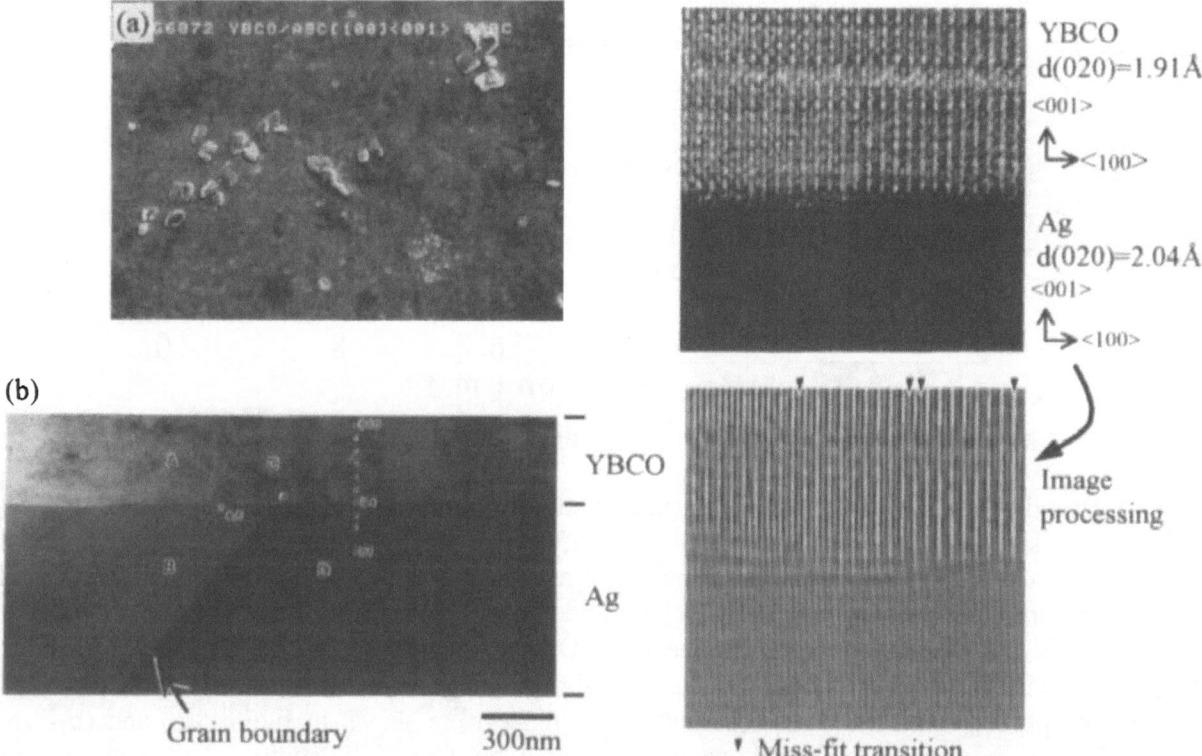

Fig.5 (a)surface morphology and (b)TEM-EDS analysis of YBCO/Ag{100}<001>

Fig.6 Result of image processing of high resolutive TEM image at interface

1. J.D.Busai, R.T.Young and B.S.Chao, Appl. Phys. Lett. **62**, 1836 (1992)
2. N.I.Sugiyama, T.Yuasa, T.Ozawa, K.Higashiyama and K.Osamura, in Adv. in Superconductivity IX, edited by S.Nakajima and M.Murakami (Springer-Verlag, Sapporo, 1996), pp.927
3. K.Onabe, S.Nagaya, Y.Iijima, N.Sadakata, T.Saitoh and O.Kohno, in Adv. in Superconductivity X, edited by K.Osamura and I.Hirabayashi (Springer-Verlag, Gifu, 1997), pp.603

INFLUENCE OF CARBON AND FLUORINE CONTAINING SOLUTION ON LIQUID PHASE EPITAXY OF $YBa_2Cu_3O_{7-\delta}$

Y. Yamada, T. Miura, Y. Koike*, I. Hirabayashi, H. Ikuta#, U. Mizutani##

Supercond. Res. Lab., ISTEC, 2-4-1, Mutsuno, Atsuta-ku, Nagoya, 456-8587, Japan
#CIRSE, Nagoya Univ., Nagoya 456-8601, Japan
##Dept. Cryst. Mater. Sci., Nagoya Univ. Nagoya 456-8603, Japan

Abstract: The influence of fluorine and carbon addition into the solution are investigated on the LPE growth of $YBa_2Cu_3O_{7-\delta}$ (YBCO) crystal. Fluorine addition into the BaO-CuO solution using BaF_2 significantly decreases the temperature range of the primary crystallization of YBCO. Superconducting transition temperature of the crystals grown from the BaF_2 added solution, however, is unchanged. Low partial pressure of oxygen further decreases the temperature of the crystallization field. Carbon containing solution is intentionally obtained by adding $BaCO_3$. The YBCO crystals are incorporated with carbon and superconducting properties are modified. Small amount of carbon is still incorporated by the equilibration of the solution with the air.

Key words: $YBa_2Cu_3O_{7-d}$,phase diagram, Liquid phase epitaxy, fluorine-doping, carbon-doping

INTRODUCTION

Deposition of $YBa_2Cu_3O_{7-d}$ (YBCO) crystalline films from high temperature solution has some advantages such as rather good crystalline perfection, high growth rate, less grain boundaries and no vacuum equipment. Liquid phase epitaxy (LPE) method, one of the high temperature solution growth methods for the formation of YBCO coated conductors, requires as low temperatures as possible since most of substrate materials can react and form other phase(s) or dissolve into the solution at the typical processing temperatures (about 900-1000°C) [1]. It has been shown that BaF_2 added BaO-CuO (BaO-CuO-Ag) solution can decrease processing temperature below 930°C. However, this temperature is still high enough for the substrate materials for a flexible YBCO coated conductor to react with the solution. To achieve further low processing temperatures, the LPE growth under low partial oxygen pressure have been investigated.

The LPE coating in the air is preferable since there is no need to use a chamber for the control of the atmosphere. However, the air contains carbon dioxide about 350ppm and it can contaminate YBCO crystal by carbon. Therefore, it is necessary to investigate the effect of the carbon incorporation in YBCO on superconducting and crystallographic properties. This paper describes the effect of low partial oxygen pressure for the BaF_2 added solution and carbon contamination of YBCO grown from an intentionally carbon doped solution.

EXPERIMENTAL

YBCO crystalline films are grown by the LPE method. The growth procedure is described in detail in elsewhere [1-3]. BaF_2 and $BaCO_3$ were used to add fluorine and carbon into the conventional BaO-CuO solution, respectively. CuO was simultaneously put in the solution to keep the Ba:Cu ratio of 3:5. Crystallization field of YBCO was determined for the solutions as a function of oxygen partial pressure with different solution compositions. The determination of the crystallization field was done by observing phase(s) deposited on a seeded substrate of MgO. When YBCO crystal grew on the seeded substrate without any other phases we determined that YBCO crystal grew as a primary phase. On the other hand, it was difficult to determine crystallization field for the carbon containing solution because carbon content in the solution decreased rapidly to the equilibrium level. YBCO crystals with various carbon content were grown during the monotonic carbon decrease at different times. Superconducting properties were measured for samples with different carbon content by a SQUID magnetometer and a typical DC four probe method. X-ray diffraction was also measured to determine lattice constant.

RESULTS AND DISCUSSION

Crystallization field of YBCO as a function of oxygen partial pressure with different solvent composition are displayed in Figure 1. As same as the conventional solution of BaO-CuO, the solution that consists of BaO-CuO-BaF$_2$(-Ag) is also influenced by oxygen partial pressure. Crystallization temperature decreased with decreasing oxygen partial pressure and the lower limit of YBCO growth was about 860°C.

Superconducting transition temperature of YBCO films grown from the fluorine containing solution was typically 90K and normal state resistivity was also in typical values. These imply that fluorine element incorporate less enough to alter electronic state.

Carbon concentration in the BaCO$_3$ added solution rapidly decreased to a finite value as time went by after the addition of BaCO$_3$. This implies that carbon was released from the solution to the ambient atmosphere and its content became in equilibrium. Holding time dependence of carbon content in the solution and that in the crystal grown from the solution are shown in Figure 2. Carbon containing YBCO crystalline films with different content were obtained by controlling holding time after the BaCO$_3$ addition. Carbon containing films show obvious decrease of transition temperature when the content exceeds about 0.06wt% in the crystals. Pinning behavior is enhanced at adequate amount of carbon and described in detail in elsewhere [4].

Remarkable splitting of lattice constant was detected by x-ray diffraction measurement. The samples annealed at 400°C

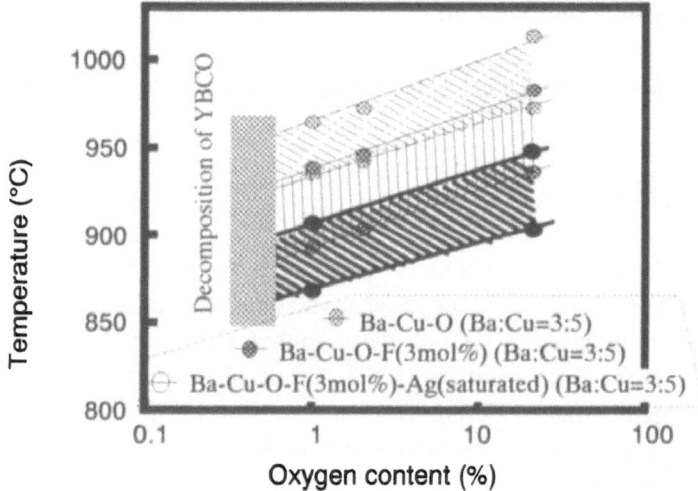

Figure 1 YBCO crystallization field versus oxygen content for the solvent composed of BaO-CuO (Ba:Cu=3:5), BaO-CuO-BaF2 (3mol% BaF2) and BaO-CuO-BaF2-Ag (3mol% BaF2 and silver saturation)

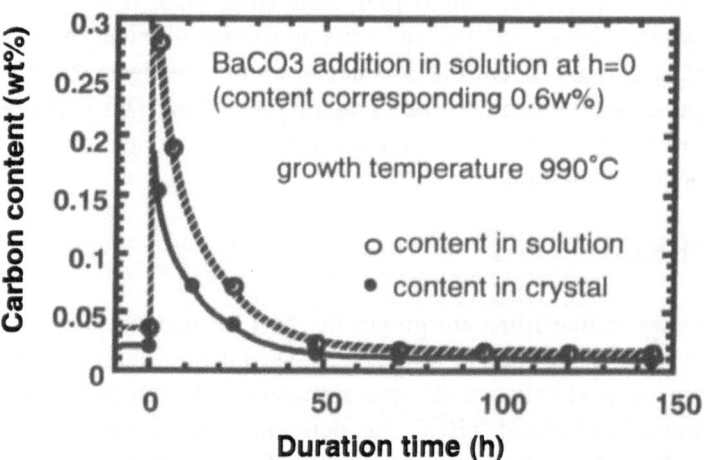

Figure 2 Holding time dependence of carbon content in the solution and in the crystal grown from the solution.

and 500°C for oxygenation show two peaks in (00*l*) diffraction. One corresponds to 11.69Å and is independent on carbon content. The other corresponds to 11.84Å at carbon concentration close to zero and decreases with increasing carbon content as shown in Figure 3. This splitting can be considered to be due to a possible phase separation in carbon incorporated YBCO crystals.

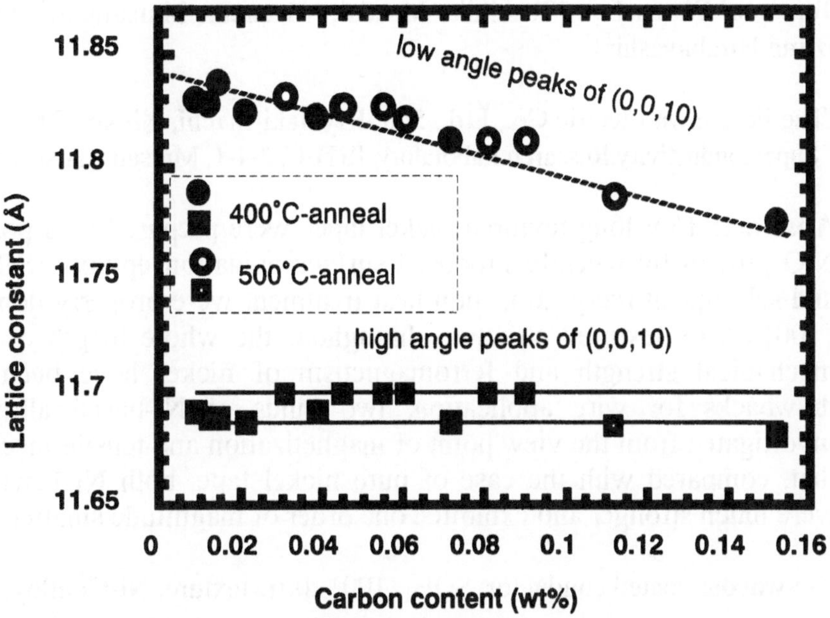

Figure 3 Relationship between c-axis lattice constant and carbon content determined from (00 10) diffraction

CONCLUSION

Effect of fluoride and carbonate added into the 3BaO-5CuO solution from which YBCO crystal grew was investigated in regard to crystallization temperature range and physical properties of YBCO crystal. Fluorine in the solution decreased crystallization temperature range of YBCO. The lowest temperature of YBCO growth reached to 860°C when the atmospheric oxygen partial pressure controlled to be 0.1%. The YBCO crystals from the fluorine containing solution showed typical resistivity in normal state and transition temperature above 90K, indicating that incorporation of fluorine in YBCO crystal does not affect its transport properties strongly.

Carbon incorporated YBCO crystals were grown from the carbon containing solution prepared by adding $BaCO_3$. Carbon released from the solution as time went by after $BaCO_3$.addition, resulting in decrease of carbon content in YBCO crystals grown from the solution. Superconducting and lattice constant were investigated by systematic variation of carbon content in YBCO crystals. Decrease of transition temperature was observed when the carbon content exceeded about 0.06wt%. Carbon content of 0.01wt%, equilibrium level with the air, did not change transition temperature from 90K. Two different (00*l*) peaks are observed in whole range of carbon content in our experiment, suggesting phase separation in carbon incorporated YBCO crystals. Interesting pinning behavior was observed and described in elsewhere [4]. According to the lower growth temperature and the unchanged transition temperature by fluorine added growth in the air, non-vacuum processing for YBCO coating on a metal substrate have been now in progress.

Acknowledgement
This work was supported by the New Energy and Industrial Technology Development Organization (NEDO) as Collaborative Research and Development of Fundamental Technologies for Superconductivity Applications.

*Present address: Chubu Electric Power CO..,INC, 20-1 Kitasekiyama, Ohdaka-cho, Midori-ku, Nagoya 459 JAPAN
1. Y. Yamada, Y. Niiori, I. Hirabayashi, S. Tanaka, Physica C 278 (1997) 180
2. Y. Yamada, Y. Niiori, I. Hirabayashi, S. Tanaka, J. Cryst. Growth 167 (1996) 566
3. Y. Niiori, Y. Yamada, I. Hirabayashi, Physica C 296 (1998) 65
4. T. Miura, Y. Yamada, Y. Koike, H. Ikuta, I. Hirabayashi, U. Mizutani, Proceeding of ISS '99.

Characterization of Biaxially Textured Ni-Based Alloy Substrates

Tomonori Watanabe[1], *Toshihiko Maeda[1], †Kaname Matsumoto[2], Naoki Uno[1], Masaru Ikeda[1] and Izumi Hirabayashi[2]

[1]The Furukawa Electric Co., Ltd., 500, Kiyotaki-machi, Nikko 321-1493, Japan
[2] Superconductivity Research Laboratory, ISTEC, 2-4-1, Mutsuno, Atsuta-ku, Nagoya 456-8587, Japan

Abstract: 15m long textured nickel tapes were prepared as a possible substrate for epitaxial NiO growth by recently proposed surface-oxidation epitaxy (SOE) method. As cold-rolling and subsequent recrystallization heat treatment were properly done, that we obtained a highly {100}<001> textured structure throughout the whole length of tapes. Since relatively low mechanical strength and ferromagnetism of nickel have been pointed out to be severe drawbacks for wire application, two kinds of Ni-based alloys, Ni-Cr and Ni-V, were investigated from the view point of magnetization and tensile mechanical property. It is shown that, compared with the case of pure nickel tape, both Ni-12wt%Cr and Ni-10wt%V tapes were much stronger and exhibited one order of magnitude smaller hysteresis loss.

Keywords: coated conductor, SOE, {100}<001>texture, Ni-Cr alloy, Ni-Cr alloy, hysteresis loss

INTRODUCTION

Several groups have reported critical current density (Jc) of over $10^6 A/cm^2$ (77K, 0T) for in-plane aligned YBCO films grown on metallic substrates [1, 2]. Recently, we have proposed a new process as "surface-oxidation epitaxy" (SOE) [3, 4]. In this method, (100)-oriented NiO layer is epitaxially grown by surface-oxidation of cube textured nickel tape. It is hopeful to be much of cost advantage because preparation of textured nickel tape and formation of textured NiO buffer layer only requires cold rolling and heat treatment process. The critical temperature (Tc) of 86-87K and Jc=4-6×$10^4 A/cm^2$ (77K, 0T) for the YBCO film on the SOE-grown NiO have been reported [4]. This paper presents the fabrication and characterization of 15m long textured pure nickel tape in order to develop a long length YBCO coated conductor. Since some problems of pure nickel for wire application, such as ferromagnetism and relatively low strength, have been known., quality of texture and magnetic and mechanical properties of pure nickel, Ni-Cr alloy and Ni-V alloy were investigated.

EXPERIMENTAL

Initial blocks of nickel, Ni-12wt%Cr alloy and Ni-10wt%V alloy were cold rolled to tapes with thickness of 0.1 – 0.3 mm. The relative reduction in thickness was over 95%. The rolled tapes were annealed to form a texture at temperatures of 900 – 1100°C. Surface oxidation of pure Ni tape was carried out for 30 – 3000 seconds at temperatures of 1000 – 1250°C in air. Preferred orientation of Ni or NiO grains were determined by x-ray diffraction (XRD). These tapes were characterized on quality of texture by x-ray pole figure, magnetization by conventional magnetization method and mechanical properties by tensile test.

RESULTS AND DISCUSSION

Pure nickel tape. The x-ray diffraction pattern at four different part of a 15m long textured nickel tape are shown in Fig. 1. For the whole length of the tape, (100) diffraction of grain was confirmed to be almost 100% parallel to the tape surface. Typical values of full width at half-maximum (FWHM) four (200) rocking curves (x-ray θ scan) was about 6 degrees. Figure 2 gives the pole figure of (111) diffraction for the textured nickel tape. Four-fold symmetry was clearly shown, which indicated that tape had the {100}<001> textured structure. The average roughness of tape surface is less than 100nm. We found no difference in roughness between the samples before and after the recrystallization heat treatment.

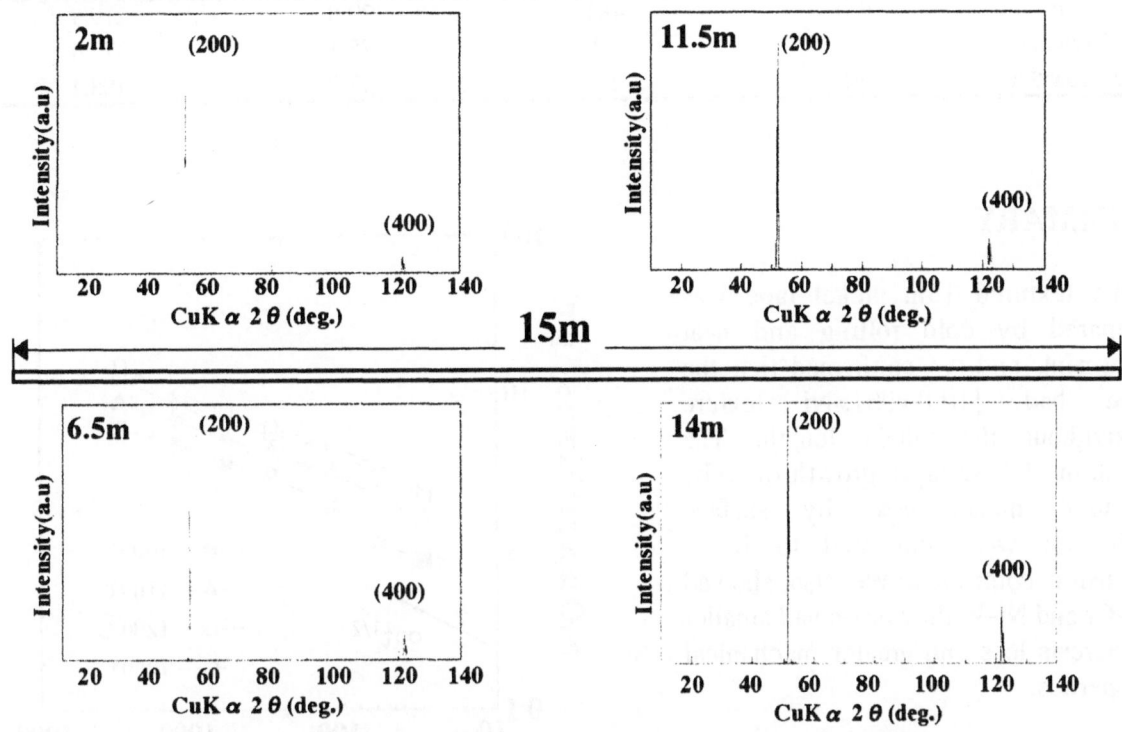

Fig. 1. X-ray diffraction patterns of 15m length textured Ni tape

NiO layer formation on textured Ni. Figure 3 gives the relationship between NiO thickness and heat treatment time during the surface-oxidation treatment of the textured nickel in air. At each temperature, the thickness of NiO varied in proportion to the square root of time, which confirmed that the reaction of NiO layer formation was controlled by diffusion. From x-ray diffraction pattern, we conclude that (100) direction was parallel to the tape surface for about 90% of NiO grains.

Ni based alloy. Although (100)Ni was almost 100% parallel to the tape for textured pure nickel as mentioned above, that for Ni-12wt%Cr alloy was slightly lower (about 95%) and that for Ni-10wt%V achieved to

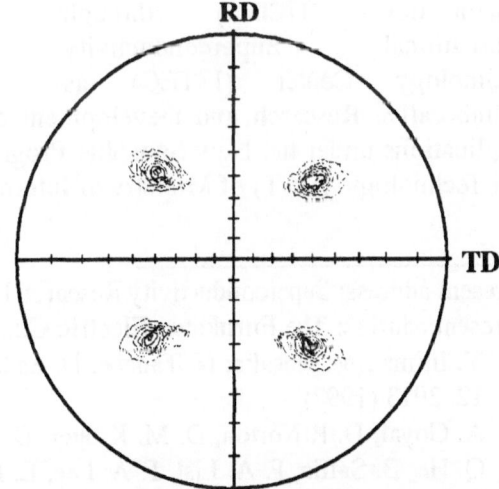

Fig. 2. X-ray pole figure of (111) peak of 15m length textured Ni tape

80%, which were determined by XRD patterns. For magnetic and mechanical properties, as shown in table 1, the hysteresis losses of alloy tapes had one order of magnitude smaller than that of pure nickel tape and proof stress and tensile strength of both of the alloy tapes were significantly higher than that of pure nickel.

Table 1. Texture quality, Hysteresis loss at 77K and Mechanical properties at room temperature of Ni, Ni-12wt%Cr alloy and Ni-10wt%V alloy

	I(200)/ Σ I(hkl) (%)	Hysteresis loss (J/m3)	Proof stress (MPa)	Tensile strength (MPa)
Ni	100	481	29.8	219.1
Ni-12wt%Cr	95	54	75.1	371.4
Ni-10wt%V	80	66	52.8	488.1

SUMMARY

Cube textured 15m nickel tape was prepared by cold rolling and heat treatment, and we confirmed that the tape had {100}<001>Ni texture throughout the whole length. The reaction of NiO layer growth on cube textured nickel tape by surface oxidation was confirmed to be a diffusion controlled. We also showed Ni-Cr and Ni-V alloy tapes had smaller hysteresis loss and greater mechanical properties.

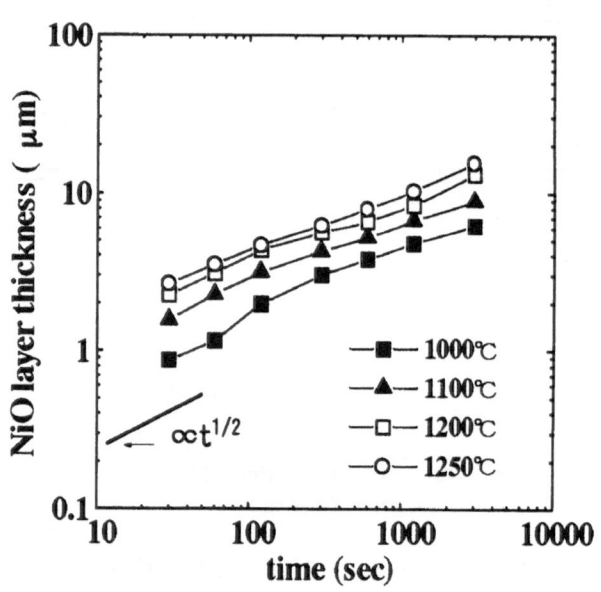

Fig. 3. Time dependence of NiO layer formation at temperature range of 1000℃-1250℃ in air

Acknowledgement. This work was supported by New Energy and Industrial Technology Development Organization (NEDO) through International Superconductivity Technology Center (ISTEC) as Collaborative Research and Development of Fundamental Technologies for Superconductivity Applications under the New Sunshine Program administrated by the Agency of Industrial Science and Technology (AIST) of Ministry of International Trade and Industry (MITI) of Japan.

*Present address: Superconductivity Research Laboratory, ISTEC
†Present address: The Furukawa Electric Co., Ltd.

1. Y. Iijima , M. Hosaka, N. Tanabe, N. Sadakata, T. Saitoh, O. Kohno, and K. Takeda, J. Mat. Res., 12, 2913 (1997)

2. A. Goyal, D. P. Norton, D. M. Kroger, D. K. Christen, M. Paranthaman, E. D. Specht, J. D. Budai, Q. He, B. Saffin, F. A. List, F. A. Lee, E. Hatfield, P. M. Martin, C. E. Klabunda, J. Mathis, and C. Park, J. Mat. Res., 12, 2924 (1997)

3. K. Matsumoto, Y. Niiori, I. Hirabayashi, N. Koshizuka, T. Watanabe, Y. Tanaka, and M. Ikeda, Advances in Superconductivity X, edited by K. Osamura and I. Hirabayashi, (Springer-Verlag, Tokyo, 1998), pp. 611-614.

4. K. Matsumoto, S. B. Kim, J. G. Wen, Y. I. Hirabayashi, T. Watanabe, N. Uno, and M. Ikeda, IEEE Trans. Appl. Supercond., 9, 1539 (1999).

Influence of Carbon on $YBa_2Cu_3O_{6+x}$ Single Crystalline Thick Film

T. Miura, Y. Yamada[#], Y. Koike[#], H. Ikuta[##], I. Hirabayashi[#], U. Mizutani

Dept of Crystalline Materials Science, Nagoya University, Furo-cho, Chikusa-ku, Nagoya 464-8603, Japan
[#]ISTEC-SRL, 2-4-1 Mutsuno, Atsuta-ku, Nagoya 456-8587, Japan
[##]CIRSE, Nagoya University, Furo-cho, Chilkusa-ku, Nagoya 464-8603, Japan

Abstract: Single crystalline thick films of $YBa_2Cu_3O_{6+x}$ including carbon were grown on (100) MgO substrates by a liquid phase epitaxy (LPE) method. Carbon was supplied by incorporating it into the solution from which crystals were grown. The carbon content decreased with increasing the holding time of the solution. The effective distribution coefficient of carbon between the crystal and solution was about 0.5 at 990°C. Transition Temperature (T_c) rapidly decreased and transition width broadened when carbon content exceeded 0.06 wt%, but the sample with 0.045 wt% carbon had a high critical current density. These results suggest that the incorporated carbon act as pinning centers.

Key words: $YBa_2Cu_3O_{6+x}$, liquid phase epitaxy (LPE) method, carbon, pinning effect

INTRODUCTION

Liquid Phase Epitaxy (LPE) method for the formation of oxide superconductor, $YBa_2Cu_3O_{6+x}$, is a promising method which can grow high J_C ($\geq 10^6$) $YBa_2Cu_3O_{6+x}$ thick films (5 μm) at high growth rates (≥ 2 μm/min) [1, 2]. In view of simplification of experimental equipment and cost, it is desirable that the LPE processing is performed in the ambient atmosphere without any gas control. In this case, however carbon may be incorporated in the $YBa_2Cu_3O_{6+x}$ crystals through the desolation of carbon dioxide (CO_2) from the ambient atmosphere to the solution from which crystals grow. Grevin *et al.* concluded that carbon locates on the Cu(1) chain site in the form of CO_3 cluster [3], while others claimed that carbon sits in between CuO chains in the form of CO_3 cluster [4]. The critical temperature (Tc) was reported to rapidly decrease when carbon content exceeds 400 ppm in weight [5]. However, most of the reported studies of the effect of carbon on $YBa_2Cu_3O_{6+x}$ are for polycrystalline samples and did not change the carbon content systematically. We grew $YBa_2Cu_3O_{6+x}$ singlecrystalline thick films containing carbon by the LPE method and quantitatively studied the influence of carbon on the superconducting properties.

EXPERIMENTAL

$YBa_2Cu_3O_{6+x}$ single crystalline films were grown by the LPE method with a modified top-seeded solution growth. Y_2BaCuO_5 and BaO-CuO (Ba:Cu = 3:5) were put into an yttria crucible and heated to 990°C. The films were grown on (100) MgO substrates whose surface were previously covered with $YBa_2Cu_3O_{6+x}$ seed layer deposited by a pulsed laser deposition method. The thickness of LPE grown $YBa_2Cu_3O_{6+x}$ was 15 μm - 25 μm. We incorporated carbon in the crystal by adding barium carbonate ($BaCO_3$) to the solution. All samples were annealed in a pure oxygen atmosphere at 400°C for 24 hours. The carbon content in the crystal was analyzed by secondary ions mass spectrometry (SIMS) and that in the solution by the combustion method and infrared absolution method. T_C was determined as a function of carbon content from magnetization measurement using a SQUID magnetometer and from resistivity measurement using the DC four-probe method. We measured the magnetization hysteresis by using a SQUID magnetometer to estimate critical current density (J_C) according to the extended Bean's critical-state model.

607

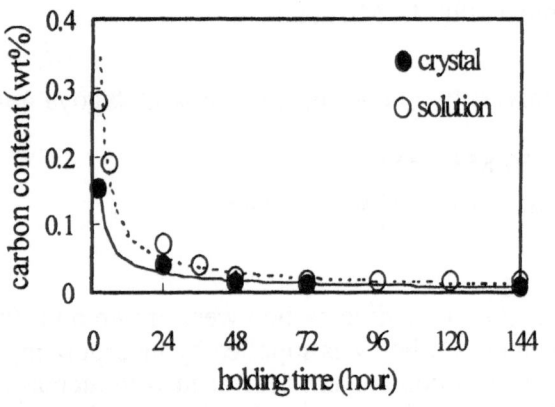

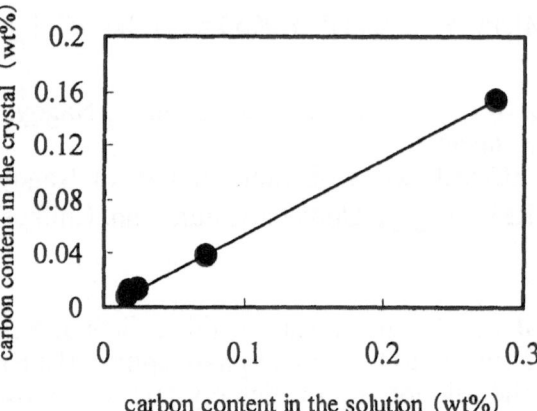

Fig. 1. The carbon content as a fuction of holding time of the solution after the addition of barium carbonate ($BaCO_3$).

Fig. 2. The relation of the carbon content in the solution and the crystal.

RESULTS AND DISCUSSION

Figure 1 shows the holding time dependence of carbon content in the solution and in the crystals. Sampling of the solution was performed just after finishing the crystal growth. The figure shows that the carbon content in both of the solution and crystal decreases with the holding time and approaches a constant value. This behavior suggests that carbon are released from the solution in the form of carbon dioxide (CO_2), and finally, the concentration of carbon in the solution will be equilibrated that in the air. In Figure 2, we replot the data of Figure 1 as the relation of the carbon content in the solution and the crystal. The slope of the line plotted in Figure 2 is about 0.5, which corresponds to the effective distribution coefficient of carbon between the crystal and solution at 990°C, 2 μm/min crystal growth rate. Note that this value may depend on the experimental condition, because the effective distribution coefficient depends on the temperature and growth velocity of the crystal. Figures 3 and 4 show the relation between T_c and carbon content. T_c rapidly decreased and the transition width is broadened when the carbon content exceeded 0.06 wt%. We measured the magnetic hysteresis curves at 50 K and 77 K by using the SQUID magnetometer and evaluated J_c which is shown in Figure 5.

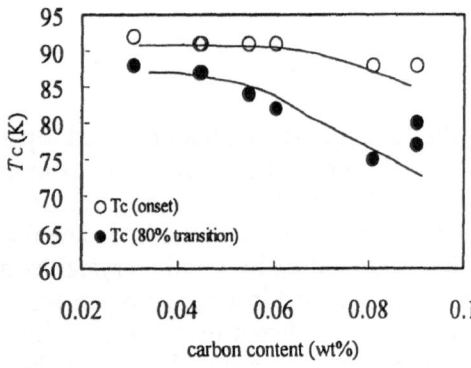

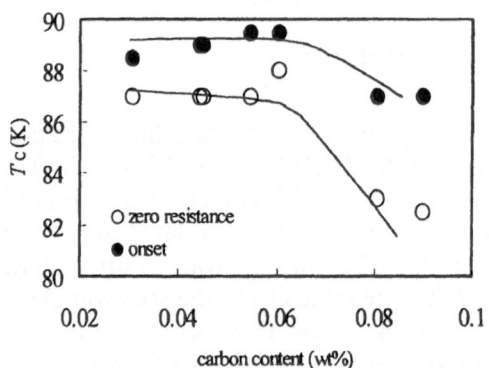

Fig. 3. The relation between T_c and carbon content. T_C was determined from magnetization measurement. The open circle represents the onset of the transition and the solid circle is the temperature where the magnetization exceeds.

Fig. 4. The relation between T_c and the carbon content. T_C was determined from resistivity mesurement. The open circle represents the zero resiststivity temperatureand the solid circle is for the onset temperature.

The sample with 0.045 wt% carbon exhibited the highest J_c value among the crystals studied here. It should be noted that the sample with 0.031 wt% carbon, which is the lowest carbon in the present work, revealed a substantically smaller J_c than the sample with 0.045 wt% carbon. This result demonstrates that appropriately incorporated carbons act as pinning centers in $YBa_2Cu_3O_{6+x}$

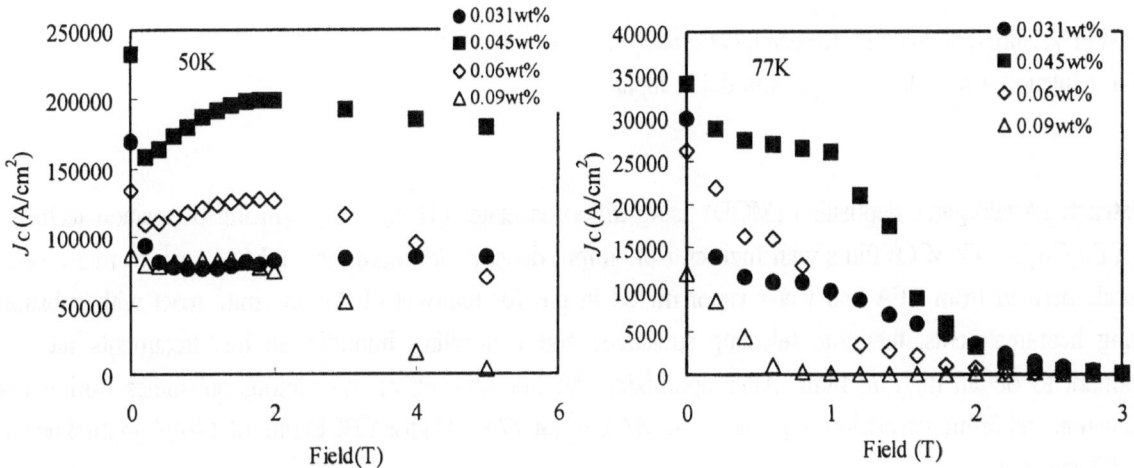

Fig.5. J_c-B curves at 50 K (left side) and 77 K (right side).

CONCLUSION

We fabricated $YBa_2Cu_3O_{6+x}$ thick films containing carbon on MgO substrates by the LPE method. The carbon content in the solution and in the crystal grown from the solution decreased with increasing the time for which the solution was hold. The effective distribution coefficient of carbon between the solution and crystal was about 0.5 at 990°C, 2 μm/min crystal growth rate. T_c rapidly decreased and the transition was broadened when the carbon content exceeded 0.06 wt%. The sample containing 0.045 wt% carbon exhibited a substantially larger J_c than the sample with less carbon content. However, J_c of the sample with the highest carbon content (0.090 wt%) showed the smallest value. Accordingly, we conclude that an appropriate amount of carbon can play the role of pinning centers, but a too much carbon incorporation in $YBa_2Cu_3O_{6+x}$ reduces T_c and J_c.

ACKNOWLEDGMENT

This work was supported by the New Energy and Industrial Technology Development Organization (NEDO) as Collaborative Research and Development of Fundamental Technologies for Superconductivity Applications.

1. M. Yoshida, T. Nakamoto, T. Kitamura, Ok. Hyun, I. Hirabayashi and S. Tanaka, Appl. Phys. Lett. 65 (1994) 1714
2. S. Miura, K. Hashimoto, F. Wang, Y. Enomoto and T. Morishita, Physica C 278 (1997) 201
3. B. Grevin, Y. Berthier, I. Monot, J. Wang, E. Weiss, Physica C 289 (1997) 77
4. F. J. Gotor, P. Odier, M. Gervais, J. Choisnet and Ph. Monod, Physica C 218 (1993) 429
5. Y. Masuda, R. Ogawa, Y. Kawate, K. Matsubara, T. tateishi and S. Sakka, J. Mater. Res 8 (1993) 693

The effects of process conditions on $YBa_2Cu_3O_{7-x}$ films by metalorganic deposition method using trifluoroacetates

T. Araki, K. Yamagiwa, S.B. Kim, K. Matsumoto and I. Hirabayashi

Division V, Superconductivity Research Laboratory, ISTEC,
2-4-1, Mutsuno, Atsuta-ku, Nagoya, 456-8587, Japan

Abstract: Metalorganic deposition (MOD) using trifluoroacetates (TFA) is a promising deposition technique for $YBa_2Cu_3O_{7-x}$ (YBCO) films with high critical current density (Jc) more than 1 MA/ cm^2. In this method, fluoride derived from TFA and water vapor mixed in gas for removal of fluorine may react with substrates during heat-treatments, therefore selecting substrates and controlling humidity in heat-treatments are very important to obtain high Jc films. After optimizing the humidity effects on various substrates during both calcination and firing, we attained high Jc of 4.6 MA/cm^2 (at 77 K, 0T) for YBCO film of 1400Å in thickness on LaAlO$_3$ substrate.

Keywords: $YBa_2Cu_3O_{7-x}$, thicker film, metalorganic deposition, trifluoroacetate

Introduction

The MOD method is to form films under heat treatments after coating on substrates with metalorganic solutions. There are many advantages for this method, such as precise controllability of composition, wide flexibility to coating objects and a low cost non-vacuum approach to ceramic films[1]. Recently Jc of 1 MA/cm^2 was reported for the YBCO films of one micron thick prepared by MOD method using TFA salts[2]. Similar result was also reported for the films prepared by electron beam process method, in which the same BaF$_2$ is contained in precursor films[3]. In the calcined films prepared by the MOD process using TFA salts, Ba component exists as BaF$_2$, that leads to carbon free films[4]. However the process requires precise humidity control in heat treatments for removal of fluorine at the final stage of the process[4-5]. Moreover, to fabricate high Jc YBCO coated conductors using this method, it is also important to find good buffer layer endurable to chemical reactions for metallic substrates.

In this study, we investigated reactions between various substrates of oxides and coated films to optimize humidity control and fabricated high Jc YBCO fimls on oxide substrates. To obtain high critical current (Ic) films, we also investigated means to obtain thicker film by multiple coatings or by controlling concentration of coating solutions in single coating.

Experimental

A trifluoroacetate solution was prepared by dissolving the acetate of Y, Ba and Cu into de-ionized water in a

1:2:3 cation ratio with stoichiometric quantity of TFA, and then solvents in the solution were removed by evaporator to yield blue glassy residue. The coating solutions were made of dissolving the residue into methanol. Various concentrations of coating solutions were prepared for experiments. The TFA gel films were coated with coating solutions onto 10 x 10 mm, (001)-oriented MgO, SrTiO₃ and LaAlO₃ single-crystalline substrates by spin coating. A spinning rate of 4000 rpm for 120 s and an acceleration time of 0.4 s were used in depositing gel films. The gel films were calcined to 400 °C for 20 hours in 0-12.1% humid oxygen, fired to 800°C for an hour in 0-12.1% humid argon mixed with 0.1% oxygen, and annealed to between 600 and 400 °C for half an hour in dry oxygen to form YBCO films.

Thermal decompositions were measured by differential thermal analysis and thermogravimeter (DTA-TG). The viscosity of coating solution was measured by corn plate viscometer. Crystalline phases into films were detected by X-ray diffraction (XRD), in-plane alignments were measured by pole figure. Surface and cross sectional morphologies were observed by scanning electron microscopy (SEM). Thickness and metal compositions were evaluated by induced coupled plasma (ICP). Tc and Ic were measured by direct current 4-probe-method.

Results and Discussion

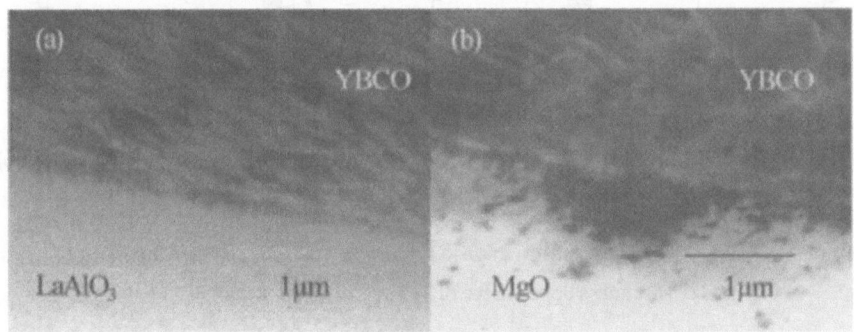

Fig. 1 Cross sectional SEM images of YBCO films on LaAlO₃ (a) and MgO (b) substrates.

In order to investigate the reactions with substrates in heat-treatments, we prepared two series of films on the MgO, SrTiO₃ and LaAlO₃ substrates with coating solution of 1.52 mol/L (M) metal concentration. One series of films were calcined in 0-12.1% humid oxygen and fired in 4.2% humid argon mixed with oxygen. Another series of films were calcined in 4.2% humid oxygen and fired in 0-12.1% humid argon mixed with oxygen. We measured the fired films on MgO, SrTiO₃ and LaAlO₃ substrates by XRD, SEM and DTA-TG. YBCO phase was confirmed by XRD on SrTiO₃ and LaAlO₃ substrates, but not confirmed on MgO substrate. Fig. 1 shows cross sectional SEM images of the fired films on MgO and LaAlO₃ substrates. Fig. 1(b) suggests chemical reaction between MgO substrate and coated film. To investigate the reaction further, we measured each MgO, SrTiO₃ and LaAlO₃ powder which was poured with coating solution, by DTA-TG between 30 and 900 °C in 4.2 % humid oxygen. Thermal decomposition behavior of only MgO powder with coating solution was different from others. Besides MgF₂ phase was detected by XRD from the residue of MgO sample after the DTA-TG measurement. Since the MgO substrate is considered to react with the coated film, finding substrates free from reactions in heat treatments, are inevitable to form YBCO by this method.

To form thicker YBCO film on this process, we attempted two ways, by controlling concentration of coating solution in single coating, and by multiple coatings. Relationship between the viscosity of coating solutions measured by corn plate viscometer and the thickness of the YBCO films, was plotted in Fig. 2. We also obtained another relationship between viscosity and concentrations of coating solutions. With these relationship, we can control the thickness of YBCO films. Several films onto LaAlO$_3$ substrates were obtained through coating, calcination and firing. One series of the films were prepared with coating solutions of 1.52M, 2.31M and 2.78M. Another film was prepared by twice coatings with coating solution of 1.52M. Jc and thickness of all films were measured and plotted in Fig. 3. It suggests that increase of YBCO film thickness causes decrease in Jc of all YBCO films. But it also shows that the decrease in Jc of twice coated YBCO film is greater than that of single coating. The reason can be understood by similar scheme of reaction between substrates and coated films, as we described. For twice coated film, in the second calcination, calcined old film was attacked by fluoride which was derived from new coated film. The correlation between Jc and ratio of a/b-axes to c-axis was observed and plotted in Fig. 4. It shows necessity for suppression of a/b-axes-oriented growth to obtain high Jc thicker films. Thus we obtained the YBCO film on LaAlO$_3$ substrate of 1400 Å thick having 4.6 MA/cm^2 of Jc (at 77K, 0T), 92.5K of Tc and 0.6K of ΔTc. We also obtained the 2900 Å thick film having 2.2 MA/cm^2 of Jc (at 77K, 0T).

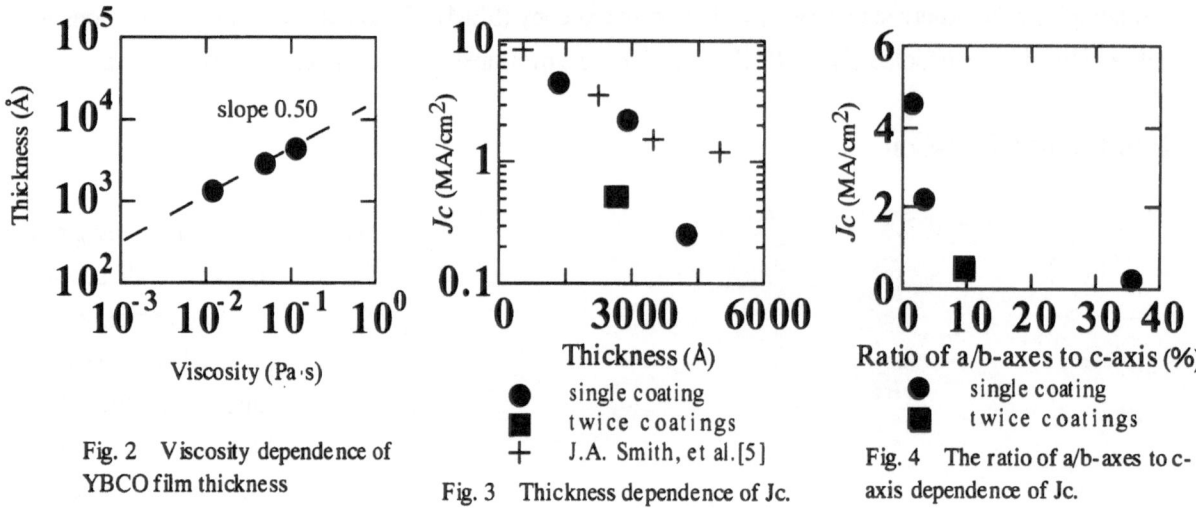

Fig. 2 Viscosity dependence of YBCO film thickness

Fig. 3 Thickness dependence of Jc.

Fig. 4 The ratio of a/b-axes to c-axis dependence of Jc.

Conclusion

YBCO films are successfully formed on SrTiO$_3$ and LaAlO$_3$ substrates by MOD using TFA. To form thicker films, controlling concentration of coating solution is superior to multiple coatings from the view point of Jc. The highest Jc of the film on LaAlO$_3$ substrate was 4.6 MA/cm^2 (at 77K, 0T).

Acknowledgment. This work is supported by the New Energy and Industrial Technology Development Organization (NEDO) as Collaborative Research and Development of Fundamental Technologies for Superconductivity Applications.

1.   P.C. McIntyre, M.J. Cima, and Man Fai Ng, J. Appl. Phys. **68** (8), 4183 (1990).

2.   J.A. Smith, M.J. Cima and N. Sonnenberg, Appl. Superconductivity, IEEE Trans. Appl. Supercon., vol. 9, 1531(1999).

3.   R. Feenstra, T.B. Lindermer, J.D. Budai, and M.D. Galloway, J. Appl. Phys. **69** (9), 6569(1991).

4.   A. Gupta, R. Jagannathan, E.I. Cooper, E.A. Giess, J.I. Landman, and B.W. Hussey, Appl. Phys. Lett. **52** (24), 2077 (1988).

5.   P.C. McIntyre, M.J. Cima, J.A. Smith, Jr., and R.B. Hallock, MP. Siegal and J.M. Philips, J. Appl. Phys. **71** (4), 1868 (1992).

PREPARATION OF LONG YBCO TAPES BY COLD-WALL MOCVD METHOD

Toyotaka Yuasa[1], Masato Hasegawa[1], Tsuneo Oyama[1], Yasuo Takahashi[1], Kaname Matsumoto[1], Hiroyuki Akata[2], Kazuthoshi Higashiyama[2], Yutaka Yoshida[3], and Izumi Hirabayashi[1]

[1]Superconductivity Research Laboratory, ISTEC, Nagoya 456-8587, Japan
[2]Hitachi Research Laboratory, Hitachi Ltd., Hitachi, Ibaraki 319-1221, Japan
[3]Department of Energy Engineering and Science, Nagoya University, Nagoya 464-8603, Japan

Abstract: We developed a reel to reel MOCVD deposition system to fabricate long YBCO coated conductor. Using this system, we deposited long YBCO film on a {100}<001> cube textured Ag tape (CUTE Ag tape) up to 1m. Homogeneous YBCO grains about 0.5 to 1 μ m in size were successfully grown on long CUTE Ag tape, and XRD pattern showed c-axis oriented. However, the in-plane orientation of YBCO was mixture of cube-on-cube and 45° rotated with respect to the underlying Ag as expected from the of lattice mismatch between YBCO and Ag. To improve in-plane alignment of the films, CeO_2 buffer layers were deposited on as-rolled CUTE Ag tapes by PLD method. By using this process, the growth of grooves at grain boundaries was suppressed. The full width at half maximum (FWHM) from the x-ray YBCO(103) ϕ scans for YBCO on CeO_2 buffered silver substrate was 11 to 12 degrees.

Keywords: MOCVD, CUTE Ag tape, CeO_2 buffer layer, surface roughness

INTRODUCTION

The fabrication of biaxially aligned YBCO films on metal substrates is necessary for coated conductor usable in liquid nitrogen. Using cube-texture (CUTE) metal such as Ni[1] and Ag[2] is the most promising technique to obtain long substrates for coated conductor because of their high processing speed. The CUTE Ag tapes is suitable for this purpose, because the tapes are very stable even at high-temperature. For the fabrication process to make long YBCO films on the tapes, MOCVD method[3] is attractive because of the ability for continuous supply of source materials.
 To make long biaxially oriented YBCO films, we have developed an apparatus and processes for preparing long YBCO films on the CUTE Ag tapes by MOCVD method.

EXPERIMENTAL

MOCVD deposition was carried out by using yttrium-tris-(2,2,6,6-tetramethylgeptane-3,5-dione) 4-t-butylpyridine-N-oxide adduct (Y(DPM)$_3$ · 4tBuPyNO),bis-dipivaroylmethanatobarium-tetranthylenepentamine adduct (Ba(DPM)$_2$ · 2tetraen) and copper-bis-(2,2,6-trimethylheptane-3,5-dione) (Cu(TMHPD)$_2$) as the MO-sources. The growth of YBCO films was deposited in a horizontal cold-wall CVD system. The MO-source vapor transported by Ar gas of 200 sccm were mixed with oxygen gas as the oxidizing agent. To achieve high deposition rate, the reactant gas was vertically introduced onto substrates heated by ceramic heater. The deposition was carried out in flow of 100 sccm O_2 at a reactor working pressure of 7 Torr. CUTE Ag tapes 5^w X 0.05^t mm in size were used as substrates. To fabricate long YBCO tapes, the substrate was moved at rate of 80 mm / hour. The samples were characterized by XRD, SEM, ICP, SQUID magnetometry and AFM.

RESULTS & DISCUSSION

Fabrication of long YBCO films on Ag tapes.
To demonstrate the ability to make long YBCO-coated conductors by our MOCVD apparatus, we deposited YBCO film on a 1m long CUTE Ag tape as shown fig.1. To verify the homogeneity of long YBCO films on a

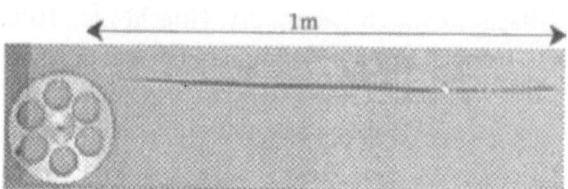

Fig. 1 The photograph of YBCO film on a 1m long CUTE Ag tape.

CUTE Ag tape, the YBCO tape was cut into 1.5 cm long at intervals of 5 cm, and the distribution of composition normalized by the content of yttrium for a 50cm long film was investigated by ICP as shown in fig. 2. Although the copper composition did not decrease as increased deposition time, the barium composition decreased. This decrease was attributed to the decrease of Ba vapor from the unstable MO-source. To get long homogeneous YBCO films, we have to optimize process condition by controlling Ba vapor. Regular θ-2θ X-ray scans for each film showed only (00l)-type reflections, indicating the c-axis orientation. The alignments of each film were inferred form X-ray ϕ scans for CUTE Ag tapes and YBCO films. Figure 3 shows typical X-ray ϕ scans of the films. The ϕ scan for (111) reflections of silver had FWHM spreads in the range of 8-9 degrees indicating very good in-plane orientation. On the other hand, the ϕ scan for (102) reflections of YBCO showed both epitaxial and in-plain 45 degrees rotated grains with the [001] YBCO rotation axes. This grain growth on Ag(100) face was explained from NCSL theory[4]. From these result, we confirmed that the epitaxial growth of YBCO film on the 50 cm long CUTE Ag tape was achieved.

YBCO deposition on CeO_2 buffered CUTE Ag tapes. To improve in-plain alignment of YBCO films, CeO_2 buffer layer was deposited on the CUTE Ag tapes by PLD method. Detailed PLD deposition conditions were described elsewhere[5]. We have alternatives in process A, CUTE Ag tapes annealed at 700℃ in ambient atmosphere for recrystallization of silver were used as substrates for

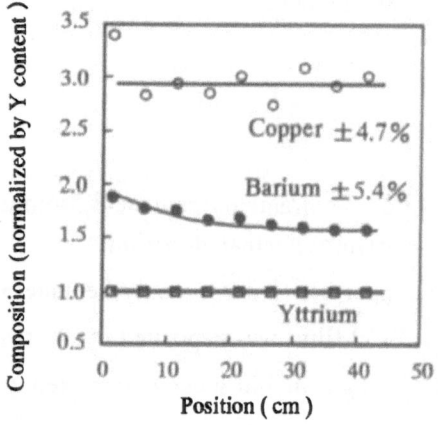

Fig. 2 The distribution of composition for YBCO films deposited on a 50cm long CUTE Ag tape.

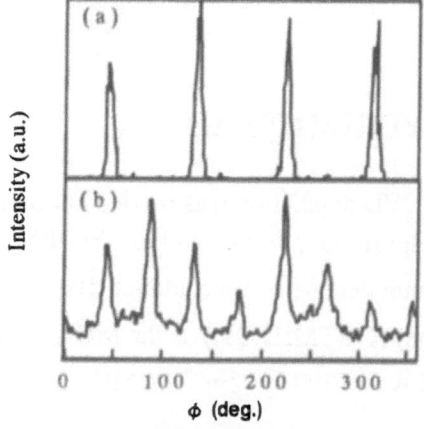

Fig. 3 ϕ scans of (a) CUTE Ag tape and (b) YBCO film on the CUTE Ag tape.

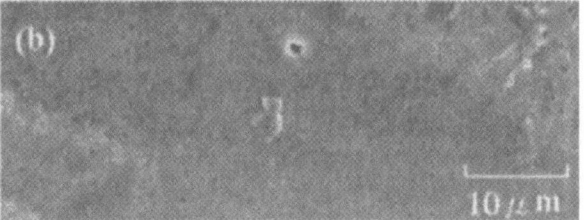

Fig. 4 SEM images of YBCO film deposited on (a) the CUTE Ag tape with CeO_2 buffer layer and (b) the as-rolled CUTE Ag tape with CeO_2 buffer layer.

CeO_2 buffer layer. On the other hand, in process B expecting to reduce surface roughness as-rolled CUTE Ag tapes were used. In this process, the recrystallization of silver was achieved during heating the substrate at 660℃ under 30 mTorr oxgen in the PLD method for CeO_2 deposition. Then, the YBCO films were deposited on the CeO_2 buffered CUTE Ag tapes by MOCVD method. Figure 4 shows SEM images of YBCO films on CeO_2 buffered CUTE Ag tapes by the process A and B respectively. Although heavy grooves over 500 nm on CUTE Ag tape by the process A were observed, the smooth surface was obtained by the process B. In addition, average roughness of AFM line scans over a 6 x 6 μm^2 area of YBCO films on CUTE Ag tape by the process B was 24 nm. Although in the process A,

the roughening at silver surface was caused by grooves formation at grain boundaries, the growth of grooves might be suppressed by coverage of the CeO_2 buffer layer during the recrystallization of silver grains in the process B. Figure 5 shows the X-ray ϕ scan for (103) reflections of the film prepared by the process B. The X-ray ϕ scan showed fourfold in-plane symmetry with FWHM of 11 to 12 degrees. At present, the Tc of the film measured by SQUID was 78 K.

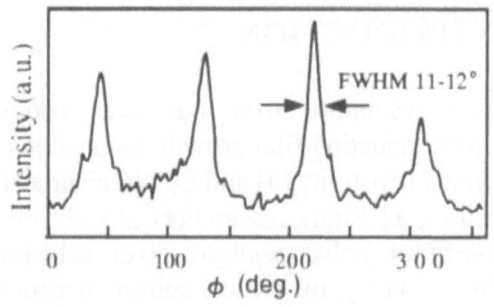

Fig. 5 ϕ scan of YBCO film on the CeO_2 buffered CUTE Ag tape.

Summary

We deposited the YBCO film on a 1m long CUTE Ag tape by MOCVD method. The YBCO films on a 50 cm long CUTE Ag tape were c-axis orientated and had both cube-on-cube and in-plain 45 degrees rotated grains. To improve in-plane alignment and surface roughness of Ag substrate, CeO_2 buffer layers were deposited on as-rolled CUTE Ag tapes by PLD method. By using this process, the growth of grooves at silver grain boundaries was suppressed. The FWHM from the x-ray ϕ scans for YBCO films on the tapes was 11 to 12 degrees.

Acknowledgments. This work was supported by the New Energy and Industrial Technology Development Organization (NEDO) as Collaborative Research and Development of Fundamental Technologies for Superconductivity Applications.

1. A. Goyal et al., Phys. Lett. 69 (12), 1795-1797
2. T. Doi et al., *Advances in Superconductivity VII*, (Springer-Verlag. Tokyo, 1995) pp. 817-820.
3. M. Hasegawa et al., *Advances in Superconductivity XI*, (Springer-Verlag. Tokyo, 1999) pp. 1031-1034
4. T. S. Ravi et al., Phys. Rev. B 42, 10141-10151(1990)
5. Y. Takahashi et al., *Advances in Superconductivity XI*, (Springer-Verlag. Tokyo, 1999) pp. 777-780

The substrate surface morphology and the $YBa_2Cu_3O_{7-x}$ film growth on polycrystalline silver

Daxiang Huang, Yasuji Yamada, and Izumi Hirabayashi

Superconductivity Research Lab., ISTEC, 2-4-1 Mutsuno, Atsuta-ku, Nagoya 456-8587, Japan

Abstract: Thin $YBa_2Cu_3O_{7-x}$ (YBCO) films were grown on the polycrystalline-silver by pulsed laser deposition (PLD). Using transmission electron microscopy (TEM), we have studied the direct relations of the YBCO film growth features with the substrate surface morphology. The obtained results show that the flat substrate surface is the only requirement for the c-axis oriented YBCO film growth. On the flat substrate surface, the YBCO film usually grows with its a-b plane simply parallel to the substrate surface plane. Any small angle change of the substrate surface plane will cause the same angle change of the c-axis orientation of the grown YBCO film. On the defective surface with valleys, however, the YBCO film can not directly grow, instead, there is an intermediate layer previously grown on the substrate surface. Concerning the surface steps and hills, the TEM observation indicates that they almost have no influence on the YBCO film growth.

Keywords: Substrate Surface Morphology, Film Microstructures, YBCO, Silver, TEM.

INTRODUCTION

Polycrystalline silver has been taken as one of the most attractive substrates for the superconducting film growth applications as it has a moderate cost, a good flexibility, [1, 2] a low contact resistivity,[3] and a good chemical compatibility with high-T_c superconducting materials as well.[4, 5] Many researchers already made efforts to grow high-quality YBCO films on single crystal or polycrystalline silver substrates by using various thin-film-growth methods.[6-10] Unfortunately, the critical current densities (J_c) for all the obtained films are just around 10^4 A/cm², which is too low for the practical high current applications.

There are many parameters which influence the quality of the YBCO films grown on Ag substrates.[8] The substrate surface roughness is an important one of them. Before the YBCO film deposition, first people usually try to treat the substrate surface using various methods to reduce the surface roughness, even for the oxide single crystal substrates such as $SrTiO_3$ et al.[11] For the soft metal, such as Ag, it is much difficult for obtaining a smooth surface comparing with the oxide single crystals. In order to clarify the detailed influence of the surface roughness on the film deposition process, in this study, we systematically investigated the change of YBCO film microstructure with the substrate surface morphology using TEM.

EXPERIMENTAL

The growth of thin 200 nm YBCO films on polycrystalline Ag substrates was carried out using single-target pulsed-laser ablation. A Lambda-Physics KrF (248 nm) excimer laser was used and operated at 50 Hz with an energy density of 2~3 J/cm². During deposition, the substrates were heated to temperatures in the range of 680 ~ 780 °C, and the oxygen partial pressure was held at about 200 mTorr. Cross-sectional TEM specimens which allowed examination of the substrate-film interface were prepared by the usual method. The TEM analyses for the substrate-film interface were performed using a JEM 2010 high-resolution TEM.

RESULTS AND DISCUSIONS

a)   On the flat substrate surface. Fig. 1 shows a good c-axis-oriented YBCO film grown on the flat substrate surface area. Seeing carefully, we can find that this flat substrate surface area is composed of many small flat surface planes with a size of about 100 ~ 300 nm. The normal directions of two neighboring small flat surface planes have a small angle difference (usually within 5°). Correspondingly, a small angle difference for the c-axis orientations of the YBCO film grown on these two neighboring small flat surface planes can be also found. At the boundary of two neighboring small flat surface planes, a YBCO grain boundary in the film side can be usually observed. The results of high resolution TEM observations further reveal that the angle between the c-axes of two neighboring YBCO grains is exact the same as that between the normal directions of two neighboring small flat surface planes on which the two YBCO grains grow. It is found that the c-axis of the deposited YBCO film is just simply parallel to the normal of substrate surface plane and changes as the normal of substrate surface plane changes.

b)   On the surface with steps. Fig. 2 shows the YBCO film growth on the Ag substrate surface with a step. The YBCO film can grow to cover the surface step. Similar to the cases on flat substrate surface, the YBCO film just grows with its a-b plane parallel to the substrate surface plane. For the two

Fig. 1.   The YBCO film grown on flat surface.

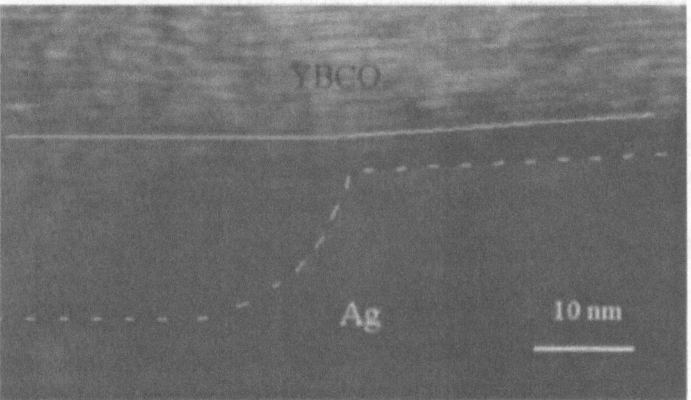

Fig. 2.   The YBCO film grown on a surface step.

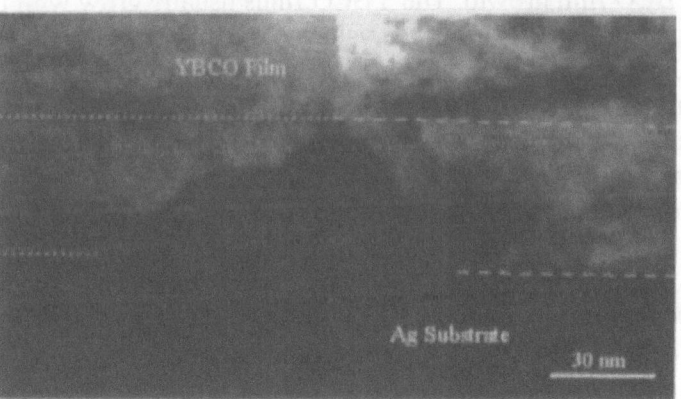

Fig. 3.   The YBCO film grown on a surface hill.

step terraces with a small angle difference of the terrace normal, the same angle   difference can be also observed for the c-axes of the YBCO grains grown on these two step terraces.

c) On the surface with hills. Fig. 3 shows the YBCO film growth on the substrate surface with a single hill surrounding by a large flat surface area. It seems that the surface hill does not influence the YBCO film growth so much. The YBCO film can simply cover this surface hill by lateral overgrowth. The grain boundary starting from the peak of the hill is also formed due to the similar reason: the small angle difference for the normal directions of the two small flat surface planes on the two sides of the surface hill.

d)On the surface with valleys. When the substrate surface has a large valley, e.g., several hundreds of nanometers in width, with a heavily-curved bottom plane as shown in Fig. 4, a very complicated structure of the intermediate layer will be formed. The intermediate layer is usually formed by several kinds of non-superconducting oxide phases such as copper oxide and barium copper oxide. Several YBCO grains, with large-angle differences of their c-axis orientations, still can nucleate and grow on the upper surface of this intermediate layer. Interestingly, the intermediate layer can play the role of a buffer layer to reduce the original surface roughness of Ag substrate and further produce a 2nd surface, with many small flat surface planes of different orientations, which is suitable for the YBCO crystal nucleation and growth. This kind of self-smooth process in the YBCO film deposition may play an important role to help the YBCO film growth on the substrate surface with valleys.

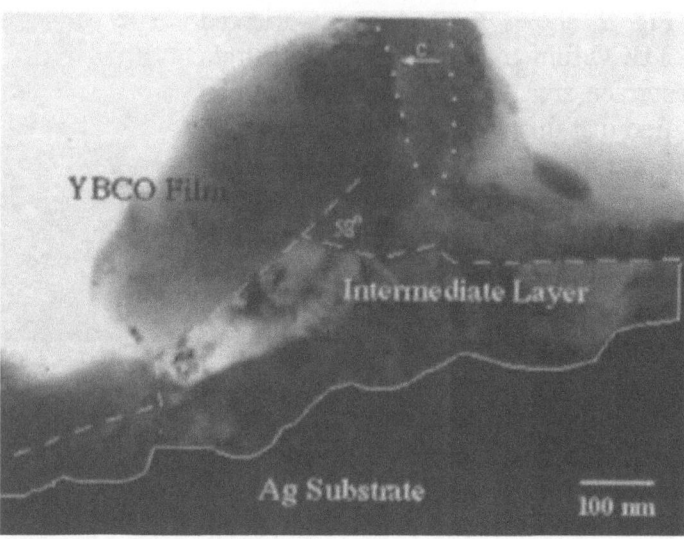

Fig. 4. The YBCO film grown on a surface valley.

CONCLUSIONS

In this study, we did a systematic investigation for the microstructure features related to the surface defect morphology in the YBCO films deposited by the PLD method on polycrystalline silver substrates. The obtained results indicate that the substrate surface morphology is a very sensitive factor to influence the film forming process strongly. The flat surface is good for the YBCO film growth. The YBCO films usually grow with their a-b planes simply parallel to the flat surface plane. The surface valley is one kind of fatal defect to strongly influence the quality of the film deposition. On the surface with valleys, usually the YBCO crystals cannot directly nucleate and grow, instead, an intermediate layer with some other oxide phases will previously form. The substrate surface steps and hills are not the suitable places for the nucleation and growth of any crystals, but they can be covered by the rapid growth of YBCO grains surrounding them and do not influence the final quality of the final-formed YBCO films so much.

Acknowledgements: This work is supported by the New Energy and Industrial Technology Development Organization (NEDO) as Collaborative Research and Development of Fundamental Technologies for Superconductivity Applications.

1. J. P. Singh, D. Shi, and D. W. Capone, Appl. Phys. Lett. 53, 237 (1988).
2. S. Jin, G. W. Kammlott, T. H. Tiefel, and S. K. Chen, Physica C 198, 333 (1992).
3. M. E. Tidjiani and R. Gronsky, Physica C 191, 260 (1992).
4. C. T. Cheung, and E. Ruckenstein, J. mater. Res. 4, 1 (1989).
5. R. Bohnenkamp-Weiss, and R. Schmid-Fetzer, Physica C 220, 396 (1994).
6. Y. Masuda, K. Matsubara, R. Ogawa, and Y. Kawate, Jpn. J. Appl. Phys. 31, 2709 (1992).
7. J. D. Budai, R. T. Young, and B. S. Chao, Appl. Phys. Lett. 62, 1836 (1993).
8. L. Chen, T. W. Piazza, et al, J. Appl. Phys. 73, 7563 (1993).
9. R. Eggenhoffner, A. Tuccio, et al, Supercond. Sci. Technol. 10, 142 (1997).
10. Y. Niiori, Y. Yamada, and I. Hirabayashi, Physica C 296, 65 (1998).
11. J. P. Contour, D. Ravelosona, et al, J. Cryst. Growth, 141, 141 (1994).

Transport characteristics in YBCO thin film superconductors

SeokBeom Kim, Yasuo Takahashi, Kaname Matsumoto, Toshiaki Takagi, Takato Machi, Naoki Koshizuka, Atsushi Ishiyama* and Izumi Hirabayashi

Superconductivity Research Laboratory, ISTEC, Nagoya, 456-8587, Japan
*Department of EEC Engineering, Waseda Univ., Tokyo, 169-8555, Japan

Abstract : We studied influences of growth temperature on superconducting characteristics in $YBa_2Cu_3O_{7-x}$ (YBCO) thin films prepared by pulsed laser deposition (PLD). The films deposited at low growth temperature (below 730°C) have a high c-axis orientation, good in-plane alignment and large amount of a-axis grains. With increasing growth temperature, the number of a-axis grains in the films were decreased, whereas superconducting properties were degraded due to the formation of misoriented grain boundaries. The crystal defects acting as barriers to supercurrent flow were observed by magneto-optical imaging (MOI) technique. Using this technique, we confirmed that the critical current densities of the films were limited by these defects. Dynamic magnetic flux behaviors in flux flow state were also observed by MOI system.

Keywords : YBCO, pulsed laser deposition, magneto-optical imaging, dynamic magnetic flux

INTRODUCTION

Recently much progress has been obtained in producing YBCO superconducting films and tapes. A pulsed laser deposition (PLD) is widely used for the fabrication of high-temperature superconducting films. One of the most important superconducting properties is the critical current density, J_c, of the films. The well-informed research on the transport characteristics is necessary in order to produce the films with high J_c. The superconducting properties of the films fabricated by PLD depend mainly on the growth temperature[1]. In this paper, the influences of growth temperature on the superconducting properties in YBCO films deposited by PLD on MgO substrate were investigated. The limitation mechanism of J_c and flux behaviors during the transition in YBCO films were also examined using magneto-optical imaging (MOI).

EXPERIMENTAL

YBCO films were grown on MgO(100) substrates by PLD method. A 248nm KrF excimer laser was used to ablate a YBCO target. In our experimental setup, the laser beam was at angle of 45° with respect to the target surface and the spot beam was focused to 0.5x6 mm², and the energy density at the target surface was about 2 J/cm². The laser repetition rate was always 25 Hz. The YBCO films were deposited at the substrate temperature range of 690-780°C, the oxygen pressure of 0.32mbar, and the target-substrate distance of 54 mm. After the film deposition, the chamber was vented with oxygen to atmospheric pressure and then the films were cooled at approximetely 30°C/min. The temperature was held at 450°C for 30 min during the cooling process .

RESULTS AND DISCUSSION

Growth temperature dependence　　Fig. 1 shows the X-ray pole figures and XRD profiles of YBCO films as a function of growth temperatures of (a)695°C, (b)720°C, (c)750°C and (d)780°C, respectively. The growth temperatures were measured by a thermocouple attached on MgO substrate surface. The temperature fluctuation during the deposition was less than 1°C. The thicknesses of the films were approximately 0.4 μm. The c-axis lengths of the films of (a)-(d) were 11.70 Å, 11.70Å, 11.69 Å and 11.69 Å, respectively. As shown in Fig. 1, the films deposited at lower growth temperatures (695-720°C) show a clear four-fold symmetry, indicating good in-plane alignment. YBCO(200) peaks shown in Figs 1(a) and 1(b) were observed in the films deposited at relatively low growth temperature. The appearence of YBCO(200) peak is undesirable factor for superconducting proper-

619

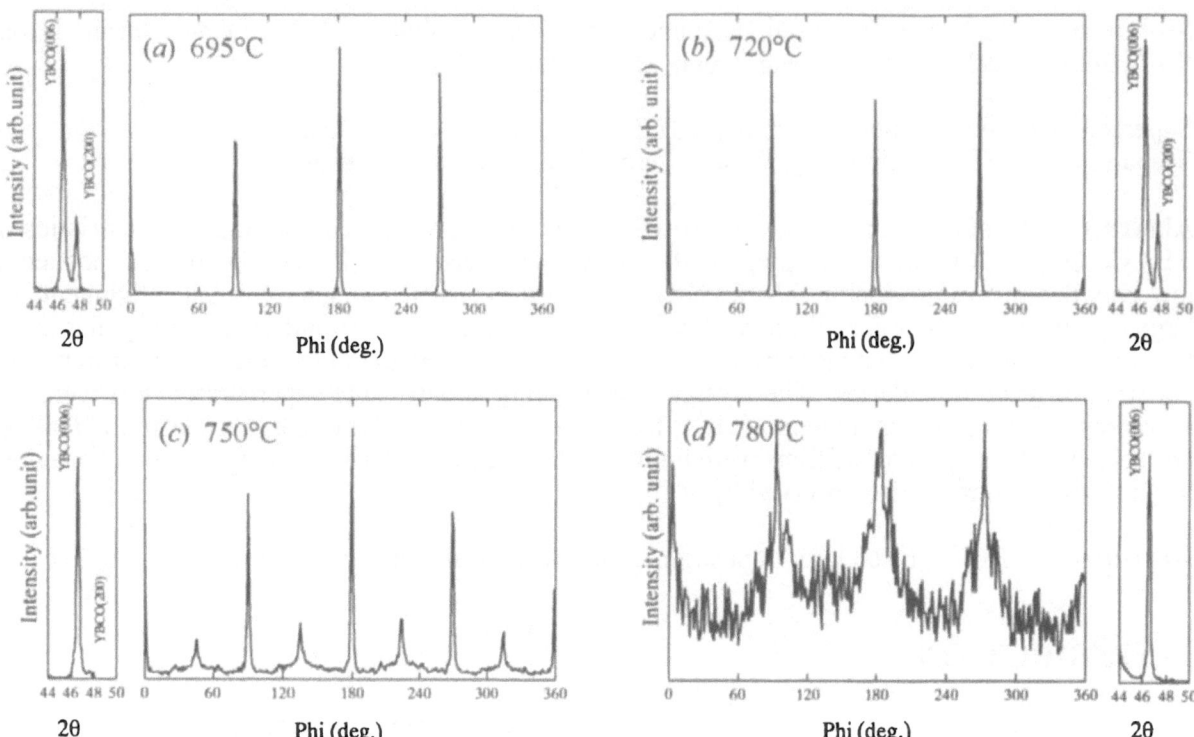

Fig. 1. The X-ray pole figures and XRD profiles of YBCO films deposited
at four different growth temperatures on MgO.

ties. The YBCO(200) peak was suppressed with increasing growth temperature, while four-fold symmetry disappeared for the films deposited around and above 750°C. J_c values of the films were 66 kA/cm² at the growth temperature of 695°C, 320 kA/cm² at 720°C and 10 kA/cm² at 750°C. The film deposited at 780°C was c-axis oriented but did not have in-plane alignment. The films with the misoriented grain boundary showed the weak superconductivity. Such a tendency was confirmed by the SEM observation of the corresponding film surfaces. In our experiment, 730°C was optimum growth temperature for high-J_c films. The width of optimum growth temperature range was 5°C.

Magneto-optical images The MOI, based on the Faraday rotation in an indicator film, is a powerful method to clarify the magnetic flux behavior in superconductors because a local and dynamic flux distribution can be observed directly[2, 3]. Our experimental setup consists of a cryostat and external magnet connected with a polarization light microscope as described in the reference list [4]. In present experiments, the films temperature was set at 20 K and the 200 G was applied perpendicular to the films surface after zero field cooling. Fig. 2 shows magneto-optical images (left) of the YBCO films and three-dimensional representations (right) of flux distribution with different J_c. The bright regions correspond to high magnetic flux density and dark regions indicate low flux density in the pictures. There is no large difference in the results of the XRD and pole figure analysis for three films used in this experiment, but the differences of magnetic field distribution by MOI were clearly observed. In the case of low J_c sample (c), MOI presents a very irregular flux density profile due to many small

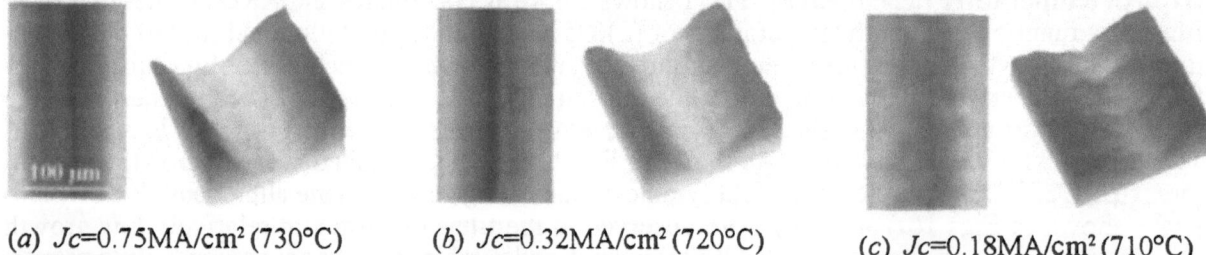

(a) Jc=0.75MA/cm² (730°C)   (b) Jc=0.32MA/cm² (720°C)   (c) Jc=0.18MA/cm² (710°C)

Fig. 2. Magneto-optically detected flux distributions for YBCO films with different Jc. And the three-dimensional representations of flux penetration taken at 20 K and 200 G after zero field cooling.

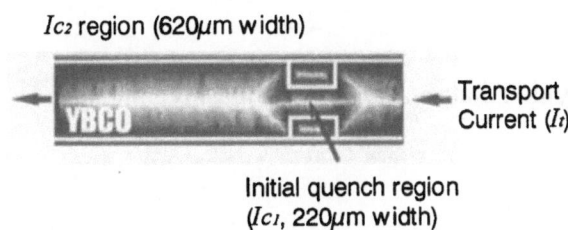

Ic₂ region (620μm width)

Transport Current (I_t)

Initial quench region
(I_{c1}, 220μm width)

Fig. 3. Schematic view of the sample configuration

defects. These defects act as barriers to current flow. On the other hand, MOI for samples (a) and (b) presents very uniform flux distribution. These films have higher J_c values than that of samlple (c). The penetration length of sample (a) is longer than that of sample (b). The result indicates that the sample (a) has the higher J_c because the strong superconducting shielding current prevents the penetration of the applied field.

Dynamic magnetic flux behaviors   Schematic view of the sample configuration to observe the dynamic magnetic flux behaviors during the transition in YBCO thin film is shown in Fig.3. The sample temperature was set at 50 K after field cooling with 0.1 T and the transport current was 3.5 A. In this experiment, the super-normal transition was caused by self-Joule heating. Initial quench region which I_{c1} is smaller than I_{c2}, was formed by making a notch on the sample as shown in Fig. 3. If the transport current (I_t) were larger than I_{c1} and smaller than the other region (I_{c2} region), the normal-zone would be generated

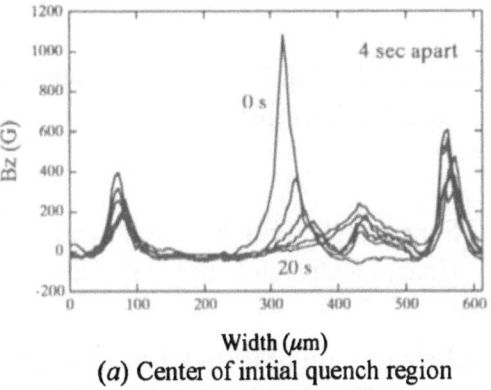

(a) Center of initial quench region

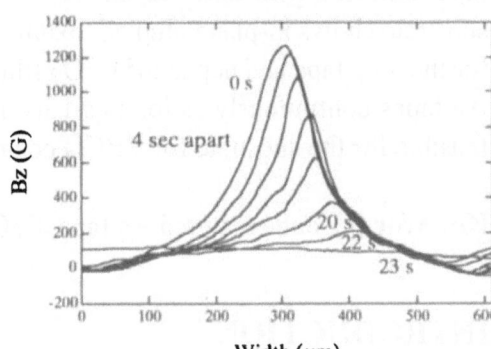

(b) 1.45mm away from the initial quench region

Fig. 4.   Measured Bz profiles for various positions and times

at the initial quench region and propagated along the sample. Fig.4 shows the time evolution of Bz profiles. As can be seen Fig. 4, Bz was decreased with times by the temperature rising according to the Joule-heat generation in the initial quench region. However, at the position 1.45 mm away from the initial quench region, the normal transition was occurred after 23 sec, while it was transited after 16 sec in the initial quench region. Thermal propagation to the longitudinal derection is very slow, because the heat propagates to sample holder made from copper through the MgO substate of which the thermal conductivity is high compared with YBCO. Therefore, Bz profiles along the sample was decreased very slowly in uniform. Unsymmetrical magnetic flux distributions during the transition was due to the effect of the Lorentz force. These experimental results agree well with the results of the theoretical calculation by Zeldov et al. [5].

SUMMARY

The limitation factors for transport characteristics of YBCO thin films deposited on MgO by PLD were studied using XRD, pole figure, SEM and MOI. For producing high J_c films, 730°C was an optimum growth temperature. The magnetic flux distribution of the YBCO films were observed by MOI, and the crystal defects act as barriers against supercurrent were confirmed. The dynamic magnetic flux behaviors in flux flow state were also observed by using the MOI.

Acknowledgements.   This work was supported by the New Energy and Industrial Technology Development Organization (NEDO) as Collaborative Research and Development of Fundamental Technologies for Superconductivity Applications.

1. M. Schieber, *et al.*, J. Crystal Growth **115**, 31-42 (1991)
2. Ch. Jooss, *et al.*, Physica C **266**, 235-252 (1996)
3. N. F. Heinig, *et al.*, IEEE Trans. on Appl. Super. **9**, 1614-1617 (1999)
4. S.Gotoh and N. Koshizuka, Physica C **176**, 300-316 (1991)
5. E. Zeldov, *et al.*, Phys. Rev. B **49**, 9802-9822 (1994)

FABRICATION OF A 1 m-LONG CUBE TEXTURED SILVER TAPE BUFFERED WITH (100) CeO$_2$ LAYER FOR YBCO COATED CONDUCTORS

Yasuo Takahashi[1], Kaname Matsumoto[1], Hiroyuki Akata[2], Kazutoshi Higashiyama[2] and Izumi Hirabayashi[1]

[1]Superconductivity Research Laboratory, Nagoya, 456-8587, Japan
[2]Hitachi Research Laboratory, Hitachi, Ltd., Ibaraki, 319-1292, Japan.

Abstract: In order to fabricate in-plane aligned YBCO films, we have studied cube textured Ag tapes buffered with CeO$_2$ layer. The CeO$_2$ films by pulsed laser deposition (PLD) on textured Ag substrate shows in-plane aligned texture with (100) orientation. We optimized deposition condition for the long tape and deposited CeO$_2$ films on the <100>{001} cube textured and the polycrystalline Ag tapes continuously as long as 1 m. These results suggest that the CeO$_2$ buffered Ag tapes are feasible for the substrate for YBCO coated superconducting wire.

Key Words: cube textured Ag tape, CeO$_2$ buffer layer, PLD, thin films

INTRODUCTION

Homogeneous biaxially aligned YBa$_2$Cu$_3$O$_{7-y}$ (YBCO) films on long metallic substrates are required for power applications. Recently the YBCO films have been deposited on the biaxially oriented substrates such as IBAD[1], RABiTS[2], and ISD[3]. We have been studying a {100}<001> textured silver tape as a substrate for coated conductors. The benign chemical interaction of silver with YBCO material is well established and the {100}<001> textures of silver tape is very stable even at high temperatures, so that the textured silver tape is an appropriate substrate for YBCO coating. Previously we investigated a fabrication of biaxially textured CeO$_2$ films on cube textured Ag tapes by PLD and found CeO$_2$ is useful as a buffer layer for YBCO orientation control on {100}<001> cube textured silver tapes[4-5]. In the present work, we investigated the deposition condition for long tape and fabricated 1 m class CeO$_2$ buffered silver tapes.

EXPERIMENTAL

The CeO$_2$ films were continuously deposited on cube textured and polycrystalline Ag tapes by PLD system. The targets were ablated by a Kr-F excimar laser with a wavelength of 248 nm. The substrate temperature was set to 500-750 ℃ under oxygen pressures of 30 mTorr. The range of tape speed during deposition was 2.5 mm/min. The energy density of laser beam was 3 J/cm^2. Structural characterization of the CeO$_2$ films was carried out by x-ray diffraction (θ-2θ scan), rocking curve and phi scan measurements using Cu-Kα source. The surface morphologies of the films were observed using a scanning electron microscope (SEM).

RESULTS AND DISCUSSION

(1) Deposition condition

The cube textured Ag substrates were prepared by hot rolling and subsequent annealing. We used both as-rolled and recrystallizing annealed Ag tape. The deposition rate of the CeO_2 films was 900 Å /min at oxygen pressure of 30 mTorr. The deposition rate was higher than those of the YSZ buffer layer deposited on Ni or Ni based alloy substrates by the IBAD [1], RABiTS [2] and that of MgO buffer layer on cube textured Ag substrate by PLD [6]. We determined the crystal orientation degrees of CeO_2 film by degree of (200) orientation; $R_{I(200)}$; [$R_{I(200)} = I(200)/\{I(200)+I(111)\} \times 100(\%)$], where I(200) and I(111) are peak intensities from CeO_2 (200) and CeO_2 (111) in x-ray θ-2θ scan. Highly (100)-oriented CeO_2 films were observed by stationary deposition, the film thickness was 1 μm deposited under 30 mTorr oxygen [4]. We controlled film thickness by the tape speed during deposition. The tape speed was 2.5 mm/min for the 1 μm CeO_2 films deposited on Ag tapes. Figure 1 represents substrate temperature dependence of the $R_{I(200)}$. The tape speed during deposition was 2.5 mm/min. The $R_{I(200)} > 90$ % were realized on the Ag substrates for the substrate temperature range of 550-650℃ at 30mTorr under oxygen atmosphere. Figure 2 shows substrate temperature dependence of FWHM for the CeO_2 (220) phi scan on the silver tape. The minimum FWHM was 13 degrees at 650℃ using as rolled tape. The optimal substrate temperature and tape speed were 650°C and 2.5 mm/min, respectively.

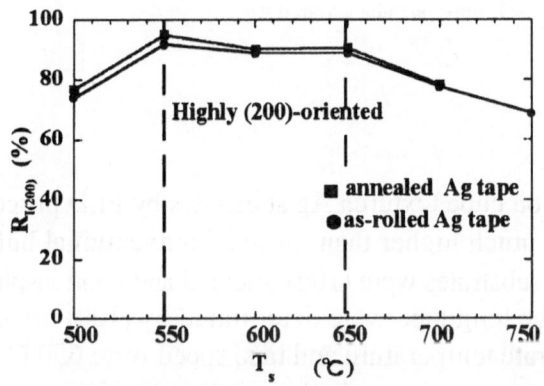

Fig.1 Substrate temperature dependence on $R_{I(200)}$ of the CeO_2 films deposited on cube textured Ag tapes.

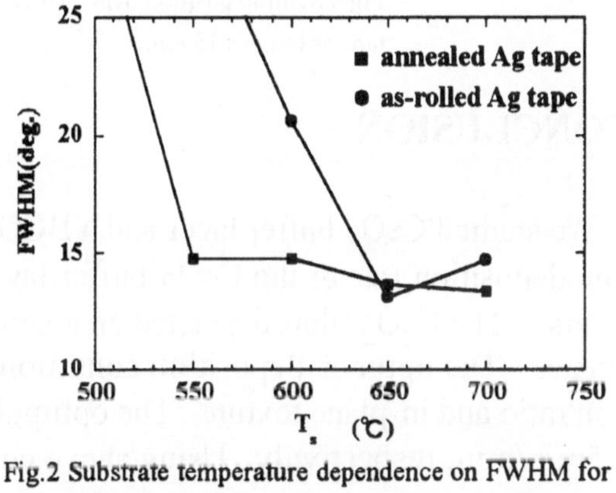

Fig.2 Substrate temperature dependence on FWHM for the rocking curve of (200) CeO2 peaks on the Ag tape.

(2) Continuous deposition of CeO_2 on the poly-crystalline and the cube textured Ag tapes

(a) CeO_2 films on the polycrystalline Ag tapes.

The CeO_2 films were deposited on the polycrystalline Ag tapes as long as 1 m. In the continuous deposition, the polycrystalline Ag tape was moved through the deposition zone at a constant speed of 2.5 mm/min. The CeO_2 films were deposited at the substrate temperature of about 650°C at oxygen pressures of 30 mTorr. The strong CeO_2 (200) peak was observed on the Ag tapes of the continuous deposition.

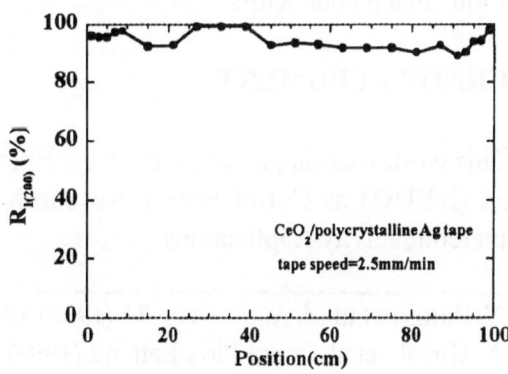

Fig.3 Sample position dependence on $R_{I(200)}$ of XRD peaks on cube textured Ag tape.

The position dependence of the $R_{I(200)}$ of the CeO_2 films is shown Fig.3. Good (100) orientation of $R_{I(200)} \approx 89\text{-}99$ % was obtained on the whole range of the 1m-long tape.

(b) CeO_2 films on the cube textured Ag tapes

The CeO_2 films were deposited on the cube textured Ag tapes as long as 1 m. The deposition condition was the same as polycrystalline Ag tapes. Fig. 4 shows a sample position dependence of x-ray pole figure of (220) CeO_2 on cube textured Ag tape of 15 cm long. $R_{I(200)}$ of 80-97 % was obtained on the tape. These films were in-plane textured. These results indicate the (100) CeO_2 buffered Ag tapes are useful for the fabrication of long YBCO coated conductors.

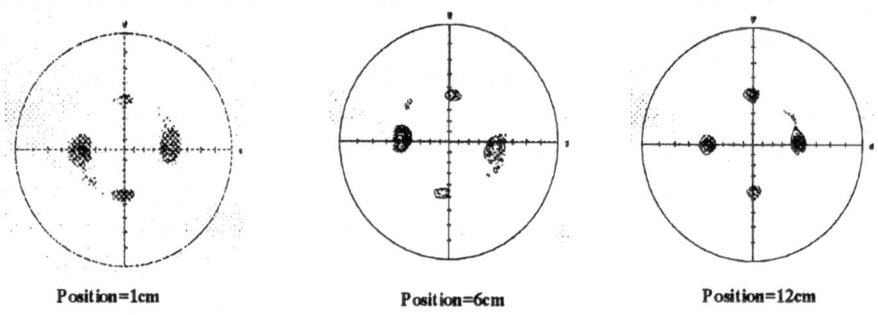

Position=1cm   Position=6cm   Position=12cm

Fig.4 Sample position dependence of (220) CeO_2 films on cube textured Ag tape as long as 15 cm.

CONCLUSION

We studied CeO_2 buffer layer and YBCO films on cube textured Ag substrates by PLD process. The deposition rate of the CeO_2 buffer layer was much higher than those of conventional buffer layers. The CeO_2 films deposited on textured Ag substrates were (100) oriented and good in-plane aligned. The optimal deposition conditions for the long tape were determined by (100) orientation ratio and in-plane texture. The optimal substrate temperature and tape speed were 650℃ and 2.5mm/min, respectively. Using above condition we continuously deposited CeO_2 films on the polycrystalline and {100}<001> cube textured Ag tape as long as 1m. These CeO_2 films were highly (100) oriented. These results indicate that the YBCO films on CeO_2 buffered Ag tapes are feasible for the coated conductors.

ACKNOWLEDGMENT

This work was supported by the New Energy and Industrial Technology Development Organization (NEDO) as Collaborative Research and Development Fundamental Technologies for Superconducivity Applications.

1. Y. Iijima , et al., J. Appl. Phys. 74:(1993) 1905.
2. A. Goyal, et al., Appl Phys Lett, 69:(1996) 1975.
3. K.Hasegawa, et al.,Advanced in Superconductivity XI,(1999) 793.
4. Y.Takahashi , et al., IEEE trans. Appl. Supercond.,9,(1999) 2272.
5. Y.Takahashi et al., Advanced in Superconductivity XI,(1999) 777.
6. Y.Niiori, et al., Physica C, 301(1998)104.

Self-field Effects in High Critical Current Density BSCCO Tapes

S. Fleshler,[1] M. Fee,[2] S. Spreafico[3] and A. P. Malozemoff[1]

[1]American Superconductor Corporation, 2 Technology Dr., Westborough MA 01581 USA
[2]Industrial Research Ltd., Gracefield Rd., P. O. Box 31-310, Lower Hutt, New Zealand
[3]Pirelli Cavi e Sistemi SpA, Viale Sarca 222, 20126 Milano, Italy

Abstract: BSCCO oxide-powder-in-tube tapes have attained average engineering critical current density J_e of 14,000 A/cm^2 (77 K, self-field) in several hundred meter production wire. At this level, self-field effects become significant. For example, an experiment to compensate the self-fields of a 134 A conductor at 77 K gives an optimally compensated I_c of 210 A, corresponding to a J_e of 25,000 A/cm^2 and a J_c of over 62,500 A/cm^2. Production tapes are thus found to have intrinsic J_c's comparable to the finer research samples, even though the latter have much higher apparent 77 K, self-field values. A variety of experiments demonstrating the self-field effect and calculations to model it are described.

Key words. BSCCO tapes, critical current density, self-field

INTRODUCTION

This paper reviews recent progress on long length BSCCO-2223 oxide-powder-in-tube tapes and addresses the strong influence self-fields exert on the electrical performance of high current conductors.

In a recent paper, Masur et al.[1,2] described production results at American Superconductor on multifilamentary BSCCO-2223 tape in several hundred meter lengths. In wires of dimension 0.168x3.1 mm^2, a maximum current density of 16,000 A/cm^2 at 77 K and 1 μV/cm was achieved in self-field. With a fill factor of about 35%, this corresponds to an average filament current density of 40,000 A/cm^2. This is the highest value reported to date for a long length tape by a significant margin.

A more recent result, described further in this paper, has been to achieve a similar maximum engineering critical current density in dimensions of 0.2x4.1 mm^2, giving a maximum current of 134 A, with a fill factor of about 40%.

These values represent the upper portion of a typical production distribution as described in reference[1]. Such levels of performance have made possible significant precommercial demonstrators. For example, coils using the wire described by Masur et. al. [1] have been delivered to the Reliance Electric laboratories of Rockwell Automation for the first 1000 hp high temperature superconducting (HTS) motor[3], a project sponsored in part by the US Department of Energy (DOE) Superconductivity Partnership Initiative (SPI). Also, production is presently underway for the wire to be delivered to Pirelli Cables and Systems for the first utility grid installation at Detroit Edison of an HTS cable[4], a project sponsored in part by the USDOE-SPI and the Electric Power Research Institute.

As the current capacity of conductors increases, the effect of self-field in suppressing the intrinsic performance of the superconducting filaments becomes more prevalent. Performance metrics must be reassessed in light of the fact that the self-field produced by the current in the conductor is increasing with improved performance. In this paper we review some results on self-field effects.

THE SELF-FIELD EFFECT: EXPERIMENTAL RESULTS

The field and temperature dependence of the critical current density has been studied extensively, and here we focus on behavior around 77 K. Early studies by Ueyama et al.[5] on multifilamentary BSCCO tapes already showed a progressive low-field flattening of $J_c(B,T)$ with increasing overall J_c. Several groups, including Grasso and Flukiger [6], Fox et al. [7], Gherardi et al.[6] and Spreafico et al.[7] recognized that this effect can be understood from the effect of local magnetic self-fields when the total current is high because, especially at 77 K, the local current density in the wire depends strongly on the local field perpendicular to the tape-plane. Simple calculations show that tapes of typical dimensions carrying 100 A can have self-fields at the surface in the range 10-20 mT; so that the self-field effect can be significant.

In reference [9], we performed several experiments to directly demonstrate the self-field effect. Sandwiching a tape tightly between two other tapes with oppositely directed currents, the perpendicular self-fields from current in the central tape could largely be compensated. Results on a 46 A tape (at 77 K, 1 μV/cm) are shown in Fig. 1a, where the central tape's I_c rises to a maximum of 70 A when a 30 A compensation current is applied in each of the outer tapes. We call this maximum I_c the "optimally compensated" value. By contrast, current applied in the same direction in all three tapes leads to a steady reduction in the central tape's I_c, as might be expected because the perpendicular fields add. Fig. 1b shows a similar experiment on a higher current (134 A) tape of similar dimensions. In this case, at the optimum compensating current in the two sandwiching tapes, the critical current I_c (77 K, 1 μV/cm) rises to 210 A, corresponding to an engineering (full cross-section) critical current density J_e of 25,000 A/cm^2 in this 0.2x4.1 mm^2 tape cross-section! This is the largest J_e level reported for BSCCO-2223 tape at 77 K, and the corresponding J_c of 62,500 A/cm^2 approaches the highest values reported in shorter research samples [10]. This experiment shows that the intrinsic performance of the tapes, i. e. their performance in the absence of self-field, is in fact much higher than the usually quoted self-field value. Since the self-field compensation is not perfect, the optimally compensated value represents only a lower limit on the intrinsic zero-field J_c (J_{co}).

Figs. 1a and 1b illustrate several less intuitive features of these data. One is that in some samples there is a single I_c maximum as a function of compensating current, and in others there are two. Also, at least in the range of I_c's from 46 to 134 A (77 K, sf = self-field), there is no obvious trend in the maximum percentage I_c increase by compensation; the values range from about 45 to 60%. This is surprising because one might expect increasingly strong self-field suppression with increasing I_c if all the intrinsic (i. e. zero-self-field) $I_c(B)$ curves scaled. However, it is important to point out that the residual magnetic field after compensation, is set to some degree by experimental geometrical factors and increases with the current in the central tape. Hence, for higher current tapes, the residual magnetic field is likely to be larger and the optimally compensated value is likely to more significantly under-estimate the intrinsic performance.

Using the compensation technique in an applied dc magnetic field, the optimally compensated I_c can be found for that particular field. Through an extensive series of such measurements, the dependence of the optimally compensated I_c on applied field was mapped out as illustrated in Fig. 2 for a 112 A (77 K, self-field) conductor. Of course, the compensation in the experimental geometry is imperfect; in particular, the parallel field component is actually enhanced, and the perpendicular field is not fully compensated to the extent that there is a gap between the center and sandwiching tapes. Nevertheless the resulting $I_c(B)$ data are a first approximation to the intrinsic behavior in the absence of self-field.

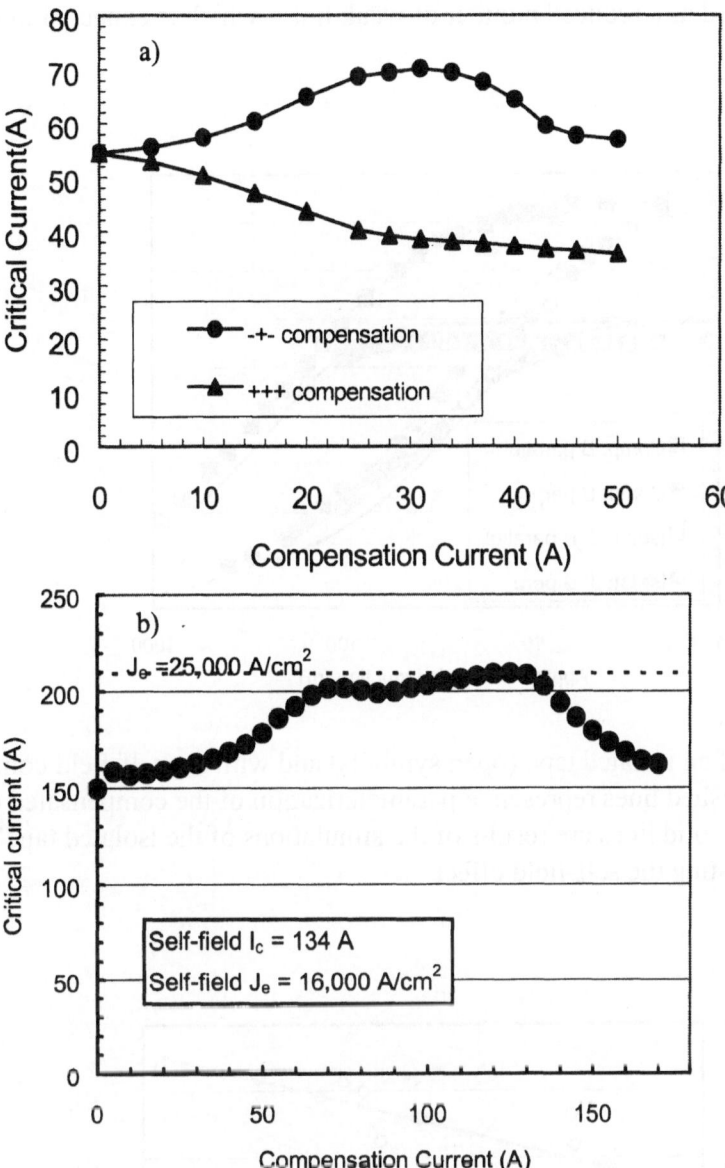

Fig. 1. I_c (77 K, 1 μV/cm) of a central BSCCO-2223 tape as a function of the current in two other tapes on top and underneath the central tape. The upper curve corresponds to opposing current in the outer tapes, the lower to all currents in the same direction. a) corresponds to a tape with I_c (77 K, sf) = 46 A, b) to I_c = 134 A.

It has been conventional practice to characterize wires by their I_c value at 77 K and self-field, and there has been a corresponding tendency to assume that performance at other temperatures and fields scales with this value. However, Fig. 3 demonstrates that such scaling is not valid for low temperatures and higher magnetic fields of the order of one Tesla. The trends with I_c(77 K, sf) can be understood if one assumes that the 77 K self-field suppression increases with I_c, while the 27 K, B_{perp} = 1 T data are unaffected by self-field and so appear to increase more rapidly relative to I_c. It remains to be understood why the same tendency of increasing self-field suppression with increasing I_c does not occur in the 77 K compensation experiments described above. The increase of the residual magnetic field leading to significantly greater under-estimation of the intrinsic performance in the compensation experiments with increasing current capacity may be one cause.

Another clue may come from Fig. 4, which compares the optimally compensated I_c(77 K, B) curves for two tapes of different I_c: the higher-I_c tape has a more gradual fall-off with field which tends to reduce the relative magnitude of the self-field suppression. The more gradual low-field fall-off can be interpreted as a reduced number of weak links, which is expected in higher performance tapes.

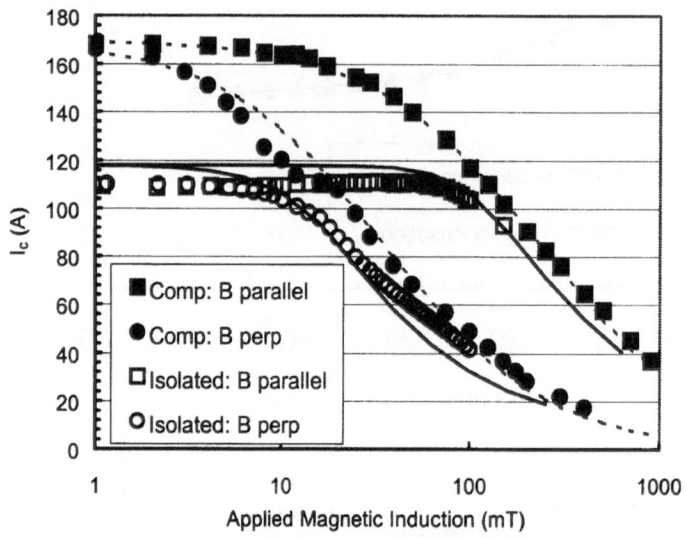

Figure 2: I_c(B) curves of an isolated tape (open symbols) and with the self-field compensation (solid symbols). The dashed lines represent a parameterization of the compensated curves using a Kim expression and the solid lines are results of the simulations of the isolated tape's response to magnetic field incorporating the self-field effect

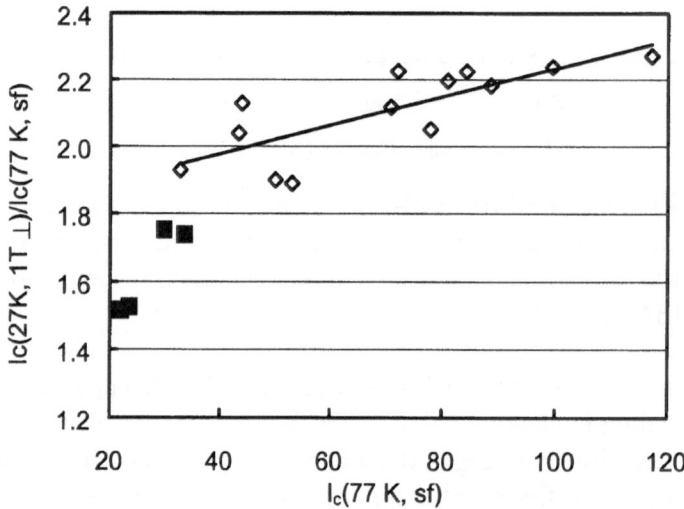

Fig. 3. Ratio of I_c at 27 K and a perpendicular field of 1 Tesla to I_c(77 K, sf), as a function of I_c(77 K, sf). The trend suggests an enhancement of self-field suppression of I_c(77 K, sf) as it increases. The solid squares correspond to data obtained for small cross-section R&D wire and the open diamonds correspond to data taken on larger production tapes of varying cross-section.

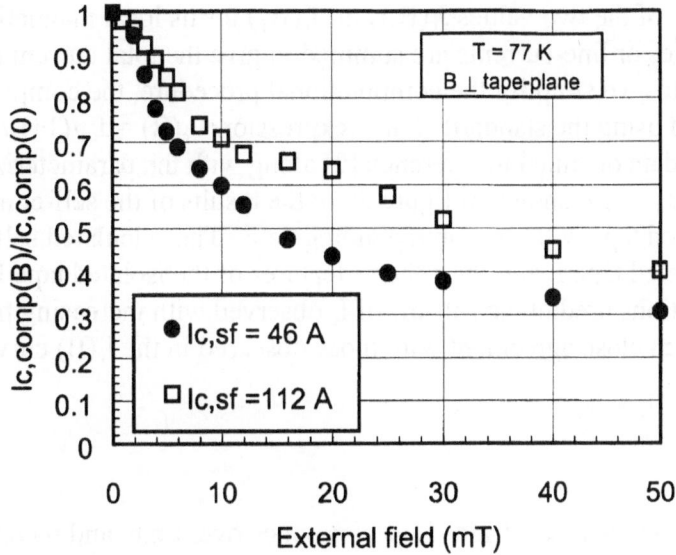

Fig. 4. Optimally compensated low-field response of I_c (normalized to the zero-field value) as a function of applied magnetic field. The higher I_c tape has a weaker drop-off, suggesting a reduced number of weak links.

Schwartzkopf et al. [11] proposed that to avoid self-field complications, which cause critical current density J_c values to depend on the size of the wire, it would be helpful to set a new convention for characterizing wires, namely, using the J_c value at 77 K and a perpendicular field value of, say, 0.1 T. The key conclusion is that the self-field J_c is not a good metric of superconductor quality, when comparing tapes of different size or architecture. One example is the comparison of two American Superconductor tapes with total cross-sections 0.20 and 0.58 mm^2. At 77 K and self-field, they showed J_c of 60 and 45 kA/cm^2, respectively, but at 27 K and a perpendicular field of 1 T, they both exhibited $J_c = 90$ kA/cm^2. The former was a short "research" sample, apparently of higher performance at 77 K, the latter a long-length production sample carrying a much higher current. A likely cause for the equivalence of the J_c's at 27 K and 1 T is that the superconducting performance is indeed similar and that the apparent difference in the 77 K J_c is due to the self-field effect! The fact that the data at 27 K, 1 T is the same speaks to the successful tech-transfer of the research results into manufacturing.

SELF-FIELD EFFECT: CALCULATIONS

To obtain further insight into the self-field effect, finite element analysis was employed to self-consistently calculate the I_c as a function of applied magnetic field including effects of non-uniform local magnetic field within the conductor produced by the current in the conductor. This was accomplished by assuming that the compensated $J_c(B)$ curves serve as a basis for the intrinsic response of the conductor in the absence of self-fields. In the calculation, the conductor was subdivided into a grid of rectangular elements and the magnetic field produced at any position in the conductor by a current within an element is computed using the Biot-Savart law for a line current. The local magnetic field at an element is the sum of the magnetic fields produced by all the other elements (the conductor self-field contribution) plus the externally applied magnetic field. The current density associated with each element in the local magnetic field computed for the position of that element is compared against the compensated $J_c(B)$ curves and incremented or decreased accordingly. The local magnetic field is decomposed into its two components parallel and perpendicular to the tape-plane, and the local current density is restricted to be the lesser of $J_c(B_\parallel)$ or $J_c(B_\perp)$. The calculation is performed iteratively until the current distribution converges to a "steady state" solution. The final solution is realized when the current density in each element equals the lesser of the two values $J_c(B_\parallel)$ and $J_c(B_\perp)$ for its local magnetic field. The elemental current densities or line-currents are summed to give the total current capacity or I_c in the applied magnetic field. To simplify the computational procedure, the compensated $J_c(B)$ curves are parameterized using the standard "Kim" expression: $J_c(B) = J_{co}/(1+B/B_o)$. Figure 2 shows the compensated data obtained in reference [9] along with the parameterized curves for a particular set of J_{co} and B_o. Also shown in figure 2 are the results of the self-consistent $I_c(B)$ calculations for an isolated tape with the corresponding data. The calculated self-field I_c is within 7% of the measured isolated tape I_c and the general features of the isolated tape $I_c(B)$ curve are reproduced. The simulations predict the roll-off of I_c observed with increasing magnetic field at values of magnetic field in close agreement with those observed in the $I_c(B)$ curves for both orientations.

CONCLUSIONS

BSCCO-2223 tapes have made great strides in performance over time, and recognition of the importance of the self-field effect in suppressing I_c makes it clear that the intrinsic J_c of these tapes is even higher than was previously supposed. Long length production tapes at American Superconductor have intrinsic J_c performance comparable to the best short-length research samples. Initial steps in characterizing and understanding the self-field effect have been made, but further work will be required to achieve a full quantitative understanding. It will be important in the future to measure critical current of conductors in applied magnetic fields that are substantially larger than the self-field to obtain an unambiguous measure of the tape performance independent of the geometric parameters.

Acknowledgments

The authors are grateful to the wire research and manufacturing teams at American Superconductor for the outstanding samples underlying this work. We thank R. Given for carrying out some of the important measurements, and G. Snitchler, L. Masur and J. Scudiere for helpful discussions. We would like to acknowledge support from the New Zealand Foundation for Research, Science and Technology.

1. L. Masur et al., to be published, International Cryogenic Materials Conference, Montreal, Quebec, Canada, July 12-16, 1999.
2. A. P. Malozemoff et al., IEEE Trans. Applied Superconductivity **9**, 2469 (1999).
3. D. Aized et al., IEEE Trans. Applied Superconductivity **9**, 1197 (1999).
4. M. Nassi, to be published, Eucas '99, Sitges, Spain, Sept. 14-17, 1999.
5. M. Ueyama et al., Jap. J. Appl. Phys. **30**, L1384 (1991).
6. G. Grasso and R. Flukiger Physica C **253**, 292(1995).
7. S. Fox et al., Physica C **257**, 332 (1996).
8. L. Gherardi et al., Mat. Sci. Eng. B **39**, 66 (1996).
9. S. Spreafico et al., IEEE Trans. Applied Superconductivity **9**, 2159 (1999).
10. Q. Li et al., IEEE Trans. Applied Superconductivity **7**, 2026 (1997).
11. L. A. Schwartzkopf et al., Appl. Phys. Lett., to be published.

Development of Bi Based Superconducting Wires

Naoki Ayai, Munetsugu Ueyama, Takeshi Kato, Shin-ichi Kobayashi, Akira Mikumo,
Tetsuyuki Kaneko, Takeshi Hikata, Kazuhiko Hayashi and Hiromi Takei

Electric Power System Technology Research Laboratories,
Sumitomo Electric Industries, LTD, 1-1-3 Shimaya, Konohana-ku, Osaka 554-0024, Japan

Abstract: Long length high-strength alloy sheathed Bi2223 tapes were developed. The tolerance for tensile stress improved to twice as high as that of the silver sheathed tape. Length and Jc at 77K of the tape achieve 800m and 30kA/cm^2 respectively. The long length tape was coated with Poly-vinyl formval, and successfully insulated along the overall length without any Jc degradation. Special samples to study AC losses were fabricated and investigated. The samples consist of successfully separated filaments without any bridging. And twisting gives an obvious loss reduction especially in the external magnetic field parallel to the tape plane.

Key words: Bi-2223, Ag sheathed tape, AC Loss

INTRODUCTION

Sumitomo Electric Industries has been developing the high temperature superconductor and studying the feasible applications using them. Among the evolving materials, the silver sheathed Bi-2223 wire by the powder in tube technique has already achieved one kilo meter class length, and possesses high Jc at even the relatively high temperature of liquid nitrogen as long as it is used in low magnetic fields. Therefore it is believed to be the most advanced wire for superconducting power applications such as power cables, fault current limiters and transformers.

Through the engineering design activities and experimental studies with available wires, the requisites for the wires have been clarified still more. Low AC loss and large current capacity are technical key issues for electric power applications in those problems. This problem should be divided into two phases from a point of a conductor design. The first phase is how to reduce the AC loss of a bundled conductor as far as the sum of that from an individual wire, in other words whether the uniform current flows in overall conductor or not. Inter strands insulation is an essential issue for the phase. We adapted an organic coat to long length Bi-2223 tapes.

The second phase is how to reduce the ac loss of a wire as far as the small level in the electro-magnetically uncoupled multi-filamentary state. We studied the relation between AC losses and some wire structures, such as the presence of inter-filament bridging points, twist pitch length and electrical resistance of matrix.

HIGH-STRENGTH ALLOY SHEATHED TAPE AND ORGANIC INSULATOR

Enameled organic insulators have advantages in electric isolating properties, heat tolerance and economy, therefore familiar as an insulator for the conventional copper coil windings. However the organic insulator burns out at high sintering temperature for Bi based superconductors. The wire has to be coated with them after sintering. And the wire should be strengthened to keep high Ic against some mechanical

631

loads in a coating line and operating conditions as the power apparatuses.

The high-strength alloy sheathed tapes have been developed. The manganese added silver alloy is arranged as the outer sheath of the wire. The composition of additive element was chosen not to degrade the Ic or workability. Fig.1 shows the current status of long length Bi-2223 tapes. On the evolving high-strength alloy sheathed tapes Jc at 77K reached 30 kA/cm^2 for 800m, which were comparable to those of conventional pure silver sheathed tapes.

The mechanical properties were investigated for the alloy sheathed tape and compared to the conventional tape on the critical stress or strain where was kept 90% Ic to unloaded wire. Fig.2 shows the relation between Ic and tensile stress or strain at 77 K load. The conventional tape has a critical tensile stress of 76 MPa, while the alloy sheathed one achieves twice as that, 150 MPa. Young's modulus decreased for the alloy sheathed tape, but that can not explain all over the

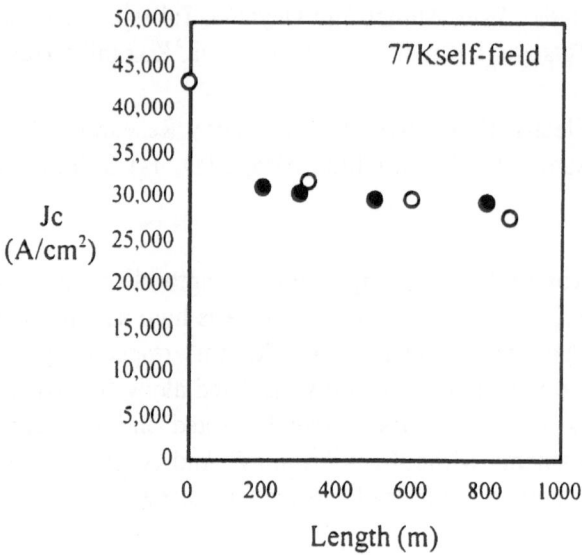

Fig.1. Jc vs. Length of Bi-2223 tapes
Open and close circles show pure Ag sheathed tapes and alloy sheathed tapes respectively.

improvement of the critical stress. In fact, the stain tolerance is also improved in the alloy sheathed tape. In Figs.2 and 3, the critical tensile strain and the bending strain are 0.17% and 0.27% on the conventional wire, while 0.26% and 0.6% on the alloy sheathed wire respectively.

Figure 4 shows the temperature dependence of the electric resistance. The alloy sheathed tape has approximately twice resistance of that of pure silver at 110 K, just above the critical temperature.

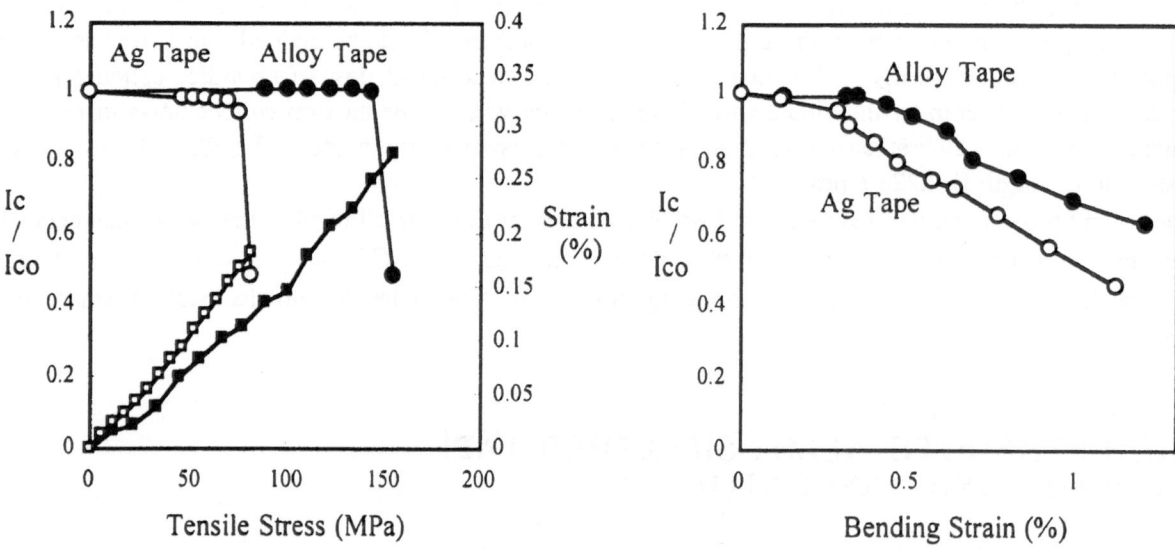

Fig.2. Ic and strain vs. tensile stress of Bi-2223 tapes
Circles and squares show Ic and strain respectively.

Fig.3. Ic vs. Bending Strain of Bi-2223 tapes

Here the alloy sheathed tape was coated with an organic insulator. Poly-vinyl formval (PVF) was chosen as the insulator because it had excellent cryogenic properties and relative a low baking temperature hardly to cause the degradation of Ic. In Fig. 5, the break down voltage (BDV) is proportional to the thickness of PVF on tape's wide plane, and approximately three times superior in 77 K to that in room temperature. For instance, the specimen with 15μm thick coating was broken down from 500V to 700V in room temperature, while about 2000V at 77 K.

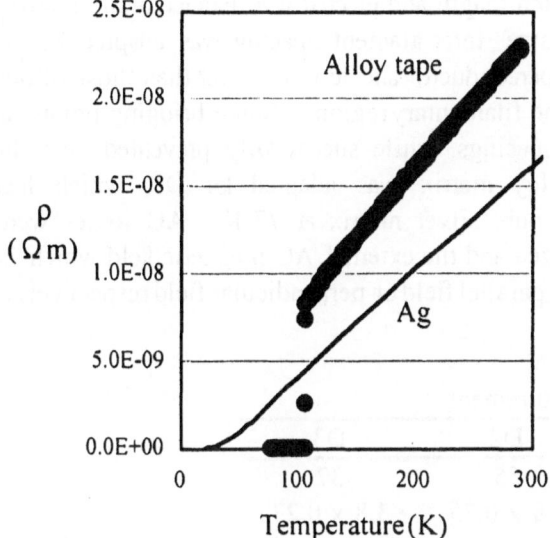

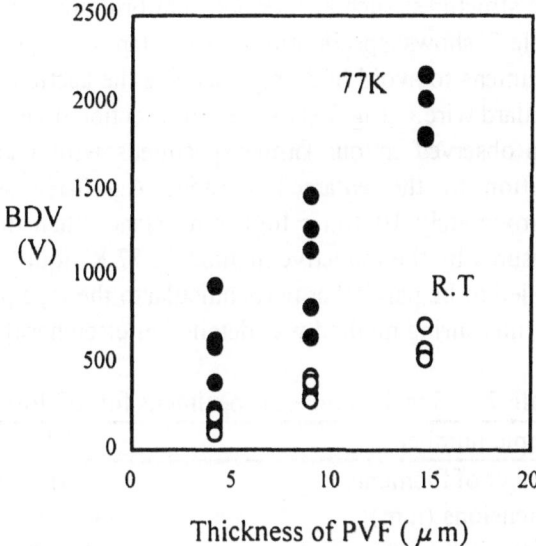

Fig.4. Resistivity of matrix vs. Temperature of the alloy sheathed Bi-2223 tapes

Fig.5. BDV vs. Thickness of PVF on the alloy sheathed Bi-2223 tapes

At last the long length alloy sheathed tape was coated with PVF and wound to a single pancake coil without any insulators except for PVF. Table 1 gives the specifications of the coil. In Table 1, Ic is that of the original bare wire before coating, continuously measured at every 4 meters length in straight shape[1]. Normal resistance and Ic of the coil agreed well with the calculation based on properties of the original bare wire. Therefore the wire surface is completely insulated along the overall length by the PVF coating alone. In addition to that, the Ic of the wire hardly degrades through the coating.

Table 1.   Specifications of the coil

Wire		Coil	
Length (m)	498	Inner diameter (mm)	317
Width (mm)	3.8	Outer diameter (mm)	522
Thickness (mm)	0.25	Number of turn	377
Ic at 77K, 0T (A)	57	Resistance (at RT. Ω)	16.4
Thickness of PVF (μm)	15	Coil Ic at 77K (A)	34

AC LOSSES

When the uniform current flow in every channels of a large conductor, the AC loss reduction on each strand will be required to be realized for a further efficient system. In some electromagnetic conditions, the silver sheathed Bi-2223 tapes often have got large AC losses corresponding to those from a likely single slab with an overall filamentary region size in spite of the multi-filamentary structure. Effect of twisting was also limited. Therefore we fabricated special tapes to study the relation between AC losses and some wire structures, such as inter filament bridging, twist pitch length and electrical resistance of the matrix. Table 2 shows specifications of the tapes. "Double size" inter-filament spacing was adapted for all specimens to avoid bridging, therefore the fraction of superconductor and Ic were lower than those of our standard wire. Fig.6 shows a cross sectional view of the filamentary region. Some bridging points had been observed in our former specimens with narrow spacings, while successfully prevented here. In addition to the enlarged spacing, Ag-10at% Au alloy matrix was adapted for D3, which has approximately 10 times higher resistance than that of pure silver matrix at 77 K. AC losses were measured by the inductive method in 77 K liquid nitrogen and the external AC magnetic field which is applied to be parallel or perpendicular to the tape plane (parallel field or perpendicular field respectively). The measuring method was detailed in elsewhere[2].

Table 2. Specifications of specimens for AC loss measurement

Sample number	D1	D2	D3
Number of filaments	61	55	37
Dimensions (mm)	4.0 x 0.26	3.8 x 0.25	3.8 x 0.23
Twist pitch (mm)	160 *	8	8
Matrix		Ag	Ag-10%atAu
Ic at 77K, 0T (A)	33	22	19

* D1 is an untwisted sample, so the twist pitch means twice of sample length for untwisted #1

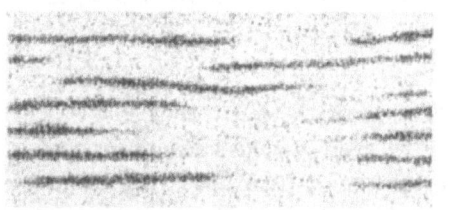

Fig.6. Cross sectional view of the Bi-2223 tapes (D2)

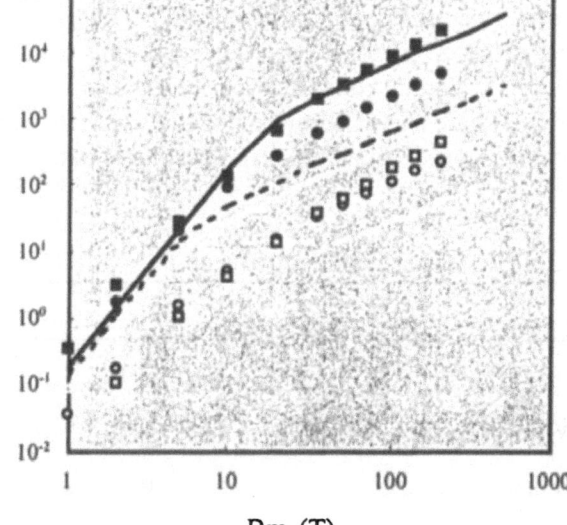

AC Losses (J/m^3/cycle)

Bm (T)

Fig.7. AC Loss vs. Bm on D2, ○: parallel 1 Hz, □: parallel 50Hz, white solid line: parallel FC, white broken line: parallel MF, ●:perpendicular 1Hz, ■:perpendicular 50Hz, black solid line: parallel FC and black broken line: parallel MF

Figure 7 shows AC losses of the sample D2 for various magnetic field amplitudes (Bm). The lines in the graph are estimated hysteresis losses based on the Been slab model, where the penetration field (Bp) are described by Bp=μ0 Jc d / 2. The solid line expresses the approximation for a filamentary region size slab corresponding to a fully coupled state (FC), while the broken line for the uncoupled multi-filamentary state (MF). In the former samples with some bridging points, the loss reductions had been observed slightly at the limited Bm around Bp of the FC prediction [2]. They are evidently observed here. In parallel field Bm higher than 20mT, the losses decrease to an intermediate level between FC and MF, while close to the MF level in lower Bm. However in perpendicular field, the remarkable improvement was observed only in low frequency of 1 Hz, scarcely seen in 50Hz.

Figure 8 shows the relation between loss and frequency in the parallel field. In 10mT, the loss of untwisted D1 increases and saturates near the FC prediction as frequency increases, while for the MF level small losses are kept on twisted D2 and D3. In large amplitude of 100mT, some frequency dependence is seen on twisted samples. However in commercial frequency around 50 Hz, the twisting reduces an FC like large loss to approximately 50% even if taking Je difference into consideration. High resistance of Ag-Au matrix hardly contributes to the reduction of the frequency dependence here.

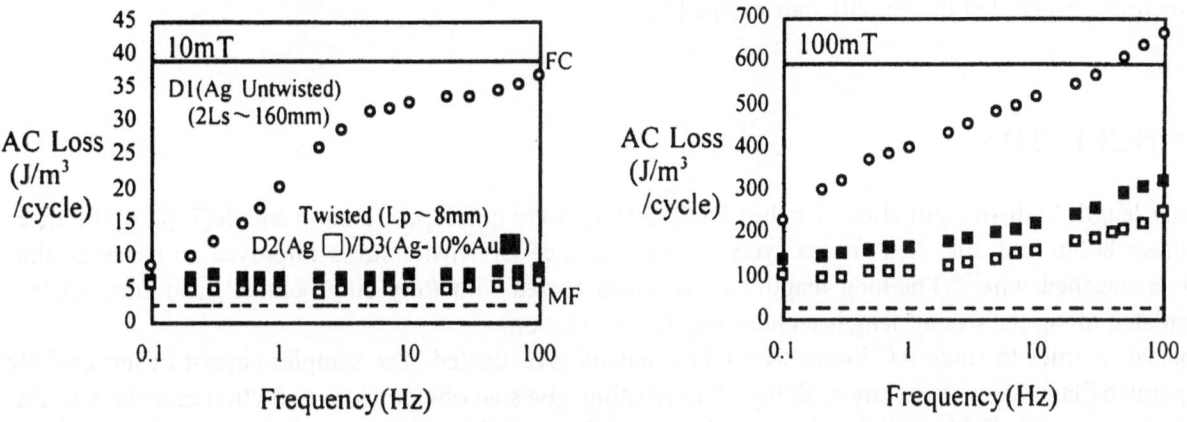

Fig.8. AC Loss vs. Frequency in parallel field

Figure 9 shows the frequency dependence in the perpendicular field. In 10mT, the peaks observed here shift to higher frequency as twisting, therefore they seem to come from resonance peaks of coupling loss components. Resonant frequency is proportional to the transverse resistivity. However, the peak of D3 with 10 times resistive matrix stays at 50Hz, only 2.5 times of D2. Those peaks stay in relative low frequency. In higher 100mT, resonant peaks are missed. Twisting or resistive matrices limit the frequency dependence and reduce absolute loss values up to the high frequency region. There gives successfully small losses close to the MF prediction in low frequency, while considerably large losses remain in high frequency even on high resistive D3.

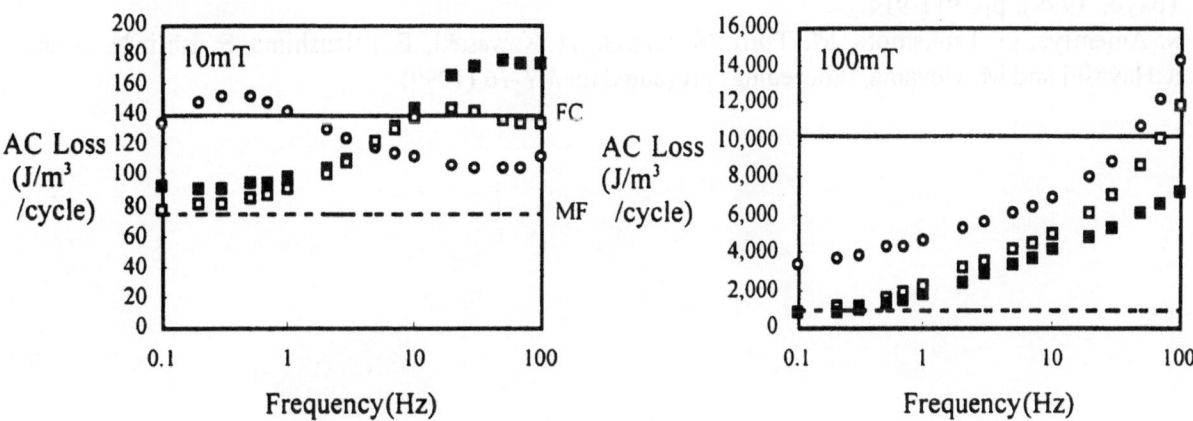

Fig.9. AC Loss vs. Frequency in perpendicular field

636

Like mentioned above, twisting gives the obvious loss reduction in tapes with completely separated filaments presented here. Twisting is more effective for separated filament wire than for inter-filament bridging wire. Especially in the parallel field, although some frequency dependence remains in large field amplitude, half losses of the untwisted tape are obtained even the commercial frequency region around 50Hz. This technique will be effective to the conductor used in mainly parallel field, such as power cables. However, additional high resistive matrix hardly contributes the loss reduction in parallel field. In the case of the perpendicular field, losses are reduced by twisting only in low frequency below a few Hz. Although an additional resistive matrix also contributes the reduced frequency dependence, considerably losses remain in the commercial frequency region. These losses will be serious to the coil windings, such as those of transformers. For scaled up power cables with a large parallel field amplitude, furthermore reduction of AC loss will be also required. Our group is involved in development of advanced tapes for AC use with an inter-filament barrier to block bridging and raise up the transverse resistivity. Recent studies find out the feasible barrier material such as Al_2O_3 and Bi2201. Some filament de-coupling effect was observed in the Bi2201 barrier tape [3].

CONCLUSION

Long length high-strength alloy sheathed Bi2223 tapes were developed. Length and Jc(77K) of the tape achieve 800m and 30kA/cm^2 respectively. The tolerance for tensile stress improved to twice as the silver sheathed wire. The long length tape was also coated with Poly-vinyl formval, and successfully insulated along the overall length without any Jc degradation.

Special samples to study AC losses were fabricated and investigated. The samples consist of successfully separated filaments without any bridging. And twisting gives an obvious loss reduction especially in the external magnetic field parallel to the tape plane. Another technique such as application of a high resistive barrier is required to reduce losses furthermore in the magnetic field perpendicular to the tape plane and for the commercial frequency region.

Acknowledgements. Part of this study was supported by MITI.

1. T. Kaneko, T. Hikata, M. Ueyama, A. Mikumo, N. Ayai, S. Kobayashi, N.Saga, K. Hayashi, K. Ohmatsu and K. Sato, *IEEE Trans. Applied Superconductivity,* vol.9 (1999), pp. 2465-2468.
2. N. Ayai, M. Ueyama, K. Hayashi and K. Sato, *Advances in Superconductivity* XI (Springer-Verlag, Tokyo, 1999), pp. 911-914.
3. N. Amemiya, O. Tsukamoto, M. Torii, M. Ciszek, H. Kawasaki, E. Mizushima, S. Ishii, N. Ayai, K Hayashi and M. Ueyama, Proceedings presented for *MT-16* (1999).

In-situ high-temperature study of Bi,Pb(2223) phase formation and stability inside Ag-sheathed tapes

Enrico Giannini, Emilio Bellingeri, Reynald Passerini, Frank Marti and René Flukiger

Dép. Phys. Mat. Cond. - Université de Genève, CH-1211 Genève 4, Switzerland

Abstract: *In-situ* diffraction techniques proved to be very powerful for the understanding of the reactions occurring inside Ag-sheathed Bi,Pb(2223) tapes. By means of *in-situ* high-temperature neutron diffraction on monofilamentary tapes, the amount of secondary cuprates was quantitatively estimated and the role of them during the reaction thermal treatment was elucidated. For the first time, a direct evidence of a partial melting occurring prior to forming the Bi,Pb(2223) was observed as a decrease of the total amount of crystalline matter in the sample and the amount of the transient liquid and the amorphous matter was quantified. Furthermore, evidence of Bi,Pb(2223) stability on cooling was observed, being the re-growth of Bi(2212) on cooling not related to any decomposition of Bi,Pb(2223). High-temperature linear expansion coefficients of Bi,Pb(2223) were measured. Preliminary results of recently performed *in-situ* synchrotron X-ray diffraction complete this work.

Keywords: BiSCCO tapes, neutron diffraction, transient liquid

INTRODUCTION

The investigation of the reactions occurring inside the silver sheath during the thermal treatment of Bi,Pb(2223)/Ag tapes is of basic importance in view of further improvements of the tape processing. In the last few years, new and interesting results have been obtained by using diffraction techniques in transmission geometry: high energy X-ray diffraction from a synchrotron source (20-100keV) [1,2] and neutron diffraction [3] have been successfully employed to study *in-situ* the reaction thermal treatment of Bi,Pb(2223)/Ag tapes. Owing to their intrinsic peculiarities, these two different diffraction techniques provide complementary results. Neutrons interact with nuclei and strongly with light elements, the neutron nuclear form factor does not decrease noticeably with scattering angle, the neutron beam is highly penetrating and it has a wave length comparable to the cell parameters, that reflects in a better scattering angle resolution and in the possibility to directly measure the absolute amount of the diffracting phases. On the other hand the synchrotron X-ray beam has a higher intensity, a strong tight collimation and a tuneable wave length. Neutron scattering is therefore preferable for a structure determination and a quantitative phase analysis, whereas high energy XRD takes the lead for local investigations and accurate texture analysis, especially in multifilamentary tapes where the neutron absorption in the Ag sheath is too strong. We performed both of these diffraction experiments, on mono- and multifilamentary Bi,Pb(2223)/Ag tapes respectively, at high temperature in air, mimicking the thermal treatment used for the standard tape processing. In this work we report mainly on the neutron diffraction experiment with a particular regard to the Bi,Pb(2223) behaviour on cooling.

EXPERIMENTAL

Measurements were done at the High Flux Reactor at the Institut Laue Langevin (ILL) in Grenoble (France), by using the D1A diffractometer, and at the European Synchrotron Radiation Facility

(ESRF) in Grenoble (France), by using the High Energy ID15 Beamline. For both the experiments dedicated beam-compatible furnaces and rotating sample holders were designed (see [3,4] for the details). A very good temperature uniformity was obtained in the vanadium furnaces employed for the neutron experiment ($\Delta T \leq 0.2°C$), whereas a bit worst but still satisfactory uniformity ($\Delta T \leq 2°C$) was obtained in the other one. Thanks to the continuous rotation of the sample along the rolling direction during neutron data acquisition the effect of the preferred orientation of the grains inside the tapes was reduced. The tapes were prepared by standard PIT technique described elsewhere [5], starting from nominal compositions $Bi_{1.72}Pb_{0.34}Sr_{1.83}Ca_{1.97}Cu_{3.13}O_{10+x}$ and $Bi_{1.8}Pb_{0.4}Sr_{2.0}Ca_{2.2}Cu_{3.0}O_{10+x}$ respectively for mono- and multifilamentary tapes. Both the thermal treatments were performed at 837°C in air and different cooling procedures were investigated. A Full Pattern Profile Refinement technique based on the Rietveld method was employed to analyse the neutron diffraction patterns: 8 phases were successfully refined and their absolute amounts were worked out during the whole treatment. Details on the refinement technique are given in [3]. Two kinds of 2D-detectors were used for hard X-rays: a CCD camera coupled to an Image Intensifier and an Image Plate. *Ex-situ* transport measurements on monofilamentary samples quenched at different times of the thermal treatment were performed by standard four-probe technique.

RESULTS AND DISCUSSION

Neutron diffraction data provided a direct *in-situ* confirmation of the Bi,Pb(2223) formation mechanism proposed earlier[7]: following to the Pb-rich phases decomposition, Pb and Ca can enter the Bi(2212) phase pushing it toward the stability edge. It corresponds to a structural change of Bi,Pb(2212) from quasi-tetragonal to orthorhombic. At this stage the Bi,Pb(2212) starts to locally melt, and from the transient liquid formed by its interaction with Ca-Cu-rich amorphous phase the Bi,Pb(2223) can nucleate. The same formation mechanism was supported by the high-energy synchrotron XRD experiment [4]. Thanks to the absolute measurement of all the present phases, it was possible to evaluate the total amount of crystalline matter inside the sample. An evidence of a partial melting and a re-crystallisation occurring at the early stages of the Bi,Pb(2223) formation is provided by the non-monotonic state of this quantity as a function of the time (Fig.1(b)). Resistivity measurements performed on monofilamentary tapes quenched in air from the same temperature range show a rise of resistivity that confirm the increase of the amorphous matter in the sample.

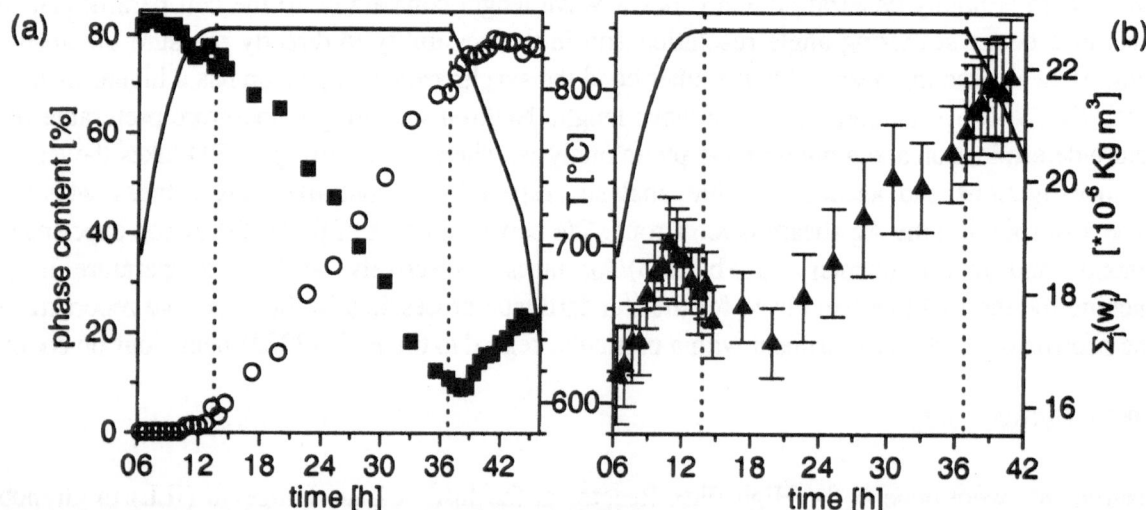

Fig. 1: (a) Bi,Pb(2223) (O) and Bi(2212) (■) weight fractions during the first thermal treatment and the slow cooling. **(b)** The total amount of crystalline matter is plotted vs. time. In both plots (a) and (b) the annealing step at constant temperature is marked by dashed lines.

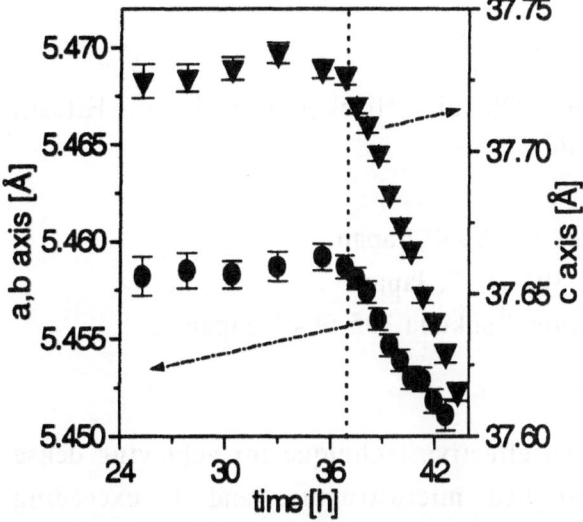

Fig.2: c (▼) and a=b (●) cell parameters of Bi,Pb(2223) on cooling. The dashed line indicates the beginning of the cooling ramp (at 0.3°C/min).

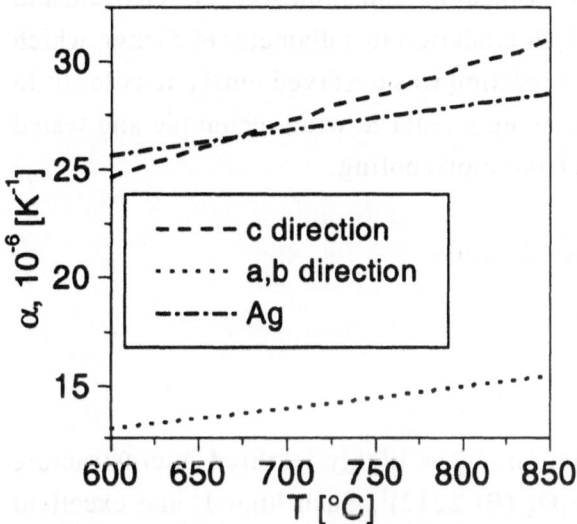

Fig.3: linear expansion coefficient α vs. T, as calculated from the cell parameters contraction on cooling.

Interesting results were found by neutron diffraction on cooling (cooling rate 0.3°C/min). As shown in Fig.1(a), at the beginning of the cooling ramp the Bi,Pb(22223) phase was still growing up and below 800°C its amount remained at a constant value. At the same stage a re-growth of Bi(2212) was observed, not related to any decomposition of the Bi,Pb(2223). A relative valuation of the phase amounts based on standard XRD spectra could not reveal such a different behaviour. A possible source for the rise of the Bi(2212) could be the decomposition of other secondary phases, such as Bi(2201) and $(Sr,Ca)_{14}Cu_{24}O_{41}$, observed at the same time [3]. The so obtained Bi(2212) should be Pb-poor, which is in good agreement with the cell parameter measurements obtained from the neutron diffraction spectra.

The *in-situ* measurement of the cell parameters of Bi(2212), Bi,Pb(2212) and Bi,Pb(2223) during the temperature variation allowed us to estimate their linear expansion coefficients. In the Fig.2 the Bi,Pb(2223) cell parameters are plotted vs. T during the slow cooling, and the corresponding linear expansion coefficient is plotted in Fig.3. An evident anisotropy of the linear expansion coefficient is shown in Fig.2 and Fig.3. Nevertheless the steeper decrease of the α parameter in the c-direction could be due to a simultaneous Pb-loss, which would not affect a and b cell parameters, as found in [8]. Analogous results were obtained for the Bi(2212) and the extrapolation down to room temperature is in good agreement with the low temperature data in the literature [9]. The value obtained for silver are in good agreement with the existing literature.

1. L. J. Wu, Y. L. Wang, W. Bian W, Y. Zhu, T. R. Thurston, P. Haldar and M. Suenaga, J. Mater. Res. **12**(11), 3055 (1997).
2. H.F. Poulsen, T. Frello, N.H. Andersen, M.D. Bentzon and M. von Zimmermann, Phys. C **298**, 265 (1998).
3. E. Giannini, E. Bellingeri, R. Passerini and R. Flükiger, Phys. C **315**, 185-197 (1999).
4. E. Giannini, E Bellingeri, R. Passerini, F. Marti, M. Dhallé, M. Ivancevic, V. Honkimäki and R. Flükiger, presented at EUCAS '99, 14-17 September 1999, Sitges, Spain.
5. G. Grasso, A. Jeremie and R. Flükiger, Supercond. Sci. Technol. **8**, 827 (1995).
6. A.P. Hammersley, ESRF Int. Report, ESRF97HA01T, FIT2D V9.129 Ref. Manual V3.1 (1998)
7. J. –C. Grivel and R. Flükiger, Supercond. Sci. Technol. **9**, 555-564 (1996)
8. F. Marti, G. Grasso, J. –C. Grivel and R. Flükiger, Supercond. Sci Technol. **11**, 485-495 (1998)
9. M. Okaji, K. Nara, H. Kato, K. Michishita and Y. Kubo, Cryogenics **34** (2), 163-165 (1994)

High J_c Bi-2212/Ag Multilayer Tape Conductor Prepared by a Coating Method

Takayo Hasegawa[1], Tsutomu Koizumi[1], Nozomu Ohtani[1], Hirokai Kumakura[2], Hitoshi Kitaguchi[2], Hanping Miao[3] and Kazumasa Togano[2,3]

[1] Showa Electric Wire & Cable Co. Ltd., Kawasaki 210-8660, Japan
[2] National Research Institute for Metals, Tsukuba 305-0047, Japan
[3] CREST, Japan Science and Technology Corporation, Tsukuba 305-0047, Japan

Abstract: The PAIR process is well known as an effective technique for achieving dense superconducting layers, resulting in highly aligned microstructure and J_c exceeding $500kA/cm^2$ at 4.2K and 10T. In order to apply this process to a long length tape conductor, many factors had to be optimized such as residual carbon content, oxide layer thickness, and intermediate rolling reduction. We successfully fabricated 100m-class Bi-2212 multilayer tapes using the PAIR process and, at 4.2K, obtained $710kA/cm^2$ and $350kA/cm^2$ in self-field and 10T, respectively. The tape could be bent without I_c degradation to a diameter of 51mm, which corresponded to a bending strain of 0.6%. No J_c degradation was observed until a tensile strain of 0.2% was applied. We fabricated three coils by using a react & wind technique and tested them in high fields at 4.2K and higher by using refrigerator-cooling.

Key words: Bi-2212 multilayer tape, PAIR process, J_c, coils

INTRODUCTION

A partial melting and slow solidification technique produces highly textured microstructure and high J_c values ($> 250kA/cm^2$) in $Bi_2Sr_2CaCu_2O_x$ (Bi-2212)[1]. Such high J_c and excellent magnetic field dependence at temperatures below 30K make Bi-2212 especially attractive as a practical wire for a refrigerator-cooled magnet and an insert magnet in high magnetic fields. However, from an industrial point of view, higher J_c is required to fabricate a compact magnet with higher coil current density. Development of a pre-annealing and intermediate rolling (PAIR) technique allowed a remarkable improvement in J_c for Bi-2212 dipcoat tapes[2]. The PAIR process was effective in reducing organic binder in the superconducting paste and promoting dense oxide layers, which resulted in homogeneous chemical reactions and ideal microstructures[3]. Dip coating is a very handy process for producing relatively low cost tape. However, tapes so produced are too brittle to make practical coils, and they are damaged by thermal cycling and operation in high magnetic fields, even if impregnated with epoxy. Lamination was useful to overcome brittleness and degradation at the surface of the superconducting layers, resulting in an ease of handling similar to that of multifilament tape. We applied the PAIR process to multilayer tapes and obtained a J_c value of 500kA/cm^2 at 4.2K and 10T in short samples. Recently, we have been developing long length Bi-2212/Ag multilayer tapes by using the PAIR process and adjusting parameters appropriate for industrial production.

In this paper, we report both the factors influencing the J_c value of 100m long multilayer tape and the superconducting and mechanical properties of the tape manufactured by industrial scale equipment. Three coils were fabricated and tested in this work in order to investigate possibilities for insert and refrigerator-cooled coils.

EXPERIMENTAL

Superconducting Bi-2212 powder with the composition of $Bi_{2.00}Sr_{2.05}Ca_{1.00}Cu_{2.00}O_x$ was prepared by a conventional ceramics technique. This powder was mixed with an organic solvent and a binder to make a slurry. Silver or AgMg alloy tapes with a thickness of 20μm, a width of 4 mm, and a length of 100 m were coated with the slurry. Four of those coated tapes were stacked together and laminated with a 20 μm thick, 10mm wide and 100m long AgMg alloy foil. The green tape was wound into a pancake shape and heat treated in a pre-annealing(PA) step at 850 ~860℃ for 10~30 hours in flowing O_2 atmosphere. After pre-annealing, the residual carbon content was analyzed by infrared spectroscopy. The carbon content was measured by using an infrared analyzer, and the total weight includes the lamination foil and substrate. The pre-annealed tape was deformed by rolling, so-called intermediate rolling (IR), at room temperature with a reduction of ~50%. Some tapes were annealed at 500℃ in flowing N_2 atmosphere to increase T_c. The variation in J_c of a 100m class tape was measured in a non-inductive coil shape at 77K using a criterion of 1μV/cm. The J_c-B characteristics at various temperatures were measured in a gas-cooled atmosphere in back-up fields up to 5T.

The microstructure was observed using a scanning electron microscope (SEM). The degradation of J_c under mechanical stress was measured as follows. A mechanical load was applied to the tape at room temperature and then removed. After unloading, the J_c was measured at 4.2K in various magnetic fields. The J_c of some samples was measured under load. The sample was mounted on a mandrel at room temperature and after soldering current leads and voltage taps onto it, the sample was cooled down to 4.2K. In the measurement of bending strain characteristics, the bending strain was defined as the calculated strain at the surface of the composite tape assuming that the neutral plane was at the center of the tape. The tensile stress-strain curves were measured using an Instron type tensile tester. In the case of measurements at 93K, a sample was placed in a cold chamber and cooled by flowing nitrogen gas. The tensile test was carried out after holding the sample for 10 minutes in the gas-cooling atmosphere. A double pancake coil was fabricated by a react & wind technique. A 100m-long PAIR processed tape was coated on both sides with a 10μm thick insulating layer of polyvinyl formval. The coil was wound using a react & wind technique and impregnated with epoxy. The coil was tested at 4.2K in magnet field. Coil I_c was determined using a criterion of $10^{-13}\Omega \cdot m$.

RESULTS AND DISCUSSION

It is well known that residual carbon causes blisters and voids in superconducting filaments in powder-in-tube processes. The lamination technique we developed has an advantage because gases can escape through a narrow gap at the seam of the lamination foil. During pre-annealing, organic binders in the slurry decompose and escape almost completely when the sample is short. However, in the case of a long sample, such as 100m, leaking through the seam is not enough to achieve a low carbon content throughout the entire length. In the first production run of a 100m long tape, the residual carbon content was 0.03wt% after pre-annealing and the J_c value was 330kA/cm^2 at 4.2K in self-field. Leaking of liquid phase was observed at the seam. Closer examination of the microstructure revealed some distortion of Bi-2212 particle alignment caused by many large Bi-free particles, resulting in a decrease in the current path[4]. There are many factors that determine the residual carbon content of multi-layer tapes in a practical process. How well we decompose the organic binders and release the gases from the heat treatment chamber during pre-annealing are very important. We optimized the annealing condition by changing the heating rate, holding time at the annealing temperature, improving the ventilation ability of the facility, and so on. In a recent process, we were able to control the residual carbon content to be 0.002wt%, and we obtained a J_c value of 710kA/cm^2 at 4.2Kin self-field. The value of residual carbon content was a convenient measure for checking the quality of pre-annealed tape, so we set the threshold value in the manufacturing process at 0.01wt %.

In the case of multifilament wires, the filament size is an important factor to obtain higher J_c. From the viewpoint of chemical reactions and homogeneity, a smaller filament size is ideal, but

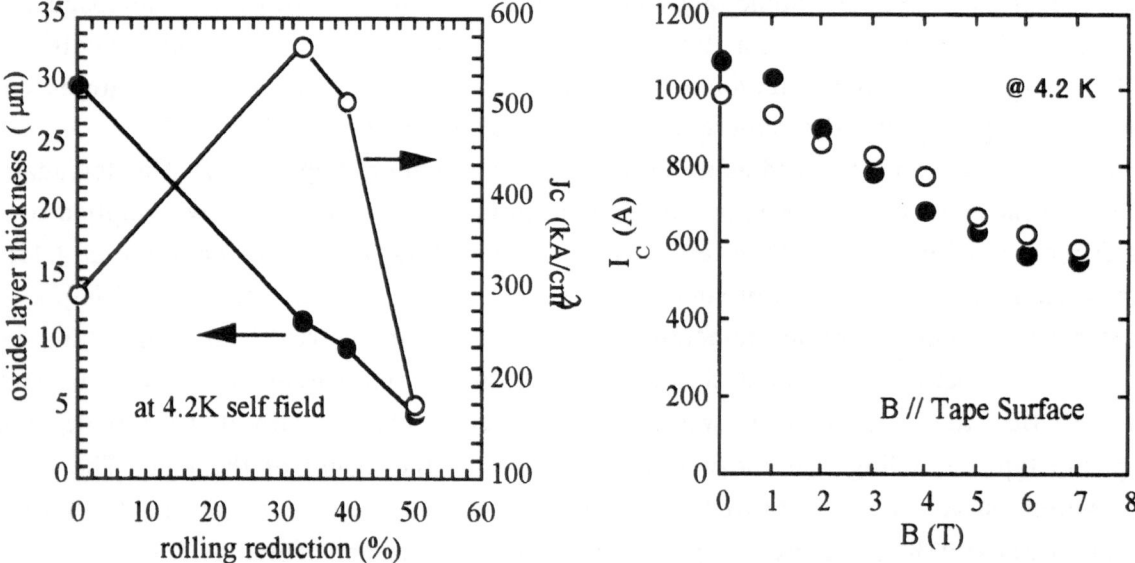

Figure1 Relationship between intermediate rolling reduction and Jc value.

Figure2 Ic-B characteristics of PAIR processed samples cut from both ends of a 100m long tape.

it is difficult to reduce the filament size in actual practice. In the multilayer tape, reducing the thickness of the coated layer in the dip coat process was not difficult in short sample, but maintaining homogeneity over an entire long length was difficult in the case of layer thicknesses below 30μm. We set the thickness of the coated layer at 50μm and did experiments to optimize the intermediate-rolling reduction.

Although we reported an optimum intermediate rolling reduction of 25%, we found that it was impossible to realize such a rolling reduction in a practical deformation process because of limitations in pay-off back tension. The results are shown in Figure 1. In the no-PAIR sample, many large voids and impurity phases were observed in the SEM photograph, which caused a lower J_c value. On the other hand, in the case of a rolling reduction of 50%, severe sausaging occurred and the oxide layer thickness varied considerably. Such inhomogeneity of geometry in multi-layer tapes led to inhomogeneity of the chemical reactions, resulting in a decrease in J_c.

Based on those results, we fabricated 100m-long multi-layer tapes using the PAIR process. The I_c-B characteristics of short samples cut from both ends of 100m long tape are shown in Figure 2. Because of an increase in the coating layer thickness in a long tape, the J_c decreased to almost 65% of short sample values in spite of having the same I_c.

Figure 3 shows the degradation of the J_c at 4.2K in self-field as a function of applied bending strain. Degradation of J_c started when the bending strain in the tape reached 0.6% (bending diameter of 30mm). The bending strain did not affect the J_c-B characteristics until the applied bending strain approached to the critical value, as shown in Figure 4. However, when the

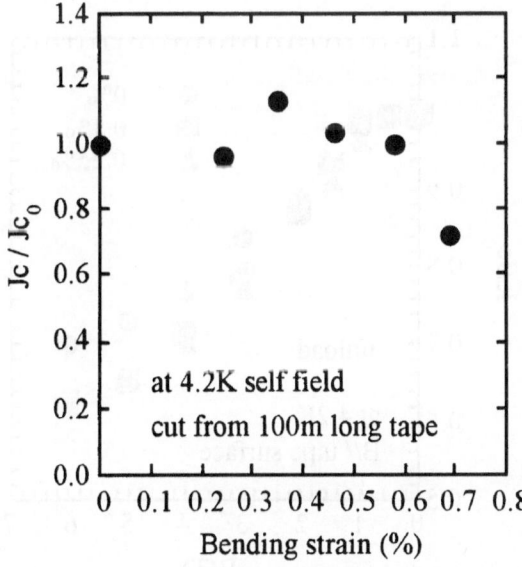

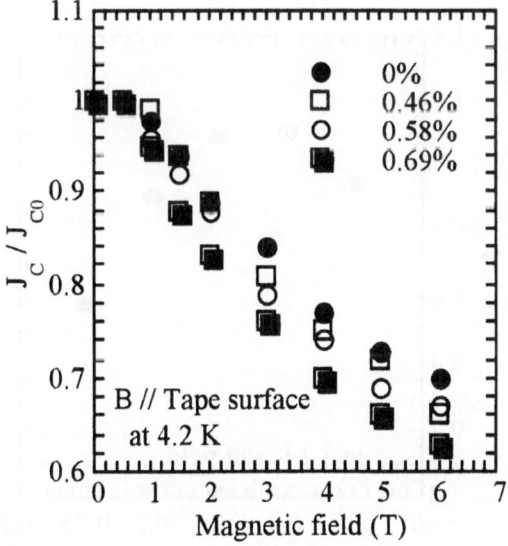

Figure 3 Bending strain characteristics of PAIR processed tapes at 4.2K in self-field.

Figure 4 Jc-B characteristics of PAIR processed tape applied bending strain.

measurement was carried out under load, the J_c value in self-field deteriorated gradually by applying magnetic fields, even when the bending strain was less than the critical value. J_c values were more sensitive against tensile strain than bending strain.

Figures 5 and 6 show the effects of tensile strain on the J_c in self-field and their magnetic field dependence, respectively. The critical value of tensile strain was 0.2%, which corresponded to a tensile stress of 40MPa according to the tensile stress-strain curves. The data in figure 6 shows that J_c degradation in the back up fields was somewhat grater than that of a virgin sample even if the applied tensile stress was lower than 0.2%. These results indicated that PAIR processed multilayer tapes exhibited mechanical strain tolerance almost equal to that of pure Ag sheathed multifilament tapes[5], which meant that it was possible to make a coil using a react & wind technique. However, the properties in magnetic fields and sensitivity to tensile strain are more matters of concern in the design of magnets and in the coiling procedure.

We fabricated two double pancake coils using a fully reacted 100m-long PAIR processed tape and tested them at 4.2K and in back up fields. The specifications of the coils are shown in Table 1. Coil #1 carried 300A and generated B_{max}=1.95T in 4T. When the backup field increased to 6T, the voltage increased rapidly before the current reached the I_c value, which meant that the coil was damaged by the hoop stress. Coil#2 wound with the reinforcement of CuAg alloy tape had an I_c value of 300A and a B_{max} of 1.35T was obtained in a back up field of 8T. The coil was not damaged after the test, demonstrating the effectiveness of the co-winding technique for PAIR-processed superconducting tapes. The performance of the coil was almost 75% that of the short sample. It is caused by the variation of I_c along a 100m long tape, the effect of handling, and by the degradation of J_c under coexisting magnetic field and loading. We also

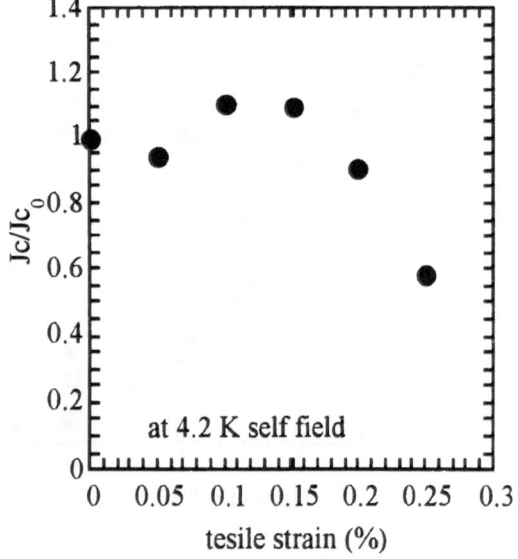

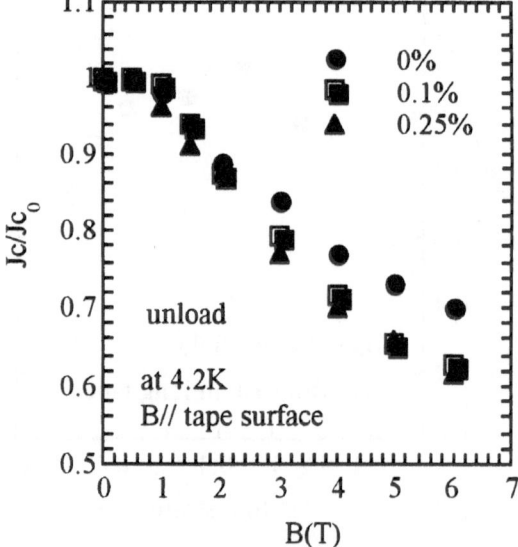

Figure 5 Tensile strain characteristics of PAIR processed tapes at 4.2K in self-field.

Figure 6 Jc-B characteristics of PAIR processed tape applied tensile strain.

fabricated a double pancake coil and measured its performance under refrigerator cooling. The coil carried 215A at 20K and generated 1.6T.

Table 1 Specifications of coils.

	Coil#1	Coil#2
Inner diameter(mm)	70	70
Outer diameter(mm)	147.2	176.6
Height(mm)	10.5	11
Total number of turns	286	243
Tape length(m)	94	91
Reinforcement	CuAg (outermost layer)	CuAg(co-winding)

CONCLUSION

The PAIR process was effective in obtaining a dense superconducting core with minimum carbon content, resulting in a small amount of impurity phases and good alignment of superconducting grains. In order to apply this process to long length conductors, we optimized critical fabrication parameters, such as heat treatment schedules for pre-annealing and rolling reduction for intermediate deformation. We successfully fabricated 100m long Bi-2212 multi-layer tapes using the PAIR process and obtained J_c values exceeding $7 \times 10^5 A/cm^2$ at 4.2K in self-field. The tape had bending and tensile strain tolerances of 0.6% and 0.2%, respectively. The coils fabricated by using 100m long PAIR processed tapes, wound with CuAg, generated 1.35T in a back up field of 8T at 4.2K.

REFERENCES

[1] M.Okada, K. Tanaka, K.Fukushima, J.Sato, H. Kitaguchi, H. Kumakura, T. Kiyoshi, K. Inoue and K. Togano, Jpn. J. Appl. Phys. Vol.35 (1996) L63

[2] H. Miao, H. Kitaguchi, H. Kumakura, K. Togano and T. Hasegawa, Physica C 301 (1998) 116

[3] T.Hasegawa, T. Koizumi, Y. Aoki, H.Kitaguchi, H. Miao, H. Kumakura and K. Togano, to be published in IEEE Trans. on Appl. Supercon. (1999)

[4] T. Koizumi, T. Hasegawa, H. Kitaguchi, H. Miao, H. Kumakura and K. Togano, Advances in Superconductivity XI (1999), 919

[5] Y. Hikichi, J. Nishioka, T. Koizumi and T. Hasegawa, Advances in Superconductivity XI (1999), 915

Strain Effect in Bi-2212/Ag PAIR Processed Tapes

Hitoshi Kitaguchi[1], Takao Takeuchi[1], Kikuo Itoh[1], Hiroaki Kumakura[1], Kazumasa Togano[1], Takayo Hasegawa[2] and Tsutomu Koizumi[2]

[1]National Research Institute for Metals, 1-2-1 Sengen, Tsukuba 305-0047, Japan
[2]Showa Electric Wire & Cable Co., Ltd., 2-1-1 Odasakae, Kawasaki 210-0843, Japan

Abstract: The influence of strain on the critical current (I_c) is investigated for Bi-2212/Ag tapes fabricated by using PAIR process. The tensile and/or compressive axial strain along tape length is successfully induced to samples by using a U-shape brass holder. The sample is soldered directly to the holder. Continuous change of the axial strain can be induced with changing the distance between both ends of the holder. Several samples are examined at 27 K in 0.5 T (B is applied perpendicular to the tape surface). While the initial I_c in zero strain state (I_{c0}) varies ranging 90~110 A, normalized I_c (I_c/I_{c0}) vs. strain relations fall on the same curve. In tensile strain tests, linear decrease of I_c is observed from as-cooled state to 0.44% strain (91% of I_{c0}). Rapid and large degradation occurs at the strain exceeding 0.44%. On the contrary, this large degradation is not observed in compressive strain tests up to -0.7% strain (75% of I_{c0}). Pre-strain caused by the thermal contraction of brass sample holder is estimated to about -0.2%. The tensile strain tolerance of PAIR tape is ~0.25%.

Keywords: Bi-2212, tape, strain effect, PAIR process, critical current density

INTRODUCTION

$Bi_2Sr_2CaCu_2O_x$ (Bi-2212) high T_c superconductors (HTS) are promising and expected for magnet applications. The conductors subject various kinds of strain (or stress) in magnet applications. For example, the conductors subject compressive strain due to thermal contraction in cooling and tensile strain due to hoop stress during operation. Another mechanical tensile strain can be considered in react & wind coil winding process. The effect of strain on the critical current (I_c) is important to design and fabricate practical HTS magnet systems. A large progress in Bi-2212/Ag conductor fabrication and a rapid progress in cryocooler engineering enhance the possibility of practical applications of Bi-2212/Ag conductors. The importance to understand the strain effects increases in these situations and a systematic study on this issue for the improved Bi-2212/Ag conductors [1] has been expected. Many efforts have been performed to clarify the effect of strain on I_c. The strain tolerance in bending deformation [2,3] and axial deformation [4-8] has been reported by many research groups. In this work, we study the effect of compressive and tensile axial strain on I_c of Bi-2212/Ag PAIR processed multilayer tape with high critical current density (J_c). It is much interested whether the large J_c enhancement (and/or the improvement of microstructure) influences the effect of strain or not. In this paper, we report I_c of Bi-2212/Ag PAIR tape under tensile and/or compressive axial strain along tape length at 27 K in the magnetic field of 0.5 T

EXPERIMENTAL

We have already reported the details of the fabrication method of Bi-2212/Ag multilayer tapes [9] and PAIR process [1]. Straight samples of ~65 mm in length were cut from 100m long PAIR tape for the strain tests. The strain tests were performed by using a U-shape sample holder. This method was established by Twente University group [7,8]. Sample setting and the principle of strain generation are summarized in Fig. 1. The U-shape holder was made of brass. The sample was fully soldered to the holder. Current leads and voltage monitor were soldered directly to the sample. Compressive or tensile strain can be induced by increasing or decreasing the pitch between both ends of the holder, respectively. The holder was attached to the movement operated with a computer-controlled actuator. Strain was calibrated by using strain gauge at liquid nitrogen temperature. In the calibration, a strain gauge was fixed by using glue directly to the holder at the center of sample position instead of mounting a sample. Strain can be controlled continuously from -1.0% (compressive) to 1.0% (tensile) in this apparatus. The probe was set to a superconducting magnet equipped with temperature controlling insert (He gas cooling). After zero field cooling to 27 K, magnetic field of 0.5 T were applied in the direction perpendicular to the sample tape surface. Sample temperature was monitored by using a resistance temperature sensor that was attached directly to the holder. I_c was determined from dc transport measurements with 1 µV/cm electric field (E) criterion. The voltage was monitored at the center part of sample that corresponds to the gauge position in the strain calibration. For the calibration of current, we monitored the voltage also at the holder itself. Current flow in the brass holder was negligible when the sample was superconducting. In I_c measurements, sample temperature was controlled within ±1% deviation.

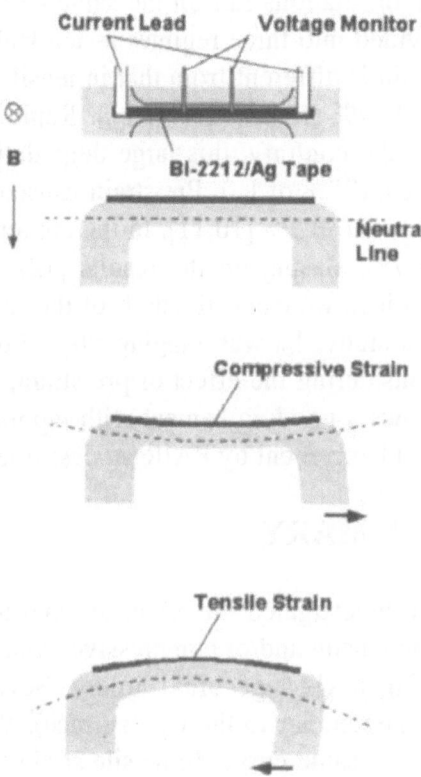

Fig. 1. Sample setting and the principle of strain generation

RESULTS AND DISCUSSION

Results for four compressive and three tensile tests at 27 K in 0.5 T are plotted together in Fig. 2. I_c/I_{c0} (I_{c0}: 0% strain I_c value) was plotted against strain in this figure. 0% strain is defined to be as-cooled state. Tensile and compressive strains are defined to be positive and negative strain, respectively. Samples subjected certain compressive pre-strain at 0% strain because of the thermal contraction in cooling from soldering temperature to 27 K. I_{c0} was ranging 90 ~ 110 A. The variation of I_c is supposed to originate in the inhomogeneity of the long tape. While I_{c0} varies, I_c/I_{c0} vs.

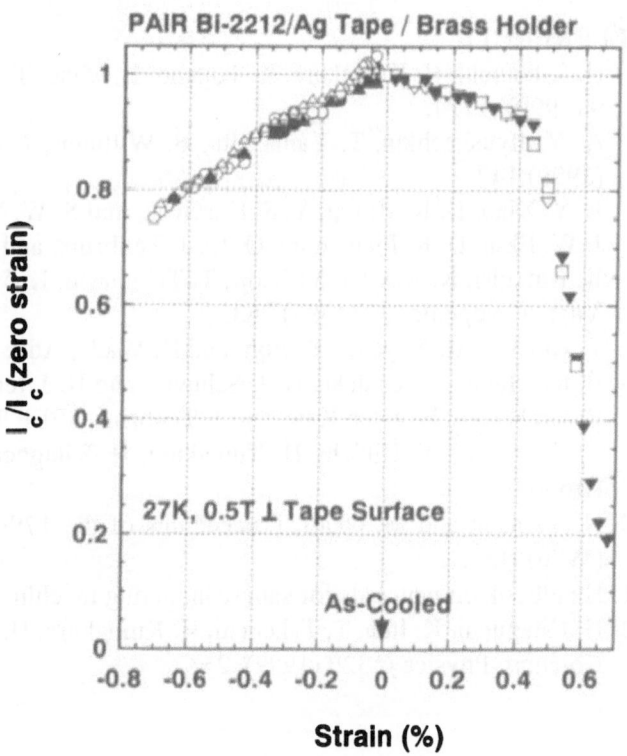

Fig. 2. Strain dependence of I_c

strain relations fall on the same curve with a very good agreement. Strain dependence of I_c can be divided into three regimes as ten Haken et al reported [7,8]. Strain dependence of I_c in compressive strain is different from that in tensile one. A linear decrease of I_c was observed from as-cooled state to 0.44% strain (91% of I_{c0}). Rapid and large degradation occurred at the strain exceeding 0.44%. On the contrary, this large degradation was not observed in compressive strain tests up to -0.67% strain (75% of I_{c0}). Pre-strain caused by the thermal contraction of brass sample holder is estimated to about -0.2% [10,11]. In the compressive strain regime, -0.2% strain corresponds to I_c degradation of 7%. Basing on the results [12] for short sample study (the same long tape was used in both studies), we can estimate I_c of the samples to be 100 ~ 130 A at 27 K in 0.5 T without pre-strain. In this study, I_{c0} was ranging 90 ~ 110 A. This degradation is attributed to the effect of pre-strain. Considering the effect of pre-strain, the tensile strain tolerance of PAIR tape is ~0.25%. This value shows a good agreement with reported strain tolerance of Bi-2212/Ag tapes [5]. This indicates that J_c enhancement by PAIR process does not degrade the strain tolerance.

SUMMARY

We investigated the effect of axial tensile and compressive strain on I_c of Bi-2212/Ag PAIR tapes. The tensile and/or compressive axial strain along tape length was successfully induced to samples by using a U-shape brass holder. Several samples were examined at 27 K in 0.5 T (B is applied perpendicular to the tape surface). While I_{c0} varies ranging 90~110 A, I_c/I_{c0} vs. strain relations fall on the same curve. In tensile strain tests, linear decrease of I_c was observed from as-cooled state to 0.44% strain (91% of I_{c0}). Rapid and large degradation occurred at the strain > 0.44%. On the contrary, this large degradation was not observed in compressive strain tests up to -0.7% strain (75% of I_{c0}). The tensile strain tolerance of PAIR tape is ~0.25% because the pre-strain caused by the thermal contraction of brass sample holder is estimated to about -0.2%.

REFERENCES

1. H. Kitaguchi, H. Kumakura, K. Togano, H. Miao, T. Hasegawa, and T. Koizumi, IEEE Trans. Supercond. 9 (1999), 1794.
2. A. Y. Ilyushechkin, T. Yamashita, B. Williams, I. D. R. Mackinnon, Supercond. Sci. Technolo. 12 (1999) 142.
3. L. Y. Xiao, D. K. Hilton, Y. S. Hascicek, and S. W. Van Sciver, Cryogenics 37 (1997) 837.
4. J. W. Ekin, D. K. Finnemore, Q. Li, J. Tenbrink, and W. Carter, Appl. Phys. Lett. 81 (1992) 858.
5. K. Katagiri, K. Kasaba, Y. Shoji, T. Takahashi, K. Watanabe, K. Noto, T. Okada, M. Hiraoka, and S, Yuya, Cryogenics 38 (1998) 283.
6. T. Kuroda, M. Yuyama, K. Itoh, and H. Wada, Adv. Cryogenic Engr. 38 (1992) 1045.
7. B. ten Haken, A. Godeke, H. J. Schuver, and H. J. ten Kate, IEEE Trans. Magn. 32 (1996) 2720.
8. B. ten Haken, H. J. ten Kate, and J. Tenbrink, IEEE Trans. Supercond. 5 (1995), 1298.
9. T. Hasegawa, Y. Hikichi, H. Kumakura, H. Kitaguchi, and K. Togano, Jpn. J. Appl. Phys. 34 (1995) L1638.
10. N. Yamada and M. Okaji, Proceedings of the 17th Japan symposium on thermophysical properties (1996), 135.
11. Handbook on materials for superconducting machinery (1977).
12. H. Kitaguchi, K. Itoh, T. Takeuchi, H. Kumakura, H. Miao, H. Wada, K. Togano, T. Hasegawa, and T. Koizumi, Physica C 320 (1999), 253.

EVALUATION OF STRESS/STRAIN DEPENDENCE OF CRITICAL CURRENT IN Ag-Mg-Ni SHEATHED Bi(2212) SUPERCONDUCTING TAPES

K. Katagiri, H.S. Shin*, I. Ishimori, Y. Shoji, K. Noto, K. Watanabe[+] and M. Okada[#]

Faculty of Eng., Iwate University, 5-4-3 Ueda, Morioka, 020-8551, Japan
*Dept. of Mech. Eng., Andong National University, Andong, 760-749, Korea
[+]MRI, Tohoku University, 2-1-1 Katahira, Aoba-ku, Sendai, 980-8577, Japan
[#]Hitachi Res. Lab., Hitachi Ltd., 1-1 Omika-cho, Hitachi, 319-12, Japan

Abstract: Stress/strain dependencies of the critical current I_c in Ag-Mg-Ni sheathed multi-filamentary Bi(2212) superconducting tapes were evaluated at 4.2K and 77K, with/without magnetic field of 0.5T. Significant variation among the strain characteristics of I_c was observed at 4.2K, while that at 77K was rather small. By soldering Ag-Mg tapes as an external reinforcing member, I_c of the tapes decreased. The I_c degradation characteristics in terms of stress were improved markedly by the reinforcement while improvement of them in terms of strain appeared less remarkable. These results are discussed from the viewpoint of monitoring sensitivity of cracking in the superconducting filaments associated with the location of them relative to the voltage-monitoring region in the tape.

Keywords: Bi(2212), Ag alloy sheath, External reinforcement, Critical current, Strain effect,

INTRODUCTION

Bi(2212) superconducting tapes are used for high-field magnets because of its high upper critical field and high I_c at 4.2K. During their operation, the windings are subjected to a high electromagnetic force. The I_c degrades irreversibly when the mechanical damage occurred at oxide superconductors. Various efforts have been made to suppress the degradation due to forces applied. The strain characteristics of I_c in short samples, however, is liable to vary in certain cases depending on the parameters such as wire construction, quality of the superconductor, measuring condition and criterion of critical current [1,2].

This paper describes the results of tensile tests and strain dependencies of I_c of Ag-Mg-Ni sheathed Bi(2212) 54 filament tapes at temperatures of 77K and 4.2K. Effects of external reinforcement with Ag-Mg tape were also investigated.

EXPERIMENTAL PROCEDURE

Specimens. The specimens used were Bi(2212) 54 filament superconducting tapes with Ag/Ag-0.025wt%Mg-0.0125wt%Ni alloy sheath prepared by a powder-in-tube method: the dimension being $4.8 \times 0.3 \times 40$ mm and Ag alloy/Ag/S.C. volume ratio, 45/35/20, respectively. In order to evaluate the effect of external reinforcement, tapes of Ag-0.5wt%Mg in the size of $4.8 \times 0.1 \times 40$ mm were soldered on one or both side of the specimens.

Apparatus. The stress-strain relation and the tensile stress/strain dependence of I_c were investigated at a temperature of 4.2K and in a magnetic field up to 14T using an apparatus described elsewhere [1]. The gage length and the voltage monitoring length were 18 and 5 mm, respectively. The criterion for I_c used was 1μ V/cm. Because the I_c of the tape is more than 200A at 4.2K even at 14T, the magnetic field of 0.5T was mainly used which minimize the effect of Lorentz force on measurements. Tests were also conducted at 77K with no filed.

RESULTS AND DISCUSSION

Effect of test temperature. Fig. 1 shows stress-strain curves of the Ag-Mg-Ni sheathed Bi(2212) tapes measured at 77K, 0T and 4.2K, 0.5T. The curves are close to each other showing no significant change in the mechanical properties, although slight decrease in the mechanical properties is seen at 4.2K due to the effect of the magnetic field intrinsic to the apparatus.

A remarkable variance in strain characteristics of I_c at 4.2K, 0.5T was found, whereas the variance was less at 77K (Fig. 2). It seems worthy of note that ε_{irr}, the strain for onset of degradation in I_c at 4.2K correspond to the point, where the stress-strain curves in Fig. 1 tends to saturate or to the one in saturated region. However, the maximum current in this experiment 150A is used as I_{c0}, I_c before straining for convenience sake here and, therefore, the value of ε_{irr} is overestimated. Although this may partly responsible for the difference, the variation in ε_{irr} is still apparent. On the other hand, ε_{irr} at 77K corresponds to near the inflection points in the steeply increasing part of the curves. The difference in the I_c vs. strain characteristics due to the testing temperature can be explained by the detecting sensitivity of cracking in the filaments, which happens to occur outside of monitoring area. The sensitivity appears to be associated with difference between the testing temperature and the critical temperature, the electric and thermal conductivity of the sheath at those temperatures. Further, there can be difference between the local strain and overall strain due to non-uniform deformation along the longitudinal direction of the tape. From a viewpoint of sensitivity determined by the criterion and the absolute value of I_c, the present result is apparently contrary to the prediction [2]. The stress dependencies of I_c, however, were almost consistent among the results both at 77K and 4.2K. The overall stress of the tape, which will induce cracking of the filaments, will not increase so much because of the low strain-hardening behavior of the sheath.

Effect of reinforcement. With increase in the amount of external reinforcement, the mechanical

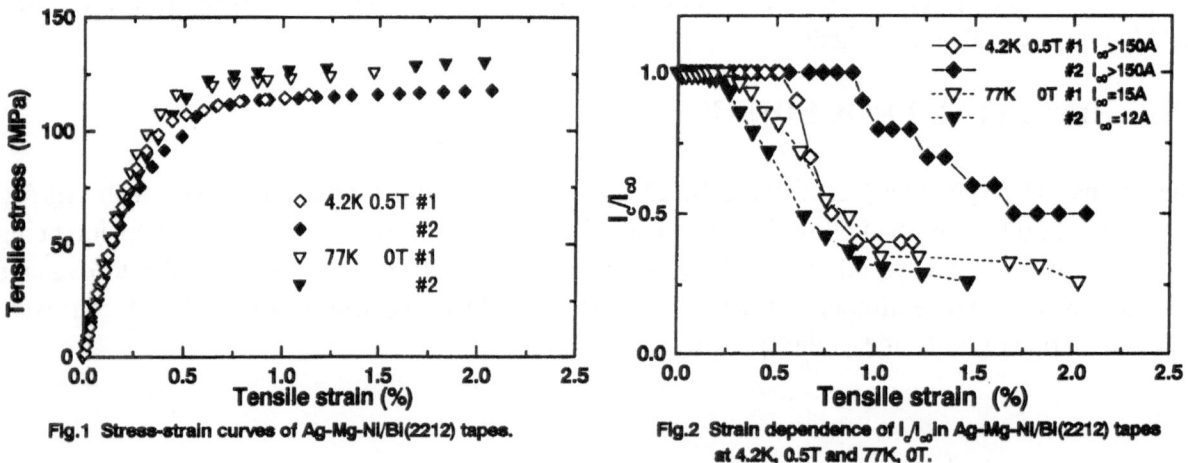

Fig.1 Stress-strain curves of Ag-Mg-Ni/Bi(2212) tapes.

Fig.2 Strain dependence of I_c/I_{c0} in Ag-Mg-Ni/Bi(2212) tapes at 4.2K, 0.5T and 77K, 0T.

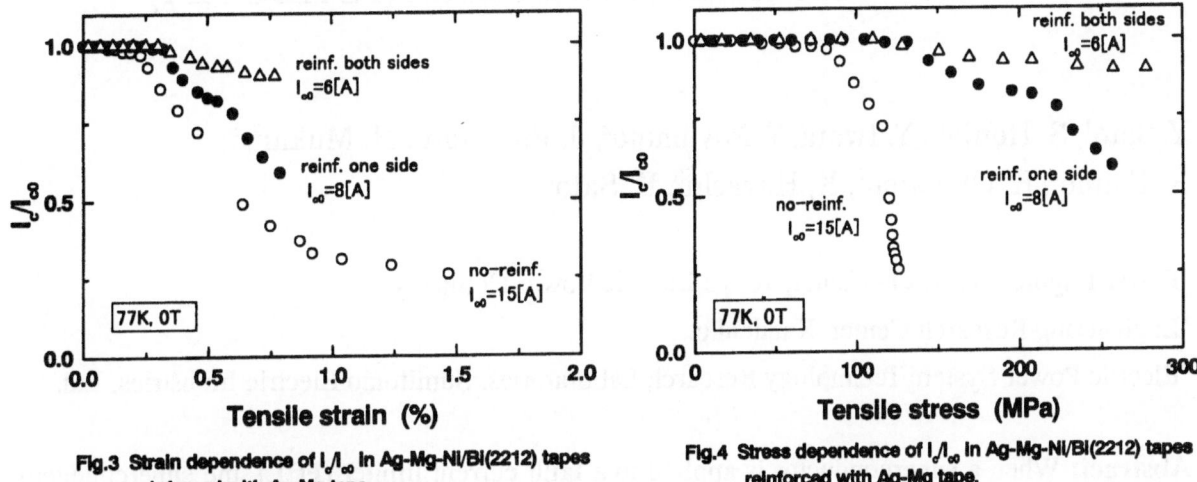

Fig.3 Strain dependence of I_c/I_{c0} in Ag-Mg-Ni/Bi(2212) tapes reinforced with Ag-Mg tape.

Fig.4 Stress dependence of I_c/I_{c0} in Ag-Mg-Ni/Bi(2212) tapes reinforced with Ag-Mg tape.

properties evaluated as stress-strain curves measured at 77K improved. The strain dependencies of I_c for samples with and without reinforcement are shown in Fig. 3. With increase in the amount of reinforcement, ε_{irr} increases. It must be noted that the value of I_{c0} drastically decreases with increase in the amount of reinforcement. This appears to be due mainly to the increase in the residual strain in the superconductor. The residual strains deduced from ε_{irr} are around 0.4 % and rough estimation obtained using the law of mixture in the present study saturates at a value of about 0.5%. According to the result by Haken [3], I_c decreases linearly to 80% of peak value with a compressive strain up to 0.6% and this cannot fully explain the present result.

The stress dependencies in reinforced tapes are shown in Fig. 4. The effect of reinforcement appears more obvious as compared to that shown as strain dependence. This is because that the degradation behavior in terms of stress in non-reinforced tape is steeper due to the reason mentioned before.

CONCLUSIONS

1. The reproducibility of strain characteristics of I_c changed depending on the testing temperature. It can be explained by the detecting sensitivity of cracking in the filaments, which is controlled by the relative location with voltage monitoring area and the temperature dependencies of material parameters related.

2. The effect of reinforcement on I_c degradation as a function of stress is obvious. This is because that the degradation in tapes without reinforcement is defined by the cracking stress of filaments irrespective of non-uniform deformation along the longitudinal direction after the cracking.

Acknowledgments. The authors wish to express their thanks to Mr. M. Ishizaki, graduate student of Iwate University for his assistance in this study. Measurements at 4.2K were carried out at the High Field Lab. for Superconducting Materials, Institute for Material Research, Tohoku University.

1. H. Shin and K.Katagiri, in *Advances in Superconductivity XI*, edited by N. Koshizuka and S. Tajima (Springer-Verlag, Tokyo, 1999), pp. 1479-1484.
2. W. Goldacker, J. Kessler, B. Ullmann, E. Mossang and M. Rikel, IEEE Trans. Appl. Supercond., **5**, 1834-1837 (1995).
3. B. ten Haken and H. ten Kate, IEEE Trans., **5**, 1298-1301 (1995).

CHARACTERISTICS OF RESISTANCE IN HIGH-Tc SUPERCONDUCTOR CARRYING CURRENTS ABOVE I_c

Y. Sato[1], S. Honjo[1], Y. Iwata, Y. Miyamoto[2], J. Fujikami[3], H. Mukai[3], K. Fujino[3], K. Ohmatsu[3], K, Hayashi[3], K. Sato[3]

[1]Power Engineering R&D Center, Tokyo Electric Power Company.
[2]Engineering Research Center, Kandenko.
[3]Electric Power System Technology Research Laboratories, Sumitomo Electric Industries, Ltd.

Abstract: When a superconductor is applied to a fault current limiter (FCL), the superconductor is exposed to currents somewhat larger than the critical current (I_c). Therefore, it is necessary to understand how the resistance behaves under the excess currents so as to evaluate temperature increase in the device and pressure rise in the vessel due to vaporization of the coolant. Resistance appearing under current above I_c was measured and investigated for high-T_c superconducting wire. For the measurements, we prepared Bi-tape wire having Ag-Au alloy for its matrix, whose resistivity is ten times larger than that of silver. The measurements were conducted at 77K under various magnetic fields. Measured data show that the resistance approaches to its free flux flow value as the current increases when H>1T, and that it strongly depends on the current for lower fields.

Keywords: Bi2223, quenching, fault current limiter, I-V characteristic

INTRODUCTION

High temperature (high-T_c) superconductors exhibit a wide dissipative region between almost zero-resistance and normal-conducting states, while low temperature (low-T_c) superconductors show a sharp transition between them. The current-voltage (I-V) characteristic in this region that dominates the dissipation in high-T_c superconductors is a critical factor in terms of application of these materials. So far, this characteristic has been studied intensively.

However, while many studies have been reported for $YBa_2Cu_3O_y$, few studies have been done for $Bi_2Sr_2Ba_2Cu_3O_y$ (Bi2223), the most prominent material for the superconducting wire used at Liquid Nitrogen (LN_2) temperature. We have measured the resistance appearing in Bi2223 tape wires in the case in which the transport current exceeds the critical current (I_c), and have found that the resistivity of Bi2223 remained as low as that of silver [1]. In this study, we measure the resistivity for I >> I_c under various magnetic fields, and investigate the appearing resistance depending on the current and the magnetic field.

EXPERIMENTAL

For the measurements, we prepared a Bi2223 tape-shaped multifilamentary wire having the specifications given in Table 1. The sheath of the wire was made of Ag10at%Au alloy, the resistivity of which is 10 times higher than that of silver. This wire is a promising candidate for a current limiting device due to the high resistivity of the sheath. The I-V characteristic was measured under the magnetic field up to 1.5 Tesla (T) applied perpendicularly to the surface of the wire. The current was induced to the sample as a pulse whose duration was ~3ms. Signals were detected and stored using a digital oscilloscope. Although the sample was immersed in LN_2, joule heating was enough to heat the sample. Due to the short duration of the current only a small temperature rise was caused (if no heat flows to LN_2 during a pulse, the temperature increase is evaluated to be less than 0.5K). The maximum value of the current was 100Apeak.

Table. 1 Specifications of the measured superconducting wire

Superconductor	$(Bi,Pb)_2Sr_2Ca_2Cu_3O_y$
Ic	35A@77K （10^{-4}V/m）
Sheath	Ag10at%Au alloy
ρ_{sheath}	$4.54 \cdot 10^{-8}\Omega$m@77K

RESULTS AND DISCUSSION

Figure 1 shows the resistance of Bi2223 region of the wire as a function of the current flowing through the Bi2223 region in various magnetic fields. The resistance and the current of the Bi2223 region (R_{sc} ans I_{sc}) were evaluated using the following equations.

$$\frac{1}{R_{sc}} = \frac{1}{R_{wire}} - \frac{1}{R_{sheath}} \qquad I_{sc} = I_{wire}\frac{R_{wire}}{R_{sc}} \qquad (1)$$

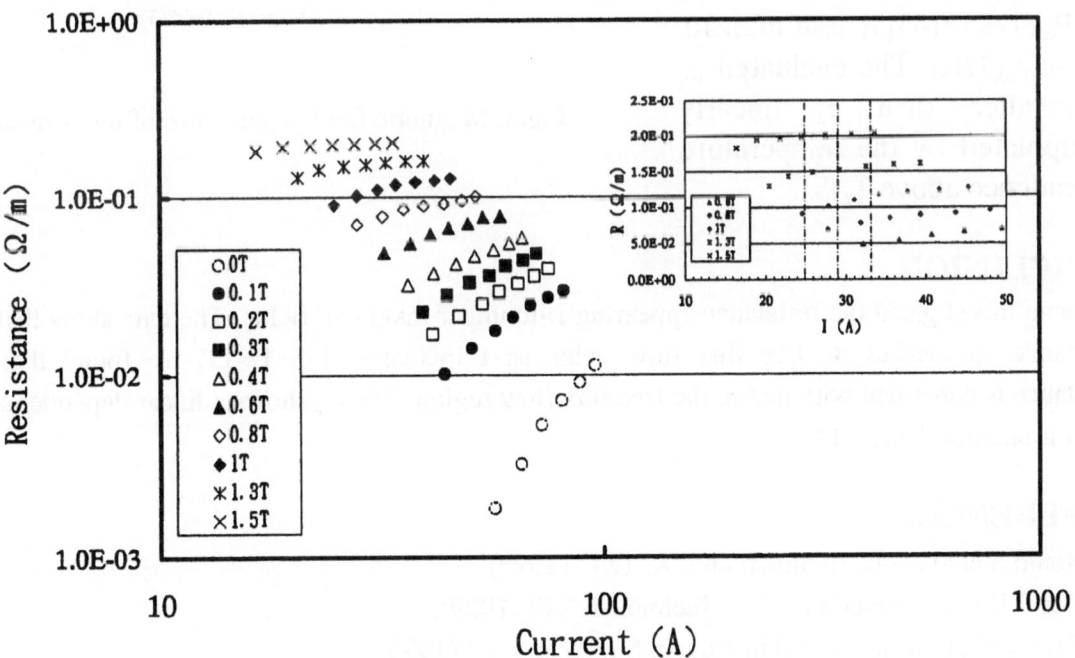

Fig. 1 Curent dependence of the resistance appearing in the wire

where, R_{wire} is the measured resistance of the wire and R_{sheath} is the resistance evaluated using Table 1. I_{wire} is the current of the entire wire.

For lower fields, R_{sc} strongly depends on the current. However, as the field becomes larger, the dependence weakens and finally, R_{sc} is almost independent of the current for B>1T. It is considered that dissipation in high-T_c superconductors has three mechanisms, thermally activated free flux motion (at a low electric filed) [2], depinning (intermediate) [3], and free flux flow (at a high electric field) [4], making it possible for the dissipative region to be divided into three according to the mechanism [4]. In the free flux flow region, the voltage and the current have a linear relation, i.e. R_{sc} is independent of the current. Therefore, the measured result suggests that the superconductor was in the free flux flow region for B>1T. Then, we observe the dependence of R_{sc} on the magnetic field. Each broken line in Fig. 1 gives the field dependence of R_{sc} for a current. The extracted dtat is given as a function of the field in Figure 2.

It should be noted that R_{sc} shows a linear dependence on the field for the higher fields. It is consistent with the Kunchur's result [4], in which the ratio of resistivities in the free flux flow region (ρ_{ff}) and normal one (ρ_n) is approximately equal to the applied field normalized by the upper critical field (B_{c2}). According to Table 1 and Fig. 2, we obtain $\rho_{ff}=5\times10^{-8}\Omega m$ for 1.7T, 3.7×10^{-8} for 1.4T, and 3.0×10^{-8} for 1.1T. The obtained ρ_{ff} and $B_{c2}(77K)\sim15T[5]$ lead to $5\times10^{-7}\Omega m$ of $\rho_n(77K)$. The evaluated ρ_n is smaller than ρ_n linearly extrapolated by the temperature dependence above $T_c[5]$.

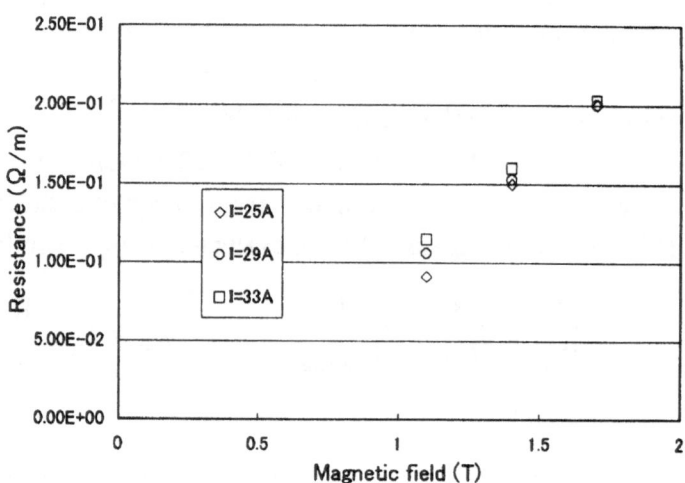

Fig.2. Magnetic field dependence of the resistance

CONCLUSION

We have investigated the resistance appearing I>Ic under magnetic fields. The data show that the resistance approaches its free flux flow value as I increases. For B>1T, we found that the resistance is consistent with that in the free flux flow region. The ρ_{ff} shows a linear dependence on B but is smaller than $\rho_n B/B_{c2}$.

REFERENCES

[1] Honjo S et al., Adv. in Supercond. X, 1243 (1998).

[2] Kes P H et al., Supercond. Sci. Technol. 1, 242 (1989).

[3] Kiss et al., to be published in Adv. in Supercond. XI (1999).

[4] Kunchur M N et al., Physical Review Lett. 70, 998 (1993).

[5]Matsubara I et al., Phys. Rev B45, 7414 (1992).

A Temperature Control System for Measuring J_c of Bi-system HTS Tapes in LN_2 Temperature Range

Huan Wu, Xinkang Teng, Yong Feng, Pingxiang Zhang, Xiaozu Wu, and Lian Zhou

Northwest Institute for Nonferrous Metal Research, P. O. Box 51, Xi'an, Shaanxi 710016, China

Abstract: In order to study practical applications and to investigate flux pinning characteristics of Bi-system HTS tapes, a temperature control system was established to measure J_c in a LN_2 temperature range. The sample to be measured was immersed in LN_2 and the desired temperature of the sample was reached by means of changing its saturated vapor pressure with a PID technique. The accuracy of the system can reach 0.1K for a long time. With this simple and effective system, J_c of Bi(2223)/Ag tapes was measured as a function of B(0-0.8T), T(77-90K), and θ(0-90°).

Keywords: temperature control system, LN_2 temperature range, J_c measurement, Bi-system HTS tapes.

INTRODUCTION

Research into J_c of Bi-system HTS tapes is useful to not only their practical applications but also detailed understandings of their flux pinning mechanisms. In general, J_c is measured in a complicated way of loading a sample with pulse measuring current in low temperature surroundings provided by a refrigerator. In this paper, a simple and effective temperature control system for measuring J_c of Bi-system HTS tapes in a LN_2 temperature range is introduced

EXPERIMENTAL

The system is made of a low resistance measuring bridge, a power regulator and a thermostat in which a thermometer, an electric heater and the sample are included. The measuring bridge specially designed for low temperature experiments has a distinguishing feature for using a four point technique and a.c. measuring current to eliminate the systematic error. PID [1] technique is used in the power regulator to adjust input power of the electric heater, so the heater can control temperature steadily, accurately and quickly. In the system designed according to a negative feedback principle, the signal of the thermometer (resistance value) and the signal of the desired temperature(resistance value) are input to the measuring bridge in which the two signals are to be compared and lead to deviation signals. Power regulator then adjusts the input power of the electric heater according to the deviation signals using PID algorithm, so the desired temperature will be reached.

Figure 1 is the schematic drawing of the thermostat. The narrow space between two brass tubes of the upper part of the thermostat is exhausted with a vacuum pump. So, the lower part has a closed thermal contact with outer surroundings, whereas the upper part will not be disturbed. The sample holder is fixed to an airtight socket located in the center of the flange through two current leads of the sample. Meanwhile, electric potential leads of the sample, thermometer leads and heating coil leads are drawn out via the airtight socket. The Pt resistance thermometer can reach an accuracy of 0.1K, not influenced by the magnetic field (B<1T). The heating coil is mounted on the lowest part of the sample holder in order to attain uniform temperature of LN_2 when it is heated. When using this system, firstly insert the thermostat in the LN_2 container, then pour LN_2 into the thermostat until the sample holder is immersed completely, finally close the flange and monitor the saturated vapor pressure. With this system, the desired temperature can be reached in 5 minutes, and the temperature fluctuation is about 0.1K. Typical results of the temperature control are as follows: heating power is 24.2W, saturated vapor pressure is 2.8 atm, and temperature can be maintained to be 86.1K. In the range of 77-90K, $J_c(B,T,\theta)$ values of Bi(2223)/Ag tapes were measured by the I-V characteristics using a standard four point technique with an electric field criterion of $1\mu V/cm$. The J_c anisotropy was measured as a function of magnetic field direction with respect to the surface of the tape($0\leq\theta\leq90^\circ$), by rotating a tape in a fixed field($0\leq B\leq0.8T$) provided by an electric magnet cooled with LN_2. The current direction was held perpendicular to the magnetic field during this measurement.

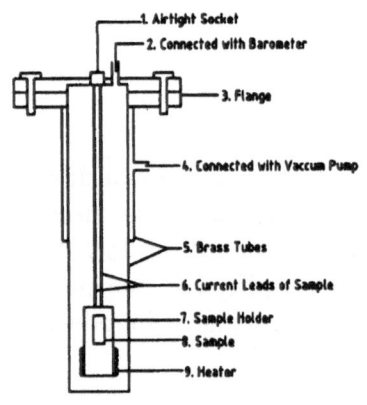

Fig. 1 Schematic drawing of the thermostat

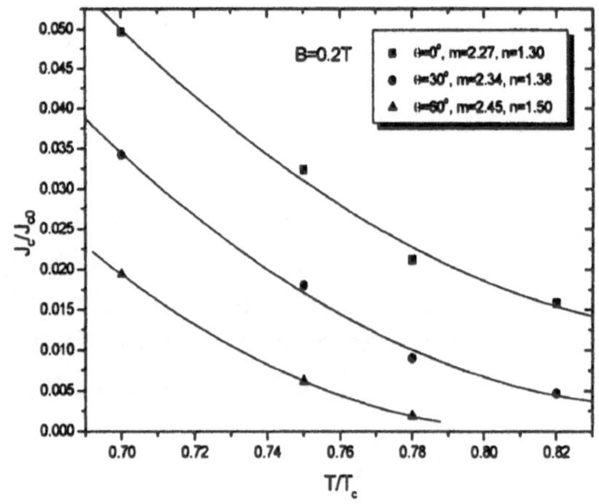

Fig. 2 J_c/J_{c0} versus T/T_c in different field directions for B=0.2T

RESULTS AND DISCUSSION

According to the flux creep model [2, 3], at low temperature($t=T/T_c<<1$), J_c is expressed as follows:

$$J_c/J_{c0}=1-mt+nt^2 \qquad (1).$$

Here, J_{c0}, m, n are parameter variables. Figure 2 plots the J_c/J_{c0} as a function of T/T_c when B=0.2T. It shows that the results are in agreement with the formula (1). It can be seen that m and n both increase because the flux creeps more and more severely when the inclined angle θ increases i.e. the component $B_\perp$ increases. Figures 3 and 4 show J_c as a function of B when B is parallel to the ab plane and J_c as a function of $Bsin\theta$ at 77K, respectively. Two figures show that the B, θ dependence of J_c can be described as the formula

$$J_c\propto exp(-\beta Bsin\theta) \qquad (2),$$

which was forwarded by Sun before [4]. Also, it indicates that the grains in our sample have a good alignment. All of the results above prove that the control system has an excellent behavior, and can meet the needs of the measurement completely.

Because a sample is immersed in LN$_2$ wholly in experiment, the uniformity of the temperature is guaranteed. Also the sample damage from heating for being loaded with strong current can be avoided, so it is needed not to load the sample with pulse current. But, the temperature dependence of the saturation vapor pressure of LN$_2$ follows an exponential relationship in the main [5]. For example, when the temperature is 110K, the pressure is about 20 atm. So, it is needed to have a better pressure resistance of the thermostat if the temperature range of LN$_2$ is to be widened.

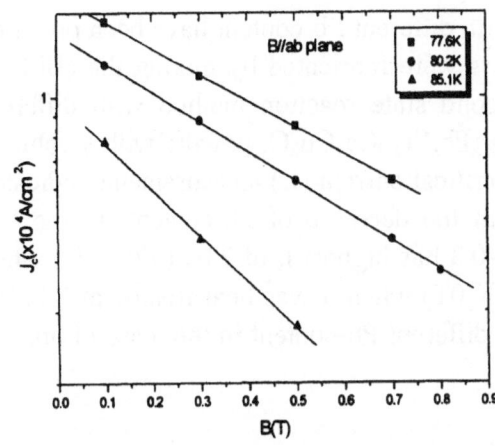

Fig. 3 J$_c$ versus B at different temperatures for B//ab plane

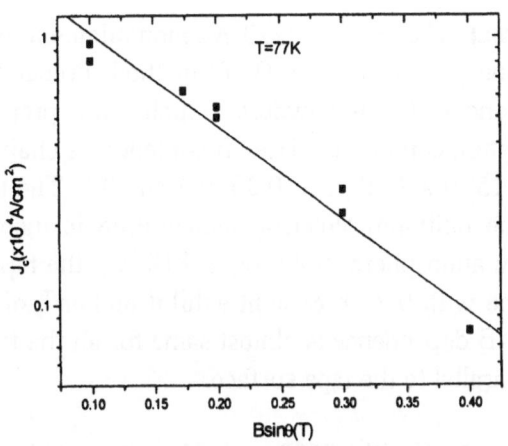

Fig. 4 Jc versus Bsinθ at 77K

CONCLUSION

By means of changing the saturation vapor pressure of LN$_2$ with a PID technique, the simple and reliable system can reach the accuracy of 0.1K for a long time in the temperature range of 77-90K, and meet the needs of measuring the J$_c$ of Bi-system HTS tapes.

Acknowledgements. We would like to thank Mr. Li Chengshan, Ms. Zheng Huiling, Mr. Duan Zhenzhong of Bi-system HTS research group of our institute for making samples available.

1. Yihui Jin. Procedure Control. Beijing, China: Tsinghua University Press, 1993
2. P. W. Anderson and Y. B. Kim. Rev. Mod. Phys. , 1964; 36: 39
3. M. Tinkham. Helv. Phys. Acta, 1988; 61: 443
4. Yuping Sun. Study of the Preparing Technique of Bi(2223)/Ag Tapes and its Transport Properties in the Mixed State. Dissertation, Hefei, China: Institute of Solid State Physics, Chinese Academy of Science, 1997: 52pp
5. Tatakoceichi, Shilu Ye et al translated. Low Temperature. Beijing, China: Editorial Department of Low Temperature Engineering, 1980: 22pp

Fabrication and Properties of (Bi,Pb)-2223 Tapes with Different Lead Content

P. X. Zhang[1,2], H. Maeda[2], A. Oota[1], R. Inada[1], W.P.Chen[2]

[1]Toyohashi University of Technology, Tempku, Toyohashi, Aichi 441, Japan
[2]Institute for Materials Research, Tohoku University, 2-1-1 Katahira, Sendai 980-8577, Japan

Abstract: The (Bi,Pb)-2223/Ag monofilamentary tapes with different Pb-content have been prepared by using powder-in tube (PIT) method. Precursor powders were fabricated by mixing the (Bi,Pb)-2212 and $CaCuO_2$ powders, which were prepared by solid state reaction method with different calcination conditions. The Pb-content was changed in $Bi_{2-x}Pb_xSr_{1.9}Ca_1Cu_2O_y$ powder with x value of 0.1, 0.15, 0.2, 0.25, 0.3, 0.35, 0.4 and 0.5. The transport critical current (I_c) measurements indicated that the optimum sintering temperature is increased with the decrease of Pb-content. In the 8% oxygen atmosphere, sintering at 818 °C, the tape with x=0.3 has highest I_c of 23A (77K,0T), while the tape with 0.2 Pb-content exhibit higher I_c of 30A (77K,0T) when it was heat treated at 828 °C. The J_c-B dependence is almost same for all the tapes with different Pb-content in the case of applied field parallel to the tape surface.

Keywords: (Bi,Pb)-2223 tape, Pb content, Critical current

INTRODUCTION

The Bi-based tape has been the most candidate for the fabrication of long HTS conductor for practical use. A big enhancement in critical current density (J_c) of (Bi,Pb)-2223 tape has been realized and the microstructural characteristics of the high J_c tape were observed [1]. Further improvements in length and performance of the tapes are still expected soon by optimization of the fabrication technique based on the understanding of phase formation mechanism and processing parameters. Partial lead substitution for bismuth was found to promote formation of high-T_c (110K) 2223 phase [2]. The influence of lead on superconducting properties of 2223 tapes with overall composition of $Bi_{1.8}Pb_xSr_2Ca_2Cu_3O_y$ was investigated recently [3], and the optimum lead contend was found to be x = 0.3. In this work, we studied the effects of lead content in 2223 tapes with overall composition of $Bi_{2-x}Pb_xSr_{1.9}Ca_2Cu_3O_y$ (x changes from 0.1 to 0.5) on processing temperature and critical current. In our case, we change the content of lead and bismuth at same time, while keep the total content of Bi and Pb in constant 2.

EXPERIMENTAL

Monofilamentary (Bi,Pb)-2223 tapes with Ag sheath were fabricated by conventional powder-in-tube method. Precursor powders with nominal composition of $Bi_{2-x}Pb_xSr_{1.9}Ca_2Cu_3O_y$ (x=0.1, 0.15, 0.2, 0.25, 0.3, 0.35, 0.4, 0.5) were prepared by mixing $Bi_{2-x}Pb_xSr_{1.9}Ca_1Cu_2O_y$ ((Bi,Pb)-2212) and $CaCuO_2$ powders. The (Bi,Pb)-2212 powder was formed by calcining a stoichiometric mixture of Bi_2O_3, PbO, $SrCO_3$, $CaCO_3$ and CuO in air and 8% oxygen atmosphere at 750-830 °C for 30 hours with two times intermediate grinding. The $CaCuO_2$ was prepared in the similar manner, except the sintering process was carried out at 800-950 °C for 40 hours. X-ray diffraction (XRD) analysis has been used to examine the phase composition of the formed powders, and DTA (differential thermal analysis) measurements were performed in air to determine the 2223 phase reaction temperature of the precursor powders. For fabrication of monofilamentary tapes, the prepared precursor powders were packed into pure silver tubes, which were then drawn to wires with diameter of 1.2 mm. Finally, two times' rolling and heat treatment process [4] was applied to make the tapes. The heat treatments were carried out at 818 - 828°C in a flowing gas containing oxygen and argon, partial pressure of oxygen is about 8%. The D.C. transport critical currents of the tapes have been measured at 77K in magnetic fields up to 1T. A standard four-probe technique and a criterion of $1\mu V/cm$ were used. The samples were mounted with tape surface parallel and applied current perpendicular to the magnetic field.

RESULTS AND DISCUSSION

The precursor powders for this study were prepared by the 'two-powder' process [5], in which the sintered intermediate powders of (Bi,Pb)-2223 and $CaCuO_2$ were mixed. X-ray diffraction analysis indicated that the main phase in all the prepared (Bi,Pb)-2212 powders with different lead content is Pb-doped 2212 phase. But, the plumbate phase and 2201 phase are increased with the enhancement of Pb-content as shown Fig.1. Although the reduced oxygen atmosphere was used for heat treatment, the content of plumbate phase is still high in the (Bi,Pb)-2212 powders with high Pb-content. The fabricated $CaCuO_2$ powder was found to be a mixture of Ca_2CuO_3 and CuO. DTA measurement of precursor powders found that the partial melting temperature, which was considered as 2223 phase formation temperature, decreases almost linearly with Pb-content (Fig.2). So the lower Pb-content tapes need higher heat treatment temperature in the same oxygen partial pressure.

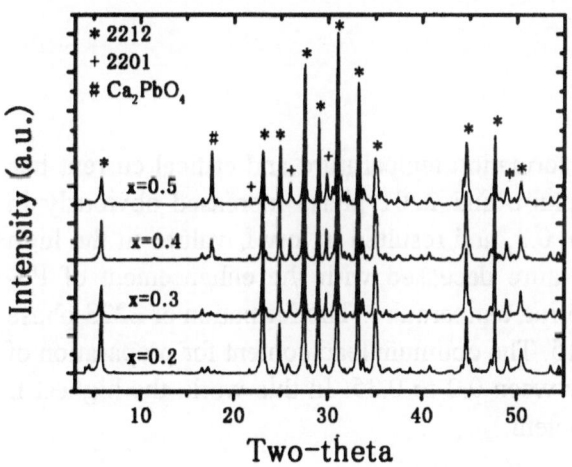

Fig.1. XRD pattern of (Bi,Pb)-2212 powders.

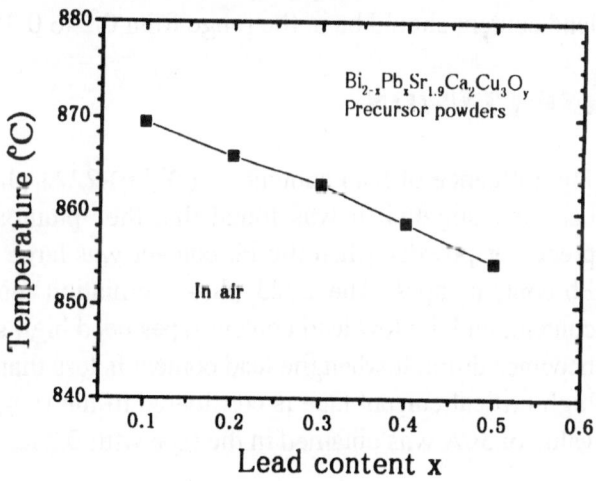

Fig.2. 2223 phase formation temperature versus x.

All the prepared tapes have same cross-section of about 0.17×2.84 mm^2. For different sintering temperature, the total heat treatment time is fixed as 200 hours. Fig.3 shows the dependence of critical currents (at 77K, self-field) on lead content for the tapes heat treated at different temperature in 8% oxygen partial pressure. When the sintering temperature increased, the I_c of lower Pb-content samples increased, and I_c of higher Pb-content tapes decreased. This is due to the decrease of 2223 phase formation temperature with increase of Pb-content. The tape with x=0.3 has the highest I_c after heat treated at 818 °C and 823 °C. A highest critical current of 30A was obtained in the tape with 0.2 Pb-content after it was sintered at higher temperature of 828 °C. Only a small critical current of 2.3A

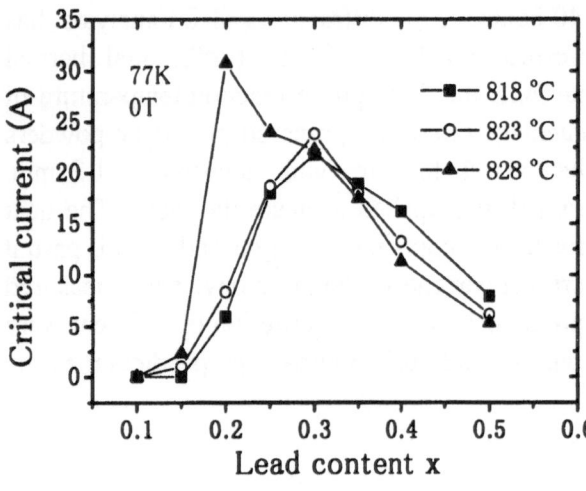

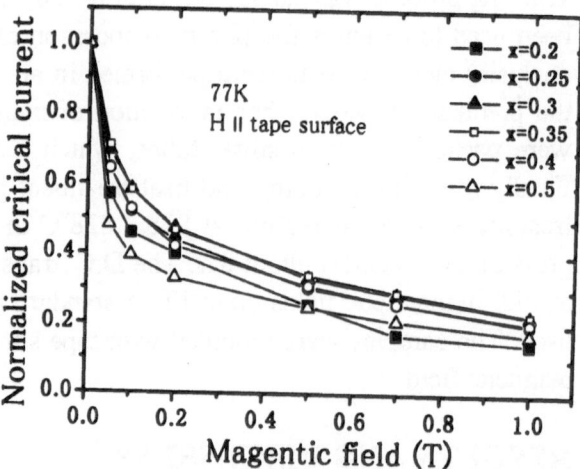

Fig.3. The dependence of I_c on Pb-content.   Fig.4. Normalized I_c on applied magnetic filed

has been obtained in the tape with x=0.15. For the tapes with 0.1 Pb-content, the transport I_c at 77K is zero. That means the formation of 2223 phase is difficult when the Pb-content is less than 0.15 in the investigated processing conditions. The tapes with higher lead content (x>0.35) exhibit lower critical current even heat treated at lower temperature. This is due to the higher content of 2201 phase and plumbate phase such as Ca_2PbO_4, which was considered to suppress the 2223 phase formation during heat treatment [4]. Although the I_c values change with lead content, the dependence of I_c on applied magnetic field is almost the same as shown in Fig.4. All the tapes have shown in Fig.4 were heat treated at 818 °C. The more drops of I_c at low field for tapes with Pb-content of 0.2 and 0.5 are due to the poor connection between 2223 grains. Based on above results, the optimum lead content should be in the range from 0.2 to 0.35.

CONCLUSION

The influence of lead content on (Bi,Pb)-2223 phase formation temperature and critical current has been investigated. It was found that the plumbate phase and 2201 phase increased obviously in precursor powder when the Pb content was large than 0.3, and resulted in low I_c values in the high Pb-content tapes. The 2223 phase formation temperature deceased with the enhancement of Pb-content, and the low lead content tapes need high sintering temperature. The formation of 2223 phase becomes difficult when the lead content is less than 0.15. The optimum lead content for preparation of high critical current tape is considered in the range between 0.2 to 0.35. In this work, the highest I_c value of 30A was obtained in the tape with 0.2 lead content.

1. T.G. Holesinger, J.F. Bingert, J.O. Willis, Q. Li, R.D. Parrella, M.D. Teplitsky, M.W. Rupich, and G.N. Riley, Jr., IEEE Transactions on Applied Superconductivity **9**, 2440 (1999).
2. S.A. Sunshine, T. Siegrist, L.F. Schneemeyer, D.W. Murphy, R.J. Cava, B. Batlogg, R.B. Dover, R.M. Fleming, S.H. Glarum, S. Nakahara, R. Farrow, J.J. Krajewski, S.M. Zahurak, J.V. Waszczak, J.H. Marshall, P. Marsh, L.W. Rupp, Jr., and W.F. Peck, Phys. Rev. B **38**, 893 (1988).
3. J.W. Anderson, S.E. Dorris, J.A. Parrell, and D.C. Larbalestier, J. Mater. Res. **14**, 340 (1999).
4. P.X. Zhang, H. Maeda, L. Zhou, C.S. Li, Z.Z. Duan, and H.L. Zheng, IEEE Transactions on Applied Superconductivity **9**, 2770 (1999).
5. S.E. Dorris, B.C. Prorok, M.T. Lanagan, S. Sinha, R.B. Poeppel, Physica C **212**, 66 (1993)

Development of Twisted Multifilamentary Tape Using AgMg Alloy

Masanao Mimura[1], Akira Takagi[1], Keizo Kosugi[1], Masahiro Sugimoto[1], Akio Kimura[1], Toru Tanigawa[1], Shoichi Honjo[2], Tomoo Mimura[2] and Yoshihisa Takahashi[2]

[1]Superconductivity Research Dept., The Furukawa Electric Co., Ltd. Kiyotaki, Nikko 321-1493, Japan
[2]Power Engineering R&D Center, Tokyo Electric Power Company, Yokohama 230-8510, Japan

Abstract: To improve over-all critical current density (J_e) and mechanical properties, and also reduce AC losses, we have developed twisted multifilamentary tapes using AgMg alloy sheath. The alloy-sheathed twisted tape achieved J_e of 5.4kA/cm^2 at 77K and 0T, with J_e uniformity over the whole length. The tape produced about twice higher strength and almost one third of particular AC loss, compared to a silver-sheathed untwisted one. A rate of inter-filament coupling in a tape was evaluated by comparing measured AC loss and calculated one.

Key words: AgMg alloy, Bi-2223, multifilamentary, twist, AC loss

INTRODUCTION

Bi-2223 multifilamentary tapes offer a relatively high overall critical current density (J_e) among high temperature superconducting wires under development for power transmission cables [1-3]. However, realizing the cables requires tapes still more improved in critical current density (J_c) and practical strength, and reduced in AC loss [4]. We reported that AgMg alloy sheathed twisted tapes had been enhanced in mechanical properties and reduced in AC loss [5]. This paper deals mainly with AC losses of the tapes superior to previously reported ones.

EXPERIMENTAL

Bi-2223 powder was processed by the powder in-tube (PIT) technique into AgMg alloy sheathed tapes with a silver ratio of about 3 and a number of filaments of 55 [5]. The alloy-sheathed tapes were twisted with 10-20mm of twist-pitches (L_p) to counteract I_c drop. For comparison, pure Ag-sheathed tapes were made by the same method. Critical currents (I_c) were taken at 77K and a self field, and defined as 1μV/cm. Mechanical properties were evaluated by the measures of I_c decline being caused by tensile stress or bending strain at room temperature. A conventional AC magnetization method was adopted to measure AC loss (Q) per cycle and unit volume of tapes, which were stacked and mutually insulated [6]. The AC magnetic field was parallel to the tape surface and its frequency (f) was 50Hz.

RESULTS AND DISCUSSION

Table 1 shows J_e values of silver-sheathed, alloy-sheathed and alloy-sheathed twisted tapes. The alloy-sheathed twisted tape achieved J_e of 5.4kA/cm^2, with J_e uniformity over the whole length of 100m. Optimizing fabrication conditions could prevent degradation of J_e by alloying and twisting. The I_c of the alloy-sheathed tape remains unchanged up to 90MPa of tensile stress or 0.3% of bending strain [5]. The improved mechanical properties of the alloy-sheathed tape are assumed to result from not only the strengthened sheath, but also some growth of the residual strain in filament [5].

Table 1. Specifications of Bi-2223 multifilamentary tapes

	Ag-Untwist	AgMg-Untwist	AgMg-Twist
Superconductor	Bi-2223	Bi-2223	Bi-2223
Matrix	Pure Ag	Pure Ag	Pure Ag
Sheath	Pure Ag	AgMg alloy	AgMg alloy
Twist pitch Lp (mm)	∞	∞	20
Je (kA/cm^2) @77K,0T	6.1	6.0	5.4

Figure 1 plots the dependence of AC losses (Q) on the applied field amplitude (B_m) for alloy- sheathed tapes untwisted and twisted with L_p=20mm, 15mm and 10mm. The solid and dotted lines in this figure denote hysteresis losses of J_e=5kA/cm^2 calculated from Bean model in electromagnetic inter-filament coupling and uncoupling, respectively [6]. The twisted tapes produced decreasing in AC losses above 3mT near the intersection of the two calculated lines. And it was confirmed that the twisted tapes had the same magnetic field dependence of J_c as that of untwisted ones. Therefore, the decrease in AC loss is considered to be due to suppression of inter-filament coupling.

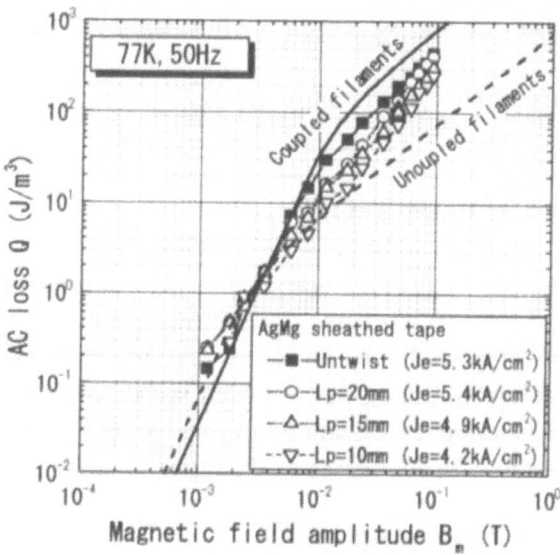

Fig. 1 AC losses of AgMg sheathed tapes

Although a tape shows low level of AC loss, the tape with low J_e is not really practical. And hysteresis loss is proportional to J_e above penetration filed. Therefore, practical AC loss suited for a power transmission cable was defined as Qf/J_e, where Q was measured at 77K, 60mT, f was 50Hz, and J_e was measured at 77K, self-field. Figure 2 shows the practical AC losses of twisted and untwisted tapes using pure silver or AgMg alloy sheaths. The silver-sheathed untwisted tapes exhibited Qf/J_e values near to 0.4mW/Am calculated in inter-filament coupling. Although some twisted tapes exhibited over 0.40mW/Am of Qf/J_e, the twisted tape with L_p=10mm achieved 0.13mW/Am of Qf/J_e.

In order to evaluate accurately the influence of twist on inter-filament coupling, a rate of

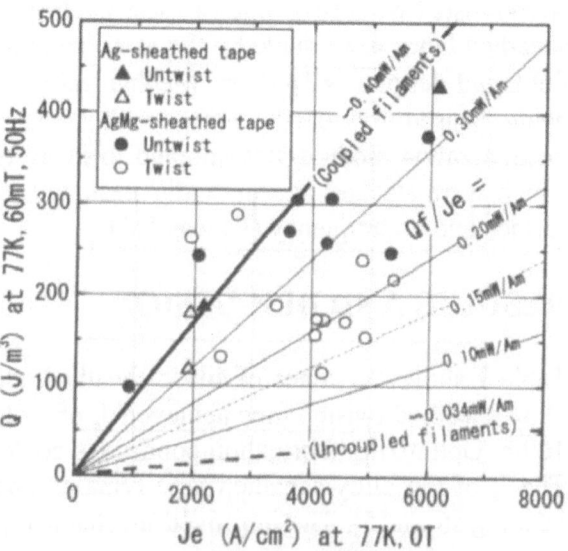

Fig. 2 Practical AC losses (Qf/Je)

inter-filament coupling (R_{fc}) was estimated from $Q=R_{fc}W_{hc}+(1-R_{fc})W_{hu}$, where W_{hc} and W_{hu} are hysteresis losses calculated in electro-magnetic inter-filament coupling and uncoupling, respectively. Figure 3 presents B_m dependence of R_{fc} for the twisted and untwisted tapes using the alloy sheath. R_{fc} tends to increase linearly with B_m. Shortening L_p leads into degrease in R_{fc}, and the twisted tape with L_p=10mm produced 22% of R_{fc} at B_m=60mT. The inter-filament coupling is considered as resulting from L_p dependent and independent factors. The dependent factor is assumed to relate to a critical length, but J_c ununiformity and mazy current paths are likely to influence the dependent factor. Also, our microscopic observation leads to the assumption that the independent factor is assigned to filament bridging.

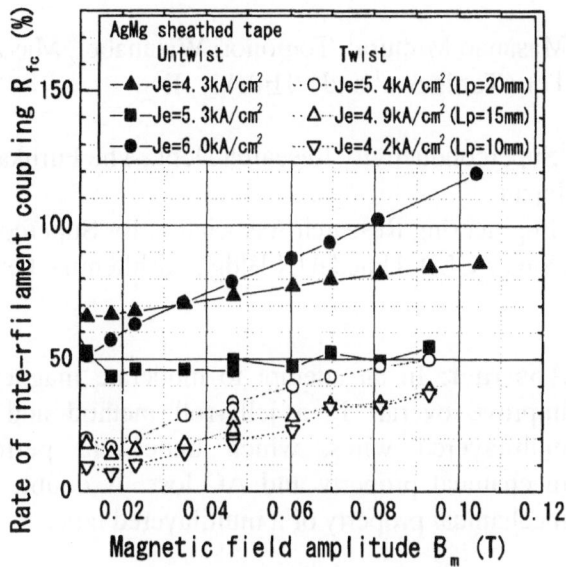

Fig.3   Rate of inter-filament coupling

CONCLUSION

The alloy-sheathed twisted multifilamentary tapes were developed in pursuit of improved mechanical properties and lessened AC losses. In the tapes, I_c remains almost unchanged till reaching 90MPa of tensile stress or 0.3% of bending strain. The tape produced 0.13mW/Am of the practical AC loss (Qf/J_e) at 77K, 60mT and 50Hz, while the silver-sheathed untwisted tape had about 0.4mW/Am of Qf/J_e. A comparison between calculated and found values leads into the supposition that the rate of inter-filament coupling in the twisted tape could have been reduced down to as low as 22%.

1. H.Kikuchi, K.Nemoto, Y.Tanaka, S.Tanaka, M.Suetsugu, T.Maeda, M.Yoshihara, H.Ishii and T.Hara.(1994) Advances in superconductivity VI, Springer-Vearlarg, Tokyo,739
2. H.Kikuchi, K.Kosugi, Y.Tanaka, S.Tanaka, M.Suetsugu, H.Ishii and T.Hara (1995) Advances in superconductivity VII, Springer-Vearlarg, Tokyo, 1199
3. M.Mimura, H.Ii, K.Kosugi, N.Uno, Y.Tanaka, H.Ishii, S.Honjyo and T.Hara (1996) 8th International Workshop on Critical Currents in Superconductors, 455
4. M.Mimura, H.Ii, A.Takagi, K.Kosugi, M.Sugimoto, A.Kimura, N.Uno, S.Mukoyama, H.Tubouti, H.Ishii, S.Honjo and Y.Iwata (1998) Advances in superconductivity X, Springer-Vearlarg, Tokyo, 747
5. M.Mimura, A.Takagi, K.Kosugi, M.Sugimoto, N.Uno, M.Ikeda, Y.Tanaka, S.Mukoyama, H.Tubouti, S.Honjo, N.Hobara and Y.Iwata (1999) Advances in superconductivity XI, Springer-Vearlarg, Tokyo, 879
6. M.Sugimoto, A.Kimura, M.Mimura, Y.Tanaka, H.Ishii, S.Honjyo, Y.Iwata (1997) Phsica C279, 225

Fabrication and Properties of Bi-Based Multilayered Wires

Masanao Mimura[1], Tomonori Watanabe[1], Masahiro Sugimoto[1], Kazutomi Miyoshi, Toru Tanigawa[1] and Hideki Ii[2]

[1]Superconductivity Research Dept., The Furukawa Electric Co., Ltd., Kiyotaki, Nikko 321-1493, Japan
[2]Engineering Research Association for Superconductive Generation Equipment and Materials (Super-GM), UmedaUN Bldg., Nishitenma, Osaka 530-0047, Japan

Abstract: In an attempt to moderate magnetic anisotropy of critical current density (J_c) and improve overall J_c, a jelly-roll method and a multi-pipe method were applied to fabricate multilayered wires, which underwent performance tests for J_c temperature dependence, mechanical property and AC loss. A double pancake coil was made on trial, considering the mechanical property of a multilayered tape.

Key Word: multilayered wire, round wire, AgMg alloy, double pancake coil, AC loss

INTRODUCTION

A multifilamentary tape by means of powder in tube (PIT) method has been under study and development in order to increase critical current density (J_c). Yet in parallel, improvement in overall J_c (J_e) and moderation in J_c anisotropy carry significant weight with its application to electric power equipment. We worked on these task points, thus contriving multilayered wires with the jelly-roll method and the multi-pipe method [1-5]. This paper describes several variant multilayered wires from the aspect of J_c temperature dependence, mechanical properties and AC loss, stretching into electrical characteristics of a double pancake coil using a tape.

EXPERIMENTAL

Multilayered wires were fabricated with the jelly-roll method in which a silver sheet and a superconducting-powder sheet rolled spirally, or the multi-pipe method in which silver pipes and superconducting powder pipes alternate coaxially [1-5]. The raw material powder was a Bi-2212 compound or Pb added Bi-2223 compound. J_c temperature dependence was measured by a four-probe method using pulse-like current for moderation in heat generation [6]. I_c decrease with bending strain was evaluated, where bending strain was applied once and relieved at room temperature before I_c measurement at 77K and self-field. AC transport loss was obtained from resistive voltage examined using a four-probe method [7]. A double pancake coil was made of a Bi-2223 multilayered tape fabricated by the multi-pipe method.

RESULTS AND DISCUSSION

Table 1 lists specifications of typical multilayered wires of round, square and tape forms. The round wires are characterized by suppression of J_c anisotropy, and the tapes feature in high J_e values due to low silver ratios.

Table 1. Specifications of typical multilayered wires

Wire name	JR round wire	MP round wire	MP square wire	MP tape wire	
Method	Jelly-roll	Multi-pipe	Multi-pipe	Multi-pipe	
Cross sectional view of multi-layered wire	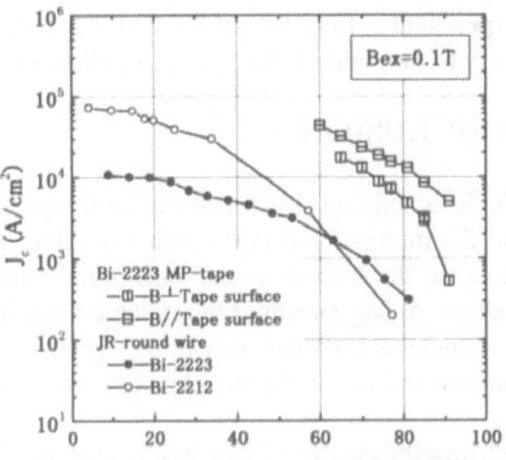				
Superconductor	Bi-2212	Bi-2223	Bi-2223	Bi-2223	
Sheath	Ag	Ag	Ag/AgMg	Ag	Ag/AgMg
Wire size (mm)	$\phi 1.5$	$\phi 1.6$	$0.84^t \times 0.84^w$	$016^t \times 2.9^w$	$0.18^t \times 0.3.9^w$
Ag/HTS ratio	4.9	1.8	2.1	1.3	2.1
I_c (A) at 0T	380 at 4.2K	22 at 77K	9.1 at 77K	43 at 77K	47 at 77K
J_e (kA/cm^2) at 0T	21.5 at 4.2K	1.1 at 77K	1.3 at 77K	9.3 at 77K	6.7 at 77K

Figure 1 plots temperature dependence of J_c at magnetic field of 0.1T, for variant multilayered wires. The Bi-2212 round wire falls short of the Bi-2223 round wire on J_c above 60K. Nevertheless, the Bi-2212 wire produces the same field dependence of J_c around 20K as a NbTi wire exhibits at 4.2K [2]. On the other hand, the Bi-2223 tape has J_c anisotropy corrective with a external magnetic field. However, the J_c of the tape exceeds that of the Bi-2223 round wire, subject to a magnetic field of 0.1T perpendicular to the tape surfaces. The J_c of the tape increases by 8% with a temperature drop of 1K around 77K.

Figure 2 shows I_c that decreases with bending strain for variant multilayered wires, where bending strains (ε_{bend}) were determined by dividing thickness of a tape and square wire, or outer diameter of a round wire by bend diameter. The Bi-2223 round wire had a bending property superior to the Bi-2212 round wire. Tape and square wires with AgMg alloy sheath have turned out to be improved in resistance against bending strain. The improvement is assumed to be assignable to an increase in residual compressive strain of superconducting layers, which arises from some difference in thermal expansion between the AgMg sheath and superconducting layers [5].

Fig. 1. Temperature dependence of J_c

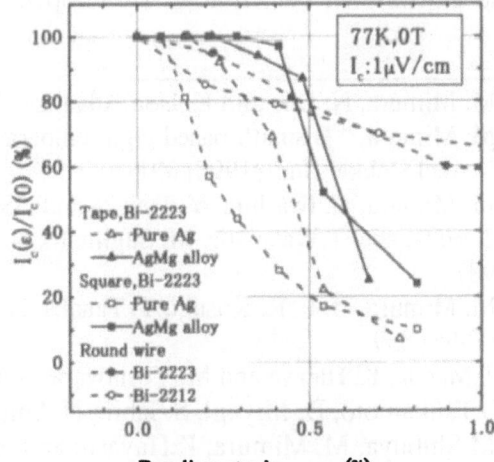

Bending strain ε_{bend} (%)

Fig.2. I_c decline due to bending strain

Figure 3 plots AC transport losses normalized by I_c for Bi-2223 multilayerd wires of round, square and tape forms. The experimental results show that the AC transport losses under self-field consists primarily of hysteresis loss, irrespective of sheath materials. The normalized losses of the tapes agreed roughly with the elliptical model of Norris [8]. On the other hand, the losses of the round and square wires were closed to the thin strip model, therefore the round and square wires are assumed to have a weak link, or intricate AC current flow.

A trial double pancake coil was made of the pure Ag-sheathed MP tape having Ic of 33A, with suppressing I_c decline to 6% by strictly controlled coil winding. The coil produced 18.3A of I_c at 77K,

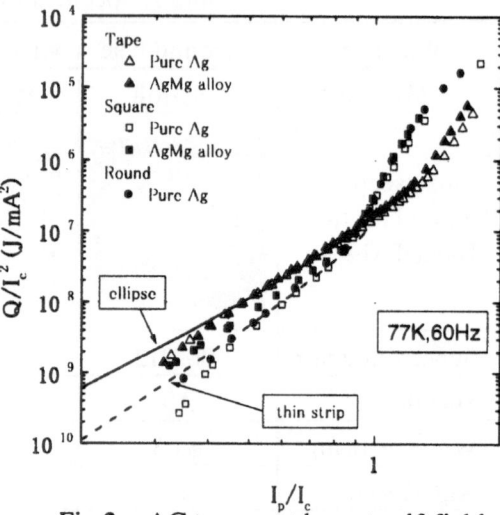

Fig.3   AC transport loss at self-field

and 42mT and 82mT of the central and maximum experienced magnetic fields, respectively, which were parallel to the tape surface. On the contrary, the maximum experienced magntic field perpendicular to the tape surface turned out to be round 25mT; therefore, it is inferred that the perpendicular field lowered the coil I_c to less than 20A. The AgMg alloy sheathed tape is more likely to moderate I_c decline by winding, enabling a coil to generate 0.1T even at 77K.

CONCLUSION

We designed and developed the multilayered round wires being free from magnetic J_c anisotropy and the multilayered tapes with low silver ratios. The tape incurs magnetic J_c anisotropy, where, however, the J_c exceeds that of a round wire, subject to a magnetic field perpendicular to the tape surface. AgMg sheathed wires have been improved in bending strain. Most of the AC transport loss under a self-field can be assigned to hysteresis loss, irrespective of sheath materials. The pancake coil of the Ag sheathed tape was made on trial, yielding central filed of 42mT at 77K.

Acknowledgment: This work was performed as a part of "R&D on Superconducting Technology for Electric Power Apparatuses" under the New Sunshine Program of Agency of Industrial Science and Technology, MITI, being consigned by New Energy and Industrial Technology Development Organization (NEDO).

1. M. Mimura, N. Uno and K. Doi, Advances in Superconductivity IV (1992) 693.
2. M. Mimura, " Bismuth-based high-temperature superconductors "ed. by H. Maeda and K. Togano, Marcel Dekker, Inc. (1996) 391.
3. M. Mimura, K. Iwashita, Y. Tanaka and N. Uno, Advances in Superconductivity VII (1995) 745.
4. T.Watanabe, T.Watanabe, M. Sugimoto, N. Uno and H. Ii, Advances in Superconductivity XI (1999) 987.
5. M. Mimura, H. Ii, K. Kosugi, T. Tanaka, N. Uno and K. Satou, Advances in Superconductivity VIII (1996) 859.
6. T.Maeda, K. Hataya and M. Yoshiwara, Advances in Superconductivity   II (1991) 149.
7. O.Tsukamoto, D. Miyagi, S. Ishii , N.Amemiya, S. Fukui, O. Kasuu, H. Ii, K. Takeda, M.Shibuya, M. Mimura, K.Hayashi and H. Yoshino, IEEE Trans. Appl. Supercond. Vol.9, No.2 (1999) 1181.
8. W.T.Norris, J.Phys. D:Appl.Phy., Vol13, (1970) 489

Development of Ag Alloy Sheathed Bi-2223 Round Wire

Yasuo Hikichi[1], Junichi Nishioka[1], Tsutomu Koizumi[1], Takayo Hasegawa[1] and Shigeo Nagaya[2]

[1]Showa Electric Wire & Cable Co., Ltd., 2-2-1, Odasakae, Kawasaki-ku, Kawasaki, Kanagawa 230-8660 Japan
[2]Chubu Electric Power Co., Ltd., 20-1, Azakitayasekiyama, Otaka-cho, Midori-ku, Nagoya, Aichi 459-8522 Japan

Abstract: We developed Ag-Mg-Sb alloy-sheathed Bi-2223 round wire with a unique design. J_c value of the round wire was 8,000A/cm^2 at 77K in self-field and had no angular dependence of backup magnetic field. The round wire had a tensile strength of 148MPa at room temperature and the J_c value remained constant up to an applied tensile stress of 140MPa. We fabricated a 1m long cable consisted with alloy-sheathed round wire and alloy round wire. The I_c value of the stranded cable was 500A at 77K in self-field.

Keywords: Bi-2223, Ag-alloy, round wire, tensile strength, stranded cable

INTRODUCTION

Recently many investigations have been carried out to determine the feasibility of practical applications using Ag-sheathed Bi-2223 wires, and many prototype model devices have been built, such as power cables, magnets, fault current limiters and transformers. Although current capacity and wire lengths have improved steadily, there are many important issues to be solved. To improve the poor mechanical strength of Ag-sheathed Bi-2223 wire, we have developed a Ag-Mg-Sb alloy with a tensile strength of 500MPa and elongation of 5.5%[1,2]. The Bi-2223 tape using the Ag-Mg-Sb alloy had a tensile strength of 230MPa at room temperature[3,4]. However, the tape shape and its angular dependence of J_c on backup magnetic field place restrictions on the design of the magnet or power cable. Therefore, there is a pressing need to develop round wire with low angular dependence of backup magnetic field. In this work we developed a Ag-Mg-Sb alloy sheathed Bi-2223 round wire with mechanical strength and no angular dependence of backup magnetic field. Moreover, we fabricated a stranded cable using the round wire.

EXPERIMENTAL

Ag-Mg-Sb alloy sheathed Bi-2223 round wire with a unique design was fabricated by using a conventional powder-in-tube technique. Powder with the stoichiometric composition of $Bi_{1.9}Pb_{0.3}Sr_{1.9}Ca_{2.1}Cu_{3.1}O_X$ was packed into a pure Ag tube and drawn into a single-filament wire. The single-filament wires were cut into pieces and bundled together into a pure Ag tube to make a 7-

filament composite wire. The 7-filament billet was drawn down to round wire and a fraction thereof rolled into a tape shape (aspect ratio = 3). The 7-firament wires and tapes were cut into pieces and stacked together into an Ag-Mg-Sb alloy tube to make a 280-filament composite round wire. The round wire was heat treated at a predetermined temperature and time. The 280-filament round wire has a novel microstructure consisting of sixteen sub-elements (four 7-filament round wires and twelve blocks with three 7-filament tapes in each block) . The tapes in neighboring blocks were arranged at right angles each other.

A multilayer 49-strand cable (7x7) was fabricated using the Bi-2223 round wires described above and Ag-Mg-Sb alloy round wires. Each of the seven 7-strand sub-cables has a Ag-Mg-Sb alloy wire at the center. The center sub-cable consists solely of a Ag-Mg-Sb alloy round wires. Each of the six outer sub-cables consists of six alloy-sheathed Bi-2223 wires around a single alloy wire. The J_c values were measured by a DC four probe method and determined using a criterion of specific resistivity of $1\mu V/cm$. Tensile stress-strain curves of the round wire were obtained using an Instron type tensile tester. J_c-tensile stress characteristics were determined by applying a range of tensile stress to the round wire and unloading them at room temperature. A transverse cross-section of the 49-strand cable is shown in Fig.1.

RESULTS AND DISCUSSION

Fig.2. shows the dependence of J_c at 77K in self-field on round wire diameter. The highest J_c value was 8,000A/cm² at 0.7mm diameter which corresponds to a filament thickness of about ~10μm. This result was similar to tape-shape wire, so we believe that a filament size of ~10μm is an optimum value.

A previous report indicated that the optimum composition for an Ag-Mg-Sb alloy sheath material was Ag-0.2wt%Mg-0.3wt%Sb, which produces a material with a tensile strength of 500MPa and elongation of 5.5%[1,2]. However, a Ag-0.1wt%Mg-0.3wt%Sb alloy with a tensile strength of

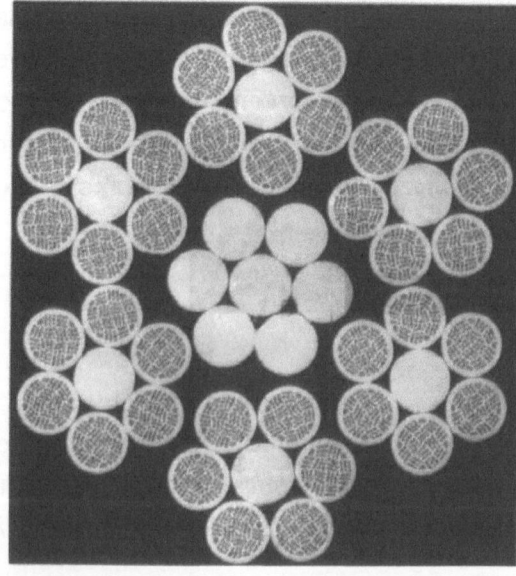

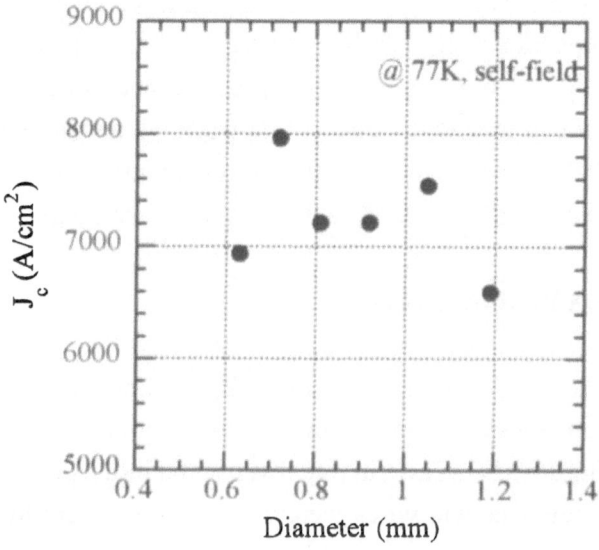

Fig. 1. A cross-section of the 49-strand cable

Fig. 2. Diameter dependence of J_c

410MPa and elongation of 11.8% was used in this study, because the round wire needs more workability. Nevertheless, the round wire had tensile strength of 148MPa and 0.2% proof stress of 120MPa at room temperature. This result was consistent with the calculated value using the 0.2% proof stress of oxidized Ag-0.1wt%Mg-0.3wt%Sb, pure Ag and Bi-2223 superconductor (filaments). The round wire kept the initial J_c value until a tensile stress of 140MPa and a tensile strain of 0.6% were applied. These results indicate that the round wire can maintain the initial J_c value until the yield point. The tensile strength tolerance was consistent with the tensile stress-strain curve.

The J_c values of round wire in magnetic fields were larger than the J_c values of tape-shape wire for the case of magnetic field applied perpendicular to the tape surface. In a 100gauss backup magnetic field the J_c value decreased to 40% of self-field values. We believe that grain alignment and linking of the Bi-2223 superconducting crystal grains need to be improved in further optimization of the preparation procedures. In order to obtain no angular dependence of backup magnetic field we designed a novel round wire with 7-filament tapes in neighboring blocks arranged at right angles to each other. Consequently, the round wire did not have any angular dependence of backup magnetic field.

We fabricated a 1m long cable twisted seven of the cables shown in Fig. 1 and tested it in liquid N_2. The final cable consisted with 252 of strand wire and 91 of AgMgSb round wire. The twist pitch of the final cable was 400mm and outer diameter of the final cable was 24mm ϕ. This final cable was heat treated after cabling and assembling terminals at the optimized temperature through the experiment for deciding the strand diameter. I_c value of the cable was 500A, which was lower than the summation value of I_c of strand wire. Shifting the optimized heat treat temperature because of the larger mass of the cable might cause such low I_c value.

CONCLUSION

We designed and developed a Ag alloy sheathed Bi-2223 round wire. This round wire had no angular dependence of backup magnetic field. The J_c value of the round wire was 8,000A/cm^2 at 77K in self-field. The J_c value remained constant up to an applied tensile stress of 140MPa and a tensile strain of 0.6%. A stranded cable was fabricated using this round wire and Ag-Mg-Sb alloy wire. The I_c value of the stranded cable was 500A at 77K in self-field.

REFERENCE

1. Y. Hikichi, T. Koizumi, J. Nishioka, T. Hirota and T. Hasegawa, Proc. of the 59th Meeting of Cryogenics and Superconductivity, 1998
2. Y. Hikichi, T. Koizumi, T. Hirota and T. Hasegawa, Proc. of the 1998 International Workshop on Superconductivity, 1998
3. Y. Hikichi, T. Koizumi, T. Hirota and T. Hasegawa, Proc. of the IEEE Trans. Mag. Technol., 1998
4. Y. Hikichi, J. Nishioka, T. Koizumi, T. Hirota and T. Hasegawa, Proc. of the 11th International Symposium on Superconductivity, 1998

Critical Current Properties and Scaling of Bi(2223)/Ag Multifilamentary Tapes at Liquid Nitrogen Temperatures

Koshichi Noto[1], Toshimi Chiba[2], Yasuo Nagai[3], Takashi Saitoh[4], Satoshi Awaji[5], Kazuo Watanabe[5], Norio Kobayashi[5] and Kaoru Yamafuji[6]

[1]Faculty of Engineering, Iwate University, Ueda 4-3-5, Morioka 020-8551, Japan
[2]Advanced Science and Technology Institute of Iwate, 3-35-2, Iioko-shinden, Morioka 020-0852, Japan
[3]Incs Ltd., 3-20-2, Nishishinjyuku, Shinjyuku, Tokyo 163-1452, Japan
[4]Fujikura Ltd., Kiba 1-5-1, Tokyo 135-8512, Japan
[5]Institute for Materials Research, Tohoku University, Sendai 980-8557, Japan
[6]Ariake National College of Technology, Ohmuta 836-8585, Japan

Abstract: Critical current properties and the scaling of the global pinning force Fp have been studied for Bi(2223)/Ag multifilamentary tapes at liquid nitrogen temperatures from 65 to 77K. It turned out that,
(1) Magnetic field dependence of Ic at low field is very strong and Ic decreases rapidly with small field application, and
(2) The global pinning force Fp shows a scaling with a peak near B/Birr ≈ 0.25 using extrapolated irreversibility fields, B_{irr}.

Keywords: High T_c, Bi(2223)/Ag Multifilamentary Tape, Critical Current, Scaling

INTRODUCTION

Recently, developmental researches on high T_c oxide superconductors are so progressed that the performance of them is almost near the practical level. Especially, critical current density characteristics of Bi(2223)/Ag tapes at liquid nitrogen temperatues (65K<T≤77K) have been progressed up to almost practical level.

We have begun to investigate superconducting properties which are important for practical applications, such as irreversibility field, B_{irr}, critical current (density), $I_c(J_c)$, and AC losses. In this paper, we report experimental results of $I_c(B)$ and scaling properties of global pinning force F_p in Bi(2223)/Ag multifilamentary tapes.

EXPERIMENTAL

Samples were prepared by Fujikura Ltd., main parameters of which are shown in Table 1. $I_c(J_c)$ was measured for a sample by transport method using a four terminal method in a liquid nitro-

Table 1 Main parameters of the samples.

Item	
SC Material	Bi(2223)
Matrix	Ag
Core number	37
Ag/SC	3.8
Thickness (mm)	0.2
Width (mm)	4.5

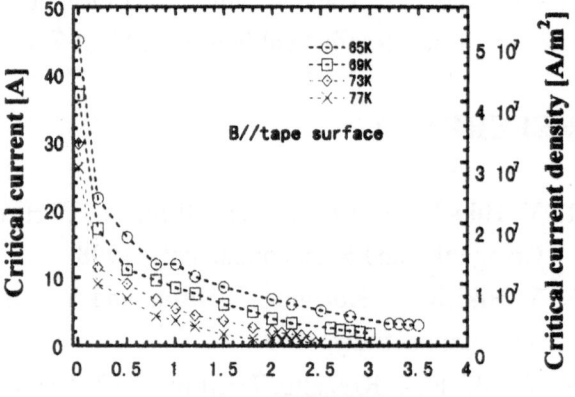

Fig. 1 Magnetic field dependence of Critical current (density) in a Bi(2223)/Ag tape sample.

670

gen bath from 65K to 77.3K. Employed criterion is $E_c=1\mu V/cm$. A 52mm bore, 5T conduction cooled superconducting magnet was used for this measurements. Transport B_{irr} was measured for other three samples by four terminal method using a criterion of $E_c=1\mu V/cm$ with a transport current of 50mA as a point of transport $J_c \cong 0$. An adiabatic cryostat and the 15T-SM superconducting magnet at HFLSM, IMR, Tohoku University were used for the B_{irr} measurements.

RESULTS AND DISCUSSION

The results of I_c measurements are shown in Fig. 1. As can be seen in this figure, $I_c(J_c)$ at B=0T increases with decreasing temperature. However, $I_c(J_c)$ decreases very rapidly with magnetic field application at low field. Fig. 2 shows the magnetic field dependence of global pinning force $F_p=Jc \times B$. As can be seen, we could not measure up to B_{irr}, where J_c becomes zero since the maximum field of our magnet is 5T. Figs. 3 and 4 show preliminary data of B_{irr} (T) and F_p/F_p^{max} vs B/B_{irr}, respectively, on other three tape samples[1]. One of the author(K.Y.) investigated theoretically the temperature dependence of B_{irr}. Although we found B_{irr} can be expressed as $B_{irr} = a(T_c-T)^b$, he can not make clear the relation between our transport B_{irr} and experimental B_{GL}[2] or theoretical prediction on B_{irr}[3, 4]. Although there is a significant scatter in Fig. 4, we can see F_p/F_p^{max} follows the following expression,

$$F_p/F_p^{max} = a \cdot \frac{B}{B_{irr}}(1-\frac{B}{B_{irr}})^3, \qquad (1)$$

which has a peak value at $B/B_{irr} \cong 0.25$. Therefore, we tried parameter fitting for the present data shown in Fig. 2 to the expression (1). The results of parameter fitting are shown in Fig. 5, and thus estimated B_{irr} is listed in Table 2. The estimated value of B_{irr} is also plotted in Fig. 3. Fig. 6 shows a plot of F_p/F_p^{max} vs. B/B_{irr} using thus estimated B_{irr}. We can see fairly good scaling characteristics.

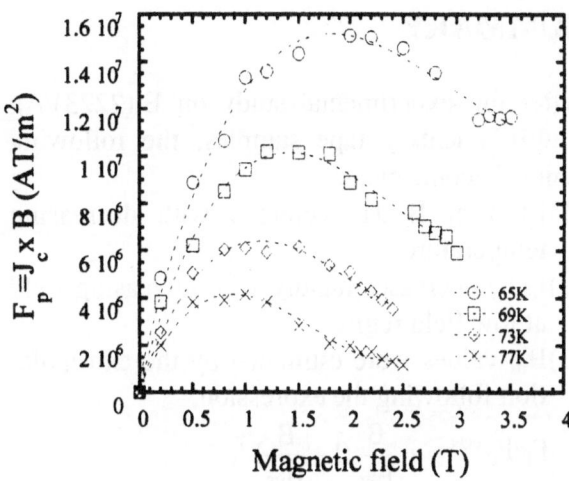

Fig. 2 Magnetic field dependence of global pinning force in Bi(2223)/Ag tape sample.

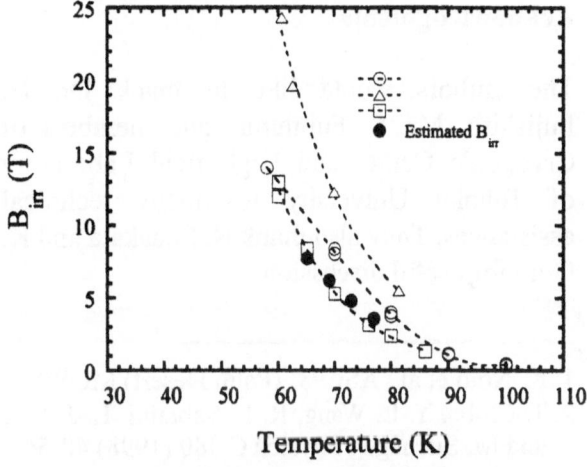

Fig. 3 Temperature dependence of B_{irr} in Bi(2223)/Ag tape sample.

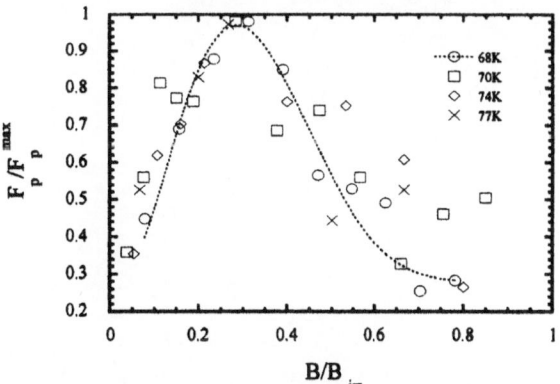

Fig. 4 F_p/F_p^{max} vs. B/B_{irr} plot for one other Bi(2223)/Ag tape sample. (previous measurement)

672

SUMMARY

After the experimental study on Bi(2223)/Ag multifilamentary tape samples, the following points become clear.

(1) $I_c(J_c)$ at B=0T increases with decreasing temperature.

(2) $I_c(J_c)$ decreases rapidly with increasing field at low field region.

(3) B_{irr} values were estimated by the extrapolation following the expression,

$$F_p/F_p{}^{max} = a \cdot \frac{B}{B_{irr}}(1-\frac{B}{B_{irr}})^3,$$

which was supposed in the previous study on F_p

(4) We can see fairly good scaling characteristics in this expression.

Acknowledgments

The authors would like to thank Dr. H. Fujishiro, Mr. S. Fujinuma, and members of Cryogenic Center and High field Laboratory of Tohoku University for many technical assistances. They also thank N. Sadakata and K. Goto for useful discussion.

1. K. Noto et al ; ASC98' (Palm Desert) MOB03
2. T. Chiba, Y.-L. Wang, R. L. Sabatini, L.-J. Wu, and M. Suenaga ; Physica C 380 (1998) 40-54
3. K. Yamafuji ; private communication.
4. K. Yamafuji and T. Kiss ; Physica C 290 (1997) 9-22

Table 2 Estimated B_{irr} values from parameter fitting.

Temperature (K)	B_{irr} (estimated) (T)
65	7.72
69	6.20
73	4.80
77	3.59

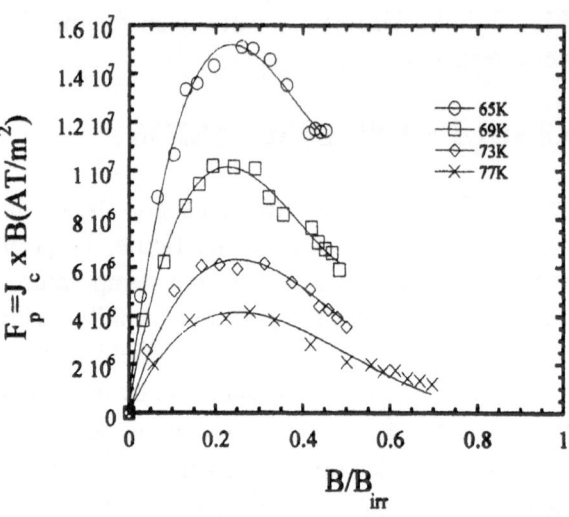

Fig. 5 B/B_{irr} dependence of F_p for a Bi(2223)/Ag tape sample.

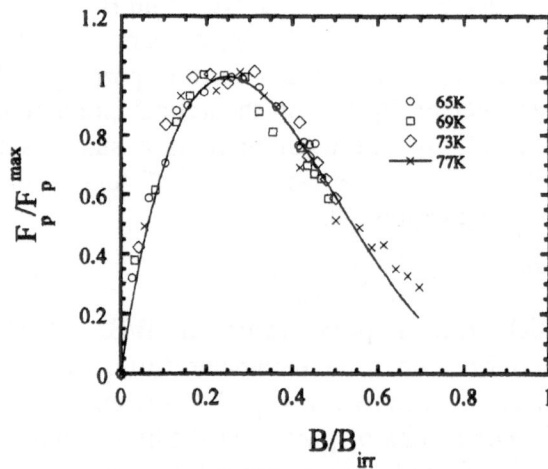

Fig. 6 $F_p/F_p{}^{max}$ vs. B/B_{irr} plot for a Bi(2223)/Ag tape sample. (present measurement)

REMANENT MAGNETIC FIELD DISTRIBUTION IN Bi-2223/Ag SUPERCONDUCTING TAPE

A.Thamizhavel, M.Sugano, N.Fukuda*, A.Sakai* and K. Osamura
Department of Materials Science and Engineering
* Mesoscopic Materials Research Center,
Faculty of Engineering, Kyoto University, Sakyo-ku, Kyoto 606 8501, Japan

Abstract: The magnetic-flux distribution inside an Ag-sheathed $(Bi,Pb)_2Sr_2Ca_2Cu_3O_x$ mono-filamentary tape is measured using Scanning Hall Probe at the liquid nitrogen temperature from the z component of the magnetic field above its surface. The spatial resolution of the imaging is 100 μm. The two dimensional self field distribution for a superconducting tape with Ic = 10.2 A is studied in a remanent state after the removal of external magnetic field normal to the tape surface. The trapped field is found to be maximum at the core of the tape and the field distribution is not uniform throughout the tape. This non-uniformity in the field distribution indicates the existence of the defects or imperfections in the oxide core which might have introduced during the tape fabrication. The spatial distribution of the magnetic field in the remanent state is related to the internal current density through the Maxwell's Equation.

Keywords: Bi-2223/Ag Tape, scanning Hall probe, remanent state

INTRODUCTION

Magnetic relaxation in high temperature superconductors is important for several reasons. With its complex dependence on temperature, field and other parameters the study of magnetic relaxation will lead to the broader understanding of the magnetic phase diagram and pinning mechanisms and to an improved determination of the thermodynamic properties of high temperature superconductors. This necessitated to study the magnetic relaxation inside a superconductor in the recent years. In order to understand the magnetic relaxation some detailed microscopic, spatially resolved probe of magnetic relaxation is necessary. Modern experimental techniques, in particular Scanning Hall Probe (SHPM) measurements [1,2] and magneto-optical (MO)[3,4] allow us to study the magnetic relaxation with rather high spatial resolution. Since in the remanent state, or at low applied magnetic fields H, any magnetic relaxation measurement integrates the contributions of regions with significant variations in properties we have studied the remanent state magnetic properties of Bi-2223/Ag tape using a scanning Hall probe microscope.

EXPERIMENTAL PROCEDURE

Tape samples of Ag-sheathed Bi-2223 have been prepared from the sol-gel derived powders using the conventional powder-in-tube (PIT) technique. The texture created by this process tends to align the c-axis of the Bi-2223 grains perpendicular to the broad face of the tape. The tape samples were cut to 20 mm length and sintered for 150 hours at 1103 K. The precise details of the fabrication process have been described elsewhere [5]. The width and thickness of the tape were 3.3 mm and 140 μm respectively and the thickness of the oxide core was 90 μm. The critical current measurement was performed at 77K with criterion of 0.1 mV/m. The transport critical current density (J_c^t) was defined as $J_c^t = I_c/S$, where the critical current (I_c) divided by the cross sectional area (S) of oxide layer. The I_c value of the tape used in the present study was 10.2 A with a Jc of approximately 10.6 KA/cm^2.

A commercial Hall IC (Toshiba THS 118) is used as the field sensing element. The size of the sensing area of this Hall IC is approximately 0.1 mm x 0.1mm which determines the spatial resolution of the Hall-probe imaging. The Hall voltage is measured with a digital voltmeter at a constant probe current of 5 mA. The Hall sensor is mounted on a pulsed *x-y* stage along with *z* axis movement also to scan along the *x* and *y* directions of the sample. The scanning area is 10 mm x 10 mm. Although the stage has a 500 nm resolution, coarse scanning with 0.1mm per step was used in all measurements considering the spatial resolution of the Hall sensor. The external magnetic field is generated by a small electro magnet. The experiment was performed by first cooling the tape specimen to the liquid nitrogen temperature. The magnetic field was applied normally to the tape surface for a predetermined time and removed off. The residual magnetic field distribution is observed under zero external field by raster scanning the hall sensor along the tape surface.

RESULTS AND DISCUSSION

Fig.1 shows the two dimensional distribution of flux density over the sample surface with a gray scale. A magnetic field of 60 G was applied perpendicular to the Bi-2223/Ag the field was removed before the sample was scanned. As can be seen from the Fig.1 the maximum amount of trapped field is at the center of the tape. The maximum value of trapped field was just above 12 G. There are no notable modulations, induced by cracks, disorders and other imperfections .

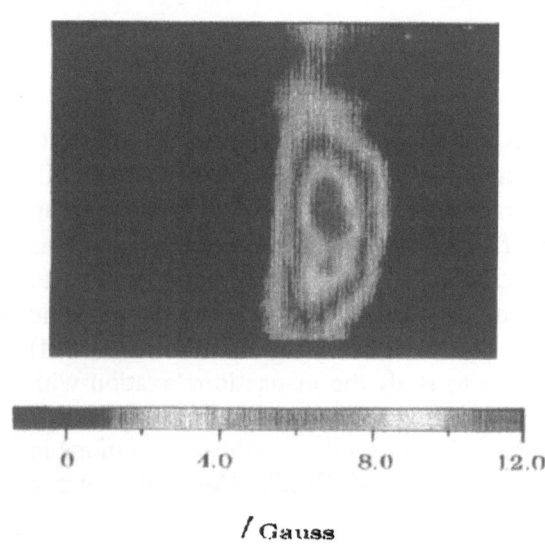

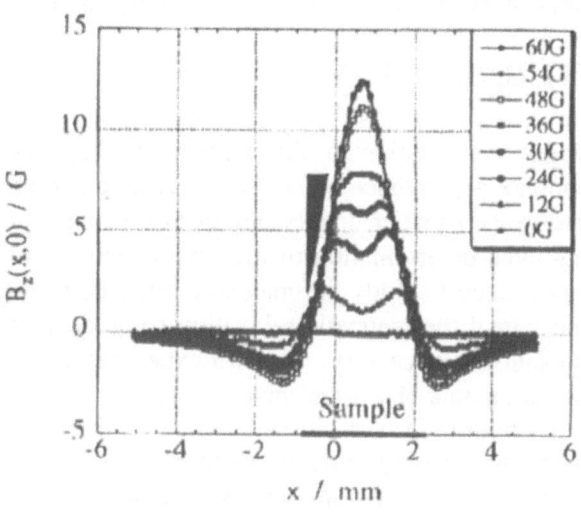

Fig.1 Two dimensional distribution of remanent magnetic flux

Fig.2 One dimensional profile of the trapped flux for different applied magnetic fields

Figure 2 shows the one dimensional profile of residual magnetic flux density at the center of the tape (y = 0) after the magnetic field was applied to a given strength indicated here. The critical state is realized above the external field of 54 G for the specimen because both data of 54 G and 60 G are overlapped. Then 2B* is considered as 54G. The observed peak of residual magnetic 12 G which is lower than B*. This discrepancy is at present not clear. Presumably it might be explained in terms of flux creep or inhomogeneous magnetic flux distribution.

In order to estimate the observed self magnetic field a transport current of 10 A is applied. When the magnetic field is induced by the current in the superconductor, the relation between the magnetic field B and the current density J is expressed by Biot-Savart Law. Here we adopt the method by Roth et al [6]. When the sample is a two dimensional thin plate the z component of magnetic field Bz (x,y,z) at the position z apart from the surface is calculated by integrating the two dimensional current distribution J(x,y). We can get the relation holding between both Fourier transforms of Bz (x,y,z) and J(x,y). Using the Fourier transform of the Maxwell equation ∇•Jc=0 the Fourier components of the current density $j_x(k_x,k_y)$, $j_y(k_x,k_y)$ are obtained as a function of $b_z (k_x, k_y, z)$. From the observed

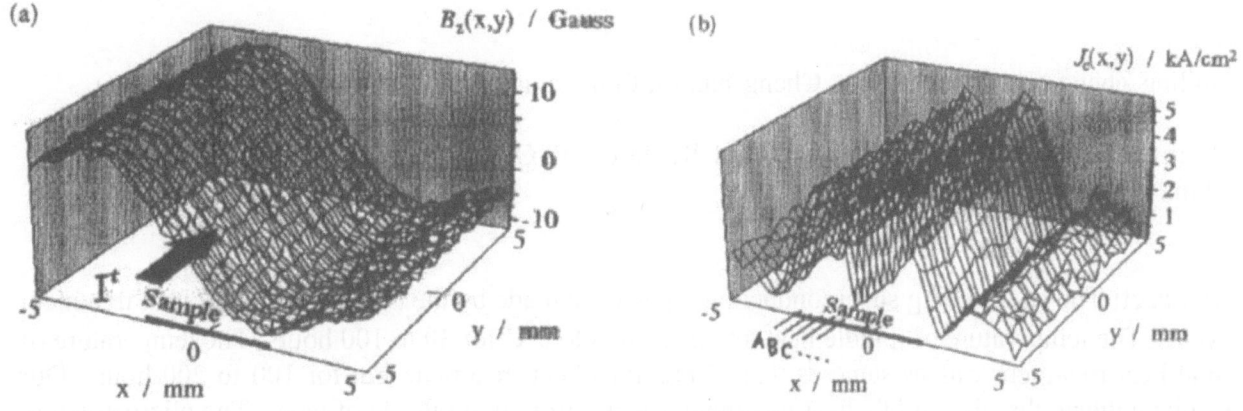

Fig.3 Two dimensional profile of (a) observed self magnetic field and (b) estimated current density

magnetic field B_z (K_x, k_y, z) the Fourier components $j_x(k_x,k_y)$ and $j_y(k_x,k_y)$ are obtained. Finally the current density distribution $J_x(x,y)$ and $J_y(x,y)$ can be analytically solved by the inverse Fourier conversion. Then the current density is defined as $J(x,y) = [J_x(x,y)^2 + J_y(x,y)^2]^{1/2}$. For a transport current of 10A, the observed self magnetic field distribution and the corresponding current density profile are shown in fig.3. The magnetic field distribution gives typically a maximum / minimum behaviour along x direction. The distribution along y direction looks to change irregularly place to place.

CONCLUSION

The spatial distribution of the trapped magnetic field inside the Ag-sheathed Bi-2223 tapes in their remanent state has been studied with scanning Hall probe microscopy. When the field is applied normally to the tape surface, the field is found to penetrate inside the tape and to give an appreciable amount of trapped field in the remanent state even when the applied field is low. The transport current re-distributed in the specimen. The critical state is realized for a field of above 54 G. The observed residual field is much less than this field and the discrepancy may be due to the inhomogenous magnetic flux distribution inside the tape.

REFERENCES

1. A. Tanihata, A. Sakai, M.Matsui, N.Nonaka and K. Osamura, Supercond. Sci. Technol. Vol.9 (1996), pp.1055-1059

2. Kozo Osamura, Ken-ichi Matsuno, Hideki Itoh and Akira Sakai, Materials Transactions, JIM, vol.38 No.8 (1997), pp.737-742

3. A.E. Pashitskii, A.Polyanskii, A.Gurevich, J.A.Parrell and D.C.Larbalestier Physica C, Vol.246 (1995) 133.

4. M.E.Gaevski, A.V.Bobyl, D.V.Shantsev, Y.M.Galperin, T.H. Johansen, M.Bazilijevich, H.Bratsberg, S.F.Karmanenko, Phys. Rev. B. Vol. 59 (1999) pp.9655-9664

5. S.Nonaka and K.Osamura, Physica C Vol. 281 (1997) pp.201-205.

6. B.J.Roth, N.G.Sepulveda and J.P.Wilswo Jr.: J. Appl. Phys. 65 (1989) pp.361-366

The Effects of Heat Treatments on Critical Current Density and Microstructure of Bi-2223/Ag Tapes

Huiling.Zheng, Zhenzhong Duan, Chengshan Li, Pingxiang Zhang, Yong Feng, and Lian Zhou

Northwest Institute For Nonferrous Metal Research, P. O. Box 51, Xi'an Shaanxi 710016, P. R. China

Abstract: The Bi-2223/Ag superconducting tapes were made by the two cold rolling / heat treatment cycles. The temperature of middle heat treatment was 835°C for 30 to 100 hours. The temperature of final heat treatment was as same as that of previous heat treatment, but for 100 to 200 hours. Our results indicate that the middle heat treatment affects strongly on the Jc of tapes. The microstructure of Bi-2223/Ag tape with different Jc values has been studied by SEM. The chemical composition of each phase has been determined by EDX in the SEM. The SEM analysis reveals larger size of 2223 grains and better connection between 2223 grains in the best tape, and more inclusions and more defects in the tapes with lower Jc value. The relationships among the heat treatment conditions, the microstructure and the transport properties of the tapes have been discussed in this work.

Keywords: heat treatment, microstructure, critical current density.

INTRODUCTION

Silver-sheathed Bi-2223 tape is one of the most important superconductors used at liquid nitrogen temperature. Some prototypes have been made from Bi-2223 tapes. A HTSC cable with capacity of 1KA was successfully constructed by using Bi-2223 tapes in China in 1998[1]. However, regarding the critical current densities in the superconducting state, the best tapes currently produced yield only about 2% of the critical current density of single crystalline thin film. The reduction of the critical current density from single crystalline films to highly textured polycrystalline material is about one order of magnitude [2]. This implies that a substantial in crease in the critical current density of the Bi(Pb)-2223 tapes would be possible if the microstructure of the tape could be improved. In this work, the microstructure of the tape was improved by heat treatment under the same conditions of rolling process.

EXPERIMENTAL DETAILS

Silver-sheathed Bi-2223/Ag tapes were produced by the powder-in-tube (PIT) method. The nominal composition of precursor powders was $Bi_{1.73}Pb_{0.34}Sr_{1.93}Ca_{2.02}Cu_{3.1}O_x$. The outer and inner diameters of the tubes used for this work were 5.2 and 4 mm, respectively. Tapes 0.30mm in thickness were obtained from wires 1.6mm in diameter by rolling. To investigate the effects of the heat treatments on the formation and microstructure of 2223 phase in the tapes, three different samples (A, B and C) were made by middle heat treatments (HT1) [3]. The Bi-2223/Ag superconducting tapes with 0.15mm in thickness were obtained by sintering at 835°C for 100~200h (HT2) after rolling. The Ic values of the final tape were measured with the standard four-probe technique with a criterion of 1 μ V/cm at 77K. Microstructure observations were made with a scanning electron microscope (SEM)(JSM5800). The local compositions of the various phases were quantified by using an energy-dispersive x-ray (EDX) detector.

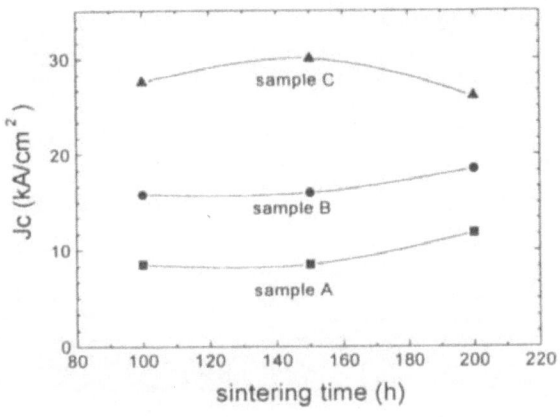

Fig. 1 shows the effects of the final sintering time on the Jc of Bi-2223/Ag tapes.

Fig. 2 is the SEM photo of sample C after first sintering (HT1) and rolling.

RESULTS AND DISCUSSIONS

Figure 1 shows the effect of the final heat treatment time on the critical current density of samples. Under the same conditions of cold-working processes, Jc values show a larger difference for different samples, which resulted from the parameters of middle heat treatments (HT1). The best Jc for different samples are dependent on the final sintering time and the conditions of middle heat treatment (HT1). The best Jc is obtained in sample C with the schedule of sintering at 835°C for 150 hours. The content and the texture degree of 2223 phase in samples A and C were studied in other work [3].

Figure 2 shows the back-scattering SEM of cross-section of sample C after middle heat treatment (HT1) and rolling. There are some kinds of phases: black, grey, pale and white. The grey particles are the Bi-2223 phase. The pale long narrow pieces are Bi-2212 phase, which likes "sandwich" between Bi-2223. The black particles are $(Ca,Sr)_2CuO_3$, $(Ca,Sr)_{14}Cu_{24}Ox$ or CuO. The white particles are mainly $(Sr,Ca)_2PbO_4$ in shapes of rectangle, triangle and circle. The 2223 phase, a number of other phases and cracks are presented as shown in Figure 2. 2223 phase had been aligned along the direction of a-b plan by rolling.

Under the same conditions of cold rolling, the characteristics of microstructure for samples A and C just caused by different schedules of heat treatments performed on tapes. Figure 3 and Figure 4 show the back-scattering SEM photos of cross-section of the final tapes A and C respectively. 2223 grains are aligned along the direction of rolling plan and connected closely each other in tape C. Size of 2223 grain in sample C is larger than that in sample A, it is about 0.6~0.9μm in thickness. For sample A, thin pieces of 2223 are dispersed over whole area of core. One 2223 layer thick is about 0.3~0.6μm. Large domains (about 50~100μm in thickness, consist of 2223 layer crystallizes) are formed in sample C. Some second phases exist in tape C, although it's Jc is up to 30000A/cm2. It indicates that the phase conversion from the 2212 to 2223 was not completely yet for tape C. $(Sr,Ca)_2PbO_4$ phases lie by big black area and lamellae of the Bi(Pb)-2212 phases (white layer) between 2223 grains in sample C are less than that in sample A. Both of the samples A and C still contain significant fractions of second phase and the crystallites contain lamellae of the Bi(Pb)-2212 phase. The significant microstructure distinguishes between sample A and sample C are summarized as following points:

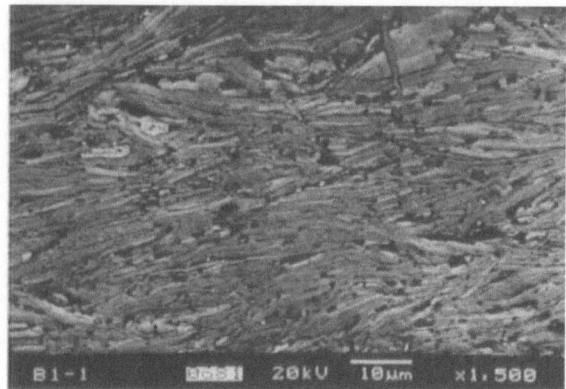 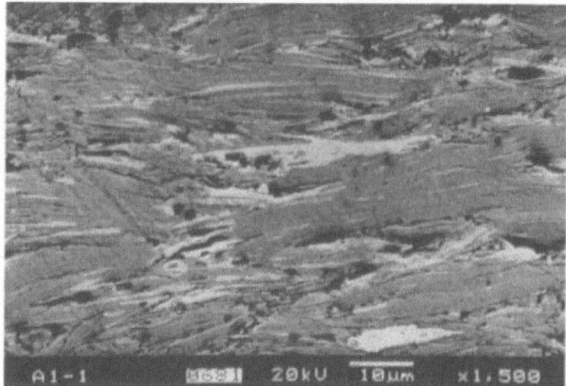

Figure 3 shows the SEM photo of sample A after the final heat treatment.

Figure 4 shows the SEM photo of sample C after final heat treatment.

(i)   Size of 2223 grain in sample C is larger than that in sample A.
(ii) Large domains have been formed of 2223 layer crystallizes in sample C, but any 2223 domains have not been formed in sample A.
(iii) Oxide core of sample C is more density than that of sample A, and more voids are in sample C.
(iv) The connection among 2223 grains in sample C is better than that in sample A.

In fact, to sample C, 2223 domains had been formed during the middle heat treatment, it can been seen from figure 2, in which more 2212 flats are clipped between 2223 grains because that the formation of 2223 phase is not complete. After middle rolling, 2223 domains were tight and aligned well, meantime some bulks of 2223 and 2212 were broken by the compression force come from rolls. However, any 2223 domains have not been formed after final heat treatments in sample A because of the short sintering time [3]. It indicates that the schedules of the middle heat treatments (HT1) affect strongly on the microstructure and the performance of tapes. On the other word, it is important to produce the precursor tapes. The purpose of the final heat treatment (HT2) is that 2223 phase should be formed further and the connections among crystals can be more closely. Meantime, the cracks come from rolling can be sealed. It is well known that 2223 phase is formed by reaction between 2212 phase and secondary phases. The formation of the 2223 phase is accompanied by a decrease in the amount of secondary phases. For high Jc value of tape, the 2223 domains must be obtained by the middle heat treatment (HT1) firstly. If there were only one large 2223 domain filled in a whole tape, the highest Jc value could be obtained surely.

CONCLUSION

We have investigated the effects of heat treatments on the microstructure and the critical current density of Bi-2223/Ag tapes. Our results indicate that the middle heat treatments (HT1) affects strongly on Jc of tapes. The final heat treatments can make 2223 grains forming completely, at the same time the cracks caused by cold rolling can be sealed. SEM analysis shows that large 2223 domains were formed in tape with high Jc values. To contrast, the flat of 2223 is thin and the distribution of 2223 is in disperse state in the tape with low Jc value, accompanying more impurity phases.

1 L Zhou, The Fifth IUMRS International Conference on Advanced Materials (IUMRS-ICAM'99)
2 O Eibl, Supercond. Sci. Technol. 8(1995) 833-861
3.H.L.Zheng et at published at ICMC'99 conference.

Effect of Intermediate Melt-Solidification Process on the Transport Property of 2223-BPSCCO/Ag Tape

May On Tjia, Darminto, Markus Diantoro, Andrivo Rusydi and Waloejo Loeksmanto

Physics Department, Bandung Institute of Technology, Jl. Ganesa 10, Bandung 40132, Indonesia.

Abstract. An experiment has been carried out to include melt-solidification process prior to the final annealing in the fabrication of $Bi_{1.8}Pb_{0.5}Sr_2Ca_2Cu_3O_x$/Ag tape by powder-in-tube (PIT) method. Measurement by 4-probe method using $1\mu V/cm$ criterion shows a remarkable enhancement of critical current density (J_c). The highest reproducible critical current densities of about 4×10^4 A/cm^2 at 77 K is obtained from the sample melted at 866°C for 5 minutes and annealed for 200 hours at 840°C. Magnetization measurement performed on the samples were analyzed by the method of Wiesinger-Sauerzopf-Weber yielding estimated J_c down to 5 K and a description of the B-dependency of J_c.

Keywords: BPSCCO-2223/Ag tape, powder in tube, melt-solidification, critical current density.

INTRODUCTION

The results of current researches on the fabrication of BSCCO/Ag composite tape by means of powder in tube (PIT) method have yielded typical values of highest critical current density of $J_c(77K,0T) \sim 10^4\text{-}10^5$ A/cm^2 and $J_c(5 K,0T) \sim 10^5\text{-}10^6$ A/cm^2 [1]. The Bi based materials employed for the tapes include both 2212 and 2223 phases incorporating Pb doping in most cases. Despite the many advances made on the studies of various practical aspects associated with wire applications [2,3], some basic problems remain to be tackled, namely those concerning the J_c-limiting factors such as material density or grain connectivity in the tape and the vortex pinning strength [1]. This study is an extension and modification of a previous experiment [4] on the fabrication of $Bi_{1.8}Pb_{0.5}Sr_2Ca_2Cu_3O_x$/Ag tape by means of PIT method with intermediate melting and post annealing processes. We have deleted in this experiment the extra rolling applied to the tape prior to its final annealing, while extended the range of post annealing time.

EXPERIMENT

The tape was fabricated by means of standard PIT method with the superconductor powder prepared by standard solid state reaction from an initial mixture of Bi_2O_3, PbO, $SrCO_3$, $CaCO_3$ and CuO and in the cation molar ratio of 1.8:0.5:2:2:3. A silver tube of 150 mm long, 3 mm inner diameter and 5 mm outer diameter, was packed with the superconductor powder at 10% filling by weight. The tube was repeatedly drawn until the outer diameter was reduced to about 2.3 mm. Subsequent multi-pass rolling and thermal annealing processes converted the tube into a tape of 1 m long, 0.15 mm thick and 3.0 mm wide. Using the material filling factor given above, the actual cross-sectional area containing the superconducting material was roughly estimated to be 0.045 mm^2. An intermediate melting was then introduced at 866 °C with different melting times of 5, 10 and 15 minutes on separate samples, followed by post annealing process at 840 °C for 80, 120, 160, 180, 200 and 230 hours. The critical current densities were measured at 77 K by the four-probe method with the measuring (voltage) electrodes spaced at a distance of 1 cm. Magnetization data were obtained by a commercial Quantum Design MPMS-5s SQUID Magnetometer and analyzed by the method of Wiesinger-Sauerzopf-Weber (WSW) [5] to yield the estimated values of J_c at 5 K and 77 K as well as their B-field dependencies.

RESULT AND DISCUSSION

The result of 4-probe J_c measurement at 77 K in zero external field is presented in Fig. 1 as functions of annealing time for samples treated with various intermediate melting times. We note that the sample treated with 5 minutes of intermediate melting exhibits dramatic enhancement of J_c after 120 hours of post annealing process, similar to our previous report [4], which tapers off for $t_s > 160$ hrs. The highest J_c attained for this sample after undergoing 200 hours of annealing is about 4×10^4 A/cm^2, corresponding to the most reproducible critical current of 18 A. Excessive annealing, on the other hand, results in degradation of J_c.

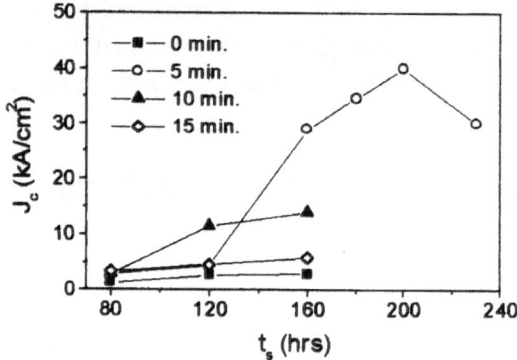

Fig. 1. Critical current densities as functions of post annealing time (t_s) on samples undergoing different intermediate melting times.

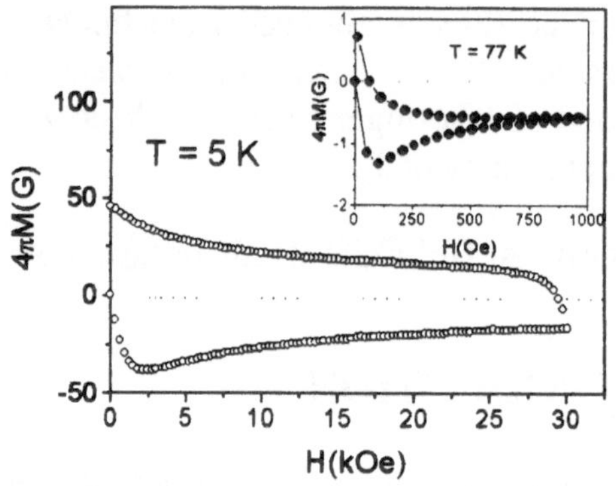

Fig. 2. Magnetization curves at 5 K and 77 K for sample prepared with 5 minutes of melting and 200 hours of annealing.

The result of magnetization measurement is given in Fig. 2 for the sample prepared with 5-minute melting and 200 hours of post annealing. It is evident from this figure and its inset that a much more effective pinning is operative at lower temperature, and hence a much higher J_c is expected at 5 K. Similar measurements were performed on the samples treated with various lengths of post annealing process.

For the analysis of magnetic data in the case where the external magnetic field is aligned parallel to the tape surface (// ab-plane), the out-of-plane J_c can be neglected compared to the in-plane critical current density (henceforth simply denoted by J_c). Consequently, we can simply write

$$J_c(B) = \frac{4}{a} \frac{|m_{irr}(B)|}{V} \qquad (1)$$

where V (= 0.85 mm^3) is the volume of the sample after being corrected with the filling factor, and a (= 6.05 mm) is the sample length perpendicular to H. Following WSW, the B-field is expressed in terms of the external H field by a transformation formula involving the field-correction function C_B which depends on the normalized applied field and the demagnetization factor D (neglected in this case due to the relatively large sample dimension parallel to H). The results of this analysis are displayed in Fig. 3a and 3b which describe the B-dependencies of J_c at 77 K and 5 K respectively. We note that J_c's at 5 K are generally much better sustained in the presence of external magnetic field than those at 77 K, which diminish to insignificant values in external magnetic field approaching 1 kG. This is apparently due to weakening of the flux pinning effect at higher temperatures. It is consistent with the more pronounced B-dependent behavior of J_c and greater differences among the differently treated samples at lower fields. It is found that both figures yield similar trends of $J_c(t_s)$ as that shown by Fig. 1, reaching the highest values of $J_c(77K,0T) \sim 6 \times 10^4$ A/cm^2 and $J_c(5K,0T) \sim 5.6 \times 10^6$ A/cm^2 for $t_s = 200$ hours. Although the numerical values are not of the same quality and significance as the transport data, the unusually large $J_c(5K,0T)/J_c(77K,0T)$ ratio (nearly two order of magnitude) for $t_s = 200$ hours, compared to the more well known ratio of ~10 as obtained for $t_s = 160$ hours, deserves nevertheless some explanation.

Fig. 3. J_c(B) curves obtained from magnetization at (a). 5 K and (b). 77 K, for samples treated with 5 minute melting and various post annealing times (t_s)

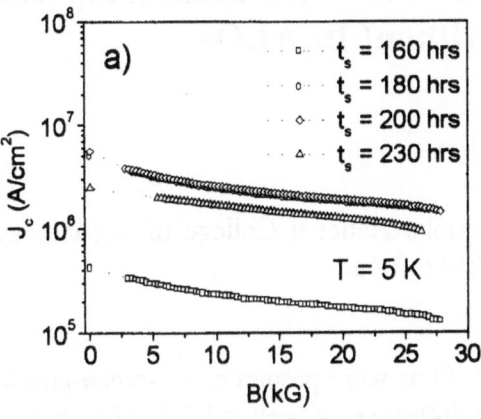

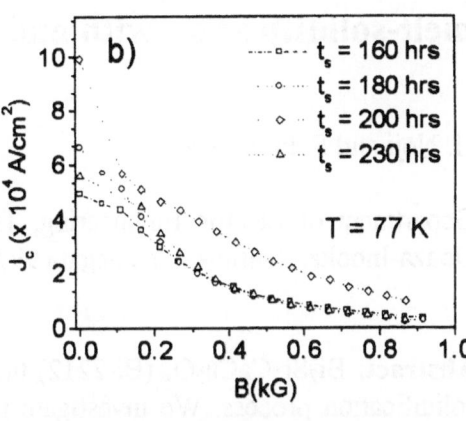

To that end, we have presented in Fig. 4 the XRD spectra for the related samples with t_s = 160 hours and 200 hours. It shows that a high purity (~ 90%) of 2223 phase is achieved around t_s = 160 hours, while further annealing leads to reconversion of 2223 to 2212 phase and resulting in a sample with the major phase (~ 55%) of 2212 for t_s = 200 hours. It is then conceivable that the large J_c(5K,0T)/J_c(77K,0T) ratio may have its origin in the stronger pinning effect in BSCCO-2212 which is less anisotropic than the 2223 phase. This suggestion of the role of pinning is also consistent with the observation made earlier on Fig. 2.

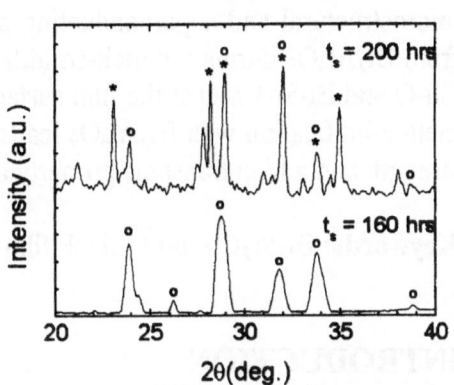

Fig. 4. XRD spectra of sample prepared with 5-minute melting showing the 2223 (o) and 2212 (*) phases.

CONCLUSION

We have shown in this work the important effects of intermediate melting and post annealing on the enhancement of J_c. It is demonstrated that a proper choice of melting and annealing times (5 minutes and 200 hours respectively) yields the highest value of J_c. It is also clear from the B-dependent behaviors of J_c that the thermal treatments have considerable effect on the pinning strength due to 2223 - 2212 phase conversion induced by prolonged annealing process.

Acknowledgement

This work is supported by RUT research grant under the contract no. 207/SP/RUT/BPPT/IV/97. We thank I.M. Sutjahja for help in magnetization measurement.

References

1. D.C. Larbalestier, et al. in *Advances in Superconductivity XI*, edited by N. Koshizuka and S. Tajima (Springer-Verlag, Tokyo, 1999) pp. 805-810.
2. S. Li, et al., Supercond. Sci. Technol. 11, 1011 (1998).
3. P. Kovac, et al., Physica C 312, 179 (1999) and references therein.
4. M.O. Tjia, et al., Technical Digest of AAPPS Seminar on Physics of Materials, IPS (1998), pp. 40-44.
5. H.P. Wiesinger, F.M. Sauerzopf and H.W. Weber, Physica C 203, 121 (1992).

Differences of microstructural and superconducting properties of Bi-2212 thick films by a variation of the maximum temperature during melt-solidification with and without $Bi_2Al_4O_9$

H. Noji and F. Furusawa

Department of Electric Engineering, Tsuruoka National College of Technology, 104 Aza-Sawada, Ooaza-Inooka, Tsuruoka, Yamagata 997-8511, Japan

Abstract: $Bi_2Sr_2CaCu_2O_x$ (Bi-2212) thick films were prepared via screen-printing method and melt-solidification process. We investigate the influence of melt-solidification process with $Bi_2Al_4O_9$ on microstructural and superconducting properties of Bi-2212 thick films. The vaporization of bismuth from $Bi_2Al_4O_9$ during the melt-solidification suppresses the growth of the impurities such as (Sr,Ca)-Cu-O and Bi-Sr-Ca-O at the film surface. The result of AC susceptibility measurements shows that the melt-solidification with $Bi_2Al_4O_9$ leads to the sample with a small scattering of intergranular-coupling strength and a high transport property at around liquid-nitrogen temperatue.

Keywords: $Bi_2Sr_2CaCu_2O_x$, thick films, $Bi_2Al_4O_9$, AC susceptibility, intergranular-coupling strength

INTRODUCTION

The enhancement of transport critical current density, J_C, of high-T_C superconducting materials is one of the most important subject for their practical applications such as coils and cables. To obtain high J_C in these materials, various problems must be overcome. One of the problems is a large number of crystals of secondary phases such as (Sr, Ca)-Cu-O and Bi-Sr-Ca-O remained at the surface of the thick films. These impurities disturb the transport current path, so that J_C values of the thick films scattered greatly. For enhancement of J_C of long films, it is necessary to reduce J_C scattering in long films by improving the homogeneity of the microstructure of the Bi-2212 thick films. Shimoyama et al successfully found that melt-solidification process with $Bi_2Al_4O_9$ is effective to decrease the Bi-poor phase, (Sr, Ca)-Cu-O, and improve the reproducibility of J_C value at 4.2 K in the Bi-2212 tapes [1]. However they have not reported that the influence of the sintering method on the transport property of the Bi-2212 tapes at around liquid-nitrogen temperature. In this study, we present results that demonstrate the influence of the melt-solidification with $Bi_2Al_4O_9$ on the microstructural and superconducting properties at around 77 K of Bi-2212 thick films.

EXPERIMENTAL

Bi-2212 thick films were prepared by applying the screen-printing method using Bi-2212 paste. The thickness of the Bi-2212 layer is 10μm. After paste deposition the thick films were presintered at 500°C for 2h in air to remove the organic binders. Then they were subjected to the melt-solidification process with or without $Bi_2Al_4O_9$. The melt-solidification process was carried out that temperature was gone up to the maximum (881°C ~ 901°C) at a rate of 300°C/h. The temperature was kept at the maximum from 1 to 5min, and subsequently cooled to 840°C at a rate of 5°C/h in air. After sintering at 840°C for 1h, the samples were cooled in the furnace to room temperature.

The microstructure of the samples was investigated by scanning electron microscopy (SEM) and X-ray diffraction analysis (XRD). An investigation of the superconducting properties such as a critical temperature and an intergranular coupling strength by AC susceptibility measurement had been conducted for 9 disks punch-pressed from the thick film. The disks were 4 mm in diameter. Complex magnetic susceptibility was measured as a function of temperature in various AC magnetic fields ranged from 0.1 to 9 Gauss (r.m.s. value) using driving frequency 333 Hz. The direction of applied AC magnetic field was always perpendicular to the disk surface.

RESULTS AND DISCUSSION

The microstructure of the surface of thick films can be sensitively dependent on the difference of amount of bismuth element in both samples. Figure 1 shows the SEM micrographs of the film surface for the samples melt-solidified (a): with $Bi_2Al_4O_9$ (b): without $Bi_2Al_4O_9$ at the maximum temperature of 891°C. On the surface of the sample sintered with $Bi_2Al_4O_9$, a very few black crystals in the form of whisker exist as shown in figure 1 (a). On the other hand, on the surface of the sample sintered without $Bi_2Al_4O_9$, a large number of impurities such as (Sr, Ca)-Cu-O and Bi-Sr-Ca-O which are distinguished as black needle-form crystals and white cubic crystals are dispersed. These results revealed that the vaporization of bismuth element from $Bi_2Al_4O_9$ during the melt-solidification process suppresses effectively the appearances of (Sr, Ca)-Cu-O and Bi-Sr-Ca-O on the surface of the samples.

Qualitative information regarding intergranular coupling properties can be investigated by measurement of the complex AC susceptibility, $\chi=\chi'-i\chi''$. We have already reported that the temperature dependence of the χ'' in the Bi-2212 screen-printed thick films without any cracks has a single loss-peak which is characterized as a contribution from the intergranular coupling region [2]. Therefore, the χ'' as a function of temperature in the sample represents the property of the intergranular coupling strength. We chose the sample sintered with $Bi_2Al_4O_9$ at the maximum temperature = 891°C for AC-susceptibility measurements and also the sample melt-solidified without $Bi_2Al_4O_9$ at the same maximum temperature for a comparison. Figure 2 shows the temperature dependence of the χ'' in these samples. AC applied field is 9 Gauss. The sample melt-solidified with $Bi_2Al_4O_9$ has a sharper loss-peak than that of the sample melt-solidified without $Bi_2Al_4O_9$. For the AC susceptibility measurements, the each sample is constructed by 9 disks which are punched out from a thick film. The loss-peak for the sample should be a superposition of intergranular coupling properties of individual disks and this peak can be sharper when the scattering of the properties of disks is smaller [2]. Therefore, the results indicate that the scattering of the intergranular coupling strength in the sample melt-solidified with $Bi_2Al_4O_9$ is smaller than that in the sample melt-solidified without $Bi_2Al_4O_9$. From the χ'' as a function of temperature, the loss-peak temperature T_M is obtained with various AC applied field. According to the Bean model [3], the flux penetrates into a middle of the sample at T_M and then the applied field values at the given temperature, $H_{AC}(T_M)$, are proportional to the transport critical current densities at this temperature, $J_C(T)$ as shown in equation (1).

$$J_C(T) \propto H_{ac}(T_M) \tag{1}$$

Therefore, T_M versus applied field characteristic indicates the transport J_C property as a function of temperature. We have already reported that the sample which has a higher intergrain flux pinning force density, i.e. stronger intergrain coupling, has a smaller reduction of T_M with increase of applied field [2]. Figure 3 shows the T_M as a function of applied field in the samples melt-solidified with and without $Bi_2Al_4O_9$. According to our previous results, the sample melt-solidified with $Bi_2Al_4O_9$ has a stronger intergrain coupling, i.e. a better transport property, than the sample melt-solidified without $Bi_2Al_4O_9$.

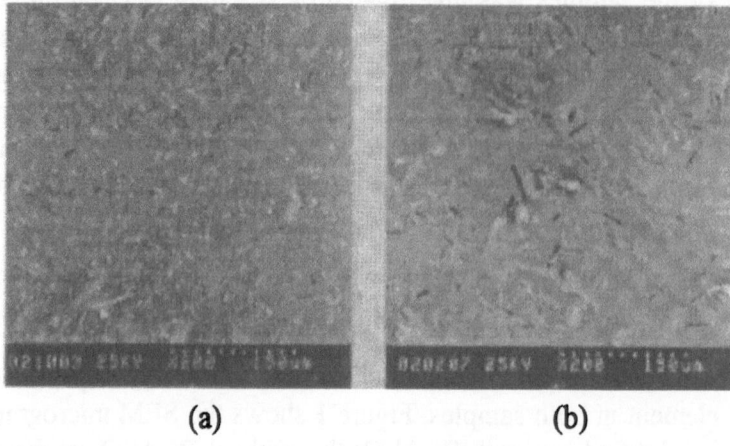

(a)   (b)

Fig. 1. SEM micrographs of the film surface for the samples melt-solidified with $Bi_2Al_4O_9$ at the maximum temperature of 891°C.

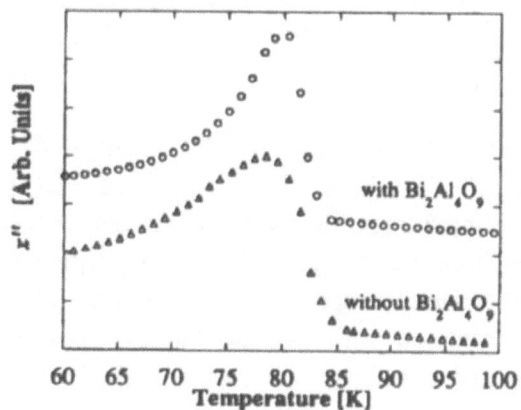

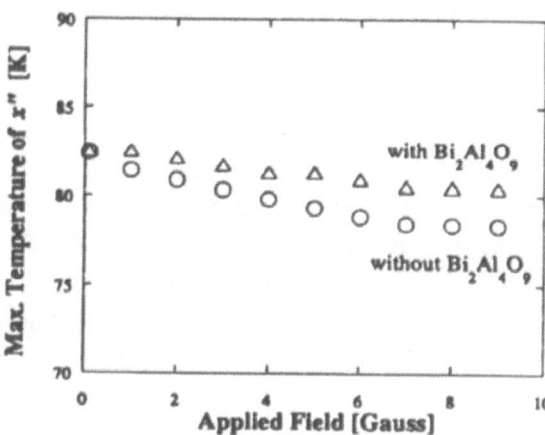

Fig. 2. Imaginary part of AC susceptibility as a function of temperature in the samples with or without $Bi_2Al_4O_9$.

Fig. 3. Maximum temperature of imaginary part of AC susceptibility as a function of AC applied field in the samples with or without $Bi_2Al_4O_9$.

CONCLUSION

The superior superconducting properties of the sample melt-solidified with $Bi_2Al_4O_9$, which has a small scattering of intergranular coupling strength and a high transport property at around liquid-nitrogen temperature, is related to the fine microstructure of this sample that has less impurity phases such as (Sr, Ca)-Cu-O and Bi-Sr-Ca-O. These characteristics is obtained by the melt-solidification with $Bi_2Al_4O_9$ due to avoid the lack of bismuth element during the sintering from the Bi-2212 thick films.

1. J.Shimoyama, N.Tomita, T.Morimoto, H.Kitaguchi, H.Kumakura, K.Togano, H.Maeda, K.Nomura and M.Seido, Japan J. Appl. Phys. **31**, L1328 (1992).
2. H.Noji, W.Zhou, B.A.Glowacki and A.Oota, Physica C **205**, 397 (1993).
3. C.P. Bean, Rev. Mod. Phys. **36**, 31 (1964).

The microstructure and the superconducting properties of Bi2212/Ag tapes with different Sr/Ca ratios

Akiyoshi Matsumoto, Hiroaki Kumakura, Hitoshi Kitaguchi and Kazumasa Togano

National Research Institute for Metals
1-2-1 Sengen, Tsukuba, Ibaraki 305-0047, Japan

Abstract: We investigated the microstructure and critical current densities (J_C) of Bi-2212/Ag tapes with Sr/Ca ratios of 1.8/1.2, 2.0/1.0 and 2.2/0.8 heat treated under various atmospheres. We fabricated Bi-2212/Ag tapes applying a dip-coating method and partial melting and subsequent slow solidification technique. The optimum heat treatment temperature T_{max} decreased with increasing Sr/Ca ratio and with decreasing oxygen partial pressure (pO_2). At the atmosphere of $pO_2 = 20\%$ and $pO_2 = 5\%$, the J_C value increased with increasing Sr/Ca ratio. At the atmosphere of $pO_2 = 100\%$ (pure oxygen), on the other hand, the J_C values of the 2.2/0.8 tapes were much lower than those of other tapes. The XRD analyses indicated that the main superconducting phase of 2.2/0.8 tapes was Bi-2201 phase. J_C increased with increasing pO_2 for the composition of 1.8/1.2 and 2.0/1.0.

Keywords: Partial-melt, Dip-coating, Transport critical current density, Microstructure, Second phase

INTRODUCTION

$Bi_2Sr_2CaCu_2O_y$ (Bi-2212) is one of the most interesting high transition temperature superconductors (HTS) for practical applications such as tape or wire. Partial melting and subsequent slow solidification process is easy and effective technique to obtain the highly c-axis grain oriented microstructure. In this technique, however, there are many factors such as heating process, atmosphere and a powder composition, which affect microstructure and critical current density [1-5]. There is still no clear understanding of all the relations among processing, microstructure and superconducting properties. In this paper, we report the relationship between microstructure and J_C of the tape conductors with different Sr/Ca ratio processed under various atmospheres.

EXPERIMENTAL

A standard calcination process with different Sr/Ca ratios of 1.8/1.2, 2.0/1.0 and 2.2/0.8 prepared Bi-2212 powders. These powders were mixed with an organic solvent and a binder to obtain slurry. A silver foil of 50 μm in thickness was dip-coated with this slurry. We applied a typical partial-melting and slow-cooling process under 100%, 20% and 5% oxygen flow. At first, the dip-coated Bi-2212/Ag tapes were heated at 500 °C for 1 hour to remove the organic materials, then heated to maximum temperature T_{max} and kept at T_{max} for 5 min, then cooled to 830 °C at 5 °C /h. The sample was kept at 830 °C for 1 hour, and finally cooled to room temperature in the furnace.

Transport I_C was measured by a four-probe method at 4.2 K in a magnetic field of 10 T parallel to the tape surface. The I_C values were determined with a criterion of 0.5μV/cm. The thickness and the width of the oxide layer were determined by microscope observation for transverse cross-section.

The free surface of the oxide layer was analyzed by X-ray diffraction (XRD) after I_C measurement. The cross section and the free surface were analyzed by scanning electron microscopy (SEM) and energy dispersive X-ray spectroscopy (EDS).

RESULTS AND DISCUSSION

Fig. 1 shows the J_C of the tapes with three different compositions for various T_{max} and atmosphere. J_C scattering was large for all compositions. However, we clearly observed that the optimum T_{max} decreased with decreasing oxygen partial pressure (pO_2).

The optimum T_{max}, which gives maximum J_C, is shown in Table 1. The optimum T_{max} decreased with increasing Sr/Ca ratio and with decreasing oxygen partial pressure (pO_2). The changes of T_{max} reflected the melting point depending on the composition and atmosphere.

Fig. 2 shows the average J_C values at 4.2 K, 10 T as a function of Sr/Ca ratios for optimum T_{max}. High J_C values were obtained for the tapes with Sr/Ca=1.8/1.2 and 2.0/1.0 fabricated under high pO_2. The J_C value increased with increasing the Sr/Ca ratio in the pO_2=5%. In the pO_2=20% and 100% the maximum J_C value were obtained for the composition of Sr/Ca=2.0/1.0. Although the 2.2/0.8 tape had high J_C values at the pO_2<20%, the J_C value droped suddenly at the pO_2=100%. We found from XRD measurements that the drop of the J_C value of 2.2/0.8 in pO_2=100% was due to the formation of Bi-2201 phase instead of Bi-2212 phase.

A second phase affects J_C value. All tapes contained second phases of Bi-free phase and Cu-free phase. Fig.3

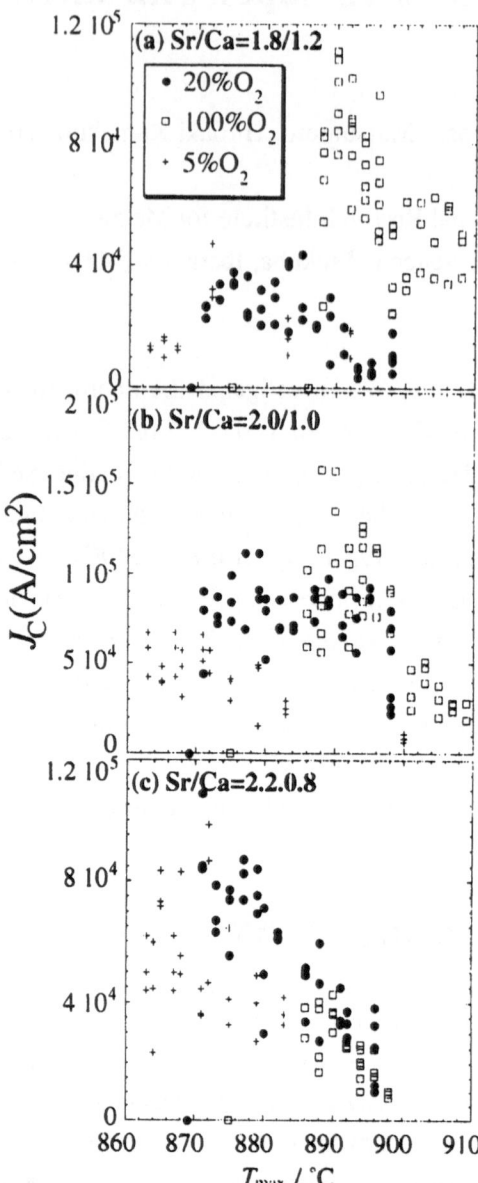

Figure 1 Critical current density J_c at 4.2 K and 10 T as a function of maximum temperature of heat treatment. Bi2212/Ag tapes of (a) Sr/Ca=1.8/1.2, (b) Sr/Ca=2.0/1.0 nd (c) Sr/Ca=2.2/0.8 were heat treated under several oxygen partial pressure pO_2.

Table 1 The optimum T_{max} of each tapes under various atmosphere.

Sr/Ca	100%O_2	20%O_2	5%O_2
1.8/1.2	892	889	872
2.0/1.0	888	885	870
2.2/0.8	884	877	865

shows the microstructure of three samples for $pO_2=20\%$. The large amount of $(Sr,Ca)CuO_x$ (1:1) phase was formed in the 1.8/1.2 tape. This (1:1) phase decreased J_C value. These results indicated that the growth of the second phase in the $pO_2=20\%$ was restrained for the larger rate of Sr/Ca. Yoshida et al.[6] reported that the Cu-free phase and Bi-free phase decreased with increasing Sr and decreasing Ca composition in the $pO_2=1\%$. We found this composition dependence was obtained in the $pO_2=5\%$ and 20%. On the other hand, the higher the oxygen partial pressure is, the smaller the size and the amount of the second phases are, except for 1.8/1.2 tape in the $pO_2=20\%$. Zhang et al. suggested the oxygen is needed for the second phase to react with the liquid to form Bi-2212[3]. High pO_2 is effective not only to decrease the size and the amount of second phase but also to improve a weak-link of intergrain[7,8]. We observed that the amplitude dependence of AC susceptibility of the 2.0/1.8 tape is smaller than that of the 1.8/1.2 tape in the $pO_2=100\%$. The highest J_C value of Sr/Ca=2.0/1.0 tapes in the $pO_2=100\%$ is due to the reduction of the second phase and the improvement of grain coupling.

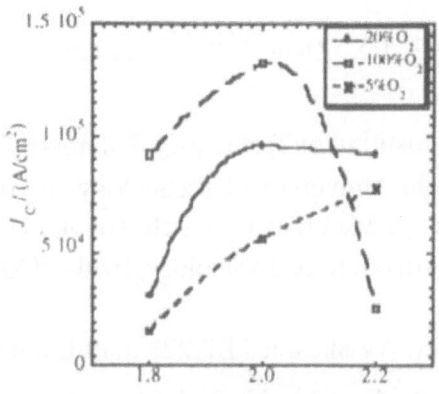

Sr composition

Figure 2 J_c value as a function of Sr composition. These tapes were heat treated at optimum T_{max} as shown by Table.

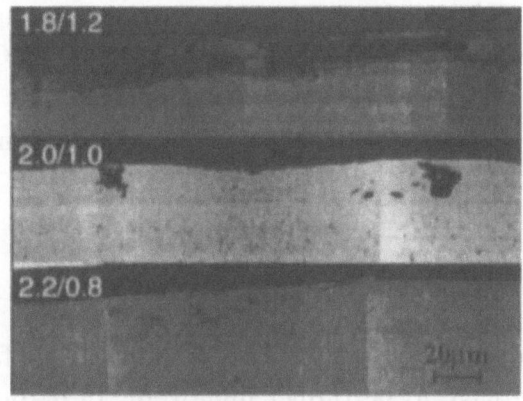

Figure 3 The back scatter images of cross sections of three tapes in the $pO_2=20\%$.

REFERENCES

(1) T. G. Holesinger, D. S. Phillips, J. Y. Coulter, J. O. Willis, and D. E. Peterson, Physica C 243 (1995) 93.

(2) J. Shimoyama, N. Tomita, T. Morimoto, H. Kitaguchi, H. Kumakura, K. Togano, H. Maeda, K. Nomura and M. Seido, Jpn. J. Appl. Phys. 31(1992) L1328.

(3) W. Zhang and E. E. Hellstrom, Phyica C 218(1993) 141.

(4) D. Buhl, Th. Lang, B. Heeb and L. J. Gauckler, Physica C 235-240(1994) 3399.

(5) D. Buhl, Th. Lang, M. Cantoni, D. Risold, B. Hallstedt and L. J. Gauckler, Physica C 257(1996) 151.

(6) M. Yoshida and A. Endo Jpn. J. Appl. Phys. 32(1993) L1509.

(7) H. Kumakura, H. Kitaguchi, K. Togano and N. Inoue, J.Appl. Phys., 80(1996) 5162.

(8) H. Fujii, H. Kumakura, H. Kitaguchi and K. Togano, Physica C 282-287(1997) 2567.

Jc properties of Ag-sheathed Bi2223 tapes with Sr-V-oxide barriers

H. Maeda[1], P.X. Zhang[2], K. Watanabe[3], T. Matsushita[4], E.S. Otabe[4]

[1]Kitami Institute of Technology, Kitami 090-8507, Japan
[2]Toyohashi University of Technology, Toyohashi 441-8580 , Japan
[3]Institute for Materials Research, Tohoku University, Sendai 980-8577, Japan
[4]Kyushu Institute of Technology, Iizuka 820-8502, Japan

Abstract: Ag-sheathed Bi2223 multifilamentary tapes with and without a layer of $Sr_6V_2O_{11}$(SVO) barrier outside and inside each filament were prepared. The SVO barrier layers have no bad effects on the critical current density Jc of the tapes. Especially, Jc increases obviously in the tapes coated with SVO layers both inside and outside each filament, due to the enhancement of Bi2223 phase formation by the reaction of the inside-coated SVO with Bi-based oxide core. This results suggest that even if the SVO barriers react with Bi oxide core due to the break of Ag sheath, the reaction have no bad effect, rather good effect on the Jc values. SVO will be one of most important candidates for barriers.

Key words: Bi2223, Ag-sheathed multifilamentary tapes, Sr-V-O barrier, critical current density

INTRODUCTION

At present, Bi2223 oxide superconducting tapes of 1 km in length with a high critical current density, Jc above 2×10^4 A/cm^2 have been fabricated. For practical use, however, the values of Jc will have to be further improved. In particular, for AC applications of use in cables , motors, transformers etc. low AC losses as well as high transport currents of the tapes will be required. In order to reach low AC losses the matrix (sheath) resistivity must be increased. From this point of view, Bi2223 multifilamentary tapes with insulating oxide barriers between filaments, such as $BaZrO_3$[1], $SrCO_3$[2] and MgO[3]have been developed.

Recently, we have reported that doping V into Bi oxide superconductors obviously enhances the formation of Bi2223 phase and improves Jc values[4], and thin plate-like precipitates of strontium vanadium oxide, such as $Sr_6V_2O_{11}$(SVO) with 1 to several μ m in diameter are distributed in the oxide core and aligned parallel to Bi2223 grains[5]. The plate-like SVO precipitates are formed and aligned during fabrication processes and give no damage to Bi2223 core matrix. The SVO is a insulator with a hexagonal structure. From these points of view the SVO will be one of the most important candidates as the barriers between filaments in Bi2223 tapes. So we prepared Ag-sheathed multifilamentary Bi2223 tapes with thin barrier layers of SVO between filaments , and investigated the structure and the superconducting properties, especially transport Jc values of the tapes.

EXPERIMENTAL

We prepared Ag-sheathed multifilamentary Bi2223 tapes with thin barrier layers of SVO by the conventional powder-in-tube method. Bi based oxide powders with a nominal composition of $Bi_{1.85}Pb_{0.35}Sr_{1.90}Ca_{2.05}Cu_{3.05}O_x$ were formed into a cylindrical bar with 8 mm in diameter.

$Sr_6V_2O_{11}$(SVO)powders,which were prepared by the conventional solid state reaction at temperatures up to 800℃, using 99.9% purity powders of $SrCO_3$ and V_2O_5, were mixed with organic binder to slurry. We prepared two types of Ag sheathed mono-core wires with about 2.0mm in diameter by drawing; (a)using the bar without SVO coating(called here "0") and (b) using the bar with SVO coating of about 200μm in thickness ("1"). Then we fabricated three types of tapes with 7 filaments by drawing and flat-rolling; (d) using "0" wire without SVO coating (called here 7V00), (e) "0" wire with SVO coating of 100μm in thickness (7V01), and (f) "1" wire with SVO coating(7V11). Finally the tapes were heat treated at 830 - 840℃ for 50h in air , then rolled to about 0.2mm in thickness. The total heat treatment time is 150h. The "1" wire is prepared to examine how the SVO barrier have an effect on the superconducting properties of tapes, imaging in case the Ag-sheath is broken and the SVO barrier is reacted with Bi based oxide core during the preparation process.

The critical current Ic of the tapes was measured by a four probe DC transport method with a criterion of 1μV/cm at 77K in magnetic fields applied parallel to the tape surface . The Jc values were calculated from the Ic's using the measured core cross sectional areas. We measured the magnetic hysteresis loops and irreversible fields Hirr by a SQUID magnetometer. The morphology of the specimens was examined by optical microscope.

RESULTS AND DISCUSSION

Fig. 1 shows the optical microscopic images of cross section of the wires before flat-rolling and the tapes after heat treated. In the round wires and heat treated tapes we can observe that the SVO thin layers are formed around almost filaments although the thickness is not uniform in the tapes. We observed the break of some filaments in the SVO coated tapes. This may be due to that uniform strain can not be applied among all the composed filaments during drawing, because slip due to low friction between filaments occurs . Especially, since we used highly dense Bi based oxide bars in the present tapes, the filament break is considered to appear. We have observed that such a filament break occurs not so frequently in case Bi based oxide powders are packed loosely into Ag tube.

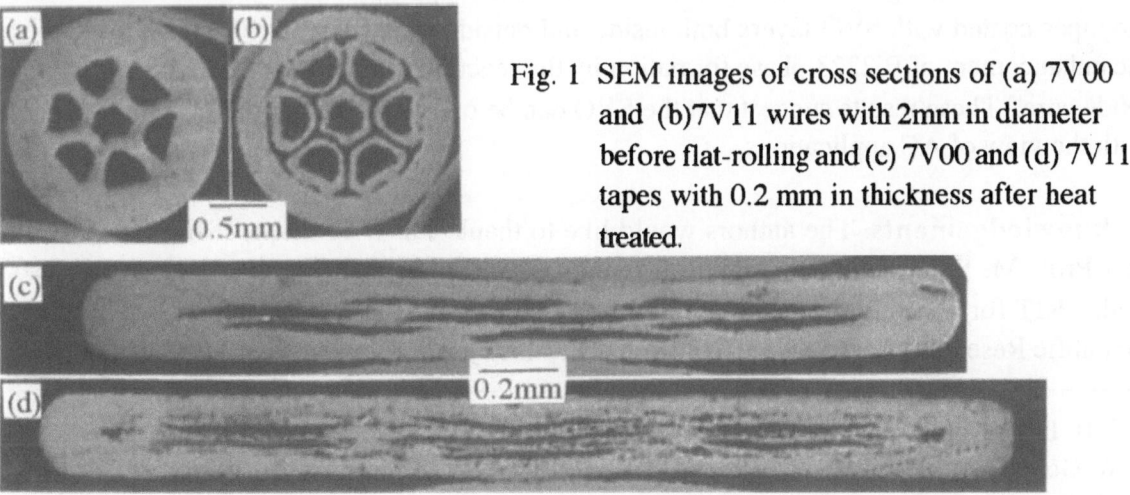

Fig. 1 SEM images of cross sections of (a) 7V00 and (b)7V11 wires with 2mm in diameter before flat-rolling and (c) 7V00 and (d) 7V11 tapes with 0.2 mm in thickness after heat treated.

Fig. 2 shows typical Ic and Jc vs. magnetic field curves at 77K for the three types of tapes heat treated at 835℃ for 150h. For all the tapes Jc values are the highest after heat treated at 835℃. The Ic and Jc values of the tapes with SVO barriers(7V01 and 7V11) are higher than those of the tapes without barriers(7V00). This may suggest that the SVO layers have some good effects on the Bi2223 formation and grain alignments. Particularly, in the SVO barrier tapes coated inside filaments(7V11) Ic and Jc values are much high. Furthermore, we observe that the irreversible fields Hirr of 7V11 tape are higher than those of 7V00. The increases in Jc and Hirr are due to that the inside SVO powders react with Bi based oxide core to promote the formation and alignment of Bi2223 grains. We have obtained such an increase of Jc in V doped Bi2223 tapes[4]. This results suggest that even if the SVO barriers react with Bi oxide core due to the break of Ag sheath, the reaction have no bad effect, rather good effect on the Jc values. Because a small amount of V doping up to 0.1 promotes the formation and alignments of Bi2223 grains[5]. From these results we consider that the SVO will be one of the important barriers for AC applications. We are now measuring AC losses for the tapes with SVO barriers.

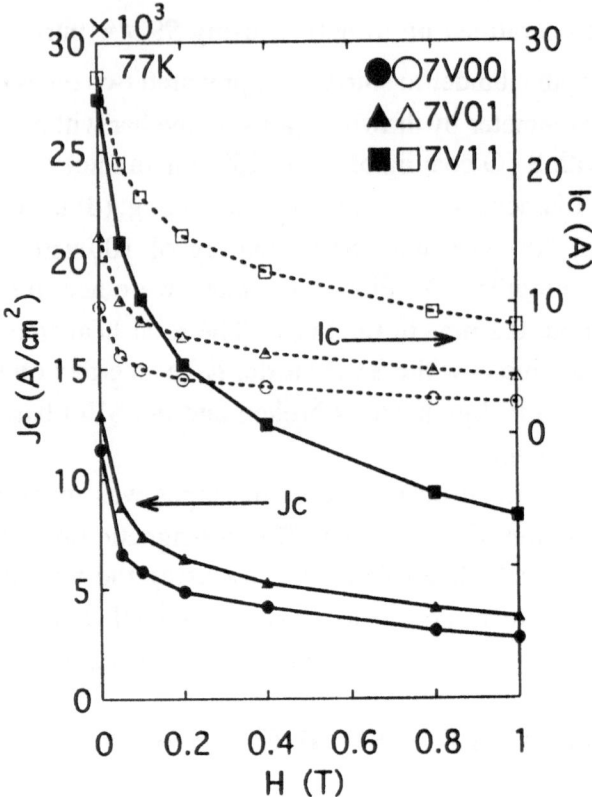

Fig. 2 Ic and Jc vs. magnetic field curves at 77K for tapes 7V00, 7V01 and 7V11

CONCLUSION

We prepared Ag-sheathed Bi2223 multifilamentary tapes with 7 filaments with and without a layer of $Sr_6V_2O_{11}$(SVO) barrier outside and inside each filament. The SVO barrier layers have no bad effects, rather some good effects on the Jc values of the tapes. Especially, Jc increases obviously in the tapes coated with SVO layers both inside and outside each filament(7V11). This can be due to the enhancement of Bi2223 phase formation by the reaction of the inside-coated SVO with Bi based oxide core. These results suggest that the SVO can be one of most important candidates for barriers with the view of AC applications.

Acknowledgements. The authors would like to thank Dr. W.P. Chen, IMR, Tohoku University and Prof. M. Sato, Kitami Institute of Technology(KIT)for valuable discussions, and also Mr. T. Inaba, KIT for assisting in some measurements. This work is supported in part by Grant-in-Aid for Scientific Research (A) No. 09355023 from the Ministry of Education, Science and Culture, Japan.

1. Y.B. Huang and R. Flukiger, Physica C294 (1998) 71

2. W. Goldacker, M. Quilitz, B. Obst and H. Eckelmann, IEEE Trans. Appl. Supercond., 9 (1999) 2155

3. N. Ayai, M. Ueyama, K. Hayashi and K. Sato, Advances in Superconductivity XI (1999) 911

4. H. Maeda, K. Kakimoto, M. Kikuchi, K, Watanabe, T. Tanaka and H. Kumakura, IEEE Trans. Appl. Supercond., 9 (1999) 2541

5. H. Maeda, K. Kakimoto, M. Kikuchi, J.O. Willis, K. Watanabe, T. Tanaka and H. Kumakura, Appl. Supercond., 5 (1997) 151

FIELD DEPENDENCE OF J_c FOR Re-DOPED FILAMENTARY Hg1212 AND Hg1223 SUPERCONDUCTORS

Tomoko Goto,[1] KenIchiro Inagaki[1*] and Kazuo Watanabe[2#]

[1]Nagoya Institute of Technology, Gokiso-cho, Showa-ku, Nagoya 466-8555, Japan
[2]Institute for Materials Research, Tohoku University, 2-1-1 Katahira, Aoba-ku, Sendai 980-8577, Japan

Abstract: Field dependence of transport J_c for Re-doped filamentary HgRe1212 superconductors was examined in comparison with the filamentary HgRe1223 superconductors. A filamentary $Ba_2CaCu_2Re_{0.2}O_x$ precursor was fabricated using a solution spinning method and reacted in an evacuated quartz tube with a pellet of $(CaHgO_2)Ba_2Cu_2Re_{0.2}O_x$ as a reactor. The reacted sample was a nearly single phase of Hg1212 with zero resistivity temperature (T_c) value of 100 K and the highest J_c value of 7000 A/cm^2 at 77 K and 0 T.
The T_c and J_c decreased slightly by post-annealing in flowing Ar, while the superconductivity at more than 77 K disappeared by post-annealing in flowing O_2. The J_c values for the as-reacted sample and the sample post-annealed in flowing Ar decreased rapidly by applying the field less than 0.5 T and the superconductivity was not observed by applying the field more than 4 T at 77 K. On the other hand, the superconductivity for the both F and Cl doped filamentary HgRe1223 was maintained at 77 K in a field of 10 T by post-annealing in flowing Ar.

Keywords: Re-doped Hg1212, Filamentary, Transport J_c, Field dependence

INTRODUCTION

It was reported that a 50-fold increase in irreversibility field was obtained by partial substitution of Re for Hg on Hg1212 phase (HgRe1212) and this was attributed to metallisation of the Hg/Re [1]. The HgRe1212 superconductors showed superior crystallographic quality and microstructure in comparison with HgRe1223 phase and demonstrated the potential of these materials for practical applications [2,3]. In this paper field dependence of transport J_c for filamentary HgRe1212 superconductors was attempted as compared with the filamentary HgRe1223 superconductors. The fabrication of filamentary HgRe1212 was already examined [4]. The filamentary (Ba,Sr)-Ca-Cu-Re-Ox precursor was reacted in an evacuated quartz tube with a pellet of (HgO)-(Ba,Sr)-Ca-Cu-Re-O. In this paper, the filamentary HgRe1212 superconductor was prepared by using CaHgO$_2$ for the vaporized materials to improve the microstructure and transport J_c [5].

EXPERIMENTAL

Long precursor $Ba_2Ca_1Cu_2Re_{0.2}O_x$ filaments were prepared by dry spinning from a homogeneous aqueous solution that contained mixed acetates of Ba, Ca and Cu, ammonium perrhenate, poly (vinyl alcohol) and organic acids as reported in a previous paper. The as-drawn filament was pyrolyzed at 450 °C at a heating rate of 25 °C /h in air and calcined at 910 °C for 1 h in flowing O$_2$. The calcined sample was vacuum-sealed in a fused quartz tube with a doped pellet of $(CaHgO_2)Ba_2Cu_2Re_{0.2}O_x$. The pellet was prepared by mixing powders of CaHgO$_2$ and Ba-Cu-Re-O.

We measured the electrical resistivity of the sample by using a standard four-probe method. The size of the sample was about 80 μm in diameter and 10 mm in length. External magnetic fields were applied in a direction normal to the filament length using helium-free 11T superconducting magnet at High Field Laboratory for Superconducting Materials, Tohoku University. Current was passed along the direction of the fiber axis and was normal to the applied magnetic field. The J_c was defined by the offset method from the point on the I-V curve at which the voltage of 1 μV appeared between the voltage terminals separated by 2 mm.

RESULT AND DISCUSSION

The filamentary sample was heated to 880 °C, partially melted for 30 min, cooled to 810 °C at a cooling rate of 30 °C/h, held for 1 h, then cooled to room temperature for 10 h. Some samples were post-annealed at 375 °C for 30 min, cooled to 250 °C at a cooling rate of 10 °C/h, followed by furnace cooling in flowing Ar. The samples were also post-annealed at 300 °C or 350 °C for 30 min in flowing O_2. A single phase of Hg1212 was detected by X-ray diffraction pattern of the as-reacted samples as same as the post-annealed samples. The highest J_c value of 7000 A/cm^2 at 77 K and 0 T was observed. Figure 1 presents a representative polished and etched longitudinal cross-section of the sample. The sample consists of semi-liquidus texture and some micro-pores are observed. The temperature dependence of the electrical resistivity of the samples is shown in Fig.2. Although the resistivity of the as-reacted sample rapidly decreases around 127 K, the zero resistivity temperature is 100 K and the width of the transition is broad. For the sample post-annealed in flowing Ar, second transition around 110 K is appeared and the zero resistivity temperature is 98K. A major diamagnetic response due to the Meissner effect below at 100 K and 90 K was observed for the samples as-reacted and post-annealed in flowing Ar, respectively. The superconductivity at more than 77 K disappeared by post-annealing in flowing O_2. The doping level of the as-reacted sample is considered to optimize and the level turns to be considerably overdoped by post-annealing in flowing O_2. A controlling of the doping level for the present filamentary HgRe1212 superconductors is difficult by post-annealing.

Figure 3 shows the field dependence of J_c for the sample as reacted and post annealed in flowing Ar. In spite of high J_c value of more than 5000 A/cm^2 at 77 K and 0 T, the J_c value decreases to two order of magnitude by applying the field less than 0.5 T. Then the J_c decreases with increasing applied field. The superconductivity at 77 K for both samples was disappeared by applying the field more than 4 T.

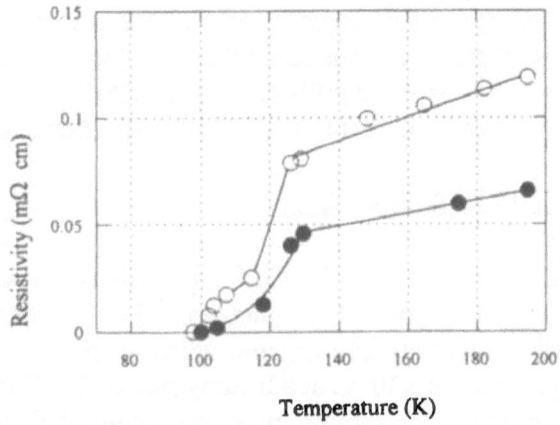

Fig.1. Polished etched surface on the longitudinal cross-section of the filamentary Hg1212 sample.

Fig.2. Resistivity as a function of temperature For the filamentary sample.: ● as-reacted, ○ post-annealed sample in flowing Ar.

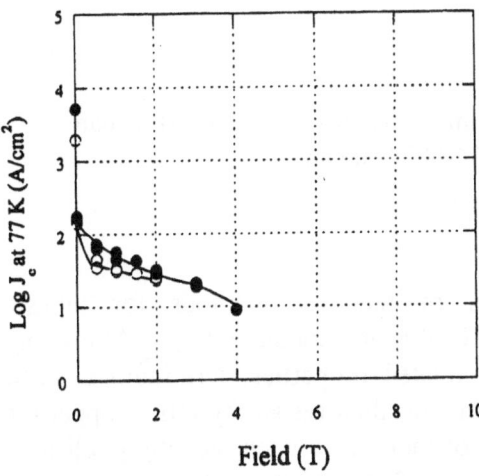

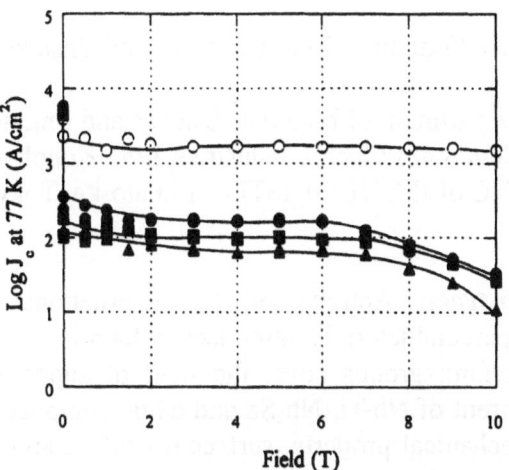

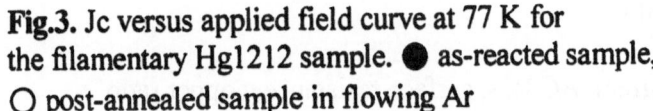

Fig.3. Jc versus applied field curve at 77 K for the filamentary Hg1212 sample. ● as-reacted sample, ○ post-annealed sample in flowing Ar

Fig.4. Jc versus applied field curve at 77 K for the filamentary Hg1223 sample post-annealed in flowing Ar.

○ 300°C, ■ 325°C, ▲ 350°C, ● 375°C

The field dependence of the J_c for filamentary HgRe1212 was poor as compared with that for filamentary HgRe1223 [5]. Field dependence of the J_c for the both F and Cl doped filamentary HgRe1223 sample partially melted with a pellet of $(HgO)_{0.8}Ba_{1.8}Ca_2Cu_3Re_{0.2}(BaF_2)_{0.2}(CuCl)_{0.03}O_x$ was examined [6]. Although the J_c value for the as-reacted sample and the sample post-annealed in flowing O_2 decreased rapidly by applying the field less than 0.5 T and the superconductivity disappeared. A J_c value exceeding 1×10^3 A/cm^2 was maintained at 77 K and in the field 10 T by post-annealing in flowing Ar. Figure 4 shows the field dependence of the J_c for the HgRe1223 sample after post-annealing in flowing Ar for various temperatures ranging from 300 °C to 375 °C. The field dependence of the J_c is dependent on the annealing temperature. It was noted that the superconductivity of the HgRe1223 sample maintained up to 10 T at 77 K. The zero resistivity temperature (T_c) for the sample annealed at 375 °C was 128 K with transition width (ΔT_c) of 8 K, while the as-reacted sample showed $T_c = 119$ K with $\Delta T_c = 4$ K. The high J_c at high field for the filamentary HgRe1223 superconductors is dependent upon the high T_c value and microstructure as well as doping effect.

On summary, the J_c of the filamentary HgRe1212 superconductors was examined in the magnetic field. As compared with the both F and Cl doped filamentary HgRe1223 superconductors, the field dependence of the J_c for filamentary HgRe1212 was poor. The superconductivity at 77 K for the HgRe1212 samples was disappeared by applying the field more than 4 T.

REFERENCES

1. J.L.Tallon, C.Bernhard, Ch.Niedermayer, J.Shimoyama, S.Hahakura, K.Yamaura, Z.Hiroi, M.Takano and K.Kishio, L.Low Temp. Phys., 105, 1379-1384 (1996).
2. A.Tsukamoto, K.Takagi, Y.Moriwaki, T.Sugano, S.Adachi and K.Tanabe, Appl.Phys.Lett., 73 990-992 (1998).
3. P.V.P.S.S.Sastry and J.Schwartz, IEEE.Trans.Appl.Supercond. 9 1684-1687 (1999).
4. T.Goto, Physica C 247 133-136 (1995).
5. T.Goto, S.Hirota and K.Watanabe, L.Low Temp. Phys., to be published.
6. T.Goto and T.Shimizu, IEEE Trans. Appl. Supercond., to be published.

Activity in International Standardization of Superconductivity

Kozo Osamura[1], Ken-ichi. Sato[2] and Yasuzo Tanaka[3]

[1] Department of Materials Science and Engineering, Kyoto University, Kyoto 606-8501, Japan
[2] Sumitomo Electric Industries, Ltd., Konohana-ku, Osaka 554-0024, Japan
[3] JNC of IEC/TC90, ISTEC, Minato-ku, Tokyo 105-0004, Japan

Abstract: Activity on the international standardization of the test methods for industrial superconductors is introduced, which is authorized to IEC/TC90 (Superconductivity). At present, 9 working groups cover the field of superconductivity and related properties; terminology, critical current of Nb-Ti, Nb_3Sn and oxide composite superconductors, residual resistivity ratio, copper ratio, mechanical property, surface resistance and AC loss. Three of them have been recently published as International Standards. Their working groups started to discuss further new working items in order to develop the systematic network for industrial use.

Keywords: International Standards, critical current, AC loss, surface resistance, copper ratio

INTRODUCTION

Industrial standard is in principle used as a common measure for the trade. For developing advanced materials, on the other hand, it provides a tool of characterizing their property. Also it is very important to establish the internationally unified standards, because it facilitates easier communication among project members for many ongoing and forthcoming international scientific and industrial projects. The activity on the international standardization of the test methods for industrial superconductors is authorized to IEC/TC90 (Superconductivity). Their activity has been reported elsewhere [1].

In the present report, a brief history of the organization of TC90 and the procedure creating the international standard are explained. At present, 9 working groups cover the field of superconductivity and related properties. Their activity and further intending development are introduced. The domestic activity supporting the international standardization is mentioned. It is also emphasized that the good liaison with groups such as VAMAS concerning the fundamental aspects of standardization is crucial to promote effective international collaboration.

PROCEDURE OF STANDARDIZATION

The technical committee 90 in IEC has been established in 1990, which engages mainly for the internal standardization of superconductivity. At present, the chairman is Dr L. Goodrich, NIST. The secretariat country is Japan and Ken Sato is her representative. Eleven countries are now registered as P member. They are China, France, Germany, Japan, Poland, USA, UK, Rumania, Russia, Korea and Italy.

In order to create a new International Standard, we need the procedure as listed in Table 1. At the proposal stage, the new work item proposal (NP) has to be submitted by National Committee of P-member country, Secretary, Liaison Organization or General Secretary. The liaison organization

Table 1. Procedure creating the international standard

Stage	Associated Documents
Preliminary Stage	PWI (Preliminary Work Item)
Proposal Stage	NP (New Work Item Proposal)
Preparatory Stage	WD (Working Draft)
Committee Stage	CD (Committee Draft)
Enquiry Stage	CDV (Committee Draft for Vote)
Approval Stage	FDIS (Final Draft for International Standard)
Publication Stage	IS (International Standard)

for TC90 is VAMAS/TWA16. When the proposal is accepted, a new international working group (WG) is set up in TC90. At present, nine WGs have been established as listed in Table 2. In each group, the content of the proposal is discussed and a working draft (WD) is prepared as a result. The WD is approved as a CD at the TC90 meeting. TC90 meeting is held approximately every one year and half. The CDV is sent to each National Committee after the CD is corrected and completed. When the majority of National Committee approves it, then it becomes FDIS. The termination of working on draft by TC90 is requested to be within 36 months.

Table 2. Present task and current stage of nine working groups.

Working Group	Present Task (1999)	Current Stage
WG1	Terms and definitions	FDIS
WG2	I_c of Cu/Nb-Ti	IS (IEC61788-1)
WG3	I_c of oxide superconductors	FDIS
WG4	RRR of Cu/Nb-Ti	WD
WG5	Tensile test of Cu/Nb-Ti	FDIS
WG6	Copper ratio of Cu/Nb-Ti	FDIS
WG7	I_c of Nb_3Sn	IS (IEC61788-2)
WG8	Surface resistance	WD
WG9	AC loss of Cu/Nb-Ti	NP

In Japan, several organizations engage to the activity for the international standardization. Japanese National Committee of IEC/TC90 settled in ISTEC takes mainly two roles; the activity for IEC/TC90 and the translation of IEC Standards to Japanese Industrial Standards. New Materials Center (NMC), OSTEC provides the proposal for the NP mainly with respect to conventional superconducting wires. The items discussed already and the advanced items are listed in Table 3. Japanese Fine Ceramics Association (JFCA) concerns mainly about oxide superconducting materials. International Superconductivity Technology Center (ISTEC) takes also part for assessing properties of thin films and bulk of oxide superconductors and creating preliminary work items as mentioned later. VAMAS is the international organization promoting the international collaboration on scientific research works. National Research Institute for Metals is the principal organization of VAMAS/TWA16.

Table 3.   Present status of standardization of test methods

Standardization Item	Cu/Nb-Ti	Cu/Cu-Ni/Nb-Ti	Nb₃Sn	Chevrel, Nb₃Al	Oxide Wires	Oxide Thin Films	Oxide Bulk
Volumetric Ratio of Matrix to SC	WG6	*	NMC	*	*		
Filament Size	*	*	*	*			
Pitch of Twisting	*	*	*	*	*		
Critical Current	WG2	NMC	WG7	*	WG3	*	*
Strain Effect			*	*	VAMAS		
AC Loss	WG9	*	*		JFCA	*	*
AC Current Limit	NMC	*	*	*			
Critical Magnetic Field	VAMAS	*	*	*			
Irreversible Magnetic Field					*		*
Coercive Force							ISTEC
Critical Temperature					JFCA	*	*
Residual Resistance Ratio	WG4	*	NMC	*	*		
Surface Resistance						WG8	
Tensile Test (R.T.)	WG5	*	NMC	*	VAMAS		*
Tensile Test (low temp.)	NMC	*	*	*	VAMAS		*
Stability		*	*		*		

* designates an item, of which standardization is requested.

ACTIVITY OF INTERNATIONAL WORKING GROUPS

Table 3 shows the overview on the present status of standardization of test methods.   First category is about the geometrical factors of volumetric ratio of matrix to superconductors, filament size and pitch of twisting.   Second category is about critical current and its strain effect.   The followings are AC characteristics, critical fields and temperature, resistivity and mechanical property.   There are a variety of superconducting materials available for industrial use.   The present majority is the supercondcuting wire consisting of Nb-Ti filaments embedded in the copper matrix.   Three layered Nb-Ti wire is designed for AC application and Nb₃Sn wire is used for high magnetic field application up to 20T.   Oxide wires indicate at present Ag sheathed Bi 2212 and 2223 tapes.   YBCO is the major material for oxide thin film and bulk applications. This table suggests that the important test methods have been taken into consideration for standardization.

(a) Critical Current: WG2 manages to evaluate the test method of the DC critical current of Cu/Nb-Ti composite superconductors.   The applicable conditions of its scope are briefly summarized in Table 4. This test method recommends the use of inductively coiled specimen.   The specimen length is larger than 310 mm. There are two criteria for determining the critical current. By electric field criterion, the critical current is defined at 10 or 100 μV/cm. The resistivity criterion is 10^{-14} or 10^{-13} Ωm. When their experimental conditions listed in Table 4 are satisfied, the COV less than 2 % is expected for the target precision[2]. WG7 treated the test method for determining the DC critical current of Nb₃Sn composite superconductors. Special care shall be paid on the mechanical brittleness and strain sensitivity of critical current[3]. The critical current of Nb₃Sn wire degrades 2 % at 12 T for a strain of 0.03%. On the other hand, the degradation for the Nb-Ti wire is 4 % at 7 T for the strain of 1%.   Reaction mandrel is selected separately from the measurement mandrel.   A comparison of thermal contraction data of Nb₃Sn with several materials, the G10 and stainless steel are suggested to be appropriate mandrel materials. WG3 manages the test method for determining DC critical current of short and straight Ag sheathed Bi2212 and Bi2223 oxide superconducting tapes. Their wires differ from metallic wires in several aspects. They are mechanically weak, exhibit weak link, and have a high degree of anisotropy. Magnetic flux flow occurs easily, and the dependency of the I_c on the magnetic field shows a large hysteresis. The approved method in WG3 can be used for short, straight, round or flat, single-core or multifilamentary Bi2212 and Bi2223 wires. The n value must be at least 5 or more, and the method is useable for currents of up to 500 A. The measurements are accurate to within 5% under self magnetic field at temperature of 4.2 K or 77 K. The strain produced by thermal shrinkage is limited to within ± 0.1% by selecting an appropriate material and design for the measurement mandrel as recommended in the text

Table 4. Comparison of test conditions for three different composite superconductors.

Item	Cu/Nb-Ti	Nb$_3$Sn	Oxide SC
Cross-Sectional Area (mm^2)	< 2	< 2	-
Cu/SC Ratio	> 1	> 0.2	-
Length (mm)	> 310	> 430	short
Shape	coiled	coiled	straight
Mandrel Diameter (mm)	> 24	25 ~ 40	-
Bending Strain (%)	< 3	< 5	-
Current Limit (A)	< 1000	< 1000	< 500
n	> 12	> 12	> 5
Magnetic Field (T)	< 0.7 B$_{c2}$	< 0.7 B$_{c2}$	-
Homogeneity of the Field (%)	< ± 2	< ± 1	< ± 2
Target Precision(%)	2	3	5

(b) Residual Resistance Ratio: The residual resistance ratio (RRR) is a physical quantity necessary for the design and operation of superconducting magnetic and is specified in commercial wires. The standardization of RRR measurements for Cu/Nb-Ti wires has been examined in WG4. RRR is defined as the resistance (R_1) at room temperature divided by the resistance (R_2) just above the critical temperature (T_c). The annealed copper is very sensitive for strain. The changing rate of RRR with bending strain is large when the initial RRR is large. The resistivity of pure copper at low temperature is appreciably influenced by bending strain. The wires is round or square but must have an RRR of 350 or less and a cross-sectional area of 2 mm^2 or less. The wires is either straight or coiled. R_1 is defined as the value at 20 °C. If the actual room temperature at the time of measurement is not 20 °C, R_1 is converted using a correction formula. Two kinds of methods for determined R_2 were proposed (i.e., the curve method and the fixed point method), and the curve method was finally adopted. The resistance is measured by reversing the current flow at low temperatures to avoid the effects of thermoelectromotive force and calculating the mean value. Using this method, measurements are accurate to within 5%. Using this measurement method, a COV of 2.4% was obtained in an international RRT performed by thirteen institutes in five countries.

(c) Surface Resistance: WG8 is responsible for the standardization of measurement method related to superconducting electronics. By the international questionnaires, the measurement of surface resistance in the microwave region of superconducting thin films was selected as the first item of standardization. The surface resistance of oxide superconductor is lower than a normal conducting metal by 2 ~ 3 order in magnitude. HTS materials can therefore be used to construct high performance microwave passive devices that have lower power consumption, higher sensitivity, lower loss and improved frequency selectivity compared to conventional devices. In the first working draft, a proposal was made to use the TE_{oml} mode of the dielectric resonance method to measure the surface resistance. This method has a measurement sensitivity of 0.1 mΩ (10GHz) in the 8 ~ 50 GHz frequency band.

(d) Mechanical Properties: The Cu/Nb-Ti superconducting composite wire is the most popular mulitfilamentary composite wire. Complicated stresses are exerted in the winding stage of the manufacturing process and large electromagnetic forces are applied during excitation. The method for testing the tensile property of Cu/Nb-Ti wires at room temperature was standardized by WG5. The wires may be either round or square and must have a cross-sectional area of 0.15 ~ 2 mm² and a copper ratio of 1.0 ~ 8.0. The measurement region of the sample must be 100 mm long. A strain gauge is used to measure the strain with an accuracy of approximately 2%. The Young modulus (loaded: E_o, unloaded; E_a), the yield strength (loaded : σ_A, unloaded: σ_B)of the composite wire when the Cu matrix yields, and the tensile strength (σ_{UTS})were selected as major properties[4]. The yield stress (σ_C) and the amount of elongation (*A*) at the point where the Nb-Ti filament yields and the composite wire breaks are described in the annex report as reference value.

(e) Geometrical structure: The copper/superconductor volumetric ratio (otherwise known as the copper ratio) of a Cu/Nb-Ti superconducting composite wire is used to calculate the critical current density (J_c) of the wire. The method for measuring the copper ratio of Cu/Nb-Ti wires was standardized by WG6. After several discussions, the working group decided that the weight method is used as the standard when the specific gravity of Nb-Ti is known. The wires may be either round or square and must have a cross-sectional area of 0.1 ~ 3 mm². The Nb-Ti filament diameter must be 2 ~ 200 μm, and the copper ratio must be 0.5 or more. Measurements are performed after removing insulating material, when it is coated. The weight of the sample should be 1 ~ 10 g. The measurement accuracy of this method is within 2 %. The working group also decided that the copper weight method should be used when the specific gravity of Nb-Ti is not known; this procedure is described in the annex report.

(f) AC loss: Even though the composite superconductors have no heat generation for a direct current, they generate AC losses in alternating electromagnetic environments. We need to pay attention to the level of AC losses, because of a penalty of refrigeration especially for the AC applications. Various types of AC losses generate in the superconducting wires. Several methods of AC loss measurements have been proposed. Two major methodologies by means of VSM and pickup coil have been employed in WG9 when the Cu/Nb-Ti wires are exposed to a transverse alternating magnetic field in DC and pulsed coils. The configuration of the transverse magnetic field is one of the most universal ones too which the wires and windings are exposed in usual coils. Two types of WD are in parallel prepared for the VSM to measure the hysteresis loss and for the pickup coil method to measure the total AC loss including the hysteresis loss and the coupling loss of Cu/Nb-Ti composite superconducting wires.

(g) Superconductivity Terminology: Definitions for superconductivity terminology and related scientific and technical terms are the first set of standards to be issued by TC90. The draft was discussed at WG1. The terms included in the draft were based on the JIS Standard (JIS H 7005) that was established in 1991. Various opinions have been expressed on the strictness of the definitions and the manner in which the terms should be arranged and supplemented. Especially, the new superconductivity-related terms arising from the recent appearance of oxide superconducting materials have been included. They are now treated as a part of the IEC glossary (IEV Chapter 815, compiled in cooperation with IEC/TC1). The standard contains three hundred and one terms categorized into eight sections: characteristics of superconductivity, superconducting substances, electromagnetic phenomena and electromagnetic physical properties, wires and conductors, manufacturing methods, superconducting magnet technology, applied technology, and testing and evaluation methods.

FURTHER ACTIVITY CREATING PRELIMINARY WORKING ITEMS

The domestic activity supporting the international standardization is supervised by the Japanese National Committee of IEC/TC90. Several TC90 working drafts have been originally created by New Material Center (NMC) and Japanese Fine Ceramics Association (JFCA). They are also active to provide preliminary working items as listed in Table 3. In the standard of I_c of Cu/Nb-Ti wires, the standard copper ratio was set at no less than one. However, wires with a small amount of stabilized copper and a copper ratio of less than one are easy to quench. NMC has therefore examined Cu/Cu-Ni/Nb-Ti three layered wires to clarify the standard for the lower limit of allowable copper and the mechanism of quench generation. The goal of WG4 also includes the standardization of RRR measurements for Nb_3Sn wires. A modified test method is necessary for Nb_3Sn wires because of technical differences. A NP will be submitted by Japan within the next two years. The methods testing the tensile strength of Nb_3Sn wires as well as oxide SC at room temperature and/or of Cu/Nb-Ti wires at low temperatures will be soon submitted to IEC/TC90. A standardization of matrix/superconductor volumetric ration for Nb_3Sn wires is now being examined in Japan. At the moment, the image processing method is believed to be the most effective because this method can be used to examine various type of wire constructions.

SUMMARY

The present status of the activities of IEC/TC90 (Superconductivity) has been overviewed. The importance of creating standards is no doubt to promote the progress of superconductivity application. In the next step, the standards of products will be settled based on the standards for test methods, while the systematic network of standards for test methods shall be completed as soon as possible.

Acknowledgments: Since its beginning, the secretariat of TC90 Japanese National Committee has been located within ISTEC International office. Several difficult times have been overcome by the strenuous efforts of Messrs. A.Negishi, H.Tanaka, Y.Furuto and Y.Tanaka as successive managers of IEC/TC90 Secretariat and Ms. M.Takezawa as secretary for IEC/TC90. Thanks to the dedication of these people, the future of IEC/TC90 is appearing brighter. On behalf of the committee members, one of the authors would like to express his appreciation for their efforts.

References:
[1] K.Osamura, K.Sato and Y.Furuto; Cryogenic Eng., 32(1997) 663.
[2] S.Tanaka, A.Murase and K.Osamura; Cryogenic Eng., 33(1998) 10.
[3] K.Itho, Y.Tanaka and K.Osamura; Proc. 6th ICEC/ICMC, Kitakyushu, (Elsevier Sci., 1996) p. 1787.
[4] K.Osamura, A.Nyilas, M.Shimada, H.Moriai, M.Hojo, T.Fuse and M.Sugano; Adv. Supercond., XI (1999) 1515.
[5] T.Kumano, M.Hiraoka and T.Shintomi; Cryogenic Eng., 33(1998) 108.

Residual Resistance Ratio Measurement Method
of Cu/Nb$_3$Sn Composite Conductors

Satoru Murase[1], Rikuo Ogawa[2], Takashi Saitoh[3], Hidezumi Moriai[4] , Teruo Matsushita[5] and Kozo Osamura[6]

[1] Tohoku University, 05 Aramaki-Aoba, Aoba-ku, Sendai 980-8579, Japan
[2] Toyohashi Institute of Technology, 1-1, Hibarigaoka, Tempaku-cho, Toyohashi 441-8122, Japan
[3] Fujikura Ltd., 1-5-1, Kiba, Koto-ku, Tokyo 135-8512, Japan
[4] Hitachi Cable, Ltd., 3550, Kidamari-cho, Tsuchiura 300-0026, Japan
[5] Kyushu Institute of Technology, 680-4, Kawazu, Iizuka 820-067, Japan
[6] Kyoto University, Yoshida-honmachi, Sakyo-ku, Kyoto 606-8501, Japan

Abstract: To submit residual resistance ratio (RRR) measurement method of Cu/Nb$_3$Sn composite conductor as one of new work items to IEC-TC90 in the near future, inter-comparison tests have been carried out. Four laboratories joined the tests using the bronze-processed Cu/Nb$_3$Sn with the external stabilizer as a sample. Laboratory W reacted and measured samples A, B and C, then they were distributed to the other three laboratories (X, Y and Z). Laboratories X, Y and Z measured samples A, B and C, respectively and those data were compared with those of W. As a result, coefficient of variation, COV, of RRR values of 6.7% was obtained. Thermoelectric voltages were also measured in the tests and it was found that it did not affect scattering of RRR measurement if we changed the polarity. If allowable target precision is extended up to around 10%, we will reach the goal of the standardization in RRR measurement method of Cu/Nb$_3$Sn in a year.

Key words: Residual resistance ratio, Cu/Nb$_3$Sn, Inter-comparison test, Standardization

INTRODUCTION

Committee draft (CD) on the residual resistance ratio (RRR) measurement method for Cu/Nb-Ti [1] was prepared and submitted to the 90th Technical Committee of the International Electrotechnical Commission (IEC/TC90). A lot of efforts in "Testing and Evaluation Studies Committee for Metallic Superconducting Materials and Wires" organized by New Material Center in Osaka, Japan, contributed to make the CD. We are now preparing the RRR measurement method for Cu/Nb$_3$Sn to be one of the next work items and carried out inter-laboratory comparison tests in order to clarify applicability of the CD for Cu/Nb-Ti. So far it was found that RRR of Cu/Nb$_3$Sn for the external Cu type was very sensitive to heat-treatment. Based on those results in recent some years, four laboratories joined and took part in the inter-comparison tests to examine the scatter of RRR values for pre-reacted samples and the effects of contamination during heat-treatment and of thermoelectric voltages on RRR.

SAMPLE PREPARATION

We focused on a bronze-processed Cu/Nb$_3$Sn conductor for an external stabilizer (No. 32) because it was widely used in the market, had simple heat-treatment process but had large scattering in the recent studies [2, 3]. The specifications are shown in Table 1. Samples were reacted on a straight base plate to form a Nb$_3$Sn layer.

Table 1. Specifications of bronze-processed Nb$_3$Sn samples

Specimen	No. 32
Wire Diameter, mm	0.82
No. of Filament	5900
Filament Diameter, μ m	3.7
Volume Ratio of Cu Stabilizer, %	58.7
Cu Stabilizer	External

MEASUREMENT METHOD

Optimized heat-treatment condition, 650 ℃ for 200 h, to obtain the highest critical current density was used to form the Nb$_3$Sn layer in laboratory W. It measured RRR for samples A, B and C and then delivered to other laboratories (X, Y and Z). Labs. X, Y and Z received those samples and measured RRR of samples A, B and C, respectively. Measured values are compared with each other for instance A and A, B and B. Beyond those samples, lab. X reacted and measured sample D to examine effects of heat-treatment conditions (650℃/200 h and 700℃/50h). The curve method by pulling up was used in all participants. Especially transport current density less than 1 A/mm^2 at room temperature was rated to prevent excessive temperature rise by joule heating. Details of measurement conditions are shown in the reference [3]. Thermoelectric voltages induced during the RRR measurements were also measured and evaluated.

RESULTS AND DISCUSSION

Inter-comparison tests. Distribution of normalized RRR, RRR values (RRR$_{XYZ}$) measured by labs. X, Y and Z divided by RRR values (RRR$_W$) measured by lab. W, is shown in Fig. 1. Average was 0.952 and coefficient of variation (COV) of 6.7% was obtained. It did not reach COV of 5% like Cu/Nb-Ti but the RRR measurement method for Cu/Nb$_3$Sn could be standardized if around 10% was allowed.

Effects of heat-treatment conditions. RRR values in short heat-treating time (700℃/50h) and long heat-treating time (650℃/200h) are about 300 and 65 - 160, respectively. RRR values for long heat-treating time are lower than those of short heat-treating time and its scattering is larger than that of it. From the above results it is considered that Sn diffuses to Cu stabilizer from the diffusion barrier and then Sn contamination results in low RRR because the barrier has small pores and/or breakage which exist at random. It brought about RRR which changed along longitudinal direction.
Effects of thermoelectric voltage. In the RRR measurements by all participants the range of

thermoelectric voltage values were $0 - 12.5$ μV just above critical temperature (T_c) and did not show any relationship with the scattering of RRR values because the polarity was changed. Thermoelectric voltage, for instance 10 μV just above T_c, increased to $15 - 19 \mu$V at 4.2 K and decreased to 5μV above T_c. The shift of thermoelectric voltages during positive and negative currents measurements is evaluated as defined by $|V_{20+} - V_{20-}|/V_2(\text{bar})$, where V_{20+} and V_{20-} are obtained at zero current in the first and second measurements under positive and negative currents, respectively, and $V_2(\text{bar})$ is the average voltage just above T_c. In all measurements $0 - 3\%$ values within acceptable limits [1] were obtained except for two of 18 data, as shown in Fig. 2.

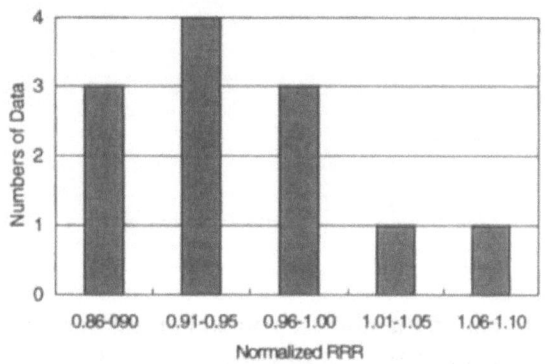

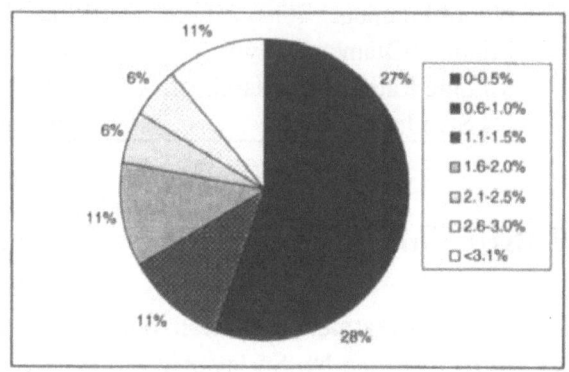

Fig. 1 Distribution of normalized RRR ($\text{RRR}_{\text{XYZ}}/\text{RRR}_{\text{W}}$)

Fig. 2 Distribution of the shift of thermoelectric voltages during positive and negative currents measurements normalized by average observed voltage ($|V_{20+} - V_{20-}|/V_2(\text{bar})$)

SUMMARY

Even when special cares were not taken, the shifts of thermoelectric voltage in the measurements are almost less than the upper limit of 3% in CD for Cu/Nb-Ti. If pre-reacted $\text{Cu/Nb}_3\text{Sn}$ is treated, COV will fit in less than 7% in the RRR measurement of $\text{Cu/Nb}_3\text{Sn}$ and the goal of the standardization in the method for $\text{Cu/Nb}_3\text{Sn}$ will be attained using extended target precision.

Acknowledgement. This work was performed as an activity of the Japanese National Committee of IEC/TC90, being conducted by the New Material Center and supported by the Agency of Industrial Science and Technology, MITI, Japan.

REFERENCES

1. IEC 61788-4/CD, Superconductivity, Part 4: Residual resistance ratio measurement --- Residual resistance ratio of Cu/Nb-Ti composite superconductors (1999).
2. S. Murase, T. Saitoh, T. Matsushita, K. Osamura Proceedings of the 16 th International Cryogenic Engineering Conference (1997) pp. 1795 – 1798.
3. S. Murase, T. Saitoh, H. Moriai, T. Matsushita, K. Osamura, in *Advances in Superconductivity XI*, edited by N. Koshizuka and S. Tajima (Springer-Verlag, Tokyo, 1999), pp. 1511 – 1514.

Inter-laboratory Comparison Test Results for Sustainable Current Measurement of AC Nb-Ti Composite SC Wires

Satoshi Fukui[1], Kazuaki Negishi[2], Osami Tsukamoto[2], Kazuya Ohmatsu[3], Hidezumi Moriai[4], Yasufumi Kasahara[5], M.Sugimoto[6], Yasuzo Tanaka[6] and Kozo Osamura[7]

[1] Niigata University, Ikarashi 8050, Niigata, 950-2181, Japan
[2] Yokohama National University, Tokiwadai 79-5 , Hodogaya, Yokohama, 240-8501, Japan
[3] Sumitomo Electric Industries, Ltd., Shimaya 1-1-3, Konohana-ku, Osaka, 554-0024, Japan
[4] Hitachi Cable Ltd., Kidayo-cho 3550, Tsuchiura, 300-0026, Japan
[5] Central Research Institute of Electric Power Industry, Tokyo, 201-0004, Japan
[6] The Furukawa Electric Co., Ltd., Kiyotaki 500, Nikko, Tochigi, 321-1493, Japan
[7] Kyoto University, Yoshidahonn-cho, Sakyo-ku, Kyoto, 606-8501, Japan

Abstract: Inter-laboratory comparison tests of sustainable current in AC NbTi composite superconducting wires have been carried out at five organizations on a tentative test method. It was found that fluctuation of the measured sustainable current defined by the power criterion was $\pm 1\%$ for intra-laboratory tests and in a range of $\pm 5\% \sim \pm 20\%$ for inter-laboratory tests. The measurement results showed that the longitudinal component of the AC external field influenced the AC sustainable current. This paper describes the test results of the AC sustainable current and discusses on the standardization of the AC sustainable current measurement for the AC wires.

Keywords: AC NbTi composite wire, AC sustainable current, AC transport current loss

INTRODUCTION

The requirement for windings of AC superconducting apparatuses is to generate necessary level of a magnetic field with high reliability and allowable AC loss value. Usually, in the study of stability of the AC wires, a maximum AC operational current is discussed based on the AC quench current. However, results of AC quench current tests are fluctuant and are not necessarily reproducible. Therefore, for the practical design of the AC apparatuses, it is necessary to establish a measure to determine the reliable operational current with an allowable AC loss level. In our previous study [1], we have proposed a method to determine the maximum current by an AC transport current loss level. In this study, to establish a standard test method, we performed round robin tests at five organizations. This paper presents the results of the AC sustainable current measurement and also discusses on the standardization of the AC sustainable current measurement for the AC wires.

TEST METHOD

An AC NbTi composite wire (S02b) and a (6+1) cable (CS021b) made of the same wire were prepared for the tests. Their specifications are listed in [1]. The sample wire and cable were wound bifilarly on a GFRP mandrel with V-shape groove under pre-tension at the room temperature. The sample coil was placed in a superconducting magnet to apply an external field of peak value B_{ex}. The external field was applied parallel to the mandrel axis. A pair of voltage taps was attached to the sample wire and cable in one side of the bifilar winding sample coil. The AC transport current losses were measured in the DC or AC external magnetic field by the direct waveform integration method [2] or by using the lock-in amplifier. The AC sustainable currents, $I_{s,AC}$ at which the AC losses were reached $25\mathrm{kW/m^3}$ and $100\mathrm{kW/m^3}$ [1] were obtained. The DC critical currents, I_c, defined at resistivity criterion of $10^{-14}\Omega\mathrm{m}$ and $10^{-13}\Omega\mathrm{m}$ under the DC external magnetic field, and the AC quench currents, $I_{q,AC}$ under the DC or AC field were also measured.

RESULTS AND DISCUSSIONS

The measurement results at the five organizations are summarized in Table 1. At the organization E, $I_{s,AC}$ and $I_{q,AC}$ were measured in the AC external magnetic field. As shown in Table 1, in some data of $I_{s,AC}$, there are the values of $I_{s,AC}$ which are higher than those of $I_{q,AC}$. These values of $I_{s,AC}$ were determined by interpolating the AC loss - transport current curves. In Table 2, the fluctuations of the measured data from the average values are presented. It is seen from these figures that the fluctuations of the measured I_c and $I_{q,AC}$ were larger than those of $I_{s,AC}$.   The measured results showed that the fluctuation of the values of $I_{s,AC}$ defined by the power criterion was ±1% for intra-laboratory tests and in a range of ±5%~±20% for inter-laboratory tests.   We consider that to define the AC sustainable current

Table 1. Summary of measurement results of AC sustainable currents.

Wire (S02b)	DC critical current I_c		AC sustainable current, $I_{s,AC}$		AC quench current, $I_{q,AC}$
	$10^{-14}\Omega$m	$10^{-13}\,\Omega$m	25kW/m³	100kW/m³	
B_{ex} (DC) (T)	Organization A				
0	52	112	90	—	94
025	36	85	88	—	104
0.5	48	67	—	—	58
1	24	32	—	—	56
B_{ex} (DC) (T)	Organization B				
0	178	Quench	—	—	88
0.2	150	Quench	—	—	99
0.5	84	151	—	—	101
0.8	53	93	—	—	109
1	39	69	—	—	97
B_{ex} (DC) (T)	Organization C				
0	173	Quench	48	83	76.3
0.2	142	Quench	46	80	75.9
0.5	80.8	157	40	67	74.6
0.8	50	96.5	35	58	59.9
1	37.5	71.8	32	53	56.9
B_{ex} (DC) (T)	Organization D				
0	—	—	—	—	93
0.2	—	—	—	—	91
0.5	—	—	—	—	60
0.8	—	—	—	—	52
1	—	—	—	—	50
B_{ex} (DC or AC) (T)	Organization E				
0	160	Quench	50	90	68
0.2	140	255	50	86	80
0.5	100	140	46	66	75
0.8	50	90	42	61	50
Cable (CS021b)	DC critical current I_c		AC sustainable current, $I_{s,AC}$		AC quench current, $I_{q,AC}$
	$10^{-14}\Omega$m	$10^{-13}\,\Omega$m	25kW/m³	100kW/m³	
B_{ex} (DC) (T)	Organization A				
0	320	570	285	585	695
025	260	530	273	520	595
0.5	200	330	238	445	560
1	115	180	284	374	374
B_{ex} (DC) (T)	Organization B				
0	730	Quench	—	—	582
0.2	660	Quench	—	—	584
0.5	427	Quench	—	—	405
0.8	260	497	—	—	412
1	197	385	—	—	382
B_{ex} (DC) (T)	Organization C				
0	713	1248	331	653	783
0.2	623	1189	294	585	776
0.5	398	780	266	516	763
0.8	249	519	242	446	667
1	207	404	227	404	563
B_{ex} (DC) (T)	Organization D				
0	—	—	—	—	—
0.2	—	—	—	—	—
0.5	—	—	—	—	490
0.8	—	—	—	—	474
1	—	—	—	—	413
B_{ex} (DC or AC) (T)	Organization E				
0	830	Over 1000A	438	762	620
0.2	720	Over 1000A	426	667	620
0.5	470	810	360	554	620
0.8	320	560	294	465	480

Table 2. Summary of fluctuation of measurement results.

B_{ex} (DC) (T)	DC critical current I_c		AC sustainable current, $I_{s,AC}$		AC quench current, $I_{q,AC}$
	$10^{-14}\Omega$m	10^{-13} Ωm	25kW/m³	100kW/m³	
	Wire (S02b)				
0	±43%	—	—	±8%	±10%
0.5	±34%	±42%	—	—	±29%
1	±22%	±35%	—	—	±37%
B_{ex} (DC) (T)	Cable (CS021b)				
0	±39%		±8%	±6%	±15%
0.5	±36%	±36%	±18%	±15%	±31%
1	±27%	±35%	±20%	±4%	±22%

by a power criterion is effective and the level of the AC transport current loss can be a practical measure to determine the maximum operational AC current of the AC wire. However, the fluctuations of the measured $I_{s,AC}$ data is still large. Therefore, the method of the inter-laboratory comparison test should be improved.

It has been revealed in the previous study [3] that the AC transport current losses of the AC wires depend by the direction and the amplitude of the longitudinal AC magnetic field component to the wires. If the AC transport current loss characteristics of the AC wire is affected by the longitudinal field components, $I_{s,AC}$ defined by the losses is also affected by them. In this measurement, when B_{ex} was applied parallel to the test coil axis, the wires in the test coils were exposed to the external longitudinal field component $B_{ex,z}$. The self longitu-

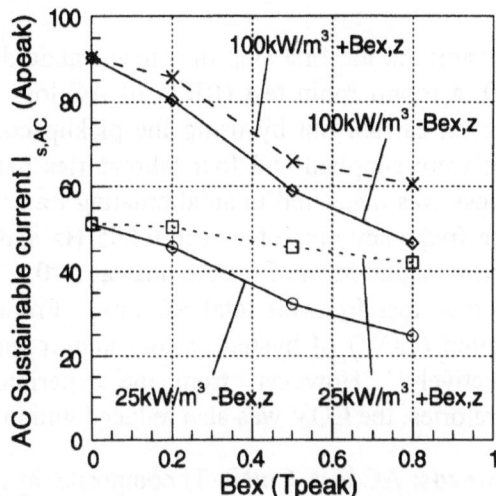

Fig.1. Influence of $B_{ex,z}$ on $I_{s,AC}$ (S02b)

dinal field component, $B_{sf,z}$, was also produced in the wire by the transport current in the twisted wire. In the following, we define that the positive direction of $B_{ex,z}$ is in same direction as that of $B_{sf,z}$. The dependence of the AC sustainable currents on the direction of the longitudinal component was evaluated. The AC transport current losses of the wire and the cable were measured for both $\pm B_{ex,z}$ at the organization E. The values of $I_{s,AC}$ determined by the measured AC transport current losses are plotted against B_{ex} in Figure 1. From the figure, it is clear that $I_{s,AC}$ decrease when the $-B_{ex,z}$ is applied. The difference of $I_{s,AC}$ of the wire is 30% at B_{ex}=0.5T$_{peak}$. The similar results were obtained for the cable and the difference of $I_{s,AC}$ is 13% at B_{ex}=0.5T$_{peak}$.

CONCLUSION

The inter-laboratory comparison tests of the AC sustainable current of the NbTi AC wire and the (6+1) cable were performed. The fluctuations of the measured AC sustainable currents defined by the power criteria were much smaller than those of the quench currents. We consider that our proposed method is effective and reasonable. However the fluctuations of the AC sustainable currents were still large. Therefore, it is necessary to improve the measurement method. It was also shown by the measurement that the longitudinal field component of the AC external field influenced the AC sustainable current. To decide the configuration of the sample coil, this influence should be taken into consideration.

Acknowledgement: This work was performed as the activity of Japan National Committee of IEC/TC 90, being conducted by the New Materials Center, Osaka Science and Technology Center and supported by Agency of Industrial Science and Technology, MITI, Japan.

1. Y.Tanaka O.Tsukamoto, S.Fukui et al, Presented at ISS '98, Adv. in Supercond. XI (1999)
2. S.Fukui et al, IEEE Trans. on Appl. Supercond., 7, 282-285 (1997)
3. S.Fukui et al, Physica C 310, 142-146 (1998)

Standardization of AC loss measurement of Cu/Nb-Ti composites exposed to alternating transverse magnetic field by pickup coil method

Kazuo Funaki[1], Hiroyasu Yumura[2], Shuma Kawabata[3], Masahiro Sugimoto[4], Kikuo Ito[5], and Kozo Osamura[6]

[1] Kyushu University, Hakozaki 6-10-1, Higashi-ku, Fukuoka 812-8581, Japan
[2] Sumitomo Electric Industries, Ltd., Shimaya 1-1-3, Konohana-ku, Osaka 554-0024, Japan
[3] Kagoshima University, Koorimoto 1-21-40, Kagoshima 890-0065, Japan
[4] The Furukawa Electric Co., Ltd., Kiyotakicho 500, Nikko 321-1493, Japan
[5] National Research Institute for Materials, Sakura 3-13, Tsukuba 305-0003, Japan
[6] Kyoto University, Yoshida-honmachi, Sakyo-ku, Kyoto 606-8501, Japan

Abstract: As the first step of a new standardization item of metallic superconductors in IEC/TC90-WG9, a round robin test (RRT) of AC loss measurement of Cu/Nb-Ti composite superconductors has been carried out by using the pickup coil method. Three kinds of Cu/NbTi multifilamentary wires were supplied and four laboratories participated in the RRT. The frequency dependence of AC loss was measured in an alternating transverse magnetic field with the amplitude of 1 T at 4.2 K in the frequency range between 0.005 Hz and 1 Hz. The hysteresis loss per cycle was obtained as an extrapolated level of the AC loss at $f = 0$. The coupling loss was also obtained by subtracting the hysteresis loss from the total AC loss. From the RRT among the 4 laboratories, the coefficients of variation (COV) of hysteresis loss and coupling time constant were within about 5 % and 10 %, respectively. However, from the experimental results over a wide range of frequency in 3 laboratories, the COV was also reduced within 5 % for the coupling time constant.

Keywords: AC loss, Cu/Nb-Ti composite, hysteresis loss, coupling loss, pickup coil method

INTRODUCTION

The 90th Technical Committee of International Electrotechnical Commission (IEC/TC90) has been intensively working on the standardization of testing and evaluation methods for superconductors to respond to the growing needs for standardization. As for the AC loss measurement of Cu/Nb-Ti composite superconductor, two types of measurement methods, VSM and pickup coil method, were proposed to IEC/TC90. The total AC loss including the hysteresis loss and the coupling loss can be measured by the pickup coil method. The first Working Draft (WD) for the pickup coil method to measure the total AC loss exposed to a transverse alternating magnetic field was proposed by the Japanese National Committee. After the proposal, a round robin test (RRT) was carried out by a working group in the Japanese National Committee to confirm the feasibility of the proposed measurement method, and to accumulate data necessary for the standardization. Four laboratories (Kyushu University, Kagoshima University, Sumitomo Electric Industries, Ltd., and The Furukawa Electric Co, Ltd.) participated in this RRT.

SAMPLES AND TEST METHOD

Three kinds of Cu/Nb-Ti wires, H1, H2 and H3, shown in Table 1 were prepared. These sample wires were the same as those that had been supplied for the 2nd VAMAS AC loss measurement intercomparison.[1] The sample wires were wound on the former in a single-layer solenoid coil. The former was made of non-metallic and non-magnetic material such as GFRP and bakelite. Since all sample wires had an insulation layer, they were wound minutely on the sample coil. The height of the sample coil was more than three times as high as that of the pickup coil to reduce geometrical error caused by the end effect of the sample coil. Both ends of a sample wire were opened and ground by emery paper of #800 - #1000 and etched with nitric acid to prevent filaments from contacting each other.

Table 1. Specifications of Cu/Nb-Ti samples

Sample	H1	H2	H3
Wire diameter	0.5mm	0.5mm	0.5mm
Cu/Nb-Ti ratio	1.0	2.6	1.6
Filament diameter	11.8 μm	3.4 μm	1.3 μm
Number of filaments	931	6517	56791
Twist pitch	9mm	9mm	9mm

A typical arrangement of pickup coils and electrical circuit for AC loss measurement is illustrated in Fig.1. The main pickup coil was arranged coaxially and adjusted concentrically outside the cancel coil, and the sample coil was arranged between both coils. The main pickup coil detects magnetic moment induced in a sample coil, and the cancel coil detects an alternating external magnetic field. In the signal from the main pickup coil, the component of the alternating external magnetic field is extremely superior to that of the magnetic moment. In the pickup coil method, the magnetization is estimated by canceling out the field component in the main pickup coil using the signal from the cancel coil. The AC loss is estimated by integrating the magnetization for the applied field through a period.

The AC loss was measured under the external alternating magnetic field of the amplitude 1 T at 4.2 K. The test conditions are listed in Table 2. The measurements in each laboratory were performed at more than 5 points of frequency in the range between 0.005 Hz and 1 Hz. The hysteresis loss was obtained as an extrapolated level of the AC loss at f =0. The level can be extrapolated in the frequency dependence of AC loss per cycle by using the least square method. The coupling loss was also obtained by subtracting the hysteresis loss from the total AC loss. The coupling time constant was calculated from the proportional coefficient of the coupling loss to the frequency.

Table 2. Test conditions with pickup coil method

Items	Test conditions
Sample coil	· a single-layer solenoid coil
	· more than three times as high as the pickup coils
	· edges ground and etched with acid
External magnetic field	· sine wave with the amplitude of 1 T
	· in the direction parallel to the coil axis
	· frequency between 0.005 Hz and 1 Hz
Temperature	· in liquid helium (around 4.2 K)

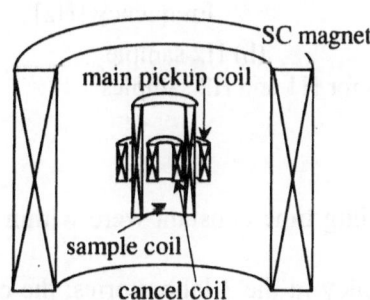 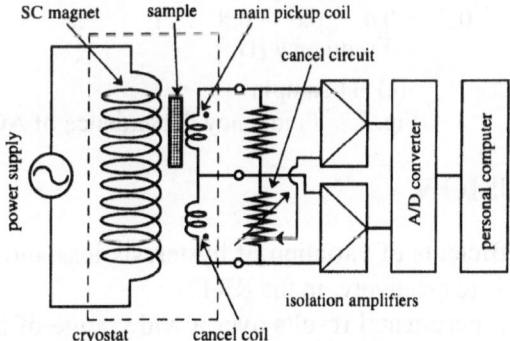

Fig.1. A typical arrangement of pickup coils and electrical circuit for AC loss measurement

RESULTS AND DISCUSSION

Results of the RRT are summarized in Table 3. The frequency dependence of AC loss for sample number H1 and H2 are given in Fig. 2. Four lines in Fig. 2 show the frequency dependence of AC loss measurement values for each laboratory calculated by using the least square method. The coefficients of variation (COV) of hysteresis loss for H1, H2 and H3 samples are 1.5 %, 5.2 %, and

708

4.0%, respectively.   The COV of coupling time constant are also estimated as 10.8 %, 4.3 % and 4.1 %, respectively.   Large COV of the coupling time constant may result from a situation in which the laboratory D obtained the frequency dependence of the AC loss in a limited lower frequency region, as seen in Fig.2.   If the results of the laboratory D are omitted, the COV attain to 3.1 %, 2.3 % and 4.6 %, respectively.   As a result of the RRT, the coupling time constant shall be calculated by the experimental results over a wide range of frequency.

Table 3.   RRT results of hysteresis loss and coupling time constant for three samples

	Hysteresis Loss			Coupling Time Constant		
Sample	H1	H2	H3	H1	H2	H3
Laboratory A	64.7 kJ/m^3	9.73 kJ/m^3	5.52 kJ/m^3	1.71 msec	2.89 msec	1.99 msec
Laboratory B	62.1 kJ/m^3	9.01 kJ/m^3	5.56 kJ/m^3	1.61 msec	2.73 msec	1.79 msec
Laboratory C	63.3 kJ/m^3	9.31 kJ/m^3	5.28 kJ/m^3	1.73 msec	2.83 msec	1.83 msec
Laboratory D	62.7 kJ/m^3	8.42 kJ/m^3	5.02 kJ/m^3	2.12 msec	3.07 msec	1.90 msec
Average	63.2 kJ/m^3	9.12 kJ/m^3	5.35 kJ/m^3	1.79 msec	2.88 msec	1.88 msec
COV	1.5 %	5.2 %	4.0 %	10.8 %	4.3%	4.1 %
Average*	63.4 kJ/m^3 *	9.35 kJ/m^3 *	5.45 kJ/m^3 *	1.68 msec*	2.82 msec*	1.87 msec*
COV*	1.7 %*	3.2 %*	2.3 %*	3.1 %*	2.3 %*	4.6 %*

* The results of the laboratory D are omitted.

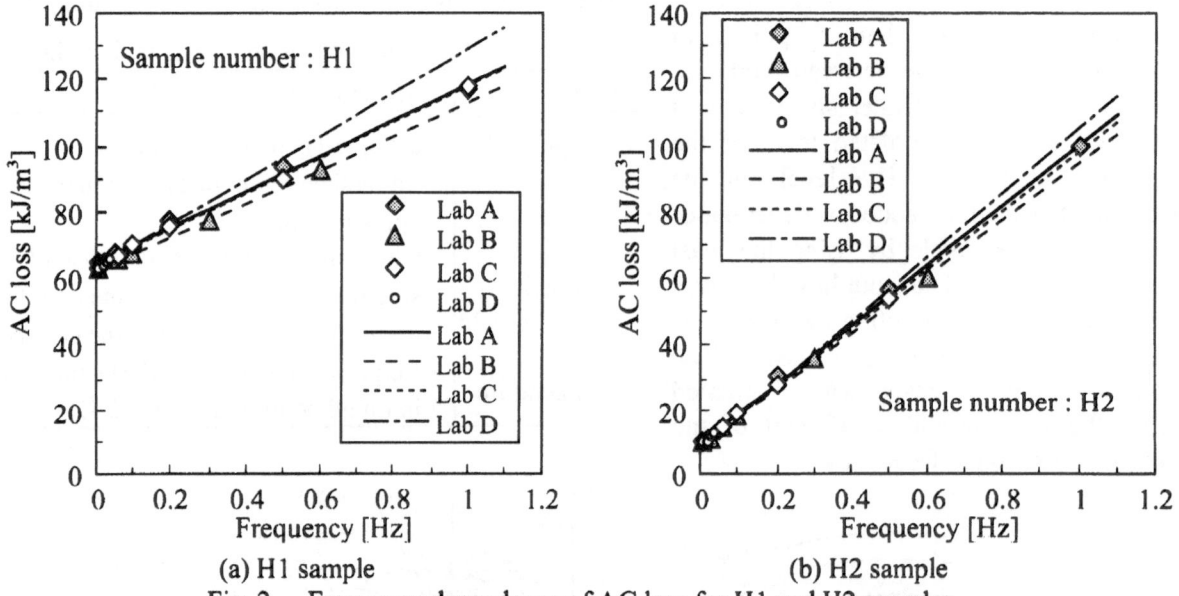

(a) H1 sample   (b) H2 sample
Fig. 2.   Frequency dependence of AC loss for H1 and H2 samples

CONCLUSION

(1) The coefficients of variation of hysteresis loss and coupling time constant were within about 5 % and 10 %, respectively, in the RRT.
(2) For the experimental results over a wide range of frequency in the 3 laboratories, the coefficients of variation were also reduced within 5 % for coupling time constant.

ACKNOWLEDGMENT

This work was performed as an activity of the Japanese National Committee of IEC/TC90, being conducted by the New Material Center of Osaka Science and Technology Center and supported by the Agency of Industrial Science and Technology, MITI, Japan.

1.   E.W. Collings, M.D. Sumption, K. Itoh, H. Wada and K. Tachikawa, *Cryogenics* 37, 49-60 (1997).

Magnetic Measurements of AC Loss in Short Multifilamentary Tapes

E.W. Collings, E. Lee and M.D. Sumption

Laboratories for Applied Superconductivity and Magnetism
Department of Materials Science and Engineering
The Ohio State University, 2041 College Rd, Columbus OH 43210, U.S.A.

Abstract: The vibrating sample magnetometer (VSM) in conjunction with an iron-cored electromagnet is suitable for magnetic AC loss measurement within a field-sweep-rate range of typically zero (point-by-point) to 4 T/min (or 0~10 mHz). In that it is capable of measuring eddy current loss over that very low range of frequencies, the VSM can be regarded as spanning the gap between the "pick-up-coil regime" and the DC limit represented by the SQUID- and extraction-type magnetometers. Presented here is an example of the use of the VSM in the measurement of low frequency AC loss. Samples for measurement are two sets of multifilamentary (MF) NbTi-base strands, with different filament numbers, that have been flattened to several aspect ratios. It is shown that for the densely filamented strands the eddy-current-measured effective transverse matrix resistivity, $\rho_\perp$, agrees well after shape-factor correction, with the standard effective-medium model. Not so for the coarsely filamented tapes which yield anomalously high apparent $\rho_\perp$s. The discrepancy is explained in terms of a reduced "eddy current activation volume" in the latter class of tape. The study demonstrates that when interpreting the results of the AC loss measurement of non-round MF strands, consideration must be given *both* to external-shape-governed demagnetization *and* to internal strand architecture.

Key words: AC loss measurement, vibrating sample magnetometer, transverse matrix resistivity, multifilamentary NbTi tapes, aspect ratio

INTRODUCTION

The vibrating sample magnetometer (VSM) in conjunction with the relatively low field sweep rate (dH/dt) achievable with an iron-core electromagnet (typically 0-4 T/min or frequency, $f = $ 0~10 mHz) can be regarded as filling an experimental niche between the strictly-AC induction (pickup-coil) magnetometer and DC instruments such as the SQUID- and extraction-coil magnetometers. The VSM magnetization (M) measurement performed as a function of field-sweep ramp rate will yield by extrapolation the "hysteretic loss", Q_h (M-H loop area Q as dH/dt$\rightarrow$0) as well as the low-frequency eddy current loss $Q_e(f)$. The presence of such an eddy current component to the total loss is indicated by an f-proportional increase in Q; or for fixed f by a Q that increases with L^2, the square of the sample length. Having determined the presence of eddy current loss it is useful to be able to quantify it in terms of an "effective transverse matrix resistivity", $\rho_\perp$. For a dense packing of round superconducting (SC) filaments in a normal-metal (NM) matrix an effective medium model for the average transverse resistivity predicts that $\rho_\perp = (1+\lambda)/(1-\lambda)^{\pm 1}\rho_{bulk}$ where λ is the SC packing fraction and the choice of "+" or "-" sign in the index is governed by whether or not there is a relatively large SC/NM interface resistance. Then for a long cylindrical twisted (to pitch, L_p) multifilamentary (MF) strand in an

709

applied transverse magnetic field Carr's "anisotropic continuum" model [1] yields

$$Q_e = \frac{L_p^2}{2x10^9 \rho_\perp} H_m^2 f \qquad (1)$$

in c.g.s units, or SI units with removal of factor 10^9, and where H_m is the field-sweep amplitude -- a well verified relationship (e.g. [2,3]) for the usual round twisted strands encountered in low-temperature superconductivity (LTSC).

Then to position ourselves for magnetic studies of Q_e and hence $\rho_\perp$ in high temperature superconducting (HTSC) tapes [4], and non-round *untwisted* strands in general, we undertook a series of measurements on LTSC strands that had been flattened to various aspect ratios. Then in order to interpret the results it was first necessary to modify Eqn.(1) to take into account the effect of strand flattening; this took place in the following two stages.

EQUATIONS FOR ASPECTED-STRAND LOSS

Eddy Current Loss in *Twisted* MF Strands. According to Campbell [5] the eddy current loss in twisted aspected strands (aspect ratio a/b, sinusoidal H perp. to a) is

$$Q_e = 2n\frac{\pi^2}{10^9} H_m^2 f \tau \qquad (2)$$

Here Q_e is the product of an "*external shape factor*", n (a function of the strand demagnetization) and some "*relaxation time*", τ. This single expression embodies the Q_e of several strand types:

$a \gg b$ *(field face-on, FO)*:	$n = \mathbf{a/b}$	$\tau = L_p^2/\rho_\perp(a/b)(7/480)$
round:	$n = \mathbf{2}$	$\tau = L_p^2/\rho_\perp(1/8\pi^2)$
$a \ll b$ *(field edge on, EO)*:	$n = \mathbf{1}$	$\tau = L_p^2/\rho_\perp(a/b)^2(1/16)$

Consider again the round strand. After substituting $n = \mathbf{2}$ and $\tau = L_p^2/\rho_\perp(1/8\pi^2)$ into Eqn.(2) we find, following Campbell:

$$Q_e = 2x2\frac{\pi^2}{10^9}H_m^2 f . \frac{L_p^2}{8\pi^2 \rho_\perp} = \frac{L_p^2}{2x10^9 \rho_\perp} H_m^2 f \qquad (3)$$

which is precisely Eqn.(1), the Carr expression. It follows that the latter includes a hidden n-factor which had been absorbed into it in the form of the factor 2. Recognizing this we re-write Eqn.(1) in the form:

$$Q_e = n . \frac{L_p^2}{4x10^9 \rho_\perp} H_m^2 f \qquad (4)$$

We are now in a position to consider eddy current loss in the **untwisted** strand.

Eddy Current Loss in *Untwisted* MF Strands. We begin with Carr's basic expression for loss in round untwisted strands of length L, viz:

$$Q_e = 0.81 \frac{(2L)^2}{2x10^9 \rho_\perp} H_m^2 f \tag{5}$$

This is comparable to the twisted case but with a prefactor 0.81 and the replacement of L_p by 2L. Just as before we must recognize the implicit presence of an n = 2 factor. Finally, after replacing field-sweep frequency, *f*, by ramp-rate, *dH/dt*, using $4H_m f = dH/dt$ we find

$$\frac{Q_e}{L^2} = n \frac{0.81 H_m}{4x10^9 \rho_\perp}(dH/dt) \tag{6}$$

which demonstrates that Q_e/L^2 versus dH/dt is linear with slope proportional to $1/\rho_\perp$. Equation (6) enables what we could refer to as an "aspect-ratio corrected $\rho_\perp$" (or n-adjusted $\rho_\perp$) to be derived from the results of a Q_e vs. dH/dt experiment. Carr has pointed out [6], and it is easily verified, that in terms of the *internal applied field* the Q_e of the untwisted strand is independent of aspect ratio. Aspect ratio enters the expression for Q_e only via the demagnetization-related prefactor n, which as before must take on the values:

flattened FO (a≫b)   **n = a/b**
round   **n = 2**
flattened EO (a≪b)   **n = 1**

It follows that in plots of Q_e vs. dH/dt, $Q_{e,EO}$ should have 1/2 the slope of $Q_{e,round}$, while the steeper $Q_{e,FO}$ curves should fan out as a/b increases, as sketched in Fig.1.

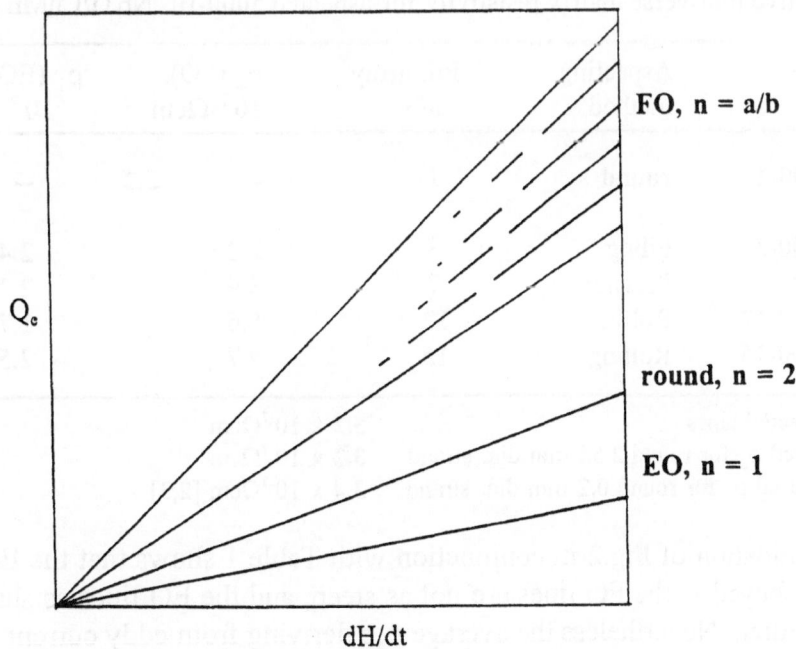

Figure 1: Eddy current loss in tapes as function of aspect ratio, a/b -- schematic

MATRIX RESISTIVITIES OF FINE MULTIFILAMENTARY TAPES

For the verification of Eqn.(6) we selected a round 2.58 mm diam. CuMn-matrix MF (5355 fils.) NbTi strand whose eddy current loss and $\rho_\perp$ (in smaller diam. form) had been the subject of several earlier measurements [2,3]; we then flattened it by filing, pressing or rolling to aspect ratios (filamentary array basis) of 3 to 15. The FO- and EO-measured Q_es are depicted in Fig.2, and the resulting slope-derived $\rho_\perp$s are presented in Table 1.

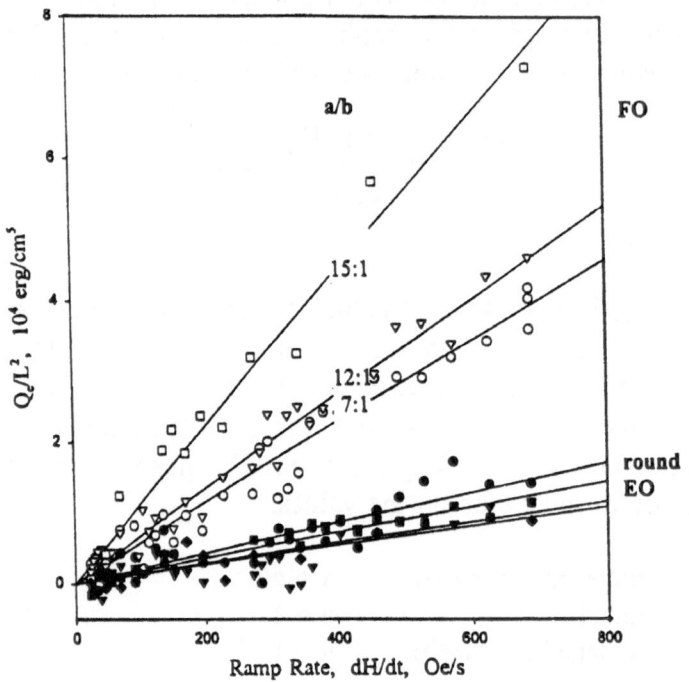

Figure 2: Eddy current loss in 5000-filament NbTi/CuMn tapes

Table 1. Effective transverse matrix resistivity for aspected 5000-fil. NbTi/CuMn

Sample code	Aspecting method	Fil. array a/b	$\rho_\perp$ (FO), 10^{-7} Ωcm	$\rho_\perp$ (EO), 10^{-7} Ωcm
NB5000-1	**round**	1	-- 3.2	--
NB5000-3	Filing	3	2.2	2.4
NB5000-7	Pressing	7	3.9	2.3
NB5000-12	Rolling	12	5.6	1.7
NB5000-15	Rolling	15	4.7	2.5

Mean $\rho_\perp$ for aspected tapes $3.2 \times 10^{-7}\Omega$cm
Presently measured $\rho_\perp$ for round 2.58 mm dia. strand $3.2 \times 10^{-7}\Omega$cm
Previously measured $\rho_\perp$ for round 0.2 mm dia. strand $3.4 \times 10^{-7}\Omega$cm [2,3]

A careful examination of Fig.2 in conjunction with Table 1 shows that the Eqn.(6) prescription is not strictly obeyed -- the FO lines are not as steep, and the EO lines as shallow as the theory (and Fig.1) predict. Nevertheless the average $\rho_\perp$s deriving from eddy current loss measurements on the aspected finely filamented MF NbTi/CuMn tapes are in reasonable agreement with those measured on the round strands, thus providing reasonable semiquantitative confirmation of Eqn.(6). But this turns out to be far from the case for coarsely filamented MF tapes.

MATRIX RESISTIVITIES OF COARSE MULTIFILAMENTARY TAPES

In order to simulate the coarse filamentary arrangement generally encountered in MF HTSC tapes (e.g. [4]), but through the use of LTSC strand material, a series of tapes was prepared by rolling to aspect ratios of from 1.5 to 30 a 54-filament NbTi/Cu strand. Figure 3 depicts the eddy current loss results displayed in the format Q_e/L^2 vs. dH/dt. In it we note that the loss data fan out both above *and* below the round-strand result. Furthermore, although the FO slopes increase in the order of increasing a/b, the rate at which they do so is unexpectedly small -- e.g. the slope corresponding to a/b = 30 is only 1.8 times greater than that of the round strand.

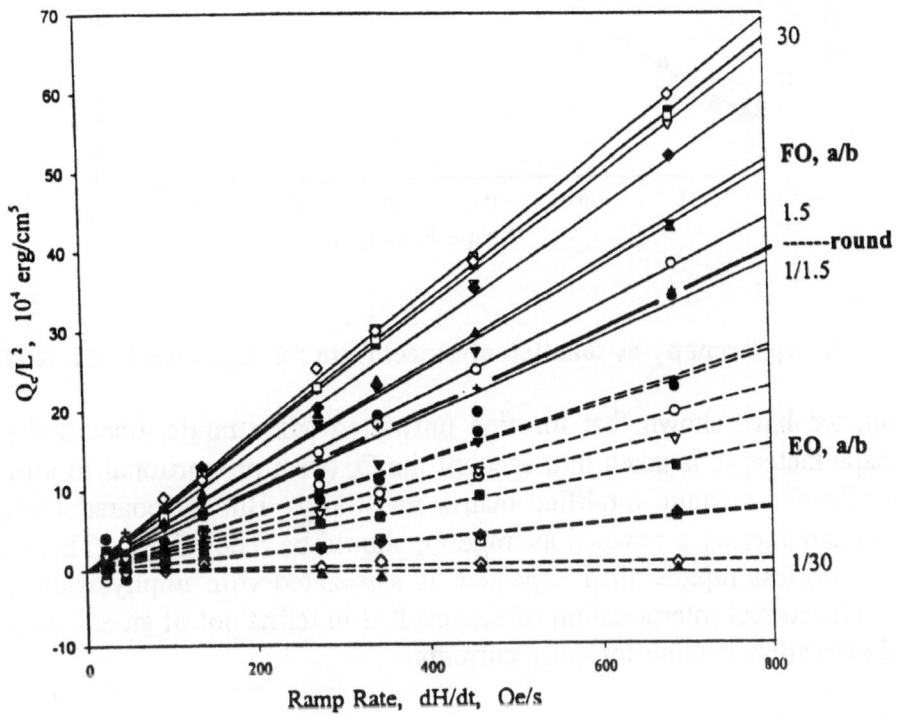

Figure 3: Eddy current loss in annealed 54-filament NbTi/Cu tapes

CONCLUDING DISCUSSION

Plotted in Fig. 4 as function of a/b are the Eqn.(6)-derived coarse-strand $\rho_\perp$s. To eliminate the effect of cold work on matrix resistivity the strands were annealed for 8h/400°C. On one of the strands, a four-terminal measurement yielded an RRR of 500, which led to an expected effective-medium ($\lambda \sim 0.7$) $\rho_\perp$ of 7.4 nΩcm. With values of up to 160 nΩcm the coarse-structured strands yield anomalously high effective $\rho_\perp$s. It seems that strand flattening spreads out the filamentary array and results in a localization of eddy current paths. Thus Q_e averaged over the entire strand volume decreases and hence $\rho_\perp$ increases. Clearly this result should not be interpreted as an intrinsic effective-medium MF resistivity. Instead, a more realistic approach stops short of a calculation of $\rho_\perp$ and recognizes that the anomalously low Q_e is the result of a restricted or confined circulation of eddy currents.

714

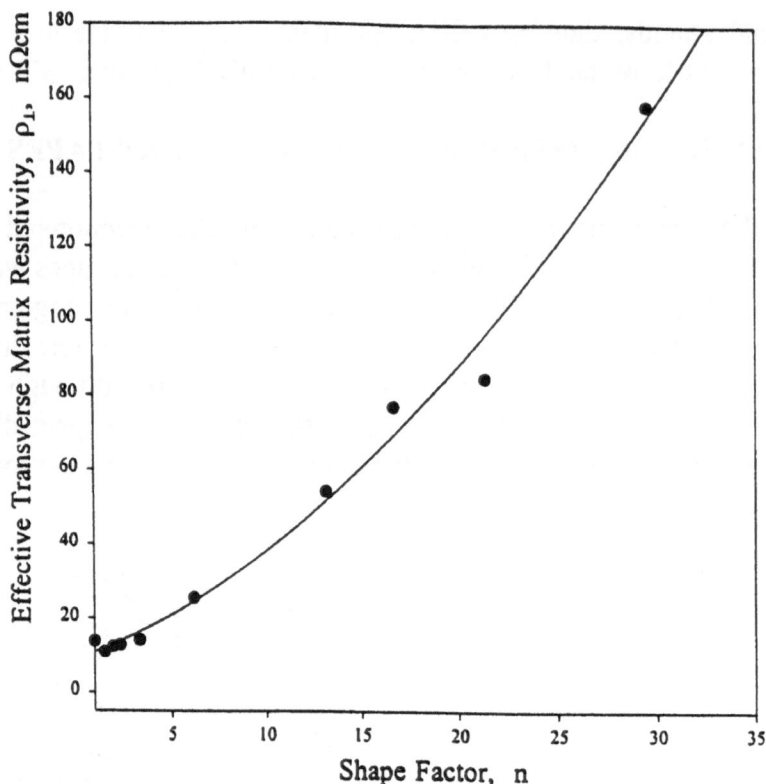

Figure 4: **Apparent $\rho_\perp$ as function of aspect ratio for annealed 54-fil. NbTi/Cu tapes**

In conclusion, we have shown that for fine untwisted MF strands, once a demagnetization-associated shape factor, n, is taken into account the Q_e (then proportional to a/b) is determined by the usual effective-medium modified matrix resistivity. But for coarse aspected arrays the possibility of encountering a reduced average Q_e should be recognized. This loss, which also increases with a/b less rapidly than expected, is associated with unphysically high values of apparent $\rho_\perp$. The correct interpretation of the result is in terms not of an enhanced $\rho_\perp$ but rather of a reduced activation volume for eddy currents.

ACKNOWLEDGEMENTS

The research was financially supported by the U.S. Department of Energy, Division of High Energy Physics, under Grant No. DE-FG02-95ER40900.

REFERENCES

1. W.J. Carr, Jr., _AC Loss and Macroscopic Theory of Superconductors_ (Gordon and Breach, London, UK, 1983).
2. M.D. Sumption, D.S. Pyun, and E.W. Collings, _Transverse and longitudinal resistivities in NbTi multifilamentary strands with Cu and CuMn matrices_, IEEE Trans. Appl. Supercond. **3**, 859-862 (1993).
3. M.D. Sumption and E.W. Collings, _Transverse resistivities of Cu-matrix and Cu-Mn matrix multifilamentary strands as functions of magnetic field and temperature_, Adv. Cryo. Eng. (Materials) **40**, 807-814 (1994).
4. M.D. Sumption and E.W. Collings, _Aspect ratio dependence of effective transverse matrix resistivity in multifilamentary HTSC/Ag strands_, Adv. in Superconductivity-XI, N. Koshizuka and S. Yajima (eds.)(Springer-Verlag, Tokyo, 1999) pp. 831-834.
5. A.M. Campbell, _General treatment of losses in multifilamentary superconductors_, Cryogenics **22**, 3-16 (1982).
6. W.J. Carr, Jr., personal communication.

ac Losses in $Bi_2Sr_2Ca_2Cu_3O_{10}$/Ag Tapes: What Are The Critical Currents to Be Used in Loss Calculations ?

M. Suenaga, T. Chiba*, S. P. Ashworth, and D. O. Welch

Department of Applied Science, Brookhaven National Laboratory, Upton, New York, 11973, USA

Abstract: Since $Bi_2Sr_2Ca_2Cu_3O_{10}$/Ag, [Bi2223/Ag], tapes have geometrically large aspect ratios, and have highly anisotropic properties along its crystallographic directions, it is not clear to what extent the ac loss theories, which are based on the critical-state model of an isotropic superconductor, are applicable to these conductors. We report on tests of these models against magnetically measured ac losses of a Bi2223/Ag tape, in magnetic environments similar to those in power devices. It is shown that the self-field critical currents are not appropriate for calculations of the losses in either parallel or perpendicular fields to the tape face. In addition, the use of the field-dependent critical currents, such as in Kim's model, is important to understand the behavior of the losses in perpendicular applied fields.

Keywords: ac losses, $Bi_2Sr_2Ca_2Cu_3O_{10}$/Ag tapes, the Bean and the Kim models

INTRODUCTION

In recent years, significant improvements in critical currents in $Bi_2Sr_2Ca_2Cu_3O_{10}$/Ag, [BSCCO/Ag], composite tapes have been made to the extent that technological utilization of these superconducting tapes for large devices in electrical power applications are seriously being considered. Also, along with such technological developments, a large number of studies on ac losses in the tapes have been reported [1]. These studies have shown that there is a sufficient understanding of the losses due to transport currents. In particular, the relationship, derived by Norris [2] for the losses and currents, appears to describe the experimental results quite well in most cases. Surprisingly, this is the case in spite of the fact that the derivation is for an isotropic superconductor, while the superconducting properties of BSCCO are highly anisotropic along the c- and a-b direction of the crystal, or the perpendicular and the parallel directions with respect to the tape face. Also, for simultaneous applications of ac transport currents and ac external magnetic fields to a single tape, an expression for the losses was found describing the losses as a function of currents at a given field, or of magnetic fields at a specific current [3]. In this particular case, the finite element numerical calculations using a power-law resistivity $\rho(J) \sim J^n$, where n is a number, can also predict the losses which are in agreement with the experimental results [4]. However, most of the studies to date were performed on a single tape rather than on a tape surrounded by other tapes as they are found in electrical power devices. Thus, studies of the losses in this more complex environment are needed to advance the capability to predict the losses in a device.

Magnetic measurements of the losses can be made for a tape in a magnetic environment which is similar to those in a device. However, we have found significant discrepancies between the calculated and the magnetically measured losses for these tapes. A factor of two or greater difference was often seen between the values of critical currents which were determined by the self field dc-transport currents and which were deduced by the application of the Bean's critical-state model to the hysteresis loss in these tapes [5]. Unfortunately, these results point out that it is difficult to calculate the losses in an electrical power device unless we understand what is (or are) appropriate measurements for I_c which can be use to estimate the losses in a device. Here we will summarize the results of our attempts to seek understanding of ac losses in these tapes such that an method(s) for the measurements of $I_c(H)$ of a tape which is appropriate for the use in an estimation of the losses in a device be found.

Hysteresis-Loss Models

Here, in order to facilitate the quantitative comparison of the observed losses with the hysteresis models for the losses in a superconductor, a brief discussion of the models is given. Also, we only consider the losses based on the Bean critical-state model and on Kim's modification to the Bean model by including the magnetic field dependence of critical currents.

Bean Model. As well known, the hysteresis loss from an infinite slab of a superconductor is given by the Bean model as [6]:

$$P = (2/3)\mu_0(H^3/H^*), \qquad H < H^* \tag{1}$$

$$= 2\mu_0 H H^*(1-2H^*/3H), \quad H > H^* \tag{2}$$

where P is in $J/m^3/cycle$, and H and H^* are the amplitude of ac applied magnetic field and the full penetration field in A/m, respectively. μ_0 is the magnetic permeability of free space. Also, $H^* = wI_c/A$ where w and A are one half width of the tape perpendicular to the direction of applied fields and the cross sectional area of a tape, respectively.

Kim's $I_c(H)$. Since the critical currents I_c of these tapes depends strongly on perpendicular applied fields, it is often expected that the values of ac losses in this geometry may be influenced by the I_c variation with the field. In order to study this possibility, we consider Kim's expression [7] for the field dependence of $I_c(H)$ for calculating the losses in perpendicular fields. There are other expressions which describe $I_c(H)$, but this expression is easier to see the physical process of the losses. Thus, we use it here and this is given as:

$$I_c(H) = I_c(0) H_o/(H + H_o) \tag{3}$$

where H_o is the characteristic field at which the value of I_c becomes one half of its value at H = 0, i.e., $I_c(0)$. Since no simple analytical expression for the losses can be obtained using this expression, we will first discuss the condition under which the use of Kim's relationship becomes important for calculating the losses in these tapes. Then, we will compare the numerically calculated losses and the measured losses at the end of this article.

From Eq. (24) in Ref. 8, the full penetration fields H^* and H^*_K, for the Bean and the Kim models, respectively, are related by an expression through the characteristic field H_o:

$$H^*_K = H_o [(2H^*/H_o + 1)^{1/2} -1]. \tag{4}$$

From this relationship, one notes that H^*_K would not be sufficiently different from H^* unless H^*/H_o is an order of one or greater. When the ratio H^*/H_o is large, the full penetration field is reduced to the logarithmic average of H^* and H_o. The range of the values for H_o for these tapes is 0.1 to 0.2 T in perpendicular fields.

Experimental Procedure

Specimens. In this study, we report the results of the study on a single multifilamentary tape of Bi(2223)/Ag having a nominal critical current of 35 A in self field, although similar measurements were made on a number of the tapes having critical currents differing by a factor of up to three. The overall dimension of the tape was 3.56 mm wide and 0.23 mm thick, and the central filamentary region was 3.37 mm by 0.154 mm.

dc Critical Currents. For comparisons of critical currents which were deduced from ac loss measurements with dc critical currents at self-field, the detailed I-V curves were measured for these tapes. For most of the cases, the critical currents of these tapes were determined using the criterion of the measured induced electric field of 100 µV/m. [Note that this I_c criterion is commonly used, but there is no justification as to why this is an appropriate measurement of I_c to compare with the ac measurements.] Also, in order to study the effects of cracks in the tapes on magnetically measured ac losses, three tapes were measured for critical currents after they were bent in two directions around pipes having diameters of 25, 38, and 50 mm. Then, the central 25 mm sections of these tapes were taken for ac loss measurements. Magnetic field dependence of critical currents for an as received tape was determined in two field directions, parallel and perpendicular, with respect to the tape face

Magnetic ac Loss Measurements. This measurement technique utilizes a pick-up coil (loss-coil), which is placed directly on a specimen inside a Cu wire wound solenoid. ac magnetic field is generated by a variable –frequency power supply. The ac magnetic field is measured by another pick-up coil (field-coil) located above the specimen in the solenoid. The wave forms from both pick-up coils are collected by a digital storage oscilloscope (Nicolet 2090-III) after the voltages have been amplified by a pair of preamplifiers (SRI Sr-560). These were then transferred to a personal computer, where the power loss P *per unit length* (W/m) along the field direction of a superconductor is calculated by the relationship:

$$P = \frac{1}{n\tau} \int_o^\tau H(t) e_f(t) dt \qquad (5)$$

Here, N is the number of turns of the loss-coil and τ is the period of the applied ac field. The voltage $e_f(t)$ and the ac magnetic field $H(t)$ are measured by the corresponding pick-up coils. It should be noted that this method is only applicable for a long uniform specimen where an infinite slab or rod configuration can be assumed.

Although a Bi(2223)/Ag tape in parallel field is a good approximation to an infinite slab, it is not at all clear whether a stack of the tapes simulates an infinite slab when the field is applied perpendicular to the tape face. In order to test this, a single loop of a # 45 Cu wire was placed around at the edge of a Bi(2223)/Ag tape (0.216 mm x 3.5 mm x 25 mm). Then the apparent losses, which were measured by this coil, were recorded as a function of the number of the tapes of the same size which were placed symmetrically on the top and the bottom of the original tape with the loss coil. The results showed that the losses at low and high fields appear to reach constant values separately and asymptotically when the number of the tapes in a stack is greater than 20. Thus, in the present study of the effects of perpendicular magnetic fields on the losses of the tape, the stack was made with 30 tapes. Also, since the thin superconducting regions in a stack are separated by the outer Ag matrix of the tapes, there will be a question about whether the internal field distribution would be close to that which is expected for the homogeneous superconductor. Fortunately, Matawari [9] calculated the field profiles in a stack of thin superconducting platelets and showed that when the ratio of the distance between the platelets in a stack and the half width of the platelets is less than 0.2, the field profile in the superconductor is essentially that for the Bean critical-state model except the small regions at the very edge and the center of the stack. Here, the ratio for the present tape is 0.16. Thus, it is assured that the stack of tapes can be treated as an infinite slab for the analysis of the losses in perpendicular fields.

Experimental results and Discussion

In Fig. 1, the critical currents as a function of applied field in both directions are shown for an as-received tape. As expected, I_c for the perpendicular field decreases very rapidly with the increasing H. In Fig. 2, ac losses at 60 Hz for the parallel fields as a function of applied magnetic field are shown for two tapes, (an as-received and one bent tape around 25 mm dia.). The vertical axis, (ac losses/$\mu_o H^2$), is the fraction of energy of applied fields which was expended in the specimen as ac losses (J/m^3) per cycle. Also shown in the figure are the calculated losses by fitting Eqs. 1 and 2 by using H* as a fitting parameter. [For clarity, the data for other bent specimens are not shown.] In all cases the Bean model appears to describe the losses very well except that the values of I_c from the dc self-field measurements and I_c from ac losses do not agree. This is shown in Table 1 where the values of the self-field critical currents and those deduced from ac losses by fitting the Bean model are compared. These discrepancies are thought to originate from two sources: (1) For the as-received tapes, the self field critical current is strongly influenced by the perpendicular component of the induced self-field. Thus, it does not reflect the *true* critical current in *purely* parallel field as in the case in the magnetically induced current in the parallel field orientation. (2) The filaments in all of the multifilamentary Bi2223/Ag tapes are electro-magnetically coupled at these frequencies and act as a mono-filament. Thus, the effects of the cracking by bending do not reflect very strongly on the losses. The first point is clearly shown by studying the self-field I_c by controlling the currents in the adjacent tapes in a three-stack tape. It was shown that I_c of the middle tape could be as high as 1.5 times of its self-field I_c when currents in two outer tapes were appropriately controlled to minimize the perpendicular field of the middle tape [10]. If one takes this fact into account, these

Table 1. Comparison of critical currents determined by transport and induced currents.

Tape I.D.	I_c^t (A)*	I_c^m (A)**
As-rec'ed A	40.7	76
As-rec'ed B	34.9	57
bent/37 mm	13.9	38
bent/25 mm	6.9	38

*Self field I_c by transport currents at E=100μV/m.
**I_c deduced by fitting the Bean model to ac losses.

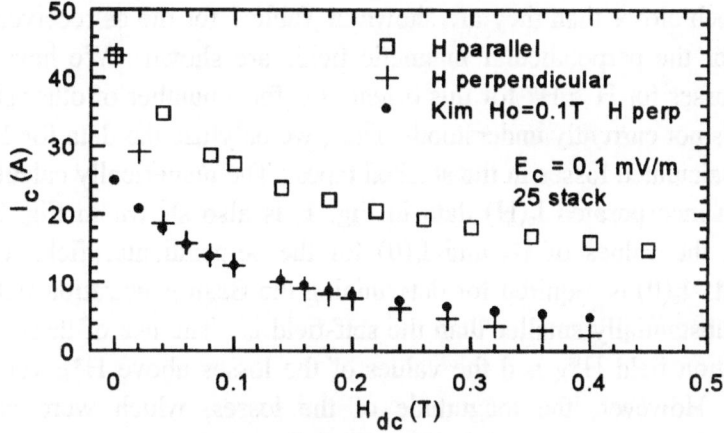

Fig. 1. Magnetic field dependence of I_c of a Bi2223/Ag tape for applied dc fields parallel and perpendicular to the tape face. Kim's expression for $I_c(H)$ is fitted to that for the perpendicular fields.

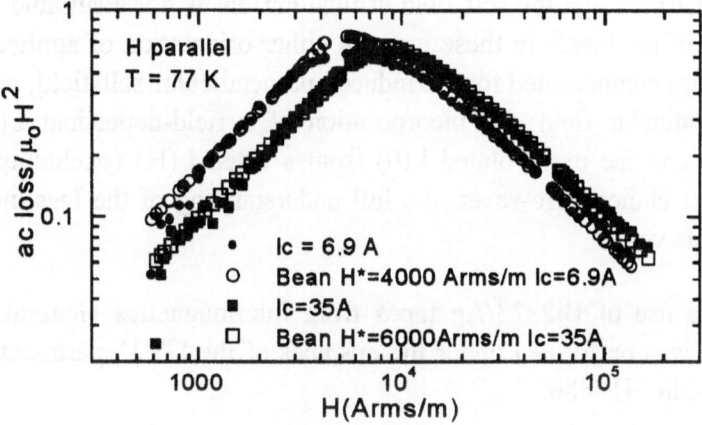

Fig. 2. ac losses for two tapes (I_c = 6.9 and 35 A) in parallel fields as a function of ac magnetic fields and the calculated ac losses from Eqs. 1 and 2 .

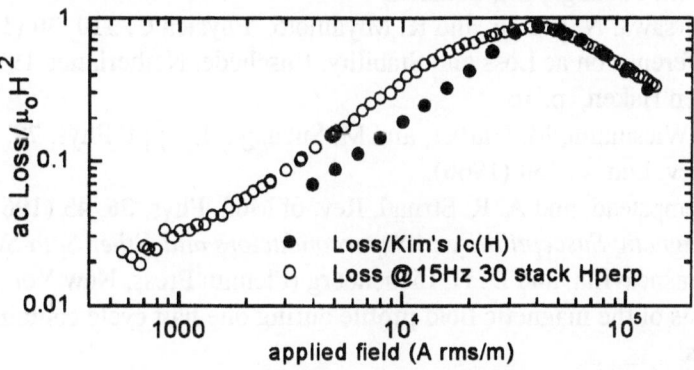

Fig. 3. ac losses as a function of applied fields for a 30 tape stack of a Bi2223/Ag tape in perpendicular fields and those calculated using the Kim's expression for $I_c(H)$.

two values of I_c are much closer than they are shown in Table 1 for the as received tapes. In Fig. 3, the losses at 15 Hz for the perpendicular magnetic fields are shown. We have observed strong frequency-dependent losses for $H > H^*$ for this orientation for a number of other stacked tapes [11] and the source of this is not currently understood. Thus, we only use the data for 15 Hz to compare the measured and the calculated losses in the stacked tapes. The numerically calculated losses using the Kim's $I_c(H)$ which incorporated $I_c(H)$ data in Fig. 1, is also shown in Fig. 3 for 15 Hz. In calculating the losses, the values of H_o and $I_c(0)$ for the perpendicular field, 0.1 T and 25 A, respectively, were used. $I_c(0)$ is required for determining the Bean penetration field H^*. Note that this value of $I_c(0)$ is substantially smaller than the self-field I_c. The use of these values makes the calculated full penetration field H^*_K and the values of the losses above H^*_K very close to that is observed in Fig. 3. However, the magnitude of the losses, which were calculated is still substantially lower than that was measured for $H < H^*_K$. This discrepancy appears to be in part due to the hump in the loss seen below the peak in Fig. 3. This implies that there is another source which reduces the apparent critical current of the tape in the perpendicular orientation below the full penetration field. This is possibly related to the manner in which the filaments couple in perpendicular ac fields.

In summary, it is demonstrated that the self-field critical current is not a suitable critical current to be used for calculations of the losses in these tapes in either orientation of applied magnetic fields. The values of I_c, which are compensated for the induced perpendicular self-field, are better values to be used. In the perpendicular field, the incorporation of a field-dependent $I_c(H)$ into the loss calculation is necessary and the extrapolated $I_c(0)$ from a fitted $I_c(H)$ (excluding the lowest field region) would be a better choice. However, the full understanding of the loss mechanisms in this orientation is not available yet.

Acknowledgment. The use of Bi2223/Ag tapes from Intermagnetics General Corp. is greatly appreciated. This work was performed under the auspices of the US Department of Energy under Contract No. DE-AC02-98CH10886.

* Present address: Iwate University, Ueda 4-3-5, Morioka, 020 Japan

1. Proc. ICMC Topical Conference on ac Loss and Stability, Enschede, Netherlands 1998 edited by H. H. J. ten Kate and B. ten Haken.
2. W. T. Norris, J. Phys. D: Appl. Phys. **3**, 49 (1970)
3. S. P. Ashworth and M. Suenaga, Unpublished.
4. N. Amemiya, S. Mursawa, N. Banno, and K. Miyamoto, Physica C, 310, 30 (1998). Also, Proc. ICMC Topical Conference on ac Loss and Stability, Enschede, Netherlands 1998, edited by H. H. J. ten Kate and B. ten Haken, p. 16.
5. Y. Fukumoto, H. J. Wiesmann, M. Garber, and M. Suenaga, J. Appl. Phys. **78**, 4584 (1995).
6. C. P. Bean, Phys. Rev. Lett. **8**, 250 (1960).
7. Y. B. Kim, C. F. Hempstead, and A. R. Strnad, Rev. of Mod. Phys. **36**, 43 (1964).
8. K.-H. Muller, in *Magnetic Susceptibility of Superconductors and Other Spin Systems*, edited by R. A. Hein, T. L. Frasncavilla, and D. H. Liebenberg (Plenum Press, New York, 1991) p. 229. Note that his analysis of the magnetic field profile during one half cycle contains a couple of typographical errors.
9. Y. Mawatari, Phys. Rev. **B54**, 13215 (1996).
10. S. Spreafico, L. Gherardi, S. Fleshler, D. Tatelbaum, J. Leone, D. Yu, and G. Snitchler, IEEE Trans. on Appl. Supercond. **9**, 2159 (1999).
11. Unpublished

AC LOSS IN HIGH-Tc SUPERCONDUCTORS

OSAMI TSUKAMOTO, MARIAN CISZEK and NAOYUKI AMEMIYA

Yokohama National University, 79-5 Tokiwadai, Hodogaya-ku, Yokohama, 240 Japan.

ABSTRACT

There are three kinds of AC losses in superconducting wires, transport current losses, magnetization losses and total losses. The first losses are caused by the AC transport currents flowing in the wires, the second caused by the AC external magnetic fields which the wires are subject to and the third caused by the combination of the transport currents and external magnetic fields. The measuring methods are divided in electric and calorimetric methods. Generally speaking, the electric methods are sensitive and precise but have some problems to measure false losses if the electromagnetic environments of the wires are not well defined. On the other hand, the calorimetric methods can measure the real losses but are not so sensitive or precise as the electric methods. In the paper, electrical methods for AC loss characterization of the HTS wires are reviewed. The AC loss data of Bi2223/Ag sheathed and YBCO tapes are also presented and the loss characteristics are studied.

Keyword: Transport current AC loss, Magnetization loss, Measurement of AC loss, YBCO tape, Bi2223/Ag sheathed tape

INTRODUCTION

High temperature superconductor (HTS) is more promising for AC power apparatuses such as power transmission cables, transformers and current limiters than low temperature superconductor (LTS) because efficiency of cooling AC losses is higher for HTS operating at 20K ~ 77K than for LTS operating at 4.2K. The knowledge of AC losses and understanding of loss mechanism in HTS wires are important to apply HTS to the AC power apparatuses and develop low AC loss HTS wires and windings. Basically the AC losses in HTS wires can be treated in the same way as those in LTS wires. However, there are complicated aspects to be considered for HTS wires. The HTS wires are usually tape-shaped and their E-J (E: the electric field and J: the current density) curves are dull, and the critical current density distribution is inhomogeneous in the wire cross section and along the wire. Those aspects influence the AC loss characteristics.

AC losses are dissipated in wires when AC transport currents and/or external AC magnetic field are applied to the wire. The loss due to the AC transport current is the AC transport current loss and the energy dissipated in the wire is supplied by a current source connected to the wire. The loss due to the external field is the AC magnetization loss which is supplied by the external field. In an AC superconducting apparatus, generally, the losses combined the transport current and magnetization losses are dissipated in the wire. Methods to measure the AC transport current losses without AC external fields and the magnetization losses without transport currents are well developed[1]. However, there are some difficulties to measure the losses due to combined transfer current and external field. In this paper, firstly AC loss measurement methods and their issues are explained focusing on electric methods. Secondly, loss characteristics of Bi2223/Ag sheathed tapes and YBCO thin film tapes are described. For the case of the Bi2223/Ag sheathed tapes, the AC loss characteristics are well studied. On the other hand for the case of YBCO tapes, there are few data of the AC losses, because YBCO tapes available for the measurements are hard to obtain.

AC LOSS MEASUREMENT

In principle, methods for AC loss measurement of HTS wires are same as those for LTS wires. However, HTS wires are usually tape-shaped with large aspect ratio and magnetic fluxes more concentrate at the edge areas than center area of the wire when a transport current and an external field are applied. Therefore, special cares are necessary for the measurement arrangement.

Transport Current Losses. A transport current loss is usually measured by measuring a resistive voltage component in the signal from voltage taps on the conductor but special cares need to be taken for arrangements of leads from the voltage taps. In commonly used rectangular leads arrangement, distance of the leads from the taps needs to be wide to make errors negligibly small[2]. However, a wide voltage lead loop induces a large inductive voltage in the voltage signal which causes errors in various manners. In a spiral leads arrangement shown in Fig. 1, the voltage leads are spirally wound on a cylindrical surface surrounding the wire. This arrangement clears problems of the rectangular arrangements[3]. The transport current loss in the section between the voltage taps can be measured correctly with this arrangement even if the wire is subject to external magnetic field. With this arrangement, the loss of a wire in tight space can be measured and inductive components in the voltage signal are remarkably suppressed.

Magnetization Loss. AC magnetization losses are usually measured by means of pick up coils by applying AC external field to the wire. From the stand pint of real applications magnetization losses of long tapes should be measured, and the pick-up coils should fully cover fluxes at the edge areas of the tape where the fluxes lines concentrate. For those purposes a pick-up coil configuration shown in Fig. 2 was developed[4].

Total Losses due to Combination of Transport Current and External Field. Energies dissipated in a tape subject to an AC external field and carrying an AC transport current are supplied by the current supply connected the tape and the external field. The energy supplied by the current supply can be measured by measuring the resistive component of the voltage across the voltage taps on the tape and the energy supplied by the AC external field can be measured by measuring the resistive component of the voltage of the pick-up coil placed on the tape. The total loss are addition of the both energies[5][6][11]. However, by this electric method the total losses can be measured only for the case that the AC transport current and external field have a same phase. By a calorimetric method method, the total AC losses of Bi2223/Ag sheathed tape can be measured even for the case that the tape is subject to the external field and carrying transfer current of different phases. However, the calorimetric method is less sensitive and less accurate than the electric methods.

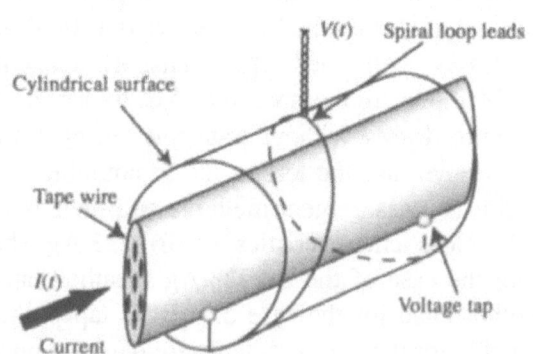

Fig. 1. Spiral voltage leads arrangement for transport current loss measurement

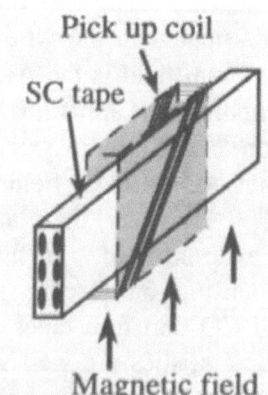

Fig. 2. Configurations of pick-up coils for AC magnetization loss measurement

AC LOSSES OF Bi/Ag SHEATHED TAPE

Transport current losses. It is commonly accepted that measured transport current losses of Bi2223/Ag sheathed tapes well agree with the Norris curve for elliptical model[1]. However, as many data have been accumulated and accuracy of a measurement system has been improved, it has been pointed out that there are some discrepancies between the measured data and the Norris curve[8][9]. Fig. 3 a) shows an example of AC transport current loss characteristics of a multifilamentary Bi2223/Ag sheathed tape which are dependent on the DC external fields B_0. In the graph the transport current loss per meter per cycle Q_t is normalized by I_c^2 where I_c is the critical current of the tape, and the amplitude of the AC transport current I_p is normalized by I_c. Fig. 3 b) shows I_c and the n values of the nth power low of E-J curves which are dependent B_0 also. I_c is defined at $E_0 = 10^{-4}$ V/m. In Fig. 3 a), a curve calculated from the Norris formula[10] for the elliptical model is also shown. As seen in Fig. 3 a), measured Q_t well agrees with the Norris curve at $B_0 = 0$ but discrepancies between Q_t and the Norris curve become obvious as B_0 increases.

In the Norris formulas, it is assumed that the current distribution follows the Bean model ($n = \infty$) and that critical current density J_c is uniform in the superconducting region[10]. The discrepancies can be explained by considering that the assumptions in the Norris formulas do not hold. We made a numerical model considering inhomogeneous J_c distribution in the cross section of the wire and the n value to calculate the transport current losses. The discrepancies can be explained by this model, provided that a proper J_c distribution where J_c is higher in the center area than in edge areas of the tape is assumed[12]. Fig. 4 shows the frequency dependence of the transport current losses of a multifilamentary Bi2223/Ag sheathed tape for 30Hz ~ 960Hz at 0T. There are small frequency dependence where at higher frequency the losses per cycle are slightly decreasing. Therefore, eddy current losses, which increase as the frequency increases and are considered to be produced in the Ag sheath, are negligibly small compared with the hysteresis losses. The slight decrease in the losses at higher frequency can be explained, we consider, by considering effect of the dull E-J curve[8].

Magnetization losses. It is fully proven that multifilamentary twisted LTS wires drastically reduce AC magnetization losses. In principle, multifilamentary twisted HTS wires can also reduce the losses. However, actual multifilamentary twisted HTS tapes developed so far are not so effective

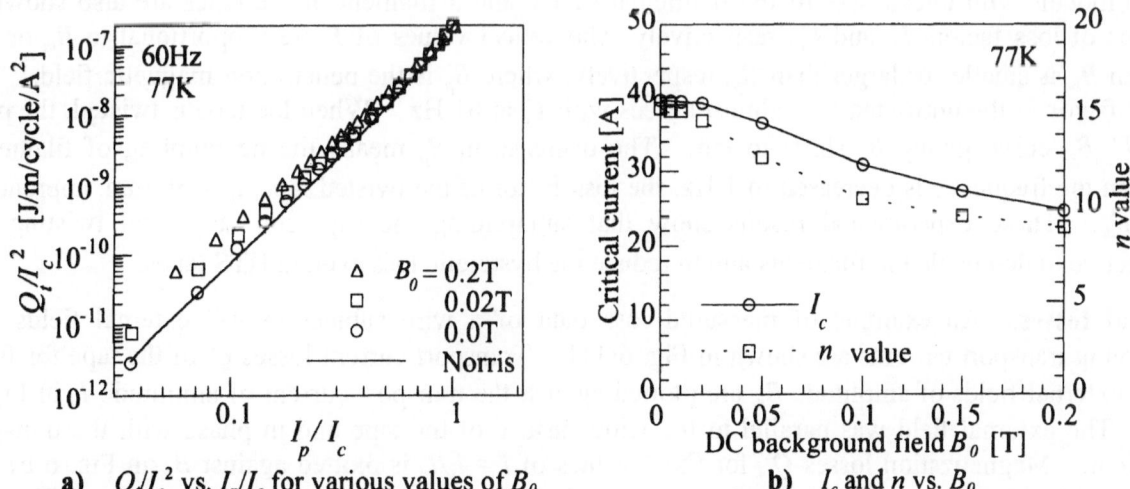

a) Q_t/I_c^2 vs. I_p/I_c for various values of B_0 **b)** I_c and n vs. B_0

Fig. 3. Example of AC transport current loss characteristics for various values of B_0 and dependence of I_c and n value on the B_0. The sample wire is a Bi2223/Ag sheathed tape of 61 filaments and 3.5×0.23 mm^2 in dimensions.

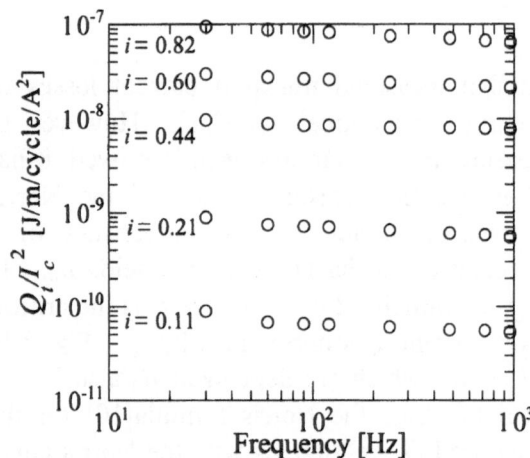

Fig. 4. Frequency dependence of Q_t/I_c^2 for various I_p/I_c. The sample wire is a multifilamentary Bi2223/Ag sheathed tape of 61 filmanets, 3.5×0.23 mm² and I_c = 50.6A at 0T 77K.

Twist pitch	Untwist	36 mm
Tape size	6.1 mm×0.39 mm	6.3 mm×0.38 mm
Number of filaments	61	61
Critical current at 0 T, 77 K	55.4 A	52 A
Critical current density	160 A/mm²	150 A/mm²

Table 1. Specifications of Bi2223 tapes with pure silver matrix

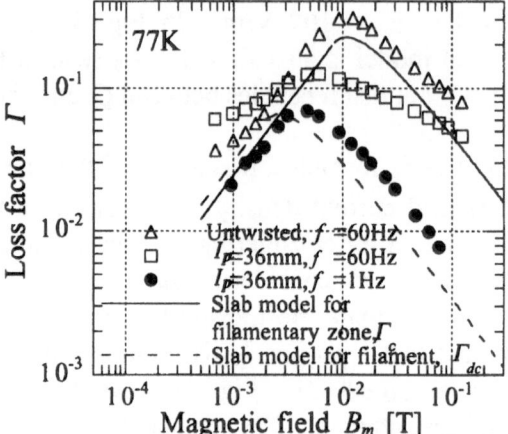

Fig. 5. Γ vs. B_m for twisted (l_p = 36 mm) and untwisted tapes. The length of the sample is ~100mm.

to reduce the losses as expected. This is because the superconducting filaments are electromagnetically coupled by superconducting bridges and whole filaments behave like a monolithic superconductor. To demonstrate that subdividing superconductor and twisting are effective to reduce the magnetization losses also in HTS wires, Bi2223/Ag sheathed tapes in which the bridging among the filaments was suppressed by taking enough space among the filaments were made[18]. The specifications of the tapes twisted and untwisted are listed in Table 1 and the magnetization losses of the tapes in terms of loss factor Γ are shown in Fig. 5 for the case that AC external field was parallel to the wide surface of the tape. Γ is defined as $\Gamma = Q_m / (2B_m^2/\mu_0)$, where Q_m and B_m are the magnetization loss per unit volume of the tape per cycle and the amplitude of the applied AC field respectively. In Fig. 5, the hysteretic magnetization losses of slab models with thicknesses of the filamentary zone and a filament of the tapes are also shown in terms of loss factors Γ_c and Γ_{dc} respectively. Theoretical values of Γ are proportional to B_m or $-B_m$ when B_m is smaller or larger than B_p, respectively, where B_p is the penetration magnetic field. The loss factor of the untwisted tape almost agrees with Γ_c at 61 Hz. When the tape is twisted, the peak of Γ- B_m curve giving B_p shifts to left. The decrease in B_p means the de-coupling of filaments. When the frequency is decreased to 1 Hz, the loss factor of the twisted tape, l_p = 36 mm, approaches to Γ_{dc}. These experimental results show that subdividing the superconductor and twisting are effective to de-couple the filaments and to reduce the hysteretic loss even in HTS tapes.

Total losses. An example of measured loss data of a wire subject to AC external fields and carrying transport currents are shown in Fig. 6[11]. Transport current losses Q_t of the tape for fixed AC external fields of amplitude B_m are plotted against the transport current of amplitude I_p in Fig. 6 a). The external field was parallel to the wide surface of the tape and in phase with the transport current. Magnetization losses Q_m for fixed values of $i = I_p/I_c$ is plotted against B_m in Fig. 6 b) and the total losses $Q_T = (Q_t + Q_m)$ for fixed values of i against B_m in Fig. 6 c). As seen from Fig. 6 a) Q_t is strongly affected by the external field especially in the region of relatively low I_p. From Fig. 6 b) dependence of Q_m on i is obvious in the region B_m < 17 mT and not obvious in the region B_m > 17 mT.   The peak of the loss factor appears at 17 mT and the penetration magnetic field is 12.2 mT.

YBCO TAPES

At the moment there are some difficulties to measure the magnetization and total losses of YBCO tapes but there are data of transport current losses[9][12][14].   An example of measured data of a YBCO thin film conductor is shown in Fig. 7[14].   The sample tape conductor was made by depositing YBCO thin layer of ~1 μm thick on a LaAlO3 single crystal substrate[15].   Dimensions of the conductor are $10 \times 1 \times 100$ mm^3 and the n value was 33 at 77 K and 0 T.   In Fig. 7, the Norris curves for the elliptical and strip models are also shown.   The data have more obvious frequency dependence compared with those of the Bi/Ag sheath tape shown in Fig. 4.   The data approximately follow the Norris curve for the strip model for $i = 0.6 \sim 1$ and the data are approximately parallel to the curve for the elliptical model for $i = 0.2 \sim 0.6$.   For $i < 0.2$, the slopes of the data are duller than the curve for the elliptical model.   Measured loss data for a given frequency can be fit to a curve numerically calculated from our model, if the J_c distribution in the conductor cross section of the conductor where J_c is higher in the edge areas than in the center area is properly assumed[14].

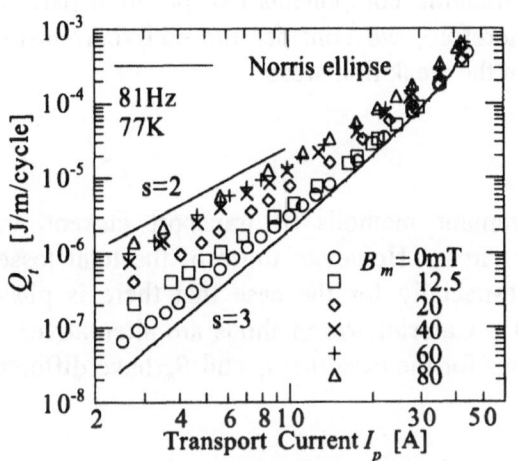

a) Q_t vs. I_p for various values of B_m. The "s" indicates the slope in log-log scale.

The frequency dependency of the AC transport current losses can be explained by assuming that Q_t has an eddy-current-like loss component Q_{ed} and a hysteretic component Q_h. Q_{ed} is considered proportional to i^2 and f and expressed as $Q_{ed} = \alpha i^2 f$ where α is a constant. Q_h is assumed equal to the analytically calculated losses based on our model to fit to the measured data of $f = 60$ Hz in Fig. 7. Q_t can be expressed as $Q_t = Q_h + \alpha i^2 f$.   In the range of low i, Q_h is low and frequency dependence of Q_t is more obvious.   Therefore, the value of α is determined as $\alpha = 9.7 \times 10^{-14}$ by fitting that equation to the measured Q_t vs. f data for a low value of i ($i = 0.11$) (Fig. 7).   In Fig. 8, Q_t vs. f curves calculated by using $\alpha = 9.7 \times 10^{-14}$ are plotted against f for

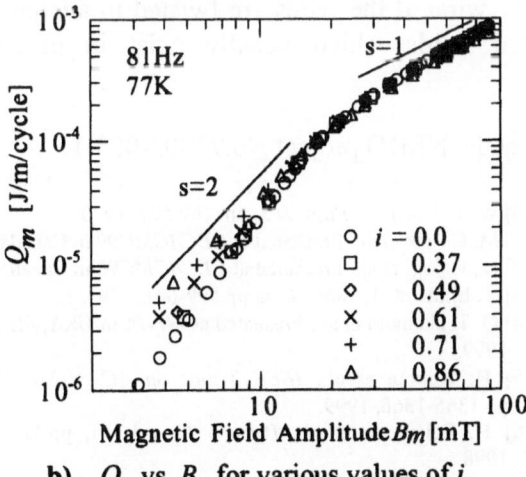

b) Q_m vs. B_m for various values of i.

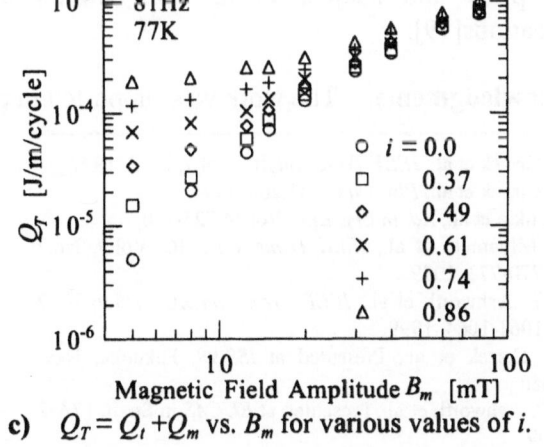

c) $Q_T = Q_t + Q_m$ vs. B_m for various values of i.

Fig. 6   Example of losses of Bi2223/Ag sheathed tape subject to AC external fields and carrying AC transport currents.   The sample wire is of 61 filaments, 3.5×0.23 mm^2 and $I_c = 40.8$ A at 0 T and 77 K.

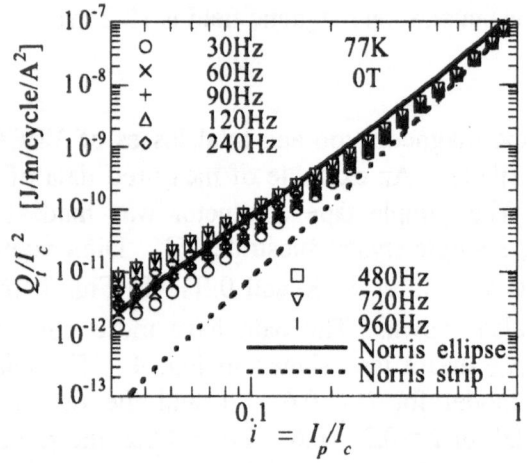

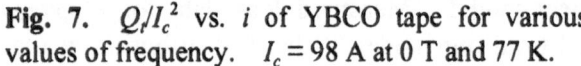

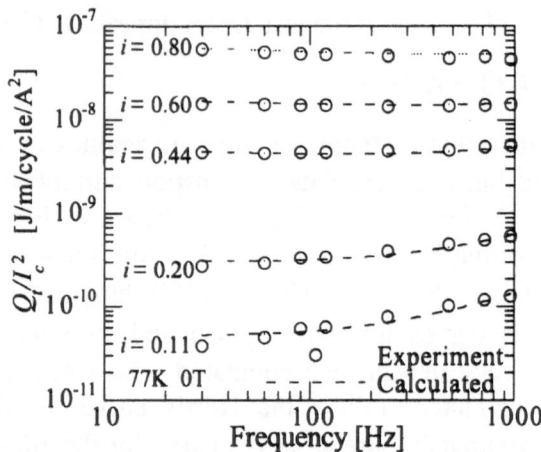

Fig. 7. Q_t/I_c^2 vs. i of YBCO tape for various values of frequency. $I_c = 98$ A at 0 T and 77 K.

Fig. 8. Comparison of frequency dependences of meausred and calculated Q_t/I_c^2's of the YBCO tape.

various i and compared with the measured data. The agreement is reasonably good. Therefore, we consider the frequency dependent AC transport current losses contain eddy-current-like loss component. The tested tape conductor has no normal metallic components except small parts of silver coating for current leads and potential taps. Therefore, we consider the eddy-current-like losses are occurring in the superconducting layer, maybe in the weak link areas.

CONCLUDING REMARKS

As to AC loss measurement system, electric measurement methods of transport current and magnetization losses are almost established for isolated wires. However, those of the total losses are not established and more works need to be done especially for the case that there is phase difference in the AC transport current and external field. Calorimetric methods are alternatives to the electric methods for the measurement of the total losses for the case that I_p and B_m have different phases but sensitivity and accuracy need to be improved.

As to AC losses of HTS wires, the losses, especially the total losses, need to be much more reduced for practical applications. Twisted multifilamentary wires with high resistive barriers among the filaments are most promising[16]-[18]. The transport current losses as well as the magnetization losses are much reduced by the twisted multifilamentary wires if the wires are twisted in adequately short pitch and subject to the longitudinal external fields which usually exist in practical applications[19].

Acknowledgments. The work was supported in part by the NEDO project No.97S03-003-1.

[1] M. Ciszek et al., *IEEE Trans. on AC.* Vol.7, No.2, 1997.

[2] M. Ciszek et al., *Physica C* 233:203, 1994.

[3] S. Fukui et al., *Ad. in Cry. Eng.* Vol.44 723-730.

[4] K. Miyamoto et al., *IEEE Trans. on . AC.* Vol.9, No.2, pp.770-773, 1999.

[5] S. P. Ashworth et al., *IEEE Trans. on AC.* Vol.9, No.2, pp.1061-1064, 1999.

[6] M. Ciszek et al., Presented at *ISS 98*, Fukuoka, Nov., Japan

[7] S. P. Ashworth et al., Presented at *EUCAS* in Spain 17A-2, 1999.

[8] D. Miyagi et al., *Physica C*, vol.310, pp.90-94, 1998.

[9] O. Tsukamoto et al., *IEEE Trans. on AC.* Vol.9, No.2, pp.1181-1184, 1999.

[10] W. T. Norris, *J. Phys. D* 3, pp. 489-507, 1970.

[11] M. Ciszek et al., Presented at *CEC/ICMC99* in Canada.

[12] D. Miyagi et al., Presented at *CEC/ICMC99* in Canada.

[13] Y. Iijima et al., *Adv. SC* X, pp.599-602, 1997.

[14] O. Tsukamoto et al., Presented at *MT-16* in USA, 4B-363, 1999.

[15] H. Kubota et al., *IEEE Trans. on AC.* Vol.9, No.2, pp.1365-1368, 1999.

[16] H. Eckelman et al., *Physica C*, vol.310, pp.122-126, 1998.

[17] M. Dhalle et al., *Physica C*, vol.310, pp.127-131, 1998.

[18] N. Amemiya et al., Presented at *MT-16* in USA, 9D-325, 1999.

[19] S. Fukui et al., *Physica C*, vol.310, pp.142-146, 1998.

Transport A.C. Losses of (Bi,Pb)-2223 Multifilamentary Tapes with Different Filament Distribution

P. X. Zhang[1,2,3], H. Maeda[2], A. Oota[1], R. Inada[1], T. Uno[1], Y. Takatori[1], S. Sakamoto[1], T. Yamamoto[1], and L. Zhou[3]

[1]Toyohashi University of Technology, Tempku, Toyohashi, Aichi 441, Japan
[2]Institute for Materials Research, Tohoku University, 2-1-1 Katahira, Sendai 980-8577, Japan
[3]Northwest Institute for Nonferrous Metal Research, P.O.Box 51, Xi'an, Shaanxi 710016, China

Abstract: The (Bi,Pb)-2223 multifilamentary wires and tapes with different filament distribution and resistive barriers have been fabricated by using powder-in tube (PIT) method. Two deformation methods were applied to prepare the wires with different filament distribution, one is ordinary drawing process using round dice, another is rectangular deformation by a "Turk's head" with four rollers. The Sr-V-O oxides, Bi-2201 and Ag-Cu alloy were used as resistive barriers to reduce the coupling between filaments. The transport a.c. losses were measured on these different wires and tapes at 77K in self-field. The results indicated that the barrier configuration and J_c distribution are important to transport a.c. losses.

Key words: (Bi,Pb)-2223 tape, Resistive barrier, Transport a.c. loss

INTRODUCTION

In recent years, many efforts have been made to understand the behaviors of a.c. losses in Bi-based tapes. The main contribution to the losses comes form hysteresis losses of superconductor [1]. The multifilamentary tape exhibits same a.c. loss behaviors as monocore tape because the filaments are fully coupled due to the low matrix resistivity and intergrowth of 2223 phase between filaments [1,2], and the results can be explained by the theory of Norris [3] for an elliptical superconductor with same critical current. The twisting process applied to Bi-based multifilamentary tapes causes slight influence on transport a.c. losses [4]. The eddy loss at low frequency (<200Hz) has been considered negligible compared with hysteresis loss and coupling loss in the matrix between filaments [5]. The hysteresis loss can be decreased by reduction of the filament dimension if the filaments are not fully coupled. So the decrease of coupling between filaments becomes very important to reduce a.c.transport losses. It was found that resistive barriers are effective to decouple the filaments in Bi-based tapes [6]. In this work, we prepared the (Bi,Pb)-2223 multifilamentary tapes with different resistive barriers, and studied the behavior of transport a.c. losses on these tapes.

EXPERIMENTAL

All the investigated (Bi,Pb)-2223 wires and tapes with Ag and Ag-alloy sheath were prepared by PIT method. Precursor powder with nominal composition of $Bi_{1.76}Pb_{0.34}Sr_{1.93}Ca_{2.02}Cu_{3.1}O_x$ were prepared by mixing the (Bi,Pb)-2212 and $CaCuO_2$ powders. The precursor powders were packed into pure silver tubes to prepare monocore wires with round, hexagonal and square shape. Pure Ag or AgMnNi alloy tubes with round or square shape were used for outer sheath to fabricate different multifilamentary wires and tapes. Two deformation techniques have been applied to the fabrication. For 19, 37, 61, 85 filaments' wires, the hexagonal monocore wires were cut and packed into a round tube, and an ordinary drawing process with round dice was used. For 40 filaments' wires, the square monocore wire was cut and packed into a sqare Ag-alloy tube, and the composite was rectangularly deformed to a wire with cross-section of 1.5×1.5 mm^2 by cold drawing using a "Turk's Head" with four rollers. Finally, the ordinary rolling process and heat treatments were applied to make all the multifilamentary tapes. The $Sr_6V_2O_{11}$ powder was introduced into 19 and 37 filaments' tapes as resistive barriers by dip-coating it on the surface of monocore wires before packing process. For the 40 filaments' wires, the filaments stacked as six layer, each layer has 7 filaments (the top and bottom layer has 6 filaments respectively). The 2201 or Ag-Cu alloy are introduced between each layer as a resistive barriers. From the sqare wire with this filament distribution, we prepared the tapes with the barrier parallel and perpendicular to the tape surface. The a.c. transport losses in self-field (at 77K, 50Hz) were measured on the samples with length of 35mm by a standard method as reported previously [1].

RESULTS AND DISCUSSION

The 19, 37, 61, and 85 filaments' tapes without resistive barrier were prepared with aspect ratio changed between 10 to 25 by ordinary PIT method. The measurements indicated that transport a.c. losses are independent on filament number, all the tapes exhibit same loss behavior as monocore tape, and can be well described by Norris theory for elliptical superconductor [3]. The filaments are almost fully coupled. So in this kind of Ag-sheathed tape, the transport a.c. losses can not be reduced by decreasing filament size and changing aspect ratio from 10 to 25. The transport a.c. losses of 19, 37 filaments' tapes with Sr-V-O barrier around each filament were reduced comparing with the no barrier tapes, as shown in Fig.1. This result means that the Sr-V-O barriers decouple the filaments to some extent, and reduce the coupling losses produced in the matrix between filaments. But the filament are far from fully decoupling, this may be due to the inhomogeneous distribution of Sr-V-O coating in the investigated samples. Actually, it is difficult to continuously cover each filament with Sr-V-O by this technique, because the thick coating would cause inhomogeneous deformation during cold-working process and slow oxygen diffusion during heat treatment, and decrease the J_c. To solve this problem, we considered other barrier configurations, and prepared 40 filamentary tapes with barriers parallel or vertical to the tape surface. Fig.2 shows the normalized transport a.c. losses per cycle versus reduced current for the 40 filaments' tapes with vertical and parallel distributed Bi-2201 barriers. The transport a.c. losses at 77K, 50Hz, are reduced significantly by the parallel Bi-2201

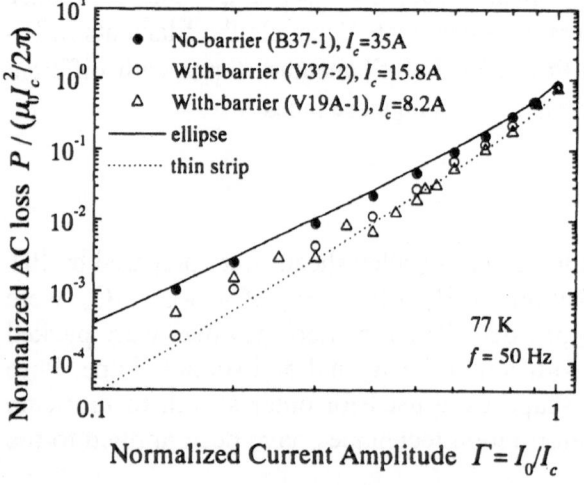

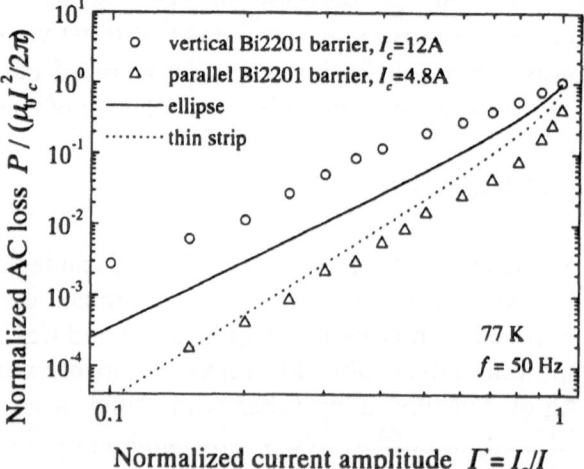

Fig.1. A.c. losses of tapes with Sr-V-O barrier.   Fig.2. A.c.losses of tapes with 2201 barriers.

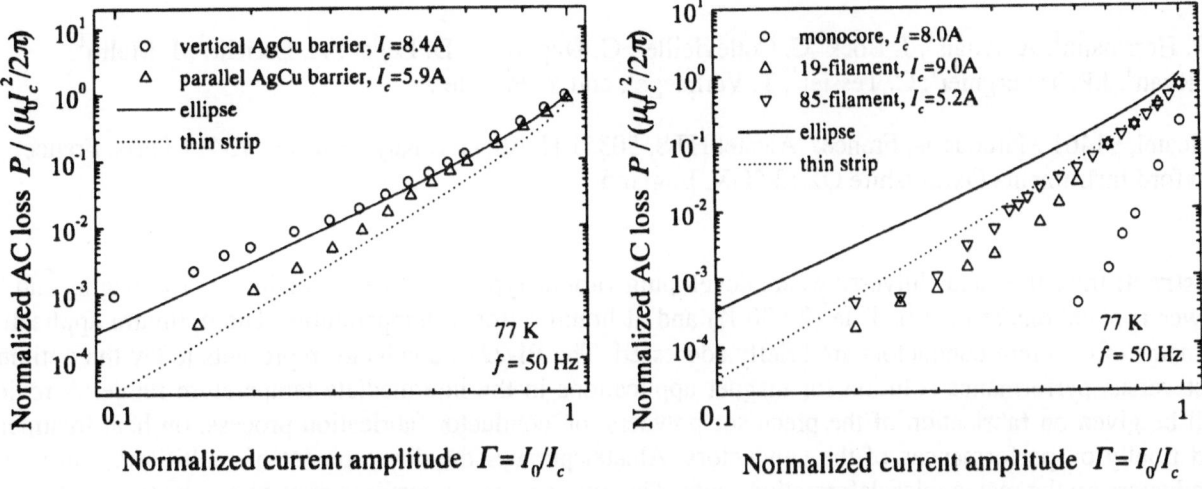

Fig.3. A.c. losses of tapes with Ag-Cu barrier.   Fig.4. A.c. losses of round wires.

barriers. Comparing with the parallel barrier, the vertical barriers are not very effective. The parallel barriers divided the tape (with cross-section 0.3×3.0 mm^2) into 6 thin strips with aspect ratio near 100. According to Norris theory, each strip has a lower a.c. loss, and the coupling between each strips are partly decoupled by the Bi-2201 barriers, so the total losses of the tapes are decreased. The similar results are obtained in the tapes with Ag-Cu barriers as shown in Fig.3. But the reduction of a.c. losses is not as significant as Bi-2201 barrier. This means the decoupling by Ag-Cu alloy is not so effective. An interesting result has been obtained on the monocore and multifilamentary round wires, which exhibit lower transport a.c. lossses as shown in Fig.4. The round filaments in multifilameantary wires are homogeneously distributed. We considered this behavior is caused by the inhomogeneous J_c distribution along the radius of the wire. The outer area near wire surface has a higher J_c than the inner area. These results mean that changing the J_c distribution in the Bi-based conductor can considerably reduce the transport a.c. losses.

CONCLUSION

The transport a.c. losses in self-field have been studied on the Bi-based wires and tapes with different filament configuration. The high transport a.c. losses in ordinary multifilamentary tapes are mainly due to the fully coupling of filaments. The resistive barrier with an appropriate configuration can decouple filaments and decrease the transport a.c. losses. Introduction of parallel resistive barriers to divide the tapes into thin strips is very effective to reduce the transport a.c. losses. The lower transport a.c. losses in the investigated round wires indicate that the J_c distribution has a big influence on a.c. loss. This result is useful to design the conductor or cable with lower a.c. losses.

1. A. Oota, T. Fukunaga, M. Matsui, S. Yuhya and M. Hiraoka, Physica C **249**, 157 (1995).
2. A.V. Volkozub, J. Everett, G. Perkins, P. Buscemi and A.D. Caplin, D. Dhallé, F.Marti, G.Grasso, Y.B. Huang and R. Flükiger, IEEE Transaction on Applied Superconductivity **9**, 2147 (1999).
3. W.T. Norris, J. Phys. D **3**, 489 (1970).
4. A. Oota, T. Fukunaga, T. Ito, Physica C **270**, 107 (1996).
5. C.M. Friend, C. Beduz, B. Dutoit, R.Navarro, E. Cereda, and J. Alonso-Lorente, IEEE Transaction on Applied Superconductivity **9**, 1165 (1999).
6. Y.B. Huang and R. Flükiger, Physica C **294**, 71 (1998).

BSCCO Based Superconductors for Magnet Applications[1]

P.F. Herrmann[1], A. Allais[1], J. Bock[2] C. Cottevieille[1], G. Duperray[1], D. Legat[1], A. Leriche[1], J. Melin[1], D. Ryan[3], J.P. Tavergnier[1], C. Tessier[1], T. Verhaege[1], and Y. Parasie[4],

[1]Alcatel, 91461 Marcoussis, France; [2]Alcatel HTS, 50351 Hürth, Germany, [4]Alcatel, 75411 Paris, France
[3]Oxford Instruments Oxfordshire OX13 5QX, England,

Abstract: Industries and Universities are developing various types of BSCCO conductors for magnet and for power applications at intermediate (20-30 K) and at liquid nitrogen temperatures. The optimum application domains of different conductors are briefly addressed. The Bi-2212 conductor represents today the optimum cost versus performance solution for magnet applications in the intermediate temperature range. A review will be given on fabrication of the precursor powders, on conductor fabrication process, on heat treatments and finally on performances of the conductors. Alcatel pursues the development of multifilamentary tape-conductors by the rectangular deformation route. The ultimate test of performances of conductors is of course their use in systems. First applications are today opening in the magnet technology domain. Some aspects on coil fabrication by the complementary R&W and W&R fabrication process are given.

INTRODUCTION

Magnets for field generation in the 20 K temperature are among the first applications which are expected for silver sheathed BiSrCaCuO superconductors. During the last years, the interest in the Bi-2212 variant has increased considerably. This is related to the larger domain of stability of the Bi-2212 phase [1,2] with respect to Bi-2223 phase [3] which explains that the conditions for synthesis can be achieved easier. A second fact is that the low temperature critical current density of the Bi-2212 phase is higher than the Bi-2223 phase. This results in several industrial R&D activities: DIP coated and PIT conductor are developed by Oxford Instruments [4] who realizes 3 to 6 m samples with J_c = 1000 A/mm^2. A round and isotropic conductor containing round filaments is developed by IGC [5] reaching a self-field current density of 2800 A/mm^2. Record current densities of 5000 A/mm^2 at 4 K and at 10 T are reached by the PAIR process at NRIM [6] and a new method for assembling anisotropic tapes to a round conductor ROSAT with almost isotropic behavior was realized by Hitachi [7]. The development at Alcatel is strongly oriented towards reproducible realization of long length Powder In Tube (PIT) conductors for W&R and R&W conductors. Green conductors have been used for realizing several W&R test coils using a novel NiO-insulating technology [8]. For AC conductors where a controlled twisting and de-coupling of filaments is necessary [9], it has been shown previously [10] that square shaped conductors are more easily twisted down to low twist pitch values. Although some approaches exists for the de-coupling of filaments in PIT conductors [11,12] there exists today no entirely satisfying conductor concept for AC applications in transversal magnetic field.

TEMPERATURE DEPENDENT REFRIGERATOR AND CONDUCTOR COST

System cost and operating cost of the superconducting devices determine the economical impact of any type of superconductor. The operating costs depend on system specific contributions and can not be assessed as a whole. One dominant contribution however is the cooling cost which is determined by the amount of losses which is generated in the system and by the Carnot efficiency. Large machines reach 30 % of the Carnot efficiency and small machines only 5%. This has the consequence that for very low temperatures the cryogenic heat load to the refrigeration system must be very small, which is in general verified by DC systems like superconducting magnets.

---

[1] Manuscript received October 18, 1999
Part of this work was supported by EU Brite/EuRam contract No BRPR-CT97-0347
Part of this work was supported by DSP-DGA contract N° 98.395

The temperature dependency of system cost can be estimated by considering the three basic components of a cryogenic system. These are weakly temperature dependent structural costs and strongly temperature dependent cost of the refrigeration equipment and the conductor costs. For estimating the optimum operating temperature, it is sufficient considering the two last contributions [13]. The refrigerator costs are known from the re-actualized empirical "Strobridge law" [14] and can be approached for a 77 K refrigerator by $C_{ref} = 125\,(P_{cp})^{0.6}$ where P_{cp} (in kW) is the refrigeration power. The temperature dependency can be estimated assuming that the cost for a given power at room temperature is constant:

$$(1) \qquad C_{ref} = 125 * \left(P_{cp} \frac{77K}{T_{op}} \right)^{0.6}$$

The device itself is typically a magnet or a power link. At the actual cost level of HTS conductors, the cost of the device is dominated by the conductor price. This contribution depends strongly on the operating temperature T_{op} and the field H_{op} through the temperature and field dependency of the critical current (see Fig 3):

$$(2) \qquad P_{Cond} = P_{kAm}(77K) \frac{I(77K, 0T)}{I(T_{op}, H_{op})}$$

The conductor price P_{Cond} in Euro/kAm is depicted in Fig.1 showing a strong temperature and field dependency. The price varies almost an order of magnitude between 77 K and low temperatures indicating that for a reference price at liquid nitrogen temperature $P_{kAm}(77K)$ of 100 E/kAm a price of 20 E/kAm is reached at 20 K for not to high fields and if $P_{kAm}(77K) = 10$ E/kAm can be reached, values as low as 2 E/kAm are possible. It is further seen that applications in magnetic above 5 Tesla remain uneconomical for temperatures above 40 K.

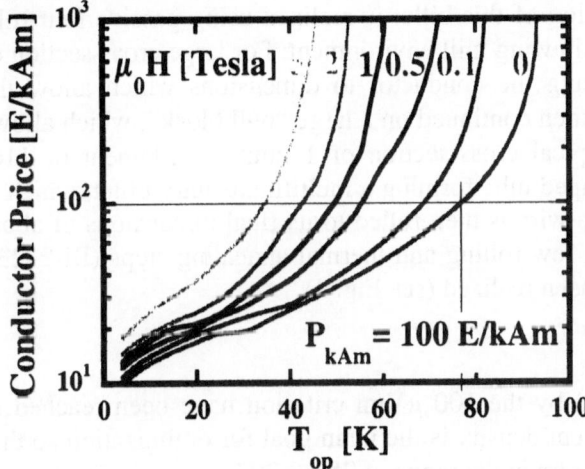
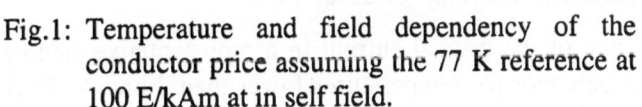

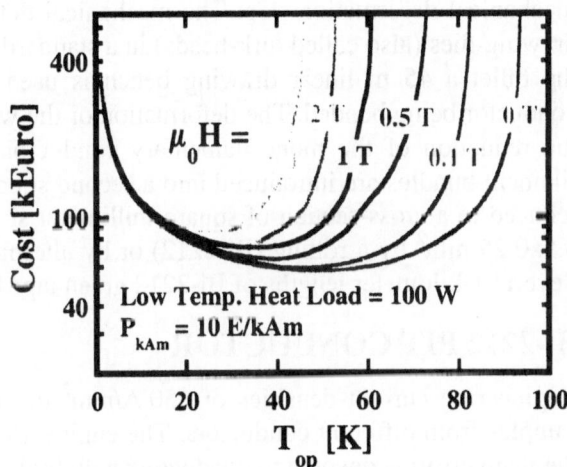

Fig.1: Temperature and field dependency of the conductor price assuming the 77 K reference at 100 E/kAm at in self field.

Fig.2: Temperature and field dependent refrigerator and conduction cost for a device generating a low temperature heat load of 100 W.

The optimum operating temperature is found (see Fig. 2) when the refrigerator cost is added to the conductor costs. For this, an application with a low temperature heat load of 100 W and which requires a conductor quantity expressed in "Ampere x length" of I x l = 5000 kAm is chosen. Figure 2 shows the evolution of this optimum. At a field of 5 Tesla, this optimum is in the 20 –30 K range even for a 77 K reference conductor price of $P_{kAm}(77K) = 10$ Euro/kAm. This indicates that for DC application especially when operated under magnetic field, the optimum operation conditions will remain in the intermediate temperature range where the Bi-2212 phase has some clear advantages over the Bi-2223 phase.

PRECURSOR POWDERS

All tapes are manufactured on the basis of precursor materials from the HTS group of Aventis Research and Technologies. This is now Alcatel High Temperature Superconductors (AHTS), a new separate legal entity located in Huerth, Germany. For the Bi-2212 material AHTS performs the unique proprietary melt casting process. This is derived from the manufacturing of bulk parts and allows to achieve homogenous and reactive precursors on an efficient route suitable for industrial scale up.

The oxides of Bi, Sr, Ca and Cu are intensively mixed and heated to about 1000°C until a homogenous melt is obtained. The melt is cast into copper moulds where solidification and cooling to room temperature takes place. There are different options for the further thermal treatment which allow to control phase composition, impurity level, and reactivity of the material. Our development showed that the high level of critical current can only be achieved if a phase mixture instead of the final Bi-2212 phase is applied. Critical for avoiding blistering during tape manufacturing are a low Carbon level (< 300 ppm) and practically now water content. Precursors used for the here reported tapes were fabricated as follows: The ingots obtained after melt casting are annealed until the desired phase content of the Bi-2212 phase is obtained and are then crushed and jet milled (under Nitrogen) to form a powder with a medium grain size (d50) of about 3μm. Dry bag pressing is used for the fabrication of rectangular precursor rods which significantly improve the efficiency of the PIT process and lead to a much a better uniformity of the tape conductors. The homogeneity of the precursor material as well as of the rods is proven by the high level of uniformity which is achieved in the Bi-2212 tapes as reported here.

CONDUCTOR FABRICATION

Conductors are realized by the rectangular deformation route which consist in introducing precursor powder bars in square shaped silver or silver-alloy tubes. The tubes are evacuated and sealed before the first mechanical deformation step. The mechanical deformation of this billet is realized using passive four-roll-drawing-dies (also called turk-heads) in a standard wire drawing mill environment: For large cross-section of the billet a 45 m linear drawing bench is used reducing the conductor to dimensions which allow the conductor being bent. The deformation of the wire is then continued on a large "bull block", which allows the reduction of the monofilamentary conductor to typical cross-section of 1 mm². 76-filament or 110-filament bundles are introduced into a second square shaped tube forming a multifilamentary billet, which is reduced to a cross-section of square-millimeter size. The wire is then rolled to its final dimensions of about 3.5x0.25 mm² by a rolling (Bi-2212) or by alternating a few rolling and thermal annealing steps (Bi-2223). Several 1-kilometer lengths of Bi-2212 green tape have been realized (see Fig. 4).

Bi-2212 PIT CONDUCTOR

Engineering current densities of 650 A/mm² determined by the 100 μV/m criterion have been reached in samples from different conductors. The engineering current density is the main goal for optimization so that the main effort is devoted to conductors with high fill-factors in the range of 25% - 30%.

The field (H ∥ to a-b plans) and temperature dependency of the critical current in a representative sample from an early 1000 m length. It shows a relatively flat behavior for temperatures lower than 30 K. At 20 K the J_c drop at 5 Tesla is roughly a factor 2 and a transport current of 100 A is still obtained. At higher temperatures the I_c drops faster until, in liquid nitrogen temperatures only applications in low field can be considered. The high field magnet laboratory NHMFL has carried out high field measurements up to 30 Tesla [13]. The measurements have been realized on a short straight sample of cross-section 0.77 mm² and a fill factor $\sigma_s = 0.29$. The Ico value was 385 A and the critical current at 30 T showed over 200 A ($J_e = 250$ A/mm²) at 4 K. This value is well above the I_c at 30 Tesla of most conductors of this type showing the potential for high field magnet applications. The relative field dependency shows that the I_c value in 30 Tesla is slightly above 50 % of the zero field value.

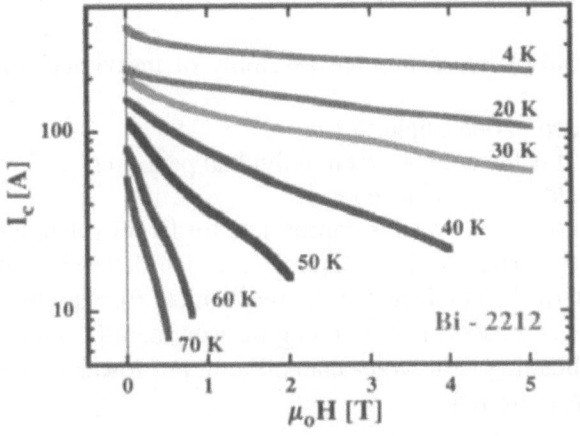

Fig.3: Field and temperature dependency of the critical current in Bi-2212 conductors. The self-field current density at 4 K is 400 A/mm^2.

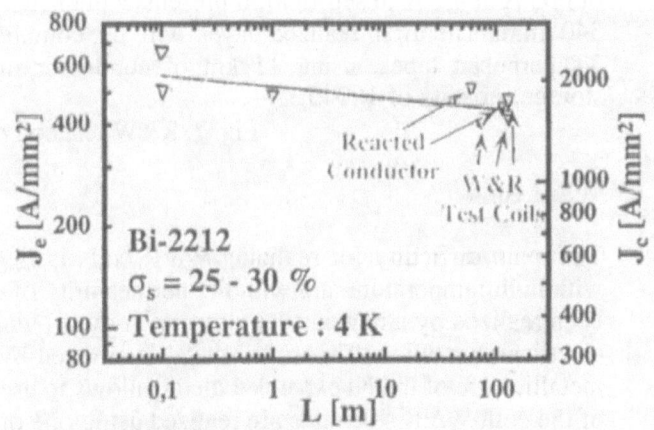

Fig.4: Alcatel PIT conductor; one-kilometer length

REACTED CONDUCTOR

In a recent 1 km length, the homogeneity was assessed on several 50 m lengths and one length of 120 m. These conductors have been submitted to the standard heat treatment in a tube furnace. After heat treatment the conductors were equipped with 20 voltage tabs measuring each a section of a length of 3 m. The distribution of critical currents of the entire 50 m or 120 m lengths are indicated in Fig. 5 which shows that the value of the critical current varies in a narrow range of 380 to 410 A. The lower value from sample 2 is related to leak of oxygen during the heat treatment. These measurements show that the performance of the kilometer lengths are rather homogeneous and that there is no strong systematic variation over the length of the conductor. A typical distribution of the I_c values of sample 1 is given in the inset of figure 5. It shows a mean value of 394 A and a halve height width of 2 x 10 A.

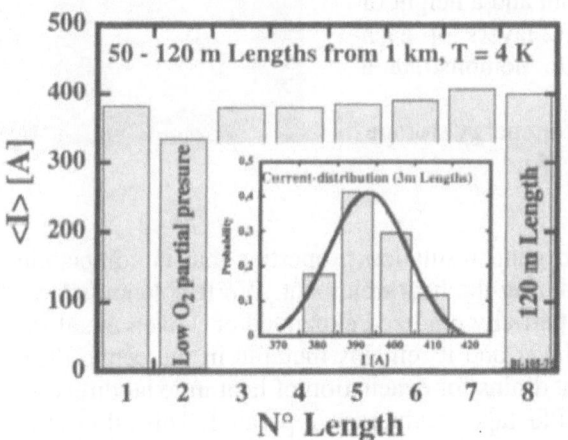

Fig.5: Reacted 50-60 m lengths from kilometer length green tape. The distribution of the critical current in 3 m section is indicated in the inset.

Fig. 6 Alcatel test coils from [PF H ASC] and 50 m and 120 m reacted lengths from kilometer length conductor.

These results are compared (Fig. 6) with the performances of a set of W&R test coils on which was reported previously [8]. The self field of these coils in the range of 1 – 2 Tesla reduces significantly their critical current. For comparison of short sample results and long length results the coil values have been corrected for field effect which allowed the estimation of the long length performance of the conductor. The comparison with the results from hectometric reacted length shows that both results are consistent and that there is no strong difference in long length and small diameter W&R coils and fully reacted lengths.

FABRICATION AND TESTING OF Bi-2212 SOLENOIDS

Magnet suppliers prefer realizing compact solenoids instead of interconnected assembly of individual pan cakes for different reasons:
1. Improved productivity of solenoids which are realized in one single step.
2. No intermediate electrical contacts are required, as it is the case between individual pancakes.
3. A higher degree of homogeneity of the magnetic field can be achieved [15].

Solenoids can be realized by React and Wind (R&W) technology is which is appropriate for larger solenoids which can not easily be heat treated as a whole in a furnace. The threshold parameter which determines if R&W technology can be used is related to the degradation of the conductor under mechanical forces which occur during the coil winding and bending of the conductor is most critical. Bending measurement [13] have shown that the Alcatel Bi-2212 conductor can be used for bending radii larger than 50 mm. For lower values of the bending radius Wind and React (W&R) technology is preferred.

R&W coils

R&W coils are made from fully reacted "ready for use" conductors. As there is no subsequent heat treatment, the size and structure of the coil can be chosen freely; in particular, classical organic spacers, insulation and resins can be used. Two types of difficulties must be overcome:
*   the heat treatment and further processing of long length conductors (see above).
*   the coil fabrication from heat treated tapes, which are mechanically weak and strain sensitive in comparison to the green tapes

In order to avoid any performance degradation, particular winding processes have been settled, where any kind of in-plane curvature is suppressed: the tape is formed in a spiral, where the radius of curvature always exceeds 50 mm, including the zones of contact to the current leads or to adjacent conductor layers. Fig. 7 shows an experimental coil made this way. It was built in the frame of a prospective study supported by the French Army (DGA), relative to energy storage. The coil is made from two parallel tapes, with a critical current of 660 A at 4.2 K in self field. It presents a diameter of 420 mm and a height of 340 mm. The first realized layer will be completed by 32 layers of each 3 superposed tapes, using 11 km of conductor, in order to demonstrate a storage capacity of 100 kJ.

Fig. 7. R&W test coil element for energy storage.

W&R coils

The main difficulty for realizing W&R coils is finding with good insulation properties that is compatible with high temperature and with the permeability of oxygen during the heat treatment. W&R solenoids have been realized by a new insulation technology [8] which uses partially oxidized expanded or woven metal for electrical insulation. The nickel metal is covered by a NiO layer and it remains metallic in the center. The metallic core of the Ni expanded metal, allows to use it as heat drains for evacuation of heat in axial direction of the coil. W&R solenoids are realized using one or several PIT tape conductors in parallel. This allows the adjustment of the coil current in the range of 100 A to about 1000 A. The oxidized expanded Ni layer is applied after each completed layer of windings. Inter-winding insulation can be applied by using appropriate NiO spacers. The solenoid (Fig. 8) is then subjected to a single step heat treatment and can be characterized without mechanical reinforcement at low fields. For the realization of the measurement in back ground field up to 5 Tesla, the coil was impregnated by Oxford Instruments. There was no change in performances of the coil before and after impregnation.

In Fig. 9 the maximum axial and perpendicular field in the test coil is compared to short sample results for parallel and for perpendicular field direction. The comparison of axial field (filled symbols) and radial field (open symbols) show that the critical current at 100 μV/m of this coil is limited by the radial field. When an external field (in the range of 1 to 5 Tesla) is applied parallel to the coil axis a field dependency smaller than the corresponding short sample field dependency is observed. This is an indication that the radial field remains the limiting factor up to the field of 6.3 Tesla.

Fig. 8: Bi-2212 solenoid with NiO insulation realized by wind & react.

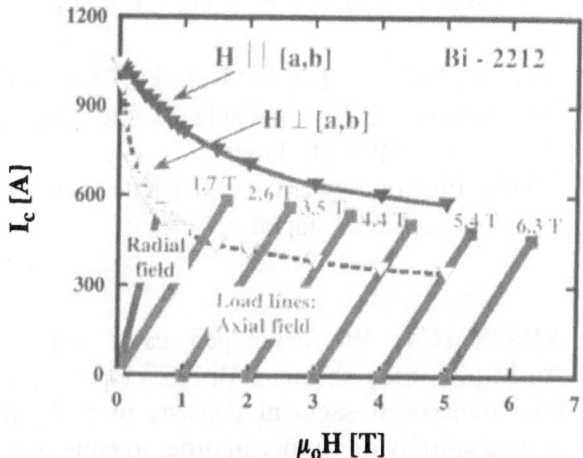

Fig. 9. Comparison of coil performances (■,□) with field dependent short sample performance at 4 K (▼,▽) in axial (■,▼) and in radial (□,▽) field.

CONCLUSIONS

A simple consideration of cost components of a superconducting device based on experimental conductor performances has been realized. It shows that for magnets, at moderate fields of a few Tesla the optimum operating conditions lays in the 20 - 30 K temperature range even for specific conductor cost as low as 10 Euro/kAm (77 K reference cost). This shows the interest of the development of Bi-2212 conductors that achieve the best performance versus cost figure for this application condition.

Bi-2212 PIT conductors are realized in kilometer length at Alcatel. The green tape can be processed either by R&W or by W&R and reproducible performances are reached on hectometric length. Critical current densities of Bi-2212 conductors lay in the range of 1800 - 2400 A/mm^2 and engineering current densities of 650 A/mm^2 are reached at 4 K. High field measurements of the conductor have been carried out showing that 200 A are reached at 30 Tesla and the critical current is reduced by less than half its value. Homogeneity studies have been carried out on reacted 50 – 120 meter lengths from a 1 km green tape, the 4 K I_c values lay all close to 400 A. These performances are reached independently which part of the km length of green tape is used. Bending tests show that the conductor looses less than 2 % for a bending radius above 50 mm.

1. E.E. Hellstrom and W. Zhang, Superconducting Glass-Ceramics in Bi-Sr-Ca-Cu-O, (World Scient. Publishing Co Pte. Ltd., Singapore, 1997).
2. L.M. Rubin et all. Physica C, 217 [1,2] 227-33, 1993.
3. J. Müller et al. Inst. Phys. Conf. Ser. No 158, 1997
4. Marken et al IEEE Trans. on Appl. Supercond. 7 (2) pp. 2211-4, 1997
5. E.W. Collings et al. Supercond. Sci. Techn. 12 (2) 87-96, 1999
6. Kitaguchi H et al IEEE Trans. on Appl. Supercond. 9 (2) pp. 1794-9, 1999
7. Okada et al. IEEE Trans. on Appl. Supercond. 7 (2) pp. 1904-7, 1998
8. P.F. Herrmann et al. IEEE Trans. on Appl. Supercond. 9 (2) pp. 2738-41, 1999
9. Février A et al IEEE Trans. Magn. MAG-25 pp 1496-9, 1989
10. P.F. Herrmann et al. IEEE Trans. on Appl. Supercond. 7 (2) pp. 2201-6, 1997
11. Hoang Y B et al EUCAS, Inst. Phys. Conf. Ser. No 158 IOP Publishing pp. 1385-8, 1997
12 Goldacker W et al, IEEE Trans. on Appl. Supercond. 9 (2) pp. 2155-8, 1999
13. P.F. Herrmann et al, to be published in proceedings of EUCAS conference held in Sitges (Barcelona), September 14-17, 1999 Paper: 8A
14. Strobridge, Design concepts for superconducting cables, EPRI report TR103631, 1994
15 T. Hase et al. "Operation of superconductivity jointed Bi-2212 solenoidal coil in persistent current mode" Cryogenics 37 (2) pp. 201-6, 1997

The development of the high-Tc superconducting cable consist of the transposed segment conductor

N. Futaki[1], S. Nagaya[3], A. Kume[1], K.Goto[1], T. Hasegawa[2] and T. Saitoh[1]

[1]Fujikura Ltd., 1-5-1, Kiba, Koto-ku Tokyo 135-8512, Japan
[2]Showa Electric Wire & Cable Co. Ltd., 2-1-1, Odasakae, Kawasaki-Ku, Kawasaki,
 Kanagawa 210-8660, Japan
[3]Chubu Electric Power Co., 20-1 Kitasekiyama, Odaka-cho, Midori-ku,
 Nagoya 459-8522, Japan

ABSTRACT: We developed cable conductors consist of transposed segments made of Ag-Mg-Sb alloy sheathed Bi-2223 tapes developed by Showa electric wire & cable co., LTD.. One transposed segment consists of 5 Ag-Mg-Sb alloy sheathed tapes. The segments were wound spirally on former in order to construct the cable conductor. At a.c. loading, a comparison of the current phase between the five tapes in a segment shows good agreement and the phase was in good agreement with the phase of the conductor. And, the amplitude between these tapes showed good agreement. From these results, the current was sharing equally to each tape in the transposed segment conductor. The a.c. loss exhibited about the cubic dependence of the current and about the proportional of the frequency.

KEY WORDS: HTS cable, Ag-Mg-Sb alloy sheathed Bi-2223 tape, transposed segment, current sharing, a.c. loss

INTRODUCTION

The power transmission system which was using the high-Tc superconducting (HTS) materials included various kinds of elements such as a cooling system [1] or structure of a connection with the part on normal temperature. In particular, various structures were proposed about the conductor, for example, the transposed conductor comprises of the round shape wire [2], and the multi-layered spiral conductor consist of the tape shape wire [3,4]. As the one, we proposed the transposed segment conductor structure [5]. Because this structure uses tape shape wire, in addition to the Ag-sheathed Bi-2223 ($(Bi, Pb)_2Sr_2Ca_2Cu_3O_x$) tape, the Y-123 ($Yba_2Cu_3O_x$) film wire can be applied. In this report, we produced the transposed segment conductor which used Ag-Mg-Sb alloy sheathed Bi-2223 tape [6] as the HTS material, and we investigated d.c. and a.c. properties of this conductor.

EXPERIMENTAL

The parameters of the transposed segment conductor were shown in Table 1. Ag-Mg-Sb sheathed Bi-2223 tapes were developed by Showa Electric Wire & Cable Co. Ltd. The conductor consisted by spirally winding with 25 transposed segments on the former of 30mm outer diameter. The transposed segment consisted of five Ag-Mg-Sb sheathed Bi-2223 tapes. The schematic of the transposed segment showed in Fig.1. The length that a tape was moved on the side tape was the crossed pitch length, which was 80mm. With the parallel part, the two tapes were moved up to vertical, and this length was 115mm by the mechanical limitation. Such side movement and vertical movement are done next in the opposite side of the segment. Such movements of tapes were repeated five times. Then, the position of the tapes became same as the start position. This length was the transposed pitch length, and it was 1950mm.

Table 1. The parameters of the transposed segment conductor

Tape: Ag-Mg-Sb alloy sheathed Bi-2223 tape	
Tape size:	$1.55mm^w \times 0.24mm^t$
Ic of the tape:	10.2A
Number of the tape (segment):	5
Number of the segment:	25
Crossed pitch length:	80mm
Transposed pitch length:	1950mm
Former size(O.D.):	30mm
Spiral pitch length:	1m
Length of the conductor:	2.5m

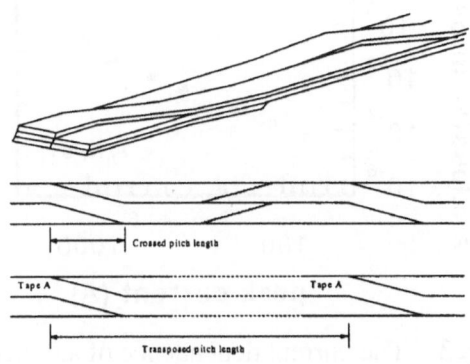

Fig. 1. The schematic of the transposed segment

The critical current (Ic) of the tapes was 10.2A on average at the liquid nitrogen (LN$_2$) temperature. And, in case of the production of the conductor, the strain was controlled within 0.01% on the edge-wise bending, and 0.3% on the flat-wise bending, and, as for this value, there was an enough margin for the mechanical characteristics of Ag-Mg-Sb alloy sheathed Bi-2223 tape. The conductor length was 2.5m, and the distance between voltage terminals was 1.95m same as the transposed pitch length. The measurement was using the 4 probe method in LN$_2$. The criterion of Ic was 1μV/cm. We used the Rogowski coil at the measurement of the current wave form at a.c. loading, about the conductor and the five tapes which constituted an arbitrary segment.

RESULT AND DISCUSSION

The current - voltage (I-V) characteristic was shown in Fig. 2. The Ic was 724A at the criterion of 1μV/cm, and it was smaller than the total Ic of the tapes. Because our conductor production technique was imperfect in the segment winding process, the arrangement disturbances of segments bred. On this account the segment which could arrange with former of 30mm O.D. became 25 with this conductor for what was 29 in ideal case. By these things, non-uniform gaps occurred between segments. In this case, the perpendicular unit of the self magnetic field was about 30-40G in a maximum.

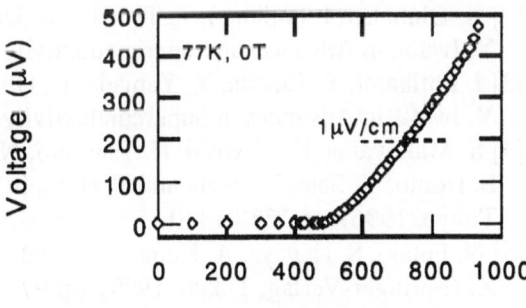

Fig. 2 1-V characteristic of the conductor

So, it was thought that Ic deteriorated by influence of this self magnetic field. We are improving the conductor production technique and possible to arrange a segment of ideally. At a.c. loading, a comparison of the current phase between the five tapes in a segment shows good agreement and the phase was in good agreement with the phase of the conductor. And the amplitude between these tapes showed good agreement. From these results, the current was sharing equally to each tape in the transposed segment conductor. The a.c. loss of this conductor was shown in Fig. 3 and Fig.4. The a.c. loss was proportional to the cubic current and the frequency.

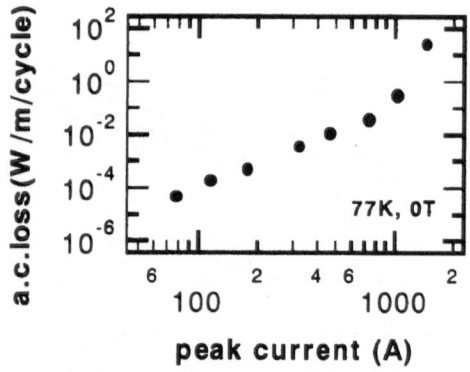

Fig. 3 The current dependence of a.c. loss of the transposed segment conductor

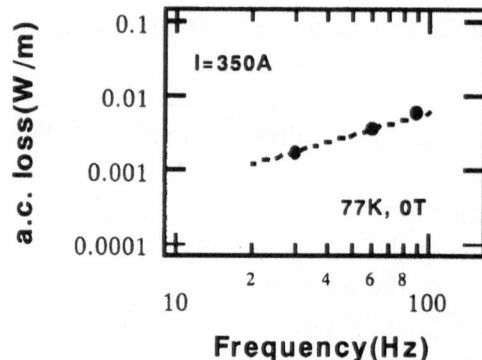

Fig. 4 The frequency dependence of a.c. loss of the transposed segment conductor

Summary

We proposed the transposed segment conductor which the Y-123 film wire could apply in addition to the Ag-sheathed tape. And, we produced the transposed segment conductor of 2.5m length using Ag-Mg-Sb sheathed Bi-2223 tapes. The Ic of this conductor was 724A at the criterion of $1\mu V/cm$. It was smaller than the total Ic of the tapes. We think that the influence of the self-magnetic field is the main reason about it. At a.c. loading, the phase of the current at the five tapes in a segment and the conductor showed good agreement. And the amplitude of these tapes also showed good agreement. The a.c. loss was proportional to the cubic current and the frequency.

[1] S. Nagaya, T. Hirasawa, Y. Nakao, T. Nakajima and T. Saitoh, in Advances in Superconductivity X (Springer-Verlag, Tokyo, 1998), pp.1271-1274

[2] K. Ohmatsu, J. Fujikami, T. Taneda, M. Ueyama, K. Hayashi, K. Sato, S. Honjo, H. Hobora and Y. Iwata, in Advances in Superconductivity XI (Springer-Verlag, Tokyo, 1999), pp. 979-982

[3] J. Fujikami, T. Taneda, Y. Yamada, K. Ohmatsu, K. Hayashi, K. Sato, S. Honjo, H. Hobora and Y. Iwata, in Advances in Superconductivity XI (Springer-Verlag, Tokyo, 1999), pp. 903-906

[4] S. Mukoyama, K. Miyoshi, H. Tsubouti, N. Ichiyanagi, M. Miura, N. Uno, Y. Tanaka, H. Ishii, S. Honjo, Y. Sato, Y. Iwata and T. Hara, in Advances in Superconductivity IX (Springer-Verlag, Tokyo, 1996), pp.1337-1340

[5] N. Futaki, S. Nagaya, A. Kume, N. Sadakata and T. Saitoh, in Advances in Superconductivity XI (Springer-Verlag, Tokyo, 1999), pp.971-974

[6] Y. Hikichi, J. Nishioka, T. Koizumi, T. Hirota, T. Hasegawa, in Advances in Superconductivity XI (Springer-Verlag, Tokyo, 1999), pp. 915-918

High-J_c Property of Round Strand for HTS Transposed Cable Conductor

T. Taneda[1], J. Fujikami[1], K. Ohmatsu[1], H. Takei[1], S. Honjo[2], T. Mimura[2], and Y. Takahashi[2]

[1]Sumitomo Electric Industries, Ltd., 1-1-3 Shimaya, Konohana-ku, Osaka, 554-0024, Japan
[2]Tokyo Electric Power Company, 4-1 Egasaki-cho, Tsurumi-ku, Yokohama, 230-8510, Japan

Abstract: Ag-sheathed Bi-2223 round strands for high temperature superconducting (HTS) transposed cable conductors were developed. The critical current density (J_c) at 77K, 0T reached $1.69 \times 10^8 A/m^2$ in short length sample and $1.0 \times 10^8 A/m^2$ in 50m-long sample. Several analyses were done to evaluate the basic properties of round strand, such as Bi-2223 volume fraction, crystal alignment, and magnetic field dependence of J_c. It was confirmed that the above properties were improved with increasing J_c.

Keywords: Bi-2223 Ag-sheathed round strand, critical current density, AC loss, transposed cable conductor

INTRODUCTION

The HTS transposed cable conductor with both large current capacity over 3kArms and low AC loss less than 1W/m can be realized by using round strand [1-6]. In the past work, our group developed round strands; J_c values were $1.05 \times 10^8 A/m^2$ in short sample [1], and $8.1 \times 10^7 A/m^2$ in 12m-long sample [2], both at 77K, 0T. However, further increases of J_c value and long length are required for practical use.

Therefore, we improved the manufacturing process of round strand and obtained high-J_c value in both short length sample and 50m-long sample. The basic properties were evaluated and analyzed in the present work. The AC transport current loss measured and AC current loading of 1m-long prototype conductors using the high-J_c round strand [3-6] were also carried out.

EXPERIMENT

Design and fabrication of strand. The specifications and cross section of round strand we have developed are shown in Table 1 and Fig. 1, respectively. To increase the overall J_c (J_e) of the strand, silver to superconductor ratio (Ag/SC ratio) was reduced in the samples No.2 and No.3. These strands were fabricated by the powder-in-tube (PIT) method. Detail conditions were described previously [1-2].

Measurement. Current-Voltage properties of round strand was measured by the conventional DC four-probe method at 77K, 0T. Critical current (I_c) was determined by $100\mu V/m$ criterion. Magnetic field dependence of J_c (J_c-B) was measured with two configuration; magnetic field applied parallel to longitudinal direction of round strand (B ∥ L), and perpendicular to longitudinal direction (B⊥L). Magnetic susceptibility was measured by SQUID magnetometer and Bi-2223 volume fraction was determined by signal ratio of Bi-2223/(Bi-2212+Bi2223) [7]. The alignment of Bi-2223 crystals in the filaments was observed by SEM.

RESULTS

The specifications of short-length samples of round strand are shown in Table 1. J_c value at 77K, 0T reached $1.69 \times 10^8 \text{A/m}^2$ in the sample No.1. The J_c of the strands were increased by reducing the Ag/SC ratio in the sample No.2 and No.3.

Table 1. Specifications of round strands.

sample No.	1	2	3	4
diameter [mm]	0.93	0.95	0.95	0.86
length [m]	4.0×10^{-3}	5.0×10^{-2}	1.0×10^{-2}	5.0×10^{1}
Ag/SC ratio	7.2	5.5	2.9	8.1
I_c [A] (77K, 0T)	14.1	16.9	17.0	6.4
J_e [10^7A/m^2]	2.06	2.40	2.4	1.1
J_c [10^8A/m^2]	1.69	1.57	0.93	1.0

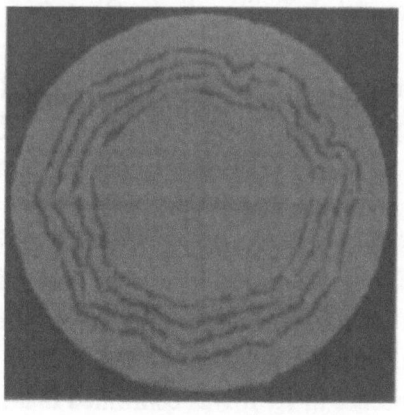

Fig. 1. Typical cross section of round strand. Bi-2223 filaments were arranged symmetrically.

The J_c -B properties of various J_c values are shown in Fig. 2. Both B ∥ L and B⊥L are expressed by average magnetic field applied to c-axis direction of Bi-2223 crystal, and the two curve fit well in the diagram. The Bi-2223 volume fractions of round strand are shown in Fig. 3. In comparison, the volume fraction data of tape wire from Ref. [7] are re-plotted in the same graph.

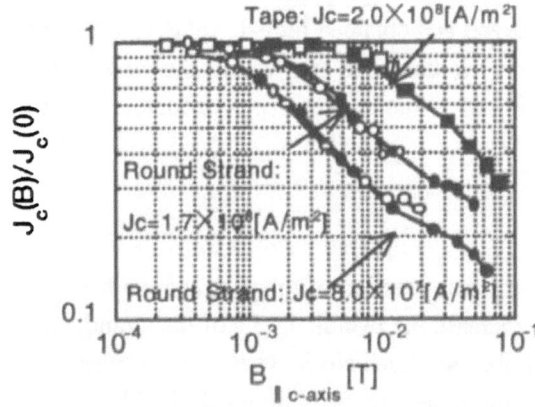

Fig. 2. The J_c -B properties of various J_c values. Both B ∥ L (white) and B⊥L (black) are expressed by average magnetic field applied to c-axis direction of Bi-2223 crystal.

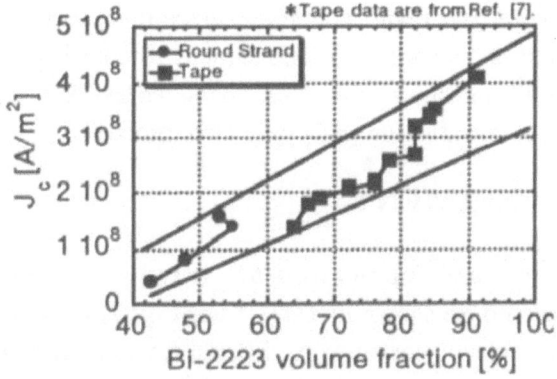

Fig. 3. The Bi-2223 volume fractions of round strand. The data of tape wire from Ref. [7] are re-plotted in the same graph.

DISCUSSION

The J_c -B analysis in Fig. 2 shows that magnetic field component parallel to c-axis of Bi-2223 crystal is dominant, and grain connectivity was improved in the lower B region. The Bi-2223 volume fraction analysis in Fig. 3 indicates that the volume fraction of round strand increases and gets closer to that of tape wire with increasing J_c. Typical SEM photographs are shown in Fig. 4 and Fig. 5, respectively. The crystal alignment near the Ag-SC interface of the filaments was improved in the higher-J_c sample.

Recently, a 1m-long transposed cable conductor was fabricated by using several sets of 50m Bi-2223/Ag-

Mn round strands. Average J_c value of these strands reached $1.0 \times 10^8 A/m^2$ at 77K. The I_c of the conductor at 77K and 63K were 2.1kA and 4.0kA, respectively. Continuous 50Hz current loadings up to 3.0kArms at 77K and 4.0kArms at 63K were performed; however, temperature rises were confirmed within 1K. At the same time, low transport current losses less than 1W/m were demonstrated when 50Hz current were lower than I_c. These results indicate that the transposed conductor with Bi-2223 round strand is promising for large capacity cable conductors.

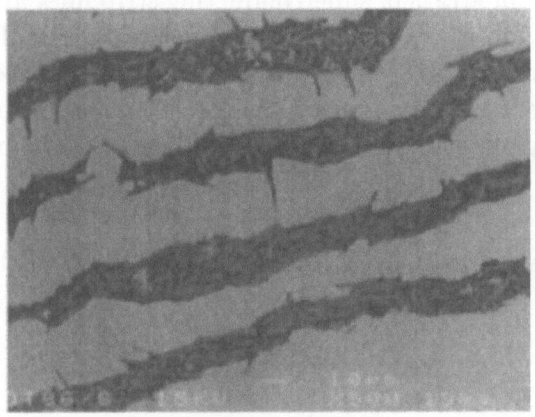

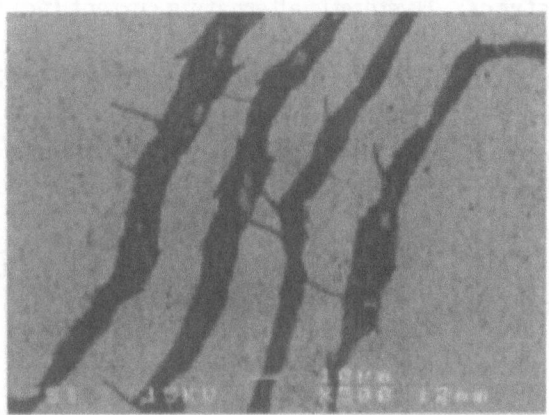

Fig. 4. SEM photograph of the cross section of low J_c sample: $J_c=5.5 \times 10^7 A/m^2$.

Fig. 5. SEM photograph of the cross section of high J_c sample: $J_c=1.60 \times 10^8 A/m^2$.

CONCLUSION

Ag-sheathed Bi-2223 round strands for HTSC transposed cable conductors were developed. The critical current density (J_c) at 77K, 0T reached $1.69 \times 10^8 A/m^2$ in short length sample and $1.0 \times 10^8 A/m^2$ in 50m-long sample. Several analyses were done to evaluate the basic properties of round strand, such as Bi-2223 volume fraction, crystal alignment, and magnetic field dependence of J_c. It was confirmed that the above properties were improved with increasing J_c.

A 1m-long transposed cable conductor was fabricated and evaluated. The I_c of the conductor at 77K and 63K were 2.1kA and 4.0kA, respectively. Continuous 50Hz current up to 3.0kArms at 77K and 4.0kArms at 63K were loaded. Temperature rises were confirmed within 1K, and low transport current losses less than 1W/m were demonstrated.

1. S. Hahakura et al., in *Advances in Superconductivity X*, edited by K. Osamura and I. Hirabayashi (Springer-Verlag, Tokyo, 1998), pp. 901-904
2. T. Taneda et al., in *Advances in Superconductivity XI*, edited by N. Koshizuka and S. Tajima (Springer-Verlag, Tokyo, 1999), pp. 983-986
3. K. Ohmatsu et al., in *Advances in Superconductivity X*, edited by K. Osamura and I. Hirabayashi (Springer-Verlag, Tokyo, 1998), pp. 905-908
4. N. Saga et al., in *Advances in Superconductivity X*, edited by K. Osamura and I. Hirabayashi (Springer-Verlag, Tokyo, 1998), pp. 893-896
5. K. Ohmatsu et al., in *Advances in Superconductivity XI*, edited by N. Koshizuka and S. Tajima (Springer-Verlag, Tokyo, 1999), pp. 817-822
6. K. Ohmatsu et al., in *Advances in Superconductivity XI*, edited by N. Koshizuka and S. Tajima (Springer-Verlag, Tokyo, 1999), pp. 979-982
7. T. Kaneko et al., in *Advances in Superconductivity IX*, edited by S. Nakajima and M. Murakami (Springer-Verlag, Tokyo, 1997), pp. 907-910

Alternating Current Loss of Strip Arrays as a Model for Resistive Fault Current Limiters

Yasunori Mawatari and Hirofumi Yamasaki

Frontier Technology Division, Electrotechnical Laboratory,
1-1-4 Umezono, Tsukuba, Ibaraki 305-8568, Japan

Abstract: Hysteretic alternating current (ac) loss P in arrays of superconducting strip lines are calculated on the bases of the critical state model. For a simplified model of a film-type fault current limiter, we consider strip arrays in which multiple strip lines are periodically arranged and are carrying bidirectional currents. The P of a stack of strip lines (Z stack) is smaller than P of an isolated strip, whereas P of a coplanar array of strip lines (X array) is larger than P of an isolated strip.

Keywords: Superconducting strip array, Critical state model, Alternating current loss, Fault current limiters

INTRODUCTION

Macroscopic electromagnetic response of a superconducting strip are well described by the critical state model for an isolated strip[1–3]. For strip arrays, the magnetic-field and current-density distributions in the critical state model are easily derived by using a transformation of these distributions for an isolated strip line[4–6]. Analytical investigation of electromagnetic response of fault current limiters (FCLs) that carry ac current may be highly complicated, because complicated meandering current paths are patterned in film type FCLs. If we ignore the effects of both ends and edges, we can simplify FCLs as a coplanar array of strips that carry bidirectional currents[6]. In the present work, we analyze ac loss of idealized FCLs, i.e., arrays of strip lines that carry bidirectional currents. We consider two types of strip arrays: a stack of strips (Z stack) and a coplanar array of strips (X array).

MAGNETIC FIELD AND CURRENT DENSITY

Consider a stack of superconducting-strip lines (Z stack) in which an infinite number of parallel strip lines are piled along the z axis. Each strip line is $2w$ wide and d thick ($d \ll 2w$), and is spaced at intervals of D along the z axis. The nth strip occupies an area where $|x| \leq w$, $|y| < \infty$, and $|z - nD| \leq d/2$ ($n = 0, \pm1, \pm2, \cdots, \pm\infty$). The strip lines carry bidirectional currents; namely, the nth strip carries a transport current $(-1)^n I_t$ along the y axis. When a Z stack carries bidirectional currents without an applied magnetic field, current-density distribution has periodicity $J_y(x, z + nD) = (-1)^n J_y(x, z)$. Using the periodicity of $J_y(x, z)$ and the symmetry of $J_y(-x, z) = J_y(x, z)$, we obtain the following relationship between the magnetic field along the z axis, $H(x) \equiv H_z(x, z = 0)$, and the mean current density along the y axis, $J(x) \equiv (1/d) \int_{-d/2}^{+d/2} J_y(x, z)dz$:

$$ H(x) = -\frac{d}{2\pi} \sum_{n=-\infty}^{+\infty} \int_{-w}^{+w} du \frac{(-1)^n (x-u) J(u)}{(x-u)^2 + (nD)^2} = -\frac{d}{D} \int_0^w du \frac{\sinh(\pi x/D)\cosh(\pi u/D) J(u)}{\sinh^2(\pi x/D) - \sinh^2(\pi u/D)}. \quad (1) $$

By introducing the variable transformation of $\tilde{x} = (D/\pi)\sinh(\pi x/D)$ and $\tilde{u} = (D/\pi)\sinh(\pi u/D)$, we can reduce Eq. (1) to $\tilde{H}(\tilde{x}) = -(d/\pi) \int_0^{\tilde{w}} d\tilde{u} \tilde{J}(\tilde{u}) \tilde{x}/(\tilde{x}^2 - \tilde{u}^2)$, where $\tilde{H}(\tilde{x})$ is the transformed magnetic field, $\tilde{J}(\tilde{x})$ is the transformed current density, and $\tilde{w} = (D/\pi)\sinh(\pi w/D)$ is the transformed strip width. Note that the relationship between $\tilde{H}(\tilde{x})$ and $\tilde{J}(\tilde{x})$ corresponds to the Biot-Savart law for an isolated strip line that carries a transport current without an applied magnetic field. In other words, by using the transformation of $x \to (D/\pi)\sinh(\pi x/D)$, we obtain $H(x)$ and $J(x)$ for Z stacks from those for an isolated strip line. This transformation method is applicable to the critical state model with a constant critical current density J_c[4–6]. When the transport current I_t is monotonically increased from zero, $H(x)$ and $J(x)$ are given as follows. In the flux-free region, $|x| < a$ (where a is the flux front), the magnetic field is shielded, $H(x) = 0$, and the current density is given by $J(x) = (2J_c/\pi)\arctan[1/\varphi(x,a)]$. In the flux-filled region, $a < |x| < w$, the current density saturates to J_c, $J(x) = J_c$, and the magnetic field is $H(x) = (J_c d/\pi)(-x/|x|)\operatorname{arctanh}[\varphi(x,a)]$. The function $\varphi(x,a)$ is defined as

$$\varphi(x,a) = \left[\frac{|\sinh^2(\pi a/D) - \sinh^2(\pi x/D)|}{\sinh^2(\pi w/D) - \sinh^2(\pi a/D)} \right]^{1/2}. \tag{2}$$

Relationship between a and I_t is determined by $I_t = d\int_{-w}^{+w} J(x)dx$.

Next we consider a coplanar array of superconducting-strip lines (X array) in which an infinite number of parallel strip lines are arranged in the xy plane. Strip lines (of width $2w$ and thickness d) is arranged along the x axis with a periodicity L ($> 2w$). The spacing between strips is given by $s = L - 2w$. The nth strip occupies an area where $|x - nL| \le w$ ($n = 0, \pm 1, \pm 2, \cdots, \pm\infty$), $|y| < \infty$, and $|z| \le d/2$. The strip lines carry bidirectional currents $(-1)^n I_t$ along the y axis. The expressions of magnetic field, current density, and ac loss for an X array are easily given by those for a Z stack by substituting $D \to iL$ [where $i = (-1)^{1/2}$][4]. In the formulas of $H(x)$ and $J(x)$ in an X array, therefore, the function $\varphi(x,a)$ should be given by the following equation instead of Eq. (2).

$$\varphi(x,a) = \left[\frac{|\sin^2(\pi a/L) - \sin^2(\pi x/L)|}{\sin^2(\pi w/L) - \sin^2(\pi a/L)} \right]^{1/2} \tag{3}$$

AC LOSS OF STRIP ARRAYS

The $H(x)$ and $J(x)$ distributions for alternating current whose amplitude is I_0 are derived from the expressions for monotonically increasing I_t[2]. Thus, the hysteretic ac loss power P of each strip per unit length in strip arrays is calculated from the magnetic-field distribution[2,6]:

$$\frac{P(a_0)}{\mu_0 \nu I_c^2} = \frac{2}{\pi w} \int_{a_0}^{w} dx \left(1 - \frac{x}{w}\right) \operatorname{arctanh}[\varphi(x,a_0)], \tag{4}$$

where ν is the frequency of alternating current. The a_0 is the flux front when the transport current reaches to the maximum value, $I_t = I_0$. Relationship between a_0 and I_0 is given by

$$\frac{I_0(a_0)}{I_c} = 1 - \frac{2}{\pi w} \int_0^{a_0} dx \arctan[\varphi(x,a_0)], \tag{5}$$

where $I_c = 2J_c wd$ is the critical current. The function $\varphi(x,a)$ in Eqs. (4) and (5) is given by Eq. (2) for Z stacks, and Eq. (3) for X arrays[6]. If we eliminate a_0 in Eqs. (4) and (5), the P can be expressed as a function of I_0. Equations (4) and (5) for $D \to \infty$ corresponds to P of an isolated strip line[1–3].

For small current limit, $I_0/I_c \to 0$, the P of a Z stack is proportional to the biquadrate of current, $P \simeq (\mu_0 \nu I_0^4/I_c^2)F_z(\pi w/D)$, where $F_z(\theta) \equiv (\pi^3/96)\theta^2/\tanh^2\theta[K(\tanh\theta)]^4$ and $K(k)$

744

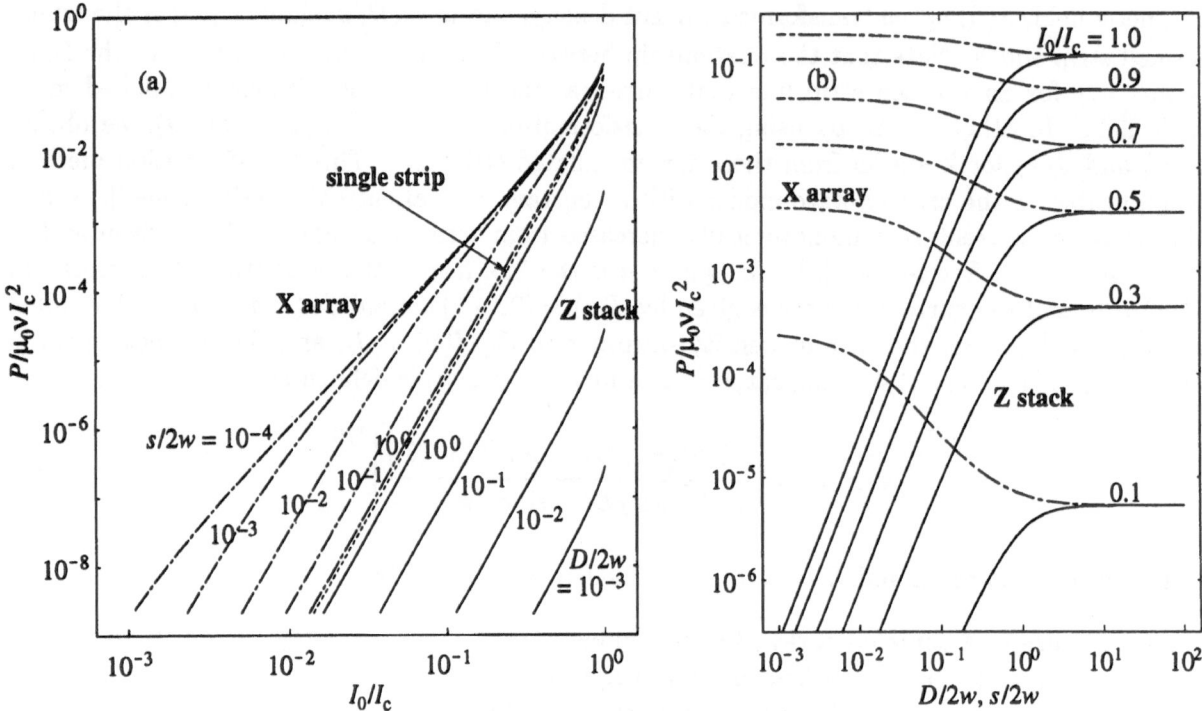

Fig. 1. AC loss power P of each strip per unit length in Z stacks (solid lines) and in X arrays (dot-dashed lines). (a) Current-amplitude I_0 dependence of P for various spacings D in Z stacks and for $s = L - 2w$ in X arrays. The P in an isolated strip line is also shown as a thin dotted line. (b) Dependence of P on periodicity of stack D in Z stacks and on spacing between strips $s = L - 2w$ in X arrays for various current amplitudes I_0.

is the complete elliptic integral[7]. The P of an X array for small current limit is[6] $P \simeq (\mu_0 \nu I_0^4 / I_c^2) F_x(\pi w/L)$, where $F_x(\theta) \equiv F_z(i\theta) = (\pi^3/24)\theta^2/\sin^2(2\theta)[K(\sin\theta)]^4$.

Figure 1(a) shows I_0 dependence of P in strip arrays. The P for small current limit is always proportional to I_0^4. For X arrays with small spacings between strips, $s/2w \ll 1$, the exponent $n \equiv \partial(\ln P)/\partial(\ln I_0)$ (i.e., $P \propto I_0^n$) can be smaller than 3. As shown in Fig. 1(b), the P becomes smaller as the periodicity D decreases, and the dependence is given as $P \propto D^2$ for $D/2w \ll 1$ in Z stacks. In X arrays, on the other hand, the P becomes larger as the spacing $s = L - 2w$ decreases, and P saturates to a finite value when $s/2w$ is small enough.

Although the configuration of strips considered here is oversimplified, the theoretical expressions of ac loss given by Eqs. (2)–(5) may be useful in the initial design phase of FCLs. In practice, arrangement of strips in FCLs must be determined by various factors, such as electric and thermal behavior during current-limiting processes.

1. W.T. Norris, J. Phys. D **3**, 489 (1970).
2. E.H. Brandt and M. Indenbom, Phys. Rev. B **48**, 12893 (1993).
3. E. Zeldov, J.R. Clem, M. McElfresh, and M. Darwin, Phys. Rev. B **49**, 9802 (1994).
4. Y. Mawatari, Phys. Rev. B **54**, 13215 (1996); IEEE Trans. Appl. Supercond. **7**, 1216 (1997).
5. Y. Mawatari, in *Advances in Superconductivity IX*, edited by S. Nakajima and M. Murakami (Springer-Verlag, Tokyo, 1997) p. 575; K.-H. Müller, Physica C **289**, 123 (1997).
6. Y. Mawatari, Appl. Phys. Lett. **75**, 406 (1999).
7. I.S. Gradshteyn and I.M. Ryzhik, *Table of Integrals, Series, and Products*, 5th ed. (Academic Press, New York, 1994) p. 908.

Stability of a Bi-2223 Refrigerator Cooled Magnet

Hiroaki Kumakura,[1] Hitoshi Kitaguchi,[1] Kazumasa Togano,[1] Hitoshi Wada,[1] Kengo Ohkura,[2] Munetugu Ueyama,[2] Kazuhiko Hayashi[2] and Ken-ichi Sato[2]

[1]National Research Institute for Metals, 1-2-1 Sengen, Tsukuba, Ibaraki 305-0047, Japan
[2]Sumitomo Electric Industries, Ltd., 1-1-3 Shimaya, Konohana-ku, Osaka 554-0024 , Japan

Abstract: A Bi-2223 pancake magnet was cooled down to various temperatures with a GM cryocooler, and was tested in order to investigate the stability of the magnet. The pancake coil at the top of the magnet which was located at a greatest distance from the cooling stage showed the lowest thermal stability of the 17 pancake coils. Below ~50K, a maximum operating current for which the magnet was stably operated was almost equal to I_c defined with $10^{-13}\Omega m$ criterion. At 20K and 30K, we repeated the excitation of the magnet. The temperature of the top pancake increased with increasing the number of cycles, but the temperature saturated at the number of cycles of 20-30 for the excitation rates of 1-3T/min.

Keywords: Pancake coil, Critical current, Magnetic separation

INTRODUCTION

Practical levels of critical current have already been obtained at ~20K for Bi-based oxide superconducting tapes. Because ~20K is efficiently obtained by a cryocooler, a Bi-based oxide superconducting magnet cooled with a cryocooler has a great potential in many technological applications. One of the interesting applications of a cryogen-free magnet is a magnet for magnetic separation. Magnetic separation has been used for many years to collect magnetic particles such as iron ore. Superconducting magnets for magnetic separation were proposed as early as 1970. In 1986, a conventional metallic superconducting magnet was installed in the High Gradient Magnetic Separator for the kaolin purification. However, conventional metallic superconducting magnets usually require liquid helium for cooling, which hinders the significant decrease of the running cost. Thus, a cryogen-free Bi-based oxide superconducting magnet is now expected to be a promising alternative for magnetic separation. In the case of a cryogen-free magnet, the thermal stability of the magnet is one of the important factors. In the previous paper, we fabricated a small $Bi_2Sr_2Ca_2Cu_3O_x$ (Bi-2223) pancake magnet using silver sheathed Bi-2223/Ag multifilamentary tapes, and tested the magnet at various temperatures using a Gifford-McMahon(GM) type cryocooler[1]. In this paper, we investigated the stability of the refrigerator-cooled magnet.

EXPERIMENTAL

The magnet consists of 17 double pancake coils whose dimension was 79mm bore and 130mm outer diameter. The detail of the magnet fabrication was described elsewhere[2]. The magnet was mounted on the cooling stage (second stage) of a GM cryocooler and the cryostat was evacuated down to 10^{-7} Torr. The detail of the cryostat was described elsewhere[1]. Electric field (E) vs. current (I) curves of the magnet were measured by a four probe resistive method at various temperatures. Voltage taps were attached to the three pancake coils located at the top, bottom and middle of the magnet, and the E-I curves of the three pancake coils were measured at a fixed cooling stage temperature. Carbon glass thermometers were also attached to these three pancakes. The time dependence of the voltage and temperature of the three pancakes was also measured at a fixed applied current for a time up to 24 hours.

For a magnetic separation system, repetition of the magnet excitation is essential, and rapid energizing and de-energizing are very important in order to attain high efficiency of the separation. At 20 and 30K, we repeated the excitation of the magnet with several excitation speeds with 20 sec interval between each energizing and de-energizing, and observed the changes of the voltages and the temperatures of the pancake coils.

RESULTS AND DISCUSSION

At the cooling stage temperature of 30K, stable operation of the magnet was obtained for an applied current of 232A, while for a current of 240A, thermal runaway of the magnet was observed. The voltage of the magnet at the current of 232A initially increased with time probably due to small heat generation at the joints. However, the voltage saturated and became constant after ~30min. When the applied current was increased to 240A, both the voltage and the temperature rapidly went up with time and finally the thermal runaway of the magnet was observed. Therefore the quenching current of the magnet should exist between 232A and 240A at the cooling stage temperature of 30K. The top pancake coil showed the lowest thermal stability of 17 pancakes. This is because the magnet was mainly cooled by the thermal contact between the bottom of the magent and the cooling stage.

hermal stability. Thus, at first, the thermal runaway of the magnet seemed to occur at the top pancake. This suggests that the improvement of heat conduction from the top of the magnet should suppress the temperature increase of the top pancake and enhances the maximum generated field of the magnet.

We measured the voltages and the temperatures of the magnet at various cooling stage temperatures for fixed operating currents. Figure 1 shows the largest applied current for stable operation and the smallest applied current which brought about thermal runaway as a function of the cooling stage temperature. I_c values of three pancake coils located at the top, middle and bottom of the magnet as well as I_c of the total magnet defined with a $10^{-13}\Omega$m criterion are also shown in the figure. At the cooling stage temperature of 20K, stable operation of the magnet was obtained for the applied current of 250A which was the maximum current capacity of our power supply. The maximum stable operating current extrapolated to 20K was ~300A which generated a field of ~1.3T at the

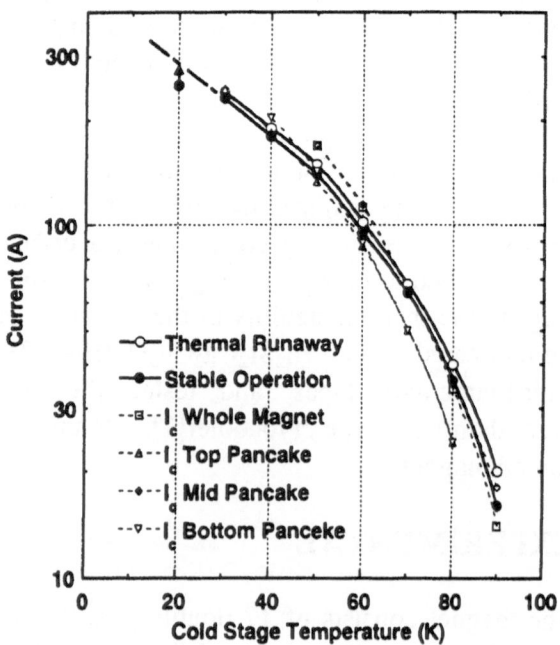

Fig. 1. The largest applied current for stable operation and the smallest applied current leading to thermal runaway of the magnet. I_c values of the whole magnet and three pancake coils are also shown in the figure for comparison.

center of the magnet. The stable operating current rapidly decreased with increasing temperature. However, operating currents larger than ~200A which seemed enough for practical applications were obtained below ~35K. Up to ~50K, the stable operating current was almost equal to I_c values of the top pancake coil having lowest stability.

In the experiments of rapid excitation cycles, the temperature change of the bottom pancake coil was as small as ~0.3K for any excitation rates. However the temperature of the top pancake increased with increasing the number of excitations. Figure 2(a) and (b) shows the temperature of the top pancake as a function of the number of excitations for cooling stage temperatures of 20K and 30K, respectively. The peak operating currents were 250A and 200A for temperatures of 20K and 30K, respectively. The applied current generated magnetic fields of 1.1T and 0.88T, respectively. The excitation speed was changed from 1 to 3T/min. Initially the temperature of the top pancake was almost equal to the cooling stage temperature. However, this temperature increased with increasing the number of cycles due to the energy dissipation in the conductors, but the temperature saturated at a certain number of cycles, and no increase of temperature was observed for larger number of cycles. The saturated temperature of the top pancake increased with increasing the excitation speed. However, we obtained stable operation even for the speed of 3T/min, although the temperature increase of the top pancake was fairly large. This result indicates that the cryocooler cooled Bi-2223 magnet is thermally stable against repetition of excitation, and is a promising candidate for magnetic separation systems.

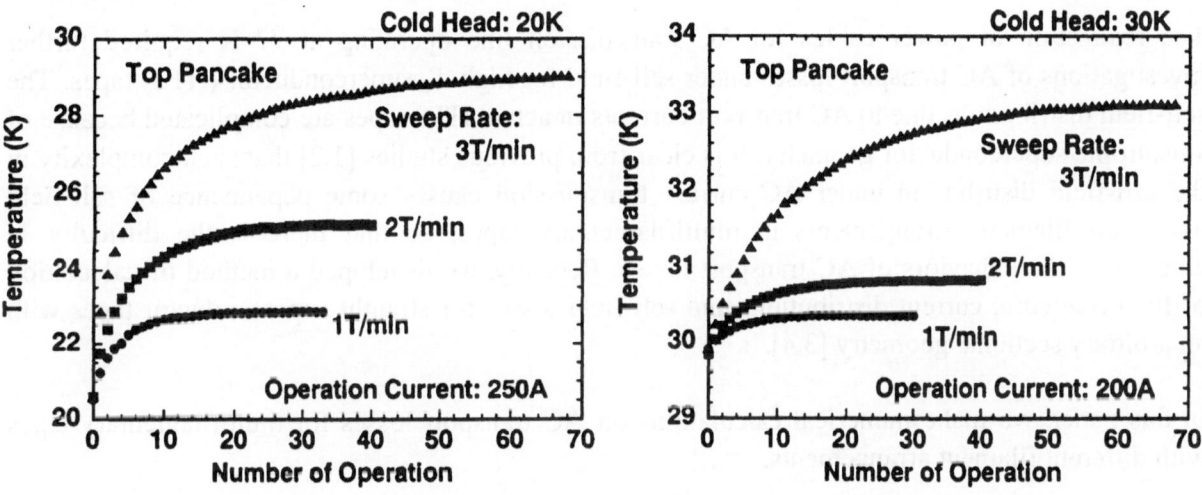

Fig. 2. The temperature increase of the top pancake during repeated excitations. The cooling stage temperature was fixed to be 20K and 30K, and peak operating current was 250A and 200A, respectively.

REFERENCES

1. H. Kumakura, H. Kitaguchi, K. Togano, H. Wada, K. Ohkura, M. Ueyama, K. Hayashi and K. Sato, Cryogenics **38**, 639-643(1998).
2. M. Ueyama, K. Ohkura, S. Kobayashi, K. Muranaka, T. Kaneko, T. Hikata, K. Hayashi, and K. Sato, in *Adv. Superconductivity* VII, edited by K. Yamafuji and T. Morishita(Springer-Verlag, Tokyo,1995), pp. 847-850.

UNDERSTANDING OF AC TRANSPORT LOSSES FOR Ag-SHEATHED Bi2223 MULTIFILAMENTARY TAPES

Ryoji INADA,[1] Tetsuya FUKUNAGA,[2] and Akio OOTA[1]

[1]Toyohaku University of Technology, Tempaku-cho, Toyohashi, Aichi 441-8580 Japan
[2]Gifu National College of Technology, Shinsei-cho, Motosu-gun, Gifu 501-0495 Japan

Abstract: AC transport losses for Ag-sheathed Bi2223 multifilamentary tapes are investigated by using numerical calculations. The measured loss values for 7-filamentary tapes can be explained neither by the prediction for an elliptic tape nor by that for a thin strip based on the theory of Norris. Furthermore, the 7-filamentary tape with sectional filament arrangements has lower loss values than that without sectional filament arrangements, which is explained by the calculated results of current distribution in these tapes. The influence of filament arrangements on the loss behaviors is suppressed as the number of filaments increases.

Keywords: AC transport losses, filament distributions, elliptic tape, thin strip, field-free core

INTRODUCTION

The realization of power cables for AC transmission line operating at 77 K required further investigations of AC transport losses under self-field for high-T_c superconductor (HTS) tapes. The self-field distributions due to AC transport currents in actual HTS tapes are complicated because of anisotropic superconductor geometry. It is clear from previous studies [1,2] that such complexity in the self-field distribution under AC current transmission causes some dependence of self-field losses on filament arrangements in multifilamentary tapes, so that there is the difficulty in elucidating the behaviors of AC transport losses. Recently, we developed a method of calculation of field-free core, current distributions and self-field losses for straight superconductor tapes with an arbitrary sectional geometry [3,4].

In this paper, we make numerical calculations on AC transport losses for multifilamentary tapes with different filament arrangements.

CALCULATIONS

Following the theory of Norris, the loss density $P_d(\mathbf{r})$ per cycle at the observation point $\mathbf{r}$ in a superconductor is calculated as $P_d(\mathbf{r}) = 4J_c\Phi(\mathbf{r})$ [5]. Here J_c is the critical current density of the superconductor and $\Phi(\mathbf{r})$ is the magnetic flux at peak current I_0 passing thorough between field-free core (FFC) and the observation point. To calculate the shape of FFC at a given current I_0 in the superconductor, we regard the superconductor core as a bundle of straight thin fibers with their sectional area dS. Under the condition I_0 less than the critical current I_c, a fiber transports a current fragment $J_c dS$ outside FFC, while it carries no current inside FFC. To determine the current distribution under $I_0 < I_c$, we first calculate $\Phi(\mathbf{r})$ under $I_0 = I_c$. With the calculated result under $I_0 = I_c$, we seek for the fiber with the minimum Φ values among all fibers. Then, we set the current of this fiber to be 0, so that the total current I carrying through the superconductor is reduced by $J_c dS$ (i.e., $I=I_c-J_c dS$). With the new current distribution, we calculate again $\Phi(r)$ and take the fiber with minimum Φ value. By setting the current of this fiber to be 0, the total current is reduced by $2J_c dS$ (i.e., $I=I_c-2J_c dS$) at this stage. We repeat these process many times until the current I reaches the value of I_0, so that we obtain the shape of FFC for a given current I_0 ($<I_c$). Using the result, we calculate the total loss value P by integrating the loss density P_d over the whole part of superconductor.

AC transport losses and current distribution for the multifilamentary tapes with different filament arrangements are calculated by using the above-mentioned method.

RESULTS AND DISCUSSION

Figure 1 shows the AC transport losses normalized to $\mu_0 I_c^2/2\pi$ as a function of normalized current amplitude $\Gamma=I_0/I_c$ for two kinds of 7-filamentary tapes S7-1 ($I_c=4.8$ A) and S7-2 ($I_c=6.6$ A) with different filament arrangements. Also shown in this figure are the results for 19-filamentary tape M19 ($I_c=14.4$ A) and 37-filamentary tape B37-1 ($I_c=35$ A). Numerical calculations of self-field losses are made by taking into account the actual size factors for these tapes, and compared with the experimental results. Note that solid and dashed lines are theoretical prediction for an elliptic tape and a thin strip by the theory of Norris [5]. As can be seen, there is a good agreement between calculated results and experimental data for all tapes. The loss value for S7-1 is smaller than that for S7-2 and this discrepancy in the loss values between these two tapes become significantly in the current range $0.4 \leq \Gamma \leq 0.8$. The loss value for S7-1 is close to a theoretical curve for a thin strip for $\Gamma > 0.5$, while it becomes larger than the theoretical value with decreasing Γ. However, the loss value for S7-2 is close to a theoretical curve for an elliptic tape for $\Gamma > 0.5$, while it becomes smaller than the theoretical value with decreasing Γ. The loss values for M19 and B37-1 are close to a theoretical curve for an elliptic tape. From these result, it is clear that the difference in the loss values between these tapes comes from the difference in the filament arrangements.

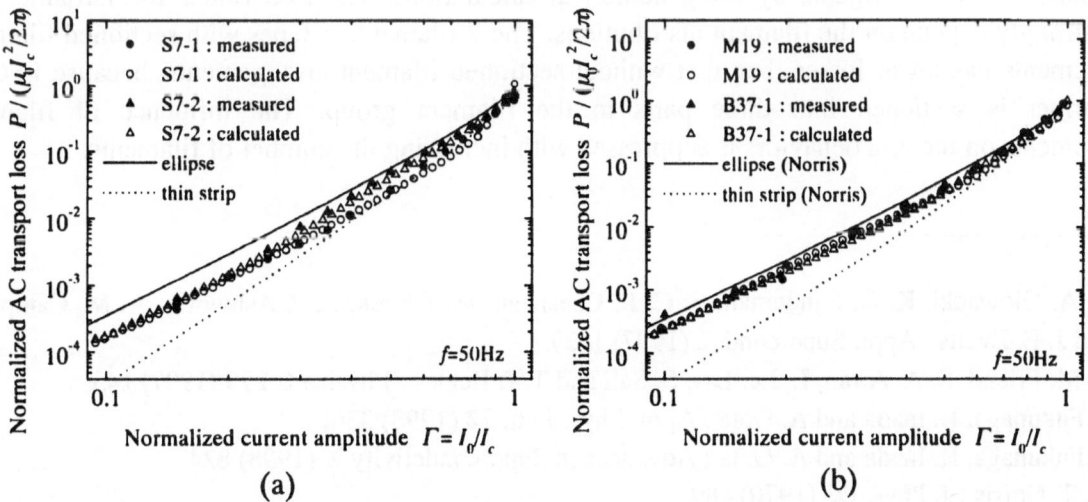

Fig. 1. AC transport losses plotted against normalized current amplitude for multifilamentary tapes with different filament arrangements (a) 7-filamentary tapes and (b) 19- and 37-filamentary tapes. Note that solid and dashed lines are theoretical predictions for an elliptic tape and a thin strip, respectively.

Figure 2 shows the calculation results for the current distributions of all tapes at $\Gamma=0.5$. The black area is the current flowing part and the gray area represents the FFC at a given current amplitude. As can be seen, transport currents in all tapes flow mainly in outer filaments, but the current distributions are quite different with each other. The shape of FFC for S7-1 is sectioned into three parts in the filament group, while that for S7-2 forms two regions connected with each other at the center filament. As the results, the former tape has lower loss values than the latter, because FFC is divided into sectional regions. The shape of FFC for M19 and B37-1 is similar to that for an elliptic tape, so that the loss behavior of these tapes is well explained by that for an elliptic tape. It is clear that the influence of filament arrangements on the self-field losses is suppressed as the number of filaments increases.

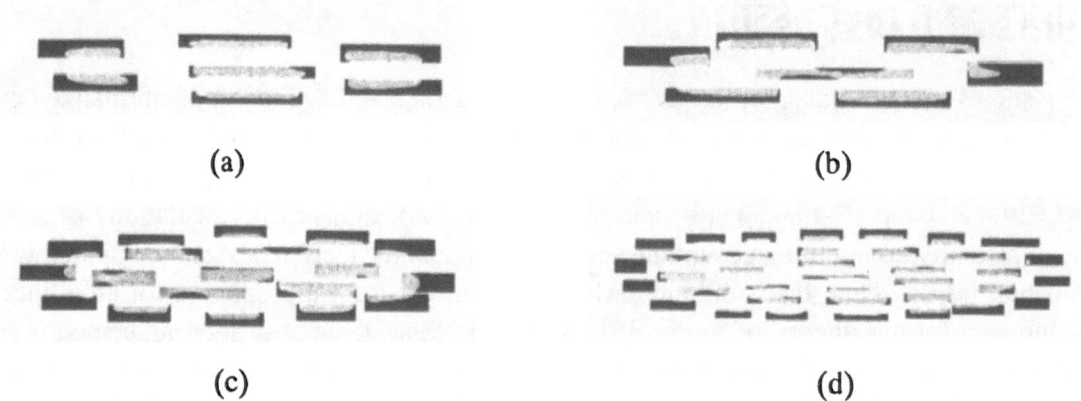

(a) (b)

(c) (d)

Fig. 2. Current distributions at $\Gamma=0.5$ for the multifilamentary tapes with different filament arrangement (a) S7-1, (b) S7-2, (c) M-19 and (d) B37-1. The black area is the current flowing part and the gray area represents the FFC at a given current amplitude.

SUMMARY

AC transport losses of Ag-sheathed Bi2223 multifilamentary tapes with different filament arrangements are investigated by using numerical calculations. The loss values for 7-filamentary tapes strongly depend on the filament distributions. The 7-filamentary tapes with sectioned filament arrangements has lower losses than that without sectioned filament arrangements, because FFC in the former is sectioned into three parts in the filament group. The influence of filament arrangements on the loss behaviors is suppressed with increasing the number of filaments.

1. B. A. Glowacki, K. G. Sandeman, E. C. L. Chesneau, M. Ciszek, S. P. Ashworth, A. M. Campbell and J. E. Evetts : Appl. Supercond. 2 (1997) 1437
2. C. M. Friend, S. A. Awan, L. Le. Lay, S. Sali and T. P. Beales : Physica C 279 (1997) 145
3. T. Fukunaga, R. Inada and A. Oota : Appl. Phys. Lett. 72 (1998) 3362
4. T. Fukunaga, R. Inada and A. Oota : Advances in Superconductivity X (1998) 824
5. W. T. Norris : J. Phys. D3 (1970) 489

AC LOSS OF HTS COIL AND ITS BEHAVIOR OF MAGNETIC FIELD

Yuji Sasaki[1], Koichiro Sawa[1], Yukikazu Iwasa[2]

[1]Department of Electrical Engineering, Keio University, Yokohama, Japan
[2]Francis Bitter Magnet Laboratory, M.I.T., Cambridge MA, U.S.A.

ABSTRACT: We obtained the AC loss of HTS coil wound with Ag sheathed Bi-2223, cooled in liquid nitrogen, from measured voltage and current waveforms. From measured data, effective resistance and inductance were calculated. They vary with amplitude and frequency of AC transport current. In order to search the cause of varying effective inductance, we also measured magnetic field generated by the coil. Behavior of amplitude of magnetic field on several measured points is very similar to that of effective inductance. It is supposed that shield current, which disturbs change of magnetic flux within the coil, might flow in the coil.

KEY WORDS: AC loss, HTS coil, effective inductance, magnetic field

INTRODUCTION

As known generally, the transport AC loss of a HTS tape under the magnetic field depends on its amplitude and angle to the surface of the tape[1,2]. When the tape was used for coil winding, magnetic field which applies on the tape is not uniform at each point of the HTS coil[3]. Then it can be thought the whole transport AC loss of the HTS coil is the sum of the transport AC loss at each point of the tape under a certain magnetic field[4]. But the property of magnetic field generated by HTS coil is not simple because of its hysteresis. Then we also measured AC magnetic field generated by the coil and discussed a relation between the behavior of AC loss and magnetic field.

EXPERIMENT

An AC loss of HTS coil P was obtained by integrating the product of current and reactance voltage directly by eq.(1). The schematic of measurement system is shown in Fig.1.

$$P = \frac{1}{T} \int_0^T V(t) \cdot I(t) \, dt \tag{1}$$

$$= I_{rms}^2 \, R_{eff} \tag{2}$$

$$= I_{rms} \cdot V_{rms} \frac{R_{eff}}{\sqrt{R_{eff}^2 + (\omega L_{eff})^2}} \tag{3}$$

Parameters of winding tape and coil are shown in Tables 1 and 2. From the data of loss, current and reactance voltage, effective resistance R_{eff} and effective inductance L_{eff} of coil were calculated from eqs.(2) and (3).

Table 1. Parameters of winding tape

Size(w×t×l)	3.5mm×0.23mm×52m
Number of filaments	61
Twist pitch	∞
Ag/SC	2.5
Ic(Ec=1 μ V/cm)	28A

Table 2. Parameters of coil

Size(od×id×h)	126mm×80mm×8mm
Number of turns	79×2
Inductance	3.40mH
Distance between voltage taps	51.5m
Ic(Ec=1 μ V/cm)	16.2A

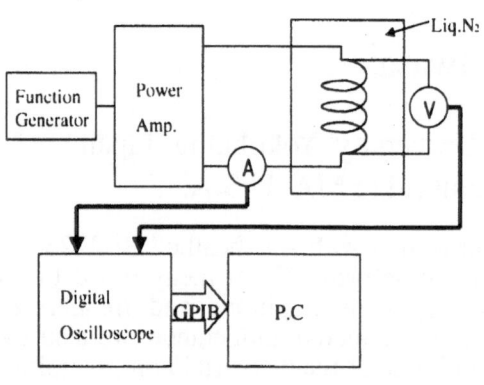

Figure 1. Schematic of measurement system

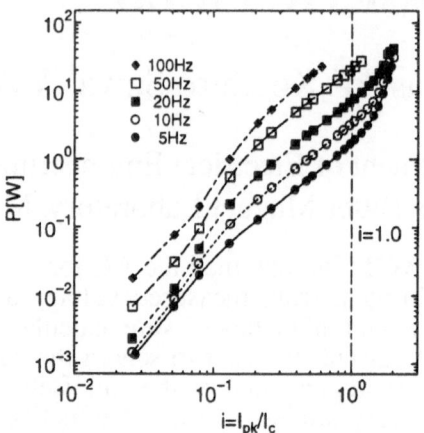

Figure 2. Characteristics of loss

RESULTS AND DISCUSSION

Fig.2 shows loss characteristics with various amplitudes and frequencies of transport current. Loss is almost in proportion to frequency at the range of $i(=I_{pk}/I_c)<1$ (I_{pk}: peak of transport current, I_c: critical current). This indicates that most loss component is hysteresis loss. Characteristics of effective resistance and inductance are shown in Figs.3 and 4. Both vary with amplitude and frequency of the transport current. At the peak points of R_{eff} curves, magnetic fields generated by the coil seemed to correspond to the value around lower critical magnetic field H_{c1}. When i becomes up to 1, curves of L_{eff} converge into around its value of at the normal state (=3.40mH).

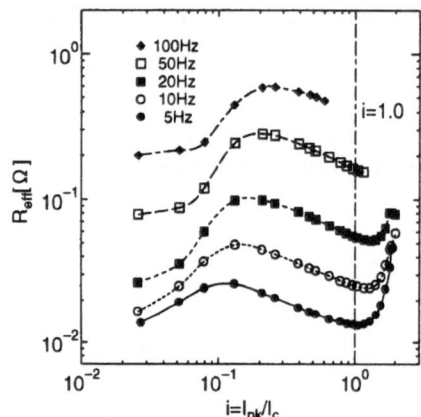

Figure 3. Effective resistance

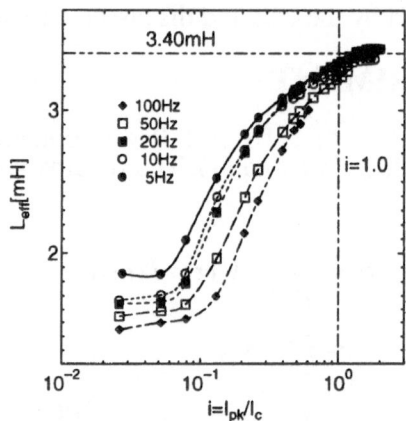

Figure 4. Effective inductance

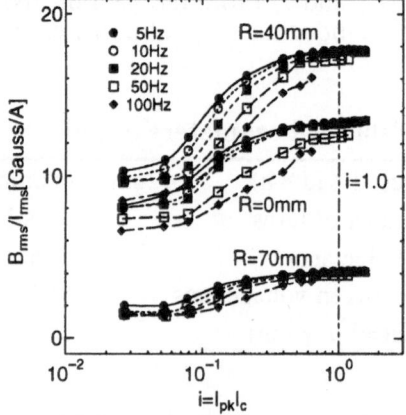

Figure 5. Magnetic field per transport current

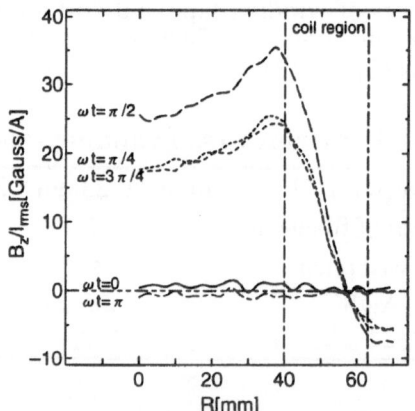

Figure 6. Field distribution of normal state coil
(5Hz, 1Arms)

EFFECTIVE INDUCTANCE AND MAGNETIC FIELD

In order to confirm the variation of L_{eff}, we also measured magnetic field (z component) generated by AC transport current along the radial direction on the top surface of the coil. Fig.5 shows rms value of magnetic field per 1Arms at several measured points.

Comparing Fig.4 and Fig.5, the behavior of effective inductance and magnetic field is so similar that we could confirm that effective inductance varies with amplitude and frequency of AC transport current. Therefore, it can be thought to be a reason of varying effective inductance that a shield current, which disturbs change of magnetic flux within the coil, might flow in the coil. Fig.6 shows a magnetic field distribution of AC transport current in the case of normal state. As shown in Fig.6, magnetic field generated by normal state coil is linear to the transport current. Field distributions generated by superconducting state coil are shown in Figs.7 and 8 with 1Arms and 6Arms. When the coil is superconducting, the field distribution is much smaller than that of normal state and not linear but very complex. But with increase of current, it becomes lager and approaches a linear characteristic. This property corresponds to the behavior of effective inductance or amplitude of the magnetic field.

From these results, it is supposed that comparatively large shield current that is induced by magnetic field might flow in the coil as if the coil is a superconducting bulk, and disturb change of magnetic flux within the coil. Then as the transport current becomes larger, the relative value of shield current becomes smaller because it approaches critical current value.

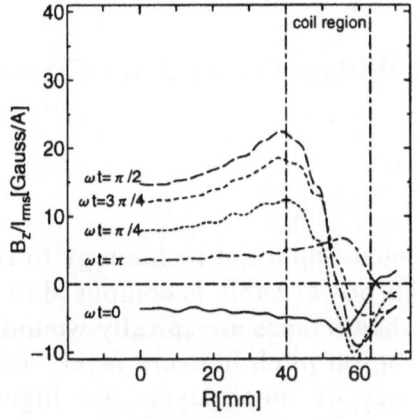

Figure 7. Field distribution of S.C. coil
(5Hz, 1Arms)

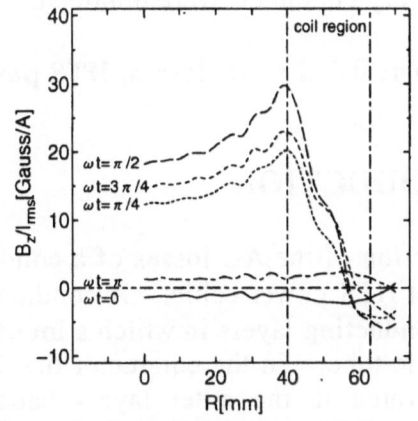

Figure 8. Field distribution of S.C. coil
(5Hz, 6Arms)

CONCLUSION

We obtained AC loss of the HTS coil from measured voltage and current waveforms, and calculated effective resistance and effective inductance from measured data. Both behaviors of effective inductance and magnetic field generated by the coil are very similar. As a reason of varying effective inductance or magnetic field, it is supposed that a shield current disturbing the change of magnetic flux within the coil might flow in the coil and causes the unique characteristics of AC loss of HTS coil or its effective resistance.

ACKNOWLEDGEMENT

The HTS coil was a courtesy of Sumitomo Electric Industries Ltd.. And we would like to appreciate the advice of members of Superconducting Research Laboratory, SEI.

REFERENCES

1 J.J.Rabbers, et al. Physica C300 (1998) 1-5
2 J.J.Rabbers, et al. Physica C310 (1998) 101-105
3 P.Fabbricatore, et al. Supercond. Sci. Technol. 11 (1998) 304-310
4 O.A.Chevchenko et al. Physica C310 (1998) 106-110

AC Losses of a Multi-layer Conductor Using Bi2223 Tapes with Twisted Filaments

Shinichi Mukoyama, Hirokazu Tsubouchi, Kazutomi Miyoshi, Toshiro Yoshida, Masanao Mimura[A], Naoki Uno[A]
Shoichi Honjo[B], Tomoo Mimura[B], Yoshihiro Iwata[B] and Yoshihisa Takahashi[B]

The Furukawa Electric Co., Ltd., 6 Yawata-Kaigandori, Ichihara, Chiba 290-8555, Japan
[A]The Furukawa Electric Co., Ltd., 500 Kiyotaki, Nikko, Tochigi 321-1493, Japan
[B]Tokyo Electric Power Company, 4-1 Egasaki-cho, Tsurumi-ku, Yokohama 230-8510, Japan

Abstract: Magnetic coupling among the filaments in multi-filament Bi2223 Ag sheath tape is sufficiently prevented by twisting filaments. Therefor, AC losses of twisted filamentary tape in which the filaments are not coupled together by external field are decreased. We produced the twisted filamentary Bi2223 AgMg sheath tapes with good mechanical properties and high critical current density. Moreover, A 1m 4-layer superconductor was built by using the tapes, and its AC losses were measured. As the result, the AC losses were decreased compared to those of the conductor that was composed of the non-twisted filamentary Ag sheath tapes. In addition, new equations for AC losses of the twisted filamentary conductor were obtained.

Keyword: Bi2223, AC losses, HTS power cable, multi-layer conductor, UCD model

INTRODUCTION

A reduction of the AC losses of a conductor is the most important technology to realize a compact HTS power cable. A conductor for an HTS power cable is composed of several superconducting layers in which a lot of Bi2223 Ag sheath tapes are spirally wound around a flexible tube. In the conductor that has the same spiral pitch in every layers, current is concentrated in the outer layers because impedances of inner layers are higher than impedances of the outer layers. As a result, AC losses of the conductor are relatively large. Meanwhile, AC losses of the uniform current distribution conductor (UCD conductor) are decreased compared with the non-uniform current distribution conductor[1], because magnetization current in the conductor is reduced by preventing magnetic coupling among layers. We thought that preventing magnetic coupling among filaments in Bi2223 tapes had also an effect to reduce AC losses of the UCD conductor.

AC LOSSES OF THE TAPE

Bi2223 Ag sheath tapes are produced in the form of filamentary composites by the PIT method; i.e. a lot of fine filaments of Bi2223 are embedded in a Ag matrix. However, the filaments are coupled magnetically and behaves as one filament for external magnetic field of the commercial frequency, because the resistance of the silver sheath is low. We thought that twisting filaments in the Ag sheath tape was the most effective method not to couple the filaments. Twisted filamentary Bi2223 AgMg sheath tapes that had a twist pitch of 9mm were developed as shown in Table 1. Fig.1 shows a cross section of the tape and one filament path. An outer sheath was made of AgMg alloy[2] to enhance mechanical strength.

Table 1 .Specifications of the Bi2223 twisted filamentary AgMg sheath tape

Tape size	0.185mm×3.3mm
The number of filaments	55
AgMg ratio	3.5
Critical current	27.4A
Critical current density	21.9kA/cm^2
Twist pitch	9mm

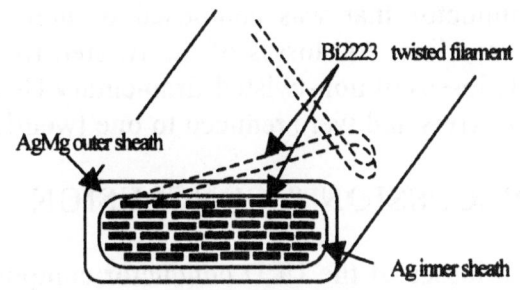

Fig.1 Conceptual cross section of the Bi2223 twisted filamentary AgMg sheath tape

AC losses of the twisted filamentary tape were measured in liquid nitrogen by a magnetization loop method[3] on condition that the tape was subjected to the external field of 50Hz parallel to the tape surface. Fig.2 shows the experimental result of AC losses of the twisted filamentary AgMg sheath tapes by circles and theoretical values of the non-twisted filamentary puer Ag sheath tapes by a solid line[2]. AC losses of the twisted filamentary tape decreased within the range of 3-100mT compared to non-twisted filamentary tapes.

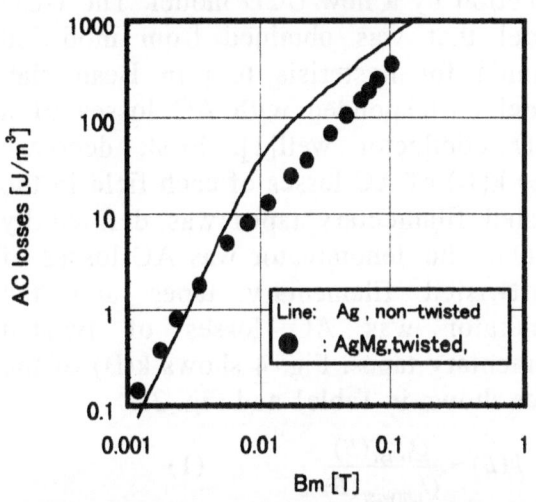

Fig.2 Hysterisis loss of AgMg twisted filamentary tapes measured by the magnetization method (50Hz,77K)

AC LOSSES OF THE CONDUCTOR

The majority of AC losses of the spiral multi-layer conductor is hysterisis loss of HTS tapes that were subjected to the magnetic field generated by transport current in the conductor. Therefore, it was expected that the AC losses of the multi-layer conductor using twisted filamentary tapes would be also decreased. A four layer conductor of 1 m length was made of the Bi2223 twisted filamentary tapes, to measure its AC losses. Table 2 shows the parameters of the conductor.

Currents in every layers of the conductor were made equal by adjusting the spiral pitch of each layer[4]. The spiral angles of the UCD conductor are shown in Table 2.

Table 2 . Parameter of 1m conductor

Diameter of flexible pipe:	φ 20.5 mm
The number of layers:	4
Diameter of the conductor:	φ 22.0 mm
Spiral angle of layers:	10°,11°,12°,25°
Critical current :	>1,300A

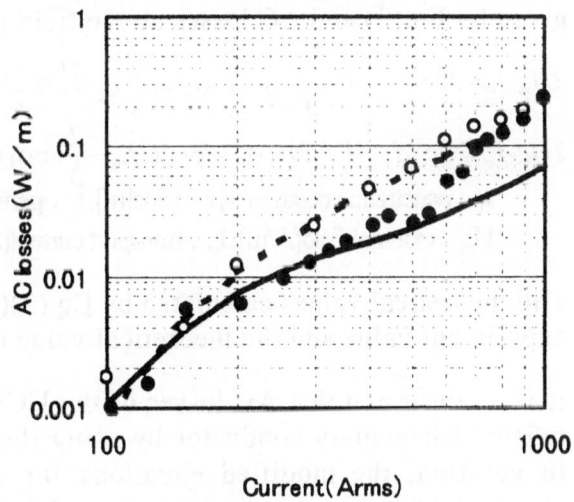

Fig.3 AC losses of the 4-layer conductor
●:AgMg twisted filamentary conductor, ○:Ag non-twisted filamentary conductor, ─ :theoretical value of twisted filamentary conductor, ---: theoretical value of non-twisted filamentary conductor

Measurement results are shown in Fig.3. The theoretical AC losses of a four layer UCD conductor that was composed of non-twisted firamentary tapes were indicated by the dotted line. AC losses of the twisted firamentary UCD conductor were reduced to half of AC losses of non-twisted firamentary UCD conductor in transport current from 200Arms to 600Arms and were reduced to one twentieth of AC losses of the non-UCD conductor.

DISCUSSION & CONCLUSION

AC losses of the UCD conductor composed of twisted filamentary tapes were decreased as described above. We tried to explain the reduction by a new UCD model. The UCD model that was obtained from modified formula for hysterisis loss in Bean slab model corresponded with AC losses of a UCD conductor well[1]. First, decrease rates k(B) of AC losses of each field in the twisted filamentary tapes was defined by Eq.(1). The denominator was AC losses of non-twisted filamentary tapes and the numerator was AC losses of twisted filamentary tapes. Fig. 4 shows k(B) of the tapes shown in Table1 and Fig. 2.

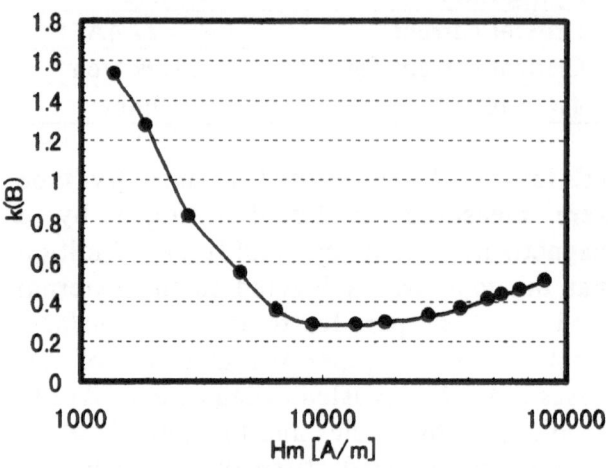

Fig.4 Decrease rates of the AgMg twisted filamentary tapes that were shown in Table1 and Fig.2.

$$k(B) = \frac{Q_{twist}(B)}{Q_{untwist}(B)} \qquad (1)$$

AC losses of the UCD conductor using twisted filamentary tapes were derived from the equations of UCD model multiplied by k(B) respectively because the transport current loss was very much less than the external field loss. AC losses below field penetration were given by Eq.(2) and AC losses above field penetration were given by Eq.(3).

$$2H_m - I_m < Jc \cdot d \qquad P = \frac{2}{3}\mu_0 f \frac{H_m^3}{Jc} k(B) + \frac{2}{3}\mu_0 f \frac{(H_m + I_m)^3}{Jc} k(B) \qquad (2)$$

$$2H_m - I_m \geqq Jc \cdot d \qquad P = 2\mu_0 \cdot t_1^2 \cdot Jc \cdot f(H_m - \frac{2}{3}Jc \cdot t_1) \cdot k(B) + 2\mu_0 \cdot t_2^2 \cdot Jc \cdot f(H_m + I_m - \frac{2}{3}Jc \cdot t_2) \cdot k(B) \qquad (3)$$

t_1 : penetration depth of tape inside[m], t_2 : penetration depth of tape outside[m], f : frequency[Hz], H_m : external field[A/m], I_m : transport current[A/m]

The theoretical values calculated by Eq.(2)(3) were shown in Fig. 3 by the solid line. The experiment value and the theoretical value of AC losses were corresponding very well.

It was confirmed that AC losses of the UCD conductor was reduced compared to the non-twisted filamentary conductor by using the filaments twisted Bi2223 AgMg sheath tapes. In addition, the modified equations for calculating AC losses of the conductor were obtained from equations of UCD model. This technology of the filament twist is very useful to realize compact HTS power cables.

[1] S. Mukoyama et. al., *Advances in Superconductivity IX*, 1997, p1337

[2] M. Mimira et. al., This Symposium, WTP-26

[3] A. Kimura et. al., Proc. of the 7th US-Japan Workshop on High-Tc Supercond., 1995, p81

[4] S. Mukoyama et. al., IEEE Trans. Appl. Supercond., VOL.9, No2, 1999, P1272

MECHANICAL PROPERTIES
OF HIGH-Tc SUPERCONDUCTOR FOR A POWER CABLE

Kazutomi Miyoshi, Shinichi Mukoyama, Hirokazu Tsubouchi, Toshiro Yoshida,
Shoichi Honjo*, Tomoo Mimura*, Yoshihisa Takahashi* and Yoshihiro Iwata*

The Furukawa ELectric Co., LTD, 6 Yawatakaigandori, Ichihara, 290-8555, Japan
Tokyo Electric Power Company, 4-1 Egasaki-cho, Tsurumi-ku, Yokohama, 230-8510, Japan

Abstract: Mechanical properties of HTS cables have been studied experimentally. The four mechanical tests were performed; (1) a thermal cycle test, (2) a bending test, (3) a tensile test, and (4) a triplex test. In the thermal cycle test, Ic was not influenced. In the bending test, Ic depended on a bending radius, and a bending radius of less than 500 mm caused a significant problem to the conducting and shielding layers. The tensile test indicated that tensile strains of up to 0.5 % did not decrease Ic. The Ic's of triplex cable's cores were not influenced by the thermal cycle and a bending radius of more than 1000 mm. These results will be applied to the HTS cable design.

Keywords: Bi-2223 tape, HTS cable, Mechanical test, Ic, Triplex conductors

INTRODUCTION

Mechanical properties of HTS cables have been studied experimentally. As the conducting layers and the shielding layers of the cables consist of HTS tapes, Ic's of the cables are influenced by mechanical strains during their production or their final use. Then, the performance of the HTS cable is limited by applied stresses. For the cable design, it is important to investigate the mechanical properties of the cables. The results of four mechanical tests are described on the HTS cable's core; (1) the thermal cycle test, (2) the bending test, (3) the tensile test, and (4) the triplex test.

EXPERIMENTALS

Structure of Core
The cable cores consist of a superconducting conducting layer (SC layer) and a superconducting shielding layer (SH layer) of spirally wound HTS tapes. The HTS tape has a width of 3.5 mm, a thickness of 0.2 mm, a ratio of non-superconductor / superconductor of 3.5, and Jc of approximately 15 kA/cm^2. There are two HTS tapes that are available for the cable production. One of them is a silver sheathed tape (Ag), and another is a silver alloy sheathed tape (Ag-Mg). Moreover, two types of formers were adopted. The one is a copper tube (Cu former), and the other is a flexible tube (FX former). So, four kinds of HTS cores, which consisted of (1) Ag tapes with Cu former, (2) Ag tapes with FX former, (3) Ag-Mg tapes with Cu former, and (4) Ag-Mg tapes with FX former, were provided for the thermal cycle test and the bending test. Table I shows the parameters of the HTS cores. For the tensile test, the core has not any shielding layer.

Table I. Parameters of Cores

testing	thermal cycle test / bending test	tensile test	triplex test
former	flexible or Cu tube (O.D. 25 mm)	SUS tube (O.D. 25 mm)	flexible (O.D. 20.5 mm)
SC laye	2 layers (Ag or Ag-Mg)	2 layers (Ag or Ag-Mg)	2 layers (Ag-Mg)
dielectric	OPPL: 36 layers	OPPL: 36 layers	OPPL: 36 layers
SH layer	2 layers (Ag or Ag-Mg)	nothing	2 layers (Ag-Mg)

Procedures of Mechanical Tests

(a) Thermal cycle test: The cores were located in a box filled with dry nitrogen gas to prevent frost. The critical current (Ic) of the SC layer and SH layer was measured after each thermal cycle between room temperature and 77 K by immersing in liquid nitrogen.

(b) Bending test: Bending the cores was performed one time for one bending radius. Bending radii were 1500 mm, 1000 mm, and 500 mm. The Ic of the bent cores was measured in liquid nitrogen. In order to measure strains of HTS tapes in the bent cores, the strain gauges were attached onto HTS tapes.

(c) Tensile test: The tensile test was carried out with an Instron-type machine in liquid nitrogen. The strains of the former and HTS tapes, and the Ic were measured under load in liquid nitrogen.

(d) Triplex (stranding three cores) bending test: Three cores were prepared for the triplex bending test. One of them consisted of SC layer and SH layer, and other two cores had only dielectric without HTS tapes. As we predicted that Ic would not decrease at stranding pitch of 1500 mm by the use of the analysis of the bending test in the single core, three cores were stranded with a wound pitch of 1500 mm. For the thermal cycle test and the bending test, Ic was measured.

RESULTS AND DISCUSSIONS

Thermal Cycle Test
Figure 1 shows the cycle dependence of the normalized Ic of SC and SH layers in two cores with Ag-Mg tapes. In the case of the core with Ag tapes, the result of thermal cycle test was similar to that of the cores with Ag-Mg tapes. The thermal cycles do not influence the critical current.

Bending Test
Figure 2 shows the result of bending tests. At bending radii of 1500 mm and 1000 mm, Ic's of SC and SH layers in the cores with the Ag-Mg tapes were not influenced. In contrast to this result, the degradation in Ic's of SC and SH layers in the cores with the Ag tapes on both formers is observed at the same bending radius. The degradation of Ic's on the Cu former with Ag-Mg tapes is greater than that of Ic's on the FX former at the same binding radius. All cores have a tendency that the degradation of Ic's increases with decreasing bending radius. As a result of the bending test, the dependence of Ic on a bending radius is found.

The strain of HTS tapes in the bent core was measured in order to investigate the relation between Ic and the strain. Figure 3 illustrates the orthogonal view of a bent core. Figure 4 and 5 show the change of strain as a function of gauge position for two different tapes with binding radius of 1000 mm. The bending strains on the Ag tapes in the bent core were higher than these on the Ag-Mg tapes. At other bending radii, the bending strains had a similar behavior. With decreasing a radius, all bending strains increased. We have proposed the two mechanisms, the sliding model and the infinite friction one, to explain degradation of Ic in the bent core [1]. According to the mechanisms, the bending test shows that the Ag-Mg tapes slide more freely than the Ag tapes in the core during bending. From these results, the Ag-Mg tapes and the flexible former are suitable for the HTS cable from the viewpoint of bending strains.

Tensile Test
Figure 6 shows the dependence of normalized Ic on former strain in the cores with the Ag tapes and the Ag-Mg tapes. In the case of the core with the Ag tapes, the degradation of Ic occurs at 0.3 %. And no degradation of Ic is observed up to 0.5 % in the core with the Ag-Mg tapes. From these results, it is found that in the tensile test the core with the Ag-Mg tapes is better than that with the Ag tapes.

Triplex (stranding three cores) test
Figure 7 shows the change of Ic on the triplex core in the thermal cycle and the bending test. the Ic did not decrease, after the cores were bent by means of triplex test. In the thermal cycle test and the bending test, no degradation of Ic was observed. This result indicates that it is possible to strand three cores without decreasing the critical current.

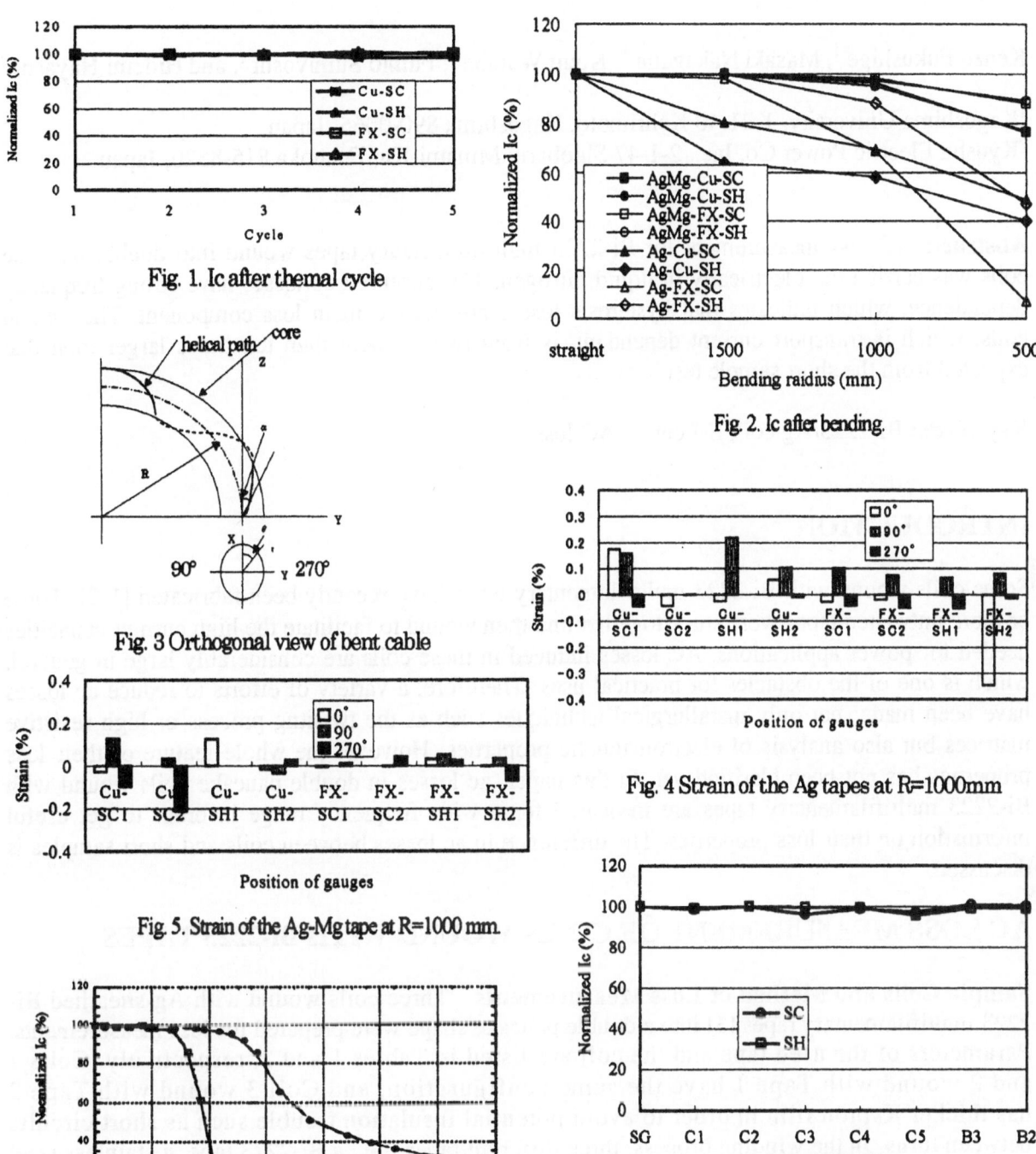

Fig. 1. Ic after themal cycle

Fig. 3 Orthogonal view of bent cable

Fig. 5. Strain of the Ag-Mg tape at R=1000 mm.

Fig. 6 Dependence of formaer strain of Ic.

Fig. 2. Ic after bending.

Fig. 4 Strain of the Ag tapes at R=1000mm

Fig. 7. Ic's of the triplex core. SG denotes Ic of a single core before triplex, and C1, C2, etc denote the thermal cycle. B3 and B2 denote bending radii, 1500 and 1000 mm respectively.

SUMMARY

Mechanical properties of HTS cables have been investigated. Thermal cycle does not influence Ic. The bending radius strongly affects Ic, and the radius is required to be larger than1000 mm for no significant degradation in Ic. A flexible former and the Ag-Mg tapes are suitable for bending and triplex HTS cables. These results will be applied to the HTS cable design.

T. K. Miyoshi, S. Mukoyama, H.Tsubouchi, T. Yoshida, M. Mimura, N. Uno, M. Ikeda, H. Ishii, S. Honjo and Y. Iwata, IEEE trans. on Appl. Super. 9, 428 - 431 (1999)

AC Losses in HTS Coils Wound with Bi-2223 Tapes

Kenzo Fukushige [1], Masaki Nakagami [1], Kouji Watabe [1], Fumio Sumiyoshi [1], and Hidemi Hayashi [2]

[1]Kagoshima University, 1-21-40 Kohrimoto, Kagoshima 890-0065, Japan
[2]Kyushu Electric Power Co. Inc., 2-1-47 Shiobaru, Minami-Ku, Fukuoka 815-8520, Japan

Abstract: AC loss measurement for Bi-2223 multifilamentary tapes wound into double pancake coils was carried out electrically in liquid nitrogen. The observed ac losses have strong frequency dependence, which indicates that hysteresis losses are not the main loss component. The loss in coils, which is transport current dependant, is from two to more than ten times larger than that expected from the short sample test.

Keywords: Bi-2223/Ag coil, E-I curve, AC loss

INTRODUCTION

Some coils wound with Bi-2223 multifilamentary tapes have recently been fabricated [1-2]. Three or more untwisted tapes were stuck together and then wound to facilitate the high current capacities needed for power applications. AC losses induced in these coils are considerably large in general, which is one of the obstacles for practical uses. Therefore, a variety of efforts to reduce ac losses have been made; not only metallurgical techniques such as the twisitng process or high resistive matrices but also analysis of electromagnetic properties. However the whole feature of their loss properties has not been clarified yet. In this paper, ac losses in double pancake coils wound with Bi-2223 multifilamentary tapes are measured for a wide frequency range in order to get useful information on their loss properties. The difference in ac losses between coils and short samples is discussed.

AC LOSS MEASUREMENT OF COILS WOUND WITH Bi-2223 TAPES

Sample Coils and Method of Loss Measurements   Three coils wound with Ag-sheathed Bi-2223 multifilamentary tapes [3] into a double pancake shape were prepared for loss measurements. Parameters of the used tape and the coil are listed in Tables 1 and 2, respectively. Coils 1 and 2 wound with Tape 1 have the same configuration, and Coil 3 wound with Tape 2 has thicker Kapton film in order to avoid potential insulation trouble such as short circuits between turns. In the winding process, three different tapes, i.e., a Bi-2223 tape, a stainless steel one and an Kapton one, were stuck together. Insulating material between single pancake coils was GFRP as well as the material of the coil bobbins. Final heat treatment of tapes was carried out before winding, and all coils were impregnated with epoxy resins. Both tapes are untwisted and their critical currents, where the measurement criterion is set at 1 micro-volt/cm, are about 35 or 39 Amps in zero magnetic fields at liquid nitrogen temperature. Preliminary measurements were made on characteristic curves of terminal voltages for dc transport currents. In spite of almost the same critical currents for coils, a clear difference among them was observed at lower currents as shown in Fig. 1. This suggests that tapes may have some pre-stress during the winding processes. AC Loss measurements for coils in liquid nitrogen were carried out electrically. In measurement circuits shown in Fig. 2, the inductive component in the terminal voltages of coils is cancelled with a high accuracy by using a transformer with concentric normal-metal coils, and

Table 1. Parameters of Bi-2223 tapes

Name	Tape 1	Tape 2
Cross section [$mm^W \times mm^t$]	3.5×0.23	3.8×0.24
Ag ratio	2.6	2.5
Filament number	61	←
Twist pitch	∞	←
Critical current [Amps]	39	35

Table 2. Parameters of coils

Name	Coil 1	Coil 2	Coil 3
Inner diameter [mm]	80	←	←
Outer diameter [mm]	114	←	122
Height [mm]	8	←	9
Inductance [mH]	1.78	←	1.74
Thickness of Kapton tape [μm]	12.5	←	70
Critical current [A] (77.3K)	22	24	←
Wound tape	Tape 1	←	Tape 2
Tape length [m]	35	←	←
Number of tape bundles	1	←	←
Number of turns	114	←	←

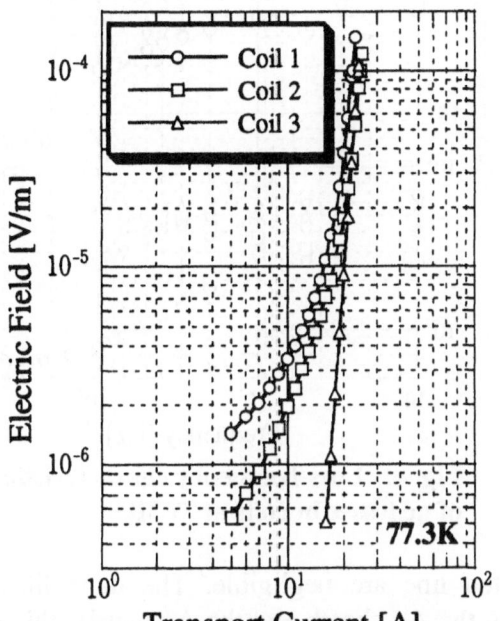

Fig. 1. E-I characteristic curves of coils.

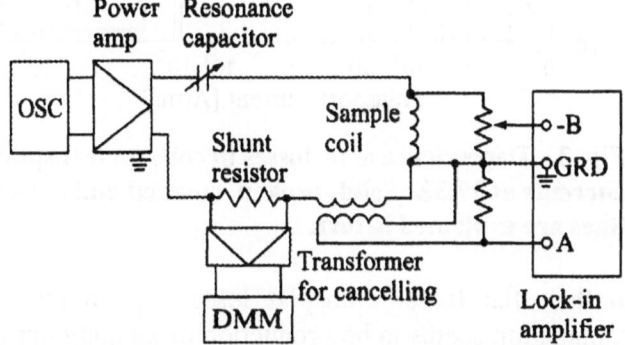

Fig. 2. Circuit for measuring ac losses in coils.

residual and extremely small loss component is processed by a Lock-in-amplifier to get loss values. The negligible error of the present method was confirmed for a wide frequency range by another preliminary measurement in which a dummy coil with the same inductance as that of the sample coil was tested in liquid helium. This dummy coil wound with NbTi ultra-fine multifilamentary wire with very small losses can be regarded as an ideal inductance, i.e., a standard free of losses.

Measured Losses in Coils The obtained ac losses for coils, dependant on ac transport currents, are shown in Fig. 3. The AC loss changes with frequency, which indicates that hysteresis losses do not predominate over the observed loss. As seen in Fig. 3, Coils 2 and 3 have similar loss properties. The data are re-plotted as loss-frequency characteristics in Fig.4, where loss values are normalised by $\mu_0 H_{\perp m}^2$, and $H_{\perp m}$ represents the maximum value of magnetic field normal to the flat face of tapes in each coil. The loss-frequency characteristic curve for low currents corresponds exactly to the typical eddy current loss curve in normal metal, and its peak frequency is about 40 Hz for Coil 1 and lower than 1 Hz for Coils 2 and 3. With increase in currents, the peak frequency shifts to higher frequency and moreover the shape of the curve is deformed.

Increases in the Eddy Current Loss for Coils As seen in Fig. 3, a large difference between the loss observed for coils and the loss expected from the short sample test exists, which can not be explained by previous studies. In Fig. 3, a solid line is measured losses in short samples of Tape 1 for simultaneous sweep of ac transport currents and transverse ac magnetic fields. The data with a weak dependence of frequency correspond to hysteresis losses and they are a little larger than magnetisation losses of tapes (a broken line) when transverse magnetic fields are applied normal

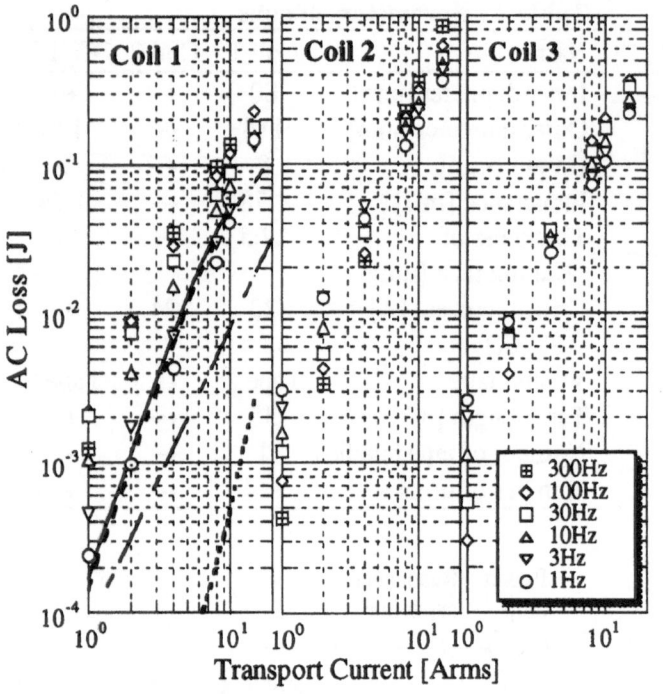

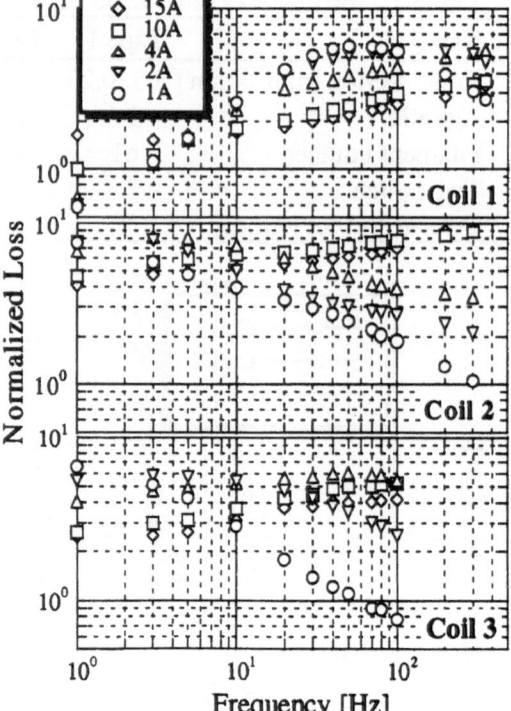

Fig. 3. Dependence of ac losses in coils on transport currents at 77.3K. Solid, broken, chained and dotted lines are explained in text.

Fig. 4. Loss-frequency characteristics of ac losses in coils at 77.3K.

to their flat faces. Transport losses shown by a dotted line are negligible. The most likely explanation seems to be production of an eddy current in the Ag sheath and the Ag matrix; this is disproved however. The calculated peak value of the loss for various frequencies for ac magnetic fields, shown by a chained line in Fig. 3, is given at the skin frequency ($\cong$ 280 Hz) and is of a lower order than observed data for coils. The loss frequency characteristics of data shown in Fig. 4 are also not explained, especially for the difference in the characteristic frequency among the three coils. Since the local magnetic field normal to the flat face of tapes in double pancake coils is relatively larger than that of solenoid coils, more detailed calculation of the eddy current loss may be required, taking the effect of the magnetisation of the superconductor on the eddy current loss into account.

CONCLUSION

Three coils wound with Bi-2223 multifilamentary tapes into double pancake shape were prepared and tested in liquid nitrogen. The observed ac losses were much larger than the loss expected from the short sample test for simultaneous sweep of ac transport currents and transverse ac magnetic fields. These extra losses are not explained by existing theories.

Acknowledgments. The authors are indebted to Messrs. K. Ohmatsu and J. Fujikami of Sumitomo Electric Industries, Ltd. Japan, where samples used in the present paper were fabricated.

1. K. Funaki, M. Iwakuma, K. Kajikawa, M. Takeo, J. Suehiro, M. Hara, K. Yamafuji, M. Konno, Y. Kasagawa, K. Okubo, Y. Yasukawa, S. Nose, M. Ueyama, K. Hayashi, and K. Sato, Cryogenics. **38**, 211-220 (1998).

2. T. Honjo, T. Hasegawa, K. Kaiho, H. Yamaguchi, K. Arai, M. Yamaguchi, S. Fukui, K. Kato, and K. Itagaki, IEEE Trans. Appl. Super. **9**, 829-832 (1999).

3. S. Kobayashi, T. Kaneko, T. Kato, J. Fujikami, and K. Sato, Physica C. **258**, 336-340 (1996).

Experimental Study on Frequency Dependence of Ac Loss in Superconducting Parallel Conductors

Hideki Tanaka[1], Masataka Iwakuma[1], Kazuhiro Kajikawa[1], Kazuo Funaki[1], Masayuki Konno[2], Shinnichi Nose[2], Kazuhiko Hayashi[3] and Kenichi Sato[3]

[1]Research Institute of Superconductivity, Kyushu University, 6-10-1 Hakozaki, Higashi-ku, Fukuoka 812-8581
[2]Fuji Electric Co.,Ltd., 1-1 Tanabeshinden, Kawasaki-ku, Kawasaki 210-8530
[3]Sumitomo Electric Ind.,Ltd., 1-1-3 Shimaya, Konohana-ku, Osaka 554-8511

Abstract: We experimentally investigated the ac loss properties in 2-strand parallel conductors. The additional ac losses due to the deviation of the transposition points from the optimum ones are expected to be inversely proportional to frequency in practical situation according to the theoretical calculation. We made sample coils with the deviation of transposition by using NbTi rectangular cross-sectional multifilamentary wires, and confirmed the theoretical predictions.

Keywords: ac loss, parallel conductor, multifilamentary wire, frequency dependence

INTRODUCTION

We introduced parallel conductors in order to enlarge the current capacity of oxide superconducting thin wires [1]. The constituent wires need to be transposed during the winding process to reduce the interlinkage flux. In the case that parallel conductors are transposed at the optimum positions, the ac loss density of parallel conductors was equal to that of the single strand [1]. However transposition point may deviate somewhat from the optimum one in practical situations. In this case, the interlinkage magnetic flux is not canceled out and the additional ac loss is induced. We reported the frequency dependence of ac losses in the parallel conductors composed of Bi2223 multifilamentary wires which had monofilamentlike electromagnetic properties [2]. It was independent of frequency. On the contrary, according to the theoretical calculation, the additional ac loss in the parallel conductors composed of multifilamentary strands has the debye-type frequency dependence. In this paper, we experimentally studied the frequency dependence of the additional ac losses in the 2-strand parallel conductors composed of NbTi rectangular cross-sectional multifilamentary strands to verify the theoretical prediction.

THEORETICAL EXPRESSION

Fig.1 shows the 2-strand parallel conductor in which the transposition points deviate from the optimum one by Δl. Strands are connected only at both ends. We applied the transverse magnetic field, $B_e = B_m \sin \omega t$, in the parallel direction to the wide surface of strands. Let us consider the loop enclosed by the E=0 lines and the contact resistances. Left band width is smaller than right hand one. Because of the unbalance in the interlinkage magnetic flux, the shielding current is induced as shown in Fig.1. A circuit equation for the induced shielding current is given by

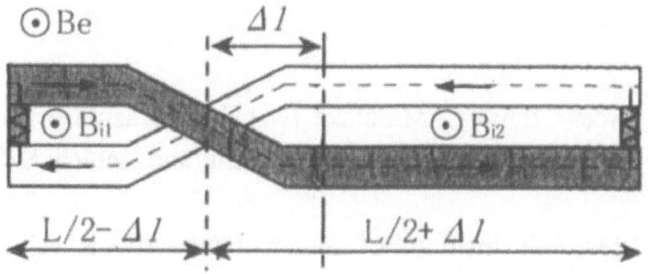

Fig.1. Projected figures of 2-strand parallel conductor

$$2RI = -d\left\{\left(\frac{L}{2}+\Delta l\right)\frac{\partial B_{i2}}{\partial t}-\left(\frac{L}{2}-\Delta l\right)\frac{\partial B_{i1}}{\partial t}\right\} \tag{1}$$

where R is the contact resistance between the strands at the terminals, I is the shielding current, d is a distance between the center lines of both strands, and L is the total length of conductor. B_{i1} and B_{i2} are the average magnetic flux densities in the left- and right- half areas of the distorted closed loop in Fig.1. These are expressed as

$$B_{i1} = B_e - \Delta B \, , \; B_{i2} = B_e + \Delta B \, , \; \Delta B = k'\left(1-\frac{t}{2d}\right)\frac{\mu_0}{w}I = k\frac{\mu_0}{w}I \tag{2}$$

where k and k' are coefficients determined by the geometrical configuration, and t and w are a thickness and a width of a strand respectively. From Eqs.(1) and (2), the shielding current, I ,is obtained and the additional ac losses is given by[1]

$$W = \frac{1}{k}\frac{d}{2t}\frac{\pi\omega\tau}{1+\omega^2\tau^2}\frac{B_m^2}{\mu_0}\left(\frac{2\Delta l}{L}\right)^2 \tag{3}$$

where τ is a time constant of the distorted loop in Fig.1. As evident from Eq.(3), in the case of $f > 1/(2\pi\tau)$, the additional ac loss is inversely proportional to frequency.

EXPERIMENTS

We made sample coils with NbTi multifilamentary wires whose characteristics are listed in Table.1. The 2-strand parallel conductor with the length of 2.5m, was wound into a one layer solenoidal coil. The transposition point deviate from the optimum position which is the middle of the conductor by Δl.

Table.1. Characteristics of the NbTi wires

Superconductor	NbTi
Matrix	CuNi
Critical current (at 4T)	50 (A)
Twist pitch	8 (mm)
Wire thickness	0.229 (mm)
Wire width	1.123 (mm)

The strands are soldered at length end by 5mm. Then we measured the ac loss densities of parallel conductors exposed to the external magnetic field.   Fig.2 shows the total ac losses observed in the parallel conductor. Then we subtracted the loss densities of the strand from those of parallel conductors to obtain the additional ac loss. The results are shown in Fig.3. The additional ac losses are proportional to B_m^2 as theoretically predicted. As frequency become lower the additional ac losses increase. So the additional ac loss is replotted in Fig.4 against the frequency. The lines represent the theoretical calculation. Here, we assume k=0.68. For any amplitude, the experimental results agreed with theoretical ones very well.

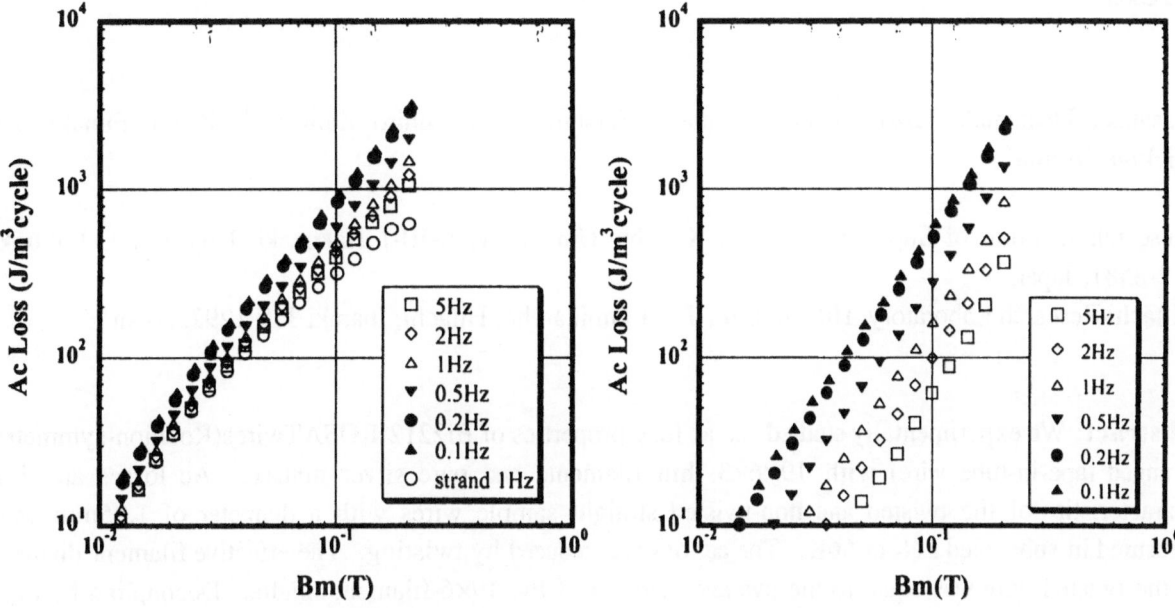

Fig.2. Total ac loss vs field amplitude

Fig.3. Additional ac loss vs field amplitude

CONCLUSION

We experimentally investigated the ac loss properties in 2-strand parallel conductors composed of multifilamentary strands. We confirmed that the additional ac losses due to the deviation of transposition points from the optimum ones are proportional to the square of field amplitude and have debye-type frequency dependence as theoretically predicted. It suggests that the ac loss densities in parallel conductors can be reduced to that of a strand in the practical situation even with the deviation of transposition points because the critical frequency, $1/(2\pi\tau)$, becomes lower with the length of conductor.

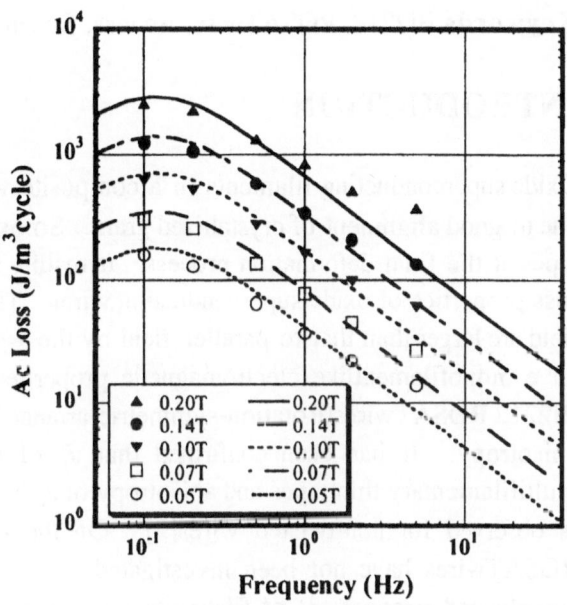

Fig.4. Additional ac loss vs frequency

1. M. Iwakuma, et al., IEEE Trans.Appl.Supercond., vol.7, pp.298-301, June 1997.

2. M. Iwakuma, et al., Proc. of MT-15, pp.1144-1147, October 1997.

Ac Losses in Bi2212 ROSATwires Exposed to an Ac Magnetic Field

Masataka Iwakuma[1], Yuzo Fukuda[1], Kentaro Matsumura[1], Kazuhiro Kajikawa[1], Kazuo Funaki[1] and Michiya Okada[2]

[1]Research Institute of Superconductivity, Kyushu University, 6-10-1 Hakozaki, Higashi-ku, Fukuoka 812-8581, Japan
[2]Hitachi Research Laboratory, Hitachi Ltd., 7-1-1 Omika-cho, Hitachi, Ibaraki 319-1292, Japan

Abstract: We experimentally studied the ac loss properties of Bi2212 ROSATwires(Rotation-symmetric arranged tape-in-tube wire) with $19 \times 6 \times 3$ thin filaments and pure silver matrix. Ac losses and I_c-B characteristics of the twisted and non-twisted straight sample wires with a diameter of 1.95mm were measured in subcooled LN_2 at 66K. The ac loss was reduced by twisting. The effective filament diameter of the twisted wire was equal to the average diameter of the 19×6-filament bundle. Decoupling between the filament bundles was confirmed. The observed coupling current loss roughly agreed with the theoretical prediction on the supposition of an ideal multifilamentary wire notwithstanding the complicated arrangement of filaments with anisotropy of J_c.

Keywords:Bi2212, ROSATwire, Ac loss, Superconducting multifilamentary wire, Twist

INTRODUCTION

Oxide superconducting filaments in a composite with Ag matrix need to be thin for the sake of high J_c due to good alignment of crystallized grain. So oxide superconducting wires are generally rolled into thin tapes at the final deformation process. It results in not only anisotropy of J_c but also anisotropy of ac loss properties of oxide superconducting wires. The ac losses in thin flat wires exposed to perpendicular field are larger than that to parallel field by the aspect ratio of the wires on the assumption that the wires have monofilamentlike electromagnetic properties due to non-twisting and/or filamentary coupling[1]. Bi2212 ROSATwires(Rotation-symmetric arranged tape-in-tube wire) is one solution for the problem of anisotropy. It has been confirmed that J_c of Bi2212 ROSATwires is as high as that of Bi2212 multifilamentary thin tapes and anisotropy of J_c is not so large though some field angular dependence of J_c is observed for non-twisted wires[2]. On the other hand, the electromagnetic properties of Bi2212 ROSATwires have not been investigated. It is difficult to predict the ac loss properties due to the complicated arrangement of filaments with anisotropy of J_c. Hence, in this paper, we experimentally studied the ac loss properties of Bi2212 ROSATwires.

SAMPLE WIRE

The cross section of a Bi2212 ROSATwire is shown in Fig.1. At first multifilamentary flat wires with 19 filaments were fabricated and, in the 2nd stuck process, 6×3 pieces of multifilamentary flat wires were assembled and inserted to a round sheath of silver. After that the wire was deformed into a final diameter by drawing with keeping a round cross section. We prepared twisted and non-twisted straight sample wires with pure silver matrix. The characteristics of sample wires are listed in Table 1.

Table 1 Characteristics of Bi2212 ROSATwires

	#A	#B
Wire diameter	1.95 mm	
Number of filament	19×6×3	
Volume fraction of filament	0.235	
Matrix	Pure silver	
Conductivity of matrix	$3.33×10^8$ $(\Omega m)^{-1}$	
Twist pitch	∞	15 mm

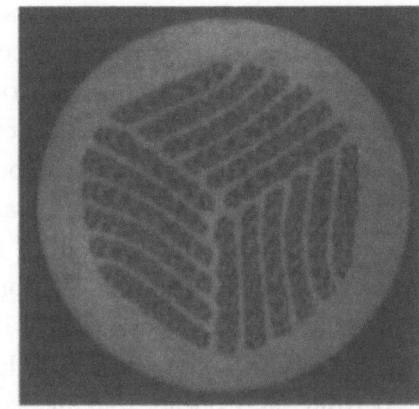

Fig.1 Cross section of sample wire.

EXPERIMENT

We measured the ac losses of sample wires in subcooled LN_2 at 66K by applying external ac magnetic field. Sample wires with a length of 50mm were arranged in parallel at an interval of 5mm and inserted to a saddle-shaped pick-up coil. Ac sinusoidal magnetic field with an amplitude of 10^{-3} to 10^{-1}T was applied perpendicularly to the wire axis. The frequency ranged from 0.5 to 60 Hz. The observed amplitude dependences of ac losses are shown in Fig.2. We can see the frequency dependences of ac losses for both of the twisted and non-twisted wires.

In order to analyze the observed ac loss properties quantitatively, we first measured the I_c-B characteristics of the sample wires in subcooled LN_2 at 66K. The observed results are shown in Fig.3. By using Kim model, they are fitted well by the following expressions; $I_c(B)=2.71/(B+0.0398)$ for the non-twisted wire and $I_c(B)=0.834/(B+0.0187)$ for the twisted wire.

Here we supposed that the sample wires were ideal multifilamentary wires in which round cross-sectional filaments were arranged uniformly and symmetrically in the cross section. For the theoretical evaluation of hysteresis loss, we introduced an effective filament diameter, d_{eff}. d_{eff} is defined as

$$d_{eff} = \frac{M}{\frac{2}{3\pi} \lambda J_c(B)} \quad (1)$$

where M is magnetization of a wire and λ is a volume fraction of filaments. Since hysteresis loss is given by $W_h = \oint MdB$, d_{eff} can be evaluated from the experimental results of ac loss and I_c-B characteristics. The broken lines in Fig.2 represent the theoretically estimated hysteresis losses in the case that d_{eff} is equal to the diameter of filamentary region, d_{fr}. As can be seen in Fig.2(a), the observed ac losses in the non-twisted wire

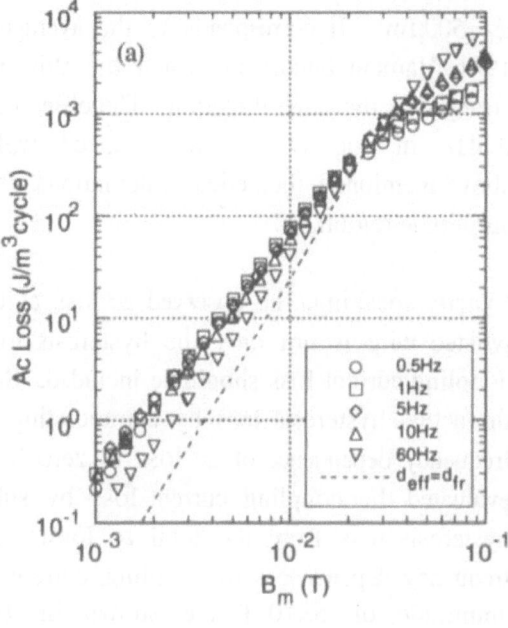

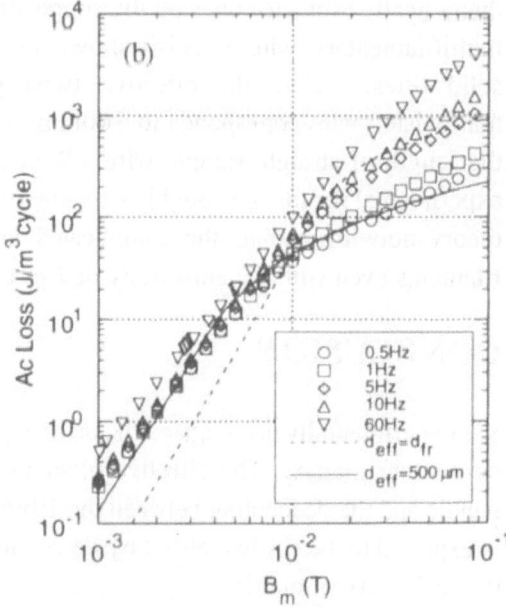

Fig.2 Amplitude dependences of ac losses in (a)non-twisted and (b)twisted wires.

for the larger amplitude than the penetration field, B_p, at 0.5 to 1Hz almost agree with the theoretical ones on the supposition of $d_{eff}=d_{fr}$. The description between the experimental results and the theoretical ones for the smaller amplitude than B_p seems to be caused by the fact that flux penetrates into the filamentary region along the wire axis due to the short length of the straight sample wires. It suggests that some decoupling between the filaments and/or the filament bundles is realized at least in the non-twisted wire. In addition, ac loss is much reduced by twisting as evident from the comparison between Figs.2(a) and (b). The solid line in Fig.2(b) represents the theoretically estimated hysteresis losses in the case of $d_{eff}=500\mu m$. It corresponds to the average diameter of 19×6-filament bundle in which the thin filaments are arranged in the same direction. The observed ac losses at 0.5Hz in the twisted wire agreed well with the above-mentioned theoretical calculations for the entire amplitude region.

Exactly speaking, the observed ac loss at 0.5Hz in the twisted wire is not only the hysteresis loss since the coupling current loss should be included. So we deduced the actual hysteresis loss by extrapolating the observed frequency dependence of ac loss to zero limit and then evaluated the coupling current loss by subtracting the hysteresis loss from the total ac loss. The obtained frequency dependences of coupling current loss for the amplitude of 5×10^{-3}T are shown in Fig.4. The theoretically estimated ones on the supposition of an ideal multifilamentary wire are also shown in Fig.4 by the solid lines. Here the effective twist pitch of the non-twisted wire corresponds to 100mm, which is twice the length of straight sample wire. We can see that the experimental results are roughly explained by the general theory notwithstanding the complicated arrangement of filaments even with the anisotropy of J_c-B characteristics.

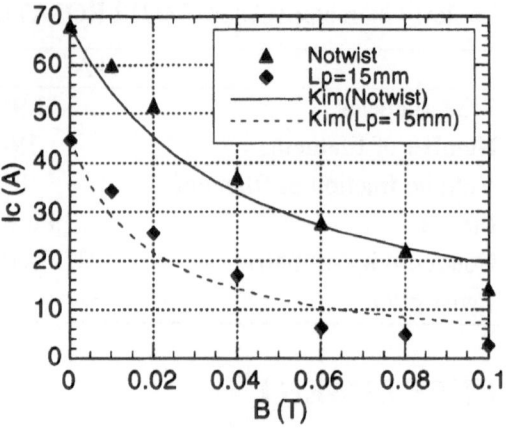

Fig.3 I_c-B characteristics.

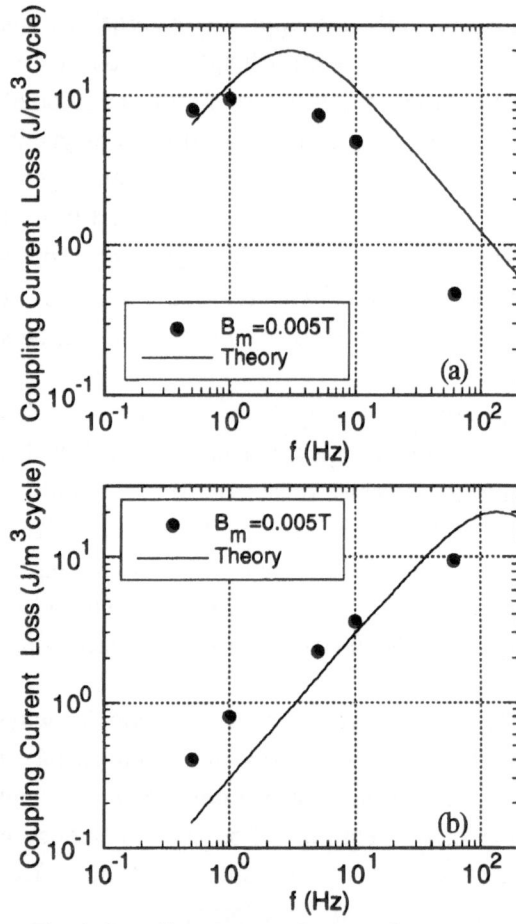

Fig.4 Coupling current loss vs frequency in (a)non-twisted and (b)twisted wires.

CONCLUSION

We experimentally investigated the ac loss properties of Bi2212 ROSATwires with 19×6×3 filaments and pure silver matrix. The effective filament diameter was equal to the average diameter of 19×6 filament bundle and the decoupling between the 19×6 filament bundles was confirmed. Further reduction of ac loss is expected to be easily realized by decreasing a final filament bundle size and a twist pitch and increasing the resistivity of matrix.

1.M.Iwakuma et al., Advances in Superconductivity X, pp.833-836 (1997)

2.M.Okada et al., IEEE Trans. Appl. Suppercond. vol.9, No.2, pp.1904-1907 (1999)

Evaluation of pancake coils made with Bi-2212 multilayer tapes Prepared by PAIR process

Nozomu Ohtani[1], Tsutomu Koizumi[1], Takayo Hasegawa[1], Hiroaki Kumakura[2], Hitoshi Kitaguchi[2], Kazumasa Togano[2] , and Hanping Miao[3]

[1]Showa Electric Wire & Cable Co. Ltd., Kawasaki 210-8660, Japan
[2] National Research Institute for Metals, Tsukuba, 305-0047, Japan*
[3]CREST, Japan Science and Technology Corporation, Tsukuba, 305-0047, Japan**

Abstract : We have successfully developed 100m-class Bi-2212 multilayer tapes prepared by the PAIR process. This conductor shows excellent Jc values of over 500 kA/cm^2 at 4.2K in 10T and the variation in Jc was $\pm 7\%$. Using this conductor we fabricated two type of high-field coils : one reinforced with CuAg alloy tape and a cryocooler-cooled coil. In backup field of 8T at 4.2K high-field coils carried more than 300A and experienced a hoop stress of 109MPa. A cryocooler-cooled coil carried an Ic of 166A, and the stable operating current was 100A for 60 minutes at 20K in self-field.

Keyword : Bi-2212 multilayer tape, PAIR process, high field coil, cryocooler-cooled coil

INTRODUCTION

Bi-2212/Ag superconductors show excellent Jc value below 20K and consequently, are expected to find application in industrial magnets. We have developed Bi-2212 multilayer tapes by applying a Pre-Annealing and Intermediate Rolling (PAIR) process [1] and successfully fabricated 100m-long tapes which exhibited Jc values of 710kA/cm^2 in self-field and 500kA/cm^2 in 10T at 4.2K[2]. When this conductor is wound for high field magnets, several support structures have to be taken into consideration. We fabricated two types of double pancake coils to evaluate the current carrying properties in high backup field. In addition a double pancake coil for cryocooler-cooling purpose was fabricated and tested.

EXPERIMENTAL

We have fabricated 100m-class Bi-2212 multilayer tapes by PAIR processing. These 0.22mm x 4.59mm tapes were heat treated in a large box furnace and were insulated with polyvinyl formal. Using 100m-class multilayer tapes we fabricated three double pancake coils by means of a react and wind method (Table 1), two type of high-field coils and a cryocooler-cooled coil. We reinforced the high-field coils against hoop stress by using CuAg alloy tapes. The CuAg alloy tape (0.16mm thick and 5.25 wide) was insulated with polyvinyl formal and shows a 0.2% proof stress of 300MPa. This reinforcement was attached as overbanding to the outer surface winding of one of the high field coils (coil A in Fig.1). Another coil (coil B) was fabricated by co-winding CuAg alloy tapes in parallel with Bi-2212 multilayer tape. The CuAg alloy tapes were fixed together to high-field coil with the stainless steel bobbin and GFRP flange. A cryocooler-cooled coil (coil C) tested in self-field had no reinforcement, so there was no need for those components. Three double pancake coils were encased and impregnated with epoxy resin.

Table 1. Parameter of double pancake coils made with Bi-2212 tapes prepared by PAIR process.

Parameter		Coil A	Coil B	Coil C
Inner diameter	(mm)	ϕ 70.0	ϕ 70.0	ϕ 60.0
Outer diameter	(mm)	ϕ 147.2	ϕ 176.6	ϕ 116.6
Height	(mm)	10.5	11.0	11.0
Number of terns		286	243	273
Inductance	(mH)	9.7	7.4	7.2
Bo	(Gauss/A)	35	28	40
Bmax	(Gauss/A)	65	45	75

We placed the high-field coil in the center of a backup magnet at the Tsukuba Magnet Laboratory, National Research Institute for Metals and evaluated their current carrying properties in field. A cryocooler-cooled coil was attached to the cooling stage on the second stage of a Gifford-McMahon type cryocooler with cooling power of 10.5W and it was time-tested at 20K. The operating temperature was measured and automatically regulated by the heater on a cooling stage. Critical currents were determined using the criterion of $10^{-13}\,\Omega\,\mathrm{m}$.

Fig. 1. High field coil.

RESULTS AND DISCUSSION

Fig. 2 shows the load line of Coil A at 4.2K. In a backup field of 4T, Coil A carried 300A and the load factor was 70% of the Ic value. In a backup field of 6T, a large voltage across the terminals appeared rapidly near a current value of 230A. The hoop stress in the coil at a current of 230A in a backup field of 6T was calculated to be 58MPa, which exceeded the tensile stress of 40MPa, but with no degradation of Jc. We checked to see if coil A was damaged by this hoop stress by checking its performance in self-field, and in fact, the inductance was reduced from 9.9mH to 8.7mH. We believe that the hoop stress damaged the insulation of coil A. The supporting structure of Coil A is not strong enough to withstand the hoop stress without internal reinforcement of the winding. Then we tested coil B fabricated by co-winding technique with CuAg alloy tape. Coil B carried more than 300A in a backup field of 8T and experienced hoop stress of over 109MPa.

A cryocooler-cooled coil (Coil C) carried 166A at 20K in self-field and generated a central field of 0.69T and a maximum field of 1.2T. Fig. 3 shows the time dependence of the temperature and the voltage across the terminals on Coil C, to which were applied two levels of current, 100A and 110A at an initial operating temperature of 20K. When the voltage gradually increased with time at 110A, the effect of the heat generation caused the temperature to rise, causing thermal runaway of Coil C. At 100A the temperature and the voltage became constant in 20 minutes. This stable operating current of 100A was 60% of the critical current of 166A. It appears necessary to improve the design of the cooling structure to maintain a low temperature in the winding.

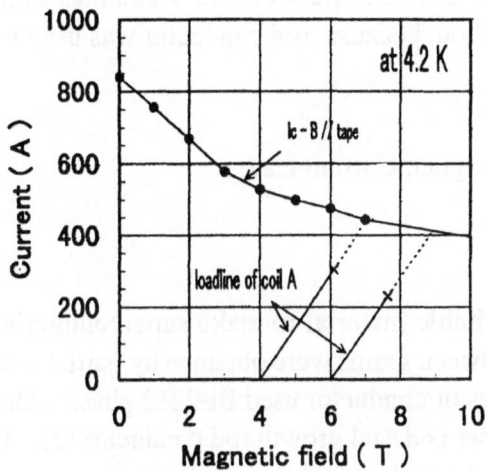

Fig. 2. Load line of high field coil.

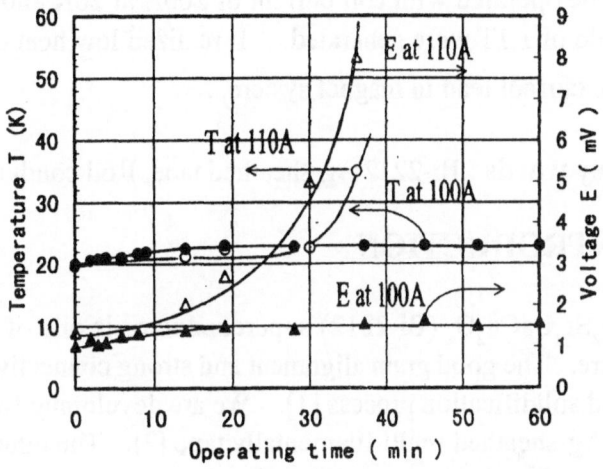

Fig. 3. Time dependence of the temperature and voltage of cryocooler-cooled coil.

CONCLUSION

We fabricated two types of high-field coils and a cryocooler-cooled coil using 100m-class Bi-2212 multilayer tapes prepared by PAIR process. One high-field coil was fabricated by winding CuAg alloy tapes in parallel with Bi-2212 multilayer tape, and it carried more than 300A at 4.2K in a backup field of 8T. The hoop stress was calculated to be over 109MPa. A cryocooler-cooled coil showed the stable operating current of 100A in 60 minutes, which was 60% of the Ic value of 166A at 20K in self-field. This Ic value was corresponded to a coil Jc of 146A/mm^2.

REFERENCES

1. H. Maio, H. Kitaguchi, K. Togano, T. Hasegawa, and Y. Hikichi, Cryogenics 38:163(1998).
2. T. Koizumi, T. Hasegawa, H. Kitaguchi, H Maio, H. Kumakura and K. Togano, Advances in Superconductivity XI, vol. 2, (1999), 919

MAGNET SYSTEM USING TAPE AND CURRENT LEAD BASED ON Bi-2212

T.Kaneko, ,K. Ohkura, K.Hayashi, H.Takei, and O.Kasuu*

Electric Power System Technology Research Laboratories,
Sumitomo Electric Industries, Ltd., Osaka, 554 Japan
*Super-GM, Osaka, 530 Japan

Abstract : We are studying both Ag-sheathed tape and rod conductor based on Bi-2212 superconductor. In the development of the Ag-sheathed tape, 1000m length tape with Jc=2,900A/cm^2 and 100 m length tape with Jc=11,000A/cm^2 at 77 K were obtained. A refrigerator cooled magnet system was assembled by using such tapes for coil part and rod conductors for current lead. The magnet was able to be operated with coil current of 200A at 20K and, center magnetic field of 0.75 T and maximum field of 1.1T were generated. It realized low heat conduction, because rod conductor was used for the current lead in magnet system. .

Key words : Bi-2212 Ag-sheathed tape, Rod conductor, Magnet, Current lead

INTRODUCTION

$Bi_2Sr_2CaCu_2O_z$ (Bi-2212) superconductor is one of the suitable material to make superconducting wire. The good grain alignment and strong connectivity between grains were obtained by partial melt and solidification process [1]. We are developing two types of conductor used Bi-2212 phase. One is Ag-sheathed multi-filamentally tape [2]. The other is laser pedestal growth rod conductor [3]. In this paper, it is reported that development of long Bi-2212 Ag-sheathed tape and magnet system which used Ag-sheathed tape in coil part and rod conductors for current lead.

Ag-SHEATHED LONG TAPES

Figure 1 shows a Ic distribution of 100 m Bi-2212 tape at 77K. As show in fig. 1, mean Ic of whole tape is about 20 A. The Ic determined from whole length I-V curve was also 20 A (criterion : 1 μ V/cm) . This Ic was correspond to Jc = 11kA/cm^2. This value is almost comparable to that of cm order short sample. There are two important points to obtain the long tape with high Ic level and uniform distribution. The tape was rolled from round wire to final tape shape by 2 steps, i.e., a round wire → 0.4mm thick tape → 0.3mm thick. This rolling process reduces filament sausaging and it makes a smooth current path. When the tape was rolled by 1 step, the filament part was strongly sausaged and the average Ic was suppressed about 15 A. The other was defect healing. In fig. 1, we can see slightly low Ic part around 35 to 40m. This part was blistering point after the first heat-treatment. The Ic of this portion was 16 A with bubble. The tape after the first heating was re-rolled and done the same heat treatment as the first one. The mean level of Ic did not change after the second heating. Only the blistering part was healed. Then, we were able to obtain a good 2212 tape with uniform Ic distribution. We tried to make further long tape. Figure 2 shows the result of 1000 m class tape. An average Ic of this tape was rather low compared to 100 m class tape in fig. 1. The Ic and Jc calculated from whole length I-V curve were 5.3 A and 2.9kA/cm^2, respectively.

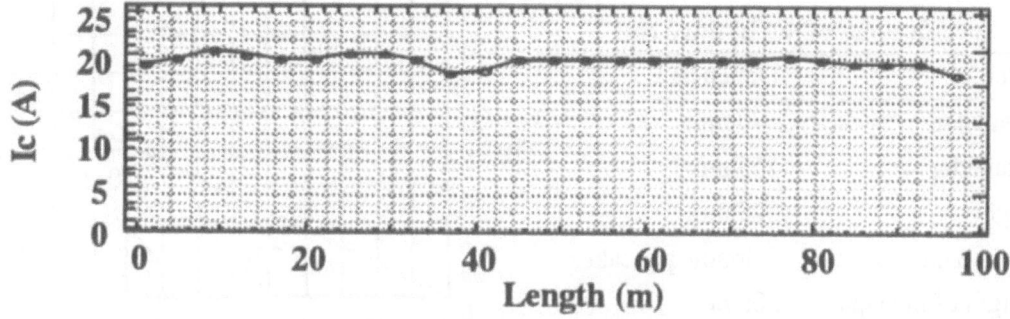

Figure 1. Distribution of Ic for100m Bi-2212 Ag-sheathed tape.

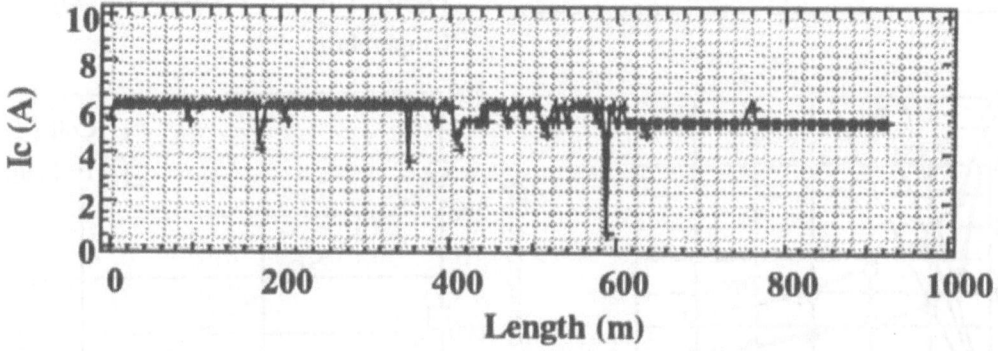

Figure 2. Distribution of Ic for 950m Bi-2212 Ag-sheathed tape.

MAGNET SYSTEM

A refrigerator cooled magnet system was assembled by using tapes mentioned above for coil part and rod conductors for current lead. The pancake coils were made by a react & wind technique. Table 1 is the coil specifications.

Figure 3 shows the I-V curves at various temperatures. The straight line in Fig. 3 is the criterion of 10^{-13}ohm • m. The temperatures in Fig. 3 are the highest and lowest value in the coil. The critical currents of coil were about 150 A, 220 A and 280 A at 26-32 K, 20-27K and 13-14 K, respectively. Figure 4 shows the coil load line at 20 K. The coil Ic was limited by the magnetic field component perpendicular to the tape surfaceat the edge of coils sets. The coil Ic of 200A was in good agreement with the experimental data in Fig. 3. The generated magnetic fields were B_{max} =1.1 T and B_0 = 0.75T at 20 K with the operation current of 200 A. At 14 K, the both of Bmax and B_0 were 1.4 T and 1.0 T, respectively. The Bi-2212 rod conductor was used for the current lead. The pair of current lead was set between the 1st stage of GM refrigirater and the coil. Figure 5 shows the initial cooling curve for the 1st stage and the coil during cooling. After 10 hr, the coil temperature was almost constant (about 15 K). The temperature of the 1st stage was 45 K. Low heat conduction was realized. The rod conductor was sufficiently applied to the current lead for the refrigerator magnet.

CONCLUSIONS

(1) Development of long Bi-2212 Ag-sheathed tape
We developed a 100m Ag-sheathed superconducting tape with a critical current of 20A and a critical current density of 11kA/cm^2 at 77 K In the case of 1000 m class tape,the critical current was rather low. However, the uniform Ic distribution of 5A was measured in whole length. This is the first result that the fabrication of 1000m tape having a filament part of Bi-2212 superconductor.

Table 1. Specifications of the Bi-2212 coil

Parameter	Dimension
Inner diameter	70 mm
Outer diameter	155 mm
Coil height	325 mm
Number of pancakes	5 double pancakes
Total length of the tape	500 m
Number of bundle	4
Turns	335

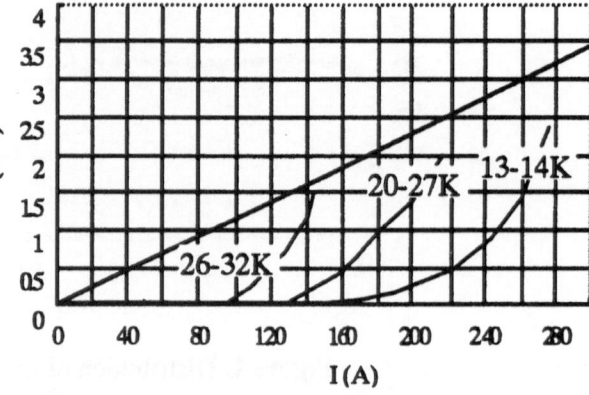

Figure 3. I-V curves of the Bi-2212 coil at various temperature

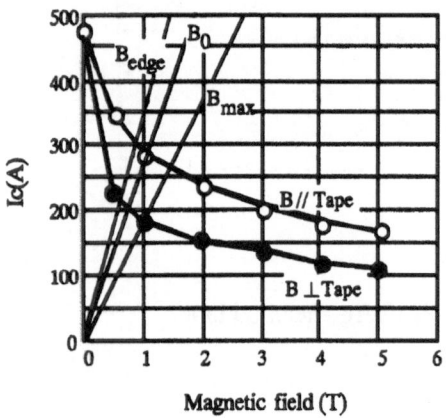

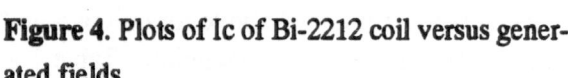

Figure 4. Plots of Ic of Bi-2212 coil versus generated fields

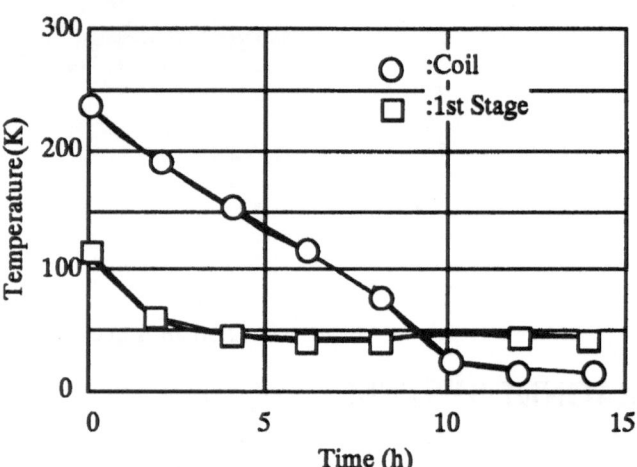

Figure 5. Temperature transition of the 1st stage and the coil during cooling.

(2) Magnet system using composite tapes

The coils were made using the tapes mentioned above. The coils were combined with a GM refrigerator for magnet system. It was possible that the operating current of 200 A generated a center magnetic field of 0.75 T and maximum magnetic field of 1.1T. These results were in good agreement with design expected from tape performances.

Rod conductor was used for the current lead in magnet system. It realized low heat conduction.

ACKNOWLEDGMENT

This study has been carried out as a part of "R&D on Superconducting Technology for Electric Power Apparatuses" under the New Sunshine Program of AIST, MITI, being consigned by NEDO.

1. Maeda H, Togano K (1996)Bismuth-Based High-Temperature Superconductors.Marcel Decker, Inc, New York, pp 369-476.

2. Kaneko T, Hayashi K, Sato K and Kasuu O, in *Advaces in Superconductivity* XI,edited by N. Koshizuka and S. Tajima (Spriger-Verlag, Tokyo,1999), pp 907-910.

3. Kasuu O, Takahashi K, Sato K and Yoshida N, in *Advaces in Superconductivity* VIII,edited by H.Hayakawa and Y. Enomoto (Spriger-Verlag, Tokyo,1996), pp 875-878.

Proposal of a New 'Low Loss-High Stability Type' Rutherford Cable

Tsuyoshi Gohda,[1] Takaaki Fukunaga,[1] Akifumi Kawagoe,[1] Fumio Sumiyoshi,[1]
Teruko Kawashima,[2] Naoki Hirano,[3] and Toshiyuki Mito[4]

[1]Kagoshima University, 1-21-40 Kohrimoto, Kagoshima, 890-0065, Japan
[2]Fukuoka Jo Gakuin University, 3-42-1 Osa, Minami-Ku, Fukuoka, 811-1313, Japan
[3]Chubu Electric Power Co., Inc., 20-1 Kitasekiyama, Ohdaka-Cho, Midori-Ku, Nagoya, 459-8522, Japan
[4]National Institute for Fusion Science, 322-6 Oroshi, Toki, 509-5292, Japan

Abstract: A new structure of Rutherford cables is proposed in which low resistive paths, enabling transfer of transport currents from one strand to another when a normal quench occurs, do not induce increase in coupling losses. This cable is designed so as to have a relatively longer strand twist pitch in relation to cable twist pitch and good crossover contacts between strands located near each edge of the cable's cross-section. The inter-strand coupling current is reduced inside the cable in changing transverse magnetic fields with a face-on orientation, which qualitatively results in decrease in coupling losses. In order to confirm quantitatively the expected loss property, two-dimensional finite element method analyses are carried out. It is found that the coupling loss of this cable can be reduced to between 1/5 and 1/10 the amount calculated in some previous cable designs in spite of the existence of the low crossover resistances between strands, required for ensuring high stability.

Keywords: Rutherford cable, coupling loss, finite element method

INTRODUCTION

Large inter-strand coupling losses are produced in superconducting Rutherford cables under changing transverse magnetic fields with a face-on orientation. Suppression of inter-strand coupling current by increasing crossover resistance between strands is thought to be effective in reducing this loss [1]. However, the high crossover resistance tends to result in low stability because transport current shearing is also suppressed when a normal quench occurs. Recently an attractive new design has been proposed; it was expected that the dilemma of low losses and high stability would be solved by adopting short circuits among strands with every length of cable twist pitches [2]. According to our numerical analyses using three-dimensional finite element method (FEM) [3], however, this design is not valid because significant reduction of losses can not be achieved due to the existence of longitudinal paths of coupling currents inside strands. The purpose of the present paper is to propose an alternative new design of Rutherford cables with both low losses and high stability. In order to confirm the validity of the design, two-dimensional FEM calculation is carried out.

PROPOSAL OF A NEW DESIGN OF RUTHERFORD CABLES

In the present design, Rutherford cables have good crossover contacts between strands located near each edge of the cable cross-section, shown in Fig. 1. Moreover, this cable is designed so as to have both relatively longer strand twist pitch in relation to cable twist pitch and the same strand twist direction as the cable twist direction. By adopting this new cable structure, potential differences between crossover strand pairs (V_c-V_s) become small and therefore induced inter-strand coupling losses are reduced. As a result, total coupling losses in the cable can be reduced in spite of increase in intra-strand coupling losses. (In Case V shown in Fig. 3 each strand is surrounded by high resistive coating, so that the loss in it corresponds to the intrinsic

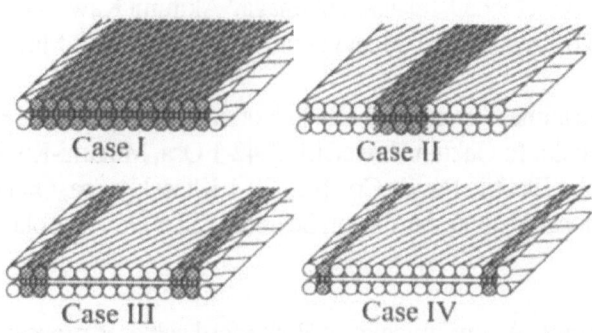

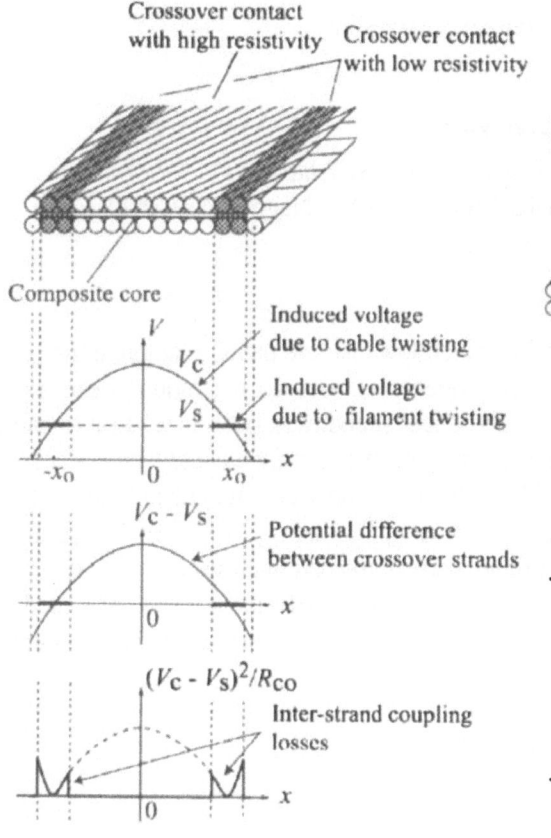

Fig. 2. Four types of Rutherford cables. Shaded area shows different low resistive section.

Fig. 1. Principle of the new design of Rutherford cables, where R_{co} is the crossover resistance between strands.

Table 1. Parameters of cables and strands used in 2D FEM analyses.

Cable	
Number of strands	30
Twist pitch (L_c)	87 mm
Thickness of core	25 μm
Core materials	Copper / Insulator
Strand	
Matrix of the filamentary region	Bronze
Matrix of the outer sheath region	Copper
Radius of strand	0.80 mm
Radius of filamentary region	0.64 mm

intra-strand coupling loss.) Characteristic features of the new design compared with the existing one are as follows: (1) Composite cores inserted in cables have the same shape and resistive pattern of thin cross-sections along the cable axis. (2) Low resistive paths enabling transfer of transport currents from one strand to another are laid at every half-length of cable twist pitches.

FEM ANALYSIS OF COUPLING LOSSES

In order to discuss properties of cables quantitatively, we shall introduce here the 'current transfer factor' ξ and the 'coupling loss factor' ζ, where ξ and ζ are defined as a number ratio of the strand pairs with low resistive contact to total strand pairs and a ratio of (optimized) coupling loss of each cable to the coupling loss of the non-cored practical cable, respectively. It must be mentioned that ζ/ξ represents the 'achievement factor of low losses and high stability'.

Four cases of Rutherford cables as shown in Fig. 2 are taken for the numerical analysis; The values of ξ vary from 0.87 for Case I to 0.13 for Case IV, which lay between 1 for the practical cable without a high resistive core and nearly zero for that with the core. Parameters of cables and strands used are listed in Table 1, and in these cables the side-by-side contact resistance between strands is assumed to be extremely large. In this analysis our original 2D FEM calculation [4] is applied only to normal metal regions, which consists of both the outer sheath region of strands and the composite core. As the boundary condition, analytic solutions are used at the outermost filament layers of each strand. These calculations take into account cable structure

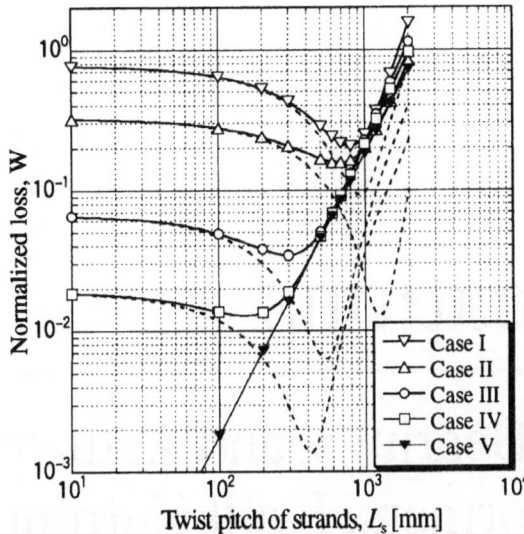

Fig. 3. Dependence of total coupling losses on twist pitches of strands, where dotted lines represent inter-strand coupling losses.

Table 2. 2D FEM results of coupling losses.

	Practical design (no core)	Practical design (cored)	Case I	Case II	Case III	Case IV
ξ	1	~ 0	0.87	0.27	0.27	0.13
L_{so}	10 mm	10 mm	800 mm	700 mm	300 mm	150 mm
W_o	7.59×10^{-1}	1.83×10^{-5}	2.06×10^{-1}	1.57×10^{-1}	3.43×10^{-1}	1.36×10^{-2}
ζ	1	2.41×10^{-5}	2.71×10^{-1}	2.07×10^{-1}	4.52×10^{-2}	1.79×10^{-2}
ζ / ξ	1	–	0.31	0.77	0.17	0.14

where the superconducting filaments inside strands are twisted and moreover whole strands are also twisted together. In Fig. 3 the obtained results are plotted for ac transverse magnetic fields with amplitude H_m, a frequency of 0.01 Hz and face-on orientation, where the vertical and the horizontal axes are the coupling loss per cycle per unit volume normalized by $\mu_0 H_m^2$, W, and the twist pitch of strands L_s, respectively. In the region of small L_s, the coupling loss tends to decrease with decreasing ξ, however there exists clear difference between Cases II and III in spite of the same value of ξ. With increase in L_s, inter-strand coupling losses shown by broken lines in Fig. 3 gradually decrease and then increase again after reaching a minimum value, where the inter-strand coupling loss is given by subtracting coupling loss for Case V from that for Cases I-IV. Since the intra-strand coupling loss is proportional to L_s^2, the characteristic curve of total coupling losses vs. L_s^2 is much deformed from that of inter-strand coupling losses at a large L_s region. The resultant minimum loss value of normalized losses W, W_o, and the optimum value of L_s, L_{so}, are shown in Table 2. The obtained value of ζ / ξ becomes smaller than 1 for Cases I-IV, and the smallest value of ζ / ξ is given as 0.14 for Case IV. This shows validity of the present design.

CONCLUSION

The validity of our new design of Rutherford cables with both low losses and high stability is clarified by 2D FEM analysis. This cable design could be useful in avoiding the deterioration of properties of multi-stranded cable caused by uncontrolled inter-strand contact resistance or the asymmetry in cable twisting.

1. F. Sumiyoshi, S. Kawabata, T. Gohda, A. Kawagoe, T. Shintomi, E.W. Collings, M.O. Sumption, and R.M. Scanlan, IEEE Trans. Appl. Super. **9**, 731-734 (1999).

2. M. Ono, T. Hamajima, M. Hiregishi, Y. Wachi, H. Maeda, and T. Fujioka, Proc. of MT-15, 457-460 (1997).

3. T. Kawashima, F. Sumiyoshi, S. Kawabata, and T. Shintomi, IEEE Trans. Appl. Super. **9**, 717-721 (1999).

4. F. Sumiyoshi, H. Kasahara, T. Kawashima, and T. Tanaka, Cryogenics. **29**, 741-747 (1989).

5 System Applications

USE OF THIN FILMS IN HIGH-TEMPERATURE SUPERCONDUCTING BEARINGS

John R. Hull and Ahmet Cansiz

Energy Technology Division, Argonne National Laboratory, Argonne, IL 60439 USA

Abstract: In a permanent magnet/high-temperature superconductor (PM/HTS) bearing, a thin-film HTS positioned above a bulk HTS has been expected to maintain the large levitation force provided by the bulk while a low rotational drag is provided by the very high current density of the film. For the low drag to be achieved, the thin film must shield the bulk from inhomogeneous magnetic fields. However, measurement of rotational drag in a PM/HTS bearing that used a combination of bulk and film HTS showed that the thin film is not effective in reducing rotational drag. Subsequent experiments, in which an AC coil was placed above the thin-film HTS and the magnetic field on the other side of the film was measured, showed that the thin film provides good shielding when the coil axis is perpendicular to the film surface but poor shielding when the coil axis is parallel to the surface. This is consistent with the lack of reduction in rotational drag being due to a horizontal magnetic moment of the permanent magnet. The poor shielding with the coil axis parallel to the film surface is attributed to the aspect ratio of the film and the three-dimensional nature of the current flow in the film for this coil orientation.

Keywords: levitation, bearings, thin films, high-temperature superconductors, bulk YBCO

INTRODUCTION

The use of bulk high-temperature superconductors (HTSs) in levitation applications such as magnetic bearings has seen considerable development during the past decade [1]. The most common configuration for an HTS bearing incorporates a rotatable permanent magnet (PM) that is stably levitated in close proximity to a stationary bulk HTS of (RE)-Ba-Cu-O (where RE denotes rare earth). More recently, interest in the use of thin-film HTSs in superconducting bearings has increased [2-6]. Although YBCO thin films often have critical current density (J_c) values of >1 MA/cm^2 at 77 K, the thickness of these films is only $\approx$1 μm and they do not provide much levitation force.

The use of high-J_c thin films in HTS bearings has long been thought to offer the potential for reduced rotational drag [7,8]. This conjecture relies on the generally accepted hysteretic loss mechanism for HTS bearings that is based on the critical-state model. Hysteretic loss is produced whenever there is a cyclic change in applied magnetic field, and the energy loss per cycle is [9]

$$E_h = K\mu_0(\Delta H)^3/J_c \ , \qquad (1)$$

where E_h is the hysteretic energy loss per unit area per cycle, K is a geometric coefficient of order unity, $\mu_0 = 4\pi \times 10^{-7}$ N/A^2 is the magnetic permeability of vacuum, $\Delta H = \Delta B/\mu_0$ is the peak-to-peak amplitude of the varying magnetic field, and J_c is the critical current density in the HTS. When a levitated PM spins over an HTS, rotational loss occurs because of azimuthal inhomogeneities of the PM's magnetic field, which causes a ΔB at a surface location on the HTS over one complete rotation of the PM. The amplitude of the ΔB increases if there is any appreciable whirl amplitude of the PM, in which case the radial gradient of the magnetic field also contributes to the ΔB. One may conceptualize the magnetics of the bearing system as the PM providing a constant magnetic field B that interacts with the HTS to provide levitation force, and a circumferentially varying magnetic field ΔB that produces the rotational loss. In general, the ΔB associated with inhomogeneity is much lower than the constant magnetic fields B that provide the levitation force.

A further consequence of the critical-state model, beyond the energy loss described in Eq. (1), is that the interior of the HTS is shielded from the varying magnetic field, so that the hysteresis losses occur only on the surface of the HTS. Thus, it was conjectured that if a thin-film HTS were interposed between the PM and the bulk HTS, the levitation force would remain approximately the same and most of the hysteretic loss would occur in the thin-film HTS. The thin-film HTS would shield the bulk HTS from the effects of the applied ΔB, so that the bulk HTS experiences only a constant applied magnetic field. Because the J_c is so much higher in the thin film, the total hysteretic loss (and therefore the rotational drag on the levitated PM) should be much lower with the HTS film installed.

When these considerations were tested in rotational drag experiments, we found that the HTS film was not effective in reducing rotational drag [10]. We then conducted magnetic shielding tests on the HTS film to help elucidate its failure to reduce the rotational drag. This article reports the results of those rotational drag and shielding tests. A theoretical interpretation of the results is also provided.

ROTATIONAL DRAG

Experimental Method. A detailed description of the experimental apparatus for rotational drag measurements has been described elsewhere [10,11]. In essence, a PM disk rotor was levitated over an HTS in a vacuum chamber, with an oil-diffusion pump reducing the pressure to <100 µPa. The HTS was inside a room-pressure cryochamber, through which liquid nitrogen flowed from a gravity feed at ≈3 kPa. Rotational loss of an HTS bearing is evidenced by the decay rate of the rotational frequency (f) and is characterized by the coefficient of friction (COF),

$$COF = -[2\pi R_\gamma^2/(gR_D)] \, df/dt \ , \qquad (2)$$

where R_γ is the radius of gyration of the rotor, g is the acceleration of gravity, t is time, and R_D is the weighted mean radius at which the drag force acts.

The height of the levitated PM was measured with a traveling telescope at intervals throughout the experiment, and no change in height from the initial levitation value was observed to within 10 µm. The levitation height in these experiments includes the thickness of the cryochamber (3.5 mm) and the gap between the PM and the top of the cryochamber. The height was recorded after the system had cooled, to avoid obfuscation by any changes due to thermal contraction.

Three configurations of HTS were used: (a) bulk HTS, (b) thin-film HTS, and (c) thin-film HTS over bulk HTS.

The bulk HTS was a melt-textured Y-Ba-Cu-O cylinder with its c-axis aligned along the vertical, J_c = 20 kA/cm^2, critical temperature T_c = 92 K, diameter of 32 mm, and thickness of 22 mm. We chose a thin-film HTS with a diameter significantly larger than that of the bulk HTS, hoping that the film would significantly shield the bulk. The disk-shaped thin-film Y-Ba-Cu-O HTS had J_c = 3.7-4.1 MA/cm^2, critical temperature T_c = 89.2-89.6 K, diameter of 51 mm, and thickness of 350 nm. The film was deposited on an La-Al-O$_3$ substrate with a thickness of 0.5 mm. When the film was used together with the bulk, they were coaxial and the film substrate was immediately above the bulk.

The same PM rotor was used in all tests; it is an axially polarized NdFeB disk with magnetization of $\mu_o M$ ≈1.1 T, diameter of 25.4 mm, height of 6.35 mm, and mass of 35.6 g. Because levitation force provided by the thin-film HTS was too low to levitate the PM, it was augmented by placing a stationary PM above the rotor in an Evershed configuration to produce an attractive force between the two PMs [12]. The distance between the two PMs was adjusted so that the upward force due to magnetic attraction was slightly lower than the gravitational force downward. Because the additional PM in the Evershed configuration does not move, it does not contribute to hysteresis loss in the thin-film HTS. The velocities of the experiment are low enough that loss due to eddy currents induced by relative velocity of the PMs is negligible compared to the hysteresis loss.

Results. In Figs. 1-3, the data sets are arranged to show the behavior for each HTS configuration at approximately the same levitation height. For all levitation heights, the resonant frequency is

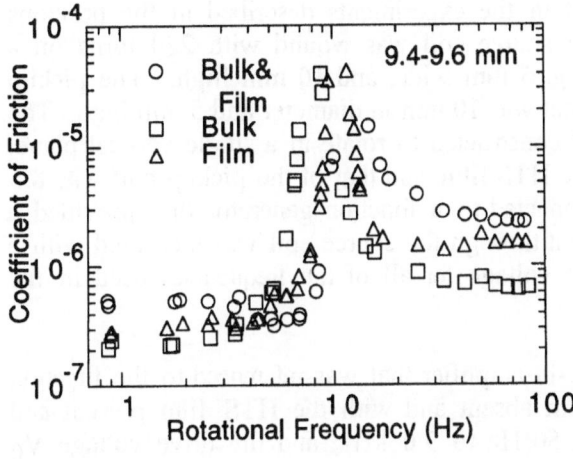

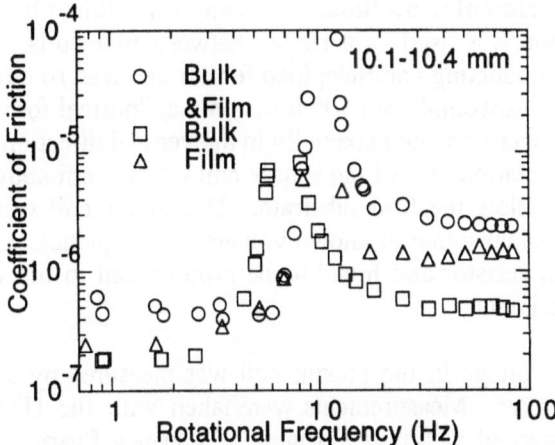

Fig. 1. COF vs. frequency at 9.4-9.6 mm height for PM levitated over HTS in various HTS configurations.

Fig. 2. COF vs. frequency at 10.1-10.4 mm height for PM levitated over HTS in various HTS configurations.

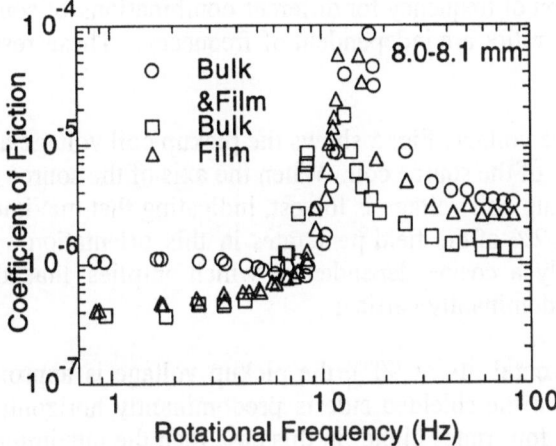

Fig. 3. COF vs. frequency at 8.0-8.1 mm height for PM levitated over HTS in various HTS configurations.

consistently lowest for the bulk alone and highest for the combination of bulk and film. For all levitation heights, both above and below the resonance, the minimum loss occurred when bulk HTS was used alone, while the maximum loss was seen when the bulk and film were used together. In these regions, the COF of the bulk and film is approximately equal to the sum of the COFs for the bulk alone and the film alone.

We assumed that the substrate on which the thin film was deposited did not contribute to the rotational loss. An experiment was performed in which a bare substrate (without an HTS film) was placed above the bulk HTS. The rotational loss in this experiment was essentially identical to that of the bulk HTS alone, consistent with our assumption.

SHIELDING

The results of the rotational loss measurements were opposite to our original expectations and motivated the shielding experiments described in this section. Although the mean magnetic moment of the PM is in the vertical direction, any inhomogeneity in the PM is likely to have a substantial amount of its magnetic moment in the horizontal direction. This horizontal moment has been previously ignored in analyzing HTS bearings.

Experimental Method. The same HTS thin film used in the experiments described in the previous section was used as a barrier between two coils. The source coil was wound with 220 turns on a nonconducting parallelepiped former that was 10 mm long, 5 mm wide, and 10 mm high. The pickup coil was wound with 100 turns on a cylindrical former that was 10 mm in diameter and 5 mm high. The coils were oriented coaxially in the vertical direction and constructed to rotate in a single vertical plane. The rotation axis of the source coil was 11 mm above the HTS film, and that of the pickup coil was 8.5 mm below the film substrate. The source coil was connected to a function generator that provided a sinusoidal signal of known voltage and frequency. Current through the source coil was measured with a shunt resistor and found to be proportional to the drive voltage for all of the frequencies used in the experiment.

The voltage in the pickup coil was measured by a lock-in amplifier that was referenced to the function generator. Measurements were taken with the HTS film absent and with the HTS film present and submerged in liquid nitrogen. Frequency f varied from 50 Hz to 3.0 kHz, and the drive voltage V_0 varied from 1.0 V to 5.0 V. At the maximum voltage, the peak amplitude of the current in the source coil was 170 mA, corresponding to a peak magnetic field amplitude of about 420 μT at the surface of the HTS film when the axis of the source coil was oriented perpendicular to the film ($\Phi_s = 0°$).

Results. Figure 4 shows the voltage on the pickup coil when the film is present, divided by the voltage with the film absent as a function of frequency for different combinations of source and pickup coil angle. Figure 4 clearly shows that the ratios are independent of frequency. These results were also independent of the coil drive voltage.

For a fixed frequency and drive voltage, Fig. 5 shows the pickup coil voltage as a function of pickup coil angle Φ_p for three orientations of the source coil. When the axis of the source coil is perpendicular to the film surface ($\Phi_s = 0$), the pickup coil voltage is lowest, indicating that maximum shielding is occurring. Figure 4 indicates that only ≈2% of the field penetrates in this orientation. As a function of Φ_p, the voltage exhibits approximately a cosine dependence, which implies that the magnetic field on the shielded side of the film is predominantly vertical.

When the source coil is horizontal ($\Phi_s = 90°$), the pickup voltage is approximately a sine function of Φ_p, indicating that the field on the shielded side is predominantly horizontal. The maximum pickup voltage (at $\Phi_p = 90°$) is about four times higher in this case than the maximum (at $\Phi_p = 0$) for $\Phi_s = 0$, despite the somewhat lower magnetic field at the film surface. This result confirms our hypothesis that the film does not shield horizontal magnetic moments very well. When the source coil is oriented at 45°, the results are a linear mixture of the two extreme cases.

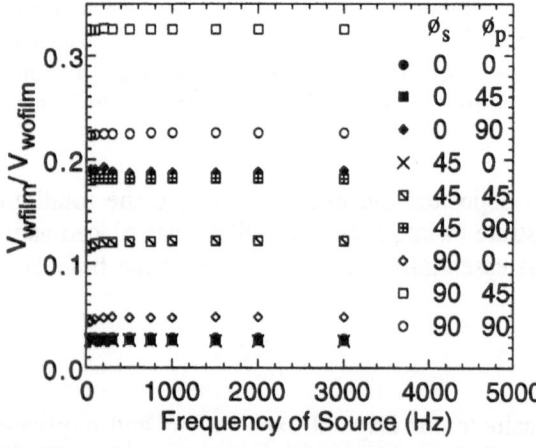

Fig. 4. Pickup coil ratio of voltage with film present divided by voltage without film for different combinations of source coil angle Φ_s and pickup coil angle Φ_p.

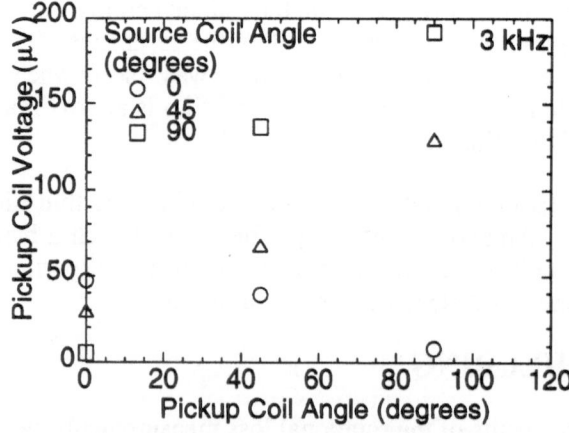

Fig. 5. Pickup coil voltage as a function of pickup coil angle Φ_p for different source coil angles Φ_s. Data from f = 3 kHz, and V_0 = 1.0 V.

The case of a thin disk in a uniform applied magnetic field has undergone considerable analysis [13-15], but the configuration of our experiment is considerably more complicated and seemingly not yet susceptible to elementary analysis. We have shown previously that if drag of magnetic flux through the film were the dominant loss mechanism, the loss in the film should be much higher [10]. We have also shown that the J_c and thickness of the film is sufficient to shield out the expected magnetic field from a vertically oriented magnetic dipole [10]. This result is consistent with the results of the experiments described in this section.

Based on the properties of the PM used in the experiments and assuming that Eq. (1) is applicable, one would expect the COF to be lower by at least a factor of 25 with the film, compared to the case of the bulk HTS alone. Because the COF is not lower for the film, plus the COF when bulk and film are used together is essentially the sum of the COFs when they are used separately, we believed that there is probably a nearly complete lack of shielding of the bulk HTS by the film. This hypothesis was verified by the results of the shielding experiments.

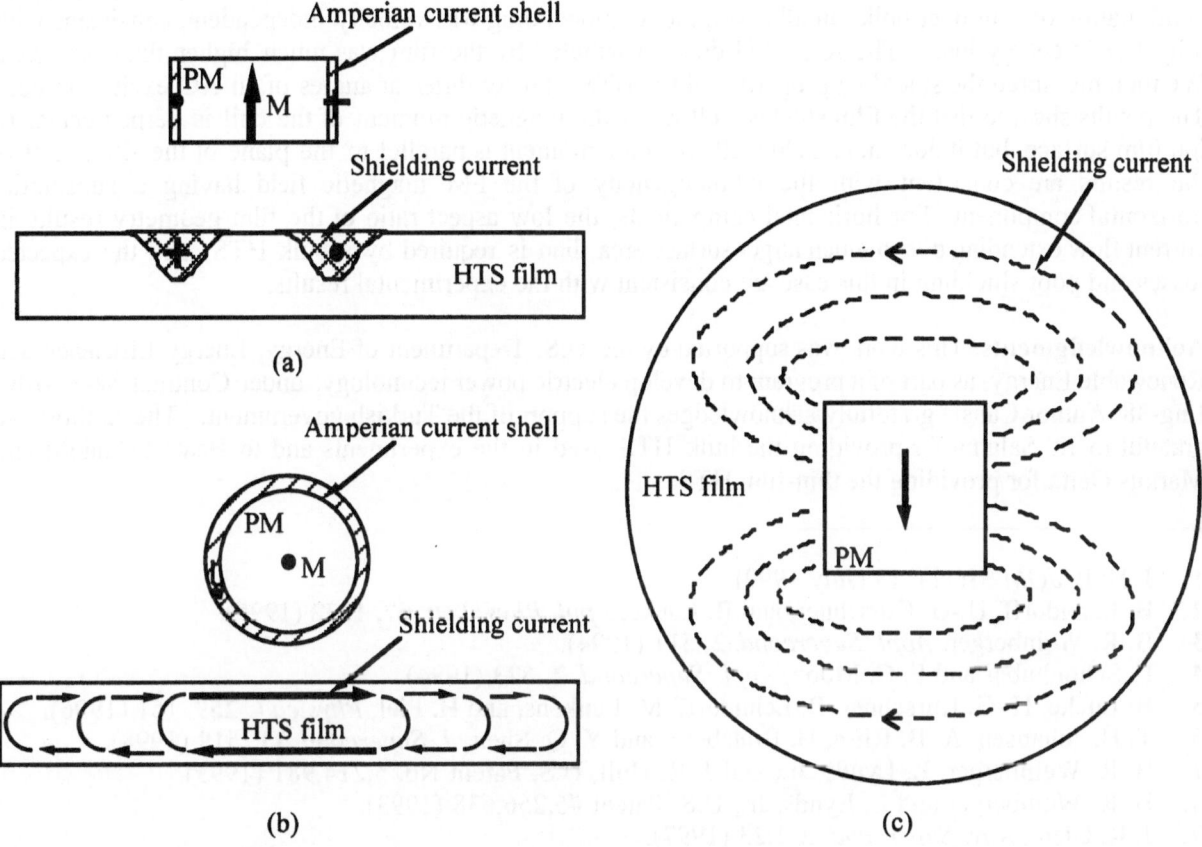

Fig. 6. Schematic diagram of hypothesized current flow in HTS film: (a) Vertical cross section, where magnetization of PM is perpendicular to surface of film. (b) Vertical cross section, where magnetization of PM is in plane of film and current flows through film thickness. (c) Top view, where magnetization of PM is in plane of film and current flows in plane of film.

These results lead us to consider the three-dimensional nature of the currents in the film. The currents under consideration are those caused by the PM inhomogeneity and are generally different from the currents associated with the levitation. The main possibilities are shown in Fig. 6. If a magnetic dipole were oriented perpendicular to the film surface, as indicated in Fig. 6(a), currents need run only in the plane of the film and mainly in the region immediately below the perimeter of the PM; therefore, we should be able to shield the field out of the film interior and below the film, which is consistent with the experimental results. However, if the dipole moment is oriented parallel to the film surface, as indicated in Figs. 6(b) and 6(c), then there are two major possible current flows. To shield the applied field in this orientation, currents at the surface of the film must flow in the same direction as the fictitious Amperian currents in the bottom arc of the PM, as indicated in Figs. 6(b) and 6(c). Because current is conserved in these systems, the currents must return within the system, either by flowing through the thickness of the film and passing along the bottom half in the opposite direction, as shown in Fig. 6(b), or in the plane of the film around the perimeter of the inhomogeneity, as shown in Fig. 6(c). In either case, the extended area over which the current must flow, compared to that shown in Fig. 6(a), is probably sufficient to account for the additional loss in the film in the rotational drag experiments. The current flow shown in Fig. 6(b) is likely to provide almost no shielding, and one would expect the ratios shown in Fig. 4 to be higher. This leaves the most likely alternative for the current flow to be that shown in Fig. 6(c). Such a current distribution would provide some shielding under the center of the film but would provide additional field under the perimeter of the film. While this seems the most likely explanation at present, final resolution of the detailed current distribution awaits further experimentation.

CONCLUSIONS

We measured rotational drag of a PM/HTS bearing that used either an HTS thin film, bulk HTS, or combination of film over bulk. In all cases, the rotational drag was velocity-independent, consistent with a hysteretic energy loss. The rotational drag contributed by the film was much higher than expected. We then measured the shielding properties of the HTS film for different angles of an AC excitation coil. The results showed that the film shields well when the magnetic moment of the coil is perpendicular to the film surface, but it does not shield well when the moment is parallel to the plane of the film. All of the results are consistent with the inhomogeneity of the PM magnetic field having a substantial horizontal component. For horizontal components, the low aspect ratio of the film geometry results in current flow extending over a much larger surface area than is required by a bulk HTS, and the expected losses and poor shielding in this case are consistent with the experimental results.

Acknowledgments. This work was supported by the U.S. Department of Energy, Energy Efficiency and Renewable Energy, as part of a program to develop electric power technology, under Contract W-31-109-Eng-38. Author Cansiz gratefully acknowledges the support of the Turkish government. The authors are grateful to K. Salama for providing the bulk HTS used in the experiments and to Beate Lehndorff and Markus Getta for providing the thin-film HTS.

1. J. Hull, *JOM* **51**(11), 13 (July 1999).
2. B. Lehndorff, H.-G. Kurschner, and B. Lucke, *Appl. Phys. Lett.* **67**, 1932 (1995).
3. B. R. Weinberger, *Appl. Supercond.* **2**, 511 (1994).
4. P. Schonhuber and F. C. Moon, *Appl. Supercond.* **2**, 523 (1994).
5. B. Lucke, H.-G. Kurschner, B. Lehndorff, M. Lenkens, and H. Piel, *Physica C* **259**, 151 (1996).
6. T. H. Johansen, A. B. Riise, H. Bratsberg, and Y. Q. Shen, *J. Supercond.* **11**, 519 (1998).
7. B. R. Weinberger, L. Lynds, Jr., and J. R. Hull, U.S. Patent No. 5,214,981 (1993).
8. B. R. Weinberger and L. Lynds, Jr., U.S. Patent #5,256,638 (1993).
9. J. R. Clem, *Adv. Supercond. X* 1:23 (1997).
10. A. Cansiz and J. R. Hull, to appear in Proc. Cryog. Eng. Conf., Montreal (July 1999).
11. J. R. Hull, T. M. Mulcahy, K. L. Uherka, R. A. Erck, and R. G. Abboud, *Appl. Supercond.* **2**, 449 (1994).
12. J. R. Hull, T. M. Mulcahy, and J. F. Labataille, *Appl. Phys. Lett.* **70**, 655 (1997).
13. E. H. Brandt, *Phys. Rev. B* 58:6506 (1998).
14. A. Rastogi, H. Yamasaki, and A. Sawa, *Adv. Supercond. X* 1:561 (1997).
15. X. N. Xu, A. M. Sun, M. J. Qin, S. Y. Ding, X. Jin, X. X. Yao, and S. L. Yan, *Physica C* 291:315 (1997).

AC LOSS OF YBCO MAGNETIC BEARING COVERED WITH HTSC THIN FILMS

Akira Miura[1], Kazuyuki Demachi[1], Ryota Shimizu[1], Tetsuya Uchimoto[1], Kenzo Miya[1] and Hiromasa Higasa[2]
[1]The Univ. of Tokyo, 2-22 Shirane Shirakata Tokai-mura Naka-gun, Ibaraki 319-1106
[2]Shikoku Res. Inst., 2108-9 Nishimachi, Yashima, Takamatsu-shi, Kagawa 761-0013

Abstract : The decreases of rotation speed and levitation force are ones of the most significant problems for the practical use of HTSC bearing. It is caused by AC field due to the inhomogeneous magnetization of the rotor's PMs. In this research, we propose the HTSC bearing covered with the HTSC thin films that have much larger Jc for suppression of rotation loss, and its effects are investigated by experimental and computational works.

Keywords: Rotation loss, YBCO thin film, HTSC Flywheel

INTRODUCTION

The HTSC flywheel has the advantage that it has no energy loss by friction which is significant in the case of the rotational energy storage system without levitation. If the inhomogeneous component of the magnetic field of the flywheel exists, however, it yields the AC magnetic field in the superconductor due to the flywheel's rotation with high speed. The AC field causes the flux flow and creep and they lead to the degradation of levitation force and of rotation speed. Such phenomena are significant for the development of HTSC flywheel and are called the rotation losses. The objectives of this research are as follows:

① Upgrade of our HTSC-FEM simulation code in order to evaluate the degradation of levitation force and rotation speed of HTSC magnetic bearing in AC field

② Preliminary experiment for observation of HTSC thin films' suppression of degradation of HTSC levitation force in AC field.

NUMERICAL SIMULATION OF ROTATION LOSS OF HTSC SMB

Numerical Method. The current vector potential method (T-method) was adopted for the electromagnetic field analysis[1-2], The governing equation is written as

$$\mu_0 \dot{T}_n + \frac{\mu_0}{4\pi} n \cdot \int_S \dot{T}_n \nabla \frac{1}{R} dS = -n \cdot \nabla \times E - n \cdot (\dot{B}_{dc} + \dot{B}_{ac}) \cdot \tag{1}$$

The thin plane assumption was applied so that the planes are perpendicular to r axis and the shielding current is assumed to flow in θ-z plane. The current density was calculated from the simulation output T_m as $J = \nabla T_n \times n$, where T_n is the normal component of current vector potential T, and n is the normal vector.

The flux flow creep model[1] was adopted as the constitutive relation between the electric field E and current density J. The electric field E was obtained using E and J as $E = (E/|J|)J$ and was substituted to eq. (1) for the calculation of next time step.

The Kim model was applied as the dependence of Jc of HTSC bulk upon the applied field: $J_c = J_{c0} \times B_0/(B_0 + |B|)$, where B_0 is 0.34T and J_{c0} is $3.0 \times 10^9 A/m^2$

The A-method was applied for the calculation of homogeneous component of the magnetic field of rotor, B_{dc}. The inhomogeneous component B_{ac} was assumed as

$$B_{ac}(r,t) = B_{dc}(r) + \gamma B_{dc} \cos(\lambda(-\omega t + \alpha(r))), \tag{2}$$

where $\gamma = |B_{ac}|/|B_{dc}|$ is set $0 < \gamma < 1$ and ω is the angular velocity and λ is wave number of B_{ac}. α is the parameter represented as $\alpha = \tan^{-1}(y/x)$ at arbitrary location $r = (x, y, z)$.

The angular velocity ω and the rotational energy P of magnetic rotor were obtained as follows,

$$I\dot{\omega} = -\int_{SC} \{r \times (J \times B_{ex})\}_z dV, \qquad P = I\omega(t)\frac{\partial \omega(t)}{\partial t}, \tag{3}$$

where I is the inertial moment of magnetic rotor.

Simulation Results. Fig. 1(a) shows the schematic drawings of HTSC radial bearing which consists of eight bulks. The analyzed region is one of the bulks. The mesh division in r, θ and z directions are 12, 24 and 7, respectively. The thickness of layeres in r direction are 3.0, 3.0, 3.0, 2.0, 2.0, 1.0 and 1.0 mm from the inner side. Fig. 1(b) shows the magnetic rotor which consists of permanent magnet rings and

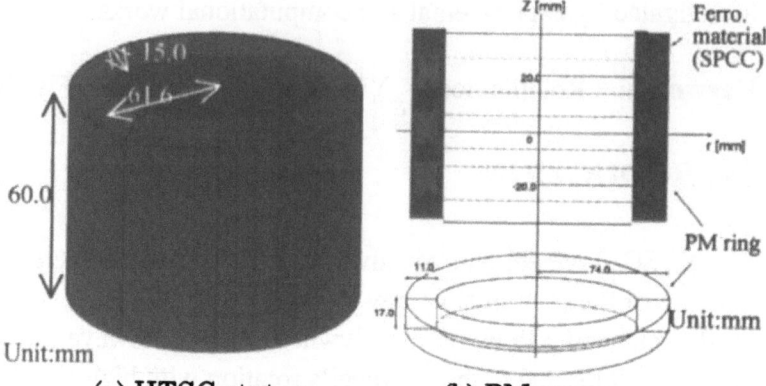

(a) HTSC stator   (b) PM rotor

Fig. 1 : Schematic drawing of SMB

ferromagnetic rings (SPCC). Fig. 2 shows the numerical results of the time-variation of rotation speed when the initial load is 200N for several cases of γ. The rotation speed decreases gradually for all cases. Table 1 shows the value of the rotation loss calculated from the results of Fig. 2. Good agreement can be found with the experimental ones[3].

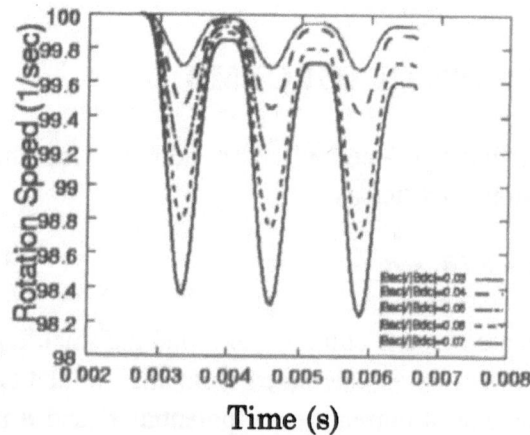

Fig. 2 : Time-variation of rotation speed of rotor when the initial load is 200N.

Table 1: Comparison of simulated rotation loss with the experimental ones[1]

F^{init} (N)	Exp. (W)	Simulated rotation loss (W)				
	γ			γ		
	~0.05	0.03	0.04	0.05	0.06	0.07
200	0.1	0.0482	0.0851	0.140	0.204	0.284
300	0.3	0.0510	0.0891	0.142	0.199	0.285

HTSC THIN FILM FOR SUPPRESSION OF ROTATION LOSS

For the technique of the suppression of rotation loss of HTSC bearing, we propose the use of YBCO thin films which have very high critical current density. We performed the preliminary experiment for observation of the suppression degradation of HTSC levitation force in AC filed with use of HTSC thin film. Fig. 3 shows the apparatus of experiment instruments. The AC field is induced on

YBCO thin films and bulk and the levitation force is measured by the strain gages attached on the cantilever. The remnant magnetization of permanent magnet is 0.34T. Fig. 4 shows the arrangement of five YBCO thin films on YBCO bulk. The thickness of YBCO thin film is 10 μm.

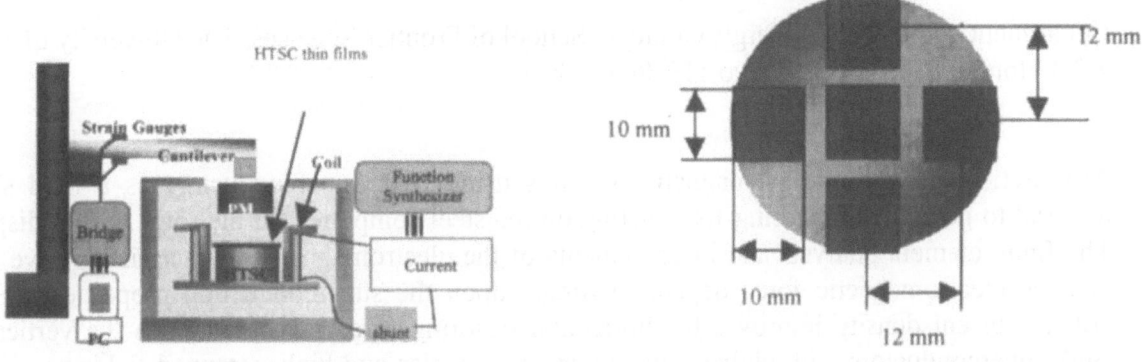

Fig.3 : Apparatus of experiment   Fig. 4 : Five YBCO thin films on an YBCO bulk

Fig. 5 shows the experimental results of degradation of levitation force with and without YBCO thin films when PM is moved from 15 mm distance to 5mm and then the AC field is induced by the coil with the amplitude of AC field is 36.4G and 182.0G. From these results, it is found that the YBCO HTSC thin films suppress degradation of levitating force in AC field for both cases.

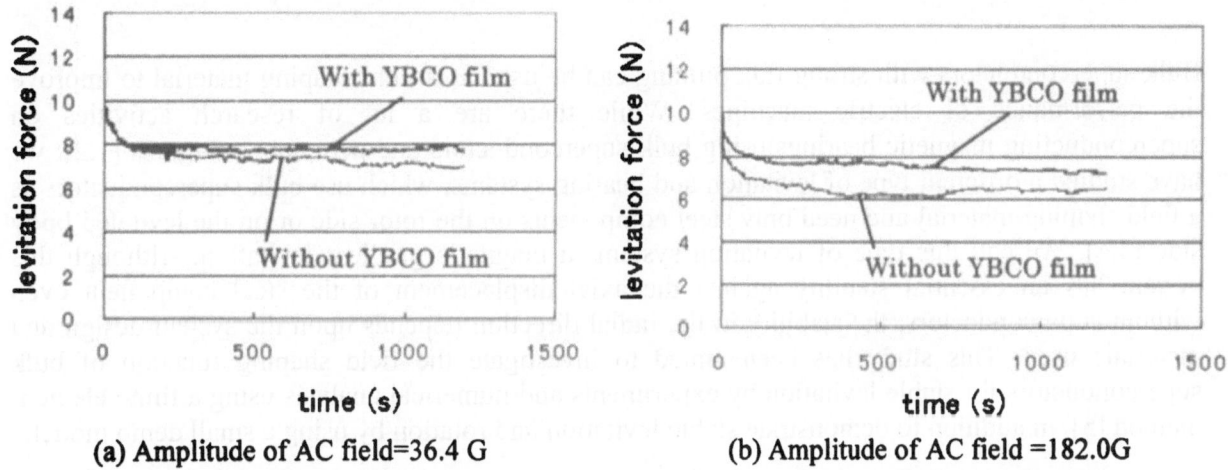

(a) Amplitude of AC field=36.4 G   (b) Amplitude of AC field =182.0G

Fig. 5 : Time evolution of degradation of levitation force with and without YBCO thin film

CONCLUSION

- Rotation loss simulation code of HTSC bearing was developed and good agreement with experiment was obtained.
- In preliminary experiment, we observed HTSC thin films suppress degradation of levitation force in AC field.

REFERENCES

[1] M. Uesaka, Y. Yoshida, N. Takeda and K. Miya, Int. J. Appl. Electromag. Mater., Vol. 4 (1993) pp. 13-25.
[2] T. Sugiura, H. Hashizume and K. Miya, Int. J. Appl. Electromagn. Mater., Vol. 2 (1991) pp. 183-196.
[3] H. Kameno, Y. Miyagawa, R. Takahata and H. Ueyama, Proc. of ASC'98, (1998) to be published.

Levitation Characteristics of Magnetic Bearings using Bulk Superconductors as a Field Shaping Material

Hiroyuki Ohsaki

Department of Advanced Energy, Graduate School of Frontier Sciences, The University of Tokyo, 7-3-1 Hongo, Bunkyo-ku, Tokyo 113,8656, Japan

Abstract: We have studied magnetic bearings using bulk superconductors as a field shaping material to generate a restoring force acting on the steel component in the rotor as it is displaced. The finite element analysis and measurements of the electromagnetic characteristics have shown that the electromagnetic force depends strongly upon the superconducting properties. A higher critical current density improves the horizontal restoring force but deteriorates the vertical one. Bulk superconductors with higher critical current densities and higher trapped fields can increase the strength of the restoring electromagnetic force both in the axial and radial directions up until the magnetic saturation of the iron occurs.

Keyword: Magnetic bearings, Bulk superconductor, Field shaping, Levitation, Magnetic gradient

INTRODUCTION

Bulk superconductors with strong flux pinning can be used as a field shaping material to improve the performance of electric machines. While there are a lot of research activities on superconducting magnetic bearings using bulk superconductors and permanent magnets [1,2], we have studied a different type of levitation and bearing systems, which use bulk superconductors as a field shaping material and need only steel components on the rotor side or on the levitated body side [3,4]. We call this type of levitation systems a magnetic gradient levitation. Although this system has an essential stability against the axial displacement of the steel component even without superconductors, the stability in the radial direction depends upon the system design and materials used. This study has been aimed to investigate the field shaping function of bulk superconductors for stable levitation by experiments and numerical analysis using a finite element method [5], in addition to demonstrate stable levitation and rotation by using a small demo model.

Fig. 1 shows an image of a flywheel supported by a mixed-μ type superconducting magnetic gradient levitation system. Vertical force is stable even without magnetic shielding or field shaping material. Radial force is stabilized by the magnetic shielding material. Magnetic flux generated by the coil is effectively used for generation of electromagnetic force.

EXPERIMENTAL

A superconducting magnetic gradient levitation system using a trapped field magnet (a bulk magnet) can be composed of only bulk superconductor and a steel component as shown in Fig. 2. After field cooling superconducting current is induced in the bulk superconductor, depending on the Jc-B characteristics. As in the case of the mixed-μ system, the bulk superconductor has a field

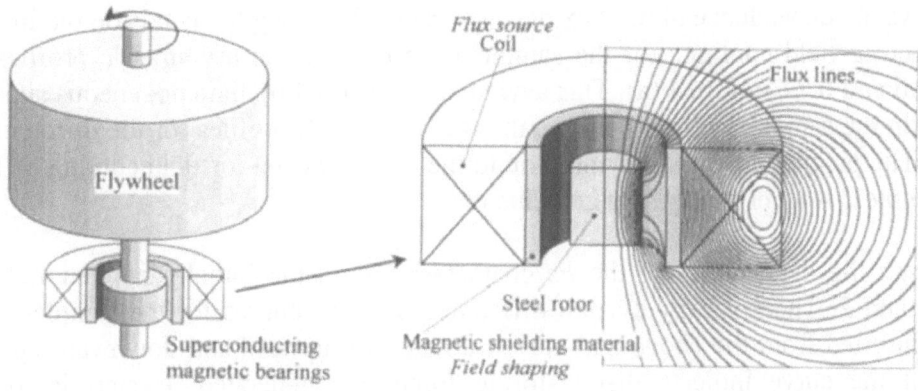

Fig. 1. Superconducting magnetic bearings based on the magnetic gradient levitation concept (mixed-μ type)

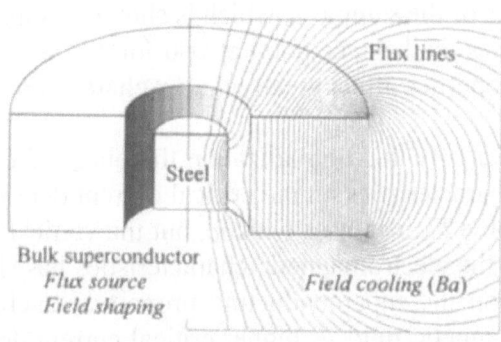

Fig. 2 Superconducting magnetic bearings using a trapped field magnet

Table 1. Specifications of the experimental system

Component	Specifications
Superconducting ring	YBCO bulk superconductor
	Inner diameter: 16mm
	Outer diameter: 46mm
	Height: 15mm
Steel cylinder	Diameter: 10mm
	Height: 10mm, 15mm, 20mm

shaping function as well for stabilization against a radial displacement. Experimental studies were made with this configuration. The specifications of the superconductors and the steel cylinder used in the experiments are listed in Table 1. The inner diameter of the YBCO superconductors is 16mm and the outer diameter is 46mm. The height is 15mm. The steel cylinders are 10mm in diameter and 10mm, 15mm or 20mm in height. In the electromagnetic measurements we used the steel cylinder of 10mm in height. Fig. 3 shows pictures of materials used in the superconducting bearing system and a steel cylinder levitated stably in the superconducting ring. We used two YBCO samples, which have a little different characteristics from each other.

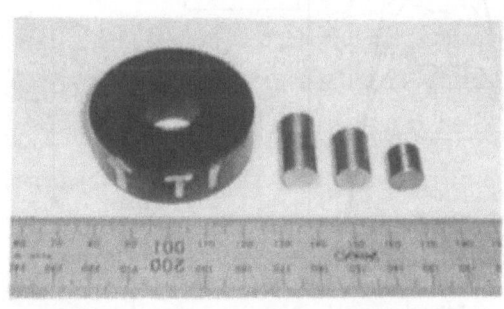

(a) Superconducting bulk ring and steel pieces

(b) Magnetic levitation of the steel cylinder

Fig. 3. Superconducting magnetic bearings using a superconducting bulk ring and a steel cylinder

Fig. 4 shows the dependence of the flux density Bz profiles along the y-axis on the initially applied field Ba during field cooling. For the sample 1 we can see the asymmetric profiles for applied fields of 0.6 and 0.7T in Fig. 4 (a). This asymmetry is caused by inhomogeneous superconducting properties in the sample. Fig. 4 (b) shows the flux density Bz profiles for the different sample, the sample 2. The profiles are almost symmetric to the axis $x=0$ even for field cooling of 1.0T. We can say the sample 2 is a fairly good superconductor.

Fig. 5 shows the dependence of the Fz characteristics on the initially applied field Ba for the samples 1 and 2. There is only a small difference between curves for the samples 1 and 2. The steel was inserted from the above and moved downward, upward and downward again. Negative gradient of the curve indicate that restoring forces are generated. Hysteresis loops are also observed. A slight shift of the point that gives $Fz=0$ is also due to the magnetic hysteresis of the superconductors. Fig. 6 shows the dependence of the electromagnetic force Fx on the initially applied field Ba for the samples 1 and 2. We can see the difference in the obtained curves between two samples. In case of the sample 1 the point giving $Fx=0$ moves with increasing Ba. This is caused by the asymmetric superconducting properties of the sample 1, which is shown in Fig. 4 (a). On the other hand the better sample, the sample 2, has the balance point at $x=0$ for all the applied fields Ba. From these figures the superconducting properties affect greatly the Fx characteristics.

Fig. 7 shows electromagnetic force normalized by Ba^2. By this normalization the obtained curves show the dependence of the electromagnetic force characteristics on the critical current density Jc. A higher Ba corresponds to a lower Jc. The radial force Fx increases with Jc, but the vertical force Fz decreases with Jc. The finite element analysis of the electromagnetic characteristics has shown that the electromagnetic force depends strongly upon the superconducting properties, such as a critical current density and its dependence on the magnetic field. A higher critical current density improves the horizontal restoring force but deteriorates the vertical one. Fig. 8 shows FEM analysis results of flux lines and superconducting current region, assuming Bean model of $Jc=0.9 \times 10^8 A/m^2$. After field cooling superconducting current is induced in the bulk superconductor, depending on the Jc-B characteristics, as shown by $+Jc$ areas in the figure. The insertion of the steel piece induces current in the area indicated by $-Jc$. As in the case of the mixed-μ system, the bulk superconductor has a field shaping function as well for stabilization against a radial displacement.

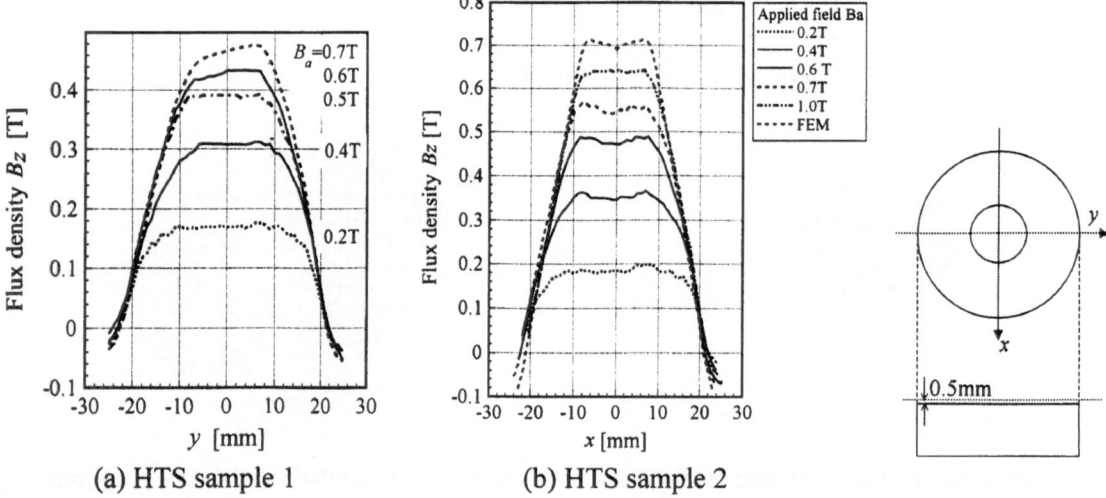

(a) HTS sample 1   (b) HTS sample 2

Fig. 4. Dependence of the flux density Bz on the initially applied field Ba during field cooling

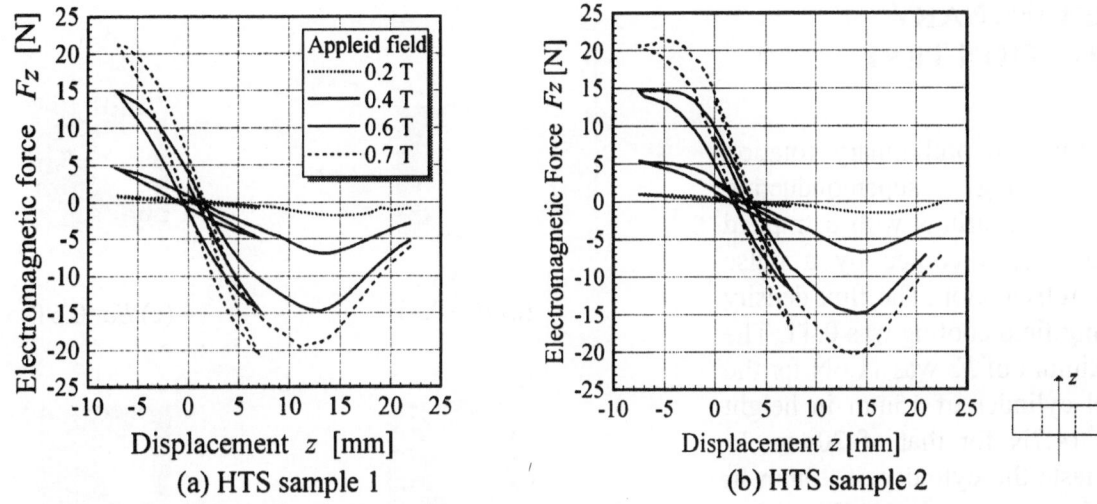

Fig. 5. Dependence of the electromagnetic force characteristics Fz on the initially applied field Ba

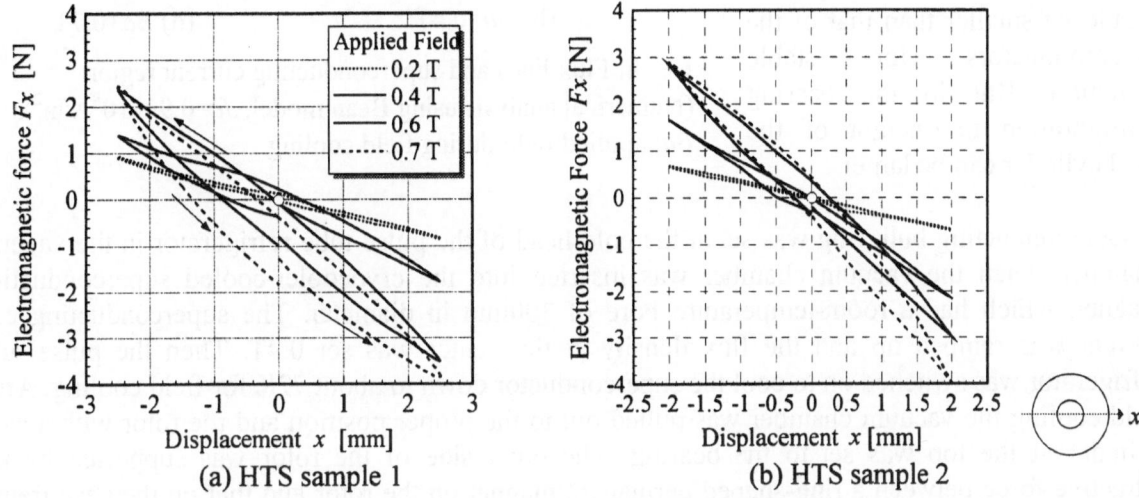

Fig. 6. Dependence of the electromagnetic force characteristics Fx on the initially applied field Ba

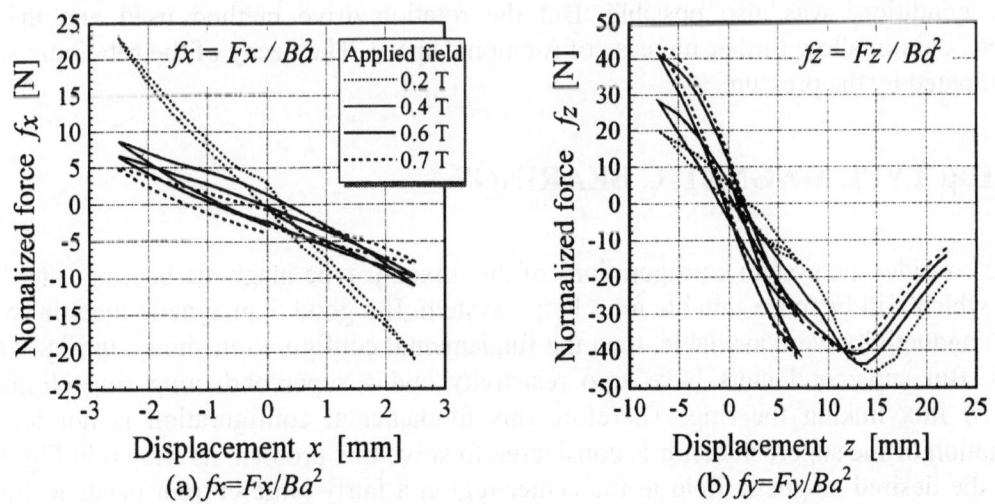

Fig. 7. Dependence of the normalized electromagnetic force characteristics on the initially applied field Ba. (Jc dependence of Fx and Fy)

PRELIMINARY ROTATION TEST

We made a preliminary rotation test using superconducting magnetic bearings with a trapped field magnet cooled by a pulse tube refrigerator. The flux density during field cooling was 0.4T. The maximum of Fz was 11.6N for the steel cylinder of 15mm in height and 14.7N for that of 20mm. In the tests the cylinder of 20mm in height was used. In the linear configuration of levitation the height of the steel component should be smaller than that of the superconductors for stable levitation. But in the bearing configuration the height of the steel cylinder can be larger.

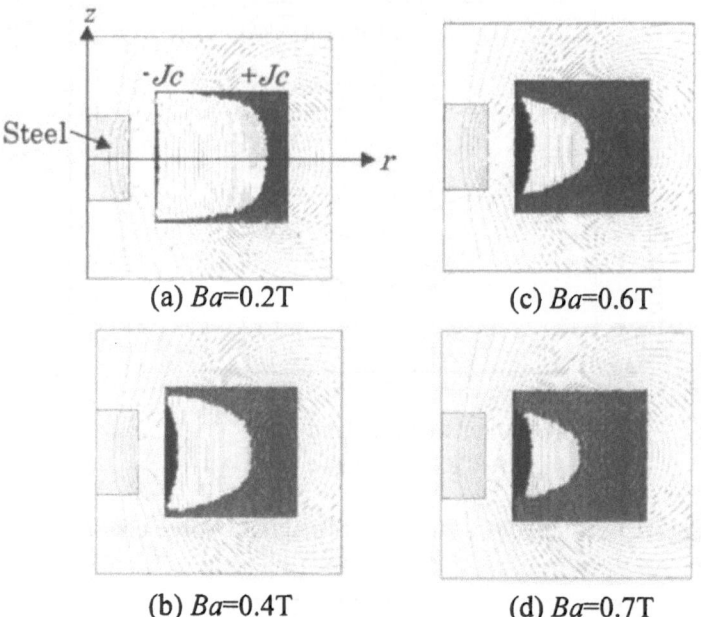

(a) Ba=0.2T   (c) Ba=0.6T

(b) Ba=0.4T   (d) Ba=0.7T

Fig. 8. Flux lines and superconducting current region (Numerical analysis using Bean model, Jc=0.9$\times$10^{8}A/m^2). Ba: applied field during field cooling.

A superconducting bulk ring was set at the cold head of the pulse tube refrigerator in the vacuum chamber. Then the vacuum chamber was inserted into the cryocooler-cooled superconducting magnet, which had a room-temperature bore of 300mm in diameter. The superconducting coil current was ramped up and the flux density at the center was set 0.4T. Then the pulse tube refrigerator was switched on to cool the superconductor down to about 72K for field cooling. After field cooling the vacuum chamber was pulled out to the proper position and the rotor with a steel cylinder at the top was set to the bearing. The other side of the rotor was supported by the attractive force between a ring-shaped permanent magnet on the rotor and that on the base frame. The weight of the rotor was 350g and the rotor was driven by compressed air. The maximum rotational speed attained in the preliminary tests was 4,400rpm. Rotation test under the reduced pressure conditions was also possible. But the rotation drive method used and the vacuum conditions did not allow further increase of rotational speed. The decay of the rotational speed was still dominated by the pressure.

MIXED-μ TYPE MAGNETIC BEARINGS

Here we consider the system configurations of the mixed-μ type magnetic bearings and levitation system, which will be more suitable for a larger system. If a good diamagnetic material with finite electric conductivity were available, then the fundamental configuration shown in Fig. 9 (a) could be used. But superconductors have zero resistivity and a superconducting ring eliminates the change of flux linking the ring. Therefore this fundamental configuration is not feasible. So segmentation of the superconductor is considered to solve this problem as shown in Fig. 9(b). But to form the desired magnetic field in the center region a fairly large current needs to flow in the superconductors. To reduce the current load of the superconductor a coil structure shown in Fig. 9 (c) is considered, which comes from the current distribution in the bulk superconductor for the bulk magnet type system (Fig. 9 (d)).

CONCLUSIONS

We studied the field shaping function of bulk superconductors for stabilization of levitation force characteristics. From the measurements and finite element analysis of electromagnetic force and flux density, we found that a higher critical current density improves the radial restoring force but deteriorates the vertical one, and homogeneous superconducting properties are necessary for good levitation characteristics. Stable levitation and rotation were demonstrated.

Acknowledgments. This work was supported in part by the Research for the Future Program, Japan Society for the Promotion of Science.

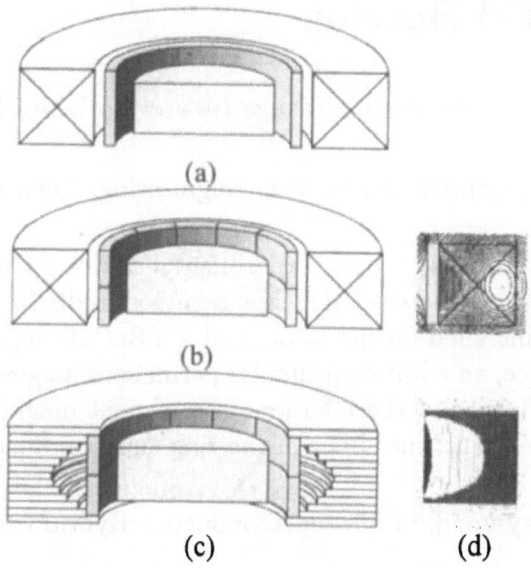

Fig. 9. System configurations of the Mixed-μ type magnetic bearings

1. J.R. Hull, IEEE SPECTRUM **34**, pp.20-25 (1997).
2. Y. Miyagawa, H. Kameno, R. Takahata and H. Ueyama, IEEE Trans. Appl. Superconductivity **9**, pp.996-999 (1999)
3. H. Ohsaki M. Takabatake and E. Masada, IEEE Trans. Appl. Superconductivity **7**, pp.908-911 (1997).
4. H. Ohsaki, Y. Fukasawa and N. Nozawa, in *Advances in Superconductivity XI*, edited by N. Koshizuka and S. Tajima (Springer-Verlag, Tokyo, 1999), pp.1333-1336.
5. H. Ohsaki and N. Nozawa, European Conference on Applied Superconductivity (EUCAS'99) (Sep. 14-17, 1999, Sitges, Spain), 4-32.

The Basic Characteristics of the Pinning Force and Flux Density Distribution of the HTSC-Permanent Magnet Hybrid Bearing

Shunsuke Ohashi, Takuya Ito and Yoshihisa Hirane

Department of Electrical Engineering, Kansai University, Suita, Osaka 564-8680, Japan

Abstract: A basic experimental device for a bearing system using high temperature superconducting material (HTSC) is introduced. In this system, a circular shaped permanent magnet is installed on the rotor, and a YBaCuO plate is used for the stator. To increase the levitation force, an additional circular permanent magnet is installed under the YBaCuO plate in the stator. Influence of the additional permanent magnet on the flux density distribution is shown. From the result, the shape of the ring superconductor is designed to increasing the lateral stability.

Keywords: Bulk superconductor, Hybrid bearing, Magnetic levitation

INTRODUCTION

The development of the superconducting materials increases their application to the various systems[1]. One of the most useful characteristics of the HTSC is the pinning effect. A levitation system using the superconducting material and permanent magnets shows good stability without gap control, the application to the bearing system is considered. But its levitation force is not so large. Although a levitation system using the repulsive force of one pair of permanent magnets has large levitation force, it is unstable for lateral direction. Thus the HTSC-permanent magnet hybrid bearing system is introduced[2]. In this system[3], a circular shaped permanent magnet is installed on the rotor, and HTSC plate (YBaCuO) is used for the stator. To increase the levitation force, an additional circular permanent magnet is installed under the HTSC plates. Influence of the additional permanent magnet on the flux density distribution is shown. From the result, the shape of the ring superconductor is designed to increase the lateral stability. The flux density trapped by the HTSC superconductor is measured, and levitation characteristics of the system is discussed.

EXPERIMENTAL DEVICE

Fig.1 shows the side view of the experimental device and the arrangement of the permanent magnets and HTSC plates. Specifications of the permanent magnet and HTSC plates are shown in Table 1. Nd-Fe-B permanent magnet is used for PM1 and PM2, and YBaCuO for a HTSC plate. The permanent magnet has three poles. PM2 is installed on the bottom of the rotor, and HTSC plates and PM1 are set in the stator filled with liquid N_2. To show the influence of the bulk shape, two kinds of bulk superconductor are used. Its diameter is 85 mm, and the thickness is 15 mm for column shape, 10 mm for ring one.

RESULTS

Fig.2 and 3 show the flux density distribution generated by PM1. Fig.4 and 5 show that by PM1 and 2. The distance between PM1 and PM2 is set at 25mm. g indicates the vertical distance from the surface of PM1. The flux density is measured by hall sensor. As PM1 and 2 is set to repulsive state, flux density with PM1 and 2 of the vertical direction becomes small. As shown in Fig.5, flux density of the lateral direction becomes small at $x \leq 12.5$ mm. Thus the inner diameter of the ring shaped superconductor is designed as 25 mm. Fig.6 and 7 show the vector of the flux density distribution. From Fig.7, we can decide the position of the HTSC plate to increase stability for the lateral direction. Fig.8 shows the flux density trapped by the HTSC plate. The gap between PM1 and HTSC plate is 8 mm.

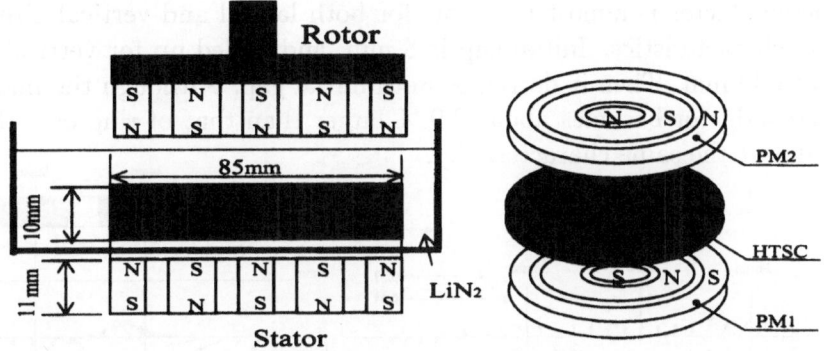

Fig. 1: HTSC-permanent magnet hybrid bearing system

Table 1: Specificatons of the permanent magnets and HTSC bulk plates

	Permanent magnet (PM1 and 2)	HTSC plate	HTSC plate
Material	Nd-Fe-B	YBaCuO	YBaCuO
Weight	300g	312.0g	506.3g
Dimension(mm)	$\phi75 \times 11$	$\phi85(25) \times 10$	$\phi85 \times 15$
	Three pole (Ring shape)	Ring shape	Column shape
Surface flux density	0.4T		

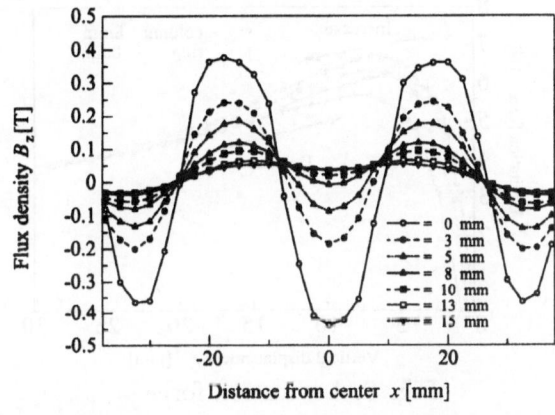

Fig. 2: Flux density distribution (PM1, vertical direction)

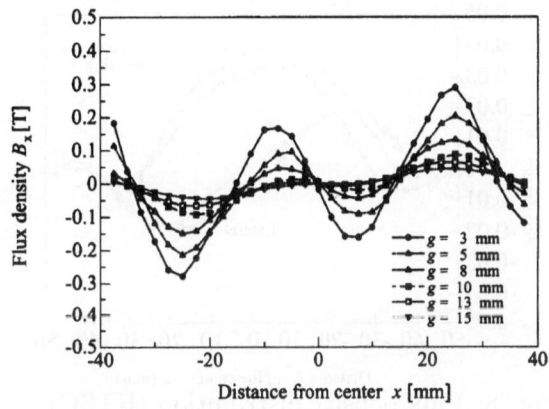

Fig. 3: Flux density distribution (PM1, lateral direction)

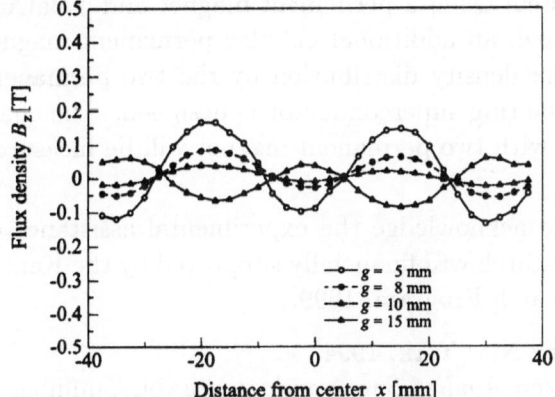

Fig. 4: Flux density distribution (PM1 and 2, vertical direction)

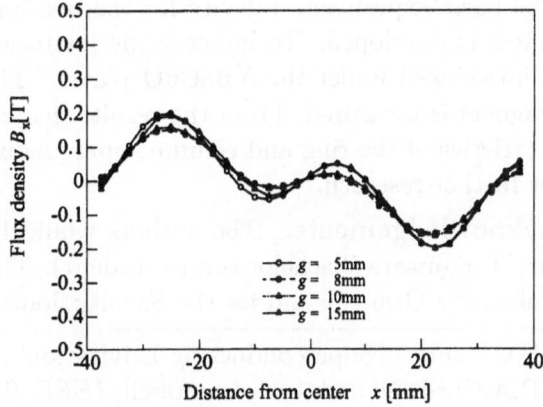

Fig. 5: Flux density distribution (PM1 and 2, lateral direction)

HTSC plates are excited and cooled keeping the gap 8 mm for an hour (field cooling). Flux density at $g=5$ mm is shown. To show the basic difference between the ring and column superconductor, permanent magnet PM2 in the rotor is not set here. Flux density distribution of the two superconductor is almost the same for both lateral and vertical direction. Fig.9 shows the levitation characteristics. Initial gap is 8 mm, and pulled up for vertical direction. Air gap increases up to 30 mm. Then it decreases until initial gap. Although the maximum force of the column superconductor becomes about 8.9 % larger than that of ring one, these two hysteresis loop shows almost the same characteristics.

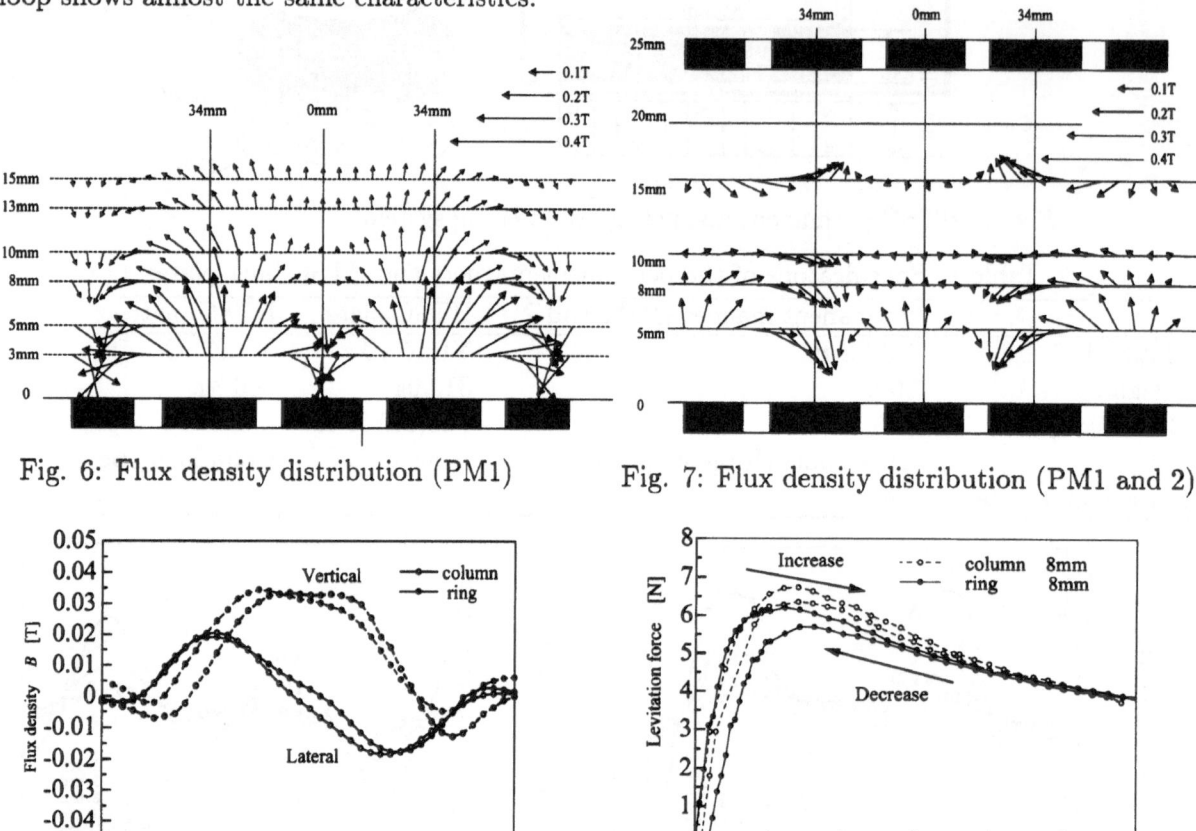

Fig. 6: Flux density distribution (PM1)

Fig. 7: Flux density distribution (PM1 and 2)

Fig. 8: Flux density distribution (HTSC)

Fig. 9: Levitation force

CONCLUSION

The basic experimental device for the bearing system using a permanent magnet and YBaCuO plates is developed. To increase the levitation force, an additional circular permanent magnet is introduced under the YBaCuO plates. The flux density distribution by the two permanent magnets is measured. From the result, shape of the ring superconductor is designed. The characteristics of the ring and column superconductor with two permanent magnet will be measured for further research.

Acknowledgements. The authors would like to acknowledge the experimental assistance of Mr. J. Fujiwara (bachelor course student). This research was financially supported by the Kansai University Grant-in-Aid for the Faculty Joint Research Program, 1999.

1. F.C. Moon, "Superconducting Leivitation", Wiley New York, 1994.

2. D.A.Cardwell and A.M. Campbell, *IEEE Trans. on Applied Superconductivity* vol.7, number 2, 1997, pp.924-927.

3. S.Ohashi, S.Tamura and Y.Hirane, *IEEE Trans. on Applied Superconductivity* vol.9, number 2, 1999, pp.988-991.

Investigation of Characteristics of Axial Gap Type Superconducting Magnetic Bearing for 1kWh Flywheel

H. Kawashima[1], S. Nagaya[2], T. Suga[2], S. Unisuga[1], K. Konno[1], M. Minami[3]

[1]Mitsubishi Heavy Industries, LTD., Takasago Machinery Works,2-1-1 Shinhama Arai-cho Takasago Hyogo Japan
[2]Chubu Electric Power Co. Inc., Electric Power R&D Ctr., 20-1 Kitasekiyama Ohdaka-cho Midori-ku Nagoya Japan
[3]Mitsubishi Heavy Industries, LTD., Takasago R&D Ctr., 2-1-1 Shinhama Arai-cho Takasago Hyogo Japan

Abstract: We have tackled with the improvement of the superconducting magnetic bearings and manufactured the newly developed facility which has possessed the cooling device with cryogenic refrigerator and the driving device for the completely floating rotational elements, and can be utilized to research and development for various type of the bearings. We have confirmed the small loss of 0.26W about the axial gap thrust bearing of 238mm diameter at speed of 3000 min^{-1} and load of 281N. We introduce the test facility and report the test results.

Keywords: Superconducting magnetic bearing, Cryogenic refrigerator, Rotating loss

INTRODUCTION

We have already achieved to manufacture the 1kWh flywheel with the superconducting magnetic bearing and stored the energy of 1.4 kWh at 20,000 min^{-1}[1]. We grasped the individual characteristics about the axially gap type superconducting magnetic bearing and we will utilize these results for the next development.

THE DEVELOPMENT OF THE TEST EQUIPMENT

The developed test equipment is shown in Fig.1. For the precise measurement of the loss we measured the decreasing rate of the speed per unit time of the rotating part and calculated loss values with the difference of inertia energies. Rotating part can be detached from the driving system with the electromagnetic clutch which has synchronous and concentrating mechanism and attached for the acceleration and deceleration without stopping. For the purpose of ignoring the windage loss and the purpose of the thermal insulation in the cryogenic part, we have made keeping the vacuum condition of below 10^{-3} Pa with the vacuum pump. We have adopted the cooling system with cryogenic refrigerator, and easily obtained the lower temperature than subcooled nitrogen. We have carried out most precise and stable feed back control for the temperature of the superconductor in a short time with directly measurement of surface temperature and additional heating control of the cold head in the cryogenic temperature. The rotating part can be lifted up with adsorption force of the electromagnetic clutch and vertically transferred with the motor to set the magnetic field condition such as null or active at initial cooling operation.

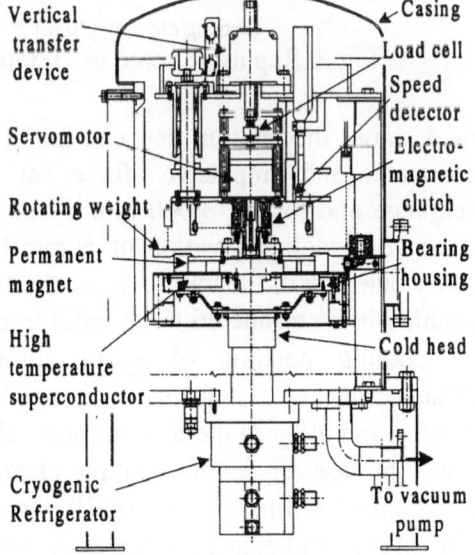

Fig.1. Test Equipment

SPECIFICATINS OF THE TEST BEARING

The maximum diameter of the permanent magnet assembled in the rotating disc is 238 mm. Its material is Pr-Fe-B. The magnetic field of its surface is 0.5 T at room temperature and the same magnetic field is gained at cold condition. The material of high temperature superconductor assembled in the stationary part is a YBCO. Its critical current density at 77K, 0T is $3 \times 10^4 A/cm^2$.

THE TEST RESULTS

We carried out the examination of the loss and repulsive force about the influence of cooling temperature, initial cooling magnetic field, bearing load and the rotating speed [2] [3]. Fig.2 shows the relation between repulsive force and bearing clearance to the cooling temperature from 80K to 50K. As for the colder condition, the repulsive force became bigger and the hysteresis became smaller. Fig.3 shows the repulsive force at the clearance of 5.5mm and the loss at rotating speed of $3,000min^{-1}$. The temperature of superconductor is 65K, and the mass of rotating part is 20kg and 29kg. The loss at the temperature of 50K was nearly one third of the temperature of 80K.

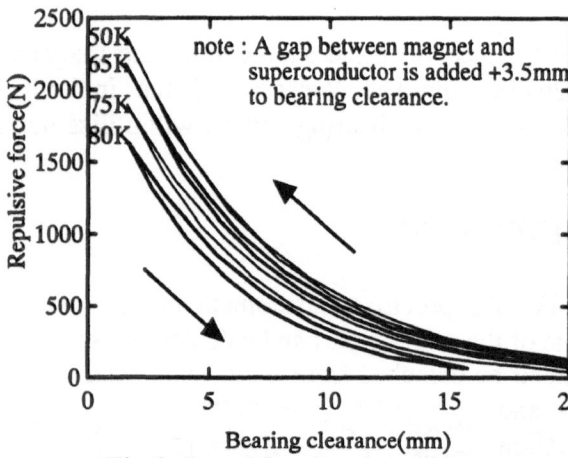

Fig.2. Repulsive force vs. Temperature

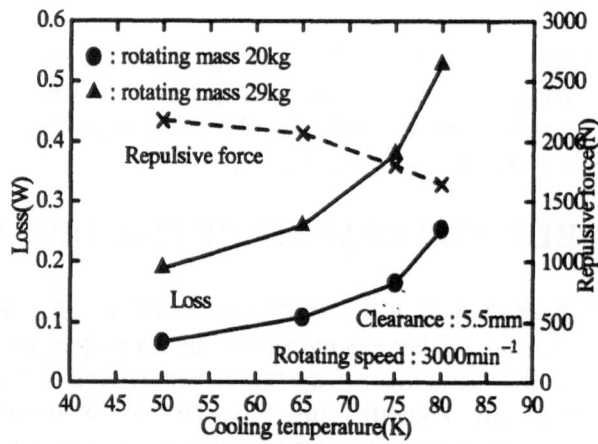

Fig.3. Loss vs. Temperature

Fig.4 shows the relation between initial cooling clearance and repulsive force at 5.5mm clearance and loss at 3,000 min^{-1}. The rotating loss at the cooling position of 50mm becomes nearly one fifth of the loss of 5mm. At the actual machine it is considered to be valid to set up at the cooling position of about 20mm. The influence of the bearing load with rotating mass of 20kg and 29kg to the loss becomes about 2.4 times in case of initial cooling clearance of 50mm and 3.8 times in case of 10 mm at 3,000 min^{-1} in spite of being 1.5 times increasing of the bearing pressure. Fig.5 shows the influence about the loss by the arrangement of the superconductor to the circumferential direction. At the rotating speed of 3,000 min^{-1} and the

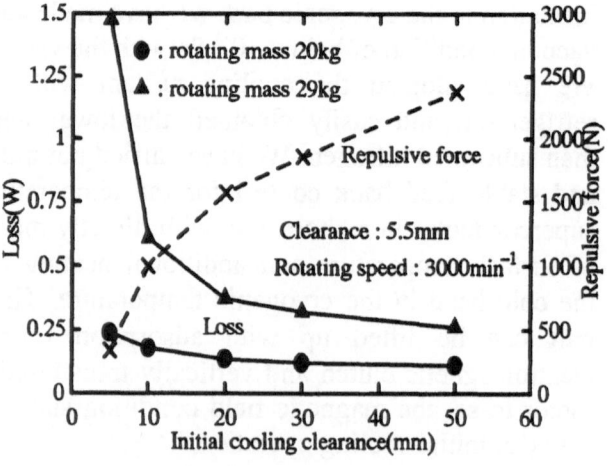

Fig.4. Loss vs. Initial cooling clearance

cooling position of 50 mm, we confirmed that the loss of three pieces of the superconductor was very high about 21 times of the loss of nine pieces of the superconductor arranged continuously to the circumferential direction. Reducing the number of superconductor to 2/3 becomes almost same condition as increasing the rotating mass to 1.5 times at the point of bearing pressure and the floating clearance which resulted 13.7 mm and 14.3 mm respectively, but as for the loss, the former became more increasing about 3 times than later. The characteristics of loss and rotating speed are shown as parameters of cooling temperatures

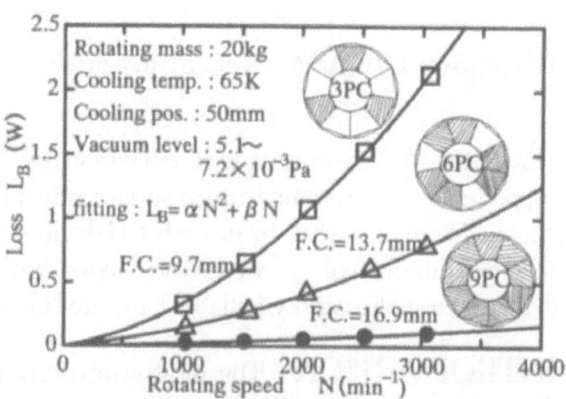

Fig.5. Loss vs. Superconductor Arrangement

in Fig.6. The test results of free running from 3,000 min^{-1} is shown in Fig.7. It has taken 68 hours and 15 minutes until stopping of the rotating part of the inertia moment of 0.434kgm^2. The change of the loss is fitted at the 1st power of the rotating speed which is represented as the hysteresis loss and square of the rotating speed which is represented as the eddy current loss. We supposed that the coefficients of each of the clauses of the 1st power and the square were constant without depending on the rotating speed. We found a coefficient in the least square method from the measurement value, and we confirmed that the good approximation was gained showing in Fig.6.

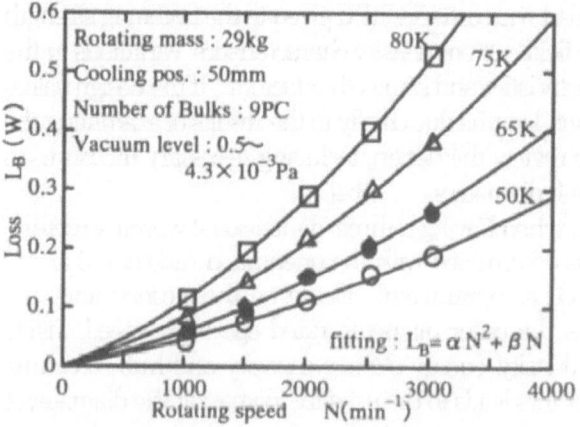

Fig.6. Loss vs. Rotating speed

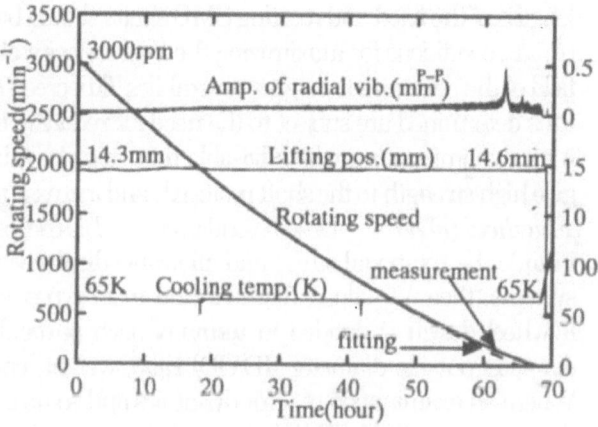

Fig.7. Free Run Trend

CONCLUSION

(1) We confirmed that the temperature of the superconductor and clearance of the bearing for the initial cooling condition were adequately each 65 K and 20 mm in the aspect of the practical use.

(2) As for the loss of the bearing we confirmed that the increasing ratio of loss was more than one of bearing load and the superconductor arrangement should be continue for the circumferential direction.

(3) About the change of loss to the speed, the good approximation could be gained with dividing into the clauses which were proportional to the 1st power and the square for the rotating speed.

References:
1.M.Minami, S.Nagaya, H.Kawashima, 4th NES&C at Osaka Univ. '99 : 349
2.H.Kameno, R.Takahata, Y.Miyagawa, H.Ueyama, 55th Meeting of Cryogenic and Superconductivity'98
3.H.Higasa, H.Kawauchi, S.Nakamura, N.Itoh, 49th Meeting of Cryogenic and Superconductivity'93

Findings for Optimal Design of Super speed Flywheel Energy Storage System with Superconducting Magnetic Bearing

Hiromasa HIGASA   Shikoku Research Institute Inc., 2109-8 Yasimanishi-machi, Takamatu, Kagawa, Japan

Abstract   The greatest technical task in developing an energy storage system using a superspeed flywheel supported by a high-temperature superconducting magnetic bearing (SMB) is to find how to optimize the system design as a whole by properly coordinating elemental and total designs to meet the practical conditions required of the system.   Conceivable topics of study involved in such total system design include the rotor dynamics, overall efficiency, initial cost, required installation space, and running cost of the system.

INTRODUCTION   The most symbolic of the rotor dynamics of the proposed system is that these dynamics are adversely affected by the superspeed capability given to the flywheel. This paper first proposes an outer rotor system as shown in Fig.1 and then discusses the relevant topics contained herein.

RELATIONSHIPS BETWEEN TOTAL AND ELEMENTAL DESIGNS INVOLVING ROTOR DYNAMICS, OVERALL EFFICIENCY AND INITIAL COST

A major factor in securing the viability of an outer rotor system is an advantage of a radial superconducting magnetic bearing (RaSMB) over an axial superconducting magnetic bearing (AxSMB).   [1-2]

SMB   Design conditions for this element include the space occupied by the SMB, its load capacity (the weight of the built-up rotor plus the required safety margin), rotation loss, time-based changes in the position of magnetic levitation, lateral stiffness, and operating speed which depends on the flywheel design. The diameters and lengths of the fixed and rotating SMB shafts should be decided with consideration given to the fastening strength of pins, conditions for maximizing the load capacity, electric field/current density characteristics, variations in the field of the permanent magnet assemblies, flux creep characteristics and some other factors. If the design values thus determined are subject to the need for reducing the shaft lengths due chiefly to the results of examining the rotor dynamics, it may be advisable that steps be taken to review the design, including necessary measures to give high strength to the shaft materials and increase the shaft diameters.   [3-4]

Flywheel (F/W)   Design conditions for this component, when having a three-dimensional woven structure, include the rotational stress and allowable displacement as determined with the orientation ratio of r, θ and Z systems, their Vf values, filament and resin types, etc. used as parameters. Based on these constraints, the flywheel design is decided in terms of such particulars as the outer diameter, rated operating speed, inside diameter-outside diameter (ID/OD) ratio, weight, energy density, energy storage capacity and hub structure. Where an examination of rotor dynamics and some other factors leads to the need for increasing the diameter of the rotating shaft, The FW design could be adjusted without involving any reduction of the rated tip speed.

Motor Generator (M/G)   Design conditions for this component include the diameters of the fixed and rotating shafts, the rated speed of the rotating shaft, the efficiency of the motor generator (including a converter/inverter unit), losses in idle running, unbalanced magnetic force and some other elements which have been almost completely determined by SMB and F/W design values. Taking these elements into account, the motor generator should be designed in respect of such particulars as the voltage, number of poles and shaft length of the M/G and the frequency of the CONV/INV carrier. To meet the requirements for reducing the M/G shaft length or downsizing the whole M/G assembly in view of rotor dynamics or the requirements for reducing losses in view of total efficiency, it may be advisable that steps be taken to make design modifications, including an increase in M/G voltage to a high level.

Active Magnetic Bearing (AMB)   Needless to say, major design conditions for this component include the capability to stably and safely pass a critical speed within the variable-speed operation limits of the flywheel by suppressing vibrations, either synchronous or asynchronous with rotation, which result from unbalanced exciting force, etc. of the rotor during its rotation. Among other important conditions is the capability to reduce rotation losses and the required control power. These design conditions are intended to decide on required AMB components, such as a solenoid for the stator and laminated magnetic steel sheets on the rotor side. The AMB's role in the area of rotor dynamics is to avoid as far as possible any change in the diameter, length or rated operating speed of the rotating shaft which has been almost completely decided by SMB, F/W and M/G designs.

DESIGN TECHNOLOGY FOR SMB ELEMENTS

The SMB with an outer rotor system as shown in Fig.1 should be so designed as to meet the design conditions stated above by adding to it the RaSMB structure described in the typified drawing in Fig.2. The load capacity characteristics of the RaSMB largely depend on the number and size of permanent magnets that are set at the maximum value for the size of the superconductor assembly used. Fig.3 shows the results of calculating these numerical values of the magnets. When the RaSMB begins to turn, the superconductor assembly is affected by a weak alternating field superimposed over a DC magnetic field, which results from such causes as the chemical composition of the magnets, uneven magnetization, rotational asymmetry of the fixed and rotating shaft assemblies, and vibrations of the rotating shaft. This phenomenon causes a rotation loss in the SMB and time-based changes in the position of magnetic levitation. Fig.4, plots the results of measuring time-based changes in the position of magnetic levitation for the RaSMB, with the bearing preloaded in the case of measurement at Load Working Points III and IV. Table 1 summarizes the results of measuring time-based changes in the position of magnetic levitation. We are now developing a technology to improve the accuracy of a projection analysis method for the nonlinear characteristics of the SMB noted above. Fig.5, shows the results of measuring the lateral stiffness of an RaSMB (in a maximum load capacity class of 1,000N). As indicated in the graph, the lateral stiffness of the SMB increases roughly in proportion to its load capacity. [5-11]

DESIGN TECHNOLOGY FOR FLYWHEEL ELEMENTS

Under the NEDO-commissioned research program, we built an experimental flywheel with a diameter of 400 mm and an ID/OD ratio of 0.8 using the multi-ring method. The flywheel attained a breaking tip speed of 1,310 m/sec and an energy storage capacity of 195 Wh/kg Fig.6, is a distribution diagram (in calculated values) of the design strength margin (strength/stress) for a CFRP flywheel with a three-dimensional woven structure. The diagram indicates it is possible to design and fabricate a flywheel with an ID/OD ratio of 0.5, a tip speed of 1,500 m/sec and an energy storage capacity of 195 Wh/kg, using Toray T-1000G commercial material with its fiber strength set at 75% of the catalog value. Should T-1500 series materials be available, we believe that building a flywheel with a tip speed of 1,800 m/sec would be no longer a mere dream. [12-13]

DESIGN TECHNOLOGY FOR M/G ELEMENTS (INCLUDING CONVERTER / INVERTER UNIT)

Table 2 compares induction motor, PM motor and coreless PM motor designs. In this paper, the coreless PM motor is considered most suitable for the M/G unit of the proposed system, based on an examination of these motor options in respect of high efficiency, losses in idle running, heat from the rotor and whether the motors involve unbalanced magnetic force.

DESIGN TECHNOLOGY FOR AMB

To secure a wide range of operating speeds (e.g., 0.05~1), therefore, we have developed commercialization technologies for a vibration stabilizer using an active magnetic bearing (AMB) with special devices that could exceed two or more critical speeds. These devices include vibration control adaptable to estimated vibration frequency and zero-power nonlinear control, of which the former has a great vibration-damping effect and involves only a small rotation loss and low control power. The vibration stabilizer is now at a prototype building stage. [14-15, 8]

Acknowledgement

For the last few decades, the author has actively participated in designing, building and testing 50-Wh and 200-Wh energy storage systems on a research contract with Shikoku Electric Power Company. Since 1995, I have been assigned to build a high-temperature superconducting magnetic bearing and design and evaluate a total system as one of the researchers engaged in a research and development program contracted out by NEDO. Under the NEDO-commissioned program, my research activities have been focused on testing and evaluating a 500-Wh energy storage system. In publishing this paper, I wish to thank all parties involved for the valuable assistance they have rendered me in carrying out my research tasks.

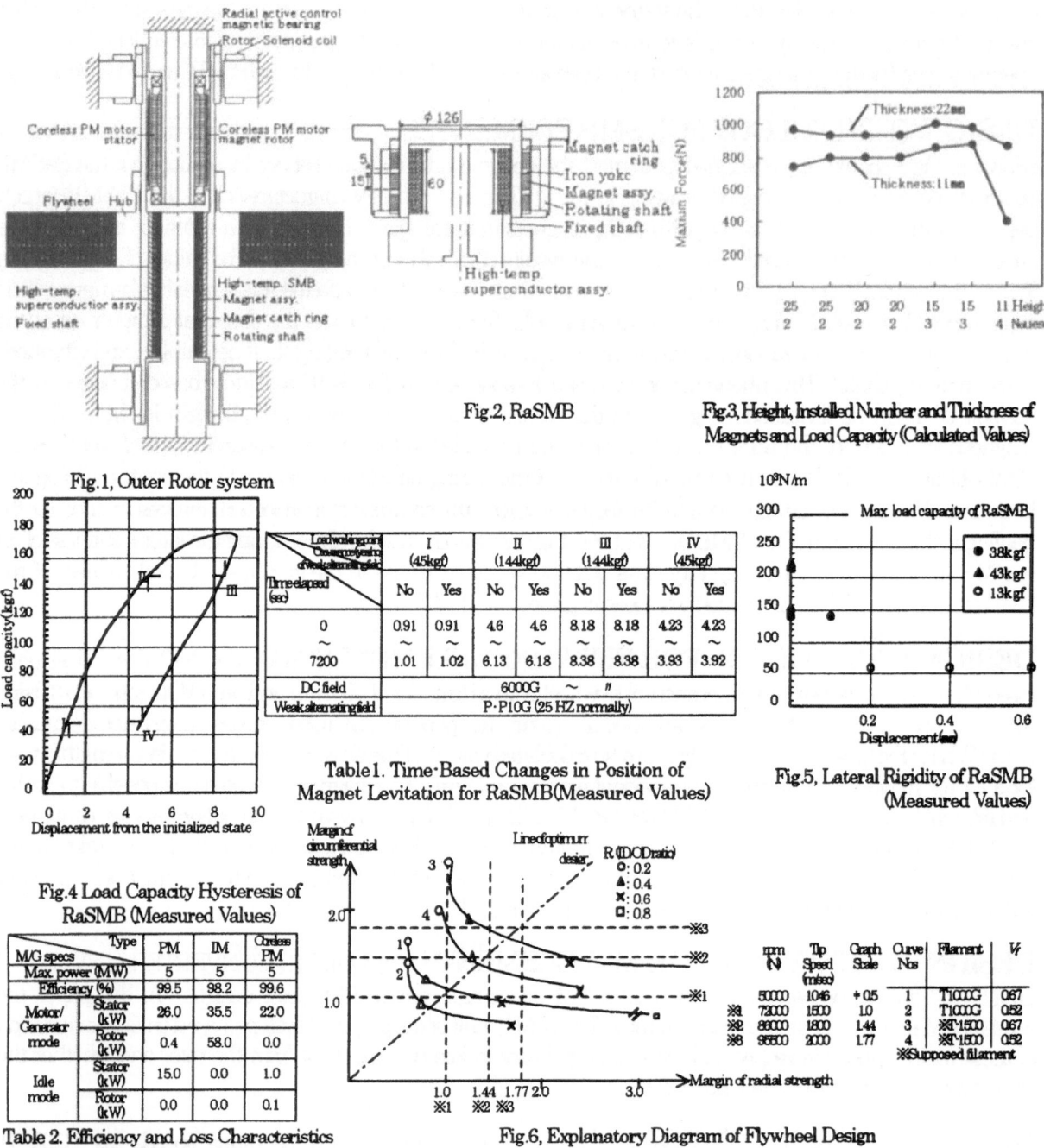

Fig.2, RaSMB

Fig.3, Height, Installed Number and Thickness of Magnets and Load Capacity (Calculated Values)

Fig.1, Outer Rotor system

Table 1. Time-Based Changes in Position of Magnet Levitation for RaSMB (Measured Values)

Fig.5, Lateral Rigidity of RaSMB (Measured Values)

Fig.4 Load Capacity Hysteresis of RaSMB (Measured Values)

Table 2. Efficiency and Loss Characteristics of M/G Units by Type of Motor

Fig.6, Explanatory Diagram of Flywheel Design

REFERENCES

[1] K.Miya, Y.Yoshida, Mathematics of Super Conductivity and it's Applications (Book), Yokendo, Tokyo, 1995
[2] H.Higasa, et al., TIEE, Japan 113-B, 7, 768-775, 1993
[3] M.Okano, Proc. 49th Cryogenic Eng. Conf., A2-12 (1993) in Japanese
[4] T.Hikihara, F.C.Moon, Phys. C250 (1995) 121
[5] M.Murakami, et al., Jpn. J. Appl. Phys.28 (1991) 11125
[6] Y.Luo, K.Miya, H.Higasa, et al., International. Sympo. Non-Linear Electro Magnetic System, Germany, 1997
[7] P.Z.Chang, F.C.Moon, J.R.Hull, J. Appl. Phys. (1990)
[8] H.Kameno, et al., ASC'98 in Palm Springs. CA. Sept., 1998
[9] H.Kameno, R.Takahata, et al., ISS'98 in Fukuoka, Japan Nov.1998
[10] K.Demachi, T.Uchimoto, K.Miya, et al., ISS'99 Oral Session SA-3
[11] J.R.Hull, ISS'99 Oral Session SA-1
[12] J.D.Eshelly, Proc. R. Soc. (London) 241A (1957) 376
[13] H.Hotta, K.Murayama, Proc. ICCSN Vol.2, (1987) 408
[14] K.Nonami, H.Higasa, et al., ISS'99, Poster Session SAP-8
[15] K.Nonami, Z.Lin, Trans. Japan Soc. M. E; C65, 639, 1999, In Press

A FEASIBILITY STUDY OF MODELING AND CONTROL FOR 10MWh CLASS ENERGY STORAGE FLYWHEEL SYSTEM USING SUPERCONDUCTING MAGNETIC BEARING

Yajun Zhang[1], Kenzo Nonami[1*] and Hiromasa Higasa[2]

[1]Department of Electronics and Mechanical Engineering, Faculty of Eng.
Chiba University 1-33 Yayoi-cho, Inage-ku, Chiba 263-8255, Japan
[2]Shikoku Research Institute Inc, 2109-8 Yashimanishi-machi, Takamatsu, Kagawa, Japan

Keywords: Flywheel, Energy Storage, Superconducting Magnetic Bearing, Stabilizing Control

INTRODUCTION

As difference of electric power required in day and night has been increasing, an energy storage and a smoothing of electric power in day and night become very important. If energy storage with high capacity could be able to realize, efficiency of electric power system will become higher. An energy storage flywheel system using superconducting magnetic bearing is one of the most useful method of energy storage[1].

This paper is concerned with the feasibility study based on computer simulation to realize the 10MWh class energy storage flywheel system. The model of system is constructed based on the results of ANSYS which is a kind of CAE software based on finite element method. Based on the eigenvalue of one dimensional finite element model is almost the same as the result of ANSYS analysis, we constructed one dimensional finite element model of flywheel system. In order to compare one dimensional finite element model with the model of ANSYS, the effects of superconducting magnetic bearing (SMB), active magnetic bearing (AMB) and coreless motor are not concerned.

Then, we check the eigenvalue of one dimensional finite element model and the result of ANSYS analysis in the range of control (within 100Hz). The eigenvalue of one dimensional finite element model and that of ANSYS's model are almost the same in the range of control. So, the one dimensional finite element model is used as basic model for control. Based on the basic model of control, we had done the stabilizing control of flywheel system taking into account the effects of SMB and AMB.

MODEL OF ANSYS AND VIBRATION ANALYSIS

Aim of vibration analysis of flywheel system is to compute the eigenvalue and natural mode of the flywheel system. A gyroscopic effect of the flywheel in vibration analysis is taken into account. When the vibration analysis of flywheel system is computed, the effects of SMB, AMB and coreless motor are not concerned.

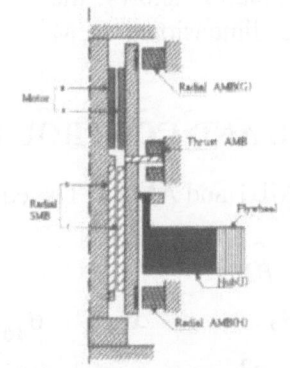

Fig.1 Cross sectional view of flywheel system

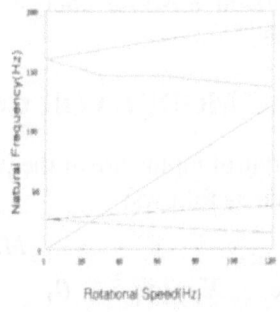

Fig.2 Realtion between Eigenvalue and rotational speed

The model used for ANSYS analysis is given in Fig.1 as a right half cross-sectional view of the right-half plane. The results without gyroscopic effect are given in Table.1. The relation between eigenvalue and rotational speed with gyroscopic effect is given in Fig.2. In Fig.2, the two natural frequencies are nutation and precession frequency. The total weight of the whole flywheel system is about 107t and the height is 6.86m.The weight of flywheel itself is approximately 45.7t and the maximum diameter of the flywheel is 4.8m. One example of ANSYS analysis is given in Fig.3.

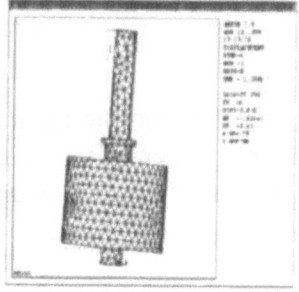

Fig.3 Cross sectional view of flywheel system

Table.1 Results of ANSYS analysis and one dimensional finite element model(Hz)

	Ansys	Without SMB	With SMB
1st rigid body mode	0	0	0.45
2nd rigid body mode	0	0	2.97
1st flexible body mode	35.82	32.03	32.03
2nd flexible body mode	145.97	151.38	151.39

One Dimensional Finite Element Model for Control System Design

There are three AMB in the system to stabilize the system against a sudden disturbance and to pass it's critical speeds. The SMB is needed for lift off the weight of flywheel system. In order to construct the model of system with the effects of SMB, AMB and coreless motor, the system is divided into ten parts.

One dimensional finite element model is given in Fig.4. AMB2 is used for the control of the axial direction. AMB1 and AMB3 are used for the control of the radial direction. Using one dimensional FEM model, the state and the output equations of control system with effect of SMB can be desired. The results of ANSYS is close to that of one dimensional finite element model. Table 1 shows the comparison ANSYS analysis with one dimensional FEM analysis.

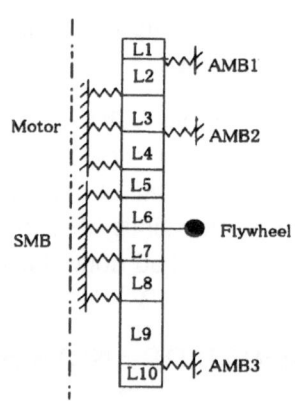

Fig.4 Model of controlled object

BASIC MODEL FOR CONTROL AND CONTROL METHOD

The control inputs are at the place of AMB1 and AMB3. The equation of motion of flywheel system is given as follows:

$$M\ddot{X} + KX = BU \tag{1}$$

Where, $X = [x_1 \quad \theta_1 \quad x_2 \quad \theta_2 \quad \cdots \quad x_{10} \quad \theta_{10} \quad x_{11} \quad \theta_{11}]$

$$B = \begin{bmatrix} 0 & 0 & 1 & \cdots & 0 & 0 & 0 & 0 \\ 0 & 0 & 0 & \cdots & 1 & 0 & 0 & 0 \end{bmatrix}$$

$$U = \begin{bmatrix} P_1 \times x_2 - P_2 \times i_{top} \\ P_1 \times x_{10} - P_2 \times i_{bottom} \end{bmatrix}$$

$$P_1 = 4 \times \frac{f_0}{x_0} \qquad\qquad P_2 = 4 \times \frac{f_0}{i_0}$$

Where, x_2 and x_{10} are the displacements. i_{top} and i_{bottom} are the control currents. f_0 is the AMB attractive force. x_0 is the gap from the equilibrium point. i_0 is the bias current.

In this paper, we only want to verify that the flywheel is controllable or not using the AMB forces and what the maximum control forces are. So, the control method of LQR with observer is used in this paper.

SIMULATIONS

It is assumed that the sensors are located at the same place of AMB1 and AMB3. The mode shape is given in Fig.5 and 6. Figure 5 is the first bending mode and Fig.6 is the second bending mode. In Fig.5 and 6, the position 5 is the place of flywheel. Because the weight of flywheel and auxiliary component is about 100t, the displacement of position 5 is close to zero.

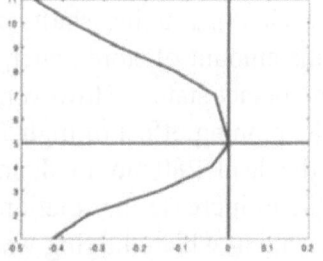

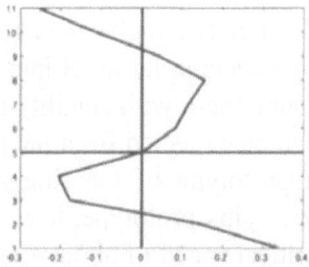

Fig.5 First flexible eigen-mode(eigenvalue 32.03Hz)

Fig.6 Second flexible eigen-mode(eigenvalue 151.38Hz)

The simulation results of stabilizing control are given in Fig.7 to Fig.9. Now, the simulation of stabilizing control is concerned with the impulse response and the response with initial value. The total force of lift off is about 1000t in Fig.9.

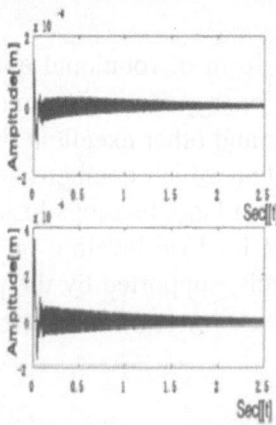

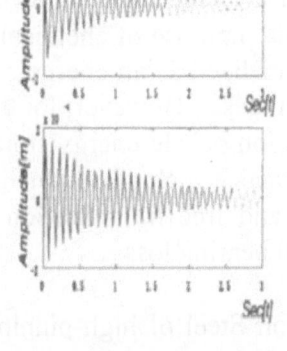

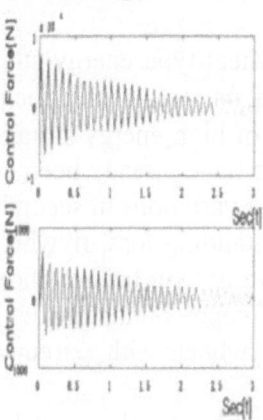

Fig.7 Impluse response

Fig.8 Response with Initial valuex3=0.2mm)

Fig.9 Control input for response with initial value

CONCLUSIONS

This paper is associated with the feasibility study of stabilizing control of 10MWh class energy storage flywheel system using superconducting magnetic bearing. From the results of simulation, it has been clarified that the flywheel system is controllable and observable system using the big control input force for AMB. It is found that the maximum control force of AMB is about 1.0×10^4 N in the worst case to stabilize again from a touch down.

Also, we had done the control of the system by influence coefficient model. The results computed by influence coefficient model is almost the same as that computed by finite element model.

1. Z,Xia,Q.Y.Chen,etc., Design of Superconducting Magnetic Bearing with High Levitating Force for Flywheel Energy Storage Systems,IEEE Transactions on Applied Superconductivity,vol.5. No.2,pp.622-625,June 1995

Ring-shaped Flywheel with Superconducting Levitation

Hidekazu Teshima[1], Satoshi Suzuka[2], and Ryuichi Shimada[2]

[1] Advanced Technology Research Laboratory, Nippon Steel Corporation, 20-1 Shintomi, Futtu-shi, Chiba 293-8511, Japan
[2] Research Laboratory for Nuclear Reactors, Tokyo Institute of Technology, 2-12-1 O-okayama, Meguro-ku, Tokyo 152-0033, Japan

Abstract: Ring-shaped flywheels without rotating shafts are superior to conventional disk-shaped flywheels in terms of increasing the amount of stored energy, but, until now, it has been difficult to rotate them with stability in a non-contact state. However, stable rotation of ring-shaped flywheels can be expected from the use of the pinning effect of high T_c bulk superconductors. Thus, we built a prototype of the ring-shaped flywheel 280mm in diameter and investigated rotation stability. With this prototype, it was possible to increase the rotational speed of the ring up to more than the critical speed of about 4.6Hz in a state in which the ring was supported by superconducting bearings alone. This level of critical speed was in good agreement with the one predicted from the one-degree-of-freedom model.

Keywords: Flywheel, Bulk superconductors, Superconducting levitation, Non-contact bearings

INTRODUCTION

The flywheel type energy storage system, which stores energy in the form of rotational energy of flywheels, is simple in principle and mechanism, compared to other energy storing technologies. Because of high energy storage density, non-use of chemical substance and other excellent features, flywheels have already been commercialized as an energy storage system which compensate short-cycle load variations in seconds or minutes. However, for the reason that large bearings loss results in large standing loss, flywheels as a long-cycle energy storage system for load leveling on a daily basis have not yet been reduced to practice. Superconducting flywheels, supported by the pinning effect of superconducting levitation and free from friction-induced bearing loss, are expected to realize flywheels with extremely small bearing loss.

Following the development by Nippon Steel of high-pinning-force bulk superconductors, such as QMG [1,2], making highly practical levels of levitation force available, expectations are running high for superconducting flywheels. Nevertheless, despite the distinct feature in superconducting levitation by the pinning effect of being able to realize non-contact bearings on a stable basis without the aid of controls, many of the prototypes heretofore reported [3,4] are in combination with the control-type magnetic bearings or machine-type bearings to compensate for low rotation stability.

This paper proposes that the ring shape of flywheels with no shaft should provide higher rotation stability when flywheels are supported by superconducting bearings. It was pointed out by Prof. Shimada's group that ring-shaped flywheels were superior to conventional disc-shaped flywheels in energy storing capacity [5], but it was difficult to achieve stable rotation of the shaft-less ring in a non-contact state. Thus, we built a small prototype ring-shaped flywheel with no shaft, in which high T_c bulk superconductors were arranged in a ring shape, and report the result of our valuation of its rotation stability.

SPECIFICATION OF THE SMALL PROTOTYPE

Figure 1 shows a photograph of the small prototype. The ring, made of iron, was 280 mm in outside diameter and 200 mm in inside diameter, and weighed 7.63 kg . For levitation and driving, Nd-Fe-B magnets were incorporated into the ring. Sixteen disk-shaped samples, 46 mm in diameter, of high T_c bulk superconductors were prepared using the improved QMG process and arranged in ring shape inside a cooling container, then cooled with liquid nitrogen. For rotary driving of the ring, the permanent magnet synchronous motor method was employed. For guiding the direction of radius vector of the ring, only the pinning effect of superconducting levitation is used with no particular control.

RESULTS AND DISCUSSION

First, the ring was made to vibrate by providing initial displacement in the horizontal direction, and time changes of the ring's vibrations were measured using a laser displacement sensor. Figure 2 shows an example of measurement at a levitation gap of 8.9 mm. The values of the spring constant, k, the vibration damping coefficient, c, and the critical speed, f_c, can be evaluated from this measurement using the one-degree-of-freedom model. For example, it can be seen from Fig.2 that under this condition, $k = 6400$ N/m, $c = 4$ Ns/m, and $f_c = 4.6$ Hz.

Next, the ring was made to rotate, and horizontal displacement during rotation was measured using the laser displacement sensor. The results of measurement are indicated with blank circles in Fig.3. Although vibration amplitude surged steeply in the vicinity of 4.6 Hz, which is considered the critical speed, the rotation speed of the ring could be increased to more than the critical speed while the ring was supported by the pinning effect of superconducting levitation alone. It was found that the ring could achieve stable rotation in the regions of rotational speed larger than the critical speed. Also, in Fig.3, calculation results that can be predicted from the one-degree-of-freedom model are shown in solid line. Because the calculation results were in good agreement with the experimental results, it was also found that vibration behaviors of ring-shaped flywheels should be predictable with the use of the one-degree-of freedom model.

Some of the conceivable reasons for the ring-shaped flywheels' ability to pass the critical speed solely on the strength of the ring's support by superconducting levitation include the structural features of the flywheel, such as position of the center of gravity being low, support being provided at a long distance from the center axis of rotation, and weight of flywheel being directly supported, compared to conventional flywheels with rotating shafts.

CONCEPTUAL DESIGN OF A 100 kWh-CLASS FLYWHEEL

Lastly, the conceptual design of a 100 kWh-class flywheel based on the results of this test is discussed. Table I shows the specification for the flywheel proper. This system, being supported by extra-low loss superconducting bearings alone, can be used as a long-cycle energy storage system, such as daily load leveling. Moreover, as this system has be no rotating shaft within the ring, refrigerators, vacuum pumps and other auxiliary machines can be arranged inside the ring, permitting efficient space utilization and compact size of the whole system.

SUMMARY

A small prototype of shaft-less ring-shaped flywheel in which high T_c bulk superconductors were arranged in ring shape was made and evaluated. The superconducting levitation type ring-shaped flywheels are a structure having high rotation stability and thus can be expected to be a promising candidate for commercialization.

Fig. 1   Prototype of ring-shaped flywheel.

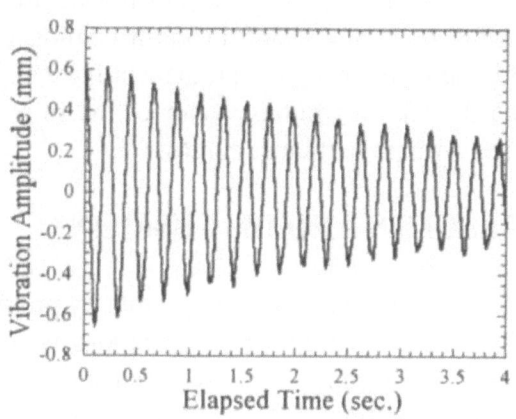

Fig.2   Vibration by providing initial displacement

Table I   100 kWh-class ring-shaped flywheel

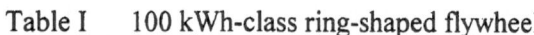

CASE	1	2	3
materials	steel	CFRP	CFRP
outside dia. (m)	3	3	2
inside dia. (m)	2.5	2.5	1.7
height (m)	1	1	1
weight (ton)	16.8	4.2	1.7
rotation speed (rpm)	1,500	3,000	7,000

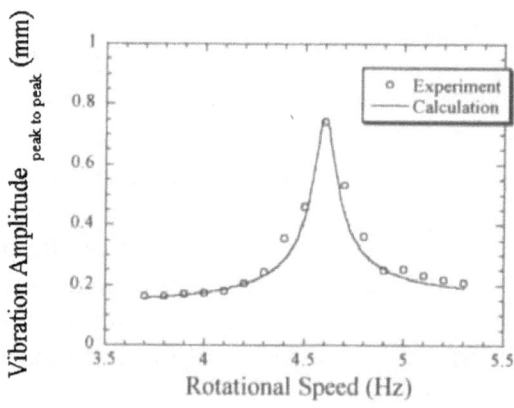

Fig.3   Vibration amplitude vs. rotational speed

1. M. Morita, K. Miyamoto, K. Doi, M. Murakami, K. Sawano, and S. Matsuda, Physica C 172, 383-387 (1990).
2. M. Morita, M. Sawamura, S. Takebayashi, K. Kimura, H. Teshima, M. Tanaka, K. Miyamoto, and M. Hashimoto, Physica C 235-240 209-212 (1994).
3. Y. Miyagawa, H. Kameno, R. Takahata, and H. Ueyama, Proceedings of *Applied Superconductivity Conference* (1998).
4. M. Minami, S. Nagaya, H. Kawashima, T. Sato, and T. Kurimura, in *Advances in Superconductivity X* (ISS'97), edited by N. Koshizuka and S. Tajima (Sringer-Verlag, Tokyo, 1998) pp.1305-1308.
5. T. Ogata, T. Takahashi, and R. Shimada, Proceedings of *Power Conversion Conference (PCC-Yokohama)* (1993) pp.587-592.

The model used for ANSYS analysis is given in Fig.1 as a right half cross-sectional view of the right-half plane. The results without gyroscopic effect are given in Table.1. The relation between eigenvalue and rotational speed with gyroscopic effect is given in Fig.2. In Fig.2, the two natural frequencies are nutation and precession frequency. The total weight of the whole flywheel system is about 107t and the height is 6.86m.The weight of flywheel itself is approximately 45.7t and the maximum diameter of the flywheel is 4.8m. One example of ANSYS analysis is given in Fig.3.

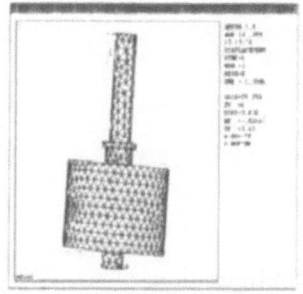

Fig.3 Cross sectional view of flywheel system

Table.1 Results of ANSYS analysis and one dimensional finite element model(Hz)

	Ansys	Without SMB	With SMB
1st rigid body mode	0	0	0.45
2nd rigid body mode	0	0	2.97
1st flexible body mode	35.82	32.03	32.03
2nd flexible body mode	145.97	151.38	151.39

One Dimensional Finite Element Model for Control System Design

There are three AMB in the system to stabilize the system against a sudden disturbance and to pass it's critical speeds. The SMB is needed for lift off the weight of flywheel system. In order to construct the model of system with the effects of SMB, AMB and coreless motor, the system is divi ded into ten parts.

One dimensional finite element model is given in Fig.4. AMB2 is used for the control of the axial direction. AMB1 and AMB3 are used for the control of the radial direction. Using one dimensional FEM model, the state and the output equations of control system with effect of SMB can be desired. The results of ANSYS is close to that of one dimensional finite element model. Table 1 shows the comparison ANSYS analysis with one dimensional FEM analysis.

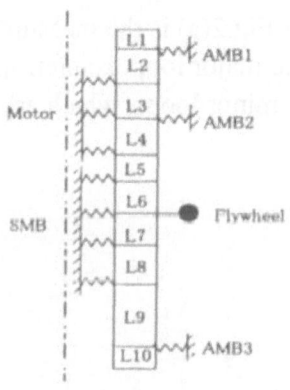

Fig.4 Model of controlled object

BASIC MODEL FOR CONTROL AND CONTROL METHOD

The control inputs are at the place of AMB1 and AMB3. The equation of motion of flywheel system is given as follows:

$$M\ddot{X} + KX = BU \qquad (1)$$

Where, $X = [x_1 \quad \theta_1 \quad x_2 \quad \theta_2 \quad \cdots \quad x_{10} \quad \theta_{10} \quad x_{11} \quad \theta_{11}]$

$$B = \begin{bmatrix} 0 & 0 & 1 & \cdots & 0 & 0 & 0 & 0 \\ 0 & 0 & 0 & \cdots & 1 & 0 & 0 & 0 \end{bmatrix} \qquad U = \begin{bmatrix} P_1 \times x_2 - P_2 \times i_{top} \\ P_1 \times x_{10} - P_2 \times i_{bottom} \end{bmatrix}$$

$$P_1 = 4 \times \frac{f_0}{x_0} \qquad P_2 = 4 \times \frac{f_0}{i_0}$$

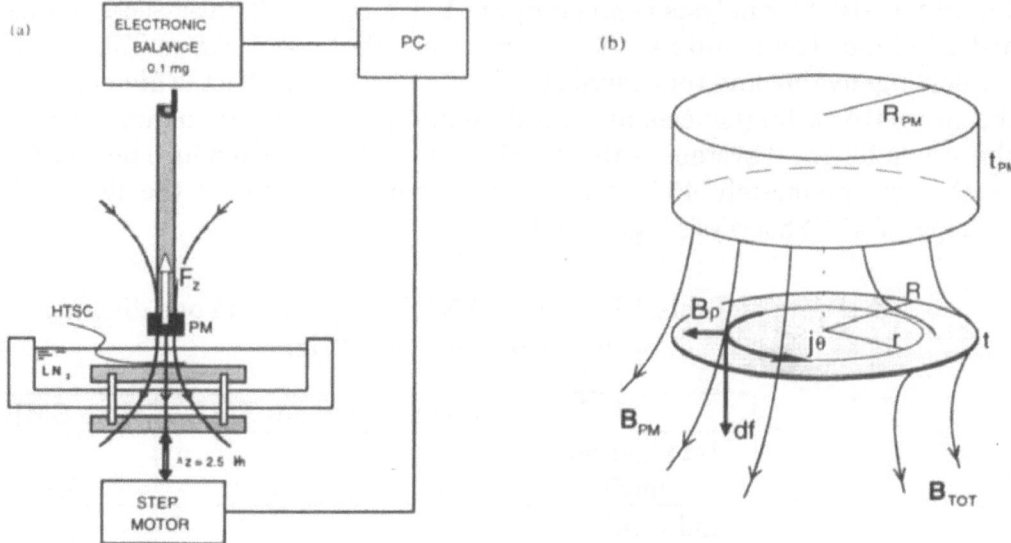

Figure 1. (a) Apparatus for measurements of levitation force. (b) Details of the configuration.

Prior to measurements the HTS film was zero-field-cooled, and then raised by a step motor towards the PM to the starting distance.

RESULTS AND DISCUSSION

Shown in Fig.2(a) is the measured levitation force as the HTS-PM distance, z, decreases. The steady approach, giving the major loop, is interrupted at equal intervals to perform minor reverse displacement. The slope of the resulting minor loops, which are essentially reversible, is the magnetic stiffness, κ_z, of the system.

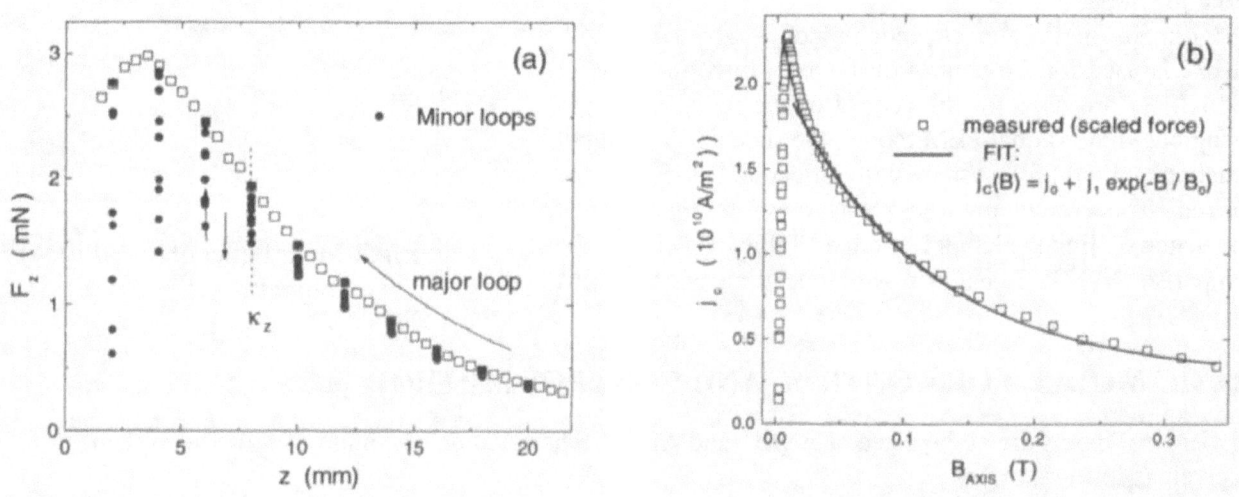

Figure 2. (a) Levitation force vs. PM-HTS film distance. (b) Inferred B-dependent critical current density. The points represent levitation force data from $z = 40$ mm and down to closest approach of 2 mm.

811

In the present configuration the force between the PM-HTS system equals $f = j_\theta \times B_{PM}$, per unit volume of the film. For the total disk it amounts to

$$F_z(z) = t \int_0^R B_\rho(r,z)\, j_\theta(r)\, 2\pi r\, dr, \tag{1}$$

$$B_\rho(r,z) = \frac{B_{rem}}{4\pi} \sum_{n=0}^{1} \int_0^{2\pi} \frac{(-1)^n R_{PM} \cos\phi\, d\phi}{\sqrt{r^2 - 2rR_{PM}\cos\phi + R_{PM}^2 + (z+nt_{PM})^2}}, \tag{2}$$

With a magnet of the present strength, the field seen by the film reaches full the penetration field already near $z = 20$ mm, hence, one may substitute j_θ by j_c . This implies that the force should follow the z-dependence of the integral over the radial field component B_ρ. An instructive plot is then a graph of the ratio $F_z(z) / 2\pi t \int^R B_\rho(z,r)\, rdr$, presumably equal to j_c, as a function of z. Fig.2(b) shows this ratio, with the horisontal axis giving the field at the film's center at the various z. The behavior consists of an initial steep rise at small fields (large z), followed by a more slow decay starting near 0.15 T. We conclude therefore that the initial rise is due to j_c filling an increasing part of the film, whereas the decaying part of the graph truly reflects the magnitude of the critical current. This part was fitted by the exponential model for $j_c(B)$, modified by an additional constant. An excellent agreement is obtained with the parameters $j_0 = 0.28 \, 10^{10}$ A/m^2 , $j_1 = 1.8 \, 10^{10}$ A/m^2 and $B_0 = 100$ mT.

The magnetic stiffness is shown in Fig.3. With increasing z it falls off close to exponentially fast, and is reduced by more than 2 decades at 20 mm. To understand this behavior one must consider how the ac-displacement transfers into force oscillations. We use here that the force on a magnetic moment, m, is the product of m and the gradient, g, of the applied field, hence the change in the force equals $\delta F = \delta m\, g + m\, \delta g$. Furthermore, one has $|\delta m| = V\chi (g/\mu_0) |\delta z|$ where V is the disk volume and χ is the susceptibility $\chi = (8/3\pi)(R/t)$. Because of the large χ for R>> t, the first term in δF dominates, and the stiffness becomes $\kappa_z(z) = (8/3\mu_0)\, R^3\, g(z)^2$. This function, is plotted together with the data in Fig.3. In view of the fact that our model has no adjustable parameters, the agreement is surprisingly good.

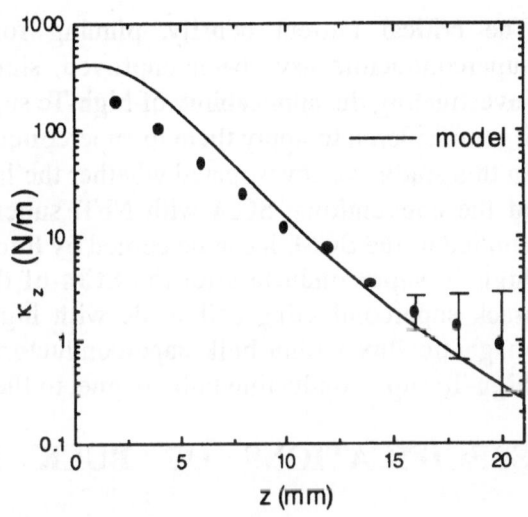

Fig.3. Magnetic stiffness as function of the PM-HTS film distance.

In summary, we have been able to explain quantitatively the magneto-mechanical behavior of a thin film HTS – PM system. The analysis also shows that force measurements can be used to infer $j_c(B)$.

Acknowledgments. This work was supported by The Norwegian Research Council.

[1] P.-Z. Chang, F. C. Moon, J. R. Hull and T. M. Mulcahy,J. Appl. Phys. 67 (1990) 4358.
[2] W. K. Chu, K. B. Ma, C. K. McMichael and M. Lamb, Appl. Supercond. 1 (1993) 1259.
[3] A. B. Riise, T. H. Johansen and H. Bratsberg, Physica C 234 (1994) 108.
[4] P. Schonhuber and F.C. Moon, Appl. Supercond, 2 (1994) 523; B. R. Weinberger, ibid 2, (1994) 511.
[5] P. Vase, Y. Shen and T. Freltoft, Physica C 180 (1991) 90.

CASE STUDY OF HIGH-Tc SUPERCONDUCTING BULK MAGNETS FOR MAGLEV

Hiroki Kamijo, Kaoru Nemoto and Hiroyuki Fujimoto

Railway Technical Research Institute, Hikari-cho 2-chome 8-38, Kokubunji-shi, Tokyo185-8540 Japan

Abstract: We investigated the applicability of a high-Tc superconducting bulk magnet to the Maglev system. For this purpose, each pole will be composed of numerous high-Tc superconductors, because their size at present is not sufficiently large. In this study, we investigated the design, weight, trapped magnetic field and other characteristics of the system required for the bulk magnet in which bulk superconductors are arranged in array per one pole, and determined it in comparison with the conventional race-track superconducting magnet of the Maglev system. We present an image of the bulk magnet for the Maglev system when 250mm square bulk superconductors with a 200mm square hole and 100mm in total thickness are arranged in four columns and two rows per one pole.

Keywords: High-Tc superconducting bulk magnet, Bulk superconductor, Maglev, Magnetic field

INTRODUCTION

The critical current density, pinning force, grain size and other characteristics of high-Tc superconductors have been improved, since they were first discovered at 1986. We have been investigating the applicability of high-Tc superconductors to the Maglev system [1,2]. For example, it is considered to apply them to superconducting magnets(SCM) and magnetic shielding of vehicles. In this study, we investigated whether the high-Tc superconductors can be used as the SCM in place of the conventional SCM with NbTi superconducting wires. If the high-Tc superconductors are applied to the SCM, it can be cooled by liquid nitrogen and the stability will be improved. In using high-Tc superconductors for the SCM of the Maglev system, it is considered to use either a race-track superconducting coil made with high-Tc superconducting wires or a bulk magnet trapping magnetic flux within bulk superconductors. In this study, we investigated the applicability of a high-Tc superconducting bulk magnet to the Maglev system.

SPECIFICATIONS OF BULK MAGNET

We consider using the bulk magnet composed of high-Tc superconductors as a substitute for the conventional SCM for the Maglev system. Therefore, the fundamental specifications of the bulk magnet are the same as those of the conventional SCM. A melt-process YBaCuO superconductor has the high critical current density at 77 K under high magnetic field. It is possible to make the superconductor of up to 150mm in diameter at present. But, if the bulk superconductor is to be used as a substitute, each pole will be composed of numerous bulk superconductors, because their size at present is not enough for the size of one pole. Then, we investigate the cases where the size of the rectangular bulk superconductors is 100mm, 167mm and 250mm square and the thickness is 100mm. For example, when the size of the bulk superconductor is 100mm square, these are arranged in ten columns and five rows per one pole. And also, we investigate the case where the bulk superconductor has a hole to make the bulk magnet light in weight.

The performance required for the bulk superconductors depends on their size. Fig.1 shows the relation between the necessary current density and the maximum empirical magnetic field, when the size of the bulk superconductors and the hole are changed and the thickness is 100mm. It is a necessary condition that the magnetic field must be the same as that of conventional SCM at the position of the ground coils. As this figure, it is clear that in the case of using 100mm square bulk superconductors, the necessary current density is larger than 50,000A/cm² in the high magnetic field, but in the case of using

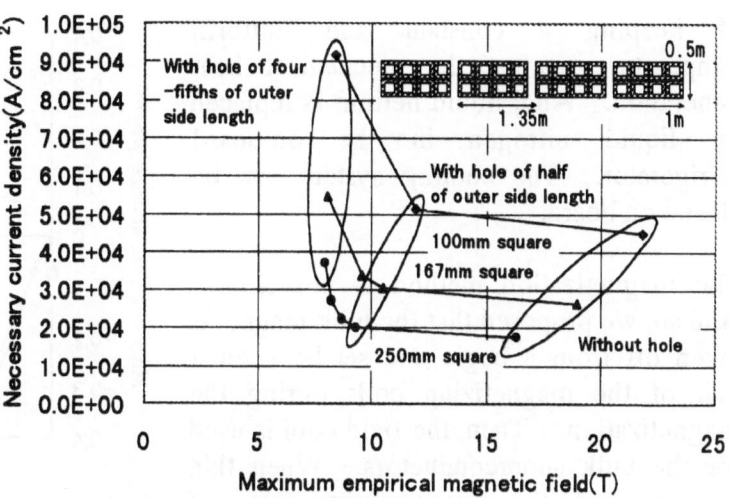

Fig.1. Relation between the necessary current density and the maximum empirical magnetic field

250mm square bulk superconductors, the necessary current density is about 30,000A/cm² and the maximum empirical magnetic field becomes less than 10T.

When the bulk magnet is used for the Maglev system in place of the conventional SCM, the bulk magnet has several advantages as follows; (1)The cooling system is simple. (2)The stability of the SCM improves. (3)The weight of the SCM is light. On the other hand, the bulk magnet has several problems as follows; (1)The magnetization method is not simple. (2)The strength of the bulk superconductors is not enough.

IMAGE OF BULK MAGNET FOR MAGLEV

We present an image of the bulk magnet for the Maglev system when 250mm square rectangular bulk superconductors with a 200mm square hole are arranged in four columns and two rows per one pole. The size of one bulk magnet corresponding to one pole is 1m×0.5m. The number of pole is 4 and pole pitch is 1.35m, that are the same as those of the conventional SCM. The thickness is 100mm, in which we put five bulk superconductors of 20mm thickness. When the current density is 37,000A/ cm², the magnetic field and the magnetic flux linkage with the ground coils of the bulk magnet is almost the same as those of the conventional SCM, as shown in Fig.2. The maximum empirical magnetic field of the bulk superconductors is about 7.9T. The weight of one bulk superconductor is 13.5 kg, if the density is 6.0g/cm². The weight of one pole is 108kg, because the bulk superconductors are arranged in four columns and two rows. This weight is a little larger than that of the conventional superconducting coil.

Fig.3 shows the image of the bulk magnet for the Maglev system. An outline of the bulk magnet is shown at the upper part and an outline of one pole is shown at the lower part. The fundamental structure of the bulk magnet is the same as that of the conventional SCM. The arranged bulk superconductors are hold in the inner vessel with a cooling channel, and cooled by liquid nitrogen. The inner vessel that is made of stainless steel acts as a cryogenic vessel in which the bulk superconductors are cooled by liquid nitrogen and as supports of the bulk superconductors to mechanical strength. The inner vessels are fixed to the outer vessel by the heat insulated supports. The outer vessel that is made of aluminum acts as mechanical support and shielding to harmonic magnetic field from the ground coils.

It is considered that the bulk magnet is cooled by the coolant of liquid nitrogen or the conductive-cooled. It is better that the bulk magnet is cooled by the coolant of liquid nitrogen at 77K, because of keeping a constant and uniform temperature against the mechanical heat generation. And, liquid helium is replaced by liquid nitrogen in the on-board refrigerator. The cooling system can be closed on board.

The magnetization method is not simple. And so, we proposed that the bulk magnet is taken off from a bogie and set between a pair of the magnetizing coils during the magnetization. Then, the field cool is used for the bulk superconductors. When this way is used, the magnetmotive force of magnetizing coils is smaller and the applied magnetic field to the bulk superconductors is nearly uniform. But in this way, the magnetizing coils should be large and have a higher magnetmotive force.

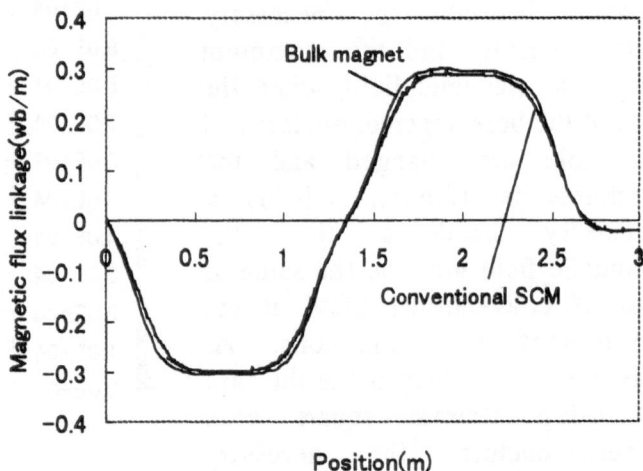

Fig.2. Magnetic flux linkage per unit length with the propulsion coil, when the balanced vertical displacement is 40mm and the current density is 37,000A/cm².

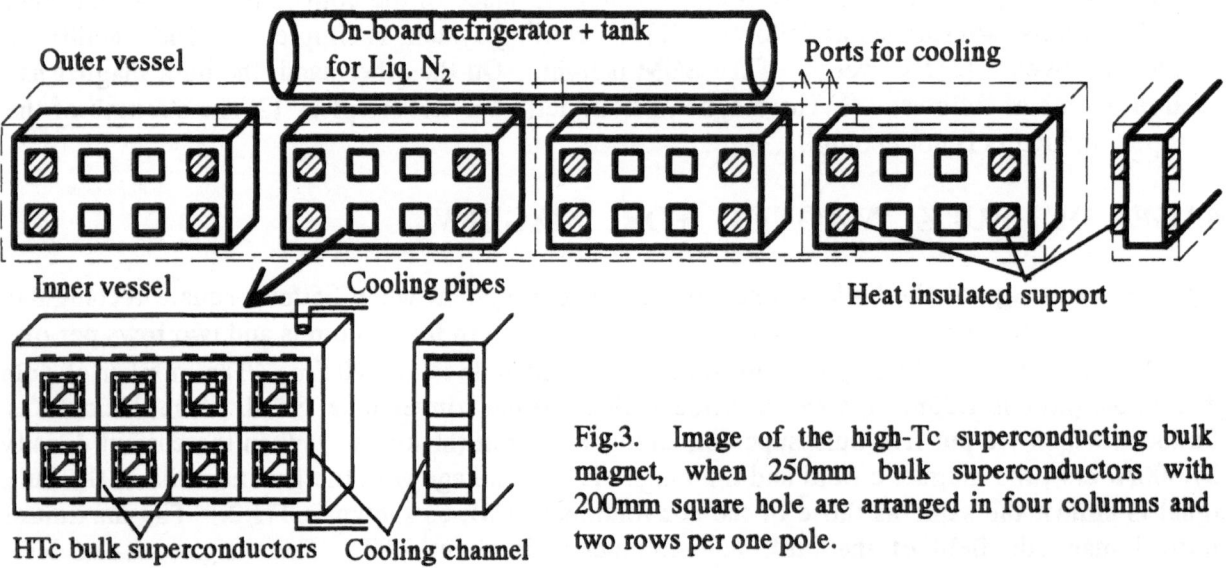

Fig.3. Image of the high-Tc superconducting bulk magnet, when 250mm bulk superconductors with 200mm square hole are arranged in four columns and two rows per one pole.

CONCLUSIONS

We investigated the applicability of high-Tc superconducting bulk magnet to the Maglev system. As the result of investigations, it is found that the bulk magnet can be made to have almost the same structure, generating magnetic field and weight as those of conventional SCM. The advantage of the bulk magnet is that the cooling system can be made simple by using only liquid nitrogen. However, it is found that the magnetizing method is not simple, because it is difficult to trap the bulk superconductors in a high magnetic flux. For applying the bulk magnet to the SCM of the Maglev system, further investigation is necessary. Especially, it is necessary to investigate the magnetization method, and improve the size and strength of the bulk superconductors.

1. H. Kamijo, H. Fujimoto, K. Nemoto, in Advances in Superconductivity IX(Springer-Verlag, Tokyo, 1997), pp.123-123.
2. H. Kamijo, T. Higuchi, H. Fujimoto, H. Ichikawa, T. Ishigohka, IEEE on transactions on applied superconductivity, Vol. 9, No. 2(1999), pp.976-979.

FEASIBILITY STUDY OF THE APPLICATION OF BULK SUPERCONDUCTORS TO THE EMS-TYPE MAGLEV VEHICLE

Mitsuyoshi Tsuchiya and Hiroyuki Ohsaki

Department of Electrical Engineering, The University of Tokyo, Bunkyo-ku, Tokyo 113-8656, Japan

Abstract: Because of the many special characteristics of bulk superconductors, they could be used as a flux source instead of electromagnets or permanent magnets in maglev systems. In this paper, we apply bulk superconductors to the EMS-type maglev vehicle. The configuration of the magnets consists of separated bulk superconductors and electromagnets. The attractive force between the magnet and an iron rail is generated by the magnetic flux pinned in the bulk superconductors. The electromagnets are controlled to produce the needed levitation force. Analysis shows that the weight of this magnetic configuration can be less than that of a conventional electromagnetic vehicle.

Keywords: Bulk superconductor, EMS-type maglev vehicle, weight of the magnet

INTRODUCTION

Magnetically levitated systems reduce noise and friction and are able to run at high speeds because it uses a non-contact drive. But when the system is very large, such as a maglev vehicle, a strong flux source is needed. Bulk superconductors can be this source. The bulk superconductor is able to generate a higher flux density than permanent magnets or electromagnets.

This research focuses on the feasibility of applying bulk superconductors to the electromagnetic suspension (EMS) type maglev vehicles[1][2]. The configuration of magnets shown in Fig.1 is considered. The two-dimensional finite element method (2D-FEM) is used to consider the electromagnetic characteristics. The bulk superconductor and the iron rail are the same size, 46mm in width and 15mm in height, and they are separated by 15mm at the steady-state value. The bulk superconductors are simplified as a single large coil with a current density in the cross section of $4 \times 10^8 \mathrm{A/m}^2$. The analysis shows that the

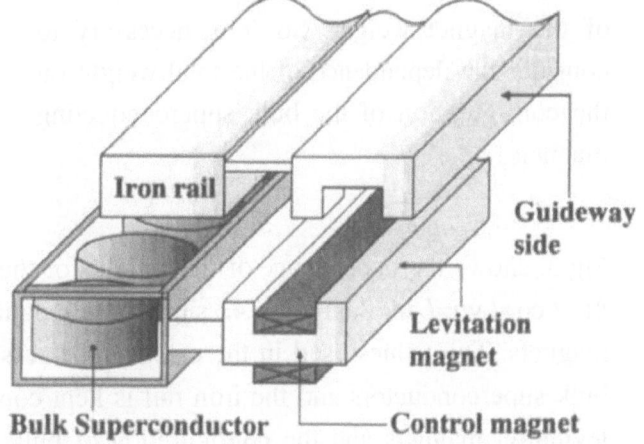

Fig. 1. Electromagnetic levitation system using bulk superconductors

attractive force between the bulk superconductors and the iron rail is enough to support the weight of the vehicles, because the weight of a conventional EMS-type maglev vehicle, HSST-100L, is estimated 1.0-1.7 ton/m and the attractive force is about 3.6 ton/m. The focus in this paper is on the dependence of the weight of the bulk superconducting magnets on its configuration.

FEASIBILITY OF LIGHTWEIGHT LEVITATION MAGNETS

If the weight of the levitation magnets shown in Fig.1 could be less than that of the magnets of a conventional electromagnetic vehicle for comparable field strengths, this would be an advantage of the application of bulk superconductors. The levitation magnets consist of the bulk superconducting magnets, the refrigerators, and electromagnets. The weight of the electromagnets is determined by the attractive force between the bulk superconductors and the iron rail. If the distance and, thus, the attractive force between them were constant, then the weight of electromagnets is unchanged, and the calculation is executed about the bulk superconducting magnets and refrigerators.

The configuration of the bulk superconducting magnets is shown in Fig.2. The bulk superconductors are cooled using liquid nitrogen, and an adiabatic vacuum is applied. The shape of supporting materials are hollow cylinders with an outer diameter of 15mm and an inner diameter of 4mm. Each set is separated by 1 m in the direction of the rail. Because the weight of refrigerators is determined by the heat loss, the configuration is very important. In order to reduce heat loss, the thickness of the cryostat is large. But it also causes an increase of the magnet weight. So it is necessary to consider the dependence of the total weight on the configuration of the bulk superconducting magnets.

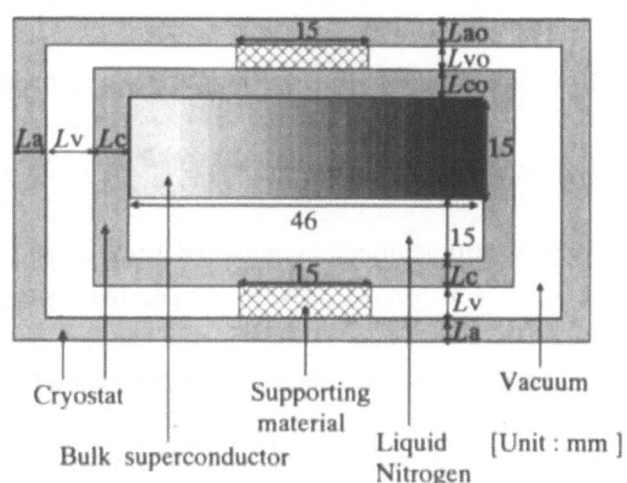

Fig. 2. Design of the bulk superconducting magnet

Fig.3. shows the dependence of the weights of the refrigerators and the cryostats on Lc or Lv when $La=Lco=Lvo=Lao=2mm$. Fig.4. shows the relationship of Lc, Lv, and the weight of levitation magnets. The values used in the calculation are shown in Table 1. Since the distance between the bulk superconductors and the iron rail is kept constant, and the relationship between the weight of levitation magnets and the configuration of bulk superconducting magnet could be illustrated, all parameters except for Lc and Lv are fixed. The weight of the electromagnets, which is determined by the electromagnetic force analyzed by the 2D-FEM, is 60kg/m. The weight of bulk superconductors is 8.0kg/m. The output of the refrigerator is estimated to be 72W-80K, and its weight including the compressor is 87kg. Fig.3 and Fig.4 indicate the weights of both the refrigerators and the cryostats greatly affect the total weight of the levitation magnets.

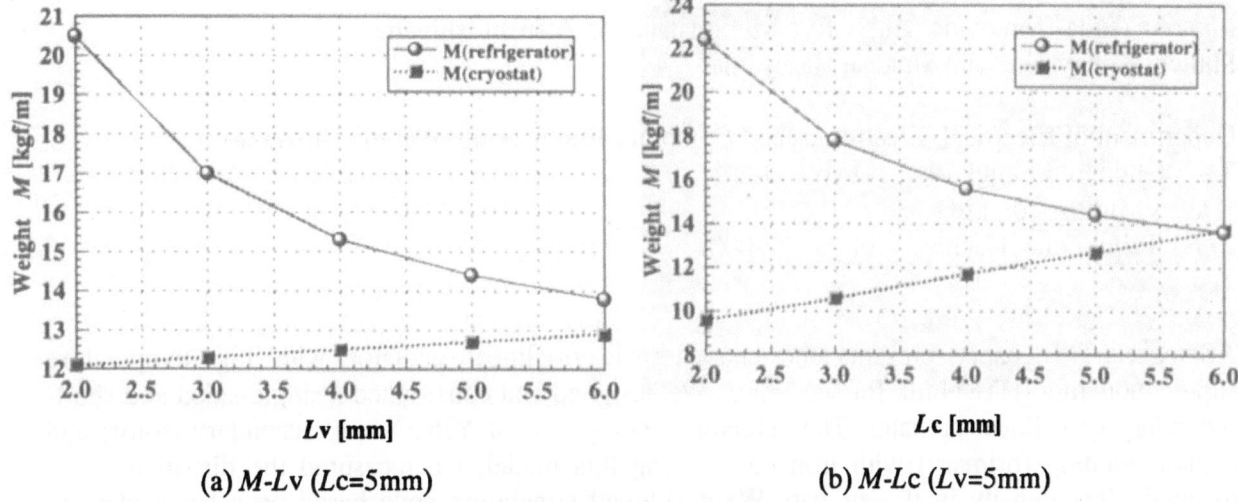

(a) *M-Lv* (*Lc*=5mm)   (b) *M-Lc* (*Lv*=5mm)

Fig. 3. Dependence of the weight of refrigerators and cryostats on the parameters, *Lv* and *Lc*

Table 1. The values used in the calculation

LN$_2$ temperature	77K
External temperature	300K
Heat conductivity Of cryostat	10.7W/m/K
Density of cryostat	8.0g/cm^3
Effective emissivity	0.12
Heat conductivity of Supporting material	0.35W/m/K

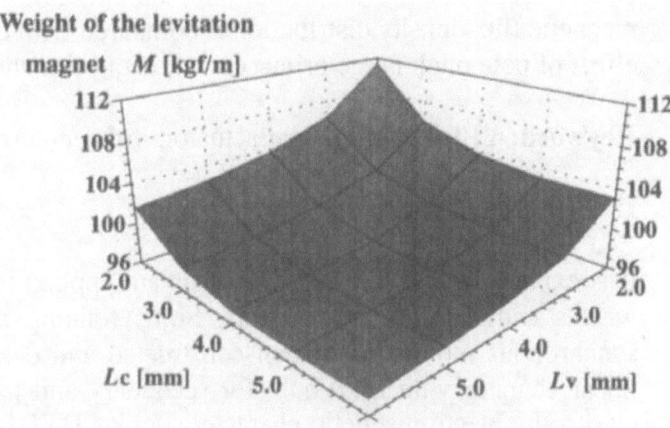

Fig. 4. Relationship of the configuration of bulk superconducting magnets and the weight of the levitation magnets

CONCLUSIONS

The feasibility of applying bulk superconductors as lightweight levitation magnets to the EMS-type maglev vehicles has been considered. Results of the calculations show the configuration of the bulk superconducting magnets greatly affects the weight of the levitation magnets.

Acknowledgments

This work was supported by the Program for Promoting Fundamental Transport Technology Research from the Corporation for Advanced Transport & Technology (CATT).

1. A. Senba, H. Kitahara, H. Ohsaki, and E. Masada, *IEEE Trans. on Magnetics*, Vol.32, pp.5049-5051, 1996
2. H.Ohsaki, M.Tokuda, and M.Tsuchiya, *The 1999 INTERMAG Conference*, DQ-12, 1999

Development of Linear Actuator with YBCO Bulk Secondary

Makoto Tsuda, Tomohide Koike, Ryo Muramatsu, and Atsushi Ishiyama
Shuichi Kohayashi* and Shuetsu Haseyama*

Department of Electrical, Electronics and Computer Engineering, Waseda University
3-4-1 Ohkubo, Shinjuku-ku, Tokyo 169-8555
*Dowa Mining Co., Ltd.,
277-1 Tobuki-cho, Hachioji, Tokyo 192-0001

Abstract: This paper presents the characteristic of linear actuator with high-temperature superconducting (HTS) bulk for secondary. We designed and constructed a single-sided and short-secondary type linear actuator. The actuator is comprised of YBCO bulk secondary (rotor) and copper winding (primary) with iron core. Using this model, we measured the distributions of magnetic flux density in the air gap. We developed simulation code based on a finite element method (FEM) taking the voltage-current (E-J) characteristic into consideration to investigate electromagnetic behavior within the bulk exposed to time-varying magnetic field. The computed magnetic flux density distribution is compared with the experiment. Numerical investigation of the effect of pole pitch in the primary on thrust is also shown.

Keyword: YBCO bulk, Linear actuator, Voltage-current characteristic, FEM

INTRODUCTION

It is expected that HTS bulk materials are applied to various electric devices such as motors, fly wheels and fault current limiters. Some rotating machines such as hysteresis, reluctance and synchronous motors have been constructed and demonstrated [1]. We have been constructing a linear actuator with HTS bulk for secondary and have been developing simulation technique to clarify the electromagnetic characteristics of HTS bulk in a realistic operational environment of electric machines. Numerical simulation based on the conventional critical state model is suited for time-independent electromagnetic behaviors in superconductors. However, it cannot represent dynamic electromagnetic phenomena in HTS bulk under realistic operation of electric machines. Therefore, we developed a simulation code of FEM based on the magnetic vector potential method taking E-J characteristic into consideration. Time-dependent electromagnetic behaviors of cylindrical HTS bulk in a fault current limiter have been evaluated by this simulation technique and good agreement between analysis and experiment has been obtained [2]. In this paper, the simulation code is applied to the numerical investigation of model linear actuator. Computed magnetic flux density is compared with experimental data and the effects of pole pitch in the primary on thrust in starting operation are investigated.

EXPERIMENTAL SYSTEM

The primary and secondary of linear actuator are composed of copper windings with iron core and YBCO bulks, respectively. The specifications of model device are shown in Table.1. As shown in Fig.1, permanent magnets, the same pole in longitudinal direction and alternating poles of N and S in transverse direction, are located at both sides of the primary windings to guide the secondary. Two types of YBCO bulk were adopted in the linear actuator model. Zero field cooled bulk is located at the center of the secondary to generate lift and thrust and field cooled bulks located above the permanent magnets play a role in guiding the secondary.

Table 1. Specifications of model device

Primary	
pole pitch	0.084 m
coil turn	50 turns
coil resistance	0.8 Ω
Secondary	
bulk length	0.084 m
bulk width	0.047 m
bulk thickness	0.003 m
Levitating	
bulk length	0.025 m
bulk width	0.025 m
bulk thickness	0.005 m

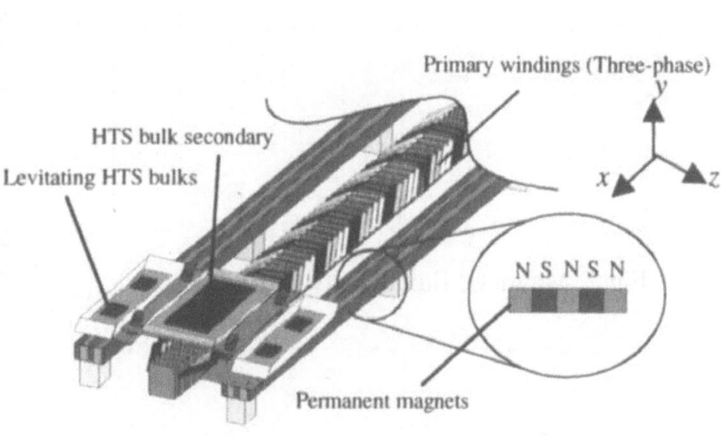

Fig.1. Overview of experimental machine.

These bulks can realize a stable levitating linear actuator, so that thrust can be obtained more effectively due to no running resistance. Magnetic flux density distribution at the top surface of the teeth in iron core was measured by Hall probe.

NUMERICAL SIMULATION

To simulate the electromagnetic behaviors of the HTS bulk exposed to time-varying magnetic field with high frequency, we developed a computer code based on the two-dimensional FEM, taking the E-J characteristic of the HTS bulk material into account. The governing equation derived from Maxwell's equation and the E-J characteristic are as follows:

$$\frac{\partial}{\partial x}\left(\frac{1}{\mu}\frac{\partial A_z}{\partial x}\right)+\frac{\partial}{\partial y}\left(\frac{1}{\mu}\frac{\partial A_z}{\partial y}\right)=J_0+J_{SC} \quad (1), \qquad E=E_C\left(\frac{J_{SC}}{J_C}\right)^n \quad (2)$$

where A_z is the magnetic vector potential in the z direction; μ is the permeability; J_0 is the exciting current density in the primary windings; J_{SC} is the supercurrent density; J_C is the critical current density; and E_C is the critical electric field that defines the critical current density, J_C. We adopt the Newton-Raphson method to solve this non-linear problem and a current sheet approximation for the primary current.

RESULTS

Magnetic flux density distributions at the top surface of the teeth in iron core are shown in Fig.2. The solid and dashed traces correspond to those of computed and experimental results, respectively. In Fig.2, YBCO bulk secondary is located at $-0.042 \leq x \leq 0.042$. Agreement between the experiment and the analysis is good. Because of shielding supercurrent within the bulk, magnetic flux density beneath the bulk is relatively small, while the larger magnetic flux density around the both edges of the bulk. The difference of the magnetic flux densities and nonsymmetry of the experimental trace may be caused by inhomogeneity of the bulk and experimental errors due to the relative location of Hall probe to iron core with teeth and slots. It can be considered that the experimental and numerical results validate our simulation model. Magnetic flux lines around the secondary of the bulk and a copper plate, as in conventional machines, at the frequency and the magnitude of the primary current of 50Hz and 1A, respectively, are shown in Fig.3. Most of the

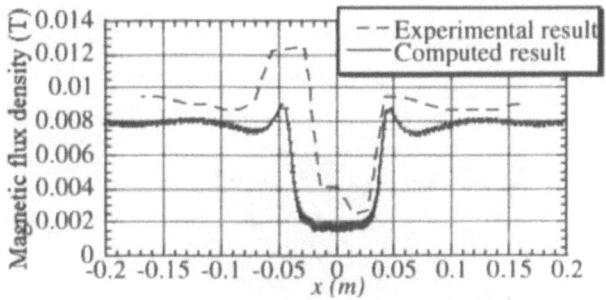

Fig.2. Magnetic flux density distributions

(a) HTS bulk
(Flux flow model)

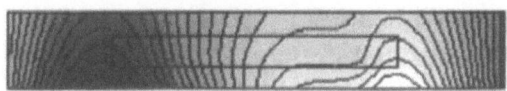

(b) Cooper plate

Fig.3. Magnetic flux lines

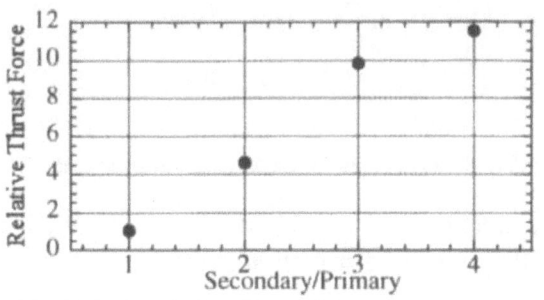

Fig.5. Pole pitch effect of the primary on thrust

Fig.4. Magnetic flux lines
(Critical state model)

magnetic flux within the bulk secondary is eliminated to the outside of the bulk, while the whole flux penetration is observed in the copper plate. This result implies that current distribution in the bulk is quite different from that of copper plate and the characteristics of linear actuator such as thrust may depend on the current distribution. The computed results based on a flux flow model taking E-J characteristic into account are compared with those of a conventional critical state model. Fig.4 shows the calculated magnetic flux lines by the critical state model. In Figs.3 and 4, the magnetic flux lines, i.e. supercurrent distributions, are different around the surfaces of the bulk. The calculated magnetic flux lines by the flux flow model varies with the frequency of the primary current, while the same distribution in the critical state model. These results confirm that the flux flow model should be adopted to evaluate the characteristics of linear actuator accompanying time-dependent electromagnetic behavior. Using our numerical model, we calculate thrust in starting operation of our constructed linear actuator at the frequency and the magnitude of the primary current of 50Hz and 1A, respectively, and investigate the pole pitch effect of the primary windings on the thrust. The computed results are shown in Fig.5. It is considered that the thrust in the starting operation can be improved by adopting longer pole pitch of the primary windings. The characteristics of the thrust, including the effects of the pole pitch and the frequency of the primary current on the thrust, will be investigated both experimentally and analytically.

SUMMARY

We designed and constructed a levitating linear actuator with YBCO bulk secondary and measured the magnetic flux distribution in the air gap. A numerical simulation code based on FEM considering the E-J characteristic was developed to evaluate the electromagnetic behaviors of the linear actuator exposed the traveling external magnetic field. It was verified that the critical state model is not suited for time-dependent electromagnetic behaviors and the remarkable characteristic of HTS bulk, such as E-J characteristic, should be taken into account in numerical approach.

1. L.K.Kovalev et al., IEEE Trans on Appl. Super., vol.9, NO.2, pp.1261-1264 (1999)
2. J. Nakatsugawa et al., IEEE Trans on Appl. Super., vol.9, NO.2, pp.1373-1376 (1999)

LATEST DEVELOPMENTS OF HTS CABLE SYSTEMS IN EUROPE AND IN THE USA

M. Nassi

Pirelli Cavi & Sistemi, Viale Sarca 222, 20126 Milano, Italy

Abstract: High Temperature Superconducting (HTS) cable systems promise a number of special opportunities in term of their high current carrying capacity, low electrical losses, compact overall dimensions, low weight, and reduced environmental impact. This paper reports the latest developments on the HTS cable system prototypes in Europe and in USA. In particular, a detailed up-date on the development stage of the application-focused HTSC projects in progress within Pirelli Cables and Systems will be provided. These projects are aimed towards the technical feasibility demonstration of HTS technology for application at medium and extra high voltage for both "warm dielectric" and "cold dielectric" cable systems. Particular attention will be devoted to the Pirelli-EPRI-DoE program for the feasibility evaluation of an HTS pipe-type retrofit cable. The program was successfully concluded in January 1999 with the demonstration of a 50m 400MVA 115kV pipe-type HTS cable system capable of carrying 2kA rms a.c. Details on the final high voltage tests and the a.c. losses measurements will be reported.

Keywords: HTS Cable, a.c. Losses

INTRODUCTION

Demand for high and bulk-power transmission is growing for systems feeding dense urban areas as well as in meshed transmission networks. In particular areas we are already approaching the upper limits of what can be achieved with conventional underground transmission cables, particularly in terms of maximum conductor size related to the current rating requirements. The prospect of adopting High Temperature Superconducting (HTS) cables therefore offers a number of opportunities and advantages, potentially overcoming constraints of conventional transmission cables.

Since the early 1990s, several studies of possible areas of application of the new generation of HTS materials operating in LN_2 have shown the potential technical and economic attractiveness of the HTS solution, particularly when a large amount of power (of the order of 1GVA or more) is bound to be transmitted in underground cables, or when the upgrade of an existing duct or pipe line to the double or more transmitting capacity is the most viable solution to the increasing energy demand.

A technical overview of the HTS cable system development projects in the USA and in Europe is here presented, with a particular focus on the Pirelli-EPRI-Doe program for the feasibility evaluation of an HTS pipe-type retrofit cable, which has recently culminated in the successful demonstration of a 50m 400MVA 115kV pipe-type HTS cable system (comprising HTS cable, joint, outdoor terminations and LN_2 cooling plant) capable of carrying $2kA_{rms}$ a.c.

HTS CABLE PROJECTS IN USA AND IN EUROPE

The technical challenge of developing and demonstrating the feasibility of a HTS cable system has been accepted by several cable manufacturers, with the involvement also of some of the more important European and American electric utilities (see table 1) and of the American Department of Energy.

Table 1. Summary of on going HTS cable projects in USA and in Europe [1].

Developer	Main Partners	Cable type	Present milestone
Alcatel	open	150 kV, CD	30 m single phase
CRPP	Brugg Rohr-Systeme	WD	5 m conductor model, 2,4 kA_{rms} @ 50 K achieved '99
NKT	NST, ELKRAFT, DTU, ELTRA, DEFU, ELTEKNIK, NESA	36 kV, WD	30 m/36 kV/2kA single phase in 1999; Field test of 30 m, 3 single phase 36 kV in '00/'01
Pirelli	DOE, EPRI, ASC	115 kV, WD	50 m complete system (cable, joint, termination, cooling),HV laboratory test completed (2 kA $_{rms}$) in Jan 1999
Pirelli	Detroit. Edison, DOE, EPRI, ASC, Lotepro	24 kV, WD	120 m, 3 single phases field test in a utility network (2.4 kA_{rms}) in '01/'02, retrofit in existing ducts of 4"
Pirelli	EdF	225 kV, CD	50 m complete system (cable, joint, termination, cooling), 2,6 kA_{rms} in '01 /'02
Pirelli	Utilities & others in Europe	110 kV, CD	100 m, 2 kA_{rms} 1-phase field trial in '01/'02
Pirelli	Edison, ENEL, CESI	132 kV, CD	30 m 1- phase, 3 kA_{rms} field trials in '02
Southwire	ORNL	12.5kV, CD	30 m, 3 single phases, 1.25 kA field trials in '99-'00

The on going development and demonstration efforts cover all the main technical issues that are necessarily to be faced if the goal is to get hold of a complete cable system, fully developed and reliable in all its components, comprising not only the cable itself (conductor, cryostat and electrical insulation) but also the cryogenic high voltage terminations, joints and refrigeration plant. Different development programs running in parallel offer a sufficiently wide choice of power capabilities, voltage ratings and geometrical arrangements to be able to match the various opportunities the market could potentially offer to an HTS cable.

Pirelli-EPRI 115 kV cable system. Thus as early as 1993, Pirelli and EPRI commenced their first feasibility evaluation of an HTS pipe-type retrofit cable that was subsequently finalised at the end of 94, under the US Department of Energy Superconductivity Partnership Initiative (SPI), to the development of a complete prototype cable system. The design of the "retrofit" cable is of the so-called "warm-dielectric" concept (see Fig. 1), where the cryostat is incorporated as an integral part of the cable core, and the electrical insulation and external protection of the cable are applied over the cryostat, and therefore operate at the external ambient temperature.

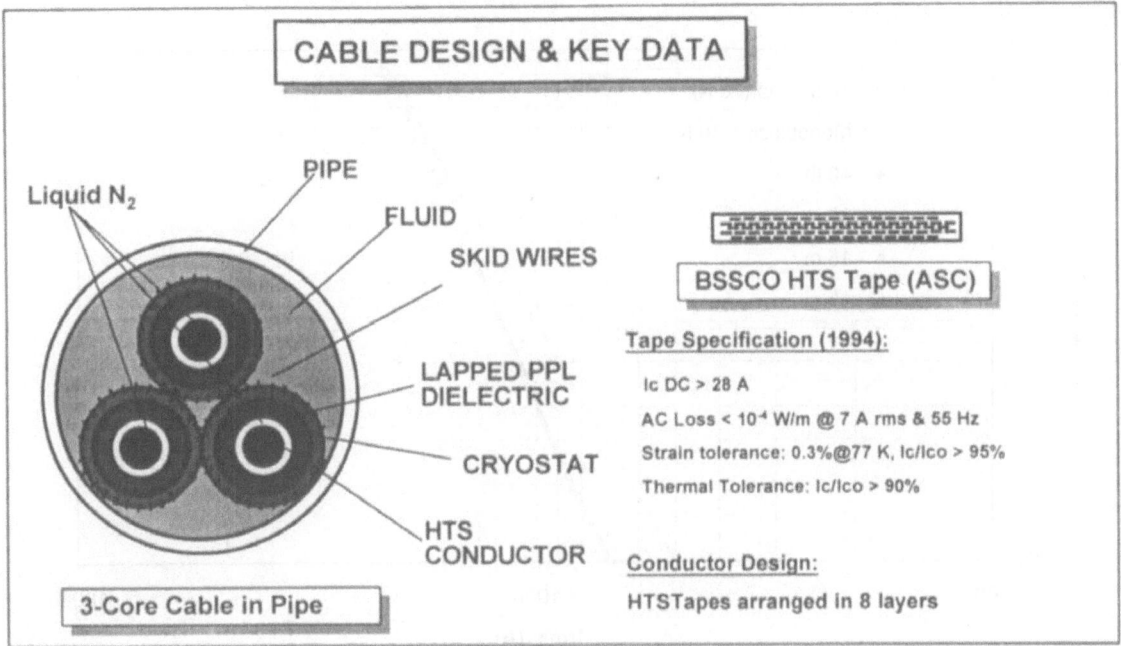

Fig. 1: 115kV Warm Dielectric cable for pipe-type retrofit application and key performance specification for tapes

This design has several advantages compared to the cold dielectric design, because it requires installation procedures similar to those used for a conventional pipe-type cable and it uses existing, well-proven dielectric materials. The conductor design was based on BSCCO2223 HTS tapes supplied by American Superconductor Corporation arranged in 8-layer. The key performance specifications of the individual tapes were as summarised in Fig. 1. The manufacture of the cable was undertaken using conventional manufacturing lines slightly modified to handle the HTS tapes with sufficient accuracy to avoid exceeding their mechanical strain limits. In order to prove the technical feasibility of incorporating the HTS cable into a transmission system, in-line splicing techniques for jointing consecutive cable lengths and also outdoor terminations were developed. The design of the termination must take into account the need for a "current lead" to bring the transmitted power up from the HTS conductor core at about 77K to a "normal ambient" exit temperature. This is achieved by use of a special thermal gradient design in which LN_2 is allowed to boil off from a chamber within the termination, to balance out the Joule heating and the heat conduction losses in the copper current lead. The concept for the jointing technology was to adopt a short copper ferrule at the centre of the joint, with large conductive cross-section, and to solder bond the HTS tapes to the ferrule on each side. Current flow thus passes through the ferrule at the joint position, but the resultant Joule heating losses are sufficiently small to be removed by the LN_2 flow through the conductor core without significantly increasing the LN_2 temperature. A vacuum-insulated cryostat tube is applied over the central part of the joint, and is hermetically welded to the cable cryostat on each side of the joint. The electrical insulation is reconstituted over the outside of the cryostat using conventional joint materials and techniques. For the a.c. loss testing the cable was laid in straight configuration, using a couple of copper braids symmetrically disposed at about half a meter away from the cable for current return, as already experimented in a previous measurement on a 13 m cable conductor section [2].

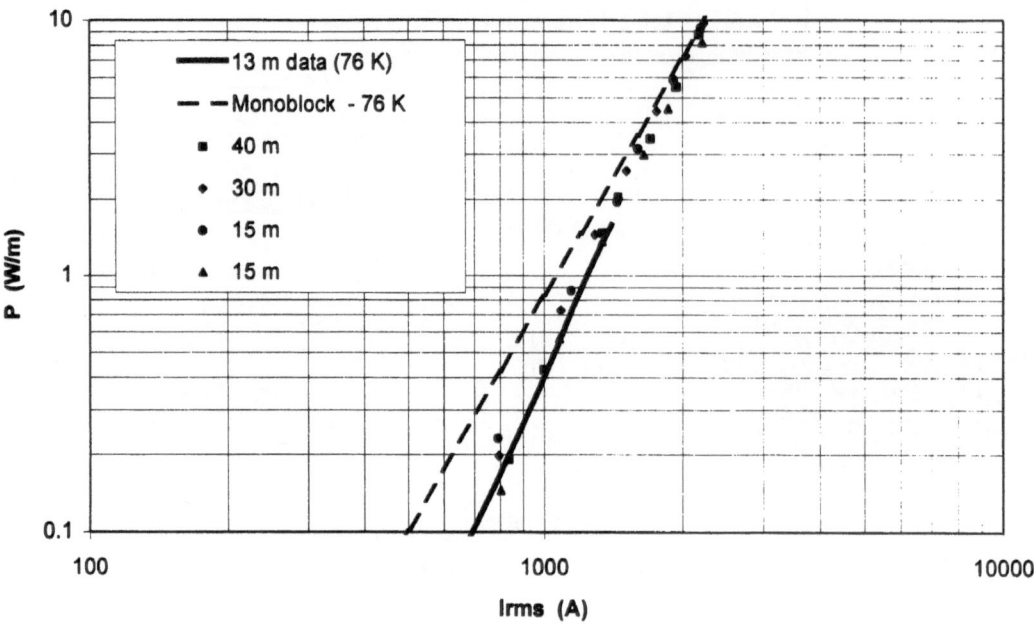

Fig. 2: Hysteretic a.c. losses at 76 K in the EPRI 50-m cable prototype compared to those previously measured on a 13 m bare conductor sample.

More than 4000 A d.c. and up to 2200 A_{rms} a.c. 50 Hz were the available power feeds. An open cycle refrigerator fed by a 12.000 litre LN2 tank provided circulating LN2 at 5 bar pressure and different temperature (72 – 83 K) at the cable's inlet. The voltage drop due to a.c. losses was measured using a lock-in amplifier for the measurement of the signal amplitude and an accurate phase meter for the phase angle with respect to a signal in phase with the current.

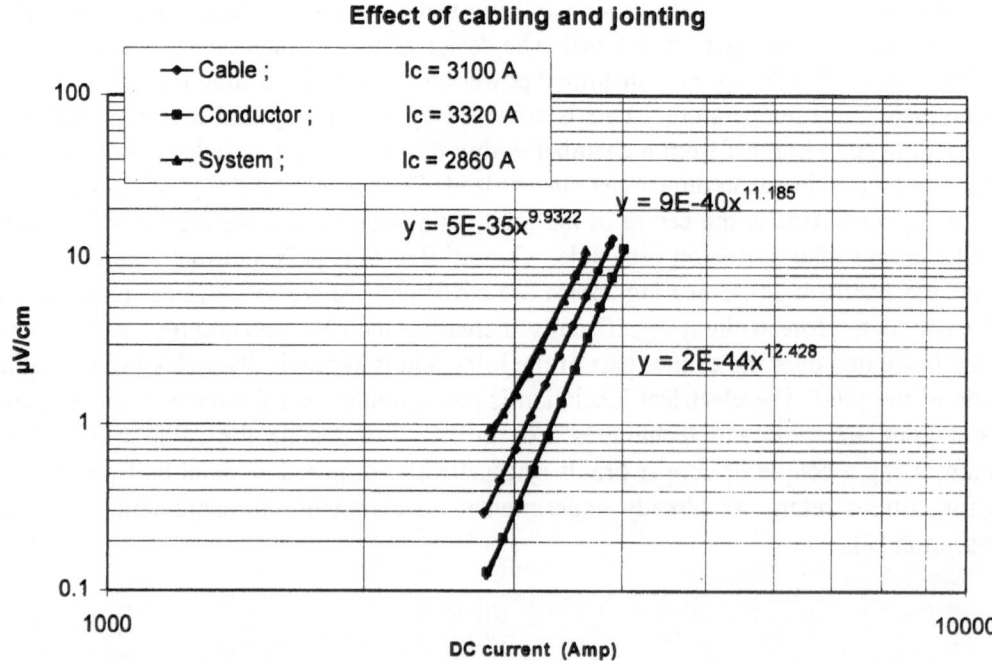

Fig. 3: Effect of cabling and jointing on HTS conductor performance.

The hysteretic losses confirmed substantially the same level of losses previously found in a 13 m bare conductor section (Fig.2).Ampacity tests were undertaken on each tape before cabling, on the completed « bare » conductor, on the cable after completion of all manufacturing stages, and on the finished cable system assembly comprising cable, terminations and one in-line joint. Before installing the joint and the terminations d.c. measurements were performed varying the operating temperature. The cable showed the same scaling law of Ic against temperature already determined for tapes. The Ic measurement confirmed that almost no degradation of the BSCCO tape performance took place through all the steps of the cable manufacturing. Fig. 3 shows results for the ampacity at 77 K for the conductor, cable, and system respectively. It can be seen that the performance of the conductor in the complete system exceeded the targeted minimum performance of Ic $\geq$ 2800A d.c.. The electrical performances of the insulation system, joint, and terminations were checked with a complete range of Type Tests, generally in accordance with AEIC CS2/90, including a.c. withstand tests to 165kV a.c. for 24 hrs, 205 kV for 1 minute, and lightning impulse tests at +/-550kV (10 pulses each).

Detroit Edison 24kV HTS Cable Project. Building on the success of the 115kV development with EPRI, in 1998 Pirelli initiated the Detroit Edison Project, which will involve the world's first HTS cable system installation in an utility site at Detroit Edison's Frisbie Station. The participants in the project are Pirelli, Detroit Edison, ASC, EPRI, and Lotepro, and the project is being supported under the US DoE's SPI initiative. This project will involve design, engineering, installation, test and routine operation of a 24kV 3-phase Warm Dielectric cable system which will replace existing conventional cables in the Frisbie station. The three HTS cables will replace nine existing copper cables, which together carry a total power of 100 MVA. Thus at 24 kV, each HTS cable will carry 2400A$_{rms}$ a.c., a level which would be inconceivable in a conventional cable of this voltage rating. The system design and engineering is under way

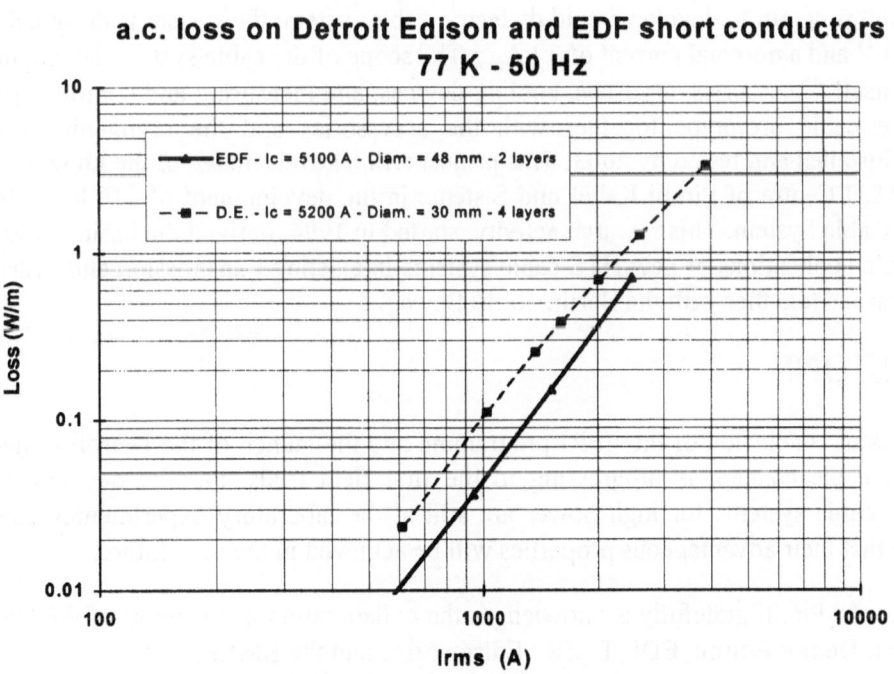

Fig. 4: a.c. loss measurements on short conductors carried out within the Detroit Edison and EDF projects

. The cable manufacture will commence in early 2000, the installation within the end of 2000, and « go-live » in the year 2001. The cable design for the Detroit project is based on the 115kV EPRI cable concept using a warm dielectric construction, which in this instance will be extruded solid insulation designed for operation at the 24kV voltage level. The cables will be installed in existing "fibre" ducts of 4 inch (~100mm) internal diameter, and will have five 90° bends in the 120m route between the 120kV/24kV transformer and the 24kV switchgear which they will interconnect. Preliminary a.c. loss measurements on short conductor samples have already been performed (see fig 4)

Pirelli-EDF HTS Cable System Development. In October 1997 Electricité de France (EDF) and Pirelli announced a joint project on high-power HTS cable link. The system is designed to carry a total power of 3000 MVA at 225 kV with 4 circuits in parallel with a total length of 10 km [3]. The main purposes of this project are specifically devoted to:

- study the technical and industrial feasibility of a high power superconducting link,
- develop a system prototype using the « cold-dielectric » cable design concept,
- estimate the investment, operating and maintenance costs of a superconducting link,
- evaluate the economic attractiveness of HTS links in a variety of network configurations.

This development focuses on CD HTS cables, potentially advantageous due to their extremely large power density, which would considerably ease route planning and installation in congested city areas. Moreover, due to their "coaxial" design they have minimal environmental impact because of their almost zero external electromagnetic and thermal interference. Several main areas of investigations were identified to characterise all the key elements of the superconducting coaxial cable system starting from the HTS and the dielectric materials to the cable, accessories and auxiliary systems. A.c. loss measurements on 1.5 long conductor samples have already been performed (fig. 4). Based on the positive results of the experimental phase, the construction of full size prototype of the system components is under way. The first set of full-size prototype test will start in 2000.

Pirelli-ENEL-Edison HTS Cable System Development. In May 1999 Edison, ENEL and Pirelli announced a joint program to develop a cold dielectric cable system; the system is designed for a service voltage of 132 kV and a nominal current of 3 kA_{rms}. The scope of the cable system development includes the HTS cable itself, accessories, cryostats, cooling devices, and measuring and control systems. The 30 m single phase cable prototype together with the accessories and the cryogenic system will be manufactured, installed and tested by 2003. This project will take advantage of the know-how developed in the Berlin R&D Centre of Pirelli Kabel und Sisteme in the development of 110 kV/400 MVA, cold dielectric HTS Cable System. This research activity, started in 1994, derived the basic design data from a study on possible applications of new HTS cable systems in existing transmission and distribution cable networks, performed together with the Berlin utility Bewag.

CONCLUSIONS

With the successful conclusion of the EPRI programme and the launch of the Detroit project the Warm Dielectric HTS cable concept is progressing to the first field trials. Parallel developments of Cold Dielectric HTS cable systems for high power are still at the laboratory experimental stage but initial results indicate that their advantageous properties will be achieved in the near future.

Acknowledgments. Pirelli gratefully acknowledges the collaboration and support of the US Department of Energy, EPRI, Detroit Edison, EDF, ENEL, Edison SpA, and the BMBF.

1.Scenet Power: Working group on power cables and high amperage conductor – 1st Report (Sept. 1999)
2. Gherardi L, Gomory F, Mele R and Coletta G, 1997 Supercond. Sci. Technol. **909-913**
3. Nassi M, Norman S, Ladiè P et al. 1999 Jicable to be published

A High-Tc Superconducting Rutherford Cable
Using Bi-2212 Oxide Superconducting Round Wire

Yuji Aoki[1], Nozomu Ohtani[1], Takayo Hasegawa[1], L. Motowidlo[2], R. S Sokolowski[2], R M Scanlan[3] and Shigeo Nagaya[4]

[1]Showa Electric Wire & Cable Co., Ltd., 2-1-1, Odasakae, Kawasaki-ku, Kawasaki, Kanagawa 210-8860 Japan.
[2]IGC-Advanced Superconductor, 1875 Thomaston Avenue, Waterbury, CT 06704, U.S.A.
[3] Lawrence Berkeley National Laboratory, 1 Cyclotron Road, Berkeley, CA MS46-16, U.S.A.
[4] Chubu Electric Power Co., Inc., 20-1, Kitasekiyama, Ohdaka-cho, Midori-ku, Nagoya, Aichi459-8522, Japan

Abstract: We developed a new robust Rutherford cable using Bi-2212 high temperature superconducting (HTS) wire, which consisted of 427 superconducting filaments and which had a critical current density (J_c) value of $2.4 \times 10^5 A/cm^2$ at 4.2K in self-field . For applications requiring a Rutherford cable with a core such as NiCr, we found that a plasma sprayed ZrO_2 film was most effective in preventing "poisoning" elements present in the core from diffusing into and poisoning the HTS wire. Rutherford cable made with this wire and a thin NiCr core protected by ZrO_2 film barrier layer was able to carry an I_c value of 3,570A. The I_c degradation in Bi-2212 AgMgSb sheathed wire was only 10%. Finally, we demonstrated the successful fabrication of an 80m-long Rutherford cable.

Keyword: Rutherford Cable, $Bi_2Sr_2CaCu_2Oy$ superconductor, AgMgSb, barrier layer, ZrO_2, Al_2O_3, 80m-class cable

INTRODUCTION

Rutherford cable using $Bi_2Sr_2CaCu_2Oy$ (Bi-2212) high critical temperature (T_c) superconducting wire is a promising candidate for large current carrying capacity conductors in electric power and high energy physics applications. Optimized heat treatment of HTS Rutherford cables demands that poisoning of the superconducting filaments by elements present in the cable core reinforcement material be prevented. In fact, we have successfully fabricated Bi-2212 Rutherford cable with a current carrying capacity of 3.5kA at 4.2K in self field by introducing a ceramic layer between the Bi-2212 superconducting wires and the Ni-alloy core used for mechanical support [1]. In this study we developed an improved double restack wire with 427 filaments (7x61), and investigated which materials would be most suitable as a barrier layer to prevent the superconducting filaments from being poisoned by elements present in the Ni alloy core.

EXPERIMENT

Seven first-stacked wires having 61 filaments were re-stacked into a AgMgSb alloy tube to make a 427-filament composite wire. This Bi-2212 round wire was fabricated by a conventional powder-in-tube method to a final diameter of 0.81mm. The ceramic barrier layers were applied to the Ni-alloy core using two methods: one by wrapping MgO ceramic tape (up to 600μm thick) around the core, and the other by depositing an Al_2O_3 or ZrO_2 film (50μm-thick) on both sides of the core by plasma spraying. Rutherford cables were fabricated by winding Bi-2212 round wires around the Ni-alloy core with different barrier layers. A partial melting and slow solidification process was employed for the heat treatment (HT). I-V traces were measured at 4.2K in self-field using a DC four probe method, and the I_c was determined using a criterion of 1μV/cm.

RESULTS AND DISCUSSION

Wire properties. It is well known that the I_c of Bi-2212 depends strongly on heat treatment conditions such as T_{max} and cooling rate. We optimized both factors using two types of round wire, 300-filament single-stack and 427-filament double-stack. The Ic values obtained under different HT conditions are summarized in Table1. The filament diameter of the 300-filament wire is larger than that of the 427-filament wire.

Table 1. Optimization of Tmax and cooling rate.

Wire	Cooling rate (°C/h)	T_{max} (°C)		
		886	884	882
Pure Ag-sheathed wire with 300 filaments	10		160.5 A	
	5	168.0A	182.3A	195.1A
	3	188.3A	213.0A	141.5A
AgMgSb-sheathed wire with 427 filaments	10		135.8A	
	5	207.8A	216.0A	85.5A
	3	252.0A	207.4A	106.5A

In AgMgSb sheathed wire the I_c value of 252A corresponds to a J_c of $2.4 \times 10^5 A/cm^2$, this value being twice that of pure Ag sheathed wire with 300 filaments. In the case of AgMgSb sheathed wire, Table 1 shows that the optimum heat treatment range is more narrow as compared with that of pure Ag sheathed wire.

Cable properties. Fig.1 shows the test results related to the development of an effective diffusion barrier to prevent poisoning of the HTS wire. Electrical performance of heat-treated wire is used as an indicator of barrier effectiveness. In the case of a core wrapped with MgO tape, the barrier effect saturates above a barrier thickness of 400μm. On the other hand, a much thinner plasma sprayed ZrO_2 film (50μm) was able to prevent poisoning to the same extent. This cable was able to carry an Ic value of 3,570A at 4.2K in self-field. We concluded that a ZrO_2 plasma sprayed film was the most effective barrier. More fundamental studies are needed to determine why these two materials differ in their effectiveness. Using MgO tape as a barrier layer for reasons of low cost and ease of manufacture, we fabricated an 80m-long Rutherford cable and cut it into four pieces, the minimum length being 17m. The I_c of each piece was measured at 4.2K in self-field (see Table 2). A minimum I_c value of 2,700A was obtained in all samples. These results give us confidence that it is possible to fabricate useful long lengths of HTS Rutherford cable.

Table2 The results of Ic value for 17m-long cable.

Sample ID	Position	Ic (A) of 17m-long cable
#1	End A	2,940
#2	End A	3,015
#3	End A	3,060
#4	End A	2,790

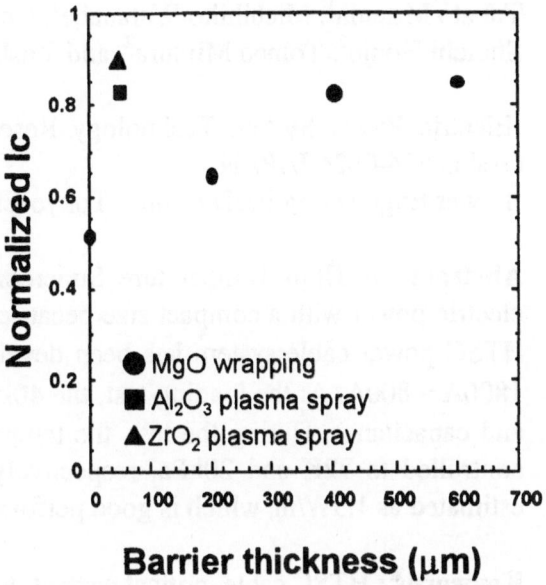

Fig.1 Effect of the ceramic barrier thickness.

CONCLUSION

Robust high-performance Bi-2212 AgMgSb alloy sheathed round HTS wire was developed for Rutherford cable. The round double restack wire consisted of 427 superconducting filaments and had a J_c value of 2.4×10^5 A/cm^2 at 4.2K in self-field. On the other hand, the evaluation of appropriate barrier layer materials indicated that a plasma sprayed ZrO$_2$ film was most effective in protecting the HTS filaments from being poisoned by elements present in the core. Rutherford cable made with 427-filament round wire and a Nichrome tape core covered with a ZrO$_2$ barrier layer was able to carry an I_c of 3,570A, the Ic degradation in Bi-2212 AgMgSb sheathed wire being only 10%. Moreover, we have successfully fabricated an 80m-long Rutherford cable with a Ni alloy reinforcement core wrapped with MgO paper.

REFERENCES

[1] Y. Aoki, T. Hasegawa, N. Aoki, L. R. Motowidlo, R. S. Sokolowski and S. Nagaya, Advances in Superconductivity XI (1999) 1377.

Development of a High Tc Superconducting Cable

Takato Masuda[1], Michihiko Watanabe[1], Chizuru Suzawa[1], Masayuki Hirose[1], Shigeki Isojima[1], Shoichi Honjo[2], Tomoo Mimura[2], and Yoshihisa Takahashi[2]

[1]Electric Power System Technology Research Laboratories, Sumitomo Electric Industries, Ltd., Osaka, 554-0024 JAPAN
[2]Power Engineering R&D Center, Tokyo Electric Power Company, Yokohama, 230-8510 JAPAN

Abstract   A High Temperature Superconducting (HTSC) cable is expected to transport large electric power with a compact size because of its high critical current density.   A 30m 66kV-1kA HTSC power cable system has been developed. The critical currents measured at 67K~80K are 1800A~800A. At the loading test, the 40kV-500A was successfully applied with a constant tan δ and capacitance. During the test, the temperature and pressure of circulating liquid nitrogen were controlled to 72K and 20kPa, respectively. Heat leak through the thermally insulated pipe was estimated as 1.5W/m, which is good performance for a long HTSC cable.

Keywords : HTSC cable, critical current, tan δ , loading tests, heat leak

INTRODUCTION

The growing electric power demand for the future will require the construction of many underground cables. However, the construction of new ducts for power cables will become increasingly difficult due to the overcrowded underground space in major cities.

A High Temperature Superconducting (HTSC) cable is expected to transport large electric power with a compact size because of its high critical current density.[1] Many key technologies, however, are needed such as reducing heat loss generated in the cable, reliable dielectric in liquid nitrogen, flexible thermally insulated pipes of minimal heat leak, a high stable and reliable cooling system, and so on. For the purpose of studying these technologies, a 30m-long 66kV-1kA HTSC cable was designed, developed and evaluated.[2]

A 30m HTSC CABLE

The cross sectional structure of the 30m cable is shown in Fig.1. The outer diameter of this cable was designed to be 130mm, applicable to existing ducts with 150mm inner diameter. The size of the cable was designed to house three cable cores together in the thermally insulated pipe.

The conductor was wound with $(Bi,Pb)_2Sr_2Ca_2Cu_3O_{10}$ silver-sheathed wires spirally around a copper pipe former to make

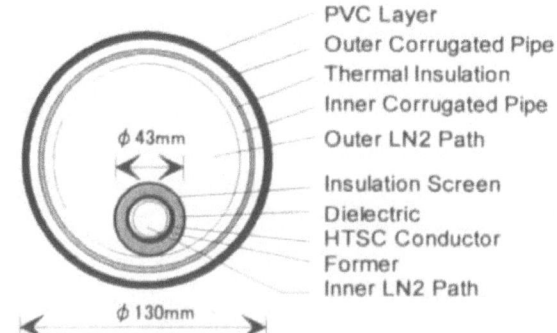

Fig.1. The structure of a 30m HTSC cable

4 layers. Each layer is insulated to cut the eddy current path between the layers for the purpose of reducing AC loss.

The conductor consists of 60 HTSC wires, taking into account the degradation of critical current (Ic) by the magnetic field and mechanical stress history. The degradation is designed 0.5 to the total

Ic calculated by Ic of wire × number of wires.

The cable is insulated with Polypropylene Laminated Paper (PPLP) impregnated with liquid nitrogen, which is similar to the dielectric structure of oil-filled cables. The target value for electrical insulation was 130kV for AC, and 385kV for impulse, which are the withstanding test voltages for 66kV class conventional oil-filled cables. The insulation thickness of the cable was designed under the condition that the electrical stress of the insulation would be free from partial discharge at 130kV AC, that is, 20kV/mm. The insulation thickness of the cable was designed to be 9mm. The thermal insulation consists of two coaxial stainless corrugated pipes and multi-layer insulation (M.L.I.) in a vacuum state.

TEST RESULTS

Test system

A schematic diagram of the HTSC cable system is shown in Fig. 2. Liquid nitrogen, cooled and pressurized by the cooling system, flows into the inner liquid nitrogen path and returns through the outer liquid nitrogen path shown in Fig.1. The liquid nitrogen is sent back to the cooling system, and is cooled and pressurized again for the next cycle.

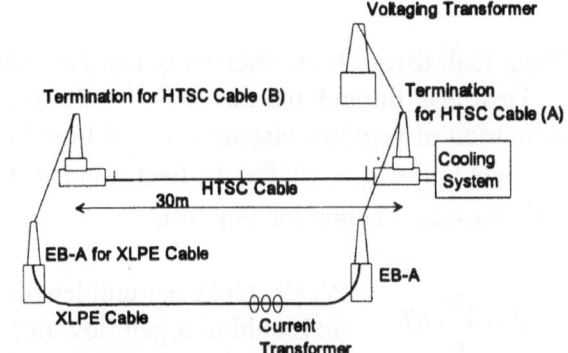

Fig. 2. Schematics diagram of HTSC cable system

The cable was cooled from room temperature to 77K using cooled nitrogen gas and liquid nitrogen. It took about 7 hours to cool the cable to 77K, after which, liquid nitrogen was circulated through the cable.

Measurements of critical current

Critical currents between 67K and 80K are measured. Figure 3 shows the voltage-current chart at each temperature and the critical current values by the definition of 10^{-12} Ωm criterion. The cable has 800A-1800A critical currents between 67K and 80K. The cable critical current dependence on the temperature agrees with those of the wires.

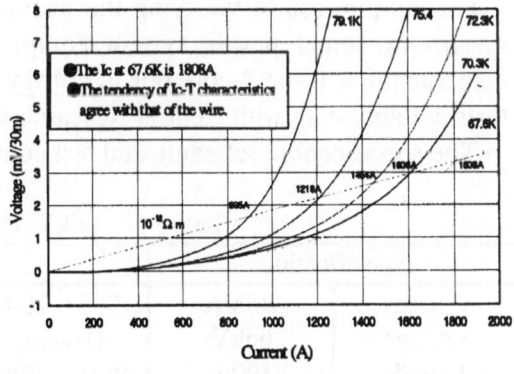

Fig. 3. Critical currents at 67K~80K

Loading tests

The initial loading test was held at 40kV-1kA for 10 minutes successfully. 40kV corresponds to the line to ground working voltage of 66kV cables.

In the next step, 40kV-500A was applied for more than 100 hours. Fig. 4 shows the pattern of the voltage and current during the test. A break of loading was needed by a day because of a supplement of liquid nitrogen to the terminations. The temperature and pressure of the circulating liquid nitrogen was controlled constantly. The tan δ and the capacitance of the cable were about constant, 0.07% and 6.7nF respectively, during the test as shown in Fig. 5.

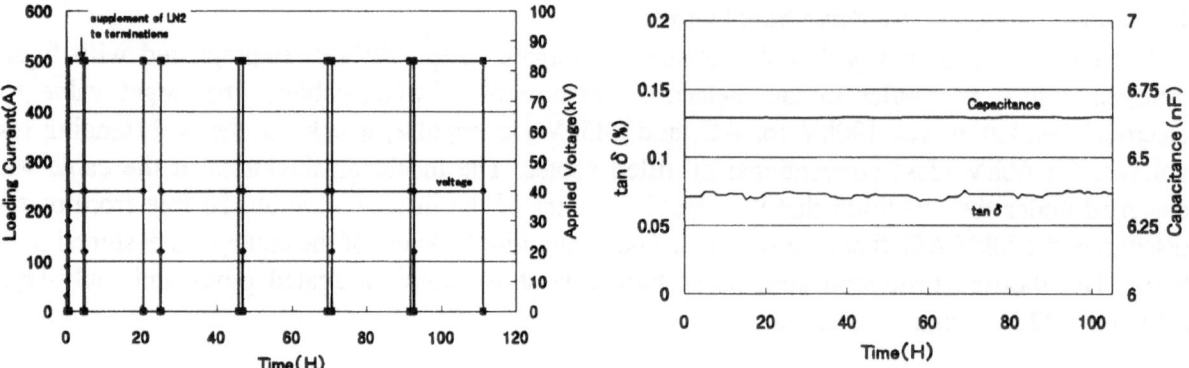

Fig. 4. 40kV-500A loading pattern

Fig. 5. Time variation of tan δ and C
(at 40kV-500A)

Heat leak through the thermally insulated pipe

Heat leak through the thermally insulated pipe is estimated by the following equation under the condition of very low viscosity loss of liquid nitrogen. In the case that the vacuum rate of thermally insulated pipe is $\sim 10^{-4}$Pa, the heat leak is calculated to be 1.5W/m at 3.8L/min LN$_2$ flow rate using $\triangle$T measured at non-loading test.

$$W = \frac{Cm}{L}\Delta T$$

W: Heat leak per unit length, C: Specific heat of liquid nitrogen,
m: Liquid nitrogen flow rate, L: Cable length
$\triangle$T: Temperature difference between entrance and exit of the cable.

CONCLUSION

A 30m long HTSC cable was developed. The cable has 800A-1800A critical currents between 67K and 80K. At the loading test, the 40kV-500A was successfully applied with a constant tan δ and capacitance. Heat leak through the thermally insulated pipe was estimated to be 1.5W/m , which is good performance for a long HTSC cable.

For the purpose of verifying the manufacturing ability and the practicability of an HTSC cable system as actual power system equipment, a new project has been started. The targets are constructing a 100m 3-core 66kV/114MVA HTSC cable system, and conducting long term loading tests collaborating with Central Research Institute of Electric Power Industry (CRIEPI).

The specification, schedule and test contents are described in Table 1.

Table 1.   66kV-1kA HTSC Cable System Prototype

Specification		Schedule	Test contents
Current	1000A	1999~2000	• Long term loading tests
Voltage	66kV	Development and construction	• Variable loading tests
Length	100m	2001~2002	• Overloading tests
Cable type	3-core	Laying , Long term tests	• Heat cycle tests etc

1   S.Kobayashi, et. al., *Progress in long length Bi-2223 tapes*, to be published in Advances in Super-conductivity X, 1998

2   T.Shibata, et. al., *Development of High Temperature Superconducting Power Cable Prototype System*, IEEE Transactions on Power Delivery, Vol.14, No.1, Jan. 1999

STUDY OF LIQUID NITROGEN PRESSURE LOSSES FOR HIGH-Tc SUPERCONDUCTING CABLE

[1]*K. Ohno, [2]*S. Nagaya*, [2]*T. Suga*, [1]*N. Futaki, and [1]*T. Saitoh

[1]*Fujikura Ltd.,Chiba 293-0043,Japan [2]*Chubu Electric Power Co.,Nagoya 459-8522,Japan

ABSTRACT Liquid nitrogen can be used as cooling medium for high-Tc superconducting cables. In order to allow liquid nitrogen to flow properly under pressure, we must consider the pressure required for electrical insulation and the pressure drop of liquid nitrogen. Pressure drop is determined by factors such as pipe shape, liquid viscosity, specific gravity, and flux. It is essential to have a good grasp of these factors in designing a cooling system.As part of preliminary study, we measured pressure drops within straight pipe and corrugated pipes. Then, we introduced a round bar into each of these pipes to determine the relationship between change in pressure drop and round bar diameter. The following makes a report on our findings.

Keyword : liquid nitrogen, straight-pipe, corrugated-pipe, pressure drop

INTRODUCTION   When a high-Tc superconducting cable is used, it is necessary to perform two tasks, namely, (1) maintaining the superconducting state and (2) allowing liquid nitrogen under pressure to flow through the pipe to ensure electrical insulation. When liquid nitrogen flows, pressure losses, determined by the pipe shape and Reynold's number, occur. Studying pressure losses is essential for reviewing high-Tc superconducting cable systems. In this experiment, we used a reduced model in which a round bar was introduced into each of various types of pipes to simulate a single conductor cable in order to measure variations in pressure loss. Based on the results of experiment, we derived a formula that would lead to variations in pressure loss with change in round bar diameter. We used this formula to evaluate pressure losses occurring in cables that are equivalent in size to actual cables. Assuming that a cable applicable in size to a pipe of 150 mm in diameter is used, we found from our calculations that the cooling length for corrugated pipe is only approximately 58% of that for straight pipe.

EXPERIMENTAL   Table 1 shows the test conditions while Table 2 and Figure 1 show the shape of the tested pipe. The pipe to be tested was provided within another pipe that was double-heat-insulated by means of vacuum. Liquid nitrogen under 500 kPa was supplied at a specific flow rate from the tank through this pipe.We collected data when the inlet and outlet pressures and temperatures stabilized.

Table 1.   Test conditions

Item	Specification
Cooling medium	LN2(79 to 80K at inlet)
Flow rate	10 to 65 liters/min
Inlet pressure	500kPa(Const.)
Effective length	25m(straight)

Table 2.   Tested pipe shape

	Straight pipe	Corrugated pipe
Inner dia. ϕD	21.4mm	22.9mm
Height	———	2.0mm
Pitch	———	7.0mm
Shape	———	Spiral
Round bar dia. ϕd	6,9,12,15mm	

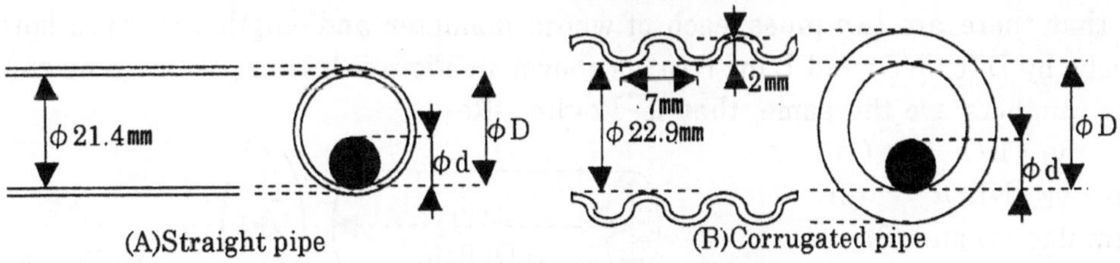

Fig.1.   Pipe shape (with round bar)

(A)Straight pipe   (B)Corrugated pipe

RESULTS & DISCUSSION

(A) Measurement results Figure 2 shows the pressure losses measured in the straight and corrugated pipes. Note that the axis of ordinates of the graph represents pressure losses in kPa/m. Two differences can be pointed out between straight and corrugated pipes.
a) Larger pressure loss occurs in the corrugated pipe than in the straight pipe.
b) Larger the round bar diameter, smaller the difference in pressure loss between pipes.

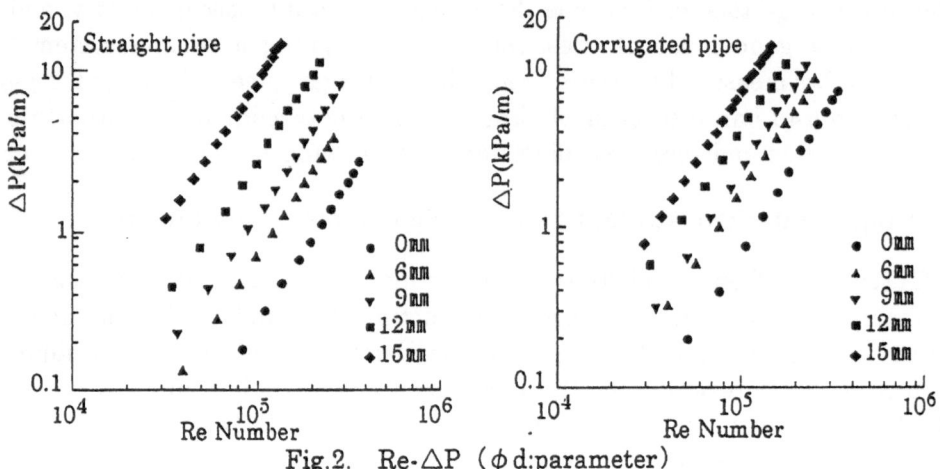

Fig.2. Re-$\triangle$P (ϕ d:parameter)

(B) Effect of ϕ d From the results shown in Figure 2, we solved for $\triangle$P when Re=10^5 to obtain Figure 3. Each of the ● and ▲ in Figure 3 represents measured values while straight lines represent the formulas that satisfy the measured values. From Figure 3, we found that $\triangle$P changes in proportion to the power of (1-n) when an object equivalent to a single conductor cable is provided within the pipe, and that the exponent for the straight pipe is different from that for the corrugated pipe.

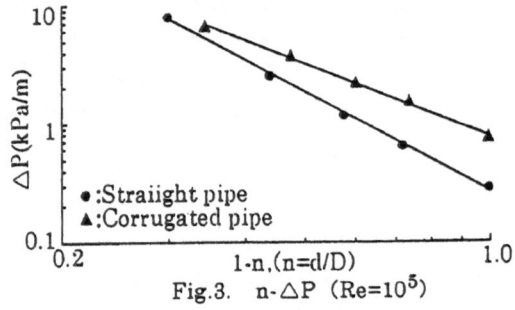

Fig.3. n-$\triangle$P (Re=10^5)

(C) Study of pressure losses occurring in pipes of actual size In the experiment, we used pipes of approximately 20 mm in inner diameter. For practical applications, however, we assume that cables applicable to pipes of 150 mm in diameter will be used. For this reason, we applied the Reynold's law of similarity to calculate pressure losses occurring in pipes of actual size based on the solution we obtained in the experiment. The following shows the concept: Suppose that there are two pipes, each of whose diameter and length are represented respectively by D (or ND) and L (or NL), as shown in Figure 4. Here, we assume that the Reynold's numbers are the same, that is, Re=Re$_1$=Re$_2$.

$\text{Re}_1 = \rho \cdot v_1 \cdot D / \mu$ ·· (1)

$\text{Re}_2 = \rho \cdot v_2 \cdot ND / \mu$ ··(2)

From formulas (1) and (2), we can derive formula (3).

$v_1 = N \cdot v_2$ ····(3)

Pressure loss $\triangle$P occurring between two ends of each pipe is as follows:

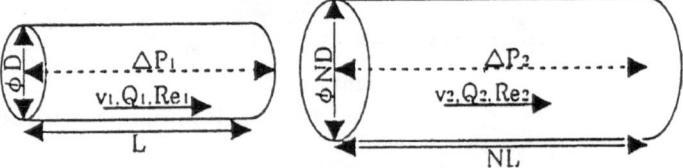

Fig.4. Geometrically similar pipes

$$\triangle P_1 = \lambda_1 \cdot L \cdot \rho \cdot v_1^2/2D \quad \cdots(4) \quad \triangle P_2 = \lambda_2 \cdot NL \cdot \rho \cdot v_2^2/2ND = \lambda_2 \cdot NL \cdot \rho \cdot (v_1/N)^2/2ND \quad \cdots(5)$$

From formulas (4) and (5), we can derive formula (6). $\quad \triangle P_2 = \lambda_2/\lambda_1 \cdot \triangle P_1/N^3 \quad \cdots(6)$

Where D:Inner diameter , L :Pipe length , $\triangle$P:Pressure drop , v:Velocity of cooling medium , Q:Value of flowing cooling medium , λ:Coefficient friction

Using formula (6), it is possible to calculate $\triangle P_2$ occurring in the pipe that is N times larger based on $\triangle P_1$ obtained in the experiment.

(D) Target model   In this study, we assumed that a single conductor cable applicable to a pipe of 150 mm in diameter was used. We assumed that the cable outer diameter was 130 mm and changed the former diameter in Figure 5 to calculate the cable length that would provide a

pressure beyond 500 kPa at the Return pipe outlet. Note that we assumed that the differences in temperature and pressure between the Go pipe inlet and the Return pipe outlet were 12 K, 1500 kPa, and that the thermal load was 2, 4, and 6 W/m.

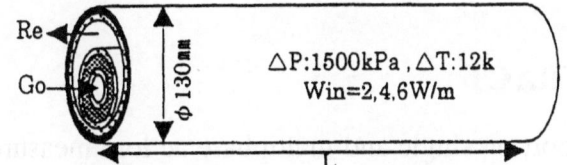

Fig.5.   Shape of the target model

(E) Calculation results   Figure 6 shows the calculation results obtained from two different Figure 5 models, one model using straight pipes for both heat-insulated and former, and the other model using corrugated pipes for both heat-insulated and former. Although the solution varies depending on the thermal load, the following tendencies have been observed:
(a) The cooling length for corrugated pipes is approximately 58% of that for straight pipes.
(b) Thermal load has large impact on cooling length in the region where thermal load is small.
(c) There is a former diameter that results in the maximum cooling length. (This is due to the fact that the outer diameter is limited.)

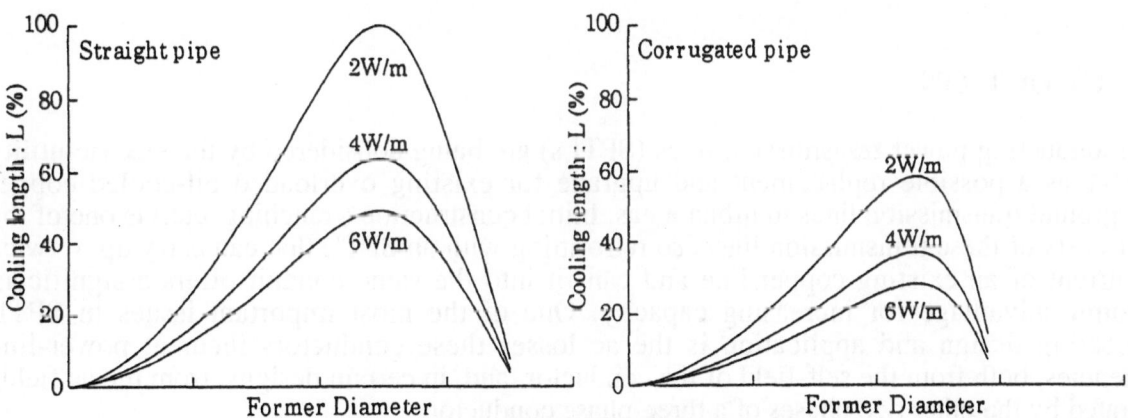

Fig.6.   Calculation results obtained from target model

Conclusion   We investigated the relationship between the round bar diameter and pressure loss by introducing a round bar into each of the reduced pipe models (straight and corrugated pipes) to simulate a single conductor cable. Based on the results obtained, we calculated pressure losses that could occur in pipes of actual size to study cooling length. As a result, we found that the cooling length for corrugated pipe is approximately 58% of that for straight pipe.We will submit 3-conductor cable to the same study to review the cable structure in terms of cooling length.

Multiphase Losses in HTS Prototype Multistrand Conductors

J.O. Willis,[1] D.E. Daney,[1*] M.P. Maley,[1] H.J. Boenig,[1] S. Fleshler,[2] R. Mele,[3] G. Coletta,[3] M. Nassi,[3] and J.R. Clem[4]

[1]MST-STC; MS-K763, Los Alamos National Laboratory, Los Alamos, NM 87545 USA
[2]American Superconductor Corp., Two Technology Drive, Westborough, MA 01581 USA
[3]Pirelli Cavi & Sistemi, Viale Sarca 222, I-20126 Milano, ITALY
[4]Department of Physics & Astronomy and Ames Laboratory, Iowa State University, A517 Physics, Ames, IA 50011-3160 USA

ABSTRACT

We report on single and multiphase ac loss measurements in four-layer prototype multi-strand conductors (PMCs) wound from HTS tape. "Two phase" losses are induced with no current flowing in the PMC but with an external ac magnetic field generated by the two normal conductors arranged at the remaining corners of an equilateral triangle forming a three-phase configuration. The losses were measured over a temperature range of 65 to 76 K, a frequency range from ≈ 10 to ≈ 200 Hz, and in magnetic fields of 0 to ≈ 5 mT. Single-phase losses are purely hysteretic in nature and exhibit power law dependence of current I ranging near 3. Two-phase loss measurements reveal two loss mechanisms with different frequency dependencies: one hysteretic in nature and the second eddy-current-like. The second term saturates near or above power line frequencies. The magnitude of the two-phase losses is proportional to the current $I^{2.6}$.

Keywords: ac losses, power transmission conductors, Bi-2223/Ag tapes

INTRODUCTION

Superconducting power transmission lines (SPTLs) are being considered by the electric-utility industry as a possible replacement and upgrade for existing overloaded oil-cooled copper underground transmission lines in urban areas. Initial construction (trenching, etc.) is one of the major costs of these transmission lines, so retrofitting with an SPTL that can carry up to twice the current of an existing copper line and can fit into the same conduit offers a significant economic advantage for increasing capacity. One of the most important issues in SPTL engineering design and application is the ac losses these conductors incur at power-line frequencies, both from the self-field of the conductor, and, in certain designs, from the ac fields generated by the other two phases of a three-phase conductor.

We report here on the "single-phase" and "two-phase" ac losses resulting from the influence of the other two phases of a three-phase SPTL on the phase under test. In a coaxial SPTL design, each superconducting phase is surrounded by a superconductive shield, thus there is no net magnetic field to influence the other phases, and the "single-phase" losses represent the total ac losses for each phase. In another design, there is no coaxial shield, thus reducing the use of HTS material significantly, but at the expense of interaction among the phases. If the conductors are relatively far apart, e.g., 20-30 cm, there is little effect of the fields from the other phase conductors.[1] However, because all three phase conductors are usually installed in a single conduit with conductor spacing on the order of 10 cm where space is at a premium, the influence of the other phases does need to be considered.

EXPERIMENTAL PROCEDURE

We use a calorimetric technique for measuring the single and multiphase ac losses, not easily determinable by voltage measurements [2]. Reference [3] contains substantial detail about the calorimeter construction and operation. Briefly, in this technique, the PMC is thermally isolated except at the two ends. Heat generated uniformly in the conductor results in a parabolic temperature profile along the conductor length. The PMC is contained in a G-10 (glass epoxy) vacuum jacket with the ends (the current leads) cooled by the surrounding liquid nitrogen. The other two phases in the three-phase configuration are normal copper conductors, with the three phases arranged at the vertices of an equilateral triangle 10 or 20 cm on a side. The current source can supply single or three-phase current at continuously variable frequencies between 10 Hz and 180 Hz at currents up to 1000 A rms and up to 3000 A rms near 60 Hz.

The PMC tested here was helically wound by Pirelli Cavi & Sistemi from Bi-2223/Ag-sheathed composite conductor HTS tape manufactured by American Superconductor Corporation. The 1-meter-long PMC was wound by hand onto a 30 mm-diameter G-10 epoxy fiberglass mandrel in four layers with alternate layers of equal and opposite pitch. The critical currents are about 6,000 A at 77.3 K with a 1 μV/cm criterion.

RESULTS AND DISCUSSION

Figure 1 shows the current dependence of the single-phase losses of the PMC scaled to 76 K. The losses at a given frequency f scale with temperature according to the critical-state model, i.e., loss $q_L \propto 1/I_c(T)$ in the measured current range. $I_c(T)$ was determined independently on a single Bi-2223/Ag tape. Figure 2 shows the results of single-phase loss measurements on PMC 4LA at 76 K and 1000 Arms as a function of frequency. The result indicates a purely hysteretic (Bean model) loss, i.e., $q_L \propto f$. [4] The filaments within each tape conductor are electromagnetically fully coupled.

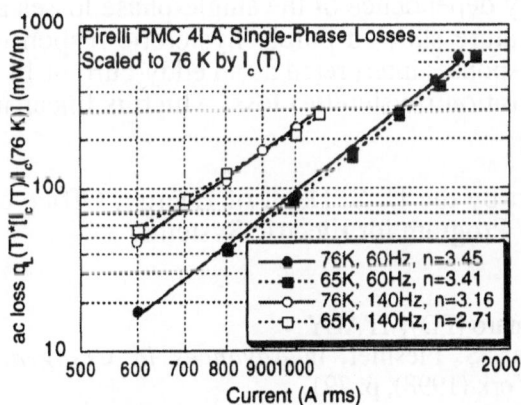
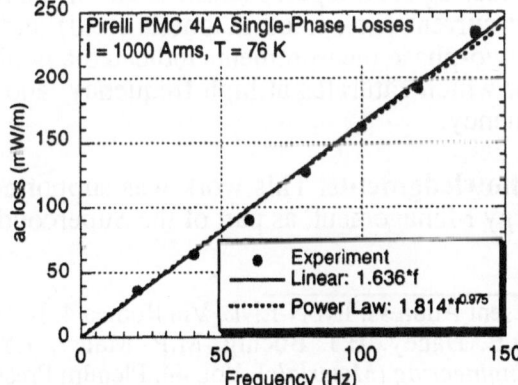

Fig. 1. Single-phase ac losses as a function of current scaled to 76 K by the critical current [$q_{l} \propto 1/I_c(T)$]. n is the exponent of the power law fit to the data.

Fig. 2. Single-phase ac losses at 76 K and 1000 A rms. The data are consistent with purely hysteretic losses represented by the good linear and power law (≈ 1) fits.

Figure 3 shows the current dependence of two-phase losses of the PMC scaled to 76 K. Here the scaling is not as good as for the single phase results. Scaling of losses with frequency ($\propto f$) is quite poor. An investigation of the two-phase losses as a function of frequency, the results of which are shown in Fig. 4, reveals the answer. The frequency dependence is very nonlinear, indicating the presence of more than one type of loss mechanism [5].

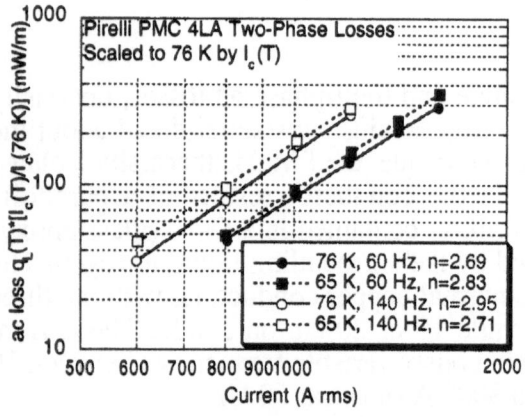
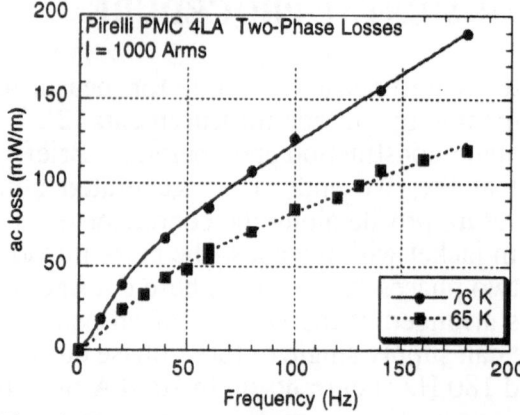

Fig. 3. Two-phase ac losses as a function of current scaled to 76 K by the critical current [$q_L \propto 1/I_c(T)$].

Fig. 4. Two-phase ac losses at 76 and 65 K and 1000 A rms. The data indicate the presence of both hysteretic and eddy-current like loss mechanisms.

In the presence of the magnetic field generated by the two normal phase conductors in the two-phase loss measurement, hysteretic, or "penetration" losses occur in the PMC. They can be characterized by the expression $q_p = (B_m^3/4\mu_0)(\omega/I_c)$, where B_m is the applied field, ω is the angular frequency, and I_c is the critical current. Currents are also forced to flow in the matrix material between the filaments and at the edges of the HTS tape. These currents also generate what are called "eddy-current" or "coupling" losses according to $q_e = (B_m^2/4\mu_0)(\omega^2\tau)/(\omega^2\tau^2+1)$, where τ characterizes the saturation behavior as a function of frequency. The lines in Fig. 4 are fits to the data that include both these loss mechanisms. Near 50-60 Hz, the eddy-current term is found to be nearly saturated, and the magnitudes of the eddy-current and penetration terms are approximately equal.

CONCLUSIONS

We have examined the single and two-phase losses in a 4-layer PMC constructed of Bi-2223/Ag HTS tapes. Measurement of the frequency dependence of the single-phase losses at fixed current (i.e., fixed magnetic field) and temperature showed purely hysteretic response. The two-phase measurements indicated a nonlinear response, interpreted as an eddy-current-like term, which saturates at high frequency, and a penetration (hysteretic) loss, which is linear in frequency.

Acknowledgments. This work was supported in part by the U. S. Dept. of Energy, Office of Energy Management, as part of the Superconductivity Program for Electric Systems.

---

* Present Address: INFN-LNL, Via Romea 4, I-35020 Legnaro (PD), ITALY
1. D.E. Daney, H.J. Boenig, M.P. Maley, J.Y. Coulter, S. Fleshler, in *Advances in Cryogenic Engineering (Materials)*, Vol. 44, Plenum Press, New York (1998), p. 791.
2. S. Fleshler, et al., *Appl. Phys. Lett.* 67, 3189 (1995).
3. D.E. Daney, H.J. Boenig, and M.P. Maley and S. Fleshler, *Cryogenics* 39, 225 (1999).
4. D.E. Daney, J.O. Willis, M.P. Maley, H.J. Boenig, R. Mele, G. Coletta, in *Advances in Cryogenic Engineering (Materials)*, Vol. 46, Plenum Press, New York (in press).
5. J.O. Willis, D.E. Daney, M.P. Maley, H.J. Boenig, R. Mele, G. Coletta, in *Advances in Cryogenic Engineering (Materials)*, Vol. 46, Plenum Press, New York (in press).

Numerical Analysis of AC Losses and Critical Current in High-Tc Superconductors with Power Law Current-Voltage Characteristics

Shoichi Honjo, Yoshibumi Sato, and Yoshihisa Takahashi

Power Engineering R&D Center, Tokyo Electric Power Company, Yokohama 230-8510, Japan

Abstract: We have started a project of fabricating a 100m, 66kV/114MVA HTS cable system prototype, aiming at certifying its practicability and extracting the necessary subjects for developing a field-level cable system. By using a numerical analysis code, which takes into account power law current-voltage characteristics of superconductors, AC losses of such conductors and their characteristics were analyzed.

Keywords: HTS cable system, power law model, AC losses, uniform current distribution

INTRODUCTION

Power demand in metropolitan areas is predicted to steadily increase in the future. HTS power cable has the potential to realize a compact and high capacity cable. Also we can expect great cost reduction if we use existing ducts of 150mm inner diameter with the HTS cables instead of the 275kV cables which require tunnels if power demand is to be increased. Our target is therefore to develop compact HTS power cables having 66kV/several 100MVA capacity.

For this purpose, TEPCO, Tokyo Electric Power Company, and SEI, Sumitomo Electric Industries, have started a project to develop a 100m, 66kV/114MVA HTS cable system prototype using Bi-2223 tapes, and conduct long term loading tests collaborating with CRIEPI, Central Research Institute for Electric Power Industries. The specifications and test conditions are listed here in after. In this project, we aim at certifying its practicability as actual power system equipment, and extracting the necessary subjects for development of a field-level HTS cable system.

Table 1   Specifications of HTS cable system prototype and schedule

◆Target	◆Schedule
· Extraction of the necessary subjects for realization of a field-level system	· Development: by the end of 03/2001
· Certification of practicability of system	· Construction at test yard: till mid 2001
◆Cable system	· Tests: mid 2001 to mid 2002
· Type: Cold dielectric type; 3-phase in thermal insulation pipe	· Project partner: TEPCO and SEI
	· Collaborator at tests: CRIEPI
· Conductor: spiral-pitch adjustment using Bi-2223 tapes for AC loss reduction	◆Test
· Capacity: 66kV/1kArms; 114MVA	· Construction, laying and initial cooling test
· Length and size: 100m having a U-type bend	· Long term rated current-voltage loading test
· Size: less than 150mm in outer diameter	· Load fluctuation test
· Cooling: closed circulation of sub-cooled LN_2	· Overloading and overvoltage withstand test
	· Heat cycle test

ANALYSIS MODEL

Estimation of AC losses is crucial for the design of HTS cable system. Though Norris' formula is conveniently used for this purpose, this formula is not accurate enough when the n-value is small, (Note: where n-value is the exponent of power law current-voltage characteristics, $E=E_0(J/J_c)^n$). Also Norris' formula is applicable only when current does not exceed the critical current. It is important to grasp AC loss characteristics around and above the critical current, as transport current may exceed the critical current in the case of overloading or fault current flowing in actual cable. This being the case, we have developed a numerical analysis code that can treat the power law characteristics to deal with these problems [1], and analyzed AC losses of such a cable.

As is listed in Table 2, the analysis model is a cylindrical conductor. Instead of actual superconducting layers, we treated them as bulk superconducting layers of uniform characteristics in the cross-section. A thin air layer was placed between each layer for insulation purposes. AC loss characteristics were compared when current distribution was not restricted among layers (mono-block model), and current was forced to equivalently flow in each layer (unified model).

RESULTS AND DISCUSSION

Current and magnetic distributions, and the AC losses of the mono-block model when $n=8$ are shown in Fig. 1(a) and 2, respectively. Where I_c denotes the critical current determined from the DC current-voltage characteristics. Current concentrates to the outer layers due to the smaller inductance, with most AC losses produced in the outermost layer. In the low current region, the slope of the graph, α, is about 2.8. The bold line is obtained from the following Rhyner's formula [2].

$$W_{ac} = \frac{\mu_0 f (\pi R^2 \tilde{J}_c)^2}{6\pi} \left(\frac{I_p}{\pi R^2 \tilde{J}_c} \right)^3 , \tilde{J}_c \approx \frac{4J_c}{3\xi(n)} \left(\frac{\mu_0 f I_p^2}{2\pi R^2 J_c E_0} \right)^{1/(1+n)} , \xi(n) = 1.33 + 3.11n^{-0.55} . \quad (1)$$

It expresses the dependence of α to n-value as $\alpha=(1+3n)/(1+n)$, which gives $\alpha=2.8$ with $n=8$. The analyzed results well coincide with the theory both in values and slope. This formula also suggests that α approaches 3 with an increase in n-value. In the high current region, the slope becomes larger and the analytical result is well represented by the bold line obtained from the next equation, where current distribution is assumed to be uniform over the cross-section of the superconducting layers.

$$W_{ac} = \int I(t) \cdot E_0 (I(t)/I_c)^n dt = (1/2) \cdot E_0 I_c (I_p/I_c)^{n+1} k(n), \ k(n) = (2/\pi) \cdot \int_0^\pi \sin^{n+1} x dx . \quad (2)$$

If we view the crossover point as AC critical current I_c^{AC}, by combining equations 1 and 2, we obtain

$$I_c^{AC} / I_c = (\mu_0 f d I_c / E_0 R)^{1/(n-1)} \cdot (\xi(n) / 2\pi k(n))^{(n+1)/n(n-1)} , \quad (3)$$

which indicates the dependence of I_c^{AC} on the n, I_c, E_0, f, and the specifications of the conductor, such as total thickness, d, and radius, R, of superconducting layers.

The effect of unifying current distribution among layers is shown in Fig. 1(b) and 3. AC losses are

Table 2   Specifications and conditions of cylindrical conductor used for analysis

◆Conductor	◆Analysis model and elements
· Inner/outer diameter, number of layers	· Size : 1/2000 cut model (0.18deg)
conducting layers = 40.00/40.1648mm, 4	length along the axis = 0.1mm
shielding layers = 80.00/80.0814mm, 2	· Division into elements of each layer
· $J_c = 1.5 \times 10^4 \text{A/cm}^2$; $I_c = 1500\text{A}$	20 divisions in radial direction
· $E_0 = 0.15 \mu\text{V/cm}$; n-value = 8	2 divisions in axial direction

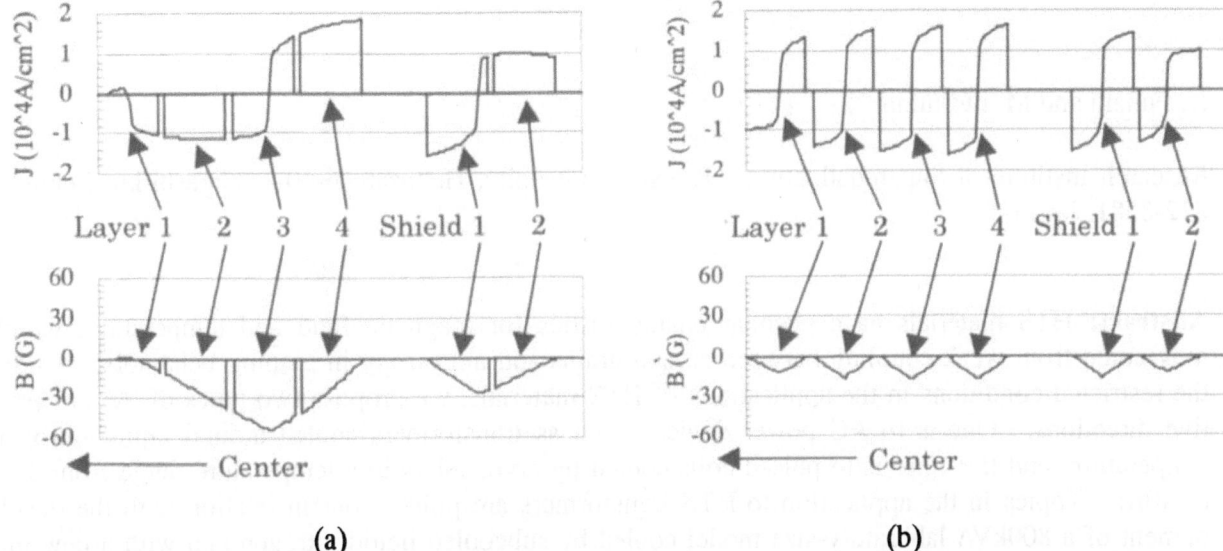

Fig. 1 Current and magnetic distributions of mono-block model (a), and unified model (b) at $I_p=0.8I_c$ and $\theta=0$, where θ is the angle of alternative current $I_p\sin\theta$. Layers are numbered from inner layers in order. Therefore, Layer 1 corresponds to the innermost layer.

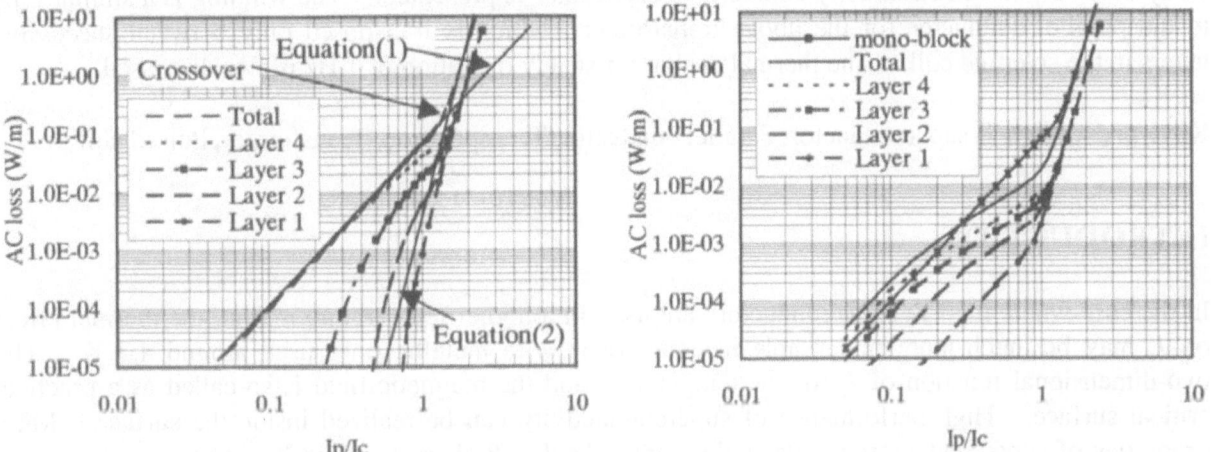

Fig. 2 AC losses of mono-block model. I_p denotes the peak transport current.

Fig. 3 AC losses of unified model.

reduced as compared with the mono-block model except low I_p region. A major portion of the total AC losses though still comes from the outermost layer. The decline of I_c^{AC} is shown in Fig. 3, which is explainable as in this case: each layer's thickness and critical current should be used as d and I_c in equation 3.

CONCLUSION

We have started a project which demonstrates a 100m, 66kV/114MVA HTS cable system prototype with the aim of certifying its practicability and extracting the necessary subjects for developing field-level HTS cable systems. AC losses of such a conductor were analyzed using a numerical analysis code. Results show that the current of inflection point in the double logarithmic graph of current vs. AC losses did not coincide with I_c, but depended on the criterion of I_c, specifications of superconductors, etc.

1. S. Honjo, et al. (1999) Advances in Superconductivity XI:1389; 2. J. Rhyner, Physica C 212 (1993)292

RECENT APPLICATIONS OF Bi-2223 TAPES TO TRANSFORMER WINDINGS AND PULSED COILS

K. Funaki and M. Iwakuma

Research Institute of Superconductivity, Kyushu University, Hakozaki 6-10-1, Higashi-ku, Fukuoka 812-8581, Japan

Abstract: HTS materials have peculiar characteristics for magnetic field and temperature, which may come from weak coupling between crystal grains and anisotropy in pinning behaviors. Under the restricted conditions in the applications of HTS materials, we propose two types of the prospective directions. One is to AC power devices such as transformers cooled around liquid nitrogen temperature, and the other is to pulsed coils cooled by cryocoolers in a temperature range from 30K to 40K. Topics in the application to HTS transformers are pointed out in relation with the development of a 800kVA laboratory-size model cooled by subcooled liquid nitrogen and with a new initiative of high-voltage model with rated levels of 22kV / 6.9kV. The key points are winding configuration of parallel conductors with low AC loss and high capacity, reliable response to excess current due to sudden short and high-voltage performance using liquid nitrogen. Secondly, recent study of a 1T pulsed coil directly cooled by a cryocooler is presented. The winding is a similar type to the parallel conductor for the above transformers, which is transposed only between successive layers in the solenoid coil. The thermal design for steady operation of 1 Hz is also discussed.

Keywords: High T_c superconductor, Parallel conductor, Transposition, Transformer, Pulsed coil

INTRODUCTION

In the case where high T_c superconductors are used in a higher temperature region, the thermal effect on J_c may be much more remarkable in comparison with metallic ones used around 4.2 K. The two-dimensional function of J_c for the temperature and the magnetic field is so-called as a practical critical surface. High performance of superconductivity can be realized inside the surface. Main properties of superconductors are dependent upon the J_c - B characteristics at operation temperature. In the case of LTS materials, the thermal effect on the J_c - B characteristic is negligible on the plane of the low-temperature region (4.2 – 20 K). On the other hand, J_c for the HTS materials have more severe degradation in the region of high magnetic field on the planes of higher temperature.

Around liquid nitrogen temperature, for example, J_c of Bi-2223 Ag-sheathed tapes decreases extremely in the magnetic field more than 1T even in the direction parallel to the tape surface. However, the efficiency of refrigeration, which is improved about 50 times as high as that around liquid helium temperature, relieves a permissive level for AC loss in HTS wires drastically. This economical merit may result in a prospective application of HTS materials to AC uses especially in lower magnetic field, such as the windings of transformers, power cables and current limiters. In the middle-temperature region from 30 and 40 K, we can utilize the advantageous aspects both at low- and high-temperature operations, namely better performance of the J_c - B characteristics and higher efficiency of refrigeration. Bi-system HTS tapes may be also applicable to the windings of high-rate pulsed coils in the middle-temperature region.

Table 1 Design parameters of parallel conductors for HTS transformers

	1-phase laboratory-test model	1-phase field-test model
capacity	0.5 MVA [0.8 MVA]	0.5 MVA [1.0 MVA]
prim./sec. voltage	6.6／3.3 kV	22／6.9 kV
prim./sec. current	72A/152A[121/242 A]	22.7/72.5A[45.4/145A]
frequency	60 Hz	60 Hz
magnetic induction	1.6 T (room temperature)	1.6 T (room temperature)
%impedance	0.67 %	2.5 % [5.0 %]
strand	Bi-2223 Ag-sheath tape	Bi-2223 Ag-sheath tape
twist pitch	no twist	no twist
matrix	pure silver	Ag/Mn
conductor type	transposed parallel	transposed parallel
strand number	3 / 6	2 / 4
layer number	2 / 2	4 / 2
coolant	liq. N2 [subcooled N2]	liq. N2 [subcooled N2]
AC withstand voltage	—	50 kV (design value)
impulse test voltage	—	100 - 150 kV (design value)

We have proposed parallel conductors of Bi-2223 Ag-sheathed tapes transposed for large-capacity HTS windings of transformers and pulsed coils. In this paper, we report main characteristics of the HTS devices designed and developed with the parallel conductors.

PARALLEL CONDUCTORS

In AC and pulsed HTS coils, one of the most important subjects is to construct large-capacity windings with low additional heat generation. This means the current distribution among the strands should be made uniform. In the case of HTS tapes with rectangular cross-section, parallel conductors are proposed as the low-AC-loss windings. [1] The uniform distribution of transport current can be designed by transposition of strands in the winding structure. We have various types of transposition for the winding structures. The strands can be transposed in each layer and/or in turning parts between layers for solenoid coils, in connection parts between single-pancakes for pancake-type coils and so on. In each case, the manner of transposition should be designed for canceling net interlinkage magnetic flux between each set of strands. In the present paper, two types of transposition are reported for the windings of transformer [2], [3] and multi-layer coil [4].

TRANSFORMER WINDINGS

800kVA laboratory-test transformer. [2] For technical applicability of the HTS parallel conductors and their high performance in AC loss characteristics, we designed and fabricated an HTS transformer with capacity of 500 kVA / 800 kVA operated in saturated liquid nitrogen of 77 K or subcooled nitrogen of 66 K, respectively. The primary / secondary voltage is 6600 V / 3300 V. Specifications of the strands and windings used for the HTS transformer of a laboratory-test device are listed in Table 1. Contents in brackets are for the operation in subcooled liquid nitrogen. The strands are Ag-sheathed Bi-2223 multifilamentary tapes that have 61 filaments with no twisting. The primary winding is a parallel conductor with three strands, that is a unit of parallel conductor, and the secondary one is composed of two units connected in parallel. In the parallel conductors, the strands are electrically insulated with each other by a glass-fiber tape and transposed five times in each layer, as shown in Fig. 1. The transposition points are set by calculating the distribution of magnetic field in the winding structure. The primary and secondary coils have two layers, respectively, that are sandwiched mutually as shown schematically in Fig. 2(a) for diminishing magnetic field and consequently AC loss in the windings. Figure 2(b) is an overview of the transformer.

The operation at 77 K included no-load, short circuit and load tests for the rated level of 500 kVA in the same manner as those for the conventional ones. The results of these steady tests show that the transformer has high performance corresponding to the designed level. Total heat loss including AC loss in the HTS windings was also measured by both of thermal and electrical methods. The efficiency of the transformer attains to 99.3 % even if the refrigeration penalty is estimated as the factor 20. Next, the transformer was operated at 66 K with a continuous-flow system of subcooled liquid nitrogen. The cryostat for the transformer and the LN_2 pump were

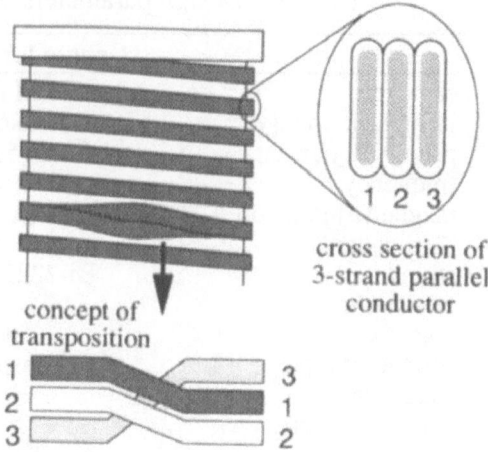

Fig. 1   Concept of transposition in layer

maintained at an atmospheric pressure. We also made use of the results of the no-load test in the saturated liquid nitrogen as those in the subcooled nitrogen. Therefore, we performed only the short-circuit test. In the maximum excitation of 121.6 A, the transformer has a capacity of 800 kVA for the primary voltage of 6600 V. Since the total thermal load to the system exceeded the cooling capacity, the thermal situation of the subcooled nitrogen was non-steady. It was, nevertheless, confirmed that the transformer has high efficiency of 99.4 % at the rated condition of 800 kVA. In this estimation, we make use of the core loss and the heat leakage obtained in the characteristic tests at 77 K in addition to the AC loss measured in the windings at the 66 K operation. The test results are listed in Table 2.

22kV/6.9kV field-test transformer. [3] A project team in the Kyushu area has designed a high-voltage-type HTS transformer with support of the New Energy and Industrial Technology Develop-

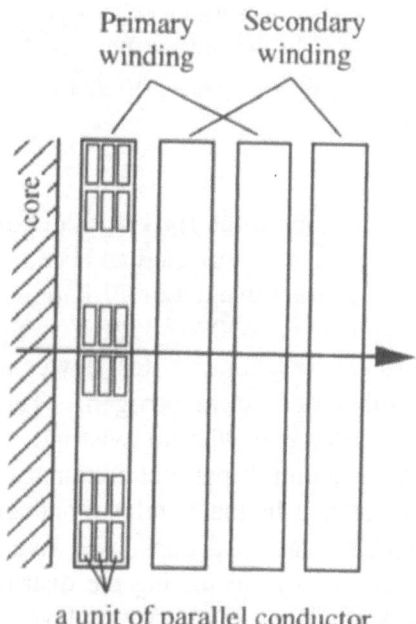

Fig. 2 (a) Structure of parallel conductor and windings, (b) Photograph of the HTS transformer

Table 2   Main results of laboratory tests for the HTS transformer

winding temperature	77 K	66 K
voltage (primary)	44.0 V	69.2 V
current (primary)	75.8 A	121.6 A
corresponding capacity	503 kVA	800 kVA
efficiency	99.1 %	99.3 %
core loss	2289 W	2289 W
a.c. loss in windings	64 W x 20*	126 W x 20*
heat leakage	51 W x 20*	51 W x 20*

* The factor of refrigeration penalty, 20, is considered

Table 3   Specifications of model coil for over-current tests

	HTS coil	copper backup coil
coil sizes		
inner diameter	200 mm	240 mm
height	497 mm	496.4 mm
winding width	1 mm	9.2 mm
conductor / winding		
strands	Bi-2223 tapes	copper tape
strand cross-section	3.4 x 0.24 mm^2	7.3 x 2.3 mm^2
parallel number	4	1
turn number	64	68
layer number	1	4
transposition	7 times	—

Fig.3   Overview of model coil for over-current tests

ment Organization.   The voltage level is equivalent to those of distribution transformers in spot networks.   The voltage is 22 kV / 6.9 kV.   The design parameters are listed in Table 2.   On the basis of the laboratory-test facility mentioned in the previous sub-section, the project is placed as a step for application to HTS transformers in underground distribution substations.

The main subjects of the field-test device are to develop advanced technology for over-current and high-voltage tests, AC withstand voltage test and lightning impulse voltage test, equivalent to those imposed to conventional oil transformers.   A model coil has been designed and fabricated for the over-current tests.   The model coil is composed of a HTS coil and a copper backup coil.   Main characteristics of the two coils are listed in Table 3.   The overview is shown in Fig. 3.   The winding of the HTS coil is composed of the same Bi-2223 wires as that of the field-test device. The winding is a 4-strand parallel conductor of Bi-2223 Mn-Ag sheathed tapes with 7-times transposition in a layer of 64 turns.   The HTS coil can be connected with the backup coil in different two ways.   One is in a normal direction and the other is in a reverse one.   In this way, the HTS winding in the model coil can be exposed to compressive or tensional excess stress in the over-current test in the same manner as the primary and secondary transformer windings.   The HTS tapes are supported from the outside with glass-fiber binding tapes mechanically in the winding structure.   Figure 4 shows the waveform of over-current and the responses of winding temperature measured with 4

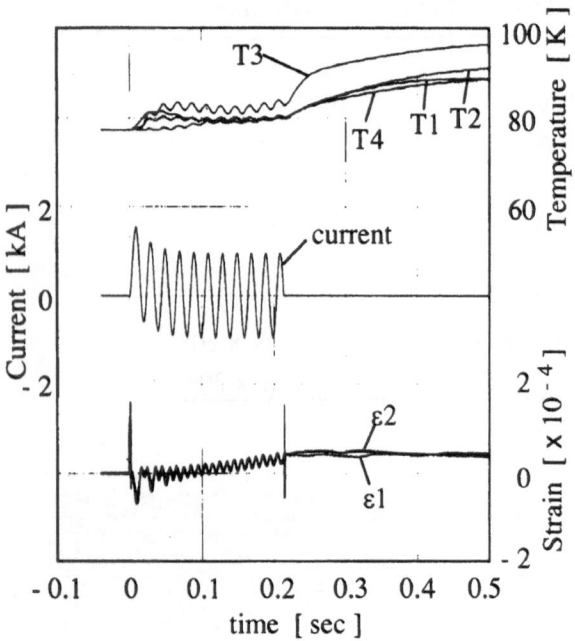

Fig. 4   Responses of winding for over-current

Table 4   Design parameters of pulsed HTS coil

inner dia.	52 mm
outer dia.	111 mm
height	120 mm
strand	Bi-2223 tape
conductor type	4-strand parallel
conductore length	123m
layer number	16
operation temperature	40 K
magnetic induction	1 T (maximum)
rated current	241 A
current waveform	triangle
frequency	1 Hz
inductance	6.89 mH
stored energy	220 J

thermal sensors and mechanical deformation with 2 strain gauges in the reverse connection.   The temperature rise due to the over-current is restricted up to 100 K at the over-current duration of 0.2 s. The thermal deformation measured is also within a permissive level.   The V-I characteristics of the winding are scarcely influenced by the over-current.   It has been confirmed from these results of the over-current test that the winding has no damage up to over-current ten times as high as the rated level.

AC withstand voltage test of 50 kV and lightning impulse voltage tests of 100 – 150 kV have been successfully completed with local model windings of the field-test transformer.   On the basis of the preliminary tests, the field-test transformer will be constructed and tested by March 2000.   The field test will be performed in a distribution grid of Kyushu Electric Power Co. near Fukuoka City.

PULSED COIL

For the designed operating current of 241 A at 40 K, we needed to adopt a 4-strand parallel conductor for a solenoid type of pulsed coil.[4]   The strands of Bi-2223 multifilamentary tapes were not twisted and had the electromagnetic properties similar to those of monofilamentary wires.   For the design of the winding structure, we estimated AC losses approximately from experimental results of short samples exposed to parallel or perpendicular ac magnetic field.   We also simulated the amount of heat conduction from the internal part of winding to the cooling head of the cryocooler using a dummy copper coil.   It was concluded from the preliminary consideration that the winding structure should be designed by the following basic concepts for the high performance of pulsed coil; i) interlayer transposition for the parallel conductor, ii) axial heat conduction by inter-layer spacers of AlN heat drain.   The interlayer transposition of the winding is schematically shown in Fig. 5. Furthermore, thermal contact between the winding and the AlN spacers is kept by epoxy resin with high thermal conductivity.   Characteristics of the coil are listed in Table 4.

The coil was uniformly cooled down between 29 K and 32 K from a room temperature with the cryocooler in 18.3 hours.   We operated the coil continuously in the rated triangular-wave mode with amplitude of 1 T and a frequency of 1 Hz shown in the insert of Fig. 6.   The changes in winding

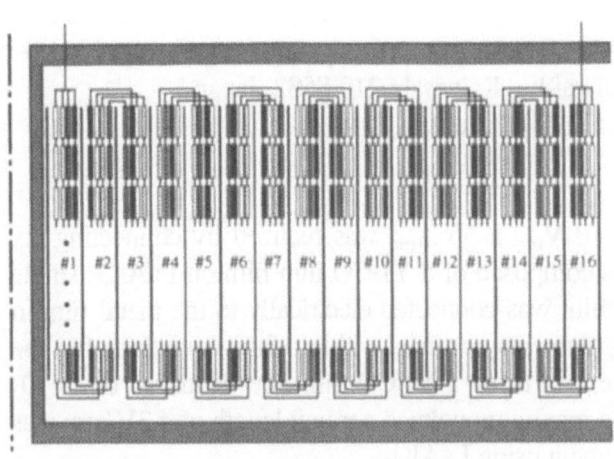

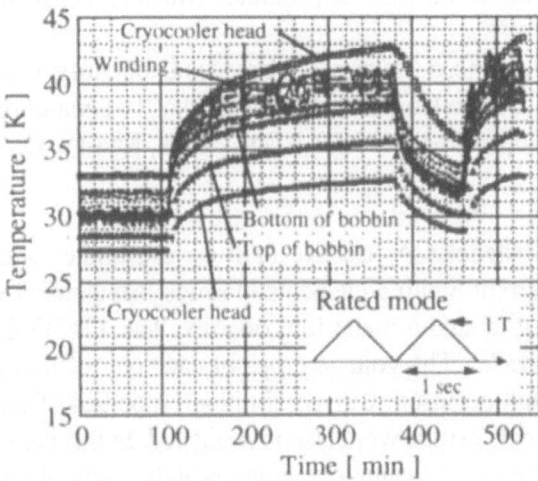

Fig. 5   Concept of interlayer transposition   Fig. 6   Overview of pulsed HTS coil

temperature with the operation, which was measured by 8 thermal sensors, are plotted in Fig. 6. The major heat input during the operation of rated mode was evaluated from the observed temperature distribution in the coil system as follows; current leads: 9 W, signal lead: 0.18 W, supporting rods of the coil from the upper flange of the cryostat: 2.99 W, the heat through the layer of superinsulation : 0.83 W.   The total heat input was about 13 W.   Adding the ac loss of 10.6 W measured electrically, the total thermal load in the rated operation was 23.6 W.   In this case, the temperature at the cryocooler head was 32.7 K.   The results of heat load estimated above seems reasonable in comparison with the nominal capacity of cryocooler of 30 W at 40 K.   The pulsed coils with a high ramp rate have AC loss as one of the major heat load.   In this case, we successfully reduced the AC loss of the winding by means of proper interlayer transposition of the strands.

CONCLUDING REMARKS

As the application of Bi-2223 tapes to large-capacity HTS conductors, parallel conductors with various types of transposition among strands are summarized.   The parallel conductors are transposed in each layer of the solenoidal windings for HTS transformers.   The applicability of parallel conductors is firstly demonstrated as the windings of transformers with low level of AC loss, where the refrigeration penalty is also accounted.   The field-test model of the HTS transformer has been designed using this type of parallel conductor and will be constructed at the beginning of 2000.   In the next example, the conductor is transposed at each turning part between layers in multi-layered pulsed solenoid coil.   The third type is also shown for pancake-type coils, where the winding is transposed at each connection part between single-pancake coils.   These types of transposition suppress electromagnetic coupling among the strands and leads to high performance of the parallel conductors in the HTS power devices.

Acknowledgments: The authors are grateful to many coworkers for their cooperative efforts to accomplish the scientific results reported in this paper.   These works are supported in part by the New Energy and Industrial Technology Development Organization, Japan (NEDO).

1. Iwakuma K et al. (1996) Proceedings of ICEC16/ICMC Part 2: 1325-1328
2. Funaki K et al. (1998) Cryogenics 38: 211-220
3. Funaki K et al. (1999) to be published in Proceedings of EUCAS99
4. Iwakuma K et al. (1998) Advances in Superconductivity XI: 963-966

400V Class Resistive Fault Current Limiter using YBCO Thin Films

Yuki Kudo[1], Hiroshi Kubota,[1] Mutsuki Yamazaki[1], Hisashi Yoshino[1], and Hidehiro Nagamura[2]

[1]Advanced Materials & Devices Laboratory, Corporate Research & Development Center,
 Toshiba Corporation, 1, Komukai Toshiba-cho, Saiwai-ku, Kawasaki 210-8582, Japan
[2]Super-GM, 5-14-10, Nishi Tenma, Kita-ku, Osaka 530-0047, Japan

Abstract : A resistive fault current limiter with 410 V_{rms} x 56 A_{rms} was realized by connecting six current limiting elements in series. An element was composed of a YBCO thin film on $LaAlO_3$ single crystal and a metal film on AlN. The YBCO thin film was connected electrically to the metal film in parallel. The voltage drops of each pair of limiting elements occurred within 0.3ms deviation after the fault occurred. Current limiting properties of an element using a YBCO thin film deposited on Al_2O_3 single crystal were also investigated. In this case, the maximum voltage per unit length of 4.3V/mm was obtained ; a value five times as high as that of an element using $LaAlO_3$.

Keywords : YBCO, Fault Current Limiter, $LaAlO_3$, Al_2O_3

INTRODUCTION

Recently, a resistive type fault current limiter (FCL) using YBCO thin films has been studied extensively[1-3]. Since an FCL of resistive type has no iron core, it is expected to be possible to construct a practical FCL that is smaller in size than an inductive type FCL. On the other hand, V_{max}, the maximum voltage drop of an FCL during current limiting operation without destroying a YBCO thin film, is low compared to a typical voltage of electric power systems at present.

Therefore, V_{max} must be improved to make a practical FCL. Besides using long YBCO films, we think that there are two methods for increasing V_{max}, connection of multiple current limiting elements in series and improvement of maximum voltage per unit length, V_{max}/L. In this paper, current limiting properties of six connected elements in series, with each element consisting of YBCO thin film deposited on $LaAlO_3$, were studied. To improve V_{max}/L, current limiting properties of an element consisted of YBCO thin film deposited on Al_2O_3 single crystal were studied. Because V_{max}/L is restricted by the temperature raising during current limiting operation, we think that V_{max}/L is improved by using YBCO thin films deposited on substrates with high thermal conductivity.

EXPERIMENTAL

Figure 1 shows the structure of one limiting element[3]. The YBCO thin film deposited on an oxide substrate is connected electrically with the metal film by indium wires at interval of 10mm. The metal film was deposited on an AlN substrate. When the $LaAlO_3$ was used as the substrate of YBCO thin film, the Ag film of 90nm thick was deposited on top of YBCO thin film, and the Ag films of 1um thick were deposited as current pad and bypass pad. The module consisted of six elements which were mounted on an FRP holder and connected in series. Critical currents of each element were slightly scattered from 82A to 92A and are listed in Table 1. On the other hand, when the Al_2O_3 was used as the substrate of YBCO thin film, the Ag films of 1um thick were deposited only as the current pad and the bypass pad. The effect of Ag film deposited on top of YBCO thin film is described later.

Measurements of current limiting properties were performed using a test circuit shown in Fig. 2. The module was cooled by liquid nitrogen. The applied voltage and the current for the module were controlled by the transformer and the resistance R_l. A half-cycle-alternating voltage was applied to the module and the current and voltage were measured by digital recorder.

Table. 1 Ic of each element

No.	Ic(A)
A	90
B	90
C	90
D	92
E	82
F	85

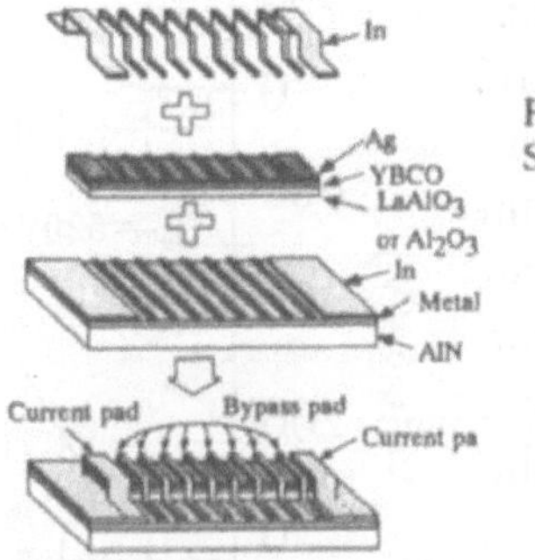

Fig. 1 Schematic drawing of an element

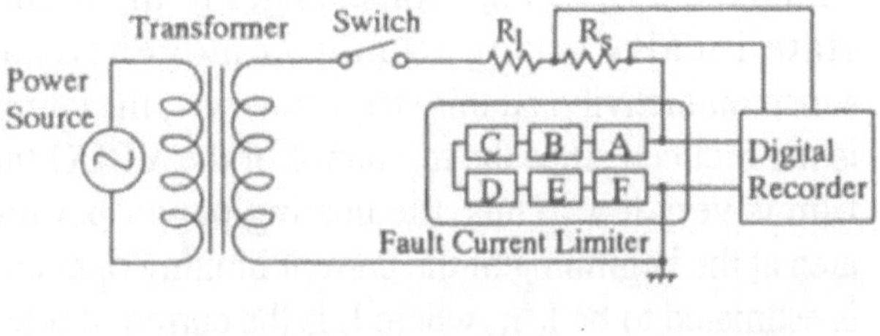

Fig. 2 Schematic drawing of test circuit

RESULTS AND DISCUSSION

Before a current limiting measurement, a current flow capacity of the module was measured. The continuous 50Hz alternating current of $56A_{rms}$ could flow in the module.

Figure 3 shows the current limiting properties of the module. When the current exceeded about 200A, the voltage increased rapidly and a prospective 2000A peak short-circuit current was reduced to a 502A peak. The maximum voltage without destroying YBCO film, V_{max}, was 580 V peak (410 V_{rms}), a value about five times as high as that of one element, which implies that V_{max} was improved by increasing the number of elements. The maximum voltage per unit length, V_{max}/L, was 0.8 V/mm. Although a recovery time was not measured electrically, it was estimated to be a few second, because the bubbling of the liquid nitrogen stopped after a few seconds.

In order to examine uniformity of the quench, the voltage for each pair of elements were also measured. As shown in Fig. 4, at near 1.2 ms, the voltage drops of three pairs increased rapidly within 0.3ms deviation, and at over 1.5ms, these increased gradually. The resistances of each pair at 1.5ms were nearly equal to the resistances at 90K. It suggests that all elements change to the normal state after 1.5ms.

Next, the current limiting properties of one element using Al_2O_3 single crystal were measured. As shown in Fig. 5, the prospective 1800A peak short-circuit current was reduced to 437A. The V_{max} was 520V

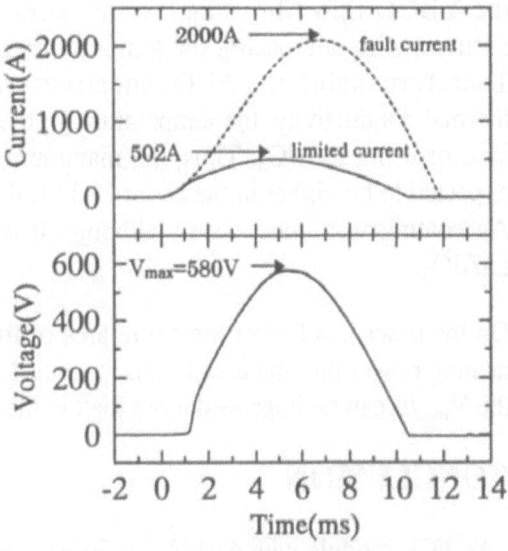

Fig. 3 Current limiting properties of a module consisting of six elements connected in series

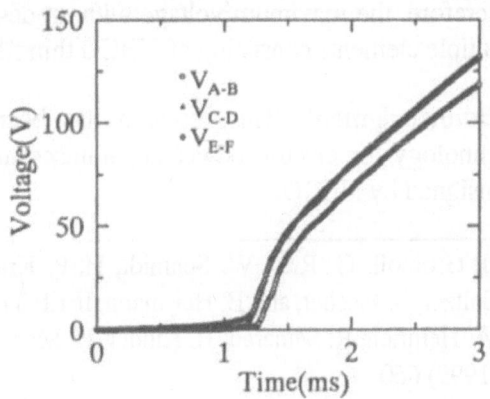

Fig. 4 Voltage drops of each pair at early stage of quench

peak (367V_{rms}) and V_{max}/L was 4.3V/mm, i.e. about five times as high as that of LaAlO$_3$.

The reason of the improvement V_{max}/L is considered as follows. When the fault current reaches a certain value, the YBCO thin film changes to the normal state locally owing to non-homogeneity of superconductivity. At this stage, however, the current is not reduced since the resistance of the YBCO thin film is very low. Thus, the heating power per unit area at the beginning of the current limiting operation is estimated to be $I_y^2 r_y$, where I_y is the current through the YBCO thin film per unit width and r_y is the resistance of the YBCO thin film per unit area. This heating power gives raise to the thermal stress in the substrate and this stress may cause the destruction of the YBCO thin film. The thermal stress can be reduced with decreasing the temperature gradient. Therefore, using the Al$_2$O$_3$ substrate with high

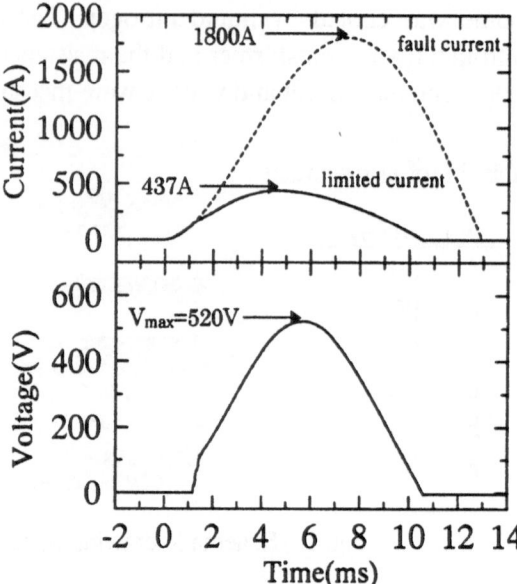

Fig. 5   Current limiting properties of one element using Al$_2$O$_3$ substrate

thermal conductivity, the temperature gradient and the thermal stress can be reduced more than in the case of using LaAlO$_3$. Thus, the maximum heating power without destroying the YBCO thin film is expected to be higher in the case of Al$_2$O$_3$ than in the case of LaAlO$_3$. And hence, in the case of Al$_2$O$_3$, Ag coating was unnecessary, although it is necessary to reduce the r_y by Ag coating in the case of LaAlO$_3$.

On the other hand, after the entire area of the YBCO thin film changes to normal state, the maximum heating power per unit area is expressed as $(V_{max}/L)^2/r_y$. Therefore, using high r_y of the YBCO thin film, the V_{max}/L can be improved more than in the case of low r_y.

CONCLUSION

An FCL module with 410 V_{rms} x 56 A_{rms} was realized by connecting six elements in series, with each element consisting of YBCO thin film on LaAlO$_3$ and metal film on AlN. The maximum voltage of 410 V_{rms} was five times as high as that of one element. Furthermore, the maximum voltage per unit length of an element using Al$_2$O$_3$ was about five times as high as that of an element using LaAlO$_3$. Therefore, the maximum voltage without destroying YBCO thin film can be improved by connecting multiple elements consisting of YBCO thin film deposited on Al$_2$O$_3$ and metal film on AlN.

Acknowledgments This research has been carried out as a part of R&D on superconducting technology for electric power apparatuses under the New Sunshine Program of AIST, MITI, being consigned by NEDO.

1. B.Gromoll, G. Ries, W. Schmidt, H.-P. Kraemer, B.Seebacher, B. Utz, R. Nies, H.-W. Neumüller, E. Baltzer, S. Fischer, and B. Heismann, IEEE Trans. on Appl. Supercond. 9 (1999) 656

2. A. Heinrich, R. Semerad, H. Kinder, H. Mosebach, and M. Lindmayer, IEEE Trans. on Appl. Supercond. 9 (1999) 660

3. H. Kubota, Y. K. Arai, M. Yamazaki, H. Yoshino, and H. Nagamura, IEEE Trans. On Appl. Supercond. 9 (1999) 1365

RECENT R&D STATUS ON 70 MW CLASS SUPERCONDUCTING GENERATORS IN SUPER-GM PROJECT

Ryukichi Takahashi[1], Masatoyo Shibuya[1], Toshimasa Shimada[1], Yoshihiro Imai[1], Ryoichi Shiobara[2], Kazuichi Suzuki[3], Kiyoshi Miyaike[4]

[1]Super-GM, Umeda UN-Bldg., 5-14-10 Nishitenma, Kita-ku, Osaka, 530-0047, Japan
[2]Hitachi Ltd., 3-1-1 Saiwai-cho, Hitachi-shi, Ibaraki, Japan
[3]Mitsubishi Elec. Corp., 1-1-2 Wadamisaki-cyo, Hyogo-ku, Kobe, Japan
[4]Toshiba Corp., 2-4 Suehiro-cho, Tsurumi-ku, Yokohama, Japan

Abstract: Super-GM has developed three types of 70 MW class superconducting generators (called model machines). On-site verification tests of the model machines were completed successfully in June 1999. World's highest generator output (79 MW), world's longest continuous operation (1500 hours) and other excellent results are obtained in the tests. Basic performance and reliability and operability of superconducting generators were verified through the comprehensive studies on the model machines. With the results, it is expected that fundamental technologies for a 200 MW class pilot machine are established.

Keywords: Superconducting generator, Reactive load capacity, Synchronous reactance

INTRODUCTION

Commissioned by NEDO as a part of the New Sunshine Program of the AIST of MITI, Super-GM has been promoting research and development on the applications of superconductive technology for electric power apparatus since 1988 [1]. The major objective of the Super-GM project is to develop three types of 70 MW class superconducting generators (called model machines) to establish fundamental technologies on design and manufacture required for a 200 MW class pilot generator.

Series of on-site verification tests of the three model machines were started in 1997 and completed in June 1999. This paper describes the current developmental status focusing on the verification test results of the model machines.

MODEL MACHINES AND THE VERIFICATION TEST SCHEMES

Two slow-response-excitation-type and one quick-response-excitation-type model machines were manufactured. The major specifications and the design features are given in Table 1. Each rotor has a different excitation control (rapid responsiveness), design concept of NbTi superconducting field winding, damper structure and thermal contraction mechanism to share the critical issues to be solved. Although the stator is a normal conducting one, challenging design concept such as air-gap armature winding with double-transposed water-cooled copper conductors is employed. The stator is common use for the three rotors.

Table 1. Major specifications of model machines

Item / Type	Slow - response - excitation - type (0.1 p.u./s)		Quick - response - excitation -type (1.0 p.u./s)
	A	B	
Specifications			
Capacity (MVA)	83	83	73
Voltage (kV)	10	10	10
Current (A)	4,792	4,792	4,215
Power Factor	0.9	0.9	0.9
Rotation speed (r.p.m.)	3,600	3,600	3,600
Synchronous reactance (p.u.)	0.35	0.35	0.45
Field current Rated / Maximum (A)	3,000/3,600	3,000/3,600	3,200/4,500
Rotor structure			
Warm damper	Single layer type	Squirrel cage type	Three layer type
Cold damper/Radiation shield	Three layer type	Single layer type	Single layer type
Thermal contraction mechanism	Double bearing	Flexible disc	Flexible support
Property of superconductor	High stability	High current density	Low AC loss
Stator structure	Air-gap winding / Double transposed copper conductor / Water cooling		

Special test facility was constructed at the Super-GM Testing Center for the on-site verification tests of the three model machines. An M-G system (pumping-back method) was employed to perform the actual load tests and the severe tests without affecting the power system. The superconducting field winding is cooled by helium coolant supplied by the refrigeration system with 100L/h liquefying capacity through HTC (helium transfer coupling).

VERIFICATION TEST RESULTS

The first model machine (type-A machine) was tested from April 1997 to December 1997. In the tests, highest superconducting generator output (79 MW) was achieved first in the world. Leading-Var operation with approximately rated capacity (82 MVar) was also demonstrated [2]. The second model machine (type-B machine) was installed in January 1998 and started the on-site tests in March 1998 and completed in September 1998. Type-B machine also made the world's highest record of generator output (79.7 MW) and world's longest continuous operation. As for the last model machine (quick-response-excitation type machine), it was installed in October 1998 and the tests were completed in April 1999. As an additional test, the last model machine was connected to a commercial power grid to study basic performance in the actual electric power system. All the on-site verification tests were completed in June 1999 successfully. Two slow-response-excitation type rotors were already disassembled for inspection, while the quick-response-excitation type rotor and the model stator are now under inspection.

The major results in the series sets of the verification tests are summarized in the following:

•**World's highest 70 MW class generation**: Demonstrated operation points are illustrated with the capability curve in Fig. 1. World's highest output of 78.7 MW and 79.7 MW were achieved for type-A and type-B machines, respectively (point C in Fig. 1; power factor=0.96, lag, terminal voltage=10 kV). Stable lagging operation (point A in Fig. 1; 0.7 MW, 47 MVar) was also confirmed. As for the quick-response-excitation type machine, stable operation with 64.5 MW (72.7 MVA, power factor=0.89, lag, terminal voltage=10 kV) was demonstrated.

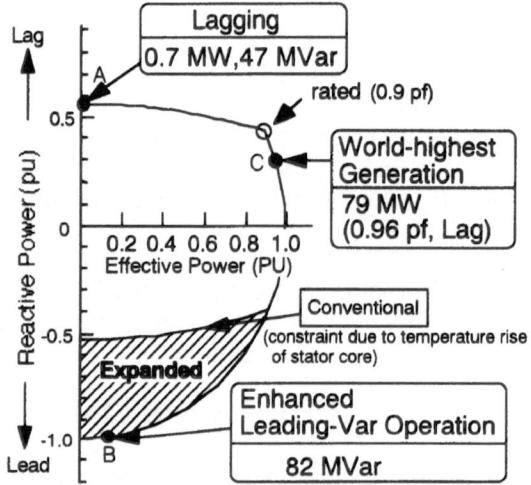

Fig. 1. Demonstrated operation
and the capability curve

•**Low synchronous reactance:** With obtained open-circuit and short-circuit characteristics, the synchronous reactance(X_d) is evaluated to be 0.41 pu for type-A machine and 0.42 pu for type-B machine. While the X_d of the quick-response-excitation type machine is 0.35 pu. The results are 1/5 to 1/3 that of the conventional generators. With such a low synchronous reactance, it is expected that the long distance transmission capacity can be increased by 20 % to 50 %.

•**Enhancement of reactive load capacity:** Demonstration of leading power factor operation with approximately equal to the rated capacity of 82 MVar (point B in Fig. 1) was also successful. Three hours' heat-run tests also revealed that both the rotor and the stator were thermally stable. These results verify that the superconducting generator can supply its rated leading-Var to the power system and can approximately double the reactive load capacity of a conventional generator. It is expected that reactors required in the midnight can be reduced by introducing superconducting generators.

•**World's longest continuous operation**: Figure 2 shows the result of long-term reliability test which was specially imposed on the type-B machine. The continuous operation with full load of 79 MW and the partial loads was performed in 814 hours. Successive 44 times DSS (Daily Start and Stop) operation including WSS (Weekly Start and Stop) operations were also carried out to confirm the operability of the superconducting generation system. The total continuous operation time reached more than 1500 hours. The generator output, heat load of the rotor and the temperature of the armature winding were stable in the long-term operation as shown in Fig. 2. The results show that the model machine has high reliability and operability under normal operation.

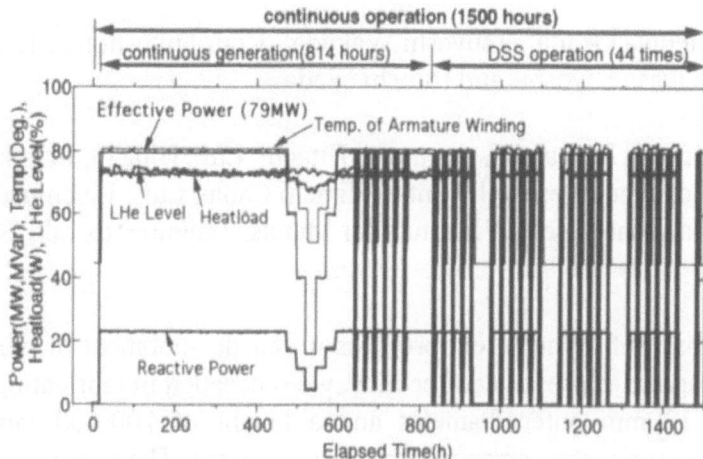

Fig. 2 Result of long-term reliability test

•**Higher generator efficiency:** The obtained generator efficiency is 98.15 % for type-A machine and 98.3 % for type-B machine. Based on the results, the efficiency of the 200 MW class pilot machine (stator air circulation type) is estimated to be 0.6 % higher than that of a conventional machine (0.5 %, if included the refrigerator power).

•**High stability against world's fastest field current change:** A fast single pulsed excitation test was imposed on the quick-response-excitation type machine. The field winding was excited from 2147 A to 3240 A with the averaged ramp rate of 3320 A/s (3.8 T/s) with no quenching and without abrupt increase of the shaft vibration. Continuous three pulsed excitation with the ramp rate of more than 3200 A/s (= 1.0 pu/s) was also performed with no quenching. With the results, it was confirmed that the superconducting field winding has enough stability and low AC losses against field current changes of 1.0 pu/s as we expected.

•**High reliability against severe operating conditions:** Severe tests such as negative-phase-sequence overcurrent test and overspeed test, etc., were performed to confirm the durability against severe operating conditions. Assumed the severest contingent fault condition in the actual power transmission line, three-phase sudden short-circuit tests were performed. The results reveal that the model machines have enough reliability and durability against severe fault conditions. It is confirmed that the superconducting generator has sufficient electrical and mechanical integrity.

•**World's first 77 kV field test:** In order to study the basic performance of the superconducting generator in the actual electric power system, the last model machine was connected to 77 kV commercial power grid of the Kansai Electric Power Co., Inc.. The model machine was operated as a synchronous rotary condenser supplying up to leading 40 MVar to the electric power system. In the tests, the model machine and the refrigeration system operated stably. The results show that the superconducting generator has effects for stabilizing voltage fluctuations and also for reducing shunt reactors in the electric power system.

CONCLUSION

Super-GM has developed three types of 70 MW class superconducting generators and performed the on-site verification tests by June 1999. We obtained superlative results such as world's highest generation, world's longest continuous operation and world's fastest pulsed operation. The field test connecting the machine with the 77kV commercial power grid was also demonstrated first in the world. With these results, it is expected that fundamental technologies on design and manufacture required for 200 MW class pilot machine are established as planned.

1. T. Ageta, in *Advances in Superconductivity X/2*, edited by K. Osamura and I. Hirabayashi (Springer-Verlag, Tokyo, 1998), pp. 1283-1288.
2. R. Takahashi, M. Shibuya, T. Shimada, Y. Imai, H. Kusafuka, R. Shiobara, K. Yamaguchi, M. Takahashi, K. Suzuki, K. Miyaike, H. Yanagi, in *Proc. of ICEC17* (1998), pp. 447-450.

Persistent Mode Operation of Bi-2212/Ag Solenoid Magnets

Michiya Okada[1], Tsuyoshi Wakuda[1], Keiji Fukushima[1], Katsumi Ohata[2], Junichi Sato[2], Tsukasa Kiyoshi[3] and Hitoshi Wada[3]

[1]Hitachi Research Laboratory, Hitachi, Ltd., Hitachi, Ibaraki 319-1292, Japan
[2]Advanced Research Center, Hitachi Cable, Ltd., Tsuchiura, Ibaraki 300-0026, Japan
[3]National Research Institute for Metals, Tsukuba, Ibaraki 305-0047, Japan

Abstract : The recent progress in our development of wind and react type Bi-2212/Ag solenoidal magnets is presented. Recently, we succeeded in fabricating solenoid magnets with 70mm clear bore, a 130mm outer diameter and a height of 100-600 mm, using Bi-2212/Ag *ROSATwire* (rotation-symmetric arranged tape-in-tube wire). The transport properties of the magnets were tested in self field conditions at 4.2K. The details of the transport property of the solenoid magnet will be discussed. Furthermore, we succeeded in fabricating a closed circuit with a PCS made by a NbTi/CuNi conductor. The closed circuits were successfully trapped a field of around 1T with a current of over 200A for over 50h at 4.2K.

Key words: Bi-2212, $Bi_2Sr_2Ca_1Cu_2O_x$, solenoid, *ROSATwire*, closed circuit

INTRODUCTION

Although persistent magnet technologies are quite common in the field of classical superconducting magnets based on NbTi and/or Nb_3Sn[1], the HTS magnets developed so far have not yet been succeeded in developing into *practical* persistent magnets. It would be of great benefit if HTS magnets could operate in a persistent mode with a practical engineering current density. However, there are still two fundamental problems, based on Physics and Engineering, that prevent the system from being made possible. One is the thermal dispassion of vortex motion and the other is the fabrication methodology of the magnet. Both problems are considered to be quite difficult to solve. However, when we use HTS in such practical application as MRI or NMR magnets, a persistent mode operation is essential. We have been developing an HTS persistent magnet system using Bi-2212/Ag conductors. In our previous study[2], 4-8 stacked double pancake magnets with HTS superconducting joints were fabricated, using silver sheathed Bi-2212/Ag multifilamentary tapes. The magnets, however, still showed poor current carrying capacities, i.e. below 50A and 0.2T after 50h operation. In early 1998, we developed a new round-shaped wire *ROSATwire* with excellent transport property, comparable to that of the tape-shaped wire[3]. One can expect that with winding *ROSATwire*, a solenoid magnet can easily be processed. In this paper, we show the first fabrication of a solenoid magnet using the *ROSATwire* and demonstrate persistent mode operation.

EXPERIMENTAL

All of the coils were constructed by means of a wind and react process (W&R) using silver sheathed Bi-2212 *ROSATwires*[3]. A powder-in-tube (PIT) method was used to fabricate the 2mm diameter, 300-400m long *ROSATwire* with heat resistive ceramic braid insulation. The *ROSATwires* were then wound on a stainless-steel bobbin to form solenoid magnets, and partially melted at around 880℃ in a flowing oxygen atmosphere. The details of the W&R process are basically the same as those for our previous study of pancake magnets[4]. An approximately 0.1mm thick SiO_2/Al_2O_3 fiber braid was used for insulation. Two solenoid magnets, A and B, with different winding heights were fabricated, as shown in Fig.1. The specifications of the magnet A and B are summarized in Table 1. After heat treatment, the magnets were impregnated wirh beeswax. Both magnets were then additionally equipped with a PCS made by NbTi/Cu-Ni. The joint between the PCS and solenoid was formed by soldering. In the case of magnet B, 6 individually wound solenoid magnets were joined together by means of butt-jointing. The transport properties of the magnets were measured at 4.2K and self field conditions.

Table 1 Specifications of the Bi-2212 solenoid magnets

	Magnet A	Magnet B
wire diameter (mm)	1.78	1.78
wire length	254m	1.2km
inner winding dia.(mm)	75	75
outer winding dia.(mm)	132	132
winding height(mm)	100	600
sheath material	pure-Ag	Ag, Ag-Mg
turn	784	4116
HTS-HTS joint	-	butt-joint x 6
HTS-NbTi joint	Soldering/1m	Soldering/1m
PCS	NbTi/CuNi	NbTi/CuNi

Fig.1　Bi-2212/Ag solenoid

RESULTS AND DISCUSSION

The dimensions of solenoid magnet B are almost comparable to those of the HTS high field insert magnet for the 1GHz-NMR spectrometer designed by NRIM[5]. Magnet A has the same outer winding diameter but is a 1/6 the height. So far, we have succeeded in fabricating both magnet A and B with the highest critical current, exceeding 400A, generating over 2.5T. These results prove that W&R processing through *partial-melting* is applicable to the practical size of Bi-2212/Ag magnets. We then tried to operate both coils in a persistent mode. The decay curve of the trapped current in the closed circuit for magnet A is shown in Fig.2. The initial current was 350A with trapping a 2.43T, but it decreased gradually. A persistent current of 200A was successfully trapped in 50hours in the closed circuit, which corresponds to a trapped field of 1.4T in 70mm clear bore. Even in the case of magnet B, the initial trapped current 200A decreased only to 179A in 2 hours. These results are almost ten times larger than those of our previous results using the stacked pancake magnet, showing the excellent homogeneity of transport J_c of the *ROSATwire* configuration in magnets wound with a long length of HTS wire. The measured decay curve of the trapped current in the HTS closed circuit can be described in the following equation [6]:

$$i(t) = \left[\left\{ i(t=0)^{1-n} + \alpha \right\} \cdot \exp\left\{ \frac{(n-1)\cdot t}{\tau} - \alpha \right\} \right]^{\frac{1}{(1-n)}}$$ (1),

where α, τ is given by $\alpha = E_c \cdot w/R \cdot i_c^n$, $\tau = L/R$, E_c is the electric field at $0.1\,\mu$V/cm, w is the conductor length, i_c is the critical current, n is the index value, and R and L are the resistance and inductance in the closed circuit, respectively. Taking $n=14.5$ and $R=70$nΩ, we observed a best fit with eq.(1), as shown in Fig.3. The analysis suggests that the good transport performance of the present solenoid magnets originated from the great index value n, which markedly improved our previous results with tape-shaped wires. Most of the residual resistance R is considered to be contributed by the two soldered joints between HTS and NbTi, which is consistent with the measured resistance of approximately $30n\Omega$ for one joint. Taking into account the conductor length of magnet B, i.e. 1.2km long Bi-2212/Ag conductor with 6 HTS-joints, the experimental results for magnet B are considered a marked progress in the development of HTS persistent magnet technology. These results are the first to prove the possibility of persistent mode operation of the practical HTS magnet system. In the case of such practical applications as NMR or MRI magnet systems, over 10km long HTS conductors are required. In this case, quite small residual resistance below $n\Omega$ originating from defects in HTS filaments, thermal and magnetic dispassion in HTS grain and resistance at joints, are also considered to exist, being the next problems to be solved. In spite of the residual resistance, the results markedly encourage us to realize future industrial NMR/MRI and/or SMES applications by using HTS persistent magnets.

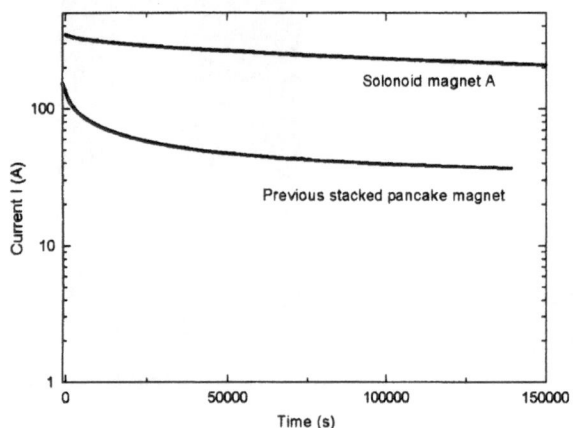

Fig.2 Current vs. time for magnet A.

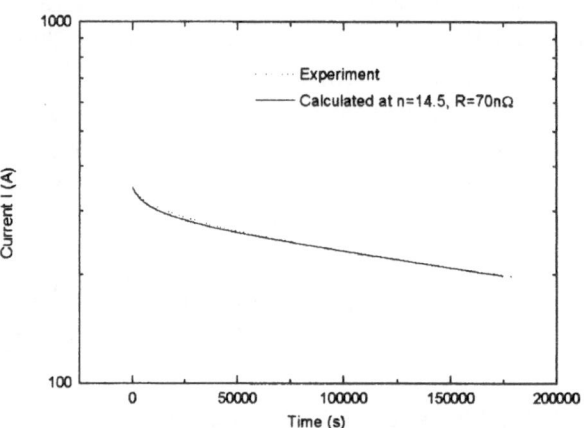

Fig.3 Comparison of calculated and experimental current decay curves.

CONCLUSION

We have succeeded in fabricating W&R type solenoid magnets with practical dimensions by using *ROSATwire*. The solenoid magnets showed excellent transport current carrying capacity for 1GHz-NMR applications. The critical current density of the solenoid magnets exceeds 400A at 4.2K and self magnetic field condition, generating 2.4T. Magnets equipped with a NbTi-PCS were also tested and succeeded in operating in a persistent mode. One magnet trapped a persistent current of 200A for 50h with a field over 1.4T. These experimental results markedly encouraged us to realize the next generation of HTS solenoid magnets with excellent field homogeneity and stability.

1. Kiyoshi T, Kosuge M, Inoue K, Maeda H, IEEE Trans. Magn.(1996) , vol.32, 2478.
2. Okada M, Fukushima K, Tanaka K, Kitaguchi H, Kumakura H, Togano K, Kiyoshi T, Inoue K, Jpn.J.Appl.Phys. (1996) vol.35, L627.
3. Okada M, Tanaka K, Wakuda T, Ohata K, Sato J, Kumakura H, Kiyoshi T, Kitaguchi H, Togano K and Wada H, Proc. 1998 Applied Superconductivity Conference, Sept 14-18, Palm Springs CA, (MIB-06), to be published.
4. Okada M, Tanaka K, Wakuda T, Ohata K, Sato J, Kiyoshi T, Kitaguchi H, Kumakura H, Togano K and Wada H, Advances in Superconductivity XI(1999) (Proc. ISS'98, Nov.16-19, Fukuoka), 851.
5. Kiyoshi T, Inoue K, Kosuge M, Itoh K, Yuyama M, Maeda H (1997): Proc. 16th Int. Cryo. Eng. Conf./ Int.Cryo. Mater. Conf. (ICEC16/ICMC), May 20-24,1996, Kitakyushu,Part 2, 1099.
6. Kiyoshi T, Inoue K, Kosuge M, Wada H and Maeda H (1997) IEEE. Trans. Appl. Super., vol.7, 877.

Mechanical and Transport Properties of Bi-2212 Coil under a Large Electromagnetic Stress State

T. Wakuda[1], M. Okada[1], S. Awaji[2] and K. Watanabe[2]

[1]Hitachi Research Laboratory, Hitachi Ltd., Hitachi 319-1292, Japan
[2]Institute for Materials Research, Tohoku University, Sendai 980-8577, Japan

Abstract: Mechanical and superconducting properties under a large electromagnetic stress state were investigated for a Bi-2212 coil consists of 5 stacked double pancake coils with a practical outer diameter of 280mm. The coils were reinforced by a co-wound Hastelloy X tape and showed a good performance against a large hoop stress of 160MPa. The obtained mechanical property of the coils was consistent with a calculation quantitatively. However, we found a stress concentration near the center of the pancake coils.

Keywords: Bi-2212 coil, Large stress state, Superconducting properties, Mechanical properties

INTRODUCTION

Since high Tc superconductors (HTS) have a large irreversible filed of far over 30T in a low temperature region, HTS magnets promise to generate a large magnetic field of over-30T. It is necessary to develop not only a superconducting wire with a large current density, but also a magnet(winding) structure against a large electromagnetic stress of over-100MPa for the high field magnets. We've developed HTS magnets consisting of stacked double pancake coils wound with a Bi-2212/Ag superconducting tape[1]. The pancake coils are reinforced by co-winding high strength tape such as Ag-Mg alloy, Hastelloy tape and so on[1,2]. The co-winding structure performed a good tolerance to the electromagnetic stress for a test coil with a relatively small dimension, but it is not obvious whether it works effectively on the 27T HTS magnet with which we are now planing to replace a water-cooling copper magnet in the hybrid magnet of Tohoku University. In this work, we fabricated a test coil consisted of stacked double pancake coils with almost the same dimension of the preliminary designed 27T magnet and investigated their mechanical strength under a large hoop stress state of over-100MPa.

EXPERIMENTAL

We prepared the W&R type Bi-2212 test coil consisted of 5 stacked double pancake coils with a dimension of ϕ 234-280mm. Specification of the test coil is shown in Table 1. The unit pancake coils were co-wound with a Bi-2212 tape, that was 5mm wide and 0.34mm thick, and a Hastelloy X (HX) tape, that was 5mm wide and 0.1mm thick. The tape surface of HX was oxidized during partial melting, forming an electric insulator between the wound tape of the pancake coil. Two types of the pancake coil, with 1 or 3 reinforcing HX tapes, were prepared to discuss a quantitative mechanical strength of the coil.

To measure an inner strain of the pancake coils caused by an electromagnetic force, 9 strain gauges were set in one pancake at intervals of 120 degrees along the circumferential direction and at equal intervals along the radial direction. They were set on a surface of the Bi-2212 tape to measure a strain of the tape directly. Three pairs of voltage taps were located at the innermost, middle and

Fig. 1. W&R Bi-2212 test coil

Table 1. Specification

Winding Inner Diameter	234mm
Winding Outer Diameter	280mm
Winding Height	70mm
Number of Pancake Coils	5
Turns	352
Conductor	Bi-2212/Ag-Mg alloy
	54 filamentary tape
	0.34mmx5mm
	total wire length: 290m
Mechanical Support	Hastelloy X
	0.1mmx5mm

outermost part of the pancake, and detect the voltage of 1-2 turns of the winding. Since they were concerned with the strain gauges located along the radial direction, the relationship between the strain of Bi-2212 tape and degradation of superconducting property was obtained.

The pancake coils were connected in series and mounted with their axes parallel to an external magnetic field of 10T, and measured in liquid Helium (4.2K). To confirm reproducibility of the transport properties of the test coils and to draw a hysteresis loop of a current vs. strain property, that is equivalent to a stress-strain property, we increased transport current up to a certain maximum value, then decreased it to zero and then increased it again to a larger value. And, this sequence was repeated until the test coils were quenched.

RESULT AND DISCUSSION

The observed typical transport characteristic of the test pancake coils is shown in Fig. 2. The critical current and the critical current density of the coil are 232A and 136A/mm^2, respectively. Each value is extremely small against expectation because of swelling of the superconducting tape occurred during a partial–melt process for the coil.

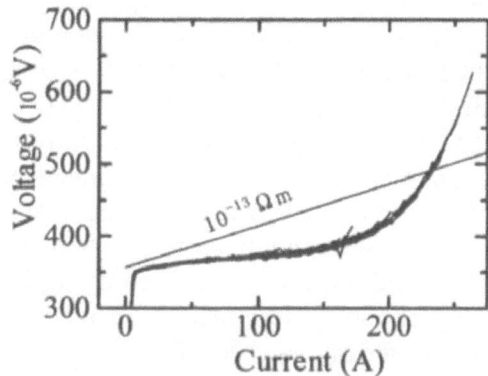

Fig. 2 The typical transport characteristic of the pancake coils.

The typical strain vs. transport current characteristics of the pancake coils with one reinforcing HX tape and three ones are shown in Fig. 3(a) and (b), respectively.
The characteristics are equivalent to the strain vs. stress characteristics through a conversion of a transport current into a hoop stress. The reinforcing winding showed a good performance up to a maximum hoop strain of 160MPa estimated by BJR when the transport current was 270A. It is obvious that the more reinforcing HX tapes can restrict the strain of the coil to the smaller level. The thick lines in the figures represent the calculated results based on the mixture rule for a Young's modulus, and they show quantitative agreement with the obtained data. Note that we can design the hoop strain with the present co-winding technique for a practical size pancake coil. However, we found that the transport property was damaged near center of the double pancake coil where the winding goes across from a lower pancake to upper. It seems that a strain caused by an electromagnetic stress or a thermal contraction concentrated on this asymmetrical part.

The estimated distribution of the hoop stress in the coil as a function of the number of applying current is shown in Fig. 4. The symbols represent the data estimated from the obtained hoop strain and a Young's modulus of the local part of winding. The broken lines represent the inner stress level

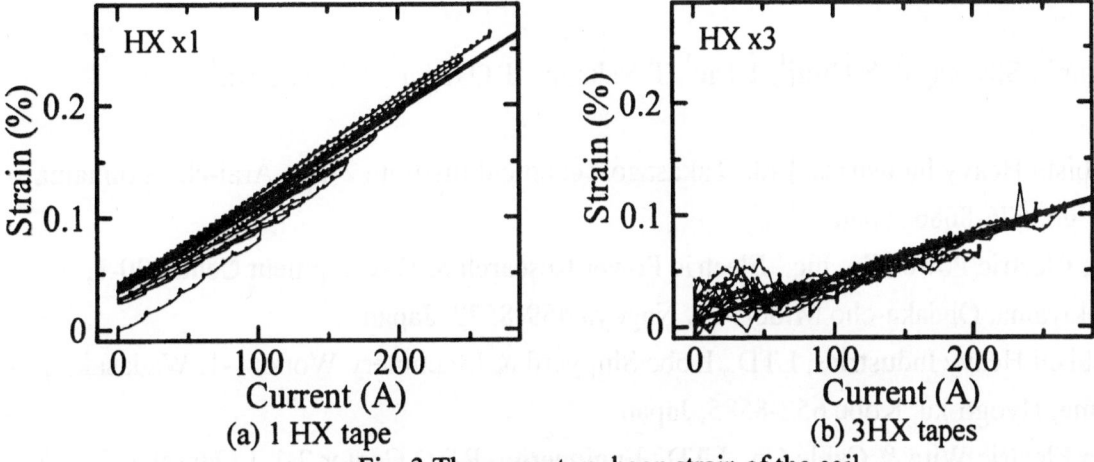

(a) 1 HX tape   (b) 3HX tapes

Fig. 3 The current vs. hoop strain of the coil

calculated assuming that the coil can be treated as an ideal elastic body. Although the obtained stress almost agrees with the level, the distribution is not cylindrically symmetric. It seems that a deal of swelling of the superconducting tape restricted a homogeneous transmission of the stress. And the distribution changes with run. The epoxy resin impregnated into the inter-winding also seemed to be destroyed gradually by the repetitious stress.

It should be concluded, from what has been said above, we established the method of reinforcing winding against a large hoop stress of over-100MPa using the co-winding technique. However, it is still immature to produce a practical pancake coil. Thus we have to improve the fabrication of a pancake coil to realize the high field HTS magnets.

In addition, a degradation of transport property of the superconducting wire as a function of a strain can be obtained from the current vs. voltage and current vs. strain characteristics. The onset of the degradation is about 0.18% and agrees with the other result we've obtained before[2]. It is effective to co-wind with a hard reinforcing tape such as a HX tape with a Young' modulus of 200GPa in order to suppress a hoop strain within the tolerance.

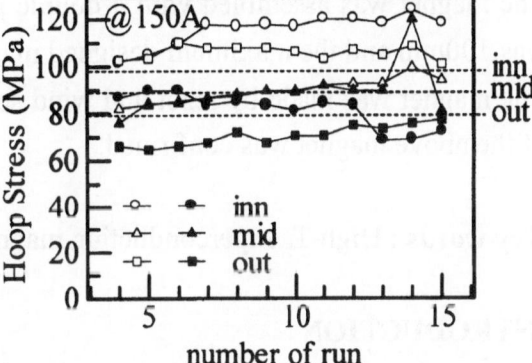

Fig. 4. The estimated stress distribution of the test coil

CONCLUSION

We fabricated a test W&R Bi-2212 coil with a large outer diameter of 280mm consisted of 5 stacked double pancake coils. The double pancake coils were co-wound with a Bi-2212/Ag-Mg tape and a Hastelloy X tape against a hoop stress. Transport and Mechanical properties of the test coil were measured in a backup field of 10T and liquid Helium(4.2K). The present reinforcing winding showed a good performance up to a maximum hoop stress of 160MPa. The mechanical property showed a good agreement with a calculation except a local concentration of the stress and a local destruction of the winding caused by repetitious load.

Acknowledgments: One of the authors K. W. thanks Grant-in-Aid for Scientific Research from the Ministry of Education, Science and Culture, Japan.

1. M. Okada *et al.*, *Advance in Superconductivity XI* (Proc. ISS'98, Nov 16-19, 1998, Fukuoka), 851
2. T.Wakuda *et al.*, Proc. 1998 Appl. Super. Conf., Sep 14-18, Palm Springs CA, to be published.

Study on High-T$_c$ Superconducting Magnet made with Bi-2212 Rutherford Cable

M.Minami[1], S.Nagaya[2], S.Hirai[1], T.Irie[1], T.Nakano[3], T.Okawa[3], T.Hasegawa[4]

[1] Mitsubishi Heavy Industries, Ltd., Takasago Technical Institute 2-1-1, Arai-cho shinhama Takasago, 676-8686 Japan

[2] Chubu Electric Power Co. Inc., Electric Power Research & Development Center 20-1, Kitasekiyama, Ohdaka-cho Midori-ku, Nagoya 459-8522, Japan

[3] Mitsubishi Heavy Industries, LTD., Kobe Shipyard & Machinery Works 1-1, Wadasaki-cho 1-chome, Hyogo-ku, Kobe 652-8585, Japan

[4] Showa Electric Wire & Cable Co., LTD., Engineering R&D Center 2-1-1 Odasakae, Kawasaki-ku, Kawasaki-shi, Kanagawa, 210-8660, Japan

Abstract : As there were no oxide superconducting wires with a large current capacity over 1kA, the high-T$_c$ superconducting magnet for large-scale magnet could not be fabricated.

In this investigation high-T$_c$ superconducting magnet using newly developed Bi-2212/Ag Rutherford cable with critical current of 3500A at 4.2K, 0T was trial manufactured as a partial model of 10T magnet.

This Rutherford cable was made to withstand to 20T class magnetic force.

The magnet was assembled with 4 double pancake coils, and the inner diameter of the magnet was 100mm and the maximum designed magnetic field was 2T.

The magnet was made by react and wind securing the electrical insulation and the performance of the above magnet was confirmed.

Key words : High-T$_c$ superconducting magnet, Bi-2212/Ag, Rutherford cable, Large current

INTRODUCTION

High-T$_c$ superconducting magnet with a large current capacity Bi-2212/Ag Rutherford cable was trial manufactured as one step of the development of large-scale magnet for high magnetic field.

DESIGN AND FABRICATION

High-T$_c$ model magnet was designed and fablicated as an elemental model of 10T magnet with a newly developed Bi-2212/Ag

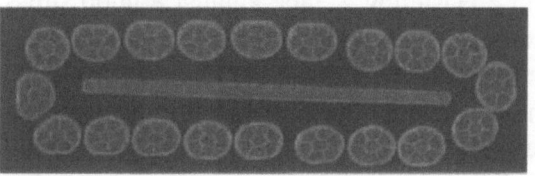

Fig.1 Bi-2212/Ag Rutherford Cable (7.8mm×2.4mm)

Rutherford cable [1][2] shown in Fig.1.

The specification of this model magnet is shown in Table 1 and Fig.2 comparing with 10T and 20T magnet. Operating this coil, the maximum hoop stress by magnetic force is 4.8MPa. This stress is very low to the tensile strength of the cable. This Rutherford cable is capable of withstanding to 20T class magnetic force. The load line of these magnets are shown in Fig.3.

This model magnet was fablicated by react and wind with glass cloth tape for electrical insulation. Its inner diameter was 100mm considering the reduction of the critical current of coil by bending strain, and it was assembled with 4 double pancake coil shown in Fig.4.

The coil was wound securing the electrical insulation and its performance between layers was confirmed.

Table 1   specification of HT$_c$-Superconducting Magnets

	Model Magnet	10T Magnet	20T Magnet
Inner Dia.	100mm	100mm	100mm
Outer Dia.	180mm	450mm	1250mm
Width	75mm	288mm	288mm
Temp.	4.2K	4.2K	4.2K
Turn×Layer	17Turn×8Layer	70Turn×36Layer	230Turn×36Layer
Coil Length	60m	2180m	17.6km
Current	1250A	1250A	1200A
Current Density	55.3A/mm^2	63A/mm^2	60A/mm^2
Magnetic Field of Coil Center	1.36T	10.0T	20.0T
Stored Energy	1.50kJ	552kJ	17.8MJ
Maximum Tensile Stress	4.8MPa	32.0MPa	61.0MPa

* Tensile Strength of the cable : 600MPa

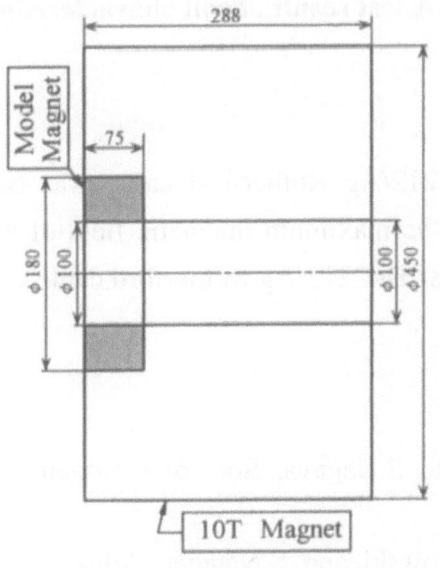

Fig.2   Dimension of Magnets

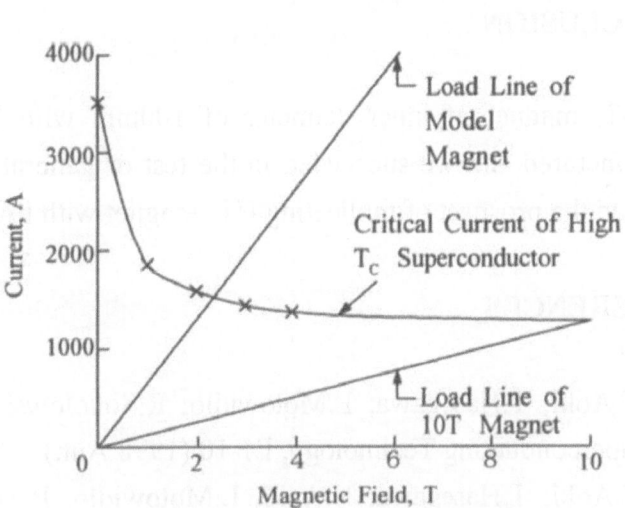

Fig.3   Characteristics of Magnets

TEST RESULTS

The performance of this magnet was tested and is shown in schematic Diagram of Fig.5.
The magnet was cooled by liquid helium and the current was supplied by DC supply unit.
It was confirmed that the coil inductance was 1.8mH and the magnetic field was the same value
as the analytical value shown in Fig.6 and the maximum magnetic field of 2T was gained and
there was no damage in the magnet under the tests of ten cycles shown in Fig.7.

Fig.4   Model Magnet

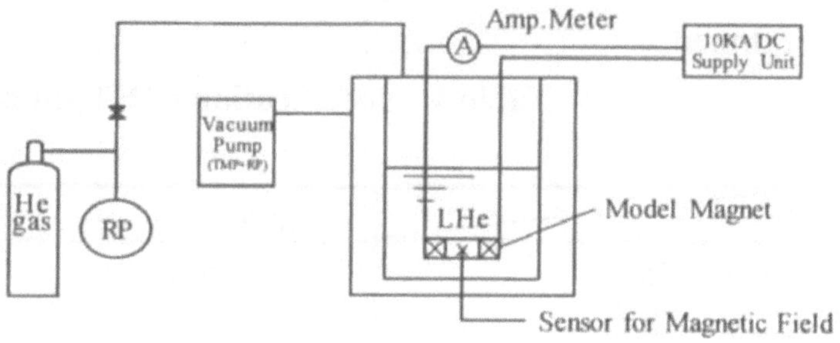

Fig.5   Schematic Diagram of Test apparatus

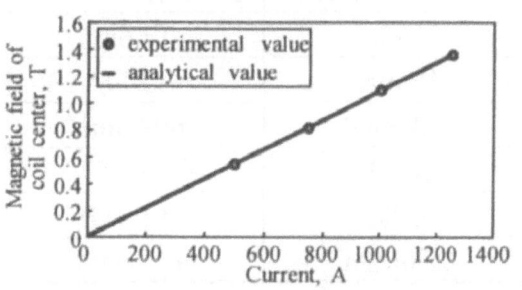

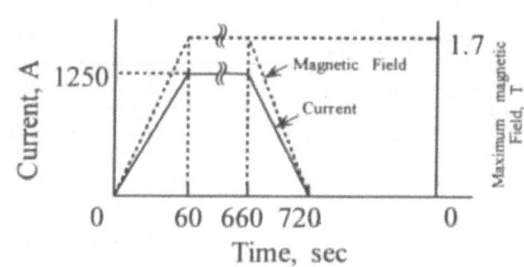

Fig.6   Characteristics of magnetic field

Fig.7   A test result of coil characteristics

CONCLUSION

High-T_c magnet of inner diameter of 100mm with Bi-2212/Ag Rutherford cable was trial
manufactured and we succeeded in the test of generating the maximum magnetic field of 2T,
and had the prospect of fablicating HT$_c$-magnet with kA class Bi-2212/Ag Rutherford cable.

REFERENCES

(1) Y.Aoki, T.Hasegawa, L.Motowidlo, R.Sokolowski and S.Nagaya, Soc. of Cryogenic &
    superconducting Technology, E1-10 (1998 Aut.)
(2) Y.Aoki, T.Hasegawa, N.Aoki, L.Motowidlo, R.Sokolowski and S.Nagaya, Advances in
    Superconductivity XI, Springer-Verlag (1998) P1377-1380.

Development of The 20kA Class HTS Current Lead for A SMES

Shinichi Mukoyama, Shinichirou Meguro, Naohiro Futaki[1], Takashi Saito[1], Tooru. Minemura[2], Kimiyuki Shinoda[2], Shirabe Akita[3], Takashi Himeno[4], MasahiroYamamoto[4] and Takashi Satow[5]

The Furukawa Electric Co., Ltd., 6, Yawata-Kaigandori, Chiba 290-8555, Japan
[1]Fujikura Ltd., 1-5-1, Kiba, Tokyo 135-8512, Japan
[2]Chubu Electric Power Co., Inc., 20-1, Kitasekiyama, Ohdaka-cho, Nagoya 459-8522, Japan
[3]CRIEPI, 2-11-1,Iwadokita,Tokyo201-8511,Japan
[4]ISTEC, Shinbashi 5-chome 34-3, Tokyo 105-0004, Japan
[5]National Institute for Fusion Science, 322-6, Oroshi-cho, Toki, Gifu 509-5292, Japan

Abstract: HTS current leads using YBCO leads for a 100kWh SMES pilot plants were developed. Moreover, a HTS lead experimental device that had a rated capacity of 20kA class and 3kV was produced and current loading tests and voltage loading tests were carried out. As a result, it was confirmed that the HTS current lead using YBCO leads had sufficient performances for practical uses through the various tests.

Key words: YBCO, current lead, SMES

INTRODUCTION

A large capacity current lead is a vital device to supply electric power to a SMES. The current lead using High-Tc superconductors (HTS current lead) can increase the efficiency of a SMES because the invasion heat and the joule heat of the current lead can be reduced by using High-Tc superconductors. Especially, a YBCO lead is a suitable material for a HTS current lead with large capacity, low heat invasion and low heat loss, because a YBCO has high critical current density, low heat conduction and an excellent magnetic property compared with a BISCCO. Fujikura manufactured YBCO leads and Furukawa fabricated a HTS current lead experimental device using the YBCO leads. Moreover, the device was carried out various experiments to verify performances of the current lead for a 100kWh SMES pilot plant in Akagi testing center of CRIEPI.

EXPERIMENTAL PROCEDURE

Producing process of the YBCO lead.[1]
Specifications of the YBCO lead used for the HTS current lead experimental device are shown in Table 1. The YBCO leads with copper electrodes are shown in Fig.1.

Table 1. Specifications of the YBCO lead

Composition:	Y123 : Y211= 10 : 4 mol	
	Ag	10 wt%
	Pt	0.5wt%
Product method:	Melt Textured Growth	
Diameter of lead:	ϕ 3mm	
Effective length of lead:	100mm	
Tc:	91.5K	
Ic (77K, 0T) :	>1kA	

The ab-planes of the YBCO have to be paralleled to a longitudinal direction of the lead to obtain high critical current density. Silver of 10wt% and platinum of 0.5wt% were mixed into Y123 and Y211 powder with a proportion of 10 mol to 4 mol.

This work was supported by the Agency of Natural Resources and Energy of MITI.

The mixed powder was shaped into a rod with about 3mm diameter by a CIP. The rod was crystallized for keeping ab-planes parallel to the lead's axis by the floating-zone-method. The (110)-pole figure of a cross section of a YBCO lead is shown in Fig. 2. Good arrays of the ab-planes are confirmed because the peak of the pole is sharp.

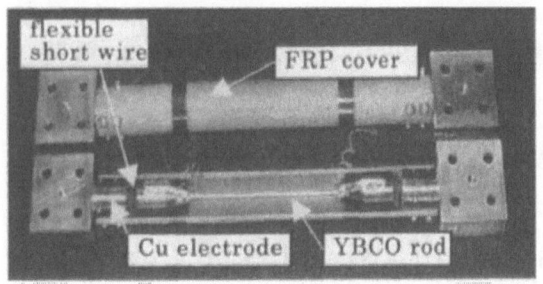

Fig.1. The YBCO lead with Cu electrodes

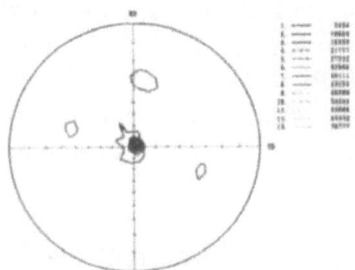

Fig.2. The (110)-pole figure of a cross section of a YBCO lead

20kA class YBCO current lead experimental device. Twenty-four YBCO leads that had a critical current of more than 1kA were connected in parallel to obtain the current capacity of the lead to 20kA. It was important to distribute current evenly to the 24 leads. Every joint resistance between the YBCO rod and the copper electrode had to be equal to each other, because the resistance determined the distribution of current. The YBCO rods were coated by thin silver film and soldered the copper electrodes to equalize each joint resistance. The YBCO lead must stand up to electromagnetic stress, and thermal stress. The YBCO rod was set in the FRP cover, and flexible short wires were fitted into both ends of the YBCO lead as shown in Fig. 1 to protect it from these mechanical stresses. The 48 YBCO leads of the plus lead and the minus lead were located in a vacuum of a cryostat as shown in Fig.3 and Fig. 4.

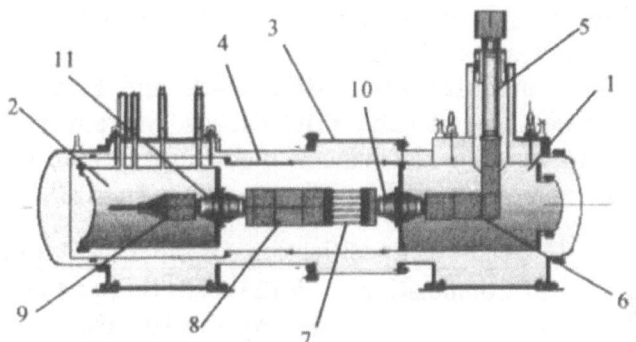

Fig.3. Schematic diagram of the experimental device. 1-LN2 can; 2-He can; 3-vacuum vessel; 4-thermal insulator, 5-Cu lead with heat exchanger; 6,8,9-Cu lead 7-YBCO leads; 10,11-feedthrough at high voltage

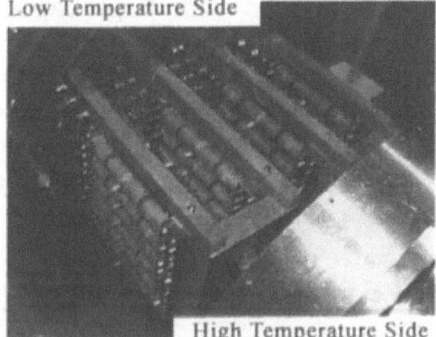

Fig.4. YBCO leads placed in the cryostat. The number of YBCO leads is 24×2.

Liquid nitrogen and liquid helium respectively cooled YBCO leads from each end of the leads by conduction. We aimed at a reduction of the refrigeration cost by placing a vacuum insulation between liquid helium and liquid nitrogen, and by cooling a high temperature part of the YBCO leads by relatively cheap liquid nitrogen.

EXPERIMENTAL RESULT

Various experiments on the 20kA class HTS current lead experimental device as shown in

Fig. 5 were carried out by using a 15kA DC power source and an helium refrigerator in Akagi Testing Center of CRIEPI.

In current loading tests, 24 YBCO leads were reduced to 18 leads to equalize current of a YBCO lead to that of the 20kA HTS current lead device. Moreover the distance between the plus and minus YBCO leads was shortened than that of the 20kA device to equalize the maximum magnetic field and the maximum magnetic force to those of the 20kA device. Current of 15kA flowed continuously for two hours without unexpected temperature raises and mechanical. The heat invasion per lead was 1.2W, and the total heat load that consisted of invasion heat and joule loss was decreased to 1/3 of heat load of the Cu current lead.

Fig.5. HTS current lead experiment device

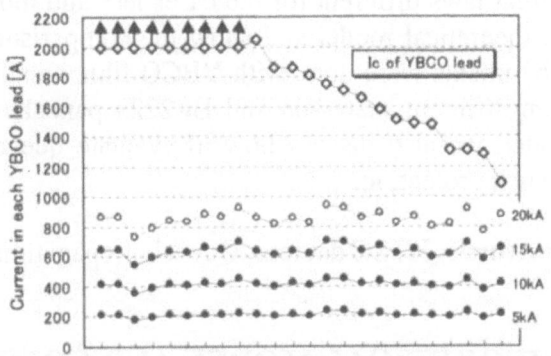

Fig.6. The currents of 24 YBCO leads at 5 to 15kA, and extrapolated value at 20kA.

Each current of 24 YBCO leads had to be equal and be less than its critical current to obtain current capacity of 20kA. Each current of the 24 leads was measured at 5 to 15kA and was almost equal to each other as shown in the Fig. 6. Moreover, each current that extrapolated to 20 kA from the experimental results was less than its critical current (Ic).

A current lead has to withstand rapid increase and decrease of current and high voltage to charge and discharge a SMES. The HTS current lead was examined by supplying the triangle-shaped wave currents with rates of 333A/s, 500A/s and 800A/s respectively for 1 hour, and by breaking current from 15kA in 1 second or less. The HTS current lead was sturdy without mechanical and thermal damage through both the experiments.

DC breakdown voltage between the leads and the cryostat grounded was measured by applying DC voltage of a maximum of 10kV to the leads. The experimental device had sufficient voltage properties because no electric breakdown was observed with keeping a high vacuum in the cryostat.

SUMMARY

It was concluded, from the results of various experiments described above, that the HTS current lead had excellent performances as the 20kA current lead for the SMES. Moreover, necessary component technologies for a power transmission system of the SMES pilot plant were established by this research.

Acknowledgments. Authors would like to thank Mr.H.Sakaki (Denryoku Tec) and members of Akagi Testing Center of CRIEPI for helps in the experiment.

1. S. Asakura et. al., Adv. in Supercond-IX pp949 (1997)

Universal Model for Quench Development in HTSC Devices

Vitaly Vysotsky,[1,2] Alexander Rakhmanov,[3] Yuri Ilyin,[1,2] Takanobu Kiss,[1] and Masakatsu Takeo [1]

[1] Graduate School of ISEE, Kyushu University, 812-8581 Fukuoka, Japan
[2] Visiting scientists from Kurchatov Institute, 123182 Moscow, Russia.
[3] SCAPE, Russian Academy of Sciences, Moscow, Russia

Abstract: We developed a new approach for quench description in HTSC devices. This approach does not rely on any ideas about normal zone propagation in HTSC. It was shown that near the quench current the time dependencies of the temperature and the electric field obey the universal scaling laws different for the cases less and more than the quench current. In this paper we outline the theoretical model and show the comparison between experimental and calculated results. The experiments were done with YBCO film, small Bi-2212 coil, Bi-2223 coils wounded onto bobbins from different materials and Bi-2223 pancake coil at different cooling conditions. Based on the model, we also discuss how to evaluate quench parameters of the HTSC devices, important for quench protection.

Keywords: HTSC devices, Quench propagation, Universal scaling law, Quench protection.

THEORETICAL MODEL AND COMPARISON WITH EXPERIMENTS

In the paper [1] we showed theoretically that temperature, T, behavior in a HTSC device obeys two universal scaling laws at transport currents less and more than quench current I_q:

$$\frac{T(t)-T_q}{T_f} = \tan\frac{t-t_q}{t_f} \quad I > I_q; \qquad \frac{T(t)-T_q}{T_f} = \frac{1+g\exp(2t/t_f)}{1-g\exp(2t/t_f)}, \quad g = \frac{T_q-T_0+T_f}{T_q-T_0-T_f} \quad I < I_q, \tag{1}$$

where t stands for time. Similar equations we can write for electric field, E, inside of conductor:

$$\frac{E-E_q}{E_f} = \tan\frac{t-t_q}{t_f} \quad I > I_q, \qquad \frac{E-E_q}{E_f} = \frac{1+g\exp(2t/t_f)}{1-g\exp(2t/t_f)} \quad I < I_q. \tag{2}$$

Supposing the voltage-current characteristic of superconductor as $E=E_0(I/I_0(T))^n$ and linear dependence of I_0 on temperature, we could derive the following analytical formulas related I_q, T_q, E_q and the other parameters:

$$T_q - T_0 = \frac{T_c - T_0}{n+1}, \quad \frac{I_q}{I_0(T_0)} = \frac{n}{n+1}\left[\frac{hP(T_c-T_0)}{nE_0 I_0(T_0)}\right]^{1/(n+1)}, \quad E_q = \frac{hPT_c}{nI_0(T_0)} \tag{3}$$

and

$$T_f = (T_c-T_0)\sqrt{\frac{2|I-I_q|}{(n+1)I_q}}, \quad E_f = nE_q\sqrt{\frac{2|I-I_q|}{(n+1)I_q}}, \quad \frac{t_q}{t_f} = \tan^{-1}\left(\frac{t_f}{2t_h}\right), \quad t_f = t_h\sqrt{\frac{2I_q}{|I-I_q|(n+1)}}, \tag{4}$$

where I_q is a quench current, I – transport current, T_q – quench temperature at $t=t_q$, T_0 – ambient temperature, T_c – critical temperature, E_q – electric field at quench point, n – resistive transition index, h – heat removal coefficient; $t_h = CA/Ph$ is the characteristic thermal time, C is specific heat, A and P are cross-section and cooling perimeter of the conductor correspondingly.

To demonstrate that theoretical model is universal and can be applied for devices made from Bi-based tape and YBCO films, we present the results of their thermal quench behavior. In Fig. 1, measured and calculated temperature traces in dimensionless form are shown for the experiments with five different objects [2,3,4]. One can see that all temperature traces measured in absolutely different experiments obey the same dimensionless curve corresponding to quench of HTSC. On the base of this very important result we may conclude, that in all experiments the quench behavior has the same nature and can be describe by the model theory of thermal quench with a good accuracy.

EVALUATION OF QUENCH PARAMETERS OF HTSC DEVICES IMPORTANT FOR QUENCH PROTECTION

From practical point of view, for quench protection design the quench current I_q and time t_q are most important parameters to know. Quench current (Eq. (3)) depends on parameters of voltage-current characteristic $I_c(T_0)$, E_0, n, critical temperature T_c and heat removal coefficient h. All parameters except heat removal coefficient are the standard parameters of voltage-current characteristic of the material used to make the winding and can be easily determined.

Heat removal coefficient h (or time t_h) – is the most difficult parameter to estimate. The best way is to determine this integral parameter from the experiments with the proper small model coil, before design of full size coil. It may also be determined in the preliminary test of the full size coil at low current [5]. On the other hand, as it is seen from Eq. (3), the heat removal coefficient h is included in the equation under $1/(n+1)$ degree. In most cases, the parameter n is more than 8-10, so the change of the heat removal coefficient by the order of value will change calculated I_q by 25% only. It is good enough for the first approximation. In Fig. 2 the calculated I_q for the pancake coil is shown versus temperature. Heat removal coefficient was determined from low current (stability) experiment [5].

Another important parameter is the characteristic time t_q in Eq. (4). This time determines the moment when rather slow rise of temperature/voltage is changed to very fast one. This time is important for design of quench protection scheme. The protection scheme must detect quench started and activate protection device (say switch, to evacuate stored energy) before time t_q, not to allow large overheating. According to Eq. (4), the time t_q is proportional to thermal time t_h and depends strongly on how close is operating current I to thermal quench current I_q. As we showed, the best way to evaluate thermal time is to find it via low-current experiment [5]. The universal dependence of normalized time t_q on normalized current is shown in Fig. 3. Time t_q is normalized on thermal time t_h and current is normalized as: $I_q/[(I-I_q)(n+1)]$. One can see that experimental curves from different HTSC devices are well coinciding with calculated one.

To protect HTSC coil most probable method should be the active one, because of very slow normal zone propagation the uniform distribution of stored energy is not achievable. The quench detector may be sensitive to the voltage as in most usual quench protection designs. In this case the parameter which should be detected is electric field E_q after which fast voltage and temperature rise starts. This parameter depends on cooling conditions and does not depend on operating current. Quench detector may be tuned for certain part of E_q to activate protection scheme in advance.

The quench detector also may be designed to check the temperature of the coil. In this case, the quench parameter to initiate protection is the temperature T_q. It depends on operating temperature,

T_c and index n only and does not depend on any other parameters. The example of calculated and experimental dependencies of T_q on T is shown in Fig. 4. Also good coincidence between theory and experiment is seen.

In summary, we showed that new theoretical model is applicable for quench description of different high-Tc devices. Similarity of the experimental curves transformed into dimensionless form proves the universality of the model and shows the same nature of the quench of different high-Tc devices. The model can be used not only to describe the experimental results but also to predict quench development in high-Tc devices that is very important at the design stage.

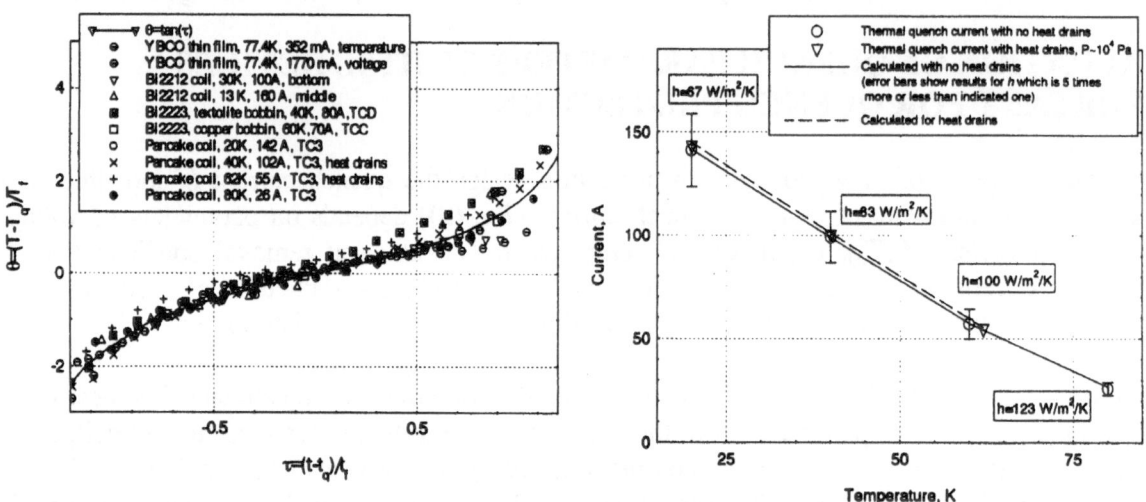

Fig.1. Dimensionless temperature vs dimensionless time for experiments with five different objects.

Fig.2. Quench current calculations for HTSC pancake coil.

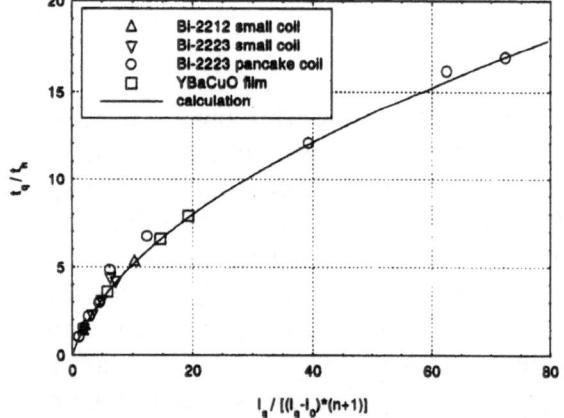

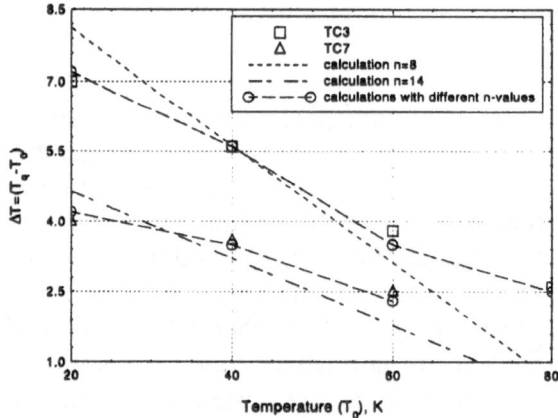

Fig.3. Relative characteristic time for devices of different kind. Symbols – experimental results.

Fig.4. Results for temperature rise before TQ happen, for two thermocouples inside the pancake coil.

1. A.L.Rakhmanov et.al., *Scaling for quench development in HTSC devices – theory*, presented at EUCAS'99, Barcelona, Spain.
2. V.S.Vystosky et.al., *Scaling for quench development in HTSC devices – comparison with the experiment*, presented at EUCAS'99, Barcelona, Spain.
3. V.S.Vysotsky et.al., *Stability and thermal quench study in HTSC pancake coil*, Advances In Superconductivity XI, Vol.2, pp. 1329-1332, 1999.
4. V.S.Vysotsky et.al., *Quench propagation in HTSC coils with bobbins from different materials*, IOP Proceedings of ICEC17, pp. 357-360, 1998.
5. T.Kiss et.al., *Heat propagation and stability in a small HTSC coil*, PHYSICA C 310, pp. 372-376, 1998.

Influence of Magnetic-Field Dependence of J_c and ρ_f on Magnetic Shielding Performance of HTS Plates

Atsushi Kamitani,[1] Kyoko Hasegawa,[1] Takafumi Yokono,[2] and Shigetoshi Ohshima [1]

[1] Faculty of Engineering, Yamagata University, Yamagata 992-8510, Japan
[2] Institute of Information Sciences and Electronics, University of Tsukuba, Ibaraki 305-0827, Japan

Abstract: The magnetic shielding performance of the high-Tc superconductor (HTS) is numerically investigated by taking account of the magnetic-field dependence of the critical current density J_c. In order to describe the B-dependence, the Kim model is employed. The numerical code for analyzing the time evolution of the shielding current density has been developed and the damping coefficients are evaluated by use of the code. The results of computations show the magnetic shielding performance of the HTS is hardly influenced by the B-dependence of J_c if the amplitude of the applied ac magnetic flux density is less than 1 T.

Keywords: magnetic shielding, YBCO, flux flow, flux creep

INTRODUCTION

In general, the critical current density J_c and the flow resistivity ρ_f of the high-Tc superconductor (HTS) strongly depend on the magnetic field. By neglecting such dependence, the magnetic shielding performance of HTS plates was numerically estimated in the previous study [1, 2]. However, the influence of such B-dependence on the magnetic shielding performance has not yet been investigated.

The purpose of the present study is to develop the numerical code for calculating the shielding current density by implementing the Kim model and the Bardeen-Stephan model, and to investigate the influence of the B-dependence of J_c on the magnetic shielding performance of HTS plates.

MATHEMATICAL FORMULATION

In this section, we explain the governing equation of the shielding current density in a disk-shaped HTS plate. By taking the symmetry axis of the plate as z-axis and by choosing its center of gravity as the origin, we use the cylindrical coordinate system (r, θ, z) throughout the present study. Let us first assume that a disk-shaped HTS plate of radius a and of thickness b is exposed to the ac homogeneous magnetic field parallel to the thickness direction. In this case, the applied magnetic flux density B_0 can be expressed as $B_0 = B_0{}^* \sin \omega t \, e_z$. Here $B_0{}^*$ and ω are constants, and e_z denotes a unit vector in the z direction. By taking account of the strong crystallographic anisotropy in J_c of the HTS, we further assume the multiple-thin-layer approximation [1-3] to modelize the HTS plate.

Under the assumptions, the behavior of the shielding current density can be expressed as follows [3]:

$$\mu_0 \frac{\partial}{\partial t} \sum_{q=1}^{M} \int_0^a S_q(r', t)\, Q^*{}_{pq}(r, r')\, r'\, \mathrm{d}r' + \frac{2M\mu_0}{b} \frac{\partial S_p}{\partial t} = -\frac{\partial}{\partial t}(B_0 \cdot e_z) - \frac{1}{r}\frac{\partial}{\partial r}(r\, E_{p\theta})\ , \qquad (1)$$

where μ_0 is a magnetic permeability of vacuum and M denotes the total number of layers. The scalar function $S_p(r, t)$ is associated with θ-component $J_{p\theta}$ of the shielding current density in the pth layer through the relation: $J_{p\theta} = -(2M/b)\partial S_p/\partial r$. In addition, the function Q^*_{pq} is defined by

$$Q^*_{pq}(r, r') = -\frac{M^2}{\pi b^2 (r r')^{1/2}} \sum_{m=0}^{1} \sum_{n=0}^{1} (-1)^{m+n} k_{pq}^{mn} K(k_{pq}^{mn}) , \qquad (2)$$

where $K(x)$ denotes a complete elliptic integral of the first kind and its parameter k_{pq}^{mn} is given by k_{pq}^{mn} = $4rr' [(r + r')^2 + \{[z_p + (-1)^m (b/2M)] - [z_q + (-1)^n (b/2M)]\}^2]^{-1}$. Here z_p denotes z-coordinate of the central plane of the pth layer and is given by $z_p \equiv b [(M-1) - 2(p-1)]/2M$. In addition, θ-component $E_{p\theta}$ of the electric field in the pth layer is determined by using the J-E constitutive relation: $E_{p\theta} = $ sgn$(J_{p\theta}) E(|J_{p\theta}|, B)$, where B is a magnitude of a magnetic flux density.

As $E(J, B)$, we adopt the flux flow creep model [3, 4] in the present study. The explicit form of $E(J, B)$ is written as $E(J, B) = 2\Theta(J_c(B) - J) \rho_c J_c(B) \sinh(\beta J/J_c(B)) \exp(-\beta) + \Theta(J - J_c(B)) \{\rho_c J_c(B) [1- \exp(-2\beta)] + \rho_f(B) (J - J_c(B))\}$, where ρ_c denotes the creep resistivity and the parameter β is defined by $\beta = U_0/(k_B T)$. Here U_0, k_B and T are the pinning potential, the Boltzmann constant and the temperature, respectively. In addition, $\Theta(x)$ denotes Heaviside's step function: it takes unity for $x > 0$ and zero for $x < 0$. Although the critical current density J_c and the flow resistivity ρ_f were regarded as constants in the previous study [1, 2], both of them are assumed as functions of B in the present study. The explicit forms of the functions are written as follows: $J_c(B) = J_{c0} B_J/(B_J + B)$ and $\rho_f(B) = \rho_{f0} (B/B_\rho)^\gamma$. Here J_{c0}, ρ_{f0}, B_J, B_ρ and γ are all constants. The former is called as the Kim model, whereas the latter becomes the Bardeen-Stephan model in case of $\gamma = 1$. Note that the former with $B_J \to \infty$ and the latter with $\gamma = 0$ are the same assumptions on J_c and ρ_f as those used in the previous study [1, 2]. As the initial and the boundary conditions to eq. (1), we assume $S_p(a, t) = S_p(r, 0) = 0$ as usual. By solving eq. (1) together with the initial and the boundary conditions, we can determine the temporal variation of the shielding current density that is used in the magnetic shielding analysis.

By discretizing eq. (1) and the boundary condition by means of the finite element method, the functions $S_1, S_2, \cdots, S_M$ at the nth time step constitute the nodal vector S^n that is a solution S of the nonlinear equation $G(S) = 0$. In order to solve the equation, the Newton method is employed. In this method, the solution S is determined iteratively. In the kth cycle, the linear equation $J \delta S = -G(S^{(k-1)})$ is solved to determine δS and, subsequently, $S^{(k)}$ is corrected by use of $S^{(k)} = S^{(k-1)} + \tau \delta S$. Here the superscript (k) and τ denote the iteration number label and the underrelaxation factor, respectively, and J represents the Jacobian matrix of G. The above cycle is repeated until the convergence is obtained.

MAGNETIC SHIELDING ANALYSIS

As a measure of the shielding performance, we define the damping coefficient by $\alpha \equiv 10 \log(\langle B^2 \rangle/ \langle B_0^2 \rangle)$. Here B denotes a magnetic flux density and the square bracket means a time average. Throughout the present analysis, the geometrical and the physical parameters are fixed as follows: $a = 20$ mm, b = 2 mm, $T = 77$K, $J_{c0} = 1.5 \times 10^6$ A/m², $U_0 = 92$ meV, $\rho_{f0} = 7.620 \times 10^{-10}$ Ωm, $\rho_c = 6.666 \times 10^{-11}$ Ωm and $B_\rho = 1$ T. The value of J_{c0} is the same as the experimental results obtained by Ohshima et al. [5] As other physical parameters, we use the same values as those used by Yoshida et al. [4] They applied the flux flow creep model to the evaluation of dynamic magnetic force in the HTS levitation system and found the good agreement between the experimental and the computed results. In order to determine the appropriate value of M, the values of α are calculated for various frequency as a function of M. The results of computations show that the values are almost constant in case of $M \geq 6$. For this reason, the layer number M is fixed as $M = 6$ throughout the present analysis.

Let us first investigate the influence of the B-dependence of ρ_f on the magnetic shielding performance.

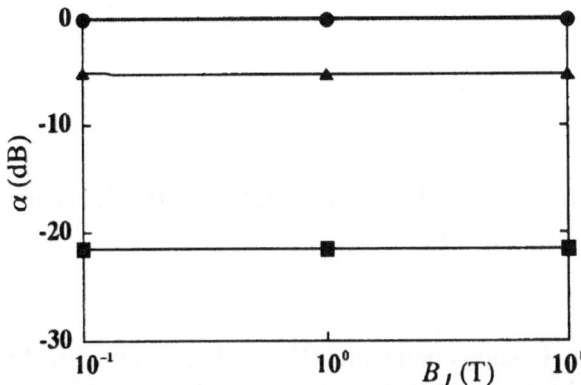

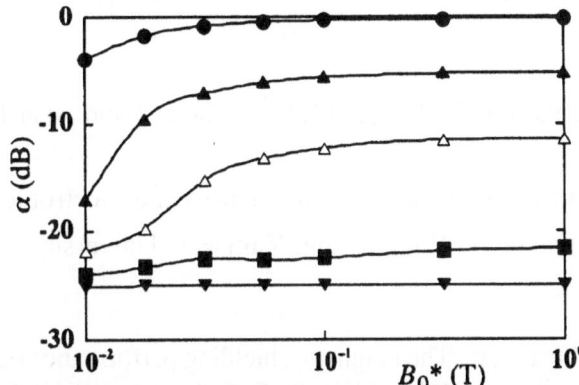

Fig. 1. Dependence of damping coefficients α on B_J for the case with $B_0^* = 1$ T. Damping coefficients are calculated at $(z/a, r/a) = (-0.06, 0.0)$. The symbols, ●, ▲ and ■, denote the values for $\omega =$ 10 Hz, 100 Hz and 1 kHz, respectively.

Fig. 2. Damping coefficients α as functions of B_0^* for the case with $B_J = 1$ T. The symbols, ●, ▲, △, ■ and ▼, denote the values of α at $(z/a, r/a) =$ $(-0.06, 0.0)$ for $\omega =$ 10 Hz, 100 Hz, 200 Hz, 1 kHz and 10kHz, respectively.

The results of computations show that, in case of $\gamma = 1$, the converged solution of eq. (1) has never been obtained. This result means that the Newton method with a constant underrelaxation factor is not applicable to the case with the Bardeen-Stephan model. For this reason, the value of γ is fixed as $\gamma = 0$ in the following. In other words, the flow resistivity is fixed as $\rho_f = 7.620 \times 10^{-10}$ Ωm.

Next, we investigate the effect of the B-dependence of J_c on the magnetic shielding performance. The damping coefficients α are calculated as functions of the parameter B_J in the Kim model and are depicted in Fig. 1. We see from this figure that the magnetic shielding performance is not affected by an increase in B_J. Although the spatial distribution of the shielding current density is expected to vary with an increase in B_J, the shielding performance does not change for the case with 0.1 T $\leq B_J \leq 1$ T.

Finally, we investigate the dependence of the shielding performance on the amplitude B_0^* of the applied magnetic flux density. In Fig. 2, we show damping coefficients as functions of B_0^*. This figure indicates that the shielding performance is independent of B_0^* in case of $\omega \gtrsim 1$ kHz. In this contrast, it is degraded with an increase in B_0^* in case of $\omega \lesssim 200$ Hz.

CONCLUSIONS

By using the Kim model, we have numerically investigated the magnetic shielding performance of the disk-shaped HTS plate against the weak ac magnetic field with $B_0^* \leq 1$ T. Conclusions obtained in the present study are summarized as follows.
1) The magnetic shielding performance is not affected by the B-dependence of J_c if 0.1 T $\leq B_J \leq 10$ T.
2) The damping coefficient α is calculated in case of $B_J = 1$ T. For the case with $\omega \gtrsim 1$ kHz, α does not change its value even when the amplitude of the applied ac magnetic field is increased. On the other hand, it increases with an increasing amplitude for the case with $\omega \lesssim 200$ Hz.

1. A. Kamitani, T. Yokono, and S. Ohshima, in Advances in Superconductivity XI, edited by N. Koshizuka and S. Tajima (Springer-Verlag, Tokyo, 1999), pp.1455-1458.
2. T. Yokono, M. Natori, S. Ohshima, and A. Kamitani, in Advances in Superconductivity XI, edited by N. Koshizuka and S. Tajima (Springer-Verlag, Tokyo, 1999), pp.1459-1462.
3. A. Kamitani and S. Ohshima, IEICE Trans. Electron. Vol. E82-C, No. 5, 766-773 (1999).
4. Y. Yoshida, M. Uesaka and K. Miya, Int. J. Appl. Electromagn. Mater. Vol. 5, 83-89 (1994).
5. S. Ohshima and K. Okuyama, Jpn. J. Appl. Phys. Vol. 29, No. 11, 2403-2406 (1990).

Numerical Investigations on Magnetic Shielding Performance of Axisymmetric HTS Plates against Several-Turn-Coil Fields

Takafumi Yokono, [1] Makoto Natori, [1] and Atsushi Kamitani [2]

[1]Institute of Information Sciences and Electronics, University of Tsukuba, Ibaraki 305-0827, Japan

[2]Faculty of Engineering, Yamagata University, Yamagata 992-8510, Japan

Abstract: The magnetic shielding performance of high-Tc superconducting (HTS) plates is numerically investigated. In order to reflect the experimental conditions accurately, a several-turn coil is assumed as the source of the magnetic field. A numerical code for analyzing the time evolution of the shielding current density has been developed and, by use of the code, the shielding performance of HTS plates is investigated. The results of computations show that the spatial distribution of the shielding current density changes drastically with the distance between the plate and the coil. The shielding current density concentrates near the edge of the plate in case of the large distance, whereas it also concentrates just below the coil in case of the sufficiently small distance.

Keywords: magnetic shielding, YBCO, flux flow, flux creep

INTRODUCTION

In the previous study [1, 2], we developed the numerical method for analyzing the magnetic shielding performance of axisymmetric high-Tc superconducting (HTS) plates. In the study, the uniformity of an externally applied magnetic field is assumed so that the shielding current density may have an updown symmetry. Consequently, the computational resources have been saved considerably. In the actual experiments, however, the several-turn coil is employed to generate the applied magnetic field. In this sense, it will be useful to investigate the magnetic shielding performance of the HTS plates against a coil fields.

The purpose of the present study is to develop the numerical code for analyze the time evolution of the shielding current density in axisymmetric HTS plates placed in the several-turn-coil field and to investigate their magnetic shielding performance by means of the code.

NUMERICAL CODE FOR MAGNETIC SHIELDING ANALYSIS

In this section, we briefly explain the governing equation of the shielding current density in the HTS plates. The method for deriving the equation has been already presented in detail in [1, 2]. In the following, we use the cylindrical coordinate system (ρ, φ, z) by taking the center of the HTS plate as the origin and the thickness direction as z-axis. In addition, we assume that the HTS plate is a disk of radius a and of thickness b and that the applied magnetic field is generated by a several-turn coil. As is well known, the critical current density of the MPMG-YBCO superconductor has the crystallographic anisotropy. By taking this experimental fact into consideration, we can adopt the multiple-thin-layer approximation [1, 2] to modelize the HTS plate. In other words, the HTS plate is assumed as a set of K pieces of thin layers. Hereafter, the pth layer denotes the region $[0, a] \times [0, 2\pi) \times [z_p - \varepsilon_p, z_p + \varepsilon_p]$.

Under the above assumptions, the shielding current density in the pth layer has only φ-component $j_{\varphi p}$ that is expressed in terms of a scalar function S_p as $j_{\varphi p} = -\varepsilon_p^{-1} \partial S_p / \partial \rho$. Here $2\varepsilon_p$ denotes the thickness of the pth layer. In addition, the scalar function S_p is governed by

$$\mu_0 \partial_t \sum_{q=1}^{K} \int_0^a Q_{pq}^* \rho' S_q \, \mathrm{d}\rho' + \frac{\mu_0}{\varepsilon_p} \partial_t S_p = -\frac{1}{\rho} \frac{\partial}{\partial \rho} (\rho E_{\varphi p}) - \partial_t \langle B_{0z} \rangle_F. \tag{1}$$

Here μ_0 is a magnetic permeability of vacuum. In addition $E_{\varphi p}$ is φ-component of the electric field that is associated with $j_{\varphi p}$ by the J-E constitutive equation. Furthermore, B_{0z} is z-component of the applied magnetic flux density and the operator $\langle \ \rangle_p$ denotes a average through the pth layer. The explicit form of the function $Q_{pq}^*(\rho, \rho')$ is given by

$$Q_{pq}^*(\rho, \rho') = \frac{-1}{4\pi \, \varepsilon_p \varepsilon_q \sqrt{\rho \rho'}} \sum_{m=0}^{1} \sum_{n=0}^{1} (-1)^{m+n} k_{pq}^{mn} \, \mathrm{K}(k_{pq}^{mn}), \tag{2}$$

where $\mathrm{K}(x)$ denotes a complete elliptic integral of the first kind and k_{pq}^{mn} is defined by $k_{pq}^{mn} \equiv 4\rho\rho'[(\rho + \rho')^2 + \{[z_p + (-1)^m \varepsilon_p] - [z_q + (-1)^n \varepsilon_q]\}^2]^{-1}$. The boundary and the initial conditions are given as follows: $S_p = 0$ on $\rho = a$ and $S_p = 0$ at $t = 0$.

The flux flow creep model is adopted to describe the characteristics of the Type-II superconductor. In the model, the function $E(j)$ of the J-E constitutive equation is given by

$$E(j) = \begin{cases} 2\rho_c j_C \sinh\left(\dfrac{U_0}{k_B T} \dfrac{j}{j_C}\right) \exp\left(-\dfrac{U_0}{k_B T}\right) & ; j \le j_C \\[2mm] \rho_c j_C \left[1 - \exp\left(-\dfrac{2U_0}{k_B T}\right)\right] + \rho_f(j - j_C) & ; j > j_C \end{cases}, \tag{3}$$

[12]. Here ρ_c, ρ_f, U_0, k_B, T and j_C are creep resistivity, flux resistivity, the pinning potential, the Boltzmann constant, the temperature and the critical current density, respectively.

In order to discretize eq. (1) and the associated boundary condition, the ordinary FEM and the θ-method are utilized. By taking account of the continuity of the electric field on the element boundary, the Hermitian elements are employed for the geometric discretization. On the other hand, the θ-method is utilized to the approximation of the integration with respect to time. Since the resulting discretized equation has a strong nonlinearity, the Newton method is employed to the solution of the equation.

Solving eq. (1) together with the initial and the boundary conditions, the temporal variation of the shielding current density is obtained. Since the total magnetic flux density is easily calculated by using the solution, we can analyze the magnetic shielding performance of the HTS plate.

RESULTS AND DISCUSSIONS

In this section, we numerically investigate the magnetic shielding performance of the HTS plate against the coil field by using the code explained above. In order to reflect the experimental conditions accurately, the applied magnetic fields are generated by a coil current $I = I_c \sin \omega_c t$. The coil of radius r_C is placed at $(\rho, z) = (0, z_C)$ so as to be parallel to the HTS plate. In addition, B_C is the amplitude of the coil field at the center of the HTS plate. Throughout the present analysis, the physical and the geometrical parameters are fixed as follows: $a = 20$ mm, $b = 1$ mm, $j_C = 1.5 \times 10^6$ A/m^2, $T = 77$ K, $U_0 = 92$ meV, $\rho_f = 7.620 \times 10^{-10}$ Ωm, $\rho_c = 6.666 \times 10^{-11}$ Ωm, $r_C = 6$ mm and $B_C = 10^{-2}$ T. Here the value of j_C is the same as the experimental result obtained by Ohshima et al. [3] On the other hand, the physical parameters

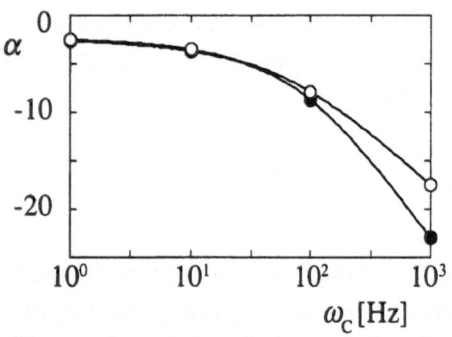

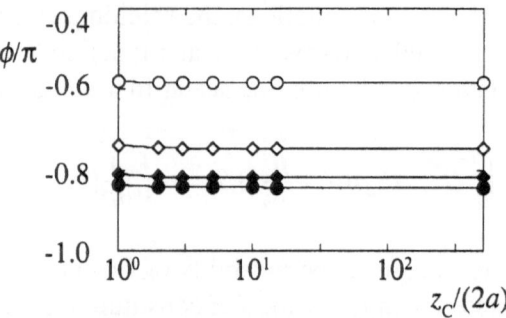

Fig. 1. Frequency dependence of the damping coefficient α at $(\rho/a, z/a) = (0, -0.051)$. The symbol $\bigcirc$ and $\bullet$ denote the values for $z_C/a = 1$ and 10^3, respectively.

Fig. 2. Phase shift ϕ as functions of $z_C/(2a)$. The symbols $\bullet$, $\blacklozenge$, $\diamondsuit$ and $\bigcirc$, denote the values of ϕ for $\omega_C = 1$Hz, 10Hz, 10^2Hz and 10^3Hz, respectively.

appeared in the eq. (3) may be the same values as used by Yoshida et al. [4]

In order to evaluate the magnetic shielding performance, the damping coefficient α is defined by $\alpha \equiv 10 \log_{10}(\langle |B_s + B_0|^2 \rangle / \langle |B_0|^2 \rangle)$ in the external region. Here, B_s is the magnetic flux density induced by the shielding current density, B_0 is the applied magnetic flux density and the operator $\langle \rangle$ means a time average.

Let us first investigate the spatial distribution of the shielding current density. The results of computations show that the distance between the plate and the coil is closely related to the distribution. In case of the large distance, the shielding current density concentrates near the edge of the HTS plate. This tendency has a close resemblance to that in the previous study. On the other hand, the shielding current density concentrates not only near the edge but also in the region just below the coil for the case with the sufficiently small distance. The damping coefficients α are evaluated and are depicted as functions of the frequency in Fig. 1. The figure indicates that α decreases with an increase in ω_C and that the shielding performance is enhanced with an increase in z_C in the high frequency region. In order to explain the reason for this, we show the phase shift ϕ as functions of z_C in Fig. 2. Here, ϕ is the phase of the dominant Fourier mode of the total shielding current in the most outer layer. This figure shows that ϕ is independent on the location of the coil and that its value is decided by the frequency ω_C only. Therefore, because the shielding current density concentrates near the edge, high shielding performance is obtained for the case with the coil placed far from the HTS plate.

CONCLUSIONS

We have investigated the magnetic shielding performance of the HTS plate against the coil field. Conclusions obtained in the present study are summarized as follows.
1) The spatial distribution of the shielding current density changes drastically with the distance between the plate and the coil.
2) When the distance is large, the shielding current density concentrates near the edge of the plate. In this contrast, it also concentrates just below the coil in case of the sufficiently small distance.

1. A. Kamitani, T. Yokono, and S. Ohshima, in Advances in Superconductivity XI, edited by N. Koshizuka and S. Tajima (Springer-Verlag, Tokyo, 1999), pp.1455-1458.
2. T. Yokono, M. Natori, S. Ohshima, and A. Kamitani, in Advances in Superconductivity XI, edited by N. Koshizuka and S. Tajima (Springer-Verlag, Tokyo, 1999), pp.1459-1462.
3. S. Ohshima and K. Okuyama, Jpn. J. Appl. Phys., vol.29, no.11, pp.2403-2406, Nov. 1990.
4. Y. Yoshida, M. Uesaka, and K. Miya, Int. J. Appl. Electromagn. Mater., vol.5, pp.83-89, 1994.

6 Films and Junctions

6.1 Film Preparation and Characterization
6.2 Film Growth and Mechanisms
6.3 Junction Fabrication and Characterization

Series production of large area $YBa_2Cu_3O_7$ – films for electronic-, microwave-, and electrical power applications

Werner Prusseit, Stefan Furtner, Ralf Nemetschek

THEVA Dünnschichttechnik GmbH, Hauptstr. 1b, D-85386 Eching-Dietersheim, Germany

Abstract: The series production of $YBa_2Cu_3O_7$ - films on a daily basis is making high demands on the reliability and reproducibility as well as the flexibility of the deposition technique. Beyond that, a strict quality management and customer support are absolute necessities to serve a market with is getting closer towards real technical applications.

Keywords: YBCO films, evaporation, protective layers

INTRODUCTION

To date a great variety of film deposition techniques has been demonstrated to yield high quality $YBa_2Cu_3O_7$ – coatings. However, quality is only a necessary but not sufficient condition for a commercial production on an everyday basis. Industrial scale production also depends to a large extend on such aspects like economic efficiency, throughput, and reliability. A mature process should guarantee high reproducibility, uniformity, and quality of the resulting films but should also provide enough flexibility for the numerous applications with respect to the substrate material and geometry as well as the film material itself. Even beyond the deposition it takes a lot more until industrial customers dare to start a real product development depending on external film supply

THERMAL EVAPORATION PROCESS

For many years the $YBa_2Cu_3O_7$ -deposition technique developed at the Technical University of Munich [1] has demonstrated its capability to yield excellent large area films [2]. From the point of economy it is rated the best candidate for commercial film production. It has turned out as a very reproducible method and due to its elaborate heater design it is very flexible as well, since it can accommodate even different substrate materials and sizes in the same deposition run and without changes in the process parameters.

The basic principle of the original turntable substrate holder is depicted in Fig. 1. The metals which build up the $YBa_2Cu_3O_7$ - lattice are evaporated from resistively heated boat sources in high vacuum ambient. The composition of the film is controlled by quartz crystal monitors individually focussed at the appropriate boat source. However, the key idea behind this deposition scheme is the spatial separation between oxidation and metal deposition. In the classic heater scheme this is realized by mounting the substrates on a turntable, whereas oxygen is introduced by an oxygen pocket which covers a sector of the whole area and is such closely spaced underneath the substrates that a pressure enhancement of 2-3 orders of magnitude can be realized against the surrounding high vacuum ambient. The turntable with its diameter of 22 cm can accommodate a large number of substrates, e.g. up to twelve 2" wafers. Since the substrates are not attached to any surface but are freely suspended and radiation heated the deposition works single as well as double sided.

Substituting the Yttrium - boat source or adding more boats allows deposition of various $REBa_2Cu_3O_7$ - films as well as in situ buffer-, cap- or contact layers.

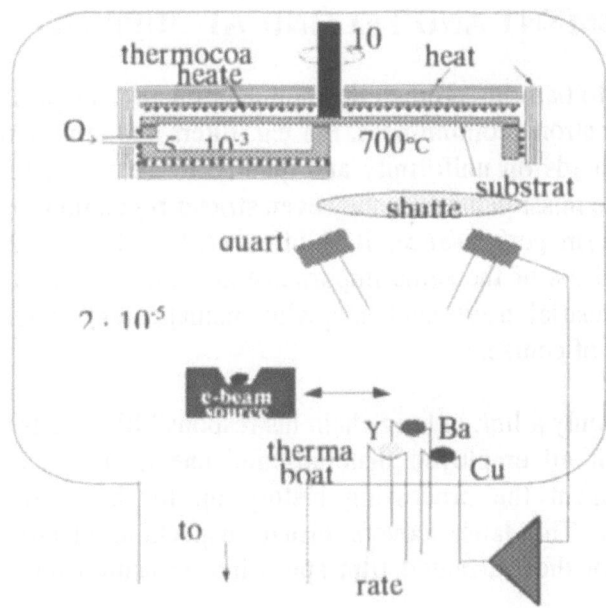

Fig. 1. YBCO - deposition principle with turntable substrate holder

APPLICATION TAILORED FILMS

The requirements on $YBa_2Cu_3O_7$ - films are as numerous and manifold as the applications. Current main stream development covers such various products as resistive fault current limiters, microwave filters for mobile or satellite communication, NMR and MRI pickup coils, Squid sensors, and superconducting electronics. While power and high frequency applications usually require highest possible critical current densities, low surface resistance, and good power handling, Squid applications need low noise figures. Superconducting electronics demand really smooth films, since YBCO is only one part in a multilayer structure where surface roughness can result in pinholes and shorts which are not tolerable. All these different and sometimes contradictory requirements cannot be met by the same $YBa_2Cu_3O_7$ - film. Rather we developed different film types optimized to certain applications as shown in Fig. 2. From 2(c) to 2(a) films are grown slightly off – stoichiometric with increasing copper and yttrium content, resulting in a distinct change in the surface morphology but also in differing superconducting properties, like e.g. surface resistance [3]. This provides a selection of films tailored for specific needs.

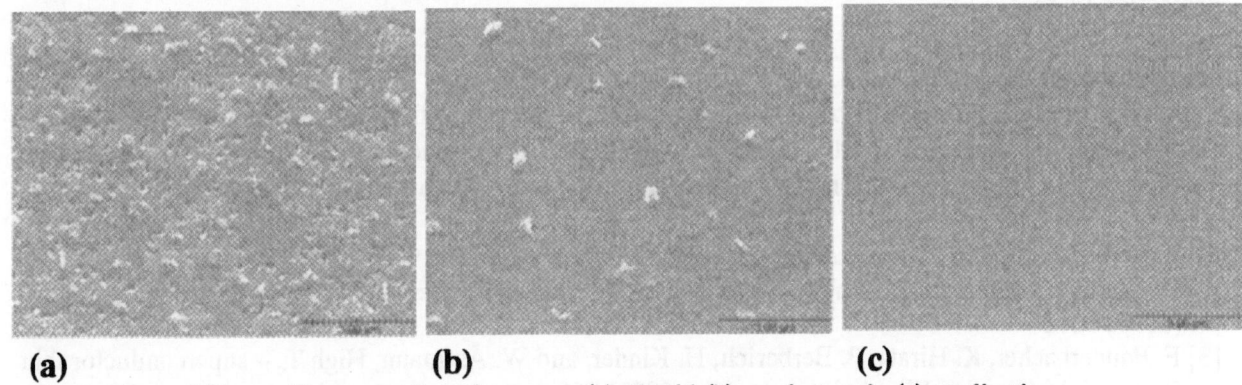

(a)   **(b)**   **(c)**

Fig. 2. Three different film types for microwave (a), Squid (b) or electronic (c) applications

QUALITY MANAGEMENT AND TECHNICAL SUPPORT

When a customer decides to base his product development on an external film supplier the producer of such films undertakes a strong commitment. For e.g. filters and fault current limiters hundreds of films with very high standards on uniformity and quality are required. Filters which are part of a platform in a manned space mission have to obey even stricter regulations, covering safety aspects as well as warranted long term performance. It is often neglected that maintaining an appropriate quality management is at least of the same importance and can even cost more time than the film fabrication itself. For industrial users such a quality management which keeps record of every fabrication step is a matter of course.

Since the film producer is only a link in long chain his responsibility starts with the inspection of the substrates and approval of all employed materials and chemicals. Records of all handling and fabrication steps complement the processing history up to the final quality inspections and certification of the films. The latter covers visual inspection, electron microscopy, and non destructive measurement of the warranted film specs like transition temperature , critical current density, and uniformity.

However, the responsibility of the film producer is not only restricted to mere fabrication. Technical consulting and support starts long before an order is placed, e.g. by selecting the appropriate substrate material and suggesting the optimum device fabrication route. It even extends to the point when long term stability becomes an issue. Long term performance is a crucial point when it comes to the decision if superconductors can be used for components in a large grid or network.

A prominent example are protective coatings which get extremely important when degradation by ambient humidity or electrochemistry may affect the device lifetime. In $YBa_2Cu_3O_7$ all cations are in their maximum oxidation state (e.g. copper in the +2 and +3 state). Consequently, $YBa_2Cu_3O_7$ is a very strong oxidizing agent and its electrochemical potential is positive and comparable to that of gold. Whenever it is brought into electrical contact with less "noble" metals, e.g. solder, in the presence of water a local galvanic cell builds up and the normal metal is oxidized. In turn, the oxidation state of Cu^{3+} is reduced, $YBa_2Cu_3O_7$ looses oxygen and becomes an insulator which is eventually destroyed by chemical reactions with water. In most practical applications it is necessary to protect the film by some kind of resin, glue, or resist against atmospheric influences. However, the requirements are very tough. The most important ones are chemical compatibility with the delicate superconductor material, curing, adhesion, water absorption and diffusion, dielectric losses, and good insulation. Screening a lot of organic compounds only a few materials like fluoropolymers, silicones etc. turned out to withstand the extreme cryogenic conditions to provide an effective protection [4-5].

The considerations above may have shown that the $YBa_2Cu_3O_7$ – film business comprises much more than the bare fabrication process. Demonstrating the reliability and potential of $YBa_2Cu_3O_7$ – films, quality management and technical support are the key elements to establish high temperature superconductors in the long run not only in a scientific but also in an engineering environment.

[1] P. Berberich, B. Utz, W. Prusseit, and H. Kinder, Physica C **219** (1993) 497

[2] H. Kinder; W. Prusseit; R. Semerad, and B. Utz, Advances in Superconductivity IX, Springer, Tokyo (1997) 1011

[3] F. Bauderbacher, K. Hirata, P. Berberich, H. Kinder, and W. Assmann, High T_c – superconductor thin films, ed. L. Correra , Elsevier, Amsterdam (1992) 365

[4] W. Prusseit, to be publ. Proc. of the EUCAS '99, Sitges, Spain (1999)

[5] Y. Nagai, N. Suzuki, M. Sato, and T. Konaka, Jap. J. Appl. Phys. **30** (1991) 2751

Preparation of NdBa$_2$Cu$_3$O$_x$ Thin Films and Their Superconducting Properties in Magnetic Fields

Yusuke Ichino, Kaname Matsumoto[1], Yasuo Takahashi[1], SeokBeom Kim[1], Hiroshi Ikuta[2], Izumi Hirabayashi[1] and Uichiro Mizutani

Department of Crystalline Materials Science, Nagoya University, Furo-cho, Chikusa-ku, Nagoya 464-8603, Japan
[1]Superconductivity Research Laboratory, ISTEC, 2-4-1 Mutsuno, Atsuta-ku, Nagoya 456-8587, Japan
[2]CIRSE, Nagoya University, Furo-cho, Chikusa-ku, Nagoya 464-8603, Japan

Abstract: We studied the superconducting properties of c-axis oriented NdBa$_2$Cu$_3$O$_x$ (Nd123) thin films prepared using the pulsed laser deposition technique. Films with highest critical temperature of 93 K and irreversibility field of 8 T at 77.3 K were successfully grown in an atmosphere of rather high oxygen pressure, 1.5 mbar. The pinning potentials of the films were estimated based on the thermally assisted flux flow model.

Keywords: NdBa$_2$Cu$_3$O$_x$ thin film, PLD, critical temperature, irreversibility field, pinning potential

INTRODUCTION

The substantially higher values of critical temperature (T_c) and irreversibility field (B_{irr}) of NdBa$_2$Cu$_3$O$_x$ (Nd123) compared to YBa$_2$Cu$_3$O$_{7-\delta}$ (Y123) make the former a potential material for high-field applications. Nevertheless, the number of works reporting superconducting properties (T_c, B_{irr}, etc) of Nd123 in the thin film form are rather limited, probably because of the difficulty of growing high quality Nd123 thin films. Previous works on Nd123 film preparation by the pulsed laser deposition (PLD) technique indicate that the deposition conditions for growing high quality films may differ from those of Y123 films. Li *et al.* [1] and Eulenburg *et al.* [2] reported that preparation of Nd123 films with T_c higher than 90 K at low oxygen pressure (less than 0.4 mbar) required a larger target-substrate distance compared to Y123, indicating that a significant clearance is necessary between the plume and substrate. The clearance depends also on the size of plume, which can be controlled by the oxygen pressure. In the present work, therefore, we studied systematically the effect of oxygen pressure and deposition temperature on T_c and B_{irr} of Nd123 films. Further, we estimated the pinning potential from the ρ-T data in magnetic field based on the thermally assisted flux flow (TAFF) model [4-6].

EXPERIMENTAL PROCEDURE

Nd123 thin films were deposited on (100) SrTiO$_3$ substrates in the presence of pure O$_2$ using KrF excimer laser (λ=248 nm) and a dense stoichiometric target. PLD method has several deposition parameters such as oxygen pressure ($P(\text{O}_2)$), substrate temperature (T_s), distance between target and substrate (d), laser energy density (D_L), laser repetition rate (f_L) and the size of the laser spot on the target (S_L). In this work, we changed $P(\text{O}_2)$ and T_s while other parameters were fixed to d=5.1 cm, D_L=2-3 J/cm^2, f_L=5 Hz, and S_L=0.52×0.07 cm^2. The number of laser shots and the thickness of the films were about 12000 and 2500 Å, respectively. T_s was measured using a pyrometer.

After the film was deposited, T_s was decreased to 700°C with a cooling rate of 20°C/min and the deposition chamber was filled with pure oxygen gas of atmospheric pressure. The heater was then switched off and the films were allowed to cool rapidly to room temperature. The crystallinity of the films was checked by x-ray diffraction.

Resistivity was measured with the conventional dc four-probe method in magnetic fields applied parallel to the crystallographic c-axis direction. Au was sputtered as electrodes on the films, which were then annealed at 400°C for 1 hour in O_2 gas flow to ensure good contacts. The T_c value and the irreversibility temperature were determined by adopting a voltage criterion of 2 μV/cm, while the current density was 100 A/cm².

RESULTS AND DISCUSSION

We first investigated the effect of changing oxygen pressure on the film quality by varying $P(O_2)$ from 0.6 to 2.0 mbar with T_s fixed to 850°C. The $P(O_2)$ dependence of the transition temperature is plotted in fig. 1, and it can be seen that the film deposited at $P(O_2)$=1.5 mbar had the highest T_c and the narrowest transition width. Further, the x-ray diffraction patterns ensured good crystallinity of the films and that c-axis is highly oriented. Therefore, we fixed $P(O_2)$ to 1.5 mbar in the following experiments, and deposited Nd123 thin films at various T_s. As a result, we succeeded to grow Nd123 thin films with T_c higher than 90 K. The top data of T_c was 93 K as shown in fig. 2. The resistivity of this film was higher than that of the film deposited at T_s=770°C. This may have been resulted due to a small amount of a-axis oriented regions which can not be detected by x-ray diffraction, because T_s=830°C is the boundary temperature above which a-axis oriented regions appear, as shown below.

Fig. 3 plots the intensity ratio of (200) to (200)+(005). Very interestingly, we found that a-axis oriented region disappears by lowering the substrate temperature in contrast to Y123. Quite recently, Zama et al. [3] have reported a similar result.

Fig. 4 shows the irreversibility lines of the Nd123 thin films deposited at various T_s ranging from 770 to 870°C. The top data of B_{irr} was 8 T at 77.3 K. T_c increased with T_s except for the film with T_s=770°C, which had a T_c comparable to that of the film deposited at 870°C. In general, higher T_s results in a film with larger grains, and this may be related to the improvement of superconducting properties. The reason of the high T_c of the film with T_s=770°C is unclear, but may be related to the disappearance of a-axis oriented region for films with low T_s, as mentioned above.

Fig. 4 shows that the slope of the irreversibility lines increased as T_c increased. This implies that the pinning force becomes larger as T_c increases. For the estimation of the pinning potential, we employed TAFF model, which describes resistivity as a function of temperature T and magnetic field B as following:

$$\rho(T,B) = \rho_0 \exp[-U(T,B)/k_B T], \quad (1a)$$

$$U(T,B) = \beta B^{-\alpha}(1-T/T_c)^n = U_{field}(1-T/T_c)^n . \quad (1b)$$

Here, ρ_0 is a prefactor related to the normal state resistivity, β is a numerical constant and U_{field} represents the field dependence of the pinning potential. n and α are expected to be 1.5 and 1 for Y123 [4,5], while 1 and 0.5 for $Bi_2Sr_2CaCu_2O_y$ (Bi2212) [6]. Fig. 5 is a plot of U_{field} of the films deposited at $P(O_2)$=1.5 mbar. U_{field} has a different field dependence

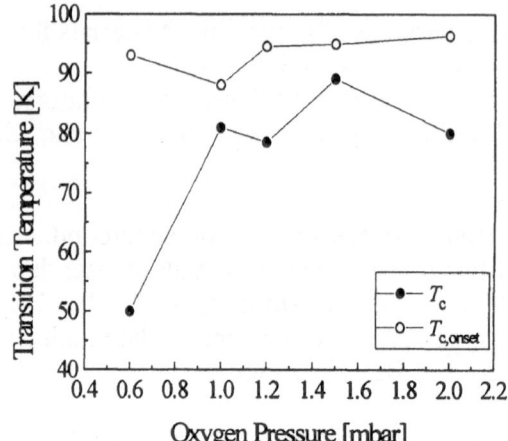

Fig. 1. $P(O_2)$ dependence of transition temperature of Nd123 films deposited at T_s=850°C.

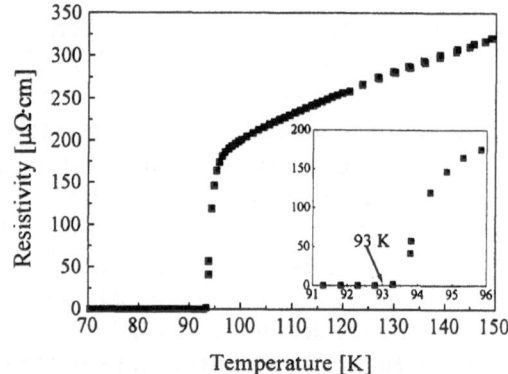

Fig. 2. The temperature dependence of resistivity of a Nd123 thin film deposited at $P(O_2)$=1.5mbar and T_s=830°C.

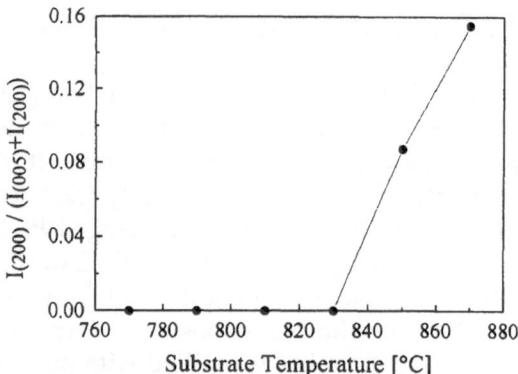

Fig. 3. The intensity ratio of (200) to (200)+(005) of the Nd123 films deposited at $P(O_2)$=1.5 mbar and various T_s.

in the lower and higher field regions, which probably comes from the difference of single vortex pinning and collective pinning. The data of U_{field} for a Y123 film reported by Wang et al. [8] is also included in fig. 5. It canbe seen that U_{field} of high quality Nd123 films is comparable to or slightly better than that of Y123.

The solid lines shown in fig. 5 give α and β, which are listed in table 1 with other characteristic values of our films. The values of n and α are smaller than that of Y123 and are rather close to those expected for Bi2212 when the film has a low T_c. This is probably reflecting the low quality of the sample. Films with T_c exceeding 90 K have n and α values close to Y123.

CONCLUSIONS

We prepared Nd123 thin films on (100) SrTiO₃ substrates using the PLD method. By growing films in a rather high oxygen pressure, $P(O_2)=1.5$ mbar, T_c exceeded 90 K, and the highest value of T_c so far is 93 K. In addition, the a-axis oriented region disappeared by lowering the substrate temperature. The irreversibility fields of the films deposited at various T_s were determined from resistivity measurements in magnetic fields. The top data of B_{irr} at 77.3 K was 8 T. In addition, we analyzed the data based on TAFF model, which revealed that the pinning potential of our Nd123 films is comparable to or slightly better than that of Y123.

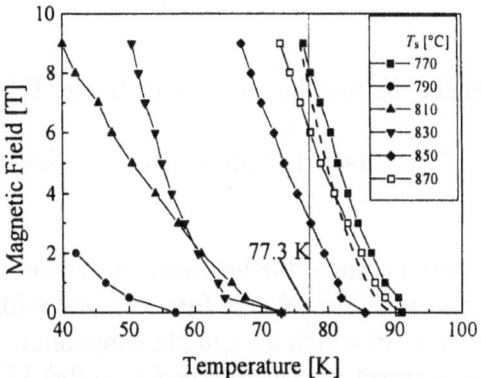

Fig. 4 The irreversibility lines of Nd123 thin films deposited at various T_s. $P(O_2)$ was fixed to 1.5 mbar. The broken line shows the irreversibility line of Y123 thin film prepared by PLD method [7].

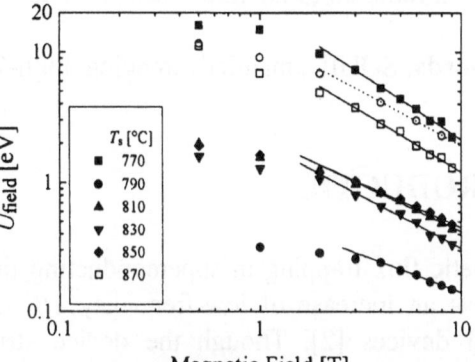

Fig. 5. The field dependence of the pinning potential. The Nd123 films were deposited at various T_s ranging from 770 to 870°C. Open circles represent the data of a Y123 film prepared by PLD method [8].

Table 1. Characteristic values of Nd123 thin films studied in this work.

T_c [K][#1]	ΔT_c [K]	T_s [°C]	c-axis [Å]	ρ_R[#2]	n	α	β [eV T$^{\alpha}$]
91	1.5	770	11.7523	2.80	1.44 ± 0.04	0.96 ± 0.04	19.74
90.5	2.8	870	11.7538	2.54	1.37 ± 0.04	0.93 ± 0.04	10.36
85.5	5.9	850	11.7487	2.48	1.12 ± 0.05	0.69 ± 0.02	2.22
73	19.9	810	11.7350	2.51	1.14 ± 0.06	0.73 ± 0.04	2.27
73	19.7	830	11.7589	2.37	1.11 ± 0.08	0.82 ± 0.05	2.07
57	35.5	790	11.7559	2.26	1.0 ± 0.1	0.58 ± 0.02	0.54

#1 T_c at 0 T   #2 $\rho_R=\rho(300\ K)/\rho(100\ K)$

Acknowledgements. This work was supported by the New Energy and Industrial Technology Development Organization (NEDO) as Collaborative Research and Development of Fundamental Technologies for Superconductivity Applications.

1. Y. Li and K. Tanabe, J. Appl. Phys. **83**, 7744-7752 (1998)
2. A. Eulenburg, E. J. Romans, Y. C. Fan and C. M. Pegrum, Physica C **312**, 91-104 (1999)
3. H. Zama, K. Ishikawa, T. Suzuki and T. Morishita, Jpn. J. Appl. Phys. (in press)
4. Y. Yeshurun and A. P. Malozemoff, Phys. Rev. Lett. **60**, 2202-2205 (1988)
5. M. Tinkham, Phys. Rev. Lett. **61**, 1658-1661 (1988)
6. J. T. Kucera, T. P. Orlando, G. Virshup and J. N. Eckstein, Phys. Rev. B **46**, 11004-11013 (1992)
7. M. Giura et al. Physica C **282-287**, 2345-2346 (1997)
8. Z. H. Wang, Supercond. Sci. Technol. **12**, 421-425 (1999)

Magnetic Imaging of High-T_c Thin Film Patterns by a Scanning SQUID Microscope

K. Tanabe, K. Suzuki, Yijie Li, N. Inoue, T. Sugano and T. Utagawa

SRL- ISTEC, 1-10-13 Shinonome, Koto-ku, Tokyo 135-0062,   Japan

Abstract: To study the behavior of magnetic flux trapping in high-T_c thin films, we have performed magnetic imaging of thin film patterns with various shapes and size by using a scanning SQUID microscope in which the sample temperature can be varied between 4 and 100 K. The images of flux quanta trapped in Nd-123 and (Hg,Re)-1212 thin film patterns are clearly observed in a wide temperature range. The number and the position of flux quanta trapped in NBCO thin films are found unchanged even at temteratures just below their T_c, which is consistent with the rather strong pinning in this material. No trapped flux is observed for the ambient field smaller than the critical field as a function of the line width predicted by Clem.

Keywords: SQUID, magnetic imaging, high-T_c films, NBCO, (Hg,Re)-1212

INTRODUCTION

Magnetic flux trapping in superconducting thin films causes serious problems for electronic devices such as an increase of low-frequency $1/f$ noise in high-T_c SQUIDs [1] and an operation error of digital devices [2]. Though the device structures with moats or holes have been employed to overcome these problems [1, 2], the behavior and mechanism of flux trapping, in particular in high-T_c thin films, have not been fully clarified yet.

Scanning SQUID microscope (SSM) has recently attracted much attention as a powerful tool to obtain magnetic field images in superconductors, magnetic materials, and electric circuits [3]. In the present study, to investigate the behavior of flux trapping in high-T_c thin films, we have taken magnetic field images of thin film patterns with various shapes and size by an SSM system in which the sample temperature can be varied between 4 and 100 K. Trapped flux quanta are clearly observed in a wide temperature range. The dependencies of flux trapping on temperature and the line width are discussed in comparison with previous results deduced from the measurements of film or device properties.

EXPERIMENTAL

150 – 200 nm thick Nd-123 (NBCO) thin films with a T_c of 89 - 92 K were prepared on $SrTiO_3(100)$ substrates by PLD (KrF excimer) with a bulk single crystal target. The details of deposition procedure and the film properties have been described elsewhere [4, 5]. Approximately 75 nm thick (Hg, Re)-1212 thin films on $SrTiO_3(100)$ with a T_c of 113 – 116 K were fabricated by the *ex situ* two-step process [6]. These films were patterned by standard photolithography and ion milling.

An SSM system with Nb-based SQUID magnetometers and a pick-up coil 10 μm in diameter (Seiko Instruments Inc.) [7] was used to take magnetic images. The spatial resolution in this system is

882

approximately 5 µm or less [7]. The system has µ-metal magnetic shields near the sample and the residual field is approximately 2 µT. The thin film patterns were cooled to a low temperature of approximately 5 K in a various field which can be controlled by a coil near the sample. Then magnetic images were taken at various temperatures by applying weak field to obtain better contrast.

RESULTS AND DISCUSSION

Fig. 1 shows the magnetic field images for NBCO and (Hg,Re)-1212 thin film patterns with a diameter of 200 µm cooled in a field of approximately 3 and 2 µT, respectively. The thin film patterns are clearly seen with a contrast to the background due to Meissner effect. Many sharp peaks are observed in the patterns. It has been confirmed that each peak corresponds to the flux quantum. These trapped flux quanta can be clearly observed even at elevated temperatures. Fig. 2 shows the images for a different NBCO pattern with a diameter of 100 µm. Though the images are less clear because of a larger gap between the pick-up coil and the sample, the number and the position of trapped flux are found unchanged even at a temperature very close to the film T_c. This seems consistent with the rather strong pinning in this material which was revealed by magnetization measurements in bulk materials [8] and thin films [5].

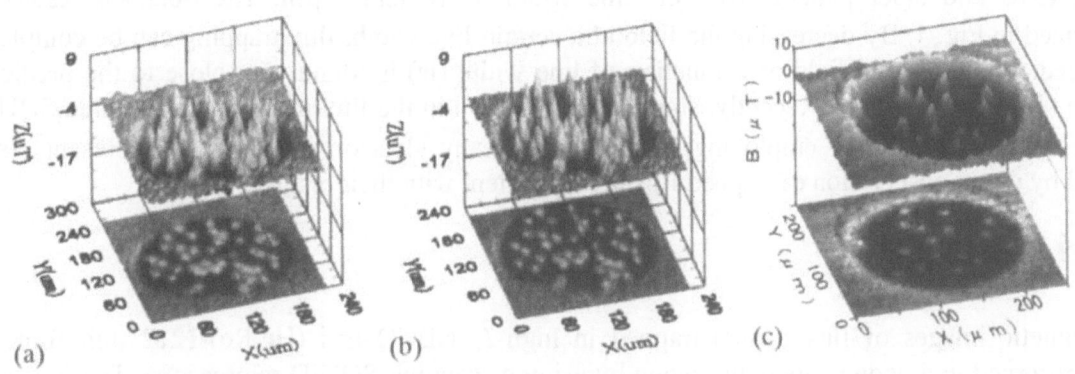

Fig. 1. Magnetic field images for NBCO and (Hg,Re)-1212 film patterns (200 µm in diameter). (a) NBCO taken at 4.8 K, (b) NBCO at 54.0 K, (c) (Hg,Re)-1212 at 58.1 K.

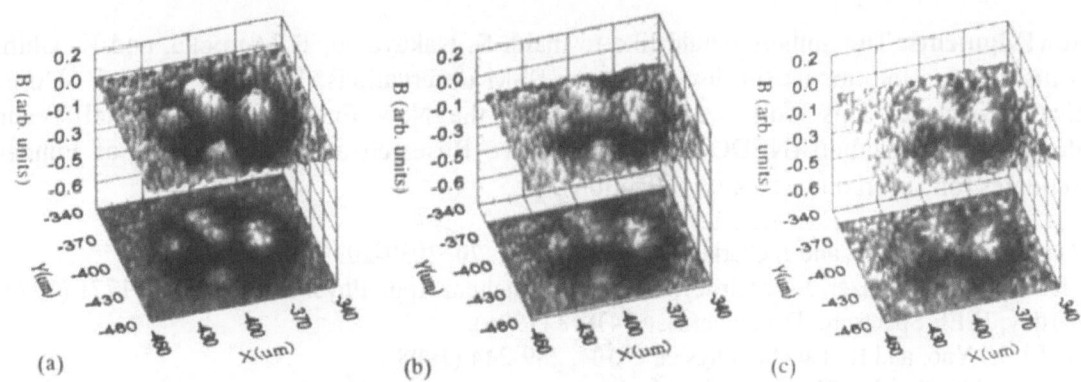

Fig. 2. Magnetic field images for an NBCO film pattern (100 µm in diameter) taken at (a) 51 K, (b) 89 K, and (c) 90 K.

Fig. 3 shows the magnetic images for NBCO patterns with a shape of SQUID washer and different line width which were cooled in an ambient field of approximately 2 µT. In these 2D images, the

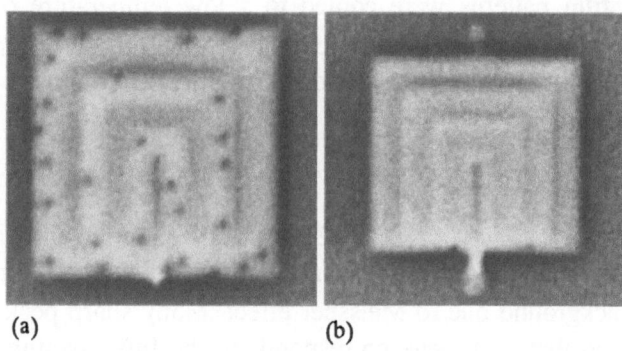
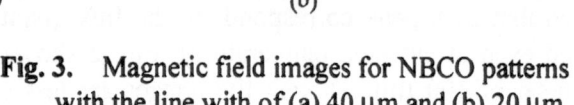

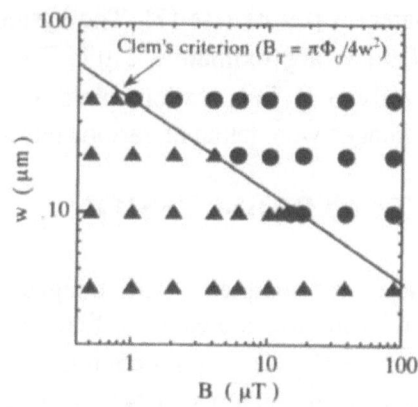

Fig. 3. Magnetic field images for NBCO patterns with the line with of (a) 40 μm and (b) 20 μm.

Fig. 4. Relation of the field with the line width giving rise to flux trapping (circles) and flux exclusion (triangles).

dark spots correspond to the flux quanta trapped in the patterns. Several flux quanta are found in the pattern with 40 μm line width, while no trapped flux is observed in the pattern with a smaller line width of 20 μm. Similar experiments were performed with changing the field during cooling for these patterns and other patterns with the line width of 10 and 4 μm. The obtained results are summarized in Fig. 4. By decreasing the field at a certain line width, flux trapping can be completely suppressed. The threshold field as a function of line width (w) is found very close to the prediction by Clem ($B_T = \pi\Phi_0/4w^2$) [1]. Recently, significant reduction in the flux noise of field-cooled SQUIDs have been demonstrated by employing washers with many slots or holes [1]. Our present results obtained by direct observation of trapped flux are consistent with their results.

CONCLUSION

The magnetic images of flux quanta trapped in high-T_c NBCO and (Hg,Re)-1212 thin films are clearly observed in a wide temperature range by using a scanning SQUID microscope. The images of trapped flux taken at temperatures near T_c support the rather strong pinning in NBCO. It is confirmed that flux trapping can be completely suppressed for the field below the threshold value as a function of line width predicted by Clem.

Acknowledgments. The authors would like to thank S. Nakayama, T. Morooka, and K. Chinone of Seiko Instruments Inc. for useful discussions on SSM observations. They also thank Y. Homani for technical assistance. This work was supported by the New Energy and Industrial Technology Development Organization (NEDO) as Collaborative Research and Development of Fundamental Technologies for Superconductivity Applications.

1. E. Dantsker, S. Tanaka, and J. Clarke, Appl. Phys. Lett. **70**, 2037-2039 (1997).
2. M. Jeffery, T. Van Duzer, J. R. Kirtley, and M. B. Ketchen, Appl. Phys. Lett. **67**, 1769-1771 (1995).
3. J. Kirtley, IEEE Spectrum, December issue 41-48 (1996).
4. Yijie Li, X. Yao, and K. Tanabe, Physica C **304**, 239-244 (1998).
5. Yijie Li and K. Tanabe, Physica C (in press).
6. Y. Moriwaki, T. Sugano, S. Adachi, K. Tanabe, and A. Tsukamoto, IEEE Appl. Supercond. **9**, 2390-2393 (1999).
7. T. Morooka, S. Nakayama, A. Odawara, M. Ikeda, S. Tanaka, and K. Chinone, IEEE Appl. Supercond. **9**, 3491-3494 (1999).
8. S. I. Yoo, N. Sakai, H. Takaichi, T. Higuchi, and M. Murakami, Appl. Phys. Lett. **65**, 633-635 (1994).

Fabrication of YBCO Thin Films by Electron Beam Deposition with Double Sources

T.Kawae, S.Kambe, and O.Ishii

Graduate School of Sience and Engineering, Yamagata-Univ. Jounan 4-3-16, Yonezawa, Yamagata 992-8510 JAPAN

We fabricated YBCO thin films by using the electron beam deposition apparatus with double hearths, in which Y_2O_3 and $Ba_2Cu_3O_X$ are contained. By changing deposition time for Y_2O_3, the composition $Y_2O_3:Ba_2Cu_3O_X$ was controlled. The prepared films were sintered at 950°C for 6 hours in air and annealed at 450°C for 5 hours in oxygen. The phase of the prepared films was analyzed by the XRD method. It was found that when the Y_2O_3 deposition time ranged from 4 to 12 minutes, the cation ratio, Y:Ba:Cu was controlled to be 1:2:3, and Tc^{on} was raised as high as 83K. When the Y_2O_3 deposition time exceeded more than 16 minutes, the amount of the Y211 phase was increased with increasing the Y_2O_3 deposition time.

INTRODUCTION

The first step of application to high Tc superconductors(HTS) is to fabricate thin films by various methods such as PVD[1], CVD[2] and PLD[3]etc. It is important to establish the reliable preparaion method of HTS thin films. Furthermore, rapid deposition rate, deposition for wide area and high productivity are also required. One of suitable methods is EB deposition. In this paper, fabrication and characterization of the YBCO thin film prepared by the EB deposition with double hearths are reported.

EXPERIMENTAL

We tried to fabricate YBCO thin films by the EB deposition with a single hearth. However, no yttrium remained in the film. So,we performed the EB deposition with two hearths.

Each target of Y_2O_3 and $Ba_2Cu_3O_X$ was pressed into a pellet and set on a hearth. MgO(100) was used as a substrate.

The pressure of the vacuum chamber

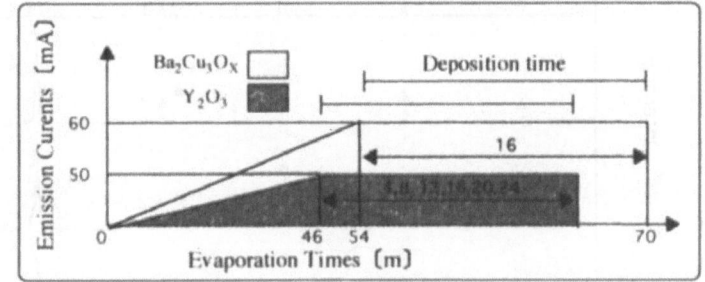

Fig. 1

was kept to be 10^{-6} Torr before deposition and under 10^{-4} Torr during the deposition. While the deposition time of the $Ba_2Cu_3O_X$ was fixed at 16 minutes, that of Y_2O_3 was changed from 4 to 24 minutes to control the composition of the film. Figure 1 shows deposition time for each target. The substrate was not heated during deposition. After the deposition, the prepared thin film was annealed in a furnace at 950°C for 6 hours. The films were cooled at 0.68°C per minutes down to 500°C and were annealed for 8 hours in flowing oxygen and then cooled down to room temperature in the furnace. The thickness was about 1-2 μm. Its structure and surface morphology were examined by X-ray diffractiometer(XRD) and scanning electron microscope(SEM), respectivery. The transition temperature was measured by standard four-probe electrical method.

RESULTS AND DISCUSSION

Figure 2(a) shows the XRD patterns of thin films prepared with various deposition time. The optimum deposition time ranged from 4 to 12 minutes for Y_2O_3. These patterns also indicate that the YBCO films are c-axis oriented.

Figure 2(b) shows the ratio between YBCO(006) and Y211(131) peak intensities calculated by

$$\frac{I_{\text{YBCO(006)}}}{I_{\text{YBCO(006)}} + I_{\text{Y211(131)}}} \times 100 \, \text{[\%]} \, .$$

Figure 2(c) shows the ratio of c-axis orientation, which was estimated by,

$$\frac{I_{\text{YBCO(006)}}}{I_{\text{YBCO(006)}} + I_{\text{YBCO(}hkl\text{)}}} \times 100 \, \text{[\%]} \, .$$

Here, the (hkl) of $I_{\text{YBCO(}hkl\text{)}}$ is either (101) or (130) peak as shown in the inlet of figure 2 (c), which are the strongest peak except YBCO (00l) peaks.

It also indicated that the deposition time between 4 and 12 minutes is the most appropriate for preparing the YBCO films.

Figure 3 shows the SEM image and morphological schemes of the prepared thin films. The composition was measured by EPMA. It was found that the Y211 phase is located like islands with the YBCO phase. We also found that the Y211 phase grows up with increasing deposition time for Y_2O_3. This is consistent with the results of Figure 2(b).

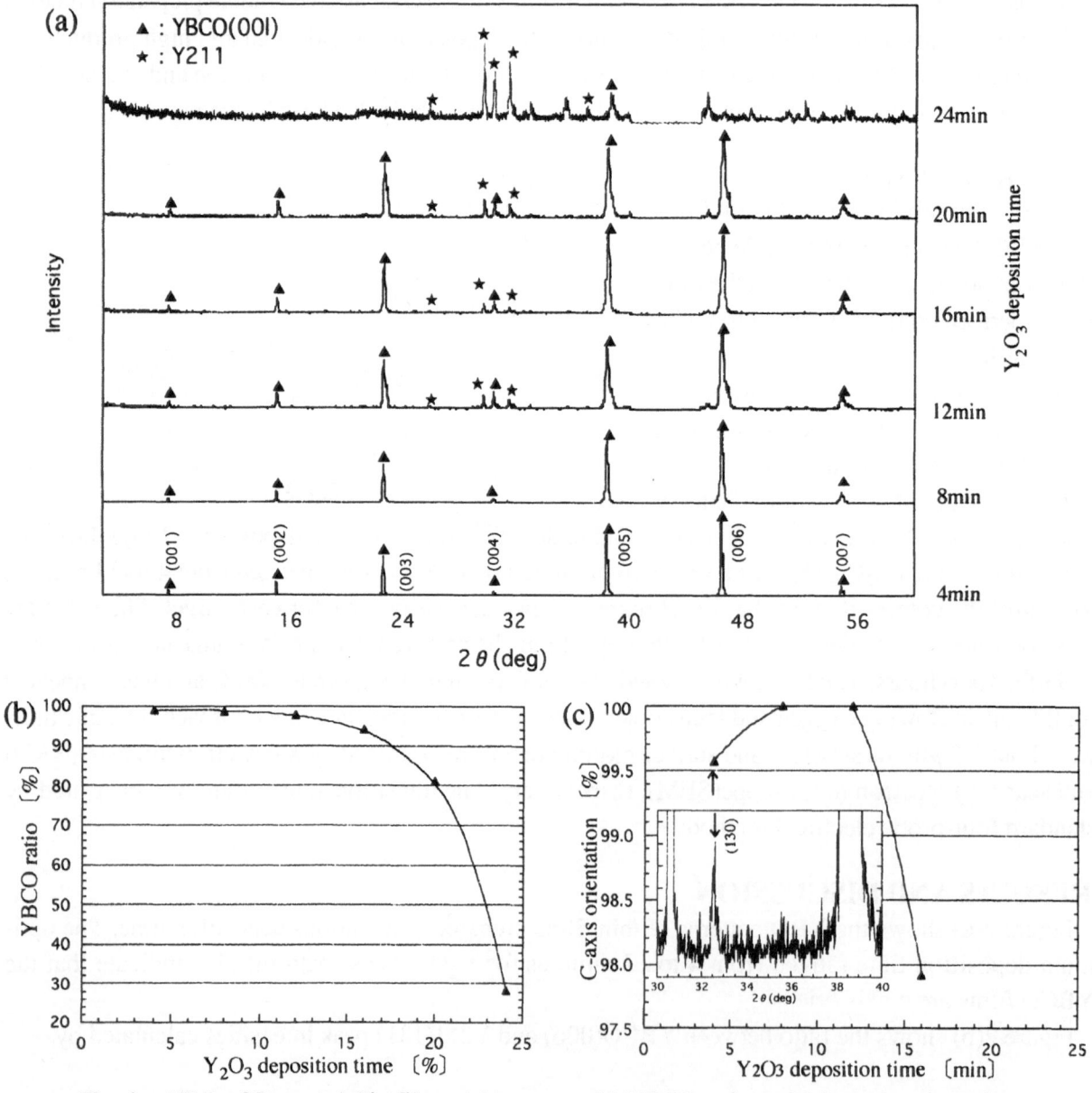

Fig. 2   XRD of Prepared thin films. (a) patterns, (b) YBCO ratio and (c) c-axis orientation.

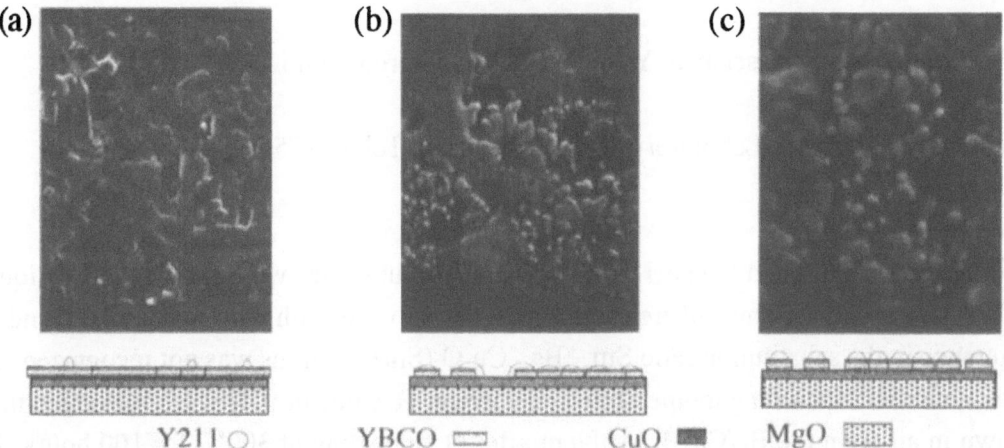

Figure 3 SEM images and schematic illustrations of the YBCO thin film prepared on the MgO substrate with changing deposition time of Y_2O_3, (a) 4 minutes, (b) 8 minutes, and (c) 12 minutes.

Figure 4 shows the R-T curves and transition temperature Tc^{on} of the prepared thin film. It was found that with increasing the deposition time, the Tc^{on} was decreased, the transport property changed into insulating, which is explained by the increase in the Y211 phase.

SUMMARY

By changing deposition time for Y_2O_3, the composition $Y_2O_3:Ba_2Cu_3O_X$ was controlled. The prepared films were sintered at 950°C for 6 hours in air and annealed at 450°C for 5 hours in oxygen. The phase of the prepared films was analyzed by the XRD method.

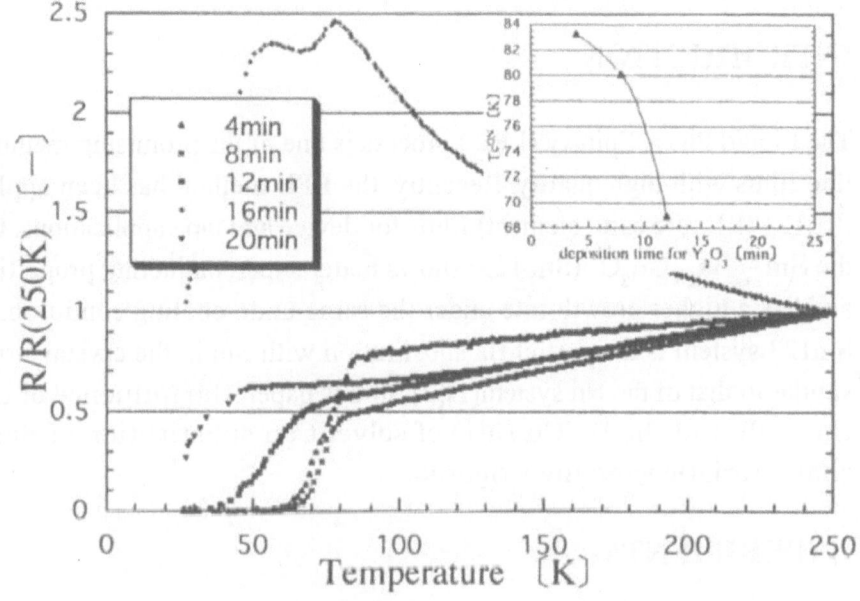

Fig. 4 R-T curves and dependence of Tc^{on} on deposition time of Y_2O_3 (inlet).

It was found that when the Y_2O_3 deposition time ranged from 4 to 12 minutes, the ratio of Y:Ba:Cu was fixed at 1:2:3, Y211 phase was minimized, the YBCO was c-axis oriented, and that Tc^{on} was raised as high as 83K when the deposition time of Y_2O_3 was 4 minutes. When the Y_2O_3 deposition time exceeded more than 16 minutes, the amount of the Y211 phase was drastically increased, and superconductivity disappeared.

REFERENCES

1) H.Kinder, R.Semerad, P.Berberich, B.Utz, and W.Prusseit, 5th International Superconductive Electronics Conference (ISEC'95) 12-3, September 18-21, (1995) Nagoya, Japan

2) H.Nagai and Y.Takai, Appl. Phys. Lett. 69 (6), 5 August (1996)

3) Masashi Mukaida and Shintaro Miyazawa, Jpn. J. Appl. Phys. Vol. 32 (1993) pp. 4521-4528

Preparation of $Sm_{1+x}Ba_{2-x}Cu_3O_y$ Film by Liquid Phase Epitaxy

Akemi Hayashi , Katsuya Hasegawa ,Yuichi Nakamura, Teruo Izumi and Yuh Shiohara

Superconductivity Research Laboratory, ISTEC, Koto-ku, Tokyo 135 ,Japan

ABSTRACT: $Sm_{1+x}Ba_{2-x}Cu_3O_y$(Sm123) films on MgO substrates were grown by the Liquid Phase Epitaxy(LPE) process in air atmosphere using BaO-CuO solvents with Ba/Cu ratio of 3/5 and 3/4. The large difference in the substitution ratio $Sm_{1+x}Ba_{2-x}Cu_3O_y$(Sm123) films was not recognized for different growth conditions including temperature and solvent. Tc value of 93K was obtained in the Sm123 films grown in air from the Ba/Cu=3/5 solvent after O_2 annealing at 300°C for 100 hours. The peak effect was observed for the Sm123 films annealed by continuous cooling heat treatment condition.

KEY WORDS: Liquid Phase Epitaxy, Sm123 film , liquid composition, Tc, peak effect

INTRODUCTION

The Liquid Phase Epitaxy (LPE) process is one of the promising methods to synthesize large crystalline films with high quality. Recently, the LPE method has been applied to fabricate $REBa_2Cu_3O_y$:RE123 (RE:rare earth element) films for device and tape applications. Comparing with the Y system, the $Sm_{1+x}Ba_{2-x}Cu_3O_y$ (Sm 123) shows better superconducting properties under high magnetic fields and has a higher growth rate under the same undercooling condition. An important feature of the Sm123 system is the partial Ba substitution with Sm in the crystal structure of the Sm123 which is similar to that of the Nd system[1][2]. In this paper, the influence of the surface temperature of the melt and the Ba/Cu ratio of solvent on substitution of Sm/Ba and superconducting characteristic were investigated.

EXPERIMENTAL

Sm123 seed films were deposited on (100) MgO ($10 \times 10 \times 0.5mm^3$) single crystal substrates by the rf plasma evaporation method. The LPE apparatus consisted of a modified vertical furnace and a crystal pulling apparatus [3][4]. $Sm_2Ba_1Cu_1O_8$ (Sm211) powder was placed at the bottom of the Sm_2O_3 crucible, and Ba-Cu-O powder with the composition ratio of Ba/Cu=3/5 or 3/4 was placed on the Sm211 layer. The powder in the crucible was heated up to 1080℃ to melt the BaO-CuO solvent completely. After complete melting was attained, the temperature was decreased to the predetermined temperatures for LPE growth. The temperature difference between the melt surface (Ts) and the crucible bottom was set to be about 10℃ during growth.

Grown Sm123 films were annealed by the following two heat treatment patterns in O_2 flow to obtain superconducting characteristics. One is the heat treatment under the constant temperatures of 300°C,400°C and 500°C for 100 hours and quenched in air. The other is the continuous cooling heat treatment from 600°C to 350°C for 100 hours. Superconducting properties were analyzed by SQUID

888

and the composition of the Sm123 films was determined by ICP.

RESULTS AND DISCUSSION

Table 1 shows the substitution value of the Sm123 LPE films grown from the different solvent compositions and the surface temperatures. As shown in this table, the Sm123 crystals had the composition close to Sm:Ba:Cu=1:2:3 for all conditions in this work. And the substitution ratio, x, was less than 0.01 for all samples, in the Nd system, the substitution ratio, x, in the Nd123 crystals

Table 1 Compositions of the Sm123 LPE films determined by ICP

	Ts	Substitution ratio (x)
Ba/Cu=3/5	1060	0.01
	1056	0.02
	1052	0.00
Ba/Cu=3/4	1060	0.00
	1055	0.01
	1050	0.00

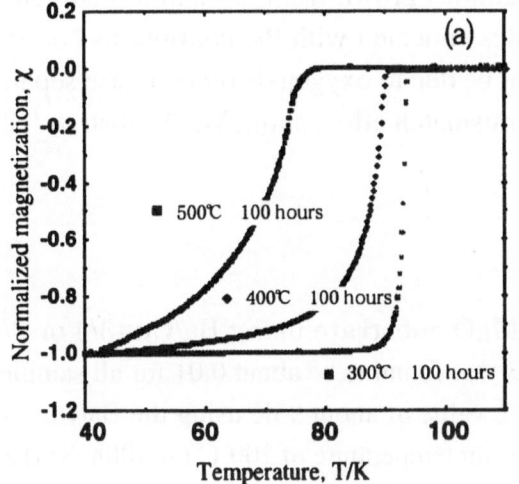
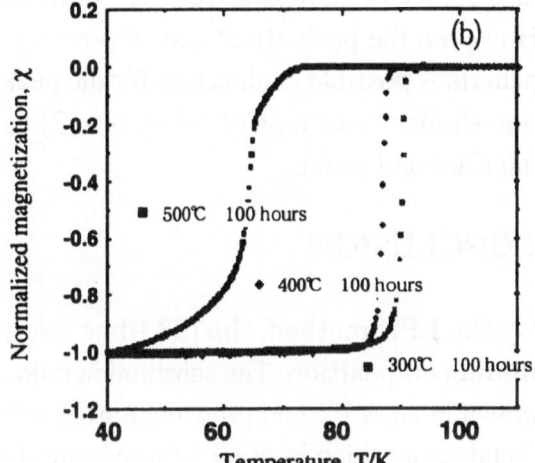

Figure 1 Temperature dependence of dc magnetization
for Sm123 LPE films using 3/5 (a) and 3/4 (b) solvents

grown in air was reported to be 0.045 for the case of Ba/Cu=3/5 solvent. [1] This suggests that the steady state growth even in air from the 3/5 liquid can be realized in the Sm-system. Figure 1 shows the annealing temperature dependence of normalized dc magnetization for Sm123 LPE films which were grown from the Ba/Cu=3/5 and Ba/Cu=3/4 solvents. As shown in figure 1, obtained Tc values were about 93K in Sm123 crystal grown from both liquid compositions by annealing at 300℃ for 100 hours. On the other hand the Tc on set value decreased with increasing annealing temperature. The decrease of Tc value would be attributed to the difference of equilibrium oxygen contents at the annealing temperature. G.Chiodelli ea al. [5] reported that the equilibrium oxygen contents at 500˚C and 300˚C are about 6.8 and 6.9 respectively. This lower oxygen contents at higher annealing temperature may lead the lower Tc value. Comparing the results of Fig.1 (a) and (b), the change of magnetic susceptibility was different between the samples grown from 3/4 and 3/5 solvent ; the transition width of the samples grown from 3/4 was narrower than that from 3/5 solvent . This suspected that the substitution ratio of the sample from 3/4 solvent is slightly smaller than that from 3/5 solvent , even though the ICP

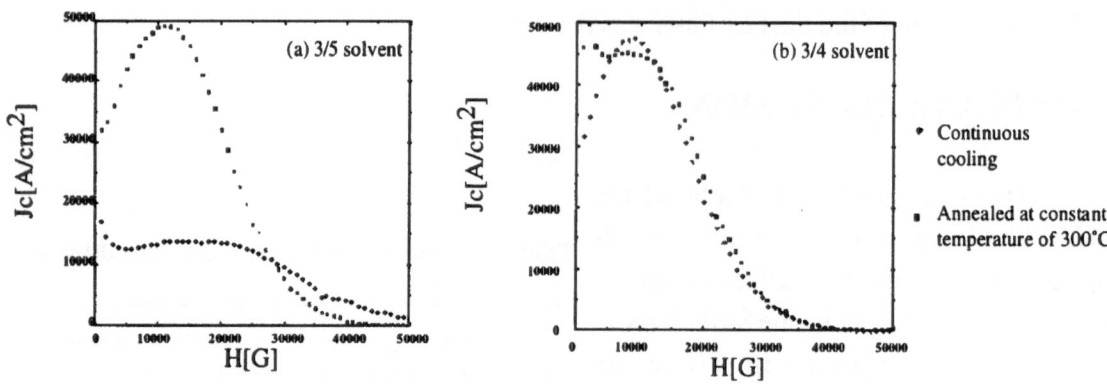

Figure 2 the Jc-H curve at 77K for the Sm123 LPE films
grown from (a) Ba/Cu=3/5 and (b)Ba/Cu=3/4 solvent

analysis showed similar value less than 0.02. Figure 2 shows the relation between the Jc-H properties and the heat pattern for oxygen annealing. When the crystal was annealed under the constant temperature of 300˚C, no peak effect was confirmed in both samples grown from 3/4 and 3/5 solvents. However, the peak effect was observed in both samples annealed with the continuous cooling pattern. A possible explanation for the peak effect might be due to oxygen disorder, phase separation similar to the case of Nd-system[2] and/or lattice mismatch effect at the MgO substrate/123 interface and so on.

CONCLUSION

By the LPE method, Sm123 films were grown on MgO substrate using Ba/Cu=3/4 or 3/5 liquid composition. The substitution ratio ,x , of Sm123 was found to be about 0.01 for all samples grown in air. We could grow the Sm123 films with the Tc value of about 93K using the Ba/Cu=3/5 liquid composition in air by followed annealing at the constant temperature of 300˚C for 100h. Sm123 LPE films for both liquid compositions followed by annealing with continuous cooling show the peak effect in the Jc-H properties.

ACKNOWLEDGMENT

This work was supported by the New Energy and Industrial Technology Development Organization (NEDO) as Collaborative Research and Development of Fundamental Technologies for Superconductivity Applications under the New Sunshine Program administered by the Agency of Industrial Science and Technology (AIST) of the Ministry of International Trade and Industry(MITI) of Japan.

1. M. Kambara X. Yao, M.Nakamura, Y.Shiohara, and T. Umeda J.Mater. Res. Vol.12 No.112866-2872
2. M. Kambara M.Tagami, E. Goodilin,X. Yao, M.Nakamura, Y.Shiohara, and T. Umeda J.Am. Ceram. Soc., 81[8] 2116-24(1998)
3. M.Nakamura, Ch.Krauns and Y.Shiohara ,Jpn. J. Appl. Phys 34(1995)6031-6035
4. Y.Kanamori and Y.Shiohara ,J.Mater. Res 11(1996)2693-2697
5. G. Chiodelli, I. Wenneker, P. Ghigna, G. Spinolo, G. Flor, M. Ferretti and E. Magnone Physica C 308 (1998)257-263

Liquid Phase Epitaxy for High Quality YBa$_2$Cu$_3$O$_{7-\delta}$ Films

Katsumi Nomura, Saburo Hoshi, Xin Yao, Kazuomi Kakimoto, Teruo Izumi and Yuh Shiohara
Superconductivity Research Laboratory, International Superconductivity Technology Center
1-10-13, Shinonome, Koto-ku, Tokyo 135-0062, Japan

Abstract: We have investigated the in-plane homogeneity of the large LPE crystals in microstructure and the growth mode in order to clarify the growth mechanism. The crystal rotation dependence on the growth rate distribution was observed in $20 \times 20\text{mm}^2$ LPE films. In the YBCO system the higher rotation rate caused a larger growth rate distribution. On the other hand, the higher rotation rate in the NdBCO system caused a smaller growth rate distribution. These phenomena could be explained by the difference in the crystal rotation effect both on the interface temperature and on the solute diffusion boundary layer thickness.

Keywords: Liquid phase epitaxy, Crystal rotation, Growth rate, Boundary layer, Interface temperature

INTRODUCTION

There have been many studies for the electronic devices using the oxide superconductors. It is necessary that the single crystal substrates have both high quality and large area for the applications to the devices. Among the film preparation methods, liquid phase epitaxy (LPE) is one of the most expected growth methods to prepare high quality and large single crystalline films. In our previous study, we have succeeded in growing the large $(20 \times 20\text{mm}^2)$ LPE films. For the large LPE films, in-plane homogeneity is one of the important qualities. In this paper, we have investigated the in-plane homogeneity of the large LPE crystals in the microstructure and the growth mode in order to clarify the growth mechanism.

EXPERIMENTAL

The MgO (100) single crystal substrates were used as the substrates for the LPE films. The polycrystalline seed films of YBa$_2$Cu$_3$O$_{7-\delta}$ (YBCO) were fabricated by the reactive thermal co-evaporation method or the pulse laser deposition method on the substrates. Both the Y$_2$BaCuO$_5$ powder and the 3BaO-5CuO powder were filled into the Y$_2$O$_3$ crucible and heated up to about 1020℃ to the melt. The supersaturated melt was prepared by decreasing the temperature lower than the peritectic temperature. The YBCO LPE film growth was initiated by making a contact of the seed film with the melt surface with a rotation rate of 20-80rpm for 30-60min. For comparison, the NdBa$_2$Cu$_3$O$_{7-\delta}$ (NdBCO) LPE films were prepared by the same way of the YBCO films as mentioned above. The microstructures of the LPE crystals were observed by optical microscopy and atomic force microscopy.

RESULTS AND DISCUSSION

Figure 1 shows the relationship between the growth rate of the YBCO crystals and the position of the LPE films. The growth rate distribution was observed in the YBCO LPE film. The growth rate of the crystal was increasing with increasing the distance from the center position of the LPE film. For the LPE films grown at a rotation rate of 20rpm, the growth rate of the film grown for 60min is almost the same of that of the film grown for 30min. It indicates that the LPE films were grown under the steady state growth mode and that the growth rate of the film grown at a rotation rate of 80rpm was larger than that of 20rpm at least for the initial 60min.

By optical microscopy with the Nomarski prism, very large hillocks were observed near the edge of the film grown under 80rpm. The bunched macrostep growth was observed at the large hillocks by atomic force microscopy. The single step growth was observed in other area. It seems that the difference of the undercooling existed at the different positions especially in the film grown under 80rpm.

It was considered that the growth rate distribution in the YBCO LPE films was caused by the difference in the solute diffusion boundary layer thickness for different locations which were enhanced by increasing rotation, due to the heat flux from the bottom of the crucible (hot region) and/or the difference of the boundary layer by the fluid flow of rotating the squared crystal. Investigation of the growth rate

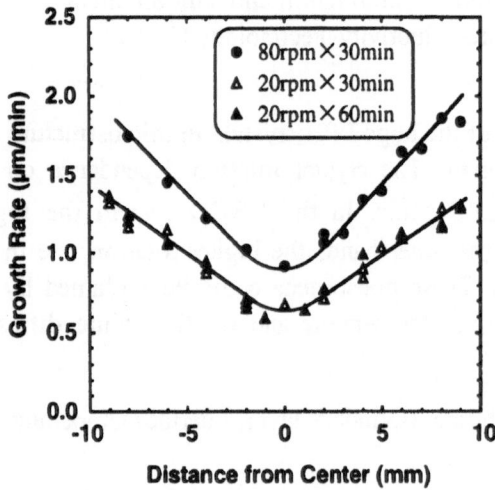

Fig. 1 Relationship between growth rate of YBCO crystals and position of LPE films with different growth conditions.

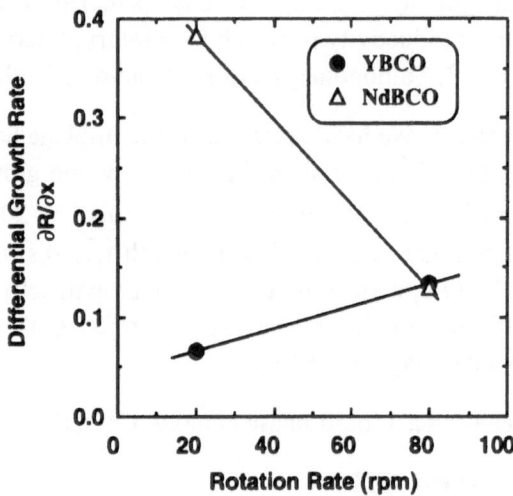

Fig. 2. Differential of growth rate as a function of rotarion rate.

Table 1. Parameters for Calculation using Equation (1)

Diffusion coefficient for YBCO	$D_Y = 2.6 \times 10^{-6}$ [cm^2/s]
Diffusion coefficient for NdBCO	$D_{Nd} = 9.6 \times 10^{-6}$ [cm^2/s]
Kinetic coefficient for YBCO	$k_Y = 1.3 \times 10^{-3}$ [cm/s]
Kinetic coefficient for NdBCO	$k_{Nd} = 1.7 \times 10^{-3}$ [cm/s]
RE concentration in REBCO crystal	$C_{123} = 16.7$ [at.%]
Growth rate of NdBCO by 20rpm at center of film	$R_{Nd}^{20-C} = 4.8$ [μm/min]
Growth rate of NdBCO by 20rpm near edge of film	$R_{Nd}^{20-E} = 8.7$ [μm/min]
Growth rate of NdBCO by 80rpm at center of film	$R_{Nd}^{80-C} = 11.3$ [μm/min]
Growth rate of NdBCO by 20rpm near edge of film	$R_{Nd}^{80-E} = 12.6$ [μm/min]

Table 2. Parameters for Calculation using Equation (2)

Reynolds number	$Re = \omega (D_{sub})^2 / 2\nu$
Prandtl number	$Pr = \nu/\alpha$
Grashof number	$Gr = g \beta \Delta T R_c^3 / \nu^2$
Rotation rate	$\omega = 2.09 - 8.37$ [rad/s]
Substrate diameter	$D_{sub} = 28$ [mm]
Crucible inner diameter	$D_{cru} = 50$ [mm]
Crucible inner radius	$R_c = (1/2) D_{cru} = 25$ [mm]
Kinematic viscosity	$\nu = 0.073$ [cm^2/s]
Thermal diffusivity of liquid	$\alpha = 6.90 \times 10^{-3}$ [cm^2/s]
Gravity	$g = 980$ [cm/s^2]
Volume expansion coefficient of liqid	$\beta = 2.60 \times 10^{-4}$ [K^{-1}]
Undercooling	$\Delta T = 20$ [K]
Temperature of crucible bottom	$T_{bot} = 1035$ [°C]
Temperature of substrate top	$T_{top} = 875$ [°C]
Interface temperature	$T_{interface}$ [°C]

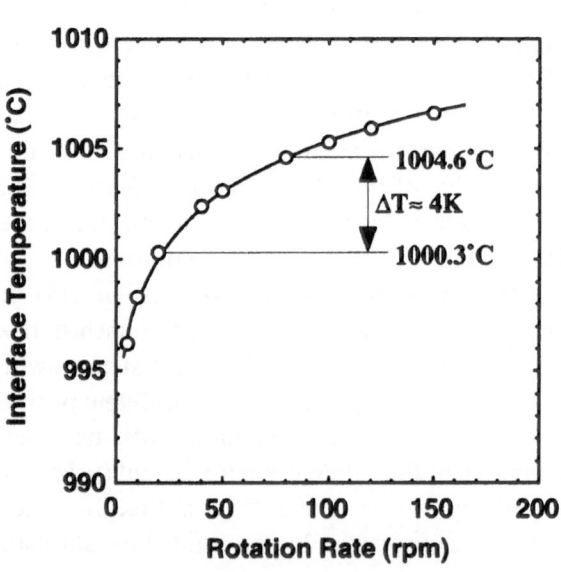

Fig. 3. Calculated results of interface temperature as a function of rotation rate.

distribution in the NdBCO system was effective in order to confirm that consideration, since the LPE growth in the NdBCO system was under the constant temperature without temperature gradient in the crucible, therefore the influence of the heat flux from the hot region was negligible.

Figure 2 shows the differential of the growth rate as a function of the rotation rate. In the YBCO system, the differential of the growth rate was increasing with increasing the rotation rate. On the other hand, in the NdBCO system, the differential of the growth rate was decreasing with increasing the rotation rate. This means that the higher rotation caused the larger growth rate distribution in the YBCO system, while in the NdBCO system the higher rotation caused the smaller growth rate distribution.

In the NdBCO system, it seems to be appropriate that only the solute diffusion boundary layer has to be considered because the influence of the heat flux from the hot region was negligible. In the steady state growth from the BPS theory for the REBCO (RE= Y, Sm, Nd), the relationship between the growth rate, the solute boundary layer, and the undercooling was derived as follows [1];

$$\frac{\delta R}{D}\frac{C_{123}-C_L(T_S)}{C_L(T_S)}+\left(\frac{R}{k}\right)^{\frac{1}{2}}=\frac{C_L(T_b)-C_L(T_S)}{C_L(T_S)} \quad \cdots\cdots (1)$$

where δ, R, D, k, C_{123}, $C_L(T_s)$ and $C_L(T_b)$ are solute boundary layer thickness, growth rate, diffusion coefficient, kinetic constant, RE concentration in the RE123 crystal, RE concentration in the liquid at the surface temperature, and RE concentration in the liquid at the temperature of the bottom of the crucible, respectively. The parameters for Eq. (1) are listed in Table 1. The thickness distributions of the solute boundary layer in the NdBCO system were calculated by Eq. (1) using the experimental results of the growth rate of the NdBCO system. On the other hand, the ratio of the solute boundary thickness for the YBCO and the NdBCO system can be obtained by the BPS theory using appropriate physical properties. Then thickness distributions of the solute boundary layer in the YBCO system were obtained by multiplying those in the NdBCO system by the ratio. Finally, the growth rate distributions in the YBCO system were estimated using the calculated solute boundary layer thickness by Eq. (1). The calculated results of the growth rate of 20rpm for the YBCO system were in good agreement with the experimental results. However, the results of 80rpm for the YBCO system were largely different from the experiment results.

To discuss the difference between the calculated results and the experimental results, the influence of the heat flux from the bottom of the crucible should be considered in the YBCO system. This phenomenon can be explained by the change of the temperature distribution in the liquid due to the substrate rotation. Namikawa et al. reported that the effect of the growth parameters for the temperature distribution and the relationship was given by the following equation [2];

$$T_{\text{interface}}=\left\{0.47Re^{0.028}Pr^{0.041}Gr^{0.019}\left(D_{\text{sub}}/D_{\text{cru}}\right)^{0.064}+0.11\right\}\left(T_{\text{bot}}-T_{\text{top}}\right)+T_{\text{top}} \quad \cdots\cdots (2)$$

where Re, Pr, and Gr mean Reynolds number, Prandtl number, and Grashof number, respectively. The parameters for Eq. (2) are listed in Table 2. The rotation rate dependence on the interface temperature was calculated using Eq. (2). The calculated results of the interface temperature were shown as a function of the rotation rate in Figure 3. The interface temperature increases with increasing the rotation. This increment value from 20 to 80rpm at the center of the film was estimated about 4K that were enough to reduce the bulk supersaturation, which resulted in suppressing the growth rate from 1.9 to 0.89µm/min at the center of the film. The calculated results for the growth rate using Eq. (1) were in good agreement with the experimental results. Consequently, this is thought to be the reason for the decreasing of the growth rate at the center of the film due to the increasing of the rotation rate.

Acknowledgment. This work was supported by the New Energy and Industrial Technology Development Organization (NEDO) as Collaborative Research and Development of Fundamental Technologies for Superconductivity Applications.

References. [1] J. A. Burton, R. C. Prim, W. P. Slichter, J. Chem. Phys. **21** (1953) 1987.
[2] Y. Namikawa, M. Egami, Y. Shiohara, J. Mater. Res. **11** (1996) 288.

TEM Observation of the Interface between (Hg,Re)-1212 Thin Film and STO Substrate

Xiao-Jing Wu, N. Inoue, T. Sugano, T. Morishita and K. Tanabe

SRL-ISTEC, 10-13 Shinonome, 1-Chome, Koto-Ku, Tokyo 135, Japan

Abstract: Two Re-doped Hg-1212 thin films prepared by the two-step process with different heating conditions were examined by TEM. Film A prepared under a temperature of 775 °C is well c-oriented. On the other hand, the quality of crystallization in film B prepared under 650 °C is rather poor. TEM observation on these films showed that some step-like regions on the surface of STO substrate appeared in film A, while such regions did not exist in film B. HRTEM observation indicated that the phase in the step-like region has a lattice parameter of about 4.10 Å. We suggest that at the interface between the film and the substrate, a chemical reaction occurred, and some of Sr and Ti were substituted by Ba and Cu, respectively.

Keywords: Re-doped Hg-1212 thin film, interface, TEM, chemical reaction

INTRODUCTION

In mercury-based superconductors, it was found that Re doping may improve the chemical stability of the phases. Recently, Moriwaki et al. succeeded in fabricating the Re-doped Hg-1212 thin films on (100) SrTiO$_3$ (STO) substrates with a Jc of about 5×10^6 A/cm^2 at 77 K in a zero field [1], being the highest J$_c$ value reported in this system. It is well-known that the physical properties of the films strongly depend on their microstructures. In this work, we report our investigation on the interface microstructures of (Hg,Re)-1212 thin films grown on the (100) STO substrates at the different annealing temperatures. The high resolution transmission electron microscope (HRTEM) observation reveals that a chemical reaction occurs in some step-like regions which appeared in the higher temperature annealed film.

EXPERIMENTAL

Amorphous precursor films with a composition of Re$_{0.1}$Ba$_2$CaCu$_2$O$_y$ were prepared by pulsed laser deposition (PLD) on the STO (100) plane using a KrF excimer laser (λ=248 nm) at room temperature. The thickness of precursor layer is about 100 nm. An HgO cap layer with a thickness of about 25 nm was deposited on top of the precursor films to protect it from moisture and CO$_2$. The precursor films were sealed in a quartz tube together with unreacted two pellets prepared by mixing monoxides with compositions of Hg:Ba:Ca:Cu=1:2:1:2 and 0:2:1:2. Two films A and B were made by annealing the precursor films under different temperatures. Film A was annealed at 725 °C and 775 °C for 5 hr and 0.25 hr, respectively, while film B was only annealed at 650 °C for 5 hr. A TEM, JEOL-4000 EX, operating at an accelerating voltage of 400 kV was employed to observe internal microstructure of the films.

RESULTS AND DISCUSSION

A low magnification micrograph taken from film A by TEM is shown in Fig. (a). The film is well c-oriented as the whole, but at the interface many defect regions can be found. Fig. (b) shows a micrograph taken from film B. Though this image has exactly the same scale as Fig. (a), we can not clearly see the

894

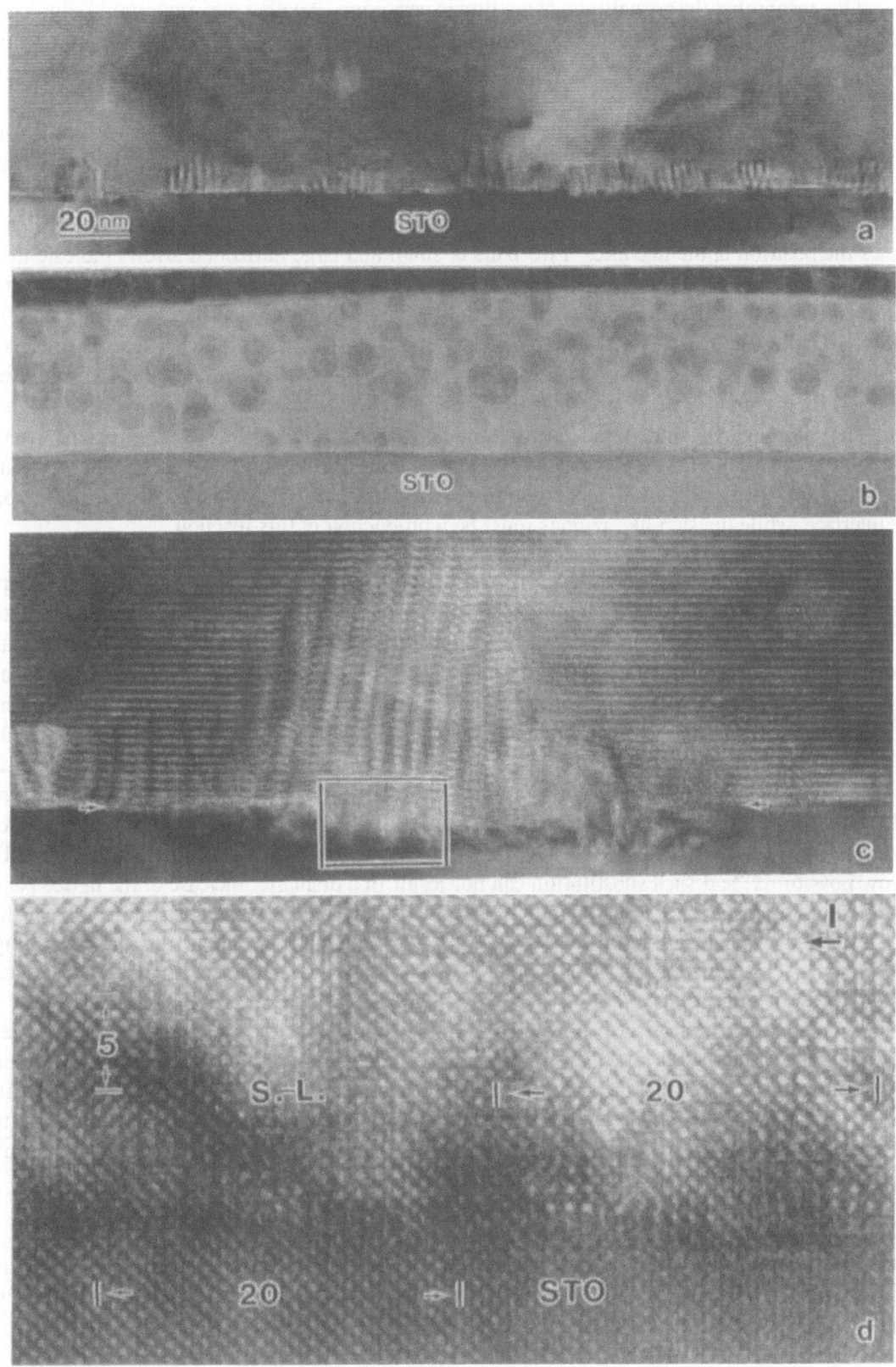

(a) Low magnification micrograph of Film A. (b) Low magnification micrograph of Film B. (c) High magnification image from a defect region in (a). (d) Enlarged image from the rectangle in (c).

lattice. This means that the crystallinity of the film is very poor. Another very important point here is that in film B there are no defect regions at the interface. The boundary between the film and STO is quite flat. By enlarging a defect region in Fig. (a), we can obtain an image as shown in Fig. (c). A pair of arrows in this image indicates the interface between the (Hg,Re)-1212 film and the STO substrate. On the top layer of STO, we can clearly see a step-like region, in which the crystal seems to have a different lattice parameter from either STO or (Hg,Re)-1212. This step-like region has a length of about 75 nm and a width (thickness) of about 6 nm. Actually, all the defect regions in Fig. (a) had a similar structure. A region indicated by a rectangle in Fig. (c) is further enlarged, as shown in Fig. (d). A large arrow, labeled as "I", indicates the position of the interface. Two pairs of small and large horizontal arrows indicate the length of 20 unit-cells in STO and the step-like region (labeled by S.-L.), respectively. On the left part of this image, a pair of vertical arrows indicates the length of 5 unit-cells in the step-like region along the vertical direction. Taking the lattice parameter of STO, 3.905 Å, as a standard, the lattice parameters in the step-like region can be determined as 4.15 and 4.13 Å along the horizontal and vertical directions, respectively, being significantly longer than that of STO. Similar measurements were carried out in several different step-like regions, and it was found that the lattice parameters in the step-like regions are variable within a range of 4.05 ~ 4.15 Å. Obviously, the phase in the step-like region must be different from either STO or (Hg,Re)-1212. It is likely that a chemical reaction between STO and the precursor film occurred at the interface, while the step-like regions must be a production of this reaction.

It can be found from Fig. (d) that the crystal structure in the step-like regions has a good epitaxial relationship with STO substrate. This implies that this phase may have a perovskite-related structure. The most possible way to form this phase is that the cations in STO were partially replaced by other cations in the precursor film. Considering the composition of precursor film, which includes Hg, Re, Ba, Ca and Cu, it is easy to recognize that Sr may be partially substituted by Ba. Ba^{2+} has a larger ionic radius than that of Sr^{2+}. Therefore, Ba substitution for Sr should cause a lattice expansion. Besides, it is considerable that the Ti site is also replaced by other elements. From structural chemistry point of view, the Ti site can not be occupied by Ca or Hg, because their ionic radii are too large to construct a perovskite-related structure. The Re content is very low in the precursor film, meanwhile the ionic radius of $Re^{4+}(VI)$ is 0.63 Å, being almost as same as that of $Ti^{4+}(VI)$, 0.61 Å [2]. These facts imply that the substitution of Re for Ti has a very low possibility, and such substitution can not result in a dramatic increase of the lattice parameter. Therefore, the only choice is Cu. The ionic radius of $Cu^{2+}(VI)$ is 0.73 Å [2], being much larger than that of Ti. According to the report from Gong et al. [3], for the tetragonal $BaCuO_{2-x}$ film, the c-axis length can reach to about 4.13 Å, while this value is very close to 4.15 Å for the present phase. The valence of Cu is always lower than that of Ti, i.e., +4, in $SrTiO_3$ structure. Therefore, Cu substituting for the Ti site must accompany with a decrease of the oxygen content in the unit cell, and then the chemical formula of this phase should be written as $(Ba_x,Sr_{1-x})(Cu_y,Ti_{1-y})O_{3-\delta}$. Accompanying with Ba and Cu substitution in STO, Sr and Ti must be released from the substrate, and diffuse into the (Hg,Re)-1212 film. Actually, in Figs. (a) or (c), we can clearly see some structural variations and long periodic fringes in the film near to such step-like regions. The long periodic fringes are so-called Moiré pattern, while the structural variations should originate from the inhomogenous diffusion of Sr and Ti in the film.

Acknowledgments. This work was supported by NEDO as Collaborative Research and Development of Fundamental Technologies for Superconductivity Applications under the New Sunshine Program administered by AIST of the Ministry of International Trade and Industry (MITI) of Japan.

1. Y. Moriwaki, et al., Advances in Superconductivity XI (Springer-Verlag Tokyo, 1999) pp1063.
2. R.D. Shannon, Acta Cryst. A 32, (1976) 751.
3. J.P. Gong, et al., Advances in Superconductivity IV (Springer-Verlag Tokyo, 1992) pp863.

FABRICATION OF (Hg,Re)-1212 THIN FILMS WITH FLAT SURFACE ON VARIOUS SUBSTRATES

Nobuyoshi Inoue[1], Tsuyoshi Sugano[1], Akira Tsukamoto[2], Xiao-Jing Wu[1], Tadashi Utagawa[1], Seiji Adachi[1] and Keiichi Tanabe[1]

[1] SRL-ISTEC, 1-10-13 Shinonome, Koto-ku, Tokyo 135-0062, Japan
[2] Advanced Research Laboratory, Hitachi, Ltd., Kokubunji, Tokyo 185-8601, Japan

Abstract: We improved the surface morphology of (Hg,Re)-1212 thin films on STO (100), LSAT(100) and LAO(100) substrates, and achieved a high value of the critical current density J_c on STO substrates. A 100-nm-thick $Re_{0.1}Ba_2CaCu_2O_z$ precursor film with an HgO protective cap layer was deposited on each substrate by pulsed laser deposition and subsequently annealed under an appropriate Hg vapor pressure in an evacuated quartz tube. The thickness of the obtained films was about 75 nm. Large pinholes and outgrowths were not conspicuously observed on the surface, indicating that homogeneous epitaxial growth was realized by employing rather thin $Re_{0.1}Ba_2CaCu_2O_z$ precursor films. As-fabricated films exhibited T_c (zero) of 109 - 117 K. The J_c values at 77 K in a self-field were 1.0×10^7, 4.5×10^6 and 2.7×10^6 A/cm^2 for STO, LSAT and LAO substrates, respectively.

Keywords: (Hg,Re)-1212 thin film, STO, LSAT, LAO, surface morphology, defect, J_c

INTRODUCTION

The preparation of high-T_c Hg-based superconducting thin films with the aim of fabricating electronic devices, such as superconducting quantum interference devices (SQUIDs) has been reported recently [1-8]. The successful fabrication of the films exhibiting J_c above 1×10^6 A/cm^2 at 77 K in a self-field was reported [2,3,6]. We have succeeded in preparing (Hg,Re)-1212 thin films on STO (SrTiO$_3$) substrates, having comparable J_c[7,8]. However, the previously reported films always contained submicron-size defects such as pinholes, outgrowths and large steps. To fabricate reliable devices, thin films with a smooth and flat surface are required. We have recently found that thickness of a precursor film is a key factor determining surface morphology of the (Hg,Re)-1212 thin films. The films fabricated from rather thin precursor films exhibited a preferable surface morphology. Here we report the successful fabrication of the smooth and flat (Hg,Re)-1212 thin films. Results of film fabrication on LSAT ((LaAlO$_3$)$_{0.3}$-(SrAl$_{0.5}$Ta$_{0.5}$O$_3$)$_{0.7}$) and LAO (LaAlO$_3$) are also described.

EXPERIMENTAL

(Hg,Re)-1212 thin films were fabricated on STO, LSAT and LAO substrates, employing the two-step process. Doping of 10-% Re to the Hg-site was carried out to improve phase stability[9]. As a precursor, approximately 100-nm-thick $Re_{0.1}Ba_2CaCu_2O_z$ film was prepared by a pulsed laser deposition (PLD) technique at room temperature in vacuum. A 25-nm-thick HgO cap layer was also deposited on top of the $Re_{0.1}Ba_2CaCu_2O_z$ film to protect it from moisture and CO$_2$. The precursor film was sealed in a quartz tube together with unreacted two pellets prepared by mixing monoxides with compositions of Hg:Ba:Ca:Cu = 1:2:1:2 and 0:2:1:2 and annealed at 650 - 800 °C for about 5 h. The structure and orientation of the obtained films were examined by x-ray diffraction (XRD). The surface and cross-sectional views were taken using a scanning electron microscope (SEM). The films were patterned into bridges with a size of 5-50 μm in width and 50-150 μm in length for electrical transport measurements.

RESULT AND DISCUSSION

Figure 1 shows the XRD θ-2θ patterns for the obtained films. Major peaks are assigned to the reflections from each substrate and the (00*l*) reflections of the 1212 phase, indicating that (Hg,Re)-1212 phase with the *c*-axis orientation perpendicular to the substrate is dominant. Only the film on LAO substrate has small peaks from *a*-axis orientation. Although peaks from impurity phases and the 1223 phase are also observed in the pattern, their intensity is quite weak. The XRD φ-scan confirmed that the greater part of (Hg,Re)-1212 grains were in-plane aligned with their *a*-axis parallel to substrate [100] direction and thus they are epitaxially grown on each substrate.

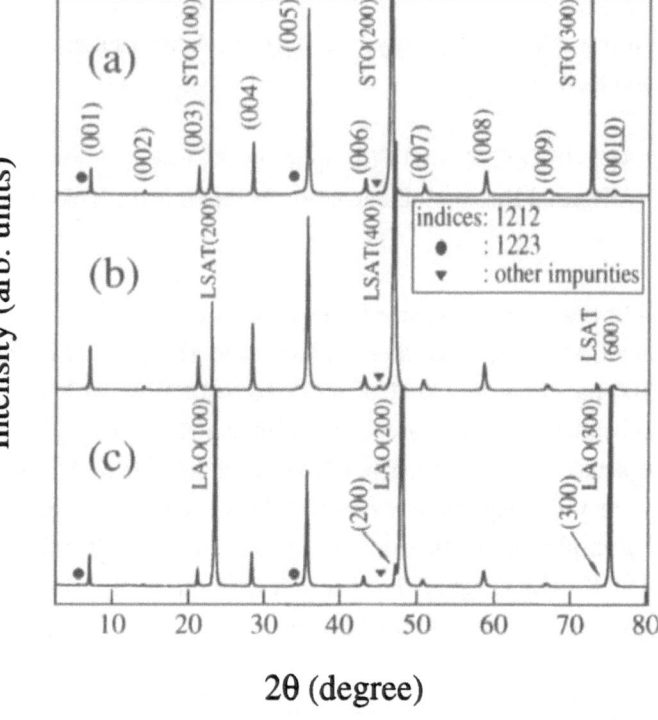

Fig. 1 X-ray diffraction θ-2θ pattern for an (Hg,Re)-1212 film on (a)STO, (b)LSAT and (c)LAO substrates.

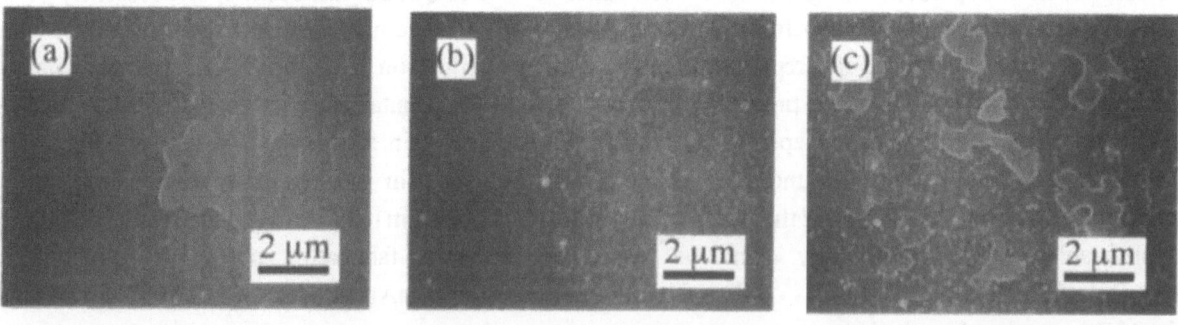

Fig.2 SEM pictures for the surface of (Hg,Re)-1212 thin films on (a)STO, (b)LSAT and (c)LAO substrates.

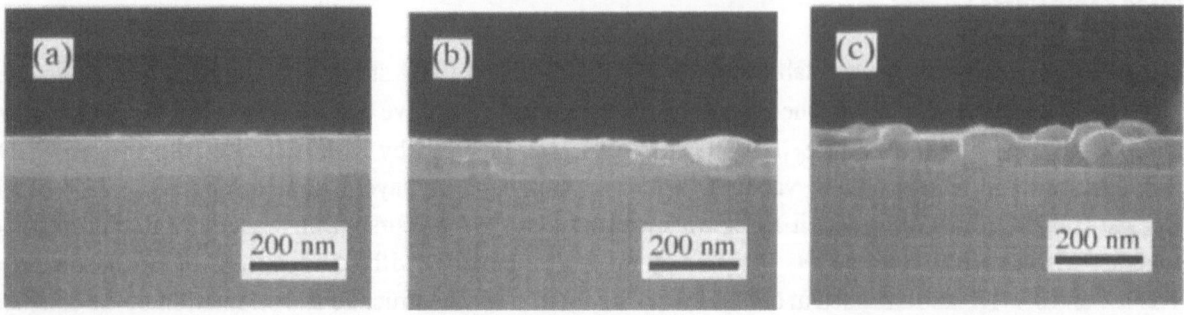

Fig.3 SEM pictures for fractured cross-section of (Hg,Re)-1212 thin films on (a)STO, (b)LSAT and (c)LAO substrates.

Figures 2 and 3 show the SEM pictures for the surfaces and the cross-sections of the films. Neither gap nor void is observed at the interface between the films and the substrates. Submicron-size defects are observed in the films on LSAT and LAO, while the film on STO has a quite smooth surface without such defects. Thin film fabrication using precursor films with different thicknesses was attempted on STO substrates. Quite homogeneous films as shown in Figs. 2 and 3 were obtained when the thickness of the precursor film was approximately 100 nm. A 150-nm-thick precursor film resulted in crystallization of randomly oriented large grains in the film. The surface morphology of the film on STO was significantly improved, comparing that of previously fabricated films[5,7,8].

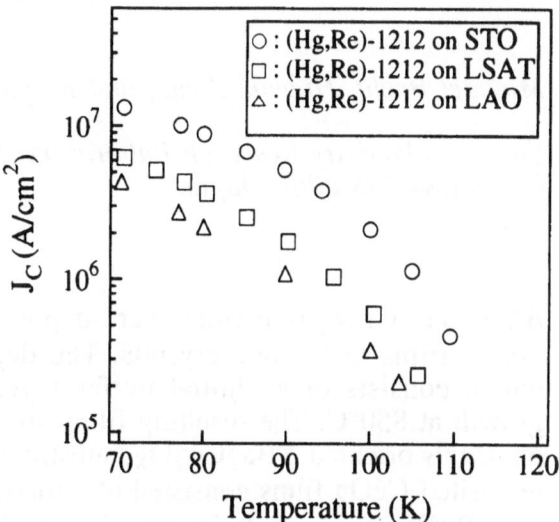

Fig. 4 Temperature dependence of transport J_c in a self-field for (Hg,Re)-1212 films on each substrate.

As-fabricated films on each substrate exhibited a typical T_c of 109 - 117 K. This is substantially lower than that of the optimally doped Hg-1212 bulk, 127 K. We attempted to optimize the carrier concentration by post-annealing, but an appreciable increase in T_c was not achieved. It is likely that the stress caused by the lattice mismatch and existence of microscopic defects suppress their T_c's. Figure 4 shows the temperature dependence of transport J_c determined with the 1 μV/mm criterion in a self-field for one of our best films on each substrate. The J_c of the film on STO substrate is as high as 2.1×10^6 A/cm^2 at 100 K and reaches 1.0×10^7 A/cm^2 at 77 K, in spite of the lowered T_c (116K). To the best of our knowledge, this is the highest J_c value at 77 K ever reported for Hg-based superconducting thin films. The J_c of the films on LSAT and LAO substrate(4.5×10^6 and 2.7×10^6 A/cm^2) are lower than that on STO substrate. Submicron-size defects on LSAT and LAO substrate are considered to hinder the superconducting current path.

Acknowledgments. This work was supported by the New Energy and Industrial Technology Development Organization (NEDO) as Collaborative Research and Development of Fundamental Technologies for Superconductivity Applications.

1. C. C. Tsuei, A. Gupta, G. Trafas, and D. Mitzi, Science **263**, 1259 (1994).
2. A. Gupta, J. Z. Sun and C. C. Tsuei, Science **265**, 1075 (1994).
3. S. H. Yun and J. Z. Wu, Appl. Phys. Lett. **68**, 862 (1996).
4. N. Khare, A. K. Gupta, S. Khare, H. K. Singh, A. K. Saxena, and O. N. srivastava, Physica C **274**, 161 (1997).
5. Y. Moriwaki, T. Sugano, A. Tsukamoto, C. Gasser, K. Nakanishi, S. Adachi, and K. Tanabe, Physica C **303**, 65 (1998).
6. S. L. Yan, Y. Y. Xie, J. Z. Wu, T. Aytug, A. A. Gapud, B. W. Kang, L. Fang, M. He, S. C. Tidrow, K. W. Kirchner, J. R. Liu, and W. K. Chu, Appl. Phys. Lett. **73**, 2989 (1998).
7. A. Tsukamoto, K. Takagi, Y. Moriwaki, T. Sugano, S. Adachi, and K. Tanabe, Appl. Phys. Lett. **73**, 990 (1998).
8. Y. Moriwaki, T. Sugano, A. Tsukamoto, S. Adachi, and K. Tanabe, in *Advances in Superconductivity XI*, edited by N. Koshizuka and S. Tajima (Springer, Tokyo, 1999) p. 1063.
9. J. Shimoyama, K. Kishio, S. Hahakura, K. Kitazawa, K. Yamaura, Z. Hiroi, and M. Takano, in *Advances in Superconductivity VII*, edited by K. Yamafuji and T. Morishita (Springer, Tokyo, 1995) p. 287.

Quality of CeO_2 Thin Films Grown on $YBa_2Cu_3O_x$ Thick Films and $YBa_2Cu_3O_x$ Single Crystals

Michael Becht, Hideaki Zama, and Keiichi Tanabe

Superconductivity Research Laboratory, Division 6, ISTEC, 1-10-13 Shinonome, Koto-ku, Tokyo 135-0062, Japan

Abstract: CeO_2 thin films were deposited by MOCVD on polished $YBa_2Cu_3O_x$ LPE grown films and single crystals. The deposition was carried out using a 2-step process which consists of an initial buffer layer deposited at 600°C followed by further film growth at 850°C. The resulting films are (100) oriented and show flat surfaces. However, on *a-axis* oriented $YBa_2Cu_3O_x$, substrates with poor surface crystallinity or polishing, the deposited CeO_2 films consisted of a mixture of CeO_2 and $BaCeO_3$. These results indicate that $BaCeO_3$ is easily formed during deposition of CeO_2 if the $YBa_2Cu_3O_x$ surface exhibits a higher density of 'free Ba' due to scratches, *a-axis* orientation, poor crystallinity, and poor polishing.

Keywords: CeO_2, $BaCeO_3$, surface morphology, MOCVD, 2-step process

INTRODUCTION

The growth of CeO_2 on $YBa_2Cu_3O_x$ (YBCO) has been studied because of the potential use of CeO_2 as an insulating layer on superconducting ground planes in junction devices[1].

The films were deposited both on mechanically polished *c-axis* oriented LPE YBCO thick films (thickness 2-5 microns) grown on MgO substrates and YBCO single crystals with *a*- or *c-axis* orientation. Metal-organic chemical vapor deposition (MOCVD) was used to deposit the films with $Ce(thd)_4$ as precursor material[2,3]. A 2-step MOCVD process has been developed earlier for the deposition of CeO_2 films (thickness = 200 nm) with a very flat surface and good electrical properties[4]. The 2-step process consists of an initial buffer layer deposition at 600°C (time = 10 min) followed by further deposition at 850°C (total deposition time of 120 min).

During our study to further improve the CeO_2 thin film morphology, we identified $BaCeO_3$ being present in some samples. The factors responsible for the formation of $BaCeO_3$ were analysed and the results are discussed in this report.

RESULTS AND DISCUSSION

Usually, we obtain shiny CeO_2 layers (film thickness of 100-300 nm), independent on the substrate material. The films on *c-axis* YBCO single crystals and LPE grown thick films are purely (100) oriented and the film surfaces are smooth with only a few outgrowths (see Fig. 1). However, in some samples, we find that close to the edges the layers are dull and rough (see Fig. 1). Sometimes, there were dull areas even in the center of the sample. In all these samples, strong reflections related to $BaCeO_3$ were observed in the X-ray diffraction pattern (see Fig. 2). By removing the 'degraded' parts from the

 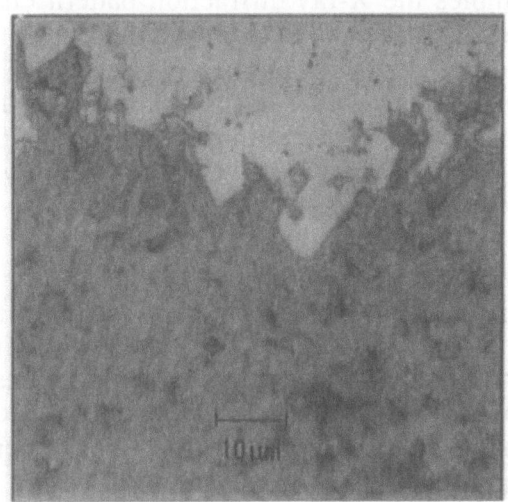

Figure 1: Surface images: average quality (left) and degraded surface (right).

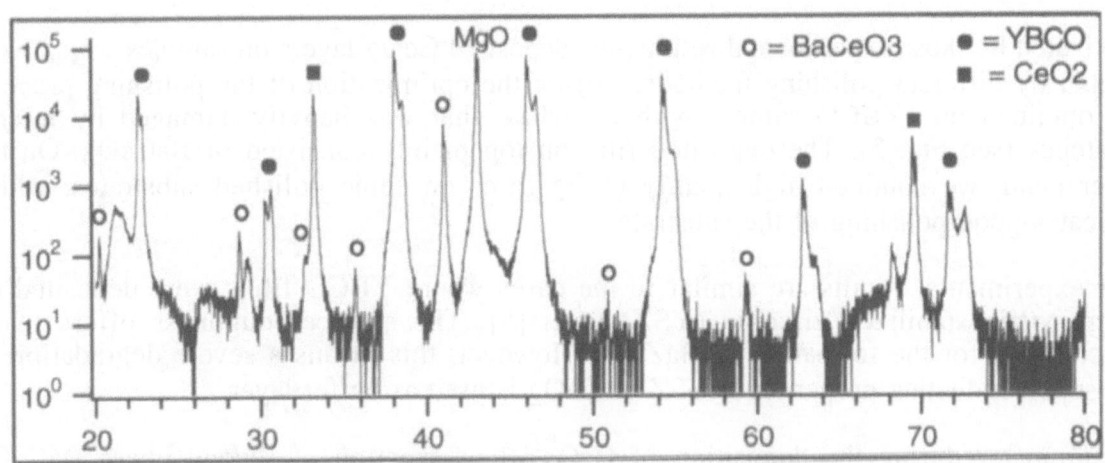

Figure 2: X-ray diffraction pattern of a sample with a dull and rough surface.

Figure 3: AFM image of a YBCO thick film with poor polishing (scratches).

3 µm

samples the X-ray diffraction pattern changed, therefore we concluded that the dull and rough parts mainly consist of $BaCeO_3$.

Furthermore, we analysed the factors which lead to the formation of $BaCeO_3$. We found that around the mechanical scratches $BaCeO_3$ is formed. One sample showed a wide area of rough surface. With X-ray topography analysis, it was found that the YBCO below the CeO_2 layer had a poor crystallinity.

Since $BaCeO_3$ must be originated from a chemical solid state reaction of CeO_2 and 'Ba', it is interesting to note, that there is no reaction if the YBCO surface is flat, smooth up to the atomic scale, and with good crystallinity. Our experimental results indicate that 'imperfections' like scratches or domains with poor crystallinity favor the formation of $BaCeO_3$. Obviously, 'free Ba' exists around these areas for the chemical reaction with CeO_2. This is clearly shown in one experiment, in which we used *a-axis* oriented YBCO single crystal as the substrate. The film surface is very degraded and a high amount of $BaCeO_3$ was detected. The chemical reaction of CeO_2 and 'Ba' is understandable, because *a-axis* oriented substrates naturally have plenty of 'Ba' at the surface. Assuming that the surface layer is not the Ba-layer in the *c-axis* polished YBCO single crystals or thick films, the density of 'Ba' is very low thus the formation of $BaCeO_3$ is highly suppressed.

Motivated by these experimental results, we deposited CeO_2 layers on samples which were treated by different polishing methods. During the optimization of the polishing process, we obtained one YBCO sample with a surface that was heavily damaged by micro-scratches (see Fig. 3). The deposited film on top mainly consisted of $BaCeO_3$. On the other hand, we obtained high quality CeO_2 films on some polished substrates, which indicates good polishing of the substrate.

Our experimental results are similar to the cases where YBCO films were deposited on rough yttria stabilized zirconia (YSZ) layers[5]. The surface roughness offers many possibilities for the formation of $BaZrO_3$. However, this means a severe degradation of the superconducting properties if YSZ or CeO_2 is used as buffer layer.

In summary, during the deposition of CeO_2, the formation of second phase $BaCeO_3$ depends on the surface quality and crystallinity of the polished YBCO thick films and single crystals. Furthermore, by using *a-axis* YBCO substrates a mixture of CeO_2 and $BaCeO_3$ was formed. These results indicate that the high density of 'Ba' on the surface of YBCO samples with many scratches and a-axis orientation strongly promotes the formation of $BaCeO_3$.

Acknowledgments. This research was supported by the New Energy and Industrial Technology Development Organization (NEDO) of Japan as Collaborative Research and development of Fundamental Technologies for Superconducting Applications.

[1] G. A. Alvarez, M. Becht, T. Utagawa, K. Toma, U. Kawabe, F. Wang, Y. Li, F. Saba, M. Sato, K. Tanabe, *IEEE Trans. Appl. Supercond.* **1998**, *9*, 3370.
[2] M. Becht, *Applied Superconductivity* **1996**, *4*, 465.
[3] M. Becht, T. Morishita, *Chem. Vap. Deposition* **1996**, *2*, 171.
[4] M. Becht, J.-G. Wen, F. M. Saba, S. Miura, K. Tanabe, in *11th International Symposium on Superconductivity (ISS `98)* (Eds: N. Koshizuka and S. Tajima), Springer Verlag Tokyo, **1998**, p.1075.
[5] J. Y. Dai, F. H. Kaatz, P. R. Markworth, D. B. Buchholz, X. Liu, W. A. Chiou, R. P. H. Chang, *J. Mater. Res.* **1998**, *13*, 1485.

Dielectric Properties for Insulator/ $NdBa_2Cu_3O_{7-\delta}$ Multi-Layer Structures

Youichi Enomoto, Michitomo Iiyama, Osami Horibe and Yasuo Oshikubo

Superconductivity Research Laboratory, ISTEC, Chiba, 270-1382, Japan

Abstract: Transmission analysis of microstrip lines shows possibility of non-metallic cuprates as insulator for Single Flux Quantum (SFQ) circuits operated at high speed above 100 GHz. Value of tan δ below 0.1 is comparable to loss of superconducting surface resistance. Value of ε_r about 15 is suitable for insulators in integrated circuits from strip line width of 10 μm order, short delay time, no excitation of surfacewave and low radiation loss. These results indicate the applicability of $Pr_{1.14}Ba_{1.86}Cu_3O_{7-\delta}$ (PBCO) for the insulator layers in multilayer electronic devices in addition to its good lattice match to superconducting $NdBa_2Cu_3O_{7-\delta}$ (NBCO).

Keywords: SFQ, multi-layer, PBCO, NBCO, microstrip line

INTRODUCTION

In integrated circuits, a role of lines increases for their speed limit. Superconducting lines are interesting due to ultra low surface resistance. As for superconducting integrated circuits, SFQ attracts researcher's interest because of its low power consumption and high speed [1]. Microstrip line structure with a ground plane is desirable to realize small and controllable SQUID inductance L_s as well as high speed signal transmission [2]. For high Tc superconductors (HTS), their longer penetration depth leads to bigger inductances from kinetic inductance part and requires the thicker line and ground plane. Consequently, insulator layers must be accomplished without producing crystallographical defects, which could introduce unintended weak-link junctions between grains and rough surfaces in the thick superconducting lines. The material used for the isolation layers must be compatible chemically with the HTS layers and must provide good matching with regards to lattice parameters. From these points, cuprates without superconductivity are candidates as the insulator. For example, crystal structure and lattice parameters of non-metallic PBCO are also known to match closely ones of superconducting NBCO [3]. PBCO worked as good tunneling barrier for SIS-type junctions in combination with NBCO [4]. The insulator layers for the multilayer integrated circuits necessitate high resistance and low dielectric constant. For the PBCO single crystal, ε_r was 15 at 100 kHz below 65 K [5]. This value is larger than that of SiO_2 used for Nb circuits, but it is lower than that of Sr_2AlNbO_6 and $La_{0.3}Sr_{0.7}Al_{0.65}Ta_{0.35}O_3$ which are frequently used for insulator layers and substrates in the HTS circuits. On the other hand, these non-superconducting cuprates were conductive at room temperature, although conductivities decrease with decrease of temperature. Below 60 K, the values of tan δ changed from +0.1 to -0.1, which were below the accuracy limit of measurement system. For such high speed switching devices as SFQ devices, the dissipative leakage currents through the insulator layers must be low for effective signal transmission. This requirement should be reconsidered for high frequency operation above 100 GHz. In this paper, we report the dielectric properties required for SFQ circuit application.

CONSTRAINTS OF DIELECTRIC PROPERTIES

To estimate dielectric properties required for transmission of SFQ signals, we used analysis and experimental results about microstrip type resonators, which have already been applied for microwave passive devices, such as filters and delay-lines [6]. Performance of the microstrip line is given by 3 parameters, attenuation a, characteristic impedance Z_0 (= $(L/C)^{1/2}$) and delay time τ_0 (= $(LC)^{1/2}$), where L and C are an inductance and a capacitance of the line for the ground plane [7]. For superconducting microstrip structures with the strip line width of w and the insulator of ε_r with thickness of t_0, L and C are described by L = $\mu_0\mu_s t_0/Kw$ and C = $\varepsilon_0\varepsilon_r Kw/t_0$ .Here, μ_s corresponding to magnetic susceptibility of the superconductor is given by

$\mu_s = 1 + (\lambda_G/t_0) \coth (t_G/\lambda_G) + (\lambda_s/t_0) \coth (t_s/\lambda_s)$,

where λ_G and λ_s are London penetration depth of the ground plane and the strip line, and t_G and t_s are thickness of them. In addition, we should consider dispersion, frequency dependence of them, over wide region due to the short SFQ pulse.

Among characteristics of the microstrip line, the line loss α is described as the following equation,

$\alpha = (\varepsilon_0\varepsilon_r /\mu_0\mu_s)^{1/2} KR_s/2t_0 + \pi f(\varepsilon_0\mu_0\varepsilon_r \mu_s)^{1/2} \tan \delta$,

where R_s is surface resistance and f is frequency and K is a fringe field factor, which is almost 1 [8]. By now, α has been neglected in integrated circuits, because of low R_s of the superconductors and low tan δ of the insulator materials at low temperatures [9]. In SFQ circuits, the voltage pulse has a narrow width of pico-second order and weak amplitude of mV. At 500 GHz corresponding to pico-second pulse width, however, R_s increases up to normal metal level according to f^2 dependence of R_s. For HTS microstrip type resonators with double-sided YBCO films, unloaded Q value, Q_u, of 20,000 was obtained at 12GHz and 77 K and was mainly determined by R_s [6]. From this value, Q_u at 500 GHz is extrapolated to be 10 using f^2 rule, although it increases for lower temperature operation. This loss level of the microstrip line corresponds to tan δ of 10^{-1} for the insulator materials. Thus, PBCO is applicable for the insulators of the microstrip line, because its contribution to the loss is comparable to that of the superconducting line.

As for Z_0, suitable value is determined by SFQ operation conditions. Impedance matching between the junction and the line requires that Z_0 is close to R_n, where R_n is normal resistance of the Josephson junction. The other condition of $(2\pi)^{1/2}R_n < Z_0$ proposed by Satchell should be satisfied to prevent hysteresis in the I-V curves caused by the additional line capacitance [10]. From the viewpoint of thermal noise stability, I_c of 0.1 mA at 4.2 K and 2 mA at 77 K are required, respectively. Consequently, Z_0 is desirable to be about 10 Ω. For the microstrip line structure, insulator thickness, t_0, is order of 0.5 μm from deposition technology. To realize Z_0 of 10 Ω using ε_r of 15, the maximum allowable linewidth is 10 μm order, resulting in suitable fabrication and designing for the circuits.

Delay time of the line, τ_0, depends on the dielectric constant of the insulator. For ideal lines,

ε_r values of 15 are sufficiently low for high speed operation. Furthermore, weak frequency dependence of L and C suppresses dispersion of the pulse. From non-ideal line with rather large R_s at high frequency, however, lower ε_r is desirable.

There are other restrictions caused by dielectric properties for high frequency operation of the microstrip line [7]. One is the frequency at which significant coupling occurs between the quasi-TEM mode and the lowest-order surface-wave spurious mode, f_T. For the microstrip line consisting of an insulator layer with t_0 of 0.5 μm and ε_r of 15, f_T is estimated to be order of 50 THz. The other is a radiation loss. This loss is estimated from unloaded Q value, Q_r, of microstrip type l/2 resonators, which may be approximated as $Q_r = 3\varepsilon_r Z_0^2 / 32f^2\mu_0 t_0^2 (\varepsilon_0\mu_0)^{1/2}$. Q_r of 10^4 order at 500 GHz indicates negligible small loss, resulting also in weak cross-talk coupling between the lines.

Finally, SFQ operation requires dc-bias for the junctions. To prevent distribution of bias current through the insulator, high resistance, which is sufficiently higher than one of resisters used for biasing, usually 10 Ω order, is necessary. For nonmetallic cuprates, dc-resistivities were above 10^6 Ω/cm below 150 K, which were sufficient high to suppress bias current leak.

CONCLUSION

Examination about constraints for SFQ circuits using HTS indicates some requirements about dielectric parameters for the insulator layers of the microstrip line structures. According to these conditions, the cuprates such as PBCO have high potential of application for the insulator in the integration circuits operated at high frequency above 100 GHz.

Acknowledgements. We would like to thank to Dr. F. M. Saba for discussion about dielectric properties of cuprates. This work was supported by the New Energy and Industrial Technology Development Organization (NEDO) as Collaborative Research and Development of Fundamental Technologies for Superconductivity Applications under the New Sunshine Program administered by the Agency of Industrial Science and Technology (AIST) of the Ministry of International Trade and Industry (MITI) of Japan.

1. K. Likharev and V. Semenov, IEEE Trans. Appl. Supercond. **1**, 3 (1991).
2. W. H. Mallison, S. J. Berkowitz, A. S. Hirahara, M. J. Neal and K. Char, Appl. Phys. Lett. **68**, 3808 (1996).
3. G. A. Alvarez, J. C. Wen, F. Wang and Y. Enomoto, Jpn. J. Appl. Phys. **35**, L1050 (1996).
4. G. A. Alvarez, T. Utagawa, and Y. Enomoto, Appl. Phys. Lett. **69**, 2743 (1996).
5. F. M. Saba, T. Utagawa, K. Tanabe and Y. Enomoto, in Advances in Superconductivity X (Springer Verlag. Tokyo, 1998) : 1069.
6. T. Yoshitake and H. Tsuge, J. Appl. Phys. **76**, 4256 (1994).
7. Inder J. Bahl, (1989) in Handbook of Microwave and Optical Components Volume 1 Microwave Passive and Antenna Components Edited by Kai Chang, New York: Weley : 27.
8. W. H. Chang, J. Appl. Phys. **50**, 8129 (1979).
9. R. L. Kautz, IEEE Trans. Magn. **MAG-15**, 566 (1979).
10. Julian Satchell, The 1996 International Workshop on Superconductivity June 24-27 Iwate, Japan : 151 (1996).

Preparation of high-T_c a-axis oriented EBCO thin films on R-sapphires with CeO$_2$\PBCO buffer layers using dc magnetron sputtering

Hironori WAKANA, Takeo HASHIMOTO, Shinji KIKUCHI and Osamu MICHIKAMI

Faculty of Engineering, Iwate University, 4-3-5 Ueda, Morioka, Iwate 020-8551, Japan

Abstract: We attempted to deposit a-axis oriented EuBa$_2$Cu$_3$O$_{7-\delta}$(EBCO) with T_{ce}'s above 85 K on Al$_2$O$_3$(1$\bar{1}$02) substrates with CeO$_2$(001) buffer layers by introducing a template layer of PrBa$_2$Cu$_3$O$_x$(PBCO). Growth conditions of substrate temperature (T_s) and deposition rate (R_d) for a-axis oriented PBCO layers were estimated. The PBCO films were prepared at a T_s of 620°C and a R_d of 35 Å/min on the CeO$_2$ buffer layers. The maximum substrate temperature where a-axis oriented EBCO films grew on the a-axis oriented PBCO layers of thicknesses above 700 Å was about 30°C higher than that on CeO$_2$(001) layers. The EBCO films with a- and b-axis orientation were deposited on R-sapphire\CeO$_2$(001) substrates with a-axis oriented 1000-Å-thick PBCO template layers at a T_s of 650°C and exhibited T_{ce}'s of about 86.7 K.

Keywords: EBCO thin film, PBCO template layer, CeO$_2$ buffer layer, sapphire substrate

INTRODUCTION

The orientation control of an oxide superconducting thin film of the YBCO system has been investigated for past ten years [1, 2]. It is possible to obtain a c-axis oriented YBCO thin film with T_{ce}'s above 90 K, but it is not easy to grow a high-quality a-axis one. Because a-axis oriented films grow at lower substrate temperatures and higher oxygen contents than c-axis oriented films, it was difficult to grow a-axis oriented films with good crystallinity. Therefore, the growth of high-quality a-axis oriented films has been investigated in many research institutes. For the preparation of a-axis oriented oxide superconducting films, it is necessary to use substrates with good lattice matching and chemical stability. SrTiO$_3$ enables us to obtain substrates with these qualities, but it is very expensive. Therefore, we attempted to use sapphire substrates. High-quality a-axis oriented YBCO thin films were deposited on LaSrGaO$_4$(100) and SrTiO$_3$ substrates using PBCO template method [3] and self-template method [4], respectively.

We have investigated the preparation of high-quality c- and a-axis oriented EBCO thin films on R-plane sapphire substrates. While c-axis oriented EBCO thin films deposited on the sapphire substrates with CeO$_2$(001) buffer layers had T_{ce}'s of 91.7 K [5], a-axis oriented EBCO thin films exhibited T_{ce}'s of 72.0 K without using self-template method [6]. In order to improve the a-axis oriented EBCO film, the PBCO film was introduced as a template layer.

In this paper, we report the epitaxial growth of PBCO films on R-plane sapphire substrates with 50-Å-CeO$_2$ buffer layers, and the effect of a PBCO buffer layer thickness on the superconducting properties and orientation of the EBCO films.

EXPERIMENTAL

A sputtering system with three magnetron cathodes was used to deposit CeO$_2$, PBCO and EBCO thin films. The position between the cathode and a sample holder can be changed with an adjusting system. Rectangular R-plane sapphires (1$\bar{1}$02) with a lattice plane (15.38×4.76 Å) were used as substrates, which have the nonsymmetry. CeO$_2$ was used as a buffer layer to prevent reaction between sapphire substrates and PBCO films. CeO$_2$ films were prepared using rf magnetron sputtering. Optimum sputtering conditions for the (00$\bar{1}$) oriented CeO$_2$ buffer layer had been clarified in previous research [5]. The 50-Å-thick CeO$_2$ films were used as the first buffer layers.

PBCO films were prepared from a sintered stoichiometric target 83 mm in diameter and 7 mm thick using dc magnetron sputtering. The PBCO deposition was carried out under the sputtering conditions of various deposition rates (R_d) and various substrate temperatures (T_s). The crystal quality of an EBCO thin film was dependent on crystallinity and orientation of the PBCO films. We investigated the influence of T_s and film thickness (t) of the PBCO film on the superconducting properties and orientation of the EBCO films.

EBCO sputtering was curried out in dc mode using a sintered stoichiometric target. The

2000-Å-thick EBCO thin films were deposited under the conditions of a mixture of Ar+20%O_2 (7 Pa), a off-center distance (D_{on-off}) of 65 mm and a target-substrate distance (H_{t-s}) of 40 mm. The structural properties were estimated by a X-ray diffractometer using CuKα radiation. The grain growth of the epitaxial films was observed using an atomic force microscope (AFM). Measurements of electrical resistance versus temperature were carried out by a dc four-probe transport method.

RESULTS AND DISCUSSION

To identify sputtering conditions where a-axis oriented PBCO films grew on R-plane sapphires with 50 Å-thick-CeO_2 buffer layers in the same atmosphere of Ar+20%O_2 at 7 Pa as a-axis oriented EBCO films[6], we examined the influence of R_d and T_s on the orientation and crystallinity. It is known that a-axis oriented EBCO films grow at lower T_s's and higher oxygen contents than c-axis oriented films. The number of oxygen ions increases with oxygen content, and it induces the result of promoting the reverse sputter of the film surface. As the result, the film surface must become rough with increasing oxygen content. The surface roughnesses of the PBCO films measured by AFM were 180 Å, 220 Å and 350 Å for 7.5%O_2, 20%O_2 and 30%O_2, respectively. The surface roughneses of PBCO and EBCO films showed the similar tendency. According to the relation of growth orientation and surface morphology to oxygen content, PBCO and EBCO films were deposited at 20%O_2.

First, we estimated sputtering conditions for a-axis oriented PBCO films. The content of a-axis oriented grains in the films depended on the sputtering conditions of T_s and R_d. The PBCO films deposited under the conditions of R_d's of 7–80 Å/min and T_s's in the vicinity of 620°C were composed of a- and c-axis oriented grains. Therefore, the content of a-axis oriented grains was estimated using a diffraction intensity ratio of $I_{200}/(I_{200}+I_{005})$, where I_{200} and I_{005} are representative diffraction peaks from a- and c-axis oriented grains, respectively. The R_d dependence of diffraction intensity ratio of the PBCO film is shown in Fig. 1. As R_d increases up to 30 Å/min, the PBCO film shifts from the c-axis orientation to the a-axis one. When the R_d are from 30 to 55 Å/min, the a-axis oriented PBCO grains grow. At R_d's above 55 Å/min, the crystallinity lowered, and I_{00l} peaks were observed again. It seems that the decline of crystallinity is caused by the collision of high energy sputter particles to the film surface. Therefore, the rise of the deposition rate induces the increase in the roughness of the film surface. Full width at half maximum (FWHM) of a (200) peak showed a

minimum value at a R_d of 35 Å/min.

We examined an optimum T_s for a-axis oriented films deposited at a R_d of 35 Å/min. The T_s dependence of orientation of the PBCO film is shown in Fig. 2. The a-axis oriented PBCO films grew at T_s's from 560 to 630°C. The c-axis oriented films grew rapidly over 630°C and the intensity ratio dropped down to 0.2 at a T_s of 680°C. At T_s's of 620°C or less, the crystallinity of a-axis oriented films were improved with a rise of T_s, and FWHM showed a minimum value at a T_s of 620°C. From these results, the conditions of a R_d of 35 Å/min and a T_s of 620°C were used in preparing a-axis oriented PBCO films.

Next, we deposited EBCO films on the a-axis oriented PBCO template layers at a T_s of 650°C. The 2000-Å-thick EBCO films were grown on the PBCO layers of different thicknesses. The orientation of the EBCO films depended significantly on the PBCO film thickness. The result is shown in Fig. 3. On the PBCO film of thicknesses above 700 Å, a-axis oriented EBCO

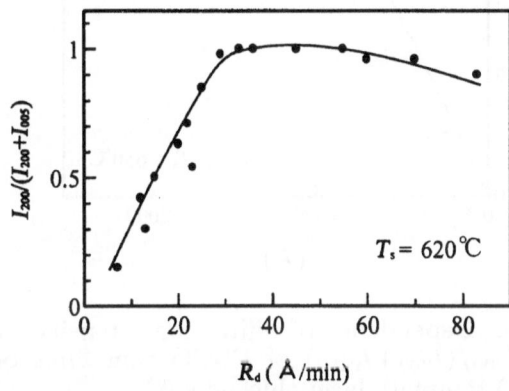

Fig. 1. Dependence of diffraction intensity ratio ($I_{200}/(I_{200}+I_{005})$) of PBCO films on deposition rate (R_d).

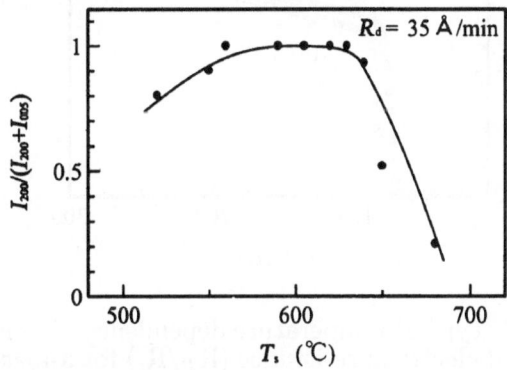

Fig. 2. Dependence of diffraction intensity ratio ($I_{200}/(I_{200}+I_{005})$) of PBCO films on substrate temperature (T_s).

films were grown. And, the upper limit of a T_s for an a-axis oriented EBCO thin film grown on 50-Å-thick CeO_2 buffer layers was 620°C [6], while the T_s for an a-axis oriented EBCO thin film deposited on the PBCO template layers was 650°C. That is, the substrate temperature increased about 30°C by using PBCO template layers. A-axis oriented EBCO films deposited directly on 50-Å-thick CeO_2 buffer layers exhibited a T_{ce} of 72.0 K (Fig. 4 (a)). A-axis oriented EBCO thin films, containing b-axis oriented grains, were successfully deposited on a-axis oriented 1000-Å-thick PBCO template layers under the condition of 7 Pa in Ar+20% O_2 and a T_s of 650°C, and exhibited T_{ce}'s of about 86.7 K. These films had the resistance ratio R_{300}/R_{100} values of about 1.50. A typical R/R_r-T curve is shown in Fig. 4(b).

The surfaces of these films were observed using AFM. The grains of EBCO and PBCO showed the similar shape. The a-axis oriented EBCO films showed the surface morphology composed of rectangular grains about 3000 Å long and 500 Å wide. The direction of the long axis of grains tilts about 45° to the $[11\bar{2}0]$ or $[\bar{1}105]$ direction of the sapphire substrates. In our previous study, the in-plane orientation relationship between $CeO_2(001)$ film and R-plane sapphire substrate was clarified. From these results, the geometrical arrangements among the R-plane of sapphire, the plane of $CeO_2(001)$, the plane of (100) oriented PBCO and the planes of (100) and (010)-oriented EBCO were $Al_2O_3[11\bar{2}0]\|CeO_2[100]\|PBCO[013]\|EBCO[013]$.

CONCLUSIONS

We attempted to deposit high-quality a-axis oriented EBCO thin films on $Al_2O_3(1\bar{1}02)$ substrates with $CeO_2\backslash PBCO$ buffer layers. A-axis oriented PBCO films were deposited using dc magnetron sputtering under conditions of 7 Pa in Ar+20% O_2, R_d=35 Å/min and T_s=620°C. As the PBCO layer thickness increased from 700 Å, a-axis oriented EBCO films grew under conditions of 7 Pa in Ar+20% O_2 and T_s=650°C. The maximum T_s for a-axis oriented EBCO films deposited on $CeO_2\backslash PBCO$ buffer layers was about 30°C higher than that on CeO_2 buffer layers. The a-axis oriented EBCO films deposited on 1000-Å-thick PBCO layers exhibited T_{ce}'s of 86.7 K.

Acknowledgement
This study was partly supported by a Grant-in-Aid for Scientific Research on Priority Area "Vortex Electronics" [A(1), No. 10142103].

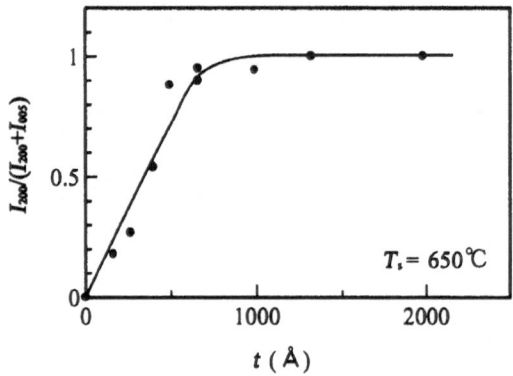

Fig. 3. Dependence of diffraction intensity ratio $(I_{200}/(I_{200}+I_{005}))$ of EBCO thin films on PBCO template layer thickness (t).

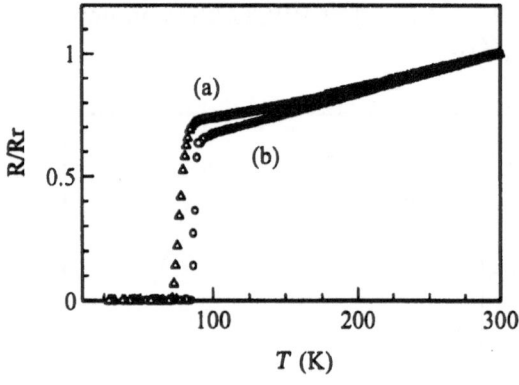

Fig. 4. Typical temperature dependence of normalized electrical resistance (R_T/R_r) for a-axis-oriented EBCO films. (a) R-$Al_2O_3\backslash CeO_2(50Å)\backslash EBCO$ (T_{ce}=72.0K, R_{300}/R_{100}=1.4) and (b) R-$Al_2O_3\backslash CeO_2(50Å)\backslash PBCO(1000Å)\backslash EBCO$ (T_{ce}=86.7 K, R_{300}/R_{100}=1.5).

REFERENCES

[1] O. Michikami and M. Asahi: Jpn. J. Appl. Phys. **30** (1991) 466.

[2] T. Kitamura, I. Hirabayashi and S. Tanaka: Appl. Phys. Lett. **68** (1996) L2002.

[3] Gun Yong Sung and Jeong Dae Suh: Appl. Phys. Lett. **67**(8) L1145 (1995).

[4] H. Ozawa, N. Terada, S. Kashiwaya, H. Takashima, M. Koyanagi and H. Ihara: IEEE TRANSACTION ON SUPERCONDUCTIVITY. **7**(2) 2161 (1997).

[5] Osamu Michikami, Atsushi Yokosawa, Hironori Wakana and Yasube Kashiwaba: Jpn. J. Appl. Phys. **36** 2646 (1997).

[6] Hironori Wakana, Atsushi Yokosawa and Osamu Michikami: Jpn. J. Appl. Phys. **38** 5857 (1999).

Preparation of high-T_c c-axis oriented EBCO thin films on Si substrates with buffer layers using DC magnetron sputtering

Takeo HASHIMOTO, Hideki MUTO, Hironori WAKANA, Hiroshi SAITO and Osamu MICHIKAMI

Faculty of Engineering, Iwate University, 4-3-5 Ueda, Morioka, Iwate 020-8551, Japan

Abstract: We investigated some buffer materials and those deposition conditions to obtain high-quality c-axis oriented $EuBa_2Cu_3O_x$ (EBCO) films on Si(001) substrates using dc magnetron sputtering. In case of the deposition of CeO_2 and Y_2O_3 on Si, it has been found difficult to grow (001) oriented films because of the existence of SiO_x films. Accordingly, (001) oriented YSZ films were deposited as the first buffer layer, and the second layer of CeO_2 or Y_2O_3 which has a small lattice mismatch with EBCO was grown epitaxially on the YSZ layer. Syntheses of the YSZ, CeO_2 and Y_2O_3 buffer layers with (001) orientation and high-quality c-axis EBCO films were examined. The c-axis oriented EBCO films with T_{ce}'s of about 90K were obtained on YSZ(200Å)\$CeO_2$(150Å) buffer layers at 600°C and on YSZ(200Å)\$Y_2O_3$(300Å) buffer layers at 650°C. The EBCO films on the YSZ\$Y_2O_3$ buffer layers were superior to those on the YSZ\$CeO_2$ buffer layers in crystallinity.

Keywords : Si substrate, YSZ film, CeO_2 film, Y_2O_3 film, c-axis EBCO films, buffer layer

INTRODUCTION

Advances in design and preparation technologies of microwave devices composed of high-T_c superconductors(HTS) have increased the necessity of large-sized c-axis superconducting films with high quality. Substrates must have the properties such as low dielectric constant, good lattice matching with HTS, chemical inertness and good crystallinity, and availability in large areas at moderate cost. $LaAlO_3$ substrates are available in large size but have a large dielectric constant and are highly expensive. On the other hand, Si and Al_2O_3, having the properties as described above, are promising. As these substrates react with HTSs, however, buffer layers are indispensable. On R-plane sapphires, we have successfully deposited the c-axis oriented EBCO films with T_{ce}'s(T_c endpoint) above 90 K by using buffer layers of CeO_2 [1]. Sapphire substrate is more expensive than Si one. Then, we attempted to grow high-quality c-axis oriented EBCO films on Si substrates with buffer layers. Buffer materials of YSZ(yttria-stabilized zirconia) [2, 3], CeO_2 [4, 5] and Y_2O_3 [6, 7, 8] were selected in view of lattice matching and growth orientation of these films for the (001) plane of Si substrates.

In this experiment, the deposition conditions for YSZ, CeO_2, Y_2O_3, YSZ\$CeO_2$, YSZ\$Y_2O_3$ buffer layers with (001) orientation on Si(001) substrates were examined. Optimum buffer structures and buffer thicknesses were determined and the high-quality c-axis oriented EBCO thin films were deposited.

EXPERIMENTAL

P-type Si(001) wafers 0.5 mm thick were used as substrates. Buffer layers were deposited on the Si substrates using electron beam evapora-

tion or rf magnetron sputtering. EBCO films were deposited at 7 Pa in Ar + 7.5% O_2 from a sintered stoichiometric target 86 mm in diameter and 6 mm in thickness using off-axis dc magnetron sputtering. Crystallinity and orientation of the films were examined using X-ray diffraction (XRD). Surface morphology was observed using atomic force microscopy(AFM). T_{ce}'s of EBCO films were measured using the dc four-probe method.

RESULTS AND DISCUSSION

To synthesize c-axis oriented EBCO films, we must grow buffer layers with suitable orientation. Then, orientation growth of CeO_2, Y_2O_3 and YSZ films on Si(001) substrates was examined. As a result, we found that CeO_2 and Y_2O_3 easily grew (111) oriented films due to the existence of a SiO_x surface layer and it was hardly possible to grow (001) oriented films. On the other hand, YSZ films grew easily with (001) orientation on Si (001) substrates in spite of the existence of SiO_x using both EB and RF sputtering. The behavior of (001) oriented YSZ film growth was similar to the previous data [9]. Therefore, optimum conditions of (001) oriented YSZ film growth were examined. YSZ films were deposited on Si substrates at various substrate temperatures(T_s) and deposition rates(R_d). The YSZ films deposited at T_s's of about 700°C were a mixture of (001) and (111) oriented grains at R_d's above 7 Å/sec. Single (001) oriented YSZ films with good crystallinity and suitable lattice constant(c_0) were deposited at a T_s of about 700°C and an R_d of about 4 Å/sec. EBCO films 1500Å thick were deposited on Si(001) substrates with (100) oriented YSZ buffer layers of various thicknesses and exhibited a maximum T_{ce} of 83 K at the layer thicknesses ranging from 300 to 500Å. However, this T_{ce} was rather low. It was probable that the low T_{ce} was caused by large mismatches (6.5%) between YSZ and EBCO.

Then, we attempted to grow epitaxially CeO_2 or Y_2O_3 films on the YSZ buffer layer as grade layers between YSZ and EBCO. The deposition conditions for CeO_2 or Y_2O_3 films with (001) orientation on (001) oriented YSZ buffer layers 200Å thick were examined. Figure 1 shows the T_s dependence of c_0 and full width at half maximum(FWHM) of the (002) peak from the CeO_2 film deposited at 80Å/sec. Though c_0 is approximately constant, FWHM decreases with increasing T_s and reaches a minimum value at T_s's of about 700°C. The CeO_2 films deposited at T_s's below 570°C did not exhibit (002) peaks. The CeO_2 films deposited at a T_s of 700°C and R_d's ranging from 3 to 300 Å/sec had the (001) orientation. The films with a minimum FWHM were obtained at R_d's above 80 Å/sec.

Figure 2 shows the T_s dependence of c_0 and FWHM of a (004) peak from the Y_2O_3 film deposited at 0.5 Å/sec. At T_s's below 600°C, c_0 and FWHM value decrease with increasing T_s, and their minimum values were obtained at T_s's

around 650°C. At T_s's below 550°C, the crystallinity of the film became worse with decreasing T_s and amorphous films were deposited at T_s's below 450°C.

The suitable thicknesses of the second buffer layers of CeO_2 and Y_2O_3 were examined by depositing the EBCO films at a T_s of 600°C. The YSZ layer thickness was fixed at 200Å. Figure 3 shows the influence of CeO_2 layer thickness on T_{ce} of the c-axis oriented EBCO film. As the CeO_2 layer thickness increases, T_{ce} increases rapidly and reaches a maximum value of 90 K at the thicknesses from about 80 to 200Å. As the layer thickness increases from 400Å, T_{ce} decreases gradually and is about 65 K at 2400Å. The result for Y_2O_3 layers is shown in Fig. 4. The T_{ce} behavior for the layer thickness is similar to that for the CeO_2 layer. However, EBCO films with T_{ce}'s above 88 K are deposited at Y_2O_3 layer thicknesses ranging from about 50 to 500Å, and those with T_{ce}'s above 90 K and a maximum R_{300}/R_{100} of about 2.8 are obtained at a buffer thickness of about 300Å.

We deposited EBCO films on the suitable buffer layers of YSZ(200Å)\$CeO_2$(150Å) and YSZ(200Å)\$Y_2O_3$(300Å) at various T_s's, and examined the optimum condition in growing

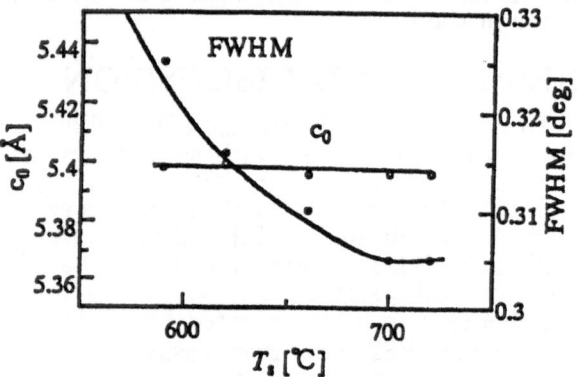

Fig. 1. T_s dependence of c_0 and FWHM obtained from CeO_2(002) reflection.

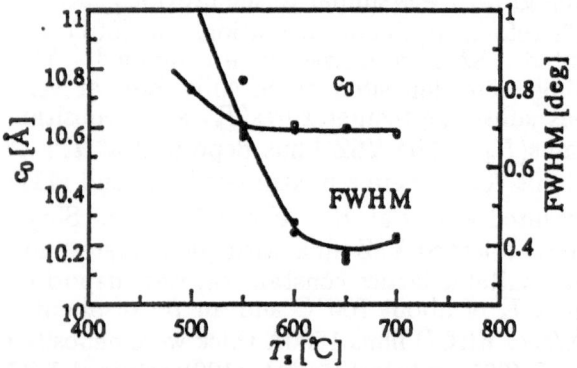

Fig. 2. T_s dependence of c_0 and FWHM obtained from Y_2O_3(004) reflection.

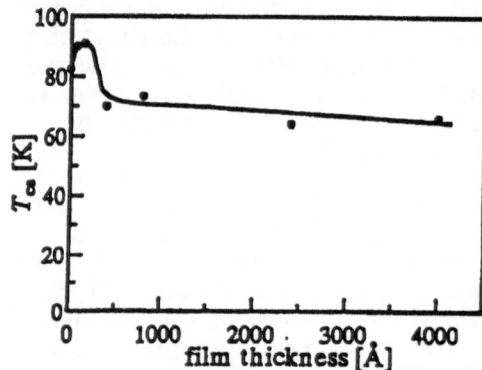

Fig. 3. Dependence of T_{ce} on thickness of CeO_2 for Si\YSZ\CeO_2\EBCO structure.

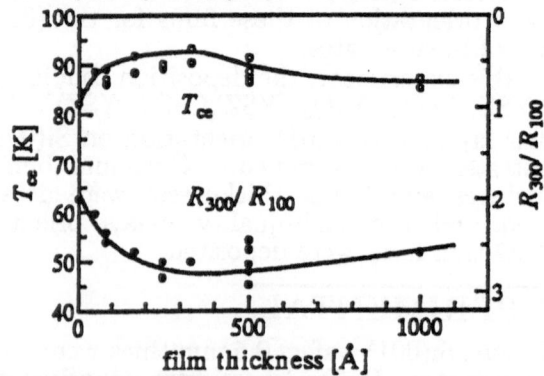

Fig. 4. Dependence of T_{ce} on thickness of Y_2O_3 for Si\YSZ\Y_2O_3\EBCO structure.

high-quality films. Figure 5 shows the T_s dependence of T_{ce} of the EBCO film deposited on the YSZ\CeO$_2$ buffer layer. The EBCO films exhibit T_{ce}'s above 88 K at T_s's between 580 and 620°C, and T_{ce}'s of about 90.7 K at T_{ce}'s of about 600°C. At T_s's above 620°C, the EBCO films deteriorated abruptly and became gray with increasing T_s. The EBCO films deposited at T_s's above 640°C did not exhibit the superconductive behavior at 4.2 K. The cause of the gray film growth is not made clear exactly. As the EBCO films deposited directly on Si substrates represented the similar gray color, however, it is probable that the efficacy of the buffer layer diminished at the high T_s and the EBCO films reacted with the Si substrate.

Figure 6 shows the T_s dependence of T_{ce} of the EBCO film deposited on the YSZ\Y$_2$O$_3$ buffer layer. The EBCO films exhibit T_{ce}'s of about 91.0 K at T_s's between 600 and 650°C.

The R_{300}/R_{100} for the film on the Y$_2$O$_3$ buffer layer is larger than that on the CeO$_2$ buffer layer. This is considered to be due to the difference in the substrate temperatures. AFM images of the EBCO films grown on Si substrates with the YSZ\CeO$_2$ and YSZ\Y$_2$O$_3$ layers exhibited slightly rougher surface morphology (R_z

$\simeq 100$Å) than that on the sapphires with CeO$_2$ layers.

CONCLUSION

The conditions where (001) oriented buffer layers of YSZ, YSZ\CeO$_2$ and YSZ\Y$_2$O$_3$ were deposited on Si(001) substrates were made clear. By optimizing the buffer layer thickness, the high-quality c-axis oriented EBCO films with T_{ce}'s above 90 K were successfully synthesized on Si substrates with YSZ(200Å)\CeO$_2$(150Å) and YSZ(200Å)\Y$_2$O$_3$(300Å) buffer layers at substrate temperatures of 600°C and 650°C, respectively. However, the surface morphology of the EBCO films was not better than that on the sapphire substrates. There is some room for further improvement in buffer materials and their deposition conditions.

Acknowledgments

This study was partly supported by Grant-in-Aid for Scientific Research on Priority Area "Vortex Electronics" [A(1), No.10142103].

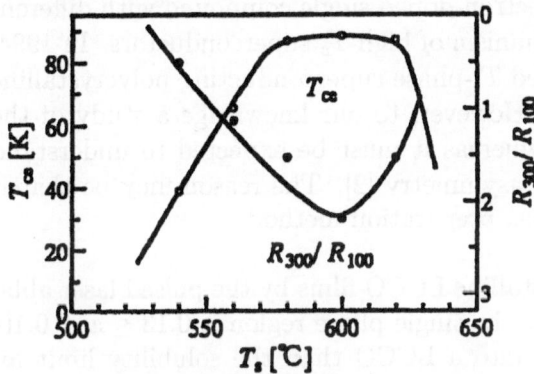

Fig. 5. T_s dependence of T_{ce} of the EBCO films deposited on the Si substrates with YSZ(200Å)\CeO$_2$(150Å) buffer layers.

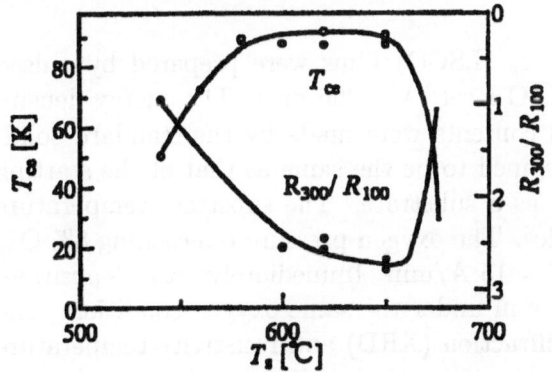

Fig. 6. T_s dependence of T_{ce} of the EBCO films deposited on the Si substrates with YSZ(200Å)\Y$_2$O$_3$(300Å) buffer layers.

REFERECES

[1] O. Michikami, A. Yokosawa, H. Wakana, Y. Kashiwaba Jpn. J. Appl. Phys. 36, (1997) 2646.

[2] D. K. Fork, D. B. Fenner, R. W. Barton, Julia M. Phillips, G. A. N. Connell, J. B. Boyce, and T. H. Geballe Appl. Phys. Lett. 57(11), (1990) 1161.

[3] H. Fukumoto, T. Imura, and Y. Osaka Jpn. J. Appl. Phys. 27(8), (1988) L1404.

[4] B. P. Chang, N. Sonnenberg, M. J. Cima, J. Z. Sun and L. S. Yu-Jahnes Appl. Phys. Lett, 67(8), (1995) 1148.

[5] T. Inoue, Y. Yamamoto and M. Satoh Proceedings of the 13th symposium on materials science and engineering research center of ion beam technology Hosei university (1994) 7.

[6] H. Myoren, Y. Nishiyama, N. Miyamoto, Y. kai, Y. Yamamoto, Y. Osaka and F. Nishiyama Jpn. J. Appl. Phys. 29(6) (1990) L955.

[7] H. Fukumoto, T. Imura, and Y. Osaka Appl. Phys. Lett. 55(4), (1989) 360.

[8] A. Rosova, S. Chromik, S. Benaka and B. Wuyts Physica C, 253, (1995) 39.

[9] H. Myoren, T. Yamashita and T. Inoue Proceedings of the 13th symposium on materials science and engineering research center of ion beam technology Hosei university (1994) 91.

FABRICATION OF $La_{2-x}Ce_xCuO_4$ THIN FILMS USING PULSED LASER ABLATION METHOD

Yasuhiro Matsuo, Kiyosi Betsuyaku, Hirosi Katayama-Yoshida and Tomoji Kawai

ISIR-Sanken, Osaka University, 8-1 Mihogaoka, Ibaraki, Osaka 567-0031 Japan

Abstract: $La_{2-x}Ce_xCuO_{4+y}$ (LCCO) films have been fabricated on $SrTiO_3$ (100) substrates using a fourth harmonic Nd:YAG laser. X-ray diffraction patterns indicate the c-axis oriented films are epitaxially grown over the solubility limit for the polycrystalline sample. Moreover, we observe that the change of c parameter is controlled in a linear dependence at $0.1 \leq x \leq 0.2$ as the case of $Nd_{2-x}Ce_xCuO_{4+y}$ and indicates the abrupt jump at $x \sim 0.075$. This result suggests that the structure of crystal changes from T-phase into T'-phase substituting four valence of ions Ce^{4+} for La^{3+} sites. The substitution of the Ce^{4+} as a dopants is confirmed.

Keywords: $La_{2-x}Ce_xCuO_4$, T'-phase, solubility limit, unstable formation

INTRODUCTION

A significant development in the field of cuprate high-T_c superconductors was the discovery of compounds of the form $La_{2-x}M_xCuO_{4-y}$ (M = Sr, Ca or Ba). The La-based cuprate superconductors has also been found in p-type (i.e. the carriers are holes) with K_2NiF_4 structure so called T-phase. Obviously comparisons between hole- and electron-doped single compound with different dopants have very important implications for the mechanism of high-T_c superconductors. In 1994, Yamada *et al.* [1] succeeded in fabricating the La-based T'-phase superconducting polycrystalline $La_{2-x}Ce_xCuO_{4+y}$ (LCCO) samples with $T_c \simeq 30$ K. However, to our knowledge a study of the system has not been done after this letter reported whereas it must be expected to understand fascinating physical properties such as a hole-electron asymmetry [2]. The reason may be that it is difficult to synthesis of LCCO because of the complex preparation method.

As a way to extend the study, we fabricated single crystalline LCCO films by the pulsed laser ablation method. In the case of the polycrystalline sample, the single phase region is $0.13 \leq x \leq 0.16$. From our composition, much more Ce can be doped into a LCCO than the solubility limit for polycrystalline sample because of their unstable formation.

EXPERIMENT

The LCCO ($0.05 \leq x \leq 0.3$) and $La_{1.85}Sr_{0.15}CuO_{4-y}$ (LSCO) films were prepared by pulsed laser ablation method using a fourth harmonic Nd:YAG laser ($\lambda = 266$ nm). The energy density was approximately 6.0 J/cm^2. Targets with various contents were made by the standard solid-state reaction method. The film composition was assumed to be the same as that of the starting material. The single crystal $SrTiO_3$ (100) was used as a substrate. The substrate temperature during film growth was kept at 730°C for both samples. The oxygen pressure (containing 8% O_3) was 8×10^{-2} mbar. Typical deposition rates were $7 \sim 15$ Å/min. Immediately after deposition, the samples were cooled to room temperature for 30 min under the same oxygen condition. The properties of the samples were evaluated by X-ray diffraction (XRD) and resistivity-temperature (ρ-T) measurements. In ρ-T measurement, the samples were patterned by photolithography into four terminal configuration with dimensions of 200 μm in width and 1 mm in length.

RESULTS AND DISCUSSION

Figure 1 shows the θ-2θ diffraction spectrum of as-deposited LCCO films. As can be seen, the (00ℓ) reflections along with the substrate reflections are visible. The LCCO peaks for $x = 0.15$ are extremely intense and narrow with a full width at half maximum of the (004) reflection equal to $0.14°$. The c parameter is found to be 12.41 Å, which is almost same as that of Ref. [1].

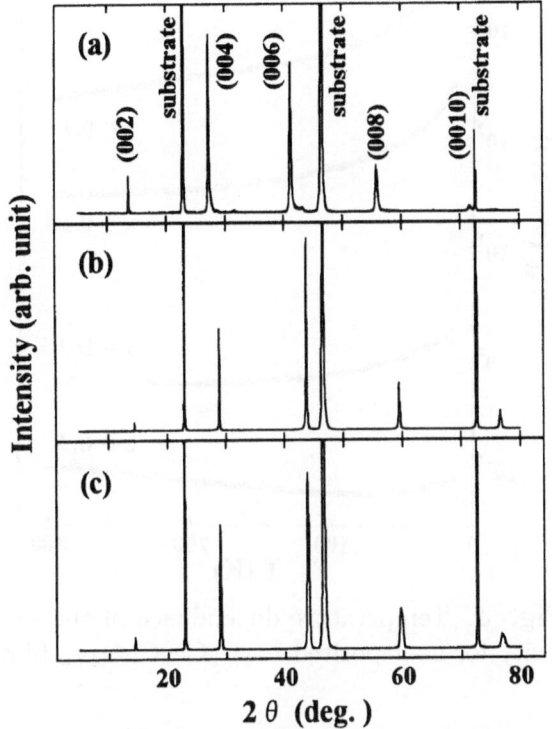

Fig. 1. Typical XRD patterns of as-deposited $La_{2-x}Ce_xCuO_{4+y}$ films for the concentration x at (a) 0.05, (b) 0.15 and (c) 0.3.

Fig. 2. Variation in the c parameter for the $La_{2-x}A_xCuO_{4+y}$ (A = Sr (open triangle), Ce (open circles)) films. We also plotted the results for $Nd_{2-x}Ce_xCuO_{4+y}$ compounds (filled squares) [3] and for $La_{2-x}Sr_xCuO_{4-y}$ compounds (filled triangles) [4]. The solid lines are guides to the eye.

We plot in Fig. 2 the changes of the c parameter as a function of x in the system of $La_{2-x}M_xCuO_{4+y}$ (M = Sr and Ce) comparing with those available in the literature [3,4]. We observe that c parameter of LCCO decrease linearly at $0.1 \leq x \leq 0.2$ with increasing the Ce content. This indicates that the Ce^{+4}, not Ce^{+3} ions are present in LCCO because the rate of decrease in c parameter is same manner as that of $Nd_{2-x}Ce_xCuO_{4+y}$ (NCCO). In order to investigate the difference in the c parameter between the samples before and after the annealing, we annealed the samples at 700°C for 12 h under O_2 (oxygenated) and Ar (reduced) gas flow with the pressure of 1 atm. As illustrated in Fig. 3(a), no remarkable difference was observed in the pattern of (006) peaks related with a c parameter for $x = 0.15$ as reported for NCCO [5]. Therefore, we suggest the structure of LCCO in the region $x \geq 0.1$ is the T'-phase. On the other hand, we observe a discontinuous jump of the c parameter at $x = 0.05$. The c parameter is 13.14 Å close to that of LSCO and is interpolated by the linear relation of that of LSCO. This means that LCCO for $x \leq 0.05$ tend to destabilize the T'-phase and form the T-phase. The phase boundary between T'- and T-phase exists at $x \sim 0.075$. Also, we find the deviation from the linear behavior at $x = 0.3$ as the overdoped region for LSCO [3]. However, the c parameter is drastically change with oxygenated or reduced manner as shown in Fig. 3(b). The origin of this behavior is not understood, but it seems likely that it arises from so much overdoping of Ce^{4+} ions giving rise to an unstable state.

To further investigate the substitution of the Ce^{4+} ions for La^{3+} site as a dopants, we measured the temperature dependence of the resistivity for as-deposited LCCO at $0.1 \leq x \leq 0.3$ as shown in Fig. 4. With increasing Ce contents x, the resistivity decreases rapidly, which is consistent with the behavior of another High-T_c cuprate. However, all samples behave as a semiconductor and do not show a superconductivity in the low temperature region.

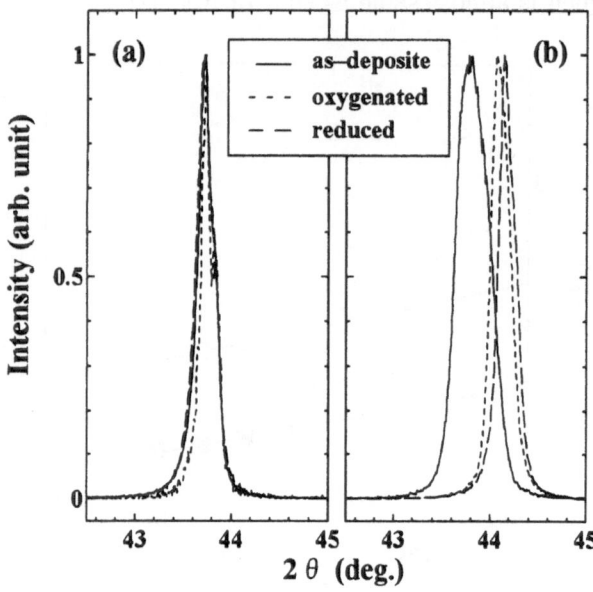

Fig. 3. XRD patterns of (006) peaks for as-deposited, oxygenated and reduced film of $La_{2-x}Ce_xCuO_{4+y}$. (a) $x = 0.15$, (b) $x = 0.3$.

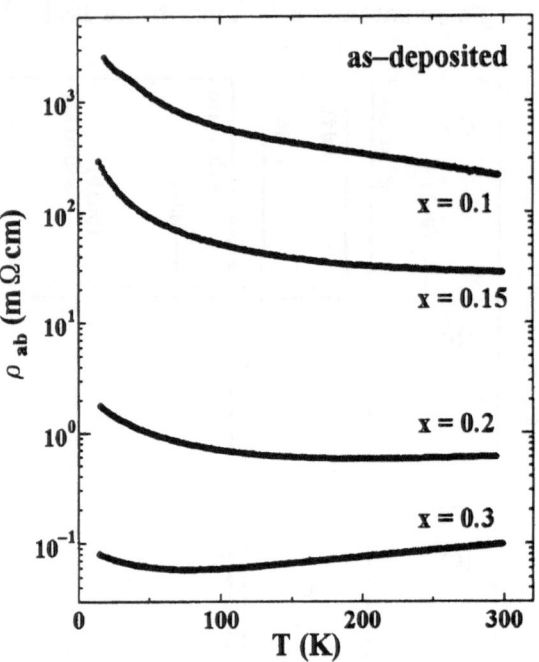

Fig. 4. Temperature dependence of the resistivity for as-deposited $La_{2-x}Ce_xCuO_{4+y}$ films.

CONCLUSION

We fabricated the c-axis oriented LCCO ($0.05 \leq x \leq 0.3$) and LSCO films with pulsed laser ablation method. The results from XRD measurement indicate that the LCCO films are epitaxially grown over the solubility limit for the polycrystalline sample. The change of c parameter obeys a linear behavior at $0.1 \leq x \leq 0.2$ and shows the abrupt jump at $x \sim 0.075$ implying the phase boundary of the crystal structure between T- and T'-phase. Also, we confirm the substitution of the Ce^{4+} ions for La^{3+} ions as a dopants.

1. T. Yamada, K. Kinoshita, and H Shibata, Jpn. J. Appl. Phys. **33**, Pt. 2, L168-L169 (1994).
2. H. A. Blackstead, and J. D. Dow, Phys. Rev. B **55**, 6605-6611 (1997).
3. E. Wang, J.-M. Tarascon, L. H. Greene, G. W. Hull, and W. R. McKinnon, Phys. Rev. B **41**, 6582-6590 (1990).
4. P. Ganguly, N. Shah, M. Phadke, V. Ramaswamy, and I. S. Mulla, Phys. Rev. B **47**, 991-995 (1993).
5. H. Takagi, S. Uchida, and Y. Tokura, Phys. Rev. Lett. **62**, 1197-1200 (1989).

Preparation of $YBa_2Cu_3O_{7-x}$ Thin Films by Nd:YAG Laser Ablation

Tetsuji UCHIYAMA and Zhen WANG

Kansai Advanced Research Center, Communications Research Laboratory, MPT, 588-2 Iwaoka, Iwaoka-cho, Nishi-Ku, Kobe 651-2492, JAPAN

ABSTRACT: We have studied the preparation of $YBa_2Cu_3O_{7-x}$ (YBCO) thin films grown on $SrTiO_3$ (100) and MgO (100) substrates using the 4th harmonics of an neodymium-doped yttrium aluminum garnet (Nd:YAG) pulse laser beam (wavelength 266 nm). The Nd:YAG laser is a solid-state laser and is safer than the excimer laser. The as-deposited YBCO thin films have been shown c-axis orientation and 90 K phase with varying the target-substrate distance and the oxygen flow rate during the ablation. We have found that the target-substrate distance is closely related to the laser-generated plume size and that the oxygen flow rate is the key preparation parameter in addition to the substrate temperature and oxygen pressure.

KEY WORDS: Nd:YAG laser, high T_c superconductor, $YBa_2Cu_3O_{7-x}$, thin film

INTRODUCTION

The pulsed laser ablation technique is one of the most promising methods for *in-situ* preparation of high T_c superconducting thin films [1]. It generates activated oxygen from molecular oxygen in the chamber due to a strong laser power. The activated oxygen accelerates to take oxygen into the films. An Excimer laser with the power of several hundred mJ/shot, such as ArF and KrF, is the most popular laser, however, it needs to make use of the dangerous and/or expensive gases. An Nd:YAG laser is a solid-state laser and is safer than the excimer laser. According to Singh *et al.* [2], the absorption coefficient of the laser-generated plume becomes stronger with shortening the wavelength. However, the laser power of the commercial 4th harmonics of an Nd:YAG laser (266 nm) is less than 100 mJ/shot, which is low to activate molecular oxygen in a chamber [3]. Therefore, an annealing treatment of oxygen after the ablation must be carried out in fabrication of superconductive YBCO thin films deposited by Nd:YAG laser [4]. In this work, we present the preparation of as-deposited superconductive YBCO thin films without an annealing treatment of oxygen deposited by the 4th harmonics of an Nd:YAG pulsed laser beam.

EXPERIMENTAL

The laser ablation was carried out as follows: The laser was operated at a repetition rate of 2 - 10 Hz with an energy density of 4 - 5 J/cm^2 and an irradiated area of about ϕ 1.5 mm. The laser beam was incident on a rotating stoichiometric YBCO target at an angle of 45°, and the evaporated material was deposited onto the rotating substrate placed at a distance of 32 - 45 mm from the on-axis target. $10 \times 10 \times 0.5$ mm^3 as-received single crystalline substrates of $SrTiO_3$ (STO) (100) and MgO (100) were used for deposition of YBCO thin films. STO has good lattice matching with YBCO, and MgO has a large lattice-mismatch, however, has an extremely low dielectric constant and a reasonably low microwave loss against STO's. The preparation parameters are shown in Table 1, where T_d, P_{O2}, F_{O2}, f_r and d_{t-s} are deposition temperature, oxygen pressure, oxygen flow rate, repetition rate, and the distance between target and substrate, respectively. The thickness of the film was fixed at around 200 nm. After ablation, oxygen gas was introduced in the chamber up to 15 Torr to prevent the film from a deoxidation, and the sample was taken after 20 minutes at about 300℃. The crystallinity of films was studied by X-ray diffraction (XRD) $2\theta/\theta$-scanning and X-ray ϕ-scanning measurements. To check the superconductivity, temperature dependence of resistance (R-T) characteristic was measured using the conventional DC four-probe method.

Table 1. Preparation parameters for STO (100) and MgO (100) substrates.

Substrate	T_d (℃)	P_{O2} (mTorr)	F_{O2} (ml/min)	f_r (Hz)	$d_{t\text{-}s}$ (mm)
STO (100)	800	350	7 – 45	1 – 10	32 – 45
MgO (100)	730	250 – 500	10 – 100	10	32

RESULTS AND DISCUSSION

Figure 1 displays the relation between $d_{t\text{-}s}$ and the zero resistance temperature (T_{c0}) for YBCO thin films deposited on STO (100) substrates under the fixed preparation parameters, F_{O2} = 40 ml/min and f_r = 10 Hz. The typical scanning electron microscope (SEM) images are inset in Fig. 1. Here, the size of the laser-generated plume was around 40 mm. It is found that T_{c0} decreases drastically with setting the sample out of the plume. However, when the sample was settled in the plume, larger outgrowths were observed by SEM images. The best position of substrate setting is decided around the top of the plume, i.e. $d_{t\text{-}s}$ is about 40 mm. In Fig. 2, the deposition rate (R_d) dependence of T_{c0} for different F_{O2} and $d_{t\text{-}s}$ fixed at 32 mm is shown. For different F_{O2}, the behavior of T_{c0} is changed at a boundary of R_d ~ 20 nm/min, that is, T_{c0} has a peak around R_d ~ 20 nm/min for F_{O2} ~ 10 - 20 ml/min, while T_{c0} increases with increasing R_d for F_{O2} ~ 40 ml/min. This is interpreted as follows: the former surface migration of the evaporated material becomes difficult with increasing R_d, and T_{c0} is suppressed. On the other hand, a reduction of the surface migration is avoided due to a faster flow of oxygen in the latter.

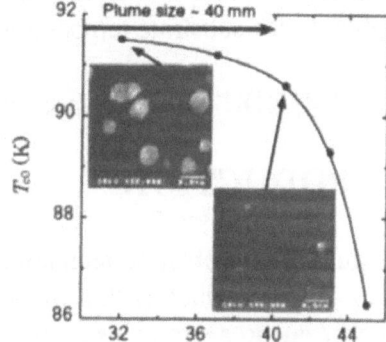

Fig. 1. Dependence of T_{c0} on $d_{t\text{-}s}$ for the films on STO (100).

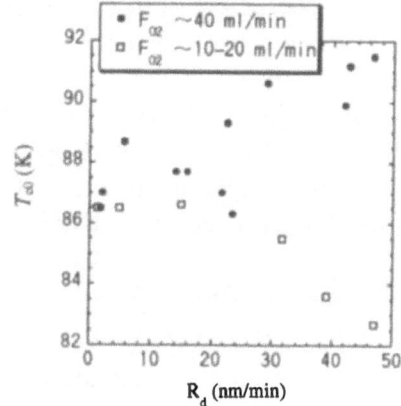

Fig. 2. Dependence of T_{c0} on R_d for the films on STO (100).

Figure 3 shows the relation between F_{O2} and T_{c0} for the films deposited on MgO (100) substrate. The P_{O2} was fixed at 300 mTorr. At F_{O2} = 40 ml/min, the highest T_{c0} ~ 86.5 K was obtained. Hence we can confirm that T_{c0} depends heavily on F_{O2} and is not given much attention as a parameter. Furthermore, the best sample on MgO (100) was obtained under the fixed preparation parameters, P_{O2} = 450 mTorr and F_{O2} = 50 ml/min. XRD $2\theta/\theta$-scanning pattern, ϕ-scanning one, and R-T characteristic for the best sample are displayed in Figs. 4-6. Figure 4 shows the highly c-axis oriented film. XRD ϕ-scanning pattern shown in Fig. 5 indicates that there is epitaxial growth with in-plane-orientation of $[100]_{MgO}$ // $[100]_{YBCO}$, i.e. " cube on cube", on MgO (100). The R-T characteristic shows T_{c0} ~ 89.2 K in Fig. 6. These results are superior to the reported as-deposited YBCO thin films prepared by Nd:YAG laser ablation [4, 5].

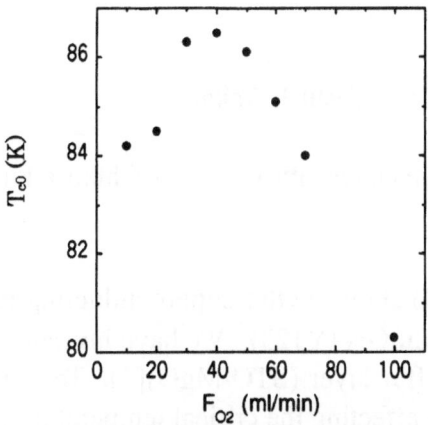

Fig. 3. Dependence of T_{c0} on F_{O2} for the films on MgO (100).

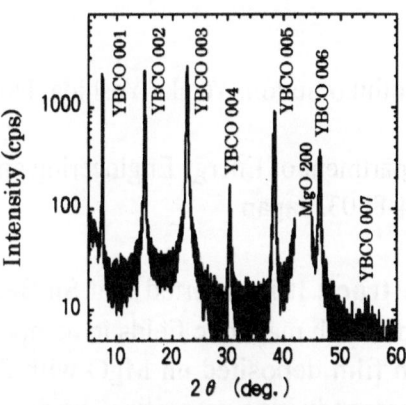

Fig. 4. XRD $2\theta/\theta$-scanning plot. 00l peaks indicate that the film is c-axis oriented.

CONCLUSION

As-deposited superconductive YBCO thin films with c-axis orientation have been prepared using the 4th harmonics of an Nd:YAG pulse laser beam. Preparation of films has been carried out by changing T_d, P_{O2}, F_{O2} and d_{t-s} for STO (100) and MgO (100) substrates. d_{t-s} is closely related to the plume size. Moreover, it is confirmed that T_{c0} depends heavily on F_{O2}, and that F_{O2} is an important dynamical preparation parameter having non-equilibrium condition.

Acknowledgment We would like to thank Dr. M. Mukaida of Yamagata University for his valuable suggestions.

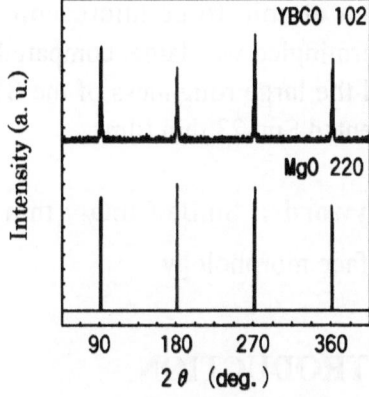

Fig. 5. XRD ϕ-scanning plots of YBCO 102 and MgO 220. Peaks appear every $n\pi/2$. (n: integer)

1. Inam A, Hedge MS, Wu XD, Venkatesan T, England P, Miccli PF, Chase EW, Chang CC, Tarascon JM, Watchman JB (1988) Appl. Phys. Lett. 53: 908
2. Singh RK, Narayan J (1990) Phys. Rev. B 41: 8843
3. Mukaida M, Miyazawa S (1993) Jpn. J. Appl. Phys. 32: 4521
4. Kiss T, Enpuku K, Matsumura T, Iriyama Y, Nakamura T, Takeo M (1996) IEICE Trans. Electron. E79-C: 1269
5. Sun Y, Strasser G, Gornik E, Seidenbusch W, Rauch W (1993) Physica C 206: 291

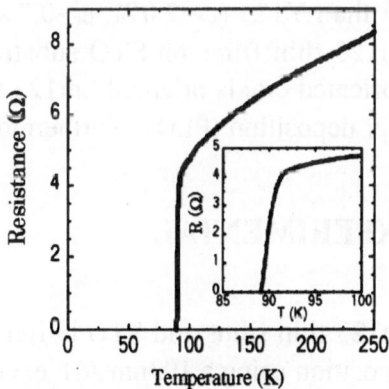

Fig. 6. Temperature dependence of the resistance on MgO (100).

Preparation of c-Axis Oriented SmBa$_2$Cu$_3$O$_{6+\delta}$ Thin Films with a Buffer Layer

Kimihiko Sudoh, Yutaka Yoshida, Noriaki Matsunami, and Yoshiaki Takai

Department of Energy Engineering and Science, Nagoya University, Furo-cho, Chikusa-ku, Nagoya, 464-8603, Japan

Abstract: It is reported that SmBa$_2$Cu$_3$O$_{6+\delta}$ (Sm123) shows better superconducting properties under high magnetic fields in comparison with YBa$_2$Cu$_3$O$_{7-y}$ (Y123). We have investigated Y123 thin film deposited on MgO with SrTiO$_3$ (STO) buffer layer (STO/MgO)[1]. To improve the superconducting properties, we discussed the factors affecting the critical temperature of Sm123 thin films deposited on MgO, STO and STO/MgO substrates. We have obtained c-axis oriented Sm123 thin films on MgO, STO and STO/MgO substrates. However, we hardly obtained c-axis oriented Sm123 thin film, when we used STO (100) buffer layer in which (110) intermingled. From atomic force microscopy, the roughness of STO (100) buffer layer in which (110) intermingled was large, compared with STO (100) buffer layer. We speculate that the orientation and the large roughness of the STO buffer layer are the reason why we could not obtain c-axis oriented Sm123 thin films.

Keywords: SmBa$_2$Cu$_3$O$_{6+\delta}$ thin film, MgO substrate, SrTiO$_3$ buffer layer, c-axis orientation, surface morphology

INTRODUCTION

SmBa$_2$Cu$_3$O$_{6+\delta}$ (Sm123) is attractive for high Tc and high Jc under magnetic field as well as NdBa$_2$Cu$_3$O$_{6+\delta}$ [2]. Sm123 thin film has a better lattice matching to SrTiO$_3$ (STO) (ε_a=1.2%, ε_b=0.5 %) than Y123 (ε_a=2.0%, ε_b=0.7%). Also, it was reported that surface growth mechanism of Sm123 thin films on STO substrate was the Stranski-Krastanov type[3]. In this report, we fabricated c-axis oriented Sm123 thin films on MgO, STO and STO/MgO substrates by pulsed laser deposition (PLD). Furthermore, we discussed the growth mechanism of Sm123 thin films.

EXPERIMENTAL

Sm123 thin films and STO buffer layers were grown on MgO (100) substrates by pulsed laser deposition using a 193nm ArF excimer laser. The targets were a sintered Sm123 and STO bulks, respectively. The energy density of the laser beam on the target surface was about 1J/cm^2 and 2J/cm^2. The ArF laser was operated at a repetition rate of 5Hz. The substrate temperature shown in this work was measured with a thermocouple attached to the substrate holder. After the deposition, oxygen was admitted to the chamber up to 20Torr and the Sm123 thin film was then cooled down at a rate of 20℃/min to 700℃, at which temperature it was kept for 15 minutes. Using the deposition conditions that c-axis oriented Sm123 thin films were grown on MgO the Sm123 thin films and STO buffer layers were determined by θ–2θ X-ray diffraction (XRD). The surface morphology was observed by atomic force microscopy (AFM).

RESULTS AND DISCUSSIONS

A: *c*-axis oriented Sm123 thin films on MgO

Figure 1 shows the peak intensity of the XRD (001) and (100) reflections vs oxygen partial pressure (Po2) with the substrate temperature (Ts). We fabricated the *c*-axis oriented Sm123 thin films at Ts of 850℃ and Po2 of 0.01 Torr. However, *c*-axis length of Sm123 thin films was too long at this condition. Therefore, in order to shorten the *c*-axis length, we deposited *c*-axis oriented Sm123 thin films at higher Po2. Figure 2 shows the peak intensity of the XRD (001) and (100) reflections vs higher Po2 from 0.1 Torr to 1 Torr together with *c*-axis length. By optimizing the position of plume to substrate by varying laser fluence, *c*-axis oriented Sm123 thin films could be fabricated at higher Po2. In addition, with rasing Po2, *c*-axis length of Sm123 thin films could be shortened.

B: Sm123 thin films on STO/MgO

There is a large lattice misfit between MgO substrates and Sm123 thin films if compared with the STO substrate. There are, however, many cracks in the Sm123 thin films deposited on the STO substrates, owing to the difference of the thermal expansion coefficients. Therefore, we have deposited Sm123 thin films on MgO substrates using STO as a buffer layer. Figure 3 shows XRD patterns and AFM images of *c*-axis oriented Sm123 thin film on STO (100) buffer layer. If we deposited Sm123 thin film on the STO (110) buffer layer in which (110) intermingled, we hardly obtained *c*-axis oriented Sm123 thin films. However, by using STO (100) buffer layer, we could

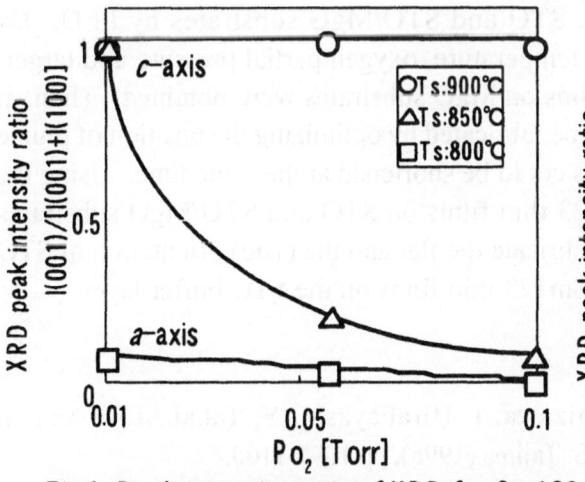

Fig.1 Peak intensity ratio of XRD for Sm123 thin films vs Po2 with Ts for 900, 850 and 800℃. (Target-substrate distance:60mm)

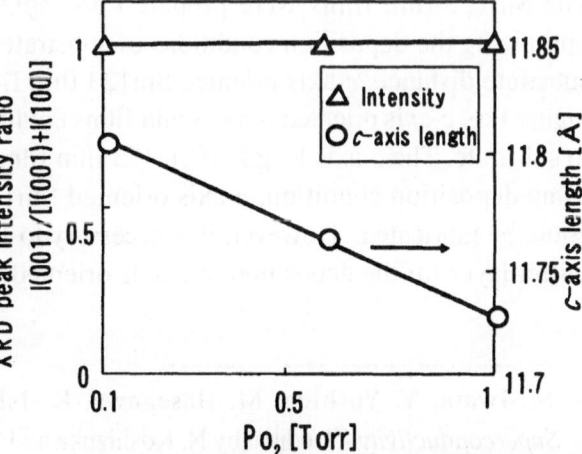

Fig.2 Peak intensity ratio of XRD and *c*-axis length for Sm123 thin films vs Po2. (Ts:850℃ Target-substrate distance:60mm)

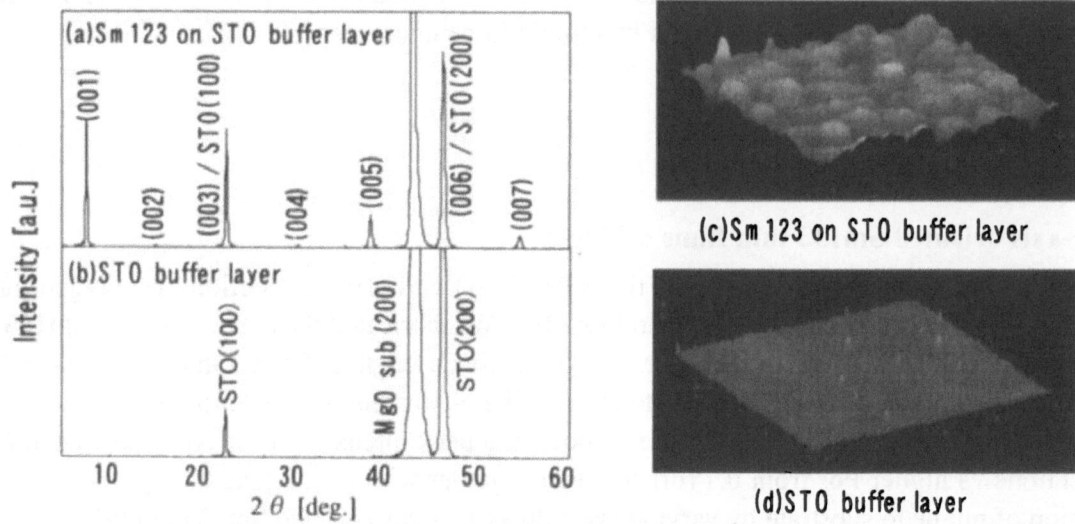

Fig.3 XRD patterns and surface morphology observed by AFM (image size:2000 by 2000 nm) of c-axis oriented Sm123 thin films on (n00) oriented STO buffer layers (a) (c) and (n00) oriented STO buffer layers (b) (d)

fabricate c-axis oriented Sm123 thin films. From the AFM images, the roughness of STO (100) buffer layer in which (110) intermingled was about 37 Å. Also, STO (100) buffer layer was grown in 2D-dimensional growth mode and the roughness was about 12 Å. Furthermore, it was confirmed the roughness of c-axis oriented Sm123 thin film on STO (100) buffer layer was about 21 Å. From the result, the orientation of Sm123 thin films is influenced by the orientation and the surface morphology of the STO buffer layer.

CONCLUSION

The Sm123 thin films were prepared on MgO, STO and STO/MgO substrates by PLD. By optimizing the deposition conditions of substrate temperature, oxygen partial pressure and target-substrate distance, c-axis oriented Sm123 thin films on MgO substrates were obtained. Then, at higher Po$_2$, c-axis oriented Sm123 thin films could be fabricated by optimizing the position of plume to substrate. The c-axis length of Sm123 thin films could be shortened at the same time. Using the same deposition condition, c-axis oriented Sm123 thin films on STO and STO/MgO substrates could be fabricated. However, it is necessary to fabricate the flat and the (100) orientation of STO buffer layer for the deposition of c-axis oriented Sm123 thin films on the STO buffer layer.

1. M. Iwata, Y. Yoshida, M. Hasegawa, K. Ishizawa, I. Hirabayashi, Y. Takai, *Advances in Superconductivity XI*, edited by N. Koshizuka and S. Tajima (1998), pp.1097-1100.

2. M. Badaye, F. Wang, Y. Kanke, K. Fukushima, T. Morishita, Appl. Phys. Lett. **66**, pp.2131-2133 (1995).

3. Q. D. Jiang, D. M. Smilgies, R. Feidenhans, M. Cardona, J. Zegenhagen, Solid State Commun **98**, No.2, pp.157-161 (1996).

The Characteristics of the YBa$_2$Cu$_3$O$_{7-x}$ Thin Films on the Metal Buffer Layer

Jouji Nishimura, Yutaka Yoshida, Noriaki Matsunami and Yoshiaki Takai

Department of Energy Engineering and Science, Nagoya University, Furo-cho, Chikusa-ku, Nagoya, 464-8603, Japan

Abstract: To understand the growth mechanism of high Tc superconductors (HTS) on the metal buffer is important for device and wire applications. Furthermore we have studied crystal orientation and superconducting properties. It was reported that the YBCO films grown directly on Pt buffer had poor superconductivity. In this report, we have prepared YBCO films with Pt buffer layers by pulsed laser deposition (PLD) method. And we have investigated the relationship with the growth mechanism of YBCO films and the metal buffer layer deposited on MgO(100) substrate. Firstly we controlled orientation and surface flatness of Pt layer. We examined the crystal orientation of Pt and YBCO on the Pt buffer. And we examine the relationship between the thickness of the films and superconducting properties. The YBCO films showed superconducting transition temperature (T_{co}) at 91K on Pt buffer layer. Therefore we conclude that Pt films can be used a noble layer for the YBCO thin films.

Keywords: YBCO, Pt buffer layer

INTRODUCTION

We have prepared the Pt films as a buffer layer and the superconducting films on the Platinum (Pt) buffer layer by the pulsed laser deposition (PLD) method. About superconducting properties, it was reported that the YBa$_2$Cu$_3$O$_{7-x}$ (YBCO) films directly deposited on metal buffer layers have poor superconducting properties[3]. If the influence of the metal buffer layer is made clear, the superconductivity of the YBCO will be able to be improved. Therefore,we controled the orientation and surface flatness of Pt, and tried to improve the superconductivity of the YBCO films by using it as the buffer layer. Firstly we examined the characteristic of Pt films. In the next, we deposited the YBCO films in various conditions, and examine their properties. In this work, we report the preparation of YBCO thin films on MgO substrates with a Pt buffer layer and investigation of various properties to understand the growth mechanism of the superconductors on the metal buffer layer.

EXPERIMENTAL

We prepared oxide superconducting YBCO and Pt thin films by pulsed laser deposition (PLD) technique on MgO(100) substrates. An ArF excimer laser (λ =193nm, pulse length τ =15ns) was operated at 10 Hz with an energy density of 0.66 to 1.32 Jcm^{-2} on the target. The substrate temperatures during the film growth were kept at room temperature to 950 °C for Pt buffer layers and 650 °C to 850 °C for the YBCO films. The oxygen pressure was 3×10^{-6} to 0.1 Torr for the deposition of Pt films and 0.2 Torr for YBCO films. The deposition rate was 20 Å /min for Pt and 90 Å /min for YBCO,respectively. We used (100) MgO single crystals ($10 \times 10 \times 0.5$ mm^3). In order to remove surface contaminants such as hydrocarbons, the MgO substrates were annealed for 1 h at 1000 °C under flowing O$_2$ gas. The crystal structure of the deposited films was examined by θ–2θ X-ray diffraction (XRD), X-ray pole figure measurements. The critical temperature (T_c) of the films was measured by four-probe resistance measurement. The surface morphology of the films was investigated by scanning electron microscopy (SEM) and atomic force microscopy (AFM). The composition of the films was examined by Rutherford backscattering spectrometry (RBS) and inductively coupled plasma emission spectrometry (ICP).

RESULTS AND DISCUSSION

Firstly, we controlled the crystallinity and the surface flatness of Pt to use the Pt films as the buffer layer. We had a conclusion that the appropriate deposite condition of Pt is such as, the substrate temperature of 350 °C, oxygen partial pressure of 1×10^{-3}Torr. The Pt films tend to orient a-axis orientation by oxgen pressure. It is observed that oxidation is combined such as PtO on the boundary of substrates. X-ray θ-2θ diffraction pattern of the Pt film on the MgO substrate is shown in figure 1(a). The obtained Pt films showed *a*-axis orientation, and the root mean square of surface roughness(Rms) of Pt film could be reduced to about 10Å by controlling the oxygen partial pressure and substrate temperature. X-ray θ-2θ diffraction pattern of the YBCO films on Pt/MgO substrate is shown in figure 1(b). Only the 00*l* reflections of YBCO are observed in the XRD pattern of YBCO. From these peaks, the *c*-axis lattice constant was calculated to be 11.69-11.71Å. Figure 2 showed a dependence of the critical temperature on the film thickness of YBCO. The critical temperature reached about 91K at a film thickness of about 3500Å. It is concluded that the roughness of the YBCO films influenced the critical temperature. The superconducting properties of YBCO on Pt/MgO was not as good as that on MgO. Therefore we investigated the surface morphology of YBCO on MgO and on Pt/MgO (Fig.3). From AFM image, surface of YBCO on Pt/MgO became rough in comparison with that on MgO. About the crystallinity, the in-

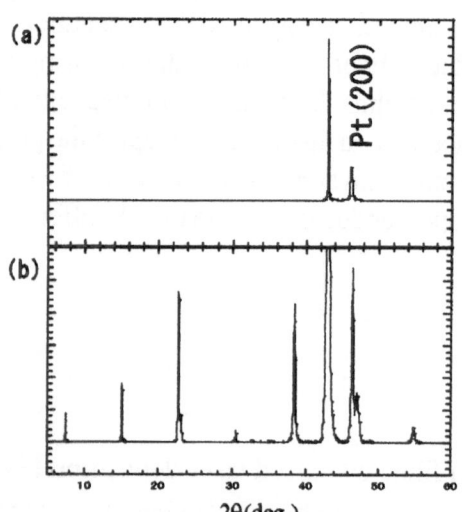

Fig. 1.X-ray diffraction patterns of Pt film on MgO(a),and YBCO film on Pt/MgO(b)

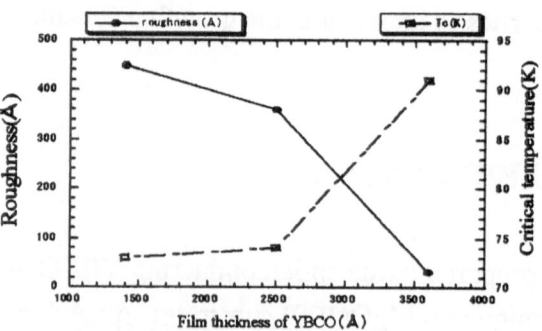

Fig.2. Dependence of the critical temperature on the film thickness of YBCO

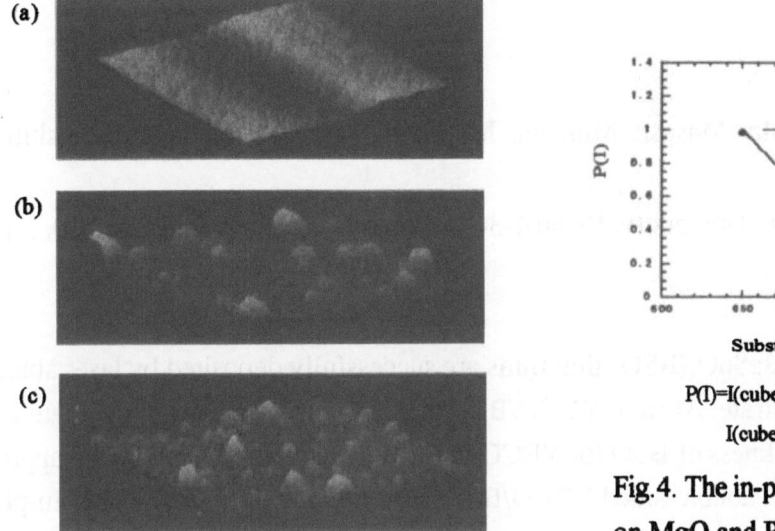

Fig.3.AFM image of Pt film (350°C) on MgO(a),YBCO film on MgO(b),and YBCO on Pt/MgO(c) (image size: 5 by 5 um)

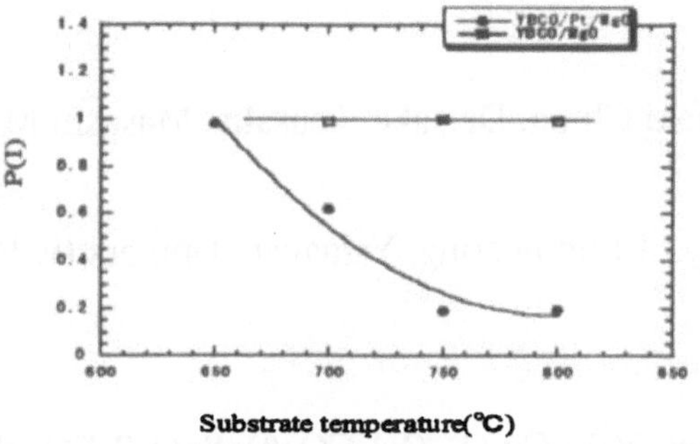

$$P(I)=I(\text{cube on cube})/$$
$$I(\text{cube on cube})+I(45°\text{misaligned grain})$$

Fig.4. The in-plane orientation ratio of YBCO on MgO and Pt/MgO

plane orientation of YBCO on Pt/MgO was easy to change to 45-rotated alignment from cube on cube. This maybe concerned with the lattice mismatch or the surface roughness which increased with the substrate temperature. From these results, in the case of using metal buffer, it was found to be very important to control the roughness and the crystallinity of Pt buffer layer, as well as YBCO.

CONCLUSION

The superconducting YBCO thin films were prepared on MgO (100) substrates with Pt(200) buffer layer by pulsed laser deposition method. The critical temperature of YBCO thin films on Pt/MgO substrate rose as the film thickness of YBCO increased, and reached 91K for the film thickness of about 3500Å. It is concluded that the superconducting property of YBCO on Pt/MgO is influenced by the orientation and the roughness of Pt buffer layer. It is more difficult to control the in-plane orientation of the YBCO films on Pt/MgO than on MgO directly. It is necessary to control the orientation and the roughness of the buffer layer using the metal buffer.

1. Masashi Mukaida et al. Jpn.J.Appl.Phys.Vol.38 (1999)pp.1945-1948
2. Shigeru Matsuno et al. Jpn.J.Appl.Phys.Vol.34 (1995)pp.2293-2299
3. P.Tiwari et al. Appl.Phys.Lett.64(5),31 January (1994)
4. S.Hontsu and J Ishii et al. Appl.Phys.Lett.67(1995)

FABRICATION OF DOUBLE-SIDED YBa$_2$Cu$_3$O$_{7-\delta}$ /BaSnO$_3$ THIN FILMS BY LASER ABLATION

Kazuaki Chiba, Daisuke Kousaka, Masashi Mukaida, Masanobu Kusunoki, Shigetoshi Ohshima

Faculty of Engineering, Yamagata University, Jonan 4-3-16, Yonezawa, Yamagata 992-8510, Japan

Abstract: YBa$_2$Cu$_3$O$_{7-\delta}$(YBCO)/BaSnO$_3$(BSO) thin films are successfully deposited by laser ablation on both sides of [100] MgO substrate. At first, YBCO/BSO thin films were grown to estimate BSO critical thickness. The critical thickness of BSO for YBCO in-plane orientation is found to be approximately 2nm. Using this condition, double-sided YBCO/BSO films are grown on both sides. In-plane orientation for both sides indicates the absence of impurity phases. The full width at half maximum (FWHM) of the YBCO (005) peak for both sides is ~0.50°. The surface resistances in both sides at 22GHz are approximately 1.0 mΩ at 35K.

Keywords: BaSnO$_3$, in-plane orientation, critical thickness, double side, surface resistance

INTRODUCTION

Superconductors have superior characteristics of quite low surface resistance (R_s), which is 2 ~ 3 orders of magnitude lower than that of normal metals in high frequency region. By using this characteristics, super low insertion loss multi-cross-coupled filters are expected to be realized. However, the usage of liquid helium for low temperature superconducting microwave devices hinders the realization. Discovery of high-T$_c$ oxide superconductors with T$_c$ higher than 77K, removed these disadvantages. Many researchers began to study superconducting microwave devices using YBCO films on MgO substrates. Using MgO as substrate, two types of grains of YBCO, which are parallel to MgO [100] and rotated 45°, are mixed in-plane at optimal fabrication temperature [1]. As this result, many 45° slant angle grain-boundaries develop and cause increase of surface resistance [2]. To reduce the surface resistance, it is necessary to control the in-plane orientation of YBCO films. By using BSO, whose lattice constant is 4.12Å and relative dielectric constant is almost 25, as buffer material, it is possible to control in-plane orientation at optimal fabrication temperature and decrease of surface resistance [3]. It is essential to fabricate YBCO/BSO films on both sides of MgO substrate to apply for microwave devices because microwave spreads both side of substrate. But at that time, there is likelihood to develop dielectric loss by BSO bacause much BSO was deposited on both side. Therefore, it is important to know the optimum thickness of BSO. In this paper we research the critical thickness of BSO to control the in-plane orientation of YBCO and fabricate double-sided YBCO/BSO thin films.

EXPERIMENTS

The thin films were grown by ArF excimer laser deposition system. At first thin films were fabricated changing BSO thickness to estimate BSO critical thickness for the control of in-plane orientation. MgO substrates were attached by silver paste to a rotating metal substrate holder which was irradiated by a lamp heater. Fabrication temperature was set to 710°C for both YBCO and BSO. Oxygen pressure was set to 40 mTorr for BSO and 400 mTorr for YBCO. YBCO film's thickness was 200 nm. Next, double-sided thin film was fabricated. As a process of double-sided thin film fabrication, YBCO/BSO was grown on first side of MgO substrate .MgO substrates was heated by lamp heater. Fabrication temperature was set to 720°C for both BSO and YBCO. Thereafter substrate was turned over at room temperature, and YBCO/BSO was grown on second side. Fabrication temperature was set to 710°C for both YBCO and BSO. BSO was fabricated with 10 nm thickness which is enough to avoid mixing other peak of in-plane orientation. Evaluation of thin film was executed as follows, XRD 2θ/θ, φ-scan and ω-scans for crystallnity and in-plane orientation, AFM for surface morphology, dielectric resonator method for characterising the microwave surface resistance.

RESULTS AND DISCUSSIONS

Figure 1 shows BSO thickness dependence of YBCO in-plane orientation. Vertical axis of this figure shows ratio of [100] in-plane orientation, which is calculated from intensity value of XRD φ-scan. From this result, covering rate of surface became wrong at a point that BSO thickness is 1.5 nm. Therefore, critical thickness of BSO is approximately 2 nm. XRD patterns of double-sided thin film are shown in Fig.2 and Fig.3. 2θ/

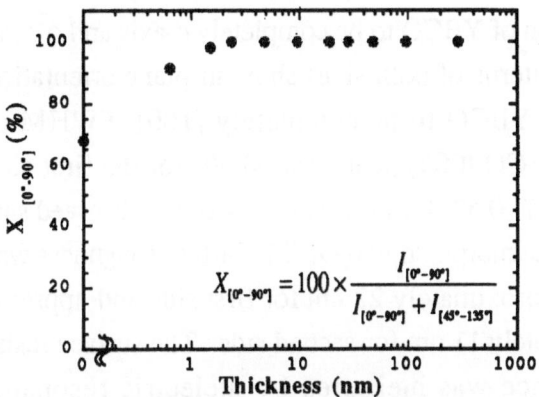

$$X_{[0°-90°]} = 100 \times \frac{I_{[0°-90°]}}{I_{[0°-90°]} + I_{[45°-135°]}}$$

Fig. 1. The BSO thickness dependence of YBCO in-plane orientation

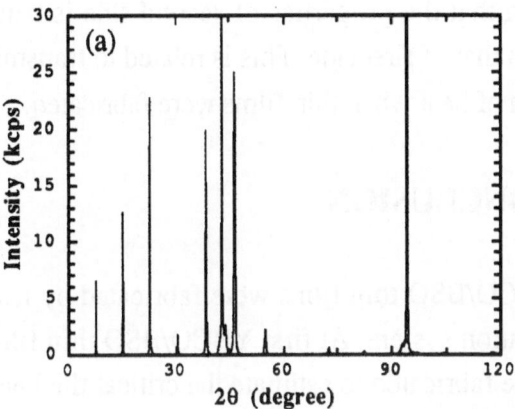

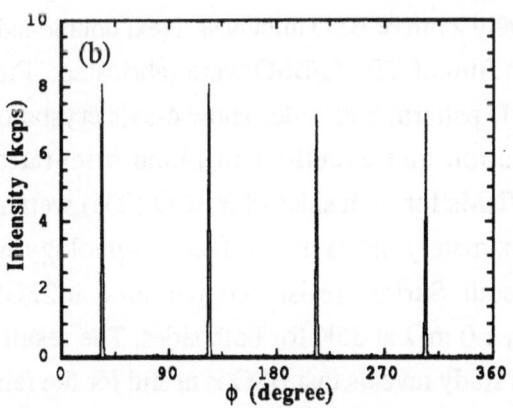

Fig. 2. XRD patrtern of YBCO /BSO thin film fabricated on first side
(a)2θ/θ (b)φ-scan

θ patterns of both sides show that crystal orientation of YBCO to be completely c-axis and φ-scan patterns of both sides show in-plane orientation of YBCO to be completely [100]. FWHM of YBCO (005) peak was ~0.60° for the first side and ~0.52° for the second. Next, we observed surface morphology by AFM. Surface roughness was approximately 22 nm for first side and approximately 11 nm for second side. The surface resistance was measured by dielectric resonator method. Measurement frequency was set to 22 GHz. Figure 4 shows the temperature dependence of surface resistance for double-sided YBCO/BSO film. Surface resistances of both side were approximately 1.0 mΩ at 35K. From these results it is seen that the properties of second side is better than that of first side. This is related to transmission of heat when thin films were fabricated.

CONCLUSION

YBCO/BSO thin films were fabricated by laser ablation system. At first YBCO/BSO thin films were fabricated to estimate the critical thickness of BSO for in-plane orientation of YBCO. In-plane orientation of YBCO was excellent for approximately 2 nm of BSO thickness. Next double-sided thin film of YBCO/BSO were fabricated. From XRD pattern, both sides show c-axis crystal orientation and excellent in-plane orientation. FWHMs for both sides of YBCO (005) were approximately the same. Surface morphology was smooth. Surface resistence measured at 22GHz was 1.0 mΩ at 35K for both sides. The result of this study reveals that BSO is useful for the fabrication of double-sided YBCO/BSO thin film.

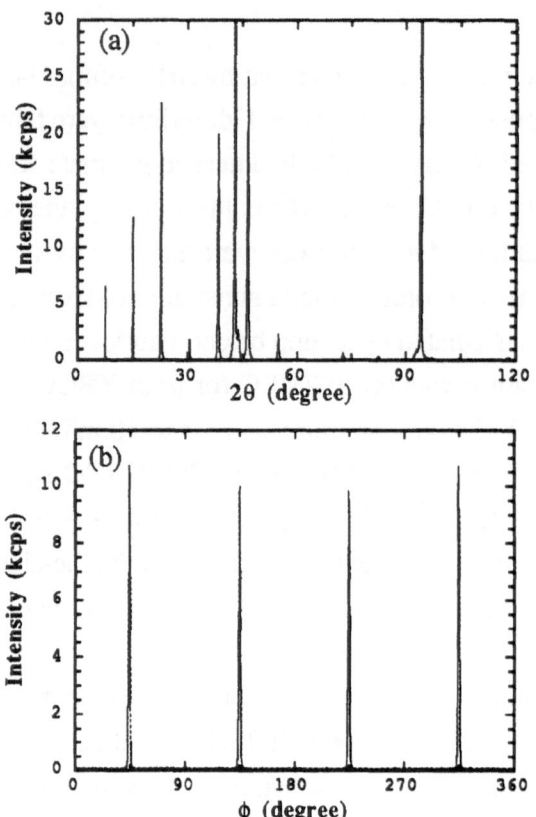

Fig. 3. XRD pattern of YBCO /BSO/MgO thin films fabricated on second side (a)2θ/θ (b)φ-scan

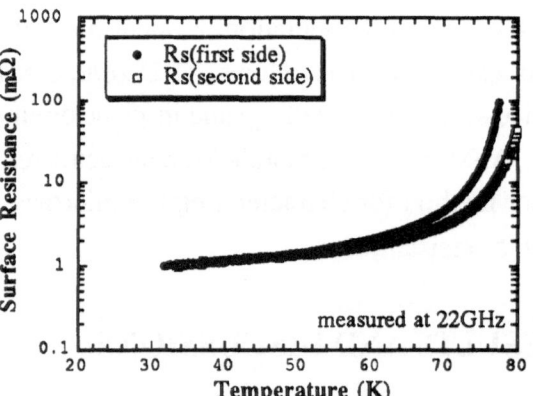

Fig. 4. Surface resistance of double-sided YBCO/ BSO thin film

[1]M.Mukaida, Y.Takano, K.Chiba, M.Kusunoki, and S.Ohshima: Jpn. J. Appl. Phys. 38 (1999)
[2]S.S.Laderman, R.C.Taber, R.D.Jacowitz, J.L.Moll, C.B.Eom, T.L.Hylton, A.F.Marshall, T.H.Geballe, and M.R.Beasley: Phys. Rev. B 43 (1991)
[3]M.Mukaida, Y.Takano, K.Chiba, T.Moriya, M.Kusunoki, and S.Ohshima: Jpn. J. Appl. Phys. 38 (1999)

Preparation of YBCO Films by CP-Process for HTS Microwave Filters

Toshiya Kumagai,[1] Takaaki Manabe,[1] Iwao Yamaguchi,[1] Susumu Nakamura,[1] Wakichi Kondo,[1] Susumu Mizuta,[1] Fumikazu Imai,[2] Kyohei Murayama,[2] Akira Shimokobe,[2] and Yoon-Myung Kang[2]

[1]National Institute of Materials and Chemical Research, Tsukuba, Ibaraki 305-8565, Japan
[2]MEC Laboratory, DAIKIN Industries, Ltd., Tsukuba, Ibaraki 305-0841, Japan

Abstract: Double-sided YBCO films were prepared on LaAlO$_3$(001) substrates ($20 \times 20 \times 0.5$ mm^3) by a coating-pyrolysis (CP) process through single annealing in a 80-mm-diameter tube furnace or by means of IR lamp heating. Superconducting transition temperatures about 90 K were obtained for the epitaxial films by dc-resistance measurement. The surface resistance at around 10 GHz was measured by the dielectric resonator method to be 0.6 mΩ at 60 K, which is comparable to those made by other high vacuum deposition techniques. Moreover, we fabricated a 5-pole microwave filter on the epitaxial c-axis YBCO films, for which an insertion loss less than 0.1 dB was obtained at 70 K.

Keywords: Coating-pyrolysis, YBCO films, LaAlO$_3$(001) substrates, Microwave filters

INTRODUCTION

Explosive growth in the wireless industry over the past several years requires new technical innovations. The use of high-temperature superconductor (HTS) microwave filters in cell site base stations is expected to upgrade the quality and reliability of the wireless transmissions because they have lower insertion loss and sharper band edge roll-off than the conventional filters using dielectrics. The HTS filters provide benefits that include improved call quality, flexibility in cell site location and increased system call capacity. For this purpose, mass production of large-area HTS films of high quality is strongly desired.

Coating-pyrolysis (CP) or metal-organic deposition is a chemical solution-based process and is noticed as being promising for fabrication of large-area HTS films with high critical current densities, e.g., J_c's $> 10^6$ A/cm^2 at 77 K [1]. In this paper, we report on the preparation of double-sided YBa$_2$Cu$_3$O$_{7-y}$ (YBCO) films on LaAlO$_3$(001) substrates through single annealing of prefired films and on fabrication of a 5-pole microwave filter. The superconducting and microwave properties of the films and filters were also investigated.

EXPERIMENTAL

A precursor solution of metal acetylacetonates with a molar ratio Y:Ba:Cu=1:2:3 was coated [2] onto one side of LaAlO$_3$(001) substrates (ε=24), $20 \times 20 \times 0.5$ mm^3 in size. The coated films were slowly heated up to about 500 °C in air to burn off most of the organic component. This coating-pyrolysis procedure was repeated several times, typically three times each, for both sides. These double-sided prefired films were annealed at about 770 °C in a 80-mm-diameter tube furnace with resistive heating, or subjected to IR lamp heating, under low oxygen partial pressure, $p(O_2)$, of about 10^{-4} atm, by flowing a gas mixture of Ar and O$_2$, followed by pure O$_2$ treatment. During the low-$p(O_2)$ annealing, the $p(O_2)$ of the outlet gas was monitored by a ZrO$_2$-type oxygen analyzer, thereby the flow rate of inlet gas was controlled. The thickness of the product films was typically 0.5-0.7 μm. The crystallinity and grain alignment of the films were assessed by x-ray diffraction

(XRD) θ-2θ scanning and ϕ-scanning at the center of each side. Some of the films were cut into pieces and subjected to dc-resistance measurement by the conventional four-probe method. Temperature dependence of surface resistance at around 10 GHz was measured by the dielectric resonator method using the TE_{011} and TE_{013} modes. A microstrip line circuit was patterned on one side (the other side being the ground plane) of the epitaxial c-axis YBCO films having smooth surfaces to make a Tchebyscheff type 5-pole filter, of which the microwave characteristics was measured by means of a network analyzer.

RESULTS AND DISCUSSION

A preliminary study showed that the following is necessary for obtaining large-area and double-sided YBCO films: (1) the precursor films should contain some organic component; (2) the $p(O_2)$ of the annealing atmosphere should be uniform and not become too low; and (3) the heating rate

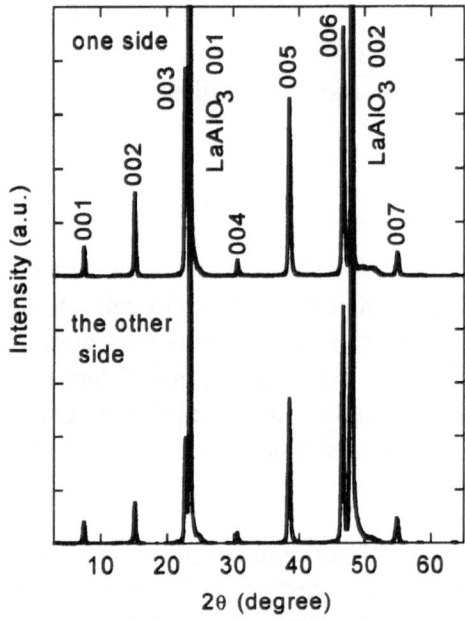

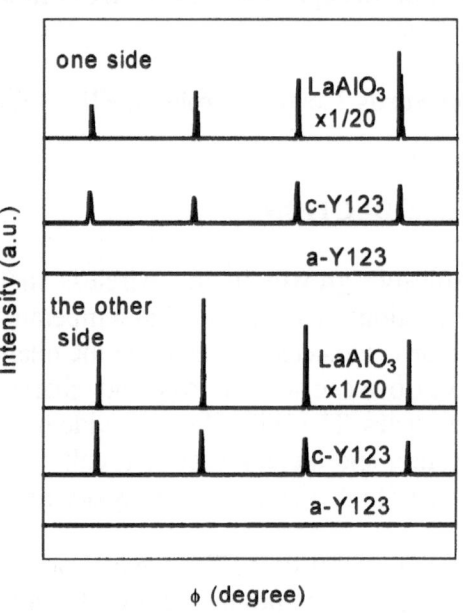

Fig. 1. XRD θ-2θ scans of a double-sided YBCO film. **Fig. 2.** XRD ϕ-scans of a double-sided YBCO film.

should be high enough for epitaxial growth of c-axis-oriented films to occur. The IR lamp heating provides a much higher heating rate (>1000 °C/min) than that of resistive heating (~20 °C/min), however, direct insertion of precursor films into the preheated furnace also provides a heating rate (~100 °C/min) high enough for the epitaxial growth of c-axis films. Thus, both heating methods resulted in epitaxial c-axis-oriented films.

Both sides of the films thus prepared were found to comprise highly c-axis-oriented YBCO grains which have grown epitaxially to the substrate surface, based on XRD θ-2θ scanning and ϕ-scanning, as shown in Figs. 1 and 2, respectively. The dc-resistance

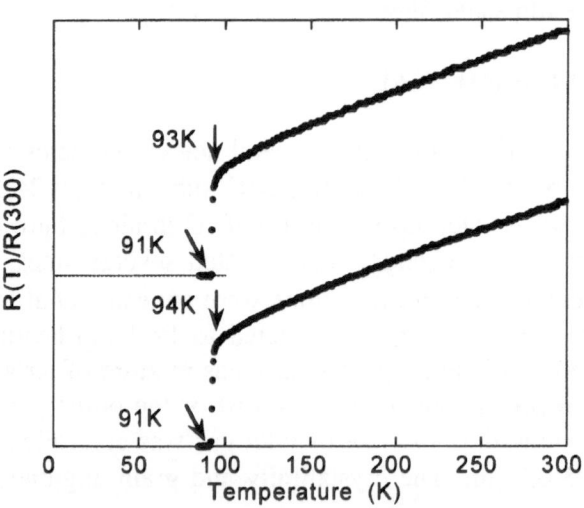

Fig. 3. Temperature dependence of dc-resistance for a double-sided epitaxial YBCO film.

measurement showed that the superconducting transition temperatures of epitaxial YBCO films were quite high, i.e., ~90 K, as typically illustrated in Fig. 3. Moreover, the surface resistance at around 10 GHz was measured to be 0.6 mΩ at 60 K and the optimum value of 0.4 mΩ was obtained at 30 K, Fig. 4. These values are lower than that of copper and comparable to the YBCO films made by other high vacuum deposition methods such as sputtering and so on. Furthermore, we

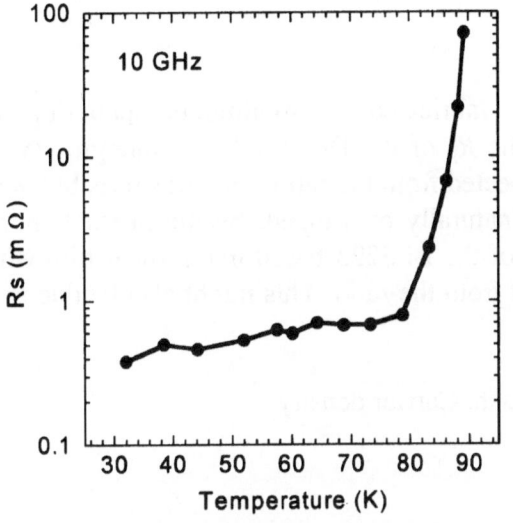

Fig. 4. Temperature dependence of surface resistance for a double-sided epitaxial YBCO film.

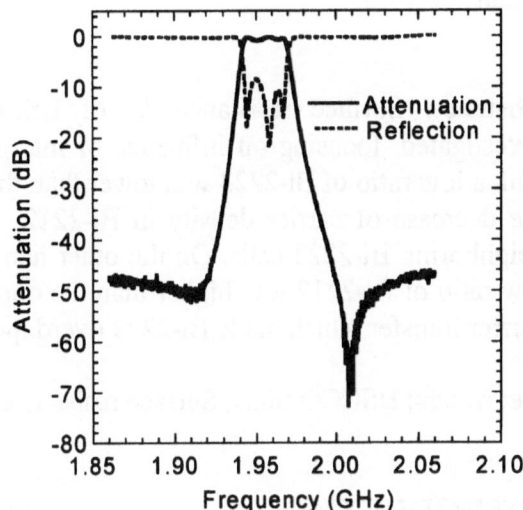

Fig. 5. Frequency response of a filter made on a double-sided epitaxial YBCO film.

fabricated a 5-pole HTS microwave filter, with a center frequency of 1.96 GHz, on the epitaxial *c*-axis YBCO films, for which an insertion loss less than 0.1 dB and a ripple less than 0.8 dB were obtained at 70 K as shown in Fig. 5. Research is in progress to fabricate double-sided YBCO films with a larger area, e.g., a 2-inch diameter, on which microwave filters are to be made.

SUMMARY

Double-sided YBCO films on $LaAlO_3(001)$ substrates ($20\times20\times0.5$ mm^3) were prepared by a CP-process through single annealing in a 80-mm-diameter tube furnace or by means of IR lamp heating. The zero-resistance temperature of the films were around 90 K. The surface resistance at around 10 GHz was measured to be 0.6 mΩ at 60 K, which is comparable to those made by other high vacuum deposition methods. Also, we fabricated a 5-pole HTS microwave filter, of which the insertion loss less than 0.1 dB was obtained at 70 K. Thus, the CP-process provides an effective means for fabricating large-area and double-sided YBCO films over other physical deposition processes in that single heat treatment be necessary at high temperatures.

1. H. Yamasaki, M. Umeda, S. Kosaka, T. Kumagai, T. Manabe, W. Kondo, and S. Mizuta, in *Advances in Superconductivity VI*, edited by T. Fujita and Y. Shiohara (Springer-Verlag, Tokyo, 1994), pp. 885-888.
2. T. Manabe, W. Kondo, S. Mizuta, and T. Kumagai, J. Mater. Res. 9, 858-865 (1994).

Surface Resistance of BSCCO Films and Influence of Intergrowth

Jun Otsuka, [1] Osamu Yamamoto, [2] Yasuaki Sugihara, [2] Toshio Senzaki, [2] Kazushige Ohbayashi, [1] Masumi Inoue, [2] Akira Fujimaki, [2] and Hisao Hayakawa [2]

[1] R & D Center, NGK Spark Plug Co., Ltd., 2808 Iwasaki, Komaki 485-8510, Japan
[2] Department of Quantum Engineering, Nagoya University, Furo-cho, Chikusa-ku, Nagoya 464-8603, Japan

Abstract: Surface resistance R_S of BSCCO films fabricated by multitarget sputtering was investigated, focusing on influence of intergrowth. The R_S of the Bi-2212-based intergrowth film with a low ratio of Bi-2223 was lower than the one expected from the ratio. This was possibly due to the decrease of carrier density in Bi-2212, which is naturally overdoped, by the interaction with neighboring Bi-2223 cells. On the other hand, the R_S of the Bi-2223-based intergrowth film with a low ratio of Bi-2212 was higher than the one expected from the ratio. This might also be due to the carrier transfer which made Bi-2223 overdoped.

Keywords: BSCCO films, Surface resistance, Intergrowth, Carrier density

INTRODUCTION

High-temperature superconducting films having excellent microwave properties are required for highly performing microwave devices such as band-pass filters. Surface resistance (R_S) is one of the most important microwave properties. It is significant to clarify the factors which cause high R_S in order to produce high quality films.

It is not easy to fabricate single-phase Bi-Sr-Ca-Cu-O (BSCCO) films such as $Bi_2Sr_2CuO_x$, $Bi_2Sr_2CaCu_2O_x$ and $Bi_2Sr_2Ca_2Cu_3O_x$ (Bi-2201, Bi-2212 and Bi-2223, respectively). We have the multitarget sputtering technique which gives BSCCO films with excellent superconducting properties and by which even single-phase Bi-2223 films can be obtained [1, 2]. However, even slight difference in chemical composition from a stoichiometric composition allows phase intergrowth with the adjacent members in the Bi-Sr-Ca-Cu-O family. Therefore, we investigated the influence of intergrowth on the R_S of BSCCO films which has not been reported although its effect on critical temperature (T_C) has been studied [3].

EXPERIMENTAL

Film preparation. BSCCO films were prepared by multitarget RF magnetron sputtering which employed $Bi_2Sr_2CuO_x$, Bi_2O_3 and $CaCuO_x$ targets so that $Bi_2Sr_2CuO_x$ and $CaCuO_x$ blocks, of which BSCCO structures consist, were deposited alternatively [1, 2]. BSCCO films were fabricated on MgO (100) substrates with a size of $20 \times 20 \times 0.5$ mm^3 at a substrate temperature of about 750°C. Bi-2201 was deposited to a thickness of about 10 nm as a buffer layer first, and Bi-2212 was subsequently deposited to a thickness of 500-600 nm. The number of $CaCuO_x$ blocks per unit cell was controlled with the RF power of the $CaCuO_x$ target in this study. The as-sputtered films were post-annealed at 800-835°C for 2 hours in O_2 in order to homogenize the oxygen content in each film.

Characterization. Crystalline phases of the prepared films were identified by the X-ray diffraction (XRD) method. Since phase intergrowth shifts XRD peaks continuously without peak splits [3, 4], the proportion of Bi-22(n - 1)n and Bi-22n(n + 1) (n = 1 – 3) were expediently determined from the peak shifts [5]. The R_S of the films was measured at about 20 GHz and 20 to 80 K by the dielectric resonator method using a sapphire single crystal as a dielectric rod [6].

RESULTS AND DISCUSSION

Typical temperature dependencies of R_S of the Bi-2212 / Bi-2223 films are shown in Fig. 1. The single phase Bi-2223 film had the most excellent R_S which was 0.5 mΩ at 30 K and 2 mΩ at 76 K at 20 GHz. This was nearly equal to the R_S of Y-Ba-Cu-O [7, 8], considering the frequency dependence of R_S [9]. The film with a higher ratio of Bi-2223 had lower R_S as shown in Fig. 1, which may be natural since Bi-2223 had lower R_S than Bi-2212. The R_S values of Bi-2212-based films are plotted against the ratio of Bi-2223 r_{23} in Fig.2. The R_S values shown as a dotted line and a broken line in this figure were calculated by a series circuit model and a parallel circuit model, respectively, using the R_S of single phase Bi-2212 and Bi-2223 films. The former model produces high R_S and the latter low R_S, of course. In the intergrowth films, Bi-2212 and Bi-2223 cells must have been connected with each other in a series-parallel mixed mode in ab plane which corresponds to the rf current direction. Although the R_S of the Bi-2212-based intergrowth films was expected to be between the both estimated R_S values, it was even lower than the R_S calculated by the parallel circuit model. The J_C also had the same tendency as the R_S and increased with an increase of r_{23} more than predicted from r_{23}. The mechanism of this improvement of R_S and J_C by the intergrowth with Bi-2223 was considered as follows.

It is generally considered that Bi-2223 is almost optimum in carrier density while Bi-2212 is naturally overdoped, which means that Bi-2223 has less holes per CuO_2 planes than Bi-2212. In the Bi-2212 / Bi-2223 intergrowth films, Bi-2212 and Bi-2223 cells were piled up randomly and mixed with each other in a manner of lattice-by-lattice. Consequently, each Bi-2223 cell was usually located next to Bi-2212 cells in the films with low r_{23}, so that the hole numbers per CuO_2 plane in Bi-2212 and Bi-2223 possibly got close to each other. Hatano et al. has had the similar opinion and reported that the T_C of BSCCO intergrowth films changed continuously with composition because

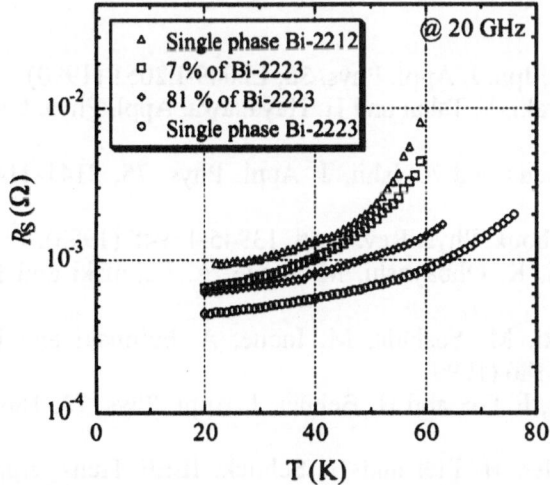

Fig. 1. Temperature dependencies of surface resistance R_S of the Bi-2212 / Bi-2223 films.

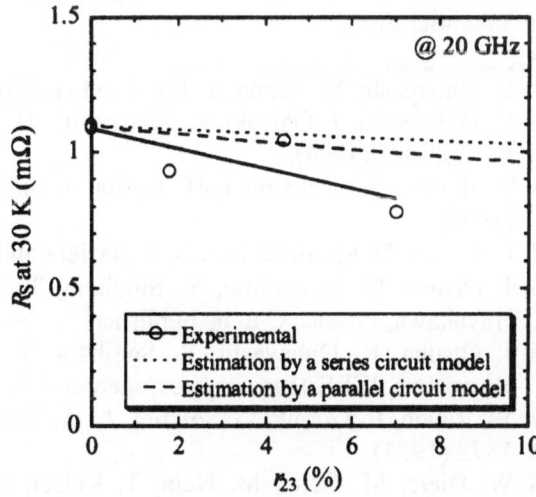

Fig. 2. Relation between the R_S of the Bi-2212-based intergrowth films and the ratio of Bi-2223 r_{23}.

the carrier concentration was averaged between the two phases in the intergrowth by the carrier diffusion [3]. Accordingly, the decrease of carrier density in Bi-2212 might result in the improvement of R_S.

Although this consideration can be confirmed by the R_S characterization over a wide range of chemical composition, it has not been sufficiently investigated yet. The verification was tried from a few data of Bi-2223-based films. The relation between the R_S of Bi-2223-based films and the ratio of Bi-2212 r_{12} is shown in Fig.3. A dotted line and a broken line in this figure are the same as those in Fig. 2. The Bi-2223-based intergrowth films had even higher R_S than the calculation by the series circuit model which estimates rather high R_S. This might also be due to the carrier transfer which made Bi-2223 overdoped. However, further investigation over a wide range of chemical composition is necessary for the confirmation of this consideration as mentioned above.

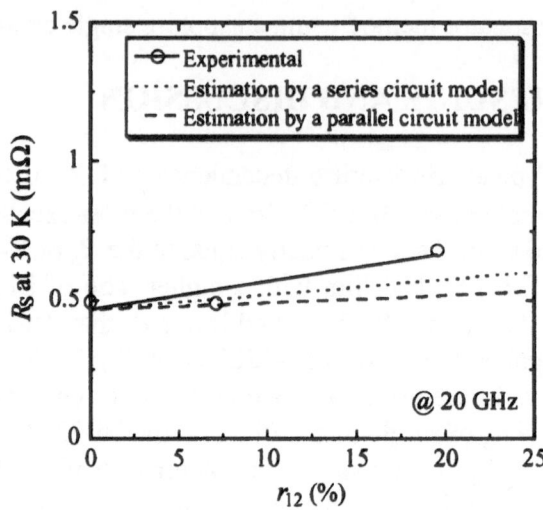

Fig. 3. Relation between the R_S of the Bi-2223-based intergrowth films and the ratio of Bi-2212 r_{12}.

CONCLUSION

Surface resistance R_S of BSCCO films fabricated by the multitarget sputtering was investigated, focusing on the effects of intergrowth on R_S. The R_S of the single phase Bi-2223 film was 0.5 mΩ at 30 K and 2 mΩ at 76 K at 20 GHz, which was nearly equal to that of Y-Ba-Cu-O. The R_S of the Bi-2212-based intergrowth film with a low ratio of Bi-2223 was lower than the one expected from the ratio. This was possibly due to the decrease of carrier density in Bi-2212, which is naturally overdoped, by the interaction with neighboring Bi-2223 cells. On the other hand, the R_S of the Bi-2223-based intergrowth film with a low ratio of Bi-2212 was higher than the one expected from the ratio. This might also be due to the carrier transfer which made Bi-2223 overdoped. However, further investigation over a wide range of chemical composition is necessary for the confirmation of this consideration.

1. K. Ohbayashi, M. Anma, Y. Takai and H. Hayakawa, Jpn. J. Appl. Phys. **35**, L2049-L2051 (1990).
2. K. Ohbayashi, T. Ohtsuki, H. Matsushita, H. Nishiwaki, Y. Takai and H. Hayakawa, Appl. Phys. Lett. **64**, 369-371 (1994).
3. T. Hatano, K. Nakamura, H. Narita, J. Sato, S. Ikeda and A. Ishii, J. Appl. Phys. **75**, 2141-2148 (1994).
4. L. Ranno, D. Martínez-García, J. Perrière and P. Barboux, Phys. Rev. B **48**, 13945-13948 (1993).
5. J. Otsuka, O. Yamamoto, Y. Sugihara, T. Senzaki, K. Ohbayashi, M. Inoue, A. Fujimaki and H. Hayakawa, Physica C to be published.
6. J. Otsuka, K. Ohbayashi, Y. Sugihara, T. Senzaki, M. Yoshida, M. Inoue, A. Fujimaki and H. Hayakawa, IEEE Trans. Appl. Supercond. **9**, 2183-2186 (1999).
7. W. Rauch, E. Gornik, G. Sölkner, A. A. Valenzuela, F. Fox and H. Behner, J. Appl. Phys. **73**, 1866-1872 (1993).
8. W. Diete, M. Getta, M. Hein, T. Kaiser, G. Müller, H. Piel and H. Schlick, IEEE Trans. Appl. Supercond. **7**, 1236-1239 (1997).
9. T. Van Duzer and C. W. Turner, "Principles of superconductive devices and circuit", Elsevier North Holland, Inc. (1981).

New Superconducting Lead Cuprates Prepared by Molecular Beam Epitaxy

Michio Naito and Shin-ichi Karimoto

NTT Basic Research Laboratories, 3-1, Morinosato-Wakamiya, Atsugi-shi, Kanagawa 243-0198, Japan

Abstract: We report the synthesis of a new superconducting lead cuprate $PbSr_2CuO_{5+\delta}$ (Pb-1201) with a pure PbO charge reservoir block using molecular beam epitaxy. The key to the successful synthesis of Pb-1201 is low-temperature growth and an appropriate choice of substrate materials. We have also been trying to synthesize higher members of this series, but this has not been straightforward partly because the growth temperature is limited by the volatility of Pb and PbO_x.

Keywords: Pb-based superconductors, MBE, epitaxy, quai-stable phase, low-temperature growth

INTRODUCTION

MBE for the next strategic material search

Since the discovery of high-T_C superconductors, most of the initial efforts to search for new superconductors have been made by means of conventional solid-state reactions. In the early 1990's, a new technique was introduced, namely high-pressure or high-oxygen-pressure synthesis, and this produced a number of new materials [1-3]. Recently, however, there have been fewer and fewer discoveries of new superconductors by bulk synthesis. So new synthetic routes are required for the next strategic material search. In 1996, we began to use molecular beam epitaxy (MBE) to search for new materials. This approach has several important advantages. The first is low temperature growth. In MBE growth, the reactants are atoms and the oxidation is initiated on a substrate, therefore the reaction temperature can be greatly lowered to ~500°C. This situation can be compared with conventional bulk synthesis where the reactants are relatively stable oxides or carbonates in the form of micron size particles which require temperatures as high as ~1000°C to initiate a chemical reaction. Low-temperature synthesis is important for cuprates because Cu-O bonds are intrinsically weak. The second advantage is epitaxy itself, which has the potential to provide quasi-stable phases. The other advantages of MBE are strong oxidation through the use of activated oxygen, and its contamination free environment which is important for chemical reactions involving Ba [4].

Pb cuprates

The crystal structure of superconducting copper oxides can be simply viewed as the stacking of two-dimensional CuO_2 planes and a charge reservoir block, and classified as a simple scheme, namely the "homologous series concept" [5]. The search for new superconductors is, in other words, the search for new charge reservoir blocks. Empirically, charge reservoir blocks containing heavy cations have high T_C's. It should be noted that Hg-, Tl-, and Bi-based superconductors are widely known, whereas Pb-based superconductors are not. Hg- and Tl-based superconductors contain toxic elements, and Bi-based superconductors have strong two-dimensionality, which is an obstacle to magnetic wire applications. This has led to the desire to develop Pb-based superconductors and there have been a large number of studies on lead cuprates. The following are important in relation to our present work. In 1990, Adachi *et al.* and Sasakura *et al.* discovered $(Pb,Cu)(Sr,La)_2CuO_{5+\delta}$ ((Pb,Cu)-1201) by conventional bulk synthesis [6,7]. This material has a $(Pb_{0.5}Cu_{0.5})$ block and contains La substituted for about half the Sr sites. In 1994, Yamauchi discovered "so-called" Pb-1212 and Pb-1223 by high-oxygen-pressure synthesis [8]. In this case, the charge reservoir blocks are reported to be $(Pb_{0.5}Cu_{0.3}Sr_{0.2})$

and ($Pb_{0.5}Cu_{0.2}Sr_{0.3}$), respectively [9]. In spite of these many studies, there has been no report on the synthesis of lead cuprates with a pure PbO block. There is one main problem as regards synthesizing lead cuprates. This is the formation of the simple perovskite $SrPbO_3$, which is a thermodynamically stable phase at high temperature [10]. As we describe later in detail, we found the choice of substrate material to be crucial in terms of avoiding this problem. The use of an $LaAlO_3$ substrate makes the quasi-stable Pb-1201 phase very easy to obtain.

EXPERIMENTAL

We grew thin films in a custom-designed UHV chamber for reactive coevaporation with precise rate control for each element. Briefly, all the elements were deposited from metal sources by multiple e-gun evaporators. The beam flux rate was controlled by electron impact emission sensors. Non-distilled ozone gas or RF activated atomic oxygen was used for oxidation. The film growth was monitored by reflection high energy electron diffraction (RHEED). The technological aspects of our MBE growth method is described in detail elsewhere [11].

RESULTS AND DISCUSSION

Synthesis of Pb-1201

Low temperature growth is a prerequisite for Pb cuprate film synthesis owing to the volatility of Pb and also PbO_x. This is unlike the growth of Bi-based superconductors, where metal Bi is volatile but BiO_x is not. Figure 1 shows the sticking coefficient of Pb for films grown at various temperatures as estimated from electron probe microanalysis (EPMA). The sticking coefficient is approximately 1 below a substrate temperature of 550 ˚C, but 0.25 at 580 ˚C and almost zero above 600 ˚C. The cation composition of films grown below 550˚C was close to the set ratios of the atomic beam fluxes. All the Pb-1201 films were prepared with atomic beam flux ratios of Pb : Sr : Cu = 1 : 2 : 1 with slightly (10~20 %) rich in Pb and Cu.

As mentioned above, the choice of substrates is crucial in terms of obtaining the Pb-1201 phase. To demonstrate this, Fig. 2 shows the X-ray diffraction (XRD) patterns for 75 nm thick films on (a) $SrTiO_3$ (001) (a_0=0.3905 nm) and (b) $LaAlO_3$ (001) (a_0=0.3788 nm). They show a striking contrast between films on $SrTiO_3$ and $LaAlO_3$. The film grown on $SrTiO_3$ was dominantly $SrPbO_3$, whereas the film

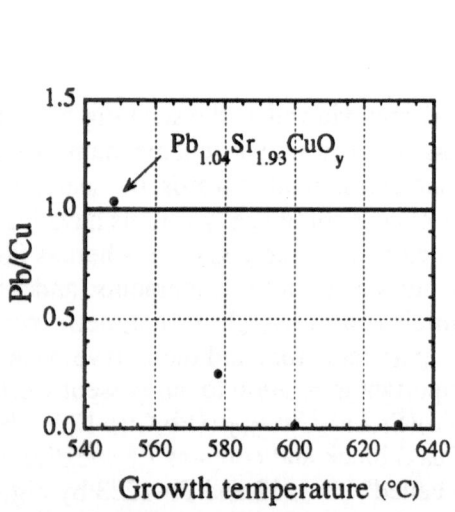

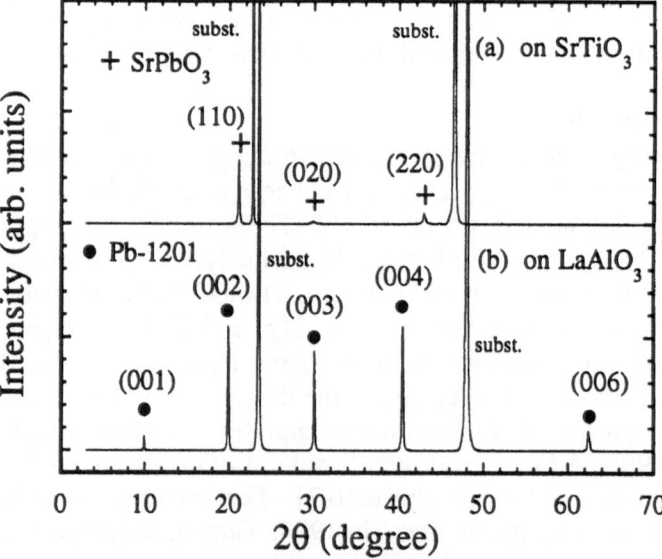

Fig. 1. Sticking coefficient of Pb for films grown at various temperatures as estimated from EPMA.

Fig. 2. Comparison of X-ray diffraction patterns for films grown on (a) $SrTiO_3$ and (b) $LaAlO_3$.

grown on LaAlO$_3$ was single-phase c-axis oriented Pb-1201 ($c_0 = 0.893$ nm). The in-plane lattice constant of Pb-1201 was 0.381 nm as estimated using a four-circle X-ray diffractometer and also a transmission electron microscope (TEM) [12]. The lattice constant of SrPbO$_3$ is 0.415 nm. The result seems to indicate the importance of lattice matching between the substrate and Pb-1201 as regards obtaining this quasi-stable phase. However, this statement is not entirely accurate.

We have examined most commercially available substrates. Figure 3 summarizes the results. It shows that Pb-1201 formed with any substrate whose lattice constant is smaller than 0.385 nm. Surprisingly this phase even formed with YAlO$_3$ substrates, for which the lattice mismatch is as large as for SrTiO$_3$ although its sign is opposite. So it may be said that substrates with lattice constants smaller than 0.385 nm prevent the growth of thermodynamically stable SrPbO$_3$, and thus promote the growth of quasi-stable Pb-1201.

Next we discuss the carrier doping mechanism of this system. Our in situ XPS showed that Pb in our Pb-1201 films was tetravalent. If the

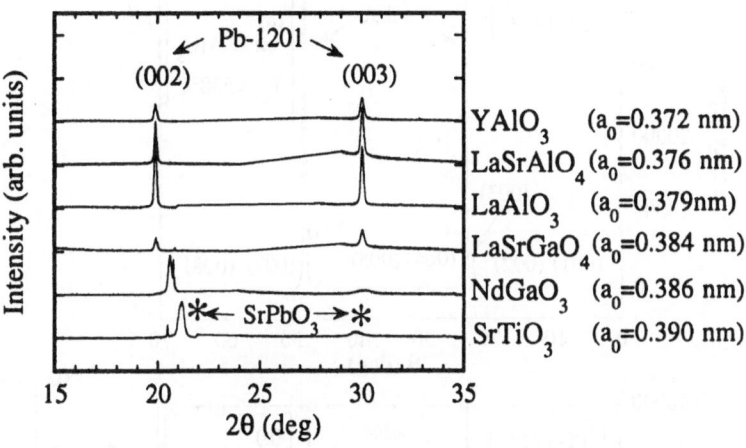

Fig. 3. X-ray diffraction patterns for films grown on various substrates. Diffraction peaks from substrates are subtracted.

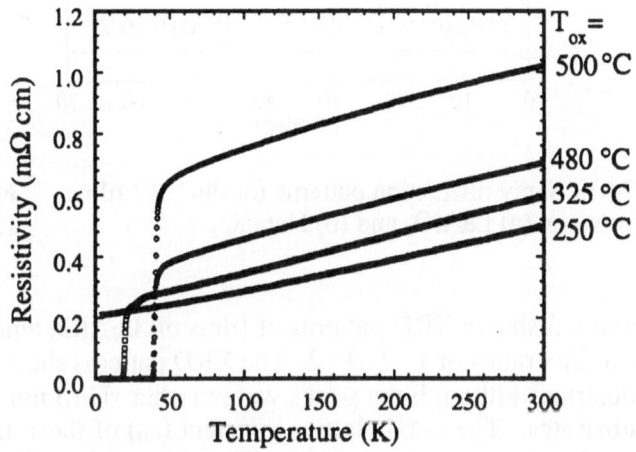

Fig. 4. Resistivity-versus-temperature curves of Pb-1201 films oxidized at various T_{ox}.

amount of excess oxygen (δ) is zero, there is no carrier in this system. So the carrier doping of this system seems to be caused by extra oxygen. We think that this extra oxygen was introduced during both the film growth and cool-down processes. In our MBE growth method, we use ozone gas or atomic oxygen to oxidize the cation elements in a high vacuum. Indeed, the Pb-1201 phase did not form when we used molecular oxygen. Ozone gas or atomic oxygen not only promotes the formation of the Pb-1201 phase with high valence Pb and Cu, but also appears to introduce excess oxygen. The amount of excess oxygen increases if the ozone gas supply is maintained during cool-down as we have already demonstrated for superconducting La$_2$CuO$_{4+\delta}$ growth [13]. This behavior is shown in Fig. 4. Here T_{ox} is defined as the temperature at which the ozone gas supply is stopped during the post-growth cooling procedure. As T_{ox} became lower, both the resistivity and T_C of films became lower, and they exhibited overdoped characteristics. The best data showed $T_C = 40.2$ K with a transition width of ~2 K, and $\rho_{(300K)} = 480$ $\mu\Omega$ cm. We also confirmed the superconductivity by DC susceptibility measurement.

Synthesis of Pb-12(n-1)n
Following our successful Pb-1201 growth, we began our attempts to synthesize higher members of this homologous series and obtained preliminary results. However, the synthesis has not been straightforward.

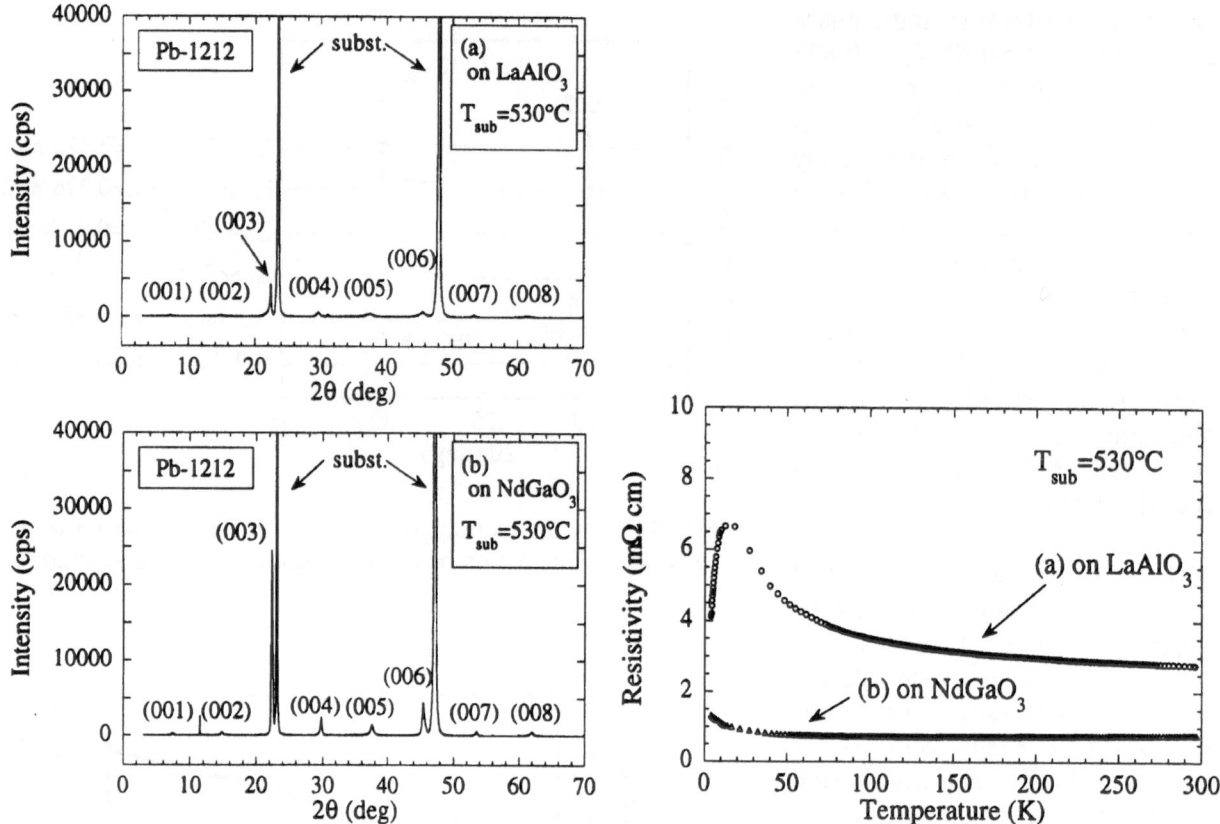

Fig. 5. X-ray diffraction patterns for Pb-1212 films grown on (a) LaAlO₃ and (b) NdGaO₃.

Fig. 6. Resistivity-versus-temperature curves of Pb-1212 films on (a) LaAlO₃ and (b) NdGaO₃.

Figure 5 shows XRD patterns of films on LaAlO₃ and NdGaO₃ substrates that were grown at 530 °C with flux ratios of 1 : 2 : 1 : 2. The XRD patterns show that the c-axis oriented Pb-1212 formed on both substrates although the peaks were neither sharp nor strong, especially for the films on the LaAlO₃ substrates. The c-axis lattice constant (c_0) of these films is 1.203 ± 0.001 nm. Figure 6 shows the resistivity-versus-temperature (ρ-T) curves of these Pb-1212 films. They were slightly semiconducting at low temperatures, and did not show zero resistance.

So far we have only examined detailed growth conditions only for films on LaAlO₃. Figure 7 shows results demonstrating the growth temperature dependence of the XRD patterns. The XRD pattern at 530 °C is the same as that shown in Fig. 5. At 570°C, the film seemed to be a-axis oriented. The a-axis lattice constant (a_0) of this film is 0.388 nm. At this temperature, the sticking coefficient of Pb was less than 1, so we doubled the Pb deposition rate. The composition of the resultant film was close to 1 : 2 : 1 : 2. At 650°C, the film lost its Pb completely, and consisted of 1D-CuO-chain material (Sr,Ca)₂CuO₃. Here the film grown at 570°C is interesting. At first we thought that this film would be a-axis oriented Pb-1212, but it was not.

Figure 8 shows planar TEM images of a-axis oriented thin film grown on LaAlO₃ substrates. The film was actually a-axis oriented with 90 degree grain boundaries. The typical grain size was about 30 nm. However, the c-axis periodicity in this film was not ordered. In this micrograph, Pb-1234 sequences are frequently observed, but Pb-1212 sequences very rarely. At present, we cannot explain the reason for the selective growth of Pb-1234 grains. We simply note that the c_0 for Pb-1234 is very close to five times the lattice constant of LaAlO₃. Generally, the ρ-T curves of these a-axis oriented Pb cuprates films are not good. Figure 9 shows a typical ρ-T curve of an a-axis oriented Pb cuprate film. The data is reminiscent of those of early a-axis oriented YBCO films. Although the onset temperature is 90 K,

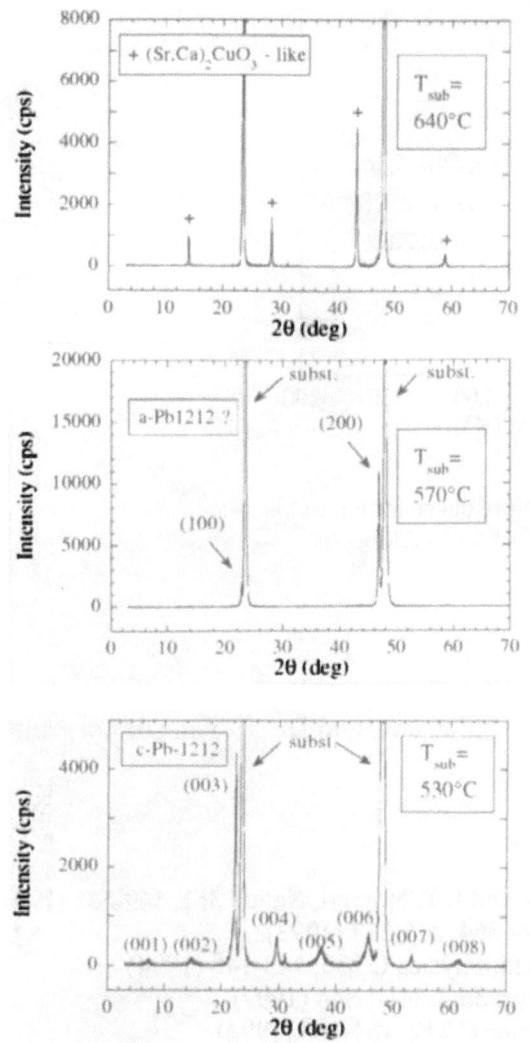

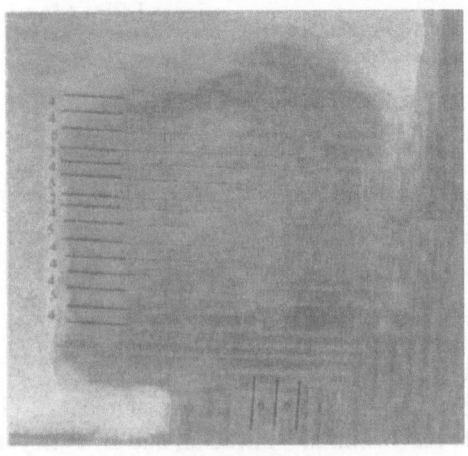

Fig. 8. Planar TEM images of an *a*-axis oriented thin film grown on LaAlO₃ at $T_s = 570$ °C. The lower panel shows a magnified image of the upper left part of the upper panel. The *c*-axis periodicity is not ordered in this film.

Fig. 7. X-ray diffraction patterns for films grown on LaAlO₃ substrates at various growth temperatures.

there is a long tail, and zero resistivity is attained below 20 K. The resistivity value is also high.

SUMMARY

We have succeeded in synthesizing the new superconductor Pb-1201 by molecular beam epitaxy. This new cuprate is thermodynamically quasi-stable phase, and has not been obtained by bulk synthesis. This material seems to have a pure PbO charge reservoir block, and should be compared with the previously reported (Pb,Cu)-1201. Pb-1201 obtained in the present work has a T_C above 40 K, whereas (Pb,Cu)-1201 has a $T_C \sim$ 30 K. This result seems to follow the empirical rule "simpler is higher". The key to obtaining this quasi-stable phase was epitaxy using a lattice-matched substrate so as to avoid the formation of the thermodynamically stable SrPbO₃ phase. The substrate field changed the chemical reaction. The synthesis of higher members of this series is under way, but is not straightforward partly because the growth temperature is limited owing to the volatility of Pb and PbOₓ. Finally, we would like to recommend the Pb homologous series as a practical material because the series is less toxic than the Hg-, Tl- series, and less anisotropic than the Bi-series.

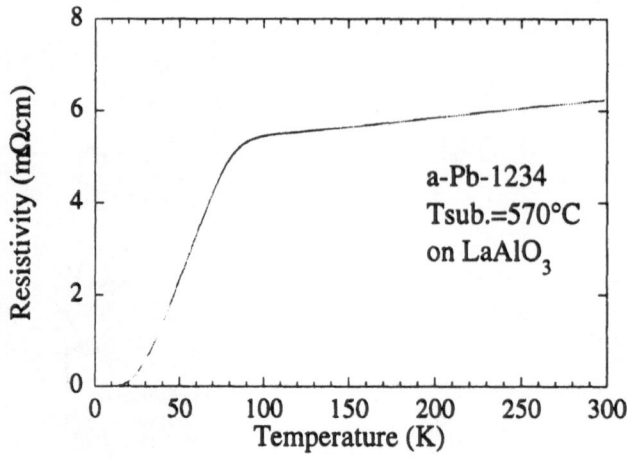

Fig. 9. Resistivity-versus-temperature curve for an *a*-axis oriented film (dominantly Pb-1234) on LaAlO$_3$.

Acknowledgments

The authors thank Dr. H. Yamamoto, Dr. H. Sato, Dr. A. Matsuda, and Dr. T. Yamada for fruitful discussions.

1. M. G. Smith, A. Manthiram, J. Zhou, J. B. Goodenough, and J. T. Markert, Nature **351**, 549-551 (1991).
2. Z. Hiroi, M. Takano, M. Azuma, and Y. Takeda, Nature **364**, 315-317 (1993).
3. T. Kawashima, Y. Matsui, and E. Takayama-Muromachi, Physica C **233**, 143-148 (1994).
4. H. Yamamoto, M. Naito, and H. Sato, Jpn. J. Appl. Phys. **36**, L341-L344 (1997)
5. S. Adachi, H. Yamauchi, S. Tanaka, and N. Mori, Physica C **212**, 164-168 (1993).
6. S. Adachi, K. Setsune, and K. Wasa, Jpn. J. Appl. Phys. **29**, L890-L892 (1990).
7. H.Sasakura, K. Nakahigashi, A. Hirose, H. Teraoka, S. Minamigawa, K. Inada, S. Noguchi, and K. Okuda, Jpn. J. Appl. Phys. **29**, L1628-L1631 (1990).
8. H. Yamauchi, T. Tamura, X.-J. Wu, S. Adachi and S. Tanaka, Jpn. J. Appl. Phys. **34**, L349-L351 (1995).
9. X.-J. Wu, T. Tamura, S. Adachi, C.-Q. Jin, T. Tatsuki, and H. Yamauchi, Physica C **247**, 96-104 (1995).
10. R. J. Cava *et al.*, Nature **336**, 211-214 (1988).
11. M. Naito, H. Sato, and H. Yamamoto, Physica C **293**, 36-43 (1997).
12. S. Karimoto, and M. Naito, Jpn. J. Appl. Phys. **38**, L283-L285 (1999).
13. H. Sato, M. Naito, and H. Yamamoto, Physica C **280**, 178-186 (1997).

Effects of Chemically Etched $YBa_2Cu_3O_x$ Surface on Homoepitaxial Growth

Hideaki Zama, Nobue Tanaka, Keiichi Tanabe, and Tadataka Morishita

Superconductivity Research Laboratory, ISTEC,
10-13 Shinonome 1-Chome, Koto-ku, Tokyo 135-0062, Japan

Abstract: We prepared homoepitaxial films on chemically etched (001)$YBa_2Cu_3O_x$ (YBCO) single-crystal substrates, examining the effect of etching applied to their surfaces. It was difficult to grow high-quality YBCO films on deeply etched YBCO substrates, because no epitaxial layer grew laterally on these substrates. By depositing one molecular layer of Cu oxide on the etched surface, the homoepitaxial films showed good crystallinity of $\chi_{min} = 4.3\%$ and surface morphology with {100} facets.

Keywords: $YBa_2Cu_3O_x$ single-crystal substrate, chemical etching, homoepitaxial growth, Cu oxide layer, atomic force microscopy, Rutherford backscattering spectrometry

INTRODUCTION

We have prepared (001)$YBa_2Cu_3O_x$ (YBCO) single-crystal substrates chemically etched to remove degradation on their surfaces caused by mechanical polishing and/or exposing in humid atmosphere for adopting them to grow homoepitaxial films. The (001) surface deeply (60sec) etched by 0.003wt%HCl/methanol shows a well-defined surface consisted of a sharp step-and-terrace feature with unit-cell-high steps. The terrace is terminated by the same kind atomic layer composing the YBCO layer structure along the c-axis. The step-and-terrace feature maintains after thermal treatments in the atmosphere corresponding to film deposition.[1,2]

In this paper, we report the results of growing Y, Ba, and Cu oxides and then homoepitaxial layers on the etched YBCO substrates.

EXPERIMENTAL

The YBCO substrates were prepared from YBCO single crystals grown by a modified top-seeded pulling method,[3] following slicing into (001) platelets and then mechanical polishing with colloidal silica. The chemical etching was applied to the substrates using 0.003wt%HCl/methanol solution as an etchant; After dipping into the etchant, the substrates were rinsed with dehydrated methanol and dried in a blow of nitrogen. The etching rate was estimated to be approximately 0.025 nm/sec at room temperature. The dipping time was 60 sec.

The metallic oxide and YBCO films were grown by metalorganic chemical vapor deposition (MOCVD). The MOCVD experimental condition is reported in our previous paper.[1]

The surface morphology was examined using an atomic force microscope (AFM). Rutherford backscattering spectrometry (RBS) was measured to evaluate the crystallinity of films using He^+ beams of 950 keV.

RESULTS AND DISCUSSION

Figure 1 shows AFM pictures of one molecular layer films after Y, Ba, and Cu were separately deposited in an oxygen atmosphere of a total pressure of 10 Torr and an oxygen partial pressure of 0.6 Torr at a substrate susceptor temperature of 825°C, which substrate surface temperature was estimated to be approximately 700°C, corresponding to YBCO film deposition, on the 60sec-etched YBCO substrates. Y and Ba oxides showed granules on terraces, while Cu oxide showed a sharp step-and-terrace feature remained, indicating that Cu oxide grew laterally on the surface.

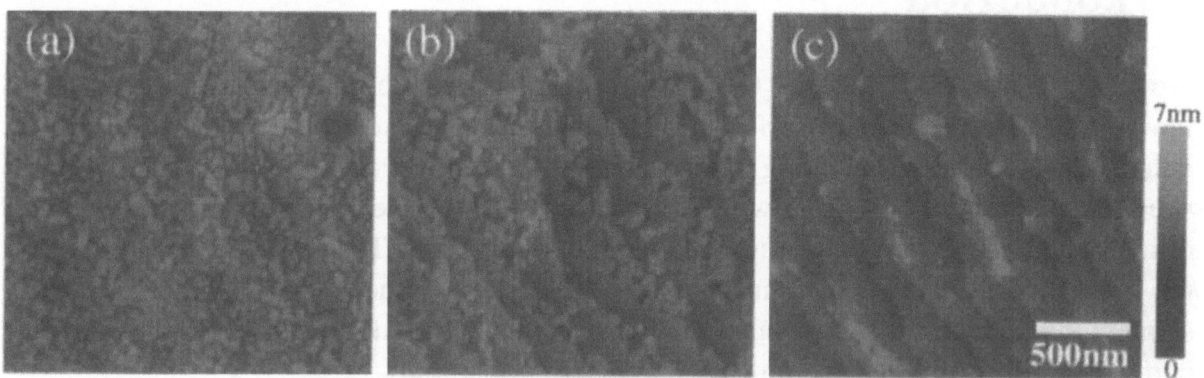

Fig. 1. AFM images of one molecular layer of (a) Y, (b) Ba, and (c) Cu oxides deposited on etched (001)YBCO single-crystal substrates.

In Fig.2(a), random nucleation and dendrite-like edges appear when a 20nm-thick YBCO film was grown on an as-polished substrate. A similar morphological image is present for the film grown on the 60sec-etched substrate as shown in Fig.2(b). The film grown on the 60sec-etched substrate had a poor crystallinity to be a χ_{min} of 10% for RBS channeling measurements.

The surface morphology and crystallinity of homoepitaxial film grown on the 60sec-etched substrate were improved by depositing the Cu oxide prior to YBCO deposition. It resulted that the {100} facets and sharp step edges were observed in Fig.2(c). The channeling yield was also improved to be a χ_{min} of 4.3%.

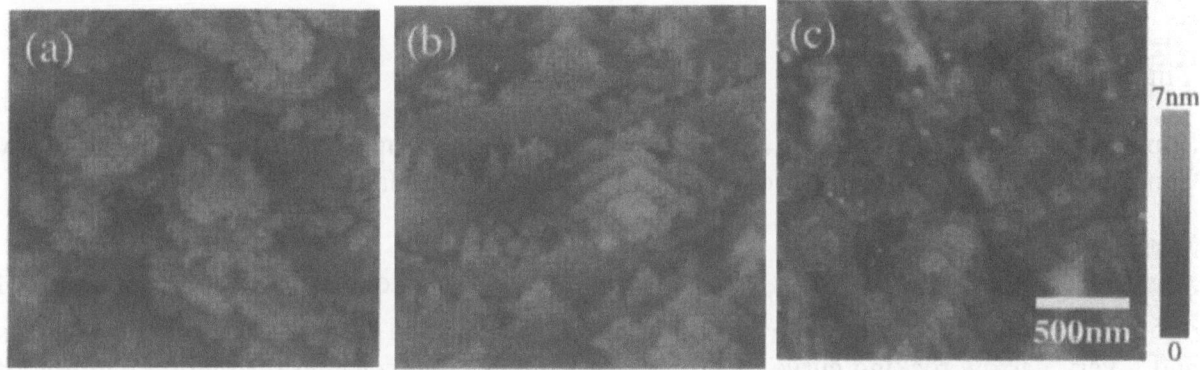

Fig. 2. AFM images of homoepitaxial films grown on (a) as-polished, (b) 60sec-etched, and (c) 60sec-etched and one molecular Cu-oxide pre-deposited surfaces of (001)YBCO single-crystal substrates.

CONCLUSIONS

The effect of chemical etching on homoepitaxial growth was examined for the (001)YBCO single-crystal substrate. One molecular layer of Cu oxide pre-deposited on the deeply 60sec-etched YBCO surfaces was effective to grow homoepitaxial layers, having better crystallinity than those grown on the 60sec-etched surfaces directly.

Acknowledgments: The authors wish to thank Y. Shiohara and S. Koyama for providing YBCO single crystals and F. Wang for RBS measurements. This work was supported by the New Energy and Industrial Technology Development Organization (NEDO) as Collaborative Research and Development of Fundamental Technologies for Superconductivity Applications under the New Sunshine Program administered by the Agency of Industrial Science and Technologies (AIST) of the Ministry of International Trade and Industry (MITI) of Japan.

1. N.Tanaka, H.Zama, T.Morishita, and H.Yamamoto, Jpn. J. Appl. Phys. **38**, L731 (1999).
2. H.Zama, N.Tanaka, K.Tanabe, and T.Morishita, *Advances in Superconductivity XI*, edited by N. Koshizuka and S. Tajima (Springer-Verlag, Tokyo, 1999), p.1113.
3. Y.Yamada and Y.Shiohara, Physica C **217**, 182 (1993).
4. H.Zama, N.Tanaka, K.Tanabe, and T.Morishita, Ext. Abst. 1999 Int. Workshop on Superconductivity, p.68.

Cation Sublattice Disorder and the Formation of Cubic Phase in $NdBa_2Cu_3O_{7-\delta}$ Thin Films

Yijie Li, X.-J. Wu, F. Wang, and K. Tanabe

Superconductivity Research Laboratory, International Superconductivity Technology Center, 10-13 Shinonome, 1-Chome, Koto-ku, Tokyo 135, Japan

Abstract: Cubic $NdBa_2Cu_3O_{7-\delta}$ (NBCO) films were deposited at substrate temperature of <650 °C. X-ray diffraction analysis shows that the deposited films have only ($h00$) peaks and a very narrow rocking curve of <0.1°. TEM images and selected diffraction patterns of the films reveal a pure cubic phase in the whole observation range. Cubic NBCO films behave semiconducting with a resistivity of $\sim10^{-2}$-10^{-1} Ωcm at 77 K.

Key words: $NdBa_2Cu_3O_{7-\delta}$ thin films, Cubic phase, Resistivity, Microstructure

INTRODUCTION

It is known that the superconducting properties and microstructures of deposited films are sensitive to the growth conditions, especially the substrate temperature. Structural disorder and cubic phase in $YBa_2Cu_3O_{7-\delta}$ (YBCO) films have been reported previously [1-3]. Because of the large difference between barium and yttrium ionic radii, it is difficult to obtain pure cubic phase YBCO films. However, in the $NdBa_2Cu_3O_{7-\delta}$ (NBCO) system, the superconductivity of NBCO samples drastically depends on the ordering of barium and neodymium ions. The substitution of the Nd^{3+} for the Ba^{2+} sites results in the disordering of cation sublattices and degrades the T_c in processed samples. In this paper, we report on the growth of cubic NBCO films by pulsed laser deposition (PLD).

EXPERIMENTAL

Epitaxial NBCO films were deposited on $SrTiO_3$ (100) and $LaAlO_3$ (100) substrates by PLD [4]. In order to study the dependence of microstructure and transport properties on the substrate temperature T_s, the T_s was changed in the range of $400<T_s<780$ °C The microstructure was characterized with x-ray diffraction (XRD) and transmission electron microscopy (TEM). The transport properties were measured by four-probe method.

RESULTS AND DISCUSSION

Fig. 1 shows XRD patterns obtained from a series of NBCO films grown on $LaAlO_3$ (100) substrates at different substrate temperatures. For the films deposited at 780 °C (optimum T_s for the growth of c-axis films), only ($00l$) peaks appear in the XRD pattern. The value of the full width at half maximum (FWHM) of the (005) rocking curve is only ~0.05° degree. As the T_s is lowered to 730-750 °C, the a- and c-axis mixed orientations are observed. The c-axis lattice parameter is slightly elongated (11.80 Å) than that of the pure *c*-axis oriented film (11.72 Å). At $T_s \leq 700$ °C, the (00*l*) peaks disappear, and only (*h*00) peaks are observed. The *a*-axis lattice constant determined from XRD is 3.90 Å. As the T_s is further reduced below 650 °C, the XRD still detects strong (*h*00) peaks. The value of FWHM of the (200) peak from the film deposited at 550 °C is as narrow as 0.1°. The calculated lattice constant is 3.90 Å. The ϕ-scan XRD patterns obtained from all NBCO films deposited at different substrate temperatures show that the films have the same *in-plane* alignment with respect to the substrates. The peaks of the films are located at the same position as those of the substrates and appear every 90°.

Fig. 2 shows the dependence of resistivity on temperature for NBCO films deposited on SrTiO$_3$ (100) substrates at different substrate temperatures. The pure *c*-axis oriented film shows the T_{c0} ~92 K. As the substrate temperature decreased below 700 °C, the T_{c0} decreases greatly with a long tail. At 650 °C, the resistivity nearly keeps flat as temperature decreases. DC SQUID measurements show that the sample is still nonsuperconducting at liquid He temperature. When the substrate temperature is further reduced below 600 °C, the normal state resistivity increases dramatically and changed from a metallic to semiconducting behavior. The value of resistivity at 100 K increases nearly two orders of magnitude comparing with one of the *c*-axis film. Since during PLD, every deposition parameter, except substrate temperature, kept unchanged, the above transport characteristics are suggested to be caused by microstructural changes in the films.

To elucidate the relationship between electrical properties and microstructures of NBCO films, cross-sectional TEM analysis was performed for films deposited at different substrate temperatures. Fig. 3 shows a TEM image for a film deposited at 600 °C. The interface between the film and the substrate is atomically flat and even the initial layer of the film has pure cubic phase. As shown in Fig. 3, in the whole view range, only pure cubic phase is observed. It should be noted that the image in Fig. 3 is only a part of the low magnification image in the range of more than 10 microns. During TEM observation by moving the specimen along the interface direction, we did not find orthorhombic phase. In addition, to identify whether the observed image belong to a-axis oriented orthorhombic phase or cubic

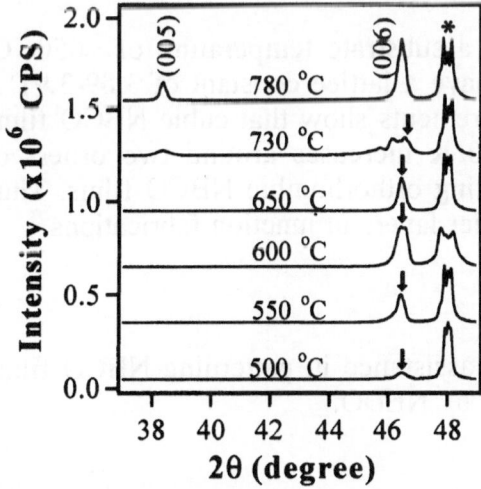

Fig. 1. XRD patterns for NBCO films deposited at different temperature. The peaks denoted by the arrow and the asterisk belong to NBCO (200) and LaAlO$_3$ (200) reflections, respectively.

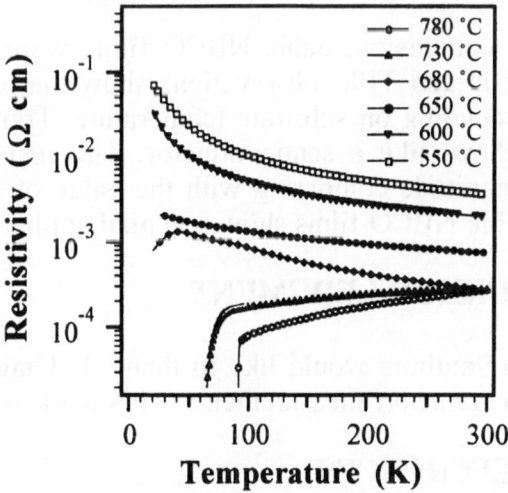

Fig. 2. Dependence of resistivity on temperature for NBCO films deposited on SrTiO$_3$ substrates at different temperatures. The NBCO film deposited at T_s=780 °C has a T_{c0}=92 K.

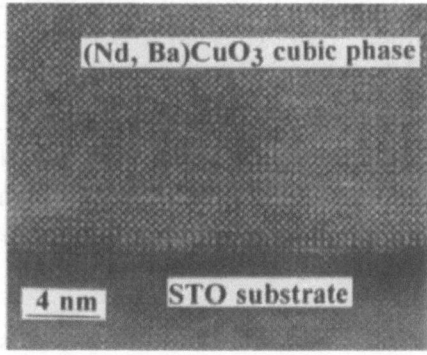

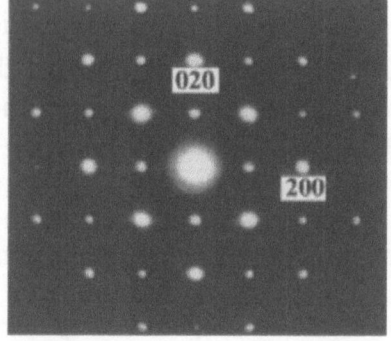

Fig. 3. TEM image for a cubic NBCO film deposited on SrTiO$_3$ substrate at 600 °C.

Fig. 4. Selected-area TEM diffraction pattern of a cubic NBCO film.

phase, the specimen was rotated and observed at different angles. However, in the selected-area electron diffraction patterns, the diffraction spots corresponding to layered structures are absent (Fig. 4). Energy dispersive x-ray analysis demonstrated that the cubic phase have the same (123) stoichiometry within the range of instrument accuracy. These results suggest that so called cubic NBCO films have a simple perovskite ABO$_3$ type unit cell, i.e., $(Nd_{1/3}Ba_{2/3})CuO_x$. Because the A cation sites are occupied by a random distribution of Nd and Ba ions, the sublattice of $NdCuO_{3-y}$ and $BaCuO_{3-z}$ have no long-rage ordering. In the XRD patterns, the peaks only corresponding to a cubic lattice parameter of 3.90 Å appear. Rutherford backscattering channeling measurements on cubic NBCO films give minimum backscattering yield of 7-9%, indicating good crystallinity of Cu sublattice. Our experimental results reveal that transport properties in NBCO system is mainly controlled by the microstructure. The superconductivity is very sensitive to the cation disordering. When the Nd and Ba sublattices become completely disordering, the films behave nonsuperconducting and a metal-to-insulator transition occurred due to abruption of Cu-O plane and thus the decrease of free electron density.

CONCLUSION

In conclusion, cubic NBCO films were prepared at a substrate temperature of <650 °C. XRD and TEM observations show that cubic films have a lattice constant of 3.89-3.91 Å depending on substrate temperature. Transport measurements show that cubic NBCO films behave like a semiconductor. The resistivity at 100 K increases around two orders of magnitude comparing with the value of superconducting orthorhombic NBCO films. Thus cubic NBCO films show potential applications as barrier layers in junction fabrications.

ACKNOWLEDGMENT

The authors would like to thank T. Utagawa for the assistance in patterning NBCO films for transport measurements. This work was supported by NEDO.

REFERENCES

1. Agostinelli JA, Chen S, Braunstein G (1991) Phys Rev B 43: 11396-11399
2. Linzen S, Krausslich J, Kohler A, Seidel P, Freitag B, Mader W (1997) Physica C 290: 323-333
3. Suh JD, Sung GY, Kang KY, Kim HS, Lee JY, Kim DK, Kim CH (1998) Physica C 308: 251-256
4. Li Yijie, Tanabe K (1998) J Appl Phys 83: 7744-7752

Analysis of Crack Formation of Y123 Single Crystal Substrate Fabricated by the Pulling Method

M. Egami, N. Itoga*, S. Koyama, T. Izumi and Y. Shiohara

Superconductivity Research Laboratory, ISTEC 1-10-13, Shinonome, Koto-ku, Tokyo 135-0062, Japan
*Japan Women's University, 2-8-1, Mejirodai, Bunkyo-ku, Tokyo 112-0015, Japan

ABSTRACT; We prepared a substrates by slicing an as-grown $YBa_2Cu_3O_{7-x}$ single crystal and a typical heating cycle was applied in the Air (750 ℃ /hour as increase/decrease rate up to 750℃) which is similar to the film vapor deposition process to the wafer, and an additional crack formation was observed. We also found that in the case of the Ar atmosphere (1.5%O_2), no additional crack was formed even at the same heating cycle. This indicated that the origin of the additional crack formation was not only the effect of the thermal distortion but also that of the lattice distortion due to the oxygen diffusion or the mixture effect.

KEY WORDS: crack formation, chemical etching, single crystal substrate

INTRODUCTION

The growth mechanism of Y123 single crystals fabricated by the pulling method was explained by the spiral growth model [1] with morphology observation of the as grown crystal surface. By using an etchant [2], these spiral dislocations and another type of dislocation was formed as an etch pit. During crystal pulling, crystals that nucleated and grown at the surface were sometimes observed. Crystallinity of these floating crystals is generally higher than that of the pulled crystal because the floating crystals are considered to growing a step flow growth mechanism. Dislocation density of the floating crystals is much less than that of the pulling crystal. We prepared two different kinds of wafer for the substrate from these two as-grown crystals, and investigated the relation between crack formation and dislocation or impurity inclusions under different atmosphere.

EXPERIMENTAL

We prepared substrates by slicing the pulled crystals and the floating crystals grown at the free solution surface. The wafer was heated up to 750℃, kept for 30 minutes and cooled to the room temperature. The cooling rates applied were 30℃/h, 100℃/h , and 750℃/h. The heat treatments were carried out either in the Air or the Ar (1.5% O_2) atmosphere.

RESULTS AND DISCUSSION

Table 1 shows the effect of the conditions of heat treatment on the crack formation. Table 2 shows the relationship between the kinds of the pre-existing defects and crack formation. We confirmed 8 different kinds of defects that exist before annealing. Figure1 shows the variety of the observed etch pit patterns observed by the optical microscopy. Deep etch pit

Table 1. Experimental results.

As Grown and slicing

High;(750,100)℃/h (to room temperature ↔ 750℃.)

Low; 30℃/h (to room temperature ↔ 750℃.)

○ : no cracks were formed.

× : cracks were formed.

	Air		Ar
	High	**Low**	**High**
Pulling crystal	×	○	○
Floating crystal	○	○	-

Table 2. Classification of pre-existing defects and the related factors.

	Types of pre-existed defects	Origin of the defects	Possibility to be the origin to new cracks due to applied heat cycle?	Removing ways
Pre-existing cracks and visualized cracks after chemical etching	Micro & macro crack	Thermal history after growth	No	Enable
	Polishing scratch	Polish	No	Enable
	Polishing remained stress	Polish	No	Enable
	Flux inclusion	Growth	?	?(growth control)
Dislocations (Appeared by the chemical etching)	Deep isolated pit	Spiral center?	Yes	Disable(other growth mechanism)
	Deep lined pit	Stacking fault	No (?)	?
	Deep assembled pit	Boundary & dislocation bundle?	?	?(growth control)
	Shallow isolated pit	?	No	?

accumulated area might correspond to the macroscopic spiral dislocation bundle or the boundary distortion region. The lined pits may be considered to be stacking faults. An isolated pit might be a center of the spiral growth hillock or an impurity inclusion. Figure 2 shows optical microscopic images of various defects that appear or disappear by chemical etching or heating cycle. High cooling rate in the Air atmosphere produced additional cracks in the case of typical pulled crystals, while no additional crack appeared in the case of the floating crystal. On the other hand, the high cooling rate in the Ar atmosphere did not cause additional crack formation even for the re-polished pulled crystal which exactly formed crack at the high cooling rate in the Air atmosphere. Additional cracks seemed to be formed originated from deep isolated pits typically as shown in Fig.2 (e). That might correspond to deep spiral holes or some impurity inclusions. The oxygen diffusivity difference at these places brought a strong lattice distortion, which might result in additional crack formation. Floating crystals may avoid this effect because of the high crystallinity. The reason for no additional crack formation in Ar might be explained by that the equilibrium oxygen concentration difference is smaller than that in Air for as-grown crystals. Another explanation might include both effects of the anisotropy of oxygen diffusion and thermal distortion. Oxygen diffusion at substrates surface will be faster if inclusions and/or dislocations may exist at the surface. In this case, in-plane carrier concentration at the surface a-b plane will increase. This means that not only an anisotropic thermal conductivity difference

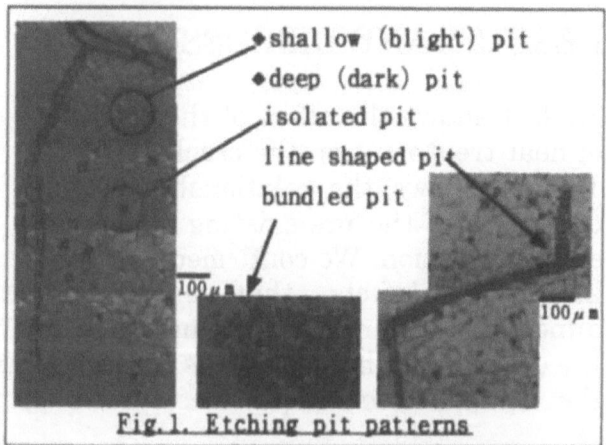

Fig.1. Etching pit patterns

shallow (blight) pit
deep (dark) pit
isolated pit
line shaped pit
bundled pit

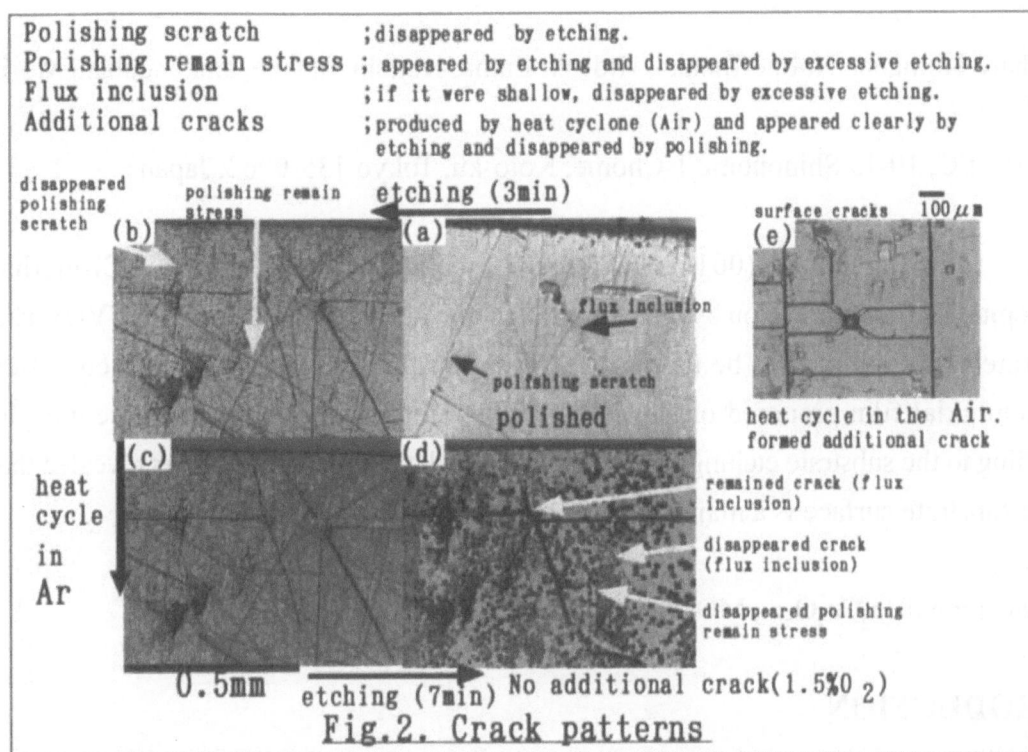

Fig.2. Crack patterns

between a-axis and c-axis increase but also increase the difference of the temperature dependence of the both axes thermal conductivity anisotropy. These effects can be enhanced in the case of higher PO_2 and may be lead to the crack formation.

CONCLUSION

Crack formation during heating cycle was not caused by distortion of thermal stress but caused by lattice distortion due to oxygen diffusion or caused by both oxygen and thermal distortion. High crystallinity can prevent from additional crack formation during the heating cycle. The additional crack formation seems to happen at the place that might consider a strong dislocation center or a flux inclusion spot according to post-etching analysis. Macroscopically crack or lined edge dislocation was not necessarily the origin of the additional crack formation for the applied heating cycle. To avoid crack formation during the heating cycle, single crystal substrates should be grown with less flux inclusions and the growth spiral center should be accumulated to one place to enlarge stress free area. Further experiments and modellings are necessary to clarify the origin of crack formation.

ACKNOWLEDGMENT

This work was supported by the New Energy and Industrial Technology Development Organization (NEDO) as Collaborative Research and Development of Fundamental Technologies for Superconductivity Applications.

REFERENCES
1.Y.Yamada and Y.Shiohara, Phys.C 217, 182(1993)
2. J.M. Rosamilia, B. Miller, L.F Schneemeyer, J.V. Waszczak and H.M. O'Brayn, J.Electrochem. Soc. 134(1987)1863.

Substrate Etching for Homoepitaxial NdBa$_2$Cu$_3$O$_{7-\delta}$ Film Deposition

Motoharu Komatsu, Nobue Tanaka, Hideaki Zama, Keiichi Tanabe, and Tadataka Morishita

SRL, ISTEC, 10-13 Shinonome 1-Chome, Koto-ku, Tokyo 135-0062, Japan

Abstract: NdBa$_2$Cu$_3$O$_{7-\delta}$ (001) substrate was etched with 0.003 wt% HCl-methanol for homoepitaxial film fabrication by metalorganic chemical vapor deposition (MOCVD). The etched substrate was examined to be thermodynamically stable on our deposition conditions. The homoepitaxial film prepared on the etched substrate showed various surface morphologies according to the substrate etching time. Etching back the homoepitaxial films revealed that defect on the substrate surface is a major cause to call the film growth mode.

Keywords: NdBa$_2$Cu$_3$O$_{7-\delta}$, MOCVD, homoepitaxy, etching

INTRODUCTION

We prepared homoepitaxial NdBa$_2$Cu$_3$O$_{7-\delta}$ (NBCO) film on as-polished (001) substrate by metalorganic chemical vapor deposition (MOCVD) [1]. The crystal growth mode in the homoepitaxial deposition was observed to depend on the vicinal angle oriented in mechanical polishing of the substrate. The report that HCl-methanol was effective to etch away deteriorated surface of YBa$_2$Cu$_3$O$_{7-\delta}$ (YBCO) (001) substrate [2] encouraged us to apply the chemical etching onto the NBCO substrate in search of yet more excellent homoepitaxial film growth.

EXPERIMENTAL

We have deposited homoepitaxial films on the NBCO (001) substrates treated with 0.003 wt% HCl-methanol for 10, 30, and 60 seconds, as well as on as-polished substrates. We started the deposition immediately after etching the substrates to keep the etched substrate surface as intact as possible in air. The precursors for MOCVD here are 2,2,6,6-tetramethyl-3,5-octanedionato (TMOD) complex of Nd, 2,2,6,6-tetramethyl-3,5-heptadionato (DPM) complex of Ba coordinated with two pentaethylenehexamines (pentaen), and DPM complex of Cu. Table1 represents the typical deposition conditions. The deposition

Table1. Typical deposition conditions.

Susceptor temperature(°C)	890
Carrier gas flow rate (sccm)	100
Oxygen partial pressure (Torr)	0.6
Total reactor pressure (Torr)	10
Deposition time (min.)	60
Nd oven temperature(°C)	135
Ba oven temperature(°C)	135
Cu oven temperature(°C)	112

948

rate was estimated as about 24 nm/hour based on inductively coupled plasma spectrometry (ICP) measurement of the cation contents in films prepared on 1 cm^2 SrTiO$_3$ substrates placed next to NBCO substrates. We grew the films for one hour. Note that the substrate temperature is 100 - 150°C lower than the susceptor temperature.

RESULTS AND DISCUSSION

We observed the surface morphology of etched NBCO substrates by atomic force microscopy (AFM) after annealing them in the temperature range of 550 - 950 °C and in the oxygen pressure range of 0.1 - 100 Torr to investigate into the thermodynamical stability of the etched substrate.

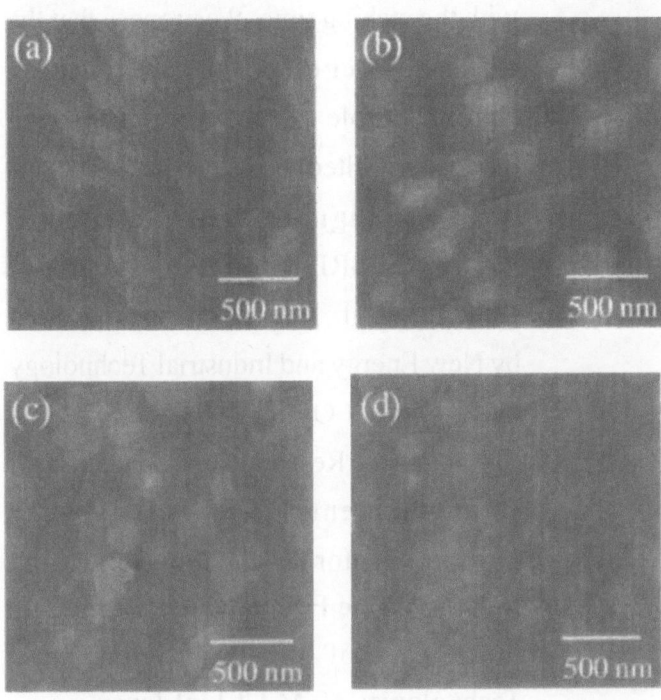

We observed none of precipitation, decomposition, nor melting at the same conditions as our deposition was carried out. This result agrees with the report of NBCO substrate surface stability examined by Rutherford backscattering spectrometry (RBS) [3]. The homoepitaxial films grown on the etched substrates had various surface morphologies depending on the etching time. In one series, for example in Fig.1, the step-flow growth appeared on as-polished substrate, then we observed a granular growth on 60s-etched substrate through the island growth on 10s- and 30s-etched substrates. In another series, the islands came on as-polished and 10s-etched substrates, then a granular growth

Fig.1. AFM images of homoepitaxial NBCO films grown on (a) as-polished, (b) 10s-, (c) 30s-, (d) 60s-etched substrates. The etching time has much influence on which film growth mode shows up.

and the step-flow growth were found on 30s- and 60s-etched substrates, respectively.

We etched back the homoepitaxial films to look into effect of substrate surface defect on the film growth mode. We place Fig.2 to demonstrate that many etch pits open on island-grown film. The etching rate was much faster than expected of bulk supposedly because the etch pits worked in favor of enhancing the film's contact with the etchant. On the other hand, as in Fig.3, the etchant needed more time with less pits to dig into step-flow-grown film than into island-grown film. With above-mentioned results, we propose that surface disorder on the substrate varies in response

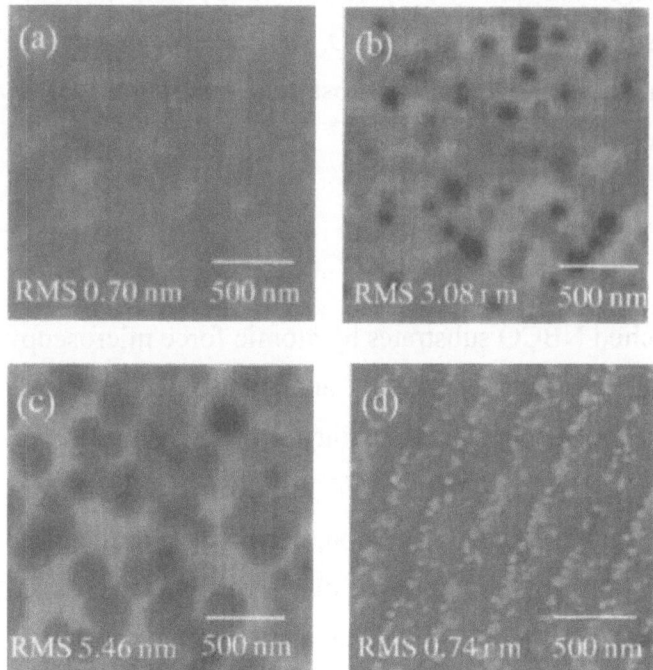

Fig.2. AFM images of an island-grown film etched back for (a) 0s, (b) 60s, (c) 120s, and (d) 240s. Note that the substrate surface emerges in (d).

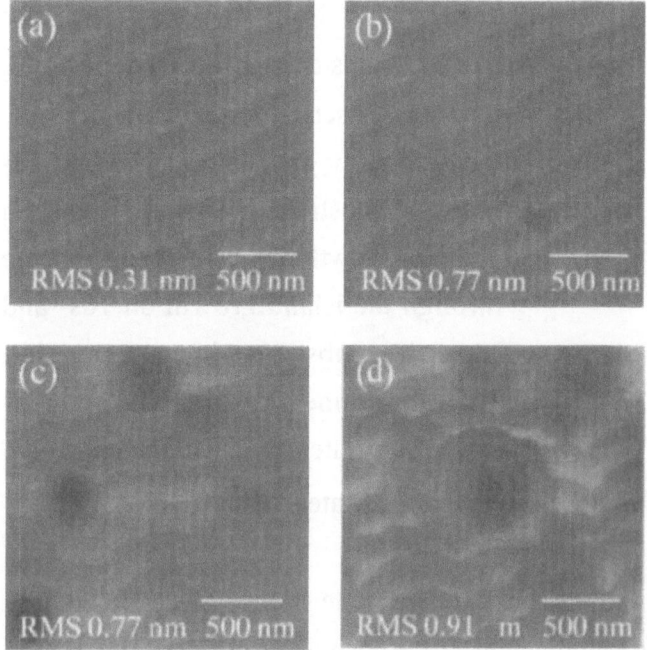

Fig.3. AFM images of a step-flow-grown film etched back for (a) 0s, (b) 60s, (c) 120s, and (d) 240s. Note that the surface doesn't roughen very much.

to the etching time to give the film growth mode the variation.

CONCLUSION

The etched NBCO substrate was thermodynamically stable on our deposition conditions. The homoepitaxial NBCO film prepared on the etched substrate varied its growth mode along with the etching time. We propose that the surface defect on the substrate plays a significant role to determine the growth mode as resulted.

Acknowledgments. We thank Mr. Koyama of SRL for his contribution of NBCO crystal. This work was supported by New Energy and Industrial Technology Development Organization (NEDO) as Collaborative Research and Development of Fundamental Technologies for Superconductor Applications under the New Sunshine Program administered by the Agency of Industrial Science and Technologies of M.I.T.I. of Japan.

[1] M. Komatsu, F. Wang, N. Tanaka, H. Zama, and T. Morishita : J. Cryst. Growth **205** (1999) 277

[2] N. Tanaka, H. Zama, T. Morishita, and H. Yamamoto : Jpn. J. Appl. Phys. **38** (1999) L731

[3] F. Wang, H. Zama, M. Sato, and T. Morishita : Advances in Superconductivity X (Proceedings of ISS '98) Springer Tokyo (1998) 106

c-Axis Oriented NdBa$_2$Cu$_3$O$_x$ Films Prepared at Substrate Temperatures Lower Than a-Axis Orientation

Kazuhiro Ishikawa[1,2], Hideaki Zama[1], Tadataka Morishita[1] and Takeo Suzuki[2]

[1] Superconductivity Research Laboratory, ISTEC
10-13 Shinonome 1-Chome, Koto-ku, Tokyo 135-0062, Japan
[2] Faculty of Engineering, Tokyo Denki University,
2-2 Kandanishiki-cho, Chiyoda-ku, Tokyo 101-8457, Japan

Abstract: The c-axis oriented NdBa$_2$Cu$_3$O$_x$ (NdBCO) film was prepared under the new growth conditions using a pulsed laser deposition. That a NdBCO film grows at higher temperature than a YBa$_2$Cu$_3$O$_x$ (YBCO) film is a drawback of composing multilayers for device applications, though NdBCO has crystalline and superconductive properties superior to those of YBCO. We successfully grew high-quality c-axis oriented NdBCO films at an oxygen pressure of 0.9Torr in the temperature range lower than that for the a-axis oriented NdBCO film, which is as low as the c-axis oriented YBCO film.

Keywords: c-axis oriented NdBa$_2$Cu$_3$O$_x$ film, pulsed laser deposition, low temperature, high oxygen pressure

INTRODUCTION

The NdBa$_2$Cu$_3$O$_x$ (NdBCO) film has the highest critical temperature (Tc(R=0)) [1] in REBa$_2$Cu$_3$O$_x$ (REBCO, RE: rare earth) system and a chemically stable and smooth surface.[2] However, the substrate temperature (Ts) at which the c-axis oriented NdBCO film grows is 80 - 100°C higher than that for the c-axis oriented YBa$_2$Cu$_3$O$_x$ (YBCO) film.[3,4] This is a serious problem in heating a substrate in the deposition system and in composing multilayers for device applications. In this paper, we propose new conditions of growth for the c-axis oriented NdBCO film; Ts is lower than what is for the a-axis oriented film and oxygen pressure (P(O$_2$)) is much higher than what has been adopted up to now.

EXPERIMENTAL

NdBCO films were fabricated by a pulsed laser deposition method. The energy density D$_L$, spot size, and pulse frequency of an ArF excimer laser were 1 J/cm^2, 1.5 x 6 mm^2, and 5 Hz, respectively. The sintered stoichiometric NdBCO disk with a density over 6g/cm^3 was used as a target. The target-substrate distance was 5 cm. The films were deposited on SrTiO$_3$(100) substrates under P(O$_2$) of 0.05 - 1.8 Torr at Ts of 720 - 935°C, which were heated by a halogen rump and calibrated by a thermocouple bonded to the substrate surface with silver paste. After deposition of 3000 shots, the films were cooled to room temperature in 700 Torr oxygen. The thickness of films ranged from 130 nm to 220 nm

951

depending on $P(O_2)$. The crystal structure of films was examined by an X-ray diffraction (XRD) θ-2θ scan using CuKα radiation. The crystallinity was evaluated by a XRD rocking curve of the (005) reflection. The critical temperature was determined by a dc four-probe method.

RESULTS AND DISCUSSION

When $P(O_2)$ was widely varied from 0.05Torr to 1.8Torr, the a-axis orientation appeared between the c-axis orientations in NdBCO films as shown by the normalized intensity ratio of XRD reflection of the a- and c-axis orientation ($R\{(h00)/(00l)\}$) in Fig.1(a). YBCO films grown at Ts of 700°C, as reported in refs. 5 and 6, showed a similar $P(O_2)$ dependence of the orientation. The Tc(R=0), however, showed a quite different $P(O_2)$ dependence (see Fig.1(b)). The c-axis oriented YBCO film grown at a low $P(O_2)$ of 0.01 Torr has a Tc(R=0) of 85K, while what is grown at a $P(O_2)$ of 0.5 Torr has a Tc(R=0) of 50K. In the case of NdBCO films both c-axis oriented films grown in $P(O_2)$ of 0.1 Torr and 1 Torr at 860°C showed Tc(R=0) above 85K.

Generally, in REBCO films the c-axis orientation appears in the temperature range higher than the range where the a-axis orientation appears in a conventional $P(O_2)$. These two c-axis oriented phases of NdBCO were investigated for samples prepared at various Ts. In Fig.2(a) the c-axis orientation does not appear on a low temperature side of the a-axis orientation at $P(O_2)$ of 0.15 Torr. At $P(O_2)$ of 0.3 Torr, the c-axis orientation appears on both sides of the a-axis orientation. At $P(O_2)$ of 0.9 Torr, the c-axis orientation grows only in the temperature range lower than the a-axis orientation. The temperature region for the a-axis orientation shifts to a higher Ts range, as $P(O_2)$ becomes higher. The purely c-axis oriented films with Tc(R=0) above 89K were obtained at Ts as low as 750 - 830°C under $P(O_2)$ of 0.9 Torr (see Fig.2(b)). These are the conditions where the new c-axis oriented NdBCO films grew with good crystallinity, $i.e.$ a

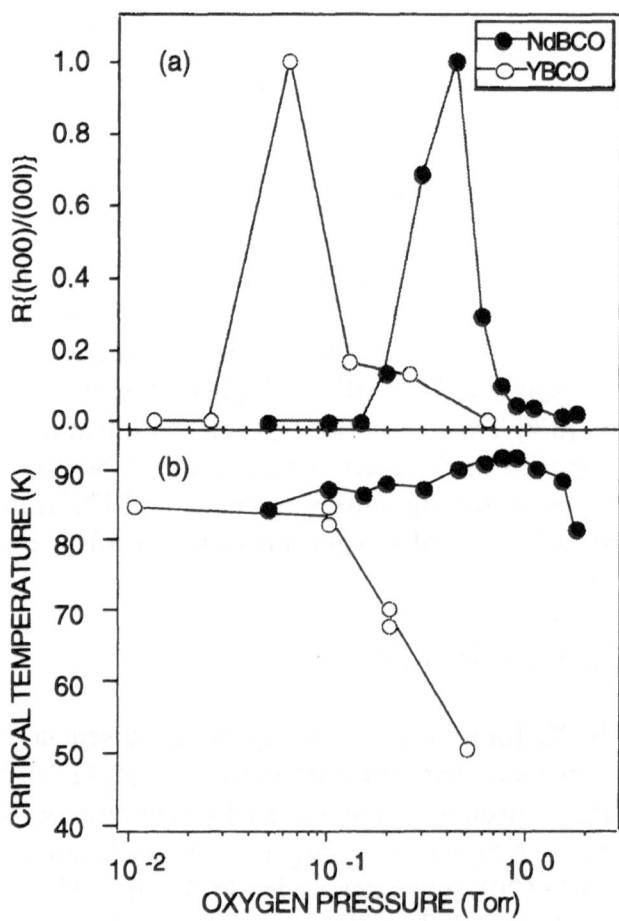

Fig. 1. $P(O_2)$ vs (a) $R\{(200)/(002)\}$ for YBCO, $R\{(200)/(005)\}$ for NdBCO and (b) Tc(R=0) for YBCO and NdBCO films.

full-width at half maximum below 0.08° for the (005) rocking curve ($\Delta \omega$). We optimized the condition to improve a surface morphology from the viewpoint of device applications by varying D_L, in fixing Ts = 790°C and $P(O_2)$ = 0.9 Torr and maintaining crystalline and superconducting properties. An optimized c-axis oriented NdBCO film grown at D_L = 0.5 J/cm^2 exhibited a root-mean-square roughness of 1.5nm averaged over a 2x2 μ m^2 area and a droplet density of 6x10^5/cm^2.[7]

CONCLUSIONS

We successfully grew the c-axis oriented NdBCO film with $\Delta \omega \sim$ 0.08° and Tc(R=0) $\sim$ 90K at Ts ranging from 750°C to 830°C and $P(O_2)$ of 0.9 Torr. This is comparable to the Ts for the c-axis oriented YBCO film.

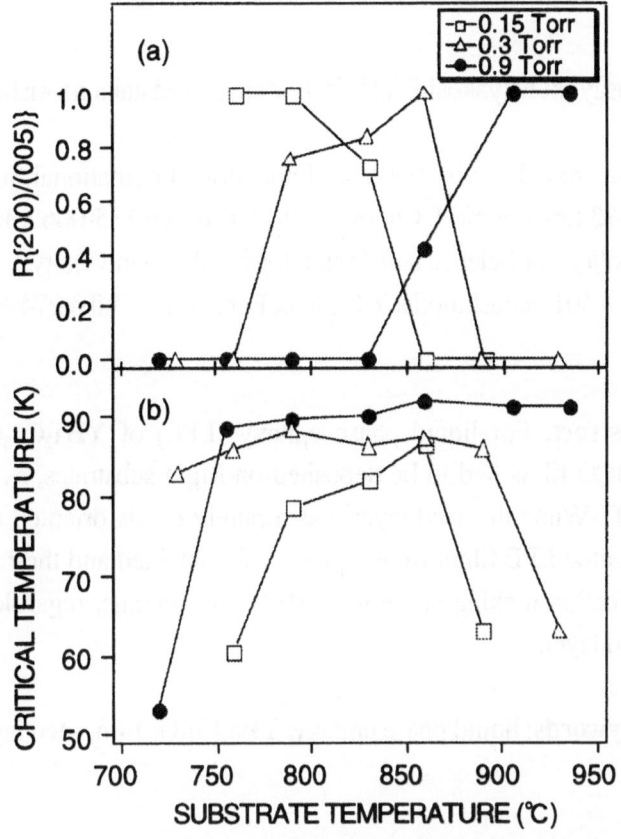

Fig. 2. Ts vs (a) R{(200)/(005)} and (b) Tc(R=0) at $P(O_2)$ of 0.15, 0.3 and 0.9 Torr for NdBCO films.

Acknowledgments: This work was supported by the New Energy and Industrial Technology Development Organization (NEDO) as Collaborative Research and Development of Fundamental Technologies for Superconductivity Applications under the New Sunshine Program administered by the Agency of Industrial Science and Technologies (AIST) of the Ministry of International Trade and Industry (MITI) of Japan.

1. M. Badaye, J.G. Wen, K. Fukushima, N. Koshizuka, T. Morishita, T. Nishimura and Y. Kido, Supercond. Sci. Technol. **10**, 825 (1997).

2. M. Badaye, W. Ting, K. Fukushima, N. Koshizuka, T. Morishita and S. Tanaka, Appl. Phys. Lett. **67**, 2155 (1995).

3. M. Badaye, F. Wang, Y. Kanke, K. Fukushima and T. Morishita, Appl. Phys. Lett. **66**, 2131 (1995).

4. Y. Li and K. Tanabe, J. Appl. Phys. **83**, 7744 (1998).

5. T. Hase, H. Takahashi, H. Izumi, K. Ohata, K. Suzuki, T. Morishita and S. Tanaka, J. Cryst. Growth **115**, 788 (1991)

6. H. Izumi : Dr. Thesis, Tokyo Institute of Technology, Tokyo, 1992.

7. H. Zama, K. Ishikawa, T. Suzuki and T. Morishita, Jpn. J. Appl. Phys. **38**, L923 (1999).

Effects of the Seed Layer for MgO Substrate on Quality of Liquid Phase Epitaxy YBa$_2$Cu$_3$O$_x$

Masayuki Miyakoshi[1,2], Hideaki Zama[1], Tadataka Morishita[1] and Hiroshi Yamamoto[2]

[1] Superconductivity Research Laboratory, International Superconductivity Technology Center,
10-13 Shinonome 1-Chome, Koto-ku, Tokyo 135-0062, Japan
[2] College of Science and Technology, Nihon University,
24-1-401 Narashinodai 7-Chome, Funabashi, Chiba 274-8501, Japan

Abstract: For liquid phase epitaxy (LPE) of YBa$_2$Cu$_3$O$_x$ (YBCO) films grown on MgO substrates, YBCO films had to be deposited on MgO substrates, as seed layers, by a vapor phase method prior to LPE. When the seed layer was a purely c-axis oriented and thicker than 24 nm, the high-quality c-axis oriented LPE films were reproducibly obtained and their full-width at half maximum of the YBCO(005) reflection rocking curve were 0.08 ° on average, regardless of the crystallinity and the thickness of the seed layer.

Keywords: liquid phase epitaxy, YBa$_2$Cu$_3$O$_x$ film, seed layer, X-ray diffraction (XRD) rocking curve

INTRODUCTION

Liquid phase epitaxy (LPE) grown YBa$_2$Cu$_3$O$_x$ (YBCO) films,[1-4] which are achieved under the conditions close to thermal equilibrium, are promising because of their exact stoichiometry and good crystallinity. In order to realize these advantages of LPE, the choice of a substrate is one of the most important factors to avoid forming cracks due to a lattice mismatch and difference in the thermal expansion coefficients between the film and substrate. MgO is the only substrate reported which has the thermal expansion coefficient larger than that of YBCO,[5] and is expected to be free from cracks not to tensile but to compressive stress during cooling from deposition temperatures. The Ba-Cu-O flux melt for LPE, with Ba : Cu = 3 : 7, shows no wetting for MgO,[6] suggesting that it is difficult to grow YBCO films epitaxially on MgO substrates using this flux. YBCO films are successfully grown on MgO substrates deposited with YBCO films from a vapor phase as a seed layer.[7,8] In this paper, we report the results examining effects of crystalline properties of the seed layer on quality of LPE films.

EXPERIMENTAL

YBCO seed films were deposited on MgO(100) substrates (20 $\times$ 20 mm^2) by an off-axis DC-RF hybrid plasma sputtering [9] at a pressure of 50 mTorr under a gas flow ratio of Ar : O$_2$ = 2 : 1 using a stoichiometric sintered target (50 mm ϕ). In order to prepare the seed films with a various thickness and graded crystallinity, we varied the deposition time from 2 min to 120 min and the substrate-heater temperature (Ts) from 750°C to 910°C. The metal composition and thickness of the seed films determined by inductively coupled plasma spectroscopy were Y : Ba : Cu = 0.99±0.03 : 2 : 2.91±0.11 and in the range from 5nm to 310nm, respectively.

The LPE experimental setup and procedure is reported in our previous paper.[7] The flux melt was 14 mol% (15 wt%) YBCO and 86 mol% (85 wt%) $Ba_3Cu_7O_{10}$. The melt was kept at 965°C at the surface and 975°C at the bottom of the Y_2O_3 crucible. The thickness of LPE films was 8.0±2.82 μm for 5 min dipping, which was determined from a cross-sectional view of an optical micrograph.

The crystalline characteristics of the seed and LPE films were examined by an X-ray diffraction (XRD) method. The XRD rocking curve of LPE films was measured using a double-crystal diffractometer.

RESULTS AND DISCUSSION

The YBCO seed layers were grown with a preferential a-axis orientation at Ts = 750°C, a- and c-axes orientations between 760°C and 790°C, and c-axis orientation above 810°C.

The full-width at half maximum (FWHM) of the (005) reflection rocking curve ($\Delta\omega$) of LPE films is plotted for the thickness of YBCO seed layers deposited at Ts = 910°C in Fig.1. When the seed layer was 5 nm and 12 nm thick, no LPE films grew on it and a bare surface of MgO substrate appeared. We can speculate the reason for this; The melting temperature of YBCO became lower in very thin films. The grains in the seed were so small that there were few nuclei larger than a critical size for a stable growth. When the seed layer was thicker than 24 nm, the c-axis oriented LPE films were obtained and $\Delta\omega$ was 0.08 ±0.01 °, regardless of the thickness of seed layers.

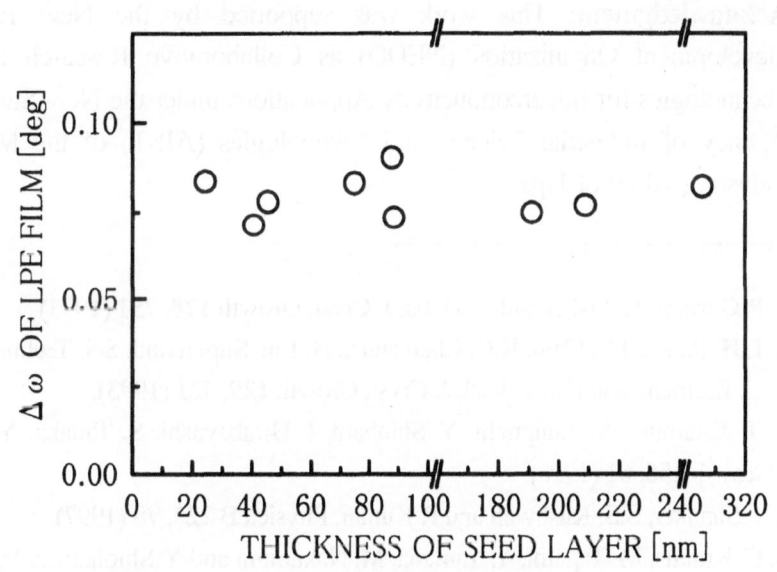

Fig. 1. $\Delta\omega$ of LPE films vs thickness of seed layers.

In the case of the a-axis oriented seed layer, even though it had a thickness of around 180 nm, we could obtain neither a- nor c-axis oriented LPE film. Only remnants of the flux clung to the substrate. The a-axis orientation needs a undercooling lower than the c-axis orientation in LPE growth.[10] The undercooling might be too small to grow the a-axis orientation in this experiment. The $\Delta\omega$ of LPE films is plotted for the intensity ratio of the (200) and (005) reflections of seed layers (R{(200)/(005)}) in

Fig.2. When we grew LPE films on the *c*-axis oriented seeds, R{(200)/(005)} = 0 in Fig.2, all the LPE films show $\Delta \omega$ below 0.08°. For the seeds containing *a*- and *c*-axes oriented grains, R{(200)/(005)} > 0 in Fig.2, $\Delta \omega$ is above 0.1°. The *a*-axis oriented grains in seed layers would have undesirable effects on the *c*-axis oriented LPE growth under the growth conditions in this experiment.

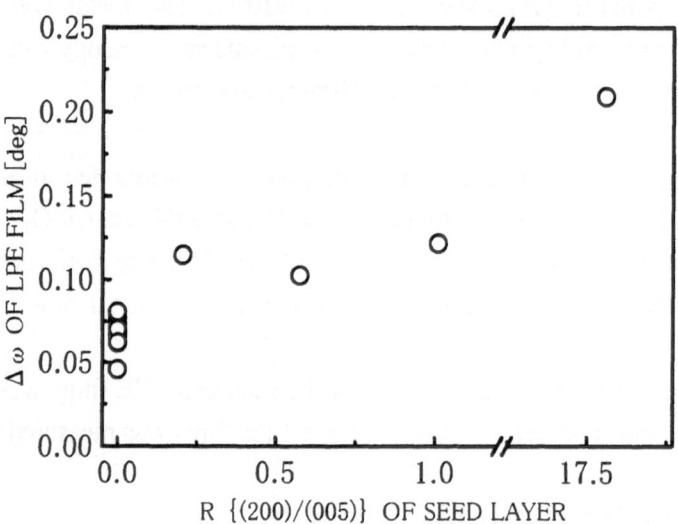

Fig. 2. $\Delta \omega$ of LPE films vs R{(200)/(005)} of seed layers.

CONCLUSIONS

We grew *c*-axis oriented LPE films with $\Delta \omega$ of 0.08° on average on purely *c*-axis oriented seed layers deposited on MgO substrates with a thickness over 24 nm. When the seed layer contained *a*-axis oriented grains, LPE films repeatedly showed a poor reproducibility of crystallinity and the value of $\Delta \omega$ above 0.1°.

Acknowledgment: This work was supported by the New Energy and Industrial Technology Development Organization (NEDO) as Collaborative Research and Development of Fundamental Technologies for Superconductivity Applications under the New Sunshine Program administered by the Agency of Industrial Science and Technologies (AIST) of the Ministry of International Trade and Industry (MITI) of Japan.

1. P. Gornert, K. Fisher and C. Dubs, J. Cryst. Growth **128**, 751 (1993).

2. L.H. Perng, T.S. Chin, K.C. Chen and C.H. Lin, Supercond. Sci. Technol. **3**, 233 (1990).

3. C. Klemenz and H.J. Scheel, J. Cryst. Growth **129**, 421 (1993).

4. T. Kitamura, S. Taniguchi, Y. Shiohara, I. Hirabayashi, S. Tanaka, Y. Sugawara and Y. Ikuhara, J. Cryst. Growth **158**, 61 (1996).

5. J. Shanker, S.S. Kushwah and P. Kumar, Physica B **223**, 78 (1997).

6. C. Krauns, M. Tagami, Y. Yamada, M. Nakamura and Y. Shiohara, J. Mater. Res. **9**, 1513 (1994).

7. S. Miura, K. Hashimoto, F. Wang, Y. Enomoto and T. Morishita, Physica C **278**, 201 (1997).

8. Y. Ishida, T. Kimura, K. Kakimoto, Y. Yamada, Z. Nakagawa, Y. Shiohara and A.B. Sawaoka, Physica C **292**, 264 (1997).

9. W. Ito, J. Mater. Res. **10**, 2216 (1995).

10. T. Aichele, S. Bornmann, C. Dubs and P. Gornert, Cryst. Res. Technol. **32**, 1145 (1997).

Effects of cation mixing on the growth of a-axis oriented NdBa$_2$Cu$_3$O$_{7-\delta}$ thin films

M. Mukaida, J. Sugimoto, S. Sato, M. Kusunoki, and S. Ohshima

Faculty of Engineering, Yamagata University, Jonan 4-3-16, Yonezawa, Yamagata 992-8510, Japan

Abstract: Effects of cation mixing on the growth of a-axis oriented NdBa$_2$Cu$_3$O$_{7-\delta}$ films grown by ArF pulsed laser deposition on LaSrGaO$_4$ (100) substrates with a buffer layer of Gd$_2$CuO$_4$ are investigated by x-ray diffractometry and cross-sectional transmission electron microscopy (TEM). Films are c-axis in-plane aligned a-axis oriented with a few amount of c-axis oriented portion. Mixture of a- and c-axis oriented regions were observed in the NdBa$_2$Cu$_3$O$_{7-\delta}$ films which are quite different from a-axis oriented YBa$_2$Cu$_3$O$_{7-\delta}$ thin films. In the a-axis oriented YBa$_2$Cu$_3$O$_{7-\delta}$ thin films, c-axis oriented portion was always observed in the vicinity of the interface between the substrates and the films. These results indicate that a preferred orientation changing of YBa$_2$Cu$_3$O$_{7-\delta}$ films from c-axis to a-axis on lattice mismatched substrates are not originated from mixing of cations of Y and Ba.

Key word: LnBa$_2$Cu$_3$O$_{7-\delta}$, a-axis orientation, growth model, cation mixing.

INTRODUCTION

Nature of preferred orientation of YBa$_2$Cu$_3$O$_{7-\delta}$ films are the most elemental problem to be solved to grow the films. Growth of a-axis oriented YBa$_2$Cu$_3$O$_{7-\delta}$ films on lattice mismatched substrates such as LaAlO$_3$ substrates has been discussed using cross-sectional TEM images. From the TEM images, the film was almost a-axis oriented, however an ultra thin c-axis oriented YBa$_2$Cu$_3$O$_{7-\delta}$ film existed in the vicinity of the substrate surface. We have proposed a quasi-lattice-matching engineering (QLME)[1] which predicts how preferred orientation of YBa$_2$Cu$_3$O$_{7-\delta}$ films are decided. This proposed model successfully explained the existence of the ultra thin c-axis oriented YBa$_2$Cu$_3$O$_{7-\delta}$ film[2] on lattice mismatched substrates such as SrTiO$_3$ and LaAlO$_3$ and a-axis oriented YBa$_2$Cu$_3$O$_{7-\delta}$ films on it.

However there is an another growth model which is originated from cation mixing. The mixing cations are Ln (lanthanide) and Ba in LnBa$_2$Cu$_3$O$_{7-\delta}$. Figure 1 shows a schematic drawing of the mixing model. In the drawing, only A-site cations are presented. The symbols "o" and "•" represent Ln and Ba, respectively. In the model Ln and Ba are some times mixed each other and coincidentally there aligned a Ba-Ln-Ba order of cations in the growing surface. The order of Ba-Ln-Ba is pointed by an arrow in the drawing. The Ba-Y-Ba ordered surface is the a-axis surface of YBa$_2$Cu$_3$O$_{7-\delta}$. Then a-axis oriented YBa$_2$Cu$_3$O$_{7-\delta}$ thin films grow on it. This is the origin of this model. However, there are two inconsistent points to be pointed out in this model. First of all, why the c-axis oriented YBa$_2$Cu$_3$O$_{7-\delta}$ films grow at the early stage of the growth. Next, why there are no preferred orientation changing after the preferred orientation of YBa$_2$Cu$_3$O$_{7-\delta}$ films changes from the c-axis to the a-axis. It is very interesting what will occur when we used Nd as Ln in LnBa$_2$Cu$_3$O$_{7-\delta}$ and grow a-axis oriented NdBa$_2$Cu$_3$O$_{7-\delta}$ films. Because Nd is easy to be mixed with Ba. Then, we can intentionally produce a cation mixed surface.

In this paper, we grow a-axis oriented NdBa$_2$Cu$_3$O$_{7-\delta}$ films and observe what will occur on the films.

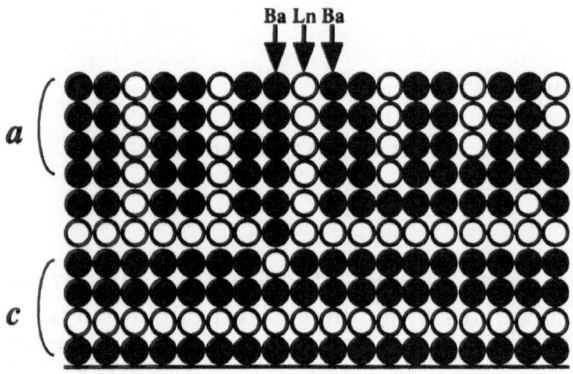

Figure 1: A cation mixing model of a-axis oriented $LnBa_2Cu_3O_{7-\delta}$ films.

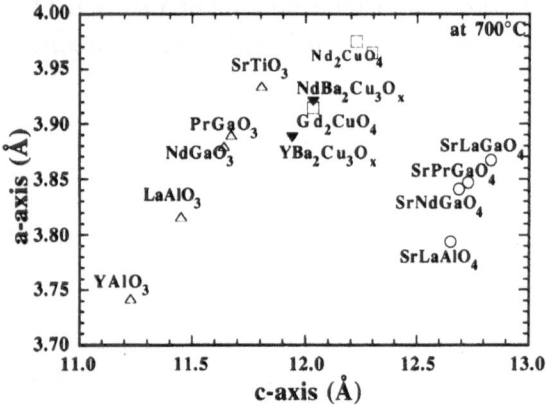

Figure 2: bc-plane lattice matching of Ln_2CuO_4 and $LnBa_2Cu_3O_{7-\delta}$.

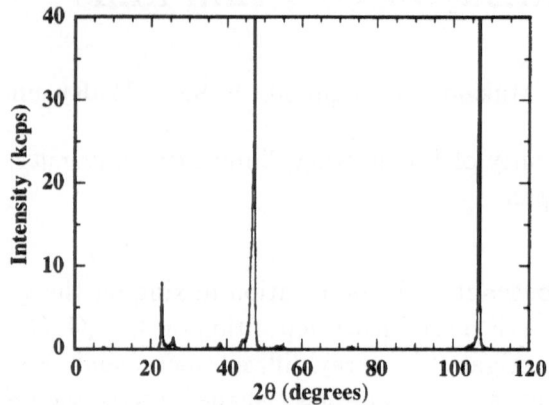

Figure 3: A typical xrd pattern of the $NdBa_2Cu_3O_{7-\delta}$ films.

$LnBa_2Cu_3O_{7-\delta}$ films are described elsewhere.[4]

The film orientation was characterized by $\theta/2\theta$ and ϕ-scan x-ray diffractometry. The microstructure of the film was observed by planar (TEM) and cross-sectional (XTEM) view transmission electron microscopy.

EXPERIMENTAL

A-axis oriented $NdBa_2Cu_3O_{7-\delta}$ films are grown by ArF pulsed laser deposition on $LaSrGaO_4$ (100) substrates with a buffer layer of Gd_2CuO_4 or Nd_2CuO_4. Because the buffer layer of Gd_2CuO_4 or Nd_2CuO_4 enables c-axis in-plane aligned a-axis oriented $LnBa_2Cu_3O_{7-\delta}$ films to grow on $LaSrGaO_4$ (100) surface[3]. Figure 2 shows a lattice matching map of a- and c-axis of materials. We have already demonstrated crystallographic orientation control of $YBa_2Cu_3O_{7-\delta}$ films and c-axis in-plane aligned a-axis oriented $YBa_2Cu_3O_{7-\delta}$ films were grown using a Gd_2CuO_4 buffer layer.[4] In the figure, the most lattice matched pair is the Nd_2CuO_4 with a K_2NiF_4 structure and $NdBa_2Cu_3O_{7-\delta}$. Therefore, we chose Nd_2CuO_4 as a buffer layer of a-axis oriented $NdBa_2Cu_3O_{7-\delta}$ films. Details of growth conditions of buffer layers and

RESULTS AND DISCUSSIONS

Figure 3 shows a typical x-ray diffraction pattern of $NdBa_2Cu_3O_{7-\delta}$ films on Gd_2CuO_4 buffer layers on $LaSrGaO_4$ (100) substrates. The x-ray diffraction pattern shows mixture of $00l$ and $h00$ peaks from $NdBa_2Cu_3O_{7-\delta}$, $h00$ peaks from Gd_2CuO_4 and peaks from the substrate. It means that the $NdBa_2Cu_3O_{7-\delta}$ film consists of mixture of a- and c-axis oriented regions. $YBa_2Cu_3O_{7-\delta}$ films grown on the Gd_2CuO_4 buffer layer on the (100) $LaSrGaO_4$ substrate result in fully a-axis oriented without c-axis regions. Then there are some effects of changing Ln from Y to Nd on the preferred orientation of $LnBa_2Cu_3O_{7-\delta}$ films.

Therefore, it is very important to observe the cation order in the film. Next, we observed cross-sectional TEM images of the film. Figure 4 is a cross-sectional TEM images of the film with the substrate. In the figure, dark lines correspond to CuO_2 planes. These dark lines indicate a film orientation. When the dark lines are parallel to the substrate surface, the film is c-axis oriented and when they are perpendicu-

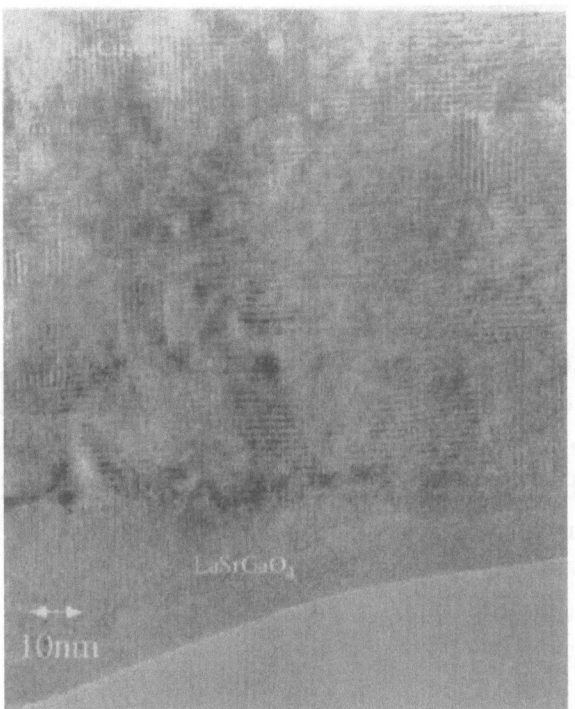

Figure 4: A typical cross sectional TEM image of an a-axis oriented $NdBa_2Cu_3O_{7-\delta}$ film on a $LaSrGaO_4$ (100) substrate.

lar to the substrate surface, the film is a-axis oriented. In the figure, it is well seen that the film consist of mixture of a- and c-axis oriented regions for every places in the growing direction. Mixture of preferred orientation for every places in the growing direction means the preferred orientation changes many times from a-axis to c-axis and c-axis to a-axis. This cross-sectional TEM image is quite different from that observed in $YBa_2Cu_3O_{7-\delta}$ films as described in literature.[5] The cross sectional TEM images[5] of $YBa_2Cu_3O_{7-\delta}$ films showed that the preferred orientation changed only once from c-axis to a-axis.

Fig. 4 is one of the answer for the questions for the cation mixing model; why there are no preferred orientation changing after the preferred orientation of $YBa_2Cu_3O_{7-\delta}$ films changes from the c-axis to the a-axis. It is clarified when the cation mixing occurs, the preferred orientation changes many times from a-axis to c-axis and c-axis to a-axis. Therefore, the growth of a-axis oriented $YBa_2Cu_3O_{7-\delta}$ films on lattice mis-

matched substrates is probably not originated from the cation mixing.

CONCLUSION

Effects of cation mixing on the growth of a-axis oriented $NdBa_2Cu_3O_{7-\delta}$ films grown by ArF pulsed laser deposition on $LaSrGaO_4$ (100) substrates with a buffer layer of Gd_2CuO_4 were investigated by cross-sectional TEM. $NdBa_2Cu_3O_{7-\delta}$ was selected for an 123-oxide superconducting material because Nd and Ba are easily to exchange their sites. Mixture of a- and c-axis oriented regions were observed in the $NdBa_2Cu_3O_{7-\delta}$ films by cross sectional TEM images, which are quite different from a-axis oriented $YBa_2Cu_3O_{7-\delta}$ thin films grown on lattice mismatched substrates as reported in a literature.[5] A preferred orientation changing of $YBa_2Cu_3O_{7-\delta}$ films from c-axis to a-axis on lattice mismatched substrates is not originated from mixing of cations of Y and Ba.

Acknowledgments. The authors would like to thank Mr. K. Aizawa for technical support through the experiment. They also acknowledge to Dr. K. D. Develos, Mr. K. Chiba, Mr. T. Suzuki, Mr. S. Oikawa, Ms. S. Makino and Mr. K. Noguchi for their valuable discussions and assistance. This research was partially done in Clean Room of Yamagata University.

References

[1] M. Mukaida, S. Miyazawa, and J. Kobayashi. In *Advances in Superconductivity V*, p. 893, Tokyo, 1993. Springer-Verlag.

[2] H. Takahashi, T. Hase, H. Izumi, K. Ohata, T. Morishita, and S. Tanaka. *Physica*, Vol. C-179, p. 291, 1991.

[3] S. Miyazawa and M. Mukaida. *Appl. Phys. Lett.*, Vol. 64, p. 2160, 1994.

[4] M. Mukaida. *Jpn. J. Appl. Phys.*, Vol. 36, p. L767, 1997.

[5] M. Mukaida and S. Miyazawa. *Jpn. J. Appl. Phys.*, Vol. 32, p. 4521, 1993.

Effects of Plasma Cleaning and Assist on Early Stage of Ca-doped Bi2201 Thin Film Growth

Masaki Tada, Akinori Hashizume, Ken-ichi Itoh, Jiro Yamada, Hisatake Uchikawa, Kohichi Ogai, Tamio Endo* and Yasuo Tsutsumi#

Faculty of Engineering, Mie University, Kamihama, Tsu, Mie 514-8507, Japan
#Department of Electrical Engineering, Akashi College of Technology, Akashi, Hyogo 674-8501, Japan

Abstract: Ca-doped Bi2201 thin films were deposited on lattice-mismatched MgO (100) substrate, and effects of oxygen plasma on early stages of the film growth were investigated. Amorphous layer grows first at interface then the 2201 crystal grows on it. However, when the substrate surface is cleaned by the oxygen plasma, the amorphous layer can be reduced then the 2201 crystal grows earlier. The plasma enhances multinucleation of 2201 crystal during the early growth.

Keywords: Ca-doped Bi2201 thin film, Plasma cleaning, Plasma assist, Early growth stage, Multinucleation

INTRODUCTION

Function-harmonized oxide devices are much attractive these days [1]. They are composed of multilayers of, say, ferroelectrics, ferromagnetics and superconductors. A key factor to develop them is to advance the multistacking techniques. In this process, it is necessary to elucidate growth mechanism of the thin films on lattice-mismatching layers. The growth mechanism of perovskite oxide thin films on the lattice-mismatched substrates has not been clarified yet and it is generally quite important in a field of "oxide electronics". In this work we deposited Ca-doped Bi2201 thin films on a lattice-mismatched MgO substrate to investigate effects of oxygen plasma on early stages of crystal growth.

EXPERIMENTAL

Thin films of Bi-Sr-Ca-Cu-O were deposited on a MgO (100) substrate at substrate temperature (T_S) of 500°C and oxygen partial pressure (P_O) of 0.75 mTorr for 1-3 hours by using ion beam sputtering (IBS). Before the deposition, MgO substrate was polished mechano-chemically using alumina powder (2.7μ) and H_3PO_4, and then cleaned by organic solvent. During the growth oxygen molecules or plasma was supplied to the substrate from a nozzle located at 11 mm above. A couple of films were deposited after cleaning the substrate surface in-situ by the oxygen plasma at comparatively low temperature of T_S= 500°C and P_O= 4 mTorr for 5 hours. Crystalline quality of the deposited films was characterized by X-ray diffraction (XRD) on θ-2θ scan using Cu Kα line.

RESULTS AND DISCUSSION

The XRD patterns are shown in Fig.1 (a) for the deposited films with supply of oxygen molecules (molecular-film) for 1-3 hours. The patterns indicate that the film with 1 hour-deposition (300 Å thickness) is amorphous, the film with 2 hour-deposition (500 Å) is $Bi_2(Sr,Ca)_2CuO_x$ (2201) crystal having broad (00ℓ) peaks, and the film with 3 hour-deposition (900 Å) is rather fine 2201 crystal having sharp (00ℓ) peaks. The film thickness does not linearly increase with the deposition time at the initial stage, because the film changes from the amorphous to the crystal structure. Moreover, the film thickness cannot be

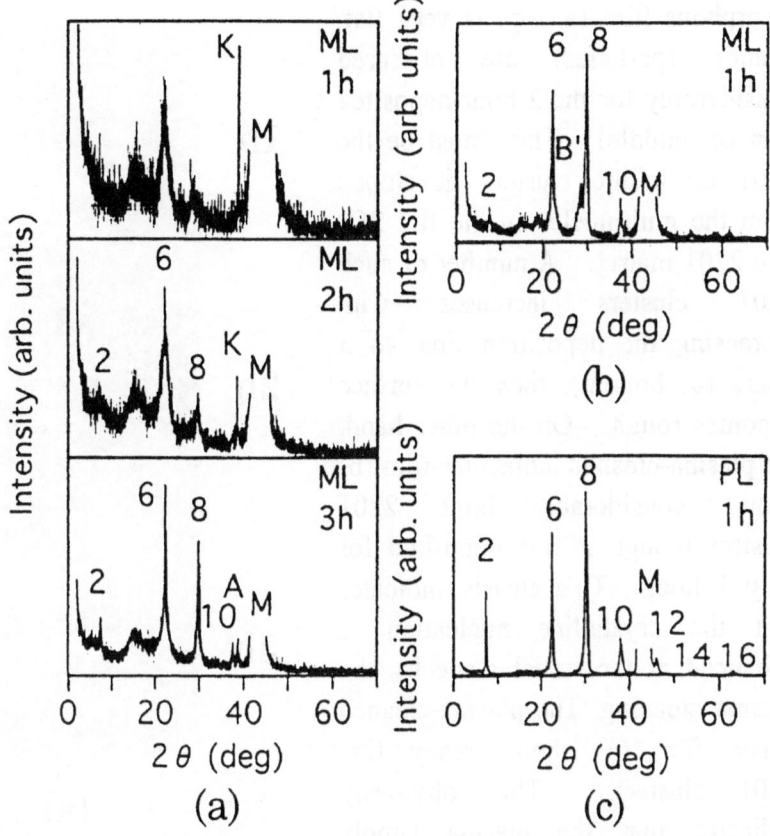

Fig.1. XRD patterns for (a) the molecular-films without the plasma cleaning, (b) the molecular-film and (c) the plasma-film with the plasma cleaning. The numbers attached to peaks are ℓ in (00ℓ). B':Bi214, A:CuO, M,K: MgO.

estimated accurately by the probe method when the surface becomes more rough. The results show that the amorphous phase is grown first as a natural buffer layer at interface directly on the mismatched substrate and then the 2201 crystal gradually grows on it. Then we tried to improve the initial growth by the plasma cleaning of the substrate surface.

Figure 1(b) shows XRD pattern for the molecular-film deposited for 1 hour after the plasma cleaning. It shows much sharper (00ℓ) peaks then the excellent 2201 crystal can be grown. A reason might be as follows. A contaminated layer on the substrate surface after the rinse can be removed by the oxygen plasma exposure then the surface is activated, resulting in improvements of adsorption of the sputtered particles and bonding to the adsorbed particles. This causes the promoted nucleations at the initial stage. We further deposited the film with supply of oxygen plasma during the growth for 1 hour after the plasma cleaning [2]. The XRD pattern for this plasma-film is shown in Fig.1 (c). The crystalline quality is much improved even compared with the plasma-cleaned molecular-film. We can suggest that Sr and Ca atoms are excited and vibrated by the plasma energy to be rearranged, then the atomic ordering is improved, and the strain energy and surface energy can be reduced [3, 4]. This causes the growth enhancement of 2201 crystal at the early stage.

Surface morphology of these films is shown in Fig.2 by SEM image. The amorphous film (a: top) is very flat. Islands (particles) are observed considerably for the 2 hour-deposited film (a: middle). They must be the 2201 crystalline clusters developed from the multinuclei on the flat 2D-like 2201 matrix. A number of such 2201 clusters increases with increasing the deposition time to 3 hours (a: bottom), then the surface becomes rough. On the other hand, the plasma-cleaned molecular-film (b) shows considerably large 2201 clusters though it was deposited for only 1 hour. This clearly indicates that the crystalline nucleation is enhanced at the initial stage by the plasma cleaning. The plasma-cleaned plasma-film (c) shows many fine 2201 clusters. This obviously indicates that the plasma supply enhances the multinucleations of 2201 crystal.

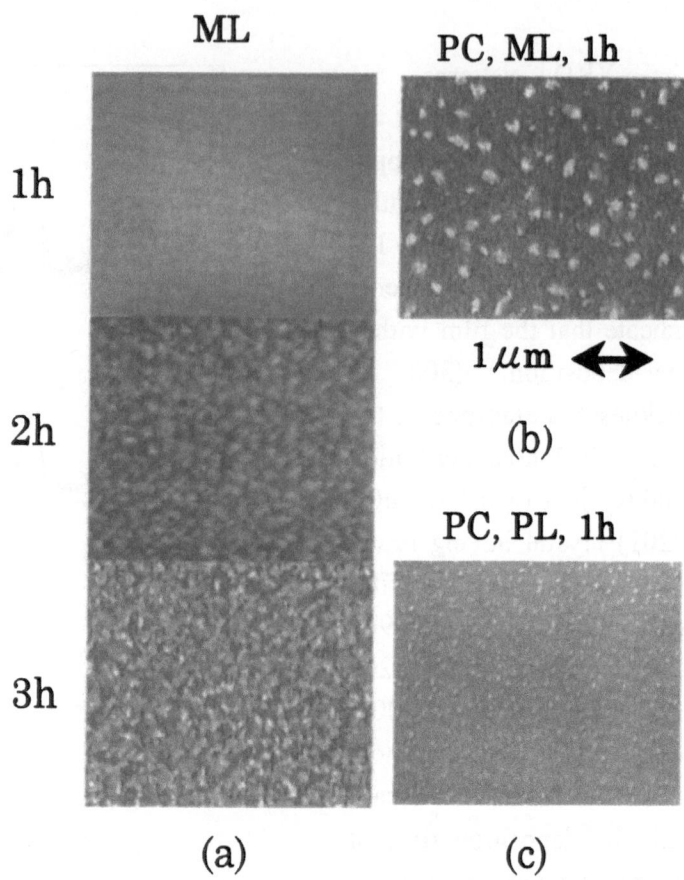

Fig.2. SEM photographs for the films shown in Fig.1.

CONCLUSION

Ca-doped Bi2201 thin films were deposited on MgO (100) substrate and the effects of plasma cleaning and plasma supply on the initial crystal growth were investigated. The plasma cleaning can promote the initial crystal nucleation. This result definitely shows that the substrate cleaning can be done at extremely low temperature as 500℃ by employing the oxygen plasma exposure. The plasma supply during the growth can enhance the multinucleation of 2201 crystal.

Acknowledgements. The authors are very grateful to TATEHO DEN-YU for supplying MgO substrates, and Professors S.Shiomi, M.Masuda and H.Kunou for XRD and SEM experiments.

*Corresponding author, E-mail: endo@cm.elec.mie-u.ac.jp

1. T. Kawai, M. Kanai, and H. Tabata, Mat. Sci. Eng. B41, 123-130 (1996).

2. T. Endo, M. Wakuta, M. Gotoh, N. Hirate, and M. Horie, Adv. Supercond. IX, 1063-1066 (1997).

3. T. Endo, M. Horie, N. Hirate, KT. Itoh, S. Yamada, M. Tada, KI. Itoh, M. Sugiyama, S. Sano, and K. Watabe, Jpn. J. Appl. Phys. 37, L886-889 (1998).

4. N. Hirate, KI. Itoh, S. Yamada, KT. Itoh, M. Tada, K. Ogai, H. Uchikawa, T. Endo, and Y. Tsutsumi, Trans. Mat. Res. Soc. Jpn. 24[1], 67-70 (1999).

A-C Phase Orientation in Thin Film Growth of YBa$_2$Cu$_3$O$_x$ in Higher Temperature Region

Tamio Endo,* Ken-ichi Itoh, Jiro Yamada, Masaki Tada, Akinori Hashizume, Morihiro Sugiyama, and Kinji Watabe

Faculty of Engineering, Mie University, Kamihama, Tsu, Mie 514-8507, Japan

Abstract: Substrate temperature (T$_S$) dependence of a-c phase orientation was investigated in growth of YBa$_2$Cu$_3$O$_x$ thin films by ion beam sputtering with supply of either oxygen molecules or plasma in T$_S$ range of 600-700°C. The growth of a-phase dominates over the c-phase at 600°C and the ratio of a-phase decreases while that of c-phase increases with increasing T$_S$. The growth of a-phase is enhanced by the plasma supply in T$_S$ <660°C. Mechanisms of a-c orientation are discussed in terms of surface migration and surface energy during the growth.

Keywords: YBa$_2$Cu$_3$O$_x$ film growth, Orientation of a-c phase, Thermal effects, Oxygen plasma

INTRODUCTION

Crystallographic orientations in growth of YBa$_2$Cu$_3$O$_x$ (YBCO) thin films on substrates are complicated because they easily depend on various growth conditions such as substrate species, substrate temperature (T$_S$), oxygen partial pressure (P$_O$) and supply of active oxygen [1-3]. It has been recognized that a "lattice matching" is the most probable mechanism for the a-c orientation. The T$_S$-dependence has been extensively investigated and it has been well known that the a-phase tends to grow at lower T$_S$ and the c-phase at higher T$_S$. Fujita et al. [3], and Mukaida and Miyazawa [2] suggested that the c-length/3 approaches a lattice constant of SrTiO$_3$ substrate in the lower T$_S$ region then the a-phase grows, while the c/3 rapidly increases and the a/b-length rather approaches to the substrate lattice constant with increasing T$_S$, then the c-phase grows in the higher T$_S$ region. This mechanism is very important when near lattice-matching substrates such as SrTiO$_3$ are employed, whereas it is not available for large mismatched substrates such as MgO. Therefore there must be some other mechanisms for the T$_S$-dependence on MgO.

We previously reported the T$_S$-dependence of a-c orientation of YBCO thin films grown by ion beam sputtering (IBS) on MgO for $400 \leq T_S \leq 650$°C [4]. The results showed that surely the a-phase is dominantly grown for T$_S$ ≤ 600°C while the c-phase for T$_S$ > 600°C, and further that the growth of a-phase is enhanced by supplying oxygen plasma instead of oxygen molecules during the growth. Owing to this enhancement effect, the a-phase could be grown at ultralow T$_S$ as 450°C. By these results we proposed the migration [5] and surface-energy mechanisms [4]. To confirm our proposals, we examined in this work the a-c orientation controls in YBCO film growth at further higher T$_S$ region in more detail using the same IBS system. The result supports our proposals again.

EXPERIMENTAL

The YBCO thin films were prepared on a MgO (100) substrate using the IBS system in the T$_S$

range from 600 to 700°C [6]. The oxygen plasma was supplied to the substrate from a plasma source located at 11 mm from the substrate during the growth in the P_O range from 0.2 to 2 mTorr. The films were also prepared by supplying oxygen molecules from the same plasma source at the same P_O without discharging. The main sample films were grown at P_O = 2 mTorr with supplying the plasma (plasma-film) or the molecules (molecular-film) at various T_S. Crystalline quality and orientation of the deposited YBCO films were investigated by X-ray diffraction (XRD) using Cu Kα line with θ-2θ scan method.

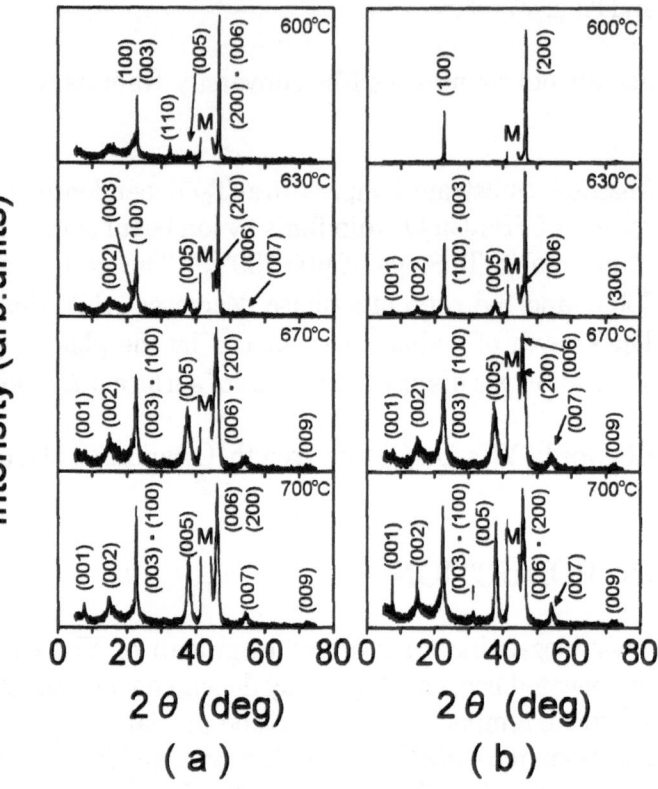

RESULTS AND DISCUSSION

The XRD patterns are shown in Figs. 1 (a) and (b) for the molecular- and plasma-films prepared at various T_S at P_O = 2 mTorr. The plasma-film at 600°C shows the single a-phase while the molecular-film at 600°C shows the major a-phase slightly mixed with the c-phase and (110)-phase. The films at 700°C clearly show the (001) and (007) peaks besides the (005) peak from the c-phase.

Fig.1. XRD Patterns for (a) the molecular-films and (b) the plasma-films prepared at various T_S at P_o =2 mTorr. M : MgO peak.

We define phase ratios γ for the a-c phases by XRD peak intensities. The γ values of a- and c-phases are plotted in Figs. 2 (a) and (b) for the molecular- and plasma-films respectively. It is clear that the ratio of a-phase decreases while that of c-phase increases with increasing T_S for the both films. The γ_a and γ_C cross each other at 640°C and 650°C for the molecular- and plasma-films respectively.

The above behaviors can be explained as follows. In the low T_S region, sputtered particles cannot migrate so long on the film surface because of shortage of thermal migration energy, then only the a-phase can grow as it needs only the short migration length [5]. Whereas as increasing T_S, the sputtered particles can get the more migration energy, then the c-phase becomes to be able to grow as it needs the long migration length [5].

The γ_a values are larger for the plasma-films than

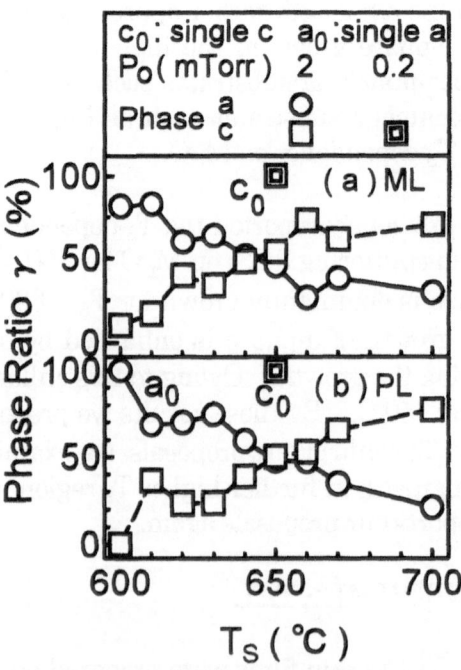

Fig.2. γ vs T_S for (a) molecular-films and (b) plasma-films prepared at 2 mTorr and 0.2 mTorr.

for the molecular-films in T_S range of 600-660°C. This indicates that the growth of a-phase is enhanced by the plasma. We propose the following mechanism [7,8]. Atomic order of Y and Ba arrangement on the growing surface is increased by utilizing the plasma energy to excite vibrations of Y and Ba atoms in their wrong sites causing the exchange of the atoms. Thus the strain energy and the surface energy can be reduced during the growth. This is supported by experimental facts that the crystallinity and surface morphology are improved by the plasma. This promotes adsorption and bonding of the sputtered particles on the surface. Thus the a-phase growth is enhanced. For $T_S > 670$ °C on the other hand, the effects of thermal migration dominates over the above plasma effects, then the c-phase grows preferentially even though the plasma is supplied. However, it should be noted from the larger γ_C values for the plasma-films than for the molecular-films that the growth of c-phase is enhanced by the plasma in this T_S range. Therefore, the plasma can assist the growth of c-phase instead if it is dominated by the thermal energy.

It was known from additional experiments that the single c-phase can be grown when P_O is reduced to 0.2 from 2 mTorr at $T_S = 650$ °C both for the molecular and plasma supplies as shown in Figs. 2 (a) and (b). Collision between the sputtered particles and the supplied oxygen species are reduced then the energy of sputtered particles are increased on the film surface compared with the higher P_O case, leading to the promotion of assist in the surface migration. This causes the enhancement of c-phase growth.

SUMMARY

Thin film orientation of YBCO on MgO were investigated. The ratio of a-phase (c-phase) decreases (increases) with increasing T_S from 600 to 700 °C because of the promotion of thermal migration. The growth of a-phase is enhanced by the plasma. The reduction of surface energy by the plasma energy is proposed for this enhancement origin. The collision effect is proposed for the enhancement of a-phase growth for the lower P_O.

Acknowledgements. The authors wish to thank Professors S. Shiomi and M. Masuda for their helps with XRD experiments, and TATEHO DEN-YU for supply of MgO substrates.

∗Corresponding author , E-mail: endo@cm.elec.mie-u.ac.jp
1. C. B. Eom, J. Z. Sun, K. Yamamoto, A. F. Marshall, K. E. Luther, T. H. Geballe, and S. S. Laderman, Appl. Phys. Lett. 55, 595-597 (1989).
2. M. Mukaida and S. Miyazawa, Jpn. J. Appl. Phys. 32, 4521-4528 (1993).
3. J. Fujita, T. Yoshitake, A. Kamijo, T. Satoh, and H. Igarashi, J. Appl. Phys. 64, 1292-1295 (1988).
4. M. Horie, KT. Itoh, S. Yamada, KI. Itoh, M. Tada, and T. Endo, Adv. Supercond. XI, 1027- 1030 (1999).
5. C. B. Eom, A. F. Marshall, S. S. Laderman, R. D. Jacowitz, and T. H. Geballe, Science 249, 1549-1552 (1990).
6. T. Endo, H. Yan, M. Wakuta, H. Nishiku, and M. Goto, Jpn. J. Appl. Phys. 35, L1260-L1263 (1996).
7. T. Endo, M. Horie, N. Hirate, KT. Itoh, S. Yamada, M. Tada, KI. Itoh, M. Sugiyama, S. Sano and K. Watabe, Jpn. J. Appl. Phys. 37, L886-L889 (1998).
8. N. Hirate, KI. Itoh, S. Yamada, KT. Itoh, M. Tada, K. Ogai, H. Uchikawa, T. Endo, and Y. Tsutsumi, Trans. Mat. Res. Soc. Jpn. 24, 67-70 (1999).

Surface morphology of $La_{1.85}Sr_{0.15}CuO_4$/$LaSrGaO_4$(100) films deposited by laser ablation at different oxygen pressures

C. Buzea[1], HB. Wang[1,2], SJ. Kim[1,2], T. Tachiki[1], Y. Uematsu [1,2], K. Nakajima[1,2], T. Yamashita[1,2,3]

[1] *Research Institute of Electrical Communication, Tohoku University, Sendai 980-8577, Japan*
[2] *CREST Japan Science and Technology Corporation (JST)*
[3] *New Industry Creation Hatchery Center, Tohoku University, Sendai 980-8579, Japan*

Abstract: The film surface morphology of LSCO films deposited by laser ablation gives information related to the velocity of the ablated species during deposition. We noticed a change in the shape of the grains as follows: at 396 mTorr oxygen pressure, the grains embody a tubular-ellipsoid form; at 230 mTorr the shape changes into a parallelepiped-type with flattened edges, one side flatter than the other; for 176 mTorr the shape becomes a "corn grain" type. These results suggest the speed of the ablated clusters is higher for low oxygen pressures, as demonstrated by the aerodynamic form of the grains.

Keywords: LSCO, grains, surface morphology, laser ablation

INTRODUCTION

One of the technological challenges in superconducting electronics is to fabricate high quality Josephson junctions using non-c axis oriented films, because of longer coherence lengths in the CuO_2 planes. The ability to increase and to control the grain size and the surface uniformity of the films is essential for optimization of junction performances. Of the several factors which influence the general nature of deposition, the oxygen pressure in the ablation chamber is particularly important. In this paper we report the surface morphology of a-axis $La_{1.85}Sr_{0.15}CuO_4$ (LSCO) grown by laser ablation at various oxygen pressures, as determined by atomic force microscopy. The grain size increases with increasing O_2 pressure.

EXPERIMENTAL DETAILS

Our LSCO films were deposited in three different oxygen pressures (397, 230 and 170 mTorr), by pulsed laser deposition (KrF excimer laser, $\lambda = 248$ nm, 20 ns pulse width, $\nu = 5$ Hz, E = 400 mJ). $LaSrGaO_4$(100) substrates were used for the deposition. In order to remove any amorphous layer, before deposition the substrates were chemically treated with 0.1% HNO_3 solution in methanol for 15 s, followed by two ultrasonic cleanings in methanol twice for 2 min each. The substrates were mounted directly on a Si heater with alumina holders. Prior to deposition the chamber was evacuated to 6×10^{-6} Torr. The oxygen pressure during deposition was measured using a capacitance manometer. The surface of the films was examined by atomic force microscopy.

DISCUSSIONS

The grains shape of the films deposited by laser ablation demonstrate the existence of an explosive target removal combined with target evaporation [1]. At high O_2 pressure, the deposited clusters grow fast, due to oxygen absorption (Fig. 1.a), giving rise to a coarse-grained film. For lower O_2 pressure, the velocity of growth of the deposited clusters is smaller, giving rise to smaller grain and a finer-grained surface (Fig. 1.b and c). The grain pattern depends also on the nature of the substrate; in Fig. 1.a, d one can notice that the grain pattern is different for films deposited on LSGO and LSAO.

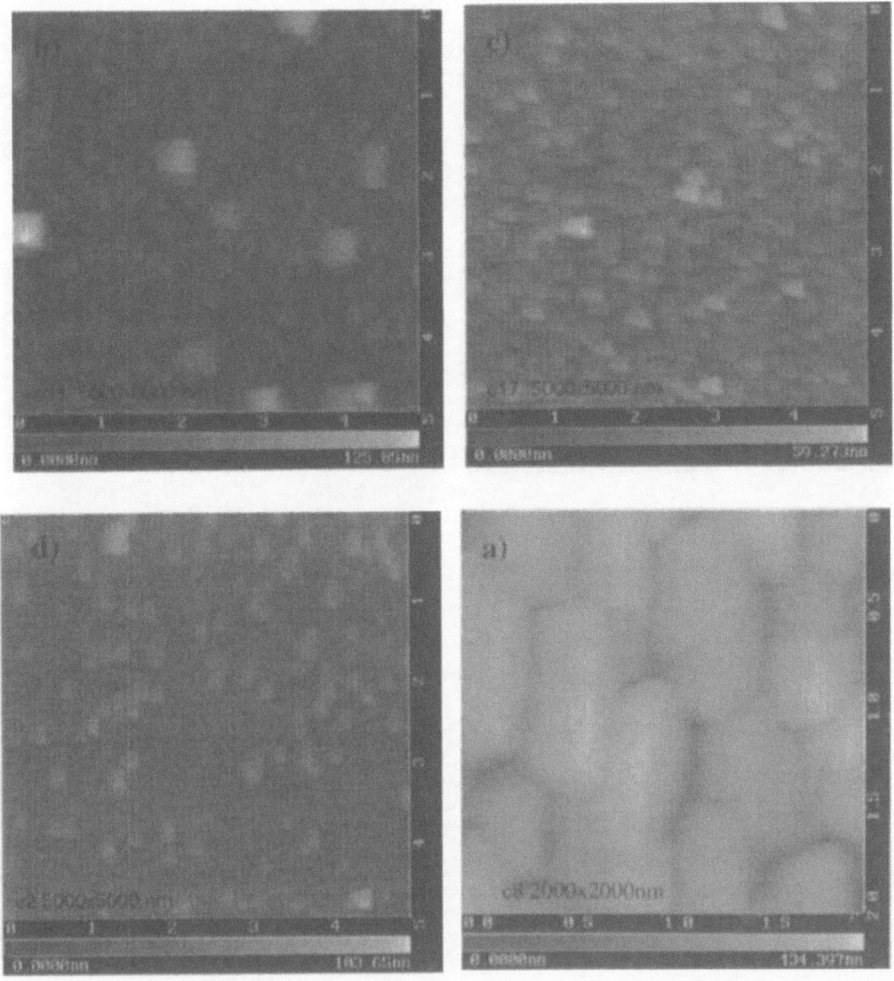

Fig. 1. *AFM images of LSCO samples deposited in an oxygen pressure of a) 397, b) 230, c) 176 mTorr on LaSrGaO₄ substrates, and of d) 398 mTorr on LaSrAlO₄ substrate (scale of 5000×5000 nm.*

As well, we noticed a change in the shape of the grains as follows: for the high O_2 pressure (397 mTorr) the grains embody a tubular-ellipsoid form (Fig. 2.a); for the lower pressure (230 mTorr) the shape changes into a parallelepiped-type with flattened edges, one side flatter than the other (Fig. 2.b); for 176 mTorr the shape becomes a "corn grain" type (Fig. 2.c). These results suggest the speed of the ablated clusters is higher for low oxygen pressure, as demonstrated by the aerodynamic form of the grains in Fig. 2.c. By increasing the pressure, the grain configuration changes, demonstrating a slightly lower speed of clusters. The above results are in agreement with the experimental observations of film surface temperature modifications during deposition by laser ablation, which provides information about plasma thermo-kinetics [1]. The shape of the grains is greatly influenced also by the nature of the substrate, as we can see in Fig. 2.d, where is the AFM image of a LSCO sample deposited on LSAO substrate.

Moreover, the dimensions of the grains are very sensitive to the oxygen pressure in the ablation chamber. At high O_2 pressures, the absorption of oxygen promotes the formation of grains with larger size than at lower pressures, as can be seen in Figs. 1,2, similar to the results for $SrTiO_3$ films [2], but opposite with the findings of Trajanovic et al. [3] for YBCO films. However, the deposition method used in ref. [3] was different from our deposition process. We kept constant the O_2 pressure during deposition, while Trajanovic et al. [3] reduced the pressure during deposition after a number of laser pulses. In the case of ref. [3], the oxygen absorption effect can be overcame by the thermo-kinetic effects occurring in the plasma plume when lowering the O_2 pressure during deposition. A possible

Fig. 2. *AFM images of LSCO samples deposited in an oxygen pressure of a) 397, b) 230, c) 176 mTorr on LaSrGaO₄ substrates, and of d) 398 mTorr on LaSrAlO₄ substrate (scale of 2000×2000 nm).*

explanation of this effect can be given according to the deposition scenario we published in ref. [1]. At the beginning of the deposition, due to an explosive target removal, clusters of different sizes are ablated. They are further heated by the incoming beam and decompose into electronically excited atoms and ions, which due to small absorption coefficient, will allow the incoming beam to reheat the target surface and explosively remove a new series of clusters in the second part of the laser pulse, which will stabilise in the expansion. After a number of consecutive laser pulses, the changes in the target morphology due to consecutive superheating of the surface during laser ablation will lead to the removal of lager clusters the first part of the laser pulse. If we keep constant the O_2 pressure, the clusters will be heated by the incoming beam and decompose again. But if at this point we decrease the O_2 pressure, the plasma expands wider, therefore a smaller number of large clusters will be heated by the incoming beam, allowing a higher number of large clusters to stabilize in the expansion and to reach the surface of the film, leading to larger grain sizes compared to the case when we keep the O_2 pressure constant.

Acknowledgements. This work was supported by CREST (Core Research for Evolutional Science and Technology) of Japan Science and Technology Corporation (JST).

1. C. Buzea, H-B. Wang, K. Nakajima, S-J. Kim, T. Yamashita, J. Appl. Phys. **86**, 2856-2864 (1999).
2. E. J. Tarsa, E. A. Hatchfeld, F. T. Quinlan, J. S. Speck, M. Eddy, Appl. Phys. Lett. **68**, 490-492 (1996).
3. Z. Trajanovic, I. Takeuchi, P. A. Warburton, C. J. Lobb, T. Venkatesan, Appl. Phys. Lett. **66**, 1536-1538 (1995).

The effect of SrTiO$_3$ buffer layer on the growth of the YBa$_2$Cu$_3$O$_{7-x}$ and SmBa$_2$Cu$_3$O$_{6+\delta}$ thin films

Kenichi Tamura[1], Yutaka Yoshida[1], Masato Hasegawa[2]*, Kimihiko Sudoh[1], Izumi Hirabayashi[2] and Yoshiaki Takai[1]

[1]Department of Energy Engineering and Science, Nagoya university, Nogoya 464-8603, Japan
[2]Surperconductivity Research Laboratory, ISTEC, Nagoya, 456-8587, Japan

ABSTRACT: We have studied YBa$_2$Cu$_3$O$_{7-x}$(Y-123) thin films on MgO substrates with SrTiO$_3$(STO) buffers using metal-organic chemical vapor deposition(MOCVD). As a result, it was indicated that the roughness of the STO buffer layer influenced the superconducting properties[1]. In this report we discussed the relationship between the growth mechanism and roughness of the buffer layer. So we determined the condition to fabricate the buffer layer with more flat surface and prepared SmBa$_2$Cu$_3$O$_{6+\delta}$(Sm–123) and Y-123 films using MOCVD on the buffer layer with various surface morphology and roughness. From the atomic force microscope image of the films, we discussed the effect of the buffer layer. In order to reveal lattice match on the buffer layer, we will also study Sm-123 films on MgO and STO substrates.

KEY WORDS: Sm-123, Y-123, STO buffer layer, MgO substrate, STO substrate, roughness, grain size

INTRODUCTION

We have prepared superconducting films by the metal-organic chemical vapor deposition (MOCVD) process using liquid-state sources and reported various issues on liquid sources. It was reported the roughness of the STO buffer layer influenced the superconducting properties[1][2]. When the roughness is about 100, the superconducting films have low Tc. On the other hand the films have high Tc on the STO buffer layer the roughness is about 20 Å . In this work, we fabricated flatter STO buffer layer on MgO substrate and prepared SmBa$_2$Cu$_3$O$_{6+\delta}$(Sm-123) and YBa$_2$Cu$_3$O$_{7-x}$(Y-123) films on the buffer layer. It is known that films is a strong substrate dependence of various properties. So we studied the effects of the buffer layer on the growth of Sm-123 thin films.

EXPERIMENTAL

We prepared Sm-123 and Y-123 thin films by MOCVD using liquid-state sources. The precursors for Y, Sm, Ba and Cu were Y(DPM)$_3$·4tBuPyNO, Sm(TMOD)$_3$, Ba(DPN)$_2$·2tetraene and Cu(TMHPD)$_2$. The source-vapors were transported by Ar gas and mixed with oxygen gas. The pressure in the reactor was adjusted to 4 Torr. The temperature of Y, Sm, Ba and Cu MO sources were kept respectively at 120, 115, 140-156 and 119°C,respectively, which were higher than the melting points of each source. The substrate temperature was set to 800-830°C. The STO buffer layer was prepared by pulsed laser deposition(PLD). The energy density was 1.05J/cm^2 by using an ArF excimer laser. The substrate temperature was varied between 850°C and 950°C. The target to

969

substrate distance was kept at 40 mm and the oxygen pressure was 0.01 Torr. The melting point of MO souces was measured by the DTA-TG scans. The crystal structure of the deposited films was examined by θ–2θ X-ray diffraction (XRD), X-ray pole figure measurements. The surface morphology of the films was investigated by atomic force microscopy (AFM).

RESULTS AND DISCUSSIONS

First, we discussed the deposition condition of Sm-123 films by MOCVD. From the result of DTA-TG of Sm(TMOD)$_3$, we found that the melting point was 105°C and the evaporation rate above the melting point was stable. So we fixed the temperature for Sm source at 115°C and fabricated Sm-123 films on MgO and STO substrates. The films on MgO and STO substrates were c-axis oriented and the length of the c-axis was about 11.71Å. Figure 1 shows the AFM image of the films on MgO substrates. Sm-123 films were grown in 3D-island-growth mode and the roughness of the film is larger than that of STO substrate because of the lattice mismatch and the roughness of the substrates. From these results, we found that the films on MgO substrates were inferior to that of STO substrates when in-plane orientation and surface morphology of the films ware compared. So we prepared STO buffer layers on MgO substrates in order to improve the properties of the films on MgO films. We prepared STO buffer layer with various surface morphology and roughness. The STO buffer layer is grown in 3D-islands-growth mode at a substrate temperature of 720°C and the roughness of the buffer layer is about 100Å. So we fabricated STO buffer layer with small roughness at a higher substrate temperature. The roughness was also controled by the film thickness of the buffer layer. Figure 2 shows X-ray θ–2θ diffraction of Sm-123 films on MgO substrate with STO buffer layer. From the result of the pattern, the STO buffer layer is oriented with (n00) in the plane perpendicular to the MgO substrate. Sm-123 films on the buffer layer is oriented with the c-axis in the plane perpendicular to the MgO substrate and the STO buffer layers. We also fabricated Y-123 films to compare with Sm-123 films. The orientation of Y-123 films was the same of that of Sm-123. We deposited Sm-123 films on STO buffer layers with various roughness to reveal the relationship between the growth mechanism and roughness of the buffer layer. Figure 3 shows the AFM image of the Sm-123 films on MgO substrate with the STO buffer layer. The surface morphology of the films on the STO buffer layer which has smaller

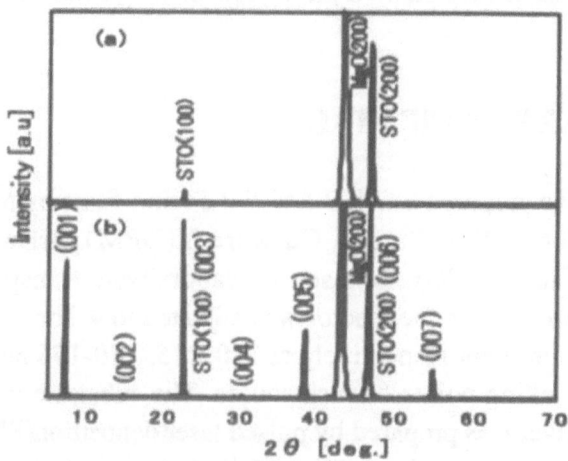

Fig. 1. AFM image of Sm-123 films on MgO substrates (5 × 5 μm)

Fig.2. XRD patterns of (a) STO buffer layer, (b) Sm-123 films on the STO buffer layer

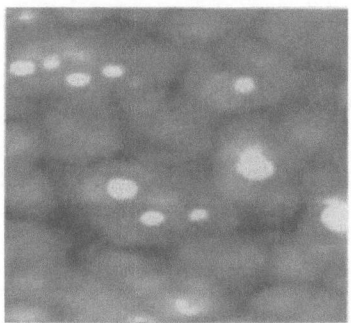 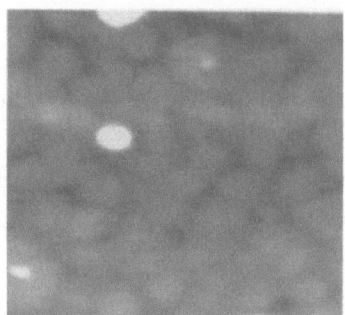

1 μ m (Rms : 30 Å)
(a)

1 μ m (Rms : 100 Å)
(b)

Fig.3. AFM image of Sm-123 on the STO buffer layer (a) the roughness of the layer is about 20 Å, (b) the roughness of the layer is about 100 Å

roughness is showed in Fig.3(a). Fig.3(b) shows the surface morphology of the films when the roughness of the buffer layer is larger than that of Fig.3(a). When comparing the surface morphology of both Sm-123 films, it is concluded that the roughness of the buffer layer influenced the surface morphology and roughness of the films. The roughness of the films in Fig.3(b) is extremely large compared with that of Fig.3(a). From the surface morphology, the grain size of Fig3.(a) is larger than that of Fig.3(b). There are larger grains and fewer grain boundaries in the films deposited on the flatter buffer layer. It is known that grain boundaries influenced Jc of the films[3]. It is important for the buffer layer not only (n00) orientation but also the flatter surface morphology.

CONCLUSION

We fabricated Sm-123 films on MgO and STO substrates by MOCVD. We found that the films on MgO substrates was inferior to that of STO substrates when various properties, for example in-plane orientation and surface morphology, was compared. Further we prepared STO buffer layers with various surface roughness and fabricated Sm-123 films on the layer. The obtained films was oriented with (n00) in the plane perpendicular to the MgO substrate. The films prepared on the STO buffer layer with small roughness have the flatter surface morphology and the larger size of the grain. It is important to fabricate the buffer layer with flatter surface for getting films with high properties .

ACKNOWLEDGMENT

This work is supported by the New Energy and Industrial Technology Development Organization (NEDO) as Collaborative Research and Development of Fundamental Technologies for Superconductivity Applications.

*Present adress : Chubu Electric Power Co., 1 Oue-cho, Minato-ku, Nagoya 455-0024, Japan

1. M. Iwata et al. 11th International symposium on superconductivity(1998) FJP-14
2. V. Boffa et al. Physica C **260**(1996), pp111
3. M. B. Field et al. Physica C **280**(1997), pp221

Surface study of the initial growth of NdBa$_2$Cu$_3$O$_{7-y}$ thin films on MgO

Koji Hattori, Shinji Santo, Yutaka Yoshida, Noriaki Matsunami and Yoshiaki Takai

Department of Energy Engineering and Science, Nagoya University, Furocho, Chikusaku, Nagoya, 464-8603, Japan

Abstract: It is reported that NdBa$_2$Cu$_3$O$_{7-y}$ (NBCO) superconductor has high critical temperature and high critical current density in a magnetic field compared with YBa$_2$Cu$_3$O$_{7-y}$ superconductor. However it is difficult to get good superconducting properties of NBCO thin films on the MgO substrate constantly. Furthermore, there are little reports about the initial growth mechanism of NBCO thin films. In this report, we discuss the initial growth mechanism of NBCO thin films fabricated by an electron beam co-evaporation method with *in-situ* reflection high energy electron diffraction (RHEED) observation. From *in-situ* RHEED observation, when *c*-axis oriented NBCO grew on the MgO substrate, it was speculated that BaO layer was the first growing layer on the surface of the MgO substrate.

Keywords: NBCO, BaO, RHEED, initial growth, in-plane orientation

INTRODUCTION

Since the discovery of High-T_c superconductors, many studies about the growth mechanisms have been made to get good superconducting properties. It is well-known about NBCO is one of the oxide superconductors which has high critical temperature ($T_c \sim$96K) and high critical current density in a magnetic field [1]. RHEED is one of the useful methods that can obtain the information on the film growth by *in-situ* observation [2]. This method has also been used to study the initial growth of the films [3].

EXPERIMENTAL

NBCO thin films were prepared by an electron beam co-evaporation method on the MgO(100) substrate. Before depositing the films, the MgO substrates were annealed at 1000°C for 1 hour in the air with O$_2$ flow. The substrate was heated to 650~850°C at a rate of 15°C/min in a flowing gas. As an oxidizer, we used O$_2$+O$_3$ (7%) mixture gas. Background gas pressure during the deposition was 2.0x10^{-5} Torr. The evaporation rate of three isolated metal sources, Neodymium, Barium and Copper, were controlled by monitoring the evaporation rate using a quartz crystal oscillator. After deposition, the films were cooled at a rate of 50°C/min in the flowing gas (2.60sccm) and took to the air out of the chamber when the substrate temperature reached 360°C. The growth rate of the film was about 0.1Å/s. RHEED technique using 15 keV electron beam was applied for monitoring the surface of the films during growth. Beam direction was parallel to the [100] and [110] direction of MgO (we describe MgO[100] and MgO[110], respectively) . The crystal structure of the deposited films was examined by θ−2θ X-ray diffraction (XRD). The critical temperature of the films was measured by the dc four-probe technique. The surface morphology of the films were investigated by atomic force microscopy (AFM).

RESULTS AND DISCUSSION

Figure 1 shows a series of RHEED patterns of the NBCO thin film growing on the MgO substrate. The direction of the electron beam is along MgO[100]. Figure 1(a) was taken from the MgO before the shutter was opened. Figure 1(b) was taken at 30 seconds after opening the shutter and this is the first change from the MgO pattern. Figures 1(c) and (d) were taken at 360 and 900 seconds after starting deposition. Figures 1(a) and (d) show that MgO[100] aligns with NBCO[100] and

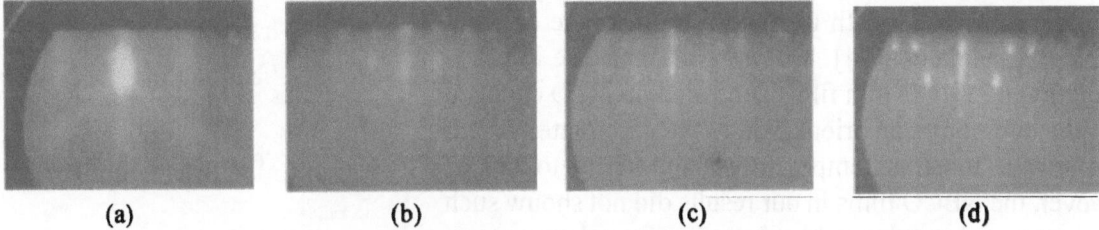

| (a) | (b) | (c) | (d) |

Fig. 1. RHEED patterns during the deposition of NBCO. (a) MgO substrate (b) after 30s deposited (c) after 360s deposited (d) after 900s deposited. (incident beam along MgO[100])

partly NBCO[110]. From figure 1(b), we can see the separation of the streaks is slightly shorter than that obtained from figures 1(c) and (d). Figure 2 was obtained when only Ba was deposited for 30 seconds on the MgO substrate. This RHEED pattern is very similar to Figure 1(b) concerning the separation and pattern of the diffraction spots, while the sharpness of the diffractions was not good in figure 2, showing a poor crystal property of BaO. This oxidized Ba is supposed to be BaO which has a rock salt structure with a lattice constant 5.52Å. Figure 2 also shows that BaO[100] rotated 45° from the MgO[100]. In the case we deposited BaO on the MgO substrate in advance before depositing NBCO, NBCO grew cube-on-cube on MgO. When Cu or Nd was firstly evapo-

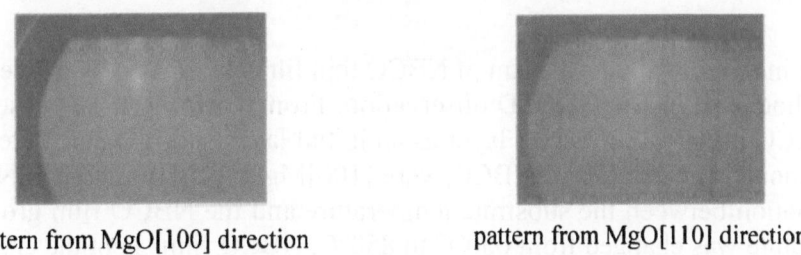

pattern from MgO[100] direction   pattern from MgO[110] direction

Fig. 2. RHEED patterns obtained by the deposition of BaO.

rated instead of Ba, the RHEED patterns, different from that shown in figure 2, were observed. From these results, we considered that when NBCO grew on MgO, BaO layer grew as a starting layer. The in-plane orientational relation among MgO, BaO and NBCO crystal lattices is speculated to be MgO[100]//BaO[110]//NBCO[100].

Figure 3 shows the surface morphology of NBCO films observed by AFM at different substrate temperatures (Ts). Figure 4 shows X-ray θ-2θ diffraction patterns of these films. Figure 4 shows that these NBCO films were all c-axis oriented NBCO. The RHEED patterns of these films indicated that these films mainly had a fourth-fold symmetric in-plane alignment on the MgO substrate while partly included 45° rotated grains. However, in the case of Ts=650°C, the weak intensity of XRD suggests that the film didn't crystallize perfectly. AFM images show a quite rough surface of

 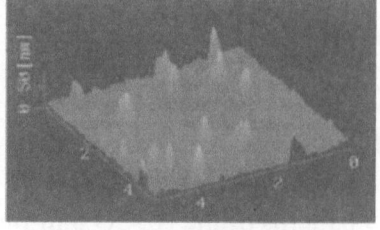 

Ts=850°C   Ts=750°C   Ts=650°C

Fig. 3. Surface morphology observed by AFM for NBCO with different Ts (image size : 5 by 5 μm).

the film grown at 650°C which is different from those grown at 750°C and 850°C. In fact, sometimes it didn't crystallize at all for Ts=650°C. With regard to the in-plane orientation, M. Mukaida et al [4]. have reported that the in-plane orientation of YBCO thin films deposited on MgO changed from the cube-on-cube orientation to the 45° rotated orientation for the substrate temperatures from 680°C to 720°C. However, the NBCO films in our results did not shouw such a temperature dependence in orientation for substrate temperatures between 750 - 850°C. Furthermore, the fact that the initial layer grown on the MgO(100) was the BaO layer was independent of the substrate temperature. The reasons for the difference between the Mukaida's results and ours may be ascribed to the difference between the ionic radius

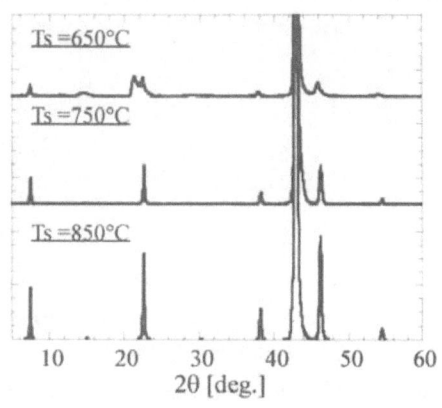

Fig. 4. X-ray θ-2θ diffraction.

of Y (0.88Å) and Nd (0.99Å) and the characteristic of an electron beam co-evaporation method, but we can't conclude from only the present experimental results.

CONCLUSION

We discussed the initial growth mechanism of NBCO thin films fabricated by an electron beam co-evaporation method with *in-situ* RHEED observation. From *in-situ* RHEED observation, the *c*-axis oriented NBCO grew with the BaO layer as an initial layer on MgO substrates. The orientational relation among MgO, BaO and NBCO were [100]MgO//[110]BaO//[100]NBCO. We also discussed the relation between the substrate temperature and the NBCO film growth. When the substrate temperature was changed from 650°C to 850°C, NBCO films kept the crystal orientation with all *c*-axis perpendicular to MgO(100). But the substrate temperature was 650°C, NBCO didn't crystallize perfectly. The in-plane orientation of these films was mainly the cube-on-cube on MgO(100), but partly 45° rotated for MgO[100] and it was independent of the substrate temperature.

1. S-I. Yoo, M. Murakami, N. Sakai, T. Higuchi and S. Tanaka: Jpn. J. Appl. Phys **33** (1994) L1000-L1003
2. T. Terashima, K. Iijima, K. Yamamoto, K. Hirata, Y. Bando and Y. Takada: Jpn. J. Appl. Phys **28** (1989) L987-L990
3. K. Hirata, F. Baudenbacker and H. Kinder: Physica C **214** (1993) 272-276
4. M. Mukaida, Y. Takano, K. Chiba, M. Kusunoki and S. Ohshima: Jpn. J. Appl. Phys **38** (1999) pp.1945-1948

The Effect of the *In-sisu* Annealing of the Growth of $Bi_2Sr_2Ca_1Cu_2O_x$ Thin Films

Koji Shinohara, Yutaka Yoshida, Noriaki Matsunami and Yoshiaki Takai

Department of Electronics, Nagoya University, Furo-cho, Chikusa-ku, Nagoya, 464-8603, Japan

Abstract: We have prepared $Bi_2Sr_2Ca_1Cu_2O_x$ (Bi-2212) thin films by a block-by-block deposition technique using pulsed laser deposition (PLD) method. We have improved the properties of Bi-2212 thin films on MgO (100) substrate by *in-situ* annealing. The stoichiometric films were prepared at 780°C, O_2 pressure of 0.2 Torr. Under the same condition the films were annealed. For comparison, we prepared Ca,Cu-rich films and annealed the films in the same way. As the annealing time increased, the intensity of XRD pattern of the stoichiometric films increased. This tendency was observed also in the Ca,Cu-rich films. These results mean that the *in-situ* annealing is effective in improvement of the crystalline perfection. Concerning the surface flatness, the stoichiometric films showed flatter surface as the annealing time increased, but the Ca,Cu-rich films didn't. We speculate that the annealing made the excess Ca and Cu atoms appear on the films surface as precipitations, and that they do not affect the quality of films inside.

Keywords: BSCCO, block-by-block deposition, *in-situ* annealing

INTRODUCTION

We have prepared superconducting Bi-Sr-Ca-Cu-O (BSCCO) thin films by a block-by-block deposition technique using a pulsed laser deposition (PLD) method with $Bi_2Sr_2CuO_x$ (BSCO) and CaCuO (CCO) targets [1]. The block-by-block deposition method was first applied for the deposition of Bi-2212 and Bi-2223 film in using an rf-magnetron sputtering technique by Ohbayashi *et al* [2]. Applying this method, we could obtain thin films of single-phase Bi-2212 and Bi-2223. About superconducting properties, Tc values of 69 and 79 K were obtained for as-deposited Bi-2212 and Bi-2223 [1]. At present, we have prepared Bi-2212 thin films for the purpose of microwave application. However, Tc's of as-deposited films we obtained were not so good as that we expected. To improve the superconducting property, we annealed the film *in-situ*. As a result, not only the superconducting property but also other properties such as surface flatness were improved. Also about the relation between the composition and the superconducting property, Egami *et al* [4] reported that superconducting properties were improved in the case of the Cu-rich composition while the surface became rough. In this work, we investigated the effect of the *in-situ* annealing on both the stoichiometric films and the Ca,Cu-rich films to obtain Bi-2212 films with good superconducting properties and also a flat surface.

EXPERIMENTAL

BSCCO films were deposited by using an ArF excimer laser (λ=193 nm) at a repetition rate of 6 Hz. The laser fluence at the surface of the target was estimated to be 0.53 J cm^{-2}. BSCO and CCO targets were selected alternately and the laser fluence was controlled for each target, if necessary. By this block-by-block deposition technique, Bi-2212 thin films were prepared. The films were deposited at 780°C in oxygen atmosphere at 0.2 Torr. The composition of the targets were $Bi_{2.0}Sr_{2.0}Cu_{1.0}O_x$ (sintered) and $Ca_{1.0}Cu_{1.0}O_x$ (sintered, but a mixed phase of Ca_2CuO_3 with CuO). MgO (100) single crystals (10 x 10 x 0.5 mm^2) were used as a substrate. In order to remove surface contaminants such as hydrocarbons, MgO substrates were annealed for 1 h at 1000°C under flowing O_2 gas. The PLD method suffers from a significant problem of particles and/or droplets grown on the film surface. Mainly they resulted from particles which come directly from the target. To solve this problem, we adopted an eclipse PLD method [3], which introduced an obstruction (a shadow mask) between the target and the substrate. In this work the target-substrate distance and the substrate-shadow mask distance were set to 50 mm and 20 mm, respectively. Buffer layer of Bi-2201 phase was deposited to a thickness of about 10 nm by employing the BSCO target. The deposited Bi-2212 film was annealed *in-situ* for 10-120 minutes at 800-900°C under the O_2 pressure of 0.002-20 Torr, then cooled to 100°C in the same atmosphere. The crystal structure of the deposited films was examined by θ-2θ X-ray diffraction (XRD) and X-ray pole figure measurements. Zero-resistance critical temperatures were measured by a four probe method. The surface roughness (Rms) of the films were investigated by atomic force microscopy (AFM).

RESULTS AND DISCUSSIONS

X-ray θ-2θ diffraction patterns of the films showed (00n) peaks of Bi-2212 phase indicating that the films oriented with the *c*-axis perpendicular to the MgO (100) substrate surface and consisted of a single phase of Bi-2212. To improve the surface flatness and the superconducting properties of films, we annealed the obtained films. At first, we tried to anneal *ex-situ* in the air, but the *ex-situ* annealing at any temperatures resulted in the film surface rougher than that before annealing. Some particles with a size of 0.2 μm were observed on the film surface by an AFM inspection. In order to improve the surface flatness of the films, we annealed the films *in-situ* where we can control O_2 pressure during annealing. By the *in-situ* annealing, various film properties such as the superconducting property, the surface flatness were improved. We annealed the films deposited at 780°C, under the following condition as at 800°C for 1 h in various O_2 pressure. We investigated the dependence of the surface morphology on the O_2 pressure during annealing. AFM images showed a flat surface. The Rms was about 20 Å when the films were annealed in O_2 pressure of 0.2 Torr. However, the film annealed at 0.002 Torr showed a wavy surface, with roughness (Rms) more than 180 Å. On the other hand, some particles with a size smaller than 0.2 mm are observed on the surfaces of the films annealed at 20 Torr. Figure 1 shows the sur-

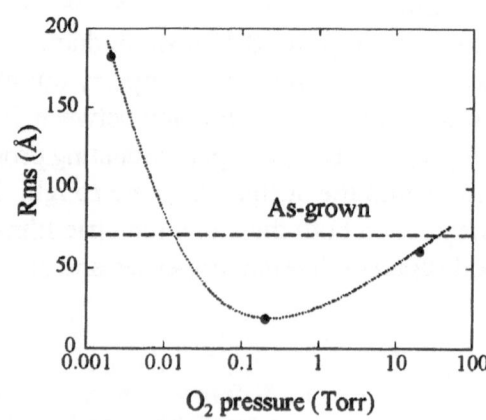

Fig.1 Dependence of the Rms measured by AFM on O_2 pressure for *in-situ* annealing

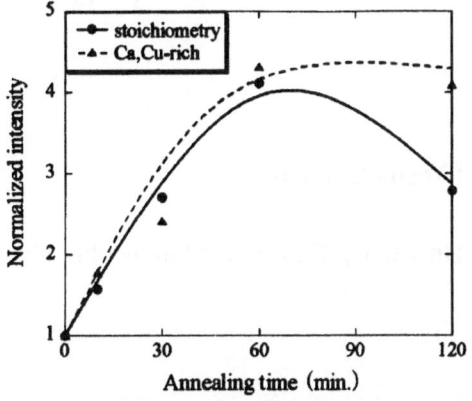

Fig.2 Dependence of normalized (002) intensity
of XRD pattern on *in-situ* annealing time

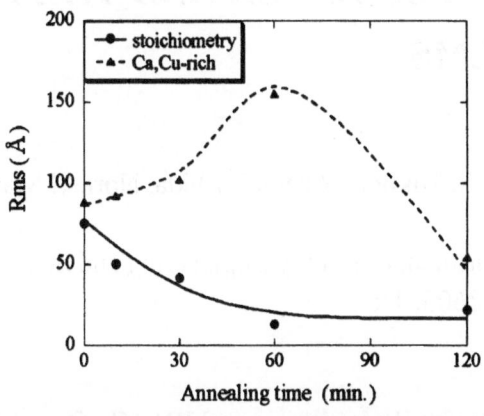

Fig.3 Dependence of rms measured by AFM of
XRD pattern on *in-situ* annealing time

face roughness (Rms) as a function of the O_2 pressure during the *in-situ* annealing. From this figure, we determined the best O_2 pressure during annealing to be 0.2 Torr. The annealed film showed a critical temperature of 80 K.

We annealed both stoichiometric films and Ca,Cu-rich films, and compared the intensity of XRD patterns and the surface roughness. Figures 2 and 3 show the annealing time dependences of the intensity of XRD pattern of the annealed film normalized for that of the un-annealed film and the surface roughness, respectively. As regards the peak intensity in figure 2, we chose the (002) peak to avoid overlapping with the peaks of the buffer layer. From figure 2, it is found that the peak intensity of both films annealed at 800°C with 0.2 Torr O_2 increased as the annealing time increased up to 60 minutes, but more than 60 minutes Ca,Cu-rich films maintained a constant intensity of (002) peak though the intensity of XRD peak of the stoichiometric films decreased. From figure 3, the surface of Ca,Cu-rich films got rougher as annealing time increased up to 60 minutes and its roughness decreased rapidly more than 60 minutes. On the other hand, Rms of stoichiometric films decreased as annealing time increased up to 60 minutes and kept constant for further annealing. It is presumed that the *in-situ* annealing makes not only the crystal growth promoted but also the excess Ca and Cu atoms appear on the films surface as precipitations, which do not affect the quality of the films inside.

CONCLUSION

We have prepared superconducting Bi-Sr-Ca-Cu-O (BSCCO) thin films by a block-by-block deposition using pulsed laser deposition (PLD) method. The properties of the films were improved by the *in-situ* annealing. We decided the best annealing condition to be 800°C for 1 h under O_2 pressure of 0.2 Torr. By the *in-situ* annealing, the surface roughness of the stoichiometric films were improved. Concerning the Ca,Cu-rich films, the *in-situ* annealing for more than 60 minutes made the surface roughness decrease but the roughness was not better than that of the stoichiometric films.

1. Asano N et al. *Supercond. Sci. Technol.* **12** (1999) 203-209
2. Ohbayashi K et al. *Appl. Phys. Lett.* **64** (1994) 369
3. Kinoshita K et al. *Jpn. J. Appl. Phys.* **33** (1994) L417
4. Egami Y et al. *Jpn. J. Appl. Phys.* **30** (1991) L478

THE GROWTH MECHANISM OF THE AMORPHOUS YSZ BUFFER LAYER AND THE PROPERTIES OF YBa$_2$Cu$_3$O$_{7-x}$ THIN FILMS

Satoshi Fukuda, Yutaka Yoshida, Noriaki Matsunami and Yoshiaki Takai

Department of Energy Engineering and Science, Nagoya University, Furo-cho, Chikusa-ku, Nagoya, 464-8603, Japan

Abstract: We have prepared YBa$_2$Cu$_3$O$_{7-x}$ (YBCO) thin films with amorphous YSZ buffer layers by a pulsed laser deposition (PLD) method. We have investigated the growth mechanism of YBCO films and the buffer layers deposited at room temperature on MgO(100) substrate and Hastelloy C-276 substrate. It was found that the YSZ film deposited at room temperature showed a weak (200) peak after post-annealing at 750℃. YBCO thin films with YSZ buffer layer on the MgO substrate showed (103) orientation at 650℃ and showed c-axis orientation at the deposition temperature higher than 700℃. In these samples, the crystallinity was improved and critical temperature of YBCO thin films increased as the substrate temperature increased. On the other hand, the crystallinity of YSZ buffer layers did not depend on the deposition temperature of YBCO thin films. YBCO thin films deposited on the Hastelloy substrates with YSZ buffer layers showed c-axis orientation at the deposition temperature lower than 750℃, but didn't show c-axis orientation at 850℃.

Keywords: YBCO, YSZ, amorphous buffer layer, Hastelloy,

INTRODUCTION

In the case of preparing YBCO thin films on the metallic substrates, buffer layer is very important to prevent mutual diffusion between the metallic substrate and the YBCO thin film[1]. Well-oriented and crystalline buffer layers grown by the techniques, such as ion-beam assisted deposition (IBAD)[2] and inclined substrate deposition (ISD)[3], are effective for the subsequent epitaxial growth of YBCO thin films. On the other hand, we have used amorphous YSZ buffer layers deposited at room temperature to improve the surface flatness. In this work, we report the growth mechanism of these buffer layers on MgO and Hastelloy substrates and investigate the properties of YBCO thin films on these buffer layers.

EXPERIMENTAL

In this study, we prepared buffer layers and YBa$_2$Cu$_3$O$_{7-x}$ thin films with film thicknesses about 2000 Å on MgO and Hastelloy substrates by a pulsed laser deposition (PLD) method with an ArF eximer laser (λ=193nm). YSZ buffer layers were deposited at room temperature. The deposition temperature of YBCO thin films on these buffer layers were 650-850°C. The laser power density was 0.66J/cm^2 and repetition rate was 10Hz. The oxygen partial pressure was maintained at 5 ×

10^{-3} Torr during deposition. After the deposition of YBCO thin films, the substrate temperature was decreased rapidly in oxygen of 20 Torr. Furthermore, for the purpose of investigating the crystallinity and surface morphology of buffer layers before YBCO deposition, we annealed YSZ single layer at 650-850°C. The crystalline properties of the obtained films were evaluated by θ-2θ mode X-ray diffraction (XRD) and X-ray pole figure measurements. The surface morphology was evaluated by atomic force microscopy (AFM). The critical temperature (T_c) was measured by four probe resistance measurement.

RESULTS AND DISCUSSIONS

XRD pattern of YSZ/MgO samples annealed at 650-850°C showed that the YSZ buffer layer crystallized weakly with a (200) orientation. From these results, it is expected that YSZ buffer layers of the YBCO/YSZ/MgO samples crystallize slightly while the substrates are heated to deposit YBCO thin films. From the AFM image, it was confirmed that the surface roughness of YSZ buffer layer annealed at 750°C is larger than that of YSZ buffer layer deposited at 750°C.

Figure 1 shows the crystallinity of YBCO thin films of the YBCO/YSZ/MgO samples as a function of the substrate temperature. YBCO films showed (103) orientation at the deposition temperature of 650°C and showed c-axis orientation at the deposition temperature higher than 700°C. The full width at half-maximum (FWHM) of the YBCO thin films grown at the deposition temperature from 650°C to 850°C indicates that the smaller values were obtained at temperatures higher than 800°C. Therefore, it was confirmed that the crystallinity of YBCO thin films was improved as the deposition temperature increased. Figure 2 shows the crystallinity of YSZ buffer layers of the YBCO/ YSZ/MgO samples. We evaluated the crystallinity of YSZ layers using the (200) intensity because of the very poor crystallinity. In this figure, YSZ (200) intensities are normalized for the YSZ (200) intensity at 650°C. Though, the crystallinity of YBCO thin films increased with the deposition temperature of YBCO thin films (Fig.1), the crystallinity of YSZ buffer layer didn't depend on the deposition temperature of YBCO thin films. Furthermore, the crystallinity of YSZ buffer layer was also same as the crystallinity of YSZ single layer annealed after deposition, for example, at 850°C as shown in Fig.2. From these results, it is concluded that the YBCO crystallizes by itself on these buffer layers.

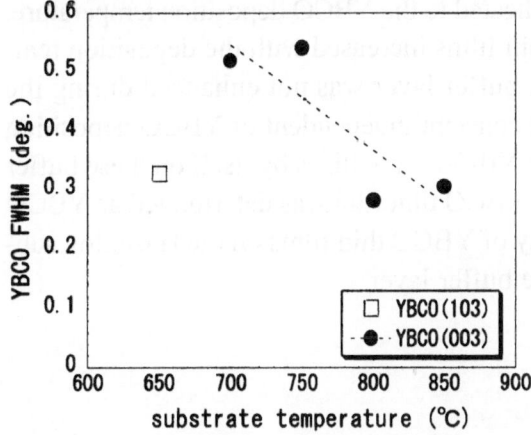

Fig.1 The dependence of the crystallinity of YBCO thin films on the deposition temperature

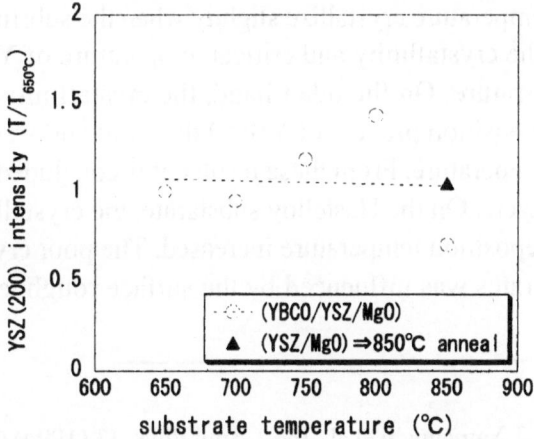

Fig.2 The dependence of the crystallinity of YSZ layer on the YBCO deposition temperature

Figure 3 shows the XRD patterns of the YBCO/YSZ/Hastelloy samples. Though the YBCO thin films deposited at 650°C and 750°C showed *c*-axis orientation, the YBCO thin film deposited at 850°C didn't show *c*-axis orientation. The peak intensity of the YBCO thin film deposited at 650°C was larger than that of the YBCO thin film deposited at 750°C. The YBCO thin film deposited at 850°C was oxidized. Figure 4 shows the surface morphology of the YSZ/Hastelloy samples annealed at 650-850°C. It can be seen from these images that the surface roughness of the YSZ buffer layers increased with the annealing temperature, for example, 90Å at 650°C, 103Å at 750°C and 150Å at 850°C. From Figs.3 and 4, it is considered that the surface roughness of buffer layer influenced the crystallinity of YBCO thin film.

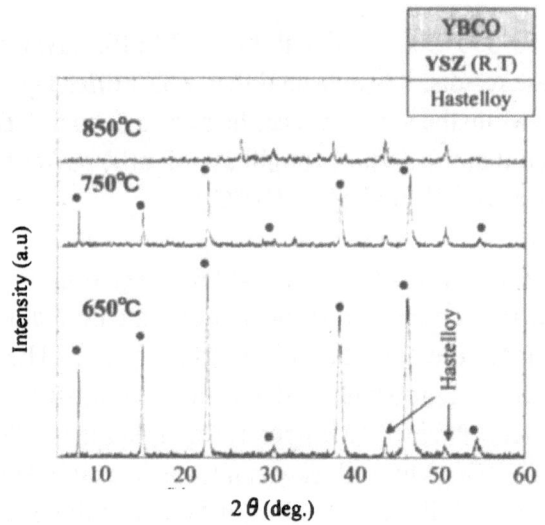

Fig.3 The dependence of the XRD patterns of the samples on the YBCO deposition temperature

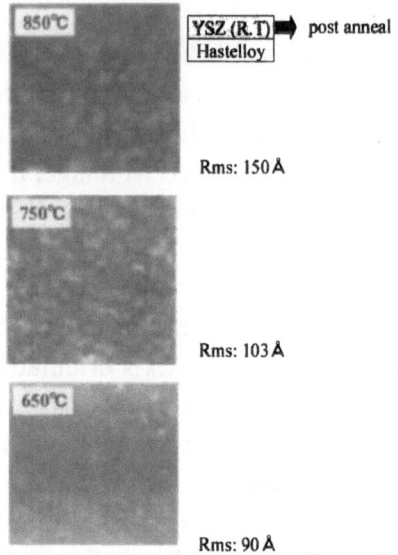

Fig.4 Surface morphology observed by AFM for YSZ layer annealed at several temperature on the Hastelloy substrate (image size: 5μm by 5μm)

CONCLUSION

The YSZ single layer deposited at room temperature crystallized weakly with orientations of (200) and (111) after post annealing. Therefore, we assume that YSZ buffer layers deposited at room temperature crystallize slightly when the substrates are heated to the YBCO deposition temperature. The crystallinity and critical temperature of YBCO thin films increased with the deposition temperature. On the other hand, the crystallinity of YSZ buffer layer was not enhanced during the deposition process of YBCO thin film and was almost constant independent of YBCO deposition temperature. From these results, it is concluded that the YBCO crystallizes by itself on these buffer layers. On the Hastelloy substarate, the crystallinity of YBCO thin film was deteriorated as YBCO deposition temperature increased. The poor crystallinity of YBCO thin films on the Hastelloy substrates was influenced by the surface roughness of the buffer layer.

1. T.Yamaguchi et al. Jpn.J.Appl.Phys. **33** (1994) 6150

2. X.D.Wu et al. IEEE Trans.Appl.Supercon. **5** (1995) 2001

3. M.Fukutomi et al. Physica C **219** (1994) 333

Growth mechanism of YBa$_2$Cu$_3$O$_{7-y}$ thin films by metal-organic chemical vapor deposition using VLS growth mode

Yutaka Yoshida[1], Masato Hasegawa[2*], Kouichi Gotou[1], Seok Beom Kim[2], Izumi Hirabayashi[2] and Yoshiaki Takai[1]

[1]Department of Energy Engineering and science, Nagoya university, Furo-cho, Chikusa-ku, Nagoya 464-8603, Japan

[2]Superconductivity Research Laboratory, Division V, International Superconductivity Technology Center, 2-4-1 Mutsuno, Atsuta-Ku, Nagoya 456-0023, Japan

Key Words: YBa$_2$Cu$_3$O$_{7-y}$, thin film, MOCVD, orientation mechanism, VLS growth

ABSTRACT: We fabricated the YBCO thin film on the MgO single crystalline substrate using VLS growth mode without YBCO seed layer. It was confirmed that the c-axis oriented YBCO film grew on top of the MgO substrate from TEM image. In addition, the *in-plane* alignment of the film was attained on MgO substrates, which had been difficult to obtain by the conventional vapor growth mode. The VLS growth enables us to control the *in-plane* alignment of YBCO films. Furthermore, superconducting properties of the film showed $T_{c(zero)}$=85K, J_c=7.0x10^{5}A/cm^2 (at 77K, 0T).

INTRODUCTION

We have studied the superconducting YBa$_2$Cu$_3$O$_{7-y}$ (YBCO) thin films deposited by metalorganic chemical vapor deposition (MOCVD) using liquid sources and reported various issues relating to liquid sources as surface morphology[1-3]. From the surface morphology, we found the existence of the liquid layer on the growing surface above a certain temperature of 750°C and proposed the possibility of the vapor-liquid-solid (VLS) growth for this observation. Furthermore we discussed about the YBCO films grown by a novel growth method using VLS mode with YBCO seed layer (Figure 1(a)). The growth rate of the films using the VLS growth mode is about five times as large as the growth rate using the conventional growth[4]. In this paper, we prepared and characterized the YBCO films using the VLS growth mode without the seed layer (Figure1 (b)). In particular, we discussed

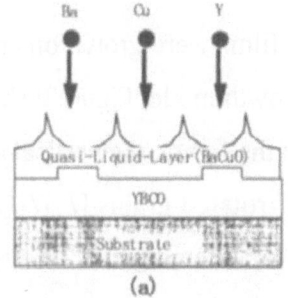

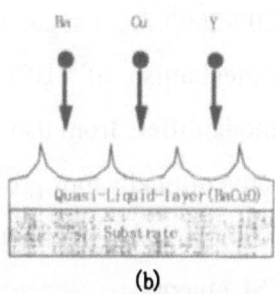

Figure1 Schmatic drawing of the model in VLS growth mode (a) with seed layer (b) without seed layer

about the orientation mechanism and the microstructure of the YBCO-VLS films.

EXPERIMENTAL

A cold-wall type MOCVD apparatus was used for preparation of the $YBa_2Cu_3O_{7-y}$ superconducting thin films. The precursors for Y, Ba and Cu were $Y(TMHD)_3 4tBPNO$, $Ba(DPM)_2$ 2tetraene and $Cu(TMHPD)_2$, respectively. The Y, Ba, and Cu sources were kept above the each melting point measured by DTA-TG. The MgO (100) single-crystalline substrates were heated at 830°C during deposition by a ceramic heater. Using the conventional vapor phase growth, the films were deposited at rate of 10-20 Å /min followed by cooling down to 200°C at 15°C /min under oxygen at 1 atm. On the other hand, when we apply the VLS growth without the YBCO seed layer, previous to depositing the YBCO films using the conventional growth mode we fabricated the liquid layer (Ba-Cu-O phase). The Ba-Cu-O liquid layer thickness were about 100-1000 Å analyzed by ICP.

RESULTS AND DISCUSSION

A. Orientation mechanism

YBCO films were grown on epitaxially MgO substrate using the conventional vapor growth and the VLS growth mode. Figure 2 shows degree of *in-plane* orientation as a function of YBCO layer thickness. In the figure, the vertical axis corresponds to the X-ray diffraction intensity ratio of cube on cube and 45° rotated grains $[I_0/(I_0+I_{45})]$ observed in X-ray ϕ -scanning. Using the conventional growth mode, 45° rotated grains ([110]YBCO) ‖ [100]MgO grains) increased as increasing the thickness of YBCO. On the other hand, using the VLS growth mode, we obtained only the cube on cube grains

([100]YBCO) ‖ [100]MgO grains) up to the thickness of 200-2000 Å . It suggest that the *in-plane* orientation mechanism of YBCO film using the VLS growth mode differs from that of YBCO film grown by the conventional vapor growth mode, which has been well explained by the Near Coincident Site Lattice (NCSL) theory.

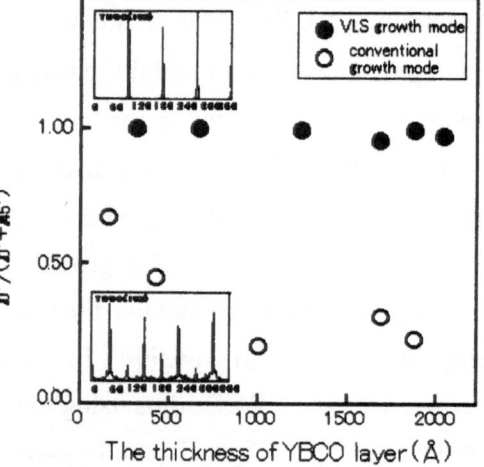

Figure2 The *in-plane* alignment versus the thickness of YBCO layer

B. Microstructure and the superconducting properties

Figure 3 is a cross sectional TEM image of the YBCO films using the VLS growth mode without seed layer.

The *in-plane* orientational relationship is [100]YBCO) ‖ [100]MgO, which is recognized from figure 2. The interface between the YBCO-VLS layer and the substrate is sharp and clean. The YBCO films grow directly on the MgO surface. There are no reaction products or intermediate layer. On the other hand, among the YBCO grains there are other phases which may be grown from the excess of the liquid layer. Although, the Ba-Cu-O layer is deposited as the initial layer of

Figure 3 Cross-sectional TEM image of the YBCO films using the VLS growth mode. The arrows shows the Ba-Cu-O phase.

the VLS growth, no extra layer is observed at this interface. It is believed that the Ba-Cu-O layer is intercorporated into the YBCO layer in the VLS growth process.

Superconducting properties of the YBCO film deposited using the VLS growth mode without seed layer showed $T_{c(zero)}$=85K, J_c=7.0x10^{5}A/cm^2 (at 77K, 0T).

CONCLUSIONS

We fabricated YBa$_2$Cu$_3$O$_{7-y}$ epitaxial thin films grown by MOCVD using the VLS growth mode without seed layer. The *in-plane* alignment of the film was attained on MgO substrates, which had been difficult to obtain by the conventional vapor growth mode. Furthermore from the TEM images there are finely distributed phases in the YBCO-VLS layers and no reaction products or intermediate layer at the interface.

Acknowledgments. This work was partly supported by the New Energy and Industrial Technology Development Organization (NEDO) as Collaborative Research and Development of Fundamental Technologies for Superconductivity Applications. We have benefited from useful comments and discussions with Prof. Yoshinori Furukawa of the Institute of Low Temperature Science of Hokkaido university.

* Present address: Chubu Electric Power Co., 1 Oue-cho, Minato-ku, Nagoya 455-0024, Japan

1. Y. Yoshida et.al.: Appl. Phys. Lett. **69**(1996) p845

2. Y. Yoshida et.al.: Jpn. J. appl. Phys. **36** (1997) p. L1376

3. M. Iwata et al : Jpn. J. appl. Phys. **37** (1998) p. L715

4. Y. Yoshida et. al.: *submitted to J. Cryst Growth*

Identification of Barrier in the Modified Interface High-T_c Josephson Junction by TEM

J.G. Wen,[1] T. Satoh,[2] M. Hidaka,[2] S. Tahara,[2] N. Koshizuka,[1] and S. Tanaka[1]

[1]Superconductivity Res. Lab., ISTEC, 1-10-13 Shinonome, Koto-ku, Tokyo 135-0062, Japan
[2]Fundamental Res. Lab., NEC Corporation, 34 Miyukigaoka, Tsukuba, Ibaraki 305-8501, Japan

Abstract: The atomic structure and composition of modified interface junctions which showed reproducible critical current I_c (I_c $1\sigma < 8\%$ for series connected 100 junctions) are investigated by transmission electron microscopy. Transmission electron microscopic observations show the existence of a thin barrier (1-2 nm) homogeneously covering the ion milled edge of the base $YBa_2Cu_3O_y$ film although there is no barrier deposition and annealing process. High-resolution electron microscopy images and energy dispersive x-ray analysis with spot size of 1 nm indicate that the barrier is a Ba-based perovskite-like structure, $(Y_{1-x}Cu_x)BaO_y$ with $x < 0.5$. A thin amorphous layer whose composition deviates from $YBa_2Cu_3O_y$ is formed due to the preferential sputtering of Cu. The amorphous layer recrystallizes into the nonequilibrium phase, $(Y_{1-x}Cu_x)BaO_y$ after heating up to the deposition temperature. The effect of La in substrate and insulation layer on the junction property was discussed.

Keywords: Josephson junction, TEM, barrier, perovskite structure, composition

INTRODUCTION

Recently, interface-engineered $YBa_2Cu_3O_y$ (YBCO) junctions (IEJ), developed by Moeckly et al. [1,2], attracted much attention since the reproducible and manufacturable process of fabrication is quite suitable for digital circuit applications. In this process, no barrier deposition is carried out: the barrier is formed just by structural modification using ion bombardment and vacuum annealing. Recently, Satoh et al. [3] modified the process, in which the edge of the base YBCO film was formed by normal Ar ion milling and then the film was heated to a deposition temperature of about 700°C in O_2 for a top YBCO layer deposition without being exposed to air. Their modified interface junctions (MIJ) also showed reproducible critical currents I_c, with a 1σ spread in I_c of less than 8% for series connected 100 junctions.

In order to understand the formation mechanism of the barrier in these junctions, it is necessary to study the atomic structure and composition of the barrier by cross-sectional transmission electron microscopy (TEM), in particular, high resolution electron microscopy (HREM) with nanometer-size element analysis. TEM investigations of Jia et al. [4] have shown that a homogenous layer of a non-superconducting, cation-disordered cubic YBCO phase can recrystallize at an interface which is made amorphous by Ar ion-beam etching. Huang et al. [5] reported their TEM observations on the microstructures of IEJs and they concluded that the barrier is probably a cubic or pseudo-cubic YBCO variant. Further study on the barrier structure of IEJs by Gustafasson et al. [6] showed that the barrier structure has a pseudo cubic unit cell which is rotated 45° around the [001] axis with respect to the underlying YBCO. No composition change of the barrier in IEJs compared to adjacent YBCO films was observed in both studies.

In this paper, the atomic structure and composition of the barrier in the MIJs are studied by TEM. We found that the barrier layer has a Ba-based perovskite-like atomic structure and composition, $(Y_{1-x}Cu_x)BaO_y$ with $x < 0.5$. The composition change was caused by the preferential sputtering of Cu during ion milling process.

EXPERIMENTAL

The base electrode YBCO layer and insulation layer $(La_{0.3}Sr_{0.7})(Al_{0.65}Ta_{0.35})O_y$ (LaSrAlTaO) (or $SrTiO_3$) were deposited on a LaSrAlTaO (or $SrTiO_3$) substrate by KrF pulsed laser deposition (PLD). In order to avoid the ion-milled YBCO edge being exposed to air prior to the top film deposition, the insulation layer was first patterned by photoresist and then the YBCO edge was fabricated using the patterned insulation layer as a mask. [7] The 200V Ar ion beam had an incidence angle of 30° to 45° to the substrate surface and the substrate was rotated during the ion-beam etching. Atomic oxygen was flowed directly to the sample to prevent oxygen depletion from the YBCO edge. The fabricated edge was directly heated to about 700°C without breaking vacuum for the further deposition of the top YBCO layer. This fabrication process has been described in detail elsewhere. [3,7]

Cross-sectional specimens for TEM are mechanically ground by tripod directly down to a uniform thickness less than 10 mm and followed by an ion milling process. More details of the sample preparation can be found elsewhere. [8] Electron microscopy for HREM was performed with a JEOL-4000EX electron microscope operating at 400 kV. Nanometer-size composition analysis was performed on a JEOL-2010 electron microscope equipped with a field-emission gun and an Oxford energy dispersive x-ray (EDX) analysis system.

RESULTS AND DISCUSSION

Two types of MIJs which exhibited resistively shunted junction (RSJ) and flux-flow (FF) type I-V characteristics were selected to carry out TEM observations. The fabrication processes for these two types of junctions are similar; the only difference is that substrate and insulation layer are LaSrAlTaO for the RSJ type junction and $SrTiO_3$ for the FF type junction. Fig.1 (a) and 1(b) show cross-sectional TEM images of MIJs exhibiting RSJ and FF type I-V characteristics, respectively. Although there was no barrier deposited, an interface can be clearly seen as a dark line in the low magnification images in the insets of Fig. 1(a) and 1(b). In Fig. 1(a), the structure of the barrier is different from YBCO, as indicated by white dashed lines. CuO layers in YBCO films appear as white lines in this image, but these white lines can not be observed in the barrier. The 2 nm thick barrier covers the base YBCO edge homogeneously. The barrier consists of many small domains with a size of 2-3 nm. The orientations of different domains are more or less the same. In contrast, one can see the continuity of CuO layers across

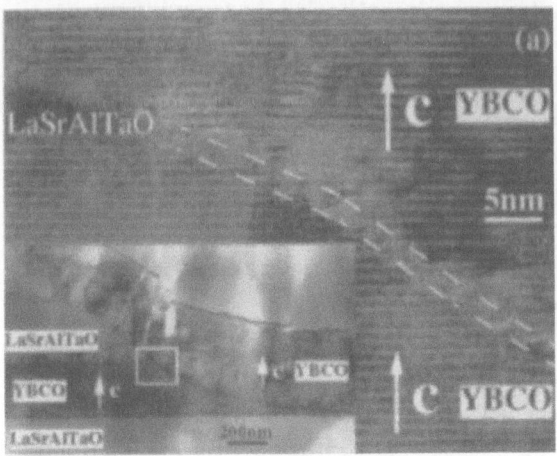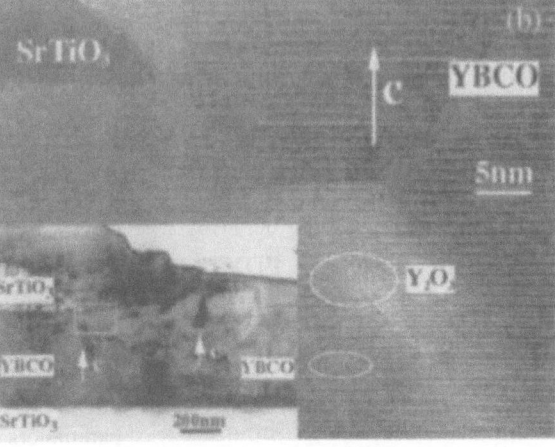

Fig. 1 Cross-sectional TEM images of MIJs which had a) RSJ type and b) Flux-flow type I-V characteristics. The barrier (about 2 nm thick) is indicated by the dashed lines in Fig. 1(a). Note the disappearance of CuO lines in the barrier of Fig. 1(a) while a continuity of CuO lines across the interface in Fig. 1(b).

the interface in Fig. 1(b). We believe this is the source of the FF type I-V characteristics. On both sides of the barrier layer, the YBCO films are c-axis oriented, indicating that the top YBCO film can grow epitaxially on the barrier although their structures are different. In both low magnification images, top YBCO films on the YBCO ion-milled edge show better crystallinity than that on LaSrAlTaO (or SrTiO₃) substrate and the insulation layer. A indistinct boundary between good and bad crystallinity films in the top YBCO film on the edge is almost parallel to substrate surface plane, indicating a step-flow growth mode of the top YBCO film on the ion-milled edge.

Fig.2(a) and 2(b) show HREM images of interfaces in MIJs exhibiting RSJ and FF type I-V characteristics, respectively. In Fig. 2(a), although steps of height of 1.2 nm can be observed on the ion-milled surface, the 2 nm thick barrier still fully covers the ion-milled surface. In contrast, one can see the continuity of CuO layers across the interface in the upper and lower part of Fig. 2(b). The lattice parameters of the barriers in Fig. 2(a) and the circled regions in Fig. 2(b) are estimated to be about 4.1 Å along both the a- and c-axes of the YBCO film, using the lattice parameters of the YBCO film as an internal standard. In some areas of the barrier, one of lattice parameters along the c-axis of the YBCO film is 4.3 Å, while the lattice parameter along the a-axis of the YBCO film is still 4.1 Å as is shown in the inset of Fig. 2(a). Due to the slightly larger lattice parameter of the barrier compared to the YBCO films, one can observe misfit dislocations at the interfaces between the YBCO films and the barrier in Fig.2(a) and 2(b).

High spatial resolution EDX analysis, with a probe size of about 1 nm, showed that the barriers in Fig. 2(a) and 2(b) have the same composition as each other but differ from YBCO, as shown in Fig. 3. The average atomic ratio of Y:Ba:Cu from 20 areas in the junction is 30: 43: 27. The EDX spectrum of the adjacent

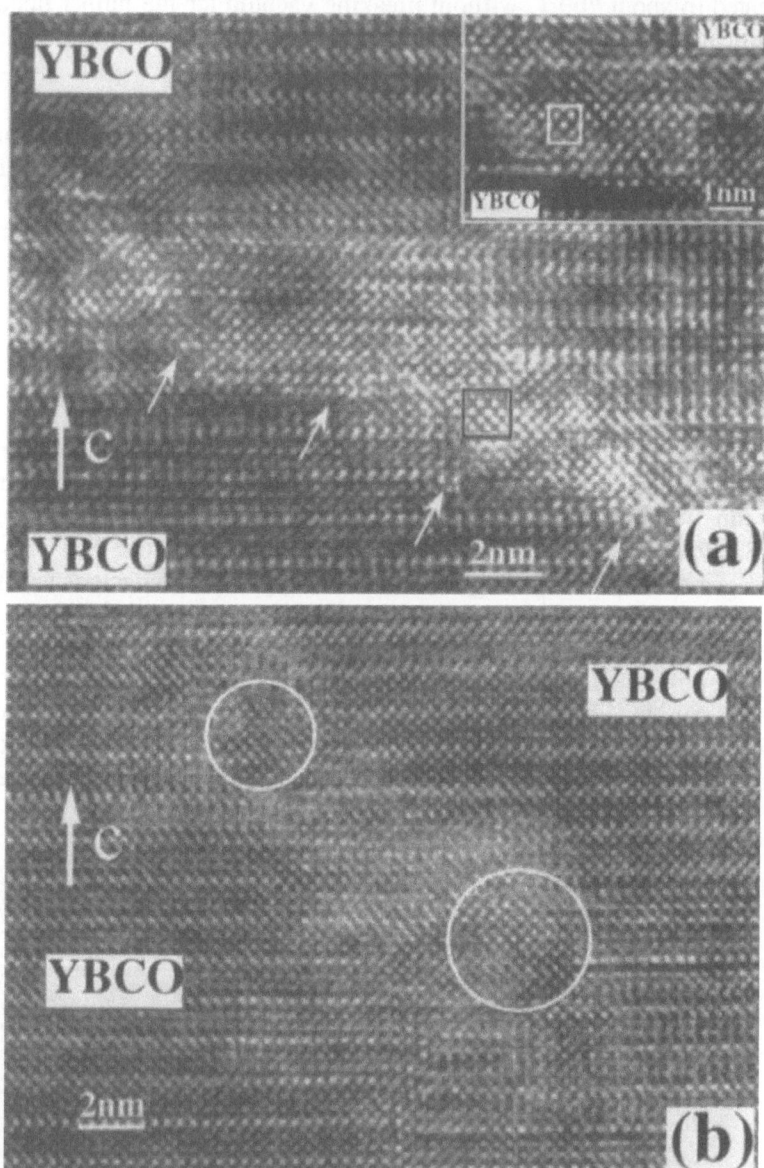

Fig. 2 HREM images of interfaces in MIJs exhibiting a) RSJ type and b) FF type I-V characteristics. In most areas the lattice parameters of the barrier are 4.1 x 4.1 Å, while in some areas, one of lattice parameters along the c-axis of YBCO film is 4.3 Å, as shown in the inset of Fig. 2(a). Steps on the bottom YBCO edge are indicated by small arrows in Fig. 2(a). Note that the continuity of CuO layers across the interface can be seen in the upper and lower part of Fig. 2(b).

YBCO film (with known atomic ratio) was measured under the same conditions for internal calibration by the Cliff-Lorimer technique. [9] Compared to the YBCO film, Cu content is low in the barrier. This can be used to distinguish whether the electron beam is on the barrier or on the adjacent YBCO film by monitoring the intensity of the Cu peak during the EDX measurements. The atomic ratio Y:Ba in the barrier might be even higher, because some x-rays from the adjacent YBCO may contribute to the EDX spectrum due to the comparable size of the electron beam and the barrier.

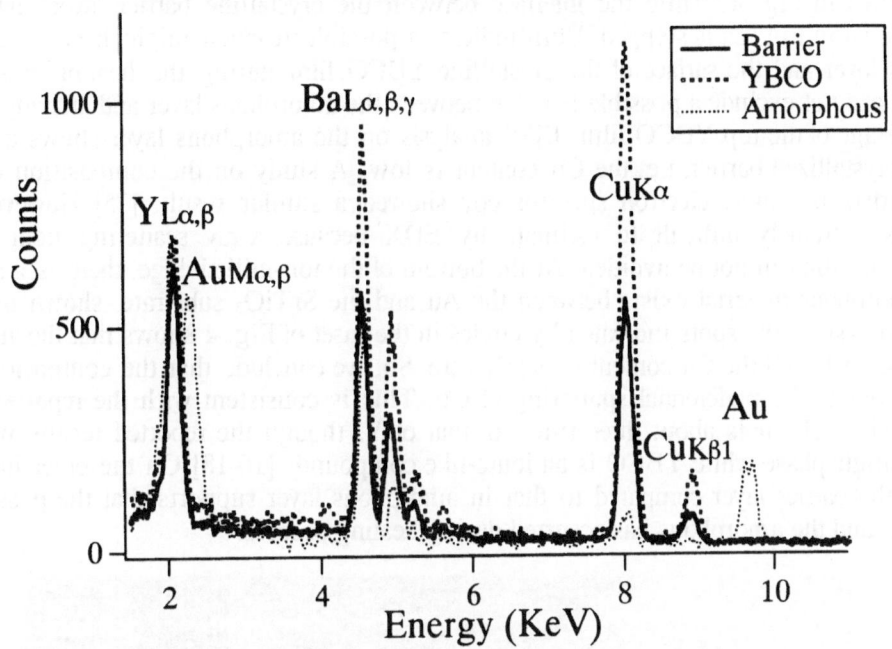

Fig. 3 EDX spectra of the barrier, the adjacent YBCO film, and the amorphous layer at circled regions in the inset of Fig. 4. The average atomic ratio of Y:Ba:Cu from 20 areas of the barrier is 30: 43: 27. The Cu content in the barrier is lower than that in the YBCO film.

From the HREM images and lattice parameters of the barrier shown in Fig. 2, the structure of the barrier is very possibly perovskite-like. All perovskite-like structures found in the YBCO system can be classified into two types according to the occupation rate at A or B sites in ABO_3. In the first type, Cu-based perovskite-like structures, Cu occupies the B site while Y and Ba occupy the A site. YBCO and cubic YBCO in which Y and Ba randomly occupy the A site are Cu-based perovskite-like structures. [4,10,11] The average lattice parameter of the Cu-based perovskite-like structure is about 3.9 Å. In the second type, Ba-based perovskite-like structures, Ba occupies the A site while Y and Cu occupy the B site. $(YCu_3)Ba_4O_y$ and $(Y_3Cu_5)Ba_8O_y$ are Ba-based perovskite-like structures. [12] The average lattice parameter of the Ba-based perovskite-like structure is about 4.1 Å, which is slightly larger than that of the Cu-based-like perovskite structure due to the large ionin radius of Ba. From HREM images, lattice parameters, and EDX analysis, one can conclude that the barrier is not a Cu-based, but a Ba-based perovskite-like structure. In all reported Ba-based perovskite-like phases, Y content is lower than Cu content. In the barrier it is the opposite, therefore, the barrier can be written as $(Y_{1-x}Cu_x)BaO_y$, with $x <$ 0.5. This phase does not exist in the thermal equilibrium YBCO phase diagram. Kwestroo et al. [13] reported a perovskite-like $Y_2Ba_2O_5$ phase with lattice parameters of a = b = 4.37 Å, and c = 11.85 Å. The lattice parameters of the primary perovskite-like unit cell are then a = b = 4.37 Å and c = 3.95 Å. The barrier which shows lattice parameters of 4.1 Å x 4.3 Å could be a $Y_2Ba_2O_5$ phase, which contains nearly no Cu. Hence, the observation of 4.1 Å x 4.3 Å in some areas may indicate an inhomogemous distribution of Cu in the barrier. Y_2O_3 phase (face-centered cubic, a = 10.5 Å) can be ruled out by the lattice parameters and the ratio of lattice parameters along the a- and c-axes of the YBCO film as shown in Fig. 1(b). [14] Y_2O_3 is always 45° in-plane tilted in order to match the YBCO lattice, so the ratio of lattice parameters in Y_2O_3 is $\sqrt{2}$ when looking along [100] of YBCO lattice.

In order to understand the composition change of the barrier which happened during the ion milling or heating process, we covered a freshly milled sample with an in-situ Au layer. The Au layer was *in-situ* deposited at room temperature by PLD to protect the surface from chemical or TEM sample preparation damage. Both substrate and insulation layer are $SrTiO_3$. Fig. 4 shows a cross-sectional TEM image of the test sample. The thickness of the thin amorphous layer is about 1 nm, which is thinner than the barrier shown in Fig. 2a. Note that the interface between the YBCO and the amorphous layer is quite smooth, as shown in Fig. 4, while the interface between the crystalline barrier layer and the bottom YBCO film shown in Fig. 2a is stepped. Both indicate a possible reaction might have occurred between the amorphous layer and the surface of the crystalline YBCO film during the heating process. On the other hand, we can not exclude a possible reaction between the amorphous layer and arriving atoms at the initial growth stage of the top YBCO film. EDX analysis on the amorphous layer shows a composition similar to the crystallized barrier, i.e. the Cu content is low. A study on the composition of ion-milled YBCO film surface by Auger electron spectroscopy showed a similar result. [15] However, the exact composition is extremely difficult to estimate by EDX because x-ray scattering from the adjacent crystalline YBCO film can not be avoided. At the bottom of the ion-milled edge, there is a small area in which only amorphous material exists between the Au and the $SrTiO_3$ substrate, shown in the inset of Fig. 4. EDX analysis at the spots indicated by circles in the inset of Fig. 4 shows that the atomic ratio of Y to Ba is close to 1, and the Cu content is nearly zero. So, we conclude that the composition deviation from YBCO is due to the preferential sputtering of Cu. This is consistent with the reported results that the sputtering yield of Cu is about three times of that of Y, though the reported results were obtained from metallic single phase while YBCO is an ionic-like compound. [16-18] On the other hand, a higher Cu content in the barrier layer compared to that in amorphous layer supports that the possible reaction between YBCO and the amorphous has occurred during heating process.

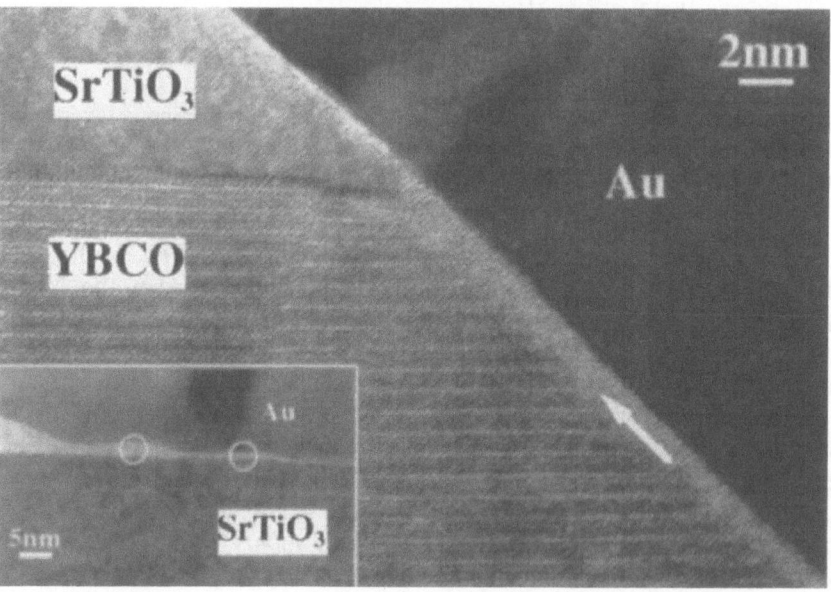

Fig. 4 Cross-sectional TEM image of a test sample in which the fresh ion-milled edge was covered in-situ by a Au layer. A thin amorphous layer of about 1 nm can be seen on the YBCO edge where there is no amorphous between $SrTiO_3$ edge and Au layer. EDX measurements at the bottom of the ramp (circled regions) show that the Cu content in the amorphous region is nearly zero.

From the above results, one can see that the atomic structure and composition of the barrier material in MIJs exhibiting RSJ and FF type I-V characteristics are similar. One difference in microstructure is that the barrier fully covers the ion-milled edge in the RSJ type MIJs while the barrier does not fully cover the edge in the FF type MIJs. The RSJ type MIJs were made with LaSrAlTaO substrates and insulators, so the existence of a small amount of La in the barrier due to redeposition during ion milling was thought as a reason to cause the different I-V characterstics. However, it was not found by EDX analysis. One possible reason for the difference is that different substrates may slightly change the effective growth temperature, such that the coverage of the barrier is different. Recent experiments confirm that an RSJ type

I-V characteristic can also be obtained on a $SrTiO_3$ substrate with a $SrTiO_3$ insulation layer over a narrow range of deposition temperatures. The coverage of the barrier is more sensitive to the deposition temperature than La impurities. After annealing the ion-milled edge at various temperatures, Fujimaki *et al.* [19] found that I_c decreases when annealing temperature decreases. Our TEM studies on MIJs with high and low I_c showed that there exist pinholes in MIJ with high I_c ($10^5 A/cm^2$) and no pinhole for MIJ with low I_c ($10^3 A/cm^2$) though RSJ-like I-V characterstics were observed for both MIJs.

Strain field caused by the mismatch between the barrier layer and the underlying YBCO may play an important role on the appearance of the nonequilibrium Ba-based perovskite phase at the ion-milled edge. On (001) YBCO surface, Y_2O_3 particles [15] and a continuous Y_2O_3 layer [5] were observed after annealing the amorphous layer. Different phases were formed on the ion-milled edge and (001) YBCO surface may result from different strain fields between phases and different orientated surfaces. Moreover, the ultra thin thickness and the homogeneity of the barrier thickness may be also governed by this strain field such that the barrier becomes unstable when thickness of the barrier is over a critical thickness.

CONCLUSION

In summary, we propose a formation mechanism of the barrier in the modified interface junctions which have no intentional barrier deposition or annealing process. A thin amorphous layer is formed on the edge of the ion-milled YBCO film whose composition deviates from YBCO due to the preferential sputtering of Cu. The barrier recrystallizes during the heating of the sample to the deposition temperature. The crystalline barrier is identified as a Ba-based perovskite-like structure, $(Y_{1-x}Cu_x)BaO_y$ with $x < 0.5$.

Acknowledgments: This work was supported by the New Energy and Industrial Technology Development Organization (NEDO) as Collaborative Research and Development of Fundamental Technologies for Superconductivity Applications.

1. B.H. Moeckly and K. Char, Appl. Phys. Lett. **71** (1997) 2526.
2. B. H. Moeckly, K. Char, Y. Huang, and K.L. Merkle, IEEE Trans. on Appl. Supercond. **9** (1999) 3358.
3. T. Satoh, M. Hidaka, and S. Tahara, IEEE Trans. on Appl. Supercond. **9** (1999) 3141.
4. C.L. Jia, M.I. Faley, U. Poppe, and K. Urban, Appl. Phys. Lett. **67** (1995) 3635.
5. Y. Huang, K.L. Merkle, B.M. Moeckly, and K. Char, Physica C **314** (1999) 36.
6. M. Gustafsson, E. Olsson, and B. Moeckly, private communication.
7. T. Satoh, M. Hidaka, and S. Tahara, IEEE Trans. on Appl. Supercond. **7** (1997) 3001.
8. J.G. Wen, in "From the Macroscopic to the Atomic Scale: Charactersation of High-T_c Superconductors by Electron Microscopy", edited by S.J. Pennycook and N.B. Browning, Cambridge Univ. Press.
9. J.I. Goldstein, in "Introduction to Analytical Electron Microscopy", edited by J.J. Hren, J.I. Goldstein, and D. C. Joy, 1979, New York.
10. J.G. Wen, T. Satoh, M. Hidaka, S. Tahara, N. Koshizuka, and S. Tanaka, to be published on Appl. Phys. Lett.
11. T. Nagano, T. Hashimoto, J. Yoshida, Physica C **265** (1996) 214.
12. D.M. De Leeuw, C.A.H.A. Mutsaers, C. Langereis, H.C.A. Smoorenburg, and P.J. Rommers, Physica C **152** (1988) 39.
13. W. Kwestroo, H.A.M. van Hal, and C. Langereis, Mat. Res. Bull. **9** (1974) 1631.
14. A. Catana, R.F. Broom, J.G. Bendnorz, J. Mannhart, and D.G. Schlom, Appl. Phys. Lett. **60** (1992) 1016.
15. S. Inoue, T. Nagano, H. Sugiyama, and J. Yoshida, extended abstract in 1999 international workshop on superconductivity by ISTEC and MRS, June 27-30, Kauai Island, Hawaii, USA.
16. G. Carter and J.S. Colligon, in "Ion Bombardment of Solids", Heinemann Educational Books Ltd, London, 1968.
17. N. Laegreid and G.K. Wehner, J. Appl. Phys. **32** (1961) 365.
18. D. Rosenburg and G.K. Wehner, J. Appl. Phys. **33** (1962) 1842.
19. A. Fujimaki, M. Horibe, K.Kawai, N. Hayashi, G. Matsuda, Y. Inagaki, and H. Hayakawa, extended abstract in 1999 international workshop on superconductivity by ISTEC and MRS, June 27-30, Kauai Island, Hawaii, USA.

Analysis of the surface-modified barrier in YBaCuO ramp-edge Josephson junctions

Yoshihisa Soutome, Tokuumi Fukazawa, Akira Tsukamoto, Yoshinobu Tarutani and Kazumasa Takagi

Advanced Research Laboratory, Hitachi, Ltd. 1-280 Higashi-Koigakubo, Kokubunji, Tokyo, 185-8601, JAPAN

Abstract: We have fabricated $YBa_2Cu_3O_{7-x}$ ramp-edge Josephson junctions with a surface-modified barrier. The surface-modified barrier was formed by Ar-ion irradiation of a $YBa_2Cu_3O_{7-x}$ ramp surface after fabrication of a ramp-edge. We investigated the composition and structure of the surface-modified barrier using XPS and TEM. Our XPS analysis suggested that the surface modified barriers were created by recrystallization of amorphous $YBa_2Cu_3O_{7-x}$ that was modified by the Ar-ion irradiation. The atomic ratio of Y:Ba:Cu in the surface-modified barrier was 1:1.2:1.1. Our TEM observations revealed no difference in the crystal structures between the barrier and the $YBa_2Cu_3O_{7-x}$ at the interface. The structure of the surface-modified barrier is appeared to be very similar to that of $YBa_2Cu_3O_{7-x}$.

Keywords: XPS, TEM, surface-modified barrier, ramp-edge junction

INTRODUCTION

Recently, ramp-edge Josephson junctions have been fabricated without deposited barriers by using interface treatments [1][2]. These junctions have exhibited excellent characteristics and small spreads in the junction parameters. We have fabricated ramp-edge Josephson junctions with a surface-modified barrier formed by Ar-ion beam irradiation and high-temperature annealing in an oxygen atmosphere [3]. The fabricated junctions had RSJ characteristics with an IcRn product of 3 mV (at 4.2 K) and a 1σ spread in Ic of 10% for 200 junctions. To improve the junction properties, though we need to better understand the nature of the surface-modified barrier. Thus, we examined the composition of the modified $YBa_2Cu_3O_{7-x}$ (YBCO) surfaces by x-ray photoelectron spectroscopy (XPS). We also investigated the lattice structure of the surface-modified barrier by cross-sectional transmission electron microscopy (TEM).

EXPERIMENTAL

The fabrication process for the junctions is described in detail elsewhere [3]. The XPS analysis was done using a KRATOS XSAM800 system equipped with a Mg ($K\alpha$,1253.6 eV) source. The XPS analysis was performed on the surface of a YBCO thin film instead of the ramp-edge surface, because the area of the ramp-edge surface was too small to be examined. We prepared three samples; an as-deposited YBCO thin film (sample A), a YBCO thin film that was modified by Ar-ion irradiation (sample B), and a YBCO thin film annealed (770℃, O_2: 0.3 Torr, 20 min) after the surface modification (sample C). In order to analyze the exposed surfaces of the samples, no surface cleaning was done before XPS analysis. The irradiation and annealing conditions were typical for the fabrication process. The cross-sectional TEM observation was done with an H-9000UHR electron microscope.

RESULTS AND DISCUSSION

Figure 1 shows Ba ($3d_{5/2}$), Cu (2p), and O (1s) XPS spectra from samples A, B, and C. The Ba ($3d_{5/2}$) spectrum of sample A has two peaks at 780 eV and 778 eV. The 778-eV peak is characteristic of the Ba-O bond in superconducting YBCO. The 780-eV peak corresponds to Ba oxides such as $BaCO_3$ and BaO. The electron escape depth of Ba ($3d_{5/2}$) is 2.1 nm [5]. The existence of the 778-eV peak indicates that superconducting YBCO remained on the surface of the as-deposited YBCO thin film. When this YBCO film was modified by the Ar-ion irradiation, the XPS spectra changed. The Ba ($3d_{5/2}$) and O (1s) spectra of sample B each had only one peak, and the Cu (2p) satellite peaks near 945 eV and 965 eV decreased compared to those of sample A. These Cu (2p) satellite peaks represent the Cu^{2+} [6]. These results suggest that the Ba-O and Cu(II)-O bonds were decomposed by the Ar-ion irradiation, and the surface of the YBCO seems to have been damaged and amorphized. The modified YBCO film was then annealed. The Ba ($3d_{5/2}$) spectrum of sample C did not change compared with that of sample B, but the Cu (2p) satellite peaks and an lower energy peak in the O (1s) spectrum were observed. These observations indicate that recrystallization occurred on the surface of the modified YBCO during the annealing. The composition on the surface of a sample can be calculated from the area of the XPS spectra from the sample. Figure 2 shows the changes in the Ba/Y and Cu/Y ratios in samples A, B, and C. The modified YBCO had a Y-rich and Cu-poor composition compared with the as-deposited YBCO film. This change in the composition was caused by differences in the sputtering of Y, Ba, and Cu. The modified YBCO had an atomic ratio of Y:Ba:Cu = 1:0.9:0.8. When the modified YBCO was annealed, the atomic ratio of Ba/Y and Cu/Y increased. We believe that the diffusion of Ba and Cu atoms from the deeper YBCO area increased the Ba/Y and Cu/Y ratios. The atomic ratio of Y:Ba:Cu in the YBCO thin film annealed after the surface modification was 1:1.2:1.1. This result is similar to that of an energy dispersive x-ray (EDX) analysis by Wen et al. [7]. Figure 3 shows a cross-sectional TEM image of a junction. The structure of the barrier, which was obviously different from YBCO, was not observed. A blurry boundary was observed at the interface between the base-YBCO and the counter-YBCO. In this image, the highlighted Cu-O layers

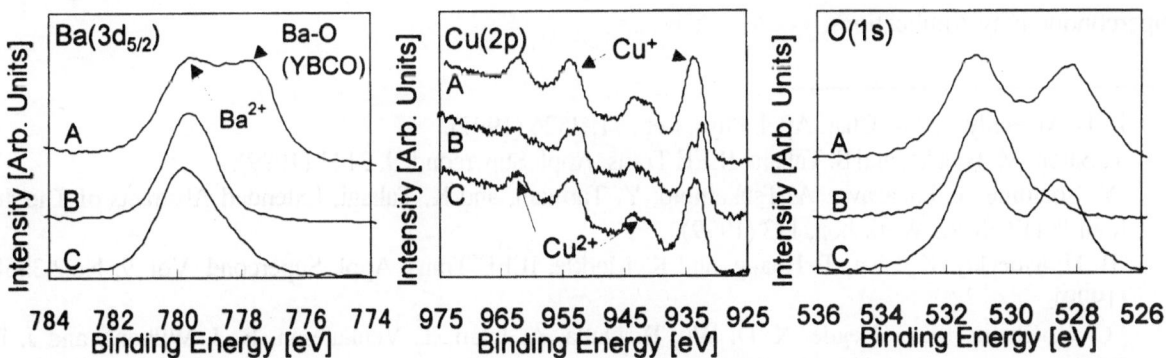

Fig. 1. Ba ($3d_{5/2}$), Cu (2p), and O (1s) XPS spectra from samples A (as-deposited YBCO), B (YBCO modified by Ar-ion irradiation), and C (modified and annealed YBCO). The take-off angle was 90°.

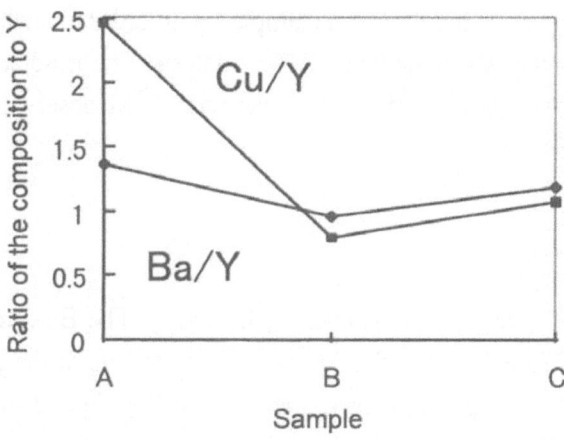

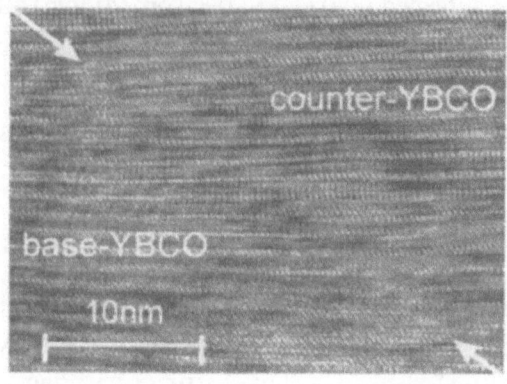

Fig. 2. Changes in the ratio of Ba/Y and Cu/Y in samples A, B, and C.

Fig. 3. Cross-sectional TEM image of the surface-modified junction at the base-YBCO/counter-YBCO interface.

often connected from the base-YBCO to the counter-YBCO. Although we observed the surface-modified barriers at several interfaces, no differences in the crystal structure between the barrier and $YBa_2Cu_3O_{7-x}$ were observed at any of the interfaces. Therefore, we think that the structure of the surface-modified barrier is very similar to that of YBCO. We also think that the thickness of the barrier is very thin ($1 \sim 2$ nm).

SUMMARY

We investigated the composition and structure of a surface-modified barrier through XPS and TEM. The XPS analysis suggested that the surface modified barriers were created by recrystallization of the amorphous YBCO that was modified by Ar-ion irradiation. The atomic ratio of Y:Ba:Cu in the surface-modified barrier was 1:1.2:1.1. The TEM observations did not reveal the structure of the barrier, which was obviously different from the YBCO. We believe, though, that the structure of the surface-modified barrier is very similar to that of YBCO, or that the thickness of the barrier is very thin ($1 \sim 2$ nm).

ACKNOWLEDGEMENT

This work was supported by New Energy and Industrial Technology Development Organization (NEDO) through ISTEC as Collaborative Research and Development of Fundamental Technologies for Superconductivity Applications.

1. B. H. Moeckly, and K. Char, Appl. Phys. Lett. 71, 2526 (1997).
2. T. Satoh, M. Hidaka, and S. Tahara, IEEE Trans. Appl. Supercond. 9, 3141 (1999).
3. Y. Soutome, T. Fukazawa, A. Tsukamoto, Y. Tarutani, and K. Takagi, Extended Abstracts of The 4th Joint ISTEC/MRS Workshop, 113 (1999).
4. B. H. Moeckly, K. Char, T. Huang, and K. Merkle, IEEE Trans. Appl. Supercond. Vol. 9, No.2,3358, (1999).
5. C. C. Chang, M. S. Hegde, X. D. Wu, B. Dutta, A. Inam, T. Venkatesan, B. J. Wilkens, and J. B. Wachtman, Jr., J. Appl. Phys. 67, 7483, (1990).
6. A. Tressaud, K. Amine, J. P. Chaminade, nad J. Etourneau, J. Appl. Phys. 68, 248, (1990).
7. J. G. Wen, T.satoh, M. Hidaka, S. Tahara, N. Koshizuka, and S. Tanaka, Appl. Phys. Lett. (to be published).

The Effect of Spatial Variation of Barrier Layer Thickness on Ramp-edge-type Josephson Junction Characteristics

Jiro Yoshida, Shinji Inoue, Toshihiko Nagano, and Hideyuki Sugiyama

Advanced Materials and Devices Laboratory, Corporate Research & Development Center, Toshiba Corporation, 1, Komukai Toshiba-cho, Saiwai-ku, Kawasaki 210-8582, Japan

Abstract: We have developed a new model for ramp-edge-type Josephson junctions with a Co-doped $PrBa_2Cu_3O_{7-x}$ barrier in which the spatial variation of the barrier layer thickness within a junction is taken into account. The reexamination of experimental data based on this new model has clarified that the barrier thickness dependence of I_c of our junctions can be explained quantitatively by a direct tunneling mechanism alone.

Keywords: Josephson junction, $PrBa_2Cu_3O_{7-x}$, Variation of barrier thickness, Tunneling effect

INTRODUCTION

In junctions with a barrier composed of oxide materials in the vicinity of metal-insulator transition, resonant tunneling via localized states in the barrier plays an important role in determining the junction characteristics[1]. However, the relevance of the localized states to Josephson coupling has not been fully understood and discussion of the subject tends to be marked by controversy[1-4]. In this paper, we examine the current transport in ramp-edge-type junctions with a Co-doped PBCO barrier using a junction model in which the spatial variation of the barrier layer thickness is taken into account.

JUNCTION CHARACTERISTICS

We have fabricated YBCO/7% Co-doped PBCO/YBCO junctions with a ramp-edge geometry. The detail of the thin-film growth and the junction fabrication process has been published elsewhere[5,6]. Josephson characteristics were observed for junctions with a Co-doped PBCO barrier ranging from 6 to 11 nm in thickness. Junctions with a thinner barrier showed flux-flow dominated characteristics, probably due to the presence of microshorts within the barrier.

The critical current (I_c) and the junction conductance (G_n) observed at 4.2 K for junctions of 4 μm in width are depicted by squares in Fig. 1 and 2 as functions of the barrier layer thickness. Both I_c and G_n exhibited a nearly exponential dependence on the barrier layer thickness with the decay parameters of 0.59 and 0.96 nm, respectively. The difference in decay parameters between I_c and G_n implies that different transport channels exist for Cooper pairs and quasiparticles, that is, Cooper pairs transfer by direct tunneling whereas quasiparticles can flow also by resonant tunneling[7]. In fact, we have confirmed that the temperature dependence of G_n with the barrier layer thickness as a parameter can be explained well by the combination of resonant tunneling and hopping conduction via a small number of localized states with a radius of around 1 nm[6].

The simple tunneling model, however, was found to be insufficient to account for the experimental data quantitatively. Theoretically, G_n due to resonant tunneling is expressed as

$$G_n = \frac{\pi e^2}{\hbar} SgaE_0 \exp\left(-\frac{d}{a}\right) , \qquad (1)$$

993

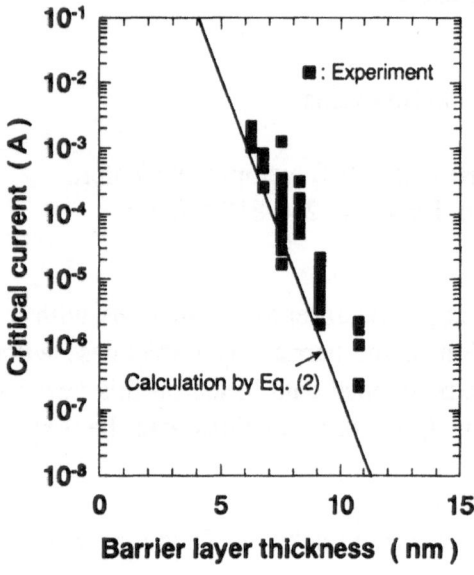

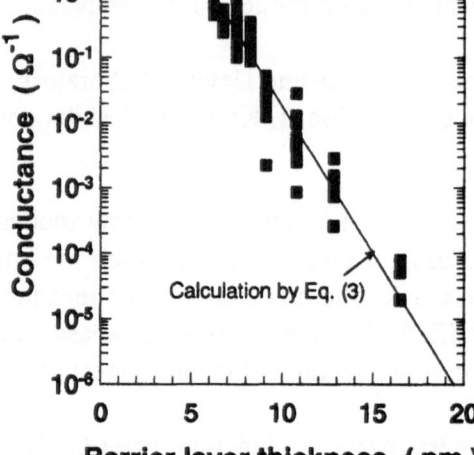

Fig. 1. Comparison between experimental and theoretical I_c versus barrier thicness.

Fig. 2. Comparison between experimental and theoretical G_n versus barrier thickness

where a is the radius of the localized states, g is the density of the localized states, E_0 is the effective potential barrier height, d is the thickness of the barrier, and S is the area of the junction[8]. If we assume that E_0 is given by the well-known expression as $E_0 = \hbar^2 / 2ma^2$, where m is the free electron mass, and that the current within a junction flows in the direction normal to the ramp-edge surface, then, we can calculate the g value to be 8.1×10^{21} $eV^{-1}cm^{-3}$ from the experimental conductance values. This value is surprisingly large and is comparable to the density of states at the Fermi surface in superconducting YBCO. It is questionable whether such a high density of localized states can actually exist in an insulator without forming an energy band.

The second problem that we noticed is that if we extrapolate the experimental I_c versus barrier layer thickness relationship to the barrier thickness of zero, the critical current density of the junction without a barrier exceeds 10^9 A/cm^2. This value is almost two orders of magnitude larger than that in bulk YBCO.

THE EFFECT OF SPATIAL VARIATION OF BARRIER THICKNESS

The discrepancy described above can be solved by taking into account the spatial variation of barrier layer thickness in the junctions. We assume that the spatial variation can be expressed by a Gaussian distribution function in which the mean value of the barrier layer thickness d_0 coincides with the nominal thickness which is experimentally determined from the deposition rate and the deposition time of the barrier layer. The standard deviation σ_d is regarded as being independent of the barrier thickness in the thickness range we are investigating. Furthermore, we assume that the Josephson and the quasiparticle currents in the junctions are entirely due to direct and resonant tunneling processes, respectively. It is straightforward to derive analytical expressions for I_c and G_n under these assumptions, as shown below

$$I_c = \frac{e\Delta}{8\pi\hbar}\frac{1}{ad}\frac{S}{\sqrt{2\pi}\sigma_d}\int_0^\infty \exp\left(-\frac{2x}{a}\right)\exp\left[-\frac{1}{2}\left(\frac{x-d_0}{\sigma_d}\right)^2\right]dx = \frac{e\Delta}{8\pi\hbar}\frac{1}{ad}S\exp\left(-\frac{2d_0}{a}\right)\exp\left(\frac{2\sigma_d^2}{a^2}\right), \quad (2)$$

$$G_n = \frac{\pi e^2}{\hbar}gaE_0\frac{S}{\sqrt{2\pi}\sigma_d}\int_0^\infty \exp\left(-\frac{x}{a}\right)\exp\left[-\frac{1}{2}\left(\frac{x-d_0}{\sigma_d}\right)^2\right] = \frac{\pi e^2}{\hbar}SgaE_0\exp\left(-\frac{d_0}{a}\right)\exp\left(\frac{\sigma_d^2}{2a^2}\right), \quad (3)$$

where Δ is the superconducting gap energy in YBCO. It should be noted that Eq. (2) is based on

995

the simple tunneling theory for isotopic materials with a parabolic band structure and the BCS approximation for Josephson current due to direct tunneling. One can see from Eqs (2) and (3) that the spatial variation of the barrier thickness enhances both I_c and G_n, but the enhancement factor differs between them. This is due to Ic and G_n having different barrier thickness dependences.

In our junctions, the nominal barrier thickness of 6 nm was the borderline below which the probability of the formation of microshorts within a junction increases rapidly. Therefore, we can suppose that $3\sigma_d$ in our junctions is around 6 nm. This gives 2nm as $1\sigma_d$. This value together with the experimentally estimated a value of around 1 nm results in the enhancements of I_c and G_n by factors of 6000 and 8.0, respectively, compared with those of spatially homogeneous junctions.

The straight lines in Figs. 1 and 2 show the calculation based on Eqs (2) and (3) using the physical parameters listed in Table I. Agreement with the experiment is satisfactory when we take the crudeness of the present model into account. The new junction model gives the density of the localized states involved in the quasiparticle transport of around 1×10^{21} eV^{-1}cm^{-3}. This value still seems large, but we think it physically permissible in oxides in the close vicinity of the metal-insulator transition like PBCO. An important point to note is that the experimental Josephson current is comparable with those expected for a direct tunneling mechanism. This indicates that the electric properties of YBCO/Co-doped PBCO/YBCO junctions can be understood well within the framework of the conventional tunneling theory.

Table I. Physical parameters characterizing Co-doped PBCO tunnel barrier

Quantity	Symbol	Value	Units
Radius of localized states	a	0.96	nm
Barrier height	E_0	41.4	meV
Density of localized states	g	1.0×10^{21}	eV^{-1}cm^{-3}
Superconducting gap energy	Δ	13.7	meV
Standard deviation of barrier thickness	σ_d	2.0	nm

CONCLUSION

We have analyzed YBCO/Co-doped PBCO/YBCO junction characteristics using a newly developed junction model in which a spatial variation of the barrier thickness is taken into account. The overall features of the junction characteristics were explained well within a framework of reasonable physical parameters. Experimental Josephson currents were found to be comparable with those expected for a direct tunneling mechanism, although quasiparticle currents were definitely governed by resonant tunneling.

Acknowledgment This work was supported by the New Energy and Industrial Technology Development Organization (NEDO) through the International Superconductivity Technology Center (ISTEC).

1. J. Yoshida and T. Nagano, Phys. Rev. B, **55**, 11860 (1997)
2. J. Halbritter, Phys. Rev. B, **46**, 14861 (1992)
3. M. A. J. Verhoeven et al., Appl. Phys. Lett., **69**, 848 (1996)
4. I. A. Devyatov and M. Yu Kupriyanov, JETP Lett., **59**, 200 (1994)
5. T. Nagano, S. Inoue, T. Hashimoto, and J. Yoshida, Physica C, **303**, 231 (1998)
6. J. Yoshida, et al, IEEE Trans. Appl. Supercond. **9**, 3366 (1999)
7. R. Gross and B. Mayer, Physica C, **180**, 235 (1991)
8. A. I. Larkin and K. A. Matveev, Sov. Phys. JETP, **66**, 580 (1987)

THE EFFECT OF PROCESS PARAMETERS ON THE ELECTRICAL PROPERTIES OF RAMP-EDGE JOSEPHSON JUNCTIONS WITH MODIFIED INTERFACES

Masahiro Horibe, Yukitoshi Inagaki, Kazuhiro Yoshida, Gen-ichiro Matsuda, Noriyoshi Hayashi, Akira Fujimaki and Hisao Hayakawa

Department of Quantum Engineering, Nagoya University,
Furo-cho, Chikusa-ku, Nagoya 464-8603, Japan

Abstract: We have studied the ramp-edge Josephson junctions with modified barrier. Modified barriers are produced in an ECR plasma etching and subsequent vacuum annealing. The electrical properties of interface modified junctions (IMJs) can be roughly controlled by process parameters. Increasing the etching time and etching voltage lead to the reduction in critical currents. On the other hand, the critical current of IMJs is increased by high temperature anneal or high pressure one. Furthermore, the deposition temperature of counter-electrodes has the strong influence on junction characteristics.

Keywords: Modified interface, etching conditions, annealing conditions, deposition conditions

INTRODUCTION

Recently, the interface modified junctions (IMJs) attract the attention [1-3]. These junctions show relatively low spread of the parameters such as I_c, R_n because of no deposited barrier. However, we have to adjust the critical current (I_c) and normal resistance (R_n) of IMJs to the suitable values for SFQ circuits. We should investigate the relationship between the process parameters and junction characteristics from the viewpoint of uniformity, reproducibility and controllability of junction characteristics. The process parameters include the etching conditions in the ramp-edge formation, deposition conditions of counter-electrodes. In this paper, we describe the effect of process parameters on the junction characteristics.

FABRICATION PROCESS

We deposit a multilayer of Au/PrGaO$_3$ (PGO)/YBa$_{1.95}$La$_{0.05}$Cu$_3$O$_{7-x}$ (La-YBCO) on an MgO(100) substrate by an rf magnetron sputtering method. YBCO film is grown with c-axis normal to substrate surface. The PGO template layers are defined by an Ar ion etching using the photoresist mask. The sample is tilted 30° to substrate surface and rotates during the etching. After removing the photoresist, the ramp-edge is formed by an Ar ion etching using PGO template layer as an etching mask[4]. In this study, we change accelerating voltages (V_{acc}) from 300 to 1100V. The etching conditions are shown in Table 1. In spite of different accelerating voltages, we observe the ramp angle of about 18° and the bulge of La-YBCO[5]. After the sample is annealed for 1 hour, the counter YBCO is deposited. Finally, the junctions are completed using Ar ion beam etching. The junction width are 5µm and every films are the thickness of 100nm. The six junctions are integrated in the area of 1.0 mm x 0.1mm. on a chip.

RESULTS AND DISCUSSION

Figure 1 shows a typical current–voltage characteristic (*IVC*) of IMJ. The resistively-shunted-junction-like (RSJ-like) *IVC*s and good characteristics are observed in an entire temperature range. The hysterisis vanishes at 30K. The characteristics at 4.2K are; I_c=115µA, R_n=10Ω, I_cR_n=1.15mV. The other samples also have the similar *IVC*s. Furthermore, the I_c is suppressed by appling magnetic fields and Fraunhofer-like modulation is observed. The modulation depth is almost 100% at 4.2K and 40K. Consequently the superconducting currents are mostly Josephson current.

We investigate systematically the dependence of accelerating voltage on the junction characteristics. We form the ramp-edge by an Ar ion etching with various V_{acc} for 20min. All junctions have RSJ-like *IVC*s. The IMJs formed in low V_{acc} have high I_c and low R_n. I_cR_n product reaches the high

Table 1 The fabrication conditions and electrical properties at 4.2K for IMJs.

Sample	A	B	C	D	E	F	G	H	I	J
V_{acc} (V)	300	500	700	900	1100	500	700	700	700	1100
t_{etch} (min)	20	20	20	20	20	30	20	20	20	20
P_{ann} (Pa)	10^{-5}	10^{-5}	10^{-5}	10^{-5}	10^{-5}	10^{-5}	1	1	1	10^{-5}
T_{ann} (°C)	680	680	680	680	680	680	680	740	800	680
T_{dep} (°C)	750	750	750	750	750	750	750	750	750	790
I_c (μA)	1200	115	72.5	34	15	10	260	280	960	1700
R_n (Ω)	2.2	10	15.5	27	41	60	4.3	4.0	1.9	1.8
I_cR_n (mV)	2.64	1.15	1.12	0.92	0.62	0.60	1.11	1.12	1.82	3.06

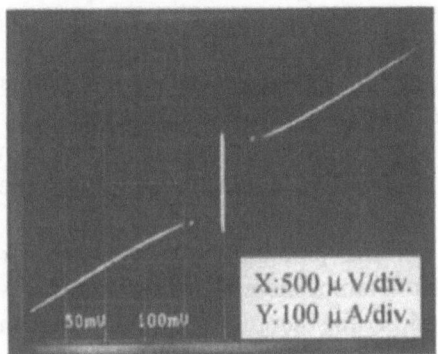

Fig. 1. A typical *IVC* of IMJ at 4.2K. This junction is Sample B.

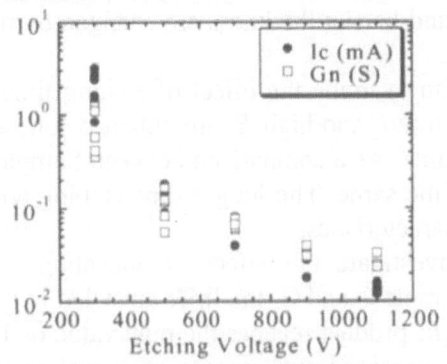

Fig. 2. The accelerating voltage dependence of junction characteristics at 4.2K.

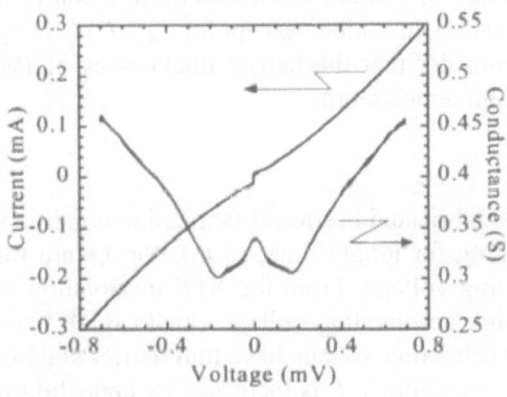

Fig. 3. The *IVC* and the differential conductance at 4.2K for Sample E

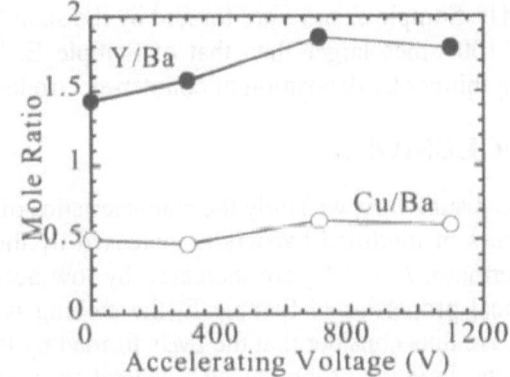

Fig. 4. The mole ratio of Y/Ba and Cu/Ba as a function of accelerating voltage by XPS.

value of 2.64mV at 4.2K for Sample A. The relationship between the accelerating voltage and the junction characteristics is shown in Figure 2. The I_c and junction conductance (G_n) are reduced with increasing V_{acc}. The accelerating voltage dependence of I_c is stronger than that of G_n, so that the I_cR_n product is reduced with increasing V_{acc}. We consider that the barrier resistivity and/or the barrier thickness is changed by the etching with high accelerating voltage.

Figure 3 shows the *IVC* and differential conductance as a function of bias voltage at 4.2K for Sample E. The junction conductances are increased with increasing bias voltage. Also, the junction conductances are increased with increasing measurement temperatures. These results suggest that the junctions with modified barriers formed in high V_{acc} have transport process via localized states for the quasiparticles. If the density of localized state in the modified barrier is increased by the etching with high V_{acc}, the barrier height and resistivity of modified barrier is reduced. However, junction resistance is increased with increasing V_{acc}. Furthermore, increasing V_{acc} is thought to cause not the highly resistive barrier but the thicker barrier.

We examine the composition of YBCO film surfaces after the etching and annealing sequence by XPS analysis. Here, etching conditions and annealing ones are the same conditions as those in formation of IMJ. The mole ratio of Y/Ba and Cu/Ba as a function of accelerating voltage are shown in Figure 4. The composition of Y to Ba is increased in the etching process. This relation becomes stronger with high V_{acc}. However, composition of Y is almost constant for V_{acc} above V_{acc} =700V. We consider that the composition on ramp-edge surfaces has the same relation as that on treated YBCO film surfaces, because of the realization of trilayer type IMJs [6]. From the XPS observation, IMJs treated at $V_{\text{acc}} \geq 700$V possibly have the barrier with the same resistivity. We, thus, speculate that increasing V_{acc} arises the thicker barrier above V_{acc}=700V. On the other hand, when V_{acc} is less than 700V, both barrier resistivity and barrier thickness are changed by the etching process. Consequently, the I_cs are enhanced by low V_{acc}.

We also examine the effect of etching time (t_{etch}) on the junction characteristics. Compared with Sample B, low I_c and high R_n are obtained for Sample F. I_cs and I_cR_n products are increased by etching of shorter time. As a comparison between Sample E and Sample F, the characteristics of these junctions are almost the same. The longer time etching and higher accelerating voltage arise the same effect on junction characteristics.

We investigate the effect of annealing conditions on the junction characteristics. Compared between Sample C and G, the IMJs annealed in a high pressure atmosphere have high I_c and low R_n. For example, I_cR_n product reaches the high value of 1.82mV at 4.2K for Sample I. On the other hand, the I_c and G_n are increased with increasing annealing temperature above the deposition temperature (T_{dep}) of the counter electrodes. Sample I have higher I_c and lower R_n than Sample H. The annealing temperature dependence of I_c is stronger than that of G_n, so that the I_cR_n product is increased with increasing T_{ann} above T_{dep}.

We also examine the effect of deposition temperature of counter electrodes on junction properties for IMJs. Sample E and J are formed by the same fabrication conditions except for T_{dep}. I_c of Sample J is about 100 times larger than that of Sample E. We consider that the barrier thicknesses of IMJs are getting thinner by deposition of counter-electrodes at high temperatures.

CONCLUSIONS

In a summary, we study the characteristics of IMJs fabricated in several fabrication conditions. The thickness of modified barriers is increased by the etching for longer time, so that the I_cs are reduced. Furthermore, I_c and G_n are increased by low accelerating voltage. From the XPS observation and the electrical properties of Sample E, the etching with high accelerating voltage causes thick barrier for IMJs. We thus consider that the IMJs formed by low accelerating voltage have thin barrier and high I_cR_n products. Junction properties also depend on annealing conditions. I_c is increased by annealing in high pressures and at high temperatures. However, I_c is independent of annealing temperature below deposition temperature of counter-0electrodes. We consider that the I_c depends on the deposition temperatures rather than annealing temperatures. However, the detail analysis such as TEM observation and discussion will be required.

1. B. H. Moeckly and K. Char, Appl. Phys. Lett., 71, 2526-2528(1997)
2. A. Fujimaki, K. Kawai, N. Hayashi, M. Horibe, M. Maruyama, and H. Hayakawa, IEEE Trans. Appl. Supercond., Vol. 9, No.2 (1999), pp. 3436-3439
3. T. Sato, M. Hidaka, and S. Tahara, IEEE Trans. Appl. Supercond., Vol. 9, No.2 (1999), pp. 3141-3144
4. T. Sato, M. Hidaka, and S. Tahara, IEICE Trans. on Electron., Vol. E81-C, No.10 (1998), pp. 1532-1537
5. M. Horibe, Y. Inagaki, G. Matsuda, N. Hayashi, K. Kawai, M. Maryuama, A. Fujimaki and H. Hayakawa, 7th International Superconductive Electronics Conference, Extended Abstract (1999), pp. 295-297
6. K. Yoshida, T. Furutani, M. Maruyama, Y. Inagaki, M. Horibe, A. Fujimaki and H. Hayakawa, in *Advances in Superconductivity XII*, (Springer-Verlag, Tokyo, 2000), to be published

Improvement of the Ramp Edge Junction Properties by Interface Treatment

Takehiko Makita, Gustavo Alvarez, Kazuyoshi Toma, and Keiichi Tanabe

Superconductivity Research Laboratory, 10-13, Shinonome 1-chome, Koto-ku, Tokyo, 135-0062, Japan

Abstract: To modify the interface of ramp edge junction and improve current-voltage characteristics, we carried out ECR (electron cyclotron resonance) ion cleaning and annealing treatments to the surface of the ramp edge before depositing a barrier layer and counter YBCO film. The surface roughness of the ramp after the interface treatment under several conditions was investigated by Atomic Force Microscope and we fabricated the junctions using PBCO as a barrier material. The junctions, which were cleaned by ECR Ar and O_2 ions and annealed under oxygen atmosphere, showed weak link characteristics, though the others showed flux flow ones. The roughness of the ramp surface for the junctions was smaller than the surface roughness after any other treatments. These results suggested that the ECR cleaning is effective to flatten the PBCO barrier on the ramp.

Keywords: YBCO, ramp edge, LSAT, ECR cleaning

INTRODUCTION

One of the most promising Josephson junction types for single flux quantum devices containing high-T_c superconductors is edge type one because of higher I_cR_n products and structural flexibility for device applications [1]. In order to improve the properties of YBCO ramp edge junctions, we recently employed $LaAlO_3$-Sr_2AlTaO_6 (LSAT) as an insulation layer and a substrate, and a resist reflow process for ramp edge fabrication. Consequently we could improve the spread of the electric properties [2], though the junction properties still have not been satisfactory probably because of poor smoothness at the ramp interface and thickness fluctuation in the barrier layer [3]. Therefore it is necessary to control roughness of the ramp surface before depositing a barrier and counter YBCO film.

In this study, we carried out ECR (electron cyclotron resonance) ion cleaning and annealing treatment to the ramp surface under several conditions to improve the smoothness of the ramp surface and investigated their influence on the electric properties of the junctions with a PBCO barrier.

EXPERIMENTAL

We deposited a 200 nm thick c-axis oriented YBCO and 300 nm LSAT bilayer on an LSAT(100) substrate as the bottom layer by off-axis rf-magnetron sputtering. The substrate temperatures for YBCO and LSAT were 755 ~ 775 ℃ and 685 ℃, respectively. We confirmed that c-axis oriented YBCO films were obtained at temperature higher than 755 ℃ by off-axis rf-magnetron sputtering as long as we investigated. Ramp edges of LSAT / YBCO bilayer structures were fabricated by the standard photolithography using a resist reflow process and Ar ion milling. After removing the resist by oxygen ashing, the sample was introduced into a vacuum chamber, cleaned by ECR ion and annealed for 10 minutes under several conditions. The samples were not exposed to the air between the ECR cleaning and annealing processes. After these processes, we evaluated the roughness of the ramp surface by atomic force microscope (AFM) in the air. We measured current-voltage (IV) characteristics of the junctions fabricated from some of these samples using a PBCO barrier.

RESULTS AND DISCUSSION

We estimated the mean roughness (R_a) of the ramp surface after several treatments from AFM images. Fig. 1 summarizes the Ra after various ECR cleaning and following annealing treatments, which is normalized by that just after fabricating the ramp structure. The Ra of the sample annealed at 775 ℃ under the same atmosphere that for the YBCO growth (partial pressure of Ar and O_2 is 192 mTorr and 8 mTorr, respectively) without ECR cleaning is the largest and the R_a of the sample annealed in vacuum is similarly relatively large.

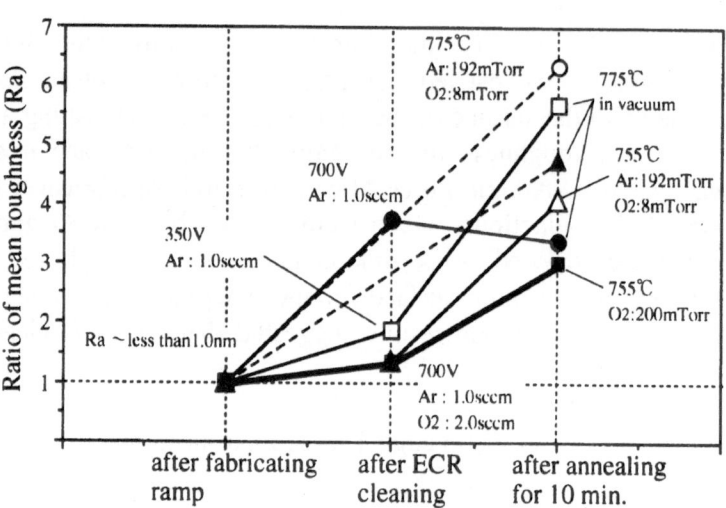

Fig. 1 R_a after various ECR cleaning and following annealing treatments for YBCO ramp structures.

The R_a after the ECR cleaning depends on the acceleration voltage of Ar ions. However, the following annealing in vacuum increases the R_a of the sample with ECR cleaning at 350 V more than that with cleaning at 700 V. Furthermore, by employing Ar + O_2 mixture in the ECR cleaning, and / or lower substrate temperature of 755 ℃ in the annealing process under pure oxygen atmosphere, the R_a is substantially suppressed. These results indicate that the optimized interface treatment to the ramp surface is ECR cleaning with Ar + O_2 mixture at a relatively high acceleration voltage and annealing under pure oxygen atmosphere.

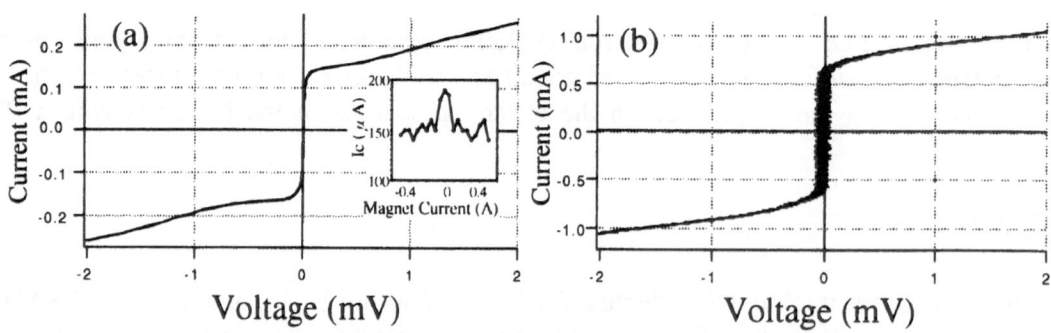

Fig. 2 IV characteristics at 4.2 K of the junctions from the samples
(a) with and (b) without the optimized treatments

We fabricated junctions using a 25nm thick PBCO barrier layer from the samples with the interface treatments. Figs. 2(a) and 2(b) show the IV characteristics at 4.2 K of the junctions fabricated from the samples with and without the optimized treatments as mentioned above. The characteristics with the optimized treatments is close to the resistively shunted junction (RSJ) type to some extent, though the modulation of critical current (I_c) with applied magnetic field was only 20 % (see the inset in Fig. 2(a)).

In order to investigate the effect of ECR cleaning to the ramp surface, we evaluated I_cs from IV characteristics of junctions subjected to the ECR cleaning under several conditions. Fig. 3 shows the dependence of I_c on the ECR acceleration voltage for the junctions with and without a 25 nm thick PBCO barrier. The I_cs of the junctions without PBCO do not depend on the acceleration voltage, while those with PBCO obviously decrease with increasing the acceleration voltage. These results suggest that the ECR cleaning with the higher acceleration voltage is effective to improve not only the flatness of the ramp surface but-also that of PBCO barrier layer without generating a significant damaged layer which is recrystallized to form insulating barrier like in interface engineered junctions [4]. Insufficient IV characteristics seem to be caused by imperfect coverage or microscopic leakage path in the PBCO barrier layer, which may still exist in spite of the improved flatness of the ramp surface.

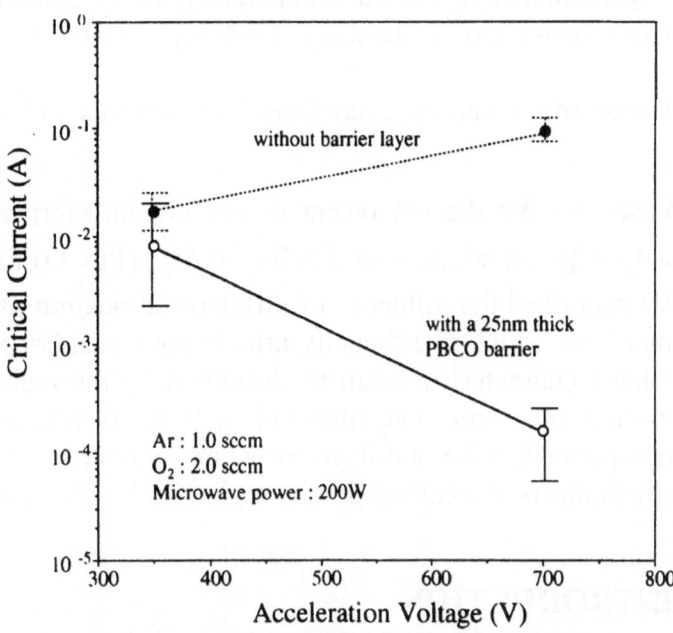

Fig. 3 Dependence of I_c on ECR acceleration voltage for the junction with and without a 25 nm thick PBCO. The junction width is 5 μm.

Summary

The roughness of the ramp surface after ECR cleaning and following annealing process under several conditions was studied by AFM observation. It was found that the optimized interface treatments to the ramp surface is ECR cleaning with Ar + O_2 mixture at a high acceleration voltage and annealing under pure oxygen atmosphere. IV characteristics of the junction with a PBCO barrier after these interface treatments were improved. The ECR cleaning with a higher acceleration voltage is effective to improve the flatness of the ramp surface and the PBCO barrier layer.

Acknowledgments. This work was supported by the New Energy and Industrial Technology Development Organization (NEDO) as Collaborative Research and Development of Fundamental Technologies for Superconductivity Applications.

1. J. Gao, W. A. M. Aarnink, G. J. Gerritsma, H. Rogalla, Physica C **171**, 126-130 (1990).
2. T. Makita, G. Alvarez, K. Toma, K. Tanabe, 1999 International Workshop on Superconductivity, pp. 105-106.
3. T. Nagano, S. Inoue, T. Hashimoto, J. Yoshida, Physica C **303**, 231-245 (1998).
4. B. H. Moeckly, K. Char, Appl. Phys. Lett. **71**, 2526 (1997)

Fabrication and Properties of $NdBa_2Cu_3O_{7-\delta}$/$PrBa_2Cu_3O_{7-\delta}$/ $NdBa_2Cu_3O_{7-\delta}$ Ramp-Edge Junctions

Gustavo A. Alvarez[1], Takehiko Makita[1], Kazuyoshi Toma[2], Tadashi Utagawa[1], Ushio Kawabe[2], Mutsumi Sato[1], and Keiichi Tanabe[1]

[1]Superconductivity Research Laboratory, 10-13 Shinonome, Koto-ku, Tokyo, 135-0062, Japan
[2]Chiba Institute of Technology, Tsudanuma, Narashino-shi, Chiba, 274-0016, Japan

Key words: Ramp-edge junctions, NBCO, PBCO, I_cR_n product.

Abstract: We discuss recent results on the fabrication process of $NdBa_2Cu_3O_{7-\delta}$ (NBCO) ramp-edge junctions with a $PrBa_2Cu_3O_{7-\delta}$ (PBCO) barrier by pulsed laser deposition (PLD). We examined the influence of different fabrication parameters on the structural quality of the interfaces which significantly affects the transport properties of the junctions. The current-voltage characteristics can be described by the resistively shunted junction model. An I_cR_n product of 1 mV was obtained at 7 K. It was confirmed that NBCO layers could be incorporated in the multilayer structure without substantial degradation. These devices exhibit I_cR_n products in a current density regime well suited for SFQ circuit fabrication.

INTRODUCTION

Although tunneling properties in the case of intrinsic Josephson junctions fabricated on $Bi_2Sr_2CaCu_2O_{8+\delta}$ [1] and planar NBCO/PBCO/NBCO quasi-homostructures [2] have been observed and reported, to date, the majority of work in the fabrication of high T_c superconducting structures has been focused on superconductor/normal-metal/superconductor (SNS) Josephson junctions, or weak links with electrical performance consistent with the resistively shunted junction (RSJ) model. Some groups have focused on the fabrication of ramp-edge type devices using high T_c superconducting oxide electrodes and normal metal layers [3]; the work in this area has involved the use of a variety of oxides such as PBCO [4]. In this paper we present recent results on the fabrication process of high T_c NBCO ramp edge junctions with PBCO barriers by PLD. Because of the close lattice match, the resulting junctions may be more ideal if their resistance is dominated by the tunneling resistance through the PBCO itself, rather than by the interface resistance

EXPERIMENTAL RESULTS AND DISCUSSION

One of the most important aspects in the ramp-edge junction fabrication process is the possibility to grow epitaxial SIS and SNS trilayer structures using c-axis orientated thin films while retaining the Josephson coupling along the ab-plane. We fabricated high T_c NBCO ramp-edge junctions with PBCO barrier by PLD. The cross section of the junction is shown schematicaly in Fig. 1. Briefly, a bilayer which consisted of a 200 nm NBCO base electrode, and a 200nm STO insulating layer was deposited in situ by PLD on $SrTiO_3$ (STO) substrates.

The films were deposited at substrate temperatures of 780 and 680 °C respectively in 200 mTorr oxygen pressure. After deposition the bilayer samples were annealed at 400° C for 5 hours in one atmosphere of oxygen and then cooled down to room temperature [5]. The NBCO films had smooth surface morphology, $T_c \sim 90$ K and $Jc > 10^6$ A/cm^2 at 77 K. The deposited STO layers on NBCO have high crystallinity and smooth surface with (100) oriented structure as shown by XRD analysis. The rocking curves give a relatively large FWHM of 0.15-0.2° comparing with 0.05° of a NBCO single layer. The base electrode was patterned by using standard photolithography and argon ion milling techniques [6]. The PBCO barrier and NBCO upper electrode layers were deposited in situ by PLD after ECR cleaning process. A 450 nm thick gold layer was evaporated and patterned in order to provide electrical contacts. The junctions were defined by a second argon ion etching step.

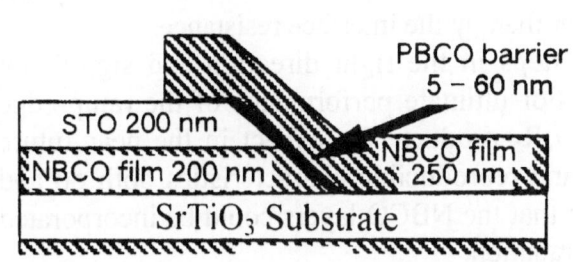

Fig. 1. Schematic cross section of a NBCO/PBCO/ NBCO ramp-edge junction.

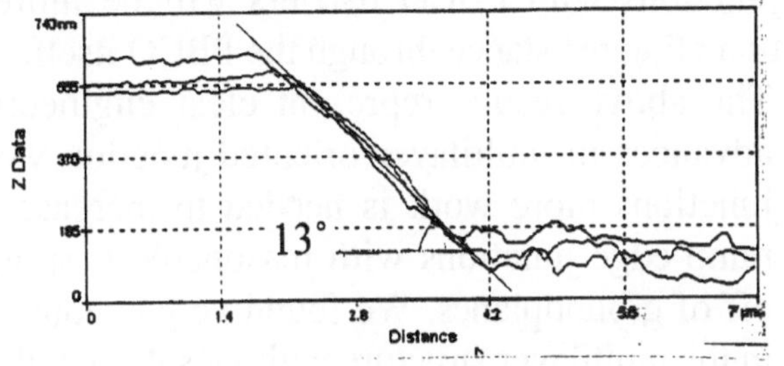

Fig. 2 AFM line scan profile of a ramp after argon ion etching.

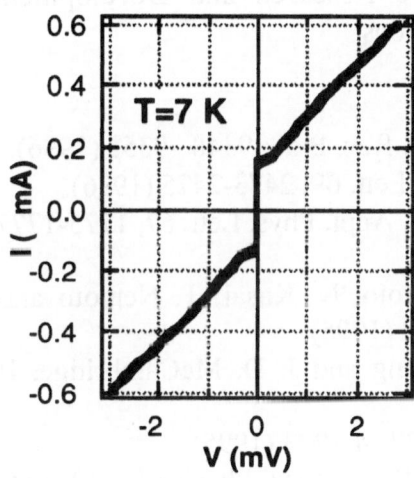

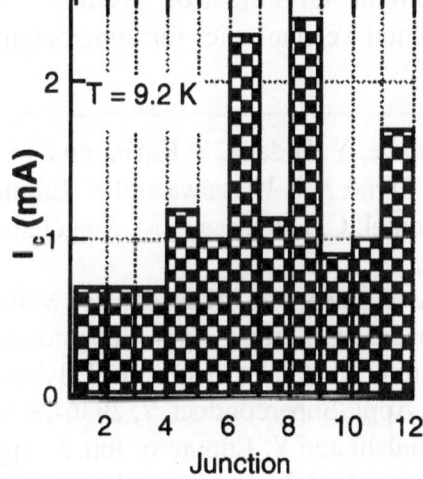

Fig. 3. I-V curve at 4.7 K for a NBCO/PBCO/NBCO ramp-edge junction. A I$_c$R$_n$ product of about 1 mV was obtained.

Fig. 4. Ic values for 5μm wide NBCO/PBCO/NBCO ramp-edge junctions on the same chip.

For the fabrication of integrated circuits, a free positioning of the junctions on the whole chip is necessary. The rotation of the sample during the ion beam etching process is essential for a sufficient design flexibility. Circuit layout considerations also make it desirable that the edge junctions properties be independent of edge orientation. Furthermore, base electrode with edge angles less than 45° are important for avoiding grain boundary formation in the counterelectrode [7]. We used a tilt angle of 60° between the ion beam and the normal plane of the substrate. Fig. 2 shows the AFM line scan of a NBCO ramp that was rotated during the argon-ion milling process. A shallow angle of 13° was obtained.

The I-V characteristics at 7 K for a ramp-edge junction are shown in Fig. 3. An $I_c R_n$ product of about 1 mV was obtained. These I-V characteristics exhibited RSJ-like behavior. Fig. 4 shows the spread in I_c for twelve junctions on the same chip with no ground plane. These junctions are 5 μm wide and have average $I_c R_n$=2.8 mV at 9.2 K. The barrier thickness for both cases was 40 nm.

SUMMARY

Because of the close lattice match between NBCO and PBCO, the resulting NBCO ramp edge junctions with PBCO barriers will be more ideal if their resistance is dominated by the tunneling resistance through the PBCO itself, rather than by the interface resistance.

The above results represent clear engineering steps in the right direction and significant advances in multilayer oriented junction work. For ultimate performance of the ramp-edge junctions more work is needed to increase the $I_c R_n$ product. We expect in the near future ramp-edge junctions with reasonable I_c spread and reproducible characteristics both on and off of groundplanes. We found very encouraging that the NBCO layers could be incorporated in the multilayer structure without substantial degradation.

Acknowledgments. This work was supported by New Energy and Industrial Technology Development Organization (NEDO) as Collaborative Research and Development of Fundamental Technologies for Superconductivity Applications

1. K. Tanabe, Y. Hidaka, S. Karimoto and M. Suzuki, Phys. Rev. B 53, 9348 –9352 (1996).

2. G. A. Alvarez, T. Utagawa and Y. Enomoto Appl. Phys. Lett. 69, 2473-2475 (1996).

3. R. Doemel, C. Horstmann, M. Siegel and A. I. Braginski, Appl. Phys. Lett. 67, 1775-1777 (1995).

4. G. A. Alvarez, M. Kuroda, M. Matsuda, K. Miyamoto, N. Kasai, T. Nemoto and M. Koyanagi IEEE Trans. Appl. Superconduct. 5, 2755-2758 (1995).

5. B. Hunt, M. G. Forrester, J. Talvacchio, R. M. Young and J. D. McCambridge, IEEE Trans. Appl. Superconduct. 7, 2936-29396 (1997).

6. M. Konishi and Y. Enomoto, Jpn. J. Appl. Phys. 34, L1271-L1273 (1995).

7. B. Hunt, M. C. Foote, W. T. Pike, J. B. Barner and R. P. Vasquez, Physica C 230, 141-152 (1994).

Fabrication and Evaluation of Sandwich-Type Interface-Treated Josephson Junctions

K.Yoshida, T.Furutani, M.Maruyama, Y.Inagaki, M.Horibe, A.Fujimaki and H.Hayakawa

Department of Quantum Engineering, Nagoya University
Furo-cho, Chikusa-ku, Nagoya, 464-8603, Japan

Abstract: We have fabricated interface-treated sandwich-type Josephson junctions without depositing an interlayer or a barrier layer. The barrier is formed during an etching process and subsequent annealing process after the base electrode is deposited. The etching conditions are various accelerating voltages and microwave power of 200W. The annealing conditions are at 680℃ in various pressures for 1hour. In case that the base electrode was etched for 15 minutes with 900V and annealed in 1Pa of an Ar+50%-O_2 mixture gas, the junctions displayed resistively-shunted-junction (RSJ) current-voltage (I-V) characteristics with hysteresis at 4.2K. The characteristic voltage (IcRn), the critical current density (Jc) and the normalized resistance (RnA) of the junction were 2.6mV, $1.5 \times 10^3 A/cm^2$ and $1.7 \times 10^{-6} \Omega cm^2$ at 4.2K, respectively. The junction showed Fraunhofer-like magnetic field modulation of the critical current (Ic), and the modulation depth of Ic is 83% at 4.2K.

Keywords: interface-treated, sandwich-type

INTRODUCTION

Interface-treated Josephson junctions have high quality because they have a thin and uniform barrier layer [1]. In recent years, they are widely studied on the ramp-edge structure [2-5]. On the other hand, vertically stacked sandwich-type Josephson junctions are suitable for digital circuit applications using high-temperature superconductors because small area of a unit cell due to the vertical structure can reduce the inductance of a SQUID loop, which is also advantage to integration. So we fabricated the interface-treated Josephson junctions by sandwich-type structure. In this paper, we report fabrication and evaluation of sandwich-type interface-treated Josephson junctions.

FABRICATION PROCESS

A bilayer of $Au/YBa_2Cu_3O_{7-x}$ (YBCO) for base electrode is deposited on an MgO(100) substrate by an rf magnetron sputtering. The Au layer is a passivation layer. The YBCO film is grown with c-axis orientation on the substrate and has the critical temperature (Tc) around 85K. After the base electrode is formed, the Au layer is removed, and the YBCO surface is etched by an Ar ion beam to form a barrier layer. The etching conditions are accelerating voltage of 700 or 900V, microwave power of 200W, incident angle of 30° and gas

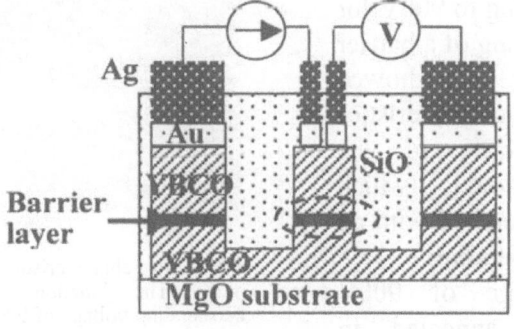

Fig.1. Schematic view of a sandwich-type interface-treated Josephson junction. A doted circle points out the area where a junction is formed. The junction size is 10 μm × 10 μm.

of Ar 1sccm. After the etching, the sample is put into the deposition chamber again. Then, the sample is annealed at 680℃ in various pressures of an Ar+50%-O$_2$ mixture gas for 1hour. After a counter-YBCO electrode with about 100nm thickness is deposited on the annealed base-YBCO films, the sample is cooled down to the room temperature in 10^3 Pa O$_2$, and an Au film is deposited on the top of the trilayer. Then, the junction structures are performed using a conventional photolithographic patterning and an Ar ion milling. The junction size is $10\,\mu$m$\times 10\,\mu$m. The junctions are measured by a four-probe method. Fig.1 shows schematic view of a fabricated junction.

EXPERIMENTAL RESULTS AND DISCUSSION

Fig.2 shows I-V characteristics at 50K of the junctions, which were etched with accelerating voltage of 700V and annealed in vacuum (about 1.0×10^{-5}Pa). The etching times of the junctions shown in Fig.2 (a) and (b) were 15min and 25min, respectively. Both junctions showed flux-flow I-V

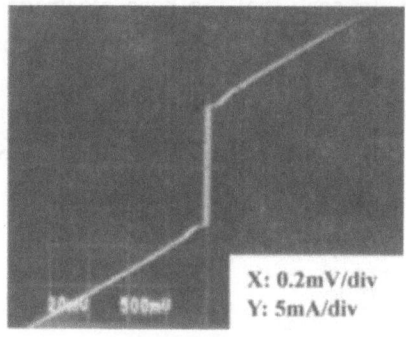

X: 0.2mV/div
Y: 5mA/div

(a) etching time 15min

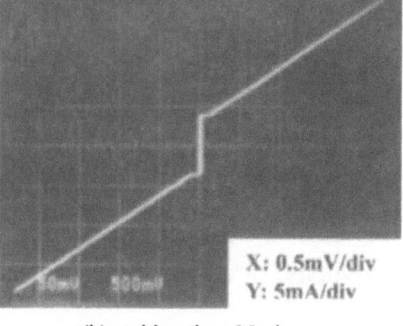

X: 0.5mV/div
Y: 5mA/div

(b) etching time 25min

Fig.2. I-V characteristics of junctions at 50K. The junctions were etched with accelerating voltage of 700V and annealed in vacuum (about 1.0×10^{-5}Pa).The junction size is $10\,\mu$m$\times 10\,\mu$m.

characteristics at 4.2K, but they showed RSJ I-V characteristics at high temperatures as shown in Fig.2. I-V characteristics showed RSJ-like characteristics at lower temperatures by lengthening the etching time. Although we thought that the properties of junctions improved more by longer etching time, we couldn't extend the etching time more because of concern the thickness of YBCO film and the etching rate, which we used.

Then, we elevated accelerating voltages of the etching to 900V for forming of a barrier layer. Fig.3 showed I-V characteristics at 4.2K of the junction, which was etched with an accelerating voltage of 900V and annealed in 1Pa. I-V characteristics

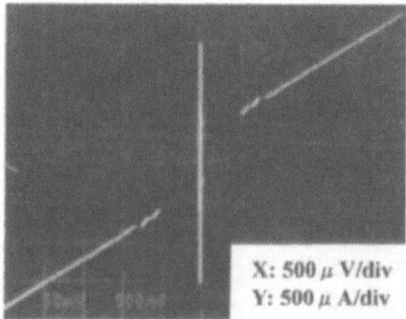

X: 500μ V/div
Y: 500μ A/div

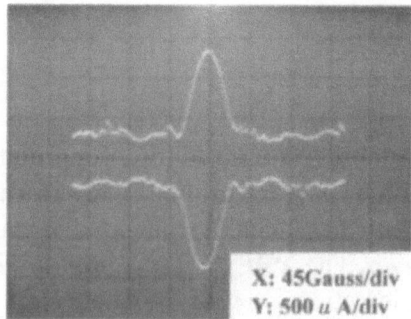

X: 45Gauss/div
Y: 500u A/div

Fig.3. I-V characteristics of a junction at 4.2K. The junction was etched with accelerating voltage of 900V and annealed in 1.0Pa.The junction size is $10\,\mu$m$\times 10\,\mu$m.

Fig.4. magnetic field modulation of Ic at 4.2K. The junction was etched with accelerating voltage of 900V and annealed in 1.0Pa.The junction size is $10\,\mu$m$\times 10\,\mu$m.

showed RSJ characteristics with hysteresis at 4.2K. IcRn, Jc and RnA were 2.6mV, 1.5×10^3A/cm^2 and $1.7 \times 10^{-6}\,\Omega$ cm^2 at 4.2K, respectively. The hysteresis of I-V characteristics vanished at 50~60K. Fig.4 showed a magnetic field modulation pattern of Ic at 4.2K. The junction showed Fraunhofer-like magnetic field modulation of Ic, and the modulation depth of Ic was 83% at 4.2K.

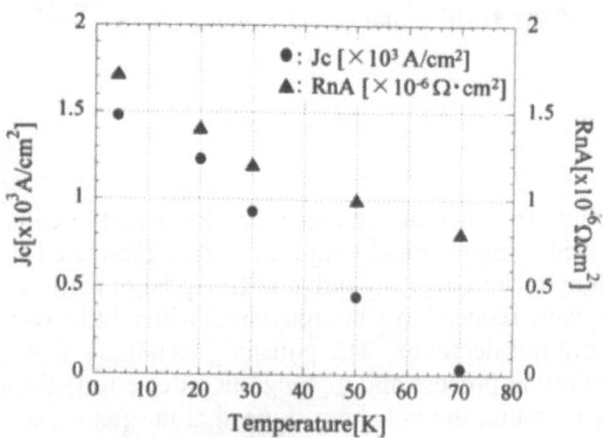 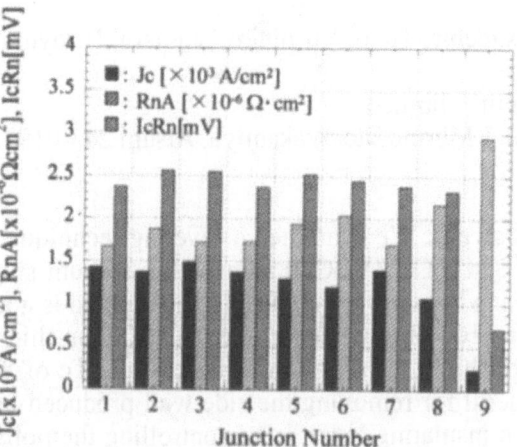

Fig.5. Temperature dependence of Jc and RnA of the junction, which was etched with accelerating voltage of 900V and annealed in 1.0Pa.The junction size is $10\,\mu\text{m} \times 10\,\mu\text{m}$.

Fig.6. The spreads of IcRn, Jc and RnA on a chip at 4.2K. The junction was etched with accelerating voltage of 900V and annealed in 1.0Pa.

Temperature dependence of Jc and RnA of the above junction was shown in Fig.5. Jc decreased linearly and RnA also decreased as the temperature increased. Fig.6 showed the spreads of IcRn, Jc and RnA on a chip at 4.2K. The 1-σ spreads of IcRn, Jc and RnA were 26%, 32% and 20%, respectively. However, the properties of Junction #9 were different from those of the others. If we ignore #9, the 1-σ spreads of IcRn, Jc and RnA were calculated to be 4%, 10% and 10%, respectively.

We once tried to form a barrier layer only by annealing without etching, but I-V characteristics showed flux-flow characteristics at an entire temperature range. Therefore we thought it was important to do etching to form a barrier layer. By extending the etching time, the properties of junctions were improved. By elevating accelerating voltage, the properties of junctions were also improved.

CONCLUSIONS

We have fabricated interface-treated sandwich-type Josephson junctions by an Ar ion beam etching and annealing without depositing a barrier layer. IcRn was 2.6mV at 4.2K, which is high value. The modulation depth of Ic was 83% at 4.2K. The 1-σ spread of IcRn was 4% at 4.2K, which is very low value. The next subject is the reproducibility and controllability of junctions.

References

1. B.H.Moeckly, K.Char, Y.Huang and K.L.Merkle, "Interface-engineered YBCO edge junctions," *IEEE Trans. Appl. Supercond.*, Vol. 9, No. 2, pp.3358-3361 (1999).

2. B.H.Moeckly and K.Char, "Properties of interface-engineered high Tc Josephson junctions," *Appl. Phys. Lett.*, Vol.71, pp.2526-2528 (1997).

3. T.Satoh, M.Hidaka, and S.Tahara, "High-Temperature Superconducting Edge-Type Josephson Junctions with Modified Interfaces," *IEEE Trans. Appl. Supercond.*, Vol. 9, No. 2, pp.3141-3144 (1999).

4. A.Fujimaki, K.Kawai, N.Hayashi, M.Horibe, M.Maruyama, and H.Hayakawa, "Preparation of Ramp-Edge Josephson Junctions with Natural Barriers," *IEEE Trans. Appl. Supercond.*, Vol. 9, No. 2, pp.3436-3439 (1999).

5. M.Horibe, Y.Inagaki, G.Matsuda, N.Hayashi, K.Kawai, M.Maruyama, A.Fujimaki, and H.Hayakawa, *7th International Superconductive Electronics Conference*, Extend Abstract (1999), pp.295-297.

Leveling Technique for High Tc Superconducting Films and Circuits

Tsunehiro Hato, Yoshihiro Ishimaru, Hiroyuki Aso, Akira Yoshida and Naoki Yokoyama

Fujitsu limited
10-1 Morinosato-Wakamiya, Atsugi 243-0197, Japan

abstract: We examined a leveling technique for high Tc superconducting (HTS) films and circuits. $YBa_2Cu_3O_{7-x}$ (YBCO) film with a smooth surface and a high critical temperature was fabricated by using leaving techniques. The method is a polishing technique using polymer fixing large particles. The 100-nm-height particles of a 200-nm-thick film were reduced to 3-nm roughness with a little over-polishing. Through this process, the Tc of 85 K did not decrease. The polishing technique is also useful for removing the side wall produced in the milling process and digging out an electrode from the insulating layer. By controlling the polishing pressure, the polishing stopped at the flat surface automatically.

Keywords: Leveling, Planarization, Polishing, Milling, Contact hole, Junction, YBCO, In_2O_3

INTRODUCTION

The $YBa_2Cu_3O_{7-x}$ (YBCO) film which has a critical temperature near 90 K tends to have precipitates on the surface even when using off-axis sputtering. To obtain a smooth film for devices, some planarization techniques have been reported [1-5]. Recently, we have been fabricating Josephson junctions and circuits, and we need a simple and high quality technique for planarization not only films but also the structures. In the circuit, side wall produced at the edge of the resist patterns in the ion-milling process makes it difficult to fabricate layer by layer structures. A large difference in level such as a deep contact hole often cuts the current line. All of these problems can be solved with planarization technique. In this paper, we report a new technique of leveling films and circuits.

EXPERIMENTAL

Figure 1 shows our polishing method. At first, diluted OMR-83 was coated using a spinner in order to hold precipitates. By using this method, we could eliminate the flaw produced by the removed precipitates. After coating 170 nm-thick resist, the film was polished with alumina of 0.03 μm diameter mixed into OS-LUBRICANT oil under a pressure of 180 g/cm². We did not use water to avoid decreasing the critical temperature. The pressure applied to the precipitates was larger than 180 g/cm² because the photo resist holding the precipitates is soft enough and has a negligibly small etching time compared with the value of YBCO. In order to keep the film parallel to the polish table, the substrate was held on the metal block by a 300 μm-thick soft tape.

Table 1 shows the polishing rate of YBCO and In_2O_3 which is used as the insulating layer of our circuits [6]. The precipitates on the YBCO surface were removed in 10 minutes even though they had a height of 100 nm. When the flat surface appeared the polishing rate was reduced to 15 A/minute. 15-minutes of polishing removed a 100 nm-thick precipitate layer and had an overetching of less than 10 nm-thick. In the case of the In_2O_3 film, the polishing stopped perfectly after the precipitate layer was removed because In_2O_3 could not be polished under 180 g/cm². This feature, which means that In_2O_3 acts as an etching stopper, was utilized in the fabrication process of our tri-layer junction.

Table 1. Polishing rate of each material

	180g/cm²	2000g/cm²
OMR-83:	1.5 μ m/min.	>3 μ m/min.
YBCO:	15A/min.	>100A/min.
In₂O₃:	0A/min.	30A/min.

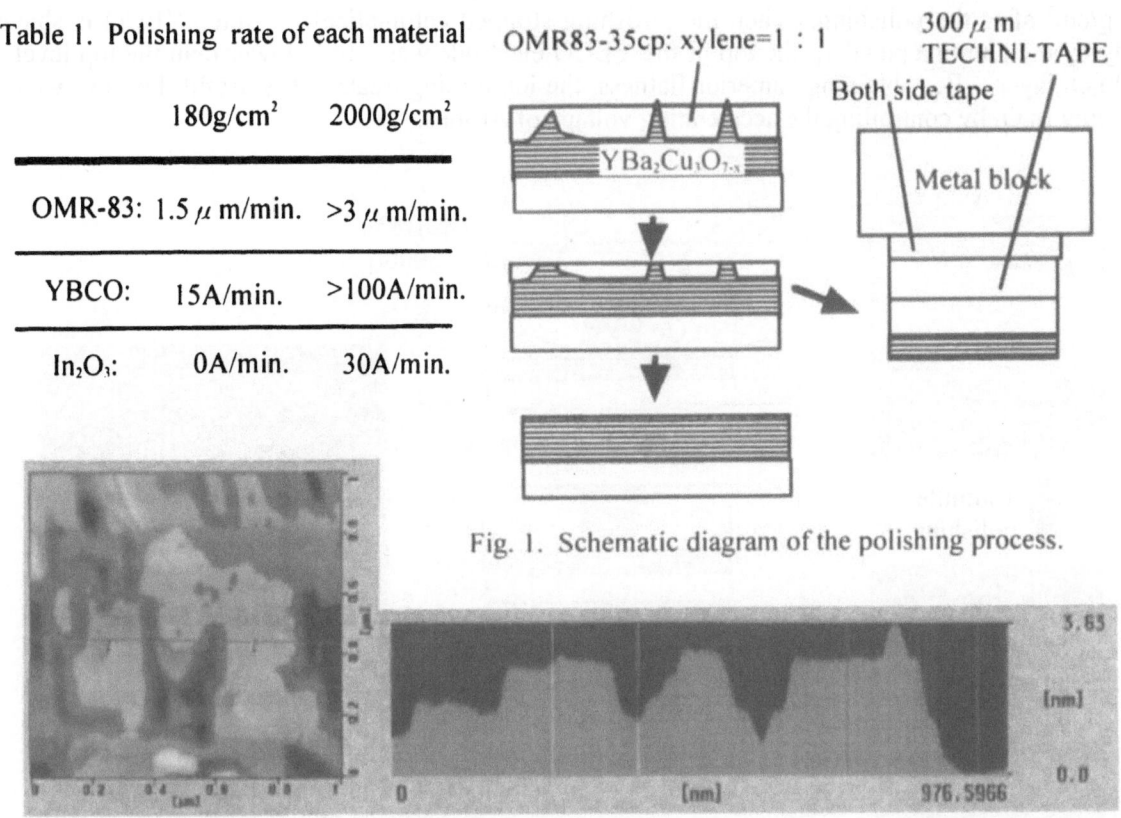

Fig. 1. Schematic diagram of the polishing process.

Fig. 2. AFM image of the polished YBCO film.

RESULTS AND DISCUSSION

The critical temperature and critical current of the polished 200 nm-thick YBCO film were 85 K and 2×10^7 A/cm² at 50 K. We could not observe any differences in the electrical characteristics, after polishing. The surface morphology observed by AFM is shown in Fig.2. The roughness was measured to be less than 3.6 nm, which means 3 units of the YBCO crystal. The terrace structure of the YBCO crystal was also observed clearly. From these results, we found that this polishing technique is useful for obtaining flat films for HTS circuits.

This polishing technique is also useful in the fabrication process of the junctions and circuits. One of the most serious issues is the side wall grown on the side of the photo resist during the ion milling process because we have no other fine patterning methods but ion milling for YBCO, In₂O₃, and other ceramics combined with YBCO. The side wall grown at the edge of a contact hole is shown in Fig. 3. The side wall, which has a height of 1 μ m, often disturbs the wiring on the contact hole from connecting the bottom layer. After holding the side wall by the resist to avoid it falling down, it was removed by only one minute polishing under 400 g to 15 mm square. The weight works as a large pressure for the side wall over 2000 g/cm², therefore the side wall was removed easily. Then, the polishing stopped automatically because the pressure was reduced to 180 g/cm² at the surface of the In₂O₃ insulating layer. Figure 4 shows a part of the fabrication process of the tri-layer junction. For junctions using the Nb superconductor, the upper wiring was contacted through a hole prepared over the upper electrode of the junction. But it is difficult to maintain superconductivity across a large difference level when we use YBCO. The planarization technique solved the issue by digging out the upper electrode from the insulating layer. In our experiment, a weight of 400 g was applied on the sample, that was equivalent to 2000 g/cm² to the mountain of In₂O₃ on the junction and equivalent to

180 g/cm² after the polishing. Then the polishing stopped automatically. Since YBCO is shaved easily by the alumina powder, the top of the YBCO electrode was a little lower than the top level of the In_2O_3 layer. For obtaining superior flatness, the ion milling treatment is useful because we can etch only In_2O_3 by controlling the accelerating voltage of Ar ions.

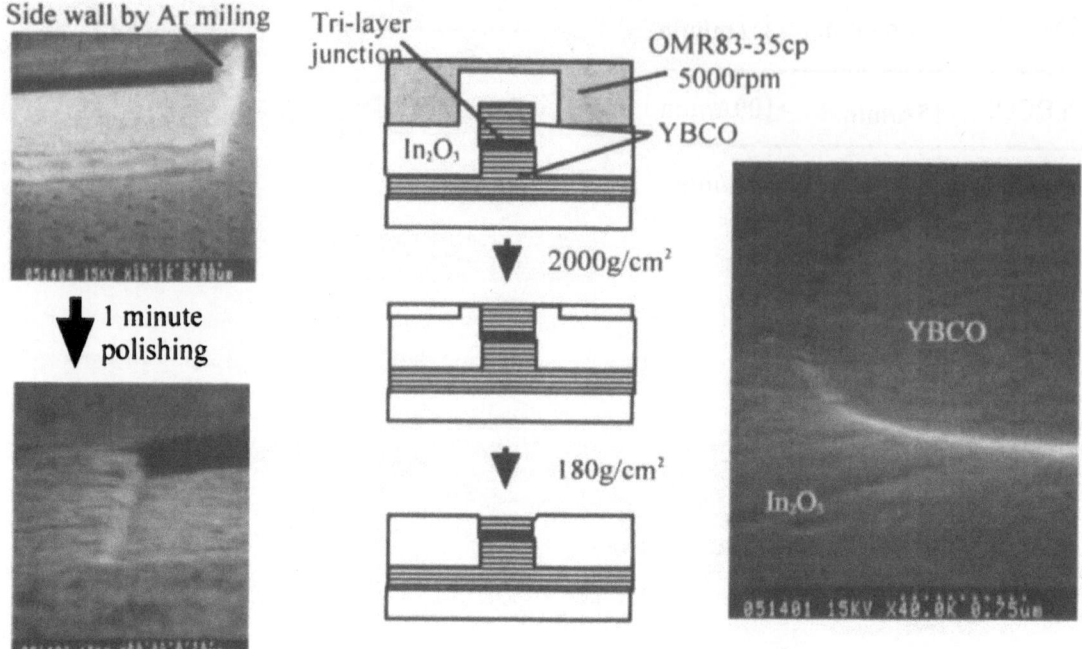

Fig. 3. Removing side wall.   Fig. 4. Digging out process of the upper electrode.

CONCLUSIONS

We examined the polishing method to solve issues in the fabrication process of the junction and circuits. The technique gave us a flat film that shows crystal steps, a pattern edge without side wall produced by the ion-milling, and a flat head of the upper electrode of the tri-layer junction. We found that the polishing method is useful for leveling the superconducting films and circuits. This method is also used as an etching method with an automatic stopping mechanism.

Acknowledgments. This work was supported by New Energy and Industrial Technology Development Organization (NEDO) through International Superconductivity Technology Center (ISTEC).

1. S. Kishida, H. Tokutaka, K. Nishimori, N. Ishihara, Y. Watanabe, and Y. Noishiki, Jpn. J. Appl. Phys., 27, L325 (1988).
2. W. Eidelloth, R. L. Sandstrom, and M. M. Plechaty, Physica C, 197, pp. 389-393 (1992).
3. A. F. Hebard, R. M. Fleming, K. T. Short, A. E. White, C. E. Rice, A. F. J. Levi, and R. H. Eike, Appl. Phys. Lett., 55, pp. 1915-1917 (1989).
4. N. Inoue, Y. Takahashi, T. Sudo, K. Sakamoto, T. Shima, and Y. Nishi, J. Appl. Phys., 71, pp. 347-349 (1992).
5. Y. Ishimaru, S. Miura, F. Wang, N. Tanaka, A. Yoshida, T. Morishita, and N. Yokoyama, in Advances in Superconductivity XI, edited by N. Koshizuka and S. Tajima (Springer-Verlag, Tokyo, 1999), pp. 1035-1038.
6. T. Hato, H. Aso, Y. Ishimaru, A. Yoshida, and N. Yokoyama, in Advances in Superconductivity XI, edited by N. Koshizuka and S. Tajima (Springer-Verlag, Tokyo, 1999), pp. 1155-1158.

Transport Properties of Grain Boundary Junctions Fabricated by Liquid Phase Epitaxy Method

T.Takagi, J.G.Wen, K.Nakao, Y.Eltsev, T.Usagawa, T.Machi, and N.Koshizuka

Superconductivity Research Laboratory, ISTEC,
1-10-13 Shinonome, Koto-ku 135-0062, Tokyo, Japan

Abstract : We succeeded in producing Y123 bicrystal films with straight GBs on MgO bicrystal substrates by LPE method. TEM images show that the GBs have straight facets with more than 30μm length. The diffraction pattern of I_c for 24° bicrystal junction shows large junction behavior with normal Fraunhopher pattern. The field suppression of I_c was beyond 85% of the maximum I_c. The large field suppression indicates small leak current through the boundary, which may provide a high performance junction. The diffraction pattern of I_c for 45° asymmetric bicrystal junction shows π junction behavior.

Keywords : Bicrystal, Straight facet, LPE, Josephson Junctions, d-wave

INTRODUCTION

The electrical transport properties across grain boundaries (GBs) in high Tc superconductors have attracted much attention not only from the physical interests but also from the applications for coated conductors[1] and electronic devices. However, the junction parameters such as Ic of the bicrystals prepared by the physical vapor deposition (PVD) in general vary widely even though they are fabricated under the same conditions[2]. TEM observations of such bicrystal junctions showed that grain boundaries (GBs) contain many microfacets[3]. Therefore, it is not so easy to understand the intrinsic relationship between Ic and misorientation angles as far as we use such bicrystal junctions. A large single facet GB is necessary for understanding the relationship between Ic and microstructures in bicrystal films.
We succeeded in preparing YBCO bicrystal films with large single facet GBs by Liquid Phase Epitaxy (LPE) method[4]. TEM images confirmed that the GBs have a single facet over about 50μm in length[5]. These bicrystal films allowed us to form striplines which enable to measure the transport properties of single facet GBs. In this paper, we report on the studies of intergrain transport properties of these samples.

EXPERIMENTS

YBCO bicrystal films with [001] tilt grain boundaries were grown by LPE method on MgO bicrystal substrates with different misorientation angles (θ = 24° and 45°). MgO bicrystal substrates with misorientation angles had symmetric GB for θ =24° and asymmetric for θ =45°, respectively. The growth conditions of LPE method are described in reference [6], [7]. The thickness of c-axis oriented YBCO films was ~3 μm. YBCO films were annealed in 100% O_2 atmosphere at 500°C for more than 80 hours. Each film was formed into a stripline with the width of 30μm by focused ion beam (FIB) etching to carry out transport measurements.

RESULTS AND DISCUSSION

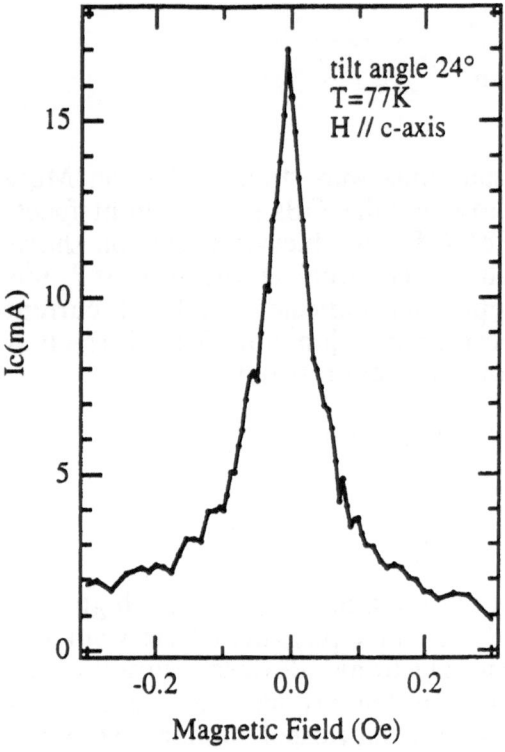

Fig.1 Magnetic field dependence of I_c for 24° tilt bicrystal junction at 77K.

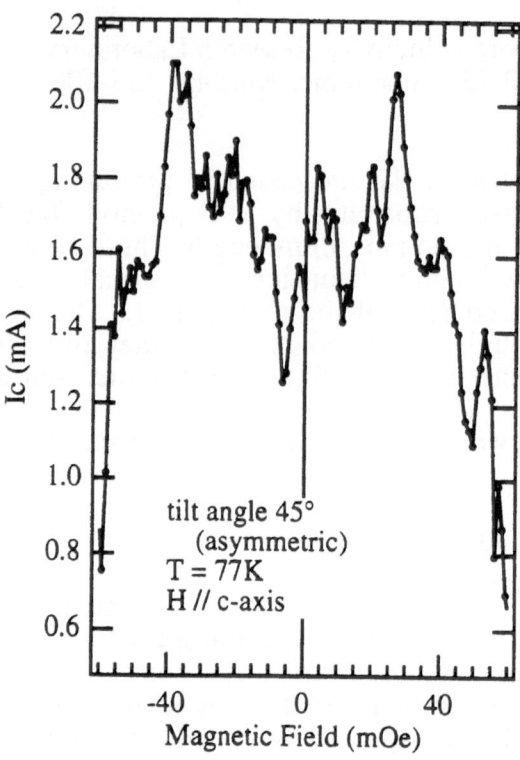

Fig.2 Magnetic field dependence of I_c for asymmetric 45° tilt bicrystal junction at 77K.

Fig.1 shows a magnetic field dependence of I_c for 24° tilt bicrystal junction at 77K. The external field was applied along the *c*-axis of YBCO film. The diffraction pattern of I_c shows the so-called large junction behavior with a normal Fraunhopher pattern. The field suppression of I_c is beyond 85% of the maximum I_c. The large field suppression indicates small leak current through the boundary, which may provide a high performance junction. The Josephson penetration depth, λ_J, is estimated to be ~2.1μm if we assume the corresponding field penetration length as $\lambda_L=0.21$μm at 77K. The ratio of the width of the junction to the Josephson penetration depth, 30μm/2.1μm ~14, is consistent with the large junction behavior in the experimental results. The field value corresponding to the first minimum, H_{c0}, is estimated to be ~80mG by extrapolation in Fig.1. This value can be explained by the flux focusing effect [8], [9]. In the geometry of our junction and electrodes, the field is concentrated into the region of the junction due to the flux focusing effect caused by shielding current of the electrodes. When we consider the focusing effect, H_{c0} is given by $H_{c0} = 2 \Phi_0 / \{ (w - 2 \lambda_L)^2 + (w + 2 \lambda_L)^2 \}$, where w is the width of the junction. From this equation, H_{c0} is estimated to be ~23mG. This value is smaller than the above estimated one, but it may be consistent with our experimental result.

Fig.2 shows a magnetic field dependence of I_c for 45° tilt bicrystal junction at 77K. Since the substrate is asymmetric, the GB consist of (110) and (100) interfaces. The diffraction pattern of I_c is similar to π junction behavior with the dip structure at zero magnetic field. This behavior suggests that high Tc superconductors have the d-wave symmetry.

SUMMARY

We succeeded in producing Y123 bicrystal films with straight GBs on MgO bicrystal substrates by LPE method. TEM images show that the GBs have straight facets with more than 30μm length. The diffraction pattern of I_c for 24° bicrystal junction shows large junction behavior with normal Fraunhofer pattern. The field suppression of I_c was beyond 85% of the maximum I_c. The large field suppression indicates small leak current through the boundary, which may provide a high performance junction. The field value corresponding to the first minimum can be explained by the flux focusing effect The diffraction pattern of I_c for 45° asymmetric bicrystal junction shows π junction behavior. This behavior suggests that high Tc superconductors have the d-wave symmetry.

ACKNOWLEDGMENT

This work was supported by the New Energy and Industrial Technology Development Organization (NEDO) as Collaborative Research and Development of Fundamental Technologies for Superconductivity Applications under the New Sunshine Program administered by Agency of Industrial Science and Technology (AIST) of the Ministry of International Trade and Industry (MITI) of Japan.

REFERENCES

[1] Y.Iijima, K.Onabe, N.Futaki, N.Tanabe, N. Sadataka, O.Kohno, Y. Ikeno, IEEE Transactions on Applied Superconductivity vol.3 no.1 March 1993 p.1510-1515

[2] D. Dimos, P. Chaudhari, J. Mannhart, F. K. LeGoues, Phys. Rev. Lett. vol.61 no.2 1988 p.219-222

[3] X. F. Zhang, D. J. Miller, J. Talvacchio, J. Mater. Res. vol.11 no.10 1996 p.2440-2449

[4] T.Takagi, J.G.Wen, T.Machi, K.Nakao and N.Koshizuka, submitted to Applied Physics Letter

[5] J.G.Wen, T.Takagi, N.Koshizuka, submitted to Phys.Rev.B

[6] S. Miura, K. Hashimoto, F. Wang, Y. Enomoto, T. Morishita, Physica C vol.278 no.1&2 1997 p.201-206

[7] K.Hashimoto, S.Miura, T.Takagi, J.G.Wen, N.Koshizuka, T.Morishita, Proceedings of 10th International Symposium on Superconductivity 1997

[8] P.A.Rosenthal, M.R.Beasley, K.Char, M.S.Colclough, and G.Zaharchuk, Appl. Phys. Lett. 59 3482 (1991)

[9] Y.Ishimaru, K.Hayashi, and Y.Enomoto, Jpn. J. Appl. Phys. 34 1123 (1995)

7 Electronic Devices

RECENT DEVELOPMENTS IN SQUID APPLICATIONS

H. Koch

Physikalisch-Technische Bundesanstalt (PTB), Abbestr. 2-12, D-10587 Berlin, Germany

Abstract: Out of the many possible SQUID applications a few illustrative examples will be presented:
- non destructive evaluation of wafer doping profiles;
- geophysical radio magnetic sounding;
- detection of binding specific immune reactions with magnetic markers;
- tracking of a gastrointestinal transit;
- determination of the blockage of a nerve signal propagation in the lumbar spine;
- biomagnetic vector magnetometer for magnetic heart signal investigations.

All applications mentioned above have been performed till now only with liquid helium cooled SQUIDs. The applicability of high-Tc SQUIDs depends first of all on its reliable performance in environmental ac and dc magnetic fields. A recently developed single layer HTS-SQUID will be described with a field noise at 1 Hz of 65 fT/$\sqrt{}$ Hz when cooled in a magnetic field of 64 μT. Such a SQUID is a good candidate for most of those applications.

Keywords: SQUID Application, Nondestructive Evaluation, Geophysics, Biomagnetism, HTS-SQUID

INTRODUCTION

Numerous laboratories in the world are contributing considerable innovations to the field of Superconducting Quantum Interference Devices (SQUIDs) and their applications. Thus, only voluminous proceedings of the big international conferences on applied superconductivity may provide a comprehensive and up-to-date overview on this fast developing field.

The following manuscript therefore does not intend to give such an overview but shall illustrate the state and diversity of SQUID technology and its applications with a few hand-picked examples. Already in this limited framework the rich diversity and the application opportunities that SQUID technology might provide become evident.

SQUID APPLICATIONS

NONDESTRUCTIVE EVALUATION OF WAFER DOPING PROFILES

With a recently developed SQUID-based, noninvasive method the investigation of doping inhomogeneities in semiconductor samples becomes feasible in a manner competitive to other established methods. It is based on the detection of photocurrents excited in the semiconductor wafer sample by a focused laser beam. A highly sensitive low-Tc SQUID system is used to detect these photocurrents via their magnetic field. The method is capable of visualizing small, growth related fluctuations of the doping level (striations) of the semiconductor with a spatial resolution of a few tens of micrometers determined by the excitation spot size. In Fig. 1. a schematic experimental setup and a photoscanning plot of a boron doped silicon sample is presented. The sample was cut from a silicon single crystal parallel to its growth direction. The magnetic field component B_x of the excited photocurrents was measured. The laser wavelength and laser power were 980 nm and 20 mW, respectively. The dynamics of the magnetic signal caused by the pulse excitation of photocurrents is related to characteristic charge carrier life times in the sample and can be measured with a high-bandwidth SQUID system. Numerical simulations of the photocurrent distribution in samples with special doping profiles have been performed and used for a forward calculation of the magnetic signal. The instrumentation of the measurement systems may still be optimized with respect to the spatial and magnetic field resolution as well as to be equipped with an automatic process control. The low-Tc SQUIDs may as well be replaced by high-Tc magnetometers in the near future [1].

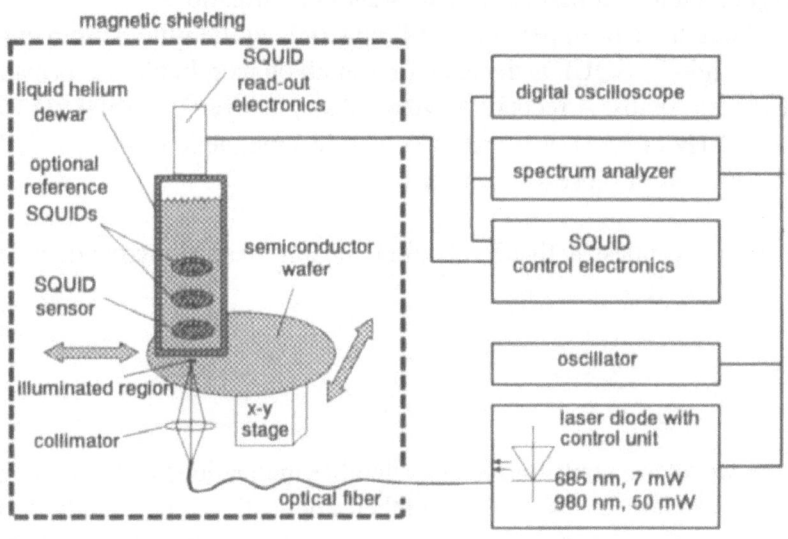

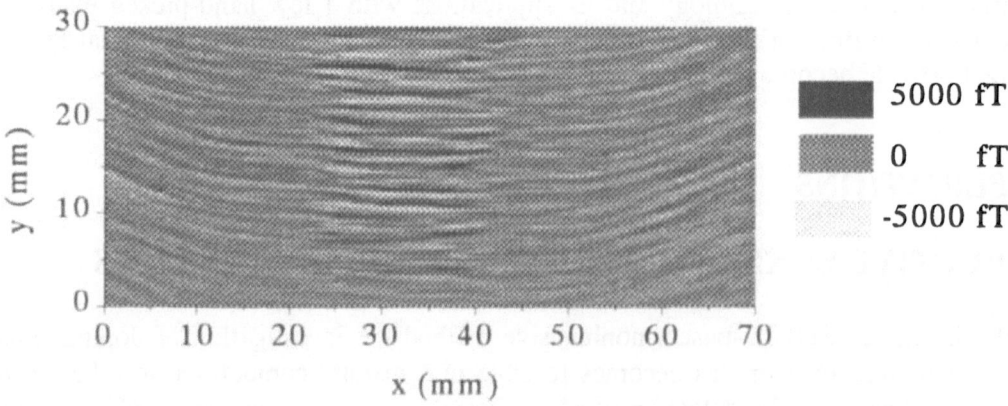

Fig. 1. Nondestructive evaluation of the doping profile of a semiconductor wafer sample: schematic experimental set-up (top) and x-y-scan of the qualitative doping concentration distribution (bottom).

GEOPHYSICAL RADIOMAGNETIC SOUNDING

Besides the more conventional geophysical exploration techniques in recent years, the interest in environmental geophysics, i.e. analysis techniques of the shallow subsoil down to some 10 m depth increased considerably. The respective geophysical methods exploit the skin-effect in the earth's crust when electromagnetic waves incident from the ionosphere induce screening currents. The skin-depth depends on the earth's resistivity and the frequency. The field of the screening current is detected electrically by electrodes and magnetically by magnetometers. With these field data measured at several sampling points the earth's impedance dependent on frequency will be derived and the 3-dimensional spatial distribution of the impedance reconstructed.

For 'shallow' investigations only a few meters below ground level one has to extend the method to higher frequencies. In the frequency range of interest the electromagnetic waves from radio transmitter stations may be exploited. Thus the respective method is called radiomagnetic sounding (RMS).

The high-frequency of the signals to be investigated (10 kHz - 2 MHz) forced a development to improve the slew-rate of SQUID-electronics. Besides the exceptionally high slew-rate requirements the homogeneity of the electromagnetic transparency of the dewar's superinsulation is of major importance [2]. Whereas low-T_c SQUID systems with their expensive cryogenic equipment are adequate to investigate the feasibility of SQUID based electromagnetic techniques, only high-T_c SQUID systems are acceptable for commercial applications.

BIOMAGNETIC APPLICATIONS IN GASTROENTEROLOGY

In gastroenterology the interest in diagnosis and therapy is focussed on phenomena of digestion, among others on the dynamics of the transit of nutrition substances through the intestinal tract under various physiological conditions. For pharmacological research and development the history of a pill, capsule or other means of oral drug delivery is of main interest, e.g. how, after what time, and where is a capsule dissolved and thus the drug released is an important question well worth to be investigated.

A novel contribution to these topics has been provided by biomagnetic methods: with a multichannel SQUID system it is possible to localize a magnetic marker and monitor its transit through the gastrointestinal tract. The magnetic marker may either be a magnetized particle coated with a protective cover or for dissolvement investigations a composition of magnetizable powder of nanoparticles fixed in a suitable matrix modeling the drug. A detailed description of such investigations may be found in [3].

IMMUNOASSAYS

Numerous medical and biochemical investigations and tests rely on binding specific antigene-antibody reactions. In clinical chemistry the respective application is performed mainly by so-called immunoassay-tests.

The detection of such molecular binding events is maintained by labeling one of the binding partners with markers that can reliably and sensitively be detected by appropriate physical methods. Such markers could be enzymes that cause a detectable color reaction, or the markers are fluorescent or radioactive and are thus detectable by optical or dosimetric means.

Recently magnetic nanoparticles have been introduced as an innovative marker in immunoassays in combination with the high sensitivity of SQUID sensors for signal detection. A very attractive feature of this method is that it measures the magnetic relaxation signal of magnetized nanoparticles.

Bound and unbound magnetically labeled makromolecules may easily discriminated due to their marked difference in relaxation time making the usually necessary separation step after immuno-reactions superfluous. This method and its instrumentation are described in depth in [4].

NERVE SIGNAL PROPAGATION

A very ambitious biomagnetic application concerns the investigation of the peripheral nerve function in man because the signals to be detected are with amplitudes of a few femtotesla extremely weak.

Thus, only with a very sensitive SQUID system these experiments are feasible. Even with PTB's multichannel system with its white noise of < 3 fT/$\sqrt{Hz}$ an averaging of 10000 stimuli was necessary to recover the relevant magnetic field signal from the noise.

Fig. 2. illustrates a typical experimental set-up for investigating the nerve signal transmission through the leg nerve bundle and entering the lumbar spinal chord. The patient lies in a prone position while his left and right foot ankles are alternately stimulated by current pulses of 6-12 mA amplitude, 0.1 – 55 ms duration and a frequency of 9 Hz for 1000 s. The SQUID system position above the region of investigation detects the spatio-temporal magnetic field evolution accompanying the propagating nerve signal [5].

The overwhelming difficulty of analysing the obtained signal data correctly stems from the poor signal-to-noise ratio. Particularly the magnetic signal from the heart activity interferes strongly and has –together with other external noise- to be eliminated by advanced methods of signal processing. Iterative weighted averaging or independent component analysis have been successfully implemented to recover the nerve signal propagation.

On the right side of Fig. 2. the reconstructed tracks of nerve pulses over the lumbar spine region is shown. In this example a clinically relevant blockade of such a signal propagation is evident for the right nerve bundle. Such a functional diagnosis cannot be provided by any other imaging technique.

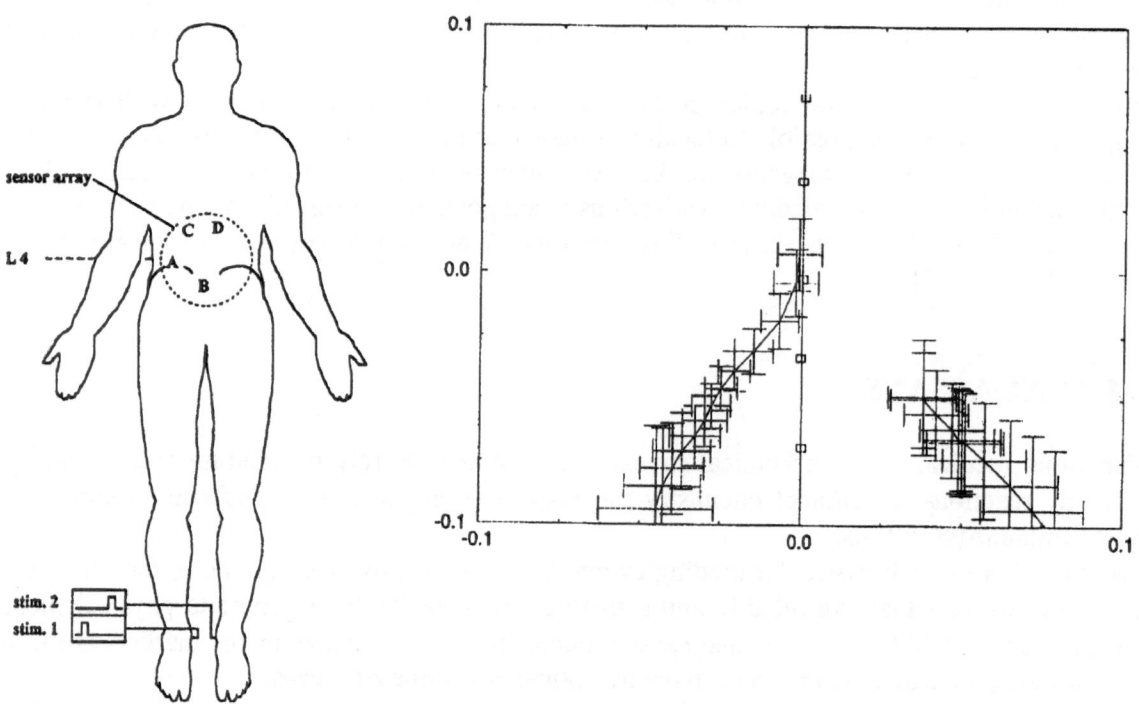

Fig. 2   Schematic experimental set-up and reconstruction of nerve signal path with blockade.

SQUID TECHNOLOGY

For optimum performance the SQUID sensors have to be specially designed according to the application requirements. One major application of SQUIDs is in biomagnetic research where an ultimate sensitivity in the low frequency range of the signal spectrum is a prerequisite. However, as shown above, SQUID sensors are also attractive for several other applications such as nondestructive evaluation, geophysics or immunoassays. These applications require sensor systems which can be operated in an unshielded or moderately shielded environment and furthermore, a wide signal bandwidth up to the MHz range is demanded in some cases. Therefore, beside the improvement of the sensitivity of the devices another design goal has to be related to the improvement of their dynamic properties as for example bandwidth and slew rate. In order to achieve reasonable results, both, SQUID design and SQUID read-out electronics have to be taken into consideration when improving the system performance.

High-Tc SQUID sensors have been developed which are suited for operation in multichannel systems in magnetically disturbed environment. Utilizing the superior properties of 30° SrTiO3 bicrystal junctions, it was possible to fabricate high-performance single-layer direct-coupled SQUID magnetometers. Operating the SQUID sensors in well shielded environment noise levels down to 24 fT Hz-1/2 and typical 1/f corners of 4 Hz were obtained for devices having a 9 mm x 9 mm pickup loop with 3 mm linewidth and a nominal SQUID inductance of 100 pH. For signal frequencies $\leq$ 50 Hz, a minimum total harmonic distortion of -118 dB was measured, dominated by nonlinearities in the read-out electronics. To protect the SQUIDs from moisture, they are encapsulated in a ceramic housing. A thick film resistor integrated on the chip carrier allows one to heat the SQUID device above Tc in order to release trapped magnetic flux. A new direct-coupled SQUID magnetometer design with improved noise performance in magnetically disturbed environment has recently been developed and characterized. These magnetometers based on 30° bicrystal junctions are composed of a 100 pH SQUID loop with a reduced linewidth of 4 μm and a modified pickup loop. The originally 3 mm wide solid pickup loop is replaced by 16 parallel loops, each 50 μm wide. The magnetic field noise down to 1 Hz did not increase when the magnetometers were exposed to ac fields with peak-to-peak amplitudes of up to 54 μT or cooled in static magnetic fields above that of the Earth. The best noise at 1 Hz of such a device cooled in 64 μT was 65 fT/√Hz. An increased noise below 1 Hz was observed when the devices were cooled in a static magnetic field. It can be quantitatively described by the measured temperature fluctuations in the liquid nitrogen bath assuming a temperature coefficient of the pickup loop area of about 1×10^{-4}/K [6].

For the operation of the low-noise high-T_c SQUIDs, a compact direct-coupled read-out electronics with a preamplifier voltage noise of 0.4 nV/√Hz and bias reversal (frequency range 0.1 kHz – 500 kHz) was developed. The maximum system bandwidth and slew rate of the bias reversal electronics are 1 MHz and about $5 \times 10^5 \Phi_0$/s, respectively.

VECTORMAGNETOMETER

The majority of the low-Tc SQUIDs fabricated at PTB will be utilized for a multichannel vectormagnetometer designed for containing 304 SQUID-chips. Contrary to existing multichannel SQUID-systems this instrument will be capable of measuring the true vector information of the biomagnetic fields and not only one field component or any other artificial gradiometric composition.

Fig: 3. shows the result of a vector field measurement at the R-peak of a heart beat. This measurement has been performed by an x-y-scan of a vectormagnetometer prototype sensor head described in [7].

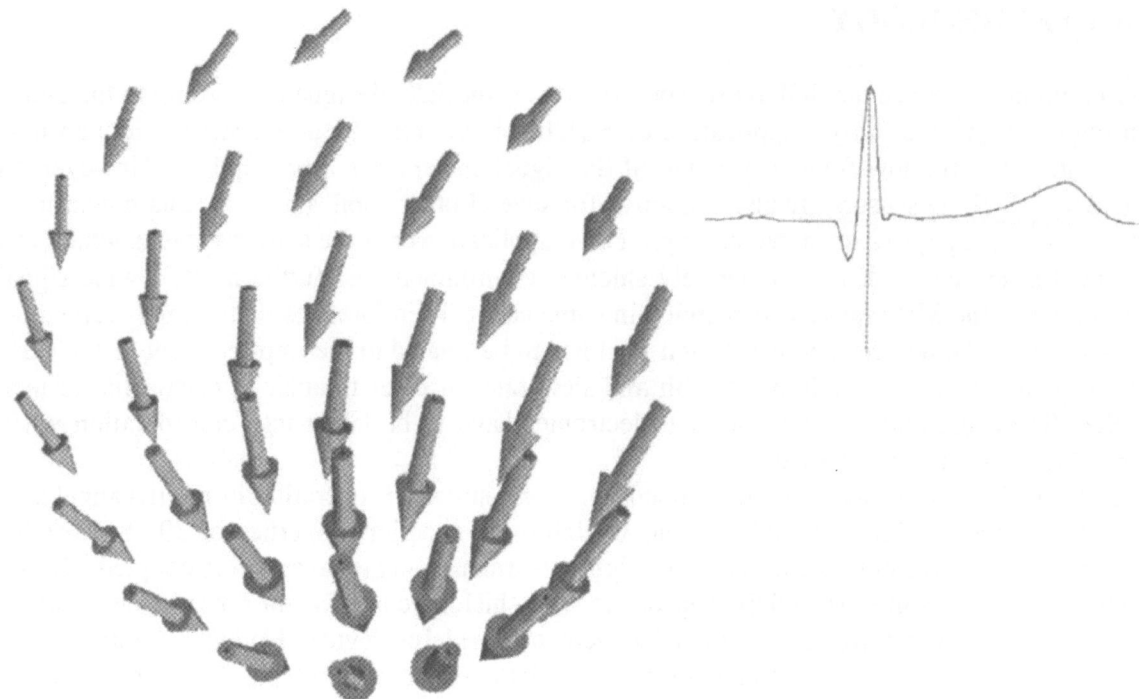

Fig. 3. Magnetocardiac vector field above a man's thorax at the instant of the heart beat's R-peak.

CONCLUDING REMARKS

The examples for SQUID applications described above should demonstrate how SQUID sensors may advantageously be employed in very different fields of application. Each of these applications require specially taylored system solutions in order to become successful. In all mentioned cases a transfer to commercially attractive products seems to be feasible but still has to be proven. In some cases the present state of high-Tc SQUID technology may already be sufficiently advanced, others require utmost sensitivity and thus demand low-Tc technology.

Acknowledgements. The author is indebted to many co-workers and partners in the R&D projects. Their respective names are listed in the cited references. The projects were partly supported by the German Federal Ministry of Education, Science, Research and Technology (BMBF), the Commission of the European Community and Deutsche Forschungsgemeinschaft (DFG).

1. J. Beyer, H. Matz, D. Drung, and Th. Schurig; Appl. Phys. Lett. **74**, 1863-1865 (1999).
2. D. Drung, T. Radic, H. Matz, H. Koch, S. Knappe, S. Menkel, and H Burkhardt; IEEE Trans. Appl. Supercond. **7**, 3283 (1997).
3. W. Weitschies, R. Kötitz, D. Cordini, L. Trahms; J. of Pharmaceut. Sci., **86** 1218-1222 (1997).
4. H. Matz, D. Drung, S. Hartwig, H. Gross, R. Kötitz, W. Müller, A. Vass, W. Weitschies, L. Trahms; Appl. Supercond. **6/10** 577-583 (1999).
5. B.-M. Mackert, G. Curio, M. Burghoff, L. Trahms, P. Marx; J. Electroencephalography and Clin. Neurophysiol. **109** 315-320 (1998).
6. F. Ludwig and D. Drung; Appl. Phys. Lett. **75**, 2821-2823 (1999)
7. M. Burghoff, H. Schleyerbach, D. Drung, L. Trahms, H. Koch; IEEE Appl. Supercond. **9** 4069-4072 (1999).

A Balanced Radio Frequency Amplifier Based on a Niobium dc SQUID with Microstrip Input Coupling

Michael Mück[1,2], Marc-Olivier André[2], John Clarke[2], Jost Gail[1] and Christoph Heiden[1]

[1]Institut für Angewandte Physik der Justus-Liebig-Universität Gießen, 35392 Gießen, Germany
[2]Department of Physics, University of California, Berkeley, CA 94720-7300, USA

Abstract: We have developed a high frequency amplifier using a dc SQUID in a new configuration in which the input coil of the SQUID is used as a microstrip resonator. By using a coil with an appropriate length, we have obtained resonant frequencies of 200 MHz to 1.3 GHz and gains of about 20 dB. We measured a minimum noise temperature of 0.12 ± 0.1 K for a frequency of 438 MHz and a bath temperature of 0.5 K. By loading the otherwise open end of the microstrip resonator with a varactor, we were able to reduce the frequency of maximum gain of such an amplifier by about 40 %. The use of two SQUID amplifiers in a balanced configuration improves the input impedance matching substantially.

Keywords: dc SQUID, rf amplifier, low noise, varactor tuning, balanced amplifier

INTRODUCTION

In order to reduce the observation time in a cryogenic axion detector [1], one requires a radio frequency amplifier which has a noise temperature of 0.1 K or less over the frequency range 0.1 - 1 GHz. Such a low noise temperature appears unachievable with currently-available cryogenic semiconductor amplifiers, which in the best case exhibit noise temperatures of 1 K or higher [2]. However, the required low noise can be achieved by means of an appropriately configured dc Superconducting Quantum Interference Device (dc SQUID). We have developed an rf amplifier based on a dc SQUID in which the input coil is used as a microstrip resonator. Unlike semiconductor amplifiers, the noise temperature of SQUID amplifiers continues to decrease linearly with the bath temperature as it is lowered to at least 0.4 K [3].

A RF AMPLIFIER BASED ON A SQUID WITH MICROSTRIP INPUT COUPLING

In contrast to a conventional dc SQUID amplifier [4,5,6] the input signal is no longer coupled to the two ends of the input coil, but rather between one end of the coil and the SQUID loop, which acts as a ground plane for the coil (see Fig. 1). A microstrip resonator is thus formed by the inductance of the input coil and its ground plane and the capacitance between them. Provided its ends are loaded with impedances greater than its characteristic impedance, the microstrip resonator is analogous to a parallel tuned circuit. At the resonant frequency the current fed into the resonator is amplified by the quality factor of the resonator Q. One selects the resonant frequency by choosing the length of the coil appropriately.

Fig. 1 Conventional SQUID amplifier with input signal coupled between both ends of the coil (left), and microstrip SQUID amplifier with signal coupled between one end of the coil and the washer (right).

Since the conventional washer SQUID is an asymmetric device (the two Josephson junctions are situated close together rather than on opposite sides of the SQUID loop), one can either ground the washer or ground the counter electrode close to the Josephson junctions. At first sight, since the washer acts as ground plane for the input coil, one would be tempted to ground the washer. However, it is also possible to ground the counter electrode and have the washer at output potential. In this case, depending on the sign of the flux-to-voltage transfer function of the SQUID, $V_\Phi = \partial V/\partial \Phi$, one can obtain either negative or positive feedback from the output to the input. In the case of positive feedback, the gain of the amplifier is increased by several dB and the quality factor Q is enhanced, thereby narrowing the bandwidth.

GAIN AND NOISE MEASUREMENTS

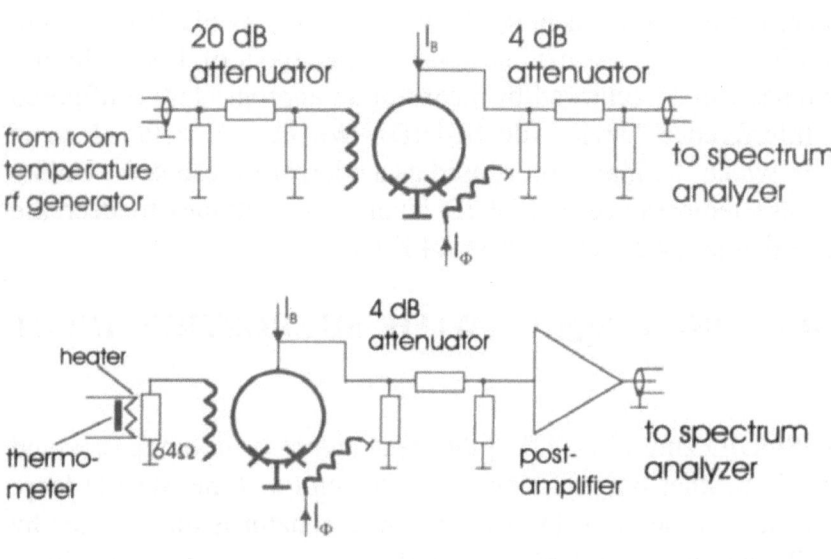

In order to measure the gain of such amplifiers, the necessary current and flux biases were supplied by batteries. We used a cold attenuator of about 20 dB between the room temperature rf generator and the input of the SQUID to prevent noise produced by the generator from saturating the SQUID (see Fig. 2, top). The attenuator also presented an impedance of 50 Ω to both the input coaxial line and the microstrip. This impedance matching largely eliminated standing

Fig. 2 Schematic diagrams of the measurement configurations for gain (top) and noise temperature (bottom).

waves on the coaxial line. It also helped to minimize errors in the measured gain due to impedance mismatch. For the same reasons, a cold 4 dB attenuator coupled the output of the

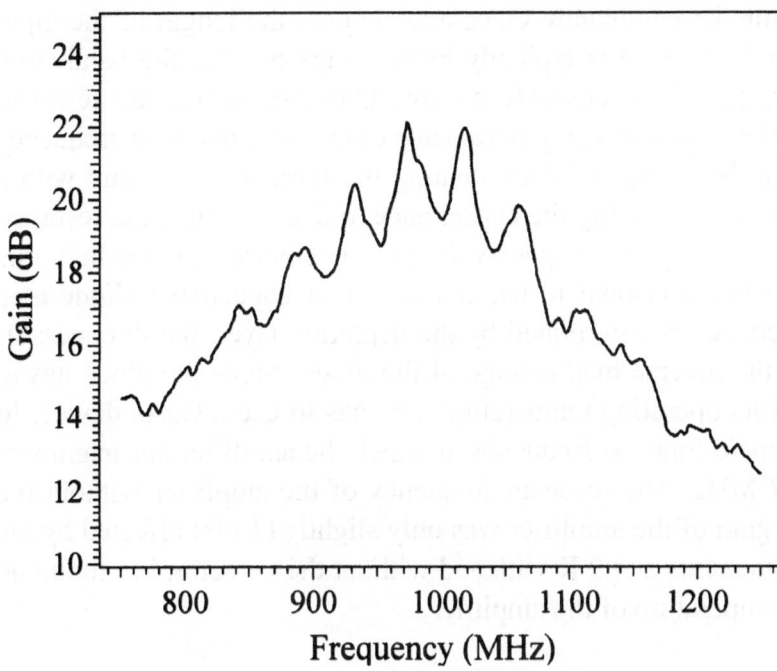

Fig. 3 Gain versus frequency of a microstrip input SQUID amplifier. The ripple on the curve is due to a slight mismatch between the coaxial line and the input of the room temperature amplifier.

SQUID to a low noise (80 K) preamplifier at room temperature. The output power of the preamplifier was measured with a spectrum analyzer. The gain of the system excluding the SQUID was calibrated by disconnecting the SQUID and connecting together the input and output attenuators. All measurements of the gain of the SQUID amplifier were referred to the baseline so obtained. Gains of the order of 20 dB in a frequency range 0.1 - 1.3 GHz were routinely obtained (see Fig. 3).

We measured the noise temperature T_N of the SQUID amplifier by connecting a 64 Ω resistor to the input of the SQUID (see Fig. 2, bottom). We could raise the temperature of this resistor by passing a current through a manganin wire wound tightly around it. By comparing the noise powers at the output of the SQUID to that of the resistor at 4.2 K and at 10 K we were able to estimate T_N. When cooled to 4.2 K, such amplifiers exhibited noise temperatures ranging from 0.5 K ± 0.3 K at 80 MHz to 1.6 K ± 1.2 K at 1 GHz. In order to reduce the contribution of the semiconductor post-amplifier to the system noise temperature, we built a single-stage cryogenic amplifier using a heterostructure field effect transistor (HFET), which we operate at 1.8 K. The noise temperature of these HFET post-amplifiers was about 4 K at 90 MHz and 14 K at 440 MHz. With the SQUID amplifiers cooled to 1.8 K and using the HFET post-amplifier, we achieved a system noise temperature of about 0.3 K ± 0.1 K for a SQUID amplifier at 250 MHz, and 0.25 K ± 0.1 K for another amplifier at 365 MHz.

To achieve an even lower noise temperature, we cooled the SQUID to a temperature of about 0.4 K using a closed cycle ^{3}He cryostat. The noise temperature of 100 mK ± 20 mK measured at 90 MHz also contained a noise contribution from the HFET post-amplifier of about 25 mK. In the case of a 438 MHz SQUID, the contribution by the post-amplifier (0.38 K) was subtracted from the measured system noise (0.5 K) to obtain a SQUID noise temperature of about 120 ± 100 mK.

TUNING THE AMPLIFIER WITH A VARACTOR DIODE

The resonant frequency of the cavity used in an axion detector is swept over a wide range of frequencies to search over a broad range of axion mass. Unfortunately, the resonant input circuit of the SQUID amplifier limits its bandwidth. Once one chooses the length of the input coil, the resonant frequency is fixed. Since Q is typically in the range 5 – 30, the bandwidth over which the gain exceeds, say, 15 dB is relatively narrow. One can reduce the resonant frequency substantially by tuning the amplifier using a varactor diode. The resonant frequency of a transmission line resonator can be changed by terminating the normally open end with a reactive load, for example a capacitor. By varying the capacitance, one can change the resonant frequency in the best case from the original frequency (zero capacitance) to one-half this frequency (infinite capacitance). It is convenient to use a varactor or capacitance diode as a variable capacitor. Here, the capacitance is determined by the depletion layer, the thickness of which can be changed by varying the reverse bias voltage of the diode. Since the diode has to be close to the SQUID and thus at its operating temperature, one has to use a GaAs device. In one of our SQUID amplifiers, we could tune the frequency at which the amplifier has maximum gain, from 195 MHz down to 117 MHz. The resonant frequency of the amplifier without the varactor was about 205 MHz. The gain of the amplifier was only slightly (1 dB) affected by the tuning. We have made noise measurements at 4.2 K with and without the varactor, but found no noticeable difference in the noise temperature of the amplifier.

A BALANCED SQUID AMPLIFIER

In the axion experiment, the amplifier has to be (critically) coupled to a high-Q resonant cavity. If the input of the amplifier is poorly matched, instabilities or even oscillations can occur. A method to improve the input matching without having to alter the amplifier is to use a cryogenic circulator between the cavity and the amplifier. For the frequency range of interest, however, a cryogenic circulator is rather bulky.

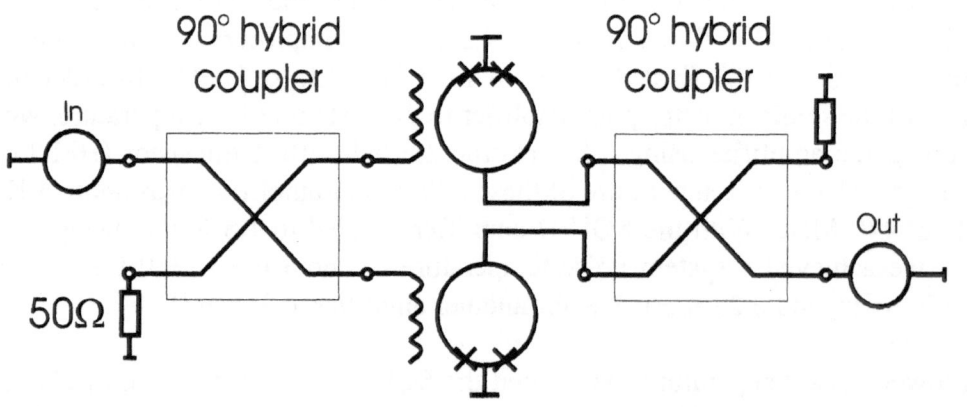

Fig. 4 Schematic diagram of a balanced SQUID amplifier.

In semiconductor amplifiers, a balanced configuration is often used to improve the input and output matching [2]. The basic circuit of a balanced SQUID amplifier is shown in Fig. 4. A so-called 90° hybrid coupler divides the input signal into two equal-amplitude components which have a 90° phase difference and which drive the two SQUIDs. The second coupler recombines the SQUID outputs. Because of the phasing properties of the 90° couplers, reflections from the individual SQUIDs cancel at the input of the coupler, resulting in an improved impedance match; a similar effect occurs at the output of the amplifier. Thermal noise generated by the termination of the fourth port of the input coupler will pass through the coupler to the SQUIDs where it is amplified but is out-of-phase at the output. In a properly balanced amplifier, this noise contribution is negligible.

There are several ways to realize 90° couplers, for example, by using quadrature hybrids or Wilkinson power dividers with an additional 90° phase shifter in one branch. We have used Lange couplers [7], which offer a relatively large bandwidth and compact design. For the projected frequency of about 200 MHz and an epoxy printed circuit board with ε ≈ 4.5, a coupler length of 180 mm results. In order to save space, the couplers were folded so that the overall length of the coupler was about 80 mm.

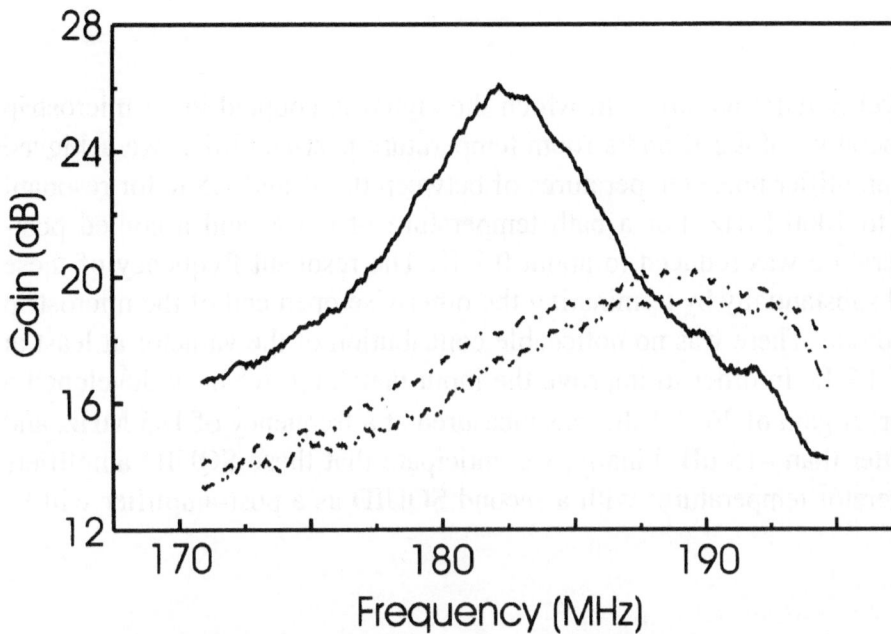

Fig. 5 Gain versus frequency of a balanced SQUID amplifier (solid line) and of individual SQUIDs used in the balanced amplifier (dashed lines).

The thickness of the substrate was chosen to be 5 mm, in order to have a width and spacing of the interdigitated wires of about 0.4 mm. For thinner boards the spacing becomes too small for a reproducible fabrication in our laboratory. The coupler exhibited an equal amplitude splitting over nearly an octave.

Figure 5 shows the gain of such a balanced amplifier versus frequency. The two SQUID amplifiers used had closely matched specifications. Their gains differed by less than 1 dB. The input coupler reduced the resonant frequency of the SQUIDs somewhat, and at the same time increased the quality factor of the amplifiers, probably because of interaction between them.

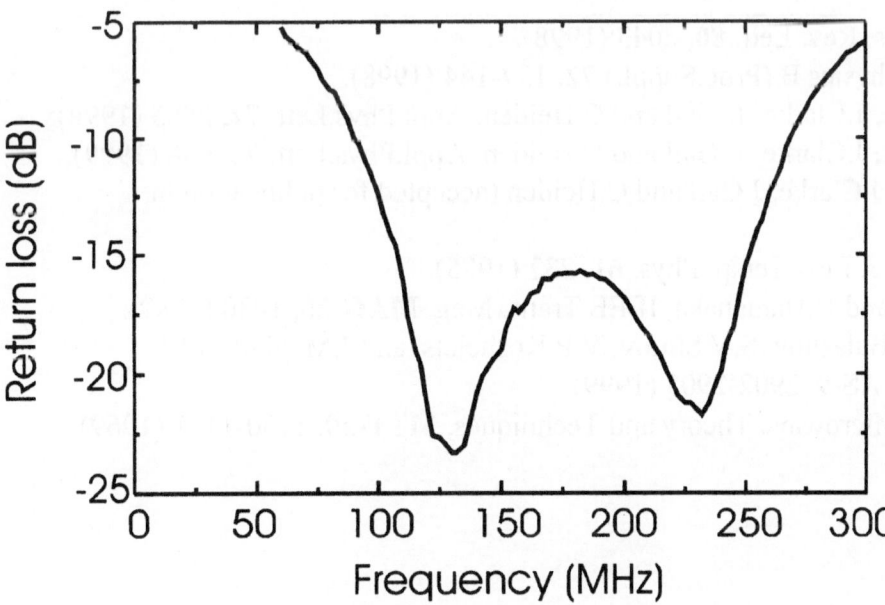

Fig. 6 Return loss at the input of a balanced SQUID amplifier versus frequency.

In Fig. 6 we plot the return loss at the input of the balanced amplifier $RL = 20 \log_{10} |\Gamma|$ dB versus frequency; Γ is the reflection coefficient. The return loss is better than -15dB ($\Gamma \leq 0.18$) over more than an octave, indicating a nearly perfect match to an input impedance of 50 Ω. A measurement of the noise temperature at 4.2 K showed no difference between the noise temperatures of the individual amplifiers ($T_N = 0.9$ K $\pm$ 0.2 K) and the balanced amplifier.

CONCLUSION

We have developed a novel SQUID amplifier in which the signal is coupled via a microstrip resonator. For a bath temperature of 4.2 K and a room temperature post-amplifier, we achieved gains of about 20 dB and amplifier noise temperatures of between 0.5 K and 1.5 K for resonant frequencies of 100 MHz to 1000 MHz. For a bath temperature of 0.4 K and a cooled post-amplifier, the noise temperature was reduced to about 0.1 K. The resonant frequency of these amplifiers could be varied substantially by terminating the otherwise open end of the microstrip resonator with a varactor diode. There was no noticeable contribution of the varactor at least at operating temperatures of 4.2 K. In order to improve the input matching, we have developed a balanced SQUID amplifier. A gain of 26 $\pm$ 1 dB was measured at a frequency of 185 MHz, and an input return loss of better than -15 dB. Finally, we anticipate that these SQUID amplifiers operated at dilution refrigerator temperatures with a second SQUID as a post-amplifier will be quantum limited.

Acknowledgments. The authors are indebted to L. Rosenberg and K. van Bibber for their ongoing encouragement, and to X. Meng for technical support. This work was supported by the National Science Foundation under grant number FD96-00014.

1. C. Hagmann *et al.*, Phys. Rev. Lett. **80**, 2043 (1998).

2. R.F.Bradley, Nuclear Physics B (Proc.Suppl.) **72**, 137-144 (1998).

3. M. Mück, M.-O. André, J.Clarke, J. Gail and C.Heiden, Appl.Phys.Lett. **72**, 2885 (1998); M.-O. André, M. Mück, J.Clarke, J. Gail and C.Heiden, Appl.Phys.Lett. **75**, 698 (1999); M.Mück, M.-O.André, J.Clarke, J.Gail and C.Heiden (accepted for publication in Appl.Phys.Lett.).

4. C.Hilbert and J.Clarke, J. Low Temp. Phys. **61**, 237 (1985).

5. T.Takami, T.Noguchi, and H.Hamanaka, IEEE Trans. Mag. **MAG-25**, 1030 (1989).

6. G.V.Prokopenko, D.V.Balashov, S.V.Shitov, V.P.Koshelets, and J.Mygind, IEEE Trans.Appl.Supercond. **AS-9**, 2902-2905 (1999).

7. J.Lange, IEEE Trans.Microwave Theory and Techniques, **MTT-20**, 1150-1151 (1969).

High-Tc dc-SQUID Magnetometers with Mesh Structure

Hiroshi Oyama[1], Koichi Yokosawa[2], Mizushi Matsuda[3] and Shinya Kuriki[1]

[1]Research Institute for Electronic Science, Hokkaido University, Sapporo, 060-0812, Japan
[2]Central Research Laboratory, Hitachi Ltd., Kokubunji, Tokyo, 185-8601, Japan
[3]Muroran Institute of Technology, Muroran, 050-8585, Japan

Abstract: In order to suppress the flux entry into $YBa_2Cu_3O_{7-x}$ (YBCO) thin film of dc-SQUIDs, we have designed and fabricated SQUID magnetometers with a mesh structure having slots and holes. The magnetometer also included flux dams within the pickup loop to suppress the flux entry driven by shielding current. In the field cooling, we observed a flux shift caused by flux trapping in the V-Φ characteristics above 10 μT, which agreed with the calculated value from the width of the YBCO film near the flux dam. When the applied field changed above 0.14 μT, the flux dams opened and the flux entered into the pickup loop. The critical current of the flux dams decreased with the applied field in a Fraunhofer-like pattern.

Keyword: SQUID magnetometer, slot structure, V-Φ characteristics, flux shift, flux dam

INTRODUCTION

Movement of flux vortices trapped in the YBCO thin film of dc-SQUID produces low frequency noise [1]. To reduce flux trapping, several methods have been proposed. Dantsker et al. showed that the entry of vortices in magnetic field cooling could be eliminated by reducing the line width of the SQUID washer [2,3]. Koch et al. reported the operation of magnetometers having flux dams, i.e., weak links in the pickup loop, which limits the shielding current flowing in the pickup loop [4,5]. In the present study, we fabricated SQUID magnetometers having slots and holes, and a flux dam structure at the grain boundary region in the pickup loop. We examined the effects of mesh structure and flux dam on the V-Φ characteristics in the magnetic field.

EXPERIMENTAL

SQUID magnetometers were fabricated by using photolithography and argon ion etching from YBCO film of 200 nm thick on a 10 mm × 10 mm $SrTiO_3$ bicrystal substrate (Fig.1). We formed many slots (45 μm × 5 μm) in the washer of pickup loop, and holes (5 μm × 5 μm) near the SQUID to divide the YBCO film into a mesh of 5 μm wide lines. The geometry of the SQUID was determined in the previous study [6]. The bias current was injected from two pads, and flowed into the SQUID symmetrically. A part of the lead connecting SQUID and pickup loop had constriction of 20 μm width that crossed the bicrystal line. This weak link would operate as flux dams with an estimated critical current of about 0.5 mA. I-V and V-Φ characteristics of the device were measured in a μ-metal cylinder located in a shielded room, where residual magnetic field in the μ-metal cylinder was less than 0.1μT. Static magnetic field was generated by a solenoid coil and applied perpendicular to the device. In the field cooling, we immersed the device into liquid nitrogen slowly in the presence of static magnetic field with a possible gradient of order of 0.5 % in the solenoid coil. In the zero field cooling, we applied the field after the device became superconducting.

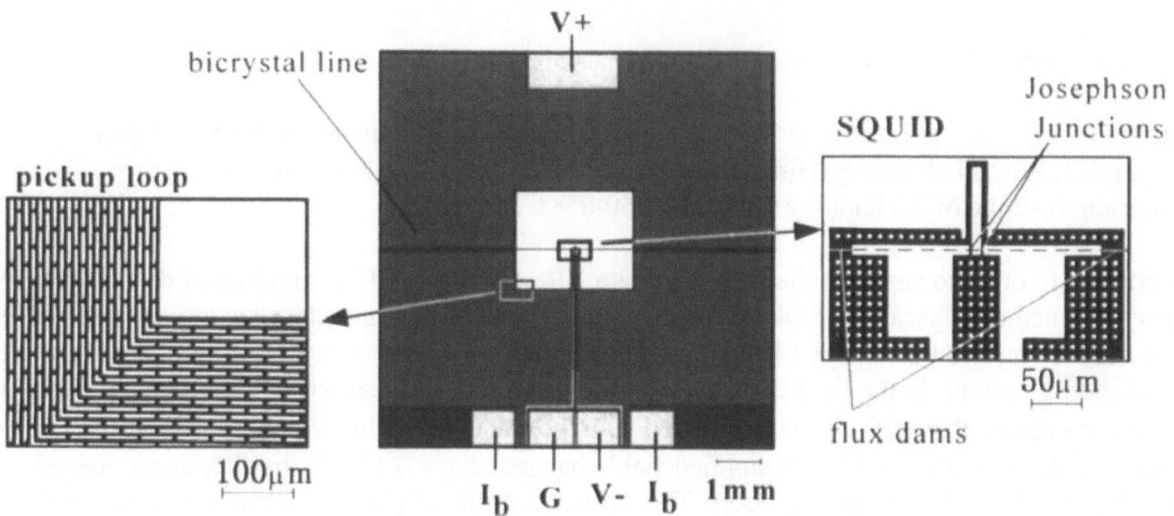

Fig. 1. Schematics of SQUID magnetometer with mesh structure.
Flux dams are located near the SQUID.

RESULTS AND DISCUSSION

After field cooling at less than about 10µT, the device was stable and operated without hysteresis of
V-Φ characteristic. Above 10 µT, flux shift occurred in the V-Φ when the bias current was changed,
suggesting flux trapping in the YBCO film. The threshold field B_{th} of flux entry may be given by B_{th}
$= \pi \Phi_0 / 4w^2$ [2,3], where w is the width of YBCO film. The largest width of the YBCO film in the
slotted washer was about 8 µm, which corresponds to B_{th} of 25 µT. However there was a part having
12 µm width, corresponding to $B_{th} \sim$ 11 µT, near the flux dam, at which flux trapping may take place.
The flux shifts could often be recovered by drawing the V-Φ characteristics repeatedly at various
bias currents. We suppose that vortices were driven out from trapping sites by the magnetic field
produced by the bias current.

Inset of Fig. 2 shows the V-Φ when we applied the offset flux after zero field cooling. Voltage
modulation disappeared at a threshold flux of $25\Phi_0$. We interpret that above the threshold the
shielding current exceeded the critical current of the flux dam. Then, the offset flux could not be
transmitted to the SQUID. The shielding current at the threshold flux of $25\Phi_0$ was estimated using
the SQUID inductance of 100 pH to be 0.5 mA. With an effective area of the device of 0.36 mm^2,
the threshold field was about 0.14 µT. When we reversed the offset flux, modulation voltage
recovered immediately. This may be due to high mobility of the vortices in the grain boundary [7];
the vortices were removed from the boundary by Lorentz force of opposite direction.

Fig. 2 shows the dependence of the estimated critical current of the flux dam on the applied field,
where Fraunhofer-like pattern was observed. The critical current of the weak link becomes
minimum at the applied field given by $B_{min}=\Phi_0/w(2\lambda_\perp+d)$, where w and $\lambda_\perp$ are the width of the
weak link and the effective penetration depth of the YBCO film, respectively. Assuming w = 20 µm,
$\lambda_\perp (=2\lambda^2/t) = 900$ nm, and neglecting the weak link length d, we obtain $B_{min} \sim$ 60 µT. This value is

much larger than the experimental value of 8 μT, suggesting enhancement of the magnetic field at the weak link. We estimated the screening current of 9.5 μA/μT, flowing near the weak link, which produces the enhanced field. Then, using the inductance of the weak link of 24pH, $B_{min} = 9$ μT is obtained, in agreement with observation.

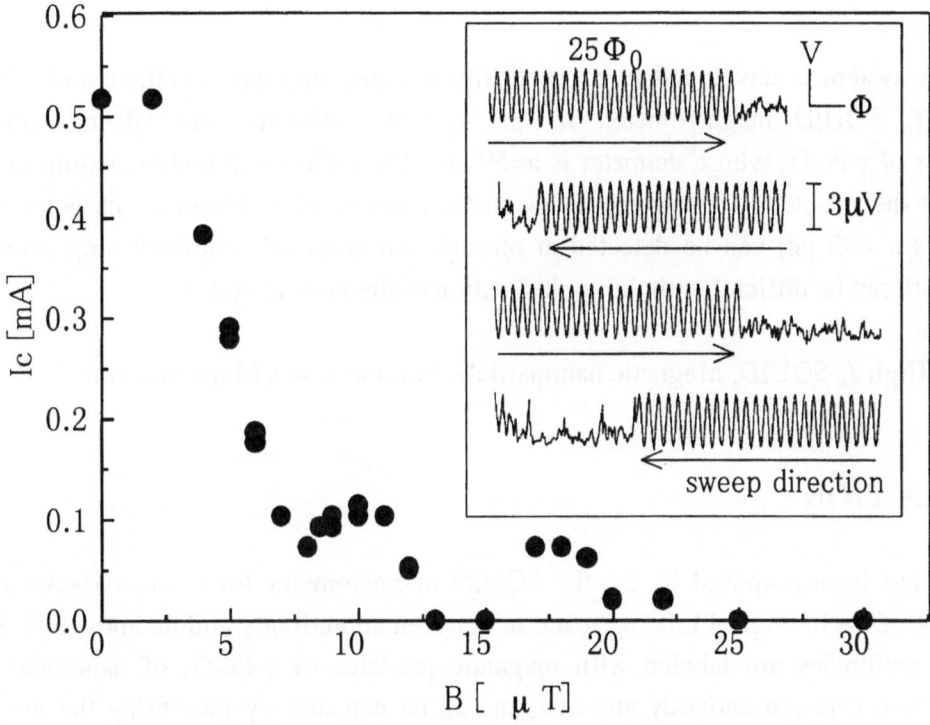

Fig. 2. Dependence of the critical current of flux dam on applied field.
Inset shows V-Φ characteristics when we applied offset flux more than 25Φ_0.

Acknowledgements This work is supported by Grant-in-Aid No.10142101 and 11359001 from the Ministry of Education, Science, Sports and Culture.

1. M. J. Ferrari, M. Johnson, F. C. Wellstood, J. J. Kingston, T. J. Shaw, and J. Clarke : J. Low Tem. Phys. **94**, 15 (1994)
2. E. Dantsker, S. Tanaka, P. Å. Nilsson, R. Kleiner, and John Clarke : Appl. Phys. Lett. **69**, 4099 (1996).
3. E. Dantsker, S. Tanaka and John Clarke : Appl. Phys. Lett. **70**, 2037 (1997) .
4. R. H. Koch, J. Z. Sun, V. Foglietta and W. J. Gallagher : Appl. Phys. Lett. **67**, 709 (1995) .
5. F. P. Milliken, S. L. Brown and R. H. Koch : Appl. Phys. Lett. **71**, 1857 (1997) .
6. S. Kuriki, H. Oyama, E. Maruyama, A. Hayashi, S. Hirano, D. Suzuki and M. Koyanagi : IEEE Trans. Appl. Supercond. **9**, 3275 (1999).
7. S. Keil, R. Straub, R. Gerber, R. P. Hauebener, D. Koelle, R. Gross and K. Barthel : IEEE Trans. Appl. Supercond. **9**, 2961 (1999).

Detection of Magnetic Nanoparticles with High T_c SQUID and Application to Biological Immunoassays

Tadashi Minotani, Takemitu Gima and Keiji Enpuku

Department of Electronic Device Engineering, Kyushu University, Fukuoka 812-8581, Japan

Abstract: A system is developed to magnetically measure biological antigen-antibody reactions with high T_c SQUID magnetometer. In this system, antibodies are labeled with magnetic nanoparticles of γ-Fe$_2$O$_3$ whose diameter is a=50 nm. The antigen-antibody reactions are measured by detecting the magnetic field from the magnetic nanoparticles. Magnetic particles as small as N=1.8 x10^6 (or 600 pg) can be detected at present. An order of magnitude improvement of the sensitivity will not be difficult with the sophistication of the present system.

Keywords: High T_c SQUID, Magnetic nanoparticle, Immunoassay, Magnetization

INTRODUCTION

Recently, it has been proposed to use the SQUID magnetometer for immunoassays, i.e., for the measurement of the biological binding reaction between an antibody and its antigen [1-2]. In this application, antibodies are labeled with magnetic particles of γ-Fe$_2$O$_3$ of nanometer size. The binding reaction between antibody and antigen can be detected by measuring the magnetic field from the magnetic nanoparticles. In this paper, we present a high T_c SQUID system developed for application to immunoassays. We describe the setup of the present system, and show the sensitivity of the present system in terms of detectable number of nanoparticles. We also discuss the ways of increasing the sensitivity of the present system.

EXPERIMENT

The measurement system is schematically shown in Fig. 1(a). The sample is an assembly of antibodies labeled with magnetic nanoparticles, as shown in Fig. 1(b). The magnetic field from the nanoparticles is measured in the following procedure. An external field B_{ex}= 8x10^{-4} T, which is parallel to the SQUID, is applied in order to magnetize the nanoparticles. Note that the field B_{ex} does not degrade SQUID performance [3], since the field is applied in parallel to the SQUID. When the particles are magnetized by B_{ex}, the z-component of the magnetic field B_z is produced by the particles. The sample is moved in the x direction with a speed of 8 mm/sec, and the SQUID measures the field B_z when the sample is passed beneath the SQUID. The distance between the SQUID and the sample is d=1.5 mm, and the diameter of the nanoparticle is a=50 nm. A large washer-type SQUID that utilizes the flux focusing effect is used, where the outer size of the washer is 4 mm. The effective area of the magnetometer is A_{eff}=0.15 mm^2 for a uniform magnetic field.

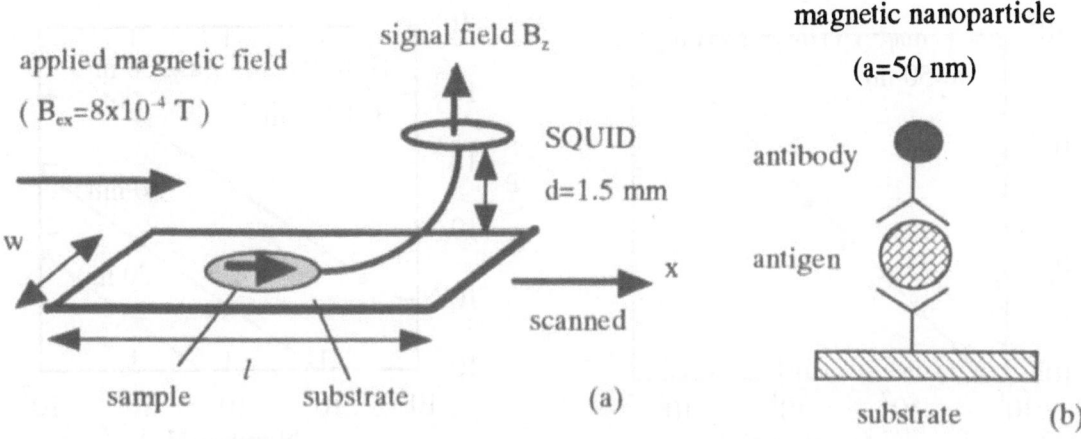

Fig. 1. (a) Schematic of the present system for detecting biological antigen-antibody reactions with the SQUID magnetometer. (b) Antibody labeled with magnetic nanoparticle.

We measured the magnetic field from a solution of antibodies which are labeled with magnetic nanoparticles (MACS, Rat Anti-Mouse IgG1). In Fig. 2, circles represent the detected signal when the amount of particles is changed. The vertical and horizontal axis represent the magnetic flux detected by the SQUID and the total number of the magnetic particles in the solution, respectively. As shown, good linearity is obtained between the magnetic signal and the concentration of the solution. In the present experiment, minimum detectable magnetic signal is limited to about 10^{-3} Φ_o. As a result, minimum detectable concentration of the particles is limited to 1.8×10^6 in terms of the number and 600 pg in terms of weight of magnetic nanoparticles. We also conducted an experiment to detect biological antigen-antibody reaction using the present system [4]. It was demonstrated that the sensitivity of the present system is better than that of the conventional optical method.

DISCUSSION

A theoretical expression for the magnetic flux Φ_s detected by the SQUID can be given by

$$\Phi_s = K(d)\,\mu_0 m\,N \quad \text{with} \quad \mu_0 m \approx 3 B_{ex}[\,4\pi(a/2)^3/3\,] \quad \text{for} \quad \mu_s >> 1, \qquad (1)$$

where N is the total number of particles and m is the magnetic moment of single particle when the external field B_{ex} is applied. The value $K(d)$ represents the efficiency of collecting the magnetic flux, which depends on the distance d as well as the geometry of the pickup coil and the sample. The moment m depends on both the applied field B_{ex} and the volume of the nanoparticles, but depends little on the relative permeability μ_s of Fe_2O_3 for $\mu_s >> 1$. In Eq. (1), we neglect the remanent magnetic moment of the nanoparticle, whose validity was checked experimentally.

In Fig. 3, the calculated result on the relation between the signal flux Φ_s and the number of particles N is show. In the calculation, we assume the parameters of $B_{ex} = 1 \times 10^{-3}$ T and $d = 1.5$ mm.

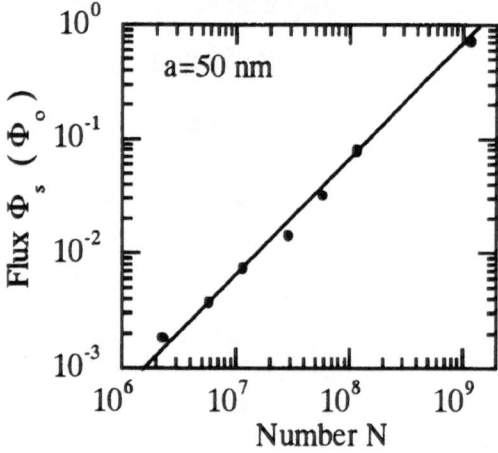

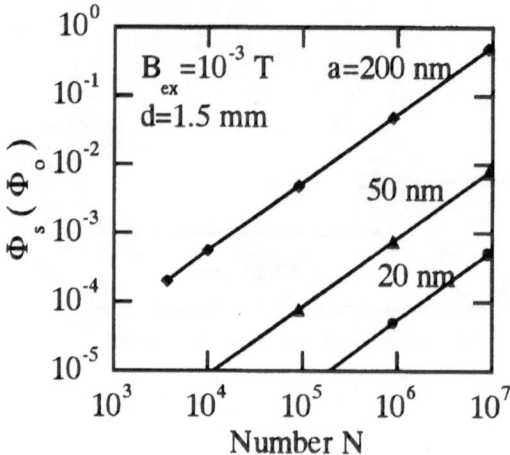

Fig. 2. Relationship between the detected flux Φ_s and the number N of particles.

Fig. 3. Calculated relationship between the signal flux Φ_s and the number N of particles

For the magnetometer, we assume 100 pH-SQUID coupled to 6 mm x 3 mm pickup coil, and the size of the sample is assumed to be 3 mm x 3 mm. As shown in Fig. 3, we can expect to detect the particles as small as $N=10^5$ for the case of a=50 nm, since the SQUID can detect magnetic flux of about 10^{-4} Φ_o. The total weight of the particle is 34 pg in this case. If the size of the particles is increased to be a=200 nm, the detectable number will become as small as $N=1.5 \times 10^3$.

It must be noted that the noise of the present system (about 10^{-3} Φ_o) is much larger than the intrinsic noise of the SQUID. Since the system noise is limited by the vibration problem, the so-called gradiometer and the method of signal averaging will be useful to decrease the system noise. When these improvements are achieved, we can realize the sensitivity shown in Fig. 3.

CONCLUSION

We developed a system to magnetically measure biological antigen-antibody reactions using the SQUID magnetometer. The present system can detect γ-Fe_2O_3 magnetic nanoparticles as small as 600 pg (or $N=1.8 \times 10^6$ for 50 nm size particle). An experiment to measure the antigen-antibody reaction revealed that the sensitivity of the present system is better than that of the conventional optical method. Since the SQUID does not limit the sensitivity of the present system, it will be easy to improve the sensitivity by more than a factor of 10 by the sophistication of the system.

REFERENCES

1. R. Kotitz et al.: IEEE Trans. Appl. Supercond. **73**, 678 (1997).
2. H. Matz et al.: Ext. Abst. 6th Int. Supercond. Electronics Conf., Berlin 1997, p.379.
3. S. Kumar et al.: Appl. Phys. Lett. **70** , 1037 (1997).
4. K. Enpuku et al.: Jpn. J. Appl. Phys., **38**, L1102 (1999).

Observation of Trapped Flux by Scanning SQUID Microscope with Novel Cooling System

Satoshi Nakayama, Toshimitsu Morooka, Akikazu Odawara, Atsushi Nagata, Masanori Ikeda, and Kazuo Chinone

Seiko Instruments Inc., Takatsukashinden 563, Matsudo-si, Chiba 2702222, Japan

Abstract: We have developed a SQUID microscope with novel cooling system. This microscope is able to observe a very weak magnetic field distribution such as a distribution generated by single flux quantum. Therefore, it is expected to be used for application and study of superconducting materials and devices. This microscope consists of a newly developed cooling system, scanning system, and controller. The cooling system is helium flow thermal conducting type which does not require pre-cooling or thermal insulation by liquid nitrogen. Consequently, the cooling time has been considerably shortened, and the sample exchange time has been shortened as well. It is possible to vary the temperature of the sample between 4K and 100K. The scanning system has wide observation area of 10mm $\times$ 10mm with a resolution of 25nm. The flux sensitivity of the micro DC-SQUID is better than $8\mu\Phi_0/\mathrm{Hz}^{-1/2}$. We present results of observing quantized flux trapped in superconducting thin film or generated in meshed superconducting thin film pattern.

Keywords: SQUID, Trapped flux, superconductor, microscope

INTRODUCTION

Recently several Scanning SQUID Microscopes (SSM) were developed for studying superconducting films and devices [1,2] . It is important to evaluate the superconducting properties of films, pinning mechanism of trapped vortex and the superconducting circuit non-destructively. Furthermore, the cycle time of measurement is an important parameter. The purpose of our work was to develop the scanning SQUID microscope system for evaluation of the superconducting material and device. Especially, High-Tc superconducting film and circuit require evaluation at various temperatures of samples. In this paper, we present the feature of scanning SQUID microscope with novel cooling system, and images of trapped vortices in Nb film and flux quantized by thin-film Nb meshed pattern.

DESIGN

SQM2000 is composed of a cooling system, a scanning system and a controller. The cooling system consists of vacuum chamber, valve box, two cold heads, turbo pump, and mechanical pump. Two cold heads are placed in the vacuum chamber. The scanning system consists of the XYZ scanning stage, magnetic shield and the coil, which are placed in the vacuum chamber. The controller consists of computer and control unit. A photograph of the scanning SQUID microscope SQM2000 is shown in Fig.1.

In order to realize easy and speedy operation and various temperature of the sample, the newly developed cooling system is a helium flow thermal conducting type. The sample and the micro DC-SQUID are placed in the chamber. Liquid helium is pumped by the mechanical pump, and supplied

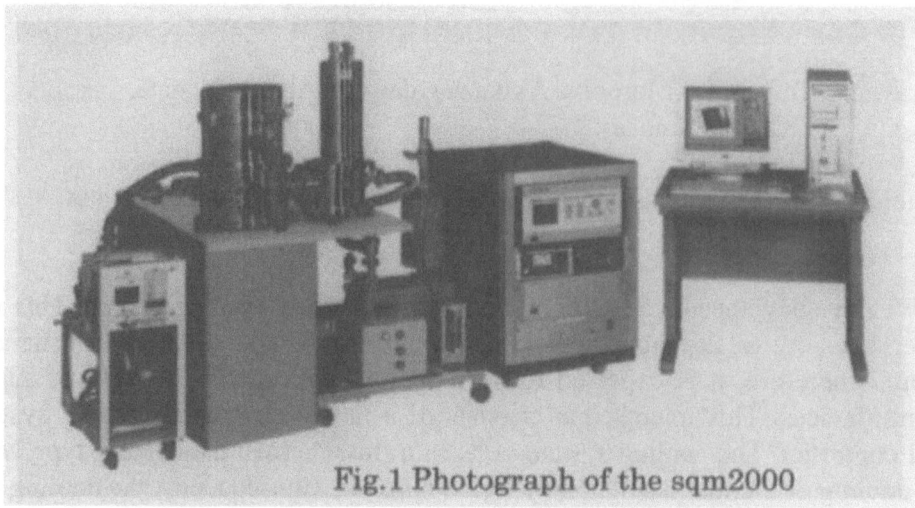

Fig.1 Photograph of the sqm2000

from helium container to the two cold heads placed in the vacuum chamber individually through the valve box. Since the cooling system dose not need the preliminary cooling by liquid nitrogen, it is easy and speedy to operate the system. The valve box controls the flow and the temperature of the helium. The SQUID is mounted on the cantilever, which is attached to a cold head. The sample is mounted on the sample holder, which is placed on the other cold head. The SQUID and the sample are cooled individually. The temperature of a sample is controlled between 3K and 100K. The SQUID is oriented about 8 degree from parallel to the sample and contacted softly with the sample.

XYZ scanning stage is placed in the vacuum chamber. The stage is driven by three stepper-motors. The stage has maximum scanning area of $10\mu m \times 10\mu m$, and minimum step is 25nm. The magnetic field is applied to sample by the coil, which is located on the cold head for placing samples. The magnetic noises radiated from the motors and environmental noises are shielded by the one-layer μ-metal.

The micro DC-SQUID is integrated on a $3mm \times 3mm$ Si chip. To realize a sufficient spatial resolution, the one-turn pick-up coil of the SQUID is $10\mu m$ in diameter [3]. The magnetic flux sensitivity of the SQUID is better than $5\mu\Phi o/Hz^{-1/2}$, where Φ_0 ($\sim 2.07 \times 10^{-15}$Wb)is the flux quantum.

SQUID controller, stage controller, temperature controller and DC power source are placed in the control unit. The computer controls the system operation and the data acquisition. 2D and 3D magnetic field images are displayed and analyzed by the software we developed.

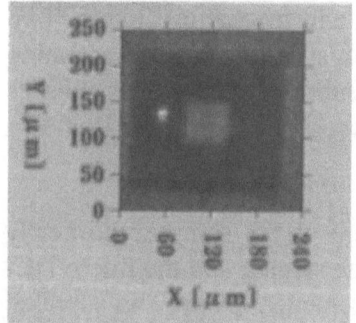

Fig.2 Magnetic image above the SQUID washer.

EXPERIMENTS

As an example of application, the magnetic image above the superconducting thin-film Nb washer SQUID was observed. The outer side of the washer was 200x200 μm^2 and the size of the washer hole was 50x50 μm^2. The washer was cooled from room temperature to 4K in the field of about $1\mu T$ applied by the coil. The physical outlines of the washer is imaged as the magnetic field screened by the Meissner effect. The magnetic image above the SQUID washer is shown in Fig.2. Only one flux vortex is trapped in the SQUID washer.

All of the images in Fig.3 were observed at 4K in the different

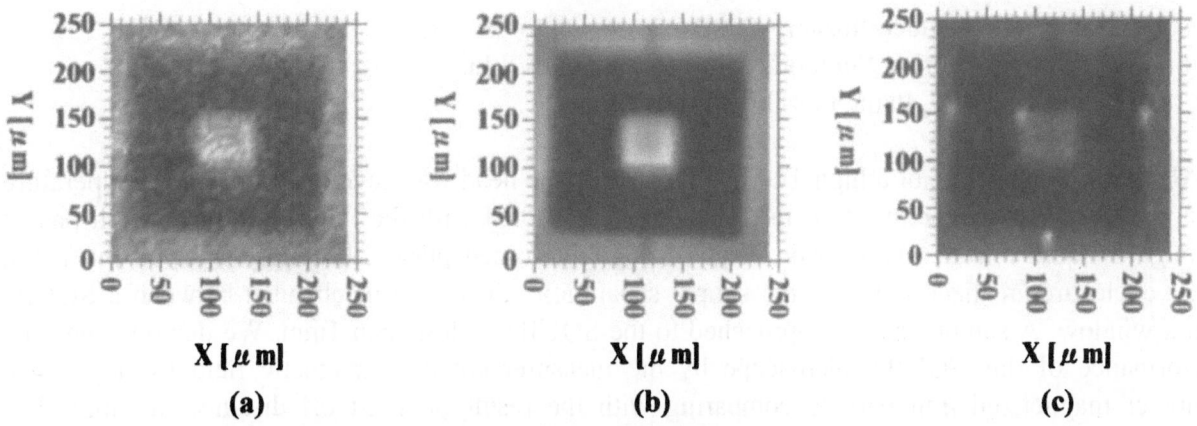

Fig.3 Magnetic images above the SQUID washer in the different field.
(a) 1μT during cooling. (b) 32μT at 4K. (c) 0T

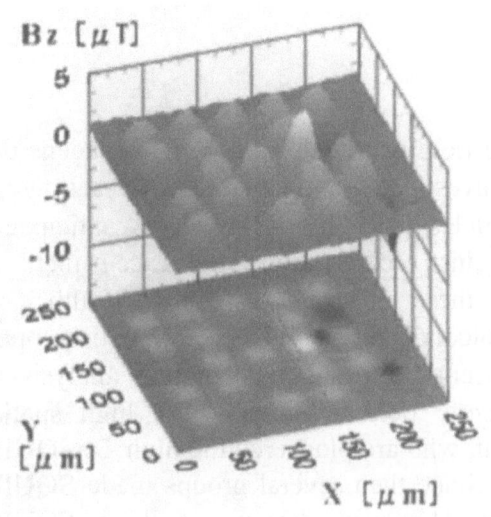

Fig.4   Magnet image above the meshed pattern

field applied by the coil continuously. The sample is the same as the thin film Nb washer SQUID mentioned above. Fig.3 (a) shows the image of the sample cooled in an additional field of about 1μT. There is no trapped vortex in the washer and weak contrast between outside and inside of the SQUID washer. It is confirmed that the background field is reduced using this coil. Fig.3 (b) shows the image of the sample measured in an additional field of about 32μT. A concentration of the magnetic flux is produced on the edge of the washer by the Meissner effect. Fig.3(c) shows the image of the sample measured in an additional field of about 0μT. Flux vortices is trapped in the edge of the SQUID washer.

Observation of the quantized flux generated in meshed Nb thin-film pattern is shown in Fig.4. The meshed pattern has 25 holes (5 rows × 5 columns). The size of each hole is 20μm × 20μm. The space between neighboring holes is 20μm. The image of the sample cooled in an additional field of about 1μT. There are 18 quantized fluxes of Φ_0, one quantized flux of $-\Phi_0$ and one quantized flux $2\Phi_0$.

CONCLUSIONS

We have developed the scanning SQUID microscope with novel cooling system operating in a temperature range of the sample 3-100K. We demonstrated to observe vortices, which are trapped in Nb thin-film in a various magnetic field generated by the coil. We will supply the scanning SQUID microscope system for evaluation of the superconducting materials and devices.

1. J.R.Kirtly, M.B.Ketchen, C.C.Tsuei, J.Z.Sun, W.J.Gallagher, L.S. Yu-Jahnes, A.Gupta, K.G.Stawiasz, and S.J.Wing, IBM J.RES.DEVELOP. **39**,1997
2. A.Y.Tzalenchuk, Z.G.Ivanov, S.Pehrson and T.Claeson,IEEE Trans. Appl. Supercond ,9, 4115-4118(1999)
3. T.Morooka, S.Nakayama, A.Odawara, M.Ikeda, S.Tanaka and K.Chinone, IEEE Trans. Appl. Supercond , **9**, 3491-3494 (1999)

High Tc SQUID microscope head for room temperature sample

Tatsuoki Nagaishi and Hideo Itozaki
Itami Research Laboratories, Sumitomo Electric Industries, Ltd.
1-1, Koya-kita, 1-Chome, Itami, 664-0016, Japan

Abstract: A new type of a high Tc SQUID microscope head was developed for room temperature sample. The sapphire thermal transfer rod is directly cooled with the liquid nitrogen which passes through a liquid nitrogen tank made of glass fiber reinforced plastic. Mutually screwed inner and outer enclosures realized bellows free simple separation adjustment mechanism between a SQUID and a window. A sample can be approached to the SQUID by less than 1mm. We demonstrated the performance of this SQUID microscope by the measurement of a magnetic field from a $70\mu m$ diameter magnetized iron particle comparing with the result of a lift off distance of 5mm. The measured magnetic field was increased by 50 times and the detection ability was enhanced drastically.

Keywords: high Tc SQUID, SQUID microscope, GFRP, spatial resolution

INTRODUCTION

Recent progress of SQUID technology especially in the field of a high Tc SQUID broadens the area of magnetic sensing applications. Matz et al. firstly investigated antigen-antibody reactions in fluid samples with the SQUID using magnetic nanoparticles [1]. This technique is anticipated because of the possibility of superiority of the SQUID against a conventional fluorescent method. Immunoassays are widely used in biology and medicine these days and the human health care industry can make progress with this technology. In the semiconductor field, Neocera Inc. developed the MAGNA-C1 prototype system which can image current flow and make failure analysis in integrated circuits [2]. In these applications high magnetic field sensitivity with high spatial resolution is required as a feature of the SQUID. Black et al. who are pioneered the high Tc SQUID microscope achieved high spatial resolution of $60\mu m$ [3]. Since then several groups made SQUID microscopes with this order of high spatial resolution. However, designs of these SQUID microscopes are very complex because of a precise separation adjustment system. In this article, we propose and introduce our new design of a high Tc SQUID microscope head which is simply made of glass fiber reinforced plastic (GFRP) except a sapphire thermal transfer rod and a sapphire window. We measured the magnetic field from a small iron particle with the SQUID microscope and compared our previous results using a capped SQUID in the liquid nitrogen Dewar.

DESIGN AND CONSTRUCTION

Fig. 1 shows a structure of a new design SQUID microscope head. The main feature of this structure is the effective heat transfer to the SQUID with direct cooling of the sapphire thermal transfer rod which is partially dipped in the liquid nitrogen. It has a simple structure with two mutually screwed different size enclosures without bellows and can adjust a separation between a SQUID and a window. The SQUID is placed in the vacuum thermal insulation layer between an inner enclosure and an outer enclosure. Since the inner and outer enclosures are screwed each other, the separation between the SQUID and the window can be adjusted by changing the screwed depth. This design leads the structure of the SQUID microscope head to be compact and simple with all GFRP made.

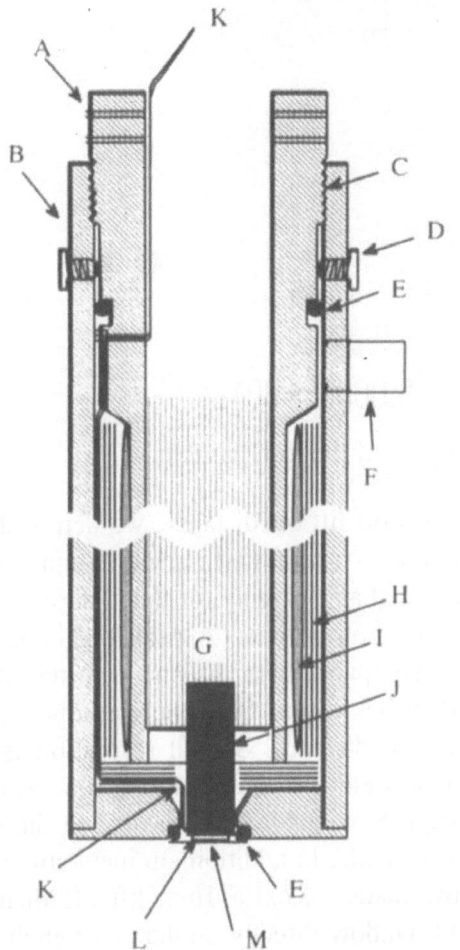

Fig. 1 Structure of a SQUID microscope. (A) inner enclosure, (B) outer enclosure, (C) screwed part, (D) stopper, (E) O-ring, (F) seal off valve, (G) liquid nitrogen, (H) super insulation, (I) activated charcoal, (J) sapphire rod, (K) wirings, (L) SQUID, (M) sapphire window

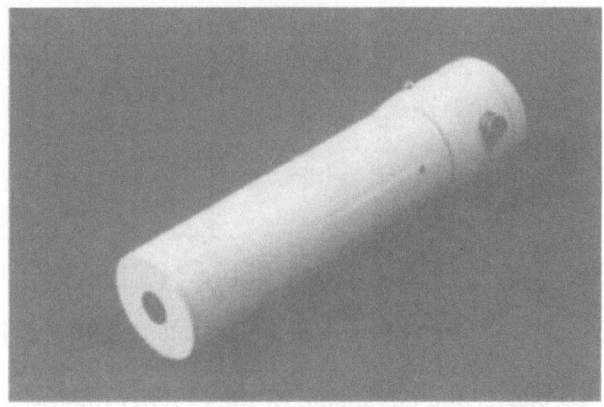

Fig. 2 Picture of SQUID microscope head
The diameter of the head is 55mm and the total length is 221 mm.

The vacuum layer between the inner (A) and outer (B) enclosure is sealed with a greased O-ring (E) to adjust the screwed depth. It is pumped through a seal off valve (F). In a vacuum region, activated charcoal (I) and ten super insulation films (H) are wound around the inner enclosure. The inner enclosure has a hole at the bottom and the sapphire thermal transfer rod (J) passes through it. The sapphire rod and the inner enclosure are glued with an epoxy resin. The SQUID chip (L) is placed at the tip of the sapphire rod with silver paste. The modulation coil is wound around the sapphire rod and four other terminal wires are also glued on it. The terminal wires are electrically connected with the SQUID at its side end. The outer enclosure also has an 11mm diameter hole and a 200μm thick sapphire window is attached with an O-ring or is fixed with the epoxy resin. The separation between the SQUID and the sapphire window is adjusted with the screwed depth of the inner and outer enclosures. Because of this structure the wiring (K) passes through the inner enclosure from atmosphere to vacuum to avoid the break during the insertion and the adjustment of the inner enclosure to the outer enclosure. The picture of the SQUID microscope head is shown in Fig. 2. The outer diameter of the head is 55mm and the total length is 221 mm. The large washer type dc SQUID can be seen in Fig.2 through the sapphire window. In this experiment, we used our conventional SQUID rather than a small SQUID chip to demonstrate the effectiveness of short lift off distance to detect a tiny magnetic field. The enclosure is pumped down to 10^{-4} Torr before the experiment and the SQUID works at least 2 hours with full volume of liquid nitrogen. The separation between the SQUID surface and the sapphire window is adjusted after pumping under the optical microscope utilizing focusing depth difference and it is set at about 110μm. A commercially available Sumitomo's flux modulation type FLL circuit [4] was used for the experiment.

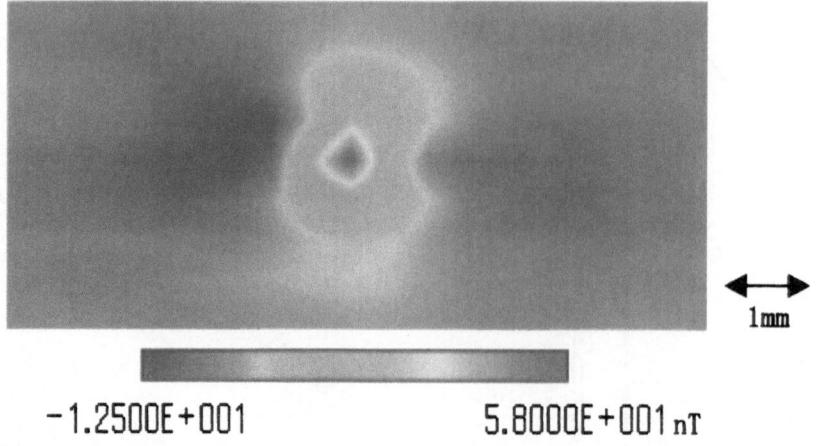

$$-1.2500E+001 \qquad\qquad 5.8000E+001\,nT$$

Fig. 3 2-dimensional magnetic field mapping of the iron particle.

To clarify the advantage of the SQUID microscope with a short lift off distance, we demonstrated the detection of the magnetic field of a magnetized iron particle. An iron particle with a diameter of $70\mu m$ magnetized under the magnetic field of 1.3Tesla was used as a sample for this demonstration. Figure 3 shows 2-dimensional magnetic field mapping of the iron particle. A peak magnetic field intensity is 5.8nT. The signal was enlarged about 50 times compared with the one with the lift off distance of 5mm [5]. The lift off distance was calculated to be 1.3mm from the formula that the magnetic field is inversely proportional to the third power of distance. Spatial resolution is also determined to be about 1.3mm from FWHM of the magnetic field distribution along the scan axis through the highest peak. Since in this experiment, the sapphire window was attached to the outer enclosure with the O-ring of the free edge, the large dent was made. In addition, an inevitable space between the sample and the edge of the sapphire window made an extra 1mm lift off distance. Renewed version of SQUID microscope head has the glued window directly on the outer enclosure, the dent of the window becomes shallow. Further improvement of sensitivity will be possible.

CONCLUSION

We developed a newly designed high Tc SQUID microscope head. The sapphire thermal transfer rod is directly cooled with liquid nitrogen for the effective heat transfer to the SQUID. The bellows free simple structure of a separation adjustment system with mutually screwed inner and outer enclosures was also designed. This design makes the high Tc SQUID microscope to be compact and simple. We measured the magnetic field of a $70\mu m$ diameter magnetized iron particle and a clear improvement of signal intensity of 50 times compared with the result of a lift off distance of 5mm. The lift off distance of the SQUID microscope was calculated to be 1.3mm. Further development with a small SQUID chip will enable high spatial resolution for applications that need spatial resolution such as failure analysis of integrated circuits, etc.

REFERENCES

1. H. Matz, D. Drung , S. Hartwig, H. Grob, R. Kotitz, W. Muller, A. Vass and L. Trahms, *Extended abstracts of 6th Int. Superconductive Electronics Conference* June 25-28, 1997, Berlin, Germany, vol. 3 A19, pp. 379-382.
2. Neocera Inc. web site : http://www.neocera.com
3. R. C. Black, A. Mathai, F. C. Wellstood, E. Dantsker, A. H. Miklich, D. T. Nemeth, J. J. Kingston and J. Clarke, *Appl. Phys. Lett.* vol. 62, pp. 2128-2130, April 1993
4. Sumitomo Electric web site : http://squid.sei.co.jp
5. T. Nagaishi H. Toyoda and H. Itozaki, *Extended abstracts of 6th Int. Superconductive Electronics Conference* June 25-28, 1997, Berlin, Germany, vol. 3 A14, pp. 364-366.

Direct flux noise measurement of $YBa_2Cu_3O_{7-\delta}$ Grain Boundary Junction

S. Hirano[1], H. Oyama[1], M. Matsuda[2], T. Morooka[3], S. Nakayama[3], and S. Kuriki[1]

[1]Research Institute for Electronic Sciences, Hokkaido University, Sapporo, 060-0812, Japan
[2]Muroran Institute of Technology, Muroran, 050-8585, Japan
[3]Seiko Instruments Inc., Matsudo, Chiba, 270-2222, Japan

Abstract: In order to study the vortex dynamics in high-Tc superconducting thin film devices, we investigated the flux noise of $YBa_2Cu_3O_{7-\delta}$ (YBCO) grain boundary junction (misorientation angle 30°, thickness 200 nm, width 84 µm) by the direct flux detection method. At about 81 K, random telegraph noise (RTN) of long time width was observed in several current regions. Zero field cooled and field cooled sample gave different I - Φ_{RMS} structures caused by the RTN, suggesting that the RTN was generated in surrounding weak links. Also, high frequency pulse trains appeared at a current slightly below dc voltage onset and disappeared with slight increase of current. The high frequency pulse trains may be related to the vortex flow at the transition to the voltage state of the current-biased grain boundary.

Keywords: Direct flux detection method, Grain boundary junction, Zero field cool, Field cool, Random telegraph noise, Surrounding weak links

INTRODUCTION

We have studied the magnetic flux noise characteristics of low-Tc and high-Tc superconducting thin film devices by the direct flux detection method [1-2]. In this method, the magnetic flux fluctuation associated with vortex motion is measured directly using a superconducting pick-up coil magnetically coupled to a SQUID. This method enables the detection of the onset of vortex motion which cannot be measured by the conventional four-terminal I -V technique and the determination of possible flux noise sources in thin film devices. In this study, we measured the magnetic flux noise of field cooled (FC) and zero-field cooled (ZFC) $YBa_2Cu_3O_{7-\delta}$ (YBCO) grain boundary junctions, and investigated the influences of the magnetic field in the cooling process to the flux noise characteristics of the high-Tc superconducting thin film devices with surrounding weak links.

EXPERIMENTAL

The $YBa_2Cu_3O_{7-\delta}$ (YBCO) thin film (thickness 200 nm) was deposited on a bicrystal $SrTiO_3$ (STO) (misorientation angle 30°, 10 mm × 10 mm) substrate (Tc = 86.20 K). Two narrow slits crossing a bicrystal line were made using photolithography and Ar ion etching techniques to form a microstrip (width 84 µm), where the thin film beside the slits was kept to detect the flux in the monopole approximation scheme [1]. Two weak links (surrounding weak links) were automatically formed in the film beside the grain boundary junction (GBJ) on the strip. A planar gradiometer (GM) was placed in such a way that inner coil of the GM was directly above the GBJ. Outer and inner coils of the GM could detect the vortex motion from all weak links and

GBJ. RMS values and real time traces of output of a SQUID coupled to the GM were measured. In field cooled (FC) measurements, static magnetic field normal to the sample surface was applied using a solenoid coil. All the measurements were made in a magnetically shielded room.

RESULTS

Fig. 1 shows the I - V, I - Φ_{RMS} characteristics, and real time traces at 81 K of zero-field cooled (ZFC) sample. The Φ_{RMS} fluctuates with time at zero bias current and low frequency RTN with two different amplitudes was observed in the time trace (A). As the current was increased, RTN of a single amplitude appeared (I = 45.6 μA, B). The fluctuation of Φ_{RMS} produced by the RTN changed, depending on the bias current and exhibited a peak structure in the I - Φ_{RMS}. At a current slightly below dc voltage onset, high frequency pulse trains appeared (I = 76.7 μA, C) and increased with the current (I = 84.7 μA, D). The pulse train disappeared as the current was increased further due to limited measurement bandwidth. At higher currents where the GBJ in the microstrip was in the voltage state, RTN of long time widths was observed (I = 99.3 μA, E) and no high frequency pulse trains appeared.

Fig. 2 shows the I - V, I - Φ_{RMS} characteristics, and time traces at 81.08 K of field cooled (FC, B= 10 μT) sample. As the current was increased, the pulse trains repeatedly appeared for short periods of time (I = 24.6 μA, A). Frequency of appearance and disappearance of the pulse trains increased with current. At a current slightly below dc voltage onset, RTN of small amplitude (I = 53.1 μA, B) was observed. The RTN changed its form to higher frequency pulse trains with increasing current. The pulse trains disappeared with slight increase of current. The high

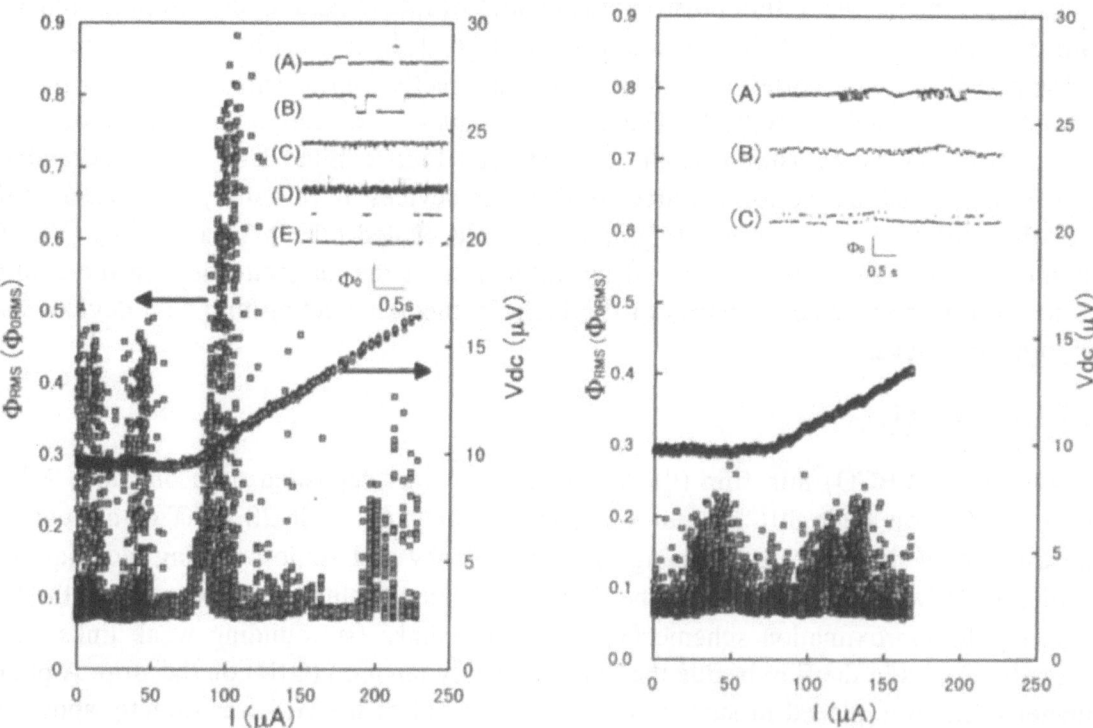

Fig. 1. I - V and I - Φ_{RMS} characteristics at 81.00 K of ZFC YBCO sample. Inset shows time traces at various bias currents.

Fig. 2. I - V and I - Φ_{RMS} characteristics at 81.08 K of FC (B = 10 μT) YBCO sample. Inset shows time traces at various currents.

frequency pulse trains were observed only within narrow current region just below the voltage onset, and RTN of long time widths (I = 118.0 μA, C) was observed at higher currents.

DISCUSSION

High frequency pulse trains observed slightly below dc voltage onset may be attributed to moving vortices in the current-biased GBJ by the following reasons: (1) the pulse trains appeared just below dc voltage onset in the I -V and were not observed in other current regions, (2) frequency of pulses increased greatly with slight increase of bias current, (3) the pulses had a unique height, indicating that the flux fluctuations were generated at a definite position of the sample. On the other hand, the RTN of long time widths had several pulse heights and appeared in several current regions, including the voltage state region of the GBJ, suggesting that the RTN was caused by vortex motion in surrounding weak links. The long time widths of the RTN reflect that the lifetime of the vortices is long compared to vortices in the current-biased GBJ. The driving force on the vortices in surrounding weak links may be the magnetic field which the bias current produces outside the microstrip, while vortices in the GBJ are directly driven by the Lorentz force exerted by bias current. Observation of the RTN at small bias currents (small driving force) is consistent with the results reported by others that vortices in the grain boundary are easier to move than in the bulk of film [3, 4]. Comparison of peak structures in the I - Φ_{RMS} between ZFC and FC sample shows strong dependence of vortex motion in surrounding weak links on the magnetic field in the cooling process.

CONCLUSION

We have investigated the flux noise characteristics of $YBa_2Cu_3O_{7-\delta}$ (YBCO) grain boundary junctions by the direct flux detection method. At about 81 K, random telegraph noise (RTN) of long time widths was observed in several current regions. Zero-field cooled and field cooled sample gave different I - Φ_{RMS} structures caused by the RTN, suggesting that the RTN of long time widths was generated in surrounding weak links. At a current slightly below dc voltage onset, high frequency pulse trains were observed. The high frequency pulse trains may be related to the transition to the voltage state of the current-biased grain boundary. It is suggested that low frequency noise characteristic of high-Tc thin film devices which contain surrounding weak links in its structure may strongly depend on the cooling environment of the device.

Acknowledgment. This work is supported by a Grant-in-Aid for Scientific Research on Priority Areas (Vortex Electronics No. 10142101) of the Ministry of Education, Science, Sports, and Culture.

1. S. Hirano, Y. Hirata, M. Matsuda, T. Morooka, S. Nakayama, and S. Kuriki, J. Appl. Phys. **85**, 7819, (1999).
2. S. Hirano, Y. Hirata, M. Matsuda, T. Morooka, S. Nakayama, and S. Kuriki, Proceedings of the 11th International Symposium on Superconductivity, Vol. 2, 1205, (1998).
3. D. Koelle, R. Gross, S. Keil, R. Straub, M. Fischer, M. Peschka, R. P. Huebener, K. Barthel, Extended Abstracts of the 7th International Superconductive Electronics Conference, 557, (1999).
4. S. Keil, R. Straub, R. Gerber, R. P. Huebener, D. Koelle, R. Gross, K. Barthel, IEEE Trans. Appl. Supercond. **9**, 2961, (1999).

Novel Cryogenic Dielectric Resonator Devices for Satellite Communication

Norbert Klein, Svetlana Vitusevych, Michael Winter and H R Yi

Forschungszentrum Jülich, Institut für Schicht- und Ionentechnik, D-52425 Jülich, Germany

Abstract: Ongoing progress in manufacturing dielectric ceramics and single crystals with high dielectric constant and   low microwave losses has turned out to be a challenge for the development of novel devices for satellite communication. From this development device performance is expected to benefit for possible device operation temperatures ranging from cryogenic temperatures around 50 - 150 K (achievable with one-stage close cycle refrigerators) over temperatures from 150 to 200K (in principal achievable with radiation cooling) towards room temperature, if novel dielectric resonator structures with lower loss contribution of the metallic housing would become available.

INTRODUCTION

Most ionic crystals exhibit a strong decrease of microwave losses upon decreasing the temperature from room temperature to liquid nitrogen temperature and below. The fundamental intrinsic loss mechanism is the absorption of electromagnetic energy by thermally excited phonons giving rise to power law dependencies of the loss tangent $\tan\delta \equiv \mathrm{Im}\{\varepsilon_r{}^*\}/\mathrm{Re}\{\varepsilon_r{}^*\}$, with $\varepsilon_r{}^*$ being the complex permittivity of the dielectric material [1,2]. This loss reduction allows for high quality factors of dielectric resonators at cryogenic temperatures if either modes with strong field confinement inside the dielectric resonator or metallic shielding cavities consisting partially of high-temperature superconducting (HTS) walls are employed. Some of the low-loss polycrystalline microwave ceramics also exhibit a significant reduction of microwave losses upon cooling and a much smaller temperature coefficient of the permittivity in comparison to single crystal materials.

Therefore, novel resonator structures utilising this material potential are worth to be investigated. If the use of cryogenics will become common in microwave communication systems due to the progress in HTS devices, the cryogenic dielectric resonators devices may be of particular relevance for frequencies above 10 GHz where the benefit of planar HTS passive devices is questionable. On the other hand, in application areas like satellite communication where the refrigerator power efficiency is an important issue, cryogenic dielectric resonator devices operating at moderate cryogenic temperatures of 100 – 200 K may be a challenging alternative to HTS systems.
In this article we describe our work on novel dielectric resonator devices with strong potential for applications at cryogenic temperatures. The first example are dielectric dual-mode filters for satellite transponders from C- to Ka-band frequencies using sapphire, single crystal lanthanum aluminate (LaAlO₃) or commercial microwave ceramics as dielectric materials [3-5]. The second example is an all-cryogenic microwave oscillator with very low phase noise at f = 23GHz based on a sapphire whispering-gallery mode resonator.

SINGLE CRYSTAL DIELECTRICS AND MICROWAVE CERAMICS AT CRYOGENIC TEMPERATURES

Among the materials usable for dielectric resonators ($\varepsilon_r \geq 10$) the following ones exhibit a strong potential for operation at cryogenic temperatures:

Al₂O₃ (sapphire and sintered alumina):   Among all dielectrics, high purity single crystals of sapphire exhibit the lowest dielectric losses at cryogenic temperatures. At 10 GHz, the loss tangent drops almost proportional to T^5 from about $7 \cdot 10^{-6}$ at room temperature to about $6 \cdot 10^{-8}$ at 77 K with a slight anisotropy for electric fields aligned along the crystallographic c- direction (ε_r = 11.4) with respect to the a,b direction (ε_r = 9.4) [7]. The frequency dependence of the loss tangent is approximately linear corresponding to a constant values of $Q \cdot f$. Sintered alumina has been optimised with respect to low losses. At 10 GHz $\tan\delta$ - values of $2 \cdot 10^{-5}$ at room temperature and $5 \cdot 10^{-6}$ at 77 K were reported [8]. Both sapphire and alumina exhibit a strong temperature coefficient of the permittivity of about 100-140 ppm at room temperature and 20 – 30 ppm at 77 K [9] resulting in stringent requirements for temperature stabilisation. On the other hand, the high thermal conductivity and the low losses of Al₂O₃ are in favour of operation at high levels of microwave power, such as transmitter filters.

TiO₂: rutile and sintered titania: Rutile is challenging because of its very high permittivity of above 100 and its relatively low loss tangent at cryogenic temperatures. The permittivity is strongly anisotropic [10,11] and exhibits an extremely large negative temperature coefficient $\partial \varepsilon_r / \partial T = - 1000$ ppm/K at $T = 77$ K. Therefore, TiO₂ is not very useful as dielectric resonator material on its own, but it can be combined with other dielectrics like sapphire in order to provide a turning point in the temperature dependence of the resonance frequency [6]. Sintered titania exhibits loss tangent values of $1.5 \cdot 10^{-5}$ at 77K and 10GHz, the temperature coefficient is similar to rutile [8]. However, titania exhibits an isotropic behaviour of permittivity and loss tangent. Similar to rutile, it is only considered to be useful in case of composite dielectric resonators.

Single crystalline lanthanum aluminate (LaAlO₃) and yttrium aluminate (YAlO₃): LaAlO₃ has become very popular as substrate material for HTS films but it is also a challenging dielectric resonator material due to its relatively high and isotropic permittivity of ε_r = 23.4. The temperature coefficient of the permittivity is similar to sapphire. As discussed in detail in one of our previous publications [12], the losses at cryogenic temperatures are governed by a relaxation peak in $\tan\delta\,(T)$ at 40-70 K. The amplitude of this peak depends on the crystal growth technique and is very small for crystals grown by the Verneuil technique. Recently, for Crochralzki grown crystals delivered from one particular crystal grower in China the peak was found to be even smaller [16]. At room temperature, $Q \cdot f =$ 500.000 GHz holds true for almost all LaAlO₃ crystals, at 77K the lowest reported loss tangent values are around $3 \cdot 10^{-6}$ at 10 GHz for Verneuil material [12]. YAlO₃ has a permittivity of 16 and the loss tangent at 77K is about 10^{-5} at 10 GHz. YAlO₃ can also be used as substrate material for HTS films.

Sintered BaMgTaO (BMT): BMT is one of the most popular commercial microwave ceramics. Its permittivity is rather high (ε_r =22-24) and its temperature coefficient is only a few ppm / K and can be optimised to be zero [14]. At room temperature, $Q \cdot f$ can be as high as 350.000 GHz for frequencies above 10 GHz (see MURATA datasheet), at lower frequencies it drops continuously according to our own measurements (e.g. $Q \cdot f =$ 150.000 GHz at $f = 4$ GHz). Fig.

1a shows our experimental results on the temperature dependence of the loss tangent for a BMT ceramic delivered by MURATA for two different frequencies. The employed dielectric resonator technique is described elsewhere [12]. The observed strong decrease of $\tan\delta$ with temperature (Fig. 1) shows that BMT is a challenging material for cryogenic dielectric resonators with Qs in the range of 10^5 at 4GHz. There are three advantages of BMT with respect to single crystal $LaAlO_3$: First, the temperature coefficient of the permittivity is much smaller from cryogenic temperatures to room temperature. Secondly, BMT is easier to machine than single crystals of $LaAlO_3$. The third advantage is that BMT is already a mass product and therefore much cheaper in comparison to $LaAlO_3$.

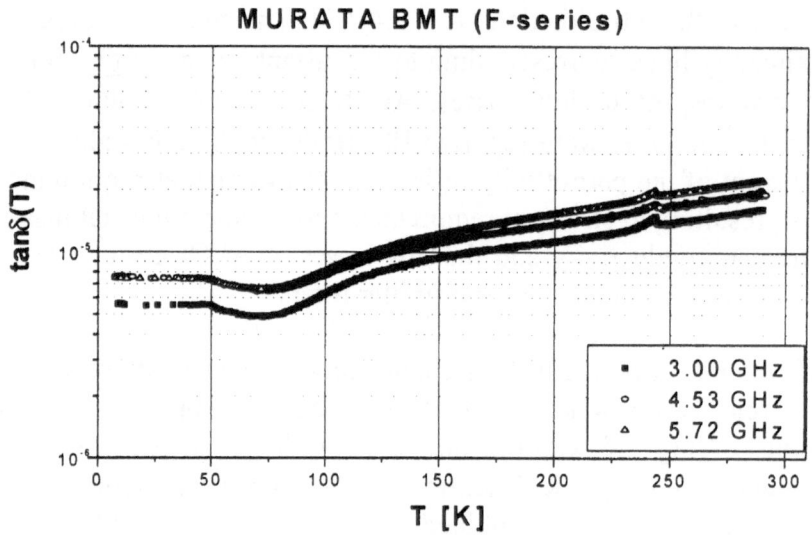

Fig.1. Measured temperature dependence of the loss tangent for a BMT sample provided by MURATA.

CRYOGENIC DIELECTRIC FILTERS FOR SATELLITE COMMUNICATION

There are several reasons to use cryogenic subsystems in satellite payloads: First, a significant weight reduction of input multiplexers (IMUX) at L and C band frequencies can be achieved if cavity or dielectric filters will be replaced by planar HTS filters to be cooled with a space qualified closed cycle refrigerator. In addition, the use of cooled preamplifiers gives rise to about 1 dB of receiver sensitivity.

For the potential use of HTS in output multiplexers (OMUX) at L- and C-band frequencies 2D planar HTS filters operated in an edge current free mode have been developed [13]. At Ku and Ka-band frequencies, the use of HTS planar filters is questionable at all, because miniaturisation is getting less important in comparison to L-and C-band. In addition, due to the quadratic frequency dependence of the surface resistance of HTS films the performance of planar HTS filters is problematic, in particular at Ka-band frequencies. On the other hand, in particular at Ka band frequencies, the performance of conventional filters is not satisfactory. Therefore, dielectric filters with improved performance are highly desired both for the IMUX and OMUX.

Our approach relies on the use of the fundamental mode of a dielectric hemisphere arranged in a metallic shielding cavity [3,4]. Originally, the design was to arrange the hemisphere on an HTS groundplane [3], however, the feedthrough of coupling and tuning elements through the ground

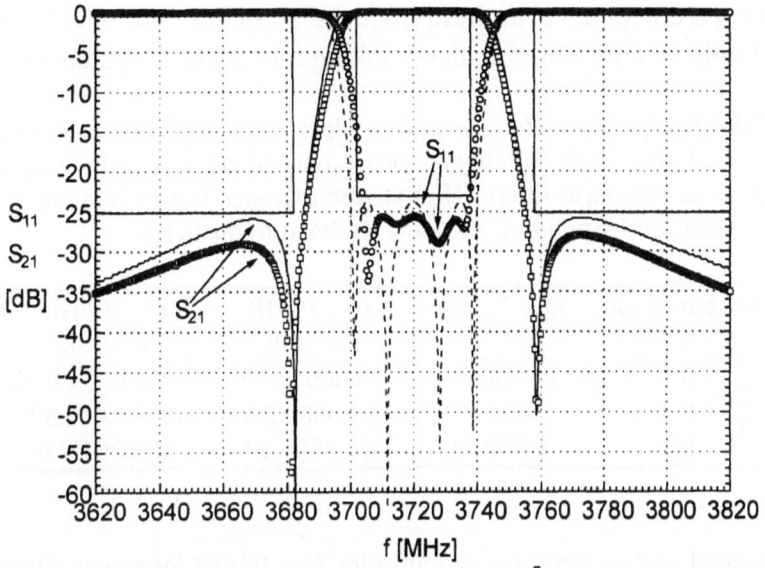

Fig.2. Measured characteristic of our quasielliptic C-band filter based on dielectric hemispheres.

plane was found to be problematic. In the current design, we are using either an HTS endplate on top or no HTS endplate all [5]. For a quasielliptic four-pole filter (see Fig.2), the in-band insertion loss was found to be (−0.04 ±0.02) dB, which is a factor of 10 lower than for conventional filters [5]. The unloaded quality factor at 77 K was found to be 111.000 with the upper HTS endplate being employed and 80.000 without HTS endplate. In both cases the hemispheres were machined from single crystal LaAlO$_3$ grown by the Verneuil technique. According to simulations of the electromagnetic fields with the computer code MAFIA [15] unloaded quality factors of 250.000 would be possible if an upper and lower HTS endplate would be used. However, already at a Q_0 of 100.000 the measured insertion loss is already dominated by losses due to the normal conducting coupling antennae. A further increase of Q_0 would therefore be only challenging for filters with smaller relative bandwidth. On the other hand, the filter with hemispheres machined from MURATA BMT ceramic exhibits a Q_0 of 27.000 at room temperature and 65000 at 77 K. These values are already very attractive for many applications.

The power handling capability of the C-band filter was found to be about 180 watts (with and without HTS endplate, the maximum rf magnetic field on the HTS film was 0.4 mT at that power level). Initially, there was a limitation due to multipacting at about 30 watts. The multipacting threshold could be increased up to about 200 W by a redesign of coupling antennae performed by our industry partner Bosch Telecom.

In table 1 the projected performance of our filter approach for the three relevant satellite bands is quoted both for room temperature operation and operation at 77K. The numbers quoted in the tables have been calculated using MAFIA simulations and experimental data for the losses of the employed dielectric materials. Note that operation at moderate cryogenic temperatures between 77K and 300K may be of some advantage from the system point of view. This holds true in particular for the OMUX, where the dissipated power in each filter is rather high. However, at T=77K the power efficiency of state-of-the art Stirling coolers is only about 4 – 5 % and increases with increasing operation temperature. According to the numbers quoted in the tables, the use of croygenic dielectric filters is challenging for Ku- and Ka band frequencies.

The additional use of HTS components at these frequencies and the use of HTS wall segments of the dielectric filters should be considered in detail taking into account system consideration.

Table 1a (OMUX). Unloaded quality factor Q_0 and dissipated power P_0 [W] of a quasielliptic 4-pole filter for an input power level of 60 watts for conventional cavity filters (conv.) in comparison to projected values for our hemispherical dielectric filters (HR) at room temperature employing BMT and at 77 K employing $LaAlO_3$ for C-band and sapphire for Ku and Ka-band frequencies.

band / downlink-frequencies [GHz]	rel. bandwidth [%]	Q_0 / P_0 conv.	Q_0 / P_0 HR $T = 300K$	Q_0 / P_0 HR $T = 77K$
C / 3,7 – 4,2	0,9	10.000 / 2,3	30.000 / 0,76	> 100.000 / 0.23
Ku / 10,95 – 12,20	0,3	9.000 / 7,7	20.000 / 3,5	80.000 / 0.87
Ka / 19,7 – 20,2	0,2	6.000 / 17	10.000 / 10	50.000 / 2,0

Table 1b (IMUX). Unloaded quality factor Q_0 and insertion loss IL[dB] for a quasielliptic 8-pole filter.

band / downlink-frequencies [GHz]	rel. Bandwidth [%]	Q_0 / IL conv.	Q_0 / IL HR $T = 300K$	Q_0 / IL HR $T = 77K$
C / 3,7 – 4,2	0,9	10.000 / 0,34	30.000 / 0.11	> 100.000 / 0.032
Ku / 10,95 – 12,20	0,3	9.000 / 1,3	20.000 / 0.5	80.000 / 0.12
Ka / 19,7 – 20,2	0,2	6.000 / 3,6	10.000 / 1.8	50.000 / 0.31

ALL CRYOGENIC Ka-BAND LOW- PHASE NOISE OSCILLATOR BASED ON A CRYOGENIC WHISPERING-GALLERY MODE RESONATOR

For high-data-rate up/down or crosslink for future multimedia satellites at Ka-band frequencies the effective date rate may be increased using higher order phase modulation schemes. Usually, the bit sequences are translated into numbers of the basis of 2^n, $n=1,2,3....$ Each of the 2^n digits of a number corresponds to one of the 2^n allocated phase shift values between $0°$ and $360°$. During data transmission, phase flips of k times $360° / 2^n$ ($k=1,2,..,2^n$) occur leading to sidebands at offset frequencies of $1/t_{switch} 2^n$, where t_{switch} is the switching time of the modulator /demodulator. For higher values of n the sidebands move closer to the carrier. The maximum number of n is limited by bit errors due to the oscillator phase noise which increases strongly with decreasing frequency offset from the carrier.

We have developed an all-cyrogenic Ka-band oscillator for f=23GHz with a three-step mechanical and electrical frequency tuning. The parts of this hybrid oscillator are as follows:

- A whispering gallery mode resonator with a mechanical tuning range of 60MHz and a piezo-mechanical fine tuning range of 50kHz. The unloaded quality factor was found to be $2 \cdot 10^6$ at 77K over the entire tuning range.
- A cryogenic two-stage HEMT amplifier including a semiconductor varactor phase shifter and a 10 dB output coupler. The amplification of the device measured between the two SMA ports was found to be 11 dB. The phase shifter allows for a phase shift of 60^0 and can be used for phase locking and fine tuning of the oscillator. The dc- power consumption of the amplifier is 200 mW. The amplifier phase noise was determined to be –135dbc/Hz at 1 kHz

frequency offset. According to the Leeson model the projected phase noise of the oscillator is expected to be -125 dBc/Hz at 1 kHz offset.

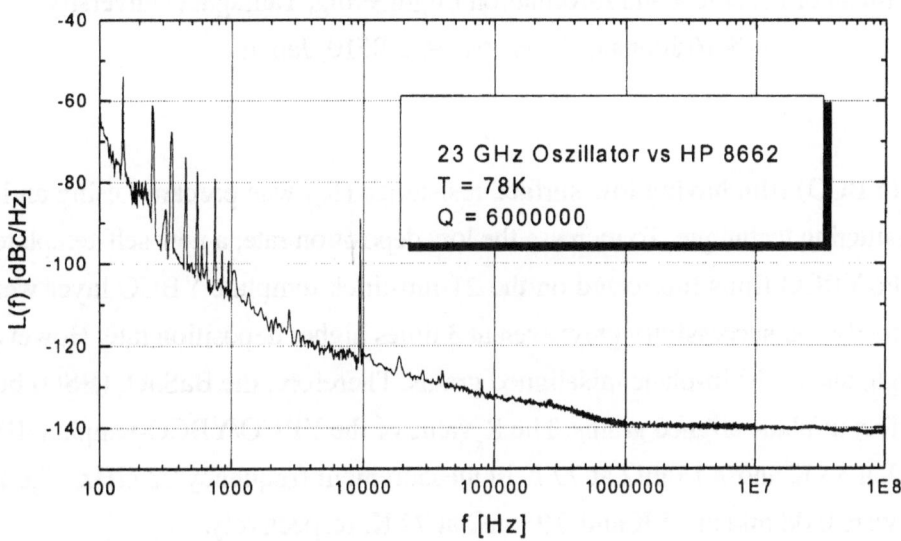

Fig. 3. Measured phase noise of our 23 GHz oscillator

Fig. 3 shows the measured phase noise of our oscillator. The measured values above offset frequencies of 1 kHz represent the sensitivity limits of our HP equipment, i.e. means the noise contribution of the quartz reference source and the downconverter. Comparative measurements of two cryogenic oscillators are in progress.

Acknowledgments. The work on the filters and the all-cryogenic low phase noise oscillator has been performed in close collaboration with Bosch Telecom in the framework of a project funded by the German ministry of research an education. The work on the experimental determination of the loss tangent of dielectric materials has been funded by the European commission in the framework of the BRITE-EURAM project "DiHiMiCo".

1. Sparks M et al. 1982, Phys. Rev. B. **26**, 6987
2. Gurevich V L and Tagantsev A K 1991, Adv. Phys. **40**, 719
3. Schornstein S et al. 1997, Inst. Phys. Conf. Ser. **158**, 267
4. Schornstein S et al. 1998S, IEEE-MTT-S Int. Microwave Symp. Digest, 1319
5. Klein N et al. 1999, IEEE Trans. on Appl. Sup. **9**, 3573
6. Hao L et al. 1999, IEEE Trans. Instr. and Meas. **48**, 99
7. Braginsky V B et al. 1987, Phys. Lett. A **120**, 300
8. Alford NMcN 1998, Mat. Res. Soc. Symp. Proc. **500**, 183
9. Dick G J et al. 1992, Proc. of the 6th European Frequency and Time Form, 35 *and* Dick G J et al. 1994, IEEE Trans IEEE Trans. UFFC **42**, 812
10. Klein N et al. 1995, J. Appl. Phys. **78**, 6683
11. Tobar M E et al. 1998, J. Appl. Phys. **83**, 1604
12. Zuccaro C et al. 1997, J.Appl. Phys. **82**, 5695
13. Baumfalk A et al 1999, IEEE Trans. on Appl. Sup. **9**, 2857
14. Wakino K et al. 1989, Ferroelectrics **91**, 69
15. Schmidt D and Weiland T 1992, IEEE Trans. Magn. **28**, 1793
16. B Aminov et al. 1999, IEEE Trans. on Appl. Sup. **9**, 4185

MICROWAVE SURFACE RESISTANCE OF $YBa_2Cu_3O_y$ THIN FILMS PREPARED BY INDUCTIVE COUPLED PLASMA SPUTTERING

T. Suzuki, S. Sato, M. Kusunoki, M. Mukaida and S. Ohshima

Department of Electrical and Information Engineering, Yamagata University,

4-3-16 Johnan, Yonezawa, 992-8510, Japan

Abstract

The $YBa_2Cu_3O_y$ (YBCO) film having low surface resistance (R_s) was prepared using an inductive coupled plasma sputtering technique. To increase the low deposition rate, a new self-template method was examined. The YBCO films fabricated on the 21-nm-thick template YBCO layer were c-axis oriented grain perfectly and successfully grew even at 3 times higher deposition rate. However, the R_s was not low enough, due to the in-plane misaligned grains. Therefore, the $BaSnO_3$ (BSO) buffer was used to obtain perfect in-plane aligned grains. The R_s value of the YBCO/YBCO-template/BSO/MgO film was 0.38 mΩ at 35 K and 1.13 mΩ at 77 K in measurement frequency 22 GHz. The R_s values scaled to 10 GHz were 0.08 mΩ at 35 K and 0.23 mΩ at 77 K, respectively.

Introduction

The trend towards practical use of high temperature superconducting (HTS) microwave devices motivated the development of deposition techniques for large area HTS thin films for wireless communication system. Various types of techniques have been successfully used so far: co-evaporation [1], planar dc high oxygen-pressure sputtering [2], off-axis sputtering [3], and pulsed laser deposition [4]. Most of these systems need complex mechanisms such as substrate motion, target motion and laser beam scanning for covering large areas to realize uniformity of the films. It is desirable to use a fabrication technique which is simple and with high cost performance. Inductive coupled plasma (ICP) sputtering [5] is a promising technique, because ICP, which generates uniform production of active species in large area [6], is intrinsically suited for large area deposition. The advantages of ICP have already been proven in the processing of more than ϕ 150 mm wafers for ultra large scale integrated (ULSI) circuits production [7]. In this letter, we present the preparation of $YBa_2Cu_3O_y$ (YBCO) films using ICP sputtering technique and evaluate the surface resistance (R_s) of the films.

Experimental and Discussion

A schematic illustration of the experimental apparatus is shown in Fig.1. A 4-turn coil is set above a target as an ICP generator. The diameter of the coil is 75 mm and length along the axis is 18 mm. A 13.56 MHz power supply is coupled to the coil. The sintered YBCO target of a 50 mm diameter is negatively biased to sustain the magnetron discharge. Substrate is set on a substrate heater positioned at the center just above the target. (100) MgO is used as the substrate in this study. It is pointed out that the low deposition rate is a disadvantage of sputtering. Furthermore, long substrate-target distance is taken in our system, to obtain the uniformity of the films and to avoid degradation by high energy minus ions. Therefore, we investigate a way to increase the deposition rate using ICP. Fig.2 is a relationship between ratio of a-axis and deposition rate. Where, I(200) and I(006) are intensities of (200) and (006) peaks of X-ray diffraction (XRD) $2\theta/\theta$ scan. This figure indicates that it is necessary to set deposition rate lower than 42 nm/h to make perfect c-axis oriented film. We supposed that the orientation of the films is decided in the beginning of the growth at the interface between substrate and film. Then the film grows keeping the orientation of under layer epitaxially [8]. Taking into account this, we developed self-template method. At first, the film was grown with deposition rate of 42 nm to obtain c-axis orientation. After that the deposition rate was increased. Figure 3 shows the dependence of a-axis ratio on thickness of template layer. The deposition rate for the template layer was 42 nm/h and that for main layer was 140 nm/h. The total film thickness was 800 nm in each film. Perfect c-axis films were successfully obtained on the template with thickness more than 21 nm. The critical temperature of the films was 86.3 K. The R_s of the film that obtained by self-template method is measured by the sapphire dielectric resonator method at 22 GHz. Open triangles in Fig.4 show the R_s values of YBCO/ YBCO-template/MgO. The R_s values seem to be relatively high. This result can be interpreted in terms of the in-plane misaligned grains in the films [8]. The mixture of 0° and 45° grains was observed by the XRD ϕ-scan. In order to prevent the 45° grain growth, BaSnO$_3$ (BSO) buffer layer [9] was used between template and MgO. The XRD ϕ-scan showed perfect in-plane alignment. Open circles in Fig.4 indicate the R_s values of the YBCO/ YBCO-template/BSO/MgO. The R_s values are

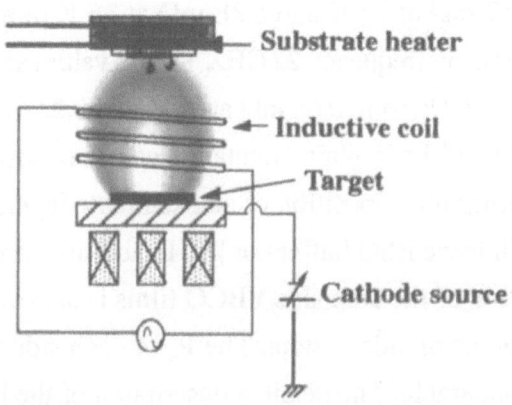

Fig1 The schematic illustration of ICP sputtering

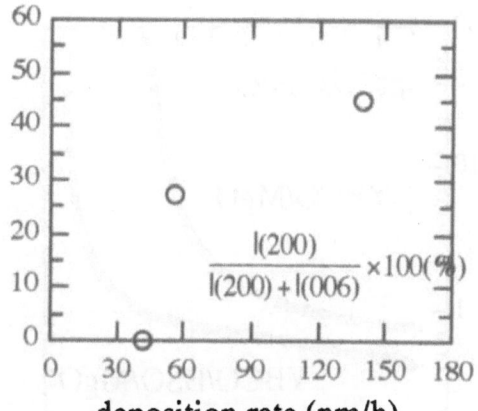

$$\frac{I(200)}{I(200)+I(006)}\times100(\%)$$

deposition rate (nm/h)

Fig2 relationship between the ratio of a-axis region and deposition rate.

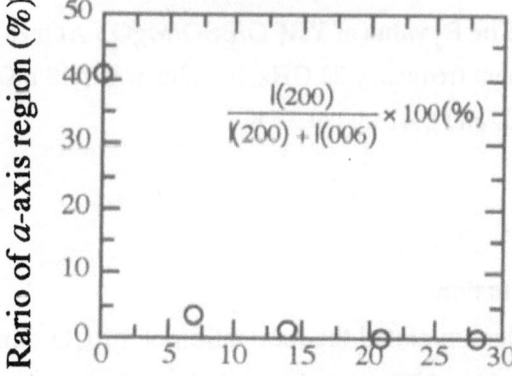

$$\frac{I(200)}{I(200)+I(006)}\times100(\%)$$

Thickness of temprate layer (nm)

Fig3 The a-axis ratio depend on temprate layer of YBCO thin films : Perfect c-axis films were successfully obtained on the template with thickness more than 21 nm.

approximately half compared with of YBCO/YBCO-template/MgO under 50K. R_s values were 1.13 mΩ at 77 K and 0.38 mΩ at 35 K in measurement frequency 22 GHz. The Rs values scaled to 10 GHz were 0.08 mΩ at 35 K and 0.23 mΩ at 77 K. If the in-plain orientation can be controlled during the deposition of self-template layer, we will leave BSO buffer out [8]. In addition, preparation of the both side YBCO films is also available using this system. The R_s of each side was comparable. The detailed description of the both side films will appear elsewhere.

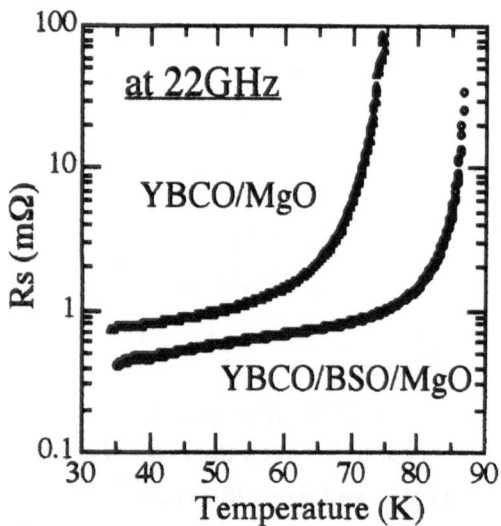

Fig4 The R_s value of YBCO/BSO/MgO : At measurement frequency 22 GHz, R_s value was 1.13 mΩ at 77 K and 0.41 mΩ at 35 K.

Conclusion

We demonstrated the fabrication of YBCO thin films using ICP sputtering technique. To avoid *a*-axis domains in the films even by higher deposition rate, self-template method was used. The *a*-axis free YBCO thin films fabricated on the 21-nm-thick template were successfully grown using 3 times higher deposition rate. However, the R_s values of the film were relatively high, which was caused by the effect of the 45° grain boundary in this film. Using the BSO buffer layer, perfect in-plane oriented YBCO films were obtained. The film showed excellent R_s values 1.13 mΩ at 77 K and 0.38 mΩ at 35 K in measurement frequency 22 GHz. The R_s values scaled to 10 GHz were 0.08 mΩ at 35 K and 0.23 mΩ at 77 K, respectively.

References

[1] B.Utz, R. Semerad, M.Bauer, W.Prusseit, P.Berberich, and H.Kinder: IEEE Trans. Appl. Supercond., 7, 1272 (1997).

[2] G.Muller, B. Aschermann, H. Chaloupka, W. Diete, M. Getta, B. Gurzinski, M. Hein, M. Jeck, T. Kaiser, S. Kolesov, H. Piel, H. Schlick, and R. Theisejans: IEEE Trans. Appl. Supercond., 7, 1287 (1997).

[3] Y. Ueno, N. Sakakibara, M. Okazaki, M. Aoki: Advances in Superconductivity, X, 1131(1997).

[4] M.Lorenz, H. Hochmuth, D. Natusch, H. Borner, G. Lippold, K. Kreher, and W. Schmitz: Appl. Phys. Lett., 68, 3332 (1996).

[5] Y.Setsuhara, H.Kamai, S.Miyake, J.Musil: Jpn. J. Appl. Phys., 36, 4568(1997).

[6] Y. Hikosaka, M. Nakamura and H. Suzuki: Jpn. J. Appl. Phys., 33, 2157(1994).

[7] H. Nogami, Y. Nakagawa, K. Mashimo, Y. Ogahara, T. Tsukada: Jpn. J. Appl. Phys., 35, 2477(1996).

[8] M. Kusunoki, Y. Takano, M. Mukaida and S. Ohshima: Physica C, 321 81 (1999).

[9]M.Mukaida, et. al. Jpn. J. Appl. Phys. vol.38 (1999) pp926~928

Characterization of double-sided NBCO thin films by measurement of Rs using sapphire rod resonator

Hajime Nakada, Norio Hasegawa, *Akira Tamaki, Youichi Enomoto, Katsumi Suzuki

Tamachi Laboratory, Superconductivity Research Laboratory, ISTEC

1-16-25 Shibaura Minato-ku Tokyo, 105-0023 Japan

*Tokyo Denki University

2-2 Kanda Nishiki-cho Chiyoda-ku Tokyo, 101-0054 Japan

Abstract: We fabricated double-sided superconducting thin films for microwave devices by pulsed laser deposition method. In the double-sided thin film fabrication, the first side film is deteriorated by the second side processing. To prevent this first side deterioration, we adopted NBCO films on the first side, and YBCO films whose deposition temperature on the second side is lower than the NBCO.

The microwave surface resistances of superconducting thin films were measured by dielectric resonator method with sapphire rod. By evaluating the microwave surface resistances double-sided NBCO/YBCO systems, it was understood that the first side deterioration was little compared with the case of double-sided thin film of YBCO/YBCO system.

Keywords: double-sided HTS thin films, NBCO, dielectric resonator, surface resistance

INTRODUCTION

The High-Tc Superconductors (HTS) are expected as efficient materials for microwave devices because of their low surface resistance (Rs). Double-sided HTS thin film is necessary to microwave devices such as microstripline filters[1]. However, in the case of double-sided YBCO($YBa_2Cu_3O_{7-\delta}$) thin film, the quality of the first side film is deteriorated because of heating process on the first film while the second side film is deposited. The first and second side films are hereafter referred to as filmA and filmB.

In this paper, we fabricated double-sided NBCO($Nd_{1+x}Ba_{2-x}Cu_3O_{7-\delta}$)/YBCO system by PLD (Pulsed Laser Deposition) method[2]. Deposition temperature and oxygen pressure of NBCO for filmA are higher and lower than those of YBCO for filmB. Moreover, NBCO thin film has the highest Tc in the $RBa_2Cu_3O_{7-\delta}$ (R:rare earth metal) system, and also has stable surface[3]. Therefore, we expected that an influence of heating process on filmA during deposition of filmB would be reduced.

EXPERIMENT

The double-sided HTS films were prepared by the on-axis PLD method on preannealed (100) MgO substrate with seize of $20{\times}20{\times}0.5mm^3$. KrF excimer laser(wave length 248nm) is reflected 90° by a reflection mirror and focused by a convex lens to a rotating target. The YBCO thin films were deposited at substrate temperature T_{sub}=740℃, and oxygen pressure Po_2=200mTorr. The NBCO thin films were deposited at T_{sub}=830℃, and

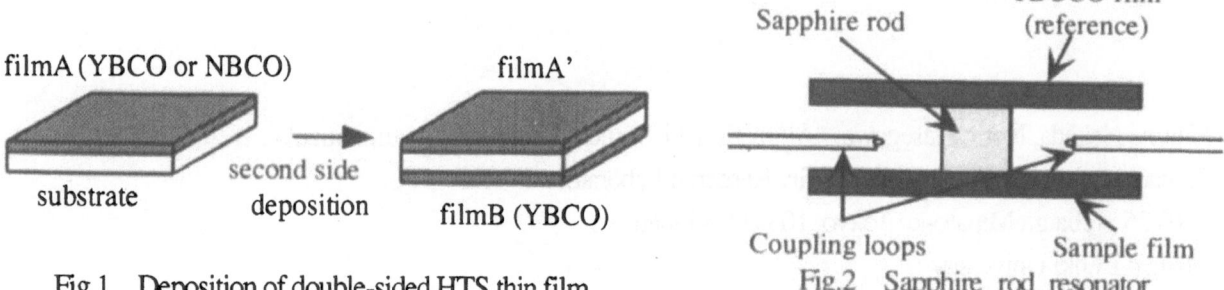

Fig.1   Deposition of double-sided HTS thin film

Fig.2   Sapphire rod resonator

mixed gas pressure P_{tot}=200mTorr(P_{O_2}:50, P_{Ar}:150). After deposition, the films were cooled in 400 Torr of oxygen. After deposition of filmA, the substrate was turned over for deposition of filmB(Fig.1). After filmB is deposited, filmA is referred to as filmA'. The film thickness was about 300nm.

The double-sided HTS films were characterized by X-ray diffraction(XRD) for crystal axis orientation, atomic force microscopy(AFM) for surface morphology, four probe method for Tc. The Rs was measured using a dielectric resonator method[4] with a sapphire rod as shown in Fig.2. The TBCCO film with higher Tc than YBCO and NBCO was used for upper part conductor as a reference. Therefore, the Rs of YBCO and NBCO thin films can be accurately measured up to their Tc.

To determine tanδ of sapphire rod in advance, we used a TE011 mode resonator of length L and a TE013 mode resonator of length 3L, and two TBCCO films.

RESULTS AND DISCUSSION

Samples of double-sided HTS films were fabricated and subsequently characterized. From XRD patterns, all the YBCO films demonstrated very sharp and high (00l)peaks which thus implied that they were highly c-axis oriented YBCO films. In the NBCO films, small other peaks could be observed. Tc of YBCO and NBCO thin films were in the range of 86~ K 88.8 and 84~86 K, respectively. Fig.3 shows the measured tanδ results for sapphire rod using two TBCCO films, together with that for two YBCO films. In the calculation of Rs, tanδ measured by two TBCCO films was used. Fig.4 shows the temperature dependence of Rs measured at 21.9GHz.

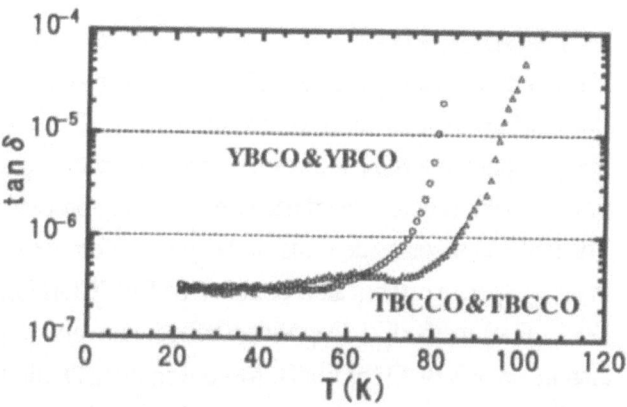

Fig3   Measured results of tanδ using two TBCCO films and YBCO films.

The Rs of YBCO drops sharply just below Tc, and is weakly temperature-dependent at a lower temperature. NBCO, on the other hand, shows slower decreases in Rs over wide temperature ranges below Tc. The Rs of filmA' increased due to the deposition process of filmB. However, deterioration of filmA of NBCO/YBCO system smaller than that of YBCO/YBCO system. The difference of Rs between filmA and filmB of YBCO/YBCO system might be due to the substrate surface temperature is different in the deposition of filmA and filmB. In the AFM observation, there is hardly a difference between the surface morphology of filmA and filmA'. From these results, deterioration of filmA might probably be due to the loss of oxygen component. Tc of NBCO thin film in the present work is lower than that of bulk samples. It is necessary to furthermore optimize the deposition parameters(mixed gas pressure, the ratio of Nd to Ba of target ,etc).

Because the deterioration of NBCO thin films to humidity is very little compared with YBCO thin films [5], NBCO thin films are preferably selected for device patterning.

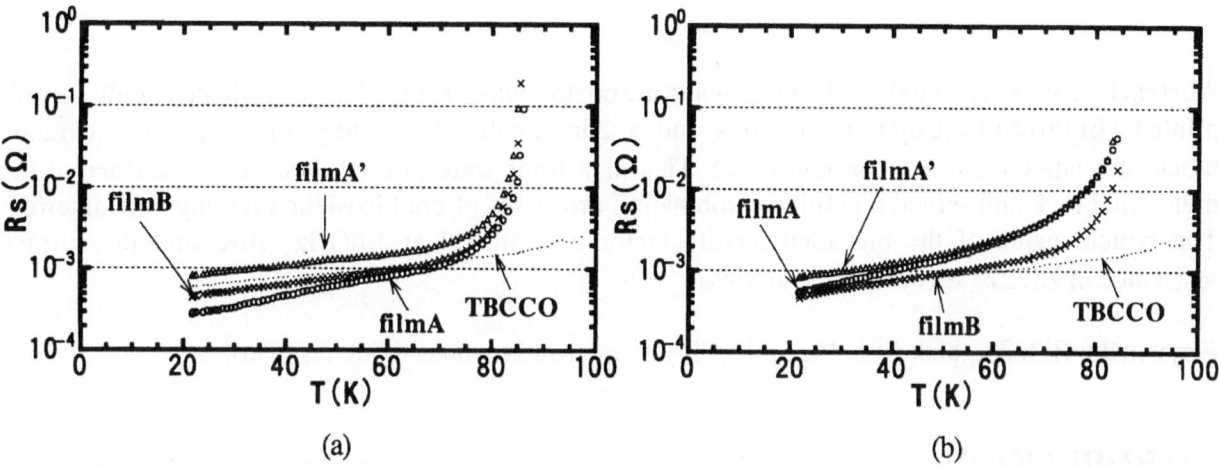

(a) (b)

Fig.4   Measured results of Rs for (a) YBCO/YBCO system and (b) NBCO/YBCO system at 21.9GHz

CONCLUSION

Double-sided YBCO/YBCO and NBCO/YBCO systems for microwave applications were fabricated on (100) MgO substrates by PLD method. The Rs of the HTS thin films were measured using dielectric resonator method with sapphire rod and TBCCO thin film. The deterioration of the first side of NBCO/YBCO system during second side(YBCO) deposition process was less than that of YBCO/YBCO system. We made it an advantage that NBCO deposit in temperature is about 100℃ higher than YBCO. Moreover, because the surface of NBCO thin film is more stable than YBCO, NBCO thin film is a promising material for the practical microwave application.

ACKNOWLEDGEMENTS

We thank Prof. Kobayashi and Mr. Hashimoto of Saitama University for their suggestions of Rs measurements. This work was supported by New Energy and Industrial Technology Development Organization(NEDO) as Collaborative Research and Development of Fundamental Technologies for Superconductivity Applications.

[1] Y.Okazaki, K.Suzuki, Y.Enomoto   Advances in Superconductivity IX 2 (1997) pp.1281-1284

[2] M.Badaye, F.Wang, Y.Kanke, K.Fukushima, and T.Morishita   Appl.Phys.Lett.66.16 (1995) pp.2131-2133

[3] S.I.Yoo, N.Sakai, H.Takaichi, and M.Murakami   Appl.Phys.Lett.65.5 (1994) pp.633-635

[4] Y.Kobayashi, T.Imai, H.Kayano   IEEE MTT-S Digest (1990) pp.281-284

[5] M.Ban, Y.Mizuno, K,Suzuki, Y.Enomoto   Physica C 270 (1996) pp129-134

Fabrication and Microwave Operation of Dual-mode Ba(Sn,Mg,Ta)O$_3$ Resonator with Bi2223 Thick Film as Superconductor Electrodes

Yuji Kintaka[1], Norifumi Matsui[1], Tsutomu Tatekawa[1], Hiroshi Tamura[1], Youhei Ishikawa[1] and Akio Oota[2]

[1]Murata Manufacturing Co. Ltd., 2-26-10, Tenjin, Nagaokakyo, Kyoto 617-8555, Japan
[2]Toyohashi University of Technology, Tempaku-cho, Toyohashi, Aichi 441-8580, Japan

Abstract: The TM$_{110}$ dual-mode microwave resonator was designed and fabricated with screen-printed (Bi,Pb)$_2$Sr$_2$Ca$_2$Cu$_3$O$_x$ thick films and a 25mm-cubic Ba(Sn,Mg,Ta)O$_3$ dielectric ceramic block of relative dielectric constant ε_r=24. The thick films were printed on every six surface of the dielectric block and served repetitive combination processes of cold isostatic pressing and sintering. The typical value of the unloaded quality factor was 40,000 at 1.8GHz, 70K, and the surface resistance of Bi2223 was estimated at 1mΩ.

Keywords: Bi2223, thick film, Ba(Sn,Mg,Ta)O$_3$, microwave, dual-mode resonator

INTRODUCTION

Passive microwave components such as filters and resonators are promising applications of high temperature superconductors (HTS) because of their small surface resistance. For its realization, much attention have been paid on the microwave properties of HTS thin films [1-3], but little on those of thick films [4,5]. To investigate the feasibility of thick films for microwave application, we have studied on the microwave properties of (Bi,Pb)$_2$Sr$_2$Ca$_2$Cu$_3$O$_x$ (Bi2223) thick films and found the crucial facts as follows: (1) a screen-printed Bi2223 thick film on Ag substrate shows low surface resistance of 1mΩ at 10.7GHz, 70K [6]; (2) Ba(Sn,Mg,Ta)O$_3$ (BSMT; ε_r=24) dielectric ceramics have low tanδ at cryogenic temperature and Bi2223 thick films can be fabricated directly on this ceramics without significant chemical reactions [7].

In order to eliminate the edge effect shown in micro-strip line resonators, planar circuit resonators (e.g. round shape resonators) have been investigated because they have the wide spread current distribution [1,2]. However, they require large size; for instance, when one uses a dielectric material of ε_r=24 as a substrate for TM$_{010}$ mode disk resonator, the diameter of the HTS electrode is needed to be 41mm for 1.8GHz operation. Moreover, they need a shielding cavity when they are mounted in a filter module.

For miniaturization of the planar electrode resonator, a TM$_{110}$ mode rectangle dielectric resonator covered with HTS electrodes on every surface (i.e. cavity resonator filled with dielectric material) is preferable. In this resonator, the size for 1.8GHz can be reduced to 25$\times$25mm (ε_r=24) and the thickness can be selected according to required value of unloaded quality factor. Moreover, this does not need a shielding cavity in a filter assembly.

In this paper, as part of the trials to make the rectangle dielectric resonator, we fabricated a 25mm-cubic BSMT dielectric resonator covered with planar Bi2223 thick film electrodes on every 6 surface, which is large but can be operated as TM$_{110}$ dual-mode resonator.

STRUCTURE OF THE DUAL-MODE RESONATOR

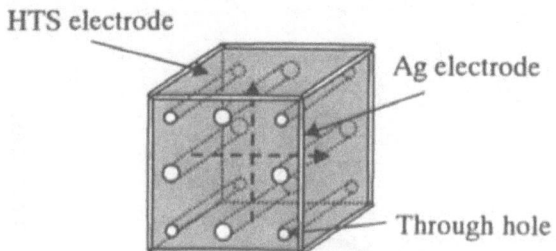

HTS electrode

Ag electrode

Through hole

Fig.1 Structure of the dual-mode resonator.
Dashed arrows indicate the propagation
direction of each mode.

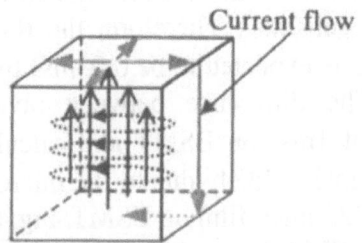

Current flow

Fig.2 Electromagnetic field (TM$_{110}$ mode) of one
mode of the dual-mode resonator: electric
filed, solid arrows; magnetic filed, dotted
arrows. Through holes are in same direction
as Fig.1 (not shown).

Fig.1 shows the structure of the dual-mode rectangle dielectric resonator, and the electromagnetic field of one mode of the TM$_{110}$ dual-mode is shown in Fig.2. This resonator consists of a 25mm cube of BSMT ceramic block with 8 through holes, Bi2223 thick film planar electrodes and Ag electrodes on side-edges of BSMT block. The through holes are for coupling of two resonant modes and used for adjustment of the resonant frequencies during construction of the microwave filter. The Ag electrodes on side-edges are to keep good connectivity between planar Bi2223 electrodes.

EXPERIMENTAL PROCEDURE

Sample preparation. The BSMT block was prepared by conventional ceramic process and every surface was mirror-polished. Ag paste was printed on every 12 side-edges of the block and fired at 850℃. After this, the Bi2223 paste was screen printed on all surfaces of BSMT block without any buffer layers. The Bi2223 thick films were heated at 400℃ to evaporate organic binders and subjected to cold isostatic pressing (CIP) at 0.2GPa. After such mechanical treatments, the sample was sintered at 830℃ for 50h. A combination process of CIP and sintering was repeated until total sintering time reached 150h. The thickness of the resultant Bi2223 films was around 10μm. The typical value of Tc for Bi2223 films on BSMT determined by a dc SQUID magnetometer was 110K.

Measurement. The dual-mode resonator with a couple of probe inserted in one of the through holes was mounted in the helium-filled gas chamber of the cryostat. The unloaded quality factors (Qu) were measured as a function of temperature between 20 and 130K by an HP8720C network analyzer with input power of –30dBm.

RESULT AND DISCUSSION

Fig.3 shows the temperature dependence of Qu and surface resistance Rs of one mode of the dual-mode resonator, where the Rs value of Bi2223 thick film is the apparent value including Rs of the side-edge Ag. Note that the data of another mode are similar to the values shown in Fig.3. For comparison, the values of Qu and Rs of the same structural resonator covered with 6 planar Ag thick film electrodes were included in Fig.3. As shown in Fig.3, the Qu value of the resonator using Bi2223 thick films reaches 40,000 at 70K, which is 4 times higher than the resonator using 6 planar Ag thick films. The Rs value of Bi2223 itself is estimated to be 1mΩ at 70K by separating the contribution of side edge Ag from apparent Rs.

As mentioned above, we obtained the Rs of 1mΩ at 10.7GHz, 70K for Bi2223 thick film on Ag substrate. Therefore, the Rs value at 1.8GHz is expected to be 0.03mΩ by using f^2 low. The difference between obtained Rs value of 1mΩ on BSMT and calculated one of 0.03mΩ is likely due to the microstructure of Bi2223 thick film on BSMT. Fig.4 shows a typical SEM image of fractured surface of Bi2223 thick film on BSMT. It is evident that the thick film has low density, poor grain growth and low degree of c-axis alignment. Consequently, it seems that so many weak links caused higher Rs value. Thus, further improvement of the microstructure of Bi2223 thick films on BSMT is required for realization of the resonator.

CONCLUSION

The 1.8GHz TM_{110} dual-mode resonator was fabricated using a 25mm cube of BSMT ceramic and 6 planar Bi2223 thick film electrodes, and showed 40,000 of Qu at 70K. Further improvement for grain growth, grain orientation and density of Bi2223 films on BSMT is our future work toward realization.

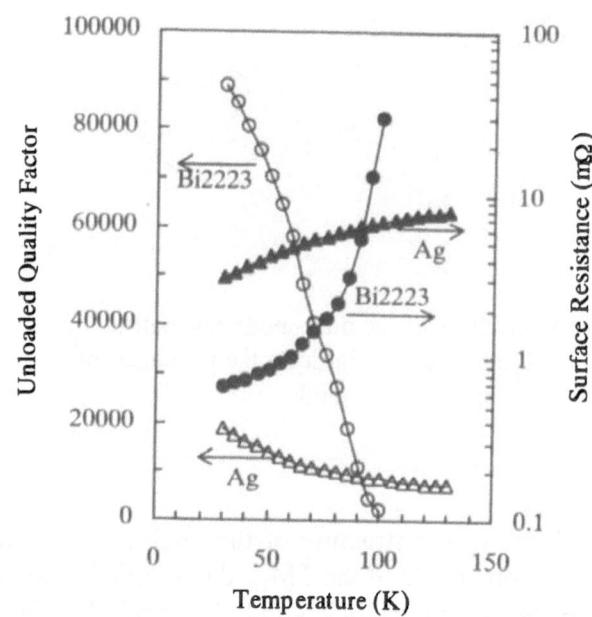

Fig.3 Temperature dependence of Qu (open symbols) and Rs (solid symbols) at 1.8GHz on a TM_{110} mode: Bi2223 electrode resonator, circles; Ag electrode resonator, triangles.

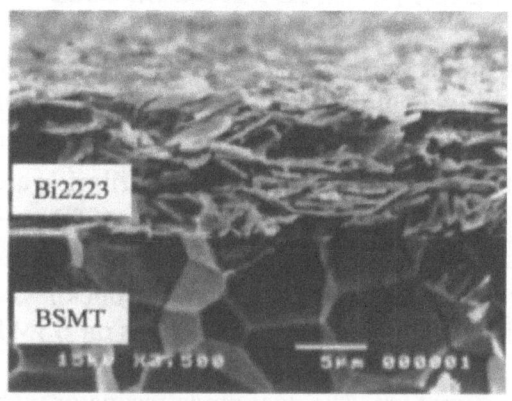

Fig.4 Typical SEM image of fractured surface of Bi2223 thick film on BSMT ceramics.

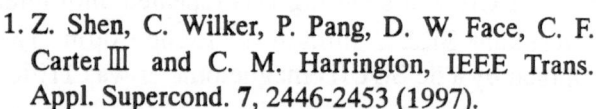

1. Z. Shen, C. Wilker, P. Pang, D. W. Face, C. F. Carter III and C. M. Harrington, IEEE Trans. Appl. Supercond. **7**, 2446-2453 (1997).
2. A. Enokihara and K. Setsune, *MWE'96 Microwave Workshop Digest*, 156-161 (1996).
3. R. B. Hammond, G. V. Negrete, L. C. Bourne, D. D. Strother, A. H. Cardona and M. M. Eddy, Appl. Phys. Lett. **57**, 825-827 (1990).
4. N. McN. Alford, T. W. Button, M. J. Adams, S. Hedges, B. Nicholson and W. A. Phillips, Nature **349**, 680-683 (1991).
5. L. Y. Su, C. R. M. Grovenor, M. J. Goringe, A. P. Jenkins and D. Dew-Hughes, Appl. Phys. Lett **66**, 1542-1544 (1995).
6. A. Oota, K. Fujikawa, K. Yamashita, M. Tanaka, T. Tatekawa, N. Matsui, Y. Kintaka and Y. Ishikawa, in *Advances in Superconductivity X*, edited by K. Osamura and I. Hirabayashi (Springer-Verlag, Tokyo, 1998), pp. 1173-1176.
7. T. Tatekawa, N. Matsui, Y. Kintaka, Y. Ishikawa, K. Fujikawa, M. Tanaka and A. Oota, IEEE Trans. Appl. Supercond. 9, 1940-1943 (1999).

Investigation of superconductive elliptic function filter

Norio Hasegawa, Hajime Nakada, Youichi Enomoto,
Akira Tamaki* and Katsumi Suzuki

Tamachi Laboratory, Superconductivity Reserch Laboratory, ISTEC
1-16-25, Shibaura, Minato-ku, Tokyo 105-0023, Japan
*Tokyo Denki University
2-2, Kanda Nishiki-cho, Chiyoda-ku, Tokyo 101-0054, Japan

Abstract : The 6-stage elliptic function filter of the microstripline structure to use $YBa_2Cu_3O_{7-\delta}$ (YBCO) was designed. The optimum configuration was obtained by corresponding Spectrum Micro-Cap which is SPICE type circuit analysis, and by Sonnet Em which is full-wave electromagnetic field analysis. In addition, the surface impedance based on two fluids model was introduced into the simulation, the microstripline filter which used superconducting material parameter was simulated.

Keywords: Elliptic function Filter, Lumped element circuit simulation, Electromagnetic field simulation, YBCO superconductor, Two fluids model

INTRODUCTION

As for high-Tc superconductors (HTS) which can keep a high unloaded Q factor, various filters were achieved in the mobile and the satellite communication [1,2]. In this field, the sharp cutting is requested as a filter characteristic. Elliptic function filter can achieve a steep attenuation characteristic compared with the maximally flat and the Tchebycheff filter by a little number of poles. Moreover, the possibility of the improvement of the group delay characteristic is expected by moving the attenuation pole [3]. Then, the achievement of the microstripline elliptic function filter was designed based on the method of a canonical filter. In addition, the surface impedance of YBCO based on two fluids model was introduced into the obtained pattern, and we investigated the temperature dependence of the transmission and the group delay characteristic.

LUMPED ELEMENT CIRCUIT SIMULATION

As a specification of the filter, (a) Elliptic function band-pass filter of 6-stage and 4 transmission zeros, (b) Passband ripple RW =0.01dB and out-of-band attenuation $SBmin$= 40dB, (c) Canonical coupling circuit and (d) Center frequency f_0 =10.5GHz and fractional bandwidth 1.8% (Δf = 0.19GHz) are given. The characteristic function of the elliptic function filter in 6-stage and 4 transmission zeros is given by

$$\varphi(s) = H \cdot \frac{s^2(s^2 + a_1^2)(s^2 + a_2^2)}{(a_1^2 s^2 + 1)(a_2^2 s^2 + 1)} \qquad (1)$$

where, s is the normalized complex frequency and a_i is the value of the pole. H is the constant decided by passband ripple and out-of-passband attenuation. Assuming (b), a_1 and a_2 are calculated by using the elliptic function by the accuracy of the effective demand 5-digit,
a_1 = 0.61950, a_2 = 0.81794 and H = 2.1911.

Next, the coupling element $m_{i,j}$ and the load n of the normalized bandpass filter of the canonical

circuit are calculated by using the expression (1) [4],
$m_{1,2}=m_{5,6}=0.82529$, $m_{2,3}=m_{4,5}=0.50974$,
$m_{3,4}=0.67058$, $m_{1,6}=0.050699$, $m_{2,5}=0.22930$
and $n=1.1549$.

When the canonical circuit is shown in the figure, 6 resonators are in the coupling relation as shown in Fig.1. The coupling coefficient $k'_{i,j}$ and the load R of the lumped element circuit are shown by using $m_{i,j}$, n and (d)

$$k'_{i,j} = \frac{m_{i,j}}{\sqrt{L_m L_n}} \cdot \frac{\Delta f}{f_0} \quad (2) \qquad R = n \frac{\Delta f}{f_0} \quad (3)$$

where, L_m and L_n are element values of the inductance of the normalized circuit which affect coupling. Moreover, lumped element circuit can be decided by using obtained the inductance L and the capacitance C according to the center frequency $f_0 = 1/(2\pi\sqrt{LC})$. Fig.2. is calculated by Micro-Cap, and the specification is satisfied with the accuracy of the effective demand 5-digit.

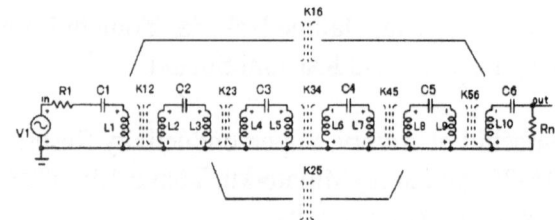

Fig.1. 6-stage lumped element canonical circuit.

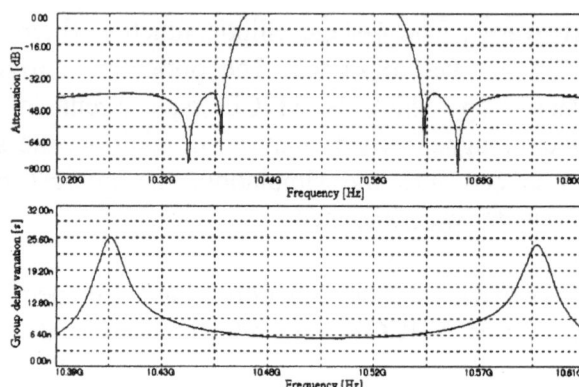

Fig.2. Calculation result of lumped element circuit. The transmission and the passband group delay characteristic respectively.

ELECTROMAGNETIC FIELD SIMULATION

To achieve the filter of the microstripline pattern, the electromagnetic field is analyzed by Sonnet Em [2,5]. Here, we propose asymmetry hairpin resonator. With asymmetry hairpin resonator, a minute adjustment of the coupling coefficient becomes possible. In addition, the selection of an external coupling can extends to input the signal from right and left in cavity. In this simulation, as a dielectric substrate is MgO, the relative permitivity of the dielectric substrate ε_r was 9.80, $tan\delta$ was 6×10^{-6} and substrate thickness was 0.5mm. The coupling coefficient of Fig.3. was calculated with asymmetry hairpin resonator. The relation between the coupling element $m_{i,j}$ and the load n obtained from the calculation and the coupling coefficient $k_{i,j}$ and the external Q factor Q_e of the microstripline circuit, are given by the following relation equation.

$$k_{i,j} = m_{i,j} \frac{\Delta f}{f_0} \quad (4) \qquad Q_e = n \frac{f_0}{\Delta f} \quad (5)$$

The configuration of asymmetry hairpin resonator was decided is Fig.4.

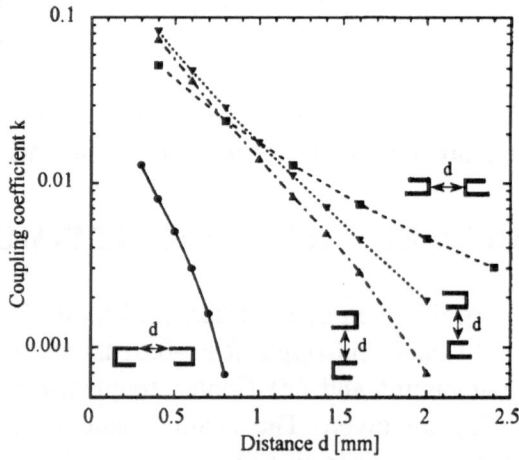

Fig.3. coupling coefficient of microstripline asymmetry hairpin resonator.

Fig.4. Geometry of 6-stage microstripline asymmetry hairpin filter.

SIMULATION RESULT AND DISCUSSION

To analyze the temperature dependence of the bandpass filter of Fig4, we use the program which built the superconductor material parameter based on two fluids model into Sonnet Em which Y.Okazaki composed [6]. In two fluids model, surface impedance Zs of a superconducting thin film and the magnetic penetration depth λ are given by

$$Zs = Rs + jXs = \frac{\omega^2 \mu_0^2 \lambda(T)^3 \sigma_n}{2} + j\omega\mu_0\lambda(T) \quad (6)$$

$$\lambda(T) = \frac{\lambda(0)}{\{1-(T/T_c)^4\}^{1/2}} \quad (7)$$

where, Rs is the surface resistance, Xs is the surface reactance, ω is the angular frequency of microwave and μ_0 is the magnetic permeability of the free space. As the parameter of YBCO, the critical temperature Tc was 85.0K, the magnetic penetration depth at 0K $\lambda(0)$ was 390nm and the conductivity of normal state at Tc σ_n was 1.00×10^6 s/m. The pattern and the grand plane were both sides of YBCO. By using the parameter, the filter was simulated at 20.0K, 77.0K and 83.0K. The temperature dependence of the transmission characteristic is shown in Fig.5. and the group delay characteristic in the passband is shown in Fig.6. The passband group delay is constant at the center frequency.

CONCLUSION

The microstripline pattern of the bandpass filter in 6-stage 4 attenuation zeros was able to be designed in asymmetry hairpin resonator which we proposed. The passband group delay as a function of tempuretures below Tc is kept constant at the center frequency.

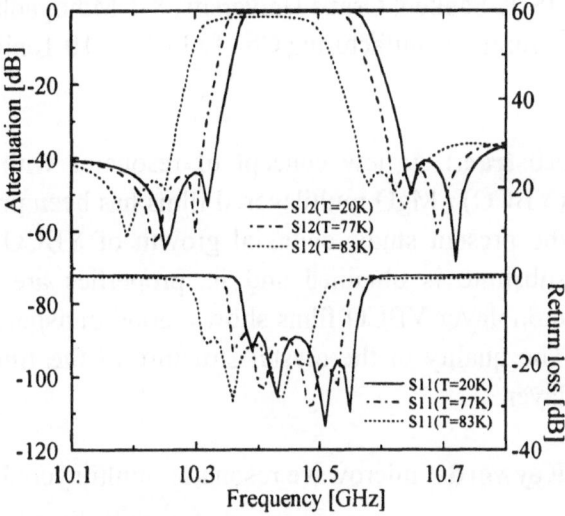

Fig.5. Transmission characteristic of 6-stage microstripline YBCO filter.

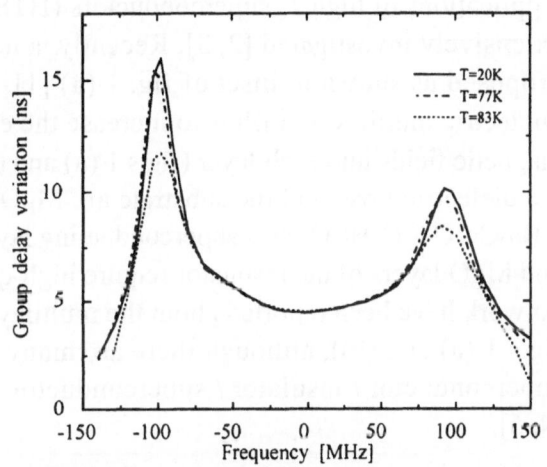

Fig.6. Passband group delay characteristic of 6-stage microstripline YBCO filter.

Acknowledgments. We would like to thank Mr. Y.Okazaki of Matsushita Electric Co.,LTD for the simulation algorithm program. This study was performed through Special Coordination Funds of the Science and Technology Agency of the Japanese Government.

1.Hong J and Lancaster J M (1998) IEEE MTT-S 46 International Microwave Symposium Digest, Baltimore pp.367-370

2.Tomiyama T, Okai D, Kusunoki M, Oshima S (1998) Advances in Superconductivity XI pp.1255-1258

3.Suginosita F and Nomomto T(1998) Asia-Pacific Microwave Conference Proceedings pp.169-172

4.A E Atia and A E Williams (1971) COMSAT Technical Review, 1, 1 pp.21-43

5.Hong J and Lancaster J M (1998) IEEE Trance. MTT46 pp.118-112

6.Okazaki Y, Suzuki K, Enomoto Y (1996) Advance in Superconductivity IX pp.1281-1284

Application of HTS / Dielectric Multilayered Thin Films for Microwave Devices

[1]Norihisa Kumagai, [1]Yasuhiro Matsuo, [1]Hitoshi Tabata, [1]Tomoji Kawai,
[2]Seiji Hidaka and [2]Katsuhiko Tanaka

[1]ISIR-Sanken, Osaka University, 8-1 Mihogaoka, Ibaraki-shi, Osaka, 567-0047 Japan
[2]Murata manufacturing Co., Ltd., 2-26-10 Tenjin, Nagaokakyou-shi, Kyoto, 617-8555 Japan

Abstract: A new concept of resonator with an electrode structure consisting of $YBa_2Cu_3O_{7-\delta}$ (YBCO) / MgO multilayered films has been proposed [1]. Toward its realization of this concept, in the present study, epitaxial growth of YBCO / MgO / YBCO trilayered film on LaAlO3 (100) substrate is obtained and its properties are compared with bi- and mono-layered films. The monolayer YBCO films show a good transport and microwave properties and a crystalline quality. The quality of the crystal structure of the film tended to degrade with increasing the number of layer.

Keywords: microwave resonator, multilayered film, epitaxial layers, YBCO, MgO, $LaAlO_3$

INTRODUCTION

Applications of high T_c superconductors (HTS) towards the fabrication of microwave devices are extensively investigated [2, 3]. Recently, a new concept for microwave disk resonator has been proposed as shown in inset of fig. 1 (a) [1]. The electrode of this resonator consists of HTS / Dielectric multilayered films to increase the effective critical current by dividing the electric and magnetic fields into each layer (fig.s 1 (a) and (b)). As suggested in ref. [1], potential candidates for the dielectric layer and the substrate are MgO and $LaAlO_3$ (LAO), respectively. And we selected $YBa_2Cu_3O_{7-\delta}$ (YBCO) as a superconducting layer because of its well-known properties. The YBCO and MgO layers of the resonator require high-quality epitaxial growth. However, to our knowledge, no work have been reported about the multilayered films having thick (≥ 100 nm) MgO layer (see fig.s 1 (a) and (b)), although there are many studies dealing with the method of preparation of superconductor / insulator / superconductor tunneling junction with thin (≤ 10nm) MgO layer [4, 5].

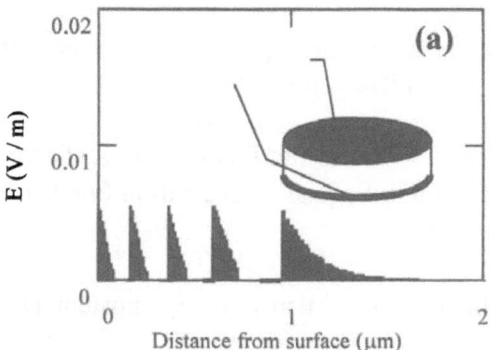
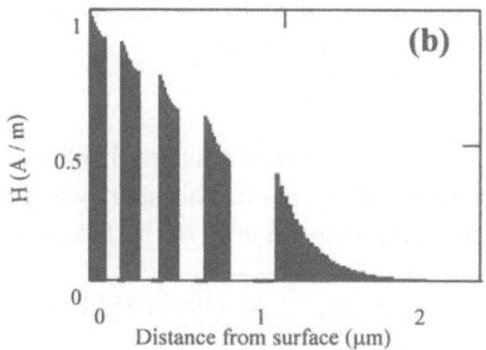

Fig. 1. Electric (a) and magnetic (b) field distribution of the multilayered film disk resonator. The inset in (a) illustrates the resonator.

In this work, we have investigated the possibility of making a resonator having thick dielectric layers. For this, we have fabricated three types of the films on the LAO (100) substrates as: i) YBCO (450 nm and 150 nm) monolayer film referred as Y1 and Y2; ii) MgO (200 nm) / YBCO (180 nm) bilayered film referred as MY and iii) YBCO (600 nm) / MgO (200 nm) / YBCO (180 nm) trilayered film referred as YMY.

EXPERIMENTAL PROCEDURE

The YBCO and MgO films were grown *in situ* by a pulsed laser deposition (PLD) method under oxygen pressure (containing 8% O_3) kept at 8 Pa. The substrates are of 7 mm diameter, both sides polished LAO. YBCO films were prepared using a fourth harmonic Nd: YAG laser ($\lambda = 266$ nm) beam operated at 0.5 Hz with a pulse energy of 6 Jcm^{-2}; deposition rate was ~1 nm / min and substrate temperature was ~750°C. MgO films were prepared with an ArF excimer ($\lambda = 193$ nm) laser operated at 2 Hz with 0.4 Jcm^{-2} / pulse. The deposition rate was 1.7 nm / min and substrate temperature was 540°C. The YBCO targets were prepared by usual solid-state reaction i.e. Y_2O_3, $BaCO_3$ and CuO powders were mixed in the ratio of Y: Ba: Cu =1: 2: 3.6, slightly enriched with Cu, and pressed into a pellet of 10 mm diameter and sintered in air initially at 900°C for 12 h then at 930°C for 24 h. Similarly, MgO target was made from powder and sintered initially at 800°C for 12 h, then at 1100°C for 24 h. Immediately after deposition, the films were cooled to room temperature in 30 min under the same oxygen condition. Ag was deposited on the other side of the substrates by conventional evaporation method in another chamber. This Ag film acts as a ground plate for the resonator. The Ag thickness was about 440 nm.

The properties of the films were investigated by unloaded-Q (Q_u) factor measurement at TM_{010} mode, temperature dependent resistivity (R-T) using a conventional four-point probe method and X-ray diffraction (XRD) measurement. Chemical etching technique with photolithography method was used in Q_u factor and R-T measurements.

RESULTS AND DISCUSSION

Figure 2 shows the temperature dependence of unloaded Q (Q_u) factor at TM_{010} mode for Y1 resonator. Similar measurement was done for Ag / LAO / Ag, a conventional metal resonator.

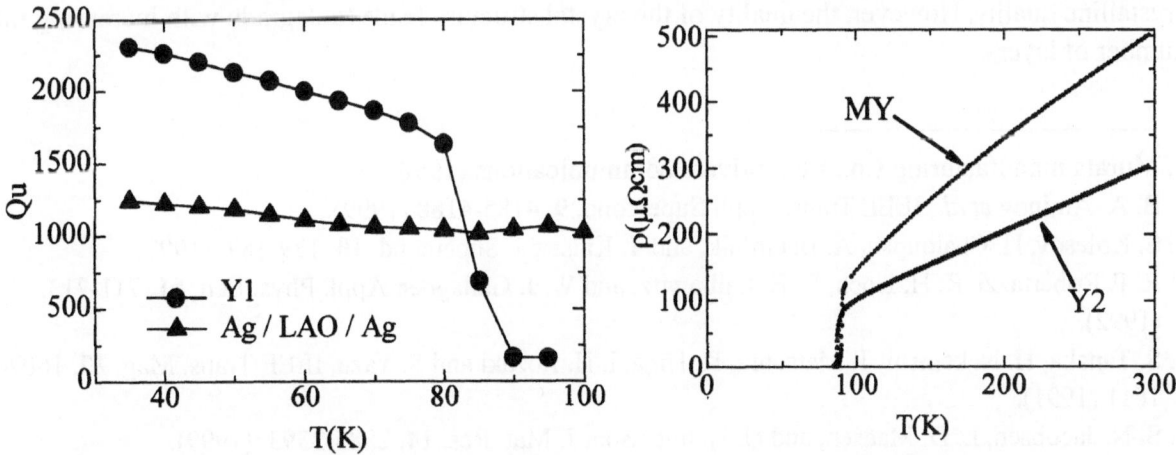

Fig. 2. Q_u factors as a function of tempera--ture for Y1 and Ag / LAO / Ag resonators.

Fig. 3. Temperature dependent resistivity for MY and Y2 films.

Below T_c, the value of the Q_u-factor for Y1 resonator is about twice as large as that of the Ag / LAO / Ag film, which is consistent with the calculated values. Figure 3 shows the temperature dependence of resistivity for MY and Y2 films. MY film shows a good superconducting transition temperature, $T_c \sim 90K$ and a sharp transition width, $\Delta T_c \sim 0.85K$. However, for this MY film the resistivity at normal state ($>T_c$) is about twice as large as that of Y2 film and proportional $\sim T^{0.6}$.

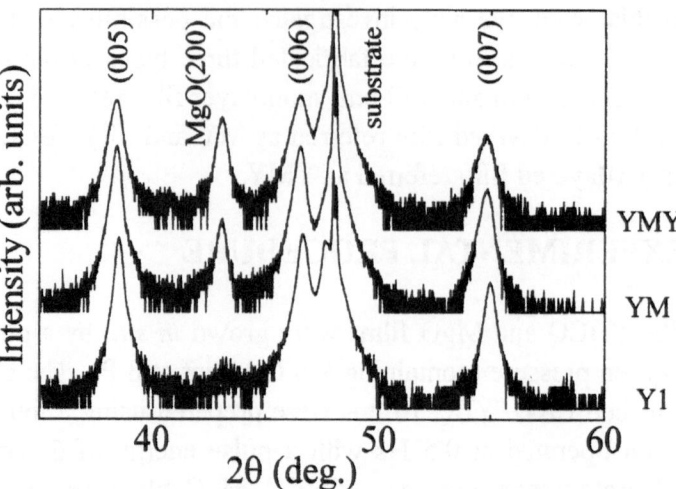

Fig. 4. X-ray diffraction pattern for Y1, YM and YMY films. The longitudinal axis is plotted in logarithmic.

The crystalline structure of the films is investigated by XRD measurement. As can be seen in Fig. 4, appearance of only (00l) peaks of YBCO and (200) peak of MgO indicates the presence of a c-axis oriented YBCO and MgO layers. The values for a full width at half maximum (FWHM) of the YBCO (005) peak of Y1, YM and YMY films are 0.22°, 0.25° and 0.36°, respectively. The c-axis lattice constants of YBCO layer present in Y1, YM and YMY films are 1.1679 nm, 1.1705 nm and 1.1699 nm, respectively. Correspondingly, the FWHM of the MgO (200) peak and the constants for YM and YMY film are 0.26°, 0.4201 nm and 0.33°, 0.4197 nm, respectively. We find that the value of FWHM and the c-axis lattice constant of YBCO tend to increase with increasing the number of layers. The structure of the films is often different from that of bulk materials because of a lattice mismatch between films and substrate or lower layer. The mismatch between YBCO and MgO may cause the deformation of the structure with increasing the number of layer. Thus it may be better to use a buffer layer such as CeO_2 [6].

CONCLUSION

We have fabricated the epitaxial YBCO monolayer, MgO / YBCO bilayered and YBCO / MgO / YBCO trilayered films on the LAO (100) substrate by PLD method using Nd: YAG laser and ArF excimer laser. The monolayer YBCO films show a good transport and microwave properties, and a crystalline quality. However, the quality of the crystal structure tends to degrade with increasing the number of layers.

1. Murata manufacturing Co., Ltd., private communications (1998).
2. B. A. Aminov *et al.*, IEEE Trans. Appl. Supercond. **9**, 4185-4188 (1999).
3. S. Kolesov, H. Chaloupka, A. Baumfalk, and T. Kaiser, J. Supercond. **10**, 179-187, (1997).
4. R. P. Robertazzi, R. H. Koch, R. B. Laibowitz, and W. J. Gallagher, Appl. Phys. Lett. **61**, 711-713, (1992).
5. S. Tanaka, H. Nakanishi, T. Matsuura, K. Higaki, H. Itozaki and S. Yazu, IEEE Trans. Mag. **27**, 1607-1611 (1991).
6. S. N. Jacobsen, L. D. Madsen, and U. Helmersson, J. Mat. Res. **14**, 2385-2393, (1999).

Effect of Cavity's Loss on the Measurement Accuracy of the Surface Resistance by Microstripline Resonator Method

Daisuke Okai, Masanobu Kusunoki, Masashi Mukaida, and Shigetoshi Ohshima

Department of Electrical and Information Engineering, Faculty of Engineering
Yamagata University, 4-3-16 Jonan, Yonezawa 992-8510, Japan

Abstract: We have examined the effect of the loss in the cavity on the measurement accuracy of surface resistance (Rs) of a superconducting thin film using a microstripline resonator. The Rs of a YBCO thin film was measured as a parameter of the height of the cavity's ceiling. The investigation revealed that the loss in the ceiling affected the measurement accuracy of Rs for a comparatively lower ceiling height. However, when the ceiling's height was sufficiently high, the effect was negligible.

Key words: Surface resistance, Cavity, Microstripline resonator

INTRODUCTION

Microstripline resonator method is a well known method of measuring the Rs of a superconducting thin film. In this method, Rs is calculated from the measured unloaded quality factor (Q_u) of a resonator. In general, Q_u of a microstripline resonator is defined by

$$\frac{1}{Q_u} = \frac{1}{Q_c} + \frac{1}{Q_d} + \frac{1}{Q_r}$$

where Q_c, Q_d and Q_r are the quality factors related to the conductor, the dielectric, and the radiation losses of the resonator, respectively. Here, the conductor loss depends on the Rs of the strip line and ground plane. The dielectric loss depends on the tanδ of the substrate. The radiation loss arises due to the radiation of electromagnetic waves from the microstrip line. When the resonator is packaged in a metal cavity, the radiated wave is reflected by the surface of the cavity's wall and returns back to the resonator. However, the radiated wave causes induced current on the surface of the cavity's wall, therefore, leading to the loss in the cavity [1]. For Rs measurement of metals, in general, we can neglect the loss in the cavity because its loss is smaller than the conductor loss of a metallic resonator. On the other hand, the Rs of a high-Tc superconducting (HTS) thin film is more than two order of magnitude smaller than that of metals at the microwave frequency range. Therefore, the conduction loss of a superconducting resonator becomes much smaller than that of the metallic resonator. For the accurate measurement of Rs of a HTS thin film, it is necessary to examine the effect of the cavity's loss on the measurement accuracy of Rs. In this study, we have investigated the effect of the loss in the cavity's ceiling on the measurement accuracy of Rs.

EXPERIMENTAL

We have measured the Rs of a HTS thin film by probe-coupling type microstripline resonator method [2]. Figure 1 shows the structures of the microstripline resonator and cavity for this method. The coupling can be varied by changing the antenna length. For the measurement of Rs, two types of cavities were used. Figure 2 shows the structures of the cavities. The cavity shown in Fig. 2 (a) was fabricated from metal (copper), where H is the height between the resonator and the ceiling of the cavity. On the other hand, the cavity shown in Fig. 2 (b) was fabricated from superconductor (YBCO) and metal (copper). The cavity's ceiling was superconductor, whereas the side walls and bottom were metal. The resonator was fabricated from a YBCO thin film on MgO substrate. The YBCO film was patterned by standard photolithography and wet etching technique. Both the strip line and ground plane were YBCO. The width and length of the strip line was 0.45 mm and 8.1 mm, respectively. The thicknesses of the strip line and ground plane were 0.62 μm each. The thickness of the substrate was 0.5 mm.

RESULTS AND DISCUSSION

Metallic ceiling. Using the cavity of metallic ceiling (m-ceiling) shown in Fig. 2 (a), the Rs of the microstripline resonator was measured with the ceiling heights of 2.5 mm, 4.5 mm and 5.5 mm. Figure 3 shows the temperature dependence of Rs of the resonator as a parameter of the ceiling's height (H). The curve measured in the condition of H = 2.5 mm was different from the curves in the conditions of H = 4.5 mm and 5.5 mm. The measured values of Rs were almost constant (0.36 mΩ) in the conditions of H = 4.5 mm and 5.5 mm at T/Tc ≒ 0.2. In the condition with H = 2.5 mm, the measured value of Rs was higher than that of the Rs in the other conditions. Here, the increase of the Rs arose from the combined conduction losses of the resonator and the cavity's ceiling. This results indicate that, in the condition where H is comparatively smaller, the loss in the cavity's ceiling becomes larger due to the excessive induced current flow along the surface of the ceiling and the measurement accuracy of Rs is strongly affected by its loss. Hence, the height of the cavity's ceiling is important for the accurate measurement of Rs.

Superconducting ceiling. Figure 4 shows the temperature dependence of Rs of the superconducting and metallic ceilings. For the superconducting ceiling (s-ceiling) of the cavity shown in Fig. 2 (b), a YBCO thin film on MgO substrate was used. The Rs of the s-ceiling was measured at 22 GHz by dielectric rod resonator method. The Rs of the m-ceiling was calculated from the resistivity of standard copper at 22

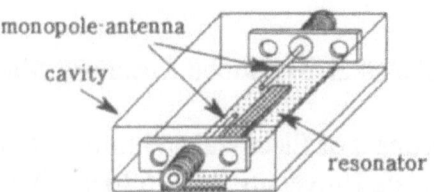

Fig. 1. The structures of the microstripline resonator and cavity for the probe-coupling type method. The RF input and output are achieved by two monopole antennas.

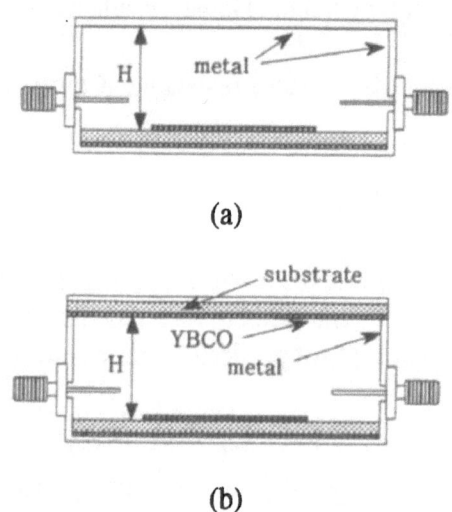

Fig. 2. The structures of the cavities. (a) Metal (copper) is used as the cavity's ceiling. (b) Superconductor (YBCO) is used as the cavity's ceiling.

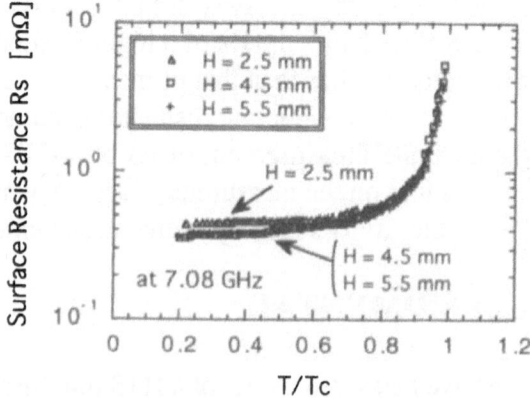

Fig. 3. The temperature dependence of Rs of the resonator as a parameter of the ceiling's height (H). The measurement of Rs was performed using the metallic (copper) ceiling shown in Fig. 2 (a).

GHz [3]. The measured Rs of the s-ceiling was 0.08 mΩ at 32 K and 0.25 mΩ at 77 K (@ 7.08 GHz). The Rs at 22 GHz was transformed into 7.08 GHz by a scaling rule of f^2, where f is frequency. The Rs of the s-ceiling was two order smaller than that of the m-ceiling at low temperature (@ 7.08 GHz). So, this results indicate that the loss in the s-ceiling is much smaller than that of the m-ceiling because the loss of the ceiling is proportional to its Rs.

Figure 5 shows the temperature dependence of Rs of the resonator measured using the cavities of superconducting and metallic ceilings shown in Fig. 2 (a) and (b). For s- and m-ceilings, the measurements of Rs of the resonator were performed in the same conditions with H = 4.5 mm. The measured curves were almost same. For s-ceiling, the measured value of Rs was almost coincided with that of m-ceiling at T/Tc ≒ 0.2. From the measurement results, it was found that, in the conditions of H = 4.5 mm, the loss in the ceiling becomes small regardless of Rs of the ceiling. This is attributed to the lesser amount of induced current flow along the surface of the ceiling. Therefore, the increase of the height of the cavity's ceiling leads to the decrease of the loss in the ceiling. So, by adjusting the height of the cavity's ceiling, we are able to obtain more accurate value of Rs of the HTS thin films.

SUMMARY

We have examined the effect of the cavity's loss on the measurement accuracy of Rs of a HTS thin film. In the condition where the height of the cavity's ceiling is comparatively lower, it was found that the accuracy of the Rs measurement is sensitive to the loss in the cavity's ceiling. In the condition where the cavity's

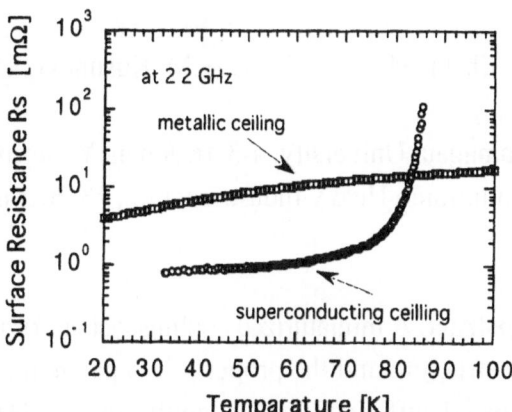

Fig. 4. The temperature dependence of Rs of the superconducting (YBCO) and metallic (copper) ceilings. The Rs of the superconducting ceiling was measured by dielectric rod resonator method. The Rs of the metallic ceiling was calculated from the resistivity of standard copper.

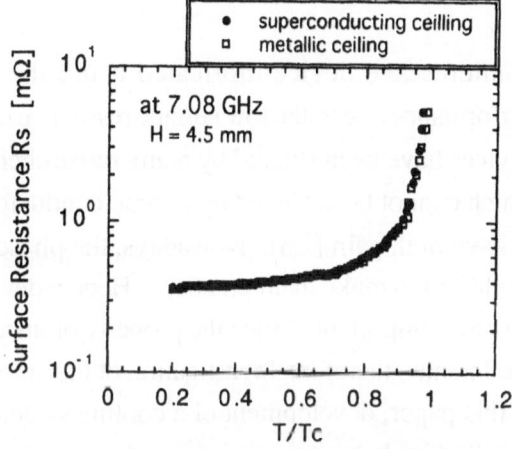

Fig. 5. The temperature dependence of Rs of the resonator measured using the cavities of superconducting (YBCO) and metallic (copper) ceilings shown in Fig. 2 (a) and (b).

ceiling is sufficiently high, the loss in the cavity's ceiling is negligible. Hence, by adjusting the height of the cavity's ceiling, we can increase the measurement accuracy of Rs of the HTS thin films. When the ceiling's heigth is relatively low, then the use of a superconducting ceiling would improve the measurement accuracy significantly.

1. Y. Ueno, M. Fuse, K Saito, and N. Sakakibara, in *Advances in Superconductivity* IX , edited by S. Nakajima and M. Murakami (Springer-Verlag, Tokyo, 1997), pp. 1261-1264.
2. D. Okai, T. Tomiyama, M. Kusunoki, M. Mukaida, and S. Ohshima, in *Advances in Superconductivity* XI , edited by N. Koshizuka and S. Tajima (Springer-Verlag, Tokyo, 1999), pp. 1259-1262.
3. H. A. Wheeler, *Proc. of the I.R.E.*, **30**, 412-424 (1942).

Development of A Miniaturized Cooling System for HTS Antennas

K. Ehata, M. I. Ali, K. Sato, M. Kusunoki, M. Mukaida, S. Ohshima, Y. Suzuki*, and K. Kanao*

Yamagata University, 4-3-16 Jonan, Yonezawa, Yamagata 992-8510, JAPAN
*Sumitomo Heavy Industries, Ltd., 63-30 Yuhigaoka, Hiratsuka, Kanagawa 254-0806, JAPAN

Abstract: A miniaturized cooling system for HTS antennas was developed using a Stirling cryocooler. Its structure and the property of a patch antenna installed in this system are reported. The system showed sufficient cooling ability to cool HTS microwave devices in spite of its portable size and weight. When a patch antenna was installed in this system, radiation from the antenna was focused to the front direction and its gain was enhanced by 11.2 dB. This fact was confirmed not only experimentally but also theoretically.

Keywords: Superconducting antenna, Patch antenna, Stirling cryocooler, Radiation pattern, Gain

INTRODUCTION

A microwave device fabricated using the High-Temperature Superconductor (HTS) has high performance due to the low surface resistance of the HTS. Therefore, a large number of HTS microwave devices have been studied by many researchers [1]. In particular, a HTS filter has excellent properties which cannot be achieved by normal conductors. On the other hand, a HTS patch antenna is expected to have high gain [2,3]. Nowadays, the phase of the research on HTS microwave devices is going to be shifted to make them practical. Hence, development of a cooling system as well as the device itself becomes important. Since the property of an antenna is affected by metals or dielectrics that surrounds the antenna, therefore, investigation of the antenna property installed in the cooling system is important. In this paper, development of a cooling system for HTS antennas and the property of a patch antenna installed in the system are reported.

EXPERIMENT

The cooling system for HTS antennas, shown in Fig. 1, was developed using a Stirling cryocooler (Sumitomo Heavy Industries, Ltd., SRS-2110). The dimensions of the system were W230 mm × D310 mm × H180 mm and its weight was 9.2 kg. A sample was cooled to 50 K by 90 min. In spite of its portable size and weight, the cooling ability of the system is enough for a HTS microwave device. Details of inside of the vacuum chamber are shown in Fig. 2. An antenna was mounted on the cold head and was thermally isolated in the vacuum chamber, which consists of a quartz glass window and a stainless steel jacket. The quartz glass window was employed for the transmission of microwaves. The diameter, thickness, and permittivity of this window were 134 mm, 8 mm, and 3.8, respectively. The distance between the window and the antenna was 8 mm. Microwave was supplied to excite the antenna through a hermetically sealed SMA connector, a Cu-Ni coaxial cable, and a K connector. Temperature of the sample stage was measured using a Pt-Co resistance thermometer. Property of the patch antenna installed in this system was measured using an Network Analyzer (Wiltron 360B) in a radiowave anechoic room. Configuration of the patch antenna is shown in Fig. 3. Superconducting (YBCO) and normal conducting (Cu) patch antennas with resonant frequency of about 5 GHz were prepared [3]. Measured results were confirmed by the 3D electromagnetic simulator, KCC Micro-Stripes.

RESULTS AND DISCUSSION

Measured radiation pattern for the YBCO and the Cu antenna installed in the system and for the Cu one in free space are shown in Fig. 4. The YBCO and the Cu antenna were measured at 60 K and at room temperature, respectively. From Fig. 4, it is found that the radiation from the Cu antenna installed in the system is focused to the front direction. This focused pattern was observed also in the YBCO antenna. Since the distance between the antenna and the vacuum chamber was less than a wave length, radiated electromagnetic field was influenced by the chamber. Simulated result of radiation pattern for a lossless patch antenna was shown in Fig. 5. The focus of the radiation shown in Fig. 4 was also obtained by the 3D electromagnetic simulation.

The measured frequency dependence of the gain for YBCO and Cu antenna is shown in Fig. 6. The gain of the Cu antenna was enhanced by 11.2 dB when it was installed in the system. This is because the radiated energy from the antenna was focused to the front direction as shown in Fig. 4 and 5. In addition, the superconducting antenna showed higher gain than normal conducting one by 2.4 dB.

The influence of the primitivity ε_r and thickness

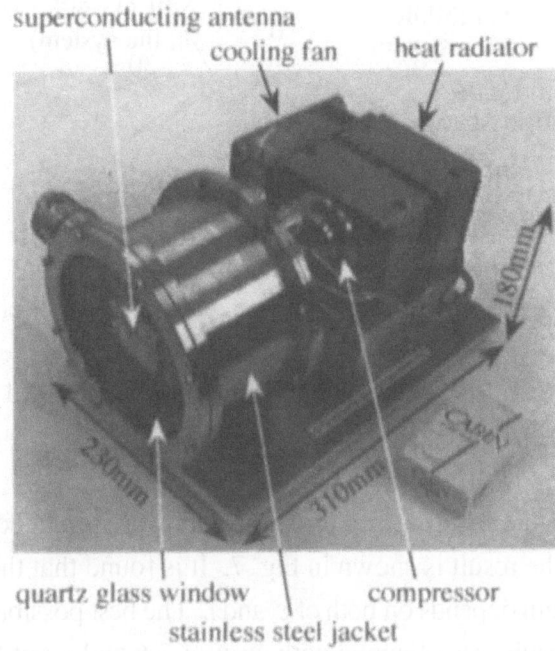

Fig.1. Miniaturized cooling system for HTS antennas

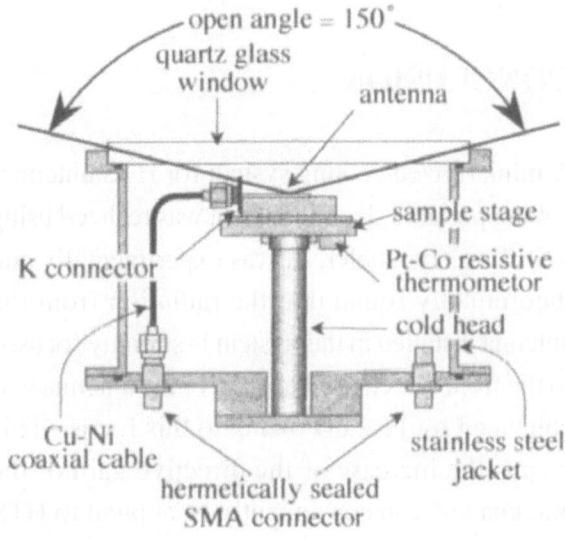

Fig.2. Details of inside of the vacuum chamber.

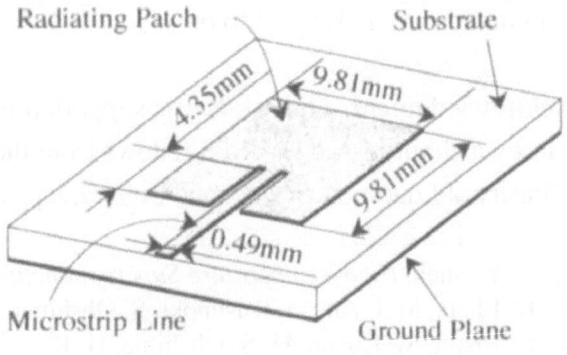

Fig.3. Patch antenna.

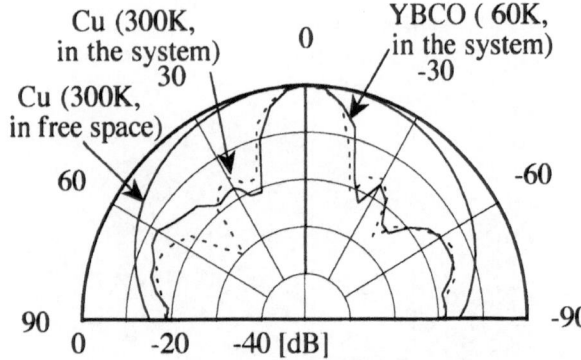

Fig.4. Measured radiation pattern of YBCO and Cu patch antenna.

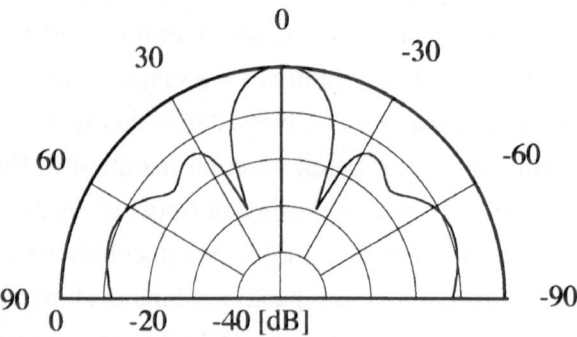

Fig.5. Simulated radiation pattern of loss less patch antenna installed in the cooling system.

of the window t on directive gain was simulated. The result is shown in Fig. 7. It is found that the gain depends on both of ε_r and t. The best possible result was obtained with an ε_r of 3.6 and a t of 8 mm. This conditions are close to the condition of the quartz glass window used in this work with an ε_r of 3.8 and a t of 8 mm.

CONCLUSION

A miniaturized cooling system for HTS antennas having portable size and weight was realized using a Stirling cryocooler. It was experimentally and theoretically found that the radiation from the antenna installed in this system is strongly focused to the front direction. The gain of the antenna was enhanced by 11.2 dB owing to this focus. This means the increase of the directive gain of the antenna and can conveniently be applied to HTS patch antennas for the systems which need narrow beam and high gain antennas such as the satellite communication, radar, and so on.

Acknowledgment: This work was supported in part by a Grant-in-Aid for JSPS fellows from the Ministry of Education, Science and Culture, Japan.

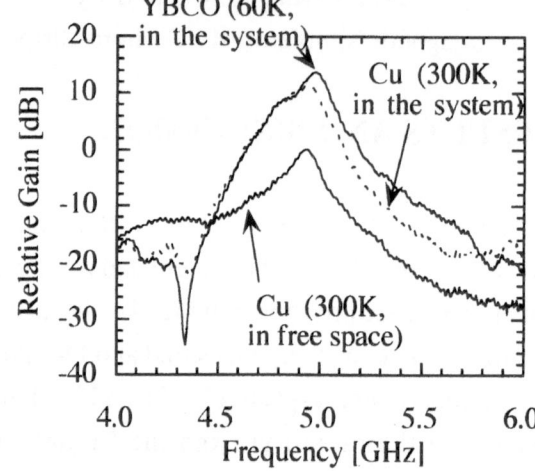

Fig.6. Measured frequency dependence of relative gain of Cu and YBCO antenna..

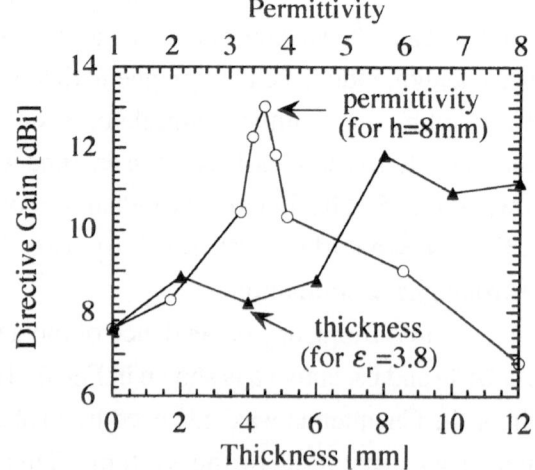

Fig.7. Relation between permittivity and thickness of window and directive gain.

1. Z. Y. Shen, *High-Temperature Superconducting Microwave Circuits*, (Artech House, Boston, 1994)
2. K. Ehata, M. I. Ali, M. Kusunoki, S. Ohshima, *IEEE Trans. Appl. Supercond.*, **9**, 3081-3085 (1999)
3. K. Ehata, M. Kusunoki, S. Ohshima, H. Kinder, *Advances in Superconductivity X* (Springer-Verlag, Tokyo, 1998), pp. 1153-1156.

Optimum Configuration of the Antenna to the Dielectric Resonator to Measure the Accurate Surface Resistance of the HTS Films

Momoko Inadomaru, Daisuke Kousaka, Masanobu Kusunoki, Masashi Mukaida, and Shigetoshi Ohshima

Department of Electrical and Information Engineering, Yamagata University, 4-3-16, Johnan, Yonezawa, 992-8510, Japan

Abstract: The configuration of the antenna of the dielectric resonator is investigated to measure the exact surface resistance (R_s) of the high temperature superconductor (HTS) thin films over a wide range of temperature. The sapphire dielectric resonator is operated at 22GHz. The loop-terminated semi-rigid cables are used as the antennas. Using a Cu resonator at room temperature, the circumference length of the loop (L), the distance from the dielectric center to the loop center (d), and the height of the loop plane from the substrate surface (h) were changed. It is found that the smaller loop set at a position halfway to the height of the dielectric rod brought the widest region of d in which exact measurements can be carried out. This result was then applied using a HTS resonator. In order to measure exact Q_u over a wide temperature region from low temperature to the critical temperature, weak coupling rather becomes a disadvantage. The best position of the antenna can be obtained when insertion loss set around –10dB at minimum temperature of the system.

Keywords: Dielectric resonator, Surface resistance, Antenna configuration, Qu

INTRODUCTION

The low surface resistance of high critical temperature superconducting (HTS) materials is a significant parameter that allows the efficient use of these materials in microwave device applications. Up to now, many kinds of techniques have been developed to fabricate high quality large-area HTS films [1]. Exact R_s measurement method of the HTS films is required to optimize [2] the manufacturing process. The dielectric resonator is useful and the most convenient method to obtain the R_s, because it has simple structure and does not require patterning of the films. Moreover, accurate formulae can be used for the calculation of R_s [3,4]. The frequency response of the resonator is measured using various configurations of the antenna in every experiment. In this paper, we investigate the suitable configuration of antenna such as distance between dielectric and antenna, height of the antenna from the plate surface and length of the loop.

EXPERIMENTAL

The structure of the dielectric resonator is shown in Fig.1. A sapphire cylindrical rod (ϕ6.5 mm, H=3.0 mm) was used as the dielectric. Oxygen-free Cu plates or YBa$_2$Cu$_3$O$_y$ films(20mm$\times$20mm) were set at both ends of the rod. The loops at the end of semi-rigid cables were used as antennas. The antennas can be freely adjusted along the horizontal axis, which is parallel to the substrate surface, and along the vertical axis, which is

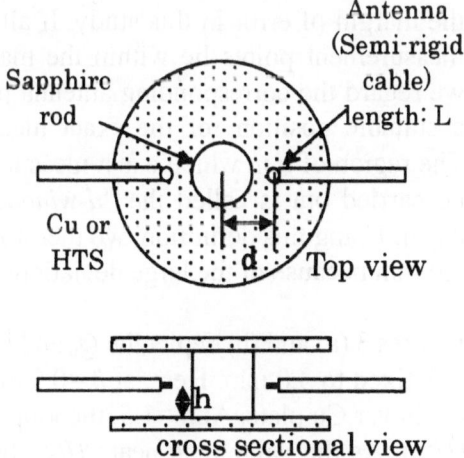

Fig.1 The structure of the dielectric resonator

parallel to the rod. The d and h denote the distance from the dielectric center to the loop center and the height of the loop plane from the substrate surface, respectively. Four kinds of loops that have different circumference lengths (L=1.6, 3.0, 4.5, 6.0mm) were prepared. The frequency response S_{21} of the resonator was measured with a vector network analyzer (HP8722D). The resonance frequency of the TE_{011} mode was 22 GHz.

RESULTS AND DISCUSSION

Figure 2 shows the dependence of the unloaded quality factor (Q_u) on d, where Cu was used as the end plates at room temperature. Open diamonds, squares, circles and triangles correspond to the Q_u values of the antennas with L=1.6, 3.0, 4.5 and 6.0 mm, respectively. The h was fixed at 0.5 mm. The measurements were repeated 10 times for each location and all of the values are plotted in the graph. The points are scattered within a relatively large d region. Moreover, the Q_u values drastically decreased within the region. The point at which Q_u begins to decrease is observed to occur at larger values of d as L is decreased. These effects are caused by background noise. The noise level arises with increasing in L. At smaller d values, Q_u also decreased due to the strong coupling between the resonator and antenna. This effect also increases as L becomes longer. From this figure, we defined exact Q_u of this resonator as 4700. We then set the condition of ±2.5 % deviation from the exact value of Q_u as the margin of error in this study. If all of the 10 measurement points lie within the margin, then we regard the corresponding antenna position as a suitable position for the exact measurement. The region of d in which exact measurement can be carried out is called the "d-window" in this

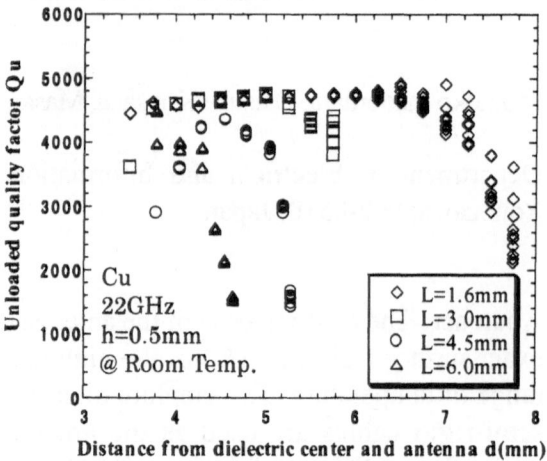

Fig.2. Dependence of Qu on d of Cu resonator. The h was fixed at 0.5 mm.

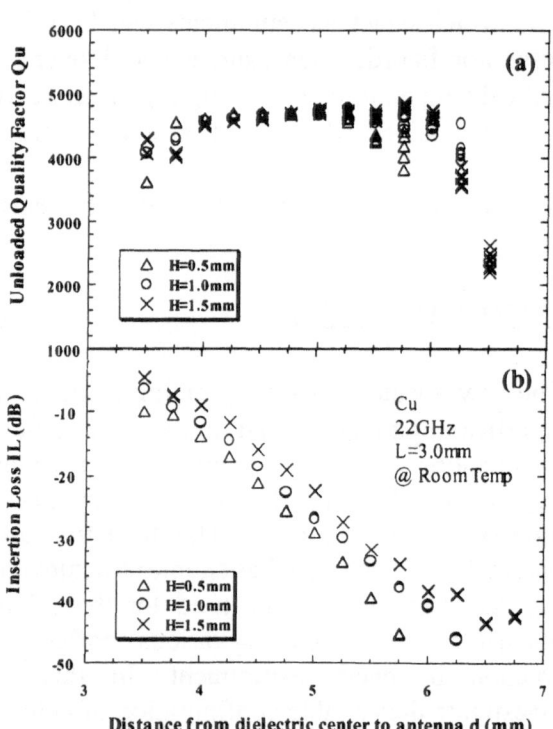

Fig.3. Relationship between Qu and d (a), Qu and IL (b). The L is fixed at 3.0 mm.

paper. Using this definition, we therefore can not obtain the d-window for antennas with L=4.5 and 6.0 mm because of the large deviations of the Q_u values.

Figures 3 (a) and (b) depict the Q_u and insertion loss (IL) as a function of d for different h values. The L is fixed to 3.0 mm. Here, at h =0.5 mm, the outer conductor of the semi-rigid cable is attached to the lower Cu plate. At h =1.5, the loop plane is positioned halfway to the height of the sapphire rod ($H/2$). As the value of h nears $H/2$, the d-window becomes wider. This implies that the coupling between the antenna and resonator becomes strong when h approaches the position at $H/2$. Taking into account that the suppression of Q_u occurs due to the overlapping of IL and the background noise level within a large d region, the stronger coupling in even larger d has an advantage of exact Q_u measurement over a wider range. As can be understood from theoretical analogy, when we take IL to be the horizontal axis and loaded Q (Q_L) as the vertical axis, then all Q_L-IL curves agree well in the range of values greater than –30 dB. It means that the best location of the antenna can be estimated from the IL values. However, sometimes other irregular modes that occur by some reasons influence the frequency response of the TE_{011} mode. In these cases, the estimation mentioned above should be ruled out.

We carried out the measurement for all possible sets of (L, h, d) in the same way as in Figs. 2 and 3. The widths of the d-windows are shown at the top line of the cells of Table 1. At the bottom line, the ranges of the IL that correspond to the d-window are shown. It is found that the smaller loop set at the height around $H/2$ brought about the widest d-window. These results can be applied to a resonator with HTS plates. When we suppose that the measurement is carried out over a wide temperature region (low temp. ~ T_c), the change of the IL with varying temperature should be taken into account. In order to obtain a high resolution of Q without the influence of the noise even at temperatures near T_c, weak coupling rather becomes a disadvantage. From Table 1, when we use the smallest 1.6-mm-long loop antenna at $h=H/2$, the range of IL that

Table 1. The widths of the d-WINDOWs and the IL that correspond to the d-WINDOW

L(mm) \ h(mm)	0.5	1.0	1.5
1.6	1.25mm 23-44dB	1.50mm 9-32 dB	2.00mm 6-36 dB
3.0	<0.25mm 29 dB	1.50mm 12-34 dB	2.00mm 9-38 dB
4.5	impossible	0.75mm 10-18 dB	1.25mm 6-23 dB
6.0	impossible	<0.25mm 19 dB	<0.25mm 19 dB

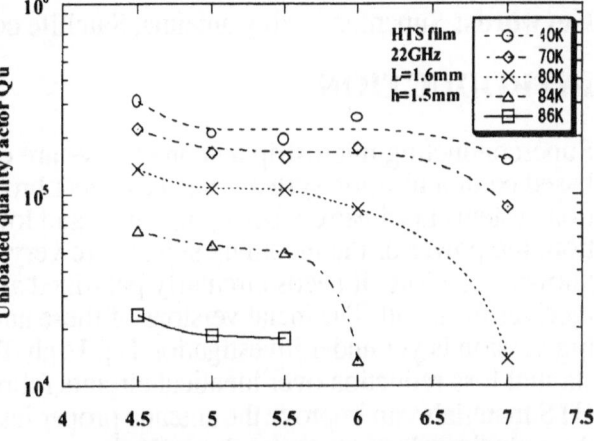

Fig.4 The relationship between Qu and d of super-conducting resonator for different temperatures.

corresponds to the d-window is from –6 dB to -36 dB. The highest resolution of Q can be obtained when the IL as a function of the temperature varies over the full region of the IL that corresponds to d-window. This is confirmed by actual experiment as shown in Fig.4. The figure shows the relationship between Q_u and d of HTS resonator for different temperatures. The condition mentioned above is satisfied from 5.0 to 6.0 mm of d in Fig.4. The exact Q_u values can not be obtained above 6.0 mm of d at higher temperatures. On the other hand, Q_u-d curves are warped at 4.5 mm of d. In this case the coupling was too strong as IL is more than –6dB.

CONCLUSION

In summary, we investigated the configuration of antenna for dielectric resonator to obtain exact Q_u. For the possible sets of (L, d, h), the smaller loop set at the height around $H/2$ brought about the widest d-window when we used Cu resonator at room temperature. The results could be applied to a resonator with HTS films. When the measurement is carried out over a wide temperature region, the change of the IL with varying temperature should be taken into account. The highest resolution of Q can be obtained when the IL as a function of the temperature varies over the full region of the IL that corresponds d-window. This is confirmed by actual experiment using HTS films. The result agreed well with the analogy.

1. H. Kinder, W. Prusseit, R. Semerad and B. Utz, Advances in Superconductivity IX (1997) 1011.
2. M. Kusunoki, Y. Takano, M. Mukaida and S. Ohshima, Physica C, 321 (1999) 81.
3. Y. Kobayashi and M. Katoh, IEEE Trans. Microwave Theory and Tech., 33 (1985) 586.
4. Z. Y. Shen, C. Wilker, P. Pang, et al., IEEE Trans. Microwave Theory Tech., 40(1992)2424.

Superconducting Circularly Polarized Antenna at 12 GHz: A Study with Single and Dual-Feeding Techniques

Mohammad I. Ali, Katsufumi Ehata, and Shigetoshi Ohshima

Faculty of Engineering, Yamagata University, 4-3-16 Jonan, Yonezawa, Yamagata 992-8510, Japan

Abstract: Superconducting microstrip antenna arrays are expected to be used in the satellite communication system which needs circularly polarized (CP) antenna with high gain and low axial ratio. We have studied the single patch CP antennas with single and dual-feed (reactive splitter coupled) structures to excite CP signals. Square patch elements with a center frequency at the proximity of 11.85 GHz and input impedance 50 Ω were used in the investigation. Measured results have been presented on the both types of antennas analyzed by a microwave simulator. Both superconducting (YBCO) and metal (Cu and Au) antennas have been considered for the comparison of antenna properties.

Keywords: Superconducting antenna, Satellite communication, CP antenna, Axial ratio, Gain.

INTRODUCTION

Superconducting microstrip antenna arrays are expected to be used in the next generation satellite-based communication systems, such as direct broadcast satellite (DBS), radar etc. These communication systems need narrow beam, high gain and low axial ratio (AR) antennas. In satellite communication, the power of the incoming signals are very low and the polarization of the signals are also unknown, therefore, it needs circularly polarized antenna with high gain to get a sensible signal in the receiver front end. The metal versions of these antennas are already in use whereas the superconducting version is yet under investigation [1]. High-Tc Superconducting (HTS) thin films provide a substantial loss reduction over identical circuits fabricated from normal metals (silver, gold, or copper). HTS materials can improve the antenna properties by reducing the losses both in the feed network and the radiating elements themselves. We have investigated the circularly polarized antenna properties with single and dual-feeds by both theoretical analysis and experimentation which may be used to fabricate the array antennas with high gain.

EXPERIMENTAL DETAILS

Antenna Design and Analysis. The antennas were analyzed by a microwave circuit simulator *Em* which can efficiently account for the microwave surface impedance of the structures under investigation. The antennas were designed to resonate with a center frequency of 11.85 GHz and input impedance of 50 Ω. Fig. 1(a) and (b) show the structures of both the antennas with square patches of length L (=W). Wmsl1 and Wmsl3 are the widths of the impedance transformers, Wmsl2 is feed line width, c is the truncation length for the single-feed antenna and λ_g is the wave length in the substrate. 15 mm x 15 mm PPO (ε_r = 10.5) substrate was used for Cu antennas whereas MgO was used for YBCO and gold antennas. To generate the CP signals by exciting orthogonal field components in phase quadrature, corner truncation was used in the single-feed type whereas a reactive splitter, with one arm quarter wave length longer than the other, was used in the dual-feeding technique. The design inputs for YBCO and metal antennas (critical temperature Tc, surface impedance etc.) and complete design procedures for both types of antennas as well as the excitation mechanism of CP signals may be found in [1-2].

Antenna Layout Fabrication and characterization. The antennas were patterned by standard photolithography and wet etching technique. The characterization includes the measurement of resonant frequency, reflection coefficient and gain. The copper antennas were measured at room temperature while the YBCO and gold antennas were measured at the cryogenic temperatures down to nearly 20 K using a Wiltron 360B network analyzer. The axial ratio was calculated from the measured antenna gains in two perpendicular planes (horizontal and vertical) and their phases. All the measurements were performed inside an anechoic chamber to avoid the incidental interference.

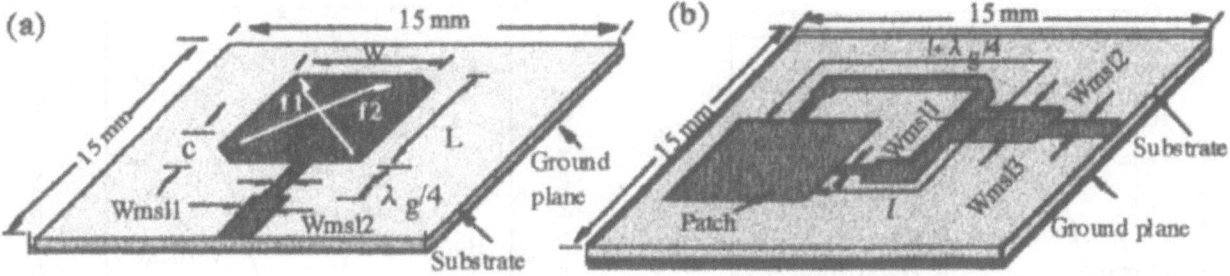

Fig. 1 Antenna structure with single and dual feed structure

RESULTS AND DISCUSSION

Fig. 2 (a) and (b) illustrates the current distributions in both single and dual-feed structures, respectively generated by *Em*. The figure shows that the orthogonal currents flow at the edges of the structures. These two orthogonal currents, in fact, produce orthogonal modes which along with the phase quadrature produce CP signals. Fig. 3 shows the reflection pattern of the single-feed Cu antenna as a function of frequency. The antenna properties with this structure depend on the truncation area which in turn depends on the unloaded Q-value of the antenna [2]. A truncation area of 1.7% of the patch area was used in this case. The figure shows that two modes are generated with a frequency difference of about 0.3 GHz which is seen both in reflection and transmission patterns (in horizontal and vertical planes) of the antenna. But the situation is completely different in the dual-feed where two modes are generated with identical frequency as seen in Fig. 4 which shows the reflection and transmission patterns of the antenna. Here, two modes coincide at a single resonant frequency. Consequently, the impedance bandwidth in this antenna is smaller than that of the single-feed one. The impedance bandwidth (vswr<2) of this antenna is about 1.8% whereas it is about 3.5% in the single-feed antenna.

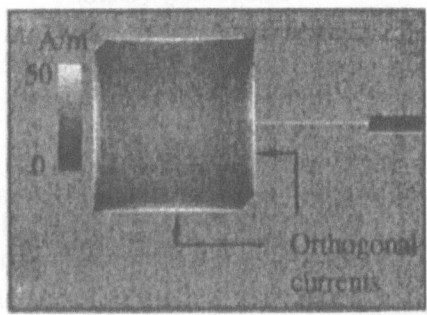

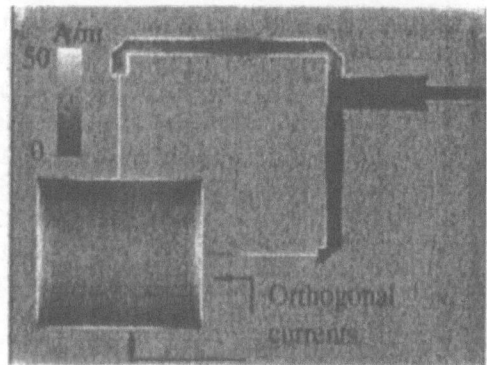

Fig. 2 Current distributions

The axial ratio was calculated from the measured transmission patterns (gain) in two perpendicular planes of the antennas and their phases. The axial ratio of the single-feed antenna was as low as 0.3 dB whereas it was about 1.4 dB at the proximity of the center frequency in the dual-feed one. The cause of higher ellipticity is believed to be due to the frequency and impedance mismatch sensitivity of the reactive power divider. Since, with the resistive power divider, the polarization purity is insensitive to the antenna mismatch, being affected only by the complex power division ratio. On the other hand, with the reactive power divider, the axial ratio is affected by the antenna mismatch which was confirmed by the theoretical analysis where the reflected signal at the antenna port was seen to affect both the realized gain and the axial ratio of the radiator. Moreover, possibile coupling between the antenna element and the feed line along with the spurious radiation from the long feed line may also influence ellipticity as well as gain. The dual-feed antenna properties were also found to be very sensitive even to the bends and feedline widths as these have influence on the mismatch causing possible amplitude and phase errors. On the other hand, the single-feed antenna was found to be less sensitive to these parameters. The AR bandwidth (3 dB) was 1.5% in the single-feed antenna while it was less than 1%

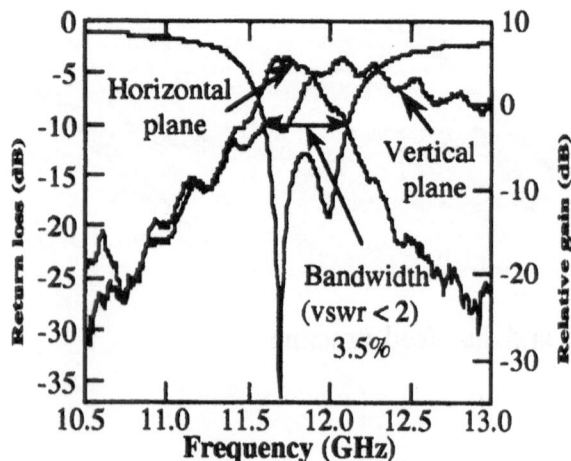

Fog. 3 Reflection and transmission patterns for single-feed antenna

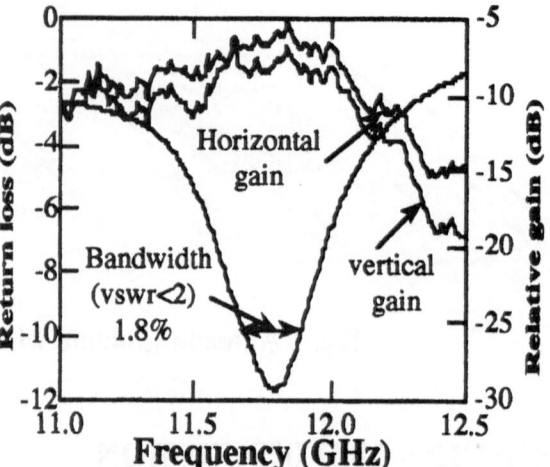

Fog. 4 Reflection and transmission patterns for dual-feed antenna

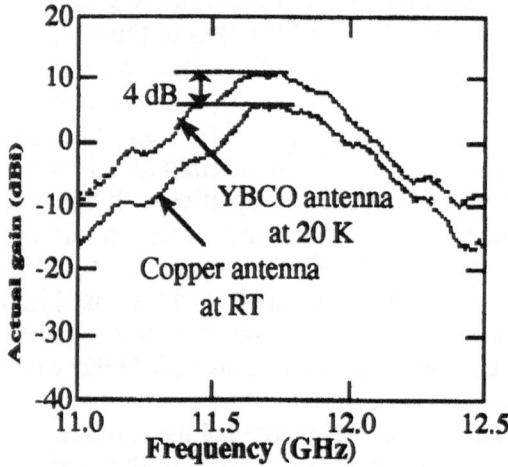

Fig. 5 Comparative gain of Cu and YBCO antenna with single-feed

in the dual-feed one. The YBCO antennas fabricated on MgO substrate had lower bandwidth than those of Cu antennas due to the higher Q-value of the radiator on MgO. The bandwidth with single and dual feed were about 1.9% and 1.2% respectively. The gold antennas also had almost the same bandwidth as those of YBCO antennas. The gain comparison between Cu (RT) and YBCO (20 K) with single-feed structure shows 4 dB improvement as shown in Fig. 5. The improvement over gold one with same structure was about 3.4 dB at 77 K. The gain improvement here agrees with the other reports [3-4]. However, the AR of this antenna was about 2 dB. The higher gain in the YBCO antenna than that of the copper is the combined effect of lower conduction and dielectric losses and focussing of the signals by the cylindrical radome which was used to cover the antenna for heat isolation [5]. In the investigation of YBCO antennas, both gain and AR were found to be temperature-dependent. The temperature dependence of gain is supposed to be due to the temperature dependence of loss factors of YBCO thin films on MgO. The change of AR is supposed to be due to the change in surface impedance with temperature causing the change in resonant frequency and input impedance affecting the phase and amplitude of the orthogonal modes.

CONCLUSION

CP antenna properties with single and dual-feed structures have been investigated by theoretical analysis and experimental results. Single-feed antenna with truncated corner was found to have superior properties than the dual-feed antenna with reactive power divider. The superconducting antennas showed a substantial improvement in gain than metal antennas. Therefore, having smaller dimension and better performance, single-feed antenna may be used to fabricate high gain superconducting array antenna.

1. M. I. Ali, K. Ehata, an S. Ohshima, IEEE Trans. Applied Superconductivity, **9**, 3077-3080 (1999).
2. R. A. Sainati, *CAD of Microstrip Antennas for Wireless Applications*, (*Artech* House, Boston, 1999).
3. X. Castel et al, Microwave and Optical Technology Letters, **13**, 255-259, (1996).
4. K. Ehata et al, *Advances in Superconductivity-XI*, edited by N. Koshizuka and S. Tajima (Springer-Verlag, Tokyo, 1999) pp. 1153-1156.
5. K. Ehata et al, "Development of a Miniaturized cooling system for HTS antenna", (to be presented in ISS '99).

A Low Linear Distortion Filter Configuration for Digital Communication Systems

Shigeki TAKEDA and Yoshinori MATSUNAGA
Kyocera Corporation R&D Center Keihanna, 3-5 Hikaridai, Seika-cho, Soraku-gun Kyoto 619-0237, Japan,

Abstract: A low linear distortion filter configuration is proposed. This filter has flat magnitude and flat group delay time in the pass band simultaneously, and has transmission zeros in the stop band. This filter is implemented on the HTS substrate, and is well applicable to software radio base stations.

Keywords: prototype low pass filter, HTS filter, group delay time, linear distortion, software radio

1.INTRODUCTION

Mobile communications are one of the most attractive communication areas in the next decade, and are now expanding rapidly. While the installation of the base stations is being promoted, energy-saving features must be implemented first. The first application of HTS filters to telecommunication was thus for energy saving, because HTS has extremely low losses, which allows us to reduce the power consumption both in the base stations and the mobile terminals.

Then, a second phase is required. The channel capacity becomes increasingly higher as our demands get more complex, and accordingly the system becomes more sophisticated such as from PDC to W-CDMA, etc.. Corresponding to this information increase, the burden to the infrastructure or the terminals becomes extremely high, both technically and economically. "Software Radio" is expected to be one of the promising solutions to overcome this difficulty. The "super conductor" becomes again an indispensable technology which supports the realization of the software radios scheme. This paper shows a low linear distortion HTS filters for software radios.

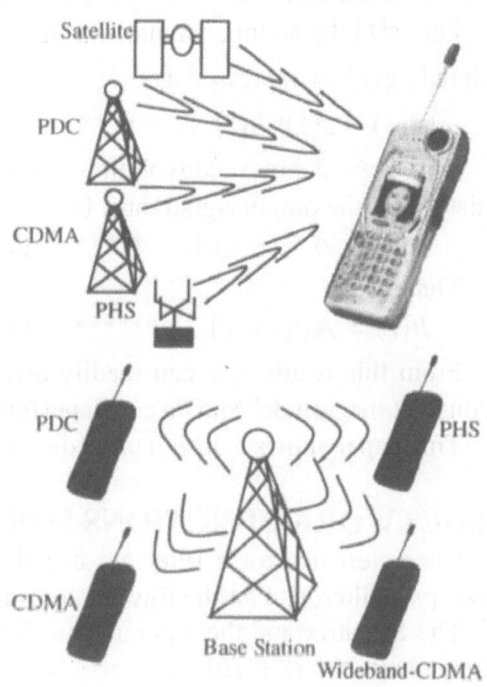

Fig.1 **Software Radio**

2.SOFTWARE RADIO

"Software Radio" is a generalized name for communication systems where a terminal or a base station is capable of operating several digital communication systems by a certain software alternation as illustrated in Fig.1 .

In the software radio scheme, we can expect especially high-speed and wide-band digital communications

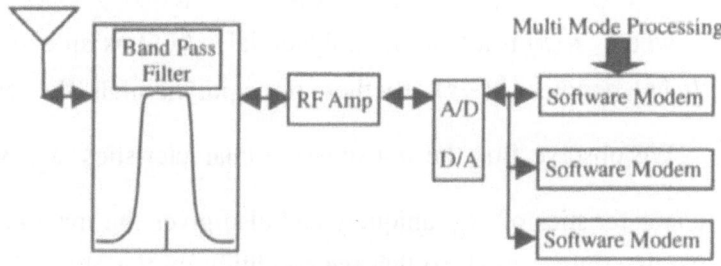

Fig.2 **Block Diagram**

which allows a variety of high speed transmission services. The key devices for that base station are a high-speed A／D converter and a low linear distortion filter as illustrated in Fig.2.

Both of them can be realized only by HTS. The filters in the base stations must keep a high channel selectivity, while they are not allowed to produce any distortion in the transmitted signals. Usually, it is very difficult to design and realize filters which satisfy these requirements simultaneously.

3.LOW LINEAR DISTORTION SYSTEM

When a signal passes through a linear system which has frequency dispersions, a signal is linearly distorted. This linear distortion, i.e. wave-form distortion, causes many undesired effects, such as degradation in B.E.R(bit error rate) in digital transmission systems. This degradation is especially critical for software radios, because the performance of digital MODEMs in those systems are affected directly by such wave-form degradation

To avoid this linear distortion, two characteristics are essential.

Let $s(t)$ be an input signal and $h(t)$ be an impulse response of a linear system, then the output signal $g(t)$ is expressed as

$$g(t) = s(t)*h(t) \quad ****** \quad (1)$$

where $*$ denotes convolution operation.. If the input signal is transmitted without linear distortion, the output signal must be

$$g(t) = A \cdot s(t - \tau) \quad ***** \quad (2)$$

That means

$$h(t) = A\delta(t - \tau) \quad ****** \quad (3)$$

From this result, one can readily derive the linear distortion-free conditions for the system, "a constant magnitude" and "a constant (flat) group delay time" in the pass band.

This paper shows a design procedure of filters which satisfy these three conditions.

4.NETWORK FUNCTIONS FOR THE FILTER

Characteristics for a filter are usually described by S-parameters. First, we derive a prototype low-pass filter, and let the filter in consideration be a reactance 2-port.

The S-matrix and the s-parameters for the filter are expressed by 3 polynomials of the complex frequency $s = \sigma + j\omega$ as

$$(S) = \begin{pmatrix} s_{11}(s) & s_{12}(s) \\ s_{21}(s) & s_{22}(s) \end{pmatrix} = \begin{pmatrix} \dfrac{h(s)}{g(s)} & \dfrac{f(s)}{g(s)} \\ \dfrac{f(s)}{g(s)} & \dfrac{-h_*(s)}{g(s)} \end{pmatrix} \quad ******* \quad (4)$$

where $g(s)$ is a Hurwitz polynomial, $f(s)$ is an even polynomial, and $h(s)$ is a polynomial. $h_*(s)$ denotes $h(-s)$. By these three polynomials, the reactance 2-port is fully described.

We observe first the transmission characteristics $s_{21}(s) = f(s)\big/g(s)$. $g(s)$ gives the phase characteristics of s_{21} uniquely and also gives the magnitude characteristics of s_{21} dominantly. So, we determine $g(s)$ so that the condition on the phase, i.e. group delay time, is satisfied, and that the magnitude approaches the prescribed values. Because $f(s)$ is an even function of s, it does not contribute to the phase characteristics, however, it can compensate the magnitude so that the remainder deviation from the end values is minimized. A pair of zeros of $f(s)$ on the imaginary axes produce a pair of transmission zeros in the stop band, and a pair of zeros on the real axes can compensate the magnitude so that the magnitude in the pass band is flattened. A pair of zeros on the real axes is realized in actual filter network in the form of a non-minimal phase shift circuit. Thus, we can determine $g(s)$ and $f(s)$ so that the linear distortion-free conditions are satisfied. The remainder polynomial $h(s)$ is uniquely determined from $g(s)$ and $f(s)$ by the unitary condition. An example of the filter characteristics is shown in Fig.3

From this prototype low-pass filter, we can derive the desired band-pass filter configuration by a frequency transform, an equivalent transform, a narrow-band approximation and some certain transform techniques.

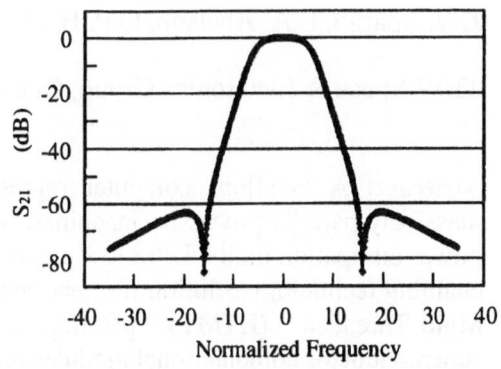

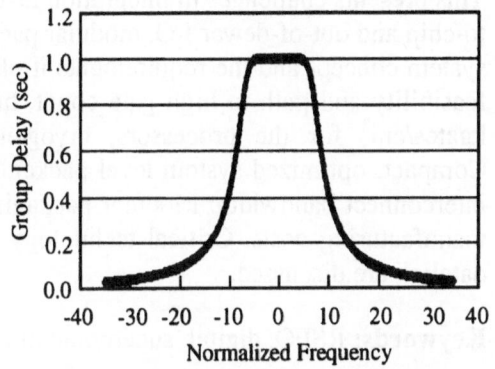

Fig.3 **Frequency Characteristics**

5.TRANSFORM TO AN ACTUAL FILTER

The prototype low-pass filter is based on 8 degree Hurwitz polynomial and a pair of transmission zeros on the real axes and the imaginary axes respectively. From this prototype low-pass filter, an equivalent band-pass filter is obtained after appropriate transforms. We see an example of such band-pass filters in Fig.4.

The filter is composed of 8 resonators coupled in cascade, and of double bridge coupling arms. Both of the bridge coupling arms corresponds to zeros of $f(s)$ on the reel axes and the imaginary axes. The configuration of this equivalent band-pass filter was so derived that the actual filter can be easily realized in the form of a planar structure. It is not our

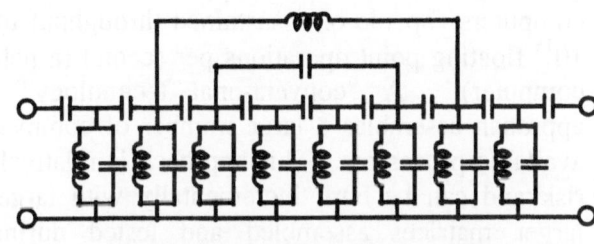

Fig.4 **Equivalent Band-pass Filter**

intention to show here an example of concrete filter patterns. However, it should be stressed here the equivalent band-pass filter is well realized in the form of an HTS planar structure.

6.CONCLUSIONS

A filter configuration which allows minimized linear distortion and enough selectivity is proposed. This filter is especially suitable for the use of software radio base stations.

Acknowledgment

This work is supported by the Japanese Ministry of International Trade and Industry.

REFERENCES

1) Matthaei. G et al., "Microwave filters impedance-matching networks and coupling structures", Artech House,1980

Superconductor Electronics for Petaflops Computing and Beyond

J.W. Spargo, L.A. Abelson, Q.P. Herr, M. Leung, G.L. Kerber, and T.S. Tighe

TRW Space & Electronics Group, One Space Park, Redondo Beach, CA, 90278, USA

Abstract: A Petaflops computer represents a thousand-fold improvement over today's largest massively-parallel-processor machines, which are susceptible to fundamental time-of-flight and power dissipation limits. Ultra-low power and ultra-high speed single-flux-quantum electronics is an enabling technology solution for near-term petaflops computing. The proposed Hybrid Technology Multi-Threaded (HTMT) petaflops-scale computer architecture includes thousands of superconductor computational modules operating at 100 GHz with an I/O throughput of 40 Petabit/s. This presents challenges in integration level of superconductor ICs, RAM size and access time, chip-to-chip and out-of-dewar I/O, modular packaging, power supply, and power dissipation. The HTMT system concept and the requirements it places on the cryogenic processing unit are described. The feasibility and path to high gate count, high clock rate SCE chips at an integration level of >100 kgates/cm^2 for the processors, cryogenic RAM, and inter-processor network are addressed. Compact, optimized system-level packaging is necessary to achieve the computational density and interconnect bandwidth. Modular packaging and automated circuit testing are required to minimize manufacturing costs. Critical technology challenges that exist for packaging, testing, and the I/O datalink are discussed.

Keywords: RSFQ, digital, supercomputing, petaflops

INTRODUCTION

Presently, there are two approaches to achieving computers capable of a sustained throughput of over 10^{15} floating point operations per second (a petaflops computer). A "conventional technology" based approach assembles a large number of commercially available processors. This approach is relatively low risk and can be built incrementally with larger and larger matrices assembled and tested during the development. Lessons learned with smaller assemblies guide the design of larger assemblies. The fastest computing engines to date, at 1-4 teraflops, have used this approach [1]. The primary problem with this approach is the difficulty in keeping processor efficiency high. As the number of lower clock rate processors increases, the number of computation problems this architecture can efficiently solve decreases. This is because latency (defined as the distance in clock ticks which separate a processor from needed data or instructions) becomes unmanageably high as processor count becomes very large. To achieve a petaflops, hundred's of thousands of processors need to be within a few thousand clock ticks of one another to achieve useful efficiencies.

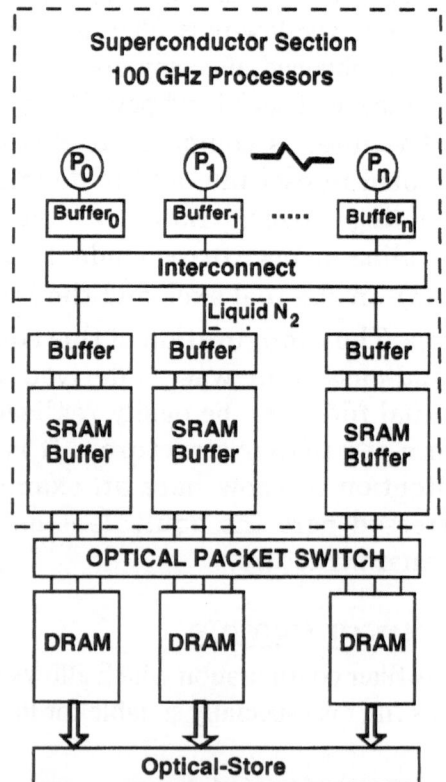

Figure 1. Hybrid Technology Multi-Threaded Architecture for Petaflops computing.

An alternative approach is enabled by tightly packed (which requires low power) processors with 50 – 100 GHz clock rates. This second approach uses a new architecture concept, multi-threaded [2] organization, to manage latency and increase processor efficiency. Only superconducting circuits, built in Nb-based fabrication technology using single flux quantum (SFQ) circuits, can simultaneously achieve the clock rates necessary to act as a petaflops processing engine for this architecture while dissipating sufficiently low power. The Hybrid-Technology Multi-Threaded Project [3] (HTMT) has been initiated to exploit Josephson digital technology (among other advanced technologies) to achieve petaflops performance over a broad class of problems. A block diagram of this innovative architecture is shown in Figure 1.

The goal of HTMT is to leapfrog conventional approaches to supercomputing in a number of areas, as is illustrated in Table 1. HTMT seeks to improve sustained performance over a broad class of applications without significant increase in power, cost, or floorspace required.

Table 1. Goals of HTMT supercomputer

	Present Capability	HTMT Goals	Improvement Factor
Performance	1 Tflops	1000 Tflops	x1000
Power	2 Mflops/W	1000 Mflops/W	x500
Cost	$250/Mflops	$0.25/Mflops	x1000
Efficiency	10%	50%	x5
Floorspace	1600 sq ft/Tflops	1 sq ft/Tflops	x1600

The performance of the superconductor-based processor is critical to the ultimate performance of the HTMT computer. While this processor is still being designed, some of the preliminary requirements are daunting, especially for a relatively young technology such as superconductor electronics (SCE). These requirements include:

- Up to 100 GHz clock rates for the processors.
- Greater than one million gates per processor element.
- Megabytes of cryogenic buffer memory (CRAM), which must be fast enough to keep the processors satisfied.
- Interconnection, I/O rates, and latencies which are commensurate with the processor speed.
- Power low enough to make cooling the processor to 4 Kelvin feasible.

Implementation of SCE processing elements (SPELL) in HTMT will require three circuit types: processor elements, cryogenic memory, and a network switch, detailed design of which has been presented elsewhere by researchers at SUNY [4]. In this paper, we will address whether it is feasible to construct a petaflops system, given the ambitious technology goals, which include: 0.8 µm Josephson junctions with J_c = 20 kA/cm^2 to achieve 100 GHz

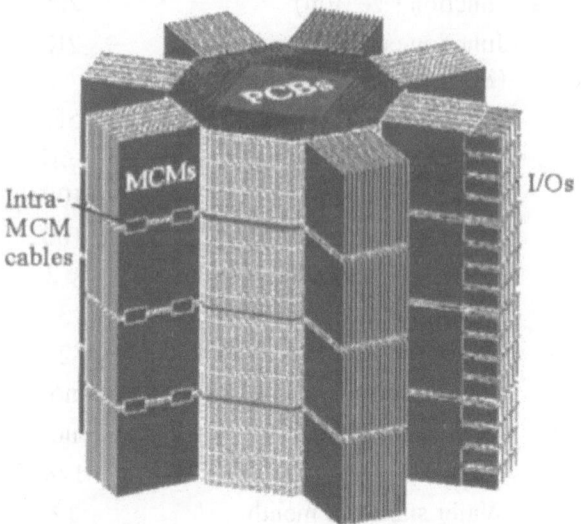

Figure. 2. Concept for HTMT superconducting processing, showing 512 MCMs connected to 160 octagonal PCBs. The MCMs, stacked four high in blocks of 16, are connected vertically with the use of short cables, and to room temperature electronics with flexible ribbon cables (I/Os).

on-chip clock speed, 300 k gates/cm^2 to achieve computational density, 1000 to 2000 pin-outs per chip, as many as 50 chips per multi-chip module (MCM) with chip-to-chip bandwidth of 30 Gbps, and up to 8M wires out of the liquid helium cryogenic region at a data rate of

8-10 Gbps. A concept for such a processing unit, comprising 4096 processors, partitioned among 512 MCMs, comprised of 37 k chips and 100 billion Josephson junctions is shown in Figure 2.

SUPERCONDUCTING INTEGRATED CIRCUIT MANUFACTURE

Scaling present day superconductor IC technology to petaflops requires more than a ten-fold increase in circuit density (to 300 k gates/cm^2), a decrease in feature size of two to three times, and additional interconnect and masking levels. An advanced IC process requires the development of a robust 0.8μm-junction technology, small-footprint resistors, planarization, and plug technology. The evolution of superconductor IC technology to achieve petaflops and beyond is shown in Table 2. Petaflops is achieved with technology corresponding to year 2004. The most dense chip will have an estimated 10 million, 0.8 μm diameter junctions and four wiring layers on a 2 cm × 2 cm die. Chip complexity exceeds the capability of existing superconductor foundries. Nevertheless, this level of integration has been available for several years in the semiconductor industry [5]. Comparison with the Semiconductor Industry Association (SIA) Roadmap suggests that 1992 semiconductor technology is comparable to that needed for petaflops. The fabrication tools, including advanced lithography, chemical mechanical planarization (CMP) and infrastructure support, are readily available [6, 7] which implies that no new technologies or tools are required to produce superconductor IC chips of equivalent density.

Table 2. IC Technology Roadmap

Year	1998	2001	2004	2007	SIA[5]1992
Minimum Feature size (μm)	1.5	0.80	0.50	0.25	0.50
Junction size (μm)	2.5	0.80	0.80	0.80	-
Junction current density (A/cm^2)	2K	20K	20K	20K	-
Gates/chip (Logic)	5K	120K	600K	2M	300K
Bits/chip (SRAM)	16K	400K	2M	6M	4M
Chip size (mm^2)	100	400	400	400	250
Wafer diameter (mm)	100	150	150	150	200
Defect density (defects /cm^2)	< 2	< 0.2	< 0.1	< 0.05	< 0.1
No. of interconnects levels	3	4-5	5-6	6	3
No. of resistor layers	2	2	2	2	-
Planarization	no	yes	yes	yes	yes
Vertical resistors	no	yes	yes	yes	-
No. of I/Os	128	2K	2-5K	2-5K	500
Wafer starts per month	12	200	1K	1K	>20K

The mature IC process as outlined in Table 2 relies extensively on metal and oxide CMP. This process has one ground plane, four wiring layers including base electrode, two resistor layers, self-aligned junction contacts, and vertical plugs for interconnection vias between wiring layers. Power lines and biasing resistors are located below the ground plane to isolate the junctions from the effect of magnetic fields and to increase circuit density. The IC process may also include an additional wiring layer, ground plane, or both, and vertical resistors that reduce the footprint of the junction shunt resistors.

Yield contributes to manufacturing cost. Circuit yield depends upon both parameter spreads and wafer fabrication yield, including number of masking levels, defect density, and chip size. Parameter spreads in superconductor circuits can cause either hard failures or soft failures in the form of increased bit-error rate (BER). Extensive data on a present day superconductor process indicates 2-4% spreads for local variations (on-chip), 5% for global variation (across-wafer) and 10% for run-to-run reproducibility [8]. As feature sizes decrease absolute critical dimension (CD) control will need to increase proportionately to maintain this level of control. Present day spreads are consistent with gate densities of 100k gates/cm^2, even while taking the low BER requirement of general computing into account [9]. The extensive modeling of defect density and its effect on die yield developed for the semiconductor industry is directly applicable to SCE. A typical estimate for present defect density in a superconductor clean foundry is 1-2 defects/cm^2. Defect density on the order of 0.1 defect/cm^2 will be needed to produce chips with yields greater than 50%.

CRYOGENIC PACKAGING AND INTEGRATION

The ultra-low power associated with SFQ logic means that a compact package may be used, enabling high computational density and interconnect bandwidth. A concept for the package that minimizes interconnect latency is shown in Fig. 2. This package occupies about 1 m^3 with a power density of 1 kW at 4 K. Chips are mounted on 512 multi-chip modules (MCM) that allow a chip-to-chip bandwidth of 30 Gbps per channel. Bisectional bandwidth into and out of the

Table 3. Packaging Technology Roadmap

Year	1998	2001	2004
MCM			
Pad size (μm)	100	75	25
Pad pitch (μm)	150	125	60
Number of pads per chip	3600	5000	12000
Pad density (cm^{-2})	1200	2000	7500
Max. no. Nb layers	5	7	9
Max. no. W layers	20	30	40
Linewidth (μm)	5	3	3
Bandwidth (Gbps/wire)	10	20	30
Chip-to-chip >30 Gbps	No	Yes	Yes
Chips per MCM	24	50	75
Backplane			
Technology	PCB	Ceramic	Ceramic/Flex
Size (cm^2)	30	30	50
Bandwidth (Gbps/wire)	2.5	5	10
I/O ribbon cables			
Line pitch (μm)	200	50	50
Cable width(cm)	2.5	5	5
Bandwidth (Gbps)	2.5	5	10
Heat load/wire (mW @ 4K)	0.4	0.2	0.1

cryostat is 32 Pbps. Each MCM has up to 50 SCE chips and eight processor units. Flip-chip bonding is used to attach the chips to the MCM providing high interconnect density (1000-2000 pinouts per chip), multi-Gbps data rates, automated assembly, and reworkability [10]. The cutoff frequency of channels between chips must be in the THz regime. The vertical MCMs are edge-mounted to 160 horizontal multi-layer printed circuit boards (PCB) which allow for MCM-to-MCM connections. Adjacent MCMs are connected to each other along their top and bottom edges with flexible ribbon cable in a stripline configuration. Such cables have been demonstrated in a cryogenic test station at data rates up to 3 Gbps. A cryogenic network (CNET) allows processor elements to communicate with minimum latency [11].

A roadmap for the packaging technology requirements, including specifications for the MCMs, PCBs, and flexible ribbon I/O cables is given in Table 3. Many techniques can be leveraged from several SCE programs currently in progress. Only marginal improvement in standard electronics packaging is required, e.g., the MCMs are expected to be commercially available. The present flex cable line density is sufficient for the MCM-to-MCM connections, but is an order of magnitude lower than required for the external I/O. The most important issues for manufacturability are minimizing parts count and system modularity. Modularity allows individual chips or boards to be easily assembled and replaced. Parts count decreases rapidly with advances in IC and packaging technology. Parts count is especially critical for the I/O data link to room temperature electronics, as discussed below.

The heat load at 4 K derives from both circuits and cabling. The cable heat load is the larger component, and could be as high as 800 W. Circuit heat load may be as small as 250 W, excluding room-temperature data link drivers. The present estimate of LHe throughput required to extract this heat (1 kW) is 1400 liter/hour; though this figure is large, such systems are in use today. A comparable system cools portions of the Tevatron accelerator at Fermilab in Illinois [12]. Systems of this size have been shown to demonstrate a refrigeration efficiency of 0.25 % [13].

CONCLUSION

Manufacture of this large-scale system will require advances in SCE integrated circuit (IC) fabrication, packaging, and high throughput, high-speed cryogenic test capabilities. Low latency, both within and between the processors, is the single most important performance issue and thus drives technology improvement on all levels. Reduced parts count dictates manufacturing complexity and cost. This is particularly critical for the data link between the cryostat and the room temperature electronics. We have examined the critical challenges in each of these areas and presented technology roadmaps showing the evolution of present technology needed to meet petaflops-scale computing using SCE in less than ten years. This is summarized in Table 4.

Table 4. System Component Challenges and Potential Approaches

Component	Challenges	Approach
SPELL chips	Clock, gate density, total current	Advanced lithography capability, integration scale
CRAM chips	Density, access time, yield	Advanced lithography capability, integration scale
Chip to MCM attach	Pin density, inductance	Solder reflow
Cryo-MCMs	Routing density, skew	Vendor fabrication
MCM-MCM attach	Density, bandwidth	Flexible ribbon cable
Data I/O	Bandwidth, number of lines	Input-Optical fiber, WDM Output-Electrical ribbon cable
Cryopackage assembly, repairability	Modularity, bandwidth, disassembly, repair	3D structure, socketed
Cooling	Power density, flow rates	Conduction cooling

There are many technical challenges in the fabrication of ICs suitable for petaflops-scale computing. SCE technology will have to improve in at least three key areas including reduction in feature size, increase in process complexity, and increase in gate density. While this level of complexity is commonplace in high-volume semiconductor fabrication plants, it represents a significant challenge, particularly in terms of yield and throughput. The leap to gate densities and feature sizes needed for petaflops cannot be achieved in one step. Sequential improvements will be needed, commensurate with the funding available for the process tools and facilities. Semiconductor technology has scaled by a factor of 0.7 in feature size and 2.5 in gate density for each generation, each of which has taken about three years and a substantial capital investment. However, much of the delay in advancing semiconductor IC technology was due to the lack of advanced process tools, especially in the area of photolithography. Now, the advanced tools and methodologies are already available.

The ultra-low power associated with SFQ logic means that a compact package may be used, which enables high computational density and interconnect bandwidth. A concept for the package that minimizes interconnection latency has been investigated. This package occupies about 1 m^3 with a power density of 1 kW at 4 K. Marginal improvement in standard electronics packaging such as the use of MCMs and flip-chip bonding die attach is required to realize this design. Modularity means that individual chips or boards may be easily repaired or replaced. Parts count decreases rapidly with advances in IC and packaging technology. Parts count is especially critical for the I/O data link to room temperature electronics.

Acknowledgments. This work was performed for the Jet Propulsion Laboratory, California Institute of Technology, and the National Aeronautics and Space Administration. Reference herein to any specific commercial product, process, or service by trade name, trademark, manufacturer, or otherwise, does not constitute or imply its endorsement by the United States Government, TRW Inc., or the Jet Propulsion Laboratory, California Institute of Technology.

1. "In pursuit of a quadrillion operations per second," T. Sterling, Insights, pp. 8-11, April 1998.
2. "Simultaneous multithreading: a platform for next generation processors," S.J. Eggers, J.S. Emer, H.M. Levy, J.L. Lo, R.L. Stamm, and D.M. Tullsen, , IEEE Micro J., vol. 17, pp. 12-19, Sept./Oct. 1997.
3. "Hybrid Technology Multithreaded Architecture," G. Gao, K.K. Likharev, P.C. Messina, and T.L. Sterling, Proceedings of Frontiers '96, pp. 98-105, October, 1996.
4. "COOL-0: Design of an RSFQ Subsystem for Petaflops Computing," M. Dorojevets, P. Bunyk, D. Zinoviev, and K. Likharev, IEEE Trans. Appl. Supercon., vol. 9, pp. 3606-3614, June 1999.
5. Semiconductor Industry Association, *National Technology Roadmap for Semiconductors, 1997 Edition*, San Jose, Calif., online at notes.scmatech.org/mcpgs/roadmap4.pdf.
6. "Fabrication technology for high-density josephson integrated circuits using mechanical polishing planarization," H. Numata, S. Nagasawa, M. Tanaka, and S. Tahara, IEEE Trans. Appl. Supercond., vol. 9, pp. 3198-3202, June 1999.
7. "Next generation Nb superconductor integrated circuit process, "L. A. Abelson, R. N. Elmadjian, and G. L. Kerber, IEEE Trans. Appl. Supercond., vol. 9, pp. 3228-3231, June 1999.
8. "LTS Josephson critical current densities for LSI applications," L. Abelson, K. Daly, N. Martinez, and A.D. Smith, IEEE Trans. Appl. Supercond., vol 5, no 2, pp. 2727-2730, June 1995.
9. "Manufacturability of Superconductor Electronics for a Petaflops-Scale Computer," L. A. Abelson, Q. P. Herr, G. L. Kerber, M. Leung, and T. Tighe, IEEE Trans. Appl. Supercond., vol. 9, pp. 3202-3207, June 1999.
10. "Robust superconducting die attach process," K. E. Yokoyama, et al., IEEE Trans. Appl. Supercond., vol. 7 , p. 2631, June 1997.
11. "CNET: Design of an RSFQ switching network for petaflops computing," D. Zinoviev, G. Sazaklis, L. Wittie, and K. Likharev, IEEE Trans. Appl. Supercond., vol. 9, pp. 4034-4039, June 1999.
12. T. Ankermann, Process Systems, Int'l, Westborough, Mass., personal communication, 1998.
13. "A 2 kW He refrigerator for SC magnet tests down to 3.3 K," F. Spath, et al., Cryogenics 32 ICEC Supplement, vol. 56, 1992.

FUNDAMENTAL RESEARCH TOWARD SFQ-INTEGRATED CIRCUITS

Akira Fujimaki, Yasutoshi Suzuki, Hiroaki Hasegawa, Futoshi Furuta, and Hisao Hayakawa

Department of Quantum Engineering, Nagoya University
Furo-cho, Chikusa-ku, Nagoya 464-8603, Japan

Abstract: We present our challenge to develop single-flux-quantum (SFQ) integrated circuits. So far, we have designed and tested several SFQ circuits to demonstrate their advantages such as high-speed operation, high throughput. Since it is difficult to detect the output signals of SFQ-DC converters over a few GHz, we prepare an on-chip testing system, in which the interface circuits between internal high-speed logics and external low-speed ones are embedded. By using the on-chip testing system, we confirm the correct operation of the shift register (SR) and that of the toggled-flip flop (T-FF) up to about 30GHz with wide bias margins. These operation frequencies are determined not by the limit of SR or T-FF but by the internal clock generator. We also describe a new logic named "SFQ with resettable latch (SFQ-RL)". The logic is suitable for a pipeline architecture that realizes high-throughput in an integrated circuit. So far, we have confirmed several circuits such as SR, M-code generator based on the SFQ-RL logic. In addition, brief comment is offered on a software radio receiver in which several advantages of using superconductor technology are found. The software radio receiver is now getting to be one of the most promising applications of superconductor electronics.

Keywords: Single-flux-quantum logic, SFQ-RL, On-chip testing, Software radio

INTRODUCTION

Single flux quantum (SFQ) logic circuits [1,2] attract attention as an alternative to the latching logic. No punch-through effect occurs at the SFQ circuits, because the circuits composed of overdamped Josephson junctions are driven by dc currents. As a result, the SFQ circuits are expected to operate up to sub-tera hertz. In addition, the return-to-zero nature of SFQ signals leads to very low power consumption.

Although the SFQ circuits have several attractive natures, there are a lot of technologies to be developed such as a fabrication process of integrated circuits, new architectures, new design tools, interface techniques for high-speed digital data access, etc. These technologies include difficult problems to be overcome. Fortunately, in Japan, several institutes including ours started to tackle the problems based on a Nb/AlOx/Nb Josephson junction technology through the project of the Science and Technology Agency of the Japanese Government. In this paper, we present our approaches to SFQ integrated circuits from various aspects.

HIGH-SPEED TESTING

To research the potential to semiconductor circuits, it is necessary to demonstrate the SFQ circuits operating above several tens of gigahertz. However, it is difficult to interchange high-frequency signals between SFQ circuits and external semicondutor circuits because of low signal level of SFQ circuits of a few hundreds of μV. Accordingly, the rate-transfer circuits should be embedded in the

SFQ circuits. On-chip testing systems [3,4] have the shift-registers (SRs) working as the rate-transfer circuits at I/Os. For the input side, the data are written in SR by providing clock pulses at a low frequency (LF), and sent to a circuit under test (CUT) by high-frequency (HF) clocks. The process is reversed for the output SRs.

Figure 1 shows the block diagrams of our on-chip testing system. To generate high frequency pulses, we adopted two ladder-type pulse generators [4] which consist of Josephson transmission lines (JTLs) with two different lengths, confluence buffers, and splitters. (See Fig. 1 (a)) One generates clock pulses of 10GHz and the other generates 25GHz. When a single SFQ pulse is provided to the ladder-type generator, the pulse is duplicated at each splitter, delayed at each JTL, and joined at each confluence buffer. As a result, a train composed of five SFQ pulses is created in each generator. The correct operation of pulse generators can be confirmed easily by SFQ/DC converter based on T-FF despite high frequency because the number of HF pulses is five that is an odd number. Increasing bias currents of JTLs, clock pulses can be generated at a higher frequency. We estimate the frequency of the pulse generators not only from the numerical simulation using JSIM[5] but also from the measurement of an average voltage of a ring oscillator which is integrated on the same chip. LF clock pulses triggered by an external signal source are generated at a DC/SFQ and joined to the outputs of other pulse generators through confluence buffers. Aa a result, we can obtain three different frequency bands in the clock pulses.

Figure 1(b) and (c) show the test systems of a SR and a T-FF, respectively. For the

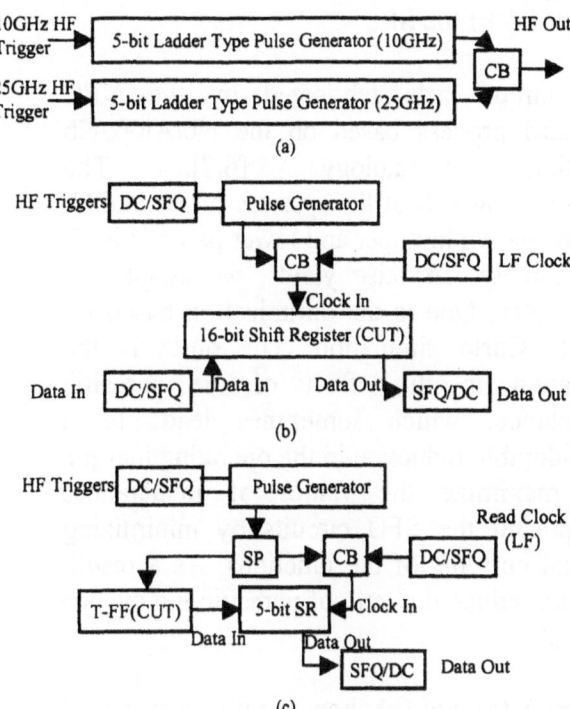

Fig. 1 (a) Pulse generator which consists of two ladder-type pulse generators. Each generates five pulses at 10GHz or 25GHz. (b) On-chip testing system for SR. (c) On-chip testing system for T-FF.

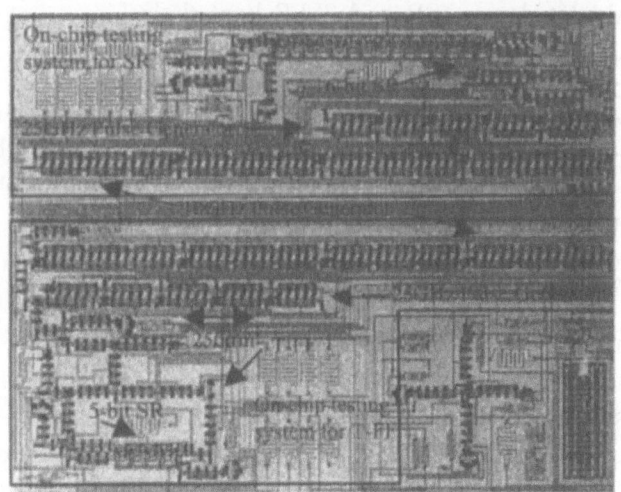

Fig. 2 Photomicrograph of the fabricated circuit. The on-chip testing systems for SR and T-FF are integrated in upper area and lower area, respectively.

high-speed testing of the SR, we jointed an input SR and an output SR to another SR of CUT. Thus, the three SRs are replaced with only one long 16-bit SR. In T-FF, the five HF clock pulses are duplicated at the splitter and sent to "clock-in" of a 5-bit output-SR and a T-FF of CUT. The output pulses from T-FF are moved to "data-in" of the SR and stored. The stored data of SR is read out by an application of five LF clock pulses to "clock-in" of the SR.

The circuit was fabricated by the NEC standard process based on the Nb/AlOx/Nb junction technology [6,7]. The photomicrograph of the systems of SR and T-FF are shown in upper and lower part of Fig. 2, respectively. To raise yields, we adopt two techniques. One is the optimization based on Monte Carlo simulation. The other is the reduction in the effect of the parasitic inductance, which sometimes leads to a considerable reduction in the operating margin. We maximize the values of inductances composing the SFQ circuits by minimizing critical currents of the junctions. As a result, we can reduce the ratio of parasitic inductance to total one.

Figure 3 (a) and (b) show the bias margins of SR and T-FF, respectively. For SR, the wide margin of ±39% was obtained experimentally and agreed well with the result of the numerical simulation. Also, relatively wide margin ±20% was obtained for T-FF, though the obtained margin was smaller than the calculated one. SR and T-FF kept their bias margins up to about 30GHz which was the upper frequency limit of the HF clock generator. This means that SR and T-FF can operate at higher frequencies.

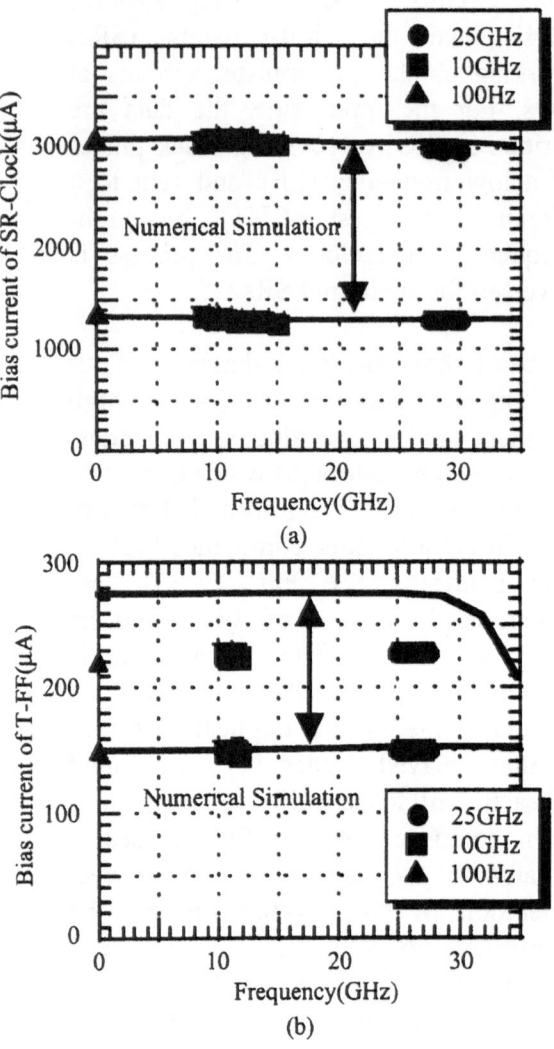

Fig. 3   Measured bias margins of (a) SR-clock and (b) T-FF.   Those obtained from numerical simulation are also plotted.

PROPOSAL OF NEW LOGIC CIRCUITS

In most SFQ circuits, the motion of an SFQ (Data) is controlled by another SFQ (Clock). To realize the coincidence of the two SFQs, the SFQ gates usually have a storage loop which can be regarded as a memory. Thus, the SFQ logic can be classified into the state machine which is composed of combinatorial logic circuits and memories. In general, the state machine should be initialized by request. Nevertheless, conventional SFQ circuits have not paid attention to such "initialization" or "reset" function. Therefore, we have proposed new logic circuits in which the initialization is adopted positively [8]. The circuits are referred to as Single Flux Quantum with Resettable Latch (SFQ-RL). The SFQ-RL logic consists of three primitive gates, "Latch-gate (L-gate)", "Copy-gate (C-gate)", and "Or-gate (O-gate)". The C-gate and O-gate exactly correspond to Splitter and Confluence Buffer in RSFQ family, respectively. The detail description concerning the equivalent circuits, margins, composition of circuits, etc. are presented in Ref.[8].

The function of the initialization or reset is strongly required for some pipeline structure embedded in CPU. If control hazard or branch hazard occur at the pipeline, the instructions stored in the pipeline should be flushed. A maximum (M)-code generator is another good example in which the initialization is an essential function. M-code generator is composed of a SR and exclusive-or (XOR) gate. The generator is a pseudo-random-bit generator and generates random codes depending on the initial data stored in the SR. Thus, the initialization is indispensable for obtaining a true code. However, the initialization cannot be achieved in a conventional SFQ logic because the generator includes a feedback loop. Once wrong data are stored in SR, the code never recovers.

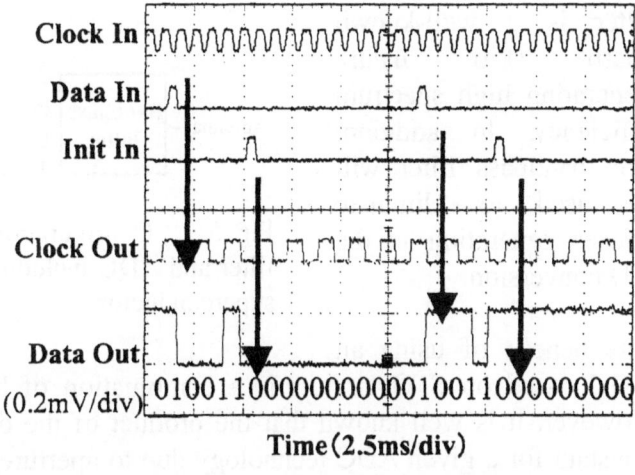

Fig. 4   Operation of the M-code generator based on the SFQ-RL logic. The data ware flushed by the "Init In" signal.

A fabricated M-code generator is composed of 5 L-gates, and both SR and XOR are made of L-gates. The length of this code is 15-bit and the correct sequence is "100110101111000". Figure 4 indicates the experimental results of the operation. The input of "Data-in" pulse triggers off the operation start. In this case, we sent the "initialization" pulse to the whole circuit after 5 clocks. The correct sequence "10011" was obtained, then the circuit was suspended. This means that the function of "initialization" is surely performed by using the SFQ-RL logic. Unfortunately, a full-code operation has not been demonstrated. The reason is still unclear and further study will be required.

To raise the integration level in SFQ circuits, we have developed an optimizer which is based on Monte Carlo method and maximize the critical margin. We also plan to develop the CAD system for SFQ circuits.

PROPOSAL OF A POSSIBLE APPLICATION

The present technology of fabricating Josephson integrated circuits (ICs) is still much immature compared to semiconductor one. Furthermore, Josephson ICs require cryocoolers or liquid-He, which cost much. These realities make circumstances surrounding Josephson technology more severe. To alleviate this situation, specific and near-term applications of SFQ circuits should be proposed and demonstrated.

One of the potential applications is the front-end hardware of a software radio receiver, which is defined as a radio receiver that has a function to be redefined in software. The Software-Defined-Radio Forum forecasts that the software radio technique will be spread quickly around the world as well as the personal computers. Our group proposed the advanced base-station in 1998 [9-10]. This is referred to as "Intelligent Super Base-Station (ISB)", and has the capability to unify the existing base-stations and future ones. The software radio technique and superconducting devices are essential for the ISB system. The basic idea of ISB is mentioned below. The similar idea that the superconducting digital electronics applies to a software radio receiver has been proposed independently by the group of SUNY at Stony Brook and Ericsson [11].

Figure 5 illustrates a configuration of a software radio receiver that is a basic component of the ISB system. A bandpass filter and analog-to-digital converter (ADC) are made of superconductors. The benefit of the bandpass filter is a well-known sharp skirt nature generating high spectrum efficiency. In addition, The bandpass filter will be used to eliminate aliases generating in the AD conversion.

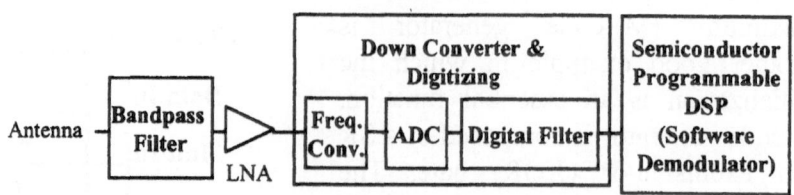

Fig. 5   Configuration of a software radio receiver. The bandpass filter and ADC including the digital (decimation) filter are made of superconductor.

The benefit of using an ADC based on SFQ circuits is a combination of high dynamic range and broad band width. However, it is well known that the product of the dynamic range and band width is considered constant for a given ADC technology due to aperture jitter [12]. At present, semiconductor ADCs seem to confront an rms aperture jitter barrier of ~0.5ps. Thus, the effective number of bit of a frontier semiconductor ADC with input band width of 25MHz is at most 13bit, which is insufficient for a software radio base-station. On the other hand, oversampling ADCs based on SFQ circuits tolerate aperture jitter [13]. High-speed nature of the SFQ circuits leads to higher oversampling ratio, resulting in higher tolerance. Furthermore, the continuous integration of the input signal on the SFQ-based ADCs proposed so far [14-16] brings additional tolerance to aperture jitter. The SFQ-based ADCs are expected to have high dynamic range and broad band width unreachable in semiconductor ADCs.

Another advantage of SFQ-based ADCs is high sensitivity. The minimum power of input signals required to drive SFQ circuits is less than $1\mu W$, while that of high-speed semiconductor ADCs is around 1mW. The nature of high sensitivity brings the reduction in the gain of a low-noise amplifier (LNA) by 30dB. Usually, the product of the gain and bandwidth is considered constant for a given amplifier. Thus, the reduction in the gain enhances the band width covered by the LNA, widening the frequency band.

We have already tackled the development of the superconducting component of the ISB. We have examined the lowpass and bandpass sigma-delta ADCs including decimation filters by means of the numerical calculations with JSIM and FFT program. The present subject is how to realize a second- or higher order ADCs to enhance the dynamic range together with broader band width.

SUMMARY

We present our approach to SFQ ICs from different aspects. We have designed and tested the SR and T-FF circuit to confirm the high-speed nature of the SFQ circuits using the on-chip test system. The circuits operate correctly up to about 30GHz with reasonable bias margins. We also tested a "flush" operation on an M-code generator based on the SFQ-RL logic. In this logic, the initialization of the whole circuit plays a positive role in the operation. The logic is thought to be suitable for a pipeline architecture that realizes high-throughput in an integrated circuit. We also describe the benefits of a superconductor technology to construction of a software radio receiver which is a promising applications of superconductor electronics.

Acknowledgment
This work is partly supported by "Research on innovative information processing using single flux quantum" under the project "Special coordination funds for promoting science and technology" and by Project for Rising Researchers from Graduate School of Engineering, Nagoya University (FY1997). The author (A. F.) would like to thank Profs. Katayama and Ogawa for their fruitful discussion.

References

1. K. K. Likharev and V. K. Semenov, IEEE Trans. Appl. Supercond., **1**, 3-28 (1991)
2. K. Nakajima, H. Mizusawa, H. Sugahara, and Y. Sawada, IEEE Trans. Appl. Supercond., **1**, 29-36 (1991)
3. Alex F. Kirichenko and Oleg A. Mukhanov, IEEE Trans. Appl. Supercond., **7**, 3438-3441 (1997)
4. Zhong J. Deng, Nobuyiki Yoshikawa, Stephen R. Whitely and Theodore Van Duzer, IEEE Trans. Appl. Supercond., **7**, 3634-3637 (1997)
5. E. S. Fang and T. Van Duzer, Extended Abstracts of ISEC'89, Tokyo, 407-410 (1989)
6. H. Numata, S. Nagasawa and S. Tahara, Extended Abstracts of ISEC'93 (Boulder) 280-281 (1993)
7. S. Nagasawa, Y. Hashimoto, H. Numata and S. Tahara, IEEE Trans. Appl. Supercond., **5**, 2447-2452 (1995)
8. F. Furuta, Y. Suzuki, E. Oya, S. Matsumoto, H. Akaike, H. Hayakawa, and Y. Takai, IEEE Trans. Applied Supercond., **9**, 3553-3556 (1999)
9. A. Fujimaki, M. Katayama, and H. Akaike, Report of Project for Rising Reserchers, *edited by Graduate School of Engineering, NagoyaUniv.* August 1998. (in Japanese)
10. A. Fujimaki, and M. Katayama, *Technical report of IEICE*, **CAS-98**, 85-92 (1999) (in Japanese)
11. E. B. Wikborg, V. K. Semenov, and K. K. Likharev, IEEE Trans. Applied Supercond., **9**, 3615-3618 (1999)
12. R. H. Walden, IEEE Commun. Mag., 96-101, Feb. (1999)
13. J. X. Przybysz, D. L. Miller, and E. H. Naviasky, IEEE Trans. Appl. Supercond., **5**, 2248-2251, (1995)
14. V. K. Semenov, Y. A. Polyakov, and T. A. Filippov, IEEE Trans. Applied Supercond., **9**, 3026-3029 (1999)
15. S. V. Rylov, D. K. Brock, D. V. Gaidarenko, A. F. Kirichenko, J. M. Vogt, and V. K. Semenov, IEEE Trans. Applied Supercond., **9**, 3016-3019 (1999)
16. A. H. Worsham, D. L. Miller, P. D. Dresselhaus, and J. X. Przybysz, IEEE Trans. Applied Supercond., **9**, 3157-3160 (1999)

Superconducting Sigma-Delta A/D Converters with a Large Dynamic Range

Tatsunori Hashimoto, Haruhiro Hasegawa, Shuichi Nagasawa, Hideo Suzuki, Kazunori Miyahara, and Youichi Enomoto

Superconductivity Research Laboratory, ISTEC, 2-1200 Musaigakuendai, Inzai 270-1382, Japan

Abstract: We have been developing a superconducting sigma-delta analog to digital converter (ADC) with a large dynamic range for software-radio applications. For an ADC with a larger dynamic range, we propose a new double-loop sigma-delta modulator utilizing single flux quantum (SFQ) pair generation. In circuit simulation, a dynamic range of 72 dB was obtained at 100 MHz for a 12 GHz clock, where the modulator was designed for the NEC Niobium standard process. Preliminary operation of a sigma-delta modulator has also been tested experimentally for a single-loop modulator to confirm our modulator design.

Keywords: A/D converter, sigma-delta modulator, software radio

INTRODUCTION

Software-radio applications utilizing superconducting analog to digital converters (ADCs) have been proposed [1], [2]. In the application, dynamic range and input frequency range of the ADC are important. Superconducting sigma-delta modulators have been studied in order to obtain a large dynamic range [3]-[5]. Several sigma-delta modulators with single- and double-loop feedback were proposed by researchers of Northrop Grumman and some of them have been tested. It was pointed out that one of the keys to obtain large dynamic range is a larger gain [4], [5].

In this paper, experimental results for a single-loop modulator are reported. Next, simulation results are shown for a new double-loop modulator designed on the basis of the single-loop modulator design, where generators of multi-flux-quanta (MFQ) are adopted to enable a large gain of 20 for a 12 GHz clock.

SINGLE-LOOP SIGMA-DELTA MODULATOR

Prior to studying double-loop modulators, we evaluated a Northrop Grumman type single-loop modulator [3] fabricated by the NEC Nb standard process [6]. Circuit parameters of the fabricated modulator are designed as follows. The integrator inductance and resistance are 80 pH and 8 mΩ, respectively, and the critical current of the comparator junction is 0.25 mA.

Figure 1 shows the measurement setup

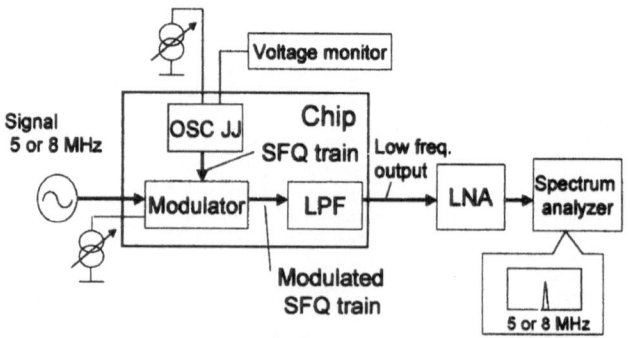

Fig. 1. Setup for testing a superconducting single-loop sigma-delta modulator. The OSC junction generates a sampling clock.

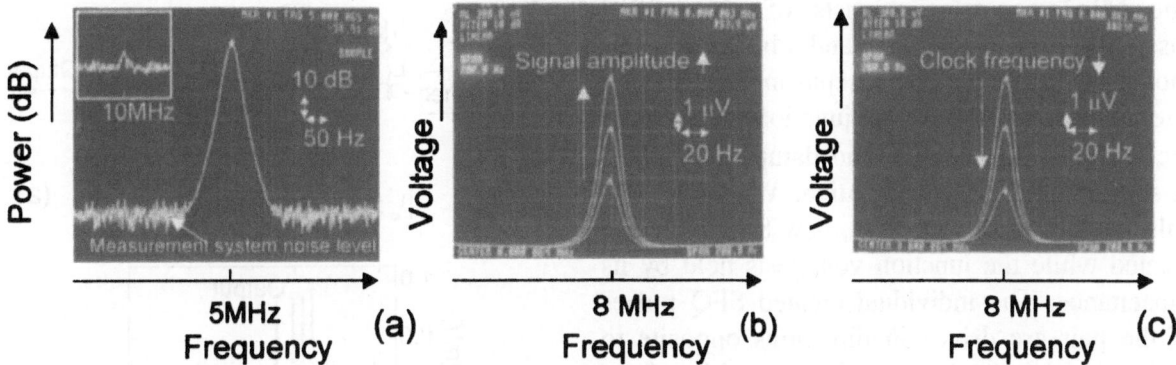

Fig. 2. Observed output spectra of the modulator. (a) Output spectrum for the input amplitude resulting in the second harmonic peak appearing as shown in the inset. (b) Input amplitude dependence. (c) Clock frequency dependence.

schematically. The modulator was operated up to a sampling clock frequency of 45 GHz for a 5 or 8 MHz sine wave signal, where the sampling clock was generated by the current-biased junction as a single flux quantum (SFQ) pulse train generator. The modulator output, which is a modulated SFQ pulse train, was averaged by the low-pass filter, and the peak corresponding to the signal was observed by the spectrum analyzer.

Figure 2 shows the observed output spectra of the modulator. As shown in Fig. 2 (a), the peak was clearly observed. Besides, the peak reasonably depended on the input amplitude and the clock frequency monitored by measuring the voltage of the clock generator junction. Consequently, fundamental modulator operation was confirmed.

DOUBLE-LOOP SIGMA-DELTA MODULATOR

To obtain a larger dynamic range, we have been studying double-loop sigma-delta modulator using L/R integrators. As researchers of Northrop Grumman have reported [3], [4], large gain for feedback to the first integrator is necessary to compensate signal attenuation in the second L/R integrator. We propose an integer-feedback-type modulator using underdamped Josephson junctions as MFQ generators.

Figure 3 shows a block diagram of the double-loop sigma-delta modulator. In this modulator, the gain is obtained by MFQ generator circuits and a splitter transferring an input SFQ pulse to six output terminals.

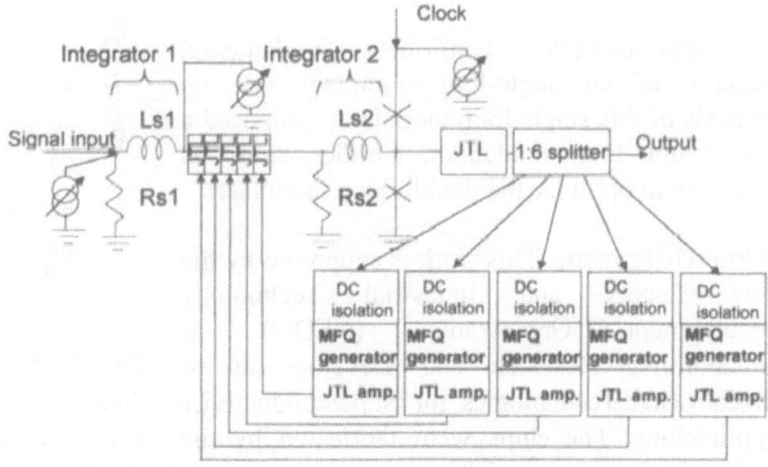

Fig. 3. Block diagram of the proposed double-loop sigma-delta modulator. The feedback to the integrator 1 is performed through the Josephson transmission lines (JTLs) including large-Ic junctions after current-amplification at the JTL amplifiers. The dc isolation circuit is a mutual-coupling JTLs including a JTL amplifier.

The MFQ generator consists of a modified Josephson transmission line and a buffer stage as shown in Fig. 4 (a). The Josephson transmission line includes one Josephson junction with a large McCumber parameter βc and damping resistors to stabilize the circuit operation. When an SFQ pulse arrives at this junction, new SFQ pairs are created while the junction voltage is held by its capacitance. The individual created SFQ pulses of the pair are driven in directions opposite to each other by the Lorentz force resulting from the bias current flow. The buffer is necessary to eliminate the SFQ pulses moving toward the input terminal. Thus, one SFQ pulse is transformed to several SFQ pulses as shown in Fig. 4 (b).

The combination of the 1:6 splitter connected to the 5 MFQ-generators and the $4\Phi_0$ MFQ generators enables a large gain of 20. By using JSIM [7], it was found that the feedback to the first integrator was completed within 70 ps. This value corresponds to around 14 GHz. Figure 5 shows the simulated power spectrum for a 12 GHz clock. The obtained dynamic range was 72 dB at 100 MHz .

CONCLUSIONS

We experimentally confirmed fundamental operation of our single-loop modulator, and, on the basis of this single-loop modulator, proposed a new double-loop modulator. Further study is necessary to optimize the double-loop modulator.

Acknowledgments. This work is supported by the New Energy and Industrial Technology Development Organization (NEDO) as Collaborative Research and Development of Fundamental Technologies for Superconductivity Applications. The chips were fabricated by the Nb standard process in NEC.

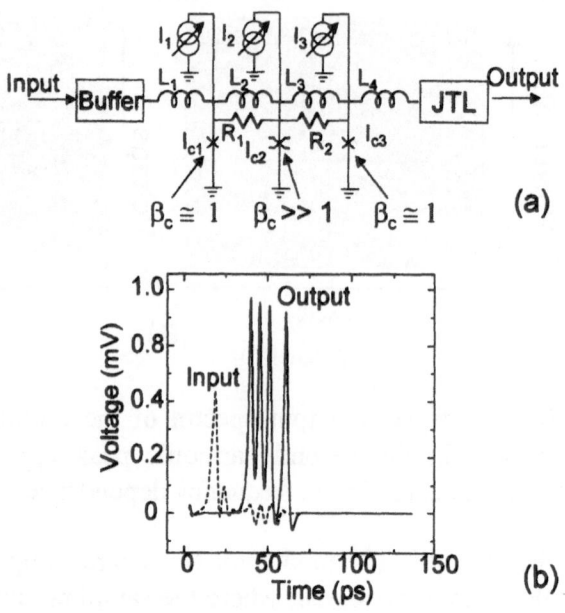

(a)

(b)

Fig. 4. MFQ generator. (a) Circuit. (b) Simulation results. The used circuit parameters for $4\Phi_0$-output case are L_1=4 pH, L_2=10 pH, L_3=8.7 pH, L_4 =2.28 pH, I_{c1}=0.28 mA, I_{c2}=0.63 mA, I_{c3}=0.17 mA, R_1=1.1 Ω, R_2=3.5 Ω, I_1=0.176 mA, I_2=0.45 mA, and I_3=0.2 mA.

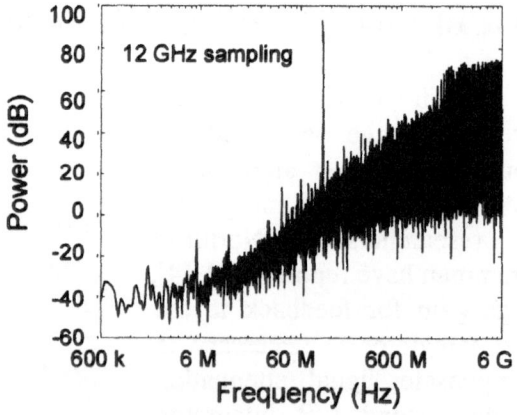

Fig. 5. Simulated power spectrum for the double-loop modulator with a gain of 20. The used parameters are L_1= L_2= 100 pH, and R_1= R_2= 8 mΩ.

1. A. Fujimaki et al., Extended Abstracts of ISEC'99, 39-41(1999).
2. E.B. Wikborg et al., IEEE Trans. Appl. Supercond. **9**, 3615-3618(1999).
3. J.X. Przybsz et al., IEEE Trans. Appl. Supercond. **3**, 2732-2735(1993).
4. A.H. Worsham, et al., IEEE Trans Appl. Supercond. **9**, 3157-3160(1999).
5. D.L. Miller et al., IEEE Trans. Appl. Supercond. **9**, 4026-4029(1999).
6. S. Nagasawa et al., IEEE Trans. Appl. Supercond. **5**, 2447-2452(1995).
7. It was written by E. Fang.

Superconducting SFQ-NOR Decoder

Shuichi Nagasawa, Haruhiro Hasegawa, Tatsunori Hashimoto, Hideo Suzuki, Kazunori Miyahara, and Youichi Enomoto

Superconductivity Research Laboratory, ISTEC, 2-1200 Musaigakuendai, Inzai 270-1382, Japan

Abstract: An SFQ-NOR decoder has developed as a main component of the 16-Kbit superconducting latch/SFQ hybrid RAM. The core circuit of the decoder is 16 SFQ-NOR circuits and a clock-distribution circuit. The SFQ-NOR circuit consists of a multi-input-OR gate and an RSFQ inverter. The OR gate is composed of serially connected SQUIDs, each including under-damping junctions. To enable operation of the OR gate with dc power, we designed it to function with a self-resetting mode. The SFQ-NOR decoder chips were fabricated using NEC's standard Nb process. The SFQ-NOR decoder functioned successfully with a large bias margin of more than ± 24%.

Keywords: Decoder, NOR, Single flux quantum device, Superconducting random access memory

INTRODUCTION

We have been developing a 16-Kbit superconducting latching/SFQ hybrid (Slash) RAM that enables high-frequency clock operation up to 10 GHz [1]. A decoder is a main component of the Slash RAM, and this decoder must meet the following requirements. First, it has to be composed mostly of circuits with dc-powered single-flux-quantum (SFQ) devices to suppress the ac-power dissipation that is a problem in conventional latching devices [2]. Second, it needs a parallel processing and pipeline architecture to enable decoder operation at a clock frequency of 10 GHz. Third, the circuit structure must match the size of the memory-cell array to enable high-density integration. To satisfy these requirements, we have designed a NOR decoder basically composed of SFQ devices.

DESIGN

The multi-input SFQ-NOR circuit, shown in Fig. 1, is a basic element of the NOR decoder. The multi-input SFQ-NOR circuit consists of a multi-input OR gate and a rapid-single-flux-quantum (RSFQ) inverter. The OR gate is composed of serially connected superconducting-quantum-interference devices (SQUIDs), each including under-damping junctions (J_{s1} and J_{s2}). The inverter circuit is almost the same as a previously reported one [3], but the circuit parameters were adjusted to obtain a larger operating margin in this NOR circuit. To enable operation of the OR gate with dc power, we designed it to function with a self-resetting

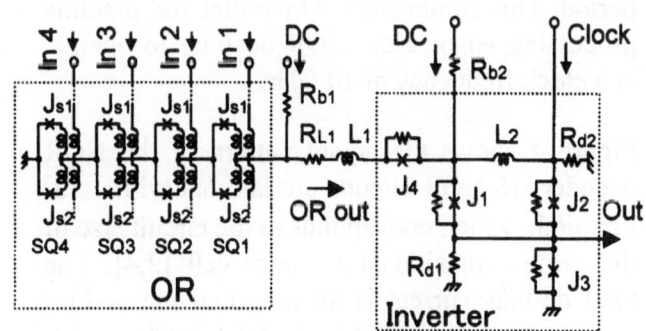

Fig. 1. Multi-input SFQ-NOR circuit. $J_{s1}= J_{s2}= 0.125$ mA, $J_1= 0.28$ mA, $J_2= 0.23$ mA, $J_3=0.3$ mA, $J_4=0.18$ mA, $L_1=10$ pH, $L_2=9.5$ pH, $R_{L1}= 1$ Ω, $R_{b1}= 57.5$ Ω, $R_{b2}= 11.5$ Ω, $R_{d1}= R_{d2}= 0.5$ Ω,

mode where the mode is determined by the value of resister RL1. As a result, the OR gate outputs a multi-flux quantum (MFQ) signal. However, the MFQ signal is converted to an SFQ signal in the inverter, because the inverter circuit ignores subsequent pulses after the arrival of the initial SFQ pulse in the MFQ pulse. An output SFQ signal is then generated on the arrival of the clock signal.

Figure 2 shows simulated waveforms of the multi-input SFQ-NOR circuit at a clock frequency of 10 GHz. In this figure, "In 1, In 2, In 3, and In 4" are input signals. "OR out" is an output signal from the OR gate. "Clock" is a clock signal for the inverter. "Out" is an output signal from the inverter. When more than one input signal are applied, the OR gate switches into the voltage state, and generates an MFQ output signal, shown as "OR out". The inverter correctly generates an SFQ signal on the arrival of the clock signal, as shown in "Out", only when there is no MFQ signal in the previous clock period. This correct NOR function was obtained with parameter margins of more than ± 30%.

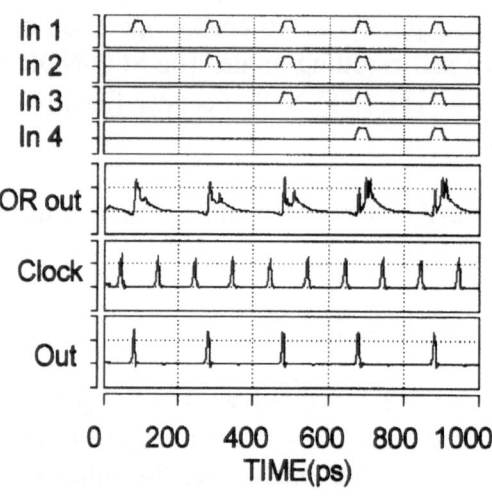

Fig. 2. Simulated waveforms of a multi-input SFQ-NOR circuit at a 10-GHz clock frequency.

Figure 3 shows half of the equivalent circuit of the NOR decoder, which outputs 16-bit decoded signals from the input of 4-bit address signals. The 16 multi-input SFQ-NOR circuits are laid out in parallel. The address signals are generated by the latching drivers, and are applied magnetically to the multi-input OR gates. A clock signal is injected simultaneously to the 16 inverters by the clock-distribution circuit. This circuit has a binary-tree structure that consists of splitters and Josephson transmission lines (JTLs). The 16-bit decoded signals are generated at the output terminals (Out 1 to Out 16) after the clock signal is injected to the inverters. Therefore, 16 NOR functions are carried out at the same time in parallel. The data stored in the inverter during a clock period are read out in the next clock period. This combination of parallel and pipeline processing enables the NOR decoder to operate at a clock frequency of 10 GHz.

Figure 4 shows the layout pattern of the NOR decoder. 16 NOR circuits are laid out with a 50-μm pitch, which corresponds to the circuit size of the vortex transitional memory cell [2,4]. The total dc-bias current is 40 mA, and the ac-bias current, which is biased to the latching drivers, is 4.5 mA. This ac-bias current is only about 1/10 that of our previous latching decoder [2]. This layout pattern was designed under the design rule of NEC's standard Nb process [4], where the critical current density of the junction is 2.5 kA/cm^2, minimum line width is 1.5 μm, and sheet resistance is 1.2 Ω/□.

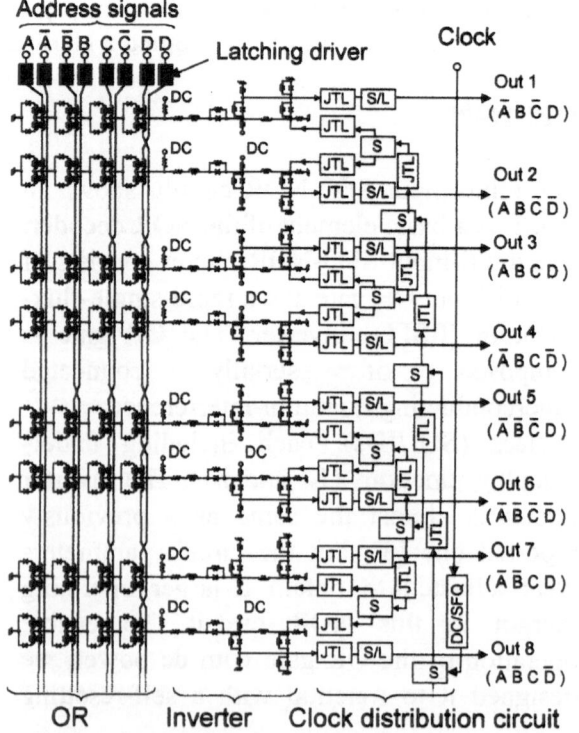

Fig. 3. Half of the equivalent circuit of the NOR decoder. S: splitter, S/L: SFQ/latch converter.

MEASUREMENTS

The decoder chips were fabricated using NEC's standard Nb process [4]. Figure 5 shows the measurement results for the SFQ-NOR decoder at a low frequency of 20 kHz. Complements for the address signals (A, B, C, and D), which are not shown in Fig. 5, were also injected. This NOR decoder has 16 outputs, but we monitored only 6 of them because test chip had only six pads that could be used for the measurements. As shown in Fig. 5, correct operation of the decoder was obtained. Table 1 shows the measured bias margins for the decoder. The NOR decoder functioned correctly with a large bias margin of more than ± 24%.

Table 1. Measured bias margins for the NOR decoder

Bias	min. ~ max.	margins
SQUID_bias	2.6 ~ 4.3 mA	±25%
NOT_bias	2.8 ~ 5.3 mA	±31%
DC/SFQ_bias	0.5 ~ 0.98 mA	±32%
Splitter_bias	14 ~ 23 mA	±24%
JTL1_bias	6.5 ~ 14 mA	±37%
JTL2_bias	1.9 ~ 7.0 mA	±57%
AC1_bias	3.5 ~ 6.1 mA	±27%
AC2_bias	3.2 ~ 5.4 mA	±26%
Clock	1.0 ~ 1.7 mA	±26%

CONCLUSION

We have designed a new NOR decoder composed of SFQ devices. The NOR decoder chips were fabricated by NEC's standard Nb process. We confirmed through the measurements that the NOR decoder functioned successfully with a large bias margin of more than ± 24%.

Acknowledgments. This work was supported by Special Coordination Funds (Research on Innovative Information Processing using Single Flux Quantum) of the Science and Technology Agency of the Japanese Government.

1. S. Nagasawa et al., ISEC'99 Extended Abstracts, PII1.16, pp. 365-367, 1999.
2. S. Nagasawa et al., IEEE Trans. Appl. Supercond., vol. 9, no. 2, pp. 3708-3713, June 1999.
3. S. Polonsky et al., IEEE Trans. Appl. Supercond., vol. 3, no. 1, pp.2566-2577, Mar. 1993.
4. S. Nagasawa et al., IEEE Trans. Appl. Supercond., vol. 5, no. 2, pp. 2447-2452, June 1995.

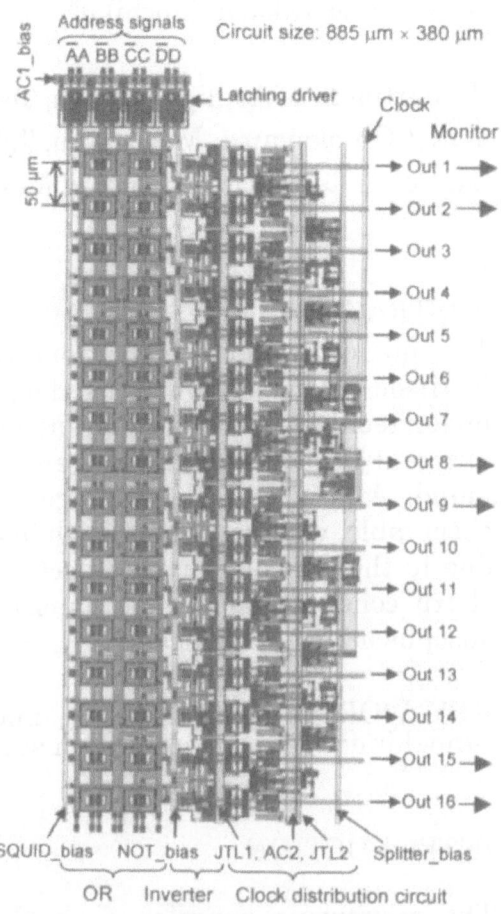

Fig. 4. Layout pattern of the NOR decoder. Junctions: 489, AC: 4.5 mA, DC: 40 mA

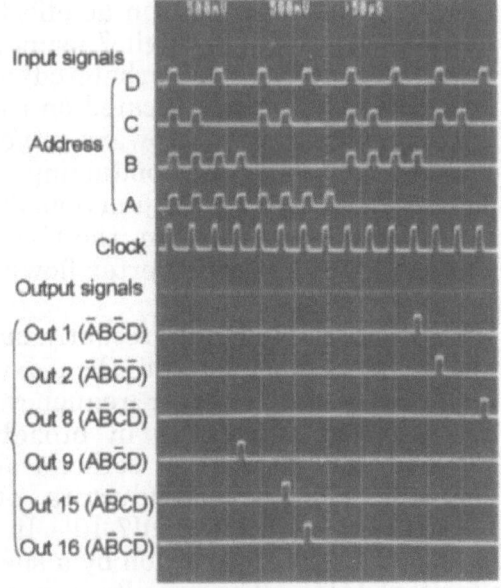

Fig. 5. Measurement results for the NOR decoder at a low frequency of 20 kHz.

Microwave Emission from Quasiparticle-injected Intrinsic Junctions

Ienari Iguchi[1,2], Kiejin Lee[2,*], Wan Wang[1], Eiji Kume[1], Masashi Tachiki[3], Kazuto Hirata[3] and Takashi Mochiku[3]

[1]Department of Applied Physics, Tokyo Institute of Technology and [2]CREST-JST, 2-12-1 Oh-okayama, Meguro-ku, Tokyo 152-8551, Japan
[3]National Research Institute of Metals, 1-2-1 Sengen, Tsukuba, Japan

ABSTRACT

The microwave emissions from the quasiparticle-injected $Bi_2Sr_2CaCu_2O_y$ (BSCCO) intrinsic Josephson junctions along the c-axis are reported. Using a superheterodyne mixer techinque, three different modes of microwave emissions corresponding to the different branches of the hysteretic $I - V$ characteristics of the intrinsic junction are found. At the low bias voltage range, the conventional Josephson self-emission is observable, whereas, at the high bias voltage range, the two novel microwave emissions due to the nonequilibrium effect are recorded; incoherent broadband emissions and a sharp coherent microwave emission. We interpret these two emissions in terms of Josephson plasma emission.

KEY WORDS: $Bi_2Sr_2CaCu_2O_y$, microwave emission, quasiparticle injection, high-Tc tunnel junction, nonequilibrium state

INTRODUCTION

The microwave emission from the low-T_c and high-T_c superconductors is a very attractive subject from both the basic study of superconducting physics and applications. It is well known that the Josephson junction biased at a finite voltage V produces microwave self-emission of frequency f satisfying the voltage-frequency relation ($2eV=nf$) due to Josephson ac effect. The Josephson self-emission has been observed for both low-T_c and high-T_c superconductors[1-3]. The power of order of lpW is detectable using a superheterodyne mixer technique. When it is detected by the Josephson junction fabricated on the same chip, the power of 8nW was detected[4]. As another means of microwave generation, there is a method utilizing the flux flow phenomenon in a superconducting Josephson junction. The flux flow oscillator has been developed for low-T_c Nb junctions[5]. With the same geometry applying to the high-T_c BSCCO intrinsic junction, the German group claimed the microwave emission due to Cherenkov radiation of vortex flow[6].

On the other hand, quite recently, our group found a novel emission of radiation from a quasiparticle injection into the c-axis of a high-T_c superconductor by a superheterodyne mixer technique (receiver frequency: 1.7GHz – 47GHz) using a tunnel-junction sample. The emission appeared in broadband spectra and was quite different from the conventional Josephson self-emission. The phenomena have been observed for cuprate oxide superconductors of $YBa_2Cu_3O_{7-y}$(YBCO), $ErBa_2Cu_3O_{7-y}$(EBCO) and $Bi_2Sr_2CaCu_2O_y$ (BSCCO)[7-10]. It has been recently confirmed that the emission extended to the THz region by a spectroscopic measurement using a high-T_c Josephson junction [11]. The results are consistently interpreted by the recent theory of

1096

Josephson plasma emission from a quasiparticle-injected nonequilibrium superconductor by Shafranjuk and Tachiki[12]. We point out that the presence of Josephson plasma has been so far demonstrated only by the resonant absorption method of microwaves by varying an external magnetic field [13,14].

In this presentation, we report a novel discovery of coherent-like radiation from a quasiparticle-injected BSCCO intrinsic junction, in addition to the nonequilibrium broadband emission[15]. The observed results are again quite different from Josephson self-emission and are also interpreted by the concept of Josephson plasma emission consistently [12]. The discovery of coherent-like emission suggests the possibility of Josephson plasma laser in the future.

EXPERIMENTAL

The BSCCO intrinsic-junction samples were prepared in the following way. The BSCCO single crystals of typical size 4mm×8mm× (50-100) μ m were grown by a traveling solvent floating zone technique. To fabricate the samples, first, a 50 nm thick Au film was deposited on the cleaved surface of the crystal by vacuum evaporation. Next, the crystal was annealed under various conditions (i.e. 400-600 ℃ for 1 to 20 hrs). Thenafter, the mesa structure was formed by photolithography and ion-milling processes. The mesa area was 50 ×50 μ m^2 and the mesa height was 30-50nm.

Fig. 1 shows the fabricated sample configuration. The quasiparticles are injected into the c-axis direction of the BSCCO single crystal. To observe the microwave emission in this system, it is necessary to fabricate the sample with low tunnel resistance so that a large amount of quasiparticles might be injected at the gap edge voltages. This could be done by controlling the annealing condition. In fact, we obtained the samples with the current density of order of 10^3A/cm^2 at the gap edge voltage, which exhibited negative resistance behavior due to intense self-injection of quasiparticles. Fig.2 and Fig.3 show the examples of the resistivity vs. temperature curves and the corresponding current-

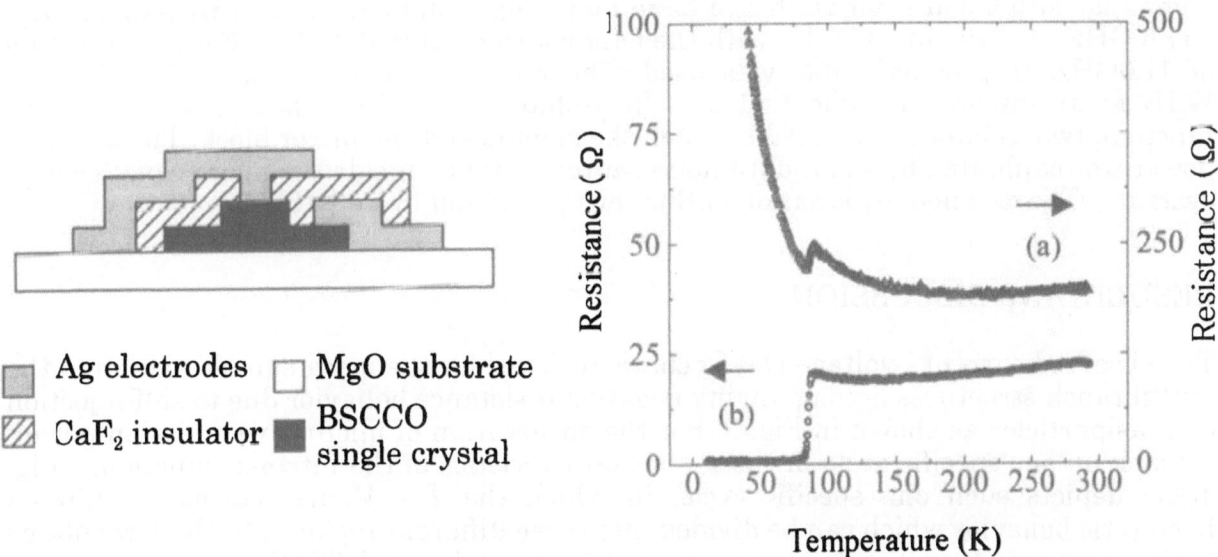

Ag electrodes ☐ MgO substrate

▨ CaF$_2$ insulator ■ BSCCO single crystal

Fig. 1 Sample configuration of an intrinsic BSCCO junction.

Fig. 2 Temperature dependence of the resistance of the intrinsic junctions.

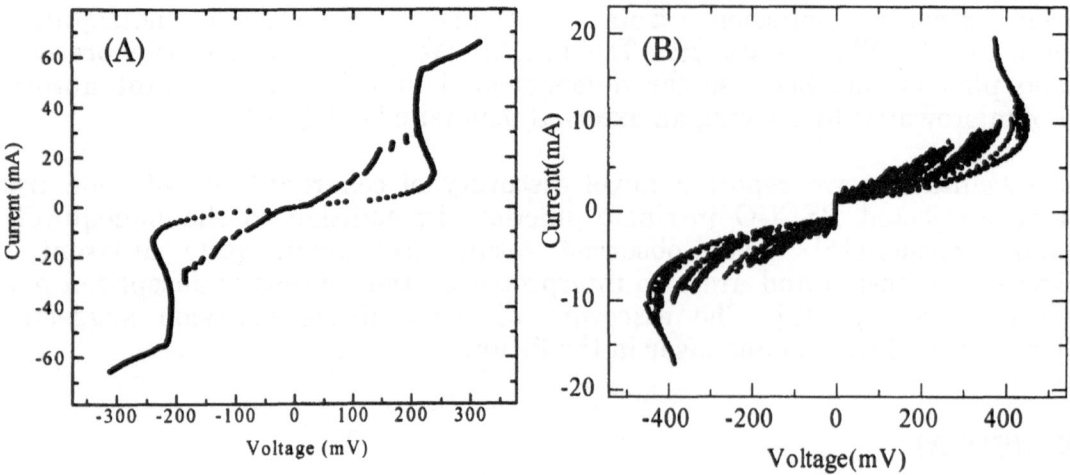

Fig. 3 $I-V$ characterisitics of the BSCCO intrinsic junctions with (B) and without (A) the Josephson current at 4.2K.

voltage $(I-V)$ characteristics. The curve (a) in Fig. 2 corresponds to the $I-V$ curve (A) in Fig. 3, while the curve (b) in Fig. 2 corresponds to the $I-V$ curve (B) in Fig. 3. Although the multi-branch structures are present and the resistance of the intrinsic junctions is small in the curve (A), no Josephson current was observable for this sample. The measurements were peformed for the sample with clear Josephson current as shown in the curve (B).

The three terminal method was used for the transport measurements with current flow along the c-axis. In the microwave measurements, the sample was mounted on the specially designed Ka-band sample holder equipped with coaxial cable which was placed at the center of a waveguide of cross section of 1x11 mm^2. This waveguide was connected to a Ka-band rectangular waveguide with a two-step quarter-wave standard matching transformer. The emitted power was detected by a superheterodyne mixer technique with a non-resonant broad-band matching system at receiver frequencies f_{REC} =11.6 GHz, 36 GHz and 47 GHz with the bandwidth of ΔB=2 GHz. For the detection of 11.6GHz, the coaxial cable was used. The receiver sensitivity was $\delta S \fallingdotseq 3 \times 10^{-24}$ W/Hz at an integrating time t =1 s. To isolate the local oscillator power from the junction, two isolators were used between the sample and the mixer block. The absolute power was calibrated by a standard noise source installed inside the microwave receiver system. The detailed explanation of this set-up is found in Ref[16].

RESULTS AND DISCUSSION

The observed current - voltage $(I-V)$ characteristics for the mesa sample exhibited the multibranch structures accompanying negative resistance behavior due to self-injection of quasiparticles as shown in Fig. 3. For the observation of microwave emission power, we pick up one specific cycle in the $I-V$ characteristics of the intrinsic junction. Fig. 4(A) depicts such one specific cycle, in which the $I-V$ characteristic exhibited hysteretic behavior which can be divided into three different regions ; (i) the low voltage region near zero bias, (ii) the quasiparticle branch region and (iii) the gap sum voltage region. The corresponding microwave emission power detected at f_{REC} =11.6 GHz with sweeping the current at 4.2 K is shown in Fig. 4(B). With increasing the current, there appeared a sharp microwave emission peak when the bias current exceeded the critical current of about 8mA. The appearance of a voltage is not clear in the $I-V$ curve since the Josephson voltage at peak point (~0.3mV) was much smaller than the voltage produced by Au/I/YBCO contact resistance (several mV). By the application of small magnetic field up to 200G, the emission peak was suppressed continuously keeping its peak voltage unchanged. Thus, it may be judged as the Josephson coherent emission peak from the series-connected phase-locked Josephson array.

When the $I - V$ characteristic jumped to one of the quasiparticle current branch, a broadband microwave emission signal was observed. A series of broadband emissions were observable from branch to branch at the high bias voltage region. When the current was increased or decreased along the quasiparticle branch, a continuous change of the emission power was observed. These emissions are qualitatively different from Josephson self-emission since it appeared at the bias voltages much greater than expected and was broadband. Besides, it appeared without applying any magnetic field.

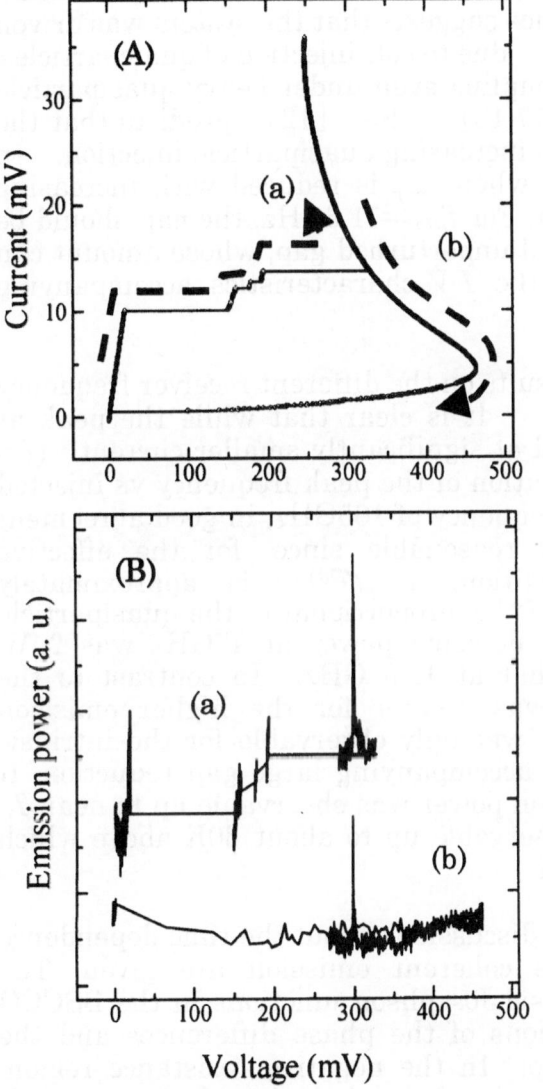

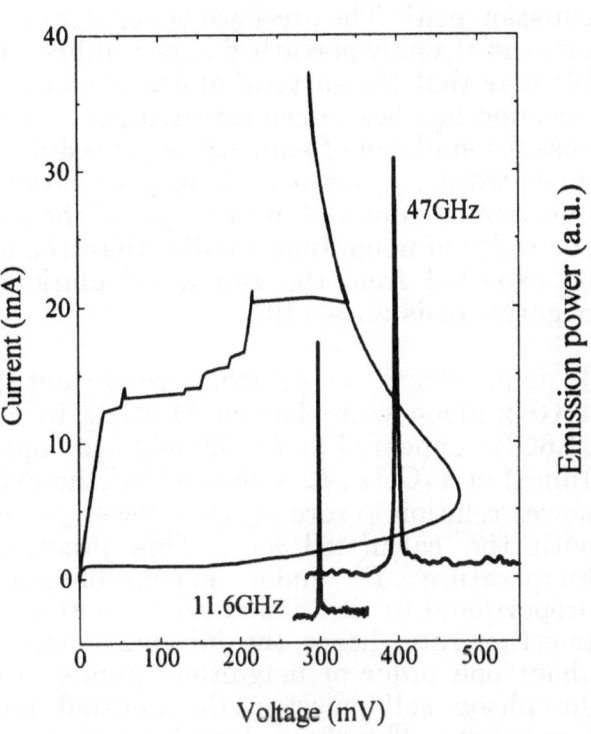

Fig. 5 $I - V$ characteristic of one specific cycle of a hysteretic junction and the detected emissions at the receiver frequency at 11.6GHz and 47GHz.

Fig. 4 (A) $I - V$ characteristic of one specific cycle of a hysteretic BSCCO junction. With sweeping the current along the path indicated by the arrows, sharp or broadband microwave emissions are detected as shown in (B). (a) corresponds to the detected emission of radiation with increasing the current and (b) corresponds to that with decreasing the current.

The observed microwave power level was, however, almost the same order of magnitude as the Josephson self-emission. The emissions were quite similar to those reported recently [7-11] and may be ascribed to the electron-plasmon interaction according to Ref. [12] . The emission of plasmons leads to microwave emission into the free space.

For further increase of current, a sharp emission peak appeared at bias point I= 30mA, V=300 mV at the gap-sum voltage portion of the I-V curve exhibiting negative resistance behavior. As shown in Fig. 4(B), this peak was observable in the processes of both increasing and decreasing the current, hence independent of the sweep direction of bias current. The half width of the peak was a few mV. This peak is again quite different from the Josephson self-emission peak since it appeared at the high bias voltage. No sharp emission peak was observable for the samples without negative resistance behavior.

We identify the sharp peak observed in Fig. 4(B) as the coherent Josephson plasma emission peak. The presence of negative resistance suggests that the system was driven into the strongly perturbed nonequilibrium state due to self-injection of quasiparticles. We note that the survival of Josephson weak coupling even under heavy quasiparticle injection has been demonstrated previously [17,18] . Ref. [12] predicts that the Josephson plasma frequency ω_p is reduced with increasing quasiparticle injection. In other words, a sharp peak may be observable when ω_p is reduced with increasing injection current and meets f_{REC} of the receiver. For f_{REC}=11.6GHz, the gap should be one order of magnitude smaller than the original unperturbed gap, whose amount can be expected from the recent calculations on the I-V characteristics accompanying negative resistance [19] .

To demonstrate this further, we present the results on the different receiver frequency 47GHz along with that on 11.6GHz in Fig. 5. It is clear that while the peak at 11.6GHz appeared at I = 30 mA, it appeared at significantly smaller current (I = 15mA) at 47GHz . It is shown that the extrapolation of the peak frequency vs injected power relation to zero injected power yields a frequency of 105GHz, in good agreement with the calculated ω_p. This procedure is reasonable since, for the effective temperature T^* under nonequilibrium condition, $\omega_p(T^*)$ is approximately proportional to $\Delta(T^*)$ and the reduction of $\Delta(T^*)$ is proportional to the quasiparticle injection rate, hence the injected power. The detected power at 47GHz was 2pW, about one order of magnitude greater than that at 11.6 GHz. In contrast to the Josephson self-emission, the emitted power was greater for the higher emission frequency. The sharp Josephson plasma peak was only observable for the intrinsic junctions exhibiting strong negative resistance accompanying large gap reduction. It was found that, while the Josephson self-emission power was observable up to near T_c, the sharp Josephson plasma peak was only observable up to about 30K above which negative resistance behavior almost disappeared.

From the theoretical point of view, the following discussions about the time dependence of the phase difference in the state of the coherent emission are given. The superconducting phase differences of the intrinsic Josephson junctions in the BSCCO crystal are obtained by solving coupled equations of the phase differences and the electromagnetic field in a nonequilibrium state. In the negative resistance region, the phase differences of all the junctions rotate in phase, inducing the same static voltage in all the junctions. The calculation shows that an oscillating phase difference with the plasma frequency is superposed to the uniformly rotating phase difference when quasiparticles are heavily doped to realize the negative resistance state. The phase difference of each junction ϕ in this case is approximately expressed as

$$\phi = (2\pi cD/\Phi_0)(E_0 t + (E_p/\omega_p)\sin\omega_p t) \tag{1}$$

and the electric field E induced in each junction is given from the Josephson relation $\partial\phi/\partial t=(2\pi cD/\Phi_0)E$ by

$$E = E_0 + E_p\cos\omega_p t,\tag{2}$$

D being the distance between the superconducting CuO_2 layers.   From Eq.(2)we see that the gap sum voltage is NDE_0, N being the total number of junctions, and the amplitude of the Josephson plasma induced in the nonequilibrium state is E_p, consistent with the experimental observation of a coherent peak.

In summary, we have reported novel microwave emissions due to tunnel injection of quasiparticles into the c-axis direction of BSCCO single crystal.   The observed microwave emissions are found to contain three different modes;   Josephson self-emission at the low bias voltage, nonequilibrium broadband emissions in the quasiparticle branches and   a sharp emission in the negative resistance region at the gap sum voltage.   From the temperature and magnetic field dependence of the emission, we identify that the broadband emission   is the incoherent plasma emission due to the elementary electron-plasmon scattering process and a sharp emission at high current and voltage is the coherent Josephson plasma emission.

* Present address: Department of Physics, Sogang University, C.P.O. Box 1142, Seoul 121-742, Korea

References
[1]   D. N. Langenberg, D. J. Scalapino, B. N. Taylor and R. E. Eck, Phys. Rev. Lett. 15, 294 (1965).
[2]   W. Reuter, M. Siegel, K. Herrmann, J. Schubert, W. Zander and A. I. Braginski, Appl. Phys. Lett. 62, 2280 (1993).
[3]   K. Lee and I. Iguchi, Appl. Phys. Lett. 66, 769 (1995).
[4]   K. Lee, I. Iguchi and K. Y. Constantinian, Physica C, 320, 65 (1999).
[5]   T. Nagatsuma, K. Enpuku, F. Irie and K. Yoshida, J. Appl. Phys. 54, 3302 (1983).
[6]   G. Hechtfischer et al.,   Phys. Rev. Lett.   79, 1365   (1997); G. Hechtfischer et al., Phys. Rev. B 55, 14638 (1997).
[7]   K. Lee, I. Iguchi, H. Arie and E. Kume, Jpn. J. Appl. Phys.37, L278 (1998).
[8]   I. Iguchi, K. Lee, H. Arie and E. Kume, *Advances in Superconductivity X* (Springer-Verlag Tokyo 1998),   p.1181.
[9]   K. Lee, H. Yamaguchi, W. Wang, E. Kume and I. Iguchi, Appl. Phys. Lett. 74, 2375 (1999).
[10]   I. Iguchi, K. Lee, E. Kume, T. Ishibashi and K. Sato, Phys. Rev. B, Jan. 1st issue (2000).
[11]   E. Kume, I. Iguchi and H. Takahashi, Appl. Phys. Lett. 75, Nov. 1 issue (1999).
[12]   S. Shafranjuk and M. Tachiki, Phys. Rev. B59, 14087 (1999).
[13]   O. K. C. Tsui, N. P. Ong, Y. Matsuda, Y. F. Yan and J. B. Peterson, Phys. Rev. Lett. 73, 724 (1995).
[14]   Y. Matsuda, M. B. Gaifullin, K. Kumagai, K. Kadowaki and T. Mochiku, Phys. Rev. Lett. 75, 4512 (1995).
[15]   I. Iguchi, K. Lee, W. Wang, E. Kume, M. Tachiki, K. Hirata and T. Mochiku, Proc. MOS'99, J. Low Temp. Phys. in press and Phys. Rev. B submitted (1999).
[16]   K. Lee and I. Iguchi, IEICE. Trans. Electron. E 78-C, 490 (1995).
[17]   T. Wong, J. T. C. Yeh and D. N. Langenberg,   Phys. Rev. Lett. 37, 150(1976).
[18]   I. Iguchi, K. Nukui, and K. Lee, Phys. Rev. B 50, 457 (1994)
[19]   M. Suzuki and K. Tanabe, Jpn. J. Appl. Phys. 35, L482 (1996).

Josephson-Junction Array Masers

Paola Barbara,[1*] Branimir Vasilić,[1] Alfred B. Cawthorne,[#] Sergey, V. Shitov,[2] and Christopher J. Lobb [1]

[1]Center for Superconductivity Research, University of Maryland, College Park, MD, 20742, USA
[2]IREE, Russian Academy of Sciences, Mokhovaya 11, 103907 Moscow, Russia

Abstract: Our recent experiments on the high-frequency properties of Josephson-junction arrays suggest that macroscopic non-identical objects, such as Josephson junctions, can synchronize through stimulated emission in the same manner as atoms in a laser.

The analogy between Josephson junctions and atoms was developed in a few theoretical works in the early 1970s' but never before confirmed by experiments. Our arrays show all the signatures of laser systems: coupling of the oscillators to a resonant cavity providing feedback, coherent emission above a pumping threshold, and synchronization up to sizes larger than the free-space radiation wavelength.

Keywords: Josephson junctions, stimulated emission, high-frequency sources.

INTRODUCTION

Josephson junctions are perfect voltage-to-frequency transducers. When a dc voltage V is applied across the junction, the current of Cooper pairs (supercurrent) oscillates at a frequency $\nu = 2eV/h$, where e is the electron charge and h is the Planck constant [1]. Early theoretical work [2] showed that a single Josephson junction can be regarded either as a radiating macroatom or as a collection of two-level atoms emitting radiation in phase. In fact, a Cooper pair tunneling from the high voltage electrode to the low voltage electrode emits a photon with energy $h\nu$ [1]. In a single Josephson junction all the radiative processes such as spontaneous emission, absorption and stimulated emission are analogous to the corresponding processes in two-level atoms [2].

While the Cooper pairs in a single junction are coherent, synchronization is not guaranteed in a collection of interconnected junctions. Josephson junctions are macroscopic non-identical objects, which usually oscillate independently from each other. The first experiments showing a correlated, single-frequency state in Josephson junction arrays were performed by Clark [3]. Tilley [4] suggested that interaction with a common radiation field could self-synchronize an array of different junctions, resulting in *superradiant* coherent emission. In this case, the emitted power is predicted to scale as the square of the number of active junctions. In general, such N^2 dependence is a signature of coherence in a system of N oscillators.

Coherent emission from Josephson-junction arrays has been previously detected [5,6] and explained in terms of classical electromagnetic coupling among the junctions. The interaction between the junctions and the radiation field was described with classical electrodynamics, where the coupling was provided by the common AC currents flowing through an external load [5,7]. It was found that synchronization could be achieved only if junctions are grouped in regions that are small with respect to the radiation wavelength.

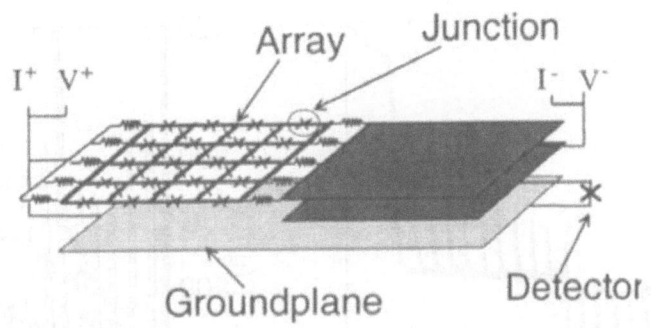

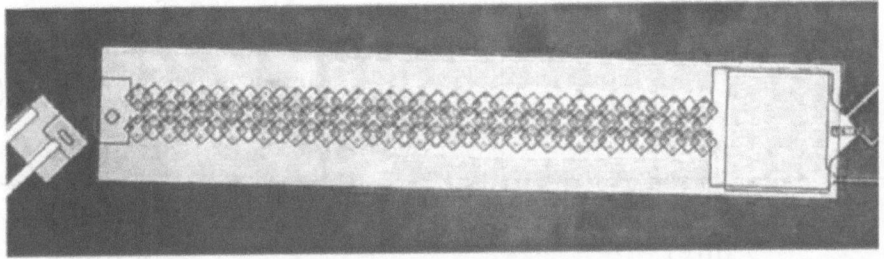

Fig.1. Sketch of a typical sample (top). Photograph of a 3 X 36 array (bottom); the length of the array is about 470 μm.

We recently reported measurements on Josephson junction arrays [8] and detected coherent radiation at 150 GHz, up to array size bigger than the free space radiation wavelength (about 2mm).

Our arrays are quite different from those previously measured. First, the junctions have very low dissipation (orders of magnitudes smaller than the junctions in Refs. [5,6]). Second, each junction is strongly coupled to a specific electromagnetic mode in the cavity formed by the array itself and a groundplane. For each array, we measured the emitted power as a function of the number of oscillating junctions. We find that, below a threshold number of oscillating junctions, no detectable power is measured. Above threshold, the array emits coherently and the power grows as the square of the number of active junctions, up to DC-to-AC conversion efficiency higher than 30% [9]. By contrast, high-dissipation arrays measured in Ref. [5,6] show DC-to-AC conversion efficiency on the order of a few percent.

The radiation output from our devices shows strong similarities to laser systems, suggesting stimulated emission as a synchronization mechanism.

EXPERIMENTAL RESULTS

A sketch of our samples is shown in Fig. 1. Junctions connected in a two-dimensional network are coupled to a two-junction detector circuit [10] through a capacitor. The current-voltage (I-V) characteristic of both the array and the detector can be independently and simultaneously measured. In the photograph, the squares are superconducting islands and the junctions are located in the overlap regions between them. The junctions are Nb/Al/AlO$_x$/Nb tunnel structures and they are spaced by 13 μm. The junction area is 16 μm^2 and the critical current density is 1kA/cm^2. These junctions are very hysteretic: when the bias current is increased above the critical value I_C, the junction switches from the zero-voltage state (supercurrent branch) to a non-zero voltage state (gap

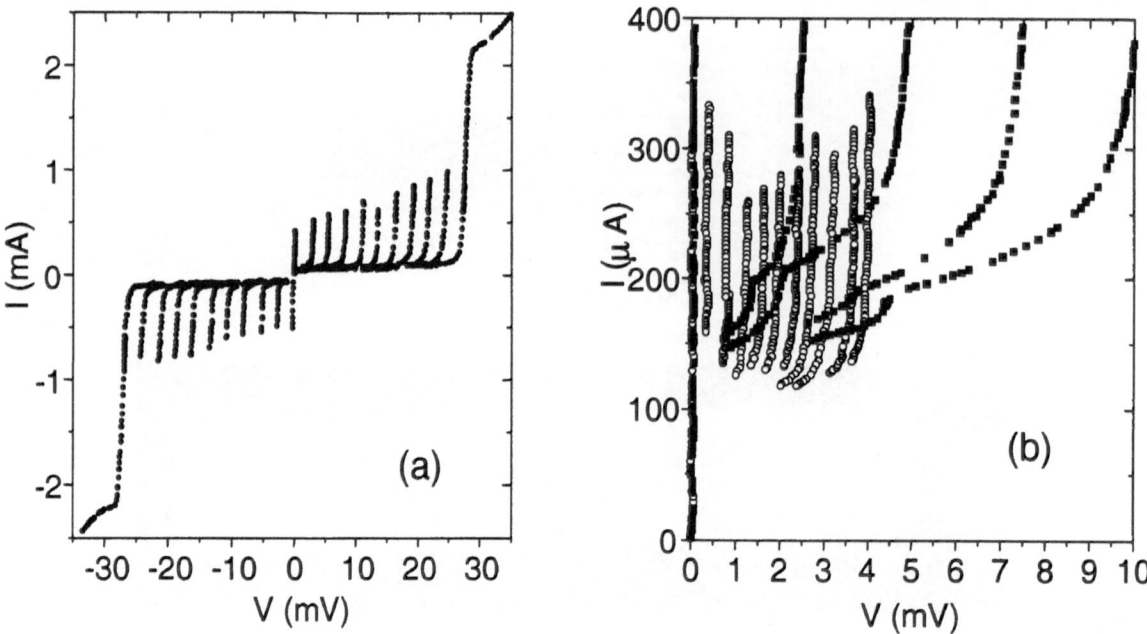

Fig. 2. (a) Current-voltage characteristic of a 10 X 10 array. (b) Enlargement of the low voltage region (solid squares). In the presence of an external magnetic field in the plane of the array, $H = 40$ Oe, sharp resonant steps appear (empty circles).

voltage). When the bias current is decreased, the junction stays trapped in the non-zero voltage state for a wide range of bias current.

Fig. 2 (a) shows a typical *IV*-curve of a 10 column by 10 row array (the rows are perpendicular to the direction of the bias current). The curve is very hysteretic and the plot is obtained by sweeping the bias current many times within the range of values shown in the graph. The 10 branches spaced by about 2.7 mV (the gap voltage of a single Nb junction) correspond to the switching of each row of junctions to the gap voltage. For example, when the array is biased on the first branch, one row of junctions is biased on the gap voltage while all the other junctions in the array are in the zero-voltage state. The next voltage branch corresponds to the switching of an additional row of junctions to the gap voltage. Eventually, at about 27 mV, all the junctions are switched.

Fig. 2 (b) is an enlargement of the low voltage region. When an external magnetic field in the plane of the array is applied, ten sharp, constant voltage steps appear on the *IV*-curve. These steps are easily distinguished from the gap structure because their voltage spacing is much smaller (about 400 µV for the array shown in Fig. 2 (b)). When the array is biased at a fixed point on the step, there is a DC voltage drop across some of the junctions, therefore they emit radiation at a frequency proportional to this voltage, according to the Josephson relation. When changing the bias current on a step, the voltage stays approximately constant, indicating that the junctions stay locked at a specific frequency for a wide range of bias current. This is a signature of a resonance in the array structure, i.e. there is a preferred radiation mode for the junction emission. A detailed investigation of the array IV-curve shows that each step corresponds to one row of junctions emitting radiation at a frequency given by the voltage spacing between the steps [8]. For example, when the array is biased on the third resonant step, three rows of junctions are emitting radiation at the resonant frequency, while all the other rows of junctions are on the zero-voltage state. In other words, the hysteresis of the IV-curve allows the number of oscillating junctions to be controlled.

In our experiment we bias the array on the resonant steps and measure the power coupled into the detector as a function of the number of oscillating rows.

The detector is made of two underdamped junctions connected in parallel by an inductance [10]. When radiation is coupled into the detector, its *IV*-curve shows an increased tunneling of quasiparticles due to photon absorption (photon-assisted tunneling) [11]. We use numerical simulations to calculate the pumped detector *IV*-curve and find the value of coupled power providing the best fit with the experimental *IV*-curve. An example is given in Fig. 3.

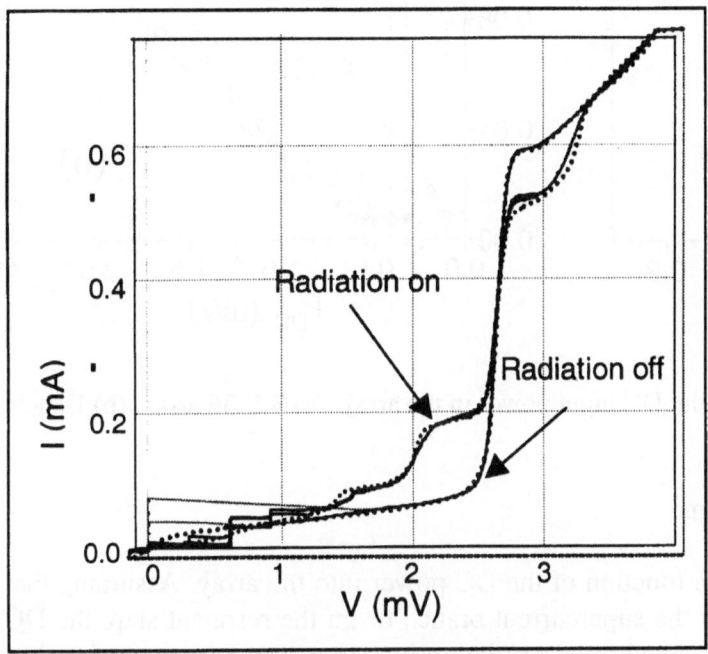

Fig. 3. Experimental (lines) and simulated (dots) current-voltage characteristics of the detector. In the simulation, $P_{AC} = 82$ nW and $\nu = 148$ GHz.

Tilley [4] predicted a quadratic dependence of the emitted AC power on the number of junctions as a signature of a coherent state. Here we use a very simple lumped circuit model to illustrate this general argument. We can model each oscillating junction in the array as an ac voltage source, V_{AC}, with a resistance in series, R_J, see Fig. 4. The junctions on the supercurrent branch do not produce any radiation, but they can dissipate the radiation produced by the other junctions because of their internal resistance R_J. In the simplest circuit model, all the junctions are connected in series and a resistive load R_L is attached to them. If all the active junctions are synchronized, the power dissipated in the load is $P_L = (N_A V_{AC})^2 R_L/(N_T R_J + R_L)^2$, where N_A is the number of oscillating junctions and N_T is the total number of junctions in the array (note that in this model V_{AC} does not depend on N_A). This simple model shows that, in case of coherent emission, the power scales like

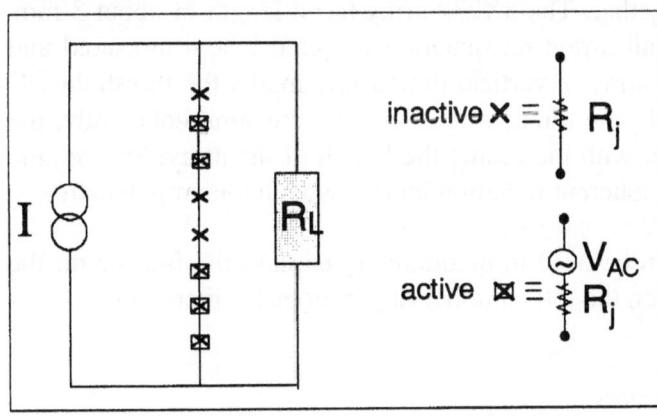

Fig. 4. Schematic of the simple circuit model used to predict the dependence of the detected AC power on the number of active junctions.

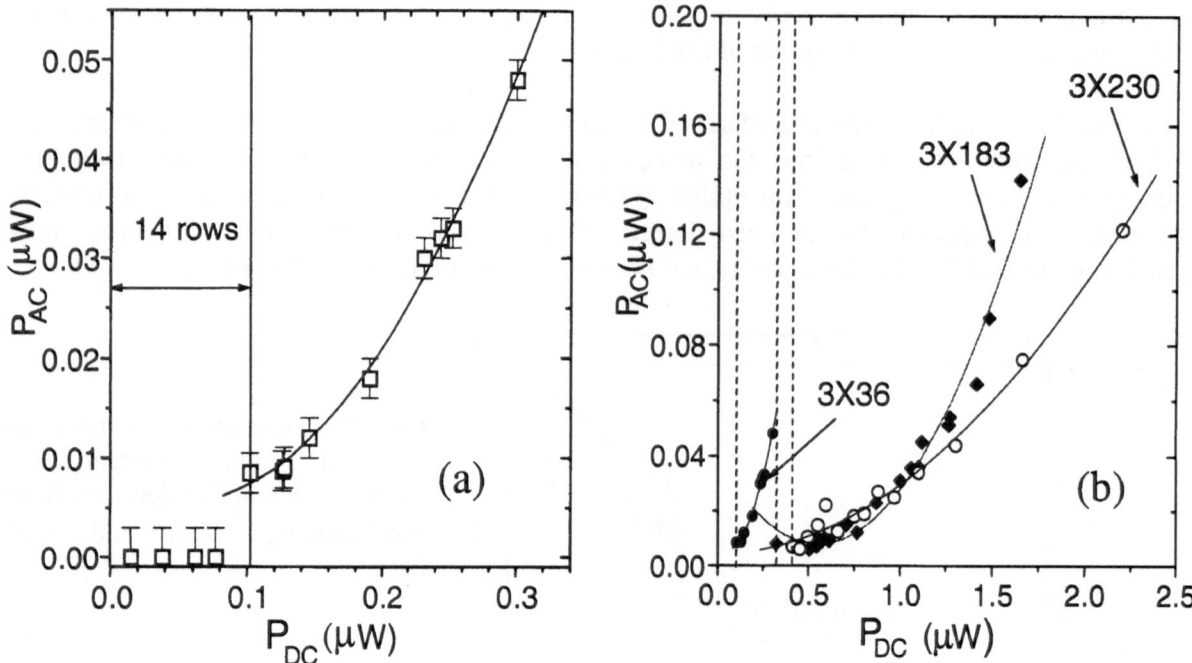

Fig. 5. Detected AC power as a function of the DC input power in the array. (a) 3 X 36 array. (b) Data from arrays with different numbers of rows.

the square of the number of active junctions.

Fig. 5 shows the measured AC power as a function of the DC power into the array. Assuming that all the junctions in the array are either on the supercurrent branch or on the resonant step, the DC power into the array is proportional to the number of oscillating junctions, i.e. it is proportional to N_A. Fig. 5 (a) shows measurements from a 3 X 36 array. Surprisingly, when biasing a few rows on the resonant state, no measurable AC power is coupled into the detector. Above a threshold number of biased rows, about 14, AC power can be detected. A further increase of the number of biased rows shows that the AC power coupled into the detector scales as the square of the number of rows, i.e. the junctions emit coherently above threshold. The maximum AC power for this array corresponds to about 17% DC-to-AC conversion efficiency.

The same experiment was performed on arrays with a larger number of junctions. The current voltage characteristic in the case of big arrays is very complex: details of the measurement technique are explained in Ref. [8]. We found a maximum DC-to-AC conversion efficiency of about 33% in a 4X36 array, corresponding to an AC power of about 0.25 μW. In Fig. 5 (b), results from arrays of different length are plotted together. The 3X230 array has a length of about 3 mm. The qualitative behavior is similar to the small array: no radiation detected below threshold and coherent emission above threshold. For each array, a vertical dotted line marks the threshold DC power. Two interesting features are worthy of note. First, when the junctions emit coherently, the coefficient of the square term becomes smaller with increasing the length of the array. Second, the threshold number of rows for the detection of coherent radiation increases with the array length.

The simple circuit model described above can be used to qualitatively explain the first point: the coefficient of the quadratic term decreases when the total number of junctions N_T increases.

The simple circuit model cannot explain the threshold because it assumes that all the junctions on the non-zero voltage state are synchronized. However, our experimental results show that, when biasing a small number of rows on the resonant step, the junctions do not emit radiation coherently. For example, in the case of the small array (Fig. 5 (a)), an extrapolation of the quadratic dependence to low values of DC power (corresponding to just a couple of rows biased on the resonant step) predicts values of AC power higher than 3 nW, which is roughly the sensitivity level of our detector. In other words, in this array, if there were any coherent radiation below threshold, it could be detected.

CONCLUSIONS

We measured a threshold from incoherent to coherent emission in two-dimensional arrays of Josephson junctions. Above threshold, the emission from all the measured arrays (up to array size larger than the free space radiation wavelength) is consistent with the theory from Tilley [4] predicting superradiance, analogously to atomic systems. The analogy between our arrays and atomic systems goes further. A threshold from the incoherent state to the coherent state is typical of laser systems. In lasers, it indicates that amplification can only be obtained with a sufficiently high population inversion. In the arrays, a certain minimum number of active junctions are necessary in order to overcome losses in the cavity and in the junctions. Analogously to lasers, the threshold in the arrays increases by increasing the array size.

In the case of Josephson-junction arrays a threshold had been predicted by Bonifacio et al. [12] by studying the analogy between free-electron lasers and Josephson junction arrays. Filatrella et al. [13] studied a lumped array of non-linear oscillators coupled through a high-Q resonator and found that the cavity plays an important role in the synchronization process and may produce a threshold. However, it is still unclear why a threshold was not observed in previous experiments [14] showing synchronization of long Josephson junctions interacting with a high-Q cavity. Further work is necessary in order to understand in detail the onset of the synchronized state in our arrays.

Acknowledgments. This work was supported by the Air Force Office of Scientific Research under grant No. F49620-98-1-0072. Samples were fabricated at Hypres, Elmsford, New York.

* E-mail: paola@squid.umd.edu
# Present address: Institute for Plasma Research, Department of Physics, University of Maryland, College Park, MD-20742.

1. B. D. Josephson, Phys. Lett. **1**, 251 (1962).
2. D. Rogovin and M. Scully, Phys. Rep. **25C**, 175 (1976) and references therein.
3. T. D. Clark, Phys. Lett. **27A**,585 (1968).
4. D. R. Tilley, Phys. Lett. **33A**, 205 (1970).
5. A. K. Jain et al., Phys, Rep. **109**, 309 (1984).
6. S. P. Benz and C. J. Burroughs, Appl. Phys. Lett. **58**, 2162 (1994).
7. K. Wiesenfeld, S. Benz and P. A. A. Booi, J. Appl. Phys. **76**, 3835 (1994).
8. P. Barbara, A. B. Cawthorne, S. V. Shitov, and C. J. Lobb, Phys. Rev. Lett. **82**, 1963 (1999).
9. B. Vasilić et al., to be published.
10. J. Zmuidzinas et al., IEEE Trans. Microwave Theory Tech. **42**, 698 (1994).
11. P. K. Tien and J. P. Gordon, Phys. Rev. **129**, 647 (1963).
12. R. Bonifacio, F. Casagrande and M. Milani, Lettere al Nuovo Cimento, 34, 520 (1982).
13. G. Filatrella, N. F. Pedersen and K Wiesenfeld, to be published.
14. R. Monaco, N. Gronbech-Jensen, and R. D. Parmentier, Phys. Lett. A **151**, 195 (1990).

Microwave Responses of BSCCO-2212 Intrinsic Josephson Junctions at Frequencies Up to 100 GHz

H. B. Wang,[1,2] T. Tachiki,[2] Y. Aruga,[2] Y. Mizugaki,[1,2] J. Chen,[1,2] K. Nakajima,[1,2] T. Yamashita,[2,3] P. H. Wu[4]

[1]CREST, Japan Science & Technology Corporation (JST), Japan
[2]Research Institute of Electrical Communication, Tohoku University, Sendai 980-8577, Japan
[3]New Industry Creation Hatchery Center, Tohoku University, Sendai, Japan
[4]Department of Electronic Science & Engineering, University of Nanjing, Nanjing 210093, China

Abstract: C-axis junction-stacks, with a-b plane sizes of 10 microns by 10 microns, were patterned on $Bi_2Sr_2CaCu_2O_8$ single crystals. We measured the current-voltage (I-V) characteristics with irradiation at a few to 100 gigahertz. At a few gigahertz and very high power level, current steps were stimulated and the voltage separations between them were independent of the microwave power. Under irradiation at 100 GHz, in addition to the suppression of critical currents, we successfully performed harmonic mixings between the 100 GHz signal and up to the 100th harmonic of a local oscillator at about 1 GHz. These experimental results show that intrinsic Josephson junctions can be good candidates for high frequency applications, and harmonic mixing may be a useful probe to investigate plasma phenomena in layered superconductors.

Keywords: intrinsic Josephson junctions, high frequency response, frequency mixing

INTRODUCTION

Since the first observation of intrinsic Josephson effects in layered high-T_C superconductors [1], much work has been focused on high frequency response of intrinsic Josephson junctions (IJJs) because of their high gap voltage (up to a few tens of mV) and characteristic voltage (up to several mV) [1, 2]. As one of the powerful tools in this connection, harmonic frequency mixings has been widely employed [3, 4]. With an intermediate frequency (f_{IF}) equal to the difference between the signal frequency (f_S) and the Nth harmonics of the local oscillator (f_{LO}), the output level at f_{IF} strongly depends on the characteristics of the Josephson junction, thus providing much useful information of the latter.

In this letter, we present experimental data on the microwave responses of stacked $Bi_2Sr_2CaCu_2O_8$ (BSCCO) intrinsic junctions at frequencies from a few to 100 GHz. In addition to the well-known dc properties of voltage jumps and resistive branches, we observed a series of remarkable microwave-induced power-independent current steps in the I-V curves when samples were subjected to a few gigahertz irradiation. At 100 GHz, we successfully performed harmonic mixings with harmonic number up to 100. Our experimental results showed the advantages in the frequency mixing due to the large dynamic resistance and possible mutual phase-locking in the junction array.

EXPERIMENTAL

The samples used in the experiments were mesa structures patterned on BSCCO single crystals. The fabrication process was already reported in detail elsewhere [5]. Measuring 10 ×10 μm^2 in the a-b plane, the number of the intrinsic junctions could be estimated from the stack depth, usually several hundred angstroms in our experiments. Electrical leads were glued onto the exposed top gold layer using silver paste. A three-probe method was used in all measurements. The linear contact resistance between the leads and the sample can easily be subtracted from the experimental data if necessary, thus it has very little effect on the microwave responses of the intrinsic junctions.

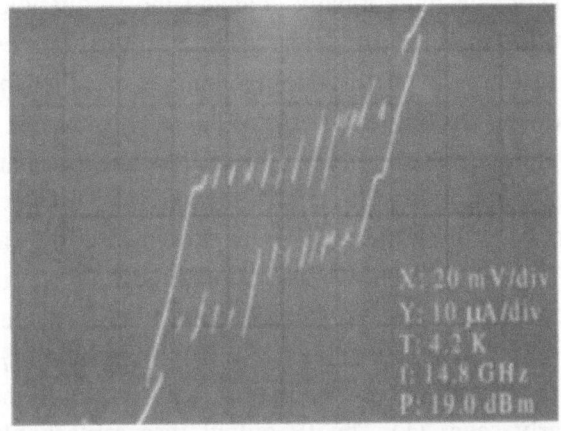

Fig. 1. Typical current-voltage curves, with microwave irradiation at 14.8 GHz, of a mesa on BSCCO single crystal. Sample sizes: 10 μm × 10 μm in the a-b plane.

Under irradiation at relatively low frequency and high power, we could often observe zero-crossing current steps appearing on the I-V curves. Shown in Fig. 1 is the typical microwave response for one sample at 14.8 GHz. Usually these steps yielded voltage intervals as large as a few millivolt, therefore, it is difficult to attribute them to Shapiro steps induced in one or many junctions. Very similar results were reported by us [6], where power-independent current steps with regular intervals of 4 mV were observed. The explanation of the step structure has been given, based on geometric resonance of the junction cavity and collective motion of rectangular vortices. For fitting the experimental data of 4 mV voltage spacing between neighboring steps, we need a velocity of $\overline{c} = 4 \times 10^7$ m/sec.. It has been noted that in a junction stack, characteristic electromagnetic wave propagation velocities can be expressed as

$$C_n = \lambda_j \omega_p \left\{ \frac{1}{1 + 2S \cos[n\pi/(N+1)]} \right\}, \qquad n = 1, 2, \ldots, N, \qquad (1)$$

where λ_j, ω_p, S, n, and N are Josephson penetration length, plasma angular frequency, coupling parameter (~-0.5), mode number, and total junction number, respectively [6, 7]. Given a critical current density of 150 A/cm^2 and a relative dielectric constant of 10 for the BSCCO samples used in our experiments, the highest mode for N=40 corresponded to a velocity of 4.5×10^6 m/sec., which is one order lower than we need. However, from Eq.1 one may get a velocity of 4×10^7 m/sec. with the assumption of N=450. As the base electrode was a few hundred microns, much thicker than 0.68 μm (=450×15 angstroms), the assumption might be reasonable, possibly implying an interesting phenomenon that the base BSCCO electrode should be taken into account for the determination of characteristic electromagnetic wave propagation velocities.

Having described the behavior of IJJs under irradiation at a few gigahertz, we now turn to the observation of microwave response at 100 GHz. In I-V curve measurements, observed was only the suppression of critical currents for most samples. To further investigate the properties of IJJs, we performed harmonic mixing. The experimental setup was reported elsewhere [8]. Keeping a Gunn oscillator (as a signal) freely running, we were able to observe the IF signal with the harmonic number up to 100. Shown in Fig. 2 is the typical IF spectrum for the 100th mixing at zero bias with a LO frequency of 1.0027456 GHz, where the signal frequency was calculated to be 100.7241 GHz. To make sure the observed spectrum was the mixing output between the 3 mm signal and the local oscillator, we intentionally changed the frequencies or the power levels of the two microwave sources, all resulting in corresponding changes in either the frequency or the magnitude of the mixing output.

In an attempt to gain a clear insight into the harmonic mixings in the IJJs, we studied the bias dependence of IF output P_{IF}. As shown in curve (a) of Fig. 3, the junctions' critical currents were strongly suppressed and most of the branches were invisible; shown as curve (b) in Fig. 3 is P_{IF}-V. On the P_{IF}-V curve a few features could be clearly seen. (1) The IF output was remarkably well above the noise level in the whole voltage range even around the gap structure; and very interesting oscillations of IF output spread over the whole measured voltage range with sharp tip-like minima. (2) IF output level was strongly dependent on where the junctions were biased. It is noticeable that IF output reached its climax in one region, implying there was either a better coupling between junctions and signal or a larger dynamic resistance due to the junction array. The oscillation of IF output was LO-power-dependent, and the intervals between its peaks or valleys were larger than those between Shapiro steps; although this was not well understood, we concluded that for large LO power, the optimum operating point moved to higher voltages where IF output was higher and more stable than for small LO power

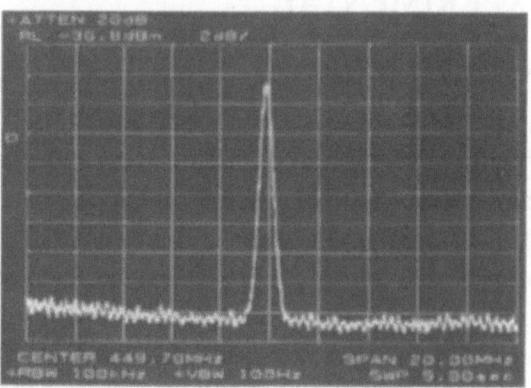

Fig. 2 Typical spectrum of the intermediate frequency (IF) output for the 100th harmonic mixing with zero bias and optimum power of local oscillator.

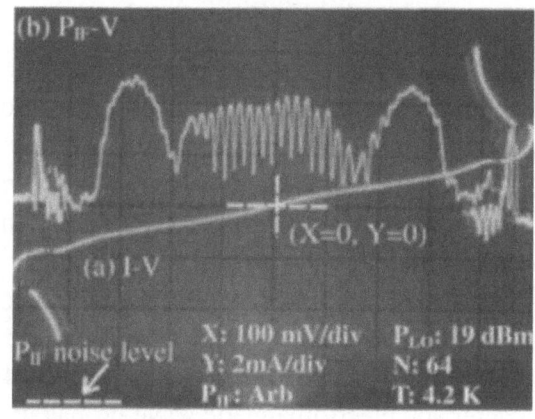

Fig. 3 Typical P_{IF} vs. V curves for the 64th harmonic mixing, and the I-V curves under irradiation from both a 3mm signal and a local oscillator.

CONCLUSION

Microwave responses of BSCCO intrinsic Josephson junctions were investigated at frequencies from a few to 100 gigahertz at 4.2K, by direct measuring the I-V curves and carrying out harmonic frequency mixing. Our experimental results demonstrated that the BSCCO intrinsic Josephson junctions can be employed as highly nonlinear devices for many applications at mm and sub-mm wavebands.

1. A. Irie, and G. Oya, *IEEE Trans. on Appl. Supercond.*, 5(2), 3267(1995).
2. W. Prusseit, M. Rapp, K. Hirata, and T. Mochiku, *Physica C*, 293, 25(1997).
3. D.C. McDonald, V.E. Kose, K.M. Evenson, J. S. Wells and J.D. Cupp, *Appl. Phys. Lett.* 15, 121(1969).
4. P.H.Wu, Y.H. Xu, and C. Heiden, *Appl. Phys. Lett.* 57, 1265(1990).
5. H. B. Wang, Y. Aruga, T. Tachiki, Y. Mizugaki, J. Chen, K. Nakajima, T. Yamashita, and P. H. Wu, *Appl. Phys. Lett.*, 74, 3693(1999).
6. R. Kleiner, *Phys. Rev. B*, 50, 6919(1994).
7. S. Sakai, A. V. Ustinov, H. Kohlstedt, A. Petraglia, and N. F. Pedersen, *Phys. Rev. B*, 50, 12905(1994).
8. H. B. Wang, Y. Aruga, T. Tachiki, Y. Mizugaki, J. Chen, K. Nakajima, T. Yamashita, and P. H. Wu, *Appl. Phys. Lett.*, 75, 2310(1999).

TERAHERTZ RADIATION PATTERNS OF YBCO THIN FILM ANTENNAS

M. Morimoto[1], H. Saijo[1], M. Yamashita[1], M. Tonouchi[1, 2], and M. Hangyo[1]

[1]Res. Ctr. Supercond. Mat. and Electron., Osaka University, Osaka 565-0871, Japan
[2]CREST, Japan Science and Technology Corporation, Osaka 565-0871, Japan

Abstract: We studied the polarization of the terahertz (THz) beam emitted from YBCO thin film patterned into dipole, bow-tie and log-periodic antennas using three sets of wire-grid polarizers. The amplitude and frequency spectra indicate that the radiation from both the dipole and the bow-tie antennas have simple polarization pattern, while the one from the log-periodic antenna shows the complicated one, which is attributed to the structural resonance.

Keywords: THz radiation, fs laser pulse, YBCO films, log-periodic antenna, polarization pattern

INTRODUCTION

THz radiation from high-T_c superconductors excited with femtosecond (fs) optical pulses has opened a new research field in microwave photonics [1,2]. The mechanism of this phenomenon is due to an ultrafast supercurrent modulation by pulsed laser excitation. Previously, we reported the radiation from YBCO bow-tie antenna, and observed the wave propagation along the antenna edges [3]. In the present work, we studied the radiation patterns emitted from YBCO thin film dipole, bow-tie, and log-periodic antennas. The polarization of the THz beam is characterized in time domain using three sets of wire-grid polarizers, and in its relation to structural resonance is discussed.

EXPERIMENTAL

A mode-locked Ti: sapphire laser was used to produce fs optical pulses with a full-width at half-maximum (FWHM) of 50 fs, a repetition rate of 82 MHz, and a center wavelength of 790nm. Using conventional photolithography and ion milling process, several 100 nm-thick thin films were patterned into dipole, bow-tie and self-complementary log-periodic toothed planar antennas. The structure of the log-periodic antenna is illustrated in Fig. 1. A hemispherical MgO lens with a diameter of 3 mm is attached on the backside of the MgO substrates of the dipole and the bow-tie antenna to increase the collection efficiency of the THz radiation. The detail of the experimental setup have been reported previously [2]. The polarization of the emitted waves is characterized using three sets of wire-grid layers as shown in Fig. 2. The 1st grid is used to pick out the polarized radiation by changing the grid angle from -90 to 90 degrees. The 2nd one is set at an angle of 45 or -45 degrees, the sign of which is the same as that of the 1st grid. The 3rd one is set at an angle of 0 degree to detect the radiation polarized parallel to the detector antenna. The 2nd one is employed to detect the wave polarized perpendicular to the detector antenna. The measurement is done at laser powers between 2 and 20 mW, bias currents between 80 and 150 mA, and a temperture of about 17K.

RESULTS AND DISCUSSION

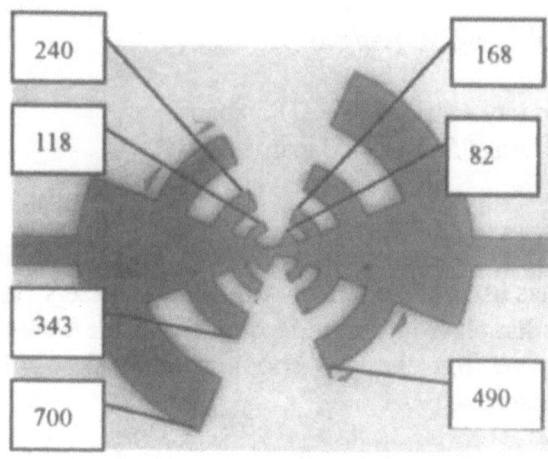

240
118
343
700
168
82
490

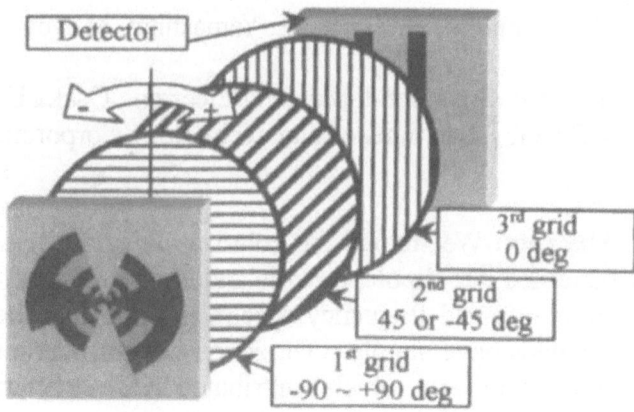

Detector

3rd grid
0 deg

2nd grid
45 or -45 deg

1st grid
-90 ~ +90 deg

Fig. 1. Structure of log-periodic antenna. The unit radius is measured in μm.

Fig. 2. Schematic of the experimental configuration

Figure 3 shows the frequency spectra of the waveforms emitted from the dipole and bow-tie antennas. The broadband spectra of the dipole and bow-tie antennas have a maximum amplitude at a frequency of about 0.26 THz and 0.21 THz, respectively, regardless of the polarizer grid angle. The corresponding polarization patterns defined at the maximum amplitude frequencies are summarized in Fig. 4. Although the bow-tie antenna has a slightly less angle dependence of the polarization than the dipole, both antennas have simple and similar patterns.

Figure 5 (a) shows the waveforms and the frequency spectra of the THz radiation emitted from log-periodic antenna detected without the MgO lens at various grid angles. It is notable that this antenna emits the radiation for more than 80 ps after the arrival of the fs pulse. This indicates that the THz radiation contains the one during the wave propagation in the antenna. The waveforms are, however, so complicated compared to those of the dipole and bow-tie antennas that we can not discuss the details such as the structural resonance. The corresponding frequency spectra are shown in Fig. 5

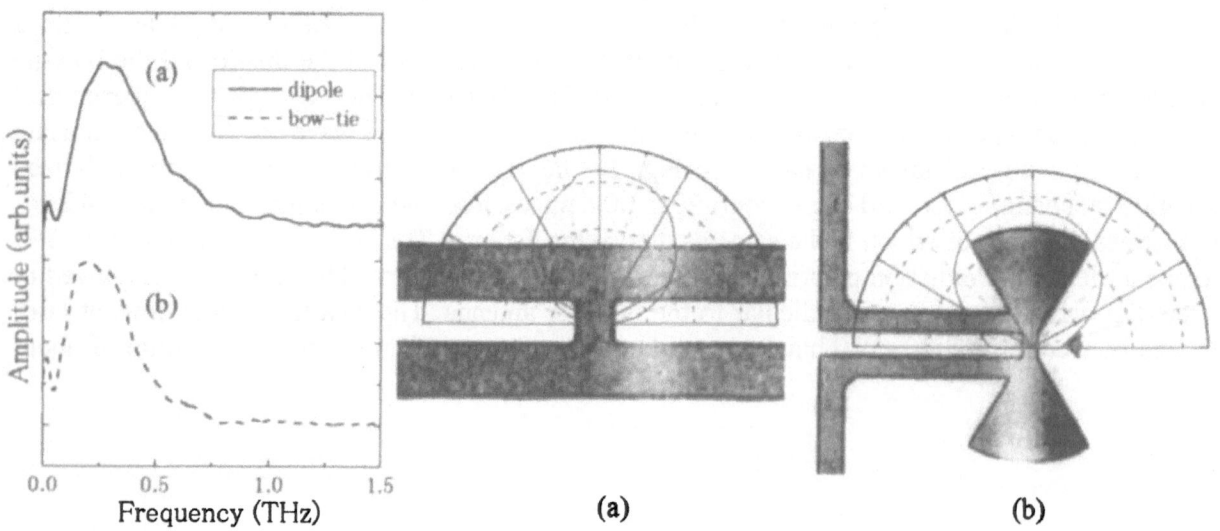

(a)
dipole
bow-tie
(b)

Amplitude (arb.units)

0.0   0.5   1.0   1.5
Frequency (THz)

(a)   (b)

Fig. 3. The frequency spectra of (a) dipole and (b) bow-tie antenna at a grid angle of 40 deg.

Fig. 4. The polarization of terahertz radiation emitted from (a) dipole and (b) bow-tie antenna.

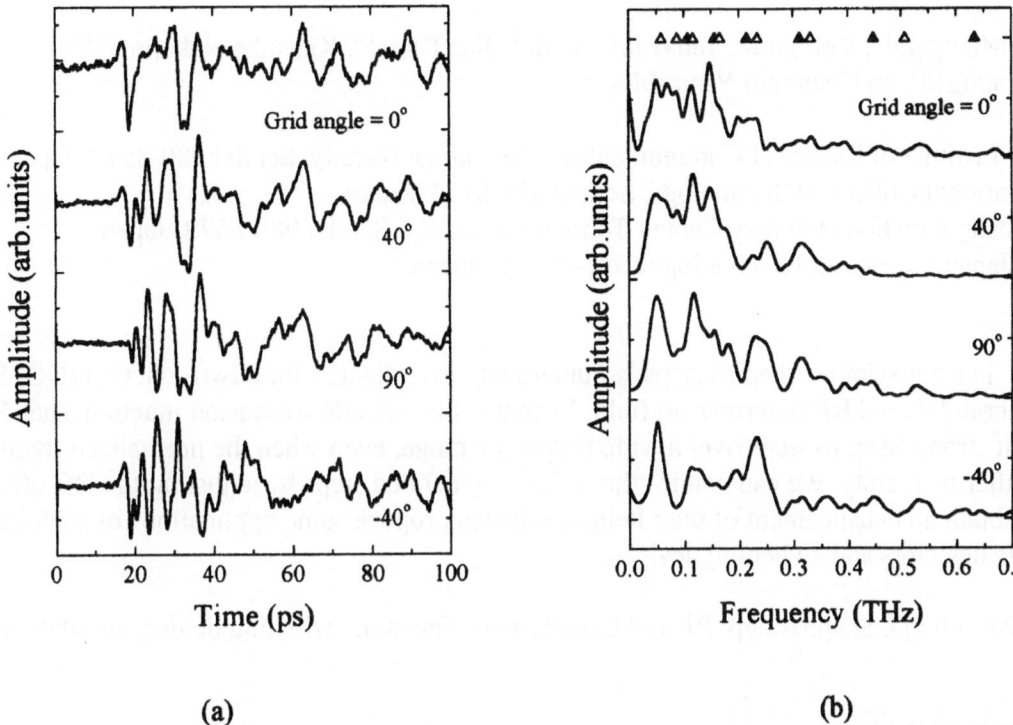

(a)

(b)

Fig. 5 (a) The THz waveforms and (b) the amplitude spectra for various grid angles detected without the MgO lens.

(b), which are also complicated. However, one can distinguish the structural resonance indicated by the triangles. The open and closed symbols correspond to the resonant lengths along the teeth edges in the direction of the circular arc, and the ones between the antenna center and the far teeth edges. It should be emphasized that although the resonant frequencies for the former should be modified by taking the dielectric constant into account, the ones for the latter should not. The polarization patterns at the resonant frequencies are qualitatively explained by the antenna structure. We can also investigate how the resonance progresses in time domain by obtaining the frequency spectrum from the waveform for the certain time period, which will be reported elsewhere.

SUMMARY

We have observed the polarization patterns of the terahertz (THz) radiation emitted from YBCO thin film patterned into dipole, bow-tie and log-periodic antennas. The polarized amplitude and frequency spectra indicated that the radiation from both the dipole and the bow-tie antennas has simple polarization pattern explained by their structure, while the one from the log-periodic antenna shows a complicated one. The time-domain-analysis of the polarized wave suggests that there co-exist a radiation directly coupled to the antenna teeth and the wave propagation along antenna edges.

Acknowledgements. This work was partially supported by a Grant-in-Aid for Scientific Research on Priority Area (A), No. 10142101, from the Ministry of Education, Science, Sports, and Culture, Japan.

References
1. M.Hangyo *et al.*, Appl. Phys. Lett., 69:2122 (1996)
2. M.Tonouch *et al.*, Jpn. J. Appl. Phys., 35, 2624 (1996)
3. Tani *et al.*, Jpn.J.Appl.Phys.,Vol.35,pp.L1184-L1187 (1996)

Two-Junction SQUID Controlled by Both DC and RF Magnetic Flux

Yoshinao Mizugaki[1,4], Kei Saito[1], Tadayuki Kondo[2], Jian Chen[1,4], Kensuke Nakajima[1,4], Alex I. Braginski[3], and Tsutomu Yamashita[3,4]

[1]Research Institute of Electrical Communication, Tohoku University, Sendai 980-8577, Japan
[2]Sendai National College of Technology, Sendai 989-3124, Japan
[3]New Industry Creation Hatchery Center, Tohoku University, Sendai 980-8579, Japan
[4]CREST, Japan Science and Technology Corporation, Japan

Abstract: The behavior of Shapiro steps is numerically investigated for a two-junction (DC) SQUID under the both DC and RF (microwave) field. In contrast to a single Josephson junction, such SQUID can exhibit strong Shapiro steps over a wide frequency range, even when the normalized frequency is much smaller than unity. We can control the order of generated steps by adjusting the DC offset field and thus obtain an enhancement of their height. We also propose some applications of such effects to microwave detectors and switching devices.

Keywords: SQUID, Shapiro step, RF-field, microwave detector, switching device, simulation

INTRODUCTION

As Lee et.al demonstrated [1], the RF-Field activated two-junction DC SQUID (RFDS) generates large Shapiro steps, the order of which can be controlled by the external magnetic field. The large steps of the RFDS are attractive for several applications, such as voltage standards [1] and digital-to-analog (D/A) converters [2]. We have also found that the terahertz response of $YBa_2Cu_3O_{7-\delta}$ grain boundary Josephson junctions can be enhanced by a DC magnetic field, and we have explained the enhancement by applicability of the RFDS model [3].

In this paper, we present numerical results on the characteristics of the RFDS, propose its microwave applications and switching applications.

BASIC PROPERITIES OF THE RFDS

Figure 1 shows the equivalent circuit of the symmetrical RFDS. The SQUID loop includes two Josephson junctions and two inductors. A DC-current source (I_{dc}) and the output line (L_{out} and R_{out}) are directly connected to the SQUID. An RF activation current ($I_{rf}\sin 2\pi ft$) and an input signal (I_{in}) are inductively coupled to the SQUID.

The operation at lower frequency is based on the quantum transitions of the SQUID [1-3]. Figure 2 shows an example of threshold characteristics of the SQUID, where n is the number of flux quanta in the SQUID loop and each operation point is represented by the horizontal line connecting two

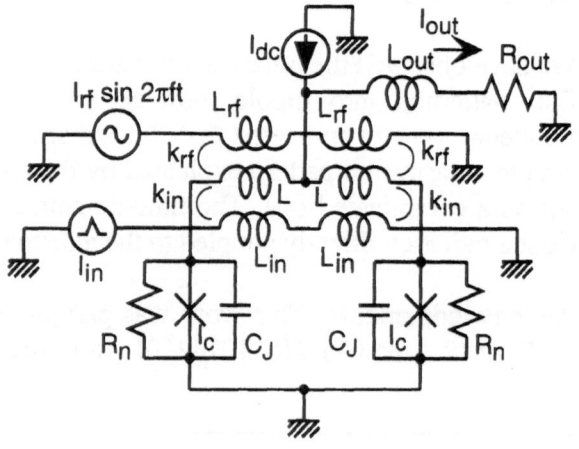

Fig. 1. Equivalent circuit of the RFDS.

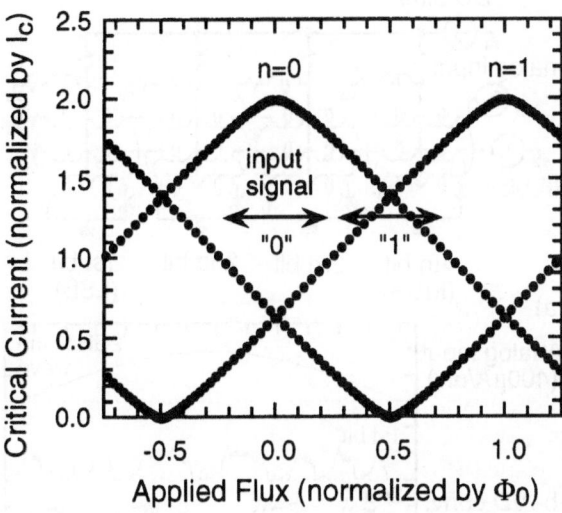

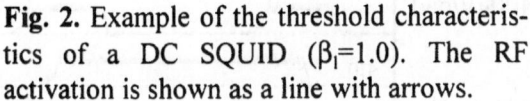

Fig. 2. Example of the threshold characteristics of a DC SQUID (β_l=1.0). The RF activation is shown as a line with arrows.

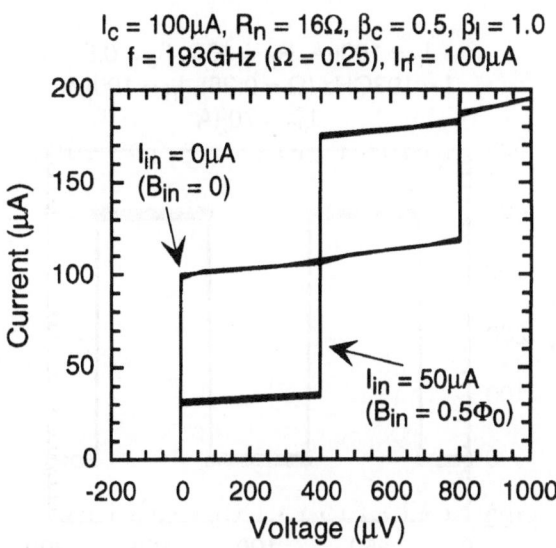

Fig. 3. Examples of the *I-V* characteristics for both I_{in} = 0µA and 50µA.

opposite arrows which indicate the peak-to-peak amplitude of the input signal. We chose the inductance parameter β_l (= LI_c/Φ_0) of 1.0. When the input (control) signal is zero and no DC flux is applied to the SQUID loop, the SQUID does not generate voltage. When the flux of the input (control) signal current (B_{in}) is $\Phi_0/2$ in the SQUID loop, the operation point moves across the threshold boundaries and the SQUID generates the first Shapiro step at the voltage $\Phi_0 f$.

Calculated *I-V* characteristics are shown in Fig. 3. In the calculation, we assumed the parameters of a YBa$_2$Cu$_3$O$_{7-\delta}$ grain boundary junction having the McCumber parameter β_c (= $2\pi I_c C_J R_n^2/\Phi_0$) of 0.5. The coupling coefficients (k_{in} and k_{rf}) equal 1. The normalized frequency of the RF-field activation is $\Omega = \Phi_0 f/I_c R_n$ = 0.25. It can be seen that the even-order steps appear for I_{in}= 0, while the odd-order steps appear for I_{in} = 50µA, which induces $\Phi_0/2$ in the SQUID. The first step appears at $\Phi_0 f$ = 400µV, and its height is as large as 70% of $2I_c$.

MICROWAVE APPLICATIONS

As shown in Fig. 3, the RFDS exhibits large Shapiro step even when the normalized frequency is much less than unity. Figure 4 shows the frequency dependence of the maximum value of the first step height (at $B_{in}=\Phi_0/2$). The results for a single junction are also plotted for reference. For the single junction, the maximum step height increases with the frequency up to Ω > 2. At Ω > 0.5, the same can be observed for the RFDS. However, the maximum step height of the RFDS maintains the value as large as 1.4I_c also in the lower frequency region. This indicates that the field-biased RFDS can work as a high-performance RF detector over a wide frequency region.

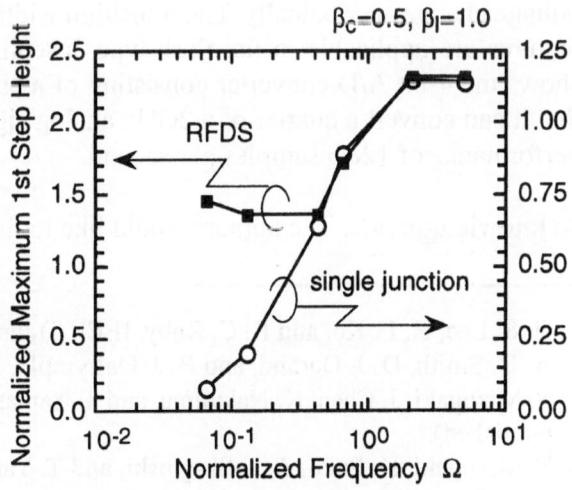

Fig. 4. Frequency dependence of the maximum first step height. The step height is normalized by I_c.

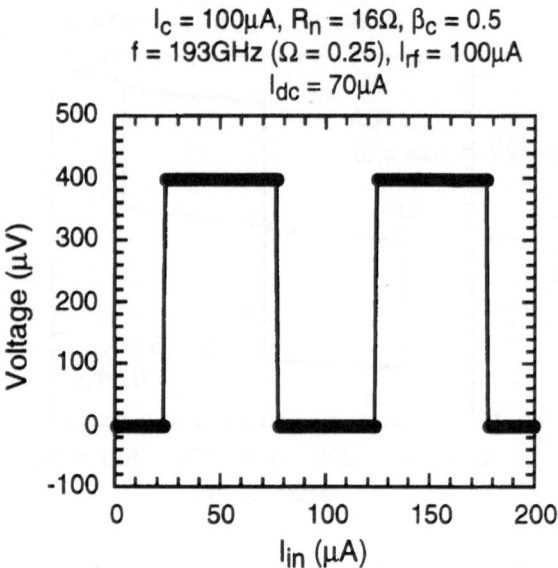

$I_c = 100\mu A$, $R_n = 16\Omega$, $\beta_c = 0.5$
$f = 193GHz$ ($\Omega = 0.25$), $I_{rf} = 100\mu A$
$I_{dc} = 70\mu A$

Fig. 5. Periodical output voltage of the RFDS as a function of I_{in}.

SWITCHING APPLICATIONS

Large current steps at the fixed voltage in the *I-V* characteristics are also attractive for the switching applications. That is, we can utilize the supercurrent for the "0" output level and the first step for the "1" output level in binary code. We have already confirmed numerically fundamental logic functions (OR, NOR, and RS flip-flop) of the RFDS [4].

Figure 5 shows the output voltage of the RFDS as a function of I_{in}. It can be seen that the output

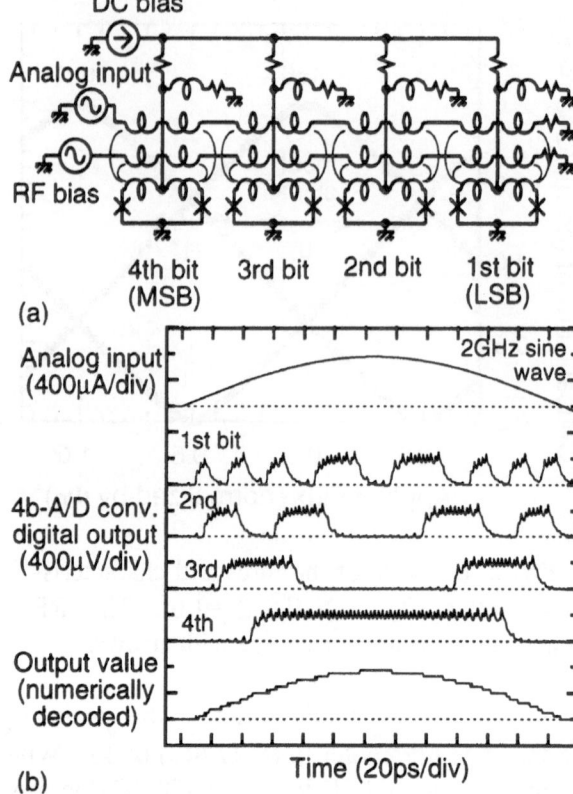

Fig. 6. Flash-type 4-bit A/D converter composed of the four RFDS's.
(a) Schematic configuration. The mutual coupling coefficient of the analog input line is set as 2^{1-m} for the *m*-th bit RFDS.
(b) Simulated waveforms. The analog input is a half period of a 2GHz sine wave. The input signal is successively reconstructed from the gray-coded digital outputs.

voltage changes periodically. The transition width for I_{in} is quite narrow, less than $1\mu A$ in Fig. 5. Such features are applicable to the flash-type Josephson analog-to-digital (A/D) converter [5]. Figure 6 shows the 4-bit A/D converter consisting of a cascade of four RFDS. The simulated results indicate that it can convert a quarter of a 2GHz analog signal into 2^4 discrete levels, which corresponds to the performance of 128G samples per second.

Acknowledgments. The authors would like to thank Dr. H. B. Wang for fruitful discussions.

1. G. S. Lee, H. L. Ko, and R. C. Ruby, IEEE Trans. Appl. Superconduct. **3**, 2740-2743 (1993).
2. A. D. Smith, D. J. Durand, and B. J. Dalrymple, IEEE Trans. Appl. Sueprconduct. **9**, 63-65 (1999).
3. Y. Mizugaki, J. Chen, K. Nakajima, and T. Yamashita, IEEE Trans. Appl. Superconduct. **9**, (1999) (to be published).
4. Y. Mizugaki, K. Saito, A. I. Braginski, and T. Yamashita, (unpublished).
5. C. A. Hamilton and F. L. Lloyd, IEEE Electron Device Lett. **EDL-1**, 92-94 (1980).

Flux Flow of Josephson Vortices in La$_{2-x}$Sr$_x$CuO$_4$ Single Crystals

T. Tachiki, [1] K. Nakajima,[1*] T. Yamashita,[2*] I. Tanaka,[3*] and H. Kojima [3*]

[1]RIEC, Tohoku University, 2-1-1 Katahira, Aoba-ku, Sendai 980-8577, Japan
[2]NICHe, Tohoku University, Aramakiaza, Aoba-ku, Sendai 980-8579, Japan
[3]Inst. of Inorganic Synthesis, Fac. of Engineering, Yamanashi University, Miyamae 7, Kofu 400-8511, Japan

Abstract: We observed a sharp up-turn in the current-voltage (*I-V*) characteristics of the intrinsic Josephson junctions fabricated from a La$_{2-x}$Sr$_x$CuO$_4$ single crystal in a magnetic field applied parallel to the *ab* plane. The *I-V* curves show a up-turn behavior with increasing the field, and become vertical for higher fields. We can explain that this up-turn is caused by matching between the velocity of Josephson vortices and the mode velocity of electromagnetic waves whose mode numbers depend on the field strength.

Keywords: Josephson vortex, La$_{2-x}$Sr$_x$CuO$_4$ single crystals

INTRODUCTION

It is recognized theoretically and experimentally that the characteristic velocities of electromagnetic waves exist in stacked Josephson junctions [1,2]. In high-temperature superconductors (HTSCs) that are naturally build in such a system, the maximum velocity of Josephson vortices can reach up to the mode velocity under currents applied along the *c* axis, if damping that give rise to viscous motion of the vortices are weak. This phenomenon is indicated by the flux-flow steps or a sharp up-turn in the current-voltage (*I-V*) characteristics of the stack such as Bi$_2$Sr$_2$CaCu$_2$O$_x$ (BSCCO) [3]. We can estimate the upper limitation of the vortex velocity from the feature in the *I-V* characteristics and it is important for application of high frequency devices based on the intrinsic Josephson effects in HTSCs. Frequency of the Josephson plasma in La$_{2-x}$Sr$_x$CuO$_4$ (LSCO) is in the range of Tera Hertz and higher than that of BSCCO. Hence it is expected that LSCO is utilized for electromagnetic wave generators in submillimeter wave band. In this paper, we report flux flow properties of stacks fabricated along the *c* axis of LSCO and discuss about the Josephson vortex dynamics in LSCO.

EXPERIMENTAL

The single crystals of La$_{2-x}$Sr$_x$CuO$_4$ (x=0.09) prepared by the traveling solvent floating zone methods were cut into rods, with about $40 \times 40 \times 500 \ \mu m^3$, where the longest dimension is parallel to the *c* axis. Four-probe geometry was formed to put Ag electrodes on both sides of the rod on a glass plate. The samples were annealed at 500 ℃ for 30 minutes in flowing oxygen to reduce contact resistance between LSCO and Ag. Micro-bridges of $20 \times 1 \ \mu m^2$ area and 1 μm long were fabricated along the *c* axis of LSCO by focused ion beam etching of Ga+ with 30kV. The micro-bridge is regarded as the stack including thousands of the intrinsic Josephson junctions. A schematic view of our sample is shown in Fig. 1.

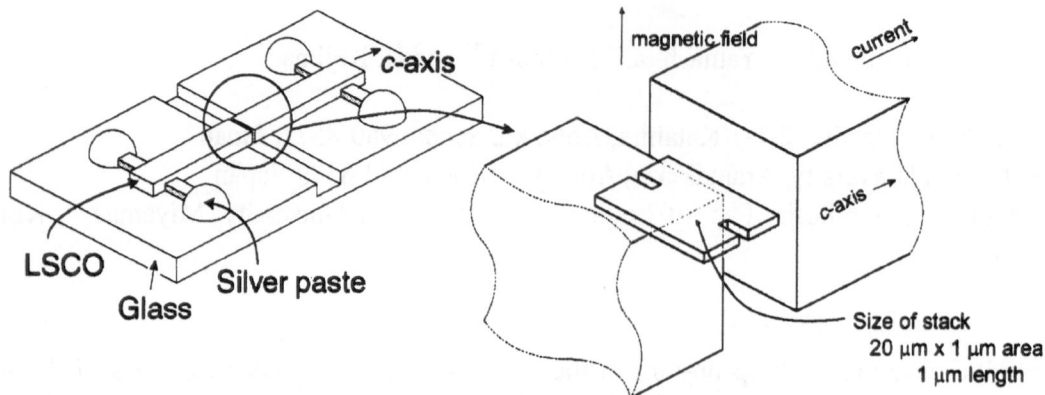

Fig. 1 A schematic view of the stack fabricated from LSCO

We recorded the voltage across the stack as a function of currents under external magnetic fields generated by a superconducting coil. The samples can be rotated in the static magnetic field with a precision of 0.05°. The external magnetic field was applied perpendicular to the longer dimension of the stack in the *ab* plane.

RESULTS AND DISCUSSION

Figure 2 shows the *I-V* characteristics of the stack fabricated from LSCO for several magnetic fields applied parallel to the *ab* plane at 4.2 K. One can see a peculiar nature that is a sharp up-turn at a high bias region in the *I-V* curves. This behavior becomes clear with increasing the magnetic field H and in particular for $\mu_0 H > 0.6$ Tesla it shows a vertical characteristic. The up-turn is very similar to so-called velocity-matching step in an underdamped junction caused by matching between the vortex velocity and the Swihart velocity that is a velocity of an electromagnetic wave propagating in the junction. The maximum voltages V_M (See Fig. 2.) versus H are plotted to investigate the origin of this behavior (Fig.3). The field dependence of V_M is approximately linear and then deviates from linearity for higher fields. For a single junction, the deviation from linearity is explained by reduction of the Swhihart velocity for higher junction voltage [4]. In a case of a close stack of Josephson junctions, it is natural to consider that the velocity of an electromagnetic wave matching the vortex motion decreases with increasing the field or with increasing the voltage across the stack. The stack consisting of N junctions has N different mode velocities c_n of electromagnetic waves. c_n is given by [2]

$$c_n = \frac{\overline{c}_S}{\sqrt{1 + 2S \cos\left(\dfrac{n\pi}{N+1}\right)}}, \tag{1}$$

where $\overline{c}_S$ is a constant corresponding to the Swihart velocity in a single junction, n the mode

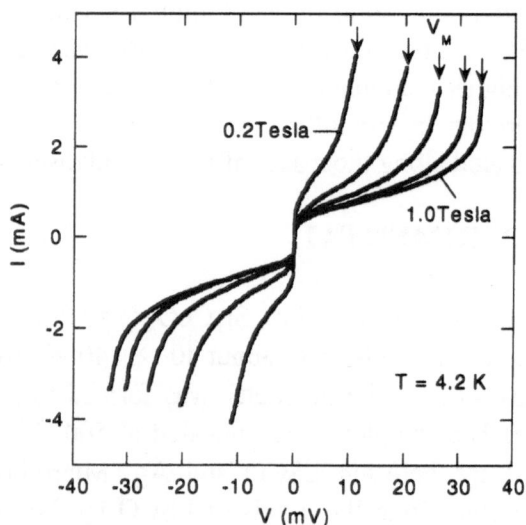

Fig. 2 *I-V* characteristics of the stack fabricated from LSCO for magnetic fields applied parallel to the *ab* plane ($\mu_0 H$ =0.2, 0.4, 0.6, 0.8, and 1.0 Tesla). The arrows indicate the maximum flux flow voltage V_M.

number, S the coupling constant between adjacent junctions. Since the CuO_2 mono-layers in LSCO are atomic layers and S is very close to -0.5, we take $S = -0.5 + 2 \times 10^{-7}$. In this case, $\bar{c}_S$ is given by $\bar{c}_S = c \cdot s/(\lambda_{ab}(2\varepsilon_r)^{1/2})$, where s is the interlayer spacing, λ_{ab} the London penetration depth of screening current in the ab plane, ε_r the dielectric constant, c the velocity of light in vacuum. In case of LSCO, $\bar{c}_S = 3.4 \times 10^4$ m/s, by substituting $\varepsilon_r = 30$, $\lambda_{ab} = 0.75\mu m$, $s = 6.6 Å$ in the above equation. We assume that the ratio $n/(N+1)$ in Eq. (1) depends on H and is approximately linear in H.

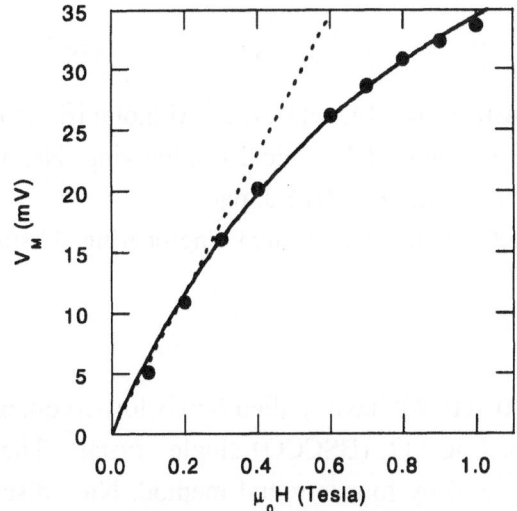

Fig. 3 Magnetic field dependence of the maximum flux-flow voltage. The solid line is a fit by Eqs. (1) and (2).

The interaction between vortices is not so strong, since H is much smaller than the characteristic field $\Phi_0/(\gamma s^2) \sim 200$ Tesla (γ is an anisotropic ratio; about 20 for LSCO which we use) where the non-linear cores of the Josephson vortices begin to overlap. Due to the weak interaction, some of the vortices are pinned, and the number of the vortices that match with the electromagnetic waves and effectively contribute to the voltage is smaller than the total number of the vortices. We calculated the maximum voltage V_M by using Eq. (1) and

$$V_M = \mu_0 H \, c_n \, \ell_{eff}, \tag{2}$$

where ℓ_{eff} is the effective length along the c axis of the area where the flux flow with the velocity c_S occurs. The results are represented in Fig. 3 by the solid line and give good agreement with the experimental data. The best fitted parameters are $n/(N+1) = 0.1$ at 0.3 Tesla and $\ell_{eff} = 0.4 \mu m$, and we obtain $c_n = 1.3 \times 10^5$ m/s from them. The maximum value of the average vortex velocity v_f estimated by using $V = \mu_0 H \, v_f \, \ell$ (ℓ is the length of the stack) is 5.8×10^4 m/s. The value of c_n estimated above is about twice of the average vortex velocity. We note that a sharp up-turn in the I-V curves (Fig.2) can indicate that the vortices move with the velocity higher than the averaged one.

CONCLUSION

We observed a sharp up-turn in the I-V curves of the stack fabricated along the c axis of $La_{2-x}Sr_xCuO_4$ single crystals. The up-turn behavior becomes clear and vertical with increasing a magnetic field. We may explain that this up-turn comes from a velocity-matching step between the vortex velocity and the mode velocity of electromagnetic waves whose mode number varies with the field. The sharp up-turn I-V curves shows that the vortex velocity reaches to the mode velocity whose value is the order of 10^5 m/s and is higher than the averaged vortex velocity.

*Also with CREST, Japan Science and Technology Corporation, Japan

1. R. Kleiner, *Phys. Rev B* **50**, pp. 6919-6922 (1994).
2. S. Sakai et al., *Phys. Rev B* **50**, pp. 12905-12914 (1994).
3. J. U. Lee et al., *Appl. Phys. Lett* **71**, pp. 1412-1414 (1997)
4. K. L. Ngai, *Phys. Rev.* **182**, pp. 555-568 (1969)

Low Frequency Noise Properties of $Bi_2Sr_2CaCu_2O_y$ Intrinsic Josephson Junctions

Atsushi Saito[1], Manabu Abe[1], Akinobu Irie[2], Gin-ichiro Oya[2] and Katsuyoshi Hamasaki[1]

[1] Department of Electrical Engineering, Nagaoka University of Technology, 1-1603 Kamitomioka, Nagaoka-shi, 940-2188 Japan

[2] Department of Electrical Engineering, Utsunomiya University, 2753 Ishii, Utsunomiya-shi, 321-8585 Japan

Abstract: We have studied c-axis low-frequency noise properties of mesa-type junctions made from $Bi_2Sr_2CaCu_2O_y$ (BSCCO) single crystals. The c-axis current vs. voltage (I-V) characteristics were measured by four-terminal method. Nine discrete-resistive-branches with hysteresis were observed in the I-V curves. We estimated the magnitude of the Hooge's 1/f noise parameter η from noise voltage spectral density in a BSCCO intrinsic Josephson junction (IJJ) when it was biased on each branch. The bias current kept constant. The value of η did not depend on the number of junctions. This result indicates that no excess noise caused by the interactions between the intrinsic junctions in the mesa.

Keywords: BSCCO intrinsic Josephson junction, mesa-type junction, 1/f noise.

INTRODUCTION

Recent observations of intrinsic Josephson tunneling in small pieces of $Bi_2Sr_2CaCu_2O_y$ (BSCCO) single crystals strongly suggest that a stack of SIS Josephson junctions is naturally formed along the c-axis in this material. Many researchers have discussed the transport mechanism of a BSCCO intrinsic Josephson junction (IJJ). [1-3] For instance, Tanabe et al. analyzed the quasiparticle current vs. voltage characteristics in a BSCCO IJJ based on a d-wave symmetry of the superconducting order parameter in the k_x-k_y space. [3]

One of the important and useful spectroscopic tools for a better understanding of the transport mechanism of a BSCCO IJJ is low-frequency noise (LFN) property. Although several researchers have studied LFN in the cuprate superconducting thin-films as YBaCuO, TlBaCaCuO and BSCCO, [4-6] the c-axis LFN property of mesa-type junctions made from BSCCO single crystal was not measured as far as we know.

In this paper, we report the first experimental study on the 1/f noise characteristics of BSCCO IJJs at 4.2K.

EXPERIMENTS AND DISCUSSION

Noise measurements were performed in superconducting Nb-can. Double shielded twisted-pair-cables were used to reduce the effect of unexpected external magnetic fields. The current was supplied from a battery-powered current source. The voltage fluctuation across the intrinsic junctions was amplified by a low-noise preamplifier (LI-75A, NF Circuit Block; Gain=100) with noise voltage V_N=-137dBV/Hz$^{1/2}$ at 50kHz, and analyzed by FFT spectrum analyzer (TR-9402, ADVANTEST; V_N=-154dBV/Hz$^{1/2}$ at 50kHz).

Figure 1 shows the schematic drawing of a mesa-type BSCCO IJJ. The BSCCO single crystal was grown by the self-flux method. [2] A 50nm thick Au layer was evaporated onto the surface of freshly cleaved BSCCO single crystals for electrical contacts. The mesa with area of 160μm $\times$ 40μm were fabricated using the Ar-ion etching and the conventional photolithography techniques. The insulating layer of hard baked Az was formed and finally an Au wiring layer was formed.

Figure 2 shows the c-axis current vs. voltage (I-V) characteristic measured by a four-terminal method at 4.2K. The junction exhibits large hysteresis with nine discrete-resistive-branches, and the voltage jump between two adjacent branches V_j is approximately 10mV. The critical current I_C of surface junction in the mesa is about 5.0mA and approximately equals to that of inner junction. The inset of Fig. 2 shows dI/dV vs. V^2 characteristic for low-bias region. The solid line shows the linear dependence of dI/dV on V^2. The observed linear relationship between dI/dV and V^2 is indicative of d-wave symmetry of the order parameter. [3] We believe that the contact resistance R_C of voltage probes is less than 10Ω. The assumption is justified by the fact that the resistances between the I_1 and V_1 electrodes (see Fig. 1) were less than 10Ω for a few samples.

For the evaluation of defect state in the mesa, 1/f noise voltage spectra S_V were measured as a function of the number of junctions, n, switched to the voltage state at 4.2K (Fig.3). The bias current was kept constant to 4mA. The closed marks in Fig. 2 show the bias points measured the noise voltage spectra S_V.

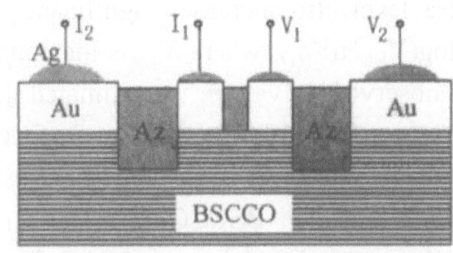

Fig. 1 Schematic draws of a mesa-type BSCCO IJJ with the dimensions of 160μm $\times$ 40μm $\times$ 13.5nm.

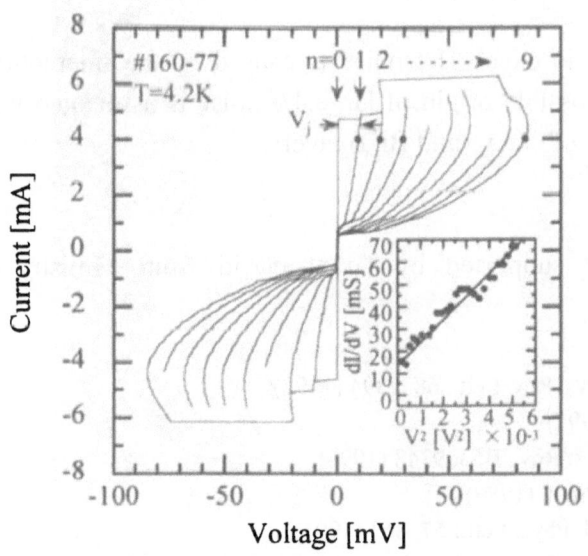

Fig. 2 Current vs. voltage characteristic of a mesa-type BSCCO IJJ at 4.2K. The inset is dI/dV-V^2 characteristic.

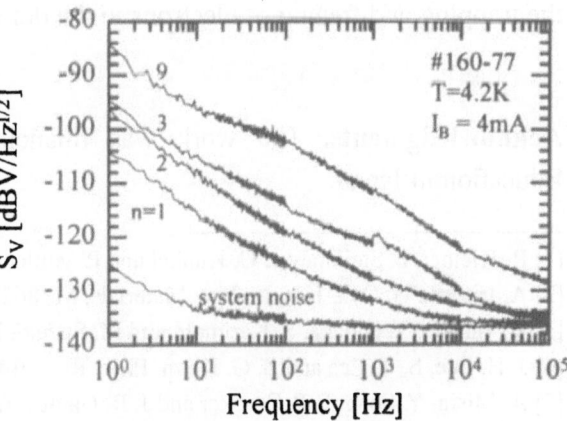

Fig. 3 Noise voltage power spectra S_V of BSCCO IJJ.

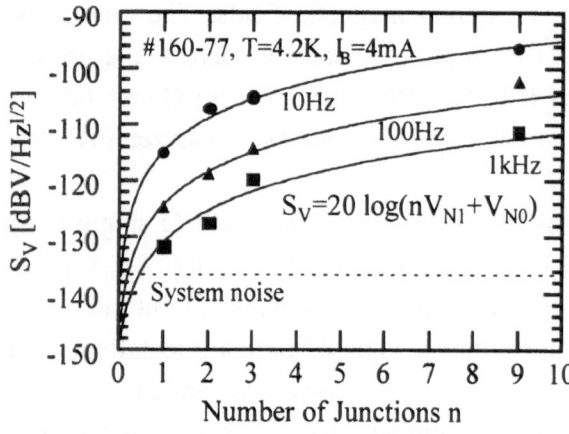 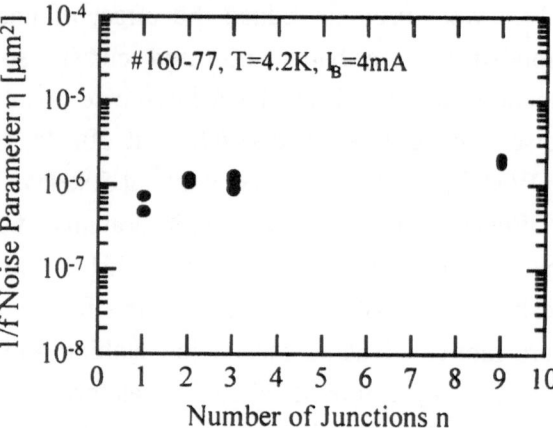

Fig. 4 S_V vs. n characteristics. The solid lines are calculated from $S_V= 20\log(V_{N0}+nV_{N1})$.

Fig. 5 Hooge's 1/f noise parameter η vs. n characteristics.

Figure 3 show the noise spectra of a BSCCO IJJ when it was biased on each branch. The measured power densities have no Lorentzian frequency dependence as expected from the random telegraph noise model.

Figure 4 shows the S_V vs. n characteristics at 10Hz, 100Hz and 1kHz. As n increases, the 1/f noise level also increases significantly. The solid lines are calculated using the $S_V[\text{dBV/Hz}^{1/2}]= 20\log(V_{N0}+nV_{N1})$, where V_{N0} is the contact noise and V_{N1} is the noise level for n=1 junction. From the observed S_V values, we estimated the 1/f noise parameter (η) using Hooge's empirical formula for investigating defect states in BSCCO IJJ: $S_V=\eta f^{-1}A^{-1}I_B{}^2R_D{}^2$, where, η is the 1/f noise parameter related to defects, A is the area of mesa, and f is the measuring frequency. [7] Figure 5 shows η vs. n characteristics in BSCCO IJJ. These results indicate that the contact noise rising from contact resistance can be almost neglected for the 1/f noise measurement. The η in BSCCO IJJ is about $10^{-6}\mu\text{m}^2$ and independent of n. This value is the same order of magnitude as for NbN/AlN/NbN tunnel junctions. [8]

In conclusion, the present data indicate that no excess 1/f noise is caused by the interactions between the intrinsic junctions. We believe the possible origin of large 1/f noise is associated with the trapping and freeing of electrons in the defects of CuO_2 (and BiO) layers.

Acknowledgments. The work was financially supported by Grant-in-Aid from Ministry of Education in Japan.

[1] R. Kleiner. F. Steinmeyer, G. Kunkel and P. Muller, Phys. Rev. Lett., **68**, 2394 (1992).

[2] A. Irie and G. Oya, J. Japan Inst. Materials, **61**, 862 (1997).

[3] K. Tanabe, Y. Hidaka, S. Karimoto and M. Suzuki, Phys. Rev., **B53**, 9348 (1996).

[4] J. H. Lee, S. C. Lee and Z. G. Khim, Phys. Rev., **B40**, 6806 (1989).

[5] A. Misra, Y. Song, P. P. Crooker and J. R. Gaines, Appl. Phys. Lett., **57**, 863 (1991).

[6] I. Shin, A. L. Li, W. W. Lam, C. X. Qiu, L. N. Phong and B. Tremblay, Can. J. Phys., **72**, 270 (1994).

[7] K. Hamasaki and A. Saito, Physics and Applications of Mesoscopic Josephson Junctions, 247 (1999).

[8] N. Sato, A. Saito, Z. Wang and K. Hamasaki, Extended Abstracts (The 60th Autumn Meeting, 1999), The Japan Society of Applied Physics, 1p-ZV-11.

Author Index

1124